PRECALCULUS

SECOND EDITION

JOHN W. COBURN

St. Louis Community College at Florissant Valley

Mc Graw Hill **Higher Education**

Boston   Burr Ridge, IL   Dubuque, IA   New York   San Francisco   St. Louis
Bangkok   Bogotá   Caracas   Kuala Lumpur   Lisbon   London   Madrid   Mexico City
Milan   Montreal   New Delhi   Santiago   Seoul   Singapore   Sydney   Taipei   Toronto

PRECALCULUS, SECOND EDITION

Published by McGraw-Hill, a business unit of The McGraw-Hill Companies, Inc., 1221 Avenue of the Americas, New York, NY 10020.
Copyright © 2010 by The McGraw-Hill Companies, Inc. All rights reserved. Previous edition © 2007. No part of this publication may be reproduced or distributed in any form or by any means, or stored in a database or retrieval system, without the prior written consent of The McGraw-Hill Companies, Inc., including, but not limited to, in any network or other electronic storage or transmission, or broadcast for distance learning.

Some ancillaries, including electronic and print components, may not be available to customers outside the United States.

This book is printed on acid-free paper.

4 5 6 7 8 9 0 DOW/DOW 10 9 8 7 6 5 4 3 2

ISBN 978–0–07–351942–5
MHID 0–07–351942–1

ISBN 978–0–07–336086–7 (Annotated Instructor's Edition)
MHID 0–07–336086–4

Editorial Director: *Stewart K. Mattson*
Sponsoring Editor: *Dawn R. Bercier*
Senior Developmental Editor: *Michelle L. Flomenhoft*
Developmental Editor: *Katie White*
Marketing Manager: *John Osgood*
Senior Project Manager: *Vicki Krug*
Senior Production Supervisor: *Sherry L. Kane*
Senior Media Project Manager: *Sandra M. Schnee*

Designer: *Laurie B. Janssen*
Cover Designer: *Christopher Reese*
(USE) Cover Image: © *Georgette Douwma/Gettyimages*
Senior Photo Research Coordinator: *John C. Leland*
Supplement Producer: *Mary Jane Lampe*
Compositor: *Aptara®, Inc.*
Typeface: *10.5/12 Times Roman*
Printer: *R. R. Donnelley Willard, OH*

Chapter 1 Opener: © Royalty-Free/CORBIS; pg. 12: NASA/RF; pg. 30: PhotoLinK/Getty Images/RF; pg. 68 top: © Brand X Pictures/PunchStock/RF; pg. 68 bottom: Photodisc Collection/Getty Images/RF. **Chapter 2** Opener: © Royalty-Free/CORBIS; pg. 139: Siede Preis/Getty Images/RF; pg. 140: The McGraw-Hill Companies, Inc./Ken Cavanagh Photographer; pg. 155: Steve Cole/Getty Images/RF; pg. 172: Alan and Sandy Carey/Getty Images/RF; pg. 183: Courtesy John Coburn; pg. 201 top: Patrick Clark/Getty Images/RF; pg. 201 bottom: © Digital Vision/PunchStock/RF. **Chapter 3** Opener: © 1997 IMS Communications Ltd./Capstone Design. All Rights Reserved/RF; pg. 238: © Adalberto Rios/Sexto Sol/Getty Images/RF; pg. 240: © Royalty-Free/CORBIS; pg. 255: © Royalty-Free/CORBIS; pg. 288: © Royalty-Free/CORBIS; pg. 314: © Royalty-Free/CORBIS; pg. 320: © Royalty-Free/CORBIS. **Chapter 4** Opener: © Andrew Ward/Life File/Getty Images/RF; pg. 364 left: © Geostock/Getty Images/RF: pg. 364 right: © Lawrence M. Sawyer/Getty Images/RF; pg. 373: Photography by G.K. Gilbert, courtesy U.S. Geological Survery; pg. 374: © Lars Niki/RF; pg. 378: © Medioimages/Superstock/RF; pg. 395: StockTrek/Getty Images/RF; pg. 415: Courtesy Simon Thomas. **Chapter 5** Opener: Digital Vision/RF; pg. 434: © Jules Frazier/Getty Images/RF; pg. 438: © Karl Weatherly/Getty Images/RF; pg. 468: © Royalty-Free/CORBIS; pg. 510: © Royalty-Free/CORBIS; pg. 527: Royalty-Free/CORBIS. **Chapter 6** Opener: © Digital Vision/Getty Images/RF; pg. 617 © John Wong/Getty Images/RF. **Chapter 7** Opener: © Royalty-Free/CORBIS/RF **Chapter 8** Opener: © Royalty-Free/CORBIS; pg. 730: © The McGraw-Hill Companies, Inc./Jill Braaten, photographer; pg. 731: © Royalty-Free/CORBIS; pg. 742: © Creatas/PunchStock/RF; pg. 751: Royalty-Free/CORBIS. **Chapter 9** Opener: © Mark Downey/Getty Images/RF; pg. 859: © Brand X Pictures/PunchStock/RF; pg. 860: © Digital Vision/Getty Images/RF; pg. 864: © H. Wiesenhofer/PhotoLink/Getty Images/RF; pg. 873 top: © Jim Wehtje/Getty Images/RF; pg. 873 middle: © Creatas/PunchStock/RF; pg. 873 bottom: © Edmond Van Hoorick/Getty Images/RF; pg. 882: © The McGraw-Hill Companies, Inc./Jill Braaten, photographer/RF; pg. 922: © PhotoLink/Getty Images/RF. **Chapter 10** Opener: © Doug Menuez/Getty Images/RF; pg. 955: Royalty-Free/CORBIS; pg. 964: © Andersen Ross/Getty Images/RF. **Chapter 11** Opener: © Royalty-Free/CORBIS. **Appendices** Pg. A-12: © Photodisc/Getty Images/RF; pg. A-50: © Glen Allison/Getty Images/RF.

Library of Congress Cataloging-in-Publication Data

Coburn, John W.
 Precalculus / John W. Coburn. —2nd ed.
  p. cm.
 Includes index.
  ISBN 978–0–07–351942–5—ISBN 0–07–351942–1 (hard copy : alk. paper) 1. Functions. 2. Trigonometry. I. Title.
 QA331.3.C63 2010
 510--dc22   2008050984

www.mhhe.com

Brief Contents

Additional Topics Online

(Visit www.mhhe.com/coburn)

Background

John Coburn grew up in the Hawaiian Islands, the seventh of sixteen children. John's mother and father were both teachers. John's mother taught English and his father, as fate would have it, held advanced degrees in physics, chemistry, and mathematics. Whereas John's father was well known, well respected, and a talented mathematician, John had to work very hard to see the connections so necessary for success in mathematics. In many ways, his writing is born of this experience.

Education

In 1979 John received a bachelor's degree in education from the University of Hawaii. After working in the business world for a number of years, John returned to his first love by accepting a teaching position in high school mathematics and in 1987 was recognized as Teacher of the Year. Soon afterward John decided to seek a master's degree, which he received two years later from the University of Oklahoma.

Teaching Experience

John is now a full professor at the Florissant Valley campus of St. Louis Community College where he has taught mathematics for the last eighteen years. During

his time there he has received numerous nominations as an outstanding teacher by the local chapter of Phi Theta Kappa, and was recognized as Post-Secondary Teacher of the Year in 2004 by Mathematics Educators of Greater St. Louis (MEGSL). John is a member of the following organizations: National Council of Teachers of Mathematics (NCTM), Missouri Council of Teachers of Mathematics (MCTM), Mathematics Educators of Greater Saint Louis (MEGSL), American Mathematical Association of two Year Colleges (AMATYC), Missouri Mathematical Association of two Year Colleges (MoMATYC), Missouri Community College Association (MCCA), and Mathematics Association of America (MAA).

Personal Interests

Some of John's other interests include body surfing, snorkeling, and beach combing whenever he gets the chance. In addition, John's loves include his family, music, athletics, games, and all things beautiful. John hopes that this love of life comes through in the writing, and serves to make the learning experience an interesting and engaging one for all students.

Dedication

To my wife and best friend Helen, whose love, support, and willingness to sacrifice never faltered.

Coral reefs support an extraordinary biodiversity as they are home to over 4000 species of tropical or reef fish. In addition, coral reefs are immensely beneficial to humans; buffering coastal regions from strong waves and storms, providing millions of people with food and jobs, and prompting advances in modern medicine.

Similar to a reef, a precalculus course is unique because of its diverse population of students. Nearly every major is represented in this course, featuring students with a wide range of backgrounds and skill sets. Just like the variety of the fish in the sea rely on the coral reefs to survive, the assortment of students in precalculus rely on succeeding in this course in order to further pursue their degree, as well as their career goals.

From the Author

This text is the result of a mighty confluence of needs, ideas, desires, and directions. This is easily understandable, as the intended audience is one of the most diverse in all of education. Our students come to us with a wide range of backgrounds, varying degrees of preparation, and interest levels that vary from apathy to excitement. In addition, our classes include those needing only a general education requirement, as well as our country's future engineers and scientists. To say our greatest challenge is meeting the needs of so diverse a population would be an understatement. In reflecting on this diversity, the image of a coral reef came to mind, and I was struck by the strength of the analogy. We have a hugely diverse population, with the reef as a common meeting place, with all the inhabitants depending on the reef for their purpose, nourishment, and direction.

Writing a text for this course has been one of the most daunting and challenging experiences in my life. Long before I began, my teaching experience left a nagging sense that most texts on the market lacked the ability to connect with so diverse an audience. In addition, they appeared to offer too scant a framework to build concepts, too terse a development to make connections, and insufficient support in their exercise sets to develop long-term retention or foster a love of mathematics. In particular, the applications seemed to lack a sense of realism, curious interest, and/or connections to a student's everyday experience.

With all of this in mind and a strong desire to write a better text, I set about the task of creating what I hoped would become a more engaging tool for students, and a more supportive tool for instructors. Drawing on the diversity of my own educational experience, and an early exposure to different cultures, views, and perspectives, I believe has contributed to the text's unique and engaging style, and I hope in the end, to more and better connections with our diverse audience. Having feedback from more than 400 people, including manuscript reviewers, focus group participants, and contributors, was invaluable to helping me hone the connections in the book. As a collateral outgrowth of this experience, I admit there was also a desire to interest and engage ourselves, the instructors—to remind us again and again, why we fell in love with mathematics in the first place.   —John Coburn

Precalculus tends to be a challenging course for many students. They don't see the connections that precalculus has to their life or why it is so critical that they take and pass this course for both technical and nontechnical careers alike. Others may enter into this course underprepared or improperly placed and with very little motivation.

Instructors are faced with several challenges as well. They are given the task of improving pass rates and student retention while energizing a classroom full of students comprised of nearly every major. Furthermore, it can be difficult to distinguish between students who are likely to succeed and students who may struggle until after the first test is given.

The goal of the Coburn series is to provide both students and instructors with tools to address these challenges, as well as the diversity of the students taking this course, so that you can experience greater success in precalculus. For instance, the comprehensive exercise sets have a range of difficulty that provides very strong support for weaker students, while advanced students are challenged to reach even further. The rest of this preface further explains the tools that John Coburn and McGraw-Hill have developed and how they can be used to *connect* students to precalculus and *connect* instructors to their students.

The Coburn Precalculus Series provides you with strong tools to achieve better outcomes in your Precalculus course as follows:

▶ *Better Student Preparedness*

▶ *Increased Student Engagement*

▶ *Solid Skill Development*

▶ *Strong Connections*

▶ Better Student Preparedness

No two students have the same strengths and weaknesses in mathematics. Typically students will enter any math course with different preparedness levels. For most students who have trouble retaining or recalling concepts learned in past courses, basic review is simply not enough to sustain them successfully throughout the course. Moreover, instructors whose main focus is to prepare students for the next course do not have adequate time in or out of class to individually help each student with review material.

ALEKS Prep uniquely assesses each student to determine their individual strengths and weaknesses and informs the student of their capabilities using a personalized pie chart. From there, students begin learning through ALEKS via a personalized learning path uniquely designed for each student. ALEKS Prep interacts with students like a private tutor and provides a safe learning environment to remediate their individual knowledge gaps of the course pre-requisite material outside of class.

ALEKS Prep is the only learning tool that empowers students by giving them an opportunity to remediate individual knowledge gaps and improve their chances for success. ALEKS Prep is especially effective when used in conjunction with ALEKS Placement and ALEKS 3.0 course-based software.

▶ Increased Student Engagement

What makes John Coburn's applications unique is that he is constantly thinking mathematically. John's applications are spawned during a trip to Chicago, a phone call with his brother or sister, or even while watching the evening news for the latest headlines. John literally takes notes on things that he sees in everyday life and connects these situations to math. This truly makes for relevant applications that are born from real-life experiences as opposed to applications that can seem fictitious or contrived.

▶ Solid Skill Development

The Coburn series intentionally relates the examples to the exercise sets so there is a strong connection between what students are learning while working through the examples in each section and the homework exercises that they complete. In turn, students who attempt to work the exercises first can surely rely on the examples to offer support as needed. Because of how well the examples and exercises are connected, key concepts are easily understood and students have plenty of help when using the book outside of class.

There are also an abundance of exercise types to choose from to ensure that homework challenges a wide variety of skills. Furthermore, John reconnects students to earlier chapter material with Mid-Chapter Checks; students have praised these exercises for helping them understand what key concepts require additional practice.

▶ Strong Connections

John Coburn's experience in the classroom and his strong connections to how students comprehend the material are evident in his writing style. This is demonstrated by the way he provides a tight weave from topic to topic and fosters an environment that doesn't just focus on procedures but illustrates the big picture, which is something that so often is sacrificed in this course. Moreover, he deploys a clear and supportive writing style, providing the students with a tool they can depend on when the teacher is not available, when they miss a day of class, or simply when working on their own.

Better Student Preparedness...

Experience Student Success!

ALEKS ALEKS is a unique online math tool that uses adaptive questioning and artificial intelligence to correctly place, prepare, and remediate students . . . all in one product! Institutional case studies have shown that **ALEKS has improved pass rates by over 20% versus traditional online homework and by over 30% compared to using a text alone.**

By offering each student an individualized learning path, ALEKS directs students to work on the math topics that they are ready to learn. Also, to help students keep pace in their course, instructors can correlate ALEKS to their textbook or syllabus in seconds.

To learn more about how ALEKS can be used to boost student performance, please visit **www.aleks.com/highered/math** or contact your McGraw-Hill representative.

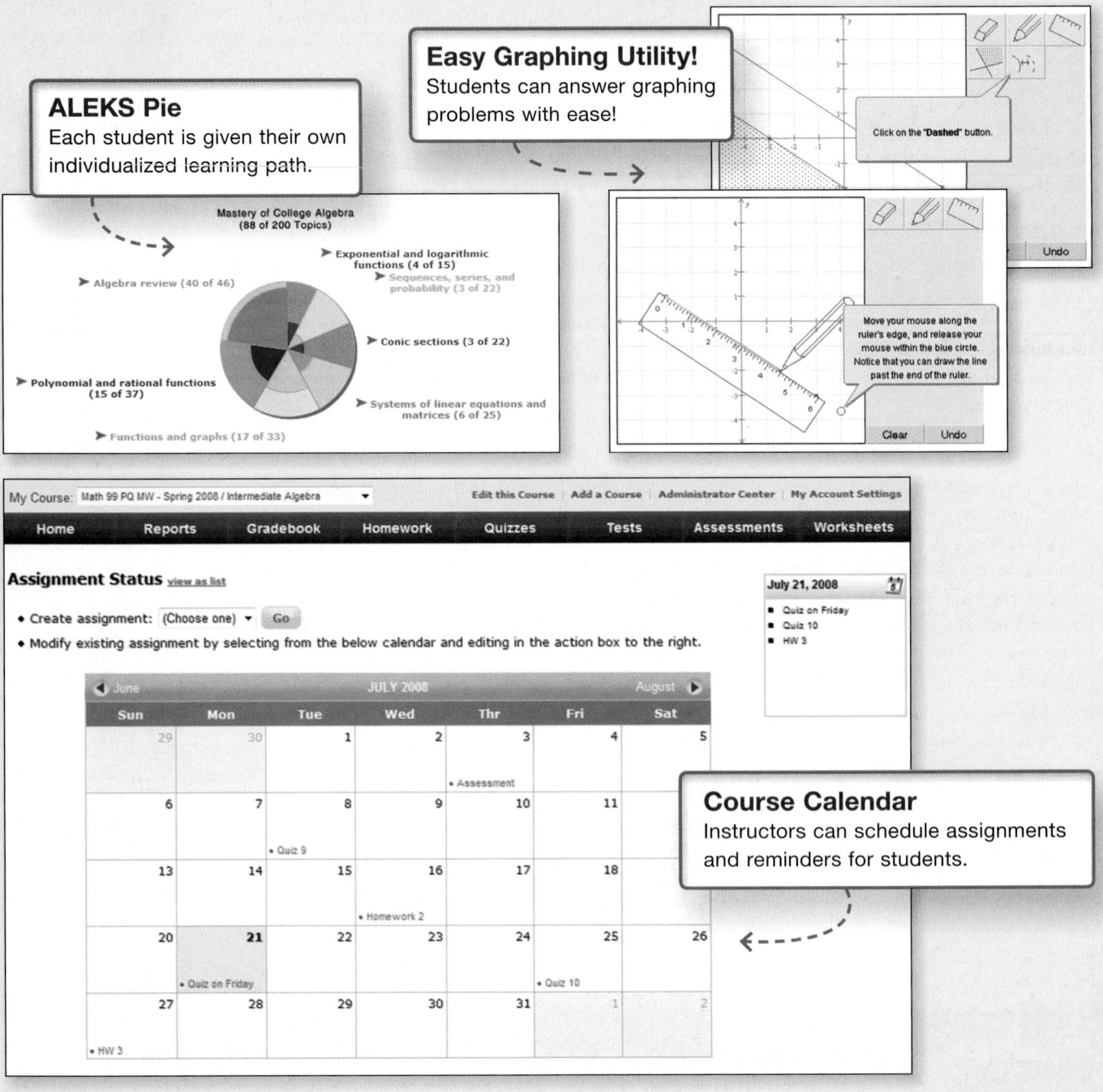

Easy Graphing Utility!
Students can answer graphing problems with ease!

ALEKS Pie
Each student is given their own individualized learning path.

Course Calendar
Instructors can schedule assignments and reminders for students.

New ALEKS Instructor Module

Enhanced Functionality and Streamlined Interface Help to Save Instructor Time

ALEKS® The new ALEKS Instructor Module features enhanced functionality and streamlined interface based on research with ALEKS instructors and homework management instructors. Paired with powerful assignment driven features, textbook integration, and extensive content flexibility, the new ALEKS Instructor Module simplifies administrative tasks and makes ALEKS more powerful than ever.

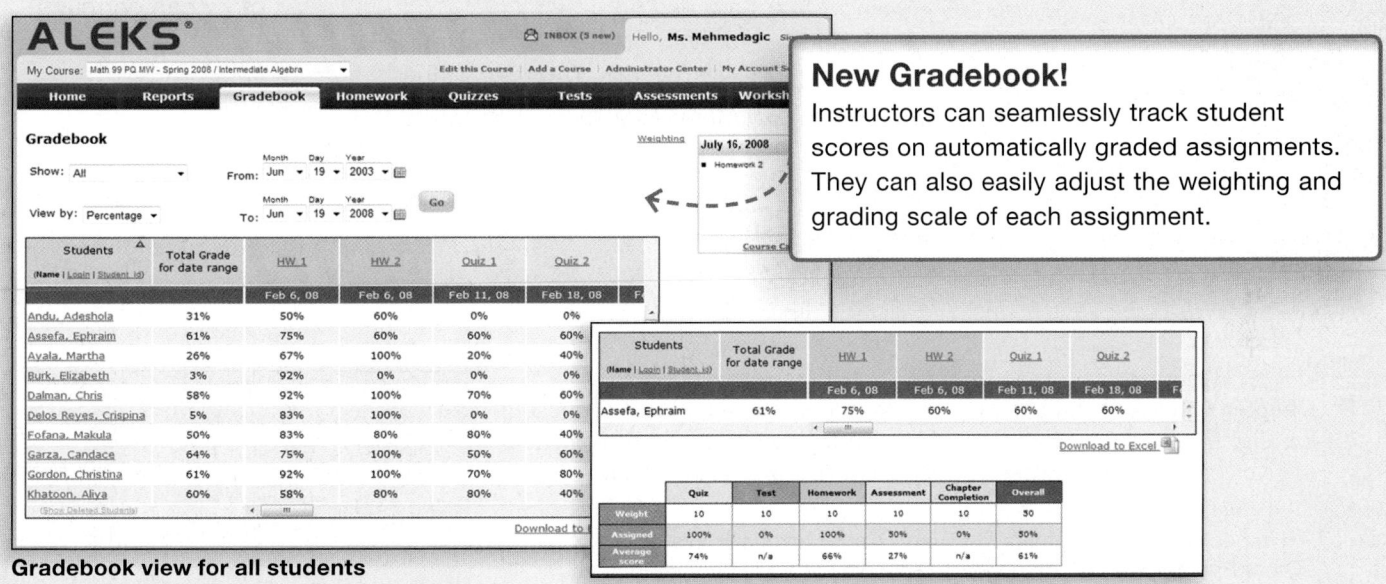

Gradebook view for all students

Gradebook view for an individual student

New Gradebook!
Instructors can seamlessly track student scores on automatically graded assignments. They can also easily adjust the weighting and grading scale of each assignment.

Track Student Progress Through Detailed Reporting
Instructors can track student progress through automated reports and robust reporting features.

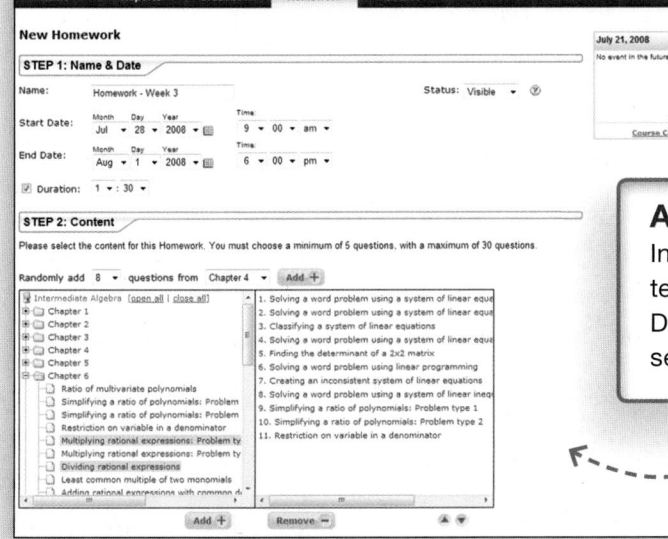

Select topics for each assignment

Automatically Graded Assignments
Instructors can easily assign homework, quizzes, tests, and assessments to all or select students. Deadline extensions can also be created for select students.

Learn more about ALEKS by visiting **www.aleks.com/highered/math** or contact your McGraw-Hill representative.

Increased Student Engagement...

Through Meaningful Applications

Making mathematics meaningful requires that students experience the connection between mathematics and its impact on the world they live in. This text is also the result of a powerful commitment to provide applications of the highest quality, having close ties to the examples, and with carefully monitored levels of difficulty.

Many of these examples were born of my own diverse life experiences, others came from a curious, lucid, and even visionary folly that allows one to seize upon the every day events of life, and see the significant or meaningful mathematics in the background. My ever-present notebook was used a thousand times to capture that casual observation, or that sudden burst of inspiration that is the genesis for outstanding applications. These were supported at home by a substantial library of reference and research books, an eye toward both history and current events, and of course our modern marvel of a research tool—the Internet. After a (sometimes long) period of thought, reflection, and research, followed by a wording and a rewording of the exercise so that it would resonate with students while filling the need, a significant and meaningful application was born. —JC

▶ **Chapter Openers** highlight Chapter Connections, an interesting application exercise from the chapter, and provide a list of other real-world connections to give context for students who wonder how math relates to them.

> **"**I especially like the depth and variety of applications in this textbook. Other Precalculus texts the department considered did not share this strength. In particular, there is a clear effort on the part of the author to include realistic examples showing how such math can be utilized in the real world.**"**   —*George Alexander, Madison Area Technical College*

▶ **Examples** throughout the text feature word problems, providing students with a starting point for how to solve these types of problems in their exercise sets.

> **"**One of this text's strongest features is the wide range of applications exercises. As an instructor, I can choose which exercises fit my teaching style as well as the student interest level.**"**
> —*Stephen Toner, Victor Valley College*

▶ **Application Exercises** at the end of each section are the hallmark of the Coburn series. Never contrived, always creative, and born out of the author's life and experiences, each application tells a story and appeals to a variety of teaching styles, disciplines, backgrounds, and interests.

> **"**[The application problems] answered the question, 'When are we ever going to use this?'**"**
> —*Student class tester at Metropolitan Community College–Longview*

▶ **Math in Action Applets,** located online, enable students to work collaboratively as they manipulate applets that apply mathematical concepts in real-world contexts.

2

CHAPTER CONNECTIONS

Relations, Functions, and Graphs

CHAPTER OUTLINE

2.1 Rectangular Coordinates; Graphing Circles and Other Relations 152
2.2 Graphs of Linear Equations 165
2.3 Linear Graphs and Rates of Change 178
2.4 Functions, Function Notation, and the Graph of a Function 190
2.5 Analyzing the Graph of a Function 206
2.6 The Toolbox Functions and Transformations 225
2.7 Piecewise-Defined Functions 240
2.8 The Algebra and Composition of Functions 254

Viewing a function in terms of an equation, a table of values, and the related graph, often brings a clearer understanding of the relationships involved. For example, the power generated by a wind turbine is often modeled by the function $P(v) = \frac{8v^3}{125}$, where P is the power in watts and v is the wind velocity in miles per hour. While the formula enables us to predict the power generated for a given wind speed, the graph offers a visual representation of this relationship, where we note a rapid growth in power output as the wind speed increases. This application appears as Exercise 107 in Section 2.6.

Check out these other real-world connections:

▶ Earthquake Area (Section 2.1, Exercise 84)
▶ Height of an Arrow (Section 2.5, Exercise 61)
▶ Garbage Collected per Number of Garbage Trucks (Section 2.2, Exercise 42)
▶ Number of People Connected to the Internet (Section 2.3, Exercise 109)

151

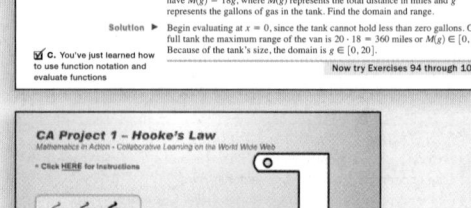

EXAMPLE 10 ▶ **Determining the Domain and Range from the Context**

Paul's 1993 Voyager has a 20-gal tank and gets 18 mpg. The number of miles he can drive (this range) depends on how much gas is in the tank. As a function we have $M(g) = 18g$, where $M(g)$ represents the total distance in miles and g represents the gallons of gas in the tank. Find the domain and range.

Solution ▶ Begin evaluating at $x = 0$, since the tank cannot hold less than zero gallons. On a full tank the maximum range of the van is $20 \cdot 18 = 360$ miles or $M(g) \in [0, 360]$. Because of the tank's size, the domain is $g \in [0, 20]$.

☑ **C.** You've just learned how to use function notation and evaluate functions

Now try Exercises 94 through 101 ▶

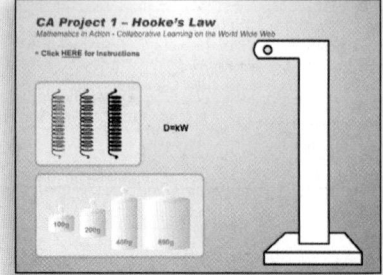

CA Project 1 – Hooke's Law
Mathematics in Action · Collaborative Learning on the World Wide Web

• Click **HERE** for instructions

D=kW

Through Timely Examples

In mathematics, it would be difficult to overstate the importance of examples that set the stage for learning. Not a few educational experiences have faltered due to an example that was too difficult, a poor fit, out of sequence, or had a distracting result. In this series, a careful and deliberate effort was made to select examples that were timely and clear, with a direct focus on the concept or skill at hand. Everywhere possible, they were further designed to link previous concepts to current ideas, and to lay the groundwork for concepts to come. As a trained educator knows, the best time to answer a question is often before it's ever asked, and a timely sequence of carefully constructed examples can go a long way in this regard, making each new idea simply the next logical, even anticipated step. When successful, the mathematical maturity of a student grows in unnoticed increments, as though it was just supposed to be that way. —JC

▶ **Titles** have been added to Examples in this edition to highlight relevant learning objectives and reinforce the importance of speaking mathematically using vocabulary.

▶ **Annotations** located to the right of the solution sequence help the student recognize which property or procedure is being applied.

▶ **"Now Try"** boxes immediately following Examples guide students to specific matched exercises at the end of the section, helping them identify exactly which homework problems coincide with each discussed concept.

▶ **Graphical Support Boxes,** located after selected examples, visually reinforce algebraic concepts with a corresponding graphing calculator example.

> **❝** The author does a great job in describing the examples and how they are to be written. In the examples, the author shows step by step ways to do just one problem . . . this makes for a better understanding of what is being done.**❞**
> —*Michael Gordon, student class tester at Navarro College*

EXAMPLE 3 ▶ Solving a System Using Substitution

Solve using substitution: $\begin{cases} 4x + y = 4 \\ y = x + 2 \end{cases}$.

Solution ▶ Since $y = x + 2$, we can replace y with $x + 2$ in the first equation.

$$4x + y = 4 \qquad \text{first equation}$$
$$4x + (x + 2) = 4 \qquad \text{substitute } x + 2 \text{ for } y$$
$$5x + 2 = 4 \qquad \text{simplify}$$
$$x = \frac{2}{5} \qquad \text{result}$$

The x-coordinate is $\frac{2}{5}$. To find the y-coordinate, substitute $\frac{2}{5}$ for x into either of the original equations. Substituting in the second equation gives

$$y = x + 2 \qquad \text{second equation}$$
$$= \frac{2}{5} + 2 \qquad \text{substitute } \frac{2}{5} \text{ for } x$$
$$= \frac{12}{5} \qquad \frac{2}{1} = \frac{10}{5}, \frac{10}{5} + \frac{2}{5} = \frac{12}{5}$$

The solution to the system is $\left(\frac{2}{5}, \frac{12}{5}\right)$. Verify by substituting $\frac{2}{5}$ for x and $\frac{12}{5}$ for y into both equations.

Now try Exercises 23 through 32 ▶

> **❝** I thought the author did a good job of explaining the content by using examples, because there was an example of every kind of problem.**❞**
> —*Brittney Pruitt, student class tester at Metropolitan Community College–Longview*

> **❝** I particularly like the 'Now Try exercises . . .' after each group of examples. I have not seen this in other texts and it is a really nice addition. I usually tell my students which examples correspond to which exercises, so this will save time and effort on my part.**❞**
> —*Scott Berthiaume, Edison State College*

GRAPHICAL SUPPORT

Graphing the lines from Example 8 as Y1 and Y2 on a graphing calculator, we note the lines do appear to be parallel (they actually *must* be since they have identical slopes). Using the ZOOM **8:ZInteger** feature of the TI-84 Plus we can quickly verify that Y2 indeed contains the point $(-6, -1)$.

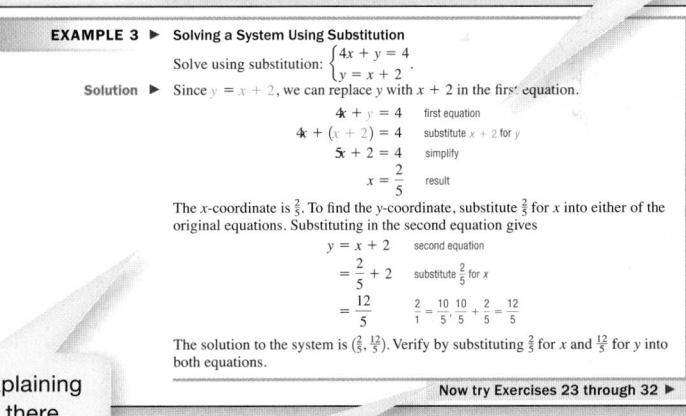

> **❝** The incorporation of technology and graphing calculator usage . . . is excellent. For the faculty that do not use the technology it is easily skipped. It is very detailed for the students or faculty that [do] use technology.**❞**
> —*Rita Marie O'Brien, Navarro College*

Solid Skill Development...

Through Exercises

I have included a wealth of exercises in support of each section's main ideas. I constructed each set with great care, in an effort to provide strong support for weaker students, while challenging advanced students to reach even further. I also designed the various exercises to support instructors in their teaching endeavors—the quantity and quality of the exercises allow for numerous opportunities to guide students through difficult calculations, and to illustrate important problem-solving techniques.—JC

Mid-Chapter Checks

Mid-Chapter Checks provide students with a good stopping place to assess their knowledge before moving on to the second half of the chapter.

End-of-Section Exercise Sets

▶ **Concepts and Vocabulary** exercises to help students recall and retain important terms.

▶ **Developing Your Skills** exercises to provide practice of relevant concepts just learned with increasing levels of difficulty.

❝Some of our instructors would mainly assign the developing your skills and working with formula problems, however, I would focus on the writing, research and decision making [in] extending the concept. The flexibility is one of the things I like about the Coburn text.❞
—*Sherry Meier, Illinois State University*

▶ **Working with Formulas** exercises to demonstrate contextual applications of well-known formulas.

▶ **Extending the Concept** exercises that require communication of topics, synthesis of related concepts, and the use of higher-order thinking skills.

▶ **Maintaining Your Skills** exercises that address skills from previous sections to help students retain previously learning knowledge.

❝He not only has exercises for skill development, but also problems for 'extending the concept' and 'maintaining your skills,' which our current text does not have. I also like the mid-chapter checks provided. All these give Coburn an advantage in my view.❞
—*Randy Ross, Morehead State University*

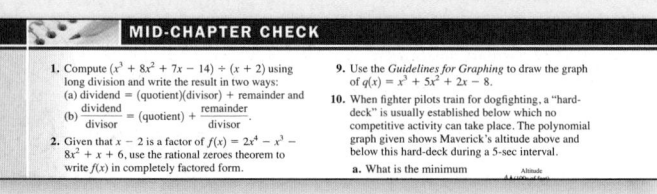

MID-CHAPTER CHECK

1. Compute $(x^3 + 8x^2 + 7x - 14) \div (x + 2)$ using long division and write the result in two ways:
   (a) dividend = (quotient)(divisor) + remainder and
   (b) $\dfrac{\text{dividend}}{\text{divisor}} = (\text{quotient}) + \dfrac{\text{remainder}}{\text{divisor}}$.
2. Given that $x - 2$ is a factor of $f(x) = 2x^4 - x^3 - 8x^2 + x + 6$, use the rational zeroes theorem to write $f(x)$ in completely factored form.
9. Use the *Guidelines for Graphing* to draw the graph of $q(x) = x^3 + 5x^2 + 2x - 8$.
10. When fighter pilots train for dogfighting, a "hard-deck" is usually established below which no competitive activity can take place. The polynomial graph given shows Maverick's altitude above and below this hard-deck during a 5-sec interval.
   a. What is the minimum

1.3 EXERCISES

▶ CONCEPTS AND VOCABULARY

Fill in the blank with the appropriate word or phrase. Carefully reread the section if needed.

1. When multiplying or dividing by a negative quantity, we _____ the inequality to maintain a true statement.
2. To write an absolute value equation or inequality in simplified form, we _____ the absolute value

4. The absolute value inequality $|3x - 6| < 12$ is true when $3x - 6 >$ _____ and $3x - 6 <$ _____.

Describe each solution set (assume $k > 0$). Justify your answer.

5. $|ax + b| < -k$

▶ DEVELOPING YOUR SKILLS

Solve each absolute value equation. Write the solution in set notation.

7. $2|m - 1| - 7 = 3$
8. $3|n - 5| - 14 = -2$
9. $-3|x + 5| + 6 = -15$
10. $-2|y + 3| - 4 = -14$
11. $2|4v + 5| - 6.5 = 10.3$
12. $7|2w + 5| + 6.3 = 11.2$
13. $-|7p - 3| + 6 = -5$
14. $-|3q + 4| + 3 = -5$

Solve each absolute value inequality. Write solutions in interval notation.

25. $|x - 2| \le 7$
26. $|y + 1| \le 3$
27. $-3|m| - 2 > 4$
28. $-2|n| + 3 > 7$
29. $\dfrac{|5v + 1|}{4} + 8 < 9$
30. $\dfrac{|3w - 2|}{2} + 6 < 8$
31. $3|p + 4| + 5 \le 8$
32. $5|q - 2| - 7 \le 8$
33. $|3b - 11| + 6 \le 9$
34. $|2c + 3| - 5 < 1$
35. $|4 - 3z| + 12 < 7$
36. $|2 - 7u| + 7 \le 4$
37. $\left|\dfrac{4x + 5}{2} - \dfrac{1}{2}\right| \le \dfrac{7}{2}$
38. $\left|\dfrac{2y - 3}{4} - \dfrac{3}{8}\right| < \dfrac{15}{16}$

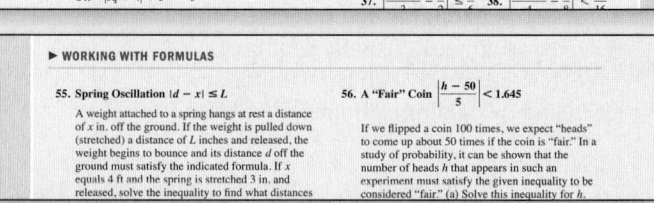

▶ WORKING WITH FORMULAS

55. **Spring Oscillation** $|d - x| \le L$

A weight attached to a spring hangs at rest a distance of x in. off the ground. If the weight is pulled down (stretched) a distance of L inches and released, the weight begins to bounce and its distance d off the ground must satisfy the indicated formula. If x equals 4 ft and the spring is stretched 3 in. and released, solve the inequality to find what distances

56. **A "Fair" Coin** $\left|\dfrac{h - 50}{5}\right| < 1.645$

If we flipped a coin 100 times, we expect "heads" to come up about 50 times if the coin is "fair." In a study of probability, it can be shown that the number of heads h that appears in such an experiment must satisfy the given inequality to be considered "fair." (a) Solve this inequality for h.

▶ EXTENDING THE CONCEPT

67. Determine the value or values (if any) that will make the equation or inequality true.
   a. $|x| + x = 8$
   b. $|x - 2| \le \dfrac{x}{2}$
   c. $x - |x| = x + |x|$
   d. $|x + 3| \ge 6x$
   e. $|2x + 1| = x - 3$

68. The equation $|5 - 2x| = |3 + 2x|$ has only one solution. Find it and explain why there is only one.

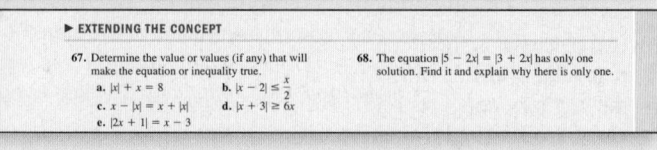

▶ MAINTAINING YOUR SKILLS

69. (R.4) Factor the expression completely:
   $18x^3 + 21x^2 - 60x$.

70. (1.1) Solve $V^2 = \dfrac{2W}{C\rho A}$ for ρ (physics).

72. (1.2) Solve the inequality, then write the solution set in interval notation:
   $-3(2x - 5) > 2(x + 1) - 7$.

❝ The strongest feature seems to be the wide variety of exercises included at the end of each section. There are plenty of drill problems along with good applications.❞
—*Jason Pallett, Metropolitan Community College–Longview*

End-of-Chapter Review Material

Exercises located at the end of the chapter provide students with the tools they need to prepare for a quiz or test. Each chapter features the following:

▶ **Chapter Summary and Concept Reviews** that present key concepts with corresponding exercises by section in a format easily used by students.

▶ **Mixed Reviews** that offer more practice on topics from the entire chapter, arranged in random order requiring students to identify problem types and solution strategies on their own.

▶ **Practice Tests** that give students the opportunity to check their knowledge and prepare for classroom quizzes, tests, and other assessments.

> **"** We always did reviews and a quiz before the actual test; it helped a lot.**"**
> —*Melissa Cowan, student class tester Metropolitan Community College–Longview*

▶ **Cumulative Reviews** that are presented at the end of each chapter help students retain previously learned skills and concepts by revisiting important ideas from earlier chapters (starting with Chapter 2).

> **"** The cumulative review is very good and is considerably better than some of the books I have reviewed/used. I have found these to be wonderful practice for the final exam.**"**
> —*Sarah Clifton, Southeastern Louisiana University*

▶ 🖩 **Graphing Calculator** icons appear next to exercises where important concepts can be supported by the use of graphing technology.

> **"** The summary and concept review was very helpful because it breaks down each section. That is what helps me the most.**"**
> —*Brittany Pratt, student class tester at Baton Rouge Community College*

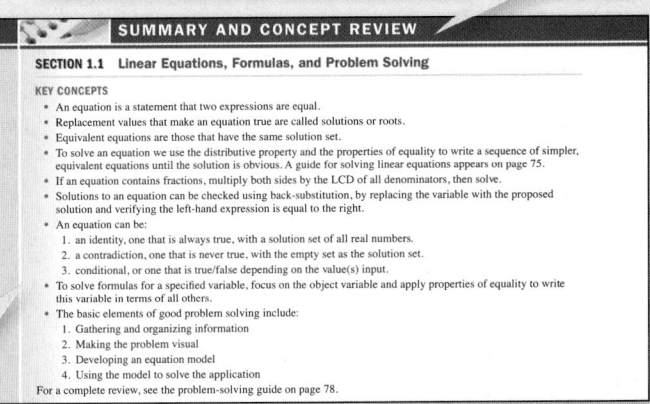

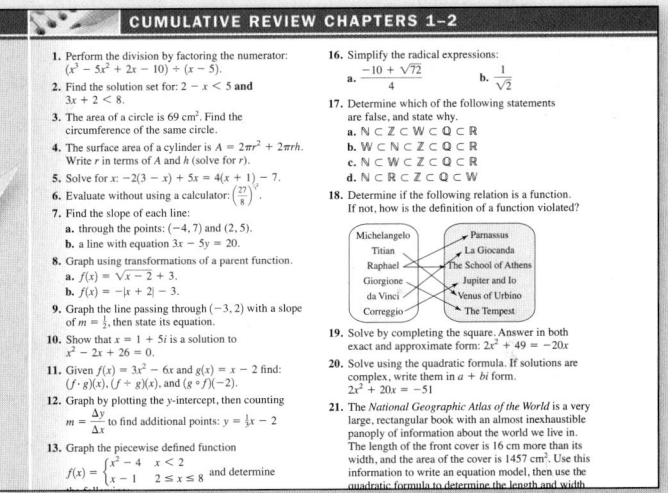

Homework Selection Guide

A list of suggested homework exercises has been provided for each section of the text (Annotated Instructor's Edition only). This feature may prove especially useful for departments that encourage consistency among many sections, or those having a large adjunct population. The feature was also designed as a convenience to instructors, enabling them to develop an inventory of exercises that is more in tune with the course as they like to teach it. The Guide provides prescreened and preselected assignments at four different levels: *Core, Standard, Extended,* and *In Depth.*

- **Core:** These assignments go right to the heart of the material, offering a minimal selection of exercises that cover the primary concepts and solution strategies of the section, along with a small selection of the best applications.
- **Standard:** The assignments at this level include the *Core* exercises, while providing for additional practice without excessive drill. A wider assortment of the possible variations on a theme are included, as well as a greater variety of applications.
- **Extended:** Assignments from the *Extended* category expand on the *Standard* exercises to include more applications, as well as some conceptual or theory-based questions. Exercises may include selected items from the *Concepts and Vocabulary, Working with Formulas,* and the *Extending* the Thought categories of the exercise sets.
- **In Depth:** The *In Depth* assignments represent a more comprehensive look at the material from each section, while attempting to keep the assignment manageable for students. These include a selection of the most popular and highest-quality exercises from each category of the exercise set, with an additional emphasis on *Maintaining Your Skills.*

Strong Connections...

Through a Conversational Writing Style

While examples and applications are arguably the most prominent features of a mathematics text, it's the writing style and readability that binds them together. It may be true that some students don't read the text, and that others open the text only when looking for an example similar to the exercise they're currently working. But when they do and for those students who do (read the text), it's important they have a text that "speaks to them," relating concepts in a form and at a level they understand and can relate to. Ideally this text will draw students in and keep their interest, becoming a positive experience and bringing them back a second and third time, until it becomes habitual. At this point, students might begin to see the true value of their text (as more that just a source of problems—pun intended), and it becomes a resource for learning on equal footing with any other form of supplemental instruction.   —JC

Conversational Writing Style

John Coburn's experience in the classroom and his strong connections to how students comprehend the material are evident in his writing style. He uses a conversational and supportive writing style, providing the students with a tool they can depend on when the teacher is not available, when they miss a day of class, or simply when working on their own. The effort John has put into the writing is representative of his unofficial mantra: "If you want more students to reach the top, you gotta put a few more rungs on the ladder."

> " The author does a fine job with his narrative. His explanations are very clear and concise. I really like his explanations better than in my current text. "
> —Tammy Potter, Gadsden State College

> " The author does an excellent job of engagement and it is easily seen that he is conscious of student learning styles. "
> —Conrad Krueger, San Antonio College

Through Student Involvement

How do you design a student-friendly textbook? We decided to get students involved by hosting two separate focus groups. During these sessions we asked students to advise us on how they use their books,

what pedagogical elements are useful, which elements are distracting and not useful, as well as general feedback on page layout. During this process there were times when we thought, "Now why hasn't anyone ever thought of that before?" Clearly these student focus groups were invaluable. Taking direct student feedback and incorporating what is feasible and doesn't detract from instructor use of the text is the best way to design a truly student-friendly text. The next two pages will highlight what we learned from students so you can see for yourself how their feedback played an important role in the development of the Coburn series.

Students said that **Learning Objectives** should clearly define the goals of each section.

Students asked for **Check Points** throughout each section to alert them when a specific learning objective has been covered and to reinforce the use of correct mathematical terms.

Described by students as one of the most useful features in a math text, **Caution Boxes** signal a student to stop and take note in order to avoid mistakes in problem solving.

Students told us that the color red should only be used for things that are really important. Also, anything significant should be included in the body of the text; marginal readings imply optional.

Examples are called out in the margins so they are easy for students to spot.

Examples are "boxed" so students can clearly see where they begin and end

Students told us they liked when the examples were linked to the exercises.

1.1 | Linear Equations, Formulas, and Problem Solving

Learning Objectives

In Section 1.1 you will learn how to:

☐ **A.** Solve linear equations using properties of equality

☐ **B.** Recognize equations that are identities or contradictions

☐ **C.** Solve for a specified variable in a formula or literal equation

☐ **D.** Use the problem-solving guide to solve various problem types

In a study of algebra, you will encounter many **families of equations,** or groups of equations that share common characteristics. Of interest to us here is the family of **linear equations in one variable,** a study that lays the foundation for understanding more advanced families. In addition to *solving* linear equations, we'll use the skills we develop to *solve for a specified variable* in a formula, a practice widely used in science, business, industry, and research.

A. Solving Linear Equations Using Properties of Equality

An **equation** is *a statement that two expressions are equal.* From the expressions $3(x - 1) + x$ and $-x + 7$, we can form the equation

$$3(x - 1) + x = -x + 7.$$

which is a **linear equation in one variable.** To solve an equation, we attempt to find a specific input or x-value that will make the equation true, meaning the left-hand expression will be equal to the right. Using

Table 1.1

x	$3(x - 1) + x$	$-x + 7$
-2	-11	9
-1	-7	8
0	-3	7
1	1	6
2	5	5

EXAMPLE 2 ▶ Solving a Linear Equation with Fractional Coefficients

Solve for n: $\frac{1}{4}(n + 8) - 2 = \frac{1}{2}(n - 6)$.

Solution ▶

$\frac{1}{4}(n + 8) - 2 = \frac{1}{2}(n - 6)$	original equation
$\frac{1}{4}n + 2 - 2 = \frac{1}{2}n - 3$	distributive property
$\frac{1}{4}n = \frac{1}{2}n - 3$	combine like terms
$4(\frac{1}{4}n) = 4(\frac{1}{2}n - 3)$	multiply both sides by LCD = 4
$n = 2n - 12$	distributive property
$-n = -12$	subtract $2n$
$n = 12$	multiply by -1

☑ **A.** You've just learned how to solve linear equations using properties of equality

Verify the solution is $n = 12$ using back-substitution.

Now try Exercises 13 through 30 ▶

$$= \frac{3}{6} = \frac{1}{2} \qquad\qquad = \frac{-4}{6} = \frac{-2}{3}$$

The slope of this line is $\frac{1}{2}$. The slope of this line is $\frac{-2}{3}$.

Now try Exercises 33 through 40 ▶

⚠ **CAUTION** ▶ When using the slope formula, try to avoid these common errors.

1. The order that the x- and y-coordinates are subtracted must be consistent, since $\frac{y_2 - y_1}{x_2 - x_1} \neq \frac{y_2 - y_1}{x_1 - x_2}$.

2. The vertical change (involving the y-values) always occurs in the numerator: $\frac{y_2 - y_1}{x_2 - x_1} \neq \frac{x_2 - x_1}{y_2 - y_1}$.

3. When x_1 or y_1 is negative, use parentheses when substituting into the formula to prevent confusing the negative sign with the subtraction operation.

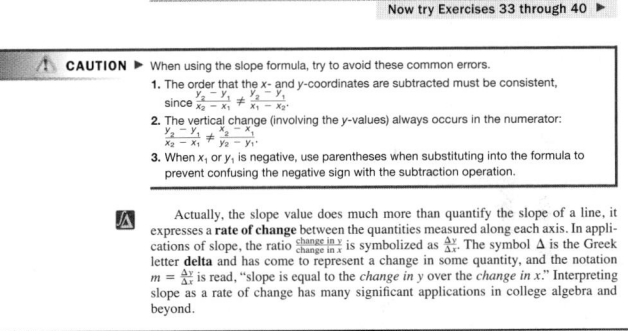

Actually, the slope value does much more than quantify the slope of a line, it expresses a **rate of change** between the quantities measured along each axis. In applications of slope, the ratio $\frac{\text{change in } y}{\text{change in } x}$ is symbolized as $\frac{\Delta y}{\Delta x}$. The symbol Δ is the Greek letter **delta** and has come to represent a change in some quantity, and the notation $m = \frac{\Delta y}{\Delta x}$ is read, "slope is equal to the *change in y* over the *change in x*." Interpreting slope as a rate of change has many significant applications in college algebra and beyond.

EXAMPLE 8 ▶ Determining the Domain of an Expression

Determine the domain of the expression $\frac{6}{x - 2}$. State the result in set notation, graphically, and using interval notation.

Solution ▶ Set the denominator equal to zero and solve: $x - 2 = 0$ yields $x = 2$. This means 2 is outside the domain and *must be excluded*.

- Set notation: $\{x | x \in \mathbb{R}, x \neq 2\}$
- Graph:
- Interval notation: $x \in (-\infty, 2) \cup (2, \infty)$

Now try Exercises 61 through 68 ▶

A second area where allowable values are a concern involves the square root operation. Recall that $\sqrt{49} = 7$ since $7 \cdot 7 = 49$. However, $\sqrt{-49}$ cannot be written as the product of two real numbers since $(-7) \cdot (-7) = 49$ and $7 \cdot 7 = 49$. In other words, $\sqrt{X}$ represents a real number only if the radicand is positive or zero. If X represents an algebraic expression, the domain of $\sqrt{X}$ is $\{X | X \geq 0\}$.

EXAMPLE 9 ▶ Determining the Domain of an Expression

Determine the domain of $\sqrt{x + 3}$. State the domain in set notation, graphically, and in interval notation.

Solution ▶ The radicand must represent a nonnegative number. Solving $x + 3 \geq 0$ gives $x \geq -3$.

- Set notation: $\{x | x \geq -3\}$
- Graph:
- Interval notation: $x \in [-3, \infty)$

Now try Exercises 69 through 76 ▶

Students told us that directions should be in bold so they are easily distinguishable from the problems.

Because students spend a lot of time in the exercise section of a text, they said that a white background is hard on their eyes…so we used a soft, off-white color for the background

Students said having a lot of icons was confusing. The graphing calculator is the only icon used in the exercise sets; no unnecessary icons are used

Solve using the zero product property. Be sure each equation is in standard form and factor out any common factors before attempting to solve. Check all answers in the original equation.

7. $22x = x^3 - 9x^2$

8. $x^3 = 13x^2 - 42x$

9. $3x^3 = -7x^2 + 6x$

10. $7x^2 + 15x = 2x^3$

11. $2x^4 - 3x^3 = 9x^2$

12. $-7x^2 = 2x^4 - 9x^3$

13. $2x^4 - 16x = 0$

14. $x^4 + 64x = 0$

15. $x^3 - 4x = 5x^2 - 20$

16. $x^3 - 18 = 9x - 2x^2$

17. $4x - 12 = 3x^2 - x^3$

18. $x - 7 = 7x^2 - x^3$

19. $2x^3 - 12x^2 = 10x - 60$

20. $9x + 81 = 27x^2 + 3x^3$

21. $x^4 - 7x^3 + 4x^2 = 28x$

22. $x^4 + 3x^3 + 9x^2 = -27x$

23. $x^4 - 81 = 0$

24. $x^4 - 1 = 0$

25. $x^4 - 256 = 0$

26. $x^4 - 625 = 0$

27. $x^6 - 2x^4 - x^2 + 2 = 0$

28. $x^6 - 3x^4 - 16x^2 + 48 = 0$

29. $x^5 - x^3 - 8x^2 + 8 = 0$

30. $x^5 - 9x^3 - x^2 + 9 = 0$

31. $x^6 - 1 = 0$

32. $x^6 - 64 = 0$

Solve each equation. Identify any extraneous roots.

33. $\frac{2}{x} + \frac{1}{x+1} = \frac{5}{x^2+x}$

34. $\frac{3}{m+3} - \frac{5}{m^2+3m} = \frac{1}{m}$

35. $\frac{21}{a+2} = \frac{3}{a-1}$

36. $\frac{4}{2y-3} = \frac{7}{3y-5}$

39. $x + \frac{14}{x-7} = 1 + \frac{2x}{x-7}$

40. $\frac{10}{x-5} + x = 1 + \frac{2x}{x-5}$

41. $\frac{6}{n+3} + \frac{20}{n^2+n-6} = \frac{5}{n-2}$

42. $\frac{7}{p+2} - \frac{1}{p^2+5p+6} = \frac{2}{p+3}$

43. $\frac{a}{2a+1} - \frac{2a^2+5}{2a^2-5a-3} = \frac{3}{a-3}$

44. $\frac{-18}{6n^2-n-1} + \frac{3n}{2n-1} = \frac{4n}{3n+1}$

Solve for the variable indicated.

45. $\frac{1}{f} = \frac{1}{f_1} + \frac{1}{f_2}$; for f

46. $\frac{1}{x} - \frac{1}{y} = \frac{1}{z}$; for z

47. $I = \frac{E}{R+r}$; for r

48. $q = \frac{pf}{p-f}$; for p

49. $V = \frac{1}{3}\pi r^2 h$; for h

50. $s = \frac{1}{2}gt^2$; for g

51. $V = \frac{4}{3}\pi r^3$; for r^3

52. $V = \frac{1}{3}\pi r^2 h$; for r^2

Solve each equation and check your solutions by substitution. Identify any extraneous roots.

53. a. $-3\sqrt{3x-5} = -9$ **b.** $x = \sqrt{3x+1} + 3$

54. a. $-2\sqrt{4x-1} = -10$ **b.** $-5 = \sqrt{5x-1} - x$

55. a. $2 = \sqrt[3]{3m-1}$ **b.** $2\sqrt[3]{7-3x} - 3 = -7$

c. $\frac{\sqrt[3]{2m+3}}{-5} + 2 = 3$ **d.** $\sqrt[3]{2x-9} = \sqrt[3]{3x+7}$

56. a. $-3 = \sqrt[3]{5p+2}$ **b.** $3\sqrt[3]{3-4x} - 7 = -4$

c. $\frac{\sqrt[3]{6x-7}}{4} - 5 = -6$

d. $3\sqrt[3]{x+3} = 2\sqrt[3]{2x+17}$

57. a. $\sqrt{x-9} + \sqrt{x} = 9$

b. $x = 3 + \sqrt{23-x}$

c. $\sqrt{x-2} - \sqrt{2x} = -2$

88. Composite figures—gelatin capsules: The gelatin capsules manufactured for cold and flu medications are shaped like a cylinder with a hemisphere on each end. The interior volume V of each capsule can be modeled by $V = \frac{4}{3}\pi r^3 + \pi r^2 h$, where h is the height of the cylindrical portion and r is its radius. If the cylindrical portion of the capsule is 8 mm long ($h = 8$ mm), what radius would give the capsule a volume that is numerically equal to 15π times this radius?

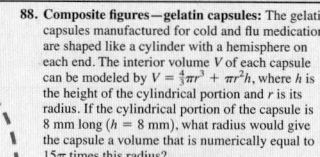

89. Running shoes: When a popular running shoe is priced at $70, The Shoe House will sell 15 pairs each week. Using a survey, they have determined that for each decrease of $2 in price, 3 additional pairs will be sold each week. What selling price will give a weekly revenue of $2250?

90. Cell phone charges: A cell phone service sells 48 subscriptions each month if their monthly fee is $30. Using a survey, they find that for each decrease of $1, 6 additional subscribers will join. What charge(s) will result in a monthly revenue of $2160?

Projectile height: In the absence of resistance, the height of an object that is projected upward can be modeled by the equation $h = -16t^2 + vt + k$, where h represents the height of the object (in feet) t sec after it has been thrown, v represents the initial velocity (in feet per second), and k represents the height of the object when $t = 0$ (before it has

velocity of 100 ft/sec and a height of 240 ft, it runs out of fuel and becomes a projectile.

a. How high is the rocket three seconds later? Four seconds later?

b. How long will it take the rocket to attain a height of 640 ft?

c. How many times is a height of 384 ft attained? When do these occur?

d. How many seconds until the rocket returns to the ground?

93. Printing newspapers: The editor of the school newspaper notes the college's new copier can complete the required print run in 20 min, while the back-up copier took 30 min to do the same amount of work. How long would it take if both copiers are used?

94. Filling a sink: The cold water faucet can fill a sink in 2 min. The drain can empty a full sink in 3 min. If the faucet were left on and the drain was left open, how long would it take to fill the sink?

95. Triathalon competition: As one part of a Mountain-Man triathalon, participants must row a canoe 5 mi down river (with the current), circle a buoy and row 5 mi back up river (against the current) to the starting point. If the current is flowing at a steady rate of 4 mph and Tom Chaney made the round-trip in 3 hr, how fast can he row in still water? (*Hint:* The time rowing down river and the time rowing up river must add up to 3 hr.)

96. Flight time: The flight distance from Cincinnati, Ohio, to Chicago, Illinois, is approximately 300 mi. On a recent round-trip between these cities in my private plane, I encountered a steady 25 mph headwind on the way to Chicago, with a 25 mph tailwind on the return trip. If my total flying time

▶ **WORKING WITH FORMULAS**

 79. Lateral surface area of a cone: $S = \pi r\sqrt{r^2 + h^2}$

The lateral surface area (surface area excluding the base) S of a cone is given by the formula shown, where r is the radius of the base and h is the height of the cone. (a) Solve the equation for h. (b) Find the surface area of a cone that has a radius of 6 m and a height of 10 m. Answer in simplest form.

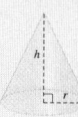

80. Painted area on a canvas: $A = \frac{4x^2 + 60x + 104}{x}$

A rectangular canvas is to contain a small painting with an area of 52 in², and requires 2-in. margins on the left and right, with 1-in. margins on the top and bottom for framing. The total area of such a canvas is given by the formula shown, where x is the height of the *painted* area.

a. What is the area A of the canvas if the height of the painting is $x = 10$ in.?

b. If the area of the canvas is $A = 120$ in², what are the dimensions of the painted area?

Connections to Calculus

Like many of you who have taught calculus, I've often been left with a strong sense that more could be done in Precalculus, to prepare our students for calculus. To this end I've included a feature called Connections to Calculus, which is designed to more closely tie their experience in PreCalculus to their future studies. This end-of-chapter feature provides a timely choice of topics from within each chapter to illustrate their use in a calculus context, giving students a more solid footing as they step up to the calculus level.   —JC

The *Connections to Calculus* feature is included at the end of Chapters 1–10 to highlight the connections between algebraic concepts presented in this course, and the calculus concepts to be learned in a later course. This feature includes exposition, examples, and exercises to reinforce the material. Each chapter opener provides a preview of the *Connection to Calculus* coming at the end of the chapter.

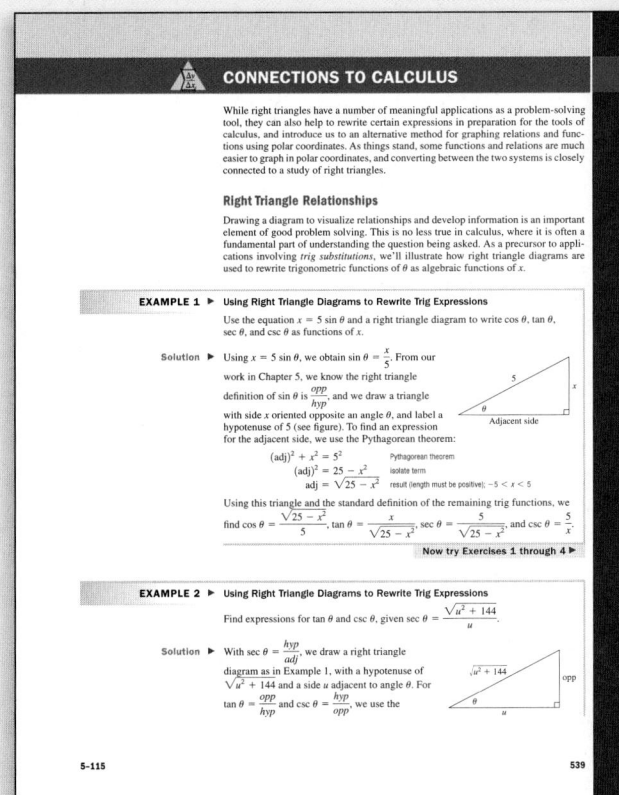

An Introduction to Trigonometric Functions

CHAPTER OUTLINE

5.1 Angle Measure, Special Triangles, and Special Angles 426
5.2 Unit Circles and the Trigonometry of Real Numbers 440
5.3 Graphs of the Sine and Cosine Functions; Cosecant and Secant Functions 454
5.4 Graphs of Tangent and Cotangent Functions 474
5.5 Transformations and Applications of Trigonometric Graphs 486
5.6 The Trigonometry of Right Triangles 499
5.7 Trigonometry and the Coordinate Plane 512

CHAPTER CONNECTIONS

While rainbows have been admired for centuries for their beauty and color, understanding the physics of rainbows is of fairly recent origin. Answers to questions regarding their seven color constitution, the order the colors appear, the circular shape of the bow, and their relationship to moisture all have answers deeply rooted in mathematics. The relationship between light and color can be understood in terms of trigonometry, with questions regarding the apparent height of the rainbow, as well as the height of other natural and man-made phenomena, found using the trigonometry of right triangles.

▶ This application appears as Exercise 85 in Section 5.6

The trigonometry of right triangles also plays an important role in a study of calculus, as we seek to simplify certain expressions, or convert from one system of graphing to another. The *Connections to Calculus* feature following Chapter 5 offers additional practice and insight in preparation for a study of calculus.

425

" I think the approach of giving them a taste of what is to come in calculus and to get them familiar with some notation is the correct one and hopefully it will motivate them to study the algebra in the chapters a little more.**"** —*Terry Hobbs, Metropolitan Community College, Maple Woods*

Making Connections...

Through New and Updated Content

New to the Second Edition

▶ An extensive reworking of the narrative and reduction of advanced concepts enhances the clarity of the exposition, improves the student's experience in the text, and decreases the overall length of the text.

▶ A modified interior design based on student and instructor feedback from focus groups features increased font size, improved exercise and example layout, more white space on the page, and the careful use of color to enhance the presentation of pedagogy.

▶ Chapter Openers based on applications bring awareness to students of the relevance of concepts presented in each chapter, and provide a preview of the connections to calculus feature at the end of the chapter.

▶ Checkpoints throughout each section alert students when a specific learning objective has been covered and reinforce the use of correct mathematical terms.

▶ The Homework Selection Guide, appearing in each exercise section in the Annotated Instructor's Edition, provides instructors with suggestions for developing core, standard, extended, and in-depth homework assignments without much prep work.

▶ A *Connections to Calculus* feature at the end of each chapter to highlight the connections between the algebraic concepts presented, and the calculus concepts to be learned in a future course.

Chapter-by-Chapter Changes

CHAPTER 1 Equations and Inequalities

• Chapter 1 now includes coverage of absolute value equations and inequalities in Section 1.3.
• Information on solving quadratics has been consolidated to a single section (1.5) and summary boxes are now used for solving linear equations, solving quadratic equations, and solution methods for quadratic equations.
• Examples and exercises employing the use of a graphing calculator have been added throughout the chapter.

CHAPTER 2 Relations, Functions, and Graphs

• The organization of Chapter 2 has changed from the first edition in an effort to concentrate the introduction of graphs and general functions.
• Coverage of the midpoint formula, the distance formula, and circles has been improved and reorganized (Section 2.1).
• Linear graphs are established early in the chapter (Sections 2.2 and 2.3) before functions are introduced.
• The section on the toolbox (basic parent) functions (2.6) now appears after analyzing graphs (Section 2.5) to improve connections among the material.
• Coverage of rates of change has been consolidated while coverage of the implied domain, distance quotient, end behavior, and even/odd functions has been expanded and improved.
• Additional applications of the floor and ceiling functions and the algebra of functions have been added.

CHAPTER 3 Polynomial and Rational Functions

• Chapter 3 has been significantly reorganized to bring focus to and provide a better bridge between general functions and polynomial functions.
• More coverage of completing the square and increased emphasis on graphing using the vertex formula is found in Section 3.1.
• Complex conjugates, zeroes of multiplicity, and number of zeroes have been realigned together in Section 3.2.
• Section 3.3 features an improved description of Descartes' rule of signs, as well as stronger connections between the fundamental theorem of algebra and the linear factorization theorem and its corollaries.
• A better introduction regarding polynomials versus nonpolynomials is found in Section 3.4, in addition to an improved discussion of end behavior.
• Section 3.6 provides better treatment of removable discontinuities and a clearer discussion of pointwise versus asymptotic continuities.
• Section 3.8 presents a clearer, stronger connection between previously covered topics and applications of variation and the toolbox functions.

CHAPTER 4 Exponential and Logarithmic Functions

• Chapter 4 now begins with coverage of one-to-one and inverse functions given their applications for exponents and logarithms.
• This section (4.1) includes examples of finding inverses of rational functions, as well as better coverage of restricting the domain to find the inverse.

- Coverage of base e as an alternative to base 10 or b is addressed in one section (4.2) as opposed to two sections as in the first edition.
- Likewise, coverage of properties of logs and log equations is found in the same section (4.4).
- A clear introduction to fundamental logarithmic properties has also been added to Section 4.4.
- Applications have been added and improved throughout the chapter.

CHAPTER 5 Introduction to Trigonometric Functions

- Section 5.1 includes improved DMS to decimal degrees conversion coverage, improved introduction to standard 45-45-90 and 30-60-90 triangles, better illustrations of longitude and latitude applications, and streamlined/clarified coverage of angular and linear velocity.
- Section 5.2 includes a table showing summary of trig functions of special angles.
- Section 5.3 has improved coverage of secant and cosecant graphs.
- Section 5.4 has a strengthened connection between $y = \tan x$ and $y = (\sin x)/\cos(x)$.
- Section 5.5 has an improved introduction to transformations, and a clearer distinction between phase angle and phase shift.
- Section 5.6 has improved coverage of co-functions, and better illustrations for angles of elevation and depression.
- Section 5.7 has improved applications, and the connection between f and f^{-1} is introduced.

CHAPTER 6 Trigonometric Identities, Inverses, and Equations

- Section 6.1 has an increased emphasis on what an identity is (the definition of an identity), as well as an additional example of quadrant and sign analysis.
- Section 6.2 has a better introduction to clarify goals, as well as an improved format for verifying identities.
- Section 6.3 has improved coverage of the co-function identities, as well as extended coverage of the sum and difference identities.
- Section 6.5 has a strengthened connection between inverse functions and drawn diagrams, improved coverage on evaluating the inverse trig functions, and more real-world applications of inverse trig functions.

CHAPTER 7 Applications of Trigonometry

- Section 7.1 has consolidated coverage of the ambiguous case.
- Section 7.2 has expanded coverage of computing areas using trig, as well as six new contextual applications of triangular area using trig.
- Section 7.3 has improved discussion, coverage, and illustrations of vector subtraction, and stronger connections between solutions using components, and solutions using the law of cosines.
- Section 7.5 has additional real-world applications of complex numbers (AC circuits).

CHAPTER 8 Systems of Equations and Inequalities

- Section 8.1 includes improved coverage of equivalent systems in addition to more examples and exercises having to do with distance and navigation.
- Section 8.2 features improved coverage of dependent and inconsistent systems.
- Section 8.3 has improved coverage of partial fractions.
- New applications of linear programming are found in Section 8.4.
- Section 8.5 features an added example of Gauss-Jordan Elimination.
- Section 8.6 includes better sequencing of examples and improved coverage of matrix properties.
- Coverage of determinants has been streamlined with more development given to determinants in Section 8.8.

CHAPTER 9 Analytical Geometry

- Section 9.1 presents a brief introduction to analytical geometry to provide a better bridge to the conic sections and show why cone/conic connection is important.
- Greater emphasis on the connection between ellipses and circles is featured in section 9.2.
- Exercises requiring the movement from graph to equation have been added throughout the chapter.
- Section 9.5 now covers non-linear systems to include parabolas.

CHAPTER 10 Additional Topics in Algebra

- The exposition has been revised throughout Chapter 10 for increased clarity and improved flow of topics.

CHAPTER 11 Bridge to Calculus: An Introduction to Limits

- New applications-based chapter on limits.
- Addresses the question of *why* limits are important.
- Close connections made between current (new) ideas and previous material.
- Information and concepts come in small, understandable increments.
- Includes extensive exercise sets, with numerous applications.
- Includes optional (online) section with an introduction to the precise definition of a limit.

Through 360° Development

McGraw-Hill's 360° Development Process is an ongoing, never-ending, market-oriented approach to building accurate and innovative print and digital products. It is dedicated to continual large-scale and incremental improvement driven by multiple customer feedback loops and checkpoints. This process is initiated during the early planning stages of our new products, intensifies during the development and production stages, and then begins again on publication, in anticipation of the next edition.

A key principle in the development of any mathematics text is its ability to adapt to teaching specifications in a universal way. The only way to do so is by contacting those universal voices—and learning from their suggestions. We are confident that our book has the most current content the industry has to offer, thus pushing our desire for accuracy to the highest standard possible. In order to accomplish this, we have moved through an arduous road to production. Extensive and open-minded advice is critical in the production of a superior text.

We engaged over 400 instructors and students to provide us guidance in the development of the second edition. By investing in this extensive endeavor, McGraw-Hill delivers to you a product suite that has been created, refined, tested, and validated to be a successful tool in your course.

Board of Advisors

A hand-picked group of trusted teachers active in the College Algebra and Precalculus course areas served as the chief advisors and consultants to the author and editorial team with regards to manuscript development. The Board of Advisors reviewed the manuscript in two drafts; served as a sounding board for pedagogical, media, and design concerns; approved organizational changes; and attended a symposium to confirm the manuscript's readiness for publication.

Bill Forrest, *Baton Rouge Community College*
Marc Grether, *University of North Texas*
Sharon Hamsa, *Metropolitan Community College –Longview*
Max Hibbs, *Blinn College*
Terry Hobbs, *Metropolitan Community College– Maple Woods*
Klay Kruczek, *Western Oregon University*
Rita Marie O'Brien's , *Navarro College*

Nancy Matthews, *University of Oklahoma*
Rebecca Muller, *Southeastern Louisiana University*
Jason Pallett, *Metropolitan Community College*
Kevin Ratliff, *Blue Ridge Community College*
Stephen Toner, *Victor Valley College*

Accuracy Panel

A selected trio of key instructors served as the chief advisors for the accuracy and clarity of the text and solutions manual. These individuals reviewed the final manuscript, the page proofs in first and revised rounds, as well as the writing and accuracy check of the instructor's solutions manuals. This trio, in addition to several other accuracy professionals, gives you the assurance of accuracy.

J.D. Herdlick, *St. Louis Community College–Meramac*
Richard A. Pescarino, *St. Louis Community College–Florissant Valley*
Nathan G. Wilson, *St. Louis Community College–Meramac*

Student Focus Groups

Two student focus groups were held at Illinois State University and Southeastern Louisiana University to engage students in the development process and provide feedback as to how the design of a textbook impacts homework and study habits in the College Algebra and Precalculus course areas.

Francisco Arceo, *Illinois State University*
Dave Cepko, *Illinois State University*
Andrea Connell, *Illinois State University*
Brian Lau, *Illinois State University*
Daniel Nathan Mielneczek, *Illinois State University*
Mingaile Orakauskaite, *Illinois State University*
Todd Michael Rapnikas, *Illinois State University*
Bethany Rollet, *Illinois State University*
Teddy Schrishuhn, *Illinois State University*
Josh Schultz, *Illinois State University*
Andy Thurman, *Illinois State University*
Candace Banos, *Southeastern Louisiana University*
Nicholas Curtis, *Southeastern Louisiana University*

M. D. "Boots" Feltenberger, *Southeastern Louisiana University*
Regina Foreman, *Southeastern Louisiana University*
Ashley Lae, *Southeastern Louisiana University*
Jessica Smith, *Southeastern Louisiana University*
Ashley Youngblood, *Southeastern Louisiana University*

Special Thanks

Sherry Meier, *Illinois State University*
Rebecca Muller, *Southeastern Louisiana University*
Anne Schmidt, *Illinois State University*

Instructor Focus Groups

Focus groups held at Baton Rouge Community College and ORMATYC provided feedback on the new Connections to Calculus feature in *Precalculus*, and shed light on the coverage of review material in this course. User focus groups at Southeastern Louisiana University and Madison Area Technical College confirmed the organizational changes planned for the second edition, provided feedback on the interior design, and helped us enhance and refine the strengths of the first edition.

Virginia Adelmann, *Southeastern Louisiana University*
George Alexander, *Madison Area Technical College*
Kenneth R. Anderson, *Chemeketa Community College*
Wayne G.Barber, *Chemeketa Community College*
Thomas Dick, *Oregon State University*
Vickie Flanders, *Baton Rouge Community College*
Bill Forrest, *Baton Rouge Community College*
Susan B. Guidroz, *Southeastern Louisiana University*
Christopher Guillory, *Baton Rouge Community College*
Cynthia Harrison, *Baton Rouge Community College*
Judy Jones, *Madison Area Technical College*
Lucyna Kabza, *Southeastern Louisiana University*
Ann Kirkpatrick, *Southeastern Louisiana University*
Sunmi Ku, *Bellevue Community College*

Pamela Larson, *Madison Area Technical College*
Jennifer Laveglia, *Bellevue Community College*
DeShea Miller, *Southeastern Louisiana University*
Elizabeth Miller, *Southeastern Louisiana University*
Rebecca Muller, *Southeastern Louisiana University*
Donna W. Newman, *Baton Rouge Community College*
Scott L. Peterson, *Oregon State University*
Ronald Posey, *Baton Rouge Community College*
Ronni Settoon, *Southeastern Louisiana University*
Jeganathan Sriskandarajah, *Madison Area Technical College*
Martha Stevens, *Bellevue Community College*
Mark J. Stigge, *Baton Rouge Community College*
Nataliya Svyeshnikova, *Southeastern Louisiana University*

John N. C. Szeto, *Southeastern Louisiana University*
Christina C. Terranova, *Southeastern Louisiana University*
Amy S. VanWey, *Clackamas Community College*
Andria Villines, *Bellevue Community College*

Jeff Weaver, *Baton Rouge Community College*
Ana Wills, *Southeastern Louisiana University*
Randall G. Wills, *Southeastern Louisiana University*
Xuezheng Wu, *Madison Area Technical College*

Developmental Symposia

McGraw-Hill conducted two symposia directly related to the development of Coburn's second edition. These events were an opportunity for editors from McGraw-Hill to gather information about the needs and challenges of instructors teaching these courses and confirm the direction of the second edition.

Rohan Dalpatadu, *University of Nevada–Las Vegas*
Franco Fedele, *University of West Florida*
Bill Forrest, *Baton Rouge Community College*
Marc Grether, *University of North Texas*
Sharon Hamsa, *Metropolitan Community College–Longview*
Derek Hein, *Southern Utah University*
Rebecca Heiskell, *Mountain View College*
Terry Hobbs, *Metropolitan Community College– Maple Woods*
Klay Kruczek, *Western Oregon University*
Nancy Matthews, *University of Oklahoma*
Sherry Meier, *Illinois State University*
Mary Ann (Molly) Misko, *Gadsden State Community College*

Rita Marie O'Brien, *Navarro College*
Jason Pallett, *Metropolitan Community College– Longview*
Christopher Parks, *Indiana University–Bloomington*
Vicki Partin, *Bluegrass Community College*
Philip Pina, *Florida Atlantic University–Boca*
Nancy Ressler, *Oakton Community College, Des Plaines Campus*
Vicki Schell, *Pensacola Junior College*
Kenan Shahla, *Antelope Valley College*
Linda Tansil, *Southeast Missouri State University*
Stephen Toner, *Victor Valley College*
Christine Walker, *Utah Valley State College*

Diary Reviews and Class Tests

Users of the first edition, Said Ngobi and Stephen Toner of Victor Valley College, provided chapter-by chapter feedback in diary form based on their experience using the text. Board of Advisors members facilitated class tests of the manuscript for a given topic. Both instructors and students returned questionnaires detailing their thoughts on the effectiveness of the text's features.

Class Tests

Instructors

Bill Forrest, *Baton Rouge Community College*
Marc Grether, *University of North Texas*
Sharon Hamsa, *Metropolitan Community College–Longview*
Rita Marie O'Brien's , *Navarro College*

Students

Cynthia Aguilar, *Navarro College*
Michalann Amoroso, *Baton Rouge Community College*
Chelsea Asbill, *Navarro College*
Sandra Atkins, *University of North Texas*
Robert Basom, *University of North Texas*
Cynthia Beasley, *Navarro College*
Michael Bermingham, *University of North Texas*
Jennifer Bickham, *Metropolitan Community College–Longview*
Rachel Brokmeier, *Baton Rouge Community College*
Amy Brugg, *University of North Texas*

Zach Burke, *University of North Texas*
Shaina Canlas, *University of North Texas*
Kristin Chambers, *University of North Texas*
Brad Chatelain, *Baton Rouge Community College*
Yu Yi Chen, *Baton Rouge Community College*
Jasmyn Clark, *Baton Rouge Community College*
Belinda Copsey, *Navarro College*
Melissa Cowan, *Metropolitan Community College–Longview*
Katlin Crooks, *Baton Rouge Community College*
Rachele Dudley, *University of North Texas*
Kevin Ekstrom, *University of North Texas*
Jade Fernberg, *University of North Texas*
Joseph Louis Fino, Jr., *Baton Rouge Community College*
Shannon M. Fleming, *University of North Texas*
Travis Flowers, *University of North Texas*
Teresa Foxx, *University of North Texas*
Michael Giulietti, *University of North Texas*
Michael Gordon, *Navarro College*

Hayley Hentzen, *University of North Texas*
Courtney Hodge, *University of North Texas*
Janice Hollaway, *Navarro College*
Weslon Hull, *Baton Rouge Community College*
Sarah James, *Baton Rouge Community College*
Georlin Johnson, *Baton Rouge Community College*
Michael Jones, *Navarro College*
Robert Koon, *Metropolitan Community College–Longview*
Ben Lenfant, *Baton Rouge Community College*
Colin Luke, *Baton Rouge Community College*
Lester Maloney, *Baton Rouge Community College*
Ana Mariscal, *Navarro College*
Tracy Ann Nguyen, *Baton Rouge Community College*
Alexandra Ortiz, *University of North Texas*
Robert T. R. Paine, *Baton Rouge Community College*
Kade Parent, *Baton Rouge Community College*
Brittany Louise Pratt, *Baton Rouge Community College*

Brittney Pruitt, *Metropolitan Community College–Longview*
Paul Rachal, *Baton Rouge Community College*
Matt Rawls, *Baton Rouge Community College*
Adam Reichert, *Metropolitan Community College–Longview*
Ryan Rodney, *Baton Rouge Community College*
Cody Scallan, *Baton Rouge Community College*
Laura Shafer, *University of North Texas*
Natina Simpson, *Navarro College*
Stephanie Sims, *Metropolitan Community College–Longview*
Cassie Snow, *University of North Texas*
Justin Stewart, *Metropolitan Community College–Longview*
Marjorie Tulana, *Navarro College*
Ashleigh Variest, *Baton Rouge Community College*
James A. Wann, *Navarro College*
Amber Wendleton, *Metropolitan Community College–Longview*
Eric Williams, *Metropolitan Community College–Longview*
Katy Wood, *Metropolitan Community College–Longview*

Developmental Editing

The manuscript has been impacted by numerous developmental editors who edited for clarity and consistency. Efforts resulted in cutting length from the manuscript, while retaining a conversational and casual narrative style. Editorial work also ensured the positive visual impact of art and photo placement.

First Edition Chapter Reviews and Manuscript Reviews

Over 200 instructors participated in postpublication single chapter reviews of the first edition and helped the team build the revision plan for the second edition. Over 100 teachers and academics from across the country reviewed the current edition text, the proposed second edition table of contents, and first-draft second edition manuscript to give feedback on reworked narrative, design changes, pedagogical enhancements, and organizational changes. This feedback was summarized by the book team and used to guide the direction of the second-draft manuscript.

Scott Adamson, *Chandler-Gilbert Community College*
Teresa Adsit, *University of Wisconsin–Green Bay*
Ebrahim Ahmadizadeh, *Northampton Community College*
George M. Alexander, *Madison Area Technical College*
Frances Alvarado, *University of Texas–Pan American*
Deb Anderson, *Antelope Valley College*
Philip Anderson, *South Plains College*
Michael Anderson, *West Virginia State University*
Jeff Anderson, *Winona State University*
Raul Aparicio, *Blinn College*
Judith Barclay, *Cuesta College*
Laurie Battle, *Georgia College and State University*
Annette Benbow, *Tarrant County College–Northwest*
Amy Benvie, *Florida Gulf Coast University*

Scott Berthiaume, *Edison State College*
Wes Black, *Illinois Valley Community College*
Arlene Blasius, *SUNY College of Old Westbury*
Caroline Maher Boulis, *Lee University*
Amin Boumenir, *University of West Georgia*
Terence Brenner, *Hostos Community College*
Gail Brooks, *McLennan Community College*
G. Robert Carlson, *Victor Valley College*
Hope Carr, *East Mississippi Community College*
Denise Chellsen, *Cuesta College*
Kim Christensen, *Metropolitan Community College–Maple Woods*
Lisa Christman, *University of Central Arkansas*
John Church, *Metropolitan Community College–Longview*
Sarah Clifton, *Southeastern Louisiana University*
David Collins, *Southwestern Illinois College*
Sarah V. Cook, *Washburn University*
Rhonda Creech, *Southeast Kentucky Community and Technical College*
Raymond L. Crownover, *Gateway College of Evangelism*
Marc Cullison, *Connors State College*
Steven Cunningham, *San Antonio College*
Callie Daniels, *St. Charles Community College*
John Denney, *Northeast Texas Community College*
Donna Densmore, *Bossier Parish Community College*
Alok Dhital, *University of New Mexico–Gallup*

James Michael *Dubrowsky Wayne Community College*
Brad Dyer, *Hazzard Community & Technical College*
Sally Edwards, *Johnson County Community College*
John Elliott, *St. Louis Community College–Meramec*
Gay Ellis, *Missouri State University*
Barbara Elzey, *Bluegrass Community College*
Dennis Evans, *Concordia University Wisconsin*
Samantha Fay, *University of Central Arkansas*
Victoria Fischer, *California State University–Monterey Bay*
Dorothy French, *Community College of Philadelphia*
Eric Garcia, *South Texas College*
Laurice Garrett, *Edison College*
Ramona Gartman, *Gadsden State Community College–Ayers Campus*
Scott Gaulke, *University of Wisconsin–Eau Claire*
Scott Gordon, *University of West Georgia*
Teri Graville, *Southern Illinois University Edwardsville*
Marc Grether, *University of North Texas*
Shane Griffith, *Lee University*
Gary Grohs, *Elgin Community College*
Peter Haberman, *Portland Community College*
Joseph Harris, *Gulf Coast Community College*
Margret Hathaway, *Kansas City Community College*
Tom Hayes, *Montana State University*
Bill Heider, *Hibbling Community College*
Max Hibbs, *Blinn College*
Terry Hobbs, *Metropolitan Community College–Maple Woods*
Sharon Holmes, *Tarrant County College–Southeast*
Jamie Holtin, *Freed-Hardeman University*
Brian Hons, *San Antonio College*
Kevin Hopkins, *Southwest Baptist University*
Teresa Houston, *East Mississippi Community College*
Keith Hubbard, *Stephen F. Austin State University*
Jeffrey Hughes, *Hinds Community College–Raymond*
Matthew Isom, *Arizona State University*
Dwayne Jennings, *Union University*
Judy Jones, *Madison Area Technical College*
Lucyna Kabza, *Southeastern Louisiana University*
Aida Kadic-Galeb, *University of Tampa*
Cheryl Kane, *University of Nebraska*
Rahim Karimpour, *Southern Illinois University Edwardsville*
Ryan Kasha, *Valencia Community College*
David Kay, *Moorpark College*
Jong Kim, *Long Beach City College*
Lynette King, *Gadsden State Community College*
Carolyn Kistner, *St. Petersburg College*
Barbara Kniepkamp, *Southern Illinois University Edwardsville*
Susan Knights, *Boise State University*
Stephanie Kolitsch, *University of Tennessee at Martin*

Louis Kolitsch, *University of Tennessee at Martin*
William Kirby, *Gadsden State Community College*
Karl Kruczek, *Northeastern State University*
Conrad Krueger, *San Antonio College*
Marcia Lambert, *Pitt Community College*
Rebecca Lanier, *Bluegrass Community College*
Marie Larsen, *Cuesta College*
Pam Larson, *Madison Area Technical College*
Jennifer Lawhon, *Valencia Community College*
John Levko, *University of Scranton*
Mitchel Levy, *Broward Community College*
John Lofberg, *South Dakota School of Mines and Technology*
Mitzi Logan, *Pitt Community College*
Sandra Maldonado, *Florida Gulf Coast University*
Robin C. Manker, *Illinois College*
Manoug Manougian, *University of South Florida*
Nancy Matthews, *University of Oklahoma*
Roger McCoach, *County College of Morris*
James McKinney, *California Polytechnic State University–Pomona*
Jennifer McNeilly, *University of Illinois Urbana Champaign*
Kathleen Miranda, *SUNY College at Old Westbury*
Mary Ann (Molly) Misko, *Gadsden State Community College*
Marianne Morea, *SUNY College of Old Westbury*
Michael Nasab, *Long Beach City College*
Said Ngobi, *Victor Valley College*
Tonie Niblett, *Northeast Alabama Community College*
Gary Nonnemacher, *Bowling Green State University*
Elaine Nye, *Alfred State College*
Rhoda Oden, *Gadsden State Community College*
Jeannette O'Rourke, *Middlesex County College*
Darla Ottman, *Elizabethtown Community & Technical College*
Jason Pallett, *Metropolitan Community College–Longview*
Priti Patel, *Tarrant County College–Southeast*
Judy Pennington-Price, *Midway College*
Susan Pfeifer, *Butler County Community College*
Margaret Poitevint, *North Georgia College & State University*
Tammy Potter, *Gadsden State Community College*
Debra Prescott, *Central Texas College*
Elise Price, *Tarrant County College*
Kevin Ratliff, *Blue Ridge Community College*
Bruce Reid, *Howard Community College*
Jolene Rhodes, *Valencia Community College*
Karen Rollins, *University of West Georgia*
Randy Ross, *Morehead State University*
Michael Sawyer, *Houston Community College*
Richard Schnackenberg, *Florida Gulf Coast University*
Bethany Seto, *Horry-Georgetown Technical College*

Delphy Shaulis, *University of Colorado–Boulder*

Jennifer Simonton, *Southwestern Illinois College*

David Slay, *McNeese State University*

David Snyder, *Texas State University at San Marcos*

Larry L. Southard, *Florida Gulf Coast University*

Lee Ann Spahr, *Durham Technical Community College*

Jeganathan Sriskandarajah, *Madison Area Technical College*

Adam Stinchcombe, *Eastern Arizona College*

Pam Stogsdill, *Bossier Parish Community College*

Eleanor Storey, *Front Range Community College*

Kathy Stover, *College of Southern Idaho*

Mary Teel, *University of North Texas*

Carlie Thompson, *Southeast Kentucky Community & Technical College*

Bob Tilidetzke, *Charleston Southern University*

Stephen Toner, *Victor Valley College*

Thomas Tunnell, *Illinois Valley Community College*

Carol Ulsafer, *University of Montana*

John Van Eps, *California Polytechnic State University–San Luis Obispo*

Andrea Vorwark, *Metropolitan Community College–Maple Woods*

Jim Voss, *Front Range Community College*

Jennifer Walsh, *Daytona State College*

Jiantian Wang, *Kean University*

Sheryl Webb, *Tennessee Technological University*

Bill Weber, *Fort Hays State University*

John Weglarz, *Kirkwood Community College*

Tressa White, *Arkansas State University–Newport*

Cheryl Winter, *Metropolitan Community College–Blue River*

Kenneth Word, *Central Texas College*

Laurie Yourk, *Dickinson State University*

Acknowledgments

I first want to express a deep appreciation for the guidance, comments and suggestions offered by all reviewers of the manuscript. I have once again found their collegial exchange of ideas and experience very refreshing and instructive, and always helping to create a better learning tool for our students.

I would especially like to thank Vicki Krug for her uncanny ability to bring innumerable pieces from all directions into a unified whole; Patricia Steele for her eagle-eyed attention to detail; Katie White and Michelle Flomenhoft for their helpful suggestions, infinite patience, tireless efforts, and steady hand in bringing the manuscript to completion; John Osgood for his ready wit and creative energies, Laurie Janssen and our magnificent design team, and Dawn Bercier, the master of this large ship, whose indefatigable spirit kept the ship on course through trial and tempest, and brought us all safely to port. In truth, my hat is off to all the fine people at McGraw-Hill for their continuing support and belief in this series. A final word of thanks must go to Rick Armstrong, whose depth of knowledge, experience, and mathematical connections seems endless; J. D. Herdlick for his friendship and his ability to fill an instant and sudden need, Anne Marie Mosher for her contributions to various features of the text, Jennifer McNeilly for her review of the Limits Chapter, Mitch Levy for his consultation on the exercise sets, Stephen Toner for his work on the videos, Rosemary Karr for her meticulous work on the solutions manuals, Donna Gerker for her work on the preformatted tests, Jay Miller and Carrie Green for their invaluable ability to catch what everyone else misses; and to Rick Pescarino, Nate Wilson, and all of my colleagues at St. Louis Community College, whose friendship, encouragement and love of mathematics makes going to work each day a joy.

Making Connections...

Through Supplements

*All online supplements are available through the book's website: www.mhhe.com/coburn.

Instructor Supplements

- **Computerized Test Bank Online:** Utilizing Brownstone Diploma® algorithm-based testing software enables users to create customized exams quickly.
- **Instructor's Solutions Manual:** Provides comprehensive, worked-out solutions to all exercises in the text.
- **Annotated Instructor's Edition:** Contains all answers to exercises in the text, which are printed in a second color, adjacent to corresponding exercises, for ease of use by the instructor.
- **PowerPoint Slides:** Fully editable slides that follow the textbook.

Student Supplements

- **Student Solutions Manual** provides comprehensive, worked-out solutions to all of the odd-numbered exercises.
- **Videos**
  - Interactive video lectures are provided for each section in the text, which explain to the students how to do key problem types, as well as highlighting common mistakes to avoid.
  - Exercise videos provide step-by-step instruction for the key exercises which students will most wish to see worked out.
  - Graphing calculator videos help students master the most essential calculator skills used in the college algebra course.
  - The videos are closed-captioned for the hearing impaired, subtitled in Spanish, and meet the Americans with Disabilities Act Standards for Accessible Design.

MathZone www.mhhe.com/coburn

McGraw-Hill's MathZone is a complete online homework system for mathematics and statistics. Instructors can assign textbook-specific content from over 40 McGraw-Hill titles as well as customize the level of feedback students receive, including the ability to have students show their work for any given exercise. Assignable content includes an array of videos and other multimedia along with algorithmic exercises, providing study tools for students with many different learning styles.

Within MathZone, a diagnostic assessment tool powered by ALEKS® is available to measure student preparedness and provide detailed reporting and personalized remediation. MathZone also helps ensure consistent assignment delivery across several sections through a course administration function and makes sharing courses with other instructors easy.

For additional study help students have access to NetTutor™, a robust online live tutoring service that incorporates whiteboard technology to communicate mathematics. The tutoring schedules are built around peak homework times to best accommodate student schedules. Instructors can also take advantage of this whiteboard by setting up a Live Classroom for online office hours or a review session with students.

For more information, visit the book's website (**www.mhhe.com/coburn**) or contact your local McGraw-Hill sales representative (**www.mhhe.com/rep**).

ALEKS® www.aleks.com

ALEKS (**A**ssessment and **LE**arning in **K**nowledge **S**paces) is a dynamic online learning system for mathematics education, available over the Web 24/7. ALEKS assesses students, accurately determines their knowledge, and then guides them to the material that they are most ready to learn. With a variety of reports, Textbook Integration Plus, quizzes, and homework assignment capabilities, ALEKS offers flexibility and ease of use for instructors.

- ALEKS uses artificial intelligence to determine exactly what each student knows and is ready to learn. ALEKS remediates student gaps and provides highly efficient learning and improved learning outcomes
- ALEKS is a comprehensive curriculum that aligns with syllabi or specified textbooks. Used in conjunction with McGraw-Hill texts, students also receive links to text-specific videos, multimedia tutorials, and textbook pages.
- Textbook Integration Plus allows ALEKS to be automatically aligned with syllabi or specified McGraw-Hill textbooks with instructor chosen dates, chapter goals, homework, and quizzes.
- ALEKS with AI-2 gives instructors increased control over the scope and sequence of student learning. Students using ALEKS demonstrate a steadily increasing mastery of the content of the course.
- ALEKS offers a dynamic classroom management system that enables instructors to monitor and direct student progress towards mastery of course objectives.

ALEKS Prep/Remediation:

- Helps instructors meet the challenge of remediating unequally prepared or improperly placed students.
- Assesses students on their pre-requisite knowledge needed for the course they are entering (i.e. Calculus students are tested on Precalculus knowledge).
- Based on the assessment, students are prescribed a unique and efficient learning path specific to address their strengths and weaknesses.
- Students can address pre-requisite knowledge gaps outside of class freeing the instructor to use class time pursuing course outcomes.

Electronic Textbook: CourseSmart is a new way for faculty to find and review eTextbooks. It's also a great option for students who are interested in accessing their course materials digitally and saving money. CourseSmart offers thousands of the most commonly adopted textbooks across hundreds of courses from a wide variety of higher education publishers. It is the only place for faculty to review and compare the full text of a textbook online, providing immediate access without the environmental impact of requesting a print exam copy. At CourseSmart, students can save up to 50% off the cost of a print book, reduce their impact on the environment, and gain access to powerful web tools for learning including full text search, notes and highlighting, and email tools for sharing notes between classmates. **www.CourseSmart.com**

Primis: You can customize this text with McGraw-Hill/Primis Online. A digital database offers you the flexibility to customize your course including material from the largest online collection of textbooks, readings, and cases. Primis leads the way in customized eBooks with hundreds of titles available at prices that save your students over 20% off bookstore prices. Additional information is available at 800-228-0634.

Contents

Preface vi

Index of Applications xxxv

CHAPTER **3** Polynomial and Rational Functions 219

CHAPTER **10**   Additional Topics in Algebra   939

CHAPTER **11**   Bridges to Calculus–An Introduction to Limits   1023

Additional Topics Online

(Visit www.mhhe.com/coburn)

Coburn's Precalculus Series

College Algebra, Second Edition

Review ◆ Equations and Inequalities ◆ Relations, Functions, and Graphs ◆ Polynomial and Rational Functions ◆ Exponential and Logarithmic Functions ◆ Systems of Equations and Inequalities ◆ Matrices ◆ Geometry and Conic Sections ◆ Additional Topics in Algebra

ISBN 0-07-351941-3, ISBN 978-0-07351941-8

College Algebra Essentials, Second Edition

Review ◆ Equations and Inequalities ◆ Relations, Functions, and Graphs ◆ Polynomial and Rational Functions ◆ Exponential and Logarithmic Functions ◆ Systems of Equations and Inequalities

ISBN 0-07-351968-5, ISBN 978-0-07351968-5

Algebra and Trigonometry, Second Edition

Review ◆ Equations and Inequalities ◆ Relations, Functions, and Graphs ◆ Polynomial and Rational Functions ◆ Exponential and Logarithmic Functions ◆ Trigonometric Functions ◆ Trigonometric Identities, Inverses and Equations ◆ Applications of Trigonometry ◆ Systems of Equations and Inequalities ◆ Matrices ◆ Geometry and Conic Sections ◆ Additional Topics in Algebra

ISBN 0-07-351952-9, ISBN 978-0-07-351952-4

Precalculus, Second Edition

Equations and Inequalities ◆ Relations, Functions, and Graphs ◆ Polynomial and Rational Functions ◆ Exponential and Logarithmic Functions ◆ Trigonometric Functions ◆ Trigonometric Identities, Inverses and Equations ◆ Applications of Trigonometry ◆ Systems of Equations and Inequalities, and Matrices ◆ Geometry and Conic Sections ◆ Additional Topics in Algebra ◆ Limits

ISBN 0-07-351942-1, ISBN 978-0-07351942-5

Trigonometry, Second Edition—Coming in 2010!

Introduction to Trigonometry ◆ Right Triangles & Static Trigonometry ◆ Radian Measure & Dynamic Trigonometry ◆ Trigonometric Graphs and Models ◆ Trigonometric Identities ◆ Inverse Functions and Trigonometric Equations ◆ Applications of Trigonometry ◆ Trigonometric Connections to Algebra

ISBN 0-07-351948-0, ISBN 978-0-07351948-7

Index of Applications

MEDICINE/NURSING/ NUTRITION/DIETETICS/HEALTH

1

Equations and Inequalities

CHAPTER OUTLINE

CHAPTER CONNECTIONS

In the world of professional sports, very strict specifications are put in place to ensure that competition is fair, with no one person or team having an unfair advantage. In professional hockey, the goal must be 48 in. high and 72 in. wide, as measured from inside the posts. In baseball, the bats used can have a maximum diameter of 69.8 mm, and a maximum length of 1006.8 mm, while in professional golf, the diameter d of a golf ball must be within 0.1 mm of the established standard of 42.7 mm. In the latter case, we can express this requirement as $|d - 42.7| < 0.1$. This application appears as Exercise 65 in Section 1.3.

Connections to Calculus

The foundation and study of calculus involves analyzing very small differences similar to the one above. The *Connections to Calculus* for Chapter 1 expands on the notation and language used in this analysis, and introduces how the absolute value concept contributes to this foundation. The need to solve a broad range of equation types is also explored.

1.1 Linear Equations, Formulas, and Problem Solving

Learning Objectives

In Section 1.1 you will learn how to:

☐ **A.** Solve linear equations using properties of equality

☐ **B.** Recognize equations that are identities or contradictions

☐ **C.** Solve for a specified variable in a formula or literal equation

☐ **D.** Use the problem-solving guide to solve various problem types

In a study of algebra, you will encounter many **families of equations,** or groups of equations that share common characteristics. Of interest to us here is the family of **linear equations in one variable,** a study that lays the foundation for understanding more advanced families. In addition to *solving* linear equations, we'll use the skills we develop to *solve for a specified variable* in a formula, a practice widely used in science, business, industry, and research.

A. Solving Linear Equations Using Properties of Equality

An **equation** is *a statement that two expressions are equal*. From the expressions $3(x - 1) + x$ and $-x + 7$, we can form the equation

$$3(x - 1) + x = -x + 7,$$

which is a **linear equation in one variable.** To solve an equation, we attempt to find a specific input or x-value that will make the equation true, meaning the left-hand expression will be equal to the right. Using Table 1.1, we find that $3(x - 1) + x = -x + 7$ is a true equation when x is replaced by 2, and is a false equation otherwise. Replacement values that make the equation true are called **solutions** or **roots** of the equation.

Table 1.1

x	$3(x - 1) + x$	$-x + 7$
-2	-11	9
-1	-7	8
0	-3	7
1	1	6
2	5	5
3	9	4
4	13	3

> ⚠ **CAUTION** ▶ From Appendix I.F, an **algebraic *expression*** is a sum or difference of algebraic terms. Algebraic expressions can be simplified, evaluated or written in an equivalent form, but cannot be "*solved,*" since we're not seeking a specific value of the unknown.

Solving equations using a table is too time consuming to be practical. Instead we attempt to write a sequence of **equivalent equations,** each one simpler than the one before, until we reach a point where the solution is obvious. Equivalent equations are those that have the same solution set, and are obtained by using the distributive property to simplify the expressions on each side of the equation, and the additive and multiplicative properties of equality to obtain an equation of the form $x =$ constant.

The Additive Property of Equality	The Multiplicative Property of Equality
If A, B, and C represent algebraic expressions and $A = B$,	If A, B, and C represent algebraic expressions and $A = B$,
then $A + C = B + C$	then $AC = BC$ and $\dfrac{A}{C} = \dfrac{B}{C}, (C \neq 0)$

In words, the additive property says that like quantities, numbers or terms can be added to both sides of an equation. A similar statement can be made for the multiplicative property. These properties are combined into a general guide for solving linear equations, which you've likely encountered in your previous studies. Note that not all steps in the guide are required to solve every equation.

Guide to Solving Linear Equations in One Variable

- Eliminate parentheses using the distributive property, then combine any like terms.
- Use the additive property of equality to write the equation with all variable terms on one side, and all constants on the other. Simplify each side.
- Use the multiplicative property of equality to obtain an equation of the form $x =$ constant.
- For applications, answer in a complete sentence and include any units of measure indicated.

For our first example, we'll use the equation $3(x - 1) + x = -x + 7$ from our initial discussion.

EXAMPLE 1 ▶ **Solving a Linear Equation Using Properties of Equality**

Solve for x: $3(x - 1) + x = -x + 7$.

Solution ▶

$3(x - 1) + x = -x + 7$	original equation
$3x - 3 + x = -x + 7$	distributive property
$4x - 3 = -x + 7$	combine like terms
$5x - 3 = 7$	add x to both sides (additive property of equality)
$5x = 10$	add 3 to both sides (additive property of equality)
$x = 2$	multiply both sides by $\frac{1}{5}$ or divide both sides by 5 (multiplicative property of equality)

As we noted in Table 1.1, the solution is $x = 2$.

Now try Exercises 7 through 12 ▶

To check a solution by substitution means we substitute the solution back into the original equation (this is sometimes called **back-substitution**), and verify the left-hand side is equal to the right. For Example 1 we have:

$3(x - 1) + x = -x + 7$	original equation
$3(2 - 1) + 2 = -2 + 7$	substitute 2 for x
$3(1) + 2 = 5$	simplify
$5 = 5\checkmark$	solution checks

If any coefficients in an equation are fractional, multiply both sides by the least common denominator (LCD) to *clear the fractions*. Since any decimal number can be written in fraction form, the same idea can be applied to decimal coefficients.

EXAMPLE 2 ▶ **Solving a Linear Equation with Fractional Coefficients**

Solve for n: $\frac{1}{4}(n + 8) - 2 = \frac{1}{2}(n - 6)$.

Solution ▶

$\frac{1}{4}(n + 8) - 2 = \frac{1}{2}(n - 6)$	original equation
$\frac{1}{4}n + 2 - 2 = \frac{1}{2}n - 3$	distributive property
$\frac{1}{4}n = \frac{1}{2}n - 3$	combine like terms
$4(\frac{1}{4}n) = 4(\frac{1}{2}n - 3)$	multiply both sides by LCD = 4
$n = 2n - 12$	distributive property
$-n = -12$	subtract $2n$
$n = 12$	multiply by -1

☑ **A.** You've just learned how to solve linear equations using properties of equality

Verify the solution is $n = 12$ using back-substitution.

Now try Exercises 13 through 30 ▶

B. Identities and Contradictions

Example 1 illustrates what is called a **conditional equation,** since the equation is true for $x = 2$, but false for all other values of x. The equation in Example 2 is also conditional. An **identity** is an equation that is *always true,* no matter what value is substituted for the variable. For instance, $2(x + 3) = 2x + 6$ is an identity with a solution set of all real numbers, written as $\{x|x \in \mathbb{R}\}$, or $x \in (-\infty, \infty)$ in interval notation. **Contradictions** are equations that are *never true,* no matter what real number is substituted for the variable. The equations $x - 3 = x + 1$ and $-3 = 1$ are contradictions. To state the solution set for a contradiction, we use the symbol "$\varnothing$" (the null set) or "$\{\ \}$" (the empty set). Recognizing these special equations will prevent some surprise and indecision in later chapters.

EXAMPLE 3 ▶   Solving an Equation That Is a Contradiction

Solve for x: $2(x - 4) + 10x = 8 + 4(3x + 1)$, and state the solution set.

Solution ▶

$$
\begin{aligned}
2(x - 4) + 10x &= 8 + 4(3x + 1) &&\text{original equation} \\
2x - 8 + 10x &= 8 + 12x + 4 &&\text{distributive property} \\
12x - 8 &= 12x + 12 &&\text{combine like terms} \\
-8 &= 12 &&\text{subtract } 12x
\end{aligned}
$$

Since -8 is never equal to 12, the original equation is a contradiction. The solution is the empty set $\{\ \}$.

Now try Exercises 31 through 36 ▶

☑ **B.** You've just learned how to recognize equations that are identities or contradictions

In Example 3, our attempt to solve for x ended with all variables being eliminated, leaving an equation that is *always false*—a contradiction (-8 is never equal to 12). There is nothing wrong with the solution process, the result is simply telling us the original equation has *no solution.* In other equations, the variables may once again be eliminated, but leave a result that is *always true*—an identity.

C. Solving for a Specified Variable in Literal Equations

A **formula** is an equation that models a known relationship between two or more quantities. A **literal equation** is simply one that has two or more variables. Formulas are a type of literal equation, but not every literal equation is a formula. For example, the formula $A = P + PRT$ models the growth of money in an account earning simple interest, where A represents the total amount accumulated, P is the initial deposit, R is the annual interest rate, and T is the number of years the money is left on deposit. To *describe* $A = P + PRT$, we might say the formula has been "solved for A" or that "A is written in terms of P, R, and T." In some cases, before using a formula it may be convenient to solve for one of the other variables, say P. In this case, P is called the **object variable.**

EXAMPLE 4 ▶   Solving for Specified Variable

Given $A = P + PRT$, write P in terms of A, R, and T (solve for P).

Solution ▶ Since the object variable occurs in more than one term, we first apply the distributive property.

$$A = P + PRT \qquad \text{focus on } \boldsymbol{P}\text{—the object variable}$$
$$A = P(1 + RT) \qquad \text{factor out } \boldsymbol{P}$$
$$\frac{A}{1 + RT} = \frac{P(1 + RT)}{(1 + RT)} \qquad \text{solve for } \boldsymbol{P} \text{ [divide by } (1 + RT)\text{]}$$
$$\frac{A}{1 + RT} = P \qquad \text{result}$$

Now try Exercises 37 through 48 ▶

We solve literal equations for a specified variable using the same methods we used for other equations and formulas. Remember that it's good practice to *focus on the object variable* to help guide you through the solution process, as again shown in Example 5.

EXAMPLE 5 ▶ Solving for a Specified Variable

Given $2x + 3y = 15$, write y in terms of x (solve for y).

Solution ▶
$$2x + 3y = 15 \qquad \text{focus on the object variable}$$
$$3y = -2x + 15 \qquad \text{subtract } 2x \text{ (isolate term with } y\text{)}$$
$$\tfrac{1}{3}(3y) = \tfrac{1}{3}(-2x + 15) \qquad \text{multiply by } \tfrac{1}{3} \text{ (solve for } y\text{)}$$
$$y = \tfrac{-2}{3}x + 5 \qquad \text{distribute and simplify}$$

Now try Exercises 49 through 54 ▶

WORTHY OF NOTE

In Example 5, notice that in the second step we wrote the subtraction of $2x$ as $-2x + 15$ instead of $15 - 2x$. For reasons that become clear later in this chapter, we generally write variable terms before constant terms.

Literal Equations and General Solutions

Solving literal equations for a specified variable can help us develop the general solution for an entire family of equations. This is demonstrated here for the family of linear equations written in the form $ax + b = c$. A side-by-side comparison with a specific linear equation demonstrates that identical ideas are used.

Specific Equation		**Literal Equation**
$2x + 3 = 15$	focus on object variable	$ax + b = c$
$2x = 15 - 3$	subtract constant	$ax = c - b$
$x = \dfrac{15 - 3}{2}$	divide by coefficient	$x = \dfrac{c - b}{a}$

Of course the solution on the left would be written as $x = 6$ and checked in the original equation. On the right we now have a general formula for all equations of the form $ax + b = c$.

EXAMPLE 6 ▶ Solving Equations of the Form $ax + b = c$ Using the General Formula

Solve $6x - 1 = -25$ using the formula just developed, and check your solution in the original equation.

Solution ▶ For this equation, $a = 6$, $b = -1$, and $c = -25$, this gives

$$x = \frac{c - b}{a}$$

$$= \frac{-25 - (-1)}{6}$$

$$= \frac{-24}{6}$$

$$= -4$$

Check: $6x - 1 = -25$

$6(-4) - 1 = -25$

$-24 - 1 = -25$

$-25 = -25$ ✓

Now try Exercises 55 through 60 ▶

☑ **C.** You've just learned how to solve for a specified variable in a formula or literal equation

D. Using the Problem-Solving Guide

Becoming a good problem solver is an evolutionary process. Over time and with continued effort, your problem-solving skills grow, as will your ability to solve a wider range of applications. Most good problem solvers develop the following characteristics:

- A positive attitude
- A mastery of basic facts
- Strong mental arithmetic skills
- Good mental-visual skills
- Good estimation skills
- A willingness to persevere

These characteristics form a solid basis for applying what we call the **Problem-Solving Guide,** which simply organizes the basic elements of good problem solving. Using this guide will help save you from two common stumbling blocks—indecision and not knowing where to start.

Problem-Solving Guide

- **Gather and organize information.**
  Read the problem several times, forming a mental picture as you read. *Highlight key phrases.* List given information, including any related formulas. *Clearly identify what you are asked to find.*

- **Make the problem visual.**
  Draw and label a diagram or create a table of values, as appropriate. This will help you see how different parts of the problem fit together.

- **Develop an equation model.**
  Assign a variable to represent what you are asked to find and build any related expressions referred to in the exercise. Write an equation model from the information given in the exercise. *Carefully reread the exercise to double-check your equation model.*

- **Use the model and given information to solve the problem.**
  Substitute given values, then simplify and solve. State the answer in sentence form, and check that the answer is reasonable. Include any units of measure indicated.

General Modeling Exercises

In Appendix I.B, we learned to translate word phrases into symbols. This skill is used to build equations from information given in paragraph form. Sometimes the variable *occurs more than once* in the equation, because two different items in the same exercise are related. If the relationship involves a comparison of size, we often use line segments or bar graphs to model the relative sizes.

EXAMPLE 7 ▶ **Solving an Application Using the Problem-Solving Guide**

The largest state in the United States is Alaska (AK), which covers an area that is 230 square miles (mi^2) more than 500 times that of the smallest state, Rhode Island (RI). If they have a combined area of 616,460 mi^2, how many square miles does each cover?

Solution ▶ Combined area is 616,460 mi^2, AK covers   gather and organize information
230 more than 500 times the area of RI.   highlight any key phrases

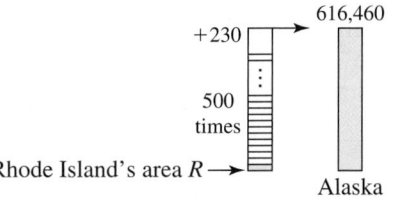

make the problem visual

Let R represent the area of Rhode Island.   assign a variable
Then $500R + 230$ represents Alaska's area.   build related expressions

Rhode Island's area + Alaska's area = Total
$$R + (500R + 230) = 616,460$$ write the equation model
$$501R = 616,230$$ combine like terms, subtract 230
$$R = 1230$$ divide by 501

Rhode Island covers an area of 1230 mi^2, while Alaska covers an area of $500(1230) + 230 = 615,230$ mi^2.

Now try Exercises 63 through 68 ▶

Consecutive Integer Exercises

Exercises involving **consecutive integers** offer excellent practice in assigning variables to unknown quantities, building related expressions, and the problem-solving process in general. We sometimes work with consecutive **odd** integers or consecutive **even** integers as well.

EXAMPLE 8 ▶ **Solving a Problem Involving Consecutive Odd Integers**

The sum of three consecutive *odd* integers is 69. What are the integers?

Solution ▶ The sum of three consecutive odd integers . . .   gather/organize information
highlight any key phrases

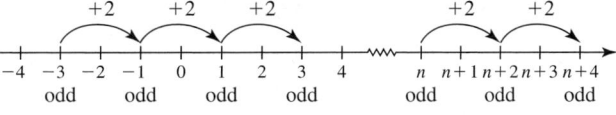

make the problem visual

Let n represent the smallest consecutive odd integer,   assign a variable
then $n + 2$ represents the second odd integer and   build related expressions
$(n + 2) + 2 = n + 4$ represents the third.

In words: first + second + third odd integer = 69   write the equation model

$$n + (n + 2) + (n + 4) = 69$$ equation model
$$3n + 6 = 69$$ combine like terms
$$3n = 63$$ subtract 6
$$n = 21$$ divide by 3

The odd integers are $n = 21$, $n + 2 = 23$, and $n + 4 = 25$.

$21 + 23 + 25 = 69$ ✓

Now try Exercises 69 through 72 ▶

WORTHY OF NOTE

The number line illustration in Example 8 shows that consecutive odd integers are *two units* apart and the related expressions were built accordingly: $n, n + 2, n + 4$, and so on. In particular, *we cannot use $n, n + 1, n + 3, \ldots$* because n and $n + 1$ are *not two units apart*. If we know the exercise involves *even* integers instead, the same model is used, since even integers are also two units apart. For *consecutive* integers, the labels are $n, n + 1, n + 2$, and so on.

Uniform Motion (Distance, Rate, Time) Exercises

Uniform motion problems have many variations, and it's important to draw a good diagram when you get started. Recall that if speed is constant, the distance traveled is equal to the rate of speed multiplied by the time in motion: $D = RT$.

EXAMPLE 9 ▶ **Solving a Problem Involving Uniform Motion**

I live 260 mi from a popular mountain retreat. On my way there to do some mountain biking, my car had engine trouble—forcing me to bike the rest of the way. If I drove 2 hr longer than I biked and averaged 60 miles per hour driving and 10 miles per hour biking, how many hours did I spend pedaling to the resort?

Solution ▶ The sum of the two distances must be 260 mi.   *gather/organize information*
The **rates** are given, and the driving time is   *highlight any key phrases*
2 hr more than biking time.   *make the problem visual*

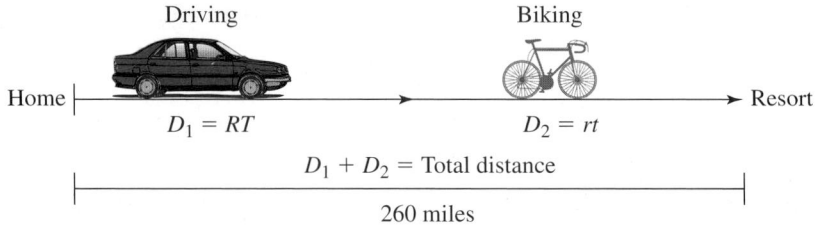

$$D_1 + D_2 = \text{Total distance}$$
$$260 \text{ miles}$$

Let t represent the biking time,   *assign a variable*
then $T = t + 2$ represents time spent driving.   *build related expressions*

$$D_1 + D_2 = 260$$ *write the equation model*
$$RT + rt = 260$$ $RT = D_1,\ rt = D_2$
$$60(t + 2) + 10t = 260$$ *substitute $t + 2$ for T, 60 for R, 10 for r*
$$70t + 120 = 260$$ *distribute and combine like terms*
$$70t = 140$$ *subtract 120*
$$t = 2$$ *divide by 70*

I rode my bike for $t = 2$ hr, after driving $t + 2 = 4$ hr.

Now try Exercises 73 through 76 ▶

Exercises Involving Mixtures

Mixture problems offer another opportunity to refine our problem-solving skills while using many elements from the problem-solving guide. They also lend themselves to a very useful mental-visual image and have many practical applications.

EXAMPLE 10 ▶ **Solving an Application Involving Mixtures**

As a nasal decongestant, doctors sometimes prescribe saline solutions with a concentration between 6% and 20%. In "the old days," pharmacists had to create different mixtures, but only needed to stock these concentrations, since any percentage in between could be obtained using a mixture. An order comes in for a 15% solution. How many milliliters (mL) of the 20% solution must be mixed with 10 mL of the 6% solution to obtain the desired 15% solution?

Solution ▶ Only 6% and 20% concentrations are available; mix a 20% solution with 10 mL of a 6% solution

gather/organize information

highlight any key phrases

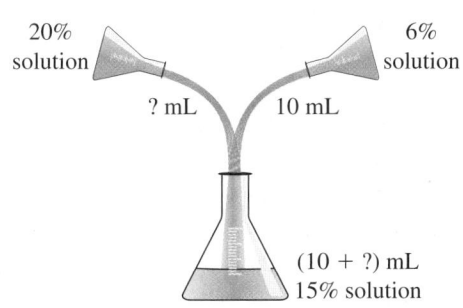

make the problem visual

Let x represent the amount of 20% solution, then $10 + x$ represents the total amount of 15% solution.

assign a variable

build related expressions

1st quantity times its concentration		2nd quantity times its concentration		1st+2nd quantity times desired concentration	
$10(0.06)$	$+$	$x(0.2)$	$=$	$(10 + x)(0.15)$	write equation model
0.6	$+$	$0.2x$	$=$	$1.5 + 0.15x$	distribute/simplify
		$0.2x$	$=$	$0.9 + 0.15x$	subtract 0.6
		$0.05x$	$=$	0.9	subtract 0.15x
		x	$=$	18	divide by 0.05

To obtain a 15% solution, 18 mL of the 20% solution must be mixed with 10 mL of the 6% solution.

Now try Exercises 77 through 84 ▶

 **D.** You've just learned how to use the problem-solving guide to solve various problem types

---

TECHNOLOGY HIGHLIGHT

Using a Graphing Calculator as an Investigative Tool

The mixture concept can be applied in a wide variety of ways, including mixing zinc and copper to get bronze, different kinds of nuts for the holidays, diversifying investments, or mixing two acid solutions in order to get a desired concentration. Whether the value of each part in the mix is monetary or a percent of concentration, the general mixture equation has this form:

Quantity 1 · Value I + Quantity 2 · Value II = Total quantity · Desired value

Graphing calculators are a great tool for exploring this relationship, because the TABLE feature enables us to test the result of various mixtures in an instant. Suppose 10 oz of an 80% glycerin solution are to be mixed with an unknown amount of a 40% solution. How much of the 40% solution is used if a 56% solution is needed? To begin, we might consider that using equal amounts of the 40% and 80% solutions would result in a 60% concentration (halfway between 40% and 80%). To illustrate, let C represent the final concentration of the mix.

$\mathbf{10}(0.8) + \mathbf{10}(0.4) = (10 + 10)C$	equal amounts
$8 + 4 = 20C$	simplify
$12 = 20C$	add
$0.6 = C$	divide by 20

Figure 1.1   **Figure 1.2**   **Figure 1.3**

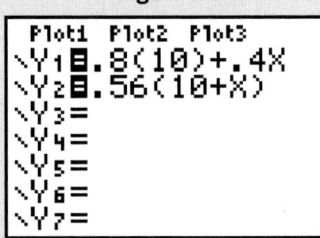

Since this is too high a concentration (a 56% = 0.56 solution is desired), we know more of the weaker solution should be used. To explore the relationship further, assume x oz of the 40% solution are used and enter the resulting equation on the [Y=] screen as $Y_1 = .8(10) + .4X$. Enter the result of the mix as $Y_2 = .56(10 + X)$ (see Figure 1.1). Next, set up a TABLE using [2nd] [WINDOW] **(TBLSET)** with **TblStart** = 10, ΔTbl = 1, and the calculator set in Indpnt: **AUTO** mode (see Figure 1.2). Finally, access the TABLE results using [2nd] [GRAPH] **(TABLE)**. The resulting screen is shown in Figure 1.3, where we note that 15 oz of the 40% solution should be used (the equation is true when X is 15: $Y_1 = Y_2$).

Exercise 1:   Use this idea to solve Exercises 81 and 82 from the Exercises.

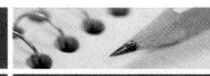

1.1 EXERCISES

▶ CONCEPTS AND VOCABULARY

Fill in each blank with the appropriate word or phrase. Carefully reread the section, if necessary.

1. A(n) _____ is an equation that is always true, regardless of the _____ value.

2. A(n) _____ is an equation that is always false, regardless of the _____ value.

3. A(n) _____ equation is an equation having _____ or more unknowns.

4. For the equation $S = 2\pi r^2 + 2\pi rh$, we can say that S is written in terms of _____ and _____.

5. Discuss/Explain the three tests used to identify a linear equation. Give examples and counterexamples in your discussion.

6. Discuss/Explain each of the four basic parts of the *problem-solving guide*. Include a solved example in your discussion.

▶ DEVELOPING YOUR SKILLS

Solve each equation. Check your answer by substitution.

7. $4x + 3(x - 2) = 18 - x$

8. $15 - 2x = -4(x + 1) + 9$

9. $21 - (2v + 17) = -7 - 3v$

10. $-12 - 5w = -9 - (6w + 7)$

11. $8 - (3b + 5) = -5 + 2(b + 1)$

12. $2a + 4(a - 1) = 3 - (2a + 1)$

Solve each equation.

13. $\frac{1}{5}(b + 10) - 7 = \frac{1}{3}(b - 9)$

14. $\frac{1}{6}(n - 12) = \frac{1}{4}(n + 8) - 2$

15. $\frac{2}{3}(m + 6) = \frac{-1}{2}$

16. $\frac{4}{5}(n - 10) = \frac{-8}{9}$

17. $\frac{1}{2}x + 5 = \frac{1}{3}x + 7$ 18. $-4 + \frac{2}{3}y = \frac{1}{2}y - 5$

19. $\frac{x + 3}{5} + \frac{x}{3} = 7$ 20. $\frac{z - 4}{6} - 2 = \frac{z}{2}$

21. $15 = -6 - \frac{3p}{8}$ 22. $-15 - \frac{2q}{9} = -21$

23. $0.2(24 - 7.5a) - 6.1 = 4.1$

24. $0.4(17 - 4.25b) - 3.15 = 4.16$

25. $6.2v - (2.1v - 5) = 1.1 - 3.7v$

26. $7.9 - 2.6w = 1.5w - (9.1 + 2.1w)$

27. $\frac{n}{2} + \frac{n}{5} = \frac{2}{3}$

28. $\frac{m}{3} - \frac{2}{5} = \frac{m}{4}$

29. $3p - \frac{p}{4} - 5 = \frac{p}{6} - 2p + 6$

30. $\frac{q}{6} + 1 - 3q = 2 - 4q + \frac{q}{8}$

Identify the following equations as an identity, a contradiction, or a conditional equation, then state the solution.

31. $-3(4z + 5) = -15z - 20 + 3z$

32. $5x - 9 - 2 = -5(2 - x) - 1$

33. $8 - 8(3n + 5) = -5 + 6(1 + n)$

34. $2a + 4(a - 1) = 1 + 3(2a + 1)$

35. $-4(4x + 5) = -6 - 2(8x + 7)$

36. $-(5x - 3) + 2x = 11 - 4(x + 2)$

Solve for the specified variable in each formula or literal equation.

37. $P = C + CM$ for C (retail)

38. $S = P - PD$ for P (retail)

39. $C = 2\pi r$ for r (geometry)

40. $V = LWH$ for W (geometry)

41. $\frac{P_1 V_1}{T_1} = \frac{P_2 V_2}{T_2}$ for T_2 (science)

42. $\frac{C}{P_2} = \frac{P_1}{d^2}$ for P_2 (communication)

43. $V = \frac{4}{3}\pi r^2 h$ for h (geometry)

44. $V = \frac{1}{3}\pi r^2 h$ for h (geometry)

45. $S_n = n\left(\frac{a_1 + a_n}{2}\right)$ for n (sequences)

46. $A = \frac{h(b_1 + b_2)}{2}$ for h (geometry)

47. $S = B + \frac{1}{2}PS$ for P (geometry)

48. $s = \frac{1}{2}gt^2 + vt$ for g (physics)

49. $Ax + By = C$ for y

50. $2x + 3y = 6$ for y

51. $\frac{5}{6}x + \frac{3}{8}y = 2$ for y

52. $\frac{2}{3}x - \frac{7}{9}y = 12$ for y

53. $y - 3 = \frac{-4}{5}(x + 10)$ for y

54. $y + 4 = \frac{-2}{15}(x + 10)$ for y

The following equations are given in $ax + b = c$ form. Solve by identifying the value of a, b, and c, then using the formula $x = \dfrac{c - b}{a}$.

55. $3x + 2 = -19$

56. $7x + 5 = 47$

57. $-6x + 1 = 33$

58. $-4x + 9 = 43$

59. $7x - 13 = -27$

60. $3x - 4 = -25$

▶ WORKING WITH FORMULAS

61. **Surface area of a cylinder:** $SA = 2\pi r^2 + 2\pi rh$

The surface area of a cylinder is given by the formula shown, where h is the height of the cylinder and r is the radius of the base. Find the height of a cylinder that has a radius of 8 cm and a surface area of 1256 cm². Use $\pi \approx 3.14$.

62. Using the equation-solving process for Exercise 61 as a model, solve the formula $SA = 2\pi r^2 + 2\pi rh$ for h.

► **APPLICATIONS**

Solve by building an equation model and using the problem-solving guidelines as needed.

General Modeling Exercises

63. Two spelunkers (cave explorers) were exploring different branches of an underground cavern. The first was able to descend 198 ft farther than twice the second. If the first spelunker descended a 1218 ft, how far was the second spelunker able to descend?

64. The area near the joining of the Tigris and Euphrates Rivers (in modern Iraq) has often been called the *Cradle of Civilization,* since the area has evidence of many ancient cultures. The length of the Euphrates River exceeds that of the Tigris by 620 mi. If they have a combined length of 2880 mi, how long is each river?

65. U.S. postal regulations require that a package can have a maximum combined length and girth (distance around) of 108 in. A shipping carton is constructed so that it has a width of 14 in., a height of 12 in., and can be cut or folded to various lengths. What is the maximum length that can be used?
    Source: www.USPS.com

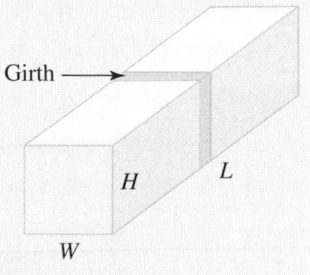
Girth →
H   L
W

66. Hi-Tech Home Improvements buys a fleet of identical trucks that cost $32,750 each. The company is allowed to depreciate the value of their trucks for tax purposes by $5250 per year. If company policies dictate that older trucks must be sold once their value declines to $6500, approximately how many years will they keep these trucks?

67. The longest suspension bridge in the world is the Akashi Kaikyo (Japan) with a length of 6532 feet. Japan is also home to the Shimotsui Straight bridge. The Akashi Kaikyo bridge is 364 ft more than twice the length of the Shimotsui bridge. How long is the Shimotsui bridge?
    Source: www.guinnessworldrecords.com

68. The Mars rover *Spirit* landed on January 3, 2004. Just over 1 yr later, on January 14, 2005, the *Huygens* probe landed on Titan (one of Saturn's moons). At their closest approach, the distance from the Earth to Saturn is 29 million mi more than 21 times the distance from the Earth to Mars. If the distance to Saturn is 743 million mi, what is the distance to Mars?

Consecutive Integer Exercises

69. Find two consecutive even integers such that the sum of twice the smaller integer plus the larger integer is one hundred forty-six.

70. When the smaller of two consecutive integers is added to three times the larger, the result is fifty-one. Find the smaller integer.

71. Seven times the first of two consecutive odd integers is equal to five times the second. Find each integer.

72. Find three consecutive even integers where the sum of triple the first and twice the second is eight more than four times the third.

Uniform Motion Exercises

73. At 9:00 A.M., Linda leaves work on a business trip, gets on the interstate, and sets her cruise control at 60 mph. At 9:30 A.M., Bruce notices she's left her briefcase and cell phone, and immediately starts after her driving 75 mph. At what time will Bruce catch up with Linda?

74. A plane flying at 300 mph has a 3-hr head start on a "chase plane," which has a speed of 800 mph. How far from the airport will the chase plane overtake the first plane?

75. Jeff had a job interview in a nearby city 72 mi away. On the first leg of the trip he drove an average of 30 mph through a long construction zone, but was able to drive 60 mph after passing through this zone. If driving time for the trip was $1\frac{1}{2}$ hr, how long was he driving in the construction zone?

76. At a high-school cross-country meet, Jared jogged 8 mph for the first part of the race, then increased his speed to 12 mph for the second part. If the race was 21 mi long and Jared finished in 2 hr, how far did he jog at the faster pace?

Mixture Exercises

Give the total amount of the mix that results and the percent concentration or worth of the mix.

77. Two quarts of 100% orange juice are mixed with 2 quarts of water (0% juice).

78. Ten pints of a 40% acid are combined with 10 pints of an 80% acid.

79. Eight pounds of premium coffee beans worth $2.50 per pound are mixed with 8 lb of standard beans worth $1.10 per pound.

80. A rancher mixes 50 lb of a custom feed blend costing $1.80 per pound, with 50 lb of cheap cottonseed worth $0.60 per pound.

Solve each application of the mixture concept.

81. To help sell more of a lower grade meat, a butcher mixes some premium ground beef worth $3.10/lb, with 8 lb of lower grade ground beef worth $2.05/lb. If the result was an intermediate grade of ground beef worth $2.68/lb, how much premium ground beef was used?

82. Knowing that the camping/hiking season has arrived, a nutrition outlet is mixing GORP (Good Old Raisins and Peanuts) for the anticipated customers. How many pounds of peanuts worth $1.29/lb, should be mixed with 20 lb of deluxe raisins worth $1.89/lb, to obtain a mix that will sell for $1.49/lb?

83. How many pounds of walnuts at 84¢/lb should be mixed with 20 lb of pecans at $1.20/lb to give a mixture worth $1.04/lb?

84. How many pounds of cheese worth 81¢/lb must be mixed with 10 lb cheese worth $1.29/lb to make a mixture worth $1.11/lb?

▶ EXTENDING THE THOUGHT

85. Look up and read the following article. Then turn in a one page summary. "Don't Give Up!," William H. Kraus, *Mathematics Teacher*, Volume 86, Number 2, February 1993: pages 110–112.

86. A chemist has four solutions of a very rare and expensive chemical that are 15% acid (cost $120 per ounce), 20% acid (cost $180 per ounce), 35% acid (cost $280 per ounce) and 45% acid (cost $359 per ounce). She requires 200 oz of a 29% acid solution. Find the combination of any two of these concentrations that will minimize the total cost of the mix.

87. P, Q, R, S, T, and U represent numbers. The arrows in the figure show the sum of the two or three numbers added in the indicated direction

(Example: $Q + T = 23$). Find $P + Q + R + S + T + U$.

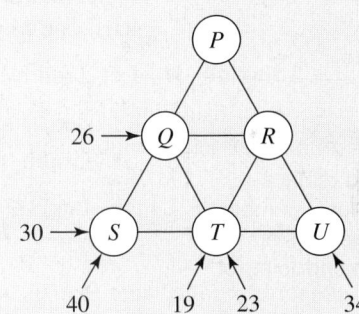

88. Given a sphere circumscribed by a cylinder, verify the volume of the sphere is $\frac{2}{3}$ that of the cylinder.

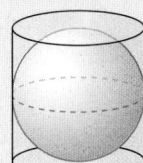

▶ MAINTAINING YOUR SKILLS

89. (R.1) Simplify the expression using the order of operations.

$-2 - 6^2 \div 4 + 8$

90. (R.3) Name the coefficient of each term in the expression:

$-3v^3 + v^2 - \frac{v}{3} + 7$

91. (R.4) Factor each expression:

  a. $4x^2 - 9$ **b.** $x^3 - 27$

92. (R.2) Identify the property illustrated:

$\frac{6}{7} \cdot 5 \cdot 21 = \frac{6}{7} \cdot 21 \cdot 5$

1.2 | Linear Inequalities in One Variable

Learning Objectives

In Section 1.2 you will learn how to:

☐ **A.** Solve inequalities and state solution sets

☐ **B.** Solve linear inequalities

☐ **C.** Solve compound inequalities

☐ **D.** Solve applications of inequalities

There are many real-world situations where the mathematical model leads to a statement of *inequality* rather than equality. Here are a few examples:

Clarice wants to buy a house costing $85,000 or less.

To earn a "B," Shanté must score more than 90% on the final exam.

To escape the Earth's gravity, a rocket must travel 25,000 mph or more.

While conditional linear equations in one variable have a single solution, linear inequalities often have an *infinite number of solutions*—which means we must develop additional methods for writing a solution set.

A. Inequalities and Solution Sets

The set of numbers that satisfy an inequality is called the **solution set.** Instead of using a simple inequality to write solution sets, we will often use (1) a form of **set notation,** (2) a **number line** graph, or (3) **interval notation.** Interval notation is a symbolic way of indicating a selected interval of the real number line. When a number acts as the **boundary point** for an interval (also called an **endpoint**), we use a left bracket "[" or a right bracket "]" to indicate **inclusion** of the endpoint. If the boundary point is **not included,** we use a left parenthesis "(" or right parenthesis ")."

EXAMPLE 1 ▶ Using Inequalities in Context

Model the given phrase using the correct inequality symbol. Then state the result in set notation, graphically, and in interval notation: "If the ball had traveled at least one more foot in the air, it would have been a home run."

Solution ▶ Let d represent additional distance: $d \geq 1$.

- Set notation: $\{d \mid d \geq 1\}$
- Graph
  $$\xleftarrow{\qquad} \underset{-2\;-1\;\;0\;\;1\;\;2\;\;3\;\;4\;\;5}{\rule{0pt}{0pt}} \xrightarrow{\qquad}$$
- Interval notation: $d \in [1, \infty)$

Now try Exercises 7 through 18 ▶

The "$\in$" symbol says the number d is *an element of the set or interval* given. The "∞" symbol represents positive infinity and indicates the interval continues forever to the right. Note that the endpoints of an interval must occur in the same order as on the number line (*smaller value on the left; larger value on the right*).

A short summary of other possibilities is given here. Many variations are possible.

Conditions ($a < b$)	Set Notation	Number Line	Interval Notation
x is greater than k	$\{x \mid x > k\}$	k	$x \in (k, \infty)$
x is less than or equal to k	$\{x \mid x \leq k\}$	k	$x \in (-\infty, k]$
x is less than b and greater than a	$\{x \mid a < x < b\}$	$a \qquad b$	$x \in (a, b)$
x is less than b and greater than or equal to a	$\{x \mid a \leq x < b\}$	$a \qquad b$	$x \in [a, b)$
x is less than a or x is greater than b	$\{x \mid x < a \text{ or } x > b\}$	$a \qquad b$	$x \in (-\infty, a) \cup (b, \infty)$

B. Solving Linear Inequalities

A linear *inequality* resembles a linear *equality* in many respects:

Linear Inequality	**Related Linear Equation**
(1) $x < 3$	$x = 3$
(2) $\frac{3}{8}p - 2 \geq -12$	$\frac{3}{8}p - 2 = -12$

A linear inequality in one variable is one that can be written in the form $ax + b < c$, where a, b, and $c \in \mathbb{R}$ and $a \neq 0$. This definition and the following properties also apply when other inequality symbols are used. Solutions to simple inequalities are easy to spot. For instance, $x = -2$ is a solution to $x < 3$ since $-2 < 3$. For more involved inequalities we use the **additive property of inequality** and the **multiplicative property of inequality.** Similar to solving equations, we solve inequalities by isolating the variable on one side to obtain a solution form such as *variable < number*.

The Additive Property of Inequality

If A, B, and C represent algebraic expressions and $A < B$,

$$\text{then} \quad A + C < B + C$$

Like quantities (numbers or terms) can be added to both sides of an inequality.

While there is little difference between the additive property of *equality* and the additive property of *inequality,* there is an *important difference* between the multiplicative property of *equality* and the multiplicative property of *inequality.* To illustrate, we begin with $-2 < 5$. Multiplying both sides by positive three yields $-6 < 15$, a true inequality. But notice what happens when we **multiply both sides by negative three:**

$$-2 < 5 \qquad \text{original inequality}$$
$$-2(-3) < 5(-3) \qquad \text{multiply by negative three}$$
$$6 < -15 \qquad \text{false}$$

This is a *false* inequality, because 6 is *to the right* of -15 on the number line. Multiplying (or dividing) an inequality by a negative quantity *reverses the order relationship between two quantities* (we say it changes the *sense* of the inequality). We must compensate for this by reversing the inequality symbol.

$$6 > -15 \qquad \text{change direction of symbol to maintain a true statement}$$

For this reason, the multiplicative property of inequality is stated in two parts.

The Multiplicative Property of Inequality

If A, B, and C represent algebraic expressions and $A < B$,	If A, B, and C represent algebraic expressions and $A < B$,
then $AC < BC$	then $AC > BC$
if C is a *positive quantity* (inequality symbol remains the same).	if C is a *negative quantity* (inequality symbol must be reversed).

EXAMPLE 2 ▶ Solving an Inequality

Solve the inequality, then graph the solution set and write it in interval notation: $\frac{-2}{3}x + \frac{1}{2} \le \frac{5}{6}$.

Solution ▶

$$\frac{-2}{3}x + \frac{1}{2} \le \frac{5}{6} \qquad \text{original inequality}$$

$$6\left(\frac{-2}{3}x + \frac{1}{2}\right) \le (6)\frac{5}{6} \qquad \text{clear fractions (multiply by LCD)}$$

$$-4x + 3 \le 5 \qquad \text{simplify}$$

$$-4x \le 2 \qquad \text{subtract 3}$$

$$x \ge -\frac{1}{2} \qquad \text{divide by } -4, \textit{reverse inequality sign}$$

WORTHY OF NOTE

As an alternative to multiplying or dividing by a negative value, the additive property of inequality can be used to ensure the variable term will be positive. From Example 2, the inequality $-4x \le 2$ can be written as $-2 \le 4x$ by adding $4x$ to both sides and subtracting 2 from both sides. This gives the solution $-\frac{1}{2} \le x$, which is equivalent to $x \ge -\frac{1}{2}$.

• Graph:

$$\begin{array}{c} -\frac{1}{2} \\ \text{—+——+——+——[——+——+——+——+——+——} \\ -3-2-101234 \end{array}$$

• Interval notation: $x \in \left[-\frac{1}{2}, \infty\right)$

Now try Exercises 19 through 28 ▶

To check a linear inequality, you often have an infinite number of choices—any number from the solution set/interval. If a test value from the solution interval results in a true inequality, all numbers in the interval are solutions. For Example 2, using $x = 0$ results in the true statement $\frac{1}{2} \le \frac{5}{6}$ ✓.

Some inequalities have all real numbers as the solution set: $\{x | x \in \mathbb{R}\}$, while other inequalities have no solutions, with the answer given as the empty set: { }.

EXAMPLE 3 ▶ Solving Inequalities

Solve the inequality and write the solution in set notation:

  a. $7 - (3x + 5) \ge 2(x - 4) - 5x$ **b.** $3(x + 4) - 5 < 2(x - 3) + x$

Solution ▶   **a.** $7 - (3x + 5) \ge 2(x - 4) - 5x$ original inequality

$$7 - 3x - 5 \ge 2x - 8 - 5x \qquad \text{distributive property}$$

$$-3x + 2 \ge -3x - 8 \qquad \text{combine like terms}$$

$$2 \ge -8 \qquad \text{add } 3x$$

Since the resulting statement is always true, the original inequality is true for all real numbers. The solution is $\{x | x \in \mathbb{R}\}$.

  b. $3(x + 4) - 5 < 2(x - 3) + x$ original inequality

$$3x + 12 - 5 < 2x - 6 + x \qquad \text{distribute}$$

$$3x + 7 < 3x - 6 \qquad \text{combine like terms}$$

$$7 < -6 \qquad \text{subtract } 3x$$

☑ **B.** You've just learned how to solve linear inequalities

Since the resulting statement is always false, the original inequality is false for all real numbers. The solution is { }.

Now try Exercises 29 through 34 ▶

C. Solving Compound Inequalities

In some applications of inequalities, we must consider more than one solution interval. These are called **compound inequalities,** and require us to take a close look at the operations of **union** "∪" and **intersection** "∩". The intersection of two sets A and B, written $A \cap B$, is the set of all elements *common to both sets*. The union of two sets A and B, written $A \cup B$, is the set of all elements *that are in either set*. When stating the union of two sets, repetitions are unnecessary.

EXAMPLE 4 ▶ **Finding the Union and Intersection of Two Sets**

For set $A = \{-2, -1, 0, 1, 2, 3\}$ and set $B = \{1, 2, 3, 4, 5\}$, determine $A \cap B$ and $A \cup B$.

Solution ▶ $A \cap B$ is the set of all elements in *both A and B:*
$A \cap B = \{1, 2, 3\}$.
$A \cup B$ is the set of all elements in *either A or B:*
$A \cup B = \{-2, -1, 0, 1, 2, 3, 4, 5\}$.

WORTHY OF NOTE

For the long term, it may help to rephrase the distinction as follows. The intersection is a *selection* of elements that are common to two sets, while the union is a *collection* of the elements from two sets (with no repetitions).

Now try Exercises 35 through 40 ▶

Notice the intersection of two sets is described using the word "and," while the union of two sets is described using the word "or." When compound inequalities are formed using these words, the solution is modeled after the ideas from Example 4. If "and" is used, the solutions must satisfy *both* inequalities. If "or" is used, the solutions can satisfy *either* inequality.

EXAMPLE 5 ▶ **Solving a Compound Inequality**

Solve the compound inequality, then write the solution in interval notation:
$-3x - 1 < -4$ **or** $4x + 3 < -6$.

Solution ▶ Begin with the statement as given:

$$-3x - 1 < -4 \qquad \text{or} \qquad 4x + 3 < -6 \qquad \text{original statement}$$
$$-3x < -3 \qquad \text{or} \qquad 4x < -9 \qquad \text{isolate variable term}$$
$$x > 1 \qquad \text{or} \qquad x < -\frac{9}{4} \qquad \text{solve for } x, \text{ reverse first inequality symbol}$$

WORTHY OF NOTE

The graphs from Example 5 clearly show the solution consists of two disjoint (disconnected) intervals. This is reflected in the "or" statement: $x < -\frac{9}{4}$ or $x > 1$, and in the interval notation. Also, note the solution $x < -\frac{9}{4}$ or $x > 1$ is not equivalent to $-\frac{9}{4} > x > 1$, as there is no single number that is both greater than 1 and less than $-\frac{9}{4}$ at the same time.

The solution $x > 1$ **or** $x < -\frac{9}{4}$ is better understood by graphing each interval separately, *then selecting both intervals (the union).*

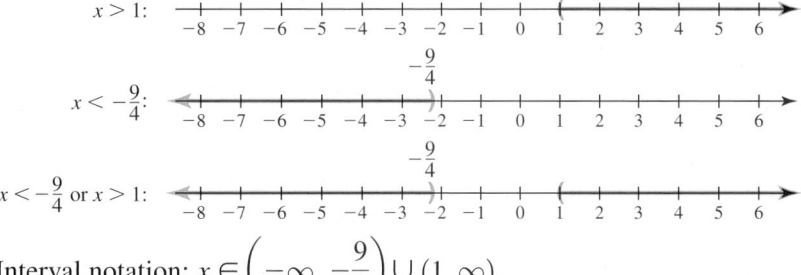

Interval notation: $x \in \left(-\infty, -\frac{9}{4}\right) \cup (1, \infty)$.

Now try Exercises 41 and 42 ▶

EXAMPLE 6 ▶ **Solving a Compound Inequality**

Solve the compound inequality, then write the solution in interval notation:
$3x + 5 > -13$ **and** $3x + 5 < -1$.

Solution ▶ Begin with the statement as given:

$$3x + 5 > -13 \qquad \text{and} \qquad 3x + 5 < -1 \qquad \text{original statement}$$
$$3x > -18 \qquad \text{and} \qquad 3x < -6 \qquad \text{subtract five}$$
$$x > -6 \qquad \text{and} \qquad x < -2 \qquad \text{divide by 3}$$

The solution $x > -6$ **and** $x < -2$ can best be understood by graphing each interval separately, then *noting where they intersect.*

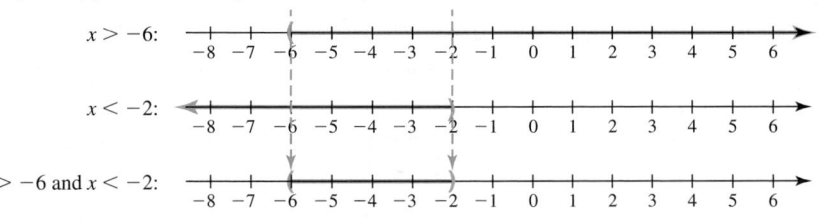

Interval notation: $x \in (-6, -2)$.

Now try Exercises 43 through 54 ▶

The solution from Example 6 consists of the single interval $(-6, -2)$, indicating the original inequality could actually be *joined* and written as $-6 < x < -2$, called a **joint** or **compound inequality** (see Worthy of Note). We solve joint inequalities in much the same way as linear inequalities, but must remember they *have three parts (left, middle, and right)*. This means operations must be applied to *all three parts* in each step of the solution process, to obtain a solution form such as *smaller number $< x <$ larger number*. The same ideas apply when other inequality symbols are used.

EXAMPLE 7 ▶ **Solving a Compound Inequality**

Solve the compound inequality, then graph the solution set and write it in interval notation: $1 > \dfrac{2x + 5}{-3} \geq -6$.

Solution ▶ $1 > \dfrac{2x + 5}{-3} \geq -6$ original inequality

$-3 < 2x + 5 \leq 18$ multiply all parts by -3; reverse the inequality symbols

$-8 < 2x \leq 13$ subtract 5 from all parts

$-4 < x \leq \dfrac{13}{2}$ divide all parts by 2

☑ **C.** You've just learned how to solve compound inequalities

• Graph:

• Interval notation: $x \in \left(-4, \dfrac{13}{2}\right]$

Now try Exercises 55 through 60 ▶

D. Applications of Inequalities

Domain and Allowable Values

One application of inequalities involves the concept of allowable values. Consider the expression $\frac{24}{x}$. As Table 1.2 suggests, we can evaluate this expression using any real number *other than zero,* since the expression $\frac{24}{0}$ is undefined. Using set notation the allowable values are written $\{x | x \in \mathbb{R}, x \neq 0\}$. To graph the solution we must be careful to exclude zero, as shown in Figure 1.4.

The graph gives us a snapshot of the solution using interval notation, which is written as a union of two **disjoint (disconnected) intervals** so as to exclude zero: $x \in (-\infty, 0) \cup (0, \infty)$. The set of allowable values is referred to as the **domain** of the expression. Allowable values are said to be "*in the domain*" of the expression; values that are not allowed are said to be "*outside the domain.*" When the denominator of a fraction contains a variable expression, values that cause a denominator of zero are outside the domain.

Table 1.2

x	$\frac{24}{x}$
6	4
-12	-2
$\frac{1}{2}$	48
0	error

Figure 1.4

EXAMPLE 8 ▶ **Determining the Domain of an Expression**

Determine the domain of the expression $\dfrac{6}{x-2}$. State the result in set notation, graphically, and using interval notation.

Solution ▶ Set the denominator equal to zero and solve: $x - 2 = 0$ yields $x = 2$. This means 2 is outside the domain and *must be excluded*.

- Set notation: $\{x | x \in \mathbb{R}, x \neq 2\}$

- Graph:

- Interval notation: $x \in (-\infty, 2) \cup (2, \infty)$

Now try Exercises 61 through 68 ▶

A second area where allowable values are a concern involves the square root operation. Recall that $\sqrt{49} = 7$ since $7 \cdot 7 = 49$. However, $\sqrt{-49}$ cannot be written as the product of two real numbers since $(-7) \cdot (-7) = 49$ and $7 \cdot 7 = 49$. In other words, $\sqrt{X}$ represents a real number only if the radicand is positive or zero. If X represents an algebraic expression, the domain of $\sqrt{X}$ is $\{X | X \geq 0\}$.

EXAMPLE 9 ▶ **Determining the Domain of an Expression**

Determine the domain of $\sqrt{x + 3}$. State the domain in set notation, graphically, and in interval notation.

Solution ▶ The radicand must represent a nonnegative number. Solving $x + 3 \geq 0$ gives $x \geq -3$.

- Set notation: $\{x | x \geq -3\}$

- Graph:

- Interval notation: $x \in [-3, \infty)$

Now try Exercises 69 through 76 ▶

Inequalities are widely used to help gather information, and to make comparisons that will lead to informed decisions. Here, the problem-solving guide is once again a valuable tool.

EXAMPLE 10 ▶ **Using an Inequality to Compute Desired Test Scores**

Justin earned scores of 78, 72, and 86 on the first three out of four exams. What score must he earn on the fourth exam to have an average of at least 80?

Solution ▶ **Gather and organize information;** highlight any key phrases.
First the scores: 78, 72, 86. An average of *at least* 80 means $A \geq 80$.
Make the problem visual.

Test 1	Test 2	Test 3	Test 4	Computed Average	Minimum
78	72	86	x	$\dfrac{78 + 72 + 86 + x}{4}$	80

Assign a variable; build related expressions.

Let x represent Justin's score on the fourth exam, then $\dfrac{78 + 72 + 86 + x}{4}$ represents his average score.

$$\frac{78 + 72 + 86 + x}{4} \geq 80 \qquad \text{average must be greater than or equal to 80}$$

Write the equation model and solve.

$$78 + 72 + 86 + x \geq 320 \qquad \text{multiply by 4}$$
$$236 + x \geq 320 \qquad \text{simplify}$$
$$x \geq 84 \qquad \text{solve for } x \text{ (subtract 236)}$$

Justin must score at least an 84 on the last exam to earn an 80 average.

Now try Exercises 79 through 86 ▶

As your problem-solving skills improve, the process outlined in the problem-solving guide naturally becomes less formal, as we work more directly toward the equation model. See Example 11.

EXAMPLE 11 ▶   **Using an Inequality to Make a Financial Decision**

As Margaret starts her new job, her employer offers two salary options. Plan 1 is base pay of $1475/mo plus 3% of sales. Plan 2 is base pay of $500/mo plus 15% of sales. What level of monthly sales is needed for her to earn more under Plan 2?

Solution ▶   Let x represent her monthly sales in dollars. The equation model for Plan 1 would be $0.03x + 1475$; for Plan 2 we have $0.15x + 500$. To find the sales volume needed for her to earn more under Plan 2, we solve the inequality

$$0.15x + 500 > 0.03x + 1475 \qquad \text{Plan 2} > \text{Plan 1}$$
$$0.12x + 500 > 1475 \qquad \text{subtract } 0.03x$$
$$0.12x > 975 \qquad \text{subtract } 500$$
$$x > 8125 \qquad \text{divide by } 0.12$$

☑ **D.** You've just learned how to solve applications of inequalities

If Margaret can generate more than $8125 in monthly sales, she will earn more under Plan 2.

Now try Exercises 87 and 88 ▶

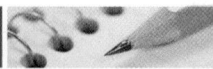

1.2 EXERCISES

▶ CONCEPTS AND VOCABULARY

Fill in each blank with the appropriate word or phrase.
Carefully reread the section, if necessary.

1. For inequalities, the three ways of writing a solution set are _____ notation, a number line graph, and _____ notation.

2. The mathematical sentence $3x + 5 < 7$ is a(n) _____ inequality, while $-2 < 3x + 5 < 7$ is a(n) _____ inequality.

3. The _____ of sets A and B is written $A \cap B$. The _____ of sets A and B is written $A \cup B$.

4. The intersection of set A with set B is the set of elements in A _____ B. The union of set A with set B is the set of elements in A _____ B.

5. Discuss/Explain how the concept of domain and allowable values relates to rational and radical expressions. Include a few examples.

6. Discuss/Explain why the inequality symbol must be reversed when multiplying or dividing by a negative quantity. Include a few examples.

▶ DEVELOPING YOUR SKILLS

Use an inequality to write a mathematical model for each statement.

7. To qualify for a secretarial position, a person must type at least 45 words per minute.

8. The balance in a checking account must remain above $1000 or a fee is charged.

9. To bake properly, a turkey must be kept between the temperatures of 250° and 450°.

10. To fly effectively, the airliner must cruise at or between altitudes of 30,000 and 35,000 ft.

Graph each inequality on a number line.

11. $y < 3$

12. $x > -2$

13. $m \leq 5$

14. $n \geq -4$

15. $x \neq 1$

16. $x \neq -3$

17. $5 > x > 2$

18. $-3 < y \leq 4$

Write the solution set illustrated on each graph in set notation and interval notation.

19.

20.

21.

22.

Solve the inequality and write the solution in set notation. Then graph the solution and write it in interval notation.

23. $5a - 11 \geq 2a - 5$

24. $-8n + 5 > -2n - 12$

25. $2(n + 3) - 4 \leq 5n - 1$

26. $-5(x + 2) - 3 < 3x + 11$

27. $\dfrac{3x}{8} + \dfrac{x}{4} < -4$ 28. $\dfrac{2y}{5} + \dfrac{y}{10} < -2$

Solve each inequality and write the solution in set notation.

29. $7 - 2(x + 3) \geq 4x - 6(x - 3)$

30. $-3 - 6(x - 5) \leq 2(7 - 3x) + 1$

31. $4(3x - 5) + 18 < 2(5x + 1) + 2x$

32. $8 - (6 + 5m) > -9m - (3 - 4m)$

33. $-6(p - 1) + 2p \leq -2(2p - 3)$

34. $9(w - 1) - 3w \geq -2(5 - 3w) + 1$

Determine the intersection and union of sets A, B, C, and D as indicated, given $A = \{-3, -2, -1, 0, 1, 2, 3\}$, $B = \{2, 4, 6, 8\}$, $C = \{-4, -2, 0, 2, 4\}$, and $D = \{4, 5, 6, 7\}$.

35. $A \cap B$ and $A \cup B$ 36. $A \cap C$ and $A \cup C$

37. $A \cap D$ and $A \cup D$ 38. $B \cap C$ and $B \cup C$

39. $B \cap D$ and $B \cup D$ 40. $C \cap D$ and $C \cup D$

Express the compound inequalities graphically and in interval notation.

41. $x < -2$ or $x > 1$ 42. $x < -5$ or $x > 5$

43. $x < 5$ and $x \geq -2$ 44. $x \geq -4$ and $x < 3$

45. $x \geq 3$ and $x \leq 1$ 46. $x \geq -5$ and $x \leq -7$

Solve the compound inequalities and graph the solution set.

47. $4(x - 1) \leq 20$ or $x + 6 > 9$

48. $-3(x + 2) > 15$ or $x - 3 \leq -1$

49. $-2x - 7 \leq 3$ and $2x \leq 0$

50. $-3x + 5 \leq 17$ and $5x \leq 0$

51. $\frac{3}{5}x + \frac{1}{2} > \frac{3}{10}$ and $-4x > 1$

52. $\frac{2}{3}x - \frac{5}{6} \leq 0$ and $-3x < -2$

53. $\dfrac{3x}{8} + \dfrac{x}{4} < -3$ or $x + 1 > -5$

54. $\dfrac{2x}{5} + \dfrac{x}{10} < -2$ or $x - 3 > 2$

55. $-3 \leq 2x + 5 < 7$ 56. $2 < 3x - 4 \leq 19$

57. $-0.5 \leq 0.3 - x \leq 1.7$

58. $-8.2 < 1.4 - x < -0.9$

59. $-7 < -\frac{3}{4}x - 1 \leq 11$

60. $-21 \leq -\frac{2}{3}x + 9 < 7$

Determine the domain of each expression. Write your answer in interval notation.

61. $\dfrac{12}{m}$

62. $\dfrac{-6}{n}$

63. $\dfrac{5}{y + 7}$

64. $\dfrac{4}{x - 3}$

65. $\dfrac{a + 5}{6a - 3}$

66. $\dfrac{m + 5}{8m + 4}$

67. $\dfrac{15}{3x - 12}$

68. $\dfrac{7}{2x + 6}$

Determine the domain for each expression. Write your answer in interval notation.

69. $\sqrt{x - 2}$

70. $\sqrt{y + 7}$

71. $\sqrt{3n - 12}$

72. $\sqrt{2m + 5}$

73. $\sqrt{b - \frac{4}{3}}$

74. $\sqrt{a + \frac{3}{4}}$

75. $\sqrt{8 - 4y}$

76. $\sqrt{12 - 2x}$

▶ WORKING WITH FORMULAS

77. Body mass index: $B = \dfrac{704W}{H^2}$

The U.S. government publishes a body mass index formula to help people consider the risk of heart disease. An index "*B*" of 27 or more means that a person is at risk. Here *W* represents weight in pounds and *H* represents height in inches. (a) Solve the formula for *W*. (b) If your height is 5′8″ what range of weights will help ensure you remain safe from the risk of heart disease?

Source: www.surgeongeneral.gov/topics.

78. Lift capacity: $75S + 125B \leq 750$

The capacity in pounds of the lift used by a roofing company to place roofing shingles and buckets of roofing nails on rooftops is modeled by the formula shown, where *S* represents packs of shingles and *B* represents buckets of nails. Use the formula to find (a) the largest number of shingle packs that can be lifted, (b) the largest number of nail buckets that can be lifted, and (c) the largest number of shingle packs that can be lifted along with three nail buckets.

▶ APPLICATIONS

Write an inequality to model the given information and solve.

79. Exam scores: Jacques is going to college on an academic scholarship that requires him to maintain at least a 75% average in all of his classes. So far he has scored 82%, 76%, 65%, and 71% on four exams. What scores are possible on his last exam that will enable him to keep his scholarship?

80. Timed trials: In the first three trials of the 100-m butterfly, Johann had times of 50.2, 49.8, and 50.9 sec. How fast must he swim the final timed trial to have an average time of 50 sec?

81. Checking account balance: If the average daily balance in a certain checking account drops below $1000, the bank charges the customer a $7.50 service fee. The table gives the daily balance for

one customer. What must the daily balance be for Friday to avoid a service charge?

Weekday	Balance
Monday	$1125
Tuesday	$850
Wednesday	$625
Thursday	$400

82. Average weight: In the National Football League, many consider an offensive line to be "small" if the average weight of the five down linemen is less than 325 lb. Using the table, what must the weight of the right tackle be so that the line will not be considered too small?

Lineman	Weight
Left tackle	318 lb
Left guard	322 lb
Center	326 lb
Right guard	315 lb
Right tackle	?

83. Area of a rectangle: Given the rectangle shown, what is the range of values for the width, in order to keep the area less than 150 m²?

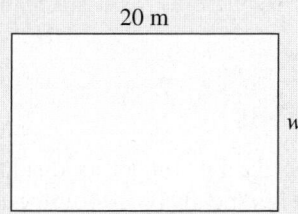

20 m

w

84. Area of a triangle: Using the triangle shown, find the height that will guarantee an area equal to or greater than 48 in².

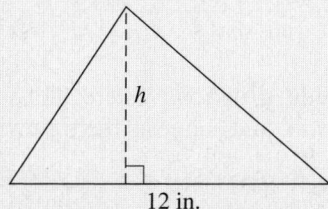

h

12 in.

85. Heating and cooling subsidies: As long as the outside temperature is over 45°F and less than 85°F ($45 < F < 85$), the city does not issue heating or cooling subsidies for low-income families. What is the corresponding range of Celsius temperatures C? Recall that $F = \frac{9}{5}C + 32$.

86. U.S. and European shoe sizes: To convert a European male shoe size "E" to an American male shoe size "A," the formula $A = 0.76E - 23$ can be used. Lillian has five sons in the U.S. military, with shoe sizes ranging from size 9 to size 14 ($9 \le A \le 14$). What is the corresponding range of European sizes? Round to the nearest half-size.

87. Power tool rentals: Sunshine Equipment Co. rents its power tools for a $20 fee, plus $4.50/hr. Kealoha's Rentals offers the same tools for an $11 fee plus $6.00/hr. How many hours h must a tool be rented to make the cost at Sunshine a better deal?

88. Moving van rentals: Davis Truck Rentals will rent a moving van for $15.75/day plus $0.35 per mile. Bertz Van Rentals will rent the same van for $25/day plus $0.30 per mile. How many miles m must the van be driven to make the cost at Bertz a better deal?

▶ EXTENDING THE CONCEPT

89. Use your local library, the Internet, or another resource to find the highest and lowest point on each of the seven continents. Express the range of altitudes for each continent as a joint inequality. Which continent has the greatest range?

90. The sum of two consecutive even integers is greater than or equal to 12 and less than or equal to 22. List all possible values for the two integers.

Place the correct inequality symbol in the blank to make the statement true.

91. If $m > 0$ and $n < 0$, then mn _____ 0.

92. If $m > n$ and $p > 0$, then mp _____ np.

93. If $m < n$ and $p > 0$, then mp _____ np.

94. If $m \le n$ and $p < 0$, then mp _____ np.

95. If $m > n$, then $-m$ _____ $-n$.

96. If $m < n$, then $\frac{1}{m}$ _____ $\frac{1}{n}$.

97. If $m > 0$ and $n < 0$, then m^2 _____ n.

98. If $m < 0$, then m^3 _____ 0.

▶ MAINTAINING YOUR SKILLS

99. (R.2) Translate into an algebraic expression: eight subtracted from twice a number.

100. (1.1) Solve: $-4(x - 7) - 3 = 2x + 1$

101. (R.3) Simplify the algebraic expression: $2(\frac{5}{9}x - 1) - (\frac{1}{6}x + 3)$.

102. (1.1) Solve: $\frac{4}{5}m + \frac{2}{3} = \frac{1}{2}$

1.3 Absolute Value Equations and Inequalities

Learning Objectives

In Section 1.3 you will learn how to:

☐ **A.** Solve absolute value equations

☐ **B.** Solve "less than" absolute value inequalities

☐ **C.** Solve "greater than" absolute value inequalities

☐ **D.** Solve applications involving absolute value

While the equations $x + 1 = 5$ and $|x + 1| = 5$ are similar in many respects, note the first has only the solution $x = 4$, while either $x = 4$ or $x = -6$ will satisfy the second. The fact there are two solutions shouldn't surprise us, as it's a natural result of how absolute value is defined.

A. Solving Absolute Value Equations

The absolute value of a number x can be thought of as its distance from zero on the number line, regardless of direction. This means $|x| = 4$ will have *two solutions*, since there are two numbers that are four units from zero: $x = -4$ and $x = 4$ (see Figure 1.5).

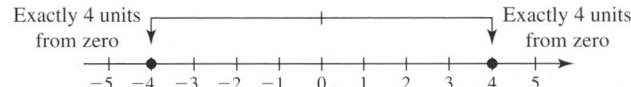

Figure 1.5

This basic idea can be extended to include situations where the quantity within absolute value bars *is an algebraic expression,* and suggests the following property.

> **Property of Absolute Value Equations**
>
> If X represents an algebraic expression and k is a positive real number,
>
> $$\text{then } |X| = k$$
>
> $$\text{implies } X = -k \text{ or } X = k$$

As the statement of this property suggests, it can only be applied *after* the absolute value expression has been isolated on one side.

WORTHY OF NOTE

Note if $k < 0$, the equation $|X| = k$ has no solutions since the absolute value of any quantity is always positive or zero. On a related note, we can verify that if $k = 0$, the equation $|X| = 0$ has only the solution $X = 0$.

EXAMPLE 1 ▶ Solving an Absolute Value Equation

Solve: $-5|x - 7| + 2 = -13$.

Solution ▶ Begin by isolating the absolute value expression.

$$\begin{aligned} -5|x - 7| + 2 &= -13 \quad \text{original equation} \\ -5|x - 7| &= -15 \quad \text{subtract 2} \\ |x - 7| &= 3 \quad \text{divide by } -5 \text{ (simplified form)} \end{aligned}$$

Now consider $x - 7$ as the variable expression "X" in the property of absolute value equations, giving

$$\begin{array}{lll} x - 7 = -3 & \text{or} & x - 7 = 3 \quad \text{apply the property of absolute value equations} \\ x = 4 & \text{or} & x = 10 \quad \text{add 7} \end{array}$$

Substituting into the original equation verifies the solution set is $\{4, 10\}$.

Now try Exercises 7 through 18 ▶

⚠ **CAUTION** ▶ For equations like those in Example 1, be careful not to treat the absolute value bars as simple grouping symbols. The equation $-5(x - 7) + 2 = -13$ has only the solution $x = 10$, and "misses" the second solution since it yields $x - 7 = 3$ in simplified form. The equation $-5|x - 7| + 2 = -13$ simplifies to $|x - 7| = 3$ and there are actually *two* solutions.

Absolute value equations come in many different forms. Always begin by isolating the absolute value expression, then apply the property of absolute value equations to solve.

EXAMPLE 2 ▶ **Solving an Absolute Value Equation**

Solve: $\left|5 - \dfrac{2}{3}x\right| - 9 = 8$

Solution ▶ $\left|5 - \dfrac{2}{3}x\right| - 9 = 8$ original equation

$\left|5 - \dfrac{2}{3}x\right| = 17$ add 9

$5 - \dfrac{2}{3}x = -17$ or $5 - \dfrac{2}{3}x = 17$ apply the property of absolute value equations

$-\dfrac{2}{3}x = -22$ or $-\dfrac{2}{3}x = 12$ subtract 5

$x = 33$ or $x = -18$ multiply by $-\frac{3}{2}$

Check ▶ For $x = 33$: $\left|5 - \dfrac{2}{3}(33)\right| - 9 = 8$ For $x = -18$: $\left|5 - \dfrac{2}{3}(-18)\right| - 9 = 8$

$|5 - 2(11)| - 9 = 8$ $|5 - 2(-6)| - 9 = 8$

$|5 - 22| - 9 = 8$ $|5 + 12| - 9 = 8$

$|-17| - 9 = 8$ $|17| - 9 = 8$

$17 - 9 = 8$ $17 - 9 = 8$

$8 = 8$ ✓ $8 = 8$ ✓

Both solutions check. The solution set is $\{-18, 33\}$.

> **WORTHY OF NOTE**
>
> As illustrated in both Examples 1 and 2, the property we use to solve absolute value equations can only be applied *after* the absolute value term has been isolated. As you will see, the same is true for the properties used to solve absolute value inequalities.

Now try Exercises 19 through 22 ▶

For some equations, it's helpful to apply the **multiplicative property of absolute value:**

Multiplicative Property of Absolute Value

If A and B represent algebraic expressions,

then $|AB| = |A||B|$.

Note that if $A = -1$ the property says $|-B| = |-1||B| = |B|$. More generally the property is applied where A is any constant.

EXAMPLE 3 ▶ **Solving Equations Using the Multiplicative Property of Absolute Value**

Solve: $|-2x| + 5 = 13$.

Solution ▶ $|-2x| + 5 = 13$ original equation

$|-2x| = 8$ subtract 5

$|-2||x| = 8$ apply multiplicative property of absolute value

$2|x| = 8$ simplify

$|x| = 4$ divide by 2

$x = -4$ or $x = 4$ apply property of absolute value equations

☑ **A. You've just learned how to solve absolute value equations**

Both solutions check. The solution set is $\{-4, 4\}$.

Now try Exercises 23 and 24 ▶

B. Solving "Less Than" Absolute Value Inequalities

Absolute value *inequalities* can be solved using the basic concept underlying the property of absolute value equalities. Whereas the equation $|x| = 4$ asks for all numbers x whose distance from zero is *equal* to 4, the inequality $|x| < 4$ asks for all numbers x whose distance from zero is *less than* 4.

Distance from zero is less than 4

Figure 1.6

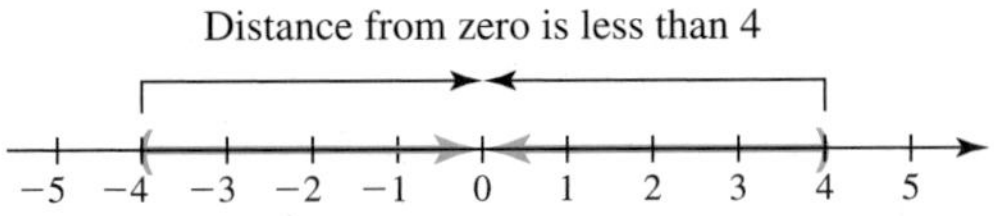

As Figure 1.6 illustrates, the solutions are $x > -4$ and $x < 4$, which can be written as the joint inequality $-4 < x < 4$. This idea can likewise be extended to include the absolute value of an algebraic expression X as follows.

WORTHY OF NOTE

Property I can also be applied when the "≤" symbol is used. Also notice that if $k < 0$, the solution is the empty set since the absolute value of any quantity is always positive or zero.

Property I: Absolute Value Inequalities

If X represents an algebraic expression and k is a positive real number,

$$\text{then } |X| < k$$

$$\text{implies } -k < X < k$$

EXAMPLE 4 ▶   Solving "Less Than" Absolute Value Inequalities

Solve the inequalities:

a. $\dfrac{|3x + 2|}{4} \leq 1$ **b.** $|2x - 7| < -5$

Solution ▶   **a.** $\dfrac{|3x + 2|}{4} \leq 1$ original inequality

$|3x + 2| \leq 4$ multiply by 4

$-4 \leq 3x + 2 \leq 4$ apply Property I

$-6 \leq 3x \leq 2$ subtract 2 from all three parts

$-2 \leq x \leq \dfrac{2}{3}$ divide all three parts by 3

The solution interval is $\left[-2, \frac{2}{3}\right]$.

b. $|2x - 7| < -5$ original inequality

Since the absolute value of any quantity is always positive or zero, the solution for this inequality is the empty set: { }.

WORTHY OF NOTE

As with the inequalities from Section 1.2, solutions to absolute value inequalities can be checked using a test value. For Example 4(a), substituting $x = 0$ from the solution interval yields:

$$\frac{1}{2} \leq 1 ✓$$

☑ **B.** You've just learned how to solve less than absolute value inequalities

Now try Exercises 25 through 38 ▶

C. Solving "Greater Than" Absolute Value Inequalities

For "greater than" inequalities, consider $|x| > 4$. Now we're asked to find all numbers x whose distance from zero is *greater than* 4. As Figure 1.7 shows, solutions are found in the interval to the left of -4, or to the right of 4. The fact the intervals are disjoint (disconnected) is reflected in this graph, in the inequalities $x < -4$ **or** $x > 4$, as well as the interval notation $x \in (-\infty, -4) \cup (4, \infty)$.

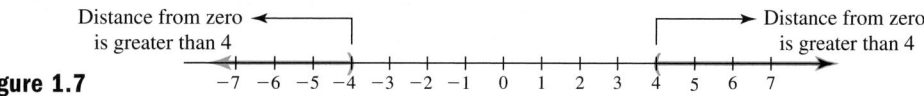

Figure 1.7

As before, we can extend this idea to include algebraic expressions, as follows:

Property II: Absolute Value Inequalities

If X represents an algebraic expression and k is a positive real number,

$$\text{then } |X| > k$$
$$\text{implies } X < -k \quad \text{or} \quad X > k$$

EXAMPLE 5 ▶ **Solving "Greater Than" Absolute Value Inequalities**

Solve the inequalities:

a. $-\dfrac{1}{3}\left|3 + \dfrac{x}{2}\right| < -2$ **b.** $|5x + 2| \geq -\dfrac{3}{2}$

Solution ▶ **a.** Note the exercise is given as a *less than* inequality, but as we multiply both sides by -3, we must *reverse the inequality symbol*.

$$-\frac{1}{3}\left|3 + \frac{x}{2}\right| < -2 \qquad \text{original inequality}$$

$$\left|3 + \frac{x}{2}\right| > 6 \qquad \text{multiply by } -3, \text{ reverse the symbol}$$

$$3 + \frac{x}{2} < -6 \quad \text{or} \quad 3 + \frac{x}{2} > 6 \qquad \text{apply Property II}$$

$$\frac{x}{2} < -9 \quad \text{or} \qquad \frac{x}{2} > 3 \qquad \text{subtract 3}$$

$$x < -18 \quad \text{or} \qquad x > 6 \qquad \text{multiply by 2}$$

Property II yields the disjoint intervals $x \in (-\infty, -18) \cup (6, \infty)$ as the solution.

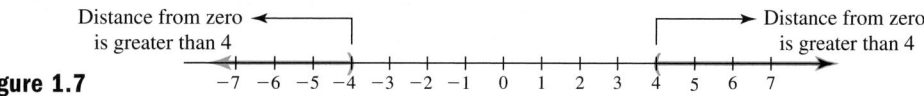

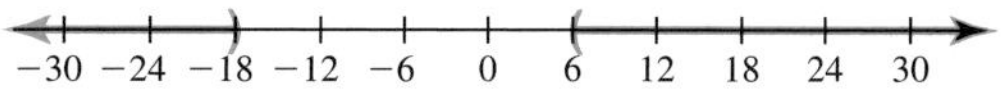

b. $|5x + 2| \geq -\dfrac{3}{2}$ original inequality

☑ **C. You've just learned how to solve greater than absolute value inequalities**

Since the absolute value of any quantity is always positive or zero, the solution for this inequality is all real numbers: $x \in \mathbb{R}$.

Now try Exercises 39 through 54 ▶

⚠ **CAUTION** ▶ Be sure you note the difference between the individual solutions of an absolute value equation, and the solution intervals that often result from solving absolute value inequalities. The solution $\{-2, 5\}$ indicates that both $x = -2$ and $x = 5$ are solutions, while the solution $[-2, 5)$ indicates that all numbers between -2 and 5, including -2, are solutions.

D. Applications Involving Absolute Value

Applications of absolute value often involve finding a range of values for which a given statement is true. Many times, the equation or inequality used must be modeled after a given description or from given information, as in Example 6.

EXAMPLE 6 ▶ **Solving Applications Involving Absolute Value Inequalities**

For new cars, the number of miles per gallon (mpg) a car will get is heavily dependent on whether it is used mainly for short trips and city driving, or primarily on the highway for longer trips. For a certain car, the number of miles per gallon that a driver can expect varies by no more than 6.5 mpg above or below its field tested average of 28.4 mpg. What range of mileage values can a driver expect for this car?

Solution ▶ Field tested average: 28.4 mpg   gather information

mileage varies by no more than 6.5 mpg   highlight key phrases

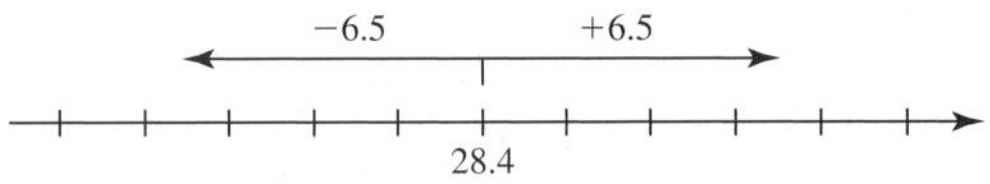

make the problem visual

Let m represent the miles per gallon a driver can expect.   assign a variable
Then the difference between m and 28.4 can be no more
than 6.5, or $|m - 28.4| \leq 6.5$.   write an equation model

$$|m - 28.4| \leq 6.5$$ equation model

$$-6.5 \leq m - 28.4 \leq 6.5$$ apply Property I

$$21.9 \leq m \leq 34.9$$ add 28.4 to all three parts

☑ **D.** You've just learned how to solve applications involving absolute value

The mileage that a driver can expect ranges from a low of 21.9 mpg to a high of 34.9 mpg.

Now try Exercises 57 through 64 ▶

TECHNOLOGY HIGHLIGHT

Absolute Value Equations and Inequalities

Graphing calculators can explore and solve inequalities in many different ways. Here we'll use a table of values and a *relational test*. To begin we'll consider the equation $2|x - 3| + 1 = 5$ by entering the left-hand side as Y_1 on the **Y=** screen. The calculator does not use absolute value bars the way they're written, and the equation is actually entered as $Y_1 = 2$ **abs** $(X - 3) + 1$ (see Figure 1.8). The **"abs("** notation is accessed by pressing **MATH**, ▶ **(NUM)** **1** (option 1 gives only the left parenthesis, you must supply the right). Preset the TABLE as in the previous Highlight (page 10). By scrolling through the table (use the up ▲ and down ▼ arrows), we find $Y_1 = 5$ when $x = 1$ or $x = 5$ (see Figure 1.9).

Although we could also solve the *inequality* $2|x - 3| + 1 \leq 5$ using the table (the solution interval is $x \in [1, 5]$), a relational test can help. Relational tests have the calculator return a "1" if a given statement is true, and a "0" otherwise. Enter $Y_2 = Y_1 \leq 5$, by accessing Y_1 using **VARS** ▶ **(Y-VARS) 1:Function** **ENTER** , and the "≤" symbol using **2nd** **MATH** **(TEST)** [the "less than or equal to" symbol is option 6]. Returning to the table shows $Y_1 \leq 5$ is true for $1 \leq x \leq 5$ (see Figure 1.9).

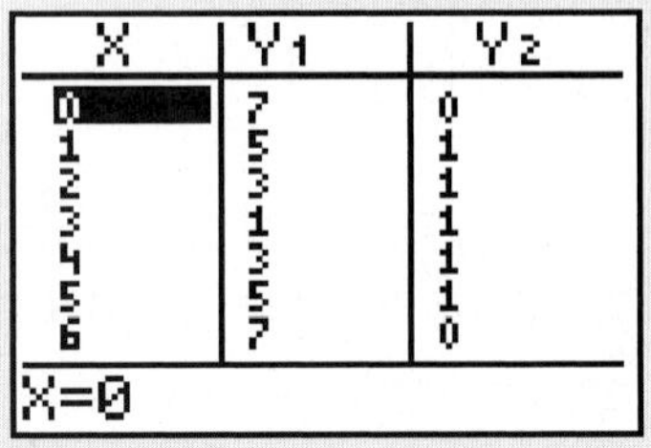

Use a table and a relational test to help solve the following inequalities. Verify the result algebraically.

Figure 1.8

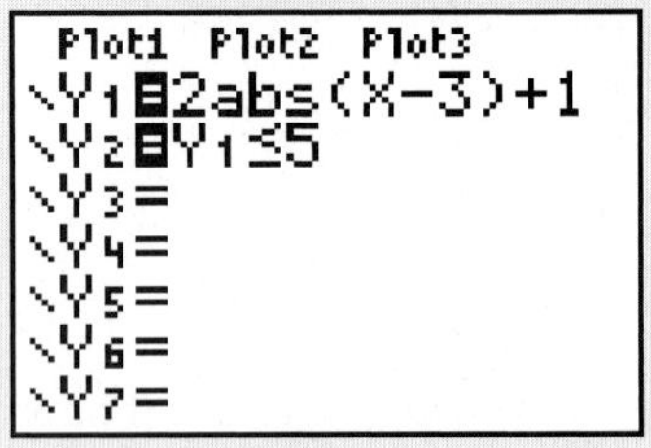

```
Plot1 Plot2 Plot3
\Y1◻2abs(X-3)+1
\Y2◻Y1≤5
\Y3=
\Y4=
\Y5=
\Y6=
\Y7=
```

Figure 1.9

X	Y1	Y2
0	7	0
1	5	1
2	3	1
3	1	1
4	3	1
5	5	1
6	7	0

X=0

Exercise 1: $3|x + 1| - 2 \geq 7$ **Exercise 2:** $-2|x + 2| + 5 \geq -1$ **Exercise 3:** $-1 \leq 4|x - 3| - 1$

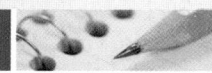

1.3 EXERCISES

▶ CONCEPTS AND VOCABULARY

Fill in the blank with the appropriate word or phrase. Carefully reread the section if needed.

1. When multiplying or dividing by a negative quantity, we _____ the inequality to maintain a true statement.

2. To write an absolute value equation or inequality in simplified form, we _____ the absolute value expression on one side.

3. The absolute value equation $|2x + 3| = 7$ is true when $2x + 3 =$ _____ or when $2x + 3 =$ _____.

4. The absolute value inequality $|3x - 6| < 12$ is true when $3x - 6 >$ _____ and $3x - 6 <$ _____.

Describe each solution set (assume $k > 0$). Justify your answer.

5. $|ax + b| < -k$

6. $|ax + b| > -k$

▶ DEVELOPING YOUR SKILLS

Solve each absolute value equation. Write the solution in set notation.

7. $2|m - 1| - 7 = 3$

8. $3|n - 5| - 14 = -2$

9. $-3|x + 5| + 6 = -15$

10. $-2|y + 3| - 4 = -14$

11. $2|4v + 5| - 6.5 = 10.3$

12. $7|2w + 5| + 6.3 = 11.2$

13. $-|7p - 3| + 6 = -5$

14. $-|3q + 4| + 3 = -5$

15. $-2|b| - 3 = -4$

16. $-3|c| - 5 = -6$

17. $-2|3x| - 17 = -5$

18. $-5|2y| - 14 = 6$

19. $-3\left|\dfrac{w}{2} + 4\right| - 1 = -4$

20. $-2\left|3 - \dfrac{v}{3}\right| + 1 = -5$

21. $8.7|p - 7.5| - 26.6 = 8.2$

22. $5.3|q + 9.2| + 6.7 = 43.8$

23. $8.7|-2.5x| - 26.6 = 8.2$

24. $5.3|1.25n| + 6.7 = 43.8$

Solve each absolute value inequality. Write solutions in interval notation.

25. $|x - 2| \leq 7$

26. $|y + 1| \leq 3$

27. $-3|m| - 2 > 4$

28. $-2|n| + 3 > 7$

29. $\dfrac{|5v + 1|}{4} + 8 < 9$

30. $\dfrac{|3w - 2|}{2} + 6 < 8$

31. $3|p + 4| + 5 < 8$

32. $5|q - 2| - 7 \leq 8$

33. $|3b - 11| + 6 \leq 9$

34. $|2c + 3| - 5 < 1$

35. $|4 - 3z| + 12 < 7$

36. $|2 - 7u| + 7 \leq 4$

37. $\left|\dfrac{4x + 5}{3} - \dfrac{1}{2}\right| \leq \dfrac{7}{6}$

38. $\left|\dfrac{2y - 3}{4} - \dfrac{3}{8}\right| < \dfrac{15}{16}$

39. $|n + 3| > 7$

40. $|m - 1| > 5$

41. $-2|w| - 5 \leq -11$

42. $-5|v| - 3 \leq -23$

43. $\dfrac{|q|}{2} - \dfrac{5}{6} \geq \dfrac{1}{3}$

44. $\dfrac{|p|}{5} + \dfrac{3}{2} \geq \dfrac{9}{4}$

45. $3|5 - 7d| + 9 \geq 15$

46. $5|2c + 7| + 1 \geq 11$

47. $|4z - 9| + 6 \geq 4$

48. $|5u - 3| + 8 > 6$

49. $4|5 - 2h| - 9 > 11$

50. $3|7 + 2k| - 11 > 10$

51. $-3.9|4q - 5| + 8.7 \leq -22.5$

52. $0.9|2p + 7| - 16.11 \geq 10.89$

53. $2 < \left|-3m + \dfrac{4}{5}\right| - \dfrac{1}{5}$

54. $4 \leq \left|\dfrac{5}{4} - 2n\right| - \dfrac{3}{4}$

▶ WORKING WITH FORMULAS

55. Spring Oscillation $|d - x| \leq L$

A weight attached to a spring hangs at rest a distance of x in. off the ground. If the weight is pulled down (stretched) a distance of L inches and released, the weight begins to bounce and its distance d off the ground must satisfy the indicated formula. If x equals 4 ft and the spring is stretched 3 in. and released, solve the inequality to find what distances from the ground the weight will oscillate between.

56. A "Fair" Coin $\left| \dfrac{h - 50}{5} \right| < 1.645$

If we flipped a coin 100 times, we expect "heads" to come up about 50 times if the coin is "fair." In a study of probability, it can be shown that the number of heads h that appears in such an experiment must satisfy the given inequality to be considered "fair." (a) Solve this inequality for h. (b) If you flipped a coin 100 times and obtained 40 heads, is the coin "fair"?

▶ APPLICATIONS

Solve each application of absolute value.

57. Altitude of jet stream: To take advantage of the jet stream, an airplane must fly at a height h (in feet) that satisfies the inequality $|h - 35{,}050| \leq 2550$. Solve the inequality and determine if an altitude of 34,000 ft will place the plane in the jet stream.

58. Quality control tests: In order to satisfy quality control, the marble columns a company produces must earn a stress test score S that satisfies the inequality $|S - 17{,}750| \leq 275$. Solve the inequality and determine if a score of 17,500 is in the passing range.

59. Submarine depth: The sonar operator on a submarine detects an old World War II submarine net and must decide to detour over or under the net. The computer gives him a depth model $|d - 394| - 20 > 164$, where d is the depth in feet that represents safe passage. At what depth should the submarine travel to go under or over the net? Answer using simple inequalities.

60. Optimal fishing depth: When deep-sea fishing, the optimal depths d (in feet) for catching a certain type of fish satisfy the inequality $28|d - 350| - 1400 < 0$. Find the range of depths that offer the best fishing. Answer using simple inequalities.

For Exercises 61 through 64, (a) develop a model that uses an absolute value inequality, and (b) solve.

61. Stock value: My stock in MMM Corporation fluctuated a great deal in 2009, but never by more than \$3.35 from its current value. If the stock is worth \$37.58 today, what was its range in 2009?

62. Traffic studies: On a given day, the volume of traffic at a busy intersection averages 726 cars per hour (cph). During rush hour the volume is much higher, during "off hours" much lighter. Find the range of this volume if it never varies by more than 235 cph from the average.

63. Physical training for recruits: For all recruits in the 3rd Armored Battalion, the average number of sit-ups is 125. For an individual recruit, the amount varies by no more than 23 sit-ups from the battalion average. Find the range of sit-ups for this battalion.

64. Computer consultant salaries: The national average salary for a computer consultant is \$53,336. For a large computer firm, the salaries offered to their employees varies by no more than \$11,994 from this national average. Find the range of salaries offered by this company.

65. According to the official rules for golf, baseball, pool, and bowling, (a) golf balls must be within 0.03 mm of $d = 42.7$ mm, (b) baseballs must be within 1.01 mm of $d = 73.78$ mm, (c) billiard balls must be within 0.127 mm of $d = 57.150$ mm, and (d) bowling balls must be within 12.05 mm of $d = 2171.05$ mm. Write each statement using an absolute value inequality, then (e) determine which sport gives the least tolerance t $\left(t = \dfrac{\text{width of interval}}{\text{average value}} \right)$ for the diameter of the ball.

66. The machines that fill boxes of breakfast cereal are programmed to fill each box within a certain tolerance. If the box is overfilled, the company loses money. If it is underfilled, it is considered unsuitable for sale. Suppose that boxes marked "14 ounces" of cereal must be filled to within 0.1 oz. Write this relationship as an absolute value inequality, then solve the inequality and explain what your answer means. Let W represent weight.

▶ EXTENDING THE CONCEPT

67. Determine the value or values (if any) that will make the equation or inequality true.

    a. $|x| + x = 8$ **b.** $|x - 2| \leq \dfrac{x}{2}$

    c. $x - |x| = x + |x|$ **d.** $|x + 3| \geq 6x$

    e. $|2x + 1| = x - 3$

68. The equation $|5 - 2x| = |3 + 2x|$ has only one solution. Find it and explain why there is only one.

▶ MAINTAINING YOUR SKILLS

69. (R.4) Factor the expression completely:
$18x^3 + 21x^2 - 60x$.

70. (1.1) Solve $V^2 = \dfrac{2W}{C\rho A}$ for ρ (physics).

71. (R.6) Simplify $\dfrac{-1}{3 + \sqrt{3}}$ by rationalizing the denominator. State the result in exact form and approximate form (to hundredths):

72. (1.2) Solve the inequality, then write the solution set in interval notation:

$-3(2x - 5) > 2(x + 1) - 7$.

MID-CHAPTER CHECK

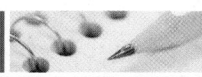

1. Solve each equation. If the equation is an identity or contradiction, so state and name the solution set.

    a. $\dfrac{r}{3} + 5 = 2$

    b. $5(2x - 1) + 4 = 9x - 7$

    c. $m - 2(m + 3) = 1 - (m + 7)$

    d. $\dfrac{1}{5}y + 3 = \dfrac{3}{2}y - 2$

    e. $\dfrac{1}{2}(5j - 2) = \dfrac{3}{2}(j - 4) + j$

    f. $0.6(x - 3) + 0.3 = 1.8$

Solve for the variable specified.

2. $H = -16t^2 + v_0 t$; for v_0

3. $S = 2\pi x^2 + \pi x^2 y$; for x

4. Solve each inequality and graph the solution set.

    a. $-5x + 16 \leq 11$ or $3x + 2 \leq -4$

    b. $\dfrac{1}{2} < \dfrac{1}{12}x - \dfrac{5}{6} \leq \dfrac{3}{4}$

5. Determine the domain of each expression. Write your answer in interval notation.

    a. $\dfrac{3x + 1}{2x - 5}$ **b.** $\sqrt{17 - 6x}$

6. Solve the following absolute value equations. Write the solution in set notation.

    a. $\dfrac{2}{3}|d - 5| + 1 = 7$ **b.** $5 - |s + 3| = \dfrac{11}{2}$

7. Solve the following absolute value inequalities. Write solutions in interval notation.

   a. $3|q + 4| - 2 < 10$

   b. $\left|\dfrac{x}{3} + 2\right| + 5 \leq 5$

8. Solve the following absolute value inequalities. Write solutions in interval notation.

   a. $3.1|d - 2| + 1.1 \geq 7.3$

   b. $\dfrac{|1 - y|}{3} + 2 > \dfrac{11}{2}$

   c. $-5|k - 2| + 3 < 4$

9. Motocross: An enduro motocross motorcyclist averages 30 mph through the first part of a 115-mi course, and 50 mph though the second part. If the rider took 2 hr and 50 min to complete the course, how long was she on the first part?

10. Kiteboarding: With the correct sized kite, a person can kiteboard when the wind is blowing at a speed w (in mph) that satisfies the inequality $|w - 17| \leq 9$. Solve the inequality and determine if a person can kiteboard with a windspeed of 9 mph.

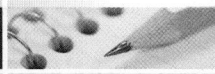

REINFORCING BASIC CONCEPTS

Using Distance to Understand Absolute Value Equations and Inequalities

In Appendix I.A we noted that for any two numbers a and b on the number line, *the distance between a and b is written* $|a - b|$ or $|b - a|$. In exactly the same way, the equation $|x - 3| = 4$ can be read, "the distance between 3 and an unknown number is equal to 4." The advantage of reading it in this way (instead of *the absolute value of x minus 3 is 4*), is that a much clearer *visualization* is formed, giving a constant reminder there are two solutions. In diagram form we have Figure 1.10.

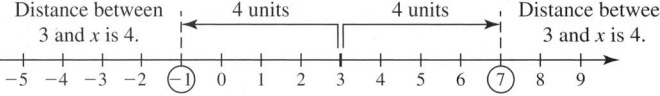

Figure 1.10

From this we note the solution is $x = -1$ or $x = 7$.

In the case of an inequality such as $|x + 2| \leq 3$, we rewrite the inequality as $|x - (-2)| \leq 3$ and read it, "the distance between -2 and an unknown number is less than or equal to 3." With some practice, visualizing this relationship mentally enables a quick statement of the solution: $x \in [-5, 1]$. In diagram form we have Figure 1.11.

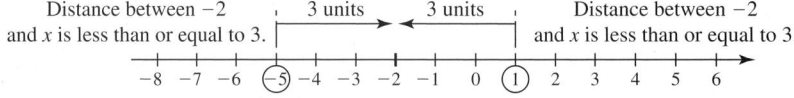

Figure 1.11

Equations and inequalities where the coefficient of x is not 1 still lend themselves to this form of conceptual understanding. For $|2x - 1| \geq 3$ we read, "the distance between 1 and twice an unknown number is greater than or equal to 3." On the number line (Figure 1.12), the number 3 units to the right of 1 is 4, and the number 3 units to the left of 1 is -2.

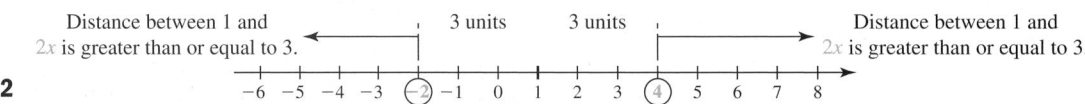

Figure 1.12

For $2x \leq -2$, $x \leq -1$, and for $2x \geq 4$, $x \geq 2$, and the solution is $x \in (-\infty, -1] \cup [2, \infty)$.

Attempt to solve the following equations and inequalities by visualizing a number line. Check all results algebraically.

Exercise 1: $|x - 2| = 5$

Exercise 2: $|x + 1| \leq 4$

Exercise 3: $|2x - 3| \geq 5$

Learning Objectives

In Section 1.4 you will learn how to:

☐ **A.** Identify and simplify imaginary and complex numbers

☐ **B.** Add and subtract complex numbers

☐ **C.** Multiply complex numbers and find powers of i

☐ **D.** Divide complex numbers

For centuries, even the most prominent mathematicians refused to work with equations like $x^2 + 1 = 0$. Using the principal of square roots gave the "solutions" $x = \sqrt{-1}$ and $x = -\sqrt{-1}$, which they found baffling and mysterious, since there is no real number whose square is -1. In this section, we'll see how this "mystery" was finally resolved.

A. Identifying and Simplifying Imaginary and Complex Numbers

The equation $x^2 = -1$ has no real solutions, since the square of any real number is positive. But if we apply the principle of square roots we get $x = \sqrt{-1}$ and $x = -\sqrt{-1}$, which seem to check when substituted into the original equation:

	$x^2 + 1 = 0$	original equation
(1)	$(\sqrt{-1})^2 + 1 = 0$	substitute $\sqrt{-1}$ for x
	$-1 + 1 = 0\checkmark$	answer "checks"
(2)	$(-\sqrt{-1})^2 + 1 = 0$	substitute $-\sqrt{-1}$ for x
	$-1 + 1 = 0\checkmark$	answer "checks"

This observation likely played a part in prompting Renaissance mathematicians to study such numbers in greater depth, as they reasoned that while these were not *real number* solutions, they must be *solutions of a new and different kind.* Their study eventually resulted in the introduction of the set of **imaginary numbers** and the **imaginary unit i,** as follows.

Imaginary Numbers and the Imaginary Unit

- Imaginary numbers are those of the form $\sqrt{-k}$, where k is a positive real number.
- The imaginary unit i represents the number whose square is -1:

$$i^2 = -1 \text{ and } i = \sqrt{-1}$$

As a convenience to understanding and working with imaginary numbers, we rewrite them in terms of i, allowing that the product property of radicals ($\sqrt{AB} = \sqrt{A}\sqrt{B}$) still applies if *only one* of the radicands is negative. For $\sqrt{-3}$, we have $\sqrt{-1 \cdot 3} = \sqrt{-1}\sqrt{3} = i\sqrt{3}$. In general, we simply state the following property.

Rewriting Imaginary Numbers

- For any positive real number k, $\sqrt{-k} = i\sqrt{k}$.

For $\sqrt{-20}$ we have:

$$\sqrt{-20} = i\sqrt{20}$$
$$= i\sqrt{4 \cdot 5}$$
$$= 2i\sqrt{5},$$

and we say the expression has been *simplified and written in terms of i.* Note that we've written the result with the unit "i" *in front of the radical* to prevent it being interpreted as being *under the radical.* In symbols, $2i\sqrt{5} = 2\sqrt{5}i \neq 2\sqrt{5i}$.

The solutions to $x^2 = -1$ also serve to illustrate that for $k > 0$, there are two solutions to $x^2 = -k$, namely, $i\sqrt{k}$ and $-i\sqrt{k}$. In other words, every negative number has two square roots, one positive and one negative. The first of these, $i\sqrt{k}$, is called the **principal square root** of $-k$.

EXAMPLE 1 ▶ **Simplifying Imaginary Numbers**

Rewrite the imaginary numbers in terms of i and simplify if possible.

a. $\sqrt{-7}$ **b.** $\sqrt{-81}$ **c.** $\sqrt{-24}$ **d.** $-3\sqrt{-16}$

Solution ▶   **a.** $\sqrt{-7} = i\sqrt{7}$ **b.** $\sqrt{-81} = i\sqrt{81}$
$$= 9i$$

c. $\sqrt{-24} = i\sqrt{24}$ **d.** $-3\sqrt{-16} = -3i\sqrt{16}$
$$= i\sqrt{4 \cdot 6}$$ $$= -3i(4)$$
$$= 2i\sqrt{6}$$ $$= -12i$$

Now try Exercises 7 through 12 ▶

EXAMPLE 2 ▶ **Writing an Expression in Terms of i**

The numbers $x = \dfrac{-6 + \sqrt{-16}}{2}$ and $x = \dfrac{-6 - \sqrt{-16}}{2}$ are not real, but are known to be solutions of $x^2 + 6x + 13 = 0$. Simplify $\dfrac{-6 + \sqrt{-16}}{2}$.

Solution ▶   Using the i notation, we have

$$\frac{-6 + \sqrt{-16}}{2} = \frac{-6 + i\sqrt{16}}{2} \qquad \text{write in } i \text{ notation}$$

$$= \frac{-6 + 4i}{2} \qquad \text{simplify}$$

$$= \frac{2(-3 + 2i)}{2} \qquad \text{factor numerator}$$

$$= -3 + 2i \qquad \text{reduce}$$

Now try Exercises 13 through 16 ▶

> **WORTHY OF NOTE**
>
> The expression $\dfrac{-6 + 4i}{2}$ from the solution of Example 2 can also be simplified by rewriting it as two separate terms, then simplifying each term:
> $$\frac{-6 + 4i}{2} = \frac{-6}{2} + \frac{4i}{2}$$
> $$= -3 + 2i.$$

The result in Example 2 contains both a **real number part** (-3) and an **imaginary part** $(2i)$. Numbers of this type are called **complex numbers.**

Complex Numbers

Complex numbers are numbers that can be written in the form $a + bi$, where a and b are real numbers and $i = \sqrt{-1}$.

The expression $a + bi$ is called the **standard form** of a complex number. From this definition we note that all real numbers are also complex numbers, since $a + 0i$ is complex with $b = 0$. In addition, all imaginary numbers are complex numbers, since $0 + bi$ is a complex number with $a = 0$.

EXAMPLE 3 ▶ **Writing Complex Numbers in Standard Form**

Write each complex number in the form $a + bi$, and identify the values of a and b.

a. $2 + \sqrt{-49}$ **b.** $\sqrt{-12}$ **c.** 7 **d.** $\dfrac{4 + 3\sqrt{-25}}{20}$

Solution ▶   **a.** $2 + \sqrt{-49} = 2 + i\sqrt{49}$ **b.** $\sqrt{-12} = 0 + i\sqrt{12}$
$$= 2 + 7i$$ $$= 0 + 2i\sqrt{3}$$
$$a = 2, b = 7$$ $$a = 0, b = 2\sqrt{3}$$

c. $7 = 7 + 0i$
$a = 7, b = 0$

d. $\dfrac{4 + 3\sqrt{-25}}{20} = \dfrac{4 + 3i\sqrt{25}}{20}$

$\qquad\qquad = \dfrac{4 + 15i}{20}$

$\qquad\qquad = \dfrac{1}{5} + \dfrac{3}{4}i$

$\qquad a = \dfrac{1}{5}, b = \dfrac{3}{4}$

Now try Exercises 17 through 24 ▶

☑ **A. You've just learned how to identify and simplify imaginary and complex numbers**

Complex numbers complete the development of our "numerical landscape." Sets of numbers and their relationships are represented in Figure 1.13, which shows how some sets of numbers are nested within larger sets and highlights the fact that complex numbers consist of a real number part (any number within the orange rectangle), and an imaginary number part (any number within the yellow rectangle).

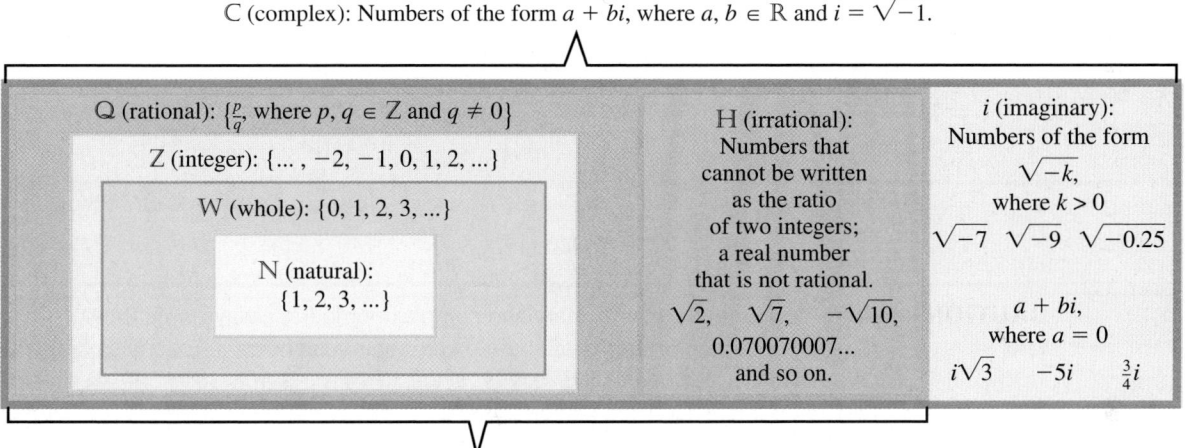

C (complex): Numbers of the form $a + bi$, where $a, b \in$ R and $i = \sqrt{-1}$.

Q (rational): $\{\frac{p}{q}, \text{ where } p, q \in$ Z and $q \neq 0\}$

Z (integer): $\{\dots, -2, -1, 0, 1, 2, \dots\}$

W (whole): $\{0, 1, 2, 3, \dots\}$

N (natural): $\{1, 2, 3, \dots\}$

H (irrational): Numbers that cannot be written as the ratio of two integers; a real number that is not rational. $\sqrt{2}$, $\sqrt{7}$, $-\sqrt{10}$, 0.070070007... and so on.

i (imaginary): Numbers of the form $\sqrt{-k}$, where $k > 0$ $\sqrt{-7}$ $\sqrt{-9}$ $\sqrt{-0.25}$ $a + bi$, where $a = 0$ $i\sqrt{3}$ $-5i$ $\frac{3}{4}i$

R (real): All rational and irrational numbers: $a + bi$, where $a \in$ R and $b = 0$.

Figure 1.13

B. Adding and Subtracting Complex Numbers

The sum and difference of two polynomials is computed by identifying and combining like terms. The sum or difference of two complex numbers is computed in a similar way, by adding the real number parts from each, and the imaginary parts from each. Notice in Example 4 that the commutative, associative, and distributive properties also apply to complex numbers.

EXAMPLE 4 ▶ **Adding and Subtracting Complex Numbers**

Perform the indicated operation and write the result in $a + bi$ form.

a. $(2 + 3i) + (-5 + 2i)$　　　　b. $(-5 - 4i) - (-2 - \sqrt{2}i)$

Solution ▶

a. $(2 + 3i) + (-5 + 2i)$　　　　　　original sum
$= 2 + 3i + (-5) + 2i$　　　　　distribute
$= 2 + (-5) + 3i + 2i$　　　　　commute terms
$= [2 + (-5)] + (3i + 2i)$　　　group like terms
$= -3 + 5i$　　　　　　　　　　result

b. $(-5 - 4i) - (-2 - \sqrt{2}i)$　　　　original difference
$= -5 - 4i + 2 + \sqrt{2}i$　　　　distribute
$= -5 + 2 + (-4i) + \sqrt{2}i$　　commute terms
$= (-5 + 2) + [(-4i) + \sqrt{2}i]$　group like terms
$= -3 + (-4 + \sqrt{2})i$　　　　result

Now try Exercises 25 through 30 ▶

B. You've just learned how to add and subtract complex numbers

C. Multiplying Complex Numbers; Powers of i

The product of two complex numbers is computed using the distributive property and the F-O-I-L process in the same way we apply these to binomials. If any result gives a factor of i^2, remember that $i^2 = -1$.

EXAMPLE 5 ▶   Multiplying Complex Numbers

Find the indicated product and write the answer in $a + bi$ form.

   a. $\sqrt{-4}\sqrt{-9}$ **b.** $\sqrt{-6}\,(2 + \sqrt{-3})$

   c. $(6 - 5i)(4 + i)$ **d.** $(2 + 3i)(2 - 3i)$

Solution ▶

a. $\sqrt{-4}\sqrt{-9} = i\sqrt{4} \cdot i\sqrt{9}$ rewrite in terms of i

$\qquad\qquad = 2i \cdot 3i$ simplify

$\qquad\qquad = 6i^2$ multiply

$\qquad\qquad = -6 + 0i$ result ($i^2 = -1$)

b. $\sqrt{-6}\,(2 + \sqrt{-3}) = i\sqrt{6}(2 + i\sqrt{3})$ rewrite in terms of i

$\qquad\qquad = 2i\sqrt{6} + i^2\sqrt{18}$ distribute

$\qquad\qquad = 2i\sqrt{6} + (-1)\sqrt{9}\sqrt{2}$ $i^2 = -1$

$\qquad\qquad = 2i\sqrt{6} - 3\sqrt{2}$ simplify

$\qquad\qquad = -3\sqrt{2} + 2i\sqrt{6}$ standard form

c. $(6 - 5i)(4 + i)$

$\quad = (6)(4) + 6i + (-5i)(4) + (-5i)(i)$ F-O-I-L

$\quad = 24 + 6i + (-20i) + (-5)i^2$ $i \cdot i = i^2$

$\quad = 24 + 6i + (-20i) + (-5)(-1)$ $i^2 = -1$

$\quad = 29 - 14i$ result

d. $(2 + 3i)(2 - 3i)$

$\quad = (2)^2 - (3i)^2$ $(A + B)(A - B) = A^2 - B^2$

$\quad = 4 - 9i^2$ $(3i)^2 = 9i^2$

$\quad = 4 - 9(-1)$ $i^2 = -1$

$\quad = 13 + 0i$ result

Now try Exercises 31 through 48 ▶

⚠ CAUTION ▶ When computing with imaginary and complex numbers, always write the square root of a negative number in terms of i before you begin, as shown in Examples 5(a) and 5(b). Otherwise we get conflicting results, since $\sqrt{-4}\,\sqrt{-9} = \sqrt{36} = 6$ if we multiply the radicands first, which is an incorrect result because the original factors were imaginary. **See Exercise 80.**

Recall that expressions $2x + 5$ and $2x - 5$ are called binomial conjugates. In the same way, $a + bi$ and $a - bi$ are called **complex conjugates.** Note from Example 5(d) that the *product* of the complex number $a + bi$ with its complex conjugate $a - bi$ is *a real number*. This relationship is useful when rationalizing expressions with a complex number in the denominator, and we generalize the result as follows:

WORTHY OF NOTE

Notice that the product of a complex number and its conjugate also gives us a method for *factoring the sum of two squares* using complex numbers! For the expression $x^2 + 4$, the factored form would be $(x + 2i)(x - 2i)$. For more on this idea, **see Exercise 79.**

Product of Complex Conjugates

For a complex number $a + bi$ and its conjugate $a - bi$, their product $(a + bi)(a - bi)$ is the real number $a^2 + b^2$;

$$(a + bi)(a - bi) = a^2 + b^2$$

Showing that $(a + bi)(a - bi) = a^2 + b^2$ is left as an exercise (see Exercise 79), but from here on, when asked to compute the product of complex conjugates, simply refer to the formula as illustrated here: $(-3 + 5i)(-3 - 5i) = (-3)^2 + 5^2$ or 34.

These operations on complex numbers enable us to verify complex solutions by substitution, in the same way we verify solutions for real numbers. In Example 2 we stated that $x = -3 + 2i$ was one solution to $x^2 + 6x + 13 = 0$. This is verified here.

EXAMPLE 6 ▶ **Checking a Complex Root by Substitution**

Verify that $x = -3 + 2i$ is a solution to $x^2 + 6x + 13 = 0$.

Solution ▶

$$x^2 + 6x + 13 = 0 \quad \text{original equation}$$
$$(-3 + 2i)^2 + 6(-3 + 2i) + 13 = 0 \quad \text{substitute } -3 + 2i \text{ for } x$$
$$(-3)^2 + 2(-3)(2i) + (2i)^2 - 18 + 12i + 13 = 0 \quad \text{square and distribute}$$
$$9 - 12i + 4i^2 + 12i - 5 = 0 \quad \text{simplify}$$
$$9 + (-4) - 5 = 0 \quad \text{combine terms } (12i - 12i = 0; \ i^2 = -1)$$
$$0 = 0 \checkmark$$

Now try Exercises 49 through 56 ▶

EXAMPLE 7 ▶ **Checking a Complex Root by Substitution**

Show that $x = 2 - i\sqrt{3}$ is a solution of $x^2 - 4x = -7$.

Solution ▶

$$x^2 - 4x = -7 \quad \text{original equation}$$
$$(2 - i\sqrt{3})^2 - 4(2 - i\sqrt{3}) = -7 \quad \text{substitute } 2 - i\sqrt{3} \text{ for } x$$
$$4 - 4i\sqrt{3} + (i\sqrt{3})^2 - 8 + 4i\sqrt{3} = -7 \quad \text{square and distribute}$$
$$4 - 4i\sqrt{3} - 3 - 8 + 4i\sqrt{3} = -7 \quad (i\sqrt{3})^2 = -3$$
$$-7 = -7 \checkmark \quad \text{solution checks}$$

Now try Exercises 57 through 60 ▶

The imaginary unit i has another interesting and useful property. Since $i = \sqrt{-1}$ and $i^2 = -1$, we know that $i^3 = i^2 \cdot i = (-1)i = -i$ and $i^4 = (i^2)^2 = 1$. We can now simplify any *higher power of i* by rewriting the expression in terms of i^4.

$$i^5 = i^4 \cdot i = i$$
$$i^6 = i^4 \cdot i^2 = -1$$
$$i^7 = i^4 \cdot i^3 = -i$$
$$i^8 = (i^4)^2 = 1$$

Notice the powers of i "cycle through" the four values i, -1, $-i$ and 1. In more advanced classes, powers of complex numbers play an important role, and next we learn to reduce higher powers using the power property of exponents and $i^4 = 1$. Essentially, we divide the exponent on i by 4, then use the remainder to compute the value of the expression. For i^{35}, $35 \div 4 = 8$ remainder 3, showing $i^{35} = (i^4)^8 \cdot i^3 = -i$.

EXAMPLE 8 ▶ **Simplifying Higher Powers of i**

Simplify:

    a. i^{22} **b.** i^{28} **c.** i^{57} **d.** i^{75}

Solution ▶   **a.** $i^{22} = (i^4)^5 \cdot (i^2)$ **b.** $i^{28} = (i^4)^7$

                    $= (1)^5(-1)$ $= (1)^7$

                    $= -1$ $= 1$

☑ **C.** You've just learned how to multiply complex numbers and find powers of i

c. $i^{57} = (i^4)^{14} \cdot i$
$= (1)^{14}i$
$= i$

d. $i^{75} = (i^4)^{18} \cdot (i^3)$
$= (1)^{18}(-i)$
$= -i$

Now try Exercises 61 and 62 ▶

D. Division of Complex Numbers

Since $i = \sqrt{-1}$, expressions like $\dfrac{3 - i}{2 + i}$ actually have a radical in the denominator. To divide complex numbers, we simply apply our earlier method of rationalizing denominators (Appendix I.F), but this time using a *complex* conjugate.

EXAMPLE 9 ▶ **Dividing Complex Numbers**

Divide and write each result in $a + bi$ form.

a. $\dfrac{2}{5 - i}$

b. $\dfrac{3 - i}{2 + i}$

c. $\dfrac{6 + \sqrt{-36}}{3 + \sqrt{-9}}$

Solution ▶

a. $\dfrac{2}{5 - i} = \dfrac{2}{5 - i} \cdot \dfrac{5 + i}{5 + i}$

$= \dfrac{2(5 + i)}{5^2 + 1^2}$

$= \dfrac{10 + 2i}{26}$

$= \dfrac{10}{26} + \dfrac{2}{26}i$

$= \dfrac{5}{13} + \dfrac{1}{13}i$

b. $\dfrac{3 - i}{2 + i} = \dfrac{3 - i}{2 + i} \cdot \dfrac{2 - i}{2 - i}$

$= \dfrac{6 - 3i - 2i + i^2}{2^2 + 1^2}$

$= \dfrac{6 - 5i + (-1)}{5}$

$= \dfrac{5 - 5i}{5} = \dfrac{5}{5} - \dfrac{5i}{5}$

$= 1 - i$

c. $\dfrac{6 + \sqrt{-36}}{3 + \sqrt{-9}} = \dfrac{6 + i\sqrt{36}}{3 + i\sqrt{9}}$ convert to i notation

$= \dfrac{6 + 6i}{3 + 3i}$ simplify

The expression can be further simplified by reducing common factors.

$= \dfrac{6(1 + i)}{3(1 + i)} = 2$ factor and reduce

Now try Exercises 63 through 68 ▶

Operations on complex numbers can be checked using inverse operations, just as we do for real numbers. To check the answer $1 - i$ from Example 9(b), we multiply it by the divisor:

$$(1 - i)(2 + i) = 2 + i - 2i - i^2$$
$$= 2 - i - (-1)$$
$$= 2 - i + 1$$
$$= 3 - i \checkmark$$

☑ **D.** You've just learned how to divide complex numbers

Several checks are asked for in the exercises.

1.4 EXERCISES

▶ **CONCEPTS AND VOCABULARY**

**Fill in each blank with the appropriate word or phrase.
Carefully reread the section, if necessary.**

1. Given the complex number $3 + 2i$, its complex conjugate is _____.

2. The product $(3 + 2i)(3 - 2i)$ gives the real number _____.

3. If the expression $\dfrac{4 + 6i\sqrt{2}}{2}$ is written in the standard form $a + bi$, then $a =$ _____ and $b =$ _____.

4. For $i = \sqrt{-1}$, $i^2 =$ ___, $i^4 =$ ___, $i^6 =$ ___, and $i^8 =$ ___, $i^3 =$ ___, $i^5 =$ ___, $i^7 =$ ___, and $i^9 =$ ___.

5. Discuss/Explain which is correct:
   a. $\sqrt{-4} \cdot \sqrt{-9} = \sqrt{(-4)(-9)} = \sqrt{36} = 6$
   b. $\sqrt{-4} \cdot \sqrt{-9} = 2i \cdot 3i = 6i^2 = -6$

6. Compare/Contrast the product $(1 + \sqrt{2})(1 - \sqrt{3})$ with the product $(1 + i\sqrt{2})(1 - i\sqrt{3})$. What is the same? What is different?

▶ **DEVELOPING YOUR SKILLS**

Simplify each radical (if possible). If imaginary, rewrite in terms of i and simplify.

7. a. $\sqrt{-16}$ b. $\sqrt{-49}$
   c. $\sqrt{27}$ d. $\sqrt{72}$

8. a. $\sqrt{-81}$ b. $\sqrt{-169}$
   c. $\sqrt{64}$ d. $\sqrt{98}$

9. a. $-\sqrt{-18}$ b. $-\sqrt{-50}$
   c. $3\sqrt{-25}$ d. $2\sqrt{-9}$

10. a. $-\sqrt{-32}$ b. $-\sqrt{-75}$
    c. $3\sqrt{-144}$ d. $2\sqrt{-81}$

11. a. $\sqrt{-19}$ b. $\sqrt{-31}$
    c. $\sqrt{\dfrac{-12}{25}}$ d. $\sqrt{\dfrac{-9}{32}}$

12. a. $\sqrt{-17}$ b. $\sqrt{-53}$
    c. $\sqrt{\dfrac{-45}{36}}$ d. $\sqrt{\dfrac{-49}{75}}$

Write each complex number in the standard form $a + bi$ and clearly identify the values of a and b.

13. a. $\dfrac{2 + \sqrt{-4}}{2}$ b. $\dfrac{6 + \sqrt{-27}}{3}$

14. a. $\dfrac{16 - \sqrt{-8}}{2}$ b. $\dfrac{4 + 3\sqrt{-20}}{2}$

15. a. $\dfrac{8 + \sqrt{-16}}{2}$ b. $\dfrac{10 - \sqrt{-50}}{5}$

16. a. $\dfrac{6 - \sqrt{-72}}{4}$ b. $\dfrac{12 + \sqrt{-200}}{8}$

17. a. 5 b. $3i$

18. a. -2 b. $-4i$

19. a. $2\sqrt{-81}$ b. $\dfrac{\sqrt{-32}}{8}$

20. a. $-3\sqrt{-36}$ b. $\dfrac{\sqrt{-75}}{15}$

21. a. $4 + \sqrt{-50}$ b. $-5 + \sqrt{-27}$

22. a. $-2 + \sqrt{-48}$ b. $7 + \sqrt{-75}$

23. a. $\dfrac{14 + \sqrt{-98}}{8}$ b. $\dfrac{5 + \sqrt{-250}}{10}$

24. a. $\dfrac{21 + \sqrt{-63}}{12}$ b. $\dfrac{8 + \sqrt{-27}}{6}$

Perform the addition or subtraction. Write the result in $a + bi$ form.

25. a. $(12 - \sqrt{-4}) + (7 + \sqrt{-9})$
    b. $(3 + \sqrt{-25}) + (-1 - \sqrt{-81})$
    c. $(11 + \sqrt{-108}) - (2 - \sqrt{-48})$

26. a. $(-7 - \sqrt{-72}) + (8 + \sqrt{-50})$
    b. $(\sqrt{3} + \sqrt{-2}) - (\sqrt{12} + \sqrt{-8})$
    c. $(\sqrt{20} - \sqrt{-3}) + (\sqrt{5} - \sqrt{-12})$

27. a. $(2 + 3i) + (-5 - i)$
    b. $(5 - 2i) + (3 + 2i)$
    c. $(6 - 5i) - (4 + 3i)$

28. a. $(-2 + 5i) + (3 - i)$
    b. $(7 - 4i) - (2 - 3i)$
    c. $(2.5 - 3.1i) + (4.3 + 2.4i)$

29. a. $(3.7 + 6.1i) - (1 + 5.9i)$

b. $\left(8 + \dfrac{3}{4}i\right) - \left(-7 + \dfrac{2}{3}i\right)$

c. $\left(-6 - \dfrac{5}{8}i\right) + \left(4 + \dfrac{1}{2}i\right)$

30. a. $(9.4 - 8.7i) - (6.5 + 4.1i)$

b. $\left(3 + \dfrac{3}{5}i\right) - \left(-11 + \dfrac{7}{15}i\right)$

c. $\left(-4 - \dfrac{5}{6}i\right) + \left(13 + \dfrac{3}{8}i\right)$

Multiply and write your answer in $a + bi$ form.

31. a. $5i \cdot (-3i)$ **b.** $(4i)(-4i)$

32. a. $3(2 - 3i)$ **b.** $-7(3 + 5i)$

33. a. $-7i(5 - 3i)$ **b.** $6i(-3 + 7i)$

34. a. $(-4 - 2i)(3 + 2i)$ **b.** $(2 - 3i)(-5 + i)$

35. a. $(-3 + 2i)(2 + 3i)$ **b.** $(3 + 2i)(1 + i)$

36. a. $(5 + 2i)(-7 + 3i)$ **b.** $(4 - i)(7 + 2i)$

For each complex number, name the complex conjugate. Then find the product.

37. a. $4 + 5i$ **b.** $3 - i\sqrt{2}$

38. a. $2 - i$ **b.** $-1 + i\sqrt{5}$

39. a. $7i$ **b.** $\frac{1}{2} - \frac{2}{3}i$

40. a. $-5i$ **b.** $\frac{3}{4} + \frac{1}{5}i$

Compute the special products and write your answer in $a + bi$ form.

41. a. $(4 - 5i)(4 + 5i)$
 b. $(7 - 5i)(7 + 5i)$

42. a. $(-2 - 7i)(-2 + 7i)$
 b. $(2 + i)(2 - i)$

43. a. $(3 - i\sqrt{2})(3 + i\sqrt{2})$
 b. $(\frac{1}{6} + \frac{2}{3}i)(\frac{1}{6} - \frac{2}{3}i)$

44. a. $(5 + i\sqrt{3})(5 - i\sqrt{3})$
 b. $(\frac{1}{2} + \frac{3}{4}i)(\frac{1}{2} - \frac{3}{4}i)$

45. a. $(2 + 3i)^2$ **b.** $(3 - 4i)^2$

46. a. $(2 - i)^2$ **b.** $(3 - i)^2$

47. a. $(-2 + 5i)^2$ **b.** $(3 + i\sqrt{2})^2$

48. a. $(-2 - 5i)^2$ **b.** $(2 - i\sqrt{3})^2$

Use substitution to determine if the value shown is a solution to the given equation.

49. $x^2 + 36 = 0; x = -6$

50. $x^2 + 16 = 0; x = -4$

51. $x^2 + 49 = 0; x = -7i$

52. $x^2 + 25 = 0; x = -5i$

53. $(x - 3)^2 = -9; x = 3 - 3i$

54. $(x + 1)^2 = -4; x = -1 + 2i$

55. $x^2 - 2x + 5 = 0; x = 1 - 2i$

56. $x^2 + 6x + 13 = 0; x = -3 + 2i$

57. $x^2 - 4x + 9 = 0; x = 2 + i\sqrt{5}$

58. $x^2 - 2x + 4 = 0; x = 1 - \sqrt{3}\,i$

59. Show that $x = 1 + 4i$ is a solution to $x^2 - 2x + 17 = 0$. Then show its complex conjugate $1 - 4i$ is also a solution.

60. Show that $x = 2 - 3\sqrt{2}\,i$ is a solution to $x^2 - 4x + 22 = 0$. Then show its complex conjugate $2 + 3\sqrt{2}\,i$ is also a solution.

Simplify using powers of i.

61. a. i^{48} **b.** i^{26} **c.** i^{39} **d.** i^{53}

62. a. i^{36} **b.** i^{50} **c.** i^{19} **d.** i^{65}

Divide and write your answer in $a + bi$ form. Check your answer using multiplication.

63. a. $\dfrac{-2}{\sqrt{-49}}$ **b.** $\dfrac{4}{\sqrt{-25}}$

64. a. $\dfrac{2}{1 - \sqrt{-4}}$ **b.** $\dfrac{3}{2 + \sqrt{-9}}$

65. a. $\dfrac{7}{3 + 2i}$ **b.** $\dfrac{-5}{2 - 3i}$

66. a. $\dfrac{6}{1 + 3i}$ **b.** $\dfrac{7}{7 - 2i}$

67. a. $\dfrac{3 + 4i}{4i}$ **b.** $\dfrac{2 - 3i}{3i}$

68. a. $\dfrac{-4 + 8i}{2 - 4i}$ **b.** $\dfrac{3 - 2i}{-6 + 4i}$

▶ WORKING WITH FORMULAS

69. Absolute value of a complex number:
$$|a + bi| = \sqrt{a^2 + b^2}$$

The absolute value of any complex number $a + bi$ (sometimes called the *modulus* of the number) is computed by taking the square root of the sum of the squares of a and b. Find the absolute value of the given complex numbers.

    a. $|2 + 3i|$ **b.** $|4 - 3i|$

    c. $|3 + \sqrt{2}\,i|$

70. Binomial cubes:
$$(A + B)^3 = A^3 + 3A^2B + 3AB^2 + B^3$$

The cube of any binomial can be found using the formula shown, where A and B are the terms of the binomial. Use the formula to compute $(1 - 2i)^3$ (note $A = 1$ and $B = -2i$).

▶ APPLICATIONS

71. Dawn of imaginary numbers: In a day when imaginary numbers were imperfectly understood, Girolamo Cardano (1501–1576) once posed the problem, "Find two numbers that have a sum of 10 and whose product is 40." In other words, $A + B = 10$ and $AB = 40$. Although the solution is routine today, at the time the problem posed an enormous challenge. Verify that $A = 5 + \sqrt{15}i$ and $B = 5 - \sqrt{15}i$ satisfy these conditions.

72. Verifying calculations using *i*: Suppose Cardano had said, "Find two numbers that have a sum of 4 and a product of 7" (see Exercise 71). Verify that $A = 2 + \sqrt{3}i$ and $B = 2 - \sqrt{3}i$ satisfy these conditions.

Although it may seem odd, imaginary numbers have several applications in the real world. Many of these involve a study of electrical circuits, in particular *alternating current* or AC circuits. Briefly, the components of an AC circuit are current I (in amperes), voltage V (in volts), and the impedance Z (in ohms). The impedance of an electrical circuit is a measure of the total opposition to the flow of current through the circuit and is calculated as $Z = R + iX_L - iX_C$ where R represents a pure resistance, X_C represents the capacitance, and X_L represents the inductance. Each of these is also measured in ohms (symbolized by Ω).

73. Find the impedance Z if $R = 7\ \Omega$, $X_L = 6\ \Omega$, and $X_C = 11\ \Omega$.

74. Find the impedance Z if $R = 9.2\ \Omega$, $X_L = 5.6\ \Omega$, and $X_C = 8.3\ \Omega$.

The voltage V (in volts) across any element in an AC circuit is calculated as a product of the current I and the impedance Z: $V = IZ$.

75. Find the voltage in a circuit with a current $I = 3 - 2i$ amperes and an impedance of $Z = 5 + 5i\ \Omega$.

76. Find the voltage in a circuit with a current $I = 2 - 3i$ amperes and an impedance of $Z = 4 + 2i\ \Omega$.

In an AC circuit, the total impedance (in ohms) is given by $Z = \dfrac{Z_1 Z_2}{Z_1 + Z_2}$, where Z represents the total impedance of a circuit that has Z_1 and Z_2 wired in parallel.

77. Find the total impedance Z if $Z_1 = 1 + 2i$ and $Z_2 = 3 - 2i$.

78. Find the total impedance Z if $Z_1 = 3 - i$ and $Z_2 = 2 + i$.

▶ EXTENDING THE CONCEPT

79. Up to this point, we've said that expressions like $x^2 - 9$ and $p^2 - 7$ are factorable:

$$x^2 - 9 = (x + 3)(x - 3) \quad \text{and}$$
$$p^2 - 7 = (p + \sqrt{7})(p - \sqrt{7}),$$

while $x^2 + 9$ and $p^2 + 7$ are prime. More correctly, we should state that $x^2 + 9$ and $p^2 + 7$

are nonfactorable *using real numbers*, since they actually *can* be factored if complex numbers are used. From $(a + bi)(a - bi) = a^2 + b^2$ we note $a^2 + b^2 = (a + bi)(a - bi)$, showing

$$x^2 + 9 = (x + 3i)(x - 3i) \quad \text{and}$$
$$p^2 + 7 = (p + i\sqrt{7})(p - i\sqrt{7}).$$

Use this idea to factor the following.

a. $x^2 + 36$ **b.** $m^2 + 3$

c. $n^2 + 12$ **d.** $4x^2 + 49$

80. In this section, we noted that the product property of radicals $\sqrt{AB} = \sqrt{A}\sqrt{B}$, can still be applied when at most one of the factors is negative. So what happens if *both* are negative? First consider the expression $\sqrt{-4 \cdot -25}$. What happens if you first multiply in the radicand, then compute the square root? Next consider the product $\sqrt{-4} \cdot \sqrt{-25}$. Rewrite each factor using the i notation, then compute the product. Do you get the same result as before? What can you say about $\sqrt{-4 \cdot -25}$ and $\sqrt{-4} \cdot \sqrt{-25}$?

81. Simplify the expression
$i^{17}(3 - 4i) - 3i^3(1 + 2i)^2$.

82. While it is a simple concept for real numbers, the square root of a complex number is much more involved due to the interplay between its real and imaginary parts. For $z = a + bi$ the square root of z can be found using the formula:

$$\sqrt{z} = \frac{\sqrt{2}}{2}(\sqrt{|z| + a} \pm i\sqrt{|z| - a}), \text{ where the sign}$$

is chosen to match the sign of b (see Exercise 69). Use the formula to find the square root of each complex number, then check by squaring.

a. $z = -7 + 24i$ **b.** $z = 5 - 12i$

c. $z = 4 + 3i$

▶ MAINTAINING YOUR SKILLS

83. (R.7) State the perimeter and area formulas for: (a) squares, (b) rectangles, (c) triangles, and (d) circles.

84. (R.1) Write the symbols in words and state True/False.

    a. $6 \notin \mathbb{Q}$ **b.** $\mathbb{Q} \subset \mathbb{R}$

    c. $103 \in \{3, 4, 5, \dots\}$ **d.** $\mathbb{R} \not\subset \mathbb{C}$

85. (1.1) John can run 10 m/sec, while Rick can only run 9 m/sec. If Rick gets a 2-sec head start, who will hit the 200-m finish line first?

86. (R.4) Factor the following expressions completely.

    a. $x^4 - 16$ **b.** $n^3 - 27$

    c. $x^3 - x^2 - x + 1$ **d.** $4n^2m - 12nm^2 + 9m^3$

1.5 | Solving Quadratic Equations

Learning Objectives

In Section 1.5 you will learn how to:

☐ **A.** Solve quadratic equations using the zero product property

☐ **B.** Solve quadratic equations using the square root property of equality

☐ **C.** Solve quadratic equations by completing the square

☐ **D.** Solve quadratic equations using the quadratic formula

☐ **E.** Use the discriminant to identify solutions

☐ **F.** Solve applications of quadratic equations

In Section 1.1 we solved the equation $ax + b = c$ for x to establish a general solution for all linear equations of this form. In this section, we'll establish a general solution for the quadratic equation $ax^2 + bx + c = 0, (a \neq 0)$ using a process known as *completing the square*. Other applications of completing the square include the graphing of parabolas, circles, and other relations from the family of *conic sections*.

A. Quadratic Equations and the Zero Product Property

A **quadratic equation** is one that can be written in the form $ax^2 + bx + c = 0$, where a, b, and c are real numbers and $a \neq 0$. As shown, the equation is written in **standard form,** meaning the terms are in decreasing order of degree and the equation is set equal to zero.

Quadratic Equations

A quadratic equation is one that can be written in the form

$$ax^2 + bx + c = 0,$$

with $a, b, c \in \mathbb{R}$, and $a \neq 0$.

Notice that a is the leading coefficient, b is the coefficient of the linear (first degree) term, and c is a constant. All quadratic equations have degree two, but can have one, two, or three terms. The equation $n^2 - 81 = 0$ is a quadratic equation with two terms, where $a = 1$, $b = 0$, and $c = -81$.

EXAMPLE 1 ▶ Determining Whether an Equation Is Quadratic

State whether the given equation is quadratic. If yes, identify coefficients a, b, and c.

a. $2x^2 - 18 = 0$　　　　　**b.** $z - 12 - 3z^2 = 0$　　　　　**c.** $\dfrac{-3}{4}x + 5 = 0$

d. $z^3 - 2z^2 + 7z = 8$　　　　**e.** $0.8x^2 = 0$

Solution ▶

	Standard Form	Quadratic	Coefficients
a.	$2x^2 - 18 = 0$	yes, deg 2	$a = 2$　$b = 0$　$c = -18$
b.	$-3z^2 + z - 12 = 0$	yes, deg 2	$a = -3$　$b = 1$　$c = -12$
c.	$\dfrac{-3}{4}x + 5 = 0$	no, deg 1	(linear equation)
d.	$z^3 - 2z^2 + 7z - 8 = 0$	no, deg 3	(cubic equation)
e.	$0.8x^2 = 0$	yes, deg 2	$a = 0.8$　$b = 0$　$c = 0$

WORTHY OF NOTE

The word *quadratic* comes from the Latin word *quadratum,* meaning square. The word historically refers to the "four sidedness" of a square, but mathematically to the *area* of a square. Hence its application to polynomials of the form $ax^2 + bx + c$— the variable of the leading term is *squared*.

Now try Exercises 7 through 18 ▶

With quadratic and other polynomial equations, we generally cannot isolate the variable on one side using only properties of equality, because the variable is raised to different powers. Instead we attempt to solve the equation by factoring and applying the **zero product property.**

Zero Product Property

If A and B represent real numbers or real valued expressions

and $A \cdot B = 0$,

then $A = 0$ or $B = 0$.

In words, the property says, *If the product of any two (or more) factors is equal to zero, then at least one of the factors must be equal to zero.* We can use this property to solve higher degree equations after rewriting them in terms of equations with lesser degree. As with linear equations, values that make the original equation true are called *solutions* or *roots* of the equation.

EXAMPLE 2 ▶ Solving Equations Using the Zero Product Property

Solve by writing the equations in factored form and applying the zero product property.

a. $3x^2 = 5x$　　　　**b.** $-5x + 2x^2 = 3$　　　　**c.** $4x^2 = 12x - 9$

Solution ▶

a.
$$3x^2 = 5x \quad \text{given equation}$$
$$3x^2 - 5x = 0 \quad \text{standard form}$$
$$x(3x - 5) = 0 \quad \text{factor}$$
$$x = 0 \quad \text{or} \quad 3x - 5 = 0 \quad \text{set factors equal to zero (zero product property)}$$
$$x = 0 \quad \text{or} \quad x = \frac{5}{3} \quad \text{result}$$

b.
$$-5x + 2x^2 = 3 \quad \text{given equation}$$
$$2x^2 - 5x - 3 = 0 \quad \text{standard form}$$
$$(2x + 1)(x - 3) = 0 \quad \text{factor}$$
$$2x + 1 = 0 \quad \text{or} \quad x - 3 = 0 \quad \text{set factors equal to zero (zero product property)}$$
$$x = -\frac{1}{2} \quad \text{or} \quad x = 3 \quad \text{result}$$

c.

$$4x^2 = 12x - 9 \qquad \text{given equation}$$

$$4x^2 - 12x + 9 = 0 \qquad \text{standard form}$$

$$(2x - 3)(2x - 3) = 0 \qquad \text{factor}$$

$$2x - 3 = 0 \quad \text{or} \quad 2x - 3 = 0 \qquad \text{set factors equal to zero (zero product property)}$$

$$x = \frac{3}{2} \quad \text{or} \quad x = \frac{3}{2} \qquad \text{result}$$

This equation has only the solution $x = \dfrac{3}{2}$, which we call a *repeated root*.

Now try Exercises 19 through 42 ▶

⚠ **CAUTION** ▶ Consider the equation $x^2 - 2x - 3 = 12$. While the left-hand side is factorable, the result is $(x - 3)(x + 1) = 12$ and finding a solution becomes a "guessing game" because the equation is not set equal to zero. If you *misapply* the zero factor property and say that $x - 3 = 12$ or $x + 1 = 12$, the "solutions" are $x = 15$ or $x = 11$, which are both incorrect! After subtracting 12 from both sides $x^2 - 2x - 3 = 12$ becomes $x^2 - 2x - 15 = 0$, giving $(x - 5)(x + 3) = 0$ with solutions $x = 5$ or $x = -3$.

☑ **A.** You've just learned how to solve quadratic equations using the zero product property

B. Solving Quadratic Equations Using the Square Root Property of Equality

The equation $x^2 = 9$ can be solved by factoring. In standard form we have $x^2 - 9 = 0$ (note $b = 0$), then $(x - 3)(x + 3) = 0$. The solutions are $x = -3$ or $x = 3$, which are simply the *positive and negative square roots of 9*. This result suggests an alternative method for solving equations of the form $X^2 = k$, known as the **square root property of equality.**

WORTHY OF NOTE

In Section R.6 we noted that for any real number a, $\sqrt{a^2} = |a|$. From Example 3(a), solving the equation by taking the square root of both sides produces $\sqrt{x^2} = \sqrt{\frac{9}{4}}$. This is equivalent to $|x| = \sqrt{\frac{9}{4}}$, again showing this equation must have two solutions, $x = -\sqrt{\frac{9}{4}}$ and $x = \sqrt{\frac{9}{4}}$.

Square Root Property of Equality

If X represents an algebraic expression

$$\text{and } X^2 = k,$$

$$\text{then } X = \sqrt{k} \text{ or } X = -\sqrt{k};$$

$$\text{also written as } X = \pm\sqrt{k}$$

EXAMPLE 3 ▶ **Solving an Equation Using the Square Root Property of Equality**

Use the square root property of equality to solve each equation.

  a. $-4x^2 + 3 = -6$ **b.** $x^2 + 12 = 0$ **c.** $(x - 5)^2 = 24$

Solution ▶ **a.**

$$-4x^2 + 3 = -6 \qquad \text{original equation}$$

$$x^2 = \frac{9}{4} \qquad \text{subtract 3, divide by } -4$$

$$x = \sqrt{\frac{9}{4}} \quad \text{or} \quad x = -\sqrt{\frac{9}{4}} \qquad \text{square root property of equality}$$

$$x = \frac{3}{2} \quad \text{or} \quad x = -\frac{3}{2} \qquad \text{simplify radicals}$$

This equation has two rational solutions.

b. $x^2 + 12 = 0$ original equation

$\qquad x^2 = -12$ subtract 12

$\qquad x = \sqrt{-12} \quad \text{or} \quad x = -\sqrt{12}$ square root property of equality

$\qquad x = 2i\sqrt{3} \quad \text{or} \quad x = -2i\sqrt{3}$ simplify radicals

This equation has two complex solutions.

☑ **B.** You've just learned how to solve quadratic equations using the square root property of equality

c. $(x - 5)^2 = 24$ original equation

$\qquad x - 5 = \sqrt{24} \quad \text{or} \quad x - 5 = -\sqrt{24}$ square root property of equality

$\qquad x = 5 + 2\sqrt{6} \qquad\qquad x = 5 - 2\sqrt{6}$ solve for x and simplify radicals

This equation has two irrational solutions.

> **Now try Exercises 43 through 58 ▶**

⚠ **CAUTION** ▶ For equations of the form $(x + d)^2 = k$ [see Example 3(c)], you should resist the temptation to expand the binomial square in an attempt to simplify the equation and solve by factoring—many times the result is nonfactorable. *Any* equation of the form $(x + d)^2 = k$ can quickly be solved using the square root property of equality.

Answers written using radicals are called **exact** or **closed form** solutions. Actually checking the exact solutions is a nice application of fundamental skills. Let's check $x = 5 + 2\sqrt{6}$ from Example 3(c).

check: $(x - 5)^2 = 24$ original equation

$\qquad\qquad\qquad (5 + 2\sqrt{6} - 5)^2 = 24$ substitute $5 + 2\sqrt{6}$ for x

$\qquad\qquad\qquad\qquad (2\sqrt{6})^2 = 24$ simplify

$\qquad\qquad\qquad\qquad\qquad 4(6) = 24$ $(2\sqrt{6})^2 = 4(6)$

$\qquad\qquad\qquad\qquad\qquad\quad 24 = 24$ ✓   result checks ($x = 5 - 2\sqrt{6}$ also checks)

C. Solving Quadratic Equations by Completing the Square

Again consider $(x - 5)^2 = 24$ from Example 3(c). If we had first expanded the binomial square, we would have obtained $x^2 - 10x + 25 = 24$, then $x^2 - 10x + 1 = 0$ in standard form. Note that this equation *cannot be solved by factoring*. Reversing this process leads us to a strategy for solving nonfactorable quadratic equations, by creating a *perfect square trinomial* from the quadratic and linear terms. This process is known as **completing the square.** To transform $x^2 - 10x + 1 = 0$ back into $x^2 - 10x + 25 = 24$ [which we would then rewrite as $(x - 5)^2 = 24$ and solve], we subtract 1 from both sides, then add 25:

$$x^2 - 10x + 1 = 0$$

$$x^2 - 10x = -1 \qquad \text{subtract 1}$$

$$x^2 - 10x + 25 = -1 + 25 \qquad \text{add 25}$$

$$(x - 5)^2 = 24 \qquad \text{factor, simplify}$$

In general, after subtracting the constant term, the number that "completes the square" is found by squaring $\frac{1}{2}$ the coefficient of the linear term: $\left[\frac{1}{2}(10)\right]^2 = 25$. **See Exercises 59 through 64 for additional practice.**

EXAMPLE 4 ▶   Solving a Quadratic Equation by Completing the Square

Solve by completing the square: $x^2 + 13 = 6x$.

Solution ▶

$$x^2 + 13 = 6x \qquad \text{original equation}$$
$$x^2 - 6x + 13 = 0 \qquad \text{standard form}$$
$$x^2 - 6x + \underline{\quad} = -13 + \underline{\quad} \qquad \text{subtract 13 to make room for new constant}$$
$$\left[\left(\tfrac{1}{2}\right)(-6)\right]^2 = 9 \qquad \text{compute } \left[\left(\tfrac{1}{2}\right)(\textit{linear coefficient})\right]^2$$
$$x^2 - 6x + 9 = -13 + 9 \qquad \text{add 9 to both sides (completing the square)}$$
$$(x - 3)^2 = -4 \qquad \text{factor and simplify}$$
$$x - 3 = \sqrt{-4} \quad \text{or} \quad x - 3 = -\sqrt{-4} \qquad \text{square root property of equality}$$
$$x = 3 + 2i \quad \text{or} \quad x = 3 - 2i \qquad \text{simplify radicals and solve for } x$$

Now try Exercises 65 through 74 ▶

The process of completing the square can be applied to any quadratic equation with a leading coefficient of 1. If the leading coefficient is not 1, we simply divide through by a before beginning, which brings us to this summary of the process.

WORTHY OF NOTE

It's helpful to note that the number you're squaring in step three, $\left[\dfrac{1}{2} \cdot \dfrac{b}{a}\right] = \dfrac{b}{2a}$, turns out to be the constant term in the factored form. From Example 4, the number we squared was $\left(\tfrac{1}{2}\right)(-6) = -3$, and the binomial square was $(x - 3)^2$.

Completing the Square to Solve a Quadratic Equation

To solve $ax^2 + bx + c = 0$ by completing the square:

1. Subtract the constant c from both sides.
2. Divide both sides by the leading coefficient a.
3. Compute $\left[\dfrac{1}{2} \cdot \dfrac{b}{a}\right]^2$ and add the result to both sides.
4. Factor left-hand side as a binomial square; simplify right-hand side.
5. Solve using the square root property of equality.

EXAMPLE 5 ▶ **Solving a Quadratic Equation by Completing the Square**

Solve by completing the square: $-3x^2 + 1 = 4x$.

Solution ▶

$$-3x^2 + 1 = 4x \qquad \text{original equation}$$
$$-3x^2 - 4x + 1 = 0 \qquad \text{standard form (nonfactorable)}$$
$$-3x^2 - 4x = -1 \qquad \text{subtract 1}$$
$$x^2 + \frac{4}{3}x + \phantom{\frac{4}{9}} = \frac{1}{3} \qquad \text{divide by } -3$$
$$x^2 + \frac{4}{3}x + \frac{4}{9} = \frac{1}{3} + \frac{4}{9} \qquad \left[\frac{1}{2}\frac{b}{a}\right]^2 = \left[\left(\frac{1}{2}\right)\left(\frac{4}{3}\right)\right]^2 = \frac{4}{9}; \text{ add } \frac{4}{9}$$
$$\left(x + \frac{2}{3}\right)^2 = \frac{7}{9} \qquad \text{factor and simplify } \left(\frac{1}{3} = \frac{3}{9}\right)$$
$$x + \frac{2}{3} = \sqrt{\frac{7}{9}} \quad \text{or} \quad x + \frac{2}{3} = -\sqrt{\frac{7}{9}} \qquad \text{square root property of equality}$$
$$x = -\frac{2}{3} + \frac{\sqrt{7}}{3} \quad \text{or} \quad x = -\frac{2}{3} - \frac{\sqrt{7}}{3} \qquad \text{solve for } x \text{ and simplify (exact form)}$$
$$x \approx 0.22 \quad \text{or} \quad x \approx -1.55 \qquad \text{approximate form (to hundredths)}$$

☑ **C.** You've just learned how to solve quadratic equations by completing the square

Now try Exercises 75 through 82 ▶

⚠ **CAUTION** ▶ For many of the skills/processes needed in a study of algebra, it's actually easier to work with the fractional form of a number, rather than the decimal form. For example, computing $\left(\frac{2}{3}\right)^2$ is easier than computing $(0.\overline{6})^2$, and finding $\sqrt{\frac{9}{16}}$ is much easier than finding $\sqrt{0.5625}$.

D. Solving Quadratic Equations Using the Quadratic Formula

In Section 1.1 we found a general solution for the linear equation $ax + b = c$ by comparing it to $2x + 3 = 15$. Here we'll use a similar idea to find a general solution for quadratic equations. In a side-by-side format, we'll solve the equations $2x^2 + 5x + 3 = 0$ and $ax^2 + bx + c = 0$ by completing the square. Note the similarities.

$2x^2 + 5x + 3 = 0$	given equations	$ax^2 + bx + c = 0$
$2x^2 + 5x + \underline{} = -3$	subtract constant term	$ax^2 + bx + \underline{} = -c$
$x^2 + \dfrac{5}{2}x + \underline{} = -\dfrac{3}{2}$	divide by lead coefficient	$x^2 + \dfrac{b}{a}x + \underline{} = -\dfrac{c}{a}$
$\left[\dfrac{1}{2}\left(\dfrac{5}{2}\right)\right]^2 = \dfrac{25}{16}$	$\left[\dfrac{1}{2}(\text{linear coefficient})\right]^2$	$\left[\dfrac{1}{2}\left(\dfrac{b}{a}\right)\right]^2 = \dfrac{b^2}{4a^2}$
$x^2 + \dfrac{5}{2}x + \dfrac{25}{16} = \dfrac{25}{16} - \dfrac{3}{2}$	add to both sides	$x^2 + \dfrac{b}{a}x + \dfrac{b^2}{4a^2} = \dfrac{b^2}{4a^2} - \dfrac{c}{a}$
$\left(x + \dfrac{5}{4}\right)^2 = \dfrac{25}{16} - \dfrac{3}{2}$	left side factors as a binomial square	$\left(x + \dfrac{b}{2a}\right)^2 = \dfrac{b^2}{4a^2} - \dfrac{c}{a}$
$\left(x + \dfrac{5}{4}\right)^2 = \dfrac{25}{16} - \dfrac{24}{16}$	determine LCDs	$\left(x + \dfrac{b}{2a}\right)^2 = \dfrac{b^2}{4a^2} - \dfrac{4ac}{4a^2}$
$\left(x + \dfrac{5}{4}\right)^2 = \dfrac{1}{16}$	simplify right side	$\left(x + \dfrac{b}{2a}\right)^2 = \dfrac{b^2 - 4ac}{4a^2}$
$x + \dfrac{5}{4} = \pm\sqrt{\dfrac{1}{16}}$	square root property of equality	$x + \dfrac{b}{2a} = \pm\sqrt{\dfrac{b^2 - 4ac}{4a^2}}$
$x + \dfrac{5}{4} = \pm\dfrac{1}{4}$	simplify radicals	$x + \dfrac{b}{2a} = \pm\dfrac{\sqrt{b^2 - 4ac}}{2a}$
$x = -\dfrac{5}{4} \pm \dfrac{1}{4}$	solve for x	$x = -\dfrac{b}{2a} \pm \dfrac{\sqrt{b^2 - 4ac}}{2a}$
$x = \dfrac{-5 \pm 1}{4}$	combine terms	$x = \dfrac{-b \pm \sqrt{b^2 - 4ac}}{2a}$
$x = \dfrac{-5 + 1}{4}$ or $x = \dfrac{-5 - 1}{4}$	solutions	$x = \dfrac{-b + \sqrt{b^2 - 4ac}}{2a}$ or $x = \dfrac{-b - \sqrt{b^2 - 4ac}}{2a}$

On the left, our final solutions are $x = -1$ or $x = -\frac{3}{2}$. The general solution is called the **quadratic formula,** which can be used to solve *any equation belonging to the quadratic family.*

Quadratic Formula

If $ax^2 + bx + c = 0$, with a, b, and $c \in \mathbb{R}$ and $a \neq 0$, then

$$x = \frac{-b + \sqrt{b^2 - 4ac}}{2a} \quad \text{or} \quad x = \frac{-b - \sqrt{b^2 - 4ac}}{2a};$$

also written $x = \dfrac{-b \pm \sqrt{b^2 - 4ac}}{2a}$.

⚠ **CAUTION** ▶ It's very important to note the values of a, b, and c come from an equation *written in standard form.* For $3x^2 - 5x = -7$, $a = 3$ and $b = -5$, but $c \neq -7$! In standard form we have $3x^2 - 5x + 7 = 0$, and note the value for use in the formula is actually $c = 7$.

EXAMPLE 6 ▶   Solving Quadratic Equations Using the Quadratic Formula

Solve $4x^2 + 1 = 8x$ using the quadratic formula. State the solution(s) in both exact and approximate form. Check one of the exact solutions in the original equation.

Solution ▶   Begin by writing the equation in standard form and identifying the values of a, b, and c.

$$4x^2 + 1 = 8x \qquad \text{original equation}$$

$$4x^2 - 8x + 1 = 0 \qquad \text{standard form}$$

$$a = 4, b = -8, c = 1$$

$$x = \frac{-(-8) \pm \sqrt{(-8)^2 - 4(4)(1)}}{2(4)} \qquad \text{substitute 4 for } a, -8 \text{ for } b, \text{ and 1 for } c$$

$$x = \frac{8 \pm \sqrt{64 - 16}}{8} = \frac{8 \pm \sqrt{48}}{8} \qquad \text{simplify}$$

$$x = \frac{8 \pm 4\sqrt{3}}{8} = \frac{8}{8} \pm \frac{4\sqrt{3}}{8} \qquad \text{rationalize the radical (see following Caution)}$$

$$x = 1 + \frac{\sqrt{3}}{2} \quad \text{or} \quad x = 1 - \frac{\sqrt{3}}{2} \qquad \text{exact solutions}$$

$$x \approx 1.87 \qquad \text{or} \qquad x \approx 0.13 \qquad \text{approximate solutions}$$

Check ▶

$$4x^2 + 1 = 8x \qquad \text{original equation}$$

$$4\left(1 + \frac{\sqrt{3}}{2}\right)^2 + 1 = 8\left(1 + \frac{\sqrt{3}}{2}\right) \qquad \text{substitute } 1 + \tfrac{\sqrt{3}}{2} \text{ for } x$$

$$4\left[1 + 2\left(\frac{\sqrt{3}}{2}\right) + \frac{3}{4}\right] + 1 = 8 + 4\sqrt{3} \qquad \text{square binomial; distribute}$$

$$4 + 4\sqrt{3} + 3 + 1 = 8 + 4\sqrt{3} \qquad \text{distribute}$$

$$8 + 4\sqrt{3} = 8 + 4\sqrt{3} \checkmark \qquad \text{result checks}$$

☑ **D.** You've just learned how to solve quadratic equations using the quadratic formula

Now try Exercises 83 through 112 ▶

⚠ **CAUTION** ▶   For $\dfrac{8 \pm 4\sqrt{3}}{8}$, be careful not to incorrectly "cancel the eights" as in $\dfrac{\overset{1}{\cancel{8}} \pm 4\sqrt{3}}{\underset{1}{\cancel{8}}} \neq 1 \pm 4\sqrt{3}$.

No! Use a calculator to verify that the results are not equivalent. Both terms in the numerator are divided by 8 and we must either rewrite the expression as separate terms (as above) or factor the numerator to see if the expression simplifies further:

$$\frac{8 \pm 4\sqrt{3}}{8} = \frac{\overset{1}{\cancel{4}}(2 \pm \sqrt{3})}{\underset{2}{\cancel{8}}} = \frac{2 \pm \sqrt{3}}{2}, \text{ which is equivalent to } 1 \pm \frac{\sqrt{3}}{2}.$$

E. The Discriminant of the Quadratic Formula

Recall that $\sqrt{X}$ represents a real number only for $X \geq 0$. Since the quadratic formula contains the radical $\sqrt{b^2 - 4ac}$, the expression $b^2 - 4ac$, called the **discriminant,** will determine the nature (real or complex) and the number of solutions to a given quadratic equation.

The Discriminant of the Quadratic Formula

For $ax^2 + bx + c = 0$, $a \neq 0$,

1. If $b^2 - 4ac = 0$, the equation has one real root.

2. If $b^2 - 4ac > 0$, the equation has two real roots.

3. If $b^2 - 4ac < 0$, the equation has two complex roots.

Further analysis of the discriminant reveals even more concerning the nature of quadratic solutions. If a, b, and c are rational and the discriminant is a perfect square, there will be two *rational* roots, which means the original equation can be solved by factoring. If the discriminant is not a perfect square, there will be two *irrational* roots that are conjugates. If the discriminant is zero there is one rational root, and the original equation is a perfect square trinomial.

EXAMPLE 7 ▶ **Using the Discriminant to Analyze Solutions**

Use the discriminant to determine if the equation given has any real root(s). If so, state whether the roots are rational or irrational, and whether the quadratic expression is factorable.

    a. $2x^2 + 5x + 2 = 0$ **b.** $x^2 - 4x + 7 = 0$ **c.** $4x^2 - 20x + 25 = 0$

Solution ▶

    a. $a = 2, b = 5, c = 2$ **b.** $a = 1, b = -4, c = 7$ **c.** $a = 4, b = -20, c = 25$

$$b^2 - 4ac = (5)^2 - 4(2)(2) \qquad b^2 - 4ac = (-4)^2 - 4(1)(7) \qquad b^2 - 4ac = (-20)^2 - 4(4)(25)$$
$$= 9 \qquad\qquad\qquad = -12 \qquad\qquad\qquad = 0$$

Since $9 > 0$,	Since $-12 < 0$,	Since $b^2 - 4ac = 0$,
→ two rational roots, factorable	→ two complex roots, nonfactorable	→ one rational root, factorable

Now try Exercises 113 through 124 ▶

In Example 7(b), $b^2 - 4ac = -12$ and the quadratic formula shows $x = \dfrac{4 \pm \sqrt{-12}}{2}$. After simplifying, we find the solutions are the complex conjugates $x = 2 + i\sqrt{3}$ or $x = 2 - i\sqrt{3}$. In general, when $b^2 - 4ac < 0$, the solutions *will be complex conjugates*.

Complex Solutions

The complex solutions of a quadratic equation with real coefficients occur in conjugate pairs.

EXAMPLE 8 ▶ **Solving Quadratic Equations Using the Quadratic Formula**

Solve: $2x^2 - 6x + 5 = 0$.

Solution ▶

With $a = 2$, $b = -6$, and $c = 5$, the discriminant becomes $(-6)^2 - 4(2)(5) = -4$, showing there will be two complex roots. The quadratic formula then yields

$$x = \frac{-b \pm \sqrt{b^2 - 4ac}}{2a} \qquad \text{quadratic formula}$$

$$x = \frac{-(-6) \pm \sqrt{-4}}{2(2)} \qquad b^2 - 4ac = -4, \text{ substitute 2 for } a, \text{ and } -6 \text{ for } b$$

$$x = \frac{6 \pm 2i}{4} \qquad \text{simplify, write in } i \text{ form}$$

$$x = \frac{3}{2} \pm \frac{1}{2}i \qquad \text{solutions are complex conjugates}$$

☑ E. You've just learned how to use the discriminant to identify solutions

Now try Exercises 125 through 130 ▶

Summary of Solution Methods for $ax^2 + bx + c = 0$

1. If $b = 0$, isolate x and use the square root property of equality.
2. If $c = 0$, factor out the GCF and solve using the zero product property.
3. If no coefficient is zero, you can attempt to solve by
   a. factoring the trinomial
   b. completing the square
   c. using the quadratic formula

F. Applications of the Quadratic Formula

A projectile is any object that is thrown, shot, or *projected* upward with no sustaining source of propulsion. The height of the projectile at time t is modeled by the equation $h = -16t^2 + vt + k$, where h is the height of the object in feet, t is the elapsed time in seconds, and v is the initial velocity in feet per second. The constant k represents the initial height of the object above ground level, as when a person releases an object 5 ft above the ground in a throwing motion. If the person were on a cliff 60 ft high, k would be 65 ft.

EXAMPLE 9 ▶   **Solving an Application of Quadratic Equations**

A person standing on a cliff 60 ft high, throws a ball upward with an initial velocity of 102 ft/sec (assume the ball is released 5 ft above where the person is standing). Find (a) the height of the object after 3 sec and (b) how many seconds until the ball hits the ground at the base of the cliff.

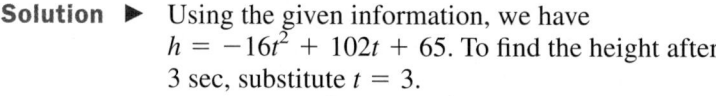

Solution ▶   Using the given information, we have $h = -16t^2 + 102t + 65$. To find the height after 3 sec, substitute $t = 3$.

a. $h = -16t^2 + 102t + 65$ original equation
    $ = -16(3)^2 + 102(3) + 65$ substitute 3 for t
    $ = 227$ result

After 3 sec, the ball is 227 ft above the ground.

b. When the ball hits the ground at the base of the cliff, it has a height of zero. Substitute $h = 0$ and solve using the quadratic formula.

$0 = -16t^2 + 102t + 65$ $a = -16, b = 102, c = 65$

$t = \dfrac{-b \pm \sqrt{b^2 - 4ac}}{2a}$ quadratic formula

$t = \dfrac{-(102) \pm \sqrt{(102)^2 - 4(-16)(65)}}{2(-16)}$ substitute -16 for a, 102 for b, 65 for c

$t = \dfrac{-102 \pm \sqrt{14{,}564}}{-32}$ simplify

Since we're trying to find the time in seconds, we go directly to the approximate form of the answer.

$t \approx -0.58$ or $t \approx 6.96$ approximate solutions

The ball will strike the base of the cliff about 7 sec later. Since t represents time, the solution $t \approx -0.58$ does not apply.

Now try Exercises 133 through 140 ▶

EXAMPLE 10 ▶   **Solving Applications Using the Quadratic Formula**

For the years 1995 to 2002, the amount A of annual international telephone traffic (in billions of minutes) can be modeled by $A = 0.3x^2 + 8.9x + 61.8$, where $x = 0$ represents the year 1995 [*Source:* Data from the *2005 Statistical Abstract of the United States,* Table 1372, page 870]. If this trend continues, in what year will the annual number of minutes reach or surpass 275 billion minutes?

Solution ▶   We are essentially asked to solve $A = 0.3x^2 + 8.9x + 61.8$, when $A = 275$.

$$275 = 0.3x^2 + 8.9x + 61.8 \qquad \text{given equation}$$
$$0 = 0.3x^2 + 8.9x - 213.2 \qquad \text{subtract 275}$$

For $a = 0.3$, $b = 8.9$, and $c = -213.2$, the quadratic formula gives

$$x = \frac{-b \pm \sqrt{b^2 - 4ac}}{2a} \qquad \text{quadratic formula}$$

$$x = \frac{-8.9 \pm \sqrt{(8.9)^2 - 4(0.3)(-213.2)}}{2(0.3)} \qquad \text{substitute known values}$$

$$x = \frac{-8.9 \pm \sqrt{335.05}}{0.6} \qquad \text{simplify}$$

$$x \approx 15.7 \quad \text{or} \quad x \approx -45.3 \qquad \text{result}$$

☑ **F. You've just learned how to solve applications of quadratic equations**

We disregard the negative solution (since x represents time), and find the annual number of international telephone minutes will reach or surpass 275 billion 15.7 years after 1995, or in the year 2010.

Now try Exercises 141 and 142 ▶

TECHNOLOGY HIGHLIGHT

The Discriminant

Quadratic equations play an important role in a study of College Algebra, forming a bridge between our previous and current studies, and the more advanced equations to come. As seen in this section, the discriminant of the quadratic formula ($b^2 - 4ac$) reveals the type and number of solutions, and whether the original equation can be solved by factoring (the discriminant is a perfect square). It will often be helpful to have this information in advance of trying to solve or graph the equation. Since this will be done for each new equation, the discriminant is a prime candidate for a short program. To begin a new program press PRGM ▶ ▶ (**NEW**) ENTER. The calculator will prompt you to name the program using the green ALPHA letters (eight letters max), then allow you to start entering program lines. In PRGM mode, pressing PRGM once again will bring up menus that contain all needed commands. For very basic programs, these commands will be in the **I/O** (Input/Output) submenu, with the most common options being **2:Prompt, 3:Disp,** and **8:CLRHOME.** As you can see, we have named our program *DISCRMNT.*

PROGRAM:DISCRMNT

:CLRHOME	Clears the home screen, places cursor in upper left corner
:DISP "DISCRIMINANT"	Displays the word *DISCRIMINANT* as user information
:DISP "B²−4AC"	Displays $B^2 - 4AC$ as user information
:DISP ""	Displays a blank line (for formatting)
:Prompt A, B, C	Prompts the user to enter the values of A, B, and C
:B²−4AC → D	Computes $B^2 - 4AC$ using given values and stores result in memory location D

—continued

:CLRHOME	Clears the home screen, places cursor in upper left corner
:DISP "DISCRIMINANT IS:"	Displays the words *DISCRIMINANT IS* as user information
:DISP D	Displays the computed value of D

Exercise 1: Run the program for $x^2 - 3x - 10 = 0$ and $x^2 + 5x - 14 = 0$ to verify that both can be solved by factoring. What do you notice?

Exercise 2: Run the program for $25x^2 - 90x + 81 = 0$ and $4x^2 + 20x + 25 = 0$, then check to see if each is a perfect square trinomial. What do you notice?

Exercise 3: Run the program for $y = x^2 + 2x + 10$ and $y = x^2 - 2x + 5$. Do these equations have real number solutions? Why or why not?

Exercise 4: Once the discriminant D is known, the quadratic formula becomes $x = \dfrac{-b \pm \sqrt{D}}{2a}$ and solutions can quickly be found. Solve the equations in Exercises 1–3 above.

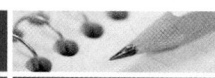

1.5 EXERCISES

▶ CONCEPTS AND VOCABULARY

Fill in each blank with the appropriate word or phrase. Carefully reread the section, if necessary.

1. A polynomial equation is in standard form when written in _____ order of degree and set equal to _____.

2. The solution $x = 2 + \sqrt{3}$ is called an _____ form of the solution. Using a calculator, we find the _____ form is $x \approx 3.732$.

3. To solve a quadratic equation by completing the square, the coefficient of the _____ term must be a _____.

4. The quantity $b^2 - 4ac$ is called the _____ of the quadratic equation. If $b^2 - 4ac > 0$, there are _____ real roots.

5. According to the summary on page 50, what method should be used to solve $4x^2 - 5x = 0$? What are the solutions?

6. Discuss/Explain why this version of the quadratic formula is incorrect:
$$x = -b \pm \frac{\sqrt{b^2 - 4ac}}{2a}$$

▶ DEVELOPING YOUR SKILLS

Determine whether each equation is quadratic. If so, identify the coefficients a, b, and c. If not, discuss why.

7. $2x - 15 - x^2 = 0$ 8. $21 + x^2 - 4x = 0$

9. $\dfrac{2}{3}x - 7 = 0$ 10. $12 - 4x = 9$

11. $\dfrac{1}{4}x^2 = 6x$ 12. $0.5x = 0.25x^2$

13. $2x^2 + 7 = 0$ 14. $5 = -4x^2$

15. $-3x^2 + 9x - 5 + 2x^3 = 0$

16. $z^2 - 6z + 9 - z^3 = 0$

17. $(x - 1)^2 + (x - 1) + 4 = 9$

18. $(x + 5)^2 - (x + 5) + 4 = 17$

Solve using the zero factor property. Be sure each equation is in standard form and factor out any common factors before attempting to solve. Check all answers in the original equation.

19. $x^2 - 15 = 2x$ 20. $z^2 - 10z = -21$

21. $m^2 = 8m - 16$ 22. $-10n = n^2 + 25$

23. $5p^2 - 10p = 0$ **24.** $6q^2 - 18q = 0$

25. $-14h^2 = 7h$ **26.** $9w = -6w^2$

27. $a^2 - 17 = -8$ **28.** $b^2 + 8 = 12$

29. $g^2 + 18g + 70 = -11$

30. $h^2 + 14h - 2 = -51$

31. $m^3 + 5m^2 - 9m - 45 = 0$

32. $n^3 - 3n^2 - 4n + 12 = 0$

33. $(c - 12)c - 15 = 30$

34. $(d - 10)d + 10 = -6$

35. $9 + (r - 5)r = 33$

36. $7 + (s - 4)s = 28$

37. $(t + 4)(t + 7) = 54$

38. $(g + 17)(g - 2) = 20$

39. $2x^2 - 4x - 30 = 0$

40. $-3z^2 + 12z + 36 = 0$

41. $2w^2 - 5w = 3$

42. $-3v^2 = -v - 2$

Solve the following equations using the square root property of equality. Write answers in exact form and approximate form rounded to hundredths. If there are no real solutions, so state.

43. $m^2 = 16$ **44.** $p^2 = 49$

45. $y^2 - 28 = 0$ **46.** $m^2 - 20 = 0$

47. $p^2 + 36 = 0$ **48.** $n^2 + 5 = 0$

49. $x^2 = \frac{21}{16}$ **50.** $y^2 = \frac{13}{9}$

51. $(n - 3)^2 = 36$ **52.** $(p + 5)^2 = 49$

53. $(w + 5)^2 = 3$ **54.** $(m - 4)^2 = 5$

55. $(x - 3)^2 + 7 = 2$ **56.** $(m + 11)^2 + 5 = 3$

57. $(m - 2)^2 = \frac{18}{49}$ **58.** $(x - 5)^2 = \frac{12}{25}$

Fill in the blank so the result is a perfect square trinomial, then factor into a binomial square.

59. $x^2 + 6x +$ _____   **60.** $y^2 + 10y +$ _____

61. $n^2 + 3n +$ _____   **62.** $x^2 - 5x +$ _____

63. $p^2 + \frac{2}{3}p +$ _____   **64.** $x^2 - \frac{3}{2}x +$ _____

Solve by completing the square. Write your answers in both exact form and approximate form rounded to the hundredths place. If there are no real solutions, so state.

65. $x^2 + 6x = -5$ **66.** $m^2 + 8m = -12$

67. $p^2 - 6p + 3 = 0$ **68.** $n^2 = 4n + 10$

69. $p^2 + 6p = -4$ **70.** $x^2 - 8x - 1 = 0$

71. $m^2 + 3m = 1$ **72.** $n^2 + 5n - 2 = 0$

73. $n^2 = 5n + 5$ **74.** $w^2 - 7w + 3 = 0$

75. $2x^2 = -7x + 4$ **76.** $3w^2 - 8w + 4 = 0$

77. $2n^2 - 3n - 9 = 0$ **78.** $2p^2 - 5p = 1$

79. $4p^2 - 3p - 2 = 0$ **80.** $3x^2 + 5x - 6 = 0$

81. $m^2 = 7m - 4$ **82.** $a^2 - 15 = 4a$

Solve each equation using the most efficient method: factoring, square root property of equality, or the quadratic formula. Write your answer in both exact and approximate form (rounded to hundredths). Check one of the exact solutions in the original equation.

83. $x^2 - 3x = 18$ **84.** $w^2 + 6w - 1 = 0$

85. $4m^2 - 25 = 0$ **86.** $4a^2 - 4a = 1$

87. $4n^2 - 8n - 1 = 0$ **88.** $2x^2 - 4x + 5 = 0$

89. $6w^2 - w = 2$ **90.** $3a^2 - 5a + 6 = 0$

91. $4m^2 = 12m - 15$ **92.** $3p^2 + p = 0$

93. $4n^2 - 9 = 0$ **94.** $4x^2 - x = 3$

95. $5w^2 = 6w + 8$ **96.** $3m^2 - 7m - 6 = 0$

97. $3a^2 - a + 2 = 0$ **98.** $3n^2 - 2n - 3 = 0$

99. $5p^2 = 6p + 3$ **100.** $2x^2 + x + 3 = 0$

101. $5w^2 - w = 1$ **102.** $3m^2 - 2 = 5m$

103. $2a^2 + 5 = 3a$ **104.** $n^2 + 4n - 8 = 0$

105. $2p^2 - 4p + 11 = 0$ **106.** $8x^2 - 5x - 1 = 0$

107. $w^2 + \frac{2}{3}w = \frac{1}{9}$ **108.** $\frac{5}{4}m^2 - \frac{8}{3}m + \frac{1}{6} = 0$

109. $0.2a^2 + 1.2a + 0.9 = 0$

110. $-5.4n^2 + 8.1n + 9 = 0$

111. $\frac{2}{7}p^2 - 3 = \frac{8}{21}p$

112. $\frac{5}{9}x^2 - \frac{16}{15}x = \frac{3}{2}$

Use the discriminant to determine whether the given equation has irrational, rational, repeated, or complex roots. Also state whether the original equation is factorable using integers, but do not solve for x.

113. $-3x^2 + 2x + 1 = 0$ **114.** $2x^2 - 5x - 3 = 0$

115. $-4x + x^2 + 13 = 0$ **116.** $-10x + x^2 + 41 = 0$

117. $15x^2 - x - 6 = 0$ **118.** $10x^2 - 11x - 35 = 0$

119. $-4x^2 + 6x - 5 = 0$ **120.** $-5x^2 - 3 = 2x$

121. $2x^2 + 8 = -9x$ **122.** $x^2 + 4 = -7x$

123. $4x^2 + 12x = -9$ **124.** $9x^2 + 4 = 12x$

Solve the quadratic equations given. Simplify each result.

125. $-6x + 2x^2 + 5 = 0$ **126.** $17 + 2x^2 = 10x$

127. $5x^2 + 5 = -5x$ **128.** $x^2 = -2x - 19$

129. $-2x^2 = -5x + 11$ **130.** $4x - 3 = 5x^2$

▶ WORKING WITH FORMULAS

131. Height of a projectile: $h = -16t^2 + vt$

If an object is projected vertically upward from ground level with no continuing source of propulsion, the height of the object (in feet) is modeled by the equation shown, where v is the initial velocity, and t is the time in seconds. Use the quadratic formula to solve for t in terms of v and h. (*Hint:* Set the equation equal to zero and identify the coefficients as before.)

132. Surface area of a cylinder: $A = 2\pi r^2 + 2\pi rh$

The surface area of a cylinder is given by the formula shown, where h is the height and r is the radius of the base. The equation can be considered a quadratic in the variable r. Use the quadratic formula to solve for r in terms of h and A. (*Hint:* Rewrite the equation in standard form and identify the coefficients as before.)

▶ APPLICATIONS

133. Height of a projectile: The height of an object thrown upward from the roof of a building 408 ft tall, with an initial velocity of 96 ft/sec, is given by the equation $h = -16t^2 + 96t + 408$, where h represents the height of the object after t seconds. How long will it take the object to hit the ground? Answer in exact form and decimal form rounded to the nearest hundredth.

134. Height of a projectile: The height of an object thrown upward from the floor of a canyon 106 ft deep, with an initial velocity of 120 ft/sec, is given by the equation $h = -16t^2 + 120t - 106$, where h represents the height of the object after t seconds. How long will it take the object to rise to the height of the canyon wall? Answer in exact form and decimal form rounded to hundredths.

135. Cost, revenue, and profit: The revenue for a manufacturer of microwave ovens is given by the equation $R = x(40 - \frac{1}{3}x)$, where revenue is in thousands of dollars and x thousand ovens are manufactured and sold. What is the minimum number of microwave ovens that must be sold to bring in a revenue of $900,000?

136. Cost, revenue, and profit: The revenue for a manufacturer of computer printers is given by the equation $R = x(30 - 0.4x)$, where revenue is in thousands of dollars and x thousand printers are manufactured and sold. What is the minimum number of printers that must be sold to bring in a revenue of $440,000?

137. Cost, revenue, and profit: The cost of raw materials to produce plastic toys is given by the cost equation $C = 2x + 35$, where x is the number of toys in hundreds. The total income (revenue) from the sale of these toys is given by $R = -x^2 + 122x - 1965$. (a) Determine the profit equation (profit = revenue − cost). During the Christmas season, the owners of the company decide to manufacture and donate as many toys as they can, without taking a loss (i.e., they break even: profit or $P = 0$). (b) How many toys will they produce for charity?

138. Cost, revenue, and profit: The cost to produce bottled spring water is given by the cost equation $C = 16x + 63$, where x is the number of bottles in thousands. The total revenue from the sale of these bottles is given by the equation $R = -x^2 + 326x - 18,463$. (a) Determine the profit equation (profit = revenue − cost). (b) After a bad flood contaminates the drinking water of a nearby community, the owners decide to bottle and donate as many bottles of water as they can, without taking a loss (i.e., they break even: profit or $P = 0$). How many bottles will they produce for the flood victims?

139. Height of an arrow: If an object is projected vertically upward from ground level with no continuing source of propulsion, its height (in feet) is modeled by the equation $h = -16t^2 + vt$, where v is the initial velocity and t is the time in seconds. Use the quadratic formula to solve for t, given an arrow is shot into the air with $v = 144$ ft/sec and $h = 260$ ft. See Exercise 131.

140. Surface area of a cylinder: The surface area of a cylinder is given by $A = 2\pi r^2 + 2\pi rh$, where h is the height and r is the radius of the base. The equation can be considered a quadratic in the variable r. Use the quadratic formula to solve for r, given $A = 4710$ cm^2 and $h = 35$ cm. See Exercise 132.

141. Cell phone subscribers: For the years 1995 to 2002, the number N of cellular phone subscribers (in millions) can be modeled by the equation $N = 17.4x^2 + 36.1x + 83.3$, where $x = 0$ represents the year 1995 [*Source:* Data from the *2005 Statistical Abstract of the United States,* Table 1372, page 870]. If this trend continued, in what year did the number of subscribers reach or surpass 3750 million?

142. U.S. international trade balance: For the years 1995 to 2003, the international trade balance B (in millions of dollars) can be approximated by the equation $B = -3.1x^2 + 4.5x - 19.9$, where $x = 0$ represents the year 1995 [*Source:* Data from the *2005 Statistical Abstract of the United States,* Table 1278, page 799]. If this trend continues, in what year will the trade balance reach a deficit of $750 million dollars or more?

143. Tennis court dimensions: A regulation tennis court for a doubles match is laid out so that its length is 6 ft more than two times its width. The area of the doubles court is 2808 ft^2. What is the length and width of the doubles court?

Exercises 143 and 144

Singles

Doubles

144. Tennis court dimensions: A regulation tennis court for a singles match is laid out so that its length is 3 ft less than three times its width. The area of the singles court is 2106 ft^2. What is the length and width of the singles court?

▶ EXTENDING THE CONCEPT

145. Using the discriminant: Each of the following equations can easily be solved by factoring, since $a = 1$. Using the discriminant, we can create factorable equations with identical values for b and c, but where $a \neq 1$. For instance, $x^2 - 3x - 10 = 0$ and $4x^2 - 3x - 10 = 0$ can both be solved by factoring. Find similar equations ($a \neq 1$) for the quadratics given here. (*Hint:* The discriminant $b^2 - 4ac$ must be a perfect square.)

  a. $x^2 + 6x - 16 = 0$

  b. $x^2 + 5x - 14 = 0$

  c. $x^2 - x - 6 = 0$

146. Using the discriminant: For what values of c will the equation $9x^2 - 12x + c = 0$ have

  a. no real roots   **b.** one rational root

  c. two real roots   **d.** two integer roots

Complex polynomials: Many techniques applied to solve polynomial equations with real coefficients can be applied to solve polynomial equations with *complex* coefficients. Here we apply the idea to carefully chosen quadratic equations, as a more general application must wait until a future course, when the square root of a complex number is fully developed. Solve each equation using the quadratic formula, noting that $\frac{1}{i} = -i$.

147. $z^2 - 3iz = -10$

148. $z^2 - 9iz = -22$

149. $4iz^2 + 5z + 6i = 0$

150. $2iz^2 - 9z + 26i = 0$

151. $0.5z^2 + (7 + i)z + (6 + 7i) = 0$

152. $0.5z^2 + (4 - 3i)z + (-9 - 12i) = 0$

► **MAINTAINING YOUR SKILLS**

153. (R.7) State the formula for the perimeter and area of each figure illustrated.

a.

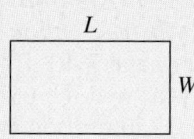

b.

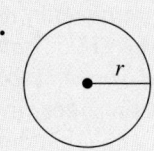

c.

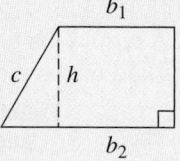

d.

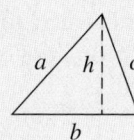

154. (1.3) Factor and solve the following equations:

    a. $x^2 - 5x - 36 = 0$ **b.** $4x^2 - 25 = 0$

    c. $x^3 + 6x^2 - 4x - 24 = 0$

155. (1.1) A total of 900 tickets were sold for a recent concert and \$25,000 was collected. If good seats were \$30 and cheap seats were \$20, how many of each type were sold?

156. (1.1) Solve for C: $P = C + Ct$.

1.6 | Solving Other Types of Equations

Learning Objectives

In Section 1.6 you will learn how to:

☐ **A.** Solve polynomial equations of higher degree

☐ **B.** Solve rational equations

☐ **C.** Solve radical equations and equations with rational exponents

☐ **D.** Solve equations in quadratic form

☐ **E.** Solve applications of various equation types

The ability to solve linear and quadratic equations is the foundation on which a large percentage of our future studies are built. Both are closely linked to the solution of other equation types, as well as to the graphs of these equations. In this section, we get our first glimpse of these connections, as we learn to solve certain polynomial, rational, radical, and other equations.

A. Polynomial Equations of Higher Degree

In standard form, linear and quadratic equations have a known number of terms, so we commonly represent their coefficients using the early letters of the alphabet, as in $ax^2 + bx + c = 0$. However, these equations belong to the larger family of **polynomial equations**. To write a general polynomial, where the number of terms is unknown, we often represent the coefficients using subscripts on a single variable, such as a_1, a_2, a_3, and so on. A *polynomial equation of degree n* has the form

$$a_n x^n + a_{n-1} x^{n-1} + \cdots + a_1 x^1 + a_0 = 0$$

where $a_n, a_{n-1}, \ldots, a_1, a_0$ are real numbers and $a_n \neq 0$. Factorable polynomials of degree 3 and higher can also be solved using the zero product property and fundamental algebra skills. As with linear equations, values that make an equation true are called *solutions* or *roots* to the equation.

EXAMPLE 1 ► **Solving Polynomials by Factoring**

Solve by factoring: $2x^3 - 20x = 3x^2$.

Solution ►

$$
\begin{aligned}
2x^3 - 20x &= 3x^2 && \text{given equation} \\
2x^3 - 3x^2 - 20x &= 0 && \text{standard form} \\
x(2x^2 - 3x - 20) &= 0 && \text{common factor is } x \\
x(2x + 5)(x - 4) &= 0 && \text{factored form}
\end{aligned}
$$

$x = 0$ or $2x + 5 = 0$ or $x - 4 = 0$ zero product property

    $x = 0$ or $x = \frac{-5}{2}$ or $x = 4$ result

Substituting these values into the original equation verifies they are solutions.

Now try Exercises 7 through 14 ►

EXAMPLE 2 ▶ **Solving Higher Degree Equations**

Solve each equation by factoring:

a. $x^3 - 7x + 21 = 3x^2$　　　　　**b.** $x^4 - 16 = 0$

Solution ▶ **a.**

$$x^3 - 7x + 21 = 3x^2 \qquad \text{given equation}$$

$$x^3 - 3x^2 - 7x + 21 = 0 \qquad \text{standard form; factor by grouping}$$

$$x^2(x - 3) - 7(x - 3) = 0 \qquad \text{remove common factors from each group}$$

$$(x - 3)(x^2 - 7) = 0 \qquad \text{factored form}$$

$$x - 3 = 0 \quad \text{or} \quad x^2 - 7 = 0 \qquad \text{zero product property}$$

$$x = 3 \quad \text{or} \quad x^2 = 7 \qquad \text{isolate variables}$$

$$x = \pm\sqrt{7} \qquad \text{square root property of equality}$$

The solutions are $x = 3$, $x = \sqrt{7}$, and $x = -\sqrt{7}$.

b.

$$x^4 - 16 = 0 \qquad \text{given equation}$$

$$(x^2 + 4)(x^2 - 4) = 0 \qquad \text{factor as a difference of squares}$$

$$(x^2 + 4)(x + 2)(x - 2) = 0 \qquad \text{factor } x^2 - 4$$

$$x^2 + 4 = 0 \quad \text{or} \quad x + 2 = 0 \quad \text{or} \quad x - 2 = 0 \qquad \text{zero product property}$$

$$x^2 = -4 \quad \text{or} \quad x = -2 \quad \text{or} \quad x = 2 \qquad \text{isolate variables}$$

$$x = \pm\sqrt{-4} \qquad \text{square root property of equality}$$

Since $\pm\sqrt{-4} = \pm 2i$, the solutions are $x = 2i$, $x = -2i$, $x = 2$, and $x = -2$.

Now try Exercises 15 through 32 ▶

In Examples 1 and 2, we were able to solve higher degree polynomials by "breaking them down" into linear and quadratic forms. This basic idea can be applied to other kinds of equations as well, by rewriting them as equivalent linear and/or quadratic equations. For future use, it will be helpful to note that for a third-degree equation in the standard form $ax^3 + bx^2 + cx + d = 0$, a solution using factoring by grouping is always possible when $ad = bc$.

☑ **A.** You've just learned how to solve polynomial equations of higher degree

B. Rational Equations

In Section 1.1 we solved linear equations using basic properties of equality. If any equation contained fractional terms, we "cleared the fractions" using the least common denominator (LCD). We can also use this idea to solve **rational equations,** or equations that contain rational *expressions*.

Solving Rational Equations

1. Identify and exclude any values that cause a zero denominator.
2. Multiply both sides by the LCD and simplify (this will eliminate all denominators).
3. Solve the resulting equation.
4. Check all solutions in the original equation.

EXAMPLE 3 ▶ **Solving a Rational Equation**

Solve for m: $\dfrac{2}{m} - \dfrac{1}{m - 1} = \dfrac{4}{m^2 - m}$.

Solution ▶ Since $m^2 - m = m(m - 1)$, the LCD is $m(m - 1)$, where $m \neq 0$ and $m \neq 1$.

$$m(m - 1)\left(\frac{2}{m} - \frac{1}{m - 1}\right) = m(m - 1)\left[\frac{4}{m(m - 1)}\right] \quad \text{multiply by LCD}$$

$$2(m - 1) - m = 4 \quad \text{simplify—denominators are eliminated}$$

$$2m - 2 - m = 4 \quad \text{distribute}$$

$$m = 6 \quad \text{solve for } m$$

Checking by substitution we have:

$$\frac{2}{m} - \frac{1}{m - 1} = \frac{4}{m^2 - m} \quad \text{original equation}$$

$$\frac{2}{(6)} - \frac{1}{(6) - 1} = \frac{4}{(6)^2 - (6)} \quad \text{substitute 6 for } m$$

$$\frac{1}{3} - \frac{1}{5} = \frac{4}{30} \quad \text{simplify}$$

$$\frac{5}{15} - \frac{3}{15} = \frac{2}{15} \quad \text{common denominator}$$

$$\frac{2}{15} = \frac{2}{15} \checkmark \quad \text{result}$$

Now try Exercises 33 through 38 ▶

Multiplying both sides of an equation by a variable sometimes introduces a solution that satisfies the *resulting equation,* but not the original equation—the one we're trying to solve. Such "solutions" are called **extraneous roots** and illustrate the need to check all apparent solutions in the original equation. In the case of rational equations, we are particularly aware that any value that causes a zero denominator is outside the domain and cannot be a solution.

EXAMPLE 4 ▶ **Solving a Rational Equation**

Solve: $x + \dfrac{12}{x - 3} = 1 + \dfrac{4x}{x - 3}$.

Solution ▶ The LCD is $x - 3$, where $x \neq 3$.

$$(x - 3)\left(x + \frac{12}{x - 3}\right) = (x - 3)\left(1 + \frac{4x}{x - 3}\right) \quad \text{multiply both sides by LCD}$$

$$x^2 - 3x + 12 = x - 3 + 4x \quad \text{simplify—denominators are eliminated}$$

$$x^2 - 8x + 15 = 0 \quad \text{set equation equal to zero}$$

$$(x - 3)(x - 5) = 0 \quad \text{factor}$$

$$x = 3 \quad \text{or} \quad x = 5 \quad \text{zero factor property}$$

Checking shows $x = 3$ is an extraneous root, and $x = 5$ is the only valid solution.

Now try Exercises 39 through 44 ▶

In many fields of study, formulas involving rational expressions are used as equation models. Frequently, we need to solve these equations for one variable in terms of others, a skill closely related to our work in Section 1.1.

EXAMPLE 5 ▶ **Solving for a Specified Variable in a Formula**

Solve for the indicated variable: $S = \dfrac{a}{1 - r}$ for r.

Solution ▶

$$S = \frac{a}{1 - r} \qquad \text{LCD is } 1 - r$$

$$(1 - r)S = (1 - r)\left(\frac{a}{1 - r}\right) \qquad \text{multiply both sides by } (1 - r)$$

$$S - Sr = a \qquad \text{simplify—denominator is eliminated}$$

$$-Sr = a - S \qquad \text{isolate term with } r$$

$$r = \frac{a - S}{-S} \qquad \text{solve for } r \text{ (divide both sides by } -S)$$

$$r = \frac{S - a}{S}; \; S \neq 0 \qquad \text{multiply numerator/denominator by } -1$$

> **WORTHY OF NOTE**
>
> Generally, we should try to write rational answers with the fewest number of negative signs possible. Multiplying the numerator and denominator in Example 5 by -1 gave $r = \frac{S - a}{S}$, a more acceptable answer.

Now try Exercises 45 through 52 ▶

☑ **B.** You've just learned how to solve rational equations

C. Radical Equations and Equations with Rational Exponents

A **radical equation** is any equation that contains terms with a variable in the radicand. To solve a radical equation, we attempt to isolate a radical term on one side, then apply the appropriate nth power to free up the radicand and solve for the unknown. This is an application of the **power property of equality**.

> **The Power Property of Equality**
>
> If $\sqrt[n]{u}$ and v are real-valued expressions and $\sqrt[n]{u} = v$,
> $$\text{then } \left(\sqrt[n]{u}\right)^n = v^n$$
> $$u = v^n$$
> for n an integer, $n \geq 2$.

Raising both sides of an equation to an *even* power can also introduce a false solution (extraneous root). Note that by inspection, the equation $x - 2 = \sqrt{x}$ has only the solution $x = 4$. But the equation $(x - 2)^2 = x$ (obtained by squaring both sides) has both $x = 4$ *and* $x = 1$ as solutions, yet $x = 1$ does not satisfy the original equation. This means we should *check all solutions of an equation where an even power is applied.*

EXAMPLE 6 ▶ **Solving Radical Equations**

Solve each radical equation:

    a. $\sqrt{3x - 2} + 12 = x + 10$ **b.** $2\sqrt[3]{x - 5} + 4 = 0$

Solution ▶   **a.** $\sqrt{3x - 2} + 12 = x + 10 \qquad \text{original equation}$

$$\sqrt{3x - 2} = x - 2 \qquad \text{isolate radical term (subtract 12)}$$

$$\left(\sqrt{3x - 2}\right)^2 = (x - 2)^2 \qquad \text{apply power property, power is even}$$

$$3x - 2 = x^2 - 4x + 4 \qquad \text{simplify; square binomial}$$

$$0 = x^2 - 7x + 6 \qquad \text{set equal to zero}$$

$$0 = (x - 6)(x - 1) \qquad \text{factor}$$

$$x - 6 = 0 \quad \text{or} \quad x - 1 = 0 \qquad \text{apply zero product property}$$

$$x = 6 \quad \text{or} \quad x = 1 \qquad \text{result, check for extraneous roots}$$

Check ▶ $x = 6$: $\qquad \sqrt{3(6) - 2} + 12 = (6) + 10$
$$\sqrt{16} + 12 = 16$$
$$16 = 16\checkmark$$

Check ▶ $x = 1$: $\qquad \sqrt{3(1) - 2} + 12 = (1) + 10$
$$\sqrt{1} + 12 = 11$$
$$13 = 11\mathbf{x}$$

The only solution is $x = 6$; $x = 1$ is extraneous.

b. $\quad 2\sqrt[3]{x - 5} + 4 = 0$ $\qquad$ original equation
$$\sqrt[3]{x - 5} = -2 \qquad \text{isolate radical term (subtract 4, divide by 2)}$$
$$(\sqrt[3]{x - 5})^3 = (-2)^3 \qquad \text{apply power property, power is odd}$$
$$x - 5 = -8 \qquad \text{simplify: } (\sqrt[3]{x - 5})^3 = x - 5$$
$$x = -3 \qquad \text{solve}$$

Substituting -3 for x in the original equation verifies it is a solution.

Now try Exercises 53 through 56 ▶

Sometimes squaring both sides of an equation still results in an equation with a radical term, but often there is *one fewer* than before. In this case, we simply repeat the process, as indicated by the flowchart in Figure 1.14.

Figure 1.14

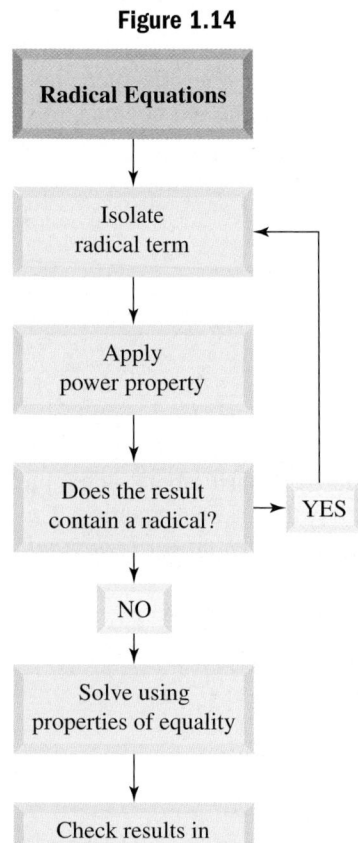

EXAMPLE 7 ▶ Solving Radical Equations

Solve the equation: $\sqrt{x + 15} - \sqrt{x + 3} = 2$.

Solution ▶ $\sqrt{x + 15} - \sqrt{x + 3} = 2$ $\qquad$ original equation

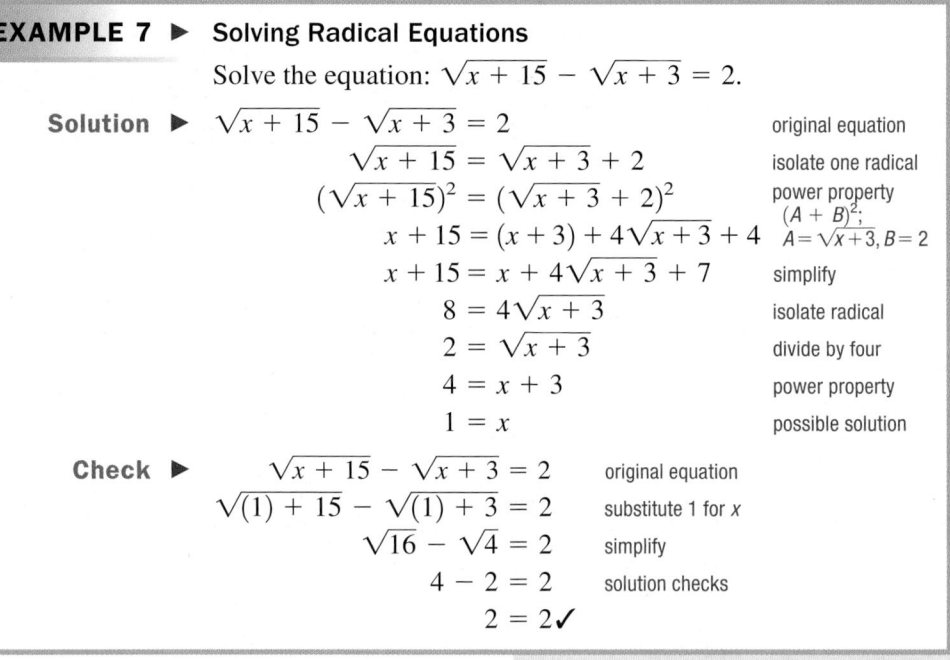

$$\sqrt{x + 15} = \sqrt{x + 3} + 2 \qquad \text{isolate one radical}$$
$$(\sqrt{x + 15})^2 = (\sqrt{x + 3} + 2)^2 \qquad \begin{array}{l}\text{power property}\\ (A + B)^2; \\ A = \sqrt{x+3}, B = 2\end{array}$$
$$x + 15 = (x + 3) + 4\sqrt{x + 3} + 4$$
$$x + 15 = x + 4\sqrt{x + 3} + 7 \qquad \text{simplify}$$
$$8 = 4\sqrt{x + 3} \qquad \text{isolate radical}$$
$$2 = \sqrt{x + 3} \qquad \text{divide by four}$$
$$4 = x + 3 \qquad \text{power property}$$
$$1 = x \qquad \text{possible solution}$$

Check ▶ $\qquad \sqrt{x + 15} - \sqrt{x + 3} = 2$ $\qquad$ original equation
$$\sqrt{(1) + 15} - \sqrt{(1) + 3} = 2 \qquad \text{substitute 1 for } x$$
$$\sqrt{16} - \sqrt{4} = 2 \qquad \text{simplify}$$
$$4 - 2 = 2 \qquad \text{solution checks}$$
$$2 = 2\checkmark$$

Now try Exercises 57 and 58 ▶

Since rational exponents are so closely related to radicals, the solution process for each is very similar. The goal is still to "undo" the radical (rational exponent) and solve for the unknown.

Power Property of Equality

For real-valued expression u and v, with positive integers m, n, and $\frac{m}{n}$ in lowest terms:

If m is odd

and $u^{\frac{m}{n}} = v$,

then $\left(u^{\frac{m}{n}}\right)^{\frac{n}{m}} = v^{\frac{n}{m}}$

$$u = v^{\frac{n}{m}}$$

If m is even

and $u^{\frac{m}{n}} = v (v > 0)$,

then $\left(u^{\frac{m}{n}}\right)^{\frac{n}{m}} = \pm v^{\frac{n}{m}}$

$$u = \pm v^{\frac{n}{m}}$$

EXAMPLE 8 ▶ **Solving Equations with Rational Exponents**

Solve each equation:

a. $3(x + 1)^{\frac{3}{4}} - 9 = 15$ **b.** $(x - 3)^{\frac{2}{3}} = 4$

Solution ▶ **a.** $3(x + 1)^{\frac{3}{4}} - 9 = 15$ original equation; $\frac{m}{n} = \frac{3}{4}$

$(x + 1)^{\frac{3}{4}} = 8$ isolate variable term (add 9, divide by 3)

$\left[(x + 1)^{\frac{3}{4}}\right]^{\frac{4}{3}} = 8^{\frac{4}{3}}$ apply power property, note m is odd

$x + 1 = 16$ simplify $\left[8^{\frac{4}{3}} = \left(8^{\frac{1}{3}}\right)^4 = 16\right]$

$x = 15$ result

Check ▶ $3(15 + 1)^{\frac{3}{4}} - 9 = 15$ substitute 15 for x in the original equation

$3\left(16^{\frac{1}{4}}\right)^3 - 9 = 15$ simplify, rewrite exponent

$3(2)^3 - 9 = 15$ $\sqrt[4]{16} = 2$

$3(8) - 9 = 15$ $2^3 = 8$

$15 = 15 ✓$ solution checks

b. $(x - 3)^{\frac{2}{3}} = 4$ original equation; $\frac{m}{n} = \frac{2}{3}$

$\left[(x - 3)^{\frac{2}{3}}\right]^{\frac{3}{2}} = \pm 4^{\frac{3}{2}}$ apply power property, note m is even

$x - 3 = \pm 8$ simplify $\left[4^{\frac{3}{2}} = \left(4^{\frac{1}{2}}\right)^3 = 8\right]$

$x = 3 \pm 8$ result

☑ **C.** You've just learned how to solve radical equations and equations with rational exponents

The solutions are $3 + 8 = 11$ and $3 - 8 = -5$.
Verify by checking both in the original equation.

Now try Exercises 59 through 64 ▶

⚠ **CAUTION** ▶ As you continue solving equations with radicals and rational exponents, be careful not to arbitrarily place the "$\pm$" sign in front of terms *given* in radical form. The expression $\sqrt{18}$ indicates the positive square root of 18, where $\sqrt{18} = 3\sqrt{2}$. The equation $x^2 = 18$ becomes $x = \pm\sqrt{18}$ after applying the power property, with solutions $x = \pm 3\sqrt{2}$ ($x = -3\sqrt{2}, x = 3\sqrt{2}$), since the square of either number produces 18.

D. Equations in Quadratic Form

In Appendix I.D we used a technique called *u-substitution* to factor expressions in quadratic form. The following equations are in quadratic form since the degree of the leading term is twice the degree of the middle term: $x^{\frac{2}{3}} - 3x^{\frac{1}{3}} - 10 = 0$, $(x^2 + x)^2 - 8(x^2 + x) + 12 = 0$ and $x - 3\sqrt{x + 4} + 4 = 0$ [*Note:* The last equation can be rewritten as $(x + 4) - 3(x + 4)^{\frac{1}{2}} = 0$]. A *u*-substitution will help to solve these equations by factoring. The first equation appears in Example 9, the other two are in Exercises 70 and 74, respectively.

EXAMPLE 9 ▶ Solving Equations in Quadratic Form

Solve using a *u*-substitution:

a. $x^{\frac{2}{3}} - 3x^{\frac{1}{3}} - 10 = 0$ **b.** $x^4 - 36 = 5x^2$

Solution ▶ **a.** This equation is in quadratic form since it can be rewritten as:
$\left(x^{\frac{1}{3}}\right)^2 - 3\left(x^{\frac{1}{3}}\right)^1 - 10 = 0$, where the degree of leading term is twice that of second term. If we let $u = x^{\frac{1}{3}}$, then $u^2 = x^{\frac{2}{3}}$ and the equation becomes $u^2 - 3u^1 - 10 = 0$ which is factorable.

$$(u - 5)(u + 2) = 0 \qquad \text{factor}$$
$$u = 5 \quad \text{or} \quad u = -2 \qquad \text{solution in terms of } u$$
$$x^{\frac{1}{3}} = 5 \quad \text{or} \quad x^{\frac{1}{3}} = -2 \qquad \text{resubstitute } x^{\frac{1}{3}} \text{ for } u$$
$$\left(x^{\frac{1}{3}}\right)^3 = 5^3 \quad \text{or} \quad \left(x^{\frac{1}{3}}\right)^3 = (-2)^3 \qquad \text{cube both sides: } \frac{1}{3}(3) = 1$$
$$x = 125 \quad \text{or} \quad x = -8 \qquad \text{solve for } x$$

Both solutions check.

b. In the standard form $x^4 - 5x^2 - 36 = 0$, we note the equation is also in quadratic form, since it can be written as $(x^2)^2 - 5(x^2)^1 - 36 = 0$. If we let $u = x^2$, then $u^2 = x^4$ and the equation becomes $u^2 - 5u - 36 = 0$, which is factorable.

$$(u - 9)(u + 4) = 0 \qquad \text{factor}$$
$$u = 9 \quad \text{or} \quad u = -4 \qquad \text{solution in terms of } u$$
$$x^2 = 9 \quad \text{or} \quad x^2 = -4 \qquad \text{resubstitute } x^2 \text{ for } u$$
$$x = \pm\sqrt{9} \quad \text{or} \quad x = \pm\sqrt{-4} \qquad \text{square root property}$$
$$x = \pm 3 \quad \text{or} \quad x = \pm 2i \qquad \text{simplify}$$

☑ **D.** You've just learned how to solve equations in quadratic form

The solutions are $x = -3$, $x = 3$, $x = -2i$, and $x = 2i$.
Verify that all solutions check.

Now try Exercises 65 through 78 ▶

E. Applications

Applications of the skills from this section come in many forms. **Number puzzles** and **consecutive integer** exercises help develop the ability to translate written information into algebraic forms **(see Exercises 81 through 84).** Applications involving **geometry** or a stated relationship between two quantities often depend on these skills, and in many scientific fields, equation models involving radicals and rational exponents are commonplace **(see Exercises 99 and 100).**

EXAMPLE 10 ▶ Solving a Geometry Application

A legal size sheet of typing paper has a length equal to 3 in. less than twice its width. If the area of the paper is 119 in², find the length and width.

Solution ▶ Let W represent the width of the paper.
Then $2W$ represents twice the width, and $2W - 3$ represents three less than twice the width: $L = 2W - 3$:

$$(\text{length})(\text{width}) = \text{area} \qquad \text{verbal model}$$
$$(2W - 3)(W) = 119 \qquad \text{substitute } 2W - 3 \text{ for length}$$

Since the equation is not set equal to zero, multiply and write the equation in standard form.

$$2W^2 - 3W = 119 \quad \text{distribute}$$
$$2W^2 - 3W - 119 = 0 \quad \text{subtract 119}$$
$$(2W - 17)(W + 7) = 0 \quad \text{factor}$$
$$W = \tfrac{17}{2} \text{ or } W = -7 \quad \text{solve}$$

We ignore $W = -7$, since the width cannot be negative. The width of the paper is $\frac{17}{2} = 8\frac{1}{2}$ in. and the length is $L = 2\left(\frac{17}{2}\right) - 3$ or 14 in.

Now try Exercises 85 and 86 ▶

EXAMPLE 11 ▶ Solving a Geometry Application

A hemispherical wash basin has a radius of 6 in. The volume of water in the basin can be modeled by $V = 6\pi h^2 - \frac{\pi}{3}h^3$, where h is the height of the water (see diagram). At what height h is the volume of water numerically equal to 15π times the height h?

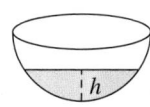

Solution ▶ We are essentially asked to solve $V = 6\pi h^2 - \frac{\pi}{3}h^3$ when $V = 15\pi h$.

The equation becomes

$$15\pi h = 6\pi h^2 - \frac{\pi}{3}h^3 \quad \text{original equation, substitute } 15\pi h \text{ for } V$$

$$\frac{\pi}{3}h^3 - 6\pi h^2 + 15\pi h = 0 \quad \text{standard form}$$

$$h^3 - 18h^2 + 45h = 0 \quad \text{multiply by } \tfrac{3}{\pi}$$
$$h(h^2 - 18h + 45) = 0 \quad \text{factor out } h$$
$$h(h - 3)(h - 15) = 0 \quad \text{factored form}$$
$$h = 0 \quad \text{or} \quad h = 3 \quad \text{or} \quad h = 15 \quad \text{result}$$

The "solution" $h = 0$ can be discounted since there would be no water in the basin, and $h = 15$ is too large for this context (the radius is only 6 in.). The only solution that fits this context is $h = 3$.

Check ▶

$$15\pi h = 6\pi h^2 - \frac{\pi}{3}h^3 \quad \text{resulting equation}$$

$$15\pi(3) = 6\pi(3)^2 - \frac{\pi}{3}(3)^3 \quad \text{substitute 3 for } h$$

$$45\pi = 6\pi(9) - \frac{\pi}{3}(27) \quad \text{apply exponents}$$

$$45\pi = 54\pi - 9\pi \quad \text{simplify}$$
$$45\pi = 45\pi \checkmark \quad \text{result checks}$$

Now try Exercises 87 and 88 ▶

In this section, we noted that extraneous roots can occur when (1) both sides of an equation are multiplied by a variable term (as when solving rational equations) and (2) when both sides of an equation are raised to an even power (as when solving certain radical equations or equations with rational exponents). Example 11 illustrates a third way that extraneous roots can occur, as when a solution checks out fine algebraically, but does not fit the context or physical constraints of the situation.

Revenue Models

In a free-market economy, we know that if the price of an item is decreased, more people will buy it. This is why stores have sales and bargain days. But if the item is sold too cheaply, revenue starts to decline because less money is coming in—even though more sales are made. This phenomenon is analyzed in Example 12, where we use the revenue formula *revenue = price · number of sales* or $R = P \cdot S$.

EXAMPLE 12 ▶ Solving a Revenue Application

When a popular printer is priced at \$300, Compu-Store will sell 15 printers per week. Using a survey, they find that for each decrease of \$8, two additional sales will be made. What price will result in weekly revenue of \$6500?

Solution ▶ Let x represent the number of times the price is decreased by \$8. Then $300 - 8x$ represents the new price. Since sales increase by 2 each time the price is decreased, $15 + 2x$ represents the total sales.

$$R = P \cdot S \qquad \text{revenue model}$$
$$6500 = (300 - 8x)(15 + 2x) \qquad R = 6500,\ P = 300 - 8x,\ S = 15 + 2x$$
$$6500 = 4500 + 600x - 120x - 16x^2 \qquad \text{multiply binomials}$$
$$0 = -16x^2 + 480x - 2000 \qquad \text{simplify and write in standard form}$$
$$0 = x^2 - 30x + 125 \qquad \text{divide by } -16$$
$$0 = (x - 5)(x - 25) \qquad \text{factor}$$
$$x = 5 \quad \text{or} \quad x = 25 \qquad \text{result}$$

Surprisingly, the store's weekly revenue will be \$6500 after 5 decreases of \$8 each (\$40 total), or 25 price decreases of \$8 each (\$200 total). The related selling prices are $300 - 5(8) = \$260$ and $300 - 25(8) = \$100$. To maximize profit, the manager of Compu-Store decides to go with the \$260 selling price.

Now try Exercises 89 and 90 ▶

Applications of rational equations can also take many forms. Work and uniform motion exercises help us develop important skills that can be used with more complex equation models. A work example follows here. For more on uniform motion, see **Exercises 95 and 96.**

EXAMPLE 13 ▶ Solving a Work Application

Lyf can clean a client's house in 5 hr, while it takes his partner Angie 4 hr to clean the same house. Both of them want to go to the Cubs' game today, which starts in $2\frac{1}{2}$ hr. If they work together, will they see the first pitch?

Solution ▶ After 1 hr, Lyf has cleaned $\frac{1}{5}$ and Angie has cleaned $\frac{1}{4}$ of the house, so together $\frac{1}{5} + \frac{1}{4} = \frac{9}{20}$ or 45% of the house has been cleaned. After 2 hr, $2\left(\frac{1}{5}\right) + 2\left(\frac{1}{4}\right)$ or $\frac{2}{5} + \frac{1}{2} = \frac{9}{10}$ or 90% of the house is clean. We can use these two illustrations to form an equation model where H represents hours worked:

$$H\left(\frac{1}{5}\right) + H\left(\frac{1}{4}\right) = 1 \text{ clean house } (1 = 100\%).$$

$$H\left(\frac{1}{5}\right) + H\left(\frac{1}{4}\right) = 1 \qquad \text{equation model}$$

$$20H\left(\frac{1}{5}\right) + 20H\left(\frac{1}{4}\right) = 1(20) \qquad \text{multiply by LCD of 20}$$

$$4H + 5H = 20 \qquad \text{simplify, denominators are eliminated}$$

$$9H = 20 \qquad \text{combine like terms}$$

$$H = \frac{20}{9} \qquad \text{solve for } H$$

It will take Lyf and Angie $2\frac{2}{9}$ hr (about 2 hr and 13 min) to clean the house. Yes! They will make the first pitch, since Wrigley Field is only 10 min away.

Now try Exercises 93 and 94 ▶

EXAMPLE 14 ▶   Solving an Application Involving a Rational Equation

In Verano City, the cost C to remove industrial waste from drinking water is given by the equation $C = \dfrac{80P}{100 - P}$, where P is the percent of total pollutants removed and C is the cost in thousands of dollars. If the City Council budgets \$1,520,000 for the removal of these pollutants, what percentage of the waste will be removed?

Solution ▶

$$C = \frac{80P}{100 - P} \qquad \text{equation model}$$

$$1520 = \frac{80P}{100 - P} \qquad \text{substitute 1520 for } C$$

$$1520(100 - P) = 80P \qquad \text{multiply by LCD of } (100 - P)$$

$$152{,}000 = 1600P \qquad \text{distribute and simplify}$$

$$95 = P \qquad \text{result}$$

☑ **E.** You've just learned how to solve applications of various equation types

On a budget of \$1,520,000, 95% of the pollutants will be removed.

Now try Exercises 97 and 98 ▶

1.6 EXERCISES

▶ CONCEPTS AND VOCABULARY

Fill in each blank with the appropriate word or phrase. Carefully reread the section, if necessary.

1. For rational equations, values that cause a zero denominator must be _____.

2. The equation or formula for revenue models is revenue = _____.

3. "False solutions" to a rational or radical equation are also called _____ roots.

4. Factorable polynomial equations can be solved using the _____ _____ property.

5. Discuss/Explain the power property of equality as it relates to rational exponents and properties of reciprocals. Use the equation $(x - 2)^{\frac{2}{3}} = 9$ for your discussion.

6. One factored form of an equation is shown. Discuss/Explain why $x = -8$ and $x = 1$ are not solutions to the equation, and what must be done to find the actual solutions: $2(x + 8)(x - 1) = -16$.

▶ **DEVELOPING YOUR SKILLS**

Solve using the zero product property. Be sure each equation is in standard form and factor out any common factors before attempting to solve. Check all answers in the original equation.

7. $22x = x^3 - 9x^2$ 8. $x^3 = 13x^2 - 42x$

9. $3x^3 = -7x^2 + 6x$ 10. $7x^2 + 15x = 2x^3$

11. $2x^4 - 3x^3 = 9x^2$ 12. $-7x^2 = 2x^4 - 9x^3$

13. $2x^4 - 16x = 0$ 14. $x^4 + 64x = 0$

15. $x^3 - 4x = 5x^2 - 20$ 16. $x^3 - 18 = 9x - 2x^2$

17. $4x - 12 = 3x^2 - x^3$ 18. $x - 7 = 7x^2 - x^3$

19. $2x^3 - 12x^2 = 10x - 60$

20. $9x + 81 = 27x^2 + 3x^3$

21. $x^4 - 7x^3 + 4x^2 = 28x$

22. $x^4 + 3x^3 + 9x^2 = -27x$

23. $x^4 - 81 = 0$

24. $x^4 - 1 = 0$

25. $x^4 - 256 = 0$

26. $x^4 - 625 = 0$

27. $x^6 - 2x^4 - x^2 + 2 = 0$

28. $x^6 - 3x^4 - 16x^2 + 48 = 0$

29. $x^5 - x^3 - 8x^2 + 8 = 0$

30. $x^5 - 9x^3 - x^2 + 9 = 0$

31. $x^6 - 1 = 0$

32. $x^6 - 64 = 0$

Solve each equation. Identify any extraneous roots.

33. $\dfrac{2}{x} + \dfrac{1}{x+1} = \dfrac{5}{x^2+x}$

34. $\dfrac{3}{m+3} - \dfrac{5}{m^2+3m} = \dfrac{1}{m}$

35. $\dfrac{21}{a+2} = \dfrac{3}{a-1}$

36. $\dfrac{4}{2y-3} = \dfrac{7}{3y-5}$

37. $\dfrac{1}{3y} - \dfrac{1}{4y} = \dfrac{1}{y^2}$

38. $\dfrac{3}{5x} - \dfrac{1}{2x} = \dfrac{1}{x^2}$

39. $x + \dfrac{14}{x-7} = 1 + \dfrac{2x}{x-7}$

40. $\dfrac{10}{x-5} + x = 1 + \dfrac{2x}{x-5}$

41. $\dfrac{6}{n+3} + \dfrac{20}{n^2+n-6} = \dfrac{5}{n-2}$

42. $\dfrac{7}{p+2} - \dfrac{1}{p^2+5p+6} = -\dfrac{2}{p+3}$

43. $\dfrac{a}{2a+1} - \dfrac{2a^2+5}{2a^2-5a-3} = \dfrac{3}{a-3}$

44. $\dfrac{-18}{6n^2-n-1} + \dfrac{3n}{2n-1} = \dfrac{4n}{3n+1}$

Solve for the variable indicated.

45. $\dfrac{1}{f} = \dfrac{1}{f_1} + \dfrac{1}{f_2}$; for f 46. $\dfrac{1}{x} - \dfrac{1}{y} = \dfrac{1}{z}$; for z

47. $I = \dfrac{E}{R+r}$; for r 48. $q = \dfrac{pf}{p-f}$; for p

49. $V = \dfrac{1}{3}\pi r^2 h$; for h 50. $s = \dfrac{1}{2}gt^2$; for g

51. $V = \dfrac{4}{3}\pi r^3$; for r^3 52. $V = \dfrac{1}{3}\pi r^2 h$; for r^2

Solve each equation and check your solutions by *substitution*. Identify any extraneous roots.

53. **a.** $-3\sqrt{3x-5} = -9$ **b.** $x = \sqrt{3x+1} + 3$

54. **a.** $-2\sqrt{4x-1} = -10$ **b.** $-5 = \sqrt{5x-1} - x$

55. **a.** $2 = \sqrt[3]{3m-1}$ **b.** $2\sqrt[3]{7-3x} - 3 = -7$

   c. $\dfrac{\sqrt[3]{2m+3}}{-5} + 2 = 3$ **d.** $\sqrt[3]{2x-9} = \sqrt[3]{3x+7}$

56. **a.** $-3 = \sqrt[3]{5p+2}$ **b.** $3\sqrt[3]{3-4x} - 7 = -4$

   c. $\dfrac{\sqrt[3]{6x-7}}{4} - 5 = -6$

   d. $3\sqrt[3]{x+3} = 2\sqrt[3]{2x+17}$

57. **a.** $\sqrt{x-9} + \sqrt{x} = 9$

   b. $x = 3 + \sqrt{23-x}$

   c. $\sqrt{x-2} - \sqrt{2x} = -2$

   d. $\sqrt{12x+9} - \sqrt{24x} = -3$

58. **a.** $\sqrt{x+7} - \sqrt{x} = 1$

   b. $\sqrt{2x+31} + x = 2$

   c. $\sqrt{3x} = \sqrt{x-3} + 3$

   d. $\sqrt{3x+4} - \sqrt{7x} = -2$

Write the equation in simplified form, then solve. Check all answers by substitution.

59. $x^{\frac{3}{5}} + 17 = 9$ **60.** $-2x^{\frac{3}{4}} + 47 = -7$

61. $0.\overline{3}x^{\frac{5}{2}} - 39 = 42$ **62.** $0.\overline{5}x^{\frac{5}{3}} + 92 = -43$

63. $2(x + 5)^{\frac{2}{3}} - 11 = 7$

64. $-3(x - 2)^{\frac{4}{5}} + 29 = -19$

Solve each equation using a *u*-substitution. Check all answers.

65. $x^{\frac{2}{3}} - 2x^{\frac{1}{3}} - 15 = 0$ **66.** $x^3 - 9x^{\frac{3}{2}} + 8 = 0$

67. $x^4 - 24x^2 - 25 = 0$ **68.** $x^4 - 37x^2 + 36 = 0$

69. $(x^2 - 3)^2 + (x^2 - 3) - 2 = 0$

70. $(x^2 + x)^2 - 8(x^2 + x) + 12 = 0$

71. $x^{-2} - 3x^{-1} - 4 = 0$

72. $x^{-2} - 2x^{-1} - 35 = 0$

73. $x^{-4} - 13x^{-2} + 36 = 0$

Use a *u*-substitution to solve each radical equation.

74. $x - 3\sqrt{x + 4} + 4 = 0$

75. $x + 4 = 7\sqrt{x + 4}$

76. $2(x + 1) = 5\sqrt{x + 1} - 2$

77. $2\sqrt{x + 10} + 8 = 3(x + 10)$

78. $4\sqrt{x - 3} = 3(x - 3) - 4$

▶ WORKING WITH FORMULAS

79. Lateral surface area of a cone: $S = \pi r \sqrt{r^2 + h^2}$

The lateral surface area (surface area excluding the base) *S* of a cone is given by the formula shown, where *r* is the radius of the base and *h* is the height of the cone. (a) Solve the equation for *h*. (b) Find the surface area of a cone that has a radius of 6 m and a height of 10 m. Answer in simplest form.

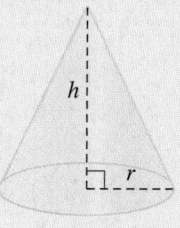

80. Painted area on a canvas: $A = \dfrac{4x^2 + 60x + 104}{x}$

A rectangular canvas is to contain a small painting with an area of 52 in², and requires 2-in. margins on the left and right, with 1-in. margins on the top and bottom for framing. The total area of such a canvas is given by the formula shown, where *x* is the height of the *painted* area.

a. What is the area *A* of the canvas if the height of the painting is $x = 10$ in.?

b. If the area of the canvas is $A = 120$ in², what are the dimensions of the painted area?

▶ APPLICATIONS

Find all real numbers that satisfy the following descriptions.

81. When the cube of a number is added to twice its square, the result is equal to 18 more than 9 times the number.

82. Four times a number decreased by 20 is equal to the cube of the number decreased by 5 times its square.

83. Find three consecutive even integers such that 4 times the largest plus the fourth power of the smallest is equal to the square of the remaining even integer increased by 24.

84. Find three consecutive integers such that the sum of twice the largest and the fourth power of the smallest is equal to the square of the remaining integer increased by 75.

85. Envelope sizes: Large mailing envelopes often come in standard sizes, with 5- by 7-in. and 9- by

12-in. envelopes being the most common. The next larger size envelope has an area of 143 in², with a length that is 2 in. longer than the width. What are the dimensions of the larger envelope?

86. Paper sizes: Letter size paper is 8.5 in. by 11 in. Legal size paper is $8\frac{1}{2}$ in. by 14 in. The next larger (common) size of paper has an area of 187 in², with a length that is 6 in. longer than the width. What are the dimensions of the Ledger size paper?

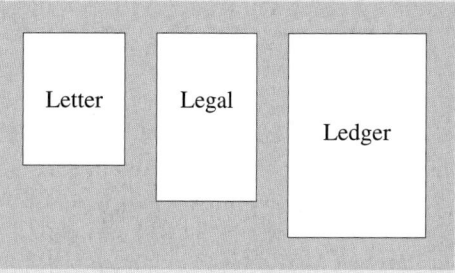

87. Composite figures—grain silos: Grain silos can be described as a hemisphere sitting atop a cylinder. The interior volume V of the silo can be modeled by $V = \frac{2}{3}\pi r^3 + \pi r^2 h$, where h is the height of a cylinder with radius r. For a cylinder 6 m tall, what radius would give the silo a volume that is numerically equal to 24π times this radius?

88. Composite figures—gelatin capsules: The gelatin capsules manufactured for cold and flu medications are shaped like a cylinder with a hemisphere on each end. The interior volume V of each capsule can be modeled by $V = \frac{4}{3}\pi r^3 + \pi r^2 h$, where h is the height of the cylindrical portion and r is its radius. If the cylindrical portion of the capsule is 8 mm long ($h = 8$ mm), what radius would give the capsule a volume that is numerically equal to 15π times this radius?

89. Running shoes: When a popular running shoe is priced at $70, The Shoe House will sell 15 pairs each week. Using a survey, they have determined that for each decrease of $2 in price, 3 additional pairs will be sold each week. What selling price will give a weekly revenue of $2250?

90. Cell phone charges: A cell phone service sells 48 subscriptions each month if their monthly fee is $30. Using a survey, they find that for each decrease of $1, 6 additional subscribers will join. What charge(s) will result in a monthly revenue of $2160?

Projectile height: In the absence of resistance, the height of an object that is projected upward can be modeled by the equation $h = -16t^2 + vt + k$, where h represents the height of the object (in feet) t sec after it has been thrown, v represents the initial velocity (in feet per second), and k represents the height of the object when $t = 0$ (before it has been thrown). Use this information to complete the following problems.

91. From the base of a canyon that is 480 feet deep (*below* ground level → -480), a slingshot is used to shoot a pebble upward toward the canyon's rim. If the initial velocity is 176 ft per second:

    a. How far is the pebble below the rim after 4 sec?

    b. How long until the pebble returns to the bottom of the canyon?

    c. What happens at $t = 5$ and $t = 6$ sec? Discuss and explain.

92. A model rocket blasts off. A short time later, at a velocity of 160 ft/sec and a height of 240 ft, it runs out of fuel and becomes a projectile.

    a. How high is the rocket three seconds later? Four seconds later?

    b. How long will it take the rocket to attain a height of 640 ft?

    c. How many times is a height of 384 ft attained? When do these occur?

    d. How many seconds until the rocket returns to the ground?

93. Printing newspapers: The editor of the school newspaper notes the college's new copier can complete the required print run in 20 min, while the back-up copier took 30 min to do the same amount of work. How long would it take if both copiers are used?

94. Filling a sink: The cold water faucet can fill a sink in 2 min. The drain can empty a full sink in 3 min. If the faucet were left on and the drain was left open, how long would it take to fill the sink?

95. Triathalon competition: As one part of a Mountain-Man triathalon, participants must row a canoe 5 mi down river (with the current), circle a buoy and row 5 mi back up river (against the current) to the starting point. If the current is flowing at a steady rate of 4 mph and Tom Chaney made the round-trip in 3 hr, how fast can he row in still water? (*Hint:* The time rowing down river and the time rowing up river must add up to 3 hr.)

96. Flight time: The flight distance from Cincinnati, Ohio, to Chicago, Illinois, is approximately 300 mi. On a recent round-trip between these cities in my private plane, I encountered a steady 25 mph headwind on the way to Chicago, with a 25 mph tailwind on the return trip. If my total flying time came to exactly 5 hr, what was my flying time to Chicago? What was my flying time back to Cincinnati? (*Hint:* The flight time between the two cities must add up to 5 hr.)

97. Pollution removal: For a steel mill, the cost C (in millions of dollars) to remove toxins from the resulting sludge is given by $C = \dfrac{92P}{100 - P}$, where P is the percent of the toxins removed. What percent can be removed if the mill spends $100,000,000 on the cleanup? Round to tenths of a percent.

98. Wildlife populations: The Department of Wildlife introduces 60 elk into a new game reserve. It is projected that the size of the herd will grow according to the equation $N = \dfrac{10(6 + 3t)}{1 + 0.05t}$, where N is the number of elk and t is the time in years. If recent counts find 225 elk, approximately how many years have passed? (See Appendix I.E, Exercise 66.)

99. Planetary motion: The time T (in days) for a planet to make one revolution around the sun is modeled by $T = 0.407R^{\frac{3}{2}}$, where R is the maximum radius of the planet's orbit in millions of miles (*Kepler's third law of planetary motion*). Use the equation to approximate the maximum radius of each orbit, given the number of days it takes for one revolution. (See Appendix I.F, Exercises 45 and 46.)

   a. Mercury: 88 days

   b. Venus: 225 days

   c. Earth: 365 days

   d. Mars: 687 days

   e. Jupiter: 4,333 days

   f. Saturn: 10,759 days

100. Wind-powered energy: If a wind-powered generator is delivering P units of power, the velocity V of the wind (in miles per hour) can be determined using $V = \sqrt[3]{\dfrac{P}{k}}$, where k is a constant that depends on the size and efficiency of the generator. Given $k = 0.004$, approximately how many units of power are being delivered if the wind is blowing at 27 miles per hour? (See Appendix I.F, Exercise 48.)

► **EXTENDING THE CONCEPT**

101. To solve the equation $3 - \dfrac{8}{x + 3} = \dfrac{1}{x}$, a student multiplied by the LCD $x(x + 3)$, simplified, and got this result: $3 - 8x = (x + 3)$. Identify and fix the mistake, then find the correct solution(s).

102. The expression $x^2 - 7$ is not factorable using *integer values*. But the expression *can be written* in the form $x^2 - (\sqrt{7})^2$, enabling us to factor it as a binomial and its conjugate: $(x + \sqrt{7})(x - \sqrt{7})$. Use this idea to solve the following equations:

   a. $x^2 - 5 = 0$ **b.** $n^2 - 19 = 0$

   c. $4v^2 - 11 = 0$ **d.** $9w^2 - 11 = 0$

Determine the values of x for which each expression represents a real number.

103. $\dfrac{\sqrt{x - 1}}{x^2 - 4}$ **104.** $\dfrac{x^2 - 4}{\sqrt{x - 1}}$

105. As an extension of working with absolute values, try the following exercises.

   Recall that for $|X| = k$, $X = -k$ or $X = k$.

   a. $|x^2 - 2x - 25| = 10$

   b. $|x^2 - 5x - 10| = 4$

   c. $|x^2 - 4| = x + 2$

   d. $|x^2 - 9| = -x + 3$

   e. $|x^2 - 7x| = -x + 7$

   f. $|x^2 - 5x - 2| = x + 5$

▶ **MAINTAINING YOUR SKILLS**

106. (1.1) Two jets take off on parallel runways going in opposite directions. The first travels at a rate of 250 mph and the second at 325 mph. How long until they are 980 miles apart?

107. (R.6) Find the missing side.

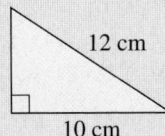

12 cm

10 cm

108. (R.3) Simplify using properties of exponents:

$$2^{-1} + (2x)^0 + 2x^0$$

109. (1.2) Graph the relation given:

$$2x - 3 < 7 \text{ and } x + 2 > 1$$

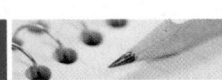

SUMMARY AND CONCEPT REVIEW

SECTION 1.1　Linear Equations, Formulas, and Problem Solving

KEY CONCEPTS

- An equation is a statement that two expressions are equal.
- Replacement values that make an equation true are called solutions or roots.
- Equivalent equations are those that have the same solution set.
- To solve an equation we use the distributive property and the properties of equality to write a sequence of simpler, equivalent equations until the solution is obvious. A guide for solving linear equations appears on page 3.
- If an equation contains fractions, multiply both sides by the LCD of all denominators, then solve.
- Solutions to an equation can be checked using back-substitution, by replacing the variable with the proposed solution and verifying the left-hand expression is equal to the right.
- An equation can be:
    1. an identity, one that is always true, with a solution set of all real numbers.
    2. a contradiction, one that is never true, with the empty set as the solution set.
    3. conditional, or one that is true/false depending on the value(s) input.
- To solve formulas for a specified variable, focus on the object variable and apply properties of equality to write this variable in terms of all others.
- The basic elements of good problem solving include:
    1. Gathering and organizing information
    2. Making the problem visual
    3. Developing an equation model
    4. Using the model to solve the application

For a complete review, see the problem-solving guide on page 6.

EXERCISES

1. Use substitution to determine if the indicated value is a solution to the equation given.

　a. $6x - (2 - x) = 4(x - 5), x = -6$ 　　**b.** $\frac{3}{4}b + 2 = \frac{5}{2}b + 16, b = -8$ 　　**c.** $4d - 2 = -\frac{1}{2} + 3d, d = \frac{3}{2}$

Solve each equation.

2. $-2b + 7 = -5$

3. $3(2n - 6) + 1 = 7$

4. $4m - 5 = 11m + 2$

5. $\frac{1}{2}x + \frac{2}{3} = \frac{3}{4}$

6. $6p - (3p + 5) - 9 = 3(p - 3)$

7. $-\frac{g}{6} = 3 - \frac{1}{2} - \frac{5g}{12}$

Solve for the specified variable in each formula or literal equation.

8. $V = \pi r^2 h$ for h **9.** $P = 2L + 2W$ for L

10. $ax + b = c$ for x **11.** $2x - 3y = 6$ for y

Use the problem-solving guidelines (page 6) to solve the following applications.

12. At a large family reunion, two kegs of lemonade are available. One is 2% sugar (too sour) and the second is 7% sugar (too sweet). How many gallons of the 2% keg, must be mixed with 12 gallons of the 7% keg to get a 5% mix?

13. A rectangular window with a width of 3 ft and a height of 4 ft is topped by a semi-circular window. Find the total area of the window.

14. Two cyclists start from the same location and ride in opposite directions, one riding at 15 mph and the other at 18 mph. If their radio phones have a range of 22 mi, how many minutes will they be able to communicate?

SECTION 1.2   Linear Inequalities in One Variable

KEY CONCEPTS

- Inequalities are solved using properties similar to those for solving equalities (see page 15). The one exception is the multiplicative property of inequality, since the truth of the resulting statement depends on whether a positive or negative quantity is used.
- Solutions to an inequality can be graphed on a number line, stated using a simple inequality, or expressed using set or interval notation.
- For two sets A and B: A intersect B ($A \cap B$) is the set of elements in both A **and** B (i.e., *elements common to both sets*). A union B ($A \cup B$) is the set of elements in either A **or** B (i.e., *all elements from either set*).
- Compound inequalities are formed using the conjunctions "and"/"or." These can be either a joint inequality as in $-3 < x \leq 5$, or a disjoint inequality, as in $x < -2$ or $x > 7$.

EXERCISES

Use inequality symbols to write a mathematical model for each statement.

15. You must be 35 yr old or older to run for president of the United States.

16. A child must be under 2 yr of age to be admitted free.

17. The speed limit on many interstate highways is 65 mph.

18. Our caloric intake should not be less than 1200 calories per day.

Solve the inequality and write the solution using interval notation.

19. $7x > 35$ **20.** $-\dfrac{3}{5}m < 6$

21. $2(3m - 2) \leq 8$ **22.** $-1 < \dfrac{1}{3}x + 2 \leq 5$

23. $-4 < 2b + 8$ and $3b - 5 > -32$ **24.** $-5(x + 3) > -7$ or $x - 5.2 > -2.9$

25. Find the allowable values for each of the following. Write your answer in interval notation.

     a. $\dfrac{7}{n - 3}$ **b.** $\dfrac{5}{2x - 3}$ **c.** $\sqrt{x + 5}$ **d.** $\sqrt{-3n + 18}$

26. Latoya has earned grades of 72%, 95%, 83%, and 79% on her first four exams. What grade must she make on her fifth and last exam so that her average is 85% or more?

SECTION 1.3   Absolute Value Equations and Inequalities

KEY CONCEPTS

- To solve absolute value equations and inequalities, begin by writing the equation in simplified form, with the absolute value isolated on one side.
- If X represents an algebraic expression and k is a nonnegative constant:
  - Absolute value equations: $|X| = k$ is equivalent to $X = -k$ or $X = k$
  - "Less than" inequalities: $|X| < k$ is equivalent to $-k < X < k$
  - "Greater than" inequalities: $|X| > k$ is equivalent to $X < -k$ or $X > k$
- These properties also apply when the symbols "$\leq$" or "$\geq$" are used.
- If the absolute value quantity has been isolated on the left, the solution to a less-than inequality will be a single interval, while the solution to a greater-than inequality will consist of two disjoint intervals.
- The multiplicative property states that for algebraic expressions A and B, $|AB| = |A||B|$.

EXERCISES

Solve each equation or inequality. Write solutions to inequalities in interval notation.

27. $7 = |x - 3|$

28. $-2|x + 2| = -10$

29. $|-2x + 3| = 13$

30. $\dfrac{|2x + 5|}{3} + 8 = 9$

31. $-3|x + 2| - 2 < -14$

32. $\left|\dfrac{x}{2} - 9\right| \leq 7$

33. $|3x + 5| = -4$

34. $3|x + 1| < -9$

35. $2|x + 1| > -4$

36. $5|m - 2| - 12 \leq 8$

37. $\dfrac{|3x - 2|}{2} + 6 \geq 10$

38. Monthly rainfall received in Omaha, Nebraska, rarely varies by more than 1.7 in. from an average of 2.5 in. per month. (a) Use this information to write an absolute value inequality model, then (b) solve the inequality to find the highest and lowest amounts of monthly rainfall for this city.

SECTION 1.4   Complex Numbers

KEY CONCEPTS

- The italicized i represents the number whose square is -1. This means $i^2 = -1$ and $i = \sqrt{-1}$.
- Larger powers of i can be simplified using $i^4 = 1$.
- For $k > 0$, $\sqrt{-k} = i\sqrt{k}$ and we say the expression has been *written in terms of i.*
- The standard form of a *complex number* is $a + bi$, where a is the *real number part* and bi is the *imaginary number part.*
- To add or subtract complex numbers, combine the like terms.
- For any complex number $a + bi$, its *complex conjugate* is $a - bi$.
- The *product* of a complex number and its conjugate is a real number.

- The commutative, associative, and distributive properties also apply to complex numbers and are used to perform basic operations.
- To multiply complex numbers, use the F-O-I-L method and simplify.
- To find a *quotient* of complex numbers, multiply the numerator and denominator by the conjugate of the denominator.

EXERCISES

Simplify each expression and write the result in standard form.

39. $\sqrt{-72}$

40. $6\sqrt{-48}$

41. $\dfrac{-10 + \sqrt{-50}}{5}$

42. $\sqrt{3}\sqrt{-6}$

43. i^{57}

Perform the operation indicated and write the result in standard form.

44. $(5 + 2i)^2$

45. $\dfrac{5i}{1 - 2i}$

46. $(-3 + 5i) - (2 - 2i)$

47. $(2 + 3i)(2 - 3i)$

48. $4i(-3 + 5i)$

Use substitution to show the given complex number and its conjugate are solutions to the equation shown.

49. $x^2 - 9 = -34$; $x = 5i$

50. $x^2 - 4x + 9 = 0$; $x = 2 + i\sqrt{5}$

SECTION 1.5   Solving Quadratic Equations

KEY CONCEPTS

- The standard form of a quadratic equation is $ax^2 + bx + c = 0$, where a, b, and c are real numbers and $a \neq 0$. In words, we say the equation is written in decreasing order of degree and set equal to zero.
- The coefficient of the squared term a is called the *leading coefficient,* b is called the *linear coefficient,* and c is called the *constant term.*
- The square root property of equality states that if $X^2 = k$, where $k \geq 0$, then $X = \sqrt{k}$ or $X = -\sqrt{k}$.
- Factorable quadratics can be solved using the zero product property, which states that if the product of two factors is zero, then one, the other, or both must be equal to zero. Symbolically, if $A \cdot B = 0$, then $A = 0$ or $B = 0$.
- Quadratic equations can also be solved by *completing the square,* or using the *quadratic formula.*
- If the discriminant $b^2 - 4ac = 0$, the equation has one real (repeated) root. If $b^2 - 4ac > 0$, the equation has two real roots; and if $b^2 - 4ac < 0$, the equation has two complex roots.

EXERCISES

51. Determine whether the given equation is quadratic. If so, write the equation in standard form and identify the values of a, b, and c.

  a. $-3 = 2x^2$ **b.** $7 = -2x + 11$ **c.** $99 = x^2 - 8x$ **d.** $20 = 4 - x^2$

52. Solve by factoring.

  a. $x^2 - 3x - 10 = 0$ **b.** $2x^2 - 50 = 0$ **c.** $3x^2 - 15 = 4x$ **d.** $x^3 - 3x^2 = 4x - 12$

53. Solve using the square root property of equality.

  a. $x^2 - 9 = 0$ **b.** $2(x - 2)^2 + 1 = 11$ **c.** $3x^2 + 15 = 0$ **d.** $-2x^2 + 4 = -46$

54. Solve by completing the square. Give real number solutions in exact and approximate form.

  a. $x^2 + 2x = 15$ **b.** $x^2 + 6x = 16$ **c.** $-4x + 2x^2 = 3$ **d.** $3x^2 - 7x = -2$

55. Solve using the quadratic formula. Give solutions in both exact and approximate form.

  a. $x^2 - 4x = -9$ **b.** $4x^2 + 7 = 12x$ **c.** $2x^2 - 6x + 5 = 0$

Solve the following quadratic applications. For 56 and 57, recall the height of a projectile is modeled by $h = -16t^2 + v_0 t + k$.

56. A projectile is fired upward from ground level with an initial velocity of 96 ft/sec. (a) To the nearest tenth of a second, how long until the object first reaches a height of 100 ft? (b) How long until the object is again at 100 ft? (c) How many seconds until it returns to the ground?

57. A person throws a rock upward from the top of an 80-ft cliff with an initial velocity of 64 ft/sec. (a) To the nearest tenth of a second, how long until the object is 120 ft high? (b) How long until the object is again at 120 ft? (c) How many seconds until the object hits the ground at the base of the cliff?

58. The manager of a large, 14-screen movie theater finds that if he charges $2.50 per person for the matinee, the average daily attendance is 4000 people. With every increase of 25 cents the attendance drops an average of 200 people. (a) What admission price will bring in a revenue of $11,250? (b) How many people will purchase tickets at this price?

59. After a storm, the Johnson's basement flooded and the water needed to be pumped out. A cleanup crew is sent out with two powerful pumps to do the job. Working alone (if one of the pumps were needed at another job), the larger pump would be able to clear the basement in 3 hr less time than the smaller pump alone. Working together, the two pumps can clear the basement in 2 hr. How long would it take the smaller pump alone?

SECTION 1.6 Solving Other Types of Equations

KEY CONCEPTS

- Certain equations of higher degree can be solved using factoring skills and the zero product property.
- To solve rational equations, clear denominators using the LCD, noting values that must be excluded.
- Multiplying an equation by a variable quantity sometimes introduces extraneous solutions. Check all results in the original equation.
- To solve radical equations, isolate the radical on one side, then apply the appropriate "nth power" to free up the radicand. Repeat the process if needed. See flowchart on page 60.
- For equations with a rational exponent $\frac{m}{n}$, isolate the variable term and raise both sides to the $\frac{n}{m}$ power. If m is even, there will be two real solutions.
- Any equation that can be written in the form $u^2 + bu + c = 0$, where u represents an algebraic expression, is said to be in quadratic form and can be solved using u-substitution and standard approaches.

EXERCISES

Solve by factoring.

60. $x^3 - 7x^2 = 3x - 21$

61. $3x^3 + 5x^2 = 2x$

62. $x^4 - 8x = 0$

63. $x^4 - \dfrac{1}{16} = 0$

Solve each equation.

64. $\dfrac{3}{5x} + \dfrac{7}{10} = \dfrac{1}{4x}$

65. $\dfrac{3h}{h + 3} - \dfrac{7}{h^2 + 3h} = \dfrac{1}{h}$

66. $\dfrac{2n}{n + 2} - \dfrac{3}{n - 4} = \dfrac{n^2 + 20}{n^2 - 2n - 8}$

67. $\dfrac{\sqrt{x^2 + 7}}{2} + 3 = 5$

68. $3\sqrt{x + 4} = x + 4$

69. $\sqrt{3x + 4} = 2 - \sqrt{x + 2}$

70. $3\left(x - \dfrac{1}{4}\right)^{-\frac{3}{2}} = \dfrac{8}{9}$

71. $-2(5x + 2)^{\frac{2}{3}} + 17 = -1$

72. $(x^2 - 3x)^2 - 14(x^2 - 3x) + 40 = 0$

73. $x^4 - 7x^2 = 18$

74. The science of *allometry* studies the growth of one aspect of an organism relative to the entire organism or to a set standard. Allometry tells us that the amount of food F (in kilocalories per day) an herbivore must eat to survive is related to its weight W (in grams) and can be approximated by the equation $F \approx 1.5W^{\frac{3}{4}}$.

    a. How many kilocalories per day are required by a 160-kg gorilla (160 kg = 160,000 g)?

    b. If an herbivore requires 40,500 kilocalories per day, how much does it weigh?

75. The area of a common stenographer's tablet, commonly called a *steno book,* is 54 in². The length of the tablet is 3 in. more than the width. Model the situation with a quadratic equation and find the dimensions of the tablet.

76. A batter has just flied out to the catcher, who catches the ball while standing on home plate. If the batter made contact with the ball at a height of 4 ft and the ball left the bat with an initial velocity of 128 ft/sec, how long will it take the ball to reach a height of 116 ft? How high is the ball 5 sec after contact? If the catcher catches the ball at a height of 4 ft, how long was it airborne?

77. Using a survey, a firewood distributor finds that if they charge $50 per load, they will sell 40 loads each winter month. For each decrease of $2, five additional loads will be sold. What selling price(s) will result in new monthly revenue of $2520?

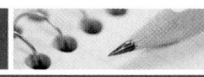

MIXED REVIEW

1. Find the allowable values for each expression. Write your response in interval notation.

   a. $\dfrac{10}{\sqrt{x-8}}$ b. $\dfrac{-5}{3x+4}$

2. Perform the operations indicated.

   a. $\sqrt{-18} + \sqrt{-50}$ b. $(1-2i)^2$

   c. $\dfrac{3i}{1+i}$ d. $(2+i\sqrt{3})(2-i\sqrt{3})$

3. Solve each equation or inequality.

   a. $-2x^3 + 4x^2 = 50x - 100$

   b. $-3x^4 - 375x = 0$ c. $-2|3x+1| = -12$

   d. $-3\left|\dfrac{x}{3} - 5\right| \le -12$ e. $v^{\frac{4}{3}} = 81$

   f. $-2(x+1)^{\frac{1}{4}} = -6$

Solve for the variable indicated.

4. $V = \dfrac{1}{3}\pi r^2 h + \dfrac{2}{3}\pi r^3$; for h 5. $3x + 4y = -12$; for y

Solve as indicated, using the method of your choice.

6. a. $-20 \le 4x + 8 < 56$

   b. $-2x + 7 \le 12$ and $3 - 4x > -5$

7. a. $5x - (2x - 3) + 3x = -4(5 + x) + 3$

   b. $\dfrac{n}{5} - 2 = 2 - \dfrac{5}{3} - \dfrac{4}{15}n$

8. $5x(x - 10)(x + 1) = 0$

9. $x^2 - 18x + 77 = 0$ 10. $3x^2 - 10 = 5 - x + x^2$

11. $4x^2 - 5 = 19$ 12. $3(x + 5)^2 - 3 = 30$

13. $25x^2 + 16 = 40x$ 14. $3x^2 - 7x + 3 = 0$

15. $2x^4 - 50 = 0$

16. a. $\dfrac{2}{x} - \dfrac{x}{5x + 12} = 0$ b. $\dfrac{1}{n-1} - \dfrac{2}{n^2 - 1} = -\dfrac{1}{2}$

   c. $\dfrac{2x}{x+3} - \dfrac{36}{x^2 - 9} = \dfrac{x}{x-3}$

17. a. $\sqrt{2v - 3} + 3 = v$

    b. $\sqrt[3]{x^2 - 9} + \sqrt[3]{x - 11} = 0$

    c. $\sqrt{x + 7} - \sqrt{2x} = 1$

18. The local Lion's Club rents out two banquet halls for large meetings and other events. The records show that when they charge $250 per day for use of the halls, there are an average of 156 bookings per year. For every increase of $20 per day, there will be three less bookings. (a) What price per day will bring in $61,950 for the year? (b) How many bookings will there be at the price from part (a)?

19. The Jefferson College basketball team has two guards who are 6′3″ tall and two forwards who are 6′7″ tall. How tall must their center be to ensure the "starting five" will have an average height of at least 6′6″?

20. The volume of an inflatable hot-air balloon can be approximated using the formulas for a hemisphere and a cone: $V = \frac{2}{3}\pi r^3 + \frac{1}{3}\pi r^2 h$. Assume the conical portion has height $h = 24$ ft. During inflation, what is the radius of the balloon at the moment the volume of air is numerically equal to 126π times this radius?

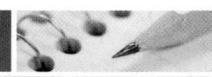

PRACTICE TEST

1. Solve each equation.

   a. $-\dfrac{2}{3}x - 5 = 7 - (x + 3)$

   b. $-5.7 + 3.1x = 14.5 - 4(x + 1.5)$

   c. $P = C + kC$; for C

   d. $2|2x + 5| - 17 = -11$

2. How much water that is 102°F must be mixed with 25 gal of water at 91°F, so that the resulting temperature of the water will be 97°F?

3. Solve each equation or inequality.

a. $-\dfrac{2}{5}x + 7 < 19$

b. $-1 < 3 - x \le 8$

c. $\dfrac{1}{2}x + 3 < 9$ or $\dfrac{2}{3}x - 1 \ge 3$

d. $\dfrac{1}{2}|x - 3| + \dfrac{5}{4} = \dfrac{7}{4}$

e. $-\dfrac{2}{3}|x + 1| - 5 < -7$

4. To make the bowling team, Jacques needs a three-game average of 160. If he bowled 141 and 162 for the first two games, what score S must be obtained in the third game so that his average is at least 160?

Solve each equation.

5. $z^2 - 7z - 30 = 0$

6. $x^2 + 25 = 0$

7. $(x - 1)^2 + 3 = 0$

8. $x^4 + 16 = 17x^2$

9. $3x^2 - 20x = -12$

10. $4x^3 + 8x^2 - 9x - 18 = 0$

11. $\dfrac{2}{x - 3} + \dfrac{2x}{x + 2} = \dfrac{x^2 + 16}{x^2 - x - 6}$

12. $\dfrac{4}{x - 3} + 2 = \dfrac{5x}{x^2 - 9}$

13. $\sqrt{x} + 1 = \sqrt{2x - 7}$

14. $(x + 3)^{\frac{-2}{3}} = \dfrac{1}{4}$

15. The Spanish Club at Rock Hill Community College has decided to sell tins of gourmet popcorn as a fundraiser. The suggested selling price is $3.00 per tin, but Maria, who also belongs to the Math Club, decides to take a survey to see if they can increase "the fruits of their labor." The survey shows it's likely that 120 tins will be sold on campus at the $3.00 price, and for each price increase of $0.10, 2 fewer tins will be sold. (a) What price per tin will bring in a revenue of $405? (b) How many tins will be sold at the price from part (a)?

16. Due to the seasonal nature of the business, the revenue of Wet Willey's Water World can be modeled by the equation $r = -3t^2 + 42t - 135$, where t is the time in months ($t = 1$ corresponds to January) and r is the dollar revenue in thousands. (a) What month does Wet Willey's open? (b) What month does Wet Willey's close? (c) Does Wet Willey's bring in more revenue in July or August? How much more?

Simplify each expression.

17. $\dfrac{-8 + \sqrt{-20}}{6}$

18. i^{39}

19. Given $x = \dfrac{1}{2} + \dfrac{\sqrt{3}}{2}i$ and $y = \dfrac{1}{2} - \dfrac{\sqrt{3}}{2}i$ find

a. $x + y$ **b.** $x - y$ **c.** xy

20. Compute the quotient: $\dfrac{3i}{1 - i}$.

21. Find the product: $(3i + 5)(5 - 3i)$.

22. Show $x = 2 - 3i$ is a solution of $x^2 - 4x + 13 = 0$.

23. Solve by completing the square.

a. $2x^2 - 20x + 49 = 0$

b. $2x^2 - 5x = -4$

24. Solve using the quadratic formula.

a. $3x^2 + 2 = 6x$ **b.** $x^2 = 2x - 10$

25. Allometric studies tell us that the necessary food intake F (in grams per day) of nonpasserine birds (birds other than song birds and other small birds) can be modeled by the equation $F \approx 0.3W^{\frac{3}{4}}$, where W is the bird's weight in grams. (a) If my Green-winged macaw weighs 1296 g, what is her anticipated daily food intake? (b) If my blue-headed pionus consumes 19.2 g per day, what is his estimated weight?

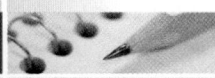

CALCULATOR EXPLORATION AND DISCOVERY

Evaluating Expressions and Looking for Patterns

These "explorations" are designed to explore the full potential of a graphing calculator, as well as to use this potential to investigate patterns and discover connections that might otherwise be overlooked. In this *Exploration and Discovery,* we point out the various ways an expression can be evaluated on a graphing calculator. Some ways seem easier, faster, and/or better than others, but each has advantages and disadvantages depending on the task at hand, and it will help to be aware of them all for future use.

One way to evaluate an expression is to use the TABLE feature of a graphing calculator, with the expression entered as Y_1 on the Y= screen. If you want the calculator to generate inputs, use the 2nd WINDOW (TBLSET) screen to indicate a starting value

(**TblStart=**) and an increment value (**ΔTbl=**), and set the calculator in **Indpnt: AUTO ASK** mode (to input specific values, the calculator should be in **Indpnt: AUTO ASK** mode). After pressing **2nd GRAPH** (**TABLE**), the calculator shows the corresponding input and output values. For help with the basic TABLE feature of the TI-84 Plus, you can visit Section R.7 at www.mhhe.com/coburn.

Expressions can also be evaluated on the home screen for a single value or a series of values. Enter the expression $-\frac{3}{4}x + 5$ on the **Y=** screen (see Figure 1.16) and use **2nd MODE** (**QUIT**) to get back to the home screen. To evaluate this expression, access Y_1 using **VARS ▶** (**Y-VARS**), and use the first option **1:Function ENTER**. This brings us to a submenu where any of the equations Y_1 through Y_0 (actually Y_{10}) can be accessed. Since the default setting is the one we need **1:Y1**, simply press **ENTER** and Y_1 appears on the home screen. To evaluate a single input, simply enclose it in parentheses. To evaluate more than one input, enter the numbers as a set of values with the set enclosed in parentheses. In Figure 1.17, Y_1 has been evaluated for $x = -4$, then simultaneously for $x = -4, -2, 0$, and 2.

A third way to evaluate expressions is using a list, with the desired inputs entered in List 1 (L1), and List 2 (L2) defined in terms of L1. For example, $L2 = -\frac{3}{4}L1 + 5$ will return the same values for inputs of $-4, -2, 0$, and 2 seen previously on the home screen (remember to clear the lists first). Lists are accessed by pressing **STAT 1:Edit.** Enter the numbers $-4, -2, 0$ and 2 in L1, then use the right arrow **▶** to move to L2. It is important to note that you *next press the up arrow key* **▲** so that the cursor overlies L2. The bottom of the screen now reads **L2=** (see Figure 1.18) and the calculator is waiting for us to define L2. After entering $L2 = -\frac{3}{4}L1 + 5$ and pressing **ENTER** we obtain the same outputs as before (see Figure 1.19).

The advantage of using the "list" method is that we can *further explore or experiment with the output values* in a search for patterns.

Exercise 1: Evaluate the expression $0.2L1 + 3$ on the list screen, using consecutive integer inputs from -6 to 6 inclusive. What do you notice about the outputs?

Exercise 2: Evaluate the expression $\sqrt{2}L1 - \sqrt{9.1}$ on the list screen, using consecutive integer inputs from -6 to 6 inclusive. We suspect there is a pattern to the output values, but this time the pattern is very difficult to see. Compute the difference between a few successive outputs from L2 [for Example $L2(1) - L2(2)$]. What do you notice?

Figure 1.16

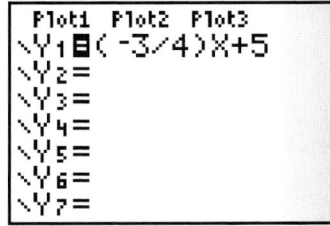

Figure 1.17

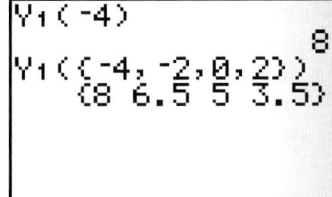

Figure 1.18

L1	■	L3	2
-4	------	------	
-2			
0			
2			

$L2 = (-3/4)L1+5$			

Figure 1.19

L1	L2	L3	3
-4	8		
-2	6.5		
0	5		
2	3.5		
------	------	------	

STRENGTHENING CORE SKILLS

An Alternative Method for Checking Solutions to Quadratic Equations

To solve $x^2 - 2x - 15 = 0$ by factoring, students will often begin by looking for two numbers whose product is -15 (the constant term) and whose sum is -2 (the linear coefficient). The two numbers are -5 and 3 since $(-5)(3) = -15$ and $-5 + 3 = -2$. In factored form, we have $(x - 5)(x + 3) = 0$ with solutions $x_1 = 5$ and $x_2 = -3$. When these solutions are compared *to the original coefficients,* we can still see the sum/product relationship, but note that while $(5)(-3) = -15$ still gives the constant term, $5 + (-3) = 2$ gives the linear coefficient *with opposite sign.* Although more difficult to accomplish,

this method can be applied to *any* factorable quadratic equation $ax^2 + bx + c = 0$ if we divide through by a, giving $x^2 + \frac{b}{a}x + \frac{c}{a} = 0$. For $2x^2 - x - 3 = 0$, we divide both sides by 2 and obtain $x^2 - \frac{1}{2}x - \frac{3}{2} = 0$, then look for two numbers whose product is $-\frac{3}{2}$ and whose sum is $-\frac{1}{2}$. The numbers are $-\frac{3}{2}$ and 1 since $\left(-\dfrac{3}{2}\right)(1) = -\dfrac{3}{2}$ and $-\dfrac{3}{2} + 1 = -\dfrac{1}{2}$, showing the

solutions are $x_1 = \dfrac{3}{2}$ and $x_2 = -1$. We again note the

product of the solutions is the constant $-\dfrac{3}{2} = \dfrac{c}{a}$, and the

sum of the solutions is the linear coefficient *with opposite*

sign: $\dfrac{1}{2} = -\dfrac{b}{a}$. No one actually promotes this method for

solving trinomials where $a \neq 1$, but it does illustrate an
important and useful concept:

If x_1 and x_2 are the two roots of $x^2 + \dfrac{b}{a}x + \dfrac{c}{a} = 0$,

then $x_1 x_2 = \dfrac{c}{a}$ and $x_1 + x_2 = -\dfrac{b}{a}$.

Justification for this can be found by taking the product

and sum of the general solutions $x_1 = \dfrac{-b}{2a} + \dfrac{\sqrt{b^2 - 4ac}}{2a}$

and $x_2 = \dfrac{-b}{2a} - \dfrac{\sqrt{b^2 - 4ac}}{2a}$. Although the computation

looks impressive, the product can be computed as a binomial times its conjugate, and the radical parts add to zero for the sum, each yielding the results as already stated.

This observation provides a useful technique for checking solutions to a quadratic equation, *even those having irrational or complex roots!* Check the solutions shown in these exercises.

Exercise 1: $2x^2 - 5x - 7 = 0$

$$x_1 = \dfrac{7}{2}$$

$$x_2 = -1$$

Exercise 2: $2x^2 - 4x - 7 = 0$

$$x_1 = \dfrac{2 + 3\sqrt{2}}{2}$$

$$x_2 = \dfrac{2 - 3\sqrt{2}}{2}$$

Exercise 3: $x^2 - 10x + 37 = 0$

$$x_1 = 5 + 2\sqrt{3}\,i$$

$$x_2 = 5 - 2\sqrt{3}\,i$$

Exercise 4: Verify this sum/product check by computing the sum and product of the general solutions.

Chapter 1 actually highlights numerous concepts and skills that transfer directly into a study of calculus. In the Chapter 1 opener, we noted that analyzing very small differences is one such skill, with this task carried out using the absolute value concept. The ability to solve a wide variety of equation types will also be a factor of your success in calculus. Here we'll explore how these concepts and skills are "connected."

Solving Various Types of Equations

The need to solve equations of various types occurs frequently in both *differential* and *integral* calculus, and the required skills will span a broad range of your algebraic experience. Here we'll solve a type of radical equation that occurs frequently in a study of *optimization* [finding the maximum or minimum value(s) of a function].

EXAMPLE 1 ▶ **Minimizing Response Time**

A boater is 70 yd away from a straight shoreline when she gets an emergency call from her home, 400 yd down shore. Knowing she can row at 200 yd/min and run at 300 yd/min, how far down shore should she land the boat to make it home in the shortest time possible?

Solution ▶ As with other forms of problem solving, drawing an accurate sketch is an important first step.

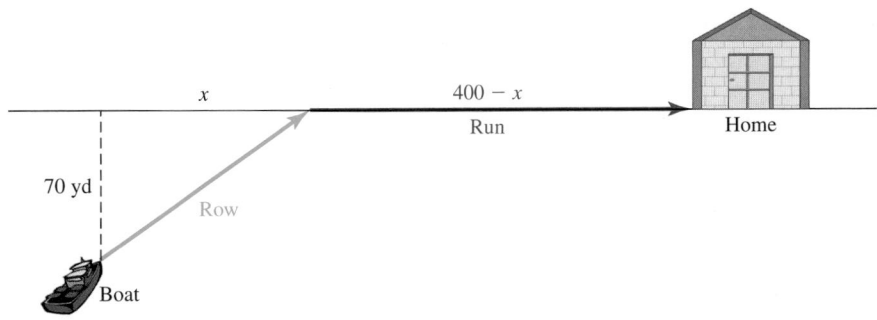

From the diagram, we note the rowing distance will be $\sqrt{x^2 + 4900}$ (using the Pythagorean theorem), and the running distance will be $400 - x$ (total minus distance downshore).

From the relationship time $= \dfrac{\text{distance}}{\text{rate}}$, we find the total time required to reach home is $t(x) = \dfrac{\sqrt{x^2 + 4900}}{200} + \dfrac{400 - x}{300}$. Using the tools of calculus it can be shown that the distance x down shore that results in the shortest possible time, is a zero of $T(x) = \dfrac{x}{200\sqrt{x^2 + 4900}} - \dfrac{1}{300}$. Find the zero(es) of $T(x)$ and state the result in both exact and approximate form.

Solution ▶ Begin by isolating the radical on one side.

$$\frac{x}{200\sqrt{x^2 + 4900}} = \frac{1}{300} \qquad \text{add } \frac{1}{300}$$

$$300x = 200\sqrt{x^2 + 4900} \qquad \text{clear denominators}$$

$$1.5x = \sqrt{x^2 + 4900} \qquad \text{divide by 200}$$

$$2.25x^2 = x^2 + 4900 \qquad \text{square both sides}$$

$$1.25x^2 = 4900 \qquad \text{subtract } x^2$$

$$x^2 = 3920 \qquad \text{divide by 1.25}$$

$$x = \sqrt{3920} \qquad \text{solve for } x; \, x > 0 \text{ (distance)}$$

$$x = 28\sqrt{5} \qquad \text{simplify radical (exact form)}$$

$$\approx 62.6 \qquad \text{approximate form}$$

The boater should row to a spot about 63 yd down shore, then run the remaining 337 yd.

Now try Exercises 1 and 2 ▶

In addition to radical equations, equations involving rational exponents are often seen in a study of calculus. Many times, solving these equations involves combining the basic properties of exponents with other familiar skills such as factoring, or in this case, *factoring least powers*.

EXAMPLE 2 ▶ **Modeling the Motion of a Particle**

Suppose the motion of an object floating in turbulent water is modeled by the function $d(t) = \sqrt{t}(t^2 - 9t + 22)$, where $d(t)$ represents the displacement (in meters) at t sec. Using the tools of calculus, it can be shown that the velocity v of the particle is given by $v(t) = \frac{5}{2}t^{\frac{3}{2}} - \frac{27}{2}t^{\frac{1}{2}} + 11t^{-\frac{1}{2}}$. Find any time(s) t when the particle is motionless ($v = 0$).

Solution ▶ Set the equation equal to zero and factor out the fraction and least power.

$$\frac{5}{2}t^{\frac{3}{2}} - \frac{27}{2}t^{\frac{1}{2}} + 11t^{-\frac{1}{2}} = 0 \qquad \text{original equation}$$

$$5t^{\frac{4}{2}}\left(\frac{1}{2}\right)t^{-\frac{1}{2}} - 27t^{\frac{2}{2}}\left(\frac{1}{2}\right)t^{-\frac{1}{2}} + 22\left(\frac{1}{2}\right)t^{-\frac{1}{2}} = 0 \qquad \text{rewrite to help factor } \frac{1}{2}t^{-\frac{1}{2}} \text{ (least power)}$$

$$\frac{1}{2}t^{-\frac{1}{2}}(5t^2 - 27t + 22) = 0 \qquad \text{common factor}$$

$$\frac{1}{2}t^{-\frac{1}{2}}(5t - 22)(t - 1) = 0 \qquad \text{factor the trinomial}$$

$$t^{-\frac{1}{2}} \neq 0; \, t = \frac{22}{5} \text{ or } t = 1 \qquad \text{result}$$

The particle is temporarily motionless at $t = 4.4$ sec and $t = 1$ sec.

Now try Exercises 3 and 4 ▶

Absolute Value Inequalities and Delta/Epsilon Form

While the terms may mean little to you now, the concept of absolute value plays an important role in the *precise definition of a limit, intervals of convergence,* and *derivatives* involving logarithmic functions. In the case of *limits,* the study of calculus concerns itself with very small differences, as in the difference between the number 3 itself, and a number very close to 3.

X	Y1			X	Y1	
2.6	5.6			3.4	6.4	
2.7	5.7			3.3	6.3	
2.8	5.8			3.2	6.2	
2.9	5.9			3.1	6.1	
2.99	5.99			3.01	6.01	
2.999	5.999			3.001	6.001	
2.9999	5.9999			3.0001	6.0001	

Y1=5.9999   Y1=6.0001

Consider the function $f(x) = \dfrac{x^2 - 9}{x - 3}$. From the implicit domain and the figures shown, we see that $f(x)$ (shown as Y_1) is not defined at 3, but is defined for any number near 3. The figures also suggest that when x is a number very close to 3, $f(x)$ is a number very close to 6. Alternatively, we might say, "if the difference between x and 3 is very small, the difference between $f(x)$ and 6 is very small." The most convenient way to express this idea and make it practical is through the use of absolute value (which allows that the difference can be either positive or negative). Using the symbols δ (delta) and ϵ (epsilon) to represent very small (and possibly unequal) numbers, we can write this phrase in *delta/epsilon form* as

$$\text{if } |x - 3| < \delta, \text{ then } |f(x) - 6| < \epsilon$$

For now, we'll simply practice translating similar relationships from words into symbols, leaving any definitive conclusions for our study of limits in Chapter 11, or a future study of calculus.

EXAMPLE 3 ▶ **Using Delta/Epsilon Form**

Use a graphing calculator to explore the value of $g(x) = \dfrac{x^2 + 3x - 10}{x - 2}$ when x is near 2, then write the relationship in delta/epsilon form.

Solution ▶ Using a graphing calculator and the approach outlined above produces the tables in the figures.

X	Y1			X	Y1	
1.6	6.6			2.4	7.4	
1.7	6.7			2.3	7.3	
1.8	6.8			2.2	7.2	
1.9	6.9			2.1	7.1	
1.99	6.99			2.01	7.01	
1.999	6.999			2.001	7.001	
1.9999	6.9999			2.0001	7.0001	

Y1=6.9999   Y1=7.0001

From these, it appears that, "if the difference between x and 2 is very small, the difference between $f(x)$ and 7 is very small." In delta/epsilon form:
$$\text{if } |x - 2| < \delta, \text{ then } |f(x) - 7| < \epsilon.$$

Now try Exercises 5 through 8 ▶

At first, modeling this relationship may seem like a minor accomplishment. But historically and in a practical sense, it is actually a major achievement as it enables us to "tame the infinite," since we can now verify that no matter how small ϵ is, there is a corresponding δ that *guarantees*

$$|f(x) - 7| < \epsilon \qquad \text{whenever} \qquad |x - 2| < \delta$$

$f(x)$ is *infinitely* close to 7 x is *infinitely* close to 2

This observation leads directly to the precise definition of a limit, the type of "limit" referred to in our *Introduction to Calculus,* found in the Preface (page 000). As noted there, such limits will enable us to find a precise formula for the instantaneous speed of the cue ball as it falls, and a precise formula for the volume of an irregular solid.

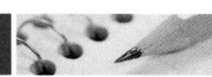

CONNECTIONS TO CALCULUS EXERCISES

Solve the following equations.

1. To find the length of a rectangle with maximum area that can be circumscribed by a circle of radius 3 in. requires that we solve $\sqrt{9 - x^2} - \dfrac{x^2}{\sqrt{9 - x^2}} = 0$, where the length of the rectangle is $2x$. To the nearest hundredth, what is the length of the rectangle?

2. To find the height of an isosceles triangle with maximum area that can be inscribed in a circle of radius $r = 5$ in. requires that we solve $\dfrac{-5x}{\sqrt{25 - x^2}} + \sqrt{25 - x^2} - \dfrac{x^2}{\sqrt{25 - x^2}} = 0$, where the height is $5 + x$. What is the height of the triangle?

3. If the motion of a particle in turbulent air is modeled by $d = \sqrt{t}(2t^2 - 9t + 18)$, the velocity of the particle is given by $v = 5t^{\frac{3}{2}} - \dfrac{27}{2}t^{\frac{1}{2}} + 9t^{-\frac{1}{2}}$ (d in meters, t in seconds). Find any time(s) t when velocity $v = 0$.

4. In order for a light source to provide maximum (circular) illumination to a workroom, the light must be hung at a certain height. While the complete development requires trigonometry, we find that maximum illumination is obtained at the solutions of the equation shown, where h is the height of the light, k is a constant, and the radius of illumination is 12 ft. Solve the equation for h by factoring the least power and simplifying the result: $k\dfrac{(h^2 + 12^2)^{\frac{3}{2}} - 3h^2(h^2 + 12^2)^{\frac{1}{2}}}{(h^2 + 12^2)^3} = 0.$

Use a graphing calculator to explore the value of the function given for values of x near the one indicated. Then write the relationship in words and in delta/epsilon form.

5. $h(x) = \dfrac{4x^2 - 9}{2x - 3}; x = \dfrac{3}{2}$

6. $v(x) = \dfrac{x^3 + 27}{x + 3}; x = -3$

7. $w(x) = \dfrac{7x^3 - 28x}{x^2 - 4}; x = 2$

8. $F(x) = \dfrac{x^2 + 7x}{x}; x = 0$

Relations, Functions, and Graphs

CHAPTER OUTLINE

CHAPTER CONNECTIONS

From the rate at which your computer can download a large file, to the rate a bacteria culture grows in the production of penicillin, science, medicine, sports, and industry all have a great interest in the rate at which change takes place. Where some measures of change have a tremendous impact on civilization (faster drying cement, stronger metal alloys, better communication), other measures of change quantify and track improvements in various areas of human endeavor. For instance, by making slight modifications in the tip or tail of an arrow, an olympic archer can alter the velocity of the arrow. Using the average rate of change formula, we can find the average velocity of the arrow for any time interval. This application appears as Exercise 61 in Section 2.5.

Connections to Calculus

In this chapter, we use the average rate of change formula to develop the difference quotient. In a calculus course, the difference quotient is combined with the concept of a limit to find the *instantaneous velocity* of the arrow at any time *t*. The *Connections to Calculus* for Chapter 2 highlights the algebraic skills necessary to make the transition from the discrete view to the instantaneous view a smooth one.

Learning Objectives

In Section 2.1 you will learn how to:

☐ **A.** Express a relation in mapping notation and ordered pair form

☐ **B.** Graph a relation

☐ **C.** Develop the equation of a circle using the distance and midpoint formulas

☐ **D.** Graph circles

WORTHY OF NOTE

From a purely practical standpoint, we note that while it is possible for two different people to share the same birthday, it is quite impossible for the same person to have two different birthdays. Later, this observation will help us mark the difference between a relation and a function.

In everyday life, we encounter a large variety of relationships. For instance, the time it takes us to get to work is related to our average speed; the monthly cost of heating a home is related to the average outdoor temperature; and in many cases, the amount of our charitable giving is related to changes in the cost of living. In each case we say that a relation exists between the two quantities.

A. Relations, Mapping Notation, and Ordered Pairs

In the most general sense, a **relation** is simply a correspondence between two sets. Relations can be represented in many different ways and may even be very "unmathematical," like the one shown in Figure 2.1 between a set of people and the set of their corresponding birthdays. If *P* represents the set of people and *B* represents the set of birthdays, we say that elements of *P* correspond to elements of *B*, or the birthday relation maps elements of *P* to elements of *B*. Using what is called **mapping notation,** we might simply write $P \to B$.

Figure 2.1

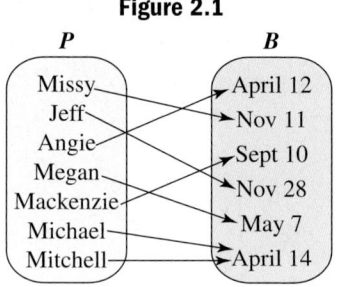

The bar graph in Figure 2.2 is also an example of a relation. In the graph, each year is related to average annual consumer spending on Internet media (music downloads, Internet radio, Web-based news articles, etc.). As an alternative to mapping or a bar graph, the relation could also be represented using **ordered pairs.** For example, the ordered pair (3, 98) would indicate that in 2003, spending per person on Internet media averaged $98 in the United States. Over a long period of time, we could collect many ordered pairs of the form (t, s), where consumer spending *s depends* on the time *t*. For this reason we often call the second coordinate of an ordered pair (in this case *s*) the **dependent variable,** with the first coordinate designated as the **independent variable.** In this form, the set of all first coordinates is called the **domain** of the relation. The set of all second coordinates is called the **range.**

Figure 2.2

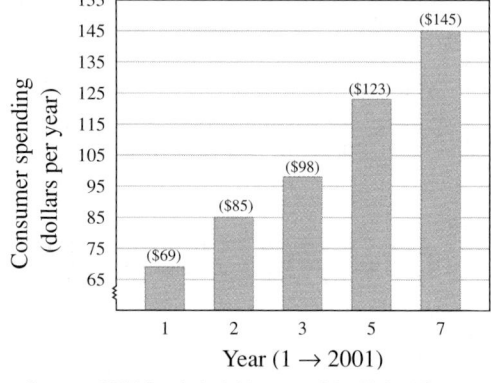

Source: 2006 Statistical Abstract of the United States

EXAMPLE 1 ▶ **Expressing a Relation as a Mapping and in Ordered Pair Form**

Represent the relation from Figure 2.2 in mapping notation and ordered pair form, then state its domain and range.

Solution ▶ Let *t* represent the year and *s* represent consumer spending. The mapping $t \to s$ gives the diagram shown. In ordered pair form we have (1, 69), (2, 85), (3, 98), (5, 123), and (7, 145). The domain is {1, 2, 3, 5, 7}, the range is {69, 85, 98, 123, 145}.

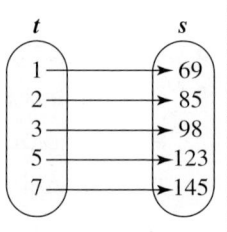

☑ **A.** You've just learned how to express a relation in mapping notation and ordered pair form

Now try Exercises 7 through 12 ▶

For more on this relation, **see Exercise 81.**

Table 2.1 $y = x - 1$

x	y
−4	−5
−2	−3
0	−1
2	1
4	3

Table 2.2 $x = |y|$

x	y
2	−2
1	−1
0	0
1	1
2	2

B. The Graph of a Relation

Relations can also be stated in **equation form.** The equation $y = x - 1$ expresses a relation where each y-value is one less than the corresponding x-value (see Table 2.1). The equation $x = |y|$ expresses a relation where each x-value corresponds to the absolute value of y (see Table 2.2). In each case, the relation is the set of all ordered pairs (x, y) that create a true statement when substituted, and a few ordered pair solutions are shown in the tables for each equation.

Relations can be expressed graphically using a **rectangular coordinate system.** It consists of a horizontal number line (the x-axis) and a vertical number line (the y-axis) intersecting at their zero marks. The point of intersection is called the *origin*. The x- and y-axes create a flat, two-dimensional surface called the **xy-plane** and divide the plane into four regions called **quadrants.** These are labeled using a capital "Q" (for quadrant) and the Roman numerals I through IV, beginning in the upper right and moving counterclockwise (Figure 2.3). The **grid lines** shown denote the integer values on each axis and further divide the plane into a **coordinate grid,** where every point in the plane corresponds to an ordered pair. Since a point at the origin has not moved along either axis, it has coordinates $(0, 0)$. To plot a point (x, y) means we place a dot at its location in the xy-plane. A few of the ordered pairs from $y = x - 1$ are plotted in Figure 2.4, where a noticeable pattern emerges—the points seem to lie along a straight line.

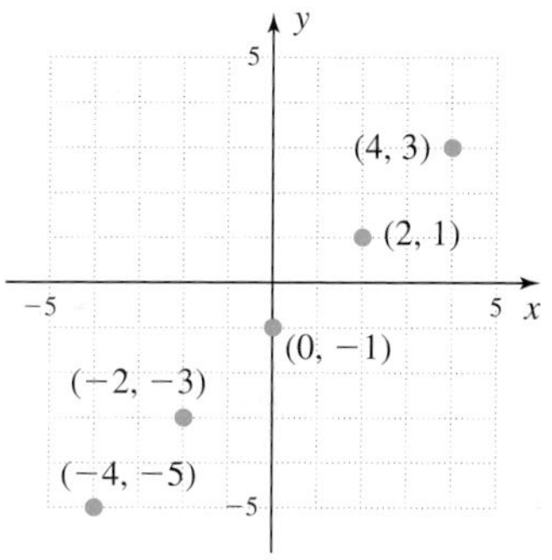

If a relation is defined by a set of ordered pairs, the graph of the relation is simply the plotted points. The graph of a relation *in equation form,* such as $y = x - 1$, is the set of *all* ordered pairs (x, y) that make the equation true. We generally use only a few select points to determine the shape of a graph, then draw a straight line or smooth curve through these points, as indicated by any patterns formed.

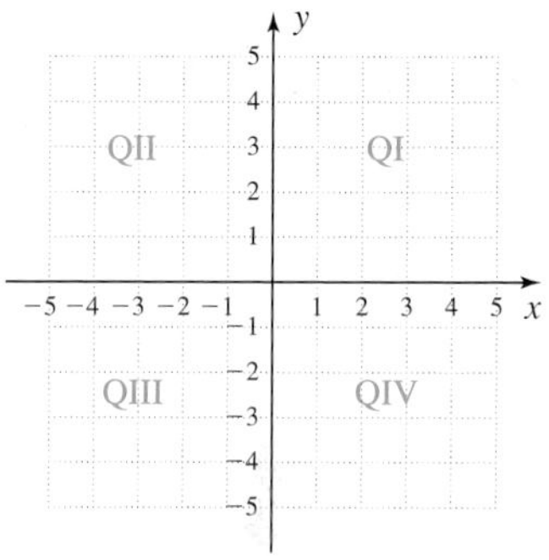

Figure 2.3

Figure 2.4

EXAMPLE 2 ▶ **Graphing Relations**

Graph the relations $y = x - 1$ and $x = |y|$ using the ordered pairs given earlier.

Solution ▶ For $y = x - 1$, we plot the points then connect them with a straight line (Figure 2.5). For $x = |y|$, the plotted points form a V-shaped graph made up of two half lines (Figure 2.6).

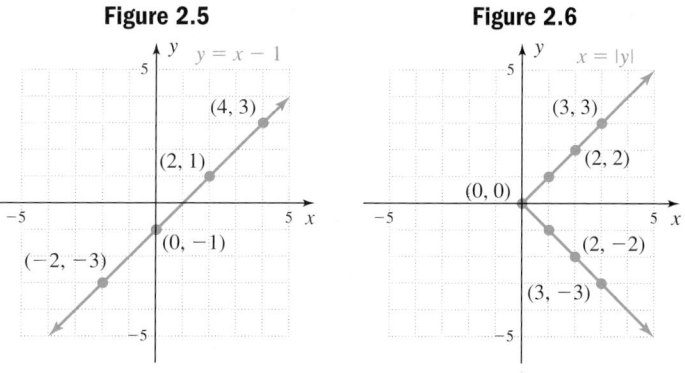

Figure 2.5   **Figure 2.6**

Now try Exercises 13 through 16 ▶

While we used only a few points to graph the relations in Example 2, they are actually made up of an *infinite number of ordered pairs* that satisfy each equation, including those that might be rational or irrational. All of these points together make these graphs **continuous,** which for our purposes means you can draw the entire graph without lifting your pencil from the paper.

Actually, a majority of graphs cannot be drawn using only a straight line or directed line segments. In these cases, we rely on a "sufficient number" of points to outline the basic shape of the graph, then connect the points with a smooth curve. As your experience with graphing increases, this "sufficient number of points" tends to get smaller as you learn to anticipate what the graph of a given relation should look like.

EXAMPLE 3 ▶ **Graphing Relations**

Graph the following relations by completing the tables given.

   a. $y = x^2 - 2x$ **b.** $y = \sqrt{9 - x^2}$ **c.** $x = y^2$

Solution ▶ For each relation, we use each x-input in turn to determine the related y-output(s), if they exist. Results can be entered in a table and the ordered pairs used to draw the graph.

a. $y = x^2 - 2x$ **Figure 2.7**

x	y	(x, y) Ordered Pairs
-4	24	$(-4, 24)$
-3	15	$(-3, 15)$
-2	8	$(-2, 8)$
-1	3	$(-1, 3)$
0	0	$(0, 0)$
1	-1	$(1, -1)$
2	0	$(2, 0)$
3	3	$(3, 3)$
4	8	$(4, 8)$

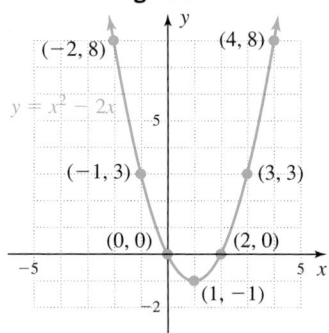

The result is a fairly common graph (Figure 2.7), called a **vertical parabola.** Although $(-4, 24)$ and $(-3, 15)$ cannot be plotted here, the arrowheads indicate an infinite extension of the graph, which will include these points.

b. $y = \sqrt{9 - x^2}$ **Figure 2.8**

x	y	(x, y) Ordered Pairs
-4	not real	—
-3	0	$(-3, 0)$
-2	$\sqrt{5}$	$(-2, \sqrt{5})$
-1	$2\sqrt{2}$	$(-1, 2\sqrt{2})$
0	3	$(0, 3)$
1	$2\sqrt{2}$	$(1, 2\sqrt{2})$
2	$\sqrt{5}$	$(2, \sqrt{5})$
3	0	$(3, 0)$
4	not real	—

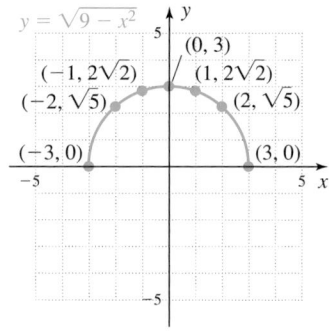

The result is the graph of a **semicircle** (Figure 2.8). The points with irrational coordinates were graphed by <u>estimating</u> their location. Note that when $x < -3$ or $x > 3$, the relation $y = \sqrt{9 - x^2}$ does not represent a real number and no points can be graphed. Also note that no arrowheads are used since the graph terminates at $(-3, 0)$ and $(3, 0)$.

   c. Similar to $x = |y|$, the relation $x = y^2$ is defined only for $x \geq 0$ since y^2 is always nonnegative ($-1 = y^2$ has no real solutions). In addition, we reason that each positive x-value will correspond to two y-values. For example, given $x = 4$, $(4, -2)$ and $(4, 2)$ are both solutions.

$x = y^2$

Figure 2.9

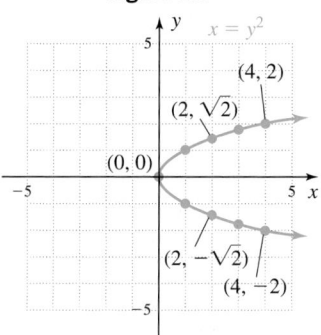

x	y	(x, y) Ordered Pairs
-2	not real	—
-1	not real	—
0	0	$(0, 0)$
1	$-1, 1$	$(1, -1)$ and $(1, 1)$
2	$-\sqrt{2}, \sqrt{2}$	$(2, -\sqrt{2})$ and $(2, \sqrt{2})$
3	$-\sqrt{3}, \sqrt{3}$	$(3, -\sqrt{3})$ and $(3, \sqrt{3})$
4	$-2, 2$	$(4, -2)$ and $(4, 2)$

☑ **B.** You've just learned how to graph a relation

This is the graph of a **horizontal parabola** (Figure 2.9).

Now try Exercises 17 through 24 ▶

C. The Equation of a Circle

Using the midpoint and distance formulas, we can develop the equation of another very important relation, that of a circle. As the name suggests, the **midpoint of a line segment** is located halfway between the endpoints. On a standard number line, the midpoint of the line segment with endpoints 1 and 5 is 3, but more important, note that 3 is the **average distance** (from zero) of 1 unit and 5 units: $\dfrac{1 + 5}{2} = \dfrac{6}{2} = 3$. This observation can be extended to find the midpoint between any two points (x_1, y_1) and (x_2, y_2). We simply find the average distance between the x-coordinates and the average distance between the y-coordinates.

The Midpoint Formula

Given any line segment with endpoints $P_1 = (x_1, y_1)$ and $P_2 = (x_2, y_2)$, the midpoint M is given by

$$M: \left(\frac{x_1 + x_2}{2}, \frac{y_1 + y_2}{2} \right)$$

The midpoint formula can be used in many different ways. Here we'll use it to find the coordinates of the center of a circle.

EXAMPLE 4 ▶ Using the Midpoint Formula

The diameter of a circle has endpoints at $P_1 = (-3, -2)$ and $P_2 = (5, 4)$. Use the midpoint formula to find the coordinates of the center, then plot this point.

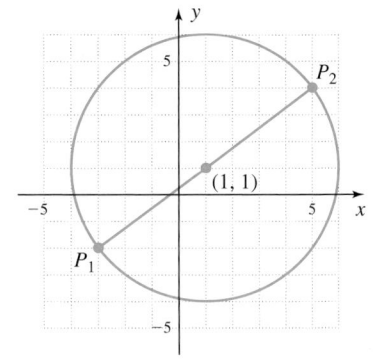

Solution ▶ Midpoint: $\left(\dfrac{x_1 + x_2}{2}, \dfrac{y_1 + y_2}{2} \right)$

$M: \left(\dfrac{-3 + 5}{2}, \dfrac{-2 + 4}{2} \right)$

$M: \left(\dfrac{2}{2}, \dfrac{2}{2} \right) = (1, 1)$

The center is at $(1, 1)$, which we graph directly on the diameter as shown.

Now try Exercises 25 through 34 ▶

Figure 2.10

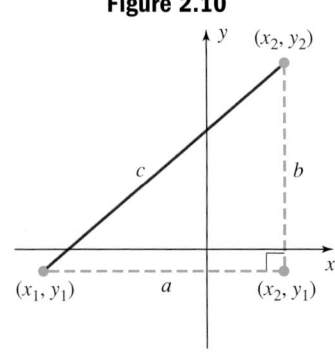

The Distance Formula

In addition to a line segment's midpoint, we are often interested in the *length* of the segment. For any two points (x_1, y_1) and (x_2, y_2) not lying on a horizontal or vertical line, a right triangle can be formed as in Figure 2.10. Regardless of the triangle's orientation, the length of side a (the horizontal segment or base of the triangle) will have length $|x_2 - x_1|$ units, with side b (the vertical segment or height) having length $|y_2 - y_1|$ units. From the Pythagorean theorem (Appendix I.F), we see that $c^2 = a^2 + b^2$ corresponds to $c^2 = (|x_2 - x_1|)^2 + (|y_2 - y_1|)^2$. By taking the square root of both sides we obtain the length of the hypotenuse, *which is identical to the distance between these two points*: $c = \sqrt{(x_2 - x_1)^2 + (y_2 - y_1)^2}$. The result is called the **distance formula,** although it's most often written using **d** for **d**istance, rather than c. Note the absolute value bars are dropped from the formula, since the square of any quantity is always nonnegative. This also means that *either* point can be used as the initial point in the computation.

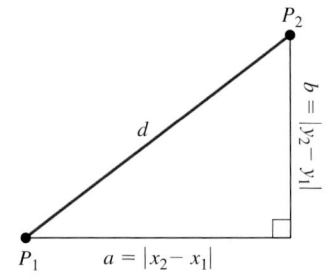

The Distance Formula

Given any two points $P_1 = (x_1, y_1)$ and $P_2 = (x_2, y_2)$, the straight line distance between them is

$$d = \sqrt{(x_2 - x_1)^2 + (y_2 - y_1)^2}$$

EXAMPLE 5 ▶ Using the Distance Formula

Use the distance formula to find the diameter of the circle from Example 4.

Solution ▶ For $(x_1, y_1) = (-3, -2)$ and $(x_2, y_2) = (5, 4)$, the distance formula gives

$$d = \sqrt{(x_2 - x_1)^2 + (y_2 - y_1)^2}$$
$$= \sqrt{[5 - (-3)]^2 + [4 - (-2)]^2}$$
$$= \sqrt{8^2 + 6^2}$$
$$= \sqrt{100} = 10$$

The diameter of the circle is 10 units long.

Now try Exercises 35 through 38 ▶

EXAMPLE 6 ▶   **Determining if Three Points Form a Right Triangle**

Use the distance formula to determine if the following points are the vertices of a right triangle: $(-8, 1)$, $(-2, 9)$, and $(10, 0)$

Solution ▶   We begin by finding the distance between each pair of points, then attempt to apply the Pythagorean theorem.

For $(x_1, y_1) = (-8, 1)$, $(x_2, y_2) = (-2, 9)$:

$$d = \sqrt{(x_2 - x_1)^2 + (y_2 - y_1)^2}$$
$$= \sqrt{[-2 - (-8)]^2 + (9 - 1)^2}$$
$$= \sqrt{6^2 + 8^2}$$
$$= \sqrt{100} = 10$$

For $(x_2, y_2) = (-2, 9)$, $(x_3, y_3) = (10, 0)$:

$$d = \sqrt{(x_3 - x_2)^2 + (y_3 - y_2)^2}$$
$$= \sqrt{[10 - (-2)]^2 + (0 - 9)^2}$$
$$= \sqrt{12^2 + (-9)^2}$$
$$= \sqrt{225} = 15$$

For $(x_1, y_1) = (-8, 1)$, $(x_3, y_3) = (10, 0)$:

$$d = \sqrt{(x_3 - x_1)^2 + (y_3 - y_1)^2}$$
$$= \sqrt{[10 - (-8)]^2 + (0 - 1)^2}$$
$$= \sqrt{18^2 + (-1)^2}$$
$$= \sqrt{325} = 5\sqrt{13}$$

Using the unsimplified form, we clearly see that $a^2 + b^2 = c^2$ corresponds to $(\sqrt{100})^2 + (\sqrt{225})^2 = (\sqrt{325})^2$, a true statement. Yes, the triangle is a right triangle.

Now try Exercises 39 through 44 ▶

A circle can be defined as the set of all points in a plane that are a *fixed distance* called the **radius,** from a *fixed point* called the **center.** Since the definition involves *distance,* we can construct the general equation of a circle using the distance formula. Assume the center has coordinates (h, k), and let (x, y) represent any point on the graph. Since the distance between these points is equal to the radius r, the distance formula yields: $\sqrt{(x - h)^2 + (y - k)^2} = r$. Squaring both sides gives the equation of a circle in **standard form:** $(x - h)^2 + (y - k)^2 = r^2$.

The Equation of a Circle

A circle of radius r with center at (h, k) has the equation $(x - h)^2 + (y - k)^2 = r^2$

If $h = 0$ and $k = 0$, the circle is centered at $(0, 0)$ and the graph is a **central circle** with equation $x^2 + y^2 = r^2$. At other values for h or k, the center is at (h, k) with no change in the radius. Note that an open dot is used for the center, as it's actually a point of reference and not a part of the actual graph.

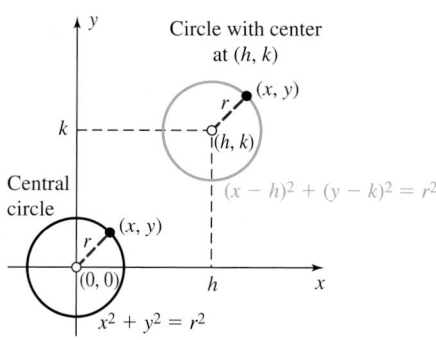

EXAMPLE 7 ▶   **Finding the Equation of a Circle**

Find the equation of a circle with center $(0, -1)$ and radius 4.

Solution ▶   Since the center is at $(0, -1)$ we have $h = 0$, $k = -1$, and $r = 4$. Using the standard form $(x - h)^2 + (y - k)^2 = r^2$ we obtain

$$(x - 0)^2 + [y - (-1)]^2 = 4^2 \quad \text{substitute 0 for } h, -1 \text{ for } k, \text{ and 4 for } r$$
$$x^2 + (y + 1)^2 = 16 \quad \text{simplify}$$

The graph of $x^2 + (y + 1)^2 = 16$ is shown in the figure.

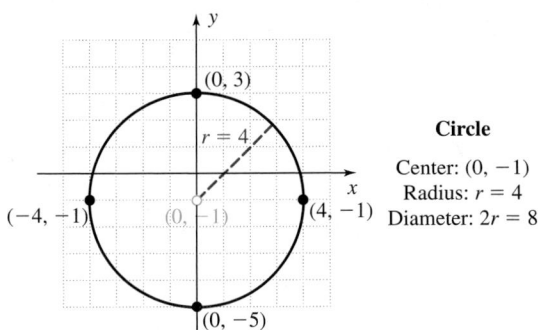

Circle
Center: $(0, -1)$
Radius: $r = 4$
Diameter: $2r = 8$

☑ **C.** You've just learned how to develop the equation of a circle using the distance and midpoint formulas

Now try Exercises 45 through 62 ▶

D. The Graph of a Circle

The graph of a circle can be obtained by first identifying the coordinates of the center and the length of the radius from the equation in standard form. After plotting the center point, we count a distance of r units left and right of center in the horizontal direction, and up and down from center in the vertical direction, obtaining four points on the circle. Neatly graph a circle containing these four points.

EXAMPLE 8 ▶ Graphing a Circle

Graph the circle represented by $(x - 2)^2 + (y + 3)^2 = 12$. Clearly label the center and radius.

Solution ▶ Comparing the given equation with the standard form, we find the center is at $(2, -3)$ and the radius is $r = 2\sqrt{3} \approx 3.5$.

$$(x - h)^2 + (y - k)^2 = r^2 \qquad \text{standard form}$$
$$(x - 2)^2 + (y + 3)^2 = 12 \qquad \text{given equation}$$
$$-h = -2 \qquad -k = 3 \qquad r^2 = 12$$
$$h = 2 \qquad k = -3 \qquad r = \sqrt{12} = 2\sqrt{3} \quad \text{radius must be positive}$$
$$\approx 3.5$$

Plot the center $(2, -3)$ and count approximately 3.5 units in the horizontal and vertical directions. Complete the circle by freehand drawing or using a compass. The graph shown is obtained.

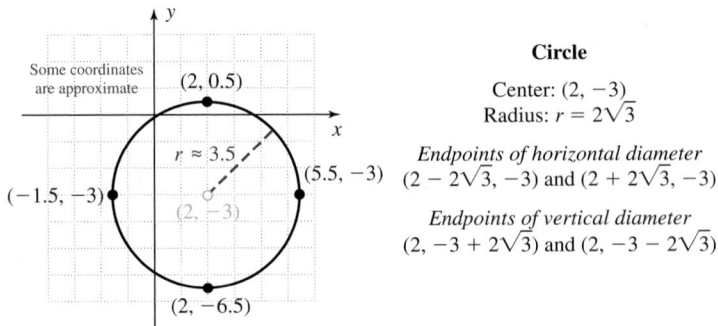

Some coordinates are approximate

Circle
Center: $(2, -3)$
Radius: $r = 2\sqrt{3}$

Endpoints of horizontal diameter
$(2 - 2\sqrt{3}, -3)$ and $(2 + 2\sqrt{3}, -3)$

Endpoints of vertical diameter
$(2, -3 + 2\sqrt{3})$ and $(2, -3 - 2\sqrt{3})$

Now try Exercises 63 through 68 ▶

WORTHY OF NOTE

After writing the equation in standard form, it is possible to end up with a constant that is zero or negative. In the first case, the graph is a single point. In the second case, no graph is possible since roots of the equation will be complex numbers. These are called *degenerate cases*. **See Exercise 91.**

In Example 8, note the equation is composed of binomial squares in both x and y. By expanding the binomials and collecting like terms, we can write the equation of the circle in the general form:

$$(x - 2)^2 + (y + 3)^2 = 12 \quad \text{standard form}$$
$$x^2 - 4x + 4 + y^2 + 6y + 9 = 12 \quad \text{expand binomials}$$
$$x^2 + y^2 - 4x + 6y + 1 = 0 \quad \text{combine like terms—general form}$$

For future reference, observe the general form contains a *sum* of second-degree terms in x and y, and that *both terms have the same coefficient* (in this case, "1").

Since this form of the equation was derived by squaring binomials, it seems reasonable to assume we can go back to the standard form by creating binomial squares in x and y. This is accomplished by *completing the square*.

EXAMPLE 9 ▶ Finding the Center and Radius of a Circle

Find the center and radius of the circle with equation $x^2 + y^2 + 2x - 4y - 4 = 0$. Then sketch its graph and label the center and radius.

Solution ▶ To find the center and radius, we complete the square in both x and y.

$$x^2 + y^2 + 2x - 4y - 4 = 0 \quad \text{given equation}$$
$$(x^2 + 2x + \_\_) + (y^2 - 4y + \_\_) = 4 \quad \text{group } x\text{-terms and } y\text{-terms; add 4}$$
$$(x^2 + 2x + 1) + (y^2 - 4y + 4) = 4 + 1 + 4 \quad \text{complete each binomial square}$$

$$(x + 1)^2 + (y - 2)^2 = 9 \quad \text{factor and simplify}$$

The center is at $(-1, 2)$ and the radius is $r = \sqrt{9} = 3$.

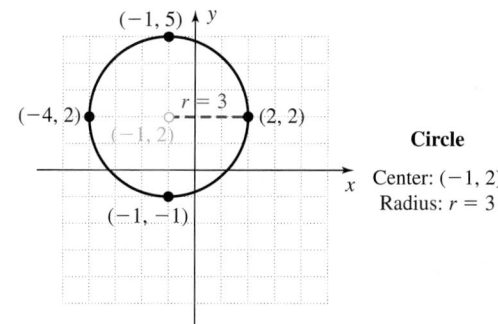

Circle
Center: $(-1, 2)$
Radius: $r = 3$

Now try Exercises 69 through 80 ▶

EXAMPLE 10 ▶ Applying the Equation of a Circle

To aid in a study of nocturnal animals, some naturalists install a motion detector near a popular watering hole. The device has a range of 10 m in any direction. Assume the water hole has coordinates $(0, 0)$ and the device is placed at $(2, -1)$.

  a. Write the equation of the circle that models the maximum effective range of the device.

  b. Use the distance formula to determine if the device will detect a badger that is approaching the water and is now at coordinates $(11, -5)$.

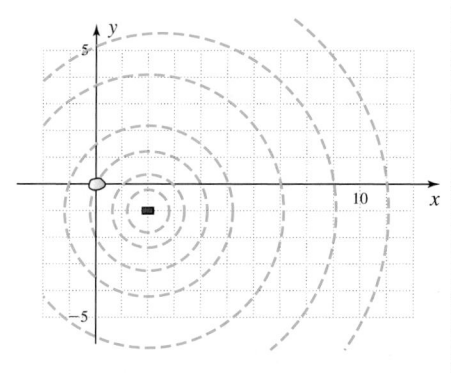

Solution ▶   **a.** Since the device is at $(2, -1)$ and the radius (or reach) of detection is 10 m, any movement in the interior of the circle defined by $(x - 2)^2 + (y + 1)^2 = 10^2$ will be detected.

b. Using the points $(2, -1)$ and $(11, -5)$ in the distance formula yields:

$$d = \sqrt{(x_2 - x_1)^2 + (y_2 - y_1)^2} \qquad \text{distance formula}$$
$$= \sqrt{(11 - 2)^2 + [-5 - (-1)]^2} \qquad \text{substitute given values}$$
$$= \sqrt{9^2 + (-4)^2} \qquad \text{simplify}$$
$$= \sqrt{81 + 16} \qquad \text{compute squares}$$
$$= \sqrt{97} \approx 9.85 \qquad \text{result}$$

☑ **D. You've just learned how to graph circles**

Since $9.85 < 10$, the badger is within range of the device and will be detected.

Now try Exercises 83 through 88 ▶

TECHNOLOGY HIGHLIGHT

The Graph of a Circle

When using a graphing calculator to study circles, it is important to keep two things in mind. First, we must modify the equation of the circle before it can be graphed using this technology. Second, most standard viewing windows have the x- and y-values preset at $[-10, 10]$ even though the calculator screen is not square. This tends to compress the y-values and give a skewed image of the graph. Consider the *relation* $x^2 + y^2 = 25$, which we know is the equation of a circle centered at (0, 0) with radius $r = 5$. To enable the calculator to graph this relation, we must define it in two pieces by solving for y:

$$x^2 + y^2 = 25 \qquad \text{original equation}$$
$$y^2 = 25 - x^2 \qquad \text{isolate } y^2$$
$$y = \pm\sqrt{25 - x^2} \qquad \text{solve for } y$$

Figure 2.11

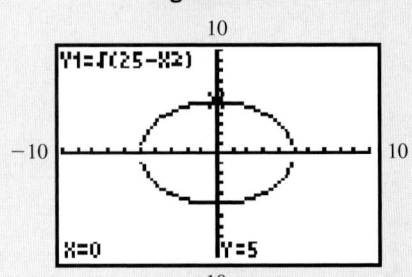

Note that we can separate this result into two parts, enabling the calculator to draw the circle: $Y_1 = \sqrt{25 - x^2}$ gives the "upper half" of the circle, and $Y_2 = -\sqrt{25 - x^2}$ gives the "lower half." Enter these on the [Y=] screen (note that $Y_2 = -Y_1$ can be used instead of reentering the entire expression: [VARS] [▶] [ENTER]). But if we graph Y_1 and Y_2 on the standard screen, the result appears more oval than circular (Figure 2.11). One way to fix this is to use the [ZOOM] **5:ZSquare** option, which places the tick marks equally spaced on both axes, instead of trying to force both to display points from -10 to 10 (see Figure 2.12). Although it is a much improved graph, the circle does not appear "closed" as the calculator lacks sufficient pixels to show the proper curvature. A second alternative is to manually set a "friendly" window. Using Xmin = -9.4, Xmax = 9.4, Ymin = -6.2, and Ymax = 6.2 will generate a better graph, which we can use to study the relation more closely. Note that we can jump between the upper and lower halves of the circle using the up [▲] or down [▼] arrows.

Figure 2.12

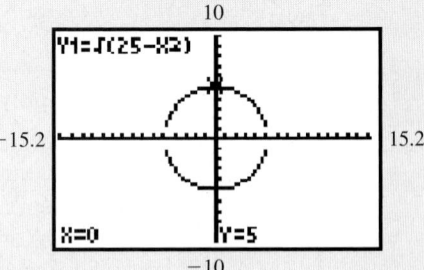

Exercise 1: Graph the circle defined by $x^2 + y^2 = 36$ using a friendly window, then use the [TRACE] feature to find the value of y when $x = 3.6$. Now find the value of y when $x = 4.8$. Explain why the values seem "interchangeable."

Exercise 2: Graph the circle defined by $(x - 3)^2 + y^2 = 16$ using a friendly window, then use the [TRACE] feature to find the value of the y-intercepts. Show you get the same intercepts by computation.

2.1 EXERCISES

▶ CONCEPTS AND VOCABULARY

Fill in each blank with the appropriate word or phrase. Carefully reread the section if needed.

1. If a relation is defined by a set of ordered pairs, the domain is the set of all _____ components, the range is the set of all _____ components.

2. For the equation $y = x + 5$ and the ordered pair (x, y), x is referred to as the input or _____ variable, while y is called the _____ or dependent variable.

3. A circle is defined as the set of all points that are an equal distance, called the _____, from a given point, called the _____.

4. For $x^2 + y^2 = 25$, the center of the circle is at _____ and the length of the radius is _____ units. The graph is called a _____ circle.

5. Discuss/Explain how to find the center and radius of the circle defined by the equation $x^2 + y^2 - 6x = 7$. How would this circle differ from the one defined by $x^2 + y^2 - 6y = 7$?

6. In Example 3b we graphed the semicircle defined by $y = \sqrt{9 - x^2}$. Discuss how you would obtain the equation of the full circle from this equation, and how the two equations are related.

▶ DEVELOPING YOUR SKILLS

Represent each relation in mapping notation, then state the domain and range.

7.

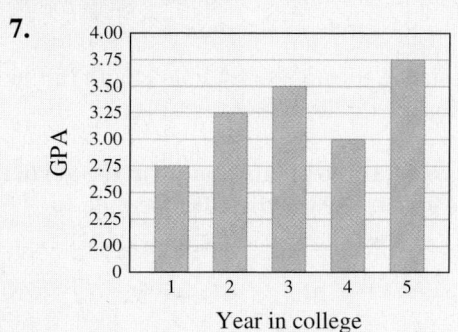

Year in college

8.

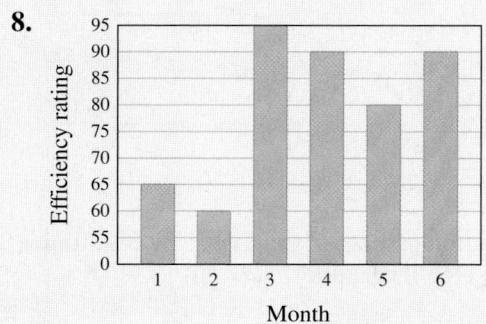

Month

State the domain and range of each relation.

9. $\{(1, 2), (3, 4), (5, 6), (7, 8), (9, 10)\}$

10. $\{(-2, 4), (-3, -5), (-1, 3), (4, -5), (2, -3)\}$

11. $\{(4, 0), (-1, 5), (2, 4), (4, 2), (-3, 3)\}$

12. $\{(-1, 1), (0, 4), (2, -5), (-3, 4), (2, 3)\}$

Complete each table using the given equation. For Exercises 15 and 16, each input may correspond to two outputs (be sure to find both if they exist). Use these points to graph the relation.

13. $y = -\dfrac{2}{3}x + 1$

x	y
-6	
-3	
0	
3	
6	
8	

14. $y = -\dfrac{5}{4}x + 3$

x	y
-8	
-4	
0	
4	
8	
10	

15. $x + 2 = |y|$

x	y
-2	
0	
1	
3	
6	
7	

16. $|y + 1| = x$

x	y
0	
1	
3	
5	
6	
7	

17. $y = x^2 - 1$

x	y
−3	
−2	
0	
2	
3	
4	

18. $y = -x^2 + 3$

x	y
−2	
−1	
0	
1	
2	
3	

19. $y = \sqrt{25 - x^2}$

x	y
−4	
−3	
0	
2	
3	
4	

20. $y = \sqrt{169 - x^2}$

x	y
−12	
−5	
0	
3	
5	
12	

21. $x - 1 = y^2$

x	y
10	
5	
4	
2	
1.25	
1	

22. $y^2 + 2 = x$

x	y
2	
3	
4	
5	
6	
11	

23. $y = \sqrt[3]{x} + 1$

x	y
−9	
−2	
−1	
0	
4	
7	

24. $y = (x - 1)^3$

x	y
−2	
−1	
0	
1	
2	
3	

Find the midpoint of each segment with the given endpoints.

25. $(1, 8), (5, -6)$ **26.** $(5, 6), (6, -8)$

27. $(-4.5, 9.2), (3.1, -9.8)$ **28.** $(5.2, 7.1), (6.3, -7.1)$

29. $\left(\dfrac{1}{5}, -\dfrac{2}{3}\right), \left(-\dfrac{1}{10}, \dfrac{3}{4}\right)$ **30.** $\left(-\dfrac{3}{4}, -\dfrac{1}{3}\right), \left(\dfrac{3}{8}, \dfrac{5}{6}\right)$

Find the midpoint of each segment.

31. 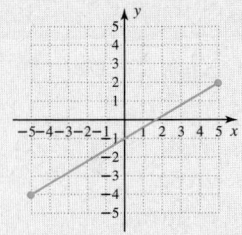   **32.**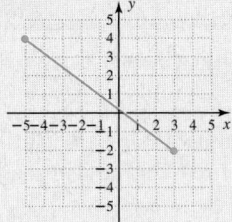

Find the center of each circle with the diameter shown.

33. 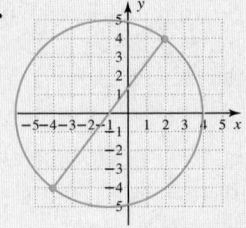   **34.**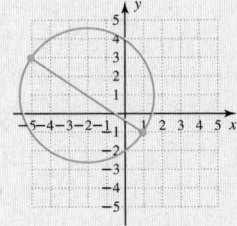

35. Use the distance formula to find the length of the line segment in Exercise 31.

36. Use the distance formula to find the length of the line segment in Exercise 32.

37. Use the distance formula to find the length of the diameter for the circle in Exercise 33.

38. Use the distance formula to find the length of the diameter for the circle in Exercise 34.

In Exercises 39 to 44, three points that form the vertices of a triangle are given. Use the distance formula to determine if any of the triangles are right triangles.

39. $(5, 2), (0, -3), (4, -4)$

40. $(7, 0), (-1, 0), (7, 4)$

41. $(-4, 3), (-7, -1), (3, -2)$

42. $(-3, 7), (2, 2), (5, 5)$

43. $(-3, 2), (-1, 5), (-6, 4)$

44. $(0, 0), (-5, 2), (2, -5)$

Find the equation of a circle satisfying the conditions given, then sketch its graph.

45. center $(0, 0)$, radius 3

46. center $(0, 0)$, radius 6

47. center $(5, 0)$, radius $\sqrt{3}$

48. center $(0, 4)$, radius $\sqrt{5}$

49. center $(4, -3)$, radius 2

50. center $(3, -8)$, radius 9

51. center $(-7, -4)$, radius $\sqrt{7}$

52. center $(-2, -5)$, radius $\sqrt{6}$

53. center $(1, -2)$, diameter 6

54. center $(-2, 3)$, diameter 10

55. center $(4, 5)$, diameter $4\sqrt{3}$

56. center $(5, 1)$, diameter $4\sqrt{5}$

57. center at $(7, 1)$, graph contains the point $(1, -7)$

58. center at $(-8, 3)$, graph contains the point $(-3, 15)$

59. center at $(3, 4)$, graph contains the point $(7, 9)$

60. center at $(-5, 2)$, graph contains the point $(-1, 3)$

61. diameter has endpoints $(5, 1)$ and $(5, 7)$

62. diameter has endpoints $(2, 3)$ and $(8, 3)$

Identify the center and radius of each circle, then graph. Also state the domain and range of the relation.

63. $(x - 2)^2 + (y - 3)^2 = 4$

64. $(x - 5)^2 + (y - 1)^2 = 9$

65. $(x + 1)^2 + (y - 2)^2 = 12$

66. $(x - 7)^2 + (y + 4)^2 = 20$

67. $(x + 4)^2 + y^2 = 81$

68. $x^2 + (y - 3)^2 = 49$

Write each equation in standard form to find the center and radius of the circle. Then sketch the graph.

69. $x^2 + y^2 - 10x - 12y + 4 = 0$

70. $x^2 + y^2 + 6x - 8y - 6 = 0$

71. $x^2 + y^2 - 10x + 4y + 4 = 0$

72. $x^2 + y^2 + 6x + 4y + 12 = 0$

73. $x^2 + y^2 + 6y - 5 = 0$

74. $x^2 + y^2 - 8x + 12 = 0$

75. $x^2 + y^2 + 4x + 10y + 18 = 0$

76. $x^2 + y^2 - 8x - 14y - 47 = 0$

77. $x^2 + y^2 + 14x + 12 = 0$

78. $x^2 + y^2 - 22y - 5 = 0$

79. $2x^2 + 2y^2 - 12x + 20y + 4 = 0$

80. $3x^2 + 3y^2 - 24x + 18y + 3 = 0$

▶ WORKING WITH FORMULAS

81. Spending on Internet media: $s = 12.5t + 59$

The data from Example 1 is closely modeled by the formula shown, where t represents the year ($t = 0$ corresponds to the year 2000) and s represents the average amount spent per person, per year in the United States. (a) List five ordered pairs for this relation using $t = 1, 2, 3, 5, 7$. Does the model give a good approximation of the actual data? (b) According to the model, what will be the average amount spent on Internet media in the year 2008? (c) According to the model, in what year will annual spending surpass \$196? (d) Use the table to graph this relation.

82. Area of an inscribed square: $A = 2r^2$

The area of a square inscribed in a circle is found by using the formula given where r is the radius of the circle. Find the area of the inscribed square shown.

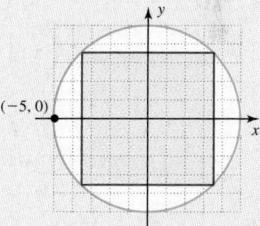

▶ APPLICATIONS

83. Radar detection: A luxury liner is located at map coordinates $(5, 12)$ and has a radar system with a range of 25 nautical miles in any direction. (a) Write the equation of the circle that models the range of the ship's radar, and (b) Use the distance formula to determine if the radar can pick up the liner's sister ship located at coordinates $(15, 36)$.

84. Earthquake range: The epicenter (point of origin) of a large earthquake was located at map coordinates $(3, 7)$, with the quake being felt up to 12 mi away. (a) Write the equation of the circle that models the range of the earthquake's effect. (b) Use the distance formula to determine if a person living at coordinates $(13, 1)$ would have felt the quake.

85. **Inscribed circle:** Find the equation for both the red and blue circles, then find the area of the region shaded in blue.

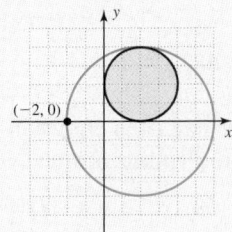

86. **Inscribed triangle:** The area of an equilateral triangle inscribed in a circle is given by the formula $A = \dfrac{3\sqrt{3}}{4}r^2$, where r is the radius of the circle. Find the area of the equilateral triangle shown.

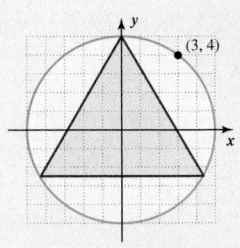

87. **Radio broadcast range:** Two radio stations may not use the same frequency if their broadcast areas *overlap.* Suppose station KXRQ has a broadcast area bounded by $x^2 + y^2 + 8x - 6y = 0$ and WLRT has a broadcast area bounded by $x^2 + y^2 - 10x + 4y = 0$. Graph the circle representing each broadcast area on the same grid to determine if both stations may broadcast on the same frequency.

88. **Radio broadcast range:** The emergency radio broadcast system is designed to alert the population by relaying an emergency signal to all points of the country. A signal is sent from a station whose broadcast area is bounded by $x^2 + y^2 = 2500$ (x and y in miles) and the signal is picked up and relayed by a transmitter with range $(x - 20)^2 + (y - 30)^2 = 900$. Graph the circle representing each broadcast area on the same grid to determine the greatest distance from the original station that this signal can be received. Be sure to scale the axes appropriately.

▶ **EXTENDING THE THOUGHT**

89. Although we use the word "domain" extensively in mathematics, it is also commonly seen in literature and heard in everyday conversation. Using a college-level dictionary, look up and write out the various meanings of the word, noting how closely the definitions given are related to its mathematical use.

90. Consider the following statement, then determine whether it is true or false and discuss why. *A graph will exhibit some form of symmetry if, given a point that is h units from the x-axis, k units from the y-axis, and d units from the origin, there is a second point*

on the graph that is a like distance from the origin and each axis.

91. When completing the square to find the center and radius of a circle, we sometimes encounter a value for r^2 that is negative or zero. These are called **degenerate cases.** If $r^2 < 0$, no circle is possible, while if $r^2 = 0$, the "graph" of the circle is simply the point (h, k). Find the center and radius of the following circles (if possible).

   a. $x^2 + y^2 - 12x + 4y + 40 = 0$
   b. $x^2 + y^2 - 2x - 8y - 8 = 0$
   c. $x^2 + y^2 - 6x - 10y + 35 = 0$

▶ **MAINTAINING YOUR SKILLS**

92. **(1.3)** Solve the absolute value inequality and write the solution in interval notation.

$$\frac{|w - 2|}{3} + \frac{1}{4} \geq \frac{5}{6}$$

93. **(R.1)** Give an example of each of the following:

   a. a whole number that is not a natural number
   b. a natural number that is not a whole number
   c. a rational number that is not an integer

   d. an integer that is not a rational number
   e. a rational number that is not a real number
   f. a real number that is not a rational number.

94. **(1.5)** Solve $x^2 + 13 = 6x$ using the quadratic equation. Simplify the result.

95. **(1.6)** Solve $1 - \sqrt{n + 3} = -n$ and check solutions by substitution. If a solution is extraneous, so state.

| **Graphs of Linear Equations**

Learning Objectives

In Section 2.2 you will learn how to:

☐ **A.** Graph linear equations using the intercept method

☐ **B.** Find the slope of a line

☐ **C.** Graph horizontal and vertical lines

☐ **D.** Identify parallel and perpendicular lines

☐ **E.** Apply linear equations in context

In preparation for sketching graphs of other relations, we'll first consider the characteristics of linear graphs. While linear graphs are fairly simple models, they have many substantive and meaningful applications. For instance, most of us are aware that music and video downloads have been increasing in popularity since they were first introduced. A close look at Example 1 of Section 2.1 reveals that spending on music downloads and Internet radio increased from $69 per person per year in 2001 to $145 in 2007 (Figure 2.13). From an investor's or a producer's point of view, there is a very high interest in the questions, How fast are sales increasing? Can this relationship be modeled mathematically to help predict sales in future years? Answers to these and other questions are precisely what our study in this section is all about.

Figure 2.13

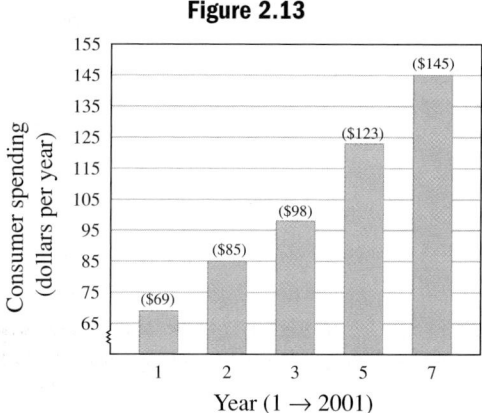

Source: 2006 SAUS

A. The Graph of a Linear Equation

A linear equation can be identified using these three tests: (1) the exponent on any variable is one, (2) no variable occurs in a denominator, and (3) no two variables are multiplied together. The equation $3y = 9$ is a linear equation in one variable, while $2x + 3y = 12$ and $y = -\frac{2}{3}x + 4$ are linear equations in two variables. In general, we have the following definition:

Linear Equations

A linear equation is one that can be written in the form

$$ax + by = c$$

where a and b are not simultaneously zero.

The most basic method for graphing a line is to simply plot a few points, then draw a straight line through the points.

EXAMPLE 1 ▶ **Graphing a Linear Equation in Two Variables**

Graph the equation $3x + 2y = 4$ by plotting points.

Solution ▶ Selecting $x = -2$, $x = 0$, $x = 1$, and $x = 4$ as inputs, we compute the related outputs and enter the ordered pairs in a table. The result is

x input	y output	(x, y) ordered pairs
-2	5	$(-2, 5)$
0	2	$(0, 2)$
1	0.5	$(1, \frac{1}{2})$
4	-4	$(4, -4)$

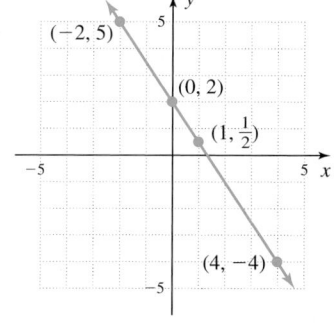

WORTHY OF NOTE

If you cannot draw a straight line through the plotted points, a computational error has been made. All points satisfying a linear equation *lie on a straight line.*

Now try Exercises 7 through 12 ▶

Note the line in Example 1 crosses the y-axis at $(0, 2)$, and this point is called the **y-intercept** of the line. In general, y-intercepts have the form $(0, y)$. Although difficult to see graphically, substituting 0 for y and solving for x shows the line crosses the x-axis at $(\frac{4}{3}, 0)$ and this point is called the **x-intercept.** In general, x-intercepts have the form $(x, 0)$. The x- and y-intercepts are usually easier to calculate than other points (since $y = 0$ or $x = 0$, respectively) and we often graph linear equations using only these two points. This is called the **intercept method** for graphing linear equations.

The Intercept Method

1. Substitute 0 for x and solve for y. This will give the y-intercept $(0, y)$.
2. Substitute 0 for y and solve for x. This will give the x-intercept $(x, 0)$.
3. Plot the intercepts and use them to graph a straight line.

EXAMPLE 2 ▶ **Graphing Lines Using the Intercept Method**

Graph $3x + 2y = 9$ using the intercept method.

Solution ▶   Substitute 0 for x (y-intercept)   Substitute 0 for y (x-intercept)

$$3(0) + 2y = 9 \qquad\qquad\qquad 3x + 2(0) = 9$$
$$2y = 9 \qquad\qquad\qquad\qquad 3x = 9$$
$$y = \frac{9}{2} \qquad\qquad\qquad\qquad x = 3$$
$$\qquad\qquad\qquad\qquad\qquad (3, 0)$$
$$\left(0, \frac{9}{2}\right)$$

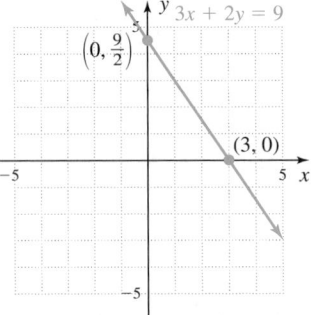

✓ **A.** You've just learned how to graph linear equations using the intercept method

Now try Exercises 13 through 32 ▶

B. The Slope of a Line

After the x- and y-intercepts, we next consider the **slope of a line.** We see applications of the concept in many diverse occupations, including the *grade* of a highway (trucking), the *pitch* of a roof (carpentry), the *climb* of an airplane (flying), the *drainage* of a field (landscaping), and the *slope* of a mountain (parks and recreation). While the general concept is an intuitive one, we seek to quantify the concept (assign it a numeric value) for purposes of comparison and decision making. In each of the preceding examples, slope is a measure of "steepness," as defined by the ratio $\frac{\text{vertical change}}{\text{horizontal change}}$. Using a line segment through arbitrary points $P_1 = (x_1, y_1)$ and $P_2 = (x_2, y_2)$, we can create the right triangle shown in Figure 2.14. The figure illustrates that the **vertical change** or the

Figure 2.14

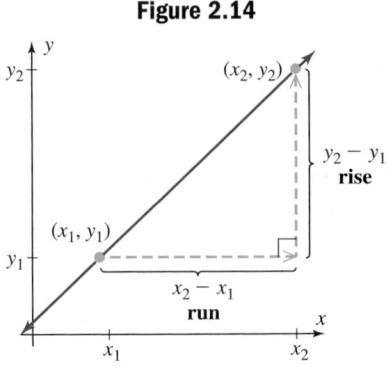

change in y (also called the **rise**) is simply the difference in y-coordinates: $y_2 - y_1$. The **horizontal change** or **change in x** (also called the **run**) is the difference in x-coordinates: $x_2 - x_1$. In algebra, we typically use the letter "m" to represent slope, giving $m = \frac{y_2 - y_1}{x_2 - x_1}$ as the $\frac{\text{change in } y}{\text{change in } x}$. The result is called the **slope formula.**

The Slope Formula

Given two points $P_1 = (x_1, y_1)$ and $P_2 = (x_2, y_2)$, the slope of any nonvertical line through P_1 and P_2 is

$$m = \frac{y_2 - y_1}{x_2 - x_1}$$

where $x_2 \neq x_1$.

EXAMPLE 3 ▶ Using the Slope Formula

Find the slope of the line through the given points.

   a. $(2, 1)$ and $(8, 4)$ **b.** $(-2, 6)$ and $(4, 2)$

Solution ▶   **a.** For $P_1 = (2, 1)$ and $P_2 = (8, 4)$,   **b.** For $P_1 = (-2, 6)$ and $P_2 = (4, 2)$,

$$m = \frac{y_2 - y_1}{x_2 - x_1} \qquad\qquad m = \frac{y_2 - y_1}{x_2 - x_1}$$

$$= \frac{4 - 1}{8 - 2} \qquad\qquad\qquad = \frac{2 - 6}{4 - (-2)}$$

$$= \frac{3}{6} = \frac{1}{2} \qquad\qquad\qquad = \frac{-4}{6} = \frac{-2}{3}$$

The slope of this line is $\frac{1}{2}$.   The slope of this line is $\frac{-2}{3}$.

Now try Exercises 33 through 40 ▶

⚠ CAUTION ▶ When using the slope formula, try to avoid these common errors.

1. The order that the x- and y-coordinates are subtracted must be consistent, since $\frac{y_2 - y_1}{x_2 - x_1} \neq \frac{y_2 - y_1}{x_1 - x_2}$.

2. The vertical change (involving the y-values) always occurs in the numerator: $\frac{y_2 - y_1}{x_2 - x_1} \neq \frac{x_2 - x_1}{y_2 - y_1}$.

3. When x_1 or y_1 is negative, use parentheses when substituting into the formula to prevent confusing the negative sign with the subtraction operation.

Actually, the slope value does much more than quantify the slope of a line, it expresses a **rate of change** between the quantities measured along each axis. In applications of slope, the ratio $\frac{\text{change in } y}{\text{change in } x}$ is symbolized as $\frac{\Delta y}{\Delta x}$. The symbol Δ is the Greek letter **delta** and has come to represent a change in some quantity, and the notation $m = \frac{\Delta y}{\Delta x}$ is read, "slope is equal to the *change in y* over the *change in x*." Interpreting slope as a rate of change has many significant applications in college algebra and beyond.

EXAMPLE 4 ▶ Interpreting the Slope Formula as a Rate of Change

Jimmy works on the assembly line for an auto parts remanufacturing company. By 9:00 A.M. his group has assembled 29 carburetors. By 12:00 noon, they have completed 87 carburetors. Assuming the relationship is linear, find the slope of the line and discuss its meaning in this context.

Solution ▶ First write the information as ordered pairs using c to represent the carburetors assembled and t to represent time. This gives $(t_1, c_1) = (9, 29)$ and $(t_2, c_2) = (12, 87)$. The slope formula then gives:

$$\frac{\Delta c}{\Delta t} = \frac{c_2 - c_1}{t_2 - t_1} = \frac{87 - 29}{12 - 9}$$

$$= \frac{58}{3} \text{ or } 19.\overline{3}$$

Here the slope ratio measures $\frac{\text{carburetors assembled}}{\text{hours}}$, and we see that Jimmy's group can assemble 58 carburetors every 3 hr, or about $19\frac{1}{3}$ carburetors per hour.

Now try Exercises 41 through 44 ▶

> **WORTHY OF NOTE**
>
> Actually, the assignment of (t_1, c_1) to $(9, 29)$ and (t_2, c_2) to $(12, 87)$ was arbitrary. The slope ratio will be the same *as long as the order of subtraction is the same.* In other words, if we reverse this assignment and use $(t_1, c_1) = (12, 87)$ and $(t_2, c_2) = (9, 29)$, we have $m = \frac{29 - 87}{9 - 12} = \frac{-58}{-3} = \frac{58}{3}$.

Positive and Negative Slope

If you've ever traveled by air, you've likely heard the announcement, "Ladies and gentlemen, please return to your seats and fasten your seat belts as we begin our descent." For a time, the descent of the airplane follows a linear path, but now the *slope of the line is negative* since the altitude of the plane is decreasing. Positive and negative slopes, as well as the rate of change they represent, are important characteristics of linear graphs. In Example 3a, the slope was a positive number ($m > 0$) and the line will slope upward from left to right since the y-values are increasing. If $m < 0$, the slope of the line is negative and the line slopes downward as you move left to right since y-values are decreasing.

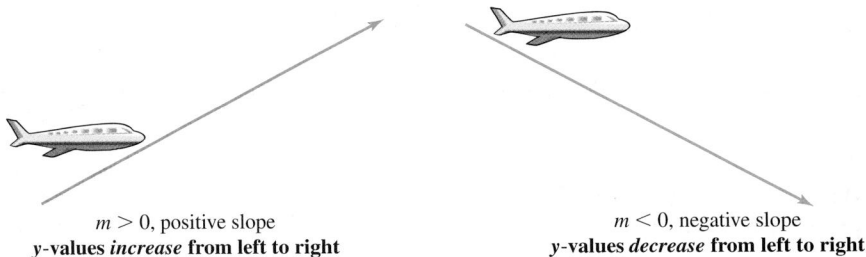

$m > 0$, positive slope
y-values *increase* **from left to right**

$m < 0$, negative slope
y-values *decrease* **from left to right**

EXAMPLE 5 ▶ **Applying Slope to Changes in Altitude**

At a horizontal distance of 10 mi after take-off, an airline pilot receives instructions to decrease altitude from their current level of 20,000 ft. A short time later, they are 17.5 mi from the airport at an altitude of 10,000 ft. Find the slope ratio for the descent of the plane and discuss its meaning in this context. Recall that 1 mi = 5280 ft.

Solution ▶ Let a represent the altitude of the plane and d its horizontal distance from the airport. Converting all measures to feet, we have $(d_1, a_1) = (52{,}800, 20{,}000)$ and $(d_2, a_2) = (92{,}400, 10{,}000)$, giving

$$\frac{\Delta a}{\Delta d} = \frac{a_2 - a_1}{d_2 - d_1} = \frac{10{,}000 - 20{,}000}{92{,}400 - 52{,}800}$$

$$= \frac{-10{,}000}{39{,}600} = \frac{-25}{99}$$

☑ **B.** You've just learned how to find the slope of a line

Since this slope ratio measures $\frac{\Delta \text{altitude}}{\Delta \text{distance}}$, we note the plane decreased 25 ft in altitude for every 99 ft it traveled horizontally.

Now try Exercises 45 through 48 ▶

C. Horizontal Lines and Vertical Lines

Horizontal and vertical lines have a number of important applications, from finding the boundaries of a given graph, to performing certain tests on nonlinear graphs. To better understand them, consider that in *one dimension*, the graph of $x = 2$ is a single point (Figure 2.15), indicating a location on the number line 2 units from zero in the positive direction. In *two dimensions*, the equation $x = 2$ represents **all points** with an x-coordinate of 2. A few of these are graphed in Figure 2.16, but since there are an infinite number, we end up with a solid *vertical line* whose equation is $x = 2$ (Figure 2.17).

Figure 2.15

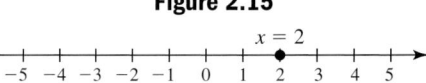

Figure 2.16 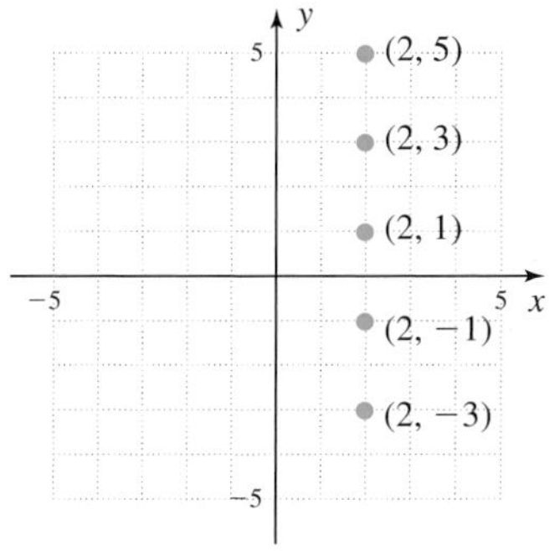   **Figure 2.17**

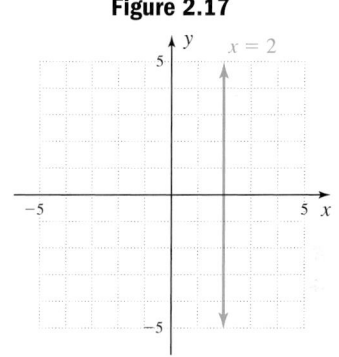

WORTHY OF NOTE

If we write the equation $x = 2$ in the form $ax + by = c$, the equation becomes $x + 0y = 2$, since the original equation has no y-variable. Notice that regardless of the value chosen for y, x will always be 2 and we end up with the set of ordered pairs $(2, y)$, which gives us a vertical line.

The same idea can be applied to horizontal lines. In *two dimensions*, the equation $y = 4$ represents *all points* with a y-coordinate of positive 4, and there are an infinite number of these as well. The result is a solid horizontal line whose equation is $y = 4$. **See Exercises 49–54.**

Vertical Lines	Horizontal Lines
The equation of a vertical line is	The equation of a horizontal line is
$x = h$	$y = k$
where $(h, 0)$ is the x-intercept.	where $(0, k)$ is the y-intercept.

So far, the slope formula has only been applied to lines that were nonhorizontal or nonvertical. So what *is* the slope of a horizontal line? On an intuitive level, we expect that a perfectly level highway would have an incline or slope of zero. In general, for any two points on a horizontal line, $y_2 = y_1$ and $y_2 - y_1 = 0$, giving a slope of $m = \frac{0}{x_2 - x_1} = 0$. For any two points on a vertical line, $x_2 = x_1$ and $x_2 - x_1 = 0$, making the slope ratio undefined: $m = \frac{y_2 - y_1}{0}$.

The Slope of a Vertical Line	The Slope of a Horizontal Line
The slope of any vertical line is undefined.	The slope of any horizontal line is zero.

EXAMPLE 6 ▶ **Calculating Slopes**

The federal minimum wage remained constant from 1997 through 2006. However, the buying power (in 1996 dollars) of these wage earners fell each year due to inflation (see Table 2.3). This decrease in buying power is approximated by the red line shown.

a. Using the data or graph, find the slope of the line segment representing the minimum wage.

b. Select two points on the line representing buying power to approximate the slope of the line segment, and explain what it means in this context.

Table 2.3

Time t (years)	Minimum wage w	Buying power p
1997	5.15	5.03
1998	5.15	4.96
1999	5.15	4.85
2000	5.15	4.69
2001	5.15	4.56
2002	5.15	4.49
2003	5.15	4.39
2004	5.15	4.28
2005	5.15	4.14
2006	5.15	4.04

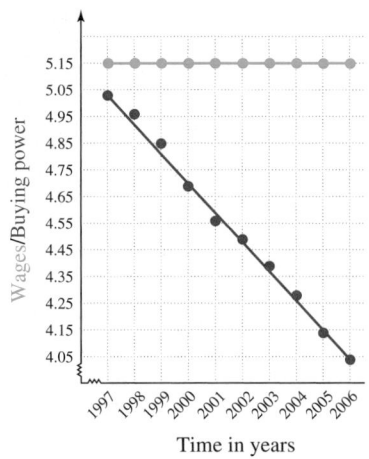

Solution ▶

a. Since the minimum wage did not increase or decrease from 1997 to 2006, the line segment has slope $m = 0$.

b. The points (1997, 5.03) and (2006, 4.04) from the table appear to be on or close to the line drawn. For buying power p and time t, the slope formula yields:

$$\frac{\Delta p}{\Delta t} = \frac{p_2 - p_1}{t_2 - t_1}$$

$$= \frac{4.04 - 5.03}{2006 - 1997}$$

$$= \frac{-0.99}{9} = \frac{-0.11}{1}$$

The buying power of a minimum wage worker decreased by 11¢ per year during this time period.

WORTHY OF NOTE

In the context of lines, try to avoid saying that a horizontal line has "no slope," since it's unclear whether a slope of zero or an undefined slope is intended.

☑ **C.** You've just learned how to graph horizontal and vertical lines

> **Now try Exercises 55 and 56 ▶**

D. Parallel and Perpendicular Lines

Two lines in the same plane that never intersect are called **parallel lines.** When we place these lines on the coordinate grid, we find that "never intersect" is equivalent to saying "the lines have equal slopes but different y-intercepts." In Figure 2.18, notice the rise and run of each line is identical, and that by counting $\frac{\Delta y}{\Delta x}$ both lines have slope $m = \frac{3}{4}$.

Figure 2.18

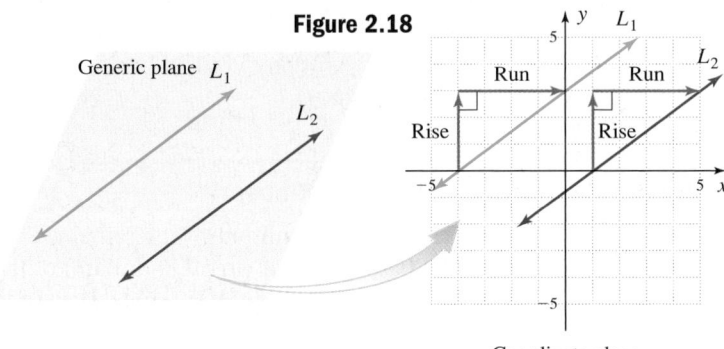

Parallel Lines

Given L_1 and L_2 are distinct, nonvertical lines with slopes of m_1 and m_2, respectively.
>   **1.** If $m_1 = m_2$, then L_1 is parallel to L_2.
>   **2.** If L_1 is parallel to L_2, then $m_1 = m_2$.
>   In symbols we write $L_1 \| L_2$.
>   *Any two vertical lines (undefined slope) are parallel.*

EXAMPLE 7A ▶ **Determining Whether Two Lines Are Parallel**

Teladango Park has been mapped out on a rectangular coordinate system, with a ranger station at $(0, 0)$. BJ and Kapi are at coordinates $(-24, -18)$ and have set a direct course for the pond at $(11, 10)$. Dave and Becky are at $(-27, 1)$ and are heading straight to the lookout tower at $(-2, 21)$. Are they hiking on parallel or nonparallel courses?

Solution ▶ To respond, we compute the slope of each trek across the park.

For BJ and Kapi:

$$m = \frac{y_2 - y_1}{x_2 - x_1}$$

$$= \frac{10 - (-18)}{11 - (-24)}$$

$$= \frac{28}{35} = \frac{4}{5}$$

For Dave and Becky:

$$m = \frac{y_2 - y_1}{x_2 - x_1}$$

$$= \frac{21 - 1}{-2 - (-27)}$$

$$= \frac{20}{25} = \frac{4}{5}$$

Since the slopes are equal, the couples are hiking on parallel courses.

Two lines in the same plane that intersect at right angles are called **perpendicular lines.** Using the coordinate grid, we note that *intersect at right angles* suggests that *their slopes are negative reciprocals.* From Figure 2.19, the ratio $\frac{\text{rise}}{\text{run}}$ for L_1 is $\frac{4}{3}$, the ratio $\frac{\text{rise}}{\text{run}}$ for L_2 is $\frac{-3}{4}$. Alternatively, we can say their **slopes have a product of -1,** since $m_1 \cdot m_2 = -1$ implies $m_1 = -\frac{1}{m_2}$.

Figure 2.19

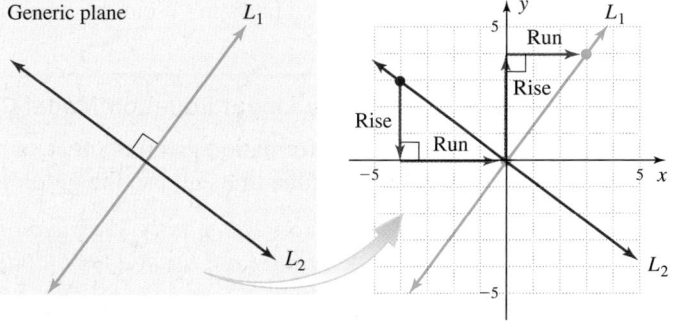

Coordinate plane

WORTHY OF NOTE

Since $m_1 \cdot m_2 = -1$ implies $m_1 = -\frac{1}{m_2}$, we can easily find the slope of a line perpendicular to a second line whose slope is given—just find the reciprocal and make it negative. For $m_1 = -\frac{3}{7}$ $m_2 = \frac{7}{3}$, and for $m_1 = -5$, $m_2 = \frac{1}{5}$.

Perpendicular Lines

Given L_1 and L_2 are distinct, nonvertical lines with slopes of m_1 and m_2, respectively.
>   **1.** If $m_1 \cdot m_2 = -1$, then L_1 is perpendicular to L_2.
>   **2.** If L_1 is perpendicular to L_2, then $m_1 \cdot m_2 = -1$.
>   In symbols we write $L_1 \perp L_2$.
>   *Any vertical line (undefined slope) is perpendicular*
>   *to any horizontal line (slope $m = 0$).*

EXAMPLE 7B ▶ **Determining Whether Two Lines Are Perpendicular**

The three points $P_1 = (5, 1)$, $P_2 = (3, -2)$, and $P_3 = (-3, 2)$ form the vertices of a triangle. Use these points to draw the triangle, then use the slope formula to determine if they form a *right* triangle.

Solution ▶ For a right triangle to be formed, two of the lines through these points must be perpendicular (forming a right angle). From Figure 2.20, it *appears* a right triangle is formed, but we must *verify* that two of the sides are perpendicular. Using the slope formula, we have:

Figure 2.20

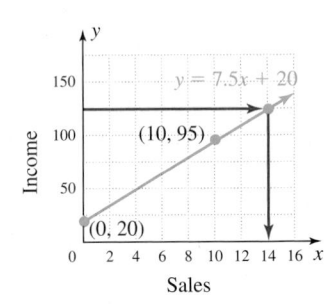

For P_1 and P_2

$$m_1 = \frac{-2 - 1}{3 - 5}$$

$$= \frac{-3}{-2} = \frac{3}{2}$$

For P_1 and P_3

$$m_2 = \frac{2 - 1}{-3 - 5}$$

$$= \frac{1}{-8}$$

For P_2 and P_3

$$m_3 = \frac{2 - (-2)}{-3 - 3}$$

$$= \frac{4}{-6} = \frac{2}{-3}$$

Since $m_1 \cdot m_3 = -1$, the triangle has a right angle and must be a right triangle.

☑ **D.** You've just learned how to identify parallel and perpendicular lines

Now try Exercises 57 through 68 ▶

E. Applications of Linear Equations

The graph of a linear equation can be used to help solve many applied problems. If the numbers you're working with are either very small or very large, **scale the axes** appropriately. This can be done by letting each tic mark represent a smaller or larger unit so the data points given will fit on the grid. Also, many applications use only nonnegative values and although points with negative coordinates may be used to graph a line, only ordered pairs in QI can be meaningfully interpreted.

EXAMPLE 8 ▶ **Applying a Linear Equation Model-Commission Sales**

Use the information given to create a linear equation model in two variables, then graph the line and use the graph to answer the question:

A salesperson gets a daily $20 meal allowance plus $7.50 for every item she sells. How many sales are needed for a daily income of $125?

Solution ▶ Let x represent sales and y represent income. This gives

verbal model: Daily income (y) equals $7.5 per sale ($x$) + $20 for meals

equation model: $y = 7.5x + 20$

Using $x = 0$ and $x = 10$, we find $(0, 20)$ and $(10, 95)$ are points on this graph. From the graph, we estimate that 14 sales are needed to generate a daily income of $125.00. Substituting $x = 14$ into the equation verifies that $(14, 125)$ is indeed on the graph:

☑ **E.** You've just learned how to apply linear equations in context

$$y = 7.5x + 20$$
$$= 7.5(14) + 20$$
$$= 105 + 20$$
$$= 125 ✓$$

Now try Exercises 71 through 74 ▶

TECHNOLOGY HIGHLIGHT

Linear Equations, Window Size, and Friendly Windows

To graph linear equations on the TI-84 Plus, we (1) solve the equation for the variable y, (2) enter the equation on the [Y=] screen, and (3) [GRAPH] the equation and adjust the [WINDOW] if necessary.

1. Solve the equation for y.
   For the equation $2x - 3y = -3$, we have

$2x - 3y = -3$	given equation
$-3y = -2x - 3$	subtract $2x$ from each side
$y = \dfrac{2}{3}x + 1$	divide both sides by -3

Figure 2.21

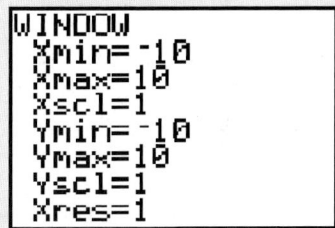

2. Enter the equation on the [Y=] screen.
   On the [Y=] screen, enter $\frac{2}{3}x + 1$. Note that for some calculators parentheses are needed to group $(2 \div 3)x$, to prevent the calculator from interpreting this term as $2 \div (3x)$.

3. [GRAPH] the equation, adjust the [WINDOW].
   Since much of our work is centered at $(0, 0)$ on the coordinate grid, the calculator's default settings have a domain of $x \in [-10, 10]$ and a range of $y \in [-10, 10]$, as shown in Figure 2.21. This is referred to as the [WINDOW] size. To graph the line in this window, it is easiest to use the [ZOOM] key and select **6:ZStandard,** which resets the window to these default settings. The graph is shown in Figure 2.22. The Xscl and Yscl

Figure 2.22

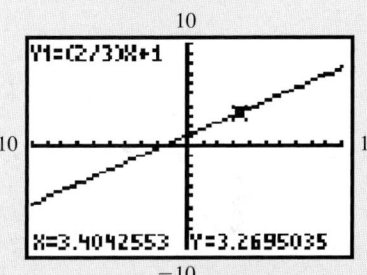

entries give the scale used on each axis, indicating that each "tic mark" represents 1 unit. Graphing calculators have many features that enable us to find ordered pairs on a line. One is the ([2nd] [GRAPH]) **(TABLE)** feature we have seen previously. We can also use the calculator's [TRACE] feature. As the name implies, this feature enables us to trace along the line by moving a blinking cursor using the left [◄] and right [►] arrow keys. The calculator simultaneously displays the coordinates of the current location of the cursor. After pressing the [TRACE] button, the cursor appears automatically— usually at the y-intercept. Moving the cursor left and right, note the coordinates changing at the bottom of the screen. The point $(3.4042553, 3.2695035)$ is on the line and satisfies the equation of the line. The calculator is displaying decimal values because the screen is exactly 95 pixels wide, 47 pixels to the left of the y-axis, and 47 pixels to the right. This means that each time you press the left or right arrow, the x-value changes by 1/47—which is *not* a nice round number. To [TRACE] through "friendlier" values, we can use the [ZOOM] **4:ZDecimal** feature, which sets Xmin $= -4.7$ and Xmax $= 4.7$, or **8:Zinteger,** which sets Xmin $= -47$ and Xmax $= 47$. Press [ZOOM] **4:ZDecimal** and the calculator will automatically regraph the line. Now when you [TRACE] the line, "friendly" decimal values are displayed.

Exercise 1: Use the [ZOOM] **4:ZDecimal** and **TRACE** features to identify the x- and y-intercepts for $Y_1 = \frac{2}{3}x + 1$.

Exercise 2: Use the [ZOOM] **8:Zinteger** and **TRACE** features to graph the line $79x - 55y = 869$, then identify the x- and y-intercepts.

2.2 EXERCISES

▶ **CONCEPTS AND VOCABULARY**

Fill in each blank with the appropriate word or phrase. Carefully reread the section if needed.

1. To find the x-intercept of a line, substitute _____ for y and solve for x. To find the y-intercept, substitute _____ for x and solve for y.

2. The slope formula is $m = $ _____ $= $ _____, and indicates a rate of change between the x- and y-variables.

3. If $m < 0$, the slope of the line is _____ and the line slopes _____ from left to right.

4. The slope of a horizontal line is _____, the slope of a vertical line is _____, and the slopes of two parallel lines are _____.

5. Discuss/Explain If $m_1 = 2.1$ and $m_2 = 2.01$, will the lines intersect? If $m_1 = \frac{2}{3}$ and $m_2 = -\frac{2}{3}$, are the lines perpendicular?

6. Discuss/Explain the relationship between the slope formula, the Pythagorean theorem, and the distance formula. Include several illustrations.

▶ **DEVELOPING YOUR SKILLS**

Create a table of values for each equation and sketch the graph.

7. $2x + 3y = 6$

x	y

8. $-3x + 5y = 10$

x	y

9. $y = \frac{3}{2}x + 4$

x	y

10. $y = \frac{5}{3}x - 3$

x	y

11. If you completed Exercise 9, verify that $(-3, -0.5)$ and $(\frac{1}{2}, \frac{19}{4})$ also satisfy the equation given. Do these points appear to be on the graph you sketched?

12. If you completed Exercise 10, verify that $(-1.5, -5.5)$ and $(\frac{11}{2}, \frac{37}{6})$ also satisfy the equation given. Do these points appear to be on the graph you sketched?

Graph the following equations using the intercept method. Plot a third point as a check.

13. $3x + y = 6$

14. $-2x + y = 12$

15. $5y - x = 5$

16. $-4y + x = 8$

17. $-5x + 2y = 6$

18. $3y + 4x = 9$

19. $2x - 5y = 4$

20. $-6x + 4y = 8$

21. $2x + 3y = -12$

22. $-3x - 2y = 6$

23. $y = -\frac{1}{2}x$

24. $y = \frac{2}{3}x$

25. $y - 25 = 50x$

26. $y + 30 = 60x$

27. $y = -\frac{2}{5}x - 2$

28. $y = \frac{3}{4}x + 2$

29. $2y - 3x = 0$

30. $y + 3x = 0$

31. $3y + 4x = 12$

32. $-2x + 5y = 8$

Compute the slope of the line through the given points, then graph the line and use $m = \frac{\Delta y}{\Delta x}$ to find two additional points on the line. Answers may vary.

33. $(3, 5), (4, 6)$

34. $(-2, 3), (5, 8)$

35. $(10, 3), (4, -5)$

36. $(-3, -1), (0, 7)$

37. $(1, -8), (-3, 7)$

38. $(-5, 5), (0, -5)$

39. $(-3, 6), (4, 2)$

40. $(-2, -4), (-3, -1)$

41. The graph shown models the relationship between the cost of a new home and the size of the home in square feet. (a) Determine the slope of the line and

interpret what the slope ratio means in this context and (b) estimate the cost of a 3000 ft² home.

Exercise 41

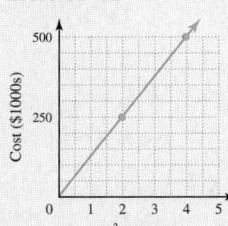

Exercise 42

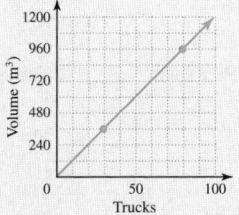

42. The graph shown models the relationship between the volume of garbage that is dumped in a landfill and the number of commercial garbage trucks that enter the site. (a) Determine the slope of the line and interpret what the slope ratio means in this context and (b) estimate the number of trucks entering the site daily if 1000 m³ of garbage is dumped per day.

43. The graph shown models the relationship between the distance of an aircraft carrier from its home port and the number of hours since departure. (a) Determine the slope of the line and interpret what the slope ratio means in this context and (b) estimate the distance from port after 8.25 hours.

Exercise 43

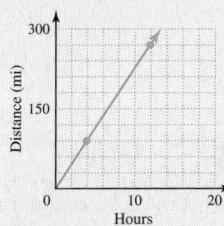

Exercise 44

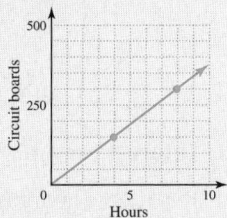

44. The graph shown models the relationship between the number of circuit boards that have been assembled at a factory and the number of hours since starting time. (a) Determine the slope of the line and interpret what the slope ratio means in this context and (b) estimate how many hours the factory has been running if 225 circuit boards have been assembled.

45. Height and weight: While there are many exceptions, numerous studies have shown a close relationship between an average height and average weight. Suppose a person 70 in. tall weighs 165 lb, while a person 64 in. tall weighs 142 lb. Assuming the relationship is linear, (a) find the slope of the line and discuss its meaning in this context and (b) determine how many pounds are added for each inch of height.

46. Rate of climb: Shortly after takeoff, a plane increases altitude at a constant (linear) rate. In 5 min the altitude is 10,000 feet. Fifteen minutes after takeoff, the plane has reached its cruising altitude of 32,000 ft. (a) Find the slope of the line and discuss its meaning in this context and (b) determine how long it takes the plane to climb from 12,200 feet to 25,400 feet.

47. Sewer line slope: Fascinated at how quickly the plumber was working, Ryan watched with great interest as the new sewer line was laid from the house to the main line, a distance of 48 ft. At the edge of the house, the sewer line was six in. under ground. If the plumber tied in to the main line at a depth of 18 in., what is the slope of the (sewer) line? What does this slope indicate?

48. Slope (pitch) of a roof: A contractor goes to a lumber yard to purchase some trusses (the triangular frames) for the roof of a house. Many sizes are available, so the contractor takes some measurements to ensure the roof will have the desired slope. In one case, the height of the truss (base to ridge) was 4 ft, with a width of 24 ft (eave to eave). Find the slope of the roof if these trusses are used. What does this slope indicate?

Graph each line using two or three ordered pairs that satisfy the equation.

49. $x = -3$ **50.** $y = 4$

51. $x = 2$ **52.** $y = -2$

Write the equation for each line L_1 and L_2 shown. Specifically state their point of intersection.

53. 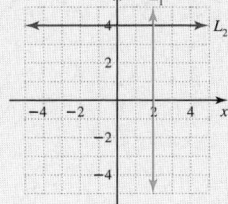   **54.**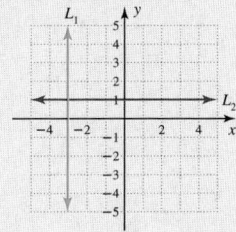

55. The table given shows the total number of justices j sitting on the Supreme Court of the United States for selected time periods t (in decades), along with the number of nonmale, nonwhite justices n for the same years. (a) Use the data to graph the linear relationship between t and j, then determine the slope of the line and discuss its meaning in this context. (b) Use the data to graph the linear relationship between t and n, then determine the slope of the line and discuss its meaning.

Exercise 55

Time t (1960 → 0)	Justices j	Nonwhite, nonmale n
0	9	0
10	9	1
20	9	2
30	9	3
40	9	4
50	9	5 (est)

56. The table shown gives the boiling temperature t of water as related to the altitude h. Use the data to graph the linear relationship between h and t, then determine the slope of the line and discuss its meaning in this context.

Exercise 56

Altitude h (ft)	Boiling Temperature t (°F)
0	212.0
1000	210.2
2000	208.4
3000	206.6
4000	204.8
5000	203.0
6000	201.2

Two points on L_1 and two points on L_2 are given. Use the slope formula to determine if lines L_1 and L_2 are parallel, perpendicular, or neither.

57. L_1: $(-2, 0)$ and $(0, 6)$
   L_2: $(1, 8)$ and $(0, 5)$

58. L_1: $(1, 10)$ and $(-1, 7)$
   L_2: $(0, 3)$ and $(1, 5)$

59. L_1: $(-3, -4)$ and $(0, 1)$
   L_2: $(0, 0)$ and $(-4, 4)$

60. L_1: $(6, 2)$ and $(8, -2)$
   L_2: $(5, 1)$ and $(3, 0)$

61. L_1: $(6, 3)$ and $(8, 7)$
   L_2: $(7, 2)$ and $(6, 0)$

62. L_1: $(-5, -1)$ and $(4, 4)$
   L_2: $(4, -7)$ and $(8, 10)$

In Exercises 63 to 68, three points that form the vertices of a triangle are given. Use the points to draw the triangle, then use the slope formula to determine if any of the triangles are right triangles. Also see Exercises 39–44 in Section 2.1.

63. $(5, 2)$, $(0, -3)$, $(4, -4)$

64. $(7, 0)$, $(-1, 0)$, $(7, 4)$

65. $(-4, 3)$, $(-7, -1)$, $(3, -2)$

66. $(-3, 7)$, $(2, 2)$, $(5, 5)$

67. $(-3, 2)$, $(-1, 5)$, $(-6, 4)$

68. $(0, 0)$, $(-5, 2)$, $(2, -5)$

▶ WORKING WITH FORMULAS

69. Human life expectancy: $L = 0.11T + 74.2$

The average number of years that human beings live has been steadily increasing over the years due to better living conditions and improved medical care. This relationship is modeled by the formula shown, where L is the average life expectancy and T is number of years since 1980. (a) What was the life expectancy in the year 2000? (b) In what year will average life expectancy reach 77.5 yr?

70. Interest earnings: $I = \left(\dfrac{7}{100}\right)(5000)T$

If $5000 dollars is invested in an account paying 7% simple interest, the amount of interest earned is given by the formula shown, where I is the interest and T is the time in years. (a) How much interest is earned in 5 yr? (b) How much is earned in 10 yr? (c) Use the two points (5 yr, interest) and (10 yr, interest) to calculate the slope of this line. What do you notice?

▶ APPLICATIONS

For exercises 71 to 74, use the information given to build a linear equation model, then use the equation to respond.

71. Business depreciation: A business purchases a copier for $8500 and anticipates it will depreciate in value $1250 per year.

   a. What is the copier's value after 4 yr of use?

   b. How many years will it take for this copier's value to decrease to $2250?

72. Baseball card value: After purchasing an autographed baseball card for $85, its value increases by $1.50 per year.

   a. What is the card's value 7 yr after purchase?

   b. How many years will it take for this card's value to reach $100?

73. Water level: During a long drought, the water level in a local lake decreased at a rate of 3 in. per month. The water level before the drought was 300 in.

a. What was the water level after 9 months of drought?

b. How many months will it take for the water level to decrease to 20 ft?

74. **Gas mileage:** When empty, a large dump-truck gets about 15 mi per gallon. It is estimated that for each 3 tons of cargo it hauls, gas mileage decreases by $\frac{3}{4}$ mi per gallon.

   a. If 10 tons of cargo is being carried, what is the truck's mileage?

   b. If the truck's mileage is down to 10 mi per gallon, how much weight is it carrying?

75. **Parallel/nonparallel roads:** Aberville is 38 mi north and 12 mi west of Boschertown, with a straight road "farm and machinery road" (FM 1960) connecting the two cities. In the next county, Crownsburg is 30 mi north and 9.5 mi west of Dower, and these cities are likewise connected by a straight road (FM 830). If the two roads continued indefinitely in both directions, would they intersect at some point?

76. **Perpendicular/nonperpendicular course headings:** Two shrimp trawlers depart Charleston Harbor at the same time. One heads for the shrimping grounds located 12 mi north and 3 mi east of the harbor. The other heads for a point 2 mi south and 8 mi east of the harbor. Assuming the harbor is at (0, 0), are the routes of the trawlers perpendicular? If so, how far apart are the boats when they reach their destinations (to the nearest one-tenth mi)?

77. **Cost of college:** For the years 1980 to 2000, the cost of tuition and fees per semester (in constant dollars) at a public 4-yr college can be approximated by the equation $y = 144x + 621$, where y represents the cost in dollars and $x = 0$

represents the year 1980. Use the equation to find: (a) the cost of tuition and fees in 2002 and (b) the year this cost will exceed $5250.

Source: 2001 New York Times Almanac, p. 356

78. **Female physicians:** In 1960 only about 7% of physicians were female. Soon after, this percentage began to grow dramatically. For the years 1980 to 2002, the percentage of physicians that were female can be approximated by the equation $y = 0.72x + 11$, where y represents the percentage (as a whole number) and $x = 0$ represents the year 1980. Use the equation to find: (a) the percentage of physicians that were female in 1992 and (b) the projected year this percentage will exceed 30%.

Source: Data from the 2004 Statistical Abstract of the United States, Table 149

79. **Decrease in smokers:** For the years 1980 to 2002, the percentage of the U.S. adult population who were smokers can be approximated by the equation $y = -\frac{7}{15}x + 32$, where y represents the percentage of smokers (as a whole number) and $x = 0$ represents 1980. Use the equation to find: (a) the percentage of adults who smoked in the year 2000 and (b) the year the percentage of smokers is projected to fall below 20%.

Source: Statistical Abstract of the United States, various years

80. **Temperature and cricket chirps:** Biologists have found a strong relationship between temperature and the number of times a cricket chirps. This is modeled by the equation $T = \frac{N}{4} + 40$, where N is the number of times the cricket chirps per minute and T is the temperature in Fahrenheit. Use the equation to find: (a) the outdoor temperature if the cricket is chirping 48 times per minute and (b) the number of times a cricket chirps if the temperature is 70°.

▶ **EXTENDING THE CONCEPT**

81. If the lines $4y + 2x = -5$ and $3y + ax = -2$ are perpendicular, what is the value of a?

82. Let m_1, m_2, m_3, and m_4 be the slopes of lines L_1, L_2, L_3, and L_4, respectively. Which of the following statements is true?

   a. $m_4 < m_1 < m_3 < m_2$

   b. $m_3 < m_2 < m_4 < m_1$

   c. $m_3 < m_4 < m_2 < m_1$

   d. $m_1 < m_3 < m_4 < m_2$

   e. $m_1 < m_4 < m_3 < m_2$

83. An *arithmetic sequence* is a sequence of numbers where each successive term is found by adding a

fixed constant, called the common difference d, to the preceding term. For instance 3, 7, 11, 15, . . . is an arithmetic sequence with $d = 4$. The formula for the "nth term" t_n of an arithmetic sequence is a linear equation of the form $t_n = t_1 + (n - 1)d$, where d is the common difference and t_1 is the first term of the sequence. Use the equation to find the term specified for each sequence.

   a. 2, 9, 16, 23, 30, . . . ; 21st term

   b. 7, 4, 1, −2, −5, . . . ; 31st term

   c. 5.10, 5.25, 5.40, 5.55, . . . ; 27th term

   d. $\frac{3}{2}, \frac{9}{4}, 3, \frac{15}{4}, \frac{9}{2}, \ldots$; 17th term

▶ **MAINTAINING YOUR SKILLS**

84. (1.1) Simplify the equation, then solve. Check your answer by substitution:
$3x^2 - 3 + 4x + 6 = 4x^2 - 3(x + 5)$

85. (R.7) Identify the following formulas:

$$P = 2L + 2W \qquad V = LWH$$
$$V = \pi r^2 h \qquad C = 2\pi r$$

86. (1.1) How many gallons of a 35% brine solution must be mixed with 12 gal of a 55% brine solution in order to get a 45% solution?

87. (1.1) Two boats leave the harbor at Lahaina, Maui, going in opposite directions. One travels at 15 mph and the other at 20 mph. How long until they are 70 mi apart?

2.3 Linear Graphs and Rates of Change

Learning Objectives

In Section 2.3 you will learn how to:

☐ **A.** Write a linear equation in slope-intercept form

☐ **B.** Use slope-intercept form to graph linear equations

☐ **C.** Write a linear equation in point-slope form

☐ **D.** Apply the slope-intercept form and point-slope form in context

The concept of slope is an important part of mathematics, because it gives us a way to measure and compare change. The value of an automobile changes with time, the circumference of a circle increases as the radius increases, and the tension in a spring grows the more it is stretched. The real world is filled with examples of how one change affects another, and slope helps us understand how these changes are related.

A. Linear Equations and Slope-Intercept Form

In Section 1.1, formulas and literal equations were written in an alternate form by solving for an object variable. The new form made using the formula more efficient. Solving for y in equations of the form $ax + by = c$ offers similar advantages to linear graphs and their applications.

EXAMPLE 1 ▶ **Solving for y in ax + by = c**

Solve $2y - 6x = 4$ for y, then evaluate at $x = 4$, $x = 0$, and $x = -\frac{1}{3}$.

Solution ▶ $\begin{aligned} 2y - 6x &= 4 && \text{given equation} \\ 2y &= 6x + 4 && \text{add } 6x \\ y &= 3x + 2 && \text{divide by 2} \end{aligned}$

Since the coefficients are integers, evaluate the function mentally. Inputs are multiplied by 3, then increased by 2, yielding the ordered pairs (4, 14), (0, 2), and $\left(-\frac{1}{3}, 1\right)$.

Now try Exercises 7 through 12 ▶

This form of the equation (where y has been written in terms of x) enables us to quickly identify what operations are performed on x in order to obtain y. For $y = 3x + 2$, *multiply inputs by 3, then add 2.*

EXAMPLE 2 ▶ **Solving for y in ax + by = c**

Solve the linear equation $3y - 2x = 6$ for y, then identify the new coefficient of x and the constant term.

Solution ▶ $3y - 2x = 6$ given equation

$\qquad 3y = 2x + 6$ add 2x

$\qquad y = \dfrac{2}{3}x + 2$ divide by 3

The new coefficient of x is $\frac{2}{3}$ and the constant term is 2.

> **Now try Exercises 13 through 18 ▶**

WORTHY OF NOTE

In Example 2, the final form can be written $y = \frac{2}{3}x + 2$ as shown (inputs are multiplied by two-thirds, then increased by 2), or written as $y = \dfrac{2x}{3} + 2$ (inputs are multiplied by two, the result divided by 3 and this amount increased by 2). The two forms are equivalent.

When the coefficient of x is rational, it's helpful to select inputs that are multiples of the denominator if the context or application requires us to evaluate the equation. This enables us to perform most operations mentally. For $y = \frac{2}{3}x + 2$, possible inputs might be $x = -9, -6, 0, 3, 6$, and so on. **See Exercises 19 through 24.**

In Section 2.2, linear equations were graphed using the intercept method. When a linear equation is written with y in terms of x, we notice a powerful connection between the graph and its equation, and one that highlights the primary characteristics of a linear graph.

EXAMPLE 3 ▶ **Noting Relationships between an Equation and Its Graph**

Find the intercepts of $4x + 5y = -20$ and use them to graph the line. Then,

a. Use the intercepts to calculate the slope of the line, then

b. Write the equation with y in terms of x and compare the calculated slope and y-intercept to the equation in this form. Comment on what you notice.

Solution ▶ Substituting 0 for x in $4x + 5y = -20$, we find the y-intercept is $(0, -4)$. Substituting 0 for y gives an x-intercept of $(-5, 0)$. The graph is displayed here.

a. By calculation or counting $\dfrac{\Delta y}{\Delta x}$, the slope is $m = -\frac{4}{5}$.

b. Solving for y:

$\qquad 4x + 5y = -20$ given equation

$\qquad 5y = -4x - 20$ subtract 4x

$\qquad y = -\dfrac{4}{5}x - 4$ divide by 5

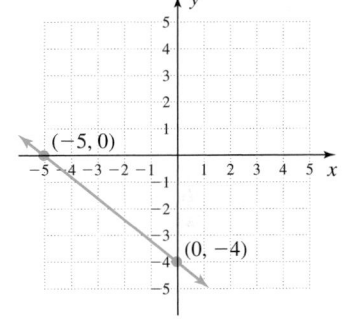

The slope value seems to be the coefficient of x, while the y-intercept is the constant term.

> **A.** You've just learned how to write a linear equation in slope-intercept form

> **Now try Exercises 25 through 30 ▶**

B. Slope-Intercept Form and the Graph of a Line

After solving a linear equation for y, an input of $x = 0$ causes the "x-term" to become zero, so the y-intercept is automatically the constant term. As Example 3 illustrates, we can also identify the slope of the line—it is the coefficient of x. In general, a linear equation of the form $y = mx + b$ is said to be in **slope-intercept form**, since the slope of the line is m and the y-intercept is $(0, b)$.

Slope-Intercept Form

For a nonvertical line whose equation is $y = mx + b$,
the slope of the line is m and the y-intercept is $(0, b)$.

EXAMPLE 4 ▶ **Finding the Slope-Intercept Form**

Write each equation in slope-intercept form and identify the slope and y-intercept of each line.

a. $3x - 2y = 9$ b. $y + x = 5$ c. $2y = x$

Solution ▶ a. $3x - 2y = 9$ b. $y + x = 5$ c. $2y = x$

$$-2y = -3x + 9 \qquad\qquad y = -x + 5 \qquad\qquad y = \frac{x}{2}$$

$$y = \frac{3}{2}x - \frac{9}{2} \qquad\qquad y = -1x + 5 \qquad\qquad y = \frac{1}{2}x$$

$$m = \frac{3}{2}, b = -\frac{9}{2} \qquad m = -1, b = 5 \qquad m = \frac{1}{2}, b = 0$$

$$y\text{-intercept}\left(0, -\frac{9}{2}\right) \qquad y\text{-intercept}\ (0, 5) \qquad y\text{-intercept}\ (0, 0)$$

Now try Exercises 31 through 38 ▶

If the slope and y-intercept of a linear equation are known or can be found, we can construct its equation by substituting these values directly into the slope-intercept form $y = mx + b$.

EXAMPLE 5 ▶ **Finding the Equation of a Line from Its Graph**

Find the slope-intercept form of the line shown.

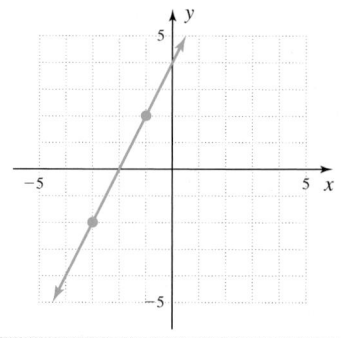

Solution ▶ Using $(-3, -2)$ and $(-1, 2)$ in the slope formula, or by simply counting $\dfrac{\Delta y}{\Delta x}$, the slope is $m = \frac{4}{2}$ or $\frac{2}{1}$.

By inspection we see the y-intercept is $(0, 4)$. Substituting $\frac{2}{1}$ for m and 4 for b in the slope-intercept form we obtain the equation $y = 2x + 4$.

Now try Exercises 39 through 44 ▶

Actually, if the slope is known and we have *any* point (x, y) on the line, we can still construct the equation since the given point *must satisfy the equation of the line*. In this case, we're treating $y = mx + b$ as a simple formula, solving for b after substituting known values for m, x, and y.

EXAMPLE 6 ▶ **Using $y = mx + b$ as a Formula**

Find the equation of a line that has slope $m = \frac{4}{5}$ and contains $(-5, 2)$.

Solution ▶ Using $y = mx + b$ as a "formula," we have $m = \frac{4}{5}$, $x = -5$, and $y = 2$.

$$y = mx + b \qquad \text{slope-intercept form}$$
$$2 = \tfrac{4}{5}(-5) + b \qquad \text{substitute } \tfrac{4}{5} \text{ for } m, -5 \text{ for } x, \text{ and } 2 \text{ for } y$$
$$2 = -4 + b \qquad \text{simplify}$$
$$6 = b \qquad \text{solve for } b$$

The equation of the line is $y = \frac{4}{5}x + 6$.

Now try Exercises 45 through 50 ▶

Writing a linear equation in slope-intercept form enables us to draw its graph with a minimum of effort, since we can easily locate the y-intercept and a second point using $m = \dfrac{\Delta y}{\Delta x}$. For instance, $\dfrac{\Delta y}{\Delta x} = \dfrac{-2}{3}$ means count down 2 and right 3 from a known point.

EXAMPLE 7 ▶ **Graphing a Line Using Slope-Intercept Form**

Write $3y - 5x = 9$ in slope-intercept form, then graph the line using the y-intercept and slope.

Solution ▶

$$3y - 5x = 9 \qquad \text{given equation}$$
$$3y = 5x + 9 \qquad \text{isolate } y \text{ term}$$
$$y = \tfrac{5}{3}x + 3 \qquad \text{divide by 3}$$

The slope is $m = \tfrac{5}{3}$ and the y-intercept is $(0, 3)$.

Plot the y-intercept, then use $\dfrac{\Delta y}{\Delta x} = \dfrac{5}{3}$ (up 5 and

right 3—shown in blue) to find another point on the line (shown in red). Finish by drawing a line through these points.

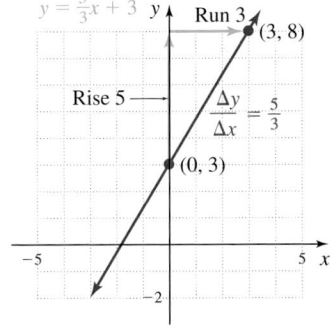

Now try Exercises 51 through 62 ▶

WORTHY OF NOTE

Noting the fraction $\tfrac{5}{3}$ is equal to $\tfrac{-5}{-3}$, we could also begin at $(0, 3)$ and count $\dfrac{\Delta y}{\Delta x} = \dfrac{-5}{-3}$ (down 5 and left 3) to find an additional point on the line: $(-3, -2)$. Also, for any negative slope $\dfrac{\Delta y}{\Delta x} = -\dfrac{a}{b}$, note $-\dfrac{a}{b} = \dfrac{-a}{b} = \dfrac{a}{-b}$.

For a discussion of what graphing method might be most efficient for a given linear equation, **see Exercises 103 and 115.**

Parallel and Perpendicular Lines

From Section 2.2 we know parallel lines have equal slopes: $m_1 = m_2$, and perpendicular lines have slopes with a product of -1: $m_1 \cdot m_2 = -1$ or $m_1 = -\dfrac{1}{m_2}$. In some applications, we need to find the equation of a second line parallel or perpendicular to a given line, through a given point. Using the slope-intercept form makes this a simple four-step process.

Finding the Equation of a Line Parallel or Perpendicular to a Given Line

1. Identify the slope m_1 of the given line.
2. Find the slope m_2 of the new line using the parallel or perpendicular relationship.
3. Use m_2 with the point (x, y) in the "formula" $y = mx + b$ and solve for b.
4. The desired equation will be $y = m_2x + b$.

EXAMPLE 8 ▶ **Finding the Equation of a Parallel Line**

Find the equation of a line that goes through $(-6, -1)$ and is parallel to $2x + 3y = 6$.

Solution ▶ Begin by writing the equation in slope-intercept form to identify the slope.

$$2x + 3y = 6 \qquad \text{given line}$$
$$3y = -2x + 6 \qquad \text{isolate } y \text{ term}$$
$$y = \tfrac{-2}{3}x + 2 \qquad \text{result}$$

The original line has slope $m_1 = \frac{-2}{3}$ and this will also be the slope of any line parallel to it. Using $m_2 = \frac{-2}{3}$ with $(x, y) \rightarrow (-6, -1)$ we have

$$y = mx + b \qquad \text{slope-intercept form}$$

$$-1 = \frac{-2}{3}(-6) + b \qquad \text{substitute } \tfrac{-2}{3} \text{ for } m, -6 \text{ for } x, \text{ and } -1 \text{ for } y$$

$$-1 = 4 + b \qquad \text{simplify}$$

$$-5 = b \qquad \text{solve for } b$$

The equation of the new line is $y = \frac{-2}{3}x - 5$.

Now try Exercises 63 through 76 ▶

GRAPHICAL SUPPORT

Graphing the lines from Example 8 as Y1 and Y2 on a graphing calculator, we note the lines do appear to be parallel (they actually *must* be since they have identical slopes). Using the ZOOM **8:ZInteger** feature of the TI-84 Plus we can quickly verify that Y2 indeed contains the point $(-6, -1)$.

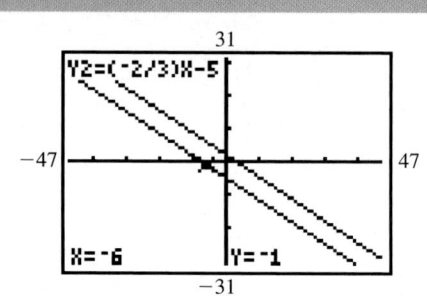

For any nonlinear graph, a straight line drawn through two points on the graph is called a **secant line.** The slope of the secant line, and lines parallel and perpendicular to this line, play fundamental roles in the further development of the rate-of-change concept.

EXAMPLE 9 ▶ **Finding Equations for Parallel and Perpendicular Lines**

A secant line is drawn using the points $(-4, 0)$ and $(2, -2)$ on the graph of the function shown. Find the equation of a line that is:

a. parallel to the secant line through $(-1, -4)$

b. perpendicular to the secant line through $(-1, -4)$.

Solution ▶ Either by using the slope formula or counting $\dfrac{\Delta y}{\Delta x}$, we find the secant line has slope

$$m = \frac{-2}{6} = \frac{-1}{3}.$$

a. For the parallel line through $(-1, -4)$, $m_2 = \dfrac{-1}{3}$.

$$y = mx + b \qquad \text{slope-intercept form}$$

$$-4 = \frac{-1}{3}(-1) + b \qquad \substack{\text{substitute } \tfrac{-1}{3} \text{ for } m, \\ -1 \text{ for } x, \text{ and } -4 \text{ for } y}$$

$$-\frac{12}{3} = \frac{1}{3} + b \qquad \text{simplify}$$

$$-\frac{13}{3} = b \qquad \text{result}$$

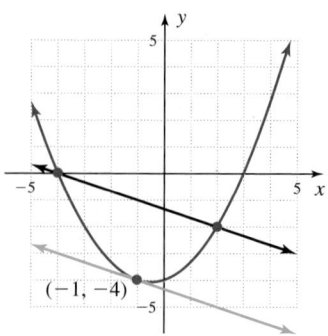

> **WORTHY OF NOTE**
>
> The word "secant" comes from the Latin word *secare,* meaning "to cut." Hence a secant line is one that cuts through a graph, as opposed to a tangent line, which touches the graph at only one point.

The equation of the parallel line (in blue) is $y = \dfrac{-1}{3}x - \dfrac{13}{3}$.

b. For the line perpendicular through $(-1, -4)$, $m_2 = 3$.

$$y = mx + b \qquad \text{slope-intercept form}$$
$$-4 = 3(-1) + b \qquad \text{substitute 3 for } m, -1 \text{ for } x, \text{ and } -4 \text{ for } y$$
$$-4 = -3 + b \qquad \text{simplify}$$
$$-1 = b \qquad \text{result}$$

The equation of the perpendicular line (in yellow) is $y = 3x - 1$.

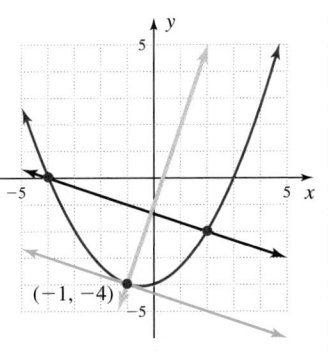

☑ **B.** You've just learned how to use the slope-intercept form to graph linear equations

> Now try Exercises 77 through 82 ▶

C. Linear Equations in Point-Slope Form

As an alternative to using $y = mx + b$, we can find the equation of the line using the slope formula $\dfrac{y_2 - y_1}{x_2 - x_1} = m$, and the fact that *the slope of a line is constant.* For a given slope m, we can let (x_1, y_1) represent a *given* point on the line and (x, y) represent *any other point* on the line, and the formula becomes $\dfrac{y - y_1}{x - x_1} = m$. Isolating the "$y$" terms on one side gives a new form for the equation of a line, called the **point-slope form:**

$$\frac{y - y_1}{x - x_1} = m \qquad \text{slope formula}$$

$$\frac{(x - x_1)}{1} \left(\frac{y - y_1}{x - x_1} \right) = m(x - x_1) \qquad \text{multiply both sides by } (x - x_1)$$

$$y - y_1 = m(x - x_1) \qquad \text{simplify} \rightarrow \text{point-slope form}$$

The Point-Slope Form of a Linear Equation

For a nonvertical line whose equation is $y - y_1 = m(x - x_1)$, the slope of the line is m and (x_1, y_1) is a point on the line.

While using $y = mx + b$ as in Example 6 may appear to be easier, both the y-intercept form and point-slope form have their own advantages and it will help to be familiar with both.

EXAMPLE 10 ▶ **Using $y - y_1 = m(x - x_1)$ as a Formula**

Find the equation of a line in point-slope form, if $m = \frac{2}{3}$ and $(-3, -3)$ is on the line. Then graph the line.

Solution ▶

$$y - y_1 = m(x - x_1) \qquad \text{point-slope form}$$

$$y - (-3) = \frac{2}{3}[x - (-3)] \qquad \text{substitute } \tfrac{2}{3} \text{ for } m; (-3, -3) \text{ for } (x_1, y_1)$$

$$y + 3 = \frac{2}{3}(x + 3) \qquad \text{simplify, point-slope form}$$

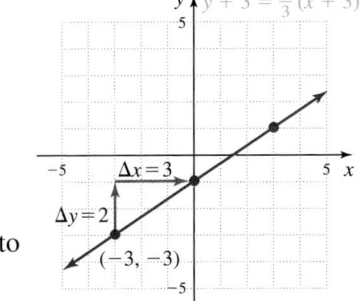

To graph the line, plot $(-3, -3)$ and use $\dfrac{\Delta y}{\Delta x} = \dfrac{2}{3}$ to find additional points on the line.

☑ **C.** You've just learned how to write a linear equation in point-slope form

> Now try Exercises 83 through 94 ▶

D. Applications of Linear Equations

As a mathematical tool, linear equations rank among the most common, powerful, and versatile. In all cases, it's important to remember that slope represents a *rate of change*. The notation $m = \dfrac{\Delta y}{\Delta x}$ literally means the quantity measured along the *y*-axis, is changing with respect to changes in the quantity measured along the *x*-axis.

EXAMPLE 11 ▶ **Relating Temperature to Altitude**

In meteorological studies, atmospheric temperature depends on the altitude according to the formula $T = -3.5h + 58.6$, where *T* represents the approximate Fahrenheit temperature at height *h* (in thousands of feet).

   a. Interpret the meaning of the slope in this context.

   b. Determine the temperature at an altitude of 12,000 ft.

   c. If the temperature is $-10°F$ what is the approximate altitude?

Solution ▶   **a.** Notice that *h* is the input variable and *T* is the output. This shows $\dfrac{\Delta T}{\Delta h} = \dfrac{-3.5}{1}$, meaning the temperature drops 3.5°F for every 1000-ft increase in altitude.

   b. Since height is in thousands, use $h = 12$.

$$
\begin{aligned}
T &= -3.5h + 58.6 && \text{original function} \\
&= -3.5(12) + 58.6 && \text{substitute 12 for } h \\
&= 16.6 && \text{result}
\end{aligned}
$$

At a height of 12,000 ft, the temperature is about 17°F.

   c. Replacing *T* with -10 and solving gives

$$
\begin{aligned}
-10 &= -3.5h + 58.6 && \text{substitute } -10 \text{ for } T \\
-68.6 &= -3.5h && \text{simplify} \\
19.6 &= h && \text{result}
\end{aligned}
$$

The temperature is $-10°F$ at a height of $19.6 \times 1000 = 19{,}600$ ft.

Now try Exercises 105 and 106 ▶

In some applications, the relationship is known to be linear but only a few points on the line are given. In this case, we can use two of the known data points to calculate the slope, then the point-slope form to find an equation model. One such application is *linear depreciation*, as when a government allows businesses to depreciate vehicles and equipment over time (the less a piece of equipment is worth, the less you pay in taxes).

EXAMPLE 12A ▶ **Using Point-Slope Form to Find an Equation Model**

Five years after purchase, the auditor of a newspaper company estimates the value of their printing press is $60,000. Eight years after its purchase, the value of the press had depreciated to $42,000. Find a linear equation that models this depreciation and discuss the slope and *y*-intercept in context.

Solution ▶ Since the value of the press depends on time, the ordered pairs have the form (time, value) or (t, v) where *time* is the input, and *value* is the output. This means the ordered pairs are (5, 60,000) and (8, 42,000).

$$m = \frac{v_2 - v_1}{t_2 - t_1} \qquad \text{slope formula}$$

$$= \frac{42,000 - 60,000}{8 - 5} \qquad (t_1, v_1) = (5, 60,000); \ (t_2, v_2) = (8, 42,000)$$

$$= \frac{-18,000}{3} = \frac{-6000}{1} \qquad \text{simplify and reduce}$$

The slope of the line is $\dfrac{\Delta \text{value}}{\Delta \text{time}} = \dfrac{-6000}{1}$, indicating the printing press loses $6000 in value with each passing year.

$$v - v_1 = m(t - t_1) \qquad \text{point-slope form}$$

$$v - 60,000 = -6000(t - 5) \qquad \text{substitute } -6000 \text{ for } m; (5, 60,000) \text{ for } (t_1, v_1)$$

$$v - 60,000 = -6000t + 30,000 \quad \text{simplify}$$

$$v = -6000t + 90,000 \quad \text{solve for } v$$

The depreciation equation is $v = -6000t + 90,000$. The v-intercept (0, 90,000) indicates the original value (cost) of the equipment was $90,000.

Once the depreciation equation is found, it represents the (time, value) relationship for all future (and intermediate) ages of the press. In other words, we can now predict the value of the press for any given year. However, note that some equation models are valid for only a set period of time, and each model should be used with care.

EXAMPLE 12B ▶ **Using an Equation Model to Gather Information**

From Example 12A,
   a. How much will the press be worth after 11 yr?
   b. How many years until the value of the equipment is less than $9,000?
   c. Is this equation model valid for $t = 18$ yr (why or why not)?

Solution ▶ **a.** Find the value v when $t = 11$:

$$v = -6000t + 90,000 \qquad \text{equation model}$$

$$v = -6000(11) + 90,000 \quad \text{substitute 11 for } t$$

$$= 24,000 \qquad\qquad\qquad \text{result (11, 24,000)}$$

After 11 yr, the printing press will only be worth $24,000.

b. "... value is less than $9000" means $v < 9000$:

$$v < 9000 \qquad \text{value at time } t$$

$$-6000t + 90,000 < 9000 \qquad \text{substitute } -6000t + 90,000 \text{ for } v$$

$$-6000t < -81,000 \quad \text{subtract 90,000}$$

$$t > 13.5 \qquad\qquad \text{divide by } -6000, \textbf{reverse inequality symbol}$$

After 13.5 yr, the printing press will be worth less than $9000.

c. Since substituting 18 for t gives a negative quantity, the equation model is not valid for $t = 18$. In the current context, the model is only valid while $v \geq 0$ and we note the domain of the function is $t \in [0, 15]$.

☑ **D.** You've just learned how to apply the slope-intercept form and point-slope form in context

Now try Exercises 107 through 112 ▶

2.3 EXERCISES

▶ CONCEPTS AND VOCABULARY

Fill in each blank with the appropriate word or phrase. Carefully reread the section if needed.

1. For the equation $y = -\dfrac{7}{4}x + 3$, the slope is_____ and the y-intercept is _____.

2. The notation $\dfrac{\Delta\text{cost}}{\Delta\text{time}}$ indicates the _____ is changing in response to changes in _____.

3. Line 1 has a slope of -0.4. The slope of any line perpendicular to line 1 is _____.

4. The equation $y - y_1 = m(x - x_1)$ is called the _____ form of a line.

5. Discuss/Explain how to graph a line using only the slope and a point on the line (no equations).

6. Given $m = -\frac{3}{5}$ and $(-5, 6)$ is on the line. Compare and contrast finding the equation of the line using $y = mx + b$ versus $y - y_1 = m(x - x_1)$.

▶ DEVELOPING YOUR SKILLS

Solve each equation for y and evaluate the result using $x = -5, x = -2, x = 0, x = 1,$ and $x = 3$.

7. $4x + 5y = 10$

8. $3y - 2x = 9$

9. $-0.4x + 0.2y = 1.4$

10. $-0.2x + 0.7y = -2.1$

11. $\frac{1}{3}x + \frac{1}{5}y = -1$

12. $\frac{1}{7}y - \frac{1}{3}x = 2$

For each equation, solve for y and identify the new coefficient of x and new constant term.

13. $6x - 3y = 9$

14. $9y - 4x = 18$

15. $-0.5x - 0.3y = 2.1$

16. $-0.7x + 0.6y = -2.4$

17. $\frac{5}{6}x + \frac{1}{7}y = -\frac{4}{7}$

18. $\frac{7}{12}y - \frac{4}{15}x = \frac{7}{6}$

Evaluate each equation by selecting three inputs that will result in integer values. Then graph each line.

19. $y = -\frac{4}{3}x + 5$

20. $y = \frac{5}{4}x + 1$

21. $y = -\frac{3}{2}x - 2$

22. $y = \frac{2}{5}x - 3$

23. $y = -\frac{1}{6}x + 4$

24. $y = -\frac{1}{3}x + 3$

Find the x- and y-intercepts for each line, then (a) use these two points to calculate the slope of the line, (b) write the equation with y in terms of x (solve for y) and compare the calculated slope and y-intercept to the equation from part (b). Comment on what you notice.

25. $3x + 4y = 12$

26. $3y - 2x = -6$

27. $2x - 5y = 10$

28. $2x + 3y = 9$

29. $4x - 5y = -15$

30. $5y + 6x = -25$

Write each equation in slope-intercept form (solve for y), then identify the slope and y-intercept.

31. $2x + 3y = 6$

32. $4y - 3x = 12$

33. $5x + 4y = 20$

34. $y + 2x = 4$

35. $x = 3y$

36. $2x = -5y$

37. $3x + 4y - 12 = 0$

38. $5y - 3x + 20 = 0$

For Exercises 39 to 50, use the slope-intercept form to state the equation of each line.

39.
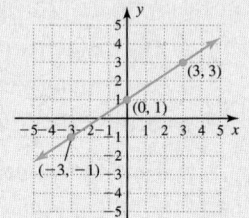

40.
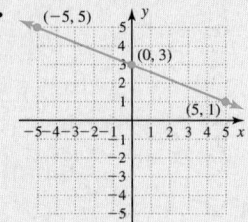

41.
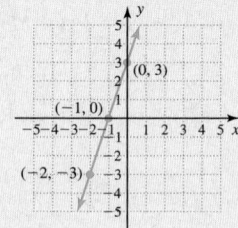

42. $m = -2$; y-intercept $(0, -3)$

43. $m = 3$; y-intercept $(0, 2)$

44. $m = -\frac{3}{2}$; y-intercept $(0, -4)$

45. 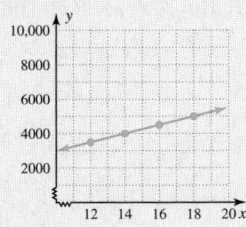   **46.**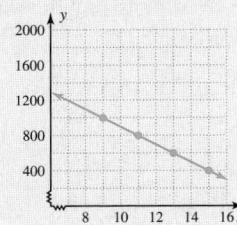

47.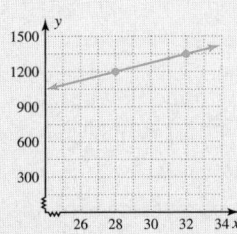

48. $m = -4; (-3, 2)$ is on the line

49. $m = 2; (5, -3)$ is on the line

50. $m = -\frac{3}{2}; (-4, 7)$ is on the line

Write each equation in slope-intercept form, then use the slope and intercept to graph the line.

51. $3x + 5y = 20$ **52.** $2y - x = 4$

53. $2x - 3y = 15$ **54.** $-3x + 2y = 4$

Graph each linear equation using the y-intercept and slope determined from each equation.

55. $y = \frac{2}{3}x + 3$ **56.** $y = \frac{5}{2}x - 1$

57. $y = \frac{-1}{3}x + 2$ **58.** $y = \frac{-4}{5}x + 2$

59. $y = 2x - 5$ **60.** $y = -3x + 4$

61. $y = \frac{1}{2}x - 3$ **62.** $y = \frac{-3}{2}x + 2$

Find the equation of the line using the information given. Write answers in slope-intercept form.

63. parallel to $2x - 5y = 10$, through the point $(-5, 2)$

64. parallel to $6x + 9y = 27$, through the point $(-3, -5)$

65. perpendicular to $5y - 3x = 9$, through the point $(6, -3)$

66. perpendicular to $x - 4y = 7$, through the point $(-5, 3)$

67. parallel to $12x + 5y = 65$, through the point $(-2, -1)$

68. parallel to $15y - 8x = 50$, through the point $(3, -4)$

69. parallel to $y = -3$, through the point $(2, 5)$

70. perpendicular to $y = -3$ through the point $(2, 5)$

Write the lines in slope-intercept form and state whether they are parallel, perpendicular, or neither.

71. $4y - 5x = 8$ **72.** $3y - 2x = 6$
$5y + 4x = -15$ $-2x + 3y = -3$

73. $2x - 5y = 20$ **74.** $5y = 11x + 135$
$4x - 3y = 18$ $11y + 5x = -77$

75. $-4x + 6y = 12$ **76.** $3x + 4y = 12$
$2x + 3y = 6$ $6x + 8y = 2$

A *secant line* is one that intersects a graph at two or more points. For each graph given, find the equation of the line (a) parallel and (b) perpendicular to the secant line, through the point indicated.

77. 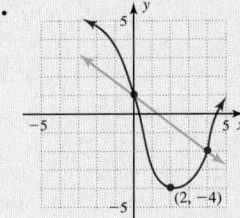   **78.**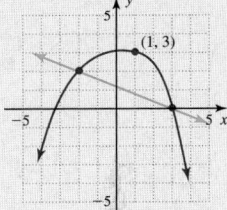

79. 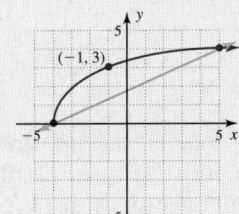   **80.**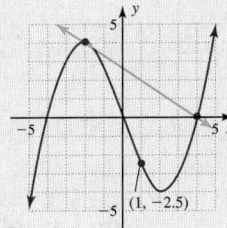

81. 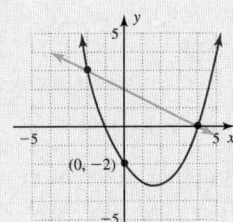   **82.**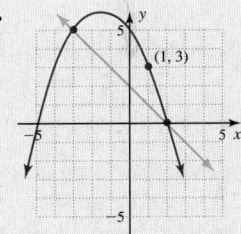

Find the equation of the line in point-slope form, then graph the line.

83. $m = 2; P_1 = (2, -5)$

84. $m = -1; P_1 = (2, -3)$

85. $P_1 = (3, -4), P_2 = (11, -1)$

86. $P_1 = (-1, 6), P_2 = (5, 1)$

87. $m = 0.5; P_1 = (1.8, -3.1)$

88. $m = 1.5; P_1 = (-0.75, -0.125)$

Find the equation of the line in point-slope form, and state the meaning of the slope in context—what information is the slope giving us?

89.
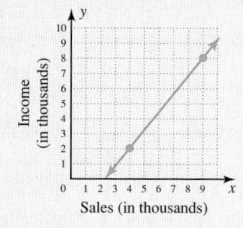

90.
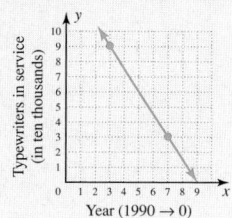

91.
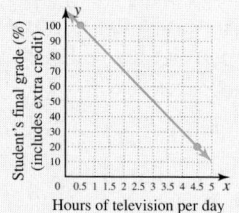

92.
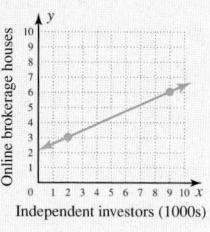

93.
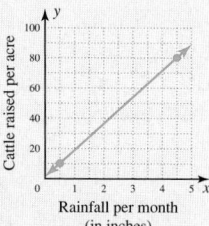

94.
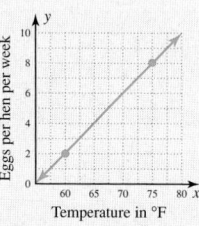

Using the concept of slope, match each description with the graph that best illustrates it. Assume time is scaled on the horizontal axes, and height, speed, or distance from the origin (as the case may be) is scaled on the vertical axis.

95. While driving today, I got stopped by a state trooper. After she warned me to slow down, I continued on my way.

96. After hitting the ball, I began trotting around the bases shouting, "Ooh, ooh, ooh!" When I saw it wasn't a home run, I began sprinting.

97. At first I ran at a steady pace, then I got tired and walked the rest of the way.

98. While on my daily walk, I had to run for a while when I was chased by a stray dog.

99. I climbed up a tree, then I jumped out.

100. I steadily swam laps at the pool yesterday.

101. I walked toward the candy machine, stared at it for a while then changed my mind and walked back.

102. For practice, the girls' track team did a series of 25-m sprints, with a brief rest in between.

▶ **WORKING WITH FORMULAS**

103. General linear equation: $ax + by = c$

The general equation of a line is shown here, where a, b, and c are real numbers, with a and b not simultaneously zero. Solve the equation for y and note the slope (coefficient of x) and y-intercept (constant term). Use these to find the slope and y-intercept of the following lines, without solving for y or computing points.

a. $3x + 4y = 8$ **b.** $2x + 5y = -15$
c. $5x - 6y = -12$ **d.** $3y - 5x = 9$

104. Intercept/Intercept form of a linear equation: $\dfrac{x}{h} + \dfrac{y}{k} = 1$

The x- and y-intercepts of a line can also be found by writing the equation in the form shown (with the equation set equal to 1). The x-intercept will be $(h, 0)$ and the y-intercept will be $(0, k)$. Find the x- and y-intercepts of the following lines using this method: (a) $2x + 5y = 10$, (b) $3x - 4y = -12$, and (c) $5x + 4y = 8$. How is the slope of each line related to the values of h and k?

▶ **APPLICATIONS**

105. Speed of sound: The speed of sound as it travels through the air depends on the temperature of the air according to the function $V = \frac{3}{5}C + 331$, where V represents the velocity of the sound waves in meters per second (m/s), at a temperature of $C°$ Celsius.

a. Interpret the meaning of the slope and y-intercept in this context.

b. Determine the speed of sound at a temperature of 20°C.

c. If the speed of sound is measured at 361 m/s, what is the temperature of the air?

106. Acceleration: A driver going down a straight highway is traveling 60 ft/sec (about 41 mph) on cruise control, when he begins accelerating at a rate of 5.2 ft/sec^2. The final velocity of the car is given by $V = \frac{26}{5}t + 60$, where V is the velocity at time t. (a) Interpret the meaning of the slope and y-intercept in this context. (b) Determine the velocity of the car after 9.4 seconds. (c) If the car is traveling at 100 ft/sec, for how long did it accelerate?

107. Investing in coins: The purchase of a "collector's item" is often made in hopes the item will increase in value. In 1998, Mark purchased a 1909-S VDB Lincoln Cent (in fair condition) for $150. By the year 2004, its value had grown to $190. (a) Use the relation (time since purchase, value) with $t = 0$ corresponding to 1998 to find a linear equation modeling the value of the coin. (b) Discuss what the slope and y-intercept indicate in this context. (c) How much will the penny be worth in 2009? (d) How many years after purchase will the penny's value exceed $250? (e) If the penny is now worth $170, how many years has Mark owned the penny?

108. Depreciation: Once a piece of equipment is put into service, its value begins to depreciate. A business purchases some computer equipment for $18,500. At the end of a 2-yr period, the value of the equipment has decreased to $11,500. (a) Use the relation (time since purchase, value) to find a linear equation modeling the value of the equipment. (b) Discuss what the slope and y-intercept indicate in this context. (c) What is the equipment's value after 4 yr? (d) How many years after purchase will the value decrease to $6000? (e) Generally, companies will sell used equipment while it still has value and use the funds to purchase new equipment. According to the function, how many years will it take this equipment to depreciate in value to $1000?

109. Internet connections: The number of households that are hooked up to the Internet (homes that are online) has been increasing steadily in recent years. In 1995, approximately 9 million homes were online. By 2001 this figure had climbed to about 51 million. (a) Use the relation (year, homes online) with $t = 0$ corresponding to 1995 to find an

equation model for the number of homes online. (b) Discuss what the slope indicates in this context. (c) According to this model, in what year did the first homes begin to come online? (d) If the rate of change stays constant, how many households will be on the Internet in 2006? (e) How many years after 1995 will there be over 100 million households connected? (f) If there are 115 million households connected, what year is it?

Source: 2004 Statistical Abstract of the United States, Table 965

110. Prescription drugs: Retail sales of prescription drugs have been increasing steadily in recent years. In 1995, retail sales hit $72 billion. By the year 2000, sales had grown to about $146 billion. (a) Use the relation (year, retail sales of prescription drugs) with $t = 0$ corresponding to 1995 to find a linear equation modeling the growth of retail sales. (b) Discuss what the slope indicates in this context. (c) According to this model, in what year will sales reach $250 billion? (d) According to the model, what was the value of retail prescription drug sales in 2005? (e) How many years after 1995 will retail sales exceed $279 billion? (f) If yearly sales totaled $294 billion, what year is it?

Source: 2004 Statistical Abstract of the United States, Table 122

111. Prison population: In 1990, the number of persons sentenced and serving time in state and federal institutions was approximately 740,000. By the year 2000, this figure had grown to nearly 1,320,000. (a) Find a linear equation with $t = 0$ corresponding to 1990 that models this data, (b) discuss the slope ratio in context, and (c) use the equation to estimate the prison population in 2007 if this trend continues.

Source: Bureau of Justice Statistics at www.ojp.usdoj.gov/bjs

112. Eating out: In 1990, Americans bought an average of 143 meals per year at restaurants. This phenomenon continued to grow in popularity and in the year 2000, the average reached 170 meals per year. (a) Find a linear equation with $t = 0$ corresponding to 1990 that models this growth, (b) discuss the slope ratio in context, and (c) use the equation to estimate the average number of times an American will eat at a restaurant in 2006 if the trend continues.

Source: The NPD Group, Inc., National Eating Trends, 2002

▶ EXTENDING THE CONCEPT

113. Locate and read the following article. Then turn in a one-page summary. "Linear Function Saves Carpenter's Time," Richard Crouse, *Mathematics Teacher,* Volume 83, Number 5, May 1990: pp. 400–401.

114. The general form of a linear equation is $ax + by = c$, where a and b are not simultaneously zero. (a) Find the x- and y-intercepts using the general form (substitute 0 for x, then 0 for y). Based on what you see, when does the intercept method work most efficiently? (b) Find the slope and *y*-intercept using the general form (solve for *y*). Based on what you see, when does the intercept method work most efficiently?.

115. Match the correct graph to the conditions stated for *m* and *b*. There are more choices than graphs.

   a. $m < 0, b < 0$ **b.** $m > 0, b < 0$

   c. $m < 0, b > 0$ **d.** $m > 0, b > 0$

   e. $m = 0, b > 0$ **f.** $m < 0, b = 0$

   g. $m > 0, b = 0$ **h.** $m = 0, b < 0$

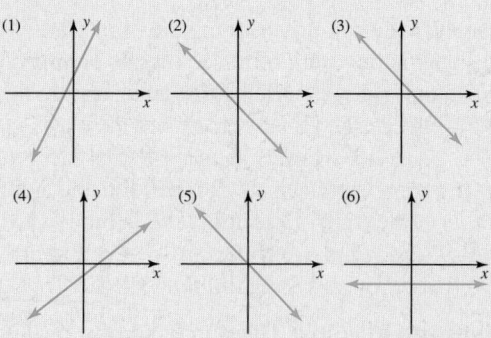

▶ MAINTAINING YOUR SKILLS

116. (2.2) Determine the domain:

   a. $y = \sqrt{2x - 5}$

   b. $y = \dfrac{5}{2x^2 + 3x - 2}$

117. (1.5) Solve using the quadratic formula. Answer in exact and approximate form: $3x^2 - 10x = 9$.

118. (1.1) Three equations follow. One is an identity, another is a contradiction, and a third has a solution. State which is which.

$$2(x - 5) + 13 - 1 = 9 - 7 + 2x$$

$$2(x - 4) + 13 - 1 = 9 + 7 - 2x$$

$$2(x - 5) + 13 - 1 = 9 + 7 + 2x$$

119. (R.7) Compute the area of the circular sidewalk shown here. Use your calculator's value of π and round the answer (only) to hundredths.

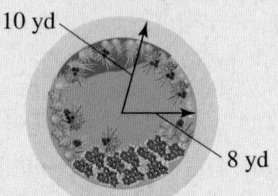

10 yd

8 yd

2.4 | Functions, Function Notation, and the Graph of a Function

Learning Objectives

In Section 2.4 you will learn how to:

☐ **A.** Distinguish the graph of a function from that of a relation

☐ **B.** Determine the domain and range of a function

☐ **C.** Use function notation and evaluate functions

☐ **D.** Apply the rate-of-change concept to nonlinear functions

In this section we introduce one of the most central ideas in mathematics—the concept of a function. Functions can model the cause-and-effect relationship that is so important to using mathematics as a decision-making tool. In addition, the study will help to unify and expand on many ideas that are already familiar.

A. Functions and Relations

There is a special type of relation that merits further attention. A **function** is a relation where each element of the domain corresponds to exactly one element of the range. In other words, for each first coordinate or input value, there is only one possible second coordinate or output.

> **Functions**
>
> A *function* is a relation that pairs each element from the *domain* with exactly one element from the *range*.

If the relation is defined by a mapping, we need only check that each element of the domain is mapped to exactly one element of the range. This is indeed the case for the mapping $P \rightarrow B$ from Figure 2.1 (page 152), where we saw that each person corresponded to only one birthday, and that it was impossible for one person to be born on two different days. For the relation $x = |y|$ shown in Figure 2.6 (page 153), each element of the domain except zero is paired with *more than one* element of the range. The relation $x = |y|$ is *not* a function.

EXAMPLE 1 ▶ **Determining Whether a Relation Is a Function**

Three different relations are given in mapping notation below. Determine whether each relation is a function.

a.   b.   c.

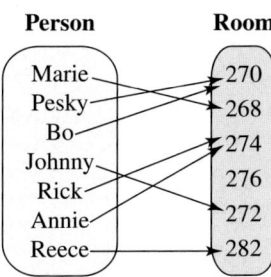   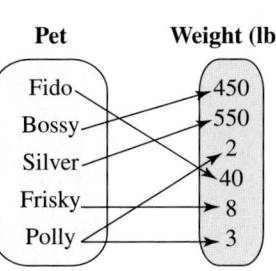   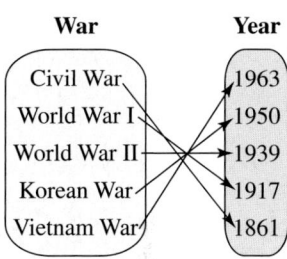

Solution ▶ Relation (a) is a function, since each person corresponds to exactly one room. This relation pairs math professors with their respective office numbers. Notice that while two people can be in one office, it is impossible for one person to physically be in two different offices. Relation (b) is not a function, since we cannot tell whether Polly the Parrot weighs 2 lb or 3 lb (one element of the domain is mapped to two elements of the range). Relation (c) is a function, where each major war is paired with the year it began.

Now try Exercises 7 through 10 ▶

If the relation is defined by a set of ordered pairs or a set of individual and distinct plotted points, we need only check that no two points have the same first coordinate with a different second coordinate.

EXAMPLE 2 ▶ **Identifying Functions**

Two relations named f and g are given; f is stated as a set of ordered pairs, while g is given as a set of plotted points. Determine whether each is a function.

f: $(-3, 0)$, $(1, 4)$, $(2, -5)$, $(4, 2)$, $(-3, -2)$, $(3, 6)$, $(0, -1)$, $(4, -5)$, and $(6, 1)$

Solution ▶ The relation f is not a function, since -3 is paired with two different outputs: $(-3, 0)$ and $(-3, -2)$.

The relation g shown in the figure *is* a function. Each input corresponds to exactly one output, otherwise one point would be directly above the other and have the same first coordinate.

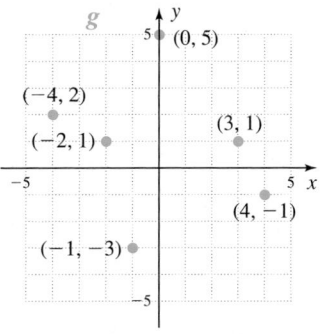

WORTHY OF NOTE

The definition of a function can also be stated in ordered pair form: *A function is a set of ordered pairs (x, y), in which each first component is paired with only one second component.*

Now try Exercises 11 through 18 ▶

The graphs of $y = x - 1$ and $x = |y|$ from Section 2.1 offer additional insight into the definition of a function. Figure 2.23 shows the line $y = x - 1$ with emphasis on the plotted points $(4, 3)$ and $(-3, -4)$. The vertical movement shown from the x-axis to a point on the graph illustrates *the pairing of a given x-value with one related y-value.* Note the vertical line shows *only one related y-value* ($x = 4$ is paired with only $y = 3$). Figure 2.24 gives the graph of $x = |y|$, highlighting the points $(4, 4)$ and $(4, -4)$. The vertical movement shown here branches in two directions, associating one x-value with more than one y-value. This shows the relation $y = x - 1$ is also a function, while the relation $x = |y|$ is not.

Figure 2.23   **Figure 2.24**

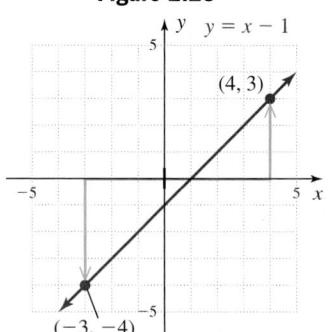   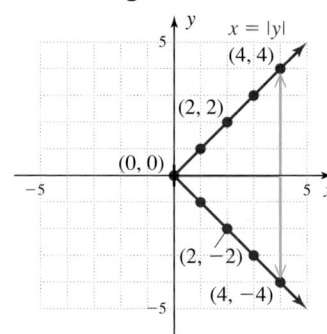

This "vertical connection" of a location on the x-axis to a point on the graph can be generalized into a **vertical line test for functions.**

Vertical Line Test

A given graph is the graph of a function, if and only if every vertical line intersects the graph in at most one point.

Applying the test to the graph in Figure 2.23 helps to illustrate that the graph of any nonvertical line is a function.

EXAMPLE 3 ▶ **Using the Vertical Line Test**

Use the vertical line test to determine if any of the relations shown (from Section 2.1) are functions.

Solution ▶ Visualize a vertical line on each coordinate grid (shown in solid blue), then mentally shift the line to the left and right as shown in Figures 2.25, 2.26, and 2.27 (dashed lines). In Figures 2.25 and 2.26, every vertical line intersects the graph only once, indicating both $y = x^2 - 2x$ and $y = \sqrt{9 - x^2}$ are functions. In Figure 2.27, a vertical line intersects the graph twice for any $x > 0$. The relation $x = y^2$ is not a function.

Figure 2.25   **Figure 2.26**   **Figure 2.27**

EXAMPLE 4 ▶ **Using the Vertical Line Test**

Use a table of values to graph the relations defined by

a. $y = |x|$ **b.** $y = \sqrt{x}$,

then use the vertical line test to determine whether each relation is a function.

Solution ▶ **a.** For $y = |x|$, using input values from $x = -4$ to $x = 4$ produces the following table and graph (Figure 2.28). Note the result is a V-shaped graph that "opens upward." The point $(0, 0)$ of this absolute value graph is called the **vertex.** Since any vertical line will intersect the graph in at most one point, this is the graph of a function.

> **WORTHY OF NOTE**
>
> For relations and functions, a good way to view the distinction is to consider a mail carrier. It is possible for the carrier to put more than one letter into the same mailbox (more than one x going to the same y), but quite impossible for the carrier to place the same letter in two different boxes (one x going to two y's).

$$y = |x|$$

x	$y = |x|$
-4	4
-3	3
-2	2
-1	1
0	0
1	1
2	2
3	3
4	4

Figure 2.28

b. For $y = \sqrt{x}$, values less than zero do not produce a real number, so our graph actually begins at $(0, 0)$ (see Figure 2.29). Completing the table for nonnegative values produces the graph shown, which appears to rise to the right and remains in the first quadrant. Since any vertical line will intersect this graph in at most one place, $y = \sqrt{x}$ is also a function.

$$y = \sqrt{x}$$

x	$y = \sqrt{x}$
0	0
1	1
2	$\sqrt{2} \approx 1.4$
3	$\sqrt{3} \approx 1.7$
4	2

Figure 2.29

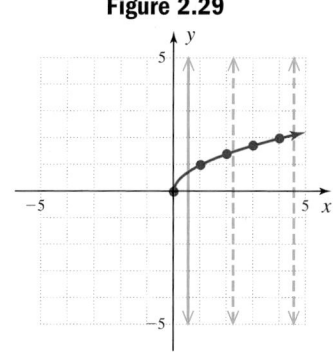

☑ **A.** You've just learned how to distinguish the graph of a function from that of a relation

Now try Exercises 31 through 34 ▶

B. The Domain and Range of a Function

Vertical Boundary Lines and the Domain

In addition to its use as a graphical test for functions, a vertical line can help determine the domain of a function from its graph. For the graph of $y = \sqrt{x}$ (Figure 2.29), a vertical line will not intersect the graph until $x = 0$, and then will intersect the graph for all values $x \geq 0$ (showing the function is defined for these values). These **vertical boundary lines** indicate the domain is $x \in [0, \infty)$. For the graph of $y = |x|$ (Figure 2.28), a vertical line will intersect the graph (or its infinite extension) for *all values* of x, and the domain is $x \in (-\infty, \infty)$. Using vertical lines in this way also affirms the domain of $y = x - 1$ (Section 2.1, Figure 2.5) is $x \in (-\infty, \infty)$ while the domain of the relation $x = |y|$ (Section 2.1, Figure 2.6) is $x \in [0, \infty)$.

Range and Horizontal Boundary Lines

The range of a relation can be found using a **horizontal "boundary line,"** since it will associate a value on the y-axis with a point on the graph (if it exists). Simply visualize a horizontal line and move the line up or down until you determine the graph will always intersect the line, or will no longer intersect the line. This will give you the boundaries of the range. Mentally applying this idea to the graph of $y = \sqrt{x}$ (Figure 2.29) shows the range is $y \in [0, \infty)$. Although shaped very differently, a horizontal boundary line shows the range of $y = |x|$ (Figure 2.28) is also $y \in [0, \infty)$.

EXAMPLE 5 ▶ **Determining the Domain and Range of a Function**

Use a table of values to graph the functions defined by

a. $y = x^2$ **b.** $y = \sqrt[3]{x}$

Then use boundary lines to determine the domain and range of each.

Solution ▶ **a.** For $y = x^2$, it seems convenient to use inputs from $x = -3$ to $x = 3$, producing the following table and graph. Note the result is a basic parabola that "opens upward" (both ends point in the positive y direction), with a vertex at $(0, 0)$. Figure 2.30 shows a vertical line will intersect the graph or its extension anywhere it is placed. The domain is $x \in (-\infty, \infty)$. Figure 2.31 shows a horizontal line will intersect the graph only for values of y that are greater than or equal to 0. The range is $y \in [0, \infty)$.

Squaring Function

x	$y = x^2$
-3	9
-2	4
-1	1
0	0
1	1
2	4
3	9

Figure 2.30

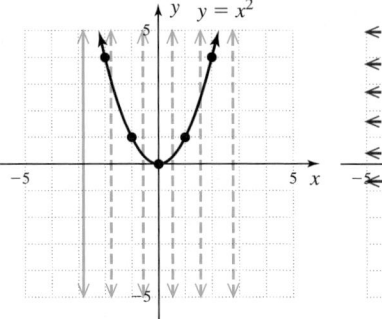

Figure 2.31

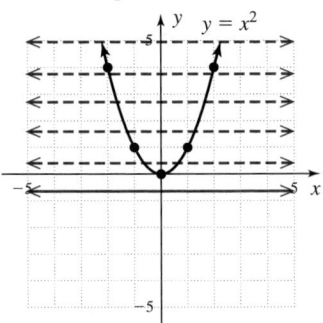

b. For $y = \sqrt[3]{x}$, we select points that are perfect cubes where possible, then a few others to round out the graph. The resulting table and graph are shown, and we notice there is a "pivot point" at $(0, 0)$ called a **point of inflection,** and the ends of the graph point in opposite directions. Figure 2.32 shows a vertical line will intersect the graph or its extension anywhere it is placed. Figure 2.33 shows a horizontal line will likewise always intersect the graph. The domain is $x \in (-\infty, \infty)$, and the range is $y \in (-\infty, \infty)$.

Cube Root Function

x	$y = \sqrt[3]{x}$
-8	-2
-4	≈ -1.6
-1	-1
0	0
1	1
4	≈ 1.6
8	2

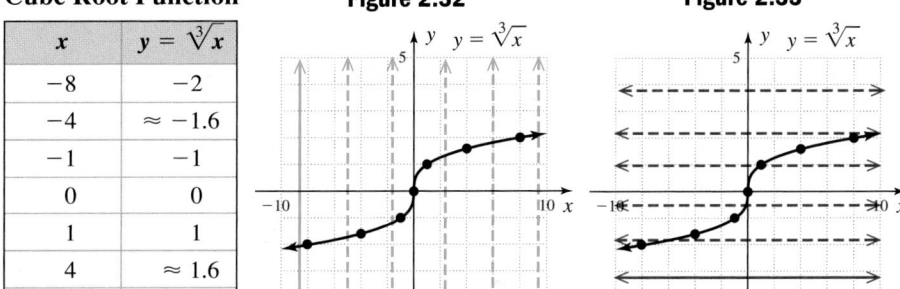

Figure 2.32

Figure 2.33

Now try Exercises 35 through 46 ▶

Implied Domains

When stated in equation form, the domain of a function is implicitly given by the expression used to define it, since the expression will dictate the allowable values (Section 1.2). The **implied domain** is the set of all real numbers for which the function represents a real number. If the function involves a rational expression, the domain will exclude any input that causes a denominator of zero. If the function involves a square root expression, the domain will exclude inputs that create a negative radicand.

EXAMPLE 6 ▶ Determining Implied Domains

State the domain of each function using interval notation.

a. $y = \dfrac{3}{x + 2}$ **b.** $y = \sqrt{2x + 3}$

c. $y = \dfrac{x - 5}{x^2 - 9}$ **d.** $y = x^2 - 5x + 7$

Solution ▶ **a.** By inspection, we note an x-value of -2 gives a zero denominator and must be excluded. The domain is $x \in (-\infty, -2) \cup (-2, \infty)$.

b. Since the radicand must be nonnegative, we solve the inequality $2x + 3 \geq 0$, giving $x \geq \frac{-3}{2}$. The domain is $x \in [\frac{-3}{2}, \infty)$.

c. To prevent division by zero, inputs of -3 and 3 must be excluded (set $x^2 - 9 = 0$ and solve by factoring). The domain is $x \in (-\infty, -3) \cup (-3, 3) \cup (3, \infty)$. Note that $x = 5$ *is in the domain* since $\frac{0}{16} = 0$ is defined.

d. Since squaring a number and multiplying a number by a constant are defined for all reals, the domain is $x \in (-\infty, \infty)$.

Now try Exercises 47 through 64 ▶

EXAMPLE 7 ▶ Determining Implied Domains

Determine the domain of each function:

a. $y = \sqrt{\dfrac{7}{x + 3}}$ **b.** $y = \dfrac{2x}{\sqrt{4x + 5}}$

Solution ▶   **a.** For $y = \sqrt{\dfrac{7}{x+3}}$, we must have $\dfrac{7}{x+3} \geq 0$ (for the radicand) **and** $x + 3 \neq 0$ (for the denominator). Since the numerator is *always* positive, we need $x + 3 > 0$, which gives $x > -3$. The domain is $x \in (-3, \infty)$.

b. For $y = \dfrac{2x}{\sqrt{4x+5}}$, we must have $4x + 5 \geq 0$ **and** $\sqrt{4x+5} \neq 0$. This indicates $4x + 5 > 0$ or $x > -\frac{5}{4}$. The domain is $x \in \left(-\frac{5}{4}, \infty\right)$.

 B. You've just learned how to determine the domain and range of a function

Now try Exercises 65 through 68 ▶

C. Function Notation

Figure 2.34

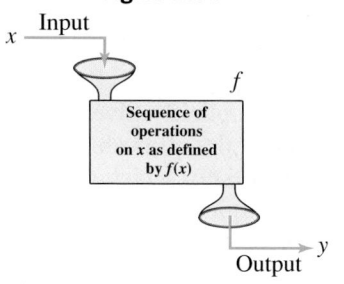

In our study of functions, you've likely noticed that the relationship between input and output values is an important one. To highlight this fact, think of a function as a simple machine, which can *process inputs* using a stated sequence of operations, then deliver a single output. The inputs are x-values, a program we'll name f performs the operations on x, and y is the resulting output (see Figure 2.34). Once again we see that "the value of y depends on the value of x," or simply "y is a function of x." Notationally, we write "y is a function of x" as $y = f(x)$ using **function notation.** You are already familiar with letting a variable represent a number. Here we do something quite different, as the letter f is used to represent *a sequence of operations to be performed on x.* Consider the function $y = \frac{x}{2} + 1$, which we'll now write as $f(x) = \frac{x}{2} + 1$ [since $y = f(x)$]. In words the function says, "divide inputs by 2, then add 1." To evaluate the function at $x = 4$ (Figure 2.35) we have:

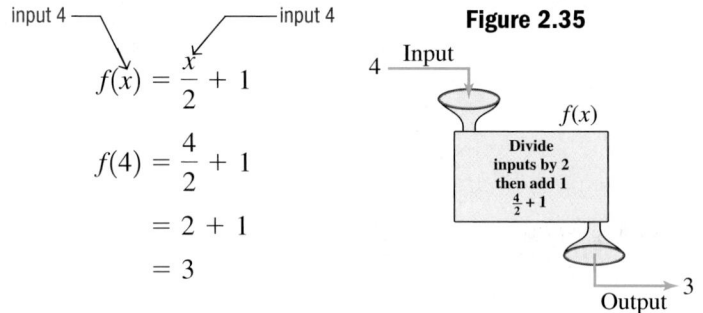

Figure 2.35

Instead of saying, ". . . when $x = 4$, the value of the function is 3," we simply say "f of 4 is 3," or write $f(4) = 3$. Note that the ordered pair $(4, 3)$ is equivalent to $(4, f(4))$.

⚠ **CAUTION** ▶ Although $f(x)$ is the favored notation for a "function of x," other letters can also be used. For example, $g(x)$ and $h(x)$ also denote functions of x, where g and h represent a different sequence of operations on the x-inputs. It is also important to remember that these represent *function values* and not the product of two variables: $f(x) \neq f \cdot (x)$.

EXAMPLE 8 ▶ **Evaluating a Function**

Given $f(x) = -2x^2 + 4x$, find

a. $f(-2)$ **b.** $f\left(\dfrac{3}{2}\right)$ **c.** $f(2a) - 2(4a^2) + 8a$ **d.** $f(a + 1)$

Solution ▶   **a.** $f(x) = -2x^2 + 4x$
$$f(-2) = -2(-2)^2 + 4(-2)$$
$$= -8 + (-8) = -16$$

b. $f(x) = -2x^2 + 4x$
$$f\left(\frac{3}{2}\right) = -2\left(\frac{3}{2}\right)^2 + 4\left(\frac{3}{2}\right)$$
$$= -\frac{9}{2} + 6 = \frac{3}{2}$$

c. $f(x) = -2x^2 + 4x$
$$f(2a) = -2(2a)^2 + 4(2a)$$
$$= -8a^2 + 8a$$

d. $f(x) = -2x^2 + 4x$
$$f(a + 1) = -2(a + 1)^2 + 4(a + 1)$$
$$= -2(a^2 + 2a + 1) + 4a + 4$$
$$= -2a^2 - 4a - 2 + 4a + 4$$
$$= -2a^2 + 2$$

> **Now try Exercises 69 through 84 ▶**

Graphs are an important part of studying functions, and learning to read and interpret them correctly is a high priority. A graph highlights and emphasizes the all-important input/output relationship that defines a function. In this study, we hope to firmly establish that the following statements are synonymous:

1. $f(-2) = 5$
2. $(-2, f(-2)) = (-2, 5)$
3. $(-2, 5)$ is on the graph of f, and
4. When $x = -2, f(x) = 5$

EXAMPLE 9A ▶   **Reading a Graph**

For the functions $f(x)$ and $g(x)$ whose graphs are shown in Figures 2.36 and 2.37
   a. State the domain of the function.
   b. Evaluate the function at $x = 2$.
   c. Determine the value(s) of x for which $y = 3$.
   d. State the range of the function.

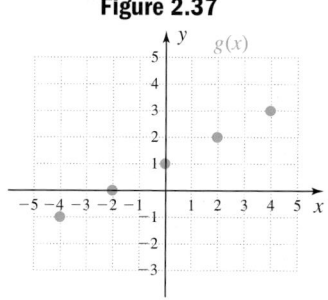

Figure 2.36   **Figure 2.37**

Solution ▶   For $f(x)$,

   a. The graph is a continuous line segment with endpoints at $(-4, -3)$ and $(5, 3)$, so we state the domain in interval notation. Using a vertical boundary line we note the smallest input is -4 and the largest is 5. The domain is $x \in [-4, 5]$.

   b. The graph shows an input of $x = 2$ corresponds to $y = 1 : f(2) = 1$ since $(2, 1)$ is a point on the graph.

   c. For $f(x) = 3$ (or $y = 3$) the input value must be $x = 5$ since $(5, 3)$ is the point on the graph.

   d. Using a horizontal boundary line, the smallest output value is -3 and the largest is 3. The range is $y \in [-3, 3]$.

For $g(x)$,
  a. Since the graph is pointwise defined, we state the domain as the set of first coordinates: $D = \{-4, -2, 0, 2, 4\}$.
  b. An input of $x = 2$ corresponds to $y = 2$: $g(2) = 2$ since $(2, 2)$ is on the graph.
  c. For $g(x) = 3$ (or $y = 3$) the input value must be $x = 4$, since $(4, 3)$ is a point on the graph.
  d. The range is the set of all second coordinates: $R = \{-1, 0, 1, 2, 3\}$.

EXAMPLE 9B ▶ **Reading a Graph**

Use the graph of $f(x)$ given to answer the following questions:
  a. What is the value of $f(-2)$?
  b. What value(s) of x satisfy $f(x) = 1$?

Solution ▶ **a.** The notation $f(-2)$ says to find the value of the function f when $x = -2$. Expressed graphically, we go to $x = -2$, locate the corresponding point on the graph of f (blue arrows), and find that $f(-2) = 4$.

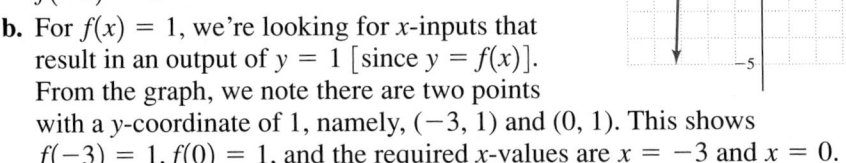

  b. For $f(x) = 1$, we're looking for x-inputs that result in an output of $y = 1$ [since $y = f(x)$]. From the graph, we note there are two points with a y-coordinate of 1, namely, $(-3, 1)$ and $(0, 1)$. This shows $f(-3) = 1, f(0) = 1$, and the required x-values are $x = -3$ and $x = 0$.

Now try Exercises 85 through 90 ▶

In many applications involving functions, the domain and range can be determined by the context or situation given.

EXAMPLE 10 ▶ **Determining the Domain and Range from the Context**

Paul's 1993 Voyager has a 20-gal tank and gets 18 mpg. The number of miles he can drive (his range) depends on how much gas is in the tank. As a function we have $M(g) = 18g$, where $M(g)$ represents the total distance in miles and g represents the gallons of gas in the tank. Find the domain and range.

Solution ▶ Begin evaluating at $x = 0$, since the tank cannot hold less than zero gallons. On a full tank the maximum range of the van is $20 \cdot 18 = 360$ miles or $M(g) \in [0, 360]$. Because of the tank's size, the domain is $g \in [0, 20]$.

☑ **C.** You've just learned how to use function notation and evaluate functions

Now try Exercises 94 through 101 ▶

D. Average Rates of Change

As noted in Section 2.3, one of the defining characteristics of a linear function is that the rate of change $m = \dfrac{\Delta y}{\Delta x}$ is constant. For nonlinear functions the rate of change is not constant, but we can use a related concept called the **average rate of change** to study these functions.

Average Rate of Change

For a function that is smooth and continuous on the interval containing x_1 and x_2, the average rate of change between x_1 and x_2 is given by

$$\frac{\Delta y}{\Delta x} = \frac{y_2 - y_1}{x_2 - x_1}$$

which is the slope of the secant line through (x_1, y_1) and (x_2, y_2)

EXAMPLE 11 ▶ **Calculating Average Rates of Change**

The graph shown displays the number of units shipped of vinyl records, cassette tapes, and CDs for the period 1980 to 2005.

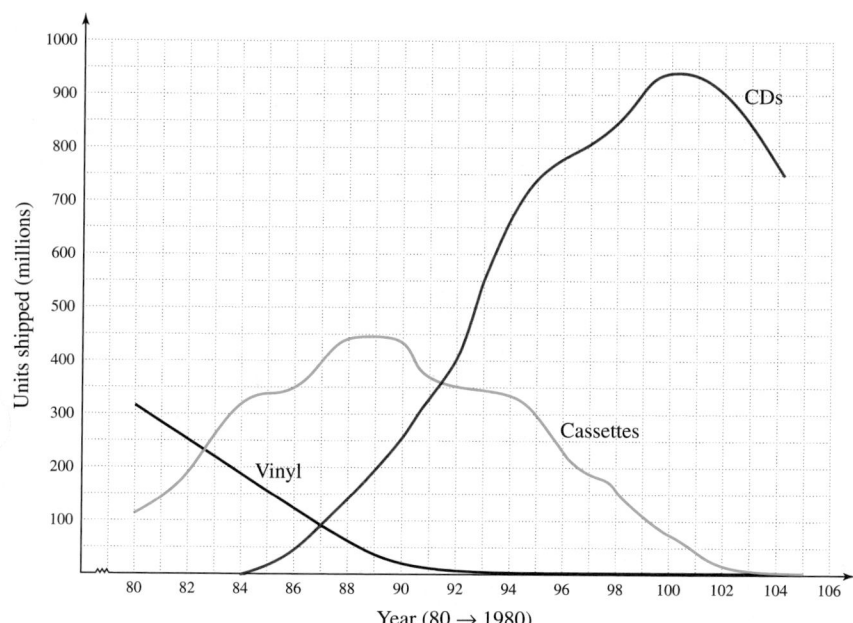

Units shipped in millions

Year	Vinyl	Cassette	CDs
1980	323	110	0
1982	244	182	0
1984	205	332	6
1986	125	345	53
1988	72	450	150
1990	12	442	287
1992	2	366	408
1994	2	345	662
1996	3	225	779
1998	3	159	847
2000	2	76	942
2004	1	5	767
2005	1	3	705

Source: Swivel.com

a. Find the average rate of change in CDs shipped and in cassettes shipped from 1994 to 1998. What do you notice?

b. Does it appear that the rate of increase in CDs shipped was greater from 1986 to 1992, or from 1992 to 1996? Compute the average rate of change for each period and comment on what you find.

Solution ▶ Using 1980 as year zero (1980 → 0), we have the following:

a.   **CDs**   **Cassettes**

1994: $(14, 662)$, 1998: $(18, 847)$ 1994: $(14, 345)$, 1998: $(18, 159)$

$$\frac{\Delta y}{\Delta x} = \frac{847 - 662}{18 - 14} \qquad\qquad \frac{\Delta y}{\Delta x} = \frac{159 - 345}{18 - 14}$$

$$= \frac{185}{4} \qquad\qquad\qquad = -\frac{186}{4}$$

$$= 46.25 \qquad\qquad\qquad = -46.5$$

The decrease in the number of cassettes shipped was roughly equal to the increase in the number of CDs shipped (about 46,000,000 per year).

b. From the graph, the secant line for 1992 to 1996 appears to have a greater slope.

1986–1992 CDs	**1992–1996 CDs**
1986: (6, 53), 1992: (12, 408)	1992: (12, 408), 1996: (16, 779)

$$\frac{\Delta y}{\Delta x} = \frac{408 - 53}{12 - 6}$$

$$= \frac{355}{6}$$

$$= 59.1\overline{6}$$

$$\frac{\Delta y}{\Delta x} = \frac{779 - 408}{16 - 12}$$

$$= \frac{371}{4}$$

$$= 92.75$$

 **D.** You've just learned how to apply the rate-of-change concept to nonlinear functions

For 1986 to 1992: $m \approx 59.2$; for 1992 to 1996: $m = 92.75$, a growth rate much higher than the earlier period.

Now try Exercises 102 and 103 ▶

2.4 EXERCISES

▶ CONCEPTS AND VOCABULARY

Fill in each blank with the appropriate word or phrase. Carefully reread the section if needed.

1. If a relation is given in ordered pair form, we state the domain by listing all of the _____ coordinates in a set.

2. A relation is a function if each element of the _____ is paired with _____ _____ element of the range.

3. The set of output values for a function is called the _____ of the function.

4. Write using function notation: The function f evaluated at 3 is negative 5: _____

5. Discuss/Explain why the relation $y = x^2$ is a function, while the relation $x = y^2$ is not. Justify your response using graphs, ordered pairs, and so on.

6. Discuss/Explain the process of finding the domain and range of a function given its graph, using vertical and horizontal boundary lines. Include a few illustrative examples.

▶ DEVELOPING YOUR SKILLS

Determine whether the mappings shown represent functions or nonfunctions. If a nonfunction, explain how the definition of a function is violated.

7.

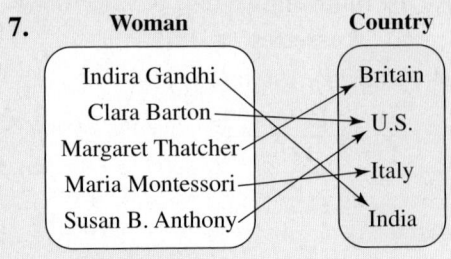

8.

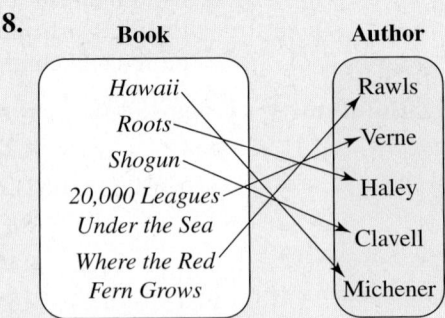

9.

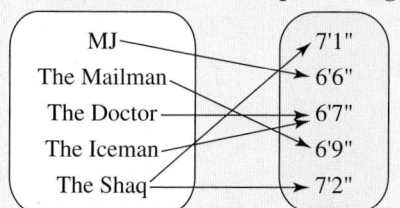

Basketball star	Reported height
MJ	7'1"
The Mailman	6'6"
The Doctor	6'7"
The Iceman	6'9"
The Shaq	7'2"

10.

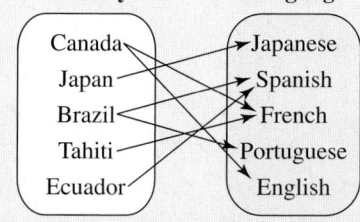

Country	Language
Canada	Japanese
Japan	Spanish
Brazil	French
Tahiti	Portuguese
Ecuador	English

Determine whether the relations indicated represent functions or nonfunctions. If the relation is a nonfunction, explain how the definition of a function is violated.

11. $(-3, 0), (1, 4), (2, -5), (4, 2), (-5, 6), (3, 6), (0, -1), (4, -5),$ and $(6, 1)$

12. $(-7, -5), (-5, 3), (4, 0), (-3, -5), (1, -6), (0, 9), (2, -8), (3, -2),$ and $(-5, 7)$

13. $(9, -10), (-7, 6), (6, -10), (4, -1), (2, -2), (1, 8), (0, -2), (-2, -7),$ and $(-6, 4)$

14. $(1, -81), (-2, 64), (-3, 49), (5, -36), (-8, 25), (13, -16), (-21, 9), (34, -4),$ and $(-55, 1)$

15.

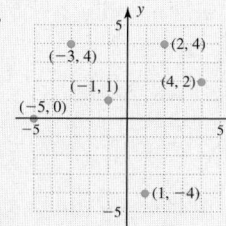

16.

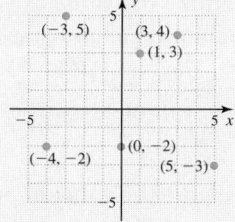

17.

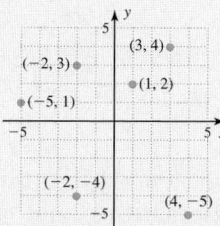

18.

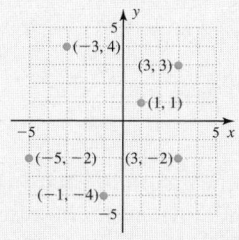

Determine whether or not the relations given represent a function. If not, explain how the definition of a function is violated.

19.

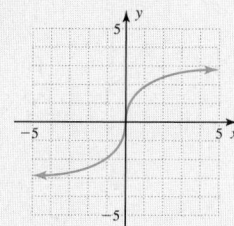

20.

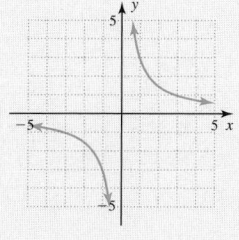

21.

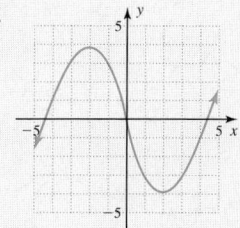

22.

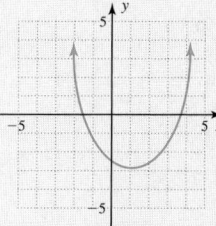

23.

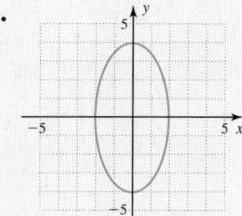

24.

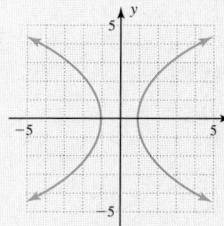

25.

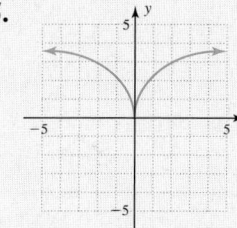

26.

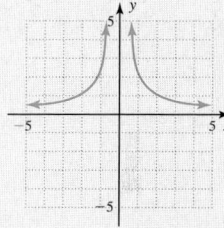

27.

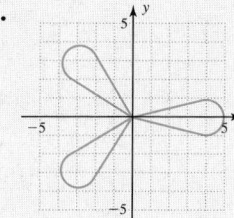

28.

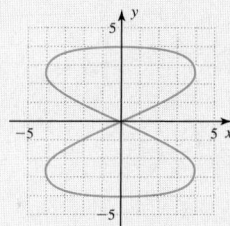

29.

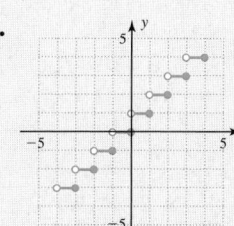

30.

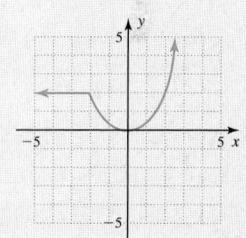

Graph each relation using a table, then use the vertical line test to determine if the relation is a function.

31. $y = x$ **32.** $y = \sqrt[3]{x}$

33. $y = (x + 2)^2$ **34.** $x = |y - 2|$

Determine whether or not the relations indicated represent a function, then determine the domain and range of each.

35.

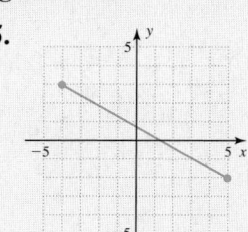

36.

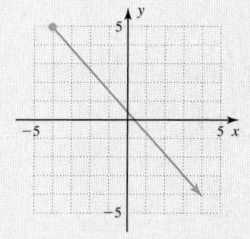

37.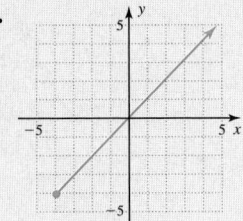

38.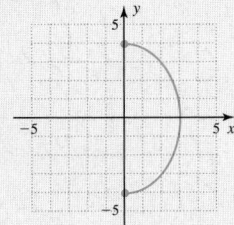

39.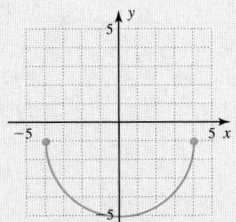

40.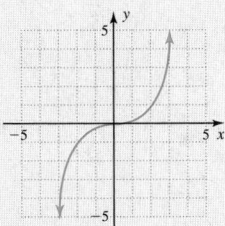

41.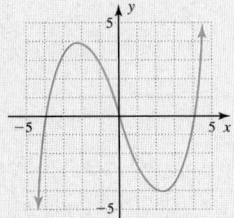

42.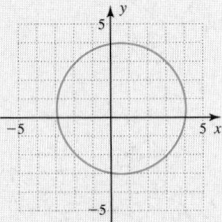

43.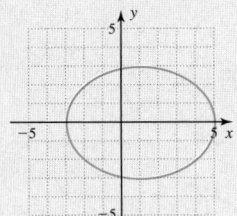

44.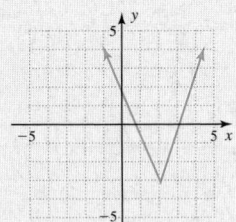

45.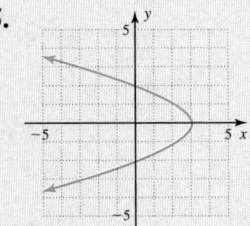

46.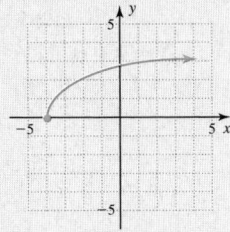

Determine the domain of the following functions.

47. $f(x) = \dfrac{3}{x - 5}$

48. $g(x) = \dfrac{-2}{3 + x}$

49. $h(a) = \sqrt{3a + 5}$

50. $p(a) = \sqrt{5a - 2}$

51. $v(x) = \dfrac{x + 2}{x^2 - 25}$

52. $w(x) = \dfrac{x - 4}{x^2 - 49}$

53. $u = \dfrac{v - 5}{v^2 - 18}$

54. $p = \dfrac{q + 7}{q^2 - 12}$

55. $y = \dfrac{17}{25}x + 123$

56. $y = \dfrac{11}{19}x - 89$

57. $m = n^2 - 3n - 10$

58. $s = t^2 - 3t - 10$

59. $y = 2|x| + 1$

60. $y = |x - 2| + 3$

61. $y_1 = \dfrac{x}{x^2 - 3x - 10}$

62. $y_2 = \dfrac{x - 4}{x^2 + 2x - 15}$

63. $y = \dfrac{\sqrt{x - 2}}{2x - 5}$

64. $y = \dfrac{\sqrt{x + 1}}{3x + 2}$

65. $f(x) = \sqrt{\dfrac{5}{x - 2}}$

66. $g(x) = \sqrt{\dfrac{-4}{3 - x}}$

67. $h(x) = \dfrac{-2}{\sqrt{4 + x}}$

68. $p(x) = \dfrac{-7}{\sqrt{5 - x}}$

Determine the value of $f(-6)$, $f(\frac{3}{2})$, $f(2c)$, and $f(c + 1)$, then simplify as much as possible.

69. $f(x) = \dfrac{1}{2}x + 3$

70. $f(x) = \dfrac{2}{3}x - 5$

71. $f(x) = 3x^2 - 4x$

72. $f(x) = 2x^2 + 3x$

Determine the value of $h(3)$, $h(-\frac{2}{3})$, $h(3a)$, and $h(a - 2)$, then simplify as much as possible.

73. $h(x) = \dfrac{3}{x}$

74. $h(x) = \dfrac{2}{x^2}$

75. $h(x) = \dfrac{5|x|}{x}$

76. $h(x) = \dfrac{4|x|}{x}$

Determine the value of $g(4)$, $g(\frac{3}{2})$, $g(2c)$, and $g(c + 3)$, then simplify as much as possible.

77. $g(r) = 2\pi r$

78. $g(r) = 2\pi rh$

79. $g(r) = \pi r^2$

80. $g(r) = \pi r^2 h$

Determine the value of $p(5)$, $p(\frac{3}{2})$, $p(3a)$, and $p(a - 1)$, then simplify as much as possible.

81. $p(x) = \sqrt{2x + 3}$

82. $p(x) = \sqrt{4x - 1}$

83. $p(x) = \dfrac{3x^2 - 5}{x^2}$

84. $p(x) = \dfrac{2x^2 + 3}{x^2}$

Use the graph of each function given to (a) state the domain, (b) state the range, (c) evaluate $f(2)$, and (d) find the value(s) x for which $f(x) = k$ (k a constant). Assume all results are integer-valued.

85. $k = 4$

86. $k = 3$

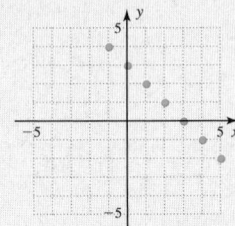

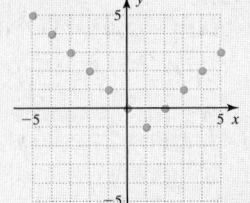

87. $k = 1$ 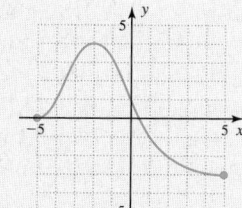   **88.** $k = -3$ 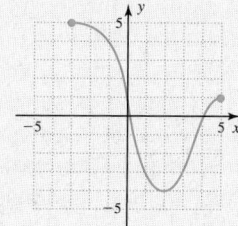   **89.** $k = 2$ 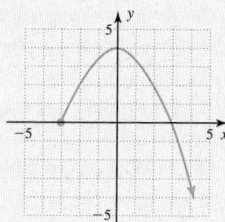   **90.** $k = -1$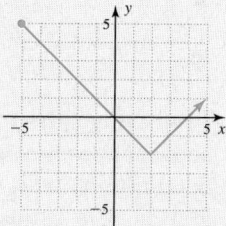

▶ WORKING WITH FORMULAS

91. Ideal weight for males: $W(H) = \frac{9}{2}H - 151$

The ideal weight for an adult male can be modeled by the function shown, where W is his weight in pounds and H is his height in inches. (a) Find the ideal weight for a male who is 75 in. tall. (b) If I am 72 in. tall and weigh 210 lb, how much weight should I lose?

92. Celsius to Fahrenheit conversions: $C = \frac{5}{9}(F - 32)$

The relationship between Fahrenheit degrees and degrees Celsius is modeled by the function shown. (a) What is the Celsius temperature if °F = 41? (b) Use the formula to solve for F in terms of C, then substitute the result from part (a). What do you notice?

93. Pick's theorem: $A = \frac{1}{2}B + I - 1$

Picks theorem is an interesting yet little known formula for computing the area of a polygon drawn in the Cartesian coordinate system. The formula can be applied as long as the vertices of the polygon are lattice points (both x and y are integers). If B represents the number of lattice points lying directly on the boundary of the polygon (including the vertices), and I represents the number of points in the interior, the area of the polygon is given by the formula shown. Use some graph paper to carefully draw a triangle with vertices at $(-3, 1)$, $(3, 9)$, and $(7, 6)$, then use Pick's theorem to compute the triangle's area.

▶ APPLICATIONS

94. Gas mileage: John's old '87 LeBaron has a 15-gal gas tank and gets 23 mpg. The number of miles he can drive is a function of how much gas is in the tank. (a) Write this relationship in equation form and (b) determine the domain and range of the function in this context.

95. Gas mileage: Jackie has a gas-powered model boat with a 5-oz gas tank. The boat will run for 2.5 min on each ounce. The number of minutes she can operate the boat is a function of how much gas is in the tank. (a) Write this relationship in equation form and (b) determine the domain and range of the function in this context.

 96. Volume of a cube: The volume of a cube depends on the length of the sides. In other words, volume is a function of the sides: $V(s) = s^3$. (a) In practical terms, what is the domain of this function? (b) Evaluate $V(6.25)$ and (c) evaluate the function for $s = 2x^2$.

97. Volume of a cylinder: For a fixed radius of 10 cm, the volume of a cylinder depends on its height. In other words, volume is a function of height:

$V(h) = 100\pi h$. (a) In practical terms, what is the domain of this function? (b) Evaluate $V(7.5)$ and (c) evaluate the function for $h = \dfrac{8}{\pi}$.

98. Rental charges: Temporary Transportation Inc. rents cars (local rentals only) for a flat fee of $19.50 and an hourly charge of $12.50. This means that cost is a function of the hours the car is rented plus the flat fee. (a) Write this relationship in equation form; (b) find the cost if the car is rented for 3.5 hr; (c) determine how long the car was rented if the bill came to $119.75; and (d) determine the domain and range of the function in this context, if your budget limits you to paying a maximum of $150 for the rental.

99. Cost of a service call: Paul's Plumbing charges a flat fee of $50 per service call plus an hourly rate of $42.50. This means that cost is a function of the hours the job takes to complete plus the flat fee. (a) Write this relationship in equation form; (b) find the cost of a service call that takes $2\frac{1}{2}$ hr; (c) find the number of hours the job took if the charge came to $262.50; and (d) determine the domain and range of the function in this context, if your insurance company has agreed to pay for all charges over $500 for the service call.

100. Predicting tides: The graph shown approximates the height of the tides at Fair Haven, New Brunswick, for a 12-hr period. (a) Is this the graph of a function? Why? (b) Approximately what time did high tide occur? (c) How high is the tide at 6 P.M.? (d) What time(s) will the tide be 2.5 m?

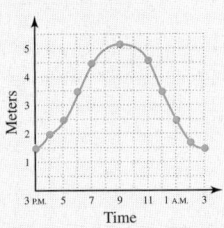

101. Predicting tides: The graph shown approximates the height of the tides at Apia, Western Samoa, for a 12-hr period. (a) Is this the graph of a function? Why? (b) Approximately what time did low tide occur? (c) How high is the tide at 2 A.M.? (d) What time(s) will the tide be 0.7 m?

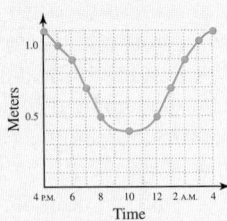

102. Weight of a fetus: The growth rate of a fetus in the mother's womb (by weight in grams) is modeled by the graph shown here, beginning with the 25th week of

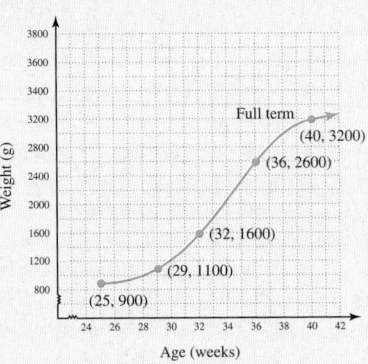

gestation. (a) Calculate the average rate of change (slope of the secant line) between the 25th week and the 29th week. Is the slope of the secant line positive or negative? Discuss what the slope means in this context. (b) Is the fetus gaining weight faster between the 25th and 29th week, or between the 32nd and 36th week? Compare the slopes of both secant lines and discuss.

103. Fertility rates: Over the years, fertility rates for women in the United States (average number of children per woman) have varied a great deal, though in the twenty-first century they've begun to level out. The graph shown models this fertility rate for most of the twentieth century. (a) Calculate the average rate of change from the years 1920 to 1940. Is the slope of the secant line positive or negative? Discuss what the slope means in this context. (b) Calculate the average rate of change from the year 1940 to 1950. Is the slope of the secant line positive or negative? Discuss what the slope means in this context. (c) Was the fertility rate increasing faster from 1940 to 1950, or from 1980 to 1990? Compare the slope of both secant lines and comment.

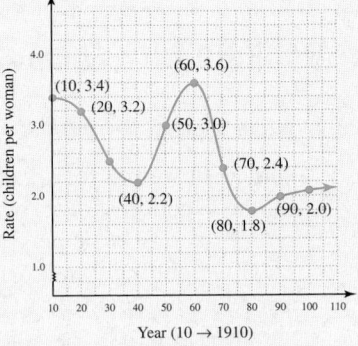

Source: Statistical History of the United States from Colonial Times to Present

▶ EXTENDING THE CONCEPT

104. A father challenges his son to a 400-m race, depicted in the graph shown here.

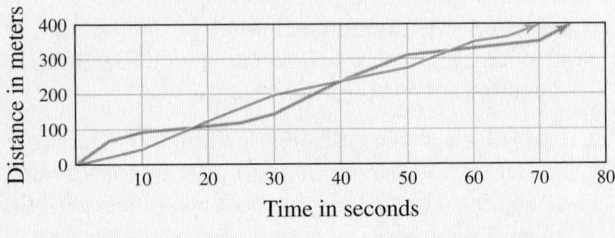

Father: —— Son: ——

   a. Who won and what was the approximate winning time?

   b. Approximately how many meters behind was the second place finisher?

   c. Estimate the number of seconds the father was in the lead in this race.

   d. How many times during the race were the father and son tied?

105. Sketch the graph of $f(x) = x$, then discuss how you could use this graph to obtain the graph of $F(x) = |x|$ without computing additional points. What would the graph of $g(x) = \dfrac{|x|}{x}$ look like?

106. Sketch the graph of $f(x) = x^2 - 4$, then discuss how you could use this graph to obtain the graph of $F(x) = |x^2 - 4|$ without computing additional points. Determine what the graph of $g(x) = \dfrac{|x^2 - 4|}{x^2 - 4}$ would look like.

107. If the equation of a function is given, the domain is implicitly defined by input values that generate real-valued outputs. But unless the graph is given or can be easily sketched, we must attempt to find the range analytically *by solving for x in terms of y*. We should note that sometimes this is an easy task, while at other times it is virtually impossible and we must rely on other methods. For the following functions, determine the implicit domain and find the range by solving for x in terms of y. **a.** $y = \frac{x-3}{x+2}$ **b.** $y = x^2 - 3$

▶ **MAINTAINING YOUR SKILLS**

108. (2.2) Which line has a steeper slope, the line through $(-5, 3)$ and $(2, 6)$, or the line through $(0, -4)$ and $(9, 4)$?

109. (R.6) Compute the sum and product indicated:
   a. $\sqrt{24} + 6\sqrt{54} - \sqrt{6}$
   b. $(2 + \sqrt{3})(2 - \sqrt{3})$

110. (1.5) Solve the equation using the quadratic formula, then check the result(s) using substitution:
$$x^2 - 4x + 1 = 0$$

111. (R.4) Factor the following polynomials completely:
   a. $x^3 - 3x^2 - 25x + 75$
   b. $2x^2 - 13x - 24$
   c. $8x^3 - 125$

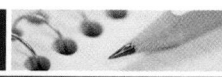

MID-CHAPTER CHECK

1. Sketch the graph of the line $4x - 3y = 12$. Plot and label at least three points.

2. Find the slope of the line passing through the given points: $(-3, 8)$ and $(4, -10)$.

3. In 2002, Data.com lost \$2 million. In 2003, they lost \$0.5 million. Will the slope of the line through these points be positive or negative? Why? Calculate the slope. Were you correct? Write the slope as a unit rate and explain what it means in this context.

4. Sketch the line passing through $(1, 4)$ with slope $m = \frac{-2}{3}$ (plot and label at least two points). Then find the equation of the line *perpendicular to this line* through $(1, 4)$.

5. Write the equation for line L_1 shown. Is this the graph of a function? Discuss why or why not.

6. Write the equation for line L_2 shown. Is this the graph of a function? Discuss why or why not.

7. For the graph of function $h(x)$ shown, (a) determine the value of $h(2)$; (b) state the domain; (c) determine the value of x for which $h(x) = -3$; and (d) state the range.

Exercises 5 and 6

Exercises 7 and 8

8. Judging from the appearance of the graph alone, compare the average rate of change from $x = 1$ to $x = 2$ to the rate of change from $x = 4$ to $x = 5$. Which rate of change is larger? How is that demonstrated graphically?

9. Find a linear function that models the graph of $F(p)$ given. Explain the slope of the line in this context, then use your model to predict the fox population when the pheasant population is 20,000.

Exercise 9

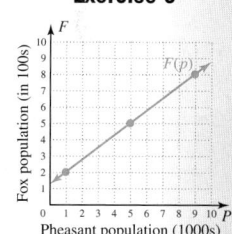

10. State the domain and range for each function below.

   a.
   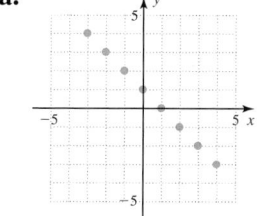

   b.
   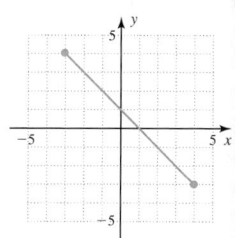

   c.
   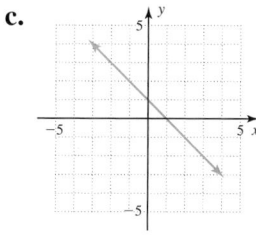

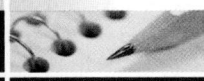

REINFORCING BASIC CONCEPTS

The Various Forms of a Linear Equation

In a study of mathematics, getting a glimpse of the "big picture" can be an enormous help. Learning mathematics is like building a skyscraper: The final height of the skyscraper ultimately depends on the strength of the foundation and quality of the frame supporting each new floor as it is built. Our work with linear functions and their graphs, while having a number of useful applications, is actually the foundation on which *much of your future work will be built*. The study of quadratic and polynomial functions and their applications all have their roots in linear equations. For this reason, it's important that you gain a certain fluency with linear functions—even to a point where things come to you effortlessly and automatically. This level of performance requires a strong desire and a sustained effort. We begin by reviewing the basic facts a student MUST know to reach this level. MUST is an acronym for <u>m</u>emorize, <u>u</u>nderstand, <u>s</u>ynthesize, and <u>t</u>each others. Don't be satisfied until you've done all four. Given points (x_1, y_1) and (x_2, y_2):

Forms and Formulas

slope formula	point-slope form	slope-intercept form	standard form
$m = \dfrac{y_2 - y_1}{x_2 - x_1}$	$y - y_1 = m(x - x_1)$	$y = mx + b$	$Ax + By = C$
given any two points on the line	given slope m and any point (x_1, y_1)	given slope m and y-intercept $(0, b)$	also used in linear systems (Chapter 6)

Characteristics of Lines

y-intercept	x-intercept	increasing	decreasing
$(0, y)$	$(x, 0)$	$m > 0$	$m < 0$
let $x = 0$, solve for y	let $y = 0$, solve for x	line slants upward from left to right	line slants downward from left to right

Practice for Speed and Accuracy

For the two points given, (a) compute the slope of the line and state whether the line is increasing or decreasing; (b) find the equation of the line using point-slope form; (c) write the equation in slope-intercept form; (d) write the equation in standard form; and (e) find the x- and y-intercepts and graph the line.

1. $P_1(0, 5); P_2(6, 7)$ **2.** $P_1(3, 2); P_2(0, 9)$ **3.** $P_1(3, 2); P_2(9, 5)$

4. $P_1(-5, -4); P_2(3, 2)$ **5.** $P_1(-2, 5); P_2(6, -1)$ **6.** $P_1(2, -7); P_2(-8, -2)$

2.5 | Analyzing the Graph of a Function

Learning Objectives

In Section 2.5 you will learn how to:

☐ **A.** Determine whether a function is even, odd, or neither

☐ **B.** Determine intervals where a function is positive or negative

☐ **C.** Determine where a function is increasing or decreasing

☐ **D.** Identify the maximum and minimum values of a function

☐ **E.** Develop a formula to calculate rates of change for any function

In this section, we'll consolidate and refine many of the ideas we've encountered related to functions. When functions and graphs are applied as real-world models, we create a numeric and visual representation that enables an informed response to questions involving *maximum* efficiency, *positive* returns, *increasing* costs, and other relationships that can have a great impact on our lives.

A. Graphs and Symmetry

While the domain and range of a function will remain dominant themes in our study, for the moment we turn our attention to other characteristics of a function's graph. We begin with the concept of symmetry.

Symmetry with Respect to the y-Axis

Figure 2.38

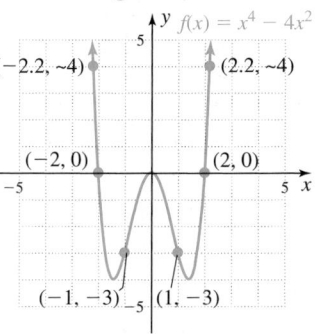

Consider the graph of $f(x) = x^4 - 4x^2$ shown in Figure 2.38, where the portion of the graph to the left of the y-axis appears to be a mirror image of the portion to the right. A function is **symmetric to the y-axis** if, given any point (x, y) on the graph, the point $(-x, y)$ is also on the graph. We note that $(-1, -3)$ is on the graph, as is $(1, -3)$, and that $(-2, 0)$ is an x-intercept of the graph, as is $(2, 0)$. Functions that are symmetric to the y-axis are also known as **even functions** and in general we have:

Even Functions: y-Axis Symmetry

A function f is an *even function* if and only if, for each point (x, y) on the graph of f, the point $(-x, y)$ is also on the graph. *In function notation*

$$f(-x) = f(x)$$

Symmetry can be a great help in graphing new functions, enabling us to plot fewer points, and to complete the graph using properties of symmetry.

EXAMPLE 1 ▶ **Graphing an Even Function Using Symmetry**

   a. The function $g(x)$ in Figure 2.39 is known to be even. Draw the complete graph (only the left half is shown).

   Figure 2.39

   b. Show that $h(x) = x^{\frac{2}{3}}$ is an even function using the arbitrary value $x = k$ [show $h(-k) = h(k)$], then sketch the complete graph using $h(0)$, $h(1)$, $h(8)$, and y-axis symmetry.

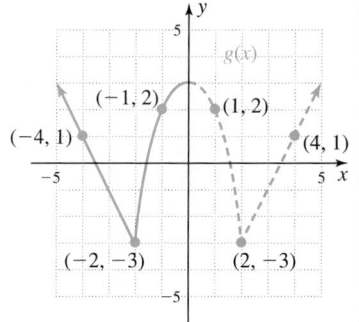

Solution ▶ a. To complete the graph of g (see Figure 2.39) use the points $(-4, 1)$, $(-2, -3)$, $(-1, 2)$, and y-axis symmetry to find additional points. The corresponding ordered pairs are $(4, 1)$, $(2, -3)$, and $(1, 2)$, which we use to help draw a "mirror image" of the partial graph given.

   b. To prove that $h(x) = x^{\frac{2}{3}}$ is an even function, we must show $h(-k) = h(k)$ for any constant k. After writing $x^{\frac{2}{3}}$ as $\left[x^2\right]^{\frac{1}{3}}$, we have:

   Figure 2.40

$$h(-k) \stackrel{?}{=} h(k) \qquad \text{first step of proof}$$

$$\left[(-k)^2\right]^{\frac{1}{3}} \stackrel{?}{=} \left[(k)^2\right]^{\frac{1}{3}} \qquad \text{evaluate } h(-k) \text{ and } h(k)$$

$$\sqrt[3]{(-k)^2} \stackrel{?}{=} \sqrt[3]{(k)^2} \qquad \text{radical form}$$

$$\sqrt[3]{k^2} = \sqrt[3]{k^2} \checkmark \qquad \text{result: } (-k)^2 = k^2$$

WORTHY OF NOTE

The proof can also be demonstrated by writing $x^{\frac{2}{3}}$ as $\left(x^{\frac{1}{3}}\right)^2$, and you are asked to complete this proof in Exercise 82.

Using $h(0) = 0$, $h(1) = 1$, and $h(8) = 4$ with y-axis symmetry produces the graph shown in Figure 2.40.

Now try Exercises 7 through 12 ▶

Symmetry with Respect to the Origin

Another common form of symmetry is known as **symmetry to the origin.** As the name implies, the graph is somehow "centered" at (0, 0). This form of symmetry is easy to see for closed figures with their center at (0, 0), like certain polygons, circles, and ellipses (these will exhibit both *y*-axis symmetry *and* symmetry to the origin). Note the relation graphed in Figure 2.41 contains the points (−3, 3) and (3, −3), along with (−1, −4) and (1, 4). But the function *f*(*x*) in Figure 2.42 also contains these points and is, in the same sense, symmetric to the origin (the paired points are on opposite sides of the *x*- and *y*-axes, and a like distance from the origin).

Figure 2.41

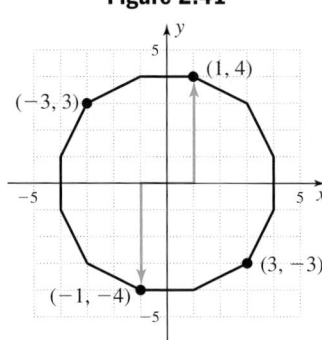

Figure 2.42

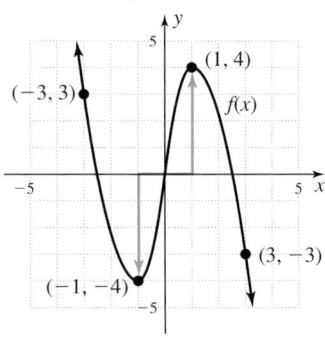

Functions symmetric to the origin are known as **odd functions** and in general we have:

Odd Functions: Symmetry about the Origin

A function *f* is an *odd function* if and only if, for each point (*x*, *y*) on the graph of *f*, the point (−*x*, −*y*) is also on the graph. *In function notation*

$$f(-x) = -f(x)$$

EXAMPLE 2 ▶   Graphing an Odd Function Using Symmetry

a. In Figure 2.43, the function *g*(*x*) given is known to be *odd*. Draw the complete graph (only the left half is shown).

b. Show that $h(x) = x^3 - 4x$ is an odd function using the arbitrary value *x* = *k* [show *h*(−*x*) = −*h*(*x*)], then sketch the graph using *h*(−2), *h*(−1), *h*(0), and odd symmetry.

Solution ▶   a. To complete the graph of *g*, use the points (−6, 3), (−4, 0), and (−2, 2) and odd symmetry to find additional points. The corresponding ordered pairs are (6, −3), (4, 0), and (2, −2), which we use to help draw a "mirror image" of the partial graph given (see Figure 2.43).

Figure 2.43

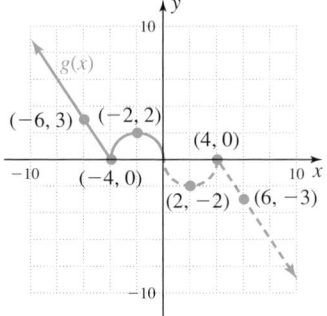

Figure 2.44

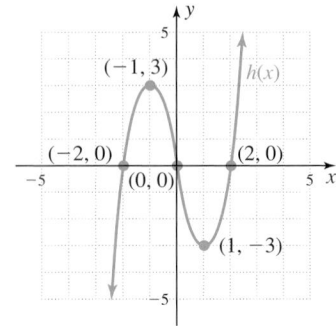

WORTHY OF NOTE

While the graph of an even function may or may not include the point (0, 0), the graph of an odd function will *always* contain this point.

b. To prove that $h(x) = x^3 - 4x$ is an odd function, we must show that $h(-k) = -h(k)$.

$$h(-k) \overset{?}{=} -h(k)$$
$$(-k)^3 - 4(-k) \overset{?}{=} -[k^3 - 4k]$$
$$-k^3 + 4k = -k^3 + 4k \checkmark$$

Using $h(-2) = 0$, $h(-1) = 3$, and $h(0) = 0$ with symmetry about the origin produces the graph shown in Figure 2.44.

☑ **A.** You've just learned how to determine whether a function is even, odd, or neither

Now try Exercises 13 through 24 ▶

B. Intervals Where a Function Is Positive or Negative

Consider the graph of $f(x) = x^2 - 4$ shown in Figure 2.45, which has x-intercepts at $(-2, 0)$ and $(2, 0)$. Since x-intercepts have the form $(x, 0)$ they are also called the **zeroes** of the function (the x-input causes an output of 0). Just as zero on the number line separates negative numbers from positive numbers, the zeroes of a function that crosses the x-axis separate x-intervals where a function is negative from x-intervals where the function is positive. Noting that outputs (y-values) are positive in Quadrants I and II, $f(x) > 0$ in intervals where its graph is *above the x-axis*. Conversely, $f(x) < 0$ in x-intervals where its graph is *below the x-axis*. To illustrate, compare the graph of f in Figure 2.45, with that of g in Figure 2.46.

Figure 2.45   **Figure 2.46**

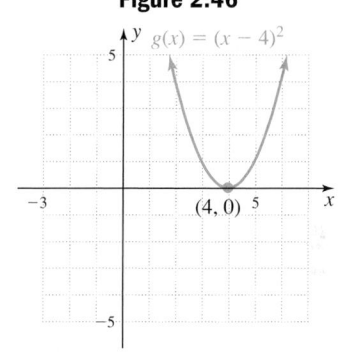

WORTHY OF NOTE

These observations form the basis for studying polynomials of higher degree, where we extend the idea to factors of the form $(x - r)^n$ in a study of **roots of multiplicity** (also see the *Calculator Exploration and Discovery* feature in this chapter).

The graph of f is a parabola, with x-intercepts of $(-2, 0)$ and $(2, 0)$. Using our previous observations, we note $f(x) \geq 0$ for $x \in (-\infty, -2] \cup [2, \infty)$ and $f(x) < 0$ for $x \in (-2, 2)$. The graph of g is also a parabola, but is entirely above or on the x-axis, showing $g(x) \geq 0$ for $x \in \mathbb{R}$. The difference is that zeroes coming from factors of the form $(x - r)$ (with degree 1) allow the graph to cross the x-axis. The zeroes of f came from $(x + 2)(x - 2) = 0$. Zeroes that come from factors of the form $(x - r)^2$ (with degree 2) cause the graph to "bounce" off the x-axis since all outputs must be nonnegative. The zero of g came from $(x - 4)^2 = 0$.

EXAMPLE 3 ▶ **Solving an Inequality Using a Graph**

Use the graph of $g(x) = x^3 - 2x^2 - 4x + 8$ given to solve the inequalities
  a. $g(x) \geq 0$
  b. $g(x) < 0$

Solution ▶ From the graph, the zeroes of g (x-intercepts) occur at $(-2, 0)$ and $(2, 0)$. a. For $g(x) \geq 0$, the graph must be on or above the x-axis, meaning the solution is $x \in [-2, \infty)$. b. For $g(x) < 0$, the graph must be below the x-axis, and the solution is $x \in (-\infty, -2)$. As we might have anticipated from the graph, factoring by grouping gives $g(x) = (x + 2)(x - 2)^2$, with the graph crossing the x-axis at -2, and bouncing off the x-axis (intersects without crossing) at $x = 2$.

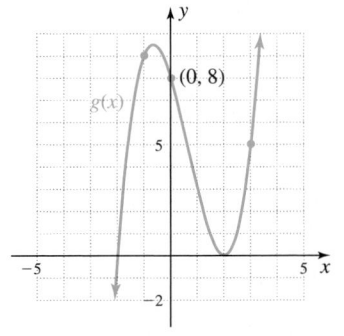

Now try Exercises 25 through 28 ▶

Even if the function is not a polynomial, the zeroes can still be used to find x-intervals where the function is positive or negative.

EXAMPLE 4 ▶ **Solving an Inequality Using a Graph**

For the graph of $r(x) = \sqrt{x + 1} - 2$ shown, solve
  a. $r(x) \leq 0$
  b. $r(x) > 0$

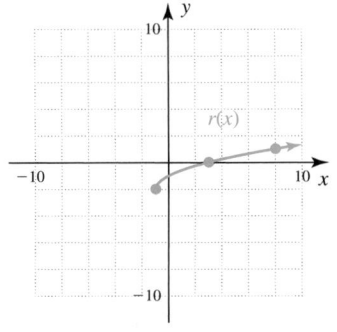

Solution ▶ **a.** The only zero of r is at $(3, 0)$. The graph is on or below the x-axis for $x \in [-1, 3]$, so $r(x) \leq 0$ in this interval.

☑ **B.** You've just learned how to determine intervals where a function is positive or negative

  b. The graph is above the x-axis for $x \in (3, \infty)$, and $r(x) > 0$ in this interval.

Now try Exercises 29 through 32 ▶

C. Intervals Where a Function Is Increasing or Decreasing

In our study of linear graphs, we said a graph was increasing if it "rose" when viewed from left to right. More generally, we say the graph of a function is increasing *on a given interval* if larger and larger x-values produce larger and larger y-values. This suggests the following tests for intervals where a function is increasing or decreasing.

Increasing and Decreasing Functions

Given an interval I that is a subset of the domain, with x_1 and x_2 in I and $x_2 > x_1$,
  1. A function is increasing on I if $f(x_2) > f(x_1)$ for all x_1 and x_2 in I (larger inputs produce larger outputs).
  2. A function is decreasing on I if $f(x_2) < f(x_1)$ for all x_1 and x_2 in I (larger inputs produce smaller outputs).
  3. A function is constant on I if $f(x_2) = f(x_1)$ for all x_1 and x_2 in I (larger inputs produce identical outputs).

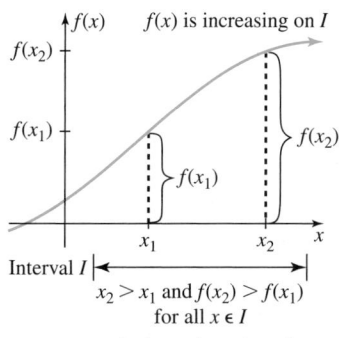

$f(x)$ is increasing on I

$x_2 > x_1$ and $f(x_2) > f(x_1)$
for all $x \in I$

graph rises when viewed
from left to right

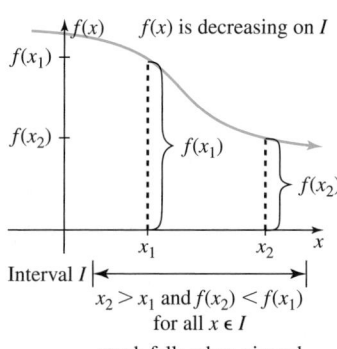

$f(x)$ is decreasing on I

$x_2 > x_1$ and $f(x_2) < f(x_1)$
for all $x \in I$

graph falls when viewed
from left to right

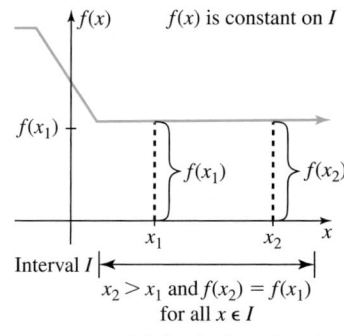

$f(x)$ is constant on I

$x_2 > x_1$ and $f(x_2) = f(x_1)$
for all $x \in I$

graph is level when viewed
from left to right

WORTHY OF NOTE

Questions about the behavior
of a function are asked with
respect to the *y* outputs:
where is the *function* positive,
where is the *function* increasing, etc. Due to the input/
output, cause/effect nature of
functions, the response is
given in terms of *x*, that is,
what is *causing* outputs to be
negative, or to be decreasing.

Consider the graph of $f(x) = -x^2 + 4x + 5$
in Figure 2.47. Since the graph opens downward
with the vertex at $(2, 9)$, the function must increase
until it reaches this maximum value at $x = 2$, and
decrease thereafter. Notationally we'll write this as
$f(x)\uparrow$ for $x \in (-\infty, 2)$ and $f(x)\downarrow$ for $x \in (2, \infty)$.
Using the interval $(-3, 2)$ shown, we see that any
larger input value from the interval will indeed
produce a larger output value, and $f(x)\uparrow$ on the
interval. For instance,

$1 > -2 \qquad\qquad x_2 > x_1$

and and

$f(1) > f(-2) \qquad f(x_2) > f(x_1)$
$8 > -7$

Figure 2.47

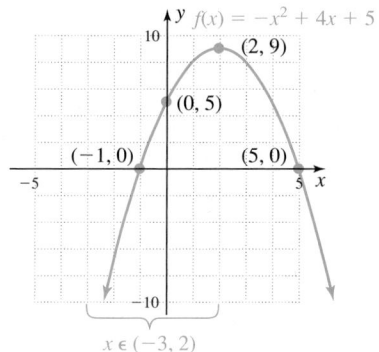

EXAMPLE 5 ▶ **Finding Intervals Where a Function Is Increasing
or Decreasing**

Use the graph of $v(x)$ given to name the interval(s)
where v is increasing, decreasing, or constant.

Solution ▶ From left to right, the graph of v increases until
leveling off at $(-2, 2)$, then it remains constant
until reaching $(1, 2)$. The graph then increases
once again until reaching a peak at $(3, 5)$ and
decreases thereafter. The result is $v(x)\uparrow$ for
$x \in (-\infty, -2) \cup (1, 3)$, $v(x)\downarrow$ for $x \in (3, \infty)$, and
$v(x)$ is constant for $x \in (-2, 1)$.

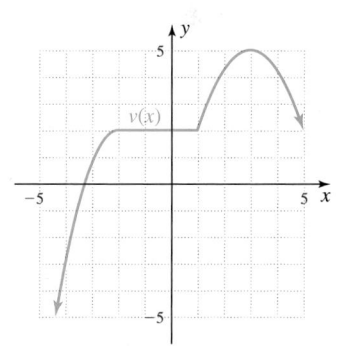

Now try Exercises 33 through 36 ▶

Notice the graph of *f* in Figure 2.47 and the graph of *v* in Example 5 have something in common. It appears that both the far left and far right branches of each graph
point downward (in the negative *y*-direction). We say that the **end behavior** of both
graphs is identical, which is the term used to describe what happens to a graph as $|x|$
becomes very large. For $x > 0$, we say a graph is, "up on the right" or "down on the
right," depending on the direction the "end" is pointing. For $x < 0$, we say the graph
is "up on the left" or "down on the left," as the case may be.

EXAMPLE 6 ▶ **Describing the End Behavior of a Graph**

The graph of $f(x) = x^3 - 3x$ is shown. Use the graph to name intervals where f is increasing or decreasing, and comment on the end-behavior of the graph.

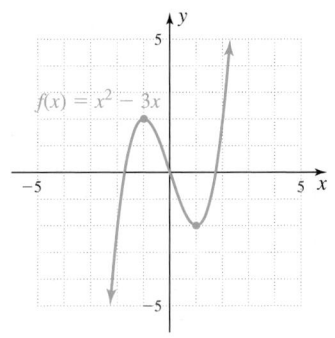

Solution ▶ From the graph we observe that: $f(x)\uparrow$ for $x \in (-\infty, -1) \cup (1, \infty)$, and $f(x)\downarrow$ for $x \in (-1, 1)$. The end behavior of the graph is down on the left, up on the right (down/up).

☑ **C. You've just learned how to determine where a function is increasing or decreasing**

Now try Exercises 37 through 40 ▶

D. More on Maximum and Minimum Values

The y-coordinate of the vertex of a parabola where $a < 0$, and the y-coordinate of "peaks" from other graphs are called **maximum values**. A **global maximum** (also called an *absolute* maximum) names the largest range value over the entire domain. A local **maximum** (also called a *relative* maximum) gives the largest range value in a specified interval; and an **endpoint maximum** can occur at an endpoint of the domain. The same can be said for the corresponding minimum values.

 We will soon develop the ability to locate maximum and minimum values for quadratic and other functions. In future courses, methods are developed to help locate maximum and minimum values for almost *any* function. For now, our work will rely chiefly on a function's graph.

EXAMPLE 7 ▶ **Analyzing Characteristics of a Graph**

Analyze the graph of function f shown in Figure 2.48. Include specific mention of
  a. domain and range,
  b. intervals where f is increasing or decreasing,
  c. maximum (max) and minimum (min) values,
  d. intervals where $f(x) \geq 0$ and $f(x) < 0$,
  e. whether the function is even, odd, or neither.

Figure 2.48

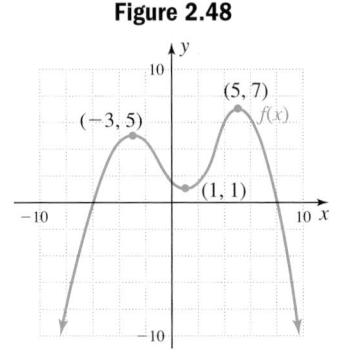

Solution ▶ **a.** Using vertical and horizontal boundary lines show the domain is $x \in \mathbb{R}$, with range: $y \in (-\infty, 7]$.
  b. $f(x)\uparrow$ for $x \in (-\infty, -3) \cup (1, 5)$ shown in blue in Figure 2.49, and $f(x)\downarrow$ for $x \in (-3, 1) \cup (5, \infty)$ as shown in **red.**
  c. From Part (b) we find that $y = 5$ at $(-3, 5)$ and $y = 7$ at $(5, 7)$ are local maximums, with a local minimum of $y = 1$ at $(1, 1)$. The point $(5, 7)$ is also a global maximum (there is no global minimum).
  d. $f(x) \geq 0$ for $x \in [-6, 8]$; $f(x) < 0$ for $x \in (-\infty, -6) \cup (8, \infty)$
  e. The function is neither even nor odd.

Figure 2.49

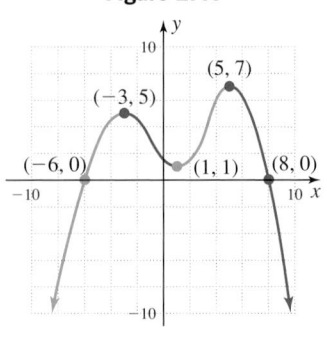

☑ **D. You've just learned how to identify the maximum and minimum values of a function**

Now try Exercises 41 through 48 ▶

The ideas presented here can be applied to functions of all kinds, including rational functions, piecewise-defined functions, step functions, and so on. There is a wide variety of applications in **Exercises 51 through 58.**

E. Rates of Change and the Difference Quotient

 We complete our study of graphs by revisiting the concept of average rates of change. In many business, scientific, and economic applications, it is this attribute of a function that draws the most attention. In Section 2.4 we computed average rates of change by selecting two points from a graph, and computing the slope of the secant line: $m = \dfrac{\Delta y}{\Delta x} = \dfrac{y_2 - y_1}{x_2 - x_1}$. With a simple change of notation, we can *use the function's equation* rather than relying on a graph. Note that y_2 corresponds to the function evaluated at x_2: $y_2 = f(x_2)$. Likewise, $y_1 = f(x_1)$. Substituting these into the slope formula yields $\dfrac{\Delta y}{\Delta x} = \dfrac{f(x_2) - f(x_1)}{x_2 - x_1}$, giving the average rate of change between x_1 and x_2 *for any function f* (assuming the function is smooth and continuous between x_1 and x_2).

Average Rate of Change

For a function f and $[x_1, x_2]$ a subset of the domain, the average rate of change between x_1 and x_2 is

$$\frac{\Delta y}{\Delta x} = \frac{f(x_2) - f(x_1)}{x_2 - x_1}, x_1 \neq x_2$$

Average Rates of Change Applied to Projectile Velocity

A projectile is any object that is thrown, shot, or cast upward, with no continuing source of propulsion. The object's height (in feet) after t sec is modeled by the function $h(t) = -16t^2 + vt + k$, where v is the initial velocity of the projectile, and k is the height of the object at contact. For instance, if a soccer ball is kicked upward from ground level ($k = 0$) with an initial speed of 64 ft/sec, the height of the ball t sec later is $h(t) = -16t^2 + 64t$. From Section 2.5, we recognize the graph will be a parabola and evaluating the function for $t = 0$ to 4 produces Table 2.4 and the graph shown in Figure 2.50. Experience tells us the ball is traveling at a faster rate immediately after being kicked, as compared to when it nears its maximum height where it momentarily stops, then begins its descent. In other words, the rate of change $\dfrac{\Delta \text{height}}{\Delta \text{time}}$ has a larger value at any time prior to reaching its maximum height. To quantify this we'll compute the average rate of change between $t = 0.5$ and $t = 1$, and compare it to the average rate of change between $t = 1$ and $t = 1.5$.

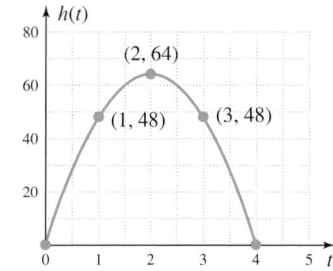

Table 2.4

Time in seconds	Height in feet
0	0
1	48
2	64
3	48
4	0

Figure 2.50

WORTHY OF NOTE

Keep in mind the graph of h represents the relationship between the soccer ball's height in feet and the elapsed time t. It does not model the actual path of the ball.

EXAMPLE 8 ▶ Calculating Average Rates of Change

For the projectile function $h(t) = -16t^2 + 64t$, find

a. the average rate of change for $t \in [0.5, 1]$

b. the average rate of change for $t \in [1, 1.5]$.

Then graph the secant lines representing these average rates of change and comment.

Solution ▶ Using the given intervals in the formula $\dfrac{\Delta h}{\Delta t} = \dfrac{h(t_2) - h(t_1)}{t_2 - t_1}$ yields

a. $\dfrac{\Delta h}{\Delta t} = \dfrac{h(1) - h(0.5)}{1 - (0.5)}$

$= \dfrac{48 - 28}{0.5}$

$= 40$

b. $\dfrac{\Delta h}{\Delta t} = \dfrac{h(1.5) - h(1)}{1.5 - 1}$

$= \dfrac{60 - 48}{0.5}$

$= 24$

For $t \in [0.5, 1]$, the average rate of change is $\frac{40}{1}$, meaning the height of the ball is increasing at an average rate of 40 ft/sec. For $t \in [1, 1.5]$, the average rate of change has slowed to $\frac{24}{1}$, and the soccer ball's height is increasing at only 24 ft/sec. The secant lines representing these rates of change are shown in the figure, where we note the line from the first interval (in **red**), has a steeper slope than the line from the second interval (in **blue**).

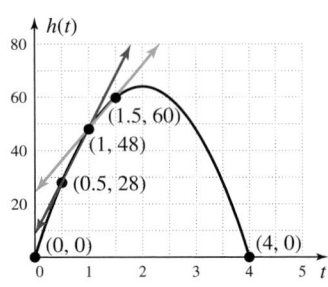

Now try Exercises 59 through 64 ▶

The approach in Example 8 works very well, but requires us to recalculate $\dfrac{\Delta y}{\Delta x}$ for each new interval. Using a slightly different approach, we can develop a general *formula* for the average rate of change. This is done by selecting a point $x_1 = x$ from the domain, then a point $x_2 = x + h$ that is very close to x. Here, $h \neq 0$ is assumed to be a small, arbitrary constant, meaning the interval $[x, x + h]$ is very small as well. Substituting $x + h$ for x_2 and x for x_1 in the rate of change formula gives $\dfrac{\Delta y}{\Delta x} = \dfrac{f(x + h) - f(x)}{(x + h) - x} = \dfrac{f(x + h) - f(x)}{h}$. The result is called the **difference quotient** and represents the average rate of change between x and $x + h$, or equivalently, the slope of the secant line for this interval.

The Difference Quotient

For a function $f(x)$ and constant $h \neq 0$,

$$\frac{f(x + h) - f(x)}{h}$$

is the difference quotient for f.

Note the formula has three parts: (1) the function f evaluated at $x + h \to f(x + h)$, (2) the function f itself, and (3) the constant h. For convenience, the expression $f(x + h)$ can be evaluated and simplified prior to its use in the difference quotient.

$$\frac{\overset{(1)}{f(x + h)} - \overset{(2)}{f(x)}}{\underset{(3)}{h}}$$

EXAMPLE 9 ▶ **Computing a Difference Quotient and Average Rates of Change**

For $f(x) = x^2 - 4x$,

　　a. Compute the difference quotient.

　　b. Find the average rate of change in the intervals [1.9, 2.0] and [3.6, 3.7].

　　c. Sketch the graph of f along with the secant lines and comment on what you notice.

Solution ▶ **a.** For $f(x) = x^2 - 4x, f(x + h) = (x + h)^2 - 4(x + h)$
$$= x^2 + 2xh + h^2 - 4x - 4h$$

Using this result in the difference quotient yields,

$$\frac{f(x + h) - f(x)}{h} = \frac{(x^2 + 2xh + h^2 - 4x - 4h) - (x^2 - 4x)}{h} \quad \text{substitute into the difference quotient}$$

$$= \frac{x^2 + 2xh + h^2 - 4x - 4h - x^2 + 4x}{h} \quad \text{eliminate parentheses}$$

$$= \frac{2xh + h^2 - 4h}{h} \quad \text{combine like terms}$$

$$= \frac{h(2x + h - 4)}{h} \quad \text{factor out } h$$

$$= 2x - 4 + h \quad \text{result}$$

b. For the interval [1.9, 2.0], $x = 1.9$ and $h = 0.1$. The slope of the secant line is $\frac{\Delta y}{\Delta x} = 2(1.9) - 4 + 0.1 = -0.1$. For the interval [3.6, 3.7], $x = 3.6$ and $h = 0.1$. The slope of this secant line is $\frac{\Delta y}{\Delta x} = 2(3.6) - 4 + 0.1 = 3.3$.

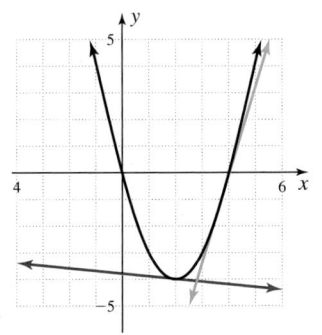

c. After sketching the graph of f and the secant lines from each interval (see the figure), we note the slope of the first line (in **red**) is negative and very near zero, while the slope of the second (in blue) is positive and very steep.

Now try Exercises 65 through 76 ▶

You might be familiar with Galileo Galilei and his studies of gravity. According to popular history, he demonstrated that unequal weights will fall equal distances in equal time periods, by dropping cannonballs from the upper floors of the Leaning Tower of Pisa. Neglecting air resistance, this distance an object falls is modeled by the function $d(t) = 16t^2$, where $d(t)$ represents the distance fallen after t sec. Due to the effects of gravity, the velocity of the object increases as it falls. In other words, the velocity or the average rate of change $\frac{\Delta \textbf{distance}}{\Delta \textbf{time}}$ is a nonconstant (increasing) rate of change. We can analyze this rate of change using the difference quotient.

EXAMPLE 10 ▶ **Applying the Difference Quotient in Context**

A construction worker drops a heavy wrench from atop the girder of new skyscraper. Use the function $d(t) = 16t^2$ to

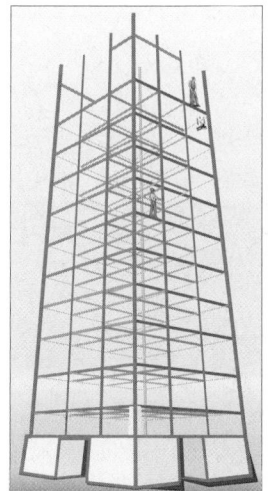

a. Compute the distance the wrench has fallen after 2 sec and after 7 sec.

b. Find a formula for the velocity of the wrench (average rate of change in distance per unit time).

c. Use the formula to find the rate of change in the intervals [2, 2.01] and [7, 7.01].

d. Graph the function and the secant lines representing the average rate of change. Comment on what you notice.

Solution ▶ **a.** Substituting $t = 2$ and $t = 7$ in the given function yields

$$d(2) = 16(2)^2 \qquad d(7) = 16(7)^2 \quad \text{evaluate } d(t) = 16t^2$$
$$= 16(4) \qquad\qquad = 16(49) \quad \text{square input}$$
$$= 64 \qquad\qquad\quad = 784 \qquad \text{multiply}$$

After 2 sec, the wrench has fallen 64 ft; after 7 sec, the wrench has fallen 784 ft.

b. For $d(t) = 16t^2$, $d(t + h) = 16(t + h)^2$, which we compute separately.

$$d(t + h) = 16(t + h)^2 \qquad\qquad \text{substitute } t + h \text{ for } t$$
$$= 16(t^2 + 2th + h^2) \qquad \text{square binomial}$$
$$= 16t^2 + 32th + 16h^2 \qquad \text{distribute 16}$$

Using this result in the difference quotient yields

$$\frac{d(t + h) - d(t)}{h} = \frac{(16t^2 + 32th + 16h^2) - 16t^2}{h} \qquad \text{substitute into the difference quotient}$$

$$= \frac{16t^2 + 32th + 16h^2 - 16t^2}{h} \qquad \text{eliminate parentheses}$$

$$= \frac{32th + 16h^2}{h} \qquad \text{combine like terms}$$

$$= \frac{h(32t + 16h)}{h} \qquad \text{factor out } h \text{ and simplify}$$

$$= 32t + 16h \qquad \text{result}$$

For any number of seconds t and h a small increment of time thereafter, the velocity of the wrench is modeled by $\dfrac{\Delta \textbf{distance}}{\Delta \textbf{time}} = \dfrac{\textbf{32}t + \textbf{16}h}{\textbf{1}}$.

c. For the interval $[t, t + h] = [2, 2.01]$, $t = 2$ and $h = 0.01$:

$$\frac{\Delta \text{distance}}{\Delta \text{time}} = \frac{32(2) + 16(0.01)}{1} \qquad \text{substitute 2 for } t \text{ and 0.01 for } h$$

$$= 64 + 0.16 = 64.16$$

Two seconds after being dropped, the velocity of the wrench is approximately 64.16 ft/sec. For the interval $[t, t + h] = [7, 7.01]$, $t = 7$ and $h = 0.01$:

$$\frac{\Delta \text{distance}}{\Delta \text{time}} = \frac{32(7) + 16(0.01)}{1} \qquad \text{substitute 7 for } t \text{ and 0.01 for } h$$

$$= 224 + 0.16 = 224.16$$

Seven seconds after being dropped, the velocity of the wrench is approximately 224.16 ft/sec (about 153 mph).

d.

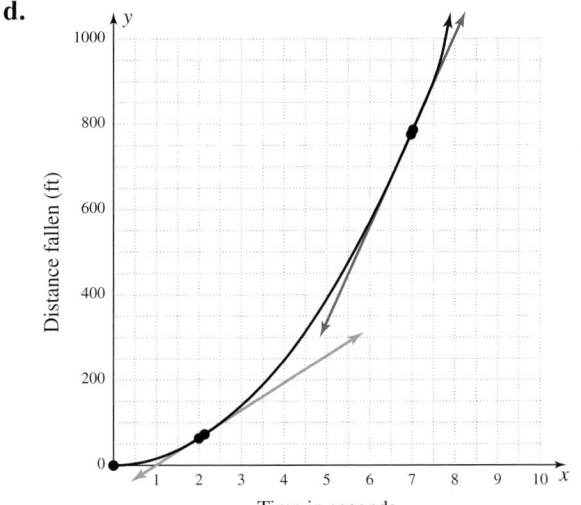

The velocity increases with time, as indicated by the steepness of each secant line.

 **E.** You've learned how to develop a formula to calculate rates of change for any function

Now try Exercises 77 and 78 ▶

TECHNOLOGY HIGHLIGHT

Locating Zeroes, Maximums, and Minimums

Graphically, the **zeroes** of a function appear as x-intercepts with coordinates $(x, 0)$. An estimate for these zeroes can easily be found using a graphing calculator. To illustrate, enter the function $y = x^2 - 8x + 9$ on the [Y=] screen and graph it using the standard window ([ZOOM] 6). We access the option for finding zeroes by pressing [2nd] [TRACE] (**CALC**), which displays the screen shown in Figure 2.51. Pressing the number "2" selects **2:zero** and returns you to the graph, where you're asked to enter a "Left Bound." The calculator is asking you to narrow the area it has to search. Select any number conveniently to the left of the x-intercept you're interested in. For this graph, we entered a left bound of "0" (press [ENTER]). The calculator marks this choice with a "▶" marker (pointing to the right), then asks you to enter a "Right Bound." Select any value to the right of the x-intercept, but be sure the value you enter *bounds only one intercept* (see Figure 2.52). For this graph, a choice of 10 would include both x-intercepts, while a choice of 3 would bound only the intercept on the left. After entering 3, the calculator asks for a "Guess." This option is used when there is more than one zero in the interval, and most of the time we'll bypass this option by pressing [ENTER] again. The calculator then finds the zero in the selected interval (if it exists), with the coordinates displayed at the bottom of the screen (Figure 2.53).

The maximum and minimum values of a function are located in the same way. Enter $y = x^3 - 3x - 2$ on the [Y=] screen and graph the function. As seen in Figure 2.54, it appears a local maximum occurs near $x = -1$. To check, we access the **CALC** **4:maximum** option, which returns you to the graph and asks you for a *Left Bound,* a *Right Bound,* and a *Guess* as before. After entering a left bound of "-3" and a right bound of "0," and

Figure 2.51

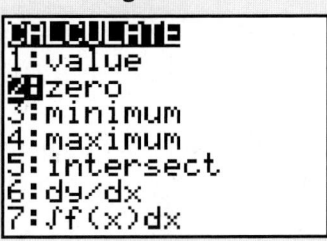

Figure 2.52

Figure 2.53

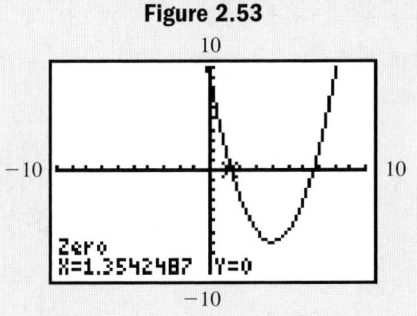

Figure 2.54

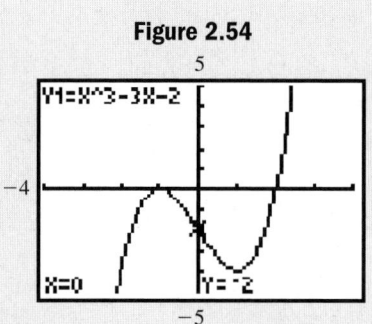

Figure 2.55

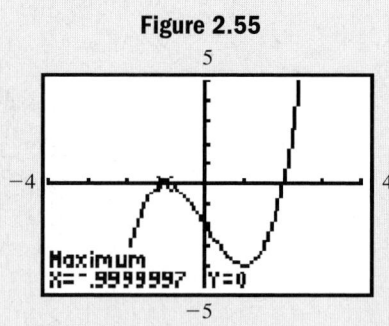

bypassing the Guess option (note the "▶" and "◀" markers), the calculator locates the maximum you selected, and again displays the coordinates. Due to the algorithm used by the calculator to find these values, a decimal number is sometimes displayed, even if the actual value is an integer (see Figure 2.55).

Use a calculator to find all zeroes and to locate the local maximum and minimum values. Round to the nearest hundredth as needed.

Exercise 1: $y = 2x^2 + 4x - 5$

Exercise 2: $y = w^3 - 3w + 1$

Exercise 3: $y = x^2 - 8x + 9$

Exercise 4: $y = x^3 - 2x^2 - 4x + 8$

Exercise 5: $y = x^4 - 5x^2 - 2x$

Exercise 6: $y = x\sqrt{x + 4}$

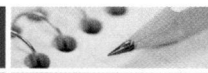

2.5 EXERCISES

▶ CONCEPTS AND VOCABULARY

Fill in each blank with the appropriate word or phrase. Carefully reread the section if needed.

1. The graph of a polynomial will cross through the x-axis at zeroes of _____ factors of degree 1, and _____ off the x-axis at the zeroes from linear factors of degree 2.

2. If $f(-x) = f(x)$ for all x in the domain, we say that f is an _____ function and symmetric to the _____ axis. If $f(-x) = -f(x)$, the function is _____ and symmetric to the _____.

3. If $f(x_2) > f(x_1)$ for $x_1 < x_2$ for all x in a given interval, the function is _____ in the interval.

4. If $f(c) \geq f(x)$ for all x in a specified interval, we say that $f(c)$ is a local _____ for this interval.

5. Discuss/Explain the following statement and give an example of the conclusion it makes. "If a function f is decreasing to the left of $(c, f(c))$ and increasing to the right of $(c, f(c))$, then $f(c)$ is either a local or a global minimum."

6. Without referring to notes or textbook, list as many features/attributes as you can that are related to analyzing the graph of a function. Include details on how to locate or determine each attribute.

▶ DEVELOPING YOUR SKILLS

The following functions are known to be even. Complete each graph using symmetry.

7.

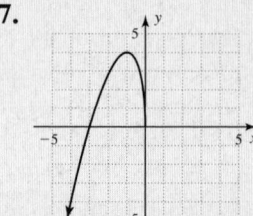

8.

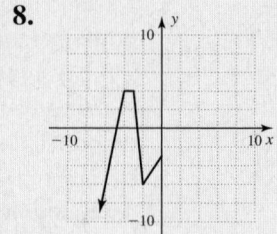

Determine whether the following functions are even:
$f(-k) = f(k)$.

9. $f(x) = -7|x| + 3x^2 + 5$ **10.** $p(x) = 2x^4 - 6x + 1$

11. $g(x) = \frac{1}{3}x^4 - 5x^2 + 1$ **12.** $q(x) = \frac{1}{x^2} - |x|$

The following functions are known to be odd. Complete each graph using symmetry.

13.

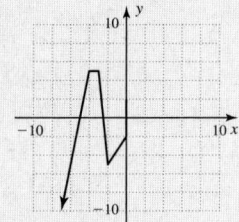

14.

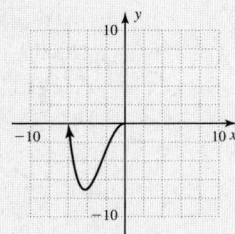

Determine whether the following functions are odd:
$f(-k) = -f(k)$.

15. $f(x) = 4\sqrt[3]{x} - x$ **16.** $g(x) = \frac{1}{2}x^3 - 6x$

17. $p(x) = 3x^3 - 5x^2 + 1$ **18.** $q(x) = \frac{1}{x} - x$

Determine whether the following functions are even, odd, or neither.

19. $w(x) = x^3 - x^2$ **20.** $q(x) = \frac{3}{4}x^2 + 3|x|$

21. $p(x) = 2\sqrt[3]{x} - \frac{1}{4}x^3$ **22.** $g(x) = x^3 + 7x$

23. $v(x) = x^3 + 3|x|$ **24.** $f(x) = x^4 + 7x^2 - 30$

Use the graphs given to solve the inequalities indicated. Write all answers in interval notation.

25. $f(x) = x^3 - 3x^2 - x + 3; f(x) \geq 0$

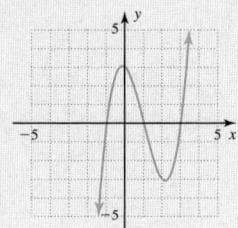

26. $f(x) = x^3 - 2x^2 - 4x + 8; f(x) > 0$

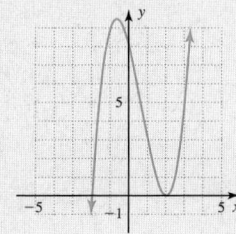

27. $f(x) = x^4 - 2x^2 + 1; f(x) > 0$

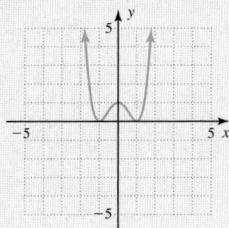

28. $f(x) = x^3 + 2x^2 - 4x - 8; f(x) \geq 0$

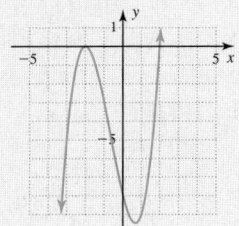

29. $p(x) = \sqrt[3]{x - 1} - 1; p(x) \geq 0$

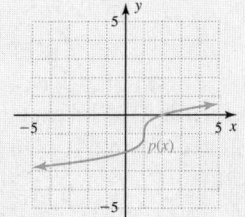

30. $q(x) = \sqrt{x + 1} - 2; q(x) > 0$

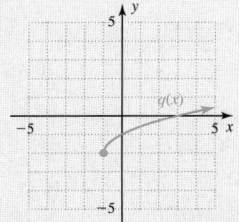

31. $f(x) = (x - 1)^3 - 1; f(x) \leq 0$

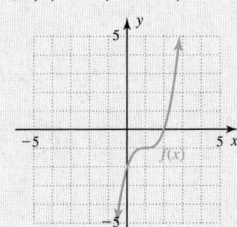

32. $g(x) = -(x + 1)^3 - 1; g(x) < 0$

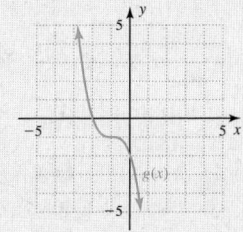

Name the interval(s) where the following functions are increasing, decreasing, or constant. Write answers using interval notation. Assume all endpoints have integer values.

33. $y = V(x)$

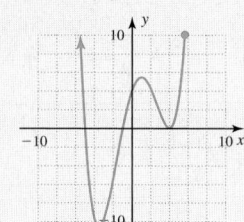

34. $y = H(x)$

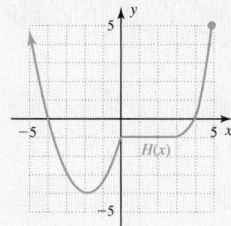

35. $y = f(x)$

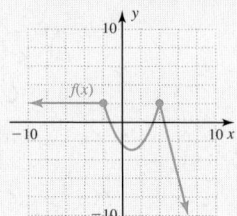

36. $y = g(x)$

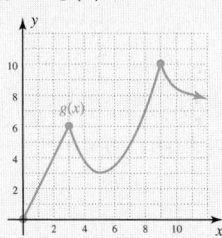

For Exercises 37 through 40, determine (a) interval(s) where the function is increasing, decreasing or constant, and (b) comment on the end behavior.

37. $p(x) = 0.5(x + 2)^3$

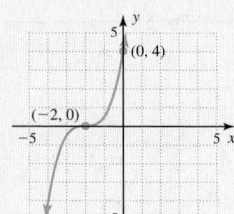

38. $q(x) = -\sqrt[3]{x + 1}$

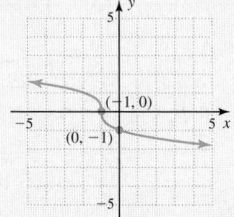

39. $y = f(x)$

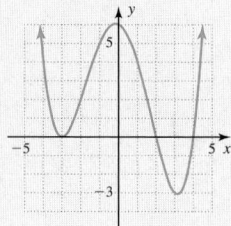

40. $y = g(x)$

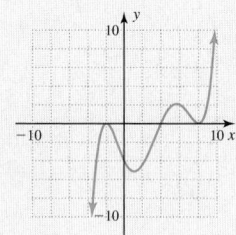

For Exercises 41 through 48, determine the following (answer in interval notation as appropriate): (a) domain and range of the function; (b) zeroes of the function; (c) interval(s) where the function is greater than or equal to zero, or less than or equal to zero; (d) interval(s) where the function is increasing, decreasing, or constant; and (e) location of any local max or min value(s).

41. $y = H(x)$

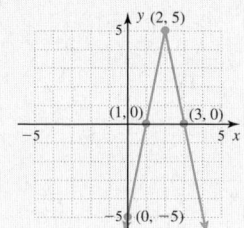

42. $y = f(x)$

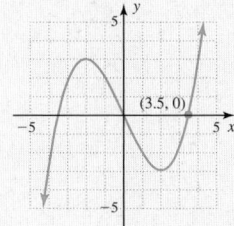

43. $y = g(x)$

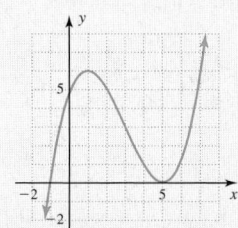

44. $y = h(x)$

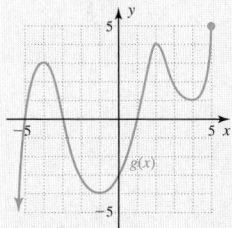

45. $y = Y_1$

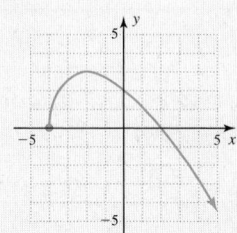

46. $y = Y_2$

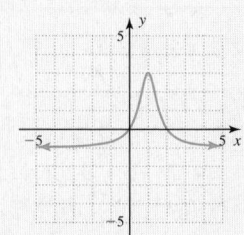

47. $p(x) = (x + 3)^3 + 1$

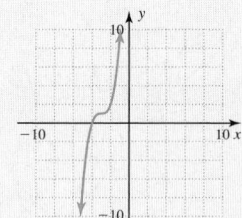

48. $q(x) = |x - 5| + 3$

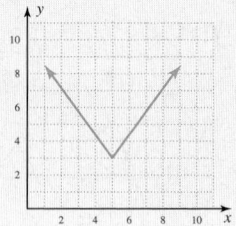

▶ **WORKING WITH FORMULAS**

49. Conic sections—hyperbola: $y = \frac{1}{3}\sqrt{4x^2 - 36}$

While the conic sections are not covered in detail until later in the course, we've already developed a number of tools that will help us understand these relations and their graphs. The equation here gives the "upper branches" of a hyperbola, as shown in the figure. Find the following by analyzing the equation: (a) the domain and range; (b) the zeroes of the relation; (c) interval(s) where y is increasing or decreasing; and (d) whether the relation is even, odd, or neither.

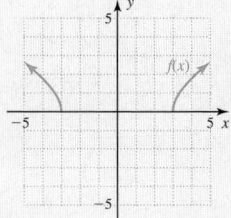

50. Trigonometric graphs: $y = \sin(x)$ and $y = \cos(x)$

The trigonometric functions are also studied at some future time, but we can apply the same tools to analyze the graphs of these functions as well. The graphs of $y = \sin x$ and $y = \cos x$ are given, graphed over the interval $x \in [-180, 360]$ degrees. Use them to find (a) the range of the functions; (b) the zeroes of the functions; (c) interval(s) where y is increasing/decreasing; (d) location of minimum/maximum values; and (e) whether each relation is even, odd, or neither.

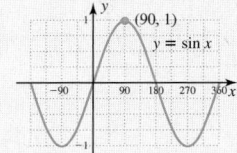 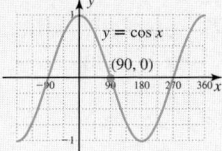

▶ APPLICATIONS

51. Catapults and projectiles: Catapults have a long and interesting history that dates back to ancient times, when they were used to launch javelins, rocks, and other projectiles. The diagram given illustrates the path of the projectile after release, which follows a parabolic arc. Use the graph to determine the following:

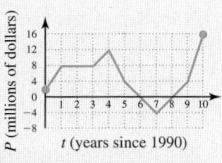

a. State the domain and range of the projectile.

b. What is the maximum height of the projectile?

c. How far from the catapult did the projectile reach its maximum height?

d. Did the projectile clear the castle wall, which was 40 ft high and 210 ft away?

e. On what interval was the height of the projectile increasing?

f. On what interval was the height of the projectile decreasing?

52. Profit and loss: The profit of DeBartolo Construction Inc. is illustrated by the graph shown. Use the graph to estimate the point(s) or the interval(s) for which the profit P was:

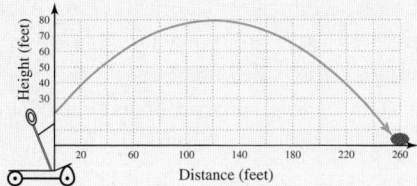

a. increasing

b. decreasing

c. constant

d. a maximum

e. a minimum

f. positive

g. negative

h. zero

53. Functions and rational exponents: The graph of $f(x) = x^{\frac{2}{3}} - 1$ is shown. Use the graph to find:

a. domain and range of the function

b. zeroes of the function

c. interval(s) where $f(x) \geq 0$ or $f(x) < 0$

d. interval(s) where $f(x)$ is increasing, decreasing, or constant

e. location of any max or min value(s)

Exercise 53   **Exercise 54**

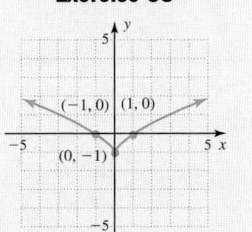 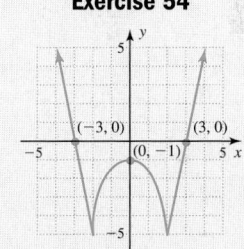

54. Analyzing a graph: Given $h(x) = |x^2 - 4| - 5$, whose graph is shown, use the graph to find:

a. domain and range of the function

b. zeroes of the function

c. interval(s) where $h(x) \geq 0$ or $h(x) \leq 0$

d. interval(s) where $f(x)$ is increasing, decreasing, or constant

e. location of any max or min value(s)

55. Analyzing interest rates: The graph shown approximates the average annual interest rates on 30-yr fixed mortgages, rounded to the nearest $\frac{1}{4}\%$. Use the graph to estimate the following (write all answers in interval notation).

a. domain and range

b. interval(s) where $I(t)$ is increasing, decreasing, or constant

c. location of the maximum and minimum values

d. the one-year period with the greatest rate of increase and the one-year period with the greatest rate of decrease

Source: 1998 Wall Street Journal Almanac, p. 446; 2004 Statistical Abstract of the United States, Table 1178

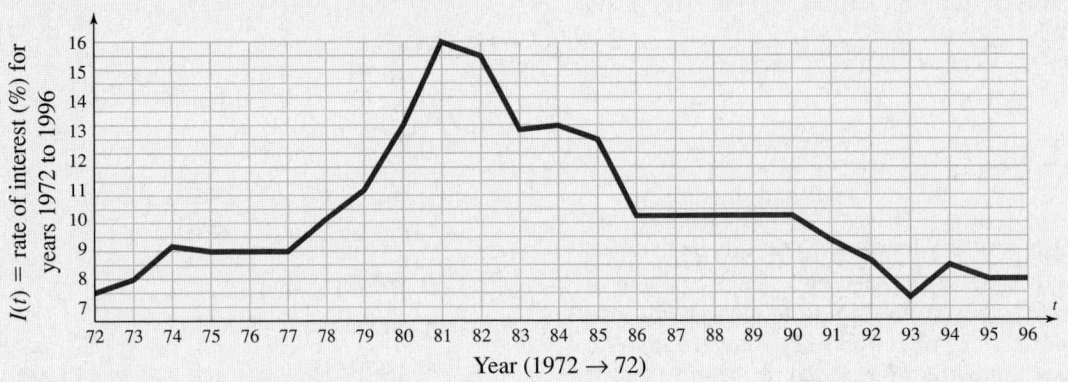

Year (1972 → 72)

56. Analyzing the deficit: The following graph approximates the federal deficit of the United States. Use the graph to estimate the following (write answers in interval notation).

a. the domain and range

b. interval(s) where $D(t)$ is increasing, decreasing, or constant

c. the location of the maximum and minimum values

d. the one-year period with the greatest rate of increase, and the one-year period with the greatest rate of decrease

Source: 2005 Statistical Abstract of the United States, Table 461

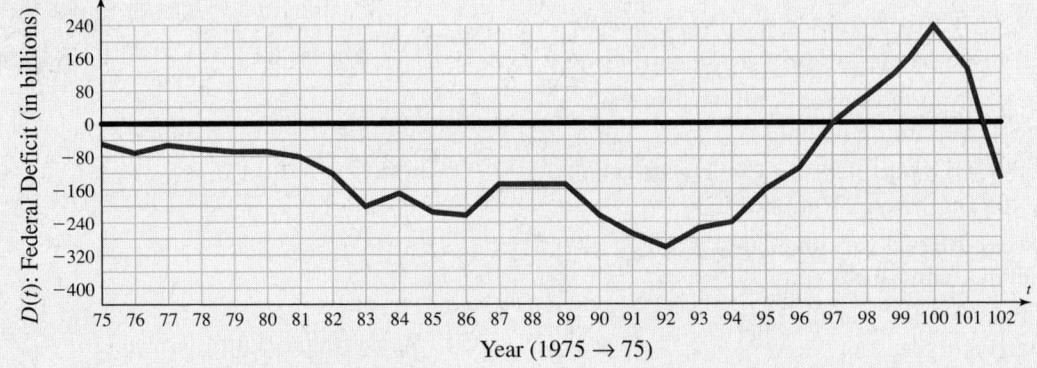

Year (1975 → 75)

57. Constructing a graph: Draw the function f that has the following characteristics, then state the zeroes and the location of all maximum and minimum values. [*Hint:* Write them as $(c, f(c))$.]

a. Domain: $x \in (-10, \infty)$

b. Range: $y \in (-6, \infty)$

c. $f(0) = 0; f(4) = 0$

d. $f(x)\uparrow$ for $x \in (-10, -6) \cup (-2, 2) \cup (4, \infty)$

e. $f(x)\downarrow$ for $x \in (-6, -2) \cup (2, 4)$

f. $f(x) \geq 0$ for $x \in [-8, -4] \cup [0, \infty)$

g. $f(x) < 0$ for $x \in (-\infty, -8) \cup (-4, 0)$

58. Constructing a graph: Draw the function g that has the following characteristics, then state the zeroes and the location of all maximum and minimum values. [*Hint:* Write them as $(c, g(c))$.]

a. Domain: $x \in (-\infty, 8)$

b. Range: $y \in [-6, \infty)$

c. $g(0) = 4.5; g(6) = 0$

d. $g(x)\uparrow$ for $x \in (-6, 3) \cup (6, 8)$

e. $g(x)\downarrow$ for $x \in (-\infty, -6) \cup (3, 6)$

f. $g(x) \geq 0$ for $x \in (-\infty, -9] \cup [-3, 8)$

g. $g(x) < 0$ for $x \in (-9, -3)$

For Exercises 59 to 64, use the formula for the average rate of change $\dfrac{f(x_2) - f(x_1)}{x_2 - x_1}$.

59. Average rate of change: For $f(x) = x^3$, (a) calculate the average rate of change for the interval $x = -2$ and $x = -1$ and (b) calculate the average rate of change for the interval $x = 1$ and $x = 2$. (c) What do you notice about the answers from parts (a) and (b)? (d) Sketch the graph of this function along with the lines representing these average rates of change and comment on what you notice.

60. Average rate of change: Knowing the general shape of the graph for $f(x) = \sqrt[3]{x}$, (a) is the average rate of change greater between $x = 0$ and $x = 1$ or between $x = 7$ and $x = 8$? Why? (b) Calculate the rate of change for these intervals and verify your response. (c) Approximately how many times greater is the rate of change?

61. Height of an arrow: If an arrow is shot vertically from a bow with an initial speed of 192 ft/sec, the height of the arrow can be modeled by the function $h(t) = -16t^2 + 192t$, where $h(t)$ represents the height of the arrow after t sec (assume the arrow was shot from ground level).

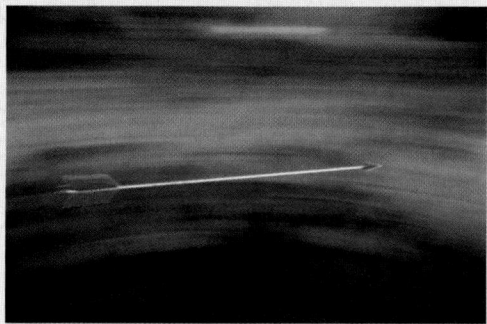

  a. What is the arrow's height at $t = 1$ sec?

  b. What is the arrow's height at $t = 2$ sec?

  c. What is the average rate of change from $t = 1$ to $t = 2$?

  d. What is the rate of change from $t = 10$ to $t = 11$? Why is it the same as (c) except for the sign?

62. Height of a water rocket: Although they have been around for decades, water rockets continue to be a popular toy. A plastic rocket is filled with water and then pressurized using a handheld pump. The rocket is then released and off it goes! If the rocket has an initial velocity of 96 ft/sec, the height of the rocket can be modeled by the function $h(t) = -16t^2 + 96t$, where $h(t)$ represents the height of the rocket after t sec (assume the rocket was shot from ground level).

  a. Find the rocket's height at $t = 1$ and $t = 2$ sec.

  b. Find the rocket's height at $t = 3$ sec.

  c. Would you expect the average rate of change to be greater between $t = 1$ and $t = 2$, or between $t = 2$ and $t = 3$? Why?

  d. Calculate each rate of change and discuss your answer.

63. Velocity of a falling object: The impact velocity of an object dropped from a height is modeled by $v = \sqrt{2gs}$, where v is the velocity in feet per second (ignoring air resistance), g is the acceleration due to gravity (32 ft/sec² near the Earth's surface), and s is the height from which the object is dropped.

  a. Find the velocity at $s = 5$ ft and $s = 10$ ft.

  b. Find the velocity at $s = 15$ ft and $s = 20$ ft.

  c. Would you expect the average rate of change to be greater between $s = 5$ and $s = 10$, or between $s = 15$ and $s = 20$?

  d. Calculate each rate of change and discuss your answer.

64. Temperature drop: One day in November, the town of Coldwater was hit by a sudden winter storm that caused temperatures to plummet. During the storm, the temperature T (in degrees Fahrenheit) could be modeled by the function $T(h) = 0.8h^2 - 16h + 60$, where h is the number of hours since the storm began. Graph the function and use this information to answer the following questions.

  a. What was the temperature as the storm began?

  b. How many hours until the temperature dropped below zero degrees?

  c. How many hours did the temperature remain below zero?

  d. What was the coldest temperature recorded during this storm?

Compute and simplify the difference quotient $\dfrac{f(x + h) - f(x)}{h}$ for each function given.

65. $f(x) = 2x - 3$ **66.** $g(x) = 4x + 1$

67. $h(x) = x^2 + 3$ **68.** $p(x) = x^2 - 2$

69. $q(x) = x^2 + 2x - 3$ **70.** $r(x) = x^2 - 5x + 2$

71. $f(x) = \dfrac{2}{x}$ **72.** $g(x) = \dfrac{-3}{x}$

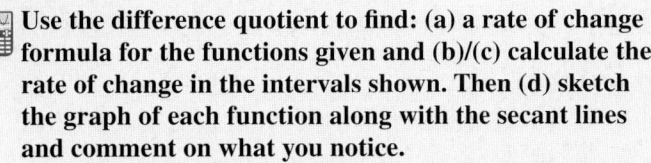 Use the difference quotient to find: (a) a rate of change formula for the functions given and (b)/(c) calculate the rate of change in the intervals shown. Then (d) sketch the graph of each function along with the secant lines and comment on what you notice.

73. $g(x) = x^2 + 2x$
    $[-3.0, -2.9], [0.50, 0.51]$

74. $h(x) = x^2 - 6x$
    $[1.9, 2.0], [5.0, 5.01]$

75. $g(x) = x^3 + 1$
    $[-2.1, -2], [0.40, 0.41]$

76. $r(x) = \sqrt{x}$ (*Hint*: Rationalize the numerator.)
    $[1, 1.1], [4, 4.1]$

77. The distance that a person can see depends on how high they're standing above level ground. On a clear day, the distance is approximated by the function $d(h) = 1.5\sqrt{h}$, where $d(h)$ represents the viewing distance (in miles) at height h (in feet). Find the average rate of change in the intervals (a) $[9, 9.01]$ and (b) $[225, 225.01]$. Then (c) graph the function along with the lines representing the average rates of change and comment on what you notice.

78. A special magnifying lens is crafted and installed in an overhead projector. When the projector is x ft from the screen, the size $P(x)$ of the projected image is x^2. Find the average rate of change for $P(x) = x^2$ in the intervals (a) $[1, 1.01]$ and (b) $[4, 4.01]$. Then (c) graph the function along with the lines representing the average rates of change and comment on what you notice.

▶ **EXTENDING THE THOUGHT**

 79. Does the function shown have a maximum value? Does it have a minimum value? Discuss/explain/justify why or why not.

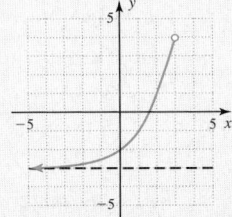

80. The graph drawn here depicts a 400-m race between a mother and her daughter. Analyze the graph to answer questions (a) through (f).

 a. Who wins the race, the mother or daughter?

 b. By approximately how many meters?

 c. By approximately how many seconds?

 d. Who was leading at $t = 40$ seconds?

 e. During the race, how many seconds was the daughter in the lead?

 f. During the race, how many seconds was the mother in the lead?

81. Draw a general function $f(x)$ that has a local *maximum* at $(a, f(a))$ and a local *minimum* at $(b, f(b))$ but with $f(a) < f(b)$.

82. Verify that $h(x) = x^{\frac{2}{3}}$ is an even function, by first rewriting h as $h(x) = (x^{\frac{1}{3}})^2$.

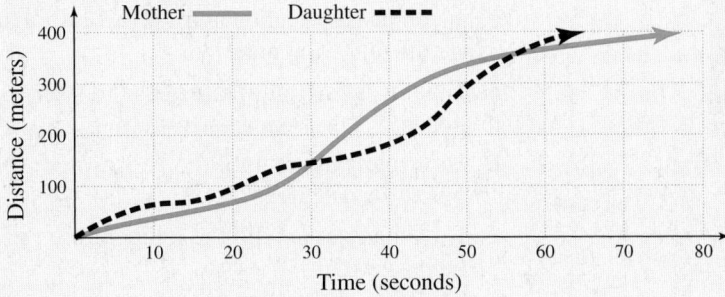

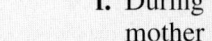 Mother ——— Daughter ▪▪▪▪

▶ **MAINTAINING YOUR SKILLS**

83. (1.5) Solve the given quadratic equation three different ways: (a) factoring, (b) completing the square, and (c) using the quadratic formula:
$x^2 - 8x - 20 = 0$

84. (R.5) Find the (a) sum and (b) product of the rational expressions $\dfrac{3}{x + 2}$ and $\dfrac{3}{2 - x}$.

85. (2.3) Write the equation of the line shown, in the form $y = mx + b$.

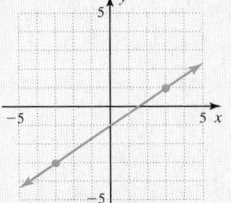

Exercise 85

86. (R.7) Find the surface area and volume of the cylinder shown.

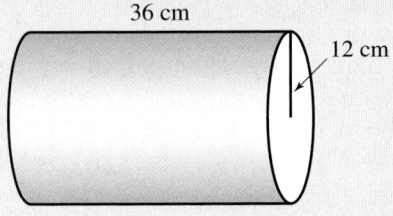

2.6 | The Toolbox Functions and Transformations

Learning Objectives

In Section 2.6 you will learn how to:

☐ **A.** Identify basic characteristics of the toolbox functions

☐ **B.** Perform vertical/horizontal shifts of a basic graph

☐ **C.** Perform vertical/horizontal reflections of a basic graph

☐ **D.** Perform vertical stretches and compressions of a basic graph

☐ **E.** Perform transformations on a general function $f(x)$

Many applications of mathematics require that we select a function known to fit the context, or build a function model from the information supplied. So far we've looked extensively at linear functions, and have introduced the absolute value, squaring, square root, cubing, and cube root functions. These are the six **toolbox functions,** so called because they give us a variety of "tools" to model the real world. In the same way a study of arithmetic depends heavily on the multiplication table, a study of algebra and mathematical modeling depends (in large part) on a solid working knowledge of these functions.

A. The Toolbox Functions

While we can accurately graph a line using only two points, most toolbox functions require more points to show all of the graph's important features. However, our work is greatly simplified in that each function belongs to a **function family,** in which all graphs from a given family share the characteristics of one basic graph, called the **parent function.** This means the number of points required for graphing will quickly decrease as we start anticipating what the graph of a given function should look like. The parent functions and their identifying characteristics are summarized here.

The Toolbox Functions

Identity function

x	$f(x) = x$
-3	-3
-2	-2
-1	-1
0	0
1	1
2	2
3	3

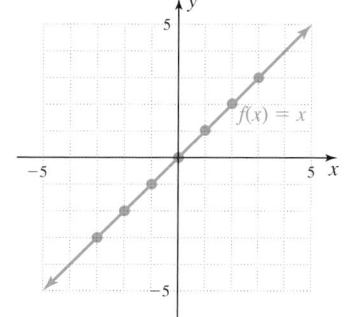

Domain: $x \in (-\infty, \infty)$, Range: $y \in (-\infty, \infty)$
Symmetry: odd
Increasing: $x \in (-\infty, \infty)$
End behavior: down on the left/up on the right

Absolute value function

x	$f(x) = |x|$
-3	3
-2	2
-1	1
0	0
1	1
2	2
3	3

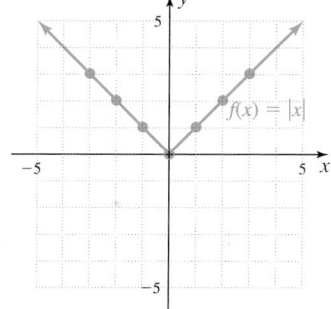

Domain: $x \in (-\infty, \infty)$, Range: $y \in [0, \infty)$
Symmetry: even
Decreasing: $x \in (-\infty, 0)$; Increasing: $x \in (0, \infty)$
End behavior: up on the left/up on the right
Vertex at $(0, 0)$

Squaring function

x	$f(x) = x^2$
-3	9
-2	4
-1	1
0	0
1	1
2	4
3	9

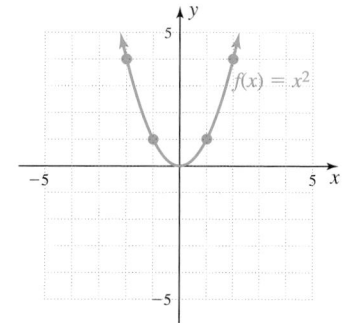

Domain: $x \in (-\infty, \infty)$, Range: $y \in [0, \infty)$
Symmetry: even
Decreasing: $x \in (-\infty, 0)$; Increasing: $x \in (0, \infty)$
End behavior: up on the left/up on the right
Vertex at $(0, 0)$

Square root function

x	$f(x) = \sqrt{x}$
-2	–
-1	–
0	0
1	1
2	≈ 1.41
3	≈ 1.73
4	2

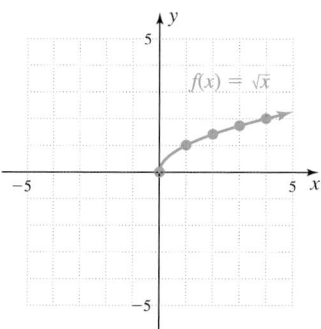

Domain: $x \in [0, \infty)$, Range: $y \in [0, \infty)$
Symmetry: neither even nor odd
Increasing: $x \in (0, \infty)$
End behavior: up on the right
Initial point at $(0, 0)$

Cubing function

x	$f(x) = x^3$
-3	-27
-2	-8
-1	-1
0	0
1	1
2	8
3	27

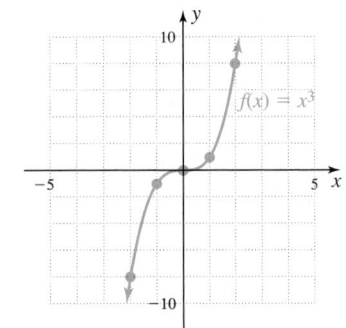

Domain: $x \in (-\infty, \infty)$, Range: $y \in (-\infty, \infty)$
Symmetry: odd
Increasing: $x \in (-\infty, \infty)$
End behavior: down on the left/up on the right
Point of inflection at $(0, 0)$

Cube root function

x	$f(x) = \sqrt[3]{x}$
-27	-3
-8	-2
-1	-1
0	0
1	1
8	2
27	3

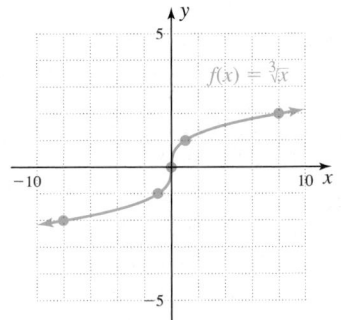

Domain: $x \in (-\infty, \infty)$, Range: $y \in (-\infty, \infty)$
Symmetry: odd
Increasing: $x \in (-\infty, \infty)$
End behavior: down on the left/up on the right
Point of inflection at $(0, 0)$

In applications of the toolbox functions, the parent graph may be altered and/or shifted from its original position, yet the graph will still retain its basic shape and features. The result is called a **transformation** of the parent graph. Analyzing the new graph (as in Section 2.5) will often provide the answers needed.

EXAMPLE 1 ▶ **Identifying the Characteristics of a Transformed Graph**

The graph of $f(x) = x^2 - 2x - 3$ is given.
Use the graph to identify each of the features
or characteristics indicated.

  a. function family

  b. domain and range

  c. vertex

  d. max or min value(s)

  e. end behavior

  f. x- and y-intercept(s)

Solution ▶   **a.** The graph is a parabola, from the squaring
function family.

  b. domain: $x \in (-\infty, \infty)$; range: $y \in [-4, \infty)$

  c. vertex: $(1, -4)$

  d. minimum value $y = -4$ at $(1, -4)$

  e. end-behavior: up/up

  f. y-intercept: $(0, -3)$; x-intercepts: $(-1, 0)$ and $(3, 0)$

Now try Exercises 7 through 34 ▶

☑ **A.** You've just learned how to identify basic characteristics of the toolbox functions

Note that we can algebraically verify the x-intercepts by substituting 0 for $f(x)$ and solving the equation by factoring. This gives $0 = (x + 1)(x - 3)$, with solutions $x = -1$ and $x = 3$. It's also worth noting that while the parabola is no longer symmetric to the y-axis, it *is* symmetric to the vertical line $x = 1$. This line is called the **axis of symmetry** for the parabola, and will always be a vertical line that goes through the vertex.

B. Vertical and Horizontal Shifts

As we study specific transformations of a graph, try to develop a *global view* as the transformations can be applied to any function. When these are applied to the toolbox functions, we rely on characteristic features of the parent function to assist in completing the transformed graph.

Vertical Translations

We'll first investigate vertical translations or vertical shifts of the toolbox functions, using the absolute value function to illustrate.

EXAMPLE 2 ▶ **Graphing Vertical Translations**

Construct a table of values for $f(x) = |x|$, $g(x) = |x| + 1$, and $h(x) = |x| - 3$ and graph the functions on the same coordinate grid. Then discuss what you observe.

Solution ▶ A table of values for all three functions is given, with the corresponding graphs shown in the figure.

x	$f(x) = |x|$	$g(x) = |x| + 1$	$h(x) = |x| - 3$
-3	3	4	0
-2	2	3	-1
-1	1	2	-2
0	0	1	-3
1	1	2	-2
2	2	3	-1
3	3	4	0

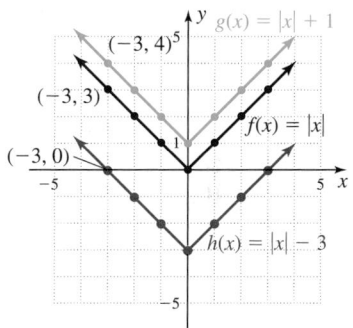

Note that outputs of $g(x)$ are one more than the outputs for $f(x)$, and that each point on the graph of f has been shifted *upward 1 unit* to form the graph of g. Similarly, each point on the graph of f has been shifted *downward 3 units* to form the graph of h. Since $h(x) = f(x) - 3$.

Now try Exercises 35 through 42 ▶

We describe the transformations in Example 2 as a **vertical shift** or **vertical translation** of a basic graph. The graph of g is the graph of f *shifted up 1 unit*, and the graph of h is the graph of f *shifted down 3 units*. In general, we have the following:

Vertical Translations of a Basic Graph

Given $k > 0$ and any function whose graph is determined by $y = f(x)$,
1. The graph of $y = f(x) + k$ is the graph of $f(x)$ shifted upward k units.
2. The graph of $y = f(x) - k$ is the graph of $f(x)$ shifted downward k units.

Horizontal Translations

The graph of a parent function can also be shifted left or right. This happens when we *alter the inputs to the basic function,* as opposed to adding or subtracting something to the basic function itself. For $Y_1 = x^2 + 2$ note that we first square inputs, then add 2, which results in a vertical shift. For $Y_2 = (x + 2)^2$, we add 2 to x *prior to squaring* and since the input values are affected, we might anticipate the graph will shift along the x-axis—horizontally.

EXAMPLE 3 ▶ **Graphing Horizontal Translations**

Construct a table of values for $f(x) = x^2$ and $g(x) = (x + 2)^2$, then graph the functions on the same grid and discuss what you observe.

Solution ▶ Both f and g belong to the quadratic family and their graphs are parabolas. A table of values is shown along with the corresponding graphs.

x	$f(x) = x^2$	$g(x) = (x + 2)^2$
-3	9	1
-2	4	0
-1	1	1
0	0	4
1	1	9
2	4	16
3	9	25

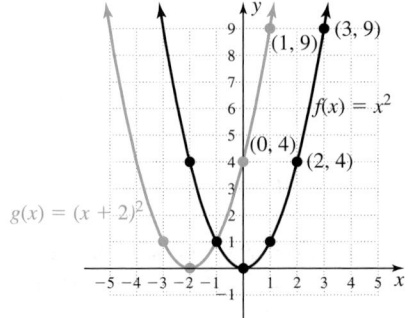

It is apparent the graphs of g and f are identical, but the graph of g has been shifted horizontally 2 units left.

Now try Exercises 43 through 46 ▶

We describe the transformation in Example 3 as a **horizontal shift** or **horizontal translation** of a basic graph. The graph of g is the graph of f, *shifted 2 units to the left.* Once again it seems reasonable that since *input* values were altered, the shift must be horizontal rather than vertical. From this example, we also learn the direction of the shift is **opposite the sign:** $y = (x + 2)^2$ is 2 units *to the left* of $y = x^2$. Although it may seem counterintuitive, the shift *opposite the sign* can be "seen" by locating the new x-intercept, which in this case is also the vertex. Substituting 0 for y gives $0 = (x + 2)^2$ with $x = -2$, as shown in the graph. In general, we have

Horizontal Translations of a Basic Graph

Given $h > 0$ and any function whose graph is determined by $y = f(x)$,
1. The graph of $y = f(x + h)$ is the graph of $f(x)$ shifted *to the left h* units.
2. The graph of $y = f(x - h)$ is the graph of $f(x)$ shifted *to the right h* units.

EXAMPLE 4 ▶ **Graphing Horizontal Translations**

Sketch the graphs of $g(x) = |x - 2|$ and $h(x) = \sqrt{x + 3}$ using a horizontal shift of the parent function and a few characteristic points (not a table of values).

Solution ▶ The graph of $g(x) = |x - 2|$ (Figure 2.56) is the absolute value function shifted 2 units to the right (shift the vertex and two other points from $y = |x|$). The graph of $h(x) = \sqrt{x + 3}$ (Figure 2.57) is a square root function, shifted 3 units to the left (shift the initial point and one or two points from $y = \sqrt{x}$).

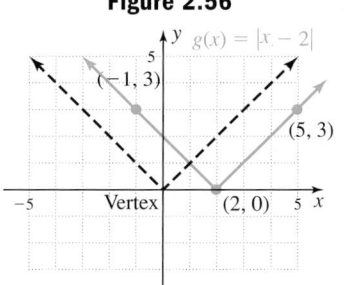

Figure 2.56

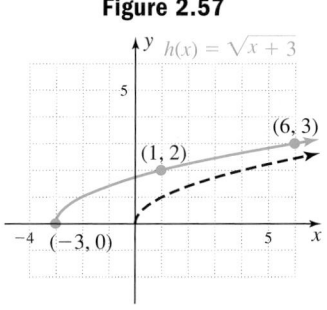

Figure 2.57

☑ **B.** You've just learned how to perform vertical/horizontal shifts of a basic graph

Now try Exercises 47 through 50 ▶

C. Vertical and Horizontal Reflections

The next transformation we investigate is called a **vertical reflection,** in which we compare the function $Y_1 = f(x)$ with the negative of the function: $Y_2 = -f(x)$.

Vertical Reflections

EXAMPLE 5 ▶ **Graphing Vertical Reflections**

Construct a table of values for $Y_1 = x^2$ and $Y_2 = -x^2$, then graph the functions on the same grid and discuss what you observe.

Solution ▶ A table of values is given for both functions, along with the corresponding graphs.

x	$Y_1 = x^2$	$Y_2 = -x^2$
-2	4	-4
-1	1	-1
0	0	0
1	1	-1
2	4	-4

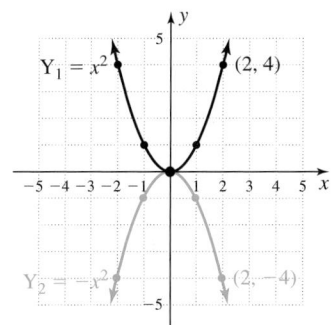

As you might have anticipated, the outputs for f and g differ only in sign. Each output is a **reflection** of the other, being an equal distance from the x-axis but on opposite sides.

Now try Exercises 51 and 52 ▶

The vertical reflection in Example 5 is called a **reflection across the x-axis.** In general,

Vertical Reflections of a Basic Graph

For any function $y = f(x)$, the graph of $y = -f(x)$ is the graph of $f(x)$ reflected across the x-axis.

Horizontal Reflections

It's also possible for a graph to be reflected horizontally *across the y-axis.* Just as we noted that $f(x)$ versus $-f(x)$ resulted in a vertical reflection, $f(x)$ versus $f(-x)$ results in a horizontal reflection.

EXAMPLE 6 ▶ **Graphing a Horizontal Reflection**

Construct a table of values for $f(x) = \sqrt{x}$ and $g(x) = \sqrt{-x}$, then graph the functions on the same coordinate grid and discuss what you observe.

Solution ▶ A table of values is given here, along with the corresponding graphs.

x	$f(x) = \sqrt{x}$	$g(x) = \sqrt{-x}$
-4	not real	2
-2	not real	$\sqrt{2} \approx 1.41$
-1	not real	1
0	0	0
1	1	not real
2	$\sqrt{2} \approx 1.41$	not real
4	2	not real

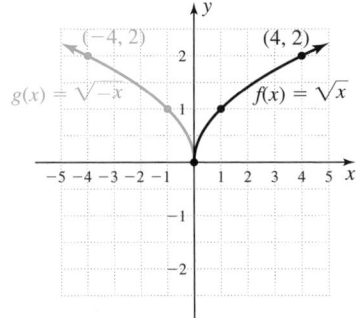

The graph of g is the same as the graph of f, but it has been reflected across the y-axis. A study of the domain shows why—f represents a real number only for nonnegative inputs, so its graph occurs to the right of the y-axis, while g represents a real number for nonpositive inputs, so its graph occurs to the left.

Now try Exercises 53 and 54 ▶

The transformation in Example 6 is called a **horizontal reflection** of a basic graph. In general,

☑ C. You've just learned how to perform vertical/horizontal reflections of a basic graph

Horizontal Reflections of a Basic Graph

For any function $y = f(x)$, the graph of $y = f(-x)$ is the graph of $f(x)$ reflected across the y-axis.

D. Vertically Stretching/Compressing a Basic Graph

As the words "stretching" and "compressing" imply, the graph of a basic function can also become elongated or flattened after certain transformations are applied. However, even these transformations preserve the key characteristics of the graph.

EXAMPLE 7 ▶ **Stretching and Compressing a Basic Graph**

Construct a table of values for $f(x) = x^2$, $g(x) = 3x^2$, and $h(x) = \frac{1}{3}x^2$, then graph the functions on the same grid and discuss what you observe.

Solution ▶ A table of values is given for all three functions, along with the corresponding graphs.

x	$f(x) = x^2$	$g(x) = 3x^2$	$h(x) = \frac{1}{3}x^2$
-3	9	27	3
-2	4	12	$\frac{4}{3}$
-1	1	3	$\frac{1}{3}$
0	0	0	0
1	1	3	$\frac{1}{3}$
2	4	12	$\frac{4}{3}$
3	9	27	3

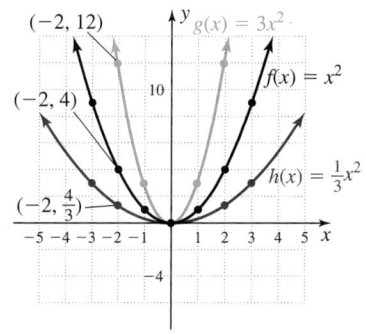

The outputs of g are triple those of f, making these outputs farther from the x-axis and *stretching* g upward (making the graph more narrow). The outputs of h are one-third those of f, and the graph of h is *compressed* downward, with its outputs closer to the x-axis (making the graph wider).

Now try Exercises 55 through 62 ▶

The transformations in Example 7 are called **vertical stretches** or **compressions** of a basic graph. In general,

---

Stretches and Compressions of a Basic Graph

For any function $y = f(x)$, the graph of $y = af(x)$ is
1. the graph of $f(x)$ stretched vertically if $|a| > 1$,
2. the graph of $f(x)$ compressed vertically if $0 < |a| < 1$.

E. Transformations of a General Function

If more than one transformation is applied to a basic graph, it's helpful to use the following sequence for graphing the new function.

General Transformations of a Basic Graph

Given a function $y = f(x)$, the graph of $y = af(x \pm h) \pm k$ can be obtained by applying the following sequence of transformations:
1. horizontal shifts   2. reflections
3. stretches or compressions   4. vertical shifts

We generally use a few characteristic points to track the transformations involved, then draw the transformed graph through the new location of these points.

EXAMPLE 8 ▶ Graphing Functions Using Transformations

Use transformations of a parent function to sketch the graphs of
  a. $g(x) = -(x + 2)^2 + 3$ **b.** $h(x) = 2\sqrt[3]{x - 2} - 1$

WORTHY OF NOTE

In a study of trigonometry, you'll find that a basic graph can also be stretched or compressed horizontally, a phenomenon known as *frequency variations*.

☑ **D.** You've just learned how to perform vertical stretches and compressions of a basic graph

Solution ▶ **a.** The graph of g is a parabola, shifted left 2 units, reflected across the x-axis, and shifted up 3 units. This sequence of transformations in shown in Figures 2.58 through 2.60.

Figure 2.58

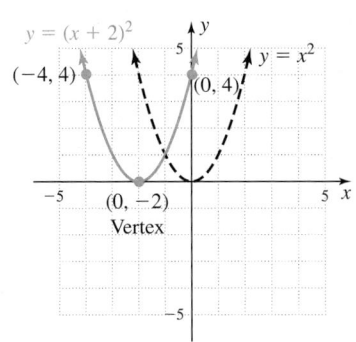

Shifted left 2 units

Figure 2.59

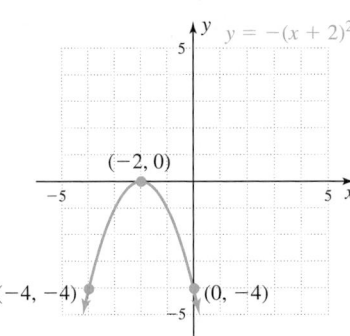

Reflected across the x-axis

Figure 2.60

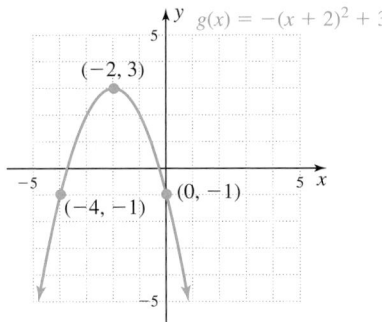

Shifted up 3

b. The graph of h is a cube root function, shifted right 2, stretched by a factor of 2, then shifted down 1. This sequence is shown in Figures 2.61 through 2.63.

Figure 2.61

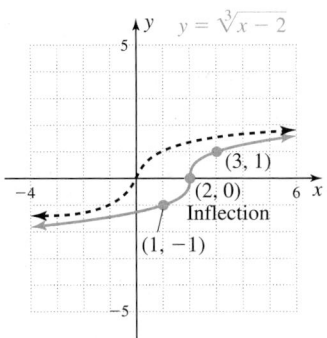

Shifted right 2

Figure 2.62

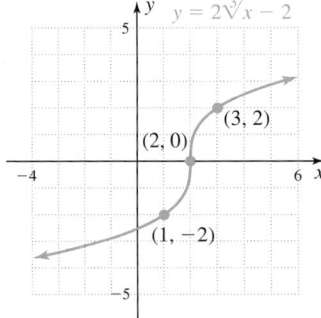

Stretched by a factor of 2

Figure 2.63

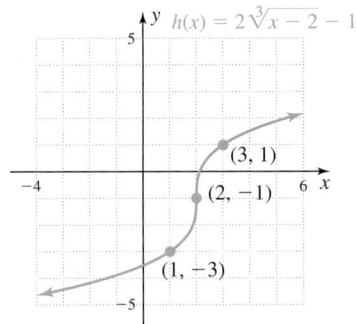

Shifted down 1

Now try Exercises 63 through 92 ▶

It's important to note that the transformations can actually be applied to *any* function, even those that are new and unfamiliar. Consider the following pattern:

Parent Function	**Transformation of Parent Function**
quadratic: $y = x^2$	$y = -2(x - 3)^2 + 1$
absolute value: $y = |x|$	$y = -2|x - 3| + 1$
cube root: $y = \sqrt[3]{x}$	$y = -2\sqrt[3]{x - 3} + 1$
general: $y = f(x)$	$y = -2f(x - 3) + 1$

In each case, the transformation involves a horizontal shift right 3, a vertical reflection, a vertical stretch, and a vertical shift up 1. Since the shifts are the same regardless of the initial function, we can generalize the results to any function $f(x)$.

General Function	**Transformed Function**
$y = f(x)$	$y = af(x \pm h) \pm k$

vertical reflections
vertical stretches and compressions

horizontal shift
h units, opposite direction of sign

vertical shift
k units, same direction as sign

Also bear in mind that the graph will be reflected across the y-axis (horizontally) if x is replaced with $-x$. Use this illustration to complete Exercise 9. Remember—if the graph of a function is shifted, the *individual points* on the graph are likewise shifted.

EXAMPLE 9 ▶ **Graphing Transformations of a General Function**

Given the graph of $f(x)$ shown in Figure 2.64, graph $g(x) = -f(x + 1) - 2$.

Solution ▶ For g, the graph of f is (1) shifted horizontally 1 unit left, (2) reflected across the x-axis, and (3) shifted vertically 2 units down. The final result is shown in Figure 2.65.

Figure 2.64

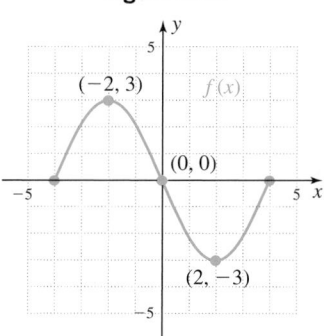

Figure 2.65

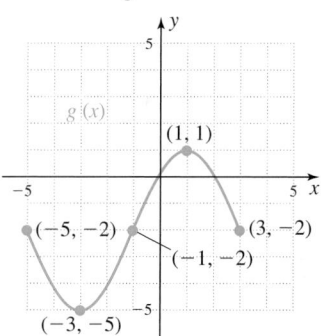

Now try Exercises 93 through 96 ▶

Using the general equation $y = af(x \pm h) \pm k$, we can identify the vertex, initial point, or inflection point of any toolbox function and sketch its graph. Given the *graph* of a toolbox function, we can likewise identify these points and reconstruct its equation. We first identify the function family and the location (h, k) of the characteristic point. By selecting one other point (x, y) on the graph, we then use the general equation as a formula (substituting h, k, and the x- and y-values of the second point) to solve for a and complete the equation.

EXAMPLE 10 ▶ **Writing the Equation of a Function Given Its Graph**

Find the equation of the toolbox function $f(x)$ shown in Figure 2.66.

Figure 2.66

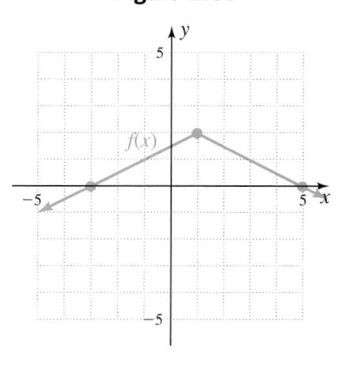

Solution ▶ The function f belongs to the absolute value family. The vertex (h, k) is at $(1, 2)$. For an additional point, choose the x-intercept $(-3, 0)$ and work as follows:

$$y = a|x - h| + k \qquad \text{general equation}$$
$$0 = a|(-3) - 1| + 2 \qquad \text{substitute 1 for } h \text{ and 2 for } k,$$
$$\qquad\qquad\qquad\qquad\quad \text{substitute } -3 \text{ for } x \text{ and 0 for } y$$
$$0 = 4a + 2 \qquad \text{simplify}$$
$$-2 = 4a \qquad \text{subtract 2}$$
$$-\frac{1}{2} = a \qquad \text{solve for } a$$

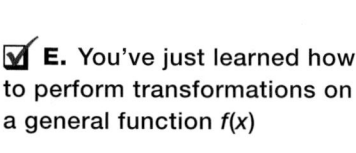

E. You've just learned how to perform transformations on a general function $f(x)$

The equation for f is $y = -\frac{1}{2}|x - 1| + 2$.

Now try Exercises 97 through 102 ▶

TECHNOLOGY HIGHLIGHT

Function Families

Graphing calculators are able to display a number of graphs simultaneously, making them a wonderful tool for studying families of functions. Let's begin by entering the function $y = |x|$ [actually $y = abs(x)$ MATH ►] as Y_1 on the Y= screen. Next, we enter different variations of the function, but always in terms of its variable name "Y_1." This enables us to simply change the basic function, and observe how the changes affect the graph. Recall that to access the function name Y_1 press VARS ► (to access the Y-VARS menu) ENTER (to access the function variables menu) and ENTER (to select Y_1). Enter the functions $Y_2 = Y_1 + 3$ and $Y_3 = Y_1 - 6$ (see Figure 2.67). Graph all three functions in the ZOOM **6:ZStandard** window. The calculator draws each graph in the order they were entered and you can always identify the functions by pressing the TRACE key and then the up arrow ▲ or down arrow ▼ keys. In the upper left corner of the window shown in Figure 2.68, the calculator identifies which function the cursor is currently on. Most importantly, note that all functions in this family maintain the same "V" shape.

Figure 2.67

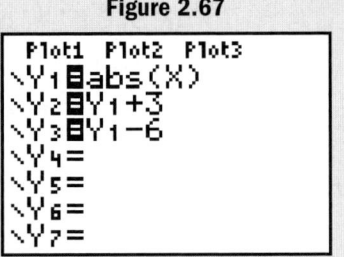

Figure 2.68

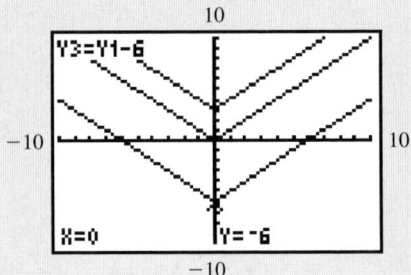

Next, change Y_1 to $Y_1 = abs(x - 3)$, leaving Y_2 and Y_3 as is. What do you notice when these are graphed again?

Exercise 1: Change Y_1 to $Y_1 = \sqrt{x}$ and graph, then enter $Y_1 = \sqrt{x - 3}$ and graph once again. What do you observe? What comparisons can be made with the translations of $Y_1 = abs(x)$?

Exercise 2: Change Y_1 to $Y_1 = x^2$ and graph, then enter $Y_1 = (x - 3)^2$ and graph once again. What do you observe? What comparisons can be made with the translations of $Y_1 = abs(x)$ and $Y_1 = \sqrt{x}$?

2.6 EXERCISES

▶ CONCEPTS AND VOCABULARY

Fill in each blank with the appropriate word or phrase. Carefully reread the section if needed.

1. After a vertical _____, points on the graph are farther from the x-axis. After a vertical _____, points on the graph are closer to the x-axis.

2. Transformations that change only the location of a graph and not its shape or form, include _____ and _____.

3. The vertex of $h(x) = 3(x + 5)^2 - 9$ is at _____ and the graph opens _____.

4. The inflection point of $f(x) = -2(x - 4)^3 + 11$ is at _____ and the end behavior is _____, _____.

5. Given the graph of a general function $f(x)$, discuss/explain how the graph of $F(x) = -2f(x + 1) - 3$ can be obtained. If $(0, 5)$, $(6, 7)$, and $(-9, -4)$ are on the graph of f, where do they end up on the graph of F?

6. Discuss/Explain why the shift of $f(x) = x^2 + 3$ is a *vertical shift* of 3 units in the *positive* direction, while the shift of $g(x) = (x + 3)^2$ is a *horizontal shift* 3 units in the *negative* direction. Include several examples linked to a table of values.

▶ DEVELOPING YOUR SKILLS

By carefully inspecting each graph given, (a) indentify the function family; (b) describe or identify the end behavior, vertex, axis of symmetry, and x- and y-intercepts; and (c) determine the domain and range. Assume required features have integer values.

For each graph given, (a) identify the function family; (b) describe or identify the end behavior, initial point, and x- and y-intercepts; and (c) determine the domain and range. Assume required features have integer values.

7. $f(x) = x^2 + 4x$ **8.** $g(x) = -x^2 + 2x$

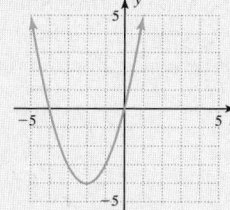 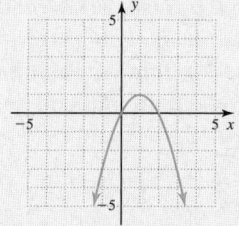

13. $p(x) = 2\sqrt{x + 4} - 2$ **14.** $q(x) = -2\sqrt{x + 4} + 2$

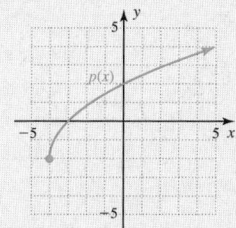 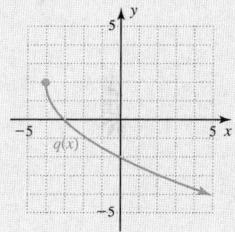
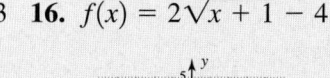

9. $p(x) = x^2 - 2x - 3$ **10.** $q(x) = -x^2 + 2x + 8$

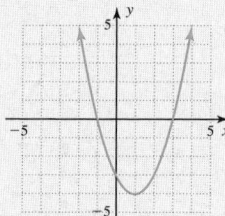 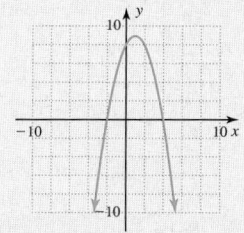

15. $r(x) = -3\sqrt{4 - x} + 3$ **16.** $f(x) = 2\sqrt{x + 1} - 4$

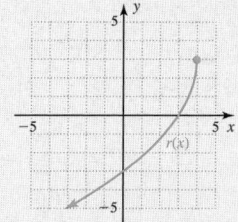 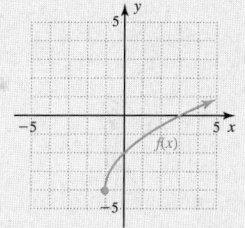
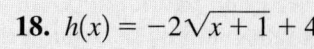

11. $f(x) = x^2 - 4x - 5$ **12.** $g(x) = x^2 + 6x + 5$

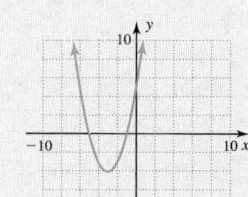 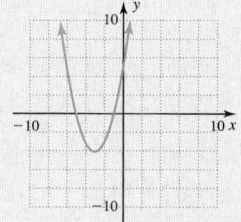

17. $g(x) = 2\sqrt{4 - x}$ **18.** $h(x) = -2\sqrt{x + 1} + 4$

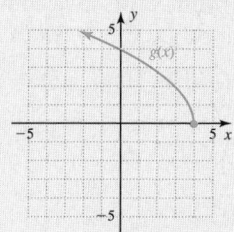 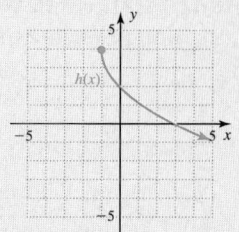

For each graph given, (a) indentify the function family; (b) describe or identify the end behavior, vertex, axis of symmetry, and x- and y-intercepts; and (c) determine the domain and range. Assume required features have integer values.

19. $p(x) = 2|x + 1| - 4$ **20.** $q(x) = -3|x - 2| + 3$

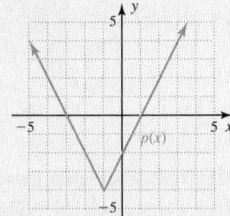
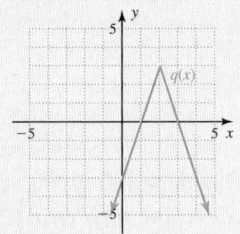

21. $r(x) = -2|x + 1| + 6$ **22.** $f(x) = 3|x - 2| - 6$

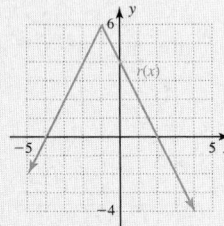
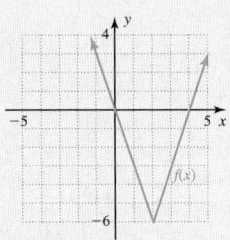
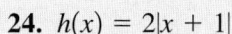

23. $g(x) = -3|x| + 6$ **24.** $h(x) = 2|x + 1|$

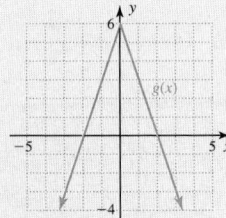
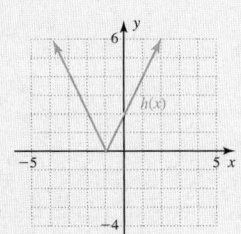

For each graph given, (a) indentify the function family; (b) describe or identify the end behavior, inflection point, and x- and y-intercepts; and (c) determine the domain and range. Assume required features have integer values. Be sure to note the scaling of each axis.

25. $f(x) = -(x - 1)^3$ **26.** $g(x) = (x + 1)^3$

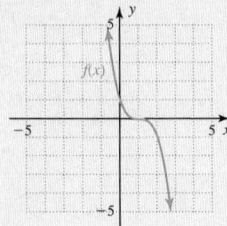
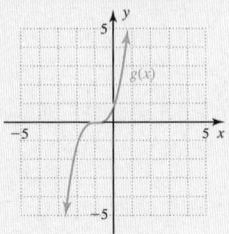

27. $h(x) = x^3 + 1$ **28.** $p(x) = -\sqrt[3]{x} + 1$

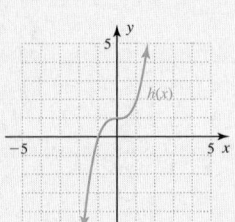
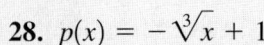
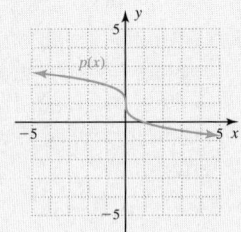

29. $q(x) = \sqrt[3]{x - 1} - 1$ **30.** $r(x) = -\sqrt[3]{x + 1} - 1$

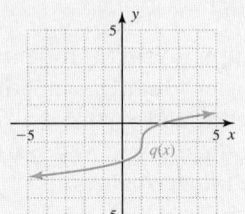
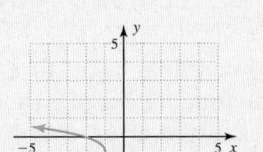

For Exercises 31–34, identify and state the characteristic features of each graph, including (as applicable) the function family, domain, range, intercepts, vertex, point of inflection, and end behavior.

31.   **32.**

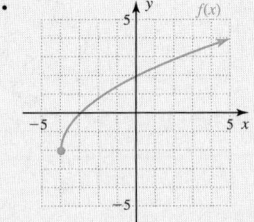
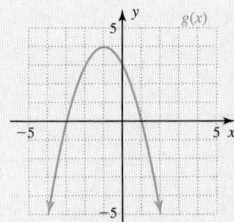

33.   **34.**

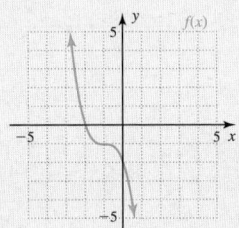
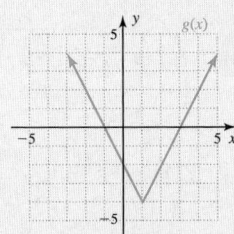

Use a table of values to graph the functions given on the same grid. Comment on what you observe.

35. $f(x) = \sqrt{x}$, $g(x) = \sqrt{x} + 2$, $h(x) = \sqrt{x} - 3$
36. $f(x) = \sqrt[3]{x}$, $g(x) = \sqrt[3]{x} - 3$, $h(x) = \sqrt[3]{x} + 1$
37. $p(x) = |x|$, $q(x) = |x| - 5$, $r(x) = |x| + 2$
38. $p(x) = x^2$, $q(x) = x^2 - 4$, $r(x) = x^2 + 1$

Sketch each graph using transformations of a parent function (without a table of values).

39. $f(x) = x^3 - 2$ **40.** $g(x) = \sqrt{x} - 4$
41. $h(x) = x^2 + 3$ **42.** $Y_1 = |x| - 3$

Use a table of values to graph the functions given on the same grid. Comment on what you observe.

43. $p(x) = x^2$, $q(x) = (x + 3)^2$

44. $f(x) = \sqrt{x}$, $g(x) = \sqrt{x + 4}$

45. $Y_1 = |x|$, $Y_2 = |x - 1|$

46. $h(x) = x^3$, $H(x) = (x - 2)^3$

Sketch each graph using transformations of a parent function (without a table of values).

47. $p(x) = (x - 3)^2$ **48.** $Y_1 = \sqrt{x - 1}$

49. $h(x) = |x + 3|$ **50.** $f(x) = \sqrt[3]{x} + 2$

51. $g(x) = -|x|$ **52.** $Y_2 = -\sqrt{x}$

53. $f(x) = \sqrt[3]{-x}$ **54.** $g(x) = (-x)^3$

Use a table of values to graph the functions given on the same grid. Comment on what you observe.

55. $p(x) = x^2$, $q(x) = 2x^2$, $r(x) = \frac{1}{2}x^2$

56. $f(x) = \sqrt{-x}$, $g(x) = 4\sqrt{-x}$, $h(x) = \frac{1}{4}\sqrt{-x}$

57. $Y_1 = |x|$, $Y_2 = 3|x|$, $Y_3 = \frac{1}{3}|x|$

58. $u(x) = x^3$, $v(x) = 2x^3$, $w(x) = \frac{1}{5}x^3$

Sketch each graph using transformations of a parent function (without a table of values).

59. $f(x) = 4\sqrt[3]{x}$ **60.** $g(x) = -2|x|$

61. $p(x) = \frac{1}{3}x^3$ **62.** $q(x) = \frac{3}{4}\sqrt{x}$

Use the characteristics of each function family to match a given function to its corresponding graph. The graphs are not scaled—make your selection based on a careful comparison.

63. $f(x) = \frac{1}{2}x^3$ **64.** $f(x) = \frac{-2}{3}x + 2$

65. $f(x) = -(x - 3)^2 + 2$ **66.** $f(x) = -\sqrt[3]{x - 1} - 1$

67. $f(x) = |x + 4| + 1$ **68.** $f(x) = -\sqrt{x + 6}$

69. $f(x) = -\sqrt{x + 6} - 1$ **70.** $f(x) = x + 1$

71. $f(x) = (x - 4)^2 - 3$ **72.** $f(x) = |x - 2| - 5$

73. $f(x) = \sqrt{x + 3} - 1$ **74.** $f(x) = -(x + 3)^2 + 5$

a.

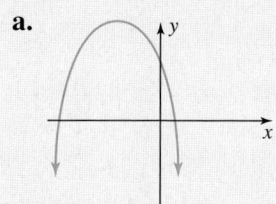

b.

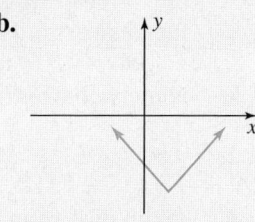

c.

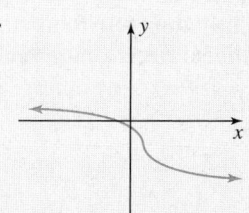

d.

e.

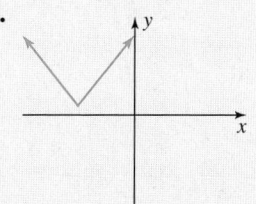

f.

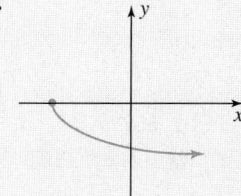

g.

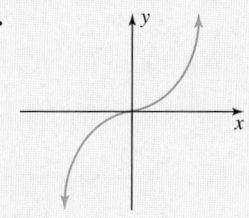

h.

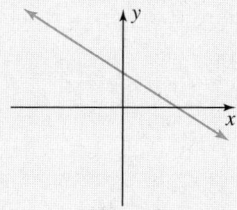

i.

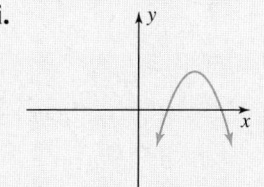

j.

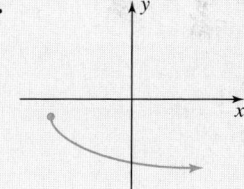

k.

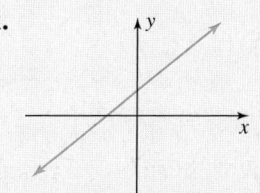

l.

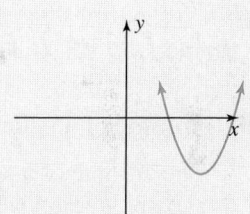

Graph each function using shifts of a parent function and a few characteristic points. *Clearly state and indicate the transformations used* and identify the location of all vertices, initial points, and/or inflection points.

75. $f(x) = \sqrt{x + 2} - 1$ **76.** $g(x) = \sqrt{x - 3} + 2$

77. $h(x) = -(x + 3)^2 - 2$ **78.** $H(x) = -(x - 2)^2 + 5$

79. $p(x) = (x + 3)^3 - 1$ **80.** $q(x) = (x - 2)^3 + 1$

81. $Y_1 = \sqrt[3]{x + 1} - 2$ **82.** $Y_2 = \sqrt[3]{x - 3} + 1$

83. $f(x) = -|x + 3| - 2$ **84.** $g(x) = -|x - 4| - 2$

85. $h(x) = -2(x + 1)^2 - 3$ **86.** $H(x) = \frac{1}{2}|x + 2| - 3$

87. $p(x) = -\frac{1}{3}(x + 2)^3 - 1$ **88.** $q(x) = 5\sqrt[3]{x + 1} + 2$

89. $Y_1 = -2\sqrt{-x - 1} + 3$ **90.** $Y_2 = 3\sqrt{-x + 2} - 1$

91. $h(x) = \frac{1}{5}(x - 3)^2 + 1$ **92.** $H(x) = -2|x - 3| + 4$

Apply the transformations indicated for the graph of the general functions given.

93.

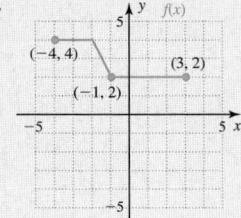

a. $f(x - 2)$
b. $-f(x) - 3$
c. $\frac{1}{2}f(x + 1)$
d. $f(-x) + 1$

94.

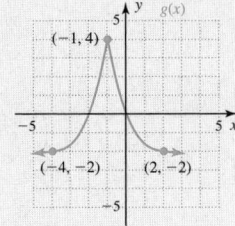

a. $g(x) - 2$
b. $-g(x) + 3$
c. $2g(x + 1)$
d. $\frac{1}{2}g(x - 1) + 2$

95.

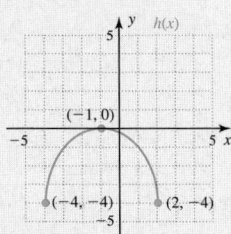

a. $h(x) + 3$
b. $-h(x - 2)$
c. $h(x - 2) - 1$
d. $\frac{1}{4}h(x) + 5$

96.

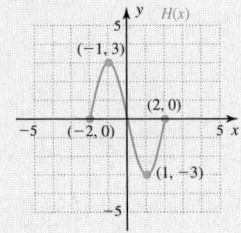

a. $H(x - 3)$
b. $-H(x) + 1$
c. $2H(x - 3)$
d. $\frac{1}{3}H(x - 2) + 1$

Use the graph given and the points indicated to determine the equation of the function shown using the general form $y = af(x \pm h) \pm k$.

97.

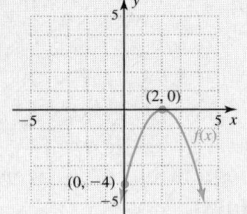

98.

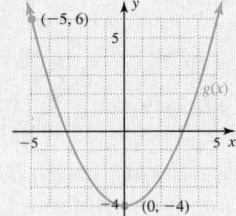

99.

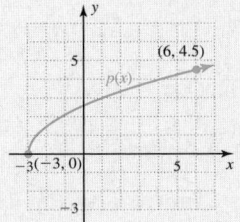

100.

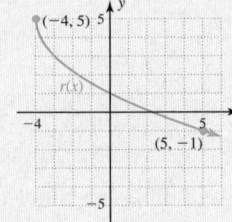

101.

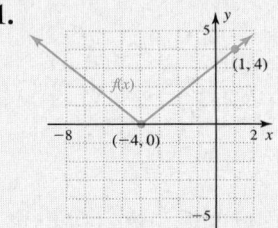

102.

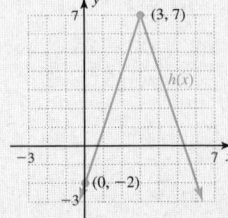

▶ WORKING WITH FORMULAS

103. Volume of a sphere: $V(r) = \frac{4}{3}\pi r^3$

The volume of a sphere is given by the function shown, where $V(r)$ is the volume in cubic units and r is the radius. Note this function belongs to the *cubic family* of functions. Approximate the value of $\frac{4}{3}\pi$ to one decimal place, then graph the function on the interval $[0, 3]$. From your *graph*, estimate the volume of a sphere with radius 2.5 in. Then compute the actual volume. Are the results close?

104. Fluid motion: $V(h) = -4\sqrt{h} + 20$

Suppose the velocity of a fluid flowing from an open tank (no top) through an opening in its side is given by the function shown, where $V(h)$ is the velocity of the fluid (in feet per second) at water height h (in feet). Note this function belongs to the *square root family* of functions. An open tank is 25 ft deep and filled to the brim with fluid. Use a table of values to graph the function on the interval $[0, 25]$. From your graph, estimate the velocity of the fluid when the water level is 7 ft, then find the actual velocity. Are the answers close? If the fluid velocity is 5 ft/sec, how high is the water in the tank?

▶ APPLICATIONS

105. Gravity, distance, time: After being released, the time it takes an object to fall x ft is given by the function $T(x) = \frac{1}{4}\sqrt{x}$, where $T(x)$ is in seconds. Describe the transformation applied to obtain the graph of T from the graph of $y = \sqrt{x}$, then sketch the graph of T for $x \in [0, 100]$. How long would it take an object to hit the ground if it were dropped from a height of 81 ft?

106. Stopping distance: In certain weather conditions, accident investigators will use the function $v(x) = 4.9\sqrt{x}$ to estimate the speed of a car (in miles per hour) that has been involved in an accident, based on the length of the skid marks x (in feet). Describe the transformation applied to obtain the graph of v from the graph of $y = \sqrt{x}$, then sketch the graph of v for $x \in [0, 400]$. If the skid marks were 225 ft long, how fast was the car traveling? Is this point on your graph?

107. Wind power: The power P generated by a certain wind turbine is given by the function $P(v) = \frac{8}{125}v^3$ where $P(v)$ is the power in watts at wind velocity v (in miles per hour). (a) Describe the transformation applied to obtain the graph of P from the graph of $y = v^3$, then sketch the graph of P for $v \in [0, 25]$ (scale the axes appropriately). (b) How much power is being generated when the wind is blowing at 15 mph? (c) Calculate the rate of change $\frac{\Delta P}{\Delta v}$ in the intervals $[8, 10]$ and $[28, 30]$. What do you notice?

108. Wind power: If the power P (in watts) being generated by a wind turbine is known, the velocity of the wind can be determined using the function $v(P) = (\frac{5}{2})\sqrt[3]{P}$. Describe the transformation applied to obtain the graph of v from the graph of $y = \sqrt[3]{P}$, then sketch the graph of v for $P \in [0, 512]$ (scale the axes appropriately). How fast is the wind blowing if 343W of power is being generated?

109. Acceleration due to gravity: The *distance* a ball rolls down an inclined plane is given by the function $d(t) = 2t^2$, where $d(t)$ represents the distance in feet after t sec. (a) Describe the transformation applied to obtain the graph of d from the graph of $y = t^2$, then sketch the graph of d for $t \in [0, 3]$. (b) How far has the ball rolled after 2.5 sec? (c) Calculate the rate of change $\frac{\Delta d}{\Delta t}$ in the intervals $[1, 1.5]$ and $[3, 3.5]$. What do you notice?

110. Acceleration due to gravity: The *velocity* of a steel ball bearing as it rolls down an inclined plane is given by the function $v(t) = 4t$, where $v(t)$ represents the velocity in feet per second after t sec. Describe the transformation applied to obtain the graph of v from the graph of $y = t$, then sketch the graph of v for $t \in [0, 3]$. What is the velocity of the ball bearing after 2.5 sec?

▶ EXTENDING THE CONCEPT

111. Carefully graph the functions $f(x) = |x|$ and $g(x) = 2\sqrt{x}$ on the same coordinate grid. From the graph, in what interval is the graph of $g(x)$ *above* the graph of $f(x)$? Pick a number (call it h) from this interval and substitute it in both functions. Is $g(h) > f(h)$? In what interval is the graph of $g(x)$ below the graph of $f(x)$? Pick a number from this interval (call it k) and substitute it in both functions. Is $g(k) < f(k)$?

112. Sketch the graph of $f(x) = -2|x - 3| + 8$ using transformations of the parent function, then determine the area of the region in quadrant I that is beneath the graph and bounded by the vertical lines $x = 0$ and $x = 6$.

113. Sketch the graph of $f(x) = x^2 - 4$, then sketch the graph of $F(x) = |x^2 - 4|$ using your intuition and the meaning of absolute value (not a table of values). What happens to the graph?

▶ MAINTAINING YOUR SKILLS

114. (2.1) Find the distance between the points $(-13, 9)$ and $(7, -12)$, and the slope of the line containing these points.

115. (R.7) Find the perimeter and area of the figure shown (note the units).

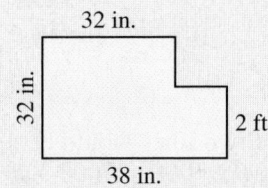

32 in.

32 in.

38 in.

2 ft

116. (1.1) Solve for x: $\frac{2}{3}x + \frac{1}{4} = \frac{1}{2}x - \frac{7}{12}$.

117. (2.5) Without graphing, state intervals where $f(x)\uparrow$ and $f(x)\downarrow$ for $f(x) = (x - 4)^2 + 3$.

Learning Objectives

In Section 2.7 you will learn how to:

☐ **A.** State the equation and domain of a piecewise-defined function

☐ **B.** Graph functions that are piecewise-defined

☐ **C.** Solve applications involving piecewise-defined functions

Most of the functions we've studied thus far have been smooth and continuous. Although "smooth" and "continuous" are defined more formally in advanced courses, for our purposes *smooth* simply means the graph has no sharp turns or jagged edges, and *continuous* means you can draw the entire graph without lifting your pencil. In this section, we study a special class of functions, called **piecewise-defined functions,** whose graphs may be various combinations of smooth/not smooth and continuous/not continuous. The absolute value function is one example (see Exercise 31). Such functions have a tremendous number of applications in the real world.

A. The Domain of a Piecewise-Defined Function

For the years 1990 to 2000, the American bald eagle remained on the nation's endangered species list, although the number of breeding pairs was growing slowly. After 2000, the population of eagles grew at a much faster rate, and they were removed from the list soon afterward. From Table 2.5 and plotted points modeling this growth (see Figure 2.69), we observe that a linear model would fit the period from 1992 to 2000 very well, but a line with greater slope would be needed for the years 2000 to 2006 and (perhaps) beyond.

Table 2.5

Year	Bald Eagle Breeding Pairs	Year	Bald Eagle Breeding Pairs
2	3700	10	6500
4	4400	12	7600
6	5100	14	8700
8	5700	16	9800

Source: www.fws.gov/midwest/eagle/population
1990 corresponds to year 0.

Figure 2.69

[Graph: Bald eagle breeding pairs (y-axis, 3,000 to 10,000) vs. t (years since 1990) (x-axis, 0 to 18). Plotted points showing increasing trend.]

WORTHY OF NOTE

For the years 1992 to 2000, we can estimate the growth in breeding pairs $\frac{\Delta\text{pairs}}{\Delta\text{time}}$ using the points (2, 3700) and (10, 6500) in the slope formula. The result is $\frac{350}{1}$, or 350 pairs per year. For 2000 to 2006, using (10, 6500) and (16, 9800) shows the rate of growth is significantly larger: $\frac{\Delta\text{pairs}}{\Delta\text{years}} = \frac{550}{1}$ or 550 pairs per year.

The combination of these two lines would be a single function that modeled the population of breeding pairs from 1990 to 2006, but it would be *defined in two pieces.* This is an example of a **piecewise-defined function.**

The notation for these functions is a large "left brace" indicating the equations it groups are part of a single function. Using selected data points and techniques from Section 2.3, we find equations that could represent each piece are $p(t) = 350t + 3000$

WORTHY OF NOTE

In Figure 2.69, note that we indicated the exclusion of $t = 10$ from the second piece of the function using an open half-circle.

for $0 \leq t \leq 10$ and $p(t) = 550t + 1000$ for $t > 10$, where $p(t)$ is the number of breeding pairs in year t. The complete function is then written:

$$
\underset{\text{function name}}{p(t)} = \underset{\text{function pieces}}{\begin{cases} 350t + 3000 \\ 550t + 1000 \end{cases}} \quad \underset{\text{domain of each piece}}{\begin{matrix} 2 \leq t \leq 10 \\ t > 10 \end{matrix}}
$$

EXAMPLE 1 ▶ **Writing the Equation and Domain of a Piecewise-Defined Function**

The linear piece of the function shown has an equation of $y = -2x + 10$. The equation of the quadratic piece is $y = -x^2 + 9x - 14$. Write the related piecewise-defined function, and state the domain of each piece by inspecting the graph.

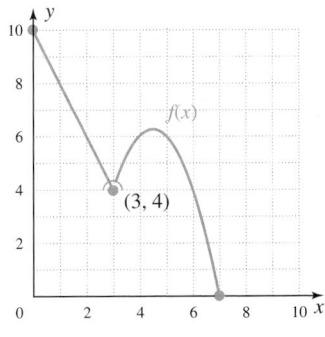

Solution ▶ From the graph we note the linear portion is defined between 0 and 3, with these endpoints included as indicated by the closed dots. The domain here is $0 \leq x \leq 3$. The quadratic portion begins at $x = 3$ *but does not include 3*, as indicated by the half-circle notation. The equation is

$$
\underset{\text{function name}}{f(x)} = \underset{\text{function pieces}}{\begin{cases} -2x + 10 \\ -x^2 + 9x - 14 \end{cases}} \quad \underset{\text{domain}}{\begin{matrix} 0 \leq x \leq 3 \\ 3 < x \leq 7 \end{matrix}}
$$

☑ **A. You've just learned how to state the equation and domain of a piecewise-defined function**

Now try Exercises 7 and 8 ▶

Piecewise-defined functions can be composed of more than two pieces, and can involve functions of many kinds.

B. Graphing Piecewise-Defined Functions

As with other functions, piecewise-defined functions can be graphed by simply plotting points. Careful attention must be paid to the domain of each piece, both to evaluate the function correctly and to consider the inclusion/exclusion of endpoints. In addition, try to keep the transformations of a basic function in mind, as this will often help graph the function more efficiently.

EXAMPLE 2 ▶ **Graphing a Piecewise-Defined Function**

Graph the function by plotting points, then state its domain and range:

$$
h(x) = \begin{cases} -x - 2 & -5 \leq x < -1 \\ 2\sqrt{x + 1} - 1 & x \geq -1 \end{cases}
$$

Solution ▶ The first piece of h is a line with negative slope, while the second is a transformed square root function. Using the endpoints of each domain specified and a few additional points, we obtain the following:

For $h(x) = -x - 2, -5 \leq x < -1$, 　　　For $h(x) = 2\sqrt{x + 1} - 1, x \geq -1$,

x	$h(x)$
-5	3
-3	1
-1	-1

x	$h(x)$
-1	-1
0	1
3	3

After plotting the points from the first piece, we connect them with a line segment noting the left endpoint is included, while the right endpoint is not (indicated using a semicircle around the point). Then we plot the points from the second piece and draw a square root graph, noting the left endpoint here *is* included, and the graph rises to the right. From the graph we note the complete domain of h is $x \in [-5, \infty)$, and the range is $y \in [-1, \infty)$.

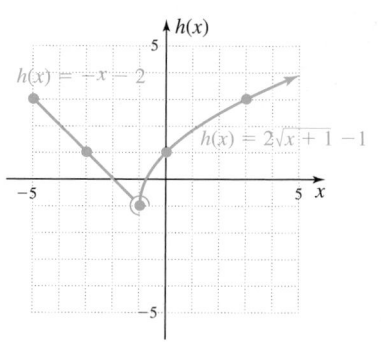

Now try Exercises 9 through 14 ▶

 As an alternative to plotting points, we can graph each piece of the function using transformations of a basic graph, then erase those parts that are outside of the corresponding domain. Repeat this procedure for each piece of the function. One interesting and highly instructive aspect of these functions is the opportunity to investigate restrictions on their domain and the ranges that result.

Piecewise and Continuous Functions

EXAMPLE 3 ▶   Graphing a Piecewise-Defined Function

Graph the function and state its domain and range:

$$f(x) = \begin{cases} -(x-3)^2 + 12 & 0 < x \le 6 \\ 3 & x > 6 \end{cases}$$

Solution ▶   The first piece of f is a basic parabola, shifted three units right, reflected across the x-axis (opening downward), and shifted 12 units up. The vertex is at $(3, 12)$ and the axis of symmetry is $x = 3$, producing the following graphs.

1. Graph first piece of f (Figure 2.70).

2. Erase portion outside domain of $0 < x \le 6$ (Figure 2.71).

Figure 2.70

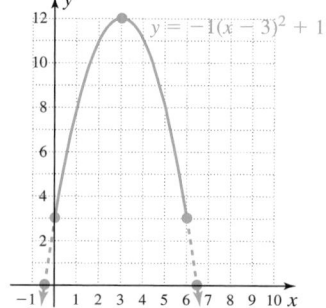

Figure 2.71

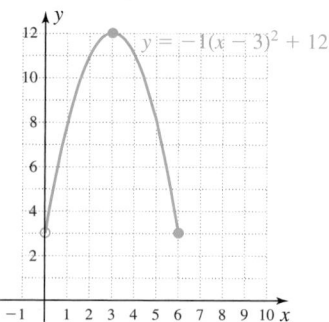

The second function is simply a horizontal line through $(0, 3)$.

3. Graph second piece of f (Figure 2.72).

4. Erase portion outside domain of $x > 6$ (Figure 2.73).

Figure 2.72

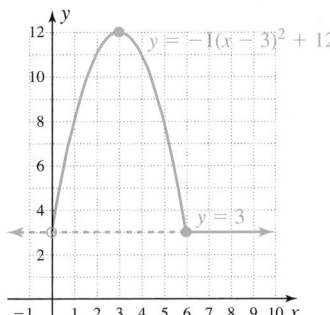

Figure 2.73

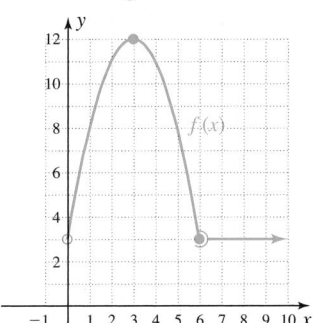

The domain of f is $x \in (0, \infty)$, and the corresponding range is $y \in [3, 12]$.

Now try Exercises 15 through 18 ▶

Piecewise and Discontinuous Functions

Notice that although the function in Example 3 was piecewise-defined, the graph was actually continuous—we could draw the entire graph without lifting our pencil. Piecewise graphs also come in the *discontinuous* variety, which makes the domain and range issues all the more important.

EXAMPLE 4 ▶ **Graphing a Discontinuous Piecewise-Defined Function**

Graph $g(x)$ and state the domain and range:

$$g(x) = \begin{cases} -\frac{1}{2}x + 6 & 0 \le x \le 4 \\ -|x - 6| + 10 & 4 < x \le 9 \end{cases}$$

Solution ▶ The first piece of g is a line, with y-intercept $(0, 6)$ and slope $\frac{\Delta y}{\Delta x} = -\frac{1}{2}$.

1. Graph first piece of g (Figure 2.74).

2. Erase portion outside domain of $0 \le x \le 4$ (Figure 2.75).

Figure 2.74

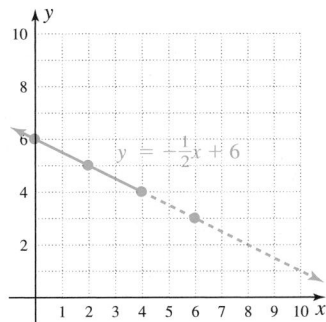

Figure 2.75

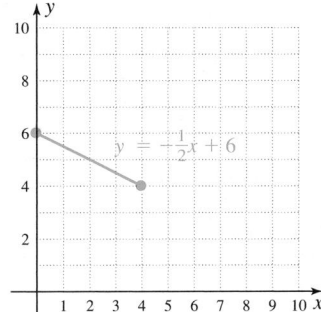

The second is an absolute value function, shifted right 6 units, reflected across the x-axis, then shifted up 10 units.

WORTHY OF NOTE

As you graph piecewise-defined functions, keep in mind that they *are* functions and the end result must pass the vertical line test. This is especially important when we are drawing each piece as a complete graph, then erasing portions outside the effective domain.

3. Graph second piece of g (Figure 2.76).

Figure 2.76

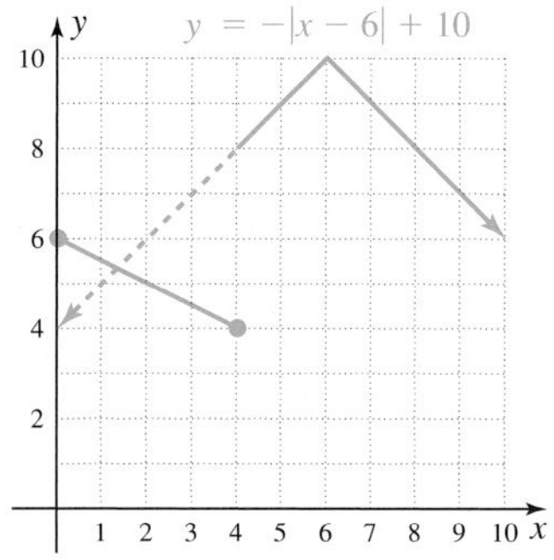

4. Erase portion outside domain of $4 < x \le 9$ (Figure 2.77).

Figure 2.77

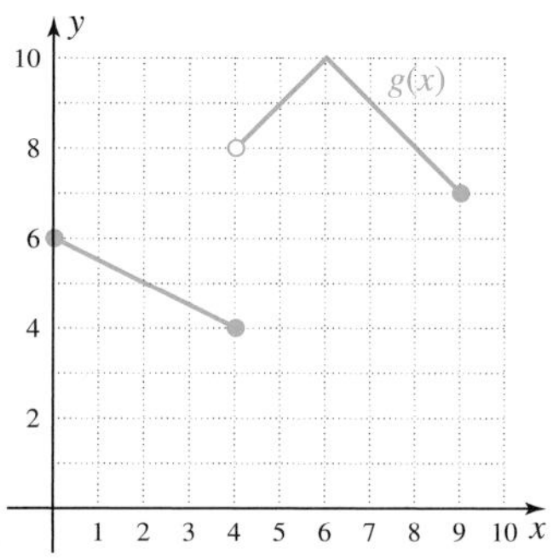

Note that the left endpoint of the absolute value portion is not included (this piece is not defined at $x = 4$), signified by the open dot. The result is a discontinuous graph, as there is no way to draw the graph other than by jumping the pencil from where one piece ends to where the next begins. Using a vertical boundary line, we note the domain of g includes all values between 0 and 9 inclusive: $x \in [0, 9]$. Using a horizontal boundary line shows the smallest y-value is 4 and the largest is 10, but no range values exist between 6 and 7. The range is $y \in [4, 6] \cup [7, 10]$.

Now try Exercises 19 through 22 ▶

EXAMPLE 5 ▶ Graphing a Discontinuous Function

The given piecewise-defined function is not continuous. Graph $h(x)$ to see why, then comment on what could be done to make it continuous.

$$h(x) = \begin{cases} \dfrac{x^2 - 4}{x - 2} & x \neq 2 \\ 1 & x = 2 \end{cases}$$

Solution ▶ The first piece of h is unfamiliar to us, so we elect to graph it by plotting points, noting $x = 2$ is outside the domain. This produces the table shown in Figure 2.78. After connecting the points, the graph of h turns out to be a straight line, but with no corresponding y-value for $x = 2$. This leaves a "hole" in the graph at (2, 4), as designated by the open dot.

Figure 2.78

x	$h(x)$
−4	−2
−2	0
0	2
2	—
4	6

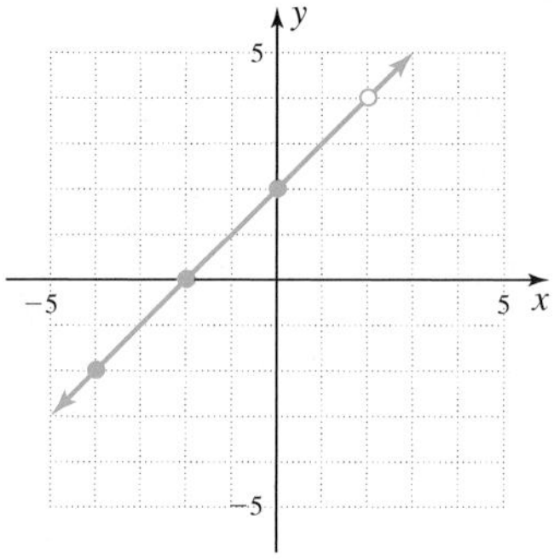

Figure 2.79

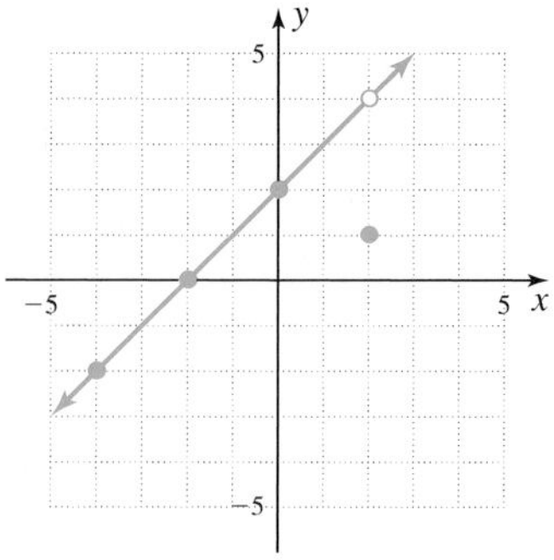

The second piece is point-wise defined, and its graph is simply the point (2, 1) shown in Figure 2.79. It's interesting to note that while the domain of h is all real numbers (h *is* defined at all points), the range is $y \in (-\infty, 4) \cup (4, \infty)$ as the function never takes on the value $y = 4$. In order for h to be continuous, we would need to redefine the second piece as $y = 4$ when $x = 2$.

Now try Exercises 23 through 26 ▶

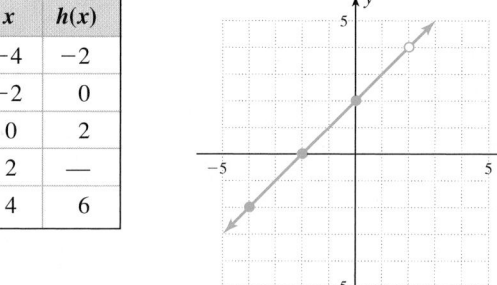

WORTHY OF NOTE

The discontinuity illustrated here is called a **removable discontinuity,** as the discontinuity can be removed by redefining a piece of the function. Note that after factoring the first piece, the denominator is a factor of the numerator, and writing the result in lowest terms gives $h(x) = \frac{(x + 2)\cancel{(x - 2)}}{\cancel{x - 2}}$ $= x + 2$, $x \neq 2$. This is precisely the equation of the line in Figure 2.78 $[h(x) = x + 2]$.

To develop these concepts more fully, it will help to practice finding the equation of a piecewise-defined function *given its graph,* a process similar to that of Example 10 in Section 2.6.

EXAMPLE 6 ▶ **Determining the Equation of a Piecewise-Defined Function**

Determine the equation of the piecewise-defined function shown, including the domain for each piece.

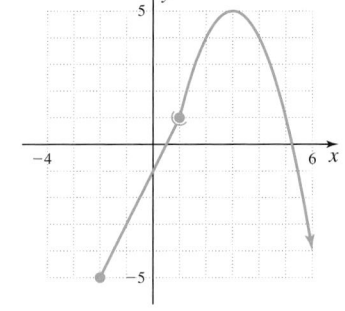

Solution ▶ By counting $\frac{\Delta y}{\Delta x}$ from $(-2, -5)$ to $(1, 1)$, we find the linear portion has slope $m = 2$, and the y-intercept must be $(0, -1)$. The equation of the line is $y = 2x - 1$. The second piece appears to be a parabola with vertex (h, k) at $(3, 5)$. Using this vertex with the point $(1, 1)$ in the general form $y = a(x - h)^2 + k$ gives

$$y = a(x - h)^2 + k \qquad \text{general form}$$
$$1 = a(1 - 3)^2 + 5 \qquad \text{substitute 1 for } x, 1 \text{ for } y, 3 \text{ for } h, 5 \text{ for } k$$
$$-4 = a(-2)^2 \qquad \text{simplify; subtract 5}$$
$$-4 = 4a \qquad (-2)^2 = 4$$
$$-1 = a \qquad \text{divide by 4}$$

The equation of the parabola is $y = -(x - 3)^2 + 5$. Considering the domains shown in the figure, the equation of this piecewise-defined function must be

$$p(x) = \begin{cases} 2x - 1 & -2 \leq x \leq 1 \\ -(x - 3)^2 + 5 & x > 1 \end{cases}$$

☑ **B.** You've just learned how to graph functions that are piecewise-defined

Now try Exercises 27 through 30 ▶

C. Applications of Piecewise-Defined Functions

The number of applications for piecewise-defined functions is practically limitless. It is actually fairly rare for a single function to accurately model a situation over a long period of time. Laws change, spending habits change, and technology can bring abrupt alterations in many areas of our lives. To accurately model these changes often requires a piecewise-defined function.

EXAMPLE 7 ▶ **Modeling with a Piecewise-Defined Function**

For the first half of the twentieth century, per capita spending on police protection can be modeled by $S(t) = 0.54t + 12$, where $S(t)$ represents per capita spending on police protection in year t (1900 corresponds to year 0). After 1950, perhaps due to the growth of American cities, this spending greatly increased: $S(t) = 3.65t - 144$. Write these as a piecewise-defined function $S(t)$, state the domain for each piece, then graph the function. According to this model, how much was spent (per capita) on police protection in 2000? How much will be spent in 2010?

Source: Data taken from the *Statistical Abstract of the United States* for various years.

Solution ▶

$$
S(t) = \begin{cases} 0.54t + 12 & 0 \le t \le 50 \\ 3.65t - 144 & t > 50 \end{cases}
$$

function name   function pieces   effective domain

Since both pieces are linear, we can graph each part using two points. For the first function, $S(0) = 12$ and $S(50) = 39$. For the second function $S(50) \approx 39$ and $S(80) = 148$. The graph for each piece is shown in the figure. Evaluating S at $t = 100$:

$$
\begin{aligned}
S(t) &= 3.65t - 144 \\
S(100) &= 3.65(100) - 144 \\
&= 365 - 144 \\
&= 221
\end{aligned}
$$

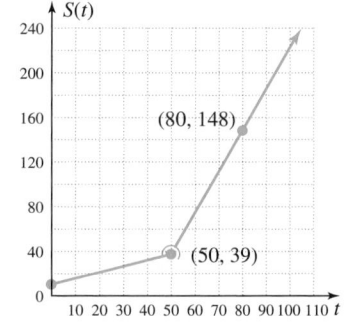

About \$221 per capita was spent on police protection in the year 2000. For 2010, the model indicates that \$257.50 per capita will be spent: $S(110) = 257.5$.

Now try Exercises 33 through 44 ▶

Step Functions

The last group of piecewise-defined functions we'll explore are the **step functions,** so called because the pieces of the function form a series of horizontal steps. These functions find frequent application in the way consumers are charged for services, and have a number of applications in number theory. Perhaps the most common is called the **greatest integer function,** though recently its alternative name, **floor function,** has gained popularity (see Figure 2.80). This is in large part due to an improvement in notation and as a better contrast to **ceiling functions.** The floor function of a real number x, denoted $f(x) = \lfloor x \rfloor$ or $[\![x]\!]$ (we will use the first), is the largest integer less than or equal to x. For instance, $\lfloor 5.9 \rfloor = 5$, $\lfloor 7 \rfloor = 7$, and $\lfloor -3.4 \rfloor = -4$.

In contrast, the ceiling function $C(x) = \lceil x \rceil$ is the smallest integer greater than or equal to x, meaning $\lceil 5.9 \rceil = 6$, $\lceil 7 \rceil = 7$, and $\lceil -3.4 \rceil = -3$ (see Figure 2.81). In simple terms, for any noninteger value on the number line, the floor function returns the integer to the left, while the ceiling function returns the integer to the right. A graph of each function is shown.

Figure 2.80

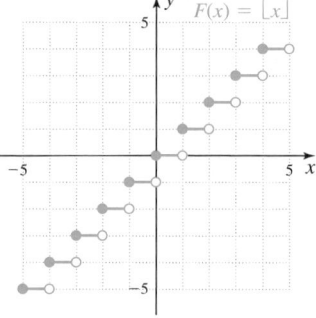

Figure 2.81

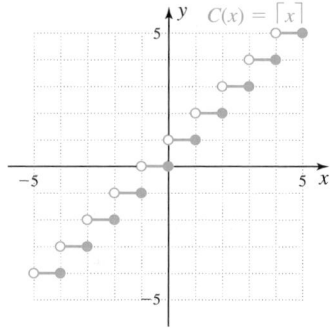

One common application of floor functions is the price of theater admission, where children 12 and under receive a discounted price. Right up until the day they're 13, they qualify for the lower price: $\lfloor 12\frac{364}{365} \rfloor = 12$. Applications of ceiling functions would include how phone companies charge for the minutes used (charging the 12-min rate for a phone call that only lasted 11.3 min: $\lceil 11.3 \rceil = 12$), and postage rates, as in Example 8.

EXAMPLE 8 ▶ **Modeling Using a Step Function**

As of May 2007, the first-class postage rate for large envelopes sent through the U.S. mail was 80¢ for the first ounce, then an additional 17¢ per ounce thereafter, up to 13 ounces. Graph the function and state its domain and range. Use the graph to state the cost of mailing a report weighing (a) 7.5 oz, (b) 8 oz, and (c) 8.1 oz in a large envelope.

Solution ▶ The 80¢ charge applies to letters weighing between 0 oz and 1 oz. Zero is not included since we have to mail *something,* but 1 is included since a large envelope and its contents weighing exactly one ounce still costs 80¢. The graph will be a horizontal line segment.

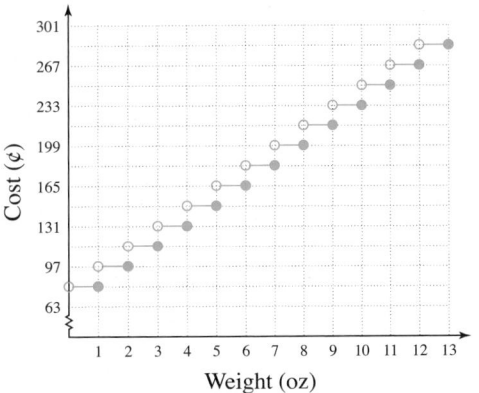

The function is defined for all weights between 0 and 13 oz, excluding zero and including 13: $x \in (0, 13]$. The range consists of single outputs corresponding to the step intervals: $R \in \{80, 97, 114, \ldots, 267, 284\}$.

a. The cost of mailing a 7.5-oz report is 199¢.

b. The cost of mailing an 8.0-oz report is still 199¢.

c. The cost of mailing an 8.1-oz report is $199 + 17 = 216$¢, since this brings you up to the next step.

☑ **C.** You've just learned how to solve applications involving piecewise-defined functions

Now try Exercises 45 through 48 ▶

TECHNOLOGY HIGHLIGHT

Piecewise-Defined Functions

Most graphing calculators are able to graph piecewise-defined functions. Consider the function f shown here:

$$f(x) = \begin{cases} x + 2 & x < 2 \\ (x - 4)^2 + 3 & x \geq 2 \end{cases}$$

Both "pieces" are well known—the first is a line with slope $m = 1$ and y-intercept $(0, 2)$. The second is a parabola that opens upward, shifted 4 units to the right and 3 units up. If we attempt to graph $f(x)$ using $Y_1 = x + 2$ and $Y_2 = (x - 4)^2 + 3$ as they stand, the resulting graph may be difficult to analyze because the pieces conflict and intersect (Figure 2.82). To graph the functions we must indicate the domain for each piece, separated by a slash and enclosed in parentheses. For instance, for the first piece we enter $Y_1 = x + 2/(x < 2)$, and for the second, $Y_2 = (x - 4)^2 + 3/(x \geq 2)$ (Figure 2.83). The slash looks like (is) the division symbol, but in this context, the calculator interprets it as a means of separating the function from the domain. The inequality symbols are accessed using the [2nd] [MATH] (TEST) keys. The graph is shown on Figure 2.84, where we see the function is linear for $x \in (-\infty, 2)$ and quadratic for $x \in [2, \infty)$. How does the calculator remind us the function is defined only for $x = 2$ on the second piece? Using the [2nd] [GRAPH] (TABLE) feature reveals the calculator will give an **ERR:** (ERROR) message for inputs outside of its domain (Figure 2.85).

We can also use the calculator to investigate endpoints of the domain. For instance, we know that $Y_1 = x + 2$ is not defined for $x = 2$, but what about numbers very close to 2? Go to [2nd] [WINDOW] (TBLSET) and place the calculator in the Indpnt: Auto [ASK] mode. With both Y_1 and Y_2 enabled, use the [2nd] [GRAPH] (TABLE) feature to evaluate the functions at numbers very near 2. Use $x = 1.9, 1.99, 1.999$, and so on.

Figure 2.82

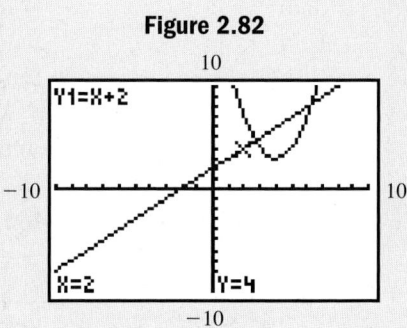

Figure 2.83

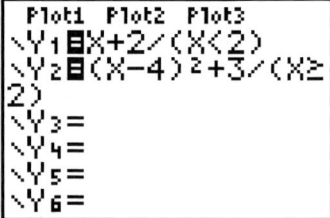

Figure 2.84

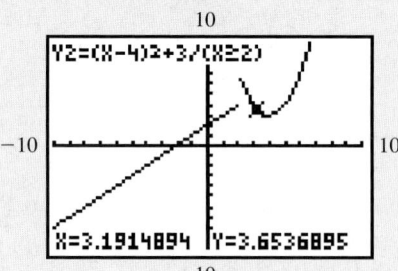

Figure 2.85

X	Y1	Y2
0	2	ERR:
.5	2.5	ERR:
1	3	ERR:
1.5	3.5	ERR:
2	ERR:	7
2.5	ERR:	5.25
3	ERR:	4

X=3

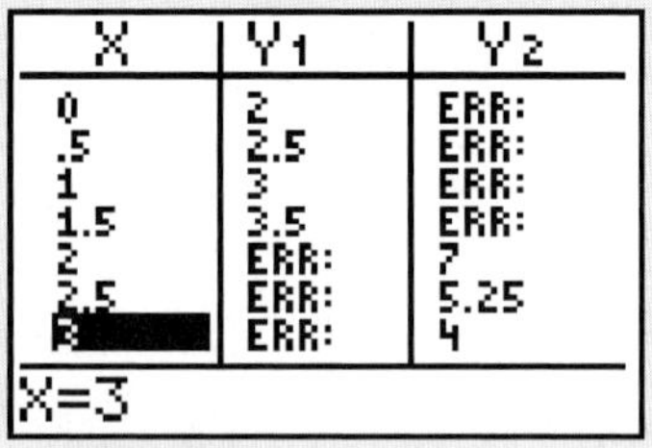

Exercise 1: What appears to be happening to the output values for Y_1? What about Y_2?

Exercise 2: What do you notice about the output values when 1.99999 is entered? Use the right arrow key [►] to move the cursor into columns Y_1 and Y_2. Comment on what you think the calculator is doing. Will Y_1 ever really have an output equal to 4?

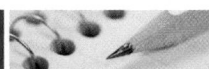

2.7 EXERCISES

▶ CONCEPTS AND VOCABULARY

Fill in each blank with the appropriate word or phrase.
Carefully reread the section if needed.

1. A function whose entire graph can be drawn without lifting your pencil is called a _____ function.

2. The input values for which each part of a piecewise function is defined is the _____ of the function.

3. A graph is called _____ if it has no sharp turns or jagged edges.

4. When graphing $2x + 3$ over a domain of $x > 0$, we leave an _____ dot at $(0, 3)$.

5. Discuss/Explain how to determine if a piecewise-defined function is continuous, without having to graph the function. Illustrate with an example.

6. Discuss/Explain how it is possible for the domain of a function to be defined for all real numbers, but have a range that is defined on more than one interval. Construct an illustrative example.

▶ DEVELOPING YOUR SKILLS

For Exercises 7 and 8, (a) use the correct notation to write them as a single piecewise-defined function, state the domain for each piece by inspecting the graph, and (b) state the range of the function.

7. $Y_1 = x^2 - 6x + 10$; $Y_2 = \frac{3}{2}x - \frac{5}{2}$

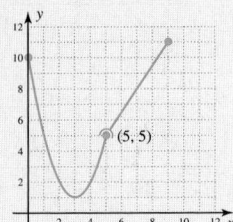

8. $Y_1 = -1.5|x - 5| + 10$; $Y_2 = -\sqrt{x - 7} + 5$

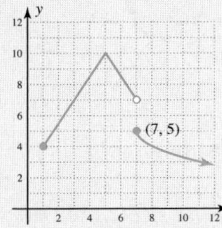

Evaluate each piecewise-defined function as indicated (if possible).

9. $h(x) = \begin{cases} -2 & x < -2 \\ |x| & -2 \le x < 3 \\ 5 & x \ge 3 \end{cases}$

$h(-5)$, $h(-2)$, $h(-\frac{1}{2})$, $h(0)$, $h(2.999)$, and $h(3)$

10. $H(x) = \begin{cases} 2x + 3 & x < 0 \\ x^2 + 1 & 0 \le x < 2 \\ 5 & x > 2 \end{cases}$

$H(-3)$, $H(-\frac{3}{2})$, $H(-0.001)$, $H(1)$, $H(2)$, and $H(3)$

11. $p(x) = \begin{cases} 5 & x < -3 \\ x^2 - 4 & -3 \le x \le 3 \\ 2x + 1 & x > 3 \end{cases}$

$p(-5)$, $p(-3)$, $p(-2)$, $p(0)$, $p(3)$, and $p(5)$

12. $q(x) = \begin{cases} -x - 3 & x < -1 \\ 2 & -1 \le x < 2 \\ -\frac{1}{2}x^2 + 3x - 2 & x \ge 2 \end{cases}$

$q(-3)$, $q(-1)$, $q(0)$, $q(1.999)$, $q(2)$, and $q(4)$

Graph each piecewise-defined function by plotting points, then state its domain and range.

13. $p(x) = \begin{cases} x + 2 & -6 \le x \le 2 \\ 2|x - 4| & x > 2 \end{cases}$

14. $q(x) = \begin{cases} \sqrt{x + 4} & -4 \le x \le 0 \\ |x - 2| & 0 < x \le 7 \end{cases}$

Graph each piecewise-defined function and state its domain and range. Use transformations of the toolbox functions where possible.

15. $g(x) = \begin{cases} -(x - 1)^2 + 5 & -2 \le x \le 4 \\ 2x - 12 & x > 4 \end{cases}$

16. $h(x) = \begin{cases} \frac{1}{2}x + 1 & x \le 0 \\ (x - 2)^2 - 3 & 0 < x \le 5 \end{cases}$

17. $p(x) = \begin{cases} \frac{1}{2}x + 1 & x \ne 4 \\ 2 & x = 4 \end{cases}$

18. $q(x) = \begin{cases} \frac{1}{2}(x - 1)^3 - 1 & x \ne 3 \\ -2 & x = 3 \end{cases}$

19. $H(x) = \begin{cases} -x + 3 & x < 1 \\ -|x - 5| + 6 & 1 \le x < 9 \end{cases}$

20. $w(x) = \begin{cases} \sqrt[3]{x + 1} & x < 1 \\ (x - 3)^2 - 2 & 1 \le x \le 6 \end{cases}$

21. $f(x) = \begin{cases} -x - 3 & x < -3 \\ 9 - x^2 & -3 \le x < 2 \\ 4 & x \ge 2 \end{cases}$

22. $h(x) = \begin{cases} -\frac{1}{2}x - 1 & x < -3 \\ -|x| + 5 & -3 \le x \le 5 \\ 3\sqrt{x - 5} & x > 5 \end{cases}$

Each of the following functions has a pointwise discontinuity. Graph the first piece of each function, then find the value of c so that a continuous function results.

23. $f(x) = \begin{cases} \dfrac{x^2 - 9}{x + 3} & x \neq -3 \\ c & x = -3 \end{cases}$

24. $f(x) = \begin{cases} \dfrac{x^2 - 3x - 10}{x - 5} & x \neq 5 \\ c & x = 5 \end{cases}$

25. $f(x) = \begin{cases} \dfrac{x^3 - 1}{x - 1} & x \neq 1 \\ c & x = 1 \end{cases}$

26. $f(x) = \begin{cases} \dfrac{4x - x^3}{x + 2} & x \neq -2 \\ c & x = -2 \end{cases}$

Determine the equation of each piecewise-defined function shown, including the domain for each piece. Assume all pieces are toolbox functions.

27.
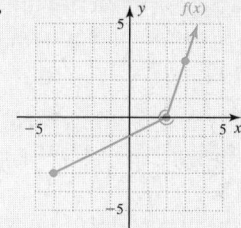

28.
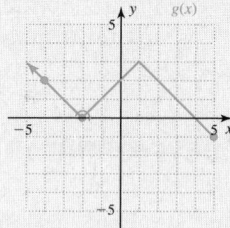

29.
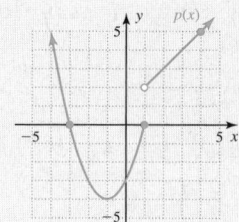

30.
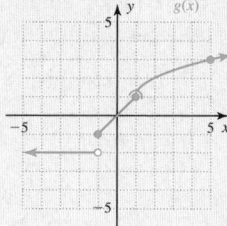

▶ WORKING WITH FORMULAS

31. **Definition of absolute value:** $|x| = \begin{cases} -x & x < 0 \\ x & x \geq 0 \end{cases}$

The absolute value function can be stated as a piecewise-defined function, a technique that is sometimes useful in graphing variations of the function or solving absolute value equations and inequalities. How does this definition ensure that the absolute value of a number is always positive? Use this definition to help sketch the graph of $f(x) = \frac{|x|}{x}$. Discuss what you notice.

32. **Sand dune function:**

$f(x) = \begin{cases} -|x - 2| + 1 & 1 \leq x < 3 \\ -|x - 4| + 1 & 3 \leq x < 5 \\ -|x - 2k| + 1 & 2k - 1 \leq x < 2k + 1, \text{ for } k \in N \end{cases}$

There are a number of interesting graphs that can be created using piecewise-defined functions, and these functions have been the basis for more than one piece of modern art. (a) Use the descriptive name and the pieces given to graph the function f. Is the function accurately named? (b) Use any combination of the toolbox functions to explore your own creativity by creating a piecewise-defined function with some interesting or appealing characteristics.

▶ APPLICATIONS

For Exercises 33 and 34, a. write the information given as a piecewise-defined function, and state the domain for each piece by inspecting the graph. b. Give the range of each.

33. Due to heavy advertising, initial sales of the Lynx Digital Camera grew very rapidly, but started to decline once the advertising blitz was over. During the advertising campaign, sales were modeled by the

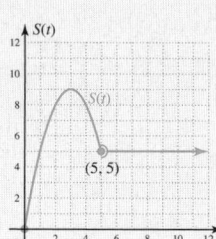

function $S(t) = -t^2 + 6t$, where $S(t)$ represents hundreds of sales in month t. However, as Lynx Inc. had hoped, the new product secured a foothold in the market and sales leveled out at a steady 500 sales per month.

34. From the turn of the twentieth century, the number of newspapers (per thousand population) grew rapidly until the 1930s, when the growth slowed down and then declined. The years 1940 to 1946 saw a "spike" in growth, but the years 1947 to 1954 saw an almost equal decline. Since 1954 the number has continued to decline, but at a slower rate. The number of papers

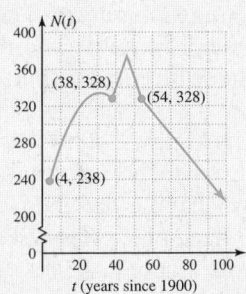

N per thousand population for each period, respectively, can be approximated by
$N_1(t) = -0.13t^2 + 8.1t + 208$,
$N_2(t) = -5.75|t - 46| + 374$, and
$N_3(t) = -2.45t + 460$.

Source: Data from the Statistical Abstract of the United States, various years; data from The First Measured Century, The AEI Press, Caplow, Hicks, and Wattenberg, 2001.

35. The percentage of American households that own publicly traded stocks began rising in the early 1950s, peaked in 1970, then began to decline until 1980 when there was a dramatic increase due to easy access over the Internet, an improved economy, and other factors. This phenomenon is modeled by the function $P(t)$, where $P(t)$ represents the percentage of households owning stock in year t, with 1950 corresponding to year 0.

$$P(t) = \begin{cases} -0.03t^2 + 1.28t + 1.68 & 0 \le t \le 30 \\ 1.89t - 43.5 & t > 30 \end{cases}$$

a. According to this model, what percentage of American households held stock in the years 1955, 1965, 1975, 1985, and 1995? If this pattern continues, what percentage held stock in 2005?

b. Why is there a discrepancy in the outputs of each piece of the function for the year 1980 ($t = 30$)? According to how the function is defined, which output should be used?

Source: 2004 Statistical Abstract of the United States, Table 1204; various other years.

36. America's dependency on foreign oil has always been a "hot" political topic, with the amount of imported oil fluctuating over the years due to political climate, public awareness, the economy, and other factors. The amount of crude oil imported can be approximated by the function given, where $A(t)$ represents the number of barrels imported in year t (in billions), with 1980 corresponding to year 0.

$$A(t) = \begin{cases} 0.047t^2 - 0.38t + 1.9 & 0 \le t < 8 \\ -0.075t^2 + 1.495t - 5.265 & 8 \le t \le 11 \\ 0.133t + 0.685 & t > 11 \end{cases}$$

a. Use $A(t)$ to estimate the number of barrels imported in the years 1983, 1989, 1995, and 2005.

b. What was the minimum number of barrels imported between 1980 and 1988?

Source: 2004 Statistical Abstract of the United States, Table 897; various other years.

37. Energy rationing: In certain areas of the United States, power blackouts have forced some counties to ration electricity. Suppose the cost is $0.09 per kilowatt (kW) for the first 1000 kW a household uses. After 1000 kW, the cost increases to 0.18 per kW: Write these charges for electricity in the form of a piecewise-defined function $C(h)$, where $C(h)$ is the cost for h kilowatt hours. State the domain for each piece. Then sketch the graph and determine the cost for 1200 kW.

38. Water rationing: Many southwestern states have a limited water supply, and some state governments try to control consumption by manipulating the cost of water usage. Suppose for the first 5000 gal a household uses per month, the charge is $0.05 per gallon. Once 5000 gal is used the charge doubles to $0.10 per gallon. Write these charges for water usage in the form of a piecewise-defined function $C(w)$, where $C(w)$ is the cost for w gallons of water and state the domain for each piece. Then sketch the graph and determine the cost to a household that used 9500 gal of water during a very hot summer month.

39. Pricing for natural gas: A local gas company charges $0.75 per therm for natural gas, up to 25 therms. Once the 25 therms has been exceeded, the charge doubles to $1.50 per therm due to limited supply and great demand. Write these charges for natural gas consumption in the form of a piecewise-defined function $C(t)$, where $C(t)$ is the charge for t therms and state the domain for each piece. Then sketch the graph and determine the cost to a household that used 45 therms during a very cold winter month.

40. Multiple births:
The number of multiple births has steadily increased in the United States during the twentieth century and beyond. Between 1985 and 1995 the number of twin births could be modeled by the function $T(x) = -0.21x^2 + 6.1x + 52$, where x is the

number of years since 1980 and T is in thousands. After 1995, the incidence of twins becomes more linear, with $T(x) = 4.53x + 28.3$ serving as a better model. Write the piecewise-defined function modeling the incidence of twins for these years, including the domain of each piece. Then sketch the graph and use the function to estimate the incidence of twins in 1990, 2000, and 2005. If this trend continues, how many sets of twins will be born in 2010?

Source: National Vital Statistics Report, Vol. 50, No. 5, February 12, 2002

41. U.S. military expenditures: Except for the year 1991 when military spending was cut drastically, the amount spent by the U.S. government on national defense and veterans' benefits rose steadily from 1980 to 1992. These expenditures can be modeled by the function $S(t) = -1.35t^2 + 31.9t + 152$, where $S(t)$ is in billions of dollars and 1980 corresponds to $t = 0$.

Source: 1992 Statistical Abstract of the United States, Table 525

From 1992 to 1996 this spending declined, then began to rise in the following years. From 1992 to 2002, military-related spending can be modeled by $S(t) = 2.5t^2 - 80.6t + 950$.

Source: 2004 Statistical Abstract of the United States, Table 492

Write $S(t)$ as a single piecewise-defined function, stating the domain for each piece. Then sketch the graph and use the function to find the projected amount the United States will spend on its military in 2005, 2008, and 2010 if this trend continues.

42. Amusement arcades: At a local amusement center, the owner has the SkeeBall machines programmed to reward very high scores. For scores of 200 or less, the function $T(x) = \frac{x}{10}$ models the number of tickets awarded (rounded to the nearest whole). For scores over 200, the number of tickets is modeled by $T(x) = 0.001x^2 - 0.3x + 40$. Write these equation models of the number of tickets awarded in the form of a piecewise-defined function and state the domain for each piece. Then sketch the graph and find the number of tickets awarded to a person who scores 390 points.

43. Phone service charges: When it comes to phone service, a large number of calling plans are available. Under one plan, the first 30 min of any phone call costs only 3.3¢ per minute. The charge increases to 7¢ per minute thereafter. Write this information in the form of a piecewise-defined function and state the domain for each piece. Then sketch the graph and find the cost of a 46-min phone call.

44. Overtime wages: Tara works on an assembly line, putting together computer monitors. She is paid $9.50 per hour for regular time (0, 40 hr], $14.25 for overtime (40, 48 hr], and when demand for computers is high, $19.00 for double-overtime (48, 84 hr]. Write this information in the form of a simplified piecewise-defined function, and state the domain for each piece. Then sketch the graph and find the gross amount of Tara's check for the week she put in 54 hr.

45. Admission prices: At Wet Willy's Water World, infants under 2 are free, then admission is charged according to age. Children 2 and older but less than 13 pay $2, teenagers 13 and older but less than 20 pay $5, adults 20 and older but less than 65 pay $7, and senior citizens 65 and older get in at the teenage rate. Write this information in the form of a piecewise-defined function and state the domain for each piece. Then sketch the graph and find the cost of admission for a family of nine which includes: one grandparent (70), two adults (44/45), 3 teenagers, 2 children, and one infant.

46. Demographics: One common use of the floor function $y = \lfloor x \rfloor$ is the reporting of ages. As of 2007, the record for longest living human is 122 yr, 164 days for the life of Jeanne Calment, formerly of France. While she actually lived $x = 122\frac{164}{365}$ years, ages are normally reported using the floor function, or the greatest integer number of years less than or equal to the actual age: $\lfloor 122\frac{164}{365} \rfloor = 122$ years. (a) Write a function $A(t)$ that gives a person's age, where $A(t)$ is the reported age at time t. (b) State the domain of the function (be sure to consider Madame Calment's record). Report the age of a person who has been living for (c) 36 years; (d) 36 years, 364 days; (e) 37 years; and (f) 37 years, 1 day.

47. Postage rates: The postal charge function from Example 8 is simply a transformation of the basic ceiling function $y = \lceil x \rceil$. Using the ideas from Section 2.6, (a) write the postal charges as a step function $C(w)$, where $C(w)$ is the cost of mailing a large envelope weighing w ounces, and (b) state the domain of the function. Then use the function to find the cost of mailing reports weighing: (c) 0.7 oz, (d) 5.1 oz, (e) 5.9 oz; (f) 6 oz, and (g) 6.1 oz.

48. Cell phone charges: A national cell phone company advertises that calls of 1 min or less do not count toward monthly usage. Calls lasting longer than 1 min are calculated normally using a ceiling function, meaning a call of 1 min, 1 sec will be counted as a 2-min call. Using the ideas from Section 2.6, (a) write the cell phone charges as a piecewise-defined function $C(m)$, where $C(m)$ is the cost of a call lasting m minutes, and include the domain of the function. Then (b) graph the function, and (c) use the graph or function to determine if a cell phone subscriber has exceeded the 30 free minutes granted by her calling plan for calls lasting 2 min 3 sec, 13 min 46 sec, 1 min 5 sec, 3 min 59 sec, 8 min 2 sec. (d) What was the actual usage in minutes and seconds?

49. Combined absolute value graphs: Carefully graph the function $h(x) = |x - 2| - |x + 3|$ using a

table of values over the interval $x \in [-5, 5]$. Is the function continuous? Write this function in piecewise-defined form and state the domain for each piece.

50. Combined absolute value graphs: Carefully graph the function $H(x) = |x - 2| + |x + 3|$ using a table of values over the interval $x \in [-5, 5]$. Is the function continuous? Write this function in piecewise-defined form and state the domain for each piece.

▶ EXTENDING THE CONCEPT

51. You've heard it said, "*any number divided by itself is one.*" Consider the functions $Y_1 = \frac{x + 2}{x + 2}$, and $Y_2 = \frac{|x + 2|}{x + 2}$. Are these functions continuous?

52. Find a linear function $h(x)$ that will make the function shown a *continuous* function. Be sure to include its domain.

$$f(x) = \begin{cases} x^2 & x < 1 \\ h(x) \\ 2x + 3 & x > 3 \end{cases}$$

▶ MAINTAINING YOUR SKILLS

53. (1.3) Solve: $\dfrac{3}{x - 2} + 1 = \dfrac{30}{x^2 - 4}$.

54. (R.5) Compute the following and write the result in lowest terms:

$$\frac{x^3 + 3x^2 - 4x - 12}{x - 3} \cdot \frac{2x - 6}{x^2 + 5x + 6} \div (3x - 6)$$

55. (R.7) For the figure shown, (a) find the length of the missing side, (b) state the area of the

triangular base, and (c) compute the volume of the prism.

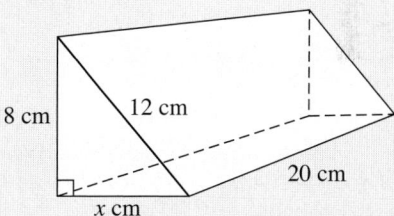

56. (2.4) Find the equation of the line perpendicular to $3x + 4y = 8$, and through the point $(0, -2)$. Write the result in slope-intercept form.

Learning Objectives

In Section 2.8 you will learn how to:

☐ **A.** Compute a sum or difference of functions and determine the domain of the result

☐ **B.** Compute a product or quotient of functions and determine the domain

☐ **C.** Compose two functions and determine the domain; decompose a function

☐ **D.** Interpret operations on functions graphically

☐ **E.** Apply the algebra and composition of functions in context

In Section 2.5, we created new functions *graphically* by applying transformations to basic functions. In this section, we'll use two (or more) functions to create new functions *algebraically*. Previous courses often contain material on the sum, difference, product, and quotient of polynomials. Here we'll combine these functions with the basic operations, noting the result is also a function that can be evaluated, graphed, and analyzed. We call these basic operations on functions the **algebra of functions.**

A. Sums and Differences of Functions

This section introduces the notation used for basic operations on functions. Here we'll note the result is also a function whose domain depends on the original functions. In general, if f and g are functions *with overlapping domains*, $f(x) + g(x) = (f + g)(x)$ and $f(x) - g(x) = (f - g)(x)$.

Sums and Differences of Functions

For functions f and g with domains P and Q respectively, the sum and difference of f and g are defined by:

	Domain of result
$(f + g)(x) = f(x) + g(x)$	$P \cap Q$
$(f - g)(x) = f(x) - g(x)$	$P \cap Q$

EXAMPLE 1A ▶ **Evaluating a Difference of Functions**

Given $f(x) = x^2 - 5x$ and $g(x) = 2x - 9$,
  a. Determine the domain of $h(x) = (f - g)(x)$.   **b.** Find $h(3)$ using the definition.

Solution ▶   **a.** Since the domain of both f and g is $\mathbb{R}$, their intersection is $\mathbb{R}$, so the domain of h is also $\mathbb{R}$.

  b.
$$
\begin{aligned}
h(x) &= (f - g)(x) && \text{given difference} \\
&= f(x) - g(x) && \text{by definition} \\
h(3) &= f(3) - g(3) && \text{substitute 3 for } x \\
&= [(3)^2 - 5(3)] - [2(3) - 9] && \text{evaluate} \\
&= [9 - 15] - [6 - 9] && \text{multiply} \\
&= -6 - [-3] && \text{subtract} \\
&= -3 && \text{result}
\end{aligned}
$$

If the function h is to be graphed or evaluated numerous times, it helps to compute a *new function rule* for h, rather than repeatedly apply the definition.

EXAMPLE 1B ▶ For the functions f, g, and h, as defined in Example 1A,
  a. Find a new function rule for h.   **b.** Use the result to find $h(3)$.

Solution ▶   **a.**
$$
\begin{aligned}
h(x) &= (f - g)(x) && \text{given difference} \\
&= f(x) - g(x) && \text{by definition} \\
&= (x^2 - 5x) - (2x - 9) && \text{replace } f(x) \text{ with } (x^2 - 5x) \text{ and } g(x) \text{ with } (2x - 9) \\
&= x^2 - 7x + 9 && \text{distribute and combine like terms}
\end{aligned}
$$

b. $h(3) = (3)^2 - 7(3) + 9$ substitute 3 for x

$\quad\quad = 9 - 21 + 9$ multiply

$\quad\quad = -3$ result

Notice the result from Part (b) is identical to that in Example 1A.

Now try Exercises 7 through 10 ▶

 CAUTION ▶ From Example 1A, note the importance of using grouping symbols with the algebra of functions. Without them, we could easily confuse the signs of g when computing the difference. Also, note that any operation applied to the functions f and g simply results in an *expression* representing a new function rule for h, and is not an *equation* that needs to be factored or solved.

EXAMPLE 2 ▶ **Evaluating a Sum of Functions**

For $f(x) = x^2$ and $g(x) = \sqrt{x - 2}$,

    a. Determine the domain of $h(x) = (f + g)(x)$.

    b. Find a new function rule for h.

    c. Evaluate $h(3)$.

    d. Evaluate $h(-1)$.

Solution ▶ **a.** The domain of f is $\mathbb{R}$, while the domain of g is $x \in [2, \infty)$. Since their intersection is $[2, \infty)$, this is the domain of the new function h.

WORTHY OF NOTE

If we *did* try to evaluate $h(-1)$, the result would be $1 + \sqrt{-3}$, which is not a real number. While it's true we could write $1 + \sqrt{-3}$ as $1 + i\sqrt{3}$ and consider it an "answer," our study here focuses on real numbers and the graphs of functions in a coordinate system where x and y are both real.

b. $h(x) = (f + g)(x)$ given sum

$\quad\quad = f(x) + g(x)$ by definition

$\quad\quad = x^2 + \sqrt{x - 2}$ substitute x^2 for $f(x)$ and $\sqrt{x-2}$ for $g(x)$ (no other simplifications possible)

c. $h(3) = (3)^2 + \sqrt{3 - 2}$ substitute 3 for x

$\quad\quad = 10$ result

d. $x = -1$ is outside the domain of h.

Now try Exercises 11 through 14 ▶

☑ A. You've just learned how to compute a sum or difference of functions and determine the domain of the result

This "intersection of domains" is illustrated in Figure 2.86 using ideas from Section 1.2.

Figure 2.86

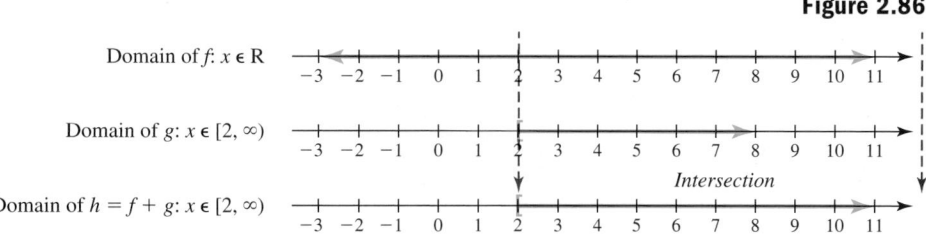

B. Products and Quotients of Functions

The product and quotient of two functions is defined in a manner similar to that for sums and differences. For example, if f and g are functions *with overlapping domains*,

$$(f \cdot g)(x) = f(x) \cdot g(x) \text{ and } \left(\frac{f}{g}\right)(x) = \frac{f(x)}{g(x)}.$$ As you might expect, for quotients we must stipulate $g(x) \neq 0$.

Products and Quotients of Functions

For functions f and g with domains P and Q, respectively, the product and quotient of f and g are defined by:

	Domain of result
$(f \cdot g)(x) = f(x) \cdot g(x)$	$P \cap Q$
$\left(\dfrac{f}{g}\right)(x) = \dfrac{f(x)}{g(x)}$	$P \cap Q$, for all $g(x) \neq 0$

EXAMPLE 3 ▶ **Computing a Product of Functions**

Given $f(x) = \sqrt{1 + x}$ and $g(x) = \sqrt{3 - x}$,

a. Determine the domian of $h(x) = (f \cdot g)(x)$.

b. Find a new function rule for h.

c. Use the result from part (b) to evaluate $h(2)$ and $h(4)$.

Solution ▶ a. The domain of f is $x \in [-1, \infty)$ and the domain of g is $x \in (-\infty, 3]$. The intersection of these domains gives $x \in [-1, 3]$, which is the domain for h.

b. $h(x) = (f \cdot g)(x)$ given product

     $= f(x) \cdot g(x)$ by definition

     $= \sqrt{1 + x} \cdot \sqrt{3 - x}$ substitute $\sqrt{1 + x}$ for f and $\sqrt{3 + x}$ for g

     $= \sqrt{3 + 2x - x^2}$ combine using properties of radicals

c. $h(2) = \sqrt{3 + 2(2) - (2)^2}$ substitute 2 for x

     $= \sqrt{3} \approx 1.732$ result

    $h(4) = \sqrt{3 + 2(4) - (4)^2}$ substitute 4 for x

     $= \sqrt{-5}$ not a real number

The second result of Part (c) is not surprising, since $x = 4$ is not in the domain of h [meaning $h(4)$ is not defined for this function].

Now try Exercises 15 through 18 ▶

In future sections, we use polynomial division as a tool for factoring, an aid to graphing, and to determine whether two expressions are equivalent. Understanding the notation and domain issues related to division will strengthen our ability in these areas.

EXAMPLE 4 ▶ **Computing a Quotient of Functions**

Given $f(x) = x^3 - 3x^2 + 2x - 6$ and $g(x) = x - 3$,

a. Determine the domain of $h(x) = \left(\dfrac{f}{g}\right)(x)$.

b. Find a new function rule for h.

c. Use the result from part (b) to evaluate $h(3)$ and $h(0)$.

Solution ▶ a. While the domain of both f and g is $\mathbb{R}$ and their intersection is also $\mathbb{R}$, we know from the definition (and past experience) *that $g(x)$ cannot be zero.* The domain of h is $x \in (-\infty, 3) \cup (3, \infty)$.

b. $h(x) = \left(\dfrac{f}{g}\right)(x)$ given quotient

     $= \dfrac{f(x)}{g(x)}$ by definition

     $= \dfrac{x^3 - 3x^2 + 2x - 6}{x - 3}$ replace f with $x^3 - 3x^2 + 2x - 6$ and g with $x - 3$

c. Recall that $x = 3$ is not in the domain of h. For $h(0)$ we have:

$$h(0) = \frac{(0)^3 - 3(0)^2 + 2(0) - 6}{(0) - 3} \qquad \text{replace } x \text{ with } 0$$

$$= \frac{-6}{-3} = 2 \qquad h(0) = 2$$

Now try Exercises 19 through 34 ▶

☑ **B.** You've just learned how to compute a product or quotient of functions and determine the domain

From our work with rational expressions in Appendix I.E, the expression that defines h can be simplified: $\dfrac{x^3 - 3x^2 + 2x - 6}{x - 3} = \dfrac{x^2(x - 3) + 2(x - 3)}{x - 3} = \dfrac{(x^2 + 2)(x\!\!\!\diagup\!\!\!-3)}{x\!\!\!\diagup\!\!\!-3} =$
$x^2 + 2$. But from the original expression, h is not defined if $g(x) = 3$, *even if the result for h is a polynomial*. In this case, we write the simplified form as $h(x) = x^2 + 2, x \neq 3$.

For additional practice with the algebra of functions, **see Exercises 35 through 46.**

C. Composition of Functions

The composition of functions is best understood by studying the "input/output" nature of a function. Consider $g(x) = x^2 - 3$. For $g(x)$ we might say, "inputs are squared, then decreased by three." In diagram form we have:

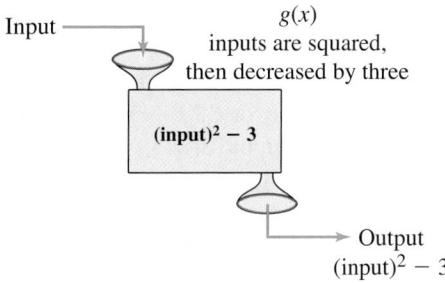

In many respects, a function box can be regarded as a very simple machine, running a simple program. It doesn't matter what the input is, this machine is going to *square the input then subtract three*.

EXAMPLE 5 ▶ **Evaluating a Function**

For $g(x) = x^2 - 3$, find

a. $g(-5)$
b. $g(5t)$
c. $g(t - 4)$

Solution ▶ **a.** $g(x) = x^2 - 3$ original function

input -5 ⟶

$g(-5) = (-5)^2 - 3$ square input, then subtract 3

$= 25 - 3$ simplify

$= 22$ result

b. $g(x) = x^2 - 3$ original function

input $5t$ ⟶

$g(5t) = (5t)^2 - 3$ square input, then subtract 3

$= 25t^2 - 3$ result

Now try Exercises 47 and 48 ▶

WORTHY OF NOTE

It's important to note that t and $t - 4$ are two different, distinct values—the number represented by t, and a number four less than t. Examples would be 7 and 3, 12 and 8, as well as -10 and -14. There should be nothing awkward or unusual about evaluating $g(t)$ versus evaluating $g(t - 4)$ as in Example 5c.

c.

$$g(x) = x^2 - 3 \qquad \text{original function}$$

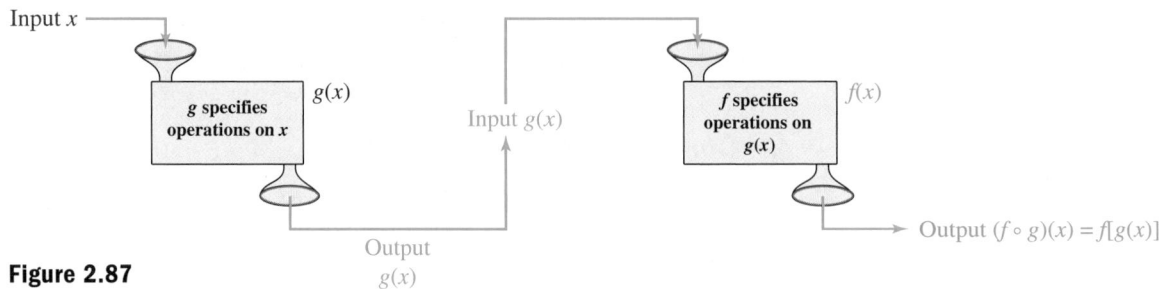

input $t - 4$

$$g(t - 4) = (t - 4)^2 - 3 \qquad \text{square input, then subtract 3}$$
$$= t^2 - 8t + 16 - 3 \qquad \text{expand binomial}$$
$$= t^2 - 8t + 13 \qquad \text{result}$$

When the input value is itself a function (rather than a single number or variable), this process is called the **composition of functions.** The evaluation method is exactly the same, we are simply using a function input. Using a general function $g(x)$ and a function diagram as before, we illustrate the process in Figure 2.87.

Figure 2.87

The notation used for the composition of f with g is an open dot "∘" placed between them, and is read, "f composed with g." The notation $(f \circ g)(x)$ indicates that $g(x)$ is an input for f: $(f \circ g)(x) = f[g(x)]$. If the order is reversed, as in $(g \circ f)(x), f(x)$ becomes the input for g: $(g \circ f)(x) = g[f(x)]$. Figure 2.87 also helps us determine the domain of a composite function, in that the first function g can operate only if x is a valid input for g, and the second function f can operate only if $g(x)$ is a valid input for f. In other words, $(f \circ g)(x)$ is defined for *all x in the domain of g, such that $g(x)$ is in the domain of f.*

⚠ **CAUTION** ▶ Try not to confuse the new "open dot" notation for the *composition* of functions, with the multiplication dot used to indicate the *product* of two functions: $(f \cdot g)(x) = (fg)(x)$ or the product of f with g; $(f \circ g)(x) = f[g(x)]$ or f composed with g.

The Composition of Functions

Given two functions f and g, the composition of f with g is defined by

$$(f \circ g)(x) = f[g(x)]$$

The domain of the composition is all x in the domain of g
for which $g(x)$ is in the domain of f.

In Figure 2.88, these ideas are displayed using mapping notation, as we consider the simple case where $g(x) = x$ and $f(x) = \sqrt{x}$.

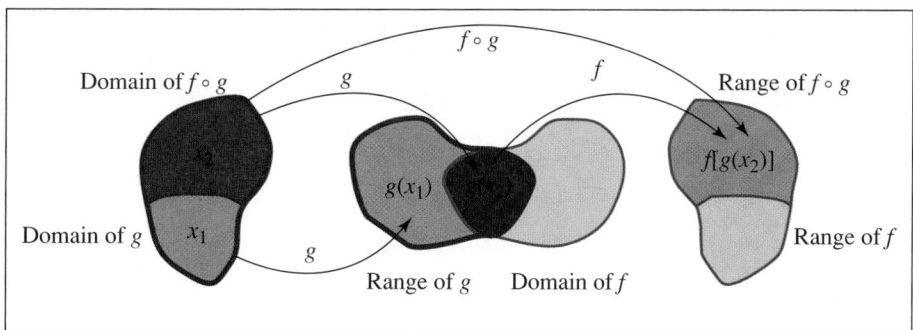

Figure 2.88

The domain of g (all real numbers) is shown within the red border, with g taking the negative inputs represented by x_1 (light red), to a like-colored portion of the range— the negative outputs $g(x_1)$. The nonnegative inputs represented by x_2 (dark red) are also mapped to a like-colored portion of the range—the nonnegative outputs $g(x_2)$. While the range of g is also all real numbers, function f can only use the nonnegative inputs represented by $g(x_2)$. This restricts the domain of $(f \circ g)(x)$ to only the inputs from g, where $g(x)$ is in the domain of f.

EXAMPLE 6 ▶ **Finding a Composition of Functions**

Given $f(x) = \sqrt{x - 4}$ and $g(x) = 3x + 2$, find

a. $(f \circ g)(x)$

b. $(g \circ f)(x)$

Also determine the domain for each.

Solution ▶ **a.** $f(x) = \sqrt{x - 4}$ says "decrease inputs by 4, and take the square root of the result."

$$(f \circ g)(x) = f[g(x)] \qquad \text{g(x) is an input for f}$$
$$= \sqrt{g(x) - 4} \qquad \text{decrease input by 4, and take the square root of the result}$$
$$= \sqrt{(3x + 2) - 4} \qquad \text{substitute } 3x + 2 \text{ for } g(x)$$
$$= \sqrt{3x - 2} \qquad \text{result}$$

While g is defined for all real numbers, f is defined only for nonnegative numbers. Since $f[g(x)] = \sqrt{3x - 2}$, we need $3x - 2 \geq 0$, $x \geq \frac{2}{3}$. In interval notation, the domain of $(f \circ g)(x)$ is $x \in [\frac{2}{3}, \infty)$.

b. The function g says "inputs are multiplied by 3, then increased by 2."

$$(g \circ f)(x) = g[f(x)] \qquad \text{f(x) is an input for g}$$
$$= 3f(x) + 2 \qquad \text{multiply input by 3, then increase by 2}$$
$$= 3\sqrt{x - 4} + 2 \qquad \text{substitute } \sqrt{x - 4} \text{ for } f(x)$$

For $g[f(x)]$, g can accept any real number input, but f can supply only those where $x \geq 4$. The domain of $(g \circ f)(x)$ is $x \in [4, \infty)$.

WORTHY OF NOTE

Example 6 shows that $(f \circ g)(x)$ is generally not equal to $(g \circ f)(x)$. On those occasions when they *are* equal, the functions have a unique relationship that we'll study in Section 4.1.

Now try Exercises 49 through 58 ▶

EXAMPLE 7 ▶ Finding a Composition of Functions

For $f(x) = \dfrac{3x}{x-1}$ and $g(x) = \dfrac{2}{x}$, analyze the domain of

a. $(f \circ g)(x)$.

b. $(g \circ f)(x)$.

c. Find the actual compositions and comment.

Solution ▶ a. $(f \circ g)(x)$: For g to be defined, $x \neq 0$ is our first restriction. Once $g(x)$ is used as the input, we have $f[g(x)] = \dfrac{3g(x)}{g(x) - 1}$, and additionally note that $g(x)$ cannot equal 1. This means $\dfrac{2}{x} \neq 1$, so $x \neq 2$. The domain of $f \circ g$ is $\{x \mid x \neq 0, x \neq 2\}$.

b. $(g \circ f)(x)$: For f to be defined, $x \neq 1$ is our first restriction. Once $f(x)$ is used as the input, we have $g[f(x)] = \dfrac{2}{f(x)}$, and additionally note that $f(x)$ cannot be 0. This means $\dfrac{3x}{x-1} \neq 0$, so $x \neq 0$. The domain of $(g \circ f)(x)$ is $\{x \mid x \neq 0, x \neq 1\}$.

c. For $(f \circ g)(x)$:

$$f[g(x)] = \frac{3g(x)}{g(x) - 1} \qquad \text{composition of } f \text{ with } g$$

$$= \frac{\left(\dfrac{3}{1}\right)\left(\dfrac{2}{x}\right)}{\left(\dfrac{2}{x}\right) - 1} \qquad \text{substitute } \frac{2}{x} \text{ for } g(x)$$

$$= \frac{\dfrac{6}{x}}{\dfrac{2-x}{x}} = \frac{6}{x} \cdot \frac{x}{2-x} \qquad \text{simplify denominator; invert and multiply}$$

$$= \frac{6}{2-x} \qquad \text{result}$$

WORTHY OF NOTE

As Example 7 illustrates, the domain of $h(x) = (f \circ g)(x)$ *cannot simply be taken from the new function rule for h*. It *must* be determined from the functions composed to obtain h.

Notice the function rule for $(f \circ g)(x)$ has an implied domain of $x \neq 2$, but does not show that g (the inner function) is undefined when $x = 0$ (see Part a). The domain of $(f \circ g)(x)$ is actually $x \neq 2$ **and** $x \neq 0$.

For $(g \circ f)(x)$ we have:

$$g[f(x)] = \frac{2}{f(x)} \qquad \text{composition of } g \text{ with } f$$

$$\frac{2}{f(x)} = \frac{2}{\dfrac{3x}{x-1}} \qquad \text{substitute } \frac{3x}{x-1} \text{ for } f(x)$$

$$= \frac{2}{1} \cdot \frac{x-1}{3x} \qquad \text{invert and multiply}$$

$$= \frac{2(x-1)}{3x} \qquad \text{result}$$

Similarly, the function rule for $(g \circ f)(x)$ has an implied domain of $x \neq 0$, but does not show that f (the inner function) is undefined when $x = 1$ (see Part a). The domain of $(g \circ f)(x)$ is actually $x \neq 0$ **and** $x \neq 1$.

Now try Exercises 59 through 64 ▶

To further explore concepts related to the domain of a composition, **see Exercises 92 through 94.**

Decomposing a Composite Function

Based on Figure 2.89, would you say that the circle is inside the square or the square is inside the circle? The decomposition of a composite function is related to a similar question, as we ask ourselves what function (of the composition) is on the "inside"—the input quantity—and what function is on the "outside." For instance, consider $h(x) = \sqrt{x - 4}$, where we see that $x - 4$ is "inside" the radical. Letting $g(x) = x - 4$ and $f(x) = \sqrt{x}$, we have $h(x) = (f \circ g)(x)$ or $f[g(x)]$.

Figure 2.89

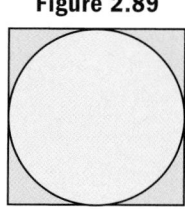

> **WORTHY OF NOTE**
>
> The decomposition of a function is not unique and can often be done in many different ways.

EXAMPLE 8 ▶ **Decomposing a Composite Function**

Given $h(x) = (\sqrt[3]{x} + 1)^2 - 3$, identify two functions f and g so that $(f \circ g)(x) = h(x)$, then check by composing the functions to obtain $h(x)$.

Solution ▶ Noting that $\sqrt[3]{x} + 1$ is inside the squaring function, we assign $g(x)$ as this inner function: $g(x) = \sqrt[3]{x} + 1$. The outer function is the squaring function decreased by 3, so $f(x) = x^2 - 3$.

> ☑ **C.** You've just learned how to compose two functions and determine the domain, and decompose a function

Check: $(f \circ g)(x) = f[g(x)]$ *$g(x)$ is an input for f*

$\qquad\qquad\quad = [g(x)]^2 - 3$ *f squares inputs, then decreases the result by 3*

$\qquad\qquad\quad = [\sqrt[3]{x} + 1]^2 - 3$ *substitute $\sqrt[3]{x} + 1$ for $g(x)$*

$\qquad\qquad\quad = h(x)$ ✓

Now try Exercises 65 through 68 ▶

D. A Graphical View of Operations on Functions

The algebra and composition of functions also has an instructive *graphical interpretation,* in which values for $f(k)$ and $g(k)$ are read from a graph (k is a given constant), with operations like $(f + g)(k) = f(k) + g(k)$ then computed and lodged. Once the value of $g(k)$ is known, $(f \circ g)(k) = f[g(k)]$ is likewise interpreted and computed (also **see Exercise 95**).

EXAMPLE 9 ▶ **Interpreting Operations on Functions Graphically**

Use the graph given to find the value of each expression:

  a. $(f + g)(-2)$

  b. $(f \circ g)(7)$

  c. $(g - f)(6)$

  d. $\left(\dfrac{g}{f}\right)(8)$

  e. $(f \cdot g)(4)$

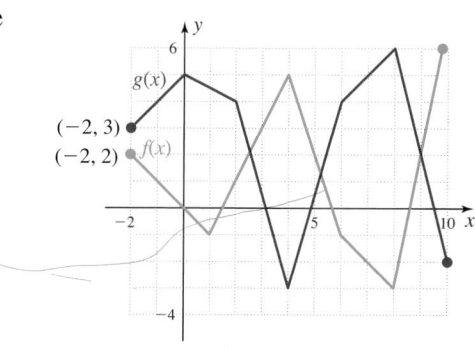

Solution ▶ Since the needed input values for this example are $x = -2, 4, 6, 7,$ and 8, we begin by reading the value of $f(x)$ and $g(x)$ at each point. From the graph, we note that $f(-2) = 2$ and $g(-2) = 3$. The other values are likewise found and appear in the table. For $(f + g)(-2)$ we have:

x	$f(x)$	$g(x)$
-2	2	3
4	5	-3
6	-1	4
7	-2	5
8	-3	6

a. $(f + g)(-2) = f(-2) + g(-2)$ definition

$\qquad\qquad = 2 + 3$ substitute 2 for $f(-2)$ and 3 for $g(-2)$

$\qquad\qquad = 5$ result

b. $(f \circ g)(7) = f[g(7)]$ definition

$\qquad\qquad = f(5)$ substitute 5 for $g(7)$

$\qquad\qquad = 2$ result read from graph: $f(5) = 2$

With some practice, the computations can be done mentally and we have

c. $(g - f)(6) = g(6) - f(6)$

$\qquad\qquad = 4 - (-1) = 5$

d. $\left(\dfrac{g}{f}\right)(8) = \dfrac{g(8)}{f(8)}$

$\qquad\qquad = \dfrac{6}{-3} = -2$

e. $(f \cdot g)(4) = f(4) \cdot g(4)$

$\qquad\qquad = 5(-3) = -15$

☑ **D.** You've just learned how to interpret operations on functions graphically

Now try Exercises 69 through 78 ▶

E. Applications of the Algebra and Composition of Functions

The algebra of functions plays an important role in the business world. For example, the cost to manufacture an item, the revenue a company brings in, and the profit a company earns are all functions of the number of items made and sold. Further, we know a company "breaks even" (making $0 profit) when the difference between their revenue R and their cost C, is zero.

EXAMPLE 10 ▶   **Applying Operations on Functions in Context**

The fixed costs to publish *Relativity Made Simple* (by N.O. Way) is $2500, and the variable cost is $4.50 per book. Marketing studies indicate the best selling price for the book is $9.50 per copy.

a. Find the cost, revenue, and profit functions for this book.

b. Determine how many copies must be sold for the company to break even.

Solution ▶ **a.** Let x represent the number of books published and sold. The cost of publishing is $4.50 per copy, plus fixed costs (labor, storage, etc.) of $2500. The cost function is $C(x) = 4.50x + 2500$. If the company charges $9.50 per book, the revenue function will be $R(x) = 9.50x$. Since profit equals revenue minus costs,

$P(x) = R(x) - C(x)$

$\qquad = 9.50x - (4.50x + 2500)$ substitute $9.50x$ for R and $4.50x + 2500$ for C

$\qquad = 9.50x - 4.50x - 2500$ distribute

$\qquad = 5x - 2500$ result

The profit function is $P(x) = 5x - 2500$.

b. When a company "breaks even," the profit is zero: $P(x) = 0$.

$$P(x) = 5x - 2500 \quad \text{profit function}$$
$$0 = 5x - 2500 \quad \text{substitute 0 for } P(x)$$
$$2500 = 5x \quad \text{add 2500}$$
$$500 = x \quad \text{divide by 5}$$

In order for the company to break even, 500 copies must be sold.

Now try Exercises 81 through 84 ▶

Suppose that due to a collision, an oil tanker is spewing oil into the open ocean. The oil is spreading outward in a shape that is roughly circular, with the radius of the circle modeled by the function $r(t) = 2\sqrt{t}$, where t is the time in minutes and r is measured in feet. How could we determine the *area* of the oil slick in terms of t? As you can see, the radius depends on the time and the area depends on the radius. In diagram form we have:

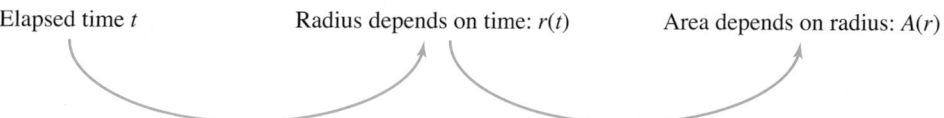

Elapsed time t Radius depends on time: $r(t)$ Area depends on radius: $A(r)$

It is possible to create a direct relationship between the elapsed time and the area of the circular spill using a composition of functions.

EXAMPLE 11 ▶   **Applying a Composition in Context**

Given $r(t) = 2\sqrt{t}$ and $A(r) = \pi r^2$,

a. Write A directly as a function of t by computing $(A \circ r)(t)$.

b. Find the area of the oil spill after 30 min.

Solution ▶   **a.** The function A squares inputs, then multiplies by π.

$$(A \circ r)(t) = A[r(t)] \quad r(t) \text{ is the input for } A$$
$$= [r(t)]^2 \cdot \pi \quad \text{square input, multiply by } \pi$$
$$= [2\sqrt{t}]^2 \cdot \pi \quad \text{substitute } 2\sqrt{t} \text{ for } r(t)$$
$$= 4\pi t \quad \text{result}$$

Since the result contains no variable r, we can now compute the area of the spill directly, given the elapsed time t (in minutes): $A(t) = 4\pi t$.

b. To find the area after 30 min, use $t = 30$.

$$A(t) = 4\pi t \quad \text{composite function}$$
$$A(30) = 4\pi(30) \quad \text{substitute 30 for } t$$
$$= 120\pi \quad \text{simplify}$$
$$\approx 377 \quad \text{result (rounded to the nearest unit)}$$

After 30 min, the area of the spill is approximately 377 ft^2.

☑ **E.** You've just learned how to apply the algebra and composition of functions in context

Now try Exercises 85 through 90 ▶

TECHNOLOGY HIGHLIGHT

Composite Functions

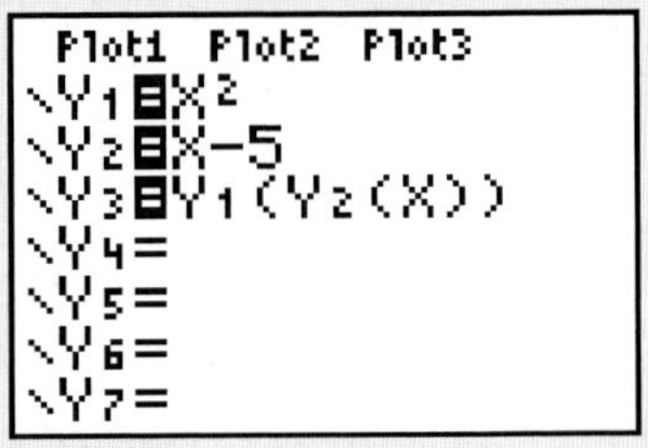

The graphing calculator is truly an amazing tool when it comes to studying composite functions. Using this powerful tool, composite functions can be graphed, evaluated, and investigated with ease. To begin, enter the functions $y = x^2$ and $y = x - 5$ as Y_1 and Y_2 on the [Y=] screen. Enter the composition $(Y_1 \circ Y_2)(x)$ as $Y_3 = Y_1(Y_2(X))$, as shown in Figure 2.90 [in our standard notation we have $f(x) = x^2$, $g(x) = x - 5$, and $h(x) = (f + g)(x) = f[g(x)]$. On the TI-84 *Plus*, we access the function variables Y_1, Y_2, Y_3, and so on by pressing [VARS]

Figure 2.90

```
Plot1  Plot2  Plot3
\Y1=X²
\Y2=X-5
\Y3=Y1(Y2(X))
\Y4=
\Y5=
\Y6=
\Y7=
```

[▶] [ENTER] and selecting the function desired. Pressing [ZOOM] **6:ZStandard** will graph all three functions in the standard window. Let's look at the relationship between Y_1 and Y_3. Deactivate Y_2 and regraph Y_1 and Y_3. What do you notice about the graphs? Y_3 is the same as the graph of Y_1, but shifted 5 units to the right! Does this have any connection to $Y_2 = x - 5$? Try changing Y_2 to $Y_2 = x + 4$, then regraph Y_1 and Y_3. Use what you notice to complete the following exercises and continue the exploration.

Exercise 1: Change Y_1 to $Y_1 = \sqrt{x}$, then experiment by changing Y_2 to $x + 3$, then to $x - 6$. Did you notice anything similar? What would happen if we changed Y_2 to $Y_2 = x + 7$?

Exercise 2: Change Y_1 to $Y_1 = x^3$, then experiment by changing Y_2 to $x + 5$, then to $x - 1$. Did the same "shift" occur? What would happen if we changed Y_1 to $Y_1 = |x|$?

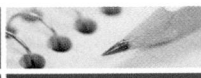

2.8 EXERCISES

▶ CONCEPTS AND VOCABULARY

**Fill in each blank with the appropriate word or phrase.
Carefully reread the section, if necessary.**

1. Given function f with domain A and function g with domain B, the sum $f(x) + g(x)$ can also be written _____. The domain of the result is _____.

2. For the product $h(x) = f(x) \cdot g(x)$, $h(5)$ can be found by evaluating f and g then multiplying the result, or multiplying $f \cdot g$ and evaluating the result. Notationally these are written _____ and _____.

3. When combining functions f and g using basic operations, the domain of the result is the _____ of the domains of f and g. For division, we further stipulate that _____ cannot equal zero.

4. When evaluating functions, if the input value is a function itself, the process is called the _____ of functions. The notation $(f \circ g)(x)$ indicates that _____ is the input value for _____, which we can also write as _____.

5. For $f(x) = 2x^3 - 50x$ and $g(x) = x - 5$, discuss/explain why the domain of $h(x) = \left(\dfrac{f}{g}\right)(x)$ must exclude $x = 5$, even though the resulting quotient is the polynomial $2x^2 + 10x$.

6. For $f(x) = \sqrt{2x + 7}$ and $g(x) = \dfrac{2}{x - 1}$, discuss/explain how the domain of $h(x) = (f \circ g)(x)$ is determined. In particular, why is $h(1)$ not defined even though $f(1) = 3$?

▶ **DEVELOPING YOUR SKILLS**

7. Given $f(x) = 2x^2 - x - 3$ and $g(x) = x^2 + 5x$,
   (a) determine the domain for $h(x) = f(x) - g(x)$ and
   (b) find $h(-2)$ using the definition.

8. Given $f(x) = 2x^2 - 18$ and $g(x) = -3x - 7$,
   (a) determine the domain for $h(x) = f(x) + g(x)$
   and (b) find $h(5)$ using the definition.

9. For the functions f, g, and h, as defined in Exercise 7,
   (a) find a new function rule for h, and (b) use the
   result to find $h(-2)$. (c) How does the result
   compare to that of Exercise 7?

10. For the functions f, g, and h as defined in
    Exercise 8, (a) find a new function rule for h, and
    (b) use the result to find $h(5)$. (c) How does the
    result compare to that in Exercise 8?

11. For $f(x) = \sqrt{x - 3}$ and $g(x) = 2x^3 - 54$,
    (a) determine the domain of $h(x) = (f + g)(x)$,
    (b) find a new function rule for h, and
    (c) evaluate $h(4)$ and $h(2)$, if possible.

12. For $f(x) = 4x^2 - 2x + 3$ and $g(x) = \sqrt{2x - 5}$,
    (a) determine the domain of $h(x) = (f - g)(x)$,
    (b) find a new function rule for h, and (c) evaluate
    $h(7)$ and $h(2)$, if possible.

13. For $p(x) = \sqrt{x + 5}$ and $q(x) = \sqrt{3 - x}$,
    (a) determine the domain of $r(x) = (p + q)(x)$,
    (b) find a new function rule for r, and
    (c) evaluate $r(2)$ and $r(4)$, if possible.

14. For $p(x) = \sqrt{6 - x}$ and $q(x) = \sqrt{x + 2}$,
    (a) determine the domain of $r(x) = (p - q)(x)$,
    (b) find a new function rule for r, and
    (c) evaluate $r(-3)$ and $r(2)$, if possible.

15. For $f(x) = \sqrt{x + 4}$ and $g(x) = 2x + 3$,
    (a) determine the domain of $h(x) = (f \cdot g)(x)$,
    (b) find a new function rule for h, and
    (c) evaluate $h(-4)$ and $h(21)$, if possible.

16. For $f(x) = -3x + 5$ and $g(x) = \sqrt{x - 7}$,
    (a) determine the domain of $h(x) = (f \cdot g)(x)$,
    (b) find a new function rule for h, and
    (c) evaluate $h(8)$ and $h(11)$, if possible.

17. For $p(x) = \sqrt{x + 1}$ and $q(x) = \sqrt{7 - x}$,
    (a) determine the domain of $r(x) = (p \cdot q)(x)$,
    (b) find a new function rule for r, and
    (c) evaluate $r(15)$ and $r(3)$, if possible.

18. For $p(x) = \sqrt{4 - x}$ and $q(x) = \sqrt{x + 4}$,
    (a) determine the domain of $r(x) = (p \cdot q)(x)$,
    (b) find a new function rule for r, and
    (c) evaluate $r(-5)$ and $r(-3)$, if possible.

For the functions f and g given, (a) determine the
domain of $h(x) = \left(\dfrac{f}{g}\right)(x)$ and (b) find a new function rule
for h in simplified form (if possible), noting the domain
restrictions along side.

19. $f(x) = x^2 - 16$ and $g(x) = x + 4$

20. $f(x) = x^2 - 49$ and $g(x) = x - 7$

21. $f(x) = x^3 + 4x^2 - 2x - 8$ and $g(x) = x + 4$

22. $f(x) = x^3 - 5x^2 + 2x - 10$ and $g(x) = x - 5$

23. $f(x) = x^3 - 7x^2 + 6x$ and $g(x) = x - 1$

24. $f(x) = x^3 - 1$ and $g(x) = x - 1$

25. $f(x) = x + 1$ and $g(x) = x - 5$

26. $f(x) = x + 3$ and $g(x) = x - 7$

For the functions p and q given, (a) determine the
domain of $r(x) = \left(\dfrac{p}{q}\right)(x)$, (b) find a new function rule
for r, and (c) use it to evaluate $r(6)$ and $r(-6)$, if possible.

27. $p(x) = 2x - 3$ and $q(x) = \sqrt{-2 - x}$

28. $p(x) = 1 - x$ and $q(x) = \sqrt{3 - x}$

29. $p(x) = x - 5$ and $q(x) = \sqrt{x - 5}$

30. $p(x) = x + 2$ and $q(x) = \sqrt{x + 3}$

31. $p(x) = x^2 - 36$ and $q(x) = \sqrt{2x + 13}$

32. $p(x) = x^2 - 6x$ and $q(x) = \sqrt{7 + 3x}$

For the functions f and g given, (a) find a new function
rule for $h(x) = \left(\dfrac{f}{g}\right)(x)$ in simplified form. (b) If $h(x)$ were
the original function, what would be its domain?
(c) Since we know $h(x) = \left(\dfrac{f}{g}\right)(x) = \dfrac{f(x)}{g(x)}$, what additional
values are excluded from the domain of h?

33. $f(x) = \dfrac{6x}{x - 3}$ and $g(x) = \dfrac{3x}{x + 2}$

34. $f(x) = \dfrac{4x}{x + 1}$ and $g(x) = \dfrac{2x}{x - 2}$

For each pair of functions f and g given, determine the
sum, difference, product, and quotient of f and g, then
determine the domain in each case.

35. $f(x) = 2x + 3$ and $g(x) = x - 2$

36. $f(x) = x - 5$ and $g(x) = 2x - 3$

37. $f(x) = x^2 + 7$ and $g(x) = 3x - 2$

38. $f(x) = x^2 - 3x$ and $g(x) = x + 4$

39. $f(x) = x^2 + 2x - 3$ and $g(x) = x - 1$

40. $f(x) = x^2 - 2x - 15$ and $g(x) = x + 3$

41. $f(x) = 3x + 1$ and $g(x) = \sqrt{x - 3}$

42. $f(x) = x + 2$ and $g(x) = \sqrt{x + 6}$

43. $f(x) = 2x^2$ and $g(x) = \sqrt{x + 1}$

44. $f(x) = x^2 + 2$ and $g(x) = \sqrt{x - 5}$

45. $f(x) = \dfrac{2}{x - 3}$ and $g(x) = \dfrac{5}{x + 2}$

46. $f(x) = \dfrac{4}{x - 3}$ and $g(x) = \dfrac{1}{x + 5}$

47. Given $f(x) = x^2 - 5x - 14$, find $f(-2), f(7)$, $f(2a)$, and $f(a - 2)$.

48. Given $g(x) = x^3 - 9x$, find $g(-3), g(2), g(3t)$, and $g(t + 1)$.

For each pair of functions below, find (a) $h(x) = (f \circ g)(x)$ and (b) $H(x) = (g \circ f)(x)$, and (c) determine the domain of each result.

49. $f(x) = \sqrt{x + 3}$ and $g(x) = 2x - 5$

50. $f(x) = x + 3$ and $g(x) = \sqrt{9 - x^2}$

51. $f(x) = \sqrt{x - 3}$ and $g(x) = 3x + 4$

52. $f(x) = \sqrt{x + 5}$ and $g(x) = 4x - 1$

53. $f(x) = x^2 - 3x$ and $g(x) = x + 2$

54. $f(x) = 2x^2 - 1$ and $g(x) = 3x + 2$

55. $f(x) = x^2 + x - 4$ and $g(x) = x + 3$

56. $f(x) = x^2 - 4x + 2$ and $g(x) = x - 2$

57. $f(x) = |x| - 5$ and $g(x) = -3x + 1$

58. $f(x) = |x - 2|$ and $g(x) = 3x - 5$

For the functions $f(x)$ and $g(x)$ given, analyze the domain of (a) $(f \circ g)(x)$ and (b) $(g \circ f)(x)$, then (c) find the actual compositions and comment.

59. $f(x) = \dfrac{2x}{x + 3}$ and $g(x) = \dfrac{5}{x}$

60. $f(x) = \dfrac{-3}{x}$ and $g(x) = \dfrac{x}{x - 2}$

61. $f(x) = \dfrac{4}{x}$ and $g(x) = \dfrac{1}{x - 5}$

62. $f(x) = \dfrac{3}{x}$ and $g(x) = \dfrac{1}{x - 2}$

63. For $f(x) = x^2 - 8$, $g(x) = x + 2$, and $h(x) = (f \circ g)(x)$, find $h(5)$ in two ways:

   a. $(f \circ g)(5)$ **b.** $f[g(5)]$

64. For $p(x) = x^2 - 8$, $q(x) = x + 2$, and $H(x) = (p \circ q)(x)$, find $H(-2)$ in two ways:

   a. $(p \circ q)(-2)$ **b.** $p[q(-2)]$

65. For $h(x) = (\sqrt{x - 2} + 1)^3 - 5$, find two functions f and g such that $(f \circ g)(x) = h(x)$.

66. For $H(x) = \sqrt[3]{x^2 - 5} + 2$, find two functions p and q such that $(p \circ q)(x) = h(x)$.

67. Given $f(x) = 2x - 1$, $g(x) = x^2 - 1$, and $h(x) = x + 4$, find $p(x) = f[g([h(x)])]$ and $q(x) = g[f([h(x)])]$.

68. Given $f(x) = 2x + 3$ and $g(x) = \dfrac{x - 3}{2}$, find

   (a) $(f \circ f)(x)$, (b) $(g \circ g)(x)$, (c) $(f \circ g)(x)$, and (d) $(g \circ f)(x)$.

69. **Reading a graph:** The graph given shows the number of cars $C(t)$ and trucks $T(t)$ sold by Ullery Used Autos for the years 2000 to 2010. Use the graph to estimate the number of

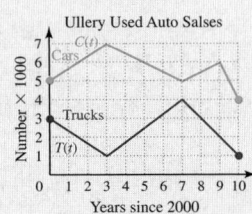
Exercise 69

   a. cars sold in 2005: $C(5)$

   b. trucks sold in 2008: $T(8)$

   c. vehicles sold in 2009: $C(9) + T(9)$

   d. In function notation, how would you determine how many more cars than trucks were sold in 2009? What was the actual number?

70. **Reading a graph:** The graph given shows a government's investment in its military $M(t)$ over time, versus its investment in public works $P(t)$, in millions of dollars. Use the graph to estimate the amount of investment in

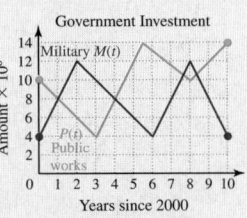
Exercise 70

   a. the military in 2002: $M(2)$

   b. public works in 2005: $P(5)$

   c. public works and the military in 2009: $M(9) + P(9)$

   d. In function notation, how would you determine how much more will be invested in public works than the military in 2010? What is the actual number?

71. Reading a graph: The graph given shows the revenue $R(t)$ and operating costs $C(t)$ of Space Travel Resources (STR), for the years 2000 to 2010. Use the graph to find the

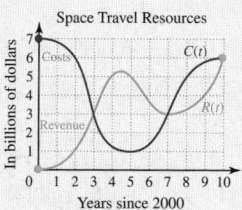

Exercise 71

a. revenue in 2002: $R(2)$

b. costs in 2008: $C(8)$

c. years STR broke even: $R(t) = C(t)$

d. years costs exceeded revenue: $C(t) > R(t)$

e. years STR made a profit: $R(t) > C(t)$

f. For the year 2005, use function notation to write the profit equation for STR. What was their profit?

72. Reading a graph: The graph given shows a large corporation's investment in research and development $R(t)$ over time, and the amount paid to investors as dividends $D(t)$, in billions of dollars. Use the graph to find the

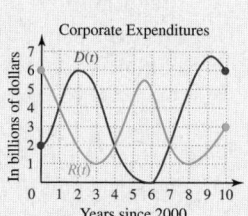

Exercise 72

a. dividend payments in 2002: $D(2)$

b. investment in 2006: $R(6)$

c. years where $R(t) = D(t)$

d. years where $R(t) > D(t)$

e. years where $R(t) < D(t)$

f. Use function notation to write an equation for the total expenditures of the corporation in year t. What was the total for 2010?

73. Reading a graph: Use the given graph to find the result of the operations indicated.

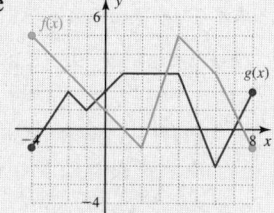

Note $f(-4) = 5$, $g(-4) = -1$, and so on.

Exercise 73

a. $(f + g)(-4)$

b. $(f \cdot g)(1)$

c. $(f - g)(4)$

d. $(f + g)(0)$

e. $\left(\dfrac{f}{g}\right)(2)$

f. $(f \cdot g)(-2)$

g. $(g \cdot f)(2)$

h. $(f - g)(-1)$

i. $(f + g)(8)$

j. $\left(\dfrac{f}{g}\right)(7)$

k. $(g \circ f)(4)$

l. $(f \circ g)(4)$

74. Reading a graph: Use the given graph to find the result of the operations indicated.

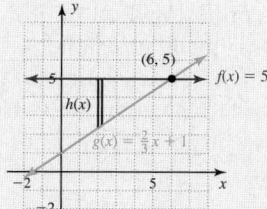

Exercise 74

Note $p(-1) = 3$, $q(5) = 6$, and so on.

a. $(p + q)(-4)$

b. $(p \cdot q)(1)$

c. $(p - q)(4)$

d. $(p + q)(0)$

e. $\left(\dfrac{p}{q}\right)(5)$

f. $(p \cdot q)(-2)$

g. $(q \cdot p)(2)$

h. $(p - q)(-1)$

i. $(p + q)(7)$

j. $\left(\dfrac{p}{q}\right)(6)$

k. $(q \circ p)(4)$

l. $(p \circ q)(-1)$

Some advanced applications require that we use the algebra of functions to find a function rule for the vertical distance between two graphs. For $f(x) = 3$ and $g(x) = -2$ (two horizontal lines), we "see" this vertical distance is 5 units, or in function form: $d(x) = f(x) - g(x) = 3 - (-2) = 5$ units. However, $d(x) = f(x) - g(x)$ also serves as a *general formula* for the vertical distance between two curves (even those that are not horizontal lines), so long as $f(x) > g(x)$ in a chosen interval. Find a function rule in simplified form, for the vertical distance $h(x)$ between the graphs of f and g shown, for the interval indicated.

75. $x \in [0, 6]$

76. $x \in [1, 7]$

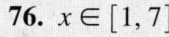
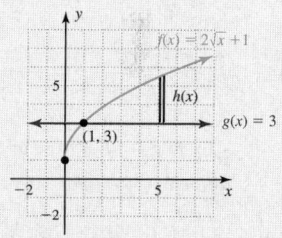

77. $x \in [0, 4]$

78. $x \in [0, 5]$

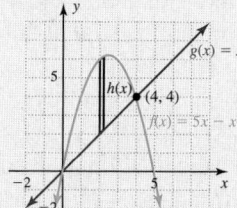
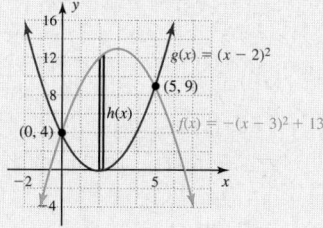

▶ WORKING WITH FORMULAS

79. Surface area of a cylinder: $A = 2\pi rh + 2\pi r^2$

If the height of a cylinder is fixed at 20 cm, the formula becomes $A = 40\pi r + 2\pi r^2$. Write this formula in factored form and find two functions $f(r)$ and $g(r)$ such that $A(r) = (f \cdot g)(r)$. Then find $A(5)$ by direct calculation and also by computing the product of $f(5)$ and $g(5)$, then comment on the results.

80. Compound annual growth: $A(r) = P(1 + r)^t$

The amount of money A in a savings account t yr after an initial investment of P dollars depends on the interest rate r. If $1000 is invested for 5 yr, find $f(r)$ and $g(r)$ such that $A(r) = (f \circ g)(r)$.

▶ APPLICATIONS

81. Boat manufacturing: Giaro Boats manufactures a popular recreational vessel, the *Revolution*. To plan for expanded production and increased labor costs, the company carefully tracks current costs and income. The fixed cost to produce this boat is $108,000 and the variable costs are $28,000 per boat. If the *Revolution* sells for $40,000, (a) find the profit function and (b) determine how many boats must be sold for the company to break even.

82. Non-profit publications: Adobe Hope, a nonprofit agency, publishes the weekly newsletter *Community Options*. In doing so, they provide useful information to the surrounding area while giving high school dropouts valuable work experience. The fixed cost for publishing the newsletter is $900 per week, with a variable cost of $0.25 per newsletter. If the newsletter is sold for $1.50 per copy, (a) find the profit function for the newsletter, (b) determine how many newsletters must be sold to break even, and (c) determine how much money will be returned to the community if 1000 newsletters are sold (to preserve their status as a nonprofit organization).

83. Cost, revenue, and profit: Suppose the total cost of manufacturing a certain computer component can be modeled by the function $C(n) = 0.1n^2$, where n is the number of components made and $C(n)$ is in dollars. If each component is sold at a price of $11.45, the revenue is modeled by $R(n) = 11.45n$. Use this information to complete the following.

  a. Find the function that represents the total profit made from sales of the components.

  b. How much profit is earned if 12 components are made and sold?

  c. How much profit is earned if 60 components are made and sold?

  d. Explain why the company is making a "negative profit" after the 114th component is made and sold.

84. Cost, revenue, and profit: For a certain manufacturer, revenue has been increasing but so has the cost of materials and the cost of employee benefits. Suppose revenue can be modeled by $R(t) = 10\sqrt{t}$, the cost of materials by $M(t) = 2t + 1$, and the cost of benefits by $C(t) = 0.1t^2 + 2$, where t represents the number of months since operations began and outputs are in thousands of dollars. Use this information to complete the following.

  a. Find the function that represents the total manufacturing costs.

  b. Find the function that represents how much more the operating costs are than the cost of materials.

  c. What was the cost of operations in the 10th month after operations began?

  d. How much less were the operating costs than the cost of materials in the 10th month?

  e. Find the function that represents the profit earned by this company.

  f. Find the amount of profit earned in the 5th month and 10th month. Discuss each result.

85. International shoe sizes: Peering inside her athletic shoes, Morgan notes the following shoe sizes: *US 8.5, UK 6, EUR 40*. The function that relates the U.S. sizes to the European (EUR) sizes is $g(x) = 2x + 23$ where x represents the U.S. size and $g(x)$ represents the EUR size. The function that relates European sizes to sizes in the United Kingdom (UK) is $f(x) = 0.5x - 14$ where x represents the EUR size and $f(x)$ represents the UK size. Find the function $h(x)$ that relates the U.S. measurement directly to the UK measurement by finding $h(x) = (f \circ g)(x)$. Find The UK size for a shoe that has a U.S. size of 13.

86. Currency conversion: On a trip to Europe, Megan had to convert American dollars to euros using the function $E(x) = 1.12x$, where x represents the number of dollars and $E(x)$ is the equivalent number of euros. Later, she converts her euros to Japanese yen using the function $Y(x) = 1061x$, where x represents the number of euros and $Y(x)$ represents the equivalent number of yen. (a) Convert 100 U.S. dollars to euros. (b) Convert the answer from part (a) into Japanese yen. (c) Express yen as a function of dollars by finding $M(x) = (Y \circ E)(x)$, then use $M(x)$ to convert 100 dollars directly to yen. Do parts (b) and (c) agree?

Source: 2005 *World Almanac*, p. 231

87. **Currency conversion:** While traveling in the Far East, Timi must convert U.S. dollars to Thai baht using the function $T(x) = 41.6x$, where x represents the number of dollars and $T(x)$ is the equivalent number of baht. Later she needs to convert her baht to Malaysian ringgit using the function $R(x) = 10.9x$. (a) Convert 100 dollars to baht. (b) Convert the result from part (a) to ringgit. (c) Express ringgit as a function of dollars using $M(x) = (R \circ T)(x)$, then use $M(x)$ to convert 100 dollars to ringgit directly. Do parts (b) and (c) agree?

Source: 2005 *World Almanac*, p. 231

88. **Spread of a fire:** Due to a lightning strike, a forest fire begins to burn and is spreading outward in a shape that is roughly circular. The radius of the circle is modeled by the function $r(t) = 2t$, where t is the time in minutes and r is measured in meters. (a) Write a function for the area burned by the fire directly as a function of t by computing $(A \circ r)(t)$. (b) Find the area of the circular burn after 60 min.

89. **Radius of a ripple:** As Mark drops firecrackers into a lake one 4th of July, each "pop" caused a circular ripple that expanded with time. The radius of the circle is a function of time t. Suppose the function is $r(t) = 3t$, where t is in seconds and r is

in feet. (a) Find the radius of the circle after 2 sec. (b) Find the area of the circle after 2 sec. (c) Express the area as a function of time by finding $A(t) = (A \circ r)(t)$ and use $A(t)$ to find the area of the circle after 2 sec. Do the answers agree?

90. **Expanding supernova:** The surface area of a star goes through an expansion phase prior to going *supernova*. As the star begins expanding, the radius becomes a function of time. Suppose this function is $r(t) = 1.05t$, where t is in days and $r(t)$ is in gigameters (Gm). (a) Find the radius of the star two days after the expansion phase begins. (b) Find the surface area after two days. (c) Express the surface area as a function of time by finding $h(t) = (S \circ r)(t)$, then use $h(t)$ to compute the surface area after two days directly. Do the answers agree?

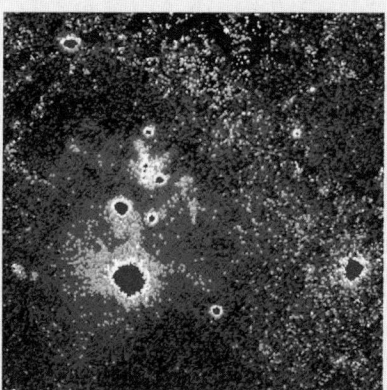

▶ **EXTENDING THE CONCEPT**

91. In a certain country, the function $C(x) = 0.0345x^4 - 0.8996x^3 + 7.5383x^2 - 21.7215x + 40$ approximates the number of Conservatives in the senate for the years 1995 to 2007, where $x = 0$ corresponds to 1995. The function $L(x) = -0.0345x^4 + 0.8996x^3 - 7.5383x^2 + 21.7215x + 10$ gives the number of Liberals for these years. Use this information to answer the following. (a) During what years did the Conservatives control the senate? (b) What was the greatest difference between the number of seats held by each faction in any one year? In what year did this occur? (c) What was the minimum number of seats held by the Conservatives? In what year? (d) Assuming no independent or third-party candidates are elected, what information does the function $T(x) = C(x) + L(x)$ give us? What information does $t(x) = |C(x) - L(x)|$ give us?

92. Given $f(x) = x^3 + 2$ and $g(x) = \sqrt[3]{x-2}$, graph each function on the same axes by plotting the points that correspond to integer inputs for $x \in [-3, 3]$. Do you notice anything? Next, find $h(x) = (f \circ g)(x)$ and $H(x) = (g \circ f)(x)$. What happened? Look closely at the functions f and g to see how they are related. Can you come up with two additional functions where the same thing occurs?

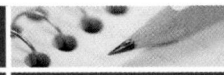

93. Given $f(x) = \sqrt{1-x}$ and $g(x) = \sqrt{x-2}$, what can you say about the domain of $(f + g)(x)$? Enter the functions as Y_1 and Y_2 on a graphing calculator, then enter $Y_3 = Y_1 + Y_2$. See if you can determine why the calculator gives an error message for Y_3, regardless of the input.

94. Given $f(x) = \dfrac{1}{x^2 - 4}$, $g(x) = \sqrt{x + 1}$, and $h(x) = (f \circ g)(x)$, (a) find the new function rule for h and (b) determine the implied domain of h. Does this *implied* domain include $x = 2$, $x = -2$, and $x = -3$ as valid inputs? (c) Determine the actual domain for $h(x) = (f \circ g)(x)$ and discuss the result.

95. Instead of calculating the result of an operation on two functions at a *specific point* as in Exercises 69–74, we can actually *graph the function* that

results from the operation. This skill, called the **addition of ordinates,** is widely applied in a study of tides and other areas. For $f(x) = (x - 3)^2 + 2$ and $g(x) = 4|x - 3| - 5$, complete a table of values like the one shown for $x \in [-2, 8]$. For the last column, remember that $(f - g)(x) = f(x) - g(x)$, and use this relation to complete the column. Finally, use the ordered pairs $(x, (f - g)(x))$ to graph the new function. Is the new function smooth? Is the new function continuous?

Exercise 95

x	$f(x)$	$g(x)$	$(f - g)(x)$
-2			
-1			
0			
1			
2			
3			
4			
5			
6			
7			
8			

▶ MAINTAINING YOUR SKILLS

96. (1.4) Find the sum and product of the complex numbers $2 + 3i$ and $2 - 3i$.

97. (2.4) Draw a sketch of the functions (a) $f(x) = \sqrt{x}$, (b) $g(x) = \sqrt[3]{x}$, and (c) $h(x) = |x|$ *from memory.*

98. (1.5) Use the quadratic formula to solve $2x^2 - 3x + 4 = 0$.

99. (2.3) Find the equation of the line perpendicular to $-2x + 3y = 9$, that also goes through the origin.

SUMMARY AND CONCEPT REVIEW

SECTION 2.1　Rectangular Coordinates; Graphing Circles and Other Relations

KEY CONCEPTS

- A relation is a collection of ordered pairs (x, y) and can be given in set or equation form.
- As a set of ordered pairs, the domain of the relation is the set of all first coordinates, and the range is the set of all corresponding second coordinates.
- A relation can be expressed in mapping notation $x \rightarrow y$, indicating an element from the domain is mapped to (corresponds to or is associated with) an element from the range.
- The graph of a relation in equation form is the set of all ordered pairs (x, y) that satisfy the equation. We plot a sufficient number of points and connect them with a straight line or smooth curve, depending on the pattern formed.
- The midpoint of a line segment with endpoints (x_1, y_1) and (x_2, y_2) is $\left(\dfrac{x_1 + x_2}{2}, \dfrac{y_1 + y_2}{2} \right)$.
- The distance between the points (x_1, y_1) and (x_2, y_2) is $d = \sqrt{(x_2 - x_1)^2 + (y_2 - y_1)^2}$.
- The equation of a circle centered at (h, k) with radius r is $(x - h)^2 + (y - k)^2 = r^2$.

EXERCISES

1. Represent the relation in mapping notation, then state the domain and range.

   $\{(-7, 3), (-4, -2), (5, 1), (-7, 0), (3, -2), (0, 8)\}$

2. Graph the relation $y = \sqrt{25 - x^2}$ by completing the table, then state the domain and range of the relation.

x	y
-5	
-4	
-2	
0	
2	
4	
5	

Mr. Northeast and Mr. Southwest live in Coordinate County and are good friends. Mr. Northeast lives at *19 East 25 North* or (19, 25), while Mr. Southwest lives at *14 West and 31 South* or $(-14, -31)$. If the streets in Coordinate County are laid out in one mile squares,

3. Use the distance formula to find how far apart they live.

4. If they agree to meet halfway between their homes, what are the coordinates of their meeting place?

5. Sketch the graph of $x^2 + y^2 = 16$.

6. Sketch the graph of $x^2 + y^2 + 6x + 4y + 9 = 0$. Clearly state the center and radius.

7. Find the equation of the circle whose diameter has the endpoints $(-3, 0)$ and $(0, 4)$.

SECTION 2.2 Graphs of Linear Equations

KEY CONCEPTS

- A linear equation can be written in the form $ax + by = c$, where a and b are not simultaneously equal to 0.

- The slope of the line through (x_1, y_1) and (x_2, y_2) is $m = \dfrac{y_2 - y_1}{x_2 - x_1}$, where $x_1 \neq x_2$.

- Other designations for slope are $m = \dfrac{\text{rise}}{\text{run}} = \dfrac{\text{change in } y}{\text{change in } x} = \dfrac{\Delta y}{\Delta x} = \dfrac{\text{vertical change}}{\text{horizontal change}}$.

- Lines with positive slope ($m > 0$) rise from left to right; lines with negative slope ($m < 0$) fall from left to right.

- The equation of a horizontal line is $y = k$; the slope is $m = 0$.

- The equation of a vertical line is $x = h$; the slope is undefined.

- Lines can be graphed using the intercept method. First determine $(x, 0)$ (substitute 0 for y and solve for x), then $(0, y)$ (substitute 0 for x and solve for y). Then draw a straight line through these points.

- Parallel lines have equal slopes ($m_1 = m_2$); perpendicular lines have slopes that are negative reciprocals $(m_1 = -\dfrac{1}{m_2}$ or $m_1 \cdot m_2 = -1)$.

EXERCISES

8. Plot the points and determine the slope, then use the ratio $\dfrac{\Delta y}{\Delta x} = \dfrac{\text{rise}}{\text{run}}$ to find an additional point on the line:

   a. $(-4, 3)$ and $(5, -2)$ and **b.** $(3, 4)$ and $(-6, 1)$.

9. Use the slope formula to determine if lines L_1 and L_2 are parallel, perpendicular, or neither:

   a. L_1: $(-2, 0)$ and $(0, 6)$; L_2: $(1, 8)$ and $(0, 5)$
   b. L_1: $(1, 10)$ and $(-1, 7)$: L_2: $(-2, -1)$ and $(1, -3)$

10. Graph each equation by plotting points: (a) $y = 3x - 2$ and (b) $y = -\frac{3}{2}x + 1$.

11. Find the intercepts for each line and sketch the graph: (a) $2x + 3y = 6$ and (b) $y = \frac{4}{3}x - 2$.

12. Identify each line as either horizontal, vertical, or neither, and graph each line.

    a. $x = 5$ **b.** $y = -4$ **c.** $2y + x = 5$

13. Determine if the triangle with the vertices given is a right triangle: $(-5, -4)$, $(7, 2)$, $(0, 16)$.

14. Find the slope and y-intercept of the line shown and discuss the slope ratio in this context.

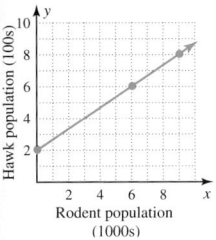

SECTION 2.3 Linear Graphs and Rates of Change

KEY CONCEPTS

- The equation of a nonvertical line in slope-intercept form is $y = mx + b$ or $f(x) = mx + b$. The slope of the line is m and the y-intercept is $(0, b)$.

- To graph a line given its equation in slope-intercept form, plot the y-intercept, then use the slope ratio $m = \dfrac{\Delta y}{\Delta x}$ to find a second point, and draw a line through these points.

- If the slope m and a point (x_1, y_1) on the line are known, the equation of the line can be written in point-slope form: $y - y_1 = m(x - x_1)$.

- A secant line is the straight line drawn through two points on a nonlinear graph.

- The notation $m = \dfrac{\Delta y}{\Delta x}$ literally means the quantity measured along the y-axis is changing with respect to changes in the quantity measured along the x-axis.

- The average rate of change on the interval containing x_1 and x_2 is the slope of the secant line through (x_1, y_1) and (x_2, y_2), or $\dfrac{\Delta y}{\Delta x} = \dfrac{y_2 - y_1}{x_2 - x_1}$.

EXERCISES

15. Write each equation in slope-intercept form, then identify the slope and y-intercept.

    a. $4x + 3y - 12 = 0$ **b.** $5x - 3y = 15$

16. Graph each equation using the slope and y-intercept.

    a. $f(x) = -\frac{2}{3}x + 1$ **b.** $h(x) = \frac{5}{2}x - 3$

17. Graph the line with the given slope through the given point.

    a. $m = \frac{2}{3};\ (1, 4)$ **b.** $m = -\frac{1}{2};\ (-2, 3)$

18. What are the equations of the horizontal line and the vertical line passing through $(-2, 5)$? Which line is the point $(7, 5)$ on?

19. Find the equation of the line passing through $(1, 2)$ and $(-3, 5)$. Write your final answer in slope-intercept form.

20. Find the equation for the line that is parallel to $4x - 3y = 12$ and passes through the point $(3, 4)$. Write your final answer in slope-intercept form.

21. Determine the slope and y-intercept of the line shown. Then write the equation of the line in slope-intercept form and interpret the slope ratio $m = \dfrac{\Delta W}{\Delta R}$ in the context of this exercise.

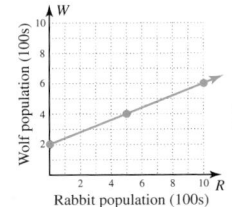

22. For the graph given, (a) find the equation of the line in point-slope form, (b) use the equation to predict the *x*- and *y*-intercepts, (c) write the equation in slope-intercept form, and (d) find *y* when *x* = 20, and the value of *x* for which *y* = 15.

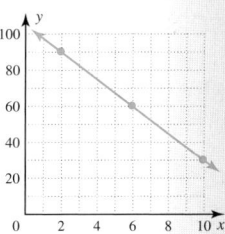

SECTION 2.4 Functions, Function Notation, and the Graph of a Function

KEY CONCEPTS

- A function is a relation, rule, or equation that pairs each element from the domain with exactly one element of the range.
- The vertical line test says that if every vertical line crosses the graph of a relation in at most one point, the relation is a function.
- On a graph, vertical boundary lines can be used to identify the domain, or the set of "allowable inputs" for a function.
- On a graph, horizontal boundary lines can be used to identify the range, or the set of *y*-values (outputs) generated by the function.
- When a function is stated as an equation, the implied domain is the set of *x*-values that yield real number outputs.
- *x*-values that cause a denominator of zero or that cause the radicand of a square root expression to be negative must be excluded from the domain.
- *The phrase* "*y* is a function of *x*," is written as $y = f(x)$. This notation enables us to evaluate functions while tracking corresponding *x*- and *y*-values.

EXERCISES

23. State the implied domain of each function:

    a. $f(x) = \sqrt{4x + 5}$ **b.** $g(x) = \dfrac{x - 4}{x^2 - x - 6}$

24. Determine $h(-2)$, $h\left(-\frac{2}{3}\right)$, and $h(3a)$ for $h(x) = 2x^2 - 3x$.

25. Determine if the mapping given represents a function. If not, explain how the definition of a function is violated.

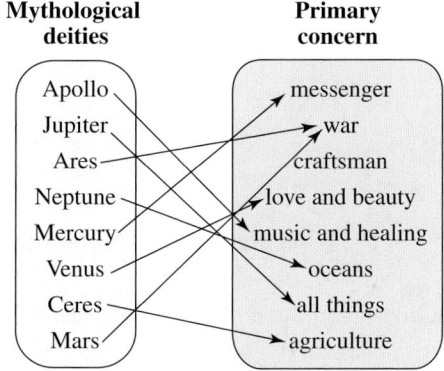

26. For the graph of each function shown, (a) state the domain and range, (b) find the value of $f(2)$, and (c) determine the value(s) of *x* for which $f(x) = 1$.

I.

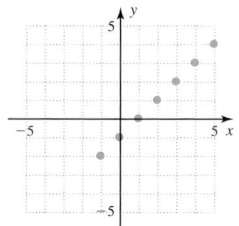

II.

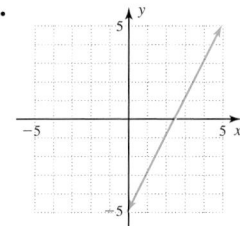

III.

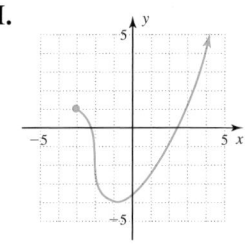

SECTION 2.5   Analyzing the Graph of a Function

KEY CONCEPTS

- A function f is even (symmetric to the y-axis), if and only if when a point (x, y) is on the graph, then $(-x, y)$ is also on the graph. In function notation: $f(-x) = f(x)$.
- A function f is odd (symmetric to the origin), if and only if when a point (x, y) is on the graph, then $(-x, -y)$ is also on the graph. In function notation: $f(-x) = -f(x)$.

Intuitive descriptions of the characteristics of a graph are given here. The formal definitions can be found within Section 2.5.

- A function is *increasing* in an interval if the graph rises from left to right (larger inputs produce larger outputs).
- A function is *decreasing* in an interval if the graph falls from left to right (larger inputs produce smaller outputs).
- A function is *positive* in an interval if the graph is above the x-axis in that interval.
- A function is *negative* in an interval if the graph is below the x-axis in that interval.
- A function is *constant* in an interval if the graph is parallel to the x-axis in that interval.
- A maximum value can be a *local* maximum, or *global* maximum. An *endpoint* maximum can occur at the endpoints of the domain. Similar statements can be made for minimum values.
- For any function f, the average rate of change in the interval $[x_1, x_2]$ is $\dfrac{f(x_2) - f(x_1)}{x_2 - x_1}$.
- The difference quotient for a function $f(x)$ is $\dfrac{f(x + h) - f(x)}{h}$.

EXERCISES

State the domain and range for each function $f(x)$ given. Then state the intervals where f is increasing or decreasing and intervals where f is positive or negative. Assume all endpoints have integer values.

27.

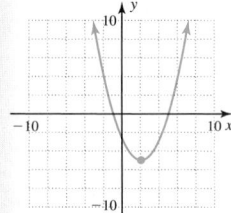

28.

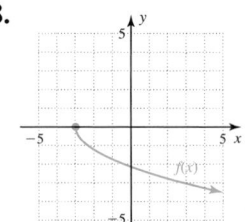

29.

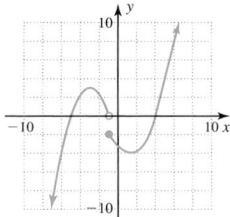

30. Determine which of the following are even $[f(-k) = f(k)]$, odd $[f(-k) = -f(k)]$, or neither.

 a. $f(x) = 2x^5 - \sqrt[3]{x}$ **b.** $g(x) = x^4 - \dfrac{\sqrt[3]{x}}{x}$

 c. $p(x) = |3x| - x^3$ **d.** $q(x) = \dfrac{x^2 - |x|}{x}$

31a. Given $f(x) = \sqrt{x + 4}$, find the average rate of change in the interval $[-3, 5]$. What does the result confirm about the graph of this toolbox function?

31b. Use the difference quotient to find a rate of change formula for the function given, then calculate the rate of change for the interval indicated: $j(x) = x^2 - x$; [2.00, 2.01].

32. Draw the function f that has all of the following characteristics, then name the zeroes of the function and the location of all maximum and minimum values. [*Hint:* Write them in the form $(c, f(c))$.]

 a. Domain: $x \in [-6, 10)$ **b.** Range: $y \in (-8, 6)$
 c. $f(0) = 0$ **d.** $f(x)\downarrow$ for $x \in (-6, -3) \cup (3, 7.5)$
 e. $f(x)\uparrow$ for $x \in (-3, 3) \cup (7.5, 10)$ **f.** $f(x) < 0$ for $x \in (-6, 0) \cup (6, 9)$
 g. $f(x) > 0$ for $x \in (0, 6) \cup (9, 10)$

SECTION 2.6   The Toolbox Functions and Transformations

KEY CONCEPTS

- The *toolbox functions* and graphs commonly used in mathematics are
  - the identity function $f(x) = x$
  - square root function: $f(x) = \sqrt{x}$
  - cubing function: $f(x) = x^3$
  - squaring function: $f(x) = x^2$, parabola
  - absolute value function: $f(x) = |x|$
  - cube root function: $f(x) = \sqrt[3]{x}$
- For a basic or parent function $y = f(x)$, the general equation of the transformed function is $y = af(x \pm h) \pm k$. For any function $y = f(x)$ and $h, k > 0$,
  - the graph of $y = f(x) + k$ is the graph of $y = f(x)$ shifted upward k units
  - the graph of $y = f(x + h)$ is the graph of $y = f(x)$ shifted left h units
  - the graph of $y = -f(x)$ is the graph of $y = f(x)$ reflected across the x-axis
  - $y = af(x)$ results in a vertical stretch when $a > 1$
  - the graph of $y = f(x) - k$ is the graph of $y = f(x)$ shifted downward k units
  - the graph of $y = f(x - h)$ is the graph of $y = f(x)$ shifted right h units
  - the graph of $y = f(-x)$ is the graph of $y = f(x)$ reflected across the y-axis
  - $y = af(x)$ results in a vertical compression when $0 < a < 1$
- Transformations are applied in the following order: (1) horizontal shifts, (2) reflections, (3) stretches or compressions, and (4) vertical shifts.

EXERCISES

Identify the function family for each graph given, then (a) describe the end behavior; (b) name the x- and y-intercepts; (c) identify the vertex, initial point, or point of inflection (as applicable); and (d) state the domain and range.

33.
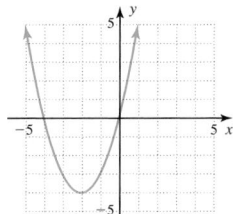

34.
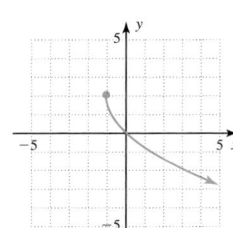

35.
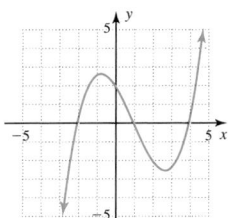

36.
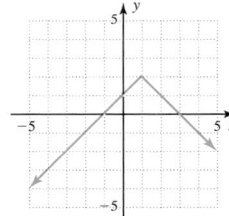

37.
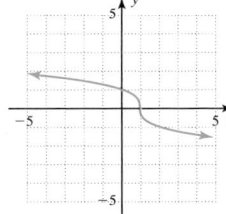

Identify each function as belonging to the linear, quadratic, square root, cubic, cube root, or absolute value family. Then sketch the graph using shifts of a parent function and a few characteristic points.

38. $f(x) = -(x + 2)^2 - 5$ **39.** $f(x) = 2|x + 3|$ **40.** $f(x) = x^3 - 1$

41. $f(x) = \sqrt{x - 5} + 2$ **42.** $f(x) = \sqrt[3]{x} + 2$

43. Apply the transformations indicated for the graph of $f(x)$ given.
  a. $f(x - 2)$
  b. $-f(x) + 4$
  c. $\frac{1}{2}f(x)$

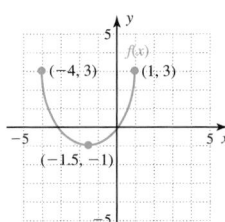

SECTION 2.7   Piecewise-Defined Functions

KEY CONCEPTS

- Each piece of a piecewise-defined function has a domain over which that piece is defined.
- To evaluate a piecewise-defined function, identify the domain interval containing the input value, then use the piece of the function corresponding to this interval.
- To graph a piecewise-defined function you can plot points, or graph each piece in its entirety, then erase portions of the graph outside the domain indicated for each piece.
- If the graph of a function can be drawn without lifting your pencil from the paper, the function is continuous.
- A pointwise discontinuity is said to be removable because we can redefine the function to "fill the hole."
- Step functions are discontinuous and formed by a series of horizontal steps.
- The floor function $\lfloor x \rfloor$ gives the first integer less than or equal to x.
- The ceiling function $\lceil x \rceil$ is the first integer greater than or equal to x.

EXERCISES

44. For the graph and functions given, (a) use the correct notation to write the relation as a single piecewise-defined function, stating the effective domain for each piece by inspecting the graph; and (b) state the range of the function: $Y_1 = 5$, $Y_2 = -x + 1$, $Y_3 = 3\sqrt{x - 3} - 1$.

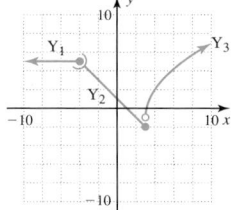

45. Use a table of values as needed to graph $h(x)$, then state its domain and range. If the function has a pointwise discontinuity, state how the second piece could be redefined so that a continuous function results.

$$h(x) = \begin{cases} \dfrac{x^2 - 2x - 15}{x + 3} & x \neq -3 \\ -6 & x = -3 \end{cases}$$

46. Evaluate the piecewise-defined function $p(x)$: $p(-4)$, $p(-2)$, $p(2.5)$, $p(2.99)$, $p(3)$, and $p(3.5)$

$$p(x) = \begin{cases} -4 & x < -2 \\ -|x| - 2 & -2 \leq x < 3 \\ 3\sqrt{x} - 9 & x \geq 3 \end{cases}$$

47. Sketch the graph of the function and state its domain and range. Use transformations of the toolbox functions where possible.

$$q(x) = \begin{cases} 2\sqrt{-x - 3} - 4 & x \leq -3 \\ -2|x| + 2 & -3 < x < 3 \\ 2\sqrt{x - 3} - 4 & x \geq 3 \end{cases}$$

48. Many home improvement outlets now rent flatbed trucks in support of customers that purchase large items. The cost is $20 per hour for the first 2 hr, $30 for the next 2 hr, then $40 for each hour afterward. Write this information as a piecewise-defined function, then sketch its graph. What is the total cost to rent this truck for 5 hr?

SECTION 2.8   The Algebra and Composition of Functions

KEY CONCEPTS

- The notation used to represent the basic operations on two functions is
  - $(f + g)(x) = f(x) + g(x)$
  - $(f - g)(x) = f(x) - g(x)$
  - $(f \cdot g)(x) = f(x) \cdot g(x)$
  - $\left(\dfrac{f}{g}\right)(x) = \dfrac{f(x)}{g(x)}$; $g(x) \neq 0$
- The result of these operations is a new function $h(x)$. The domain of h is the intersection of domains for f and g, excluding values that make $g(x) = 0$ for $h(x) = \left(\dfrac{f}{g}\right)(x)$.

- The composition of two functions is written $(f \circ g)(x) = f[g(x)]$ (g is an input for f).
- The domain of $f \circ g$ is all x in the domain of g, such that $g(x)$ is in the domain of f.
- To evaluate $(f \circ g)(2)$, we find $(f \circ g)(x)$ then substitute $x = 2$. Alternatively, we can find $g(2) = k$, then find $f(k)$.
- A composite function $h(x) = (f \circ g)(x)$ can be "decomposed" into individual functions by identifying functions f and g such that $(f \circ g)(x) = h(x)$. The decomposition is not unique.

EXERCISES

For $f(x) = x^2 + 4x$ and $g(x) = 3x - 2$, find the following:

49. $(f + g)(a)$ **50.** $(f \cdot g)(3)$ **51.** the domain of $\left(\dfrac{f}{g}\right)(x)$

Given $p(x) = 4x - 3$, $q(x) = x^2 + 2x$, and $r(x) = \dfrac{x + 3}{4}$ find:

52. $(p \circ q)(x)$ **53.** $(q \circ p)(3)$ **54.** $(p \circ r)(x)$ and $(r \circ p)(x)$

For each function here, find functions $f(x)$ and $g(x)$ such that $h(x) = f[g(x)]$:

55. $h(x) = \sqrt{3x - 2} + 1$ **56.** $h(x) = x^{\frac{2}{3}} - 3x^{\frac{1}{3}} - 10$

57. A stone is thrown into a pond causing a circular ripple to move outward from the point of entry. The radius of the circle is modeled by $r(t) = 2t + 3$, where t is the time in seconds. Find a function that will give the area of the circle directly as a function of time. In other words, find $A(t)$.

58. Use the graph given to find the value of each expression:

    a. $(f + g)(-2)$

    b. $(g \circ f)(5)$

    c. $(g - f)(7)$

    d. $\left(\dfrac{g}{f}\right)(10)$

    e. $(f \cdot g)(3)$

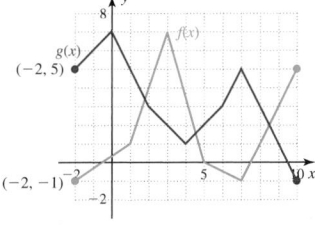

MIXED REVIEW

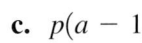

1. Write the given equation in slope-intercept form:
$4x + 3y = 12$

2. Find the equation of the line perpendicular to $x - 2y = 8$ that passes through $(1, 3)$.

3. Find the implied domain of:

    a. $f(x)\dfrac{x + 1}{x^2 - 5x + 4}$ **b.** $g(x) = \dfrac{1}{\sqrt{2x - 3}}$

4. Given $p(x) = -x^2 + 3x - 1$, find

    a. $p\left(\dfrac{-1}{3}\right)$ **b.** $p(3a)$ **c.** $p(a - 1)$

5. State the equation of the line shown, in slope-intercept form.

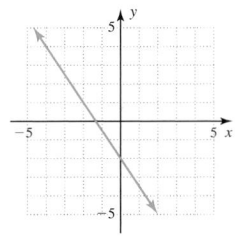

6. For the function g whose graph is given, find (a) domain, (b) $g(2)$, and (c) k if $g(k) = -3$.

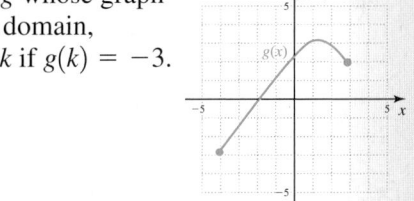

7. The following three points form a right triangle: $(-3, 7)$, $(2, 2)$ and $(5, 5)$. Use the distance formula to help determine which point is at the vertex of the right angle. Then find the equation of the smallest circle, centered at that point, that encloses the triangle.

8. Discuss the end behavior of $F(x)$ and name the vertex, axis of symmetry, and all intercepts.

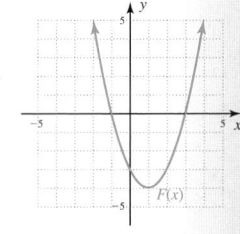

9. Graph by plotting the y-intercept, then counting
   $m = \dfrac{\Delta y}{\Delta x}$ to find additional points:

   $y = \dfrac{3}{5}x - 2$

10. Solve the inequality using the graph provided:

    $f(x) = 4x - \dfrac{4}{3}x^2;\ f(x) < 0.$

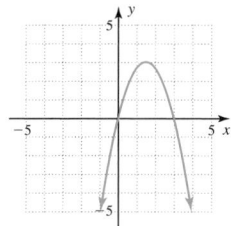

11. **a.** Graph the function $p(x) = -2x^2 + 8x$. By observing the graph, is the average rate of change positive or negative in the interval $[-2, -1]$? Why? Do you expect the rate of change in $[1, 2]$ to be greater or less than the rate of change in $[-2, -1]$? Calculate the average rate of change in each interval and comment.

     b. If $1000 is deposited in an account paying 7% interest compounded continuously, the function model is $A(t) = 1000e^{0.07t}$. Use the average rate of change formula to determine if the amount of interest added to the account exceeds $200 per year $\left(\dfrac{\Delta A}{\Delta t} > 200\right)$ in the 10th, 15th, or 20th year. Use the intervals $[10, 10.01]$, $[15, 15.01]$, and $[20, 20.01]$.

Given $f(x) = \dfrac{3}{x^2 - 1}$ and $g(x) = 3x - 2$, find

12. $\dfrac{g}{f}\left(\dfrac{1}{2}\right)$

13. $(f \circ g)(x)$ and its domain

14. Sketch the function h as defined.

    $h(x) = \begin{cases} 5 & 0 \le x < 8 \\ x - 3 & 8 \le x \le 15 \\ -2x + 40 & x > 15 \end{cases}$

15. Given $f(x) = x^2 + 1$ and $g(x) = 3x - 2$, calculate the difference quotient for each function and use the results to estimate the value of x for which their rates of change are equal.

16. Identify the function family for the function $g(x) = -2|x + 3| + 4$. Then sketch the graph using transformations of a parent function and a few characteristic points.

17. For the graph shown, determine
    a. the domain and range of g,
    b. intervals where g is increasing, decreasing, or constant,
    c. intervals where g is positive or negative,
    d. any maximum or minimum values for g.

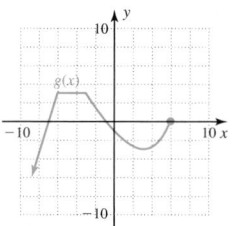

18. Draw a function f that has the following characteristics, then write the zeroes of the function and the location of all maximum and minimum values.

    a. domain: $x \in [0, 30]$
    b. range: $y \in [-10, 12]$
    c. $f(2) = f(10) = 0$
    d. $f(x)\downarrow$ from $x \in (0, 5) \cup (15, 20)$
    e. $f(x)\uparrow$ for $x \in (5, 15)$
    f. $f(x) < 0$ for $x \in (2, 10)$
    g. $f(x) > 0$ for $x \in (0, 2) \cup (10, 30)$
    h. $f(x) = 5$ for $x \in [20, 30]$

19. Find the equation of the function $f(x)$ whose graph is given.

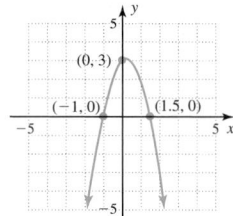

20. Since 1975, the number of deaths in the United States due to heart disease has been declining at a rate that is close to linear. Find an equation model if there were 431 thousand deaths in 1975 and 257 thousand deaths in 2000 (let x represent years since 1975 and $f(x)$ deaths in thousands). How many deaths due to heart disease does the model predict for 2008?

Source: 2004 Statistical Abstract of the United States, Table 102

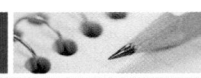

PRACTICE TEST

1. Two relations here are functions and two are not. Identify the nonfunctions (justify your response).
   a. $x = y^2 + 2y$ **b.** $y = \sqrt{5 - 2x}$
   c. $|y| + 1 = x$ **d.** $y = x^2 + 2x$

2. Determine if the lines are parallel, perpendicular, or neither:
   L_1: $2x + 5y = -15$ and L_2: $y = \frac{2}{5}x + 7$.

3. Graph the line using the slope and y-intercept: $x + 4y = 8$

4. Find the center and radius of the circle defined by $x^2 + y^2 - 4x + 6y - 3 = 0$, then sketch its graph.

5. Find the equation of the line parallel to $6x + 5y = 3$, containing the point $(2, -2)$. Answer in slope-intercept form.

6. My partner and I are at coordinates $(-20, 15)$ on a map. If our destination is at coordinates $(35, -12)$, (a) what are the coordinates of the rest station located halfway to our destination? (b) How far away is our destination? Assume that each unit is 1 mi.

7. Write the equations for lines L_1 and L_2 shown.

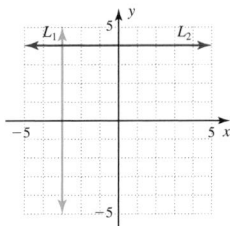

8. State the domain and range for the relations shown on graphs 8(a) and 8(b).

Exercise 8(a)

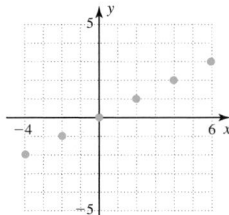

Exercise 8(b)

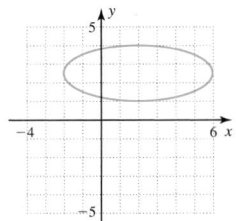

9. For the linear function shown,
   a. Determine the value of $W(24)$ from the graph.
   b. What input h will give an output of $W(h) = 375$?
   c. Find a linear function for the graph.

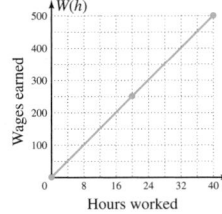

d. What does the slope indicate in this context?

e. State the domain and range of h.

10. Each function graphed here is from a toolbox function family. For each graph, (a) identify the function family, (b) state the domain and range, (c) identify x- and y-intercepts, (d) discuss the end behavior, and (e) solve the inequality $f(x) > 0$, and (f) solve $f(x) < 0$.

I.

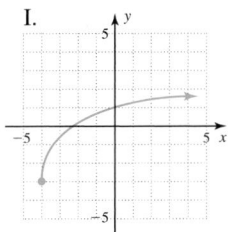

II.

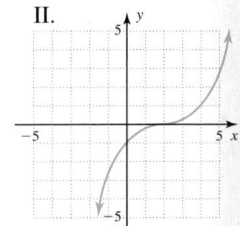

III.

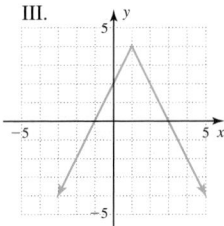

IV.

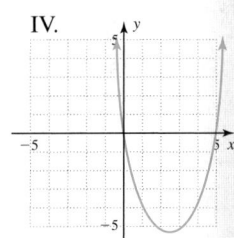

11. Given $f(x) = \dfrac{2 - x^2}{x^2}$, evaluate and simplify:
   (a) $f\left(\frac{2}{3}\right)$ (b) $f(a + 3)$ (c) $f(1 + 2i)$

12. Given $f(x) = x^2 + 2$ and $g(x) = \sqrt{3x - 1}$, determine $(f \circ g)(x)$ and its domain.

13. Monthly sales volume for a successful new company is modeled by $S(t) = 2t^2 - 3t$, where $S(t)$ represents sales volume in thousands in month t ($t = 0$ corresponds to January 1).
   (a) Would you expect the average rate of change from May to June to be greater than that from June to July? Why? (b) Calculate the rates of change in these intervals to verify your answer. (c) Calculate the difference quotient for $S(t)$ and use it to estimate the sales volume rate of change after 10, 18, and 24 months.

Sketch each graph using a transformation.

14. $f(x) = |x - 2| + 3$

15. $g(x) = -(x + 3)^2 - 2$

16. A snowball increases in size as it rolls downhill. The snowball is roughly spherical with a radius that can be modeled by the function $r(t) = \sqrt{t}$, where t

is time in seconds and r is measured in inches. The volume of the snowball is given by the function $V(r) = \frac{4}{3}\pi r^3$. Use a composition to (a) write V directly as a function of t and (b) find the volume of the snowball after 9 sec.

17. Determine the following from the graph shown.
    a. the domain and range
    b. estimate the value of $f(-1)$
    c. interval(s) where $f(x)$ is negative or positive
    d. interval(s) where $f(x)$ is increasing, decreasing, or constant
    e. an equation for $f(x)$

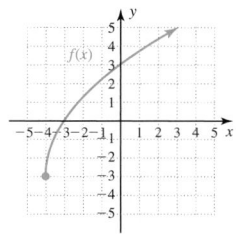

18. Given $h(x) = \begin{cases} 4 & x < -2 \\ 2x & -2 \le x \le 2 \\ x^2 & x > 2 \end{cases}$
    a. Find $h(-3)$, $h(-2)$, and $h\left(\frac{5}{2}\right)$
    b. Sketch the graph of h. Label important points.

For the function $h(x)$ whose partial graph is given,

19. complete the graph if h is known to be even.

20. complete the graph if h is known to be odd.

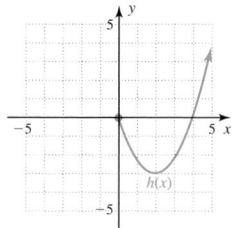

CALCULATOR EXPLORATION AND DISCOVERY

Using a Simple Program to Explore Transformations

In Section 2.6 we studied transformations of the toolbox functions. On page 231, an organized sequence for applying these transformations was given. Since the transformations are identical regardless of the function used, a simple program is an efficient way to explore these transformations further. As a good programming practice, clear all functions on the [Y =] screen, and preset the graphing window to [ZOOM] 6:ZStandard. Begin by pressing the [PRGM] key, then the [▶] key twice to name the new program. At the prompt, enter TRANSFRM. The specific functions we use for programming are all accessed in sub menus of the [PRGM] key. Recall that the relational operators ($=$, $<$, $>$, $\le$, $\ge$) are accessed using [2nd] [MATH] (TEST).

PROGRAM: TRANSFRM

:ClrHome
:FnOff 1,2,3,4,5,6,7,8,9
:Disp "FUNCTION FAMILY"
:Disp "1:SQUARING"
:Disp "2:SQUARE ROOT"
:Disp "3:ABSOLUTE VALUE"
:Disp "4:CUBING"
:Disp "5:CUBE ROOT"
:Input T

:If T=1:"X^2"→Y1
:If T=2:"√(X)"→Y1
:If T=3:"abs(X)"→Y1
:If T=4:"X^3"→Y1
:If T=5:"X^(1/3)"→Y1
:DispGraph:Pause
:ClrHome
:Disp "HORIZONTAL SHIFT"
:Disp "ENTER 0 IF NONE"
:Prompt H
:"Y1(X + H)"→Y2
:DispGraph
:FnOff 1
:DispGraph:Pause
:ClrHome
:Disp "STRETCH FACTOR A"
:Disp "(A>0)"
:Disp "ENTER 1 IF A=1"
:Input A
:"A*Y2"→Y3
:DispGraph
:FnOff 2
:DispGraph:Pause

:ClrHome
:Disp "REFLECTIONS?"
:Disp "0:NONE"
:Disp "1:ACROSS X-AXIS"
:Disp "2:ACROSS Y-AXIS"
:Disp "3:ACROSS BOTH"
:Input B
:If B=0:"Y3"→Y4
:If B=1:"-Y3"→Y4
:If B=2:"Y3(-X)"→Y4
:If B=3:"-Y3(-X)"→Y4
:DispGraph
:FnOff 3
:DispGraph:Pause
:ClrHome
:Disp "VERTICAL SHIFT"
:Disp "ENTER 0 IF NONE"
:Prompt V
:"Y4 + V"→Y5
:DispGraph

:FnOff 4
:DispGraph:Pause
:Stop

Enter the TRANSFRM program into your calculator. Note that as you are writing or editing a program:

1. The "FnOff" command is located at $\boxed{\text{VARS}}$ Y–VARS 4:On/Off.

2. The "ClrHome" command is located at $\boxed{\text{PRGM}}$ CTL 8.

3. The "Pause" command is located at $\boxed{\text{PRGM}}$ I/O 8.

All other needed commands are visible as Options 1 through 7 on the CTL and I/O menus.

Exercise 1: Use the TRANSFRM program to apply the following transformations to $y = x^2$: (1) shift left 4 units, (2) stretch by a factor of 5, (3) reflect across the x-axis, (4) shift up 6 units. What is the equation of the final graph? Where is the vertex located?

Exercise 2: Use TRANSFRM to graph the function $y = -4\sqrt[3]{x} - 2 + 3$. Where is the point of inflection? Estimate the y-intercept from the graph, then compare the estimate to the computed value.

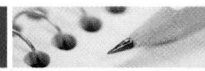

STRENGTHENING CORE SKILLS

Transformations via Composition

Historically, many of the transformations studied in this chapter played a fundamental role in the development of modern algebra. To make the connection, we note that many transformations can be viewed as a composition of functions. For instance, for $f(x) = x^2 + 2$ (a parabola shifted two units up) and $g(x) = (x - 3)$, the composition $h(x) = f[g(x)]$ yields $(x - 3)^2 + 2$, a parabola shifted 2 units up *and* 3 units right. Enter $f(x)$ as Y_1 and $h(x)$ as Y_2 on your graphing calculator, then graph and inspect the results. As you see, we do obtain the same parabola shifted 3 units to the right (see figure). But now, notice what happens when we compose using $g(x) = x + 2$. After simplification, the result is $h(x) = x^2 - 9$ or *a quadratic function whose zeroes can easily be solved by taking square roots*, since the linear term is eliminated. The zeroes of h (the shifted quadratic) are $x = -3$ and $x = 3$, which means the zeroes of f (the original function) can be found by shifting two units *right*, returning them to their original position. The zeroes are $x = -3 + 2 = -1$ and $x = 3 + 2 = 5$ (verify by factoring). Transformations of this type are especially insightful when the zeroes of a

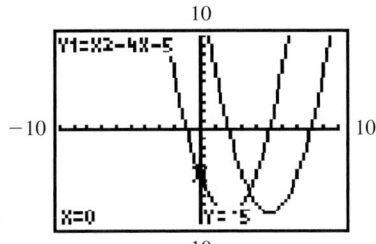

quadratic equation are irrational, since it enables us to find the radical portion by taking square roots, and the rational portion by addition. The key is to shift the quadratic function $y = ax^2 + bx + c$ using $x - \dfrac{b}{2a}$. Let's find the zeroes of $f(x) = x^2 + 6x - 11$ in this way. We find that $\dfrac{b}{2a} = 3$, giving $g(x) = x - 3$. This gives $h(x) = f[g(x)] = (x - 3)^2 + 6(x - 3) - 11$, which simplifies to $h(x) = x^2 - 20$. The zeroes of h are $x = -2\sqrt{5}$ and $x = 2\sqrt{5}$, so the solutions to the original equation must be $x = -2\sqrt{5} - 3$ and $x = 2\sqrt{5} - 3$. For Exercises 1–3, use this method to: (a) find such functions $h(x)$, and (b) use the zeroes of h to find the zeroes of f. Verify each solution using a calculator.

Exercise 1: $f(x) = x^2 - 8x - 12$
Exercise 2: $f(x) = x^2 + 4x + 5$
Exercise 3: $f(x) = 2x^2 - 10x + 11$

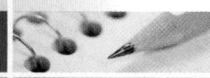

CUMULATIVE REVIEW CHAPTERS 1–2

1. Perform the division by factoring the numerator: $(x^3 - 5x^2 + 2x - 10) \div (x - 5)$.

2. Find the solution set for: $2 - x < 5$ **and** $3x + 2 < 8$.

3. The area of a circle is 69 cm². Find the circumference of the same circle.

4. The surface area of a cylinder is $A = 2\pi r^2 + 2\pi rh$. Write r in terms of A and h (solve for r).

5. Solve for x: $-2(3 - x) + 5x = 4(x + 1) - 7$.

6. Evaluate without using a calculator: $\left(\frac{27}{8}\right)^{\frac{-2}{3}}$.

7. Find the slope of each line:
   a. through the points: $(-4, 7)$ and $(2, 5)$.
   b. a line with equation $3x - 5y = 20$.

8. Graph using transformations of a parent function.
   a. $f(x) = \sqrt{x - 2} + 3$.
   b. $f(x) = -|x + 2| - 3$.

9. Graph the line passing through $(-3, 2)$ with a slope of $m = \frac{1}{2}$, then state its equation.

10. Show that $x = 1 + 5i$ is a solution to $x^2 - 2x + 26 = 0$.

11. Given $f(x) = 3x^2 - 6x$ and $g(x) = x - 2$ find: $(f \cdot g)(x)$, $(f \div g)(x)$, and $(g \circ f)(-2)$.

12. Graph by plotting the y-intercept, then counting $m = \dfrac{\Delta y}{\Delta x}$ to find additional points: $y = \frac{1}{3}x - 2$

13. Graph the piecewise defined function
$f(x) = \begin{cases} x^2 - 4 & x < 2 \\ x - 1 & 2 \le x \le 8 \end{cases}$ and determine the following:
   a. the domain and range
   b. the value of $f(-3), f(-1), f(1), f(2)$, and $f(3)$
   c. the zeroes of the function
   d. interval(s) where $f(x)$ is negative/positive
   e. location of any max/min values
   f. interval(s) where $f(x)$ is increasing/decreasing

14. Given $f(x) = x^2$ and $g(x) = x^3$, use the formula for average rate of change to determine which of these functions is increasing faster in the intervals:
   a. $[0.5, 0.6]$ b. $[1.5, 1.6]$.

15. Add the rational expressions:
   a. $\dfrac{-2}{x^2 - 3x - 10} + \dfrac{1}{x + 2}$
   b. $\dfrac{b^2}{4a^2} - \dfrac{c}{a}$

16. Simplify the radical expressions:
   a. $\dfrac{-10 + \sqrt{72}}{4}$ b. $\dfrac{1}{\sqrt{2}}$

17. Determine which of the following statements are false, and state why.
   a. $\mathbb{N} \subset \mathbb{Z} \subset \mathbb{W} \subset \mathbb{Q} \subset \mathbb{R}$
   b. $\mathbb{W} \subset \mathbb{N} \subset \mathbb{Z} \subset \mathbb{Q} \subset \mathbb{R}$
   c. $\mathbb{N} \subset \mathbb{W} \subset \mathbb{Z} \subset \mathbb{Q} \subset \mathbb{R}$
   d. $\mathbb{N} \subset \mathbb{R} \subset \mathbb{Z} \subset \mathbb{Q} \subset \mathbb{W}$

18. Determine if the following relation is a function. If not, how is the definition of a function violated?

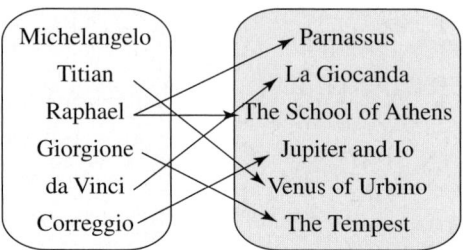

19. Solve by completing the square. Answer in both exact and approximate form: $2x^2 + 49 = -20x$

20. Solve using the quadratic formula. If solutions are complex, write them in $a + bi$ form.
$2x^2 + 20x = -51$

21. The *National Geographic Atlas of the World* is a very large, rectangular book with an almost inexhaustible panoply of information about the world we live in. The length of the front cover is 16 cm more than its width, and the area of the cover is 1457 cm². Use this information to write an equation model, then use the quadratic formula to determine the length and width of the Atlas.

22. Compute as indicated:
   a. $(2 + 5i)^2$ b. $\dfrac{1 - 2i}{1 + 2i}$

23. Solve by factoring:
   a. $6x^2 - 7x = 20$
   b. $x^3 + 5x^2 - 15 = 3x$

24. A theorem from elementary geometry states, "A line tangent to a circle is perpendicular to the radius at the point of tangency." Find the equation of the tangent line for the circle and radius shown.

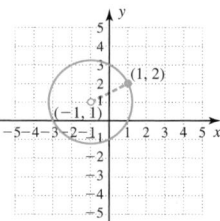

25. A triangle has its vertices at $(-4, 5)$, $(4, -1)$, and $(0, 8)$. Find the perimeter of the triangle and determine whether or not it is a *right* triangle.

In Section 2.5, we noted the difference quotient $\dfrac{f(x + h) - f(x)}{h}$ was actually a formula for the *average rate of change* of a function f. In calculus, our study of very small differences (Chapter 1 *Connections*) investigates what happens to this quotient as the difference between h and 0 becomes very small (as h approaches 0: $h \rightarrow 0$). As noted in the Chapter 2 opener, this investigation is closely linked to the concept of instantaneous rates of change. Also from our work in Chapter 2, we'll briefly look at how transformations of basic functions and the area under a curve play significant roles in our future studies.

Rates of Change and the Difference Quotient

In Exercise 61 of Section 2.5, the height of an arrow shot vertically into the air was modeled by the function $f(t) = -16t^2 + 192t$ (the general function notation is used here). Calculating the average rate of change using $\dfrac{f(t_2) - f(t_1)}{t_2 - t_1}$, we can find the average velocity of the arrow for any specified interval of time (see Exercise 61c). To consider the velocity of the arrow at a *precise instant*, we apply the difference quotient $\dfrac{f(t + h) - f(t)}{h}$ to this function, and investigate what happens as h becomes very small ($h \rightarrow 0$) using ideas similar to those employed in the Chapter 1 *Connections*. In this process, we note that while h may become infinitely *close* to zero, it never takes on a value of 0.

EXAMPLE 1 ▶ **Applying the Difference Quotient to a Polynomial Function**

For the function $f(t) = -16t^2 + 192t$ modeling the height of an arrow, apply the difference quotient to $f(t)$ and simplify the result. Then approximate the velocity of the arrow at the moment $t = 2$, by investigating what happens as $h \rightarrow 0$.

Solution ▶ For $f(t) = -16t^2 + 192t, f(t + h) = -16(t + h)^2 + 192(t + h)$.

$$= -16(t^2 + 2th + h^2) + 192t + 192h$$

$$= -16t^2 - 32th - 16h^2 + 192t + 192h.$$

The resulting difference quotient is

$$\frac{f(t + h) - f(t)}{h} = \frac{(-16t^2 - 32th - 16h^2 + 192t + 192h) - (-16t^2 + 192t)}{h}$$

$$= \frac{-16t^2 - 32th - 16h^2 + 192t + 192h + 16t^2 - 192t}{h}$$

$$= \frac{-32th - 16h^2 + 192h}{h}$$

$$= \frac{h(-32t - 16h + 192)}{h}$$

For any value of $h \neq 0$, however small, $\dfrac{h}{h} = 1$ and we obtain

$$= -32t - 16h + 192$$

When $t = 2$, the expression becomes $-32(2) - 16h + 192 = \mathbf{128 - 16h,}$ and our investigation shows that as $h \rightarrow 0$, the velocity of the arrow approaches 128 ft/sec.

Now try Exercises 1 through 4 ▶

Other real-world applications rely heavily on the ability to apply the difference quotient to a variety of functions, and simplify the result. In all cases, remember that the difference quotient represents a *rate of change,* which appears graphically as the slope of a secant line.

EXAMPLE 2 ▶ **Applying the Difference Quotient to a Radical Function**

Apply the difference quotient to $f(x) = \sqrt{x}$ and simplify the result. Then consider the slope of a line drawn tangent to the graph of f at $x = 2$, by investigating what happens as $h \to 0$.

Solution ▶ For $f(x) = \sqrt{x}, f(x + h) = \sqrt{x + h}$, giving the difference quotient $\dfrac{\sqrt{x + h} - \sqrt{x}}{h}$.

Note that multiplying the numerator and denominator by $\sqrt{x + h} + \sqrt{x}$ will eliminate the radicals in the numerator, since $(A + B)(A - B) = A^2 - B^2$. This will "free up" the terms in the numerator and we then have

$$\frac{(\sqrt{x + h} - \sqrt{x})}{h} \frac{(\sqrt{x + h} + \sqrt{x})}{(\sqrt{x + h} + \sqrt{x})} = \frac{(x + h) - x}{h(\sqrt{x + h} + \sqrt{x})} \qquad (A - B)(A + B) = A^2 - B^2$$

$$= \frac{h}{h(\sqrt{x + h} + \sqrt{x})} \qquad \text{simplify}$$

For any value of $h \neq 0$, however small, $\dfrac{h}{h} = 1$ and we obtain

$$= \frac{1}{\sqrt{x + h} + \sqrt{x}} \qquad \frac{h}{h} = 1, h \neq 0$$

When $x = 2$, the expression becomes $\dfrac{1}{\sqrt{2 + h} + \sqrt{2}}$ and our investigation shows

that as $h \to 0$, the slope of the tangent line approaches $\dfrac{1}{2\sqrt{2}} \approx 0.35$. The graph of $y = \sqrt{x}$ along with an approximation for the tangent line at $x = 2$ are shown in the figure. Note that for values of x close to zero, the tangent line will be much steeper. Verify this using $x = 1, x = 0.1$, and $x = 0.01$.

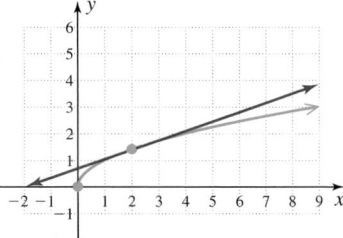

Now try Exercises 5 through 8 ▶

In Example 1, we simplified the difference quotient using a binomial square and the distributive property. In Example 2, the needed skills involved working with radicals and rationalizing the numerator. In other cases, the algebra needed may involve complex fractions, binomial powers, trigonometric identities, or other skills.

Transformations and the Area Under a Curve

The transformations of basic graphs studied in Chapter 2 can help to illustrate some important topics in calculus. One of these involves a simple computing of the area between the graph of a function and the x-axis, between stipulated boundaries.

EXAMPLE 3 ▶ **Using the Area Under a Graph to Determine Distance Covered**

Consider a jogger who is running at a steady pace of 600 ft/min (about 7 mph). Represent this graphically and use the result to determine the distance run if she continues this pace for 5 min.

Solution ▶ Graphically her running speed is given by the horizontal line $v(t) = 600$, where $v(t)$ represents the velocity at time t in minutes, with minutes scaled on the horizontal axis. Note this creates a rectangular shape with an area that is numerically the same as the distance run ($A = LW$ corresponds to $D = RT$), which turns out to be 3000 ft . This is not a coincidence, and in fact, this area represents the total distance run *even when the velocity is not constant*. If these were the last 5 min of the race, she might increase her speed (begin her final kick) in order to overtake the runners ahead of her. If she increases her velocity at a rate of 50 ft/min, the equation becomes $v(t) = 50t + 600$, and the graphical model for her velocity is a line with slope 50 and y-intercept $(0, 600)$. Since her velocity is increasing, it seems reasonable that she covers a greater distance in 5 min, and sure enough we find she covers a distance of 3625 ft (the area of the trapezoid is numerically the same as the distance run).

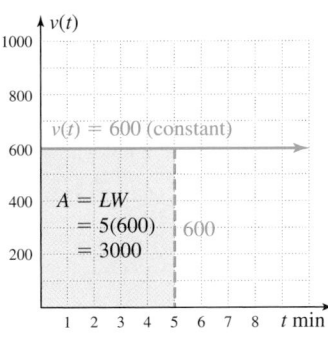
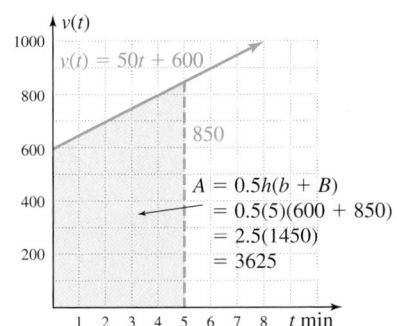

Now try Exercises 9 through 12 ▶

This very simple idea has a number of powerful real-world applications, because the principle holds *even when the velocity function is nonlinear*. Even more importantly, this principle holds for other rate of change relationships as well. For example, power consumption is the rate of change of energy, and if we have a graph representing the power consumption in an office building, the area under this curve will represent the total amount of energy consumed!

Connections to Calculus Exercises

For Exercises 1 through 8, (a) apply the difference quotient to the functions given and simplify the result, then (b) find the value that the slope of the tangent line approaches when $x = 2$, by investigating what happens as $h \to 0$.

1. $f(x) = -3x + 5$

2. $g(x) = \dfrac{2}{3}x - 7$

3. $d(x) = x^2 - 3x$

4. $r(x) = -2x^2 + 3x + 7$

5. $f(x) = \dfrac{1}{x}$

6. $g(x) = \dfrac{3}{x + 1}$

7. $d(x) = \dfrac{1}{2x^2}$

8. $r(x) = x^3 - 2x - 2$

To elude a hungry bird, a crawling insect may vary its speed. Suppose the velocity of the insect (in ft/sec) for an 8-sec time period ($t = 0$ to $t = 8$) is modeled by the following functions. Draw an accurate graph (use transformations of a basic graph as needed) in each case, and use the graph and basic geometry to determine the total distsance traveled by the insect. For Exercise 12, the area of a right parabolic segment is required, and the formula is given here.

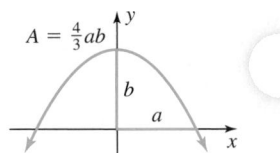

9. $v(t) = 3$

10. $v(t) = t + 3$

11. $v(t) = -|t - 4| + 7$

12. $v(t) = -\dfrac{1}{2}(t - 4)^2 + 11$

Polynomial and
Rational Functions

CHAPTER OUTLINE

CHAPTER CONNECTIONS

For the most part, the growth of a civilization depends on a variety of improvements, such as better infrastructure, more superior tools, advances in medical care, minimizing waste, and maximizing efficiency, to name a few. When the performance of any of these is modeled using a function and its graph, the question becomes one of finding maximum and minimum values. For instance, health agencies might be very interested in modeling the population of disease-carrying insects, to stock up on certain vaccines during the peak season. County governments use traffic flow patterns to aid repair work, opting to fill potholes when traffic in a certain area is at a minimum. A solid understanding of functions and graphs is a valuable asset in both cases. These applications appear as Exercises 85 and 86 in Section 3.4.

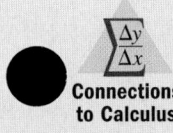

Connections to Calculus

In Chapter 2, we were able to find the maximum or minimum value of a parabola and other elementary functions. In a calculus course, certain techniques are developed that enable us to find these extreme values for almost any function. The ability to graph functions will play an important role in this development, as well as the skills necessary to solve a variety of equation types. These are explored in the *Connections to Calculus* feature for Chapter 3.

Learning Objectives

In Section 3.1 you will learn how to:

☐ **A.** Graph quadratic functions by completing the square

☐ **B.** Graph quadratic functions using the vertex formula

☐ **C.** Find the equation of a quadratic function from its graph

☐ **D.** Solve applications involving extreme values

As our knowledge of functions grows, our ability to apply mathematics in new ways likewise grows. In this section, we'll build on the foundation laid in Chapter 2, as we introduce additional function families and the tools needed to apply them effectively. We begin with the family of quadratic functions.

A. Graphing Quadratic Functions by Completing the Square

The squaring function $f(x) = x^2$ is actually a member of the family of **quadratic functions,** defined as follows.

Quadratic Functions

A quadratic function is one of the form

$$f(x) = ax^2 + bx + c,$$

where a, b, and c are real numbers and $a \neq 0$.

As shown in Figure 3.1, the function is written in **standard form.** For $f(x) = x^2$, $a = 1$ with b and c equal to 0. The function $f(x) = 2x^2 + x - 3$ is also quadratic, with $a = 2$, $b = 1$ and $c = -3$. Our earlier work suggests the graph of *any* quadratic function will be a parabola. Figure 3.1 provides a summary of the characteristic features of this graph. As pictured, the parabola opens upward with the vertex at (h, k), so k is a global minimum. Since the vertex is below the x-axis, the graph has two x-intercepts. The axis of symmetry goes through the vertex, and has equation $x = h$. The y-intercept is $(0, c)$, since $f(0) = c$.

Figure 3.1 $f(x) = ax^2 + bx + c$

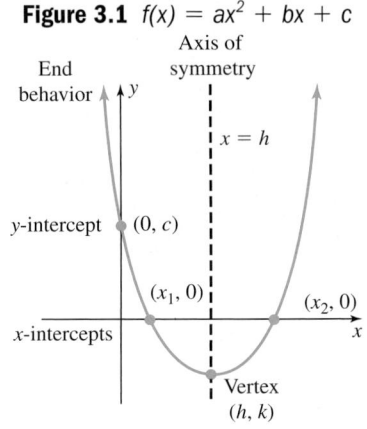

In Section 2.6, we graphed transformations of $f(x) = x^2$, using $y = a(x \pm h)^2 \pm k$. Here, we'll show that by completing the square, we can graph *any* quadratic function as a transformation of this basic graph.

When completing the square on a quadratic *equation* (Section 1.5), we applied the standard properties of equality to both sides of the equation. When completing the square on a *quadratic function,* the process is altered slightly in that we operate on only one side.

EXAMPLE 1 ▶ **Graphing a Quadratic Function by Completing the Square**

Given $g(x) = x^2 - 6x + 5$, complete the square to rewrite g as a transformation of $f(x) = x^2$, then graph the function.

Solution ▶ To begin we note the leading coefficient is $a = 1$.

$$g(x) = x^2 - 6x + 5 \qquad \text{given function}$$

$$= 1(x^2 - 6x + \underline{\quad}) + 5 \qquad \text{group variable terms, note } a = 1$$

$$= 1(\underbrace{x^2 - 6x + 9}_{\text{adds } 1 \cdot 9 = 9}) \underbrace{- 9}_{\text{subtract 9}} + 5 \qquad \left[\left(\frac{1}{2}\right)(-6)\right]^2 = 9$$

$$= (x - 3)^2 - 4 \qquad \text{factor and simplify}$$

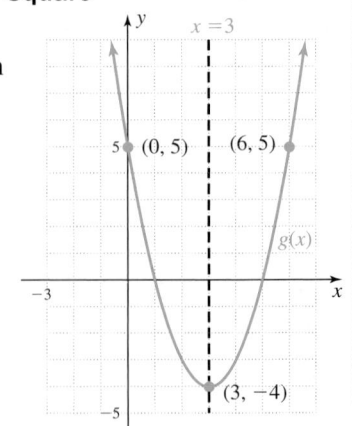

The graph of g is the graph of f shifted 3 units right, and 4 units down. The graph opens upward ($a > 0$) with the vertex at $(3, -4)$, and axis of symmetry $x = 3$. From the original equation we find $g(0) = 5$, giving a y-intercept of $(0, 5)$. The point $(6, 5)$ was obtained using the axis of symmetry. The graph is shown in the figure.

Now try Exercises 7 through 10 ▶

Note that by **adding 9** and simultaneously **subtracting 9** (essentially adding "0"), we changed only the *form* of the function, not its value. In other words, the resulting expression is equivalent to the original. If the leading coefficient is not 1, we factor it out from the variable terms, but take it into account when we add the constant needed to maintain an equivalent expression.

EXAMPLE 2 ▶ **Graphing a Quadratic Function by Completing the Square**

Given $p(x) = -2x^2 - 8x - 3$, complete the square to rewrite p as a transformation of $f(x) = x^2$, then graph the function.

Solution ▶

$$p(x) = -2x^2 - 8x - 3 \qquad \text{given function}$$
$$= (-2x^2 - 8x + \underline{\quad}) - 3 \qquad \text{group variable terms}$$
$$= -2(x^2 + 4x + \underline{\quad}) - 3 \qquad \text{factor out } a = -2 \text{ (notice sign change)}$$
$$= -2(x^2 + 4x + 4) - (-8) - 3 \qquad \left[\left(\tfrac{1}{2}\right)(4)\right]^2 = 4$$
$$\underbrace{}_{\text{adds } -2 \cdot 4 = -8} \quad \overset{\text{subtract } -8}{}$$
$$= -2(x + 2)^2 + 8 - 3 \qquad \text{factor trinomial, simplify}$$
$$= -2(x + 2)^2 + 5 \qquad \text{result}$$

The graph of p is a parabola, shifted 2 units left, stretched by a factor of 2, reflected across the x-axis (opens downward), and shifted up 5 units. The vertex is $(-2, 5)$, and the axis of symmetry is $x = -2$. From the original function, the y-intercept is $(0, -3)$. The point $(-4, -3)$ was obtained using the axis of symmetry. The graph is shown in the figure.

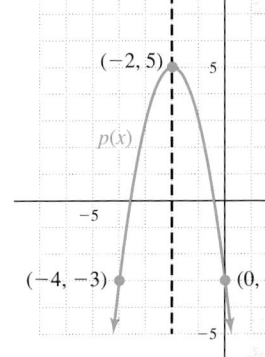

Now try Exercises 11 through 14 ▶

By adding 4 to the variable terms within parentheses, we actually added $-2 \cdot 4 = -8$ to the value of the function. To adjust for this we subtracted -8. The basic ideas are summarized here.

WORTHY OF NOTE

In cases like $f(x) = 3x^2 - 10x + 5$, where the linear coefficient has no integer factors of a, we factor out 3 and *simultaneously divide the linear coefficient by 3*. This yields

$$h(x) = 3\left(x^2 - \frac{10}{3}x + \underline{\quad}\right) + 5,$$

and the process continues as before: $\left[\left(\tfrac{1}{2}\right)\left(\tfrac{10}{3}\right)\right]^2 = \left(\tfrac{5}{3}\right)^2 = \tfrac{25}{9}$, and so on. For more on this idea, **see Exercises 15 through 20.**

Graphing $f(x) = ax^2 + bx + c$ by Completing the Square

1. Group the variable terms apart from the constant c.
2. Factor out the leading coefficient a.
3. Compute $\left[\tfrac{1}{2}\left(\tfrac{b}{a}\right)\right]^2$ and add the result to the grouped terms,

   then subtract $a \cdot \left[\tfrac{1}{2}\left(\tfrac{b}{a}\right)\right]^2$ to maintain an equivalent expression.
4. Factor the grouped terms as a binomial square and simplify.
5. Graph using transformations of $f(x) = x^2$.

✓ **A.** You've just learned how to graph quadratic functions by completing the square

B. Graphing Quadratic Functions Using the Vertex Formula

When the process of completing the square is applied to $f(x) = ax^2 + bx + c$, we obtain a very useful result. Notice the close similarities to Example 2.

$$f(x) = ax^2 + bx + c \qquad \text{quadratic function}$$

$$= (ax^2 + bx + \underline{\quad}) + c \qquad \text{group variable terms apart from the constant } c$$

$$= a\left(x^2 + \frac{b}{a}x + \underline{\quad}\right) + c \qquad \text{factor out } a$$

$$= a\left(x^2 + \frac{b}{a}x + \frac{b^2}{4a^2}\right) - a\left(\frac{b^2}{4a^2}\right) + c \qquad \left[\left(\frac{1}{2}\right)\left(\frac{b}{a}\right)\right]^2 = \frac{b^2}{4a^2}$$

$$= a\left(x + \frac{b}{2a}\right)^2 - \frac{b^2}{4a} + c \qquad \text{factor the trinomial, simplify}$$

$$= a\left(x + \frac{b}{2a}\right)^2 + \frac{4ac - b^2}{4a} \qquad \text{result}$$

By comparing this result with previous transformations, we note the x-coordinate of the vertex is $h = \dfrac{-b}{2a}$ (since the graph shifts horizontally "opposite the sign"). While we could use the expression $\dfrac{4ac - b^2}{4a}$ to find k, we find it easier to substitute $\dfrac{-b}{2a}$ back into the function: $k = f\left(\dfrac{-b}{2a}\right)$. The result is called the **vertex formula.**

Vertex Formula

For the quadratic function $f(x) = ax^2 + bx + c$, the coordinates of the vertex are
$$(h, k) = \left(\frac{-b}{2a}, f\left(\frac{-b}{2a}\right)\right)$$

Since all characteristic features of the graph (end-behavior, vertex, axis of symmetry, x-intercepts, and y-intercept) can now be determined using the original equation, we'll rely on these features to sketch quadratic graphs, rather than having to complete the square.

EXAMPLE 3 ▶ **Graphing a Quadratic Function Using the Vertex Formula**

Graph $f(x) = 2x^2 + 8x + 3$ using the vertex formula and other features of a quadratic graph.

Solution ▶ The graph will open upward since $a > 0$.
The y-intercept is $(0, 3)$.
The vertex formula gives

$$h = \frac{-b}{2a} \qquad \text{x-coordinate of vertex}$$

$$= \frac{-8}{2(2)} \qquad \text{substitute 2 for } a \text{ and 8 for } b$$

$$= -2 \qquad \text{simplify}$$

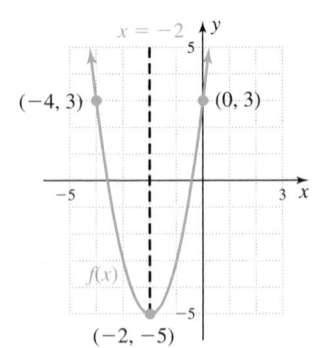

Computing $f(-2)$ to find the y-coordinate of the vertex yields

$$\begin{aligned}
f(-2) &= 2(-2)^2 + 8(-2) + 3 && \text{substitute } -2 \text{ for } x \\
&= 2(4) - 16 + 3 && \text{multiply} \\
&= 8 - 13 && \text{simplify} \\
&= -5 && \text{result}
\end{aligned}$$

☑ **B.** You've just learned how to graph quadratic functions using the vertex formula

The vertex is $(-2, -5)$. The graph is shown in the figure, with the point $(-4, 3)$ obtained using symmetry.

> **Now try Exercises 21 through 32 ▶**

C. Finding the Equation of a Quadratic Function from Its Graph

While most of our emphasis so far has centered on graphing quadratic functions, it would be hard to overstate the importance of the reverse process—determining the equation of the function from its graph (as in Section 2.6). This reverse process, which began with our study of lines, will be a continuing theme each time we consider a new function.

EXAMPLE 4 ▶ Finding the Equation of a Quadratic Function

The graph shown is a transformation of $f(x) = x^2$. What function defines this graph?

Solution ▶ Compared to the graph of $f(x) = x^2$, the vertex has been shifted left 1 and up 2, so the function will have the form $F(x) = a(x + 1)^2 + 2$. Since the graph opens downward, we know a will be negative. As before, we select one additional point on the graph and substitute to find the value of a. Using $(x, y) \to (1, 0)$ we obtain

> **WORTHY OF NOTE**
> It helps to remember that any point (x, y) on the parabola can be used. To verify this, try the calculation again using $(-3, 0)$.

$$\begin{aligned}
F(x) &= a(x + 1)^2 + 2 && \text{transformation} \\
0 &= a(1 + 1)^2 + 2 && \text{substitute 1 for } x \text{ and 0} \\
&&& \text{for } F(x)\text{: } (x, y) \to (1, 0) \\
0 &= 4a + 2 && \text{simplify} \\
-2 &= 4a && \text{subtract 2} \\
-\frac{1}{2} &= a && \text{solve for } a
\end{aligned}$$

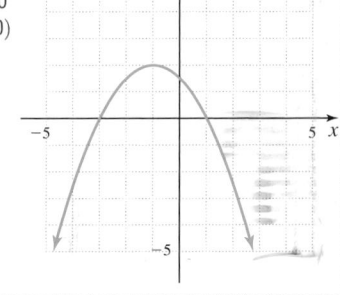

☑ **C.** You've just learned how to find the equation of a quadratic function from its graph

The equation of this function is

$$F(x) = -\frac{1}{2}(x + 1)^2 + 2.$$

> **Now try Exercises 33 through 38 ▶**

D. Quadratic Functions and Extreme Values

If $a > 0$, the parabola opens upward, and the y-coordinate of the vertex is a global minimum, the smallest value attained by the function anywhere in its domain. Conversely, if $a < 0$ the parabola opens downward and the vertex yields a global maximum. These greatest and least points are known as **extreme values** and have a number of significant applications.

EXAMPLE 5 ▶ Applying a Quadratic Model to Manufacturing

An airplane manufacturer can produce up to 15 planes per month. The profit made from the sale of these planes is modeled by $P(x) = -0.2x^2 + 4x - 3$, where $P(x)$ is the profit in hundred-thousands of dollars per month, and x is the number of planes sold. Based on this model,

a. Find the y-intercept and explain what it means in this context.

b. How many planes should be made and sold to maximize profit?

c. What is the maximum profit?

Solution ▶

a. $P(0) = -3$, which means the manufacturer loses $300,000 each month if the company produces no planes.

b. Since $a < 0$, we know the graph opens downward and has a maximum value. To find the required number of sales needed to "maximize profit," we use the vertex formula with $a = -0.2$ and $b = 4$:

$$x = \frac{-b}{2a} \qquad \text{vertex formula}$$

$$= \frac{-4}{2(-0.2)} \qquad \text{substitute } -0.2 \text{ for } a \text{ and } 4 \text{ for } b$$

$$= 10 \qquad \text{result}$$

The result shows 10 planes should be sold each month for maximum profit.

c. Evaluating $P(10)$ we find that a maximum profit of 17 "hundred thousand dollars" will be earned ($1,700,000).

Now try Exercises 41 through 45 ▶

Note that if the leading coefficient is positive and the vertex is below the x-axis ($k < 0$), the graph will have two x-intercepts (see Figure 3.2). If $a > 0$ and the vertex is above the x-axis ($k > 0$), the graph will not cross the x-axis (Figure 3.3). Similar statements can be made for the case where a is negative.

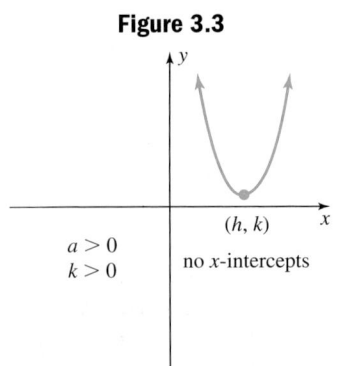

Figure 3.2

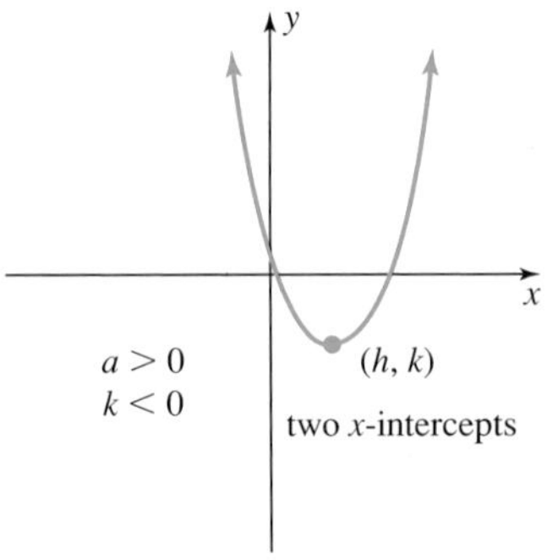

$a > 0$
$k < 0$
two x-intercepts
(h, k)

Figure 3.3

$a > 0$
$k > 0$
no x-intercepts
(h, k)

In some applications of quadratic functions, our interest includes the x-intercepts of the graph. Drawing on our previous work, we note that the following statements are equivalent, meaning if any one statement is true, then all four statements are true.

- $(r, 0)$ is an **x-intercept** of the graph of $f(x)$.
- $x = r$ is a **solution** or **root** of the equation $f(x) = 0$.
- $(x - r)$ is a **factor** of $f(x)$.
- r is a **zero** of $f(x)$.

When the quadratic function is in standard form, our primary tool for finding the zeroes is the quadratic formula. If the function is expressed as a transformation, we will often solve for x using inverse operations.

EXAMPLE 6 ▶   **Modeling the Height of a Projectile**

In the 1976 Pro Bowl, NFL punter Ray Guy of the Oakland Raiders kicked the ball so high it hit the scoreboard hanging from the roof of the New Orleans SuperDome. If we let $h(t)$ represent the height of the football (in feet) after t sec, the function $h(t) = -22t^2 + 132t + 1$ models the relationship (time, height of ball).

a. What does the y-intercept of this function represent?

b. After how many seconds did the football reach its maximum height?

c. What was the maximum height of this kick?

d. To the nearest hundredth of a second, how long until the ball returns to the ground (what was the hang time)?

Solution ▶ **a.** $h(0) = 1$, meaning the ball was 1 ft off the ground when Ray Guy kicked it.

b. Since $a < 0$, we know the graph opens downward and has a maximum value. To find the time needed to reach the maximum height, we use the vertex formula with $a = -22$ and $b = 132$:

$$t = \frac{-b}{2a} \qquad \text{vertex formula}$$

$$= \frac{-132}{2(-22)} \qquad \text{substitute } -22 \text{ for } a \text{ and } 132 \text{ for } b$$

$$= 3 \qquad \text{result}$$

The ball reached its maximum height after 3 sec.

c. To find the maximum height, we substitute 3 for t [evaluate $h(3)$]:

$$h(t) = -22t^2 + 132t + 1 \qquad \text{given function}$$

$$h(3) = -22(3)^2 + 132(3) + 1 \qquad \text{substitute 3 for } t$$

$$= 199 \qquad \text{result}$$

The ball reached a maximum height of 199 ft.

 d. When the ball returns to the ground it has a height of 0 ft. Substituting 0 for $h(t)$ gives $0 = -22t^2 + 132t + 1$, which we solve using the quadratic formula.

$$t = \frac{-b \pm \sqrt{b^2 - 4ac}}{2a} \qquad \text{quadratic formula}$$

$$= \frac{-132 \pm \sqrt{132^2 - 4(-22)(1)}}{2(-22)} \qquad \text{substitute } -22 \text{ for } a, 132 \text{ for } b, \text{ and 1 for } c$$

$$= \frac{-132 \pm \sqrt{17512}}{-44} \qquad \text{simplify}$$

$$t \approx -0.01 \quad \text{or} \quad t \approx 6.01$$

☑ **D.** You've just learned how to solve applications involving extreme values

The punt had a hang time of just over 6 sec.

Now try Exercises 46 through 49 ▶

TECHNOLOGY HIGHLIGHT

Estimating Irrational Zeroes

Once a function is entered into a graphing calculator, an estimate for irrational zeroes can easily be found. Enter the function $y = x^2 - 8x + 9$ on the Y= screen and graph using the standard window (ZOOM 6). Pressing 2nd TRACE (CALC) displays the screen in Figure 3.4. Pressing the number "2" selects the **2:zero** option and returns you to the graph, where you are asked to enter a "Left Bound." The calculator is asking you to narrow down the area it has to search for the x-intercept. Select any number that is conveniently to the left of the x-intercept you're interested in. For this graph, we entered a left bound of "0" (press ENTER), which the calculator indicates with a "▶" marker. It then asks you to enter a "Right Bound."

Figure 3.4

```
CALCULATE
1:value
2:zero
3:minimum
4:maximum
5:intersect
6:dy/dx
7:∫f(x)dx
```

—continued

Figure 3.5

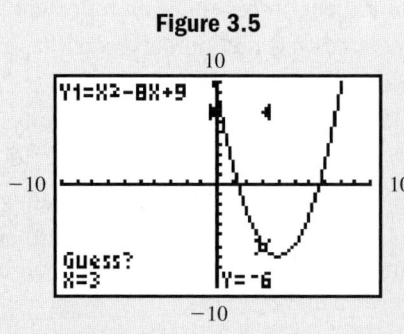

Figure 3.6

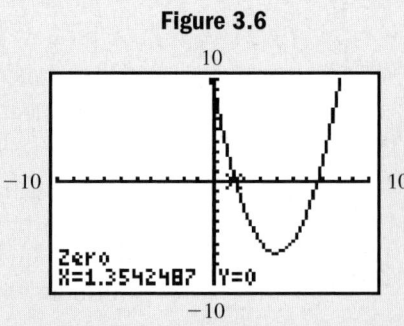

Select any value to the right of this x-intercept, but be sure the value *bounds only one intercept* (see Figure 3.5). For this graph, a choice of 10 would include both x-intercepts, while a choice of 3 would bound only the x-intercept on the left. After entering 3, the calculator asks for a "guess." This option is used only when there are many different zeroes close by or if you entered a large interval. Most of the time we'll simply bypass this option by pressing ENTER . The cursor will be located at the zero you chose, with the coordinates displayed at the bottom of the screen (see Figure 3.6). The x-value is an approximation of the irrational zero. Find the zeroes of these functions using the 2nd   TRACE (CALC) **2:Zero** feature.

Exercise 1: $y = x^2 - 8x + 9$ **Exercise 2:** $y = 3a^2 - 5a - 6$

Exercise 3: $y = 2x^2 + 4x - 5$ **Exercise 4:** $y = 9w^2 + 6w - 1$

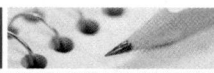

3.1 EXERCISES

▶ CONCEPTS AND VOCABULARY

Fill in each blank with the appropriate word or phrase.
Carefully reread the section if needed.

1. Fill in the blank to complete the square, given $f(x) = -2x^2 - 10x - 7$:
   $f(x) = -2(x^2 + 5x + \frac{25}{4}) - 7 + $ _____ .

2. The maximum and minimum values are called _____ values and can be found using the _____ formula.

3. To find the zeroes of $f(x) = ax^2 + bx + c$, we substitute _____ for _____ and solve.

4. If the leading coefficient is positive and the vertex (h, k) is in Quadrant IV, the graph will have _____ x-intercepts.

5. Compare/Contrast how to complete the square on an *equation,* versus how to complete the square on a function. Use the equation $2x^2 + 6x - 3 = 0$ and the function $f(x) = 2x^2 + 6x - 3 = 0$ to illustrate.

6. Discuss/Explain why the graph of a quadratic function has no x-intercepts if a and k [vertex (h, k)] have like signs. Under what conditions will the function have a single real root?

▶ DEVELOPING YOUR SKILLS

Graph each function using end behavior, intercepts, and completing the square to write the function in shifted form. Clearly state the transformations used to obtain the graph, and label the vertex and all intercepts (if they exist). Use the quadratic formula to find the x-intercepts.

7. $f(x) = x^2 + 4x - 5$ 8. $g(x) = x^2 - 6x - 7$ 9. $h(x) = -x^2 + 2x + 3$ 10. $H(x) = -x^2 + 8x - 7$

11. $Y_1 = 3x^2 + 6x - 5$

12. $Y_2 = 4x^2 - 24x + 15$

13. $f(x) = -2x^2 + 8x + 7$

14. $g(x) = -3x^2 + 12x - 7$

15. $p(x) = 2x^2 - 7x + 3$

16. $q(x) = 4x^2 - 9x + 2$

17. $f(x) = -3x^2 - 7x + 6$

18. $g(x) = -2x^2 + 9x - 7$

19. $p(x) = x^2 - 5x + 2$

20. $q(x) = x^2 + 7x + 4$

Graph each function using the vertex formula and other features of a quadratic graph. Label all important features.

21. $f(x) = x^2 + 2x - 6$ **22.** $g(x) = x^2 + 8x + 11$

23. $h(x) = -x^2 + 4x + 2$

24. $H(x) = -x^2 + 10x - 19$

25. $Y_1 = 0.5x^2 + 3x + 7$ **26.** $Y_2 = 0.2x^2 - 2x + 8$

27. $Y_1 = -2x^2 + 10x - 7$ **28.** $Y_2 = -2x^2 + 8x - 3$

29. $f(x) = 4x^2 - 12x + 3$ **30.** $g(x) = 3x^2 + 12x + 5$

31. $p(x) = \frac{1}{2}x^2 + 3x - 5$ **32.** $q(x) = \frac{1}{3}x^2 - 2x - 4$

State the equation of the function whose graph is shown.

33. 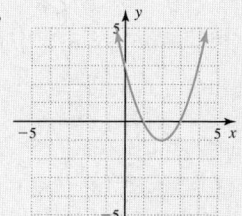   **34.**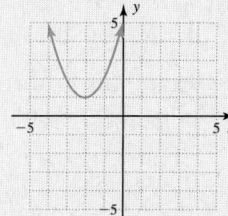

35. 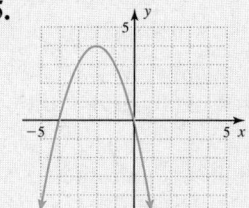   **36.**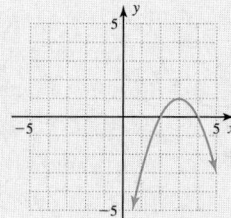

37. 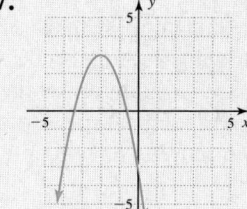   **38.**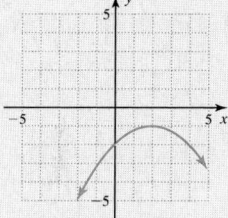

▶ WORKING WITH FORMULAS

39. Vertex/intercept formula: $x = h \pm \sqrt{-\dfrac{k}{a}}$

As an alternative to using the quadratic formula *prior* to completing the square, the x-intercepts can more easily be found using the vertex/intercept formula *after* completing the square, when the coordinates of the vertex are known. (a) Beginning with the shifted form $y = a(x - h)^2 + k$, substitute 0 for y and solve for x to derive the formula, and (b) use the formula to find zeroes, real or complex, of the following functions.

  i. $y = (x + 3)^2 - 5$ **ii.** $y = -(x - 4)^2 + 3$

iii. $y = 2(x + 4)^2 - 7$ **iv.** $y = -3(x - 2)^2 + 6$

  v. $s(t) = 0.2(t + 0.7)^2 - 0.8$

vi. $r(t) = -0.5(t - 0.6)^2 + 2$

40. Surface area of a rectangular box with square ends: $S = 2h^2 + 4Lh$

The surface area of a rectangular box with square ends is given by the formula shown, where h is the height and width of the square ends, and L is the length of the box. (a) If L is 3 ft and the box must have a surface area of 32 ft^2, find the dimensions of the square ends. (b) Solve for L, then find the length if the height is 1.5 ft and surface area is 22.5 ft^2.

▶ APPLICATIONS

41. Maximum profit: An automobile manufacturer can produce up to 300 cars per day. The profit made from the sale of these vehicles can be modeled by the function $P(x) = -10x^2 + 3500x - 66,000$, where $P(x)$ is the profit in dollars and x is the number of automobiles made and sold. Based on this model:

  a. Find the y-intercept and explain what it means in this context.

  b. Find the x-intercepts and explain what they mean in this context.

c. How many cars should be made and sold to maximize profit?

d. What is the maximum profit?

42. **Maximum profit:** The profit for a manufacturer of collectible grandfather clocks is given by the function shown here, where $P(x)$ is the profit in dollars and x is the number of clocks made and sold. Answer the following questions based on this model: $P(x) = -1.6x^2 + 240x - 375$.

   a. Find the y-intercept and explain what it means in this context.

   b. Find the x-intercepts and explain what they mean in this context.

   c. How many clocks should be made and sold to maximize profit?

   d. What is the maximum profit?

43. **Depth of a dive:** As it leaves its support harness, a minisub takes a deep dive toward an underwater exploration site. The dive path is modeled by the function $d(x) = x^2 - 12x$, where $d(x)$ represents the depth of the minisub in hundreds of feet at a distance of x mi from the surface ship.

   a. How far from the mother ship did the minisub reach its deepest point?

   b. How far underwater was the submarine at its deepest point?

   c. At $x = 4$ mi, how deep was the minisub explorer?

   d. How far from its entry point did the minisub resurface?

44. **Optimal pricing strategy:** The director of the Ferguson Valley drama club must decide what to charge for a ticket to the club's performance of *The Music Man*. If the price is set too low, the club will lose money; and if the price is too high, people won't come. From past experience she estimates that the profit P from sales (in hundreds) can be approximated by $P(x) = -x^2 + 46x - 88$, where x is the cost of a ticket and $0 \le x \le 50$.

   a. Find the lowest cost of a ticket that would allow the club to break even.

   b. What is the highest cost that the club can charge to break even?

   c. If the theater were to close down before any tickets are sold, how much money would the club lose?

   d. How much should the club charge to maximize their profits? What is the maximum profit?

45. **Maximum profit:** A kitchen appliance manufacturer can produce up to 200 appliances per day. The profit made from the sale of these machines can be modeled by the function $P(x) = -0.5x^2 + 175x - 3300$, where $P(x)$ is the profit in dollars, and x is the number of appliances made and sold. Based on this model,

   a. Find the y-intercept and explain what it means in this context.

   b. Find the x-intercepts and explain what they mean in this context.

   c. Determine the domain of the function and explain its significance.

   d. How many should be sold to maximize profit? What is the maximum profit?

The projectile function: $h(t) = -16t^2 + vt + k$ applies to any object projected upward with an initial velocity v, from a height k but not to objects under propulsion (such as a rocket). Consider this situation and answer the questions that follow.

46. **Model rocketry:** A member of the local rocketry club launches her latest rocket from a large field. At the moment its fuel is exhausted, the rocket has a velocity of 240 ft/sec and an altitude of 544 ft (t is in seconds).

   a. Write the function that models the height of the rocket.

   b. How high is the rocket at $t = 0$? If it took off from the ground, why is it this high at $t = 0$?

   c. How high is the rocket 5 sec after the fuel is exhausted?

   d. How high is the rocket 10 sec after the fuel is exhausted?

   e. How could the rocket be at the same height at $t = 5$ and at $t = 10$?

   f. What is the maximum height attained by the rocket?

   g. How many seconds was the rocket airborne *after* its fuel was exhausted?

47. **Height of a projectile:** A projectile is thrown upward with an initial velocity of 176 ft/sec. After t sec, its height $h(t)$ above the ground is given by the function $h(t) = -16t^2 + 176t$.

   a. Find the projectile's height above the ground after 2 sec.

   b. Sketch the graph modeling the projectile's height.

   c. What is the projectile's maximum height? What is the value of t at this height?

   d. How many seconds after it is thrown will the projectile strike the ground?

48. Height of a projectile: In the movie *The Court Jester* (1956; Danny Kaye, Basil Rathbone, Angela Lansbury, and Glynis Johns), a catapult is used to toss the nefarious adviser to the king into a river. Suppose the path flown by the king's adviser is modeled by the function $h(d) = -0.02d^2 + 1.64d + 14.4$, where $h(d)$ is the height of the adviser in feet at a distance of d ft from the base of the catapult.

   a. How high was the release point of this catapult?

   b. How far from the catapult did the adviser reach a maximum altitude?

   c. What was this maximum altitude attained by the adviser?

   d. How far from the catapult did the adviser splash into the river?

49. Blanket toss competition: The Fraternities at Steele Head University are participating in a blanket toss competition, an activity borrowed from the whaling villages of the Inuit Eskimos. If the person being tossed is traveling at 32 ft/sec as he is projected into the air, and the Frat members are holding the canvas blanket at a height of 5 ft,

   a. Write the function that models the height at time t of the person being tossed.

   b. How high is the person when (i) $t = 0.5$, (ii) $t = 1.5$?

   c. From part (b) what do you know about *when* the maximum height is reached?

   d. To the nearest tenth of a second, when is the maximum height reached?

   e. To the nearest one-half foot, what was the maximum height?

   f. To the nearest tenth of a second, how long was this person airborne?

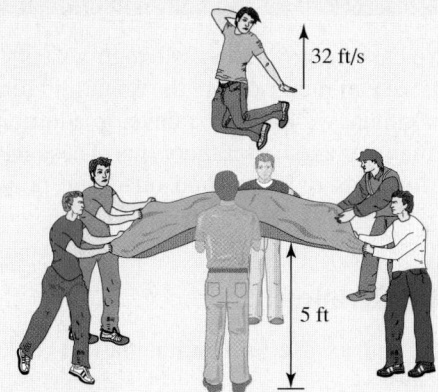

32 ft/s

5 ft

50. Cost of production: The cost of producing a plastic toy is given by the function $C(x) = 2x + 35$, where x is the number of hundreds of toys. The revenue from toy sales is given by $R(x) = -x^2 + 122x - 365$. Since profit = revenue − cost, the profit function must be $P(x) = -x^2 + 120x - 400$ (verify). How many toys sold will produce the maximum profit? What is the maximum profit?

51. Cost of production: The cost to produce bottled spring water is given by $C(x) = 16x - 63$, where x is the number of thousands of bottles. The total income (revenue) from the sale of these bottles is given by the function $R(x) = -x^2 + 326x - 7463$. Since profit = revenue − cost, the profit function must be $P(x) = -x^2 + 310x - 7400$ (verify). How many bottles sold will produce the maximum profit? What is the maximum profit?

52. Fencing a backyard: Tina and Imai have just purchased a purebred German Shepherd, and need to fence in their backyard so the dog can run. What is the maximum rectangular area they can enclose with 200 ft of fencing, if (a) they use fencing material along all four sides? What are the dimensions of the rectangle? (b) What is the maximum area if they use the house as one of the sides? What are the dimensions of *this* rectangle?

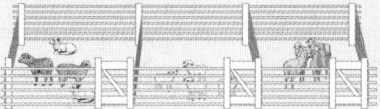

53. Building sheep pens: It's time to drench the sheep again, so Chance and Chelsey Lou are fencing off a large rectangular area to build some temporary holding pens. To prep the males, females, and kids, they are separated into three smaller and equal-size pens partitioned within the large rectangle. If 384 ft of fencing is available and the maximum area is desired, what will be (a) the dimensions of the larger, outer rectangle? (b) the dimensions of the smaller holding pens?

▶ EXTENDING THE CONCEPT

54. Use the general solutions from the quadratic formula to show that the average value of the x-intercepts is $\dfrac{-b}{2a}$. Explain/Discuss why the result is valid even if the roots are complex.

$$x_1 = \frac{-b + \sqrt{b^2 - 4ac}}{2a} \qquad x_2 = \frac{-b - \sqrt{b^2 - 4ac}}{2a}$$

55. Write the equation of a quadratic function whose x-intercepts are given by $x = 2 \pm 3i$.

56. Write the equation for the parabola given.

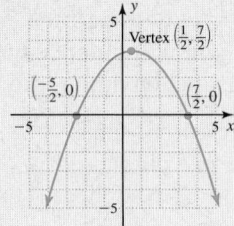

57. Referring to Exercise 39, discuss the nature (real or complex, rational or irrational) and number of zeroes (0, 1, or 2) given by the vertex/intercept formula if (a) a and k have like signs, (b) a and k have unlike signs, (c) k is zero, (d) the ratio $-\dfrac{k}{a}$ is positive and a perfect square, and (e) the ratio $-\dfrac{k}{a}$ is positive and not a perfect square.

▶ MAINTAINING YOUR SKILLS

58. (2.3) Identify the slope and y-intercept for $-4x + 3y = 9$. Do not graph.

59. (R.5) Multiply: $\dfrac{x^2 - 4x + 4}{x^2 + 3x - 10} \cdot \dfrac{x^2 - 25}{x^2 - 10x + 25}$

60. (2.8) Given $f(x) = \sqrt[3]{x + 3}$ and $g(x) = x^3 - 3$, find $(f \circ g)(x)$ and $(g \circ f)(x)$.

61. (2.7) Given $f(x) = 3x^2 + 7x - 6$, solve $f(x) \le 0$ using the x-intercepts and concavity of f.

3.2 | Synthetic Division; the Remainder and Factor Theorems

Learning Objectives

In Section 3.2 you will learn how to:

☐ **A.** Divide polynomials using long division and synthetic division

☐ **B.** Use the remainder theorem to evaluate polynomials

☐ **C.** Use the factor theorem to factor and build polynomials

☐ **D.** Solve applications using the remainder theorem

To find the zero of a linear function, we can use properties of equality to isolate x. To find the zeroes of a quadratic function, we can factor or use the quadratic formula. To find the zeroes of higher degree polynomials, we must first develop additional tools, including synthetic division and the remainder and factor theorems. These will help us to write a higher degree polynomial in terms of linear and quadratic polynomials, whose zeroes can easily be found.

A. Long Division and Synthetic Division

To help understand **synthetic division** and its use as a mathematical tool, we first review the process of **long division.**

Long Division

Polynomial long division closely resembles the division of whole numbers, with the main difference being that *we group each partial product* in parentheses to prevent errors in subtraction.

EXAMPLE 1 ▶ **Dividing Polynomials Using Long Division**

Divide $x^3 - 4x^2 + x + 6$ by $x - 1$.

Solution ▶ The divisor is $(x - 1)$ and the dividend is $(x^3 - 4x^2 + x + 6)$. To find the first multiplier, we compute *the ratio of leading terms* from each expression. Here the ratio $\dfrac{x^3 \text{ from dividend}}{x \text{ from divisor}}$ shows our first multiplier will be "x^2," with $x^2(x - 1) = x^3 - x^2$.

$$
\begin{array}{r}
x^2 \\
x - 1 \overline{)x^3 - 4x^2 + x + 6} \\
\underline{-(x^3 - x^2)} \quad \text{subtraction}
\end{array}
\qquad\longrightarrow\qquad
\begin{array}{r}
x^2 \\
x - 1 \overline{)x^3 - 4x^2 + x + 6} \\
\underline{-x^3 + x^2} \quad \text{algebraic addition} \\
-3x^2 + x
\end{array}
$$

At each stage, after writing the subtraction as algebraic addition (distributing the negative) we compute the sum in each column and "bring down" the next term.

Each following multiplier is found as before, using the ratio $\dfrac{ax^k \text{ next leading term}}{x \text{ from divisor}}$.

$$
\begin{array}{r}
x^2 - 3x - 2 \\
x - 1 \overline{)x^3 - 4x^2 + x + 6} \\
\underline{-(x^3 - x^2)} \\
-3x^2 + x \\
\underline{-(-3x^2 + 3x)} \\
-2x + 6 \\
\underline{-(-2x + 2)} \\
4
\end{array}
$$

next multiplier: $\frac{-3x^2}{x} = -3x$

(ratio of leading terms)

next multiplier: $\frac{-2x}{x} = -2$

subtract $-3x(x - 1) = -3x^2 + 3x$

algebraic addition, bring down next term

subtract $-2(x - 1) = -2x + 2$

algebraic addition, remainder is 4

The result shows $\dfrac{x^3 - 4x^2 + x + 6}{x - 1} = x^2 - 3x - 2 + \dfrac{4}{x - 1}$, or after multiplying both sides by $x - 1$, $x^3 - 4x^2 + x + 6 = (x - 1)(x^2 - 3x - 2) + 4$.

Now try Exercises 7 through 12 ▶

The process illustrated is called the **division algorithm,** and like the division of whole numbers, the final result can be checked by multiplication.

$$
\begin{array}{llll}
\overset{\text{dividend}}{} & \overset{\text{divisor}}{} & \overset{\text{quotient}}{} & \overset{\text{remainder}}{} \\
\end{array}
$$

check: $\begin{aligned}
x^3 - 4x^2 + x + 6 &= (x - 1)(x^2 - 3x - 2) + 4 \\
&= (x^3 - 3x^2 - 2x - x^2 + 3x + 2) + 4 \quad \text{divisor · quotient} \\
&= (x^3 - 4x^2 + x + 2) + 4 \quad \text{combine like terms} \\
&= x^3 - 4x^2 + x + 6 \checkmark \quad \text{add remainder}
\end{aligned}$

In general, the division algorithm for polynomials says

Division of Polynomials

Given polynomials $p(x)$ and $d(x) \neq 0$, there exists unique polynomials $q(x)$ and $r(x)$ such that

$$p(x) = d(x)q(x) + r(x),$$

where $r(x) = 0$ or the degree of $r(x)$ is less than the degree of $d(x)$.
Here, $d(x)$ is called the *divisor,* $q(x)$ is the *quotient,* and $r(x)$ is the *remainder.*

In other words, "a polynomial of greater degree can be divided by a polynomial of equal or lesser degree to obtain a quotient and a remainder." As with whole numbers, if the remainder is zero, the divisor is a factor of the dividend.

Synthetic Division

As the word "synthetic" implies, synthetic division *simulates* the long division process, but condenses it and makes it more efficient when the divisor is linear. The process works by capitalizing on the repetition found in the division algorithm. First, the polynomials involved are written in decreasing order of degree, so the variable part of each term is unnecessary as we can let the *position of each coefficient* indicate the degree of the term. For the dividend from Example 1, $1 \quad -4 \quad 1 \quad 6$ would represent the polynomial $1x^3 - 4x^2 + 1x + 6$. Also, each stage of the algorithm involves a product of the divisor with the next multiplier, followed by a subtraction. These can likewise be computed using the coefficients only, as the degree of each term is still determined by its position. Here is the division from Example 1 in the synthetic division format. Note that we must use the *zero of the divisor* (as in $x = \frac{3}{2}$ for a divisor of $2x - 3$, or in this case, "1" from $x - 1 = 0$) and the coefficients of the dividend in the following format:

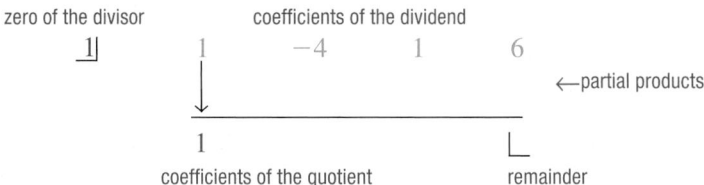

As this template indicates, the quotient and remainder will be read from the last row.

The arrow indicates we begin by "dropping the leading coefficient into place." We then multiply this coefficient by the "divisor," and place the result in the next column and add. Note that using the zero of the divisor enables us to *add in each column directly,* rather than subtracting then changing to algebraic addition as in long division.

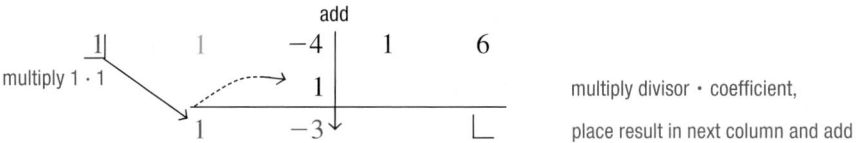

In a sense, we "multiply in the diagonal direction," and "add in the vertical direction." Repeat the process until the division is complete.

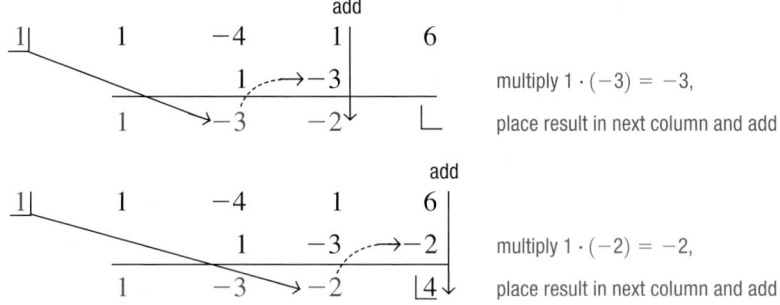

The quotient is read from the last row by noting the remainder is 4, leaving the coefficients $1 \quad -3 \quad -2$, which translates back into the polynomial $x^2 - 3x - 2$. The final result is identical to that in Example 1, but the new process is more efficient, since all stages are actually computed on a single template as shown here:

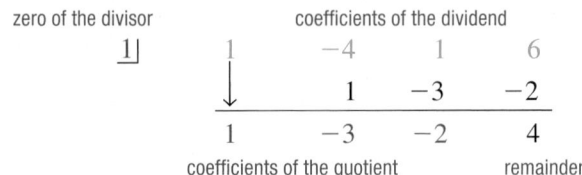

EXAMPLE 2 ▶ Dividing Polynomials Using Synthetic Division

Compute the quotient of $(x^3 + 3x^2 - 4x - 12)$ and $(x + 2)$, then check your answer.

Solution ▶ Using -2 as our "divisor" (from $x + 2 = 0$), we set up the synthetic division template and begin.

use -2 as a "divisor"
$$
\begin{array}{r|rrrr}
-2 & 1 & 3 & -4 & -12 \\
   & \downarrow & -2 & -2 & 12 \\
\hline
   & 1 & 1 & -6 & 0
\end{array}
$$
drop lead coefficient into place;
multiply by divisor, place result
in next column and add

The result shows $\dfrac{x^3 + 3x^2 - 4x - 12}{x + 2} = x^2 + x - 6$, with no remainder.

Check ▶
$$x^3 + 3x^2 - 4x - 12 = (x + 2)(x^2 + x - 6)$$
$$= (x^3 + x^2 - 6x + 2x^2 + 2x - 12)$$
$$= x^3 + 3x^2 - 4x - 12\ ✓$$

Now try Exercises 13 through 20 ▶

Since the division process is so dependent on the place value (degree) of each term, polynomials such as $2x^3 + 3x + 7$, which has no term of degree 2, must be written using a zero *placeholder:* $2x^3 + \mathbf{0}x^2 + 3x + 7$. This ensures that like place values "line up" as we carry out the division.

EXAMPLE 3 ▶ Dividing Polynomials Using a Zero Placeholder

Compute the quotient $\dfrac{2x^3 + 3x + 7}{x - 3}$ and check your answer.

Solution ▶ use 3 as a "divisor"
$$
\begin{array}{r|rrrr}
3 & 2 & 0 & 3 & 7 \\
  & \downarrow & 6 & 18 & 63 \\
\hline
  & 2 & 6 & 21 & 70
\end{array}
$$
note place holder $\mathbf{0}x^2$ for "x^2" term

The result shows $\dfrac{2x^3 + 3x + 7}{x - 3} = 2x^2 + 6x + 21 + \dfrac{70}{x - 3}$. Multiplying by $x - 3$ gives

$$2x^3 + 3x + 7 = (2x^2 + 6x + 21)(x - 3) + 70$$

Check ▶
$$2x^3 + 3x + 7 = (x - 3)(2x^2 + 6x + 21) + 70$$
$$= (2x^3 + 6x^2 + 21x - 6x^2 - 18x - 63) + 70$$
$$= 2x^3 + 3x + 7\ ✓$$

Now try Exercises 21 through 30 ▶

WORTHY OF NOTE

Many corporations now pay their employees monthly to save on payroll costs. If your monthly salary was $2037/mo, but you received a check for only $237, would you complain? Just as placeholder zeroes ensure the correct value of each digit, they also ensure the correct valuation of each term in the division process.

As noted earlier, for synthetic division the divisor must be a linear polynomial and the zero of this divisor is used. This means for the quotient $\dfrac{2x^3 - 3x^2 - 8x + 12}{2x - 3}, \dfrac{3}{2}$ would be used for synthetic division **(see Exercises 43 and 44).** If the divisor is nonlinear, long division must be used.

EXAMPLE 4 ▶ **Division with a Nonlinear Divisor**

Compute the quotient: $\dfrac{2x^4 + x^3 - 7x^2 + 3}{x^2 - 2}$.

Solution ▶ Write the dividend as $2x^4 + x^3 - 7x^2 + \mathbf{0}x + 3$, and the divisor as $x^2 + \mathbf{0}x - 2$.

The quotient of leading terms gives $\dfrac{2x^4 \text{ from dividend}}{x^2 \text{ from divisor}} = 2x^2$ as our first multiplier.

$$
\begin{array}{r}
2x^2 + x - 3 \\
x^2 + 0x - 2 \overline{)2x^4 + x^3 - 7x^2 + 0x + 3} \\
\end{array}
$$

divisor → $x^2 + 0x - 2$)$2x^4 + x^3 - 7x^2 + 0x + 3$

Multiply $2x^2(x^2 + 0x - 2)$ $-(2x^4 + 0x^3 - 4x^2)$ subtract (algebraic addition)

           $x^3 - 3x^2 + 0x$ bring down next term

Multiply $x(x^2 + 0x - 2)$ $-(x^3 + 0x^2 - 2x)$ subtract (algebraic addition)

           $-3x^2 + 2x + 3$ bring down next term

Multiply $-3(x^2 + 0x - 2)$ $-(-\mathbf{3x^2} + \mathbf{0}x + \mathbf{6})$ subtract

           $2x - 3$ remainder is $2x - 3$

☑ **A.** You've just learned how to divide polynomials using long division and synthetic division

Since the degree of $2x - 3$ (degree 1) is less than the degree of the divisor (degree 2), the process is complete.

$$\frac{2x^4 + x^3 - 7x^2 + 3}{x^2 - 2} = (2x^2 + x - 3) + \frac{(2x - 3)}{x^2 - 2}$$

Now try Exercises 31 through 34 ▶

B. The Remainder Theorem

In Example 2, we saw that $(x^3 + 3x^2 - 4x - 12) \div (x + 2) = x^2 + x - 6$, with no remainder. Similar to whole number division, this means $x + 2$ must be a factor of $x^3 + 3x^2 - 4x - 12$, a fact made clear as we checked our answer: $x^3 + 3x^2 - 4x - 12 = (x + 2)(x^2 + x - 6)$. To help us find the factors of higher degree polynomials, we combine synthetic division with a relationship known as the **remainder theorem.** Consider the functions $p(x) = x^3 + 5x^2 + 2x - 8$, $d(x) = x + 3$, and their quotient $\dfrac{p(x)}{d(x)} = \dfrac{x^3 + 5x^2 + 2x - 8}{x + 3}$. Using -3 as the divisor in synthetic division gives

$$
\begin{array}{r}
\text{use } -3 \text{ as a "divisor"} \quad \underline{-3|} \quad
\begin{array}{rrrr}
1 & 5 & 2 & -8 \\
& -3 & -6 & 12 \\
\hline
1 & 2 & -4 & \underline{|4}
\end{array}
\end{array}
$$

This shows $x + 3$ is *not* a factor of $P(x)$, since it didn't "divide evenly." However, from the result $p(x) = (x + 3)(x^2 + 2x - 4) + 4$, we make a remarkable observation—if we evaluate $p(-3)$, *the quotient portion becomes zero*, showing $p(-3) = 4$—*which is the remainder*.

$$p(-3) = (-3 + 3)\left[(-3)^2 + 2(-3) - 4\right] + 4$$
$$= (\mathbf{0})(-1) + 4$$
$$= 4$$

This can also be seen by evaluating $p(-3)$ in its original form:

$$p(x) = x^3 + 5x^2 + 2x - 8$$
$$p(-3) = (-3)^3 + 5(-3)^2 + 2(-3) - 8$$
$$= -27 + 45 + (-6) - 8$$
$$= 4$$

The result is no coincidence, and illustrates the conclusion of the remainder theorem.

The Remainder Theorem

If a polynomial $p(x)$ is divided by $(x - c)$ using synthetic division,
the remainder is equal to $p(c)$.

Proof of the Remainder Theorem

From our previous work, any number c used in synthetic division will occur as the factor $(x - c)$ when written as (quotient)(divisor) + remainder:

$$p(x) = (x - c)\, q(x) + r.$$

Here, $q(x)$ represents the quotient polynomial and r is a constant. Evaluating $p(c)$ gives

$$p(x) = (x - c)q(x) + r$$
$$p(c) = (c - c)q(c) + r$$
$$= 0 \cdot q(c) + r$$
$$= r \checkmark$$

This gives us a powerful tool for evaluating polynomials. Where a direct evaluation involves powers of numbers and a long series of calculations, synthetic division reduces the process to simple products and sums.

EXAMPLE 5 ▶ **Using the Remainder Theorem to Evaluate Polynomials**

Use the remainder theorem to find $p(-5)$ for $p(x) = x^4 + 3x^3 - 8x^2 + 5x - 6$. Verify the result using a substitution.

Solution ▶

$$
\begin{array}{r|rrrrr}
\text{use } -5 \text{ as a "divisor"} \quad -5 & 1 & 3 & -8 & 5 & -6 \\
 & & -5 & 10 & -10 & 25 \\
\hline
 & 1 & -2 & 2 & -5 & \underline{|19} \\
\end{array}
$$

The result shows $p(-5) = 19$, which we verify directly:

$$p(-5) = (-5)^4 + 3(-5)^3 - 8(-5)^2 + 5(-5) - 6$$
$$= 625 - 375 - 200 - 25 - 6$$
$$= 19 \checkmark$$

☑ **B.** You've just learned how to use the Remainder Theorem to evaluate polynomials

Now try Exercises 35 through 44 ▶

C. The Factor Theorem

As a consequence of the remainder theorem, when $p(x)$ is divided by $x - c$ and the remainder is 0, $p(c) = 0$, and c is a zero of the polynomial. The relationship between $x - c$, c, and $p(c) = 0$ are summarized into the **factor theorem.**

WORTHY OF NOTE

Since $p(-5) = 19$, we know $(-5, 19)$ must be a point of the graph of $p(x)$. The ability to quickly evaluate polynomial functions using the remainder theorem will be used extensively in the sections that follow.

The Factor Theorem

For a polynomial $p(x)$,

    1. If $p(c) = 0$, then $x - c$ is a factor of $p(x)$.
    2. If $x - c$ is a factor of $p(x)$, then $p(c) = 0$.

Proof of the Factor Theorem

  1. Consider a polynomial p written in the form $p(x) = (x - c)q(x) + r$. From the remainder theorem we know $p(c) = r$, and substituting $p(c)$ for r in the equation shown gives:

$$p(x) = (x - c)q(x) + p(c),$$
$$\text{showing } x - c \text{ is a factor of } p(x), \text{ if } p(c) = 0.$$
$$p(x) = (x - c)q(x) \checkmark$$

2. The steps from part 1 can be reversed, since for any factor $(x - c)$ of $p(x)$, we have $p(x) = (x - c)q(x)$. Evaluating at $x = c$ produces a result of zero:

$$p(c) = (c - c)q(c)$$
$$= 0 \checkmark$$

The remainder and factor theorems often work together to help us find factors of higher degree polynomials.

EXAMPLE 6 ▶ **Using the Factor Theorem to Find Factors of a Polynomial**

Use the factor theorem to determine if

a. $x - 2$ **b.** $x + 1$

are factors of $p(x) = x^4 + x^3 - 10x^2 - 4x + 24$.

Solution ▶ **a.** If $x - 2$ is a factor, then $p(2)$ must be 0. Using the remainder theorem we have

$$
\begin{array}{r|rrrrr}
2) & 1 & 1 & -10 & -4 & 24 \\
  & \downarrow & 2 & 6 & -8 & -24 \\
\hline
  & 1 & 3 & -4 & -12 & \underline{|0}
\end{array}
$$

Since the remainder is zero, we know $p(2) = 0$ (remainder theorem) and $(x - 2)$ is a factor (factor theorem).

b. Similarly, if $x + 1$ is a factor, then $p(-1)$ must be 0.

$$
\begin{array}{r|rrrrr}
-1) & 1 & 1 & -10 & -4 & 24 \\
  & \downarrow & -1 & 0 & 10 & -6 \\
\hline
  & 1 & 0 & -10 & 6 & \underline{|18}
\end{array}
$$

Since the remainder is not zero, $(x + 1)$ is not a factor of p.

Now try Exercises 45 through 56 ▶

EXAMPLE 7 ▶ **Building a Polynomial Using the Factor Theorem**

A polynomial $p(x)$ has the zeroes 3, $\sqrt{2}$, and $-\sqrt{2}$. Use the factor theorem to find the polynomial.

Solution ▶ Using the factor theorem, the factors of $p(x)$ must be $(x - 3)$, $(x - \sqrt{2})$, and $(x + \sqrt{2})$. Computing the product will yield the polynomial.

$$p(x) = (x - 3)(x - \sqrt{2})(x + \sqrt{2})$$
$$= (x - 3)(x^2 - 2)$$
$$= x^3 - 3x^2 - 2x + 6$$

Now try Exercises 57 through 64 ▶

As the following *Graphical Support* feature shows, the result obtained in Example 7 is not unique, since any polynomial of the form $a(x^3 - 3x^2 - 2x + 6)$ will also have the same three roots for $a \in \mathbb{R}$.

GRAPHICAL SUPPORT

A graphing calculator helps to illustrate there are actually many different polynomials that have the three roots required by Example 7. Figure 3.7 shows the graph of $Y_1 = p(x)$, as well as graph of $Y_2 = 2p(x)$. The only difference is $2p(x)$ has been vertically stretched. Likewise, the graph of $-1p(x)$ would be a vertical reflection, *but still with the same zeroes.*

Figure 3.7

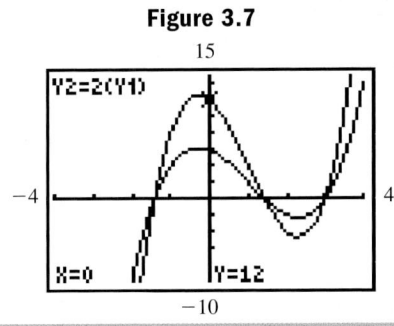

EXAMPLE 8 ▶ **Finding Zeroes Using the Factor Theorem**

Given that 2 is a zero of $p(x) = x^4 + x^3 - 10x^2 - 4x + 24$, use the factor theorem to help find all other zeroes.

Solution ▶ Using synthetic division gives:

$$
\text{use 2 as a "divisor"} \quad
\begin{array}{r|rrrrr}
2 & 1 & 1 & -10 & -4 & 24 \\
  & \downarrow & 2 & 6 & -8 & -24 \\
\hline
  & 1 & 3 & -4 & -12 & \underline{|0}
\end{array}
$$

Since the remainder is zero, $(x - 2)$ is a factor and p can be written:

$$x^4 + x^3 - 10x^2 - 4x + 24 = (x - 2)(x^3 + 3x^2 - 4x - 12)$$

Note the quotient polynomial can be factored by grouping to find the remaining factors of p.

$$
\begin{aligned}
x^4 + x^3 - 10x^2 - 4x + 24 &= (x - 2)(x^3 + 3x^2 - 4x - 12) && \text{group terms (in color)} \\
&= (x - 2)[x^2(x + 3) - 4(x + 3)] && \text{remove common factors from each group} \\
&= (x - 2)[(x + 3)(x^2 - 4)] && \text{factor common binomial} \\
&= (x - 2)(x + 3)(x + 2)(x - 2) && \text{factor difference of squares} \\
&= (x + 3)(x + 2)(x - 2)^2 && \text{completely factored form}
\end{aligned}
$$

The final result shows $(x - 2)$ is actually a repeated factor, and the remaining zeroes of p are -3 and -2.

WORTHY OF NOTE

In Appendix I.D we noted a third degree polynomial $ax^3 + bx^2 + cx + d$ is factorable if $ad = bc$. In Example 8, $1(-12) = 3(-4)$ and the polynomial is factorable.

☑ **C.** You've just learned how to use the factor theorem to factor and build polynomials

Now try Exercises 65 through 78 ▶

D. Applications

While the factor and remainder theorems are valuable tools for factoring higher degree polynomials, each has applications that extend beyond this use.

EXAMPLE 9 ▶ **Using the Remainder Theorem to Solve a Discharge Rate Application**

The *discharge rate* of a river is a measure of the river's water flow as it empties into a lake, sea, or ocean. The rate depends on many factors, but is primarily influenced by the precipitation in the surrounding area and is often seasonal. Suppose the discharge rate of the Shimote River was modeled by

$$D(m) = -m^4 + 22m^3 - 147m^2 + 315m + 150$$

where $D(m)$ represents the discharge rate in thousands of cubic meters of water per second in month m ($m = 1 \rightarrow$ Jan).

a. What was the discharge rate in June (summer heat)?

b. Is the discharge rate higher in February (winter runoff) or October (fall rains)?

Solution ▶ a. To find the discharge rate in June, we evaluate D at $m = 6$.
Using the remainder theorem gives

$$
\begin{array}{r|rrrrr}
6| & -1 & 22 & -147 & 317 & 150 \\
  & \downarrow & -6 & 96 & -306 & 66 \\
\hline
  & -1 & 16 & -51 & 11 & |216
\end{array}
$$

In June, the discharge rate is 216,000 m³/sec.

b. For the discharge rates in February ($m = 2$) and October ($m = 10$), we have

$$
\begin{array}{r|rrrrr}
2| & -1 & 22 & -147 & 317 & 150 \\
  & \downarrow & -2 & 40 & -214 & 206 \\
\hline
  & -1 & 20 & -107 & 103 & |356
\end{array}
\qquad
\begin{array}{r|rrrrr}
10| & -1 & 22 & -147 & 317 & 150 \\
  & \downarrow & -10 & 120 & -270 & 470 \\
\hline
  & -1 & 12 & -27 & 47 & |620
\end{array}
$$

☑ **D.** You've just learned how to solve applications using the remainder theorem

The discharge rate during the fall rains in October is much higher.

Now try Exercises 81 through 84 ▶

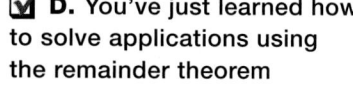

3.2 EXERCISES

▶ **CONCEPTS AND VOCABULARY**

Fill in each blank with the appropriate word or phrase. Carefully reread the section if needed.

1. For _____ division, we use the _____ of the divisor to begin.

2. If the _____ is zero after division, then the _____ is a factor of the dividend.

3. If polynomial $P(x)$ is divided by a linear divisor of the form $x - c$, the remainder is identical to _____. This is a statement of the _____ theorem.

4. If $P(c) = 0$, then _____ must be a factor of $P(x)$. Conversely, if _____ is a factor of $P(x)$, then $P(c) = 0$. These are statements from the _____ theorem.

5. Discuss/Explain how to write the quotient and remainder using the last line from a synthetic division.

6. Discuss/Explain why (a, b) is a point on the graph of P, given b was the remainder after P was divided by a using synthetic division.

Divide using long division. Write the result as dividend = (divisor)(quotient) + remainder.

7. $\dfrac{x^3 - 5x^2 - 4x + 23}{x - 2}$ 8. $\dfrac{x^3 + 5x^2 - 17x - 26}{x + 7}$

9. $(2x^3 + 5x^2 + 4x + 17) \div (x + 3)$

10. $(3x^3 + 14x^2 - 2x - 37) \div (x + 4)$

11. $(x^3 - 8x^2 + 11x + 20) \div (x - 5)$

12. $(x^3 - 5x^2 - 22x - 16) \div (x + 2)$

Divide using synthetic division. Write answers in two ways: (a) $\frac{dividend}{divisor}$ = quotient + $\frac{remainder}{divisor}$, and (b) dividend = (divisor)(quotient) + remainder. For Exercises 13–18, check answers using multiplication.

13. $\dfrac{2x^2 - 5x - 3}{x - 3}$ 14. $\dfrac{3x^2 + 13x - 10}{x + 5}$

15. $(x^3 - 3x^2 - 14x - 8) \div (x + 2)$

16. $(x^3 - 6x^2 - 25x - 17) \div (x + 1)$

17. $\dfrac{x^3 - 5x^2 - 4x + 23}{x - 2}$ 18. $\dfrac{x^3 + 12x^2 + 34x - 7}{x + 7}$

19. $(2x^3 - 5x^2 - 11x - 17) \div (x - 4)$

20. $(3x^3 - x^2 - 7x + 27) \div (x - 1)$

Divide using synthetic division. Note that some terms of a polynomial may be "missing." Write answers as dividend = (divisor)(quotient) + remainder.

21. $(x^3 + 5x^2 + 7) \div (x + 1)$

22. $(x^3 - 3x^2 - 37) \div (x - 5)$

23. $(x^3 - 13x - 12) \div (x - 4)$

24. $(x^3 - 7x + 6) \div (x + 3)$

25. $\dfrac{3x^3 - 8x + 12}{x - 1}$ 26. $\dfrac{2x^3 + 7x - 81}{x - 3}$

27. $(n^3 + 27) \div (n + 3)$ 28. $(m^3 - 8) \div (m - 2)$

29. $(x^4 + 3x^3 - 16x - 8) \div (x - 2)$

30. $(x^4 + 3x^2 + 29x - 21) \div (x + 3)$

Compute each indicated quotient. Write answers in the form $\frac{dividend}{divisor}$ = quotient + $\frac{remainder}{divisor}$.

31. $\dfrac{2x^3 + 7x^2 - x + 26}{x^2 + 3}$ 32. $\dfrac{x^4 + 3x^3 + 2x^2 - x - 5}{x^2 - 2}$

33. $\dfrac{x^4 - 5x^2 - 4x + 7}{x^2 - 1}$ 34. $\dfrac{x^4 + 2x^3 - 8x - 16}{x^2 + 5}$

▶ **DEVELOPING YOUR SKILLS**

Use the remainder theorem to evaluate $P(x)$ as given.

35. $P(x) = x^3 - 6x^2 + 5x + 12$
    a. $P(-2)$ b. $P(5)$

36. $P(x) = x^3 + 4x^2 - 8x - 15$
    a. $P(-2)$ b. $P(3)$

37. $P(x) = 2x^3 - x^2 - 19x + 4$
    a. $P(-3)$ b. $P(2)$

38. $P(x) = 3x^3 - 8x^2 - 14x + 9$
    a. $P(-2)$ b. $P(4)$

39. $P(x) = x^4 - 4x^2 + x + 1$
    a. $P(-2)$ b. $P(2)$

40. $P(x) = x^4 + 3x^3 - 2x - 4$
    a. $P(-2)$ b. $P(2)$

41. $P(x) = 2x^3 - 7x + 33$
    a. $P(-2)$ b. $P(-3)$

42. $P(x) = -2x^3 + 9x^2 - 11$
    a. $P(-2)$ b. $P(-1)$

43. $P(x) = 2x^3 + 3x^2 - 9x - 10$
    a. $P\left(\frac{3}{2}\right)$ b. $P\left(-\frac{5}{2}\right)$

44. $P(x) = 3x^3 + 11x^2 + 2x - 16$
    a. $P\left(\frac{1}{3}\right)$ b. $P\left(-\frac{8}{3}\right)$

Use the factor theorem to determine if the factors given are factors of $f(x)$.

45. $f(x) = x^3 - 3x^2 - 13x + 15$
    a. $(x + 3)$ b. $(x - 5)$

46. $f(x) = x^3 + 2x^2 - 11x - 12$
    a. $(x + 4)$ b. $(x - 3)$

47. $f(x) = x^3 - 6x^2 + 3x + 10$
    a. $(x + 2)$ b. $(x - 5)$

48. $f(x) = x^3 + 2x^2 - 5x - 6$
    a. $(x - 2)$ b. $(x + 4)$

49. $f(x) = -x^3 + 7x - 6$
    a. $(x + 3)$ b. $(x - 2)$

50. $f(x) = -x^3 + 13x - 12$
    a. $(x + 4)$ b. $(x - 3)$

Use the factor theorem to show the given value is a zero of $P(x)$.

51. $P(x) = x^3 + 2x^2 - 5x - 6$
$x = -3$

52. $P(x) = x^3 + 3x^2 - 16x + 12$
$x = -6$

53. $P(x) = x^3 - 7x + 6$
$x = 2$

54. $P(x) = x^3 - 13x + 12$
$x = -4$

55. $P(x) = 9x^3 + 18x^2 - 4x - 8$
$x = \dfrac{2}{3}$

56. $P(x) = 5x^3 + 13x^2 - 9x - 9$
$x = -\dfrac{3}{5}$

A polynomial P with integer coefficients has the zeroes and degree indicated. Use the factor theorem to write the function in factored form and standard form.

57. $-2, 3, -5$; degree 3

58. $1, -4, 2$; degree 3

59. $-2, \sqrt{3}, -\sqrt{3}$; degree 3

60. $\sqrt{5}, -\sqrt{5}, 4$; degree 3

61. $-5, 2\sqrt{3}, -2\sqrt{3}$; degree 3

62. $4, 3\sqrt{2}, -3\sqrt{2}$; degree 3

63. $1, -2, \sqrt{10}, -\sqrt{10}$; degree 4

64. $\sqrt{7}, -\sqrt{7}, 3, -1$; degree 4

In Exercises 65 through 70, a known zero of the polynomial is given. Use the factor theorem to write the polynomial in completely factored form.

65. $P(x) = x^3 - 5x^2 - 2x + 24$; $x = -2$

66. $Q(x) = x^3 - 7x^2 + 7x + 15$; $x = 3$

67. $p(x) = x^4 + 2x^3 - 12x^2 - 18x + 27$; $x = -3$

68. $q(x) = x^4 + 4x^3 - 6x^2 - 4x + 5$; $x = 1$

69. $f(x) = 2x^3 + 11x^2 - x - 30$; $x = \frac{3}{2}$

70. $g(x) = 3x^3 + 2x^2 - 75x - 50$; $x = -\frac{2}{3}$

If $p(x)$ is a polynomial with rational coefficients and a leading coefficient of $a = 1$, the rational zeroes of p (if they exist) *must be factors of the constant term*. Use this property of polynomials with the factor and remainder theorems to factor each polynomial completely.

71. $p(x) = x^3 - 3x^2 - 9x + 27$

72. $p(x) = x^3 - 4x^2 - 16x + 64$

73. $p(x) = x^3 - 6x^2 + 12x - 8$

74. $p(x) = x^3 - 15x^2 + 75x - 125$

75. $p(x) = (x^2 - 6x + 9)(x^2 - 9)$

76. $p(x) = (x^2 - 1)(x^2 - 2x + 1)$

77. $p(x) = (x^3 + 4x^2 - 9x - 36)(x^2 + x - 12)$

78. $p(x) = (x^3 - 3x^2 + 3x - 1)(x^2 - 3x + 2)$

▶ **WORKING WITH FORMULAS**

Volume of an open box: $V(x) = 4x^3 - 84x^2 + 432x$

An open box is constructed by cutting square corners from a 24 in. by 18 in. sheet of cardboard and folding up the sides. Its volume is given by the formula shown, where x represents the size of the square cut.

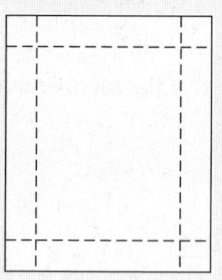

79. Given a volume of 640 in³, use synthetic division and the

remainder theorem to determine if the squares were 2-, 3-, 4-, or 5-in. squares and state the dimensions of the box. (*Hint:* Write as a function $v(x)$ and use synthetic division.)

80. Given the volume is 357.5 in³, use synthetic division and the remainder theorem to determine if the squares were 5.5-, 6.5-, or 7.5-in. squares and state the dimensions of the box. (*Hint:* Write as a function $v(x)$ and use synthetic division.)

▶ **APPLICATIONS**

81. Tourist population: During the 12 weeks of summer, the population of tourists at a popular beach resort is modeled by the polynomial

$P(w) = -0.1w^4 + 2w^3 - 14w^2 + 52w + 5$, where $P(w)$ is the tourist population (in 1000s) during week w. Use the remainder theorem to help answer the following questions.

a. Were there more tourists at the resort in week 5 ($w = 5$) or week 10? How many more tourists?

b. Were more tourists at the resort one week after opening ($w = 1$) or one week before closing ($w = 11$). How many more tourists?

c. The tourist population peaked (reached its highest) between weeks 7 and 10. Use the remainder theorem to determine the peak week.

82. Debt load: Due to a fluctuation in tax revenues, a county government is projecting a deficit for the next 12 months, followed by a quick recovery and the repayment of all debt near the end of this period. The projected debt can be modeled by the polynomial $D(m) = 0.1m^4 - 2m^3 + 15m^2 - 64m - 3$, where $D(m)$ represents the amount of debt (in millions of dollars) in month m. Use the remainder theorem to help answer the following questions.

a. Was the debt higher in month 5 ($m = 5$) or month 10 of this period? How much higher?

b. Was the debt higher in the first month of this period (one month into the deficit) or after the eleventh month (one month before the expected recovery)? How much higher?

c. The total debt reached its maximum between months 7 and 10. Use the remainder theorem to determine which month.

83. Volume of water: The volume of water in a rectangular, in-ground, swimming pool is given by $V(x) = x^3 + 11x^2 + 24x$, where $v(x)$ is the volume in cubic feet when the water is x ft high. (a) Use

the remainder theorem to find the volume when $x = 3$ ft. (b) If the volume is 100 ft^3 of water, what is the height x? (c) If the maximum capacity of the pool is 1000 ft^3, what is the maximum depth (to the nearest integer)?

84. Amusement park attendance: Attendance at an amusement park depends on the weather. After opening in spring, attendance rises quickly, slows during the summer, soars in the fall, then quickly falls with the approach of winter when the park closes. The model for attendance is given by $A(m) = -\frac{1}{4}m^4 + 6m^3 - 52m^2 + 196m - 260$, where $A(m)$ represents the number of people attending in month m (in thousands). (a) Did more people go to the park in April ($m = 4$) or June ($m = 6$)? (b) In what month did maximum attendance occur? (c) When did the park close?

In these applications, synthetic division is applied in the usual way, treating k as an unknown constant.

85. Find a value of k that will make $x = -2$ a zero of $f(x) = x^3 - 3x^2 - 5x + k$.

86. Find a value of k that will make $x - 3$ a factor of $g(x) = x^3 + 2x^2 - 7x + k$.

87. For what value(s) of k will $x - 2$ be a factor of $p(x) = x^3 - 3x^2 + kx + 10$?

88. For what value(s) of k will $x + 5$ be a factor of $q(x) = x^3 + 6x^2 + kx + 50$?

▶ **EXTENDING THE CONCEPT**

89. To investigate whether the remainder and factor theorems can be applied when the coefficients or zeroes of a polynomial are complex, try using the factor theorem to find a polynomial with degree 3, whose zeroes are $x = 2i$, $x = -2i$, and $x = 3$. Then see if the result can be verified using the remainder theorem and these zeroes. What does the result suggest?

90. Since we use a base-10 number system, numbers like 1196 can be written in polynomial form as $p(x) = 1x^3 + 1x^2 + 9x + 6$, where $x = 10$. Divide $p(x)$ by $x + 3$ using synthetic division and write your answer as $\frac{x^3 + x^2 + 9x + 6}{x + 3}$ = quotient + $\frac{remainder}{divisor}$. For $x = 10$, what is the value of quotient + $\frac{remainder}{divisor}$? What is the result of dividing 1196 by $10 + 3 = 13$? What can you conclude?

91. The sum of the first n perfect cubes is given by the formula $S = \frac{1}{4}(n^4 + 2n^3 + n^2)$. Use the remainder theorem on S to find the sum of (a) the first three

perfect cubes (divide by $n - 3$) and (b) the first five perfect cubes (divide by $n - 5$). Check results by adding the perfect cubes manually. To avoid working with fractions you can initially ignore the $\frac{1}{4}$ (use $n^4 + 2n^3 + n^2 + 0n + 0$), as long as you divide the remainder by 4.

92. Though not a direct focus of this course, the remainder and factor theorems, as well as synthetic division, *can also be applied using complex numbers*. Use the remainder theorem to show the value given is a zero of $P(x)$.

a. $P(x) = x^3 - 4x^2 + 9x - 36$; $x = 3i$

b. $P(x) = x^4 + x^3 + 2x^2 + 4x - 8$; $x = -2i$

c. $P(x) = -x^3 + x^2 - 3x - 5$; $x = 1 + 2i$

d. $P(x) = x^3 + 2x^2 + 16x + 32$; $x = -4i$

e. $P(x) = x^4 + x^3 - 5x^2 + x - 6$; $x = i$

f. $P(x) = -x^3 + x^2 - 8x - 10$; $x = 1 + 3i$

MAINTAINING YOUR SKILLS

93. (1.1) John and Rick are out orienteering. Rick finds the last marker first and is heading for the finish line, 1275 yd away. John is just seconds behind, and after locating the last marker tries to overtake Rick, who by now has a 250-yd lead. If Rick runs at 4 yd/sec and John runs at 5 yd/sec, will John catch Rick before they reach the finish line?

94. (1.5) Solve for w: $-2(3w^2 + 5) + 3 = -7w + w^2 - 7$

95. (2.3) The profit of a small business increased linearly from $5000 in 2005 to $12,000 in 2010. Find a linear function $G(t)$ modeling the growth of the company's profit (let $t = 0$ correspond to 2005).

96. (2.7) Given $f(x) = x^2 - 4x$, use the average rate of change formula to find $\frac{\Delta y}{\Delta x}$ in the interval $x \in [1.0, 1.1]$.

3.3 The Zeroes of Polynomial Functions

Learning Objectives

In Section 3.3 you will learn how to:

☐ **A.** Apply the fundamental theorem of algebra and the linear factorization theorem

☐ **B.** Locate zeroes of a polynomial using the intermediate value theorem

☐ **C.** Find rational zeroes of a polynomial using the rational zeroes theorem

☐ **D.** Use Descartes' rule of signs and the upper/lower bounds theorem

☐ **E.** Solve applications of polynomials

This section represents one of the highlights in the college algebra curriculum, because it offers a look at what many call *the big picture*. The ideas presented are the result of a cumulative knowledge base developed over a long period of time, and give a fairly comprehensive view of the study of polynomial functions.

A. The Fundamental Theorem of Algebra

From Section 1.4, we know that real numbers are a subset of the complex numbers: $\mathbb{R} \subset \mathbb{C}$. Because complex numbers are the "larger" set (containing all other number sets), properties and theorems about complex numbers are more powerful and far reaching than theorems about real numbers. In the same way, real polynomials are a subset of the complex polynomials, and the same principle applies.

Complex Polynomial Functions

A complex polynomial of degree n has the form
$$P(x) = a_n x^n + a_{n-1} x^{n-1} + \cdots + a_1 x^1 + a_0,$$
where $a_n, a_{n-1}, \cdots, a_1, a_0$ are complex numbers and $a_n \neq 0$.

Notice that real polynomials have the same form, but here $a_n, a_{n-1}, \ldots, a_1, a_0$ *represent complex numbers*. In 1797, Carl Friedrich Gauss (1777–1855) proved that *all* polynomial functions have zeroes, and that the number of zeroes is equal to the degree of the polynomial. The proof of this statement is based on a theorem that is the bedrock for a complete study of polynomial functions, and has come to be known as the **fundamental theorem of algebra.**

The Fundamental Theorem of Algebra

Every complex polynomial of degree $n \geq 1$ has at least one complex zero.

Although the statement may seem trivial, it allows us to draw two important conclusions. The first is that our search for a solution will not be fruitless or wasted—zeroes for *all* polynomial equations exist. Second, the fundamental theorem combined with the factor theorem allows us to state the **linear factorization theorem.**

WORTHY OF NOTE

Quadratic functions also belong to the larger family of **complex polynomial functions.** Since quadratics have a known number of terms, it is common to write the general form using the early letters of the alphabet: $P(x) = ax^2 + bx + c = 0$. For higher degree polynomials, the number of terms is unknown or unspecified, and the general form is written using subscripts on a single letter.

The Linear Factorization Theorem

If $p(x)$ is a polynomial function of degree $n \geq 1$, then p has exactly n linear factors and can be written in the form,

$$p(x) = a(x - c_1)(x - c_2) \cdot \cdots \cdot (x - c_n)$$

where $a \neq 0$ and $c_1, c_2, \ldots, c_n$ are (not necessarily distinct) complex numbers.

Proof of the Linear Factorization Theorem

Given $p(x) = a_n x^n + a_{n-1} x^{n-1} + \cdots + a_1 x^1 + a_0$ is a complex polynomial, the fundamental theorem of algebra establishes that $p(x)$ has a least one complex zero, call it c_1. The factor theorem stipulates $(x - c_1)$ must be a factor of P, giving

$$p(x) = (x - c_1)q_1(x)$$

where $q_1(x)$ is a complex polynomial of degree $n - 1$.

Since $q_1(x)$ is a complex polynomial in its own right, it too must also have a complex zero, call it c_2. Then $(x - c_2)$ must be a factor of $q_1(x)$, giving

$$p(x) = (x - c_1)(x - c_2)q_2(x)$$

where $q_2(x)$ is a complex polynomial of degree $n - 2$.

Repeating this rationale n times will cause $p(x)$ to be rewritten in the form

$$p(x) = (x - c_1)(x - c_2) \cdot \cdots \cdot (x - c_n)q_n(x)$$

where $q_n(x)$ has a degree of $n - n = 0$, a nonzero constant typically called a_n.

The result is $p(x) = a_n(x - c_1)(x - c_2) \cdot \cdots \cdot (x - c_n)$.

In other words, every complex polynomial of degree n can be rewritten as the product of a nonzero constant and exactly n linear factors.

EXAMPLE 1 ▶ **Writing Polynomials as a Product of Linear Factors**

Rewrite $P(x) = x^4 - 8x^2 - 9$ as a product of linear factors, and find its zeroes.

Solution ▶ From its given form, we know $a = 1$. Since P has degree 4, the factored form must be $P(x) = (x - c_1)(x - c_2)(x - c_3)(x - c_4)$. Noting that P is in quadratic form, we substitute u for x^2 and u^2 for x^4 and attempt to factor:

$$x^4 - 8x^2 - 9 \rightarrow u^2 - 8u - 9 \qquad \text{substitute } u \text{ for } x^2; u^2 \text{ for } x^4$$
$$= (u - 9)(u + 1) \qquad \text{factor in terms of } u$$
$$= (x^2 - 9)(x^2 + 1) \qquad \text{rewrite in terms of } x \text{ (substitute } x^2 \text{ for } u)$$

We know $x^2 - 9$ will factor since it is a difference of squares. From our work with complex numbers (Section 1.4), we know $(a + bi)(a - bi) = a^2 + b^2$, and the factored form of $x^2 + 1$ must be $(x + i)(x - i)$. The completely factored form is

$$P(x) = (x + 3)(x - 3)(x + i)(x - i), \text{ and}$$

the zeroes of P are $-3, 3, -i$, and i.

WORTHY OF NOTE

While polynomials with complex coefficients are not the focus of this course, interested students can investigate the wider application of these theorems by completing **Exercise 115**.

Now try Exercises 7 through 10 ▶

EXAMPLE 2 ▶ **Writing Polynomials as a Product of Linear Factors**

Rewrite $P(x) = x^3 + 2x^2 - 4x - 8$ as a product of linear factors and find its zeroes.

Solution ▶ We observe that $a = 1$ and P has degree 3, so the factored form must be $P(x) = (x - c_1)(x - c_2)(x - c_3)$. Noting that $ad = bc$ (Appendix I.D), we start with factoring by grouping.

$$P(x) = x^3 + 2x^2 - 4x - 8 \qquad \text{group terms (in color)}$$
$$= x^2(x + 2) - 4(x + 2) \qquad \text{remove common factors (note sign change)}$$
$$= (x + 2)(x^2 - 4) \qquad \text{factor common binomial}$$
$$= (x + 2)(x + 2)(x - 2) \qquad \text{factor difference of squares}$$

The zeroes of P are -2, -2, and 2.

Now try Exercises 11 through 14 ▶

Note the polynomial in Example 2 has three zeroes, but the zero -2 was repeated two times. In this case we say -2 is a zero of multiplicity two, and a zero of **even multiplicity.** It is also possible for a zero to be repeated three or more times, with those repeated an odd number of times called zeroes of **odd multiplicity** [the factor $(x - 2) = (x - 2)^1$ also gives a zero of odd multiplicity]. In general, repeated factors are written in exponential form and we have

Zeroes of Multiplicity

If p is a polynomial function with degree $n \geq 1$, and $(x - c)$ occurs as a factor of p exactly m times, then c is a zero of multiplicity m.

EXAMPLE 3 ▶ **Identifying the Multiplicity of a Zero**

Factor the given function completely, writing repeated factors in exponential form. Then state the multiplicity of each zero: $P(x) = (x^2 + 8x + 16)(x^2 - x - 20)(x - 5)$

Solution ▶ $P(x) = (x^2 + 8x + 16)(x^2 - x - 20)(x - 5) \qquad \text{given polynomial}$
$= (x + 4)(x + 4)(x - 5)(x + 4)(x - 5) \qquad \text{trinomial factoring}$
$= (x + 4)^3(x - 5)^2 \qquad \text{exponential form}$

For function P, -4 is a zero of multiplicity 3 (odd multiplicity), and 5 is a zero of multiplicity 2 (even multiplicity).

Now try Exercises 15 through 18 ▶

WORTHY OF NOTE

When reconstructing a polynomial P having complex zeroes, it is often more efficient to determine the irreducible quadratic factors of P separately, as shown here. For the zeroes $2 \pm \sqrt{3}i$ we have

$$x = 2 \pm i\sqrt{3}$$
$$x - 2 = \pm i\sqrt{3}$$
$$(x - 2)^2 = (\pm i\sqrt{3})^2$$
$$x^2 - 4x + 4 = -3$$
$$x^2 - 4x + 7 = 0.$$

The quadratic factor is $(x^2 - 4x + 7)$.

These examples help illustrate three important consequences of the linear factorization theorem. From Example 1, if the coefficients of P are real, the polynomial can be factored into linear and quadratic factors using real numbers only $[(x + 3)(x - 3)(x^2 + 1)]$, where the quadratic factors have no real zeroes. Quadratic factors of this type are said to be **irreducible.**

Corollary I: Irreducible Quadratic Factors

If p is a polynomial with real coefficients, p can be factored into a product of linear factors (which are not necessarily distinct) and irreducible quadratic factors having real coefficients.

Closely related to this corollary and our previous study of quadratic functions, complex zeroes of the irreducible factors must occur in conjugate pairs.

Corollary II: Complex Conjugates

If p is a polynomial with real coefficients, complex zeroes must occur in conjugate pairs. If $a + bi$, $b \neq 0$ is a zero, then $a - bi$ will also be a zero.

Finally, the polynomial in Example 1 has degree 4 with 4 zeroes (two real, two complex), and the polynomial in Example 2 has degree 3 with 3 zeroes (three real, one repeated). While not shown explicitly, the polynomial in Example 3 has degree 5, and there were 5 zeroes (one repeated twice, one repeated three times). This suggests our final corollary.

> **Corollary III: Number of Zeroes**
>
> If p is a polynomial function with degree $n \geq 1$, then p has exactly n zeroes (real or complex), where zeroes of multiplicity m are counted m times.

These corollaries help us gain valuable information about a polynomial, when only partial information is given or known. For a proof of Corollary II, see Appendix V.

EXAMPLE 4 ▶ **Constructing a Polynomial from Its Zeroes**

A polynomial P of degree 3 with real coefficients has zeroes of -1 and $2 + i\sqrt{3}$. Find the polynomial (assume $a = 1$).

Solution ▶ Using the factor theorem, two of the factors are $(x + 1)$ and $x - (2 + i\sqrt{3})$. From Corollary II, $2 - i\sqrt{3}$ must also be a zero and $x - (2 - i\sqrt{3})$ is also a factor of P. This gives

$$
\begin{aligned}
P(x) &= (x + 1)[x - (2 + i\sqrt{3})][x - (2 - i\sqrt{3})] \\
&= (x + 1)[(x - 2) - i\sqrt{3}][(x - 2) + i\sqrt{3}] && \text{associative property} \\
&= (x + 1)[(x^2 - 4x + 4) + 3] && (a + bi)(a - bi) = a^2 + b^2 \\
&= (x + 1)(x^2 - 4x + 7) && \text{simplify} \\
&= x^3 - 3x^2 + 3x + 7 && \text{result}
\end{aligned}
$$

The polynomial is $P(x) = x^3 - 3x^2 + 3x + 7$, which can be verified using the remainder theorem and any of the original zeroes.

Now try Exercises 19 through 22 ▶

EXAMPLE 5 ▶ **Building a Polynomial from Its Zeroes**

Find a fourth degree polynomial P with real coefficients, if 3 is the only real zero and $2i$ is also a zero of P.

Solution ▶ Since complex zeroes must occur in conjugate pairs, $-2i$ is also a zero, but this accounts for only three zeroes. Since P has degree 4, 3 must be a *repeated* zero, and the factors of P are $(x - 3)(x - 3)(x - 2i)(x + 2i)$.

☑ **A.** You've just learned how to apply the fundamental theorem of algebra and the linear factorization theorem

$$
\begin{aligned}
P(x) &= (x - 3)(x - 3)(x - 2i)(x + 2i) && \text{factored form} \\
&= (x^2 - 6x + 9)(x^2 + 4) && \text{multiply binomials, } (a + bi)(a - bi) = a^2 + b^2 \\
&= x^4 - 6x^3 + 13x^2 - 24x + 36 && \text{result}
\end{aligned}
$$

The polynomial is $P(x) = x^4 - 6x^3 + 13x^2 - 24x + 36$, which can be verified using the remainder theorem and any of the original zeroes.

Now try Exercises 23 through 28 ▶

B. Real Polynomials and the Intermediate Value Theorem

The fundamental theorem of algebra is called an **existence theorem,** as it affirms the *existence* of the zeroes but does not tell us where or how to find them. Because polynomial graphs are continuous (there are no holes or breaks in the graph), the **intermediate value theorem (IVT)** can be used for this purpose.

The Intermediate Value Theorem

Given P is a polynomial with real coefficients, if $P(a)$ and $P(b)$ have opposite signs, there is *at least* one value c between a and b such that $P(c) = 0$.

Figure 3.8

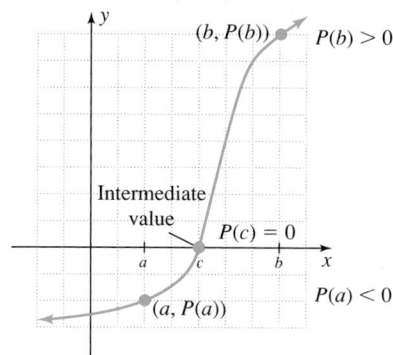

EXAMPLE 6 ▶ **Finding Zeroes Using the Intermediate Value Theorem**

Use the intermediate value theorem to show $P(x) = x^3 - 9x + 6$ has at least one zero in the interval given:

    a. $[-4, -3]$ **b.** $[0, 1]$

Solution ▶   **a.** Begin by evaluating P at $x = -4$ and $x = -3$.

$$P(-4) = (-4)^3 - 9(-4) + 6 \qquad P(-3) = x^3 - 9x + 6$$
$$= -64 + 36 + 6 \qquad\qquad = -27 + 27 + 6$$
$$= -22 \qquad\qquad\qquad = 6$$

Since $P(-4) < 0$ and $P(-3) > 0$, there must be at least one number c_1 between -4 and -3 where $P(c_1) = 0$. The graph must cross the x-axis in this interval.

    b. Evaluate P at $x = 0$ and $x = 1$.

$$P(0) = (0)^3 - 9(0) + 6 \qquad P(1) = (1)^3 - 9(1) + 6$$
$$= 0 - 0 + 6 \qquad\qquad = 1 - 9 + 6$$
$$= 6 \qquad\qquad\qquad = -2$$

☑ **B.** You've just learned how to locate zeroes of a real polynomial function using the intermediate value theorem

Since $P(0) > 0$ and $P(1) < 0$, there must be at least one number c_2 between 0 and 1 where $P(c_2) = 0$.

Now try Exercises 29 through 32 ▶

C. The Rational Zeroes Theorem

The fundamental theorem of algebra tells us that zeroes of a polynomial function *exist*. The intermediate value theorem tells us how to *locate* zeroes within an interval. Our next theorem gives us the information we need to actually *find* certain zeroes of a polynomial. Recall that if c is a zero of P, then $P(c) = 0$, and when $P(x)$ is divided by $x - c$ using synthetic division, the remainder is zero (from the remainder and factor theorems).

    To find *divisors that give a remainder of zero,* we make the following observations. To solve $3x^2 - 11x - 20 = 0$ by factoring, a beginner might write out all possible binomial pairs where the **F**irst term in the F-O-I-L process multiplies to $3x^2$ and the **L**ast term multiplies to 20. The six possibilities are shown here:

$$(3x \quad 1)(x \quad 20) \qquad (3x \quad 20)(x \quad 1) \qquad (3x \quad 2)(x \quad 10) \qquad (3x \quad 10)(x \quad 2)$$

$$(3x \quad 4)(x \quad 5) \qquad\qquad (3x \quad 5)(x \quad 4)$$

If $3x^2 - 11x - 20$ is factorable using integers, the factors *must be somewhere in this list*. Also, the first coefficient in each binomial must be a factor of the leading coefficient, and the second coefficient must be a factor of the constant term. This means that regardless of which factored form is correct, the solution will be a rational number whose numerator comes from the factors of 20, and whose denominator comes from the factors of 3. The correct factored form is shown here, along with the solution:

$$3x^2 - 11x - 20 = 0$$
$$(3x + 4)(x - 5) = 0$$
$$3x + 4 = 0 \qquad\qquad x - 5 = 0$$

$$x = \frac{-4}{3} \quad \begin{array}{l} \leftarrow \text{from the factors of 20} \\ \leftarrow \text{from the factors of 3} \end{array} \qquad\qquad x = \frac{5}{1} \quad \begin{array}{l} \leftarrow \text{from the factors of 20} \\ \leftarrow \text{from the factors of 3} \end{array}$$

This same principle also applies to polynomials of higher degree, and these observations suggest the following theorem.

The Rational Zeroes Theorem

Given polynomial P with integer coefficients, and $\frac{p}{q}$ a rational number in lowest terms, the rational zeroes of P (if they exist) must be of the form $\frac{p}{q}$, where p is a factor of the constant term, and q is a factor of the leading coefficient.

Note that if the leading coefficient is 1, the possible rational zeroes are limited to factors of the constant term: $\frac{p}{1} = p$. If the leading coefficient is not "1" and the constant term has a large number of factors, the set of possible rational zeroes becomes rather large. To list these possibilities, it helps to begin with all factor *pairs* of the constant a_0, then divide each of these by the factors of a_n as shown in Example 7.

EXAMPLE 7 ▶ **Identifying the Possible Rational Zeroes of a Polynomial**

List all possible rational zeroes for $3x^4 + 14x^3 - x^2 - 42x - 24 = 0$, but do not solve.

Solution ▶ All rational zeroes must be of the form $\frac{p}{q}$, where p is a factor of $a_0 = -24$ and q is a factor of $a_n = 3$. The factor pairs of -24 are: ± 1, ± 24, ± 2, ± 12, ± 3, ± 8, ± 4 and ± 6. Dividing each by ± 1 and ± 3 (the factor pairs of 3), we note division by ± 1 will not change any of the previous values, while division by ± 3 gives $\pm\frac{1}{3}$, $\pm\frac{2}{3}$, $\pm\frac{8}{3}$, $\pm\frac{4}{3}$ as additional possibilities. Any rational zeroes must be in the set $\{\pm 1, \pm 24, \pm 2, \pm 12, \pm 3, \pm 8, \pm 4, \pm 6, \pm\frac{1}{3}, \pm\frac{2}{3}, \pm\frac{8}{3}, \pm\frac{4}{3}\}$.

Now try Exercises 33 through 40 ▶

WORTHY OF NOTE

To test for -1, only the sign of terms with odd degree must be changed, since $(-1)^{\text{even\#}} = 1$, while $(-1)^{\text{odd\#}} = -1$. The method simply gives a shortcut for evaluating $P(1)$ and $P(-1)$, which often helps to break down a higher degree polynomial.

The actual solutions to the equation in Example 7 are $x = \sqrt{3}$, $x = -\sqrt{3}$, $x = -\frac{2}{3}$, and $x = -4$. Although the *rational* zeroes are indeed in the set noted, it's apparent we need a way to narrow down the number of possibilities (we don't want to try all 24 possible zeroes). If we're able to find even one factor easily, we can rewrite the polynomial using this factor and the quotient polynomial, with the hope of factoring further using trinomial factoring or factoring by grouping. Many times testing to see if 1 or -1 are zeroes will help.

Tests to Determine If 1 or −1 is a Zero of *P*

For any polynomial P with real coefficients,

1. *If the sum of all coefficients is zero, then 1 is a root and $(x - 1)$ is a factor.*
2. *After changing the sign of all terms with odd degree, if the sum of the coefficients is zero, then -1 is a root and $(x + 1)$ is a factor.*

EXAMPLE 8 ▶ **Finding the Rational Zeroes of a Polynomial**

Find all rational zeroes of $P(x) = 3x^4 - x^3 - 8x^2 + 2x + 4$, and use them to write the function in completely factored form. Then use the factored form to name all zeroes of P.

Solution ▶ Instead of listing all possibilities using the rational zeroes theorem, we first test for 1 and -1, then see if we're able to complete the factorization using other means. The sum of the coefficients is: $3 - 1 - 8 + 2 + 4 = 0$, which means 1 is a zero and $x - 1$ is a factor. By changing the sign on terms of odd degree, we have $3x^4 + x^3 - 8x^2 - 2x + 4$ and $3 + 1 - 8 - 2 + 4 = -2$, showing -1 is *not* a zero. Using $x = 1$ and the factor theorem, we have

$$\text{use 1 as a "divisor"}\quad 1\underline{|}\ \begin{array}{rrrrr} 3 & -1 & -8 & 2 & 4 \\ & 3 & 2 & -6 & -4 \\ \hline 3 & 2 & -6 & -4 & \underline{|0} \end{array}$$

and we write P as $P(x) = (x - 1)(\mathbf{3x^3 + 2x^2 - 6x - 4})$. Noting the quotient polynomial can be factored by grouping ($ad = bc$), we need not continue with synthetic division or the factor theorem.

$$\begin{array}{ll} P(x) = (x - 1)(\underline{3x^3 + 2x^2} - \underline{6x - 4}) & \text{group terms} \\ = (x - 1)[x^2(3x + 2) - 2(3x + 2)] & \text{factor common terms} \\ = (x - 1)(3x + 2)(x^2 - 2) & \text{factor common binomial} \\ = (x - 1)(3x + 2)(x + \sqrt{2})(x - \sqrt{2}) & \text{completely factored form} \end{array}$$

The zeroes of P are 1, $\frac{-2}{3}$, and $\pm\sqrt{2}$.

WORTHY OF NOTE

In the second to last line of Example 8, we factored $x^2 - 2$ as $(x + \sqrt{2})(x - \sqrt{2})$. As discussed in Appendix I.D, this is an application of factoring the difference of two squares: $a^2 - b^2 = (a + b)(a - b)$. By mentally rewriting $x^2 - 2$ as $x^2 - (\sqrt{2})^2$, we obtain the result shown. Also **see Exercise 113.**

Now try Exercises 41 through 62 ▶

In cases where the quotient polynomial is not easily factored, we continue with synthetic division and other possible zeroes, until the remaining zeroes can be determined.

EXAMPLE 9 ▶ **Finding the Zeroes of a Polynomial**

Find all zeroes of $P(x) = x^5 - 3x^4 + 3x^3 - 5x^2 + 12$.

Solution ▶ Using the rational zeroes theorem, the possibilities are: $\{\pm1, \pm12, \pm2, \pm6, \pm3, \pm4\}$. The test for 1 shows 1 is not a zero. After changing the signs of all terms with odd degree, we have $-1 - 3 - 3 - 5 + 12 = 0$, and find -1 *is* a zero. Using -1 with the factor theorem, we continue our search for additional factors. Noting that P is missing a linear term, we include a place-holder zero:

$$\text{use } -1 \text{ as a "divisor"}\quad -1\underline{|}\ \begin{array}{rrrrrr} 1 & -3 & 3 & -5 & 0 & 12 \\ & -1 & 4 & -7 & 12 & -12 \\ \hline 1 & -4 & 7 & -12 & 12 & \underline{|0} \end{array}\ \begin{array}{l}\text{coefficients of } P \\ \\ \text{coefficients of } q_1(x)\end{array}$$

Here the quotient polynomial $q_1(x) = x^4 - 4x^3 + 7x^2 - 12x + 12$ is not easily factored, so we next try 2, *using the quotient polynomial:*

$$\text{use 2 as a "divisor" on } \boldsymbol{q_1(x)}\quad 2\underline{|}\ \begin{array}{rrrrr} 1 & -4 & 7 & -12 & 12 \\ & 2 & -4 & 6 & -12 \\ \hline 1 & -2 & 3 & -6 & \underline{|0} \end{array}\ \begin{array}{l}\text{coefficients of } q_1(x) \\ \\ \text{coefficients of } q_2(x)\end{array}$$

If you miss the fact that $q_2(x)$ is actually factorable ($ad = bc$), the process would continue using -2 and the current quotient.

$$\text{use } -2 \text{ as a "divisor"}\quad -2\underline{|}\ \begin{array}{rrrr} 1 & -2 & 3 & -6 \\ & -2 & 8 & -22 \\ \hline 1 & -4 & 11 & \underline{|-28} \end{array}\ \begin{array}{l}\text{coefficients of } q_2(x) \\ \\ -2 \text{ is not a zero}\end{array}$$

We find -2 is not a zero, and in fact, trying *all other possible zeroes* will show that *none* of them are zeroes. As there must be five zeroes, we are reminded of three things:

1. This process can only find *rational zeros* (the remaining zeroes may be irrational or complex),
2. This process cannot find irreducible quadratic factors (unless they appear as the quotient polynomial), and
3. Some of the zeroes *may have multiplicities greater than 1!*

Testing the zero 2 for a second time using $q_2(x)$ gives

$$\begin{array}{r} \text{use 2 as a ``divisor''} \quad \underline{2\,\rfloor}\ \ 1 \quad -2 \quad\ \ 3 \quad -6 \qquad \text{coefficients of } q_2(x) \\ \ \ \ 2 \quad\ \ 0 \quad\ \ \ 6 \\ \hline 1 \quad\ \ \ \ 0 \quad\ \ \ 3 \quad\ \ |0 \qquad \text{2 is a } \textit{repeated} \text{ zero} \end{array}$$

✅ **C.** You've just learned how to find rational zeroes of a real polynomial function using the rational zeroes theorem

and we see that 2 is actually a zero of multiplicity two, and the final quotient is the irreducible quadratic factor $x^2 + 3$. Using this information produces the factored form $P(x) = (x + 1)(x - 2)^2(x^2 + 3) = (x + 1)(x - 2)^2(x + i\sqrt{3})(x - i\sqrt{3})$, and the zeroes of P are $-i\sqrt{3}, i\sqrt{3}, -1$, and 2 with multiplicity two.

Now try Exercises 63 through 82 ▶

D. Descartes' Rule of Signs and Upper/Lower Bounds

Testing $x = 1$ and $x = -1$ is one way to reduce the number of possible rational zeroes, but unless we're very lucky, factoring the polynomial can still be a challenge. **Descartes' rule of signs** and the **upper and lower bounds property** offer additional assistance.

Descartes' Rule of Signs

Given the real polynomial equation $P(x) = 0$,

1. The number of positive real zeroes is equal to the number of variations in sign for $P(x)$, or an even number less.
2. The number of negative real zeroes is equal to the number of variations in sign for $P(-x)$, or an even number less.

EXAMPLE 10 ▶ **Finding the Zeroes of a Polynomial**

For $P(x) = 2x^5 - 5x^4 + x^3 + x^2 - x + 6$,

    a. Use the rational zeroes theorem to list all possible rational zeroes.

    b. Apply Descartes' rule to count the number of possible positive, negative, and complex roots.

    c. Use this information and the tools of this section to find all zeroes of P.

Solution ▶ **a.** The factors of 2 are $\{\pm 1, \pm 2\}$ and the factors of 6 are $\{\pm 1, \pm 6, \pm 2, \pm 3\}$. The possible rational zeroes for P are $\{\pm 1, \pm 6, \pm 2, \pm 3, \pm\frac{1}{2}, \pm\frac{3}{2}\}$.

    b. For Descartes' rule, we organize our work in a table. Since P has degree 5, there must be a total of five zeroes. For this illustration, positive terms are in **blue** and negative terms in **red**: $P(x) = 2x^5 - 5x^4 + x^3 + x^2 - x + 6$. The terms change sign a total of four times, meaning there are four, two, or zero positive roots. For the negative roots, recall that $P(-x)$ will change the sign of *all odd-degree terms,* giving $P(-x) = -2x^5 - 5x^4 - x^3 + x^2 + x + 6$. This time there is only one sign change (from negative to positive) showing there is exactly one negative root, a fact that is highlighted in the following table.

possible positive zeroes	known negative zeroes	possibilities for complex roots	total number *must be 5*
4	1	0	5
2	1	2	5
0	1	4	5

c. Testing 1 and -1 shows $x = 1$ is not a root, but $x = -1$ *is,* and using -1 in synthetic division gives:

use -1 as a "divisor" $\underline{-1|}$ 2 -5 1   1 -1 6   coefficients of $P(x)$

-2 7 -8 7 -6

2   -7 8 -7 6   0 $q_1(x)$ is not easily factored

Since there is *only one* negative root, we need only check the remaining positive zeroes. The quotient $q_1(x)$ is not easily factored, so we continue with synthetic division using the next larger positive root, $x = 2$.

use 2 as a "divisor" $\underline{2|}$ 2 -7 8 -7 6   coefficients of $q_1(x)$

4   -6 4 -6

2   -3 2 -3 0 $q_2(x)$ is easily factored

The partially factored form is $P(x) = (x + 1)(x - 2)(2x^3 - 3x^2 + 2x - 3)$, which we can complete using factoring by grouping. The factored form is

$$
\begin{aligned}
P(x) &= (x + 1)(x - 2)(\underline{2x^3 - 3x^2} + \underline{2x - 3}) && \text{group terms} \\
&= (x + 1)(x - 2)[x^2(2x - 3) + 1(2x - 3)] && \text{factor common terms} \\
&= (x + 1)(x - 2)(2x - 3)(x^2 + 1) && \text{factor out common binomial} \\
&= (x + 1)(x - 2)(2x - 3)(x + i)(x - i) && \text{completely factored form}
\end{aligned}
$$

The zeroes of P are $-1, 2, \frac{3}{2}, -i$ and i, with two positive, one negative, and two complex zeroes.

Now try Exercises 83 through 96 ▶

One final idea that helps reduce the number of possible zeroes is the **upper and lower bounds property**. A number b is an **upper bound** on the positive zeroes of a function if no positive zero is greater than b. In the same way, a number a is a **lower bound** on the negative zeroes if no negative zero is less than a.

Upper and Lower Bounds Property

Given $P(x)$ is a polynomial with real coefficients.

1. If $P(x)$ is divided by $x - b$ ($b > 0$) using synthetic division and all coefficients in the quotient row are either positive or zero, then b is an upper bound on the zeroes of P.

2. If $P(x)$ is divided by $x - a$ ($a < 0$) using synthetic division and all coefficients in the quotient row alternate in sign, then a is a lower bound on the zeroes of P.

For both 1 and 2, zero coefficients can be either positive or negative as needed.

☑ **D.** You just learned how to gain more information on the zeroes of real polynomials using Descartes' rule of signs and upper/lower bounds

While this test certainly helps narrow the possibilities, we gain the additional benefit of knowing the property actually places boundaries on *all* real zeroes of the polynomial, both rational and irrational. In Part (c) of Example 10, the quotient row of the first division alternates in sign, showing $x = -1$ is both a zero and a lower bound on the real zeroes of P. For more on the upper and lower bounds property, **see Exercise 111.**

E. Applications of Polynomial Functions

Polynomial functions can be very accurate models of real-world phenomena, though we often must restrict their domain, as illustrated in Example 11.

EXAMPLE 11 ▶ **Using the Remainder Theorem to Solve an Oceanography Application**

As part of an environmental study, scientists use radar to map the ocean floor from the coastline to a distance 12 mi from shore. In this study, ocean trenches appear as negative values and underwater mountains as positive values, as measured from the surrounding ocean floor. The terrain due west of a particular island can be modeled by $h(x) = x^4 - 25x^3 + 200x^2 - 560x + 384$, where $h(x)$ represents the height in feet, x mi from shore ($0 < x \le 12$).

  a. Use the remainder theorem to find the "height of the ocean floor" 10 mi out.

  b. Use the tools developed in this section to find the number of times the ocean floor has height $h(x) = 0$ in this interval, given this occurs 12 mi out.

Solution ▶ **a.** For part (a) we simply evaluate $h(10)$ using the remainder theorem.

use 10 as a "divisor" $10\rfloor$ 1 -25 200 -560 384   coefficients of $h(x)$

                                 10   -150 500 -600

                         1   -15 50 -60 $\lfloor-216$ remainder is -216

Ten miles from shore, there is an ocean trench 216 ft deep.

  b. For part (b), we know 12 is zero, so we again use the remainder theorem and work with the quotient polynomial.

use 12 as a "divisor" $12\rfloor$ 1 -25 200 -560 384   coefficients of $h(x)$

                                 12   -156 528 -384

                         1   -13 44 -32 $\lfloor 0$ $q_1(x)$

The quotient is $q_1(x) = x^3 - 13x^2 + 44x - 32$. Since $a = 1$, we know the remaining zeroes must be factors of -32: $\{\pm1, \pm32, \pm2, \pm16, \pm4, \pm8\}$. Using $x = 1$ gives

use 1 as a "divisor" $1\rfloor$ 1 -13 44 -32 coefficients of $q_1(x)$

                                 1   -12 32

                         1   -12 32 $\lfloor 0$ $q_2(x)$

☑ **E.** You've just learned how to solve an application of polynomial functions

The function can now be written as $h(x) = (x - 12)(x - 1)(x^2 - 12x + 32)$ and in completely factored form $h(x) = (x - 12)(x - 1)(x - 4)(x - 8)$. The ocean floor has height zero at distances of 1, 4, 8, and 12 mi from shore.

Now try Exercises 99 through 110 ▶

GRAPHICAL SUPPORT

The graph of $h(x)$ is shown here using a window size of X ∈ [0, 13] and Y ∈ [−450, 450]. The graphs shows a great deal of variation in the ocean floor, but the zeroes occurring at 1, 4, 8, and 12 mi out are clearly evident.

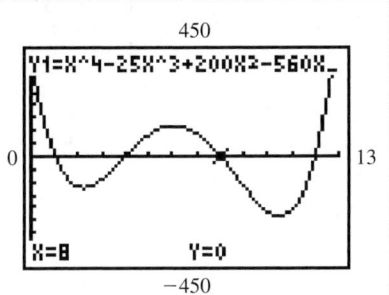

TECHNOLOGY HIGHLIGHT

The Intermediate Value Theorem and Split Screen Viewing

Graphical support for the results of Example 6 is shown in Figure 3.9 using the window $x \in [-5, 5]$ and $y \in [-10, 20]$. The zero of P between 0 and 1 is highlighted, and the zero between $x = -4$ and $x = -3$ is clearly seen. Note there is also a third zero between 2 and 3.

The TI 84 Plus (and other models) offer a useful feature called *split screen viewing,* that enables us to view a table of values and the graph of a function at the same time. To illustrate, enter the function $y = x^3 - 9x + 6$ for Y₁ on the [Y=] screen. Press the [ZOOM] **4:ZDecimal** keys to view the graph, then adjust the viewing window as needed to get a comprehensive view. Set up your table in **AUTO** mode with ΔTbl = 1 [use [2nd] [WINDOW] (TBLSET)]. Use the table of values ([2nd] [GRAPH]) to locate any real zeroes of f [look for where $f(x)$ changes in sign]. To support this concept we can view *both the graph and table at the same time.* Press the [MODE] key and notice the second-to-last entry on this screen reads: **Full** (for full screen viewing), **Horiz** for splitting the screen horizontally with the graph above a reduced home screen, and **G-T**, which represents **Graph-Table** and splits the screen vertically. In the **G-T** mode, the graph appears on the left and the table of values on the right. Navigate the cursor to the **G-T** mode and press [ENTER]. Pressing the [GRAPH] key at this point should give you a screen similar to Figure 3.10. Use this feature to complete the following exercises.

Figure 3.9

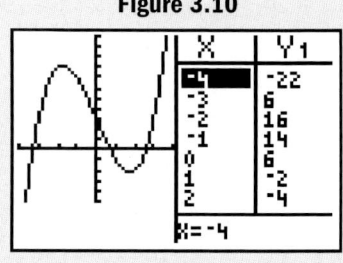

Figure 3.10

Exercise 1: What do the graph, table, and the IVT tell you about the zeroes of this function?

Exercise 2: Go to TBLSET and reset TblStart = −4 and ΔTbl = 0.1. Use [2nd] [GRAPH] to walk through the table values. Does this give you a better idea about where the zeroes are located?

Exercise 3: Press the [TRACE] key. What happens to the table as you trace through the points on Y₁?

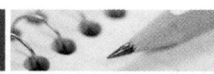

3.3 EXERCISES

▶ **CONCEPTS AND VOCABULARY**

Fill in each blank with the appropriate word or phrase. Carefully reread the section if needed.

1. A complex polynomial is one where one or more _____ are complex numbers.

2. A polynomial function of degree n will have exactly _____ zeroes, real or _____, where zeroes of multiplicity m are counted m times.

3. If $a + bi$ is a complex zero of polynomial P with real coefficients, then _____ is also a zero.

4. According to Descartes' rule of signs, there are as many _____ real roots as changes in sign from term to term, or an _____ number less.

5. Which of the following values is *not* a possible root of $f(x) = 6x^3 - 2x^2 + 5x - 12$:
   a. $x = \frac{4}{3}$ **b.** $x = \frac{3}{4}$ **c.** $x = \frac{1}{2}$

   Discuss/Explain why.

6. Discuss/Explain each of the following:
   (a) irreducible quadratic factors, (b) factors that are complex conjugates, (c) zeroes of multiplicity m, and (d) upper bounds on the zeroes of a polynomial.

▶ DEVELOPING YOUR SKILLS

Rewrite each polynomial as a product of linear factors, and find the zeroes of the polynomial.

7. $P(x) = x^4 + 5x^2 - 36$

8. $Q(x) = x^4 + 21x^2 - 100$

9. $Q(x) = x^4 - 16$

10. $P(x) = x^4 - 81$

11. $P(x) = x^3 + x^2 - x - 1$

12. $Q(x) = x^3 - 3x^2 - 9x + 27$

13. $Q(x) = x^3 - 5x^2 - 25x + 125$

14. $P(x) = x^3 + 4x^2 - 16x - 64$

Factor each polynomial completely. Write any repeated factors in exponential form, then name all zeroes and their multiplicity.

15. $p(x) = (x^2 - 10x + 25)(x^2 + 4x - 45)(x + 9)$

16. $q(x) = (x^2 + 12x + 36)(x^2 + 2x - 24)(x - 4)$

17. $P(x) = (x^2 - 5x - 14)(x^2 - 49)(x + 2)$

18. $Q(x) = (x^2 - 9x + 18)(x^2 - 36)(x - 3)$

Find a polynomial $P(x)$ having real coefficients, with the degree and zeroes indicated. Assume the lead coefficient is 1. Recall $(a + bi)(a - bi) = a^2 + b^2$.

19. degree 3, $x = 3, x = 2i$

20. degree 3, $x = -5, x = -3i$

21. degree 4, $x = -1, x = 2, x = i$

22. degree 4, $x = -1, x = 3, x = -2i$

23. degree 4, $x = 3, x = 2i$

24. degree 4, $x = -2, x = -3i$

25. degree 4, $x = -1, x = 1 + 2i$

26. degree 4, $x = -1, x = 1 - 3i$

27. degree 4, $x = -3, x = 1 + i\sqrt{2}$

28. degree 4, $x = -2, x = 1 + i\sqrt{3}$

Use the intermediate value theorem to verify the given polynomial has at least one zero "c_i" in the intervals specified. Do not find the zeroes.

29. $f(x) = x^3 + 2x^2 - 8x - 5$

    a. $[-4, -3]$ **b.** $[2, 3]$

30. $g(x) = x^4 - 2x^2 + 6x - 3$

    a. $[-3, -2]$ **b.** $[0, 1]$

31. $h(x) = 2x^3 + 13x^2 + 3x - 36$

    a. $[1, 2]$ **b.** $[-3, -2]$

32. $H(x) = 2x^4 + 3x^3 - 14x^2 - 9x + 8$

    a. $[-4, -3]$ **b.** $[-2, -1]$

List all possible rational zeroes for the polynomials given, but do not solve.

33. $f(x) = 4x^3 - 19x - 15$

34. $g(x) = 3x^3 - 2x + 20$

35. $h(x) = 2x^3 - 5x^2 - 28x + 15$

36. $H(x) = 2x^3 - 19x^2 + 37x - 14$

37. $p(x) = 6x^4 - 2x^3 + 5x^2 - 28$

38. $q(x) = 7x^4 + 6x^3 - 49x^2 + 36$

39. $Y_1 = 32t^3 - 52t^2 + 17t + 3$

40. $Y_2 = 24t^3 + 17t^2 - 13t - 6$

Use the rational zeroes theorem to write each function in factored form and find all zeroes. Note $a = 1$.

41. $f(x) = x^3 - 13x + 12$

42. $g(x) = x^3 - 21x + 20$

43. $h(x) = x^3 - 19x - 30$

44. $H(x) = x^3 - 28x - 48$

45. $p(x) = x^3 - 2x^2 - 11x + 12$

46. $q(x) = x^3 - 4x^2 - 7x + 10$

47. $Y_1 = x^3 - 6x^2 - x + 30$

48. $Y_2 = x^3 - 4x^2 - 20x + 48$

49. $Y_3 = x^4 - 15x^2 + 10x + 24$

50. $Y_4 = x^4 - 23x^2 - 18x + 40$

51. $f(x) = x^4 + 7x^3 - 7x^2 - 55x - 42$

52. $g(x) = x^4 + 4x^3 - 17x^2 - 24x + 36$

Find all rational zeroes of the functions given and use them to write the function in factored form. Use the factored form to state *all* zeroes of f. Begin by applying the tests for 1 and −1.

53. $f(x) = 4x^3 - 7x + 3$

54. $g(x) = 9x^3 - 7x - 2$

55. $h(x) = 4x^3 + 8x^2 - 3x - 9$

56. $H(x) = 9x^3 + 3x^2 - 8x - 4$

57. $Y_1 = 2x^3 - 3x^2 - 9x + 10$

58. $Y_2 = 3x^3 - 14x^2 + 17x - 6$

59. $p(x) = 2x^4 + 3x^3 - 9x^2 - 15x - 5$

60. $q(x) = 3x^4 + x^3 - 11x^2 - 3x + 6$

61. $r(x) = 3x^4 - 5x^3 + 14x^2 - 20x + 8$

62. $s(x) = 2x^4 - x^3 + 17x^2 - 9x - 9$

Find the zeroes of the polynomials given using any combination of the rational zeroes theorem, testing for 1 and −1, and/or the remainder and factor theorems.

63. $f(x) = 2x^4 - 9x^3 + 4x^2 + 21x - 18$

64. $g(x) = 3x^4 + 4x^3 - 21x^2 - 10x + 24$

65. $h(x) = 3x^4 + 2x^3 - 9x^2 + 4$

66. $H(x) = 7x^4 + 6x^3 - 49x^2 + 36$

67. $p(x) = 2x^4 + 3x^3 - 24x^2 - 68x - 48$

68. $q(x) = 3x^4 - 19x^3 + 6x^2 + 96x - 32$

69. $r(x) = 3x^4 - 20x^3 + 34x^2 + 12x - 45$

70. $s(x) = 4x^4 - 15x^3 + 9x^2 + 16x - 12$

71. $Y_1 = x^5 + 6x^2 - 49x + 42$

72. $Y_2 = x^5 + 2x^2 - 9x + 6$

73. $P(x) = 3x^5 + x^4 + x^3 + 7x^2 - 24x + 12$

74. $P(x) = 2x^5 - x^4 - 3x^3 + 4x^2 - 14x + 12$

75. $Y_1 = x^4 - 5x^3 + 20x - 16$

76. $Y_2 = x^4 - 10x^3 + 90x - 81$

77. $r(x) = x^4 + 2x^3 - 5x^2 - 4x + 6$

78. $s(x) = x^4 + x^3 - 5x^2 - 3x + 6$

79. $p(x) = 2x^4 - x^3 + 3x^2 - 3x - 9$

80. $q(x) = 3x^4 + x^3 + 13x^2 + 5x - 10$

81. $f(x) = 2x^5 - 7x^4 + 13x^3 - 23x^2 + 21x - 6$

82. $g(x) = 4x^5 + 3x^4 + 3x^3 + 11x^2 - 27x + 6$

Gather information on each polynomial using (a) the rational zeroes theorem, (b) testing for 1 and −1, (c) applying Descartes' rule of signs, and (d) using the upper and lower bounds property. Respond explicitly to each.

83. $f(x) = x^4 - 2x^3 + 4x - 8$

84. $g(x) = x^4 + 3x^3 - 7x - 6$

85. $h(x) = x^5 + x^4 - 3x^3 + 5x + 2$

86. $H(x) = x^5 + x^4 - 2x^3 + 4x - 4$

87. $p(x) = x^5 - 3x^4 + 3x^3 - 9x^2 - 4x + 12$

88. $q(x) = x^5 - 2x^4 - 8x^3 + 16x^2 + 7x - 14$

89. $r(x) = 2x^4 + 7x^2 + 11x - 20$

90. $s(x) = 3x^4 - 8x^3 - 13x - 24$

⊞ **Use Descartes' rule of signs to determine the possible combinations of real and complex zeroes for each polynomial. Then graph the function on the standard window of a graphing calculator and adjust it as needed until you're certain all real zeroes are in clear view. Use this screen and a list of the possible rational zeroes to factor the polynomial and find all zeroes (real and complex).**

91. $f(x) = 4x^3 - 16x^2 - 9x + 36$

92. $g(x) = 6x^3 - 41x^2 + 26x + 24$

93. $h(x) = 6x^3 - 73x^2 + 10x + 24$

94. $H(x) = 4x^3 + 60x^2 + 53x - 42$

95. $p(x) = 4x^4 + 40x^3 - 97x^2 - 10x + 24$

96. $q(x) = 4x^4 - 42x^3 - 70x^2 - 21x - 36$

▶ **WORKING WITH FORMULAS**

97. **The absolute value of a complex number**
$z = a + bi$: $|z| = \sqrt{a^2 + b^2}$

The absolute value of a complex number z, denoted $|z|$, represents the distance between the origin and the point (a, b) in the complex plane. Use the formula to find $|z|$ for the complex numbers given (also see Section 1.4, Exercise 69): (a) $3 + 4i$, (b) $-5 + 12i$, and (c) $1 + \sqrt{3}\,i$.

98. **The square root of $z = a + bi$:**
$\sqrt{z} = \frac{\sqrt{2}}{2}\left(\sqrt{|z| + a} \pm i\sqrt{|z| - a}\right)$

The square roots of a complex number are given by the relations shown, where $|z|$ represents the absolute value of z and the sign is chosen to match the sign of b. Use the formula to find the square root of each complex number from Exercise 97, then check your answer by squaring the result (also see Section 1.4, Exercise 82).

▶ **APPLICATIONS**

99. Maximum and minimum values: To locate the maximum and minimum values of $F(x) = x^4 - 4x^3 - 12x^2 + 32x + 15$ requires finding the zeroes of $f(x) = 4x^3 - 12x^2 - 24x + 32$. Use the rational zeroes theorem and synthetic division to find the zeroes of f, then graph $F(x)$ on a calculator and see if the graph tends to support your calculations—do the maximum and minimum values occur at the zeroes of f?

100. Graphical analysis: Use the rational zeroes theorem and synthetic division to find the zeroes of $F(x) = x^4 - 4x^3 - 12x^2 + 32x + 15$ (see Exercise 99).

101. Maximum and minimum values: To locate the maximum and minimum values of $G(x) = x^4 - 6x^3 + x^2 + 24x - 20$ requires finding the zeroes of $g(x) = 4x^3 - 18x^2 + 2x + 24$. Use the rational zeroes theorem and synthetic division to find the zeroes of g, then graph $G(x)$ on a calculator and see if the graph tends to support your calculations—do the maximum and minimum values occur at the zeroes of g?

102. Graphical analysis: Use the rational zeroes theorem and synthetic division to find the zeroes of $G(x) = x^4 - 6x^3 + x^2 + 24x - 20$ (see Exercise 101).

Geometry: The volume of a cube is $V = x \cdot x \cdot x = x^3$, where x represents the length of the edges. If a slice 1 unit thick is removed from the cube, the remaining volume is $v = x \cdot x \cdot (x - 1) = x^3 - x^2$. Use this information for Exercises 103 and 104.

103. A slice 1 unit in thickness is removed from one side of a cube. Use the rational zeroes theorem and synthetic division to find the original dimensions of the cube, if the remaining volume is (a) 48 cm³ and (b) 100 cm³.

104. A slice 1 unit in thickness is removed from one side of a cube, then a second slice of the same thickness is removed from a different side (not the opposite side). Use the rational zeroes theorem and synthetic division to find the original dimensions of the cube, if the remaining volume is (a) 36 cm³ and (b) 80 cm³.

Geometry: The volume of a rectangular box is $V = LWH$. For the box to satisfy certain requirements, its length must be twice the width, and its height must be two inches less than the width. Use this information for Exercises 105 and 106.

105. Use the rational zeroes theorem and synthetic division to find the dimensions of the box if it must have a volume of 150 in³.

106. Suppose the box must have a volume of 64 in³. Use the rational zeroes theorem and synthetic division to find the dimensions required.

Government deficits: Over a 14-yr period, the balance of payments (deficit versus surplus) for a certain county government was modeled by the function $f(x) = \frac{1}{4}x^4 - 6x^3 + 42x^2 - 72x - 64$, where $x = 0$ corresponds to 1990 and $f(x)$ is the deficit or surplus in tens of thousands of dollars. Use this information for Exercises 107 and 108.

107. Use the rational zeroes theorem and synthetic division to find the years when the county "broke even" (debt = surplus = 0) from 1990 to 2004. How many years did the county run a surplus during this period?

108. The deficit was at the $84,000 level $[f(x) = -84]$, four times from 1990 to 2004. Given this occurred in 1992 and 2000 ($x = 2$ and $x = 10$), use the rational zeroes theorem, synthetic division, and the remainder theorem to find the other two years the deficit was at $84,000.

109. Drag resistance on a boat: In a scientific study on the effects of drag against the hull of a sculling boat, some of the factors to consider are displacement, draft, speed, hull shape, and length, among others. If the first four are held constant and we assume a flat, calm water surface, length becomes the sole variable (as length changes, we adjust the beam by a uniform scaling to keep a constant displacement). For a fixed sculling speed of 5.5 knots, the relationship between drag and length can be modeled by $f(x) = -0.4192x^4 + 18.9663x^3 - 319.9714x^2 + 2384.2x - 6615.8$, where $f(x)$ is the efficiency rating of a boat with length x ($8.7 < x < 13.6$). Here, $f(x) = 0$ represents an *average* efficiency rating. (a) Under these conditions, what lengths (to the nearest hundredth) will give the boat an average rating? (b) What length will maximize the efficiency of the boat? What is this rating?

110. Comparing densities: Why is it that when you throw a rock into a lake, it sinks, while a wooden ball will float half submerged, but the bobber on your fishing line floats on the surface? It all depends on the density of the object compared to the density of water ($d = 1$). For uniformity, we'll consider spherical objects of various densities, each with a radius of 5 cm. When placed into water, the depth that the sphere will sink beneath the surface (while still floating) is modeled by the polynomial $p(x) = \frac{\pi}{3}x^3 - 5\pi x^2 + \frac{500\pi}{3}d$, where d is the density of the object and the smallest positive zero of p is the depth of the sphere below the surface (in centimeters). How far submerged is the sphere if it's made of (a) balsa wood, $d = 0.17$;

(b) pine wood, $d = 0.55$; (c) ebony wood, $d = 1.12$; (d) a large bobber made of lightweight plastic, $d = 0.05$?

▶ **EXTENDING THE CONCEPT**

111. In the figure, $P(x) = 0.02x^3 - 0.24x^2 - 1.04x + 2.68$ is graphed on the standard screen ($-10 \leq x \leq 10$), which shows two real zeroes. Since P has degree 3, there must be one more real zero but is it negative or positive? Use the upper/lower bounds property (a) to see if -10 is a lower bound and (b) to see if 10 is an upper bound. (c) Then use your calculator to find the remaining zero.

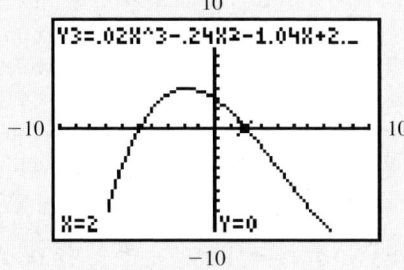

112. From Example 11, (a) what is the significance of the y-intercept? (b) If the domain were extended to include $0 < x \leq 13$, what happens when x is approximately 12.8?

113A. It is often said that while the difference of two squares is factorable, $a^2 - b^2 = (a + b)(a - b)$, the sum of two squares is prime. To be 100% correct, we should say the sum of two squares cannot be factored *using real numbers*. If complex numbers are used, $(a^2 + b^2) = (a + bi)(a - bi)$. Use this idea to factor the following binomials.

a. $p(x) = x^2 + 25$ **b.** $q(x) = x^2 + 9$
c. $r(x) = x^2 + 7$

113B. It is often said that while $x^2 - 16$ is factorable as a difference of squares, $a^2 - b^2 = (a + b)(a - b)$, $x^2 - 17$ is not. To be 100% correct, we should say that $x^2 - 17$ is not factorable *using integers*. Since $(\sqrt{17})^2 = 17$, it can actually be factored in the same way: $x^2 - 17 = (x + \sqrt{17})(x - \sqrt{17})$. Use this idea to solve the following equations.

a. $x^2 - 7 = 0$ **b.** $x^2 - 12 = 0$ **c.** $x^2 - 18 = 0$

114. Every general cubic equation $aw^3 + bw^2 + cw + d = 0$ can be written in the form $x^3 + px + q = 0$ (where the squared term has been "depressed"), using the transformation $w = x - \dfrac{b}{3}$. Use this transformation to solve the following equations.

a. $w^3 - 3w^2 + 6w - 4 = 0$
b. $w^3 - 6w^2 + 21w - 26 = 0$

Note: It is actually very rare that the transformation produces a value of $q = 0$ for the "depressed" cubic $x^3 + px + q = 0$, and general solutions must be found using what has become known as *Cardano's formula*. For a complete treatment of cubic equations and their solutions, visit our website at www.mhhe.com/coburn. Here we'll focus on the primary root of selected cubics.

115. For each of the following complex polynomials, one of its zeroes is given. Use this zero to help write the polynomial in completely factored form. (*Hint:* Synthetic division and the quadratic formula can be applied to *all polynomials,* even those with complex coefficients.)

a. $C(z) = z^3 + (1 - 4i)z^2 + (-6 - 4i)z + 24i;$
$z = 4i$

b. $C(z) = z^3 + (5 - 9i)z^2 + (4 - 45i)z - 36i;$
$z = 9i$

c. $C(z) = z^3 + (-2 - 3i)z^2 + (5 + 6i)z - 15i;$
$z = 3i$

d. $C(z) = z^3 + (-4 - i)z^2 + (29 + 4i)z - 29i;$
$z = i$

e. $C(z) = z^3 + (-2 - 6i)z^2 + (4 + 12i)z - 24i;$
$z = 6i$

f. $C(z) = z^3 + (-6 + 4i)z^2 + (11 - 24i)z + 44i;$
$z = -4i$

g. $C(z) = z^3 + (-2 - i)z^2 + (5 + 4i)z + (-6 + 3i);$
$z = 2 - i$

h. $C(z) = z^3 - 2z^2 + (19 + 6i)z + (-20 + 30i);$
$z = 2 - 3i$

▶ **MAINTAINING YOUR SKILLS**

116. (2.6) Graph the piecewise-defined function and find the value of $f(-3), f(2)$, and $f(5)$.

$$f(x) = \begin{cases} 2 & x \le -1 \\ |x - 1| & -1 < x < 5 \\ 4 & x \ge 5 \end{cases}$$

117. (3.1) For a county fair, officials need to fence off a large rectangular area, then subdivide it into three equal (rectangular) areas. If the county provides 1200 ft of fencing, (a) what dimensions will maximize the area of the larger (outer) rectangle? (b) What is the area of each smaller rectangle?

118. (2.7) Use the graph given to (a) state intervals where $f(x) \ge 0$, (b) locate local maximum and minimum values, and (c) state intervals where $f(x)\uparrow$ and $f(x)\downarrow$.

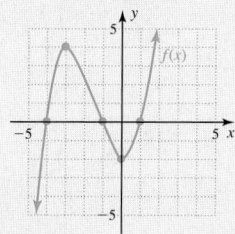

119. (2.5) Write the equation of the function shown.

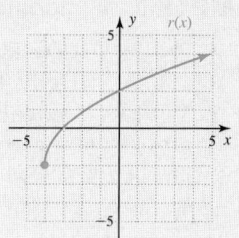

3.4 Graphing Polynomial Functions

Learning Objectives

In Section 3.4 you will learn how to:

☐ **A.** Identify the graph of a polynomial function and determine its degree

☐ **B.** Describe the end behavior of a polynomial graph

☐ **C.** Discuss the attributes of a polynomial graph with zeroes of multiplicity

☐ **D.** Graph polynomial functions in standard form

☐ **E.** Solve applications of polynomials

As with linear and quadratic functions, understanding graphs of *polynomial* functions will help us apply them more effectively as mathematical models. Since all real polynomials can be written in terms of their linear and quadratic factors (Section 3.3), these functions provide the basis for our continuing study.

A. Identifying the Graph of a Polynomial Function

Consider the graphs of $f(x) = x + 2$ and $g(x) = (x - 1)^2$, which we know are smooth, continuous curves. The graph of f is a straight line with positive slope, that crosses the x-axis at -2. The graph of g is a parabola, opening upward, shifted 1 unit to the right, and touching the x-axis at $x = 1$. When f and g are "combined" into the single function $P(x) = (x + 2)(x - 1)^2$, the behavior of the graph at these zeroes is still evident. In Figure 3.11, the graph of P crosses the x-axis at $x = -2$, "bounces" off the x-axis at $x = 1$, and is still a smooth, continuous curve. This observation could be

Figure 3.11

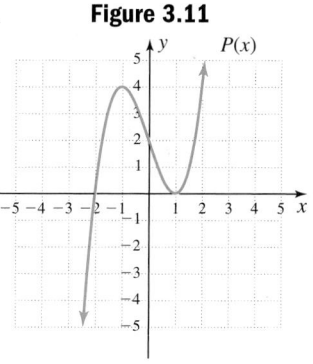

WORTHY OF NOTE

While defined more precisely in a future course, we will take "smooth" to mean the graph has no sharp turns or jagged edges, and "continuous" to mean the entire graph can be drawn without lifting your pencil.

extended to include additional linear or quadratic factors, and helps affirm that the graph of a polynomial function is a *smooth, continuous curve.*

Further, after the graph of P crosses the axis at $x = -2$, it must "turn around" at some point to reach the zero at $x = 1$, then turn again as it touches the x-axis without crossing. By combining this observation with our work in Section 3.3, we can state the following:

Polynomial Graphs and Turning Points

1. If $P(x)$ is a polynomial function of degree n, then the graph of P has at most $n - 1$ turning points.
2. If the graph of a function P has $n - 1$ turning points, then the degree of $P(x)$ is at least n.

EXAMPLE 1 ▶ **Identifying Polynomial Graphs**

Determine whether each graph could be the graph of a polynomial. If not, discuss why. If so, use the number of turning points and zeroes to identify the least possible degree of the function.

a.

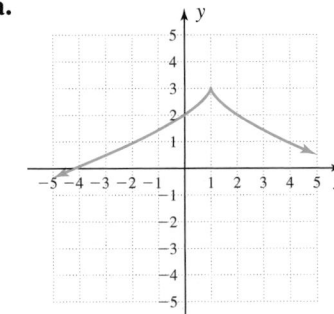

b.

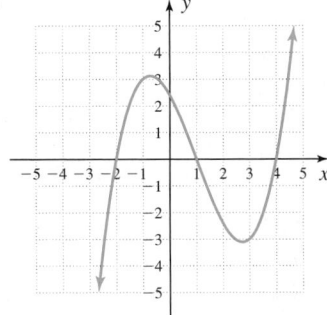

c.

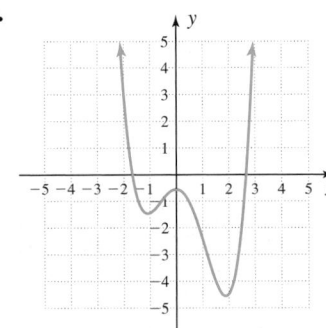

d.

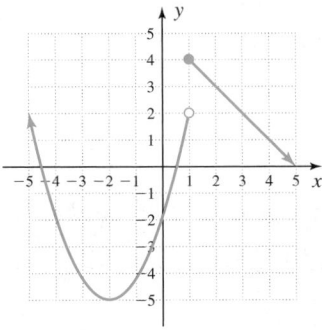

Solution ▶ **a.** This is not a polynomial graph, as it has a sharp turn (called a **cusp**) at $(1, 3)$. A polynomial graph is always smooth.

b. This graph is smooth and continuous, and could be that of a polynomial. With two turning points and three zeroes, the function is at least degree 3.

c. This graph is smooth and continuous, and could be that of a polynomial. With three turning points and two zeroes, the function is at least degree 4.

d. This is not a polynomial graph, as it has a break (discontinuity) at $x = 1$. A polynomial graph is always continuous.

☑ **A.** You've just learned how to identify the graph of a polynomial function and determine its degree

Now try Exercises 7 through 12 ▶

B. The End Behavior of a Polynomial Graph

Once the graph of a function has "made its last turn" and crossed or touched its last real zero, it will continue to increase or decrease without bound as $|x|$ becomes large. As before, we refer to this as the **end behavior** of the graph. In previous sections we noted that quadratic functions (degree 2) with a positive leading coefficient ($a > 0$), had the end behavior "up on the left" and "up on the right (up/up)." If the leading coefficient was negative ($a < 0$), end behavior was "down on the left" and "down on the right (down/down)." These descriptions were also applied to the graph of a linear function $y = mx + b$ (degree 1). A positive leading coefficient ($m > 0$) indicates the graph will be down on the left, up on the right (down/up), and so on. All polynomial graphs exhibit some form of end behavior, which can be likewise described.

EXAMPLE 2 ▶ **Identifying the End Behavior of a Graph**

State the end behavior of each graph shown:

    a. $f(x) = x^3 - 4x + 1$ **b.** $g(x) = -2x^5 + 7x^3 - 4x$ **c.** $h(x) = -2x^4 + 5x^2 + x - 1$

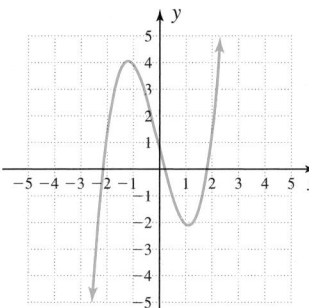 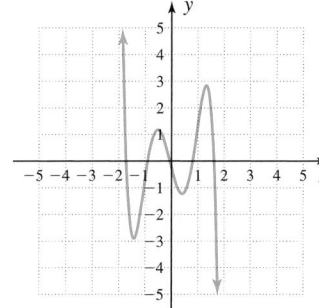 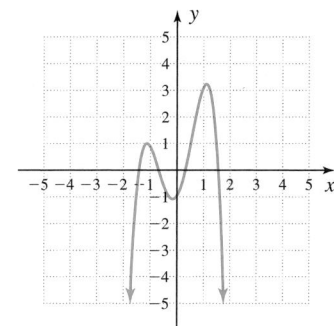

Solution ▶   **a.** down on the left, up on the right

           b. up on the left, down on the right

           c. down on the left, down on the right

Now try Exercises 13 through 16 ▶

WORTHY OF NOTE

As a visual aid to end behavior, it might help to picture a signalman using semaphore code as illustrated here. As you view the end behavior of a polynomial graph, there is a striking resemblance.

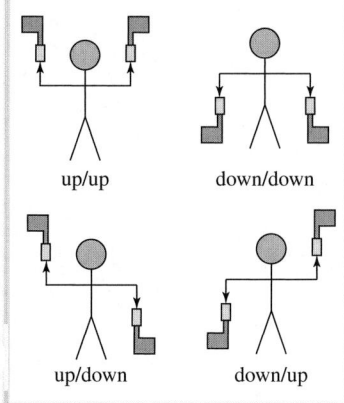

   up/up   down/down

  up/down   down/up

The leading term ax^n of a polynomial function is said to be the **dominant term,** because for large values of $|x|$, the value of ax^n is much larger than all other terms combined. This means that like linear and quadratic graphs, polynomial end behavior can be predicted in advance by analyzing this term alone.

1. For ax^n when n is even, any nonzero number raised to an even power is positive, so the ends of the graph must point in the same direction. If $a > 0$, both point upward. If $a < 0$, both point downward.
2. For ax^n when n is odd, any number raised to an odd power has the same sign as the input value, so the ends of the graph must point in opposite directions. If $a > 0$, end behavior is down on the left, up on the right. If $a < 0$, end behavior is up on the left, down on the right.

From this we find that end behavior depends on two things: *the degree of the function* (even or odd) and the *sign of the leading coefficient* (positive or negative). In more formal terms, this is described in terms of how the graph "behaves" for large values of x. For end behavior that is "up on the right," we mean that as x becomes a large positive number, y becomes a large positive number. This is indicated using the notation: as $x \to \infty$, $y \to \infty$. Similar notation is used for the other possibilities. These facts are summarized in Table 3.1. The interior portion of each graph is dashed since the actual number of turning points may vary, although a polynomial of odd degree will have an even number of turning points, and a polynomial of even degree will have an odd number of turning points.

Table 3.1
Polynomial End Behavior

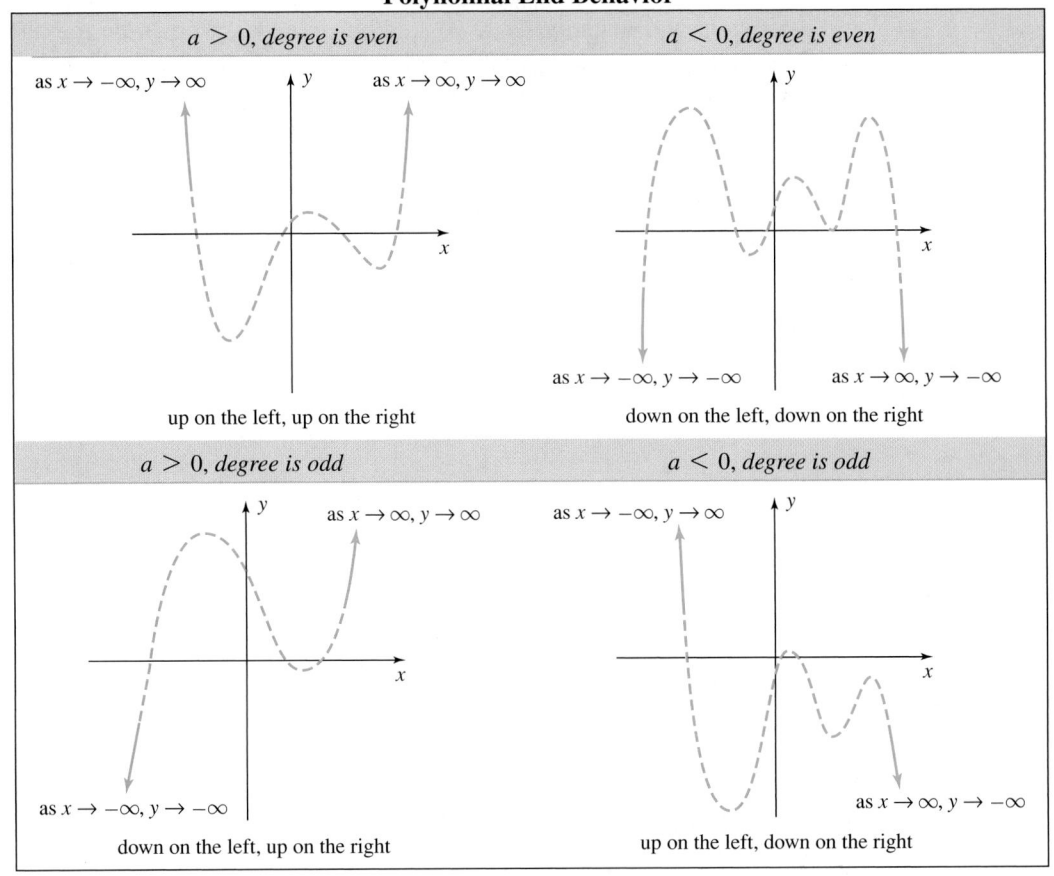

Note the end behavior of $y = mx$ can be used as a representative of all odd degree functions, and the end behavior of $y = ax^2$ as a representative of all even degree functions.

The End Behavior of a Polynomial Graph

Given a polynomial $P(x)$ with leading term ax^n and $n \geq 1$.

If n is **even,** ends will point in the **same direction,**
1. for $a > 0$: up on the left, up on the right (*as with* $y = x^2$);
$$\text{as } x \to -\infty, y \to \infty; \qquad \text{as } x \to \infty, y \to \infty$$
2. for $a < 0$: down on the left, down on the right (*as with* $y = -x^2$);
$$\text{as } x \to -\infty, y \to {}^-\infty; \qquad \text{as } x \to \infty, y \to {}^-\infty$$

If n is **odd,** the ends will point in **opposite directions,**
1. for $a > 0$: down on the left, up on the right (*as with* $y = x$);
$$\text{as } x \to -\infty, y \to {}^-\infty; \qquad \text{as } x \to \infty, y \to \infty$$
2. for $a < 0$: up on the left, down on the right (*as with* $y = -x$);
$$\text{as } x \to -\infty, y \to \infty; \qquad \text{as } x \to \infty, y \to {}^-\infty$$

EXAMPLE 3 ▶ Identifying the End Behavior of a Function

State the end behavior of each function, without actually graphing.

a. $f(x) = 0.5x^4 + 3x^3 - 5x + 6$ **b.** $g(x) = -2x^5 - 5x^3 - 3$

Solution ▶   **a.** The function has degree 4 (even), and the ends will point in the same direction. The leading coefficient is positive, so end behavior is up/up.

b. The function has degree 5 (odd), and the ends will point in opposite directions. The leading coefficient is negative, so the end behavior is up/down.

☑ **B.** You've just learned how to describe the end behavior of a polynomial graph

Now try Exercises 17 through 22 ▶

C. Attributes of Polynomial Graphs with Zeroes of Multiplicity

Another important aspect of polynomial functions is the behavior of a graph near its zeroes. In the simplest case, consider the functions $f(x) = x$ and $g(x) = x^3$. Both have odd degree, like end behavior (down/up), and a zero at $x = 0$. But the zero of f has multiplicity 1, while the zero from g has multiplicity 3. Notice the graph of g is vertically compressed near $x = 0$ and seems to approach this zero "more gradually."

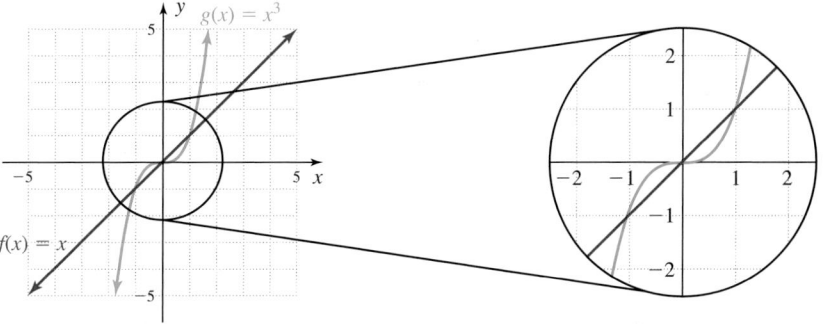

This behavior can be explained by noting that for $x = -1$ and 1, $f(x) = g(x)$. But for $|x| < 1$, the graph of g *will be closer to the x-axis* since the cube of a fractional number is smaller than the fraction itself. We further note that for $|x| > 1$, g increases much faster than f, and $|g(x)| > |f(x)|$. Similar observations can be made regarding $f(x) = x^2$ and $g(x) = x^4$. Both functions have even degree, a zero at $x = 0$, and $f(x) = g(x)$ for $x = -1$ and 1. But for $|x| < 1$, the function with higher degree is once again closer to the x-axis.

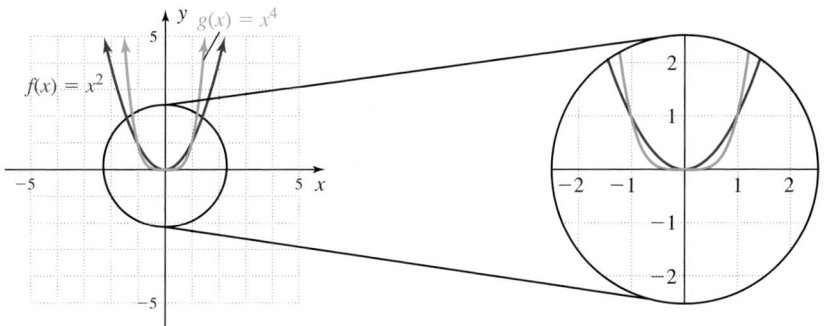

These observations can be generalized and applied to all real zeroes of a function.

Polynomial Graphs and Zeroes of Multiplicity

Given $P(x)$ is a polynomial with factors of the form $(x - c)^m$, with c a real number,
- If m is odd, the graph will cross through the x-axis.
- If m is even, the graph will bounce off the x-axis (touching at just one point).
In each case, the graph will be more compressed (flatter) near c for larger values of m.

To illustrate, compare the graph of $P(x) = (x + 2)(x - 1)^2$ from page 257, with the graph of $p(x) = (x + 2)^3(x - 1)^4$ shown, noting the increased multiplicity of each zero.

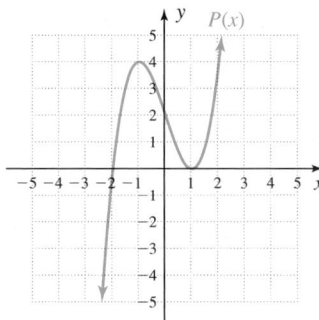 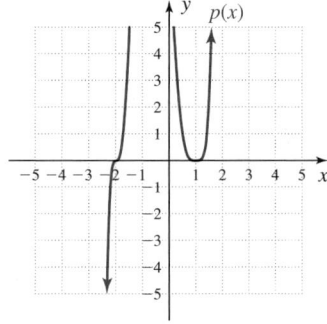

Both graphs show the expected zeroes at $x = -2$ and $x = 1$, but the graph of $p(x)$ is flatter near $x = -2$ and $x = 1$, due to the increased multiplicity of each zero. We lose sight of the graph of $p(x)$ between $x = -2$ and $x = 0$, since the increased multiplicities produce larger values than the original grid could display.

EXAMPLE 4 ▶ **Naming Attributes of a Function from Its Graph**

The graph of a polynomial $f(x)$ is shown.
  a. State whether the degree of f is even or odd.
  b. Use the graph to name the zeroes of f, then state whether their multiplicity is even or odd.
  c. State the minimum possible degree of f.
  d. State the domain and range of f.

Solution ▶ **a.** Since the ends of the graph point in opposite directions, the degree of the function must be odd.

  b. The graph crosses the x-axis at $x = -3$ and is compressed near -3, meaning it must have odd multiplicity with $m > 1$. The graph bounces off the x-axis at $x = 2$ and 2 must be a zero of even multiplicity.

  c. The minimum possible degree of f is 5, as in $f(x) = a(x - 2)^2(x + 3)^3$.

  d. $x \in \mathbb{R}, y \in \mathbb{R}$.

Now try Exercises 23 through 28 ▶

To find the degree of a polynomial from its factored form, add the exponents on all linear factors, then add 2 for each irreducible quadratic factor (the degree of any quadratic factor is 2). The sum gives the degree of the polynomial, from which end behavior can be determined. To find the y-intercept, substitute 0 for x as before, noting this is equivalent to applying the exponent to the constant from each factor.

EXAMPLE 5 ▶ **Naming Attributes of a Function from Its Factored Form**

State the degree of each function, then describe the end behavior and name the y-intercept of each graph.
  a. $f(x) = (x + 2)^3(x - 3)$ **b.** $g(x) = -(x + 1)^2(x^2 + 3)(x - 6)$

Solution ▶ **a.** The degree of f is $3 + 1 = 4$. With even degree and positive leading coefficient, end behavior is up/up. For $f(0) = (2)^3(-3) = -24$, the y-intercept is $(0, -24)$.

  b. The degree of g is $2 + 2 + 1 = 5$. With odd degree and negative leading coefficient, end behavior is up/down. For $g(0) = -1(1)^2(3)(-6) = 18$, the y-intercept is $(0, 18)$.

Now try Exercises 29 through 36 ▶

EXAMPLE 6 ▶ Matching Graphs to Functions Using Zeroes of Multiplicity

The following functions all have zeroes at $x = -2, -1$, and 1. Match each function to the corresponding graph *using its degree and the multiplicity of each zero.*

a. $y = (x + 2)(x + 1)^2(x - 1)^3$ **b.** $y = (x + 2)(x + 1)(x - 1)^3$

c. $y = (x + 2)^2(x + 1)^2(x - 1)^3$ **d.** $y = (x + 2)^2(x + 1)(x - 1)^3$

Solution ▶ The functions in Figures 3.12 and 3.14 must have even degree due to end behavior, so each corresponds to (a) or (d). At $x = -1$ the graph in Figure 3.12 "crosses," while the graph in Figure 3.14 "bounces." This indicates Figure 3.12 matches equation (d), while Figure 3.14 matches equation (a).

The graphs in Figures 3.13 and 3.15 must have odd degree due to end behavior, so each corresponds to (b) or (c). Here, one graph "bounces" at $x = -2$, while the other "crosses." The graph in Figure 3.13 matches equation (c), the graph in Figure 3.15 matches equation (b).

Figure 3.12

Figure 3.13

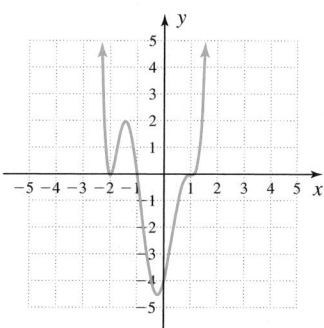

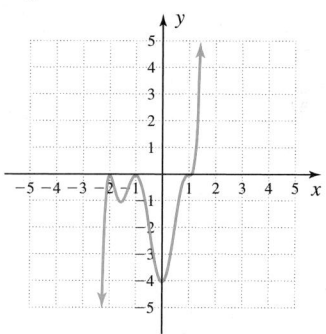

Figure 3.14

Figure 3.15

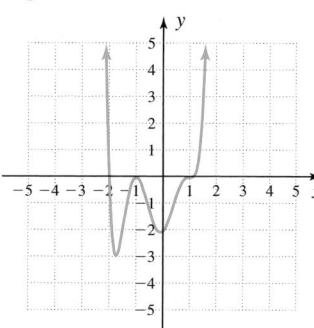

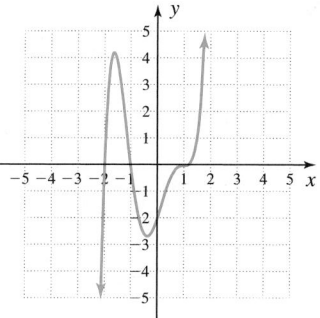

Now try Exercises 37 through 42 ▶

Using the ideas from Examples 5 and 6, we're able to draw a fairly accurate graph given the factored form of a polynomial. Convenient values between two zeroes, called **mid-interval points,** can be used to help complete the graph.

EXAMPLE 7 ▶ Graphing a Function Given the Factored Form

Sketch the graph of $f(x) = (x - 2)(x - 1)^2(x + 1)^3$ using end behavior; the x- and y-intercepts, and zeroes of multiplicity.

Solution ▶ Adding the exponents of each factor, we find that f is a function of degree 6 with a positive lead coefficient, so end behavior will be up/up. Since $f(0) = -2$, the y-intercept is $(0, -2)$. The graph will bounce off the x-axis at $x = 1$ (even multiplicity), and cross the axis at $x = -1$ and 2 (odd multiplicities). The graph will "flatten out" near $x = -1$ because of its higher multiplicity. To help "round-out" the graph we evaluate f at $x = 1.5$, giving $(-0.5)^2(0.5)^3(2.5) \approx -1.95$ (note scaling of the x- and y-axes).

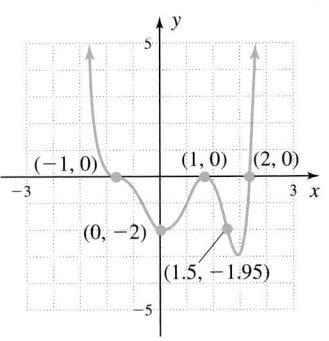

☑ **C. You've just learned how to discuss the attributes of a polynomial graph with zeroes of multiplicity**

> **Now try Exercises 43 through 56** ▶

D. The Graph of a Polynomial Function

Using the cumulative observations from this and previous sections, a general strategy emerges for the graphing of polynomial functions.

Guidelines for Graphing Polynomial Functions

1. Determine the end behavior of the graph.
2. Find the y-intercept $(0, a_0)$
3. Find the zeroes using any combination of the rational zeroes theorem, the factor and remainder theorems, tests for 1 and -1 (p. 310), factoring, and the quadratic formula.
4. Use the y-intercept, end behavior, the multiplicity of each zero, and midinterval points as needed to sketch a smooth, continuous curve.

    Additional tools include (a) polynomial zeroes theorem, (b) complex conjugates theorem, (c) number of turning points, (d) Descartes' rule of signs, (e) upper and lower bounds, and (f) symmetry.

EXAMPLE 8 ▶ **Graphing a Polynomial Function**

Sketch the graph of $g(x) = -x^4 + 9x^2 - 4x - 12$.

Solution ▶
1. End behavior: The function has degree 4 (even) with a negative leading coefficient, so end behavior is *down on the left, down on the right.*
2. Since $g(0) = -12$, the y-intercept is $(0, -12)$.
3. Zeroes: Using the test for $x = 1$ gives $-1 + 9 - 4 - 12 = -8$, showing $x = 1$ is not a zero but $(1, -8)$ is a point on the graph. Using the test for $x = -1$ gives $-1 + 9 + 4 - 12 = 0$, so -1 is a zero and $(x + 1)$ is a factor. Using $x = -1$ with the factor theorem yields

$$
\begin{array}{r|rrrrr}
-1 & -1 & 0 & 9 & -4 & -12 \\
   &   & 1 & -1 & -8 & 12 \\
\hline
   & -1 & 1 & 8 & -12 & 0
\end{array}
$$

The quotient polynomial is not easily factorable so we continue with synthetic division. Using the rational zeroes theorem, the possible rational zeroes are $\{\pm 1, \pm 12, \pm 2, \pm 6, \pm 3, \pm 4\}$, so we try $x = 2$.

use 2 as a "divisor" on
the quotient polynomial

$$
\begin{array}{r|rrrr}
2 & -1 & 1 & 8 & -12 \\
  &   & -2 & -2 & 12 \\
\hline
  & -1 & -1 & 6 & 0
\end{array}
$$

This shows $x = 2$ is a zero, $x - 2$ is a factor, and the function can now be written as

$$g(x) = (x + 1)(x - 2)(-x^2 - x + 6).$$

Factoring -1 from the trinomial gives

$$\begin{aligned} g(x) &= -1(x + 1)(x - 2)(x^2 + x - 6) \\ &= -1(x + 1)(x - 2)(x + 3)(x - 2) \\ &= -1(x + 1)(x - 2)^2(x + 3) \end{aligned}$$

The zeroes of g are $x = -1$ and -3, both with multiplicity 1, and $x = 2$ with multiplicity 2.

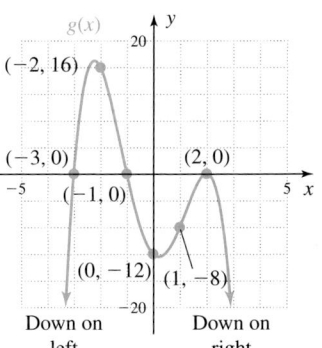

4. To help "round-out" the graph we evaluate the midinterval point $x = -2$ using the remainder theorem, which shows that $(-2, 16)$ is also a point on the graph.

$$\text{use } -2 \text{ as a "divisor"} \quad \underline{-2|} \quad \begin{array}{rrrrr} -1 & 0 & 9 & -4 & -12 \\ & 2 & -4 & -10 & 28 \\ \hline -1 & 2 & 5 & -1 & \underline{|16} \end{array}$$

The final result is the graph shown.

> **Now try Exercises 57 through 72 ▶**

⚠ **CAUTION** ▶ Sometimes using a midinterval point to help draw a graph will give the illusion that a maximum or minimum value has been located. This is rarely the case, as demonstrated in the figure in Example 8, where the maximum value in Quadrant II is actually closer to $(-2.22, 16.95)$.

EXAMPLE 9 ▶ **Using the Guidelines to Sketch a Polynomial Graph**

Sketch the graph of $h(x) = x^7 - 4x^6 + 7x^5 - 12x^4 + 12x^3$.

Solution ▶ 1. End behavior: The function has degree 7 (odd) and the ends will point in opposite directions. The leading coefficient is positive and the end behavior will be *down on the left* and *up on the right*.

2. y-intercept: Since $h(0) = 0$, the y-intercept is $(0, 0)$.

3. Zeroes: Testing 1 and -1 shows neither are zeroes but $(1, 4)$ and $(-1, -36)$ are points on the graph. Factoring out x^3 produces $h(x) = x^3(x^4 - 4x^3 + 7x^2 - 12x + 12)$, and we see that $x = 0$ is a zero of multiplicity 3. We next use synthetic division with $x = 2$ on the fourth-degree polynomial:

$$\text{use 2 as a "divisor"} \quad \underline{2|} \quad \begin{array}{rrrrr} 1 & -4 & 7 & -12 & 12 \\ & 2 & -4 & 6 & -12 \\ \hline 1 & -2 & 3 & -6 & \underline{|0} \end{array}$$

This shows $x = 2$ is a zero and $x - 2$ is a factor. At this stage, it appears the quotient can be factored by grouping. From $h(x) = x^3(x - 2)(x^3 - 2x^2 + 3x - 6)$, we obtain $h(x) = x^3(x - 2)(x^2 + 3)(x - 2)$ after factoring and

$$h(x) = x^3(x - 2)^2(x^2 + 3)$$

as the completely factored form. We find that $x = 2$ is a zero of multiplicity 2, and the remaining two zeroes are complex.

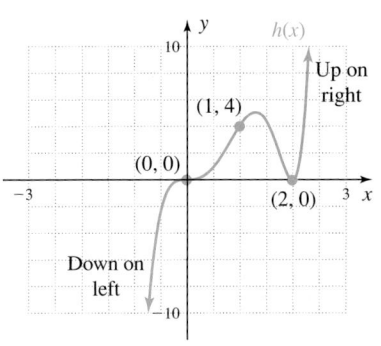

4. Using this information produces the graph shown in the figure.

> **Now try Exercises 73 through 76 ▶**

☑ **D. You've just learned how to graph polynomial functions in standard form**

For practice with these ideas using a graphing calculator, **see Exercises 77 through 80.** Similar to our work in previous sections, **Exercises 81 and 82** ask you to reconstruct the complete equation of a polynomial from its given graph.

E. Applications of Polynomials

EXAMPLE 10 ▶ **Modeling the Value of an Investment**

In the year 2000, Marc and his wife Maria decided to invest some money in precious metals. As expected, the value of the investment fluctuated over the years, sometimes being worth more than they paid, other times less. Through 2008, the value of the investment was modeled by $v(t) = t^4 - 11t^3 + 38t^2 - 40t$, where $v(t)$ represents the gain or loss (in hundreds of dollars) in year t ($t = 0 \rightarrow 2000$).

 a. Use the rational zeroes theorem to find the years when their gain/loss was zero.
 b. Sketch the graph of the function.
 c. In what years was the investment worth less than they paid?
 d. What was their gain or loss in 2008?

Solution ▶ a. Writing the function as $v(t) = t(t^3 - 11t^2 + 38t - 40)$, we note $t = 0$ shows no gain or loss on purchase, and attempt to find the remaining zeroes. Testing for 1 and -1 shows neither is a zero, but $(1, -12)$ and $(-1, 90)$ are points on the graph of v. Next we try $t = 2$ with the factor theorem and the cubic polynomial.

$$\underline{2|} \quad \begin{array}{rrrr} 1 & -11 & 38 & -40 \\ & 2 & -18 & 40 \\ \hline 1 & -9 & 2 & \underline{|0} \end{array}$$

We find that 2 is a zero and write $v(t) = t(t - 2)(t^2 - 9t + 20)$, then factor to obtain $v(t) = t(t - 2)(t - 4)(t - 5)$. Since $v(t) = 0$ for $t = 0, 2, 4,$ and 5, they "broke even" in years 2000, 2002, 2004, and 2005.

 b. With even degree and a positive leading coefficient, the end behavior is up/up. All zeroes have multiplicity 1. As an additional midinterval point we find $v(3) = 6$:

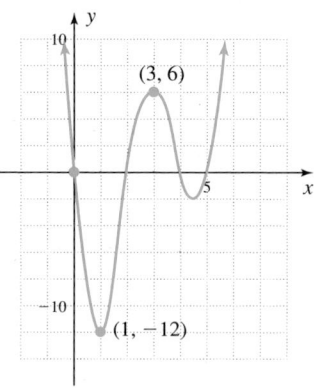

$$\underline{3|} \quad \begin{array}{rrrrr} 1 & -11 & 38 & -40 & 0 \\ & 3 & -24 & 42 & 6 \\ \hline 1 & -8 & 14 & 2 & \underline{|6} \end{array}$$

The complete graph is shown.

 c. The investment was worth less than what they paid (outputs are negative) from 2000 to 2002 and 2004 to 2005.

 d. In 2008, they were "sitting pretty," as their investment had gained $576.

$$\underline{8|} \quad \begin{array}{rrrrr} 1 & -11 & 38 & -40 & 0 \\ & 8 & -24 & 112 & 576 \\ \hline 1 & -3 & 14 & 72 & \underline{|576} \end{array}$$

WORTHY OF NOTE

Due to the context, the domain of $v(t)$ in Example 10 actually begins at $x = 0$, which we could designate with a point at $(0, 0)$. In addition, note there are three sign changes in the terms of $v(t)$, indicating there will be 3 or 1 positive roots (we found 3).

☑ **E. You've just learned how to solve an application of polynomials**

Now try Exercises 85 through 88 ▶

3.4 EXERCISES

▶ CONCEPTS AND VOCABULARY

Fill in each blank with the appropriate word or phrase. Carefully reread the section if needed.

1. For a polynomial with factors of the form $(x - c)^m$, c is called a _____ of multiplicity _____.

2. A polynomial function of degree n has _____ zeroes and at most _____ "turning points."

3. The graphs of $Y_1 = (x - 2)^2$ and $Y_2 = (x - 2)^4$ both _____ at $x = 2$, but the graph of Y_2 is _____ than the graph of Y_1 at this point.

4. Since $x^4 > 0$ for all x, the ends of its graph will always point in the _____ direction. Since

$x^3 > 0$ when $x > 0$ and $x^3 < 0$ when $x < 0$, the ends of its graph will always point in the _____ direction.

5. In your own words, explain/discuss how to find the degree and y-intercept of a function that is given in factored form. Use $f(x) = (x + 1)^3(x - 2)(x + 4)^2$ to illustrate.

6. Name all of the "tools" at your disposal that play a role in the graphing of polynomial functions. Which tools are indispensable and always used? Which tools are used only as the situation merits?

▶ DEVELOPING YOUR SKILLS

Determine whether each graph is the graph of a polynomial function. If yes, state the least possible degree of the function. If no, state why.

7.

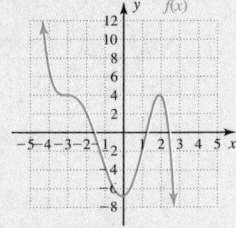

8.

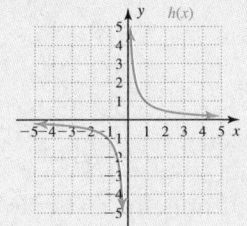

9.

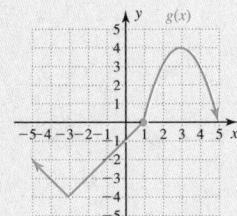

10.

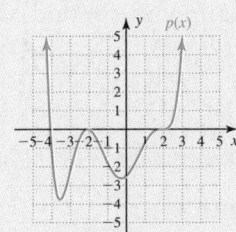

11.

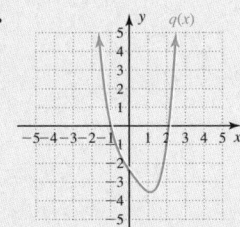

12.

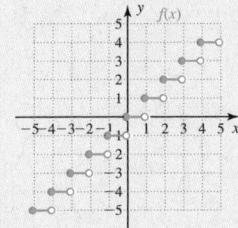

State the end behavior of the functions given.

13. $f(x)$

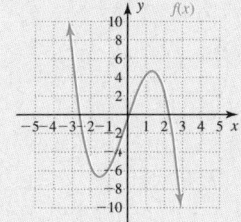

14. $g(x)$

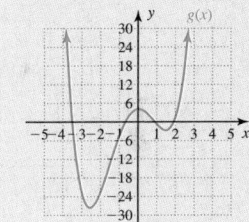

15. $H(x)$

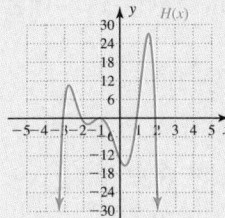

16. $h(x)$

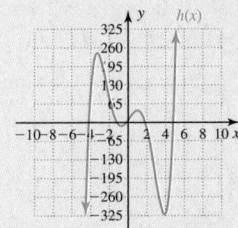

State the end behavior and y-intercept of the functions given. Do not graph.

17. $f(x) = x^3 + 6x^2 - 5x - 2$

18. $g(x) = x^4 - 4x^3 - 2x^2 + 16x - 12$

19. $p(x) = -2x^4 + x^3 + 7x^2 - x - 6$

20. $q(x) = -2x^3 - 18x^2 + 7x + 3$

21. $Y_1 = -3x^5 + x^3 + 7x^2 - 6$

22. $Y_2 = -x^6 - 4x^5 + 4x^3 + 16x - 12$

For each polynomial graph, (a) state whether the degree of the function is even or odd; (b) use the graph to name the zeroes of f, then state whether their multiplicity is even or odd; (c) state the minimum possible degree of f and write it in factored form; and (d) estimate the domain and range. Assume all zeroes are real.

23.

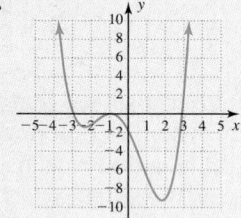

24.

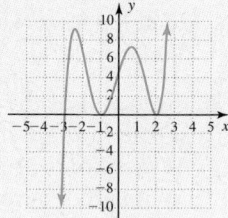

25.

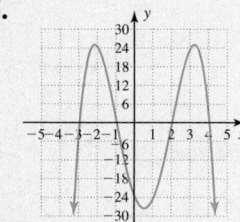

26.

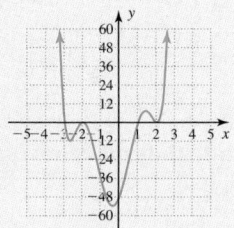

27.

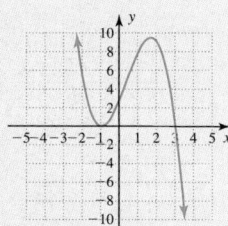

28.

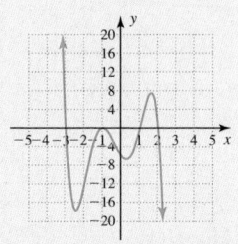

State the degree of each function, the end behavior, and y-intercept of its graph.

29. $f(x) = (x - 3)(x + 1)^3(x - 2)^2$

30. $g(x) = (x + 2)^2(x - 4)(x + 1)$

31. $Y_1 = -(x + 1)^2(x - 2)(2x - 3)(x + 4)$

32. $Y_2 = -(x + 1)(x - 2)^3(5x - 3)$

33. $r(x) = (x^2 + 3)(x + 4)^3(x - 1)$

34. $s(x) = (x + 2)^2(x - 1)^2(x^2 + 5)$

35. $h(x) = (x^2 + 2)(x - 1)^2(1 - x)$

36. $H(x) = (x + 2)^2(2 - x)(x^2 + 4)$

Every function in Exercises 37 through 42 has the zeroes $x = -1$, $x = -3$, and $x = 2$. Match each to its corresponding graph using degree, end behavior, and the multiplicity of each zero.

37. $f(x) = (x + 1)^2(x + 3)(x - 2)$

38. $F(x) = (x + 1)(x + 3)^2(x - 2)$

39. $g(x) = (x + 1)(x + 3)(x - 2)^3$

40. $G(x) = (x + 1)^3(x + 3)(x - 2)$

41. $Y_1 = (x + 1)^2(x + 3)(x - 2)^2$

42. $Y_2 = (x + 1)^3(x + 3)(x - 2)^2$

a.

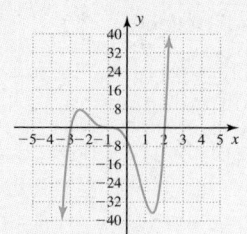

b.

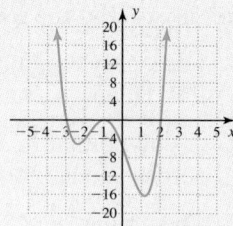

c.

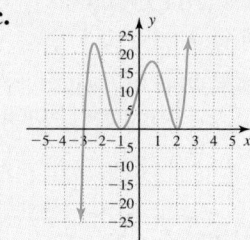

d.

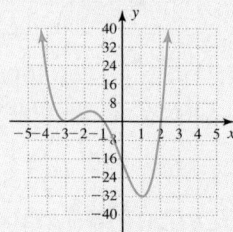

e.

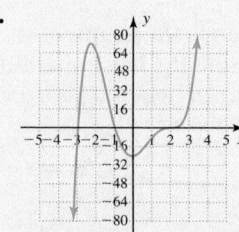

f.

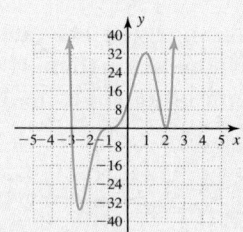

Sketch the graph of each function using the degree, end behavior, x- and y-intercepts, zeroes of multiplicity, and a few midinterval points to round-out the graph. Connect all points with a smooth, continuous curve.

43. $f(x) = (x + 3)(x + 1)(x - 2)$

44. $g(x) = (x + 2)(x - 4)(x - 1)$

45. $p(x) = -(x + 1)^2(x - 3)$

46. $q(x) = -(x + 2)(x - 2)^2$

47. $Y_1 = (x + 1)^2(3x - 2)(x + 3)$

48. $Y_2 = (x + 2)(x - 1)^2(5x - 2)$

49. $r(x) = -(x + 1)^2(x - 2)^2(x - 1)$

50. $s(x) = -(x - 3)(x - 1)^2(x + 1)^2$

51. $f(x) = (2x + 3)(x - 1)^3$

52. $g(x) = (3x - 4)(x + 1)^3$

53. $h(x) = (x + 1)^3(x - 3)(x - 2)$

54. $H(x) = (x + 3)(x + 1)^2(x - 2)^2$

55. $Y_3 = (x + 1)^3(x - 1)^2(x - 2)$

56. $Y_4 = (x - 3)(x - 1)^3(x + 1)^2$

Use the *Guidelines for Graphing Polynomial Functions* to graph the polynomials.

57. $y = x^3 + 3x^2 - 4$

58. $y = x^3 - 13x + 12$

59. $f(x) = x^3 - 3x^2 - 6x + 8$

60. $g(x) = x^3 + 2x^2 - 5x - 6$

61. $h(x) = -x^3 - x^2 + 5x - 3$

62. $H(x) = -x^3 - x^2 + 8x + 12$

63. $p(x) = -x^4 + 10x^2 - 9$

64. $q(x) = -x^4 + 13x^2 - 36$

65. $r(x) = x^4 - 9x^2 - 4x + 12$

66. $s(x) = x^4 - 5x^3 + 20x - 16$

67. $Y_1 = x^4 - 6x^3 + 8x^2 + 6x - 9$

68. $Y_2 = x^4 - 4x^3 - 3x^2 + 10x + 8$

69. $Y_3 = 3x^4 + 2x^3 - 36x^2 + 24x + 32$

70. $Y_4 = 2x^4 - 3x^3 - 15x^2 + 32x - 12$

71. $F(x) = 2x^4 + 3x^3 - 9x^2$

72. $G(x) = 3x^4 + 2x^3 - 8x^2$

73. $f(x) = x^5 + 4x^4 - 16x^2 - 16x$

74. $g(x) = x^5 - 3x^4 + x^3 - 3x^2$

75. $h(x) = x^6 - 2x^5 - 4x^4 + 8x^3$

76. $H(x) = x^6 + 3x^5 - 4x^4$

In preparation for future course work, it becomes helpful to recognize the most common square roots in mathematics: $\sqrt{2} \approx 1.414$, $\sqrt{3} \approx 1.732$, and $\sqrt{6} \approx 2.449$. Graph the following polynomials *on a graphing calculator,* and use the calculator to locate the maximum/minimum values and all zeroes. Use the zeroes to write the polynomial in factored form, then verify the *y*-intercept from the factored form and polynomial form.

77. $h(x) = x^5 + 4x^4 - 9x - 36$

78. $H(x) = x^5 + 5x^4 - 4x - 20$

79. $f(x) = 2x^5 + 5x^4 - 10x^3 - 25x^2 + 12x + 30$

80. $g(x) = 3x^5 + 2x^4 - 24x^3 - 16x^2 + 36x + 24$

Use the graph of each function to construct its equation in factored form and in polynomial form. Be sure to check the *y*-intercept and adjust the lead coefficient if necessary.

81.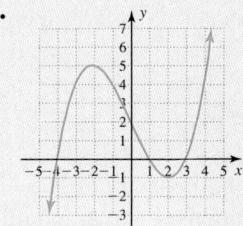

82.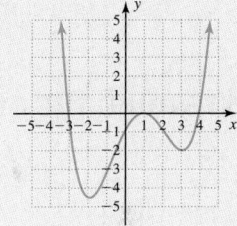

► **WORKING WITH FORMULAS**

83. Root tests for quartic polynomials: $ax^4 + bx^3 + cx^2 + dx + e = 0$

If u, v, w, and z represent the roots of a quartic polynomial, then the following relationships are true: (a) $u + v + w + z = -b$, (b) $u(v + z) + v(w + z) + w(u + z) = c$, (c) $u(vw + wz) + v(uz + wz) = -d$, and (d) $u \cdot v \cdot w \cdot z = e$. Use these tests to verify that $x = -3, -1, 2, 4$ are the solutions to $x^4 - 2x^3 - 13x^2 + 14x + 24 = 0$,

then use these zeroes and the factored form to write the equation in polynomial form to confirm results.

84. It is worth noting that the root tests in Exercise 83 still apply when the roots are irrational and/or complex. Use these tests to verify that $x = -\sqrt{3}, \sqrt{3}, 1 + 2i$, and $1 - 2i$ are the solutions to $x^4 - 2x^3 + 2x^2 + 6x - 15 = 0$, then use these zeroes and the factored form to write the equation in polynomial form to confirm results.

► **APPLICATIONS**

85. Traffic volume: Between the hours of 6:00 A.M. and 6.00 P.M., the volume of traffic at a busy intersection can be modeled by the polynomial $v(t) = -t^4 + 25t^3 - 192t^2 + 432t$, where $v(t)$ represents the number of vehicles above/below average, and t is number of hours past 6:00 A.M. (6:00 A.M. $\rightarrow$ 0). (a) Use the remainder theorem to find the volume of traffic during rush hour (8:00 A.M.), lunch time (12 noon), and the trip home (5:00 P.M.). (b) Use the rational zeroes theorem to find the times when the volume of

traffic is at its average $[v(t) = 0]$. (c) Use this information to graph $v(t)$, then use the graph to estimate the maximum and minimum flow of traffic and the time at which each occurs.

86. Insect population: The population of a certain insect varies dramatically with the weather, with spring-like temperatures causing a population boom and extreme weather (summer heat and winter cold) adversely affecting the population. This phenomena can be modeled by the polynomial $p(m) = -m^4 + 26m^3 - 217m^2 + 588m$, where $p(m)$

represents the number of live insects (in hundreds of thousands) in month m ($m = 1 \rightarrow$ Jan). (a) Use the remainder theorem to find the population of insects during the cool of spring (March) and the fair weather of fall (October). (b) Use the rational zeroes theorem to find the times when the population of insects becomes dormant $[p(m) = 0]$. (c) Use this information to graph $p(m)$, then use the graph to estimate the maximum and minimum population of insects, and the month at which each occurs.

87. **Balance of payments:** The graph shown represents the balance of payments (surplus versus deficit) for a large county over a 9-yr period. Use it to answer the following:

   a. What is the minimum possible degree polynomial that can model this graph?

   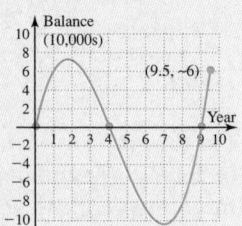

   b. How many years did this county run a deficit?

   c. Construct an equation model in factored form and in polynomial

form, adjusting the lead coefficient as needed. How large was the deficit in year 8?

88. **Water supply:** The graph shown represents the water level in a reservoir (above and below normal) that supplies water to a metropolitan area, over a 6-month period. Use it to answer the following:

   a. What is the minimum possible degree polynomial that can model this graph?

   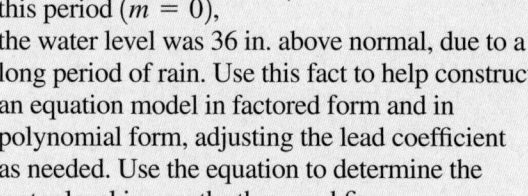

   b. How many months was the water level below normal in this 6-month period?

   c. At the beginning of this period ($m = 0$), the water level was 36 in. above normal, due to a long period of rain. Use this fact to help construct an equation model in factored form and in polynomial form, adjusting the lead coefficient as needed. Use the equation to determine the water level in months three and five.

▶ **EXTENDING THE CONCEPT**

89. As discussed in this section, the study of end behavior looks at what happens to the graph of a function as $|x| \rightarrow \infty$. Notice that as $|x| \rightarrow \infty$, both $\frac{1}{x}$ and $\frac{1}{x^2}$ approach zero. This fact can be used to study the end behavior of polynomial graphs.

   a. For $f(x) = x^3 + x^2 - 3x + 6$, factoring out x^3 gives the expression

   $$f(x) = x^3\left(1 + \frac{1}{x} - \frac{3}{x^2} + \frac{6}{x^3}\right).$$ What happens to the value of the expression as $x \rightarrow \infty$? As $x \rightarrow -\infty$?

   $$x^4 \qquad g(x) = x^4 + 3x^3 - 4x^2 +$$

   b. Factor out   from $5x - 1$. What happens to the value of the expression as $x \rightarrow \infty$? As $x \rightarrow -\infty$? How does this affirm the end behavior must be up/up?

90. For what value of c will three of the four real roots of $x^4 + 5x^3 + x^2 - 21x + c = 0$ be shared by the polynomial $x^3 + 2x^2 - 5x - 6 = 0$?

Show that the following equations have no rational roots.

91. $x^5 - x^4 - x^3 + x^2 - 2x + 3 = 0$

92. $x^5 - 2x^4 - x^3 + 2x^2 - 3x + 4 = 0$

▶ **MAINTAINING YOUR SKILLS**

93. (2.8) Given $f(x) = x^2 - 2x$ and $g(x) = \frac{1}{x}$, find the compositions $h(x) = (f \circ g)(x)$ and $H(x) = (g \circ f)(x)$, then state the domain of each.

94. (1.5) By direct substitution, verify that $x = 1 - 2i$ is a solution to $x^2 - 2x + 5 = 0$ and name the second solution.

95. (1.1/1.6) Solve each of the following equations.

   a. $-(2x + 5) - (6 - x) + 3 = x - 3(x + 2)$
   b. $\sqrt{x + 1} + 3 = \sqrt{2x + 2}$

   c. $\dfrac{2}{x - 3} + 5 = \dfrac{21}{x^2 - 9} + 4$

96. (2.2) Determine if the relation shown is a function. If not, explain how the definition of a function is violated.

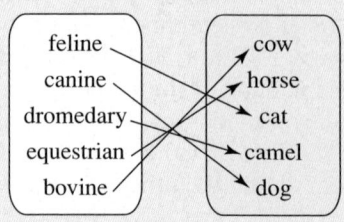

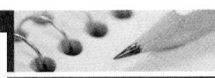

MID-CHAPTER CHECK

1. Compute $(x^3 + 8x^2 + 7x - 14) \div (x + 2)$ using long division and write the result in two ways:
   (a) dividend = (quotient)(divisor) + remainder and
   (b) $\dfrac{\text{dividend}}{\text{divisor}} = (\text{quotient}) + \dfrac{\text{remainder}}{\text{divisor}}$.

2. Given that $x - 2$ is a factor of $f(x) = 2x^4 - x^3 - 8x^2 + x + 6$, use the rational zeroes theorem to write $f(x)$ in completely factored form.

3. Use the remainder theorem to evaluate $f(-2)$, given $f(x) = -3x^4 + 7x^2 - 8x + 11$.

4. Use the factor theorem to find a third-degree polynomial having $x = -2$ and $x = 1 + i$ as roots.

5. Use the intermediate value theorem to show that $g(x) = x^3 - 6x - 4$ has a root in the interval $(2, 3)$.

6. Use the rational zeroes theorem, tests for -1 and 1, synthetic division, and the remainder theorem to write $f(x) = x^4 + 5x^3 - 20x - 16$ in completely factored form.

7. Find all the zeroes of h, real and complex: $h(x) = x^4 + 3x^3 + 10x^2 + 6x - 20$.

8. Sketch the graph of p using its degree, end behavior, y-intercept, zeroes of multiplicity, and any midinterval points needed, given $p(x) = (x + 1)^2(x - 1)(x - 3)$.

9. Use the *Guidelines for Graphing* to draw the graph of $q(x) = x^3 + 5x^2 + 2x - 8$.

10. When fighter pilots train for dogfighting, a "hard-deck" is usually established below which no competitive activity can take place. The polynomial graph given shows Maverick's altitude above and below this hard-deck during a 5-sec interval.

    a. What is the minimum possible degree polynomial that could form this graph? Why?

    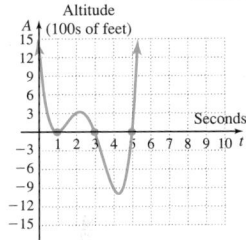

    b. How many seconds (total) was Maverick below the hard-deck for these 5 sec of the exercise?

    c. At the beginning of this time interval ($t = 0$), Maverick's altitude was 1500 ft above the hard-deck. Use this fact and the graph given to help construct an equation model in factored form and in polynomial form, adjusting the lead coefficient if needed. Use the equation to determine Maverick's altitude in relation to the hard-deck at $t = 2$ and $t = 4$.

REINFORCING BASIC CONCEPTS

Approximating Real Zeroes

Consider the equation $x^4 + x^3 + x - 6 = 0$. Using the rational zeroes theorem, the possible rational zeroes are $\{\pm 1, \pm 6, \pm 2, \pm 3\}$. The tests for 1 and -1 indicate that neither is a zero: $f(1) = -3$ and $f(-1) = -7$. Descartes' rule of signs reveals there must be one positive real zero since the coefficients of $f(x)$ change sign one time: $f(x) = x^4 + x^3 + x - 6$, and one negative real zero since $f(-x)$ also changes sign one time: $f(-x) = x^4 - x^3 - x - 6$. The remaining two zeroes must be complex. Using $x = 2$ with synthetic division shows 2 is not a zero, but the coefficients in the quotient row are all positive, so 2 is an upper bound:

$$\begin{array}{r|rrrrr} 2 & 1 & 1 & 0 & 1 & -6 \quad \text{coefficients of } f(x) \\ & & 2 & 6 & 12 & 26 \\ \hline & 1 & 3 & 6 & 13 & 20 \quad q(x) \end{array}$$

Using $x = -2$ shows that -2 is a zero *and a lower bound* for all other zeroes (quotient row alternates in sign):

$$\begin{array}{r|rrrrr} -2 & 1 & 1 & 0 & 1 & -6 \quad \text{coefficients of } f(x) \\ & & -2 & 2 & -4 & 6 \\ \hline & 1 & -1 & 2 & -3 & 0 \quad q_1(x) \end{array}$$

This means the remaining real zero must be a positive irrational number less than 2 (all other possible rational zeroes were eliminated). The quotient polynomial $q_1(x) = x^3 - x^2 + 2x - 3$ is not factorable, yet we're left with the challenge of finding this final zero. While there are many advanced techniques available for approximating irrational zeroes, at this level either technology or a technique called **bisection** is commonly used. The bisection method combines the intermediate value theorem with successively smaller intervals of the input variable, to narrow down the location of the irrational zero. Although "bisection" implies halving the interval each time, any number within the interval will do. The bisection method may be most efficient using a succession of short input/output tables as shown, with the number of tables increased if greater accuracy is desired. Since $f(1) = -3$ and $f(2) = 20$, the intermediate value theorem tells us the zero must be in the interval $[1, 2]$. We begin our search here, rounding noninteger outputs to the nearest 100th. As a visual aid, positive outputs are in blue, negative outputs in red.

x	$f(x)$	Conclusion
1	-3	← Zero is here, use $x = 1.25$ next
1.5	3.94	
2	20	

x	$f(x)$	Conclusion
1	-3	Zero is here, ← use $x = 1.30$ next
1.25	-0.36	
1.5	3.94	

x	$f(x)$	Conclusion
1.25	-0.36	← Zero is here, use $x = 1.275$ next
1.30	0.35	
1.5	3.94	

A reasonable estimate for the zero appears to be $x = 1.275$. Evaluating the function at this point gives $f(1.275) \approx 0.0098$, which is very close to zero.

Naturally, a closer approximation is obtained using the capabilities of a graphing calculator. To seven decimal places the zero is $x \approx 1.2756822$.

Exercise 1: Use the intermediate value theorem to show that $f(x) = x^3 - 3x + 1$ has a zero in the interval $[1, 2]$, then use bisection to locate the zero to three decimal place accuracy.

Exercise 2: The function $f(x) = x^4 + 3x - 15$ has two real zeroes in the interval $[-5, 5]$. Use the intermediate value theorem to locate the zeroes, then use bisection to find the zeroes accurate to three decimal places.

3.5 | Graphing Rational Functions

Learning Objectives

In Section 3.5 you will learn how to:

☐ **A.** Identify horizontal and vertical asymptotes

☐ **B.** Find the domain of a rational function

☐ **C.** Apply the concept of "multiplicity" to rational graphs

☐ **D.** Find the horizontal asymptotes of a rational function

☐ **E.** Graph general rational functions

☐ **F.** Solve applications of rational functions

In this section we introduce an entirely new kind of relation, called a **rational function.** While we've already studied a variety of functions, we still lack the ability to model a large number of important situations. For example, functions that model the amount of medication remaining in the bloodstream over time, the relationship between altitude and weightlessness, and the relationship between predator and prey populations are all rational functions.

A. Rational Functions and Asymptotes

Just as a rational number is the ratio of two integers, a **rational function** is the ratio of two polynomials. In general,

Rational Functions

A rational function $V(x)$ is one of the form

$$V(x) = \frac{p(x)}{d(x)},$$

where p and d are polynomials and $d(x) \neq 0$.
The domain of $V(x)$ is all real numbers, *except the zeroes of d.*

The simplest rational functions are the reciprocal function $y = \frac{1}{x}$ and the reciprocal square function $y = \frac{1}{x^2}$, as both have a constant numerator and a single term in the denominator, with the domain of both excluding $x = 0$.

The Reciprocal Function: $y = \dfrac{1}{x}$

The reciprocal function takes any input (other than zero) and gives its reciprocal as the output. This means large inputs produce small outputs and vice versa. A table of values (Table 3.2) and the resulting graph (Figure 3.16) are shown.

Table 3.2

x	y	x	y
-1000	$-1/1000$	$1/1000$	1000
-5	$-1/5$	$1/3$	3
-4	$-1/4$	$1/2$	2
-3	$-1/3$	1	1
-2	$-1/2$	2	$1/2$
-1	-1	3	$1/3$
$-1/2$	-2	4	$1/4$
$-1/3$	-3	5	$1/5$
$-1/1000$	-1000	1000	$1/1000$
0	undefined		

Figure 3.16

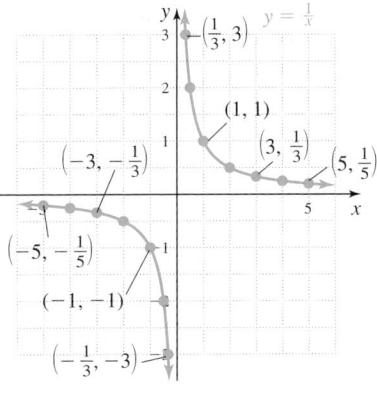

WORTHY OF NOTE

The notation used for graphical behavior always begins by describing what is happening to the x-values, and the resulting effect on the y-values. Using Figure 3.17, visualize that for a point (x, y) on the graph of $y = \frac{1}{x}$, as x gets larger, y must become smaller, particularly since their product must always be 1 ($y = \frac{1}{x} \Rightarrow xy = 1$).

Figure 3.17

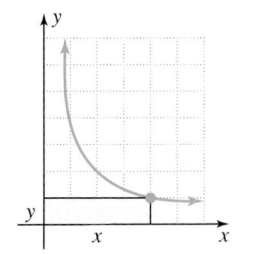

Table 3.2 and Figure 3.16 reveal some interesting features. First, the graph passes the vertical line test, verifying $y = \frac{1}{x}$ is indeed a function. Second, since division by zero is undefined, there can be no corresponding point on the graph, *creating a break at $x = 0$*. In line with our definition of rational functions, the domain is $x \in (-\infty, 0) \cup (0, \infty)$. Third, this is an odd function, with a "branch" of the graph in the first quadrant and one in the third quadrant, as the reciprocal of any input maintains its sign. Finally, we note in QI that as x becomes an infinitely large positive number, y gets closer and closer to zero. It seems convenient to symbolize this end behavior using the notation adopted in Section 3.4, and we write as $x \to \infty, y \to 0$. Graphically, the curve becomes very close to, or *approaches the x-axis.*

We also note that as x approaches zero from the right, y becomes an infinitely large positive number: as $x \to 0^{+}, y \to \infty$. Note a superscript $+$ or $-$ sign is used to indicate the *direction of the approach,* meaning *from the positive side* (right) or *from the negative side* (left).

EXAMPLE 1 ▶ **Describing the End Behavior of Rational Functions**

For $y = \frac{1}{x}$ in QIII,

    a. Describe the end behavior of the graph.

    b. Describe what happens as x approaches zero.

Solution ▶ Similar to the graph's behavior in QI, we have

    a. In words: As x becomes an infinitely large negative number, y approaches zero. In notation: As $x \to -\infty, y \to 0$.

    b. In words: As x approaches zero from the left, y becomes an infinitely large negative number. In notation: As $x \to 0^{-}, y \to -\infty$.

Now try Exercises 7 and 8 ▶

The Reciprocal Square Function: $y = \dfrac{1}{x^2}$

From our previous work, we anticipate this graph will also have a break at $x = 0$. But since the square of any negative number is positive, the branches of the **reciprocal square function** are both *above the x-axis.* Note the result is the graph of an even function. See Table 3.3 and Figure 3.18.

Table 3.3

x	y	x	y
-1000	1/1,000,000	1/1000	1,000,000
-5	1/25	1/3	9
-4	1/16	1/2	4
-3	1/9	1	1
-2	1/4	2	1/4
-1	1	3	1/9
$-1/2$	4	4	1/16
$-1/3$	9	5	1/25
$-1/1000$	1,000,000	1000	1/1,000,000
0	undefined		

Figure 3.18

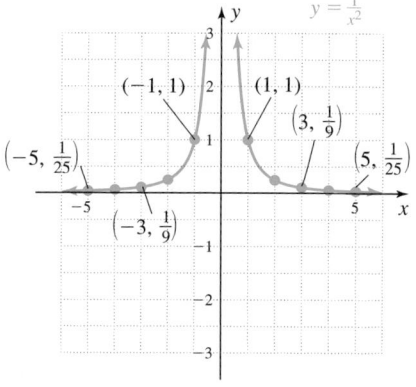

Similar to $y = \frac{1}{x}$, large positive inputs generate small, positive outputs: as $x \to \infty$, $y \to 0$. This is one indication of **asymptotic behavior** in the horizontal direction, and we say the line $y = 0$ is a **horizontal asymptote** for the reciprocal and reciprocal square functions. In general,

Horizontal Asymptotes

Given a constant k, the line $y = k$ is a horizontal asymptote for a function V if as x increases without bound, $V(x)$ approaches k:

$$\text{as } x \to -\infty, \ V(x) \to k \qquad \text{or} \qquad \text{as } x \to \infty, \ V(x) \to k$$

Figure 3.19 shows a horizontal asymptote at $y = 1$, which suggests the graph of $f(x)$ is the graph of $y = \frac{1}{x}$ shifted up 1 unit. Figure 3.20 shows a horizontal asymptote at $y = -2$, which suggests the graph of $g(x)$ is the graph of $y = \frac{1}{x^2}$ shifted down 2 units.

Figure 3.19

Figure 3.20

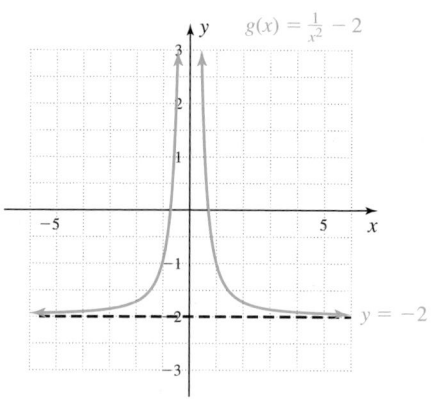

WORTHY OF NOTE

As seen in Figures 3.19 and 3.20, asymptotes appear graphically as dashed lines that seem to "guide" the branches of the graph.

EXAMPLE 2 ▶ **Describing the End Behavior of Rational Functions**

For the graph in Figure 3.20, use mathematical notation to

a. Describe the end behavior of the graph.

b. Describe what happens as x approaches zero.

Solution ▶ a. as $x \to -\infty$, $g(x) \to -2$ b. as $x \to 0^-$, $g(x) \to \infty$

as $x \to \infty$, $g(x) \to -2$ as $x \to 0^+$, $g(x) \to \infty$

Now try Exercises 9 and 10 ▶

From Example 2b, we note that as x becomes *smaller and close to 0*, g becomes very large and *increases without bound*. This is an indication of asymptotic behavior in the vertical direction, and we say the line $x = 0$ is a **vertical asymptote** for g ($x = 0$ is also a vertical asymptote for f). In general,

---

Vertical Asymptotes

Given a constant h, the line $x = h$ is a vertical asymptote for a function V if as x approaches h, $V(x)$ increases or decreases without bound:

$$\text{as } x \to h^{+},\ V(x) \to \pm\infty \qquad \text{or} \qquad \text{as } x \to h^{-},\ V(x) \to \pm\infty$$

Identifying these asymptotes is useful because the graphs of $y = \frac{1}{x}$ and $y = \frac{1}{x^2}$ can be transformed *in exactly the same way as the toolbox functions*. When their graphs shift—the vertical and horizontal asymptotes shift with them and can be used as guides to redraw the graph. In shifted form, $f(x) = \dfrac{a}{(x \pm h)} \pm k$ for the reciprocal function, and $g(x) = \dfrac{a}{(x \pm h)^2} \pm k$ for the reciprocal square function.

EXAMPLE 3 ▶ **Writing the Equation of a Basic Rational Function, Given Its Graph**

Identify the function family for the graph given, then use the graph to write the equation of the function in "shifted form." Assume $|a| = 1$.

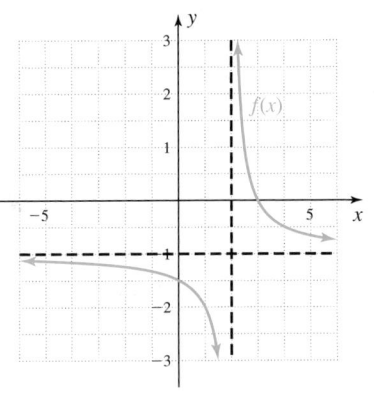

Solution ▶ The graph appears to be from the reciprocal function family, and has been shifted 2 units right (the vertical asymptote is at $x = 2$), and 1 unit down (the horizontal asymptote is at $y = -1$). From $y = \frac{1}{x}$, we obtain $f(x) = \frac{1}{x - 2} - 1$ as the shifted form.

☑ **A.** You've just learned how to identify and name horizontal and vertical asymptotes

Now try Exercises 11 through 22 ▶

WORTHY OF NOTE

In Section 2.7, we studied special cases of $\frac{p(x)}{d(x)}$, where p and d shared a common factor, creating a "hole" in the graph. In this section, we'll assume the functions are given in simplest form (the numerator and denominator have no common factors).

B. Vertical Asymptotes and the Domain

Much of what we know about these basic functions can be generalized and applied to general rational functions. The graphs in Figures 3.21 through 3.24 show that rational graphs come in many shapes, often in "pieces," and exhibit asymptotic behavior.

Figure 3.21
$$f(x) = \frac{1}{x + 2}$$

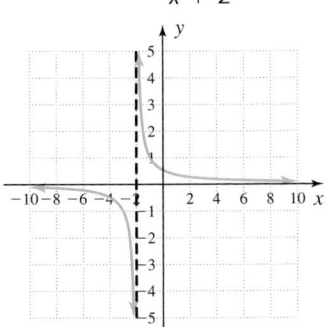

Figure 3.22
$$g(x) = \frac{2x}{x^2 - 1}$$

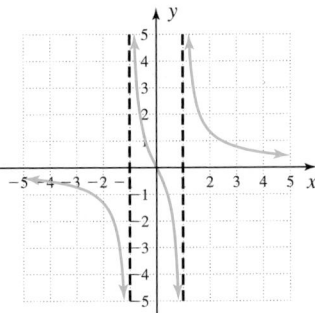

Figure 3.23
$$w(x) = \frac{3}{x^2 + 1}$$

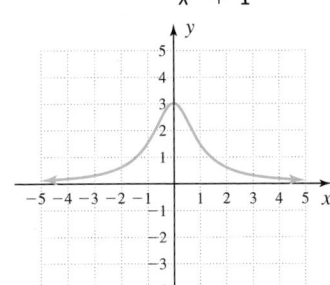

Figure 3.24
$$H(x) = \frac{x^2}{x^2 - 2x - 3}$$

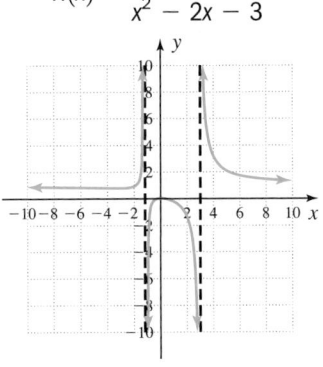

For $y = \frac{1}{x}$ and $y = \frac{1}{x^2}$, a vertical asymptote occurred at the zero of the denominator. This actually applies to all rational functions *in simplified form.* For $V(x) = \frac{p(x)}{d(x)}$, if c is a zero of $d(x)$, the function can be evaluated at every point near c, but not *at c.* This creates a **break** or **discontinuity** in the graph, resulting in the asymptotic behavior.

Vertical Asymptotes of a Rational Function

Given $V(x) = \frac{p(x)}{d(x)}$ is a rational function in simplest form, vertical asymptotes will occur at the real zeroes of d.

EXAMPLE 4 ▶ Finding Vertical Asymptotes

Locate the vertical asymptote(s) of each function given, then state its domain.

a. $f(x) = \dfrac{2x}{x^2 - 1}$ **b.** $g(x) = \dfrac{3}{x^2 + 1}$ **c.** $v(x) = \dfrac{x^2}{x^2 - 2x - 3}$

Solution ▶ **a.** Setting the denominator equal to zero gives $x^2 - 1 = 0$, so vertical asymptotes will occur at $x = -1$ and $x = 1$. The domain of f is $x \in (-\infty, -1) \cup (-1, 1) \cup (1, \infty)$.

b. Since the equation $x^2 + 1 = 0$ has no real zeroes, there are no vertical asymptotes and the domain of g is unrestricted: $x \in R$.

c. Solving $x^2 - 2x - 3 = 0$ gives $(x + 1)(x - 3) = 0$, with solutions $x = -1$ and $x = 3$. There are vertical asymptotes at $x = -1$ and $x = 3$, and the domain of v is $x \in (-\infty, -1) \cup (-1, 3) \cup (3, \infty)$.

☑ **B.** You've just learned how to find the domain of a rational function

Now try Exercises 23 through 30 ▶

C. Vertical Asymptotes and Multiplicities

The "cross" and "bounce" concept used for polynomial graphs can also be applied to rational graphs, particularly when viewed in terms of sign changes in the dependent variable. As you can see in Figures 3.25 to 3.27, the function $f(x) = \dfrac{1}{x + 2}$ changes sign at the asymptote $x = -2$ (negative on one side, positive on the other), and the denominator has multiplicity 1 (odd). The function $g(x) = \dfrac{1}{(x - 1)^2}$ does not change sign at the asymptote $x = 1$ (positive on both sides), and its denominator has multiplicity 2 (even). As with our earlier study of multiplicities, when these two are combined into the single function $v(x) = \dfrac{1}{(x + 2)(x - 1)^2}$, the function still changes sign at $x = -2$, and does not change sign at $x = 1$.

Figure 3.25

$f(x) = \dfrac{1}{x + 2}$

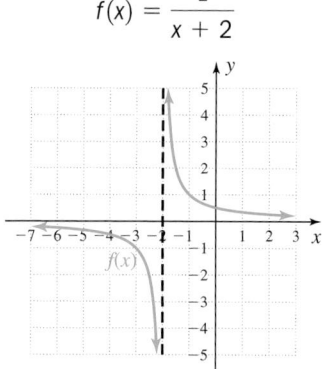

Figure 3.26

$g(x) = \dfrac{1}{(x - 1)^2}$

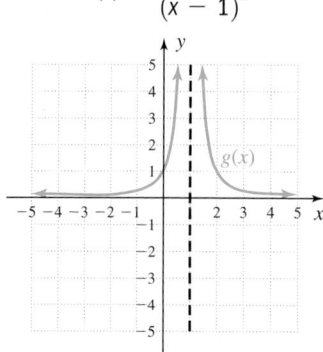

Figure 3.27

$h(x) = \dfrac{1}{(x + 2)(x - 1)^2}$

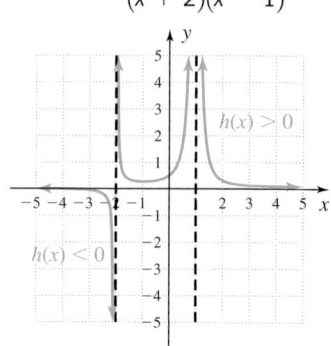

EXAMPLE 5 ▶ Finding Sign Changes at Vertical Asymptotes

Locate the vertical asymptotes of each function and state whether the function will change sign from one side of the asymptote(s) to the other.

a. $f(x) = \dfrac{x^2 - 4x + 4}{x^2 - 2x - 3}$ **b.** $g(x) = \dfrac{x^2 + 2}{x^2 + 2x + 1}$

Solution ▶ **a.** Factoring the denominator of f and setting it equal to zero gives $(x + 1)(x - 3) = 0$, and vertical asymptotes will occur at $x = -1$ and $x = 3$ (both multiplicity 1). The function will change sign at each asymptote (see Figure 3.28).

b. Factoring the denominator of g and setting it equal to zero gives $(x + 1)^2 = 0$. There will be a vertical asymptote at $x = -1$, but the function will not change sign since it's a zero of even multiplicity (see Figure 3.29).

Figure 3.28

$$f(x) = \dfrac{x^2 - 4x + 4}{x^2 - 2x - 3}$$

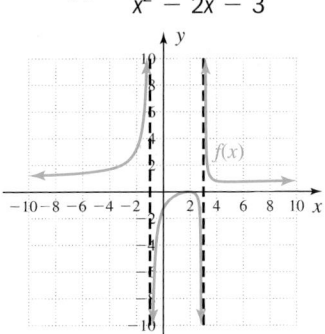

Figure 3.29

$$g(x) = \dfrac{x^2 + 2}{x^2 + 2x + 1}$$

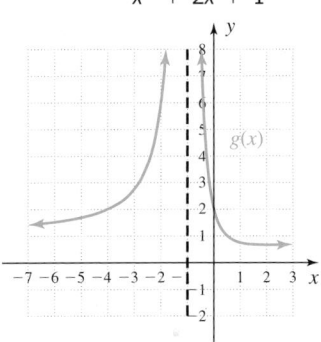

☑ **C.** You've just learned how to apply the concept of "multiplicity" to rational graphs

Now try Exercises 31 through 36 ▶

D. Finding Horizontal Asymptotes

A study of horizontal asymptotes is closely related to our study of "dominant terms" in Section 3.4. Recall the highest degree term in a polynomial tends to dominate all other terms as $|x| \to \infty$. For $v(x) = \dfrac{2x^2 + 4x + 3}{x^2 + 2x + 1}$, both polynomials *have the same degree*, so $\dfrac{2x^2 + 4x + 3}{x^2 + 2x + 1} \approx \dfrac{2x^2}{x^2} = 2$ for large values of x: as $|x| \to \infty$, $y \to 2$ and $y = 2$ is a horizontal asymptote for v. When the degree of the numerator is *smaller* than the degree of the denominator, our earlier work with $y = \frac{1}{x}$ and $y = \frac{1}{x^2}$ showed there was a horizontal asymptote at $y = 0$ (the x-axis), since as $|x| \to \infty$, $y \to 0$. In general,

LOOKING AHEAD

In Section 3.6 we will explore two additional kinds of asymptotic behavior, (1) oblique (slant) asymptotes and (2) asymptotes that are nonlinear.

Horizontal Asymptotes

Given $V(x) = \dfrac{p(x)}{d(x)}$ is a rational function in lowest terms, where the leading term of p is ax^n and the leading term of d is bx^m (polynomial p has degree n, polynomial d has degree m).

  I. If $n < m$, there is a horizontal asymptote at $y = 0$ (the x-axis).
 II. If $n = m$, there is a horizontal asymptote at $y = \frac{a}{b}$.
III. If $n > m$, the graph has no horizontal asymptote.

Finally, while the graph of a rational function can never "cross" the vertical asymptote $x = h$ (since the function simply cannot be evaluated at h), it is possible for a graph to cross the horizontal asymptote $y = k$ (some do, others do not). To find out which is the case, we set the function equal to k and solve.

EXAMPLE 6 ▶ **Locating Horizontal Asymptotes**

Locate the horizontal asymptote for each function, if one exists. Then determine if the graph will cross the asymptote.

    a. $f(x) = \dfrac{3x}{x^2 + 2}$ **b.** $g(x) = \dfrac{x^2 - 4}{x^2 - 1}$ **c.** $v(x) = \dfrac{3x^2 - x - 6}{x^2 + x - 6}$

Solution ▶ **a.** For $f(x)$, the degree of the numerator $<$ degree of the denominator, indicating a horizontal asymptote at $y = 0$. Solving $f(x) = 0$, we find $x = 0$ is the only solution and the graph will cross the horizontal asymptote at $(0, 0)$ (see Figure 3.30).

    b. For $g(x)$, the degree of the numerator and the denominator are equal. This means $g(x) \approx \dfrac{x^2}{x^2} = 1$ for large values of x, and there is a horizontal asymptote at $y = 1$. Solving $g(x) = 1$ gives

$$\dfrac{x^2 - 4}{x^2 - 1} = 1 \qquad\qquad y = 1 \rightarrow \text{horizontal asymptote}$$
$$x^2 - 4 = x^2 - 1 \qquad \text{multiply by } x^2 - 1$$
$$-4 = -1 \qquad\qquad \text{no solution}$$

The graph will not cross the asymptote (see Figure 3.31).

    c. For $v(x)$, the degree of the numerator and denominator are once again equal, so $v(x) \approx \dfrac{3x^2}{x^2} = 3$ and there is a horizontal asymptote at $y = 3$. Solving $v(x) = 3$ gives

$$\dfrac{3x^2 - x - 6}{x^2 + x - 6} = 3 \qquad\qquad y = 3 \rightarrow \text{horizontal asymptote}$$
$$3x^2 - x - 6 = 3(x^2 + x - 6) \qquad \text{multiply by } x^2 + x - 6$$
$$3x^2 - x - 6 = 3x^2 + 3x - 18 \qquad \text{distribute}$$
$$-4x + 12 = 0 \qquad\qquad \text{simplify}$$
$$x = 3 \qquad\qquad\qquad \text{result}$$

☑ **D. You've just learned how to find the horizontal asymptotes of a rational function**

The graph will cross its asymptote at $x = 3$ (see Figure 3.32).

Figure 3.30

$f(x) = \dfrac{3x}{x^2 + 2}$

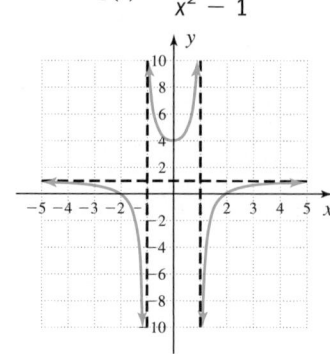

Figure 3.31

$g(x) = \dfrac{x^2 - 4}{x^2 - 1}$

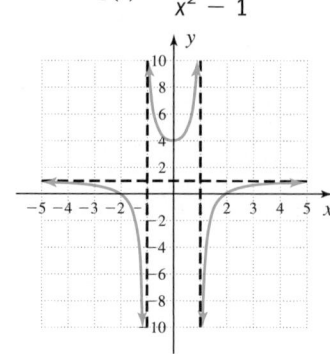

Figure 3.32

$v(x) = \dfrac{3x^2 - x - 6}{x^2 + x - 6}$

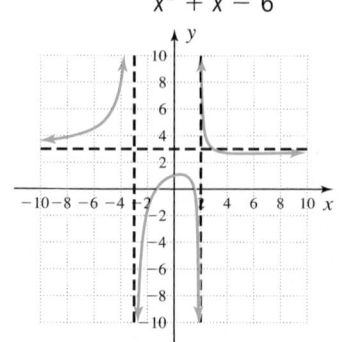

Now try Exercises 37 through 42 ▶

WORTHY OF NOTE

It's helpful to note that all nonvertical asymptotes and whether they cross the graph can actually be found using long division. The quotient $q(x)$ gives the equation of the asymptote, and the zeroes of the remainder $r(x)$ will indicate if or where the two will cross. From Example 6c, long division gives $q(x) = 3$ and $r(x) = -4x + 12$ (verify this), showing there is a horizontal asymptote at $y = 3$, which the graph crosses at $x = 3$ [the zero of $r(x)$].

E. The Graph of a Rational Function

Our observations to this point lead us to this general strategy for graphing rational functions. Not all graphs require every step, but together they provide an effective approach.

Guidelines for Graphing Rational Functions

Given $V(x) = \frac{p(x)}{d(x)}$, $d(x) \neq 0$, is a rational function in lowest terms,

1. Find the y-intercept at $V(0)$.
2. Find vertical asymptotes (if any) at $d(x) = 0$.
3. Find x-intercepts at $p(x) = 0$.
4. Locate the horizontal asymptote (if any).
5. Determine if the graph will cross the horizontal asymptote.
6. If needed, compute "midinterval" points to help complete the graph.
7. Draw the asymptotes, plot the intercepts and additional points, and use intervals where $V(x)$ changes sign to complete the graph.

EXAMPLE 7 ▶ **Graphing Rational Functions**

Graph each function given.

a. $f(x) = \dfrac{x^2 - x - 6}{x^2 + x - 6}$ **b.** $g(x) = \dfrac{2x^2 - 4x + 2}{x^2 - 7}$

Solution ▶ **a.** Begin by writing f in factored form: $f(x) = \dfrac{(x + 2)(x - 3)}{(x + 3)(x - 2)}$.

1. y-intercept: $f(0) = \dfrac{(2)(-3)}{(3)(-2)} = 1$, so the y-intercept is $(0, 1)$.

2. Vertical asymptote(s): Setting the denominator equal to zero gives $(x + 3)(x - 2) = 0$, showing there will be vertical asymptotes at $x = -3$, $x = 2$.

3. x-intercepts: Setting the numerator equal to zero gives $(x + 2)(x - 3) = 0$, showing the x-intercepts will be $(-2, 0)$ and $(3, 0)$.

4. Horizontal asymptote: Since the degree of the numerator and the degree of the denominator are equal, $y = \dfrac{x^2}{x^2} = 1$ is a horizontal asymptote.

5. Solving $\dfrac{x^2 - x - 6}{x^2 + x - 6} = 1$ $f(x) = 1 \rightarrow$ horizontal asymptote

           $x^2 - x - 6 = x^2 + x - 6$ multiply by $x^2 + x - 6$

                       $-2x = 0$ simplify

                         $x = 0$ solve

The graph will cross the horizontal asymptote at $(0, 1)$.

    The information from steps 1 through 5 is shown in Figure 3.33, and indicates we have no information about the graph in the interval $(-\infty, -3)$. Since rational functions are defined for all real numbers except the zeroes of d, we know there must be a "piece" of the graph in this interval.

6. Selecting $x = -4$ to compute one additional point, we find $f(-4) = \dfrac{(-2)(-7)}{(-1)(-6)} = \dfrac{14}{6} = \dfrac{7}{3}$. The point is $\left(-4, \dfrac{7}{3}\right)$.

7. All factors of f are linear, so function values will alternate sign in the intervals created by x-intercepts and vertical asymptotes. The y-intercept $(0, 1)$ shows $f(x)$ is positive in the interval containing 0. To meet all necessary conditions, we complete the graph, as shown in Figure 3.34.

Figure 3.33

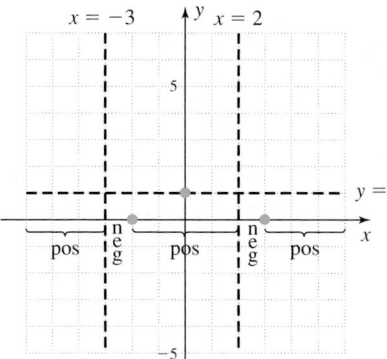

Figure 3.34

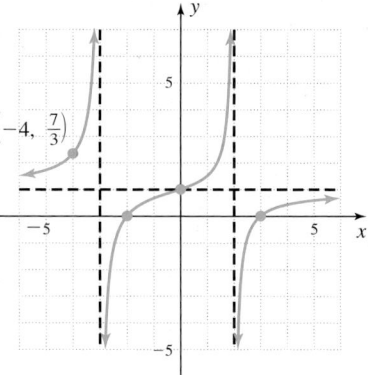

b. Writing g in factored form gives $g(x) = \dfrac{2(x^2 - 2x + 1)}{x^2 - 7} = \dfrac{2(x - 1)^2}{(x + \sqrt{7})(x - \sqrt{7})}$.

1. y-intercept: $g(0) = \dfrac{2(-1)^2}{(\sqrt{7})(-\sqrt{7})} = -\dfrac{2}{7}$. The y-intercept is $(0, -\dfrac{2}{7})$.

2. Vertical asymptote(s): Setting the denominator equal to zero gives $(x + \sqrt{7})(x - \sqrt{7}) = 0$, showing there will be asymptotes at $x = -\sqrt{7}, x = \sqrt{7}$.

3. x-intercept(s): Setting the numerator equal to zero gives $2(x - 1)^2 = 0$, with $x = 1$ a zero of multiplicity 2. The x-intercept is $(1, 0)$.

4. Horizontal asymptote: The degree of the numerator is equal to the degree of denominator, so $y = \dfrac{2x^2}{x^2} = 2$ is a horizontal asymptote.

5. Solve $\dfrac{2x^2 - 4x + 2}{x^2 - 7} = 2$ $g(x) = 2 \rightarrow$ horizontal asymptote

$2x^2 - 4x + 2 = 2x^2 - 14$ multiply by $x^2 - 7$

$-4x = -16$ simplify

$x = 4$ solve

The graph will cross its horizontal asymptote at $(4, 2)$. The information from steps 1 to 5 is shown in Figure 3.35, and indicates we have no information about the graph in the interval $(-\infty, -\sqrt{7})$.

Figure 3.35

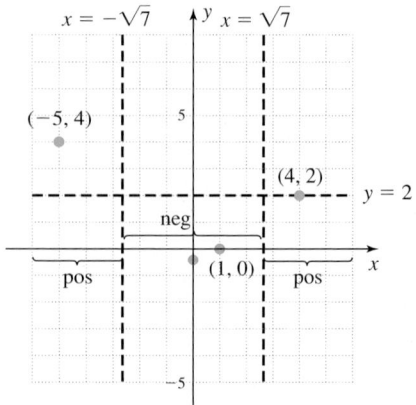

6. Selecting $x = -5$, $g(-5) = \dfrac{2(-5 - 1)^2}{(-5)^2 - 7}$

$= \dfrac{2(-6)^2}{25 - 7}$

$= \dfrac{2(36)}{18}$

$= 4$

The point $(-5, 4)$ is on the graph.

7. Since factors of the denominator have odd multiplicity, function values will alternate sign on either side of the asymptotes. The factor in the numerator has even multiplicity, so the graph will "bounce off" the x-axis at $x = 1$ (no change in sign). The y-intercept $(0, -\frac{2}{7})$ shows the function is negative in the interval containing 0. This information and the completed graph are shown in Figure 3.36.

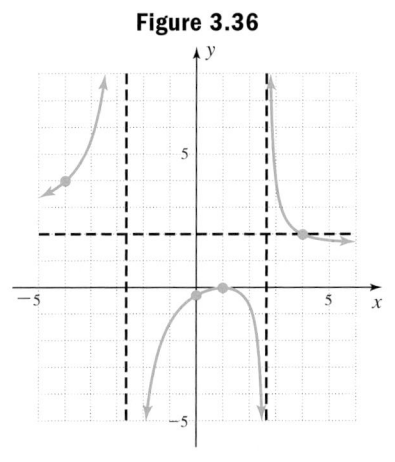

Figure 3.36

Now try Exercises 43 through 66 ▶

Examples 6 and 7 demonstrate that graphs of rational functions come in a large variety. Once the components of the graph have been found, completing the graph presents an intriguing and puzzle-like challenge as we attempt to sketch a graph that meets all conditions. As we've done with other functions, can you reverse this process? That is, given the <u>graph</u> of a rational function, can you construct its equation?

EXAMPLE 8 ▶ **Finding the Equation of a Rational Function from Its Graph**

Use the graph of $f(x)$ shown to construct its equation.

Solution ▶ The x-intercepts are $(-1, 0)$ and $(4, 0)$, so the numerator must contain the factors $(x + 1)$ and $(x - 4)$. The vertical asymptotes are $x = -2$ and $x = 3$, so the denominator must have the factors $(x + 2)$ and $(x - 3)$. So far we have:

$$f(x) = \frac{a(x + 1)(x - 4)}{(x + 2)(x - 3)}$$

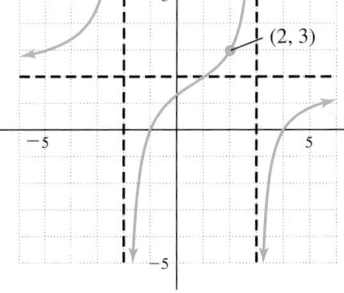

Since $(2, 3)$ is on the graph, we substitute 2 for x and 3 for $f(x)$ to solve for a:

$$3 = \frac{a(2 + 1)(2 - 4)}{(2 + 2)(2 - 3)} \quad \text{substitute 3 for } f(x) \text{ and 2 for } x$$

$$3 = \frac{3a}{2} \quad \text{simplify}$$

$$2 = a \quad \text{solve}$$

The result is $f(x) = \dfrac{2(x + 1)(x - 4)}{(x + 2)(x - 3)} = \dfrac{2x^2 - 6x - 8}{x^2 - x - 6}$, with a horizontal asymptote at $y = 2$ and a y-intercept of $(0, \frac{4}{3})$, which fits the graph very well.

Now try Exercises 67 through 70 ▶

F. Applications of Rational Functions

In many applications of rational functions, the coefficients can be rather large and the graph should be scaled appropriately.

EXAMPLE 9 ▶   Modeling the Cost to Remove Chemical Waste

For a large urban-centered county, the cost to remove chemical waste from a local river is modeled by $C(p) = \frac{180p}{100 - p}$, where $C(p)$ represents the cost (in thousands of dollars) to remove p percent of the pollutants.

  a. Find the cost to remove 25%, 50%, and 75% of the pollutants and comment.

  b. Graph the function using an appropriate scale.

  c. In mathematical notation, state what happens if the county attempts to remove 100% of the pollutants.

Solution ▶   **a.** Evaluating the function for the values indicated, we find $C(25) = 60$, $C(50) = 180$, and $C(75) = 540$. The cost is escalating rapidly. The change from 25% to 50% brought a $120,000 increase, but the change from 50% to 75% brought *a $360,000 increase!*

b. From $C(p) = \frac{180p}{100 - p}$, we see that C has a y-intercept at $(0, 0)$ and a vertical asymptote at $p = 100$. Since the degree of the numerator and denominator are equal, there is a horizontal asymptote at $y = \frac{180p}{-p} = -180$. From the context we need only graph the portion from $0 \le p < 100$, producing the following graph:

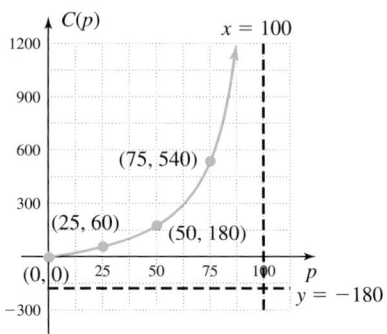

☑ **F. You've just learned how to solve an application of rational functions**

c. As the percentage of the pollutants removed approaches 100%, the cost of the cleanup skyrockets. Notationally, as $p \to 100^-$, $C \to \infty$.

Now try Exercises 73 through 84 ▶

TECHNOLOGY HIGHLIGHT

Rational Functions and Appropriate Domains

In Example 9, portions of the graph were ignored due to the context of the application. To see the full graph, we use the fact that a second branch of C occurs on the opposite side of the vertical and horizontal asymptotes, and set a window size like the one shown in Figure 3.37. After entering $C(p)$ as Y_1 on the [Y=] screen and pressing [GRAPH], the full graph shown in Figure 3.38 appears (the horizontal asymptote was drawn using $Y_2 = -180$).

Figure 3.37

```
WINDOW
 Xmin=0
 Xmax=200
 Xscl=20
 Ymin=-2000
 Ymax=2000
 Yscl=200
 Xres=1
```

Figure 3.38

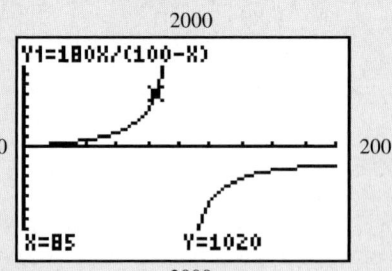

Exercise 1:   Use the [TRACE] feature to verify that as $p \to 100^-$, $C \to \infty$. Approximately how much money must be spent to remove 95% of the pollutants? What happens when you [TRACE] to 100%? Past 100%?

Exercise 2:   Calculate the rate of change $\frac{\Delta C}{\Delta p}$ for the intervals [60, 65], [85, 90], and [90, 95] (use the *Technology Extension* from Chapter 3 at www.mhhe.com/coburn if desired). Comment on what you notice.

Exercise 3:   Reset the window size changing only Xmax to 100 and Ymin to 0 for a more relevant graph. How closely does it resemble the graph from Example 9?

3.5 EXERCISES

▶ CONCEPTS AND VOCABULARY

Fill in each blank with the appropriate word or phrase. Carefully reread the section if needed.

1. Write the following in direction/approach notation. *As x becomes an infinitely large negative number, y approaches 2.* _____

2. For any constant k, the notation as $|x| \to +\infty$, $y \to k$ is an indication of a _____ asymptote, while $x \to k$, $|y| \to +\infty$ indicates a _____ asymptote.

3. Vertical asymptotes are found by setting the _____ equal to zero. The x-intercepts are found by setting the _____ equal to zero.

4. If the degree of the numerator is equal to the degree of the denominator, a horizontal asymptote occurs at $y = \frac{a}{b}$, where $\frac{a}{b}$ represents the ratio of the _____ _____.

5. Use the function $g(x) = \dfrac{3x^2 - 2x}{2x^2 - 3}$ and a table of values to discuss the concept of horizontal asymptotes. At what positive value of x is the graph of g within 0.01 of its horizontal asymptote?

6. Name all of the "tools" at your disposal that play a role in the graphing of rational functions. Which tools are indispensable and always used? Which are used only as the situation merits?

▶ DEVELOPING YOUR SKILLS

For each graph given, (a) use mathematical notation to describe the end behavior of each graph and (b) describe what happens as x approaches 1.

7. $V(x) = \dfrac{1}{(x - 1)} + 2$ 8. $v(x) = \dfrac{1}{(x - 1)} - 2$

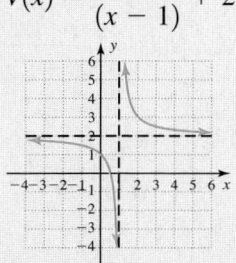   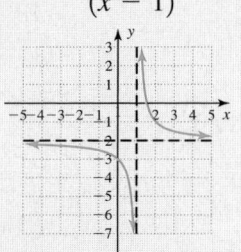

For each graph given, (a) use mathematical notation to describe the end behavior of each graph, and (b) describe what happens as x approaches -2.

9. $Q(x) = \dfrac{1}{(x + 2)^2} + 1$ 10. $q(x) = \dfrac{-1}{(x + 2)^2} + 2$

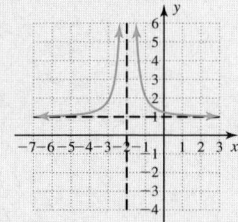   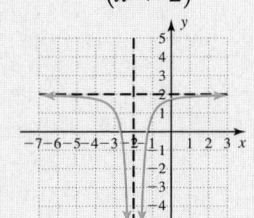

Identify the parent function for each graph given, then use the graph to construct the equation of the function in shifted form. Assume $|a| = 1$.

11.   12.

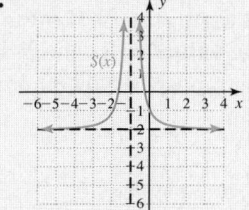   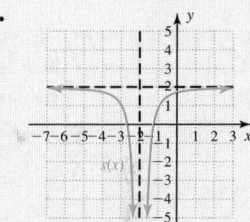

13.   14.

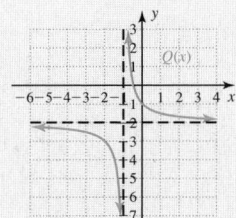   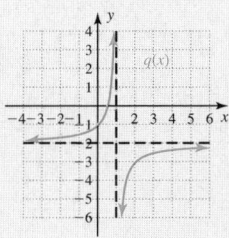

15.   16.

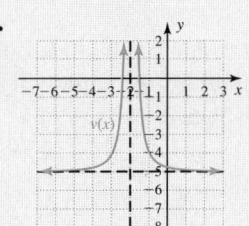   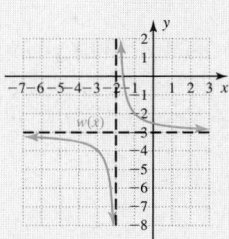

Use the graph shown to complete each statement using the direction/approach notation.

Exercises 17 through 22

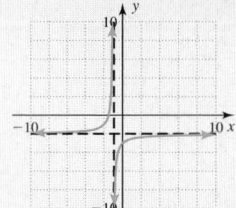

17. As $x \to -\infty$, y _____.

18. As $x \to \infty$, y _____.

19. As $x \to -1^+$, y _____.

20. As $x \to -1^-$, y _____.

21. The line $x = -1$ is a vertical asymptote, since: as $x \to$ _____, $y \to$ _____.

22. The line $y = -2$ is a horizontal asymptote, since: as $x \to$ _____, $y \to$ _____.

Give the location of the vertical asymptote(s) if they exist, and state the function's domain.

23. $f(x) = \dfrac{x + 2}{x - 3}$

24. $F(x) = \dfrac{4x}{2x - 3}$

25. $g(x) = \dfrac{3x^2}{x^2 - 9}$

26. $G(x) = \dfrac{x + 1}{9x^2 - 4}$

27. $h(x) = \dfrac{x^2 - 1}{2x^2 + 3x - 5}$

28. $H(x) = \dfrac{x - 5}{2x^2 - x - 3}$

29. $p(x) = \dfrac{2x + 3}{x^2 + x + 1}$

30. $q(x) = \dfrac{2x^3}{x^2 + 4}$

Give the location of the vertical asymptote(s) if they exist, and state whether function values will change sign (positive to negative or negative to positive) from one side of the asymptote to the other.

31. $Y_1 = \dfrac{x + 1}{x^2 - x - 6}$

32. $Y_2 = \dfrac{2x + 3}{x^2 - x - 20}$

33. $r(x) = \dfrac{x^2 + 3x - 10}{x^2 - 6x + 9}$

34. $R(x) = \dfrac{x^2 - 2x - 15}{x^2 - 4x + 4}$

35. $Y_1 = \dfrac{x}{x^3 + 2x^2 - 4x - 8}$

36. $Y_2 = \dfrac{-2x}{x^3 + x^2 - x - 1}$

For the functions given, (a) determine if a horizontal asymptote exists and (b) determine if the graph will cross the asymptote, and if so, where it crosses.

37. $Y_1 = \dfrac{2x - 3}{x^2 + 1}$

38. $Y_2 = \dfrac{4x + 3}{2x^2 + 5}$

39. $r(x) = \dfrac{4x^2 - 9}{x^2 - 3x - 18}$

40. $R(x) = \dfrac{2x^2 - x - 10}{x^2 + 5}$

41. $p(x) = \dfrac{3x^2 - 5}{x^2 - 1}$

42. $P(x) = \dfrac{3x^2 - 5x - 2}{x^2 - 4}$

Give the location of the *x*- and *y*-intercepts (if they exist), and discuss the behavior of the function (bounce or cross) at each *x*-intercept.

43. $f(x) = \dfrac{x^2 - 3x}{x^2 - 5}$

44. $F(x) = \dfrac{2x - x^2}{x^2 + 2x - 3}$

45. $g(x) = \dfrac{x^2 + 3x - 4}{x^2 - 1}$

46. $G(x) = \dfrac{x^2 + 7x + 6}{x^2 - 2}$

47. $h(x) = \dfrac{x^3 - 6x^2 + 9x}{4 - x^2}$

48. $H(x) = \dfrac{4x + 4x^2 + x^3}{x^2 - 1}$

Use the *Guidelines for Graphing Rational Functions* to graph the functions given.

49. $f(x) = \dfrac{x + 3}{x - 1}$

50. $g(x) = \dfrac{x - 4}{x + 2}$

51. $F(x) = \dfrac{8x}{x^2 + 4}$

52. $G(x) = \dfrac{-12x}{x^2 + 3}$

53. $p(x) = \dfrac{-2x^2}{x^2 - 4}$

54. $P(x) = \dfrac{3x^2}{x^2 - 9}$

55. $q(x) = \dfrac{2x - x^2}{x^2 + 4x - 5}$

56. $Q(x) = \dfrac{x^2 + 3x}{x^2 - 2x - 3}$

57. $h(x) = \dfrac{-3x}{x^2 - 6x + 9}$

58. $H(x) = \dfrac{2x}{x^2 - 2x + 1}$

59. $Y_1 = \dfrac{x - 1}{x^2 - 3x - 4}$

60. $Y_2 = \dfrac{1 - x}{x^2 - 2x}$

61. $s(x) = \dfrac{4x^2}{2x^2 + 4}$

62. $S(x) = \dfrac{-2x^2}{x^2 + 1}$

63. $Y_1 = \dfrac{x^2 - 4}{x^2 - 1}$

64. $Y_2 = \dfrac{x^2 - x - 6}{x^2 + x - 6}$

65. $v(x) = \dfrac{-2x}{x^3 + 2x^2 - 4x - 8}$

66. $V(x) = \dfrac{3x}{x^3 + x^2 - x - 1}$

Use the vertical asymptotes, x-intercepts, and their multiplicities to construct an equation that corresponds to each graph. Be sure the y-intercept estimated from the graph matches the value given by your equation for $x = 0$. Check work on a graphing calculator.

67.

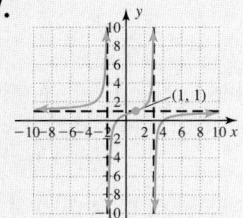

68.

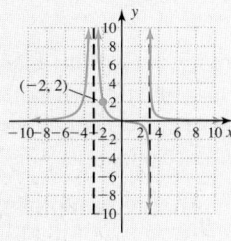

69.

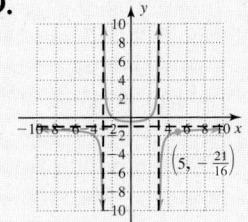

70.

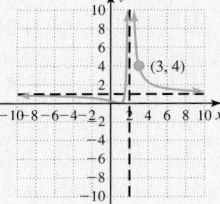

▶ WORKING WITH FORMULAS

71. Population density: $D(x) = \dfrac{ax}{x^2 + b}$

The population density of urban areas (in people per square mile) can be modeled by the formula shown, where a and b are constants related to the overall population and sprawl of the area under study, and $D(x)$ is the population density (in hundreds), x mi from the center of downtown.

Graph the function for $a = 63$ and $b = 20$ over the interval $x \in [0, 25]$, and then use the graph to answer the following questions.

  a. What is the significance of the *horizontal asymptote* (what does it mean in this context)?

  b. How far from downtown does the population density fall below 525 people per square mile? How far until the density falls below 300 people per square mile?

  c. Use the graph and a table to determine how far from downtown the population density reaches a maximum? What is this maximum?

72. Cost of removing pollutants: $C(x) = \dfrac{kx}{100 - x}$

Some industries resist cleaner air standards because the cost of removing pollutants rises dramatically as higher standards are set. This phenomenon can be modeled by the formula given, where $C(x)$ is the cost (in thousands of dollars) of removing $x\%$ of the pollutant and k is a constant that depends on the type of pollutant and other factors.

Graph the function for $k = 250$ over the interval $x \in [0, 100]$, and then use the graph to answer the following questions.

  a. What is the significance of the *vertical asymptote* (what does it mean in this context)?

  b. If new laws are passed that require 80% of a pollutant to be removed, while the existing law requires only 75%, how much will the new legislation cost the company? Compare the cost of the 5% increase from 75% to 80% with the cost of the 1% increase from 90% to 91%.

  c. What percent of the pollutants can be removed if the company budgets 2250 thousand dollars?

▶ APPLICATIONS

73. For a certain coal-burning power plant, the cost to remove pollutants from plant emissions can be modeled by $C(p) = \dfrac{80p}{100 - p}$, where $C(p)$ represents the cost (in thousands of dollars) to remove p percent of the pollutants. (a) Find the cost to remove 20%, 50%, and 80% of the

pollutants, then comment on the results; (b) graph the function using an appropriate scale; and (c) use the direction/approach notation to state what happens if the power company attempts to remove 100% of the pollutants.

74. A large city has initiated a new recycling effort, and wants to distribute recycling bins for use in separating various recyclable materials. City planners anticipate the cost of the program can be modeled by the function $C(p) = \dfrac{220p}{100 - p}$, where $C(p)$ represents the cost (in \$10,000) to distribute the bins to p percent of the population. (a) Find the cost to distribute bins to 25%, 50%, and 75% of the population, then comment on the results; (b) graph the function using an appropriate scale; and (c) use the direction/approach notation to state what happens if the city attempts to give recycling bins to 100% of the population.

75. The concentration C of a certain medicine in the bloodstream h hours after being injected into the shoulder is given by the function: $C(h) = \dfrac{2h^2 + h}{h^3 + 70}$. Use the given graph of the function to answer the following questions.

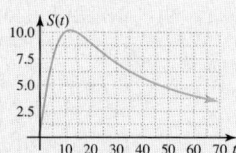

   a. Approximately how many hours after injection did the maximum concentration occur? What was the maximum concentration?

   b. Use $C(h)$ to *compute* the rate of change for the intervals $h = 8$ to $h = 10$ and $h = 20$ to $h = 22$. What do you notice?

   c. Use the direction/approach notation to state what happens to the concentration C as the number of hours becomes infinitely large. What role does the h-axis play for this function?

76. In response to certain market demands, manufacturers will quickly get a product out on the market to take advantage of consumer interest. Once the product is released, it is not uncommon for sales to initially skyrocket, taper off and then gradually decrease as consumer interest wanes. For a certain product, sales can be modeled by the function $S(t) = \dfrac{250t}{t^2 + 150}$, where $S(t)$ represents the daily sales (in \$10,000) t days after the product has debuted. Use the given graph of the function to answer the following questions.

   a. Approximately how many days after the product came out did sales reach a maximum? What was the maximum sales?

   b. Use $S(t)$ to compute the rate of change for the intervals $t = 7$ to $t = 8$ and $t = 60$ to $t = 62$. What do you notice?

   c. Use the direction/approach notation to state what happens to the daily sales S as the number of days becomes infinitely large. What role does the t-axis play for this function?

Memory retention: Due to their asymptotic behavior, rational functions are often used to model the mind's ability to retain information over a long period of time—the "use it or lose it" phenomenon.

77. A large group of students is asked to memorize a list of 50 Italian words, a language that is unfamiliar to them. The group is then tested regularly to see how many of the words are retained over a period of time. The average number of words retained is modeled by the function $W(t) = \dfrac{6t + 40}{t}$, where $W(t)$ represents the number of words remembered after t days.

   a. Graph the function over the interval $t \in [0, 40]$. How many days until only half the words are remembered? How many days until only one-fifth of the words are remembered?

   b. After 10 days, what is the average number of words retained? How many days until only 8 words can be recalled?

   c. What is the significance of the horizontal asymptote (what does it mean in this context)?

78. A similar study asked students to memorize 50 Hawaiian words, a language that is both unfamiliar and phonetically foreign to them (see Exercise 77). The average number of words retained is modeled by the function $W(t) = \dfrac{4t + 20}{t}$, where $W(t)$ represents the number of words after t days.

   a. Graph the function over the interval $t \in [0, 40]$. How many days until only half the words are remembered? How does this compare to Exercise 77? How many days until only one-fifth of the words are remembered?

   b. After 7 days, what is the average number of words retained? How many days until only 5 words can be recalled?

   c. What is the significance of the horizontal asymptote (what does it mean in this context)?

Concentration and dilution: When antifreeze is mixed with water, it becomes diluted—less than 100% antifreeze. The more water added, the less concentrated the antifreeze becomes, with this process continuing until a desired concentration is met. This application and many similar to it can be modeled by rational functions.

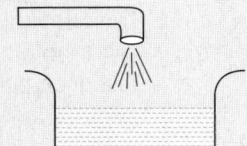

79. A 400-gal tank currently holds 40 gal of a 25% antifreeze solution. To raise the concentration of the antifreeze in the tank, x gal of a 75% antifreeze solution is pumped in.

   a. Show the formula for the resulting concentration is $C(x) = \dfrac{40 + 3x}{160 + 4x}$ after simplifying, and graph the function over the interval $x \in [0, 360]$.

   b. What is the concentration of the antifreeze in the tank after 10 gal of the new solution are added? After 120 gal have been added? How much liquid is now in the tank?

   c. If the concentration level is now at 65%, how many gallons of the 75% solution have been added? How many gallons of liquid are in the tank now?

   d. What is the maximum antifreeze concentration that can be attained in a tank of this size? What is the maximum concentration that can be attained in a tank of "unlimited" size?

80. A sodium chloride solution has a concentration of 0.2 oz (weight) per gallon. The solution is pumped into an 800-gal tank currently holding 40 gal of pure water, at a rate of 10 gal/min.

   a. Find a function $A(t)$ modeling the amount of liquid in the tank after t min, and a function $S(t)$ for the amount of sodium chloride in the tank after t min.

   b. The concentration $C(t)$ in ounces per gallon is measured by the ratio $\dfrac{S(t)}{A(t)}$, a rational function. Graph the function on the interval $t \in [0, 100]$. What is the concentration level (in ounces per gallon) after 6 min? After 28 min? How many gallons of liquid are in the tank at this time?

   c. If the concentration level is now 0.184 oz/gal, how long have the pumps been running? How many gallons of liquid are in the tank now?

   d. What is the maximum concentration that can be attained in a tank of this size? What is the maximum concentration that can be attained in a tank of "unlimited" size?

Average cost of manufacturing an item: The cost "C" to manufacture an item depends on the relatively fixed costs "K" for remaining in business (utilities, maintenance, transportation, etc.) and the actual cost "c" of manufacturing the item (labor and materials). For x items the cost is $C(x) = K + cx$. The average cost "A" of manufacturing an item is then $A(x) = \dfrac{C(x)}{x}$.

81. A company that manufactures water heaters finds their fixed costs are normally $50,000 per month, while the cost to manufacture each heater is $125. Due to factory size and the current equipment, the company can produce a maximum of 5000 water heaters per month during a good month.

   a. Use the average cost function to find the average cost if 500 water heaters are manufactured each month. What is the average cost if 1000 heaters are made?

   b. What level of production will bring the average cost down to $150 per water heater?

   c. If the average cost is currently $137.50, how many water heaters are being produced that month?

   d. What's the significance of the horizontal asymptote for the average cost function (what does it mean in this context)? Will the company ever break the $130 average cost level? Why or why not?

82. An enterprising company has finally developed a disposable diaper that is biodegradable. The brand becomes wildly popular and production is soaring. The fixed cost of production is $20,000 per month, while the cost of manufacturing is $6.00 per case (48 diapers). Even while working three shifts around-the-clock, the maximum production level is 16,000 cases per month. The company figures it will be profitable if it can bring costs down to an average of $7 per case.

   a. Use the average cost function to find the average cost if 2000 cases are produced each month. What is the average cost if 4000 cases are made?

   b. What level of production will bring the average cost down to $8 per case?

   c. If the average cost is currently $10 per case, how many cases are being produced?

   d. What's the significance of the horizontal asymptote for the average cost function (what does it mean in this context)? Will the company ever reach its goal of $7/case at its maximum production? What level of production would help them meet their goal?

Test averages and grade point averages: To calculate a test average we sum all test points P and divide by the number of tests N: $\dfrac{P}{N}$. To compute

the score or scores needed on future tests to raise the average grade to a desired grade G, we add the number of additional tests n to the denominator, and the number of additional tests times the projected grade g on each test to the numerator:

$G(n) = \dfrac{P + ng}{N + n}$. The result is a rational function with some "eye-opening" results.

83. After four tests, Bobby Lou's test average was an 84. [*Hint:* $P = 4(84) = 336$.]

   a. Assume that she gets a 95 on all remaining tests ($g = 95$). Graph the resulting function on a calculator using the window $n \in [0, 20]$ and $G(n) \in [80$ to $100]$. Use the calculator to determine how many tests are required to lift her grade to a 90 under these conditions.

   b. At some colleges, the range for an "A" grade is 93–100. How many tests would Bobby Lou have to score a 95 on, to raise her average to higher than 93? Were you surprised?

c. Describe the significance of the horizontal asymptote of the average grade function. Is a test average of 95 possible for her under these conditions?

d. Assume now that Bobby Lou scores 100 on all remaining tests ($g = 100$). Approximately how many more tests are required to lift her grade average to higher than 93?

84. At most colleges, $A \to 4$ grade points, $B \to 3$, $C \to 2$, and $D \to 1$. After taking 56 credit hours, Aurelio's GPA is 2.5. [*Hint:* In the formula given, $P = 2.5(56) = 140$.]

   a. Assume Aurelio is determined to get A's (4 grade points or $g = 4$), for all remaining credit hours. Graph the resulting function on a calculator using the window $n \in [0, 60]$ and $G(n) \in [2, 4]$. Use the calculator to determine the number of credit hours required to lift his GPA to over 2.75 under these conditions.

   b. At some colleges, scholarship money is available only to students with a 3.0 average or higher. How many (perfect 4.0) credit hours would Aurelio have to earn, to raise his GPA to 3.0 or higher? Were you surprised?

   c. Describe the significance of the horizontal asymptote of the GPA function. Is a GPA of 4.0 possible for him under these conditions?

▶ **EXTENDING THE CONCEPT**

85. In addition to determining *if* a function has a vertical asymptote, we are often interested in *how fast* the graph approaches the asymptote. As in previous investigations, this involves the function's rate of change over a small interval. Exercise 72 describes the rising cost of removing pollutants from the air. As noted there, the rate of increase in the cost changes as higher requirements are set. To quantify this change, we'll compute the rate of change

$\dfrac{\Delta C}{\Delta x} = \dfrac{C(x_2) - C(x_1)}{x_2 - x_1}$ for $C(x) = \dfrac{250x}{100 - x}$.

   a. Find the rate of change of the function in the following intervals:

      $x \in [60, 61]$ $x \in [70, 71]$

      $x \in [80, 81]$ $x \in [90, 91]$

   b. What do you notice? How much did the rate increase from the first interval to the second? From the second to the third? From the third to the fourth?

   c. Recompute parts (a) and (b) using the function $C(x) = \dfrac{350x}{100 - x}$. Comment on what you notice.

86. Consider the function $f(x) = \dfrac{ax^2 + k}{bx^2 + h}$, where a, b, k, and h are constants and $a, b > 0$.

   a. What can you say about asymptotes and intercepts of this function if $h, k > 0$?

   b. Now assume $k < 0$ and $h > 0$. How does this affect the asymptotes? The intercepts?

   c. If $b = 1$ and $a > 1$, how does this affect the results from part (b)?

   d. How is the graph affected if $k > 0$ and $h < 0$?

   e. Find values of a, b, h, and k that create a function with a horizontal asymptote at $y = \frac{3}{2}$, x-intercepts at $(-2, 0)$ and $(2, 0)$, a y-intercept of $(0, -4)$, and no vertical asymptotes.

87. The horizontal asymptotes of a rational function, and whether or not a graph crosses this asymptote, can be found using long division. The quotient polynomial $q(x)$ gives the equation of the asymptote, and the zeroes of the remainder $r(x)$ will indicate if and where the graph crosses it. Use this idea to help graph these functions.

   a. $V(x) = \dfrac{3x^2 - 16x - 20}{x^2 - 3x - 10}$

   b. $v(x) = \dfrac{-2x^2 + 4x + 13}{x^2 - 2x - 3}$

▶ MAINTAINING YOUR SKILLS

88. (R.1/1.4) Describe/Define each set of numbers: complex C, rational Q, and integers Z.

89. (2.3) Find the equation of a line that is perpendicular to $3x - 4y = 12$ and contains the point $(2, -3)$.

90. (1.5) Solve the following equation using the quadratic formula, then write the equation in factored form: $12x^2 + 55x - 48 = 0$.

91. (3.2) Use synthetic division and the remainder theorem to find the value of $f(4)$, $f(\frac{3}{2})$, and $f(2)$: $f(x) = 2x^3 - 7x^2 + 5x + 3$.

3.6 | Additional Insights into Rational Functions

Learning Objectives

In Section 3.6 you will learn how to:

❑ **A.** Graph rational functions with removable discontinuities

❑ **B.** Graph rational functions with oblique or non-linear asymptotes

❑ **C.** Solve applications involving rational functions

In Section 3.5, we saw that rational graphs can have both a horizontal and vertical asymptote. In this section, we'll study functions with asymptotes that are *neither* horizontal nor vertical. In addition, we'll further explore the "break" we saw in graphs of certain piecewise-defined functions, that of a simple "hole" created when the numerator and denominator share a common variable factor.

A. Rational Functions and Removable Discontinuities

In Example 5 of Section 2.7, we graphed the piecewise-defined function $h(x) = \begin{cases} \dfrac{x^2 - 4}{x - 2} & x \neq 2 \\ 1 & x = 2 \end{cases}$.

Figure 3.39

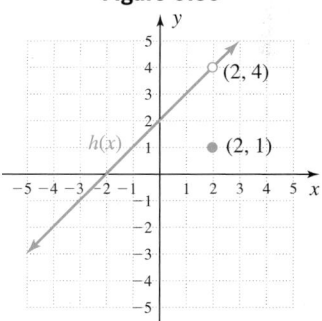

The second piece is simply the point $(2, 1)$. The first piece is a rational function, but instead of a vertical asymptote at $x = 2$ (the zero of the denominator), its graph was actually the line $y = x + 2$ with a "hole" at $(2, 4)$, called a **removable discontinuity** (Figure 3.39). As the name implies, we can *remove* or fix this break by redefining the second piece as $h(x) = 4$, when $x = 2$. This would create a new and continuous function,

$$H(x) = \begin{cases} \dfrac{x^2 - 4}{x - 2} & x \neq 2 \\ 4 & x = 2 \end{cases}$$ (Figure 3.40). It's possible

for a rational graph to have more than one removable discontinuity, or to be nonlinear with a removable discontinuity. For cases where we elect to repair the break, we will adopt the convention of using the corresponding upper case letter to name the new function, as we did here.

Figure 3.40

WORTHY OF NOTE

The graph of $f(x) = \dfrac{1}{x - 2}$ also has a break at $x = 2$, but this time the result is a *vertical asymptote*. The difference is the numerator and denominator of $h(x) = \dfrac{x^2 - 4}{x - 2}$ share a common factor, and canceling these factors leaves $y = x + 2$, which is a continuous function. However, the *original function* is not defined at $x = 2$, so we must remove the single point $(2, 4)$ from the domain of $y = x + 2$ (Figure 3.39).

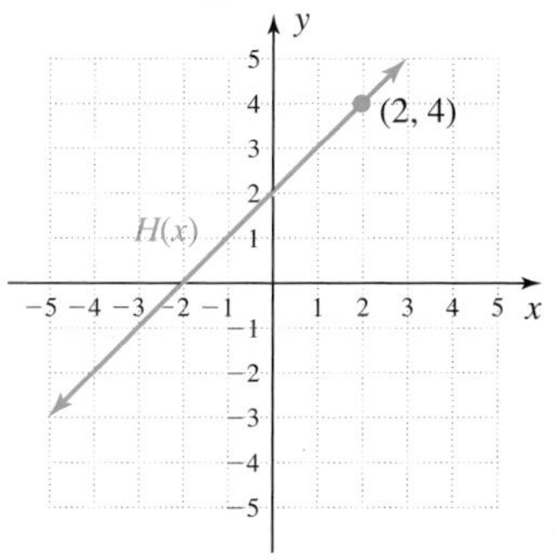

EXAMPLE 1 ▶ **Graphing Rational Functions with Removable Discontinuities**

Graph the function $t(x) = \dfrac{x^3 + 8}{x + 2}$. If there is a removable discontinuity, repair the break using an appropriate piecewise-defined function.

Solution ▶ Note the domain of t does not include $x = -2$. We begin by factoring as before to identify zeroes and asymptotes, but find the numerator and denominator share a common factor, which we remove.

$$t(x) = \frac{x^3 + 8}{x + 2}$$

$$= \frac{(x + 2)(x^2 - 2x + 4)}{x + 2}$$

$$= x^2 - 2x + 4; \text{ where } x \neq -2$$

The graph of t will be the same as $y = x^2 - 2x + 4$ *for all values except $x = -2$.* Here we have a parabola, opening upward, with y-intercept $(0, 4)$. From the vertex formula, the x-coordinate of the vertex will be $\dfrac{-b}{2a} = \dfrac{-(-2)}{2(1)} = 1$, giving $y = 3$ after substitution. The vertex is $(1, 3)$. Evaluating $t(-1)$ we find $(-1, 7)$ is on the graph, giving the point $(3, 7)$ using the axis of symmetry. We draw a parabola through these points, noting the original function is not defined at -2, and there will be a "hole" in the graph at $(-2, y)$. The value of y is found by substituting -2 for x in the simplified form: $(-2)^2 - 2(-2) + 4 = 12$. This information produces the graph shown. We can repair the break using the function

$$T(x) = \begin{cases} \dfrac{x^3 + 8}{x + 2} & x \neq -2 \\ 12 & x = -2 \end{cases}$$

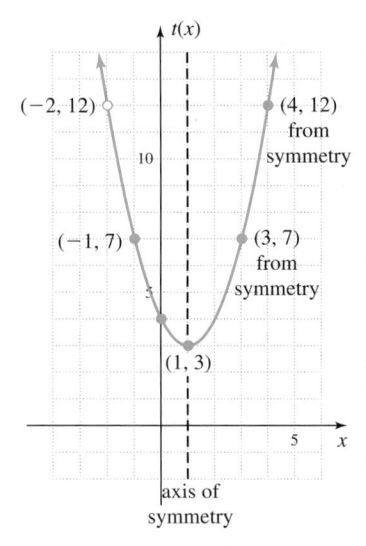

WORTHY OF NOTE

For more on removable discontinuities, see the *Technology Highlight* feature on page 370.

☑ **A.** You've just learned how to graph rational functions with removable discontinuities

Now try Exercises 7 through 18 ▶

B. Rational Functions with Oblique and Nonlinear Asymptotes

In Section 3.5, we found that for $V(x) = \dfrac{p(x)}{d(x)}$, the location of nonvertical asymptotes was determined by comparing the degree of p with the degree of d. As review, for $p(x)$ with leading term ax^n and $d(x)$ with leading term degree bx^m,

- If $n < m$, the line $y = 0$ is a horizontal asymptote.
- If $n = m$, the line $y = \frac{a}{b}$ is a horizontal asymptote.

But what happens if the degree of the numerator is *greater than* the degree of the denominator? To investigate, consider the functions f, g, and h in Figures 3.41 to 3.43, whose only difference is the degree of the numerator.

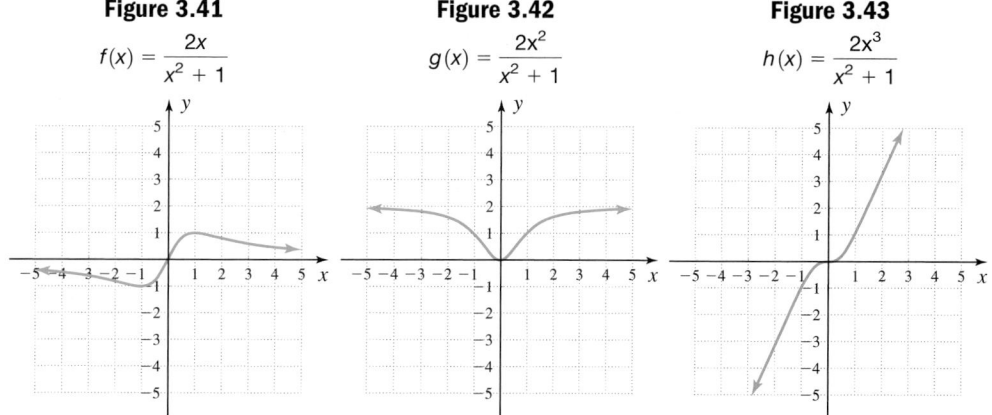

Figure 3.41

$f(x) = \dfrac{2x}{x^2 + 1}$

Figure 3.42

$g(x) = \dfrac{2x^2}{x^2 + 1}$

Figure 3.43

$h(x) = \dfrac{2x^3}{x^2 + 1}$

The graph of f has a horizontal asymptote at $y = 0$ since the denominator is of larger degree (as $|x| \to \infty$, $y \to 0$). As we might have anticipated, the horizontal asymptote for g is $y = 2$, the ratio of leading coefficients (as $|x| \to \infty$, $y \to 2$). The graph of h has no horizontal asymptote, yet appears to be asymptotic to some slanted line. The table in Figure 3.44 suggests that as $|x| \to \infty$, $y \to 2x = Y_2$. To see why, note the function $h(x) = \dfrac{2x^3}{x^2 + 1}$ can be considered an "improper fraction," similar to how we apply this designation to the fraction $\frac{3}{2}$. To write h in "proper" form, we use long division, writing the dividend as $2x^3 + 0x^2 + 0x + 0$, and the divisor as $x^2 + 0x + 1$.

Figure 3.44

$Y_2 = 2X$

X	Y1	Y2
0	0	0
25	49.92	50
50	99.96	100
75	149.97	150
100	199.98	200
125	249.98	250
150	299.99	300

X=0

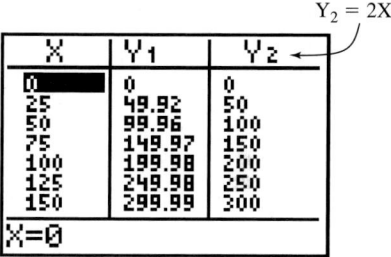

The ratio $\dfrac{2x^3}{x^2}$ (from dividend / from divisor) shows $\mathbf{2x}$ will be our first multiplier.

$$\begin{array}{r} 2x \\ \text{divisor} \rightarrow x^2 + 0x + 1 \overline{)2x^3 + 0x^2 + 0x + 0} \\ \underline{-(2x^3 + 0x^2 + 2x)} \\ -2x \end{array}$$

multiply $2x(x^2 + 0x + 1)$

subtract, next term is 0

The result shows $h(x) = 2x + \dfrac{-2x}{x^2 + 1}$. Note as $|x| \to \infty$, the term $\dfrac{-2x}{x^2 + 1}$ becomes very small and closer to zero, so $h(x) \approx 2x$ for large x. This is an example of an **oblique asymptote.** In general,

Oblique and Nonlinear Asymptotes

Given $V(x) = \frac{p(x)}{d(x)}$ is a rational function in simplest form, where the degree of p is greater than the degree of d, the graph will have an oblique or nonlinear asymptote as determined by $q(x)$, where $q(x)$ is the quotient polynomial after division.

We conclude that an oblique or slant asymptote occurs when the degree of the numerator is one more than the degree of the denominator, and a nonlinear asymptote occurs when its degree is larger by two or more.

EXAMPLE 2 ▶ **Graphing a Rational Function with an Oblique Asymptote**

Graph the function $f(x) = \dfrac{x^2 - 1}{x}$.

Solution ▶ Using the *Guidelines*, we find $f(x) = \dfrac{(x + 1)(x - 1)}{x}$ and proceed:

1. *y*-intercept: The graph has no *y*-intercept.

2. Vertical asymptote(s): $x = 0$ with multiplicity 1. The function will change sign at $x = 0$.

3. *x*-intercepts: From $(x + 1)(x - 1) = 0$, the *x*-intercepts are $(-1, 0)$ and $(1, 0)$. Since both have multiplicity 1, the graph will cross the *x*-axis and the function will change sign at these points

4. Horizontal/oblique asymptote: Since the degree of numerator $>$ the degree of denominator, we rewrite f using division. Using term-by-term division (the denominator is a monomial) produces $f(x) = \dfrac{x^2 - 1}{x} = \dfrac{x^2}{x} - \dfrac{1}{x} = x - \dfrac{1}{x}$. The quotient polynomial is $q(x) = x$ and the graph has the oblique asymptote $y = x$.

5. To determine if the function will cross the asymptote, we solve

$$\frac{x^2 - 1}{x} = x \qquad \text{\small } q(x) = x \text{ is the slant asymptote}$$

$$x^2 - 1 = x^2 \qquad \text{\small multiply by } x$$

$$-1 = 0 \qquad \text{\small no solutions possible}$$

The graph will not cross the oblique asymptote.

The information from steps 1 through 5 is displayed in Figure 3.45. While this is sufficient to complete the graph, we select $x = -4$ and 4 to compute additional points and find $f(-4) = -\frac{15}{4}$ and $f(4) = \frac{15}{4}$. To meet all necessary conditions, we complete the graph as shown in Figure 3.46.

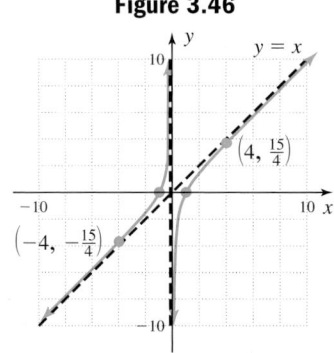

Figure 3.45

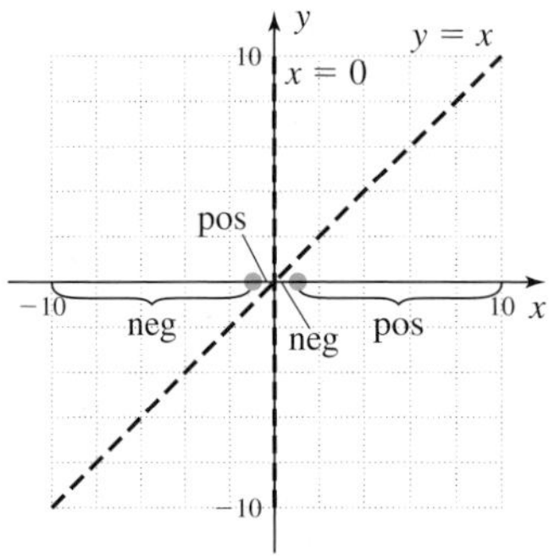

Figure 3.46

Now try Exercises 19 through 24 ▶

EXAMPLE 3 ▶ **Graphing a Rational Function with an Oblique Asymptote**

Graph the function: $h(x) = \dfrac{x^2}{x - 1}$

Solution ▶ The function is already in "factored form."

1. *y*-intercept: Since $h(0) = 0$, the *y*-intercept is $(0, 0)$.

2. Vertical asymptote: Solving $x - 1 = 0$ gives $x = 1$ with multiplicity one. There is a vertical asymptote at $x = 1$ and the function will change sign here.

3. x-intercept: $(0, 0)$; From, $x^2 = 0$, we have $x = 0$ with multiplicity two. The x-intercept is $(0, 0)$ and the function will not change sign here.

4. Horizontal/oblique asymptote: Since the degree of numerator $>$ the degree of denominator, we rewrite h using division. The denominator is linear so we use synthetic division:

$$\text{use 1 as a "divisor"}\quad \underline{1\rvert}\quad \begin{array}{ccc} 1 & 0 & 0 \\ & \downarrow 1 & 1 \\ \hline 1 & 1 & 1 \end{array}\quad \begin{array}{l}\text{coefficients of dividend} \\ \\ \text{quotient and remainder}\end{array}$$

Since $q(x) = x + 1$ the graph has an oblique asymptote at $y = x + 1$.

5. To determine if the function crosses the asymptote, we solve

$$\frac{x^2}{x - 1} = x + 1 \qquad q(x) = x + 1 \text{ is the slant asymptote}$$
$$x^2 = x^2 - 1 \qquad \text{cross multiply}$$
$$0 = -1 \qquad \text{no solutions possible}$$

The graph will not cross the slant asymptote.

The information gathered in steps 1 through 5 is shown Figure 3.47, and is actually sufficient to complete the graph. If you feel a little unsure about how to "puzzle" out the graph, find additional points in the first and third quadrants: $h(2) = 4$ and $h(-2) = -\frac{4}{3}$. Since the graph will "bounce" at $x = 0$ and output values must change sign at $x = 1$, all conditions are met with the graph shown in Figure 3.48.

Figure 3.47

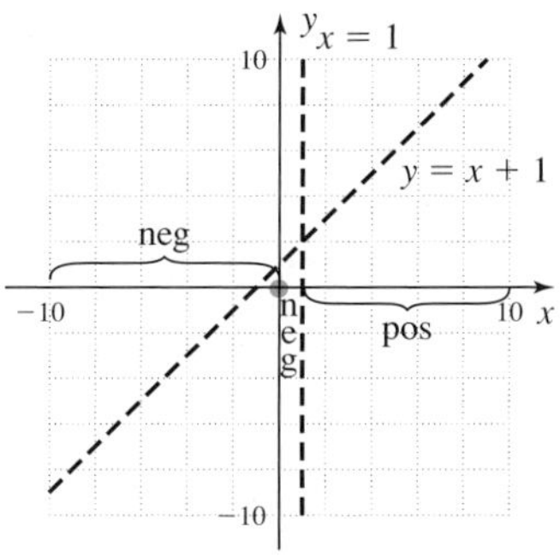

Figure 3.48

Now try Exercises 25 through 46 ▶

Finally, it would be a mistake to think that all asymptotes are linear. In fact, when the degree of the numerator is two more than the degree of the denominator, a parabolic asymptote results. Functions of this type often occur in applications of rational functions, and are used to minimize cost, materials, distances, or other considerations of great importance to business and industry. For $f(x) = \dfrac{x^4 + 1}{x^2}$, term-by-term division

Figure 3.49

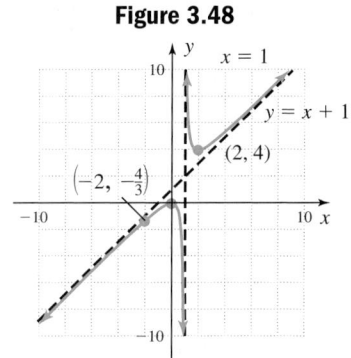

gives $x^2 + \dfrac{1}{x^2}$ and the quotient $q(x) = x^2$ is a nonlinear,

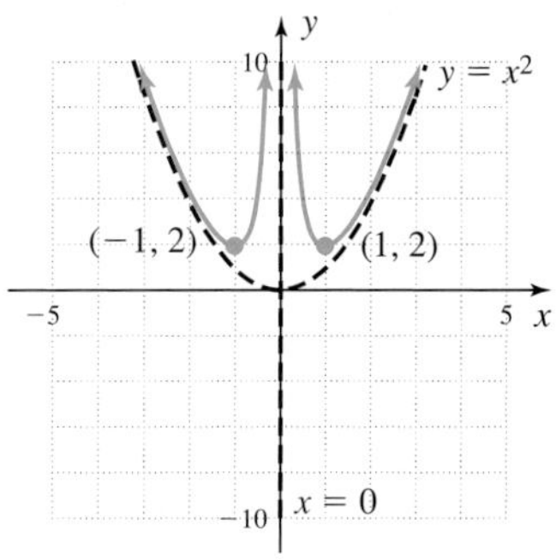

☑ **B.** You've just learned how to graph rational functions with oblique or nonlinear asymptotes

parabolic asymptote (see Figure 3.49). For more on nonlinear asymptotes, **see Exercises 47 through 50.**

C. Applications of Rational Functions

Rational functions have applications in a wide variety of fields, including environmental studies, manufacturing, and various branches of medicine. In most practical applications, only the values from Quadrant I have meaning since inputs and outputs must often be positive (**see Exercises 51 and 52**). Here we investigate an application involving manufacturing and average cost.

EXAMPLE 4 ▶ **Solving an Application of Rational Functions**

Suppose the cost (in thousands of dollars) of manufacturing x thousand of a given item is modeled by the function $C(x) = x^2 + 4x + 3$. The *average cost* of each item would then be expressed by

$$A(x) = \frac{x^2 + 4x + 3}{x} = \frac{\text{total cost}}{\text{number of items}}$$

a. Graph the function $A(x)$.

b. Find how many thousand items are manufactured when the average cost is $8.

c. Determine how many thousand items should be manufactured to minimize the average cost (use the graph to estimate this minimum average cost).

Solution ▶ **a.** The function is already in simplest form.

1. y-intercept: none [$A(0)$ is undefined]

2. Vertical asymptote: $x = 0$, multiplicity one; the function will change sign at $x = 0$.

3. x-intercept(s): After factoring we obtain $(x + 3)(x + 1) = 0$, and the zeroes of the numerator are $x = -1$ and $x = -3$, both with multiplicity one. The graph will cross the x-axis at each intercept.

4. Horizontal/oblique asymptote: The degree of numerator > the degree of denominator, so we divide using term-by-term division:

$$\frac{x^2 + 4x + 3}{x} = \frac{x^2}{x} + \frac{4x}{x} + \frac{3}{x}$$

$$= x + 4 + \frac{3}{x}$$

The line $q(x) = x + 4$ is an oblique asymptote.

5. Solve

$$\frac{x^2 + 4x + 3}{x} = x + 4 \qquad \text{$q(x) = x + 4$ is a slant asymptote}$$

$$x^2 + 4x + 3 = x^2 + 4x \qquad \text{cross multiply}$$

$$3 = 0 \qquad \text{no solutions possible}$$

The graph will not cross the slant asymptote.

The function changes sign at both x-intercepts and at the asymptote $x = 0$. The information from steps 1 through 5 is shown in Figure 3.50 and perhaps an additional point in Quadrant I would help to complete the graph: $A(1) = 8$. The point $(1, 8)$ is on the graph, showing A is positive in the interval containing 1. Since output values will alternate in sign as stipulated above, all conditions are met with the graph shown in Figure 3.51.

Figure 3.50

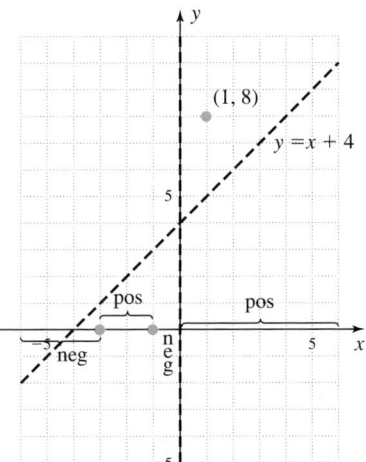

Figure 3.51

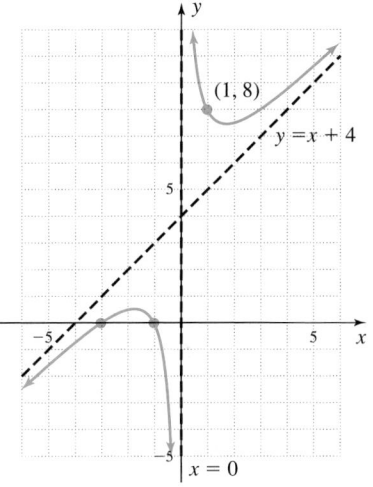

b. To find the number of items manufactured when average cost is $8, we replace $A(x)$ with 8 and solve: $\dfrac{x^2 + 4x + 3}{x} = 8$:

$$x^2 + 4x + 3 = 8x$$
$$x^2 - 4x + 3 = 0$$
$$(x - 1)(x - 3) = 0$$
$$x = 1 \quad \text{or} \quad x = 3$$

The average cost is $8 when 1000 items or 3000 items are manufactured.

c. From the graph, it appears that the minimum average cost is close to $7.50, when approximately 1500 to 1800 items are manufactured.

Now try Exercises 55 and 56 ▶

GRAPHICAL SUPPORT

In the *Technology Highlight* from Section 2.5, we saw how a graphing calculator can be used to locate the extreme values of a function. Applying this technology to the graph from Example 4 we find that the minimum average cost is approximately $7.46, when about 1732 items are manufactured.

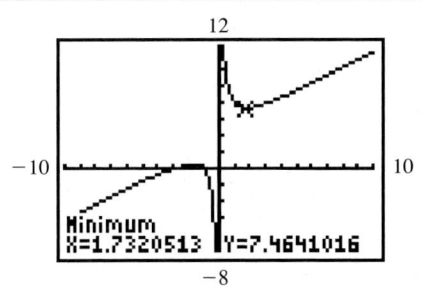

In some applications, the functions we use are initially defined in *two variables* rather than just one, as in $H(x, y) = (x - 50)(y - 80)$. However, in the solution process a substitution is used to rewrite the relationship as a rational function in one variable and we can proceed as before.

EXAMPLE 5 ▶ **Using a Rational Function to Solve a Layout Application**

 The building codes in a new subdivision require that a rectangular home be built at least 20 ft from the street, 40 ft from the neighboring lots, and 30 ft from the house to the rear fence line.

a. Find a function $A(x, y)$ for the area of the lot, and a function $H(x, y)$ for the area of the home (the inner rectangle).

b. If a new home is to have a floor area of 2000 ft², $H(x, y) = 2000$. Substitute 2000 for $H(x, y)$ and solve for y, then substitute the result in $A(x, y)$ to write the area A as a function of x alone (simplify the result).

c. Graph $A(x)$ on a calculator, using the window $X \in [-50, 150]$; $Y \in [-30{,}000, 30{,}000]$. Then graph $y = 80x + 2000$ on the same screen. How are these two graphs related?

d. Use the graph of $A(x)$ in Quadrant I to determine the minimum dimensions of a lot that satisfies the subdivision's requirements (to the nearest tenth of a foot). Also state the dimensions of the house.

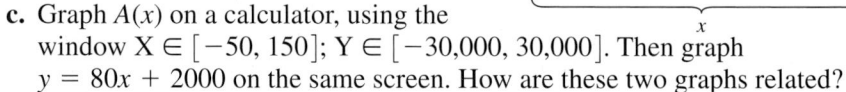

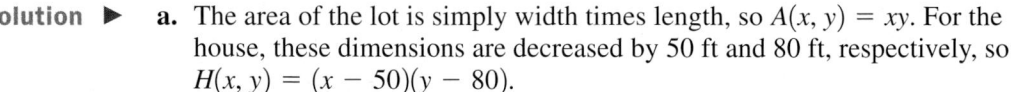

Solution ▶ a. The area of the lot is simply width times length, so $A(x, y) = xy$. For the house, these dimensions are decreased by 50 ft and 80 ft, respectively, so $H(x, y) = (x - 50)(y - 80)$.

b. Given $H(x, y) = 2000$ produces the equation $2000 = (x - 50)(y - 80)$, and solving for y gives

$$2000 = (x - 50)(y - 80) \quad \text{given equation}$$

$$\frac{2000}{x - 50} = y - 80 \quad \text{divide by } x - 50$$

$$\frac{2000}{x - 50} + 80 = y \quad \text{add 80}$$

$$\frac{2000}{x - 50} + \frac{80(x - 50)}{x - 50} = y \quad \text{find LCD}$$

$$\frac{80x - 2000}{x - 50} = y \quad \text{combine terms}$$

Substituting this expression for y in $A(x, y) = xy$ produces

$$A(x) = x\left(\frac{80x - 2000}{x - 50}\right) \quad \text{substitute } \frac{80x - 2000}{x - 50} \text{ for } y$$

$$= \frac{80x^2 - 2000x}{x - 50} \quad \text{multiply}$$

 c. The graph of $Y1 = A(x)$ appears in Figure 3.52 using the prescribed window. $Y_2 = 80x + 2000$ appears to be an oblique asymptote for A, which can be verified using synthetic division.

Figure 3.52

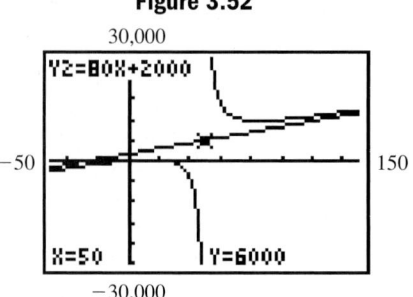

d. Using the [2nd] [TRACE] (**CALC**)
3:minimum feature of a calculator, the
minimum width is $x \approx 85.4$ ft.
Substituting 85.4 for x in
$y = \dfrac{80x - 2000}{x - 50}$, gives the length
$y \approx 136.5$ ft. The dimensions of the
house must be $85.4 - 50 = 35.4$ ft,
by $136.5 - 80 = 56.5$ ft
(see Figure 3.53).
    As expected, the area of the house will be $(35.4)(56.5) \approx 2000$ ft².

Figure 3.53

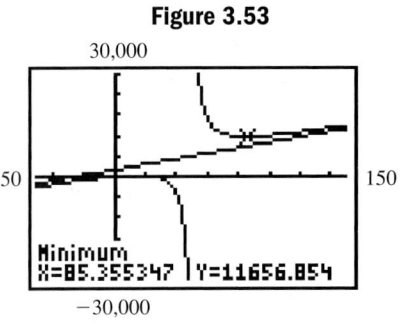

☑ **C.** You've just learned how
to solve applications involving
rational functions

Now try Exercises 57 through 60 ▶

TECHNOLOGY HIGHLIGHT

Removable Discontinuities

Graphing calculators offer both numerical and visual
representations of removable discontinuities. For instance, enter
the function $r(x) = \dfrac{x^2 - 4x + 3}{x - 1}$ on the [Y=] screen, then use the
[TBLSET] feature to set up the table as shown in Figure 3.54.
Pressing [2nd] [GRAPH] displays the expected table, which shows
the function cannot be evaluated at $x = 1$ (see Figure 3.55). Now
change the [TBLSET] screen so that **ΔTbl = 0.01**. Note again that
the function is defined for all values except $x = 1$. Reset the table
to **ΔTbl = 0.001** and investigate further.

Figure 3.54

    We can actually see the gap or hole in the graph using a "friendly window." Since the screen of the
TI-84 Plus is 95 pixels wide and 63 pixels high, multiples of 4.7 for Xmin and Xmax, and multiples of 3.1
for Ymin and Ymax, display what happens at integer (and other) values (see Figure 3.56). Pressing [GRAPH]
gives Figure 3.57, which shows a noticeable gap at $(1, -2)$. With the [TRACE] feature, move the cursor over
to the gap and notice what happens.
    Use these ideas to view the discontinuities in the following rational functions. State the ordered
pair location of each discontinuity.

Figure 3.55

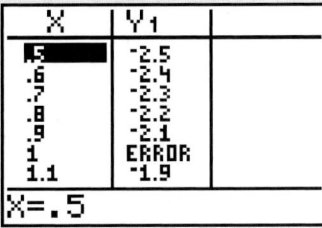

Figure 3.56

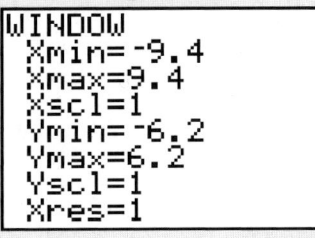

Figure 3.57

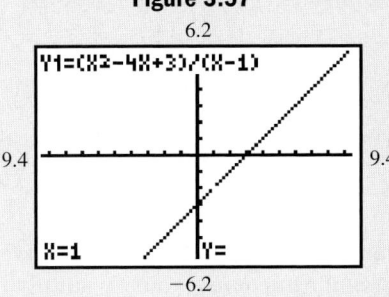

Exercise 1: $r(x) = \dfrac{x^2 - 4}{x + 2}$

Exercise 2: $f(x) = \dfrac{x^2 - 2x - 3}{x + 1}$

Exercise 3: $r(x) = \dfrac{x^3 + 1}{x + 1}$

Exercise 4: $f(x) = \dfrac{x^3 - 7x + 6}{x^2 + x - 6}$

3.6 EXERCISES

▶ CONCEPTS AND VOCABULARY

Fill in each blank with the appropriate word or phrase. Carefully reread the section if needed.

1. The discontinuity in the graph of $y = \dfrac{1}{(x+3)^2}$ is called a _____ discontinuity, since it cannot be "repaired."

2. If the degree of the numerator is greater than the degree of the denominator, the graph will have an _____ or _____ asymptote.

3. If the degree of the numerator is _____ more than the degree of the denominator, the graph will have a parabolic asymptote.

4. If the denominator is a _____, use term by term division to find the quotient. Otherwise _____ or long division must be used.

5. Discuss/Explain how you would create a function with a parabolic asymptote and two vertical asymptotes.

6. Complete Exercise 7 in expository form. That is, work this exercise out completely, discussing each step of the process as you go.

▶ DEVELOPING YOUR SKILLS

Graph each function. If there is a removable discontinuity, repair the break using an appropriate piecewise-defined function.

7. $f(x) = \dfrac{x^2 - 4}{x + 2}$

8. $f(x) = \dfrac{x^2 - 9}{x + 3}$

9. $g(x) = \dfrac{x^2 - 2x - 3}{x + 1}$

10. $g(x) = \dfrac{x^2 - 3x - 10}{x - 5}$

11. $h(x) = \dfrac{3x - 2x^2}{2x - 3}$

12. $h(x) = \dfrac{4x - 5x^2}{5x - 4}$

13. $p(x) = \dfrac{x^3 - 8}{x - 2}$

14. $p(x) = \dfrac{8x^3 - 1}{2x - 1}$

15. $q(x) = \dfrac{x^3 - 7x - 6}{x + 1}$

16. $q(x) = \dfrac{x^3 - 3x + 2}{x + 2}$

17. $r(x) = \dfrac{x^3 + 3x^2 - x - 3}{x^2 + 2x - 3}$

18. $r(x) = \dfrac{x^3 - 2x^2 - 4x + 8}{x^2 - 4}$

Graph each function using the *Guidelines for Graphing Rational Functions,* which is simply modified to include nonlinear asymptotes. Clearly label all intercepts and asymptotes and any additional points used to sketch the graph.

19. $Y_1 = \dfrac{x^2 - 4}{x}$

20. $Y_2 = \dfrac{x^2 - x - 6}{x}$

21. $v(x) = \dfrac{3 - x^2}{x}$

22. $V(x) = \dfrac{7 - x^2}{x}$

23. $w(x) = \dfrac{x^2 + 1}{x}$

24. $W(x) = \dfrac{x^2 + 4}{2x}$

25. $h(x) = \dfrac{x^3 - 2x^2 + 3}{x^2}$

26. $H(x) = \dfrac{x^3 + x^2 - 2}{x^2}$

27. $Y_1 = \dfrac{x^3 + 3x^2 - 4}{x^2}$

28. $Y_2 = \dfrac{x^3 - 3x^2 + 4}{x^2}$

29. $f(x) = \dfrac{x^3 - 3x + 2}{x^2}$

30. $F(x) = \dfrac{x^3 - 12x - 16}{x^2}$

31. $Y_3 = \dfrac{x^3 - 5x^2 + 4}{x^2}$

32. $Y_4 = \dfrac{x^3 + 5x^2 - 6}{x^2}$

33. $r(x) = \dfrac{x^3 - x^2 - 4x + 4}{x^2}$

34. $R(x) = \dfrac{x^3 - 2x^2 - 9x + 18}{x^2}$

35. $g(x) = \dfrac{x^2 + 4x + 4}{x + 3}$ **36.** $G(x) = \dfrac{x^2 - 2x + 1}{x - 2}$

37. $f(x) = \dfrac{x^2 + 1}{x + 1}$ **38.** $F(x) = \dfrac{x^2 + x + 1}{x - 1}$

39. $Y_3 = \dfrac{x^2 - 4}{x + 1}$ **40.** $Y_4 = \dfrac{x^2 - x - 6}{x - 1}$

41. $v(x) = \dfrac{x^3 - 4x}{x^2 - 1}$ **42.** $V(x) = \dfrac{9x - x^3}{x^2 - 4}$

43. $w(x) = \dfrac{16x - x^3}{x^2 + 4}$ **44.** $W(x) = \dfrac{x^3 - 7x + 6}{2 + x^2}$

45. $Y_1 = \dfrac{x^3 - 3x + 2}{x^2 - 9}$ **46.** $Y_2 = \dfrac{x^3 - x^2 - 12x}{x^2 - 7}$

47. $p(x) = \dfrac{x^4 + 4}{x^2 + 1}$ **48.** $P(x) = \dfrac{x^4 - 5x^2 + 4}{x^2 + 2}$

49. $q(x) = \dfrac{10 + 9x^2 - x^4}{x^2 + 5}$ **50.** $Q(x) = \dfrac{x^4 - 2x^2 + 3}{x^2}$

 Graph each function and its nonlinear asymptote on the same screen, using the window specified. Then locate the minimum value of f in the first quadrant.

51. $f(x) = \dfrac{x^3 + 500}{x}$;

$x \in [-24, 24], y \in [-500, 500]$

52. $f(x) = \dfrac{2\pi x^3 + 750}{x}$;

$x \in [-12, 12], y \in [-750, 750]$

 ▶ **WORKING WITH FORMULAS**

53. Area of a first quadrant triangle:

$$A(a) = \frac{1}{2}\left(\frac{ka^2}{a - h}\right)$$

The area of a right triangle in the first quadrant, formed by a line with negative slope through the point (h, k) and legs that lie along the positive axes is given by the formula shown, where a represents the x-intercept of the resulting line $(h < a)$. The area of the triangle varies with the slope of the line. Assume the line contains the point $(5, 6)$.

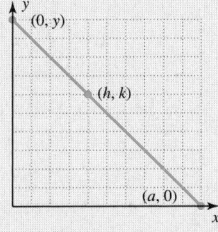

a. Find the equation of the vertical and slant asymptotes.

b. Find the area of the triangle if it has an x-intercept of $(11, 0)$.

 c. Use a graphing calculator to graph the function on an appropriate window. Does the shape of the graph look familiar? Use the calculator to find the value of a that minimizes $A(a)$. That is, find the x-intercept that results in a triangle with the smallest possible area.

54. Surface area of a cylinder with fixed volume:

$$S = \frac{2\pi r^3 + 2V}{r}$$

It's possible to construct many different cylinders that will hold a specified volume, by changing the radius and height. This is critically important to producers who want to minimize the cost of packing canned goods and marketers who want to present an attractive product. The surface area of the cylinder can be found using the formula shown, where the radius is r and $V = \pi r^2 h$ is known. Assume the fixed volume is 750 cm³.

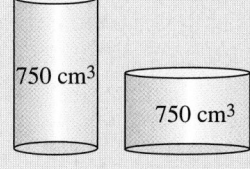

a. Find the equation of the vertical asymptote. How would you describe the nonlinear asymptote?

b. If the radius of the cylinder is 2 cm, what is its surface area?

 c. Use a graphing calculator to graph the function on an appropriate window, and use it to find the value of r that minimizes $S(r)$. That is, find the radius that results in a cylinder with the smallest possible area, while still holding a volume of 750 cm³.

▶ **APPLICATIONS**

 Costs of manufacturing: As in Example 4, the cost $C(x)$ of manufacturing is sometimes nonlinear and can increase dramatically with each item. For the average cost function $A(x) = \dfrac{C(x)}{x}$, consider the following.

55. Assume the monthly cost of manufacturing custom-crafted storage sheds is modeled by the function $C(x) = 4x^2 + 53x + 250$.

a. Write the average cost function and state the equation of the vertical and oblique asymptotes.

b. Enter the cost function $C(x)$ as Y_1 on a graphing calculator, and the average cost function $A(x)$ as Y_2. Using the TABLE feature, find the cost and average cost of making 1, 2, and 3 sheds.

c. Scroll down the table to where it appears that average cost is a minimum. According to the table, how many sheds should be made each month to minimize costs? What is the minimum cost?

d. Graph the average cost function and its asymptotes, using a window that shows the entire function. Use the graph to confirm the result from part (c).

56. Assume the monthly cost of manufacturing playground equipment that combines a play house, slides, and swings is modeled by the function $C(x) = 5x^2 + 94x + 576$. The company has projected that they will be profitable if they can bring their average cost down to $200 per set of playground equipment.

a. Write the average cost function and state the equation of the vertical and oblique asymptotes.

b. Enter the cost function $C(x)$ as Y_1 on a graphing calculator, and the average cost function $A(x)$ as Y_2. Using the TABLE feature, find the cost and average cost of making 1, 2, and 3 playground equipment combinations. Why would the average cost fall so dramatically early on?

c. Scroll down the table to where it appears that average cost is a minimum. According to the table, how many sets of equipment should be made each month to minimize costs? What is the minimum cost? Will the company be profitable under these conditions?

d. Graph the average cost function and its asymptotes, using a window that shows the entire function. Use the graph to confirm the result from part (c).

Minimum cost of packaging: Similar to Exercise 54, manufacturers can minimize their costs by shipping merchandise in packages that use a minimum amount of material. After all, rectangular boxes come in different sizes and there are many combinations of length, width, and height that will hold a specified volume.

57. A clothing manufacturer wishes to ship lots of 12 ft^3 of clothing in boxes with square ends and rectangular sides.

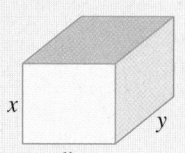

a. Find a function $S(x, y)$ for the surface area of the box, and a function $V(x, y)$ for the volume of the box.

b. Solve for y in $V(x, y) = 12$ (volume is 12 ft^3) and use the result to write the surface area as a function $S(x)$ in terms of x alone (simplify the result).

c. On a graphing calculator, graph the function $S(x)$ using the window $x \in [-8, 8]$; $y \in [-100, 100]$. Then graph $y = 2x^2$ on the same screen. How are these two graphs related?

d. Use the graph of $S(x)$ in Quadrant I to determine the dimensions that will minimize the surface area of the box, yet still hold 12 ft^3 of clothing. Clearly state the values of x and y, *in terms of feet and inches*, rounded to the nearest $\frac{1}{2}$ in.

58. A maker of packaging materials needs to ship 36 ft^3 of foam "peanuts" to his customers across the country, using boxes with the dimensions shown.

a. Find a function $S(x, y)$ for the surface area of the box, and a function $V(x, y)$ for the volume of the box.

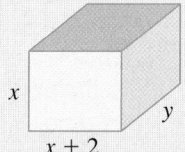

b. Solve for y in $V(x, y) = 36$ (volume is 36 ft^3), and use the result to write the surface area as a function $S(x)$ in terms of x alone (simplify the result).

c. On a graphing calculator, graph the function $S(x)$ using the window $x \in [-10, 10]$; $y \in [-200, 200]$. Then graph $y = 2x^2 + 4x$ on the same screen. How are these two graphs related?

d. Use the graph of $S(x)$ in Quadrant I to determine the dimensions that will minimize the surface area of the box, yet still hold the foam peanuts. Clearly state the values of x and y, *in terms of feet and inches*, rounded to the nearest $\frac{1}{2}$ in.

Printing and publishing: In the design of magazine pages, posters, and other published materials, an effort is made to maximize the usable area of the page while maintaining an attractive border, or minimizing the page size that will hold a certain amount of print or art work.

59. An editor has a story that requires 60 in^2 of print. Company standards require a 1-in. border at the top and bottom of a page, and 1.25-in. borders along both sides.

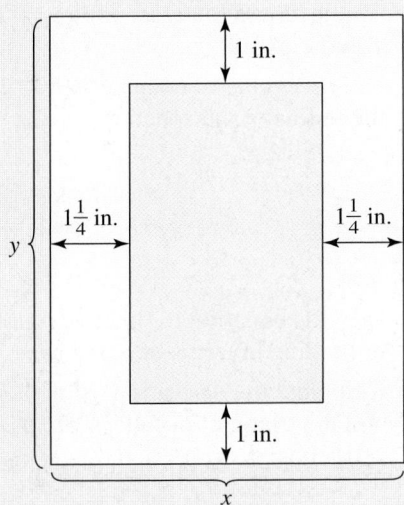

a. Find a function $A(x, y)$ for the area of the page, and a function $R(x, y)$ for the area of the inner rectangle (the printed portion).

b. Solve for y in $R(x, y) = 60$, and use the result to write the area from part (a) as a function $A(x)$ in terms of x alone (simplify the result).

c. On a graphing calculator, graph the function $A(x)$ using the window $x \in [-30, 30]$; $y \in [-100, 200]$. Then graph $y = 2x + 60$ on the same screen. How are these two graphs related?

d. Use the graph of $A(x)$ in Quadrant I to determine the page of minimum size that satisfies these border requirements and holds the necessary print. Clearly state the values of x and y, rounded to the nearest hundredth of an inch.

60. *The Poster Shoppe* creates posters, handbills, billboards, and other advertising for business customers. An order comes in for a poster with 500 in² of usable area, with margins of 2 in. across the top, 3 in. across the bottom, and 2.5 in. on each side.

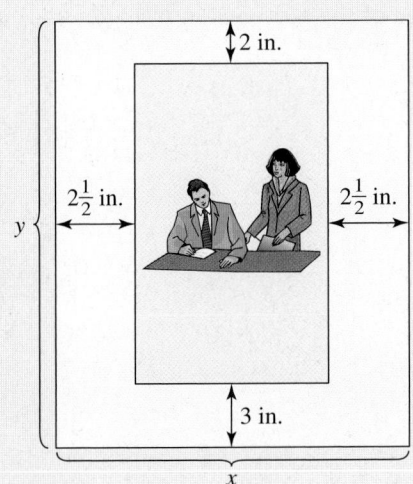

a. Find a function $A(x, y)$ for the area of the page, and a function $R(x, y)$ for the area of the inner rectangle (the usable area).

b. Solve for y in $R(x, y) = 500$, and use the result to write the area from part (a) as a function $A(x)$ in terms of x alone (simplify the result).

c. On a graphing calculator, graph $A(x)$ using the window $x \in [-100, 100]$; $y \in [-800, 1600]$. Then graph $y = 5x + 500$ on the same screen. How are these two graphs related?

d. Use the graph of $A(x)$ in Quadrant I to determine the poster of minimum size that satisfies these border requirements and has the necessary usable area. Clearly state the values of x and y, rounded to the nearest hundredth of an inch.

61. The formula from Exercise 54 has an interesting derivation. The volume of a cylinder is $V = \pi r^2 h$, while the surface area is given by $S = 2\pi r^2 + 2\pi rh$ (the circular top and bottom + the area of the side).

a. Solve the volume formula for the variable h.

b. Substitute the resulting expression for h into the surface area formula and simplify.

c. Combine the resulting two terms using the least common denominator, and the result is the formula from Exercise 54.

d. Assume the volume of a can must be 1200 cm³. Use a calculator to graph the function S using an appropriate window, then use it to find the radius r and height h that will result in a cylinder with the smallest possible area, while still holding a volume of 1200 cm³. Also see Exercise 62.

62. The surface area of a spherical cap is given by $S = 2\pi rh$, where r is the radius of the sphere and h is the perpendicular distance from the sphere's surface to the plane intersecting the sphere, forming the cap. The volume of the cap is $V = \frac{1}{3}\pi h^2(3r - h)$. Similar to Exercise 61, a formula can be found that will minimize the area of a cap that holds a specified volume.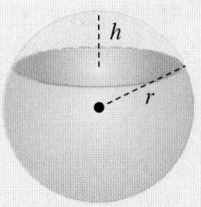

a. Solve the volume formula for the variable r.

b. Substitute the resulting expression for r into the surface area formula and simplify. The result is a formula for surface area given solely in terms of the volume V and the height h.

c. Assume the volume of the spherical cap is 500 cm³. Use a graphing calculator to graph the resulting function on an appropriate window, and use the graph to find the height h that will result in a spherical cap with the

smallest possible area, while still holding a volume of 500 cm³.

d. Use this value of h and $V = 500$ cm³ to find the radius of the sphere.

▶ EXTENDING THE CONCEPT

63. Consider rational functions of the form $f(x) = \dfrac{x^2 - a}{x - b}$. Use a graphing calculator to explore cases where $a = b^2 + 1$, $a = b^2$, and $a = b^2 - 1$. What do you notice? Explain/Discuss why the graphs differ. It's helpful to note that when graphing functions of this form, the "center" of the graph will be at $(b, b^2 - a)$, and the window size can be set accordingly for an optimal view. Do some investigation on this function and determine/explain why the "center" of the graph is at $(b, b^2 - a)$.

64. The formula from Exercise 53 also has an interesting derivation, and the process involves this sequence:

a. Use the points $(a, 0)$ and (h, k) to find the slope of the line, and the point-slope formula to find the equation of the line in terms of y.

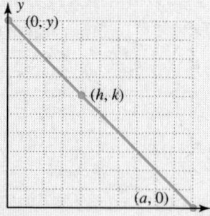

b. Use this equation to find the x- and y-intercepts of the line in terms of a, k, and h.

c. Complete the derivation using these intercepts and the triangle formula $A = \frac{1}{2}BH$.

d. If the lines goes through $(4, 4)$ the area formula becomes $A = \dfrac{1}{2}\left(\dfrac{4a^2}{a - 4}\right)$. Find the minimum value of this rational function. What can you say about the triangle with minimum area through (h, k), where $h = k$? Verify using the points $(5, 5)$, and $(6, 6)$.

65. Referring to Exercises 54 and 61, suppose that instead of a closed cylinder, with both a top and bottom, we needed to manufacture *open cylinders,* like tennis ball cans that use a lid made from a different material. Derive the formula that will minimize the surface area of an open cylinder, and use it to find the cylinder with minimum surface area that will hold 90 in³ of material.

▶ MAINTAINING YOUR SKILLS

66. (1.4) Compute the quotient $\dfrac{5i}{1 + 2i}$, then check your answer using multiplication.

67. (2.3) Write the equation of the line in slope intercept form and state the slope and y-intercept: $-3x + 4y = -16$.

68. (1.5) Given $f(x) = ax^2 + bx + c$, for what real values of a, b, and c will the function have: (a) two, real/rational roots, (b) two, real/irrational roots, (c) one real and rational root, (d) one real/irrational root, (e) one complex root, and (f) two complex roots?

69. (R.2/1.5) For triangle ABC as shown, (a) find the perimeter; (b) find the length of $\overline{CD}$, given $(\overline{CB})^2 = \overline{AB} \cdot \overline{DB}$; (c) find the area; and (d) find the area of the two smaller triangles.

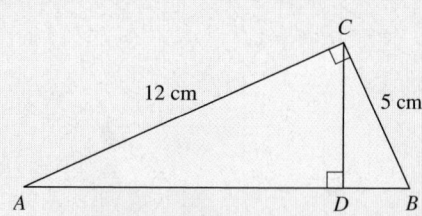

3.7 Polynomial and Rational Inequalities

Learning Objectives

In Section 3.7 you will learn how to:

☐ **A.** Solve quadratic inequalities

☐ **B.** Solve polynomial inequalities

☐ **C.** Solve rational inequalities

☐ **D.** Use interval tests to solve inequalities

☐ **E.** Solve applications of inequalities

The study of polynomial and rational inequalities is simply an extension of our earlier work in analyzing functions (Section 2.5). While we've developed the ability to graph a variety of new functions, solution sets will still be determined by analyzing the behavior of the function at its zeroes, and in the case of rational functions, on either side of any vertical asymptotes. The key idea is to recognize the following statements are synonymous:

1. $f(x) > 0$.   **2.** Outputs are positive.   **3.** The graph is *above the x-axis*.

Similar statements can be made using the other inequality symbols.

A. Quadratic Inequalities

Solving a quadratic inequality only requires that we (a) locate any real zeroes of the function and (b) determine whether the graph opens upward or downward. If there are no x-intercepts, the graph is entirely above the x-axis (output values are positive), or entirely below the x-axis (output values are negative), making the solution either all real numbers or the empty set.

EXAMPLE 1 ▶ Solving a Quadratic Inequality

For $f(x) = x^2 + x - 6$, solve $f(x) > 0$.

Solution ▶ The graph of f will open upward since $a > 0$. Factoring gives $f(x) = (x + 3)(x - 2)$, with zeroes at -3 and 2. Using a the x-axis alone (since graphing the function is not our focus), we plot $(-3, 0)$ and $(2, 0)$ and visualize a parabola opening upward through these points (Figure 3.58).

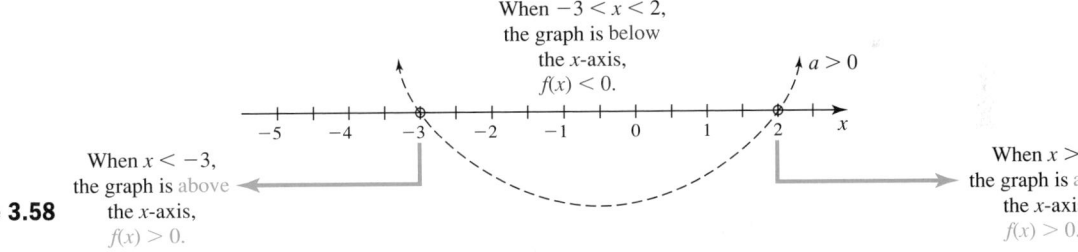

When $-3 < x < 2$, the graph is below the x-axis, $f(x) < 0$.

When $x < -3$, the graph is above the x-axis, $f(x) > 0$.

When $x > 2$, the graph is above the x-axis, $f(x) > 0$.

Figure 3.58

Figure 3.59

The diagram clearly shows the graph is *above* the x-axis (outputs are positive) when $x < -3$ or when $x > 2$. The solution is $x \in (-\infty, -3) \cup (2, \infty)$. For reference only, the complete graph is given in Figure 3.59.

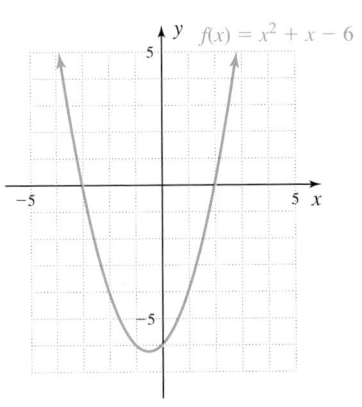

$f(x) = x^2 + x - 6$

Now try Exercises 7 through 18 ▶

When solving general inequalities, zeroes of multiplicity continue to play a role. In Example 1, the zeroes of f were both of multiplicity 1, and the graph crossed the x-axis at these points. In other cases, the zeroes may have even multiplicity.

EXAMPLE 2 ▶ **Solving a Quadratic Inequality**

Solve the inequality $-x^2 + 6x \leq 9$.

Solution ▶ Begin by writing the inequality in standard form: $-x^2 + 6x - 9 \leq 0$. Note this is equivalent to $g(x) \leq 0$ for $g(x) = -x^2 + 6x - 9$. Since $a < 0$, the graph of g will open downward. The factored form is $g(x) = -(x - 3)^2$, showing 3 is a zero with multiplicity 2. Using the x-axis, we plot the point $(3, 0)$ and visualize a parabola opening downward through this point.

Figure 3.60

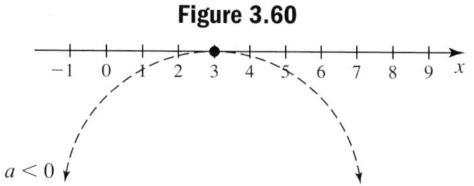

Figure 3.61

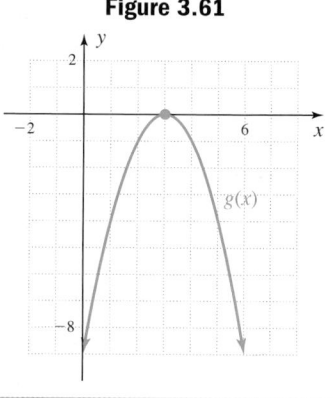

WORTHY OF NOTE

Since $x = 3$ was a zero of multiplicity 2, the graph "bounced off" the x-axis at this point, with no change of sign for g. The graph is entirely below the x-axis, except at the vertex $(3, 0)$.

Figure 3.60 shows the graph is *below* the x-axis (outputs are negative) for *all values* of x except $x = 3$. But since this is a less than *or equal to* inequality, the solution is $x \in \mathbb{R}$. For reference only, the complete graph is given in Figure 3.61.

☑ **A.** You've just learned how to solve quadratic inequalities

Now try Exercises 19 through 36 ▶

B. Polynomial Inequalities

The reasoning in Examples 1 and 2 transfers seamlessly to inequalities involving higher degree polynomials. After writing the polynomial in standard form, find the zeroes, plot them on the x-axis, and determine the solution set using end behavior and the behavior at each zero (cross—sign change; or bounce—no change in sign). In this process, any irreducible quadratic factors can be ignored, as they have no effect on the solution set. In summary,

---

Solving Polynomial Inequalities

Given $f(x)$ is a polynomial in standard form,

1. Write f in completely factored form.
2. Plot real zeroes on the x-axis, noting their multiplicity.
   - If the multiplicity is odd the function will **change** sign.
   - If the multiplicity is even, there will be **no change** in sign.
3. Use the end behavior to determine the sign of f in the outermost intervals, then label the other intervals as $f(x) < 0$ or $f(x) > 0$ by analyzing the multiplicity of neighboring zeroes.
4. State the solution in interval notation.

EXAMPLE 3 ▶ **Solving a Polynomial Inequality**

Solve the inequality $x^3 - 18 < -4x^2 + 3x$.

Solution ▶ In standard form we have $x^3 + 4x^2 - 3x - 18 < 0$, which is equivalent to $f(x) < 0$ where $f(x) = x^3 + 4x^2 - 3x - 18$. The polynomial cannot be factored by grouping and testing 1 and -1 shows neither is a zero. Using $x = 2$ and synthetic division gives

$$
\text{use 2 as a ``divisor''} \quad
\begin{array}{r|rrrr}
2 & 1 & 4 & -3 & -18 \\
  & \downarrow & 2 & 12 & 18 \\
\hline
  & 1 & 6 & 9 & 0,
\end{array}
$$

with a quotient of $x^2 + 6x + 9$ and a remainder of zero.

1. The factored form is $f(x) = (x - 2)(x^2 + 6x + 9) = (x - 2)(x + 3)^2$.

2. The graph will bounce off the x-axis at $x = -3$ (f will not change sign), and cross the x-axis at $x = 2$ (f will change sign). This is illustrated in Figure 3.62, which uses open dots due to the strict inequality.

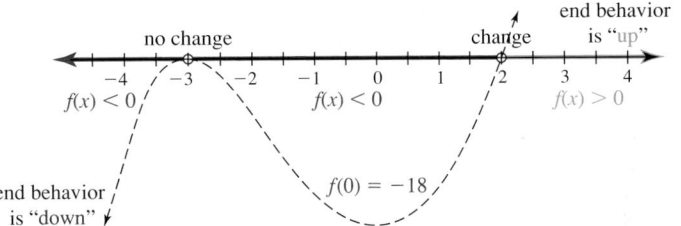

Figure 3.62

3. The polynomial has odd degree with a positive lead coefficient, so end behavior is down/up, which we note in the outermost intervals. Working from the left, f will not change sign at $x = -3$, showing $f(x) < 0$ in the left and middle intervals. This is supported by the y-intercept $(0, -18)$. See Figure 3.63.

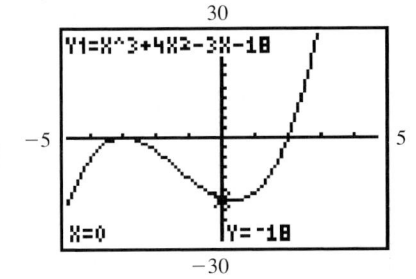

Figure 3.63   end behavior is "down"

4. From the diagram, we see that $f(x) < 0$ for $x \in (-\infty, -3) \cup (-3, 2)$, which must also be the solution interval for $x^3 - 18 < -4x^2 + 3x$.

Now try Exercises 37 through 48 ▶

GRAPHICAL SUPPORT

The results from Example 3 can easily be verified using a graphing calculator. The graph shown here is displayed using a window of $X \in [-5, 5]$ and $Y \in [-30, 30]$, and definitely shows the graph is below the x-axis $[f(x) < 0]$ from $-\infty$ to 2, except at $x = -3$ where the graph touches the x-axis without crossing.

EXAMPLE 4 ▶ Solving a Polynomial Inequality

Solve the inequality $x^4 + 4x \le 9x^2 - 12$.

Solution ▶ Writing the polynomial in standard form gives $x^4 - 9x^2 + 4x + 12 \le 0$. The equivalent inequality is $f(x) \le 0$. Testing 1 and -1 shows $x = 1$ is not a zero, but $x = -1$ is. Using synthetic division with $x = -1$ gives

use -1 as a "divisor"

$$
\begin{array}{r|rrrrr}
-1 & 1 & 0 & -9 & 4 & 12 \\
   &   & -1 & 1 & 8 & -12 \\
\hline
   & 1 & -1 & -8 & 12 & \underline{|0}
\end{array}
$$

with a quotient of $q_1(x) = x^3 - x^2 - 8x + 12$ and a remainder of zero. As $q_1(x)$ is not easily factored, we continue with synthetic division using $x = 2$.

use 2 as a "divisor"

$$
\begin{array}{r|rrrr}
2 & 1 & -1 & -8 & 12 \\
  &   & 2 & 2 & -12 \\
\hline
  & 1 & 1 & -6 & \underline{|0}
\end{array}
$$

The result is $q_2(x) = x^2 + x - 6$ with a remainder of zero.

1. The factored form is
$$f(x) = (x + 1)(x - 2)(x^2 + x - 6) = (x + 1)(x - 2)^2(x + 3).$$

2. The graph will "cross" at $x = -1$ and -3, and f will change sign. The graph will bounce at $x = 2$ and f will not change sign. This is illustrated in Figure 3.65 which uses closed dots since $f(x)$ can be equal to zero. See Figure 3.64.

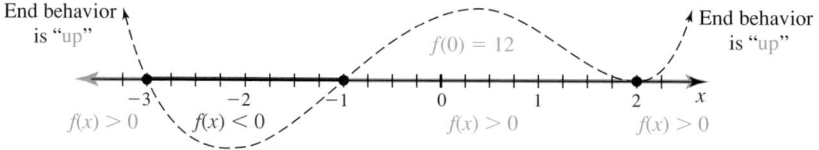

Figure 3.64

3. With even degree and positive lead coefficient, the end behavior is up/up. Working from the leftmost interval, $f(x) > 0$, the function must change sign at $x = -3$ (going below the x-axis), and again at $x = -1$ (going above the x-axis). This is supported by the y-intercept $(0, 12)$. The graph then "bounces" at $x = 2$, remaining above the x-axis (no sign change). This produces the sketch shown in Figure 3.65.

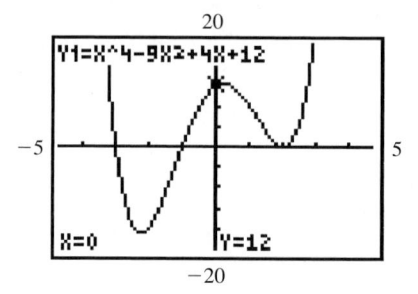

Figure 3.65

☑ **B.** You've just learned how to solve polynomial inequalities

4. From the diagram, we see that $f(x) \le 0$ for $x \in [-3, -1]$, and at the single point $x = 2$. This shows the solution for $x^4 + 4x \le 9x^2 - 12$ is $x \in [-3, -1] \cup \{2\}$.

Now try Exercises 49 through 54 ▶

GRAPHICAL SUPPORT

As with Example 3, the results from Example 4 can be confirmed using a graphing calculator. The graph shown here is displayed using $X \in [-5, 5]$ and $Y \in [-20, 20]$. The graph is below or touching the x-axis $[f(x) < 0]$ from -3 to -1 and at $x = 2$.

C. Rational Inequalities

In general, the solution process for polynomial and rational inequalities is virtually identical, once we recognize that vertical asymptotes also break the x-axis into intervals where function values may change sign. However, for rational functions it's more efficient to begin the analysis using the y-intercept or a test point, rather than end behavior, although either will do.

EXAMPLE 5 ▶ Solving a Rational Inequality

Solve $\dfrac{x^2 - 9}{x^3 - x^2 - x + 1} \le 0$.

Solution ▶ In function form, $v(x) = \dfrac{x^2 - 9}{x^3 - x^2 - x + 1}$ and we want the solution for $v(x) \le 0$.

The numerator and denominator are in standard form. The numerator factors easily, and the denominator can be factored by grouping.

1. The factored form is $v(x) = \dfrac{(x - 3)(x + 3)}{(x - 1)^2(x + 1)}$.

2. $v(x)$ will change sign at $x = 3, -3$, and -1 as all have odd multiplicity, but will not change sign at $x = 1$ (even multiplicity). Note that zeroes of the denominator will always be indicated by open dots (Figure 3.66) as they are excluded from any solution set.

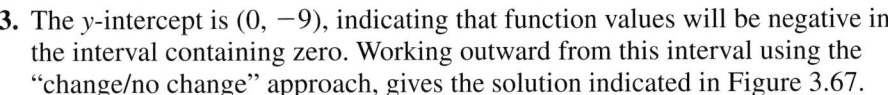

Figure 3.66

3. The y-intercept is $(0, -9)$, indicating that function values will be negative in the interval containing zero. Working outward from this interval using the "change/no change" approach, gives the solution indicated in Figure 3.67.

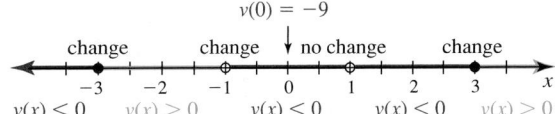

Figure 3.67

4. For $v(x) \le 0$, the solution is $x \in (-\infty, -3] \cup (-1, 1) \cup (1, 3]$.

Now try Exercises 55 through 66 ▶

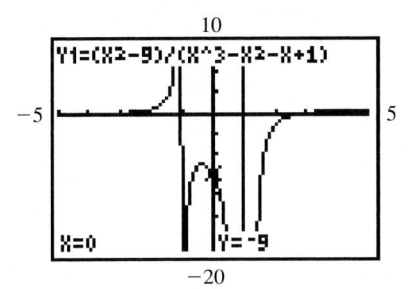

WORTHY OF NOTE

End behavior can also be used to analyze rational inequalities, although using the y-intercept may be more efficient. For the function $v(x)$ from Example 5 we have $\dfrac{x^2 - 9}{x^3 - x^2 - x + 1} \approx \dfrac{x^2}{x^3} = \dfrac{1}{x}$ for large values of x, indicating $v(x) > 0$ to the far right and $v(x) < 0$ to the far left. The analysis of each interval can then begin from either side.

GRAPHICAL SUPPORT

Sometimes finding a window that clearly displays all features of rational function can be difficult. In these cases, we can investigate each piece separately to confirm solutions. For Example 5, most of the features of $v(x)$ can be seen using a window $X \in [-5, 5]$ and $Y \in [-20, 10]$, and we note the graph displayed strongly tends to support our solution.

If the rational inequality is not given in function form or is composed of more than one term, start by writing the inequality with zero on one side, then combine terms into a single expression.

EXAMPLE 6 ▶ **Solving a Rational Inequality**

Solve $\dfrac{x-2}{x-3} \le \dfrac{1}{x+3}$.

Solution ▶ Rewrite the inequality with zero on one side: $\dfrac{x-2}{x-3} - \dfrac{1}{x+3} \le 0$. This is

equivalent to $v(x) \le 0$, where $v(x) = \dfrac{x-2}{x-3} - \dfrac{1}{x+3}$. Combining the expressions
on the right, we have

$$v(x) = \frac{(x-2)(x+3) - 1(x-3)}{(x+3)(x-3)} \qquad \text{LCD is } (x+3)(x-3)$$

$$= \frac{x^2 + x - 6 - x + 3}{(x+3)(x-3)} \qquad \text{multiply}$$

$$= \frac{x^2 - 3}{(x+3)(x-3)} \qquad \text{simplify}$$

1. The factored form is $v(x) = \dfrac{(x+\sqrt{3})(x-\sqrt{3})}{(x+3)(x-3)}$. $x^2 - k = (x+\sqrt{k})(x-\sqrt{k})$

2. $v(x)$ will change sign at $x = -\sqrt{3}, \sqrt{3}, -3$, and 3, as all have odd
   multiplicity (Figure 3.68).

Figure 3.68
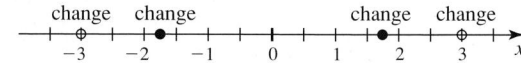

3. Since $v(0) = \frac{1}{3}$ (verify this), function values will be positive in the interval
   containing zero. Working outward from this interval produces the diagram
   shown in Figure 3.69.

Figure 3.69
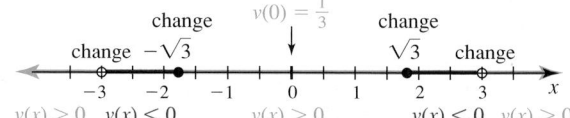

4. The solution for $\dfrac{x-2}{x-3} \le \dfrac{1}{x+3}$ is $x \in (-3, -\sqrt{3}] \cup [\sqrt{3}, 3)$.

☑ **C.** You've just learned how
to solve rational inequalities

Now try Exercises 67 through 82 ▶

GRAPHICAL SUPPORT

To check the solutions to $\dfrac{x-2}{x-3} \le \dfrac{1}{x+3}$, we
subtract $\dfrac{1}{x+3}$ and graph $Y_1 = \dfrac{x-2}{x-3} - \dfrac{1}{x+3}$
to look for intervals where the graph is below
the x-axis. The graph is shown here using the
window $X \in [-5, 5]$ and $y \in [-10, 10]$, and
verifies our solution.

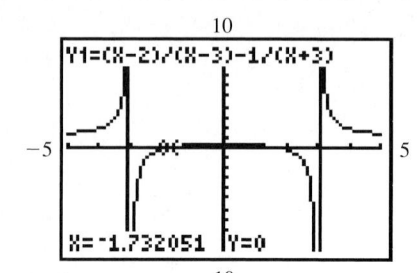

D. Solving Function Inequalities Using Interval Tests

As an alternative to the "zeroes method," an **interval test method** can be used to solve polynomial and rational inequalities. The x-intercepts and vertical asymptotes (in the case of rational functions) are noted on the x-axis, then a test number is selected from each interval. Since polynomial and rational functions are continuous over their entire domain, the sign of the function at these test values will be the sign of the function for all values of x in the chosen interval.

EXAMPLE 7 ▶ **Solving a Polynomial Inequality**

Solve the inequality $x^3 + 8 \leq 5x^2 - 2x$.

Solution ▶ Writing the relationship in function form gives $p(x) = x^3 - 5x^2 + 2x + 8$, with solutions needed to $p(x) \leq 0$. The tests for 1 and -1 show $x = -1$ is a root, and using -1 with synthetic division gives

use -1 as a "divisor" $\underline{-1|}$ $\begin{array}{cccc} 1 & -5 & 2 & 8 \\ \downarrow & -1 & 6 & -8 \\ \hline 1 & -6 & 8 & \underline{|0} \end{array}$

The quotient is $q(x) = x^2 - 6x + 8$, with a remainder of 0.

The factored form is $p(x) = (x + 1)(x^2 - 6x + 8) = (x + 1)(x - 2)(x - 4)$. The x-intercepts are $(-1, 0)$, $(2, 0)$, and $(4, 0)$. Plotting these intercepts creates four intervals on the x-axis (Figure 3.70).

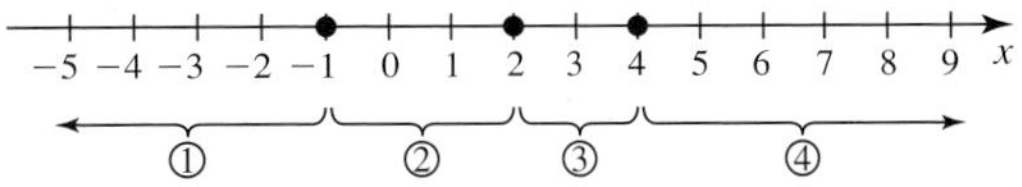

Figure 3.70

Selecting a test value from each interval gives Figure 3.71.

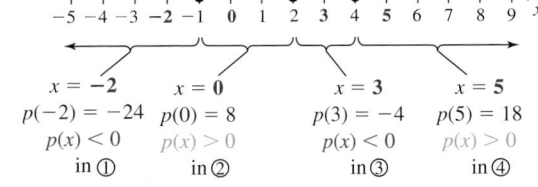

Figure 3.71

$x = -2$	$x = 0$	$x = 3$	$x = 5$
$p(-2) = -24$	$p(0) = 8$	$p(3) = -4$	$p(5) = 18$
$p(x) < 0$	$p(x) > 0$	$p(x) < 0$	$p(x) > 0$
in ①	in ②	in ③	in ④

The interval tests show $x^3 + 8 \leq 5x^2 - 2x$ for $x \in (-\infty, -1] \cup [2, 4]$.

☑ **D.** You've just learned how to use interval tests to solve inequalities

Now try Exercises 83 through 90 ▶

E. Applications of Inequalities

Applications of inequalities come in many varieties. In addition to stating the solution algebraically, these exercises often compel us to consider the *context of each application* as we state the solution set.

EXAMPLE 8 ▶ **Solving Applications of Inequalities**

The velocity of a particle (in feet per second) as it floats through air turbulence is given by $V(t) = t^5 - 10t^4 + 35t^3 - 50t^2 + 24t$, where t is the time in seconds and $0 < t < 4.5$. During what intervals of time is the particle moving in the positive direction $[V(t) > 0]$?

Solution ▶ Begin by writing V in factored form. Testing 1 and -1 shows $t = 1$ is a root. Factoring out t gives $V(t) = t(t^4 - 10t^3 + 35t^2 - 50t + 24)$, and using $t = 1$ with synthetic division yields

use 1 as a "divisor" $\underline{1|}$ $\begin{array}{ccccc} 1 & -10 & 35 & -50 & 24 \\ \downarrow & 1 & -9 & 26 & -24 \\ \hline 1 & -9 & 26 & -24 & \underline{|0} \end{array}$

The quotient is $q_1(t) = t^3 - 9t^2 + 26t - 24$. Using $t = 2$, we continue with the division on $q_1(t)$ which gives:

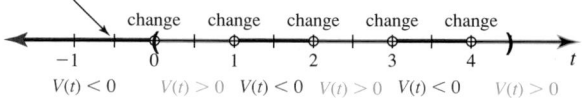

use 2 as a "divisor" $2\rfloor$

$$\begin{array}{r|rrrr} & 1 & -9 & 26 & -24 \\ \downarrow & & 2 & -14 & 24 \\ \hline & 1 & -7 & 12 & \underline{0} \end{array}$$

This shows $V(t) = t(t - 1)(t - 2)(t^2 - 7t + 12)$.

1. The completely factored form is $V(t) = t(t - 1)(t - 2)(t - 3)(t - 4)$.

2. All zeroes have odd multiplicity and function values will change sign.

3. With odd degree and a positive leading coefficient, end behavior is down/up.

Function values will be negative in the far left interval and alternate in sign thereafter. The solution diagram is shown in the figure.

Since end behavior is down/up, function values
are negative in this interval, and will alternate thereafter.

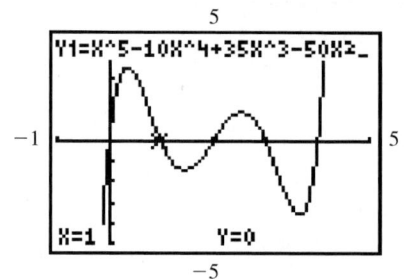

change   change   change   change   change

-1 0   1   2   3   4 t

$V(t) < 0$ $V(t) > 0$ $V(t) < 0$ $V(t) > 0$ $V(t) < 0$ $V(t) > 0$

☑ **E.** You've just learned how to solve applications of inequalities

4. For $V(t) > 0$, the solution is $t \in (0, 1) \cup (2, 3) \cup (4, 4.5)$. The particle is moving in the positive direction in these time intervals.

> **Now try Exercises 93 through 100** ▶

GRAPHICAL SUPPORT

To verify our analysis of Example 8, we graph $V(t)$ using the window $X \in [-1, 5]$ and $Y \in [-5, 5]$. As the graph shows, function values are positive (graph is above the x-axis) when $t \in (0, 1) \cup (2, 3) \cup (4, 4.5)$. Also see the *Technology Highlight* for this section.

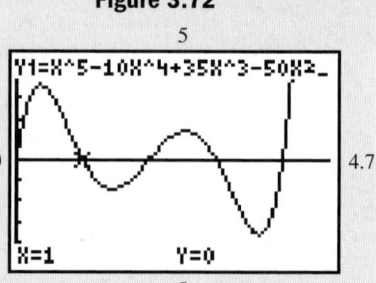

TECHNOLOGY HIGHLIGHT

Polynomial and Rational Inequalities

Consider the results from Example 8, where we solved the inequality $V(t) > 0$ for $V(t) = t^5 - 10t^4 + 35t^3 - 50t^2 + 24t$. To emphasize that we are seeking intervals where the function is above the x-axis (the horizontal line $y = 0$), we can have the calculator *shade these areas*. Begin by entering $V(t)$ as Y_1 on the
| Y = | screen, and the line $y = 0$ as Y_2. Using $x \in [0, 4.7]$ and $y \in [-5, 5]$ (a "friendly" window) and | GRAPH | ing the functions produces Figure 3.72. To shade all portions of the graph that are above the x-axis, go to the home screen, and press | 2nd | | PRGM |

(DRAW) 7:Shade. This feature requires six arguments, all separated by commas. These are (in order): *lower function, upper function, left endpoint, right endpoint, pattern choice,* and *density.* The calculator will then shade the area between the lower and upper functions,

Figure 3.72

between the left and right endpoints, using the pattern and density chosen. The patterns are (1) vertical lines, (2) horizontal lines, (3) lines with negative slope, and (4) lines with positive slope. There are eight density settings, from every pixel (1), to every eight pixels (8). Figure 3.73 shows the options we've selected, with the resulting graph shown in Figure 3.74. The friendly window makes it easy to investigate the inequality further using the [TRACE] feature.

Use these ideas to visually study and explore the solution to the following inequality.

Exercise 1: Use window size $x \in [-4.7, 4.7]; y \in [-10, 20]$, **(DRAW) 7:Shade**, and [TRACE] to solve $P(x) < 0$ for $P(x) = x^4 + 1.1x^3 - 9.37x^2 - 4.523x + 16.4424$.

Figure 3.73

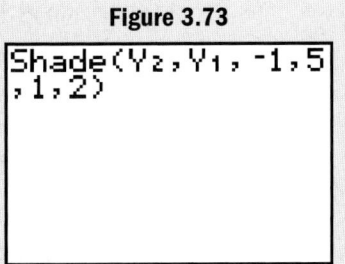

Figure 3.74

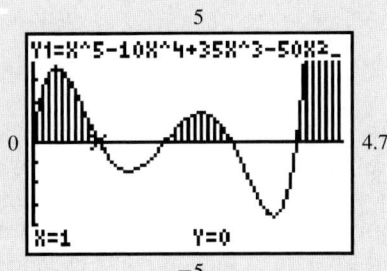

3.7 EXERCISES

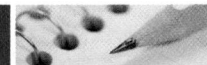

▶ CONCEPTS AND VOCABULARY

Fill in each blank with the appropriate word or phrase. Carefully reread the section if needed.

1. To solve a polynomial or rational inequality, begin by plotting the location of all zeroes and _____ asymptotes (if they exist), then consider the _____ of each.

2. For strict inequalities, the zeroes are _____ from the solution set. For nonstrict inequalities, zeroes are _____. The values at which vertical asymptotes occur are always _____.

3. If the graph of a quadratic function $g(x)$ opens downward with a vertex at $(5, -1)$, the solution set for $g(x) > 0$ is _____.

4. To solve a polynomial/rational inequality, it helps to find the sign of f in some interval. This can quickly be done using the _____ _____ or _____ of the function.

5. Compare/Contrast the process for solving $x^2 - 3x - 4 \geq 0$ with $\dfrac{1}{x^2 - 3x - 4} \geq 0$. Are there similarities? What are the differences?

6. Compare/Contrast the process for solving $(x + 1)(x - 3)(x^2 + 1) > 0$ with $(x + 1)(x - 3) > 0$. Are there similarities? What are the differences?

▶ DEVELOPING YOUR SKILLS

Solve each quadratic inequality by locating the x-intercept(s) (if they exist), and noting the end behavior of the graph. Begin by writing the inequality in function form as needed.

7. $f(x) = -x^2 + 4x; f(x) > 0$

8. $g(x) = x^2 - 5x; g(x) < 0$

9. $h(x) = x^2 + 4x - 5; h(x) \geq 0$

10. $p(x) = -x^2 + 3x + 10; p(x) \leq 0$

11. $q(x) = 2x^2 - 5x - 7; q(x) < 0$

12. $r(x) = -2x^2 - 3x + 5; r(x) > 0$

13. $7 \geq x^2$

14. $x^2 \leq 13$

15. $x^2 + 3x \leq 6$

16. $x^2 - 2 \leq 5x$

17. $3x^2 \geq -2x + 5$

18. $4x^2 \geq 3x + 7$

19. $s(x) = x^2 - 8x + 16; s(x) \geq 0$

20. $t(x) = x^2 - 6x + 9; t(x) \geq 0$

21. $r(x) = 4x^2 + 12x + 9; r(x) < 0$

22. $f(x) = 9x^2 - 6x + 1; f(x) < 0$

23. $g(x) = -x^2 + 10x - 25; g(x) < 0$

24. $h(x) = -x^2 + 14x - 49; h(x) < 0$

25. $-x^2 > 2$ **26.** $x^2 < -4$

27. $x^2 - 2x > -5$ **28.** $-x^2 + 3x < 3$

29. $2x^2 \geq 6x - 9$ **30.** $5x^2 \geq 4x - 4$

Recall that for a square root expression to represent a real number, the radicand must be greater than or equal to zero. Applying this idea results in an inequality that can be solved using the skills from this section. Determine the domain of the following radical functions.

31. $h(x) = \sqrt{x^2 - 25}$ **32.** $p(x) = \sqrt{25 - x^2}$

33. $q(x) = \sqrt{x^2 - 5x}$ **34.** $r(x) = \sqrt{6x - x^2}$

35. $t(x) = \sqrt{-x^2 + 3x - 4}$

36. $Y_1 = \sqrt{x^2 - 6x + 9}$

Solve the inequality indicated using a number line and the behavior of the graph at each zero. Write all answers in interval notation.

37. $(x + 3)(x - 5) < 0$ **38.** $(x - 2)(x + 7) < 0$

39. $(x + 1)^2(x - 4) \geq 0$ **40.** $(x + 6)(x - 1)^2 \leq 0$

41. $(x + 2)^3(x - 2)^2(x - 4) \geq 0$

42. $(x - 1)^3(x + 2)^2(x - 3) \leq 0$

43. $x^2 + 4x + 1 < 0$ **44.** $x^2 - 6x + 4 > 0$

45. $x^3 + x^2 - 5x + 3 \leq 0$

46. $x^3 + x^2 - 8x - 12 \geq 0$

47. $x^3 - 7x + 6 > 0$ **48.** $x^3 - 13x + 12 > 0$

49. $x^4 - 10x^2 > -9$ **50.** $x^4 + 36 < 13x^2$

51. $x^4 - 9x^2 > 4x - 12$ **52.** $x^4 - 16 > 5x^3 - 20x$

53. $x^4 - 6x^3 \leq -8x^2 - 6x + 9$

54. $x^4 - 3x^2 + 8 \leq 4x^3 - 10x$

55. $f(x) = \dfrac{x + 3}{x - 2}; f(x) \leq 0$

56. $F(x) = \dfrac{x - 4}{x + 1}; F(x) \geq 0$

57. $g(x) = \dfrac{x + 1}{x^2 + 4x + 4}; g(x) < 0$

58. $G(x) = \dfrac{x - 3}{x^2 - 2x + 1}; G(x) > 0$

59. $\dfrac{2 - x}{x^2 - x - 6} \geq 0$ **60.** $\dfrac{1 - x}{x^2 - 2x - 8} \leq 0$

61. $\dfrac{2x - x^2}{x^2 + 4x - 5} < 0$ **62.** $\dfrac{x^2 + 3x}{x^2 - 2x - 3} > 0$

63. $\dfrac{x^2 - 4}{x^3 - 13x + 12} \geq 0$ **64.** $\dfrac{x^2 + x - 6}{x^3 - 7x + 6} \leq 0$

65. $\dfrac{x^2 + 5x - 14}{x^3 + x^2 - 5x + 3} > 0$

66. $\dfrac{x^2 + 2x - 8}{x^3 + 5x^2 + 3x - 9} < 0$

67. $\dfrac{2}{x - 2} \leq \dfrac{1}{x}$ **68.** $\dfrac{5}{x + 3} \geq \dfrac{3}{x}$

69. $\dfrac{x - 3}{x + 17} > \dfrac{1}{x - 1}$ **70.** $\dfrac{1}{x + 5} < \dfrac{x - 2}{x - 7}$

71. $\dfrac{x + 1}{x - 2} \geq \dfrac{x + 2}{x + 3}$ **72.** $\dfrac{x - 3}{x - 6} \leq \dfrac{x + 1}{x + 4}$

73. $\dfrac{x + 2}{x^2 + 9} > 0$ **74.** $\dfrac{x^2 + 4}{x - 3} < 0$

75. $\dfrac{x^3 + 1}{x^2 + 1} > 0$ **76.** $\dfrac{x^2 + 4}{x^3 - 8} < 0$

77. $\dfrac{x^4 - 5x^2 - 36}{x^2 - 2x + 1} > 0$ **78.** $\dfrac{x^4 - 3x^2 - 4}{x^2 - x - 20} < 0$

79. $x^2 - 2x \geq 15$ **80.** $x^2 + 3x \geq 18$

81. $x^3 \geq 9x$ **82.** $x^3 \leq 4x$

83. $-4x + 12 < -x^3 + 3x^2$

84. $x^3 + 8 < 5x^2 - 2x$ **85.** $\dfrac{x^2 - x - 6}{x^2 - 1} \geq 0$

86. $\dfrac{x^2 - 4x - 21}{x - 3} < 0$

Match the correct solution with the inequality and graph given.

87. $f(x) < 0$

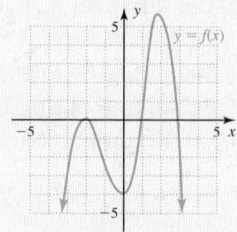

a. $x \in (-5, -2) \cup (3, 5)$

b. $x \in (-\infty, -2) \cup (-2, 1) \cup (3, \infty)$

c. $x \in (-\infty, -2) \cup (3, \infty)$

d. $x \in (-\infty, -2) \cup (-2, 1] \cup [3, \infty)$

e. none of these

88. $g(x) \geq 0$

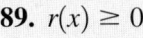

   a. $x \in (-4, -0.5) \cup (4, \infty)$

   b. $x \in [-0.5, 4] \cup [4, 5]$

   c. $x \in (-\infty, -4) \cup (-0.5, 4)$

   d. $x \in [-4, -0.5] \cup [4, \infty)$

   e. none of these

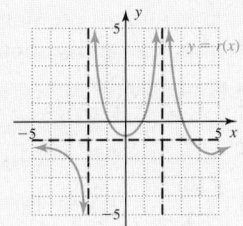

89. $r(x) \geq 0$

   a. $x \in (-\infty, -2) \cup [-1, 1] \cup [3, \infty)$

   b. $x \in (-2, -1] \cup [1, 2) \cup (2, 3]$

   c. $x \in (-\infty, -2) \cup (2, \infty)$

   d. $x \in (-2, -1) \cup (1, 2) \cup (2, 3]$

   e. none of these

90. $R(x) \leq 0$

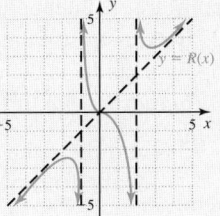

   a. $x \in (-\infty, -1) \cup (0, 2)$

   b. $x \in [0, 1] \cup (2, \infty)$

   c. $x \in [-5, -1] \cup [2, 5]$

   d. $x \in (-\infty, -1) \cup [0, 2)$

   e. none of these

▶ WORKING WITH FORMULAS

91. Discriminant of the reduced cubic
$$x^3 + px + q = 0: D = -(4p^3 + 27q^2)$$
The discriminant of a cubic equation is less well known than that of the quadratic, but serves the same purpose. The discriminant of the reduced cubic is given by the formula shown, where p is the linear coefficient and q is the constant term. If $D > 0$, there will be three real and distinct roots. If $D = 0$, there are still three real roots, but one is a repeated root (multiplicity two). If $D < 0$, there are one real and two complex roots. Suppose we wish to study the family of cubic equations where $q = p + 1$.

   a. Verify the resulting discriminant is
$$D = -(4p^3 + 27p^2 + 54p + 27).$$

   b. Determine the values of p and q for which this family of equations has a repeated real root. In other words, solve the equation $-(4p^3 + 27p^2 + 54p + 27) = 0$ using the rational zeroes theorem and synthetic division to write D in completely factored form.

   c. Use the factored form from part (b) to determine the values of p and q for which this family of equations has three real and distinct roots. In other words, solve $D > 0$.

    d. Verify the results of parts (b) and (c) on a graphing calculator.

92. Coordinates for the folium of Descartes:
$$\begin{cases} a = \dfrac{3kx}{1 + x^3} \\[2mm] b = \dfrac{3kx^2}{1 + x^3} \end{cases}$$

The interesting relation shown here is called the folium (leaf) of Descartes. The folium is most often graphed using what are called *parametric equations,* in which the coordinates a and b are expressed in terms of the parameter x

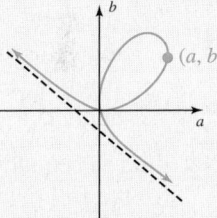

Folium of Descartes

("k" is a constant that affects the size of the leaf). Since each is an individual function, the x- and y-coordinates can be investigated individually in rectangular coordinates using $F(x) = \dfrac{3x}{1 + x^3}$ and

$G(x) = \dfrac{3x^2}{1 + x^3}$ (assume $k = 1$ for now).

   a. Graph each function using the techniques from this section.

   b. According to your graph, for what values of x will the x-*coordinate* of the folium be positive? In other words, solve $F(x) = \dfrac{3x}{1 + x^3} > 0$.

   c. For what values of x will the y-*coordinate* of the folium be positive? Solve
$$G(x) = \dfrac{3x^2}{1 + x^3} > 0.$$

   d. Will $F(x)$ ever be equal to $G(x)$? If so, for what values of x?

► **APPLICATIONS**

Deflection of a beam: The amount of deflection in a rectangular wooden beam of length L ft can be approximated by $d(x) = k(x^3 - 3L^2x + 2L^3)$, where k is a constant that depends on the characteristics of the wood and the force applied, and x is the *distance from the unsupported end* of the beam ($x < L$).

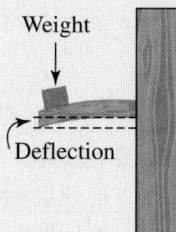

Weight

Deflection

93. Find the equation for a beam 8 ft long and use it for the following:

   a. For what distances x is the quantity $\dfrac{d(x)}{k}$ less than 189 units?

   b. What is the amount of deflection 4 ft from the unsupported end ($x = 4$)?

   c. For what distances x is the quantity $\dfrac{d(x)}{k}$ greater than 475 units?

   d. If safety concerns prohibit a deflection of more than 648 units, what is the shortest distance from the end of the beam that the force can be applied?

94. Find the equation for a beam 9 ft long and use it for the following:

   a. For what distances x is the quantity $\dfrac{d(x)}{k}$ less than 216 units?

   b. What is the amount of deflection 4 ft from the unsupported end ($x = 4$)?

   c. For what distances x is the quantity $\dfrac{d(x)}{k}$ greater than 550 units?

    d. Compare the answer to 93b with the answer to 94b. What can you conclude?

Average speed for a round-trip: Surprisingly, the average speed of a round-trip is *not* the sum of the average speed in each direction divided by two. For a fixed distance D, consider rate r_1 in time t_1 for one direction, and rate r_2 in time t_2 for the other, giving $r_1 = \dfrac{D}{t_1}$ and $r_2 = \dfrac{D}{t_2}$. The average speed for the round-trip is $R = \dfrac{2D}{t_1 + t_2}$.

95. The distance from St. Louis, Missouri, to Springfield, Illinois, is approximately 80 mi. Suppose that Sione, due to the age of his vehicle, made the round-trip with an average speed of 40 mph.

   a. Use the relationships stated to verify that $$r_2 = \frac{20r_1}{r_1 - 20}.$$

   b. Discuss the meaning of the horizontal and vertical asymptotes in this context.

   c. Verify algebraically the speed returning would be greater than the speed going for $20 < r_1 < 40$. In other words, solve the inequality $\dfrac{20r_1}{r_1 - 20} > r_1$ using the ideas from this section.

96. The distance from Boston, Massachusetts, to Hartford, Connecticut, is approximately 100 mi. Suppose that Stella, due to excellent driving conditions, made the round-trip with an average speed of 60 mph.

   a. Use the relationships above to verify that $$r_2 = \frac{30r_1}{r_1 - 30}.$$

   b. Discuss the meaning of the horizontal and vertical asymptotes in this context.

   c. Verify algebraically the speed returning would be greater than the speed going for $30 < r_1 < 60$. In other words, solve the inequality $\dfrac{30r_1}{r_1 - 30} > r_1$ using the ideas from this section.

Electrical resistance and temperature: The amount of electrical resistance R in a medium depends on the temperature, and for certain materials can be modeled by the equation $R(t) = 0.01t^2 + 0.1t + k$, where $R(t)$ is the resistance (in ohms Ω) at temperature t ($t \geq 0°$) in degrees Celsius, and k is the resistance at $t = 0°C$.

97. Suppose $k = 30$ for a certain medium. Write the resistance equation and use it to answer the following.

   a. For what temperatures is the resistance less than 42 Ω?

   b. For what temperatures is the resistance greater than 36 Ω?

   c. If it becomes uneconomical to run electricity through the medium for resistances greater than 60 Ω, for what temperatures should the electricity generator be shut down?

98. Suppose $k = 20$. Write the resistance equation and solve the following.

    a. For what temperatures is the resistance less than 26 Ω?

    b. For what temperatures is the resistance greater than 40 Ω?

    c. If it becomes uneconomical to run electricity through the medium for resistances greater than 50 Ω, for what temperatures should the electricity generator be shut down?

99. Sum of consecutive squares: The sum of the first n squares $1^2 + 2^2 + 3^2 + \cdots + n^2$ is given by the formula $S(n) = \dfrac{2n^3 + 3n^2 + n}{6}$. Use the equation to solve the following inequalities.

    a. For what number of consecutive squares is $S(n) \geq 30$?

    b. For what number of consecutive squares is $S(n) \leq 285$?

    c. What is the maximum number of consecutive squares that can be summed without the result exceeding three digits?

100. Sum of consecutive cubes: The sum of the first n cubes $1^3 + 2^3 + 3^3 + \cdots + n^3$ is given by the formula $S(n) = \dfrac{n^4 + 2n^3 + n^2}{4}$. Use the equation to solve the following inequalities.

    a. For what number of consecutive cubes is $S(n) \geq 100$?

    b. For what number of consecutive cubes is $S(n) \leq 784$?

    c. What is the maximum number of consecutive cubes that can be summed without the result exceeding three digits?

▶ EXTENDING THE CONCEPT

101. (a) Is it possible for the solution set of a polynomial inequality to be all real numbers? If not, discuss why. If so, provide an example. (b) Is it possible for the solution set of a rational inequality to be all real numbers? If not, discuss why. If so, provide an example.

102. The domain of radical functions: As in Exercises 31–36, if n is an even number, the expression $\sqrt[n]{A}$ represents a real number only if $A \geq 0$. Use this idea to find the domain of the following functions.

    a. $f(x) = \sqrt{2x^3 - x^2 - 16x + 15}$

    b. $g(x) = \sqrt[4]{2x^3 + x^2 - 22x + 24}$

    c. $p(x) = \sqrt[4]{\dfrac{x + 2}{x^2 - 2x - 35}}$

    d. $q(x) = \sqrt{\dfrac{x^2 - 1}{x^2 - x - 6}}$

103. Find one polynomial inequality and one rational inequality that have the solution $x \in (-\infty, -2) \cup (0, 1) \cup (1, \infty)$.

104. Using the tools of calculus, it can be shown that $f(x) = x^4 - 4x^3 - 12x^2 + 32x + 39$ is increasing in the intervals where $F(x) = x^3 - 3x^2 - 6x + 8$ is positive. Solve the inequality $F(x) > 0$ using the ideas from this section, then verify $f(x)\uparrow$ in these intervals by graphing f on a graphing calculator and using the $\boxed{\text{TRACE}}$ feature.

105. Using the tools of calculus, it can be shown that $r(x) = \dfrac{x^2 - 3x - 4}{x - 8}$ is decreasing in the intervals where $R(x) = \dfrac{x^2 - 16x + 28}{(x - 8)^2}$ is negative. Solve the inequality $R(x) < 0$ using the ideas from this section, then verify $r(x)\downarrow$ in these intervals by graphing r on a graphing calculator and using the $\boxed{\text{TRACE}}$ feature.

▶ MAINTAINING YOUR SKILLS

106. (2.5) Use the graph of $f(x)$ given to sketch the graph of $y = f(x + 2) - 3$.

Exercise 106

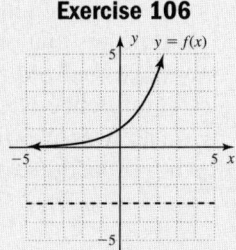

107. (3.5) Graph the function $f(x) = \dfrac{x^2 + 2x - 8}{x + 4}$. If there is a removable discontinuity,

repair the break using an appropriate piecewise-defined function.

108. (1.6) Solve the equation $\dfrac{1}{2}\sqrt{16 - x} - \dfrac{x}{2} = 2$. Check solutions in the original equation.

109. (1.2/3.7) Graph the solution set for the relation: $3x + 1 < 10$ *and* $x^2 - 3 < 1$.

Learning Objectives

In Section 3.8 you will learn how to:

☐ **A.** Solve direct variations

☐ **B.** Solve inverse variations

☐ **C.** Solve joint variations

A study of direct and inverse variation offers perhaps our clearest view of how mathematics is used to model real-world phenomena. While the basis of our study is elementary, involving only the toolbox functions, the applications are at the same time elegant, powerful, and far reaching. In addition, these applications unite some of the most important ideas in algebra, including functions, transformations, rates of change, and graphical analysis, to name a few.

A. Toolbox Functions and Direct Variation

If a car gets 24 miles per gallon (mpg) of gas, we could express the distance d it could travel as $d = 24g$. Table 3.4 verifies the distance traveled by the car changes in *direct* or *constant proportion* to the number of gallons used, and here we say, "distance traveled *varies directly* with gallons used." The equation $d = 24g$ is called a **direct variation,** and the coefficient 24 is called the **constant of variation.**

Using the rate of change notation, $\dfrac{\Delta\text{distance}}{\Delta\text{gallons}} = \dfrac{\Delta d}{\Delta g} = \dfrac{24}{1}$, and we

note this is actually a *linear equation* with slope $m = 24$. When working with variations, the constant k is preferred over m, and in general we have the following:

Table 3.4

g	d
1	24
2	48
3	72
4	96

Direct Variation

y varies directly with x, or *y is directly proportional to x,*

if there is a nonzero constant k such that

$$y = kx.$$

k is called the *constant of variation.*

EXAMPLE 1 ▶ **Writing a Variation Equation**

Write the variation equation for these statements:

    a. Wages earned varies directly with the number of hours worked.
    b. The value of an office machine varies directly with time.
    c. The circumference of a circle varies directly with the length of the diameter.

Solution ▶   **a.** **W**ages varies directly with **h**ours worked: $W = kh$
    b. The **V**alue of an office machine varies directly with **t**ime: $V = kt$
    c. The **C**ircumference varies directly with the **d**iameter: $C = kd$

Now try Exercises 7 through 10 ▶

Once we determine the relationship between two variables is a direct variation, we try to find the value of k and develop a general equation model for the relationship indicated. Note that "varies directly" indicates that one value is a constant multiple of the other. In Example 1(c), you may have realized that for $C = kd$, $k = \pi$ since $\pi = \frac{C}{d}$ and the formula for a circle's circumference is $C = \pi d$. The connection helps illustrate the procedure for finding k, as it shows that only *one known relationship is needed!* This suggests the following procedure:

Solving Applications of Variation

1. Write the information given as an equation, using k as the constant multiple.
2. Substitute the first relationship (pair of values) given and solve for k.
3. Substitute this value for k in the original equation to obtain the variation equation.
4. Use the variation equation to complete the application.

EXAMPLE 2 ▶ **Solving an Application of Direct Variation**

The weight of an astronaut on the surface of another planet **varies directly** with their weight on Earth. An astronaut weighing 140 lb on Earth weighs only 53.2 lb on Mars. How much would a 170-lb astronaut weigh on Mars?

Solution ▶
1. $M = kE$ "Mars weight **varies directly** with Earth weight"
2. $53.2 = k(140)$ substitute 53.2 for M and 140 for E
   $k = 0.38$ solve for k (constant of variation)

Substitute this value of k in the original equation to obtain the variation equation, then find the weight of a 170-lb astronaut that landed on Mars.

3. $M = 0.38E$ variation equation
4. $ = 0.38(170)$ substitute 170 for E
   $ = 64.6$ result

An astronaut weighing 170 lb on Earth weighs only 64.6 lb on Mars.

Now try Exercises 11 through 14 ▶

The toolbox function from Example 2 was a line with slope $k = 0.38$, or $k = \frac{19}{50}$ as a fraction in simplest form. As a rate of change, $k = \frac{\Delta M}{\Delta E} = \frac{19}{50}$, and we see that for every 50 additional pounds on Earth, the weight of an astronaut would increase by only 19 lb on Mars.

EXAMPLE 3 ▶ **Making Estimates from the Graph of a Variation**

The scientists at NASA are planning to send additional probes to the red planet (Mars), that will weigh from 250 to 450 lb. Graph the variation equation from Example 2, then *use the graph* to estimate the corresponding range of weights on Mars. Check your estimate using the variation equation.

Solution ▶ After selecting an appropriate scale, begin at (0, 0) and count off the slope $k = \frac{\Delta M}{\Delta E} = \frac{19}{50}$. This gives the points (50, 19), (100, 38), (200, 76), and so on. From the graph (see dashed arrows), it appears the weights corresponding to 250 lb and 450 lb on Earth are near 95 lb and 170 lb on Mars. Using the equation gives

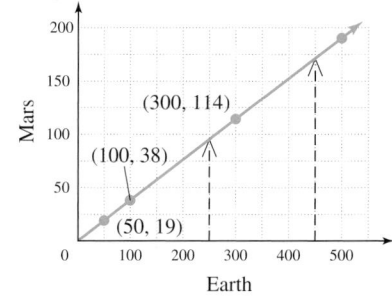

$M = 0.38E$ variation equation
$ = 0.38(250)$ substitute 250 for E
$ = 95$, and
$M = 0.38E$ variation equation
$ = 0.38(450)$ substitute 450 for E
$ = 171$, very close to our estimate from the graph.

Now try Exercises 15 and 16 ▶

When a toolbox function is used to model a variation, our knowledge of their graphs and defining characteristics strengthens a contextual understanding of the application. Consider Examples 4 and 5, where the squaring function is used.

EXAMPLE 4 ▶ **Writing Variation Equations**

Write the variation equation for these statements:

a. In free fall, the distance traveled by an object varies directly with the square of the time.

b. The area of a circle varies directly with the square of its radius.

Solution ▶ **a.** Distance varies directly with the square of the time: $D = kt^2$.

b. Area varies directly with the square of the radius: $A = kr^2$.

Now try Exercises 17 through 20 ▶

Both variations in Example 4 use the squaring function, where k represents the amount of stretch or compression applied, and whether the graph will open upward or downward. However, regardless of the function used, the four-step solution process remains the same.

EXAMPLE 5 ▶ **Solving an Application of Direct Variation**

The range of a projectile varies directly with the square of its initial velocity. As part of a circus act, Bailey the Human Bullet is shot out of a cannon with an initial velocity of 80 feet per second (ft/sec), into a net 200 ft away.

a. Find the constant of variation and write the variation equation.

b. Graph the equation and *use the graph* to estimate how far away the net should be placed if initial velocity is increased to 95 ft/sec.

c. Determine the accuracy of the estimate from (b) using the variation equation.

Solution ▶ **a. 1.** $R = kv^2$ "Range varies directly with the square of the velocity"

    2. $200 = k(80)^2$ substitute 200 for R and 80 for v

        $k = 0.03125$ solve for k (constant of variation)

    3. $R = 0.03125v^2$ variation equation (substitute 0.03125 for k)

b. Since velocity and distance are positive, we again use only QI. The graph is a parabola that opens upward, with the vertex at (0, 0). Selecting velocities from 50 to 100 ft/s, we have:

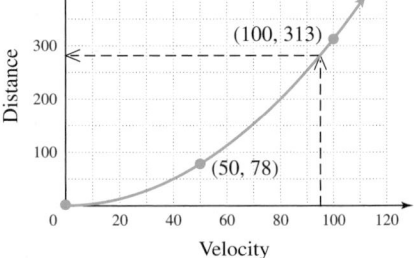

     $R = 0.03125v^2$ variation equation

       $= 0.03125(50)^2$ substitute 50 for v

       $= 78.125$ result

Likewise substituting 100 for v gives $R = 312.5$ ft. Scaling the axes and using (0, 0), (50, 78), and (100, 313) produces the graph shown. At 95 ft/s (dashed lines), it appears the net should be placed about 280 ft away.

c. Using the variation equation gives:

    4. $R = 0.03125v^2$ variation equation

       $= 0.03125(95)^2$ substitute 95 for v

     $R = 282.03125$ result

Our estimate was off by about 2 ft. The net should be placed about 282 ft away.

☑ **A. You've just learned how to solve direct variations**

Now try Exercises 21 through 26 ▶

Note: For Examples 6 to 8, the four steps of the solution process are used in sequence, but are not numbered.

B. Inverse Variation

Table 3.5

Price (dollars)	Demand (1000s)
8	288
9	144
10	96
11	72
12	57.6

Numerous studies have been done that relate the price of a commodity to the demand—the willingness of a consumer to pay that price. For instance, if there is a sudden increase in the price of a popular tool, hardware stores know there will be a corresponding decrease in the demand for that tool. The question remains, "What is this rate of decrease?" Can it be modeled by a linear function with a negative slope? A parabola that opens downward? Some other function? Table 3.5 shows some (simulated) data regarding price versus demand. It appears that a linear function is not appropriate because the rate of change in the number of tools sold is not constant. Likewise a quadratic model seems inappropriate, since we don't expect demand to suddenly start rising again as the price continues to increase. This phenomenon is actually an example of an **inverse variation,** modeled by a transformation of the reciprocal function $y = \frac{k}{x}$. We will often rewrite the equation as $y = k\left(\frac{1}{x}\right)$ to clearly see the inverse relationship. In the case at hand, we might write $D = k\left(\frac{1}{P}\right)$, where k is the constant of variation, D represents the demand for the product, and P the price of the product. In words, we say that "demand *varies inversely* as the price." In other applications of inverse variation, one quantity may vary inversely as the *square* of another, and in general we have

Inverse Variation

y varies inversely with x, or *y is inversely proportional to x,*
if there is a nonzero constant k such that

$$y = k\left(\frac{1}{x}\right).$$

k is called the *constant of variation.*

EXAMPLE 6 ▶   Writing Inverse Variation Equations

Write the variation equation for these statements:

a. In a closed container, pressure varies inversely with the volume of gas.

b. The intensity of light varies inversely with the square of the distance from the source.

Solution ▶   a. Pressure varies inversely with the *Volume* of gas: $P = k\left(\frac{1}{V}\right)$.

b. Intensity of light varies inversely with the square of the distance: $I = k\frac{1}{(d^2)}$.

Now try Exercises 27 through 30 ▶

EXAMPLE 7 ▶   Solving an Application of Inverse Variation

Boyle's law tells us that in a closed container with constant temperature, the pressure of a gas varies inversely with its volume (see illustration on page 393). Suppose the air pressure in a closed cylinder is 60 pounds per square inch (psi) when the volume of the cylinder is 50 in^3.

a. Find the constant of variation and write the variation equation.

b. Use the equation to find the pressure, if volume is compressed to 30 in^3.

Illustration of Boyle's Law

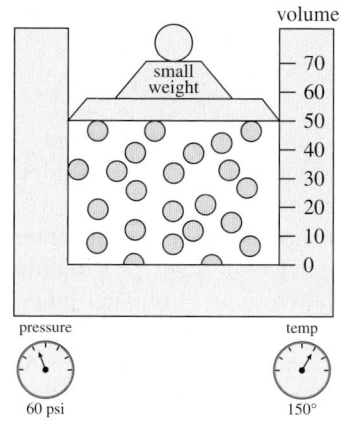

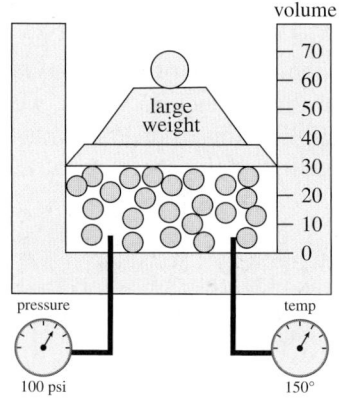

☑ **B. You've just learned how to solve inverse variations**

Solution ▶ **a.** $P = k\left(\dfrac{1}{V}\right)$ "Pressure varies inversely with the volume"

$60 = k\left(\dfrac{1}{50}\right)$ substitute 60 for *P* and 50 for *V.*

$k = 3000$ constant of variation

$P = 3000\left(\dfrac{1}{V}\right)$ variation equation (substitute 3000 for *k*)

b. Using the variation equation we have:

$P = 3000\left(\dfrac{1}{V}\right)$ variation equation

$= 3000\left(\dfrac{1}{30}\right)$ substitute 30 for *V*

$= 100$ result

When the volume is decreased to 30 in³, the pressure increases to 100 psi.

Now try Exercises 31 through 34 ▶

C. Joint or Combined Variations

Just as some decisions might be based on many considerations, often the relationship between two variables depends on a combination of factors. Imagine a wooden plank laid across the banks of a stream for hikers to cross the streambed (see Figure 3.75). The amount of weight the plank will support depends on the type of wood, the width and height of the plank's cross section, and the distance between the supported ends (**see Exercises 59 and 60**). This is an example of a **joint variation,** which can combine any number of variables in different ways. Two general possibilities are: (1) *y varies jointly with the product of x and p:* $y = kxp$; and (2) *y varies jointly with the product of x and p, and inversely with the square of q:* $y = kxp\left(\dfrac{1}{q^2}\right)$. For practice writing joint variations as an equation model, **see Exercises 35 through 40.**

Figure 3.75

EXAMPLE 8 ▶ Solving an Application of Joint Variation

The amount of fuel used by a ship traveling at a uniform speed varies jointly with the distance it travels and the square of the velocity. If 200 barrels of fuel are used to travel 10 mi at 20 nautical miles per hour, how far does the ship travel on 500 barrels of fuel at 30 nautical miles per hour?

Solution ▶

$F = kdv^2$ "fuel use *varies jointly* with distance and velocity squared"

$200 = k(10)(20)^2$ substitute 200 for *F*, 10 for *d*, and 20 for *v*

$200 = 4000k$ simplify and solve for *k*

$0.05 = k$ constant of variation

$F = 0.05dv^2$ equation of variation

To find the distance traveled at 30 nautical miles per hour using 500 barrels of fuel, substitute 500 for F and 30 for v:

$$F = 0.05dv^2 \qquad \text{equation of variation}$$
$$500 = 0.05d(30)^2 \qquad \text{substitute 500 for } F \text{ and 30 for } v$$
$$500 = 45d \qquad \text{simplify}$$
$$11.\overline{1} = d \qquad \text{result}$$

If 500 barrels of fuel are consumed while traveling 30 nautical miles per hour, the ship covers a distance of just over 11 mi.

> **Now try Exercises 41 through 44 ▶**

It's interesting to note that the ship covers just over one additional mile, but consumes over 2.5 times the amount of fuel. The additional speed requires a great deal more fuel.

☑ **C. You've just learned how to solve joint variations**

There is a variety of additional applications in the Exercise Set. **See Exercises 47 through 55.**

3.8 EXERCISES

▶ CONCEPTS AND VOCABULARY

Fill in each blank with the appropriate word or phrase. Carefully reread the section if needed.

1. The phrase "y varies directly with x" is written $y = kx$, where k is called the _____ of variation.

2. If more than two quantities are related in a variation equation, the result is called a _____ variation.

3. The statement "y varies inversely with the square of x" is written _____.

4. For $y = kx$, $y = kx^2$, $y = kx^3$, and $y = k\sqrt{x}$, it is true that as $x \to \infty$, $y \to \infty$ (functions increase). One important difference among the functions is $\frac{\Delta y}{\Delta x}$, or their _____ of _____.

5. Discuss/Explain the general procedure for solving applications of variation. Include references to keywords, and illustrate using an example.

6. The basic percent formula is *amount equals percent times base*, or $A = PB$. In words, write this out as a direct variation with B as the constant of variation, then as an inverse variation with the amount A as the constant of variation.

▶ DEVELOPING YOUR SKILLS

Write the variation equation for each statement.

7. distance traveled varies directly with rate of speed

8. cost varies directly with the quantity purchased

9. force varies directly with acceleration

10. length of a spring varies directly with attached weight

For Exercises 11 and 12, find the constant of variation and write the variation equation. Then use the equation to complete the table.

11. y varies directly with x; $y = 0.6$ when $x = 24$.

x	y
500	
	16.25
750	

12. w varies directly with v; $w = \frac{1}{3}$ when $v = 5$.

v	w
291	
	21.8
339	

13. Wages earned varies directly with the number of hours worked. Last week I worked 37.5 hr and my gross pay was $344.25. Write the variation equation and determine how much I will gross this week if I work 35 hr. What does the value of k represent in this case?

14. The thickness of a paperback book varies directly as the number of pages. A book 3.2 cm thick has 750 pages. Write the variation equation and approximate the thickness of *Roget's 21st Century Thesaurus* (paperback—2nd edition), which has 957 pages.

15. The number of stairs in the stairwell of tall buildings and other structures varies directly as the height of the structure. The base and pedestal for the Statue of Liberty are 47 m tall, with 192 stairs from ground level to the observation deck at the top of the pedestal (at the statue's feet). (a) Find the constant of variation and write the variation equation, (b) graph the variation equation, (c) use the graph to estimate the number of stairs from ground level to the observation deck in the statue's crown 81 m above ground level, and (d) use the equation to check this estimate. Was it close?

16. The height of a projected image varies directly as the distance of the projector from the screen. At a distance of 48 in., the image on the screen is 16 in. high. (a) Find the constant of variation and write the variation equation, (b) graph the variation equation, (c) use the graph to estimate the height of the image if the projector is placed at a distance of 5 ft 3 in., and (d) use the equation to check this estimate. Was it close?

Write the variation equation for each statement.

17. Surface area of a cube varies directly with the square of a side.

18. Potential energy in a spring varies directly with the square of the distance the spring is compressed.

19. Electric power varies directly with the square of the current (amperes).

20. Manufacturing cost varies directly as the square of the number of items made.

For Exercises 21 through 26, find the constant of variation and write the variation equation. Then use the equation to complete the table or solve the application.

21. p varies directly with the square of q; $p = 280$ when $q = 50$

q	p
45	
	338.8
70	

22. n varies directly with m squared; $n = 24.75$ when $m = 30$

m	n
40	
	99
88	

23. The surface area of a cube varies directly as the square of one edge. A cube with edges of $14\sqrt{3}$ cm has a surface area of 3528 cm². Find the surface area in square meters of the spaceships used by the Borg Collective in *Star Trek—The Next Generation*, cubical spacecraft with edges of 3036 m.

24. The area of an equilateral triangle varies directly as the square of one side. A triangle with sides of 50 yd has an area of 1082.5 yd². Find the area in mi² of the region bounded by straight lines connecting the cities of Cincinnati, Ohio, Washington, D.C., and Columbia, South Carolina, which are each approximately 400 mi apart.

25. The distance an object falls varies directly as the square of the time it has been falling. The cannonballs dropped by Galileo from the Leaning Tower of Pisa fell about 169 ft in 3.25 sec. (a) Find the constant of variation and write the variation equation, (b) graph the variation equation, (c) use the graph to estimate how long it would take a hammer, accidentally dropped from a height of 196 ft by a bridge repair crew, to splash into the water below, and (d) use the equation to check this estimate. Was it close? (e) According to the equation, if a camera accidentally fell out of the *News 4 Eye-in-the-Sky* helicopter from a height of 121 ft, how long until it strikes the ground?

26. When a child blows small soap bubbles, they come out in the form of a sphere because the surface tension in the soap seeks to minimize the surface area. The surface area of any sphere varies directly with the square of its radius. A soap bubble with a $\frac{3}{4}$ in. radius has a surface area of approximately 7.07 in². (a) Find the constant of variation and write the variation equation, (b) graph the variation equation, (c) use the graph to estimate the radius of a seventeenth-century cannonball that has a surface area of 113.1 in^2, and (d) use the equation to check this estimate. Was it close? (e) According to the equation, what is the surface area of an orange with a radius of $1\frac{1}{2}$ in.?

Write the variation equation for each statement.

27. The force of gravity varies inversely as the square of the distance between objects.

28. Pressure varies inversely as the area over which it is applied.

29. The safe load of a beam supported at both ends varies inversely as its length.

30. The intensity of sound varies inversely as the square of its distance from the source.

For Exercises 31 through 34, find the constant of variation and write the variation equation. Then use the equation to complete the table or solve the application.

31. Y varies inversely as the square of Z; $Y = 1369$ when $Z = 3$

Z	Y
37	
	2.25
111	

32. A varies inversely with B; $A = 2450$ when $B = 0.8$

B	A
140	
	6.125
560	

33. The effect of Earth's gravity on an object (its weight) varies inversely as the square of its distance from the center of the planet (assume the Earth's radius is 6400 km). If the weight of an astronaut is 75 kg on Earth (when $r = 6400$), what would this weight be at an altitude of 1600 km *above the surface* of the Earth?

34. The demand for a popular new running shoe varies inversely with the cost of the shoes. When the wholesale price is set at $45, the manufacturer ships 5500 orders per week to retail outlets. Based on this information, how many orders would be shipped if the wholesale price rose to $55?

Write the variation equation for each statement.

35. Interest earned varies jointly with the rate of interest and the length of time on deposit.

36. Horsepower varies jointly as the number of cylinders in the engine and the square of the cylinder's diameter.

37. The area of a trapezoid varies jointly with its height and the sum of the bases.

38. The area of a triangle varies jointly with its height and the length of the base.

39. The volume of metal in a circular coin varies directly with the thickness of the coin and the square of its radius.

40. The electrical resistance in a wire varies directly with its length and inversely as the cross-sectional area of the wire.

For Exercises 41–44, find the constant of variation and write the related variation equation. Then use the equation to complete the table or solve the application.

41. C varies directly with R and inversely with S squared, and $C = 21$ when $R = 7$ and $S = 1.5$.

R	S	C
120		22.5
200	12.5	
	15	10.5

42. J varies directly with P and inversely with the square root of Q, and $J = 19$ when $P = 4$ and $Q = 25$.

P	Q	J
47.5		118.75
112	31.36	
	44.89	66.5

43. Kinetic energy: Kinetic energy (energy attributed to motion) varies jointly with the mass of the object and the square of its velocity. Assuming a unit mass of $m = 1$, an object with a velocity of 20 m per sec (m/s) has kinetic energy of 200 J. How much energy is produced if the velocity is increased to 35 m/s?

44. Safe load: The load that a horizontal beam can support varies jointly as the width of the beam, the square of its height, and inversely as the length of the beam. A beam 4 in. wide and 8 in. tall can safely support a load of 1 ton when the beam has a length of 12 ft. How much could a similar beam 10 in. tall safely support?

▶ WORKING WITH FORMULAS

45. Required interest rate: $R(A) = \sqrt[3]{A} - 1$

To determine the simple interest rate R that would be required for each dollar ($1) left on deposit for 3 yr to grow to an amount A, the formula $R(A) = \sqrt[3]{A} - 1$ can be applied. To what function family does this formula belong? Complete the table using a calculator, then use the table to estimate the interest rate required for each $1 to grow to $1.17. Compare your estimate to the value you get by evaluating $R(1.17)$.

Amount A	Rate R
1.0	
1.05	
1.10	
1.15	
1.20	
1.25	

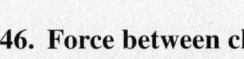 **46. Force between charged particles:** $F = k\dfrac{Q_1 Q_2}{d^2}$

The force between two charged particles is given by the formula shown, where F is the force (in joules—J), Q_1 and Q_2 represent the electrical charge on each particle (in coulombs—C), and d is the distance between them (in meters). If the particles have a like charge, the force is repulsive; if the charges are unlike, the force is attractive. (a) Write the variation equation in words. (b) Solve for k and use the formula to find the electrical constant k, given $F = 0.36$ J, $Q_1 = 2 \times 10^{-6}$ C, $Q_2 = 4 \times 10^{-6}$ C, and $d = 0.2$ m. Express the result in scientific notation.

▶ APPLICATIONS

Find the constant of variation "k" and write the variation equation, then use the equation to solve.

47. Cleanup time: The time required to pick up the trash along a stretch of highway varies inversely as the number of volunteers who are working. If 12 volunteers can do the cleanup in 4 hr, how many volunteers are needed to complete the cleanup in just 1.5 hr?

48. Wind power: The wind farms in southern California contain wind generators whose power production varies directly with the cube of the wind's speed. If one such generator produces 1000 W of power in a 25 mph wind, find the power it generates in a 35 mph wind.

49. Pull of gravity: The weight of an object on the moon varies directly with the weight of the object on Earth. A 96-kg object on Earth would weigh only 16 kg on the moon. How much would a fully suited 250-kg astronaut weigh on the moon?

50. Period of a pendulum: The time that it takes for a simple pendulum to complete one period (swing over and back) varies directly as the square root of its length. If a pendulum 20 ft long has a period of 5 sec, find the period of a pendulum 30 ft long.

51. Stopping distance: The stopping distance of an automobile varies directly as the square root of its speed when the brakes are applied. If a car requires

108 ft to stop from a speed of 25 mph, estimate the stopping distance if the brakes were applied when the car was traveling 45 mph.

52. Supply and demand: A chain of hardware stores finds that the demand for a special power tool varies inversely with the advertised price of the tool. If the price is advertised at $85, there is a monthly demand for 10,000 units at all participating stores. Find the projected demand if the price were lowered to $70.83.

53. Cost of copper tubing: The cost of copper tubing varies jointly with the length and the diameter of the tube. If a 36-ft spool of $\frac{1}{4}$-in.-diameter tubing costs $76.50, how much does a 24-ft spool of $\frac{3}{8}$-in.-diameter tubing cost?

54. Electrical resistance: The electrical resistance of a copper wire varies directly with its length and inversely with the square of the diameter of the wire. If a wire 30 m long with a diameter of 3 mm has a resistance of 25 Ω, find the resistance of a wire 40 m long with a diameter of 3.5 mm.

55. Volume of phone calls: The number of phone calls per day between two cities varies directly as the product of their populations and inversely as the square of the distance between them. The city of Tampa, Florida (pop. 300,000), is 430 mi from the city of Atlanta, Georgia (pop. 420,000).

Telecommunications experts estimate there are about 300 calls per day between the two cities. Use this information to estimate the number of daily phone calls between Amarillo, Texas (pop. 170,000), and Denver, Colorado (pop. 550,000), which are also separated by a distance of about 430 mi. Note: Population figures are for the year 2000 and rounded to the nearest ten-thousand.

Source: 2005 World Almanac, p. 626.

56. **Internet commerce:** The likelihood of an eBay® item being sold for its "Buy it Now®" price P, varies directly with the feedback rating of the seller, and inversely with the cube of $\frac{P}{MSRP}$, where MSRP represents the manufacturer's suggested retail price. A power eBay® seller with a feedback rating of 99.6%, knows she has a 60% likelihood of selling an item at 90% of the MSRP. What is the likelihood a seller with a 95.3% feedback rating can sell the same item at 95% of the MSRP?

57. **Volume of an egg:** The volume of an egg laid by an average chicken varies jointly with its length and the square of its width. An egg measuring 2.50 cm wide and 3.75 cm long has a volume of 12.27 cm³. A Barret's Blue Ribbon hen can lay an egg measuring 3.10 cm wide and 4.65 cm long. (a) What is the volume of this egg? (b) As a percentage, how much greater is this volume than that of an average chicken?

58. **Athletic performance:** Researchers have estimated that a sprinter's time in the 100-m dash varies directly as the square root of her age and inversely as the number of hours spent training each week. At 20 yr old, Gail trains 10 hr per week (hr/wk) and has an average time of 11 sec. (a) Assuming she continues to train 10 hr/wk, what will her average time be at 30 yr old? (b) If she wants to keep her average time at 11 sec, how many hours per week should she train?

59. **Maximum safe-load:** The maximum safe load M that can be placed on a uniform horizontal beam supported at both ends varies directly as the width w and the square of the height h of the beam's cross section, and inversely as its length L (width and height are assumed to be in inches, and length in feet). (a) Write the variation equation. (b) If a beam 18 in. wide, 2 in. high, and 8 ft long can safely support 270 lb, what is the safe load for a beam of like dimensions with a length of 12 ft?

60. **Maximum safe load:** Suppose a 10-ft wooden beam with dimensions 4 in. by 6 in. is made from the same material as the beam in Exercise 59 (the same k value can be used). (a) What is the maximum safe load if the beam is placed so that width is 6 in. and height is 4 in.? (b) What is the maximum safe load if the beam is placed so that width is 4 in. and height is 6 in.?

▶ **EXTENDING THE CONCEPT**

61. In function form, the variations $Y_1 = k\frac{1}{x}$ and $Y_2 = k\frac{1}{x^2}$ become $f(x) = k\frac{1}{x}$ and $g(x) = k\frac{1}{x^2}$. Both graphs appear similar in Quadrant I and both may "fit" a scatter-plot fairly well, but there is a big difference between them—they decrease as x gets larger, but *they decrease at very different rates*. Assume $k = 1$ and use the ideas from Section 2.5 to compute the rate of change for f and g for the interval from $x = 0.5$ to $x = 0.6$. Were you surprised? In the interval $x = 0.7$ to $x = 0.8$, will the rate of decrease for each function be greater or less than in the interval $x = 0.5$ to $x = 0.6$? Why?

62. The gravitational force F between two celestial bodies varies jointly as the product of their masses and inversely as the square of the distance d between them. The relationship is modeled by Newton's law of universal gravitation: $F = k\frac{m_1 m_2}{d^2}$.

Given that $k = 6.67 \times 10^{-11}$, what is the gravitational force exerted by a 1000-kg sphere on another identical sphere that is 10 m away?

63. The intensity of light and sound both vary inversely as the square of their distance from the source.

    a. Suppose you're relaxing one evening with a copy of *Twelfth Night* (Shakespeare), and the reading light is placed 5 ft from the surface of the book. At what distance would the intensity of the light be twice as great?

    b. *Tamino's Aria* (*The Magic Flute*—Mozart) is playing in the background, with the speakers 12 ft away. At what distance from the speakers would the intensity of sound be three times as great?

▶ **MAINTAINING YOUR SKILLS**

64. (R.3) Evaluate: $\left(\dfrac{2x^4}{3x^3y}\right)^{-2}$

65. (1.5) Find all zeroes, real and complex:
$x^3 + 4x^2 + 8x = 0$.

66. (2.2) State the domains of f and g given here:

a. $f(x) = \dfrac{x-3}{x^2 - 16}$ **b.** $g(x) = \dfrac{x-3}{\sqrt{x^2 - 16}}$

67. (2.5) Graph by using transformations of the parer function and plotting a minimum number of points:
$f(x) = -2|x-3| + 5$.

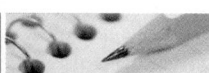

SUMMARY AND CONCEPT REVIEW

SECTION 3.1   Quadratic Functions and Applications

KEY CONCEPTS

- A quadratic function is one of the form $f(x) = ax^2 + bx + c$; $a \neq 0$. The simplest quadratic is the squaring function $f(x) = x^2$, where $a = 1$ and $b, c = 0$.
- The graph of a quadratic function is a parabola. Parabolas have three distinctive features: (1) like end behavior on the left and right, (2) an axis of symmetry, (3) a highest or lowest point called the vertex.
- For a quadratic function in the standard form $y = ax^2 + bx + c$,
  - End behavior: graph opens upward if $a > 0$, opens downward if $a < 0$
  - Zeroes/x-intercepts: substitute 0 for y and solve for x (if they exist)
  - y-intercept: substitute 0 for $x \to (0, c)$
  - Vertex: (h, k), where $h = \dfrac{-b}{2a}$, $k = f\left(\dfrac{-b}{2a}\right)$
  - Maximum value: If the parabola opens downward, $y = k$ is the maximum value of f.
  - Minimum value: If the parabola opens upward, $y = k$ is the minimum value of f.
  - Line of symmetry: $x = h$ is the line (or axis) of symmetry [if $(h + c, y)$ is on the graph, then $(h - c, y)$ is also on the graph].
- By completing the square, $f(x) = ax^2 + bx + c$ can be written as the transformation $f(x) = a(x + h)^2 \pm k$, and graphed using transformations of $y = x^2$.

EXERCISES

Graph $f(x)$ by completing the square and using transformations of the parent function. Graph $g(x)$ and $h(x)$ using the vertex formula and y-intercept. Find the x-intercepts (if they exist) for all functions.

1. $f(x) = x^2 + 8x + 15$ **2.** $g(x) = -x^2 + 4x - 5$ **3.** $h(x) = 4x^2 - 12x + 3$

4. Height of a superball: A teenager tries to see how high she can bounce her superball by throwing it downward on her driveway. The height of the ball (in feet) at time t (in seconds) is given by $h(t) = -16t^2 + 96t$. (a) How high is the ball at $t = 0$? (b) How high is the ball after 1.5 sec? (c) How long until the ball is 135 ft high? (d) What is the maximum height attained by the ball? At what time t did this occur?

SECTION 3.2   Synthetic Division; the Remainder and Factor Theorems

KEY CONCEPTS

- Synthetic division is an abbreviated form of long division. Only the coefficients of the dividend are used, since "standard form" ensures like place values are aligned. Zero placeholders are used for "missing" terms. The "divisor" must be linear with leading coefficient 1.
- To divide a polynomial by $x - c$, use c in the synthetic division; to divide by $x + c$, use $-c$.
- After setting up the synthetic division template, drop the leading coefficient of the dividend into place, then multiply in the diagonal direction, place the product in the next column, and add in the vertical direction, continuing to the last column.
- The final sum is the remainder r, the numbers preceding it are the coefficients of $q(x)$.
- Remainder theorem: If $p(x)$ is divided by $x - c$, the remainder is equal to $p(c)$. The theorem can be used to evaluate polynomials at $x = c$.
- Factor theorem: If $p(c) = 0$, then c is a zero of p and $(x - c)$ is a factor. Conversely, if $(x - c)$ is a factor of p, then $p(c) = 0$. The theorem can be used to factor a polynomial or build a polynomial from its zeroes.
- The remainder and factor theorems also apply when c is a complex number.

EXERCISES

Divide using long division and clearly identify the quotient and remainder:

5. $\dfrac{x^3 + 4x^2 - 5x - 6}{x - 2}$

6. $\dfrac{x^3 + 2x - 4}{x^2 - x}$

7. Use the factor theorem to show that $x + 7$ is a factor of $2x^4 + 13x^3 - 6x^2 + 9x + 14$.

8. Complete the division and write $h(x)$ as $h(x) = d(x)q(x) + r(x)$, given $\dfrac{h(x)}{d(x)} = \dfrac{x^3 - 4x + 5}{x - 2}$.

9. Use the factor theorem to help factor $p(x) = x^3 + 2x^2 - 11x - 12$ completely.

10. Use the factor and remainder theorems to factor h, given $x = 4$ is a zero: $h(x) = x^4 - 3x^3 - 4x^2 - 2x + 8$.

Use the remainder theorem:

11. Show $x = \frac{1}{2}$ is a zero of V: $V(x) = 4x^3 + 8x^2 - 3x - 1$.

12. Show $x = 3i$ is a zero of W: $W(x) = x^3 - 2x^2 + 9x - 18$.

13. Find $h(-7)$ given $h(x) = x^3 + 9x^2 + 13x - 10$.

Use the factor theorem:

14. Find a degree 3 polynomial in standard form with zeroes $x = 1$, $x = -\sqrt{5}$, and $x = \sqrt{5}$.

15. Find a fourth-degree polynomial in standard form with one real zero, given $x = 1$ and $x = -2i$ are zeroes.

16. Use synthetic division and the remainder theorem to answer: At a busy shopping mall, customers are constantly coming and going. One summer afternoon during the hours from 12 o'clock noon to 6 in the evening, the number of customers in the mall could be modeled by $C(t) = 3t^3 - 28t^2 + 66t + 35$, where $C(t)$ is the number of customers (in tens), t hours after 12 noon. (a) How many customers were in the mall at noon? (b) Were more customers in the mall at 2:00 or at 3:00 P.M.? How many more? (c) Was the mall busier at 1:00 P.M. (after lunch) or 6:00 P.M. (around dinner time)?

SECTION 3.3   Zeroes of Polynomial Functions

KEY CONCEPTS

- Fundamental theorem of algebra: Every complex polynomial of degree $n \geq 1$ has at least one complex zero.
- Linear factorization theorem: Every complex polynomial of degree $n \geq 1$ has exactly n linear factors, and can be written in the form $p(x) = a(x - c_1)(x - c_2) \ldots (x - c_n)$, where $a \neq 0$ and $c_1, c_2, \ldots, c_n$ are (not necessarily distinct) complex numbers.

- For a polynomial p in factored form with repeated factors $(x - c)^m$, c is a zero of multiplicity m. If m is odd, c is a zero of odd multiplicity; if m is even, c is a zero of even multiplicity.
- Corollaries to the linear factorization theorem:
  - I. If p is a polynomial with real coefficients, p can be factored into linear factors (not necessarily distinct) and irreducible quadratic factors having real coefficients.
  - II. If p is a polynomial with real coefficients, the complex zeroes of p must occur in conjugate pairs. If $a + bi$ $(b \neq 0)$, is a zero, then $a - bi$ is also a zero.
  - III. If p is a polynomial with degree $n \geq 1$, then p will have exactly n zeroes (real or complex), where zeroes of multiplicity m are counted m times.
- Intermediate value theorem: If p is a polynomial with real coefficients where $p(a)$ and $p(b)$ have opposite signs, then there is at least one c between a and b such that $p(c) = 0$.
- Rational zeroes theorem: If a real polynomial has integer coefficients, rational zeroes must be of the form $\frac{p}{q}$, where p is a factor of the constant term and q is a factor of the leading coefficient.
- Descartes' rule of signs, upper and lower bounds property, tests for -1 and 1, and graphing technology can all be used with the rational zeroes theorem to factor, solve, and graph polynomial functions.

EXERCISES

Using the tools from this section,

17. List all possible rational zeroes of
$p(x) = 4x^3 - 16x^2 + 11x + 10$.

18. Find all rational zeroes of
$p(x) = 4x^3 - 16x^2 + 11x + 10$.

19. Write $P(x) = 2x^3 - 3x^2 - 17x - 12$
in completely factored form.

20. Prove that $h(x) = x^4 - 7x^2 - 2x + 3$
has no rational zeroes.

21. Identify two intervals (of those given) that contain a zero of $P(x) = x^4 - 3x^3 - 8x^2 + 12x + 6$: $[-2, -1]$, $[1, 2]$, $[2, 3]$, $[4, 5]$. Then verify your answer using a graphing calculator.

22. Discuss the number of possible positive, negative, and complex zeroes for $g(x) = x^4 + 3x^3 - 2x^2 - x - 30$. Then identify which combination is correct using a graphing calculator.

SECTION 3.4   Graphing Polynomial Functions

KEY CONCEPTS

- All polynomial graphs are smooth, continuous curves.
- A polynomial of degree n has *at most $n - 1$* turning points. The precise location of these turning points are the local maximums or local minimums of the function.
- If the degree of a polynomial is odd, the ends of its graph will point in opposite directions (like $y = mx$). If the degree is even, the ends will point in the same direction (like $y = ax^2$). The sign of the lead coefficient determines the actual behavior.
- The "behavior" of a polynomial graph near its zeroes is determined by the multiplicity of the zero. For any factor $(x - c)^m$, the graph will "cross through" the x-axis if m is odd and "bounce off" the x-axis (touching at just one point) if m is even. The larger the value of m, the flatter (more compressed) the graph will be near c.
- To "round-out" a graph, additional *midinterval points* can be found between known zeroes.
- These ideas help to establish the *Guidelines for Graphing Polynomial Functions*. See page 327.

EXERCISES

State the degree, end behavior, and y-intercept, but do not graph.

23. $f(x) = -3x^5 + 2x^4 + 9x - 4$

24. $g(x) = (x - 1)(x + 2)^2(x - 2)$

Graph using the *Guidelines for Graphing Polynomials.*

25. $p(x) = (x + 1)^3(x - 2)^2$ **26.** $q(x) = 2x^3 - 3x^2 - 9x + 10$ **27.** $h(x) = x^4 - 6x^3 + 8x^2 + 6x - 9$

28. For the graph of $P(x)$ shown, (a) state whether the degree of P is even or odd, (b) use the graph to locate the zeroes of P and state whether their multiplicity is even or odd, and (c) find the minimum possible degree of P and write it in factored form. Assume all zeroes are real.

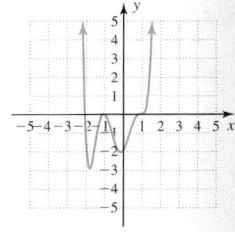

SECTION 3.5   Graphing Rational Functions

KEY CONCEPTS

• A rational function is one of the form $V(x) = \dfrac{p(x)}{d(x)}$, where p and d are polynomials and $d(x) \neq 0$.

• The domain of V is all real numbers, except the zeroes of d.
• If zero is in the domain of V, substitute 0 for x to find the y-intercept.
• The zeroes of V (if they exist), are solutions to $p(x) = 0$.
• The line $y = k$ is a horizontal asymptote of V if as $|x|$ increases without bound, $V(x)$ approaches k.

• If $\dfrac{p(x)}{d(x)}$ is in simplest form, vertical asymptotes will occur at the zeroes of d.

• The line $x = h$ is a vertical asymptote of V if as x approaches h, $V(x)$ increases/decreases without bound.
• If the degree of p is less than the degree of d, $y = 0$ (the x-axis) is a horizontal asymptote. If the degree of p is

  equal to the degree of d, $y = \dfrac{a}{b}$ is a horizontal asymptote, where a is the leading coefficient of p, and b is the

  leading coefficient of d.
• The *Guidelines for Graphing Rational Functions* can be found on page 342.

EXERCISES

29. For the function $V(x) = \dfrac{x^2 - 9}{x^2 - 3x - 4}$, state the following but do not graph: (a) domain (in set notation),

   (b) equations of the horizontal and vertical asymptotes, (c) the x- and y-intercept(s), and (d) the value of $V(1)$.

30. For $v(x) = \dfrac{(x + 1)^2}{x + 2}$, will the function change sign at $x = -1$? Will the function change sign at $x = -2$? Justify your responses.

Graph using the *Guidelines for Graphing Rational Functions*.

31. $v(x) = \dfrac{x^2 - 4x}{x^2 - 4}$

32. $t(x) = \dfrac{2x^2}{x^2 - 5}$

33. Use the vertical asymptotes, x-intercepts, and their multiplicities to construct an equation that corresponds to the given graph. Be sure the y-intercept on the graph matches the value given by your equation. Assume these features are integer-valued. Check your work on a graphing calculator.

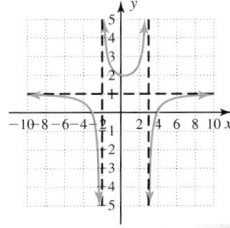

34. The average cost of producing a popular board game is given by the function

   $A(x) = \dfrac{5000 + 15x}{x}$; $x \geq 1000$. (a) Identify the horizontal asymptote of the function and

   explain its meaning in this context. (b) To be profitable, management believes the average cost must be below $17.50. What levels of production will make the company profitable?

SECTION 3.6   Additional Insights into Rational Functions

KEY CONCEPTS

- If $V = \dfrac{p(x)}{d(x)}$ is not in simplest form, with p and d sharing factors of the form $x - c$, the graph will have a removable discontinuity (a hole or gap) at $x = c$. The discontinuity can be "removed" (repaired) by redefining V using a piecewise-defined function.

- If $V = \dfrac{p(x)}{d(x)}$ is in simplest form, and the degree of p is greater than the degree of d, the graph will have an oblique or nonlinear asymptote, as determined by the quotient polynomial after division. If the degree of p is greater by 1, the result is a slant (oblique) asymptote. If the degree of p is greater by 2, the result is a parabolic asymptote.

EXERCISES

35. Determine if the graph of h will have a vertical asymptote or a removable discontinuity, then graph the function

$$h(x) = \frac{x^3 - 2x^2 - 9x + 18}{x - 2}.$$

36. Sketch the graph of $h(x) = \dfrac{x^2 - 3x - 4}{x + 1}$. If there is a removable discontinuity, repair the break by redefining h using an appropriate piecewise-defined function.

Graph the functions using the *Guidelines for Graphing Rational Functions.*

37. $h(x) = \dfrac{x^2 - 2x}{x - 3}$

38. $t(x) = \dfrac{x^3 - 7x + 6}{x^2}$

39. The cost to make x thousand party favors is given by $C(x) = x^2 - 2x + 6$, where $x \geq 1$ and C is in thousands of dollars. For the average cost of production $A(x) = \dfrac{x^2 - 2x + 6}{x}$, (a) graph the function, (b) use the graph to estimate the level of production that will make average cost a minimum, and (c) state the average cost of a single party favor at this level of production.

SECTION 3.7   Polynomial and Rational Inequalities

KEY CONCEPTS

- To solve polynomial inequalities, write $P(x)$ in factored form and note the multiplicity of each real zero.
- Plot real zeroes on a number line. The graph will cross the x-axis at zeroes of odd multiplicity (P will change sign), and bounce off the axis at zeroes of even multiplicity (P will not change sign).
- Use the end behavior, y-intercept, or a test point to determine the sign of P in a given interval, then label all other intervals as $P(x) > 0$ or $P(x) < 0$ by analyzing the multiplicity of neighboring zeroes. Use the resulting diagram to state the solution.
- The solution process for rational inequalities and polynomial inequalities is virtually identical, considering that vertical asymptotes also create intervals where function values may change sign, depending on their multiplicity.
- Polynomial and rational inequalities can also be solved using an interval test method. Since polynomials and rational functions are continuous on their domains, the sign of the function at any one point in an interval will be the same as for all other points in that interval.

EXERCISES

Solve each inequality indicated using a number line and the behavior of the graph at each zero.

40. $x^3 + x^2 > 10x - 8$

41. $\dfrac{x^2 - 3x - 10}{x - 2} \geq 0$

42. $\dfrac{x}{x - 2} \leq \dfrac{-1}{x}$

SECTION 3.8   Variation: Function Models in Action

KEY CONCEPTS

- *Direct variation:* If there is a nonzero constant k such that $y = kx$, we say, "y varies directly with x" or "y is directly proportional to x" (k is called the constant of variation).

- *Inverse variation:* If there is a nonzero constant k such that $y = k\left(\dfrac{1}{x}\right)$ we say, "y varies inversely with x" or y is inversely proportional to x.

- In some cases, direct and inverse variations work simultaneously to form a *joint variation.*

- The process for solving variation equations can be found on page 380.

EXERCISES

Find the constant of variation and write the equation model, then use this model to complete the table.

43. y varies directly as the cube root of x; $y = 52.5$ when $x = 27$.

x	y
216	
	12.25
729	

44. z varies directly as v and inversely as the square of w; $z = 1.62$ when $w = 8$ and $v = 144$.

v	w	z
196	7	
	1.25	17.856
24		48

45. Given t varies jointly with u and v, and inversely as w, if $t = 30$ when $u = 2$, $v = 3$, and $w = 5$, find t when $u = 8$, $v = 12$, and $w = 15$.

46. The time that it takes for a simple pendulum to complete one period (swing over and back) is directly proportional to the square root of its length. If a pendulum 16 ft long has a period of 3 sec, find the time it takes for a 36-ft pendulum to complete one period.

MIXED REVIEW

1. Find the equation of the function whose graph is shown.

2. Complete the square to write each function as a transformation. Then graph each function, clearly labeling the vertex and all intercepts (if they exist).
   a. $f(x) = 2x^2 + 8x + 3$ **b.** $g(x) = -x^2 - 4x$

3. A computer components manufacturer produces external 2.5″ hard drives. Their sizes range from 20 GB to 200 GB. The cost of producing a hard drive can be modeled by the function

$$C(s) = \frac{1}{180}s^2 - \frac{8}{9}s + \frac{680}{9},$$

where s is the size of the hard drive, in gigabytes. Find the hard drive size that has the lowest cost of production. What is the cost of production?

4. Divide using long division and name the quotient and remainder: $\dfrac{x^3 + 3x^2 - 5x - 7}{x + 3}$.

5. Divide using synthetic division and name the quotient and remainder: $\dfrac{x^4 - 3x^2 + 5x - 1}{x + 2}$.

Use synthetic division and the remainder theorem to complete Exercises 6 and 7.

6. State which of the following *are not factors* of $x^3 - 9x^2 + 2x + 48$: (a) $(x + 6)$, (b) $(x - 8)$, (c) $(x - 12)$, (d) $(x - 4)$, (e) $(x + 2)$.

7. Given $P(x) = 6x^3 - 23x^2 - 40x + 31$, find (a) $P(-1)$, (b) $P(1)$, and (c) $P(5)$.

8. Use the factor theorem.
   a. Find a real polynomial of degree 3 with roots $x = 3$ and $x = -5i$.
   b. Find a real polynomial of degree 2 with $x = 2 - 3i$ as one of the roots.

9. Use the rational zeroes theorem.
   a. Which of the following *cannot be* roots of
      $6x^3 + x^2 - 20x - 12 = 0$?
      $x = 9 \quad x = -3 \quad x = \frac{3}{2} \quad x = \frac{8}{3} \quad x = -\frac{2}{3}$
   b. Write P in completely factored form. Then state
      all zeroes of P, real and complex.
      $P(x) = x^4 - x^3 + 7x^2 - 9x - 18$.

10. Graph using the *Guidelines for Graphing Polynomials.*
    a. $f(x) = x^3 - 13x + 12$
    b. $g(x) = x^4 - 10x^2 + 9$
    c. $h(x) = (x - 1)^3 (x + 2)^2 (x + 1)$

Graph using the *Guidelines for Graphing Rational Functions.*

11. $p(x) = \dfrac{x^2 - 2x}{x^2 - 2x + 1}$ 12. $q(x) = \dfrac{x^2 - 4}{x^2 - 3x - 4}$

13. $r(x) = \dfrac{x^3 - 13x + 12}{x^2}$ 14. $y = \dfrac{x^2 - 4x}{x - 3}$
    (see Exercise 10a)

Solve each inequality.

15. $x^3 - 4x < 12 - 3x^2$ 16. $\dfrac{4}{x + 2} \geq \dfrac{3}{x}$

17. An open, rectangular box is to
    be made from a 24-in. by 16-in.
    piece of sheet metal, by cutting
    a square from each corner and
    folding up the sides.
    a. Show that the resulting
       volume is given by
       $V(x) = 4x^3 - 80x^2 + 384x$.

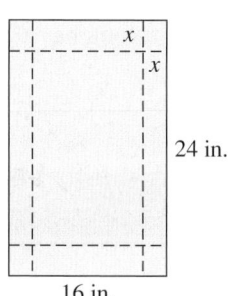
24 in.

16 in.

b. Show that for a desired volume of 512 in³, the
   height "x" of the box can be found by solving
   $x^3 - 20x^2 + 96x - 128 = 0$.
c. According to the rational roots theorem *and the
   context of this application,* what are the possible
   rational zeroes for this equation?
d. Find the *rational zero x* (the height) that gives the
   box a volume of 512 in³.
e. Use the zero from part (d) and synthetic division
   to help find the *irrational zero x* that also gives
   the box a volume of 512 in³. Round the solution
   to hundredths.

Write the variation equation for each statement.

18. The volume of metal in a circular coin varies
    directly with the thickness of the coin and the square
    of its radius.

19. The electrical resistance in a wire varies directly
    with its length and inversely as the cross-sectional
    area of the wire.

20. **Cost of copper tubing:** The cost of copper tubing
    varies jointly with the length and the diameter of the
    tube. If a 36-ft spool of $\frac{1}{4}$-in. diameter tubing costs
    $76.50, how much does a 24-ft spool of $\frac{3}{8}$-in.
    diameter tubing cost?

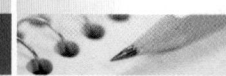

PRACTICE TEST

1. Complete the square to write each function as a
   transformation. Then graph each function and label
   the vertex and all intercepts (if they exist).
   a. $f(x) = -x^2 + 10x - 16$
   b. $g(x) = \dfrac{1}{2}x^2 + 4x + 16$

2. The graph of a quadratic function has a vertex of
   $(-1, -2)$, and passes through the origin. Find the
   other intercept, and the equation of the graph in
   standard form.

3. Suppose the function $d(t) = t^2 - 14t$ models the
   depth of a scuba diver at time t, as she dives

underwater from a steep shoreline, reaches a certain
depth, and swims back to the surface.
a. What is her depth after 4 sec? After 6 sec?
b. What was the maximum depth of the dive?
c. How many seconds was the diver beneath the
   surface?

4. Compute the quotient using long division:
   $\dfrac{x^3 - 3x^2 + 5x - 2}{x^2 + 2x + 1}$.

5. Find the quotient and remainder using synthetic
   division: $\dfrac{x^3 + 4x^2 - 5x - 20}{x + 2}$.

6. Use the remainder theorem to show $(x + 3)$ is a factor of $x^4 - 15x^2 - 10x + 24$.

7. Given $f(x) = 2x^3 + 4x^2 - 5x + 2$, find the value of $f(-3)$ using synthetic division and the remainder theorem.

8. Given $x = 2$ and $x = 3i$ are two zeroes of a real polynomial $P(x)$ with degree 3. Use the factor theorem to find $P(x)$.

9. Factor the polynomial and state the multiplicity of each zero: $Q(x) = (x^2 - 3x + 2)(x^3 - 2x^2 - x + 2)$.

10. Given $C(x) = x^4 + x^3 + 7x^2 + 9x - 18$, (a) use the rational zeroes theorem to list all possible rational zeroes; (b) apply Descartes' rule of signs to count the number of possible positive, negative, and complex zeroes; and (c) use this information along with the tests for 1 and -1, synthetic division, and the factor theorem to factor C completely.

11. Over a 10-yr period, the balance of payments (deficit versus surplus) for a small county was modeled by the function $f(x) = \frac{1}{2}x^3 - 7x^2 + 28x - 32$, where $x = 0$ corresponds to 1990 and $f(x)$ is the deficit or surplus in millions of dollars. (a) Use the rational roots theorem and synthetic division to find the years the county "broke even" (debt = surplus = 0) from 1990 to 2000. (b) How many years did the county run a surplus during this period? (c) What was the surplus/deficit in 1993?

12. Sketch the graph of $f(x) = (x - 3)(x + 1)^3(x + 2)^2$ using the degree, end behavior, x- and y-intercepts, zeroes of multiplicity, and a few "midinterval" points.

13. Use the *Guidelines for Graphing Polynomials* to graph $g(x) = x^4 - 9x^2 - 4x + 12$.

14. Use the *Guidelines for Graphing Rational Functions* to graph $h(x) = \dfrac{x - 2}{x^2 - 3x - 4}$.

15. Suppose the cost of cleaning contaminated soil from a dump site is modeled by $C(x) = \dfrac{300x}{100 - x}$, where $C(x)$ is the cost (in $1000s) to remove $x\%$ of the contaminants. Graph using $x \in [0, 100]$, and use the graph to answer the following questions.

   a. What is the significance of the *vertical asymptote* (what does it mean in this context)?

b. If EPA regulations are changed so that 85% of the contaminants must be removed, instead of the 80% previously required, how much additional cost will the new regulations add? Compare the cost of the 5% increase from 80% to 85% with the cost of the 5% increase from 90% to 95%. What do you notice?

   c. What percent of the pollutants can be removed if the company budgets $2,200,000?

16. Graph using the *Guidelines for Graphing Rational Functions.*

   a. $r(x) = \dfrac{x^3 - x^2 - 9x + 9}{x^2}$

   b. $R(x) = \dfrac{x^3 + 7x - 6}{x^2 - 4}$

17. Find the level of production that will minimize the average cost of an item, if production costs are modeled by $C(x) = 2x^2 + 25x + 128$, where $C(x)$ is the cost to manufacture x hundred items.

18. Solve each inequality

   a. $x^3 - 13x \le 12$ b. $\dfrac{3}{x - 2} < \dfrac{2}{x}$

19. Suppose the concentration of a chemical in the bloodstream of a large animal h hr after injection into muscle tissue is modeled by the formula

   $C(h) = \dfrac{2h^2 + 5h}{h^3 + 55}$.

   a. Sketch a graph of the function for the intervals $x \in [-5, 20]$, $y \in [0, 1]$.

   b. Where is the vertical asymptote? Does it play a role in this context?

   c. What is the concentration after 2 hr? After 8 hr?

   d. How long does it take the concentration to fall below 20% $[C(h) < 0.2]$?

   e. When does the maximum concentration of the chemical occur? What is this maximum?

   f. Describe the significance of the horizontal asymptote in this context.

20. The maximum load that can be supported by a rectangular beam varies jointly with its width and its height squared and inversely with its length. If a beam 10 ft long, 3 in. wide, and 4 in. high can support 624 lb, how many pounds could a beam support with the same dimensions but 12 ft long?

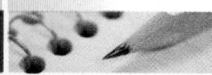

CALCULATOR EXPLORATION AND DISCOVERY

Complex Zeroes, Repeated Zeroes, and Inequalities

This *Calculator Exploration and Discovery* will explore the relationship between the solution of a polynomial (or rational) inequality and the complex zeroes and repeated zeroes of the related function. After all, if complex zeroes can never create an x-intercept, how do they affect the function? And if a zero of even multiplicity never crosses the x-axis (always bounces), can it still affect a nonstrict (*less than or equal to* or *greater than or equal to*) inequality? These are interesting and important questions, with numerous avenues of exploration. To begin, consider the function $Y_1 = (x + 3)^2(x^3 - 1)$. In completely factored form $Y_1 = (x + 3)^2(x - 1)(x^2 + x + 1)$. This is a polynomial function of degree 5 with two real zeroes (one repeated), two complex zeroes (the quadratic factor is irreducible), and after viewing the graph on Figure 3.76, four turning points. From the graph (or by analysis), we have $Y_1 \leq 0$ for $x \leq 1$. Now let's consider $Y_2 = (x + 3)^2(x - 1)$, the same function as Y_1, less the quadratic factor. Since complex zeroes never "cross the x-axis" anyway, the removal of this factor *cannot affect the solution set of the inequality!* But how does it affect the function? Y_2 is now a function of degree three, with three real zeroes (one repeated) and only two turning points (Figure 3.77). But even so, the solution to $Y_2 \leq 0$ is the same as for $Y_1 \leq 0$: $x \leq 1$. Finally, let's look at $Y_3 = x - 1$, the same function as Y_2 but with the repeated zero removed. The key here is to notice that since $(x - 3)^2$

will be nonnegative for any value of x, it too does not change the solution set of the "less than or equal to inequality," only the shape of the graph. Y_3 is a function of degree 1, with one real zero and no turning points, *but the solution interval for $Y_3 \leq 0$ is the same solution interval as Y_2 and Y_1: $x \leq 1$* (see Figure 3.78).

Explore these relationships further using the following exercises and a "greater than or equal to" inequality. Begin by writing Y_1 in completely factored form.

Exercise 1: $Y_1 = (x^3 - 6x^2 + 32)(x^2 + 1)$
$\qquad\quad\ Y_2 = x^3 - 6x^2 + 32$
$\qquad\quad\ Y_3 = x + 2$

Exercise 2: $Y_1 = (x + 3)^2(x^3 - 2x^2 + x - 2)$
$\qquad\quad\ Y_2 = (x + 3)^2(x - 2)$
$\qquad\quad\ Y_3 = x - 2$

Exercise 3: Based on what you've noticed, comment on how the irreducible quadratic factors of a polynomial affect its graph. What role do they play in the solution of inequalities?

Exercise 4: How do zeroes of even multiplicity affect the solution set of nonstrict inequalities (less/greater than or equal to)?

For more on these ideas, see the *Strengthening Core Skills* feature from this chapter.

Figure 3.76

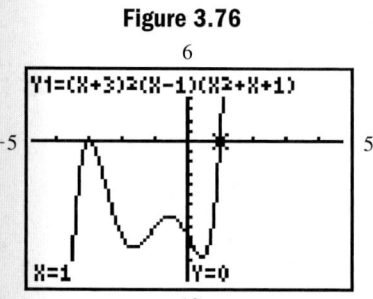

Figure 3.77

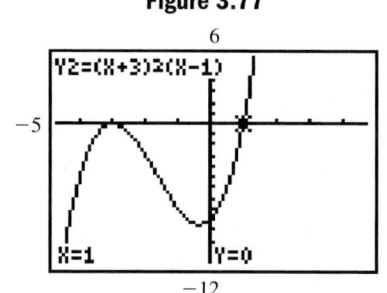

Figure 3.78

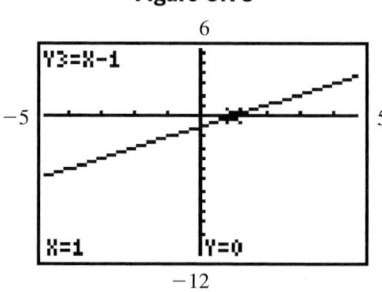

STRENGTHENING CORE SKILLS

Solving Inequalities Using the Push Principle

The most common method for solving polynomial inequalities involves finding the zeroes of the function and checking the sign of the function in the intervals between these zeroes. In Section 3.7, we relied on the end behavior of the graph, the sign of the function at the y-intercept, and the multiplicity of the zeroes to determine the solution. There is a third method that is more conceptual in nature,

but in many cases highly efficient. It is based on two very simple ideas, the first involving only order relations and the number line:

A. Given any number x and constant $k > 0$: $x > x - k$ and $x < x + k$.

$$x - 4 < x \qquad x < x + 3$$

This statement simply reinforces the idea that if a is left of b on the number line, then $a < b$. As shown in the diagram, $x - 4 < x$ and $x < x + 3$, from which $x - 4 < x + 3$ for any x.

B. The second idea reiterates well-known ideas regarding the multiplication of signed numbers. For any number of factors:

if there are an even number of negative factors, the result is positive;

if there are an odd number of negative factors, the result is negative.

These two ideas work together to solve inequalities using what we'll call the *push principle*. Consider the inequality $x^2 - x - 12 > 0$. The factored form is $(x - 4)(x + 3) > 0$ and we want the product of these two factors to be positive. From (A), both factors will be positive if $(x - 4)$ is positive, since it's the smaller of the two; and both factors will be negative if $x + 3 < 0$, since it's the larger. The solution set is found by solving these two simple inequalities: $x - 4 > 0$ gives $x > 4$ and $x + 3 < 0$ gives $x < -3$. If the inequality were $(x - 4)(x + 3) < 0$ instead, we require one negative factor and one positive factor. Due to order relations and the number line, the larger factor must be the positive one: $x + 3 > 0$ so $x > -3$. The smaller factor must be the negative one: $x - 4 < 0$ and $x < 4$. This gives the solution $-3 < x < 4$ as can be verified using any alternative method. Solutions to all other polynomial and rational inequalities are an extension of these two cases.

Illustration 1 ▶ Solve $x^3 - 7x + 6 < 0$ using the push principle.

Solution ▶ The polynomial can be factored using the tests for 1 and -1 and synthetic division. The factors are $(x - 2)(x - 1)(x + 3) < 0$, which we've conveniently written in increasing order. For the product of three factors to be negative we require: (1) three negative factors or (2) one negative and two positive factors. The first condition is met

by simply making the largest factor negative, as it will ensure the smaller factors are also negative: $x + 3 < 0$ so $x < -3$. The second condition is met by making the smaller factor negative and the "middle" factor positive: $x - 2 < 0$ *and* $x - 1 > 0$. The second solution interval is $x < 2$ and $x > 1$, or $1 < x < 2$.

Note the push principle does not require the testing of intervals between the zeroes, nor the "cross/bounce" analysis at the zeroes and vertical asymptotes (of rational functions). In addition, irreducible quadratic factors can still be ignored as they contribute nothing to the solution of real inequalities, and factors of even multiplicity can be overlooked precisely because there is no sign change at these roots.

Illustration 2 ▶ Solve $(x^2 + 1)(x - 2)^2(x + 3) \geq 0$ using the push principle.

Solution ▶ Since the factor $(x^2 + 1)$ does not affect the solution set, this inequality will have the same solution as $(x - 2)^2(x + 3) \geq 0$. Further, since $(x - 2)^2$ will be nonnegative for all x, the original inequality *has the same solution set as* $(x + 3) \geq 0$! The solution is $x \geq -3$.

With some practice, the push principle can be a very effective tool. Use it to solve the following exercises. Check all solutions by graphing the function on a graphing calculator.

Exercise 1: $x^3 - 3x - 18 \leq 0$

Exercise 2: $\dfrac{x + 1}{x^2 - 4} > 0$

Exercise 3: $x^3 - 13x + 12 < 0$

Exercise 4: $x^3 - 3x + 2 \geq 0$

Exercise 5: $x^4 - x^2 - 12 > 0$

Exercise 6: $(x^2 + 5)(x^2 - 9)(x + 2)^2(x - 1) \geq 0$

CUMULATIVE REVIEW CHAPTERS 1–3

1. Solve for R: $\dfrac{1}{R} = \dfrac{1}{R_1} + \dfrac{1}{R_2}$

2. Solve for x: $\dfrac{2}{x + 1} + 1 = \dfrac{5}{x^2 - 1}$

3. Factor the expressions:
   a. $x^3 - 1$ **b.** $x^3 - 3x^2 - 4x + 12$

4. Solve using the quadratic formula. Write answers in both exact and approximate form:
$2x^2 + 4x + 1 = 0$.

5. Solve the following inequality: $x + 3 < 5$ *or* $5 - x < 4$.

6. Name the eight toolbox functions, give their equations, then draw a sketch of each.

7. Use substitution to verify that $x = 2 - 3i$ is a solution to $x^2 - 4x + 13 = 0$.

8. Solve the rational inequality:
$\dfrac{x + 4}{x - 2} < 3$.

9. As part of a study on traffic conditions, the mayor of a small city tracks her driving time to work each day for six months and finds a linear and increasing relationship. On day 1, her drive time was 17 min. By day 61 the drive time had increased to 28 min. Find a linear function that models the drive time and use it to estimate the drive time on day 121, if the trend continues. Explain what the slope of the line means in this context.

10. Does the relation shown represent a function? If not, discuss/explain why not.

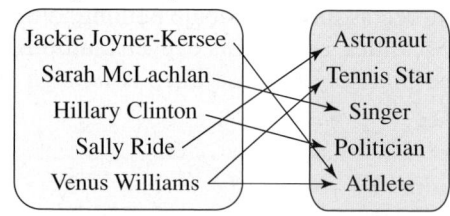

11. The data given shows the profit of a new company for the first 6 months of business, and is closely modeled by the function $p(m) = 1.18x^2 - 10.99x + 4.6$; where $p(m)$ is the profit earned in month m. Assuming this trend continues, use this function to find the first month a profit will be earned ($p > 0$).

Exercise 11

Month	Profit (1000s)
1	−5
2	−13
3	−18
4	−20
5	−21
6	−19

12. Graph the function

$$g(x) = \frac{-1}{(x+2)^2} + 3 \text{ using}$$

transformations of a basic function.

13. Find $f^{-1}(x)$, given $f(x) = \sqrt[3]{2x - 3}$, then use composition to verify your inverse is correct.

14. Graph $f(x) = x^2 - 4x + 7$ by completing the square, then state intervals where:
  a. $f(x) \geq 0$ **b.** $f(x)\uparrow$

15. Given the graph of a general function $f(x)$, graph $F(x) = -f(x + 1) + 2$.

Exercise 15

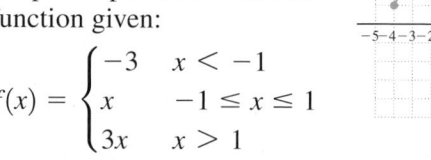

16. Graph the piecewise-defined function given:

$$f(x) = \begin{cases} -3 & x < -1 \\ x & -1 \leq x \leq 1 \\ 3x & x > 1 \end{cases}$$

17. Y varies directly with X and inversely with the square of Z. If $Y = 10$ when $X = 32$ and $Z = 4$, find X when $Z = 15$ and $Y = 1.4$.

18. Use the rational zeroes theorem and synthetic division to find all zeroes (real and complex) of $f(x) = x^4 - 2x^2 + 16x - 15$.

19. Sketch the graph of $f(x) = x^3 - 3x^2 - 6x + 8$.

20. Sketch the graph of $h(x) = \dfrac{x-1}{x^2 - 4}$ and use the zeroes and vertical asymptotes to solve $h(x) \geq 0$.

CONNECTIONS TO CALCULUS

In a calculus course, many of the graphing techniques demonstrated in Chapters 2 and 3 are applied to a wider variety of functions, but with little change in the basic approach. The graph of a function will always depend on its domain, end behavior, y-intercept, x-intercept(s), midinterval points, maximum and minimum values, intervals where the function is positive or negative, and so on. In addition, these elements remain in play even when the function is not polynomial or rational, as demonstrated in Example 1. Further, locating the maximum and minimum values of a function is a vital component of calculus, as in its application we often seek the maximum illumination, lowest cost, greatest efficiency, and the like. These ideas behind locating extreme values are addressed in Examples 2 and 3.

Graphing Techniques

EXAMPLE 1 ▶ **Graphing a Radical Function**

Graph the function $f(x) = x\sqrt{2 + x}$.

Solution ▶ From the radical factor we note the domain is $x \in [-2, \infty)$. The y-intercept is $(0, 0)$, with x-intercepts at $(0, 0)$ and $(-2, 0)$. From the given expression, we see that as $x \to \infty$, $y \to \infty$, and the end behavior will be "up on the right." Evaluating at the "midinterval point" $x = -1$ gives $f(-1) = -1$, showing $f(x) < 0$ in the interval $(-2, 0)$, and $f(x) > 0$ for $x > 0$. Using this information and connecting the given points with a smooth curve produces the graph shown.

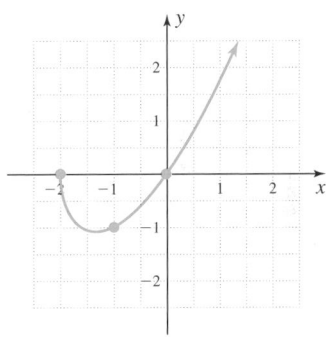

Now try Exercises 1 through 4 ▶

From the graph of $f(x) = x\sqrt{2 + x}$ in Example 1, the point $(-1, -1)$ appears to be near the minimum value of the function. However, as we noted in Chapter 3, maximum and minimum values of a function rarely occur at these "midinterval points," and finding them often requires the tools of calculus and good algebra skills.

EXAMPLE 2 ▶ **Locating Maximum or Minimum Values**

Using calculus, it can be shown that the minimum value of $f(x) = x\sqrt{2 + x}$ actually occurs at the zero of $f(x) = x\left(\dfrac{1}{2\sqrt{2 + x}}\right) + \sqrt{2 + x}$. Find the zero and locate this minimum value.

Solution ▶ Substituting 0 for $f(x)$ produces the following sequence:

$$0 = \frac{x}{2\sqrt{2 + x}} + \sqrt{2 + x}$$

$$-\sqrt{2 + x} = \frac{x}{2\sqrt{2 + x}} \qquad \text{subtract } \sqrt{2 + x}$$

$$2 + x = \frac{x^2}{4(2 + x)} \qquad \text{square both sides}$$

$$4(2 + x)^2 = x^2 \qquad \text{multiply by } 4(2+x)$$

$$16 + 16x + 4x^2 = x^2 \qquad \text{expand binomial square and distribute 4}$$
$$3x^2 + 16x + 16 = 0 \qquad \text{set equal to zero}$$
$$(3x + 4)(x + 4) = 0 \qquad \text{factor}$$
$$x = -\frac{4}{3} \quad x = -4 \qquad \text{result}$$

Since -4 is outside the domain of f, the minimum value must occur at $x = -\frac{4}{3}$.

Note the zero(s) of f will tell us *where* the minimum value occurs, but to find this minimum we must use this zero *in the original function*. A minimum value of $f\left(-\frac{4}{3}\right) \approx -1.1$ will occur at $\left(-\frac{4}{3}, f\left(-\frac{4}{3}\right)\right) \approx \left(-\frac{4}{3}, -1.1\right)$.

Now try Exercises 7 through 10 ▶

In Exercise 86 of Section 3.4, the function model for a population of insects was given as $p(m) = -m^4 + 26m^3 - 217m^2 + 588m$, where $p(m)$ represents the insect population (in hundreds of thousands) for month m. As mentioned in the chapter opener, health officials may be very interested in the month(s) that this population reaches a peak, as they prepare vaccines or attempt to raise public awareness. A graph of the function provides a visual representation of population growth, and helps to identify what months need to be targeted.

EXAMPLE 3 ▶   Locating Maximum and Minimum Values

The graph of

$$p(m) = -m^4 + 26m^3 - 217m^3 + 588m$$

(modeling the population of insects) is shown. Using the tools of calculus, it can be shown that the zeroes of

$$p(m) = -4m^3 + 78m^2 - 434m + 588$$

give the *location* of the maximum and minimum values of p:

a. Find these zeroes.

b. Then substitute these zeroes into $p(m)$ to find the actual maximum and minimum values.

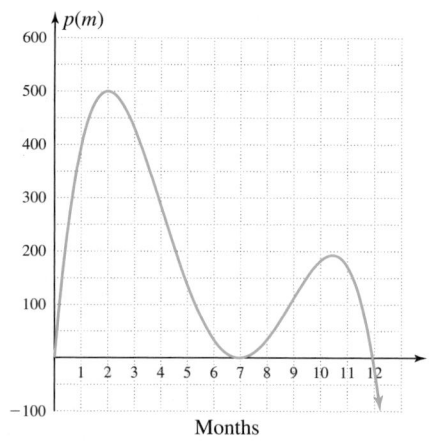

Months

Solution ▶   a. To begin, we set the equation equal to zero and factor out -2:
$$0 = -2(2m^3 - 39m^2 + 217m - 294)$$
The tests for 1 and -1 show neither is a zero. Using $x = 2$ and the remainder theorem to help factor p gives

$$\begin{array}{r|rrrr} 2 & 2 & -39 & 217 & -294 \\ & & 4 & -70 & 294 \\ \hline & 2 & -35 & 147 & \underline{0} \end{array}$$

This shows $m = 2$ is a zero and $m - 2$ is a factor, resulting in the equation $0 = -2(m - 2)(2m^2 - 35m + 147)$. Using the quadratic formula (or trial-and-error factoring) shows the remaining zeroes of p are $m = 7$ and $m = \frac{21}{2}$.

b. By inspecting the graph of p it indeed appears the maximum/minimum values (peaks and valleys) occur at $m = 2, 7$, and 10.5, and it only remains to find the actual values by substituting these into $p(m)$.

$$p(2) = -(2)^4 + 26(2)^3 - 217(2)^2 + 588(2)$$
$$= -16 + 208 - 868 + 1176$$
$$= 500$$
$$p(7) = -(7)^4 + 26(7)^3 - 217(7)^2 + 588(7)$$
$$= -2401 + 8918 - 10{,}633 + 4116$$
$$= 0$$
$$p(10.5) = -(10.5)^4 + 26(10.5)^3 - 217(10.5)^2 + 588(10.5)$$
$$= 192.9375$$

The maximum infestation of 50,000,000 insects occurs at $m = 2$ (February), and a minimum population of 0 insects occurs at $m = 7$ (July). The population of insects then peaks at a lower level of about 19,300,000 insects at $m = 10.5$ (mid-October).

> **Now try Exercises 5 and 6 ▶**

In Examples 2 and 3 it's important to note that a maximum or minimum value is an *output value* (from the range), and not an ordered pair. Considering the "cause and effect" nature of a function, the cause of a higher or lower insect population is the time of year, the effect is the population of insects at that time.

Connections to Calculus Exercises

State the domain, end behavior, and y-intercept of each function. Then find the x-intercepts and use any midinterval points or other graphing "tools" needed to complete the graph.

1. $f(x) = x\sqrt{3 - x}$ **2.** $g(x) = 2x^2\sqrt{x} - 6x\sqrt{x}$

3. $h(x) = x\sqrt{9 - x^2}$ **4.** $v(x) = \sqrt[3]{2x}\sqrt{x + 5}$

5. The anxiety level of an Olympic skater is modeled by the function $p(x) = x^3 + 2x^2 - 5x - 6$, where $p(x)$ is his anxiety level at time x in minutes, and $x = 0$ is the start time of his skating routine $[p < 0 \rightarrow$ calm and relaxed, $p > 0 \rightarrow$ anxious, $p = 0 \rightarrow$ normal]. State the domain, end behavior, and y-intercept of p, then find the x-intercepts and use any midinterval points or other graphing "tools" needed to complete the graph.

6. From a height of 1200 ft, an ASARI plane (advanced sonar and radar imaging) flies over a canyon to ascertain the depth of a channel cut by a river through the canyon. The main point of interest is the location where a tall hill remains in the middle of the channel. Using the data gathered, the operators determine the terrain can be modeled by the function $h(x) = x^4 - 10x^2 + 9$, where $h(x)$ represents the height (in hundreds of feet) at a distance of x hundred feet from the top of the hill. State the domain, end behavior, and y-intercept of h, then find the x-intercepts and use any midinterval points or other graphing "tools" needed to complete the graph.

Using calculus, it can be shown that the maximum or minimum value of the functions from 1 through 6 (designated by lower case letters), actually occur at the zeroes of the corresponding functions in 7 through 12 (designated by identical letter names in a different font). Find and clearly state the maximum or minimum value(s) for each function, and where they will occur. Round to two decimal places as needed.

7. $f(x) = \sqrt{3 - x} - \dfrac{x}{2\sqrt{3 - x}}$

8. $g(x) = 5x^{\frac{3}{2}} - 9x^{\frac{1}{2}}$

9. $h(x) = \dfrac{-x^2}{\sqrt{9 - x^2}} + \sqrt{9 - x^2}$

10. $v(x) = \dfrac{(2x)^{\frac{1}{3}}}{2(x + 5)^{\frac{1}{2}}} + \dfrac{2(x + 5)^{\frac{1}{2}}}{3(2x)^{\frac{2}{3}}}$

11. Using the tools of calculus, it can be shown that the maximum and minimum values of p (Exercise 5) occur at the zeroes of $p(x) = 3x^2 + 4x - 5$. Determine the maximum level of anxiety felt by the skater and when it occurs. What is the minimum level of anxiety? When does it occur? [Note: In realistic terms we might consider the domain of p to be $-3 \leq x \leq 3$.]

12. Using the tools of calculus, it can be shown that the maximum and minimum values of h (Exercise 6) occur at the zeroes of $h(x) = 4x^3 - 20x$. Determine the maximum height, and where this maximum occurs, then find the depth of the first channel the plane flies over. How far from the top of the hill does this occur? What do you notice about the depth of the second channel? (Note: In realistic terms we might consider the domain of h to be about $-3.1 \leq x \leq 3.1$.)

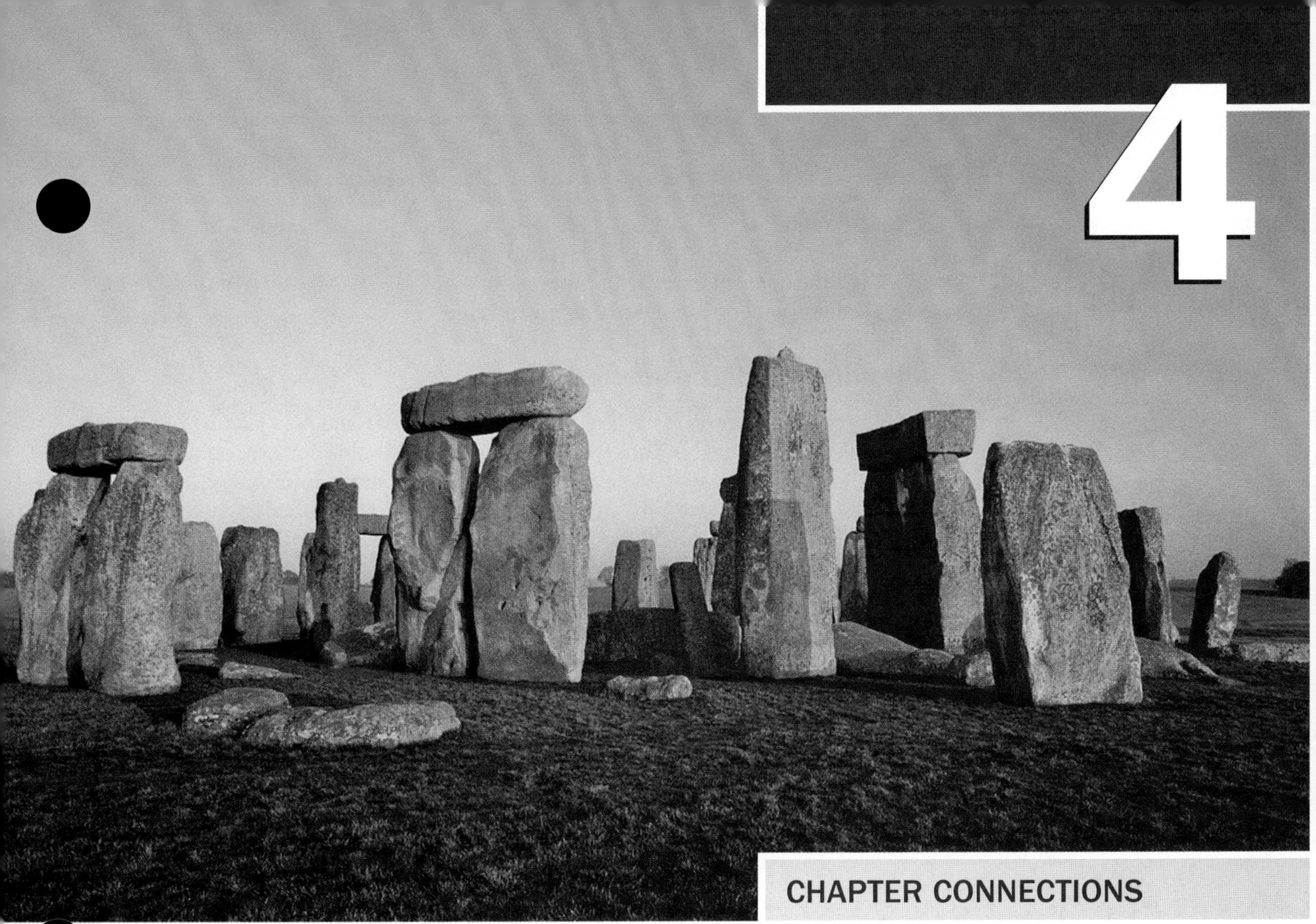

4

Exponential and Logarithmic Functions

CHAPTER OUTLINE

CHAPTER CONNECTIONS

The power and importance of exponential and logarithmic functions would be hard to overstate. From ecology and economics to environmental studies and atomic research, we simply couldn't do without them. While many important applications may be of interest only to scientists or investment firms, their use actually extends in many directions, and often serves to broaden our understanding of history and civilization. For instance, using bits of organic material from the immediate area of the famous Stonehenge site (in southern England), scientists were able to estimate the date that the area was last inhabited, even though it was thousands of years ago. This application appears as Exercise 62 in Section 4.5

While exponential and logarithmic functions play a vital role in modeling real-world phenomena, they are also invaluable tools in a study of calculus. The properties of logarithms help simplify some otherwise difficult computations, while many new and important functions are defined in terms of base-*e* exponential functions. The *Connections to Calculus* feature for Chapter 4 offers additional work with exponential and logarithmic properties in preparation for their use in a calculus course.

Learning Objectives

In Section 4.1 you will learn how to:

☐ **A.** Identify one-to-one functions

☐ **B.** Explore inverse functions using ordered pairs

☐ **C.** Find inverse functions using an algebraic method

☐ **D.** Graph a function and its inverse

☐ **E.** Solve applications of inverse functions

Consider the function $f(x) = 2x - 3$. If $f(x) = 7$, the equation becomes $2x - 3 = 7$, and the corresponding value of x can be found using *inverse operations*. In this section, we introduce the concept of an *inverse function,* which can be viewed as a formula for finding x-values that correspond to *any* given value of $f(x)$.

A. Identifying One-to-One Functions

The graphs of $y = 2x$ and $y = x^2$ are shown in Figures 4.1 and 4.2. The dashed, vertical lines clearly indicate both are functions, with each x-value corresponding to only one y. But the points on $y = 2x$ have one characteristic those from $y = x^2$ do not— *each y-value also corresponds to only one x* (for $y = x^2$, 4 corresponds to both -2 and 2). If each element from the range of a function corresponds to only one element of the domain, the function is said to be **one-to-one.**

Figure 4.1

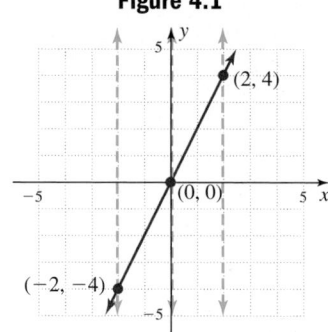

Figure 4.2

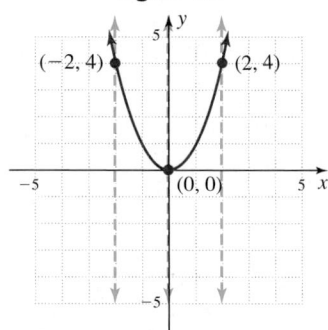

One-to-One Functions

A function f is one-to-one if every element in the range corresponds to only one element of the domain.

In symbols, if $f(x_1) = f(x_2)$ then $x_1 = x_2$, or
if $x_1 \neq x_2$, then $f(x_1) \neq f(x_2)$.

From this definition we note the graph of a one-to-one function must not only pass a vertical line test (to show each x corresponds to only one y), but also pass a **horizontal line test** (to show each y corresponds to only one x).

Horizontal Line Test

If every horizontal line intersects the graph of a function in at most one point, the function is one-to-one.

Notice the graph of $y = 2x$ (Figure 4.3) passes the horizontal line test, while the graph of $y = x^2$ (Figure 4.4) does not.

Figure 4.3

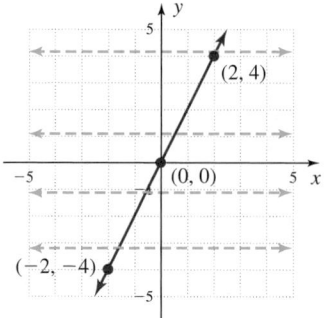

Figure 4.4

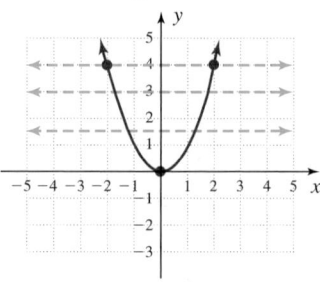

EXAMPLE 1 ▶ Identifying One-to-One Functions

Use the horizontal line test to determine whether each graph is the graph of a one-to-one function.

a.

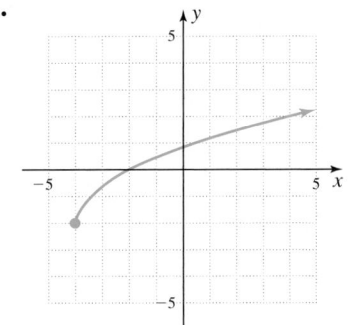

b.

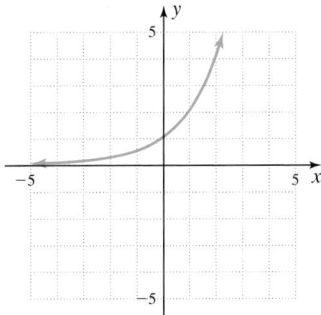

c.

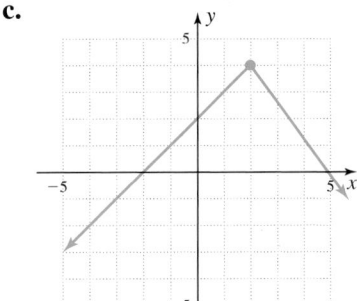

d.

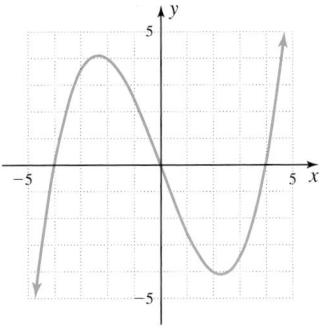

Solution ▶ A careful inspection shows all four graphs depict a function, since each passes the vertical line test. Only (a) and (b) pass the horizontal line test and are *one-to-one* functions.

Now try Exercises 7 through 28 ▶

☑ **A.** You've just learned how to identify a one-to-one function

If the function is given in ordered pair form, we simply check to see that no given second coordinate is paired with more than one first coordinate.

B. Inverse Functions and Ordered Pairs

Table 4.1

x	$f(x)$
−3	−9
0	−3
2	1
5	7
8	13

Consider the function $f(x) = 2x - 3$ and the solutions shown in Table 4.1. Figure 4.5 shows this function in diagram form (in blue), and illustrates that for each element of the domain, we *multiply by 2, then subtract 3*. An **inverse function** for f is one that takes the result of these operations (elements of the range), and returns the original domain element. Figure 4.6 shows that function F achieves this by "undoing" the operations in reverse order: *add 3, then divide by 2* (in red). A table of values for $F(x)$ is shown (Table 4.2).

Table 4.2

x	$F(x)$
−9	−3
−3	0
1	2
7	5
13	8

Figure 4.5

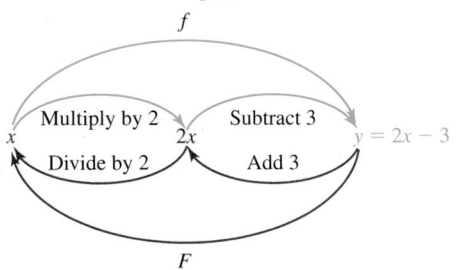

Figure 4.6

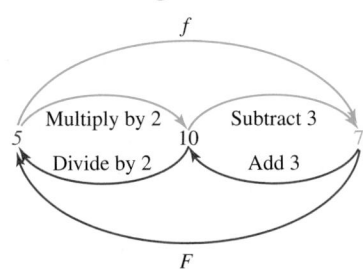

From this illustration we make the following observations regarding an inverse function, which we actually denote as $f^{-1}(x)$.

Inverse Functions

If f is a one-to-one function with ordered pairs (a, b),
1. $f^{-1}(x)$ is a one-to-one function with ordered pairs (b, a).
2. The range of f will be the domain of $f^{-1}(x)$.
3. The domain of f will be the range of $f^{-1}(x)$.

⚠ **CAUTION** ▶ The notation $f^{-1}(x)$ is simply a way of denoting an inverse function and has nothing to do with exponential properties. In particular, $f^{-1}(x)$ does *not* mean $\dfrac{1}{f(x)}$.

EXAMPLE 2 ▶ **Finding the Inverse of a Function**

Find the inverse of each one-to-one function given:
a. $f(x) = \{(-4, 13), (-1, 7), (0, 5), (2, 1), (5, -5), (8, -11)\}$
b. $p(x) = -3x + 2$

Solution ▶ a. When a function is defined as a set of ordered pairs, the inverse function is found by simply interchanging the x- and y-coordinates:
$f^{-1}(x) = \{(13, -4), (7, -1), (5, 0), (1, 2), (-5, 5), (-11, 8)\}$.

b. Using diagrams similar to Figures 4.5 and 4.6, we reason that $p^{-1}(x)$ will subtract 2, then divide the result by -3: $p^{-1}(x) = \dfrac{x - 2}{-3}$. As a test, we find that $(-2, 8)$, $(0, 2)$, and $(3, -7)$ are solutions to $p(x)$, and note that $(8, -2)$, $(2, 0)$, and $(-7, 3)$ are indeed solutions to $p^{-1}(x)$.

☑ **B. You've just learned how to explore inverse functions using ordered pairs**

Now try Exercises 29 through 40 ▶

C. Finding Inverse Functions Using an Algebraic Method

WORTHY OF NOTE

If a function is *not* one-to-one, no inverse function exists since interchanging the x- and y-coordinates will result in a nonfunction. For instance, interchanging the coordinates of $(-2, 4)$ and $(2, 4)$ from $y = x^2$ results in $(4, -2)$ and $(4, 2)$, and we have one x-value being mapped to two y-values, in violation of the function definition.

The fact that interchanging x- and y-values helps determine an inverse function can be generalized to develop an **algebraic method** for finding inverses. Instead of interchanging *specific x- and y-values*, we actually interchange the x- and y-*variables,* then solve the equation for y. The process is summarized here.

Finding an Inverse Function

1. Use y instead of $f(x)$.
2. Interchange x and y.
3. Solve the equation for y.
4. The result gives the inverse function: substitute $f^{-1}(x)$ for y.

In this process, it might seem like we're using the *same y* to represent two different functions. To see why there is actually no contradiction, **see Exercise 103.**

EXAMPLE 3 ▶ **Finding Inverse Functions Algebraically**

Use the algebraic method to find the inverse function for

a. $f(x) = \sqrt[3]{x + 5}$ b. $g(x) = \dfrac{2x}{x + 1}$

Solution ▶   **a.** $f(x) = \sqrt[3]{x + 5}$ given function

$y = \sqrt[3]{x + 5}$ use y instead of $f(x)$

$x = \sqrt[3]{y + 5}$ interchange x and y

$x^3 = y + 5$ cube both sides

$x^3 - 5 = y$ solve for y

$x^3 - 5 = f^{-1}(x)$ the result is $f^{-1}(x)$

For $f(x) = \sqrt[3]{x + 5}$, $f^{-1}(x) = x^3 - 5$.

b. $g(x) = \dfrac{2x}{x + 1}$ given function

$y = \dfrac{2x}{x + 1}$ use y instead of $f(x)$

$x = \dfrac{2y}{y + 1}$ interchange x and y

$xy + x = 2y$ multiply by $y + 1$ and distribute

$x = 2y - xy$ gather terms with y

$x = y(2 - x)$ factor

$\dfrac{x}{2 - x} = y$ solve for y

$\dfrac{x}{2 - x} = g^{-1}(x)$ the result is $g^{-1}(x)$

For $g(x) = \dfrac{2x}{x + 1}$, $g^{-1}(x) = \dfrac{x}{2 - x}$.

Now try Exercises 41 through 48 ▶

In cases where a given function is *not* one-to-one, we can sometimes restrict the domain to create a function that *is,* and then determine an inverse. The restriction we use is arbitrary, and only requires that the result still produce all possible range values. For the most part, we simply choose a limited domain that seems convenient or reasonable.

EXAMPLE 4 ▶   **Restricting the Domain to Create a One-to-One Function**

Given $f(x) = (x - 4)^2$, restrict the domain to create a one-to-one function, then find $f^{-1}(x)$. State the domain and range of both resulting functions.

Solution ▶   The graph of f is a parabola, opening upward with the vertex at $(4, 0)$. Restricting the domain to $x \geq 4$ (see figure) leaves only the "right branch" of the parabola, creating a one-to-one function without affecting the range, $y \in [0, \infty)$. For $f(x) = (x - 4)^2$, $x \geq 4$, we have

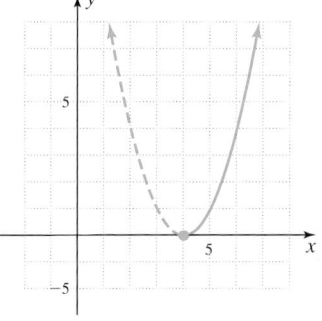

$f(x) = (x - 4)^2$ given function

$y = (x - 4)^2$ use y instead of $f(x)$

$x = (y - 4)^2$ interchange x and y

$\pm\sqrt{x} = y - 4$ take square roots

$\sqrt{x} + 4 = y$ solve for y, use $\sqrt{x}$ since $x \geq 4$

The result shows $f^{-1}(x) = \sqrt{x} + 4$, with domain $x \in [0, \infty)$ and range $y \in [4, \infty)$ (the domain of f becomes the range of f^{-1}, and the range of f becomes the domain of f^{-1}).

Now try Exercises 49 through 54 ▶

While we now have the ability to *find* the inverse of a function, we still lack a definitive method of *verifying* the inverse is correct. Actually, the diagrams in Figures 4.5 and 4.6 suggest just such a method. If we use the function f itself as an input for f^{-1}, or the function f^{-1} as an input for f, the end result should simply be x, as each function "undoes" the operations of the other. From Section 2.8 this is called a composition of functions and using the notation for composition we have,

Verifying Inverse Functions

If f is a one-to-one function, then the function f^{-1} exists, where
$$(f \circ f^{-1})(x) = x \qquad \text{and} \qquad (f^{-1} \circ f)(x) = x$$

EXAMPLE 5 ▶ **Finding and Verifying an Inverse Function**

Use the algebraic method to find the inverse function for $f(x) = \sqrt{x + 2}$. Then verify the inverse you found is correct.

Solution ▶ Since the graph of f is the graph of $y = \sqrt{x}$ shifted 2 units left, we know f is one-to-one with domain $x \in [-2, \infty)$ and range $y \in [0, \infty)$. This is important since the *domain and range values will be interchanged for the inverse function.* The domain of f^{-1} will be $x \in [0, \infty)$ and its range $y \in [-2, \infty)$.

$$f(x) = \sqrt{x + 2} \qquad \text{given function; } x \geq -2$$
$$y = \sqrt{x + 2} \qquad \text{use } y \text{ instead of } f(x)$$
$$x = \sqrt{y + 2} \qquad \text{interchange } x \text{ and } y$$
$$x^2 = y + 2 \qquad \text{solve for } y \text{ (square both sides)}$$
$$x^2 - 2 = y \qquad \text{subtract 2}$$
$$f^{-1}(x) = x^2 - 2 \qquad \text{the result is } f^{-1}(x); \text{ } D{:}\ x \in [0, \infty), \text{ } R{:}\ y \in [-2, \infty)$$

Verify ▶
$$(f \circ f^{-1})(x) = f[f^{-1}(x)] \qquad f^{-1}(x) \text{ is an input for } f$$
$$= \sqrt{f^{-1}(x) + 2} \qquad f \text{ adds 2 to inputs, then takes the square root}$$
$$= \sqrt{(x^2 - 2) + 2} \qquad \text{substitute } x^2 - 2 \text{ for } f^{-1}(x)$$
$$= \sqrt{x^2} \qquad \text{simplify}$$
$$= x \checkmark \qquad \text{since the domain of } f^{-1}(x) \text{ is } x \in [0, \infty)$$

Verify ▶
$$(f^{-1} \circ f)(x) = f^{-1}[f(x)] \qquad f(x) \text{ is an input for } f^{-1}$$
$$= [f(x)]^2 - 2 \qquad f^{-1} \text{ squares inputs, then subtracts 2}$$
$$= [\sqrt{x + 2}]^2 - 2 \qquad \text{substitute } \sqrt{x + 2} \text{ for } f(x)$$
$$= x + 2 - 2 \qquad \text{simplify}$$
$$= x \checkmark \qquad \text{result}$$

☑ **C.** You've just learned how to find inverse functions using an algebraic method

Now try Exercises 55 through 80 ▶

D. The Graph of a Function and Its Inverse

Graphing a function and its inverse on the same axes reveals an interesting and useful relationship—the graphs are reflections across the line $y = x$ (the identity function).

Consider the function $f(x) = 2x + 3$, and its inverse $f^{-1}(x) = \dfrac{x - 3}{2} = \dfrac{1}{2}x - \dfrac{3}{2}$. In Figure 4.7, the points $(1, 5)$, $(0, 3)$, $(-\frac{3}{2}, 0)$, and $(-4, -5)$ from f (see Table 4.3) are graphed in blue, with the points $(5, 1)$, $(3, 0)$, $(0, -\frac{3}{2})$, and $(-5, -4)$ (see Table 4.4) from f^{-1} graphed in red (note the x- and y-values are reversed). Graphing both lines illustrates this symmetry (Figure 4.8).

Table 4.3

x	$f(x)$
1	5
0	3
$-\dfrac{3}{2}$	0
-4	-5

Figure 4.7

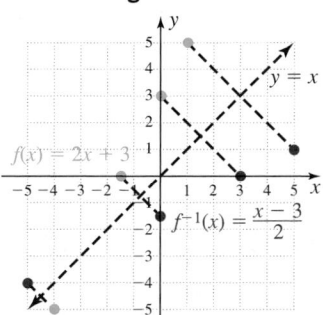

Figure 4.8

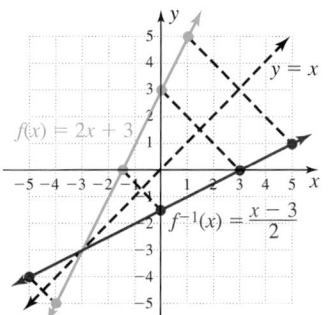

Table 4.4

x	$f^{-1}(x)$
5	1
3	0
0	$-\dfrac{3}{2}$
-5	-4

EXAMPLE 6 ▶ Graphing a Function and Its Inverse

In Example 5, we found the inverse function for $f(x) = \sqrt{x + 2}$ was $f^{-1}(x) = x^2 - 2, x \geq 0$. Graph these functions on the same axes and comment on how the graphs are related.

Solution ▶ The graph of f is a square root function with initial point $(-2, 0)$, a y-intercept of $(0, \sqrt{2})$, and an x-intercept of $(-2, 0)$ (Figure 4.9 in blue). The graph of $x^2 - 2, x \geq 0$ is the right-hand branch of a parabola, with y-intercept at $(0, -2)$ and an x-intercept at $(\sqrt{2}, 0)$ (Figure 4.9 in red).

Figure 4.9

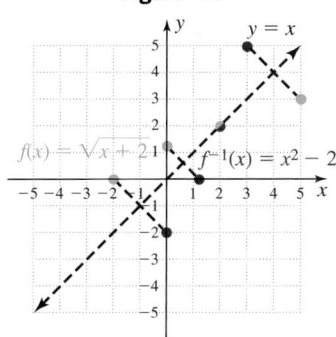

Figure 4.10

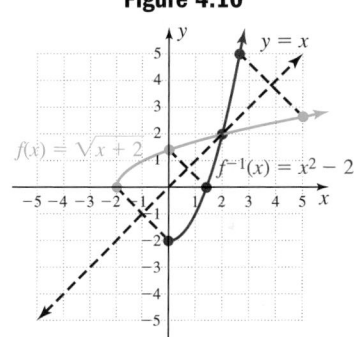

Connecting these points with a smooth curve indeed shows their graphs are symmetric to the line $y = x$ (Figure 4.10).

Now try Exercises 81 through 88 ▶

EXAMPLE 7 ▶ Graphing a Function and Its Inverse

Given the graph shown in Figure 4.11, use the grid in Figure 4.12 to draw a graph of the inverse function.

Figure 4.11

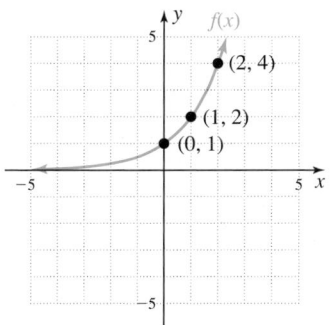

Figure 4.12

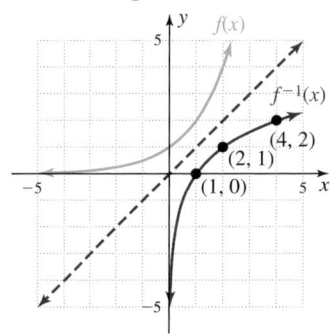

Solution ▶ From the graph, the domain of f appears to be $x \in \mathbb{R}$ and the range is $y \in (0, \infty)$. This means the domain of f^{-1} will be $x \in (0, \infty)$ and the range will be $y \in \mathbb{R}$. To sketch f^{-1}, draw the line $y = x$, interchange the x- and y-coordinates of the selected points, then plot these points and draw a smooth curve using the domain and range boundaries as a guide.

> **Now try Exercises 89 through 94** ▶

A summary of important points is given here followed by their application in Example 8.

☑ **D.** You've just learned how to graph a function and its inverse

> ### Functions and Inverse Functions
>
> 1. If the graph of a function passes the horizontal line test, the function is one-to-one.
> 2. If a function f is one-to-one, the function f^{-1} exists.
> 3. The domain of f is the range of f^{-1}, and the range of f is the domain of f^{-1}.
> 4. For a function f and its inverse f^{-1}, $(f \circ f^{-1})(x) = x$ and $(f^{-1} \circ f)(x) = x$.
> 5. The graphs of f and f^{-1} are symmetric with respect to the line $y = x$.

E. Applications of Inverse Functions

Our final example illustrates one of the many ways that inverse functions can be applied.

EXAMPLE 8 ▶ **Using Volume to Understand Inverse Functions**

The volume of an equipoise cylinder (height equal to diameter) is given by $v(x) = 2\pi x^3$ (since $h = d = 2r$), where $v(x)$ represents the volume in units cubed and x represents the radius of the cylinder.

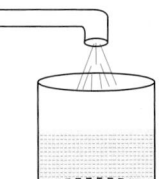

  a. Find the volume of such a cylinder if $x = 10$ ft.
  b. Find $v^{-1}(x)$, and discuss what the input and output variables represent.
  c. If a volume of 1024π ft^3 is required, which formula would be easier to use to find the radius? What is this radius?

Solution ▶ **a.** $\quad v(x) = 2\pi x^3 \qquad$ given function
$\qquad v(10) = 2\pi(10)^3 \qquad$ substitute 10 for x
$\qquad\quad\ = 2000\pi \qquad$ $10^3 = 1000$, exact form

With a radius of 10 ft, the volume of the cylinder would be 2000π ft^3.

b. $\qquad v(x) = 2\pi x^3 \qquad$ given function
$\qquad\quad y = 2\pi x^3 \qquad$ use y instead of $v(x)$
$\qquad\quad x = 2\pi y^3 \qquad$ interchange x and y

$\qquad \dfrac{x}{2\pi} = y^3 \qquad$ solve for y

$\qquad \sqrt[3]{\dfrac{x}{2\pi}} = y \qquad$ result

The inverse function is $v^{-1}(x) = \sqrt[3]{\dfrac{x}{2\pi}}$. In this case, the input x is a given volume, the output $v^{-1}(x)$ is the radius of an equipoise cylinder that will hold this volume.

c. Since the volume is known and we need the radius, using $v^{-1}(x) = \sqrt[3]{\dfrac{x}{2\pi}}$ would be more efficient.

$$v^{-1}(1024\pi) = \sqrt[3]{\dfrac{1024\pi}{2\pi}} \quad \text{substitute } 1024\pi \text{ for } x \text{ in } v^{-1}(x)$$

$$= \sqrt[3]{512} \qquad \dfrac{2\pi}{2\pi} = 1$$

$$= 8 \qquad \text{result}$$

☑ **E.** You've just learned how to solve an application of inverse functions

The radius of the cylinder would be 8 ft.

Now try Exercises 97 through 102 ▶

TECHNOLOGY HIGHLIGHT

Investigating Inverse Functions

Many important ideas from this section can be illustrated using a graphing calculator. To begin, enter the function

$Y_1 = 2\sqrt[3]{x-2}$ and $Y_2 = \dfrac{x^3}{8} + 2$ (which appear to be inverse

functions) on the Y= screen, then press ZOOM
5:ZSquare. The graphs seem to be reflections across the line $y = x$ (Figure 4.13). To verify, use the TABLE feature with inputs $x = -2, -1, 0, 1, 2,$ and 3. As shown in Figure 4.14, the points $(1, -2)$, $(2, 0)$, and $(3, 2)$ are on Y_1, and the points $(-2, 1)$, $(0, 2)$, and $(2, 3)$ are all on Y_2. While this seems convincing (the x- and y-coordinates are interchanged), the technology can actually *compose the two functions* to verify an inverse relationship. Function names Y_1 and Y_2 can be accessed using the VARS and ▶ keys, then pressing ENTER . After entering $Y_3 = Y_1(Y_2)$ and $Y_4 = Y_2(Y_1)$ on the Y= screen, we observe whether one function "undoes" the other using the TABLE feature (Figure 4.15).

Figure 4.13

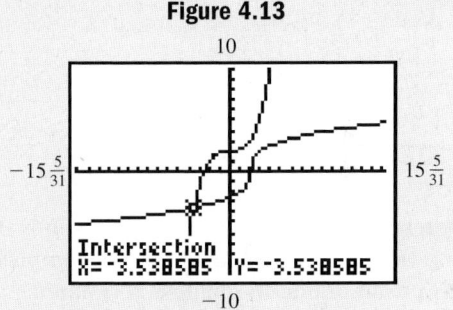

Figure 4.14

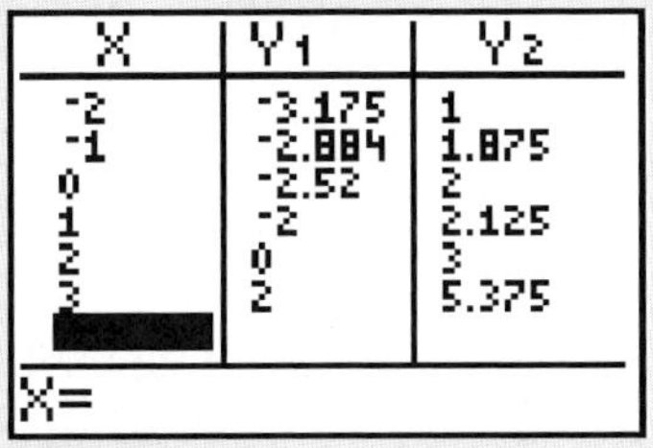

X	Y₁	Y₂
-2	-3.175	1
-1	-2.884	1.875
0	-2.52	2
1	-2	2.125
2	0	3
3	2	5.375

X=

Figure 4.15

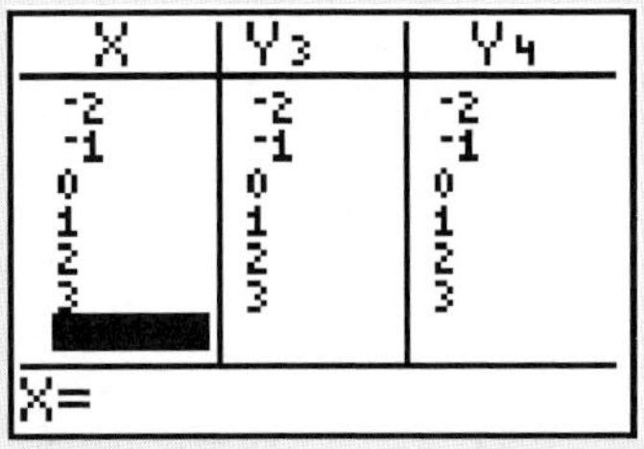

X	Y₃	Y₄
-2	-2	-2
-1	-1	-1
0	0	0
1	1	1
2	2	2
3	3	3

X=

For the functions given, (a) find $f^{-1}(x)$, then use your calculator to verify they are inverses by (b) using ordered pairs, (c) composing the functions, and (d) showing their graphs are symmetric to $y = x$.

Exercise 1: $f(x) = 2x + 1$ **Exercise 2:** $g(x) = x^2 + 1; x \geq 0$

Exercise 3: $h(x) = \dfrac{x}{x+1}$

4.1 EXERCISES

▶ **CONCEPTS AND VOCABULARY**

Fill in each blank with the appropriate word or phrase. Carefully reread the section if needed.

1. A function is one-to-one if each _____ coordinate corresponds to exactly _____ first coordinate.

2. If every _____ line intersects the graph of a function in at most _____ point, the function is one-to-one.

3. A certain function is defined by the ordered pairs $(-2, -11), (0, -5), (2, 1),$ and $(4, 19)$. The inverse function is _____ .

4. To find f^{-1} using the algebraic method, we (1) use _____ instead of $f(x)$, (2) _____ x and y, (3) _____ for y and replace y with $f^{-1}(x)$.

5. State true or false and explain why: *To show that g is the inverse function for f, simply show that* $(f \circ g)(x) = x$. Include an example in your response.

6. Discuss/Explain why no inverse function exists for $f(x) = (x + 3)^2$ and $g(x) = \sqrt{4 - x^2}$. How would the domain of each function have to be restricted to allow for an inverse function?

▶ **DEVELOPING YOUR SKILLS**

Determine whether each graph given is the graph of a one-to-one function. If not, give examples of how the definition of one-to-oneness is violated.

7.
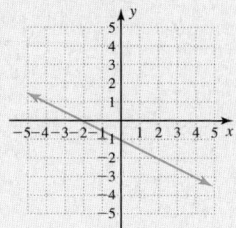

8.
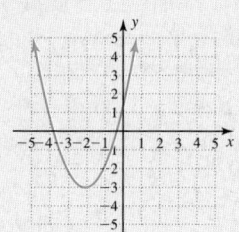

9.
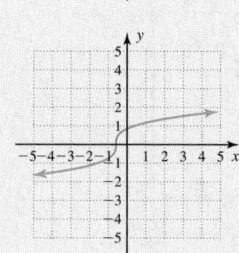

10.
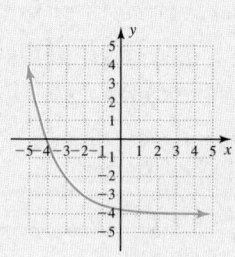

11.
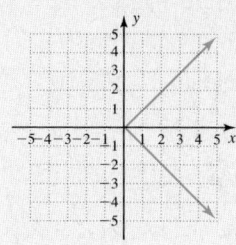

12.
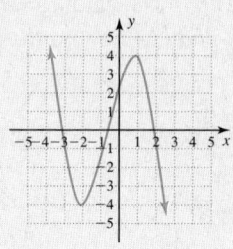

13.
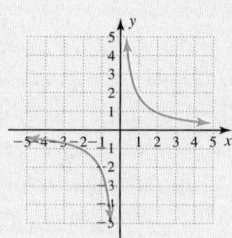

14.
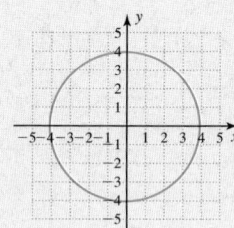

15.
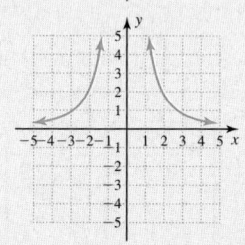

Determine whether the functions given are one-to-one. If not, state why.

16. $\{(-7, 4), (-1, 9), (0, 5), (-2, 1), (5, -5)\}$

17. $\{(9, 1), (-2, 7), (7, 4), (3, 9), (2, 7)\}$

18. $\{(-6, 1), (4, -9), (0, 11), (-2, 7), (-4, 5), (8, 1)\}$

19. $\{(-6, 2), (-3, 7), (8, 0), (12, -1), (2, -3), (1, 3)\}$

Determine if the functions given are one-to-one by noting the function family to which each belongs and mentally picturing the shape of the graph. If a function is not one-to-one, discuss how the definition of one-to-oneness is violated.

20. $f(x) = 3x - 5$ **21.** $g(x) = (x + 2)^3 - 1$

22. $h(x) = -|x - 4| + 3$ **23.** $p(t) = 3t^2 + 5$

24. $s(t) = \sqrt{2t - 1} + 5$ **25.** $r(t) = \sqrt[3]{t + 1} - 2$

26. $y = 3$ **27.** $y = -2x$ **28.** $y = x$

For Exercises 29 to 32, find the inverse function of the one-to-one functions given.

29. $f(x) = \{(-2, 1), (-1, 4), (0, 5), (2, 9), (5, 15)\}$

30. $g(x) = \{(-2, 30), (-1, 11), (0, 4), (1, 3), (2, 2)\}$

31. $v(x)$ is defined by the ordered pairs shown.

X	Y₁
-4	3
-3	2
0	1
5	0
12	-1
21	-2
	-3

X=32

32. $w(x)$ is defined by the ordered pairs shown.

X	Y₁
-6	4
-5	2.5
-2	-2
0	-5
3	-9.5
4	-11
	-15.5

X=7

Find the inverse function using diagrams similar to those illustrated in Example 2. Check the result using three test points.

33. $f(x) = x + 5$ **34.** $g(x) = x - 4$

35. $p(x) = -\dfrac{4}{5}x$ **36.** $r(x) = \dfrac{3}{4}x$

37. $f(x) = 4x + 3$ **38.** $g(x) = 5x - 2$

39. $Y_1 = \sqrt[3]{x - 4}$ **40.** $Y_2 = \sqrt[3]{x + 2}$

Find each function $f(x)$ given, (a) find any three ordered pair solutions (a, b), then (b) algebraically compute $f^{-1}(x)$, and (c) verify the ordered pairs (a, b) satisfy $f^{-1}(x)$.

41. $f(x) = \sqrt[3]{x - 2}$ **42.** $f(x) = \sqrt[3]{x + 3}$

43. $f(x) = x^3 + 1$ **44.** $f(x) = x^3 - 2$

45. $f(x) = \dfrac{8}{x + 2}$ **46.** $f(x) = \dfrac{12}{x - 1}$

47. $f(x) = \dfrac{x}{x + 1}$ **48.** $f(x) = \dfrac{x + 2}{1 - x}$

The functions given in Exercises 49 through 54 are not one-to-one. (a) Determine a domain restriction that preserves all range values, then state this domain and range. (b) Find the inverse function and state its domain and range.

49. $f(x) = (x + 5)^2$ **50.** $g(x) = x^2 + 3$

51. $v(x) = \dfrac{8}{(x - 3)^2}$ **52.** $V(x) = \dfrac{4}{x^2} + 2$

53. $p(x) = (x + 4)^2 - 2$ **54.** $q(x) = \dfrac{4}{(x - 2)^2} + 1$

For each function $f(x)$ given, prove (using a composition) that $g(x) = f^{-1}(x)$.

55. $f(x) = -2x + 5$, $g(x) = \dfrac{x - 5}{-2}$

56. $f(x) = 3x - 4$, $g(x) = \dfrac{x + 4}{3}$

57. $f(x) = \sqrt[3]{x} + 5$, $g(x) = x^3 - 5$

58. $f(x) = \sqrt[3]{x} - 4$, $g(x) = x^3 + 4$

59. $f(x) = \frac{2}{3}x - 6$, $g(x) = \frac{3}{2}x + 9$

60. $f(x) = \frac{4}{5}x + 6$, $g(x) = \frac{5}{4}x - \frac{15}{2}$

61. $f(x) = x^2 - 3$; $x \geq 0$, $g(x) = \sqrt{x + 3}$

62. $f(x) = x^2 + 8$; $x \geq 0$, $g(x) = \sqrt{x - 8}$

Find the inverse of each function $f(x)$ given, then prove (by composition) your inverse function is correct. Note the domain of f is all real numbers.

63. $f(x) = 3x - 5$ **64.** $f(x) = 5x + 4$

65. $f(x) = \dfrac{x - 5}{2}$ **66.** $f(x) = \dfrac{x + 4}{3}$

67. $f(x) = \frac{1}{2}x - 3$ **68.** $f(x) = \frac{2}{3}x + 1$

69. $f(x) = x^3 + 3$ **70.** $f(x) = x^3 - 4$

71. $f(x) = \sqrt[3]{2x + 1}$ **72.** $f(x) = \sqrt[3]{3x - 2}$

73. $f(x) = \dfrac{(x - 1)^3}{8}$ **74.** $f(x) = \dfrac{(x + 3)^3}{-27}$

Find the inverse of each function, then prove (by composition) your inverse function is correct. State the implied domain and range as you begin, and use these to state the domain and range of the inverse function.

75. $f(x) = \sqrt{3x + 2}$ **76.** $g(x) = \sqrt{2x - 5}$

77. $p(x) = 2\sqrt{x - 3}$ **78.** $q(x) = 4\sqrt{x + 1}$

79. $v(x) = x^2 + 3$; $x \geq 0$ **80.** $w(x) = x^2 - 1$; $x \geq 0$

Graph each function $f(x)$ and its inverse $f^{-1}(x)$ on the same grid and "dash-in" the line $y = x$. Note how the graphs are related. Then verify the "inverse function" relationship using a composition.

81. $f(x) = 4x + 1; f^{-1}(x) = \dfrac{x - 1}{4}$

82. $f(x) = 2x - 7; f^{-1}(x) = \dfrac{x + 7}{2}$

83. $f(x) = \sqrt[3]{x + 2}; f^{-1}(x) = x^3 - 2$

84. $f(x) = \sqrt[3]{x - 7}; f^{-1}(x) = x^3 + 7$

85. $f(x) = 0.2x + 1; f^{-1}(x) = 5x - 5$

86. $f(x) = \dfrac{2}{9}x + 4; f^{-1}(x) = \dfrac{9}{2}x - 18$

87. $f(x) = (x + 2)^2; x \geq -2; f^{-1}(x) = \sqrt{x} - 2$

88. $f(x) = (x - 3)^2; x \geq 3; f^{-1}(x) = \sqrt{x} + 3$

Determine the domain and range for each function whose graph is given, and use this information to state the domain and range of the inverse function. Then sketch in the line $y = x$, estimate the location of two or more points on the graph, and use these to graph $f^{-1}(x)$ on the same grid.

89.
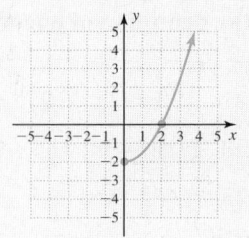

90.
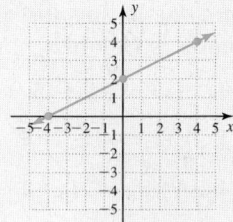

91.
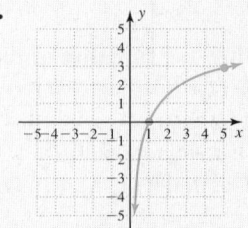

92.
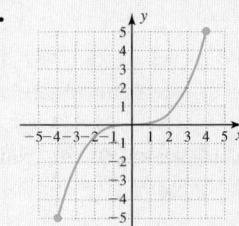

93.
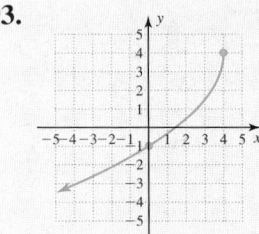

94.
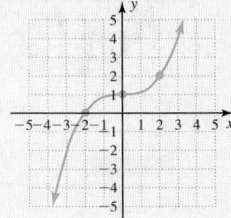

▶ WORKING WITH FORMULAS

95. The height of a projected image: $f(x) = \frac{1}{2}x - 8.5$

The height of an image projected on a screen by a projector is given by the formula shown, where $f(x)$ represents the actual height of the image on the projector (in centimeters) and x is the distance of the projector from the screen (in centimeters). (a) When the projector is 80 cm from the screen, how large is the image? (b) Show that the inverse function is $f^{-1}(x) = 2x + 17$, then input your answer from part (a) and comment on the result. What information does the inverse function give?

96. The radius of a sphere: $r(x) = \sqrt[3]{\dfrac{3x}{4\pi}}$

In generic form, the radius of a sphere is given by the formula shown, where $r(x)$ represents the radius and x represents the volume of the sphere in cubic units. (a) If a weather balloon that is roughly spherical holds 14,130 in^3 of air, what is the radius of the balloon (use $\pi \approx 3.14$)? (b) Show that the inverse function is $r^{-1}(x) = \frac{4}{3}\pi x^3$, then input your answer from part (a) and comment on the result. What information does the inverse function give?

▶ APPLICATIONS

97. Temperature and altitude: The temperature (in degrees Fahrenheit) at a given altitude can be approximated by the function $f(x) = -\frac{7}{2}x + 59$, where $f(x)$ represents the temperature and x represents the altitude in thousands of feet. (a) What is the approximate temperature at an altitude of 35,000 ft (normal cruising altitude for commercial airliners)? (b) Find $f^{-1}(x)$, and state what the independent and dependent variables represent. (c) If the temperature outside a weather balloon is $-18°$F, what is the approximate altitude of the balloon?

98. Fines for speeding: In some localities, there is a set formula to determine the amount of a fine for exceeding posted speed limits. Suppose the amount of the fine for exceeding a 50 mph speed limit was given by the function $f(x) = 12x - 560$ ($x > 50$) where $f(x)$ represents the fine in dollars for a speed of x mph. (a) What is the fine for traveling 65 mph through this speed zone? (b) Find $f^{-1}(x)$, and state what the independent and dependent variables represent. (c) If a fine of $172 were assessed, how fast was the driver going through this speed zone?

99. **Effect of gravity:** Due to the effect of gravity, the distance an object has fallen after being dropped is given by the function $f(x) = 16x^2; x \geq 0$, where $f(x)$ represents the distance in feet after x sec. (a) How far has the object fallen 3 sec after it has been dropped? (b) Find $f^{-1}(x)$, and state what the independent and dependent variables represent. (c) If the object is dropped from a height of 784 ft, how many seconds until it hits the ground (stops falling)?

100. **Area and radius:** In generic form, the area of a circle is given by $f(x) = \pi x^2$, where $f(x)$ represents the area in square units for a circle with radius x. (a) A pet dog is tethered to a stake in the backyard. If the tether is 10 ft long, how much area does the dog have to roam (use $\pi \approx 3.14$)? (b) Find $f^{-1}(x)$, and state what the independent and dependent variables represent. (c) If the owners want to allow the dog 1256 ft^2 of area to live and roam, how long a tether should be used?

101. **Volume of a cone:** In generic form, the volume of an equipoise cone (height equal to radius) is given by $f(x) = \frac{1}{3}\pi x^3$, where $f(x)$ represents the volume

in units^3 and x represents the height of the cone. (a) Find the volume of such a cone if $r = 30$ ft (use $\pi \approx 3.14$). (b) Find $f^{-1}(x)$, and state what the independent and dependent variables represent. (c) If the volume of water in the cone is 763.02 ft^3, how deep is the water at its deepest point?

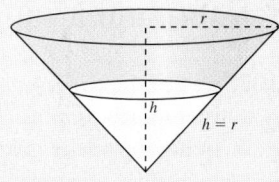

102. **Wind power:** The power delivered by a certain wind-powered generator can be modeled by the function $f(x) = \frac{x^3}{2500}$, where $f(x)$ is the horsepower (hp) delivered by the generator and x represents the speed of the wind in miles per hour. (a) Use the model to determine how much horsepower is generated by a 30 mph wind. (b) The person monitoring the output of the generators (wind generators are usually erected in large numbers) would like a function that gives the wind speed based on the horsepower readings on the gauges in the monitoring station. For this purpose, find $f^{-1}(x)$ and state what the independent and dependent variables represent. (c) If gauges show 25.6 hp is being generated, how fast is the wind blowing?

▶ **EXTENDING THE CONCEPT**

103. For a deeper understanding of the algebraic method for finding an inverse, suppose a function f is defined as $f(x): \{(x, y)|y = 3x - 6\}$. We can then define the inverse as $f^{-1}: \{(x, y)|x = 3y - 6\}$, having interchanged x and y in the equation portion. The equation for f^{-1} is not in standard form, but (x, y) still represents all ordered pairs satisfying either equation. Solving for y gives $f^{-1}: \left\{(x, y)|y = \frac{x}{3} + 2\right\}$, and demonstrates the role of steps 2, 3, and 4 of the method. (a) Find five ordered pairs that satisfy the equation for f, then (b) interchange their coordinates and show they satisfy the equation for f^{-1}.

104. The function $f(x) = \frac{1}{x}$ is one of the few functions that is its own inverse. This means the ordered pairs (a, b) and (b, a) must satisfy both f and f^{-1}. (a) Find f^{-1} using the algebraic method to verify that $f(x) = f^{-1}(x) = \frac{1}{x}$. (b) Graph the function $f(x) = \frac{1}{x}$ using a table of integers from -4 to 4. Note that for any ordered pair (a, b) on f, the

ordered pair (b, a) is also on f. (c) State where the graph of $y = x$ will intersect the graph of this function and discuss why.

105. By inspection, which of the following is the inverse function for $f(x) = \frac{2}{3}\left(x - \frac{1}{2}\right)^5 + \frac{4}{5}$?

   a. $f^{-1}(x) = \sqrt[5]{\frac{1}{2}\left(x - \frac{2}{3}\right)} - \frac{4}{5}$

   b. $f^{-1}(x) = \frac{3}{2}\sqrt[5]{(x - 2)} - \frac{5}{4}$

   c. $f^{-1}(x) = \frac{3}{2}\sqrt[5]{\left(x + \frac{1}{2}\right)} - \frac{5}{4}$

   d. $f^{-1}(x) = \sqrt[5]{\frac{3}{2}\left(x - \frac{4}{5}\right)} + \frac{1}{2}$

106. Suppose a function is defined as $f(x) =$ *the exponent that goes on 9 to obtain* x. For example, $f(81) = 2$ since 2 is the exponent that goes on 9 to obtain 81, and $f(3) = \frac{1}{2}$ since $\frac{1}{2}$ is the exponent that goes on 9 to obtain 3. Determine the value of each of the following:

   a. $f(1)$ **b.** $f(729)$ **c.** $f^{-1}(2)$ **d.** $f^{-1}\left(\frac{1}{2}\right)$

▶ **MAINTAINING YOUR SKILLS**

107. (2.5) Given $f(x) = x^2 - x - 2$, solve the inequality $f(x) \leq 0$ using the x-intercepts and end behavior of the graph.

108. (2.4) For the function $y = 2\sqrt{x} + 3$, find the average rate of change between $x = 1$ and $x = 2$, and between $x = 4$ and $x = 5$. Which is greater? Why?

109. (R.7) Write as many of the following formulas as you can from memory:

   a. perimeter of a rectangle

   b. area of a circle

   c. volume of a cylinder

   d. volume of a cone

   e. circumference of a circle

   f. area of a triangle

   g. area of a trapezoid

   h. volume of a sphere

   i. Pythagorean theorem

110. (1.3) Solve the following cubic equations by factoring:

   a. $x^3 - 5x = 0$

   b. $x^3 - 7x^2 - 4x + 28 = 0$

   c. $x^3 - 3x^2 = 0$

   d. $x^3 - 3x^2 - 4x = 0$

4.2 | Exponential Functions

Learning Objectives

In Section 4.2 you will learn how to:

☐ **A.** Evaluate an exponential function

☐ **B.** Graph general exponential functions

☐ **C.** Graph base-e exponential functions

☐ **D.** Solve exponential equations and applications

Demographics is the statistical study of human populations. In this section, we introduce the family of *exponential functions,* which are widely used to model population growth or decline with additional applications in science, engineering, and many other fields. As with other functions, we begin with a study of the graph and its characteristics.

A. Evaluating Exponential Functions

In the boomtowns of the old west, it was not uncommon for a town to double in size every year (at least for a time) as the lure of gold drew more and more people westward. When this type of growth is modeled using mathematics, exponents play a lead role. Suppose the town of Goldsboro had 1000 residents when gold was first discovered. After 1 yr the population doubled to 2000 residents. The next year it doubled again to 4000, then

again to 8000, then to 16,000 and so on. You probably recognize the digits in blue as powers of two (indicating the population is *doubling*), with each one multiplied by 1000 (the initial population). This suggests we can model the relationship using

$$P(x) = 1000 \cdot 2^x$$

where $P(x)$ is the population after x yr. Further, we can evaluate this function, called an **exponential function,** for *fractional parts of a year* using rational exponents. The population of Goldsboro one-and-a-half years after the gold rush was

$$P\left(\frac{3}{2}\right) = 1000 \cdot 2^{\frac{3}{2}}$$

$$= 1000 \cdot (\sqrt{2})^3$$

$$\approx 2828 \text{ people}$$

WORTHY OF NOTE

To properly understand the exponential function and its graph requires that we evaluate $f(x) = 2^x$ even when x is *irrational*. For example, what does $2^{\sqrt{5}}$ mean? While the technical details require calculus, it can be shown that successive approximations of $2^{\sqrt{5}}$ as in $2^{2.2360}$, $2^{2.23606}$, $2^{2.23236067}$, ... approach a unique real number, and $f(x) = 2^x$ exists for all real numbers x.

In general, exponential functions are defined as follows.

Exponential Functions

For $b > 0$, $b \neq 1$, and all real numbers x,

$$f(x) = b^x$$

defines the base b exponential function.

Limiting b to positive values ensures that outputs will be real numbers, and the restriction $b \neq 1$ is needed since $y = 1^x$ is a constant function (1 raised to *any* power is still 1). Specifically note the domain of an exponential function is *all real numbers*, and that all of the familiar properties of exponents still hold. A summary of these properties follows. For a complete review, see Appendix I.C.

Exponential Properties

For real numbers a, b, m, and n, with $a, b > 0$,

$$b^m \cdot b^n = b^{m+n} \qquad \frac{b^m}{b^n} = b^{m-n} \qquad (b^m)^n = b^{mn}$$

$$(ab)^n = a^n \cdot b^n \qquad b^{-n} = \frac{1}{b^n} \qquad \left(\frac{b}{a}\right)^{-n} = \left(\frac{a}{b}\right)^n$$

EXAMPLE 1 ▶ **Evaluating Exponential Functions**

Evaluate each exponential function for $x = 2$, $x = -1$, $x = \frac{1}{2}$, and $x = \pi$. Use a calculator for $x = \pi$, rounding to five decimal places.

    a. $f(x) = 4^x$ **b.** $g(x) = \left(\frac{4}{9}\right)^x$

Solution ▶ **a.** For $\quad f(x) = 4^x$,   **b.** For $\quad g(x) = \left(\frac{4}{9}\right)^x$,

$$f(2) = 4^2 = 16$$
$$f(-1) = 4^{-1} = \frac{1}{4}$$
$$f\left(\frac{1}{2}\right) = 4^{\frac{1}{2}} = \sqrt{4} = 2$$
$$f(\pi) = 4^\pi \approx 77.88023$$

$$g(2) = \left(\frac{4}{9}\right)^2 = \frac{16}{81}$$
$$g(-1) = \left(\frac{4}{9}\right)^{-1} = \frac{9}{4}$$
$$g\left(\frac{1}{2}\right) = \left(\frac{4}{9}\right)^{\frac{1}{2}} = \sqrt{\frac{4}{9}} = \frac{2}{3}$$
$$g(\pi) = \left(\frac{4}{9}\right)^\pi \approx 0.07827$$

☑ **A.** You've just learned how to evaluate an exponential function

Now try Exercises 7 through 12 ▶

B. Graphing Exponential Functions

To gain a better understanding of exponential functions, we'll graph examples of $y = b^x$ and note some of the characteristic features. Since $b \neq 1$, it seems reasonable that we graph one exponential function where $b > 1$ and one where $0 < b < 1$.

EXAMPLE 2 ▶ **Graphing Exponential Functions with $b > 1$**

Graph $y = 2^x$ using a table of values.

Solution ▶ To get an idea of the graph's shape we'll use integer values from -3 to 3 in our table, then draw the graph as a continuous curve, since the function is defined for all real numbers.

x	$y = 2^x$
-3	$2^{-3} = \frac{1}{8}$
-2	$2^{-2} = \frac{1}{4}$
-1	$2^{-1} = \frac{1}{2}$
0	$2^0 = 1$
1	$2^1 = 2$
2	$2^2 = 4$
3	$2^3 = 8$

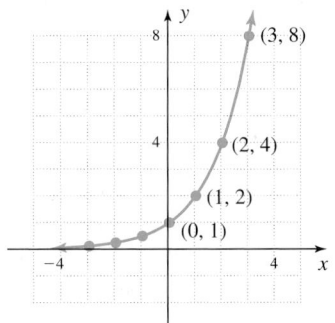

Now try Exercises 13 and 14 ▶

> **WORTHY OF NOTE**
>
> As in Example 2, functions that are increasing for all $x \in D$ are said to be **monotonically increasing** or simply **monotonic functions**. The function in Example 3 is monotonically decreasing.

Several important observations can now be made. First note the x-axis (the line $y = 0$) is a horizontal asymptote for the function, because as $x \to -\infty$, $y \to 0$. Second, the function is increasing over its entire domain, giving the function a range of $y \in (0, \infty)$.

EXAMPLE 3 ▶ **Graphing Exponential Functions with $0 < b < 1$**

Graph $y = \left(\frac{1}{2}\right)^x$ using a table of values.

Solution ▶ Using properties of exponents, we can write $\left(\frac{1}{2}\right)^x$ as $\left(\frac{2}{1}\right)^{-x} = 2^{-x}$. Again using integers from -3 to 3, we plot the ordered pairs and draw a continuous curve.

x	$y = 2^{-x}$
-3	$2^{-(-3)} = 2^3 = 8$
-2	$2^{-(-2)} = 2^2 = 4$
-1	$2^{-(-1)} = 2^1 = 2$
0	$2^0 = 1$
1	$2^{-1} = \frac{1}{2}$
2	$2^{-2} = \frac{1}{4}$
3	$2^{-3} = \frac{1}{8}$

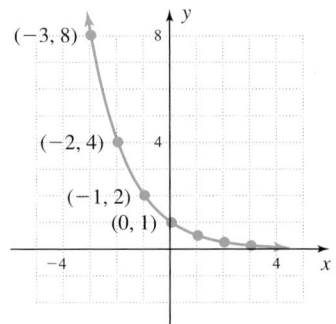

Now try Exercises 15 and 16 ▶

We note this graph is also asymptotic to the x-axis, but *decreasing on its domain.* In addition, both $y = 2^x$ and $y = 2^{-x} = \left(\frac{1}{2}\right)^x$ are one-to-one, and have a y-intercept of $(0, 1)$—which we expect since any base to the zero power is 1. Finally, observe that $y = b^{-x}$ is *a reflection of $y = b^x$ across the y-axis,* a property that suggests these basic graphs might also be transformed in other ways, as were the toolbox functions. The characteristics of exponential functions are summarized here:

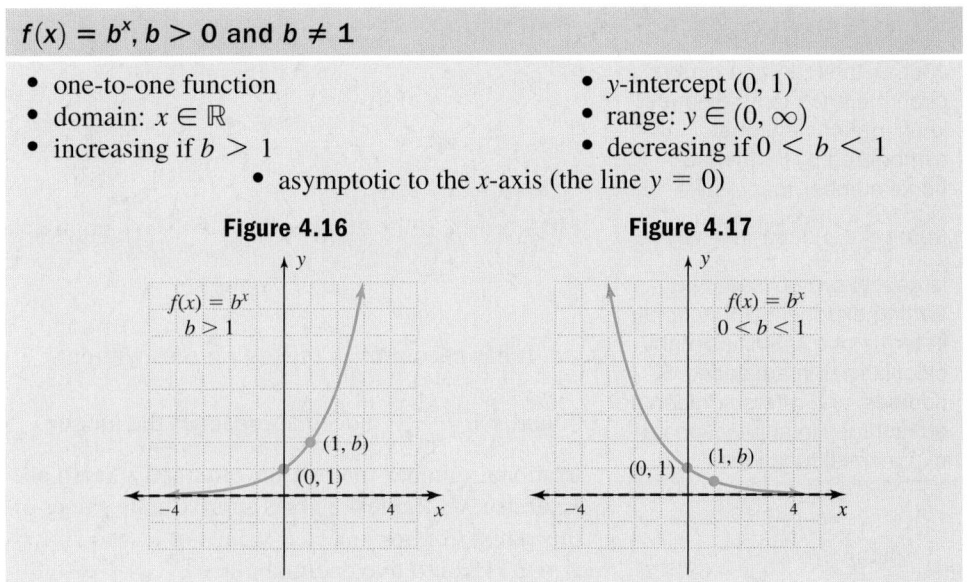

$f(x) = b^x, b > 0$ and $b \neq 1$

- one-to-one function
- domain: $x \in \mathbb{R}$
- increasing if $b > 1$
- y-intercept $(0, 1)$
- range: $y \in (0, \infty)$
- decreasing if $0 < b < 1$
- asymptotic to the x-axis (the line $y = 0$)

Figure 4.16

$f(x) = b^x$
$b > 1$

$(1, b)$
$(0, 1)$

Figure 4.17

$f(x) = b^x$
$0 < b < 1$

$(0, 1)$ $(1, b)$

WORTHY OF NOTE

When an exponential function is increasing, it can be referred to as a "growth function." When decreasing, it is often called a "decay function." Each of the graphs shown in Figures 4.16 and 4.17 should now be added to your repertoire of basic functions, to be sketched from memory and analyzed or used as needed.

Just as the graph of a quadratic function maintains its parabolic shape regardless of the transformations applied, exponential functions will also maintain their general shape and features. Any sum or difference applied to the basic function ($y = b^x \pm k$ vs. $y = b^x$) will cause a vertical shift in the same direction as the sign, and any change to input values ($y = b^{x+h}$ vs. $y = b^x$) will cause a horizontal shift in a direction opposite the sign.

EXAMPLE 4 ▶ **Graphing Exponential Functions Using Transformations**

Graph $F(x) = 2^{x-1} + 2$ using transformations of the basic function (not by simply plotting points). Clearly name the parent function and state what transformations are applied.

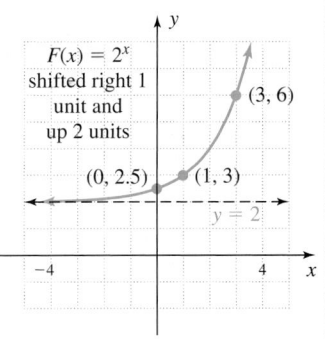

Solution ▶ The graph of F is that of the basic function $y = 2^x$ shifted 1 unit right and 2 units up. With this in mind the horizontal asymptote shifts from $y = 0$ to $y = 2$ and $(0, 1)$ shifts to $(1, 3)$. The y-intercept of F is at $(0, 2.5)$ and to help sketch a more accurate graph, the additional point $(3, 6)$ is used: $F(3) = 6$.

☑ **B.** You've just learned how to graph general exponential functions

Now try Exercises 17 through 38 ▶

C. The Base-e Exponential Function: $f(x) = e^x$

In nature, exponential growth occurs when the rate of change in a population's growth, is in constant proportion to its current size. Using the rate of change notation, $\dfrac{\Delta P}{\Delta t} = kP$, where k is a constant. For the city of Goldsboro, we know the population at time t is given by $P(t) = 1000 \cdot 2^t$, but have no information on this value of k **(see Exercise 96).** We can actually rewrite this function, and other exponential functions, using a base that gives the value of k directly and without having to apply the difference quotient. This new base is an irrational number, symbolized by the letter e and defined as follows.

The Number e

For $x > 0$,

$$\text{as } x \to \infty, \left(1 + \frac{1}{x}\right)^x \to e$$

In words, e is the number that $\left(1 + \dfrac{1}{x}\right)^x$ approaches as x becomes infinitely large.

It has been proven that as x grows without bound, $\left(1 + \dfrac{1}{x}\right)^x$ indeed approaches the unique, irrational number that we have named e (**also see Exercise 97).** Table 4.5 gives approximate values of the expression for selected values of x, and shows $e \approx 2.71828$ to five decimal places.

The result is the base-e **exponential function:** $f(x) = e^x$, also called the **natural exponential function.** Instead of having to enter a decimal approximation when computing with e, most calculators have an "e^x" key, usually as the 2nd function for the key marked LN . To find the value of e^2, use the keystrokes 2nd LN 2) ENTER , and the calculator display should read 7.389056099. Note the calculator supplies the left parenthesis for the exponent, and you must supply the right.

Table 4.5

x	$(1 + \frac{1}{x})^x$
1	2
10	2.59
100	2.705
1000	2.7169
10,000	2.71815
100,000	2.718268
1,000,000	2.7182804
10,000,000	2.71828169

EXAMPLE 5 ▶ **Evaluating the Natural Exponential Function**

Use a calculator to evaluate $f(x) = e^x$ for the values of x given. Round to six decimal places.

a. $f(3)$ **b.** $f(1)$ **c.** $f(0)$ **d.** $f(\frac{1}{2})$

Solution ▶ **a.** $f(3) = e^3 \approx 20.085537$ **b.** $f(1) = e^1 \approx 2.718282$

c. $f(0) = e^0 = 1$ (exactly) **d.** $f(\frac{1}{2}) = e^{\frac{1}{2}} \approx 1.648721$

Now try Exercises 39 through 46 ▶

Although e is an irrational number, the graph of $y = e^x$ behaves in exactly the same way and has the same characteristics as other exponential graphs. Figure 4.18 shows this graph on the same grid as $y = 2^x$ and $y = 3^x$. As we might expect, all three graphs are increasing, have an asymptote at $y = 0$, and contain the point $(0, 1)$, with the graph of $y = e^x$ "between" the other two. The domain for all three functions, as with all basic exponential functions, is $x \in (-\infty, \infty)$ with range $y \in (0, \infty)$. The same transformations applied earlier can also be applied to the graph of $y = e^x$.

Figure 4.18

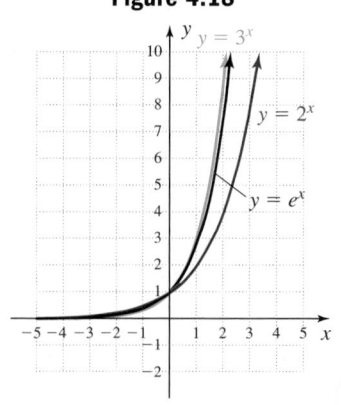

EXAMPLE 6 ▶ **Graphing Exponential Functions Using a Transformation**

Graph $f(x) = e^{x+1} - 2$ using transformations of $y = e^x$. Clearly state the transformations applied.

Solution ▶ The graph of f is the same as $y = e^x$, shifted 1 unit left and 2 units down. The point $(0, 1)$ becomes $(-1, -1)$, and the horizontal asymptote becomes $y = -2$. As the basic shape of the graph is known, we compute $f(1) \approx 5.4$, and complete the graph as shown.

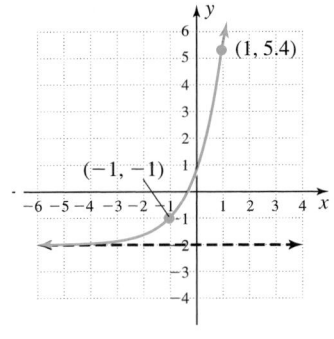

☑ **C.** You've just learned how to graph base-e exponential functions

Now try Exercises 47 through 52 ▶

D. Solving Exponential Equations Using the Uniqueness Property

Since exponential functions are one-to-one, we can solve equations where each side is an exponential term with the identical base. This is because one-to-oneness guarantees a unique solution to the equation.

Exponential Equations and the Uniqueness Property

For all real numbers m, n, and b, where $b > 0$ and $b \neq 1$,

$$\text{If } b^m = b^n,$$

$$\text{then } m = n.$$

Equal bases imply exponents are equal.

The equation $2^x = 32$ can be written as $2^x = 2^5$, and we note $x = 5$ is a solution. Although $3^x = 32$ can be written as $3^x = 2^5$, the bases are not alike and the solution to this equation must wait until additional tools are developed in Section 4.4.

EXAMPLE 7 ▶ **Solving Exponential Equations**

Solve the exponential equations using the uniqueness property.

a. $3^{2x-1} = 81$ **b.** $25^{-2x} = 125^{x+7}$

c. $\left(\frac{1}{6}\right)^{-3x-2} = 36^{x+1}$ **d.** $e^x e^2 = \dfrac{e^4}{e^{x+1}}$

Solution ▶ **a.** $3^{2x-1} = 81$ given

$3^{2x-1} = 3^4$ rewrite using base 3

$\Rightarrow 2x - 1 = 4$ uniqueness property

$x = \dfrac{5}{2}$ solve for x

Check ▶ $3^{2x-1} = 81$ given

$3^{2\left(\frac{5}{2}\right)-1} = 81$ substitute $\frac{5}{2}$ for x

$3^{5-1} = 81$ simplify

$3^4 = 81$ result checks

$81 = 81$

The remaining checks are left to the student.

b.
$$25^{-2x} = 125^{x+7} \qquad \text{given}$$
$$(5^2)^{-2x} = (5^3)^{x+7} \qquad \text{rewrite using base 5}$$
$$5^{-4x} = 5^{3x+21} \qquad \text{power property of exponents}$$
$$\Rightarrow -4x = 3x + 21 \qquad \text{uniqueness property}$$
$$x = -3 \qquad \text{solve for } x$$

c.
$$\left(\frac{1}{6}\right)^{-3x-2} = 36^{x+1} \qquad \text{given}$$
$$(6^{-1})^{-3x-2} = (6^2)^{x+1} \qquad \text{rewrite using base 6}$$
$$6^{3x+2} = 6^{2x+2} \qquad \text{power property of exponents}$$
$$\Rightarrow 3x + 2 = 2x + 2 \qquad \text{uniqueness property}$$
$$x = 0 \qquad \text{solve for } x$$

d.
$$e^x e^2 = \frac{e^4}{e^{x+1}} \qquad \text{given}$$
$$e^{x+2} = e^{4-(x+1)} \qquad \text{product property; quotient property}$$
$$e^{x+2} = e^{3-x} \qquad \text{simplify}$$
$$\Rightarrow x + 2 = 3 - x \qquad \text{uniqueness property}$$
$$2x = 1 \qquad \text{add } x, \text{ subtract 2}$$
$$x = \frac{1}{2} \qquad \text{solve for } x$$

Now try Exercises 53 through 72 ▶

One very practical application of the natural exponential function involves **Newton's law of cooling.** This law or formula models the temperature of an object as it cools down, as when a pizza is removed from the oven and placed on the kitchen counter. The function model is

$$T(x) = T_R + (T_0 - T_R)e^{kx}, \, k < 0$$

where T_0 represents the initial temperature of the object, T_R represents the temperature of the room or surrounding medium, $T(x)$ is the temperature of the object x min later, and k is the cooling rate as determined by the nature and physical properties of the object.

EXAMPLE 8 ▶ **Applying an Exponential Function—Newton's Law of Cooling**

A pizza is taken from a 425°F oven and placed on the counter to cool. If the temperature in the kitchen is 75°F, and the cooling rate for this type of pizza is $k = -0.35$,

a. What is the temperature (to the nearest degree) of the pizza 2 min later?

b. To the nearest minute, how long until the pizza has cooled to a temperature below 90°F?

c. If Zack and Raef like to eat their pizza at a temperature of about 110°F, how many minutes should they wait to "dig in"?

Solution ▶ Begin by substituting the given values to obtain the equation model:

$$T(x) = T_R + (T_0 - T_R)e^{kx} \qquad \text{general equation model}$$
$$= 75 + (425 - 75)e^{-0.35x} \qquad \text{substitute 75 for } T_R, \text{ 425 for } T_0 \text{ and } -0.35 \text{ for } k$$
$$= 75 + 350e^{-0.35x} \qquad \text{simplify}$$

For part (a) we simply find $T(2)$:

a. $T(2) = 75 + 350e^{-0.35(2)} \qquad \text{substitute 2 for } x$
$$\approx 249 \qquad \text{result}$$

Two minutes later, the temperature of the pizza is near 249°.

b. Using the TABLE feature of a graphing calculator shows the pizza reaches a temperature of just under 90° after 9 min: $T(9) \approx 90°F$.

c. We elect to use the intersection of graphs method (see the *Technology Highlight* on page 432). After setting an appropriate window, we enter $Y_1 = 75 + 350e^{-0.35x}$ and $Y_2 = 110$, then press 2nd CALC option **5: intersect**. After pressing ENTER three times, the coordinates of the point of intersection appear at the bottom of the screen: $x \approx 6.6$, $y = 110$. It appears the boys should wait about $6\frac{1}{2}$ min for the pizza to cool.

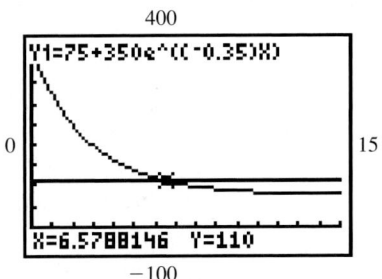

Now try Exercises 75 and 76 ▶

EXAMPLE 9 ▶ **Applications of Exponential Functions—Depreciation**

For insurance purposes, it is estimated that large household appliances lose $\frac{1}{5}$ of their value each year. The current value can then be modeled by the function $V(t) = V_0(\frac{4}{5})^t$, where V_0 is the initial value and $V(t)$ represents the value after t years. How many years does it take a washing machine that cost $625 new, to depreciate to a value of $256?

Solution ▶ For this exercise, $V_0 = \$625$ and $V(t) = \$256$. The formula yields

$$V(t) = V_0\left(\frac{4}{5}\right)^t \qquad \text{given}$$

$$256 = 625\left(\frac{4}{5}\right)^t \qquad \text{substitute known values}$$

$$\frac{256}{625} = \left(\frac{4}{5}\right)^t \qquad \text{divide by 625}$$

$$\left(\frac{4}{5}\right)^4 = \left(\frac{4}{5}\right)^t \qquad \text{equate bases } \frac{256}{625} = \left(\frac{4}{5}\right)^4$$

$$\Rightarrow 4 = t \qquad \text{Uniqueness Property}$$

✓ **D.** You've just learned how to solve exponential equations and applications

After 4 yr, the washing machine's value has dropped to $256.

Now try Exercises 77 through 90 ▶

TECHNOLOGY HIGHLIGHT

Solving Exponential Equations Graphically

In this section, we showed that the exponential function $f(x) = b^x$ was defined for all real numbers. This is important because it establishes that equations like $2^x = 7$ must have a solution, even if x is not rational. In fact, since $2^2 = 4$ and $2^3 = 8$, the following inequalities indicate the solution must be between 2 and 3

$$4 < 7 < 8 \qquad \text{7 is between 4 and 8}$$
$$2^2 < 2^x < 2^3 \qquad \text{replace 4 with } 2^2, \text{ 8 with } 2^3$$
$$2 < x < 3 \qquad x \text{ must be between 2 and 3}$$

—continued

Until we develop an inverse for exponential functions, we are unable to solve many of these equations in exact form. We can, however, get a very close approximation using a graphing calculator. For the equation $2^x = 7$, enter $Y_1 = 2^x$ and $Y_2 = 7$ on the Y= screen. Then press ZOOM 6 to graph both functions (see Figure 4.19). To find the point of intersection, press 2nd TRACE (CALC) and select option **5: intersect** and press ENTER *three* times (to identify the intersecting functions and bypass "Guess"). The *x*- and *y*-coordinates of the point of intersection will appear at the bottom of the screen, with the *x*-coordinate being the solution. As you can see, *x* is indeed between 2 and 3. Solve the following equations. First estimate the answer by bounding it between two integers, then solve the equation graphically. Adjust the viewing window as needed.

Figure 4.19

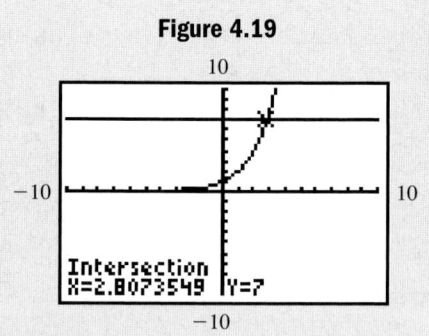

Exercise 1: $3^x = 22$	**Exercise 2:** $2^x = 0.125$
Exercise 3: $e^{x-1} = 9$	**Exercise 4:** $e^{0.5x} = 0.1x^3$

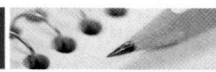

4.2 EXERCISES

▶ CONCEPTS AND VOCABULARY

Fill in each blank with the appropriate word or phrase. Carefully reread the section if needed.

1. An exponential function is one of the form $y = $ _____, where _____ > 0, _____ $\neq 1$, and _____ is any real number.

2. The domain of $y = b^x$ is all _____, and the range is $y \in$ _____. Further, as $x \to -\infty$, y _____.

3. For exponential functions of the form $y = ab^x$, the *y*-intercept is $(0, $ _____$)$, since $b^0 = $ _____ for any real number b.

4. If each side of an equation can be written as an exponential term with the same base, the equation can be solved using the _____ _____.

5. State true or false and explain why: $y = b^x$ is always increasing if $0 < b < 1$.

6. Discuss/Explain the statement, "For $k > 0$, the *y*-intercept of $y = ab^x + k$ is $(0, a + k)$."

▶ DEVELOPING YOUR SKILLS

Use a calculator (as needed) to evaluate each function as indicated. Round answers to thousandths.

7. $P(t) = 2500 \cdot 4^t$;
   $t = 2, t = \frac{1}{2}, t = \frac{3}{2}$,
   $t = \sqrt{3}$

8. $Q(t) = 5000 \cdot 8^t$;
   $t = 2, t = \frac{1}{3}, t = \frac{5}{3}$,
   $t = 5$

9. $f(x) = 0.5 \cdot 10^x$;
   $x = 3, x = \frac{1}{2}, x = \frac{2}{3}$,
   $x = \sqrt{7}$

10. $g(x) = 0.8 \cdot 5^x$;
    $x = 4, x = \frac{1}{4}, x = \frac{4}{5}$,
    $x = \pi$

11. $V(n) = 10{,}000(\frac{2}{3})^n$;
    $n = 0, n = 4, n = 4.7$,
    $n = 5$

12. $W(m) = 3300(\frac{4}{5})^m$;
    $m = 0, m = 5, m = 7.2$,
    $m = 10$

Graph each function using a table of values and integer inputs between −3 and 3. Clearly label the y-intercept and one additional point, then indicate whether the function is increasing or decreasing.

13. $y = 3^x$

14. $y = 4^x$

15. $y = \left(\frac{1}{3}\right)^x$

16. $y = \left(\frac{1}{4}\right)^x$

Graph each of the following functions by *translating the basic function* $y = b^x$, sketching the asymptote, and strategically plotting a few points to round out the graph. Clearly state the basic function and what shifts are applied.

17. $y = 3^x + 2$

18. $y = 3^x - 3$

19. $y = 3^{x+3}$

20. $y = 3^{x-2}$

21. $y = 2^{-x}$

22. $y = 3^{-x}$

23. $y = 2^{-x} + 3$

24. $y = 3^{-x} - 2$

25. $y = 2^{x+1} - 3$

26. $y = 3^{x-2} + 1$

27. $y = \left(\frac{1}{3}\right)^x + 1$

28. $y = \left(\frac{1}{3}\right)^x - 4$

29. $y = \left(\frac{1}{3}\right)^{x-2}$

30. $y = \left(\frac{1}{3}\right)^{x+2}$

31. $f(x) = \left(\frac{1}{3}\right)^x - 2$

32. $g(x) = \left(\frac{1}{3}\right)^x + 2$

Match each graph to the correct exponential equation.

33. $y = 5^{-x}$

34. $y = 4^{-x}$

35. $y = 3^{-x+1}$

36. $y = 3^{-x} + 1$

37. $y = 2^{x+1} - 2$

38. $y = 2^{x+2} - 1$

a.

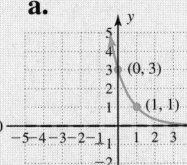

b.

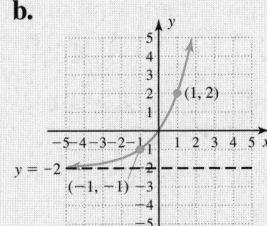

c.

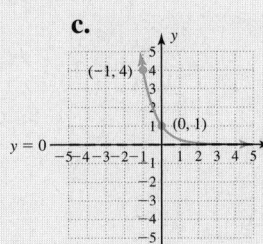

d.

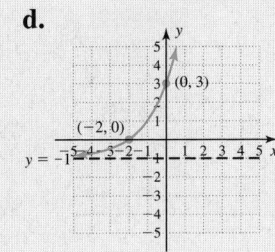

e.

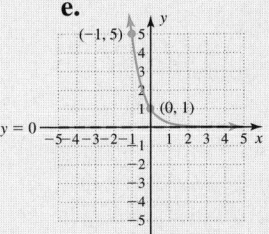

f.

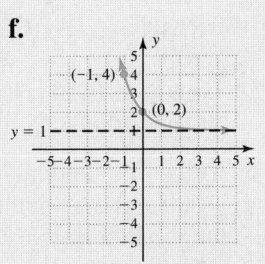

Use a calculator to evaluate each expression, rounded to six decimal places.

39. e^1

40. e^0

41. e^2

42. e^5

43. $e^{1.5}$

44. $e^{-3.2}$

45. $e^{\sqrt{2}}$

46. e^{π}

Graph each exponential function.

47. $f(x) = e^{x+3} - 2$

48. $g(x) = e^{x-2} + 1$

49. $r(t) = -e^t + 2$

50. $s(t) = -e^{t+2}$

51. $p(x) = e^{-x+2} - 1$

52. $q(x) = e^{-x-1} + 2$

Solve each exponential equation and check your answer by substituting into the original equation.

53. $10^x = 1000$

54. $144 = 12^x$

55. $25^x = 125$

56. $81 = 27^x$

57. $8^{x+2} = 32$

58. $9^{x-1} = 27$

59. $32^x = 16^{x+1}$

60. $100^{x+2} = 1000^x$

61. $\left(\frac{1}{5}\right)^x = 125$

62. $\left(\frac{1}{4}\right)^x = 64$

63. $\left(\frac{1}{3}\right)^{2x} = 9^{x-6}$

64. $\left(\frac{1}{2}\right)^{3x} = 8^{x-2}$

65. $\left(\frac{1}{9}\right)^{x-5} = 3^{3x}$

66. $2^{-2x} = \left(\frac{1}{32}\right)^{x-3}$

67. $25^{3x} = 125^{x-2}$

68. $27^{2x+4} = 9^{4x}$

69. $\dfrac{e^4}{e^{2-x}} = e^3 e$

70. $e^x(e^x + e) = \dfrac{e^x + e^{3x}}{e^{-x}}$

71. $\left(e^{2x-4}\right)^3 = \dfrac{e^{x+5}}{e^2}$

72. $e^x e^{x+3} = \left(e^{x+2}\right)^3$

▶ WORKING WITH FORMULAS

73. The growth of a bacteria population:
$P(t) = 1000 \cdot 3^t$

If the initial population of a common bacterium is 1000 and the population triples every day, its population is given by the formula shown, where $P(t)$ is the total population after t days. (a) Find the total population 12 hr, 1 day, $1\frac{1}{2}$ days, and 2 days later. (b) Do the outputs show the population is tripling every 24 hr (1 day)? (c) Explain why this is an increasing function. (d) Graph the function using an appropriate scale.

74. Games involving a spinner with numbers 1 through 4: $P(x) = (\frac{1}{4})^x$

Games that involve moving pieces around a board using a fair spinner are fairly common. If the spinner has the numbers 1 through 4, the probability that any one number is spun repeatedly is given by the formula shown, where x represents the number of spins and $P(x)$ represents the probability the same number results x times. (a) What is the probability that the first player spins a 2? (b) What is the probability that all four players spin a 2? (c) Explain why this is a decreasing function.

▶ APPLICATIONS

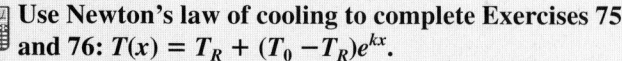 Use Newton's law of cooling to complete Exercises 75 and 76: $T(x) = T_R + (T_0 - T_R)e^{kx}$.

75. Cold party drinks: Janae was late getting ready for the party, and the liters of soft drinks she bought were still at room temperature (73°F) with guests due to arrive in 15 min. If she puts these in her freezer at $-10°F$, will the drinks be cold enough (35°F) for her guests? Assume $k \approx -0.031$.

76. Warm party drinks: Newton's law of cooling applies equally well if the "cooling is negative," meaning the object is taken from a colder medium and placed in a warmer one. If a can of soft drink is taken from a 35°F cooler and placed in a room where the temperature is 75°F, how long will it take the drink to warm to 65°F? Assume $k \approx -0.031$.

77. Depreciation: The financial analyst for a large construction firm estimates that its heavy equipment loses one-fifth of its value each  year. The current value of the equipment is then modeled by the function $V(t) = V_0(\frac{4}{5})^t$, where V_0 represents the initial value, t is in years, and $V(t)$ represents the value after t years. (a) How much is a large earthmover worth after 1 yr if it cost $125 thousand new? (b) How many years does it take for the earthmover to depreciate to a value of $64 thousand?

78. Depreciation: Photocopiers have become a critical part of the operation of many businesses, and due to their heavy use they can depreciate in value very quickly. If a copier loses $\frac{3}{8}$ of its value each year,

the current value of the copier can be modeled by the function $V(t) = V_0(\frac{5}{8})^t$, where V_0 represents the initial value, t is in years, and $V(t)$ represents the value after t yr. (a) How much is this copier worth after one year if it cost $64 thousand new? (b) How many years does it take for the copier to depreciate to a value of $25 thousand?

79. Depreciation: Margaret Madison, DDS, estimates that her dental equipment loses one-sixth of its value each year. (a) Determine the value of an x-ray machine after 5 yr if it cost $216 thousand new, and (b) determine how long until the machine is worth less than $125 thousand.

80. Exponential decay: The groundskeeper of a local high school estimates that due to heavy usage by the baseball and softball teams, the pitcher's mound loses one-fifth of its height every month. (a) Determine the height of the mound after 3 months if it was 25 cm to begin, and (b) determine how long until the pitcher's mound is less than 16 cm high (meaning it must be rebuilt).

81. Exponential growth: Similar to a small town doubling in size after a discovery of gold, a business that develops a product in high demand has the potential for doubling its revenue each year for a number of years. The revenue would be modeled by the function $R(t) = R_0 2^t$, where R_0 represents the initial revenue, and $R(t)$ represents the revenue after t years. (a) How much revenue is being generated after 4 yr, if the company's initial revenue was $2.5 million? (b) How many years does it take for the business to be generating $320 million in revenue?

82. Exponential growth: If a company's revenue grows at a rate of 150% per year (rather than doubling as in Exercise 81), the revenue would be modeled by the function $R(t) = R_0 (\frac{3}{2})^t$, where R_0 represents the initial revenue, and $R(t)$ represents the revenue after t years. (a) How much revenue is being generated after 3 yr, if the company's initial revenue was $256 thousand? (b) How long until the business is generating $1944 thousand in revenue? (*Hint:* Reduce the fraction.)

Photochromatic sunglasses: Sunglasses that darken in sunlight (photochromatic sunglasses) contain millions of molecules of a substance known as *silver halide*. The molecules are transparent indoors in the absence of ultraviolet (UV) light. Outdoors, UV light from the sun causes the molecules to change shape, darkening the lenses in response to the intensity of the UV light. For certain lenses, the function $T(x) = 0.85^x$ models the transparency of the lenses (as a percentage) based on a UV index x. Find the transparency (to the nearest percent), if the lenses are exposed to

83. sunlight with a UV index of 7 (a high exposure).

84. sunlight with a UV index of 5.5 (a moderate exposure).

85. Given that a UV index of 11 is very high and most individuals should stay indoors, what is the minimum transparency percentage for these lenses?

86. Use trial-and-error to determine the UV index when the lenses are 50% transparent.

Modeling inflation: Assuming the rate of inflation is 5% per year, the predicted price of an item can be modeled by the function $P(t) = P_0(1.05)^t$, where P_0 represents the initial price of the item and t is in years. Use this information to solve Exercises 87 and 88.

87. What will the price of a new car be in the year 2010, if it cost $20,000 in the year 2000?

88. What will the price of a gallon of milk be in the year 2010, if it cost $2.95 in the year 2000? Round to the nearest cent.

Modeling radioactive decay: The half-life of a radioactive substance is the time required for half an initial amount of the substance to disappear through decay. The amount of the substance remaining is given by the formula $Q(t) = Q_0(\frac{1}{2})^{\frac{t}{h}}$, where h is the half-life, t represents the elapsed time, and $Q(t)$ represents the amount that remains (t and h must have the same unit of time). Use this information to solve Exercises 89 and 90.

89. Some isotopes of the substance known as thorium have a half-life of only 8 min. (a) If 64 grams are initially present, how many grams (g) of the substance remain after 24 min? (b) How many minutes until only 1 gram (g) of the substance remains?

90. Some isotopes of sodium have a half-life of about 16 hr. (a) If 128 g are initially present, how many grams of the substance remain after 2 days (48 hr)? (b) How many hours until only 1 g of the substance remains?

▶ **EXTENDING THE CONCEPT**

91. The formula $f(x) = (\frac{1}{2})^x$ gives the probability that "x" number of flips result in heads (or tails). First determine the probability that 20 flips results in *20 heads in a row*. Then use the Internet or some other resource to determine the probability of winning a state lottery (expressed as a decimal). Which has the greater probability? Were you surprised?

92. If $10^{2x} = 25$, what is the value of 10^{-x}?

93. If $5^{3x} = 27$, what is the value of 5^{2x}?

94. If $3^{0.5x} = 5$, what is the value of 3^{x+1}?

95. If $\left(\frac{1}{2}\right)^{x+1} = \frac{1}{3}$, what is the value of $\left(\frac{1}{2}\right)^{-x}$?

The growth rate constant that governs an exponential function was introduced on page 427.

96. In later sections, we will easily be able to find the growth constant k for Goldsboro, where

$P(t) = 1000 \cdot 2^t$. For now we'll approximate its value using the rate of change formula on a very small interval of the domain. From the definition of an exponential function, $\dfrac{\Delta P}{\Delta t} = kP(t)$. Since k is constant, we can choose any value of t, say $t = 4$. For $h = 0.0001$, we have

$$\frac{1000 \cdot 2^{4+0.0001} - 1000 \cdot 2^4}{0.0001} = k \cdot P(4)$$

(a) Use the equation shown to solve for k (round to thousandths). (b) Show that k is constant by completing the same exercise for $t = 2$ and $t = 6$. (c) Verify that $P(t) = 1000 \cdot 2^t$ and $P(t) = 1000e^{kt}$ give approximately the same results.

97. As we analyze the expression $\left(1 + \dfrac{1}{x}\right)^x$, we notice a battle (of sorts) takes place between the base $\left(1 + \dfrac{1}{x}\right)$ and the exponent x. As $x \to \infty$, $\dfrac{1}{x}$ becomes infinitely small, but the exponent becomes

infinitely large. So what happens? The answer is best understood by computing a series of *average rates of change,* using the intervals given here. Using the tools of Calculus, it can be shown that this rate of change becomes infinitely small, and that the "battle" ends at the irrational number e. In other words, e is an upper bound on the value of this expression, regardless of how large x becomes.

a. Use a calculator to find the average rate of change for $y = \left(1 + \dfrac{1}{x}\right)^x$ in these intervals:
[1, 1.01], [4, 4.01], [10, 10.01], and [20, 20.01]. What do you notice?

b. What is the smallest integer value for x that gives the value of e correct to four decimal places?

c. Use a graphing calculator to graph this function on a window size of $x \in [0, 25]$ and $y \in [0, 3]$. Does the graph seem to support the statements above?

▶ **MAINTAINING YOUR SKILLS**

98. (2.2) Given $f(x) = 2x^2 - 3x$, determine:

$f(-1), \quad f(\tfrac{1}{3}), \quad f(a), \quad f(a + h)$

99. (3.3) Graph $g(x) = \sqrt{x + 2} - 1$ using a shift of the parent function. Then state the domain and range of g.

100. (1.3) Solve the following equations:

a. $-2\sqrt{x - 3} + 7 = 21$

b. $\dfrac{9}{x + 3} + 3 = \dfrac{12}{x - 3}$

101. (R.7) Identify each formula:

a. $\tfrac{4}{3}\pi r^3$

b. $\tfrac{1}{2}bh$

c. lwh

d. $a^2 + b^2 = c^2$

4.3 Logarithms and Logarithmic Functions

Learning Objectives

In Section 4.3 you will learn how to:

☐ **A.** Write exponential equations in logarithmic form

☐ **B.** Find common logarithms and natural logarithms

☐ **C.** Graph logarithmic functions

☐ **D.** Find the domain of a logarithmic function

☐ **E.** Solve applications of logarithmic functions

A **transcendental function** is one whose solutions are beyond or *transcend* the methods applied to polynomial functions. The exponential function and its inverse, called the logarithmic function, are transcendental functions. In this section, we'll use the concept of an inverse to develop an understanding of the logarithmic function, which has numerous applications that include measuring pH levels, sound and earthquake intensities, barometric pressure, and other natural phenomena.

A. Exponential Equations and Logarithmic Form

While exponential functions have a large number of significant applications, we can't appreciate their full value until we develop the inverse function. Without it, we're unable to solve all but the simplest equations, of the type encountered in Section 4.2. Using the fact that $f(x) = b^x$ is one-to-one, we have the following:

1. The function $f^{-1}(x)$ must exist.
2. We can graph $f^{-1}(x)$ by interchanging the x- and y-coordinates of points from $f(x)$.
3. The domain of $f(x)$ will become the range of $f^{-1}(x)$.
4. The range of $f(x)$ will become the domain of $f^{-1}(x)$.
5. The graph of $f^{-1}(x)$ will be a reflection of $f(x)$ across the line $y = x$.

Table 4.6 contains selected values for $f(x) = 2^x$. The values for $f^{-1}(x)$ in Table 4.7 were found by interchanging x- and y-coordinates. Both functions were then graphed using these values.

Table 4.6

$f(x): y = 2^x$

x	y
-3	$\frac{1}{8}$
-2	$\frac{1}{4}$
-1	$\frac{1}{2}$
0	1
1	2
2	4
3	8

Table 4.7

$f^{-1}(x): x = 2^y$

x	(x)
$\frac{1}{8}$	-3
$\frac{1}{4}$	-2
$\frac{1}{2}$	-1
1	0
2	1
4	2
8	3

Figure 4.20

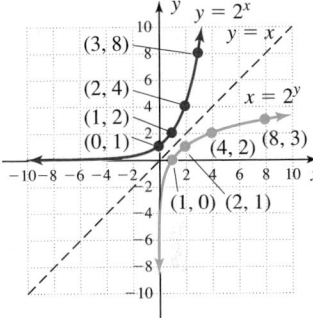

The interchange of x and y and the graphs in Figure 4.20 show that $f^{-1}(x)$ has an x-intercept of $(1, 0)$, a vertical asymptote at $x = 0$, a domain of $x \in (0, \infty)$, and a range of $y \in (-\infty, \infty)$. To find *an equation* for $f^{-1}(x)$, we'll attempt to use the algebraic approach employed previously. For $f(x) = 2^x$,

1. use y instead of $f(x)$: $y = 2^x$.
2. interchange x and y: $x = 2^y$.

At this point we have an *implicit* equation for the inverse function, but no algebraic operations that enable us to solve *explicitly* for y in terms of x. Instead, we write $x = 2^y$ in function form by noting that "y is the exponent that goes on base 2 to obtain x."

In the language of mathematics, this phrase is represented by $y = \log_2 x$ and is called a **logarithmic function** with base 2. For $y = b^x$, $x = b^y \rightarrow y = \log_b x$ is the inverse function, and is read, "y is the logarithm base b of x." For this new function, we must always keep in mind what y *represents*—y is an exponent. In fact, y *is the exponent that goes on base b to obtain x*: $y = \log_b x$.

Logarithmic Functions

For positive numbers x and b, with $b \neq 1$,

$$y = \log_b x \text{ if and only if } x = b^y$$

The function $f(x) = \log_b x$ is a logarithmic function with base b. The expression $\log_b x$ is simply called a logarithm, and represents the exponent on b that yields x.

Finally, note the equations $x = b^y$ and $y = \log_b x$ are equivalent. We say that $x = b^y$ is the **exponential form** of the equation, whereas $y = \log_b x$ is written in **logarithmic form**.

EXAMPLE 1 ▶ **Converting from Logarithmic Form to Exponential Form**

Write each equation in words, then in exponential form.

a. $3 = \log_2 8$ **b.** $1 = \log_{10} 10$ **c.** $0 = \log_e 1$ **d.** $-2 = \log_3\left(\frac{1}{9}\right)$

Solution ▶   **a.** $3 = \log_2 8 \rightarrow 3$ is the exponent on base 2 for 8: $2^3 = 8$.

b. $1 = \log_{10} 10 \rightarrow 1$ is the exponent on base 10 for 10: $10^1 = 10$.

c. $0 = \log_e 1 \rightarrow 0$ is the exponent on base e for 1: $e^0 = 1$.

d. $-2 = \log_3\left(\frac{1}{9}\right) \rightarrow -2$ is the exponent on base 3 for $\frac{1}{9}$: $3^{-2} = \frac{1}{9}$.

Now try Exercises 7 through 22 ▶

To convert from exponential form to logarithmic form, note the exponent on the base and read from there. For $5^3 = 125$, "3 is the exponent that goes on base 5 for 125," or *3 is the logarithm base 5 of 125*: $3 = \log_5 125$.

EXAMPLE 2 ▶ **Converting from Exponential Form to Logarithmic Form**

Write each equation in words, then in logarithmic form.

a. $10^3 = 1000$ **b.** $2^{-1} = \frac{1}{2}$ **c.** $e^2 \approx 7.389$ **d.** $9^{\frac{3}{2}} = 27$

Solution ▶   **a.** $10^3 = 1000 \rightarrow 3$ is the exponent on base 10 for 1000, or
3 is the logarithm base 10 of 1000: $3 = \log_{10} 1000$.

b. $2^{-1} = \frac{1}{2} \rightarrow -1$ is the exponent on base 2 for $\frac{1}{2}$, or
-1 is the logarithm base 2 of $\frac{1}{2}$: $-1 = \log_2\left(\frac{1}{2}\right)$.

c. $e^2 \approx 7.389 \rightarrow 2$ is the exponent on base e for 7.389, or
2 is the logarithm base e of 7.389: $2 \approx \log_e 7.389$.

d. $9^{\frac{3}{2}} = 27 \rightarrow \frac{3}{2}$ is the exponent on base 9 for 27, or
$\frac{3}{2}$ is the logarithm base 9 of 27: $\frac{3}{2} = \log_9 27$.

☑ **A.** You've just learned how to write exponential equations in logarithmic form

Now try Exercises 23 through 38 ▶

B. Finding Common Logarithms and Natural Logarithms

Of all possible bases for $\log_b x$, the most common are base 10 (likely due to our base-10 number system), and base e *(due to the advantages it offers in advanced courses)*. The expression $\log_{10} x$ is called a **common logarithm,** and we simply write $\log x$ for $\log_{10} x$. The expression $\log_e x$ is called a **natural logarithm,** and is written in abbreviated form as $\ln x$.

Some logarithms are easy to evaluate. For example, $\log 100 = 2$ since $10^2 = 100$, and $\log \frac{1}{100} = -2$ since $10^{-2} = \frac{1}{100}$. But what about the expressions $\log 850$ and $\ln 4$? Because logarithmic functions are continuous on their domains, a value exists for $\log 850$ and the equation $10^x = 850$ must have a solution. Further, the inequalities

$$\log 100 < \log 850 < \log 1000$$

$$2 < \log 850 < 3$$

tell us that $\log 850$ must be between 2 and 3. Fortunately, modern calculators can compute base-10 and base-e logarithms instantly, often with nine-decimal-place accuracy. For $\log 850$, press ⌞**LOG**⌟, then input 850 and press ⌞**ENTER**⌟. The display should read 2.929418926. We can also use the calculator to verify $10^{2.929418926} = 850$ (see Figure 4.21). For $\ln 4$, press the ⌞**LN**⌟ key, then input 4 and press ⌞**ENTER**⌟ to obtain 1.386294361. Figure 4.22 verifies that $e^{1.386294361} = 4$.

WORTHY OF NOTE

We do something similar with square roots. Technically, the "square root of x" should be written $\sqrt[2]{x}$. However, square roots are so common we often leave off the two, assuming that if no index is written, an index of two is intended.

Figure 4.21

```
log(850)
            2.929418926
10^Ans
                      850
```

Figure 4.22

```
ln(4)
           1.386294361
e^(Ans)
                     4
```

EXAMPLE 3 ▶ **Finding the Value of a Logarithm**

Determine the value of each logarithm without using a calculator:

a. $\log_2 8$ b. $\log_5(\frac{1}{25})$ c. $\log_e e$ d. $\log_{10}\sqrt{10}$

Solution ▶ a. $\log_2 8$ represents the exponent on 2 for 8: $\log_2 8 = 3$, since $2^3 = 8$.
b. $\log_5(\frac{1}{25})$ represents the exponent on 5 for $\frac{1}{25}$: $\log_5\frac{1}{25} = -2$, since $5^{-2} = \frac{1}{25}$.
c. $\log_e e$ represents the exponent on e for e: $\log_e e = 1$, since $e^1 = e$.
d. $\log_{10}\sqrt{10}$ represents the exponent on 10 for $\sqrt{10}$: $\log_{10}\sqrt{10} = \frac{1}{2}$, since $10^{\frac{1}{2}} = \sqrt{10}$.

Now try Exercises 39 through 50 ▶

EXAMPLE 4 ▶ **Using a Calculator to Find Logarithms**

Use a calculator to evaluate each logarithmic expression. Verify the result.

a. $\log 1857$ b. $\log 0.258$ c. $\ln 3.592$

Solution ▶ a. $\log 1857 = 3.268811904$,
$10^{3.268811904} = 1857$ ✓

b. $\log 0.258 = -0.588380294$,
$10^{-0.588380294} = 0.258$ ✓

c. $\ln 3.592 \approx 1.27870915$
$e^{1.27870915} \approx 3.592$ ✓

☑ **B. You've just learned how to find common logarithms and natural logarithms**

Now try Exercises 51 through 58 ▶

C. Graphing Logarithmic Functions

For convenience and ease of calculation, our first examples of logarithmic graphs are done using base-2 logarithms. However, the basic shape of a logarithmic graph remains unchanged regardless of the base used, and transformations can be applied to $y = \log_b(x)$ for any value of b. For $y = a\log(x \pm h) \pm k$, a continues to govern stretches, compressions, and vertical reflections, the graph will shift horizontally h units opposite the sign, and shift k units vertically in the same direction as the sign. Our earlier graph of $y = \log_2 x$ was completed using $x = 2^y$ as the inverse function for $y = 2^x$ (Figure 4.20). For reference, the graph is repeated in Figure 4.23.

WORTHY OF NOTE

As with the basic graphs we studied in Section 2.6, logarithmic graphs maintain the same characteristics when transformations are applied, and these graphs *should be added to your collection of basic functions,* ready for recall or analysis as the situation requires.

Figure 4.23

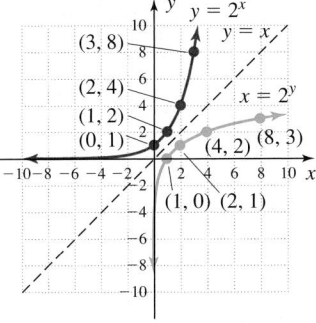

EXAMPLE 5 ▶ **Graphing Logarithmic Functions Using Transformations**

Graph $f(x) = \log_2(x - 3) + 1$ using transformations of $y = \log_2 x$ (not by simply plotting points). Clearly state what transformations are applied.

Solution ▶ The graph of f is the same as that of $y = \log_2 x$, shifted 3 units right and 1 unit up. The vertical asymptote will be at $x = 3$ and the point $(1, 0)$ from the basic graph becomes $(1 + 3, 0 + 1) = (4, 1)$. Knowing the graph's basic shape, we compute one additional point using $x = 7$:

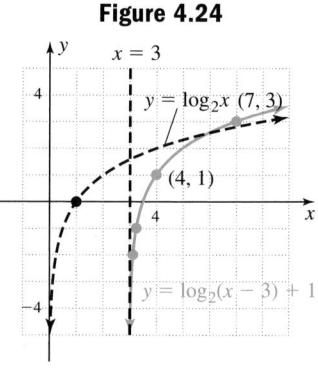

Figure 4.24

$$f(7) = \log_2(7 - 3) + 1$$
$$= \log_2 4 + 1$$
$$= 2 + 1$$
$$= 3$$

The point $(7, 3)$ is on the graph, shown in Figure 4.24.

Now try Exercises 59 through 62 ▶

As with the exponential functions, much can be learned from graphs of logarithmic functions and a summary of important characteristics is given here.

$f(x) = \log_b x, b > 0$ and $b \neq 1$

- one-to-one function
- domain: $x \in (0, \infty)$
- increasing if $b > 1$

- x-intercept $(1, 0)$
- range: $y \in \mathbb{R}$
- decreasing if $0 < b < 1$

- asymptotic to the y-axis (the line $x = 0$)

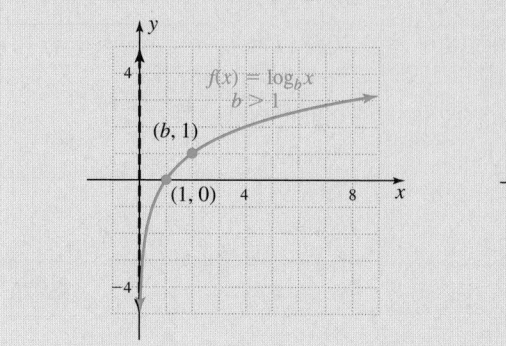

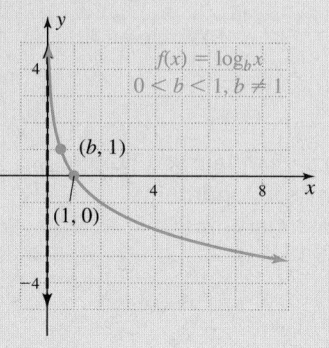

EXAMPLE 6 ▶ **Graphing Logarithmic Functions Using Transformations**

Graph $g(x) = -\ln(x + 2)$ using transformations of $y = \ln x$ (not by simply plotting points). Clearly state what transformations are applied.

Solution ▶ The graph of g is the same as $y = \ln x$, shifted 2 units left, then reflected across the x-axis. The vertical asymptote will be at $x = -2$, and the point $(1, 0)$ from the basic function becomes $(-1, 0)$. To complete the graph we compute $f(6)$:

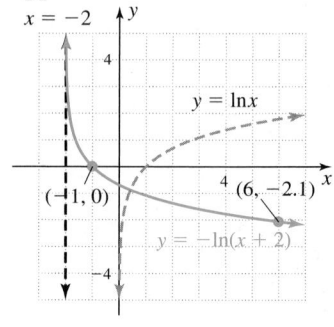

WORTHY OF NOTE

Accurate graphs can actually be drawn for logarithms of *any base* using what is called the base-change formula, introduced in Section 4.4.

$$f(6) = -\ln(6 + 2)$$
$$= -\ln 8$$
$$\approx -2.1 \text{ (using a calculator)}.$$

☑ **C.** You've just learned how to graph logarithmic functions

The point $(6, -2.1)$ is on the graph shown in the figure.

Now try Exercises 63 through 72 ▶

D. Finding the Domain of a Logarithmic Function

Examples 5 and 6 illustrate how the domain of a logarithmic function can change when certain transformations are applied. Since the domain consists of *positive* real numbers, the argument of a logarithmic function must be greater than zero. This means finding the domain often consists of solving various inequalities, which can be done using the skills acquired in Sections 2.5 and 3.7.

EXAMPLE 7 ▶ **Finding the Domain of a Logarithmic Function**

Determine the domain of each function.

a. $p(x) = \log_2(2x + 3)$ **b.** $q(x) = \log_5(x^2 - 2x)$

c. $r(x) = \log\left(\dfrac{3 - x}{x + 3}\right)$ **d.** $f(x) = \ln|x - 2|$

Solution ▶ Begin by writing the argument of each logarithmic function as a greater than inequality.

a. Solving $2x + 3 > 0$ for x gives $x > -\frac{3}{2}$, and the domain of p is $x \in (-\frac{3}{2}, \infty)$.

b. For $x^2 - 2x > 0$, we note $y = x^2 - 2x$ is a parabola, opening upward, with zeroes at $x = 0$ and $x = 2$ (see Figure 4.25). This means $x^2 - 2x$ will be positive for $x < 0$ and $x > 2$. The domain of q is $x \in (-\infty, 0) \cup (2, \infty)$.

Figure 4.25

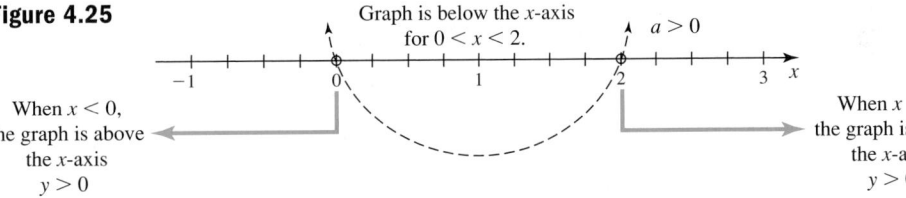

c. For $\dfrac{3 - x}{x + 3} > 0$, we note $y = \dfrac{3 - x}{x + 3}$ has a zero at $x = 3$, and a vertical asymptote at $x = -3$. Outputs are positive when $x = 0$ (see Figure 4.26), so y is positive in the interval $(-3, 3)$ and negative elsewhere. The domain of r is $x \in (-3, 3)$.

Figure 4.26

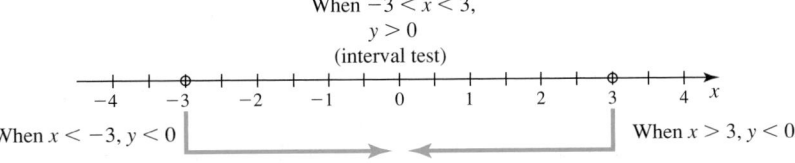

d. For $|x - 2| > 0$, we note $y = |x - 2|$ is the graph of $y = |x|$ shifted 2 units right, with its vertex at $(2, 0)$. The graph is positive for all x, except at $x = 2$. The domain of f is $x \in (-\infty, 2) \cup (2, \infty)$.

☑ **D. You've just learned how to find the domain of a logarithmic function**

Now try Exercises 73 through 78 ▶

GRAPHICAL SUPPORT

The domain for $r(x) = \log_{10}\left(\dfrac{3 - x}{x + 3}\right)$ from Example 6c can be confirmed using the LOG key on a graphing calculator. Use the key to enter the equation as Y_1 on the Y= screen, then graph the function using the ZOOM 4:ZDecimal option. Both the graph and TABLE feature help to confirm the domain is $x \in (-3, 3)$.

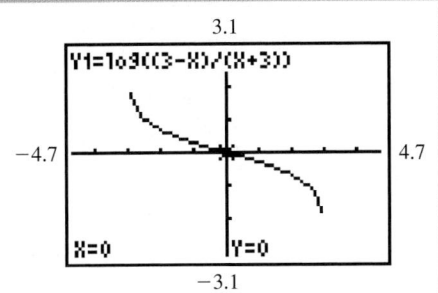

E. Applications of Logarithms

As we use mathematics to model the real world, there are times when the range of outcomes is so large that using a linear scale would be hard to manage. For example, compared to a whisper—the scream of a jet engine may be up to *ten billion times* louder. Similar ranges exist in the measurement of earthquakes, light, acidity, and voltage. In lieu of a linear scale, logarithms are used where each whole number increase in magnitude represents a tenfold increase in the intensity. For earthquake intensities (a measure of the wave energy produced by the quake), units called **magnitudes** (or **Richter values**) are used. Earthquakes with a magnitude of 6.5 or more often cause significant damage, while the slightest earthquakes have magnitudes near 1 and are barely perceptible. The magnitude of the intensity $M(I)$ is given by $M(I) = \log\left(\dfrac{I}{I_0}\right)$ where I is the measured intensity and I_0 represents a minimum or **reference intensity.** The value of I is often given as a multiple of this reference intensity.

WORTHY OF NOTE

The **decibel** (dB) is the reference unit for sound, and is based on the faintest sound a person can hear, called the **threshold of audibility.** It is a base-10 logarithmic scale, meaning a sound 10 times more intense is one decibel louder.

EXAMPLE 8A ▶ **Finding the Magnitude of an Earthquake**

Find the magnitude of an earthquake (rounded to hundredths) with the intensities given.

     a. $I = 4000I_0$ **b.** $I = 8{,}252{,}000I_0$

Solution ▶   **a.** $M(I) = \log\left(\dfrac{I}{I_0}\right)$ magnitude equation

$$M(4000I_0) = \log\left(\dfrac{4000I_0}{I_0}\right)$$ substitute $4000I_0$ for I

$$= \log 4000$$ simplify

$$\approx 3.60$$ result

The earthquake had a magnitude of 3.6.

   b. $M(I) = \log\left(\dfrac{I}{I_0}\right)$ magnitude equation

$$M(8{,}252{,}000I_0) = \log\left(\dfrac{8252000I_0}{I_0}\right)$$ substitute $8{,}252{,}000\,I_0$ for I

$$= \log 8{,}252{,}000$$ simplify

$$\approx 6.92$$ result

The earthquake had a magnitude of about 6.92.

EXAMPLE 8B ▶ **Comparing Earthquake Intensity to the Reference Intensity**

How many times more intense than the reference intensity I_0 is an earthquake with a magnitude of 6.7?

Solution ▶ $M(I) = \log\left(\dfrac{I}{I_0}\right)$ magnitude equation

$$6.7 = \log\left(\dfrac{I}{I_0}\right)$$ substitute 6.7 for $M(I)$

$$10^{6.7} = \left(\dfrac{I}{I_0}\right)$$ exponential form

$$I = 10^{6.7}I_0$$ solve for I

$$I = 5{,}011{,}872I_0$$ $10^{6.7} \approx 5{,}011{,}872$

An earthquake of magnitude 6.7 is over 5 million times more intense than the reference intensity.

EXAMPLE 8C ▶ **Comparing Earthquake Intensities**

The Great San Francisco Earthquake of 1906 left over 800 dead, did \$80,000,000 in damage (see photo), and had an estimated magnitude of 7.7. The 2004 Indian Ocean earthquake, which had a magnitude of approximately 9.2, triggered a series of deadly tsunamis and was responsible for nearly 300,000 casualties. How much more intense was the 2004 quake?

Solution ▶ To find the intensity of each quake, substitute the given magnitude for $M(I)$ and solve for I:

$$M(I) = \log\left(\frac{I}{I_0}\right) \quad \text{magnitude equation} \qquad M(I) = \log\left(\frac{I}{I_0}\right)$$

$$7.7 = \log\left(\frac{I}{I_0}\right) \quad \text{substitute for } M(I) \qquad 9.2 = \log\left(\frac{I}{I_0}\right)$$

$$10^{7.7} = \left(\frac{I}{I_0}\right) \quad \text{exponential form} \qquad 10^{9.2} = \left(\frac{I}{I_0}\right)$$

$$10^{7.7}I_0 = I \quad \text{solve for } I \qquad 10^{9.2}I_0 = I$$

Using these intensities, we find that the Indian Ocean quake was $\dfrac{10^{9.2}}{10^{7.7}} = 10^{1.5} \approx 31.6$ times more intense.

Now try Exercises 81 through 90 ▶

A second application of logarithmic functions involves the relationship between altitude and barometric pressure. The altitude or height above sea level can be determined by the formula $H = (30T + 8000) \ln\left(\dfrac{P_0}{P}\right)$, where H is the altitude in meters for a temperature T in degrees Celsius, P is the barometric pressure at a given altitude in units called **centimeters of mercury** (cmHg), and P_0 is the barometric pressure at sea level: 76 cmHg.

EXAMPLE 9 ▶ **Using Logarithms to Determine Altitude**

Hikers at the summit of Mt. Shasta in northern California take a pressure reading of 45.1 cmHg at a temperature of 9°C. How high is Mt. Shasta?

Solution ▶ For this exercise, $P_0 = 76$, $P = 45.1$, and $T = 9$. The formula yields

$$H = (30T + 8000) \ln\left(\frac{P_0}{P}\right) \qquad \text{given formula}$$

$$= [30(9) + 8000] \ln\left(\frac{76}{45.1}\right) \qquad \text{substitute given values}$$

$$= 8270 \ln\left(\frac{76}{45.1}\right) \qquad \text{simplify}$$

$$\approx 4316 \qquad \text{result}$$

Mt. Shasta is about 4316 m high.

Now try Exercises 91 through 94 ▶

Our final application shows the versatility of logarithmic functions, and their value as a real-world model. Large advertising agencies are well aware that after a new ad campaign, sales will increase rapidly as more people become aware of the product. Continued advertising will give the new product additional market share, but once the "newness" wears off and the competition begins responding, sales tend to taper off—regardless of any additional amount spent on ads. This phenomenon can be modeled by the function

$$S(d) = k + a \ln d,$$

where $S(d)$ is the number of expected sales after d dollars are spent, and a and k are constants related to product type and market size **(see Exercises 95 and 96).**

EXAMPLE 10 ▶ **Using Logarithms for Marketing Strategies**

Market research has shown that sales of the MusicMaster, a new system for downloading and playing music, can be approximated by the equation $S(d) = 2500 + 250 \ln d$, where $S(d)$ is the number of sales after d thousand dollars is spent on advertising.

a. What sales volume is expected if the advertising budget is $40,000?

b. If the company needs to sell 3500 units to begin making a profit, how much should be spent on advertising?

c. To gain a firm hold on market share, the company is willing to continue spending on advertising up to a point where only 3 additional sales are gained for each $1000 spent, in other words, $\dfrac{\Delta S}{\Delta d} = \dfrac{3}{1}$. Verify that spending between $83,200 and $83,300 puts them very close to this goal.

Solution ▶   **a.** For sales volume, we simply evaluate the function for $d = 40$ (d in thousands):

$$
\begin{aligned}
S(d) &= 2500 + 250 \ln d & \text{given equation} \\
S(40) &= 2500 + 250 \ln 40 & \text{substitute 40 for } d \\
&\approx 2500 + 922 & 250 \ln 40 \approx 922 \\
&= 3422
\end{aligned}
$$

Spending $40,000 on advertising will generate approximately 3422 sales.

b. To find the advertising budget needed, we substitute number of sales and solve for d.

$$
\begin{aligned}
S(d) &= 2500 + 250 \ln d & \text{given equation} \\
3500 &= 2500 + 250 \ln d & \text{substitute 2500 for } S(d) \\
1000 &= 250 \ln d & \text{subtract 2500} \\
4 &= \ln d & \text{divide by 250} \\
e^4 &= d & \text{exponential form} \\
54.598 &\approx d & e^4 \approx 54.598
\end{aligned}
$$

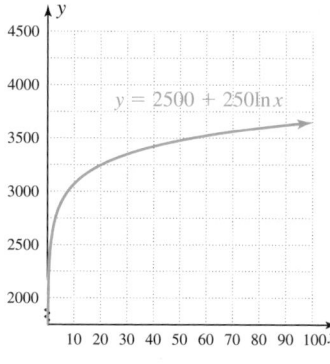

About $54,600 should be spent in order to sell 3500 units.

c. To verify, we calculate the average rate of change on the interval [83.2, 83.3].

$$\frac{\Delta S}{\Delta d} = \frac{S(d_2) - S(d_1)}{d_2 - d_1} \qquad \text{formula for average rate of change}$$

$$= \frac{S(83.3) - S(83.2)}{83.3 - 83.2} \qquad \text{substitute 83.3 for } d_2 \text{ and 83.2 for } d_1$$

$$\approx \frac{3605.6 - 3605.3}{0.1} \qquad \text{evaluate } S(83.3) \text{ and } S(83.2)$$

$$= 3$$

☑ **E.** You've just learned how to solve applications of logarithmic functions

The average rate of change in this interval is very close to $\frac{3}{1}$.

Now try Exercises 95 and 96 ▶

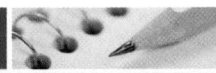

4.3 EXERCISES

▶ **CONCEPTS AND VOCABULARY**

Fill in each blank with the appropriate word or phrase. Carefully reread the section if needed.

1. A logarithmic function is of the form $y =$ _____, where _____ > 0, _____ $\neq 1$ and inputs are _____ than zero.

2. The range of $y = \log_b x$ is all _____ _____, and the domain is $x \in$ _____. Further, as $x \to 0$, $y \to$ ___.

3. For logarithmic functions of the form $y = \log_b x$, the x-intercept is _____, since $\log_b 1 =$ _____.

4. The function $y = \log_b x$ is an increasing function if _____, and a decreasing function if _____.

5. What number does the expression $\log_2 32$ represent? Discuss/Explain how $\log_2 32 = \log_2 2^5$ justifies this fact.

6. Explain how the graph of $Y = \log_b(x - 3)$ can be obtained from $y = \log_b x$. Where is the "new" x-intercept? Where is the new asymptote?

▶ **DEVELOPING YOUR SKILLS**

Write each equation in exponential form.

7. $3 = \log_2 8$

8. $2 = \log_3 9$

9. $-1 = \log_7 \frac{1}{7}$

10. $-3 = \log_e \frac{1}{e^3}$

11. $0 = \log_9 1$

12. $0 = \log_e 1$

13. $\frac{1}{3} = \log_8 2$

14. $\frac{1}{2} = \log_{81} 9$

15. $1 = \log_2 2$

16. $1 = \log_e e$

17. $\log_7 49 = 2$

18. $\log_4 16 = 2$

19. $\log_{10} 100 = 2$

20. $\log_{10} 10,000 = 4$

21. $\log_e(54.598) \approx 4$

22. $\log_{10} 0.001 = -3$

Write each equation in logarithmic form.

23. $4^3 = 64$

24. $e^3 \approx 20.086$

25. $3^{-2} = \frac{1}{9}$

26. $2^{-3} = \frac{1}{8}$

27. $e^0 = 1$

28. $8^0 = 1$

29. $\left(\frac{1}{3}\right)^{-3} = 27$

30. $\left(\frac{1}{5}\right)^{-2} = 25$

31. $10^3 = 1000$

32. $e^1 = e$

33. $10^{-2} = \frac{1}{100}$

34. $10^{-5} = \frac{1}{100,000}$

35. $4^{\frac{3}{2}} = 8$

36. $e^{\frac{3}{4}} \approx 2.117$

37. $4^{\frac{-3}{2}} = \frac{1}{8}$

38. $27^{\frac{-2}{3}} = \frac{1}{9}$

Determine the value of each logarithm without using a calculator.

39. $\log_4 4$

40. $\log_9 9$

41. $\log_{11} 121$

42. $\log_{12} 144$

43. $\log_e e$ **44.** $\log_e e^2$

45. $\log_4 2$ **46.** $\log_{81} 9$

47. $\log_7 \frac{1}{49}$ **48.** $\log_9 \frac{1}{81}$

49. $\log_e \frac{1}{e^2}$ **50.** $\log_e \frac{1}{\sqrt{e}}$

I.

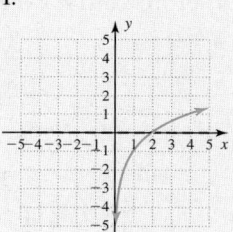

II.

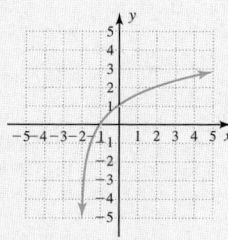

Use a calculator to evaluate each expression, rounded to four decimal places.

51. $\log 50$ **52.** $\log 47$

53. $\ln 1.6$ **54.** $\ln 0.75$

55. $\ln 225$ **56.** $\ln 381$

57. $\log \sqrt{37}$ **58.** $\log 4\pi$

III.

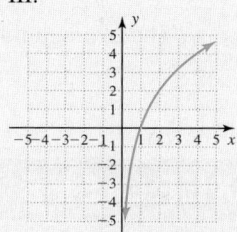

IV.

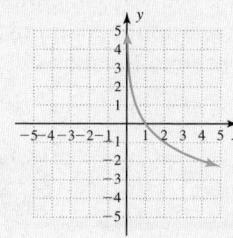

Graph each function *using transformations* of $y = \log_b x$ and strategically plotting a few points. Clearly state the transformations applied.

59. $f(x) = \log_2 x + 3$ **60.** $g(x) = \log_2(x - 2)$

61. $h(x) = \log_2(x - 2) + 3$ **62.** $p(x) = \log_3 x - 2$

63. $q(x) = \ln(x + 1)$ **64.** $r(x) = \ln(x + 1) - 2$

65. $Y_1 = -\ln(x + 1)$ **66.** $Y_2 = -\ln x + 2$

V.

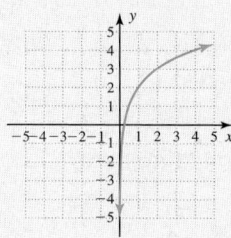

VI.

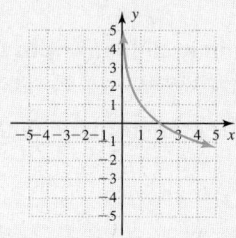

Use the transformation equation $y = af(x \pm h) \pm k$ and the asymptotes and intercept(s) of the parent function to match each equation to one of the graphs given.

67. $y = \log_b(x + 2)$ **68.** $y = 2\log_b x$

69. $y = 1 - \log_b x$ **70.** $y = \log_b x - 1$

71. $y = \log_b x + 2$ **72.** $y = -\log_b x$

Determine the domain of the following functions.

73. $y = \log_6\left(\dfrac{x + 1}{x - 3}\right)$ **74.** $y = \ln\left(\dfrac{x - 2}{x + 3}\right)$

75. $y = \log_5 \sqrt{2x - 3}$ **76.** $y = \ln\sqrt{5 - 3x}$

77. $y = \log(9 - x^2)$ **78.** $y = \ln(9x - x^2)$

▶ **WORKING THE FORMULAS**

79. pH level: $f(x) = -\log_{10} x$

The pH level of a solution indicates the concentration of hydrogen (H^+) ions in a unit called *moles per liter.* The pH level $f(x)$ is given by the formula shown, where x is the ion concentration (given in scientific notation). A solution with pH < 7 is called an acid (lemon juice: pH ≈ 2), and a solution with pH > 7 is called a base (household ammonia: pH ≈ 11). Use the formula to determine the pH level of tomato juice if $x = 7.94 \times 10^{-5}$ moles per liter. Is this an acid or base solution?

80. Time required for an investment to double:

$$T(r) = \frac{\log 2}{\log(1 + r)}$$

The time required for an investment to double in value is given by the formula shown, where r represents the interest rate (expressed as a decimal) and $T(r)$ gives the years required. How long would it take an investment to double if the interest rate were (a) 5%, (b) 8%, (c) 12%?

🔼 ▶ **APPLICATIONS**

Earthquake intensity: Use the information provided in Example 8 to answer the following.

81. Find the value of $M(I)$ given

   a. $I = 50{,}000I_0$ and   **b.** $I = 75{,}000I_0$.

82. Find the intensity I of the earthquake given

   a. $M(I) = 3.2$ and   **b.** $M(I) = 8.1$.

83. Earthquake intensity: On June 25, 1989, an earthquake with magnitude 6.2 shook the southeast side of the Island of Hawaii (near Kalapana), causing some $1,000,000 in damage. On October 15, 2006, an earthquake measuring 6.7 on the Richter scale shook the northwest side of the island, causing over $100,000,000 in damage. How much more intense was the 2006 quake?

84. Earthquake intensity: The most intense earthquake of the modern era occurred in Chile on May 22, 1960, and measured 9.5 on the Richter scale. How many times more intense was this earthquake, than the quake that hit Northern Sumatra (Indonesia) on March 28, 2005, and measured 8.7?

Brightness of a star: The brightness or intensity I of a star as perceived by the naked eye is measured in units called *magnitudes*. The brightest stars have magnitude 1 $[M(I) = 1]$ and the dimmest have magnitude 6 $[M(I) = 6]$. The magnitude of a star is given by the equation $M(I) = 6 - 2.5 \cdot \log\left(\dfrac{I}{I_0}\right)$, where I is the actual intensity of light from the star and I_0 is the faintest light visible to the human eye, called the reference intensity. The intensity I is often given as a multiple of this reference intensity.

85. Find the value of $M(I)$ given

   a. $I = 27I_0$ and   **b.** $I = 85I_0$.

86. Find the intensity I of a star given

   a. $M(I) = 1.6$ and   **b.** $M(I) = 5.2$.

Intensity of sound: The intensity of sound as perceived by the human ear is measured in units called decibels (dB). The loudest sounds that can be withstood without damage to the eardrum are in the 120- to 130-dB range, while a whisper may measure in the 15- to 20-dB range. Decibel measure is given by the equation $D(I) = 10 \log\left(\dfrac{I}{I_0}\right)$, where I is the actual intensity of the sound and I_0 is the faintest sound perceptible by the human ear— called the reference intensity. The intensity I is often given as a multiple of this reference intensity, but often the constant 10^{-16} (watts per cm^2; W/cm^2) is used as the threshold of audibility.

87. Find the value of $D(I)$ given

   a. $I = 10^{-14}$ and   **b.** $I = 10^{-4}$.

88. Find the intensity I of the sound given

   a. $D(I) = 83$ and   **b.** $D(I) = 125$.

89. Sound intensity of a hair dryer: Every morning (it seems), Jose is awakened by the mind-jarring, ear-jamming sound of his daughter's hair dryer (75 dB). He knew he was exaggerating, but told her (many times) of how it reminded him of his railroad days, when the air compressor for the pneumatic tools was running (110 dB). In fact, how many times more intense was the sound of the air compressor compared to the sound of the hair dryer?

90. Sound intensity of a busy street: The decibel level of noisy, downtown traffic has been estimated at 87 dB, while the laughter and banter at a loud party might be in the 60 dB range. How many times more intense is the sound of the downtown traffic?

The *barometric equation* $H = (30T + 8000)\ln\left(\dfrac{P_0}{P}\right)$ was discussed in Example 9.

91. Temperature and atmospheric pressure: Determine the height of Mount McKinley (Alaska), if the temperature at the summit is $-10°$C, with a barometric reading of 34 cmHg.

92. Temperature and atmospheric pressure: A large passenger plane is flying cross-country. The instruments on board show an air temperature of 3°C, with a barometric pressure of 22 cmHg. What is the altitude of the plane?

93. Altitude and atmospheric pressure: By definition, a mountain pass is a low point between two mountains. Passes may be very short with steep slopes, or as large as a valley between two peaks. Perhaps the highest drivable pass in the world is the Semo La pass in central Tibet. At its highest elevation, a temperature reading of 8°C was taken, along with a barometer reading of 39.3 cmHg. (a) Approximately how high is the Semo La pass? (b) While traveling up to this pass, an elevation marker is seen. If the barometer reading was 47.1 cmHg at a temperature of 12°C, what height did the marker give?

94. **Altitude and atmospheric pressure:** Hikers on Mt. Everest take successive readings of 35 cmHg at 5°C and 30 cmHg at −10°C. (a) How far up the mountain are they at each reading? (b) Approximate the height of Mt. Everest if the temperature at the summit is −27°C and the barometric pressure is 22.2 cmHg.

95. **Marketing budgets:** An advertising agency has determined the number of items sold by a certain client is modeled by the equation $N(A) = 1500 + 315 \ln A$, where $N(A)$ represents the number of sales after spending A thousands of dollars on advertising. Determine the approximate number of items sold on an advertising budget of (a) $10,000; (b) $50,000. (c) Use the TABLE feature of a calculator to estimate how large a budget is needed (to the nearest $500 dollars) to sell 3000 items. (d) This company is willing to continue advertising as long as eight additional sales are gained for every $1000 spent: $\dfrac{\Delta N}{\Delta A} = \dfrac{8}{1}$. Show this occurs by spending between $39,300 to $39,400.

96. **Sports promotions:** The accountants for a major boxing promoter have determined that the number of pay-per-view subscriptions sold to their championship bouts can be modeled by the function $N(d) = 15,000 + 5850 \ln d$, where $N(d)$ represents the number of subscriptions sold after spending d thousand dollars on promotional activities. Determine the number of subscriptions sold if (a) $50,000 and (b) $100,000 is spent. (c) Use the TABLE feature of a calculator to estimate how much should be spent (to the nearest $1000 dollars) to sell over 50,000 subscriptions. (d) This promoter is willing to continue promotional spending as long as 14 additional subscriptions are sold for every $1000 spent: $\dfrac{\Delta N}{\Delta d} = \dfrac{14}{1}$. Show this occurs by spending between $417,800 and $417,900.

97. **Home ventilation:** In the construction of new housing, there is considerable emphasis placed on correct ventilation. If too little outdoor air enters a home, pollutants can sometimes accumulate to levels that pose a health risk. For homes of various sizes, ventilation requirements have been established and are based on floor area and the number of bedrooms. For a three-bedroom home,

the relationship can be modeled by the function $C(x) = 42 \ln x - 270$, where $C(x)$ represents the number of cubic feet of air per minute (cfm) that should be exchanged with outside air in a home with floor area x (in square feet). (a) How many cfm of exchanged air are needed for a three-bedroom home with a floor area of 2500 ft^2? (b) If a three-bedroom home is being mechanically ventilated by a system with 40 cfm capacity, what is the square footage of the home, assuming it is built to code?

98. **Runway takeoff distance:** Many will remember the August 27, 2006, crash of a commuter jet at Lexington's Blue Grass Airport, that was mistakenly trying to take off on a runway that was just too short. Forty-nine lives were lost. The minimum required length of a runway depends on the maximum allowable takeoff weight (mtw) of a specific plane. This relationship can be approximated by the function

$L(x) = 2085 \ln x - 14,900$, where $L(x)$ represents the required length of a runway in feet, for a plane with x mtw in pounds.

   a. The Airbus-320 has a 169,750 lb mtw. What minimum runway length is required for takeoff?

   b. A Learjet 30 model requires a runway of 5550 ft to takeoff safely. What is its mtw?

Memory retention: Under certain conditions, a person's retention of random facts can be modeled by the equation $P(x) = 95 - 14 \log_2 x$, where $P(x)$ is the percentage of those facts retained after x number of days. Find the percentage of facts a person might retain after:

99. **a.** 1 day   **b.** 4 days   **c.** 16 days

100. **a.** 32 days   **b.** 64 days   **c.** 78 days

101. **pH level:** Use the formula given in Exercise 79 to determine the pH level of black coffee if $x = 5.1 \times 10^{-5}$ moles per liter. Is black coffee considered an acid or base solution?

102. The length of time required for an amount of money to *triple* is given by the formula $T(r) = \dfrac{\log 3}{\log(1 + r)}$ (refer to Exercise 80). Construct a table of values to help estimate what interest rate is needed for an investment to triple in nine years.

▶ EXTENDING THE CONCEPT

103. Many texts and reference books give estimates of the noise level (in decibels dB) of common sounds. Through reading and research, try to locate or approximate where the following sounds would fall along this scale. In addition, determine at what point pain or ear damage begins to occur.

     a. threshold of audibility   **b.** lawn mower

     c. whisper   **d.** loud rock concert

     e. lively party   **f.** jet engine

104. Determine the value of x that makes the equation true: $\log_3[\log_3(\log_3 x)] = 0$.

105. Find the value of each expression without using a calculator.

     a. $\log_{64}\frac{1}{16}$ **b.** $\log_{\frac{4}{9}}\frac{27}{8}$ **c.** $\log_{0.25}32$

106. Suppose you and I represent two different numbers. Is the following cryptogram true or false? *The log of me base me is one and the log of you base you is one, but the log of you base me is equal to the log of me base you turned upside down.*

▶ MAINTAINING YOUR SKILLS

107. (3.3) Graph $g(x) = \sqrt[3]{x + 2} - 1$ by shifting the parent function. Then state the domain and range of g.

108. (R.4) Factor the following expressions:

     a. $x^3 - 8$ **b.** $a^2 - 49$

     c. $n^2 - 10n + 25$ **d.** $2b^2 - 7b + 6$

109. (3.4/3.7) For the graph shown, write the solution set for $f(x) < 0$. Then write the equation of the graph in factored form and in polynomial form.

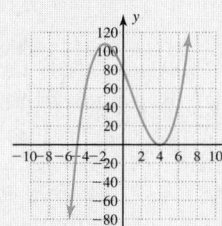

110. (2.2) A function $f(x)$ is defined by the ordered pairs shown in the table. Is the function (a) linear? (b) increasing? Justify your answers.

x	y
-10	0
-9	-2
-8	-8
-6	-18
-5	-50
-4	-72

MID-CHAPTER CHECK

1. Write the following in logarithmic form.

     a. $27^{\frac{2}{3}} = 9$ **b.** $81^{\frac{5}{4}} = 243$

2. Write the following in exponential form.

     a. $\log_8 32 = \frac{5}{3}$ **b.** $\log_{1296}6 = 0.25$

3. Solve each equation for the unknown:

     a. $4^{2x} = 32^{x-1}$ **b.** $\left(\frac{1}{3}\right)^{4b} = 9^{2b-5}$

4. Solve each equation for the unknown:

     a. $\log_{27}x = \frac{1}{3}$ **b.** $\log_b 125 = 3$

 5. The homes in a popular neighborhood are growing in value according to the formula $V(t) = V_0(\frac{9}{8})^t$, where t is the time in years, V_0 is the purchase price of the home, and $V(t)$ is the current value of the home. (a) In 3 yr, how much will a $50,000 home be worth? (b) Use the TABLE feature of your calculator to estimate how many years (to the nearest year) until the home doubles in value.

6. The graph of the function $f(x) = 5^x$ has been shifted right 3 units, up 2 units, and stretched by a factor of 4. What is the equation of the resulting function?

7. State the domain and range for $f(x) = \sqrt{x - 3} + 1$, then find $f^{-1}(x)$ and state its domain and range. Verify the inverse relationship using composition.

8. Write the following equations in logarithmic form, then verify the result on a calculator.

     a. $81 = 3^4$ **b.** $e^4 \approx 54.598$

9. Write the following equations in exponential form, then verify the result on a calculator.

   a. $\dfrac{2}{3} = \log_{27} 9$

   b. $1.4 \approx \ln 4.0552$

10. On August 15, 2007, an earthquake measuring 8.0 on the Richter scale struck coastal Peru. On October 17, 1989, right before Game 3 of the World Series between the Oakland A's and the San Francisco Giants, the Loma Prieta earthquake, measuring 7.1 on the Richter scale, struck the San Francisco Bay area. How much more intense was the Peruvian earthquake?

REINFORCING BASIC CONCEPTS

Linear and Logarithm Scales

The use of logarithmic scales as a tool of measurement is primarily due to the range of values for the phenomenon being measured. For instance, time is generally measured on a linear scale, and for short periods a linear scale is appropriate. For the time line in Figure 4.27, each tick-mark *represents 1 unit,* and the time line can display a period of 10 yr. However, the scale would be useless in a study of world history or geology. If we scale the number line logarithmically, each tick-mark *represents a power of 10* (Figure 4.28) and a scale of the same length can now display a time period of 10 billion years.

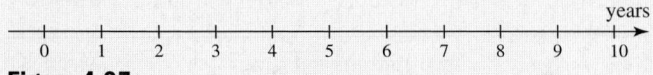

Figure 4.27

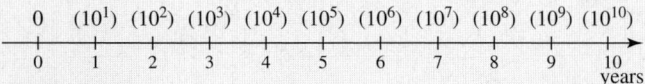

Figure 4.28

In much the same way, logarithmic measures are needed in a study of sound and earthquake intensity, as the scream of a jet engine is over 1 billion times more intense than the threshold of hearing, and the most destructive earthquakes are billions of times stronger than the slightest earth movement that can be felt. Figures 4.29 and 4.30 show logarithmic scales for measuring sound in decibels (1 bel = 10 decibels) and earthquake intensity in Richter values (or magnitudes).

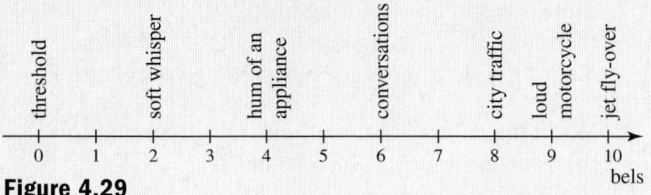

Figure 4.29

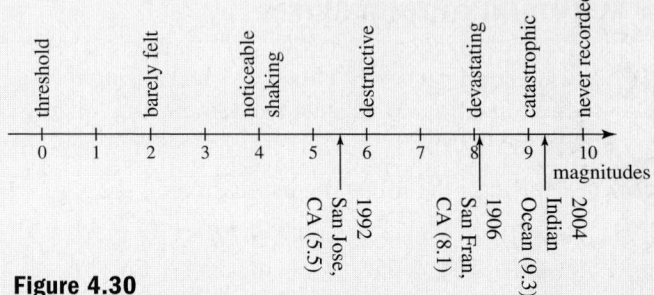

Figure 4.30

As you view these scales, remember that each unit increase represents a power of 10. For instance, the 1906 San Francisco earthquake was $8.1 - 5.5 = 2.6$ magnitudes greater than the San Jose quake of 1992, meaning it was $10^{2.6} \approx$ *398 times more intense.* Use this information to complete the following exercises. Determine how many times more intense the first sound is compared to the second.

Exercise 1: jet engine: 14 bels
             rock concert: 11.8 bels

Exercise 2: pneumatic hammer: 11.2 bels
             heavy lawn mower: 8.5 bels

Exercise 3: train horn: 7.5 bels
             soft music: 3.4 bels

Determine how many times more intense the first quake was compared to the second.

Exercise 4: Great Chilean quake (1960): magnitude 9.5
             Kobe, Japan, quake (1995): magnitude 6.9

Exercise 5: Northern Sumatra (2004): magnitude 9.1
             Southern, Greece (2008): magnitude 4.5

Properties of Logarithms; Solving Exponential/ Logarithmic Equations

Learning Objectives

In Section 4.4 you will learn how to:

☐ **A.** Solve logarithmic equations using the fundamental properties of logarithms

☐ **B.** Apply the product, quotient, and power properties of logarithms

☐ **C.** Solve general logarithmic and exponential equations

☐ **D.** Solve applications involving logistic, exponential, and logarithmic functions

In this section, we develop the ability to solve logarithmic and exponential equations of any base. A **logarithmic equation** has at least one term that involves the logarithm of a variable. Likewise, an **exponential equation** has at least one term that involves a variable exponent. In the same way that we might *square both sides* or *divide both sides* of an equation in the solution process, we'll show that we can also *exponentiate both sides* or *take logarithms of both sides* to help obtain a solution.

A. Solving Equations Using the Fundamental Properties of Logarithms

In Section 4.3, we converted expressions from exponential form to logarithmic form using the basic definition: $x = b^y \Leftrightarrow y = \log_b x$. This relationship reveals the following four properties:

Fundamental Properties of Logarithms

For any base $b > 0$, $b \neq 1$,

  I. $\log_b b = 1$, since $b^1 = b$
  II. $\log_b 1 = 0$, since $b^0 = 1$
  III. $\log_b b^x = x$, since $b^x = b^x$ (exponential form)
  IV. $b^{\log_b x} = x$, since $\log_b x = \log_b x$ (logarithmic form)

To see the verification of Property IV more clearly, again note that for $y = \log_b x$, $b^y = x$ is the exponential form, and substituting $\log_b x$ for y yields $b^{\log_b x} = x$. Also note that Properties III and IV demonstrate that $y = \log_b x$ and $y = b^x$ are inverse functions. In common language, "a base-b logarithm *undoes* a base-b exponential," and "a base-b exponential *undoes* a base-b logarithm." For $f(x) = \log_b x$ and $f^{-1}(x) = b^x$, using a composition verifies the inverse relationship:

$$(f \circ f^{-1})(x) = f[f^{-1}(x)] \qquad (f^{-1} \circ f)(x) = f^{-1}[f(x)]$$
$$= \log_b b^x \qquad\qquad\qquad = b^{\log_b x}$$
$$= x \qquad\qquad\qquad\qquad = x$$

These properties can be used to solve basic equations involving logarithms and exponentials.

EXAMPLE 1 ▶ **Solving Basic Logarithmic Equations**

Solve each equation by applying fundamental properties. Answer in exact form and approximate form using a calculator (round to 1000ths).

  a. $\ln x = 2$ **b.** $-0.52 = \log x$

Solution ▶ **a.** $\ln x = 2$ \qquad given
  $e^{\ln x} = e^2$ \qquad exponentiate both sides
  $x = e^2$ \qquad Property IV
  ≈ 7.389 \qquad result

  b. $-0.52 = \log x$ \qquad given
  $10^{-0.52} = 10^{\log x}$ \qquad exponentiate both sides
  $10^{-0.52} = x$ \qquad Property IV
  $0.302 \approx x$ \qquad result

Now try Exercises 7 through 10 ▶

Note that exponentiating both sides of the equation produced the same result as simply writing the original equation in exponential form, and either approach can be used.

EXAMPLE 2 ▶ **Solving Basic Exponential Equations**

Solve each equation by applying fundamental properties. Answer in exact form and approximate form using a calculator (round to 1000ths).

a. $e^x = 167$ b. $10^x = 8.223$

Solution ▶ a. $e^x = 167$ given

$\ln e^x = \ln 167$ take natural log of both sides

$x = \ln 167$ Property III

$x \approx 5.118$ result

b. $10^x = 8.223$ given

$\log 10^x = \log 8.223$ take common log of both sides

$x = \log 8.223$ Property III

$x \approx 0.915$ result

Now try Exercises 11 through 14 ▶

Similar to our previous observation, taking the logarithm of both sides produced the same result as writing the original equation in logarithmic form, and either approach can be used.

If an equation has a single logarithmic or exponential term (base 10 or base e), the equation can be solved by isolating this term and applying one of the fundamental properties.

EXAMPLE 3 ▶ **Solving Exponential Equations**

Solve each equation. Write answers in exact form and approximate form to four decimal places.

a. $10^x - 29 = 51$ b. $3e^{x+1} - 5 = 7$

Solution ▶ a. $10^x - 29 = 51$ given

$10^x = 80$ add 29

Since the left-hand side is base 10, we apply a common logarithm.

$\log 10^x = \log 80$ take the common log of both sides

$x = \log 80$ Property III (exact form)

≈ 1.9031 approximate form

b. $3e^{x+1} - 5 = 7$ given

$3e^{x+1} = 12$ add 5

$e^{x+1} = 4$ divide by 3

Since the left-hand side is base e, we apply a natural logarithm.

$\ln e^{x+1} = \ln 4$ take the natural log of both sides

$x + 1 = \ln 4$ Property III

$x = \ln 4 - 1$ solve for x (exact form)

≈ 0.3863 approximate form

Now try Exercises 15 through 20 ▶

WORTHY OF NOTE

To check solutions using a calculator, we can $\boxed{\text{STO▸}}$ (store) the exact result in storage location $\boxed{x, T, \theta, n}$ (the function variable x) and simply enter the original equation on the home screen. The figure shows this verification for Example 3b.

```
ln(4)-1→X
          .3862943611
3e^(X+1)-5
                   7
```

EXAMPLE 4 ▶   **Solving Logarithmic Equations**

Solve each equation. Write answers in exact form and approximate form to four decimal places.

a. $2 \log(7x) + 1 = 4$ **b.** $-4 \ln(x + 1) - 5 = 7$

Solution ▶   **a.**

$2 \log(7x) + 1 = 4$	given
$2 \log(7x) = 3$	subtract 1
$\log(7x) = \dfrac{3}{2}$	divide by 2
$7x = 10^{\frac{3}{2}}$	exponential form
$x = \dfrac{10^{\frac{3}{2}}}{7}$	divide by 7 (exact form)
≈ 4.5175	approximate form

b.

$-4 \ln(x + 1) - 5 = 7$	given
$-4 \ln(x + 1) = 12$	add 5
$\ln(x + 1) = -3$	divide by -4
$x + 1 = e^{-3}$	exponential form
$x = e^{-3} - 1$	subtract 1 (exact form)
≈ -0.9502	approximate form

☑ **A.** You've just learned how to solve logarithmic equations using the fundamental properties of logarithms

Now try Exercises 21 through 26 ▶

GRAPHICAL SUPPORT

Solutions can be also checked using the intersection of graphs method (Technology Highlight, page 432). For Example 4a, enter $Y_1 = 2 \log(7x) + 1$ and $Y_2 = 4$ on the $\boxed{Y=}$ screen. From the domain of the function and the expected answer, we set a window that includes only Quadrant I. Use the keystrokes $\boxed{\text{2nd}}$ $\boxed{\text{TRACE}}$ **(CALC) 5:intersect**, and identify each graph by pressing $\boxed{\text{ENTER}}$ 3 times. The calculator will find the point of intersection and display it at the bottom of the screen.

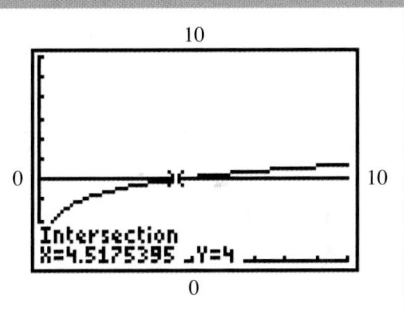

B. The Product, Quotient, and Power Properties of Logarithms

Generally speaking, equation solving involves simplifying the equation, isolating a variable term on one side, and applying an inverse to solve for the unknown. For logarithmic equations such as $\log x + \log(x + 3) = 1$, we must find a way to combine the terms on the left, before we can work toward a solution. This requires a further exploration of logarithmic properties.

Due to the close connection between exponents and logarithms, their properties are very similar. To illustrate, we'll use terms that can all be written in the form 2^x, and write the equations $8 \cdot 4 = 32$, $\frac{8}{4} = 2$, and $8^2 = 64$ in both exponential form and logarithmic form.

The exponents from a product are added:	exponential form:	$2^3 \cdot 2^2 = 2^{3+2}$
	logarithmic form:	$\log_2(8 \cdot 4) = \log_2 8 + \log_2 4$
The exponents from a quotient are subtracted:	exponential form:	$\dfrac{2^3}{2^2} = 2^{3-2}$
	logarithmic form:	$\log_2\left(\dfrac{8}{4}\right) = \log_2 8 - \log_2 4$

The exponents from a power are multiplied:

exponential form: $(2^3)^2 = 2^{3 \cdot 2}$

logarithmic form: $\log_2 8^2 = 2 \cdot \log_2 8$

Each illustration can be generalized and applied with any base b.

Properties of Logarithms

Give M, N, and $b \neq 1$ are *positive* real numbers, and *any* real number p.

Product Property

$\log_b(MN) = \log_b M + \log_b N$

The log of a product is a sum of logarithms.

Quotient Property

$\log_b\left(\dfrac{M}{N}\right) = \log_b M - \log_b N$

The log of a quotient is a difference of logarithms.

Power Property

$\log_b M^p = p\log_b M$

The log of a quantity to a power is the power times the log of the quantity.

Proof of the Product Property

Given M, N, and $b \neq 1$ are positive real numbers, $\log_b(MN) = \log_b M + \log_b N$.

For $P = \log_b M$ and $Q = \log_b N$, we have $b^P = M$ and $b^Q = N$ in exponential form. It follows that

$$\begin{aligned}
\log_b(MN) &= \log_b(b^P b^Q) && \text{substitute } b^P \text{ for } M \text{ and } b^Q \text{ for } N \\
&= \log_b(b^{P+Q}) && \text{properties of exponents} \\
&= P + Q && \text{log property 3} \\
&= \log_b M + \log_b N && \text{substitute } \log_b M \text{ for } P \text{ and } \log_b N \text{ for } Q
\end{aligned}$$

Proof of the Quotient Property

Given M, N, and $b \neq 1$ are positive real numbers, $\log_b\left(\dfrac{M}{N}\right) = \log_b M - \log_b N$.

For $P = \log_b M$ and $Q = \log_b N$, we have $b^P = M$ and $b^Q = N$ in exponential form. It follows that

$$\begin{aligned}
\log_b\left(\frac{M}{N}\right) &= \log_b\left(\frac{b^P}{b^Q}\right) && \text{substitute } b^P \text{ for } M \text{ and } b^Q \text{ for } N \\
&= \log_b(b^{P-Q}) && \text{properties of exponents} \\
&= P - Q && \text{log property 3} \\
&= \log_b M - \log_b N && \text{substitute } \log_b M \text{ for } P \text{ and } \log_b N \text{ for } Q
\end{aligned}$$

Proof of the Power Property

Given M and $b \neq 1$ are positive real numbers and any real number x, $\log_b M^x = x \log_b M$.

For $P = \log_b M$, we have $b^P = M$ in exponential form. It follows that

$$\begin{aligned}
\log_b(M)^x &= \log_b(b^P)^x && \text{substitute } b^P \text{ for } M \\
&= \log_b(b^{Px}) && \text{properties of exponents} \\
&= Px && \text{log property 3} \\
&= (\log_b M)x && \text{substitute } \log_b M \text{ for } P \\
&= x \log_b M && \text{rewrite factors}
\end{aligned}$$

CAUTION ▶ It's very important that you read and understand these properties correctly. In particular, note that $\log_b(M + N) \neq \log_b M + \log_b N$, and $\log_b\left(\dfrac{M}{N}\right) \neq \dfrac{\log_b M}{\log_b N}$. In the first case, it might help to compare the statement with $f(x + 3)$, which represents a horizontal shift of the graph 3 units left, and in particular, $f(x + 3) \neq f(x) + f(3)$.

In many cases, these properties are applied to consolidate logarithmic terms in preparation for equation solving.

EXAMPLE 5 ▶ **Rewriting Expressions Using Logarithmic Properties**

Use the properties of logarithms to write each expression as a single term.
 a. $\log_2 7 + \log_2 5$ **b.** $2 \ln x + \ln(x + 6)$ **c.** $\ln(x + 2) - \ln x$

Solution ▶ **a.** $\log_2 7 + \log_2 5 = \log_2(7 \cdot 5)$ product property
         $= \log_2 35$ simplify
 b. $2 \ln x + \ln(x + 6) = \ln x^2 + \ln(x + 6)$ power property
          $= \ln[x^2(x + 6)]$ product property
          $= \ln[x^3 + 6x^2]$ simplify
 c. $\ln(x + 2) - \ln x = \ln\left(\dfrac{x + 2}{x}\right)$ quotient property

Now try Exercises 27 through 42 ▶

EXAMPLE 6 ▶ **Rewriting Logarithmic Expressions Using the Power Property**

Use the power property of logarithms to rewrite each term as a product.
 a. $\ln 5^x$ **b.** $\log 32^{x+2}$ **c.** $\log \sqrt{x}$

Solution ▶ **a.** $\ln 5^x = x \ln 5$ power property
 b. $\log 32^{x+2} = (x + 2)\log 32$ power property (note use of parentheses)
 c. $\log \sqrt{x} = \log x^{\frac{1}{2}}$ write radical using a rational exponent

    $= \dfrac{1}{2}\log x$ power property

Now try Exercises 43 through 50 ▶

 CAUTION ▶ Note from Example 6b that parentheses *must be used* whenever the exponent is a sum or difference. There is a huge difference between $(x + 2)\log 32$ and $x + 2\log 32$.

In other cases, these properties help rewrite an expression so that certain procedures can be applied more easily. Example 7 actually lays the foundation for more advanced work in mathematics.

EXAMPLE 7 ▶ **Rewriting Expressions Using Logarithmic Properties**

Use the properties of logarithms to write the following expressions as a sum or difference of simple logarithmic terms.

 a. $\log(x^2 z)$ **b.** $\ln \sqrt{\dfrac{x}{x + 5}}$ **c.** $\ln\left[\dfrac{e\sqrt{x^2 + 1}}{(2x + 5)^3}\right]$

Solution ▶   **a.** $\log(x^2 z) = \log x^2 + \log z$ product property

$= 2 \log x + \log z$ power property

b. $\ln \sqrt{\dfrac{x}{x+5}} = \ln\left(\dfrac{x}{x+5}\right)^{\frac{1}{2}}$ write radical using a rational exponent

$= \dfrac{1}{2} \ln\left(\dfrac{x}{x+5}\right)$ power property

$= \dfrac{1}{2}[\ln x - \ln(x+5)]$ quotient property

c. $\ln\left[\dfrac{e\sqrt{x^2+1}}{(2x+5)^3}\right] = \ln\left[\dfrac{e(x^2+1)^{\frac{1}{2}}}{(2x+5)^3}\right]$ write radical using a rational exponent

$= \ln[e(x^2+1)^{\frac{1}{2}}] - \ln(2x+5)^3$ quotient property

$= \ln e + \ln(x^2+1)^{\frac{1}{2}} - \ln(2x+3)^3$ product property

$= 1 + \dfrac{1}{2}\ln(x^2+1) - 3\ln(2x+3)$ power property

Now try Exercises 51 through 60 ▶

Although base-10 and base-e logarithms dominate the mathematical landscape, there are many practical applications that use other bases. Fortunately, a formula exists that will convert any given base into either base 10 or base e. It's called the **change-of-base formula.**

Change-of-Base Formula

For the positive real numbers M, a, and b, with $a, b \neq 1$,

$$\log_b M = \frac{\log M}{\log b} \qquad\qquad \log_b M = \frac{\ln M}{\ln b} \qquad\qquad \log_b M = \frac{\log_a M}{\log_a b}$$

       base 10   base e arbitrary base a

Proof of the Change-of-Base Formula:

For $y = \log_b M$, we have $b^y = M$ in exponential form. It follows that

$$\log_a(b^y) = \log_a M \qquad \text{take base-}a\text{ logarithm of both sides}$$

$$y \log_a b = \log_a M \qquad \text{power property of logarithms}$$

$$y = \frac{\log_a M}{\log_a b} \qquad \text{divide by } \log_a b$$

$$\log_b M = \frac{\log_a M}{\log_a b} \qquad \text{substitute } \log_b M \text{ for } y$$

EXAMPLE 8 ▶   **Using the Change-of-Base Formula to Evaluate Expressions**

Find the value of each expression using the change-of-base formula. Answer in exact form and approximate form using nine digits, then *verify the result* using the original base.

     a. $\log_3 29$ **b.** $\log_5 3.6$

Solution ▶   **a.** $\log_3 29 = \dfrac{\log 29}{\log 3}$ **b.** $\log_5 3.6 = \dfrac{\log 3.6}{\log 5}$

$= 3.065044752$ $= 0.795888947$

Check: $3^{3.065044752} = 29$ ✓   **Check:** $5^{0.795888947} = 3.6$ ✓

Now try Exercises 61 through 72 ▶

The change-of-base formula can also be used to study and graph logarithmic functions of *any* base. For $y = \log_b x$, the right-hand expression is simply rewritten using the formula and the equivalent function is $y = \dfrac{\log x}{\log b}$. The new function can then be evaluated as in Example 8, or used to study the graph of $y = \log_b x$ for any base b.

✓ **B.** You've just learned how to apply the product, quotient, and power properties of logarithms

C. Solving Logarithmic Equations

One of the most common mistakes in solving exponential and logarithmic equations is to apply the inverse function too early—before the equation has been simplified. In addition, since the domain of $y = \log_b x$ is $x > 0$, logarithmic equations can sometimes produce **extraneous roots,** and checking all answers is a good practice. We'll illustrate by solving the equation mentioned earlier: $\log x + \log (x + 3) = 1$.

EXAMPLE 9 ▶ Solving a Logarithmic Equation

Solve for x and check your answer: $\log x + \log (x + 3) = 1$.

Solution ▶

$$
\begin{aligned}
\log x + \log (x + 3) &= 1 && \text{original equation} \\
\log [x(x + 3)] &= 1 && \text{product property} \\
x^2 + 3x &= 10^1 && \text{exponential form, distribute } x \\
x^2 + 3x - 10 &= 0 && \text{set equal to 0} \\
(x + 5)(x - 2) &= 0 && \text{factor} \\
x = -5 \text{ or } x &= 2 && \text{result}
\end{aligned}
$$

Check: The "solution" $x = -5$ is outside the domain and is discarded. For $x = 2$,

$$
\begin{aligned}
\log x + \log (x + 3) &= 1 && \text{original equation} \\
\log 2 + \log (2 + 3) &= 1 && \text{substitute 2 for } x \\
\log 2 + \log 5 &= 1 && \text{simplify} \\
\log (2 \cdot 5) &= 1 && \text{product property} \\
\log 10 &= 1 && \text{Property I}
\end{aligned}
$$

Now try Exercises 73 through 80 ▶

As an alternative check, you could also use a calculator to verify $\log 2 + \log 5 = 1$ directly.

If the simplified form of an equation yields a logarithmic term on both sides, the **uniqueness property of logarithms** provides an efficient way to work toward a solution. Since logarithmic functions are one-to-one, we have

The Uniqueness Property of Logarithms

For positive real numbers m, n, and $b \neq 1$,

$$\text{If } \log_b m = \log_b n, \quad \text{then } m = n$$

Equal bases imply equal arguments.

EXAMPLE 10 ▶ Solving Logarithmic Equations Using the Uniqueness Property

Solve each equation using the uniqueness property.

a. $\log (x + 2) = \log 7 + \log x$ **b.** $\ln 87 - \ln x = \ln 29$

Solution ▶ **a.** $\log(x + 2) = \log 7 + \log x$

$\log(x + 2) = \log 7x$ properties of logarithms

$x + 2 = 7x$ uniqueness property

$2 = 6x$ solve for x

$\dfrac{1}{3} = x$ result

b. $\ln 87 - \ln x = \ln 29$

$\ln\left(\dfrac{87}{x}\right) = \ln 29$

$\dfrac{87}{x} = 29$

$87 = 29x$

$3 = x$

The checks are left to the student.

> **WORTHY OF NOTE**
>
> The uniqueness property can also be viewed as exponentiating both sides using the appropriate base, then applying Property IV.

Now try Exercises 81 through 86 ▶

Often the solution may depend on using a variety of algebraic skills in addition to logarithmic or exponential properties.

EXAMPLE 11 ▶ **Solving Logarithmic Equations**

Solve each equation and check your answers.

a. $\ln(x + 7) - 2\ln 5 = 0.9$ **b.** $\log(x + 12) - \log x = \log(x + 9)$

Solution ▶ **a.** $\ln(x + 7) - 2\ln 5 = 0.9$ given

$\ln(x + 7) - \ln 5^2 = 0.9$ power property

$\ln\left(\dfrac{x + 7}{25}\right) = 0.9$ quotient property

$\dfrac{x + 7}{25} = e^{0.9}$ exponential form

$x + 7 = 25e^{0.9}$ clear denominator

$x = 25e^{0.9} - 7$ solve for x (exact form)

≈ 54.49 approximate form (to 100ths)

Check: $\ln(x + 7) - 2\ln 5 = 0.9$ original equation

$\ln(54.49 + 7) - 2\ln 5 \approx 0.9$ substitute 54.49 for x

$\ln 61.49 - 2\ln 5 \approx 0.9$ simplify

$0.9 \approx 0.9$ ✓   result checks

b. $\log(x + 12) - \log x = \log(x + 9)$ given equation

$\log\left(\dfrac{x + 12}{x}\right) = \log(x + 9)$ quotient property

$\dfrac{x + 12}{x} = x + 9$ uniqueness property

$x + 12 = x^2 + 9x$ clear denominator

$0 = x^2 + 8x - 12$ set equal to 0

The equation is not factorable, and the quadratic formula must be used.

$x = \dfrac{-b \pm \sqrt{b^2 - 4ac}}{2a}$ quadratic formula

$= \dfrac{-8 \pm \sqrt{(8)^2 - 4(1)(-12)}}{2(1)}$ substitute 1 for a, 8 for b, -12 for c

$= \dfrac{-8 \pm \sqrt{112}}{2} = \dfrac{-8 \pm 4\sqrt{7}}{2}$ simplify

$= -4 \pm 2\sqrt{7}$ result

> **WORTHY OF NOTE**
>
> If all digits given by your calculator are used in the check, a calculator will generally produce "exact" answers. Try using the solution $x = 54.49007778$ in Example 11a by substituting directly, or by storing the result of the original computation and using your home screen.

Substitution shows $x = -4 + 2\sqrt{7}$ ($x \approx 1.29150$) checks, but substituting $-4 - 2\sqrt{7}$ for x gives $\log(2.7085) - \log(-9.2915) = \log(-0.2915)$ and two of the three terms do not represent real numbers ($x = -4 - 2\sqrt{7}$ is an extraneous root).

> **Now try Exercises 87 through 102 ▶**

GRAPHICAL SUPPORT

Logarithmic equations can also be checked using the intersection of graphs method. For Example 11b, we first enter $\log(x + 12) - \log x$ as Y_1 and $\log(x + 9)$ as Y_2 on the ｜**Y=**｜ screen. Using ｜**2nd**｜ ｜**TRACE**｜ (CALC) 5:intersect, we find the graphs intersect at $x = 1.2915026$, and that *this is the only solution* (knowing the graph's basic shape, we conclude they cannot intersect again).

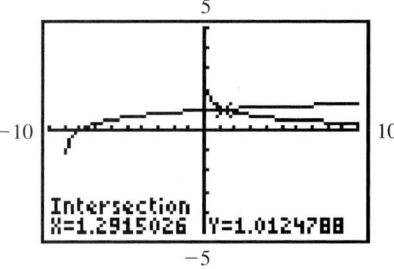

⚠ **CAUTION ▶** Be careful not to dismiss or discard a possible solution simply because it's negative. For the equation $\log(-6 - x) = 1$, $x = -16$ is the solution (the domain here allows negative numbers: $-6 - x > 0$ yields $x < -6$ as the domain). In general, when a logarithmic equation has multiple solutions, all solutions should be checked.

Solving exponential equations likewise involves isolating an exponential term on one side, or writing the equation where exponential terms of like base occur on each side. The latter case can be solved using the uniqueness property. If the exponential base is neither 10 nor e, logarithms of either base can be used along with the Power Property to solve the equation.

EXAMPLE 12 ▶   **Solving an Exponential Equation Using Base 10 or Base e**

Solve the exponential equation. Answer in both exact form, and approximate form to four decimal places: $4^{3x} - 1 = 8$

Solution ▶

$4^{3x} - 1 = 8$ given equation

$4^{3x} = 9$ add 1

WORTHY OF NOTE

The equation $\log 4^{3x} = \log 9$ from Example 12, can actually be solved using the change-of-base property, by taking logarithms base 4 of both sides.

$\log_4 4^{3x} = \log_4 9$ logarithms base 4

$3x = \dfrac{\log 9}{\log 4}$ Property III; change-of-base property

$x = \dfrac{\log 9}{3 \log 4}$ divide by 3

The left-hand side is neither base 10 or base e, so the choice is arbitrary. Here we chose base 10 to solve.

$$\log 4^{3x} = \log 9 \quad \text{take logarithm base 10 of both sides}$$

$$3x \log 4 = \log 9 \quad \text{power property}$$

$$x = \frac{\log 9}{3 \log 4} \quad \text{divide by 3 log 4 (exact form)}$$

$$x \approx 0.5283 \quad \text{approximate form}$$

> **Now try Exercises 103 through 106 ▶**

In some cases, two exponential terms with *unlike* bases may be involved. Here again, either common logs or natural logs can be used, but be sure to distinguish between constant terms like *ln 5* and variable terms like *x ln 5*. As with all equations, the goal is to isolate the *variable terms* on one side, with all constant terms on the other.

EXAMPLE 13 ▶ **Solving an Exponential Equation with Unlike Bases**

Solve the exponential equation $5^{x+1} = 6^{2x}$.

Solution ▶ $5^{x+1} = 6^{2x}$ original equation

Begin by taking the natural log of both sides:

$$\ln(5^{x+1}) = \ln(6^{2x}) \qquad \text{apply base-}e \text{ logarithms}$$
$$(x+1)\ln 5 = 2x\ln 6 \qquad \text{power property}$$
$$x\ln 5 + \ln 5 = 2x\ln 6 \qquad \text{distribute}$$
$$\ln 5 = 2x\ln 6 - x\ln 5 \qquad \text{variable terms to one side}$$
$$\ln 5 = x(2\ln 6 - \ln 5) \qquad \text{factor out } x$$
$$\frac{\ln 5}{2\ln 6 - \ln 5} = x \qquad \text{solve for } x \text{ (exact form)}$$
$$0.8153 \approx x \qquad \text{approximate form}$$

☑ **C. You've just learned how to solve general logarithmic and exponential equations**

The solution can be checked on a calculator.

Now try Exercises 107 through 110 ▶

D. Applications of Logistic, Exponential, and Logarithmic Functions

Applications of exponential and logarithmic functions take many different forms and it would be impossible to illustrate them all. As you work through the exercises, try to adopt a "big picture" approach, applying the general principles illustrated here to other applications. Some may have been introduced in previous sections. The difference here is that we can now *solve for the independent variable,* instead of simply evaluating the relationships.

In applications involving the **logistic growth** of animal populations, the initial stage of growth is virtually exponential, but due to limitations on food, space, or other resources, growth slows and at some point it reaches a limit. In business, the same principle applies to the logistic growth of sales or profits, due to market saturation. In these cases, the exponential term appears in the denominator of a quotient, and we "clear denominators" to begin the solution process.

EXAMPLE 14 ▶ **Solving a Logistics Equation**

A small business makes a new discovery and begins an aggressive advertising campaign, confident they can capture 66% of the market in a short period of time. They anticipate their market share will be modeled by the function

$M(t) = \dfrac{66}{1 + 10e^{-0.05t}}$, where $M(t)$ represents the percentage after t days. Use this function to answer the following.

a. What was the company's initial market share ($t = 0$)? What was their market share 30 days later?

b. How long will it take the company to reach a 60% market share?

Solution ▶ **a.** $M(t) = \dfrac{66}{1 + 10e^{-0.05t}}$ given

$M(0) = \dfrac{66}{1 + 10e^{-0.05(0)}}$ substitute 0 for t

$= \dfrac{66}{11}$ simplify

$= 6$ result

The company originally had only a 6% market share.

$M(30) = \dfrac{66}{1 + 10e^{-0.05(30)}}$ substitute 30 for t

$= \dfrac{66}{1 + 10e^{-1.5}}$ simplify

≈ 20.4 result

After 30 days, they held a 20.4% market share.

b. For Part b, we replace $M(t)$ with 60 and solve for t.

$60 = \dfrac{66}{1 + 10e^{-0.05t}}$ given

$60(1 + 10e^{-0.05t}) = 66$ multiply by $1 + 10e^{-0.05t}$

$1 + 10e^{-0.05t} = 1.1$ divide by 60

$10e^{-0.05t} = 0.1$ subtract 1

$e^{-0.05t} = 0.01$ divide by 10

$\ln e^{-0.05t} = \ln 0.01$ apply base-e logarithms

$-0.05t = \ln 0.01$ Property III

$t = \dfrac{\ln 0.01}{-0.05}$ solve for t (exact form)

≈ 92 approximate form

According to this model, the company will reach a 60% market share in about 92 days.

Now try Exercises 111 through 116 ▶

Earlier we used the barometric equation $H = (30T + 8000) \ln\left(\dfrac{P_0}{P}\right)$ to find an altitude H, given a temperature and the atmospheric (barometric) pressure in centimeters of mercury (cmHg). Using the tools from this section, we are now able to find the atmospheric pressure for a given altitude and temperature.

EXAMPLE 15 ▶ **Using Logarithms to Determine Atmospheric Pressure**

Suppose a group of climbers has just scaled Mt. Rainier, the highest mountain of the Cascade Range in western Washington State. If the mountain is about 4395 m high and the temperature at the summit is $-22.5°C$, what is the atmospheric pressure at this altitude? The pressure at sea level is $P_0 = 76$ cmHg.

Solution ▶

$$H = (30T + 8000) \ln\left(\frac{P_0}{P}\right) \qquad \text{given}$$

$$4395 = [30(-22.5) + 8000] \ln\left(\frac{76}{P}\right) \qquad \text{substitute 4395 for } H, 76 \text{ for } P_0, \text{ and } -22.5 \text{ for } T$$

$$4395 = 7325 \ln\left(\frac{76}{P}\right) \qquad \text{simplify}$$

$$0.6 = \ln\left(\frac{76}{P}\right) \qquad \text{divide by 7325}$$

$$e^{0.6} = \frac{76}{P} \qquad \text{exponential form}$$

$$Pe^{0.6} = 76 \qquad \text{multiply by } P$$

$$P = \frac{76}{e^{0.6}} \qquad \text{divide by } e^{0.6} \text{ (exact form)}$$

$$\approx 41.7 \qquad \text{approximate form}$$

 D. You've just learned how to solve applications involving logistic, exponential and logarithmic functions

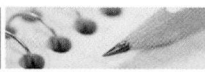 Under these conditions and at this altitude, the atmospheric pressure would be 41.7 cmHg.

Now try Exercises 117 through 120 ▶

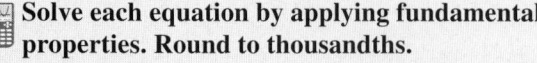

4.4 EXERCISES

▶ **CONCEPTS AND VOCABULARY**

Fill in each blank with the appropriate word or phrase. Carefully reread the section if needed.

1. For $e^{-0.02x+1} = 10$, the solution process is most efficient if we apply a base _____ logarithm to both sides.

2. To solve $3 \ln x - \ln(x + 3) = 0$, we can combine terms using the _____ property, or add $\ln(x + 3)$ to both sides and use the _____ property.

3. Since logarithmic functions are not defined for all real numbers, we should check all "solutions" for _____ roots.

4. The statement $\log_e 10 = \dfrac{\log 10}{\log e}$ is an example of the _____ -of- _____ property.

5. Solve the equation here, giving a step-by-step discussion of the solution process: $\ln(4x + 3) + \ln(2) = 3.2$

6. Describe the difference between *evaluating* the equation below given $x = 9.7$ and *solving* the equation given $y = 9.7$: $y = 3 \log_2(x - 1.7) - 2.3$.

▶ **DEVELOPING YOUR SKILLS**

Solve each equation by applying fundamental properties. Round to thousandths.

7. $\ln x = 3.4$ 8. $\ln x = \frac{1}{2}$

9. $\log x = \frac{1}{4}$ 10. $\log x = 1.6$

11. $e^x = 9.025$ 12. $e^x = 0.343$

13. $10^x = 18.197$ 14. $10^x = 0.024$

Solve each equation. Write answers in exact form and in approximate form to four decimal places.

15. $4e^{x-2} + 5 = 70$ 16. $2 - 3e^{0.4x} = -7$

17. $10^{x+5} - 228 = -150$ 18. $10^{2x} + 27 = 190$

19. $-150 = 290.8 - 190e^{-0.75x}$

20. $250e^{0.05x+1} + 175 = 1175$

Solve each equation. Write answers in exact form and in approximate form to four decimal places.

21. $3 \ln(x + 4) - 5 = 3$ **22.** $-15 = -8 \ln(3x) + 7$

23. $-1.5 = 2 \log(5 - x) - 4$

24. $-4 \log(2x) + 9 = 3.6$

25. $\frac{1}{2} \ln(2x + 5) + 3 = 3.2$

26. $\frac{3}{4} \ln(4x) - 6.9 = -5.1$

Use properties of logarithms to write each expression as a single term.

27. $\ln(2x) + \ln(x - 7)$ **28.** $\ln(x + 2) + \ln(3x)$

29. $\log(x + 1) + \log(x - 1)$

30. $\log(x - 3) + \log(x + 3)$

31. $\log_3 28 - \log_3 7$ **32.** $\log_6 30 - \log_6 10$

33. $\log x - \log(x + 1)$ **34.** $\log(x - 2) - \log x$

35. $\ln(x - 5) - \ln x$ **36.** $\ln(x + 3) - \ln(x - 1)$

37. $\ln(x^2 - 4) - \ln(x + 2)$

38. $\ln(x^2 - 25) - \ln(x + 5)$

39. $\log_2 7 + \log_2 6$ **40.** $\log_9 2 + \log_9 15$

41. $\log_5(x^2 - 2x) + \log_5 x^{-1}$

42. $\log_3(3x^2 + 5x) - \log_3 x$

Use the power property of logarithms to rewrite each term as the product of a constant and a logarithmic term.

43. $\log 8^{x+2}$ **44.** $\log 15^{x-3}$

45. $\ln 5^{2x-1}$ **46.** $\ln 10^{3x+2}$

47. $\log \sqrt{22}$ **48.** $\log \sqrt[3]{34}$

49. $\log_5 81$ **50.** $\log_7 121$

Use the properties of logarithms to write the following expressions as a sum or difference of simple logarithmic terms.

51. $\log(a^3 b)$ **52.** $\log(m^2 n)$

53. $\ln(x \sqrt[4]{y})$ **54.** $\ln(\sqrt[3]{pq})$

55. $\ln\left(\dfrac{x^2}{y}\right)$ **56.** $\ln\left(\dfrac{m^2}{n^3}\right)$

57. $\log\left(\sqrt{\dfrac{x - 2}{x}}\right)$ **58.** $\log\left(\sqrt[3]{\dfrac{3 - v}{2v}}\right)$

59. $\ln\left(\dfrac{7x\sqrt{3 - 4x}}{2(x - 1)^3}\right)$ **60.** $\ln\left(\dfrac{x^4\sqrt{x^2 - 4}}{\sqrt[3]{x^2 + 5}}\right)$

Evaluate each expression using the change-of-base formula and either base 10 or base e. Answer in exact form and in approximate form using nine decimal places, then verify the result using the original base.

61. $\log_7 60$ **62.** $\log_8 92$

63. $\log_5 152$ **64.** $\log_6 200$

65. $\log_3 1.73205$ **66.** $\log_2 1.41421$

67. $\log_{0.5} 0.125$ **68.** $\log_{0.2} 0.008$

Use the change-of-base formula to write an equivalent function, then evaluate the function as indicated (round to four decimal places). Investigate and discuss any patterns you notice in the output values, then determine the next input that will continue the pattern.

69. $f(x) = \log_3 x; f(5), f(15), f(45)$

70. $g(x) = \log_2 x; g(5), g(10), g(20)$

71. $h(x) = \log_9 x; h(2), h(4), h(8)$

72. $H(x) = \log_\pi x; H(\sqrt{2}), H(2), H(\sqrt{2^3})$

Solve each equation and check your answers.

73. $\log 4 + \log(x - 7) = 2$

74. $\log 5 + \log(x - 9) = 1$

75. $\log(2x - 5) - \log 78 = -1$

76. $\log(4 - 3x) - \log 145 = -2$

77. $\log(x - 15) - 2 = -\log x$

78. $\log x - 1 = -\log(x - 9)$

79. $\log(2x + 1) = 1 - \log x$

80. $\log(3x - 13) = 2 - \log x$

Solve each equation using the uniqueness property of logarithms.

81. $\log(5x + 2) = \log 2$

82. $\log(2x - 3) = \log 3$

83. $\log_4(x + 2) - \log_4 3 = \log_4(x - 1)$

84. $\log_3(x + 6) - \log_3 x = \log_3 5$

85. $\ln(8x - 4) = \ln 2 + \ln x$

86. $\ln(x - 1) + \ln 6 = \ln(3x)$

Solve each logarithmic equation using any appropriate method. Clearly identify any extraneous roots. If there are no solutions, so state.

87. $\log(2x - 1) + \log 5 = 1$

88. $\log(x - 7) + \log 3 = 2$

89. $\log_2(9) + \log_2(x + 3) = 3$

90. $\log_3(x - 4) + \log_3(7) = 2$

91. $\ln(x + 7) + \ln 9 = 2$

92. $\ln 5 + \ln(x - 2) = 1$

93. $\log(x + 8) + \log x = \log(x + 18)$

94. $\log(x + 14) - \log x = \log(x + 6)$

95. $\ln(2x + 1) = 3 + \ln 6$

96. $\ln 21 = 1 + \ln(x - 2)$

97. $\log(-x - 1) = \log(5x) - \log x$

98. $\log(1 - x) + \log x = \log(x + 4)$

99. $\ln(2t + 7) = \ln 3 - \ln(t + 1)$

100. $\ln 6 - \ln(5 - r) = \ln(r + 2)$

101. $\log(x - 1) - \log x = \log(x - 3)$

102. $\ln x + \ln(x - 2) = \ln 4$

103. $7^{x+2} = 231$ **104.** $6^{x+2} = 3589$

105. $5^{3x-2} = 128{,}965$ **106.** $9^{5x-3} = 78{,}462$

107. $2^{x+1} = 3^x$ **108.** $7^x = 4^{2x-1}$

109. $5^{2x+1} = 9^{x+1}$ **110.** $\left(\dfrac{1}{5}\right)^{x-1} = \left(\dfrac{1}{2}\right)^{3-x}$

111. $\dfrac{250}{1 + 4e^{-0.06x}} = 200$ **112.** $\dfrac{80}{1 + 15e^{-0.06x}} = 50$

▶ WORKING WITH FORMULAS

113. Logistic growth: $P(t) = \dfrac{C}{1 + ae^{-kt}}$

For populations that exhibit logistic growth, the population at time t is modeled by the function shown, where C is the carrying capacity of the population (the maximum population that can be supported over a long period of time), k is the growth constant, and $a = \frac{C - P(0)}{P(0)}$. Solve the formula for t, then use the result to find the value of t given $C = 450$, $a = 8$, $P = 400$, and $k = 0.075$.

114. Forensics—estimating time of death:

$$h = -3.9 \cdot \ln\left(\frac{T - T_R}{T_0 - T_R}\right)$$

Using the formula shown, a forensic expert can compute the approximate time of death for a person found recently expired, where T is the body temperature when it was found, T_R is the (constant) temperature of the room, T_0 is the body temperature at the time of death ($T_0 = 98.6°F$), and h is the number of hours since death. If the body was discovered at 9:00 A.M. with a temperature of 86.2°F, in a room at 73°F, at approximately what time did the person expire? (Note this formula is a version of Newton's law of cooling.)

▶ APPLICATIONS

115. Stocking a lake: A farmer wants to stock a private lake on his property with catfish. A specialist studies the area and depth of the lake, along with other factors, and determines it can support a maximum population of around 750 fish, with growth modeled by the function $P(t) = \dfrac{750}{1 + 24e^{-0.075t}}$, where $P(t)$ gives the current population after t months.
(a) How many catfish did the farmer initially put in the lake? (b) How many months until the population reaches 300 fish?

116. Increasing sales: After expanding their area of operations, a manufacturer of small storage buildings believes the larger area can support sales of 40 units per month. After increasing the advertising budget and enlarging the sales force, sales are expected to grow according to the model

$S(t) = \dfrac{40}{1 + 1.5e^{-0.08t}}$, where $S(t)$ is the expected number of sales after t months. (a) How many sales were being made each month, prior to the expansion? (b) How many months until sales reach 25 units per month?

Use the *barometric equation* $H = (30T + 8000) \ln\left(\dfrac{P_0}{P}\right)$
for exercises 117 and 118. Recall that $P_0 = 76$ cmHg.

117. Altitude and temperature: A sophisticated spy plane is cruising at an altitude of 18,250 m. If the temperature at this altitude is −75°C, what is the barometric pressure?

118. Altitude and temperature: A large weather balloon is released and takes altitude, pressure, and temperature readings as it climbs, and radios the information back to Earth. What is the pressure reading at an altitude of 5000 m, given the temperature is $-18°C$?

Use *Newton's law of cooling* $T = T_R + (T_0 - T_R)e^{kh}$ to complete Exercises 119 and 120. Recall that water freezes at 32°F and use $k = -0.012$. Refer to Section 4.2, page 430 as needed.

119. Making popsicles: On a hot summer day, Sean and his friends mix some Kool-Aid® and decide to freeze it in an ice tray to make popsicles. If the water used for the Kool-Aid® was 75°F and the freezer has a temperature of $-20°F$, how long will they have to wait to enjoy the treat?

120. Freezing time: Suppose the current temperature in Esconabe, Michigan, was 47°F when a 5°F arctic cold front moved over the state. How long would it take a puddle of water to freeze over?

Depreciation/appreciation: As time passes, the value of certain items decrease (appliances, automobiles, etc.), while the value of other items increase (collectibles, real estate, etc.). The time T in years for an item to reach a future value can be modeled by the formula $T = k \ln\left(\dfrac{V_n}{V_f}\right)$, where V_n is the purchase price when new, V_f is its future value, and k is a constant that depends on the item.

121. Automobile depreciation: If a new car is purchased for $28,500, find its value 3 yr later if $k = 5$.

122. Home appreciation: If a new home in an "upscale" neighborhood is purchased for $130,000, find its value 12 yr later if $k = -16$.

Drug absorption: The time required for a certain percentage of a drug to be *absorbed* by the body depends on the drug's absorption rate. This can be modeled by the function $T(p) = \dfrac{-\ln p}{k}$, where p represents the percent of the drug that *remains unabsorbed* (expressed as a decimal), k is the absorption rate of the drug, and $T(p)$ represents the elapsed time.

123. For a drug with an absorption rate of 7.2%, (a) find the time required (to the nearest hour) for the body to *absorb* 35% of the drug, and (b) find the percent of this drug (to the nearest half percent) that remains unabsorbed after 24 hr.

124. For a drug with an absorption rate of 5.7%, (a) find the time required (to the nearest hour)

for the body to *absorb* 50% of the drug, and (b) find the percent of this drug (to the nearest half percent) that remains unabsorbed after 24 hr.

Spaceship velocity: In space travel, the change in the velocity of a spaceship V_s (in km/sec) depends on the mass of the ship M_s (in tons), the mass of the fuel which has been burned M_f (in tons) and the escape velocity of the exhaust V_e (in km/sec). Disregarding frictional forces, these are related by the equation

$$V_s = V_e \ln\left(\frac{M_s}{M_s - M_f}\right).$$

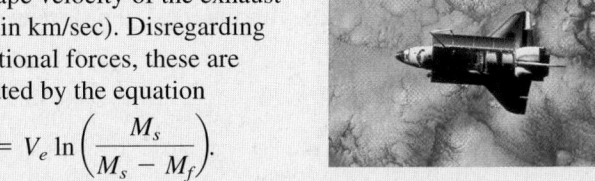

125. For the Jupiter VII rocket, find the mass of the fuel M_f that has been burned if $V_s = 6$ km/sec when $V_e = 8$ km/sec, and the ship's mass is 100 tons.

126. For the Neptune X satellite booster, find the mass of the ship M_s if $M_f = 75$ tons of fuel has been burned when $V_s = 8$ km/sec and $V_e = 10$ km/sec.

Learning curve: The job performance of a new employee when learning a repetitive task (as on an assembly line) improves very quickly at first, then grows more slowly over time. This can be modeled by the function $P(t) = a + b \ln t$, where a and b are constants that depend on the type of task and the training of the employee.

127. The number of toy planes an employee can assemble from its component parts depends on the length of time the employee has been working. This output is modeled by $P(t) = 5.9 + 12.6 \ln t$, where $P(t)$ is the number of planes assembled daily after working t days. (a) How many planes is an employee making after 5 days on the job? (b) How many days until the employee is able to assemble 34 planes per day?

128. The number of circuit boards an associate can assemble from its component parts depends on the length of time the associate has been working. This output is modeled by $B(t) = 1 + 2.3 \ln t$, where $B(t)$ is the number of boards assembled daily after working t days. (a) How many boards is an employee completing after 9 days on the job? (b) How long will it take until the employee is able to complete 10 boards per day?

▶ EXTENDING THE CONCEPT

Use prime factors, properties of logs, and the values given to evaluate each expression without a calculator. Check each result using the change-of-base formula:

129. $\log_3 4 = 1.2619$ and $\log_3 5 = 1.4649$:

    a. $\log_3 20$ **b.** $\log_3 \dfrac{4}{5}$ **c.** $\log_3 25$

130. $\log_5 2 \approx 0.4307$ and $\log_5 3 \approx 0.6826$:

    a. $\log_5 \dfrac{9}{2}$

    b. $\log_5 216$

    c. $\log_5 \sqrt[3]{6}$

131. Match each equation with the most appropriate solution strategy, and justify/discuss why.

    a. $e^{x+1} = 25$ _____   apply base-10 logarithm to both sides

    b. $\log(2x + 3) = \log 53$ _____   rewrite and apply uniqueness property for exponentials

    c. $\log(x^2 - 3x) = 2$ _____   apply uniqueness property for logarithms

    d. $10^{2x} = 97$ _____   apply either base-10 or base-e logarithm

    e. $2^{5x-3} = 32$ _____   apply base-e logarithm

    f. $7^{x+2} = 23$ _____   write in exponential form

Solve the following equations. Note that equations Exercises 132 and 133 are in quadratic form.

132. $2e^{2x} - 7e^x = 15$

133. $3e^{2x} - 4e^x - 7 = -3$

134. $\log_2(x + 5) = \log_4(21x + 1)$

135. Show that $g(x) = f^{-1}(x)$ by composing the functions.

    a. $f(x) = 3^{x-2}$; $g(x) = \log_3 x + 2$

    b. $f(x) = e^{x-1}$; $g(x) = \ln x + 1$

136. Use the algebraic method to find the inverse function.

    a. $f(x) = 2^{x+1}$ **b.** $y = 2\ln(x - 3)$

137. Use properties of logarithms and/or exponents to show

    a. $y = 2^x$ is equivalent to $y = e^{x\ln 2}$.

    b. $y = b^x$ is equivalent to $y = e^{rx}$, where $r = \ln b$.

138. To understand the formula for the half-life of radioactive material, consider that for each time increment, a constant proportion of mass m is lost. In symbols; $m(t + 1) - m(t) = -km(t)$. (a) Solve for $m(t + 1)$ and factor the right-hand side. (b) Evaluate the new equation for $t = 0, 1, 2,$ and 3, to show that $m(t) = m(0)(1 - k)^t$. (c) For any half-life h, we have $m(h) = m(0)(1 - k)^h = \frac{1}{2}m(0)$. Solve for $1 - k$, raise both sides to the power t, and substitute to show $m(t) = m(0)(\frac{1}{2})^{\frac{t}{h}}$.

139. Use test values for p and q to demonstrate that the following relationships are *false*, then state the correct property and use the same test value to verify the property.

    a. $\ln(pq) = \ln p \ln q$

    b. $\ln\left(\dfrac{p}{q}\right) = \dfrac{\ln p}{\ln q}$

    c. $\ln p + \ln q = \ln(p + q)$

140. Verify that $\ln x = (\ln 10)(\log x)$, and discuss *why* they're equal. Then use the relationship to find the value of $\ln e$, $\ln 10$, and $\ln 2$.

▶ MAINTAINING YOUR SKILLS

141. **(2.4)** Match the graph shown with its correct equation, without actually graphing the function.

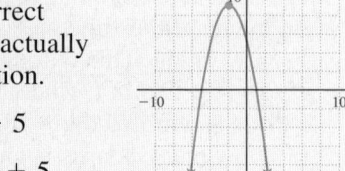

    a. $y = x^2 + 4x - 5$

    b. $y = -x^2 - 4x + 5$

    c. $y = -x^2 + 4x + 5$

    d. $y = x^2 - 4x - 5$

142. **(3.3)** State the domain and range of the functions.

    a. $y = \sqrt{2x + 3}$ **b.** $y = |x + 2| - 3$

143. **(4.6)** Graph the function $r(x) = \dfrac{x^2 - 4}{x - 1}$. Label all intercepts and asymptotes.

144. **(3.6)** Suppose the maximum load (in tons) that can be supported by a cylindrical post varies directly with its diameter raised to the fourth power and inversely as the square of its height. A post 8 ft high and 2 ft in diameter can support 6 tons. How many tons can be supported by a post 12 ft high and 3 ft in diameter?

4.5 | Applications from Business, Finance, and Science

Learning Objectives

In Section 4.5 you will learn how to:

☐ **A.** Calculate simple interest and compound interest

☐ **B.** Calculate interest compounded continuously

☐ **C.** Solve applications of annuities and amortization

☐ **D.** Solve applications of exponential growth and decay

Would you pay $750,000 for a home worth only $250,000? Surprisingly, when a conventional mortgage is repaid over 30 years, this is not at all rare. Over time, the accumulated interest on the mortgage is easily more than two or three times the original value of the house. In this section we explore how interest is paid or charged, and look at other applications of exponential and logarithmic functions from business, finance, as well as the physical and social sciences.

A. Simple and Compound Interest

Simple interest is an amount of interest that is computed only once during the lifetime of an investment (or loan). In the world of finance, the initial deposit or base amount is referred to as the **principal p,** the **interest rate r** is given as a percentage and stated as an annual rate, with the term of the investment or loan most often given as *time t* in years. Simple interest is merely an application of the basic percent equation, with the additional element of time coming into play: *interest = principal × rate × time*, or $I = prt$. To find the total amount A that has accumulated (for deposits) or is due (for loans) after t years, we merely add the accumulated interest to the initial principal: $A = p + prt$.

> **WORTHY OF NOTE**
>
> If a loan is kept for only a certain number of months, weeks, or days, the time t should be stated as a fractional part of a year so the time period for the rate (years) matches the time period over which the loan is repaid.

> **Simple Interest Formula**
>
> If principal p is deposited or borrowed at interest rate r for a period of t years, the simple interest on this account will be
>
> $$I = prt$$
>
> The total amount A accumulated or due after this period will be:
>
> $$A = p + prt \quad \text{or} \quad A = p(1 + rt)$$

EXAMPLE 1 ▶ **Solving an Application of Simple Interest**

Many finance companies offer what have become known as *PayDay Loans*—a small $50 loan to help people get by until payday, usually no longer than 2 weeks. If the cost of this service is $12.50, determine the annual rate of interest charged by these companies.

Solution ▶ The interest charge is $12.50, the initial principal is $50.00, and the time period is 2 weeks or $\frac{2}{52} = \frac{1}{26}$ of a year. The simple interest formula yields

$$I = prt \qquad \text{simple interest formula}$$

$$12.50 = 50r\left(\frac{1}{26}\right) \qquad \text{substitute \$12.50 for } I, \text{ \$50.00 for } p, \text{ and } \tfrac{1}{26} \text{ for } t$$

$$6.5 = r \qquad \text{solve for } r$$

The annual interest rate on these loans is a whopping 650%!

Now try Exercises 7 through 16 ▶

Compound Interest

Many financial institutions pay **compound interest** on deposits they receive, which is interest paid on previously accumulated interest. The most common compounding periods are yearly, semiannually (two times per year), quarterly (four times per year), monthly (12 times per year), and daily (365 times per year). Applications of compound interest typically involve exponential functions. For convenience, consider $1000 in principal, deposited at 8% for 3 yr. The simple interest calculation shows $240 in interest is earned and there will be $1240 in the account: $A = 1000[1 + (0.08)(3)] = \$1240$. If the interest is *compounded each year* $(t = 1)$ instead of once at the start of the 3-yr period, the interest calculation shows

$A_1 = 1000(1 + 0.08) = 1080$ in the account at the end of year 1,
$A_2 = 1080(1 + 0.08) = 1166.40$ in the account at the end of year 2,
$A_3 = 1166.40(1 + 0.08) \approx 1259.71$ in the account at the end of year 3.

The account has earned an additional $19.71 interest. More importantly, notice that we're multiplying by $(1 + 0.08)$ each compounding period, meaning results can be computed more efficiently by simply applying the factor $(1 + 0.08)^t$ to the initial principal p. For example,

$$A_3 = 1000(1 + 0.08)^3 \approx \$1259.71.$$

In general, for interest compounded yearly the **accumulated value** is $A = p(1 + r)^t$. Notice that solving this equation for p will tell us the amount we need to deposit *now*, in order to accumulate A dollars in t years: $p = \frac{A}{(1 + r)^t}$. This is called the **present value equation.**

Interest Compounded Annually

If a principal p is deposited at interest rate r and compounded yearly for a period of t yr, the *accumulated value* is

$$A = p(1 + r)^t$$

If an accumulated value A is desired after t yr, and the money is deposited at interest rate r and compounded yearly, the *present value* is

$$p = \frac{A}{(1 + r)^t}$$

EXAMPLE 2 ▶ **Finding the Doubling Time of an Investment**

An initial deposit of $1000 is made into an account paying 6% compounded yearly. How long will it take for the money to double?

Solution ▶ Using the formula for interest compounded yearly we have

$A = p(1 + r)^t$	given
$2000 = 1000(1 + 0.06)^t$	substitute 2000 for A, 1000 for p, and 0.06 for r
$2 = 1.06^t$	isolate variable term
$\ln 2 = t \ln 1.06$	apply base-e logarithms; power property
$\dfrac{\ln 2}{\ln 1.06} = t$	solve for t
$11.9 \approx t$	approximate form

The money will double in just under 12 yr.

Now try Exercises 17 through 22 ▶

If interest is compounded monthly (12 times each year), the bank will divide the interest rate by 12 (the number of compoundings), but then pay you interest 12 times per year (interest is *compounded*). The net effect is an increased gain in the interest you earn, and the final compound interest formula takes this form:

$$\text{total amount} = \text{principal}\left(1 + \frac{\text{interest rate}}{\text{compoundings per year}}\right)^{(\text{years} \times \text{compoundings per year})}$$

> **Compounded Interest Formula**
>
> If principal p is deposited at interest rate r and compounded n times per year for a period of t yr, the *accumulated value* will be:
>
> $$A = p\left(1 + \frac{r}{n}\right)^{nt}$$

EXAMPLE 3 ▶ **Solving an Application of Compound Interest**

Macalyn won $150,000 in the Missouri lottery and decides to invest the money for retirement in 20 yr. Of all the options available here, which one will produce the most money for retirement?

    a. A certificate of deposit paying 5.4% compounded yearly.

    b. A money market certificate paying 5.35% compounded semiannually.

    c. A bank account paying 5.25% compounded quarterly.

    d. A bond issue paying 5.2% compounded daily.

Solution ▶   **a.** $A = \$150{,}000\left(1 + \dfrac{0.054}{1}\right)^{(20\times1)}$

          $\approx \$429{,}440.97$

  b. $A = \$150{,}000\left(1 + \dfrac{0.0535}{2}\right)^{(20\times2)}$

          $\approx \$431{,}200.96$

  c. $A = \$150{,}000\left(1 + \dfrac{0.0525}{4}\right)^{(20\times4)}$

          $\approx \$425{,}729.59$

  d. $A = \$150{,}000\left(1 + \dfrac{0.052}{365}\right)^{(20\times365)}$

☑ **A.** You've just learned how to calculate simple interest and compound interest

          $\approx \$424{,}351.12$

The best choice is (b), semiannual compounding at 5.35% for 20 yr.

Now try Exercises 23 through 30 ▶

B. Interest Compounded Continuously

It seems natural to wonder what happens to the interest accumulation as n (the number of compounding periods) becomes very large. It appears the interest rate becomes very small (because we're dividing by n), but the exponent becomes very large (since we're multiplying by n). To see the result of this interplay more clearly, it will help to rewrite the compound interest formula $A = p(1 + \frac{r}{n})^{nt}$ using the substitution $n = xr$. This gives $\frac{r}{n} = \frac{1}{x}$, and by direct substitution (xr for n and $\frac{1}{x}$ for $\frac{r}{n}$) we obtain the form

$$A = p\left[\left(1 + \frac{1}{x}\right)^{x}\right]^{rt}$$

by regrouping. This allows for a more careful study of the "denominator versus exponent" relationship using $(1 + \frac{1}{x})^{x}$, *the same expression we used in Section 4.2 to define the number e* (also **see Section 4.2 Exercise 97**). Once again, note what happens as $x \to \infty$ (meaning the number of compounding periods increase without bound).

x	1	10	100	1000	10,000	100,000	1,000,000
$\left(1 + \dfrac{1}{x}\right)^x$	2	2.56374	2.70481	2.71692	2.71815	2.71827	2.71828

As before, we have, as $x \to \infty$, $(1 + \frac{1}{x})^x \to e$. The net result of this investigation is a formula for **interest compounded continuously,** derived by replacing $(1 + \frac{1}{x})^x$ with the number e in the formula for compound interest, where

$$A = p\left[\left(1 + \frac{1}{x}\right)^x\right]^{rt} = pe^{rt}$$

Interest Compounded Continuously

If a principal p is deposited at interest rate r and compounded continuously for a period of t years, the *accumulated value* will be

$$A = pe^{rt}$$

EXAMPLE 4 ▶ **Solving an Application of Interest Compounded Continuously**

Jaimin has $10,000 to invest and wants to have at least $25,000 in the account in 10 yr for his daughter's college education fund. If the account pays interest compounded continuously, what interest rate is required?

Solution ▶ In this case, $P = \$10,000$, $A = \$25,000$, and $t = 10$.

$A = pe^{rt}$	given
$25,000 = 10,000e^{10r}$	substitute 25,000 for A, 10,000 for p, and 10 for t
$2.5 = e^{10r}$	isolate variable term
$\ln 2.5 = 10r \ln e$	apply base-e logarithms ($\ln e = 1$); power property
$\dfrac{\ln 2.5}{10} = r$	solve for r
$0.092 \approx r$	approximate form

Jaimin will need an interest rate of about 9.2% to meet his goal.

Now try Exercises 31 through 40 ▶

☑ **B.** You've just learned how to calculate interest compounded continuously

GRAPHICAL SUPPORT

To check the result from Example 4, use $Y_1 = 10{,}000e^{10x}$ and $Y_2 = 25{,}000$, then look for their point of intersection. We need only set an appropriate window size to ensure the answer will appear in the viewing window. Since 25,000 is the goal, $y \in [0, 30{,}000]$ seems reasonable for y. Although 12% interest ($x = 0.12$) is too good to be true, $x \in [0, 0.12]$ leaves a nice frame for the x-values. Verify that the calculator's answer is equal to $\frac{\ln 2.5}{10}$.

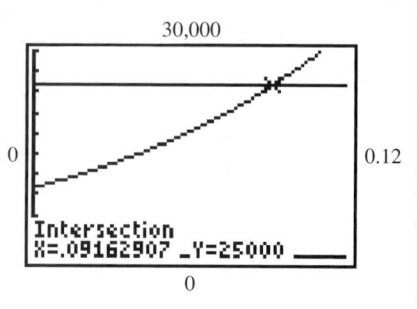

WORTHY OF NOTE

It is often assumed that the first payment into an annuity is made *at the end of a compounding period,* and hence earns no interest. This is why the first $100 deposit is not multiplied by the interest factor. These terms are actually the terms of a **geometric sequence,** which we will study later in Section 8.3.

C. Applications Involving Annuities and Amortization

Our previous calculations for simple and compound interest involved a single (lump) deposit (the principal) that accumulated interest over time. Many savings and investment plans involve a regular schedule of deposits (monthly, quarterly, or annual deposits) over the life of the investment. Such an investment plan is called an **annuity.**

Suppose that for 4 yr, $100 is deposited annually into an account paying 8% compounded yearly. Using the compound interest formula we can track the accumulated value A in the account:

$$A = 100 + 100(1.08)^1 + 100(1.08)^2 + 100(1.08)^3$$

To develop an annuity formula, we multiply the annuity equation by 1.08, then subtract the original equation. This leaves only the first and last terms, since the other (interior) terms add to zero:

$$1.08A = 100(1.08) + 100(1.08)^2 + 100(1.08)^3 + 100(1.08)^4 \quad \text{multiply by 1.08}$$
$$\underline{-A = -[100 + 100(1.08)^1 + 100(1.08)^2 + 100(1.08)^3]} \quad \text{original equation}$$
$$1.08A - A = 100(1.08)^4 - 100 \quad \text{subtract ("interior terms" sum to zero)}$$
$$0.08A = 100[(1.08)^4 - 1] \quad \text{factor out 100}$$
$$A = \frac{100[(1.08)^4 - 1]}{0.08} \quad \text{solve for } A$$

This result can be generalized for any periodic payment $\mathcal{P}$, interest rate r, number of compounding periods n, and number of years t. This would give

$$A = \frac{\mathcal{P}\left[\left(1 + \dfrac{r}{n}\right)^{nt} - 1\right]}{\dfrac{r}{n}}$$

The formula can be made less formidable using $R = \frac{r}{n}$, where R is the interest rate per compounding period.

Accumulated Value of an Annuity

If a periodic payment $\mathcal{P}$ is deposited n times per year at an *annual interest rate r* with interest compounded n times per year for t years, the accumulated value is given by

$$A = \frac{\mathcal{P}}{R}[(1 + R)^{nt} - 1], \text{ where } R = \frac{r}{n}$$

This is also referred to as the **future value** of the account.

EXAMPLE 5 ▶ **Solving an Application of Annuities**

Since he was a young child, Fitisemanu's parents have been depositing $50 each month into an annuity that pays 6% annually and is compounded monthly. If the account is now worth $9875, how long has it been open?

Solution ▶ In this case $\mathcal{P} = 50$, $r = 0.06$, $n = 12$, $R = 0.005$, and $A = 9875$. The formula gives

$$A = \frac{\mathcal{P}}{R}[(1 + R)^{nt} - 1] \qquad \text{future value formula}$$

$$9875 = \frac{50}{0.005}[(1.005)^{(12)(t)} - 1] \qquad \text{substitute 9875 for } A, \text{ 50 for } \mathcal{P}, \text{ 0.005 for } R, \text{ and 12 for } n$$

$$1.9875 = 1.005^{12t} \qquad \text{simplify and isolate variable term}$$

$$\ln(1.9875) = 12t(\ln 1.005) \qquad \text{apply base-}e\text{ logarithms; power property}$$

$$\frac{\ln(1.9875)}{12\ln(1.005)} = t \qquad \text{solve for } t \text{ (exact form)}$$

$$11.5 \approx t \qquad \text{approximate form}$$

The account has been open approximately 11.5 yr.

Now try Exercises 41 through 44 ▶

The periodic payment required to meet a future goal or obligation can be computed by solving for $\mathcal{P}$ in the future value formula: $\mathcal{P} = \dfrac{AR}{[(1 + R)^{nt} - 1]}$. In this form, $\mathcal{P}$ is referred to as a **sinking fund.**

EXAMPLE 6 ▶ **Solving an Application of Sinking Funds**

Sheila is determined to stay out of debt and decides to save $20,000 to pay cash for a new car in 4 yr. The best investment vehicle she can find pays 9% compounded monthly. If $300 is the most she can invest each month, can she meet her "4-yr" goal?

Solution ▶ Here we have $\mathcal{P} = 300$, $A = 20,000$, $r = 0.09$, $n = 12$, and $R = 0.0075$. The sinking fund formula gives

$$\mathcal{P} = \frac{AR}{[(1 + R)^{nt} - 1]} \qquad \text{sinking fund}$$

$$300 = \frac{(20,000)(0.0075)}{(1.0075)^{12t} - 1} \qquad \begin{array}{l}\text{substitute 300 for } \mathcal{P}, \text{ 20,000 for } A, \\ \text{0.0075 for } R, \text{ and 12 for } n\end{array}$$

$$300(1.0075^{12t} - 1) = 150 \qquad \text{multiply in numerator, clear denominators}$$

$$1.0075^{12t} = 1.5 \qquad \text{isolate variable term}$$

$$12t\ln(1.0075) = \ln 1.5 \qquad \text{apply base-}e\text{ logarithms; power property}$$

$$t = \frac{\ln(1.5)}{12\ln(1.0075)} \qquad \text{solve for } t \text{ (exact form)}$$

$$\approx 4.5 \qquad \text{approximate form}$$

No. She is close, but misses her original 4-yr goal.

Now try Exercises 45 and 46 ▶

☑ **C. You've just learned how to solve applications of annuities and amortization**

For Example 6, we could have substituted 4 for t and left $\mathcal{P}$ unknown, to see if a payment of $300 per month would be sufficient. You can verify the result would be $\mathcal{P} \approx \$347.70$, which is what Sheila would need to invest to meet her 4-yr goal exactly.

For additional practice with the formulas for interest earned or paid, the *Working with Formulas* portion of this Exercise Set has been expanded. See **Exercises 47 through 54.**

WORTHY OF NOTE

Notice the formula for exponential growth is virtually identical to the formula for interest compounded continuously. In fact, both are based on the same principles. If we let $A(t)$ represent the amount in an account after t years and A_0 represent the initial deposit (instead of P), we have: $A(t) = A_0e^{rt}$ versus $Q(t) = Q_0e^{rt}$ and the two cannot be distinguished.

D. Applications Involving Exponential Growth and Decay

Closely related to interest compounded continuously are applications of **exponential growth** and **exponential decay.** If Q (quantity) and t (time) are variables, then Q grows exponentially as a function of t if $Q(t) = Q_0e^{rt}$ for positive constants Q_0 and r. Careful studies have shown that population growth, whether it be humans, bats, or bacteria, can be modeled by these "base-e" exponential growth functions. If $Q(t) = Q_0e^{-rt}$, then we say Q decreases or **decays exponentially** over time. The constant r determines how rapidly a quantity grows or decays and is known as the **growth rate** or **decay rate** constant.

EXAMPLE 7 ▶ Solving an Application of Exponential Growth

Because fruit flies multiply very quickly, they are often used in a study of genetics. Given the necessary space and food supply, a certain population of fruit flies is known to double every 12 days. If there were 100 flies to begin, find (a) the growth rate r and (b) the number of days until the population reaches 2000 flies.

Solution ▶ **a.** Using the formula for exponential growth with $Q_0 = 100$, $t = 12$, and $Q(t) = 200$, we can solve for the growth rate r.

$$Q(t) = Q_0e^{rt} \qquad \text{exponential growth function}$$
$$200 = 100e^{12r} \qquad \text{substitute 200 for } Q(t) \text{ 100 for } Q_0, \text{ and 12 for } t$$
$$2 = e^{12r} \qquad \text{isolate variable term}$$
$$\ln 2 = 12r \ln e \qquad \text{apply base-}e \text{ logarithms; power property}$$
$$\frac{\ln 2}{12} = r \qquad \text{solve for } r \text{ (exact form)}$$
$$0.05776 \approx r \qquad \text{approximate form}$$

The growth rate is approximately 5.78%.

b. To find the number of days until the fly population reaches 2000, we substitute 0.05776 for r in the exponential growth function.

$$Q(t) = Q_0e^{rt} \qquad \text{exponential growth function}$$
$$2000 = 100e^{0.05776t} \qquad \text{substitute 2000 for } Q(t), \text{ 100 for } Q_0, \text{ and 0.05776 for } r$$
$$20 = e^{0.05776t} \qquad \text{isolate variable term}$$
$$\ln 20 = 0.05776t \ln e \qquad \text{apply base-}e \text{ logarithms; power property}$$
$$\frac{\ln 20}{0.05776} = t \qquad \text{solve for } t \text{ (exact form)}$$
$$51.87 \approx t \qquad \text{approximate form}$$

The fruit fly population will reach 2000 on day 51.

Now try Exercises 55 and 56 ▶

WORTHY OF NOTE

Many population growth models assume an unlimited supply of resources, nutrients, and room for growth. When this is not the case, a logistic growth model often results. See the *Modeling with Technology* feature following this chapter.

Perhaps the best-known examples of exponential decay involve radioactivity. Ever since the end of World War II, common citizens have been aware of the existence of **radioactive elements** and the power of atomic energy. Today, hundreds of additional applications have been found for these materials, from areas as diverse as biological research, radiology, medicine, and archeology. Radioactive elements decay of their own accord by emitting radiation. The rate of decay is measured using the **half-life** of the substance, which is the time required for a mass of radioactive material to decay until only one-half of its original mass remains. This half-life is used to find the rate of decay r, first mentioned in Section 4.4. In general, if h represents the half-life of the substance, one-half the initial amount remains when $t = h$.

$$Q(t) = Q_0e^{-rt} \qquad \text{exponential decay function}$$
$$\frac{1}{2}Q_0 = Q_0e^{-rh} \qquad \text{substitute } \tfrac{1}{2}Q_0 \text{ for } Q(t), h \text{ for } t$$
$$\frac{1}{2} = \frac{1}{e^{rh}} \qquad \text{divide by } Q_0; \text{ rewrite expression}$$
$$2 = e^{rh} \qquad \text{property of ratios}$$
$$\ln 2 = rh \ln e \qquad \text{apply base-}e \text{ logarithms; power property}$$
$$\frac{\ln 2}{h} = r \qquad \text{solve for } r$$

> ### Radioactive Rate of Decay
>
> If h represents the half-life of a radioactive substance per unit time, the nominal rate of decay per a like unit of time is given by
>
> $$r = \frac{\ln 2}{h}$$

The rate of decay for known radioactive elements varies greatly. For example, the element carbon-14 has a half-life of about 5730 yr, while the element lead-211 has a half-life of only about 3.5 min. Radioactive elements can be detected in extremely small amounts. If a drug is "labeled" (mixed with) a radioactive element and injected into a living organism, its passage through the organism can be traced and information on the health of internal organs can be obtained.

EXAMPLE 8 ▶   **Solving a Rate of Decay Application**

The radioactive element potassium-42 is often used in biological experiments, since it has a half-life of only about 12.4 hr. How much of a 2-g sample will remain after 18 hr and 45 min?

Solution ▶   To begin we must find the nominal rate of decay r and use this value in the exponential decay function.

$$r = \frac{\ln 2}{h} \qquad \text{radioactive rate of decay}$$

$$r = \frac{\ln 2}{12.4} \qquad \text{substitute 12.4 for } h$$

$$r \approx 0.056 \qquad \text{result}$$

The rate of decay is approximately 5.6%. To determine how much of the sample remains after 18.75 hr, we use $r = 0.056$ in the decay function and evaluate it at $t = 18.75$.

$$Q(t) = Q_0 e^{-rt} \qquad \text{exponential decay function}$$

$$Q(18.75) = 2e^{(-0.056)(18.75)} \qquad \text{substitute 2 for } Q_0, \text{ 0.056 for } r, \text{ and 18.75 for } t$$

$$Q(18.75) \approx 0.7 \qquad \text{evaluate}$$

☑ **D.** You've just learned how to solve applications of exponential growth and decay

After 18 hr and 45 min, only 0.7 g of potassium-42 will remain.

Now try Exercises 57 through 62 ▶

TECHNOLOGY HIGHLIGHT

Exploring Compound Interest

The graphing calculator is an excellent tool for exploring mathematical relationships, particularly when many variables work simultaneously to produce a single result. For example, the formula $A = P\left(1 + \frac{r}{n}\right)^{nt}$ has five different unknowns. In Example 2, we asked how long it would take $1000 to double if it were compounded yearly at 6% ($n = 1$, $r = 0.06$). What if we deposited $5000 instead of $1000? Compounded daily instead of quarterly? Or invested at 12% rather than 10%? There are many ways a graphing calculator can be used to answer such questions. In this exercise, we make use of the calculator's "alpha constants." Most graphing calculators can use any of the 26 letters of the English alphabet (and even a few other symbols) to store constant values. We can use them to write a formula

—continued

Figure 4.31	Figure 4.32	Figure 4.33
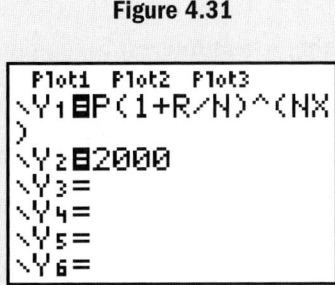	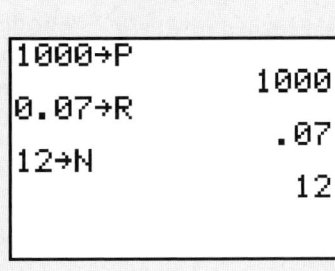	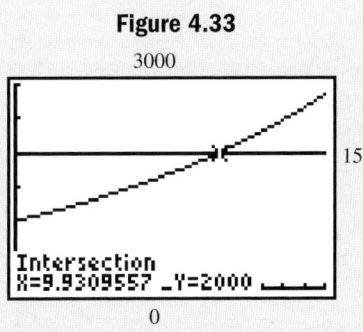

on the [Y=] screen, then change any constant on the home screen to see how other values are affected. On the TI-84 Plus, these alpha constants are shown in green and accessed by pressing the [ALPHA] key. Suppose we wanted to study the relationship between an interest rate r and the time t required for a deposit to double. Using Y_1 in place of A as output variable, and x in place of t, enter $A = P\left(1 + \frac{r}{n}\right)^{nt}$ as Y_1 on the [Y=] screen (Figure 4.31). Let's start with a deposit of $1000 at 7% interest compounded monthly. The keystrokes are: 1000 [STO →] [ALPHA] [8] [ENTER], 0.07 [STO →] [ALPHA] [×] [ENTER], and 12 [STO →] [ALPHA] [LOG] [ENTER] (Figure 4.32). After setting an appropriate window size (perhaps Xmax = 15 and Ymax = 3000), and entering $Y_2 = 2000$ we can graph both functions and use the intersection of graphs method to find the doubling time. This produces the result in Figure 4.33, where we note it will take about 9.9 yr. Return to the home screen ([2nd] [MODE]), change the interest rate to 10%, and graph the functions again. This time the point of intersection is just less than 7 (yr). Experiment with other rates and compounding periods.

Exercise 1: With $P = \$1000$, and $r = 0.08$, investigate the "doubling time" for interest compounded quarterly, monthly, daily, and hourly.

Exercise 2: With $P = \$1000$, investigate "doubling time" for rates of 6%, 8%, 10%, and 12%, and $n = 4$, $n = 12$, and $n = 365$. Which had a more significant impact, more compounding periods, or a greater interest rate?

Exercise 3: Will a larger principal cause the money to double faster? Investigate and respond.

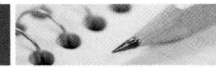

4.5 EXERCISES

▶ CONCEPTS AND VOCABULARY

Fill in each blank with the appropriate word or phrase. Carefully reread the section if needed.

1. _____ interest is interest paid to you on previously accumulated interest.

2. The formula for interest compounded _____ is $A = pe^{rt}$, where e is approximately _____.

3. Given constants Q_0 and r, and that Q decays exponentially as a function of t, the equation model is $Q(t) =$ _____.

4. Investment plans calling for regularly scheduled deposits are called _____. The annuity formula gives the _____ value of the account.

5. Explain/Describe the difference between the future value and present value of an annuity. Include an example.

6. Describe/Explain how you would find the rate of growth r, given that a population of ants grew from 250 to 3000 in 6 weeks.

▶ **DEVELOPING YOUR SKILLS**

For simple interest accounts, the interest earned or due depends on the principal p, interest rate r, and the time t in years according to the formula $I = prt$.

7. Find p given $I = \$229.50$, $r = 6.25\%$, and $t = 9$ months.

8. Find r given $I = \$1928.75$, $p = \$8500$, and $t = 3.75$ yr.

9. Larry came up a little short one month at bill-paying time and had to take out a title loan on his car at Check Casher's, Inc. He borrowed \$260, and 3 weeks later he paid off the note for \$297.50. What was the annual interest rate on this title loan? (*Hint:* How much *interest* was charged?)

10. Angela has \$750 in a passbook savings account that pays 2.5% simple interest. How long will it take the account balance to hit the \$1000 mark at this rate of interest, if she makes no further deposits? (*Hint:* How much *interest* will be paid?)

For simple interest accounts, the amount A accumulated or due depends on the principal p, interest rate r, and the time t in years according to the formula $A = p(1 + rt)$.

11. Find p given $A = \$2500$, $r = 6.25\%$, and $t = 31$ months.

12. Find r given $A = \$15,800$, $p = \$10,000$, and $t = 3.75$ yr.

13. Olivette Custom Auto Service borrowed \$120,000 at 4.75% simple interest to expand their facility from three service bays to four. If they repaid \$149,925, what was the term of the loan?

14. Healthy U sells nutritional supplements and borrows \$50,000 to expand their product line. When the note is due 3 yr later, they repay the lender \$62,500. If it was a simple interest note, what was the annual interest rate?

15. **Simple interest:** The owner of Paul's Pawn Shop loans Larry \$200.00 using his Toro riding mower as collateral. Thirteen weeks later Larry comes back to get his mower out of pawn and pays Paul \$240.00. What was the annual simple interest rate on this loan?

16. **Simple interest:** To open business in a new strip mall, Laurie's Custom Card Shoppe borrows \$50,000 from a group of investors at 4.55% simple interest. Business booms and blossoms, enabling Laurie to repay the loan fairly quickly. If Laurie repays \$62,500, how long did it take?

For accounts where interest is compounded annually, the amount A accumulated or due depends on the principal p, interest rate r, and the time t in years according to the formula $A = p(1 + r)^t$.

17. Find t given $A = \$48,428$, $p = \$38,000$, and $r = 6.25\%$.

18. Find p given $A = \$30,146$, $r = 5.3\%$, and $t = 7$ yr.

19. How long would it take \$1525 to triple if invested at 7.1%?

20. What interest rate will ensure a \$747.26 deposit will be worth \$1000 in 5 yr?

For accounts where interest is compounded annually, the principal P needed to ensure an amount A has been accumulated in the time period t when deposited at interest rate r is given by the formula $P = \dfrac{A}{(1 + r)^t}$.

21. The Stringers need to make a \$10,000 balloon payment in 5 yr. How much should be invested now at 5.75%, so that the money will be available?

22. Morgan is 8 yr old. If her mother wants to have \$25,000 for Morgan's first year of college (in 10 yr), how much should be invested now if the account pays a 6.375% fixed rate?

For compound interest accounts, the amount A accumulated or due depends on the principal p, interest rate r, number of compoundings per year n, and the time t in years according to the formula $A = p\left(1 + \frac{r}{n}\right)^{nt}$.

23. Find t given $A = \$129,500$, $p = \$90,000$, and $r = 7.125\%$ compounded weekly.

24. Find r given $A = \$95,375$, $p = \$65,750$, and $t = 15$ yr with interest compounded monthly.

25. How long would it take a \$5000 deposit to double, if invested at a 9.25% rate and compounded daily?

26. What principal should be deposited at 8.375% compounded monthly to ensure the account will be worth \$20,000 in 10 yr?

27. **Compound interest:** As a curiosity, David decides to invest \$10 in an account paying 10% interest compounded 10 times per year for 10 yr. Is that enough time for the \$10 to triple in value?

28. **Compound interest:** As a follow-up experiment (see Exercise 27), David invests \$10 in an account paying 12% interest compounded 10 times per year

for 10 yr, and another $10 in an account paying 10% interest compounded 12 times per year for 10 yr. Which produces the better investment—more compounding periods or a higher interest rate?

29. **Compound interest:** Due to demand, Donovan's Dairy (Wisconsin, USA) plans to double its size in 4 yr and will need $250,000 to begin development. If they invest $175,000 in an account that pays 8.75% compounded semiannually, (a) will there be sufficient funds to break ground in 4 yr? (b) If not, find the *minimum interest rate* that will allow the dairy to meet its 4-yr goal.

30. **Compound interest:** To celebrate the birth of a new daughter, Helyn invests 6000 Swiss francs in a college savings plan to pay for her daughter's first year of college in 18 yr. She estimates that 25,000 francs will be needed. If the account pays 7.2% compounded daily, (a) will she meet her investment goal? (b) If not, find the *minimum rate of interest* that will enable her to meet this 18-yr goal.

For accounts where interest is compounded continuously, the amount *A* accumulated or due depends on the principal *p*, interest rate *r*, and the time *t* in years according to the formula $A = pe^{rt}$.

31. Find t given $A = 2500, $p = 1750, and $r = 4.5\%$.

32. Find r given $A = $325,000$, $p = $250,000$, and $t = 10$ yr.

33. How long would it take $5000 to double if it is invested at 9.25%? Compare the result to Exercise 25.

34. What principal should be deposited at 8.375% to ensure the account will be worth $20,000 in 10 yr? Compare the result to Exercise 26.

35. **Interest compounded continuously:** Valance wants to build an addition to his home outside Madrid (Spain) so he can watch over and help his parents in their old age. He hopes to have 20,000 euros put aside for this purpose within 5 yr. If he invests 12,500 euros in an account paying 8.6% interest compounded continuously, (a) will he meet his investment goal? (b) If not, find the *minimum rate of interest* that will enable him to meet this 5-yr goal.

36. **Interest compounded continuously:** Minh-Ho just inherited her father's farm near Mito (Japan), which badly needs a new barn. The estimated cost of the barn is 8,465,000 yen and she would like to begin construction in 4 yr. If she invests 6,250,000 yen in

an account paying 6.5% interest compounded continuously, (a) will she meet her investment goal? (b) If not, find the *minimum rate of interest* that will enable her to meet this 4-yr goal.

37. **Interest compounded continuously:** William and Mary buy a small cottage in Dovershire (England), where they hope to move after retiring in 7 yr. The cottage needs about 20,000 euros worth of improvements to make it the retirement home they desire. If they invest 12,000 euros in an account paying 5.5% interest compounded continuously, (a) will they have enough to make the repairs? (b) If not, find the *minimum amount they need to deposit* that will enable them to meet this goal in 7 yr.

38. **Interest compounded continuously:** After living in Oslo (Norway) for 20 years, Zirkcyt and Shybrt decide to move inland to help operate the family ski resort. They hope to make the move in 6 yr, after they have put aside 140,000 kroner. If they invest 85,000 kroner in an account paying 6.9% interest compounded continuously, (a) will they meet their 140,000 kroner goal? (b) If not, find the *minimum amount they need to deposit* that will allow them to meet this goal in 6 yr.

The length of time *T* (in years) required for an initial principal *P* to grow to an amount *A* at a given interest rate *r* is given by $T = \frac{1}{r} \ln\left(\frac{A}{P}\right)$.

39. **Investment growth:** A small business is planning to build a new $350,000 facility in 8 yr. If they deposit $200,000 in an account that pays 5% interest compounded continuously, will they have enough for the new facility in 8 yr? If not, what amount should be invested on these terms to meet the goal?

40. **Investment growth:** After the twins were born, Sasan deposited $25,000 in an account paying 7.5% compounded continuously, with the goal of having $120,000 available for their college education 20 yr later. Will Sasan meet the 20-yr goal? If not, what amount should be invested on these terms to meet the goal?

Ordinary annuities: If a periodic payment $\mathcal{P}$ is deposited *n* times per year, with annual interest rate *r* also compounded *n* times per year for *t* years, the future value of the account is given by $A = \frac{\mathcal{P}[(1 + R)^{nt} - 1]}{R}$, where $R = \frac{r}{n}$ (if the rate is 9% compounded monthly, $R = \frac{0.09}{12} = 0.0075$).

41. **Saving for a rainy day:** How long would it take Jasmine to save $10,000 if she deposits $90/month at an annual rate of 7.75% compounded monthly?

42. Saving for a sunny day: What quarterly investment amount is required to ensure that Larry can save $4700 in 4 yr at an annual rate of 8.5% compounded quarterly?

43. Saving for college: At the birth of their first child, Latasha and Terrance opened an annuity account and have been depositing $50 per month in the account ever since. If the account is now worth $30,000 and the interest on the account is 6.2% compounded monthly, how old is the child?

44. Saving for a bequest: When Cherie (Brandon's first granddaughter) was born, he purchased an annuity account for her and stipulated that she should receive the funds (in trust, if necessary) upon his death. The quarterly annuity payments were $250 and interest on the account was 7.6% compounded quarterly. The account balance of $17,500 was recently given to Cherie. How much longer did Brandon live?

45. Saving for a down payment: Tae-Hon is tired of renting and decides that within the next 5 yr he must save $22,500 for the down payment on a home. He finds an investment company that offers 8.5% interest compounded monthly and begins depositing $250 each month in the account. (a) Is this monthly amount sufficient to help him meet his 5 yr goal? (b) If not, find the *minimum amount he needs to deposit each month* that will allow him to meet his goal in 5 yr.

46. Saving to open a business: Madeline feels trapped in her current job and decides to save $75,000 over the next 7 yr to open up a Harley Davidson franchise. To this end, she invests $145 every week in an account paying $7\frac{1}{2}\%$ interest compounded weekly. (a) Is this weekly amount sufficient to help her meet the seven-year goal? (b) If not, find the *minimum amount she needs to deposit each week* that will allow her to meet this goal in 7 yr?

▶ WORKING WITH FORMULAS

Solve for the indicated unknowns.

47. $A = p + prt$
   a. solve for t
   b. solve for p

48. $A = p(1 + r)^t$
   a. solve for t
   b. solve for r

49. $A = P\left(1 + \dfrac{r}{n}\right)^{nt}$
   a. solve for r
   b. solve for t

50. $A = pe^{rt}$
   a. solve for p
   b. solve for r

51. $Q(t) = Q_0 e^{rt}$
   a. solve for Q_0
   b. solve for t

52. $p = \dfrac{AR}{[(1 + R)^{nt} - 1]}$
   a. solve for A
   b. solve for n

 53. Amount of a mortgage payment:

$$\mathcal{P} = \dfrac{AR}{1 - (1 + R)^{-nt}}$$

The mortgage payment required to pay off (or amortize) a loan is given by the formula shown, where $\mathcal{P}$ is the payment amount, A is the original amount of the loan, t is the time in years, r is the annual interest rate, n is the number of payments per year, and $R = \frac{r}{n}$. Find the *monthly payment* required to amortize a $125,000 home, if the interest rate is 5.5%/year and the home is financed over 30 yr.

54. Total interest paid on a home mortgage:

$$I = \left[\dfrac{prt}{1 - \left(\dfrac{1}{1 + 0.08\overline{3}r}\right)^{12t}}\right] - p$$

The total interest I paid in t years on a home mortgage of p dollars is given by the formula shown, where r is the interest rate on the loan (note that $0.08\overline{3} = \frac{1}{12}$). If the original mortgage was $198,000 at an interest rate of 6.5%, (a) how much interest has been paid in 10 yr? (b) Use a table of values to determine how many years it will take for the interest paid to exceed the amount of the original mortgage.

▶ APPLICATIONS

55. Exponential growth: As part of a lab experiment, Luamata needs to grow a culture of 200,000 bacteria, which are known to double in number in 12 hr. If he begins with 1000 bacteria, (a) find the growth rate r and (b) find how many hours it takes for the culture to produce the 200,000 bacteria.

56. Exponential growth: After the wolf population was decimated due to overhunting, the rabbit population in the Boluhti Game Reserve began to double every 6 months. If there were an estimated 120 rabbits to begin, (a) find the growth rate r and (b) find the number of months required for the population to reach 2500.

57. Radioactive decay: The radioactive element iodine-131 has a half-life of 8 days and is often used to help diagnose patients with thyroid problems. If a certain thyroid procedure requires 0.5 g and is scheduled to take place in 3 days, what is the minimum amount that must be on hand now (to the nearest hundredth of a gram)?

58. Radioactive decay: The radioactive element sodium-24 has a half-life of 15 hr and is used to help locate obstructions in blood flow. If the procedure requires 0.75 g and is scheduled to take place in 2 days (48 hr), what minimum amount must be on hand *now* (to the nearest hundredth of a gram)?

59. Radioactive decay: The radioactive element americium-241 has a half-life of 432 yr and although extremely small amounts are used (about 0.0002 g), it is the most vital component of standard household smoke detectors. How many years will it take a 10-g mass of americium-241 to decay to 2.7 g?

60. Radioactive decay: Carbon-14 is a radioactive compound that occurs naturally in all living

organisms, with the amount in the organism constantly renewed. After death, no new carbon-14 is acquired and the amount in the organism begins to decay exponentially. If the half-life of carbon-14 is 5730 yr, how old is a mummy having only 30% of the normal amount of carbon-14?

Carbon-14 dating: If the percentage p of carbon-14 that remains in a fossil can be determined, the formula $T = -8267 \ln p$ can be used to estimate the number of years T since the organism died.

61. Dating the Lascaux Cave Dwellers: Bits of charcoal from Lascaux Cave (home of the prehistoric Lascaux Cave Paintings) were used to estimate that the fire had burned some 17,255 yr ago. What percent of the original amount of carbon-14 remained in the bits of charcoal?

62. Dating Stonehenge: Using organic fragments found near Stonehenge (England), scientists were able to determine that the organism that produced the fragments lived about 3925 yr ago. What percent of the original amount of carbon-14 remained in the organism?

► EXTENDING THE CONCEPT

63. Many claim that inheritance taxes are put in place simply to prevent a massive accumulation of wealth by a select few. Suppose that in 1890, your great-grandfather deposited $10,000 in an account paying 6.2% compounded continuously. If the account were to pass to you untaxed, what would it be worth in 2010? Do some research on the inheritance tax laws in your state. In particular, what amounts can be inherited untaxed (i.e., before the inheritance tax kicks in)?

64. In Section 4.2, we noted that one important characteristic of exponential functions is their rate of growth is in constant proportion to the population at time t: $\frac{\Delta P}{\Delta t} = kP$. This rate of growth can also be applied to finance and biological models, as well as the growth of tumors, and is of great value in studying these applications. In Exercise 96 of Section 4.2, we computed the value of k for the Goldsboro model ($P = 1000 \cdot 2^t$) using the difference quotient. If we rewrite this model in terms of base e ($P = 1000 \cdot e^{kt}$), the value of k is given directly. The following sequence shows how

this is done, and you are asked to supply the reason or justification for each step.

$P = b^t$	base-b exponential
$\ln P = \ln b^t$	_____
$\ln P = t \ln b$	_____
$P = e^{t \ln b}$	_____
$P = e^{kt}$	$k = \ln b$

The last step shows the growth rate constant is equal to the natural log of the given base b: $k = \ln b$.

a. Use this result to verify the growth rate constant for Goldsboro is 0.6931472.

b. After the Great Oklahoma Land Run of 1890, the population of the state grew rapidly for the next 2 decades. For this time period, population growth could be approximated by $P = 260(1.10^t)$. Find the growth rate constant for this model, and use it to write the base-e population equation. Use the TABLE feature of a graphing calculator to verify that the equations are equivalent.

65. If you have not already completed Exercise 30, please do so. For *this* exercise, *solve the compound interest equation for r* to find the exact rate of interest that will allow Helyn to meet her 18-yr goal.

 66. If you have not already completed Exercise 43, please do so. Suppose the final balance of the account was $35,100 with interest again being compounded monthly. For *this* exercise, use a graphing calculator to find *r*, the exact rate of interest the account would have been earning.

▶ **MAINTAINING YOUR SKILLS**

67. (2.1) In an effort to boost tourism, a trolley car is being built to carry sightseers from a strip mall to the top of Mt. Vernon, 1580-m high. Approximately how long will the trolley cables be?

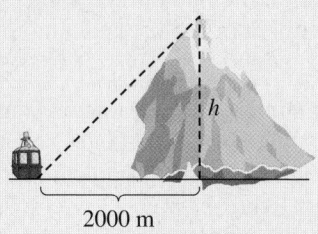

2000 m

68. (2.2) Is the following relation a function? If not, state how the definition of a function is violated.

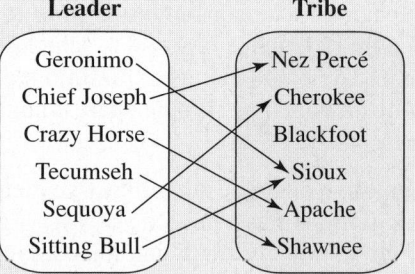

Leader	Tribe
Geronimo	Nez Percé
Chief Joseph	Cherokee
Crazy Horse	Blackfoot
Tecumseh	Sioux
Sequoya	Apache
Sitting Bull	Shawnee

69. (4.3) A polynomial is known to have the zeroes $x = 3$, $x = -1$, and $x = 1 + 2i$. Find the equation of the polynomial, given it has degree 4 and a y-intercept of $(0, -15)$.

70. (2.2/3.8) Name the toolbox functions that are (a) one-to-one, (b) even, (c) increasing for $x \in R$, and (d) asymptotic.

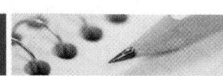 **SUMMARY AND CONCEPT REVIEW**

SECTION 4.1 One-to-One and Inverse Functions

KEY CONCEPTS

- A function is one-to-one if each element of the range corresponds to a unique element of the domain.
- If every horizontal line intersects the graph of a function in at most one point, the function is one-to-one.
- If f is a one-to-one function with ordered pairs (a, b), then the inverse of f exists and is that one-to-one function f^{-1} with ordered pairs of the form (b, a).
- The range of f becomes the domain of f^{-1}, and the domain of f becomes the range of f^{-1}.
- To find f^{-1} using the algebraic method:
  1. Use y instead of $f(x)$.   2. Interchange x and y.
  3. Solve the equation for y.   4. Substitute $f^{-1}(x)$ for y.
- If f is a one-to-one function, the inverse f^{-1} exists, where $(f \circ f^{-1})(x) = x$ and $(f^{-1} \circ f)(x) = x$.
- The graphs of f and f^{-1} are symmetric with respect to the identity function $y = x$.

EXERCISES

Determine whether the functions given are one-to-one by noting the function family to which each belongs and mentally picturing the shape of the graph.

1. $h(x) = -|x - 2| + 3$ **2.** $p(x) = 2x^2 + 7$ **3.** $s(x) = \sqrt{x - 1} + 5$

Find the inverse of each function given. Then show graphically and using composition that your inverse function is correct. State any necessary restrictions.

4. $f(x) = -3x + 2$ **5.** $f(x) = x^2 - 2, x \geq 0$ **6.** $f(x) = \sqrt{x - 1}$

Determine the domain and range for each function whose graph is given, and use this information to state the domain and range of the inverse function. Then sketch in the line $y = x$, estimate the location of three points on the graph, and use these to graph $f^{-1}(x)$ on the same grid.

7. 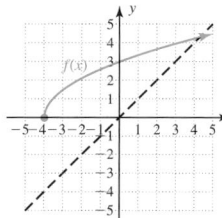   **8.** 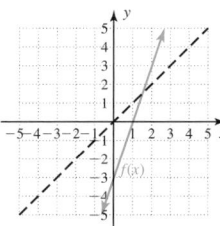   **9.**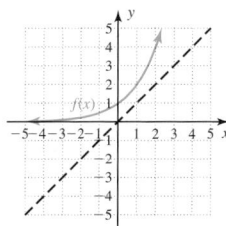

10. Fines for overdue material: Some libraries have set fees and penalties to discourage patrons from holding borrowed materials for an extended period. Suppose the fine for overdue DVDs is given by the function $f(t) = 0.15t + 2$, where $f(t)$ is the amount of the fine t days after it is due. (a) What is the fine for keeping a DVD seven (7) extra days? (b) Find $f^{-1}(t)$, then input your answer from part (a) and comment on the result. (c) If a fine of \$3.80 was assessed, how many days was the DVD overdue?

SECTION 4.2   Exponential Functions

KEY CONCEPTS

- An exponential function is defined as $f(x) = b^x$, where $b > 0$, $b \neq 1$, and b, x are real numbers.
- The natural exponential function is $f(x) = e^x$, where $e \approx 2.71828182846$.
- For exponential functions, we have
  - one-to-one function
  - y-intercept $(0, 1)$
  - domain: $x \in \mathbb{R}$
  - range: $y \in (0, \infty)$
  - increasing if $b > 1$
  - decreasing if $0 < b < 1$
  - asymptotic to x-axis
- The graph of $y = b^{x \pm h} \pm k$ is a translation of the basic graph of $y = b^x$, horizontally h units opposite the sign and vertically k units in the same direction as the sign.
- If an equation can be written with like bases on each side, we solve it using the uniqueness property: If $b^m = b^n$, then $m = n$ (equal bases imply equal exponents).
- All previous properties of exponents also apply to exponential functions.

EXERCISES

Graph each function using *transformations of the basic function,* then strategically plot a few points to check your work and round out the graph. Draw and label the asymptote.

11. $y = 2^x + 3$ **12.** $y = 2^{-x} - 1$ **13.** $y = -e^{x+1} - 2$

Solve using the uniqueness property.

14. $3^{2x-1} = 27$ **15.** $4^x = \frac{1}{16}$ **16.** $e^x \cdot e^{x+1} = e^6$

17. A ballast machine is purchased new for \$142,000 by the AT & SF Railroad. The machine loses 15% of its value each year and must be replaced when its value drops below \$20,000. How many years will the machine be in service?

SECTION 4.3   Logarithms and Logarithmic Functions

KEY CONCEPTS

- A logarithm is an exponent. For $x, b > 0$, and $b \neq 1$, the expression $\log_b x$ represents the exponent that goes on base b to obtain x: If $y = \log_b x$, then $b^y = x \Rightarrow b^{\log_b x} = x$ (by substitution).

- The equations $x = b^y$ and $y = \log_b x$ are equivalent. We say $x = b^y$ is the *exponential* form and $y = \log_b x$ is the *logarithmic* form of the equation.
- The value of $\log_b x$ can sometimes be determined by writing the expression in exponential form. If $b = 10$ or $b = e$, the value of $\log_b x$ can be found directly using a calculator.
- A logarithmic *function* is defined as $f(x) = \log_b x$, where $x, b > 0$, and $b \neq 1$.
  - $y = \log_{10} x = \log x$ is called a *common* logarithmic function.
  - $y = \log_e x = \ln x$ is called a *natural* logarithmic function.
- For $f(x) = \log_b x$ as defined we have
  - one-to-one function
  - x-intercept $(1, 0)$
  - domain: $x \in (0, \infty)$
  - range: $y \in \mathbb{R}$
  - increasing if $b > 1$
  - decreasing if $0 < b < 1$
  - asymptotic to y-axis
- The graph of $y = \log_b(x \pm h) \pm k$ is a translation of the graph of $y = \log_b x$, horizontally h units opposite the sign and vertically k units in the same direction as the sign.

EXERCISES

Write each expression in *exponential* form.

18. $\log_3 9 = 2$ **19.** $\log_5 \frac{1}{125} = -3$ **20.** $\ln 43 \approx 3.7612$

Write each expression in *logarithmic* form.

21. $5^2 = 25$ **22.** $e^{-0.25} \approx 0.7788$ **23.** $3^4 = 81$

Find the value of each expression without using a calculator.

24. $\log_2 32$ **25.** $\ln(\frac{1}{e})$ **26.** $\log_9 3$

Graph each function using *transformations of the basic function,* then strategically plot a few points to check your work and round out the graph. Draw and label the asymptote.

27. $f(x) = \log_2 x$ **28.** $f(x) = \log_2(x + 3)$ **29.** $f(x) = 2 + \ln(x - 1)$

Find the domain of the following functions.

30. $f(x) = \ln(x^2 - 6x)$ **31.** $g(x) = \log \sqrt{2x + 3}$

32. The magnitude of an earthquake is given by $M(I) = \log\dfrac{I}{I_0}$, where I is the intensity and I_0 is the reference intensity. (a) Find $M(I)$ given $I = 62{,}000I_0$ and (b) find the intensity I given $M(I) = 7.3$.

SECTION 4.4   Properties of Logarithms; Solving Exponential and Logarithmic Equations

KEY CONCEPTS

- The basic definition of a logarithm gives rise to the following properties: For any base $b > 0, b \neq 1$,
  1. $\log_b b = 1$ (since $b^1 = b$)
  2. $\log_b 1 = 0$ (since $b^0 = 1$)
  3. $\log_b b^x = x$ (since $b^x = b^x$)
  4. $b^{\log_b x} = x$
- Since a logarithm is an exponent, they have properties that parallel those of exponents.

Product Property	**Quotient Property**	**Power Property**
like base and multiplication, add exponents:	like base and division, subtract exponents:	exponent raised to a power, multiply exponents:
$\log_b(MN) = \log_b M + \log_b N$	$\log_b\left(\dfrac{M}{N}\right) = \log_b M - \log_b N$	$\log_b M^p = p\log_b M$

- The logarithmic properties can be used to expand an expression: $\log(2x) = \log 2 + \log x$.
- The logarithmic properties can be used to contract an expression: $\ln(2x) - \ln(x + 3) = \ln\left(\dfrac{2x}{x + 3}\right)$.

- To evaluate logarithms with bases other than 10 or e, use the change-of-base formula:

$$\log_b M = \frac{\log M}{\log b} = \frac{\ln M}{\ln b}$$

- If an equation can be written with like bases on each side, we solve it using the uniqueness property: if $\log_b m = \log_b n$, then $m = n$ (equal bases imply equal arguments).
- If a single exponential or logarithmic term can be isolated on one side, then for any base b:

$$\text{If } b^x = k, \text{ then } x = \frac{\log k}{\log b} \qquad\qquad \text{If } \log_b x = k, \text{ then } x = b^k.$$

EXERCISES

33. Solve each equation by applying fundamental properties.

    a. $\ln x = 32$ **b.** $\log x = 2.38$ **c.** $e^x = 9.8$ **d.** $10^x = \sqrt{7}$

34. Solve each equation. Write answers in exact form and in approximate form to four decimal places.

    a. $15 = 7 + 2e^{0.5x}$ **b.** $10^{0.2x} = 19$ **c.** $-2\log(3x) + 1 = -5$ **d.** $-2\ln x + 1 = 6.5$

35. Use the product or quotient property of logarithms to write each sum or difference as a single term.

    a. $\ln 7 + \ln 6$ **b.** $\log_9 2 + \log_9 15$ **c.** $\ln(x+3) - \ln(x-1)$ **d.** $\log x + \log(x+1)$

36. Use the power property of logarithms to rewrite each term as a product.

    a. $\log_5 9^2$ **b.** $\log_7 4^2$ **c.** $\ln 5^{2x-1}$ **d.** $\ln 10^{3x+2}$

37. Use the properties of logarithms to write the following expressions as a sum or difference of simple logarithmic terms.

    a. $\ln(x\sqrt[4]{y})$ **b.** $\ln(\sqrt[3]{pq})$ **c.** $\log\left(\dfrac{\sqrt[3]{x^5 \cdot y^4}}{\sqrt{x^5 y^3}}\right)$ **d.** $\log\left(\dfrac{4\sqrt[3]{p^5 q^4}}{\sqrt{p^3 q^2}}\right)$

38. Evaluate using a change-of-base formula. Answer in exact form and approximate form to thousandths.

    a. $\log_6 45$ **b.** $\log_3 128$ **c.** $\ln_2 124$ **d.** $\ln_5 0.42$

Solve each equation.

39. $2^x = 7$ **40.** $3^{x+1} = 5$ **41.** $e^{x-2} = 3^x$

42. $\ln(x+1) = 2$ **43.** $\log x + \log(x-3) = 1$ **44.** $\log_{25}(x+2) - \log_{25}(x-3) = \frac{1}{2}$

45. The rate of decay for radioactive material is related to its half-life by the formula $R(h) = \frac{\ln 2}{h}$, where h represents the half-life of the material and $R(h)$ is the rate of decay expressed as a decimal. The element radon-222 has a half-life of approximately 3.9 days. (a) Find its rate of decay to the nearest hundredth of a percent. (b) Find the half-life of thorium-234 if its rate of decay is 2.89% per day.

46. The *barometric equation* $H = (30T + 8000)\ln\left(\frac{P_0}{P}\right)$ relates the altitude H to atmospheric pressure P, where $P_0 = 76$ cmHg. Find the atmospheric pressure at the summit of Mount Pico de Orizaba (Mexico), whose summit is at 5657 m. Assume the temperature at the summit is $T = 12°C$.

SECTION 4.5   Applications from Investment, Finance, and Physical Science

KEY CONCEPTS

- Simple interest: $I = prt$; p is the initial principal, r is the interest rate per year, and t is the time in years.
- Amount in an account after t years: $A = p + prt$ or $A = p(1 + rt)$.
- Interest compounded n times per year: $A = p\left(1 + \dfrac{r}{n}\right)^{nt}$; p is the initial principal, r is the interest rate per year, t is the time in years, and n is the times per year interest is compounded.
- Interest compounded continuously: $A = pe^{rt}$; p is the initial principal, r is the interest rate per year, and t is the time in years.

- If a loan or savings plan calls for a regular schedule of deposits, the plan is called an annuity.
- For periodic payment P, deposited or paid n times per year, at annual interest rate r, with interest compounded or calculated n times per year for t years, and $R = \dfrac{r}{n}$:

  - The accumulated value of the account is $A = \dfrac{p}{R}[(1 + R)^{nt} - 1]$.

  - The payment required to meet a future goal is $P = \dfrac{AR}{[(1 + R)^{nt} - 1]}$.

  - The payment required to amortize an amount A is $P = \dfrac{AR}{1 - (1 + R)^{-nt}}$.

  - The general formulas for exponential growth and decay are $Q(t) = Q_0 e^{rt}$ and $Q(t) = Q_0 e^{-rt}$, respectively.

EXERCISES

Solve each application.

47. Jeffery borrows $600.00 from his dad, who decides it's best to charge him interest. Three months later Jeff repays the loan plus interest, a total of $627.75. What was the annual interest rate on the loan?

48. To save money for her first car, Cheryl invests the $7500 she inherited in an account paying 7.8% interest compounded monthly. She hopes to buy the car in 6 yr and needs $12,000. Is this possible?

49. To save up for the vacation of a lifetime, Al-Harwi decides to save $15,000 over the next 4 yr. For this purpose he invests $260 every month in an account paying $7\frac{1}{2}\%$ interest compounded monthly. (a) Is this monthly amount sufficient to meet the four-year goal? (b) If not, find the *minimum amount he needs to deposit each month* that will allow him to meet this goal in 4 yr.

50. Eighty prairie dogs are released in a wilderness area in an effort to repopulate the species. Five years later a statistical survey reveals the population has reached 1250 dogs. Assuming the growth was exponential, approximate the growth rate to the nearest tenth of a percent.

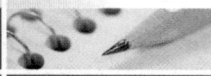

MIXED REVIEW

1. Evaluate each expression using the change-of-base formula.

   a. $\log_2 30$ **b.** $\log_{0.25} 8$

   c. $\log_8 2$

2. Solve each equation using the uniqueness property.

   a. $10^{4x-5} = 1000$ **b.** $5^{3x-1} = \sqrt{5}$

   c. $2^x \cdot 2^{0.5x} = 64$

3. Use the power property of logarithms to rewrite each expression as a product.

   a. $\log_{10} 20^2$ **b.** $\log 10^{0.05x}$

   c. $\ln 2^{x-3}$

Graph each of the following functions by shifting the basic function, then strategically plotting a few points to check your work and round out the graph. Graph and label the asymptote.

4. $y = -e^x + 15$ **5.** $y = 5 \cdot 2^{-x}$

6. $y = \ln(x + 5) + 7$ **7.** $y = \log_2(-x) - 4$

8. Use the properties of logarithms to write the following expressions as a sum or difference of simple logarithmic terms.

   a. $\ln\left(\dfrac{x^3}{2y}\right)$ **b.** $\log(10a\sqrt[3]{a^2 b})$

   c. $\log_2\left(\dfrac{8x^4\sqrt{x}}{3\sqrt{y}}\right)$

9. Write the following expressions in exponential form.

   a. $\log_5 625 = 4$ **b.** $\ln 0.15x = 0.45$

   c. $\log(0.1 \times 10^8) = 7$

10. Write the following expressions in logarithmic form.

   a. $343^{1/3} = 7$ **b.** $256^{3/4} = 64$

   c. $2^{-3} = \frac{1}{8}$

11. For $g(x) = \sqrt{x - 1} + 2$, (a) state the domain and range, (b) find $g^{-1}(x)$ and state its domain and range, and (c) compute at least three ordered pairs (a, b) for g and show the order pairs (b, a) are solutions to g^{-1}.

Solve the following equations. State answers in exact form.

12. $\log_5(4x + 7) = 0$ **13.** $10^{x-4} = 200$

14. $e^{x+1} = 3^x$

15. $\log_2(2x - 5) + \log_2(x - 2) = 4$

16. $\log(3x - 4) - \log(x - 2) = 1$

Solve each application.

17. The magnitude of an earthquake is given by

$M(I) = \log\left(\dfrac{I}{I_0}\right)$, where I is the intensity of the quake

and I_0 is the reference intensity 2×10^{11} (energy released from the smallest detectable quake). On October 23, 2004, the Niigata region of Japan was hit by an earthquake that registered 6.5 on the Richter scale. Find the intensity of this earthquake by solving the following equation for

I: $6.5 = \log\left(\dfrac{I}{2 \times 10^{11}}\right)$.

18. Serene is planning to buy a house. She has $6500 to invest in a certificate of deposit that compounds interest quarterly at an annual rate of 4.4%. (a) Find how long it will take for this account to grow to the $12,500 she will need for a 10% down payment for a $125,000 house. Round to the nearest tenth of a year. (b) Suppose instead of investing an initial $6500, Serene deposits $500 a quarter in an account paying 4% each quarter. Find how long it will take for this account to grow to $12,500. Round to the nearest tenth of a year.

19. British artist Simon Thomas designs sculptures he calls hypercones. These sculptures involve rings of exponentially decreasing radii rotated through space. For one sculpture, the radii follow the model  $r(n) = 2(0.8)^n$, where n counts the rings (outer-most first) and $r(n)$ is radii in meters. Find the radii of the six largest rings in the sculpture. Round to the nearest hundredth of a meter.

Source: http://www.plus.maths.org/issue8/features/art/

20. Ms. Chan-Chiu works for MediaMax, a small business that helps other companies purchase advertising in publications. Her model for the benefits of advertising is $P(a) = 1000(1.07)^a$, where P represents the number of potential customers reached when a dollars (in thousands) are invested in advertising.

a. Use this model to predict (to the nearest thousand) how many potential customers will be reached when $50,000 is invested in advertising.

b. Use this model to determine how much money a company should expect to invest in advertising (to the nearest thousand), if it wants to reach 100,000 potential customers.

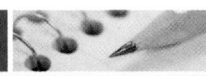 **PRACTICE TEST**

1. Write the expression $\log_3 81 = 4$ in exponential form.

2. Write the expression $25^{1/2} = 5$ in logarithmic form.

3. Write the expression $\log_b\left(\dfrac{\sqrt{x^5}y^3}{z}\right)$ as a sum or difference of logarithmic terms.

4. Write the expression $\log_b m + \left(\frac{3}{2}\right)\log_b n - \frac{1}{2}\log_b p$ as a single logarithm.

Solve for x using the uniqueness property.

5. $5^{x-7} = 125$ **6.** $2 \cdot 4^{3x} = \dfrac{8^x}{16}$

Given $\log_a 3 \approx 0.48$ and $\log_a 5 \approx 1.72$, evaluate the following without the use of a calculator:

7. $\log_a 45$ **8.** $\log_a 0.6$

Graph using transformations of the parent function. Verify answers using a graphing calculator.

9. $g(x) = -2^{x-1} + 3$ **10.** $h(x) = \log_2(x - 2) + 1$

11. Use the change-of-base formula to evaluate. Verify results using a calculator.

a. $\log_3 100$ **b.** $\log_6 0.235$

12. State the domain and range of $f(x) = (x - 2)^2 - 3$ and determine if f is a one-to-one function. If so, find its inverse. If not, restrict the domain of f to create a one-to-one function, then find the inverse of this new function, including the domain and range.

Solve each equation.

13. $3^{x-1} = 89$

14. $\log_5 x + \log_5(x + 4) = 1$

15. A copier is purchased new for $8000. The machine loses 18% of its value each year and must be replaced when its value drops below $3000. How many years will the machine be in service?

16. How long would it take $1000 to double if invested at 8% annual interest compounded daily?

17. The number of ounces of unrefined platinum drawn from a mine is modeled by $Q(t) = -2600 + 1900 \ln(t)$, where $Q(t)$ represents the number of ounces mined in t months. How many months did it take for the number of ounces mined to exceed 3000?

18. Septashi can invest his savings in an account paying 7% compounded semi-annually, or in an account paying 6.8% compounded daily. Which is the better investment?

19. Jacob decides to save $4000 over the next 5 yr so that he can present his wife with a new diamond ring for their 20th anniversary. He invests $50 every month in an account paying $8\frac{1}{4}\%$ interest compounded monthly. (a) Is this amount sufficient to meet the 5-yr goal? (b) If not, find the *minimum amount he needs to save monthly* that will enable him to meet this goal.

20. Chaucer is a typical Welsh Corgi puppy. During his first year of life, his weight very closely follows the model $W(t) = 6.79 \ln t - 11.97$, where $W(t)$ is his weight in pounds after t weeks and $8 \leq t \leq 52$.

   a. How much will Chaucer weigh when he is 6 months old (to the nearest one-tenth pound)?

   b. To the nearest week, how old is Chaucer when he weighs 8 lb?

CALCULATOR EXPLORATION AND DISCOVERY

Investigating Logistic Equations

As we saw in Section 4.4, logistics models have the form

$$P(t) = \frac{c}{1 + ae^{-bt}},$$ where a, b, and c are constants and $P(t)$ represents the population at time t. For populations modeled by a logistics curve (sometimes called an "S" curve) growth is very rapid at first (like an exponential function), but this growth begins to slow down and level off due to various factors. This *Calculator Exploration and Discovery* is designed to investigate the effects that a, b, and c have on the resulting graph.

I. From our earlier observation, as t becomes larger and larger, the term ae^{-bt} becomes smaller and smaller (approaching 0) because it is a decreasing function: as $t \to \infty$, $ae^{-bt} \to 0$. If we allow that the term eventually becomes so small it can be disregarded, what remains is $P(t) = \frac{c}{1}$ or c. This is why c is called the capacity constant and the population can get no larger than c. In Figure 4.34, the graph of

Figure 4.34

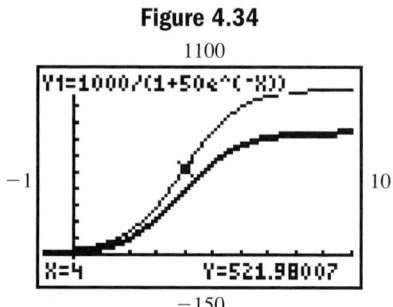

$$P(t) = \frac{1000}{1 + 50e^{-1x}} \ (a = 50, b = 1, \text{ and } c = 1000) \text{ is}$$

shown using a lighter line, while the graph of

$$P(t) = \frac{750}{1 + 50e^{-1x}} \ (a = 50, b = 1, \text{ and } c = 750), \text{ is}$$

given in bold. The window size is indicated in Figure 4.35.

Figure 4.35

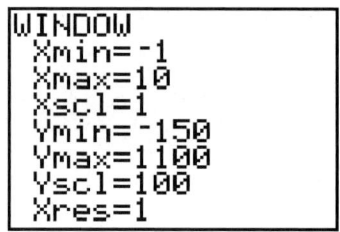

```
WINDOW
 Xmin=-1
 Xmax=10
 Xscl=1
 Ymin=-150
 Ymax=1100
 Yscl=100
 Xres=1
```

Also note that if a is held constant, smaller values of c cause the "interior" of the S curve to grow at a slower rate than larger values, a concept studied in some detail in a Calculus I class.

II. If $t = 0$, $ae^{-bt} = ae^0 = a$, and we note the ratio $P(0) = \frac{c}{1 + a}$ represents the *initial population*. This also means for constant values of c, larger values of a make the ratio $\frac{c}{1 + a}$ smaller; while smaller values of a make the ratio $\frac{c}{1 + a}$ larger. From this we conclude that a primarily affects the initial population. For the screens shown next, $P(t) = \dfrac{1000}{1 + 50e^{-1x}}$ (from I) is graphed using a lighter line. For comparison, the graph of $P(t) = \dfrac{1000}{1 + 5e^{-1x}}$ ($a = 5, b = 1$, and $c = 1000$) is shown in bold in Figure 4.36, while the graph of $P(t) = \dfrac{1000}{1 + 500e^{-1x}}$ ($a = 500, b = 1$, and $c = 1000$) is shown in bold in Figure 4.37.

Figure 4.36

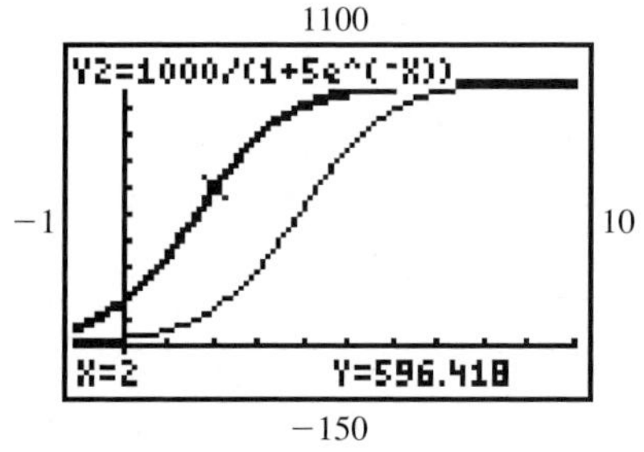

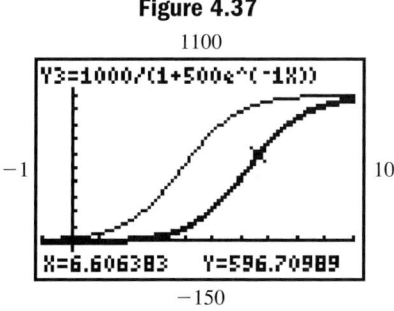

Figure 4.37

Note that changes in a appear to have no effect on the rate of growth in the interior of the S curve.

III. As for the value of b, we might expect that it affects the rate of growth in much the same way as the growth rate r does for exponential functions $Q(t) = Q_0 e^{-rt}$. Sure enough, we note from the graphs shown that b has no effect on the initial value or the eventual capacity, but causes the population to approach this capacity more quickly for larger values of b, and more slowly for smaller values of b. For the screens shown, $P(t) = \dfrac{1000}{1 + 50e^{-1x}}$ ($a = 50$, $b = 1$, and $c = 1000$) is graphed using a lighter line. For comparison, the graph of $P(t) = \dfrac{1000}{1 + 50e^{-1.2x}}$ ($a = 50$, $b = 1.2$, and $c = 1000$) is shown in bold in Figure 4.38, while the graph of $P(t) =$

Figure 4.38

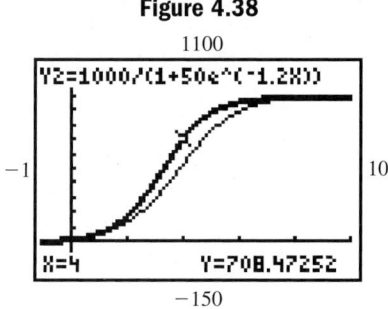

Figure 4.39

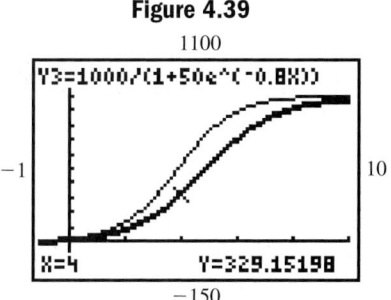

$\dfrac{1000}{1 + 50e^{-0.8x}}$ ($a = 50, b = 0.8$, and $c = 1000$) is shown in bold in Figure 4.39.

The following exercises are based on the population of an ant colony, modeled by the logistic function $P(t) = \dfrac{2500}{1 + 25e^{-0.5x}}$. Respond to Exercises 1 through 6 without the use of a calculator.

Exercise 1: Identify the values of $a, b,$ and c for this logistics curve.

Exercise 2: What was the approximate initial population of the colony?

Exercise 3: Which gives a larger initial population: (a) $c = 2500$ and $a = 25$ or (b) $c = 3000$ and $a = 15$?

Exercise 4: What is the maximum population capacity for this colony?

Exercise 5: Would the population of the colony surpass 2000 more quickly if $b = 0.6$ or if $b = 0.4$?

Exercise 6: Which causes a slower population growth: (a) $c = 2000$ and $a = 25$ or (b) $c = 3000$ and $a = 25$?

Exercise 7: Verify your responses to Exercises 2 through 6 using a graphing calculator.

STRENGTHENING CORE SKILLS

Understanding Properties of Logarithms

To effectively use the properties of logarithms as a mathematical tool, a student must attain some degree of comfort and fluency in their application. Otherwise we are resigned to using them as a template or formula, leaving little room for growth or insight. This feature is divided into two parts. The first is designed to promote an understanding of the product and quotient properties of logarithms, which play a role in the solution of logarithmic and exponential equations.

We begin by looking at some logarithmic expressions that are obviously true:

$$\log_2 2 = 1 \qquad \log_2 4 = 2 \qquad \log_2 8 = 3$$
$$\log_2 16 = 4 \qquad \log_2 32 = 5 \qquad \log_2 64 = 6$$

Next, we view the same expressions with their value *understood mentally,* illustrated by the numbers in the background, rather than expressly written.

$$\log_2 2 \quad \log_2 4 \quad \log_2 8 \quad \log_2 16 \quad \log_2 32 \quad \log_2 64$$

This will make the product and quotient properties of equality much easier to "see." Recall the product property states: $\log_b M + \log_b N = \log_b(MN)$ and the quotient property states: $\log_b M - \log_b N = \log_b\left(\dfrac{M}{N}\right)$. Consider the following.

$$\log_2 4 + \log_2 8 = \log_2 32 \qquad \log_2 64 - \log_2 32 = \log_2 2$$

which is the same as saying

$$\log_2 4 + \log_2 8 = \log_2(4 \cdot 8)$$
$$(\text{since } 4 \cdot 8 = 32)$$

which is the same as saying

$$\log_2 64 - \log_2 32 = \log_2\left(\tfrac{64}{32}\right)$$
$$(\text{since } \tfrac{64}{32} = 2)$$

$$\log_b M + \log_b N = \log_b(MN) \qquad \log_b M - \log_b N = \log_b\left(\dfrac{M}{N}\right)$$

Exercise 1: Repeat this exercise using logarithms of base 3 and various sums and differences.

Exercise 2: Use the basic concept behind these exercises to combine these expressions: (a) $\log(x) + \log(x + 3)$, (b) $\ln(x + 2) + \ln(x - 2)$, and (c) $\log(x) - \log(x + 3)$.

The second part is similar to the first, but highlights the power property: $\log_b M^x = x \log_b M$. For instance, knowing that $\log_2 64 = 6$, $\log_2 8 = 3$, and $\log_2 2 = 1$, consider the following:

$\log_2 8$ can be written as $\log_2 2^3$ (since $2^3 = 8$). Applying the power property gives $3 \cdot \log_2 2 = 3$.
$\log_2 64$ can be written as $\log_2 2^6$ (since $2^6 = 64$). Applying the power property gives $6 \cdot \log_2 2 = 6$.

$$\log_b M^x = x \log_b M$$

Exercise 3: Repeat this exercise using logarithms of base 3 and various powers.

Exercise 4: Use the basic concept behind these exercises to rewrite each expression as a product: (a) $\log 3^x$, (b) $\ln x^5$, and (c) $\ln 2^{3x-1}$.

CUMULATIVE REVIEW CHAPTERS 1–4

Use the quadratic formula to solve for x.

1. $x^2 - 4x + 53 = 0$ 2. $6x^2 + 19x = 36$

3. Use substitution to show that $4 + 5i$ is a zero of $f(x) = x^2 - 8x + 41$.

4. Graph using transformations of a basic function: $y = 2\sqrt{x + 2} - 3$.

5. Find $(f \circ g)(x)$ and $(g \circ f)(x)$ and comment on what you notice: $f(x) = x^3 - 2$; $g(x) = \sqrt[3]{x + 2}$.

6. State the domain of $h(x)$ in interval notation:
$$h(x) = \frac{\sqrt{x + 3}}{x^2 + 6x + 8}.$$

7. According to the 2002 *National Vital Statistics Report* (Vol. 50, No. 5, page 19) there were 3100 sets of triplets born in the United States in 1991, and 6740 sets of triplets born in 1999. Assuming the relationship (year, sets of triplets) is linear: (a) find the equation of the line, (b) explain the meaning of the slope in this context, and (c) use the equation to estimate the number of sets born in 1996, and to project the number of sets that will be born in 2007 if this trend continues.

8. State the following geometric formulas:
   a. area of a circle
   b. Pythagorean theorem
   c. perimeter of a rectangle
   d. area of a trapezoid

9. Graph the following piecewise-defined function and state its domain, range, and intervals where it is increasing and decreasing.

$$h(x) = \begin{cases} -4 & -10 \le x < -2 \\ -x^2 & -2 \le x < 3 \\ 3x - 18 & x \ge 3 \end{cases}$$

10. Solve the inequality and write the solution in interval notation: $\dfrac{2x + 1}{x - 3} \ge 0$.

11. Use the rational roots theorem to find all zeroes of $f(x) = x^4 - 3x^3 - 12x^2 + 52x - 48$.

12. Given $f(c) = \dfrac{9}{5}c + 32$, find k, where $k = f(25)$.

    Then find the inverse function using the algebraic method, and verify that $f^{-1}(k) = 25$.

13. Solve the formula $V = \dfrac{1}{2}\pi b^2 a$ (the volume of a paraboloid) for the variable b.

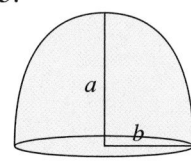

14. Use the *Guidelines for Graphing* to graph
    a. $p(x) = x^3 - 4x^2 + x + 6$.
    b. $r(x) = \dfrac{5x^2}{x^2 + 4}$.

15. For $f(x) = \dfrac{2x + 3}{5}$, (a) find f^{-1}, (b) graph both functions and verify they are symmetric to the line $y = x$, and (c) show they are inverses using composition.

16. Solve for x: $10 = -2e^{-0.05x} + 25$.

17. Solve for x: $\ln(x + 3) + \ln(x - 2) = \ln(24)$.

18. Once in orbit, satellites are often powered by radioactive isotopes. From the natural process of radioactive decay, the power output declines over a period of time. For an initial amount of 50 g, suppose the power output is modeled by the function $p(t) = 50e^{-0.002t}$, where $p(t)$ is the power output in watts, t days after the satellite has been put into service. (a) Approximately how much power remains 6 *months* later? (b) How many *years* until only one-fourth of the original power remains?

19. Simon and Christine own a sport wagon and a minivan. The sport wagon has a power curve that is closely modeled by $H(r) = 123 \ln r - 897$, where $H(r)$ is the horsepower at r rpm, with $2200 \le r \le 5600$. The power curve for the minivan is $h(r) = 193 \ln r - 1464$, for $2600 < r < 5800$.

    a. How much horsepower is generated by each engine at 3000 rpm?

    b. At what rpm are the engines generating the same horsepower?

    c. If Christine wants the maximum horsepower available, which vehicle should she drive? What is the maximum horsepower?

20. Wilson's disease is a hereditary disease that causes the body to retain copper. Radioactive copper, ^{64}Cu, has been used extensively to study and understand this disease. ^{64}Cu has a relatively short half-life of 12.7 hr. How many hours will it take for a 5-g mass of ^{64}Cu to decay to 1 g?

While exponents and logarithms will continue to help us solve noteworthy applications in the calculus sequence, the *properties* of logarithms and exponentials also play a significant role outside of their use in the equation-solving process. Here we'll see how these properties are applied in the context of exercises seen in a first-semester calculus course. We'll also explore the concept of an **area function,** a simple idea with some profound consequences in a study of calculus. The area functions we'll use, while not logarithmic, can also be applied to logarithmic functions, where it provides what we might call a "missing link."

Properties of Logarithms

In calculus, the properties of logarithms play an important role in *logarithmic differentiation*, the *derivative of composite functions,* and other areas. Regarding the first, some calculus concepts are more easily applied to sums and differences, rather than products or quotients, making the properties of logarithms an invaluable tool.

EXAMPLE 1 ▶ **Rewriting Expressions Using the Properties of Logarithms**

Use properties of logarithms to rewrite the expression $\ln\left[\dfrac{x^{\frac{2}{3}}\sqrt{x^2+2}}{(2x+3)^4}\right]$ as the sum or difference of simple logarithmic terms.

Solution ▶ Applying the quotient, product, and power properties, respectively, gives

$$\ln\left[\frac{x^{\frac{2}{3}}\sqrt{x^2+2}}{(2x+3)^4}\right] = \ln\left(x^{\frac{2}{3}}\sqrt{x^2+2}\right) - \ln(2x+3)^4 \qquad \text{quotient property}$$

$$= \ln x^{\frac{2}{3}} + \ln(x^2+2)^{\frac{1}{2}} - \ln(2x+3)^4 \qquad \text{product property}$$

$$= \frac{2}{3}\ln x + \frac{1}{2}\ln(x^2+2) - 4\ln(2x+3) \qquad \text{power property}$$

Now try Exercises 1 through 4 ▶

Area Functions

Figure C2C 4.1

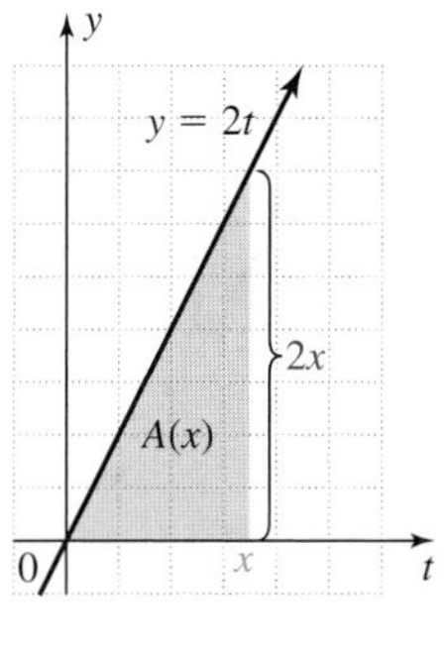

An area function $A(x)$ is defined to be the area under the graph of a function $f(t)$, between a fixed point a and a variable point x on the t-axis (for now we'll assume $x > a$). The basic ideas are illustrated in Figure C2C 4.1 for the function $y = 2t$, using $a = 0$. At this point, the primary goal is to collect values of $A(x)$ for various x-values to see if we can detect a pattern and actually come up with an expression for $A(x)$. For this illustration, a triangle $\left(A = \dfrac{1}{2}bh\right)$ is formed at each value of x, where $b = x$ and $h = 2x$.

EXAMPLE 2 ▶ **Developing an Area Function**

For the area function illustrated in Figure C2C4.1, evaluate $A(x)$ for integer values 1 through 5, then try to name an expression for $A(x)$.

Solution ▶ For clarity, we'll organize the information as follows:

Base of Triangle	Height of Triangle	Area of Triangle $= \dfrac{1}{2}bh$
1	2	$\dfrac{1}{2}(1)(2) = 1$
2	4	$\dfrac{1}{2}(2)(4) = 4$
3	6	$\dfrac{1}{2}(3)(6) = 9$
4	8	$\dfrac{1}{2}(4)(8) = 16$
5	10	$\dfrac{1}{2}(5)(10) = 25$

We note the results are all perfect squares, and we find that the area function for $y = 2t$ is $A(x) = x^2$ for any x.

Now try Exercises 5 and 6 ▶

Expressions Involving e^x

Because of some extraordinary properties, the function $y = e^x$ is one of the most commonly seen functions in calculus and its applications. Here we'll simplify expressions containing e^x to gain some facility for their application in a calculus context. Essentially, it will be a reminder that the standard properties of exponents also apply to these base-e exponential expressions.

EXAMPLE 3 ▶ **Simplifying Expressions Involving e**

A function called the *hyperbolic cotangent*
$(y = \coth x)$ is defined by $f(x) = \dfrac{e^x + e^{-x}}{e^x - e^{-x}}$. Using

the tools of calculus, it can be shown that the slope of a line drawn tangent to the curve at x, is given by

$$f(x) = \frac{(e^x - e^{-x})(e^x - e^{-x}) - (e^x + e^{-x})(e^x + e^{-x})}{(e^x - e^{-x})^2}.$$

Simplify the expression to show that slope of the tangent line is always negative.

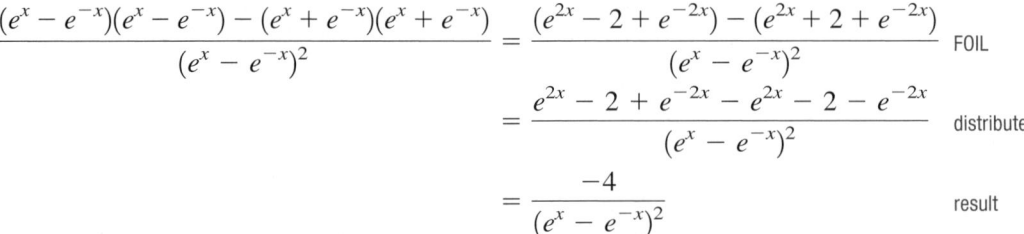

Solution ▶ The products in the numerator can be expanded using the FOIL process, giving

$$\frac{(e^x - e^{-x})(e^x - e^{-x}) - (e^x + e^{-x})(e^x + e^{-x})}{(e^x - e^{-x})^2} = \frac{(e^{2x} - 2 + e^{-2x}) - (e^{2x} + 2 + e^{-2x})}{(e^x - e^{-x})^2} \quad \text{FOIL}$$

$$= \frac{e^{2x} - 2 + e^{-2x} - e^{2x} - 2 - e^{-2x}}{(e^x - e^{-x})^2} \quad \text{distribute}$$

$$= \frac{-4}{(e^x - e^{-x})^2} \quad \text{result}$$

Notice that $x = 0$ is not in the domain of f. Since the square of any quantity is nonnegative and the numerator is a negative constant, the slope of the tangent line will always be negative (see graph).

Now try Exercises 7 through 10 ▶

Connections to Calculus Exercises

Rewrite the following expressions using properties of logarithms.

1. $\ln\left(e^{5x}x^{\frac{3}{2}}\right)$

2. $\log_b\left[\dfrac{x^3\sqrt{y}}{(z+1)^4}\right]$

3. $\log\left(\dfrac{x^5}{y^3\sqrt{z}}\right)$

4. $\ln\left[\dfrac{x^2\sqrt[3]{x-1}}{\sqrt{(x+1)^3}}\right]$

5. Find the area function $A(x)$ for the function $y=t$ with $a=0$. Include a graph with your solution.

6. Find the area function $A(x)$ for the function $y=2t+1$ with $a=0$. Include a graph with your solution.

The expressions and equations that follow are taken from the area of calculus noted in *italics*. Simplify, solve, or verify as indicated.

7. *Identities:* Verify that

$$\dfrac{\dfrac{1}{2}(e^x+e^{-x})+\dfrac{1}{2}(e^x-e^{-x})}{\dfrac{1}{2}(e^x+e^{-x})-\dfrac{1}{2}(e^x-e^{-x})}=e^{2x}$$

8. *Hyperbolic functions:* Verify that

$$\left(\dfrac{e^x+e^{-x}}{2}\right)^2-\left(\dfrac{e^x-e^{-x}}{2}\right)^2=1$$

9. *Inverse hyperbolic functions:* Solve for t:
$$xe^{2t}+x=e^{2t}-1$$

10. *Graphing:* It can be shown that the maximum value of $f(x)=\sqrt{1+xe^{-3x}}$ is located at the zero of

$$f'(x)=\dfrac{1}{2}(1+xe^{-3x})^{-\frac{1}{2}}[x(-3e^{-3x})+e^{-3x}].$$

Find the x-coordinate of the maximum value.

5

An Introduction to Trigonometric Functions

CHAPTER OUTLINE

CHAPTER CONNECTIONS

While rainbows have been admired for centuries for their beauty and color, understanding the physics of rainbows is of fairly recent origin. Answers to questions regarding their seven-color constitution, the order the colors appear, the circular shape of the bow, and their relationship to moisture all have answers deeply rooted in mathematics. The relationship between light and color can be understood in terms of trigonometry, with questions regarding the apparent height of the rainbow, as well as the height of other natural and man-made phenomena, found using the trigonometry of right triangles. This application appears as Exercise 85 in Section 5.6

Connections to Calculus

The trigonometry of right triangles also plays an important role in a study of calculus, as we seek to simplify certain expressions, or convert from one system of graphing to another. The *Connections to Calculus* feature following Chapter 5 offers additional practice and insight in preparation for a study of calculus.

Learning Objectives

In Section 5.1 you will learn how to:

☐ **A.** Use the vocabulary associated with a study of angles and triangles

☐ **B.** Find fixed ratios of the sides of special triangles

☐ **C.** Use radians for angle measure and compute circular arc length and area using radians

☐ **D.** Convert between degrees and radians for nonstandard angles

☐ **E.** Solve applications involving angular velocity and linear velocity using radians

Trigonometry, like her sister science geometry, has its origins deeply rooted in the practical use of measurement and proportion. In this section, we'll look at the fundamental concepts on which trigonometry is based, which we hope will lead to a better understanding and a greater appreciation of the wonderful study that trigonometry has become.

A. Angle Measure in Degrees

Beginning with the common notion of a straight line, a **ray** is a half line, or all points extending from a single point, in a single direction. An **angle** is the joining of two rays at a common endpoint called the **vertex.** Arrowheads are used to indicate the half lines continue forever and can be extended if necessary. Angles can be named using a single letter at the vertex, the letters from the rays forming the sides, or by a single Greek letter, with the favorites being **alpha** α, **beta** β, **gamma** γ, and **theta** θ. The symbol $\angle$ is often used to designate an angle (see Figure 5.1).

Figure 5.1

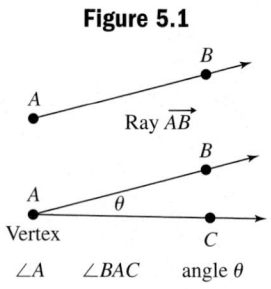

Euclid (325–265 B.C.), often thought of as the *father of geometry,* described an angle as "the inclination of one to another of two lines which meet in a plane." This *amount of inclination* gives rise to the common notion of angle measure in degrees, often measured with a semicircular **protractor** like the one shown in Figure 5.2. The notation for degrees is the ° symbol. By definition $1°$ is $\frac{1}{360}$ of a full rotation, so this

Figure 5.2

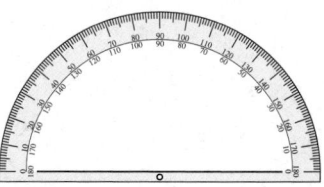

protractor can be used to measure any angle from $0°$ (where the two rays are coincident), to $180°$ (where they form a straight line). An angle measuring $180°$ is called a **straight angle,** while an angle that measures $90°$ is called a **right angle.** Two angles that sum to $90°$ are said to be **complementary,** while two that sum to $180°$ are **supplementary** angles. Recall the "⌐" symbol represents a $90°$ angle.

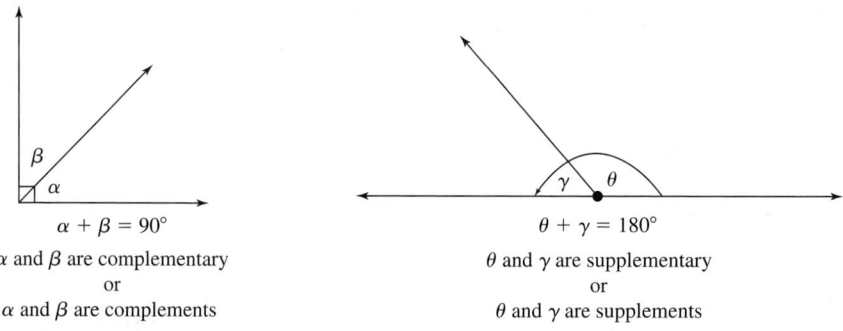

EXAMPLE 1 ▶ Finding the Complement and Supplement of an Angle

Determine the measure of each angle described.

    a. the complement of a $57°$ angle   **b.** the supplement of a $132°$ angle

    c. the measure of angle θ shown in the figure

Solution ▶   **a.** The complement of $57°$ is $33°$ since $90 - 57 = 33° \Rightarrow 33 + 57 = 90°$.

    b. The supplement of $132°$ is $48°$ since $180 - 132 = 48° \Rightarrow 48 + 132 = 180°$.

    c. Since θ and $39°$ are complements, $\theta = 90 - 39 = 51°$.

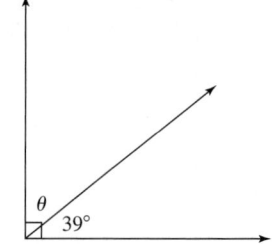

Now try Exercises 7 through 10 ▶

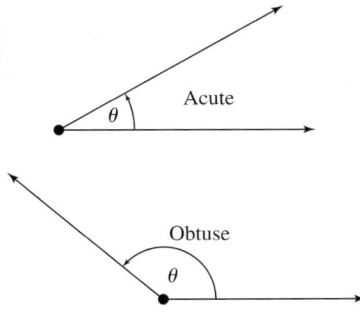

Acute

Obtuse

In the study of trigonometry, it helps to further classify the various angles we encounter. An angle greater than 0° but less than 90° is called an **acute** angle. An angle greater than 90° but less than 180° is called an **obtuse** angle. For very fine measurements, each degree is divided into 60 smaller parts called **minutes,** and each minute into 60 smaller parts called **seconds.** This means that a minute is $\frac{1}{60}$ of a degree, while a second is $\frac{1}{3600}$ of a degree. The angle whose measure is "sixty-one degrees, eighteen minutes, and forty-five seconds" is written as 61° 18′ 45″. The degrees-minutes-seconds (DMS) method of measuring angles is commonly used in aviation and navigation, while in other areas **decimal degrees** such as 61.3125° are preferred. You will sometimes be asked to convert between the two.

EXAMPLE 2 ▶ **Converting Between Decimal Degrees and Degrees/Minutes/Seconds**

Convert as indicated.

    a. 61° 18′ 45″ to decimal degrees

    b. 142.2075° to DMS

Solution ▶   **a.** Since $1' = \frac{1}{60}$ of a degree and $1'' = \frac{1}{3600}$ of a degree, we have

$$61° \ 18' \ 45'' = \left[61 + 18\left(\frac{1}{60}\right) + 45\left(\frac{1}{3600}\right) \right]^{\circ}$$

$$= 61.3125°$$

    b. For the conversion to DMS we write the fractional part separate from the whole number part to compute the number of degrees and minutes represented, then repeat the process to find the number of degrees, minutes, and seconds:

$142.2075° = 142° + 0.2075°$	separate fractional part from the whole
$= 142° + 0.2075(60)'$	$0.2075° = 0.2075 \cdot 1°$; substitute 60′ for 1°
$= 142° \ 12.45'$	result in degrees and minutes
$= 142° \ 12' + 0.45'$	separate fractional part from the whole
$= 142° \ 12' + 0.45(60)''$	$0.45' = 0.45 \cdot 1'$; substitute 60″ for 1′
$= 142° \ 12' \ 27''$	result in degrees, minutes, and seconds

☑ **A.** You've just learned how to use the vocabulary associated with a study of angles and triangles

Now try Exercises 11 through 26 ▶

B. Triangles and Properties of Triangles

A triangle is a closed plane figure with three straight sides and three angles. It is customary to name each angle using a capital letter and the side opposite the angle using the corresponding lowercase letter. Regardless of their size or orientation, triangles have the following properties.

Properties of Triangles

Given triangle ABC with sides a, b, and c respectively,

    I. The sum of the angles is 180°:

        $A + B + C = 180°$

    II. The combined length of any two sides exceeds that of the third side:

        $a + b > c$, $a + c > b$, and $b + c > a$.

    III. Larger angles are opposite larger sides:

        If $C > B$, then $c > b$.

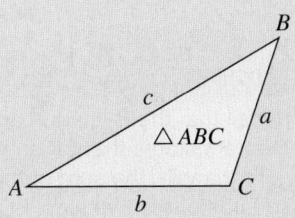

Two triangles are **similar triangles** if corresponding angles are equal, meaning for $\triangle ABC$ and $\triangle DEF$, $A = D$, $B = E$, and $C = F$. Since antiquity it's been known that *if two triangles are similar, corresponding sides are proportional* (corresponding sides are those opposite the equal angles from each triangle). This relationship, used extensively by the engineers of virtually all ancient civilizations, is very important to our study of trigonometry. Example 3 illustrates how proportions and similar triangles are often used.

EXAMPLE 3 ▶ **Using Similar Triangles to Find Heights**

To estimate the height of a flagpole, Carrie reasons that $\triangle ABC$ formed by her height and shadow must be similar to $\triangle DEF$ formed by the flagpole. She is 5 ft 6 in. tall and casts an 8-ft shadow, while the shadow of the flagpole measures 44 ft. How tall is the pole?

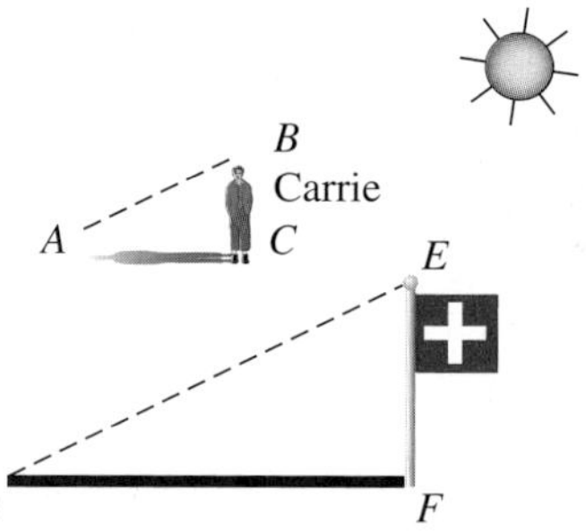

Solution ▶ Let H represent the height of the flagpole.

$$\frac{\text{Height}}{\text{Shadow Length}} : \frac{5.5}{8} = \frac{H}{44} \qquad \text{original proportion, } 5'\,6'' = 5.5 \text{ ft}$$

$$8H = 242 \qquad \text{cross multiply}$$

$$H = 30.25 \qquad \text{result}$$

The flagpole is $30\frac{1}{4}$ ft tall (30 ft 3 in.).

Now try Exercises 27 through 34 ▶

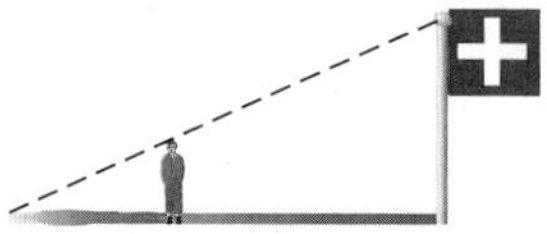

Figure 5.3 shows Carrie standing along the shadow of the flagpole, again illustrating the proportional relationships that exist. Early mathematicians recognized the power of these relationships, realizing that if the triangles were similar and the related fixed proportions were known, they had the ability to find mountain heights, the widths of lakes, and even the ability to estimate the distance to the Sun and Moon. What was needed was an accurate and systematic method of finding these "fixed proportions" for various angles, so they could be applied more widely. In support of this search, two special triangles were used. These triangles, commonly called **45-45-90** and **30-60-90** triangles, are *special* because no estimation or interpolation is needed to find the relationships between their sides. For the first, consider an isosceles right triangle—a right triangle with two equal sides and two 45° angles (Figure 5.4). After naming the equal sides x and the hypotenuse h, we can apply the Pythagorean theorem to find a relationship between the sides and the hypotenuse in terms of x.

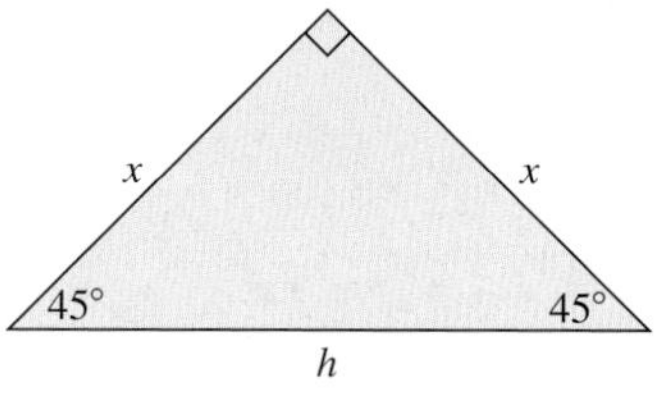

Figure 5.3

Figure 5.4

$$c^2 = a^2 + b^2 \qquad \text{Pythagorean theorem}$$

$$h^2 = x^2 + x^2 \qquad \text{substitute } x \text{ for } a, x \text{ for } b, \text{ and } h \text{ for } c$$

$$= 2x^2 \qquad \text{combine like terms}$$

$$h = \sqrt{2}x \qquad \text{solve for } h\,(h > 0)$$

This important result is summarized in the following box.

WORTHY OF NOTE

Recall that the Pythagorean theorem states that for any right triangle, the sum of the squares of the two legs, is equal to the square of the hypotenuse: $a^2 + b^2 = c^2$.

45-45-90 Triangles

Given a 45-45-90 triangle with one side of length x, the relationship between corresponding sides is:

$$1x : 1x : \sqrt{2}\,x.$$

(1) The two legs are equal
(2) The hypotenuse is $\sqrt{2}$ times the length of either leg.

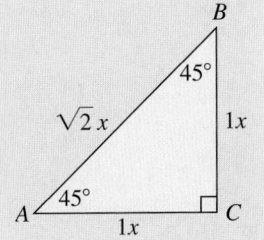

The proportional relationship for a 30-60-90 triangle is developed in **Exercise 110,** and the result is stated here.

30-60-90 Triangles

Given a 30-60-90 triangle with the shortest side of length x, the relationship between corresponding sides is:

$$1x : \sqrt{3}\,x : 2x.$$

(1) The hypotenuse is 2 times the length of the shorter leg,
(2) The longer leg is $\sqrt{3}$ times the length of the shorter leg.

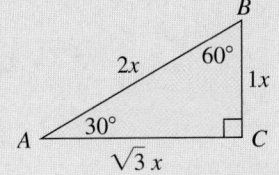

EXAMPLE 4 ▶ Applications of 45-45-90 Triangles: The Height of a Cliff

A group of campers has pitched their tent some distance from the base of a tall cliff. The evening's conversation turns to a discussion of the cliff's height, and they all lodge an estimate. Then one of them says, "Wait . . . how will we know who's closest?" Describe how a 45-45-90 triangle can help determine a winner.

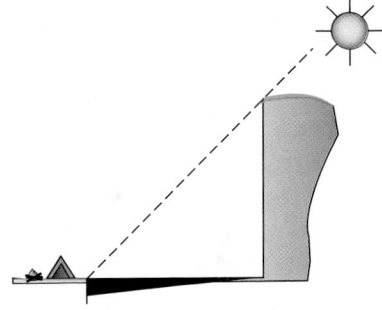

Solution ▶ In the morning, cut a pole equal in height to any of the campers. Then follow the shadow of the cliff as it moves, with the selected camper occasionally laying the pole at her feet and checking her shadow's length against the length of the pole. At the moment her shadow is equal to the pole's length, the sun is shining at a 45° angle and the campers can use the pole to measure the shadow cast by the cliff (by counting the number of pole lengths needed to reach it), which will be equal to its height since a 45-45-90 triangle is formed.

☑ **B.** You've just learned how to find fixed ratios of the sides of special triangles

Now try Exercises 35 and 36 ▶

C. Angle Measure in Radians; Arc Length and Area

Figure 5.5

Counter-clockwise

Clockwise

As an alternative to viewing angles as "the amount of inclination" between two rays, angle measure can also be considered as the *amount of rotation* from a fixed ray called the **initial side,** to a rotated ray called the **terminal side.** This enables angle measure to be free from the context of a triangle, and allows for positive or negative angles, depending on the direction of rotation. Angles formed by a counterclockwise rotation are considered **positive angles,** and angles formed by a clockwise rotation are **negative angles** (see Figure 5.5). We can then name an angle of any size, including those greater than 360° where the amount of rotation exceeds one revolution. See Figures 5.6 through 5.10.

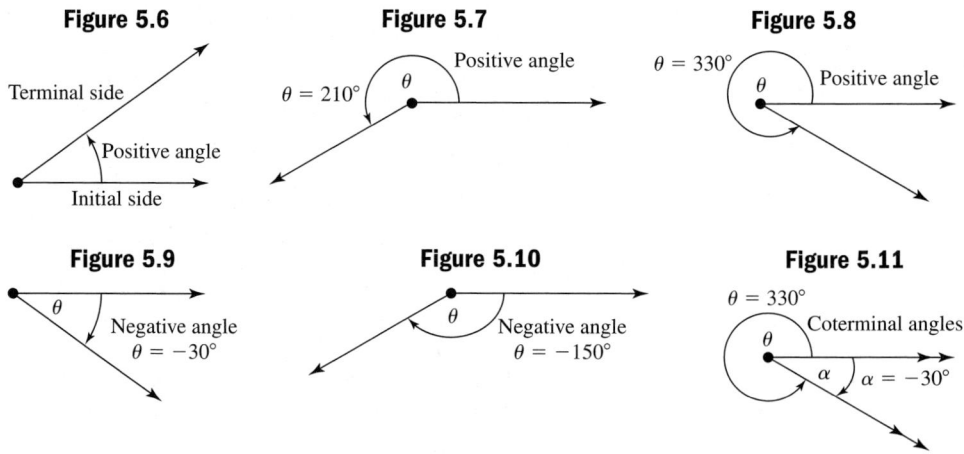

Figure 5.6   Figure 5.7   Figure 5.8

Figure 5.9   Figure 5.10   Figure 5.11

Note in Figure 5.11 that angle $\theta = 330°$ and angle $\alpha = -30°$ share the same initial and terminal sides and are called **coterminal angles.** Coterminal angles will always differ by 360°, meaning for that any integer k, angles θ and $\theta + 360k$ will be coterminal.

EXAMPLE 5 ▶ **Finding Coterminal Angles**

Find two positive angles and two negative angles that are coterminal with 60°.

Solution ▶   For $k = -2, 60° + 360(-2) = -660°$.
For $k = -1, 60° + 360(-1) = -300°$.
For $k = 1, 60° + 360(1) = 420°$.
For $k = 2, 60° + 360(2) = 780°$.

Note that many other answers are possible.

Now try Exercises 37 through 40 ▶

WORTHY OF NOTE

Using the properties of ratios, we note that since both r (radius) and s (arc length) are measured in like units, the units actually "cancel" making radians a unitless measure:

$$\theta = \frac{s \text{ units}}{r \text{ units}} = \frac{s}{r}.$$

An angle is said to be in **standard position** in the rectangular coordinate system if its vertex is at the origin and the initial side is along the positive x-axis. In standard position, the terminal sides of 90°, 180°, 270°, and 360° angles coincide with one of the axes and are called **quadrantal angles.** To help develop these ideas further, we use a **central circle,** that is, a circle in the xy-plane with its center at the origin. A **central angle** is an angle whose vertex is at the center of the circle. For central angle θ intersecting the circle at points B and C, we say circular arc BC, denoted $\overparen{BC}$, **subtends** $\angle BAC$, as shown in Figure 5.12. The letter s is commonly used to represent arc length, and if we define **1 radian** (abbreviated *rad*) to be the measure of an angle subtended by an arc equal in length to the radius, then $\theta = 1$ rad when $s = r$ (see Figure 5.13). We can then find the radian measure of any central angle by dividing the length of the subtended arc by r: $\frac{s}{r} = \theta$ radians.

Multiplying both sides by r gives a formula for the length of any arc subtended on a circle of radius r: $s = r\theta$ if θ is in radians.

Figure 5.12

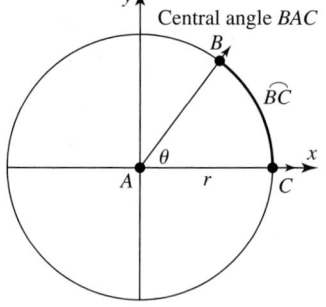

Central angle BAC

Figure 5.13

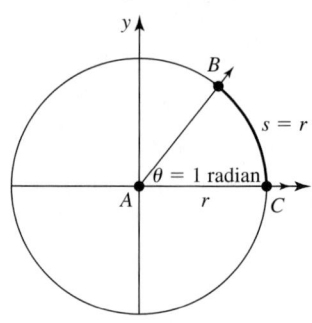

Radians

If central angle θ is subtended by an arc that is equal in length to the radius, then
$$\theta = 1 \; radian.$$

Arc Length

If θ is a central angle in a circle of radius r, then the length of the subtended arc s is
$$s = r\theta,$$
provided θ is expressed in radians.

EXAMPLE 6 ▶ **Using the Formula for Arc Length**

If the circle in Figure 5.13 has radius $r = 10$ cm, what is the length of the arc subtended by an angle of 3.5 rad?

Solution ▶ Using the formula $s = r\theta$ with $r = 10$ and $\theta = 3.5$ gives

$$s = 10(3.5) \quad \text{substitute 10 for } r \text{ and 3.5 for } \theta$$
$$s = 35 \qquad \text{result}$$

The subtended arc has a length of 35 cm.

Now try Exercises 41 through 52 ▶

Using a central angle θ measured in radians, we can also develop a formula for the **area of a circular sector** (a pie slice) using a proportion. Recall the circumference of a cicle is $C = 2\pi r$. While you may not have considered this before, note the formula can be written as $C = 2\pi \cdot r$, which implies that the radius, or an arc of length r, can be wrapped around the circumference of the circle $2\pi \approx 6.28$ times, as illustrated in Figure 5.14. This shows the radian measure of a full 360° rotation is 2π: $2\pi \; rad = 360°$. This can be verified as before, using the relation $\theta \; \text{radians} = \dfrac{s}{r} = \dfrac{2\pi r}{r} = 2\pi$. The ratio of the area of a sector to the total area will be identical to the ratio of the subtended angle to one full rotation. Using $\mathcal{A}$ to represent the area of the sector, we have $\dfrac{\mathcal{A}}{\pi r^2} = \dfrac{\theta}{2\pi}$ and solving for $\mathcal{A}$ gives $\mathcal{A} = \dfrac{1}{2}r^2\theta$.

Figure 5.14

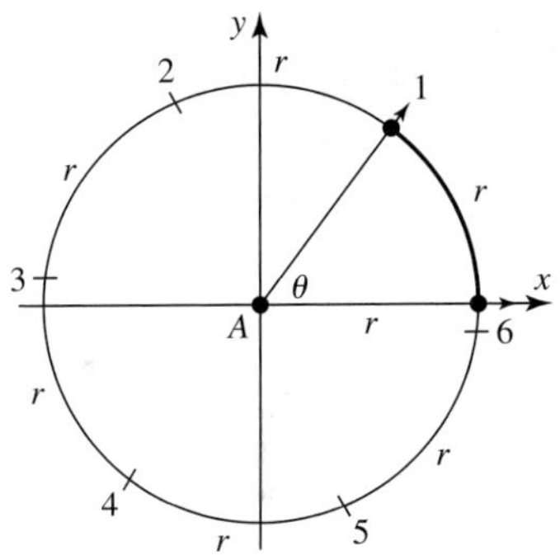

Area of a Sector

If θ is a central angle in a circle of radius r, the area of the sector formed is

$$\mathcal{A} = \frac{1}{2}r^2\theta,$$

provided θ is expressed in radians.

EXAMPLE 7 ▶ **Using the Formula for the Area of a Sector**

What is the area of the circular sector formed by a central angle of $\dfrac{3\pi}{4}$, if the radius of the circle is 72 ft? Round to tenths.

Solution ▶ Using the formula $\mathcal{A} = \frac{1}{2}r^2\theta$ we have

☑ **C.** You've just learned how to use radians for angle measure and to compute circular arc length and area using radians

$$\mathcal{A} = \left(\frac{1}{2}\right)(72)^2\left(\frac{3\pi}{4}\right) \quad \text{substitute 72 for } r, \frac{3\pi}{4} \text{ for } \theta$$

$$= 1944\pi \text{ ft}^2 \quad \text{result}$$

The area of this sector is approximately 6107.3 ft².

Now try Exercises 53 through 64 ▶

D. Converting Between Degrees and Radians

In addition to its use in developing formulas for arc length and the area of a sector, the relation 2π rad $= 360°$ enables us to state the radian measures of the standard angles using a simple division. For π rad $= 180°$ we have

division by 2: $\frac{\pi}{2} = 90°$ division by 3: $\frac{\pi}{3} = 60°$

division by 4: $\frac{\pi}{4} = 45°$ division by 6: $\frac{\pi}{6} = 30°$.

See Figure 5.15. The radian measures of these standard angles play a major role in this chapter, and you are encouraged to become very familiar with them. Additional conversions can quickly be found using multiples of these four. For example, multiplying both sides of $\frac{\pi}{3} = 60°$ by two gives $\frac{2\pi}{3} = 120°$. The relationship $\pi = 180°$ also gives the factors needed for converting from degrees to radians or from radians to degrees, even if θ is a nonstandard angle. Dividing by π we have $1 = \frac{180°}{\pi}$, while division by 180° shows $1° = \frac{\pi}{180°}$. Multiplying a given angle by the appropriate conversion factor gives the equivalent measure.

Figure 5.15

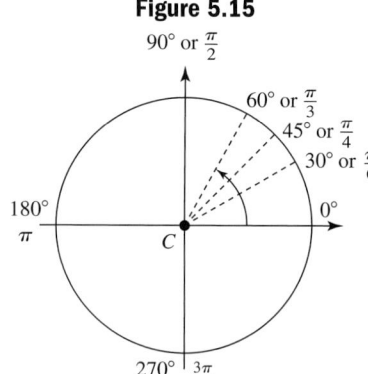

Degrees/Radians Conversion Factors

To convert from radians to degrees: multiply by $\dfrac{180°}{\pi}$.

To convert from degrees to radians: multiply by $\dfrac{\pi}{180°}$.

EXAMPLE 8 ▶ **Converting Between Radians and Degrees**

Convert each angle as indicated:

a. 75° to radians.   **b.** $\dfrac{\pi}{24}$ to degrees.

Solution ▶ **a.** For degrees to radians, use the conversion factor $\dfrac{\pi}{180°}$.

$$75° = 75° \cdot \frac{\pi}{180°} = \frac{5\pi}{12} \qquad \frac{75}{180} \quad \frac{5}{12}$$

b. For radians to degrees, use the conversion factor $\dfrac{180°}{\pi}$:

$$\frac{\pi}{24} = \frac{\pi}{24} \cdot \left(\frac{180°}{\pi}\right) = 7.5° \qquad \frac{\pi}{\pi} = 1, \frac{180}{24} = 7.5$$

Now try Exercises 65 through 92 ▶

One example where these conversions are useful is in applications involving longitude and latitude (see Figure 5.16). The **latitude** of a fixed point on the Earth's surface tells how many degrees north or south of the equator the point is, as measured from the center of the Earth. The **longitude** of a fixed point on the Earth's surface tells how many degrees east or west of the Prime Meridian (through Greenwich, England) the point is, as measured along the equator to the longitude line going through the point. For example, the city of New Orleans, Louisiana, is located at 30° N latitude, 90° W longitude (see Figure 5.16).

Figure 5.16

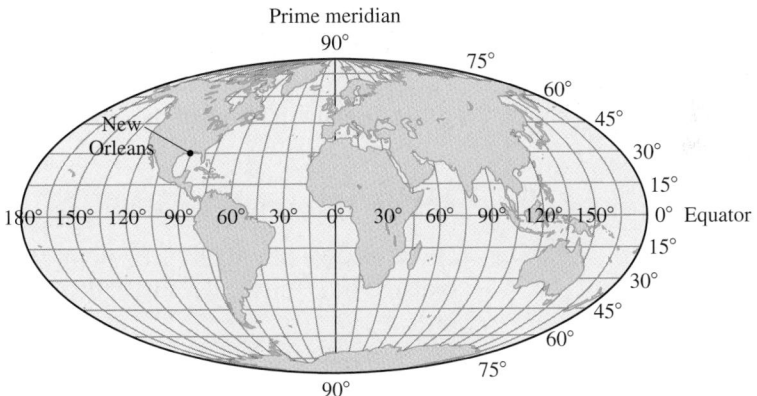

EXAMPLE 9 ▶ **Applying the Arc Length Formula: Distances Between Cities**

The cities of Quito, Ecuador, and Macapá, Brazil, both lie very near the equator, at a latitude of 0°. However, Quito is at approximately 78° west longitude, while Macapá is at 51° west longitude (see Figure 5.16). Assuming the Earth has a radius of 3960 mi, how far apart are these cities?

Solution ▶ First we note that $(78 - 51)° = 27°$ of longitude separate the two cities. Using the conversion factor $1° = \dfrac{\pi}{180}$ we find the equivalent radian measure is $27\left(\dfrac{\pi}{180}\right) = \dfrac{3\pi}{20}$. The arc length formula gives

$$\begin{aligned} s &= r\theta & \text{arc length formula; } \theta \text{ in radians} \\ &= 3960\left(\frac{3\pi}{20}\right) & \text{substitute 3960 for } r \text{ and } \frac{3\pi}{20} \text{ for } \theta \\ &= 594\pi & \text{result} \end{aligned}$$

Quito and Macapá are approximately 1866 mi apart (see *Worthy of Note* in the margin).

> **WORTHY OF NOTE**
>
> Note that $r = 3960$ mi was used because Quito and Macapá *are both on the equator.* For other cities sharing the same longitude but not on the equator, the radius of the Earth *at that longitude* must be used. See Section 5.2, Exercise 91.

☑ **D.** You've just learned how to convert between degrees and radians for nonstandard angles

Now try Exercises 95 through 98 ▶

E. Angular and Linear Velocity

The **angular velocity** of an object is defined as the *amount of rotation* per unit time. Here, we often use the symbol ω (omega) to represent the angular velocity, and θ to represent the angle through which the terminal side has rotated, measured in radians: $\omega = \dfrac{\theta}{t}$. For instance, a Ferris wheel turning at 10 revolutions per minute has an angular velocity of

$$\omega = \frac{10 \text{ revolutions}}{1 \text{ min}} \qquad \omega = \frac{\theta}{t}$$

$$= \frac{10(2\pi)}{1 \text{ min}} \qquad \text{substitute } 2\pi \text{ for 1 revolution}$$

$$= \frac{20\pi \text{ rad}}{1 \text{ min}} \qquad 10(2) = 20$$

WORTHY OF NOTE

Generally speaking, the *velocity* of an object is its change in position per unit time, and can be either positive or negative. The *rate* or *speed* of an object is the magnitude of the velocity, regardless of direction.

The **linear velocity** of an object is defined as a *change of position* or *distance traveled* per unit time. In the context of angular motion, we consider the distance traveled by a point on the circumference of the Ferris wheel, *which is equivalent to the length of the resulting arc s*. This relationship is expressed as $V = \dfrac{s}{t}$, a formula that can be written directly in terms of the angular velocity since $s = r\theta$: $V = \dfrac{r\theta}{t} = r\left(\dfrac{\theta}{t}\right) = r\omega$.

Angular and Linear Velocity

Given a circle of radius r with point P on the circumference, and central angle θ in radians with P on the terminal side. If P moves along the circumference at a uniform rate:

1. The rate at which θ changes is called the *angular velocity* ω,

$$\omega = \frac{\theta}{t}.$$

2. The rate at which the position of P changes is called the *linear velocity* V,

$$V = \frac{r\theta}{t} \quad \Rightarrow \quad V = r\omega.$$

EXAMPLE 10 ▶ **Using Angular Velocity to Determine Linear Velocity**

The wheels on a racing bicycle have a radius of 13 in. How fast is the cyclist traveling in miles per hour, if the wheels are turning at 300 rpm?

Solution ▶ Note that $\omega = \dfrac{300 \text{ rev}}{1 \text{ min}} = \dfrac{300(2\pi)}{1 \text{ min}} = \dfrac{600\pi}{1 \text{ min}}$.

Using the formula $V = r\omega$ gives a linear velocity of

$$V = (13 \text{ in.})\frac{600\pi}{1 \text{ min}} \approx \frac{24{,}504.4 \text{ in.}}{1 \text{ min}}.$$

To convert this to miles per hour we convert minutes to hours (1 hr = 60 min) and inches to miles (1 mi = 5280 × 12 in.):

$$\left(\frac{24{,}504.4 \text{ in.}}{1 \text{ min}}\right)\left(\frac{60 \text{ min}}{1 \text{ hr}}\right)\left(\frac{1 \text{ mi}}{63{,}360 \text{ in.}}\right) \approx 23.2 \text{ mph}.$$

The bicycle is traveling about 23.2 mph.

Now try Exercises 99 through 102 ▶

To help understand the relationship between angular velocity and linear velocity, consider two large rollers with a radius of 1.6 ft, used to move an industrial conveyor belt. The rollers have a circumference of $C = 2\pi(1.6 \text{ ft}) \approx 10.05 \text{ ft}$, meaning that for each revolution of the rollers, an object on the belt will move 10.05 ft (from P_1 to P_2).

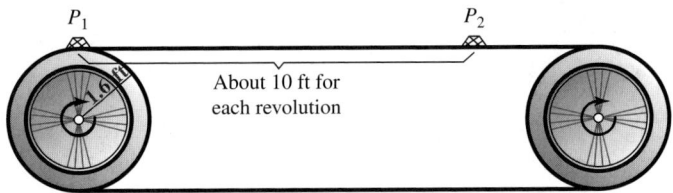

If the rollers are rotating at 20 revolutions per minute (rpm), an object on the belt (or a point on the circumference of a roller), will be moving at a rate of $20 \cdot 10.05 = 201$ ft/min (about 2.3 miles per hour). In other words,

$$\omega = \frac{20 \text{ revolutions}}{1 \text{ min}} \qquad \omega = \frac{\theta}{t}$$

$$= \frac{20 \cdot 2\pi}{1 \text{ min}} = \frac{40\pi}{1 \text{ min}} \qquad \text{substitute } 2\pi \text{ for 1 revolution}$$

$$V = r\omega \qquad \text{formula for velocity}$$

$$= (1.6 \text{ ft})\frac{40\pi}{1 \text{ min}} \qquad \text{substitute 1.6 ft for } r, 40\pi \text{ for } \omega$$

$$\approx 201 \text{ ft per min} \qquad \text{result}$$

☑ **E.** You've just learned how to solve applications involving angular velocity and linear velocity using radians

---

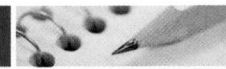

5.1 EXERCISES

▶ CONCEPTS AND VOCABULARY

Fill in each blank with the appropriate word or phrase. Carefully reread the section if needed.

1. _____ angles sum to 90°. Supplementary angles sum to _____°. Acute angles are _____ than 90°. Obtuse angles are _____ than 90°.

2. The expression "theta equals two degrees" is written _____ using the ""°"" notation. The expression, "theta equals two radians" is simply written _____.

3. The formula for arc length is $s = $ _____. The area of a sector is $A = $ _____. For both formulas, θ must be in _____.

4. If θ is not a special angle, multiply by _____ to convert radians to degrees. To convert degrees to radians, multiply by _____.

5. Discuss/Explain the difference between angular velocity and linear velocity. In particular, why does one depend on the radius while the other does not?

6. Discuss/Explain the difference between 1° and 1 radian. Exactly what is a radian? Without any conversions, explain why an angle of 4 rad terminates in QIII.

▶ DEVELOPING YOUR SKILLS

Determine the measure of the angle described.

7. **a.** The complement of a 12.5° angle
   b. The supplement of a 149.2° angle

8. **a.** The complement of a 62.4° angle
   b. The supplement of a 74.7° angle

9. The measure of angle α

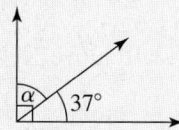

10. The measure of angle β

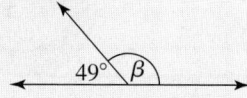

Convert from DMS (degree/minute/seconds) notation to decimal degrees.

11. $42°30'$

12. $125°45'$

13. $67°33'18''$

14. $9°15'36''$

15. $285°00'09''$

16. $312°00'54''$

17. $45°45'45''$

18. $30°30'27''$

Convert the angles from decimal degrees to DMS (degree/minute/sec) notation.

19. $20.25°$

20. $40.75°$

21. $67.307°$

22. $83.516°$

23. $275.33°$

24. $330.45°$

25. $5.4525°$

26. $12.3275°$

27. Is the triangle shown possible? Why/why not?

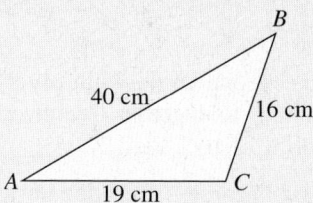

28. Is the triangle below possible? Why/why not?

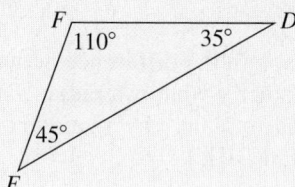

Determine the measure of the angle indicated.

29. angle α

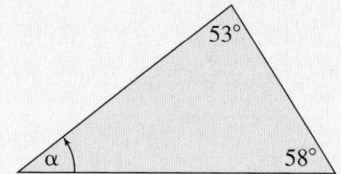

30. angle β

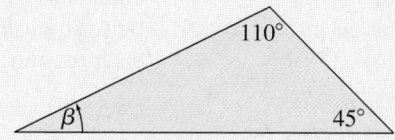

31. $\angle A$

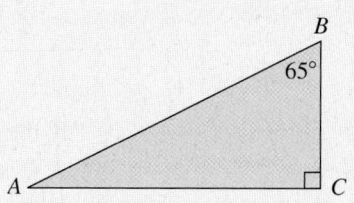

32. $\angle B$

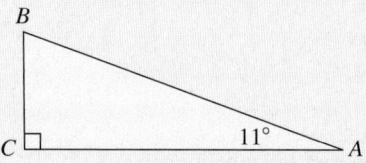

33. Similar triangles: A helicopter is hovering over a crowd of people watching a police standoff in a parking garage across the street. Stewart notices the shadow of the helicopter is lagging approximately 50 m behind a point directly below the helicopter. If he is 2 m tall and casts a shadow of 1.6 m at this time, what is the altitude of the helicopter?

34. Similar triangles: Near Fort Macloud, Alberta (Canada), there is a famous cliff known as *Head Smashed in Buffalo Jump*. The area is now a Canadian National Park, but at one time the Native Americans hunted buffalo by steering a part of the herd over the cliff. While visiting the park late one afternoon, Denise notices that its shadow reaches 201 ft from the foot of the cliff, at the same time she is casting a shadow of $12'1''$. If Denise is $5'4''$. tall, what is the height of the cliff?

Solve using special triangles. Answer in both exact and approximate form.

35. Special triangles: A ladder-truck arrives at a high-rise apartment complex where a fire has broken out. If the maximum length the ladder extends is 82 ft and the angle of inclination is $45°$, how high up the side of the building does the ladder reach? Assume the ladder is mounted atop a 10 ft high truck.

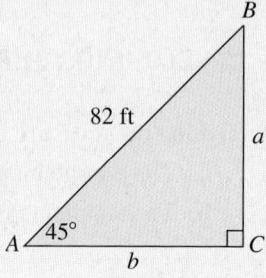

36. Special triangles: A heavy-duty ramp is used to winch heavy appliances from street level up to a warehouse loading

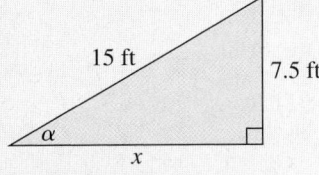

dock. If the ramp is 7.5 ft high and the incline is 15 ft long, (a) what angle α does the dock make with the street? (b) How long is the base of the ramp?

Find two positive angles and two negative angles that are coterminal with the angle given. Answers may vary.

37. $\theta = 75°$ **38.** $\theta = 225°$

39. $\theta = -45°$ **40.** $\theta = -60°$

Use the formula for arc length to find the value of the unknown quantity: $s = r\theta$.

41. $\theta = 3.5; r = 280$ m

42. $\theta = 2.3; r = 129$ cm

43. $s = 2007$ mi; $r = 2676$ mi

44. $s = 4435.2$ km; $r = 12{,}320$ km

45. $\theta = \dfrac{3\pi}{4}; s = 4146.9$ yd

46. $\theta = \dfrac{11\pi}{6}; s = 28.8$ nautical miles

47. $\theta = \dfrac{4\pi}{3}; r = 2$ mi

48. $\theta = \dfrac{3\pi}{2}; r = 424$ in.

49. $s = 252.35$ ft; $r = 980$ ft

50. $s = 942.3$ mm; $r = 1800$ mm

51. $\theta = 320°; s = 52.5$ km

52. $\theta = 220.5°; s = 7627$ m

Use the formula for area of a circular sector to find the value of the unknown quantity: $A = \frac{1}{2}r^2\theta$.

53. $\theta = 5; r = 6.8$ km

54. $\theta = 3; r = 45$ mi

55. $A = 1080$ mi^2; $r = 60$ mi

56. $A = 437.5$ cm^2; $r = 12.5$ cm

57. $\theta = \dfrac{7\pi}{6}; A = 16.5$ m^2

58. $\theta = \dfrac{19\pi}{12}; A = 753$ cm^2

Find the angle, radius, arc length, and/or area as needed, until all values are known.

59.

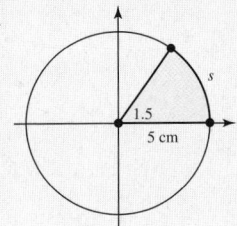

60.

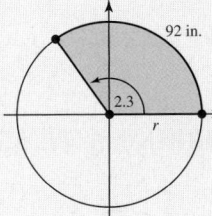

61.

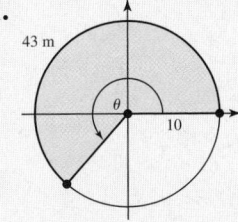

62.

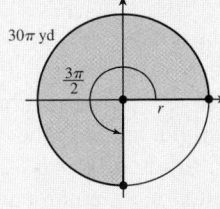

63.

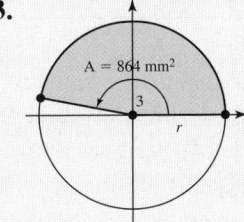

64.

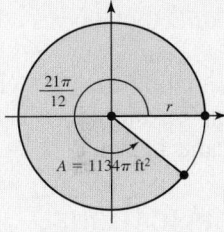

Convert the following degree measures to radians in exact form, without the use of a calculator.

65. $\theta = 360°$ **66.** $\theta = 180°$

67. $\theta = 45°$ **68.** $\theta = 30°$

69. $\theta = 210°$ **70.** $\theta = 330°$

71. $\theta = -120°$ **72.** $\theta = -225°$

Convert each degree measure to radians. Round to the nearest ten-thousandth.

73. $\theta = 27°$ **74.** $\theta = 52°$

75. $\theta = 227.9°$ **76.** $\theta = 154.4°$

Convert each radian measure to degrees, without the use of a calculator.

77. $\theta = \dfrac{\pi}{3}$ **78.** $\theta = \dfrac{\pi}{4}$

79. $\theta = \dfrac{\pi}{6}$ **80.** $\theta = \dfrac{\pi}{2}$

81. $\theta = \dfrac{2\pi}{3}$ **82.** $\theta = \dfrac{5\pi}{6}$

83. $\theta = 4\pi$ **84.** $\theta = 6\pi$

Convert each radian measure to degrees. Round to the nearest tenth.

85. $\theta = \dfrac{11\pi}{12}$ **86.** $\theta = \dfrac{17\pi}{36}$

87. $\theta = 3.2541$ **88.** $\theta = 1.0257$

89. $\theta = 3$ **90.** $\theta = 5$

91. $\theta = -2.5$ **92.** $\theta = -3.7$

▶ WORKING WITH FORMULAS

93. Relationships in a right triangle: $h = \dfrac{ab}{c}$, $m = \dfrac{b^2}{c}$, and $n = \dfrac{a^2}{c}$

Given $\angle C$ is a right angle, and h is the altitude of $\triangle ABC$, then h, m, and n can all be expressed directly in terms of a, b, and c by the relationships shown here. Compute the value of h, m, and n for a right triangle with sides of 8, 15, and 17 cm.

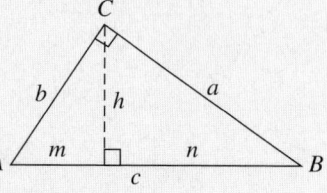

94. The height of an equilateral triangle: $H = \dfrac{\sqrt{3}}{2} S$

Given an equilateral triangle with sides of length S, the height of the triangle is given by the formula shown. Once the height is known the area of the triangle can easily be found (also see Exercise 93). The Gateway Arch in St. Louis, Missouri, is actually composed of stainless steel sections that are equilateral triangles. At the base of the arch the length of the sides is 54 ft. The smallest cross section at the top of the arch has sides of 17 ft. Find the area of these cross sections.

▶ APPLICATIONS

95. Arc length: The city of Pittsburgh, Pennsylvania, is directly north of West Palm Beach, Florida. Pittsburg is at 40.3° north latitude, while West Palm Beach is at 26.4° north latitude. Assuming the Earth has a radius of 3960 mi, how far apart are these cities?

96. Arc length: Both Libreville, Gabon, and Jamame, Somalia, lie near the equator, but on opposite ends of the African continent. If Libreville is at 9.3° east longitude and Jamame is 42.5° east longitude, how wide is the continent of Africa at the equator?

97. Area of a sector: A water sprinkler is set to shoot a stream of water a distance of 12 m and rotate through an angle of 40°. (a) What is the area of the lawn it waters? (b) For $r = 12$ m, what angle is required to water twice as much area? (c) For $\theta = 40°$, what range for the water stream is required to water twice as much area?

98. Area of a sector: A motion detector can detect movement up to 25 m away through an angle of 75°. (a) What area can the motion detector monitor? (b) For $r = 25$ m, what angle is required to monitor 50% more area? (c) For $\theta = 75°$, what range is required for the detector to monitor 50% more area?

99. Riding a round-a-bout: At the park two blocks from our home, the kids' round-a-bout has a radius of 56 in. About the time the kids stop screaming, "Faster, Daddy, faster!" I estimate the round-a-bout is turning at $\frac{3}{4}$

revolutions per second. (a) What is the related angular velocity? (b) What is the linear velocity (in miles per hour) of Eli and Reno, who are "hanging on for dear life" at the rim of the round-a-bout?

100. Carnival rides: At carnivals and fairs, the *Gravity Drum* is a popular ride. People stand along the wall of a circular drum with radius 12 ft, which begins spinning very fast, pinning them against the wall. The drum is then turned on its side by an armature, with the riders screaming and squealing with delight. As the drum is raised to a near-vertical position, it is spinning at a rate of 35 rpm. (a) What is the angular velocity in radians? (b) What is the linear velocity (in miles per hour) of a person on this ride?

101. Speed of a winch: A winch is being used to lift a turbine off the ground so that a tractor-

trailer can back under it and load it up for transport. The winch drum has a radius of 3 in. and is turning at 20 rpm. Find (a) the angular velocity of the drum in radians, (b) the linear velocity of the turbine in feet per second as it is being raised, and (c) how long it will take to get the load to the desired height of 6 ft (ignore the fact that the cable may wind over itself on the drum).

102. **Speed of a current:** An instrument called a flowmeter is used to measure the speed of flowing water, like that in a river or stream. A cruder method involves placing a paddle wheel in the current, and using the wheel's radius and angular velocity to calculate the speed of water flow. If the paddle wheel has a radius of 5.6 ft and is turning at 30 rpm, find (a) the angular velocity of the wheel in radians and (b) the linear velocity of the water current in miles per hour.

On topographical maps, each closed figure represents a fixed elevation (a vertical change) according to a given *contour interval*. The *measured distance* on the map from point *A* to point *B* indicates the horizontal distance or the horizontal change between point *A* and a location directly beneath point *B*, according to a given *scale of distances*.

Exercise 103 and 104

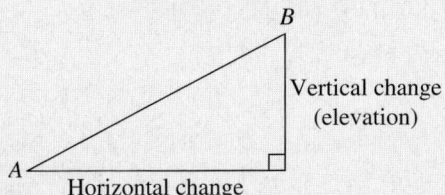

103. **Special triangles:** In the figure shown, the *contour interval* is 1:250 (each figure indicates a change of 250 m in elevation) and the scale of distances is 1 cm = 625 m. (a) Find the change of elevation from *A* to *B*; (b) use a proportion to find the horizontal distance between points *A* and *B* if the measured distance on the map is 1.6 cm; and (c) Draw the corresponding

Exercise 103

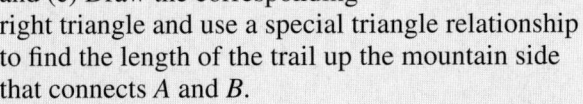

right triangle and use a special triangle relationship to find the length of the trail up the mountain side that connects *A* and *B*.

104. **Special triangles:** As part of park maintenance, the 2 by 4 handrail alongside a mountain trail leading

to the summit of Mount Marilyn must be replaced. In the figure, the *contour interval* is 1:200 (each figure indicates a change of 200 m in elevation) and the scale of distances is 1 cm = 400 m. (a) Find the change of elevation from *A* to *B*; (b) use a proportion to find the horizontal distance between *A* and *B* if the measured distance on the map is 4.33 cm; and (c) draw the corresponding right triangle and use a special triangle relationship to find the length needed to replace the handrail (recall that $\sqrt{3} \approx 1.732$).

Exercise 104

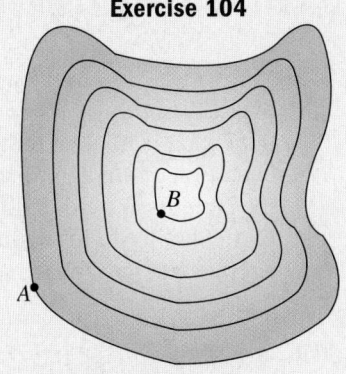

105. **Special triangles:** Two light planes are flying in formation at 100 mph, doing some reconnaissance work. At a designated instant, one pilot breaks to the left at an angle of 90° to the other plane. Assuming they keep the same altitude and continue to fly at 100 mph, use a special triangle to find the distance between them after 0.5 hr.

106. **Special triangles:** Two ships are cruising together on the open ocean at 10 nautical miles per hour. One of them turns to make a 90° angle with the first and increases speed, heading for port. Assuming the first ship continues traveling at 10 knots, use a special triangle to find the speed of the other ship if they are 20 mi apart after 1 hr.

107. **Angular and linear velocity:** The planet Jupiter's largest moon, Ganymede, rotates around the planet at a distance of about 656,000 miles, in an orbit that is perfectly circular. If the moon completes one rotation about Jupiter in 7.15 days, (a) find the angle θ that the moon moves through in 1 day, in both degrees and radians, (b) find the angular velocity of the moon in radians per hour, and (c) find the moon's linear velocity in miles per second as it orbits Jupiter.

108. **Angular and linear velocity:** The planet Neptune has an orbit that is nearly circular. It orbits the Sun at a distance of 4497 million kilometers and completes one revolution every 165 yr. (a) Find the angle θ that the planet moves through in one year in both degrees and radians and (b) find the linear velocity (km/hr) as it orbits the Sun.

▶ **EXTENDING THE CONCEPT**

109. Many methods have been used for angle measure over the centuries, some more logical or meaningful than what is popular today. Do some research on the evolution of angle measure, and compare/contrast the benefits and limitations of each method. In particular, try to locate information on the history of degrees, radians, mils, and gradients, and identify those still in use.

110. Ancient geometers knew that a hexagon (six sides) could be inscribed in a circle by laying out six consecutive chords equal in length to the radius ($r = 10$ cm for illustration). After connecting the diagonals of the

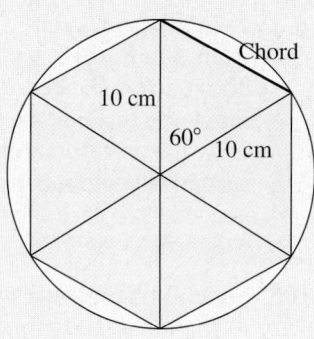

hexagon, six equilateral triangles are formed with sides of 10 cm. Use the diagram given to develop the fixed ratios for the sides of a 30-60-90 triangle. (*Hint:* Use a perpendicular bisector.)

111. The Duvall family is out on a family bicycle ride around Creve Couer Lake. The adult bikes have a pedal sprocket with a 4-in. radius, wheel sprocket with 2-in. radius, and tires with a 13-in. radius. The kids' bikes have pedal sprockets with a 2.5-in. radius, wheel sprockets with 1.5-in. radius, and tires with a 9-in. radius. (a) If adults and kids all pedal at 50 rpm, how far ahead (in yards) are the adults after 2 min? (b) If adults pedal at 50 rpm, how fast do the kids have to pedal to keep up?

▶ **MAINTAINING YOUR SKILLS**

112. (2.6) Describe how the graph of $g(x) = -2\sqrt{x+3} - 1$ can be obtained from transformations of $y = \sqrt{x}$.

113. (4.5) Find the interest rate required for $1000 to grow to $1500 if the money is compounded monthly and remains on deposit for 5 yr.

114. (2.2) Given a line segment with endpoints $(-2, 3)$ and $(6, -1)$, find the equation of the

line that bisects and is perpendicular to this segment.

115. (3.1) Find the equation of the function whose graph is shown.

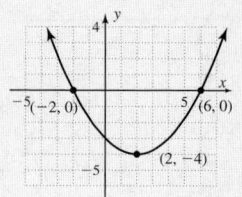

5.2 | Unit Circles and the Trigonometry of Real Numbers

Learning Objectives

In Section 5.2 you will learn how to:

☐ **A.** Locate points on a unit circle and use symmetry to locate other points

☐ **B.** Use special triangles to find points on a unit circle and locate other points using symmetry

☐ **C.** Define the six trig functions in terms of a point on the unit circle

☐ **D.** Define the six trig functions in terms of a real number t

☐ **E.** Find the real number t corresponding to given values of sin t, cos t, and tan t

In this section, we introduce the **trigonometry of real numbers,** a view of trigonometry that can exist free of its historical roots in a study of right triangles. In fact, the ultimate value of these functions is not in their classical study, but in the implications and applications that follow from understanding them as functions of a real number, rather than simply as functions of a given angle.

A. The Unit Circle

A circle is defined as the set of all points in a plane that are a *fixed distance* called the **radius** from a *fixed point* called the **center.** Since the definition involves distance, we can construct the general equation of a circle using the distance formula. Assume the center has coordinates (h, k) and let (x, y) represent any point on the graph. Since the distance between these points is the radius r, the distance formula yields $\sqrt{(x - h)^2 + (y - k)^2} = r$. Squaring both sides gives $(x - h)^2 + (y - k)^2 = r^2$. For central circles both h and k are zero, and the result is the equation for a **central circle** of radius r: $x^2 + y^2 = r^2 \,(r > 0)$. The **unit circle** is defined as a central circle with radius 1 unit: $x^2 + y^2 = 1$. As such, the figure can easily be graphed by drawing a circle through the four **quadrantal points** $(1, 0)$, $(-1, 0)$, $(0, 1)$, and $(0, -1)$ as in Figure 5.17. To find other points on the circle, we simply select any value of x, where $|x| < 1$, then substitute and solve for y; or any value of y, where $|y| < 1$, then solve for x.

Figure 5.17

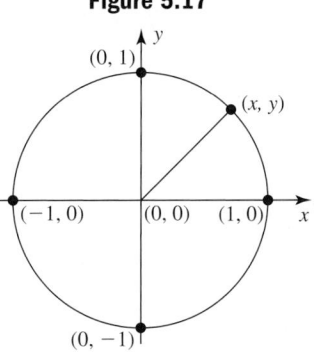

EXAMPLE 1 ▶ **Finding Points on a Unit Circle**

Find a point on the unit circle given $y = \frac{1}{2}$ with (x, y) in QII.

Solution ▶ Using the equation of a unit circle, we have

$$x^2 + y^2 = 1 \qquad \text{unit circle equation}$$

$$x^2 + \left(\frac{1}{2}\right)^2 = 1 \qquad \text{substitute } \frac{1}{2} \text{ for } y$$

$$x^2 + \frac{1}{4} = 1 \qquad \left(\frac{1}{2}\right)^2 = \frac{1}{4}$$

$$x^2 = \frac{3}{4} \qquad \text{subtract } \frac{1}{4}$$

$$x = \pm\frac{\sqrt{3}}{2} \qquad \text{result}$$

With (x, y) in QII, we choose $x = -\frac{\sqrt{3}}{2}$. The point is $\left(-\frac{\sqrt{3}}{2}, \frac{1}{2}\right)$.

Now try Exercises 7 through 18 ▶

Additional points on the unit circle can be found using symmetry. The simplest examples come from the quadrantal points, where $(1, 0)$ and $(-1, 0)$ are on opposite sides of the y-axis, and $(0, 1)$ and $(0, -1)$ are on opposite sides of the x-axis. In general, if a and b are positive real numbers and (a, b) is on the unit circle, then $(-a, b)$, $(a, -b)$, and $(-a, -b)$ are also on the circle *because a circle is symmetric to both axes and the origin*! For the point $\left(-\frac{\sqrt{3}}{2}, \frac{1}{2}\right)$ from Example 1, three other points are $\left(-\frac{\sqrt{3}}{2}, -\frac{1}{2}\right)$ in QIII, $\left(\frac{\sqrt{3}}{2}, -\frac{1}{2}\right)$ in QIV, and $\left(\frac{\sqrt{3}}{2}, \frac{1}{2}\right)$ in QI. See Figure 5.18.

Figure 5.18

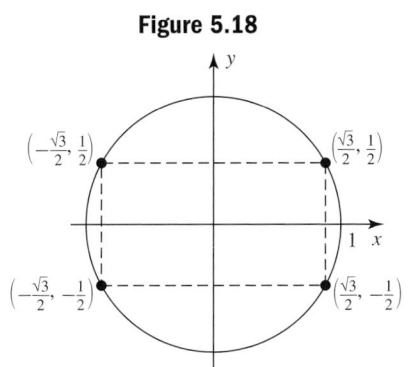

EXAMPLE 2 ▶ **Using Symmetry to Locate Points on a Unit Circle**

Name the quadrant containing $\left(-\frac{3}{5}, -\frac{4}{5}\right)$ and verify it's on a unit circle. Then use symmetry to find three other points on the circle.

Solution ▶ Since both coordinates are negative, $\left(-\frac{3}{5}, -\frac{4}{5}\right)$ is in QIII. Substituting into the equation for a unit circle yields

$$x^2 + y^2 = 1 \qquad \text{unit circle equation}$$

$$\left(\frac{-3}{5}\right)^2 + \left(\frac{-4}{5}\right)^2 \overset{?}{=} 1 \qquad \text{substitute } \tfrac{-3}{5} \text{ for } x \text{ and } \tfrac{-4}{5} \text{ for } y$$

$$\frac{9}{25} + \frac{16}{25} \overset{?}{=} 1 \qquad \text{simplify}$$

$$\frac{25}{25} = 1 \qquad \text{result checks}$$

☑ **A.** You've just learned how to locate points on a unit circle and use symmetry to locate other points

Since $\left(\frac{-3}{5}, \frac{-4}{5}\right)$ is on the unit circle, $\left(\frac{3}{5}, \frac{-4}{5}\right)$, $\left(\frac{-3}{5}, \frac{4}{5}\right)$, and $\left(\frac{3}{5}, \frac{4}{5}\right)$ are also on the circle due to symmetry (see figure).

Now try Exercises 19 through 26 ▶

B. Special Triangles and the Unit Circle

The special triangles from Section 5.1 can also be used to find points on a unit circle. As usually written, the triangles state a proportional relationship between their sides after assigning a value of 1 to the shortest side. However, precisely due to this proportional relationship, *we can divide all sides by the length of the hypotenuse,* giving *it* a length of 1 unit (see Figures 5.19 and 5.20).

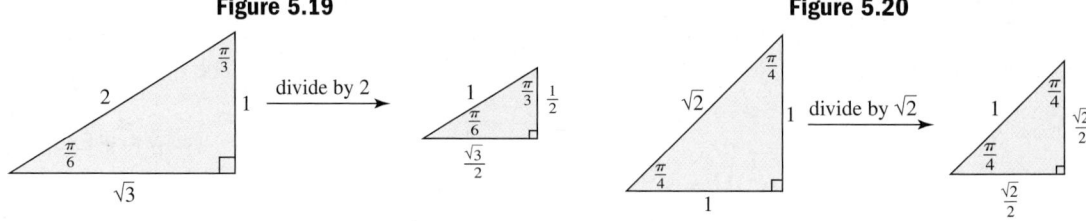

Figure 5.19   **Figure 5.20**

We then place the triangle within the unit circle, and reflect it from quadrant to quadrant to find additional points. We use *the sides* of the triangle to determine the absolute value of each coordinate, and *the quadrant* to give each coordinate the appropriate sign. Note the angles in these special triangles are now expressed in radians.

EXAMPLE 3 ▶ **Using a Special Triangle and Symmetry to Locate Points on a Unit Circle**

Use the $\frac{\pi}{4} : \frac{\pi}{4} : \frac{\pi}{2}$ triangle from Figure 5.20 to find four points on the unit circle.

Solution ▶ Begin by superimposing the triangle in QI, noting it gives the point $\left(\frac{\sqrt{2}}{2}, \frac{\sqrt{2}}{2}\right)$ shown in Figure 5.21. By reflecting the triangle into QII, we find the additional point $\left(-\frac{\sqrt{2}}{2}, \frac{\sqrt{2}}{2}\right)$ on this circle. Realizing we can simply apply the circle's remaining symmetries, we obtain the two additional points $\left(-\dfrac{\sqrt{2}}{2}, -\dfrac{\sqrt{2}}{2}\right)$ and $\left(\dfrac{\sqrt{2}}{2}, -\dfrac{\sqrt{2}}{2}\right)$ shown in Figure 5.22.

Figure 5.21   **Figure 5.22**

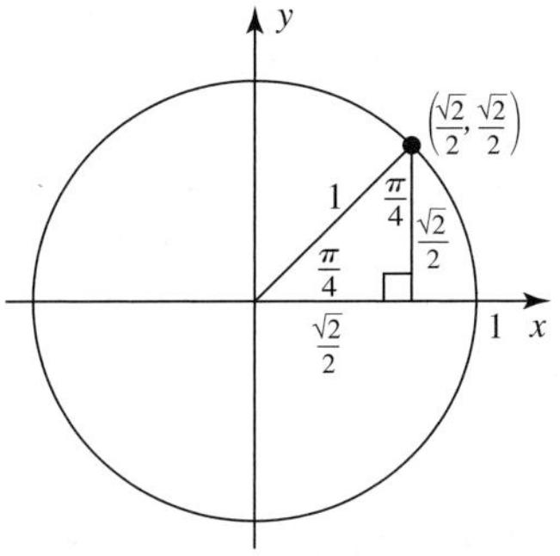

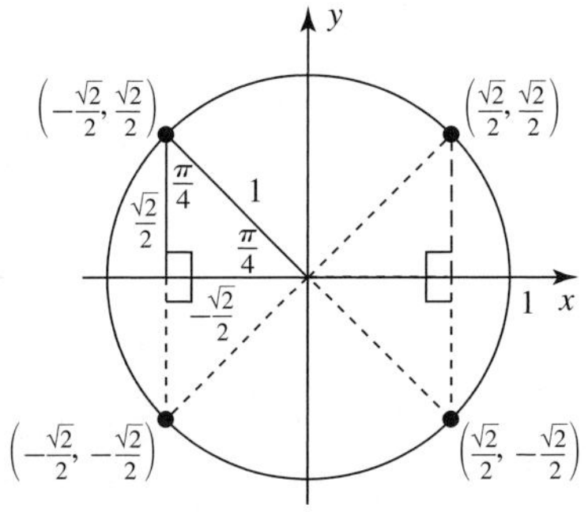

Now try Exercises 27 and 28 ▶

Applying the same idea to a $\dfrac{\pi}{6} : \dfrac{\pi}{3} : \dfrac{\pi}{2}$ triangle would give the points $\left(\dfrac{\sqrt{3}}{2}, \dfrac{1}{2}\right)$, $\left(-\dfrac{\sqrt{3}}{2}, \dfrac{1}{2}\right), \left(-\dfrac{\sqrt{3}}{2}, -\dfrac{1}{2}\right)$ and $\left(\dfrac{\sqrt{3}}{2}, -\dfrac{1}{2}\right)$, *the same points we found in Example 1.*

Figure 5.23

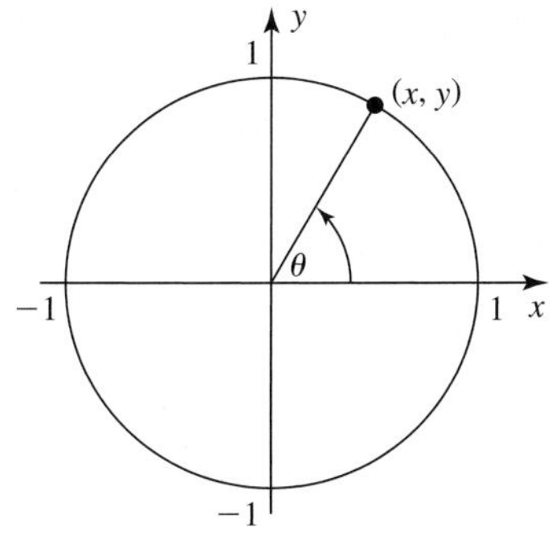

When a central angle θ is viewed as a rotation, each rotation can be associated with a unique point (x, y) on the terminal side, where it intersects the unit circle (see Figure 5.23). For the quadrantal angles $\dfrac{\pi}{2}, \pi, \dfrac{3\pi}{2}$, and 2π, we associate the points $(0, 1), (-1, 0), (0, -1)$, and $(0, 0)$, respectively. When this rotation results in a special angle θ, the association can be found using a special triangle in a manner similar to Example 3. Figure 5.24 shows we associate the point $\left(\dfrac{\sqrt{3}}{2}, \dfrac{1}{2}\right)$ with $\theta = \dfrac{\pi}{6}$, $\left(\dfrac{\sqrt{2}}{2}, \dfrac{\sqrt{2}}{2}\right)$ with $\theta = \dfrac{\pi}{4}$, and by reorienting the $\dfrac{\pi}{6} : \dfrac{\pi}{3} : \dfrac{\pi}{2}$ triangle, $\left(\dfrac{1}{2}, \dfrac{\sqrt{3}}{2}\right)$ is associated with a rotation of $\theta = \dfrac{\pi}{3}$.

Figure 5.24

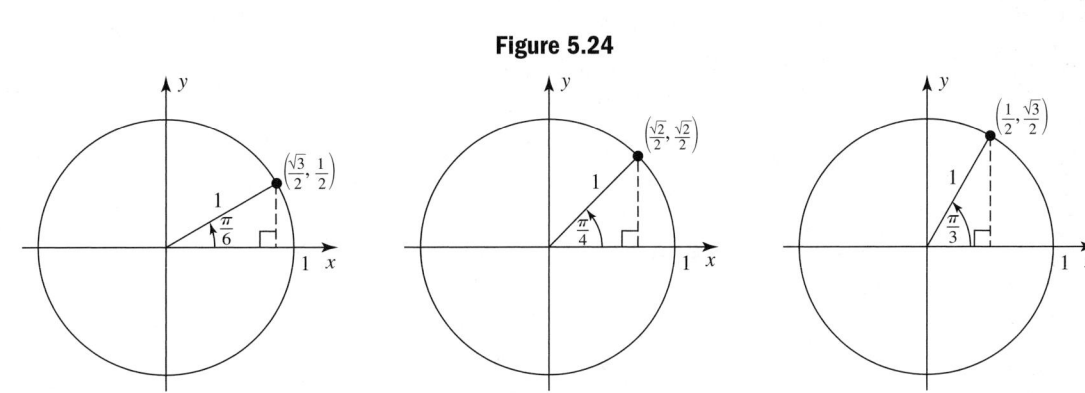

For standard rotations from $\theta = 0$ to $\theta = \dfrac{\pi}{2}$ we have the following:

Rotation θ	0	$\dfrac{\pi}{6}$	$\dfrac{\pi}{4}$	$\dfrac{\pi}{3}$	$\dfrac{\pi}{2}$
Associated point (x, y)	(0, 0)	$\left(\dfrac{\sqrt{3}}{2}, \dfrac{1}{2}\right)$	$\left(\dfrac{\sqrt{2}}{2}, \dfrac{\sqrt{2}}{2}\right)$	$\left(\dfrac{1}{2}, \dfrac{\sqrt{3}}{2}\right)$	(0, 1)

Each of these points give rise to three others using the symmetry of the circle. By defining a reference angle θ_r, we can associate these points with the related rotation $\theta > \dfrac{\pi}{2}$.

Reference Angles

For any angle θ in standard position, the acute angle θ_r formed by the terminal side and the x-axis is called the *reference angle* for θ.

Several examples of the reference angle concept are shown in Figure 5.25 for $\theta > 0$ in radians.

Figure 5.25

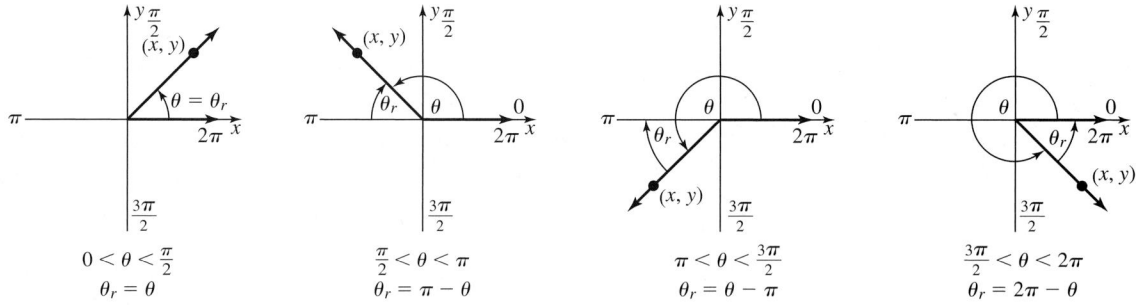

$$0 < \theta < \dfrac{\pi}{2}$$
$$\theta_r = \theta$$
$$\dfrac{\pi}{2} < \theta < \pi$$
$$\theta_r = \pi - \theta$$
$$\pi < \theta < \dfrac{3\pi}{2}$$
$$\theta_r = \theta - \pi$$
$$\dfrac{3\pi}{2} < \theta < 2\pi$$
$$\theta_r = 2\pi - \theta$$

Due to the symmetries of the circle, reference angles of $\dfrac{\pi}{6}, \dfrac{\pi}{4}$, and $\dfrac{\pi}{3}$ serve to fix the absolute value of the coordinates for x and y, and we simply *use the appropriate sign for each coordinate* (r is always positive). As before this depends solely on the quadrant of the terminal side.

EXAMPLE 4 ▶ **Finding Points on a Unit Circle Associated with a Rotation θ**

Determine the reference angle for each rotation given, then find the associated point (x, y) on the unit circle.

a. $\theta = \dfrac{5\pi}{6}$ **b.** $\theta = \dfrac{4\pi}{3}$ **c.** $\theta = \dfrac{7\pi}{4}$

Figure 5.26

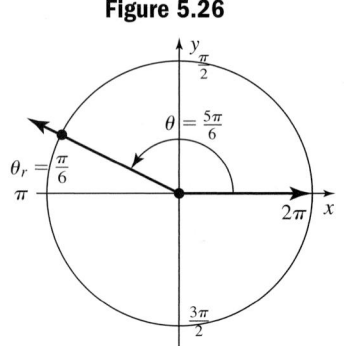

Solution ▶ **a.** A rotation of $\dfrac{5\pi}{6}$ terminates in QII:

$\theta_r = \pi - \dfrac{5\pi}{6} = \dfrac{\pi}{6}$. The associated point is $\left(-\dfrac{\sqrt{3}}{2}, \dfrac{1}{2}\right)$ since $x < 0$ in QII. See Figure 5.26.

Figure 5.27

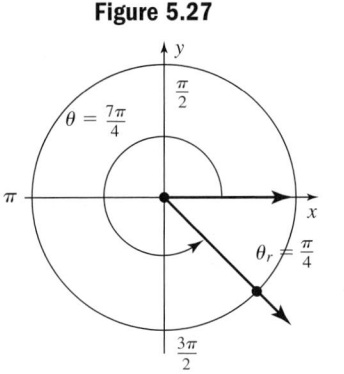

b. A rotation of $\dfrac{4\pi}{3}$ terminates in QIII:

$\theta_r = \dfrac{4\pi}{3} - \pi = \dfrac{\pi}{3}$. The associated point is

$\left(-\dfrac{1}{2}, -\dfrac{\sqrt{3}}{2}\right)$ since $x < 0$ and $y < 0$ in QIII.

c. A rotation of $\dfrac{7\pi}{4}$ terminates in QIV:

$\theta_r = 2\pi - \dfrac{7\pi}{4} = \dfrac{\pi}{4}$. The associated point is

$\left(\dfrac{\sqrt{2}}{2}, -\dfrac{\sqrt{2}}{2}\right)$ since $y < 0$ in QIV. See

Figure 5.27.

✓ **B.** You've just learned how to use special triangles to find points on a unit circle and locate other points using symmetry

Now try Exercises 29 through 36 ▶

C. Trigonometric Functions and Points on the Unit Circle

We can now define the six trigonometric functions in terms of a point (x, y) on the unit circle, with the use of right triangles fading from view. For this reason they are sometimes called the **circular functions.**

The Circular Functions

For any rotation θ and point $P(x, y)$ on the unit circle associated with θ,

$$\cos\theta = x \qquad\qquad \sin\theta = y \qquad\qquad \tan\theta = \dfrac{y}{x}; x \neq 0$$

$$\sec\theta = \dfrac{1}{x}; x \neq 0 \qquad\qquad \csc\theta = \dfrac{1}{y}; y \neq 0 \qquad\qquad \cot\theta = \dfrac{x}{y}; y \neq 0$$

Note that once $\sin\theta$, $\cos\theta$, and $\tan\theta$ are known, the values of $\csc\theta$, $\sec\theta$, and $\cot\theta$ follow automatically since a number and its reciprocal always have the same sign. See Figure 5.28.

Figure 5.28

QII $x < 0, y > 0$ (only y is positive)	QI $x > 0, y > 0$ (both x and y are positive)
$\sin\theta$ is positive $\tan\theta$ is positive	All functions are positive $\cos\theta$ is positive
QIII $x < 0, y < 0$ (both x and y are negative)	QIV $x > 0, y < 0$ (only x is positive)

EXAMPLE 5 ▶ **Evaluating Trig Functions for a Rotation θ**

Evaluate the six trig functions for $\theta = \dfrac{5\pi}{4}$.

Solution ▶ A rotation of $\dfrac{5\pi}{4}$ terminates in QIII, so

$\theta_r = \dfrac{5\pi}{4} - \pi = \dfrac{\pi}{4}$. The associated point is

$\left(-\dfrac{\sqrt{2}}{2}, -\dfrac{\sqrt{2}}{2}\right)$ since $x < 0$ and $y < 0$ in QIII.

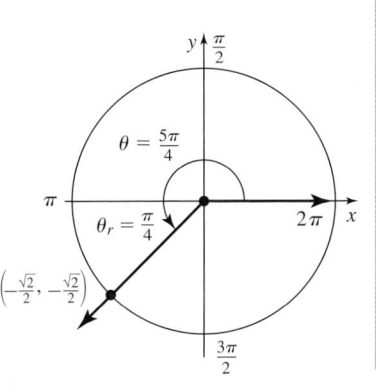

This yields

$$\cos\left(\frac{5\pi}{4}\right) = -\frac{\sqrt{2}}{2} \qquad \sin\left(\frac{5\pi}{4}\right) = -\frac{\sqrt{2}}{2} \qquad \tan\left(\frac{5\pi}{4}\right) = 1$$

Noting the reciprocal of $-\dfrac{\sqrt{2}}{2}$ is $-\sqrt{2}$ after rationalizing, we have

$$\sec\left(\frac{5\pi}{4}\right) = -\sqrt{2} \qquad \csc\left(\frac{5\pi}{4}\right) = -\sqrt{2} \qquad \cot\left(\frac{5\pi}{4}\right) = 1$$

☑ **C.** You've just learned how to define the six trig functions in terms of a point on the unit circle

Now try Exercises 37 through 40 ▶

D. The Trigonometry of Real Numbers

Defining the trig functions in terms of a point on the unit circle is precisely what we needed to work with them as functions of real numbers. This is because when $r = 1$ and θ is in radians, *the length of the subtended arc is numerically the same as the measure of the angle:* $s = (1)\theta \Rightarrow s = \theta$! This means we can view any function of θ as a like function of arc length s, where $s \in \mathbb{R}$ (see the *Reinforcing Basic Concepts* feature following this section.). As a compromise the variable t is commonly used, with t representing *either* the amount of rotation *or* the length of the arc. As such we will assume t is a unitless quantity, although there are other reasons for this assumption. In Figure 5.29, a rotation of $\theta = \dfrac{3\pi}{4}$ is subtended by an arc length of $s = \dfrac{3\pi}{4}$ (about 2.356 units). The reference angle for θ is $\dfrac{\pi}{4}$, which we will now refer to as a **reference arc.** As you work through the remaining examples and the exercises that follow, it will often help to draw a quick sketch similar to that in Figure 5.29 to determine the quadrant of the terminal side, the reference arc, and the sign of each function.

Figure 5.29

$\left(-\frac{\sqrt{2}}{2}, \frac{\sqrt{2}}{2}\right)$

$s = \frac{3\pi}{4}$

$s_r = \frac{\pi}{4}$

$\theta = \frac{3\pi}{4}$

EXAMPLE 6 ▶ **Evaluating Trig Functions for a Real Number t**

Evaluate the six trig functions for the given value of t.

a. $t = \dfrac{11\pi}{6}$ **b.** $t = \dfrac{3\pi}{2}$

Solution ▶ **a.** For $t = \dfrac{11\pi}{6}$, the arc terminates in QIV where $x > 0$ and $y < 0$. The reference arc is $\dfrac{\pi}{6}$ and from our previous work we know the corresponding point (x, y) is $\left(\dfrac{\sqrt{3}}{2}, -\dfrac{1}{2}\right)$. This gives

$$\cos\left(\frac{11\pi}{6}\right) = \frac{\sqrt{3}}{2} \qquad \sin\left(\frac{11\pi}{6}\right) = -\frac{1}{2} \qquad \tan\left(\frac{11\pi}{6}\right) = -\frac{\sqrt{3}}{3}$$

$$\sec\left(\frac{11\pi}{6}\right) = \frac{2\sqrt{3}}{3} \qquad \csc\left(\frac{11\pi}{6}\right) = -2 \qquad \cot\left(\frac{11\pi}{6}\right) = -\sqrt{3}$$

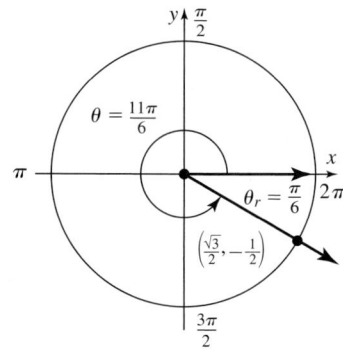

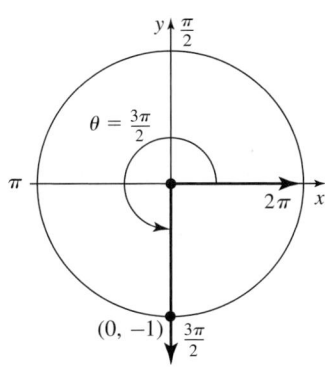

b. $t = \dfrac{3\pi}{2}$ is a quadrantal angle and the associated point is $(0, -1)$.

This yields

$$\cos\left(\frac{3\pi}{2}\right) = 0 \qquad \sin\left(\frac{3\pi}{2}\right) = -1 \qquad \tan\left(\frac{3\pi}{2}\right) = \text{undefined}$$

$$\sec\left(\frac{3\pi}{2}\right) = \text{undefined} \qquad \csc\left(\frac{3\pi}{2}\right) = -1 \qquad \cot\left(\frac{3\pi}{2}\right) = 0$$

Now try Exercises 41 through 44 ▶

As Example 6(b) indicates, as functions of a real number the concept of domain comes into play. From their definition it is apparent there are no restrictions on the domain of cosine and sine, but the domains of the other functions must be restricted to exclude division by zero. For functions with x in the denominator, we cast out the odd multiples of $\dfrac{\pi}{2}$, since the x-coordinate of the related quadrantal points is zero: $\dfrac{\pi}{2} \to (0, 1), \dfrac{3\pi}{2} \to (0, -1)$, and so on. The excluded values can be stated as $t \neq \dfrac{\pi}{2} + \pi k$ for all integers k. For functions with y in the denominator, we cast out all multiples of π ($t \neq \pi k$ for all integers k) since the y-coordinate of these points is zero: $0 \to (1, 0), \pi \to (-1, 0), 2\pi \to (1, 0)$, and so on.

The Domains of the Trig Functions as Functions of a Real Number

For $t \in \mathbb{R}$ and $k \in \mathbb{Z}$, the domains of the trig functions are:

$$\cos t = x \qquad\qquad \sin t = y \qquad\qquad \tan t = \frac{y}{x}\,;\, x \neq 0$$

$$t \in \mathbb{R} \qquad\qquad t \in \mathbb{R} \qquad\qquad t \neq \frac{\pi}{2} + \pi k$$

$$\sec t = \frac{1}{x}\,;\, x \neq 0 \qquad \csc t = \frac{1}{y}\,;\, y \neq 0 \qquad \cot t = \frac{x}{y}\,;\, y \neq 0$$

$$t \neq \frac{\pi}{2} + \pi k \qquad\qquad t \neq \pi k \qquad\qquad t \neq \pi k$$

For a given point (x, y) on the unit circle associated with the real number t, the value of each function at t can still be determined even if t is unknown.

EXAMPLE 7 ▶ **Finding Function Values Given a Point on the Unit Circle**

Given $\left(\frac{-7}{25}, \frac{24}{25}\right)$ is a point on the unit circle corresponding to a real number t, find the value of all six trig functions of t.

Solution ▶ Using the definitions from the previous box we have $\cos t = \frac{-7}{25}$, $\sin t = \frac{24}{25}$, and $\tan t = \frac{\sin t}{\cos t} = \frac{24}{-7}$. The values of the reciprocal functions are then $\sec t = \frac{25}{-7}$, $\csc t = \frac{25}{24}$, and $\cot t = \frac{-7}{24}$.

☑ D. You've just learned how to define the six trig functions in terms of a real number t

Now try Exercises 45 through 70 ▶

E. Finding a Real Number *t* Whose Function Value Is Known

In Example 7, we were able to determine the values of the trig functions even though *t* was unknown. In many cases, however, we need to *find* the value of *t*. For instance, what is the value of *t* given $\cos t = -\dfrac{\sqrt{3}}{2}$ with *t* in QII? Exercises of this type fall into two broad categories: (1) you recognize the given number as one of the special values: $\pm\left\{0, \dfrac{1}{2}, \dfrac{\sqrt{2}}{2}, \dfrac{\sqrt{3}}{2}, \dfrac{\sqrt{3}}{3}, \sqrt{3}, 1\right\}$; or (2) you don't. If you recognize a special value, you can often name the real number *t* after a careful consideration of the related quadrant and required sign.

Figure 5.30

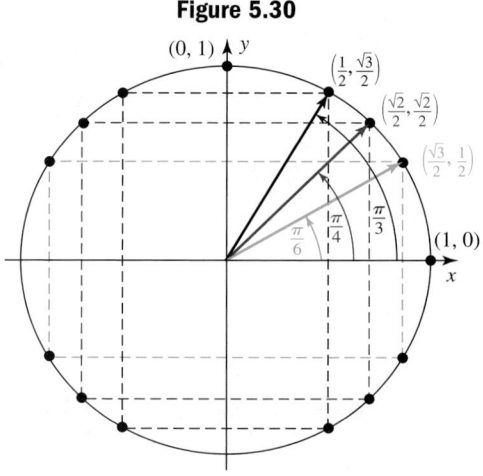

The diagram in Figure 5.30 reviews these special values for $0 \le t \le \dfrac{\pi}{2}$ but remember—all other special values can be found using reference arcs and the symmetry of the circle.

EXAMPLE 8 ▶ **Finding *t* for Given Values and Conditions**

Find the value of *t* that corresponds to the given function values.

　　a. $\cos t = -\dfrac{\sqrt{2}}{2}$; *t* in QII　　　**b.** $\tan t = \sqrt{3}$; *t* in QIII

Solution ▶　**a.** The cosine function is negative in QII and QIII, where $x < 0$. We recognize $-\dfrac{\sqrt{2}}{2}$ as a standard value for sine and cosine, related to certain multiples of $t = \dfrac{\pi}{4}$. In QII, we have $t = \dfrac{3\pi}{4}$.

　　　　b. The tangent function is positive in QI and QIII, where *x* and *y* have like signs. We recognize $\sqrt{3}$ as a standard value for tangent and cotangent, related to certain multiples of $t = \dfrac{\pi}{3}$. For tangent in QIII, we have $t = \dfrac{4\pi}{3}$.

Now try Exercises 71 through 94 ▶

If the given function value is not one of the special values, properties of the inverse trigonometric functions must be used to find the associated value of *t*. The inverse functions are developed in Section 6.5.

Using radian measure and the unit circle is much more than a simple convenience to trigonometry and its applications. Whether the unit is 1 cm, 1 m, 1 km, or even 1 light-year, using 1 unit designations serves to simplify a great many practical applications, including those involving the arc length formula, $s = r\theta$. **See Exercises 97 through 104.**

The following table summarizes the relationship between a special arc *t* (*t* in QI) and the value of each trig function at *t*. Due to the frequent use of these relationships, students are encouraged to commit them to memory.

t	$\sin t$	$\cos t$	$\tan t$	$\csc t$	$\sec t$	$\cot t$
0	0	1	0	undefined	1	undefined
$\dfrac{\pi}{6}$	$\dfrac{1}{2}$	$\dfrac{\sqrt{3}}{2}$	$\dfrac{1}{\sqrt{3}} = \dfrac{\sqrt{3}}{3}$	2	$\dfrac{2}{\sqrt{3}} = \dfrac{2\sqrt{3}}{3}$	$\sqrt{3}$
$\dfrac{\pi}{4}$	$\dfrac{\sqrt{2}}{2}$	$\dfrac{\sqrt{2}}{2}$	1	$\sqrt{2}$	$\sqrt{2}$	1
$\dfrac{\pi}{3}$	$\dfrac{\sqrt{3}}{2}$	$\dfrac{1}{2}$	$\sqrt{3}$	$\dfrac{2}{\sqrt{3}} = \dfrac{2\sqrt{3}}{3}$	2	$\dfrac{1}{\sqrt{3}} = \dfrac{\sqrt{3}}{3}$
$\dfrac{\pi}{2}$	1	0	undefined	1	undefined	0

☑ **E.** You've just learned how to find the real number t corresponding to given values of $\sin t$, $\cos t$, and $\tan t$

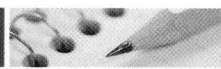

5.2 EXERCISES

▶ CONCEPTS AND VOCABULARY

Fill in each blank with the appropriate word or phrase. Carefully reread the section if needed.

1. A central circle is symmetric to the _____ axis, the _____ axis and to the _____.

2. Since $(\frac{5}{13}, -\frac{12}{13})$ is on the unit circle, the point _____ in QII is also on the circle.

3. On a unit circle, $\cos t =$ _____, $\sin t =$ _____, and $\tan t =$ _____; while $\dfrac{1}{x} =$ _____, $\dfrac{1}{y} =$ _____, and $\dfrac{x}{y} =$ _____.

4. On a unit circle with θ in radians, the length of a(n) _____ is numerically the same as the measure of the _____, since for $s = r\theta$, $s = \theta$ when $r = 1$.

5. Discuss/Explain how knowing only one point on the unit circle, actually gives the location of four points. Why is this helpful to a study of the circular functions?

6. A student is asked to find t using a calculator, given $\sin t \approx 0.5592$ with t in QII. The answer submitted is $t = \sin^{-1} 0.5592 \approx 34°$. Discuss/Explain why this answer is not correct. What is the correct response?

▶ DEVELOPING YOUR SKILLS

Given the point is on a unit circle, complete the ordered pair (x, y) for the quadrant indicated. For Exercises 7 to 14, answer in radical form as needed. For Exercises 15 to 18, round results to four decimal places.

7. $(x, -0.8)$; QIII

8. $(-0.6, y)$; QII

9. $\left(\dfrac{5}{13}, y\right)$; QIV

10. $\left(x, -\dfrac{8}{17}\right)$; QIV

11. $\left(\dfrac{\sqrt{11}}{6}, y\right)$; QI

12. $\left(x, -\dfrac{\sqrt{13}}{7}\right)$; QIII

13. $\left(-\dfrac{\sqrt{11}}{4}, y\right)$; QII

14. $\left(x, \dfrac{\sqrt{6}}{5}\right)$; QI

15. $(x, -0.2137)$; QIII

16. $(0.9909, y)$; QIV

17. $(x, 0.1198)$; QII

18. $(0.5449, y)$; QI

Verify the point given is on a unit circle, then use symmetry to find three more points on the circle. Results for Exercises 19 to 22 are exact, results for Exercises 23 to 26 are approximate.

19. $\left(-\dfrac{\sqrt{3}}{2}, \dfrac{1}{2}\right)$

20. $\left(\dfrac{\sqrt{7}}{4}, -\dfrac{3}{4}\right)$

21. $\left(\dfrac{\sqrt{11}}{6}, -\dfrac{5}{6}\right)$

22. $\left(-\dfrac{\sqrt{6}}{3}, -\dfrac{\sqrt{3}}{3}\right)$

23. $(0.3325, 0.9431)$ **24.** $(0.7707, -0.6372)$

25. $(0.9937, -0.1121)$ **26.** $(-0.2029, 0.9792)$

27. Use a $\dfrac{\pi}{6}:\dfrac{\pi}{3}:\dfrac{\pi}{2}$ triangle with a hypotenuse of length 1 to verify that $\left(\dfrac{1}{2}, \dfrac{\sqrt{3}}{2}\right)$ is a point on the unit circle.

28. Use the results from Exercise 27 to find three additional points on the circle and name the quadrant of each point.

Find the reference angle associated with each rotation, then find the associated point (x, y) on the unit circle.

29. $\theta = \dfrac{5\pi}{4}$ **30.** $\theta = \dfrac{5\pi}{3}$

31. $\theta = -\dfrac{5\pi}{6}$ **32.** $\theta = -\dfrac{7\pi}{4}$

33. $\theta = \dfrac{11\pi}{4}$ **34.** $\theta = \dfrac{11\pi}{3}$

35. $\theta = \dfrac{25\pi}{6}$ **36.** $\theta = \dfrac{39\pi}{4}$

Without the use of a calculator, state the exact value of the trig functions for the given angle. A diagram may help.

37. a. $\sin\left(\dfrac{\pi}{4}\right)$ **b.** $\sin\left(\dfrac{3\pi}{4}\right)$

c. $\sin\left(\dfrac{5\pi}{4}\right)$ **d.** $\sin\left(\dfrac{7\pi}{4}\right)$

e. $\sin\left(\dfrac{9\pi}{4}\right)$ **f.** $\sin\left(-\dfrac{\pi}{4}\right)$

g. $\sin\left(-\dfrac{5\pi}{4}\right)$ **h.** $\sin\left(-\dfrac{11\pi}{4}\right)$

38. a. $\tan\left(\dfrac{\pi}{3}\right)$ **b.** $\tan\left(\dfrac{2\pi}{3}\right)$

c. $\tan\left(\dfrac{4\pi}{3}\right)$ **d.** $\tan\left(\dfrac{5\pi}{3}\right)$

e. $\tan\left(\dfrac{7\pi}{3}\right)$ **f.** $\tan\left(-\dfrac{\pi}{3}\right)$

g. $\tan\left(-\dfrac{4\pi}{3}\right)$ **h.** $\tan\left(-\dfrac{10\pi}{3}\right)$

39. a. $\cos\pi$ **b.** $\cos 0$

c. $\cos\left(\dfrac{\pi}{2}\right)$ **d.** $\cos\left(\dfrac{3\pi}{2}\right)$

40. a. $\sin\pi$ **b.** $\sin 0$

c. $\sin\left(\dfrac{\pi}{2}\right)$ **d.** $\sin\left(\dfrac{3\pi}{2}\right)$

Use the symmetry of the circle and reference arcs as needed to state the exact value of the trig functions for the given real number, without the use of a calculator. A diagram may help.

41. a. $\cos\left(\dfrac{\pi}{6}\right)$ **b.** $\cos\left(\dfrac{5\pi}{6}\right)$

c. $\cos\left(\dfrac{7\pi}{6}\right)$ **d.** $\cos\left(\dfrac{11\pi}{6}\right)$

e. $\cos\left(\dfrac{13\pi}{6}\right)$ **f.** $\cos\left(-\dfrac{\pi}{6}\right)$

g. $\cos\left(-\dfrac{5\pi}{6}\right)$ **h.** $\cos\left(-\dfrac{23\pi}{6}\right)$

42. a. $\csc\left(\dfrac{\pi}{6}\right)$ **b.** $\csc\left(\dfrac{5\pi}{6}\right)$

c. $\csc\left(\dfrac{7\pi}{6}\right)$ **d.** $\csc\left(\dfrac{11\pi}{6}\right)$

e. $\csc\left(\dfrac{13\pi}{6}\right)$ **f.** $\csc\left(-\dfrac{\pi}{6}\right)$

g. $\csc\left(-\dfrac{11\pi}{6}\right)$ **h.** $\csc\left(-\dfrac{17\pi}{6}\right)$

43. a. $\tan\pi$ **b.** $\tan 0$

c. $\tan\left(\dfrac{\pi}{2}\right)$ **d.** $\tan\left(\dfrac{3\pi}{2}\right)$

44. a. $\cot\pi$ **b.** $\cot 0$

c. $\cot\left(\dfrac{\pi}{2}\right)$ **d.** $\cot\left(\dfrac{3\pi}{2}\right)$

Given (x, y) is a point on a unit circle corresponding to t, find the value of all six circular functions of t.

45.

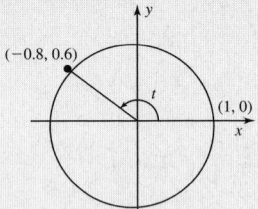

46.

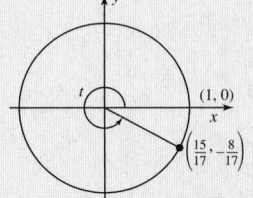

47.

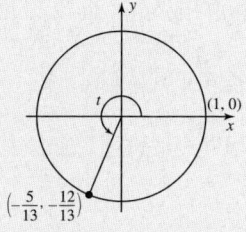

$\left(-\frac{5}{13}, -\frac{12}{13}\right)$

48.

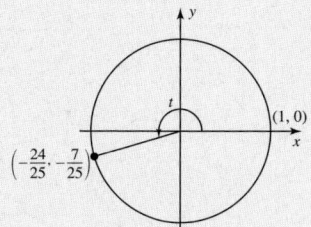

$\left(-\frac{24}{25}, -\frac{7}{25}\right)$

49.

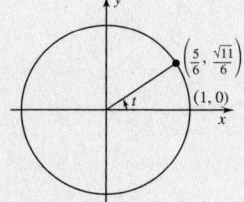

$\left(\frac{5}{6}, \frac{\sqrt{11}}{6}\right)$

50.

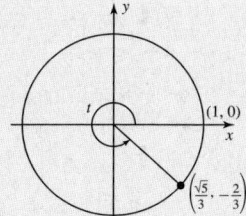

$\left(\frac{\sqrt{5}}{3}, -\frac{2}{3}\right)$

51. $\left(-\dfrac{2}{5}, \dfrac{\sqrt{21}}{5}\right)$ **52.** $\left(\dfrac{\sqrt{7}}{4}, -\dfrac{3}{4}\right)$

53. $\left(-\dfrac{1}{3}, -\dfrac{2\sqrt{2}}{3}\right)$ **54.** $\left(-\dfrac{2\sqrt{6}}{5}, -\dfrac{1}{5}\right)$

55. $\left(\dfrac{1}{2}, \dfrac{\sqrt{3}}{2}\right)$ **56.** $\left(\dfrac{\sqrt{3}}{2}, \dfrac{1}{2}\right)$

57. $\left(-\dfrac{\sqrt{2}}{2}, \dfrac{\sqrt{2}}{2}\right)$ **58.** $\left(\dfrac{\sqrt{2}}{3}, -\dfrac{\sqrt{7}}{3}\right)$

On a unit circle, the real number t can represent either the amount of rotation or the *length of the arc* when we associate t with a point (x, y) on the circle. In the circle diagram shown, the real number t in radians is marked off along the circumference. For Exercises 59 through 70, name the quadrant in which t terminates and use the figure to estimate function values to one decimal place (use a straightedge). Check results using a calculator.

Exercises 59 to 70

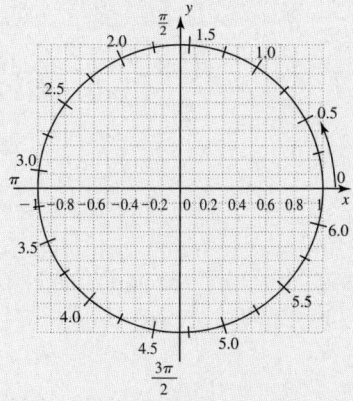

59. sin 0.75   **60.** cos 2.75

61. cos 5.5   **62.** sin 4.0

63. tan 0.8   **64.** sec 3.75

65. csc 2.0   **66.** cot 0.5

67. $\cos\left(\dfrac{5\pi}{8}\right)$ **68.** $\sin\left(\dfrac{5\pi}{8}\right)$

69. $\tan\left(\dfrac{8\pi}{5}\right)$ **70.** $\sec\left(\dfrac{8\pi}{5}\right)$

Without using a calculator, find the value of t in $[0, 2\pi)$ that corresponds to the following functions.

71. $\sin t = \dfrac{\sqrt{3}}{2}$; t in QII

72. $\cos t = \dfrac{1}{2}$; t in QIV

73. $\cos t = -\dfrac{\sqrt{3}}{2}$; t in QIII

74. $\sin t = -\dfrac{1}{2}$; t in QIV

75. $\tan t = -\sqrt{3}$; t in QII

76. $\sec t = -2$; t in QIII

77. $\sin t = 1$; t is quadrantal

78. $\cos t = -1$; t is quadrantal

Without using a calculator, find the two values of t (where possible) in $[0, 2\pi)$ that make each equation true.

79. $\sec t = -\sqrt{2}$ **80.** $\csc t = -\dfrac{2}{\sqrt{3}}$

81. $\tan t$ undefined   **82.** $\csc t$ undefined

83. $\cos t = -\dfrac{\sqrt{2}}{2}$ **84.** $\sin t = \dfrac{\sqrt{2}}{2}$

85. $\sin t = 0$ **86.** $\cos t = -1$

87. Given $\left(\frac{3}{4}, -\frac{4}{5}\right)$ is a point on the unit circle that corresponds to t. Find the coordinates of the point corresponding to (a) $-t$ and (b) $t + \pi$.

88. Given $\left(-\frac{7}{25}, \frac{24}{25}\right)$ is a point on the unit circle that corresponds to t. Find the coordinates of the point corresponding to (a) $-t + \pi$ and (b) $t - \pi$.

Find an additional value of t in $[0, 2\pi)$ that makes the equation true.

89. $\sin 0.8 \approx 0.7174$

90. $\cos 2.12 \approx -0.5220$

91. $\cos 4.5 \approx -0.2108$

92. $\sin 5.23 \approx -0.8690$

93. $\tan 0.4 \approx 0.4228$

94. $\sec 5.7 \approx 1.1980$

▶ WORKING WITH FORMULAS

95. From Pythagorean triples to points on the unit circle: $(x, y, r) \rightarrow \left(\dfrac{x}{r}, \dfrac{y}{r}, 1\right)$

While not strictly a "formula," dividing a Pythagorean triple by r is a simple algorithm for rewriting any Pythagorean triple as a triple with hypotenuse 1. This enables us to identify certain points on a unit circle, and to evaluate the six trig functions of the related acute angle. Rewrite each triple as a triple with hypotenuse 1, verify $\left(\dfrac{x}{r}, \dfrac{y}{r}\right)$ is a point on the unit circle, and evaluate the six trig functions using this point.

   a. (5, 12, 13)   **b.** (7, 24, 25)
   c. (12, 35, 37)   **d.** (9, 40, 41)

96. The sine and cosine of $(2k + 1)\dfrac{\pi}{4}; k \in \mathbb{Z}$

In the solution to Example 8(a), we mentioned $\pm\dfrac{\sqrt{2}}{2}$ were standard values for sine and cosine, "related to certain multiples of $\dfrac{\pi}{4}$." Actually, we meant "odd multiples of $\dfrac{\pi}{4}$." The odd multiples of $\dfrac{\pi}{4}$ are given by the "formula" shown, where k is any integer. (a) What multiples of $\dfrac{\pi}{4}$ are generated by $k = -3, -2, -1, 0, 1, 2, 3$? (b) Find similar formulas for Example 8(b), where $\sqrt{3}$ is a standard value for tangent and cotangent, "related to certain multiples of $\dfrac{\pi}{6}$."

▶ APPLICATIONS

97. Laying new sod: When new sod is laid, a heavy roller is used to press the sod down to ensure good contact with the ground beneath. The radius of the roller is 1 ft. (a) Through what angle (in radians) has the roller turned after being pulled across 5 ft of yard? (b) What angle must the roller turn through to press a length of 30 ft?

1 ft

98. Cable winch: A large winch with a radius of 1 ft winds in 3 ft of cable. (a) Through what angle (in radians) has it turned? (b) What angle must it turn through in order to winch in 12.5 ft of cable?

Exercise 98

99. Wiring an apartment: In the wiring of an apartment complex, electrical wire is being pulled from a spool with radius 1 decimeter (1 dm = 10 cm). (a) What length (in decimeters) is removed as the spool turns through 5 rad? (b) How many decimeters are removed in one complete turn ($t = 2\pi$) of the spool?

100. Barrel races: In the barrel races popular at some family reunions, contestants stand on a hard rubber barrel with a radius of 1 cubit (1 cubit = 18 in.), and try to "walk the barrel" from the start line to the finish line without falling. (a) What distance (in cubits) is traveled as the barrel is walked through an angle of 4.5 rad? (b) If the race is 25 cubits long, through what angle will the winning barrel walker walk the barrel?

Interplanetary measurement: In the year 1905, astronomers began using astronomical units or AU to study the distances between the celestial bodies of our solar system. One AU represents the average distance between the Earth and the Sun, which is about 93 million miles. Pluto is roughly 39.24 AU from the Sun.

101. If the Earth travels through an angle of 2.5 rad about the Sun, (a) what distance in astronomical units (AU) has it traveled? (b) How many AU does it take for one complete orbit around the Sun?

102. If you include the dwarf planet Pluto, Jupiter is the middle (fifth of nine) planet from the Sun. Suppose astronomers had decided to use *its* average distance from the Sun as 1 AU. In this case, 1 AU would be 480 million miles. If Jupiter travels through an angle of 4 rad about the Sun, (a) what distance in the "new" astronomical units (AU) has it traveled? (b) How many of the new AU does it take to complete one-half an orbit about the Sun? (c) What distance in the new AU is the dwarf planet Pluto from the Sun?

103. Compact disk circumference: A standard compact disk has a radius of 6 cm. Call this length "1 unit." Mark a starting point on any large surface, then carefully roll the compact disk along this line without slippage, through one full revolution (2π rad) and mark this spot. Take an accurate measurement of the resulting line segment. Is the result close to 2π "units" ($2\pi \times 6$ cm)?

104. Verifying $s = r\theta$: On a protractor, carefully measure the distance from the middle of the protractor's eye to the edge of the protractor along the 0° mark, to the nearest half-millimeter. Call this length "1 unit." Then use a ruler to draw a straight line on a blank sheet of paper, and with the protractor on edge, start the zero degree mark at one end of the line, carefully roll the protractor until it reaches 1 radian (57.3°), and mark this spot. Now measure the length of the line segment created. Is it very close to 1 "unit" long?

Exercise 104

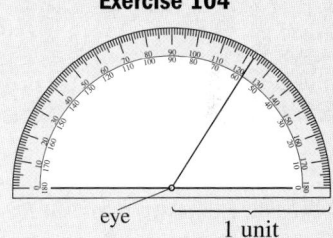

► EXTENDING THE CONCEPT

105. In this section, we discussed the *domain* of the circular functions, but said very little about their *range*. Review the concepts presented here and determine the range of $y = \cos t$ and $y = \sin t$. In other words, what are the smallest and largest output values we can expect?

106. Since $\tan t = \dfrac{\sin t}{\cos t}$, what can you say about the range of the tangent function?

Use the radian grid given with Exercises 59–70 to answer Exercises 107 and 108.

107. Given $\cos(2t) = -0.6$ with the terminal side of the arc in QII, (a) what is the value of $2t$? (b) What quadrant is t in? (c) What is the value of $\cos t$? (d) Does $\cos(2t) = 2\cos t$?

108. Given $\sin(2t) = -0.8$ with the terminal side of the arc in QIII, (a) what is the value of $2t$? (b) What quadrant is t in? (c) What is the value of $\sin t$? (d) Does $\sin(2t) = 2\sin t$?

▶ MAINTAINING YOUR SKILLS

109. (2.1) Given the points $(-3, -4)$ and $(5, 2)$ find
    a. the distance between them
    b. the midpoint between them
    c. the slope of the line through them.

110. (4.3) Use a calculator to find the value of each expression, then explain the results.
    a. $\log 2 + \log 5 =$ _____
    b. $\log 20 - \log 2 =$ _____

111. (1.3) Solve each equation:
    a. $2|x + 1| - 3 = 7$
    b. $2\sqrt{x + 1} - 3 = 7$

112. (3.2) Use the rational zeroes theorem to solve the equation completely, given $x = -3$ is one root.
$$x^4 + x^3 - 3x^2 + 3x - 18 = 0$$

5.3 | Graphs of the Sine and Cosine Functions; Cosecant and Secant Functions

Learning Objectives

In Section 5.3 you will learn how to:

☐ **A.** Graph $f(t) = \sin t$ using special values and symmetry

☐ **B.** Graph $f(t) = \cos t$ using special values and symmetry

☐ **C.** Graph sine and cosine functions with various amplitudes and periods

☐ **D.** Investigate graphs of the reciprocal functions $f(t) = \csc (Bt)$ and $f(t) = \sec (Bt)$

☐ **E.** Write the equation for a given graph

As with the graphs of other functions, trigonometric graphs contribute a great deal toward the understanding of each trig function and its applications. For now, our primary interest is the general shape of each basic graph and some of the transformations that can be applied. We will also learn to analyze each graph, and to capitalize on the features that enable us to apply the functions as real-world models.

A. Graphing $f(t) = \sin t$

Consider the following table of values (Table 5.1) for $\sin t$ and the special angles in *QI*.

Table 5.1

t	0	$\dfrac{\pi}{6}$	$\dfrac{\pi}{4}$	$\dfrac{\pi}{3}$	$\dfrac{\pi}{2}$
$\sin t$	0	$\dfrac{1}{2}$	$\dfrac{\sqrt{2}}{2}$	$\dfrac{\sqrt{3}}{2}$	1

Observe that in this interval, sine values are increasing from 0 to 1. From $\dfrac{\pi}{2}$ to π (QII), special values taken from the unit circle show sine values are decreasing from 1 to 0, *but through the same output values as in QI*. See Figures 5.31 through 5.33.

Figure 5.31

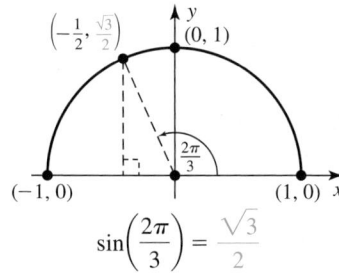

$$\sin\left(\frac{2\pi}{3}\right) = \frac{\sqrt{3}}{2}$$

Figure 5.32

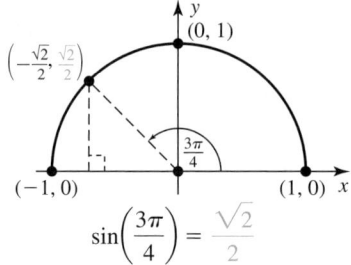

$$\sin\left(\frac{3\pi}{4}\right) = \frac{\sqrt{2}}{2}$$

Figure 5.33

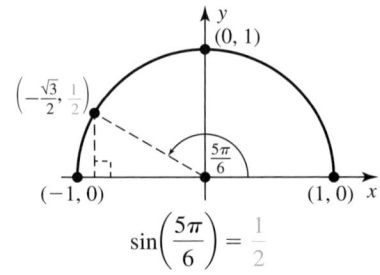

$$\sin\left(\frac{5\pi}{6}\right) = \frac{1}{2}$$

With this information we can extend our table of values through π, noting that $\sin \pi = 0$ (see Table 5.2).

Table 5.2

t	0	$\dfrac{\pi}{6}$	$\dfrac{\pi}{4}$	$\dfrac{\pi}{3}$	$\dfrac{\pi}{2}$	$\dfrac{2\pi}{3}$	$\dfrac{3\pi}{4}$	$\dfrac{5\pi}{6}$	π
$\sin t$	0	$\dfrac{1}{2}$	$\dfrac{\sqrt{2}}{2}$	$\dfrac{\sqrt{3}}{2}$	1	$\dfrac{\sqrt{3}}{2}$	$\dfrac{\sqrt{2}}{2}$	$\dfrac{1}{2}$	0

Using the symmetry of the circle and the fact that y is negative in QIII and QIV, we can complete the table for values between π and 2π.

EXAMPLE 1 ▶ **Finding Function Values Using Symmetry**

Use the symmetry of the unit circle and reference arcs of special values to complete Table 5.3. Recall that y is negative in QIII and QIV.

Table 5.3

t	π	$\dfrac{7\pi}{6}$	$\dfrac{5\pi}{4}$	$\dfrac{4\pi}{3}$	$\dfrac{3\pi}{2}$	$\dfrac{5\pi}{3}$	$\dfrac{7\pi}{4}$	$\dfrac{11\pi}{6}$	2π
$\sin t$									

Solution ▶ Symmetry shows that for any odd multiple of $t = \dfrac{\pi}{4}$, $\sin t = \pm \dfrac{\sqrt{2}}{2}$ depending on the quadrant of the terminal side. Similarly, for any reference arc of $\dfrac{\pi}{6}$, $\sin t = \pm \dfrac{1}{2}$, while any reference arc of $\dfrac{\pi}{3}$ will give $\sin t = \pm \dfrac{\sqrt{3}}{2}$. The completed table is shown in Table 5.4.

Table 5.4

t	π	$\dfrac{7\pi}{6}$	$\dfrac{5\pi}{4}$	$\dfrac{4\pi}{3}$	$\dfrac{3\pi}{2}$	$\dfrac{5\pi}{3}$	$\dfrac{7\pi}{4}$	$\dfrac{11\pi}{6}$	2π
$\sin t$	0	$-\dfrac{1}{2}$	$-\dfrac{\sqrt{2}}{2}$	$-\dfrac{\sqrt{3}}{2}$	-1	$-\dfrac{\sqrt{3}}{2}$	$-\dfrac{\sqrt{2}}{2}$	$-\dfrac{1}{2}$	0

Now try Exercises 7 and 8 ▶

Noting that $\dfrac{1}{2} = 0.5$, $\dfrac{\sqrt{2}}{2} \approx 0.71$, and $\dfrac{\sqrt{3}}{2} \approx 0.87$, we plot these points and connect them with a smooth curve to graph $y = \sin t$ in the interval $[0, 2\pi]$. The first five plotted points are labeled in Figure 5.34.

Figure 5.34

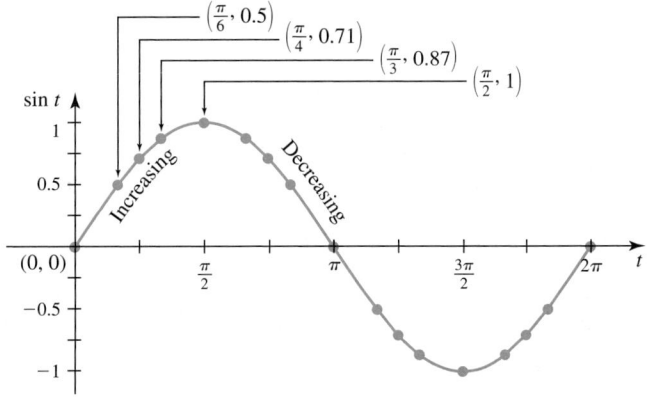

Expanding the table from 2π to 4π using reference arcs and the unit circle shows that function values begin to repeat. For example, $\sin\left(\dfrac{13\pi}{6}\right) = \sin\left(\dfrac{\pi}{6}\right)$ since $\theta_r = \dfrac{\pi}{6}$; $\sin\left(\dfrac{9\pi}{4}\right) = \sin\left(\dfrac{\pi}{4}\right)$ since $\theta_r = \dfrac{\pi}{4}$, and so on. Functions that cycle through a set pattern of values are said to be **periodic functions.**

Periodic Functions

A function f is said to be periodic if there is a positive number P such that $f(t + P) = f(t)$ for all t in the domain. The smallest number P for which this occurs is called the **period** of f.

For the sine function we have $\sin t = \sin(t + 2\pi)$, as in $\sin\left(\dfrac{13\pi}{6}\right) = \sin\left(\dfrac{\pi}{6} + 2\pi\right)$ and $\sin\left(\dfrac{9\pi}{4}\right) = \sin\left(\dfrac{\pi}{4} + 2\pi\right)$, with the idea extending to all other real numbers t: $\sin t = \sin(t + 2\pi k)$ for all integers k. The sine function is periodic with period $P = 2\pi$.

Although we initially focused on positive values of t in $[0, 2\pi]$, $t < 0$ and $k < 0$ are certainly possibilities and we note the graph of $y = \sin t$ extends infinitely in both directions (see Figure 5.35).

Figure 5.35

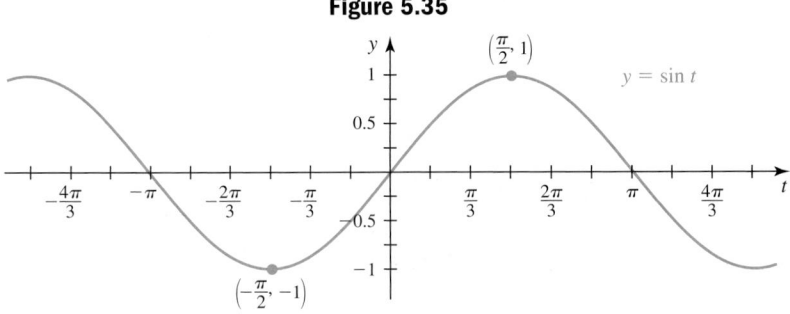

Finally, both the graph and the unit circle confirm that the range of $y = \sin t$ is $[-1, 1]$, and that $y = \sin t$ is an odd function. In particular, the graph shows $\sin\left(-\dfrac{\pi}{2}\right) = -\sin\left(\dfrac{\pi}{2}\right)$, and the unit circle shows (Figure 5.36) $\sin t = y$, and $\sin(-t) = -y$, from which we obtain $\sin(-t) = -\sin t$ by substitution. As a handy reference, the following box summarizes the main characteristics of $y = \sin t$.

Figure 5.36

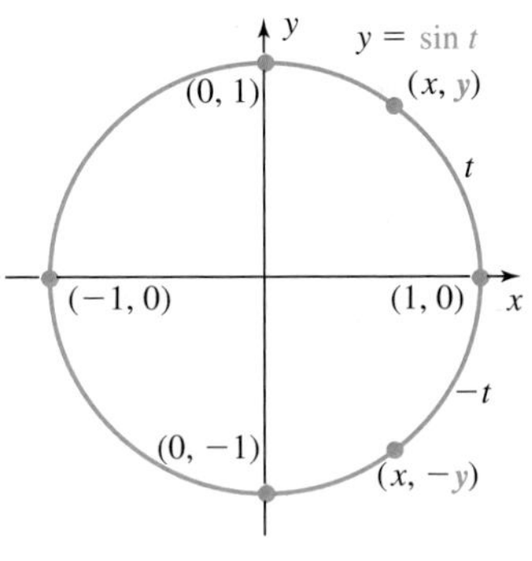

Characteristics of $f(t) = \sin t$

For all real numbers t and integers k,

Domain	**Range**	**Period**
$(-\infty, \infty)$	$[-1, 1]$	2π
Symmetry	**Maximum value**	**Minimum value**
odd	$\sin t = 1$	$\sin t = -1$
$\sin(-t) = -\sin t$	at $t = \dfrac{\pi}{2} + 2\pi k$	at $t = \dfrac{3\pi}{2} + 2\pi k$
Increasing	**Decreasing**	**Zeroes**
$\left(0, \dfrac{\pi}{2}\right) \cup \left(\dfrac{3\pi}{2}, 2\pi\right)$	$\left(\dfrac{\pi}{2}, \dfrac{3\pi}{2}\right)$	$t = k\pi$

EXAMPLE 2 ▶ **Using the Period of sin t to Find Function Values**

Use the characteristics of $f(t) = \sin t$ to match the given value of t to the correct value of $\sin t$.

a. $t = \left(\dfrac{\pi}{4} + 8\pi\right)$ **b.** $t = -\dfrac{\pi}{6}$ **c.** $t = \dfrac{17\pi}{2}$ **d.** $t = 21\pi$ **e.** $t = \dfrac{11\pi}{2}$

I. $\sin t = 1$ **II.** $\sin t = -\dfrac{1}{2}$ **III.** $\sin t = -1$ **IV.** $\sin t = \dfrac{\sqrt{2}}{2}$ **V.** $\sin t = 0$

Solution ▶ **a.** Since $\sin\left(\dfrac{\pi}{4} + 8\pi\right) = \sin\dfrac{\pi}{4}$, the correct match is (IV).

b. Since $\sin\left(-\dfrac{\pi}{6}\right) = -\sin\dfrac{\pi}{6}$, the correct match is (II).

c. Since $\sin\left(\dfrac{17\pi}{2}\right) = \sin\left(\dfrac{\pi}{2} + 8\pi\right) = \sin\dfrac{\pi}{2}$, the correct match is (I).

d. Since $\sin(21\pi) = \sin(\pi + 20\pi) = \sin\pi$, the correct match is (V).

e. Since $\sin\left(\dfrac{11\pi}{2}\right) = \sin\left(\dfrac{3\pi}{2} + 4\pi\right) = \sin\left(\dfrac{3\pi}{2}\right)$, the correct match is (III).

Now try Exercises 9 and 10 ▶

Many of the transformations applied to algebraic graphs can also be applied to trigonometric graphs. These transformations may stretch, reflect, or translate the graph, but it will still retain its basic shape. In numerous applications it will help if you're able to draw a quick, accurate sketch of the transformations involving $f(t) = \sin t$. To assist this effort, we'll begin with the interval $[0, 2\pi]$, combine the characteristics just listed with some simple geometry, and offer the following four-step process. Steps I through IV are illustrated in Figures 5.37 through 5.40.

Figure 5.37

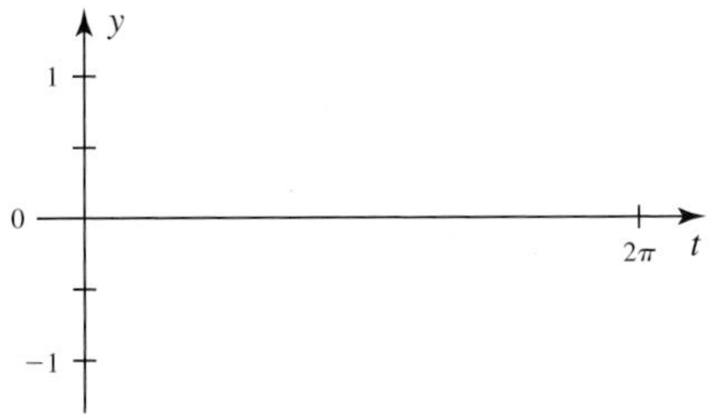

Figure 5.38

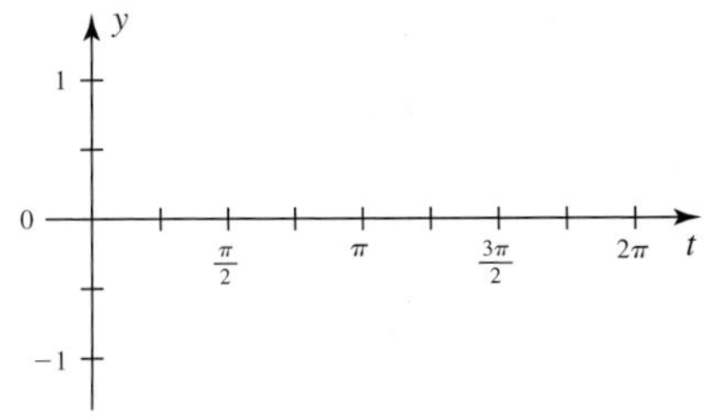

Figure 5.39   **Figure 5.40**

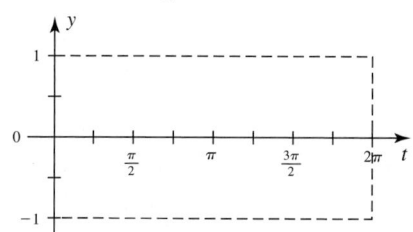   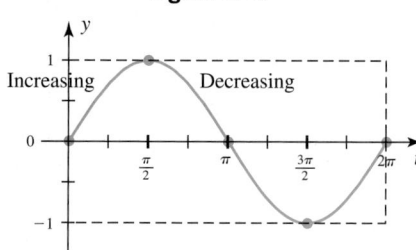

Step I: Draw the *y*-axis, mark zero halfway up, with -1 and 1 an equal distance from this zero. Then draw an extended *t*-axis and tick mark 2π to the extreme right (Figure 5.37).

Step II: On the *t*-axis, mark halfway between 0 and 2π and label it "π," mark halfway between π on either side and label the marks $\dfrac{\pi}{2}$ and $\dfrac{3\pi}{2}$. Halfway between these you can draw additional tick marks to represent the remaining multiples of $\dfrac{\pi}{4}$ (Figure 5.38).

Step III: Next, lightly draw a rectangular frame, which we'll call the **reference rectangle,** $P = 2\pi$ units wide and 2 units tall, centered on the *t*-axis and with the *y*-axis along one side (Figure 5.39).

Step IV: Knowing $y = \sin t$ is positive and increasing in QI, that the range is $[-1, 1]$, that the zeroes are 0, π, and 2π, and that maximum and minimum values *occur halfway between the zeroes* (since there is no horizontal shift), we can draw a reliable graph of $y = \sin t$ by partitioning the rectangle into four equal parts to locate these values (note **bold** tick-marks). We will call this partitioning of the reference rectangle the **rule of fourths,** since we are then scaling the *t*-axis in increments of $\dfrac{P}{4}$ (Figure 5.40).

EXAMPLE 3 ▶ Graphing $y = \sin t$ Using a Reference Rectangle

Use steps I through IV to draw a sketch of $y = \sin t$ for the interval $\left[-\dfrac{\pi}{2}, \dfrac{3\pi}{2}\right]$.

Solution ▶ Start by completing steps I and II, then extend the *t*-axis to include $-\dfrac{\pi}{2}$. Beginning at $-\dfrac{\pi}{2}$, draw a reference rectangle 2π units wide and 2 units tall, centered on the *x*-axis $\left(\text{ending at } \dfrac{3\pi}{2}\right)$. After applying the rule of

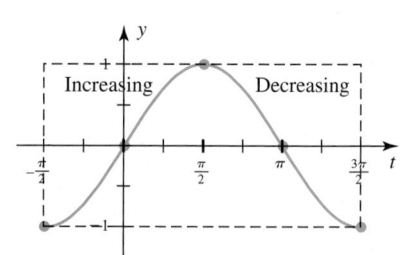

fourths, we note the zeroes occur at $t = 0$ and $t = \pi$, with the max/min values spaced equally between and on either side. Plot these points and connect them with a smooth curve (see the figure).

☑ **A.** You've just learned how to graph $f(t) = \sin t$ using special values and symmetry

Now try Exercises 11 and 12 ▶

B. Graphing $f(t) = \cos t$

With the graph of $f(t) = \sin t$ established, sketching the graph of $f(t) = \cos t$ is a very natural next step. First, note that when $t = 0$, $\cos t = 1$ so the graph of $y = \cos t$ will begin at (0, 1) in the interval $[0, 2\pi]$. Second, we've seen $\left(\pm\dfrac{1}{2}, \pm\dfrac{\sqrt{3}}{2} \right)$, $\left(\pm\dfrac{\sqrt{3}}{2}, \pm\dfrac{1}{2} \right)$ and $\left(\pm\dfrac{\sqrt{2}}{2}, \pm\dfrac{\sqrt{2}}{2} \right)$ are all points on the unit circle since they satisfy $x^2 + y^2 = 1$. Since $\cos t = x$ and $\sin t = y$, the equation $\cos^2 t + \sin^2 t = 1$ can be obtained by direct substitution. This means if $\sin t = \pm\dfrac{1}{2}$, then $\cos t = \pm\dfrac{\sqrt{3}}{2}$ and vice versa, with the signs taken from the appropriate quadrant. The table of values for cosine then becomes a simple variation of the table for sine, as shown in Table 5.5 for $t \in [0, \pi]$.

Table 5.5

t	0	$\dfrac{\pi}{6}$	$\dfrac{\pi}{4}$	$\dfrac{\pi}{3}$	$\dfrac{\pi}{2}$	$\dfrac{2\pi}{3}$	$\dfrac{3\pi}{4}$	$\dfrac{5\pi}{6}$	π
$\sin t$	0	$\dfrac{1}{2} = 0.5$	$\dfrac{\sqrt{2}}{2} \approx 0.71$	$\dfrac{\sqrt{3}}{2} \approx 0.87$	1	$\dfrac{\sqrt{3}}{2} \approx 0.87$	$\dfrac{\sqrt{2}}{2} \approx 0.71$	$\dfrac{1}{2} = 0.5$	0
$\cos t$	1	$\dfrac{\sqrt{3}}{2} \approx 0.87$	$\dfrac{\sqrt{2}}{2} \approx 0.71$	$\dfrac{1}{2} = 0.5$	0	$-\dfrac{1}{2} = -0.5$	$-\dfrac{\sqrt{2}}{2} \approx -0.71$	$-\dfrac{\sqrt{3}}{2} \approx -0.87$	-1

The same values can be taken from the unit circle, but this view requires much less effort and easily extends to values of t in $[\pi, 2\pi]$. Using the points from Table 5.5 and its extension through $[\pi, 2\pi]$, we can draw the graph of $y = \cos t$ in $[0, 2\pi]$ and identify where the function is increasing and decreasing in this interval. See Figure 5.41.

Figure 5.41

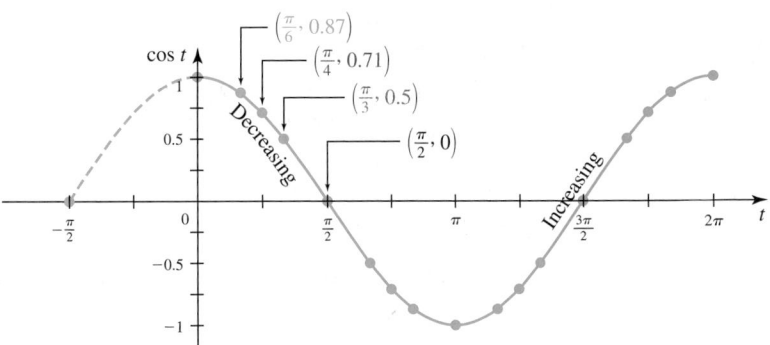

The function is decreasing for t in $(0, \pi)$, and increasing for t in $(\pi, 2\pi)$. The end result appears to be the graph of $y = \sin t$ shifted to the left $\dfrac{\pi}{2}$ units, a fact more easily seen if we extend the graph to $-\dfrac{\pi}{2}$ as shown. This is in fact the case, and is a relationship we will later prove in Chapter 6. Like $y = \sin t$, the function $y = \cos t$ is periodic with period $P = 2\pi$, with the graph extending infinitely in both directions.

Finally, we note that cosine is an **even function,** meaning $\cos(-t) = \cos t$ for all t in the domain. For instance, $\cos\left(-\dfrac{\pi}{2}\right) = \cos\left(\dfrac{\pi}{2}\right) = 0$ (see Figure 5.41). Here is a summary of important characteristics of the cosine function.

Characteristics of $f(t) = \cos t$		
For all real numbers t and integers k,		
Domain	**Range**	**Period**
$(-\infty, \infty)$	$[-1, 1]$	2π
Symmetry	**Maximum value**	**Minimum value**
even	$\cos t = 1$	$\cos t = -1$
$\cos(-t) = \cos t$	at $t = 2\pi k$	at $t = \pi + 2\pi k$
Increasing	**Decreasing**	**Zeroes**
$(\pi, 2\pi)$	$(0, \pi)$	$t = \dfrac{\pi}{2} + \pi k$

EXAMPLE 4 ▶ **Graphing $y = \cos t$ Using a Reference Rectangle**

Draw a sketch of $y = \cos t$ for t in $\left[-\pi, \dfrac{3\pi}{2}\right]$.

Solution ▶ After completing steps I and II, extend the negative x-axis to include $-\pi$. Beginning at $-\pi$, draw a reference rectangle 2π units wide and 2 units tall, centered on the x-axis. After applying the rule of fourths, we note the zeroes will occur at $t = -\pi/2$ and $t = \pi/2$, with

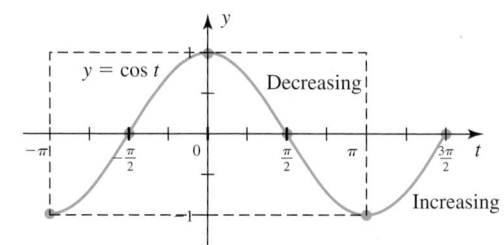

☑ **B.** You've just learned how to graph $f(t) = \cos t$ using special values and symmetry

the max/min values spaced equally between these zeroes and on either side (at $t = -\pi$, $t = 0$, and $t = \pi$). Finally, we extend the graph to include $3\pi/2$.

Now try Exercises 13 and 14 ▶

WORTHY OF NOTE

Note that the equations $y = A \sin t$ and $y = A \cos t$ both indicate y is a function of t, with no reference to the unit circle definitions $\cos t = x$ and $\sin t = y$.

C. Graphing $y = A \sin(Bt)$ and $y = A \cos(Bt)$

In many applications, trig functions have maximum and minimum values other than 1 and -1, and periods other than 2π. For instance, in tropical regions the maximum and minimum temperatures may vary by no more than $20°$, while for desert regions this difference may be $40°$ or more. This variation is modeled by the *amplitude* of sine and cosine functions.

Amplitude and the Coefficient A (assume B = 1)

For functions of the form $y = A \sin t$ and $y = A \cos t$, let M represent the *M*aximum value and m the *m*inimum value of the functions. Then the quantity $\dfrac{M + m}{2}$ gives the **average value** of the function, while $\dfrac{M - m}{2}$ gives the **amplitude** of the function.

Amplitude is the maximum displacement from the average value in the positive or negative direction. It is represented by $|A|$, with A playing a role similar to that seen for algebraic graphs [$Af(t)$ vertically stretches or compresses the graph of f, and reflects it across the t-axis if $A < 0$]. Graphs of the form $y = \sin t$ (and $y = \cos t$) can quickly be sketched with any amplitude by noting (1) the *zeroes of the function remain fixed* since $\sin t = 0$ implies $A \sin t = 0$, and (2) the *maximum and minimum values are A and $-A$*, respectively, since $\sin t = 1$ or -1 implies $A \sin t = A$ or $-A$. Note this implies the reference rectangle will be $2A$ units tall and P units wide. Connecting the points that result with a smooth curve will complete the graph.

EXAMPLE 5 ▶ **Graphing $y = A \sin t$ Where $A \neq 1$**

Draw a sketch of $y = 4 \sin t$ in the interval $[0, 2\pi]$.

Solution ▶ With an amplitude of $|A| = 4$, the reference rectangle will be $2(4) = 8$ units tall, by 2π units wide. Using the rule of fourths, the zeroes are still $t = 0$, $t = \pi$, and $t = 2\pi$, with the max/min values spaced equally between. The maximum value is $4 \sin\left(\dfrac{\pi}{2}\right) = 4(1) = 4$, with a minimum value of $4 \sin\left(\dfrac{3\pi}{2}\right) = 4(-1) = -4$.

Connecting these points with a "sine curve" gives the graph shown ($y = \sin t$ is also shown for comparison).

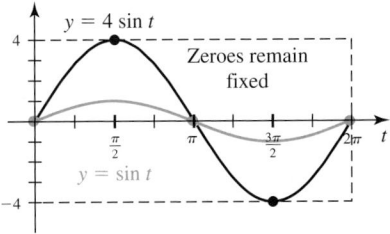

Now try Exercises 15 through 20 ▶

Period and the Coefficient B

While basic sine and cosine functions have a period of 2π, in many applications the period may be very long (tsunami's) or very short (electromagnetic waves). For the equations $y = A \sin(Bt)$ and $y = A \cos(Bt)$, the period depends on the value of B. To see why, consider the function $y = \cos(2t)$ and Table 5.6. Multiplying input values by 2 means each cycle will be completed twice as fast. The table shows that $y = \cos(2t)$ completes a full cycle in $[0, \pi]$, giving a period of $P = \pi$ (Figure 5.42, red graph).

Table 5.6

t	0	$\dfrac{\pi}{4}$	$\dfrac{\pi}{2}$	$\dfrac{3\pi}{4}$	π
$2t$	0	$\dfrac{\pi}{2}$	π	$\dfrac{3\pi}{2}$	2π
$\cos(2t)$	1	0	-1	0	1

Dividing input values by 2 (or multiplying by $\frac{1}{2}$) will cause the function to complete a cycle only half as fast, doubling the time required to complete a full cycle. Table 5.7 shows $y = \cos\left(\frac{1}{2}t\right)$ completes only one-half cycle in 2π (Figure 5.42, blue graph).

Table 5.7
(values in blue are approximate)

t	0	$\dfrac{\pi}{4}$	$\dfrac{\pi}{2}$	$\dfrac{3\pi}{4}$	π	$\dfrac{5\pi}{4}$	$\dfrac{3\pi}{2}$	$\dfrac{7\pi}{4}$	2π
$\dfrac{1}{2}t$	0	$\dfrac{\pi}{8}$	$\dfrac{\pi}{4}$	$\dfrac{3\pi}{8}$	$\dfrac{\pi}{2}$	$\dfrac{5\pi}{8}$	$\dfrac{3\pi}{4}$	$\dfrac{7\pi}{8}$	π
$\cos\left(\dfrac{1}{2}t\right)$	1	0.92	$\dfrac{\sqrt{2}}{2}$	0.38	0	-0.38	$-\dfrac{\sqrt{2}}{2}$	-0.92	-1

The graphs of $y = \cos t$, $y = \cos(2t)$, and $y = \cos\left(\frac{1}{2}t\right)$ shown in Figure 5.42 clearly illustrate this relationship and how the value of B affects the period of a graph.

To find the period for arbitrary values of B, the formula $P = \dfrac{2\pi}{|B|}$ is used. Note for $y = \cos(2t)$, $B = 2$ and $P = \dfrac{2\pi}{2} = \pi$, as shown. For $y = \cos\left(\dfrac{1}{2}t\right)$, $|B| = \dfrac{1}{2}$, and $P = \dfrac{2\pi}{1/2} = 4\pi$.

Figure 5.42

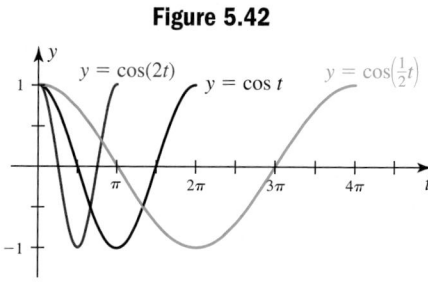

Period Formula for Sine and Cosine

For B a real number and functions $y = A\sin(Bt)$ and $y = A\cos(Bt)$,

$$P = \frac{2\pi}{|B|}.$$

To sketch these functions for periods other than 2π, we still use a reference rectangle of height $2A$ and length P, then break the enclosed t-axis in four equal parts to help draw the graph. In general, if the period is "very large" one full cycle is appropriate for the graph. If the period is very small, graph at least two cycles.

Note the value of B in Example 6 includes a factor of π. This actually happens quite frequently in applications of the trig functions.

EXAMPLE 6 ▶ **Graphing $y = A\cos(Bt)$, Where A, $B \neq 1$**

Draw a sketch of $y = -2\cos(0.4\pi t)$ for t in $[-\pi, 2\pi]$.

Solution ▶ The amplitude is $|A| = 2$, so the reference rectangle will be $2(2) = 4$ units high. Since $A < 0$ the *graph will be vertically reflected across the t-axis*. The period is

$$P = \frac{2\pi}{0.4\pi} = 5 \text{ (note the factors of } \pi \text{ reduce to 1), so the reference rectangle will}$$

be 5 units in length. Breaking the t-axis into four parts within the frame (rule of fourths) gives $\left(\dfrac{1}{4}\right)5 = \dfrac{5}{4}$ units, indicating that we should scale the t-axis in multiples of $\dfrac{1}{4}$. Note the zeroes occur at $\dfrac{5\pi}{4}$ and $\dfrac{15\pi}{4}$, with a maximum value at $\dfrac{10\pi}{4}$. In cases where the π factor reduces, we scale the t-axis as a "standard" number line, and *estimate the location of multiples of π*. For practical reasons, we first draw the unreflected graph (shown in blue) for guidance in drawing the reflected graph, which is then extended to fit the given interval.

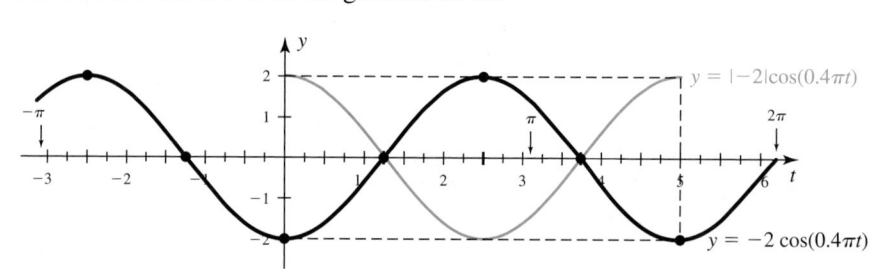

☑ **C. You've just learned how to graph sine and cosine functions with various amplitudes and periods**

Now try Exercises 21 through 32 ▶

D. Graphs of $y = \csc(Bt)$ and $y = \sec(Bt)$

The graphs of these reciprocal functions follow quite naturally from the graphs of $y = A\sin(Bt)$ and $y = A\cos(Bt)$, by using these observations: (1) you cannot divide by zero, (2) the reciprocal of a very small number is a very large number (and vice versa), and (3) the reciprocal of ± 1 is ± 1. Just as with rational functions, division by zero creates a vertical asymptote, so the graph of $y = \csc t = \dfrac{1}{\sin t}$ will have a vertical asymptote at every point where $\sin t = 0$. This occurs at $t = \pi k$, where k is an integer $(\ldots -2\pi, -\pi, 0, \pi, 2\pi, \ldots)$. Further, when $\csc(Bt) = \pm 1$, $\sin(Bt) = \pm 1$ since the reciprocal of 1 and -1 are still 1 and -1, respectively. Finally, due to observation 2, the graph of the cosecant function will be increasing when the sine function is decreasing, and decreasing when the sine function is increasing. In most cases, we graph $y = \csc(Bt)$ by drawing a sketch of $y = \sin(Bt)$, then using these observations as demonstrated in Example 7. In doing so, we discover that the period of the cosecant function is also 2π and that $y = \csc(Bt)$ is an odd function.

EXAMPLE 7 ▶ **Graphing a Cosecant Function**

Graph the function $y = \csc t$ for $t \in [0, 4\pi]$.

Solution ▶ The related sine function is $y = \sin t$, which means we'll draw a rectangular frame $2A = 2$ units high. The period is $P = \dfrac{2\pi}{1} = 2$, so the reference frame will be 2π units in length. Breaking the t-axis into four parts within the frame means each tick mark will be $\left(\dfrac{1}{4}\right)\left(\dfrac{2\pi}{1}\right) = \dfrac{\pi}{2}$ units apart, with the asymptotes occurring at 0, π, and 2π. A partial table and the resulting graph are shown.

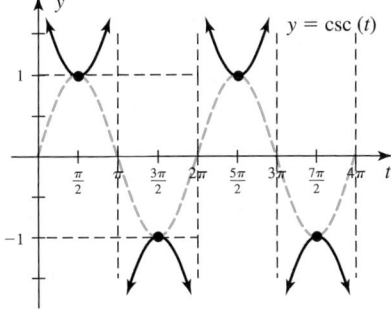

- Vertical asymptotes where sine is zero
- When $\csc(Bt) = \pm 1$, $\sin(Bt) = \pm 1$
- Output values are reciprocated

t	$\sin t$	$\csc t$
0	0	$\dfrac{1}{0} \to$ undefined
$\dfrac{\pi}{6}$	$\dfrac{1}{2} = 0.5$	$\dfrac{2}{1} = 2$
$\dfrac{\pi}{4}$	$\dfrac{\sqrt{2}}{2} \approx 0.71$	$\dfrac{2}{\sqrt{2}} \approx 1.41$
$\dfrac{\pi}{3}$	$\dfrac{\sqrt{3}}{2} \approx 0.87$	$\dfrac{2}{\sqrt{3}} \approx 1.15$
$\dfrac{\pi}{2}$	1	1

Now try Exercises 33 and 34 ▶

☑ **D. You've just learned how to investigate graphs of the reciprocal functions** $f(t) = \csc(Bt)$ **and** $f(t) = \sec(Bt)$

Similar observations can be made regarding $y = \sec(Bt)$ and its relationship to $y = \cos(Bt)$ (see **Exercises 8, 35, and 36**). The most important characteristics of the cosecant and secant functions are summarized in the following box. For these functions, there is no discussion of amplitude, and no mention is made of their zeroes since neither graph intersects the t-axis.

Characteristics of $f(t) = \csc t$ and $f(t) = \sec t$

For all real numbers t and integers k,

$y = \csc t$		
Domain	**Range**	**Asymptotes**
$t \neq k\pi$	$(-\infty, -1] \cup [1, \infty)$	$t = k\pi$

Period	**Symmetry**
2π	odd
	$\csc(-t) = -\csc t$

$y = \sec t$		
Domain	**Range**	**Asymptotes**
$t \neq \dfrac{\pi}{2} + \pi k$	$(-\infty, -1] \cup [1, \infty)$	$t = \dfrac{\pi}{2} + \pi k$

Period	**Symmetry**
2π	even
	$\sec(-t) = \sec t$

E. Writing Equations from Graphs

Mathematical concepts are best reinforced by working with them in both "forward and reverse." Where graphs are concerned, this means we should attempt to find the equation of a given graph, rather than only using an equation to sketch the graph. Exercises of this type require that you become very familiar with the graph's basic characteristics and how each is expressed as part of the equation.

EXAMPLE 8 ▶   **Determining the Equation of a Given Graph**

The graph shown here is of the form $y = A \sin(Bt)$. Find the value of A and B.

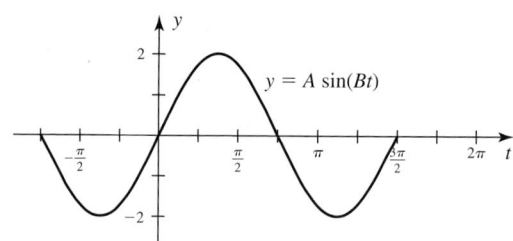

Solution ▶   By inspection, the graph has an amplitude of $A = 2$ and a period of $P = \dfrac{3\pi}{2}$.

To find B we used the period formula $P = \dfrac{2\pi}{|B|}$, substituting $\dfrac{3\pi}{2}$ for P and solving.

$$P = \frac{2\pi}{|B|} \quad \text{period formula}$$

$$\frac{3\pi}{2} = \frac{2\pi}{B} \quad \text{substitute } \frac{3\pi}{2} \text{ for } P; B > 0$$

$$3\pi B = 4\pi \quad \text{multiply by } 2B$$

$$B = \frac{4}{3} \quad \text{solve for } B$$

☑ **E.** You've just learned how to write the equation for a given graph

The result is $B = \frac{4}{3}$, which gives us the equation $y = 2 \sin\left(\frac{4}{3}t\right)$.

Now try Exercises 37 through 58 ▶

There are a number of interesting applications of this "graph to equation" process in the exercise set. **See Exercises 61 to 72.**

TECHNOLOGY HIGHLIGHT

Exploring Amplitudes and Periods

In practice, trig applications offer an immense range of coefficients, creating amplitudes that are sometimes very large and sometimes extremely small, as well as periods ranging from nanoseconds, to many years. This *Technology Highlight* is designed to help you use the calculator more effectively in the study of these functions. To begin, we note that many calculators offer a preset [ZOOM] option that automatically sets a window size convenient to many trig graphs. The resulting [WINDOW] after pressing [ZOOM] **7:ZTrig** on a TI-84 Plus is shown in Figure 5.43 for a calculator set in **Radian** [MODE].

In Section 5.3 we noted that a change in amplitude will not change the location of the zeroes or max/min values. On the [Y=] screen, enter $Y_1 = \frac{1}{2}\sin x$, $Y_2 = \sin x$, $Y_3 = 2\sin x$, and $Y_4 = 4\sin x$, then use [ZOOM] **7:ZTrig** to graph the functions. As you see in Figure 5.44, each graph rises to the expected amplitude at the expected location, while "holding on" to the zeroes.

To explore concepts related to the coefficient B and the period of a trig function, enter $Y_1 = \sin\left(\frac{1}{2}x\right)$ and $Y_2 = \sin(2x)$ on the [Y=] screen and graph using [ZOOM] **7:ZTrig**. While the result is "acceptable," the graphs are difficult to read and compare, so we manually change the window size to obtain a better view (Figure 5.45).

A true test of effective calculator use comes when the amplitude or period is a very large or very small number. For instance, the tone you hear while pressing "5" on your telephone is actually a combination of the tones modeled by $Y_1 = \sin[2\pi(770)t]$ and $Y_2 = \sin[2\pi(1336)t]$. Graphing these functions requires a careful analysis of the period, otherwise the graph can appear garbled, misleading, or difficult to read —try graphing Y_1 on the [ZOOM] **7:ZTrig** or [ZOOM] **6:ZStandard** screens (see Figure 5.46). First note $A = 1$, and $P = \frac{2\pi}{2\pi770}$ or $\frac{1}{770}$. With a period this short, even graphing the function from Xmin = −1 to Xmax = 1 gives a distorted graph. Setting Xmin to −1/770, Xmax to 1/770, and Xscl to (1/770)/10 gives the graph in Figure 5.47, which can be used to investigate characteristics of the function.

Exercise 1: Graph the second tone $Y_2 = \sin[2\pi(1336)t]$ and find its value at $t = 0.00025$ sec.

Exercise 2: Graph the function $Y_1 = 950\sin(0.005t)$ on a "friendly" window and find the value at $x = 550$.

Figure 5.43

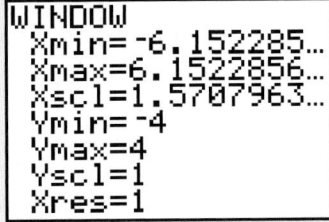

Figure 5.44

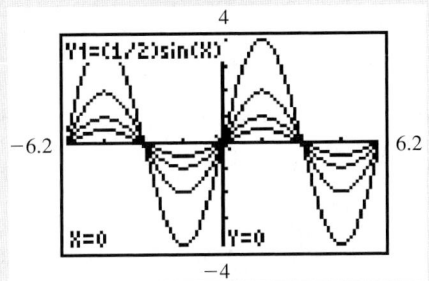

Figure 5.45

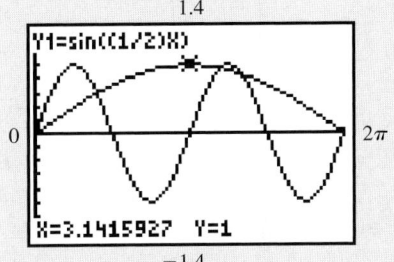

Figure 5.46

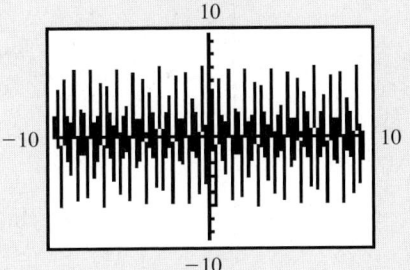

Figure 5.47

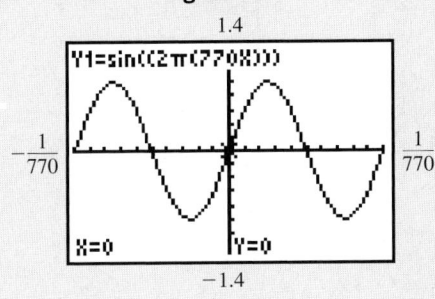

5.3 EXERCISES

▶ CONCEPTS AND VOCABULARY

Fill in each blank with the appropriate word or phrase. Carefully reread the section if needed.

1. For the sine function, output values are _____ in the interval $\left[0, \dfrac{\pi}{2}\right]$.

2. For the cosine function, output values are _____ in the interval $\left[0, \dfrac{\pi}{2}\right]$.

3. For the sine and cosine functions, the domain is _____ and the range is _____.

4. The amplitude of sine and cosine is defined to be the maximum _____ from the _____ value in the positive and negative directions.

5. Discuss/Describe the four-step process outlined in this section for the graphing of basic trig functions. Include a worked-out example and a detailed explanation.

6. Discuss/Explain how you would determine the domain and range of $y = \sec x$. Where is this function undefined? Why? Graph $y = 2 \sec(2t)$ using $y = 2 \cos(2t)$. What do you notice?

▶ DEVELOPING YOUR SKILLS

7. Use the symmetry of the unit circle and reference arcs of standard values to complete a table of values for $y = \cos t$ in the interval $t \in [\pi, 2\pi]$.

8. Use the standard values for $y = \cos t$ for $t \in [\pi, 2\pi]$ to create a table of values for $y = \sec t$ on the same interval.

Use the characteristics of $f(t) = \sin t$ to match the given value of t to the correct value of $\sin t$.

9. **a.** $t = \left(\dfrac{\pi}{6} + 10\pi\right)$ **b.** $t = -\dfrac{\pi}{4}$

   c. $t = \dfrac{-15\pi}{4}$ **d.** $t = 13\pi$

   e. $t = \dfrac{21\pi}{2}$ **I.** $\sin t = 0$

   II. $\sin t = \dfrac{1}{2}$ **III.** $\sin t = 1$

   IV. $\sin t = \dfrac{\sqrt{2}}{2}$ **V.** $\sin t = -\dfrac{\sqrt{2}}{2}$

10. **a.** $t = \left(\dfrac{\pi}{4} - 12\pi\right)$ **b.** $t = \dfrac{11\pi}{6}$

    c. $t = \dfrac{23\pi}{2}$ **d.** $t = -19\pi$

    e. $t = -\dfrac{25\pi}{4}$ **I.** $\sin t = -\dfrac{1}{2}$

   II. $\sin t = -\dfrac{\sqrt{2}}{2}$ **III.** $\sin t = 0$

   IV. $\sin t = \dfrac{\sqrt{2}}{2}$ **V.** $\sin t = -1$

Use steps I through IV given in this section to draw a sketch of each graph.

11. $y = \sin t$ for $t \in \left[-\dfrac{3\pi}{2}, \dfrac{\pi}{2}\right]$

12. $y = \sin t$ for $t \in [-\pi, \pi]$

13. $y = \cos t$ for $t \in \left[-\dfrac{\pi}{2}, 2\pi\right]$

14. $y = \cos t$ for $t \in \left[-\dfrac{\pi}{2}, \dfrac{5\pi}{2}\right]$

Use a reference rectangle and the *rule of fourths* to draw an accurate sketch of the following functions through two complete cycles—one where $t > 0$, and one where $t < 0$. Clearly state the amplitude and period as you begin.

15. $y = 3 \sin t$ 16. $y = 4 \sin t$

17. $y = -2 \cos t$ 18. $y = -3 \cos t$

19. $y = \dfrac{1}{2} \sin t$ 20. $y = \dfrac{3}{4} \sin t$

21. $y = -\sin(2t)$

22. $y = -\cos(2t)$

23. $y = 0.8 \cos(2t)$

24. $y = 1.7 \sin(4t)$

25. $f(t) = 4 \cos\left(\frac{1}{2}t\right)$

26. $y = -3 \cos\left(\frac{3}{4}t\right)$

27. $f(t) = 3 \sin(4\pi t)$

28. $g(t) = 5 \cos(8\pi t)$

29. $y = 4 \sin\left(\frac{5\pi}{3}t\right)$

30. $y = 2.5 \cos\left(\frac{2\pi}{5}t\right)$

31. $f(t) = 2 \sin(256\pi t)$

32. $g(t) = 3 \cos(184\pi t)$

Draw the graph of each function by first sketching the related sine and cosine graphs, and applying the observations made in this section.

33. $y = 3 \csc t$

34. $g(t) = 2 \csc(4t)$

35. $y = 2 \sec t$

36. $f(t) = 3 \sec(2t)$

Clearly state the amplitude and period of each function, then match it with the corresponding graph.

37. $y = -2 \cos(4t)$

38. $y = 2 \sin(4t)$

39. $y = 3 \sin(2t)$

40. $y = -3 \cos(2t)$

41. $y = 2 \csc\left(\frac{1}{2}t\right)$

42. $y = 2 \sec\left(\frac{1}{4}t\right)$

43. $f(t) = \frac{3}{4}\cos(0.4t)$

44. $g(t) = \frac{7}{4} \cos(0.8t)$

45. $y = \sec(8\pi t)$

46. $y = \csc(12\pi t)$

47. $y = 4 \sin(144\pi t)$

48. $y = 4 \cos(72\pi t)$

a.

b.
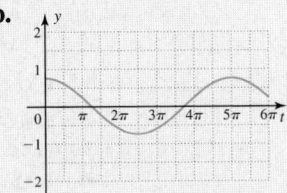

c.
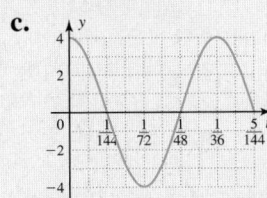

d.
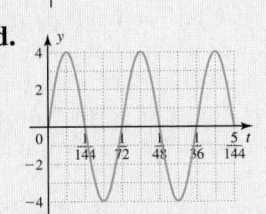

e.
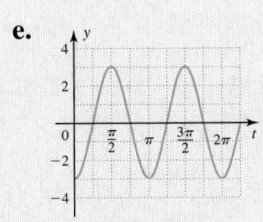

f.
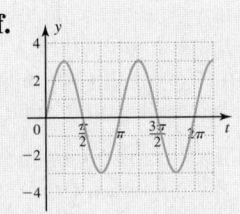

g.
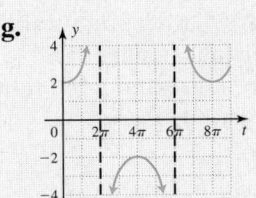

h.
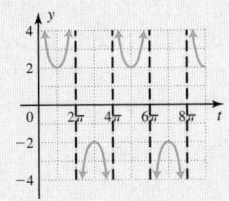

i.
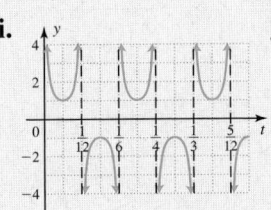

j.
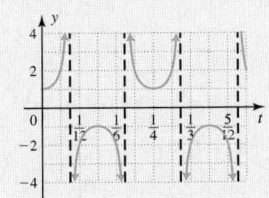

k.
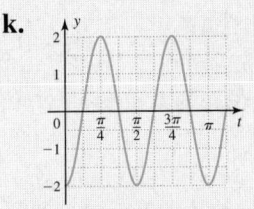

l.
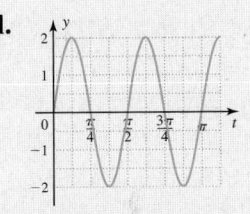

The graphs shown are of the form $y = A \cos(Bt)$ or $y = A \csc(Bt)$. Use the characteristics illustrated for each graph to determine its equation.

49.
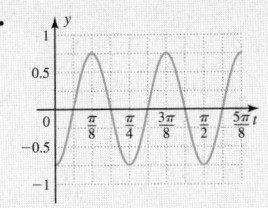

50.
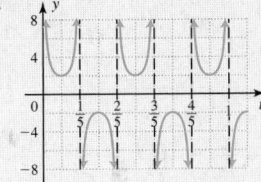

51.
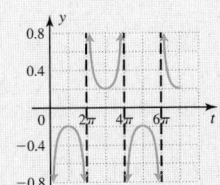

52.
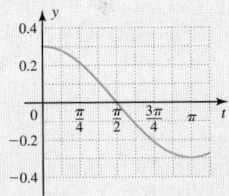

53.
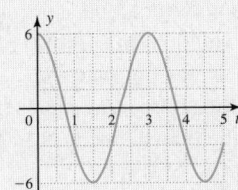

54.
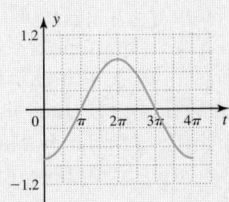

Match each graph to its equation, then graphically estimate the points of intersection. Confirm or contradict your estimate(s) by substituting the values into the given equations using a calculator.

55. $y = -\cos x$; 1
$y = \sin x$

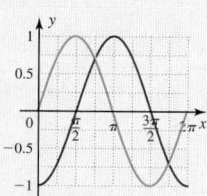

56. $y = -\cos x$;
$y = \sin(2x)$

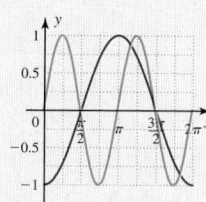

57. $y = -2\cos x$;
$y = 2\sin(3x)$

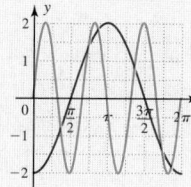

58. $y = 2\cos(2\pi x)$;
$y = -2\sin(\pi x)$

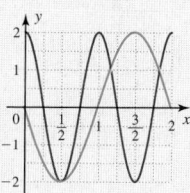

▶ **WORKING WITH FORMULAS**

59. The Pythagorean theorem in trigonometric form: $\sin^2\theta + \cos^2\theta = 1$

The formula shown is commonly known as a Pythagorean identity and is introduced more formally in Chapter 6. It is derived by noting that on a unit circle, $\cos t = x$ and $\sin t = y$, while $x^2 + y^2 = 1$. Given that $\sin t = \frac{15}{113}$, use the formula to find the value of $\cos t$ in Quadrant I. What is the Pythagorean triple associated with these values of x and y?

60. Hydrostatics, surface tension, and contact angles: $y = \dfrac{2\gamma \cos \theta}{kr}$

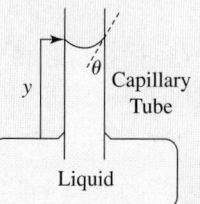

The height that a liquid will rise in a capillary tube is given by the formula shown, where r is the radius of the tube, θ is the contact angle of the liquid (the meniscus), γ is the surface tension of the liquid-vapor film, and k is a constant that depends on the weight-density of the liquid. How high will the liquid rise given that the surface tension $\gamma = 0.2706$, the tube has radius $r = 0.2$ cm, the contact angle $\theta = 22.5°$, and $k = 1.25$?

▶ **APPLICATIONS**

Tidal waves: Tsunamis, also known as tidal waves, are ocean waves produced by earthquakes or other upheavals in the Earth's crust and can move through the water undetected for hundreds of miles at great speed. While traveling in the open ocean, these waves can be represented by a sine graph with a very long wavelength (period) and a very small amplitude. Tsunami waves only attain a monstrous size as they approach the shore, and represent a very different phenomenon than the ocean swells created by heavy winds over an extended period of time.

61. A graph modeling a tsunami wave is given in the figure. (a) What is the height of the tsunami wave (from crest to trough)? Note that $h = 0$ is considered the level of a calm ocean. (b) What is the tsunami's wavelength? (c) Find the equation for this wave.

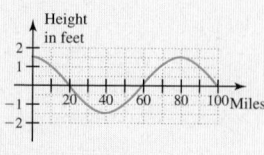

62. A heavy wind is kicking up ocean swells approximately 10 ft high (from crest to trough), with wavelengths of 250 ft. (a) Find the equation that models these swells. (b) Graph the equation. (c) Determine the height of a wave measured 200 ft from the trough of the previous wave.

Sinusoidal models: The sine and cosine functions are of great importance to meteorological studies, as when modeling the temperature based on the time of day, the illumination of the Moon as it goes through its phases, or even the prediction of tidal motion.

63. The graph given shows the deviation from the average daily temperature for the hours of a given day, with $t = 0$

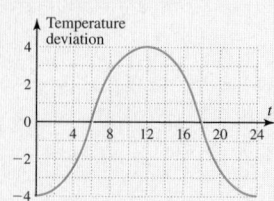

corresponding to 6 A.M. (a) Use the graph to determine the related equation. (b) Use the equation to find the deviation at $t = 11$ (5 P.M.) and confirm that this point is on the graph. (c) If the average temperature for this day was 72°, what was the temperature at midnight?

64. The equation $y = 7\sin\left(\dfrac{\pi}{6}t\right)$ models the height of the tide along a certain coastal area, as compared to average sea level. Assuming $t = 0$ is midnight, (a) graph this function over a 12-hr period. (b) What will the height of the tide be at 5 A.M.? (c) Is the tide rising or falling at this time?

Sinusoidal movements: Many animals exhibit a wavelike motion in their movements, as in the tail of a shark as it swims in a straight line or the wingtips of a large bird in flight. Such movements can be modeled by a sine or cosine function and will vary depending on the animal's size, speed, and other factors.

65. The graph shown models the position of a shark's tail at time t, as measured to the left (negative) and right (positive) of a straight line along its length. (a) Use the graph to determine the related equation. (b) Is the tail to the right, left, or at center when $t = 6.5$ sec? How far? (c) Would you say the shark is "swimming leisurely," or "chasing its prey"? Justify your answer.

66. The State Fish of Hawaii is the *humuhumunukunukuapua'a*, a small colorful fish found abundantly in coastal waters. Suppose the tail motion of an adult fish is modeled by the equation $d(t) = \sin(15\pi t)$ with $d(t)$ representing the position of the fish's tail at time t, as measured in inches to the left (negative) or right (positive) of a straight line along its length. (a) Graph the equation over two periods. (b) Is the tail to the left or right of center at $t = 2.7$ sec? How far? (c) Would you say this fish is "swimming leisurely," or "running for cover"? Justify your answer.

Kinetic energy: The kinetic energy a planet possesses as it orbits the Sun can be modeled by a cosine function. When the planet is at its apogee (greatest distance from the Sun), its kinetic energy is at its lowest point as it slows down and "turns around" to head back toward the Sun. The kinetic energy is at its highest when the planet "whips around the Sun" to begin a new orbit.

67. Two graphs are given here. (a) Which of the graphs could represent the kinetic energy of a planet

orbiting the Sun if the planet is at its perigee (closest distance to the Sun) when $t = 0$? (b) For what value(s) of t does this planet possess 62.5% of its maximum kinetic energy with the kinetic energy increasing? (c) What is the orbital period of this planet?

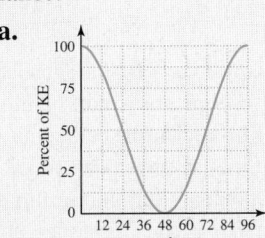
a.

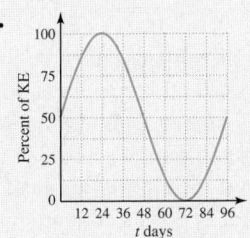
b.

68. The *potential energy* of the planet is the antipode of its kinetic energy, meaning when kinetic energy is at 100%, the potential energy is 0%, and when kinetic energy is at 0% the potential energy is at 100%. (a) How is the graph of the kinetic energy related to the graph of the potential energy? In other words, what transformation could be applied to the kinetic energy graph to obtain the potential energy graph? (b) If the kinetic energy is at 62.5% and increasing [as in Graph 67(b)], what can be said about the potential energy in the planet's orbit at this time?

Visible light: One of the narrowest bands in the electromagnetic spectrum is the region involving visible light. The wavelengths (periods) of visible light vary from 400 nanometers (purple/violet colors) to 700 nanometers (bright red). The approximate wavelengths of the other colors are shown in the diagram.

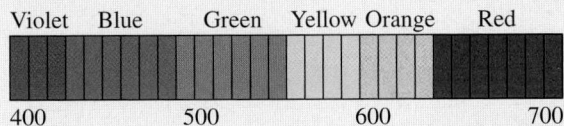

69. The equations for the colors in this spectrum have the form $y = \sin(\gamma t)$, where $\dfrac{2\pi}{\gamma}$ gives the length of the sine wave. (a) What color is represented by the equation $y = \sin\left(\dfrac{\pi}{240}t\right)$? (b) What color is represented by the equation $y = \sin\left(\dfrac{\pi}{310}t\right)$?

70. Name the color represented by each of the graphs (a) and (b) here and write the related equation.

a.

b.

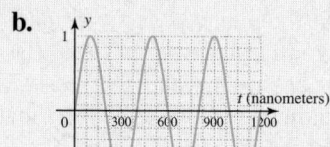

Alternating current: Surprisingly, even characteristics of the electric current supplied to your home can be modeled by sine or cosine functions. For alternating current (AC), the amount of current I (in amps) at time t can be modeled by $I = A \sin(\omega t)$, where A represents the maximum current that is produced, and ω is related to the frequency at which the generators turn to produce the current.

71. Find the equation of the household current modeled by the graph, then use the equation to determine I when $t = 0.045$ sec. Verify that the resulting ordered pair is on the graph.

Exercise 71

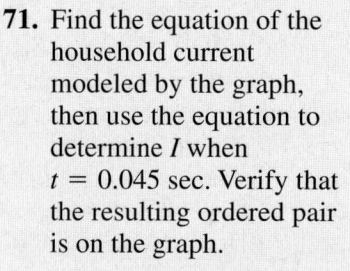

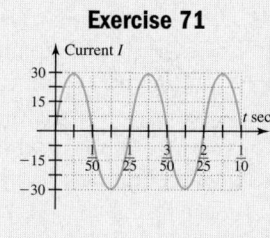

72. If the *voltage* produced by an AC circuit is modeled by the equation $E = 155 \sin(120\pi t)$, (a) what is the period and amplitude of the related graph? (b) What voltage is produced when $t = 0.2$?

▶ **EXTENDING THE CONCEPT**

73. For $y = A \sin(Bx)$ and $y = A \cos(Bx)$, the expression $\dfrac{M + m}{2}$ gives the average value of the function, where M and m represent the maximum and minimum values, respectively. What was the average value of every function graphed in this section? Compute a table of values for $y = 2 \sin t + 3$, and note its maximum and minimum values. What is the average value of this function? What transformation has been applied to change the average value of the function? Can you name the average value of $y = -2 \cos t + 1$ by inspection?

74. To understand where the period formula $P = \dfrac{2\pi}{B}$ came from, consider that if $B = 1$, the graph of $y = \sin(Bt) = \sin(1t)$ completes one cycle from $1t = 0$ to $1t = 2\pi$. If $B \neq 1$, $y = \sin(Bt)$ completes one cycle from $Bt = 0$ to $Bt = 2\pi$. Discuss how this observation validates the period formula.

75. The tone you hear when pressing the digit "9" on your telephone is actually a combination of two separate tones, which can be modeled by the functions $f(t) = \sin[2\pi(852)t]$ and $g(t) = \sin[2\pi(1477)t]$. Which of the two functions has the shortest period? By carefully scaling the axes, graph the function having the shorter period using the steps I through IV discussed in this section.

▶ **MAINTAINING YOUR SKILLS**

76. (5.2) Given $\sin 1.12 \approx 0.9$, find an additional value of t in $[0, 2\pi)$ that makes the equation $\sin t \approx 0.9$ true.

77. (5.1) Use a special triangle to calculate the distance from the ball to the pin on the seventh hole, given the ball is in a straight line with the 100-yd plate, as shown in the figure.

Exercise 77

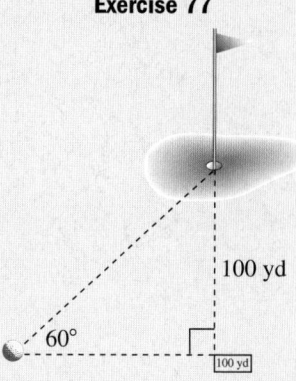

78. (5.1) Invercargill, New Zealand, is at 46°14′24″ south latitude. If the Earth has a radius of 3960 mi, how far is Invercargill from the equator?

79. (1.4) Given $z_1 = 1 + i$ and $z_2 = 2 - 5i$, compute the following:

   a. $z_1 + z_2$

   b. $z_1 - z_2$

   c. $z_1 z_2$

   d. $\dfrac{z_2}{z_1}$

5.4 | Graphs of Tangent and Cotangent Functions

Learning Objectives

In Section 5.4 you will learn how to:

☐ **A.** Graph $y = \tan t$ using asymptotes, zeroes, and the ratio $\dfrac{\sin t}{\cos t}$

☐ **B.** Graph $y = \cot t$ using asymptotes, zeroes, and the ratio $\dfrac{\cos t}{\sin t}$

☐ **C.** Identify and discuss important characteristics of $y = \tan t$ and $y = \cot t$

☐ **D.** Graph $y = A \tan(Bt)$ and $y = A \cot(Bt)$ with various values of A and B

☐ **E.** Solve applications of $y = \tan t$ and $y = \cot t$

Unlike the other four trig functions, tangent and cotangent have no maximum or minimum value on any open interval of their domain. However, it is precisely this unique feature that adds to their value as mathematical models. Collectively, the six functions give scientists the tools they need to study, explore, and investigate a wide range of phenomena, extending our understanding of the world around us.

A. The Graph of $y = \tan t$

Like the secant and cosecant functions, tangent is defined in terms of a ratio, creating asymptotic behavior at the zeroes of the denominator. In terms of the unit circle, $\tan t = \dfrac{y}{x}$, which means in $[-\pi, 2\pi]$, vertical asymptotes occur at $t = -\dfrac{\pi}{2}, t = \dfrac{\pi}{2}$, and $\dfrac{3\pi}{2}$, since the x-coordinate on the unit circle is zero (see Figure 5.48). We further note $\tan t = 0$ when the y-coordinate is zero, so the function will have t-intercepts at $t = -\pi, 0, \pi$, and 2π in the same interval. This produces the framework for graphing the tangent function shown in Figure 5.49.

Figure 5.48

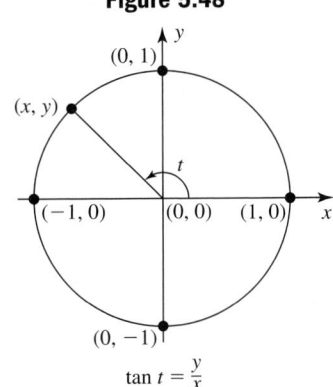

$$\tan t = \frac{y}{x}$$

Figure 5.49

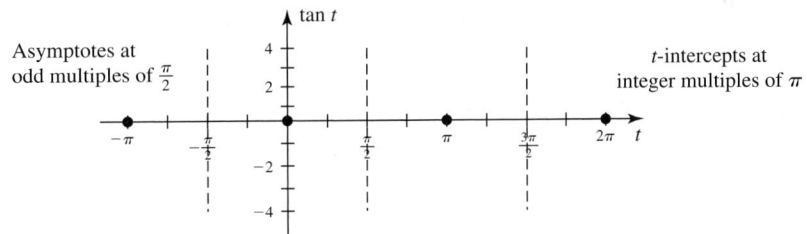

Knowing the graph must go through these zeroes and approach the asymptotes, we are left with determining the *direction of the approach*. This can be discovered by noting that in QI, the y-coordinates of points on the unit circle start at 0 and increase, while the x-values start at 1 and decrease. This means the ratio $\dfrac{y}{x}$ defining $\tan t$ is increasing, and in fact becomes infinitely large as t gets very close to $\dfrac{\pi}{2}$. A similar observation can be made for a negative rotation of t in QIV. Using the additional points provided by $\tan\left(-\dfrac{\pi}{4}\right) = -1$ and $\tan\left(\dfrac{\pi}{4}\right) = 1$, we find the graph of $\tan t$ is increasing throughout the interval $\left(-\dfrac{\pi}{2}, \dfrac{\pi}{2}\right)$ and that the function has a period of π. We also note $y = \tan t$ is an odd function (symmetric about the origin), since $\tan(-t) = -\tan t$ as evidenced by the two points just computed. The completed graph is shown in Figure 5.50 with the primary interval in red.

Figure 5.50

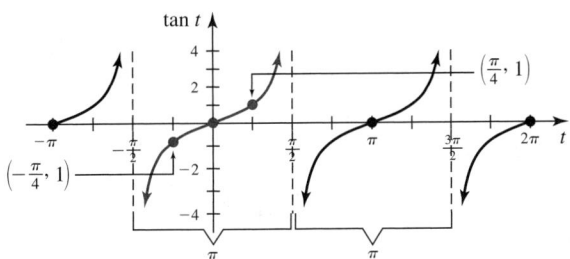

The graph can also be developed by noting $\sin t = y$, $\cos t = x$, and $\tan t = \dfrac{y}{x}$.

This gives $\tan t = \dfrac{\sin t}{\cos t}$ by direct substitution and we can quickly complete a table of values for $\tan t$, as shown in Example 1. These and other relationships between the trig functions will be fully explored in Chapter 6.

EXAMPLE 1 ▶ **Constructing a Table of Values for $f(t) = \tan t$**

Complete Table 5.8 shown for $\tan t = \dfrac{y}{x}$ using the values given for $\sin t$ and $\cos t$, then graph the function by plotting points.

Table 5.8

t	0	$\dfrac{\pi}{6}$	$\dfrac{\pi}{4}$	$\dfrac{\pi}{3}$	$\dfrac{\pi}{2}$	$\dfrac{2\pi}{3}$	$\dfrac{3\pi}{4}$	$\dfrac{5\pi}{6}$	π
$\sin t = y$	0	$\dfrac{1}{2}$	$\dfrac{\sqrt{2}}{2}$	$\dfrac{\sqrt{3}}{2}$	1	$\dfrac{\sqrt{3}}{2}$	$\dfrac{\sqrt{2}}{2}$	$\dfrac{1}{2}$	0
$\cos t = x$	1	$\dfrac{\sqrt{3}}{2}$	$\dfrac{\sqrt{2}}{2}$	$\dfrac{1}{2}$	0	$-\dfrac{1}{2}$	$-\dfrac{\sqrt{2}}{2}$	$-\dfrac{\sqrt{3}}{2}$	-1
$\tan t = \dfrac{y}{x}$									

Solution ▶ For the noninteger values of x and y, the "twos will cancel" each time we compute $\dfrac{y}{x}$. This means we can simply list the ratio of numerators. The resulting points are shown in Table 5.9, along with the plotted points. The graph shown in Figure 5.51 was completed using symmetry and the previous observations.

Table 5.9

t	0	$\dfrac{\pi}{6}$	$\dfrac{\pi}{4}$	$\dfrac{\pi}{3}$	$\dfrac{\pi}{2}$	$\dfrac{2\pi}{3}$	$\dfrac{3\pi}{4}$	$\dfrac{5\pi}{6}$	π
$\sin t = y$	0	$\dfrac{1}{2}$	$\dfrac{\sqrt{2}}{2}$	$\dfrac{\sqrt{3}}{2}$	1	$\dfrac{\sqrt{3}}{2}$	$\dfrac{\sqrt{2}}{2}$	$\dfrac{1}{2}$	0
$\cos t = x$	1	$\dfrac{\sqrt{3}}{2}$	$\dfrac{\sqrt{2}}{2}$	$\dfrac{1}{2}$	0	$-\dfrac{1}{2}$	$-\dfrac{\sqrt{2}}{2}$	$-\dfrac{\sqrt{3}}{2}$	-1
$\tan t = \dfrac{y}{x}$	0	$\dfrac{1}{\sqrt{3}} \approx 0.58$	1	$\sqrt{3} \approx 1.7$	undefined	$-\sqrt{3}$	-1	$-\dfrac{1}{\sqrt{3}}$	0

Figure 5.51

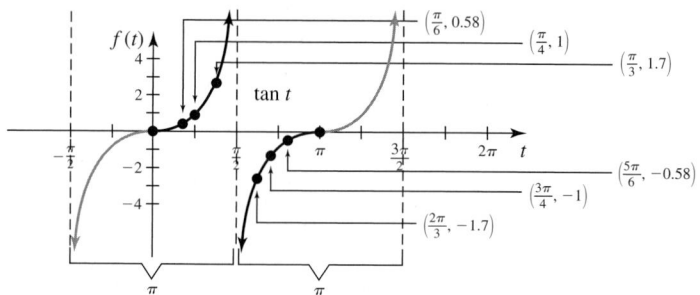

Now try Exercises 7 and 8 ▶

☑ **A.** You've just learned
how to graph $y = \tan t$ using
asymptotes, zeroes, and the
ratio $\dfrac{\sin t}{\cos t}$

Additional values can be found using a calculator as needed. For future use and reference, it will help to recognize the approximate decimal equivalent of all special values and radian angles. In particular, note that $\sqrt{3} \approx 1.73$ and $\dfrac{1}{\sqrt{3}} \approx 0.58$. **See Exercises 9 through 14.**

B. The Graph of $y = \cot t$

Since the cotangent function is also defined in terms of a ratio, it too displays asymptotic behavior at the zeroes of the denominator, with t-intercepts at the zeroes of the numerator. Like the tangent function, $\cot t = \dfrac{x}{y}$ can be written in terms of $\cos t = x$ and $\sin t = y$: $\cot t = \dfrac{\cos t}{\sin t}$, and the graph obtained by plotting points.

EXAMPLE 2 ▶ **Constructing a Table of Values for $f(t) = \cot t$**

Complete a table of values for $\cot t = \dfrac{x}{y}$ for t in $[0, \pi]$ using its ratio relationship with $\cos t$ and $\sin t$. Use the results to graph the function for t in $(-\pi, 2\pi)$.

Solution ▶ The completed table is shown here. In this interval, the cotangent function has asymptotes at 0 and π since $y = 0$ at these points, and has a t-intercept at $\dfrac{\pi}{2}$ since $x = 0$. The graph shown in Figure 5.52 was completed using the period $P = \pi$.

t	0	$\dfrac{\pi}{6}$	$\dfrac{\pi}{4}$	$\dfrac{\pi}{3}$	$\dfrac{\pi}{2}$	$\dfrac{2\pi}{3}$	$\dfrac{3\pi}{4}$	$\dfrac{5\pi}{6}$	π
$\sin t = y$	0	$\dfrac{1}{2}$	$\dfrac{\sqrt{2}}{2}$	$\dfrac{\sqrt{3}}{2}$	1	$\dfrac{\sqrt{3}}{2}$	$\dfrac{\sqrt{2}}{2}$	$\dfrac{1}{2}$	0
$\cos t = x$	1	$\dfrac{\sqrt{3}}{2}$	$\dfrac{\sqrt{2}}{2}$	$\dfrac{1}{2}$	0	$-\dfrac{1}{2}$	$-\dfrac{\sqrt{2}}{2}$	$-\dfrac{\sqrt{3}}{2}$	-1
$\cot t = \dfrac{x}{y}$	undefined	$\sqrt{3}$	1	$\dfrac{1}{\sqrt{3}}$	0	$-\dfrac{1}{\sqrt{3}}$	-1	$-\sqrt{3}$	undefined

Figure 5.52

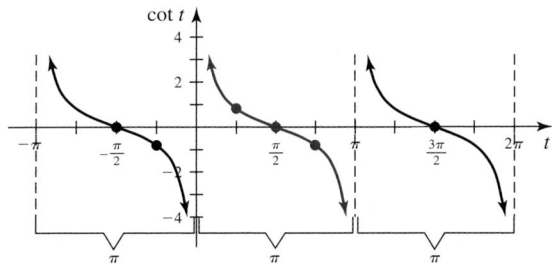

Now try Exercises 15 and 16 ▶

☑ **B.** You've just learned
how to graph $y = \cot t$ using
asymptotes, zeroes, and the
ratio $\dfrac{\cos t}{\sin t}$

C. Characteristics of $y = \tan t$ and $y = \cot t$

The most important characteristics of the tangent and cotangent functions are summarized in the following box. There is no discussion of amplitude, maximum, or minimum values, since maximum or minimum values do not exist. For future use and reference, perhaps the most significant characteristic distinguishing tan t from cot t is that *tan t increases,* while *cot t decreases* over their respective domains. Also note that due to symmetry, the zeroes of each function are always located halfway between the asymptotes.

Characteristics of $f(t) = \tan t$ and $f(t) = \cot t$

For all real numbers t and integers k,

$y = \tan t$			$y = \cot t$		
Domain	**Range**	**Asymptotes**	**Domain**	**Range**	**Asymptotes**
$t \neq \dfrac{\pi}{2} + \pi k$	R	$t = \dfrac{\pi}{2} + \pi k$	$t \neq k\pi$	R	$t = k\pi$
Period	**Behavior**	**Symmetry**	**Period**	**Behavior**	**Symmetry**
π	increasing	odd $\tan(-t) = -\tan t$	π	decreasing	odd $\cot(-t) = -\cot t$

EXAMPLE 3 ▶ **Using the Period of $f(t) = \tan t$ to Find Additional Points**

Given $\tan\left(\dfrac{\pi}{6}\right) = \dfrac{1}{\sqrt{3}}$, what can you say about $\tan\left(\dfrac{7\pi}{6}\right)$, $\tan\left(\dfrac{13\pi}{6}\right)$, and $\tan\left(-\dfrac{5\pi}{6}\right)$?

Solution ▶ Each value of t differs from $\dfrac{\pi}{6}$ by a multiple of π: $\tan\left(\dfrac{7\pi}{6}\right) = \tan\left(\dfrac{\pi}{6} + \pi\right)$, $\tan\left(\dfrac{13\pi}{6}\right) = \tan\left(\dfrac{\pi}{6} + 2\pi\right)$ and $\tan\left(-\dfrac{5\pi}{6}\right) = \tan\left(\dfrac{\pi}{6} - \pi\right)$. Since the period of the tangent function is $P = \pi$, all of these expressions have a value of $\dfrac{1}{\sqrt{3}}$.

Now try Exercises 17 through 22 ▶

Since the tangent function is more common than the cotangent, many needed calculations will first be done using the tangent function and its properties, then reciprocated. For instance, to evaluate $\cot\left(-\dfrac{\pi}{6}\right)$ we reason that cot t is an odd function, so $\cot\left(-\dfrac{\pi}{6}\right) = -\cot\left(\dfrac{\pi}{6}\right)$. Since cotangent is the reciprocal of tangent and $\tan\left(\dfrac{\pi}{6}\right) = \dfrac{1}{\sqrt{3}}$, $-\cot\left(\dfrac{\pi}{6}\right) = -\sqrt{3}$. **See Exercises 23 and 24.**

☑ **C.** You've just learned how to identify and discuss important characteristics of $y = \tan t$ and $y = \cot t$

D. Graphing $y = A \tan(Bt)$ and $y = A \cot(Bt)$

The Coefficient A: Vertical Stretches and Compressions

For the tangent and cotangent functions, the role of coefficient A is best seen through an analogy from basic algebra (the concept of amplitude is foreign to these functions). Consider the graph of $y = x^3$ (Figure 5.53). Comparing the parent function $y = x^3$ with functions $y = Ax^3$, the graph is stretched vertically if $|A| > 1$ (see Figure 5.54) and compressed if $0 < |A| < 1$. In the latter case the graph becomes very "flat" near the zeroes, as shown in Figure 5.55.

Figure 5.53
$y = x^3$

Figure 5.54
$y = 4x^3; A = 4$

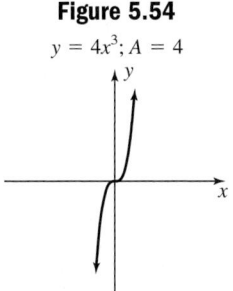

Figure 5.55
$y = \frac{1}{4}x^3; A = \frac{1}{4}$

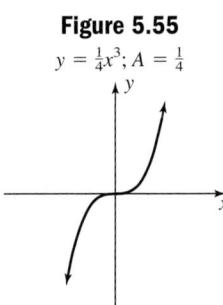

While **cubic functions are not asymptotic**, they are a good illustration of A's effect on the tangent and cotangent functions. Fractional values of A ($|A| < 1$) compress the graph, flattening it out near its zeroes. Numerically, this is because a fractional part of a small quantity is an even smaller quantity. For instance, compare $\tan\left(\frac{\pi}{6}\right)$ with $\frac{1}{4}\tan\left(\frac{\pi}{6}\right)$. To two decimal places, $\tan\left(\frac{\pi}{6}\right) = 0.57$, while $\frac{1}{4}\tan\left(\frac{\pi}{6}\right) = 0.14$, so the graph must be "nearer the t-axis" at this value.

EXAMPLE 4 ▶ **Comparing the Graph of $f(t) = \tan t$ and $g(t) = A \tan t$**

Draw a "comparative sketch" of $y = \tan t$ and $y = \frac{1}{4}\tan t$ on the same axis and discuss similarities and differences. Use the interval $[-\pi, 2\pi]$.

Solution ▶ Both graphs will maintain their essential features (zeroes, asymptotes, period, increasing, and so on). However, the graph of $y = \frac{1}{4}\tan t$ is vertically compressed, causing it to flatten out near its zeroes and changing how the graph approaches its asymptotes in each interval.

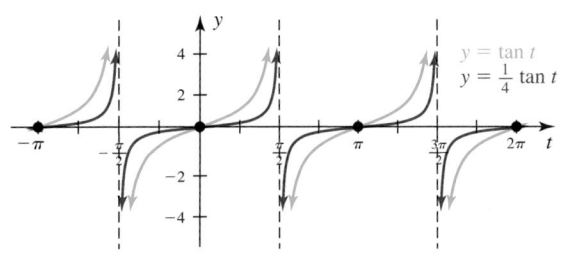

Now try Exercises 25 through 28 ▶

The Coefficient B: The Period of Tangent and Cotangent

WORTHY OF NOTE

It may be easier to interpret the phrase "twice as fast" as $2P = \pi$ and "one-half as fast" as $\frac{1}{2}P = \pi$. In each case, solving for P gives the correct interval for the period of the new function.

Like the other trig functions, the value of B has a material impact on the period of the function, and with the same effect. The graph of $y = \cot(2t)$ completes a cycle twice as fast as $y = \cot t \left(P = \frac{\pi}{2} \text{ versus } P = \pi\right)$, while $y = \cot\left(\frac{1}{2}t\right)$ completes a cycle one-half as fast ($P = 2\pi$ versus $P = \pi$).

This reasoning leads us to a **period formula** for tangent and cotangent, namely, $P = \frac{\pi}{|B|}$, where B is the coefficient of the input variable.

Similar to the four-step process used to graph sine and cosine functions, we can graph tangent and cotangent functions using a rectangle $P = \dfrac{\pi}{B}$ units in length and $2A$ units high, centered on the primary interval. After dividing the length of the rectangle into fourths, the t-intercept will always be the halfway point, with y-values of $|A|$ occurring at the $\frac{1}{4}$ and $\frac{3}{4}$ marks. See Example 5.

EXAMPLE 5 ▶ Graphing $y = A \cot(Bt)$ for $A, B, \neq 1$

Sketch the graph of $y = 3 \cot(2t)$ over the interval $[-\pi, \pi]$.

Solution ▶ For $y = 3 \cot(2t)$, $|A| = 3$ which results in a vertical stretch, and $|B| = 2$ which gives a period of $\dfrac{\pi}{2}$. The function is still undefined at $t = 0$ and is asymptotic there, then at all integer multiples of $P = \dfrac{\pi}{2}$. We also know the graph is decreasing, with zeroes of the function halfway between the asymptotes. The inputs $t = \dfrac{\pi}{8}$ and $t = \dfrac{3\pi}{8}$ $\left(\text{the } \dfrac{1}{4} \text{ and } \dfrac{3}{4} \text{ marks between 0 and } \dfrac{\pi}{2}\right)$ yield the points $\left(\dfrac{\pi}{8}, 3\right)$ and $\left(\dfrac{3\pi}{8}, -3\right)$, which we'll use along with the period and symmetry of the function to complete the graph:

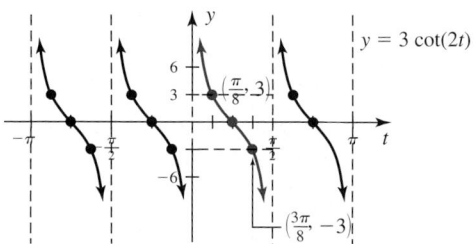

Now try Exercises 29 through 40 ▶

As with the trig functions from Section 5.3, it is possible to determine the equation of a tangent or cotangent function from a given graph. Where previously we used the amplitude, period, and max/min values to obtain our equation, here we first determine the period of the function by calculating the "distance" between asymptotes, then choose any convenient point on the graph (other than a t-intercept) and substitute in the equation to solve for A.

EXAMPLE 6 ▶ Constructing the Equation for a Given Graph

Find the equation of the graph, given it's of the form $y = A \tan(Bt)$.

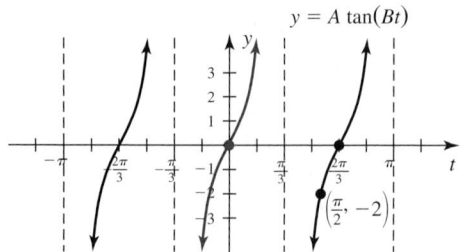

Solution ▶ Using the primary interval and the asymptotes at $t = -\dfrac{\pi}{3}$ and $t = \dfrac{\pi}{3}$, we find the

period is $P = \dfrac{\pi}{3} - \left(-\dfrac{\pi}{3}\right) = \dfrac{2\pi}{3}$. To find the value of B we substitute $\dfrac{2\pi}{3}$ for P in

$P = \dfrac{\pi}{B}$ and find $B = \dfrac{3}{2}$ (verify). This gives the equation $y = A\tan\left(\dfrac{3}{2}t\right)$.

To find A, we take the point $\left(\dfrac{\pi}{2}, -2\right)$ shown, and use $t = \dfrac{\pi}{2}$ with $y = -2$ to
solve for A:

$$y = A\tan\left(\dfrac{3}{2}t\right) \qquad \text{substitute } \tfrac{3}{2} \text{ for } B$$

$$-2 = A\tan\left[\left(\dfrac{3}{2}\right)\left(\dfrac{\pi}{2}\right)\right] \qquad \text{substitute } -2 \text{ for } y \text{ and } \tfrac{\pi}{2} \text{ for } t$$

$$-2 = A\tan\left(\dfrac{3\pi}{4}\right) \qquad \text{multiply}$$

$$A = \dfrac{-2}{\tan\left(\dfrac{3\pi}{4}\right)} \qquad \text{solve for } A$$

$$= 2 \qquad \text{result}$$

☑ **D.** You've just learned how
to graph $y = A\tan(Bt)$ and
$y = A\cot(Bt)$ with various
values of A and B

The equation of the graph is $y = 2\tan(\tfrac{3}{2}t)$.

Now try Exercises 41 through 46 ▶

E. Applications of Tangent and Cotangent Functions

We end this section with one example of how tangent and cotangent functions can be applied. Numerous others can be found in the exercise set.

EXAMPLE 7 ▶ **Applications of $y = A\tan(Bt)$: Modeling the Movement of a Light Beam**

One evening, in port during a *Semester at Sea*, Richard is debating a project choice for his Precalculus class. Looking out his porthole, he notices a revolving light turning at a constant speed near the corner of a long warehouse. The light throws its beam along the length of the warehouse, then disappears into the air, and then returns time and time again. Suddenly—Richard has his project. He notes the time it takes the beam to traverse the warehouse wall is very close to 4 sec, and in the morning he measures the wall's length at 127.26 m. His project? Modeling the distance of the beam from the corner of the warehouse as a function of time using a tangent function. Can you help?

Solution ▶ The equation model will have the form $D(t) = A\tan(Bt)$, where $D(t)$ is the distance (in meters) of the beam from the corner after t sec. The distance along the wall is measured in positive values so we're using only $\tfrac{1}{2}$ the period of the function, giving $\tfrac{1}{2}P = 4$ (the beam "disappears" at $t = 4$) so $P = 8$. Substitution in the period formula gives $B = \dfrac{\pi}{8}$ and the equation $D = A\tan\left(\dfrac{\pi}{8}t\right)$.

Knowing the beam travels 127.26 m in about 4 sec (when it disappears into infinity), we'll use $t = 3.9$ and $D = 127.26$ in order to solve for A and complete our equation model (see note following this example).

$$A \tan\left(\frac{\pi}{8}t\right) = D \qquad \text{equation model}$$

$$A \tan\left[\frac{\pi}{8}(3.9)\right] = 127.26 \qquad \text{substitute 127.26 for } D \text{ and 3.9 for } t$$

$$A = \frac{127.26}{\tan\left[\frac{\pi}{8}(3.9)\right]} \qquad \text{solve for } A$$

$$\approx 5 \qquad \text{result}$$

One equation approximating the distance of the beam from the corner of the warehouse is $D(t) = 5 \tan\left(\frac{\pi}{8}t\right)$.

Now try Exercises 49 through 52 ▶

☑ E. You've just learned how to solve applications of $y = \tan t$ and $y = \cot t$

For Example 7, we should note the choice of 3.9 for t was arbitrary, and while we obtained an "acceptable" model, different values of A would be generated for other choices. For instance, $t = 3.95$ gives $A \approx 2.5$, while $t = 3.99$ gives $A \approx 0.5$. The true value of A depends on the distance of the light from the corner of the warehouse wall. In any case, it's interesting to note that at $t = 2$ sec (one-half the time it takes the beam to disappear), the beam has traveled only 5m from the corner of the building: $D(2) = 5 \tan\left(\frac{\pi}{4}\right) = 5$ m. Although the light is rotating at a constant angular speed, the speed of the beam along the wall increases *dramatically* as t gets close to 4 sec.

TECHNOLOGY HIGHLIGHT

Zeroes, Asymptotes, and the Tangent/Cotangent Functions

In this *Technology Highlight* we'll explore the tangent and cotangent functions from the perspective of their ratio definition. While we could easily use $Y_1 = \tan x$ to generate and explore the graph, we would miss an opportunity to note the many important connections that emerge from a ratio definition perspective. To begin, enter $Y_1 = \sin x$, $Y_2 = \cos x$, and $Y_3 = \dfrac{Y_1}{Y_2}$, as shown in Figure 5.56 [recall that function variables are accessed using VARS   ▶ **(Y-VARS)** ENTER **(1:Function)]**. Note that Y_2 has been disabled by overlaying the cursor on the equal sign and pressing ENTER . In addition, note the slash next to Y_1 is more **bold** than the other slashes. The TI-84 Plus offers options that help distinguish between graphs when more than one is being displayed, and we selected a **bold** line for Y_1 by moving the cursor to the far left position and repeatedly pressing ENTER until the desired option appeared. Pressing ZOOM **7:ZTrig** at this point produces the screen shown in Figure 5.57, where we note that tan x is zero everywhere that sin x

Figure 5.56

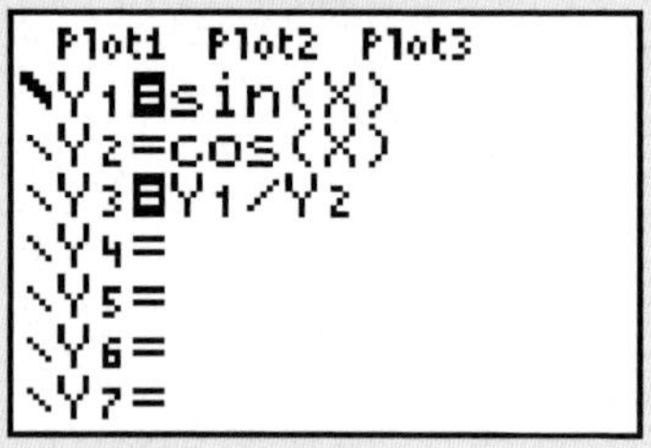

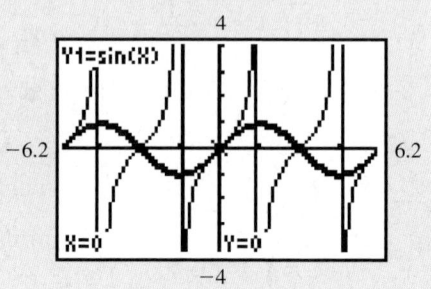

Figure 5.57

is zero. This is hardly surprising since $\tan x = \dfrac{\sin x}{\cos x}$, but is a point that is often overlooked. Going back to the Y= screen and disabling Y_1 while enabling Y_2 will produce the graph shown in Figure 5.58.

Exercise 1: What do you notice about the zeroes of cos x as they relate to the graph of $Y_3 = \tan x$?

Exercise 2: Go to the Y= screen and change Y_3 from $\dfrac{Y_1}{Y_2}$ (tangent) to $\dfrac{Y_2}{Y_1}$ (cotangent), then repeat the previous investigation regarding $y = \sin x$ and $y = \cos x$.

Figure 5.58

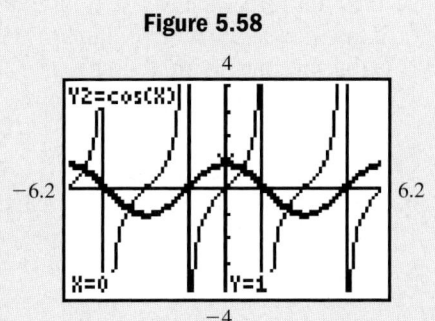

5.4 EXERCISES

▶ CONCEPTS AND VOCABULARY

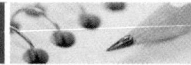

Fill in each blank with the appropriate word or phrase. Carefully reread the section if needed.

1. The period of $y = \tan t$ and $y = \cot t$ is _____. To find the period of $y = \tan(Bt)$ and $y = \cot(Bt)$, the formula _____ is used.

2. The function $y = \tan t$ is _____ everywhere it is defined. The function $y = \cot t$ is _____ everywhere it is defined.

3. Tan t and cot t are _____ functions, so $f(-t) =$ _____. If $\tan\left(-\dfrac{11\pi}{12}\right), \approx 0.268$, then $\tan\left(\dfrac{11\pi}{12}\right) \approx$ _____.

4. The asymptotes of $y =$ _____ are located at odd multiples of $\dfrac{\pi}{2}$. The asymptotes of $y =$ _____ are located at integer multiples of π.

5. Discuss/Explain how you can obtain a table of values for $y = \cot t$ (a) given the values for $y = \sin t$ and $y = \cos t$, and (b) given the values for $y = \tan t$.

6. Explain/Discuss how the zeroes of $y = \sin t$ and $y = \cos t$ are related to the graphs of $y = \tan t$ and $y = \cot t$. How can these relationships help graph functions of the form $y = A\tan(Bt)$ and $y = A\cot(Bt)$?

▶ DEVELOPING YOUR SKILLS

Use the values given for sin t and cos t to complete the tables.

7.

t	π	$\dfrac{7\pi}{6}$	$\dfrac{5\pi}{4}$	$\dfrac{4\pi}{3}$	$\dfrac{3\pi}{2}$
$\sin t = y$	0	$-\dfrac{1}{2}$	$-\dfrac{\sqrt{2}}{2}$	$-\dfrac{\sqrt{3}}{2}$	-1
$\cos t = x$	-1	$-\dfrac{\sqrt{3}}{2}$	$-\dfrac{\sqrt{2}}{2}$	$-\dfrac{1}{2}$	0
$\tan t = \dfrac{y}{x}$					

8.

t	$\dfrac{3\pi}{2}$	$\dfrac{5\pi}{3}$	$\dfrac{7\pi}{4}$	$\dfrac{11\pi}{6}$	2π
$\sin t = y$	-1	$-\dfrac{\sqrt{3}}{2}$	$-\dfrac{\sqrt{2}}{2}$	$-\dfrac{1}{2}$	0
$\cos t = x$	0	$\dfrac{1}{2}$	$\dfrac{\sqrt{2}}{2}$	$\dfrac{\sqrt{3}}{2}$	1
$\tan t = \dfrac{y}{x}$					

9. Without reference to a text or calculator, attempt to name the decimal equivalent of the following values to one decimal place.

$$\frac{\pi}{2} \quad \frac{\pi}{4} \quad \frac{\pi}{6} \quad \sqrt{2} \quad \frac{\sqrt{2}}{2} \quad \frac{2}{\sqrt{3}}$$

10. Without reference to a text or calculator, attempt to name the decimal equivalent of the following values to one decimal place.

$$\frac{\pi}{3} \quad \pi \quad \frac{3\pi}{2} \quad \sqrt{3} \quad \frac{\sqrt{3}}{2} \quad \frac{1}{\sqrt{3}}$$

11. State the value of each expression without the use of a calculator.

   a. $\tan\left(-\frac{\pi}{4}\right)$ b. $\cot\left(\frac{\pi}{6}\right)$

   c. $\cot\left(\frac{3\pi}{4}\right)$ d. $\tan\left(\frac{\pi}{3}\right)$

12. State the value of t without the use of a calculator.

   a. $\cot\left(\frac{\pi}{2}\right)$ b. $\tan \pi$

   c. $\tan\left(-\frac{5\pi}{4}\right)$ d. $\cot\left(-\frac{5\pi}{6}\right)$

13. State the value of t without the use of a calculator, given $t \in [0, 2\pi)$ terminates in the quadrant indicated.

   a. $\tan t = -1$, t in QIV
   b. $\cot t = \sqrt{3}$, t in QIII
   c. $\cot t = -\frac{1}{\sqrt{3}}$, t in QIV
   d. $\tan t = -1$, t in QII

14. State the value of each expression without the use of a calculator, given $t \in [0, 2\pi)$ terminates in the quadrant indicated.

   a. $\cot t = 1$, t in QI
   b. $\tan t = -\sqrt{3}$, t in QII
   c. $\tan t = \frac{1}{\sqrt{3}}$, t in QI
   d. $\cot t = 1$, t in QIII

Use the values given for sin t and cos t to complete the tables.

15.

t	π	$\frac{7\pi}{6}$	$\frac{5\pi}{4}$	$\frac{4\pi}{3}$	$\frac{3\pi}{2}$
$\sin t = y$	0	$-\frac{1}{2}$	$-\frac{\sqrt{2}}{2}$	$-\frac{\sqrt{3}}{2}$	-1
$\cos t = x$	-1	$-\frac{\sqrt{3}}{2}$	$-\frac{\sqrt{2}}{2}$	$-\frac{1}{2}$	0
$\cot t = \frac{x}{y}$					

16.

	$\frac{3\pi}{2}$	$\frac{5\pi}{3}$	$\frac{7\pi}{4}$	$\frac{11\pi}{6}$	2π
$\sin t = y$	-1	$-\frac{\sqrt{3}}{2}$	$-\frac{\sqrt{2}}{2}$	$-\frac{1}{2}$	0
$\cos t = x$	0	$\frac{1}{2}$	$\frac{\sqrt{2}}{2}$	$\frac{\sqrt{3}}{2}$	1
$\cot t = \frac{x}{y}$					

17. Given $t = \frac{11\pi}{24}$ is a solution to $\tan t \approx 7.6$, use the period of the function to name three additional solutions. Check your answer using a calculator.

18. Given $t = \frac{7\pi}{24}$ is a solution to $\cot t \approx 0.77$, use the period of the function to name three additional solutions. Check your answer using a calculator.

19. Given $t \approx 1.5$ is a solution to $\cot t = 0.07$, use the period of the function to name three additional solutions. Check your answers using a calculator.

20. Given $t \approx 1.25$ is a solution to $\tan t = 3$, use the period of the function to name three additional solutions. Check your answers using a calculator.

Verify the value shown for t is a solution to the equation given, then use the period of the function to name all real roots. Check two of these roots on a calculator.

21. $t = \frac{\pi}{10}$; $\tan t \approx 0.3249$

22. $t = -\frac{\pi}{16}$; $\tan t \approx -0.1989$

23. $t = \frac{\pi}{12}$; $\cot t = 2 + \sqrt{3}$

24. $t = \frac{5\pi}{12}$; $\cot t = 2 - \sqrt{3}$

Graph each function over the interval indicated, noting the period, asymptotes, zeroes, and value of A. Include a comparative sketch of $y = \tan t$ or $y = \cot t$ as indicated.

25. $f(t) = 2 \tan t;\ [-2\pi, 2\pi]$

26. $g(t) = \dfrac{1}{2} \tan t;\ [-2\pi, 2\pi]$

27. $h(t) = 3 \cot t;\ [-2\pi, 2\pi]$

28. $r(t) = \dfrac{1}{4} \cot t;\ [-2\pi, 2\pi]$

Graph each function over the interval indicated, noting the period, asymptotes, zeroes, and value of A and B.

29. $y = \tan(2t);\ \left[-\dfrac{\pi}{2}, \dfrac{\pi}{2}\right]$

30. $y = \tan\left(\dfrac{1}{4}t\right);\ [-4\pi, 4\pi]$

31. $y = \cot(4t);\ \left[-\dfrac{\pi}{4}, \dfrac{\pi}{4}\right]$

32. $y = \cot\left(\dfrac{1}{2}t\right);\ [-2\pi, 2\pi]$

33. $y = 2\tan(4t);\ \left[-\dfrac{\pi}{4}, \dfrac{\pi}{4}\right]$

34. $y = 4\tan\left(\dfrac{1}{2}t\right);\ [-2\pi, 2\pi]$

35. $y = 5\cot\left(\dfrac{1}{3}t\right);\ [-3\pi, 3\pi]$

36. $y = \dfrac{1}{2}\cot(2t);\ \left[-\dfrac{\pi}{2}, \dfrac{\pi}{2}\right]$

37. $y = 3\tan(2\pi t);\ \left[-\dfrac{1}{2}, \dfrac{1}{2}\right]$

38. $y = 4\tan\left(\dfrac{\pi}{2}t\right);\ [-2, 2]$

39. $f(t) = 2\cot(\pi t);\ [-1, 1]$

40. $p(t) = \dfrac{1}{2}\cot\left(\dfrac{\pi}{4}t\right);\ [-4, 4]$

Find the equation of each graph, given it is of the form $y = A\tan(Bt)$.

41.

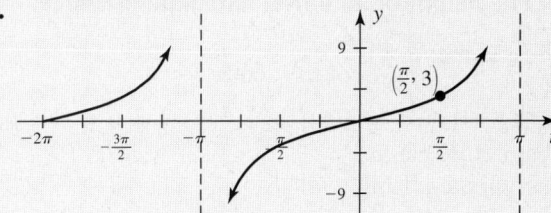

42.

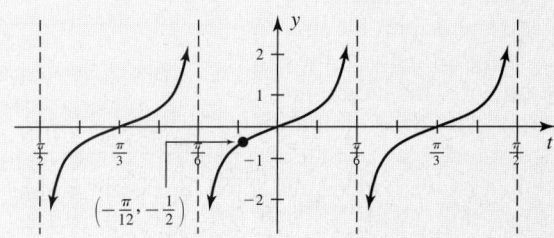

Find the equation of each graph, given it is of the form $y = A\cot(Bt)$.

43.

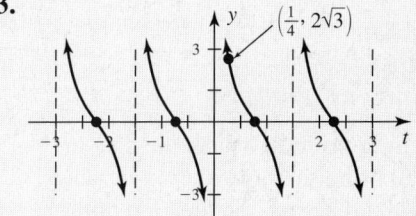

44.

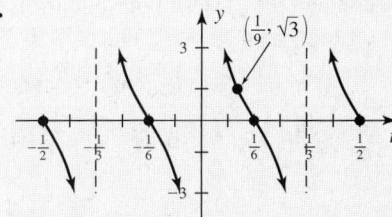

45. Given that $t = -\dfrac{\pi}{8}$ and $t = -\dfrac{3\pi}{8}$ are solutions to $\cot(3t) = \tan t$, use a graphing calculator to find two additional solutions in $[0, 2\pi]$.

46. Given $t = \dfrac{1}{6}$ is a solution to $\tan(2\pi t) = \cot(\pi t)$, use a graphing calculator to find two additional solutions in $[-1, 1]$.

▶ WORKING WITH FORMULAS

47. The height of an object calculated from a

distance: $h = \dfrac{d}{\cot u - \cot v}$

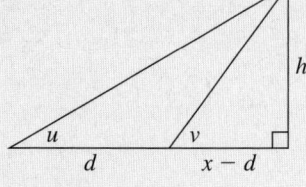

The height h of a tall structure can be computed using two angles of elevation measured some distance apart along a straight line with the object. This height is given by the formula shown, where d is the distance between the two points from which angles u and v were measured. Find the height h of a building if $u = 40°$, $v = 65°$, and $d = 100$ ft.

48. Position of an image reflected from a spherical

lens: $\tan \theta = \dfrac{h}{s - k}$

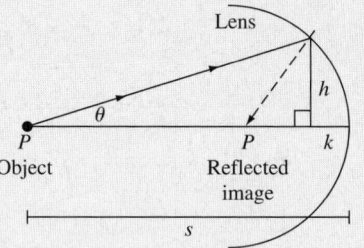

The equation shown is used to help locate the position of an image reflected by a spherical mirror, where s is the distance of the object from the lens along a horizontal axis, θ is the angle of elevation from this axis, h is the altitude of the right triangle indicated, and k is distance from the lens to the foot of altitude h. Find the distance k given $h = 3$ mm, $\theta = \dfrac{\pi}{24}$, and that the object is 24 mm from the lens.

▶ APPLICATIONS

Tangent function data models: Model the data in Exercises 49 and 50 using the function $y = A\tan(Bx)$. State the period of the function, the location of the asymptotes, the value of A, and name the point (x, y) used to calculate A (answers may vary). Use your equation model to evaluate the function at $x = -2$ and $x = 2$. What observations can you make? Also see Exercise 58.

49.

Input	Output	Input	Output
−6	−∞	1	1.4
−5	−20	2	3
−4	−9.7	3	5.2
−3	−5.2	4	9.7
−2	−3	5	20
−1	−1.4	6	∞
0	0		

50.

Input	Output	Input	Output
−3	−∞	0.5	6.4
−2.5	−91.3	1	13.7
−2	−44.3	1.5	23.7
−1.5	−23.7	2	44.3
−1	−13.7	2.5	91.3
−0.5	−6.4	3	∞
0	0		

51. As part of a lab setup, a laser pen is made to swivel on a large protractor as illustrated in the figure. For their lab project, students are asked to take the instrument to one end of a long hallway and measure the distance of the projected beam relative to the angle the pen is being held, and collect the data in a table. Use the data to find a function of the form $y = A\tan(B\theta)$. State the period of the function, the location of the asymptotes, the value of A, and name the point (θ, y) you used to calculate A (answers may vary). Based on the result, can you approximate the length of the laser pen? Note that in degrees, the period formula for tangent is $P = \dfrac{180°}{B}$.

Exercise 51

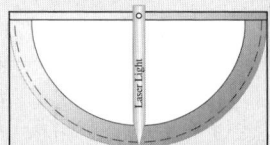

θ (degrees)	Distance (cm)
0	0
10	2.1
20	4.4
30	6.9
40	10.1
50	14.3
60	20.8
70	33.0
80	68.1
89	687.5

52. Use the equation model obtained in Exercise 51 to compare the values given by the equation with the actual data. As a percentage, what was the largest deviation between the two?

53. Circumscribed polygons:
The *perimeter* of a regular polygon circumscribed about a circle of radius r is given by $P = 2nr \tan\left(\dfrac{\pi}{n}\right)$, where n is the number of sides ($n \geq 3$) and r is the radius of the circle. Given $r = 10$ cm, (a) What is the circumference of the circle? (b) What is the perimeter of the polygon when $n = 4$? Why? (c) Calculate the perimeter of the polygon for $n = 10, 20, 30$, and 100. What do you notice?

Exercise 53

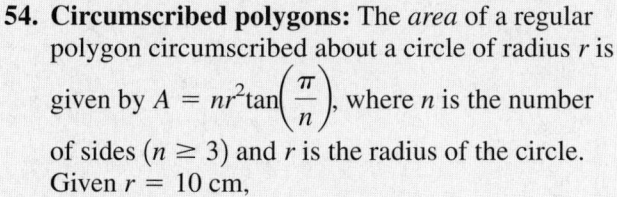

54. Circumscribed polygons: The *area* of a regular polygon circumscribed about a circle of radius r is given by $A = nr^2 \tan\left(\dfrac{\pi}{n}\right)$, where n is the number of sides ($n \geq 3$) and r is the radius of the circle. Given $r = 10$ cm,

  a. What is the area of the circle?

  b. What is the area of the polygon when $n = 4$? Why?

  c. Calculate the area of the polygon for $n = 10, 20, 30$, and 100. What do you notice?

Coefficients of friction:
Pulling someone on a sled is much easier during the winter than in the summer, due to a phenomenon known as the *coefficient of friction*. The friction between the sled's skids and the snow is much lower than the friction between the skids and the dry ground or pavement. Basically, the coefficient of friction is defined by the relationship $\mu = \tan \theta$, where θ is the angle at which a block composed of one material will slide down an inclined plane made of another material, with a constant velocity. Coefficients of friction have been established experimentally for many materials and a short list is shown here.

Material	Coefficient
steel on steel	0.74
copper on glass	0.53
glass on glass	0.94
copper on steel	0.68
wood on wood	0.5

55. Graph the function $\mu = \tan \theta$, with θ in degrees over the interval $[0°, 60°]$ and use the graph to estimate solutions to the following. Confirm or contradict your estimates using a calculator.

  a. A block of copper is placed on a sheet of steel, which is slowly inclined. Is the block of copper moving when the angle of inclination is 30°? At what angle of inclination will the copper block be moving with a constant velocity down the incline?

b. A block of copper is placed on a sheet of cast-iron. As the cast-iron sheet is slowly inclined, the copper block begins sliding at a constant velocity when the angle of inclination is approximately 46.5°. What is the coefficient of friction for copper on cast-iron?

c. Why do you suppose coefficients of friction greater than $\mu = 2.5$ are extremely rare? Give an example of two materials that likely have a high μ-value.

56. Graph the function $\mu = \tan \theta$ with θ in radians over the interval $\left[0, \dfrac{5\pi}{12}\right]$ and use the graph to estimate solutions to the following. Confirm or contradict your estimates using a calculator.

  a. A block of glass is placed on a sheet of glass, which is slowly inclined. Is the block of glass moving when the angle of inclination is $\dfrac{\pi}{4}$? What is the smallest angle of inclination for which the glass block will be moving with a constant velocity down the incline (rounded to four decimal places)?

  b. A block of Teflon is placed on a sheet of steel. As the steel sheet is slowly inclined, the Teflon block begins sliding at a constant velocity when the angle of inclination is approximately 0.04. What is the coefficient of friction for Teflon on steel?

  c. Why do you suppose coefficients of friction less than $\mu = 0.04$ are extremely rare for two solid materials? Give an example of two materials that likely have a very low μ value.

57. Tangent lines: The actual definition of the word *tangent* comes from the Latin *tangere*, meaning "to touch." In mathematics, a tangent line touches the graph of a circle at only one point and function values for $\tan \theta$ are obtained from the length of the line segment tangent to a unit circle.

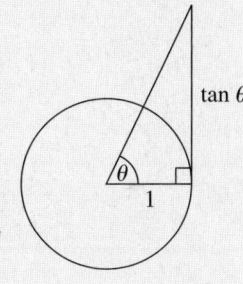

  a. What is the length of the line segment when $\theta = 80°$?

  b. If the line segment is 16.35 units long, what is the value of θ?

  c. Can the line segment ever be greater than 100 units long? Why or why not?

  d. How does your answer to (c) relate to the asymptotic behavior of the graph?

▶ EXTENDING THE CONCEPT

 58. Rework Exercises 49 and 50, obtaining a new equation for the data using a different ordered pair to compute the value of A. What do you notice? Try yet another ordered pair and calculate A once again for another equation Y_2. Complete a table of values using the given inputs, with the outputs of the three equations generated (original, Y_1, and Y_2). Does any one equation seem to model the data better than the others? Are all of the equation models "acceptable"? Please comment.

 59. Regarding Example 7, we can use the standard distance/rate/time formula $D = RT$ to compute the average velocity of the beam of light along the wall in any interval of time: $R = \dfrac{D}{T}$. For example, using

$D(t) = 5 \tan\left(\dfrac{\pi}{8}t\right)$, the average velocity in the

interval $[0, 2]$ is $\dfrac{D(2) - D(0)}{2 - 0} = 2.5$ m/sec.

Calculate the average velocity of the beam in the time intervals $[2, 3]$, $[3, 3.5]$, and $[3.5, 3.8]$ sec. What do you notice? How would the average velocity of the beam in the interval $[3.9, 3.99]$ sec compare?

▶ MAINTAINING YOUR SKILLS

60. **(5.1)** A lune is a section of surface area on a sphere, which is subtended by an angle θ at the circumference. For θ *in radians,* the surface area of a lune is $A = 2r^2\theta$, where r is the radius of the sphere. Find the area of a lune on the surface of the Earth which is subtended by an angle of 15°. Assume the radius of the Earth is 6373 km.

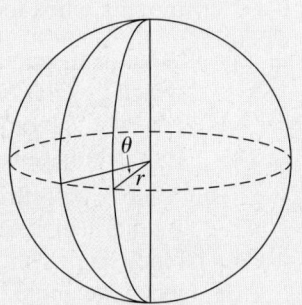

61. **(3.4/3.5)** Find the y-intercept, x-intercept(s), and all asymptotes of each function, but do not graph.

a. $h(x) = \dfrac{3x^2 - 9x}{2x^2 - 8}$ **b.** $t(x) = \dfrac{x + 1}{x^2 - 4x}$

c. $p(x) = \dfrac{x^2 - 1}{x + 2}$

62. **(5.2)** State the points on the unit circle that correspond to $t = 0, \dfrac{\pi}{4}, \dfrac{\pi}{2}, \pi, \dfrac{3\pi}{4}, \dfrac{3\pi}{2}$, and 2π.

What is the value of $\tan\left(\dfrac{\pi}{2}\right)$? Why?

63. **(4.1)** The radioactive element potassium-42 is sometimes used as a tracer in certain biological experiments, and its decay can be modeled by the formula $Q(t) = Q_0 e^{-0.055t}$, where $Q(t)$ is the amount that remains after t hours. If 15 grams (g) of potassium-42 are initially present, how many hours until only 10 g remain?

MID-CHAPTER CHECK

1. The city of Las Vegas, Nevada, is located at 36°06′36″ north latitude, 115°04′48″ west longitude. (a) Convert both measures to decimal degrees. (b) If the radius of the Earth is 3960 mi, how far north of the equator is Las Vegas?

Exercise 2

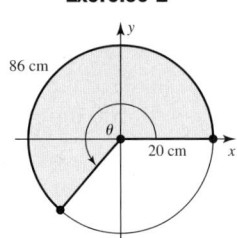

2. Find the angle subtended by the arc shown in the figure, then determine the area of the sector.

3. Evaluate without using a calculator: (a) $\cot 60°$ and (b) $\sin\left(\dfrac{7\pi}{4}\right)$.

4. Evaluate using a calculator: (a) $\sec\left(\dfrac{\pi}{12}\right)$ and (b) $\tan 83.6°$.

5. Complete the ordered pair indicated on the unit circle in the figure and find the value of all six trigonometric functions at this point.

Exercise 5

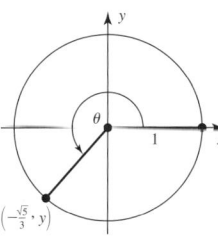

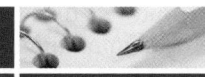 6. For the point on the unit circle in Exercise 5, find the related angle t in both degrees (to tenths) and radians (to ten-thousandths).

7. Name the location of the asymptotes and graph $y = 3 \tan\left(\frac{\pi}{2}t\right)$ for $t \in [-2\pi, 2\pi]$.

8. Clearly state the amplitude and period, then sketch the graph: $y = -3 \cos\left(\frac{\pi}{2}t\right)$.

9. On a unit circle, if arc t has length 5.94, (a) in what quadrant does it terminate? (b) What is its reference arc? (c) Of $\sin t$, $\cos t$, and $\tan t$, which are negative for this value of t?

10. For the graph given here, (a) clearly state the amplitude and period; (b) find the equation of the graph; (c) graphically find $f(\pi)$ and then confirm/contradict your estimation using a calculator.

Exercise 10

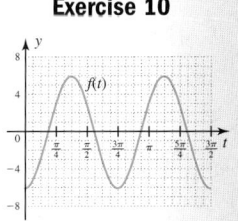

REINFORCING BASIC CONCEPTS

Trigonometry of the Real Numbers and the Wrapping Function

The circular functions are sometimes discussed in terms of what is called a *wrapping function,* in which the real number line is literally wrapped around the unit circle. This approach can help illustrate how the trig functions can be seen as functions of the real numbers, and apart from any reference to a right triangle. Figure 5.59 shows (1) a unit circle with the location of certain points on the circumference clearly marked and (2) a number line that has been marked in multiples of $\frac{\pi}{12}$ to coincide with the length of the special arcs (integers are shown in the background). Figure 5.60 shows this same number line wrapped counterclockwise around the unit circle in the positive direction. Note how the resulting diagram confirms that an arc of length $t = \frac{\pi}{4}$ is associated with the point $\left(\frac{\sqrt{2}}{2}, \frac{\sqrt{2}}{2}\right)$ on the unit circle: $\cos\frac{\pi}{4} = \frac{\sqrt{2}}{2}$ and $\sin\frac{\pi}{4} = \frac{\sqrt{2}}{2}$; while an arc of length of $t = \frac{5\pi}{6}$ is associated with the point $\left(-\frac{\sqrt{3}}{2}, \frac{1}{2}\right)$: $\cos\frac{5\pi}{6} = -\frac{\sqrt{3}}{2}$ and $\sin\frac{5\pi}{6} = \frac{1}{2}$. Use this information to complete the exercises given.

Figure 5.59

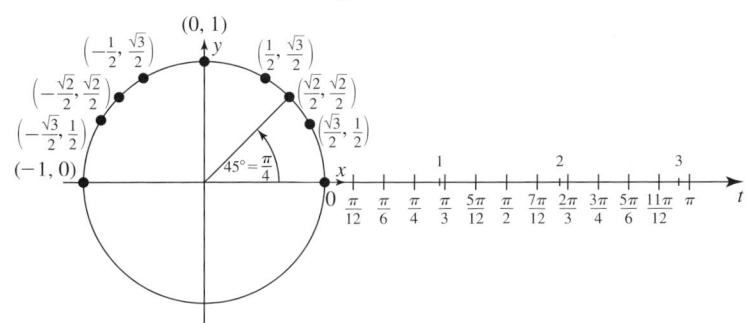

Figure 5.60

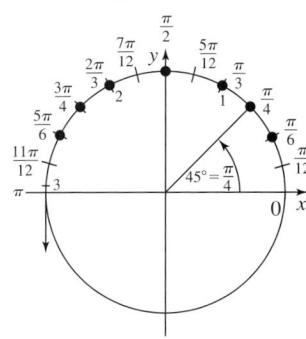

1. What is the ordered pair associated with an arc length of $t = \dfrac{2\pi}{3}$? What is the value of cos t? sin t?

2. What arc length t is associated with the ordered pair $\left(-\dfrac{\sqrt{3}}{2}, \dfrac{1}{2}\right)$? Is cos t positive or negative? Why?

3. If we continued to wrap this number line all the way around the circle, in what quadrant would an arc length of $t = \dfrac{11\pi}{6}$ terminate? Would sin t be positive or negative?

4. Suppose we wrapped a number line with negative values clockwise around the unit circle. In what quadrant would an arc length of $t = -\dfrac{5\pi}{3}$ terminate? What is cos t? sin t? What positive rotation terminates at the same point?

5.5 | Transformations and Applications of Trigonometric Graphs

Learning Objectives

In Section 5.5 you will learn how to:

☐ **A.** Apply vertical translations in context

☐ **B.** Apply horizontal translations in context

☐ **C.** Solve applications involving harmonic motion

From your algebra experience, you may remember beginning with a study of linear graphs, then moving on to quadratic graphs and their characteristics. By combining and extending the knowledge you gained, you were able to investigate and understand a variety of polynomial graphs—along with some powerful applications. A study of trigonometry follows a similar pattern, and by "combining and extending" our understanding of the basic trig graphs, we'll look at some powerful applications in *this* section.

A. Vertical Translations: $y = A \sin(Bt) + D$

On any given day, outdoor temperatures tend to follow a **sinusoidal pattern,** or a pattern that can be modeled by a sine function. As the sun rises, the morning temperature begins to warm and rise until reaching its high in the late afternoon, then begins to cool during the early evening and nighttime hours until falling to its nighttime low just prior to sunrise. Next morning, the cycle begins again. In the northern latitudes where the winters are very cold, it's not unreasonable to assume an average daily temperature of 0°C (32°F), and a temperature graph in degrees Celsius that looks like the one in Figure 5.61. For the moment, we'll assume that

Figure 5.61

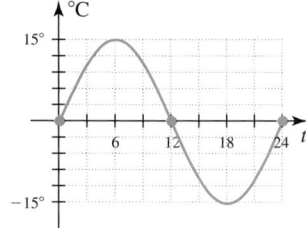

$t = 0$ corresponds to 12:00 noon. Note that $|A| = 15$ and $P = 24$, yielding $24 = \dfrac{2\pi}{B}$ or $B = \dfrac{\pi}{12}$.

If you live in a more temperate area, the daily temperatures still follow a sinusoidal pattern, but the average temperature could be much higher. This is an example of a **vertical shift,** and is the role D plays in the equation $y = A \sin(Bt) + D$. All other aspects of a graph remain the same; it is simply shifted D units up if $D > 0$ and D units down if $D < 0$. As in Section 5.3, for maximum value M and minimum value m, $\dfrac{M - m}{2}$ gives the amplitude A of a sine curve, while $\dfrac{M + m}{2}$ gives the **average value** D.

EXAMPLE 1 ▶   **Modeling Temperature Using a Sine Function**

On a fine day in Galveston, Texas, the high temperature might be about 85°F with an overnight low of 61°F.

   a. Find a sinusoidal equation model for the daily temperature.

   b. Sketch the graph.

   c. Approximate what time(s) of day the temperature is 65°F. Assume $t = 0$ corresponds to 12:00 noon.

Solution ▶   **a.** We first note the period is still $P = 24$, so $B = \dfrac{\pi}{12}$, and the equation model will have the form $y = A \sin\left(\dfrac{\pi}{12}t\right) + D$. Using $\dfrac{M + m}{2} = \dfrac{85 + 61}{2}$, we find the *average value* $D = 73$, with amplitude $A = \dfrac{85 - 61}{2} = 12$. The resulting equation is $y = 12 \sin\left(\dfrac{\pi}{12}t\right) + 73$.

   b. To sketch the graph, use a reference rectangle $2A = 24$ units tall and $P = 24$ units wide, along with the *rule of fourths* to locate zeroes and max/min values (see Figure 5.62). Then lightly sketch a sine curve through these points and within the rectangle as shown. This is the graph of $y = 12 \sin\left(\dfrac{\pi}{12}t\right) + 0$.

Using an appropriate scale, shift the rectangle and plotted points vertically upward 73 units and carefully draw the finished graph through the points and within the rectangle (see Figure 5.63).

Figure 5.62

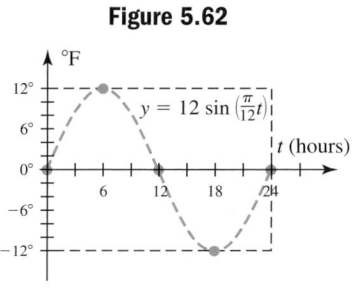

Figure 5.63

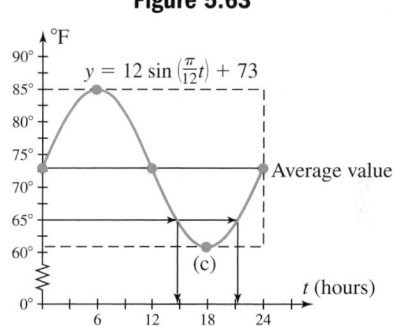

This gives the graph of $y = 12 \sin\left(\dfrac{\pi}{12}t\right) + 73$. Note the brokenline notation "≶" in Figure 5.63 indicates that certain values along an axis are unused (in this case, we skipped 0° to 60°), and we began scaling the axis with the values needed.

   c. As indicated in Figure 5.63, the temperature hits 65° twice, at about 15 and 21 hr after 12:00 noon, or at 3:00 A.M. and 9:00 A.M. Verify by computing $f(15)$ and $f(21)$.

Now try Exercises 7 through 18 ▶

> **WORTHY OF NOTE**
>
> Recall from Section 5.5 that transformations of any function $y = f(x)$ remain consistent regardless of the function f used. For the sine function, the transformation $y = af(x \pm h) \pm k$ is more commonly written $y = A \sin(t \pm C) \pm D$, and $|A|$ gives a vertical stretch or compression, C is a horizontal shift opposite the sign, and D is a vertical shift, as seen in Example 1.

Sinusoidal graphs actually include both sine and cosine graphs, the difference being that sine graphs begin at the average value, while cosine graphs begin at the maximum value. Sometimes it's more advantageous to use one over the other, but equivalent forms can easily be found. In Example 2, a cosine function is used to model an animal population that fluctuates sinusoidally due to changes in food supplies.

EXAMPLE 2 ▶ **Modeling Population Fluctuations Using a Cosine Function**

The population of a certain animal species can be modeled by the function

$$P(t) = 1200 \cos\left(\frac{\pi}{5}t\right) + 9000,$$ where $P(t)$ represents the population in year t.

Use the model to

a. Find the period of the function.
b. Graph the function over one period.
c. Find the maximum and minimum values.
d. Estimate the number of years the population is less than 8000.

Solution ▶ a. Since $B = \frac{\pi}{5}$, the period is $P = \frac{2\pi}{\pi/5} = 10$, meaning the population of this species fluctuates over a 10-yr cycle.

b. Use a reference rectangle ($2A = 2400$ by $P = 10$ units) and the *rule of fourths* to locate zeroes and max/min values, then sketch the unshifted graph

$$y = 1200 \cos\left(\frac{\pi}{5}t\right).$$ With $P = 10$, these occur at $t = 0, 2.5, 5, 7.5,$ and 10

(see Figure 5.64). Shift this graph upward 9000 units (using an appropriate scale) to obtain the graph of $P(t)$ shown in Figure 5.65.

Figure 5.64

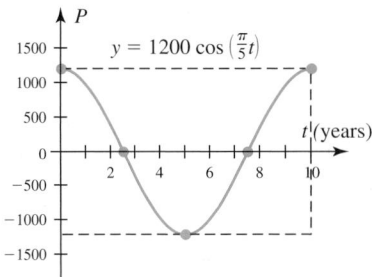

Figure 5.65

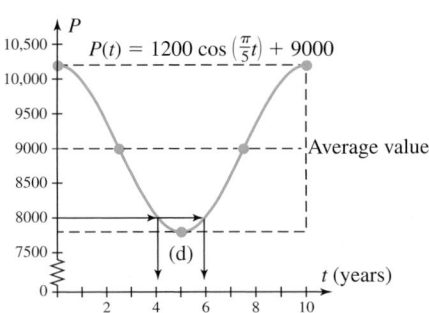

☑ **A.** You've just learned how to apply vertical translations in context

c. The maximum value is $9000 + 1200 = 10,200$ and the minimum value is $9000 - 1200 = 7800$.

d. As determined from the graph, the population drops below 8000 animals for approximately 2 yr. Verify by computing $P(4)$ and $P(6)$.

Now try Exercises 19 and 20 ▶

B. Horizontal Translations: $y = A \sin(Bt + C) + D$

In some cases, scientists would rather "benchmark" their study of sinusoidal phenomena by placing the average value at $t = 0$ instead of a maximum value (as in Example 2), or by placing the maximum or minimum value at $t = 0$ instead of the average value (as in Example 1). Rather than make additional studies or recompute using available data, *we can simply shift these graphs using a horizontal translation.* To help understand how, consider the graph of $y = x^2$. The graph is a parabola, concave up, with a vertex at the origin. Comparing this function with $y_1 = (x - 3)^2$ and

$y_2 = (x + 3)^2$, we note y_1 is simply the parent graph shifted 3 units right, and y_2 is the parent graph shifted 3 units left ("opposite the sign"). See Figures 5.66 through 5.68.

While *quadratic functions have no maximum value if A > 0*, these graphs are a good reminder of how a basic graph can be horizontally shifted. We simply *replace the independent variable x with $(x \pm h)$ or t with $(t \pm h)$,* where h is the desired shift and the sign is chosen depending on the direction of the shift.

Figure 5.66	Figure 5.67	Figure 5.68
$y = x^2$	$y_1 = (x - 3)^2$	$y_2 = (x + 3)^2$
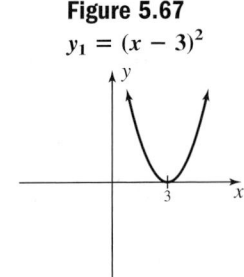		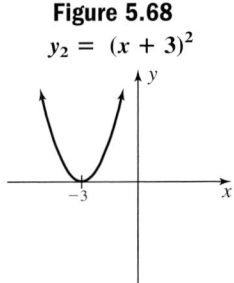

EXAMPLE 3 ▶ **Investigating Horizontal Shifts of Trigonometric Graphs**

Use a horizontal translation to shift the graph from Example 2 so that the average population begins at $t = 0$. Verify the result on a graphing calculator, then find a sine function that gives the same graph as the shifted cosine function.

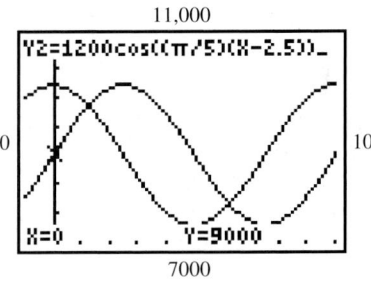

Solution ▶ For $P(t) = 1200 \cos\left(\dfrac{\pi}{5}t\right) + 9000$ from Example 2, the average value first occurs at $t = 2.5$. For the average value to occur at $t = 0$, we must shift the graph to the right 2.5 units. Replacing t with $(t - 2.5)$ gives $P(t) = 1200 \cos\left[\dfrac{\pi}{5}(t - 2.5)\right] + 9000$.

A graphing calculator shows the desired result is obtained (see figure). The new graph appears to be a sine function with the same amplitude and period, and the equation is $y = 1200 \sin\left(\dfrac{\pi}{5}t\right) + 9000$.

Now try Exercises 21 and 22 ▶

WORTHY OF NOTE

When the function

$P(t) = 1200 \cos\left[\dfrac{\pi}{5}(t - 2.5)\right]$
$+ 9000$ is written in standard form as $P(t) = 1200$ $\cos\left[\dfrac{\pi}{5}t - \dfrac{\pi}{2}\right] + 9000$, we can easily see why they are equivalent to $P(t) = 1200$ $\sin\left(\dfrac{\pi}{5}t\right) + 9000$. Using the cofunction relationship, $\cos\left[\dfrac{\pi}{5}t - \dfrac{\pi}{2}\right] = \sin\left(\dfrac{\pi}{5}t\right)$.

Equations like $P(t) = 1200 \cos\left[\dfrac{\pi}{5}(t - 2.5)\right] + 9000$ from Example 3 are said to be written in **shifted form,** since we can easily tell the magnitude and direction of the shift. To obtain the **standard form** we *distribute the value of B:* $P(t) = 1200 \cos\left(\dfrac{\pi}{5}t - \dfrac{\pi}{2}\right) + 9000$. In general, the *standard form* of a sinusoidal equation (using *either* a cosine or sine function) is written $y = A \sin(Bt \pm C) + D$, with the *shifted form* found by factoring out B from $Bt \pm C$:

$$y = A \sin(Bt \pm C) + D \rightarrow y = A \sin\left[B\left(t \pm \dfrac{C}{B}\right)\right] + D$$

In either case, C gives what is known as the **phase angle** of the function, and is used in a study of AC circuits and other areas, to discuss how far a given function is "out of phase" with a reference function. In the latter case, $\dfrac{C}{B}$ is simply the horizontal shift (or phase shift) of the function and gives the magnitude and direction of this shift (opposite the sign).

Characteristics of Sinusoidal Models

Transformations of the graph of $y = \sin t$ are written as $y = A \sin(Bt)$, where

1. $|A|$ gives the *amplitude* of the graph, or the maximum displacement from the average value.
2. B is related to the *period P* of the graph according to the ratio $P = \dfrac{2\pi}{B}$ (the interval required for one complete cycle). Translations of $y = A \sin(Bt)$ can be written as follows:

    Standard form

    $$y = A \sin(Bt \pm C) + D$$

    Shifted form

    $$y = A \sin\left[B\left(t \pm \dfrac{C}{B} \right) \right] + D$$

3. In either case, C is called the *phase angle* of the graph, while $\pm \dfrac{C}{B}$ gives the magnitude and direction of the *horizontal shift* (opposite the given sign).

4. D gives the *vertical shift* of the graph, and the location of the average value. The shift will be in the same direction as the given sign.

Knowing where each cycle begins and ends is a helpful part of sketching a graph of the equation model. The **primary interval** for a sinusoidal graph can be found by solving the inequality $0 \le Bt \pm C < 2\pi$, with the reference rectangle and *rule of fourths* giving the zeroes, max/min values, and a sketch of the graph in this interval. The graph can then be extended in either direction, and shifted vertically as needed.

EXAMPLE 4 ▶ **Analyzing the Transformation of a Trig Function**

Identify the amplitude, period, horizontal shift, vertical shift (average value), and endpoints of the primary interval.

$$y = 2.5 \sin\left(\dfrac{\pi}{4}t + \dfrac{3\pi}{4} \right) + 6$$

Solution ▶ The equation gives an amplitude of $|A| = 2.5$, with an average value of $D = 6$. The maximum value will be $y = 2.5(1) + 6 = 8.5$, with a minimum of $y = 2.5(-1) + 6 = 3.5$. With $B = \dfrac{\pi}{4}$, the period is $P = \dfrac{2\pi}{\pi/4} = 8$. To find the horizontal shift, we factor out $\dfrac{\pi}{4}$ to write the equation in shifted form: $\left(\dfrac{\pi}{4}t + \dfrac{3\pi}{4} \right) = \dfrac{\pi}{4}(t + 3)$. The horizontal shift is 3 units left. For the endpoints of the primary interval we solve $0 \le \dfrac{\pi}{4}(t + 3) < 2\pi$, which gives $-3 \le t < 5$.

Now try Exercises 23 through 34 ▶

GRAPHICAL SUPPORT

The analysis of $y = 2.5 \sin\left[\dfrac{\pi}{4}(t + 3)\right] + 6$ from

Example 4 can be verified on a graphing calculator. Enter the function as Y_1 on the $\boxed{Y=}$ screen and set an appropriate window size using the information gathered. Press the $\boxed{TRACE}$ key and -3 $\boxed{ENTER}$ and the calculator gives the average value $y = 6$ as output. Repeating this for $x = 5$ shows one complete cycle has been completed.

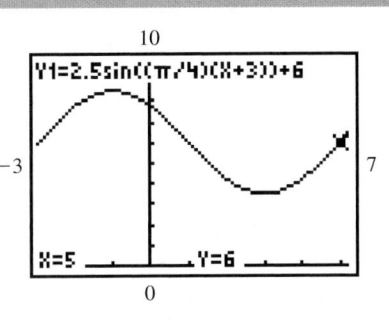

To help gain a better understanding of sinusoidal functions, their graphs, and the role the coefficients A, B, C, and D play, it's often helpful to reconstruct the equation of a given graph.

EXAMPLE 5 ▶ **Determining the Equation of a Trig Function from Its Graph**

Determine the equation of the given graph using a sine function.

Solution ▶ From the graph it is apparent the maximum value is 300, with a minimum of 50. This gives a value

of $\dfrac{300 + 50}{2} = 175$ for D and $\dfrac{300 - 50}{2} = 125$

for A. The graph completes one cycle from $t = 2$

to $t = 18$, showing $P = 18 - 2 = 16$ and $B = \dfrac{\pi}{8}$.

The average value first occurs at $t = 2$, so the basic graph has been shifted to the right 2 units.

The equation is $y = 125 \sin\left[\dfrac{\pi}{8}(t - 2)\right] + 175$.

☑ **B. You've just learned how to apply horizontal translations in context**

Now try Exercises 35 through 44 ▶

C. Simple Harmonic Motion: $y = A \sin(Bt)$ or $y = A \cos(Bt)$

The periodic motion of springs, tides, sound, and other phenomena all exhibit what is known as **harmonic motion**, which can be modeled using sinusoidal functions.

Harmonic Models—Springs

Consider a spring hanging from a beam with a weight attached to one end. When the weight is at rest, we say it is in **equilibrium**, or has zero displacement from center. Stretching the spring and then releasing it causes the weight to "bounce up and down," with its displacement from center neatly modeled over time by a sine wave (see Figure 5.69).

Figure 5.69

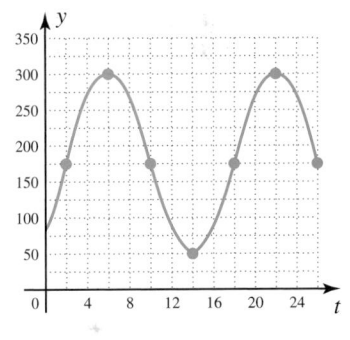

At rest   Stretched   Released

For objects in harmonic *motion* (there are other harmonic models), the input variable t is always a time unit (seconds, minutes, days, etc.), so in addition to the period of the sinusoid, we are very interested in its **frequency**—the number of cycles it completes per unit time (see Figure 5.70). Since the period gives the time required to complete one cycle, the frequency f is given by $f = \dfrac{1}{P} = \dfrac{B}{2\pi}$.

Figure 5.70

Harmonic motion

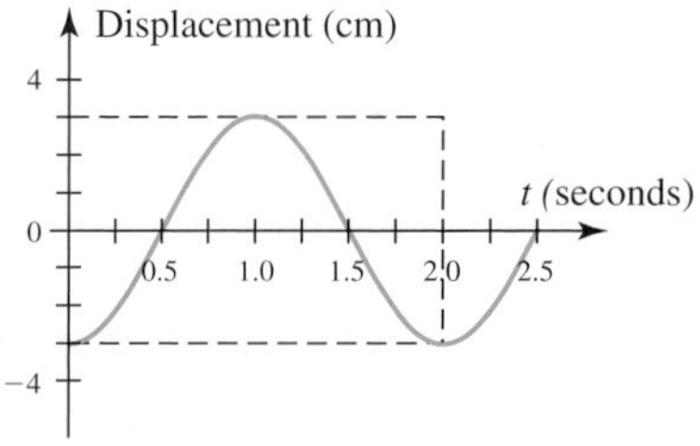

EXAMPLE 6 ▶ **Applications of Sine and Cosine: Harmonic Motion**

For the harmonic motion modeled by the sinusoid in Figure 5.70,

a. Find an equation of the form $y = A\cos(Bt)$.

b. Determine the frequency.

c. Use the equation to find the position of the weight at $t = 1.8$ sec.

Solution ▶ **a.** By inspection the graph has an amplitude $|A| = 3$ and a period $P = 2$. After substitution into $P = \dfrac{2\pi}{B}$, we obtain $B = \pi$ and the equation $y = -3\cos(\pi t)$.

b. Frequency is the reciprocal of the period so $f = \dfrac{1}{2}$, showing one-half a cycle is completed each second (as the graph indicates).

c. Evaluating the model at $t = 1.8$ gives $y = -3\cos[\pi(1.8)] \approx -2.43$, meaning the weight is 2.43 cm below the equilibrium point at this time.

Now try Exercises 47 through 50 ▶

Harmonic Models—Sound Waves

A second example of harmonic motion is the production of sound. For the purposes of this study, we'll look at musical notes. The vibration of matter produces a **pressure wave** or **sound energy,** which in turn vibrates the eardrum. Through the intricate structure of the middle ear, this sound energy is converted into mechanical energy and sent to the inner ear where it is converted to nerve impulses and transmitted to the brain. If the sound wave has a high frequency, the eardrum vibrates with greater frequency, which the brain interprets as a "high-pitched" sound. The *intensity* of the sound wave can also be transmitted to the brain via these mechanisms, and if the arriving sound wave has a high amplitude, the eardrum vibrates more forcefully and the sound is interpreted as "loud" by the brain. These characteristics are neatly modeled using $y = A\sin(Bt)$. For the moment we will focus on the frequency, keeping the amplitude constant at $A = 1$.

The musical note known as A_4 or "the A above middle C" is produced with a frequency of 440 vibrations per second, or 440 hertz (Hz) (this is the note most often used in the tuning of pianos and other musical instruments). For any given note, the same note one octave higher will have double the frequency, and the same note one octave lower will have one-half the frequency. In addition, with $f = \dfrac{1}{P}$ the value of $B = 2\pi\left(\dfrac{1}{P}\right)$ can always be expressed as $B = 2\pi f$, so A_4 has the equation $y = \sin[440(2\pi t)]$ (after rearranging the factors). The same note one octave lower is A_3

and has the equation $y = \sin[220(2\pi t)]$, with one-half the frequency. To draw the representative graphs, we must scale the t-axis in very small increments (seconds $\times$ 10^{-3}) since $P = \dfrac{1}{440} \approx 0.0023$ for A_4, and

$P = \dfrac{1}{220} \approx 0.0045$ for A_3. Both are graphed in Figure 5.71, where we see that the higher note completes two cycles in the same interval that the lower note completes one.

Figure 5.71

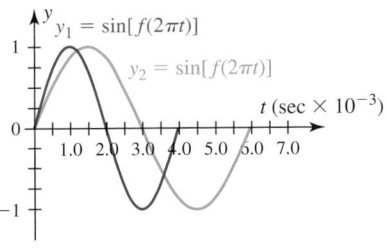

$A_4 \rightarrow y = \sin[440(2\pi t)]$
$A_3 \rightarrow y = \sin[220(2\pi t)]$

$t\,(\text{sec} \times 10^{-3})$

EXAMPLE 7 ▶ **Applications of Sine and Cosine: Sound Frequencies**

The table here gives the frequencies for three octaves of the 12 "chromatic" notes with frequencies between 110 Hz and 840 Hz. Two of the 36 notes are graphed in the figure. Which two?

$y_1 = \sin[f(2\pi t)]$

$y_2 = \sin[f(2\pi t)]$

$t\,(\text{sec} \times 10^{-3})$

	Frequency by Octave		
Note	Octave 3	Octave 4	Octave 5
A	110.00	220.00	440.00
A#	116.54	233.08	466.16
B	123.48	**246.96**	493.92
C	130.82	261.64	523.28
C#	138.60	277.20	554.40
D	146.84	293.68	587.36
D#	155.56	311.12	622.24
E	**164.82**	329.24	659.28
F	174.62	349.24	698.48
F#	185.00	370.00	740.00
G	196.00	392.00	784.00
G#	207.66	415.32	830.64

Solution ▶ Since amplitudes are equal, the only difference is the frequency and period of the notes. It appears that y_1 has a period of about 0.004 sec, giving a frequency of

$\dfrac{1}{0.004} = 250$ Hz—very likely a B_4 (in bold). The graph of y_2 has a period of about

0.006, for a frequency of $\dfrac{1}{0.006} \approx 167$ Hz—probably an E_3 (also in bold).

✓ **C. You've just learned how to solve applications involving harmonic motion**

Now try Exercises 51 through 54 ▶

TECHNOLOGY HIGHLIGHT

Locating Zeroes, Roots, and x-Intercepts

As you know, the zeroes of a function are *input* values that cause an *output* of zero. Graphically, these show up as x-intercepts and once a function is graphed they can be located (if they exist) using the [2nd] [CALC] **2:zero** feature. This feature is similar to the **3:minimum** and **4:maximum** features, in that we have the calculator search a specified interval by giving a **left bound** and a **right bound**. To illustrate, enter $Y_1 = 3 \sin\left(\frac{\pi}{2}x\right) - 1$ on the [Y=] screen and graph it using the [ZOOM] **7:ZTrig** option. The resulting graph shows there are six zeroes in this interval and we'll locate the first negative root. Knowing the **7:Trig** option uses tick marks that are spaced every $\frac{\pi}{2}$ units, this root is in the interval $\left(-\pi, -\frac{\pi}{2}\right)$. After pressing [2nd]

Figure 5.72

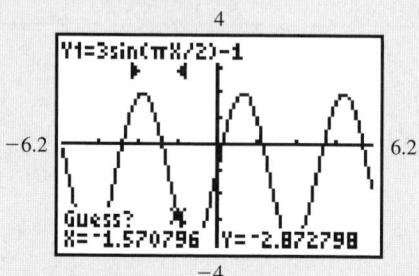

[CALC] **2:zero** the calculator returns you to the graph, and requests a "Left Bound," (see Figure 5.72). We enter $-\pi$ (press [ENTER]) and the calculator marks this choice with a "▶" marker (pointing to the right), then asks for a "Right Bound." After entering $-\frac{\pi}{2}$, the calculator marks this with a "◀" marker and asks for a "Guess." Bypass this option by pressing [ENTER] once again (see Figure 5.73). The calculator searches the interval until it locates a zero (Figure 5.74) or displays an error message indicating it was unable to comply (no zeroes in the interval). Use these ideas to locate the zeroes of the following functions in $[0, \pi]$.

Figure 5.73

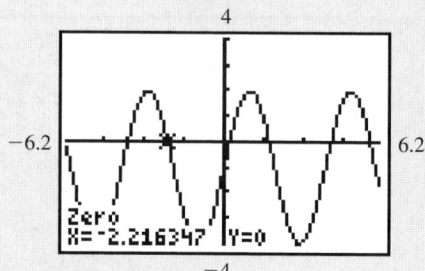

Figure 5.74

Exercise 1: $y = -2\cos(\pi t) + 1$

Exercise 2: $y = 0.5 \sin[\pi(t - 2)]$

Exercise 3: $y = \frac{3}{2}\tan(2x) - 1$

Exercise 4: $y = x^3 - \cos x$

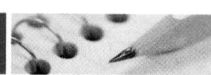

5.5 EXERCISES

▶ CONCEPTS AND VOCABULARY

Fill in each blank with the appropriate word or phrase. Carefully reread the section if needed.

1. A sinusoidal wave is one that can be modeled by functions of the form _____ or _____.

2. The graph of $y = \sin x + k$ is the graph of $y = \sin x$ shifted _____ k units. The graph of $y = \sin(x - h)$ is the graph of $y = \sin x$ shifted _____ h units.

3. To find the primary interval of a sinusoidal graph, solve the inequality _____.

4. Given the period P, the frequency is _____, and given the frequency f, the value of B is _____.

5. Explain/Discuss the difference between the *standard form* of a sinusoidal equation, and the *shifted form*. How do you obtain one from the other? For what benefit?

6. Write out a step-by-step procedure for sketching the graph of $y = 30 \sin\left(\dfrac{\pi}{2}t - \dfrac{1}{2}\right) + 10$. Include use of the reference rectangle, primary interval, zeroes, max/mins, and so on. Be complete and thorough.

▶ DEVELOPING YOUR SKILLS

Use the graphs given to (a) state the amplitude A and period P of the function; (b) estimate the value at $x = 14$; and (c) estimate the interval in $[0, P]$ where $f(x) \geq 20$.

7.
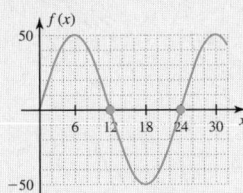

8.
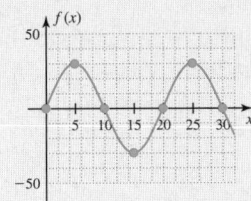

Use the graphs given to (a) state the amplitude A and period P of the function; (b) estimate the value at $x = 2$; and (c) estimate the interval in $[0, P]$, where $f(x) \leq -100$.

9.
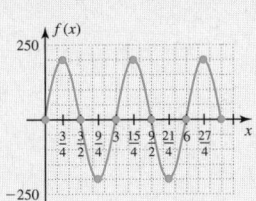

10.
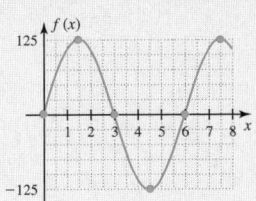

Use the information given to write a sinusoidal equation and sketch its graph. Recall $B = \dfrac{2\pi}{P}$.

11. Max: 100, min: 20, $P = 30$

12. Max: 95, min: 40, $P = 24$

13. Max: 20, min: 4, $P = 360$

14. Max: 12,000, min: 6500, $P = 10$

Use the information given to write a sinusoidal equation, sketch its graph, and answer the question posed.

15. In Geneva, Switzerland, the daily temperature in January ranges from an average high of 39°F to an average low of 29°F. (a) Find a sinusoidal equation model for the daily temperature; (b) sketch the graph; and (c) approximate the time(s) each January day the temperature reaches the freezing point (32°F). Assume $t = 0$ corresponds to noon.

Source: 2004 Statistical Abstract of the United States, Table 1331.

16. In Nairobi, Kenya, the daily temperature in January ranges from an average high of 77°F to an average low of 58°F. (a) Find a sinusoidal equation model for the daily temperature; (b) sketch the graph; and (c) approximate the time(s) each January day the temperature reaches a comfortable 72°F. Assume $t = 0$ corresponds to noon.

Source: 2004 Statistical Abstract of the United States, Table 1331.

17. In Oslo, Norway, the number of hours of daylight reaches a low of 6 hr in January, and a high of nearly 18.8 hr in July. (a) Find a sinusoidal equation model for the number of daylight hours each month; (b) sketch the graph; and (c) approximate the number of *days* each year there are more than 15 hr of daylight. Use 1 month $\approx$ 30.5 days. Assume $t = 0$ corresponds to January 1.

Source: www.visitnorway.com/templates.

18. In Vancouver, British Columbia, the number of hours of daylight reaches a low of 8.3 hr in January, and a high of nearly 16.2 hr in July. (a) Find a sinusoidal equation model for the number of daylight hours each month; (b) sketch the graph; and (c) approximate the number of *days* each year there are more than 15 hr of daylight. Use 1 month $\approx$ 30.5 days. Assume $t = 0$ corresponds to January 1.

Source: www.bcpassport.com/vital/temp.

19. Recent studies seem to indicate the population of North American porcupine (*Erethizon dorsatum*) varies sinusoidally with the solar (sunspot) cycle due to its effects on Earth's ecosystems. Suppose the population of this species in a certain locality is modeled by the function $P(t) = 250 \cos\left(\dfrac{2\pi}{11}t\right) + 950$, where $P(t)$ represents the population of porcupines in year t. Use the model to (a) find the period of the function; (b) graph the function over one period; (c) find the maximum and minimum values; and

(d) estimate the number of years the population is less than 740 animals.

Source: Ilya Klvana, McGill University (Montreal), Master of Science thesis paper, November 2002.

20. The population of mosquitoes in a given area is primarily influenced by precipitation, humidity, and temperature. In tropical regions, these tend to fluctuate sinusoidally in the course of a year. Using trap counts and statistical projections, fairly accurate estimates of a mosquito population can be obtained. Suppose the population in a certain region was modeled by the function

$$P(t) = 50 \cos\left(\frac{\pi}{26}t\right) + 950, \text{ where } P(t) \text{ was the}$$

mosquito population (in thousands) in week t of the year. Use the model to (a) find the period of the function; (b) graph the function over one period; (c) find the maximum and minimum population values; and (d) estimate the number of weeks the population is less than 915,000.

 21. Use a horizontal translation to shift the graph from Exercise 19 so that the average population of the North American porcupine begins at $t = 0$. Verify results on a graphing calculator, then find a sine function that gives the same graph as the shifted cosine function.

 22. Use a horizontal translation to shift the graph from Exercise 20 so that the average population of mosquitoes begins at $t = 0$. Verify results on a graphing calculator, then find a sine function that gives the same graph as the shifted cosine function.

Identify the amplitude (A), period (P), horizontal shift (HS), vertical shift (VS), and endpoints of the primary interval (PI) for each function given.

23. $y = 120 \sin\left[\frac{\pi}{12}(t - 6)\right]$

24. $y = 560 \sin\left[\frac{\pi}{4}(t + 4)\right]$

25. $h(t) = \sin\left(\frac{\pi}{6}t - \frac{\pi}{3}\right)$

26. $r(t) = \sin\left(\frac{\pi}{10}t - \frac{2\pi}{5}\right)$

27. $y = \sin\left(\frac{\pi}{4}t - \frac{\pi}{6}\right)$

28. $y = \sin\left(\frac{\pi}{3}t + \frac{5\pi}{12}\right)$

29. $f(t) = 24.5 \sin\left[\frac{\pi}{10}(t - 2.5)\right] + 15.5$

30. $g(t) = 40.6 \sin\left[\frac{\pi}{6}(t - 4)\right] + 13.4$

31. $g(t) = 28 \sin\left(\frac{\pi}{6}t - \frac{5\pi}{12}\right) + 92$

32. $f(t) = 90 \sin\left(\frac{\pi}{10}t - \frac{\pi}{5}\right) + 120$

33. $y = 2500 \sin\left(\frac{\pi}{4}t + \frac{\pi}{12}\right) + 3150$

34. $y = 1450 \sin\left(\frac{3\pi}{4}t + \frac{\pi}{8}\right) + 2050$

Find the equation of the graph given. Write answers in the form $y = A \sin(Bt + C) + D$.

35.
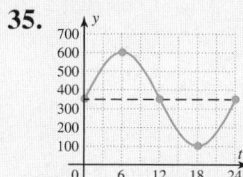

36.
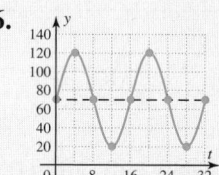

37.
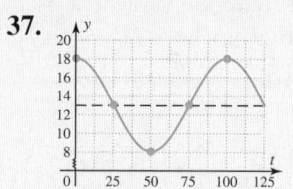

38.
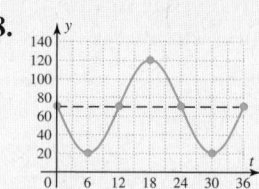

39.
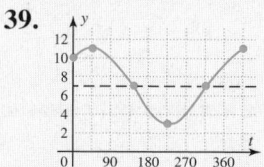

40.
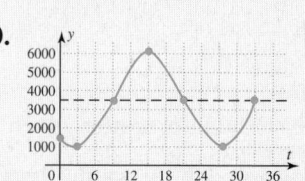

Sketch one complete period of each function.

41. $f(t) = 25 \sin\left[\frac{\pi}{4}(t - 2)\right] + 55$

42. $g(t) = 24.5 \sin\left[\frac{\pi}{10}(t - 2.5)\right] + 15.5$

43. $h(t) = 3 \sin(4t - \pi)$

44. $p(t) = -2 \cos\left(3t - \frac{\pi}{2}\right)$

▶ WORKING WITH FORMULAS

45. The relationship between the coefficient B, the frequency f, and the period P

In many applications of trigonometric functions, the equation $y = A \sin(Bt)$ is written as $y = A \sin[(2\pi f)t]$, where $B = 2\pi f$. Justify the new equation using $f = \dfrac{1}{P}$ and $P = \dfrac{2\pi}{B}$. In other words, explain how $A \sin(Bt)$ becomes $A \sin[(2\pi f)t]$, as though you were trying to help another student with the ideas involved.

46. Number of daylight hours:

$$D(t) = \frac{K}{2} \sin\left[\frac{2\pi}{365}(t - 79) \right] + 12$$

The number of daylight hours for a particular day of the year is modeled by the formula given, where $D(t)$ is the number of daylight hours on day t of the year and K is a constant related to the total variation of daylight hours, latitude of the location, and other factors. For the city of Reykjavik, Iceland, $K \approx 17$, while for Detroit, Michigan, $K \approx 6$. How many hours of daylight will each city receive on June 30 (the 182nd day of the year)?

▶ APPLICATIONS

47. Harmonic motion: A weight on the end of a spring is oscillating in harmonic motion. The equation model for the oscillations is

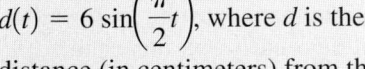

$d(t) = 6 \sin\left(\dfrac{\pi}{2}t\right)$, where d is the

distance (in centimeters) from the equilibrium point in t sec.

  a. What is the period of the motion? What is the frequency of the motion?

  b. What is the displacement from equilibrium at $t = 2.5$? Is the weight moving toward the equilibrium point or away from equilibrium at this time?

  c. What is the displacement from equilibrium at $t = 3.5$? Is the weight moving toward the equilibrium point or away from equilibrium at this time?

  d. How far does the weight move between $t = 1$ and $t = 1.5$ sec? What is the average velocity for this interval? Do you expect a greater or lesser velocity for $t = 1.75$ to $t = 2$? Explain why.

48. Harmonic motion: The bob on the end of a 24-in. pendulum is oscillating in harmonic motion. The equation model for the oscillations is $d(t) = 20 \cos(4t)$, where d is the distance (in inches) from the equilibrium point, t sec after being released from one side.

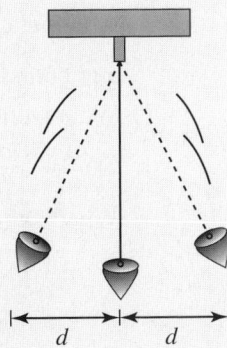

  a. What is the period of the motion? What is the frequency of the motion?

  b. What is the displacement from equilibrium at $t = 0.25$ sec? Is the weight moving toward the equilibrium point or away from equilibrium at this time?

  c. What is the displacement from equilibrium at $t = 1.3$ sec? Is the weight moving toward the equilibrium point or away from equilibrium at this time?

  d. How far does the bob move between $t = 0.25$ and $t = 0.35$ sec? What is its average velocity for this interval? Do you expect a greater velocity for the interval $t = 0.55$ to $t = 0.6$? Explain why.

49. Harmonic motion: A simple pendulum 36 in. in length is oscillating in harmonic motion. The bob at the end of the pendulum swings through an arc of 30 in. (from the far left to the far right, or one-half cycle) in about 0.8 sec. What is the equation model for this harmonic motion?

50. Harmonic motion: As part of a study of wave motion, the motion of a floater is observed as a series of uniform ripples of water move beneath it. By careful observation, it is noted that the floater bobs up and down through a distance of 2.5 cm every $\dfrac{1}{3}$ sec. What is the equation model for this harmonic motion?

51. Sound waves: Two of the musical notes from the chart on page 493 are graphed in the figure. Use the graphs given to determine which two.

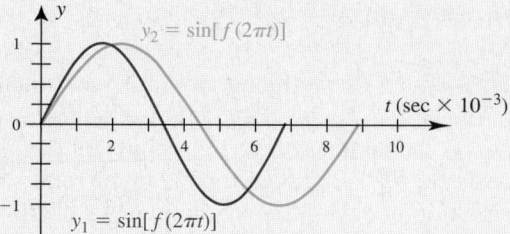

52. Sound waves: Two chromatic notes *not on the chart from page* 493 are graphed in the figure. Use the graphs and the discussion regarding octaves to determine which two. Note the scale of the *t*-axis *has been changed* to hundredths of a second.

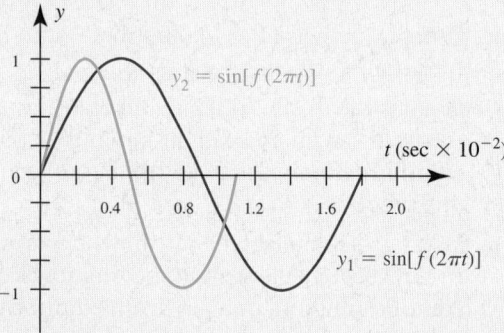

Sound waves: Use the chart on page 493 to write the equation for each note in the form $y = \sin[f(2\pi t)]$ and clearly state the period of each note.

53. notes D_3 and G_4 **54.** the notes A_5 and $C\#_3$

Daylight hours model: Solve using a graphing calculator and the formula given in Exercise 46.

55. For the city of Caracas, Venezuela, $K \approx 1.3$, while for Tokyo, Japan, $K \approx 4.8$.

   a. How many hours of daylight will each city receive on January 15th (the 15th day of the year)?

   b. Graph the equations modeling the hours of daylight on the same screen. Then determine (i) what days of the year these two cities will have the same number of hours of daylight, and (ii) the number of days each year that each city receives 11.5 hr or less of daylight.

56. For the city of Houston, Texas, $K \approx 3.8$, while for Pocatello, Idaho, $K \approx 6.2$.

   a. How many hours of daylight will each city receive on December 15 (the 349th day of the year)?

   b. Graph the equations modeling the hours of daylight on the same screen. Then determine (i) how many days each year Pocatello receives more daylight than Houston, and (ii) the number of days each year that each city receives 13.5 hr or more of daylight.

▶ EXTENDING THE CONCEPT

57. The formulas we use in mathematics can sometimes seem very mysterious. We know they "work," and we can graph and evaluate them—but where did they come from? Consider the formula for the number of daylight hours from Exercise 46:

$$D(t) = \frac{K}{2}\sin\left[\frac{2\pi}{365}(t - 79)\right] + 12.$$

   a. We know that the addition of 12 represents a vertical shift, but what does a vertical shift of 12 mean *in this context*?

   b. We also know the factor $(t - 79)$ represents a phase shift of 79 to the right. But what does a horizontal (phase) shift of 79 mean *in this context*?

   c. Finally, the coefficient $\frac{K}{2}$ represents a change in amplitude, but what does a change of amplitude mean *in this context*? Why is the coefficient bigger for the northern latitudes?

58. Use a graphing calculator to graph the equation

$$f(x) = \frac{3x}{2} - 2\sin(2x) - 1.5.$$

   a. Determine the interval between each peak of the graph. What do you notice?

   b. Graph $g(x) = \frac{3x}{2} - 1.5$ on the same screen and comment on what you observe.

   c. What would the graph of

$$f(x) = -\frac{3x}{2} + 2\sin(2x) + 1.5 \text{ look like?}$$

What is the *x*-intercept?

▶ MAINTAINING YOUR SKILLS

59. (5.1) In what quadrant does the arc $t = 3.7$ terminate? What is the reference arc?

60. (3.1) Given $f(x) = -3(x + 1)^2 - 4$, name the vertex and solve the inequality $f(x) > 0$.

61. (1.4) Compute the sum, difference, product and quotient of $-1 + i\sqrt{5}$ and $-1 - i\sqrt{5}$.

62. (5.3/5.4) Sketch the graph of (a) $y = \cos t$ in the interval $[0, 2\pi)$ and (b) $y = \tan t$ in the interval $\left(-\frac{\pi}{2}, \frac{3\pi}{2}\right)$.

5.6 The Trigonometry of Right Triangles

Learning Objectives

In Section 5.6 you will learn how to:

- ☐ **A.** Find values of the six trigonometric functions from their ratio definitions
- ☐ **B.** Solve a right triangle given one angle and one side
- ☐ **C.** Solve a right triangle given two sides
- ☐ **D.** Use cofunctions and complements to write equivalent expressions
- ☐ **E.** Solve applications involving angles of elevation and depression
- ☐ **F.** Solve general applications of right triangles

Over a long period of time, what began as a study of chord lengths by Hipparchus, Ptolemy, Aryabhata, and others became a systematic application of the ratios of the sides of a right triangle. In this section, we develop the sine, cosine, and tangent functions from a right triangle perspective, and explore certain relationships that exist between them. This view of the trig functions also leads to a number of significant applications.

A. Trigonometric Ratios and Their Values

In Section 5.1, we looked at applications involving 45-45-90 and 30-60-90 triangles, using the fixed ratios that exist between their sides. To apply this concept more generally using other right triangles, each side is given a specific name using its location relative to a specified angle. For the 30-60-90 triangle in Figure 5.75(a), the side **opposite (opp)** and the side **adjacent (adj)** are named with respect to the 30° angle, with the **hypotenuse (hyp)** always across from the right angle. Likewise for the 45-45-90 triangle in Figure 5.75(b).

Figure 5.75

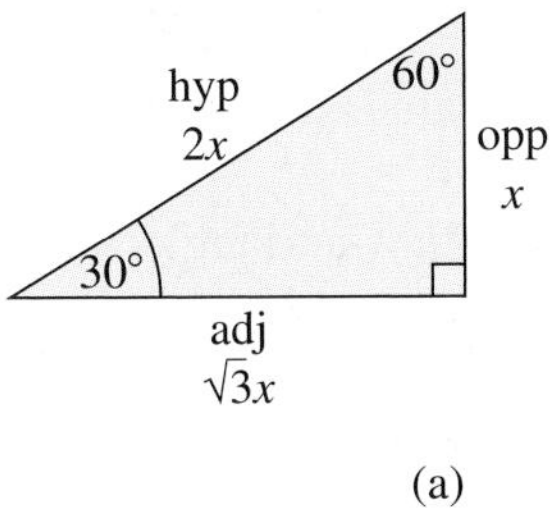

for 30°

$\frac{\text{opp}}{\text{hyp}} = \frac{1}{2}$

$\frac{\text{adj}}{\text{hyp}} = \frac{\sqrt{3}}{2}$

$\frac{\text{opp}}{\text{adj}} = \frac{1}{\sqrt{3}} = \frac{\sqrt{3}}{3}$

(a)

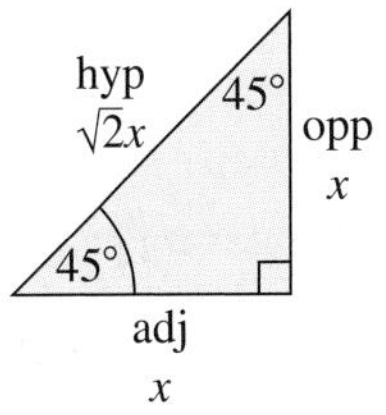

for 45°

$\frac{\text{opp}}{\text{hyp}} = \frac{1}{\sqrt{2}} = \frac{\sqrt{2}}{2}$

$\frac{\text{adj}}{\text{hyp}} = \frac{1}{\sqrt{2}} = \frac{\sqrt{2}}{2}$

$\frac{\text{opp}}{\text{adj}} = \frac{1}{1}$

(b)

Using these designations to define the various trig ratios, we can now develop a systematic method for applying them. Note that the x's "cancel" in each ratio, reminding us the ratios are independent of the triangle's size (if two triangles are similar, the ratio of corresponding sides is constant).

Ancient mathematicians were able to find values for the ratios corresponding to *any acute angle* in a right triangle, and realized that *naming* each ratio would be helpful. These names are $\frac{\text{opp}}{\text{hyp}} \rightarrow$ **sine,** $\frac{\text{adj}}{\text{hyp}} \rightarrow$ **cosine,** and $\frac{\text{opp}}{\text{adj}} \rightarrow$ **tangent.** Since each ratio depends on the measure of an acute angle θ, they are often referred to as **functions of an acute angle** and written in function form.

$$\text{sine } \theta = \frac{\text{opp}}{\text{hyp}} \qquad \text{cosine } \theta = \frac{\text{adj}}{\text{hyp}} \qquad \text{tangent } \theta = \frac{\text{opp}}{\text{adj}}$$

The reciprocal of these ratios, for example, $\dfrac{\text{hyp}}{\text{opp}}$ instead of $\dfrac{\text{opp}}{\text{hyp}}$, also play a significant role in this view of trigonometry, and are likewise given names:

$$\text{cosecant } \theta = \dfrac{\text{hyp}}{\text{opp}} \qquad \text{secant } \theta = \dfrac{\text{hyp}}{\text{adj}} \qquad \text{cotangent } \theta = \dfrac{\text{adj}}{\text{opp}}$$

The definitions hold regardless of the triangle's orientation or which of the acute angles is used.

In actual use, each function name is written in abbreviated form as sin θ, cos θ, tan θ, csc θ, sec θ, and cot θ respectively. Note that based on these designations, we have the following reciprocal relationships:

$$\sin \theta = \dfrac{1}{\csc \theta} \qquad \cos \theta = \dfrac{1}{\sec \theta} \qquad \tan \theta = \dfrac{1}{\cot \theta}$$

$$\csc \theta = \dfrac{1}{\sin \theta} \qquad \sec \theta = \dfrac{1}{\cos \theta} \qquad \cot \theta = \dfrac{1}{\tan \theta}$$

In general:

WORTHY OF NOTE

Over the years, a number of memory tools have been invented to help students recall these ratios correctly. One such tool is the acronym SOH CAH TOA, from the first letter of the function and the corresponding ratio. It is often recited as, "Sit On a Horse, Canter Away Hurriedly, To Other Adventures." Try making up a memory tool of your own.

Trigonometric Functions of an Acute Angle

$$\sin \alpha = \dfrac{a}{c} \qquad\qquad \sin \beta = \dfrac{b}{c}$$

$$\cos \alpha = \dfrac{b}{c} \qquad\qquad \cos \beta = \dfrac{a}{c}$$

$$\tan \alpha = \dfrac{a}{b} \qquad\qquad \tan \beta = \dfrac{b}{a}$$

Now that these ratios have been formally named, we can state values of all six functions given sufficient information about a right triangle.

EXAMPLE 1 ▶ **Finding Function Values Using a Right Triangle**

Given $\sin \theta = \frac{4}{7}$, find the values of the remaining trig functions.

Solution ▶ For $\sin \theta = \dfrac{4}{7} = \dfrac{\text{opp}}{\text{hyp}}$, we draw a triangle with a side of 4 units opposite a designated angle θ, and label a hypotenuse of 7 (see the figure). Using the Pythagorean theorem we find the length of the adjacent side: $\text{adj} = \sqrt{7^2 - 4^2} = \sqrt{33}$. The ratios are

$$\sin \theta = \dfrac{4}{7} \qquad \cos \theta = \dfrac{\sqrt{33}}{7} \qquad \tan \theta = \dfrac{4}{\sqrt{33}}$$

$$\csc \theta = \dfrac{7}{4} \qquad \sec \theta = \dfrac{7}{\sqrt{33}} \qquad \cot \theta = \dfrac{\sqrt{33}}{4}$$

Now try Exercises 7 through 12 ▶

☑ **A.** You've just learned how to find values of the six trigonometric functions from their ratio definitions

Note that due to the properties of similar triangles, identical results would be obtained using any ratio of sides that is equal to $\frac{4}{7}$. In other words, $\frac{2}{3.5} = \frac{4}{7} = \frac{8}{14} = \frac{16}{28}$ and so on, will all give the same value for sin θ.

B. Solving Right Triangles Given One Angle and One Side

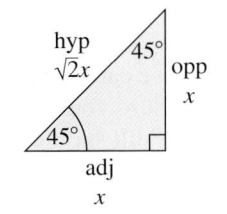

Example 1 gave values of the trig functions for an *unknown angle* θ. Using the special triangles, we can state the value of each trig function for 30°, 45°, and 60° based on the related ratio (see Table 5.10). These values are used extensively in a study of trigonometry and must be committed to memory.

Table 5.10

θ	$\sin\theta$	$\cos\theta$	$\tan\theta$	$\csc\theta$	$\sec\theta$	$\cot\theta$
30°	$\dfrac{1}{2}$	$\dfrac{\sqrt{3}}{2}$	$\dfrac{1}{\sqrt{3}}=\dfrac{\sqrt{3}}{3}$	2	$\dfrac{2}{\sqrt{3}}=\dfrac{2\sqrt{3}}{3}$	$\sqrt{3}$
45°	$\dfrac{\sqrt{2}}{2}$	$\dfrac{\sqrt{2}}{2}$	1	$\sqrt{2}$	$\sqrt{2}$	1
60°	$\dfrac{\sqrt{3}}{2}$	$\dfrac{1}{2}$	$\sqrt{3}$	$\dfrac{2}{\sqrt{3}}=\dfrac{2\sqrt{3}}{3}$	2	$\dfrac{1}{\sqrt{3}}=\dfrac{\sqrt{3}}{3}$

To **solve a right triangle** means to find the measure of all three angles and all three sides. This is accomplished using combinations of the Pythagorean theorem, the properties of triangles, and the trigonometric ratios. We will adopt the convention of naming each angle with a capital letter at the vertex or using a Greek letter on the interior. Each side is labeled using the related lowercase letter from the angle opposite. The complete solution should be organized in table form as in Example 2. Note the quantities shown in **bold** were given, and the remaining values were found using the techniques mentioned.

EXAMPLE 2 ▶ **Solving a Right Triangle**

Solve the triangle given.

Solution ▶ Applying the sine ratio (since the side opposite 30° is given), we have: $\sin 30° = \dfrac{\text{opp}}{\text{hyp}}$.

For side c: $\sin 30° = \dfrac{17.9}{c}$ $\sin 30° = \dfrac{\text{opposite}}{\text{hypotenuse}}$

$c\sin 30° = 17.9$ multiply by c

$c = \dfrac{17.9}{\sin 30°}$ divide by $\sin 30° = \dfrac{1}{2}$

$= 35.8$ result

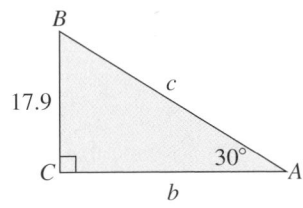

Using the Pythagorean theorem shows $b \approx 31$, and since $\angle A$ and $\angle B$ are complements, $B = 60°$. Note the results would have been identical if the special ratios from the 30-60-90 triangle were applied. The hypotenuse is twice the shorter side: $c = 2(17.9) = 35.8$, and the longer side is $\sqrt{3}$ times the shorter: $b = 17.9(\sqrt{3}) \approx 31$.

Angles	Sides
$A = 30°$	$a = 17.9$
$B = 60°$	$b \approx 31$
$C = 90°$	$c = 35.8$

Now try Exercises 13 through 16 ▶

Prior to the widespread availability of handheld calculators, a table of values was used to find $\sin\theta$, $\cos\theta$, and $\tan\theta$ for nonstandard angles. Table 5.11 shows the sine of 49° 30' is approximately 0.7604.

Table 5.11
$\sin \theta$

θ	0′	10′	20′	30′
45°	0.7071	0.7092	0.7112	0.7133
46	0.7193	0.7214	0.7234	0.7254
47	0.7314	0.7333	0.7353	0.7373
48	0.7431	0.7451	0.7470	0.7490
49	0.7547	0.7566	0.7585	0.7604

Today these trig values are programmed into your calculator and we can retrieve them with the push of a button (or two). To find the sine of 48°, make sure your calculator is in degree ⎡MODE⎤, then press the ⎡SIN⎤ key, 48, and ⎡ENTER⎤. The result should be very close to 0.7431 as the table indicates.

EXAMPLE 3 ▶ **Solving a Right Triangle**

Solve the triangle shown in the figure.

Solution ▶ We know $\angle B = 58°$ since $A + B = 90°$. We can find length b using the tangent function:

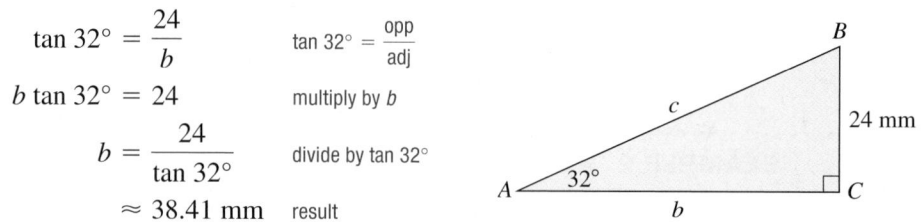

$$\tan 32° = \frac{24}{b} \qquad \tan 32° = \frac{\text{opp}}{\text{adj}}$$

$$b \tan 32° = 24 \qquad \text{multiply by } b$$

$$b = \frac{24}{\tan 32°} \qquad \text{divide by } \tan 32°$$

$$\approx 38.41 \text{ mm} \qquad \text{result}$$

We can find the length c by simply applying the Pythagorean theorem, or by using another trig ratio and a known angle.

For side c: $\quad \sin 32° = \frac{24}{c} \qquad \sin 32° = \frac{\text{opp}}{\text{hyp}}$

$$c \sin 32° = 24 \qquad \text{multiply by } c$$

$$c = \frac{24}{\sin 32°} \qquad \text{divide by } \sin 32°$$

$$\approx 45.29 \text{ mm} \qquad \text{result}$$

Angles	Sides
$A = 32°$	$a = 24$
$B = 58°$	$b \approx 38.41$
$C = 90°$	$c \approx 45.29$

The complete solution is shown in the table.

Now try Exercises 17 through 22 ▶

When solving a right triangle, any of the triangle relationships can be employed: (1) angles must sum to 180°, (2) Pythagorean theorem, (3) special triangles, and (4) the trigonometric functions of an acute angle. However, the resulting equation must have only one unknown or it cannot be used. For the triangle shown in Figure 5.76, we cannot begin with the Pythagorean theorem since sides a and b are unknown, and tan 51° is unusable for the same reason. Since the hypotenuse is given, we could begin with $\cos 51° = \frac{b}{152}$ and solve for b, or with $\sin 51° = \frac{a}{152}$ and solve for a, then work out a complete solution. Verify that $a \approx 118.13$ ft and $b \approx 95.66$ ft.

Figure 5.76

☑ **B.** You've just learned how to solve a right triangle given one angle and one side

C. Solving Right Triangles Given Two Sides

The partial table for $\sin \theta$ given earlier was also used in times past to find an angle whose sine was known, meaning if $\sin \theta \approx 0.7604$, then θ must be $49.5°$ (see the last line of Table 5.11). The modern notation for "an angle whose sine is known" is $\theta = \mathbf{sin^{-1}}\,x$ or $\theta = \mathbf{arcsin}\,x$, where x is the known value for $\sin \theta$. The values for the acute angles $\theta = \sin^{-1}x$, $\theta = \cos^{-1}x$, and $\theta = \tan^{-1}x$ are also programmed into your calculator and are generally accessed using the INV or 2nd keys with the related SIN, COS, or TAN key. With these we are completely equip to find all six measures of a right triangle, given at least one side and any two other measures.

EXAMPLE 4 ▶ **Solving a Right Triangle**

Solve the triangle given in the figure.

Solution ▶ Since the hypotenuse is unknown, we cannot begin with the sine or cosine ratios. The opposite and adjacent

sides for α *are* known, so we use $\tan \alpha$. For $\tan \alpha = \dfrac{17}{25}$

we find $\alpha = \tan^{-1}\left(\dfrac{17}{25}\right) \approx 34.2°$ [verify that

$\tan(34.2°) = 0.6795992982 \approx \dfrac{17}{25}$]. Since α and β are

complements, $\beta \approx 90 - 34.2 = 55.8°$. The Pythagorean theorem shows the hypotenuse is about 30.23 m.

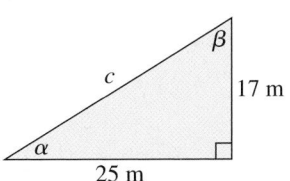

Angles	Sides
$\alpha \approx 34.2°$	$a = 17$
$\beta \approx 55.8°$	$b = 25$
$\gamma = 90°$	$c \approx 30.23$

✓ **C.** You've just learned how to solve a right triangle given two sides

Now try Exercises 23 through 54 ▶

D. Using Cofunctions and Complements to Write Equivalent Expressions

In Figure 5.77, $\angle \alpha$ and $\angle \beta$ must be complements since we have a right triangle, and the sum of the three angles must be $180°$. The complementary angles in a right triangle have a unique relationship that is often used. Specifically $\alpha + \beta = 90°$ means

$\beta = 90° - \alpha$. Note that $\sin \alpha = \dfrac{a}{c}$ and $\cos \beta = \dfrac{a}{c}$. This means

$\sin \alpha = \cos \beta$ or $\sin \alpha = \cos(90° - \alpha)$ by substitution. In words, "The sine of an angle is equal to the cosine of its complement." For this reason sine and cosine are called **cofunctions** (hence the name **co**sine), as are secant/cosecant, and tangent/cotangent. As a test, we use a calculator to check the statement $\sin 52.3° = \cos(90 - 52.3)°$

Figure 5.77

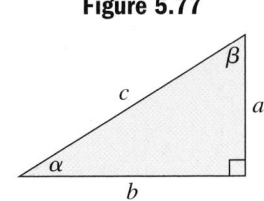

WORTHY OF NOTE

The word *cosine* is actually a shortened form of the words "*complement of sine*," a designation suggested by Edmund Gunter around 1620 since the sine of an angle is equal to the cosine of its complement $[\sin(\theta) = \cos(90° - \theta)]$.

$$\sin 52.3° \overset{?}{=} \cos 37.7°$$
$$0.791223533 = 0.791223533 \checkmark$$

To verify the cofunction relationship for $\sec \theta$ and $\csc \theta$, recall their reciprocal relationship to cosine and sine, respectively.

$$\sec 52.3° \overset{?}{=} \csc \overset{?}{=} 37.7°$$
$$\dfrac{1}{\cos 52.3°} \overset{?}{=} \dfrac{1}{\sin 37.7°}$$
$$1.635250666 = 1.635250666 \checkmark$$

The cofunction relationship for $\tan \theta$ and $\cot \theta$ can similarly be verified.

Summary of Cofunctions

sine and cosine	tangent and cotangent	secant and cosecant
$\sin\theta = \cos(90 - \theta)$	$\tan\theta = \cot(90 - \theta)$	$\sec\theta = \csc(90 - \theta)$
$\cos\theta = \sin(90 - \theta)$	$\cot\theta = \tan(90 - \theta)$	$\csc\theta = \sec(90 - \theta)$

For use in Example 5 and elsewhere in the text, note the expression $\tan^2 15°$ is simply a more convenient way of writing $(\tan 15°)^2$.

EXAMPLE 5 ▶ Applying the Cofunction Relationship

Given $\cot 75° = 2 - \sqrt{3}$ in exact form, find the exact value of $\tan^2 15°$ using a cofunction. Check the result using a calculator.

Solution ▶ Using $\cot 75° = \tan(90° - 75°) = \tan 15°$ gives

$$\cot^2 75° = \tan^2 15° \qquad \text{cofunctions}$$
$$= (2 - \sqrt{3})^2 \qquad \text{substitute known value}$$
$$= 4 - 4\sqrt{3} + 3 \qquad \text{square as indicated}$$
$$= 7 - 4\sqrt{3} \qquad \text{result}$$

☑ **D.** You've just learned how to use cofunctions and complements to write equivalent expressions

Using a calculator, we verify $\tan^2 15° \approx 0.0717967697 \approx 7 - 4\sqrt{3}$.

Now try Exercises 55 through 68 ▶

E. Applications Using Angles of Elevation/Depression

While the name seems self-descriptive, in more formal terms an **angle of elevation** is defined to be the acute angle formed by a **horizontal line of orientation** (parallel to level ground) and the line of sight (see Figure 5.78). An **angle of depression** is likewise defined but involves a line of sight that is below the horizontal line of orientation (Figure 5.79).

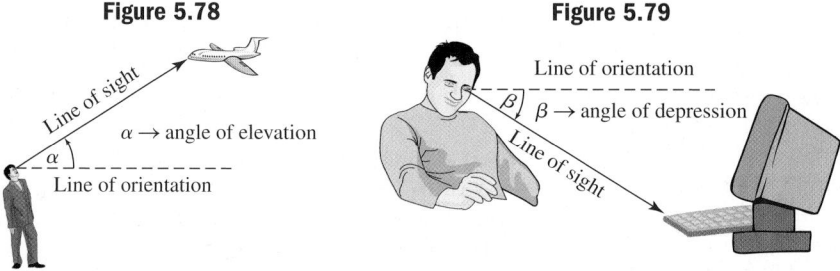

Figure 5.78

Line of sight
$\alpha \rightarrow$ angle of elevation
α
Line of orientation

Figure 5.79

Line of orientation
β $\beta \rightarrow$ angle of depression
Line of sight

Angles of elevation/depression make distance and length computations of all sizes a relatively easy matter and are extensively used by surveyors, engineers, astronomers, and even the casual observer who is familiar with the basics of trigonometry.

EXAMPLE 6 ▶ Applying Angles of Elevation

In Example 4 from Section 5.1, a group of campers used a 45-45-90 triangle to estimate the height of a cliff. It was a time consuming process as they had to wait until mid-morning for the shadow of the cliff to make the needed 45° angle. If the campsite was 250 yd from the base of the cliff and the angle of elevation was 40° at that point, how tall is the cliff?

Solution ▶ As described we want to know the height of the opposite side, given the adjacent side, so we use the tangent function.

For height h:

$$\tan 40° = \frac{h}{250} \quad \tan 40° = \frac{\text{opp}}{\text{adj}}$$

$$250 \tan 40° = h \qquad \text{multiply by 250}$$

$$209.8 \approx h \qquad \text{result } (\tan 40° \approx 0.8391)$$

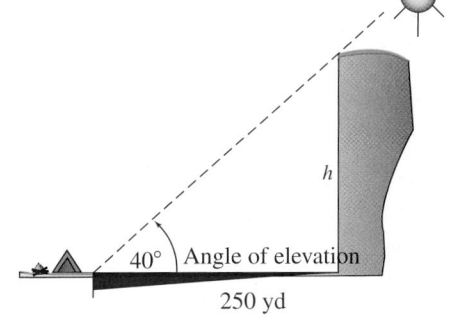

☑ **E. You've just learned how to solve applications involving angles of elevation and depression**

The cliff is approximately 209.8 yd high (about 629 ft).

Now try Exercises 71 through 76 ▶

F. Additional Applications of Right Triangles

In their widest and most beneficial use, the trig functions of acute angles are used with other problem-solving skills, such as drawing a diagram, labeling unknowns, working the solution out in stages, and so on. Example 7 serves to illustrate some of these combinations.

EXAMPLE 7 ▶ **Applying Angles of Elevation and Depression**

From his hotel room window on the sixth floor, Singh notices some window washers high above him on the hotel across the street. Curious as to their height above ground, he quickly estimates the buildings are 50 ft apart, the angle of elevation to the workers is about 80°, and the angle of depression to the base of the hotel is about 50°.

   a. How high above ground is the window of Singh's hotel room?

   b. How high above ground are the workers?

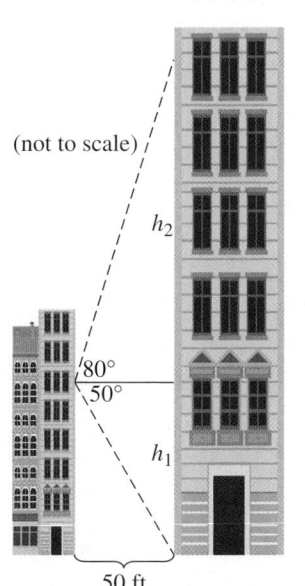

(not to scale)

Solution ▶ **a.** Begin by drawing a diagram of the situation (see figure). To find the height of the window we'll use the tangent ratio, since the adjacent side of the angle is known, and the opposite side is the height we desire.

For the height h_1: $\qquad \tan 50° = \frac{h_1}{50} \quad \tan 50° = \frac{\text{opp}}{\text{adj}}$

$$50 \tan 50° = h_1 \qquad \text{solve for } h_1$$

$$59.6 \approx h_1 \qquad \text{result } (\tan 50° \approx 1.1918)$$

The window is approximately 59.6 ft above ground.

   b. For the height h_2: $\qquad \tan 80° = \frac{h_2}{50} \quad \tan 80° = \frac{\text{opp}}{\text{adj}}$

$$50 \tan 80° = h_2 \qquad \text{solve for } h_2$$

$$283.6 \approx h_2 \qquad \text{result } (\tan 80° \approx 5.6713)$$

The workers are approximately $283.6 + 59.6 = 343.2$ ft above ground.

Now try Exercises 77 through 80 ▶

☑ **F. You've just learned how to solve general applications of right triangles**

There are a number of additional, interesting applications in the exercise set.

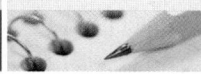

5.6 EXERCISES

▶ CONCEPTS AND VOCABULARY

Fill in each blank with the appropriate word or phrase. Carefully reread the section if needed.

1. The phrase, "an angle whose tangent is known," is written notationally as _____.

2. Given $\sin \theta = \frac{7}{24}$, $\csc \theta =$ _____ because they are _____.

3. The sine of an angle is the ratio of the _____ side to the _____.

4. The cosine of an angle is the ratio of the _____ side to the _____.

5. Discuss/Explain exactly what is meant when you are asked to "solve a triangle." Include an illustrative example.

6. Given an acute angle and the length of the adjacent leg, which four (of the six) trig functions could be used to begin solving the triangle?

▶ DEVELOPING YOUR SKILLS

Use the function value given to determine the value of the other five trig functions of the acute angle θ. Answer in exact form (a diagram will help).

7. $\cos \theta = \dfrac{5}{13}$ 8. $\sin \theta = \dfrac{20}{29}$

9. $\tan \theta = \dfrac{84}{13}$ 10. $\sec \theta = \dfrac{53}{45}$

11. $\cot \theta = \dfrac{2}{11}$ 12. $\cos \theta = \dfrac{2}{3}$

Solve each triangle using trig functions of an acute angle θ. Give a complete answer (in table form) using exact values.

13.

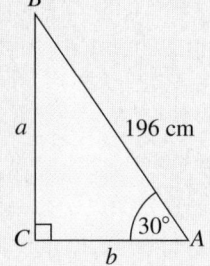

14.

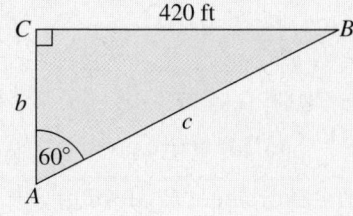

15.

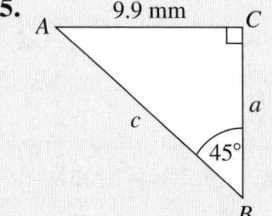

16.

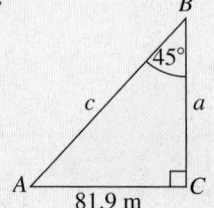

Solve the triangles shown and write answers in table form. Round sides to the nearest 100th of a unit. Verify that angles sum to 180° and that the three sides satisfy (approximately) the Pythagorean theorem.

17.

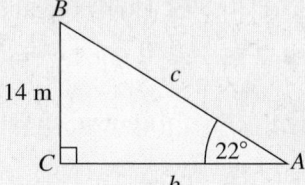

18.

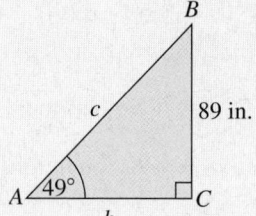

19.

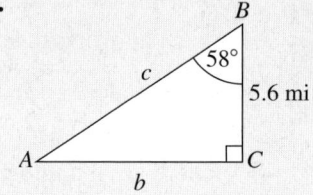

20.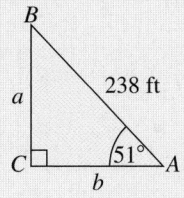

21.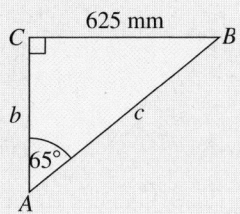

22.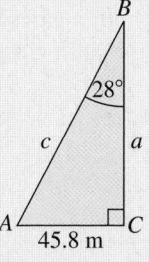

Use a calculator to find the value of each expression, rounded to four decimal places.

23. $\sin 27°$ **24.** $\cos 72°$

25. $\tan 40°$ **26.** $\cot 57.3°$

27. $\sec 40.9°$ **28.** $\csc 39°$

29. $\sin 65°$ **30.** $\tan 84.1°$

Use a calculator to find the acute angle whose corresponding ratio is given. Round to the nearest 10th of a degree. For Exercises 31 through 38, use Exercises 23 through 30 to answer.

31. $\sin A = 0.4540$ **32.** $\cos B = 0.3090$

33. $\tan \theta = 0.8391$ **34.** $\cot A = 0.6420$

35. $\sec B = 1.3230$ **36.** $\csc \beta = 1.5890$

37. $\sin A = 0.9063$ **38.** $\tan B = 9.6768$

39. $\tan \alpha = 0.9896$ **40.** $\cos \alpha = 0.7408$

41. $\sin \alpha = 0.3453$ **42.** $\tan \alpha = 3.1336$

Select an appropriate function to find the angle indicated (round to 10ths of a degree).

43.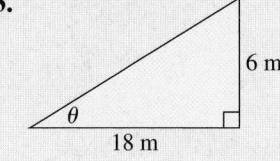

44.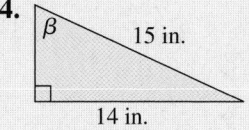

45.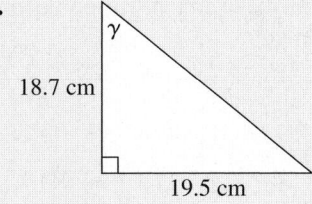

46.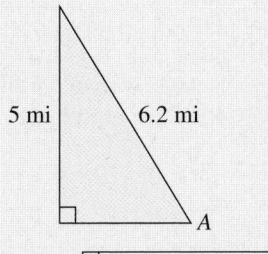

47.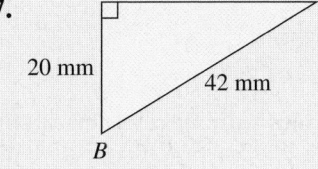

48.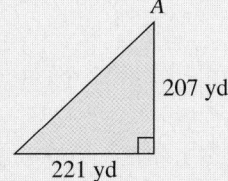

Draw a right triangle ABC as shown, using the information given. Then select an appropriate ratio to find the side indicated. Round to the nearest 100th.

Exercises 49 to 54

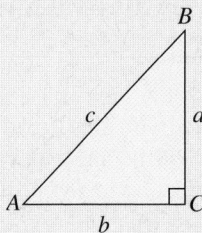

49. $\angle A = 25°$ **50.** $\angle B = 55°$

$c = 52$ mm $b = 31$ ft

find side a find side c

51. $\angle A = 32°$ **52.** $\angle B = 29.6°$

$a = 1.9$ mi $c = 9.5$ yd

find side b find side a

53. $\angle A = 62.3°$ **54.** $\angle B = 12.5°$

$b = 82.5$ furlongs $a = 32.8$ km

find side c find side b

Use a calculator to evaluate each pair of functions and comment on what you notice.

55. sin 25°, cos 65°

56. sin 57°, cos 33°

57. tan 5°, cot 85°

58. sec 40°, csc 50°

Based on your observations in Exercises 55 to 58, fill in the blank so that the functions given are equal.

59. sin 47°, cos ___

60. cos ___, sin 12°

61. cot 69°, tan ___

62. csc 17°, sec ___

Complete the following tables without referring to the text or using a calculator.

63.

θ	$\sin\theta$	$\cos\theta$	$\tan\theta$	$\sin(90-\theta)$
30°				
$\cos(90-\theta)$	$\tan(90-\theta)$	$\csc\theta$	$\sec\theta$	$\cot\theta$

64.

θ	$\sin\theta$	$\cos\theta$	$\tan\theta$	$\sin(90-\theta)$
45°				
$\cos(90-\theta)$	$\tan(90-\theta)$	$\csc\theta$	$\sec\theta$	$\cot\theta$

Evaluate the following expressions without a calculator, using the cofunction relationship and the following exact forms: sec 75° = $\sqrt{6}+\sqrt{2}$; tan 75° = $2+\sqrt{3}$.

65. $\sqrt{6}$ csc 15°

66. $\csc^2 15°$

67. $\cot^2 15°$

68. $\sqrt{3}$ cot 15°

▶ **WORKING WITH FORMULAS**

69. The sine of an angle between two sides of a triangle: $\sin\theta = \dfrac{2A}{ab}$

If the area A and two sides a and b of a triangle are known, the sine of the angle between the two sides is given by the formula shown. Find the angle θ for the triangle below given $A \approx 38.9$, and use it to solve the triangle. (*Hint:* Apply the same concept to angle γ or β.)

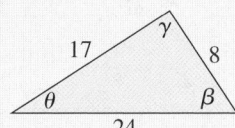

70. Illumination of a surface: $E = \dfrac{I\cos\theta}{d^2}$

The illumination E of a surface by a light source is a measure of the luminous flux per unit area that reaches the surface. The value of E [in lumens (lm) per square foot] is given by the formula shown, where d is the distance from the light source (in feet), I is the intensity of the light [in candelas (cd)], and θ is the angle the light source makes with the vertical. For reading a book, an illumination E of at least 18 lm/ft² is recommended. Assuming the open book is lying on a horizontal surface, how far away should a light source be placed if it has an intensity of 90 cd (about 75 W) and the light flux makes an angle of 65° with the book's surface (i.e., $\theta = 25°$)?

90 cd (about 75 W)

65°

► **APPLICATIONS**

71. Angle of elevation: For a person standing 100 m from the center of the base of the Eiffel Tower, the angle of elevation to the top of the tower is 71.6°. How tall is the Eiffel Tower?

72. Angle of depression: A person standing near the top of the Eiffel Tower notices a car wreck some distance from the tower. If the angle of depression from the person's eyes to the wreck is 32°, how far away is the accident from the base of the tower? See Exercise 71.

73. Angle of elevation: In 2001, the tallest building in the world was the Petronas Tower I in Kuala Lumpur, Malaysia. For a person standing 25.9 ft from the base of the tower, the angle of elevation to the top of the tower is 89°. How tall is the Petronas tower?

74. Angle of depression: A person standing on the top of the Petronas Tower I looks out across the city and pinpoints her residence. If the angle of depression from the person's eyes to her home is 5°, how far away (in feet and in miles) is the residence from the base of the tower? See Exercise 73.

75. Crop duster's speed: While standing near the edge of a farmer's field, Johnny watches a crop duster dust the farmer's field for insect control. Curious as to the plane's speed during each drop, Johnny attempts an estimate using the angle of rotation from one end of the field to the other, while standing 50 ft from one corner. Using a stopwatch he finds the plane makes each pass in 2.35 sec. If the angle of rotation was 83°, how fast (in miles per hour) is the plane flying as it applies the insecticide?

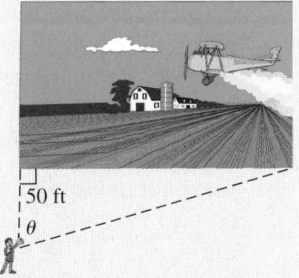

76. Train speed: While driving to their next gig, Josh and the boys get stuck in a line of cars at a railroad crossing as the gates go down. As the sleek, speedy express train approaches, Josh decides to pass the time estimating its speed. He spots a large oak tree beside the track some distance away, and figures the angle of rotation from the crossing to the tree is about 80°. If their car is 60 ft from the crossing and it takes the train 3 sec to reach the tree, how fast is the train moving in miles per hour?

77. Height of a climber: A local Outdoors Club has just hiked to the south rim of a large canyon, when they spot a climber attempting to scale the taller northern face. Knowing the distance between the sheer walls of the northern and southern faces of the canyon is approximately 175 yd, they attempt to compute the distance remaining for the climbers to reach the top of the northern rim. Using a homemade transit, they sight an angle of depression of 55° to the bottom of the north face, and angles of elevation of 24° and 30° to the climbers and top of the northern rim respectively. (a) How high is the southern rim of the canyon? (b) How high is the northern rim? (c) How much farther until the climber reaches the top?

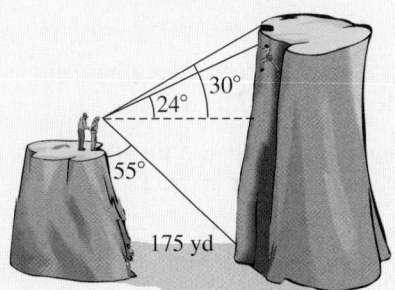

78. Observing wildlife: From her elevated observation post 300 ft away, a naturalist spots a troop of baboons high up in a tree. Using the small transit attached to her telescope, she finds the angle of depression to the bottom of this tree is 14°, while the angle of elevation to the top of the tree is 25°. The angle of elevation to the troop of baboons is 21°. Use this information to find (a) the height of the observation post, (b) the height of the baboons' tree, and (c) the height of the baboons above ground.

79. Angle of elevation: The tallest free-standing tower in the world is the CNN Tower in Toronto, Canada. The tower includes a rotating restaurant high above the ground. From a distance of 500 ft the angle of elevation to the pinnacle of the tower is 74.6°. The angle of elevation to the restaurant from the same vantage point is 66.5°. How tall is the CNN Tower? How far below the pinnacle of the tower is the restaurant located?

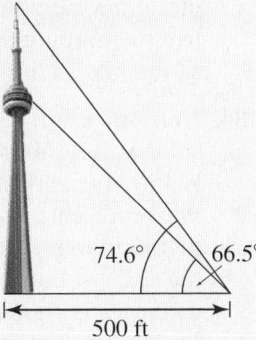

80. Angle of elevation: In August 2004, Taipei 101 captured the record as the world's tallest building, according to the Council on Tall Buildings and Urban Habitat [Source: www.ctbuh.org]. Measured at a point 108 m from its base, the angle of elevation to the top of the spire is 78°. From a distance of about 95 m, the angle of elevation to the top of the roof is also 78°. How tall is Taipei 101 from street level to the top of the spire? How tall is the spire itself?

Alternating current: In AC (alternating current) applications, the relationship between measures known as the impedance (Z), resistance (R), and the phase angle (θ) can be demonstrated using a right triangle. Both the resistance and the impedance are measured in ohms (Ω).

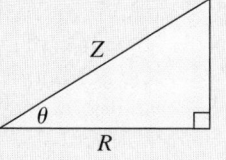

81. Find the impedance Z if the phase angle θ is 34°, and the resistance R is 320 Ω.

82. Find the phase angle θ if the impedance Z is 420 Ω, and the resistance R is 290 Ω.

83. Contour maps: In the figure shown, the *contour interval* is 175 m (each concentric line represents an increase of 175 m in elevation), and the scale of horizontal distances is 1 cm = 500 m. (a) Find the vertical change from A to B (the increase in elevation); (b) use a proportion to find the horizontal change between points A and B if the measured distance on the map is 2.4 cm; and (c) draw the corresponding right triangle and use it to estimate the length of the trail up the mountain side that connects A and B, then use trig to compute the approximate angle of incline as the hiker climbs from point A to point B.

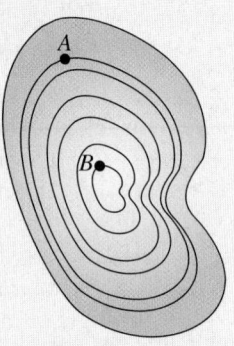

84. Contour maps: In the figure shown, the *contour interval* is 150 m (each concentric line represents an increase of 150 m in elevation), and the scale of horizontal distances is 1 cm = 250 m. (a) Find the vertical change from A to B (the increase in elevation); (b) use a proportion to find the

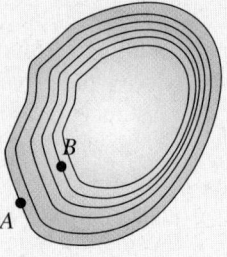

horizontal change between points A and B if the measured distance on the map is 4.5 cm; and (c) draw the corresponding right triangle and use it to estimate the length of the trail up the mountain side that connects A and B, then use trig to compute the approximate angle of incline as the hiker climbs from point A to point B.

85. Height of a rainbow: While visiting the Lapahoehoe Memorial on the island of Hawaii, Bruce and Carma see a spectacularly vivid rainbow arching over the bay. Bruce speculates the rainbow is 500 ft away, while Carma estimates the angle of elevation to the highest point of the rainbow is about 42°. What was the approximate height of the rainbow?

86. High-wire walking: As part of a circus act, a high-wire walker not only "walks the wire," she walks a wire that is *set at an incline of* 10° to the horizontal! If the length of the (inclined) wire is 25.39 m, (a) how much higher is the wire set at the destination pole than at the departure pole? (b) How far apart are the poles?

87. Diagonal of a cube: A cubical box has a diagonal measure of 35 cm. (a) Find the dimensions of the box and (b) the angle θ that the diagonal makes at the lower corner of the box.

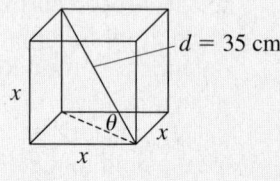

88. Diagonal of a rectangular parallelepiped: A rectangular box has a width of 50 cm and a length of 70 cm. (a) Find the height h that ensures the diagonal across the middle of the box will be 90 cm and (b) the angle θ that the diagonal makes at the lower corner of the box.

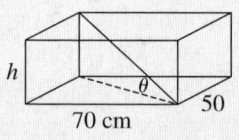

▶ **EXTENDING THE CONCEPT**

89. The formula $h = \dfrac{d}{\cot u - \cot v}$ can be used to calculate the height h of a building when distance x is unknown but distance d is known (see the diagram). Use the ratios for $\cot u$ and $\cot v$ to derive the formula (note x is "absent" from the formula).

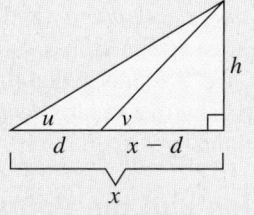

90. Use the diagram given to derive a formula for the height h of the taller building in terms of the height x of the shorter building and the ratios for $\tan u$ and $\tan v$. Then use the formula to find h given the shorter building is 75 m tall with $u = 40°$ and $v = 50°$.

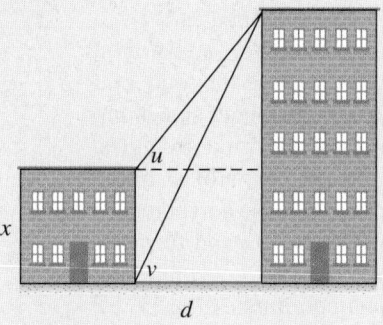

91. The radius of the Earth at the equator (0° N latitude) is approximately 3960 mi. Beijing, China, is located at 39.5° N latitude, 116° E longitude. Philadelphia, Pennsylvania, is located at the same latitude, but at 75° W longitude. (a) Use the diagram given and a cofunction relationship to find the radius r of the Earth (parallel to the equator) at this latitude; (b) use the arc length formula to compute the *shortest distance* between these two cities along this latitude; and (c) if the supersonic Concorde flew a direct flight between Beijing and Philadelphia along this latitude, approximate the flight time assuming a cruising speed of 1250 mph. Note: The shortest distance is actually traversed by heading northward, using the arc of a "great circle" that goes through these two cities.

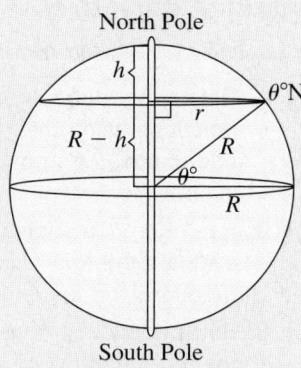

▶ **MAINTAINING YOUR SKILLS**

92. (1.5) Solve by factoring:
 a. $g^2 - 9g = 0$
 b. $g^2 - 9 = 0$
 c. $g^2 - 9g - 10 = 0$
 d. $g^2 + 9g - 10 = 0$
 e. $g^3 - 9g^2 - 10g + 90 = 0$

93. (2.5) For the graph of $T(x)$ given, (a) name the local maximums and minimums, (b) the zeroes of T, (c) intervals where $T(x)\downarrow$ and $T(x)\uparrow$, and (d) intervals where $T(x) > 0$ and $T(x) < 0$.

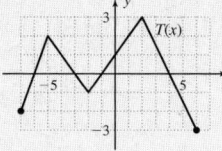

94. (5.1) The armature for the rear windshield wiper has a length of 24 in., with a rubber wiper blade that is 20 in. long. What area of my rear windshield is cleaned as the armature swings back-and-forth through an angle of 110°?

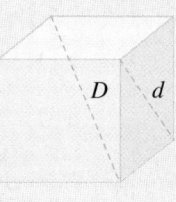

95. (5.1) The boxes used to ship some washing machines are perfect cubes with edges measuring 38 in. Use a special triangle to find the length of the diagonal d of one side, and the length of the interior diagonal D (through the middle of the box).

Learning Objectives

In Section 5.7 you will learn how to:

☐ **A.** Define the trigonometric functions using the coordinates of a point in QI

☐ **B.** Use reference angles to evaluate the trig functions for any angle

☐ **C.** Solve applications using the trig functions of any angle

This section tends to bridge the study of *static trigonometry* and the angles of a right triangle, with the study of *dynamic trigonometry* and the unit circle. This is accomplished by noting that the domain of the trig functions (unlike a triangle point of view) *need not be restricted to acute angles*. We'll soon see that the domain can be extended to include trig functions of *any* angle, a view that greatly facilitates our work in Chapter 7, where many applications involve angles greater than 90°.

A. Trigonometric Ratios and the Point $P(x, y)$

Regardless of where a right triangle is situated or how it is oriented, each trig function can be defined as a given ratio of sides with respect to a given angle. In this light, consider a 30-60-90 triangle placed in the first quadrant with the 30° angle at the origin and the longer side along the x-axis. From our previous review of similar triangles, the trig ratios will have the same value regardless of the triangle's size so for convenience, we'll use a hypotenuse of 10. This gives sides of 5, $5\sqrt{3}$, and 10, and from the diagram in Figure 5.80 we note the point (x, y) marking the vertex of the 60° angle has coordinates $(5\sqrt{3}, 5)$.

Figure 5.80

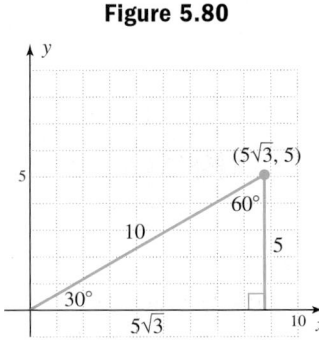

Further, the diagram shows that sin 30°, cos 30°, and tan 30° can all be expressed in terms of these coordinates since $\dfrac{\text{opp}}{\text{hyp}} = \dfrac{5}{10} = \dfrac{y}{r}$(sine), $\dfrac{\text{adj}}{\text{hyp}} = \dfrac{5\sqrt{3}}{10} = \dfrac{x}{r}$(cosine), and

$\dfrac{\text{opp}}{\text{adj}} = \dfrac{5}{5\sqrt{3}} = \dfrac{y}{x}$(tangent), where r is the length of the hypotenuse. Each result

reduces to the more familiar values seen earlier: $\sin 30° = \dfrac{1}{2}$, $\cos 30° = \dfrac{\sqrt{3}}{2}$, and

$\tan 30° = \dfrac{1}{\sqrt{3}} = \dfrac{\sqrt{3}}{3}$. This suggests we can define the six trig functions in terms

of x, y, and r, where $r = \sqrt{x^2 + y^2}$.

Consider that the slope of the line coincident with the hypotenuse is $\dfrac{\text{rise}}{\text{run}} = \dfrac{5}{5\sqrt{3}} = \dfrac{\sqrt{3}}{3}$, and since the line goes through the origin its equation must be

$y = \dfrac{\sqrt{3}}{3}x$. Any point (x, y) on this line will be at the 60° vertex of a right triangle

formed by drawing a perpendicular line from the point (x, y) to the x-axis. As Example 1 shows, we obtain the special values for sin 30°, cos 30°, and tan 30° regardless of the point chosen.

EXAMPLE 1 ▶ Evaluating Trig Functions Using x, y, and r

Pick an arbitrary point in QI that satisfies $y = \dfrac{\sqrt{3}}{3}x$,

then draw the corresponding right triangle and

evaluate sin 30°, cos 30°, and tan 30°.

Solution ▶ The coefficient of x has a denominator of 3, so we choose a multiple of 3 for convenience. For $x = 6$

we have $y = \dfrac{\sqrt{3}}{3}(6) = 2\sqrt{3}$. As seen in the figure,

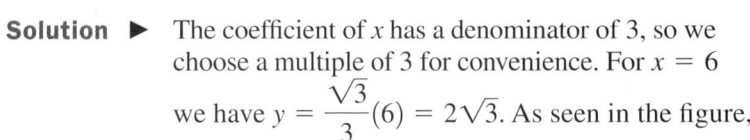

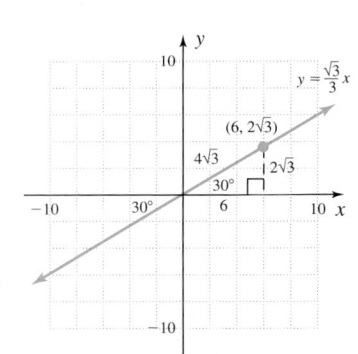

the point $(6, 2\sqrt{3})$ is on the line and at the vertex of the 60° angle. Evaluating the trig functions at 30°, we obtain:

$$\sin 30° = \frac{y}{r} = \frac{2\sqrt{3}}{4\sqrt{3}} \qquad \cos 30° = \frac{x}{r} = \frac{6}{4\sqrt{3}} \qquad \tan 30° = \frac{y}{x} = \frac{2\sqrt{3}}{6}$$

$$= \frac{1}{2} \qquad\qquad = \frac{6\sqrt{3}}{4\sqrt{3}\sqrt{3}} = \frac{\sqrt{3}}{2} \qquad\qquad = \frac{\sqrt{3}}{3}$$

Now try Exercises 7 and 8 ▶

In general, consider *any* two points (x, y) and (X, Y) on an arbitrary line $y = kx$, at corresponding distances r and R from the origin (Figure 5.81). Because the triangles formed are similar, we have $\dfrac{y}{x} = \dfrac{Y}{X}, \dfrac{x}{r} = \dfrac{X}{R}$, and so on, and we conclude that the value of the trig functions are indeed independent of the point chosen.

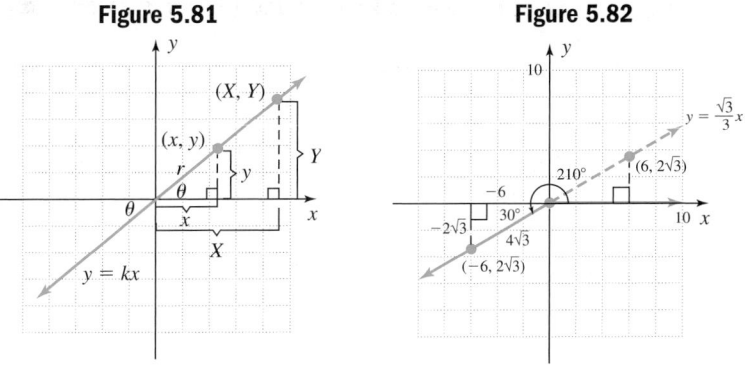

Figure 5.81   **Figure 5.82**

Viewing the trig functions in terms of x, y, and r produces significant results. In Figure 5.82, we note the line $y = \dfrac{\sqrt{3}}{3}x$ from Example 1 also extends into QIII, and *creates another* 30° *angle whose vertex is at the origin* (since vertical angles are equal). The sine, cosine, and tangent functions can still be evaluated for this angle, but in QIII both x and y are negative. If we consider the angle in QIII to be a positive rotation of 210° (180° + 30°), we can evaluate the trig functions using the values of x, y, and r from any point on the terminal side, since these are fixed by the 30° angle created and are the same as those in QI except for their sign:

$$\sin 210° = \frac{y}{r} = \frac{-2\sqrt{3}}{4\sqrt{3}} \qquad \cos 210° = \frac{x}{r} = \frac{-6}{4\sqrt{3}} \qquad \tan 210° = \frac{y}{x} = \frac{-2\sqrt{3}}{-6}$$

$$= -\frac{1}{2} \qquad\qquad = -\frac{\sqrt{3}}{2} \qquad\qquad = \frac{\sqrt{3}}{3}$$

For *any* rotation θ and a point (x, y) on the terminal side, the distance r can be found using $r = \sqrt{x^2 + y^2}$ and the six trig functions likewise evaluated. Note that evaluating them correctly depends on the quadrant of the terminal side, since this will dictate the signs for x and y. Students are strongly encouraged to make these quadrant and sign observations the *first step* in any solution process. In summary, we have

Trigonometric Functions of Any Angle

Given $P(x, y)$ is any point on the terminal side of angle θ in standard position, with $r = \sqrt{x^2 + y^2} \,(r > 0)$ the distance from the origin to (x, y). The six trigonometric functions of θ are

$$\sin \theta = \frac{y}{r} \qquad \cos \theta = \frac{x}{r} \qquad \tan \theta = \frac{y}{x}$$

$$x \neq 0$$

$$\csc \theta = \frac{r}{y} \qquad \sec \theta = \frac{r}{x} \qquad \cot \theta = \frac{x}{y}$$

$$y \neq 0 \qquad\qquad x \neq 0 \qquad\qquad y \neq 0$$

EXAMPLE 2 ▶ **Evaluating Trig Functions Given the Terminal Side is on $y = mx$**

Given that $P(x, y)$ is a point on the terminal side of angle θ in standard position, find the value of $\sin \theta$ and $\cos \theta$, if

  a. The terminal side is in QII and coincident with the line $y = -\frac{12}{5}x$,
  b. The terminal side is in QIV and coincident with the line $y = -\frac{12}{5}x$.

Solution ▶ **a.** Select any convenient point in QII that satisfies this equation. We select $x = -5$ since x is negative in QII, which gives $y = 12$ and the point $(-5, 12)$.

Solving for r gives $r = \sqrt{(-5)^2 + (12)^2} = 13$. The ratios are

$$\sin \theta = \frac{y}{r} = \frac{12}{13} \qquad \cos \theta = \frac{x}{r} = \frac{-5}{13}$$

  b. In QIV we select $x = 10$ since x is positive in QIV, giving $y = -24$ and the point $(10, -24)$. Solving for r gives $r = \sqrt{(10)^2 + (-24)^2} = 26$. The ratios are

$$\sin \theta = \frac{y}{r} = \frac{-24}{26} \qquad \cos \theta = \frac{x}{r} = \frac{10}{26}$$

$$= -\frac{12}{13} \qquad\qquad = \frac{5}{13}$$

Now try Exercises 9 through 12 ▶

In Example 2, note the ratios are the same in QII and QIV *except for their sign.* We will soon use this observation to great advantage.

EXAMPLE 3 ▶ **Evaluating Trig Functions Given a Point P**

Find the value of the six trigonometric functions given $P(-5, 5)$ is on the terminal side of angle θ in standard position.

Solution ▶ For $P(-5, 5)$ we have $x < 0$ and $y > 0$ so the terminal side is in QII. Solving for r yields $r = \sqrt{(-5)^2 + (5)^2} = \sqrt{50} = 5\sqrt{2}$. For $x = -5$, $y = 5$, and $r = 5\sqrt{2}$, we obtain

$$\sin \theta = \frac{y}{r} = \frac{5}{5\sqrt{2}} \qquad \cos \theta = \frac{x}{r} = \frac{-5}{5\sqrt{2}} \qquad \tan \theta = \frac{y}{x} = \frac{5}{-5}$$

$$= \frac{\sqrt{2}}{2} \qquad\qquad = -\frac{\sqrt{2}}{2} \qquad\qquad = -1$$

The remaining functions can be evaluated using reciprocals.

$$\csc\theta = \frac{2}{\sqrt{2}} = \sqrt{2} \qquad \sec\theta = -\frac{2}{\sqrt{2}} = -\sqrt{2} \qquad \cot\theta = -1$$

Note the connection between these results and the special values for $\theta = 45°$.

Now try Exercises 13 through 28 ▶

Figure 5.83

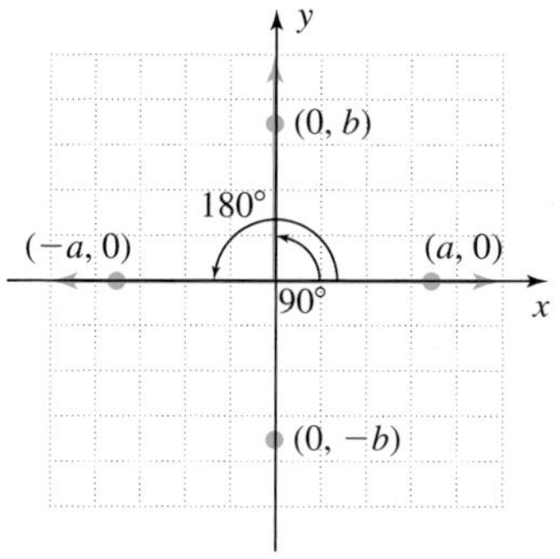

Now that we've defined the trig functions in terms of ratios involving x, y, and r, the question arises as to their value at the quadrantal angles. For 90° and 270°, any point on the terminal side of the angle has an *x-value* of zero, meaning $\tan 90°$, $\sec 90°$, $\tan 270°$, and $\sec 270°$ are all undefined since $x = 0$ is in the denominator. Similarly, at 180° and 360°, the *y*-value of any point on the terminal side is zero, so $\cot 180°$, $\csc 180°$, $\cot 360°$, and $\csc 360°$ are likewise undefined (see Figure 5.83).

EXAMPLE 4 ▶ **Evaluating the Trig Functions for $\theta = 90°k$, k an Integer**

Evaluate the six trig functions for $\theta = 270°$.

Solution ▶ Here, θ is the quadrantal angle whose terminal side separates QIII and QIV. Since the evaluation is independent of the point chosen on this side, we choose $(0, -1)$ for convenience, giving $r = 1$. For $x = 0$, $y = -1$, and $r = 1$ we obtain

$$\sin\theta = \frac{-1}{1} = -1 \qquad \cos\theta = \frac{0}{-1} = 0 \qquad \tan\theta = \frac{-1}{0}\ (\textit{undefined})$$

The remaining ratios can be evaluated using reciprocals.

$$\csc\theta = -1 \qquad \sec\theta = \frac{-1}{0}\ (\textit{undefined}) \qquad \cot\theta = \frac{0}{-1} = 0$$

Now try Exercises 29 and 30 ▶

Results for the quadrantal angles are summarized in Table 5.12.

Table 5.12

θ	$\sin\theta = \dfrac{y}{r}$	$\cos\theta = \dfrac{x}{r}$	$\tan\theta = \dfrac{y}{x}$	$\csc\theta = \dfrac{r}{y}$	$\sec\theta = \dfrac{r}{x}$	$\cot\theta = \dfrac{x}{y}$
$0° \to (1, 0)$	0	1	0	undefined	1	undefined
$90° \to (0, 1)$	1	0	undefined	1	undefined	0
$180° \to (-1, 0)$	0	-1	0	undefined	-1	undefined
$270° \to (0, -1)$	-1	0	undefined	-1	undefined	0

☑ **A.** You've just learned how to define the trigonometric functions using the coordinates of a point in QI

B. Reference Angles and the Trig Functions of Any Angle

Recall that for any angle θ in standard position, the acute angle θ_r formed by the terminal side and the *x*-axis is called the **reference angle**. Several examples of this definition are illustrated in Figures 5.84 through 5.87 for $\theta > 0$ in degrees.

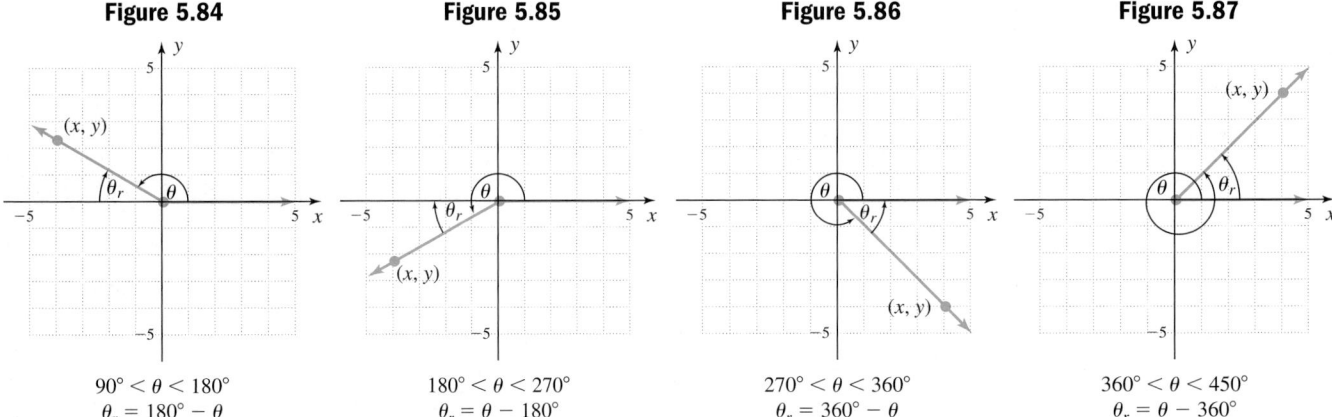

Figure 5.84

$90° < \theta < 180°$
$\theta_r = 180° - \theta$

Figure 5.85

$180° < \theta < 270°$
$\theta_r = \theta - 180°$

Figure 5.86

$270° < \theta < 360°$
$\theta_r = 360° - \theta$

Figure 5.87

$360° < \theta < 450°$
$\theta_r = \theta - 360°$

EXAMPLE 5 ► **Finding Reference Angles**

Determine the reference angle for

    a. 315°   **b.** 150°   **c.** −121°   **d.** 425°

Solution ► Begin by mentally visualizing each angle and the quadrant where it terminates.

    a. 315° is a QIV angle: **c.** −121° is a QIII angle:
      $\theta_r = 360° - 315° = 45°$ $\theta_r = 180° - 121° = 59°$

    b. 150° is a QII angle: **d.** 425° is a QI angle:
      $\theta_r = 180° - 150° = 30°$ $\theta_r = 425° - 360° = 65°$

Now try Exercises 31 through 42 ►

The reference angles from Examples 5(a) and 5(b) were special angles, which means we automatically know the absolute value of the trig ratios using θ_r. The best way to remember the signs of the trig functions is to keep in mind that sine is associated with y, cosine with x, and tangent with both x and y (r is always positive). In addition, there are several mnemonic devices (memory tools) to assist you. One is to use the first letter of the function that is positive in each quadrant and create a catchy acronym. For instance **ASTC** → **A**ll **S**tudents **T**ake **C**lasses (see Figure 5.88). Note that a trig function and its reciprocal function will always have the same sign.

Figure 5.88

Quadrant II	Quadrant I
Sine is positive	All are positive
Tangent is positive	Cosine is positive
Quadrant III	Quadrant IV

EXAMPLE 6 ► **Evaluating Trig Functions Using θ_r**

Use a reference angle to evaluate sin θ, cos θ, and tan θ for $\theta = 315°$.

Solution ► The terminal side is in QIV where x is positive and y is negative. With $\theta_r = 45°$, we have:

$$\sin 315° = -\frac{\sqrt{2}}{2} \quad \cos 315° = \frac{\sqrt{2}}{2}$$

$$\tan 315° = -1$$

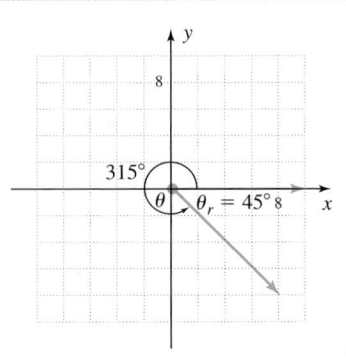

Now try Exercises 43 through 54 ►

EXAMPLE 7 ▶ Finding Function Values Using a Quadrant and Sign Analysis

Given $\sin \theta = \dfrac{5}{13}$ and $\cos \theta < 0$, find the value of the other ratios.

Solution ▶ Always begin with a quadrant and sign analysis: $\sin \theta$ is positive in QI and QII, while $\cos \theta$ is negative in QII and QIII. Both conditions are satisfied in QII only. For $r = 13$ and $y = 5$, the Pythagorean theorem shows $x = \sqrt{13^2 - 5^2} = \sqrt{144} = 12$.

With θ in QII this gives $\cos \theta = \dfrac{-12}{13}$ and $\tan \theta = \dfrac{5}{-12}$. The reciprocal values are

$\csc \theta = \dfrac{13}{5}$, $\sec \theta = \dfrac{13}{-12}$, and $\cot \theta = \dfrac{-12}{5}$.

Now try Exercises 55 through 62 ▶

In our everyday experience, there are many actions and activities where angles greater than or equal to 360° are applied. Some common instances are a professional basketball player who "does a three-sixty" (360°) while going to the hoop, a diver who completes a "two-and-a-half" (900°) off the high board, and a skater who executes a perfect triple axel ($3\frac{1}{2}$ turns or 1260°). As these examples suggest, angles greater than 360° must still terminate on a quadrantal axis, or in one of the four quadrants, allowing a reference angle to be found and the functions to be evaluated for any angle *regardless of size*. Figure 5.89 illustrates that $\alpha = 135°$, $\beta = -225°$, and $\theta = 495°$ are all coterminal, with *each having a reference angle of 45°.*

Figure 5.89

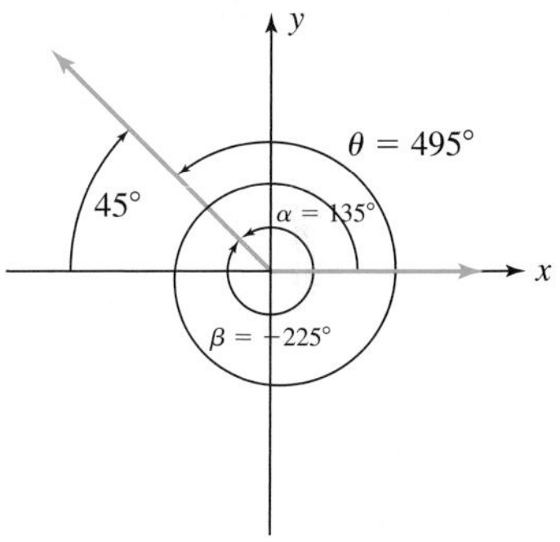

EXAMPLE 8 ▶ Evaluating Trig Functions of Any Angle

Evaluate $\sin 135°$, $\cos (-225°)$, and $\tan 495°$.

Solution ▶ The angles are coterminal and terminate in QII, where $x < 0$ and $y > 0$. With $\theta_r = 45°$ we have $\sin 135° = \dfrac{\sqrt{2}}{2}$, $\cos (-225°) = -\dfrac{\sqrt{2}}{2}$, and $\tan 495° = -1$.

Now try Exercises 63 through 74 ▶

☑ **B. You've just learned how to use reference angles to evaluate the trig functions for any angle**

Since 360° is one full rotation, all angles $\theta + 360°k$ will be coterminal for any integer k. For angles with a very large magnitude, we can find the quadrant of the terminal side by subtracting as many integer multiples of 360° as needed from the angle. For $\alpha = 1908°$, $\dfrac{1908}{360} = 5.3$ and $1908 - 360(5) = 108°$. This angle is in QII with $\theta_r = 72°$. **See Exercises 75 through 90.**

C. Applications of the Trig Functions of Any Angle

One of the most basic uses of coterminal angles is determining all values of θ that satisfy a stated relationship. For example, by now you are aware that if $\sin \theta = \dfrac{1}{2}$

Figure 5.90

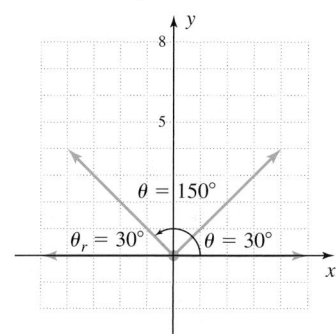

(positive one-half), then $\theta = 30°$ or $\theta = 150°$ (see Figure 5.90). But this is also true for all angles coterminal with these two, and we would write the solutions as $\theta = 30° + 360°k$ and $\theta = 150° + 360°k$ for all integers k.

EXAMPLE 9 ▶ **Finding All Angles that Satisfy a Given Equation**

Find all angles satisfying the relationship given. Answer in degrees.

a. $\cos \theta = -\dfrac{\sqrt{2}}{2}$ **b.** $\tan \theta = -1.3764$

Solution ▶ **a.** Cosine is negative in QII and QIII. Recognizing $\cos 45° = \dfrac{\sqrt{2}}{2}$, we reason $\theta_r = 45°$ and two solutions are $\theta = 135°$ from QII and $\theta = 225°$ from QIII. For all values of θ satisfying the relationship, we have $\theta = 135° + 360°k$ and $\theta = 225° + 360°k$. See Figure 5.91.

Figure 5.91

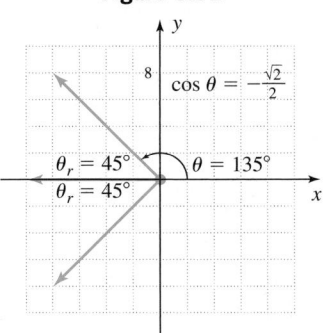

Figure 5.92

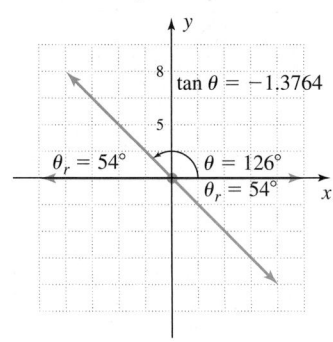

b. Tangent is negative in QII and QIV. For -1.3764 we find θ_r using a calculator:

[2nd] [TAN] (**tan^{-1}**) -1.3764 [ENTER] shows $\tan^{-1}(-1.3764) \approx -54$, so $\theta_r = 54°$.

Two solutions are $\theta = 180° - 54° = 126°$ from QII, and in QIV $\theta = 360° - 54° = 306°$. The result is $\theta = 126° + 360°k$ and $\theta = 306° + 360°k$. Note these can be combined into the single statement $\theta = 126° + 180°k$. See Figure 5.92.

Now try Exercises 93 through 100 ▶

We close this section with an additional application of the concepts related to trigonometric functions of any angle.

EXAMPLE 10 ▶ **Applications of Coterminal Angles: Location on Radar**

A radar operator calls the captain over to her screen saying, "Sir, we have an unidentified aircraft heading 20° (20° east of due north or a standard 70° rotation). I think it's a UFO." The captain asks, "What makes you think so?" To which the sailor replies, "Because it's at 5000 ft and not moving!" Name all angles for which the UFO causes a "blip" to occur on the radar screen.

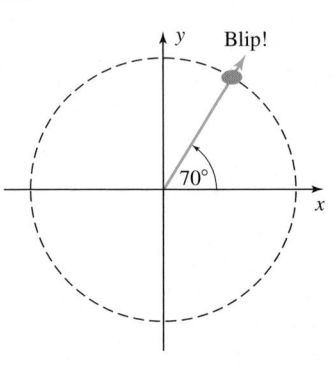

Solution ▶ Since radar typically sweeps out a 360° angle, a blip will occur on the screen for all angles $\theta = 70° + 360°k$, where k is an integer.

Now try Exercises 101 through 106 ▶

☑ **C.** You've just learned how to solve applications using the trig functions of any angle

5.7 EXERCISES

▶ CONCEPTS AND VOCABULARY

Fill in each blank with the appropriate word or phrase. Carefully reread the section if needed.

1. An angle is in standard position if its vertex is at the _____ and the initial side is along the _____ .

2. A(n) _____ angle is one where the _____ side is coincident with one of the coordinate axes.

3. Angles formed by a counterclockwise rotation are _____ angles. Angles formed by a _____ rotation are negative angles.

4. For any angle θ, its reference angle θ_r is the positive _____ angle formed by the _____ side and the nearest x-axis.

5. Discuss the similarities and differences between the trigonometry of right triangles and the trigonometry of *any* angle.

6. Let $T(x)$ represent any one of the six basic trig functions. Explain why the equation $T(x) = k$ will always have exactly two solutions in $[0, 2\pi)$ if x is not a quadrantal angle.

▶ DEVELOPING YOUR SKILLS

7. Draw a 30-60-90 triangle with the 60° angle at the origin and the short side along the positive x-axis. Determine the slope and equation of the line coincident with the hypotenuse, then pick any point on this line and evaluate sin 60°, cos 60°, and tan 60°. Comment on what you notice.

8. Draw a 45-45-90 triangle with a 45° angle at the origin and one side along the positive x-axis. Determine the slope and equation of the line coincident with the hypotenuse, then pick any point on this line and evaluate sin 45°, cos 45°, and tan 45. Comment on what you notice.

Graph each linear equation and state the quadrants it traverses. Then pick one point on the line from each quadrant and evaluate the functions sin θ, cos θ and tan θ using these points.

9. $y = \dfrac{3}{4}x$ 10. $y = \dfrac{5}{12}x$

11. $y = -\dfrac{\sqrt{3}}{3}x$ 12. $y = -\dfrac{\sqrt{3}}{2}x$

Find the value of the six trigonometric functions given $P(x, y)$ is on the terminal side of angle θ, with θ in standard position.

13. $(8, 15)$ 14. $(7, 24)$

15. $(-20, 21)$ 16. $(-3, -1)$

17. $(7.5, -7.5)$ 18. $(9, -9)$

19. $(4\sqrt{3}, 4)$ 20. $(-6, 6\sqrt{3})$

21. $(2, 8)$ 22. $(6, -15)$

23. $(-3.75, -2.5)$ 24. $(6.75, 9)$

25. $\left(-\dfrac{5}{9}, \dfrac{2}{3}\right)$ 26. $\left(\dfrac{3}{4}, -\dfrac{7}{16}\right)$

27. $\left(\dfrac{1}{4}, -\dfrac{\sqrt{5}}{2}\right)$ 28. $\left(-\dfrac{\sqrt{3}}{5}, \dfrac{22}{25}\right)$

29. Evaluate the six trig functions in terms of x, y, and r for $\theta = 90°$.

30. Evaluate the six trig functions in terms of x, y, and r for $\theta = 180°$.

Name the reference angle θ_r for the angle θ given.

31. $\theta = 120°$ 32. $\theta = 210°$

33. $\theta = 135°$ 34. $\theta = 315°$

35. $\theta = -45°$ 36. $\theta = -240°$

37. $\theta = 112°$ 38. $\theta = 179°$

39. $\theta = 500°$ 40. $\theta = 750°$

41. $\theta = -168.4°$ 42. $\theta = -328.2°$

State the quadrant of the terminal side of θ, using the information given.

43. $\sin \theta > 0, \cos \theta < 0$

44. $\cos \theta < 0, \tan \theta < 0$

45. $\tan \theta < 0, \sin \theta > 0$

46. $\sec \theta > 0, \tan \theta > 0$

Find the exact value of $\sin \theta$, $\cos \theta$, and $\tan \theta$ using reference angles.

47. $\theta = 330°$

48. $\theta = 390°$

49. $\theta = -45°$

50. $\theta = -120°$

51. $\theta = 240°$

52. $\theta = 315°$

53. $\theta = -150°$

54. $\theta = -210°$

For the information given, find the values of x, y, and r. Clearly indicate the quadrant of the terminal side of θ, then state the values of the six trig functions of θ.

55. $\cos \theta = \dfrac{4}{5}$ and $\sin \theta < 0$

56. $\tan \theta = -\dfrac{12}{5}$ and $\cos \theta > 0$

57. $\csc \theta = -\dfrac{37}{35}$ and $\tan \theta > 0$

58. $\sin \theta = -\dfrac{20}{29}$ and $\cot \theta < 0$

59. $\csc \theta = 3$ and $\cos \theta > 0$

60. $\csc \theta = -2$ and $\cos \theta > 0$

61. $\sin \theta = -\dfrac{7}{8}$ and $\sec \theta < 0$

62. $\cos \theta = \dfrac{5}{12}$ and $\sin \theta < 0$

Find two positive and two negative angles that are coterminal with the angle given. Answers will vary.

63. $52°$

64. $12°$

65. $87.5°$

66. $22.8°$

67. $225°$

68. $175°$

69. $-107°$

70. $-215°$

Evaluate in exact form as indicated.

71. $\sin 120°, \cos -240°, \tan 480°$

72. $\sin 225°, \cos 585°, \tan -495°$

73. $\sin -30°, \cos -390°, \tan -690°$

74. $\sin 210°, \cos 570°, \tan -150°$

Find the exact value of $\sin \theta$, $\cos \theta$, and $\tan \theta$ using reference angles.

75. $\theta = 600°$

76. $\theta = 480°$

77. $\theta = -840°$

78. $\theta = -930°$

79. $\theta = 570°$

80. $\theta = 495°$

81. $\theta = -1230°$

82. $\theta = 3270°$

For each exercise, state the quadrant of the terminal side and the sign of the function in that quadrant. Then evaluate the expression using a calculator. Round to four decimal places.

83. $\sin 719°$

84. $\cos 528°$

85. $\tan -419°$

86. $\sec -621°$

87. $\csc 681°$

88. $\tan 995°$

89. $\cos 805°$

90. $\sin 772°$

▶ WORKING WITH FORMULAS

91. The area of a parallelogram: $A = ab \sin \theta$

The area of a parallelogram is given by the formula shown, where a and b are the lengths of the sides and θ is the angle between them. Use the formula to complete the following: (a) find the area of a parallelogram with sides $a = 9$ and $b = 21$ given $\theta = 50°$. (b) What is the smallest integer value of θ where the area is greater than 150 units^2? (c) State what happens when $\theta = 90°$. (d) How can you find the area of a triangle using this formula?

92. The angle between two intersecting lines:

$$\tan \theta = \frac{m_2 - m_1}{1 + m_2 m_1}$$

Given line 1 and line 2 with slopes m_1 and m_2, respectively, the angle between the two lines is given by the formula shown. Find the angle θ if the equation of line 1 is $y_1 = \frac{3}{4}x + 2$ and line 2 has equation $y_2 = -\frac{2}{3}x + 5$.

▶ APPLICATIONS

Find all angles satisfying the stated relationship. For standard angles, express your answer in exact form. For nonstandard values, use a calculator and round function values to tenths.

93. $\cos \theta = \dfrac{1}{2}$ **94.** $\sin \theta = \dfrac{\sqrt{2}}{2}$

95. $\sin \theta = -\dfrac{\sqrt{3}}{2}$ **96.** $\tan \theta = -\dfrac{\sqrt{3}}{1}$

97. $\sin \theta = 0.8754$ **98.** $\cos \theta = 0.2378$

99. $\tan \theta = -2.3512$ **100.** $\cos \theta = -0.0562$

101. Nonacute angles: At a recent carnival, one of the games on the midway was played using a large spinner that turns clockwise. On Jorge's spin the number 25 began at the 12 o'clock (top/center) position, returned to this position five times during the spin and stopped at the 3 o'clock position. What angle θ did the spinner spin through? Name all angles that are coterminal with θ.

Exercise 101

102. Nonacute angles: One of the four blades on a ceiling fan has a decal on it and begins at a designated "12 o'clock" position. Turning the switch on and then immediately off, causes the blade to make over three complete, counterclockwise rotations, with the blade stopping at the 8 o'clock position. What angle θ did the blade turn through? Name all angles that are coterminal with θ.

103. High dives: As part of a diving competition, David executes a perfect reverse two-and-a-half flip. Does he enter the water feet first or head first? Through what angle did he turn from takeoff until the moment he entered the water?

Exercise 103

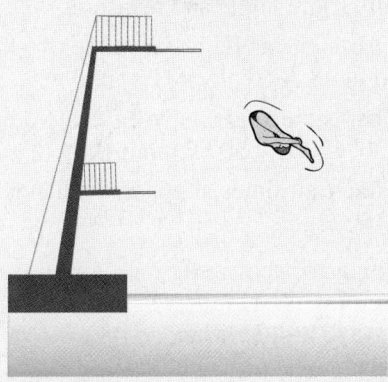

104. Gymnastics: While working out on a trampoline, Charlene does three complete, forward flips and then belly-flops on the trampoline before returning to the upright position. What angle did she turn through from the start of this maneuver to the moment she belly-flops?

105. Spiral of Archimedes: The graph shown is called the spiral of Archimedes. Through what angle θ has the spiral turned, given the spiral terminates at $(6, -2)$ as indicated?

Exercise 105

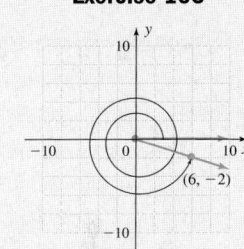

106. Involute of a circle: The graph shown is called the involute of a circle. Through what angle θ has the involute turned, given the graph terminates at $(-4, -3.5)$ as indicated?

Exercise 106

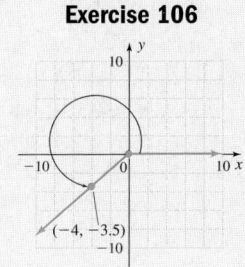

Area bounded by chord and circumference: Find the area of the shaded region, rounded to the nearest 100th. Note the area of a triangle is one-half the area of a parallelogram (see Exercise 91).

107.

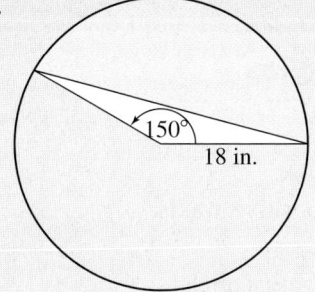

108.

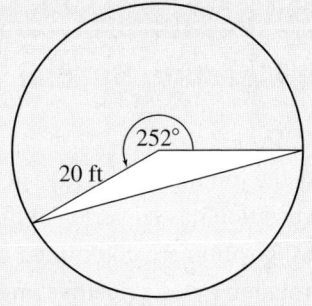

► EXTENDING THE CONCEPT

109. In an elementary study of trigonometry, the hands of a clock are often studied because of the angle relationship that exists between the hands. For example, at 3 o'clock, the angle between the two hands is a right angle and measures 90°.

    a. What is the angle between the two hands at 1 o'clock? 2 o'clock? Explain why.

    b. What is the angle between the two hands at 6:30? 7:00? 7:30? Explain why.

    c. Name four times at which the hands will form a 45° angle.

110. In the diagram shown, the indicated ray is of arbitrary length. (a) Through what additional angle α would the ray have to be rotated to create triangle ABC? (b) What will be the length of side AC once the triangle is complete?

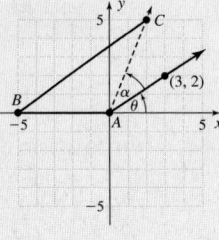

111. Referring to Exercise 102, suppose the fan blade had a radius of 20 in. and is turning at a rate of 12 revolutions per second. (a) Find the angle the blade turns through in 3 sec. (b) Find the circumference of the circle traced out by the tip of the blade. (c) Find the total distance traveled by the blade tip in 10 sec. (d) Find the speed, in miles per hour, that the tip of the blade is traveling.

► MAINTAINING YOUR SKILLS

112. (5.1) For emissions testing, automobiles are held stationary while a heavy roller installed in the floor allows the wheels to turn freely. If the large wheels of a customized pickup have a radius of 18 in. and are turning at 300 revolutions per minute, what speed is the odometer of the truck reading in miles per hour?

113. (5.2) Jazon is standing 117 ft from the base of the Washington Monument in Washington, D.C. If his eyes are 5 ft above level ground and he must hold his head at a 78° angle from horizontal to see the top of the monument (the angle of elevation of 78°), estimate the height of the monument. Answer to the nearest tenth of a foot.

114. (4.4) Solve for t. Answer in both exact and approximate form:

$$-250 = -150e^{-0.05t} - 202.$$

115. (2.3) Find the equation of the line perpendicular to $4x - 5y = 15$ that contains the point $(4, -3)$.

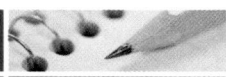

SUMMARY AND CONCEPT REVIEW

SECTION 5.1   Angle Measure, Special Triangles, and Special Angles

KEY CONCEPTS

- An angle is defined as the joining of two rays at a common endpoint called the vertex.
- An angle in standard position has its vertex at the origin and its initial side on the positive x-axis.
- Two angles in standard position are coterminal if they have the same terminal side.
- A counterclockwise rotation gives a positive angle, a clockwise rotation gives a negative angle.

- One (1°) degree is defined to be $\frac{1}{360}$ of a full revolution. One (1) radian is the measure of a central angle subtended by an arc equal in length to the radius.
- Degrees can be divided into a smaller unit called minutes: $1° = 60'$; minutes can be divided into a smaller unit called seconds: $1' = 60''$. This implies $1° = 3600''$.
- Two angles are complementary if they sum to 90° and supplementary if they sum to 180°.
- Properties of triangles: (I) the sum of the angles is 180°; (II) the combined length of any two sides must exceed that of the third side and; (III) larger angles are opposite larger sides.
- Given two triangles, if all three corresponding angles are equal, the triangles are said to be similar. If two triangles are similar, then corresponding sides are in proportion.
- In a 45-45-90 triangle, the sides are in the proportion $1x: 1x: \sqrt{2}x$.
- In a 30-60-90 triangle, the sides are in the proportion $1x: \sqrt{3}x: 2x$.
- The formula for arc length: $s = r\theta$, θ in radians.
- The formula for the area of a circular sector: $\mathcal{A} = \frac{1}{2}r^2\theta$, θ in radians.
- To convert degree measure to radians, multiply by $\frac{\pi}{180°}$; for radians to degrees, multiply by $\frac{180°}{\pi}$.
- Special angle conversions: $30° = \frac{\pi}{6}$, $45° = \frac{\pi}{4}$, $60° = \frac{\pi}{3}$, $90° = \frac{\pi}{2}$.
- A location north or south of the equator is given in degrees latitude; a location east or west of the Greenwich Meridian is given in degrees longitude.
- Angular velocity is a rate of rotation per unit time: $\omega = \frac{\theta}{t}$.
- Linear velocity is a change in position per unit time: $V = \frac{\theta r}{t}$ or $V = r\omega$.

EXERCISES

1. Convert $147°36'48''$ to decimal degrees.
2. Convert $32.87°$ to degrees, minutes, and seconds.
3. All of the right triangles given are similar. Find the dimensions of the largest triangle.

Exercise 3

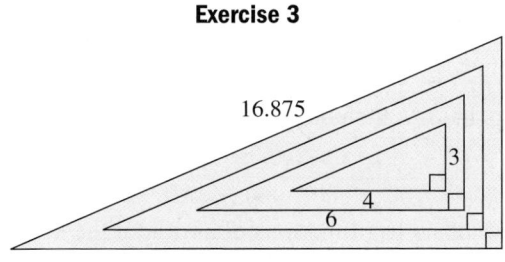

Exercise 4

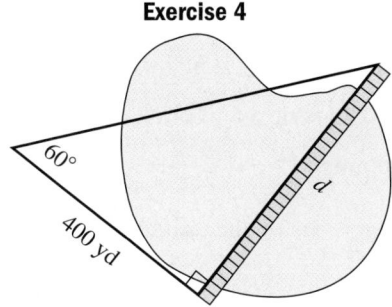

4. Use special angles/special triangles to find the length of the bridge needed to cross the lake shown in the figure.

5. Convert to degrees: $\frac{2\pi}{3}$.

6. Convert to radians: 210°.

7. Find the arc length if $r = 5$ and $\theta = 57°$.

8. Evaluate without using a calculator:
$$\sin\left(\frac{7\pi}{6}\right).$$

Find the angle, radius, arc length, and/or area as needed, until all values are known.

9.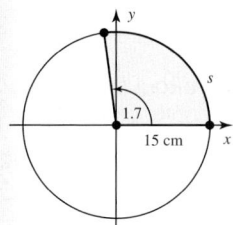

10.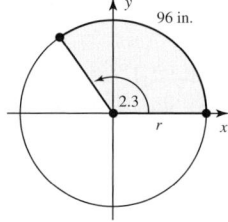

11.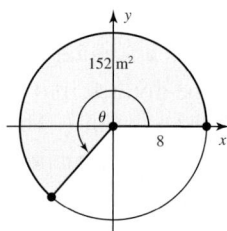

12. With great effort, 5-year-old Mackenzie has just rolled her bowling ball down the lane, and it is traveling painfully slow. So slow, in fact, that you can count the number of revolutions the ball makes using the finger holes as a reference. (a) If the ball is rolling at 1.5 revolutions per second, what is the angular velocity? (b) If the ball's radius is 5 in., what is its linear velocity in feet per second? (c) If the distance to the first pin is 60 feet and the ball is true, how many seconds until it hits?

SECTION 5.2 Unit Circles and the Trigonometry of Real Numbers

KEY CONCEPTS

• A central unit circle is a circle with radius 1 unit having its center at the origin.
• A central circle is symmetric to both axes and the origin. This means that if (a, b) is a point on the circle, then $(-a, b)$, $(-a, -b)$, and $(a, -b)$ are also on the circle and satisfy the equation of the circle.
• On a unit circle with θ in radians, the length of a subtended arc is numerically the same as the subtended angle, making the arc a "circular number line" and associating any given rotation with a unique real number.
• A reference angle is defined to be the acute angle formed by the terminal side of a given angle and the x-axis. For functions of a real number we refer to a reference arc rather than a reference angle.
• For any real number t and a point on the unit circle associated with t, we have:

$$\cos t = x \qquad \sin t = y \qquad \tan t = \frac{y}{x} \qquad \sec t = \frac{1}{x} \qquad \csc t = \frac{1}{y} \qquad \cot t = \frac{x}{y}$$
$$x \neq 0 \qquad\quad x \neq 0 \qquad\quad y \neq 0 \qquad\quad y \neq 0$$

• Given the specific value of any function, the related real number t or angle θ can be found using a reference arc/angle, or the $\sin^{-1}$, $\cos^{-1}$, or $\tan^{-1}$ features of a calculator.

EXERCISES

13. Given $\left(\dfrac{\sqrt{13}}{7}, y\right)$ is on a unit circle, find y if the point is in QIV, then use the symmetry of the circle to locate three other points.

14. Given $\left(\dfrac{3}{4}, -\dfrac{\sqrt{7}}{4}\right)$ is on the unit circle, find the value of all six trig functions of t without the use of a calculator.

15. Without using a calculator, find two values in $[0, 2\pi)$ that make the equation true: $\csc t = \dfrac{2}{\sqrt{3}}$.

16. Use a calculator to find the value of t that corresponds to the situation described: $\cos t = -0.7641$ with t in QII.

17. A crane used for lifting heavy equipment has a winch-drum with a 1-yd radius. (a) If 59 ft of cable has been wound in while lifting some equipment to the roof-top of a building, what radian angle has the drum turned through? (b) What angle must the drum turn through to wind in 75 ft of cable?

SECTION 5.3   Graphs of the Sine and Cosine Functions; Cosecant and Secant Functions

KEY CONCEPTS

- Graphing sine and cosine functions using the special values from the unit circle results in a periodic, wavelike graph with domain $(-\infty, \infty)$.

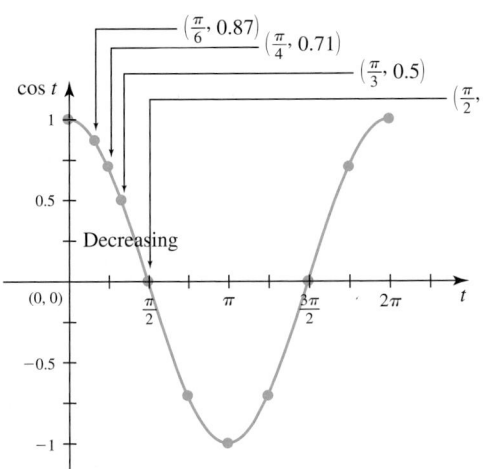
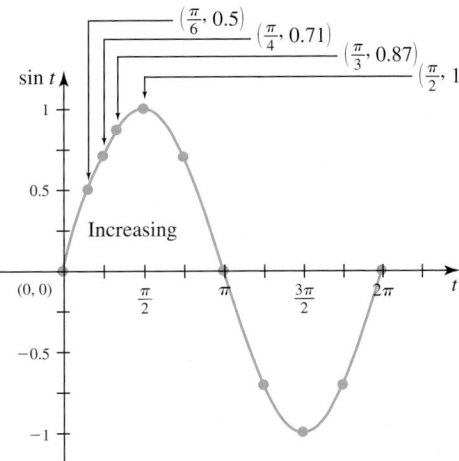

- The characteristics of each graph play a vital role in their contextual application, and these are summarized on pages 457 and 460.
- The amplitude of a sine or cosine graph is the maximum displacement from the average value. For $y = A\sin(Bt)$ and $y = A\cos(Bt)$, the amplitude is $|A|$.
- The period of a periodic function is the smallest interval required to complete one cycle.

  For $y = A\sin(Bt)$ and $y = A\cos(Bt)$, $P = \dfrac{2\pi}{B}$ gives the period.

- If $|A| > 1$, the graph is vertically stretched, if $0 < |A| < 1$ the graph is vertically compressed, and if $A < 0$ the graph is reflected across the x-axis.
- If $B > 1$, the graph is horizontally compressed (the period is smaller/shorter); if $B < 1$ the graph is horizontally stretched (the period is larger/longer).
- To graph $y = A\sin(Bt)$ or $A\cos(Bt)$, draw a reference rectangle $2A$ units high and $P = \dfrac{2\pi}{B}$ units wide,

  centered on the x-axis, then use the *rule of fourths* to locate zeroes and max/min values. Connect these points with a smooth curve.

- The graph of $y = \sec t = \dfrac{1}{\cos t}$ will be asymptotic everywhere $\cos t = 0$, increasing where $\cos t$ is decreasing, and decreasing where $\cos t$ is increasing.

- The graph of $y = \csc t = \dfrac{1}{\sin t}$ will be asymptotic everywhere $\sin t = 0$, increasing where $\sin t$ is decreasing, and decreasing where $\sin t$ is increasing.

EXERCISES

Use a reference rectangle and the *rule of fourths* to draw an accurate sketch of the following functions through at least one full period. Clearly state the amplitude (as applicable) and period as you begin.

18. $y = 3\sin t$

19. $y = 3\sec t$

20. $y = -\cos(2t)$

21. $y = 1.7\sin(4t)$

22. $f(t) = 2\cos(4\pi t)$

23. $g(t) = 3\sin(398\pi t)$

The given graphs are of the form $y = A \sin(Bt)$ and $y = A \csc(Bt)$. Determine the equation of each graph.

24.

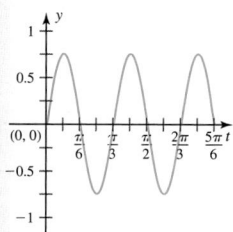

25.

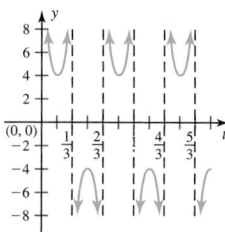

26. Referring to the chart of colors visible in the electromagnetic spectrum (page 469), what color is represented by the equation $y = \sin\left(\dfrac{\pi}{270}t\right)$? By $y = \sin\left(\dfrac{\pi}{320}t\right)$?

SECTION 5.4   Graphs of Tangent and Cotangent Functions

KEY CONCEPTS

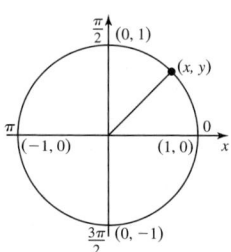

- Since $\tan t$ is defined in terms of the ratio $\dfrac{y}{x}$, the graph will be asymptotic everywhere $x = 0$ on the unit circle, meaning all odd multiples of $\dfrac{\pi}{2}$.

- Since $\cot t$ is defined in terms of the ratio $\dfrac{x}{y}$, the graph will be asymptotic everywhere $y = 0$ on the unit circle, meaning all integer multiples of π.

- The graph of $y = \tan t$ is increasing everywhere it is defined; the graph of $y = \cot t$ is decreasing everywhere it is defined.

- The characteristics of each graph play a vital role in their contextual application, and these are summarized on page 474.

- For the more general tangent and cotangent graphs $y = A \tan(Bt)$ and $y = A \cot(Bt)$, if $|A| > 1$, the graph is vertically stretched, if $0 < |A| < 1$ the graph is vertically compressed, and if $A < 0$ the graph is reflected across the x-axis.

- If $B > 1$, the graph is horizontally compressed (the period is smaller/shorter); if $B < 1$ the graph is horizontally stretched (the period is larger/longer).

- To graph $y = A \tan(Bt)$, note $A \tan(Bt)$ is zero at $t = 0$. Compute the period $P = \dfrac{\pi}{B}$ and draw asymptotes a distance of $\dfrac{P}{2}$ on either side of the y-axis. Plot zeroes halfway between the asymptotes and use symmetry to complete the graph.

- To graph $y = A \cot(Bt)$, note it is asymptotic at $t = 0$. Compute the period $P = \dfrac{\pi}{B}$ and draw asymptotes a distance P on either side of the y-axis. Plot zeroes halfway between the asymptotes and use symmetry to complete the graph.

EXERCISES

27. State the value of each expression without the aid of a calculator:

$$\tan\left(\frac{7\pi}{4}\right) \qquad \cot\left(\frac{\pi}{3}\right)$$

28. State the value of each expression without the aid of a calculator, given that t terminates in QII.

$$\tan^{-1}(-\sqrt{3}) \qquad \cot^{-1}\left(-\frac{1}{\sqrt{3}}\right)$$

29. Graph $y = 6 \tan\left(\dfrac{1}{2}t\right)$ in the interval $[-2\pi, 2\pi]$.   **30.** Graph $y = \dfrac{1}{2}\cot(2\pi t)$ in the interval $[-1, 1]$.

31. Use the period of $y = \cot t$ to name three additional solutions to $\cot t = 0.0208$, given $t = 1.55$ is a solution. Many solutions are possible.

32. Given $t = 0.4444$ is a solution to $\cot^{-1}(t) = 2.1$, use an analysis of signs and quadrants to name an additional solution in $[0, 2\pi)$.

33. Find the approximate height of Mount Rushmore, using $h = \dfrac{d}{\cot u - \cot v}$ and the values shown.

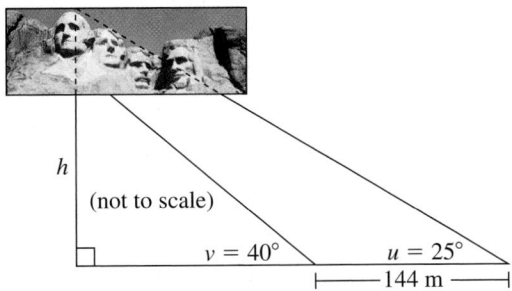

34. Model the data in the table using a tangent function. Clearly state the period, the value of A, and the location of the asymptotes.

Input	Output	Input	Output
−6	−∞	1	1.4
−5	−19.4	2	3
−4	−9	3	5.2
−3	−5.2	4	9
−2	−3	5	19.4
1	−1.4	6	∞
0	0		

SECTION 5.5   Transformations and Applications of Trigonometric Graphs

KEY CONCEPTS

- Many everyday phenomena follow a sinusoidal pattern, or a pattern that can be modeled by a sine or cosine function (e.g., daily temperatures, hours of daylight, and more).

- To obtain accurate equation models of sinusoidal phenomena, vertical and horizontal shifts of a basic function are used.

- The equation $y = A \sin(Bt \pm C) + D$ is called the *standard form* of a general sinusoid. The equation

  $y = A \sin\left[B\left(t \pm \dfrac{C}{B}\right)\right] + D$ is called the *shifted form* of a general sinusoid.

- In either form, D represents the average value of the function and a vertical shift D units upward if $D > 0$, D units downward if $D < 0$. For a maximum value M and minimum value m, $\dfrac{M + m}{2} = D$, $\dfrac{M - m}{2} = A$.

- The shifted form $y = A \sin\left[B\left(t \pm \dfrac{C}{B}\right)\right] + D$ enables us to quickly identify the horizontal shift of the function:

  $\dfrac{C}{B}$ units in a direction opposite the given sign.

- To graph a shifted sinusoid, locate the primary interval by solving $0 \le Bt + C < 2\pi$, then use a reference rectangle along with the *rule of fourths* to sketch the graph in this interval. The graph can then be extended as needed, then shifted vertically D units.

- One basic application of sinusoidal graphs involves phenomena in harmonic motion, or motion that can be modeled by functions of the form $y = A \sin(Bt)$ or $y = A \cos(Bt)$ (with no horizontal or vertical shift).
- If the period P and critical points (X, M) and (x, m) of a sinusoidal function are known, a model of the form $y = A \sin(Bx + C) + D$ can be obtained:

$$B = \frac{2\pi}{P} \qquad A = \frac{M - m}{2} \qquad D = \frac{M + m}{2} \qquad C = \frac{3\pi}{2} - Bx$$

EXERCISES

For each equation given, (a) identify/clearly state the amplitude, period, horizontal shift, and vertical shift; then (b) graph the equation using the primary interval, a reference rectangle, and *rule of fourths*.

35. $y = 240 \sin\left[\dfrac{\pi}{6}(t - 3)\right] + 520$ **36.** $y = 3.2 \cos\left(\dfrac{\pi}{4}t + \dfrac{3\pi}{2}\right) + 6.4$

For each graph given, identify the amplitude, period, horizontal shift, and vertical shift, and give the equation of the graph.

37. 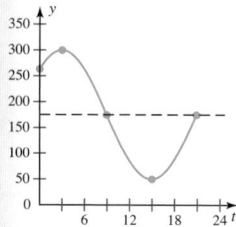   **38.**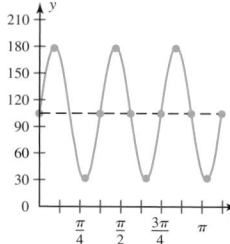

39. Monthly precipitation in Cheyenne, Wyoming, can be modeled by a sine function, by using the average precipitation for July (2.26 in.) as a maximum (actually slightly higher in May), and the average precipitation for February (0.44 in.) as a minimum. Assume $t = 0$ corresponds to March. (a) Use the information to construct a sinusoidal model, and (b) use the model to estimate the inches of precipitation Cheyenne receives in August ($t = 5$) and December ($t = 9$).

Source: 2004 Statistical Abstract of the United States, Table 380.

SECTION 5.6   The Trigonometry of Right Triangles

KEY CONCEPTS

- The sides of a right triangle can be named relative to their location with respect to a given angle.

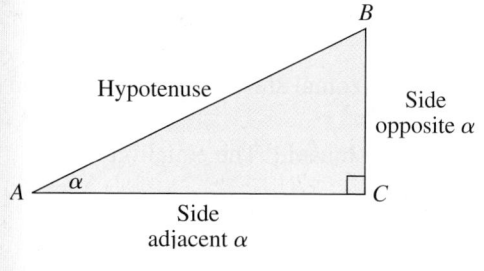   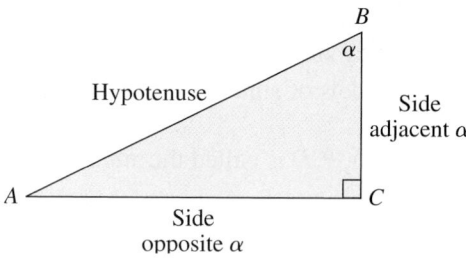

- The ratios of two sides with respect to a given angle are named as follows:

$$\sin \alpha = \frac{\text{opp}}{\text{hyp}} \qquad \cos \alpha = \frac{\text{adj}}{\text{hyp}} \qquad \tan \alpha = \frac{\text{opp}}{\text{adj}}$$

- The reciprocal of the ratios above play a vital role and are likewise given special names:

$$\csc \alpha = \frac{\text{hyp}}{\text{opp}} \qquad \sec \alpha = \frac{\text{hyp}}{\text{adj}} \qquad \cot \alpha = \frac{\text{adj}}{\text{opp}}$$

$$\csc \alpha = \frac{1}{\sin \alpha} \qquad \sec \alpha = \frac{1}{\cos \alpha} \qquad \cot \alpha = \frac{1}{\tan \alpha}$$

- Each function of α is equal to the cofunction of its complement. For instance, the complement of sine is *co*sine and $\sin \alpha = \cos(90° - \alpha)$.
- To solve a right triangle means to apply any combination of the trig functions, along with the triangle properties, until all sides and all angles are known.
- An angle of elevation is the angle formed by a horizontal line of sight (parallel to level ground) and the true line of sight. An angle of depression is likewise formed, but with the line of sight below the line of orientation.

EXERCISES

40. Use a calculator to solve for A:

    a. $\cos 37° = A$

    b. $\cos A = 0.4340$

41. Rewrite each expression in terms of a cofunction.

    a. $\tan 57.4°$

    b. $\sin(19°30'15'')$

Solve each triangle. Round angles to the nearest tenth and sides to the nearest hundredth.

42.

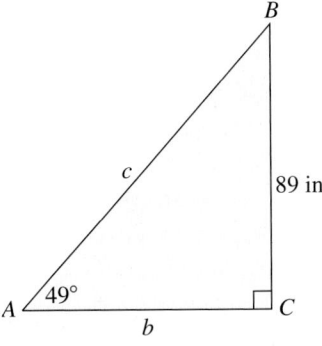

43.

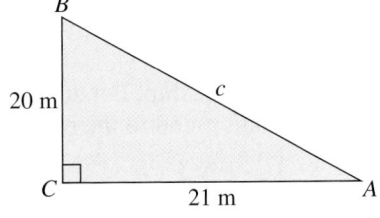

44. Josephine is to weld a vertical support to a 20-m ramp so that the incline is exactly 15°. What is the height h of the support that must be used?

45. From the observation deck of a seaside building 480 m high, Armando sees two fishing boats in the distance. The angle of depression to the nearer boat is 63.5°, while for the boat farther away the angle is 45°. (a) How far out to sea is the nearer boat? (b) How far apart are the two boats?

46. A slice of bread is roughly 14 cm by 10 cm. If the slice is cut diagonally in half, what acute angles are formed?

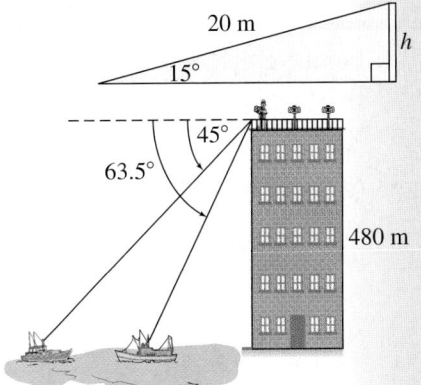

SECTION 5.7 Trigonometry and the Coordinate Plane

KEY CONCEPTS

- In standard position, the terminal sides of 0°, 90°, 180°, 270°, and 360° angles coincide with one of the axes and are called quadrantal angles.
- By placing a right triangle in the coordinate plane with one acute angle at the origin and one side along the x-axis, we note the trig functions can be defined in terms of a point $P(x, y)$ on the hypotenuse.
- Given $P(x, y)$ is any point on the terminal side of an angle θ in standard position. Then $r = \sqrt{x^2 + y^2}$ is the distance from the origin to this point. The six trigonometric functions of θ are defined as

$$\sin \theta = \frac{y}{r} \qquad \cos \theta = \frac{x}{r} \qquad \tan \theta = \frac{y}{x} \qquad \csc \theta = \frac{r}{y} \qquad \sec \theta = \frac{r}{x} \qquad \cot \theta = \frac{x}{y}$$

$$x \neq 0 \qquad\qquad y \neq 0 \qquad\qquad x \neq 0 \qquad\qquad y \neq 0$$

- A reference angle θ_r is defined to be the acute angle formed by the terminal side of a given angle θ and the x-axis.
- Reference angles can be used to evaluate the trig functions of any nonquadrantal angle, since the values are fixed by the ratio of sides and the signs are dictated by the quadrant of the terminal side.

- If the value of a trig function and the quadrant of the terminal side are known, the related angle θ can be found using a reference arc/angle, or the $\sin^{-1}$, $\cos^{-1}$, or $\tan^{-1}$ features of a calculator.
- If θ is a solution to $\sin \theta = k$, then $\theta + 360°k$ is also a solution for any integer k.

EXERCISES

47. Find two positive angles and two negative angles that are coterminal with $\theta = 207°$.

48. Name the reference angle for the angles given: $\theta = -152°$ $\theta = 521°$ $\theta = 210°$

49. Find the value of the six trigonometric functions, given $P(x, y)$ is on the terminal side of angle θ in standard position.

    a. $P(-12, 35)$ **b.** $(12, -18)$

50. Find the values of x, y, and r using the information given, and state the quadrant of the terminal side of θ. Then state the values of the six trig functions of θ.

    a. $\cos \theta = \dfrac{4}{5}$; $\sin \theta < 0$ **b.** $\tan \theta = -\dfrac{12}{5}$; $\cos \theta > 0$

51. Find all angles satisfying the stated relationship. For standard angles, express your answer in exact form. For nonstandard angles, use a calculator and round to the nearest tenth.

    a. $\tan \theta = -1$ **b.** $\cos \theta = \dfrac{\sqrt{3}}{2}$ **c.** $\tan \theta = 4.0108$ **d.** $\sin \theta = -0.4540$

MIXED REVIEW

1. For the graph of periodic function f given, state the (a) amplitude, (b) average value, (c) period, and (d) value of $f(4)$.

Exercise 1

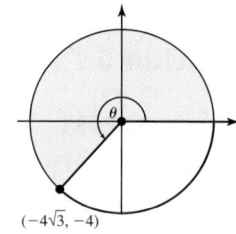
(Exercise 1 graph: $y = f(x)$)

2. Name two values in $[0, 2\pi)$ where $\tan t = 1$.

3. Name two values in $[0, 2\pi)$ where $\cos t = -\dfrac{1}{2}$.

4. Given $\sin \theta = \dfrac{8}{\sqrt{185}}$ with θ in QII, state the value of the other five trig functions.

5. Convert to DMS form: $220.81\overline{38}°$.

6. Find two negative angles and two positive angles that are coterminal with (a) $57°$ and (b) $135°$.

7. To finish the top row of the tile pattern on our bathroom wall, $12''$ by $12''$ tiles must be cut diagonally. Use a standard triangle to find the length of each cut and the width of the wall covered by tiles.

Exercise 7

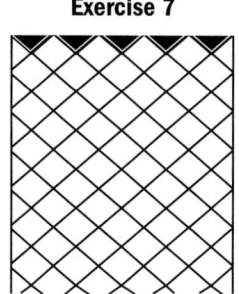

8. The service door into the foyer of a large office building is $36''$ wide by $78''$ tall. The building manager has ordered a large wall painting $85''$ by $85''$ to add some atmosphere to the foyer area. (a) Can the painting be brought in the service door? (b) If so, at what two integer-valued angles (with respect to level ground) could the painting be tilted?

9. Find the arc length and area of the shaded sector.

Exercise 9

10. Monthly precipitation in Minneapolis, Minnesota, can be modeled by a sine function, by using the average precipitation for August (4.05 in.) as a maximum (actually slightly higher in June), and the average precipitation for February (0.79 in.) as a minimum. Assume $t = 0$ corresponds to April. (a) Use the information to construct a sinusoidal model, and (b) Use the model to approximate the inches of precipitation Minneapolis receives in July ($t = 3$) and December ($t = 8$).

Source: 2004 Statistical Abstract of the United States, Table 380

11. Convert from DMS to decimal degrees: $86° \, 54' \, 54''$.

12. Name the reference angle θ_r for the angle θ given.

    a. $735°$ **b.** $-135°$ **c.** $\dfrac{5\pi}{6}$ **d.** $-\dfrac{5\pi}{3}$

13. Find the value of all six trig functions of θ, given the point $(15, -8)$ is on the terminal side.

14. Verify that $\left(-\dfrac{\sqrt{2}}{2}, \dfrac{\sqrt{2}}{2}\right)$ is a point on the unit circle and find the value of all six trig functions at this point.

15. On your approach shot to the ninth green, the Global Positioning System (GPS) your cart is equipped with tells you the pin is 115.47 yd away. The distance plate states the straight line distance to the hole is 100 yd (see the diagram). Relative to a straight line between the plate and the hole, at what acute angle θ should you hit the shot?

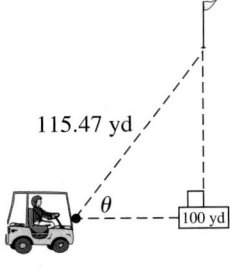

115.47 yd

θ

100 yd

16. The electricity supply lines to the top of Lone Eagle Plateau must be replaced, and the new lines will be run in conduit buried slightly beneath the surface. The scale of

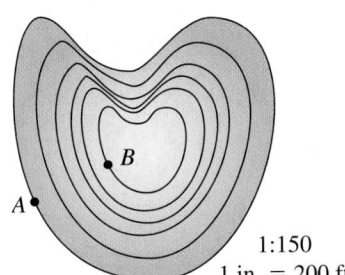

B

A

1:150
1 in. = 200 ft

elevation is 1:150 (each closed figure indicates an increase in 150 ft of elevation), and the scale of horizontal distance is 1 in. = 200 ft. (a) Find the increase in elevation from point A to point B, (b) use a proportion to find the horizontal distance from A to B if the measured distance on the map is $2\frac{1}{4}$ in., (c) draw the corresponding right triangle and use it to estimate the length of conduit needed from A to B and the angle of incline the installers will experience while installing the conduit.

17. A salad spinner consists of a colander basket inside a large bowl, and is used to wash and dry lettuce and other salad ingredients. The spinner is turned at about 3 revolutions per second. (a) Find the angular velocity and (b) find the linear velocity of a point of the circumference if the basket has a 20 cm radius.

18. Solve each equation in $[0, 2\pi)$ without the use of a calculator. If the expression is undefined, so state.

    a. $x = \sin\left(-\dfrac{\pi}{4}\right)$ **b.** $\sec x = \sqrt{2}$

    c. $\cot\left(\dfrac{\pi}{2}\right) = x$ **d.** $\cos \pi = x$

    e. $\csc x = \dfrac{2\sqrt{3}}{3}$ **f.** $\tan\left(\dfrac{\pi}{2}\right) = x$

19. State the amplitude, period, horizontal shift, vertical shift, and endpoints of the primary interval (as applicable), then sketch the graph using a reference rectangle and the rule of fourths.

    a. $y = 5\cos(2t) - 8$ **b.** $y = \dfrac{7}{2}\sin\left[\dfrac{\pi}{2}(x - 1)\right]$

    c. $y = 2\tan\left(\dfrac{1}{4}t\right)$ **d.** $y = 3\sec\left(x - \dfrac{\pi}{2}\right)$

20. Virtually everyone is familiar with the Statue of Liberty in New York Bay, but fewer know that America is home to a second "Statue of Liberty" standing proudly atop the iron dome of the Capitol Building. From a distance of 600 ft, the angle of elevation from ground level to the top of the statue (from the east side) is 25.60°. The angle of elevation to the base of the statue is 24.07°. How tall is the statue *Freedom* (the name sculptor Thomas Crawford gave this statue)?

H

25.6°

600 ft

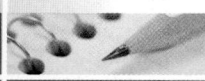

PRACTICE TEST

1. State the complement and supplement of 35°.

2. Name the reference angle of each angle given.

    a. 225°   **b.** −510°

    c. $\dfrac{7\pi}{6}$ **d.** $\dfrac{25\pi}{3}$

3. Find two negative angles and two positive angles that are coterminal with $\theta = 30°$. Many solutions are possible.

4. Convert from DMS to decimal degrees or decimal degrees to DMS as indicated.

    a. 100°45′18″ to decimal degrees

    b. 48.2125° to DMS

5. Four Corners USA is the point at which Utah, Colorado, Arizona, and New Mexico meet. The southern border of Colorado, the western border of Kansas, and the point P where Colorado, Nebraska, and Kansas meet, very nearly approximates a 30-60-90 triangle. If the western border of Kansas is 215 mi long, (a) what is the distance from Four Corners USA to point P? (b) How long is Colorado's southern border?

Exercise 5

6. Complete the table from memory using exact values. If a function is undefined, so state.

t	$\sin t$	$\cos t$	$\tan t$	$\csc t$	$\sec t$	$\cot t$
0						
$\dfrac{2\pi}{3}$						
$\dfrac{7\pi}{6}$						
$\dfrac{5\pi}{4}$						
$\dfrac{5\pi}{3}$						
$\dfrac{13\pi}{6}$						

7. Given $\cos\theta = \dfrac{2}{5}$ and $\tan\theta < 0$, find the value of the other five trig functions of θ.

8. Verify that $\left(\dfrac{1}{3}, -\dfrac{2\sqrt{2}}{3}\right)$ is a point on the unit circle, then find the value of all six trig functions associated with this point.

9. In order to take pictures of a dance troupe as it performs, a camera crew rides in a cart on tracks that trace a circular arc. The radius of the arc is 75 ft, and from end to end the cart sweeps out an angle of 172.5° in 20 seconds. Use this information to find (a) the length of the track in feet and inches, (b) the angular velocity of the cart, and (c) the linear velocity of the cart in both ft/sec and mph.

10. Solve the triangle shown. Answer in table form.

Exercise 10

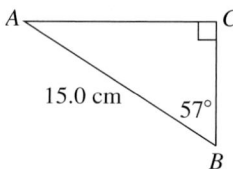

11. The "plow" is a yoga position in which a person lying on their back brings their feet up, over, and behind their head and touches them to the floor. If distance from hip to shoulder (at the right angle) is 57 cm and from hip to toes is 88 cm, find the distance from shoulders to toes and the angle formed at the hips.

Exercise 11

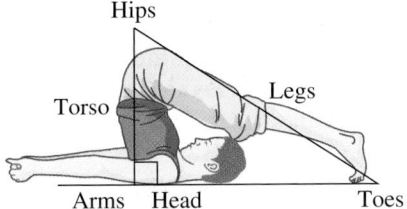

Hips

Torso   Legs

Arms   Head   Toes

12. While doing some night fishing, you round a peninsula and a tall light house comes into view. Taking a sighting, you find the angle of elevation to the top of the lighthouse is 25°. If the lighthouse is known to be 27 m tall, how far from the lighthouse are you?

Exercise 12

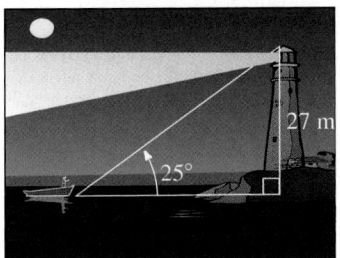

27 m

25°

13. Find the value of $t \in [0, 2\pi]$ satisfying the conditions given.

a. $\sin t = -\dfrac{1}{2}$, t in QIII

b. $\sec t = \dfrac{2\sqrt{3}}{3}$, t in QIV

c. $\tan t = -1$, t in QII

14. In arid communities, daily water usage can often be approximated using a sinusoidal model. Suppose water consumption in the city of Caliente del Sol reaches a maximum of 525,000 gallons in the heat of the day, with a minimum usage of 157,000 gallons in the cool of the night. Assume $t = 0$ corresponds to 6:00 A.M. (a) Use the information to construct a sinusoidal model, and (b) Use the model to approximate water usage at 4:00 P.M. and 4:00 A.M.

15. State the domain, range, period, and amplitude (if it exists), then graph the function over 1 period.

a. $y = 2\sin\left(\dfrac{\pi}{5}t\right)$ **b.** $y = \sec t$

c. $y = 2\tan(3t)$

16. State the amplitude, period, horizontal shift, vertical shift, and endpoints of the primary interval. Then sketch the graph using a reference rectangle and the rule of fourths:

$$y = 12\sin\left(3t - \dfrac{\pi}{4}\right) + 19.$$

17. An athlete throwing the shot-put begins his first attempt facing due east, completes three and one-half turns and launches the shot facing due west. What angle did his body turn through?

18. State the domain, range, and period, then sketch the graph in $[0, 2\pi)$.

a. $y = \tan(2t)$ **b.** $y = \cot\left(\dfrac{1}{2}t\right)$

19. Due to tidal motions, the depth of water in Brentwood Bay varies sinusoidally as shown in the diagram, where time is in hours and depth is in feet. Find an equation that models the depth of water at time t.

Exercise 19

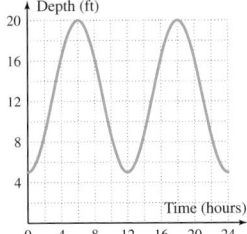

Depth (ft)

Time (hours)

20. Find the value of t satisfying the given conditions.

a. $\sin t = -0.7568$; t in QIII

b. $\sec t = -1.5$; t in QII

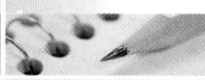

CALCULATOR EXPLORATION AND DISCOVERY

Variable Amplitudes and Modeling the Tides

Tidal motion is often too complex to be modeled by a single sine function. In this *Exploration and Discovery*, we'll look at a method that combines two sine functions to help model a tidal motion with variable amplitude. In the process, we'll use much of what we know about the amplitude, horizontal shifts and vertical shifts of a sine function, helping to reinforce these important concepts and broaden our understanding about how they can be applied. The graph in Figure 5.93 shows three days of tidal motion for Davis Inlet, Canada.

Figure 5.93

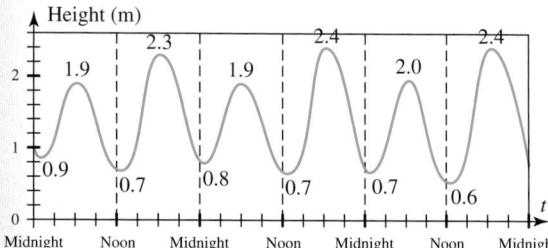

As you can see, the amplitude of the graph varies, and there is no *single* sine function that can serve as a model. However, notice that the amplitude *varies predictably*, and that the high tides and low tides can independently be modeled by a sine function. To simplify our exploration, we will use the assumption that tides have an exact 24-hr period (close, but no), that variations between high and low tides takes place every 12 hr (again close but not exactly true), and the variation between the "low-high" (1.9 m) and the "high-high" (2.4 m) is uniform. A similar assumption is made for the low tides. The result is the graph in Figure 5.94.

Figure 5.94

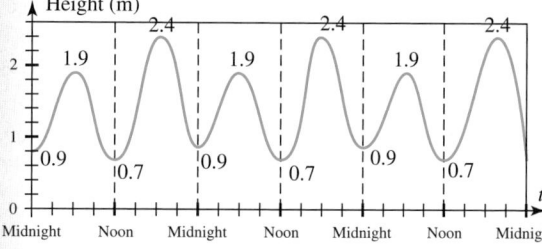

First consider the high tides, which vary from a maximum of 2.4 to a minimum of 1.9. Using the ideas from Section 5.7 to construct an equation model gives

$$A = \frac{2.4 - 1.9}{2} = 0.25 \text{ and } D = \frac{2.4 + 1.9}{2} = 2.15. \text{ With}$$

a period of $P = 24$ hr we obtain the equation $Y_1 = 0.25 \sin\left(\frac{\pi}{12}x\right) + 2.15$. Using 0.9 and 0.7 as the

maximum and minimum low tides, similar calculations yield the equation $Y_2 = 0.1 \sin\left(\frac{\pi}{12}x\right) + 0.8$ (verify this).

Graphing these two functions over a 24-hr period yields the graph in Figure 5.95, where we note the high and low values are correct, but the two functions are in phase with each other. As can be determined from Figure 5.94, we want the high tide model to start at the average value and decrease, and the low tide equation model to start at high-low and decrease. Replacing x with $x - 12$ in Y_1 and x with $x + 6$ in Y_2 accomplishes this result (see Figure 5.96). Now comes the fun part! Since Y_1 represents the low/high maximum values for high tide, and Y_2 represents the low/high minimum values for low tide, *the amplitude and average value for the tidal motion at Davis Inlet*

Figure 5.95

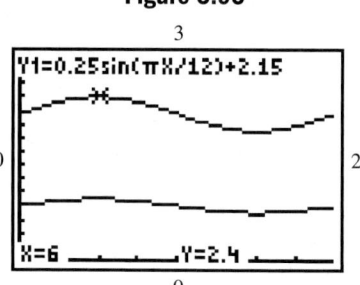

Figure 5.96

Figure 5.97

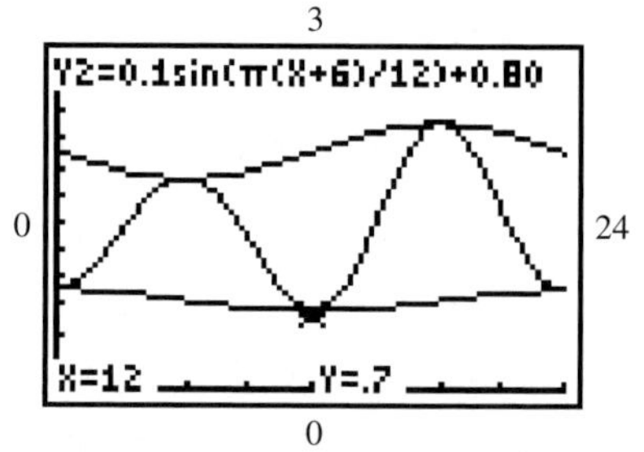

are $A = \dfrac{Y_1 - Y_2}{2}$

and $D = \dfrac{Y_1 + Y_2}{2}$!

By entering $Y_3 = \dfrac{Y_1 - Y_2}{2}$ and

$Y_4 = \dfrac{Y_1 + Y_2}{2}$, the equation for the tidal

motion (with its variable amplitude) will have the form $Y_5 = Y_3 \sin(Bx \pm C) + Y_4$, where the value of B and C must be determined. The key here is to note there is only a 12-hr difference between the changes in amplitude, so $P = 12$ (instead of 24) and $B = \dfrac{\pi}{6}$ for this function.

function. Also, from the graph (Figure 5.94) we note the tidal motion begins at a minimum and increases, indicating a shift of 3 units to the right is required. Replacing x with $x - 3$ gives the equation modeling these tides, and the final

equation is $Y_5 = Y_3 \sin\left[\dfrac{\pi}{6}(x - 3)\right] + Y_4$. Figure 5.97

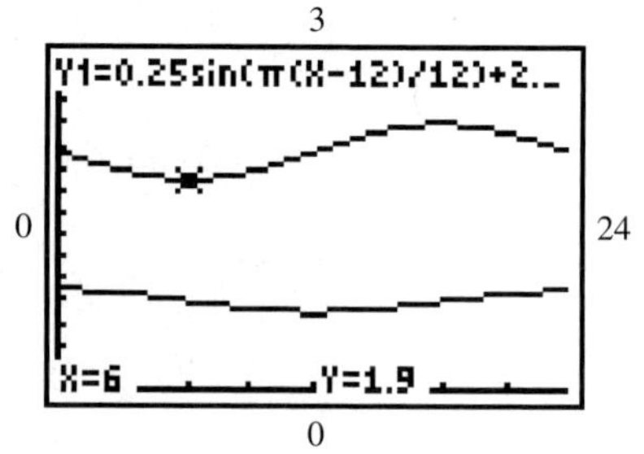

gives a screen shot of Y_1, Y_2, and Y_5 in the interval [0, 24]. The tidal graph from Figure 5.94 is shown in Figure 5.98 with Y_3 and Y_4 superimposed on it.

Figure 5.98

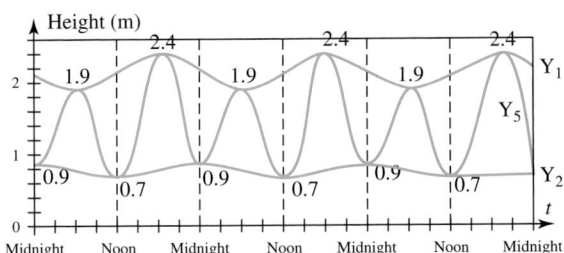

Exercise 1: The website www.tides.com/tcpred.htm offers both *t*ide and *c*urrent *pred*ictions for various locations around the world, in both numeric and graphical form. In addition, data for the "two" high tides and "two" low tides are clearly highlighted. Select a coastal area where tidal motion is similar to that of Davis Inlet, and repeat this exercise. Compare your model to the actual data given on the website. How good was the fit?

STRENGTHENING CORE SKILLS

Standard Angles, Reference Angles, and the Trig Functions

A review of the main ideas discussed in this chapter indicates there are four of what might be called "core skills." These are skills that (a) play a fundamental part in the acquisition of concepts, (b) hold the overall structure together as we move from concept to concept, and (c) are ones we return to again and again throughout our study. The first of these is *(1) knowing the standard angles and standard values.* These values are "standard" because no estimation, interpolation, or special methods are required to name their value, and each can be expressed as a single factor. This gives them a great advantage in that further conceptual development can take place without the main points being obscured by large expressions or decimal approximations. Knowing the value of the trig functions for each standard angle will serve you very well throughout this study. *Know* the chart on page ••• and the ideas that led to it.

The standard angles/values brought us to the trigonometry of any angle, forming a strong bridge to the second core skill:
(2) using reference angles to determine the value of the trig functions in each quadrant. For review, a 30-60-90 triangle will always have sides that are in the proportion $1x : \sqrt{3}x : 2x$, regardless of its size. This means for any angle θ, where $\theta_r = 30°$, $\sin\theta = \frac{1}{2}$ or $\sin\theta = -\frac{1}{2}$ since the *ratio is fixed* but the sign *depends on the quadrant of θ*: $\sin 30° = \frac{1}{2}$

Figure 5.99

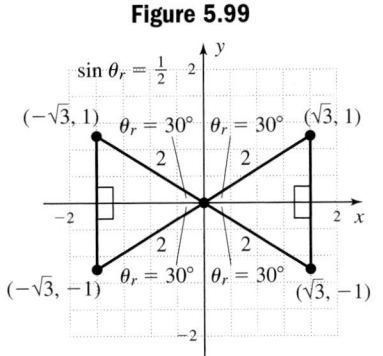

[QI], $\quad \sin 150° = \frac{1}{2}$ [QII], $\quad \sin 210° = -\frac{1}{2}$ [QIII], $\sin 330° = -\frac{1}{2}$ [QII], and so on (see Figure 5.99).

In turn, the reference angles led us to a third core skill, helping us realize that if θ was not a quadrantal angle, *(3) equations like*
$$\cos(\theta) = -\frac{\sqrt{3}}{2} \text{ must have}$$
two solutions in [0, 360°). From the standard angles and standard values we learn to recognize that for
$$\cos\theta = -\frac{\sqrt{3}}{2}, \ \theta_r = 30°,$$

Figure 5.100

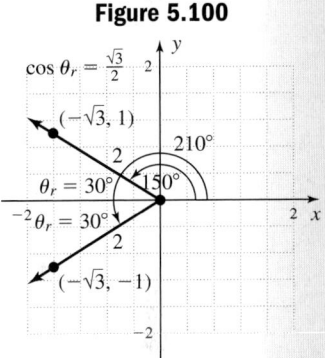

which will occur as a reference angle in the two quadrants where cosine is negative, QII and QIII. The solutions in [0, 360°) are $\theta = 150°$ and $\theta = 210°$ (see Figure 5.100).

Of necessity, this brings us to the fourth core skill, *(4) effective use of a calculator.* The standard angles are a wonderful vehicle for introducing the basic ideas of trigonometry, and actually occur quite frequently in real-world applications. But by far, most of the values we encounter will be nonstandard values where θ_r must be found using a calculator. However, once θ_r is found, the reason and reckoning inherent in these ideas can be directly applied.

The *Summary and Concept Review Exercises*, as well as the *Practice Test* offer ample opportunities to refine these skills, so that they will serve you well in future chapters as we continue our attempts to explain and understand the world around us in mathematical terms.

Exercise 1: Fill in the table from memory.

t	0	$\dfrac{\pi}{6}$	$\dfrac{\pi}{4}$	$\dfrac{\pi}{3}$	$\dfrac{\pi}{2}$
$\sin t = y$					
$\cos t = x$					
$\tan t = \dfrac{y}{x}$					

$\dfrac{2\pi}{3}$	$\dfrac{3\pi}{4}$	$\dfrac{5\pi}{6}$	π	$\dfrac{7\pi}{6}$	$\dfrac{5\pi}{4}$

Exercise 2: Solve each equation in $[0, 2\pi)$ without the use of a calculator.

a. $2 \sin t + \sqrt{3} = 0$

b. $-3\sqrt{2} \cos t + 4 = 1$

c. $-\sqrt{3} \tan t + 2 = 1$

d. $\sqrt{2} \sec t + 1 = 3$

Exercise 3: Solve each equation in $[0, 2\pi)$ using a calculator and rounding answers to four decimal places.

a. $\sqrt{6} \sin t - 2 = 1$

b. $-3\sqrt{2} \cos t + \sqrt{2} = 0$

c. $3 \tan t + \dfrac{1}{2} = -\dfrac{1}{4}$

d. $2 \sec t = -5$

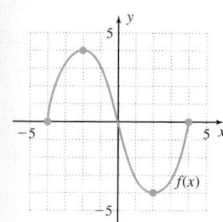

CUMULATIVE REVIEW CHAPTERS 1–5

1. Solve the inequality given: $2|x + 1| - 3 < 5$

2. Find the domain of the function: $y = \sqrt{x^2 - 2x - 15}$

3. Given that $\tan \theta = \dfrac{80}{39}$, draw a right triangle that corresponds to this ratio, then use the Pythagorean theorem to find the length of the missing side. Finally, find the two acute angles.

4. Without a calculator, what values in $[0, 2\pi)$ make the equation true: $\sin t = -\dfrac{\sqrt{3}}{2}$?

5. Given $\left(\dfrac{3}{4}, -\dfrac{\sqrt{7}}{4}\right)$ is a point on the unit circle corresponding to t, find all six trig functions of t.

State the domain and range of each function shown:

6. $y = f(x)$

7. **a.** $f(x) = \sqrt{2x - 3}$

   b. $g(x) = \dfrac{2x}{x^2 - 49}$

8. $y = T(x)$

x	$T(x)$
0	-7
1	-5
2	-3
3	-1
4	1
5	3
6	5

9. Analyze the graph of the function in Exercise 6, including: (a) maximum and minimum values; (b) intervals where $f(x) \geq 0$ and $f(x) < 0$; (c) intervals where f is increasing or decreasing; and (d) any symmetry noted. Assume the features you are to describe have integer values.

10. The attractive force that exists between two magnets varies inversely as the square of the distance between them. If the attractive force is 1.5 newtons (N) at a distance of 10 cm, how close are the magnets when the attractive force reaches 5 N?

11. The world's tallest indoor waterfall is in Detroit, Michigan, in the lobby of the International Center Building. Standing 66 ft from the base of the falls, the angle of elevation is 60°. How tall is the waterfall?

12. It's a warm, lazy Saturday and Hank is watching a county maintenance crew mow the park across the street. He notices the mower takes 29 sec to pass through 77° of rotation from one end of the park to the other. If the corner of the park is 60 ft directly across the street from his house, (a) how wide is the park? (b) How fast (in mph) does the mower travel as it cuts the grass?

13. Graph using transformations of a parent function:
$$f(x) = \frac{1}{x + 1} - 2.$$

14. Graph using transformations of a parent function:
$$g(x) = e^{x-1} - 2.$$

15. Find $f(\theta)$ for all six trig functions, given the point $P(-9, 40)$ is a point on the terminal side of the angle. Then find the angle θ in degrees, rounded to tenths.

16. Given $t = 5.37$, (a) in what quadrant does the arc terminate? (b) What is the reference arc? (c) Find the value of $\sin t$ rounded to four decimal places.

17. A jet-stream water sprinkler shoots water a distance of 15 m and turns back-and-forth through an angle of $t = 1.2$ rad. (a) What is the length of the arc that the sprinkler reaches? (b) What is the area in m² of the yard that is watered?

18. Determine the equation of graph shown given it is of the form $y = A \tan(Bt)$.

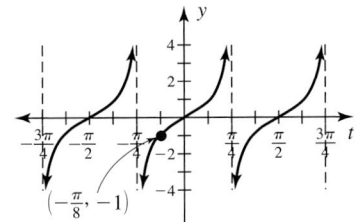

19. Determine the equation of the graph shown given it is of the form $y = A \sin(Bt \pm C) + D$.

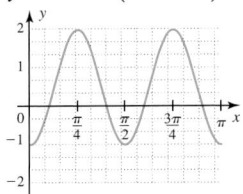

20. In London, the average temperatures on a summer day range from a high of 72°F to a low of 56°F (Source: 2004 Statistical Abstract of the United States, Table 1331). Use this information to write a sinusoidal equation model, assuming the low temperature occurs at 6:00 A.M. Clearly state the amplitude, average value, period, and horizontal shift.

21. The graph of a function $f(x)$ is given. Sketch the graph of $f^{-1}(x)$.

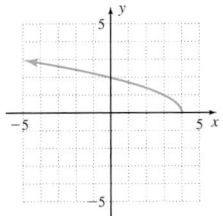

22. The volume of a spherical cap is given by
$$V = \frac{\pi h^2}{3}(3r - h).$$ Solve for r in terms of V and h.

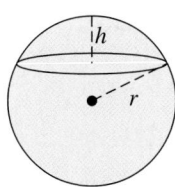

23. Find the slope and y-intercept: $3x - 4y = 8$.

24. Solve by factoring: $4x^3 - 8x^2 - 9x + 18 = 0$.

25. At what interest rate will $1000 grow to $2275 in 12 yr if compounded continuously?

CONNECTIONS TO CALCULUS

While right triangles have a number of meaningful applications as a problem-solving tool, they can also help to rewrite certain expressions in preparation for the tools of calculus, and introduce us to an alternative method for graphing relations and functions using polar coordinates. As things stand, some functions and relations are much easier to graph in polar coordinates, and converting between the two systems is closely connected to a study of right triangles.

Right Triangle Relationships

Drawing a diagram to visualize relationships and develop information is an important element of good problem solving. This is no less true in calculus, where it is often a fundamental part of understanding the question being asked. As a precursor to applications involving *trig substitutions*, we'll illustrate how right triangle diagrams are used to rewrite trigonometric functions of θ as algebraic functions of x.

EXAMPLE 1 ▶ **Using Right Triangle Diagrams to Rewrite Trig Expressions**

Use the equation $x = 5 \sin \theta$ and a right triangle diagram to write $\cos \theta$, $\tan \theta$, $\sec \theta$, and $\csc \theta$ as functions of x.

Solution ▶ Using $x = 5 \sin \theta$, we obtain $\sin \theta = \dfrac{x}{5}$. From our

work in Chapter 5, we know the right triangle

definition of $\sin \theta$ is $\dfrac{opp}{hyp}$, and we draw a triangle

with side x oriented opposite an angle θ, and label a
hypotenuse of 5 (see figure). To find an expression
for the adjacent side, we use the Pythagorean theorem:

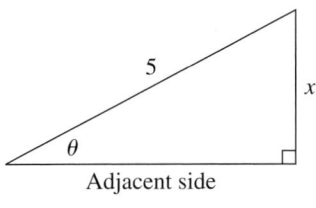

$$(\text{adj})^2 + x^2 = 5^2 \qquad \text{Pythagorean theorem}$$
$$(\text{adj})^2 = 25 - x^2 \qquad \text{isolate term}$$
$$\text{adj} = \sqrt{25 - x^2} \qquad \text{result (length must be positive); } -5 < x < 5$$

Using this triangle and the standard definition of the remaining trig functions, we

find $\cos \theta = \dfrac{\sqrt{25 - x^2}}{5}$, $\tan \theta = \dfrac{x}{\sqrt{25 - x^2}}$, $\sec \theta = \dfrac{5}{\sqrt{25 - x^2}}$, and $\csc \theta = \dfrac{5}{x}$.

Now try Exercises 1 through 4 ▶

EXAMPLE 2 ▶ **Using Right Triangle Diagrams to Rewrite Trig Expressions**

Find expressions for $\tan \theta$ and $\csc \theta$, given $\sec \theta = \dfrac{\sqrt{u^2 + 144}}{u}$.

Solution ▶ With $\sec \theta = \dfrac{hyp}{adj}$, we draw a right triangle

diagram as in Example 1, with a hypotenuse of
$\sqrt{u^2 + 144}$ and a side u adjacent to angle θ. For

$\tan \theta = \dfrac{opp}{hyp}$ and $\csc \theta = \dfrac{hyp}{opp}$, we use the

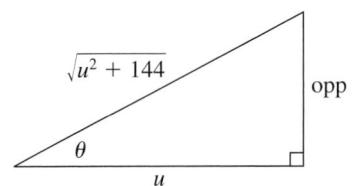

Pythagorean theorem to find an expression for the opposite side:

$$\text{opp}^2 + u^2 = (\sqrt{u^2 + 144})^2 \qquad \text{Pythagorean theorem}$$
$$\text{opp}^2 + u^2 = u^2 + 144 \qquad \text{square radical}$$
$$\text{opp}^2 = 144 \qquad \text{subtract } u^2$$
$$\text{opp} = 12 \qquad \text{result (length must be positive)}$$

With an opposite side of 12 units, the figure shows $\tan \theta = \dfrac{12}{u}$ and
$\csc \theta = \dfrac{\sqrt{u^2 + 144}}{12}$.

Now try Exercises 5 and 6 ▶

Converting from Rectangular Coordinates to Trigonometric (Polar) Form

Using the equations $x = r \cos \theta$, $y = r \sin \theta$, and $x^2 + y^2 = r^2$ from Section 5.2, we can rewrite functions of x given in rectangular form as equations of θ in trigonometric (*polar*) form. In Chapter 9 we'll see how this offers us certain advantages. Note that while algebraic equations are often written with y in terms of x, polar equations are written with r in terms of θ.

EXAMPLE 3 ▶ **Converting from Rectangular to Polar Form**

Rewrite the equation $2x + 3y = 6$ in trigonometric form using the substitutions indicated and solving for r. Note the given equation is that of a line with x-intercept $(3, 0)$ and y-intercept $(0, 2)$.

Solution ▶ Using $x = r \cos \theta$ and $y = r \sin \theta$ we proceed as follows:

$$2x + 3y = 6 \qquad \text{given}$$
$$2(r \cos \theta) + 3(r \sin \theta) = 6 \qquad \text{substitute } r \cos \theta \text{ for } x, r \sin \theta \text{ for } y$$
$$r[2 \cos \theta + 3 \sin \theta] = 6 \qquad \text{factor out } r$$
$$r = \frac{6}{2 \cos \theta + 3 \sin \theta} \qquad \text{solve for } r$$

This is the equation of the same line, but in trigonometric form.

Now try Exercises 7 through 10 ▶

While it is somewhat simplistic (there are other subtleties involved), we can verify the equation obtained in Example 3 produces the same line as $2x + 3y = 6$ by evaluating the equation at $\theta = 0°$ to find a point on the positive x-axis, $\theta = 90°$ to find a point of the positive y-axis, and $\theta = 135°$ to find a point in QII.

For $\theta = 0°$:
$$r = \frac{6}{2 \cos 0 + 3 \sin 0}$$
$$= \frac{6}{2(1) + 3(0)} = \frac{6}{2}$$
$$= 3$$

For $\theta = 90°$:
$$r = \frac{6}{2 \cos 90 + 3 \sin 90}$$
$$= \frac{6}{2(0) + 3(1)} = \frac{6}{3}$$
$$= 2$$

For $\theta = 135°$: $r = \dfrac{6}{2 \cos 135 + 3 \sin 135}$

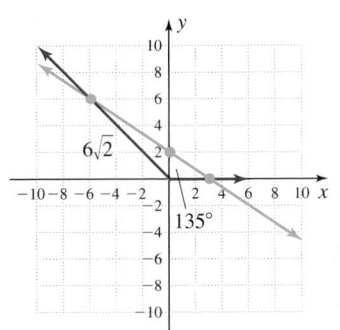

$$= \dfrac{6}{2\left(-\dfrac{\sqrt{2}}{2}\right) + 3\left(\dfrac{\sqrt{2}}{2}\right)} = \dfrac{6}{\dfrac{\sqrt{2}}{2}}$$

$$= \dfrac{12}{\sqrt{2}} = 6\sqrt{2} \approx 8.5$$

Using a distance r from the origin at each θ given, we note that all three points are on the line $2x + 3y = 6$, as shown in the figure.

EXAMPLE 4 ▶ **Converting from Polar to Rectangular Form**

Rewrite the equation $r = 4 \cos \theta$ in rectangular form using the relationships $x = r \cos \theta$, $y = r \sin \theta$, and/or $x^2 + y^2 = r^2$. Identify the resulting equation.

Solution ▶ From $x = r \cos \theta$ we have $\cos \theta = \dfrac{x}{r}$. Substituting into the given equation we have

$$r = 4 \cos \theta \quad \text{given}$$
$$r = 4\left(\dfrac{x}{r}\right) \quad \text{substitute } \dfrac{x}{r} \text{ for } \cos \theta$$
$$r^2 = 4x \quad \text{multiply by } r$$
$$x^2 + y^2 = 4x \quad \text{substitute } x^2 + y^2 \text{ for } r^2$$
$$x^2 - 4x + y^2 = 0 \quad \text{set equal to 0}$$
$$(x^2 - 4x + 4) + y^2 = 4 \quad \text{complete the square in } x$$
$$(x - 2)^2 + y^2 = 4 \quad \text{standard form}$$

The result shows $r = 4 \cos \theta$ is a trigonometric form for the equation of a circle with center at $(2, 0)$ and radius $r = 2$.

Now try Exercises 11 through 14 ▶

Connections to Calculus Exercises

Use the diagram given to find the remaining side of the triangle. Then write the six trig functions as functions of x.

1.

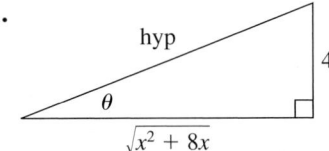

2.

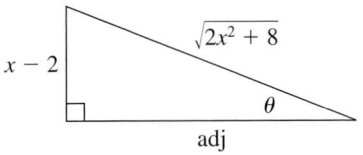

Use the equation given and a sketch of the corresponding right triangle diagram to write the remaining five trig functions as functions of x.

3. $x = 4 \tan \theta$

4. $x = 5 \sec \theta$

5. Find expressions for $\cot \theta$ and $\sec \theta$, given
$$\csc \theta = \dfrac{\sqrt{u^2 + 169}}{u}.$$

6. Find expressions for $\sin \theta$ and $\cos \theta$, given
$$\cot \theta = \dfrac{2\sqrt{5}}{x}.$$

Use $x = r \cos \theta$, $y = r \sin \theta$, and $x^2 + y^2 = r^2$ to rewrite the expressions in trigonometric form.

7. $y = 2$

8. $y = \dfrac{1}{4}x^2$

9. $y = -2x + 3$

10. $(x - 2)^2 + y^2 = 4$

Use $x = r \cos \theta$, $y = r \sin \theta$, and $x^2 + y^2 = r^2$ to rewrite the expressions in rectangular form, then identify the equation as that of a line, circle, vertical parabola, or horizontal parabola.

11. $r = 5 \sin \theta$

12. $r = \dfrac{4}{1 + \sin \theta}$ $\quad \left(Hint\text{: } \sin \theta = \dfrac{y}{r}. \right)$

13. $r(3 \cos \theta - 2 \sin \theta) = 6$

14. $r = \dfrac{4}{2 - 2 \cos \theta}$ $\quad \left(Hint\text{: } \cos \theta = \dfrac{x}{r}. \right)$

6

Trigonometric Identities, Inverses, and Equations

CHAPTER OUTLINE

CHAPTER CONNECTIONS

Have you ever noticed that people who arrive early at a movie tend to choose seats about halfway up the theater's incline and in the middle of a row? More than likely, this is due to a phenomenon called the *optimal viewing angle*, or the angle formed by the viewer's eyes and the top and bottom of the screen. Seats located in this area maximize the viewing angle, with the measure of the angle depending on factors such as the distance from the floor to the bottom of the screen, the height of the screen, the location of a seat, and the incline of the auditorium. Here, trigonometric functions and identities play an important role. This application appears as Exercise 59 of Section 6.2.

Connections to Calculus

Trigonometric equations, identities, and substitutions also play a vital role in a study of calculus, helping to simplify complex expressions, or rewrite an expression in a form more suitable for the tools of calculus. These connections are explored in the *Connections to Calculus* feature following Chapter 6.

Learning Objectives

In Section 6.1 you will learn how to:

☐ **A.** Use fundamental identities to help understand and recognize identity "families"

☐ **B.** Verify other identities using the fundamental identities and basic algebra skills

☐ **C.** Use fundamental identities to express a given trig function in terms of the other five

☐ **D.** Use counterexamples and contradictions to show an equation is not an identity

In this section we begin laying the foundation necessary to work with identities successfully. The cornerstone of this effort is a healthy respect for the fundamental identities and vital role they play. Students are strongly encouraged to do more than memorize them—they should be *internalized,* meaning they must become a natural and instinctive part of your core mathematical knowledge.

A. Fundamental Identities and Identity Families

An **identity** is an equation that is true for all elements in the domain. In trigonometry, some identities result directly from the way the trig functions are defined. For instance, since $\sin\theta = \dfrac{y}{r}$ and $\csc\theta = \dfrac{r}{y}$, $\dfrac{1}{\csc\theta} = \dfrac{y}{r}$, and the identity $\sin\theta = \dfrac{1}{\csc\theta}$ immediately follows. We call identities of this type *fundamental identities.* Successfully working with *other* identities will depend a great deal on your mastery of these fundamental types. For convenience, the definition of the trig functions are reviewed here, followed by the fundamental identities that result.

Given point $P(x, y)$ on the unit circle, and the central angle θ associated with P, we have $\sqrt{x^2 + y^2} = 1$ and

$$\cos\theta = x \qquad \sin\theta = y \qquad \tan\theta = \frac{y}{x}; x \neq 0$$

$$\sec\theta = \frac{1}{x}; x \neq 0 \qquad \csc\theta = \frac{1}{y}; y \neq 0 \qquad \cot\theta = \frac{x}{y}; y \neq 0$$

WORTHY OF NOTE

The word *fundamental* itself means, "a basis or foundation supporting an essential structure or function" (Merriam Webster).

Fundamental Trigonometric Identities

Reciprocal identities	Ratio identities	Pythagorean identities	Identities due to symmetry
$\sin\theta = \dfrac{1}{\csc\theta}$	$\tan\theta = \dfrac{\sin\theta}{\cos\theta}$	$\sin^2\theta + \cos^2\theta = 1$	$\sin(-\theta) = -\sin\theta$
$\cos\theta = \dfrac{1}{\sec\theta}$	$\tan\theta = \dfrac{\sec\theta}{\csc\theta}$	$\tan^2\theta + 1 = \sec^2\theta$	$\cos(-\theta) = \cos\theta$
$\tan\theta = \dfrac{1}{\cot\theta}$	$\cot\theta = \dfrac{\cos\theta}{\sin\theta}$	$1 + \cot^2\theta = \csc^2\theta$	$\tan(-\theta) = -\tan\theta$

These identities seem to naturally separate themselves into the four groups or families listed, with each group having additional relationships that can be inferred from the definitions. For instance, since $\sin\theta$ is the reciprocal of $\csc\theta$, $\csc\theta$ must be the reciprocal of $\sin\theta$. Similar statements can be made regarding $\cos\theta$ and $\sec\theta$ as well as $\tan\theta$ and $\cot\theta$. Recognizing these additional "family members" enlarges the number of identities you can work with, and will help you use them more effectively. In particular, since they *are* reciprocals: $\sin\theta\csc\theta = 1$, $\cos\theta\sec\theta = 1$, $\tan\theta\cot\theta = 1$. **See Exercises 7 and 8.**

EXAMPLE 1 ▶ **Identifying Families of Identities**

Use algebra to write four additional identities that belong to the Pythagorean family.

Solution ▶ Starting with $\sin^2\theta + \cos^2\theta = 1$,

$$\sin^2\theta + \cos^2\theta = 1 \qquad \text{original identity}$$
$$\bullet \quad \sin^2\theta = 1 - \cos^2\theta \qquad \text{subtract } \cos^2\theta$$
$$\bullet \quad \sin\theta = \pm\sqrt{1 - \cos^2\theta} \qquad \text{take square root}$$
$$\sin^2\theta + \cos^2\theta = 1 \qquad \text{original identity}$$
$$\bullet \quad \cos^2\theta = 1 - \sin^2\theta \qquad \text{subtract } \sin^2\theta$$
$$\bullet \quad \cos\theta = \pm\sqrt{1 - \sin^2\theta} \qquad \text{take square root}$$

For the identities involving a radical, the choice of sign will depend on the quadrant of the terminal side.

Now try Exercises 9 and 10 ▶

The four additional Pythagorean identities are marked with a "•" in Example 1. The fact that each of them represents an equality gives us more options when attempting to verify or prove more complex identities. For instance, since $\cos^2\theta = 1 - \sin^2\theta$, we can replace $\cos^2\theta$ with $1 - \sin^2\theta$, or replace $1 - \sin^2\theta$ with $\cos^2\theta$, *any time they occur in an expression.* Note there are many other members of this family, since similar steps can be performed on the other Pythagorean identities. In fact, each of the fundamental identities can be similarly rewritten and there are a variety of exercises at the end of this section for practice.

☑ **A.** You've just learned how to use fundamental identities to help understand and recognize identity "families"

B. Verifying an Identity Using Algebra

Note that we cannot *prove* an equation is an identity by repeatedly substituting input values and obtaining a true equation. This would be an infinite exercise and we might easily miss a value or even a range of values for which the equation is false. Instead we attempt to rewrite one side of the equation until we obtain a match with the other side, so there can be no doubt. As hinted at earlier, this is done using basic algebra skills combined with the fundamental identities and the substitution principle. For now we'll focus on verifying identities by using algebra. In Section 6.2 we'll introduce some guidelines and ideas that will help you verify a wider range of identities.

EXAMPLE 2 ▶ **Using Algebra to Help Verify an Identity**

Use the distributive property to verify that $\sin\theta(\csc\theta - \sin\theta) = \cos^2\theta$ is an identity.

Solution ▶ Use the distributive property to simplify the left-hand side.

$$\sin\theta(\csc\theta - \sin\theta) = \sin\theta\csc\theta - \sin^2\theta \qquad \text{distribute}$$
$$= 1 - \sin^2\theta \qquad \text{substitute 1 for } \sin\theta\csc\theta$$
$$= \cos^2\theta \qquad 1 - \sin^2\theta = \cos^2\theta$$

Since we were able to transform the left-hand side into a duplicate of the right, there can be no doubt the original equation is an identity.

Now try Exercises 11 through 20 ▶

Often we must *factor* an expression, rather than multiply, to begin the verification process.

EXAMPLE 3 ▶ **Using Algebra to Help Verify an Identity**

Verify that $1 = \cot^2 x \sec^2 x - \cot^2 x$ is an identity.

Solution ▶ The left side is as simple as it gets. The terms on the right side have a common factor and we begin there.

$$\cot^2 x \sec^2 x - \cot^2 x = \cot^2 x (\sec^2 x - 1) \quad \text{factor out } \cot^2 x$$
$$= \cot^2 x \tan^2 x \quad \text{substitute } \tan^2 x \text{ for } \sec^2 x - 1$$
$$= (\cot x \tan x)^2 \quad \text{power property of exponents}$$
$$= 1^2 = 1 \quad \cot x \, \tan x = 1$$

Now try Exercises 21 through 28 ▶

Examples 2 and 3 show you can begin the verification process on either the left or right side of the equation, whichever seems more convenient. Example 4 shows how the special products $(A + B)(A - B) = A^2 - B^2$ and/or $(A \pm B)^2 = A^2 \pm 2AB + B^2$ can be used in the verification process.

EXAMPLE 4 ▶ Using a Special Product to Help Verify an Identity

Use a special product and fundamental identities to verify that $(\sin x - \cos x)^2 = 1 + 2\sin(-x)\cos x$ is an identity.

Solution ▶ Begin by squaring the left-hand side, in hopes of using a Pythagorean identity.

$$(\sin x - \cos x)^2 = \sin^2 x - 2\sin x \cos x + \cos^2 x \quad \text{binomial square}$$
$$= \sin^2 x + \cos^2 x - 2\sin x \cos x \quad \text{rewrite terms}$$
$$= 1 - 2\sin x \cos x \quad \text{substitute 1 for } \sin^2 x + \cos^2 x$$

At this point we appear to be off by a sign, but quickly recall that sine is on odd function and $-\sin x = \sin(-x)$. By writing $1 - 2\sin x \cos x$ as $1 + 2(-\sin x)(\cos x)$, we can complete the verification:

$$= 1 + 2(-\sin x)(\cos x) \quad \text{rewrite expression to obtain } -\sin x$$
$$= 1 + 2\sin(-x)\cos x \checkmark \quad \text{substitute } \sin(-x) \text{ for } -\sin x$$

Now try Exercises 29 through 34 ▶

☑ **B.** You've just learned how to verify other identities using the fundamental identities and basic algebra skills

Another common method used to verify identities is simplification by combining terms, using the model $\dfrac{A}{B} \pm \dfrac{C}{D} = \dfrac{AD \pm BC}{BD}$. For $\sec u = \dfrac{\sin^2 u}{\cos u} + \cos u$, the right-hand side immediately becomes $\dfrac{\sin^2 u + \cos^2 u}{\cos u}$, which gives $\dfrac{1}{\cos u} = \sec u$. **See Exercises 35 through 40.**

C. Writing One Function in Terms of Another

Any one of the six trigonometric functions can be written in terms of any of the other functions using fundamental identities. The process involved offers practice in working with identities, highlights how each function is related to the other, and has practical applications in verifying more complex identities.

EXAMPLE 5 ▶ Writing One Trig Function in Terms of Another

Write the function $\cos x$ in terms of the tangent function.

Solution ▶ Begin by noting these functions share "common ground" via $\sec x$, since $\sec^2 x = 1 + \tan^2 x$ and $\cos x = \dfrac{1}{\sec x}$. Starting with $\sec^2 x$,

$$\sec^2 x = 1 + \tan^2 x \quad \text{Pythagorean identity}$$
$$\sec x = \pm\sqrt{1 + \tan^2 x} \quad \text{square roots}$$

We can now substitute $\pm\sqrt{1 + \tan^2 x}$ for sec x in $\cos x = \dfrac{1}{\sec x}$.

$$\cos x = \dfrac{1}{\pm\sqrt{1 + \tan^2 x}} \qquad \text{substitute } \pm\sqrt{1 + \tan^2 x} \text{ for sec } x$$

Now try Exercises 41 through 46 ▶

Example 5 also reminds us of a very important point — the sign we choose for the final answer is dependent on the terminal side of the angle. If the terminal side is in QI or QIV we chose the positive sign since $\cos x > 0$ in those quadrants. If the angle terminates in QII or QIII, the final answer is negative since $\cos x < 0$ in those quadrants.

Similar to our work in Chapter 5, given the value of cot t and the quadrant of t, the fundamental identities enable us to find the value of the other five functions at t. In fact, this is generally true for any given trig function and real number or angle t.

EXAMPLE 6 ▶ **Using a Known Value and Quadrant Analysis to Find Other Function Values**

Given $\cot t = \dfrac{-9}{40}$ with t in QIV, find the value of the other five functions at t.

Solution ▶ The function value $\tan t = -\dfrac{40}{9}$ follows immediately, since cotangent and tangent are reciprocals. The value of sec t can be found using $\sec^2 t = 1 + \tan^2 t$.

$$\sec^2 t = 1 + \tan^2 t \qquad \text{Pythagorean identity}$$

$$= 1 + \left(-\dfrac{40}{9}\right)^2 \qquad \text{substitute } -\dfrac{40}{9} \text{ for tan } t$$

$$= \dfrac{81}{81} + \dfrac{1600}{81} \qquad \text{square } -\dfrac{40}{9}, \text{ substitute } \dfrac{81}{81} \text{ for 1}$$

$$= \dfrac{1681}{81} \qquad \text{combine terms}$$

$$\sec t = \pm\dfrac{41}{9} \qquad \text{take square roots}$$

Since sec t is positive in QIV, we have $\sec t = \dfrac{41}{9}$. This automatically gives

$\cos t = \dfrac{9}{41}$ (reciprocal identities), and we find $\sin t = -\dfrac{40}{41}$ using $\sin^2 t = 1 - \cos^2 t$

or the ratio identity $\tan t = \dfrac{\sin t}{\cos t}$ (verify).

☑ **C.** You've just learned how to use fundamental identities to express a given trig function in terms of the other five

Now try Exercises 47 through 55 ▶

D. Showing an Equation Is Not an Identity

To show an equation is *not* an identity, we need only find a single value for which the functions involved are defined but the equation is *false*. This can often be done by trial and error, or even by inspection. To illustrate the process, we'll use two common misconceptions that arise in working with identities.

EXAMPLE 7 ▶ **Showing an Equation is Not an Identity**

Show the equations given are *not* identities.

  a. $\sin(2x) = 2\sin x$ **b.** $\cos(\alpha + \beta) = \cos \alpha + \cos \beta$

Solution ▶   **a.** The assumption here seems to be that we can factor out the coefficient from the argument. By inspection we note the amplitude of $\sin(2x)$ is $A = 1$, while the amplitude of $2\sin x$ is $A = 2$. This means they cannot possibly be equal for all values of x, although they are equal for integer multiples of π. Verify they are not equivalent using $x = \dfrac{\pi}{6}$ or other standard values.

GRAPHICAL SUPPORT

While not a definitive method of proof, a graphing calculator can be used to investigate whether an equation is an identity. Since the left and right members of the equation must be equal for all values (where they are defined), their graphs must be identical. Graphing the functions from Example 7(a) as Y_1 and Y_2 shows the equation $sin(2x) = 2\,sin\,x$ is definitely *not* an identity.

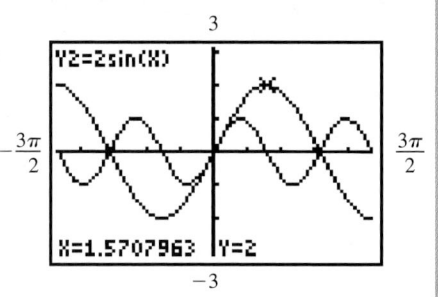

---

  b. The assumption here is that we can distribute function values. This is similar to saying $\sqrt{x + 4} = \sqrt{x} + 2$, a statement obviously false for all values except $x = 0$. Here we'll substitute convenient values to prove the equation false, namely, $\alpha = \dfrac{3\pi}{4}$ and $\beta = \dfrac{\pi}{4}$.

$$\cos\left(\frac{3\pi}{4} + \frac{\pi}{4}\right) = \cos\left(\frac{3\pi}{4}\right) + \cos\left(\frac{\pi}{4}\right) \qquad \text{substitute } \tfrac{\pi}{3} \text{ for } \alpha \text{ and } \tfrac{\pi}{4} \text{ for } \beta$$

$$\cos \pi = -\frac{\sqrt{2}}{2} + \frac{\sqrt{2}}{2} \qquad \text{simplify}$$

$$-1 \neq 0 \qquad \text{result is false}$$

☑ **D. You've just learned how to use counterexamples and contradictions to show an equation is not an identity**

Now try Exercises 56 through 62 ▶

6.1 EXERCISES

▶ CONCEPTS AND VOCABULARY

Fill in each blank with the appropriate word or phrase. Carefully reread the section if needed.

1. Three fundamental ratio identities are $\tan\theta = \dfrac{?}{\cos\theta}$, $\tan\theta = \dfrac{?}{\csc\theta}$, and $\cot\theta = \dfrac{?}{\sin\theta}$.

2. When applying identities due to symmetry, $\sin(-x)\tan x = $ _____ and $\cos(-x)\cot x = $ _____.

3. To show an equation is *not an identity*, we must find at least ____ value(s) where both sides of the equation are defined, but which makes the equation _____.

4. Using a calculator we find $\sec^2 45° = $ ____ and $3\tan 45° - 1 = $ ____. We also find $\sec^2 225° = $ ____ and $3\tan 225° - 1 = $ ____. Is the equation $\sec^2\theta = 3\tan\theta - 1$ an identity?

5. Use the pattern $\dfrac{A}{B} \pm \dfrac{C}{D} = \dfrac{AD \pm BC}{BD}$ to add the following terms, and comment on this process versus "finding a common denominator:" $\dfrac{\cos x}{\sin x} - \dfrac{\sin x}{\sec x}$.

6. Name at least four algebraic skills that are used with the fundamental identities in order to rewrite a trigonometric expression. Use algebra to quickly rewrite $(\sin x + \cos x)^2$.

▶ DEVELOPING YOUR SKILLS

Starting with the ratio identity given, use substitution and fundamental identities to write four new identities belonging to the ratio family. Answers may vary.

7. $\tan x = \dfrac{\sin x}{\cos x}$ **8.** $\cot x = \dfrac{\cos x}{\sin x}$

Starting with the Pythagorean identity given, use algebra to write four additional identities belonging to the Pythagorean family. Answers may vary.

9. $1 + \tan^2 x = \sec^2 x$ **10.** $1 + \cot^2 x = \csc^2 x$

Verify the equation is an identity using multiplication and fundamental identities.

11. $\sin x \cot x = \cos x$ **12.** $\cos x \tan x = \sin x$

13. $\sec^2 x \cot^2 x = \csc^2 x$ **14.** $\csc^2 x \tan^2 x = \sec^2 x$

15. $\cos x (\sec x - \cos x) = \sin^2 x$

16. $\tan x (\cot x + \tan x) = \sec^2 x$

17. $\sin x (\csc x - \sin x) = \cos^2 x$

18. $\cot x (\tan x + \cot x) = \csc^2 x$

19. $\tan x (\csc x + \cot x) = \sec x + 1$

20. $\cot x (\sec x + \tan x) = \csc x + 1$

Verify the equation is an identity using factoring and fundamental identities.

21. $\tan^2 x \csc^2 x - \tan^2 x = 1$

22. $\sin^2 x \cot^2 x + \sin^2 x = 1$

23. $\dfrac{\sin x \cos x + \sin x}{\cos x + \cos^2 x} = \tan x$

24. $\dfrac{\sin x \cos x + \cos x}{\sin x + \sin^2 x} = \cot x$

25. $\dfrac{1 + \sin x}{\cos x + \cos x \sin x} = \sec x$

26. $\dfrac{1 + \cos x}{\sin x + \cos x \sin x} = \csc x$

27. $\dfrac{\sin x \tan x + \sin x}{\tan x + \tan^2 x} = \cos x$

28. $\dfrac{\cos x \cot x + \cos x}{\cot x + \cot^2 x} = \sin x$

Verify the equation is an identity using special products and fundamental identities.

29. $\dfrac{(\sin x + \cos x)^2}{\cos x} = \sec x + 2 \sin x$

30. $\dfrac{(1 + \tan x)^2}{\sec x} = \sec x + 2 \sin x$

31. $(1 + \sin x)[1 + \sin(-x)] = \cos^2 x$

32. $(\sec x + 1)[\sec(-x) - 1] = \tan^2 x$

33. $\dfrac{(\csc x - \cot x)(\csc x + \cot x)}{\tan x} = \cot x$

34. $\dfrac{(\sec x + \tan x)(\sec x - \tan x)}{\csc x} = \sin x$

Verify the equation is an identity using fundamental identities and $\dfrac{A}{B} \pm \dfrac{C}{D} = \dfrac{AD \pm BC}{BD}$ to combine terms.

35. $\dfrac{\cos^2 x}{\sin x} + \dfrac{\sin x}{1} = \csc x$

36. $\dfrac{\sec \alpha}{1} - \dfrac{\tan^2 \alpha}{\sec \alpha} = \cos \alpha$

37. $\dfrac{\tan x}{\csc x} - \dfrac{\sin x}{\cos x} = \dfrac{\sin x - 1}{\cot x}$

38. $\dfrac{\cot x}{\sec x} - \dfrac{\cos x}{\sin x} = \dfrac{\cos x - 1}{\tan x}$

39. $\dfrac{\sec x}{\sin x} - \dfrac{\csc x}{\sec x} = \tan x$ **40.** $\dfrac{\csc x}{\cos x} - \dfrac{\sec x}{\csc x} = \cot x$

Write the given function entirely in terms of the second function indicated.

41. $\tan x$ in terms of $\sin x$ **42.** $\tan x$ in terms of $\sec x$

43. $\sec x$ in terms of $\cot x$ **44.** $\sec x$ in terms of $\sin x$

45. $\cot x$ in terms of $\sin x$ **46.** $\cot x$ in terms of $\csc x$

For the function $f(\theta)$ and the quadrant in which θ terminates, state the value of the other five trig functions.

47. $\cos \theta = -\dfrac{20}{29}$ with θ in QII

48. $\sin \theta = \dfrac{12}{37}$ with θ in QII

49. $\tan \theta = \dfrac{15}{8}$ with θ in QIII

50. $\sec \theta = \dfrac{45}{27}$ with θ in QIV

51. $\cot \theta = \dfrac{x}{5}$ with θ in QI

52. $\csc\theta = \dfrac{7}{x}$ with θ in QII

53. $\sin\theta = -\dfrac{7}{13}$ with θ in QIII

54. $\cos\theta = \dfrac{23}{25}$ with θ in QIV

55. $\sec\theta = -\dfrac{9}{7}$ with θ in QII

Show that the following equations *are not identities*.

56. $\sin\left(\theta + \dfrac{\pi}{3}\right) = \sin\theta + \sin\left(\dfrac{\pi}{3}\right)$

57. $\cos\left(\dfrac{\pi}{4}\right) + \cos\theta = \cos\left(\dfrac{\pi}{4} + \theta\right)$

58. $\cos(2\theta) = 2\cos\theta$

59. $\tan(2\theta) = 2\tan\theta$

60. $\tan\left(\dfrac{\theta}{4}\right) = \dfrac{\tan\theta}{\tan 4}$

61. $\cos^2\theta - \sin^2\theta = -1$

62. $\sqrt{\sin^2 x - 9} = \sin x - 3$

▶ WORKING WITH FORMULAS

63. The illuminance of a point on a surface by a source of light: $E = \dfrac{I\cos\theta}{r^2}$

The illuminance E (in lumens/m²) of a point on a horizontal surface is given by the formula shown, where I is the intensity of the light source in lumens, r is the distance in meters from the light source to the point, and θ is the complement of the angle α (in degrees) made by the light source and the horizontal surface. Calculate the illuminance if $I = 800$ lumens, and the flashlight is held so that the distance r is 2 m while the angle α is 40°.

Exercise 63

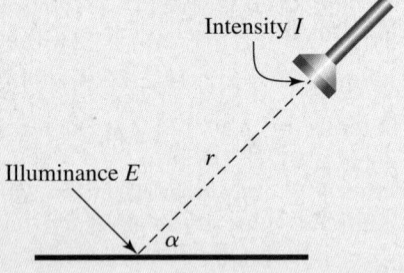

Intensity I

Illuminance E

r

α

64. The area of regular polygon: $A = \left(\dfrac{nx^2}{4}\right)\dfrac{\cos(\frac{\pi}{n})}{\sin(\frac{\pi}{n})}$

The area of a regular polygon is given by the formula shown, where n represents the number of sides and x is the length of each side.

  a. Rewrite the formula in terms of a single trig function.

  b. Verify the formula for a square with sides of 8 m.

  c. Find the area of a dodecagon (12 sides) with 10-in. sides.

▶ APPLICATIONS

Writing a given expression in an alternative form is an idea used at all levels of mathematics. In future classes, it is often helpful to decompose a power into smaller powers (as in writing A^3 as $A \cdot A^2$) or to rewrite an expression using known identities so that it can be factored.

  65. Show that $\cos^3 x$ can be written as $\cos x(1 - \sin^2 x)$.

  66. Show that $\tan^3 x$ can be written as $\tan x(\sec^2 x - 1)$.

  67. Show that $\tan x + \tan^3 x$ can be written as $\tan x(\sec^2 x)$.

68. Show that $\cot^3 x$ can be written as $\cot x(\csc^2 x - 1)$.

69. Show $\tan^2 x \sec x - 4\tan^2 x$ can be factored into $(\sec x - 4)(\sec x - 1)(\sec x + 1)$.

70. Show $2\sin^2 x \cos x - \sqrt{3}\sin^2 x$ can be factored into $(1 - \cos x)(1 + \cos x)(2\cos x - \sqrt{3})$.

71. Show $\cos^2 x \sin x - \cos^2 x$ can be factored into $-1(1 + \sin x)(1 - \sin x)^2$.

72. Show $2\cot^2 x \csc x + 2\sqrt{2}\cot^2 x$ can be factored into $2(\csc x + \sqrt{2})(\csc x - 1)(\csc x + 1)$.

▣ **Many applications of fundamental identities involve geometric figures, as in Exercises 73 and 74.**

73. **Area of a polygon:** The area of a regular polygon that has been circumscribed about a circle of radius r (see figure) is given by the formula $A = nr^2 \dfrac{\sin(\frac{\pi}{n})}{\cos(\frac{\pi}{n})}$, where n represents the number of sides. (a) Rewrite the formula in terms of a single trig function; (b) verify the formula for a square circumscribed about a circle with radius 4 m; and (c) find the area of a dodecagon (12 sides) circumscribed about the same circle.

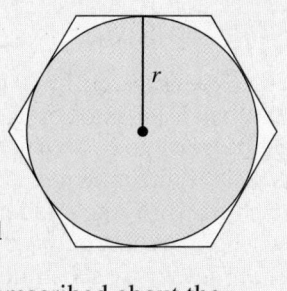

74. **Perimeter of a polygon:** The perimeter of a regular polygon circumscribed about a circle of radius r is given by the formula $P = 2nr \dfrac{\sin(\frac{\pi}{n})}{\cos(\frac{\pi}{n})}$,

where n represents the number of sides. (a) Rewrite the formula in terms of a single trig function; (b) verify the formula for a square circumscribed about a circle with radius 4 m; and (c) Find the perimeter of a dodecagon (12 sides) circumscribed about the same circle.

75. **Angle of intersection:** At their point of intersection, the angle θ between any two nonparallel lines satisfies the relationship $(m_2 - m_1)\cos \theta = \sin \theta + m_1 m_2 \sin \theta$, where m_1 and m_2 represent the slopes of the two lines. Rewrite the equation in terms of a single trig function.

76. **Angle of intersection:** Use the result of Exercise 75 to find the angle between the lines $Y_1 = \dfrac{2}{5}x - 3$ and $Y_2 = \dfrac{7}{3}x + 1$.

77. **Angle of intersection:** Use the result of Exercise 75 to find the angle between the lines $Y_1 = 3x - 1$ and $Y_2 = -2x + 7$.

▶ **EXTENDING THE CONCEPT**

78. The word *tangent* literally means "to touch," which in mathematics we take to mean *touches in only and exactly one point.* In the figure, the circle has a radius of 1 and the vertical line is

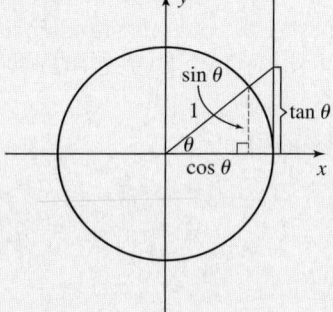

"tangent" to the circle at the x-axis. The figure can be used to verify the Pythagorean identity for sine and cosine, as well as the ratio identity for tangent. Discuss/Explain how.

79. Use factoring and fundamental identities to help find the x-intercepts of f in $[0, 2\pi)$.
$$f(\theta) = -2 \sin^4\theta + \sqrt{3} \sin^3\theta + 2 \sin^2\theta - \sqrt{3} \sin\theta.$$

▶ **MAINTAINING YOUR SKILLS**

80. (4.6) Solve for x:
$$2351 = \frac{2500}{1 + e^{-1.015x}}$$

81. (5.6) Standing 265 ft from the base of the Strastosphere Tower in Las Vegas, Nevada, the angle of elevation to the top of the tower is about 77°. Approximate the height of the tower to the nearest foot.

82. (3.3) Use the rational zeroes theorem and other "tools" to find all zeroes of the function $f(x) = 2x^4 + 9x^3 - 4x^2 - 36x - 16$.

83. (5.3) Use a reference rectangle and the *rule of fourths* to sketch the graph of $y = 2 \sin(2t)$ for t in $[0, 2\pi)$.

6.2 | Constructing and Verifying Identities

Learning Objectives

In Section 6.2 you will learn how to:

☐ **A.** Create and verify a new identity

☐ **B.** Verify general identities

In Section 6.1, our primary goal was to illustrate how basic algebra skills are used to help rewrite trigonometric expressions. In this section, we'll sharpen and refine these skills so they can be applied more generally, as we develop the ability to verify a much wider range of identities.

A. Creating and Verifying Identities

In Example 2 of Section 6.1, we showed $\sin\theta\,(\csc\theta - \sin\theta) = \cos^2\theta$ was an identity by transforming the left-hand side into $\cos^2\theta$. There, the instructions were very specific: "Use the distributive property to . . ." When verifying identities, one of the biggest issues students face is that the directions are deliberately vague—because there is no single, fail-proof approach for verifying an identity. This sometimes leaves students feeling they don't know where to start, or what to do first. To help overcome this discomfort, we'll first *create an identity* by substituting fundamental identities into a given expression, then reverse these steps to get back the original expression. This return to the original illustrates the essence of verifying identities, namely, if two things are equal, one can be substituted for the other at any time. The process may seem arbitrary (actually—it *is*), and the steps could vary. But try to keep the underlying message in mind, rather than any specific steps. When working with identities, there is actually *no right place* to start, and the process begins by using the substitution principle to create an equivalent expression as you work toward the expression you're trying to match.

EXAMPLE 1 ▶ **Creating and Verifying an Identity**

Starting with the expression $\csc x + \cot x$, use fundamental identities to rewrite the expression and create a new identity. Then verify the identity by reversing the steps.

Solution ▶
$$\csc x + \cot x \qquad \text{original expression}$$

$$= \frac{1}{\sin x} + \frac{\cos x}{\sin x} \qquad \text{substitute reciprocal and ratio identities}$$

$$= \frac{1 + \cos x}{\sin x} \qquad \text{write as a single term}$$

Resulting identity ▶
$$\csc x + \cot x = \frac{1 + \cos x}{\sin x}$$

Verify identity ▶ Working with the right-hand side, we reverse each step with a view toward the original expression.

$$\frac{1 + \cos x}{\sin x} = \frac{1}{\sin x} + \frac{\cos x}{\sin x} \qquad \text{rewrite as individual terms}$$

$$= \csc x + \cot x \qquad \text{substitute reciprocal and ratio identities}$$

Now try Exercises 7 through 9 ▶

In actual practice, all you'll see is this instruction, "Verify the following is an identity: $\csc x + \cot x = \dfrac{1 + \cos x}{\sin x}$," and it will be up to you to employ the algebra and fundamental identities needed.

EXAMPLE 2 ▶ Creating and Verifying an Identity

Starting with the expression $2 \tan x \sec x$, use fundamental identities to rewrite the expression and create a new identity. Then verify the identity by reversing the steps.

Solution ▶ $2 \tan x \sec x$ original expression

$$= 2 \cdot \frac{\sin x}{\cos x} \cdot \frac{1}{\cos x} \qquad \text{substitute ratio and reciprocal identities}$$

$$= \frac{2 \sin x}{\cos^2 x} \qquad \text{multiply}$$

$$= \frac{2 \sin x}{1 - \sin^2 x} \qquad \text{substitute } 1 - \sin^2 x \text{ for } \cos^2 x$$

Resulting identity ▶ $2 \tan x \sec x = \dfrac{2 \sin x}{1 - \sin^2 x}$ identity

Verify identity ▶ Working with the right-hand side, we reverse each step with a view toward the original expression.

$$\frac{2 \sin x}{1 - \sin^2 x} = \frac{2 \sin x}{\cos^2 x} \qquad \text{substitute } \cos^2 x \text{ for } 1 - \sin^2 x$$

$$= 2 \cdot \frac{\sin x}{\cos x} \cdot \frac{1}{\cos x} \qquad \text{substitute } \cos x \cdot \cos x \text{ for } \cos^2 x$$

$$= 2 \tan x \sec x \qquad \text{substitute ratio and reciprocal identities}$$

☑ **A.** You've just learned how to create and verify a new identity

Now try Exercises 10 through 12 ▶

B. Verifying Identities

We're now ready to put these ideas, and the ideas from Section 6.1, to work for us. When verifying identities we attempt to mold, change, or rewrite one side of the equality until we obtain a match with the other side. What follows is a collection of the ideas and methods we've observed so far, which we'll call the *Guidelines for Verifying Identities*. But remember, there really is no *right* place to start. Think things over for a moment, then attempt a substitution, simplification, or operation and see where it leads. If you hit a dead end, that's okay! Just back up and try something else.

WORTHY OF NOTE

When verifying identities, it is actually permissible to work on each side of the equality *independently,* in the effort to create a "match." But properties of equality can never be used, since we cannot assume an equality exists.

Guidelines for Verifying Identities

1. As a general rule, work on only one side of the identity.
   • We cannot assume the equation is true, so properties of equality cannot be applied.
   • We verify the identity by changing the form of one side until we get a match with the other.
2. Work with the more complex side, as it is easier to reduce/simplify than to "build."
3. If an expression contains more than one term, it is often helpful to combine terms using $\dfrac{A}{B} \pm \dfrac{C}{D} = \dfrac{AD \pm BC}{BD}$.
4. Converting all functions to sines and cosines can be helpful.
5. Apply other algebra skills as appropriate: distribute, factor, multiply by a conjugate, and so on.
6. *Know the fundamental identities inside out, upside down, and backward—they are the key!*

Note how these ideas are employed in Examples 3 through 5, particularly the frequent use of fundamental identities.

EXAMPLE 3 ▶ **Verifying an Identity**

Verify the identity: $\sin^2\theta \tan^2\theta = \tan^2\theta - \sin^2\theta$.

Solution ▶ As a general rule, the side with the greater number of terms or the side with rational terms is considered "more complex," so we begin with the right-hand side.

$$\tan^2\theta - \sin^2\theta = \frac{\sin^2\theta}{\cos^2\theta} - \sin^2\theta \qquad \text{substitute } \tfrac{\sin^2\theta}{\cos^2\theta} \text{ for } \tan^2\theta$$

$$= \frac{\sin^2\theta}{1} \cdot \frac{1}{\cos^2\theta} - \sin^2\theta \qquad \text{decompose rational term}$$

$$= \sin^2\theta \sec^2\theta - \sin^2\theta \qquad \text{substitute } \sec^2\theta \text{ for } \tfrac{1}{\cos^2\theta}$$

$$= \sin^2\theta (\sec^2\theta - 1) \qquad \text{factor out } \sin^2\theta$$

$$= \sin^2\theta \tan^2\theta \qquad \text{substitute } \tan^2\theta \text{ for } \sec^2\theta - 1$$

Now try Exercises 13 through 18 ▶

Example 3 involved *factoring* out a common expression. Just as often, we'll need to *multiply* numerators and denominators by a common expression, as in Example 4.

EXAMPLE 4 ▶ **Verifying an Identity by Multiplying Conjugates**

Verify the identity: $\dfrac{\cos t}{1 + \sec t} = \dfrac{1 - \cos t}{\tan^2 t}$.

Solution ▶ Both sides of the identity have a single term and one is really no more complex than the other. As a matter of choice we begin with the left side. Noting the denominator on the left has the term sec t, with a corresponding term of $\tan^2 t$ to the right, we reason that multiplication by a conjugate might be productive.

$$\frac{\cos t}{1 + \sec t} = \left(\frac{\cos t}{1 + \sec t}\right)\left(\frac{1 - \sec t}{1 - \sec t}\right) \qquad \text{multiply above and below by the conjugate}$$

$$= \frac{\cos t - 1}{1 - \sec^2 t} \qquad \text{distribute: } \cos t \sec t = 1,\ (A + B)(A - B) = A^2 - B^2$$

$$= \frac{\cos t - 1}{-\tan^2 t} \qquad \begin{array}{l}\text{substitute } -\tan^2 t \text{ for } 1 - \sec^2 t \\ (1 + \tan^2 t = \sec^2 t \Rightarrow 1 - \sec^2 t = -\tan^2 t)\end{array}$$

$$= \frac{1 - \cos t}{\tan^2 t} \qquad \text{multiply above and below by } -1$$

Now try Exercises 19 through 22 ▶

Example 4 highlights the need to be very familiar with families of identities. To replace $1 - \sec^2 t$, we had to use $-\tan^2 t$, not simply $\tan^2 t$, since the related Pythagorean identity is $1 + \tan^2 t = \sec^2 t$.

As noted in the *Guidelines,* combining rational terms is often helpful. At this point, students are encouraged to work with the pattern $\dfrac{A}{B} \pm \dfrac{C}{D} = \dfrac{AD \pm BC}{BD}$ as a means of combing rational terms quickly and efficiently.

EXAMPLE 5 ▶ **Verifying an Identity by Combining Terms**

Verify the identity: $\dfrac{\sec x}{\sin x} - \dfrac{\sin x}{\sec x} = \dfrac{\tan^2 x + \cos^2 x}{\tan x}$.

Solution ▶ We begin with the left-hand side.

$$\dfrac{\sec x}{\sin x} - \dfrac{\sin x}{\sec x} = \dfrac{\sec^2 x - \sin^2 x}{\sin x \sec x}$$
  combine terms: $\dfrac{A}{B} - \dfrac{C}{D} = \dfrac{AD - BC}{BD}$

$$= \dfrac{(1 + \tan^2 x) - (1 - \cos^2 x)}{\left(\dfrac{\sin x}{1}\right)\left(\dfrac{1}{\cos x}\right)}$$
  substitute $1 + \tan^2 x$ for $\sec^2 x$, $1 - \cos^2 x$ for $\sin^2 x$, $\dfrac{1}{\cos x}$ for $\sec x$,

$$= \dfrac{\tan^2 x + \cos^2 x}{\tan x}$$
  simplify numerator, substitute $\tan x$ for $\dfrac{\sin x}{\cos x}$

✓ **B.** You've just learned how to verify general identities

Now try Exercises 23 through 28 ▶

Identities come in an infinite variety and it would be impossible to illustrate all variations. Using the general ideas and skills presented should prepare you to verify any of those given in the exercise set, as well as those you encounter in your future studies. See **Exercises 29 through 58.**

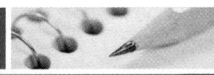

6.2 EXERCISES

▶ CONCEPTS AND VOCABULARY

Fill in each blank with the appropriate word or phrase. Carefully reread the section if needed.

1. If two expressions are equal, then one may be _____ for the other at any time and the result will be equivalent.

2. We verify an identity by changing the form of one side, working until we _____ the other side.

3. To verify an identity, always begin with the more _____ expression, since it is easier to _____ than to _____.

4. Converting all terms to functions of _____ and _____ may help verify an identity.

5. Discuss/Explain why you must not add, subtract, multiply, or divide both sides of the equation when verifying identities.

6. Discuss/Explain the difference between operating on both sides of an equation (see Exercise 5) and working on each side independently.

▶ DEVELOPING YOUR SKILLS

Using algebra and the fundamental identities, rewrite each given expression to create a new identity relationship. Then verify your identity by reversing the steps. Answers will vary.

7. $\sec x + \tan x$

8. $(\cos x + \sin x)^2$

9. $(1 - \sin^2 x)\sec x$

10. $2\cot x \csc x$

11. $\dfrac{\sin x - \sin x \cos x}{\sin^2 x}$

12. $(\cos x + \sin x)(\cos x - \sin x)$

Verify that the following equations are identities.

13. $\cos^2 x \tan^2 x = 1 - \cos^2 x$

14. $\sin^2 x \cot^2 x = 1 - \sin^2 x$

15. $\tan x + \cot x = \sec x \csc x$

16. $\cot x \cos x = \csc x - \sin x$

17. $\dfrac{\cos x}{\tan x} = \csc x - \sin x$

18. $\dfrac{\sin x}{\cot x} = \sec x - \cos x$

19. $\dfrac{\cos \theta}{1 - \sin \theta} = \sec \theta + \tan \theta$

20. $\dfrac{\sin \theta}{1 - \cos \theta} = \csc \theta + \cot \theta$

21. $\dfrac{1 - \sin x}{\cos x} = \dfrac{\cos x}{1 + \sin x}$

22. $\dfrac{1 - \cos x}{\sin x} = \dfrac{\sin x}{1 + \cos x}$

23. $\dfrac{\csc x}{\cos x} - \dfrac{\cos x}{\csc x} = \dfrac{\cot^2 x + \sin^2 x}{\cot x}$

24. $\dfrac{1}{\cos^2 x} + \dfrac{1}{\sin^2 x} = \csc^2 x \sec^2 x$

25. $\dfrac{\sin x}{1 + \sin x} - \dfrac{\sin x}{1 - \sin x} = -2 \tan^2 x$

26. $\dfrac{\cos x}{1 + \cos x} - \dfrac{\cos x}{1 - \cos x} = -2 \cot^2 x$

27. $\dfrac{\cot x}{1 + \csc x} - \dfrac{\cot x}{1 - \csc x} = 2 \sec x$

28. $\dfrac{\tan x}{1 + \sec x} - \dfrac{\tan x}{1 - \sec x} = 2 \csc x$

29. $\dfrac{\sec^2 x}{1 + \cot^2 x} = \tan^2 x$ 30. $\dfrac{\csc^2 x}{1 + \tan^2 x} = \cot^2 x$

31. $\sin^2 x (\cot^2 x - \csc^2 x) = -\sin^2 x$

32. $\cos^2 x (\tan^2 x - \sec^2 x) = -\cos^2 x$

33. $\cos x \cot x + \sin x = \csc x$

34. $\sin x \tan x + \cos x = \sec x$

35. $\dfrac{\sec x}{\cot x + \tan x} = \sin x$

36. $\dfrac{\csc x}{\cot x + \tan x} = \cos x$

37. $\dfrac{\sin x - \csc x}{\csc x} = -\cos^2 x$

38. $\dfrac{\cos x - \sec x}{\sec x} = -\sin^2 x$

39. $\dfrac{1}{\csc x - \sin x} = \tan x \sec x$

40. $\dfrac{1}{\sec x - \cos x} = \cot x \csc x$

41. $\dfrac{1 + \sin x}{1 - \sin x} = (\tan x + \sec x)^2$

42. $\dfrac{1 - \cos x}{1 + \cos x} = (\csc x - \cot x)^2$

43. $\dfrac{\cos x - \sin x}{1 - \tan x} = \dfrac{\cos x + \sin x}{1 + \tan x}$

44. $\dfrac{1 - \cot x}{1 + \cot x} = \dfrac{\sin x - \cos x}{\sin x + \cos x}$

45. $\dfrac{\tan^2 x - \cot^2 x}{\tan x - \cot x} = \csc x \sec x$

46. $\dfrac{\cot x - \tan x}{\cot^2 x - \tan^2 x} = \sin x \cos x$

47. $\dfrac{\cot x}{\cot x + \tan x} = 1 - \sin^2 x$

48. $\dfrac{\tan x}{\cot x + \tan x} = 1 - \cos^2 x$

49. $\dfrac{\sec^4 x - \tan^4 x}{\sec^2 x + \tan^2 x} = 1$

50. $\dfrac{\csc^4 x - \cot^4 x}{\csc^2 x + \cot^2 x} = 1$

51. $\dfrac{\cos^4 x - \sin^4 x}{\cos^2 x} = 2 - \sec^2 x$

52. $\dfrac{\sin^4 x - \cos^4 x}{\sin^2 x} = 2 - \csc^2 x$

53. $(\sec x + \tan x)^2 = \dfrac{(\sin x + 1)^2}{\cos^2 x}$

54. $(\csc x + \cot x)^2 = \dfrac{(\cos x + 1)^2}{\sin^2 x}$

55. $\dfrac{\cos x}{\sin x} + \dfrac{\sin x}{\cos x} + \dfrac{\csc x}{\sec x} = \dfrac{\sec x + \cos x}{\sin x}$

56. $\dfrac{\cos x}{\sin x} + \dfrac{\sin x}{\cos x} + \dfrac{\sec x}{\csc x} = \dfrac{\csc x + \sin x}{\cos x}$

57. $\dfrac{\sin^4 x - \cos^4 x}{\sin^3 x + \cos^3 x} = \dfrac{\sin x - \cos x}{1 - \sin x \cos x}$

58. $\dfrac{\sin^4 x - \cos^4 x}{\sin^3 x - \cos^3 x} = \dfrac{\sin x + \cos x}{1 + \sin x \cos x}$

▶ **WORKING WITH FORMULAS**

59. Distance to top of movie screen:
$$d^2 = (20 + x \cos \theta)^2 + (20 - x \sin \theta)^2$$

At a theater, the optimum viewing angle depends on a number of factors, like the height of the screen, the incline of the auditorium, the location of a seat, the height of your eyes while seated, and so on. One of the measures needed to find the "best" seat is the distance from your eyes to the top of the screen. For a theater with the dimensions shown, this distance is given by the formula here (x is the diagonal distance from the horizontal floor to your seat).
(a) Show the formula is equivalent to $800 + 40x(\cos \theta - \sin \theta) + x^2$. (b) Find the distance d if $\theta = 18°$ and you are sitting in the eighth row with the rows spaced 3 ft apart.

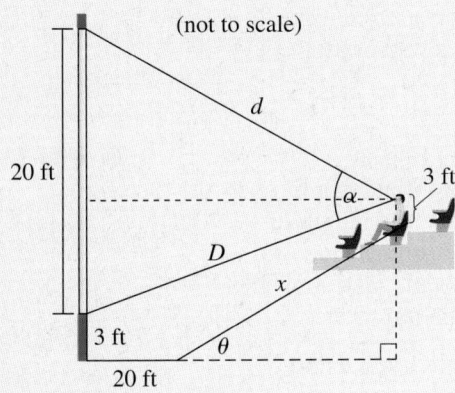

60. The area of triangle ABC: $A = \dfrac{c^2 \sin A \sin B}{2 \sin C}$

If one side and three angles of a triangle are known, its area can be computed using this formula, where side c is opposite angle C. Find the area of the triangle shown in the diagram.

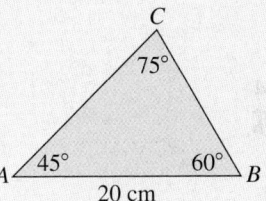

▶ **APPLICATIONS**

61. Pythagorean theorem: For the triangle shown, (a) find an expression for the length of the hypotenuse in terms of $\tan x$ and $\cot x$, then determine the length of the hypotenuse when $x = 1.5$ rad; (b) show the expression you found in part (a) is equivalent to $h = \sqrt{\csc x \sec x}$ and recompute the length of the hypotenuse using this expression. Did the answers match?

62. Pythagorean theorem: For the triangle shown, (a) find an expression for the area of the triangle in terms of $\cot x$ and $\cos x$, then determine its area given $x = \dfrac{\pi}{6}$;

(b) show the expression you found in part (a) is equivalent to $A = \dfrac{1}{2}(\csc x - \sin x)$ and recompute the area using this expression. Did the answers match?

Exercise 62

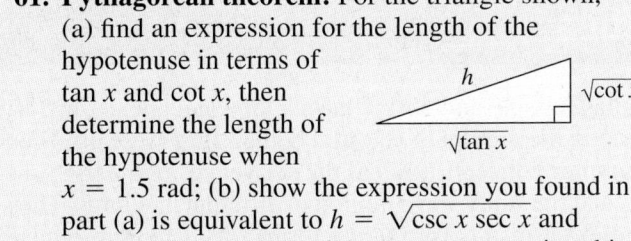

63. Viewing distance: Referring to Exercise 59, find a formula for D—the distance from this patron's eyes to the *bottom* of the movie screen. Simplify the result using a Pythagorean identity, then find the value of D.

64. Viewing angle: Referring to Exercises 59 and 63, once d and D are known, the viewing angle α (the angle subtended by the movie screen and the viewer's eyes) can be found using the formula $\cos \alpha = \dfrac{d^2 + D^2 - 20^2}{2dD}$. Find the value of $\cos \alpha$ for this particular theater, person, and seat.

65. Intensity of light: In a study of the luminous intensity of light, the expression
$$\sin \alpha = \frac{I_1 \cos \theta}{\sqrt{(I_1 \cos \theta)^2 + (I_2 \sin \theta)^2}}$$
can occur. Simplify the equation for the moment $I_1 = I_2$.

66. Intensity of light: Referring to Exercise 65, find the angle θ given $I_1 = I_2$ and $\alpha = 60°$.

▶ **EXTENDING THE CONCEPT**

67. Just as the points $P(x, y)$ on the unit circle $x^2 + y^2 = 1$ are used to name the circular trigonometric functions, the points $P(x, y)$ on the unit hyperbola $x^2 - y^2 = 1$ are used to name what are called the **hyperbolic trigonometric functions.** The hyperbolic functions are used extensively in many of the applied sciences. The identities for these functions have many similarities to those for the circular functions, but also have some significant

differences. Using the Internet or the resources of a library, do some research on the functions sinh t, cosh t, and tanh t, where t is any real number. In particular, see how the Pythagorean identities compare/contrast between the two forms of trigonometry.

68. Verify the identity $\dfrac{\sin^6 x - \cos^6 x}{\sin^4 x - \cos^4 x} = 1 - \sin^2 x \cos^2 x.$

69. Use factoring to show the equation is an identity: $\sin^4 x + 2 \sin^2 x \cos^2 x + \cos^4 x = 1.$

▶ **MAINTAINING YOUR SKILLS**

70. **(3.5)** Graph the rational function given.

$$h(x) = \frac{x - 1}{x^2 - 4}$$

71. **(5.2)** Verify that $\left(\dfrac{\sqrt{7}}{4}, \dfrac{3}{4}\right)$ is a point on the unit circle, then state the values of sin t, cos t, and tan t associated with this point.

72. **(5.7)** Use an appropriate trig ratio to find the length of the bridge needed to cross the lake shown in the figure.

Exercise 72

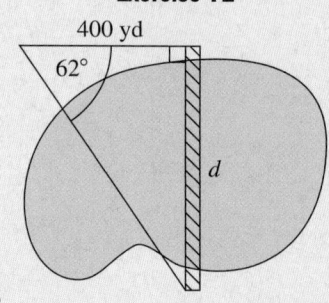

73. **(2.5)** Graph using transformations of a basic function: $f(x) = -2|x - 3| + 6$

6.3 | The Sum and Difference Identities

Learning Objectives

In Section 6.3 you will learn how to:

☐ **A.** Develop and use sum and difference identities for cosine

☐ **B.** Use the cofunction identities to develop the sum and difference identities for sine and tangent

☐ **C.** Use the sum and difference identities to verify other identities

The sum and difference formulas for sine and cosine have a long and ancient history. Originally developed to help study the motion of celestial bodies, they were used centuries later to develop more complex concepts, such as the derivatives of the trig functions, complex number theory, and the study wave motion in different mediums. These identities are also used to find exact results (in radical form) for many nonstandard angles, a result of great importance to the ancient astronomers and still of notable mathematical significance today.

A. The Sum and Difference Identities for Cosine

On a unit circle with center C, consider the point A on the terminal side of angle α, and point B on the terminal side of angle β, as shown in Figure 6.1. Since $r = 1$, the coordinates of A and B are $(\cos \alpha, \sin \alpha)$ and $(\cos \beta, \sin \beta)$, respectively. Using the distance formula, we find that $\overline{AB}$ is equal to

Figure 6.1

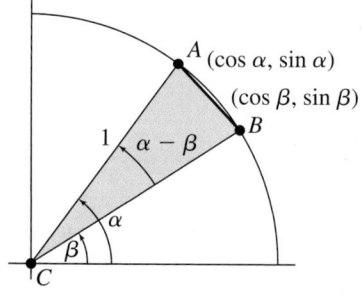

$$\overline{AB} = \sqrt{(\cos \alpha - \cos \beta)^2 + (\sin \alpha - \sin \beta)^2} \qquad \text{binomial}$$
$$= \sqrt{\cos^2\alpha - 2\cos\alpha\cos\beta + \cos^2\beta + \sin^2\alpha - 2\sin\alpha\sin\beta + \sin^2\beta} \qquad \text{squares}$$
$$= \sqrt{(\cos^2\alpha + \sin^2\alpha) + (\cos^2\beta + \sin^2\beta) - 2\cos\alpha\cos\beta - 2\sin\alpha\sin\beta} \qquad \text{regroup}$$
$$= \sqrt{2 - 2\cos\alpha\cos\beta - 2\sin\alpha\sin\beta} \qquad \cos^2 u + \sin^2 u = 1$$

With no loss of generality, we can rotate sector ACB clockwise, until side $\overline{CB}$ coincides with the x-axis. This creates new coordinates of $(1, 0)$ for B, and new coordinates of $(\cos(\alpha - \beta), \sin(\alpha - \beta))$ for A, *but the distance $\overline{AB}$ remains unchanged!* (see Figure 6.2). Recomputing the distance gives

Figure 6.2

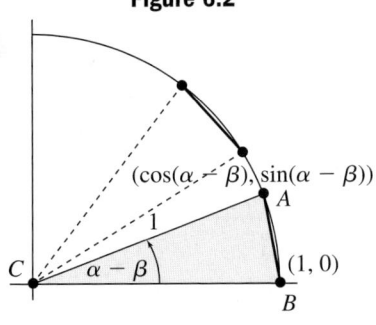

$$\overline{AB} = \sqrt{[\cos(\alpha - \beta) - 1]^2 + [\sin(\alpha - \beta) - 0]^2}$$
$$= \sqrt{\cos^2(\alpha - \beta) - 2\cos(\alpha - \beta) + 1 + \sin^2(\alpha - \beta)}$$
$$= \sqrt{[\cos^2(\alpha - \beta) + \sin^2(\alpha - \beta)] - 2\cos(\alpha - \beta) + 1}$$
$$= \sqrt{2 - 2\cos(\alpha - \beta)}$$

Since both expressions represent the same distance, we can set them equal to each other and solve for $\cos(\alpha - \beta)$.

$\sqrt{2 - 2\cos(\alpha - \beta)} = \sqrt{2 - 2\cos\alpha\cos\beta - 2\sin\alpha\sin\beta}$	$\overline{AB} = \overline{AB}$
$2 - 2\cos(\alpha - \beta) = 2 - 2\cos\alpha\cos\beta - 2\sin\alpha\sin\beta$	property of radicals
$-2\cos(\alpha - \beta) = -2\cos\alpha\cos\beta - 2\sin\alpha\sin\beta$	subtract 2
$\cos(\alpha - \beta) = \cos\alpha\cos\beta + \sin\alpha\sin\beta$	divide both sides by -2

The result is called the **difference identity for cosine.** The **sum identity for cosine** follows immediately, by substituting $-\beta$ for β.

$\cos(\alpha - \beta) = \cos\alpha\cos\beta + \sin\alpha\sin\beta$	difference identity
$\cos(\alpha - [-\beta]) = \cos\alpha\cos(-\beta) + \sin\alpha\sin(-\beta)$	substitute $-\beta$ for β
$\cos(\alpha + \beta) = \cos\alpha\cos\beta - \sin\alpha\sin\beta$	$\cos(-\beta) = \cos\beta; \sin(-\beta) = -\sin\beta$

The sum and difference identities can be used to find exact values for the trig functions of certain angles (values written in nondecimal form using radicals), simplify expressions, and to establish additional identities.

EXAMPLE 1 ▶ **Finding Exact Values for Non-Standard Angles**

Use the sum and difference identities for cosine to find exact values for
   a. $\cos 15° = \cos(45° - 30°)$ **b.** $\cos 75° = \cos(45° + 30°)$
Check results on a calculator.

Solution ▶ Each involves a direct application of the related identity, and uses special values.

a.
$\cos(\alpha - \beta) = \cos\alpha\cos\beta + \sin\alpha\sin\beta$	difference identity
$\cos(45° - 30°) = \cos 45°\cos 30° + \sin 45°\sin 30°$	$\alpha = 45°, \beta = 30°$
$= \left(\dfrac{\sqrt{2}}{2}\right)\left(\dfrac{\sqrt{3}}{2}\right) + \left(\dfrac{\sqrt{2}}{2}\right)\left(\dfrac{1}{2}\right)$	standard values
$\cos 15° = \dfrac{\sqrt{6} + \sqrt{2}}{4}$	combine terms

> **WORTHY OF NOTE**
>
> Be aware that $\cos(60° + 30°) \neq \cos 60° + \cos 30°$
> $\left(0 \neq \dfrac{1}{2} + \dfrac{\sqrt{3}}{2}\right)$ and in general $f(a + b) \neq f(a) + f(b)$.

To 10 decimal places, $\cos 15° = 0.9659258263$.

b.
$\cos(\alpha + \beta) = \cos\alpha\cos\beta - \sin\alpha\sin\beta$	sum identity
$\cos(45° + 30°) = \cos 45°\cos 30° - \sin 45°\sin 30°$	$\alpha = 45°, \beta = 30°$
$= \left(\dfrac{\sqrt{2}}{2}\right)\left(\dfrac{\sqrt{3}}{2}\right) - \left(\dfrac{\sqrt{2}}{2}\right)\left(\dfrac{1}{2}\right)$	standard values
$\cos 75° = \dfrac{\sqrt{6} - \sqrt{2}}{4}$	combine terms

To 10 decimal places, $\cos 75° = 0.2588190451$.

Now try Exercises 7 through 12 ▶

These identities are listed here using the "$\pm$" and "$\mp$" notation to avoid needless repetition. In their application, use both upper symbols or both lower symbols depending on whether you're evaluating the cosine of a sum or difference of two angles. As with the other identities, these can be rewritten to form other members of the identity family, as when they are used to consolidate a larger expression. This is shown in Example 2.

The Sum and Difference Identities for Cosine

> **cosine family:** $\cos(\alpha \pm \beta) = \cos\alpha\cos\beta \mp \sin\alpha\sin\beta$ functions repeat, signs alternate
>
> $\cos\alpha\cos\beta \mp \sin\alpha\sin\beta = \cos(\alpha \pm \beta)$ can be used to *expand* or *contract*

EXAMPLE 2 ▶ **Using a Sum/Difference Identity to Simplify an Expression**

Write as a single expression in cosine and evaluate: $\cos 57° \cos 78° - \sin 57° \sin 78°$

Solution ▶ Since the functions repeat and are expressed as a difference, we use the sum identity for cosine to rewrite the difference as a single expression.

$$\cos\alpha\cos\beta - \sin\alpha\sin\beta = \cos(\alpha + \beta) \qquad \text{sum identity for cosine}$$
$$\cos 57° \cos 78° - \sin 57° \sin 78° = \cos(57° + 78°) \qquad \alpha = 57°, \beta = 78°$$

The expression is equal to $\cos 135° = -\dfrac{\sqrt{2}}{2}$.

> **Now try Exercises 13 through 16** ▶

The sum and difference identities can be used to evaluate the cosine of the sum of two angles, even when they are not adjacent, or even expressed in terms of cosine.

EXAMPLE 3 ▶ **Computing the Cosine of a Sum**

Given $\sin\alpha = \frac{5}{13}$ with the terminal side in QI, and $\tan\beta = -\frac{24}{7}$ with the terminal side in QII. Compute the value of $\cos(\alpha + \beta)$.

Solution ▶ To use the sum formula we need the value of $\cos\alpha$, $\sin\alpha$, $\cos\beta$, and $\sin\beta$. Using the given information about the quadrants along with the Pythagorean theorem, we draw the triangles shown in Figures 6.3 and 6.4, yielding the values that follow.

Figure 6.3

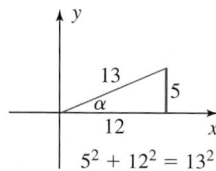

$5^2 + 12^2 = 13^2$

Figure 6.4

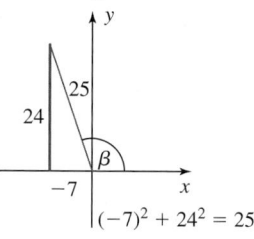

$(-7)^2 + 24^2 = 25^2$

$$\cos\alpha = \frac{12}{13}\,(\text{QI}), \sin\alpha = \frac{5}{13}\,(\text{QI}), \cos\beta = -\frac{7}{25}\,(\text{QII}), \text{ and } \sin\beta = \frac{24}{25}\,(\text{QII})$$

Using $\cos(\alpha + \beta) = \cos\alpha\cos\beta - \sin\alpha\sin\beta$ gives this result:

$$\cos(\alpha + \beta) = \left(\frac{12}{13}\right)\left(-\frac{7}{25}\right) - \left(\frac{5}{13}\right)\left(\frac{24}{25}\right)$$
$$= -\frac{84}{325} - \frac{120}{325}$$
$$= -\frac{204}{325}$$

> **Now try Exercises 17 and 18** ▶

B. The Sum and Difference Identities for Sine and Tangent

☑ **A. You've just learned how to develop and use sum and difference identities for cosine**

The cofunction identities were actually introduced in Section 5.1, using the complementary angles in a right triangle. In this section we'll *verify* that $\cos\left(\dfrac{\pi}{2} - \theta\right) = \sin\theta$

and $\sin\left(\dfrac{\pi}{2} - \theta\right) = \cos\theta$. For the first, we use the difference identity for cosine to obtain

$$\cos\left(\frac{\pi}{2} - \theta\right) = \cos\frac{\pi}{2}\cos\theta + \sin\frac{\pi}{2}\sin\theta$$

$$= (0)\cos\theta + (1)\sin\theta$$

$$= \sin\theta$$

For the second, we use $\cos\left(\dfrac{\pi}{2} - \theta\right) = \sin\theta$, and replace θ with the real number $\dfrac{\pi}{2} - t$. This gives

$$\cos\left(\frac{\pi}{2} - \theta\right) = \sin\theta \qquad \text{cofunction identity for cosine}$$

$$\cos\left(\frac{\pi}{2} - \left[\frac{\pi}{2} - t\right]\right) = \sin\left(\frac{\pi}{2} - t\right) \qquad \text{replace } \theta \text{ with } \frac{\pi}{2} - t$$

$$\cos t = \sin\left(\frac{\pi}{2} - t\right) \qquad \text{result, note } \left[\frac{\pi}{2} - \left(\frac{\pi}{2} - t\right)\right] = t$$

This establishes the cofunction relationship for sine: $\sin\left(\dfrac{\pi}{2} - t\right) = \cos t$ for any real number t. Both identities can be written in terms of the real number t. **See Exercises 19 through 24.**

The Cofunction Identities

$$\cos\left(\frac{\pi}{2} - t\right) = \sin t \qquad \sin\left(\frac{\pi}{2} - t\right) = \cos t$$

The sum and difference identities for sine can easily be developed using cofunction identities. Since $\sin t = \cos\left(\dfrac{\pi}{2} - t\right)$, we need only rename t as the sum $(\alpha + \beta)$ or the difference $(\alpha - \beta)$ and work from there.

$$\sin t = \cos\left(\frac{\pi}{2} - t\right) \qquad \text{cofunction identity}$$

$$\sin(\alpha + \beta) = \cos\left[\frac{\pi}{2} - (\alpha + \beta)\right] \qquad \text{substitute } (\alpha + \beta) \text{ for } t$$

$$= \cos\left[\left(\frac{\pi}{2} - \alpha\right) - \beta\right] \qquad \text{regroup argument}$$

$$= \cos\left(\frac{\pi}{2} - \alpha\right)\cos\beta + \sin\left(\frac{\pi}{2} - \alpha\right)\sin\beta \qquad \begin{array}{l}\text{apply difference identity}\\ \text{for cosine}\end{array}$$

$$\sin(\alpha + \beta) = \sin\alpha\cos\beta + \cos\alpha\sin\beta \qquad \text{result}$$

The difference identity for sine is likewise developed. The sum and difference identities for tangent can be derived using ratio identities and their derivation is left as an exercise **(see Exercise 78).**

The Sum and Difference Identities for Sine and Tangent

sine family: $\sin(\alpha \pm \beta) = \sin\alpha\cos\beta \pm \cos\alpha\sin\beta$ functions alternate, signs repeat

$\sin\alpha\cos\beta \pm \cos\alpha\sin\beta = \sin(\alpha \pm \beta)$ can be used to *expand* or *contract*

tangent family: $\tan(\alpha \pm \beta) = \dfrac{\tan\alpha \pm \tan\beta}{1 \mp \tan\alpha\tan\beta}$ signs match original in numerator
signs alternate in denominator

$\dfrac{\tan\alpha \pm \tan\beta}{1 \mp \tan\alpha\tan\beta} = \tan(\alpha \pm \beta)$ can be used to *expand* or *contract*

EXAMPLE 4A ▶ **Simplifying Expressions Using Sum/Difference Identities**

Write as a single expression in sine: $\sin(2t)\cos t + \cos(2t)\sin t$.

Solution ▶ Since the functions in each term alternate and the expression is written as a sum, we use the sum identity for sine:

$$\sin\alpha\cos\beta + \cos\alpha\sin\beta = \sin(\alpha + \beta) \quad \text{sum identity for sine}$$

$$\sin(2t)\cos t + \cos(2t)\sin t = \sin(2t + t) \quad \text{substitute } 2t \text{ for } \alpha \text{ and } t \text{ for } \beta$$

The expression is equal to $\sin(3t)$.

EXAMPLE 4B ▶ **Simplifying Expressions Using Sum/Difference Identities**

Use the sum or difference identity for tangent to find the exact value of $\tan\dfrac{11\pi}{12}$.

Solution ▶ Since an exact value is requested, $\dfrac{11\pi}{12}$ must be the sum or difference of two

standard angles. A casual inspection reveals $\dfrac{11\pi}{12} = \dfrac{2\pi}{3} + \dfrac{\pi}{4}$. This gives

$$\tan(\alpha + \beta) = \frac{\tan\alpha + \tan\beta}{1 - \tan\alpha\tan\beta} \quad \text{sum identity for tangent}$$

$$\tan\left(\frac{2\pi}{3} + \frac{\pi}{4}\right) = \frac{\tan\left(\dfrac{2\pi}{3}\right) + \tan\left(\dfrac{\pi}{4}\right)}{1 - \tan\left(\dfrac{2\pi}{3}\right)\tan\left(\dfrac{\pi}{4}\right)} \quad \alpha = \frac{2\pi}{3}, \beta = \frac{\pi}{4}$$

$$= \frac{-\sqrt{3} + 1}{1 - (-\sqrt{3})(1)} \quad \tan\left(\frac{2\pi}{3}\right) = -\sqrt{3}, \tan\left(\frac{\pi}{4}\right) = 1$$

$$= \frac{1 - \sqrt{3}}{1 + \sqrt{3}} \quad \text{simplify expression}$$

☑ **B.** You've just learned how to use the cofunction identities to develop the sum and difference identities for sine and tangent

Now try Exercises 25 through 54 ▶

C. Verifying Other Identities

Once the sum and difference identities are established, we can simply add these to the tools we use to verify other identities.

EXAMPLE 5 ▶ **Verifying an Identity**

Verify that $\tan\left(\theta - \dfrac{\pi}{4}\right) = \dfrac{\tan\theta - 1}{\tan\theta + 1}$ is an identity.

Solution ▶ Using a direct application of the difference formula for tangent we obtain

$$\tan\left(\theta - \frac{\pi}{4}\right) = \frac{\tan\theta - \tan\dfrac{\pi}{4}}{1 + \tan\theta \tan\dfrac{\pi}{4}} \qquad \alpha = \theta, \beta = \frac{\pi}{4}$$

$$= \frac{\tan\theta - 1}{1 + \tan\theta} = \frac{\tan\theta - 1}{\tan\theta + 1} \qquad \tan\left(\frac{\pi}{4}\right) = 1$$

Now try Exercises 55 through 60 ▶

EXAMPLE 6 ▶ **Verifying an Identity**

Verify that $\sin(\alpha + \beta)\sin(\alpha - \beta) = \sin^2\alpha - \sin^2\beta$ is an identity.

Solution ▶ Using the sum and difference formulas for sine we obtain

$$\sin(\alpha + \beta)\sin(\alpha - \beta) = (\sin\alpha\cos\beta + \cos\alpha\sin\beta)(\sin\alpha\cos\beta - \cos\alpha\sin\beta)$$

$$= \sin^2\alpha\cos^2\beta - \cos^2\alpha\sin^2\beta \qquad (A+B)(A-B) = A^2 - B^2$$

$$= \sin^2\alpha\,(1 - \sin^2\beta) - (1 - \sin^2\alpha)\sin^2\beta \qquad \begin{array}{l}\text{use } \cos^2 x = 1 - \sin^2 x \text{ to write the}\\ \text{expression solely in terms of sine}\end{array}$$

☑ **C.** You've just learned how to use the sum and difference identities to verify other identities

$$= \sin^2\alpha - \sin^2\alpha\sin^2\beta - \sin^2\beta + \sin^2\alpha\sin^2\beta \qquad \text{distribute}$$

$$= \sin^2\alpha - \sin^2\beta \qquad \text{simplify}$$

Now try Exercises 61 through 68 ▶

6.3 EXERCISES

▶ CONCEPTS AND VOCABULARY

Fill in each blank with the appropriate word or phrase. Carefully reread the section if needed.

1. Since $\tan 45° + \tan 60° > 1$, we know $\tan 45° + \tan 60° = \tan 105°$ is _____ since $\tan\theta < 0$ in _____.

2. To find an exact value for $\tan 105°$, use the sum identity for tangent with $\alpha =$ _____ and $\beta =$ _____.

3. For the cosine sum/difference identities, the functions _____ in each term, with the _____ sign between them.

4. For the sine sum/difference identities, the functions _____ in each term, with the _____ sign between them.

5. Discuss/Explain how we know the exact value for
$$\cos\frac{11\pi}{12} = \cos\left(\frac{2\pi}{3} + \frac{\pi}{4}\right)$$ will be negative, prior to applying any identity.

6. Discuss/Explain why $\tan(\alpha - \beta) = \dfrac{\sin(\alpha - \beta)}{\cos(\beta - \alpha)}$ is an identity, even though the arguments of cosine have been reversed. Then verify the identity.

▶ **DEVELOPING YOUR SKILLS**

Find the exact value of the expression given using a sum or difference identity. Some simplifications may involve using symmetry and the formulas for negatives.

7. $\cos 105°$ **8.** $\cos 135°$

9. $\cos\left(\dfrac{7\pi}{12}\right)$ **10.** $\cos\left(-\dfrac{5\pi}{12}\right)$

Use sum/difference identities to verify that both expressions give the same result.

11. a. $\cos(45° + 30°)$ **b.** $\cos(120° - 45°)$

12. a. $\cos\left(\dfrac{\pi}{6} - \dfrac{\pi}{4}\right)$ **b.** $\cos\left(\dfrac{\pi}{4} - \dfrac{\pi}{3}\right)$

Rewrite as a single expression in cosine.

13. $\cos(7\theta)\cos(2\theta) + \sin(7\theta)\sin(2\theta)$

14. $\cos\left(\dfrac{\theta}{3}\right)\cos\left(\dfrac{\theta}{6}\right) - \sin\left(\dfrac{\theta}{3}\right)\sin\left(\dfrac{\theta}{6}\right)$

Find the exact value of the given expressions.

15. $\cos 183° \cos 153° + \sin 183° \sin 153°$

16. $\cos\left(\dfrac{7\pi}{36}\right)\cos\left(\dfrac{5\pi}{36}\right) - \sin\left(\dfrac{7\pi}{36}\right)\sin\left(\dfrac{5\pi}{36}\right)$

17. For $\sin \alpha = -\dfrac{4}{5}$ with terminal side in QIV and

$\tan \beta = -\dfrac{5}{12}$ with terminal side in QII, find $\cos(\alpha + \beta)$.

18. For $\sin \alpha = \dfrac{112}{113}$ with terminal side in QII and

$\sec \beta = -\dfrac{89}{39}$ with terminal side in QII, find $\cos(\alpha - \beta)$.

Use a cofunction identity to write an equivalent expression.

19. $\cos 57°$ **20.** $\sin 18°$ **21.** $\tan\left(\dfrac{5\pi}{12}\right)$

22. $\sec\left(\dfrac{\pi}{10}\right)$ **23.** $\sin\left(\dfrac{\pi}{6} - \theta\right)$ **24.** $\cos\left(\dfrac{\pi}{3} + \theta\right)$

Rewrite as a single expression.

25. $\sin(3x)\cos(5x) + \cos(3x)\sin(5x)$

26. $\sin\left(\dfrac{x}{2}\right)\cos\left(\dfrac{x}{3}\right) - \cos\left(\dfrac{x}{2}\right)\sin\left(\dfrac{x}{3}\right)$

27. $\dfrac{\tan(5\theta) - \tan(2\theta)}{1 + \tan(5\theta)\tan(2\theta)}$

28. $\dfrac{\tan\left(\dfrac{x}{2}\right) + \tan\left(\dfrac{x}{8}\right)}{1 - \tan\left(\dfrac{x}{2}\right)\tan\left(\dfrac{x}{8}\right)}$

Find the exact value of the given expressions.

29. $\sin 137° \cos 47° - \cos 137° \sin 47°$

30. $\sin\left(\dfrac{11\pi}{24}\right)\cos\left(\dfrac{5\pi}{24}\right) + \cos\left(\dfrac{11\pi}{24}\right)\sin\left(\dfrac{5\pi}{24}\right)$

31. $\dfrac{\tan\left(\dfrac{11\pi}{21}\right) - \tan\left(\dfrac{4\pi}{21}\right)}{1 + \tan\left(\dfrac{11\pi}{21}\right)\tan\left(\dfrac{4\pi}{21}\right)}$

32. $\dfrac{\tan\left(\dfrac{3\pi}{20}\right) + \tan\left(\dfrac{\pi}{10}\right)}{1 - \tan\left(\dfrac{3\pi}{20}\right)\tan\left(\dfrac{\pi}{10}\right)}$

33. For $\cos \alpha = -\dfrac{7}{25}$ with terminal side in QII and

$\cot \beta = \dfrac{15}{8}$ with terminal side in QIII, find

a. $\sin(\alpha + \beta)$ **b.** $\tan(\alpha + \beta)$

34. For $\csc \alpha = \dfrac{29}{20}$ with terminal side in QI and

$\cos \beta = -\dfrac{12}{37}$ with terminal side in QII, find

a. $\sin(\alpha - \beta)$ **b.** $\tan(\alpha - \beta)$

Find the exact value of the expression given using a sum or difference identity. Some simplifications may involve using symmetry and the formulas for negatives.

35. $\sin 105°$ **36.** $\sin(-75°)$

37. $\sin\left(\dfrac{5\pi}{12}\right)$ **38.** $\sin\left(\dfrac{11\pi}{12}\right)$

39. $\tan 150°$ **40.** $\tan 75°$

41. $\tan\left(\dfrac{2\pi}{3}\right)$ **42.** $\tan\left(-\dfrac{\pi}{12}\right)$

Use sum/difference identities to verify that both expressions give the same result.

43. a. $\sin(45° - 30°)$

b. $\sin(135° - 120°)$

44. a. $\sin\left(\dfrac{\pi}{3} - \dfrac{\pi}{4}\right)$

b. $\sin\left(\dfrac{\pi}{4} - \dfrac{\pi}{6}\right)$

45. Find $\sin 255°$ given $150° + 105° = 255°$. See Exercises 7 and 35.

46. Find $\cos\left(\dfrac{19\pi}{12}\right)$ given $2\pi - \dfrac{5\pi}{12} = \dfrac{19\pi}{12}$. See Exercises 10 and 37.

47. Given α and β are acute angles with $\sin \alpha = \dfrac{12}{13}$ and $\tan \beta = \dfrac{35}{12}$, find

   a. $\sin(\alpha + \beta)$ **b.** $\cos(\alpha - \beta)$ **c.** $\tan(\alpha + \beta)$

48. Given α and β are acute angles with $\cos \alpha = \dfrac{8}{17}$ and $\sec \beta = \dfrac{25}{7}$, find

   a. $\sin(\alpha + \beta)$ **b.** $\cos(\alpha - \beta)$ **c.** $\tan(\alpha + \beta)$

49. Given α and β are obtuse angles with $\sin \alpha = \dfrac{28}{53}$ and $\cos \beta = -\dfrac{13}{85}$, find

   a. $\sin(\alpha - \beta)$ **b.** $\cos(\alpha + \beta)$ **c.** $\tan(\alpha - \beta)$

50. Given α and β are obtuse angles with $\tan \alpha = -\dfrac{60}{11}$ and $\sin \beta = \dfrac{35}{37}$, find

   a. $\sin(\alpha - \beta)$ **b.** $\cos(\alpha + \beta)$ **c.** $\tan(\alpha - \beta)$

51. Use the diagram indicated to compute the following:

   a. $\sin A$ **b.** $\cos A$ **c.** $\tan A$

Exercise 51   **Exercise 52**

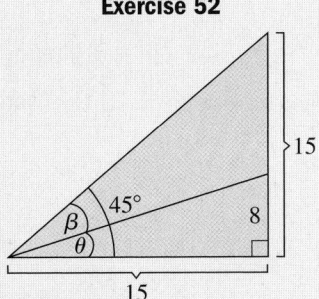

52. Use the diagram indicated to compute the following:

   a. $\sin \beta$ **b.** $\cos \beta$ **c.** $\tan \beta$

53. For the figure indicated, show that $\theta = \alpha + \beta$ and compute the following:

   a. $\sin \theta$ **b.** $\cos \theta$ **c.** $\tan \theta$

Exercise 53

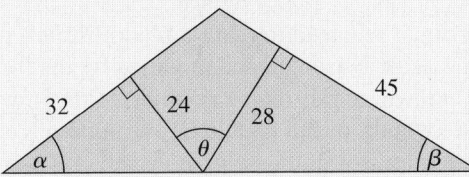

54. For the figure indicated, show that $\theta = \alpha + \beta$ and compute the following:

   a. $\sin \theta$ **b.** $\cos \theta$ **c.** $\tan \theta$

Exercise 54

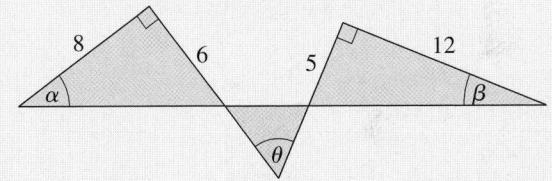

Verify each identity.

55. $\sin(\pi - \alpha) = \sin \alpha$

56. $\cos(\pi - \alpha) = -\cos \alpha$

57. $\cos\left(x + \dfrac{\pi}{4}\right) = \dfrac{\sqrt{2}}{2}(\cos x - \sin x)$

58. $\sin\left(x + \dfrac{\pi}{4}\right) = \dfrac{\sqrt{2}}{2}(\sin x + \cos x)$

59. $\tan\left(x + \dfrac{\pi}{4}\right) = \dfrac{1 + \tan x}{1 - \tan x}$

60. $\tan\left(x - \dfrac{\pi}{4}\right) = \dfrac{\tan x - 1}{\tan x + 1}$

61. $\cos(\alpha + \beta) + \cos(\alpha - \beta) = 2 \cos \alpha \cos \beta$

62. $\sin(\alpha + \beta) + \sin(\alpha - \beta) = 2 \sin \alpha \sin \beta$

63. $\cos(2t) = \cos^2 t - \sin^2 t$

64. $\sin(2t) = 2 \sin t \cos t$

65. $\sin(3t) = -4 \sin^3 t + 3 \sin t$

66. $\cos(3t) = 4 \cos^3 t - 3 \cos t$

67. Use a difference identity to show

   $\cos\left(x - \dfrac{\pi}{4}\right) = \dfrac{\sqrt{2}}{2}(\cos x + \sin x)$.

68. Use sum/difference identities to show

   $\sin\left(x + \dfrac{\pi}{4}\right) + \sin\left(x - \dfrac{\pi}{4}\right) = \sqrt{2} \sin x$.

▶ WORKING WITH FORMULAS

69. Force required to maintain equilibrium using a screw jack: $F = \dfrac{Wk}{c} \tan(p - \theta)$

The force required to maintain equilibrium when a screw jack is used can be modeled by the formula shown, where p is the pitch angle of the screw, W is the weight of the load, θ is the angle of friction, with k and c being constants related to a particular jack. Simplify the formula using the

difference formula for tangent given $p = \dfrac{\pi}{6}$ and $\theta = \dfrac{\pi}{4}$.

70. Brewster's law: reflection and refraction of unpolarized light: $\tan \theta_p = \dfrac{n_2}{n_1}$

Brewster's law of optics states that when unpolarized light strikes a dielectric surface, the transmitted light rays and the reflected light rays are perpendicular to each other. The proof of Brewster's law involves the expression $n_1 \sin \theta_p = n_2 \sin\left(\dfrac{\pi}{2} - \theta_p\right)$. Use the difference identity for sine to verify that this expression leads to Brewster's law.

▶ APPLICATIONS

71. AC circuits: In a study of AC circuits, the equation $R = \dfrac{\cos s \cos t}{\omega C \sin(s + t)}$ sometimes arises. Use a sum identity and algebra to show this equation is equivalent to $R = \dfrac{1}{\omega C(\tan s + \tan t)}$.

72. Fluid mechanics: In studies of fluid mechanics, the equation $\gamma_1 V_1 \sin \alpha = \gamma_2 V_2 \sin(\alpha - \beta)$ sometimes arises. Use a difference identity to show that if $\gamma_1 V_1 = \gamma_2 V_2$, the equation is equivalent to $\cos \beta - \cot \alpha \sin \beta = 1$.

73. Art and mathematics: When working in two-point geometric perspective, artists must scale their work to fit on the paper or canvas they are using. In doing so, the equation $\dfrac{A}{B} = \dfrac{\tan \theta}{\tan(90° - \theta)}$ arises. Rewrite the expression on the right in terms of sine and cosine, then use the difference identities to show the equation can be rewritten as $\dfrac{A}{B} = \tan^2\theta$.

74. Traveling waves: If two waves of the same frequency, velocity, and amplitude are traveling along a string in opposite directions, they can be represented by the equations $Y_1 = A \sin(kx - \omega t)$ and $Y_2 = A \sin(kx + \omega t)$. Use the sum and difference formulas for sine to show the result $Y_R = Y_1 + Y_2$ of these waves can be expressed as $Y_R = 2A \sin(kx)\cos(\omega t)$.

75. Pressure on the eardrum: If a frequency generator is placed a certain distance from the ear, the pressure on the eardrum can be modeled by the function $P_1(t) = A \sin(2\pi ft)$, where f is the frequency and t is the time in seconds. If a second frequency generator with identical settings is placed slightly closer to the ear, its pressure on the eardrum could be represented by $P_2(t) = A \sin(2\pi ft + C)$, where C is a constant. Show that if $C = \dfrac{\pi}{2}$, the total pressure on the eardrum $[P_1(t) + P_2(t)]$ is $P(t) = A[\sin(2\pi ft) + \cos(2\pi ft)]$.

76. Angle between two cables: Two cables used to steady a radio tower are attached to the tower at heights of 5 ft and 35 ft, with both secured to a stake 12 ft from the tower (see figure). Find the value of $\cos \theta$, where θ is the angle between the upper and lower cables.

Exercise 76

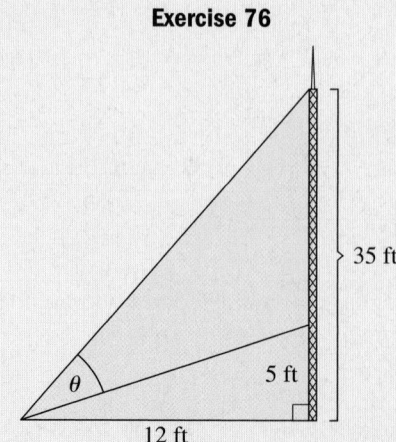

77. Difference quotient: Given $f(x) = \sin x$, show that the difference quotient results in the expression

$$\sin x \frac{\cos h - 1}{h} + \cos x \left(\frac{\sin h}{h} \right).$$

78. Difference identity: Derive the difference identity for tangent using $\tan(\alpha - \beta) = \dfrac{\sin(\alpha - \beta)}{\cos(\alpha - \beta)}$.

(*Hint:* After applying the difference identities, divide the numerator and denominator by $\cos \alpha \cos \beta$.)

▶ EXTENDING THE CONCEPT

A family of identities called the *angle reduction formulas,* will be of use in our study of complex numbers and other areas. These formulas use the period of a function to reduce large angles to an angle in $[0, 360°)$ or $[0, 2\pi)$ having an equivalent function value: (1) $\cos(t + 2\pi k) = \cos t$; (2) $\sin(t + 2\pi k) = \sin t$. Use the reduction formulas to find values for the following functions (note the formulas can also be expressed in degrees).

Exercise 83

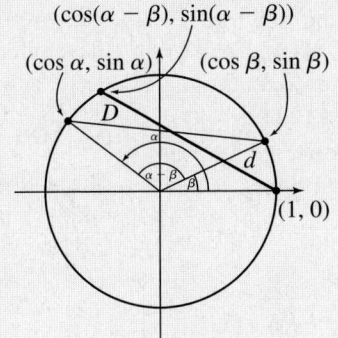

$(\cos(\alpha - \beta), \sin(\alpha - \beta))$

79. $\cos 1665°$ **80.** $\cos\left(\dfrac{91\pi}{6} \right)$

81. $\sin\left(\dfrac{41\pi}{6} \right)$ **82.** $\sin 2385°$

83. An alternative method of proving the difference formula for cosine uses a unit circle and the fact that equal arcs are subtended by equal chords ($D = d$ in the diagram). Using a combination of algebra, the distance formula, and a Pythagorean identity, show that $\cos(\alpha - \beta) = \cos \alpha \cos \beta + \sin \alpha \sin \beta$ (start by computing D^2 and d^2). Then discuss/explain how the sum identity can be found using the fact that $\beta = -(-\beta)$.

84. A proof without words: Verify the Pythagorean theorem for each right triangle in the diagram, then discuss/explain how the diagram offers a proof of the sum identities for sine and cosine. Be detailed and thorough.

Exercise 84

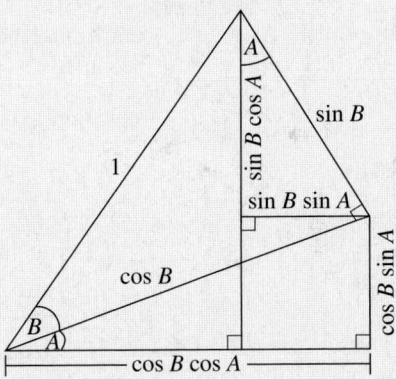

▶ MAINTAINING YOUR SKILLS

85. (5.3/5.4) State the period of the functions given:

  a. $y = 3 \sin\left(\dfrac{\pi}{8}x - \dfrac{\pi}{3} \right)$

  b. $y = 4 \tan\left(2x + \dfrac{\pi}{4} \right)$

86. (2.7) Graph the piecewise-defined function given:

$$f(x) = \begin{cases} 3 & x < -1 \\ x^2 & -1 \le x \le 1 \\ x & x > 1 \end{cases}$$

87. (5.2) Clarence the Clown is about to be shot from a circus cannon to a safety net on the other side of the main tent. If the cannon is 30 ft long and must be aimed at 40° for Clarence to hit the net, the end of the cannon must be how high from ground level?

88. (2.3) Find the equation of the line parallel to $2x + 5y = -10$, containing the point $(5, -2)$. Write your answer in standard form.

Learning Objectives

In Section 6.4 you will learn how to:

☐ **A.** Derive and use the double-angle identities for cosine, tangent, and sine

☐ **B.** Develop and use the power reduction and half-angle identities

☐ **C.** Derive and use the product-to-sum and sum-to-product identities

☐ **D.** Solve applications using these identities

The derivation of the sum and difference identities in Section 6.3 was a "watershed event" in the study of identities. By making various substitutions, they lead us very naturally to many new identity families, giving us a heightened ability to simplify expressions, solve equations, find exact values, and model real-world phenomena. In fact, many of the identities are applied in very practical ways, as in a study of projectile motion and the conic sections (Chapter 10). In addition, one of the most profound principles discovered in the eighteenth and nineteenth centuries was that electricity, light, and sound could all be studied using sinusoidal waves. These waves often interact with each other, creating the phenomena known as reflection, diffraction, superposition, interference, standing waves, and others. The product-to-sum and sum-to-product identities play a fundamental role in the investigation and study of these phenomena.

A. The Double-Angle Identities

The double-angle identities for sine, cosine, and tangent can all be derived using the related sum identities with two equal angles ($\alpha = \beta$). We'll illustrate the process here for the cosine of twice an angle.

$\cos(\alpha + \beta) = \cos \alpha \cos \beta - \sin \alpha \sin \beta$ sum identity for cosine

$\cos(\alpha + \alpha) = \cos \alpha \cos \alpha - \sin \alpha \sin \alpha$ assume $\alpha = \beta$ and substitute α for β

$\cos(2\alpha) = \cos^2\alpha - \sin^2\alpha$ simplify—double-angle identity for cosine

Using the Pythagorean identity $\cos^2\alpha + \sin^2\alpha = 1$, we can easily find two additional members of this family, which are often quite useful. For $\cos^2\alpha = 1 - \sin^2\alpha$ we have

$\cos(2\alpha) = \cos^2\alpha - \sin^2\alpha$ double-angle identity for cosine

$= (1 - \sin^2\alpha) - \sin^2\alpha$ substitute $1 - \sin^2\alpha$ for $\cos^2\alpha$

$\cos(2\alpha) = 1 - 2\sin^2\alpha$ double-angle in terms of sine

Using $\sin^2\alpha = 1 - \cos^2\alpha$ we obtain an additional form:

$\cos(2\alpha) = \cos^2\alpha - \sin^2\alpha$ double-angle identity for cosine

$= \cos^2\alpha - (1 - \cos^2\alpha)$ substitute $1 - \cos^2\alpha$ for $\sin^2\alpha$

$\cos(2\alpha) = 2\cos^2\alpha - 1$ double-angle in terms of cosine

The derivations of $\sin(2\alpha)$ and $\tan(2\alpha)$ are likewise developed and are asked for in **Exercise 103.** The double-angle identities are collected here for your convenience.

The Double-Angle Identities

cosine: $\cos(2\alpha) = \cos^2\alpha - \sin^2\alpha$ **sine:** $\sin(2\alpha) = 2\sin\alpha\cos\alpha$

$= 1 - 2\sin^2\alpha$

$= 2\cos^2\alpha - 1$

tangent: $\tan(2\alpha) = \dfrac{2\tan\alpha}{1 - \tan^2\alpha}$

EXAMPLE 1 ▶ **Using a Double-Angle Identity to Find Function Values**

Given $\sin \alpha = \dfrac{5}{8}$, find the value of $\cos(2\alpha)$.

Solution ▶ Using the double-angle identity for cosine interms of sine, we find

$$\cos(2\alpha) = 1 - 2\sin^2\alpha \qquad \text{double-angle in terms of sine}$$

$$= 1 - 2\left(\frac{5}{8}\right)^2 \qquad \text{substitute } \frac{5}{8} \text{ for } \sin \alpha$$

$$= 1 - \frac{25}{32} \qquad 2\left(\frac{5}{8}\right)^2 = \frac{25}{32}$$

$$= \frac{7}{32} \qquad \text{result}$$

If $\sin \alpha = \dfrac{5}{8}$, then $\cos(2\alpha) = \dfrac{7}{32}$.

Now try Exercises 7 through 20 ▶

Like the fundamental identities, the double-angle identities can be used to verify or develop others. In Example 2, we explore one of many **multiple-angle identities,** verifying that $\cos(3\theta)$ can be rewritten as $4\cos^3\theta - 3\cos\theta$ (in terms of powers of $\cos\theta$).

EXAMPLE 2 ▶ **Verifying a Multiple Angle Identity**

Verify that $\cos(3\theta) = 4\cos^3\theta - 3\cos\theta$ is an identity.

Solution ▶ Use the sum identity for cosine, with $\alpha = 2\theta$ and $\beta = \theta$. Note that our goal is an expression using cosines only, with no multiple angles.

$$\cos(\alpha + \beta) = \cos\alpha\cos\beta - \sin\alpha\sin\beta \qquad \text{sum identity for cosine}$$

$$\cos(2\theta + \theta) = \cos(2\theta)\cos\theta - \sin(2\theta)\sin\theta \qquad \text{substitute } 2\theta \text{ for } \alpha \text{ and } \theta \text{ for } \beta$$

$$\cos(3\theta) = (2\cos^2\theta - 1)\cos\theta - (2\sin\theta\cos\theta)\sin\theta \qquad \text{substitute for } \cos(2\theta) \text{ and } \sin(2\theta)$$

$$= 2\cos^3\theta - \cos\theta - 2\cos\theta\sin^2\theta \qquad \text{multiply}$$

$$= 2\cos^3\theta - \cos\theta - 2\cos\theta(1 - \cos^2\theta) \qquad \text{substitute } 1 - \cos^2\theta \text{ for } \sin^2\theta$$

$$= 2\cos^3\theta - \cos\theta - 2\cos\theta + 2\cos^3\theta \qquad \text{multiply}$$

$$= 4\cos^3\theta - 3\cos\theta \qquad \text{combine terms}$$

Now try Exercises 21 and 22 ▶

EXAMPLE 3 ▶ **Using a Double-Angle Formula to Find Exact Values**

Find the exact value of $\sin 22.5° \cos 22.5°$.

Solution ▶ A product of sines and cosines having the same argument hints at the double-angle identity for sine. Using $\sin(2\alpha) = 2\sin\alpha\cos\alpha$ and dividing by 2 gives

$$\sin \alpha \cos \alpha = \frac{\sin(2\alpha)}{2} \qquad \text{double-angle identity for sine}$$

$$\sin 22.5° \cos 22.5° = \frac{\sin(2[22.5°])}{2} \qquad \text{replace } \alpha \text{ with } 22.5°$$

$$= \frac{\sin 45°}{2} \qquad \text{multiply}$$

$$= \frac{\frac{\sqrt{2}}{2}}{2} = \frac{\sqrt{2}}{4} \qquad \sin 45° = \frac{\sqrt{2}}{2}$$

☑ **A.** You've just learned how to derive and use the double-angle identities for cosine, tangent, and sine

Now try Exercises 23 through 30 ▶

B. The Power Reduction and Half-Angle Identities

Expressions having a trigonometric function raised to a power occur quite frequently in various applications. We can rewrite even powers of these trig functions in terms of an expression containing only cosine to the power 1, using what are called the **power reduction identities.** This makes the expression easier to use and evaluate. It can legitimately be argued that the power reduction identities are actually members of the double-angle family, as all three are a direct consequence. To find identities for $\cos^2 x$ and $\sin^2 x$, we solve the related double-angle identity involving $\cos(2x)$.

$$1 - 2\sin^2\alpha = \cos(2\alpha) \qquad \cos(2\alpha) \text{ in terms of sine}$$

$$-2\sin^2\alpha = \cos(2\alpha) - 1 \qquad \text{subtract 1, then divide by } -2$$

$$\sin^2\alpha = \frac{1 - \cos(2\alpha)}{2} \qquad \text{power reduction identity for sine}$$

Using the same approach for $\cos^2\alpha$ gives $\cos^2\alpha = \dfrac{1 + \cos(2\alpha)}{2}$. The identity for $\tan^2\alpha$ can be derived from $\tan(2\alpha) = \dfrac{2\tan\alpha}{1 - \tan^2\alpha}$ **(see Exercise 104),** but in this case it's easier to use the identity $\tan^2 u = \dfrac{\sin^2 u}{\cos^2 u}$. The result is $\dfrac{1 - \cos(2\alpha)}{1 + \cos(2\alpha)}$.

The Power Reduction Identities

$$\cos^2\alpha = \frac{1 + \cos(2\alpha)}{2} \qquad \sin^2\alpha = \frac{1 - \cos(2\alpha)}{2} \qquad \tan^2\alpha = \frac{1 - \cos(2\alpha)}{1 + \cos(2\alpha)}$$

EXAMPLE 4 ▶   **Using a Power Reduction Formula**

Write $8\sin^4 x$ in terms of an expression containing only cosines to the power 1.

Solution ▶
$$8\sin^4 x = 8(\sin^2 x)^2 \qquad \text{original expression}$$

$$= 8\left[\frac{1 - \cos(2x)}{2}\right]^2 \qquad \text{substitute } \frac{1 - \cos(2x)}{2} \text{ for } \sin^2 x$$

$$= 2[1 - 2\cos(2x) + \cos^2(2x)] \qquad \text{multiply}$$

$$= 2\left[1 - 2\cos(2x) + \frac{1 + \cos(4x)}{2}\right] \qquad \text{substitute } \frac{1 + \cos(4x)}{2} \text{ for } \cos^2(2x)$$

$$= 2 - 4\cos(2x) + 1 + \cos(4x) \qquad \text{multiply}$$

$$= 3 - 4\cos(2x) + \cos(4x) \qquad \text{result}$$

Now try Exercises 31 through 36 ▶

The half-angle identities follow directly from those above, using algebra and a simple change of variable. For $\cos^2\alpha = \dfrac{1 + \cos(2\alpha)}{2}$, we first take square roots and obtain $\cos\alpha = \pm\sqrt{\dfrac{1 + \cos(2\alpha)}{2}}$. Using the substitution $u = 2\alpha$ gives $\alpha = \dfrac{u}{2}$, and making these substitutions results in the half-angle identity for cosine: $\cos\left(\dfrac{u}{2}\right) = \pm\sqrt{\dfrac{1 + \cos u}{2}}$, where the radical's sign depends on the quadrant in which $\dfrac{u}{2}$ terminates. Using the same substitution for sine gives $\sin\left(\dfrac{u}{2}\right) = \pm\sqrt{\dfrac{1 - \cos u}{2}}$, and for the tangent identity, $\tan\left(\dfrac{u}{2}\right) = \pm\sqrt{\dfrac{1 - \cos u}{1 + \cos u}}$. In the case of $\tan\left(\dfrac{u}{2}\right)$, we can actually develop identities that are free of radicals by rationalizing the denominator or numerator. We'll illustrate the former, leaving the latter as an exercise (**see Exercise 102**).

$$\tan\left(\frac{u}{2}\right) = \pm\sqrt{\frac{(1 - \cos u)(1 - \cos u)}{(1 + \cos u)(1 - \cos u)}} \qquad \text{multiply by the conjugate}$$

$$= \pm\sqrt{\frac{(1 - \cos u)^2}{1 - \cos^2 u}} \qquad \text{rewrite}$$

$$= \pm\sqrt{\frac{(1 - \cos u)^2}{\sin^2 u}} \qquad \text{Pythagorean identity}$$

$$= \pm\left|\frac{1 - \cos u}{\sin u}\right| \qquad \sqrt{x^2} = |x|$$

Since $1 - \cos u > 0$ and $\sin u$ has the same sign as $\tan\left(\dfrac{u}{2}\right)$ for all u in its domain, the relationship can simply be written $\tan\left(\dfrac{u}{2}\right) = \dfrac{1 - \cos u}{\sin u}$.

The Half-Angle Identities

$$\cos\left(\frac{u}{2}\right) = \pm\sqrt{\frac{1 + \cos u}{2}} \qquad \sin\left(\frac{u}{2}\right) = \pm\sqrt{\frac{1 - \cos u}{2}} \qquad \tan\left(\frac{u}{2}\right) = \pm\sqrt{\frac{1 - \cos u}{1 + \cos u}}$$

$$\tan\left(\frac{u}{2}\right) = \frac{1 - \cos u}{\sin u} \qquad \tan\left(\frac{u}{2}\right) = \frac{\sin u}{1 + \cos u}$$

EXAMPLE 5 ▶ **Using Half-Angle Formulas to Find Exact Values**

Use the half-angle identities to find exact values for (a) $\sin 15°$ and (b) $\tan 15°$.

Solution ▶ Noting that $15°$ is one-half the standard angle $30°$, we can find each value by applying the respective half-angle identity with $u = 30°$ in Quadrant I.

a. $\sin\left(\dfrac{30}{2}\right) = \sqrt{\dfrac{1 - \cos 30}{2}}$

$= \sqrt{\dfrac{1 - \dfrac{\sqrt{3}}{2}}{2}}$

$\sin 15° = \dfrac{\sqrt{2 - \sqrt{3}}}{2}$

b. $\tan\left(\dfrac{30}{2}\right) = \dfrac{1 - \cos 30}{\sin 30}$

$\tan 15° = \dfrac{1 - \dfrac{\sqrt{3}}{2}}{\dfrac{1}{2}} = 2 - \sqrt{3}$

Now try Exercises 37 through 48 ▶

EXAMPLE 6 ▶   **Using Half-Angle Formulas to Find Exact Values**

For $\cos \theta = -\dfrac{7}{25}$ and θ in QIII, find exact values of $\sin\left(\dfrac{\theta}{2}\right)$ and $\cos\left(\dfrac{\theta}{2}\right)$.

Solution ▶   With θ in QIII $\to \pi < \theta < \dfrac{3\pi}{2}$, we know $\dfrac{\theta}{2}$ must be in QII $\to \dfrac{\pi}{2} < \dfrac{\theta}{2} < \dfrac{3\pi}{4}$ and

we choose our signs accordingly: $\sin\left(\dfrac{\theta}{2}\right) > 0$ and $\cos\left(\dfrac{\theta}{2}\right) < 0$.

$$\sin\left(\dfrac{\theta}{2}\right) = \sqrt{\dfrac{1 - \cos \theta}{2}} \qquad \cos\left(\dfrac{\theta}{2}\right) = -\sqrt{\dfrac{1 + \cos \theta}{2}}$$

$$= \sqrt{\dfrac{1 - \left(-\dfrac{7}{25}\right)}{2}} \qquad = -\sqrt{\dfrac{1 + \left(-\dfrac{7}{25}\right)}{2}}$$

$$= \sqrt{\dfrac{16}{25}} = \dfrac{4}{5} \qquad = -\sqrt{\dfrac{9}{25}} = -\dfrac{3}{5}$$

☑ **B. You've just learned how to develop and use the power reduction and half-angle identities**

Now try Exercises 49 through 64 ▶

C. The Product-to-Sum Identities

As mentioned in the introduction, the product-to-sum and sum-to-product identities are of immense importance to the study of any phenomenon that travels in waves, like light and sound. In fact, the tones you hear as you dial a telephone are actually the sum of two sound waves interacting with each other. Each derivation of a product-to-sum identity is very similar **(see Exercise 105),** and we illustrate by deriving the identity for $\cos \alpha \cos \beta$. Beginning with the sum and difference identities for cosine, we have

$$\cos \alpha \cos \beta + \sin \alpha \sin \beta = \cos(\alpha - \beta) \qquad \text{cosine of a difference}$$

$$+ \ \underline{\cos \alpha \cos \beta - \sin \alpha \sin \beta = \cos(\alpha + \beta)} \qquad \text{cosine of a sum}$$

$$2 \cos \alpha \cos \beta = \cos(\alpha - \beta) + \cos(\alpha + \beta) \qquad \text{combine equations}$$

$$\cos \alpha \cos \beta = \dfrac{1}{2}[\cos(\alpha - \beta) + \cos(\alpha + \beta)] \qquad \text{divide by 2}$$

The identities from this family are listed here.

The Product-to-Sum Identities

$$\cos\alpha\cos\beta = \frac{1}{2}[\cos(\alpha - \beta) + \cos(\alpha + \beta)] \qquad \sin\alpha\sin\beta = \frac{1}{2}[\cos(\alpha - \beta) - \cos(\alpha + \beta)]$$

$$\sin\alpha\cos\beta = \frac{1}{2}[\sin(\alpha + \beta) + \sin(\alpha - \beta)] \qquad \cos\alpha\sin\beta = \frac{1}{2}[\sin(\alpha + \beta) - \sin(\alpha - \beta)]$$

EXAMPLE 7 ▶ **Rewriting a Product as an Equivalent Sum Using Identities**

Write the product $2\cos(27t)\cos(15t)$ as the sum of two cosine functions.

Solution ▶ This is a direct application of the product-to-sum identity, with $\alpha = 27t$ and $\beta = 15t$.

$$\cos\alpha\cos\beta = \frac{1}{2}[\cos(\alpha - \beta) + \cos(\alpha + \beta)] \qquad \text{product-to-sum identity}$$

$$2\cos(27t)\cos(15t) = 2\left(\frac{1}{2}\right)[\cos(27t - 15t) + \cos(27t + 15t)] \qquad \text{substitute}$$

$$= \cos(12t) + \cos(42t) \qquad \text{result}$$

Now try Exercises 65 through 73 ▶

There are times we find it necessary to "work in the other direction," writing a sum of two trig functions as a product. This family of identities can be derived from the product-to-sum identities using a change of variable. We'll illustrate the process for $\sin u + \sin v$. You are asked for the derivation of $\cos u + \cos v$ in **Exercise 106.** To begin, we use $2\alpha = u + v$ and $2\beta = u - v$. This creates the sum $2\alpha + 2\beta = 2u$ and the difference $2\alpha - 2\beta = 2v$, yielding $\alpha + \beta = u$ and $\alpha - \beta = v$, respectively. Dividing the original expressions by 2 gives $\alpha = \dfrac{u + v}{2}$ and $\beta = \dfrac{u - v}{2}$, which all together make the derivation a matter of direct substitution. Using these values in any product-to-sum identity gives the related sum-to-product, as shown here.

$$\sin\alpha\cos\beta = \frac{1}{2}[\sin(\alpha + \beta) + \sin(\alpha - \beta)] \qquad \text{product-to-sum identity (sum of sines)}$$

$$\sin\left(\frac{u + v}{2}\right)\cos\left(\frac{u - v}{2}\right) = \frac{1}{2}(\sin u + \sin v) \qquad \text{substitute } \frac{u + v}{2} \text{ for } \alpha, \frac{u - v}{2} \text{ for } \beta,$$

$$\text{substitute } u \text{ for } \alpha + \beta \text{ and } v \text{ for } \alpha - \beta$$

$$2\sin\left(\frac{u + v}{2}\right)\cos\left(\frac{u - v}{2}\right) = \sin u + \sin v \qquad \text{multiply by 2}$$

The sum-to-product identities follow.

The Sum-to-Product Identities

$$\cos u + \cos v = 2\cos\left(\frac{u + v}{2}\right)\cos\left(\frac{u - v}{2}\right) \qquad \sin u + \sin v = 2\sin\left(\frac{u + v}{2}\right)\cos\left(\frac{u - v}{2}\right)$$

$$\sin u - \sin v = 2\cos\left(\frac{u + v}{2}\right)\sin\left(\frac{u - v}{2}\right) \qquad \cos u - \cos v = -2\sin\left(\frac{u + v}{2}\right)\sin\left(\frac{u - v}{2}\right)$$

EXAMPLE 8 ▶ **Rewriting a Sum as an Equivalent Product Using Identities**

Given $y_1 = \sin(12\pi t)$ and $y_2 = \sin(10\pi t)$, express $y_1 + y_2$ as a product of trigonometric functions.

Solution ▶ This is a direct application of the sum-to-product identity $\sin u + \sin v$, with $u = 12\pi t$ and $v = 10\pi t$.

$$\sin u + \sin v = 2\sin\left(\frac{u+v}{2}\right)\cos\left(\frac{u-v}{2}\right) \qquad \text{sum-to-product identity}$$

☑ **C.** You've just learned how to derive and use the product-to-sum and sum-to-product identities

$$\sin(12\pi t) + \sin(10\pi t) = 2\sin\left(\frac{12\pi t + 10\pi t}{2}\right)\cos\left(\frac{12\pi t - 10\pi t}{2}\right) \qquad \begin{array}{l}\text{substitute } 12\pi t \text{ for}\\ u \text{ and } 10\pi t \text{ for } v\end{array}$$

$$= 2\sin(11\pi t)\cos(\pi t) \qquad \text{substitute}$$

Now try Exercises 74 through 82 ▶

For a mixed variety of identities, **see Exercises 83–100.**

D. Applications of Identities

 In more advanced mathematics courses, rewriting an expression using identities enables the extension or completion of a task that would otherwise be very difficult (or even impossible). In addition, there are a number of practical applications in the physical sciences.

Projectile Motion

A projectile is any object that is thrown, shot, kicked, dropped, or otherwise given an initial velocity, but lacking a continuing source of propulsion. If air resistance is ignored, the range of the projectile depends only on its initial velocity v and the angle θ at which it is propelled. This phenomenon is modeled by the function $r(\theta) = \frac{1}{16}v^2\sin\theta\cos\theta$.

EXAMPLE 9 ▶ **Using Identities to Solve an Application**

a. Use an identity to show $r(\theta) = \frac{1}{16}v^2\sin\theta\cos\theta$ is equivalent to

$$r(\theta) = \frac{1}{32}v^2\sin(2\theta).$$

b. If the projectile is thrown with an initial velocity of $v = 96$ ft/sec, how far will it travel if $\theta = 15°$?

c. From the result of part (a), determine what angle θ will give the maximum range for the projectile.

Solution ▶ **a.** Note that we can use a double-angle identity if we rewrite the coefficient. Writing $\frac{1}{16}$ as $2\left(\frac{1}{32}\right)$ and commuting the factors gives

$$r(\theta) = \left(\frac{1}{32}\right)v^2(2\sin\theta\cos\theta) = \left(\frac{1}{32}\right)v^2\sin(2\theta).$$

b. With $v = 96$ ft/sec and $\theta = 15°$, the formula gives $r(15°) = \left(\frac{1}{32}\right)(96)^2\sin 30°$.

Evaluating the result shows the projectile travels a horizontal distance of 144 ft.

c. For any initial velocity v, $r(\theta)$ will be maximized when $\sin(2\theta)$ is a maximum. This occurs when $\sin(2\theta) = 1$, meaning $2\theta = 90°$ and $\theta = 45°$. The maximum range is achieved when the projectile is released at an angle of $45°$.

Now try Exercises 109 and 110 ▶

GRAPHICAL SUPPORT

The result in Example 9(c) can be verified graphically by assuming an initial velocity of 96 ft/sec and entering the function

$r(\theta) = \dfrac{1}{32}(96)^2\sin(2\theta) = 288\sin(2\theta)$ as Y_1 on a

graphing calculator. With an amplitude of 288 and results confined to the first quadrant, we set an appropriate window, graph the function, and use the 2nd   TRACE **(CALC 4:maximum)** feature. As shown in the figure, the max occurs at $\theta = 45°$.

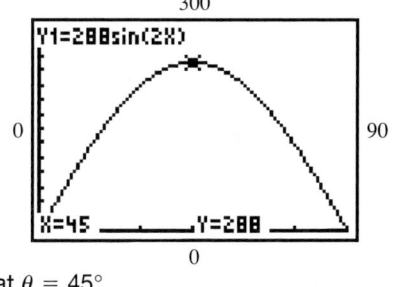

Sound Waves

Each tone you hear on a touch-tone phone is actually the combination of precisely two sound waves with different frequencies (frequency f is defined as $f = \dfrac{B}{2\pi}$). This is why the tones you hear sound identical, regardless of what phone you use. The sum-to-product and product-to-sum formulas help us to understand, study, and use sound in very powerful and practical ways, like sending faxes and using other electronic media.

EXAMPLE 10 ▶ **Using an Identity to Solve an Application**

On a touch-tone phone, the sound created by pressing 5 is produced by combining a sound wave with frequency 1336 cycles/sec, with another wave having frequency 770 cycles/sec. Their respective equations are $y_1 = \cos(2\pi\,1336t)$ and $y_2 = \cos(2\pi\,770t)$, with the resultant wave being $y = y_1 + y_2$ or $y = \cos(2672\pi t) + \cos(1540\pi t)$. Rewrite this sum as a product.

1	2	3	← 697 cps
4	5	6	← 770 cps
7	8	9	← 852 cps
*	0	#	← 941 cps

↑ 1209 cps   ↑ 1336 cps   ↑ 1477 cps

Solution ▶ This is a direct application of the sum-to-product identity, with $u = 2672\pi t$ and $v = 1540\pi t$. Computing one-half the sum/difference of u and v gives

$$\frac{2672\pi t + 1540\pi t}{2} = 2106\pi t \text{ and } \frac{2672\pi t - 1540\pi t}{2} = 566\pi t.$$

$$\cos u + \cos v = 2\cos\left(\frac{u+v}{2}\right)\cos\left(\frac{u-v}{2}\right) \qquad \text{sum-to-product identity}$$

$$\cos(2672\pi t) + \cos(1540\pi t) = 2\cos(2106\pi t)\cos(566\pi t) \qquad \begin{array}{l}\text{substitute } 2672\pi t \text{ for } u\\ \text{and } 1540\pi t \text{ for } v\end{array}$$

Now try Exercises 111 and 112 ▶

Note we can identify the button pressed when the wave is written as a sum. If we have only the resulting wave (written as a product), the product-to-sum formula must be used to identify which button was pressed.

Additional applications requiring the use of identities can be found in **Exercises 113 through 117.**

☑ **D.** You've just learned how to solve applications using identities

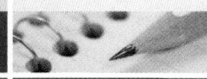

6.4 EXERCISES

► CONCEPTS AND VOCABULARY

Fill in each blank with the appropriate word or phrase. Carefully reread the section if needed.

1. The double-angle identities can be derived using the _____ identities with $\alpha = \beta$. For $\cos(2\theta)$ we expand $\cos(\alpha + \beta)$ using _____.

2. If θ is in QIII then $180° < \theta < 270°$ and $\dfrac{\theta}{2}$ must be in _____ since _____ $< \dfrac{\theta}{2} <$ _____.

3. Multiple-angle identities can be derived using the sum and difference identities. For $\sin(3x)$ use $\sin$ (_____ + _____).

4. For the half-angle identities the sign preceding the radical depends on the _____ in which $\dfrac{u}{2}$ _____.

5. Explain/Discuss how the three different identities for $\tan\left(\dfrac{u}{2}\right)$ are related. Verify that
$$\frac{1 - \cos x}{\sin x} = \frac{\sin x}{1 + \cos x}.$$

6. In Example 6, we were given $\cos\theta = -\dfrac{7}{25}$ and θ in QIII. Discuss how the result would differ if we stipulate that θ is in QII instead.

► DEVELOPING YOUR SKILLS

Find exact values for $\sin(2\theta)$, $\cos(2\theta)$, and $\tan(2\theta)$ using the information given.

7. $\sin\theta = \dfrac{5}{13}$; θ in QII

8. $\cos\theta = -\dfrac{21}{29}$; θ in QII

9. $\cos\theta = -\dfrac{9}{41}$; θ in QII

10. $\sin\theta = -\dfrac{63}{65}$; θ in QIII

11. $\tan\theta = \dfrac{13}{84}$; θ in QIII

12. $\sec\theta = \dfrac{53}{28}$; θ in QI

13. $\sin\theta = \dfrac{48}{73}$; $\cos\theta < 0$

14. $\cos\theta = -\dfrac{8}{17}$; $\tan\theta > 0$

15. $\csc\theta = \dfrac{5}{3}$; $\sec\theta < 0$

16. $\cot\theta = -\dfrac{80}{39}$; $\cos\theta > 0$

Find exact values for $\sin\theta$, $\cos\theta$, and $\tan\theta$ using the information given.

17. $\sin(2\theta) = \dfrac{24}{25}$; 2θ in QII

18. $\sin(2\theta) = -\dfrac{240}{289}$; 2θ in QIII

19. $\cos(2\theta) = -\dfrac{41}{841}$; 2θ in QII

20. $\cos(2\theta) = \dfrac{120}{169}$; 2θ in QIV

21. Verify the following identity:
$\sin(3\theta) = 3\sin\theta - 4\sin^3\theta$

22. Verify the following identity:
$\cos(4\theta) = 8\cos^4\theta - 8\cos^2\theta + 1$

Use a double-angle identity to find exact values for the following expressions.

23. $\cos 75° \sin 75°$

24. $\cos^2 15° - \sin^2 15°$

25. $1 - 2\sin^2\left(\dfrac{\pi}{8}\right)$

26. $2\cos^2\left(\dfrac{\pi}{12}\right) - 1$

27. $\dfrac{2\tan 22.5°}{1 - \tan^2 22.5°}$

28. $\dfrac{2\tan\left(\frac{\pi}{12}\right)}{1 - \tan^2\left(\frac{\pi}{12}\right)}$

29. Use a double-angle identity to rewrite $9\sin(3x)\cos(3x)$ as a single function.
[*Hint:* $9 = \frac{9}{2}(2)$.]

30. Use a double-angle identity to rewrite $2.5 - 5\sin^2 x$ as a single term.
[*Hint:* Factor out a constant.]

Rewrite in terms of an expression containing only cosines to the power 1.

31. $\sin^2 x \cos^2 x$

32. $\sin^4 x \cos^2 x$

33. $3 \cos^4 x$

34. $\cos^4 x \sin^4 x$

35. $2 \sin^6 x$

36. $4 \cos^6 x$

Use a half-angle identity to find exact values for $\sin\theta$, $\cos\theta$, and $\tan\theta$ for the given value of θ.

37. $\theta = 22.5°$

38. $\theta = 75°$

39. $\theta = \dfrac{\pi}{12}$

40. $\theta = \dfrac{5\pi}{12}$

41. $\theta = 67.5°$

42. $\theta = 112.5°$

43. $\theta = \dfrac{3\pi}{8}$

44. $\theta = \dfrac{11\pi}{12}$

Use the results of Exercises 37–40 and a half-angle identity to find the exact value.

45. $\sin 11.25°$

46. $\tan 37.5°$

47. $\sin\left(\dfrac{\pi}{24}\right)$

48. $\cos\left(\dfrac{5\pi}{24}\right)$

Use a half-angle identity to rewrite each expression as a single, nonradical function.

49. $\sqrt{\dfrac{1 + \cos 30°}{2}}$

50. $\sqrt{\dfrac{1 - \cos 45°}{2}}$

51. $\sqrt{\dfrac{1 - \cos(4\theta)}{1 + \cos(4\theta)}}$

52. $\dfrac{1 - \cos(6x)}{\sin(6x)}$

53. $\dfrac{\sin(2x)}{1 + \cos(2x)}$

54. $\dfrac{\sqrt{2(1 + \cos x)}}{1 + \cos x}$

Find exact values for $\sin\left(\dfrac{\theta}{2}\right)$, $\cos\left(\dfrac{\theta}{2}\right)$, and $\tan\left(\dfrac{\theta}{2}\right)$ using the information given.

55. $\sin\theta = \dfrac{12}{13}$; θ is obtuse

56. $\cos\theta = -\dfrac{8}{17}$; θ is obtuse

57. $\cos\theta = -\dfrac{4}{5}$; θ in QII

58. $\sin\theta = -\dfrac{7}{25}$; θ in QIII

59. $\tan\theta = -\dfrac{35}{12}$; θ in QII

60. $\sec\theta = -\dfrac{65}{33}$; θ in QIII

61. $\sin\theta = \dfrac{15}{113}$; θ is acute

62. $\cos\theta = \dfrac{48}{73}$; θ is acute

63. $\cot\theta = \dfrac{21}{20}$; $\pi < \theta < \dfrac{3\pi}{2}$

64. $\csc\theta = \dfrac{41}{9}$; $\dfrac{\pi}{2} < \theta < \pi$

Write each product as a sum using the product-to-sum identities.

65. $\sin(-4\theta)\sin(8\theta)$

66. $\cos(15\alpha)\sin(-3\alpha)$

67. $2\cos\left(\dfrac{7t}{2}\right)\cos\left(\dfrac{3t}{2}\right)$

68. $2\sin\left(\dfrac{5t}{2}\right)\sin\left(\dfrac{9t}{2}\right)$

69. $2\cos(1979\pi t)\cos(439\pi t)$

70. $2\cos(2150\pi t)\cos(268\pi t)$

Find the exact value using product-to-sum identities.

71. $2\cos 15°\sin 135°$

72. $\sin\left(\dfrac{7\pi}{8}\right)\cos\left(\dfrac{\pi}{8}\right)$

73. $\sin\left(\dfrac{7\pi}{12}\right)\sin\left(-\dfrac{\pi}{12}\right)$

Write each sum as a product using the sum-to-product identities.

74. $\cos(9h) + \cos(4h)$

75. $\sin(14k) + \sin(41k)$

76. $\sin\left(\dfrac{11x}{8}\right) - \sin\left(\dfrac{5x}{8}\right)$

77. $\cos\left(\dfrac{7x}{6}\right) - \cos\left(\dfrac{5x}{6}\right)$

78. $\cos(697\pi t) + \cos(1447\pi t)$

79. $\cos(852\pi t) + \cos(1209\pi t)$

Find the exact value using sum-to-product identities.

80. $\cos 75° + \cos 15°$

81. $\sin\left(\dfrac{17\pi}{12}\right) - \sin\left(\dfrac{13\pi}{12}\right)$

82. $\sin\left(\dfrac{11\pi}{12}\right) + \sin\left(\dfrac{7\pi}{12}\right)$

Verify the following identities.

83. $\dfrac{2\sin x \cos x}{\cos^2 x - \sin^2 x} = \tan(2x)$

84. $\dfrac{1 - 2\sin^2 x}{2\sin x \cos x} = \cot(2x)$

85. $(\sin x + \cos x)^2 = 1 + \sin(2x)$

86. $(\sin^2 x - 1)^2 = \sin^4 x + \cos(2x)$

87. $\cos(8\theta) = \cos^2(4\theta) - \sin^2(4\theta)$

88. $\sin(4x) = 4 \sin x \cos x (1 - 2 \sin^2 x)$

89. $\dfrac{\cos(2\theta)}{\sin^2\theta} = \cot^2\theta - 1$

90. $\csc^2\theta - 2 = \dfrac{\cos(2\theta)}{\sin^2\theta}$

91. $\tan(2\theta) = \dfrac{2}{\cot\theta - \tan\theta}$

92. $\cot\theta - \tan\theta = \dfrac{2\cos(2\theta)}{\sin(2\theta)}$

93. $\tan x + \cot x = 2\csc(2x)$

94. $\csc(2x) = \dfrac{1}{2}\csc x \sec x$

95. $\cos^2\left(\dfrac{x}{2}\right) - \sin^2\left(\dfrac{x}{2}\right) = \cos x$

96. $1 - 2\sin^2\left(\dfrac{x}{4}\right) = \cos\left(\dfrac{x}{2}\right)$

97. $1 - \sin^2(2\theta) = 1 - 4\sin^2\theta + 4\sin^4\theta$

98. $2\cos^2\left(\dfrac{x}{2}\right) - 1 = \cos x$

99. $\dfrac{\sin(120\pi t) + \sin(80\pi t)}{\cos(120\pi t) - \cos(80\pi t)} = -\cot(20\pi t)$

100. $\dfrac{\sin m + \sin n}{\cos m + \cos n} = \tan\left(\dfrac{m+n}{2}\right)$

101. Show $\sin^2\alpha + (1 - \cos\alpha)^2 = \left[2\sin\left(\dfrac{\alpha}{2}\right)\right]^2$.

102. Show that $\tan\left(\dfrac{u}{2}\right) = \pm\sqrt{\dfrac{1 - \cos u}{1 + \cos u}}$ is equivalent to $\dfrac{\sin u}{1 + \cos u}$ by rationalizing the numerator.

103. Derive the identity for $\sin(2\alpha)$ and $\tan(2\alpha)$ using $\sin(\alpha + \beta)$ and $\tan(\alpha + \beta)$, where $\alpha = \beta$.

104. Derive the identity for $\tan^2(\alpha)$ using $\tan(2\alpha) = \dfrac{2\tan(\alpha)}{1 - \tan^2(\alpha)}$. *Hint:* Solve for $\tan^2\alpha$ and work in terms of sines and cosines.

105. Derive the product-to-sum identity for $\sin\alpha \sin\beta$.

106. Derive the sum-to-product identity for $\cos u + \cos v$.

▶ WORKING WITH FORMULAS

 107. Supersonic speeds, the sound barrier, and Mach numbers: $\mathcal{M} = \csc\left(\dfrac{\theta}{2}\right)$

The speed of sound varies with temperature and altitude. At 32°F, sound travels about 742 mi/hr at sea level. A jet-plane flying faster than the speed of sound (called supersonic speed) has "broken the sound barrier." The plane projects three-dimensional sound waves about the nose of the craft that form the shape of a cone. The cone intersects the Earth along a hyperbolic path, with a sonic boom being heard by anyone along this path. The ratio of the plane's speed to the speed of sound is

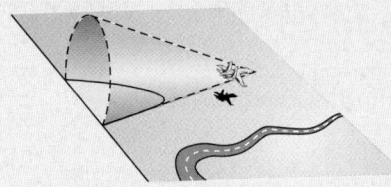

called its Mach number $\mathcal{M}$, meaning a plane flying at $\mathcal{M} = 3.2$ is traveling 3.2 times the speed of sound. This Mach number can be determined using the formula given here, where θ is the vertex angle of the cone described. For the following exercises, use the formula to find $\mathcal{M}$ or θ as required. For parts (a) and (b), answer in exact form (using a half-angle identity) and approximate form.

   a. $\theta = 30°$ **b.** $\theta = 45°$ **c.** $\mathcal{M} = 2$

108. Malus's law: $I = I_0 \cos^2\theta$

When a beam of plane-polarized light with intensity I_0 hits an analyzer, the intensity I of the transmitted beam of light can be found using the formula shown, where θ is the angle formed between the transmission axes of the polarizer and the analyzer. Find the intensity of the beam when $\theta = 15°$ and $I_0 = 300$ candelas (cd). Answer in exact form (using a power reduction identity) and approximate form.

▶ **APPLICATIONS**

Range of a projectile: Exercises 109 and 110 refer to Example 9. In Example 9, we noted that the range of a projectile was maximized at $\theta = 45°$. If $\theta > 45°$ or $\theta < 45°$, the projectile falls short of its maximum potential distance. In Exercises 109 and 110 assume that the projectile has an initial velocity of 96 ft/sec.

109. Compute how many feet short of maximum the projectile falls if (a) $\theta = 22.5°$ and (b) $\theta = 67.5°$. Answer in both exact and approximate form.

110. Use a calculator to compute how many feet short of maximum the projectile falls if (a) $\theta = 40°$ and $\theta = 50°$ and (b) $\theta = 37.5°$ and $\theta = 52.5°$. Do you see a pattern? Discuss/explain what you notice and experiment with other values to confirm your observations.

Touch-tone phones: The diagram given in Example 10 shows the various frequencies used to create the tones for a touch-tone phone. One button is randomly pressed and the resultant wave is modeled by $y(t)$ shown. Use a product-to-sum identity to write the expression as a sum and determine the button pressed.

111. $y(t) = 2\cos(2150\pi t)\cos(268\pi t)$

112. $y(t) = 2\cos(1906\pi t)\cos(512\pi t)$

113. Clock angles: Kirkland City has a large clock atop city hall, with a minute hand that is 3 ft long. Claire and Monica independently attempt to devise a function that will track the distance between the tip of the

minute hand at t minutes between the hours, and the tip of the minute hand when it is in the vertical position as shown. Claire finds the function $d(t) = \left| 6\sin\left(\dfrac{\pi t}{60}\right) \right|$, while Monica devises $d(t) = \sqrt{18\left[1 - \cos\left(\dfrac{\pi t}{30}\right)\right]}$. Use the identities from this section to show the functions are equivalent.

114. Origami: The Japanese art of origami involves the repeated folding of a single piece of paper to create various art forms. When the upper right corner of

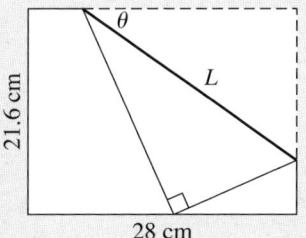

a rectangular 21.6-cm by 28-cm piece of paper is folded down until the corner is flush with the other side, the length L of the fold is related to the angle θ by $L = \dfrac{10.8}{\sin\theta\cos^2\theta}$. (a) Show this is equivalent to $L = \dfrac{21.6\sec\theta}{\sin(2\theta)}$, (b) find the length of the fold if $\theta = 30°$, and (c) find the angle θ if $L = 28.8$ cm.

115. Machine gears: A machine part involves two gears. The first has a radius of 2 cm and the second a radius of 1 cm, so the smaller gear turns twice as fast as the larger gear. Let θ represent the angle of rotation in the larger gear, measured from a vertical and downward starting position. Let P be a point on the circumference

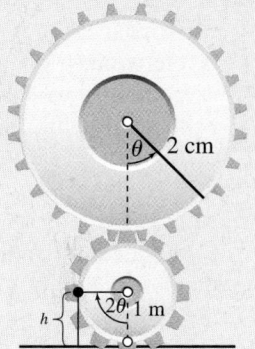

of the smaller gear, starting at the vertical and downward position. Four engineers working on an improved design for this component devise functions that track the height of point P above the horizontal plane shown, for a rotation of $\theta°$ by the larger gear. The functions they develop are: Engineer A: $f(\theta) = \sin(2\theta - 90°) + 1$; Engineer B: $g(\theta) = 2\sin^2\theta$; Engineer C: $k(\theta) = 1 + \sin^2\theta - \cos^2\theta$; and Engineer D: $h(\theta) = 1 - \cos(2\theta)$. Use any of the identities you've learned so far to show these four functions are equivalent.

116. Working with identities: Compute the value of $\sin 15°$ two ways, first using the half-angle identity for sine, and second using the difference identity for sine. (a) Find a decimal approximation for each to show the results are equivalent and (b) verify algebraically that they are equivalent. (*Hint:* Square both sides.)

117. Working with identities: Compute the value of $\cos 15°$ two ways, first using the half-angle identity for cosine, and second using the difference identity for cosine. (a) Find a decimal approximation for each to show the results are equivalent and (b) verify algebraically that they are equivalent. (*Hint:* Square both sides.)

▶ EXTENDING THE CONCEPT

 118. Can you find three distinct, real numbers whose sum is equal to their product? A little known fact from trigonometry stipulates that for any triangle, the sum of the tangents of the angles is equal to the products of their tangents. Use a calculator to test this statement, recalling the three angles must sum to 180°. Our website at www.mhhe.com/coburn shows a method that enables you to verify the statement using tangents that are all rational values.

119. A proof without words: From elementary geometry we have the following: (a) an angle inscribed in a semicircle is a right angle; and (b) the measure of an

Exercise 119

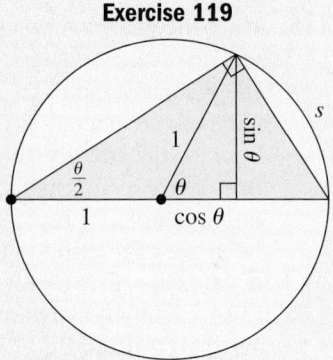

inscribed angle (vertex on the circumference) is one-half the measure of its intercepted arc. Discuss/explain how the unit-circle diagram offers a proof that $\tan\left(\dfrac{x}{2}\right) = \dfrac{\sin x}{1 + \cos x}$. Be detailed and thorough.

120. Using $\theta = 30°$ and repeatedly applying the half-angle identity for cosine, show that $\cos 3.75°$ is equal to $\dfrac{\sqrt{2 + \sqrt{2 + \sqrt{2 + \sqrt{3}}}}}{2}$. Verify the result using a calculator, then use the patterns noted to write the value of $\cos 1.875°$ in closed form (also verify this result). As θ becomes very small, what appears to be happening to the value of $\cos \theta$?

▶ MAINTAINING YOUR SKILLS

121. (3.3) Use the rational roots theorem to find all zeroes of $x^4 + x^3 - 8x^2 - 6x + 12 = 0$.

122. (5.1) The hypotenuse of a certain right triangle is twice the shortest side. Solve the triangle.

123. (5.3) Verify that $\left(\frac{16}{65}, \frac{63}{65}\right)$ is on the unit circle, then find $\tan \theta$ and $\sec \theta$ to verify $1 + \tan^2\theta = \sec^2\theta$.

124. (5.5) Write the equation of the function graphed in terms of a sine function of the form $y = A \sin(Bx + C) + D$.

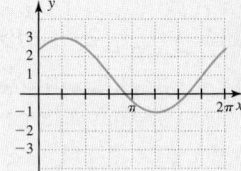

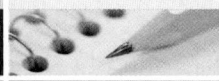 ## MID-CHAPTER CHECK

1. Verify the identity using a multiplication:
$\sin x(\csc x - \sin x) = \cos^2 x$

2. Verify the identity by factoring:
$\cos^2 x - \cot^2 x = -\cos^2 x \cot^2 x$

3. Verify the identity by combining terms:
$\dfrac{2 \sin x}{\sec x} - \dfrac{\cos x}{\csc x} = \cos x \sin x$

4. Show the equation given is not an identity.
$1 + \sec^2 x = \tan^2 x$

5. Verify each identity.

  a. $\dfrac{\sin^3 x + \cos^3 x}{\sin x + \cos x} = 1 - \sin x \cos x$

  b. $\dfrac{1 + \sec x}{\csc x} - \dfrac{1 + \cos x}{\cot x} = 0$

6. Verify each identity.

  a. $\dfrac{\sec^2 x - \tan^2 x}{\sec^2 x} = \cos^2 x$

  b. $\dfrac{\cot x - \tan x}{\csc x \sec x} = \cos^2 x - \sin^2 x$

7. Given α and β are obtuse angles with $\sin \alpha = \dfrac{56}{65}$ and $\tan \beta = -\dfrac{80}{39}$, find

  a. $\sin(\alpha - \beta)$

  b. $\cos(\alpha + \beta)$

  c. $\tan(\alpha - \beta)$

8. Use the diagram shown to compute $\sin A$, $\cos A$, and $\tan A$.

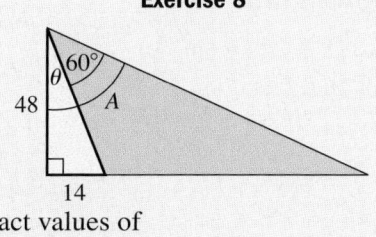

Exercise 8

10. Given $\sin \alpha = -\dfrac{7}{25}$ with α in QIII, find the value of $\sin(2\alpha)$, $\cos(2\alpha)$, and $\tan(2\alpha)$.

9. Given $\cos \theta = -\dfrac{15}{17}$ and θ in QII, find exact values of $\sin\left(\dfrac{\theta}{2}\right)$ and $\cos\left(\dfrac{\theta}{2}\right)$.

REINFORCING BASIC CONCEPTS

Identities—Connections and Relationships

It is a well-known fact that information is retained longer and used more effectively when it is organized, sequential, and connected. In this *Strengthening Core Skills (SCS)*, we attempt to do just that with our study of identities. In flowchart form we'll show that the entire range of identities has only two tiers, and that the fundamental identities and the sum and difference identities are really the keys to the entire range of identities. Beginning with the right triangle definition of sine, cosine, and tangent, the **reciprocal identities** and **ratio identities** are more semantic (word related) than mathematical, and the **Pythagorean identities** follow naturally from the properties of right triangles. These form the first tier.

Basic Definitions

$$\sin \theta = \frac{\text{opp}}{\text{hyp}} \qquad \cos \theta = \frac{\text{adj}}{\text{hyp}} \qquad \tan \theta = \frac{\text{opp}}{\text{adj}}$$

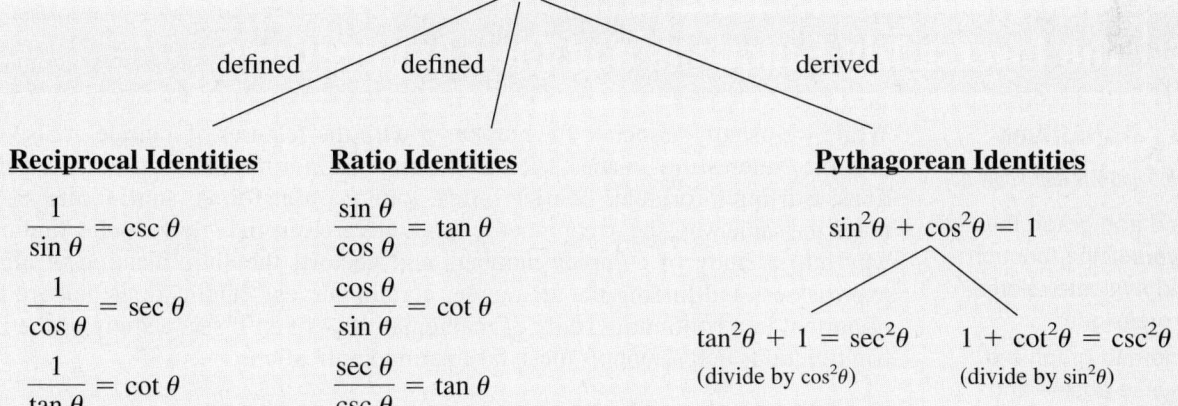

The reciprocal and ratio identities are actually *defined*, while the Pythagorean identities are *derived* from these two families. In addition, the identity $\sin^2\theta + \cos^2\theta = 1$ is the only Pythagorean identity we actually need to memorize; the other two follow by division of $\cos^2\theta$ and $\sin^2\theta$ as indicated.

In virtually the same way, the sum and difference identities for sine and cosine are the only identities that need to be memorized, as all other identities in the second tier flow from these.

Sum/Difference Identities

$$\cos(\alpha \pm \beta) = \cos\alpha\cos\beta \mp \sin\alpha\sin\beta$$
$$\sin(\alpha \pm \beta) = \sin\alpha\cos\beta \pm \cos\alpha\sin\beta$$

Double-Angle Identities
use $\alpha = \beta$
in sum identities

Power Reduction Identities
solve for $\cos^2\alpha$, $\sin^2\alpha$ in
related $\cos(2\alpha)$ identity

Half-Angle Identities
solve for $\cos\alpha$, $\sin\alpha$
and use $\alpha = u/2$ in the
power reduction identities

Product-to-Sum Identities
combine various
sum/difference identities

$$\sin(2\alpha) = 2\sin\alpha\cos\alpha \qquad \cos^2\alpha = \frac{1 + \cos(2\alpha)}{2} \qquad \cos\left(\frac{u}{2}\right) = \pm\sqrt{\frac{1 + \cos u}{2}} \qquad \text{see Section 6.4}$$

$$\cos(2\alpha) = \cos^2\alpha - \sin^2\alpha \qquad \sin^2\alpha = \frac{1 - \cos(2\alpha)}{2} \qquad \sin\left(\frac{u}{2}\right) = \pm\sqrt{\frac{1 - \cos u}{2}} \qquad \text{see Section 6.4}$$

$$\cos(2\alpha) = 2\cos^2\alpha - 1 \qquad\qquad \cos(2\alpha) = 1 - 2\sin^2\alpha$$
(use $\sin^2\alpha = 1 - \cos^2\alpha$) $\qquad\qquad$ (use $\cos^2\alpha = 1 - \sin^2\alpha$)

Exercise 1: Starting with the identity $\sin^2\alpha + \cos^2\alpha = 1$, derive the other two Pythagorean identities.

Exercise 2: Starting with the identity $\cos(\alpha \pm \beta) = \cos\alpha\cos\beta \mp \sin\alpha\sin\beta$, derive the double-angle identities for cosine.

6.5 | The Inverse Trig Functions and Their Applications

Learning Objectives

In Section 6.5 you will learn how to:

☐ **A.** Find and graph the inverse sine function and evaluate related expressions

☐ **B.** Find and graph the inverse cosine and tangent functions and evaluate related expressions

☐ **C.** Apply the definition and notation of inverse trig functions to simplify compositions

☐ **D.** Find and graph inverse functions for sec x, csc x, and cot x

☐ **E.** Solve applications involving inverse functions

While we usually associate the number π with the features of a circle, it also occurs in some "interesting" places, such as the study of normal (bell) curves, Bessel functions, Stirling's formula, Fourier series, Laplace transforms, and infinite series. In much the same way, the trigonometric functions are surprisingly versatile, finding their way into a study of complex numbers and vectors, the simplification of algebraic expressions, and finding the area under certain curves—applications that are hugely important in a continuing study of mathematics. As you'll see, a study of the inverse trig functions helps support these fascinating applications.

A. The Inverse Sine Function

In Section 4.1 we established that only one-to-one functions have an inverse. All six trig functions fail the horizontal line test and are not one-to-one as given. However, by suitably restricting the domain, a one-to-one function can be defined that makes finding an inverse possible. For the sine function, it seems natural to choose the interval $\left[-\frac{\pi}{2}, \frac{\pi}{2}\right]$ since it is centrally located and the sine function attains all possible range values in this interval. A graph of $y = \sin x$ is shown in Figure 6.5, with the portion corresponding to this interval colored in red. Note the range is still $[-1, 1]$ (Figure 6.6).

Figure 6.5

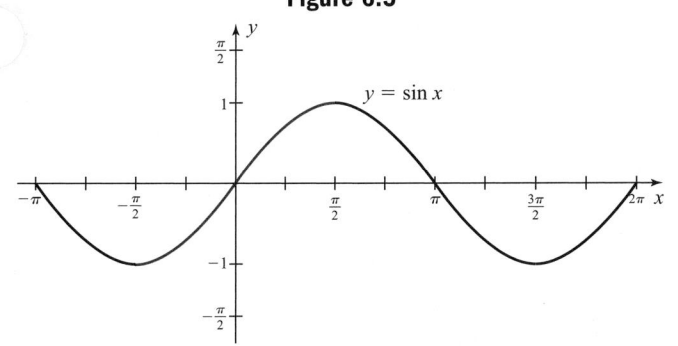

Figure 6.6   **Figure 6.7**

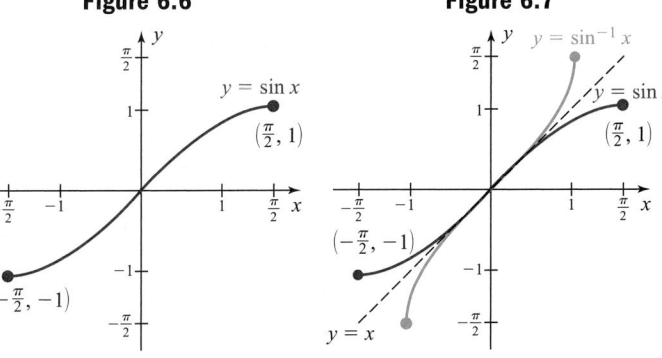

In Example 4 of Section 4.1, we noted that by suitably restricting the domain of $y = x^2$, a one-to-one function could be defined that made finding an inverse function possible. Specifically, for $f(x) = x^2; x \geq 0, f^{-1}(x) = \sqrt{x}$.

We can obtain an implicit equation for the inverse of $y = \sin x$ by interchanging x- and y-values, obtaining $x = \sin y$. By accepted convention, the *explicit* form of the inverse sine function is written $y = \sin^{-1}x$ or $y = \arcsin x$. Since domain and range values have been interchanged, the domain of $y = \sin^{-1}x$ is $[-1, 1]$ and the range is $\left[-\frac{\pi}{2}, \frac{\pi}{2}\right]$. The graph of $y = \sin^{-1}x$ can be found by reflecting the portion in red across the line $y = x$ and using the endpoints of the domain and range (see Figure 6.7).

The Inverse Sine Function

For $y = \sin x$ with domain $\left[-\frac{\pi}{2}, \frac{\pi}{2}\right]$ and range $[-1, 1]$,

the inverse sine function is

$$y = \sin^{-1}x \text{ or } y = \arcsin x,$$

with domain $[-1, 1]$ and range $\left[-\frac{\pi}{2}, \frac{\pi}{2}\right]$.

$$y = \sin^{-1}x \text{ if and only if } \sin y = x$$

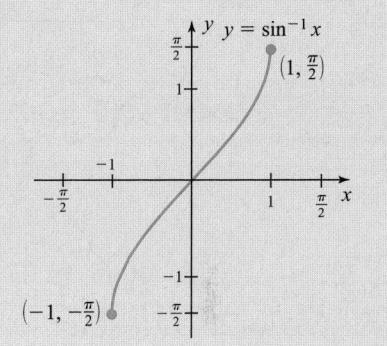

From the implicit form $x = \sin y$, we learn to interpret the inverse function as, "y is the number or angle whose sine is x." Learning to read and interpret the explicit form in this way will be helpful. That is, $y = \sin^{-1}x$ means "y is the number or angle whose sine is x."

$$y = \sin^{-1}x \Leftrightarrow x = \sin y \qquad x = \sin y \Leftrightarrow y = \sin^{-1}x$$

EXAMPLE 1 ▶ **Evaluating $y = \sin^{-1}x$ Using Special Values**

Evaluate the inverse sine function for the values given:

   a. $y = \sin^{-1}\left(\frac{\sqrt{3}}{2}\right)$ **b.** $y = \arcsin\left(-\frac{1}{2}\right)$ **c.** $y = \sin^{-1}2$

Solution ▶ For x in $[-1, 1]$ and y in $\left[-\frac{\pi}{2}, \frac{\pi}{2}\right]$,

   a. $y = \sin^{-1}\left(\frac{\sqrt{3}}{2}\right)$: y is the number or angle whose sine is $\frac{\sqrt{3}}{2}$

   $\Rightarrow \sin y = \frac{\sqrt{3}}{2}$, so $\sin^{-1}\left(\frac{\sqrt{3}}{2}\right) = \frac{\pi}{3}$.

b. $y = \arcsin\left(-\dfrac{1}{2}\right)$: y is the arc or angle whose sine is $-\dfrac{1}{2}$

$\Rightarrow \sin y = -\dfrac{1}{2}$, so $\arcsin\left(-\dfrac{1}{2}\right) = -\dfrac{\pi}{6}$.

c. $y = \sin^{-1}(2)$: y is the number or angle whose sine is 2
$\Rightarrow \sin y = 2$. Since 2 is not in $[-1, 1]$, $\sin^{-1}(2)$ is undefined.

Now try Exercises 7 through 12 ▶

Table 6.1

x	$\sin x$
$-\dfrac{\pi}{2}$	-1
$-\dfrac{\pi}{3}$	$-\dfrac{\sqrt{3}}{2}$
$-\dfrac{\pi}{4}$	$-\dfrac{\sqrt{2}}{2}$
$-\dfrac{\pi}{6}$	$-\dfrac{1}{2}$
0	0
$\dfrac{\pi}{6}$	$\dfrac{1}{2}$
$\dfrac{\pi}{4}$	$\dfrac{\sqrt{2}}{2}$
$\dfrac{\pi}{3}$	$\dfrac{\sqrt{3}}{2}$
$\dfrac{\pi}{2}$	1

In Examples 1a and 1b, note that the equations $\sin y = \dfrac{\sqrt{3}}{2}$ and $\sin y = -\dfrac{1}{2}$ each have an infinite number of solutions, but only one solution in $\left[-\dfrac{\pi}{2}, \dfrac{\pi}{2}\right]$.

When x is one of the standard values $\left(0, \dfrac{1}{2}, \dfrac{\sqrt{3}}{2}, 1, \text{ and so on}\right)$, $y = \sin^{-1}x$ can be evaluated by reading a standard table "in reverse." For $y = \arcsin(-1)$, we locate the number -1 in the right-hand column of Table 6.1, and note the "number or angle whose sine is -1," is $-\dfrac{\pi}{2}$. If x is between -1 and 1 but is not a standard value, we can use the $\sin^{-1}$ function on a calculator, which is most often the [2nd] or [INV] function for [SIN].

EXAMPLE 2 ▶ **Evaluating $y = \sin^{-1}x$ Using a Calculator**

Evaluate each inverse sine function twice. First in radians rounded to four decimal places, then in degrees to the nearest tenth.
a. $y = \sin^{-1}0.8492$ **b.** $y = \arcsin(-0.2317)$

Solution ▶ For x in $[-1, 1]$, we evaluate $y = \sin^{-1}x$.

a. $y = \sin^{-1}0.8492$: With the calculator in radian [MODE], use the keystrokes [2nd] [SIN] 0.8492 [ )] [ENTER]. We find $\sin^{-1}(0.8492) \approx 1.0145$ radians. In degree [MODE], the same sequence of keystrokes gives $\sin^{-1}(0.8492) \approx 58.1°$ (note that 1.0145 rad $\approx 58.1°$).

b. $y = \arcsin(-0.2317)$: In radian [MODE], we find $\sin^{-1}(-0.2317) \approx -0.2338$ rad. In degree [MODE], $\sin^{-1}(-0.2317) \approx -13.4°$.

Now try Exercises 13 through 16 ▶

WORTHY OF NOTE

The $\sin^{-1}x$ notation for the inverse sine function is a carryover from the $f^{-1}(x)$ notation for a general inverse function, and likewise has nothing to do with the reciprocal of the function. The arcsin x notation derives from our work in radians on the unit circle, where $y = \arcsin x$ can be interpreted as "y is an arc whose sine is x."

From our work in Section 4.1, we know that if f and g are inverses, $(f \circ g)(x) = x$ and $(g \circ f)(x) = x$. This suggests the following properties.

Inverse Function Properties for Sine

For $f(x) = \sin x$ and $g(x) = \sin^{-1}x$:

    I. $(f \circ g)(x) = \sin(\sin^{-1}x) = x$ for x in $[-1, 1]$

and

    II. $(g \circ f)(x) = \sin^{-1}(\sin x) = x$ for x in $\left[-\dfrac{\pi}{2}, \dfrac{\pi}{2}\right]$

EXAMPLE 3 ▶ **Evaluating Expressions Using Inverse Function Properties**

Evaluate each expression and verify the result on a calculator.

a. $\sin\left[\sin^{-1}\left(\dfrac{1}{2}\right)\right]$ **b.** $\arcsin\left[\sin\left(\dfrac{\pi}{4}\right)\right]$ **c.** $\sin^{-1}\left[\sin\left(\dfrac{5\pi}{6}\right)\right]$

Solution ▶ **a.** $\sin\left[\sin^{-1}\left(\dfrac{1}{2}\right)\right] = \dfrac{1}{2}$, since $\dfrac{1}{2}$ is in $[-1, 1]$ Property I

b. $\arcsin\left[\sin\left(\dfrac{\pi}{4}\right)\right] = \dfrac{\pi}{4}$, since $\dfrac{\pi}{4}$ is in $\left[-\dfrac{\pi}{2}, \dfrac{\pi}{2}\right]$ Property II

c. $\sin^{-1}\left[\sin\left(\dfrac{5\pi}{6}\right)\right] \neq \dfrac{5\pi}{6}$, since $\dfrac{5\pi}{6}$ is not in $\left[-\dfrac{\pi}{2}, \dfrac{\pi}{2}\right]$.

This doesn't mean the expression cannot be evaluated, only that we cannot use

Property II. Since $\sin\left(\dfrac{5\pi}{6}\right) = \sin\left(\dfrac{\pi}{6}\right)$, $\sin^{-1}\left[\left(\sin\dfrac{5\pi}{6}\right)\right] = \sin^{-1}\left[\sin\left(\dfrac{\pi}{6}\right)\right] = \dfrac{\pi}{6}$.

The calculator verification for each is shown in Figures 6.8 and 6.9. Note

$\dfrac{\pi}{6} \approx 0.5236$ and $\dfrac{\pi}{4} \approx 0.7854$.

Figure 6.8
Parts (a) and (b)

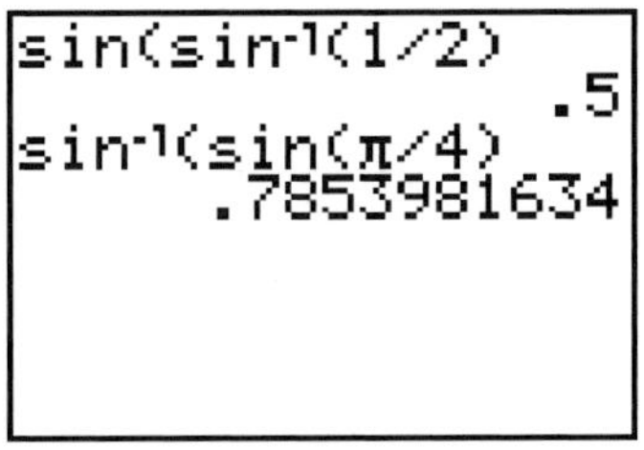

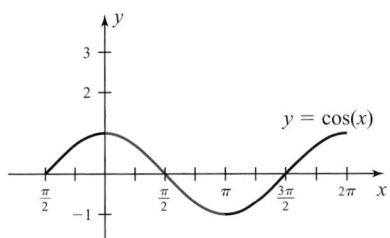

Figure 6.9
Part (c)

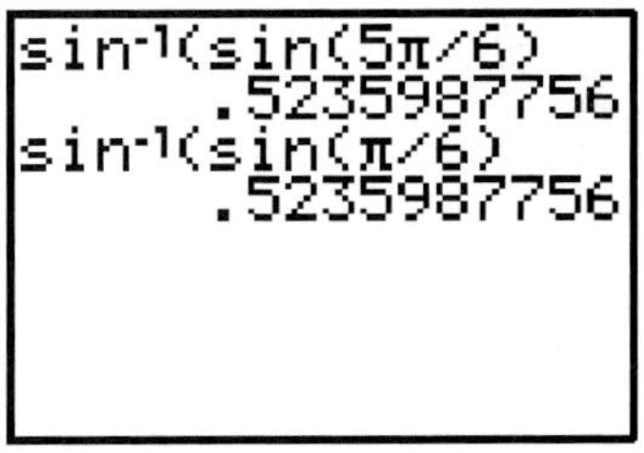

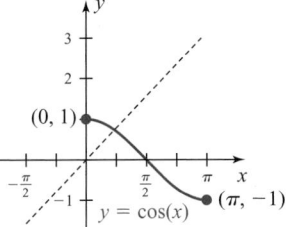

✓ **A.** You've just learned how to find and graph the inverse sine function and evaluate related expressions

Now try Exercises 17 through 24 ▶

B. The Inverse Cosine and Inverse Tangent Functions

Like the sine function, the cosine function is not one-to-one and its domain must also be restricted to develop an inverse function. For convenience we choose the interval $x \in [0, \pi]$ since it is again somewhat central and takes on all of its range values in this interval. A graph of the cosine function, with the interval corresponding to this interval shown in red, is given in Figure 6.10. Note the range is still $[-1, 1]$ (Figure 6.11).

Figure 6.10

Figure 6.11

Figure 6.12

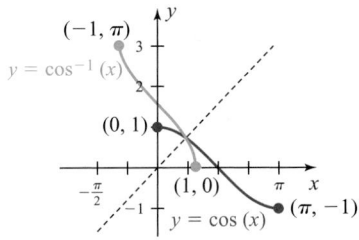

For the implicit equation of inverse cosine, $y = \cos x$ becomes $x = \cos y$, with the corresponding explicit forms being $y = \cos^{-1} x$ or $y = \arccos x$. By reflecting the graph of $y = \cos x$ across the line $y = x$, we obtain the graph of $y = \cos^{-1} x$ shown in Figure 6.12.

The Inverse Cosine Function

For $y = \cos x$ with domain $[0, \pi]$ and range $[-1, 1]$, the inverse cosine function is

$$y = \cos^{-1}x \text{ or } y = \arccos x$$

with domain $[-1, 1]$ and range $[0, \pi]$.

$$y = \cos^{-1}x \text{ if and only if } \cos y = x$$

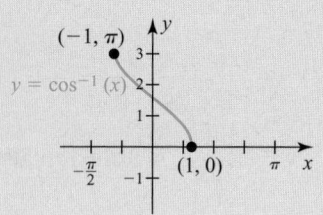

EXAMPLE 4 ▶ **Evaluating $y = \cos^{-1}x$ Using Special Values**

Evaluate the inverse cosine for the values given:

    a. $y = \cos^{-1}0$ **b.** $y = \arccos\left(-\dfrac{\sqrt{3}}{2}\right)$ **c.** $y = \cos^{-1}\pi$

Solution ▶ For x in $[-1, 1]$ and y in $[0, \pi]$,

    a. $y = \cos^{-1}0$: y is the number or angle whose cosine is $0 \Rightarrow \cos y = 0$.

    This shows $\cos^{-1}0 = \dfrac{\pi}{2}$.

    b. $y = \arccos\left(-\dfrac{\sqrt{3}}{2}\right)$: y is the arc or angle whose cosine is

    $-\dfrac{\sqrt{3}}{2} \Rightarrow \cos y = -\dfrac{\sqrt{3}}{2}$. This shows $\arccos\left(-\dfrac{\sqrt{3}}{2}\right) = \dfrac{5\pi}{6}$.

    c. $y = \cos^{-1}\pi$: y is the number or angle whose cosine is $\pi \Rightarrow \cos y = \pi$. Since $\pi \notin [-1, 1]$, $\cos^{-1}\pi$ is undefined.

Now try Exercises 25 through 34 ▶

Knowing that $y = \cos x$ and $y = \cos^{-1}x$ are inverse functions enables us to state inverse function properties similar to those for sine.

Inverse Function Properties for Cosine

For $f(x) = \cos x$ and $g(x) = \cos^{-1}x$:

$$\text{I. } (f \circ g)(x) = \cos(\cos^{-1}x) = x \text{ for } x \text{ in } [-1, 1]$$

and

$$\text{II. } (g \circ f)(x) = \cos^{-1}(\cos x) = x \text{ for } x \text{ in } [0, \pi]$$

EXAMPLE 5 ▶ **Evaluating Expressions Using Inverse Function Properties**

Evaluate each expression.

    a. $\cos[\cos^{-1}(0.73)]$ **b.** $\arccos\left[\cos\left(\dfrac{\pi}{12}\right)\right]$ **c.** $\cos^{-1}\left[\cos\left(\dfrac{4\pi}{3}\right)\right]$

Solution ▶   **a.** $\cos[\cos^{-1}(0.73)] = 0.73$, since 0.73 is in $[-1, 1]$ Property I

    b. $\arccos\left[\cos\left(\dfrac{\pi}{12}\right)\right] = \dfrac{\pi}{12}$, since $\dfrac{\pi}{12}$ is in $[0, \pi]$ Property II

c. $\cos^{-1}\left[\cos\left(\dfrac{4\pi}{3}\right)\right] \neq \dfrac{4\pi}{3}$, since $\dfrac{4\pi}{3}$ is not in $[0, \pi]$.

This expression cannot be evaluated using Property II. Since

$$\cos\left(\dfrac{4\pi}{3}\right) = \cos\left(\dfrac{2\pi}{3}\right), \cos^{-1}\left[\cos\left(\dfrac{4\pi}{3}\right)\right] = \cos^{-1}\left[\cos\left(\dfrac{2\pi}{3}\right)\right] = \dfrac{2\pi}{3}.$$

The results can also be verified using a calculator.

Now try Exercises 35 through 42 ▶

For the tangent function, we likewise restrict the domain to obtain a one-to-one function, with the most common choice being $\left(-\dfrac{\pi}{2}, \dfrac{\pi}{2}\right)$. The corresponding range is $\mathbb{R}$. The *implicit* equation for the inverse tangent function is $x = \tan y$ with the explicit forms $y = \tan^{-1}x$ or $y = \arctan x$. With the domain and range interchanged, the domain of $y = \tan^{-1}x$ is $\mathbb{R}$, and the range is $\left(-\dfrac{\pi}{2}, \dfrac{\pi}{2}\right)$. The graph of $y = \tan x$ for x in $\left(-\dfrac{\pi}{2}, \dfrac{\pi}{2}\right)$ is shown in red (Figure 6.13), with the inverse function $y = \tan^{-1}x$ shown in blue (Figure 6.14).

Figure 6.13

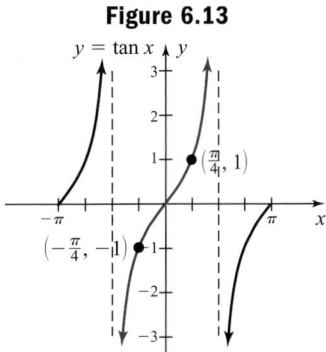

Figure 6.14

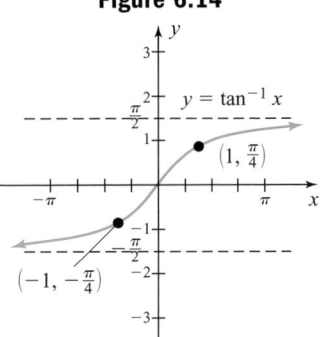

The Inverse Tangent Function

For $y = \tan x$ with domain $\left(-\dfrac{\pi}{2}, \dfrac{\pi}{2}\right)$ and range $\mathbb{R}$, the inverse tangent function is

$$y = \tan^{-1}x \text{ or } y = \arctan x,$$

with domain $\mathbb{R}$ and range $\left(-\dfrac{\pi}{2}, \dfrac{\pi}{2}\right)$.

$y = \tan^{-1}x$ if and only if $\tan y = x$

Inverse Function Properties for Tangent

For $f(x) = \tan x$ and $g(x) = \tan^{-1}x$:

I. $(f \circ g)(x) = \tan(\tan^{-1}x) = x$ for x in $\mathbb{R}$

and

II. $(g \circ f)(x) = \tan^{-1}(\tan x) = x$ for x in $\left(-\dfrac{\pi}{2}, \dfrac{\pi}{2}\right)$.

EXAMPLE 6 ▶ Evaluating Expressions Involving Inverse Tangent

Evaluate each expression.

a. $\tan^{-1}(-\sqrt{3})$ **b.** $\arctan[\tan(-0.89)]$

Solution ▶ For x in $\mathbb{R}$ and y in $\left(-\dfrac{\pi}{2}, \dfrac{\pi}{2}\right)$,

a. $\tan^{-1}(-\sqrt{3}) = -\dfrac{\pi}{3}$, since $\tan\left(-\dfrac{\pi}{3}\right) = -\sqrt{3}$

☑ B. You've just learned how to find and graph the inverse cosine and tangent functions and evaluate related expressions

b. $\arctan[\tan(-0.89)] = -0.89$, since -0.89 is in $\left(-\dfrac{\pi}{2}, \dfrac{\pi}{2}\right)$ Property II

Now try Exercises 43 through 52 ▶

C. Using the Inverse Trig Functions to Evaluate Compositions

In the context of angle measure, the expression $y = \sin^{-1}\left(-\dfrac{1}{2}\right)$ represents an angle—the *angle* y whose sine is $-\dfrac{1}{2}$. It seems natural to ask, "What happens if we take the tangent of this angle?" In other words, what does the expression $\tan\left[\sin^{-1}\left(-\dfrac{1}{2}\right)\right]$ mean? Similarly, if $y = \cos\left(\dfrac{\pi}{3}\right)$ represents a real number between -1 and 1, how do we compute $\sin^{-1}\left[\cos\left(\dfrac{\pi}{3}\right)\right]$? Expressions like these occur in many fields of study.

EXAMPLE 7 ▶ **Simplifying Expressions Involving Inverse Trig Functions**

Simplify each expression:

a. $\tan\left[\arcsin\left(-\dfrac{1}{2}\right)\right]$ **b.** $\sin^{-1}\left[\cos\left(\dfrac{\pi}{3}\right)\right]$

Solution ▶ **a.** In Example 1 we found $\arcsin\left(-\dfrac{1}{2}\right) = -\dfrac{\pi}{6}$. Substituting $-\dfrac{\pi}{6}$ for $\arcsin\left(-\dfrac{1}{2}\right)$

gives $\tan\left(-\dfrac{\pi}{6}\right) = -\dfrac{\sqrt{3}}{3}$, showing $\tan\left[\arcsin\left(-\dfrac{1}{2}\right)\right] = -\dfrac{\sqrt{3}}{3}$.

b. For $\sin^{-1}\left[\cos\left(\dfrac{\pi}{3}\right)\right]$, we begin with the inner function $\cos\left(\dfrac{\pi}{3}\right) = \dfrac{1}{2}$.

Substituting $\dfrac{1}{2}$ for $\cos\left(\dfrac{\pi}{3}\right)$ gives $\sin^{-1}\left(\dfrac{1}{2}\right)$. With the appropriate checks

satisfied we have $\sin^{-1}\left(\dfrac{1}{2}\right) = \dfrac{\pi}{6}$, showing $\sin^{-1}\left[\cos\left(\dfrac{\pi}{3}\right)\right] = \dfrac{\pi}{6}$.

Now try Exercises 53 through 64 ▶

WORTHY OF NOTE

To verify the result of Example 8, we can actually find the value of $\sin^{-1}\left(-\dfrac{8}{17}\right)$ on a calculator, then take the tangent of the result. See the figure.

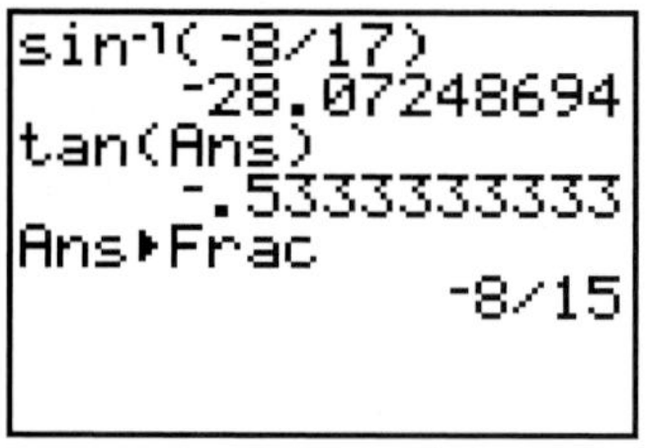

```
sin⁻¹(-8/17)
         -28.07248694
tan(Ans)
         -.5333333333
Ans▶Frac
                -8/15
```

If the argument is not a special value and we need the answer in exact form, we can draw the triangle described by the inner expression using the definition of the trigonometric functions as ratios. In other words, for either y or $\theta = \sin^{-1}\left(\dfrac{8}{17}\right)$, we draw a triangle with hypotenuse 17 and side 8 opposite θ to model the statement, "an angle whose sine is $\dfrac{8}{17} = \dfrac{\text{opp}}{\text{hyp}}$," (see Figure 6.15). Using the Pythagorean theorem, we find the adjacent side is 15 and can now name any of the other trig functions.

Figure 6.15

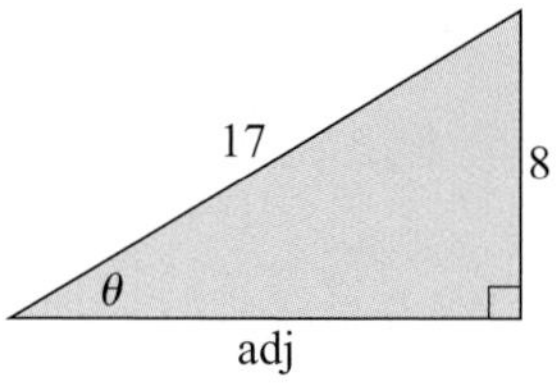

EXAMPLE 8 ▶ **Using a Diagram to Evaluate an Expression Involving Inverse Trig Functions**

Evaluate the expression $\tan\left[\sin^{-1}\left(-\dfrac{8}{17}\right)\right]$.

Solution ▶ The expression $\tan\left[\sin^{-1}\left(-\dfrac{8}{17}\right)\right]$ is equivalent to $\tan\theta$,

where $\theta = \sin^{-1}\left(-\dfrac{8}{17}\right)$ with θ in

$\left[-\dfrac{\pi}{2}, \dfrac{\pi}{2}\right]$ (QIV or QI). For

$\sin\theta = -\dfrac{8}{17}$ ($\sin\theta < 0$), θ must be in

QIII or QIV. To satisfy both, θ must be in QIV. From the figure

we note $\tan\theta = -\dfrac{8}{15}$, showing $\tan\left[\sin^{-1}\left(-\dfrac{8}{17}\right)\right] = -\dfrac{8}{15}$.

Now try Exercises 65 through 72 ▶

 These ideas apply even when one side of the triangle is unknown. In other words,

we can still draw a triangle for $\theta = \cos^{-1}\left(\dfrac{x}{\sqrt{x^2 + 16}}\right)$, since "$\theta$ is an angle whose

cosine is $\dfrac{x}{\sqrt{x^2 + 16}} = \dfrac{\text{adj}}{\text{hyp}}$."

EXAMPLE 9 ▶ **Using a Diagram to Evaluate an Expression Involving Inverse Trig Functions**

Evaluate the expression $\tan\left[\cos^{-1}\left(\dfrac{x}{\sqrt{x^2 + 16}}\right)\right]$. Assume $x > 0$ and the inverse

function is defined for the expression given.

Solution ▶ Rewrite $\tan\left[\cos^{-1}\left(\dfrac{x}{\sqrt{x^2 + 16}}\right)\right]$ as $\tan\theta$, where

$\theta = \cos^{-1}\left(\dfrac{x}{\sqrt{x^2 + 16}}\right)$. Draw a triangle with

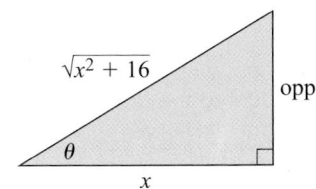

side x adjacent to θ and a hypotenuse of

$\sqrt{x^2 + 16}$. The Pythagorean theorem gives

$x^2 + \text{opp}^2 = (\sqrt{x^2 + 16})^2$, which leads to

$\text{opp}^2 = (x^2 + 16) - x^2$ giving $\text{opp} = \sqrt{16} = 4$. This shows

$\tan\theta = \tan\left[\cos^{-1}\left(\dfrac{x}{\sqrt{x^2 + 16}}\right)\right] = \dfrac{4}{x}$ (see the figure).

☑ **C.** You've just learned how to apply the definition and notation of inverse trig functions to simplify compositions

Now try Exercises 73 through 76 ▶

D. The Inverse Functions for Secant, Cosecant, and Cotangent

As with the other functions, we restrict the domain of the secant, cosecant, and cotangent functions to obtain a one-to-one function that is invertible (an inverse can be found). Once again the choice is arbitrary, and some domains are easier to work with than others in more advanced mathematics. For $y = \sec x$, we've chosen the "most

Figure 6.16
$y = \sec x$

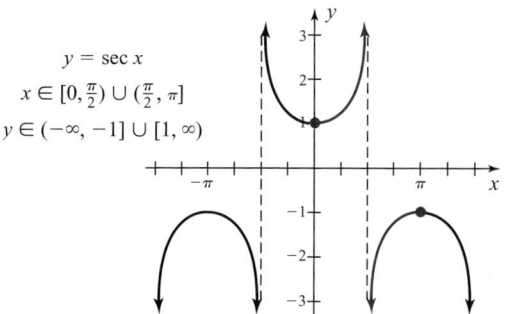

Figure 6.17
$y = \sec^{-1} x$

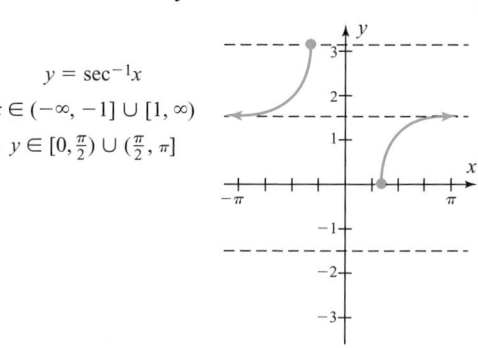

intuitive" restriction, one that seems more centrally located (nearer the origin). The graph of $y = \sec x$ is reproduced here, along with its inverse function (see Figures 6.16 and 6.17). The domain, range, and graphs of the functions $y = \csc^{-1}x$ and $y = \cot^{-1}x$ are asked for in the Exercises **(see Exercise 100).**

The functions $y = \sec^{-1}x$, $y = \csc^{-1}x$, and $y = \cot^{-1}x$ can be evaluated by noting their relationship to $y = \cos^{-1}x$, $y = \sin^{-1}x$, and $y = \tan^{-1}x$, respectively. For $y = \sec^{-1}x$, we have

<table>
<tr><td>$\sec y = x$</td><td>definition of inverse function</td></tr>
<tr><td>$\dfrac{1}{\sec y} = \dfrac{1}{x}$</td><td>property of reciprocals</td></tr>
<tr><td>$\cos y = \dfrac{1}{x}$</td><td>reciprocal ratio</td></tr>
<tr><td>$y = \cos^{-1}\left(\dfrac{1}{x}\right)$</td><td>rewrite using inverse function notation</td></tr>
<tr><td>$\sec^{-1}x = \cos^{-1}\left(\dfrac{1}{x}\right)$</td><td>substitute $\sec^{-1}x$ for y</td></tr>
</table>

In other words, to find the value of $y = \sec^{-1}x$, evaluate $y = \cos^{-1}\left(\dfrac{1}{x}\right)$, $|x| \geq 1$. Similarly, the expression $\csc^{-1}x$ can be evaluated using $\sin^{-1}\left(\dfrac{1}{x}\right)$, $|x| \geq 1$. The expression $\cot^{-1}x$ can likewise be evaluated using an inverse tangent function: $\cot^{-1}x = \tan^{-1}\left(\dfrac{1}{x}\right)$.

> **WORTHY OF NOTE**
>
> While the domains of $y = \cot^{-1}x$ and $y = \tan^{-1}x$ both include all real numbers, evaluating $\cot^{-1}x$ using $\tan^{-1}\left(\dfrac{1}{x}\right)$ involves the restriction $x \neq 0$. To maintain consistency, the equation $\cot^{-1}x = \dfrac{\pi}{2} - \tan^{-1}x$ is often used. The graph of $y = \dfrac{\pi}{2} - \tan^{-1}x$ is that of $y = \tan^{-1}x$ reflected across the x-axis and shifted $\dfrac{\pi}{2}$ units up, with the result identical to the graph of $y = \cot^{-1}x$.

EXAMPLE 10 ▶ **Evaluating an Inverse Trig Function**

Evaluate using a calculator only if necessary:

a. $\sec^{-1}\left(\dfrac{2}{\sqrt{3}}\right)$ **b.** $\cot^{-1}\left(\dfrac{\pi}{12}\right)$

Solution ▶ **a.** From our previous discussion, for $\sec^{-1}\left(\dfrac{2}{\sqrt{3}}\right)$, we evaluate $\cos^{-1}\left(\dfrac{\sqrt{3}}{2}\right)$.

Since this is a standard value, no calculator is needed and the result is $30°$.

b. For $\cot^{-1}\left(\dfrac{\pi}{12}\right)$, find $\tan^{-1}\left(\dfrac{12}{\pi}\right)$ on a calculator:

$$\cot^{-1}\left(\frac{\pi}{12}\right) = \tan^{-1}\left(\frac{12}{\pi}\right) \approx 1.3147.$$

☑ **D.** You've just learned how to find and graph inverse functions for sec x, csc x, and cot x

Now try Exercises 77 through 86 ▶

A summary of the highlights from this section follows.

Summary of Inverse Function Properties and Compositions

1. For $\sin x$ and $\sin^{-1}x$,

$\sin(\sin^{-1}x) = x$, for any x in the interval $[-1, 1]$

$\sin^{-1}(\sin x) = x$, for any x in the interval $\left[-\dfrac{\pi}{2}, \dfrac{\pi}{2}\right]$

3. For $\tan x$ and $\tan^{-1}x$,

$\tan(\tan^{-1}x) = x$, for any real number x

$\tan^{-1}(\tan x) = x$, for any x in the interval $\left(-\dfrac{\pi}{2}, \dfrac{\pi}{2}\right)$

2. For $\cos x$ and $\cos^{-1}x$,

$\cos(\cos^{-1}x) = x$, for any x in the interval $[-1, 1]$

$\cos^{-1}(\cos x) = x$, for any x in the interval $[0, \pi]$

4. To evaluate $\sec^{-1}x$, use $\cos^{-1}\left(\dfrac{1}{x}\right)$, $|x| \geq 1$,

$\csc^{-1}x$, use $\sin^{-1}\left(\dfrac{1}{x}\right)$, $|x| \geq 1$

$\cot^{-1}x$, use $\dfrac{\pi}{2} - \tan^{-1}x$, for all real numbers x

E. Applications of Inverse Trig Functions

We close this section with one example of the many ways that inverse functions can be applied.

EXAMPLE 11 ▶ **Using Inverse Trig Functions to Find Viewing Angles**

Believe it or not, the drive-in movie theaters that were so popular in the 1950s are making a comeback! If you arrive early, you can park in one of the coveted "center spots," but if you arrive late, you might have to park very close and strain your neck to watch the movie. Surprisingly, the maximum viewing angle (not the most comfortable viewing angle in this case) is actually very close to the front. Assume the base of a 30-ft screen is 10 ft above eye level (see Figure 6.18).

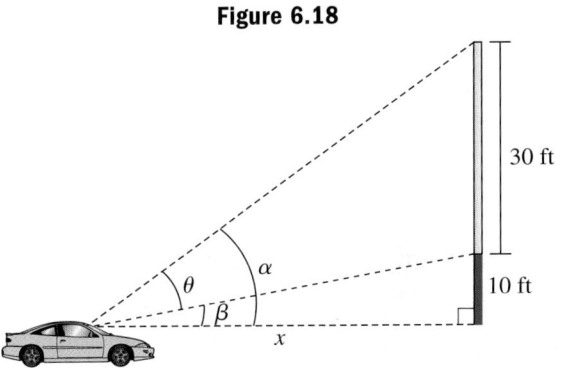

Figure 6.18

30 ft

10 ft

a. Use the inverse function concept to find expressions for angle α and angle β.

b. Use the result of Part (a) to find an expression for the *viewing angle* θ.

c. Use a calculator to find the viewing angle θ (to tenths of a degree) for distances of 15, 25, 35, and 45 ft, then to determine the distance x (to tenths of a foot) that maximizes the viewing angle.

Solution ▶ **a.** The side opposite β is 10 ft, and we want to know x — the adjacent side. This suggests we use $\tan \beta = \dfrac{10}{x}$, giving $\beta = \tan^{-1}\left(\dfrac{10}{x}\right)$. In the same way, we find that $\alpha = \tan^{-1}\left(\dfrac{40}{x}\right)$.

b. From the diagram we note that $\theta = \alpha - \beta$, and substituting for α and β directly gives $\theta = \tan^{-1}\left(\dfrac{40}{x}\right) - \tan^{-1}\left(\dfrac{10}{x}\right)$.

c. After we enter $Y_1 = \tan^{-1}\left(\dfrac{40}{x}\right) - \tan^{-1}\left(\dfrac{10}{x}\right)$, a graphing calculator gives

approximate viewing angles of
35.8°, 36.2°, 32.9°, and 29.1°, for $x = 15$,
25, 35, and 45 ft, respectively. From these
data, we note the distance x that makes θ a
maximum must be between 15 and 35 ft,
and using [2nd] [TRACE] (CALC)
4:maximum shows θ is a maximum of
36.9° at a distance of 20 ft from the screen
(see Figure 6.19).

Figure 6.19

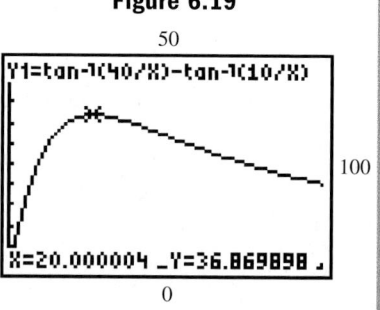

☑ **E.** You've just learned
how to solve applications
involving inverse functions

Now try Exercises 89 through 95 ▶

TECHNOLOGY HIGHLIGHT

More on Inverse Functions

The domain and range of the inverse functions for sine, cosine, and
tangent are preprogrammed into most graphing calculators, making
them an ideal tool for reinforcing the concepts involved. In particular,
$\sin x = y$ implies that $\sin^{-1}y = x$ only if $-90° \le y \le 90°$ and
$-1 \le x \le 1$. For a stark reminder of this fact we'll use the TABLE
feature of the grapher. Begin by using the TBLSET screen ([2nd]
[WINDOW]) to set TblStart = 90 with ΔTbl = −30. After placing the
calculator in degree [MODE], go to the [Y=] screen and input
$Y_1 = \sin x$, $Y_2 = \sin^{-1}x$, and $Y_3 = Y_2(Y_1)$ (the composition $Y_2 \circ Y_1$).
Then disable Y_2 [turn it off—Y_3 will read it anyway) so that both Y_1 and
Y_3 will be displayed simultaneously on the TABLE screen. Pressing
[2nd] [GRAPH] brings up the TABLE shown in Figure 6.20, where we
note the inputs are standard angles, the outputs in Y_1 are the
(expected) standard values, and the outputs in Y_3 return the original
standard values. Now scroll upward until 180° is at the top of the
X column (Figure 6.21), and note that Y_3 continues to return standard
angles from the interval $[-90°, 90°]$—a stark reminder that while the
expression $\sin 150° = 0.5$, $\sin^{-1}(\sin 150°) \ne 150°$. Once again we note that while $\sin^{-1}(\sin 150°)$ can be
evaluated, it cannot be evaluated directly using the inverse function properties. Use these ideas to
complete the following exercises.

Figure 6.20

X	Y₁	Y₃
90	1	90
60	.86603	60
30	.5	30
0	0	0
-30	-.5	-30
-60	-.866	-60
-90	-1	-90

X=90

Figure 6.21

X	Y₁	Y₃
180	0	0
150	.5	30
120	.86603	60
90	1	90
60	.86603	60
30	.5	30
0	0	0

X=180

Exercise 1:   Go through an exercise similar to the one here using $Y_1 = \cos x$ and $Y_2 = \cos^{-1}x$. Remember to modify the TBLSET to accommodate the restricted domain for cosine.

Exercise 2:   Complete parts (a) and (b) using the TABLE from Exercise 1. Complete parts (c) and (d) without a calculator.

a. $\cos^{-1}(\cos 150°)$

b. $\cos^{-1}(\cos 210°)$

c. $\cos^{-1}(\cos 120°)$

d. $\cos^{-1}(\cos 240°)$

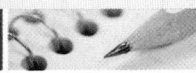

6.5 EXERCISES

► CONCEPTS AND VOCABULARY

Fill in each blank with the appropriate word or phrase. Carefully reread the section if needed.

1. All six trigonometric functions fail the _____ _____ test and therefore are not _____ to _____.

2. The two most common ways of writing the inverse function for $y = \sin x$ are _____ and _____.

3. The domain for the inverse sine function is _____ and the range is _____.

4. The domain for the inverse cosine function is _____ and the range is _____.

5. Most calculators do not have a key for evaluating an expression like $\sec^{-1}5$. Explain how it is done using the COS key.

6. Discuss/Explain what is meant by the *implicit form* of an inverse function and the *explicit form*. Give algebraic and trigonometric examples.

► DEVELOPING YOUR SKILLS

The tables here show values of $\sin \theta$, $\cos \theta$, and $\tan \theta$ for $\theta \in [-180°$ to $210°]$. The restricted domain used to develop the inverse functions is shaded. Use the information from these tables to complete the exercises that follow.

$y = \sin \theta$

θ	$\sin \theta$	θ	$\sin \theta$
$-180°$	0	$30°$	$\frac{1}{2}$
$-150°$	$-\frac{1}{2}$	$60°$	$\frac{\sqrt{3}}{2}$
$-120°$	$-\frac{\sqrt{3}}{2}$	$90°$	1
$-90°$	-1	$120°$	$\frac{\sqrt{3}}{2}$
$-60°$	$-\frac{\sqrt{3}}{2}$	$150°$	$\frac{1}{2}$
$-30°$	$-\frac{1}{2}$	$180°$	0
0	0	$210°$	$-\frac{1}{2}$

$y = \cos \theta$

θ	$\cos \theta$	θ	$\cos \theta$
$-180°$	-1	$30°$	$\frac{\sqrt{3}}{2}$
$-150°$	$-\frac{\sqrt{3}}{2}$	$60°$	$\frac{1}{2}$
$-120°$	$-\frac{1}{2}$	$90°$	0
$-90°$	0	$120°$	$-\frac{1}{2}$
$-60°$	$\frac{1}{2}$	$150°$	$-\frac{\sqrt{3}}{2}$
$-30°$	$\frac{\sqrt{3}}{2}$	$180°$	-1
0	1	$210°$	$-\frac{\sqrt{3}}{2}$

$y = \tan \theta$

θ	$\tan \theta$	θ	$\tan \theta$
$-180°$	0	$30°$	$\frac{\sqrt{3}}{3}$
$-150°$	$\frac{\sqrt{3}}{3}$	$60°$	$\sqrt{3}$
$-120°$	$\sqrt{3}$	$90°$	$-$
$-90°$	$-$	$120°$	$-\sqrt{3}$
$-60°$	$-\sqrt{3}$	$150°$	$-\frac{\sqrt{3}}{3}$
$-30°$	$-\frac{\sqrt{3}}{3}$	$180°$	0
0	0	$210°$	$\sqrt{3}$

Use the preceding tables to fill in each blank (principal values only).

7.

$\sin 0 = 0$	$\sin^{-1} 0 = $ _____
$\sin\left(\frac{\pi}{6}\right) = $ _____	$\arcsin\left(\frac{1}{2}\right) = \frac{\pi}{6}$
$\sin\left(-\frac{5\pi}{6}\right) = -\frac{1}{2}$	$\sin^{-1}\left(-\frac{1}{2}\right) = $ _____
$\sin\left(-\frac{\pi}{2}\right) = -1$	$\sin^{-1}(-1) = $ _____

8.

$\sin \pi = 0$	$\sin^{-1} 0 = $ _____
$\sin 120° = \frac{\sqrt{3}}{2}$	$\sin^{-1}\left(\frac{\sqrt{3}}{2}\right) = $ _____
$\sin(-60°) = -\frac{\sqrt{3}}{2}$	$\arcsin\left(-\frac{\sqrt{3}}{2}\right) = $ _____
$\sin 180° = $ _____	$\arcsin 0 = 0$

Evaluate without the aid of calculators or tables, *keeping the domain and range of each function in mind.* **Answer in radians.**

9. $\sin^{-1}\left(\dfrac{\sqrt{2}}{2}\right)$ 　　10. $\arcsin\left(\dfrac{\sqrt{3}}{2}\right)$

11. $\sin^{-1}1$ 　　12. $\arcsin\left(-\dfrac{1}{2}\right)$

Evaluate using a calculator, *keeping the domain and range of each function in mind.* **Answer in radians to the nearest ten-thousandth** *and* **in degrees to the nearest tenth.**

13. $\arcsin 0.8892$ 　　14. $\arcsin\left(\dfrac{7}{8}\right)$

15. $\sin^{-1}\left(\dfrac{1}{\sqrt{7}}\right)$ 　　16. $\sin^{-1}\left(\dfrac{1-\sqrt{5}}{2}\right)$

Evaluate each expression.

17. $\sin\left[\sin^{-1}\left(\dfrac{\sqrt{2}}{2}\right)\right]$ 　　18. $\sin\left[\arcsin\left(\dfrac{\sqrt{3}}{2}\right)\right]$

19. $\arcsin\left[\sin\left(\dfrac{\pi}{3}\right)\right]$ 　　20. $\sin^{-1}(\sin 30°)$

21. $\sin^{-1}(\sin 135°)$ 　　22. $\arcsin\left[\sin\left(\dfrac{-2\pi}{3}\right)\right]$

23. $\sin(\sin^{-1} 0.8205)$ 　　24. $\sin\left[\arcsin\left(\dfrac{3}{5}\right)\right]$

Use the tables given prior to Exercise 7 to fill in each blank (principal values only).

25.

$\cos 0 = 1$	$\cos^{-1}1 = $ _____
$\cos\left(\dfrac{\pi}{6}\right) = $ _____	$\arccos\left(\dfrac{\sqrt{3}}{2}\right) = \dfrac{\pi}{6}$
$\cos 120° = -\dfrac{1}{2}$	$\arccos\left(-\dfrac{1}{2}\right) = $ _____
$\cos \pi = -1$	$\cos^{-1}(-1) = $ _____

26.

$\cos(-60°) = \dfrac{1}{2}$	$\cos^{-1}\left(\dfrac{1}{2}\right) = $ _____
$\cos\left(-\dfrac{\pi}{6}\right) = \dfrac{\sqrt{3}}{2}$	$\cos^{-1}\left(\dfrac{\sqrt{3}}{2}\right) = $ _____
$\cos(-120°) = $ _____	$\arccos\left(-\dfrac{1}{2}\right) = 120°$
$\cos(2\pi) = 1$	$\cos^{-1}1 = $ _____

Evaluate without the aid of calculators or tables. Answer in radians.

27. $\cos^{-1}\left(\dfrac{1}{2}\right)$ 　　28. $\arccos\left(-\dfrac{\sqrt{3}}{2}\right)$

29. $\cos^{-1}(-1)$ 　　30. $\arccos(0)$

Evaluate using a calculator. Answer in radians to the nearest ten-thousandth, degrees to the nearest tenth.

31. $\arccos 0.1352$ 　　32. $\arccos\left(\dfrac{4}{7}\right)$

33. $\cos^{-1}\left(\dfrac{\sqrt{5}}{3}\right)$ 　　34. $\cos^{-1}\left(\dfrac{\sqrt{6}-1}{5}\right)$

Evaluate each expression.

35. $\arccos\left[\cos\left(\dfrac{\pi}{4}\right)\right]$ 　　36. $\cos^{-1}(\cos 60°)$

37. $\cos(\cos^{-1} 0.5560)$ 　　38. $\cos\left[\arccos\left(-\dfrac{8}{17}\right)\right]$

39. $\cos\left[\cos^{-1}\left(-\dfrac{\sqrt{2}}{2}\right)\right]$ 　　40. $\cos\left[\arccos\left(\dfrac{\sqrt{3}}{2}\right)\right]$

41. $\cos^{-1}\left[\cos\left(\dfrac{5\pi}{4}\right)\right]$ 　　42. $\arccos(\cos 44.2°)$

Use the tables presented before Exercise 7 to fill in each blank. Convert from radians to degrees as needed.

43.

$\tan 0 = 0$	$\tan^{-1}0 = $ _____
$\tan\left(-\dfrac{\pi}{3}\right) = $ _____	$\arctan(-\sqrt{3}) = -\dfrac{\pi}{3}$
$\tan 30° = \dfrac{\sqrt{3}}{3}$	$\arctan\left(\dfrac{\sqrt{3}}{3}\right) = $ _____
$\tan\left(\dfrac{\pi}{3}\right) = $ _____	$\tan^{-1}(\sqrt{3}) = $ _____

44.

$\tan(-150°) = \dfrac{\sqrt{3}}{3}$	$\tan^{-1}\left(\dfrac{\sqrt{3}}{3}\right) = $ _____
$\tan \pi = 0$	$\tan^{-1}0 = $ _____
$\tan 120° = -\sqrt{3}$	$\arctan(-\sqrt{3}) = $ _____
$\tan\left(\dfrac{\pi}{4}\right) = $ _____	$\arctan 1 = \dfrac{\pi}{4}$

Evaluate without the aid of calculators or tables.

45. $\tan^{-1}\left(-\dfrac{\sqrt{3}}{3}\right)$ 　　46. $\arctan(-1)$

47. $\arctan(\sqrt{3})$ 　　48. $\tan^{-1}0$

Evaluate using a calculator, *keeping the domain and range of each function in mind.* **Answer in radians to the nearest ten-thousandth** *and* **in degrees to the nearest tenth.**

49. $\tan^{-1}(-2.05)$ **50.** $\tan^{-1}(0.3267)$

51. $\arctan\left(\dfrac{29}{21}\right)$ **52.** $\arctan(-\sqrt{6})$

Simplify each expression without using a calculator.

53. $\sin^{-1}\left[\cos\left(\dfrac{2\pi}{3}\right)\right]$ **54.** $\cos^{-1}\left[\sin\left(-\dfrac{\pi}{3}\right)\right]$

55. $\tan\left[\arccos\left(\dfrac{\sqrt{3}}{2}\right)\right]$ **56.** $\sec\left[\arcsin\left(\dfrac{1}{2}\right)\right]$

57. $\csc\left[\sin^{-1}\left(\dfrac{\sqrt{2}}{2}\right)\right]$ **58.** $\cot\left[\cos^{-1}\left(-\dfrac{1}{2}\right)\right]$

59. $\arccos[\sin(-30°)]$ **60.** $\arcsin(\cos 135°)$

Explain why the following expressions are not defined.

61. $\tan(\sin^{-1} 1)$ **62.** $\cot(\arccos 1)$

63. $\sin^{-1}\left[\csc\left(\dfrac{\pi}{4}\right)\right]$ **64.** $\cos^{-1}\left[\sec\left(\dfrac{2\pi}{3}\right)\right]$

Use the diagrams below to write the value of: (a) sin θ, (b) cos θ, and (c) tan θ.

65. 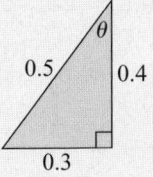   **66.**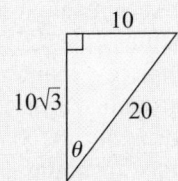

67. 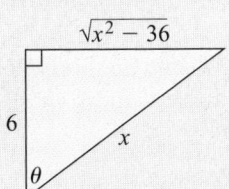   **68.**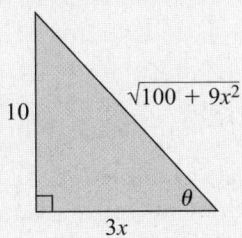

Evaluate each expression by drawing a right triangle and labeling the sides.

69. $\sin\left[\cos^{-1}\left(-\dfrac{7}{25}\right)\right]$ **70.** $\cos\left[\sin^{-1}\left(-\dfrac{11}{61}\right)\right]$

71. $\sin\left[\tan^{-1}\left(\dfrac{\sqrt{5}}{2}\right)\right]$ **72.** $\tan\left[\cos^{-1}\left(\dfrac{\sqrt{23}}{12}\right)\right]$

73. $\cot\left[\arcsin\left(\dfrac{3x}{5}\right)\right]$ **74.** $\tan\left[\arcsec\left(\dfrac{5}{2x}\right)\right]$

75. $\cos\left[\sin^{-1}\left(\dfrac{x}{\sqrt{12+x^2}}\right)\right]$

76. $\tan\left[\sec^{-1}\left(\dfrac{\sqrt{9+x^2}}{x}\right)\right]$

Use the tables given prior to Exercise 7 to help fill in each blank.

77.

$\sec 0 = 1$	$\sec^{-1} 1 = $ _____
$\sec\left(\dfrac{\pi}{3}\right) = $ _____	$\arcsec 2 = \dfrac{\pi}{3}$
$\sec(-30°) = \dfrac{2}{\sqrt{3}}$	$\arcsec\left(\dfrac{2}{\sqrt{3}}\right) = $ _____
$\sec(\pi) = $ _____	$\sec^{-1}(-1) = \pi$

78.

$\sec(-60°) = 2$	$\arcsec 2 = $ _____
$\sec\left(\dfrac{7\pi}{6}\right) = -\dfrac{2}{\sqrt{3}}$	$\arcsec\left(-\dfrac{2}{\sqrt{3}}\right) = $ _____
$\sec(-360°) = 1$	$\arcsec 1 = $ _____
$\sec(60°) = $ _____	$\sec^{-1} 2 = 60°$

Evaluate using a calculator only as necessary.

79. $\arccsc 2$ **80.** $\csc^{-1}\left(-\dfrac{2}{\sqrt{3}}\right)$

81. $\cot^{-1}\sqrt{3}$ **82.** $\arccot(-1)$

83. $\arcsec 5.789$ **84.** $\cot^{-1}\left(-\dfrac{\sqrt{7}}{2}\right)$

85. $\sec^{-1}\sqrt{7}$ **86.** $\arccsc 2.9875$

▶ **WORKING WITH FORMULAS**

87. **The force normal to an object on an inclined plane:** $F_N = mg\cos\theta$

When an object is on an inclined plane, the **normal force** is the force acting perpendicular to the plane and away from the force of gravity, and is measured in a unit called **newtons (N).** The magnitude of this force depends on the angle of incline of the plane according to the formula

above, where m is the mass of the object in kilograms and g is the force of gravity (9.8 m/sec²). Given $m = 225$ g, find (a) F_N for $\theta = 15°$ and $\theta = 45°$ and (b) θ for $F_N = 1$ N and $F_N = 2$ N.

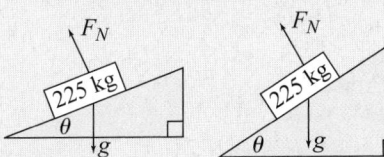

88. Heat flow on a cylindrical pipe: $T = (T_0 - T_R)\sin\left(\dfrac{y}{\sqrt{x^2 + y^2}}\right) + T_R; \ y \geq 0$

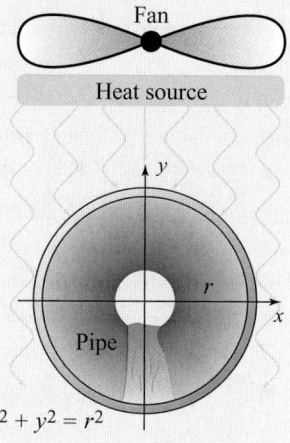

Fan

Heat source

When a circular pipe is exposed to a fan-driven source of heat, the temperature of the air reaching the pipe is greatest at the point nearest to the source (see diagram). As you move around the circumference of the pipe away from the source, the temperature of the air reaching the pipe gradually decreases. One

possible model of this phenomenon is given by the formula shown, where T is the temperature of the air at a point (x, y) on the circumference of a pipe with outer radius $r = \sqrt{x^2 + y^2}$, T_0 is the temperature of the air at the source, and T_R is the surrounding room temperature. Assuming $T_0 = 220°F$, $T_R = 72°$ and $r = 5$ cm: (a) Find the temperature of the air at the points $(0, 5)$, $(3, 4)$, $(4, 3)$, $(4.58, 2)$, and $(4.9, 1)$. (b) Why is the temperature decreasing for this sequence of points? (c) Simplify the formula using $r = 5$ and use it to find two points on the pipe's circumference where the temperature of the air is $113°$.

▶ **APPLICATIONS**

89. Snowcone dimensions: *Made in the Shade Snowcones* sells a colossal size cone that uses a conical cup holding 20 oz of ice and liquid. The cup is 20 cm tall and has a radius of 5.35 cm. Find the angle θ formed by a cross-section of the cup.

Exercise 89

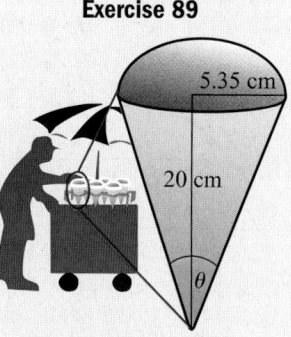

5.35 cm

20 cm

θ

90. Avalanche conditions: Winter avalanches occur for many reasons, one being the slope of the mountain. Avalanches seem to occur most often for slopes between 35° and 60° (snow gradually slides off steeper slopes). The slopes at a local ski resort have an average rise of 2000 ft for each horizontal run of 2559 ft. Is this resort prone to avalanches? Find the angle θ and respond.

Exercise 90

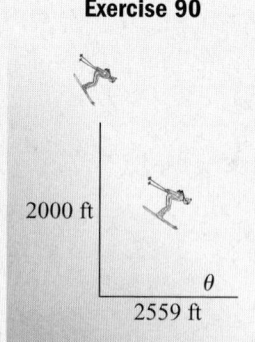

2000 ft

θ

2559 ft

91. Distance to hole: A popular story on the PGA Tour has Gerry Yang, Tiger Woods' teammate at Stanford and occasional caddie, using the Pythagorean theorem to find the distance Tiger needed to reach a particular hole. Suppose you notice a marker in the ground stating that the straight line distance from the marker to the hole (H) is 150 yd. If your ball B is 48 yd from the marker (M) and angle BMH is a right angle, determine the angle θ and *your* straight line distance from the hole.

Exercise 91

H

150 yd

M

B θ Marker

48 yd

92. Ski jumps: At a waterskiing contest on a large lake, skiers use a ramp rising out of the water that is 30 ft long and 10 ft high at the high end. What angle θ does the ramp make with the lake?

Exercise 92

θ 10 ft

30 ft

93. Viewing angles for advertising: A 25-ft-wide billboard is erected perpendicular to a straight highway, with the closer edge 50 ft away (see figure). Assume the advertisement on the billboard is most easily read when the viewing angle is 10.5° or more. (a) Use inverse functions to find an expression for the viewing angle θ. (b) Use a calculator to help determine the distance d (to tenths of a foot) for which the viewing angle is greater than 10.5°. (c) What distance d maximizes this viewing angle?

Exercise 93

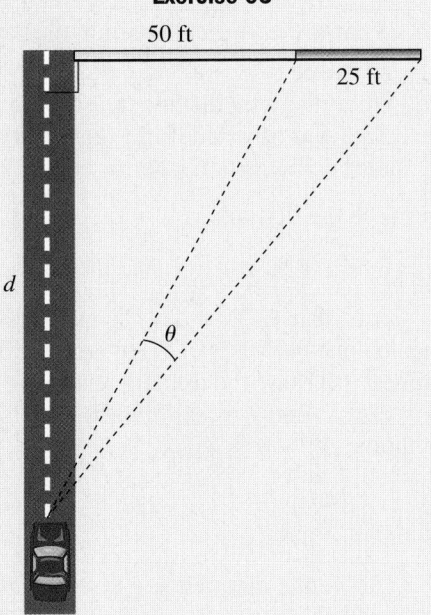

94. Viewing angles at an art show: At an art show, a painting 2.5 ft in height is hung on a wall so that its base is 1.5 ft above the eye level of an average viewer (see figure). (a) Use inverse functions to find expressions for angles α and β. (b) Use the result to find an expression for the *viewing angle θ*. (c) Use a

Exercise 94

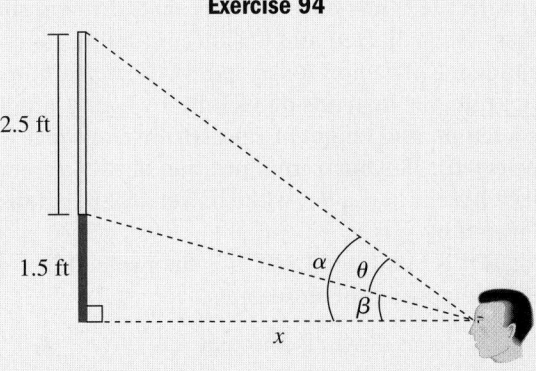

calculator to help determine the distance x (to tenths of a foot) that maximizes this viewing angle.

95. Shooting angles and shots on goal: A soccer player is on a breakaway and is dribbling just inside the right sideline toward the opposing goal (see figure). As the defense closes in, she has just a few seconds to decide when to shoot. (a) Use inverse functions to find an expression for the shooting angle θ. (b) Use a calculator to help determine the distance d (to tenths of a foot) that will maximize the shooting angle for the dimensions shown.

Exercise 95

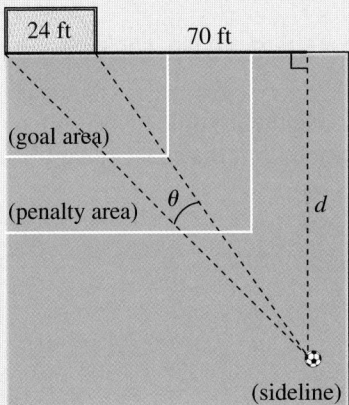

▶ **EXTENDING THE CONCEPT**

Consider a satellite orbiting at an altitude of x mi above the Earth. The distance d from the satellite to the horizon and the length s of the corresponding arc of the Earth are shown in the diagram.

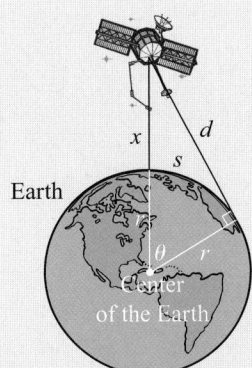

96. To find the distance d we use the formula $d = \sqrt{2rx + x^2}$. (a) Show how this formula was developed using the

Pythagorean theorem. (b) Find a formula for the angle θ in terms of r and x, then a formula for the arc length s.

97. If the Earth has a radius of 3960 mi and the satellite is orbiting at an altitude of 150 mi, (a) what is the measure of angle θ? (b) how much longer is d than s?

A projectile is any object that is shot, thrown, slung, or otherwise projected and has no continuing source of propulsion. The horizontal and vertical position of the projectile depends on its initial velocity, angle of projection, and height of release (air resistance is neglected). The horizontal position of the projectile is given by $x = v_0 \cos \theta \, t$, while its vertical position is modeled by $y = y_0 + v_0 \sin \theta \, t - 16t^2$, where y_0 is the height it is projected from, θ is the projection angle, and t is the elapsed time in seconds.

98. A circus clown is shot out of a specially made cannon at an angle of 55°, with an initial velocity of 85 ft/sec, and the end of the cannon is 10 ft high.

a. Find the position of the safety net (distance from the cannon and height from the ground) if the clown hits the net after 4.3 sec.

b. Find the angle at which the clown was shot if the initial velocity was 75 ft/sec and the clown hits a net that is placed 175.5 ft away after 3.5 sec.

99. A winter ski jumper leaves the ski-jump with an initial velocity of 70 ft/sec at an angle of 10°. Assume the jump-off point has coordinates (0, 0).

Exercise 99

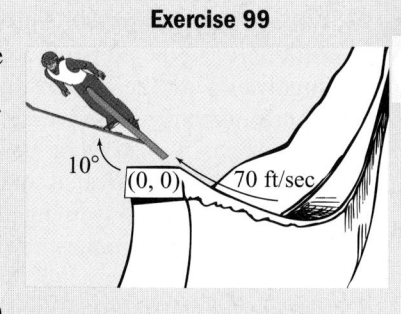

a. What is the horizontal position of the skier after 6 sec?

b. What is the vertical position of the skier after 6 sec?

c. What diagonal distance (down the mountain side) was traveled if the skier touched down after being airborne for 6 sec?

100. Suppose the domain of $y = \csc x$ was restricted to $x \in \left[-\dfrac{\pi}{2}, 0 \right) \cup \left(0, \dfrac{\pi}{2} \right]$, and the domain of $y = \cot x$ to $x \in (0, \pi)$. **(a)** Would these functions then be one-to-one? **(b)** What are the corresponding ranges? **(c)** State the domain and range of $y = \csc^{-1} x$ and $y = \cot^{-1} x$. **(d)** Graph each function.

▶ MAINTAINING YOUR SKILLS

101. (6.4) Use the triangle given with a double-angle identity to find the exact value of $\sin(2\theta)$.

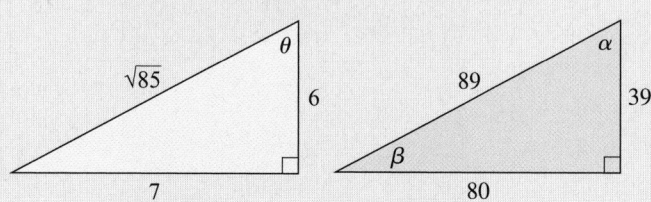

102. (6.3) Use the triangle given with a sum identity to find the exact value of $\sin(\alpha + \beta)$.

103. (3.7) Solve the inequality $f(x) \leq 0$ using zeroes and end behavior given $f(x) = x^3 - 9x$.

104. (2.3) In 2000, Space Tourists Inc. sold 28 low-orbit travel packages. By 2005, yearly sales of the low-orbit package had grown to 105. Assuming the growth is linear, (a) find the equation that models this growth ($2000 \rightarrow t = 0$), (b) discuss the meaning of the slope in this context, and (c) use the equation to project the number of packages that will be sold in 2010.

Solving Basic Trig Equations

Learning Objectives

In Section 6.6 you will learn how to:

- ☐ **A.** Use a graph to gain information about principal roots, roots in $[0, 2\pi)$, and roots in $\mathbb{R}$
- ☐ **B.** Use inverse functions to solve trig equations for the principal root
- ☐ **C.** Solve trig equations for roots in $[0, 2\pi)$ or $[0, 360°)$
- ☐ **D.** Solve trig equations for roots in $\mathbb{R}$
- ☐ **E.** Solve trig equations using fundamental identities
- ☐ **F.** Solve trig equations using graphing technology

In this section, we'll take the elements of basic equation solving and use them to help solve **trig equations,** or equations containing trigonometric functions. All of the algebraic techniques previously used can be applied to these equations, including the properties of equality and all forms of factoring (common terms, difference of squares, etc.). As with polynomial equations, we continue to be concerned with the *number of solutions* as well as with the *solutions themselves,* but there is one major difference. There is no "algebra" that can transform a function like $\sin x = \frac{1}{2}$ into $x = solution$. For that we rely on the inverse trig functions from Section 6.5.

A. The Principal Root, Roots in $[0, 2\pi)$, and Real Roots

In a study of polynomial equations, making a connection between the degree of an equation, its graph, and its possible roots, helped give insights as to the number, location, and nature of the roots. Similarly, keeping graphs of basic trig functions *constantly* in mind helps you gain information regarding the solutions to trig equations. When solving trig equations, we refer to the solution found using $\sin^{-1}$, $\cos^{-1}$, and $\tan^{-1}$ as the **principal root.** You will alternatively be asked to find (1) the principal root, (2) solutions in $[0, 2\pi)$ or $[0°, 360°)$, or (3) solutions from the set of real numbers $\mathbb{R}$. For convenience, graphs of the basic sine, cosine, and tangent functions are repeated in Figures 6.22 through 6.24. Take a mental snapshot of them and keep them close at hand.

Figure 6.22

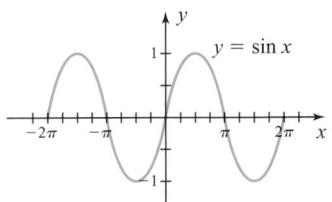

Figure 6.23

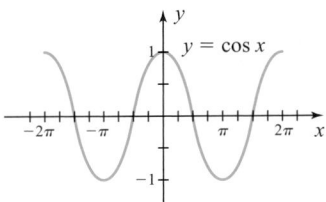

Figure 6.24

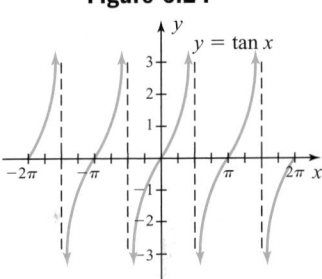

EXAMPLE 1 ▶ **Visualizing Solutions Graphically**

Consider the equation $\sin x = \frac{2}{3}$. Using a graph of $y = \sin x$ and $y = \frac{2}{3}$,

  a. State the quadrant of the principal root.

  b. State the number of roots in $[0, 2\pi)$ and their quadrants.

  c. Comment on the number of real roots.

Solution ▶ We begin by drawing a quick sketch of $y = \sin x$ and $y = \frac{2}{3}$, noting that solutions will occur where the graphs intersect.

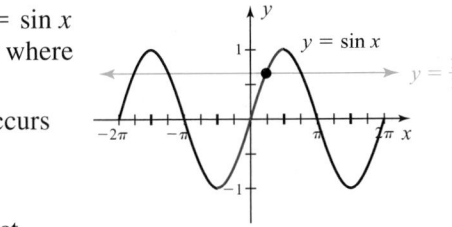

  a. The sketch shows the principal root occurs between 0 and $\dfrac{\pi}{2}$ in QI.

  b. For $[0, 2\pi)$ we note the graphs intersect twice and there will be two solutions in this interval.

  c. Since the graphs of $y = \sin x$ and $y = \frac{2}{3}$ extend infinitely in both directions, they will intersect an infinite number of times—*but at regular intervals!* Once a root is found, adding integer multiples of 2π (the period of sine) to this root will give the location of additional roots.

Now try Exercises 7 through 10 ▶

WORTHY OF NOTE

Note that we refer to $\left(0, \dfrac{\pi}{2}\right)$ as Quadrant I or QI, regardless of whether we're discussing the unit circle or the graph of the function. In Example 1b, the solutions correspond to those found in QI and QII on the unit circle, where $\sin x$ is also positive.

When this process is applied to the equation $\tan x = -2$, the graph shows the principal root occurs between $-\dfrac{\pi}{2}$ and 0 in QIV (see Figure 6.25). In the interval $[0, 2\pi)$ the graphs intersect twice, in QII and QIV where $\tan x$ is negative (graphically—below the x-axis). As in Example 1, the graphs continue infinitely and will intersect an infinite number of times—*but again at regular intervals!* Once a root is found, adding integer multiples of π (the period of tangent) to this root will give the location of other roots.

Figure 6.25

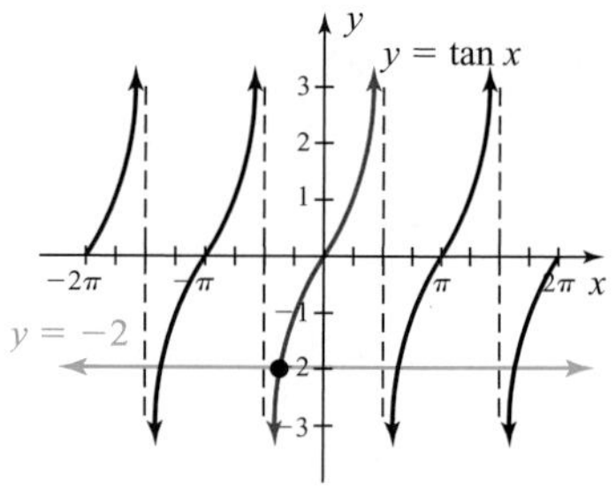

☑ **A.** You've just learned how to use a graph to gain information about principal roots, roots in $[0, 2\pi)$, and roots in $\mathbb{R}$

B. Inverse Functions and Principal Roots

To solve equations having a single variable term, the basic goal is to isolate the variable term and apply the inverse function or operation. This is true for algebraic equations like $2x - 1 = 0$, $2\sqrt{x} - 1 = 0$, or $2x^2 - 1 = 0$, and for trig equations like $2\sin x - 1 = 0$. In each case we would add 1 to both sides, divide by 2, then apply the appropriate inverse. When the inverse trig functions are applied, the result is only the principal root and other solutions may exist depending on the interval under consideration.

EXAMPLE 2 ▶ **Finding Principal Roots**

Find the principal root of $\sqrt{3} \tan x - 1 = 0$.

Solution ▶ We begin by isolating the variable term, then apply the inverse function.

$$\sqrt{3} \tan x - 1 = 0 \qquad \text{given equation}$$

$$\tan x = \frac{1}{\sqrt{3}} \qquad \text{add 1 and divide by } \sqrt{3}$$

$$\tan^{-1}(\tan x) = \tan^{-1}\left(\frac{1}{\sqrt{3}}\right) \qquad \text{apply inverse tangent to both sides}$$

$$x = \frac{\pi}{6} \qquad \text{result (exact form)}$$

☑ **B.** You've just learned how to use inverse functions to solve trig equations for the principal root

Now try Exercises 11 through 28 ▶

Table 6.2

θ	$\sin \theta$	$\cos \theta$
0	0	1
$\dfrac{\pi}{6}$	$\dfrac{1}{2}$	$\dfrac{\sqrt{3}}{2}$
$\dfrac{\pi}{4}$	$\dfrac{\sqrt{2}}{2}$	$\dfrac{\sqrt{2}}{2}$
$\dfrac{\pi}{3}$	$\dfrac{\sqrt{3}}{2}$	$\dfrac{1}{2}$
$\dfrac{\pi}{2}$	1	0
$\dfrac{2\pi}{3}$	$\dfrac{\sqrt{3}}{2}$	$-\dfrac{1}{2}$
$\dfrac{3\pi}{4}$	$\dfrac{\sqrt{2}}{2}$	$-\dfrac{\sqrt{2}}{2}$
$\dfrac{5\pi}{6}$	$\dfrac{1}{2}$	$-\dfrac{\sqrt{3}}{2}$
π	0	-1

Equations like the one in Example 2 demonstrate the need to be *very* familiar with the functions of a special angle. They are frequently used in equations and applications to ensure results don't get so messy they obscure the main ideas. For convenience, the values of $\sin \theta$ and $\cos \theta$ are repeated in Table 6.2 for $x \in [0, \pi]$. Using symmetry and the appropriate sign, the table can easily be extended to all values in $[0, 2\pi)$. Using the reciprocal and ratio relationships, values for the other trig functions can also be found.

C. Solving Trig Equations for Roots in $[0, 2\pi)$ or $[0°, 360°)$

To find multiple solutions to a trig equation, we simply take the reference angle of the principal root, and *use this angle to find all solutions* within a specified range. A mental image of the graph still guides us, and the standard table of values (also held in memory) allows for a quick solution to many equations.

EXAMPLE 3 ▶ **Finding Solutions in [0, 2π)**

For $2 \cos \theta + \sqrt{2} = 0$, find all solutions in $[0, 2\pi)$.

Solution ▶ Isolate the variable term, then apply the inverse function.

$$2 \cos \theta + \sqrt{2} = 0 \qquad \text{given equation}$$

$$\cos \theta = -\frac{\sqrt{2}}{2} \qquad \text{subtract } \sqrt{2} \text{ and divide by 2}$$

$$\cos^{-1}(\cos \theta) = \cos^{-1}\left(-\frac{\sqrt{2}}{2}\right) \qquad \text{apply inverse cosine to both sides}$$

$$\theta = \frac{3\pi}{4} \qquad \text{result}$$

> **WORTHY OF NOTE**
>
> Note how the graph of a trig function displays the information regarding *quadrants*. From the graph of $y = \cos x$ we "read" that cosine is negative in QII and QIII [the lower "hump" of the graph is below the *x*-axis in $(\pi/2, 3\pi/2)$] and positive in QI and QIV [the graph is above the *x*-axis in the intervals $(0, \pi/2)$ and $(3\pi/2, 2\pi)$].

With $\frac{3\pi}{4}$ as the principal root, we know $\theta_r = \frac{\pi}{4}$. Since $\cos x$ is negative in QII and QIII, the second solution is $\frac{5\pi}{4}$. The second solution could also have been found from memory, recognition, or symmetry on the unit circle. Our (mental) graph verifies these are the only solutions in $[0, 2\pi)$.

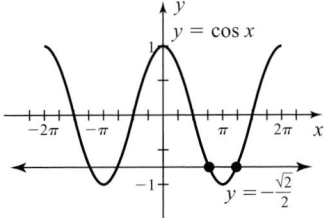

Now try Exercises 29 through 34 ▶

EXAMPLE 4 ▶ **Finding Solutions in [0, 2π)**

For $\tan^2 x - 1 = 0$, find all solutions in $[0, 2\pi)$.

Solution ▶ As with the other equations having a single variable term, we try to isolate this term or attempt a solution by factoring.

$$\tan^2 x - 1 = 0 \qquad \text{given equation}$$

$$\sqrt{\tan^2 x} = \pm\sqrt{1} \qquad \text{add 1 to both sides and take square roots}$$

$$\tan x = \pm 1 \qquad \text{result}$$

The algebra gives $\tan x = 1$ or $\tan x = -1$ and we solve each equation independently.

$$\tan x = 1 \qquad\qquad \tan x = -1$$

$$\tan^{-1}(\tan x) = \tan^{-1}(1) \qquad \tan^{-1}(\tan x) = \tan^{-1}(-1) \qquad \text{apply inverse tangent}$$

$$x = \frac{\pi}{4} \qquad\qquad x = -\frac{\pi}{4} \qquad \text{principal roots}$$

Of the principal roots, only $x = \frac{\pi}{4}$ is in the specified interval. With $\tan x$ positive in QI and QIII, a second solution is $\frac{5\pi}{4}$. While $x = -\frac{\pi}{4}$ is not in the interval, we still use it as a reference angle in QII and QIV (for $\tan x = -1$) and find the solutions $x = \frac{3\pi}{4}$ and $\frac{7\pi}{4}$. The four solutions are $x = \frac{\pi}{4}, \frac{3\pi}{4}, \frac{5\pi}{4}$, and $\frac{7\pi}{4}$, which is supported by the graph shown.

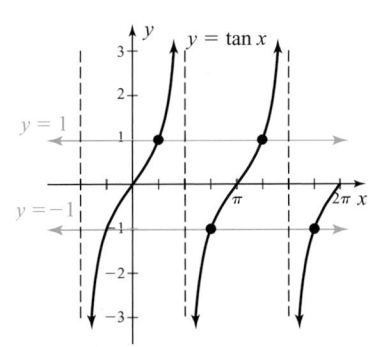

Now try Exercises 35 through 42 ▶

For any trig function that is not equal to a standard value, we can use a calculator to approximate the principal root or leave the result in exact form, and apply the same ideas to this root to find all solutions in the interval.

EXAMPLE 5 ▶ **Finding Solutions in [0, 360°)**

Find all solutions in $[0°, 360°)$ for $3\cos^2\theta + \cos\theta - 2 = 0$.

Solution ▶ Use a u-substitution to simplify the equation and help select an appropriate strategy. For $u = \cos\theta$, the equation becomes $3u^2 + u - 2 = 0$ and factoring seems the best approach. The factored form is $(u + 1)(3u - 2) = 0$, with solutions $u = -1$ and $u = \frac{2}{3}$. Re-substituting $\cos\theta$ for u gives

$$\cos\theta = -1 \qquad\qquad\qquad \cos\theta = \frac{2}{3} \qquad\qquad \text{equations from factored form}$$

$$\cos^{-1}(\cos\theta) = \cos^{-1}(-1) \qquad \cos^{-1}(\cos\theta) = \cos^{-1}\left(\frac{2}{3}\right) \qquad \text{apply inverse cosine}$$

$$\theta = 180° \qquad\qquad\qquad \theta \approx 48.2° \qquad\qquad \text{principal roots}$$

Both principal roots are in the specified interval. The first is quadrantal, the second was found using a calculator and is approximately 48.2°. With $\cos x$ positive in QI and QIV, a second solution is $(360 - 48.2)° = 311.8°$. The three solutions are 48.2°, 180°, and 311.8° although only $\theta = 180°$ is exact.

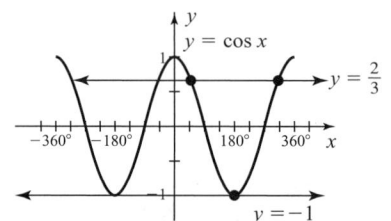

☑ **C. You've just learned how to solve trig equations for roots in [0, 2π) or [0, 360°)**

Now try Exercises 43 through 50 ▶

D. Solving Trig Equations for All Real Roots (ℝ)

As we noted, the intersections of a trig function with a horizontal line occur at regular, *predictable* intervals. This makes finding solutions from the set of real numbers a simple matter of extending the solutions we found in $[0, 2\pi)$ or $[0°, 360°)$. To illustrate, consider the solutions to Example 3. For $2\cos\theta + \sqrt{2} = 0$, we found the solutions $\theta = \frac{3\pi}{4}$ and $\theta = \frac{5\pi}{4}$. For solutions in ℝ, we note the "predictable interval" between roots *is identical to the period of the function*. This means all real solutions will be represented by $\theta = \frac{3\pi}{4} + 2\pi k$ and $\theta = \frac{5\pi}{4} + 2\pi k$, $k \in \mathbb{Z}$ (k is an integer). Both are illustrated in Figures 6.26 and 6.27 with the primary solution indicated with a "*."

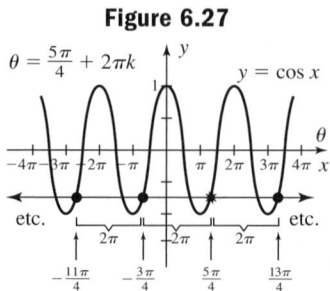

Figure 6.26

Figure 6.27

EXAMPLE 6 ▶ **Finding Solutions in** $\mathbb{R}$

Find all real solutions to
$\sqrt{3}\tan x - 1 = 0$.

Solution ▶ In Example 2 we found the

principal root was $x = \dfrac{\pi}{6}$. Since the

tangent function has a period of π, adding integer multiples of π to this root will identify all solutions:

$x = \dfrac{\pi}{6} + \pi k, k \in \mathbb{Z}$, as illustrated here.

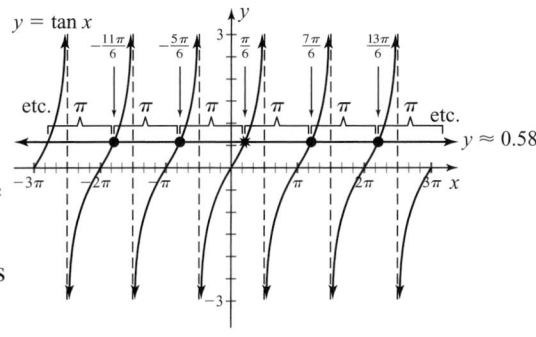

Now try Exercises 51 through 56 ▶

These fundamental ideas can be extended to many different situations. When asked to find *all real solutions,* be sure you find all roots in a stipulated interval before naming solutions by applying the period of the function. For instance, $\cos x = 0$ has two solutions in $[0, 2\pi)$ $\left[x = \dfrac{\pi}{2} \text{ and } x = \dfrac{3\pi}{2}\right]$, which we can quickly extend to find all real roots. But using $x = \cos^{-1}0$ or a calculator limits us to the single (principal) root $x = \dfrac{\pi}{2}$, and we'd miss all solutions stemming from $\dfrac{3\pi}{2}$. Note that solutions involving multiples of an angle (or fractional parts of an angle) should likewise be "handled with care," as in Example 7.

EXAMPLE 7 ▶ **Finding Solutions in** $\mathbb{R}$

Find all real solutions to $2\sin(2x)\cos x - \cos x = 0$.

Solution ▶ Since we have a common factor of $\cos x$, we begin by rewriting the equation as $\cos x[2\sin(2x) - 1] = 0$ and solve using the zero factor property. The resulting equations are $\cos x = 0$ and $2\sin(2x) - 1 = 0 \rightarrow \sin(2x) = \frac{1}{2}$.

$$\cos x = 0 \qquad \sin(2x) = \dfrac{1}{2} \quad \text{equations from factored form}$$

WORTHY OF NOTE

When solving trig equations that involve arguments other than a single variable, a *u*-substitution is sometimes used. For Example 7, substituting *u* for 2*x* gives the equation $\sin u = \dfrac{1}{2}$, making it "easier to see" that $u = \dfrac{\pi}{6}$ (since $\dfrac{1}{2}$ is a special value), and therefore $2x = \dfrac{\pi}{6}$ and $x = \dfrac{\pi}{12}$.

In $[0, 2\pi)$, $\cos x = 0$ has solutions $x = \dfrac{\pi}{2}$ and $x = \dfrac{3\pi}{2}$, giving $x = \dfrac{\pi}{2} + 2\pi k$ and $x = \dfrac{3\pi}{2} + 2\pi k$ as solutions in $\mathbb{R}$. Note these can actually be combined and written as $x = \dfrac{\pi}{2} + \pi k, k \in \mathbb{Z}$. The solution process for $\sin(2x) = \dfrac{1}{2}$ yields $2x = \dfrac{\pi}{6}$ and $2x = \dfrac{5\pi}{6}$. Since we seek all real roots, *we first extend each solution by $2\pi k$ before dividing by 2,* otherwise multiple solutions would be overlooked.

$$2x = \dfrac{\pi}{6} + 2\pi k \qquad\qquad 2x = \dfrac{5\pi}{6} + 2\pi k \quad \text{solutions from } \sin(2x) = \frac{1}{2}; k \in \mathbb{Z}$$

$$x = \dfrac{\pi}{12} + \pi k \qquad\qquad x = \dfrac{5\pi}{12} + \pi k \quad \text{divide by 2}$$

☑ **D.** You've just learned how to solve trig equations for roots in $\mathbb{R}$

Now try Exercises 57 through 66 ▶

E. Trig Equations and Trig Identities

In the process of solving trig equations, we sometimes employ fundamental identities to help simplify an equation, or to make factoring or some other method possible.

EXAMPLE 8 ▶ **Solving Trig Equations Using an Identity**

Find all solutions in $[0°, 360°)$ for $\cos(2\theta) + \sin^2\theta - 3\cos\theta = 1$.

Solution ▶ With a mixture of functions, exponents, and arguments, the equation is almost impossible to solve as it stands. But we can eliminate the sine function using the identity $\cos(2\theta) = \cos^2\theta - \sin^2\theta$, leaving a quadratic equation in $\cos x$.

$$\cos(2\theta) + \sin^2\theta - 3\cos\theta = 1 \quad \text{given equation}$$
$$\cos^2\theta - \sin^2\theta + \sin^2\theta - 3\cos\theta = 1 \quad \text{substitute } \cos^2\theta - \sin^2\theta \text{ for } \cos(2\theta)$$
$$\cos^2\theta - 3\cos\theta = 1 \quad \text{combine like terms}$$
$$\cos^2\theta - 3\cos\theta - 1 = 0 \quad \text{subtract 1}$$

Let's substitute u for $\cos\theta$ to give us a simpler view of the equation. This gives $u^2 - 3u - 1 = 0$, which is clearly not factorable over the integers. Using the quadratic formula with $a = 1$, $b = -3$, and $c = -1$ gives

$$u = \frac{3 \pm \sqrt{(-3)^2 - 4(1)(-1)}}{2(1)} \quad \text{quadratic formula in } u$$

$$= \frac{3 \pm \sqrt{13}}{2} \quad \text{simplified}$$

To four decimal places we have $u = 3.3028$ and $u = -0.3028$. To answer in terms of the original variable we re-substitute $\cos\theta$ for u, realizing that $\cos\theta \approx 3.3028$ has no solution, so solutions in $[0°, 360°)$ must be provided by $\cos\theta \approx -0.3028$ and occur in QII and QIII. The solutions are $\theta = \cos^{-1}(-0.3028) = 107.6°$ and $360° - 107.6° = 252.4°$ to the nearest tenth of a degree.

☑ **E. You've just learned how to solve trig equations using fundamental identities**

Now try Exercises 67 through 82 ▶

F. Trig Equations and Graphing Technology

A majority of the trig equations you'll encounter in your studies can be solved using the ideas and methods presented here. But there are some equations that cannot be solved using standard methods because they mix polynomial functions (linear, quadratic, and so on) that can be solved using algebraic methods, with what are called **transcendental functions** (trigonometric, logarithmic, and so on). By definition, transcendental functions are those that *transcend* the reach of standard algebraic methods. These kinds of equations serve to highlight the value of graphing and calculating technology to today's problem solvers.

EXAMPLE 9 ▶ **Solving Trig Equations Using Technology**

Use a graphing calculator in radian mode to find all real roots of $2\sin x + \dfrac{3x}{5} - 2 = 0$. Round solutions to four decimal places.

Solution ▶ When using graphing technology our initial concern is the size of the viewing window. After carefully entering the equation on the ▣ Y= screen, we note the term $2\sin x$ will never be larger than 2 or less than -2 for any real number x. On the other hand, the term $\dfrac{3x}{5}$ becomes larger for larger values of x, which would seem to cause $2\sin x + \dfrac{3x}{5}$ to "grow" as x gets larger. We conclude the standard window is a good place to start, and the resulting graph is shown in Figure 6.28.

Figure 6.28

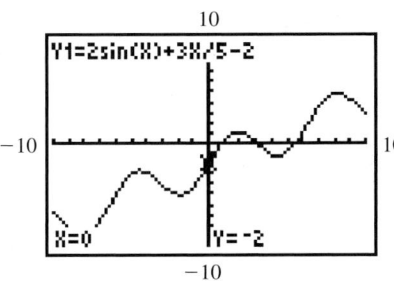

Figure 6.29

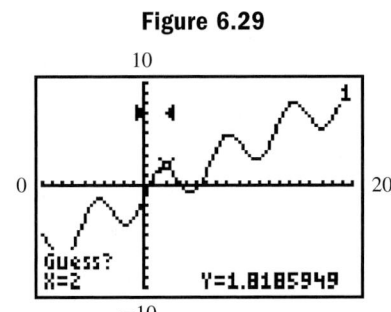

From this screen it appears there are three real roots, but to be sure none are hidden to the right, we extend the Xmax value to 20 (Figure 6.29). Using [2nd] [TRACE] CALC **2:zero,** we follow the prompts and enter a left bound of 0 (a number to the left of the zero) and a right bound of 2 (a number to the right of the zero—see Figure 6.29). If you can visually approximate the root, the calculator prompts you for a GUESS, otherwise just bypass the request by pressing [ENTER]. The smallest root is approximately $x = 0.8435$. Repeating this sequence we find the other roots are $x \approx 3.0593$ and $x \approx 5.5541$.

☑ **F.** You've just learned how to solve trig equations using graphing technology

Now try Exercises 83 through 88 ▶

TECHNOLOGY HIGHLIGHT

Solving Equations Graphically

Some equations are very difficult to solve analytically, and even with the use of a graphing calculator, a strong combination of analytical skills with technical skills is required to state the solution set. Consider the equation $5 \sin\left(\frac{1}{2}x\right) + 5 = \cot\left(\frac{1}{2}x\right)$ and solutions in $[-2\pi, 2\pi)$. There appears to be no quick analytical solution, and the first attempt at a graphical solution holds some hidden surprises. Enter $Y_1 = 5 \sin\left(\frac{1}{2}x\right) + 5$ and

$Y_2 = \dfrac{1}{\tan(\frac{1}{2}x)}$ on the [Y=] screen. Pressing [ZOOM] **7:ZTrig** gives

the screen in Figure 6.30, where we note there are at least two and possibly three solutions, depending on how the sine graph intersects the cotangent graph. We are also uncertain as to whether the graphs intersect again between $-\dfrac{\pi}{2}$ and $\dfrac{\pi}{2}$.

Increasing the maximum Y-value to Ymax = 8 shows they do indeed. But once again, are there now three or four solutions? In situations like this it may be helpful to use the **zeroes method** for solving graphically. On the [Y=] screen, disable Y_1 and Y_2 and enter Y_3 as $Y_1 - Y_2$. Pressing [ZOOM] **7:ZTrig** at this point clearly shows that there are four solutions in this interval (Figure 6.31), which can easily be found using [2nd] [CALC] **2:zero:** $x \approx -5.7543$, -4.0094, -3.1416, and 0.3390. Use these ideas to find solutions in $[-2\pi, 2\pi)$ for the exercises that follow.

Figure 6.30

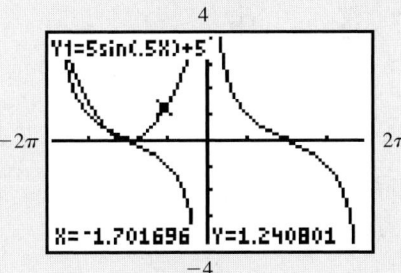

Figure 6.31

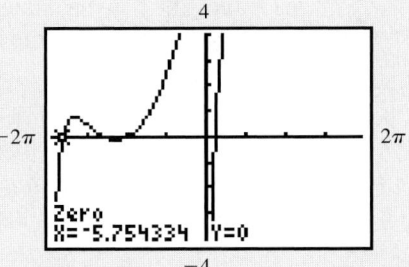

Exercise 1: $(1 + \sin x)^2 + \cos(2x) = 4 \cos x(1 + \sin x)$

Exercise 2: $4 \sin x = 2 \cos^2\left(\dfrac{x}{2}\right)$

6.6 EXERCISES

▶ CONCEPTS AND VOCABULARY

Fill in each blank with the appropriate word or phrase. Carefully reread the section if necessary.

1. For simple equations, a mental graph will tell us the quadrant of the _____ root, the number of roots in _____, and show a pattern for all _____ roots.

2. Solving trig equations is similar to solving algebraic equations, in that we first _____ the variable term, then apply the appropriate _____ function.

3. For $\sin x = \dfrac{\sqrt{2}}{2}$ the principal root is _____, solutions in $[0, 2\pi)$ are _____ and _____, and an expression for all real roots is _____ and _____; $k \in \mathbb{Z}$.

4. For $\tan x = -1$, the principal root is _____, solutions in $[0, 2\pi)$ are _____ and _____, and an expression for all real roots is _____.

5. Discuss/Explain/Illustrate why $\tan x = \dfrac{3}{4}$ and $y = \cos x$ have two solutions in $[0, 2\pi)$, even though the period of $y = \tan x$ is π, while the period of $y = \cos x$ is 2π.

6. The equation $\sin^2 x = \dfrac{1}{2}$ has four solutions in $[0, 2\pi)$. Explain how these solutions can be viewed as the vertices of a square inscribed in the unit circle.

▶ DEVELOPING YOUR SKILLS

7. For the equation $\sin x = -\dfrac{3}{4}$ and the graphs of $y = \sin x$ and $y = -\dfrac{3}{4}$ given, state (a) the quadrant of the principal root and (b) the number of roots in $[0, 2\pi)$.

Exercise 7

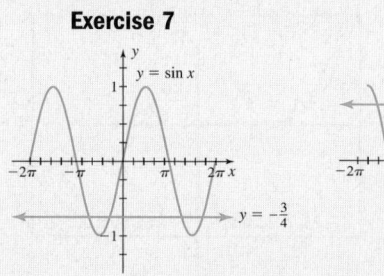

Exercise 8

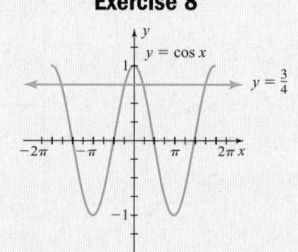

8. For the equation $\cos x = \dfrac{3}{4}$ and the graphs of $y = \cos x$ and $y = \dfrac{3}{4}$ given, state (a) the quadrant of the principal root and (b) the number of roots in $[0, 2\pi)$.

9. Given the graph $y = \tan x$ shown here, draw the horizontal line $y = -1.5$ and then for $\tan x = -1.5$, state (a) the quadrant of the principal root and (b) the number of roots in $[0, 2\pi)$.

Exercise 9

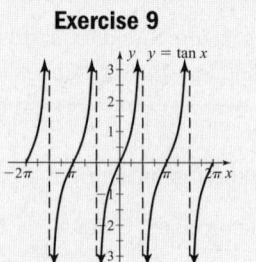

Exercise 10

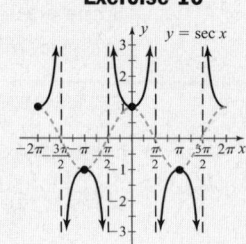

10. Given the graph of $y = \sec x$ shown, draw the horizontal line $y = \dfrac{5}{4}$ and then for $\sec x = \dfrac{5}{4}$, state (a) the quadrant of the principal root and (b) the number of roots in $[0, 2\pi)$.

11. The table that follows shows θ in multiples of $\dfrac{\pi}{6}$ between 0 and $\dfrac{4\pi}{3}$, with the values for $\sin \theta$ given.

Complete the table without a calculator or references using your knowledge of the unit circle, the signs of $f(\theta)$ in each quadrant, memory/recognition, $\tan \theta = \dfrac{\sin \theta}{\cos \theta}$, and so on.

Exercise 11

θ	$\sin\theta$	$\cos\theta$	$\tan\theta$
0	0		
$\dfrac{\pi}{6}$	$\dfrac{1}{2}$		
$\dfrac{\pi}{3}$	$\dfrac{\sqrt{3}}{2}$		
$\dfrac{\pi}{2}$	1		
$\dfrac{2\pi}{3}$	$\dfrac{\sqrt{3}}{2}$		
$\dfrac{5\pi}{6}$	$\dfrac{1}{2}$		
π	0		
$\dfrac{7\pi}{6}$	$-\dfrac{1}{2}$		
$\dfrac{4\pi}{3}$	$-\dfrac{\sqrt{3}}{2}$		

Exercise 12

θ	$\sin\theta$	$\cos\theta$	$\tan\theta$
0		1	
$\dfrac{\pi}{4}$		$\dfrac{\sqrt{2}}{2}$	
$\dfrac{\pi}{2}$		0	
$\dfrac{3\pi}{4}$		$-\dfrac{\sqrt{2}}{2}$	
π		-1	
$\dfrac{5\pi}{4}$		$-\dfrac{\sqrt{2}}{2}$	
$\dfrac{3\pi}{2}$		0	
$\dfrac{7\pi}{4}$		$\dfrac{\sqrt{2}}{2}$	
2π		1	

12. The table shows θ in multiples of $\dfrac{\pi}{4}$ between 0 and 2π, with the values for $\cos\theta$ given. Complete the table without a calculator or references using your knowledge of the unit circle, the signs of $f(\theta)$ in each quadrant, memory/recognition, $\tan\theta = \dfrac{\sin\theta}{\cos\theta}$, and so on.

Find the principal root of each equation.

13. $2\cos x = \sqrt{2}$ **14.** $2\sin x = -1$

15. $-4\sin x = 2\sqrt{2}$ **16.** $-4\cos x = 2\sqrt{3}$

17. $\sqrt{3}\tan x = 1$ **18.** $-2\sqrt{3}\tan x = 2$

19. $2\sqrt{3}\sin x = -3$ **20.** $-3\sqrt{2}\csc x = 6$

21. $-6\cos x = 6$ **22.** $4\sec x = -8$

23. $\dfrac{7}{8}\cos x = \dfrac{7}{16}$ **24.** $-\dfrac{5}{3}\sin x = \dfrac{5}{6}$

25. $2 = 4\sin\theta$ **26.** $\pi\tan x = 0$

27. $-5\sqrt{3} = 10\cos\theta$ **28.** $4\sqrt{3} = 4\tan\theta$

Find all solutions in $[0, 2\pi)$.

29. $9\sin x - 3.5 = 1$ **30.** $6.2\cos x + 4 = 7.1$

31. $8\tan x + 7\sqrt{3} = -\sqrt{3}$

32. $\dfrac{1}{2}\sec x - \dfrac{3}{4} = -\dfrac{7}{4}$ **33.** $\dfrac{2}{3}\cot x - \dfrac{5}{6} = -\dfrac{3}{2}$

34. $-110\sin x = -55\sqrt{3}$ **35.** $4\cos^2 x = 3$

36. $4\sin^2 x = 1$ **37.** $-7\tan^2 x = -21$

38. $3\sec^2 x = 6$ **39.** $-4\csc^2 x = -8$

40. $6\sqrt{3}\cos^2 x = 3\sqrt{3}$ **41.** $4\sqrt{2}\sin^2 x = 4\sqrt{2}$

42. $\dfrac{2}{3}\cos^2 x + \dfrac{5}{6} = \dfrac{4}{3}$

Solve the following equations by factoring. State all real solutions in radians using the exact form where possible and rounded to four decimal places if the result is not a standard value.

43. $3\cos^2\theta + 14\cos\theta - 5 = 0$

44. $6\tan^2\theta - 2\sqrt{3}\tan\theta = 0$

45. $2\cos x\sin x - \cos x = 0$

46. $2\sin^2 x + 7\sin x = 4$ **47.** $\sec^2 x - 6\sec x = 16$

48. $2\cos^3 x + \cos^2 x = 0$ **49.** $4\sin^2 x - 1 = 0$

50. $4\cos^2 x - 3 = 0$

Find all real solutions. Note that identities are not required to solve these exercises.

51. $-2\sin x = \sqrt{2}$ **52.** $2\cos x = 1$

53. $-4\cos x = 2\sqrt{2}$ **54.** $4\sin x = 2\sqrt{3}$

55. $\sqrt{3}\tan x = -\sqrt{3}$ **56.** $2\sqrt{3}\tan x = 2$

57. $6\cos(2x) = -3$ **58.** $2\sin(3x) = -\sqrt{2}$

59. $\sqrt{3}\tan(2x) = -\sqrt{3}$ **60.** $2\sqrt{3}\tan(3x) = 6$

61. $-2\sqrt{3}\cos\left(\dfrac{1}{3}x\right) = 2\sqrt{3}$

62. $-8\sin\left(\dfrac{1}{2}x\right) = -4\sqrt{3}$

63. $\sqrt{2}\cos x\sin(2x) - 3\cos x = 0$

64. $\sqrt{3}\sin x\tan(2x) - \sin x = 0$

65. $\cos(3x)\csc(2x) - 2\cos(3x) = 0$

66. $\sqrt{3}\sin(2x)\sec(2x) - 2\sin(2x) = 0$

Solve each equation using calculator and inverse trig functions to determine the principal root (not by graphing). Clearly state (a) the principal root and (b) all real roots.

67. $3\cos x = 1$ **68.** $5\sin x = -2$

69. $\sqrt{2}\sec x + 3 = 7$ **70.** $\sqrt{3}\csc x + 2 = 11$

71. $\dfrac{1}{2}\sin(2\theta) = \dfrac{1}{3}$ **72.** $\dfrac{2}{5}\cos(2\theta) = \dfrac{1}{4}$

73. $-5\cos(2\theta) - 1 = 0$ **74.** $6\sin(2\theta) - 3 = 2$

Solve the following equations using an identity. State all real solutions in radians using the exact form where possible and rounded to four decimal places if the result is not a standard value.

75. $\cos^2 x - \sin^2 x = \dfrac{1}{2}$

76. $4 \sin^2 x - 4 \cos^2 x = 2\sqrt{3}$

77. $2 \cos\left(\dfrac{1}{2}x\right)\cos x - 2 \sin\left(\dfrac{1}{2}x\right)\sin x = 1$

78. $\sqrt{2} \sin(2x)\cos(3x) + \sqrt{2} \sin(3x)\cos(2x) = 1$

79. $(\cos \theta + \sin \theta)^2 = 1$ **80.** $(\cos \theta + \sin \theta)^2 = 2$

81. $\cos(2\theta) + 2 \sin^2\theta - 3 \sin \theta = 0$

82. $3 \sin(2\theta) - \cos^2(2\theta) - 1 = 0$

Find all roots in $[0, 2\pi)$ using a graphing calculator. State answers in radians rounded to four decimal places.

83. $5 \cos x - x = 3$ **84.** $3 \sin x + x = 4$

85. $\cos^2(2x) + x = 3$ **86.** $\sin^2(2x) + 2x = 1$

87. $x^2 + \sin(2x) = 1$ **88.** $\cos(2x) - x^2 = -5$

▶ WORKING WITH FORMULAS

89. Range of a projectile: $R = \dfrac{5}{49}v^2 \sin(2\theta)$

The distance a projectile travels is called its range and is modeled by the formula shown, where R is the range in meters, v is the initial velocity in meters per second, and θ is the angle of release. Two friends are standing 16 m apart playing catch. If the first throw has an initial velocity of 15 m/sec, what *two* angles will insure the ball travels the 16 m between the friends?

90. Fine-tuning a golf swing:
(club head to shoulder)^2 = (club length)^2 + (arm length)^2 − 2 (club length)(arm length)cos θ

A golf pro is taking specific measurements on a client's swing to help improve her game. If the angle θ is too small, the ball is hit late and "too thin" (you *top the ball*). If θ is too large, the ball is hit early and "too fat" (you *scoop the ball*). Approximate the angle θ formed by the club and the extended (left) arm using the given measurements and formula shown.

37 in.
39 in.
27 in.

▶ APPLICATIONS

Acceleration due to gravity: When a steel ball is released down an inclined plane, the rate of the ball's acceleration depends on the angle of incline. The acceleration can be approximated by the formula $A(\theta) = 9.8 \sin \theta$, where θ is in degrees and the acceleration is measured in meters per second/per second. To the nearest tenth of a degree,

91. What angle produces an acceleration of 0 m/sec^2 when the ball is released? Explain why this is reasonable.

92. What angle produces an acceleration of 9.8 m/sec^2? What does this tell you about the acceleration due to gravity?

93. What angle produces an acceleration of 5 m/sec^2? Will the angle be larger or smaller for an acceleration of 4.5 m/sec^2?

94. Will an angle producing an acceleration of 2.5 m/sec^2 be one-half the angle required for an acceleration of 5 m/sec^2? Explore and discuss.

Snell's law states that when a ray of light passes from one medium into another, the sine of the angle of incidence α *varies directly with* the sine of the angle of refraction β (see the figure). This phenomenon is modeled by the formula $\sin \alpha = k \sin \beta$, where k is called the **index of refraction.** Note the angle θ is the angle at which the light strikes the surface, so that $\alpha = 90° - \theta$. Use this information to work Exercises 95 to 98.

Exercises 95 to 98

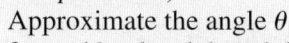
light   incidence | reflection
α α'
θ θ
new medium β refraction
$\sin(\alpha) = k \sin(\beta)$

95. A ray of light passes from air into water, striking the water at an angle of 55°. Find the angle of incidence α and the angle of refraction β, if the index of refraction for water is $k = 1.33$.

96. A ray of light passes from air into a diamond, striking the surface at an angle of 75°. Find the angle of incidence α and the angle of refraction β, if the index of refraction for a diamond is $k = 2.42$.

97. Find the index of refraction for ethyl alcohol if a beam of light strikes the surface of this medium at an angle of 40° and produces an angle of refraction $\beta = 34.3°$. Use this index to find the angle of incidence if a second beam of light created an angle of refraction measuring 15°.

98. Find the index of refraction for rutile (a type of mineral) if a beam of light strikes the surface of this medium at an angle of 30° and produces an angle of refraction $\beta = 18.7°$. Use this index to find the angle of incidence if a second beam of light created an angle of refraction measuring 10°.

99. Roller coaster design: As part of a science fair project, Hadra builds a scale model of a roller coaster using the equation $y = 5 \sin\left(\dfrac{1}{2}x\right) + 7$, where y is the height of the model in inches and x is the distance from the "loading platform" in inches. (a) How high is the platform? (b) What distances from the platform does the model attain a height of 9.5 in.?

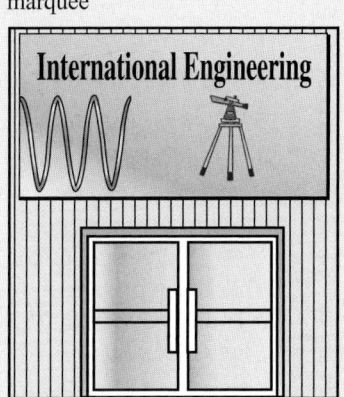

100. Company logo: Part of the logo for an engineering firm was modeled by a cosine function. The logo was then manufactured in steel and installed on the entrance marquee of the home office. The position and size of the logo is modeled by the function $y = 9 \cos x + 15$, where y is the height of the graph above the base of the marquee in inches and x represents the distance from the edge of the marquee. Assume the graph begins flush with the edge. (a) How far above the base is the beginning of the cosine graph? (b) What distances from the edge does the graph attain a height of 19.5 in.?

Geometry applications: Solve Exercises 101 and 102 graphically using a calculator. For Exercise 101, give θ in radians rounded to four decimal places. For Exercise 102, answer in degrees to the nearest tenth of a degree.

101. The area of a circular segment (the shaded portion shown) is given by the formula $A = \dfrac{1}{2} r^2(\theta - \sin \theta)$, where θ is in radians. If the circle has a radius of 10 cm, find the angle θ that gives an area of 12 cm².

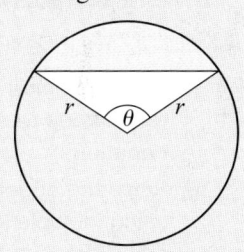

Exercise 102

102. The perimeter of a trapezoid with parallel sides B and b, altitude h, and base angles α and β is given by the formula $P = B + b + h(\csc \alpha + \csc \beta)$. If $b = 30$ m, $B = 40$ m, $h = 10$ m, and $\alpha = 45°$, find the angle β that gives a perimeter of 105 m.

▶ **EXTENDING THE CONCEPT**

103. Find all real solutions to $5 \cos x - x = -x$ in two ways. First use a calculator with $Y_1 = 5 \cos x - x$ and $Y_2 = -x$ to determine the regular intervals between points of intersection. Second, simplify by adding x to both sides, and draw a quick sketch of the result to locate x-intercepts. Explain why both methods give the same result, even though the first presents you with a very different graph.

104. Once the fundamental ideas of solving a given family of equations is understood and practiced, a student usually begins to generalize them—making the numbers or symbols used in the equation irrelevant. (a) Use the inverse sine function to find the principal root of $y = A \sin(Bx - C) + D$, by solving for x in terms of y, A, B, C, and D. (b) Solve the following equation using the techniques addressed in this section, and then using the "formula" from Part (a) $5 = 2 \sin\left(\dfrac{1}{2}x + \dfrac{\pi}{4}\right) + 3$. Do the results agree?

▶ **MAINTAINING YOUR SKILLS**

105. (1.4) Use a substitution to show that $x = 2 + i$ is a zero of $f(x) = x^2 - 4x + 5$.

106. (3.1) Currently, tickets to productions of the Shakespeare Community Theater cost $10.00, with an average attendance of 250 people. Due to market research, the theater director believes that for each $0.50 reduction in price, 25 more people will attend. What ticket price will maximize the theater's revenue? What will the average attendance projected to become at that price?

107. (6.5) Evaluate without using a calculator:

a. $\tan\left[\sin^{-1}\left(-\dfrac{1}{2}\right)\right]$ **b.** $\sin[\tan^{-1}(-1)]$

108. (5.1) The largest Ferris wheel in the world, located in Yokohama, Japan, has a radius of 50 m. To the nearest hundredth of a meter, how far does a seat on the rim travel as the wheel turns through $\theta = 292.5°$?

6.7 | General Trig Equations and Applications

Learning Objectives

In Section 6.7 you will learn how to:

☐ **A.** Use additional algebraic techniques to solve trig equations

☐ **B.** Solve trig equations using multiple angle, sum and difference, and sum-to-product identities

☐ **C.** Solve trig equations of the form $A\sin(Bx + C) + D = k$

☐ **D.** Use a combination of skills to model and solve a variety of applications

At this point you're likely beginning to understand the true value of trigonometry to the scientific world. Essentially, any phenomenon that is cyclic or periodic is beyond the reach of polynomial (and other) functions, and may require trig for an accurate understanding. And while there is an abundance of trig applications in oceanography, astronomy, meteorology, geology, zoology, and engineering, their value is not limited to the hard sciences. There are also rich applications in business and economics, and a growing number of modern artists are creating works based on attributes of the trig functions. In this section, we try to place some of these applications within your reach, with the exercise set offering an appealing variety from many of these fields.

A. Trig Equations and Algebraic Methods

We begin this section with a follow-up to Section 6.6, by introducing trig equations that require slightly more sophisticated methods to work out a solution.

EXAMPLE 1 ▶ **Solving a Trig Equation by Squaring Both Sides**

Find all solutions in $[0, 2\pi)$: $\sec x + \tan x = \sqrt{3}$.

Solution ▶ Our first instinct might be to rewrite the equation in terms of sine and cosine, but that simply leads to a similar equation that still has two different functions $[\sqrt{3}\cos x - \sin x = 1]$. Instead, we *square both sides* and see if the Pythagorean identity $1 + \tan^2 x = \sec^2 x$ will be of use. Prior to squaring, we separate the functions on opposite sides to avoid the mixed term $2\tan x \sec x$.

$$\sec x + \tan x = \sqrt{3} \qquad \text{given equation}$$
$$(\sec x)^2 = (\sqrt{3} - \tan x)^2 \qquad \text{subtract tan } x \text{ and square}$$
$$\sec^2 x = 3 - 2\sqrt{3}\tan x + \tan^2 x \qquad \text{result}$$

Since $\sec^2 x = 1 + \tan^2 x$, we substitute directly and obtain an equation in tangent alone.

$$1 + \tan^2 x = 3 - 2\sqrt{3}\tan x + \tan^2 x \qquad \text{substitute } 1 + \tan^2 x \text{ for sec}^2 x$$
$$-2 = -2\sqrt{3}\tan x \qquad \text{simplify}$$
$$\frac{1}{\sqrt{3}} = \tan x \qquad \text{solve for tan } x$$
$$\tan x > 0 \text{ in QI and QIII}$$

The proposed solutions are $x = \dfrac{\pi}{6}$ [QI] and $\dfrac{7\pi}{6}$ [QIII]. Since squaring an equation sometimes introduces extraneous roots, both should be checked in the original equation. The check shows only $x = \dfrac{\pi}{6}$ is a solution.

Now try Exercises 7 through 12 ▶

Here is one additional example that uses a factoring strategy commonly employed when an equation has more than three terms.

EXAMPLE 2 ▶   Solving a Trig Equation by Factoring

Find all solutions in $[0°, 360°)$: $8 \sin^2\theta \cos\theta - 2 \cos\theta - 4 \sin^2\theta + 1 = 0$.

Solution ▶   The four terms in the equation share no common factors, so we attempt to factor by grouping. We could factor $2 \cos\theta$ from the first two terms but instead elect to group the $\sin^2\theta$ terms and begin there.

$$8 \sin^2\theta \cos\theta - 2 \cos\theta - 4 \sin^2\theta + 1 = 0 \qquad \text{given equation}$$
$$(8 \sin^2\theta \cos\theta - 4 \sin^2\theta) - (2 \cos\theta - 1) = 0 \qquad \text{rearrange and group terms}$$
$$4 \sin^2\theta(2 \cos\theta - 1) - 1(2 \cos\theta - 1) = 0 \qquad \text{remove common factors}$$
$$(2 \cos\theta - 1)(4 \sin^2\theta - 1) = 0 \qquad \text{remove common binomial factors}$$

Using the zero factor property, we write two equations and solve each independently.

$2 \cos\theta - 1 = 0$	$4 \sin^2\theta - 1 = 0$ resulting equations
$2 \cos\theta = 1$	$\sin^2\theta = \dfrac{1}{4}$ isolate variable term
$\cos\theta = \dfrac{1}{2}$	$\sin\theta = \pm\dfrac{1}{2}$ solve
$\cos\theta > 0$ in QI and QIV	$\sin\theta > 0$ in QI and QII
$\theta = 60°, 300°$	$\sin\theta < 0$ in QIII and QIV
	$\theta = 30°, 150°, 210°, 330°$ solutions

☑ A. You've just learned how to use additional algebraic techniques to solve trig equations

Initially factoring $2 \cos\theta$ from the first two terms and proceeding from there would have produced the same result.

Now try Exercises 13 through 16 ▶

B. Solving Trig Equations Using Various Identities

To solve equations effectively, a student should strive to develop *all* of the necessary "tools." Certainly the underlying concepts and graphical connections are of primary importance, as are the related algebraic skills. But to solve *trig* equations effectively we must also have a ready command of commonly used identities. Observe how Example 3 combines a double-angle identity with factoring by grouping.

EXAMPLE 3 ▶   Using Identities and Algebra to Solve a Trig Equation

Find all solutions in $[0, 2\pi)$: $3 \sin(2x) + 2 \sin x - 3 \cos x = 1$. Round solutions to four decimal places as necessary.

Solution ▶   Noting that one of the terms involves a double angle, we attempt to replace that term to make factoring a possibility. Using the double identity for sine, we have

$$3(2 \sin x \cos x) + 2 \sin x - 3 \cos x = 1 \quad \text{substitute } 2 \sin x \cos x \text{ for } \sin(2x)$$

$$(6 \sin x \cos x + 2 \sin x) - (3 \cos x + 1) = 0 \quad \text{set equal zero and group terms}$$

$$2 \sin x(3 \cos x + 1) - 1(3 \cos x + 1) = 0 \quad \text{factor using } 3 \cos x + 1$$

$$(3 \cos x + 1)(2 \sin x - 1) = 0 \quad \text{common binomial factor}$$

Use the zero factor property to solve each equation independently.

$$3 \cos x + 1 = 0 \qquad\qquad 2 \sin x - 1 = 0 \quad \text{resulting equations}$$

$$\cos x = -\frac{1}{3} \qquad\qquad \sin x = \frac{1}{2} \quad \text{isolate variable term}$$

$$\cos x < 0 \text{ in QII and QIII} \qquad \sin x > 0 \text{ in QI and QII}$$

$$x \approx 1.9106, 4.3726 \qquad\qquad x = \frac{\pi}{6}, \frac{5\pi}{6} \quad \text{solutions}$$

☑ **B.** You've just learned how to solve trig equations using multiple angle, sum and difference, and sum-to-product identities

Should you prefer the exact form, the solutions from the cosine equation could be written as $x = \cos^{-1}\left(-\frac{1}{3}\right)$ and $x = 2\pi - \cos^{-1}\left(-\frac{1}{3}\right)$.

Now try Exercises 17 through 26 ▶

C. Solving Equations of the Form $A \sin (Bx \pm C) \pm D = k$

You may remember equations of this form from Section 5.7. They actually occur quite frequently in the investigation of many natural phenomena and in the modeling of data from a periodic or seasonal context. Solving these equations requires a good combination of algebra skills with the fundamentals of trig.

EXAMPLE 4 ▶   **Solving Equations That Involve Transformations**

Given $f(x) = 160 \sin\left(\frac{\pi}{3}x + \frac{\pi}{3}\right) + 320$ and $x \in [0, 2\pi)$, for what real numbers x is $f(x)$ less than 240?

Solution ▶   We reason that to find values where $f(x) < 240$, we should begin by finding values where $f(x) = 240$. The result is

$$160 \sin\left(\frac{\pi}{3}x + \frac{\pi}{3}\right) + 320 = 240 \quad \text{equation}$$

$$\sin\left(\frac{\pi}{3}x + \frac{\pi}{3}\right) = -0.5 \quad \text{subtract 320 and divide by 160; isolate variable term}$$

At this point we elect to use a u-substitution for $\left(\frac{\pi}{3}x + \frac{\pi}{3}\right) = \frac{\pi}{3}(x + 1)$ to obtain a "clearer view."

$$\sin u = -0.5 \quad \text{substitute } u \text{ for } \frac{\pi}{3}(x + 1)$$

$$\sin u < 0 \text{ in QIII and QIV}$$

$$u = \frac{7\pi}{6} \qquad u = \frac{11\pi}{6} \quad \text{solutions in } u$$

To complete the solution we re-substitute $\frac{\pi}{3}(x + 1)$ for u and solve.

$$\frac{\pi}{3}(x + 1) = \frac{7\pi}{6} \qquad \frac{\pi}{3}(x + 1) = \frac{11\pi}{6} \quad \text{re-substitute } \frac{\pi}{3}(x + 1) \text{ for } u$$

$$x + 1 = \frac{7}{2} \qquad\qquad x + 1 = \frac{11}{2} \quad \text{multiply both sides by } \frac{3}{\pi}$$

$$x = 2.5 \qquad\qquad\qquad x = 4.5 \quad \text{solutions}$$

We now know $f(x) = 240$ when $x = 2.5$ and $x = 4.5$ but when will $f(x)$ be *less*

than 240? By analyzing the equation, we find the function has period of

$P = \dfrac{2\pi}{\frac{\pi}{3}} = 6$ and is shifted to the left $\dfrac{\pi}{3}$ units. This would indicate the graph peaks

early in the interval $[0, 2\pi)$ with a "valley" in the interior. We conclude

$f(x) < 240$ in the interval $(2.5, 4.5)$.

☑ **C.** You've just learned how to solve trig equations of the form $A \sin(Bx + C) + D = k$

Now try Exercises 27 through 30 ▶

GRAPHICAL SUPPORT

Support for the result in Example 4 can be
obtained by graphing the equation over the
specified interval. Enter

$Y_1 = 160 \sin\left(\dfrac{\pi}{3}x + \dfrac{\pi}{3}\right) + 320$ on the ⬚ Y= ⬚

screen, then $Y_2 = 240$. After locating points of
intersection, we note the graphs indeed verify

that in the interval $[0, 2\pi)$, $f(x) < 240$ for $x \in (2.5, 4.5)$.

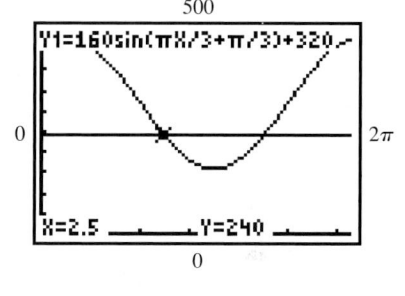

There is a mixed variety of equation types in **Exercises 31 through 40.**

D. Applications Using Trigonometric Equations

Figure 6.32

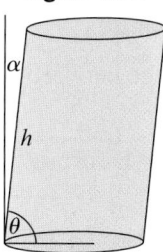

Using characteristics of the trig functions, we can often generalize and extend many
of the formulas that are familiar to you. For example, the formulas for the volume of
a right circular cylinder and a right circular cone are well known, but what about the
volume of a nonright figure (see Figure 6.32)? Here, trigonometry provides the answer,
as the most general volume formula is $V = V_0 \sin \theta$, where V_0 is a "standard" volume
formula and θ is the complement of angle of deflection **(see Exercises 43 and 44).**

As for other applications, consider the following from the environmental sciences.
Natural scientists are very interested in the discharge rate of major rivers, as this gives
an indication of rainfall over the inland area served by the river. In addition, the discharge rate has a large impact on the freshwater and saltwater zones found at the river's
estuary (where it empties into the sea).

EXAMPLE 5 ▶ Solving an Equation Modeling the Discharge Rate of a River

For May through December, the discharge rate of the Ganges River (Bangladesh)

can be modeled by $D(t) = 16{,}580 \sin\left(\dfrac{\pi}{3}t - \dfrac{2\pi}{3}\right) + 17{,}760$ where $t = 1$

represents May 1, and $D(t)$ is the discharge rate in m³/sec.

Source: Global River Discharge Database Project; www.rivdis.sr.unh.edu.

a. What is the discharge rate in mid-October?

b. For what months (within this interval) is the discharge rate over 26,050 m³/sec?

Solution ▶ **a.** To find the discharge rate in mid-October we simply evaluate the function at $t = 6.5$:

$$D(t) = 16{,}580 \sin\left(\frac{\pi}{3}t - \frac{2\pi}{3}\right) + 17{,}760 \qquad \text{given function}$$

$$D(6.5) = 16{,}580 \sin\left[\frac{\pi}{3}(6.5) - \frac{2\pi}{3}\right] + 17{,}760 \qquad \text{substitute 6.5 for } t$$

$$= 1180 \qquad \text{compute result on a calculator}$$

In mid-October the discharge rate is 1180 m³/sec.

b. We first find when the rate is *equal* to 26,050 m³/sec: $D(t) = 26{,}050$.

$$26{,}050 = 16{,}580 \sin\left(\frac{\pi}{3}t - \frac{2\pi}{3}\right) + 17{,}760 \qquad \text{substitute 26,050 for } D(t)$$

$$0.5 = \sin\left(\frac{\pi}{3}t - \frac{2\pi}{3}\right) \qquad \text{subtract 17,760; divide by 16,580}$$

Using a u-substitution for $\left(\dfrac{\pi}{3}t - \dfrac{2\pi}{3}\right)$ we obtain the equation

$$0.5 = \sin u$$
$$\sin u > 0 \text{ in QI and QII}$$

$$u = \frac{\pi}{6} \qquad\qquad u = \frac{5\pi}{6} \qquad \text{solutions in } u$$

To complete the solution we re-substitute $\left(\dfrac{\pi}{3}t - \dfrac{2\pi}{3}\right) = \dfrac{\pi}{3}(t - 2)$ for u and solve.

$$\frac{\pi}{3}(t - 2) = \frac{\pi}{6} \qquad\qquad \frac{\pi}{3}(t - 2) = \frac{5\pi}{6} \qquad \text{re-substitute } \frac{\pi}{3}(t-2) \text{ for } u$$

$$t - 2 = 0.5 \qquad\qquad t - 2 = 2.5 \qquad \text{multiply both sides by } \frac{3}{\pi}$$

$$t = 2.5 \qquad\qquad\qquad t = 4.5 \qquad \text{solutions}$$

The Ganges River will have a flow rate of over 26,050 m³/sec between mid-June (2.5) and mid-August (4.5).

Now try Exercises 45 through 48 ▶

GRAPHICAL SUPPORT

To obtain a graphical view of the solution to Example 5, enter

$Y_1 = 16{,}580 \sin\left(\dfrac{\pi}{3}t - \dfrac{2\pi}{3}\right) + 17{,}760$ on the [**Y =**] screen, then $Y_2 = 26{,}050$. To set

an appropriate window, note the amplitude is

16,580 and that the graph has been vertically

shifted by 17,760. Also note the x-axis

represents months 5 through 12. After locating

points of intersection, we note the graphs verify

that in the interval $[1, 9]$ $D(t) > 26{,}050$ for

$t \in (2.5, 4.5)$.

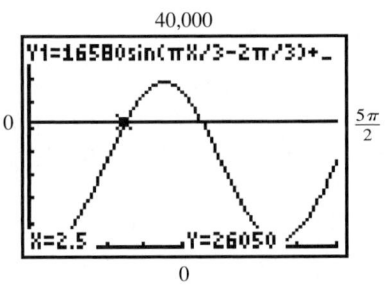

☑ **D.** You've just learned how to use a combination of skills to model and solve a variety of applications

There is a variety of additional exercises in the Exercise Set. **See Exercises 49 through 54.**

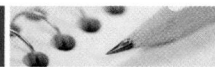 **6.7 EXERCISES**

▶ **CONCEPTS AND VOCABULARY**

Fill in each blank with the appropriate word or phrase. Carefully reread the section if needed.

1. The three Pythagorean identities are _____, _____, and _____.

2. When an equation contains two functions from a Pythagorean identity, sometimes _____ both sides will lead to a solution.

3. One strategy to solve equations with four terms and no common factors is _____ by _____.

4. To combine two sine or cosine terms with different arguments, we can use the _____ to _____ formulas.

5. Regarding Example 5, discuss/explain how to determine the months of the year the discharge rate is *under* 26,050 m³/sec, using the solution set given.

6. Regarding Example 6, discuss/explain how to determine the months of the year the revenue projection is *under* $1250 using the solution set given.

▶ **DEVELOPING YOUR SKILLS**

Solve each equation in $[0, 2\pi)$ using the method indicated. Round nonstandard values to four decimal places.

• **Squaring both sides**

7. $\sin x + \cos x = \dfrac{\sqrt{6}}{2}$ 8. $\cot x - \csc x = \sqrt{3}$

9. $\tan x - \sec x = -1$ 10. $\sin x + \cos x = \sqrt{2}$

11. $\cos x + \sin x = \dfrac{4}{3}$ 12. $\sec x + \tan x = 2$

• **Factor by grouping**

13. $\cot x \csc x - 2 \cot x - \csc x + 2 = 0$

14. $4 \sin x \cos x - 2\sqrt{3} \sin x - 2 \cos x + \sqrt{3} = 0$

15. $3 \tan^2 x \cos x - 3 \cos x + 2 = 2 \tan^2 x$

16. $4\sqrt{3} \sin^2 x \sec x - \sqrt{3} \sec x + 2 = 8 \sin^2 x$

• **Using identities**

17. $\dfrac{1 + \cot^2 x}{\cot^2 x} = 2$ 18. $\dfrac{1 + \tan^2 x}{\tan^2 x} = \dfrac{4}{3}$

19. $3 \cos(2x) + 7 \sin x - 5 = 0$

20. $3 \cos(2x) - \cos x + 1 = 0$

21. $2 \sin^2\left(\dfrac{x}{2}\right) - 3 \cos\left(\dfrac{x}{2}\right) = 0$

22. $2 \cos^2\left(\dfrac{x}{3}\right) + 3 \sin\left(\dfrac{x}{3}\right) - 3 = 0$

23. $\cos(3x) + \cos(5x)\cos(2x) + \sin(5x)\sin(2x) - 1 = 0$

24. $\sin(7x)\cos(4x) + \sin(5x) - \cos(7x)\sin(4x) + \cos x = 0$

25. $\sec^4 x - 2 \sec^2 x \tan^2 x + \tan^4 x = \tan^2 x$

26. $\tan^4 x - 2 \sec^2 x \tan^2 x + \sec^4 x = \cot^2 x$

State the period P of each function and find all solutions in $[0, P)$. Round to four decimal places as needed.

27. $250 \sin\left(\dfrac{\pi}{6}x + \dfrac{\pi}{3}\right) - 125 = 0$

28. $-75\sqrt{2} \sec\left(\dfrac{\pi}{4}x + \dfrac{\pi}{6}\right) + 150 = 0$

29. $1235 \cos\left(\dfrac{\pi}{12}x - \dfrac{\pi}{4}\right) + 772 = 1750$

30. $-0.075 \sin\left(\dfrac{\pi}{2}x + \dfrac{\pi}{3}\right) - 0.023 = -0.068$

• **Using any appropriate method to solve.**

31. $\cos x - \sin x = \dfrac{\sqrt{2}}{2}$

32. $5\sec^2 x - 2\tan x - 8 = 0$

33. $\dfrac{1 - \cos^2 x}{\tan^2 x} = \dfrac{\sqrt{3}}{2}$

34. $5\csc^2 x - 5\cot x - 5 = 0$

35. $\csc x + \cot x = 1$ **36.** $\dfrac{1 - \sin^2 x}{\cot^2 x} = \dfrac{\sqrt{2}}{2}$

37. $\sec x \cos\left(\dfrac{\pi}{2} - x\right) = -1$

38. $\sin\left(\dfrac{\pi}{2} - x\right)\csc x = \sqrt{3}$

39. $\sec^2 x \tan\left(\dfrac{\pi}{2} - x\right) = 4$

40. $2\tan\left(\dfrac{\pi}{2} - x\right)\sin^2 x = \dfrac{\sqrt{3}}{2}$

▶ WORKING WITH FORMULAS

41. The equation of a line in trigonometric form:

$$y = \dfrac{D - x\cos\theta}{\sin\theta}$$

Exercise 41

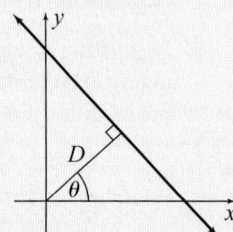

The trigonometric form of a linear equation is given by the formula shown, where D is the perpendicular distance from the origin to the line and θ is the angle between the perpendicular segment and the x-axis. For each pair of perpendicular lines given, (a) find the point (a, b) of their intersection; (b) compute the distance $D = \sqrt{a^2 + b^2}$ and the angle $\theta = \tan^{-1}\left(\dfrac{b}{a}\right)$, and give the equation of the line in trigonometric form; and (c) use the GRAPH or the 2nd GRAPH TABLE feature of a graphing calculator to verify that both equations name the same line.

I. $L_1: y = -x + 5$ **II.** $L_1: y = -\dfrac{1}{2}x + 5$

$L_2: y = x$ $L_2: y = 2x$

III. $L_1: y = -\dfrac{\sqrt{3}}{3}x + \dfrac{4\sqrt{3}}{3}$

$L_2: y = \sqrt{3}x$

42. Rewriting $y = a\cos x + b\sin x$ as a single function: $y = k\sin(x + \theta)$

Linear terms of sine and cosine can be rewritten as a single function using the formula shown, where $k = \sqrt{a^2 + b^2}$ and $\theta = \sin^{-1}\left(\dfrac{a}{k}\right)$. Rewrite the equations given using these relationships and verify they are equivalent using the GRAPH or the 2nd GRAPH TABLE feature of a graphing calculator:

a. $y = 2\cos x + 2\sqrt{3}\sin x$

b. $y = 4\cos x + 3\sin x$

The ability to rewrite a trigonometric equation in simpler form has a tremendous number of applications in graphing, equation solving, working with identities, and solving applications.

▶ APPLICATIONS

43. Volume of a cylinder: The volume of a cylinder is given by the formula $V = \pi r^2 h\sin\theta$, where r is the radius and h is the height of the cylinder, and θ is the indicated complement of the angle of deflection α. Note that when

Exercise 43

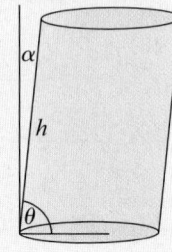

$\theta = \dfrac{\pi}{2}$, the formula becomes that of a right circular cylinder (if $\theta \neq \dfrac{\pi}{2}$, then h is called the *slant height or lateral height* of the cylinder). An old farm silo is built in the form of a right circular cylinder with a radius of 10 ft and a

height of 25 ft. After an earthquake, the silo became tilted with an angle of deflection $\alpha = 5°$. (a) Find the volume of the silo before the earthquake. (b) Find the volume of the silo after the earthquake. (c) What angle θ is required to bring the original volume of the silo down 2%?

44. Volume of a cone: The volume of a cone is given by the formula $V = \dfrac{1}{3}\pi r^2 h\sin\theta$, where r is the radius and h is the height of the cone, and θ is the indicated complement of the angle of deflection α.

Note that when $\theta = \dfrac{\pi}{2}$, the formula becomes that of a right circular cone (if $\theta \neq \dfrac{\pi}{2}$, then h is called the *slant height or lateral height* of the cone). As part of a sculpture exhibit, an artist is constructing three such structures each with a radius of 2 m and a slant height of 3 m. (a) Find the volume of the sculptures if the angle of deflection is $\alpha = 15°$. (b) What angle θ was used if the volume of each sculpture is 12 m^3?

Exercise 44

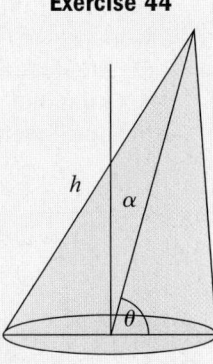

45. **River discharge rate:** For June through February, the discharge rate of the La Corcovada River (Venezuela) can be modeled by the function

$$D(t) = 36 \sin\left(\frac{\pi}{4}t - \frac{9}{4}\right) + 44,$$ where t represents

the months of the year with $t = 1$ corresponding to June, and $D(t)$ is the discharge rate in cubic meters per second. (a) What is the discharge rate in mid-September? (b) For what months of the year is the discharge rate over 50 m^3/sec?

Source: Global River Discharge Database Project; www.rivdis.sr.unh.edu.

46. **River discharge rate:** For February through June, the average monthly discharge of the Point Wolfe River (Canada) can be modeled by the function

$$D(t) = 4.6 \sin\left(\frac{\pi}{2}t + 3\right) + 7.4,$$ where t represents

the months of the year with $t = 1$ corresponding to February, and $D(t)$ is the discharge rate in cubic meters/second. (a) What is the discharge rate in mid-March ($t = 2.5$)? (b) For what months of the year is the discharge rate less than 7.5 m^3/sec?

Source: Global River Discharge Database Project; www.rivdis.sr.unh.edu.

47. **Seasonal sales:** Hank's Heating Oil is a very seasonal enterprise, with sales in the winter far exceeding sales in the summer. Monthly sales for the company can be modeled by

$$S(x) = 1600 \cos\left(\frac{\pi}{6}x - \frac{\pi}{12}\right) + 5100,$$ where $S(x)$

is the average sales in month x ($x = 1 \rightarrow$ January). (a) What is the average sales amount for July? (b) For what months of the year are sales less than \$4000?

48. **Seasonal income:** As a roofing company employee, Mark's income fluctuates with the seasons and the availability of work. For the past several years his average monthly income could be approximated by

the function $$I(m) = 2100 \sin\left(\frac{\pi}{6}m - \frac{\pi}{2}\right) + 3520,$$

where $I(m)$ represents income in month m ($m = 1 \rightarrow$ January). (a) What is Mark's average

monthly income in October? (b) For what months of the year is his average monthly income over \$4500?

49. **Seasonal ice thickness:** The average thickness of the ice covering an arctic lake can be modeled by

the function $$T(x) = 9 \cos\left(\frac{\pi}{6}x\right) + 15,$$ where $T(x)$

is the average thickness in month x ($x = 1 \rightarrow$ January). (a) How thick is the ice in mid-March? (b) For what months of the year is the ice at most 10.5 in. thick?

50. **Seasonal temperatures:** The function

$$T(x) = 19 \sin\left(\frac{\pi}{6}x - \frac{\pi}{2}\right) + 53$$ models the average

monthly temperature of the water in a mountain stream, where $T(x)$ is the temperature (°F) of the water in month x ($x = 1 \rightarrow$ January). (a) What is the temperature of the water in October? (b) What two months are most likely to give a temperature reading of 62°F? (c) For what months of the year is the temperature below 50°F?

51. **Coffee sales:** Coffee sales fluctuate with the weather, with a great deal more coffee sold in the winter than in the summer. For Joe's Diner, assume

the function $$G(x) = 21 \cos\left(\frac{2\pi}{365}x + \frac{\pi}{2}\right) + 29$$

models daily coffee sales (for non-leap years), where $G(x)$ is the number of gallons sold and x represents the days of the year ($x = 1 \rightarrow$ January 1). (a) How many gallons are projected to be sold on March 21? (b) For what days of the year are more than 40 gal of coffee sold?

52. **Park attendance:** Attendance at a popular state park varies with the weather, with a great deal more visitors coming in during the summer months. Assume daily attendance at the park can be modeled

by the function $$V(x) = 437 \cos\left(\frac{2\pi}{365}x - \pi\right) + 545$$

(for non-leap years), where $V(x)$ gives the number of visitors on day x ($x = 1 \rightarrow$ January 1). (a) Approximately how many people visited the park on November 1 ($11 \times 30.5 = 335.5$)? (b) For what days of the year are there more than 900 visitors?

53. Exercise routine: As part of his yearly physical, Manu Tuiosamoa's heart rate is closely monitored during a 12-min, cardiovascular exercise routine. His heart rate in beats per minute (bpm) is modeled by the function $B(x) = 58 \cos\left(\dfrac{\pi}{6}x + \pi\right) + 126$

where x represents the duration of the workout in minutes. (a) What was his resting heart rate? (b) What was his heart rate 5 min into the workout? (c) At what times during the workout was his heart rate over 170 bpm?

54. Exercise routine: As part of her workout routine, Sara Lee programs her treadmill to begin at a slight initial grade (angle of incline), gradually increase to a maximum grade, then gradually decrease back to the original grade. For the duration of her workout, the grade is modeled by the function

$G(x) = 3 \cos\left(\dfrac{\pi}{5}x - \pi\right) + 4$, where $G(x)$ is the

percent grade x minutes after the workout has

begun. (a) What is the initial grade for her workout? (b) What is the grade at $x = 4$ min? (c) At $G(x) = 4.9\%$, how long has she been working out? (d) What is the duration of the treadmill workout?

▶ EXTENDING THE CONCEPT

 55. As we saw in Chapter 6, cosine is the cofunction of sine and each can be expressed in terms of the other: $\cos\left(\dfrac{\pi}{2} - \theta\right) = \sin\theta$ and $\sin\left(\dfrac{\pi}{2} - \theta\right) = \cos\theta$.

This implies that either function can be used to model the phenomenon described in this section by adjusting the phase shift. By experimentation, (a) find a model using cosine that will produce results identical to the sine function in Exercise 50 and (b) find a model using sine that will produce results identical to the cosine function in Exercise 51.

56. Use multiple identities to find all real solutions for the equation given: $\sin(5x) + \sin(2x)\cos x + \cos(2x)\sin x = 0$.

57. A rectangular parallelepiped with square ends has 12 edges and six surfaces. If the sum of all edges is 176 cm and the total surface area is 1288 cm², find (a) the length of the diagonal of the parallelepiped (shown in bold) and (b) the angle the diagonal makes with the base (two answers are possible).

Exercise 57

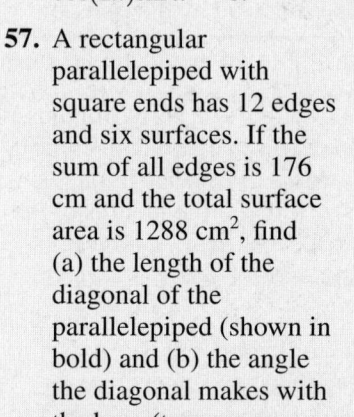

▶ MAINTAINING YOUR SKILLS

58. (5.7) Find $f(\theta)$ for all six trig functions, given $P(-51, 68)$ is on the terminal side.

59. (3.4) Sketch the graph of f by locating its zeroes and using end behavior: $f(x) = x^4 - 3x^3 + 4x$.

60. (4.3) Use a calculator and the change-of-base formula to find the value of $\log_5 279$.

61. (5.6) The Sears Tower in Chicago, Illinois, remains one of the tallest structures in the world. The top of the roof reaches 1450 ft above the street below and the antenna extends an additional 280 ft

into the air. Find the viewing angle θ for the antenna from a distance of 1000 ft (the angle formed from the base of the antenna to its top).

Exercise 61

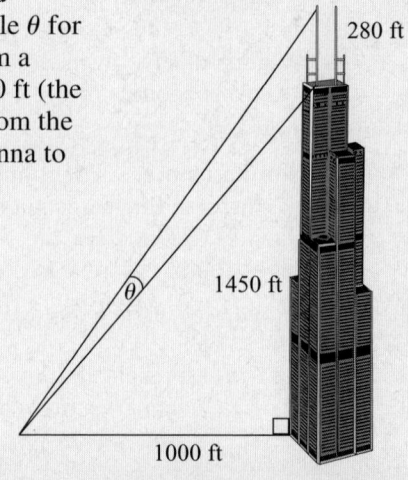

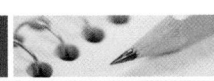

SUMMARY AND CONCEPT REVIEW

SECTION 6.1   Fundamental Identities and Families of Identities

KEY CONCEPTS

- The fundamental identities include the *reciprocal, ratio, and Pythagorean identities.*
- A given identity can algebraically be rewritten to obtain other identities in an identity "family."
- Standard algebraic skills like distribution, factoring, combining terms, and special products play an important role in working with identities.
- The pattern $\dfrac{A}{B} \pm \dfrac{C}{D} = \dfrac{AD \pm BC}{BD}$ gives an efficient method for combining rational terms.
- Using fundamental identities, a given trig function can be expressed in terms of any other trig function.
- Once the value of a given trig function is known, the value of the other five can be uniquely determined using fundamental identities, *if the quadrant of the terminal side is known.*
- To show an equation is not an identity, find any one value where the expressions are defined but the equation is false, or graph both functions on a calculator to see if the graphs are identical.

EXERCISES

Verify using the method specified and fundamental identities.

1. multiplication

$\sin x(\csc x - \sin x) = \cos^2 x$

2. factoring

$\dfrac{\tan^2 x \csc x + \csc x}{\sec^2 x} = \csc x$

3. special products

$\dfrac{(\sec x - \tan x)(\sec x + \tan x)}{\csc x} = \sin x$

4. combine terms using

$\dfrac{A}{B} \pm \dfrac{C}{D} = \dfrac{AD \pm BC}{BD}$

$\dfrac{\sec^2 x}{\csc x} - \sin x = \dfrac{\tan^2 x}{\csc x}$

Find the value of all six trigonometric functions using the information given.

5. $\cos \theta = -\dfrac{12}{37}$; θ in QIII

6. $\sec \theta = \dfrac{25}{23}$; θ in QIV

SECTION 6.2   Constructing and Verifying Identities

KEY CONCEPTS

- The steps used to verify an identity must be reversible.
- If two expressions are equal, one may be substituted for the other and the result will be equivalent.
- To verify an identity we mold, change, substitute, and rewrite one side until we "match" the other side.
- Verifying identities often involves a combination of algebraic skills with the fundamental trig identities.
  A collection and summary of the *Guidelines for Verifying Identities* can be found on page 553.

EXERCISES

Rewrite each expression to create a new identity, then verify the identity by reversing the steps.

7. $\csc x + \cot x$

8. $\dfrac{\cos x - \sin x \cos x}{\cos^2 x}$

Verify that each equation is an identity.

9. $\dfrac{\csc^2 x(1 - \cos^2 x)}{\tan^2 x} = \cot^2 x$

10. $\dfrac{\cot x}{\sec x} - \dfrac{\csc x}{\tan x} = \cot x(\cos x - \csc x)$

11. $\dfrac{\sin^4 x - \cos^4 x}{\sin x \cos x} = \tan x - \cot x$

12. $\dfrac{(\sin x + \cos x)^2}{\sin x \cos x} = \csc x \sec x + 2$

SECTION 6.3   The Sum and Difference Identities

KEY CONCEPTS

The sum and difference identities can be used to
- Find exact values for nonstandard angles that are a sum or difference of two standard angles.
- Verify the cofunction identities and to rewrite a given function in terms of its cofunction.
- Find coterminal angles in $[0, 360°)$ for very large angles (the angle reduction formulas).
- Evaluate the difference quotient for $\sin x$, $\cos x$, and $\tan x$.
- Rewrite a sum as a single expression: $\cos \alpha \cos \beta + \sin \alpha \sin \beta = \cos(\alpha - \beta)$.

The sum and difference identities for sine and cosine can be remembered by noting
- For $\cos(\alpha \pm \beta)$, the function repeats and the signs alternate: $\cos(\alpha \pm \beta) = \cos \alpha \cos \beta \mp \sin \alpha \sin \beta$
- For $\sin(\alpha \pm \beta)$ the signs repeat and the functions alternate: $\sin(\alpha \pm \beta) = \sin \alpha \cos \beta \pm \cos \alpha \sin \beta$

EXERCISES

Find exact values for the following expressions using sum and difference formulas.

13. a. $\cos 75°$ **b.** $\tan\left(\dfrac{\pi}{12}\right)$ **14. a.** $\tan 15°$ **b.** $\sin\left(-\dfrac{\pi}{12}\right)$

Evaluate exactly using sum and difference formulas.

15. a. $\cos 109° \cos 71° - \sin 109° \sin 71°$ **b.** $\sin 139° \cos 19° - \cos 139° \sin 19°$

Rewrite as a single expression using sum and difference formulas.

16. a. $\cos(3x)\cos(-2x) - \sin(3x)\sin(-2x)$ **b.** $\sin\left(\dfrac{x}{4}\right)\cos\left(\dfrac{3x}{8}\right) + \cos\left(\dfrac{x}{4}\right)\sin\left(\dfrac{3x}{8}\right)$

Evaluate exactly using sum and difference formulas, by reducing the angle to an angle in $[0, 360°)$ or $[0, 2\pi)$.

17. a. $\cos 1170°$ **b.** $\sin\left(\dfrac{57\pi}{4}\right)$

Use a cofunction identity to write an equivalent expression for the one given.

18. a. $\cos\left(\dfrac{x}{8}\right)$ **b.** $\sin\left(x - \dfrac{\pi}{12}\right)$

19. Verify that both expressions yield the same result using sum and difference formulas. $\tan 15° = \tan(45° - 30°)$ and $\tan 15° = \tan(135° - 120°)$.

20. Use sum and difference formulas to verify the following identity.
$$\cos\left(x + \dfrac{\pi}{6}\right) + \cos\left(x - \dfrac{\pi}{6}\right) = \sqrt{3} \cos x$$

SECTION 6.4   The Double-Angle, Half-Angle, and Product-to-Sum Identities

KEY CONCEPTS

- When multiple angle identities (identities involving $n\theta$) are used to find exact values, the terminal side of θ must be determined so the appropriate sign can be used.
- The power reduction identities for $\cos^2 x$ and $\sin^2 x$ are closely related to the double-angle identities, and can be derived directly from $\cos(2x) = 2 \cos^2 x - 1$ and $\cos(2x) = 1 - 2 \sin^2 x$.

- The half-angle identities can be developed from the power reduction identities by using a change of variable and taking square roots. The sign is then chosen based on the quadrant of the half angle.
- The product-to-sum and sum-to-product identities can be derived using the sum and difference formulas, and have important applications in many areas of science.

EXERCISES

Find exact values for $\sin(2\theta)$, $\cos(2\theta)$, and $\tan(2\theta)$ using the information given.

21. a. $\cos\theta = \dfrac{13}{85}$; θ in QIV

b. $\csc\theta = -\dfrac{29}{20}$; θ in QIII

Find exact values for $\sin\theta$, $\cos\theta$, and $\tan\theta$ using the information given.

22. a. $\cos(2\theta) = -\dfrac{41}{841}$; θ in QII

b. $\sin(2\theta) = -\dfrac{336}{625}$; θ in QII

Find exact values using the appropriate double-angle identity.

23. a. $\cos^2 22.5° - \sin^2 22.5°$

b. $1 - 2\sin^2\left(\dfrac{\pi}{12}\right)$

Find exact values for $\sin\theta$ and $\cos\theta$ using the appropriate half-angle identity.

24. a. $\theta = 67.5°$

b. $\theta = \dfrac{5\pi}{8}$

Find exact values for $\sin\left(\dfrac{\theta}{2}\right)$ and $\cos\left(\dfrac{\theta}{2}\right)$ using the given information.

25. a. $\cos\theta = \dfrac{24}{25}$; $0° < \theta < 360°$; θ in QIV

b. $\csc\theta = -\dfrac{65}{33}$; $-90° < \theta < 0$; θ in QIV

26. Verify the equation is an identity.

$$\dfrac{\cos(3\alpha) - \cos\alpha}{\cos(3\alpha) + \cos\alpha} = \dfrac{2\tan^2\alpha}{\sec^2\alpha - 2}$$

27. Solve using a sum-to-product formula.

$$\cos(3x) + \cos x = 0$$

28. The area of an isosceles triangle (two equal sides) is given by the formula $A = x^2\sin\left(\dfrac{\theta}{2}\right)\cos\left(\dfrac{\theta}{2}\right)$, where the equal sides have length x and the vertex angle measures $\theta°$. (a) Use this formula and the half-angle identities to find the area of an isosceles triangle with vertex angle $\theta = 30°$ and equal sides of 12 cm. (b) Use substitution and a double-angle identity to verify that $x^2\sin\left(\dfrac{\theta}{2}\right)\cos\left(\dfrac{\theta}{2}\right) = \dfrac{1}{2}x^2\sin\theta$, then recompute the triangle's area. Do the results match?

SECTION 6.5   The Inverse Trig Functions and Their Applications

KEY CONCEPTS

- In order to create one-to-one functions, the domains of $y = \sin t$, $y = \cos t$, and $y = \tan t$ are restricted as follows:

(a) $y = \sin t$, $t \in \left[-\dfrac{\pi}{2}, \dfrac{\pi}{2}\right]$; (b) $y = \cos t$, $t \in [0, \pi]$; and (c) $y = \tan t$; $t \in \left(-\dfrac{\pi}{2}, \dfrac{\pi}{2}\right)$.

- For $y = \sin x$, the inverse function is given implicitly as $x = \sin y$ and explicitly as $y = \sin^{-1} x$ or $y = \arcsin x$.
- The expression $y = \sin^{-1} x$ is read, "y is the angle or real number whose sine is x." The other inverse functions are similarly read/understood.
- For $y = \cos x$, the inverse function is given implicitly as $x = \cos y$ and explicitly as $y = \cos^{-1} x$ or $y = \arccos x$.
- For $y = \tan x$, the inverse function is given implicitly as $x = \tan y$ and explicitly as $y = \tan^{-1} x$ or $y = \arctan x$.

- The domains of $y = \sec x$, $y = \csc x$, and $y = \cot x$ are likewise restricted to create one-to-one functions:
  (a) $y = \sec t;\ t \in \left[0, \dfrac{\pi}{2}\right) \cup \left(\dfrac{\pi}{2}, \pi\right]$; (b) $y = \csc t,\ t \in \left[-\dfrac{\pi}{2}, 0\right) \cup \left(0, \dfrac{\pi}{2}\right]$; and (c) $y = \cot t,\ t \in (0, \pi)$.

- In some applications, inverse functions occur in a composition with other trig functions, with the expression best evaluated by drawing a diagram using the ratio definition of the trig functions.

- To evaluate $y = \sec^{-1}t$, we use $y = \cos^{-1}\left(\dfrac{1}{t}\right)$; for $y = \cot^{-1}t$, use $\tan^{-1}\left(\dfrac{1}{t}\right)$; and so on.

- Trigonometric substitutions can be used to simplify certain algebraic expressions.

EXERCISES

Evaluate without the aid of calculators or tables. State answers in both radians and degrees in exact form.

29. $y = \sin^{-1}\left(\dfrac{\sqrt{2}}{2}\right)$

30. $y = \csc^{-1}2$

31. $y = \arccos\left(-\dfrac{\sqrt{3}}{2}\right)$

 Evaluate the following using a calculator, *keeping the domain and range of each function in mind.* Answer in radians to the nearest ten-thousandth *and* in degrees to the nearest tenth. Some may be undefined.

32. $y = \tan^{-1}4.3165$

33. $y = \sin^{-1}0.8892$

34. $f(x) = \arccos\left(\dfrac{7}{8}\right)$

Evaluate the following without the aid of a calculator. Some may be undefined.

35. $\sin\left[\sin^{-1}\left(\dfrac{1}{2}\right)\right]$

36. $\operatorname{arcsec}\left[\sec\left(\dfrac{\pi}{4}\right)\right]$

37. $\cos(\cos^{-1}2)$

Evaluate the following using a calculator. Some may be undefined.

38. $\sin^{-1}(\sin 1.0245)$

39. $\arccos[\cos(-60°)]$

40. $\cot^{-1}\left[\cot\left(\dfrac{11\pi}{4}\right)\right]$

Evaluate each expression by drawing a right triangle and labeling the sides.

41. $\sin\left[\cos^{-1}\left(\dfrac{12}{37}\right)\right]$

42. $\tan\left[\operatorname{arcsec}\left(\dfrac{7}{3x}\right)\right]$

43. $\cot\left[\sin^{-1}\left(\dfrac{x}{\sqrt{81 + x^2}}\right)\right]$

Use an inverse function to solve the following equations for θ in terms of x.

44. $x = 5\cos\theta$

45. $7\sqrt{3}\sec\theta = x$

46. $x = 4\sin\left(\theta - \dfrac{\pi}{6}\right)$

SECTION 6.6 Solving Basic Trig Equations

KEY CONCEPTS

- When solving trig equations, we often consider either the principal root, roots in $[0, 2\pi)$, or all real roots.
- Keeping the graph of each function in mind helps to determine the desired solution set.
- After isolating the trigonometric term containing the variable, we solve by applying the appropriate inverse function, realizing the result is only the principal root.
- Once the principal root is found, roots in $[0, 2\pi)$ or all real roots can be found using reference angles and the period of the function under consideration.
- Trig identities can be used to obtain an equation that can be solved by factoring or other solution methods.

EXERCISES

Solve each equation without the aid of a calculator (all solutions are standard values). Clearly state (a) the principal root; (b) all solutions in the interval $[0, 2\pi)$; and (c) all real roots.

47. $2\sin x = \sqrt{2}$

48. $3\sec x = -6$

49. $8\tan x + 7\sqrt{3} = -\sqrt{3}$

Solve using a calculator and the inverse trig functions (not by graphing). Clearly state (a) the principal root; (b) solutions in $[0, 2\pi)$; and (c) all real roots. Answer in radians to the nearest ten-thousandth as needed.

50. $9 \cos x = 4$ **51.** $\dfrac{2}{5} \sin(2\theta) = \dfrac{1}{4}$ **52.** $\sqrt{2} \csc x + 3 = 7$

 53. The area of a circular segment (the shaded portion shown in the diagram) is given by the

formula $A = \dfrac{1}{2}r^2(\theta - \sin \theta)$, where θ is in radians. If the circle has a radius of 10 cm, find

the angle θ that gives an area of 12 cm^2.

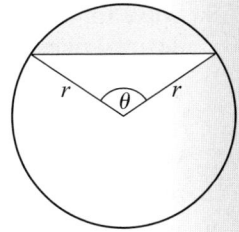

SECTION 6.7   General Trig Equations and Applications

KEY CONCEPTS

- In addition to the basic solution methods from Section 6.6, additional strategies include squaring both sides, factoring by grouping, and using the full range of identities to simplify an equation.
- Many applications result in equations of the form $A\sin(Bx + C) + D = k$. To solve, isolate the factor $\sin(Bx + C)$ (subtract D and divide by A), then apply the inverse function.
- Once the principal root is found, roots in $[0, 2\pi)$ or all real roots can be found using reference angles and the period of the function under consideration.

EXERCISES

Find solutions in $[0, 2\pi)$ using the method indicated. Round nonstandard values to four decimal places.

54. squaring both sides

$\sin x + \cos x = \dfrac{\sqrt{6}}{2}$

55. using identities

$3 \cos(2x) + 7 \sin x - 5 = 0$

56. factor by grouping

$4 \sin x \cos x - 2\sqrt{3} \sin x - 2 \cos x + \sqrt{3} = 0$

57. using any appropriate method

$\csc x + \cot x = 1$

State the period P of each function and find all solutions in $[0, P)$. Round to four decimal places as needed.

58. $-750 \sin\left(\dfrac{\pi}{6}x + \dfrac{\pi}{2}\right) + 120 = 0$ **59.** $80 \cos\left(\dfrac{\pi}{3}x + \dfrac{\pi}{4}\right) - 40\sqrt{2} = 0$

60. The revenue earned by Waipahu Joe's Tanning Lotions fluctuates with the seasons, with a great deal more lotion sold in the summer than in the winter. The function $R(x) = 15 \sin\left(\dfrac{\pi}{6}x - \dfrac{\pi}{2}\right) + 30$ models the monthly sales of lotion nationwide, where $R(x)$ is the revenue in thousands of dollars and x represents the months of the year ($x = 1 \rightarrow$ Jan). (a) How much revenue is projected for July? (b) For what months of the year does revenue exceed $37,000?

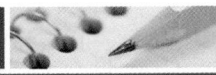 **MIXED REVIEW**

Find the value of all six trig functions using the information given.

Find the exact value of each expression using a sum or difference identity.

1. $\csc \theta = \dfrac{\sqrt{117}}{6}$; θ in QII   **2.** $\tan^{-1}\left(\dfrac{4}{3}\right) = \theta$

3. $\tan 255°$ **4.** $\cos\left(\dfrac{19\pi}{12}\right)$

Evaluate each expression by drawing a right triangle and labeling the sides appropriately.

5. $\tan\left[\operatorname{arccsc}\left(\dfrac{10}{x}\right)\right]$ **6.** $\sin\left[\sec^{-1}\left(\dfrac{\sqrt{64+x^2}}{x}\right)\right]$

7. Solve for x in the interval $[0, 2\pi)$. Round to four decimal places as needed:

$$-100\sin\left(\dfrac{\pi}{4}x - \dfrac{\pi}{6}\right) + 80 = 100$$

8. Without the aid of a calculator, find: (a) the principal roots, (b) all solutions in $[0, 2\pi)$ and (c) all real solutions: $(\cos x - 1)[2\cos^2(x) - 1] = 0$

9. The horizontal distance R that an object will travel when it is projected at angle θ with initial velocity v is given by the equation $R = \dfrac{1}{16}v^2 \sin\theta \cos\theta$.

   a. Use an identity to show this equation can be

     written as $R = \dfrac{1}{32}v^2 \sin(2\theta)$.

   b. Use this equation to show why the horizontal distance traveled by the object is the same for any two complementary angles.

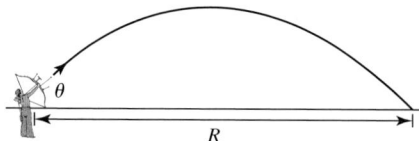

10. The profits of Red-Bud Nursery can be modeled by a sinusoid, with profit peaking twice each year. Given profits reach a yearly low of \$4000 in mid-January (month 1.5), and a yearly high of \$14,000 in mid-April (month 4.5). (a) Construct an equation for their yearly profits. (b) Use the model to find their profits for August. (c) Name the other month at which profit peaks.

11. Find the exact value of $2\cos^2\left(\dfrac{\pi}{12}\right) - 1$ using an

appropriate identity.

Verify the following identity.

12. $\dfrac{1 - \cos^2\theta + \sin^2\theta}{\tan^2\theta} = 1 + \cos(2\theta)$

13. $\dfrac{(\cos t + \sin t)^2}{\tan t} = \cot t + 2\cos^2 t$

14. Find exact values for $\sin\left(\dfrac{x}{2}\right)$ and $\cos\left(\dfrac{x}{2}\right)$ using the

information given.

   a. $\sin x = \dfrac{-6}{7.5}$; $540° < x < 630°$

   b. $\sec x = \dfrac{11.7}{4.5}$; $0 < x < \dfrac{\pi}{2}$

Evaluate without the aid of calculator or tables. Answer in both radians and degrees.

15. $y = \operatorname{arcsec}(-\sqrt{2})$ **16.** $y = \sin^{-1}0$

17. $y = \tan^{-1}\sqrt{3}$

18. Verify the following identities *using a sum formula*.

   a. $\sin(2x) = 2\sin x \cos x$

   b. $\cos(2x) = \cos^2 x - \sin^2 x$

Use an inverse function to solve each equation for θ in terms of x.

19. $\dfrac{x}{10} = \tan\theta$ **20.** $2\sqrt{2}\csc\left(\theta - \dfrac{\pi}{4}\right) = x$

21. On a large clock, the distance from the tip of the hour hand to the base of the "12" can be approxi-

mated by the function $D(t) = \left| 8\sin\left(\dfrac{\pi t}{12}\right)\right| + 2,$

where $D(t)$ is this distance in feet at time t in hours. Use this function to approximate (a) the time of day when the hand is 6 ft from the 12 and 10 ft from the 12 and (b) the distance between the tip and the 12 at 4:00. Check your answer graphically.

22. The figure shows a smaller pentagon inscribed within a larger pentagon. Find the measure of angle θ using the diagram and equation given:
$3.2^2 = 11^2 + 9.4^2 - 2(11)(9.4)\cos\theta$

23. Find the value of each expression using sum-to-product and half-angle identities (without using a calculator).

   a. $\sin 172.5° - \sin 52.5°$

   b. $\cos 172.5° + \cos 52.5°$

24. Given $100\sin t = 70$, use a calculator to find (a) the principal root, (b) all solutions in $[0, 2\pi]$, and (c) all real solutions. Round to the nearest ten-thousandth.

25. Use the product-to-sum formulas to find the exact value of

   a. $\sin\left(\dfrac{13\pi}{24}\right)\cos\left(\dfrac{7\pi}{24}\right)$ **b.** $\sin\left(\dfrac{13\pi}{24}\right)\sin\left(\dfrac{7\pi}{24}\right)$

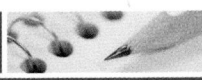

PRACTICE TEST

Verify each identity using fundamental identities and the method specified.

1. special products
$$\frac{(\csc x - \cot x)(\csc x + \cot x)}{\sec x} = \cos x$$

2. factoring $\dfrac{\sin^3 x - \cos^3 x}{1 + \cos x \sin x} = \sin x - \cos x$

3. Find the value of all six trigonometric functions given $\cos \theta = \dfrac{48}{73}$; θ in QIV

4. Find the exact value of $\tan 15°$ using a sum or difference formula.

5. Rewrite as a single expression and evaluate: $\cos 81° \cos 36° + \sin 81° \sin 36°$

6. Evaluate $\cos 1935°$ exactly using an angle reduction formula.

7. Use sum and difference formulas to verify
$$\sin\left(x + \frac{\pi}{4}\right) - \sin\left(x - \frac{\pi}{4}\right) = \sqrt{2} \cos x.$$

8. Find exact values for $\sin \theta$, $\cos \theta$, and $\tan \theta$ given
$$\cos(2\theta) = -\frac{161}{289}; \theta \text{ in QI}$$

9. Use a double-angle identity to evaluate $2\cos^2 75° - 1$.

10. Find exact values for $\sin\left(\dfrac{\theta}{2}\right)$ and $\cos\left(\dfrac{\theta}{2}\right)$ given $\tan \theta = \dfrac{12}{35}$; θ in QI

11. The area of a triangle is given geometrically as $A = \dfrac{1}{2}$ base $\cdot$ height. The trigonometric formula for

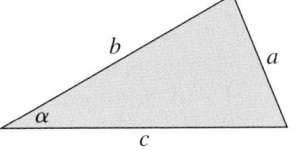

the triangle's area is $A = \dfrac{1}{2}bc \sin \alpha$, where α is the angle formed by the sides b and c. In a certain triangle, $b = 8$, $c = 10$, and $\alpha = 22.5°$. Use the formula for A given here and a half-angle identity to find the area of the triangle in exact form.

12. The equation $Ax^2 + Bxy + Cy^2 = 0$ can be written in an alternative form that makes it easier to graph. This is done by eliminating the mixed xy-term using the relation $\tan(2\theta) = \dfrac{B}{A - C}$ to find θ. We can then

find values for $\sin \theta$ and $\cos \theta$, which are used in a conversion formula. Find $\sin \theta$ and $\cos \theta$ for $17x^2 + 5\sqrt{3}xy + 2y^2 = 0$, assuming 2θ in QI.

13. Evaluate without the aid of calculators or tables.
  a. $y = \tan^{-1}\left(\dfrac{1}{\sqrt{3}}\right)$ **b.** $y = \sin\left[\sin^{-1}\left(\dfrac{1}{2}\right)\right]$
  c. $y = \arccos(\cos 30°)$

14. Evaluate the following. Use a calculator for part (a), give exact answers for part (b), and find the value of the expression in part (c) without using a calculator. Some may be undefined.
  a. $y = \sin^{-1} 0.7528$ **b.** $y = \arctan(\tan 78.5°)$
  c. $y = \sec^{-1}\left[\sec\left(\dfrac{7\pi}{24}\right)\right]$

Evaluate the expressions by drawing a right triangle and labeling the sides.

15. $\cos\left[\tan^{-1}\left(\dfrac{56}{33}\right)\right]$

16. $\cot\left[\cos^{-1}\left(\dfrac{x}{\sqrt{25 + x^2}}\right)\right]$

17. Solve without the aid of a calculator (all solutions are standard values). Clearly state (a) the principal root, (b) all solutions in the interval $[0, 2\pi)$, and (c) all real roots.
  I. $8 \cos x = -4\sqrt{2}$ **II.** $\sqrt{3} \sec x + 2 = 4$

18. Solve each equation using a calculator and inverse trig functions to find the principal root (not by graphing). Then state (a) the principal root, (b) all solutions in the interval $[0, 2\pi)$, and (c) all real roots.
  I. $\dfrac{2}{3}\sin(2x) = \dfrac{1}{4}$ **II.** $-3\cos(2x) - 0.8 = 0$

19. Solve the equations graphically in the indicated interval using a graphing calculator. State answers in radians rounded to the nearest ten-thousandth.
  a. $3\cos(2x - 1) = \sin x; x \in [-\pi, \pi]$
  b. $2\sqrt{x} - 1 = 3\cos^2 x; x \in [0, 2\pi)$

20. Solve the following equations for $x \in [0, 2\pi)$ using a combination of identities and/or factoring. State solutions in radians using the exact form where possible.
  a. $2\sin x \sin(2x) + \sin(2x) = 0$
  b. $(\cos x + \sin x)^2 = \dfrac{1}{2}$

Solve each equation in $[0, 2\pi)$ by squaring both sides, factoring, using identities or by using any appropriate method. Round nonstandard values to four decimal places.

21. $3\sin(2x) + \cos x = 0$

22. $\dfrac{2}{3}\sin\left(2x - \dfrac{\pi}{6}\right) + \dfrac{3}{2} = \dfrac{5}{6}$

23. The revenue for Otake's Mower Repair is very seasonal, with business in the summer months far exceeding business in the winter months. Monthly revenue for the company can be modeled by the function $R(x) = 7.5\cos\left(\dfrac{\pi}{6}x + \dfrac{4\pi}{3}\right) + 12.5$, where $R(x)$ is the average revenue (in thousands of dollars) for month x ($x = 1 \rightarrow$ Jan). (a) What is the average revenue for September? (b) For what months of the year is revenue at least $12,500?

24. The lowest temperature on record for the even months of the year are given in the table for the city of Denver, Colorado. The equation $y = 35.223\sin(0.576x - 2.589) + 6$ is a fairly accurate model for this data. Use the equation to estimate the record low temperature for the odd numbered months.

Month (Jan → 1)	Low Temp. (°F)
2	−30
4	−2
6	30
8	41
10	3
12	−25

Source: 2004 Statistical Abstract of the United States, Table 379.

25. Write the product as a sum using a product-to-sum identity: $2\cos(1979\pi t)\cos(439\pi t)$.

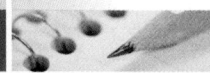

CALCULATOR EXPLORATION AND DISCOVERY

Seeing the Beats as the Beats Go On

When two sound waves of slightly different frequencies are combined, the resultant wave varies periodically in amplitude over time. These amplitude pulsations are called **beats**. In this *Exploration and Discovery*, we'll look at ways to "see" the beats more clearly on a graphing calculator, by representing sound waves very simplistically as $Y_1 = \cos(mt)$ and $Y_2 = \cos(nt)$ and noting a relationship between m, n, and the number of beats in $[0, 2\pi]$. Using a sum-to-product formula, we can represent the resultant wave as a single term. For $Y_1 = \cos(12t)$ and $Y_2 = \cos(8t)$ the result is

Figure 6.33

WINDOW
Xmin=0
Xmax=6.2831853...
Xscl=1.5707963...
Ymin=-3
Ymax=3
Yscl=1
Xres=1

Figure 6.34

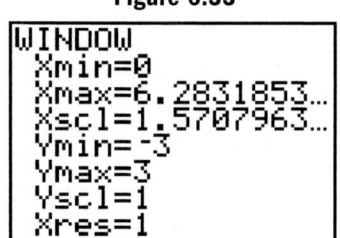

$$\cos(12t) + \cos(8t) = 2\cos\left(\frac{12t + 8t}{2}\right)\cos\left(\frac{12t - 8t}{2}\right)$$

$$= 2\cos(10t)\cos(2t)$$

The window used and resulting graph are shown in Figures 6.33 and 6.34, and it appears that "silence" occurs four times in this interval— where the graph of the combined waves is tangent to (bounces off of) the x-axis. This indicates a total of four beats. Note the number of beats is equal to the difference $m - n$: $12 - 8 = 4$. Further experimentation will show this is not a coincidence, and this enables us to construct two additional functions that will *frame these pulsations* and make them easier to see. Since the maximum amplitude of the resulting wave is 2, we use functions of the form $\pm 2\cos\left(\dfrac{k}{2}x\right)$ to construct the frame, where k is the number of beats in the interval ($m - n = k$). For $Y_1 = \cos(12t)$ and $Y_2 = \cos(8t)$, we have $k = \dfrac{12 - 8}{2} = 2$ and the functions we use will be $Y_2 = 2\cos(2x)$ and $Y_3 = -2\cos(2x)$ as shown in Figure 6.35. The result is shown in Figure 6.36, where the

Figure 6.35

Plot1 Plot2 Plot3
\Y1◻2cos(10X)cos(2X)
\Y2◻2cos(2X)
\Y3◻-2cos(2X)
\Y4=
\Y5=
\Y6=

Figure 6.36

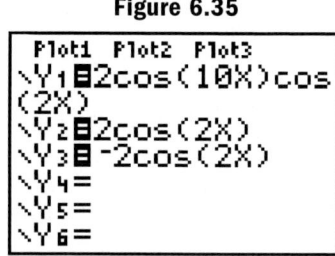

frame clearly shows the four beats or more precisely, the four moments of silence.

For each exercise, (a) express the sum $Y_1 + Y_2$ as a product, (b) graph Y_R on a graphing calculator for $x \in [0, 2\pi]$ and identify the number of beats in this interval, and (c) determine what value of k in $\pm 2 \cos\left(\dfrac{k}{2}x\right)$

would be used to frame the resultant Y_R, then enter these as Y_2 and Y_3 to check the result.

Exercise 1: $Y_1 = \cos(14t)$; $Y_2 = \cos(8t)$
Exercise 2: $Y_1 = \cos(12t)$; $Y_2 = \cos(9t)$
Exercise 3: $Y_1 = \cos(14t)$; $Y_2 = \cos(6t)$
Exercise 4: $Y_1 = \cos(11t)$; $Y_2 = \cos(10t)$

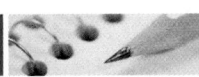

STRENGTHENING CORE SKILLS

Trigonometric Equations and Inequalities

The ability to draw a quick graph of the trigonometric functions is a tremendous help in understanding equations and inequalities. A basic sketch can help reveal the number of solutions in $[0, 2\pi)$ and the quadrant of each solution. For nonstandard angles, the value given by the inverse function can then be used as a basis for stating the solution set for all real numbers. We'll illustrate the process using a few simple examples, then generalize our observations to solve more realistic applications. Consider the function $f(x) = 2 \sin x + 1$, a sine wave with amplitude 2, and a vertical translation of $+1$. To find intervals in $[0, 2\pi)$ where $f(x) > 2.5$, we reason that f has a maximum of $3 = 2(1) + 1$ and a minimum of $-1 = 2(-1) + 1$, since $-1 \le \sin x \le 1$. With no phase shift and a standard period of 2π, we can easily draw a quick sketch of f by vertically translating x-intercepts and max/min points 1 unit up. After drawing the line $y = 2.5$ (see Figure 6.37), it appears there are two intersections in the interval, one in QI and one in QII. More importantly, it is clear that $f(x) > 2.5$ *between these two solutions.* Substituting 2.5 for $f(x)$ in $f(x) = 2 \sin x + 1$, we solve for $\sin x$ to obtain $\sin x = 0.75$, which we use to state the solution in exact form: $f(x) > 2.5$ for $x \in (\sin^{-1}0.75, \pi - \sin^{-1}0.75)$. In approximate form the solution interval is $x \in (0.85, 2.29)$.

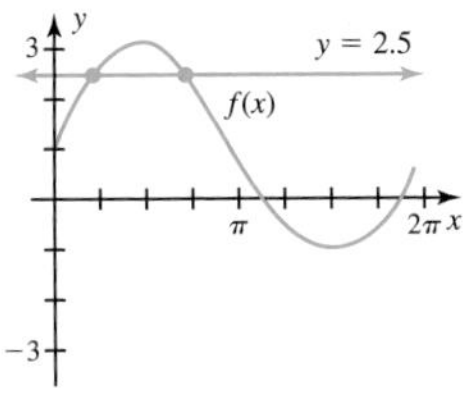

If the function involves a horizontal shift, the graphical analysis will reveal which intervals should be chosen to satisfy the given inequality.

Figure 6.37

y = 2.5
f(x)

unit down and draw a sine wave through the points (see Figure 6.38). This sketch along with the graph of $y = -1.2$ is sufficient to reveal that solutions to $g(x) = -1.2$ occur in QI and QIII, with solutions to $g(x) \le -1.2$ *outside this interval.* Substituting -1.2 for $g(x)$ and isolating the sine function we obtain $\sin\left(x - \dfrac{\pi}{4}\right) = -\dfrac{1}{15}$,

Figure 6.38

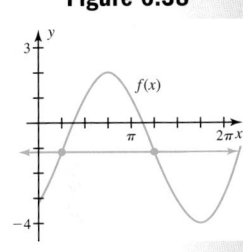

f(x)

then $x = \sin^{-1}\left(-\dfrac{1}{15}\right) + \dfrac{\pi}{4}$ after taking the inverse sine of both sides. This is the QI solution, with $x = \left[\pi - \sin^{-1}\left(-\dfrac{1}{15}\right)\right] + \dfrac{\pi}{4}$ being the solution in QIII. In approximate form the solution interval is $x \in [0, 0.72] \cup [3.99, 2\pi]$.

The basic ideas remain the same regardless of the complexity of the equation. Remember—our current goal is not a supremely accurate graph, just a sketch that will guide us to the solution using the inverse functions and the correct quadrants. Perhaps that greatest challenge is recalling that when $B \ne 1$, the horizontal shift is $-\dfrac{C}{B}$, but other than this a fairly accurate sketch can quickly be obtained.

Practice with these ideas by solving the following inequalities within the intervals specified.

Exercise 1: $f(x) = 3 \sin x + 2$; $f(x) > 3.7$; $x \in [0, 2\pi)$

Exercise 2: $g(x) = 4 \sin\left(x - \dfrac{\pi}{3}\right) - 1$;

$g(x) \le -2$; $x \in [0, 2\pi)$

Exercise 3: $h(x) = 125 \sin\left(\dfrac{\pi}{6}x - \dfrac{\pi}{2}\right) + 175$;

$h(x) \le 150$; $x \in [0, 12)$

Exercise 4: $f(x) = 15,750 \sin\left(\dfrac{2\pi}{360}x - \dfrac{\pi}{4}\right) + 19,250$;

$f(x) > 25,250$; $x \in [0, 360)$

Illustration 1 ▶ Given $g(x) = 3 \sin\left(x - \dfrac{\pi}{4}\right) - 1$, solve $g(x) \le -1.2$ for $x \in [0, 2\pi)$.

Solution ▶ Plot the x-intercepts and maximum/minimum values for a standard sine wave with amplitude 3, then shift these points $\dfrac{\pi}{4}$ units to the right. Then shift each point one

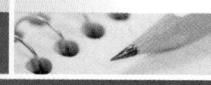

CUMULATIVE REVIEW CHAPTERS 1–6

1. Find $f(\theta)$ for all six trig functions, given $P(-13, 84)$ is on the terminal side with θ in QII.

Exercise 2

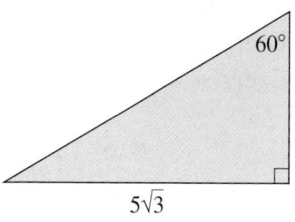

2. Find the lengths of the missing sides.

3. Verify that $x = 2 + \sqrt{3}$ is a zero of $g(x) = x^2 - 4x + 1$.

4. Determine the domain of $r(x) = \sqrt{9 - x^2}$. Answer in interval notation.

5. Standing 5 mi (26,400 ft) from the base of Mount Logan (Yukon) the angle of elevation to the summit is $36° \, 56'$. How much taller is Mount McKinley (Alaska) which stands at 20,320 ft high?

6. Use the *Guidelines for Graphing Polynomial Functions* to sketch the graph of $f(x) = x^3 + 3x^2 - 4$.

7. Use the *Guidelines for Graphing Rational Functions* to sketch the graph of $h(x) = \dfrac{x - 1}{x^2 - 4}$

8. The Petronas Towers in Malaysia are two of the tallest structures in the world. The top of the roof reaches 1483 ft above the street below and the stainless steel pinnacles extend an additional 241 ft into the air (see figure). Find the viewing angle θ for the pinnacles from a distance of 1000 ft (the angle formed from the base of the antennae to its top).

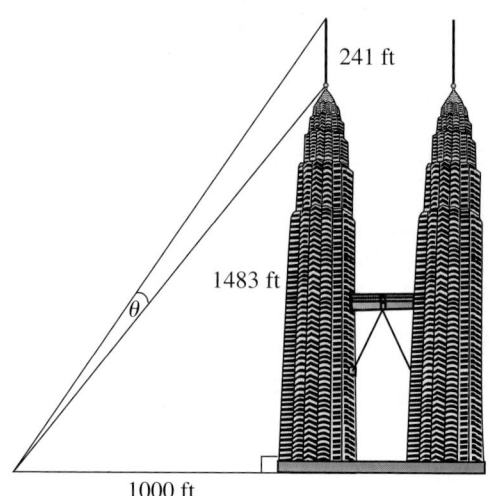

241 ft

1483 ft

θ

1000 ft

9. A wheel with radius 45 cm is turning at 5 revolutions per second. Find the linear velocity of a point on the rim in kilometers per hour, rounded to the nearest 10th of a kilometer.

10. Solve for x: $2(x - 3)^{\frac{3}{4}} + 1 = 55$.

11. Solve for x: $-3|x - \dfrac{1}{2}| + 5 \geq -10$

12. The Earth has a radius of 3960 mi. Tokyo, Japan, is located at 35.4° N latitude, very near the 139° E latitude line. Adelaide, Australia, is at 34.6° S latitude, and also very near 139° E latitude. How many miles separate the two cities?

13. Since 1970, sulphur dioxide emissions in the United States have been decreasing at a nearly linear rate. In 1970, about 31 million tons were emitted into the atmosphere. In 2000, the amount had decreased to approximately 16 million tons. (a) Find a linear equation that models sulphur dioxide emissions. (b) Discuss the meaning of the slope ratio in this context. (c) Use the equation model to estimate the emissions in 1985, and project the emission for 2010. *Source: 2004 Statistical Abstract of the United States, Table 360.*

14. List the three Pythagorean identities and three identities equivalent to $\cos(2\theta)$.

15. For $f(x) = 325 \cos\left(\dfrac{\pi}{6}x - \dfrac{\pi}{2}\right) + 168$, what values of x in $[0, 2\pi)$ satisfy $f(x) > 330.5$?

16. Write as a single logarithmic expression in simplest form: $\log(x^2 - 9) + \log(x + 1) - \log(x^2 - 2x - 3)$.

17. After doing some market research, the manager of a sporting goods store finds that when a four-pack of premium tennis balls are priced at $9 per pack, 20 packs per day are sold. For each decrease of $0.25, 1 additional pack per day will be sold. Find the price at which four-packs of tennis balls should be sold in order to maximize the store's revenue on this item.

Exercise 18

18. Write the equation of the function whose graph is given, in terms of a sine function.

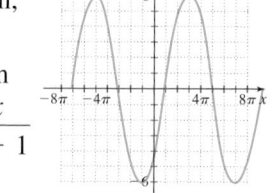

19. Verify that the following is an identity: $\dfrac{\cos x + 1}{\tan^2 x} = \dfrac{\cos x}{\sec x - 1}$

20. The graph of a function $f(x)$ is shown. Given the zeroes are $x = \pm 4$ and $x = \pm \sqrt{2}$, estimate the following:

   a. the domain and range of the function

   b. intervals where $f(x) > 0$ and $f(x) \le 0$

   c. intervals where $f(x)\downarrow$ and $f(x)\uparrow$

   d. name the location of all local maximums and minimums

21. Use the triangle shown to find the exact value of $\sin(2\theta)$.

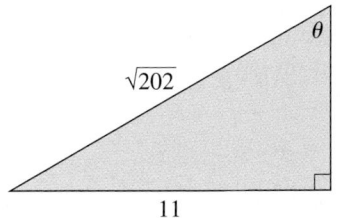

22. Use the triangle shown to find the exact value of $\sin(\alpha + \beta)$.

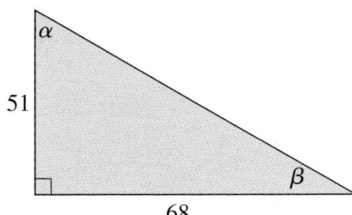

23. The amount of waste product released by a manufacturing company varies according to its production schedule, which is much heavier during the summer months and lighter in the winter. Waste product amount reaches a maximum of 32.5 tons in the month of July, and falls to a minimum of 21.7 tons in January ($t = 1$). (a) Use this information to build a sinusoidal equation that models the amount of waste produced each month. (b) During what months of the year does output exceed 30 tons?

24. At what interest rate will $2500 grow to $3500 if it's left on deposit for 6 yr and interest is compounded continuously?

25. Identify each geometric formula:

   a. $y = \pi r^2 h$ **b.** $y = LWH$

   c. $y = 2\pi r$ **d.** $y = \dfrac{1}{2}bh$

In calculus, as in college algebra, we often encounter expressions that are difficult to use in their given form, and so attempt to write the expression in an alternative form more suitable to the task at hand. Often, we see algebra and trigonometry working together to achieve this goal.

Simplifying Expressions Using a Trigonometric Substitution

For instance, it is difficult to apply certain concepts from calculus to the equation $y = \dfrac{x}{\sqrt{9 - x^2}}$, and we attempt to rewrite the expression using a trig substitution and a Pythagorean Identity. When doing so, we're careful to ensure the substitution used represents a one-to-one function, and that the substitution maintains the integrity of the domain.

EXAMPLE 1 ▶ **Simplifying Algebraic Expressions Using Trigonometry**

Simplify $y = \dfrac{x}{\sqrt{9 - x^2}}$ using the substitution $x = 3 \sin \theta$ for $-\dfrac{\pi}{2} < \theta < \dfrac{\pi}{2}$, then verify that the result is equivalent to the original function.

Solution ▶ $y = \dfrac{3 \sin \theta}{\sqrt{9 - (3 \sin \theta)^2}}$ substitute $3 \sin \theta$ for x

$= \dfrac{3 \sin \theta}{\sqrt{9 - 9 \sin^2 \theta}}$ $(3 \sin \theta)^2 = 9 \sin^2 \theta$

$= \dfrac{3 \sin \theta}{\sqrt{9(1 - \sin^2 \theta)}}$ factor

$= \dfrac{3 \sin \theta}{\sqrt{9 \cos^2 \theta}}$ substitute $\cos^2 \theta$ for $1 - \sin^2 \theta$

$= \dfrac{3 \sin \theta}{3 \cos \theta}$ $\sqrt{9 \cos^2 \theta} = 3 \cos \theta$ since $-\dfrac{\pi}{2} \leq \theta \leq \dfrac{\pi}{2}$

$= \tan \theta$ result

Now try Exercises 1 through 6 ▶

Using the notation for inverse functions, we can rewrite $y = \tan \theta$ as a function of x and use a calculator to compare it with the original function. For $x = 3 \sin \theta$ we obtain $\dfrac{x}{3} = \sin \theta$ or $\theta = \sin^{-1}\left(\dfrac{x}{3}\right)$. Substituting $\sin^{-1}\left(\dfrac{x}{3}\right)$ for θ in $y = \tan \theta$ gives $y = \tan\left[\sin^{-1}\left(\dfrac{x}{3}\right)\right]$. With the calculator in radian radian **MODE**, enter $Y_1 = \dfrac{x}{\sqrt{9 - x^2}}$ and $Y_2 = \tan\left[\sin^{-1}\left(\dfrac{x}{3}\right)\right]$ on the **Y=** screen. Using TblStart $= -3$ (due to the domain), the resulting table seems to indicate that the functions are indeed equivalent (see the figure).

X	Y₁	Y₂
-3	ERROR	ERROR
-2.9	-3.775	-3.775
-2.8	-2.6	-2.6
-2.7	-2.065	-2.065
-2.6	-1.737	-1.737
-2.5	-1.508	-1.508
	-1.333	-1.333

X=-2.4

Trigonometric Identities and Equations

While the tools of calculus are very powerful, it is the application of rudimentary concepts that makes them work. Earlier, we saw how basic algebra skills were needed to simplify expressions that resulted from applications of calculus. Here we illustrate the use of basic trigonometric skills combined with basic algebra skills to achieve the same end.

EXAMPLE 2 ▶ **Finding Maximum and Minimum Values**

Using the tools of calculus, it can be shown that the maximum and/or minimum values of $f(\theta) = \dfrac{1 + \cot \theta}{\csc \theta}$ will occur at the zero(s) of the function

$$f(\theta) = \frac{\csc \theta(-\csc^2\theta) - (1 + \cot\theta)(-\csc \theta \cot \theta)}{\csc^2\theta}.$$

Simplify the right-hand side and use the result to find the location of any maximum or minimum values that occur in the interval $[0, 2\pi)$.

Solution ▶ Begin by simplifying the numerator.

$$f(\theta) = \frac{\csc \theta(-\csc^2\theta) - (-\csc \theta \cot \theta - \csc \theta \cot^2\theta)}{\csc^2\theta} \qquad \text{distribute}$$

$$= \frac{-\csc^3\theta + \csc \theta \cot \theta + \csc \theta \cot^2\theta}{\csc^2\theta} \qquad \text{simplify}$$

$$= \frac{\csc \theta(\cot^2\theta + \cot \theta - \csc^2\theta)}{\csc^2\theta} \qquad \text{factor, commute terms}$$

$$= \frac{(\csc^2\theta - 1) + \cot \theta - \csc^2\theta}{\csc \theta} \qquad \text{simplify, substitute } \csc^2\theta - 1 \text{ for } \cot^2\theta$$

$$= \frac{\cot \theta - 1}{\csc \theta} \qquad \text{result}$$

This shows that $f(\theta) = 0$ when $\cot \theta = 1$, or when $\theta = \dfrac{\pi}{4} + \pi k, k \in \mathbb{Z}$. In the interval $[0, 2\pi)$, this gives $\theta = \dfrac{\pi}{4}$ and $\dfrac{5\pi}{4}$. The function f has a maximum value of $\sqrt{2}$ at $\dfrac{\pi}{4}$, with a minimum value of $-\sqrt{2}$ at $\dfrac{5\pi}{4}$.

Now try Exercises 7 through 10 ▶

Connections to Calculus Exercises

For the functions given, (a) use the substitution indicated to find an equivalent function of θ, (b) rewrite the resulting function in terms of x using an inverse trig function, and (c) use the $\boxed{\text{TABLE}}$ feature of a graphing calculator to verify the two functions are equivalent for $x \neq 0$.

1. $y = \dfrac{\sqrt{169 + x^2}}{x}$;

$x = 13 \tan \theta, \theta \in \left(-\dfrac{\pi}{2}, \dfrac{\pi}{2}\right)$

2. $y = \dfrac{\sqrt{144 - x^2}}{x}$;

$x = 12 \sin \theta, \theta \in \left(-\dfrac{\pi}{2}, \dfrac{\pi}{2}\right)$

Rewrite the following expressions using the substitution indicated.

3. $\dfrac{x^2}{\sqrt{16 - x^2}}$; $x = 4 \sin \theta$

4. $\dfrac{\sqrt{81 - x^2}}{x}$; $x = 9 \sin \theta$

5. $\dfrac{x}{\sqrt{9 + x^2}}$; $x = 3 \tan \theta$

6. $\dfrac{8}{(x^2 - 4)^{\frac{3}{2}}}$; $x = 2 \sec \theta$

Using the tools of calculus, it can be shown that for each function $f(x)$ given, the zeroes of $f'(x)$ give the location of any maximum and/or minimum values. Find the location of these values in the interval $[0, 2\pi)$, using trig identities as needed to solve $f'(x) = 0$. Verify solutions using a graphing calculator.

7. $f(x) = \dfrac{1 + \cos x}{\sec x}$;

$f'(x) = \dfrac{\sec x(-\sin x) - (1 + \cos x)\sec x \tan x}{\sec^2 x}$

8. $f(x) = \sin x \tan x$;

$f'(x) = \sin x \sec^2 x + \tan x \cos x$

9. $f(x) = 2 \sin x \cos x$;

$f'(x) = 2 \sin x(-\sin x) + 2 \cos x \cos x$

10. $f(x) = \dfrac{\cos x}{2 + \sin x}$;

$f'(x) = \dfrac{(2 + \sin x)(-\sin x) - \cos x \cos x}{(2 + \sin x)^2}$

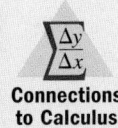

Chapter

7

Applications of Trigonometry

CHAPTER OUTLINE

CHAPTER CONNECTIONS

When an airline pilot charts a course, it's not as simple as pointing the airplane in the right direction. Wind currents must be taken into consideration, and compensated for by additional thrust or a change of heading that will help equalize the force of the wind and keep the plane flying in the desired direction. The effect of these forces working together can be modeled using a carefully drawn *vector diagram*, and with the aid of trigonometry, a pilot can easily determine any adjustments in navigation needed. This application appears as Exercise 85 in Section 7.3.

$\dfrac{\Delta y}{\Delta x}$

Connections to Calculus

As with other forms of problem solving, drawing an accurate sketch or diagram of the relationships involved has a large impact on our ability to understand vector applications and other applications of trigonometry. This is certainly no less true in a calculus course and is the subject of the Chapter 7 *Connections to Calculus*.

Learning Objectives

In Section 7.1 you will learn how to:

☐ **A.** Develop the law of sines, and use it to solve ASA and AAS triangles

☐ **B.** Solve SSA triangles (the ambiguous case) using the law of sines

☐ **C.** Use the law of sines to solve applications

Many applications of trigonometry involve *oblique triangles*, or triangles that do not have a 90° angle. For example, suppose a trolley carries passengers from ground level up to a mountain chateau, as shown in Figure 7.1. Assuming the cable could be held taut, what is its approximate length? Can we also determine the slant height of the mountain? To answer questions like these, we'll develop techniques that enable us to solve acute and obtuse triangles using fundamental trigonometric relationships.

Figure 7.1

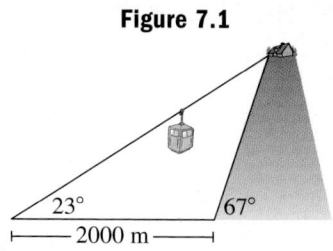

A. The Law of Sines and Unique Solutions

Consider the oblique triangle *ABC* pictured in Figure 7.2. Since it is not a right triangle, it seems the trigonometric ratios studied earlier cannot be applied. But if we draw the altitude *h* (from vertex *B*), two right triangles are formed that *share a common side*. By applying the sine ratio to angles *A* and *C*, we can develop a relationship that will help us solve the triangle.

Figure 7.2

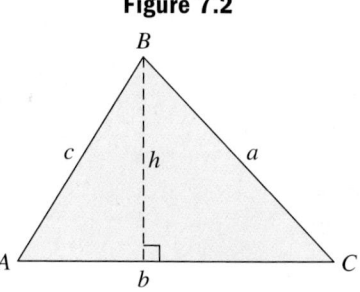

For $\angle A$ we have $\sin A = \dfrac{h}{c}$ or $h = c \sin A$. For $\angle C$ we have $\sin C = \dfrac{h}{a}$ or $h = a \sin C$. Since both products are equal to *h*, the transitive property gives $c \sin A = a \sin C$, which leads to

$$c \sin A = a \sin C \qquad \text{since } h = h$$

$$\frac{c \sin A}{ac} = \frac{a \sin C}{ac} \qquad \text{divide by } ac$$

$$\frac{\sin A}{a} = \frac{\sin C}{c} \qquad \text{simplify}$$

Using the same triangle and the altitude drawn from *C* (Figure 7.3), we note a similar relationship involving angles *A* and *B*: $\sin A = \dfrac{h}{b}$ or $h = b \sin A$, and $\sin B = \dfrac{h}{a}$ or $h = a \sin B$. As before, we can then write $\dfrac{\sin A}{a} = \dfrac{\sin B}{b}$. If $\angle A$ is obtuse, the altitude *h* actually falls outside the triangle, as shown in Figure 7.4. In this case, consider that $\sin(180° - \alpha) = \sin \alpha$ from the difference formula for sines (Exercise 55, Section 6.3). In the figure we note $\sin(180° - \alpha) = \dfrac{h}{c} = \sin \alpha$, yielding $h = c \sin \alpha$

WORTHY OF NOTE

As with right triangles, solving an oblique triangle involves determining the lengths of all three sides and the measures of all three angles.

Figure 7.3

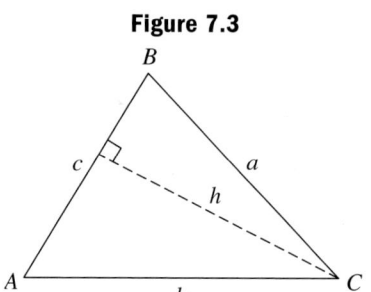

Figure 7.4

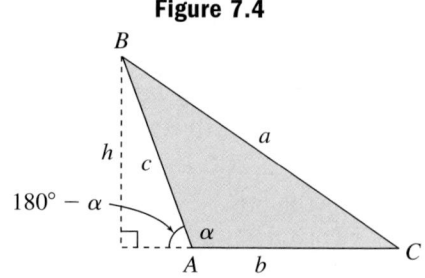

and the preceding relationship can now be stated using any pair of angles and corresponding sides. The result is called the **law of sines,** which is usually stated by combing the three possible proportions.

The Law of Sines

For any triangle ABC, the ratio of the sine of an angle to the side opposite that angle is constant:

$$\frac{\sin A}{a} = \frac{\sin B}{b} = \frac{\sin C}{c}$$

As a proportional relationship, the law requires that we have three parts in order to solve for the fourth. This suggests the following possibilities:

1. two angles and an included side (ASA)
2. two angles and a side opposite one of these angles (AAS)
3. two sides and a angle opposite one of these sides (SSA)
4. two sides and an included angle (SAS)
5. three sides (SSS)

Each of these possibilities is diagrammed in Figures 7.5 through 7.9.

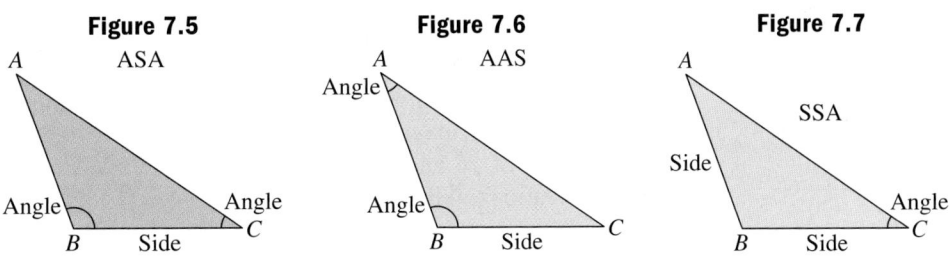

Figure 7.5 ASA   **Figure 7.6** AAS   **Figure 7.7** SSA

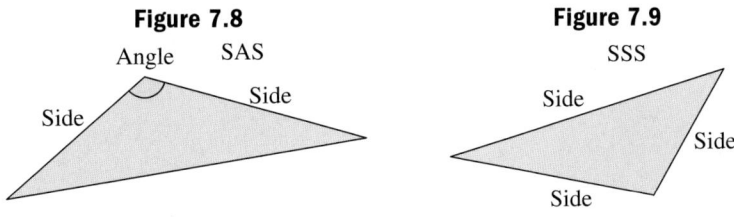

Figure 7.8 SAS   **Figure 7.9** SSS

> **WORTHY OF NOTE**
>
> When working with triangles, keeping these basic properties in mind will prevent errors and assist in their solution:
>
> 1. The angles must sum to 180°.
> 2. The combined length of any two sides must exceed the length of the third side.
> 3. Longer sides will be opposite larger angles.
> 4. This sine of an angle cannot be greater than 1.
> 5. For $y \in (0, 1)$, the equation $y = \sin \theta$ has two solutions in $(0°, 180°)$ that are supplements.

Since applying the law of sines requires we have a given side opposite a known angle, it cannot be used in the case of SAS or SSS triangles. These require the law of cosines, which we will develop in Section 7.2. In the case of ASA and AAS triangles, a unique triangle is formed since the measure of the third angle is fixed by the two angles given (they must sum to 180°) and the remaining sides must be of fixed length.

EXAMPLE 1 ▶ Solving a Triangle Using the Law of Sines

Solve the triangle shown, and state your answer using a table.

Solution ▶ This is *not* a right triangle, so the standard ratios cannot be used. Since $\angle B$ and $\angle C$ are given, we know $\angle A = 180° - (110° + 32°) = 38°$. With $\angle A$ and side a, we have

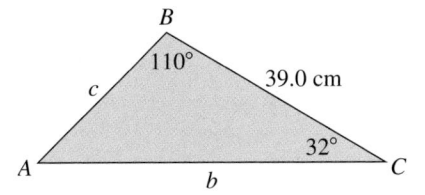

$$\frac{\sin A}{a} = \frac{\sin B}{b} \qquad \text{law of sines applied to } \angle A \text{ and } \angle B$$

$$\frac{\sin 38°}{39} = \frac{\sin 110°}{b} \qquad \text{substitute given values}$$

$$b \sin 38° = 39 \sin 110° \qquad \text{multiply by } 39b$$

$$b = \frac{39 \sin 110°}{\sin 38°} \qquad \text{divide by } \sin 38°$$

$$b \approx 59.5 \qquad \text{result}$$

Repeating this procedure using $\dfrac{\sin A}{a} = \dfrac{\sin C}{c}$ shows side $c \approx 33.6$ cm. In table form we have

Angles	Sides (cm)
$A = 38°$	$a = 39.0$
$B = 110°$	$b \approx 59.5$
$C = 32°$	$c \approx 33.6$

☑ **A.** You've just learned to develop the law of sines and use it to solve ASA and AAS triangles

Now try Exercises 7 through 24 ▶

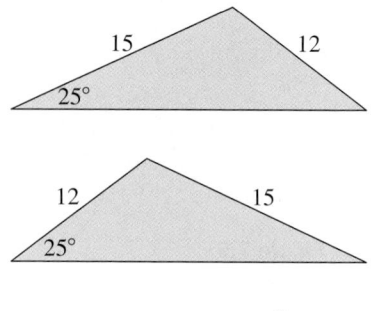

Figure 7.10

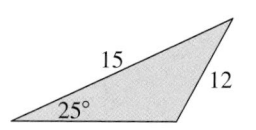

B. Solving SSA Triangles—The Ambiguous Case

To understand the concept of unique and nonunique solutions regarding the law of sines, consider an instructor who asks a large group of students to draw a triangle with sides of 15 and 12 units, and a nonincluded 25° angle. Unavoidably, three different solutions will be offered (see Figure 7.10). For the SSA case, there is some doubt as to the number of solutions possible, or whether a solution even exists.

To further understand why, consider a triangle with side $c = 30$ cm, $\angle A = 30°$, and side a opposite the 30° angle (Figure 7.11—note the length of side b is yet to be determined). From our work with 30-60-90 triangles, we know if $a = 15$ cm, it is exactly the length needed to form a right triangle (Figure 7.12).

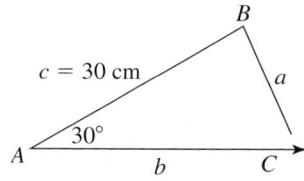

Figure 7.11

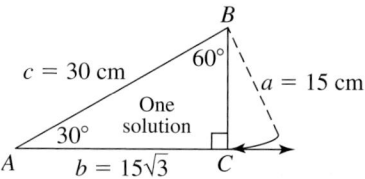

Figure 7.12

By varying the length of side a, we note three other possibilities. If side $a < 15$ cm, no triangle is possible since a is too short to contact side b (Figure 7.13), while if 15 cm $<$ side $a < 30$ cm, two triangles are possible since side a will then intersect side b at two points, C_1 and C_2 (Figure 7.14).

For future use, note that when two triangles are possible, angles C_1 and C_2 must be supplements since an-isosceles triangle is formed. Finally, if side $a > 30$ cm, it will

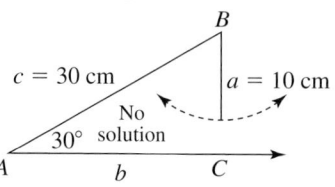

Figure 7.13

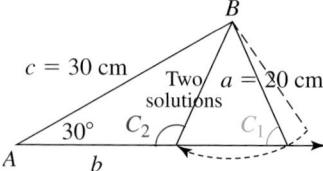

Figure 7.14

WORTHY OF NOTE

The case where three angles are known (AAA) is not considered since we then have a family of similar triangles, with infinitely many solutions.

Figure 7.15

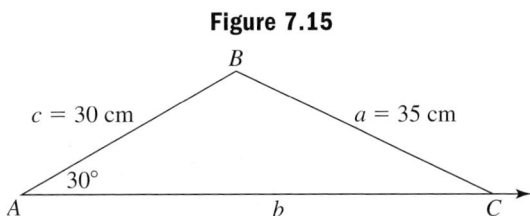

intersect side b only once, forming the obtuse triangle shown in Figure 7.15, where we've assumed $a = 35$ cm. Since the final solution is in doubt until we do further work, the SSA case is called the **ambiguous case** of the law of sines.

EXAMPLE 2 ▶ **Analyzing the Ambiguous Case of the Law of Sines**

Given triangle ABC with $\angle A = 45°$ and side $c = 100\sqrt{2}$ mm,

  a. What length for side a will produce a right triangle where $\angle C = 90°$?

  b. How many triangles can be formed if side $a = 90$ mm?

  c. If side $a = 120$ mm, how many triangles can be formed?

  d. If side $a = 145$ mm, how many triangles can be formed?

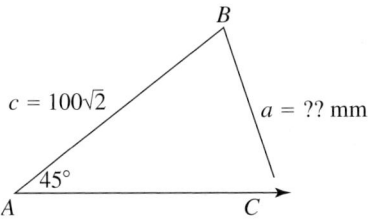

Solution ▶   **a.** Recognizing the sides of a 45-45-90 triangle are in proportion according to $1x:1x: \sqrt{2}x$, side a must be 100 mm for a right triangle to be formed.

  b. If $a = 90$ mm, it will be too short to contact side b and no triangle is possible.

  c. As shown in Figure 7.16, if $a = 120$ mm, it will contact side b in two distinct places and two triangles are possible.

  d. If $a = 145$ mm, it will contact side b only once, since it is longer than side c and will "miss" side b as it pivots around $\angle B$ (see Figure 7.17). One triangle is possible.

Figure 7.16   **Figure 7.17**

Now try Exercises 25 and 26 ▶

For a better understanding of the SSA (ambiguous) case, scaled drawings can initially be used along with a metric ruler and protractor. Begin with a horizontal line segment of undetermined length to represent the third (unknown) side, and use the protractor to draw the given angle on either the left or right side of this segment (we chose the left). Then use the metric ruler to draw an adjacent side of appropriate length, *choosing a scale that enables a complete diagram.* For instance, if the given sides are 3 ft and 5 ft, use 3 cm and 5 cm instead (1 cm = 1 ft). If the sides are 80 mi and 120 mi, use 8 cm and 12 cm (1 cm = 10 mi), and so on. Once the adjacent side is drawn, start at the free endpoint and draw a vertical segment to represent the remaining side. A careful sketch will often indicate whether none, one, or two triangles are possible (see the *Reinforcing Basic Concepts* feature on page 673).

EXAMPLE 3 ▶ **Solving the Ambiguous Case of the Law of Sines**

Solve the triangle with side $b = 100$ ft, side $c = 60$ ft, and $\angle C = 28.0°$.

Solution ▶ Two sides and an angle opposite are given (SSA), and we draw a diagram to help determine the possibilities. Draw the horizontal segment of some length and use a protractor to mark $\angle C = 28°$. Then draw a segment 10 cm long (to represent $b = 100$ ft) as the adjacent side of the angle, with a vertical segment 6 cm long from the free end of b (to represent $c = 60$ ft). It seems apparent that side c will intersect the horizontal side in two places (see figure), and two triangles are possible. We apply the law of sines to solve the first triangle, whose features we'll note with a subscript of 1.

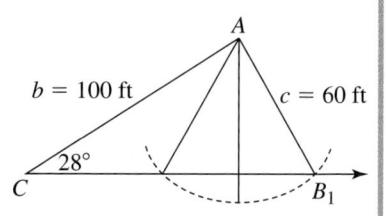

$$\frac{\sin B_1}{b} = \frac{\sin C}{c} \qquad \text{law of sines}$$

$$\frac{\sin B_1}{100} = \frac{\sin 28°}{60} \qquad \text{substitute}$$

$$\sin B_1 = \frac{5}{3} \sin 28° \qquad \text{solve for } \sin B_1$$

$$B_1 \approx 51.5° \qquad \text{apply arcsine}$$

WORTHY OF NOTE

In Example 3, we found $\angle B_2$ using the property that states the angles in a triangle must sum to 180°. We could also view B_1 as a QI reference angle, which also gives a QII solution of $(180 - 51.5)° = 128.5°$.

Since $\angle B_1 + \angle B_2 = 180°$, we know $\angle B_2 = 128.5°$. These values give 100.5° and 23.5° as the measures of $\angle A_1$ and $\angle A_2$, respectively. By once again applying the law of sines to each triangle, we find side $a_1 \approx 125.7$ ft and $a_2 \approx 51.0$ ft. See Figure 7.18.

Angles	Sides (ft)
$A_1 \approx 100.5°$	$a_1 \approx 125.7$
$B_1 \approx 51.5°$	$b = 100$
$C = 28°$	$c = 60$

Angles	Sides (ft)
$A_2 \approx 23.5°$	$a_2 \approx 51.0$
$B_2 \approx 128.5°$	$b = 100$
$C = 28°$	$c = 60$

Figure 7.18

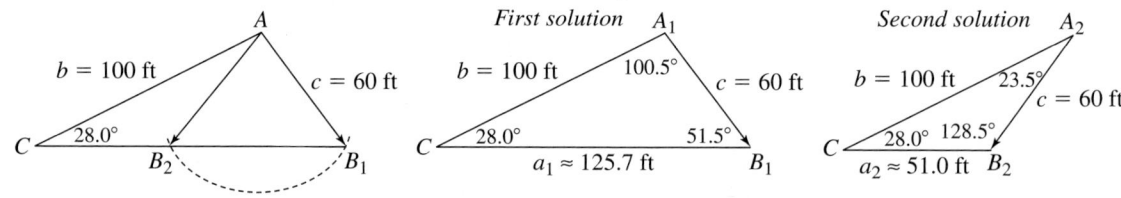

Now try Exercises 27 through 32 ▶

Admittedly, the scaled drawing approach has some drawbacks—it takes time to draw the diagrams and is of little use if the situation is a close call. It does, however, offer a deeper understanding of the subtleties involved in solving the SSA case. Instead of a scaled drawing, we can use a simple sketch *as a guide*, while keeping in mind the properties mentioned in the *Worthy of Note* on page 635.

EXAMPLE 4 ▶ **Solving the Ambiguous Case of the Law of Sines**

Solve the triangle with side $a = 220$ ft, side $b = 200$ ft, and $\angle A = 40°$.

Solution ▶ The information given is again SSA, and we apply the law of sines with this in mind.

$$\frac{\sin A}{a} = \frac{\sin B}{b} \qquad \text{law of sines}$$

$$\frac{\sin 40°}{220} = \frac{\sin B}{200} \qquad \text{substitute}$$

$$\sin B = \frac{200 \sin 40°}{220} \qquad \text{solve for } \sin B$$

$$B_1 \approx 35.7° \qquad \text{apply arcsine}$$

This is the solution from Quadrant I. The QII solution is about $(180 - 35.7)° = 144.3°$. At this point our solution tables have this form:

Angles	Sides (ft)
$A = 40°$	$a = 220$
$B_1 \approx 35.7°$	$b = 200$
$C_1 =$	$c_1 =$

Angles	Sides (ft)
$A = 40°$	$a = 220$
$B_2 \approx 144.3°$	$b = 200$
$C_2 =$	$c_2 =$

☑ **B.** You've just learned how to solve SSA triangles (the ambiguous case) using the law of sines

It seems reasonable to once again find the remaining angles and finish by reapplying the law of sines, but observe that the sum of the two angles from the second solution *already exceeds 180°*: $40° + 144.3° = 188.3°$! This means no second solution is possible (side a is too long). We find that $C_1 \approx 104.3°$, and applying the law of sines gives a value of $c_1 \approx 331.7$ ft.

Now try Exercises 33 through 44 ▶

C. Applications of the Law of Sines

As "ambiguous" as it is, the ambiguous case has a number of applications in engineering, astronomy, physics, and other areas. Here is an example from astronomy.

EXAMPLE 5 ▶ **Solving an Application of the Ambiguous Case—Planetary Distance**

The planet Venus can be seen from Earth with the naked eye, but as the diagram indicates, the position of Venus is uncertain (we are unable to tell if Venus is in the near position or the far position). Given the Earth is 93 million miles from the Sun and Venus is 67 million miles from the Sun, determine the closest and farthest possible distances that separate the planets in this alignment. Assume a viewing angle of $\theta \approx 18°$ and that the orbits of both planets are roughly circular.

Solution ▶ A close look at the information and diagram shows a SSA case. Begin by applying the law of sines where $E \to$ Earth, $V \to$ Venus, and $S \to$ Sun.

$$\frac{\sin E}{e} = \frac{\sin V}{v} \qquad \text{law of sines}$$

$$\frac{\sin 18°}{67} = \frac{\sin V}{93} \qquad \text{substitute given values}$$

$$\sin V = \frac{93 \sin 18°}{67} \qquad \text{solve for } \sin V$$

$$V \approx 25.4° \qquad \text{apply arcsine}$$

This is the angle V_1 formed when Venus is farthest away. The angle V_2 at the closer distance is $180° - 25.4° = 154.6°$. At this point, our solution tables have this form:

Angles	Sides (10^6 mi)
$E = 18°$	$e = 67$
$V_1 \approx 25.4°$	$v = 93$
$S_1 =$	$s_1 =$

Angles	Sides (10^6 mi)
$E = 18°$	$e = 67$
$V_2 = 154.6°$	$v = 93$
$S_2 =$	$s_2 =$

For S_1 and S_2 we have $S_1 \approx 180 - (18 + 25.4°) = 136.6°$ (larger angle) and $S_2 \approx 180 - (18 + 154.6°) = 7.4°$ (smaller angle). Re-applying the law of sines for s_1 shows the farther distance between the planets is about 149 million miles. Solving for s_2 shows that the closer distance is approximately 28 million miles.

> **Now try Exercises 47 and 48 ▶**

EXAMPLE 6 ▶ **Solving an Application of the Ambiguous Case—Radar Detection**

As shown in Figure 7.19, a radar ship is 30.0 mi off shore when a large fleet of ships leaves port at an angle of 43.0°.

a. If the maximum range of the ship's radar is 20.0 mi, will the departing fleet be detected?

b. If the maximum range of the ship's radar is 25.0 mi, how far from port is the fleet when it is first detected?

Figure 7.19

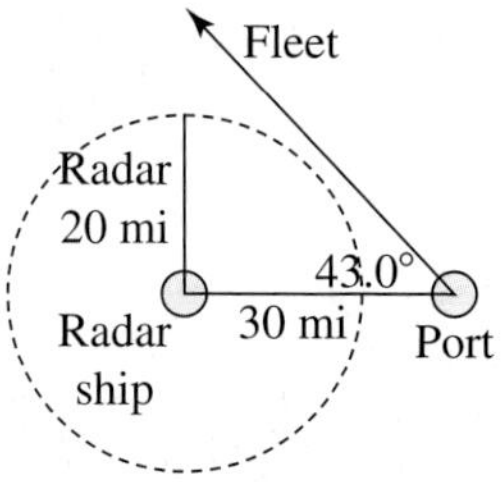

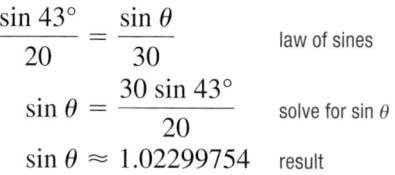

Solution ▶ **a.** This is again the SSA (ambiguous) case. Applying the law of sines gives

$$\frac{\sin 43°}{20} = \frac{\sin \theta}{30} \qquad \text{law of sines}$$

$$\sin \theta = \frac{30 \sin 43°}{20} \qquad \text{solve for } \sin \theta$$

$$\sin \theta \approx 1.02299754 \qquad \text{result}$$

No triangle is possible and the departing fleet will not be detected.

b. If the radar has a range of 25.0 mi, the radar beam will intersect the projected course of the fleet in two places.

$$\frac{\sin 43°}{25} = \frac{\sin \theta}{30} \qquad \text{law of sines}$$

$$\sin \theta = \frac{30 \sin 43°}{25} \qquad \text{solve for } \sin \theta$$

$$\theta \approx 54.9° \qquad \text{apply arcsine}$$

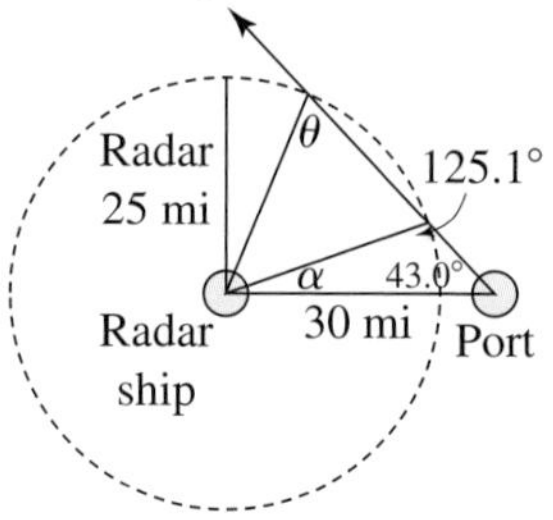

This is the acute angle θ related to the *farthest point* from port at which the fleet could be detected (see Figure 7.20). For the second triangle, we have $180° - 54.9° = 125.1°$ (the obtuse angle) giving a measure of $180° - (125.1° + 43°) = 11.9°$ for angle α. For d as the side opposite α we have

Figure 7.20

$$\frac{\sin 43°}{25} = \frac{\sin 11.9°}{d} \qquad \text{law of sines}$$

$$d = \frac{25 \sin 11.9°}{\sin 43°} \qquad \text{solve for } d$$

$$\approx 7.6 \qquad \text{simplify}$$

This shows the fleet is first detected about 7.6 mi from port.

☑ C. You've just learned how to use the law of sines to solve applications

> **Now try Exercises 49 and 50 ▶**

There are a number of additional, interesting applications in the exercise set (see **Exercises 51 through 70**).

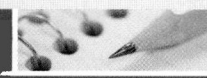

7.1 EXERCISES

▶ CONCEPTS AND VOCABULARY

Fill in each blank with the appropriate word or phrase. Carefully reread the section if needed.

1. For the law of sines, if two sides and an angle opposite one side are given, this is referred to as the _____ case, since the solution is in doubt until further work.

2. Two inviolate properties of a triangle that can be used to help solve the ambiguous case are: (a) the angles must sum to _____ and (b) no sine ratio can exceed _____.

3. For positive k, the equation $\sin \theta = k$ has two solutions, one in Quadrant _____ and the other in Quadrant _____.

4. After a triangle is solved, you should always check to ensure that the _____ side is opposite the _____ angle.

5. In your own words, explain why the AAS case results in a unique solution while the SSA case does not. Give supporting diagrams.

6. Explain why no triangle is possible in each case:
   a. $A = 34°, B = 73°, C = 52°,$
      $a = 14', b = 22', c = 18'$
   b. $A = 42°, B = 57°, C = 81°,$
      $a = 7'', b = 9'', c = 22''$

▶ DEVELOPING YOUR SKILLS

Solve each of the following equations for the unknown part (if possible). Round sides to the nearest hundredth and degrees to the nearest tenth.

7. $\dfrac{\sin 32°}{15} = \dfrac{\sin 18.5°}{a}$

8. $\dfrac{\sin 52°}{b} = \dfrac{\sin 30°}{12}$

9. $\dfrac{\sin 63°}{21.9} = \dfrac{\sin C}{18.6}$

10. $\dfrac{\sin B}{3.14} = \dfrac{\sin 105°}{6.28}$

11. $\dfrac{\sin C}{48.5} = \dfrac{\sin 19°}{43.2}$

12. $\dfrac{\sin 38°}{125} = \dfrac{\sin B}{190}$

Solve each triangle using the law of sines. If the law of sines cannot be used, state why. Draw and label a triangle or label the triangle given before you begin.

13. side $a = 75$ cm
    $\angle A = 38°$
    $\angle B = 64°$

14. side $b = 385$ m
    $\angle B = 47°$
    $\angle A = 108°$

15. side $b = 10\sqrt{3}$ in.
    $\angle A = 30°$
    $\angle B = 60°$

16.
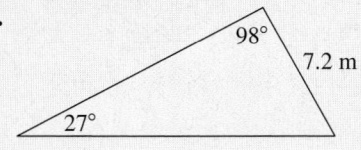

17.
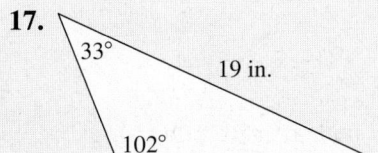

18.
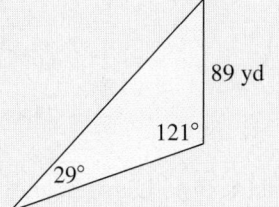

19. $\angle A = 45°$
    $\angle B = 45°$
    side $c = 15\sqrt{2}$ mi

20. $\angle A = 20.4°$
    side $c = 12.9$ mi
    $\angle B = 63.4°$

21. $\angle B = 103.4°$
    side $a = 42.7$ km
    $\angle C = 19.6°$

22.
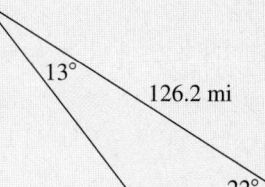

23.
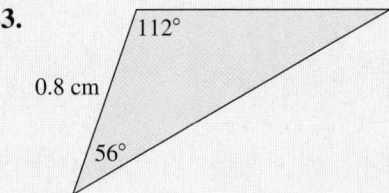

24.
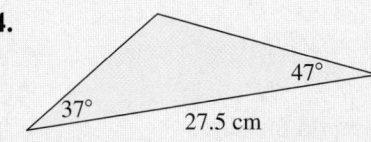

Answer each question and justify your response using a diagram, but do not solve.

25. Given $\triangle ABC$ with $\angle A = 30°$ and side $c = 20$ cm, (a) what length for side a will produce a right triangle? (b) How many triangles can be formed if side $a = 8$ cm? (c) If side $a = 12$ cm, how many triangles can be formed? (d) If side $a = 25$ cm, how many triangles can be formed?

26. Given $\triangle ABC$ with $\angle A = 60°$ and side $c = 6\sqrt{3}$ m, (a) what length for side a will produce a right triangle? (b) How many triangles can be formed if side $a = 8$ m? (c) If side $a = 10$ m, how many triangles can be formed? (d) If side $a = 15$ m, how many triangles can be formed?

Solve using the law of sines and a scaled drawing. If two triangles exist, solve both completely.

27. side $b = 385$ m
    $\angle B = 67°$
    side $a = 490$ m

28. side $a = 36.5$ yd
    $\angle B = 67°$
    side $b = 12.9$ yd

29. side $c = 25.8$ mi
    $\angle A = 30°$
    side $a = 12.9$ mi

30. side $c = 10\sqrt{3}$ in.
    $\angle A = 60°$
    side $a = 15$ in.

31. side $c = 58$ mi
    $\angle C = 59°$
    side $b = 67$ mi

32. side $b = 24.9$ km
    $\angle B = 45°$
    side $a = 32.8$ km

Use the law of sines to determine if no triangle, one triangle, or two triangles can be formed from the diagrams given (diagrams *may not be to scale*), then solve. If two solutions exist, solve both completely. Note the arrowhead marks the side of undetermined length.

33.
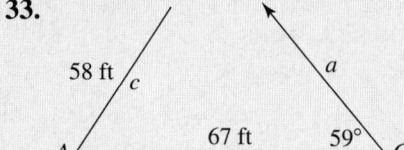

34.
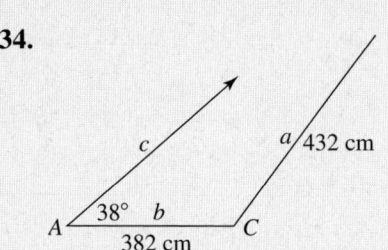

35.
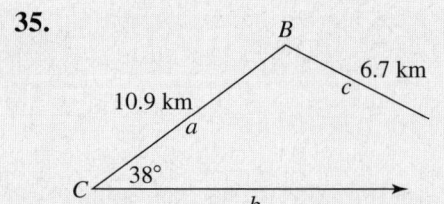

36.
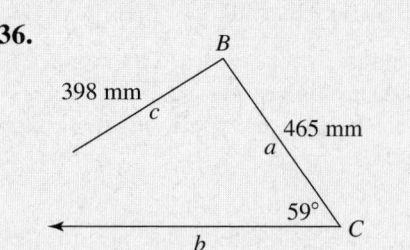

37.
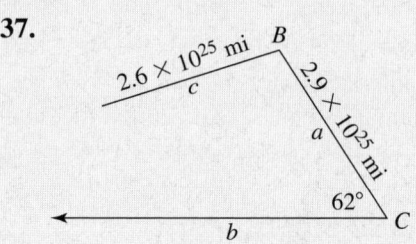

38.
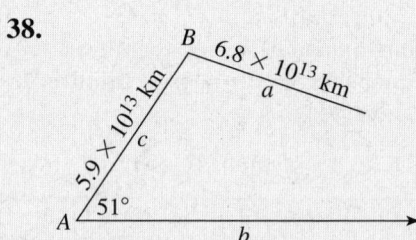

For Exercises 39 to 44, assume the law of sines is being applied to solve a triangle. Solve for the unknown angle (if possible), then determine if a second angle ($0° < \theta < 180°$) exists that also satisfies the proportion.

39. $\dfrac{\sin A}{12} = \dfrac{\sin 48°}{27}$

40. $\dfrac{\sin 60°}{32} = \dfrac{\sin B}{9}$

41. $\dfrac{\sin 57°}{35.6} = \dfrac{\sin C}{40.2}$

42. $\dfrac{\sin B}{5.2} = \dfrac{\sin 65°}{4.9}$

43. $\dfrac{\sin A}{280} = \dfrac{\sin 15°}{52}$

44. $\dfrac{\sin 29°}{121} = \dfrac{\sin B}{321}$

▶ **WORKING WITH FORMULAS**

45. **Triple angle formula for sine:**
    $\sin(3\theta) = 3\sin\theta - 4\sin^3\theta$

    Most students are familiar with the double angle formula for sine: $\sin(2\theta) = 2\sin\theta\cos\theta$. The triple angle formula for sine is given here. Use the formula to find an exact value for $\sin 135°$, then verify the result using a reference angle.

46. Radius of a circumscribed circle: $R = \dfrac{b}{2 \sin B}$

Given $\triangle ABC$ is circumscribed by a circle of radius R, the radius of the circle can be found using the formula shown, where side b is opposite angle B. Find the radius of the circle shown.

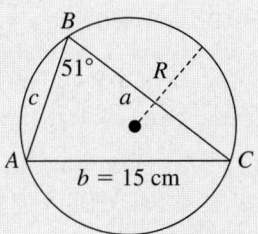

▶ **APPLICATIONS**

47. Planetary distances: In a solar system that parallels our own, the planet Sorus can be seen from a Class M planet with the naked eye, but as the diagram indicates, the position of Sorus is uncertain. Assume the orbits of both planets are roughly circular and that the viewing angle θ is about 20°. If the Class M planet is 82 million miles from its sun and Sorus is 51 million miles from this sun, determine the closest and farthest possible distances that separate the planets in this alignment.

Exercise 47

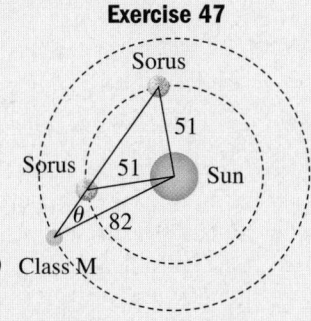

48. Planetary distances: In a solar system that parallels our own, the planet Cirrus can be seen from a Class M planet with the naked eye, but as the diagram indicates, the position of Cirrus is uncertain. Assume the orbits of both planets are roughly circular and that the viewing angle θ is about 15°. If the Class M planet is 105 million miles from its sun and Cirrus is 70 million miles from this sun, determine the closest and farthest possible distances that separate the planets in this alignment.

Exercise 48

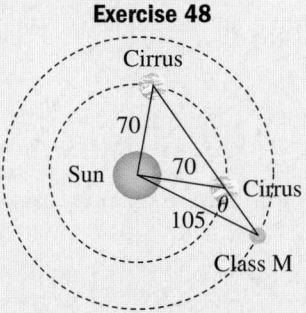

49. Radar detection: A radar ship is 15.0 mi off shore from a major port when a large fleet of ships leaves the port at the 35.0° angle shown. (a) If the maximum range of the ship's radar is 8.0 mi, will the departing fleet be detected? (b) If the maximum range of the ship's radar is 12 mi, how far from port is the fleet when it is first detected?

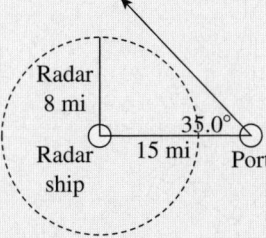

50. Motion detection: To notify environmentalists of the presence of big game, motion detectors are installed 200 yd from a watering hole. A pride of lions has just visited the hole and is leaving the area at the 29.0° angle shown. (a) If the maximum range of the motion detector is 90 yd, will the pride be detected? (b) If the maximum range of the motion detector is 120 yd, how far from the watering hole is the pride when first detected?

Exercise 50

51. Distance between cities: The cities of Van Gogh, Rembrandt, Pissarro, and Seurat are situated as shown in the diagram. Assume that triangle RSP is isosceles and use the law of sines to find the distance between Van Gogh and Seurat, and between Van Gogh and Pissarro.

Exercise 51

52. Distance between cities: The cities of Mozart, Rossini, Offenbach, and Verdi are situated as shown in the diagram. Assume that triangle ROV is isosceles and use the law of sines to find the distance between Mozart and Verdi, and between Mozart and Offenbach.

Exercise 52

53. Distance to target: To practice for a competition, an archer stands as shown in the diagram and attempts to hit a moving target. (a) If the archer has a maximum effective range of about 180 ft, can the target be hit?

Exercise 53

(b) What is the shortest range the archer can have and still hit the target? (c) If the archer's range is 215 ft and the target is moving at 10 ft/sec, how many seconds is the target within range?

54. **Distance to target:** As part of an All-Star competition, a quarterback stands as shown in the diagram and attempts to hit a moving target with a football. (a) If the quarterback has a maximum effective range of about 35 yd, can the target be hit? (b) What is the shortest range the quarterback can have and still hit the target? (c) If the quarterback's range is 45 yd and the target is moving at 5 yd/sec, how many seconds is the target within range?

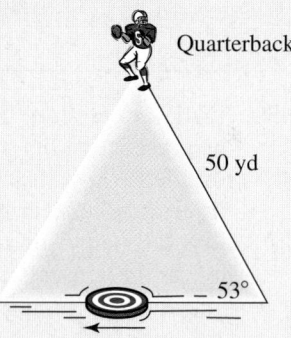

In Exercises 55 and 56, three rods are attached via pivot joints so the rods can be manipulated to form a triangle. How many triangles can be formed if angle *B* must measure 26°? If one triangle, solve it. If two, solve both. Diagrams are not drawn to scale.

55.

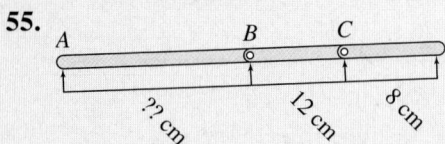

56.

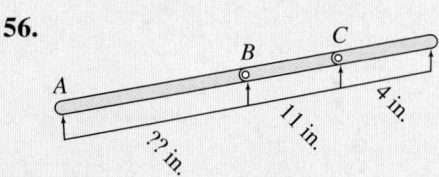

In the diagrams given, the measure of angle *C* and the length of sides *a* and *c* are fixed. Side *c* can be rotated at pivot point *B*. Solve any triangles that can be formed. (*Hint:* Begin by using the grid to find lengths *a* and *c*, then find angle *C*.)

57.

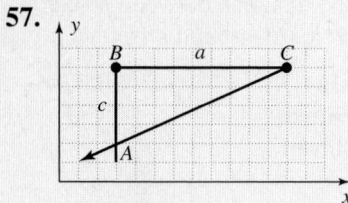

58.

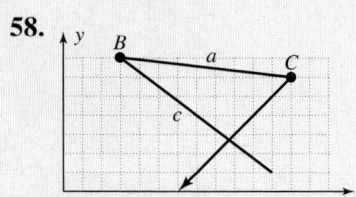

Length of a rafter: Determine the length of both roof rafters in the diagrams given.

59.

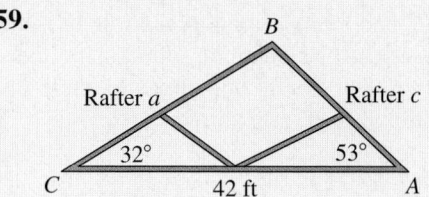

60.

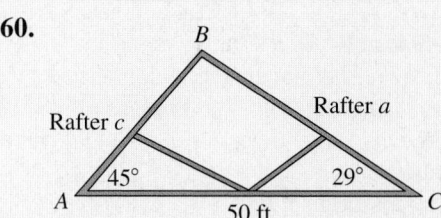

61. **Map distance:** A cartographer is using aerial photographs to prepare a map for publication. The distance from Sexton to Rhymes is known to be 27.2 km. Using a protractor, the map maker measures an angle of 96° from Sexton to Tarryson (a newly developed area) and an angle of 58° from Rhymes to Tarryson. Compute each unknown distance.

62. **Height of a fortress:** An ancient fortress is built on a steep hillside, with the base of the fortress walls making a 102° angle with the hill. At the moment the fortress casts a 112-ft shadow, the angle of elevation from the tip of the shadow to the top of the wall is 32°. What is the distance from the base of the fortress to the top of the tower?

Exercise 62

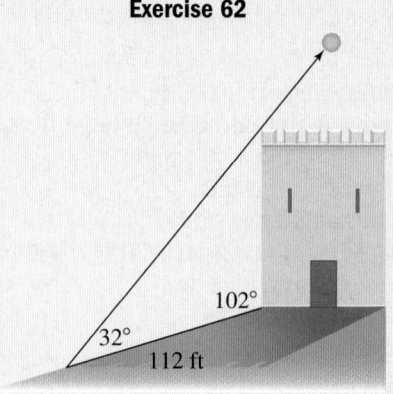

63. Distance to a fire: In
Yellowstone Park, a fire is
spotted by park rangers
stationed in two towers that
are known to be 5 mi apart.
Using the line between
them as a baseline, tower A
reports the fire is at an
angle of 39°, while tower B reports an angle of 58°.
How far is the fire from the closer tower?

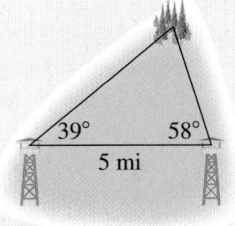

**64. Width of a
canyon:** To find
the distance across
Waimea Canyon
(on the island of
Kauai), a surveyor
marks a 1000-m
baseline along the
southern rim.
Using a transit, she
sights on a large
rock formation on
the north rim, and
finds the angles indicated. How wide is the canyon
from point *B* to point *C*?

Exercise 64

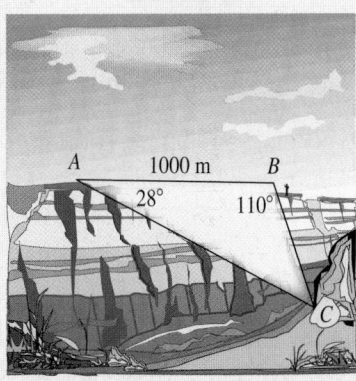

65. Height of a blimp:
When the Good-
Year Blimp is
viewed from the
field-level bleachers
near the southern
end-zone of a
football stadium, the
angle of elevation is
62°. From the field-
level bleachers near
the northern end-
zone, the angle of elevation is 70°. Find the height of
the blimp if the distance from the southern bleachers
to the northern bleachers is 145 yd.

Exercise 65

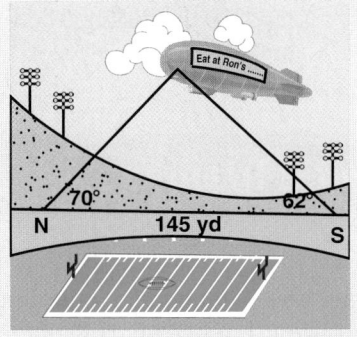

66. Height of a blimp: The rock-n-roll group *Pink
Floyd* just finished their most recent tour and has
moored their touring blimp at a hangar near the
airport in Indianapolis, Indiana. From an unknown
distance away, the angle of elevation is measured at
26.5°. After moving 110 yd closer, the angle of

elevation has become 48.3°. At what height is the
blimp moored?

**67. Circumscribed
triangles:** A triangle is
circumscribed within
the upper semicircle
drawn in the figure. Use
the law of sines to solve
the triangle given the
measures shown. What
is the diameter of the
circle? What do you notice about the triangle?

Exercises 67 and 68

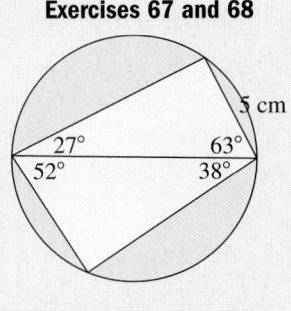

68. Circumscribed triangles: A triangle is
circumscribed within the lower semicircle shown.
Use the law of sines to solve the triangle given the
measures shown. How long is the longer chord?
What do you notice about the triangle?

69. Height of a mountain:
Approaching from the
west, a group of
hikers notes the angle
of elevation to the
summit of a steep
mountain is 35° at a
distance of 1250
meters. Arriving at the base of the mountain, they
estimate this side of the mountain has an average
slope of 48°. (a) Find the slant height of the
mountain's west side. (b) Find the slant height of
the east side of the mountain, if the east side has an
average slope of 65°. (c) How tall is the mountain?

Exercise 69

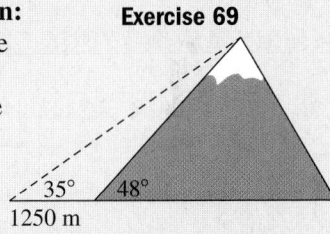

70. Distance on a map: Coffeyville and Liberal,
Kansas, lie along the state's southern border and
are roughly 298 miles apart. Olathe, Kansas, is
very near the
state's eastern
border at an
angle of 23°
with Liberal
and 72° with
Coffeyville
(using the
southern border as one side of the angle).
(a) Compute the distance between these cities.
(b) What is the shortest (straight line) distance
from Olathe to the southern border of Kansas?

Exercise 70

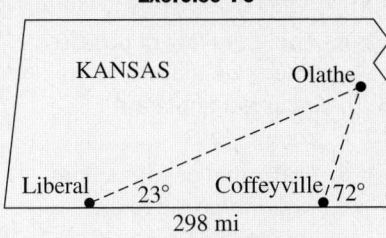

▶ **EXTENDING THE CONCEPT**

71. Solve the triangle shown in three ways—first by
using the law of sines, second using right triangle
trigonometry, and third using the standard 30-60-90
triangle. Was one method "easier" than the others?
Use these connections to express the irrational
number $\sqrt{3}$ as a quotient of two trigonometric

functions of an
angle. Can you
find a similar
expression
for $\sqrt{2}$?

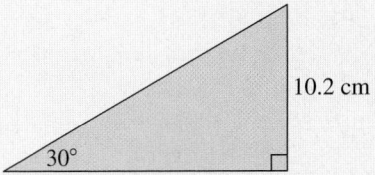

72. Use the law of sines and any needed identities to solve the triangles shown.

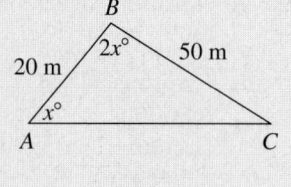

73. Similar to the law of sines, there is a *law of tangents*. The law says for any triangle

$$ABC, \frac{a + b}{a - b} = \frac{\tan\left[\frac{1}{2}(A + B)\right]}{\tan\left[\frac{1}{2}(A - B)\right]}.$$

Use the law of tangents to solve the triangle shown.

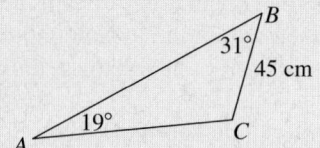

74. Lines L_1 and L_2 shown are parallel. The three triangles between these lines all share the same base (in bold). Explain why all three triangles must have the same area.

Exercise 74

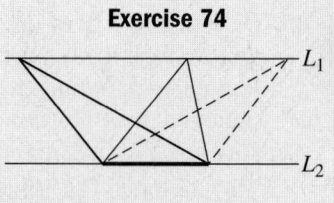

75. A UFO is sighted on a direct line between the towns of Batesville and Cave City, sitting stationary in the sky. The towns are 13 mi apart as the crow flies. A student in Batesville calls a friend in Cave City and both take measurements of the angle of elevation: 35° from Batesville and 42° from Cave City. Suddenly the UFO zips across the sky at a level altitude heading directly for Cave City, then stops and hovers long enough for an additional measurement from Batesville: 24°. If the UFO was in motion for 1.2 sec, at what average speed (in mph) did it travel?

▶ **MAINTAINING YOUR SKILLS**

76. (6.7) Find all solutions to the equation $2 \sin x = \cos(2x)$

77. (6.2) Prove the given identity:
$\tan^2 x - \sin^2 x = \tan^2 x \sin^2 x$

78. (3.3) Write an equation for the real polynomial with smallest degree, possible, having the solutions $x = 2, x = -1,$ and $x = 1 + 2i$.

79. (2.3) Given the points $(-5, -3)$ and $(4, 2)$, find (a) the equation of the line containing these points and (b) the distance between these points.

7.2 The Law of Cosines; the Area of a Triangle

Learning Objectives

In Section 7.2 you will learn how to:

- ☐ **A.** Apply the law of cosines when two sides and an included angle are known (SAS)
- ☐ **B.** Apply the law of cosines when three sides are known (SSS)
- ☐ **C.** Solve applications using the law of cosines
- ☐ **D.** Use trigonometry to find the area of a triangle

The distance formula $d = \sqrt{(x_2 - x_1)^2 + (y_2 - y_1)^2}$ is traditionally developed by placing two arbitrary points on a rectangular coordinate system and using the Pythagorean theorem. The relationship known as the *law of cosines* is developed in much the same way, but this time by using *three* arbitrary points (the vertices of a triangle). After giving the location of one vertex in trigonometric form, we obtain a formula that enables us to solve SSS and SAS triangles, which cannot be solved using the law of sines alone.

A. The Law of Cosines and SAS Triangles

In situations where all three sides are known (but no angles), the law of sines cannot be applied. The same is true when two sides and the angle between them are known, since we must have an angle opposite one of the sides. In these two cases (Figure 7.21), side-side-side (**SSS**) and side-angle-side (**SAS**), we use the **law of cosines.**

Figure 7.21
Law of Sines cannot be applied.

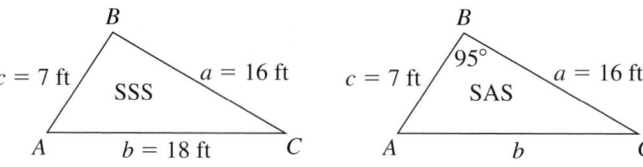

Keep in mind that the sum of any two sides of a triangle must be greater than the remaining side. For example, if $a = 7$, $B = 20$, and $C = 12$, no triangle is possible (see the figure).

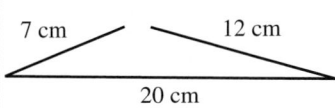

To solve these cases, it's evident we need additional insight on the unknown angles. Consider a general triangle ABC on the rectangular coordinate system conveniently placed with vertex A at the origin, side c along the x-axis, and the vertex C at some point (x, y) in QI (Figure 7.22). Note $\cos \theta = \dfrac{x}{b}$ giving $x = b \cos \theta$, and $\sin \theta = \dfrac{y}{b}$ or $y = b \sin \theta$. This means we can write the point (x, y) as $(b \cos \theta, b \sin \theta)$ as shown, and use the Pythagorean theorem with side $x - c$ to find the length of side a of the exterior, right triangle. It follows that

Figure 7.22

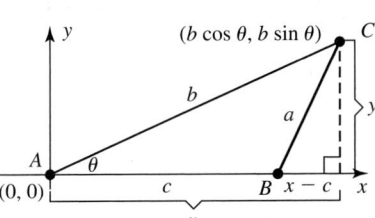

$$
\begin{aligned}
a^2 &= (x - c)^2 + y^2 & \text{Pythagorean theorem} \\
&= (b \cos \theta - c)^2 + (b \sin \theta)^2 & \text{substitute } b \cos \theta \text{ for } x \text{ and } b \sin \theta \text{ for } y \\
&= b^2 \cos^2 \theta - 2bc \cos \theta + c^2 + b^2 \sin^2 \theta & \text{square binomial, square term} \\
&= b^2 \cos^2 \theta + b^2 \sin^2 \theta + c^2 - 2bc \cos \theta & \text{rearrange terms} \\
&= b^2 (\cos^2 \theta + \sin^2 \theta) + c^2 - 2bc \cos \theta & \text{factor out } b^2 \\
&= b^2 + c^2 - 2bc \cos \theta & \text{substitute 1 for } \cos^2 \theta + \sin^2 \theta
\end{aligned}
$$

We now have a formula relating all three sides and an included angle. Since the naming of the angles is purely arbitrary, the formula can be used in any of the three forms shown. For the derivation of the formula where $\angle B$ is acute, **see Exercise 61.**

The Law of Cosines

For any triangle ABC and corresponding sides a, b, and c,

$$a^2 = b^2 + c^2 - 2bc \cos A$$
$$b^2 = a^2 + c^2 - 2ac \cos B$$
$$c^2 = a^2 + b^2 - 2ab \cos C$$

Note the relationship between the indicated angle and the squared term.

In words, the law of cosines says that the square of any side is equal to the sums of the squares of the other two sides, minus twice their product times the cosine of the included angle. It is interesting to note that if the included angle is 90°, the formula reduces to the Pythagorean theorem since $\cos 90° = 0$.

EXAMPLE 1 ▶ **Verifying the Law of Cosines**

For the triangle shown, verify:

a. $c^2 = a^2 + b^2 - 2ab \cos C$

b. $b^2 = a^2 + c^2 - 2ac \cos B$

Solution ▶ Note the included angle C is a right angle.

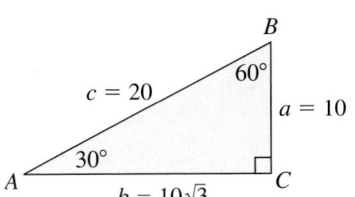

a. $c^2 = a^2 + b^2 - 2ab \cos C$

$20^2 = 10^2 + (10\sqrt{3})^2 - 2(10\sqrt{3})(10)\cos 90°$

$400 = 100 + 300 - 0$

$\quad\quad = 400 ✓$

b. $\quad b^2 = a^2 + c^2 - 2ac \cos B$

$(10\sqrt{3})^2 = 10^2 + 20^2 - 2(10)(20)\cos 60°$

$300 = 100 + 400 - 400\left(\dfrac{1}{2}\right)$

$\quad\quad = 500 - 200$

$\quad\quad = 300 ✓$

Now try Exercises 7 through 14 ▶

> ⚠ **CAUTION** ▶ When evaluating the law of cosines, a common error is to combine the coefficient of cos θ with the squared terms (the terms shown in blue): $a^2 = b^2 + c^2 - 2bc \cos A$. Be sure to use the correct order of operations when simplifying the expression.

Once additional information about the triangle is known, the law of sines can be used to complete the solution.

EXAMPLE 2 ▶ **Solving a Triangle Using the Law of Cosines—SAS**

Solve the triangle shown. Write the solution in table form.

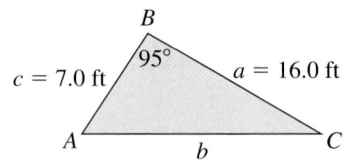

Solution ▶ The given information is SAS. Apply the law of cosines with respect to side b and $\angle B$:

$b^2 = a^2 + c^2 - 2ac \cos B$ law of cosines with respect to b

$b^2 = (16)^2 + (7)^2 - 2(16)(7)\cos 95°$ substitute known values

$\quad \approx 324.522886$ simplify

$b \approx 18.0$ $\sqrt{324.522886} \approx 18.0$

WORTHY OF NOTE

After using the law of cosines, we often use the law of sines to complete a solution. With a little foresight, we can avoid the ambiguous case—since the ambiguous case occurs only if θ *could be* obtuse (the largest angle of the triangle). After calculating the third side of a SAS triangle using the law of cosines, use the law of sines to find the smallest angle, since it cannot be obtuse. For SSS triangles, using the law of cosines to find the largest angle will ensure that when the second angle is found using the law of sines, it cannot be obtuse.

We now have side b opposite $\angle B$, and complete the solution using the law of sines, selecting the smaller angle to avoid the ambiguous case (we *could* apply the law of cosines again, if we chose).

$\dfrac{\sin C}{c} = \dfrac{\sin B}{b}$ law of sines applied to $\angle C$ and $\angle B$

$\dfrac{\sin C}{7} \approx \dfrac{\sin 95°}{18}$ substitute given values

$\sin C \approx 7 \cdot \dfrac{\sin 95°}{18}$ solve for $\sin C$

$C \approx \sin^{-1}\left(\dfrac{7 \sin 95°}{18}\right)$ apply $\sin^{-1}$

$\quad \approx 22.8°$ result

For the remaining angle, $\angle C$: $180° - (95° + 22.8°) = 62.2°$. The finished solution is shown in the table (given information is in bold).

Angles	Sides (ft)
$A \approx 62.2°$	$a = \mathbf{16.0}$
$B = \mathbf{95.0°}$	$b \approx 18.0$
$C \approx 22.8°$	$c = \mathbf{7.0}$

✅ **A.** You've just learned how to apply the law of cosines when two sides and an included angle are known (SAS)

Now try Exercises 15 through 26 ▶

B. The Law of Cosines and SSS Triangles

When three sides of a triangle are given, we use the law of cosines to find any one of the three angles. As a good practice, we first find the *largest* angle, or the angle opposite the largest side. This will ensure that the remaining two angles are acute, avoiding the ambiguous case if the law of sines is used to complete the solution.

EXAMPLE 3 ▶ **Solving a Triangle Using the Law of Cosines—SSS**

Solve the triangle shown. Write the solution in table form, with angles rounded to tenths of a degree.

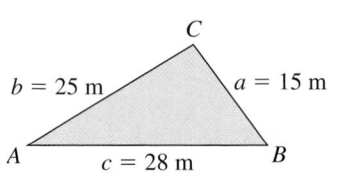

Solution ▶ The information is given as SSS. Since side c is the longest side, we apply the law of cosines with respect to side c and $\angle C$:

$$c^2 = a^2 + b^2 - 2ab \cos C \qquad \text{law of cosines with respect to } c$$
$$28^2 = (15)^2 + (25)^2 - 2(15)(25)\cos C \qquad \text{substitute known values}$$
$$784 = 850 - 750 \cos C \qquad \text{simplify}$$
$$-66 = -750 \cos C \qquad \text{isolate variable term}$$
$$0.088 = \cos C \qquad \text{divide}$$
$$\cos^{-1} 0.088 = C \qquad \text{solve for } C$$
$$85.0 \approx C \qquad \text{result}$$

We now have side c opposite $\angle C$ and finish up using the law of sines.

$$\frac{\sin A}{a} = \frac{\sin C}{c} \qquad \text{law of sines applied to } \angle A \text{ and } \angle C$$

$$\frac{\sin A}{15} \approx \frac{\sin 85°}{28} \qquad \text{substitute given values}$$

$$\sin A \approx 15 \cdot \frac{\sin 85°}{28} \qquad \text{solve for } \sin A$$

$$\approx 0.5336757311 \qquad \text{simplify}$$

$$A \approx \sin^{-1} 0.5336757311 \qquad \text{solve for } A$$

$$\approx 32.3° \qquad \text{result}$$

☑ **B. You've just learned how to apply the law of cosines when three sides are known (SSS)**

Since the remaining angle must be acute, we compute it directly. $\angle B$: $180° - (85° + 32.3°) = 62.7°$. The finished solution is shown in the table, with the information originally given shown in bold.

Angles	Sides (m)
$A \approx 32.3°$	$a = 15$
$B \approx 62.7°$	$b = 25$
$C \approx 85°$	$c = 28$

Now try Exercises 27 through 34 ▶

C. Applications Using the Law of Cosines

As with the law of sines, the law of cosines has a large number of applications from very diverse fields including geometry, navigation, surveying, and astronomy, as well as being put to use in solving recreational exercises **(see Exercises 37 through 40).**

EXAMPLE 4 ▶ **Solving an Application of the Law of Cosines—Geological Surveys**

A volcanologist needs to measure the distance across the base of an active volcano. Distance AB is measured at 1.5 km, while distance AC is 3.2 km. Using a theodolite (a sighting instrument used by surveyors), angle BAC is found to be 95.7°. What is the distance across the base?

Solution ▶ The information is given as SAS. To find the distance BC across the base of the volcano, we apply the law of cosines with respect to $\angle A$.

$$a^2 = b^2 + c^2 - 2bc \cos A \qquad \text{law of cosines with respect to } a$$
$$= (1.5)^2 + (3.2)^2 - 2(1.5)(3.2)\cos 95.7° \qquad \text{substitute known values}$$
$$\approx 13.44347 \qquad \text{simplify}$$

☑ **C. You've just learned how to solve applications using the law of cosines**

$$a \approx 3.7 \qquad \text{solve for } a$$

The volcano is approximately 3.7 km wide at its base.

Now try Exercises 41 through 52 ▶

A variety of additional applications can be found in the exercise set (**see Exercises 45 through 52**).

D. Trigonometry and the Area of a Triangle

While you're likely familiar with the most common formula for a triangle's area, $A = \frac{1}{2}bh$, there are actually over 20 formulas for computing this area. Many involve basic trigonometric ideas, and we'll use some of these ideas to develop three additional formulas here.

For $A = \frac{1}{2}bh$, recall that b represents the length of a designated base, and h represents the length of the altitude drawn to that base (see Figure 7.23). If the height h is unknown, but sides a and b with angle C between them are known, h can be found using $\sin C = \dfrac{h}{a}$, giving $h = a \sin C$. Figure 7.24 indicates the same result is obtained if C is obtuse, since $\sin(180° - C) = \sin C$. Substituting for in the formula $A = \frac{1}{2}b$ gives $A = \frac{1}{2}b$, or $A = \frac{1}{2}ab \sin C$ in more common form. Since naming the angles in a triangle is arbitrary, the formulas $A = \frac{1}{2}bc \sin A$ and $A = \frac{1}{2}ac \sin B$ can likewise be obtained.

Figure 7.23

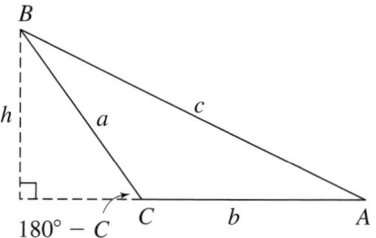

Figure 7.24

Area Given Two Sides and an Included Angle (SAS)

1. $A = \dfrac{1}{2} ab \sin C$ **2.** $A = \dfrac{1}{2} bc \sin A$ **3.** $A = \dfrac{1}{2} ac \sin B$

In words, the formulas say the area of a triangle is equal to one-half the product of two sides times the sine of the angle between them.

EXAMPLE 5 ▶ **Finding the Area of a Nonright Triangle**

Find the area of $\triangle ABC$, if $a = 16.2$ cm, $b = 25.6$ cm, and $C = 28.3°$.

Solution ▶ Since sides a and b and angle C are given, we apply the first formula.

$$A = \frac{1}{2} ab \sin C \qquad \text{area formula}$$

$$= \frac{1}{2}(16.2)(25.6) \sin 28.3° \qquad \text{substitute 15.2 for } a, \text{ 25.6 for } b, \text{ and 28.3° for } C$$

$$\approx 98.3 \text{ cm}^2 \qquad \text{result}$$

The area of this triangle is approximately 98.3 cm².

Now try Exercises 53 and 54 ▶

Using these formulas, a second formula type requiring two angles and one side (AAS or ASA) can be developed. Solving for b in $\mathcal{A} = \frac{1}{2}bc \sin A$ gives $b = \frac{2\mathcal{A}}{c \sin A}$. Likewise, solving for a in $\mathcal{A} = \frac{1}{2}ac \sin B$ yields $a = \frac{2\mathcal{A}}{c \sin B}$. Substituting these for b and a in $\mathcal{A} = \frac{1}{2}ab \sin C$ gives

$$\mathcal{A} = \frac{1}{2} \cdot a \cdot b \cdot \sin C \qquad \text{given formula}$$

$$2\mathcal{A} = \frac{2\mathcal{A}}{c \sin B} \cdot \frac{2\mathcal{A}}{c \sin A} \cdot \sin C \qquad \text{substitute } \frac{2\mathcal{A}}{c \sin B} \text{ for } a, \frac{2\mathcal{A}}{c \sin A} \text{ for } b; \text{ multiply by 2}$$

$$c^2 \sin A \cdot \sin B = 2\mathcal{A} \cdot \sin C \qquad \text{multiply by } c \sin A \cdot c \sin B; \text{ divide by } 2\mathcal{A}$$

$$\frac{c^2 \sin A \sin B}{2 \sin C} = \mathcal{A} \qquad \text{solve for } \mathcal{A}$$

As with the previous formula, versions relying on side a or side b can also be found.

Area Given Two Angles and Any Side (AAS/ASA)

1. $\mathcal{A} = \dfrac{c^2 \cdot \sin A \cdot \sin B}{2 \sin C}$ **2.** $\mathcal{A} = \dfrac{a^2 \cdot \sin B \cdot \sin C}{2 \sin A}$ **3.** $\mathcal{A} = \dfrac{b^2 \cdot \sin A \cdot \sin C}{2 \sin B}$

EXAMPLE 6 ▶   **Finding the Area of a Nonright Triangle**

Find the area of $\triangle ABC$ if $a = 34.5$ ft, $B = 87.9°$, and $C = 29.3°$.

Solution ▶   Since side a is given, we apply the second version of the formula. First we find the measure of angle A, then make the appropriate substitutions:

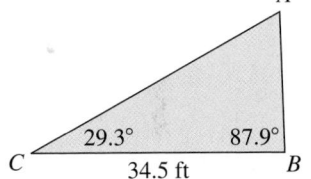

$A = 180° - (87.9 + 29.3)° = 62.8°$

$\mathcal{A} = \dfrac{a^2 \sin B \sin C}{2 \sin A}$ area formula—side a

$\quad = \dfrac{(34.5)^2 \sin 87.9° \sin 29.3°}{2 \sin 62.8°}$ substitute 34.5 for a, 87.9° for B, 29.3° for C, and 62.8° for A

$\quad \approx 327.2 \text{ ft}^2$ simplify

The area of this triangle is approximately 327.2 ft^2.

Now try Exercises 55 and 56 ▶

Our final formula for a triangle's area is a useful addition to the other two, as it requires only the lengths of the three sides. The development of the formula requires only a Pythagorean identity and solving for the angle C in the law of cosines, as follows.

$$a^2 + b^2 - 2ab \cos C = c^2 \qquad \text{law of cosines}$$

$$a^2 + b^2 - c^2 = 2ab \cos C \qquad \text{add } 2ab \cos C, \text{ subtract } c^2$$

$$\frac{a^2 + b^2 - c^2}{2ab} = \cos C \qquad \text{divide by } 2ab$$

Beginning with our first area formula, we then have

$$\mathcal{A} = \frac{1}{2} ab \sin C \qquad \text{previous area formula}$$

$$= \frac{1}{2} ab \sqrt{1 - \cos^2 C} \qquad \sin^2 C + \cos^2 C = 1 \rightarrow \sin C = \sqrt{1 - \cos^2 C}$$

$$= \frac{1}{2} ab \sqrt{1 - \left(\frac{a^2 + b^2 - c^2}{2ab}\right)^2} \qquad \text{substitute } \frac{a^2 + b^2 + c^2}{2ab} \text{ for cos } C$$

and can find the area of any triangle given its three sides. While the formula certainly serves this purpose, it is not so easy to use. By working algebraically and using the perimeter of the triangle, we can derive a more elegant version.

$$\mathcal{A}^2 = \frac{1}{4} a^2 b^2 \left[1 - \left(\frac{a^2 + b^2 - c^2}{2ab}\right)^2\right] \qquad \text{square both sides}$$

$$= \frac{1}{4} a^2 b^2 \left[1 + \left(\frac{a^2 + b^2 - c^2}{2ab}\right)\right]\left[1 - \left(\frac{a^2 + b^2 - c^2}{2ab}\right)\right] \qquad \begin{array}{l}\text{factor as a difference of} \\ \text{squares}\end{array}$$

$$= \frac{1}{4} a^2 b^2 \left[\frac{2ab + a^2 + b^2 - c^2}{2ab}\right]\left[\frac{2ab - a^2 - b^2 + c^2}{2ab}\right] \qquad 1 = \frac{2ab}{2ab}; \text{combine terms}$$

$$= \frac{1}{4} \left[\frac{(a^2 + 2ab + b^2) - c^2}{2}\right]\left[\frac{-(a^2 - 2ab + b^2) + c^2}{2}\right] \qquad \begin{array}{l}\text{rewrite/regroup} \\ \text{numerator; cancel } a^2 b^2\end{array}$$

$$= \frac{1}{16} \left[(a + b)^2 - c^2\right]\left[c^2 - (a - b)^2\right] \qquad \text{factor (binomial squares)}$$

$$= \frac{1}{16} (a + b + c)(a + b - c)(c + a - b)(c - a + b) \qquad \text{factor (difference of squares)}$$

For the perimeter $p = a + b + c$, we note the following relationships:

$$a + b - c = p - 2c \qquad c + a - b = p - 2b \qquad c - a + b = p - 2a$$

and making the appropriate substitutions gives

$$= \frac{1}{16} p(p - 2c)(p - 2b)(p - 2a) \qquad \text{substitute}$$

While this would provide a usable formula for the area in terms of the perimeter, we can refine it further using the *semi*perimeter $s = \dfrac{a + b + c}{2} = \dfrac{p}{2}$. Since $\dfrac{1}{16} = \left(\dfrac{1}{2}\right)^4$, we can write the expression as

$$= \frac{p}{2}\left(\frac{p - 2c}{2}\right)\left(\frac{p - 2b}{2}\right)\left(\frac{p - 2a}{2}\right) \qquad \text{rewrite expression}$$

$$= \frac{p}{2}\left(\frac{p}{2} - c\right)\left(\frac{p}{2} - b\right)\left(\frac{p}{2} - a\right) \qquad \text{simplify}$$

$$= s(s - c)(s - b)(s - a) \qquad \text{substitute } s \text{ for } \frac{p}{2}$$

Taking the square root of each side produces what is known as **Heron's formula.**

$$\mathcal{A} = \sqrt{s(s - a)(s - b)(s - c)} \qquad \text{Heron's formula}$$

Heron's Formula

Given $\triangle ABC$ with sides a, b, and c and semiperimeter $s = \dfrac{a + b + c}{2}$, the area of the triangle is

$$\mathcal{A} = \sqrt{s(s - a)(s - b)(s - c)}$$

EXAMPLE 7 ▶ Solving an Application of Heron's Formula—Construction Planning

A New York City developer wants to build condominiums on the triangular lot formed by Greenwich, Watts, and Canal Streets. How many square meters does the developer have to work with if the frontage along each street is approximately 34.1 m, 43.5 m, and 62.4 m, respectively?

Solution ▶ The perimeter of the lot is $p = 34.1 + 43.5 + 62.4 = 140$ m, so $s = 70$ m. By direct substitution we obtain

$$\mathcal{A} = \sqrt{s(s - a)(s - b)(s - c)} \qquad \text{Heron's formula}$$
$$= \sqrt{70(70 - 34.1)(70 - 43.5)(70 - 62.4)} \qquad \text{substitute known values}$$
$$= \sqrt{70(35.9)(26.5)(7.6)} \qquad \text{simplify}$$
$$= \sqrt{506{,}118.2} \qquad \text{multiply}$$
$$\approx 711.4 \qquad \text{result}$$

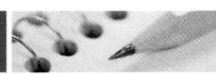 **D.** You've just learned how to use trigonometry to find the area of a triangle

The developer has about 711.4 m^2 of land to work with.

Now try Exercises 57 and 58 ▶

For a derivation of Heron's formula that does not depend on trigonometry, see Appendix V.

7.2 EXERCISES

▶ CONCEPTS AND VOCABULARY

Fill in each blank with the appropriate word or phrase. Carefully reread the section if needed.

1. When the information given is SSS or SAS, the law of _____ is used to solve the triangle.

2. Fill in the blank so that the law of cosines is complete: $c^2 = a^2 + b^2 - $ _____ $\cos C$

3. If the law of cosines is applied to a right triangle, the result is the same as the _____ theorem, since $\cos 90° = 0$.

4. Write out which version of the law of cosines you would use to begin solving the triangle shown: _____

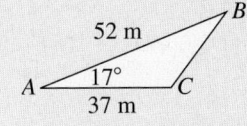

5. Solve the triangle in Exercise 4 using only the law of cosines, then by using the law of cosines followed by the law of sines. Which method was more efficient?

6. Begin with $a^2 = b^2 + c^2 - 2bc \cos A$ and write $\cos A$ in terms of a, b, and c (solve for $\cos A$). Why must $b^2 + c^2 - a^2 < 2bc$ hold in order for a solution to exist?

▶ DEVELOPING YOUR SKILLS

Determine whether the law of cosines can be used to begin the solution process for each triangle.

7.

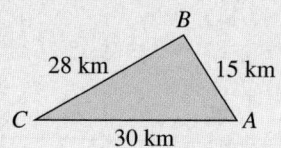

8.

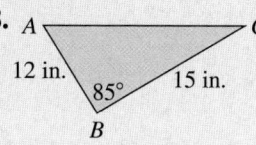

9.

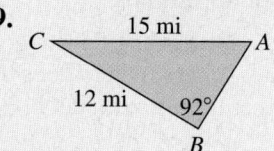

10.

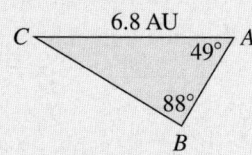

11.

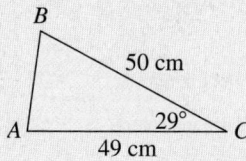

12.

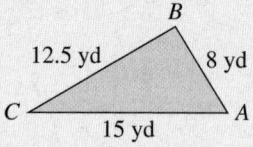

For each triangle, verify all three forms of the law of cosines.

13.

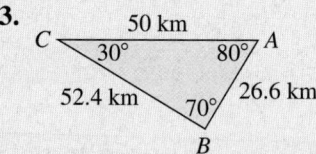

14.

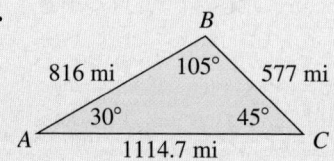

Solve each of the following equations for the unknown part.

15. $4^2 = 5^2 + 6^2 - 2(5)(6)\cos B$

16. $12.9^2 = 15.2^2 + 9.8^2 - 2(15.2)(9.8)\cos C$

17. $a^2 = 9^2 + 7^2 - 2(9)(7)\cos 52°$

18. $b^2 = 3.9^2 + 9.5^2 - 2(3.9)(9.5)\cos 30°$

19. $10^2 = 12^2 + 15^2 - 2(12)(15)\cos A$

20. $202^2 = 182^2 + 98^2 - 2(182)(98)\cos B$

Solve each triangle using the law of cosines.

21. side $a = 75$ cm
$\angle C = 38°$
side $b = 32$ cm

22. side $b = 385$ m
$\angle C = 67°$
side $a = 490$ m

23. side $c = 25.8$ mi
$\angle B = 30°$
side $a = 12.9$ mi

Solve using the law of cosines (if possible). Label each triangle appropriately before you begin.

24.

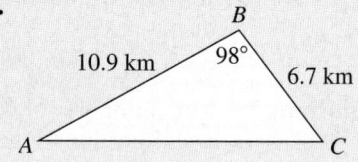

25.

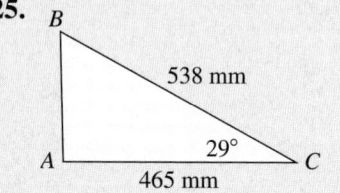

26.

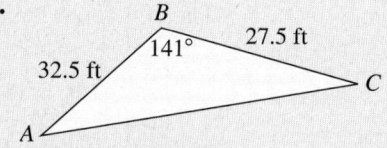

27. side $c = 10\sqrt{3}$ in.
side $b = 6\sqrt{3}$ in.
side $a = 15\sqrt{3}$ in.

28. side $a = 282$ ft
side $b = 129$ ft
side $c = 300$ ft

29. side $a = 32.8$ km
side $b = 24.9$ km
side $c = 12.4$ km

30.

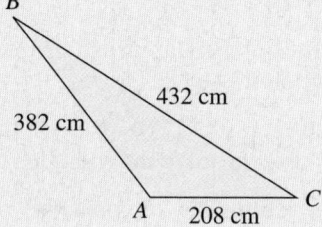

31.

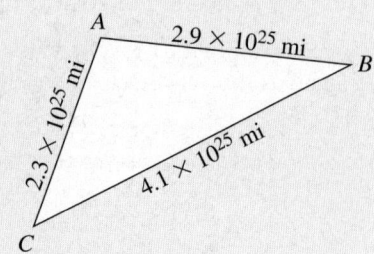

32.

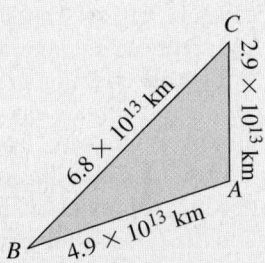

33. side $a = 12\sqrt{3}$ yd
side $b = 12.9$ yd
side $c = 9.2$ yd

34. side $a = 36.5$ AU
side $b = 12.9$ AU
side $c = 22$ AU

▶ **WORKING WITH FORMULAS**

35. Alternative form for the law of cosines:

$$\cos A = \frac{b^2 + c^2 - a^2}{2bc}$$

By solving the law of cosines for the cosine of the angle, the formula can be written as shown. Derive this formula (solve for $\cos \theta$), beginning from $a^2 = b^2 + c^2 - 2bc \cos A$, then use this form to begin the solution of the triangle given.

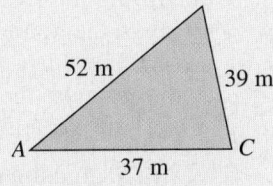

36. The Perimeter of a Trapezoid:
$$P = a + b + h(\csc \alpha + \csc \beta)$$

The perimeter of a trapezoid can be found using the formula shown, where a and b represent the lengths of the parallel sides, h is the height of the trapezoid, and α and β are the base angles. Find the perimeter of Trapezoid Park (to the nearest foot) if $a = 5000$ ft, $b = 7500$ ft, and $h = 2000$ ft, with base angles $\alpha = 42°$ and $\beta = 78°$.

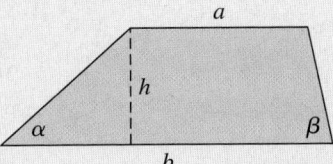

▶ **APPLICATIONS**

37. Distance between cities: The satellite Mercury II measures its distance from Portland and from Green Bay using radio waves as shown. Using an on-board sighting device, the satellite determines that $\angle M$ is 99°. How many miles is it from Portland to Green Bay?

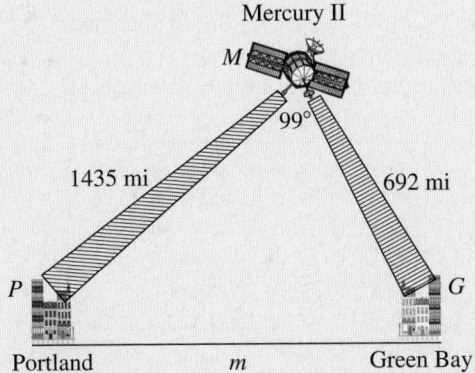

38. Distance between cities: Voyager VII measures its distance from Los Angeles and from San Francisco using radio waves as shown. Using an on-board sighting device, the satellite determines $\angle V$ is 95°. How many kilometers separate Los Angeles and San Francisco?

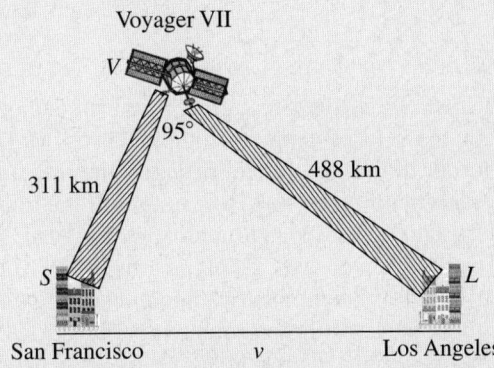

39. Trip planning: A business executive is going to fly the corporate jet from Providence to College Cove.

Exercise 39

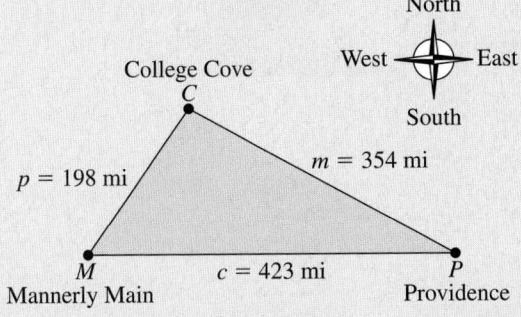

WORTHY OF NOTE

In navigation, there are two basic methods for defining a course. *Headings* are understood to be the amount of rotation from due north in the clockwise direction $(0 \le \theta < 360°)$. *Bearings* give the number of degrees East or West from a due North or due South orientation, hence the angle indicated is always less than 90°. For instance, the bearing N 25° W and a heading of 335° would indicate the same direction.

She calculates the distances shown using a map, with Mannerly Main for reference since it is due east of Providence. What is the measure of angle *P*? What heading should she set for this trip?

40. **Trip planning:** A troop of Scouts is planning a hike from *M*ontgomery to *P*attonville. They calculate the distances shown using a map, using *B*radleyton for reference since it is due east of *M*ontgomery. What is the measure of angle *M*? What heading should they set for this trip?

Exercise 40

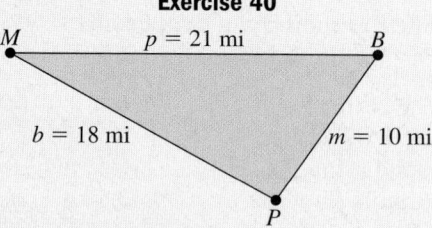

41. **Runway length:** Surveyors are measuring a large, marshy area outside of the city as part of a feasibility study for the construction of a new airport. Using a theodolite and the markers shown gives the information indicated. If the main runway must be at least 11,000 ft long, and environmental concerns are satisfied, can the airport be constructed at this site (recall that 1 mi = 5280 ft)?

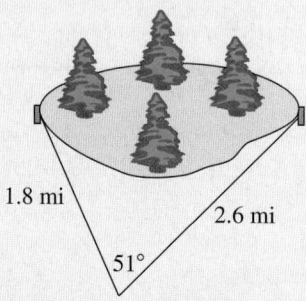

42. **Tunnel length:** An engineering firm decides to bid on a proposed tunnel through Harvest Mountain. In order to find the tunnel's length, the measurements shown are taken. (a) How long will the tunnel be? (b) Due to previous tunneling experience, the firm estimates a cost of $5000 per foot for boring through this type of rock and constructing the tunnel according to required specifications. If management insists on a 25% profit, what will be their minimum bid to the nearest hundred?

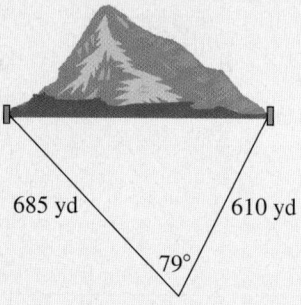

43. **Aerial distance:** Two planes leave Los Angeles International Airport at the same time. One travels due west (at heading 270°) with a cruising speed of 450 mph, going to Tokyo, Japan, with a group that seeks tranquility at the foot of Mount Fuji. The other travels at heading 225° with a cruising speed of 425 mph, going to Brisbane, Australia, with a group seeking adventure in the Great Outback. Approximate the distance between the planes after 5 hr of flight.

44. **Nautical distance:** Two ships leave Honolulu Harbor at the same time. One travels 15 knots (nautical miles per hour) at heading 150°, and is going to the Marquesas Islands (*Crosby, Stills, and Nash*). The other travels 12 knots at heading 200°, and is going to the Samoan Islands (*Samoa, le galu a tu*). How far apart are the two ships after 10 hr?

45. **Geoboard geometry:** A rubber band is placed on a geoboard (a board with all pegs 1 cm apart) as shown. Approximate the perimeter of the triangle formed by the rubber band *and* the angle formed at each vertex. (*Hint:* Use a standard triangle to find ∠*A* and length $\overline{AB}$.)

Exercise 45

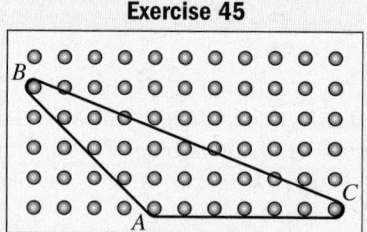

46. **Geoboard geometry:** A rubber band is placed on a geoboard as shown. Approximate the perimeter of the triangle formed by the rubber band *and* the angle formed at each vertex. (*Hint:* Use a Pythagorean triple, then find angle *A*.)

Exercise 46

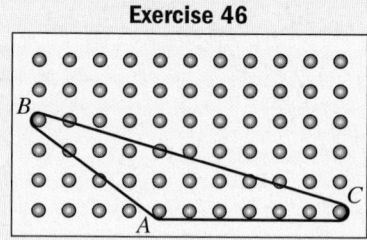

In Exercises 47 and 48, three rods are attached via pivot joints so the rods can be manipulated to form a triangle. Find the three angles of the triangle formed.

47.

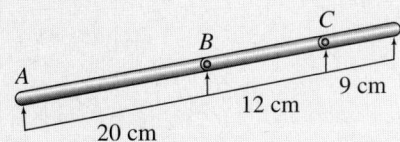

48.

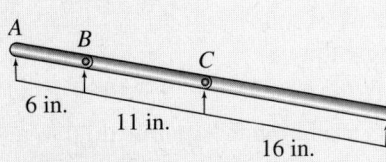

49. Pentagon perimeter: Find the perimeter of a regular *pentagon* that is circumscribed by a circle with radius $r = 10$ cm.

Exercise 49

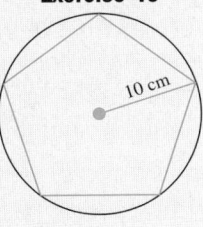

50. Hexagon perimeter: Find the perimeter of a regular *hexagon* that is circumscribed by a circle with radius $r = 15$ cm.

Solve the following triangles. Round sides and angles to the nearest tenth. (*Hint:* Use Pythagorean triples.)

51.

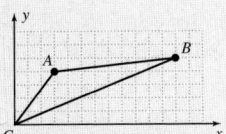

52.

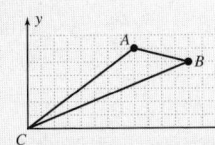

53. Billboard design: Creative Designs iNc. has designed a flashy, new billboard for one of its clients. Using a rectangular highway billboard measuring 20 ft by 30 ft, the primary advertising area is a triangle formed using the diagonal of the billboard as one side, and one-half the base as another (see figure). Use the dimensions given to find the angle α formed at the corner, then compute the area of the triangle using two sides and this included angle.

Exercise 53

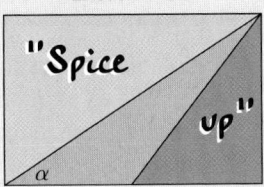

54. Area caught by surveillance camera: A stationary surveillance camera is set up to monitor activity in the parking lot of a shopping mall. If the camera has a 38° field of vision, how many square feet of the parking lot can it tape using the dimensions given?

Exercise 54

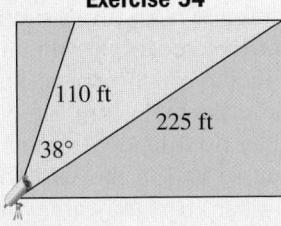

55. Pricing for undeveloped lots: Undeveloped land in a popular resort area is selling for $3,000,000/acre. Given the dimensions of the lot shown, (a) find what percent of a full acre is being purchased (to the nearest whole percent), and (b) compute the cost of the lot. Recall that 1 acre $= 43,560$ ft^2.

Exercise 55

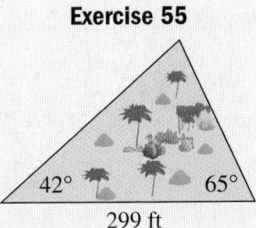

56. Area of the Nile River Delta: The Nile River Delta is one of the world's largest. The delta begins slightly up river from the Egyptian capitol (Cairo) and stretches along the Mediterranean from Alexandria in the west to Port Said in the east (over 240 km). Approximate the area of this rich agricultural region using the two triangles shown.

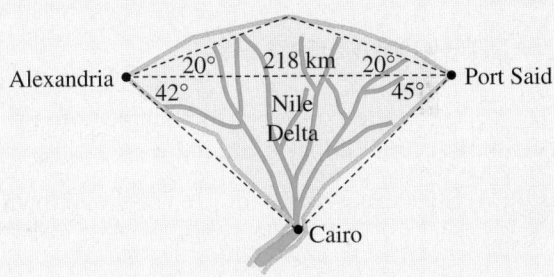

57. Area of the Yukon Territory: The Yukon Territory in northwest Canada is roughly triangular in shape with sides of 1289 km, 1063 km, and 922 km. What is the approximate area covered by this territory?

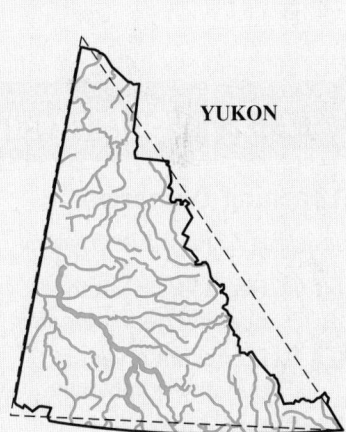

58. Alternate method for computing area: Referring to Exercise 53, since the dimensions of the billboard are known, all three sides of the triangle can actually be determined. Find the length of the sides rounded to the nearest whole, then use Heron's formula to find the area of the triangle. How close was your answer to that in Exercise 53?

▶ EXTENDING THE CONCEPT

59. No matter how hard I try, I cannot solve the triangle shown. Why?

Exercise 59

Sputnik 10

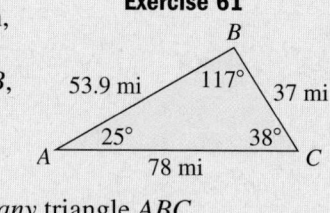

387 mi

502 mi

A   902 mi   *C*

60. In Figure 7.22 (page 647, note that if the *x*-coordinate of vertex *B* is greater than the *x*-coordinate of vertex *C*, ∠*B* becomes acute, and ∠*C* obtuse. How does this change the relationship between *x* and *c*? Verify the law of cosines remains unchanged.

61. For the triangle shown, verify that
$c = b \cos A + a \cos B$,
then use two different forms of the law of cosines to show this relationship holds for *any* triangle *ABC*.

Exercise 61

B

53.9 mi   117°   37 mi

25°   38°

A   78 mi   *C*

62. Most students are familiar with this double-angle formula for cosine: $\cos(2\theta) = \cos^2\theta - \sin^2\theta$. The *triple* angle formula for cosine is $\cos(3\theta) = 4\cos^3\theta - 3\cos\theta$. Use the formula to find an exact value for $\cos 135°$. Show that you get the same result as when using a reference angle.

▶ MAINTAINING YOUR SKILLS

63. (4.4) Write the expression as a single term in simplest form: $2\log_2 4 + 2\log_2 3 - 2\log_2 6$

64. (5.4) State exact forms for each of the following:
$\sin\left(\dfrac{\pi}{6}\right)$, $\cos\left(\dfrac{7\pi}{6}\right)$, and $\tan\left(\dfrac{\pi}{3}\right)$.

65. (5.7) Use fundamental identities to find the values of all six trig functions that satisfy the conditions.
$\sin x = -\dfrac{5}{13}$ and $\cos x > 0$.

66. (3.2) Use synthetic division to show $f(-2) > 0$ for $f(x) = -x^4 - x^3 + 7x^2 + x - 6$.

7.3 | Vectors and Vector Diagrams

Learning Objectives

In Section 7.3 you will learn how to:

☐ **A.** Represent a vector quantity geometrically

☐ **B.** Represent a vector quantity graphically

☐ **C.** Perform defined operations on vectors

☐ **D.** Represent a vector quantity algebraically and find unit vectors

☐ **E.** Use vector diagrams to solve applications

The study of vectors is closely connected to the study of force, motion, velocity, and other related phenomena. Vectors enable us to quantify certain characteristics of these phenomena and to physically represent their magnitude and direction with a simple model. To quantify something means we assign it a relative numeric value for purposes of study and comparison. While very uncomplicated, this model turns out to be a powerful mathematical tool.

A. The Notation and Geometry of Vectors

Measurements involving time, area, volume, energy, and temperature are called **scalar measurements** or **scalar quantities** because each can be adequately described by their magnitude alone and the appropriate unit or "scale." The related real number is simply called a **scalar.** Concepts that require more than a single quantity to describe their attributes are called **vector quantities.** Examples might include force, velocity, and displacement, which require knowing a magnitude *and* direction to describe them completely.

To begin our study, consider two identical airplanes flying at 300 mph, on a parallel course and in the same direction. Although we don't know how far apart they are, what direction they're flying, or if one is "ahead" of the other, we can still model, "300 mph on a parallel course," using **directed line segments** (Figure 7.25). Drawing these segments parallel with the arrowheads pointing the same way models the direction of

Figure 7.25

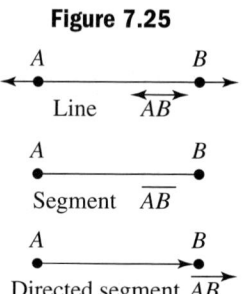

A   *B*
Line $\overleftrightarrow{AB}$

A   *B*
Segment $\overline{AB}$

A   *B*
Directed segment $\overrightarrow{AB}$

flight, while drawing segments the *same length* indicates the velocities are equal. The directed segment used to represent a vector quantity is simply called a **vector.** In this case the length of the vector models the **magnitude** of the velocity, while the arrowhead indicates the **direction** of travel. The origin of the segment is called the **initial point,** with the arrowhead pointing to the **terminal point.** Both are labeled using capital letters as shown in Figure 7.26 and we call this a *geometric representation* of the vectors.

Figure 7.26

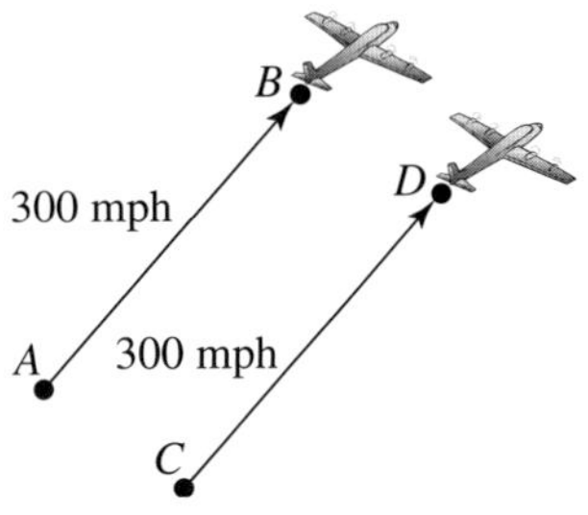

Vectors can be named using the initial and terminal points that define them (initial point first) as in $\overrightarrow{AB}$ and $\overrightarrow{CD}$, or using a bold, small case letter with the favorites being **v** (first letter of the word vector) and **u.** Other small case, bold letters can be used and subscripted vector names ($\mathbf{v}_1$, $\mathbf{v}_2$, $\mathbf{v}_3$, . . .) are also common. Two **vectors are equal** if they have the same magnitude and direction. For $\mathbf{u} = \overrightarrow{AB}$ and $\mathbf{v} = \overrightarrow{CD}$, we can say $\mathbf{u} = \mathbf{v}$ or $\overrightarrow{AB} = \overrightarrow{CD}$ since both airplanes are flying at the same speed and in the same direction (Figure 7.26).

Based on these conventions, it seems reasonable to represent an airplane flying at 600 mph with a vector that is twice as long as **u** and **v,** and one flying at 150 mph with a vector that is half as long. If all planes are flying in the same direction on a parallel course, we can represent them geometrically as shown in Figure 7.27, and state that $\mathbf{w} = 2\mathbf{v}$, $\mathbf{x} = \frac{1}{2}\mathbf{v}$, and $\mathbf{w} = 4\mathbf{x}$. The multiplication of a vector by a constant is called **scalar multiplication,** since the product changes only the scale or size of the vector and not its direction.

Figure 7.27

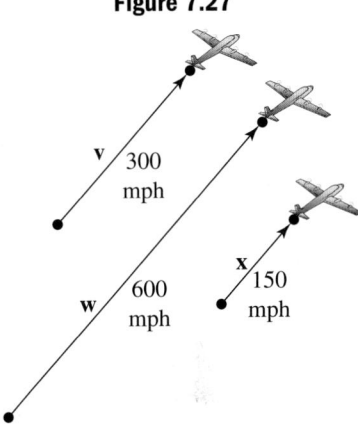

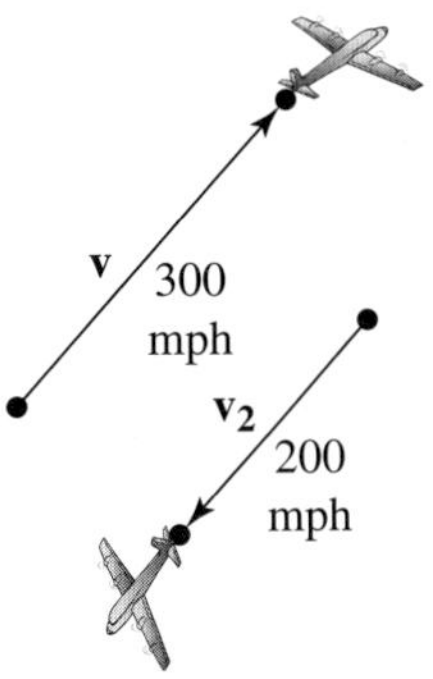

Finally, consider the airplane represented by vector $\mathbf{v}_2$, flying at 200 mph on a parallel course *but in the opposite direction* (see Figure 7.28). In this case, the directed segment will be $\frac{200}{300} = \frac{2}{3}$ as long as **v** and point in the opposite or "negative" direction. In perspective we can now state: $\mathbf{v}_2 = -\frac{2}{3}\mathbf{v}$, $\mathbf{v}_2 = -\frac{1}{3}\mathbf{w}$, $\mathbf{v}_2 = -\frac{4}{3}\mathbf{x}$, or any equivalent form of these equations.

Figure 7.28

EXAMPLE 1 ▶ **Using Geometric Vectors to Model Forces Acting on a Point**

Two tugboats are attempting to free a barge that is stuck on a sand bar. One is pulling with a force of 2000 newtons (N) in a certain direction, the other is pulling with a force of 1500 N in a direction that is *perpendicular to the first.* Represent the situation geometrically using vectors.

Solution ▶ We could once again draw a vector of arbitrary length and let it represent the 2000-N force applied by the first tugboat. For better perspective, we can actually use a ruler and choose a convenient length, say 6 cm. We then represent the pulling force of the second tug with a vector that is $\frac{1500}{2000} = \frac{3}{4}$ as long (4.5 cm), drawn at a 90° angle with relation to the first. Note that many correct solutions are possible, depending on the direction of the first vector drawn.

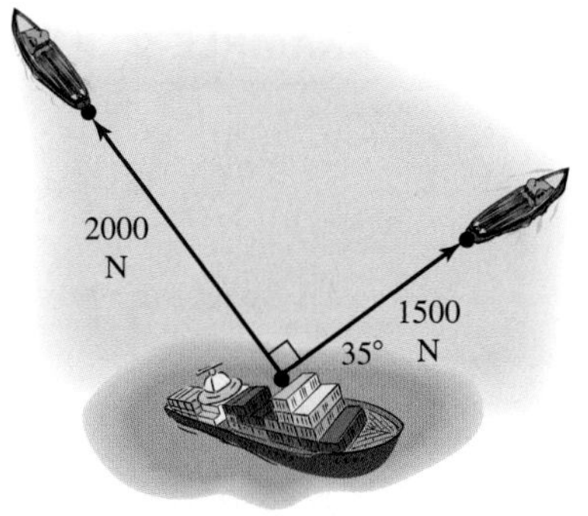

☑ **A.** You've just learned how to represent a vector quantity geometrically

Now try Exercises 7 through 12 ▶

B. Vectors and the Rectangular Coordinate System

Representing vectors geometrically (with a directed line segment) is fine for simple comparisons, but many applications involve numerous vectors acting on a single point or changes in a vector quantity over time. For these situations, a graphical representation in the coordinate plane helps to analyze this interaction. The only question is *where* to place the vector on the grid, and the answer is—it really doesn't matter. Consider the three vectors shown in Figure 7.29. From the initial point of each, counting four units in the vertical direction, then three units in the horizontal direction, puts us at the terminal point. This shows the vectors are all 5 units long (since a 3-4-5 triangle is formed) and are all parallel (since slopes are equal: $\dfrac{\Delta y}{\Delta x} = \dfrac{4}{3}$). In other words, they are **equivalent vectors.**

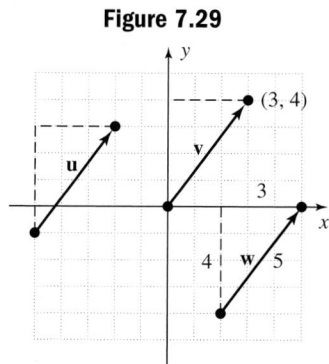

Figure 7.29

Since a vector's location is unimportant, we can replace any given vector with a unique and equivalent vector whose initial point is (0, 0), called the **position vector.**

WORTHY OF NOTE

For vector **u**, the initial and terminal points are $(-5, -1)$ and $(-2, 3)$, respectively, yielding the position vector $\langle -2 - (-5), 3 - (-1)\rangle = \langle 3, 4\rangle$ as before.

Position Vectors

For a vector **v** with initial point (x_1, y_1) and terminal point (x_2, y_2), the position vector for **v** is

$$\mathbf{v} = \langle x_2 - x_1, y_2 - y_1 \rangle,$$

an equivalent vector with initial point (0, 0) and terminal point $(x_2 - x_1, y_2, -y_1)$.

For instance, the initial and terminal points of vector **w** in Figure 7.29 are $(2, -4)$ and $(5, 0)$, respectively, with $(5 - 2, 0 - (-4)) = (3, 4)$. Since $(3, 4)$ is also the terminal point of **v** (whose initial point is at the origin), **v** is the position vector for **u** and **w**. This observation also indicates that every geometric vector in the *xy*-plane corresponds to a unique ordered pair of real numbers (a, b), with a as the **horizontal component** and b as the **vertical component** of the vector. As indicated, we denote the vector in **component form** as $\langle a, b \rangle$, using the new notation to prevent confusing vector $\langle a, b \rangle$ with the ordered pair (a, b). Finally, while each of the vectors in Figure 7.29 has a component form of $\langle 3, 4 \rangle$, the horizontal and vertical components can be read directly only from $\mathbf{v} = \langle 3, 4 \rangle$, giving it a distinct advantage.

EXAMPLE 2 ▶   **Verifying the Components of a Position Vector**

Vector $\mathbf{v} = \langle 12, -5 \rangle$ has initial point $(-4, 3)$.

a. Find the coordinates of the terminal point.

b. Verify the position vector for **v** is also $\langle 12, -5 \rangle$ and find its length.

Solution ▶   **a.** Since **v** has a horizontal component of 12 and a vertical component of -5, we add 12 to the *x*-coordinate and -5 to the *y*-coordinate of the initial point. This gives a terminal point of $(12 + (-4), -5 + 3) = (8, -2)$.

b. To verify we use the initial and terminal points to compute $\langle x_2 - x_1, y_2, -y_1 \rangle$, giving a position vector of $\langle 8 - (-4), -2 - 3 \rangle = \langle 12, -5 \rangle$. To find its length we can use either the Pythagorean theorem or simply note that a 5-12-13 Pythagorean triple is formed. Vector **v** has a length of 13 units.

Now try Exercises 13 through 20 ▶

For the remainder of this section, vector $\mathbf{v} = \langle a, b \rangle$ will refer to the unique position vector for all those equivalent to $\mathbf{v}$. Upon considering the graph of $\langle a, b \rangle$ (shown in QI for convenience in Figure 7.30), several things are immediately evident. The length or **magnitude** of the vector, which is denoted $|\mathbf{v}|$, can be determined using the Pythagorean theorem: $|\mathbf{v}| = \sqrt{a^2 + b^2}$. In addition, basic trigonometry shows the horizontal component can be

Figure 7.30

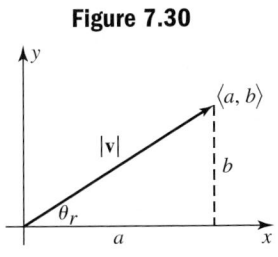

found using $\cos\theta = \dfrac{a}{|\mathbf{v}|}$ or $a = |\mathbf{v}|\cos\theta$, with the vertical component being

$\sin\theta = \dfrac{b}{|\mathbf{v}|}$ or $b = |\mathbf{v}|\sin\theta$. Finally, we note the angle θ can be determined using

$\tan\theta = \left(\dfrac{b}{a}\right)$, or $\theta_r = \tan^{-1}\left(\dfrac{b}{a}\right)$ and the quadrant of $\mathbf{v}$.

Vector Components in Trig Form

For a position vector $\mathbf{v} = \langle a, b \rangle$ and angle θ, we have

horizontal component: $a = |\mathbf{v}|\cos\theta$ vertical component: $b = |\mathbf{v}|\sin\theta$,

where $\theta_r = \tan^{-1}\left(\dfrac{b}{a}\right)$ and

$$|\mathbf{v}| = \sqrt{a^2 + b^2}$$

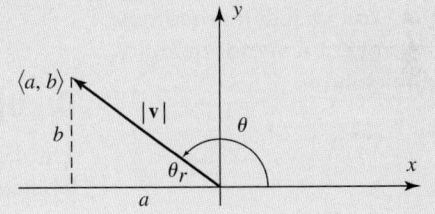

The ability to model characteristics of a vector using these equations is a huge benefit to solving applications, since we must often work out solutions using only the partial information given.

EXAMPLE 3 ▶   **Finding the Magnitude and Direction Angle of a Vector**

For $\mathbf{v}_1 = \langle -2.5, -6 \rangle$ and $\mathbf{v}_2 = \langle 3\sqrt{3}, 3 \rangle$,

  a. Graph each vector and name the quadrant where it is located.

  b. Find their magnitudes.

  c. Find the angle θ for each vector (round to tenths of a degree as needed).

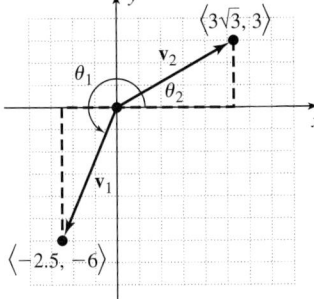

Solution ▶   **a.** The graphs of $\mathbf{v}_1$ and $\mathbf{v}_2$ are shown in the figure. Using the signs of each coordinate, we note that $\mathbf{v}_1$ is in QIII, and $\mathbf{v}_2$ is in QI.

  b. $\begin{aligned} |\mathbf{v}_1| &= \sqrt{(-2.5)^2 + (-6)^2} \\ &= \sqrt{6.25 + 36} \\ &= \sqrt{42.25} \\ &= 6.5 \end{aligned}$ $\begin{aligned} |\mathbf{v}_2| &= \sqrt{(3\sqrt{3})^2 + (3)^2} \\ &= \sqrt{27 + 9} \\ &= \sqrt{36} \\ &= 6 \end{aligned}$

  c. For $\mathbf{v}_1$: $\theta_r = \tan^{-1}\left(\dfrac{-6}{-2.5}\right)$ For $\mathbf{v}_2$: $\theta_r = \tan^{-1}\left(\dfrac{3}{3\sqrt{3}}\right)$

                $= \tan^{-1}(2.4) \approx 67.4°$ $= \tan^{-1}\left(\dfrac{\sqrt{3}}{3}\right) = 30°$

   In QIII, $\theta \approx 247.4°$.   In QI, $\theta = 30°$.

Now try Exercises 21 through 24 ▶

EXAMPLE 4 ▶ **Finding the Horizontal and Vertical Components of a Vector**

The vector $\mathbf{v} = \langle a, b \rangle$ is in QIII, has a magnitude of $|\mathbf{v}| = 21$, and forms an angle of 25° with the negative *x*-axis (Figure 7.31). Find the horizontal and vertical components of the vector, rounded to tenths.

Figure 7.31

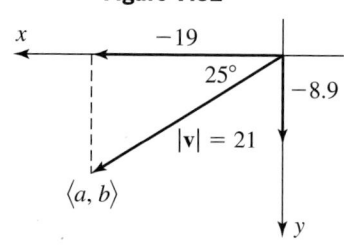

Solution ▶ Begin by graphing the vector and setting up the equations for its components. For $\theta_r = 25°$, $\theta = 205°$.

Figure 7.32

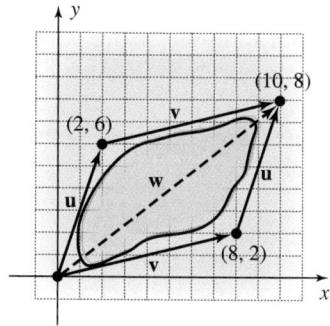

For the horizontal component:

$$a = |\mathbf{v}|\cos\theta$$
$$= 21\cos 205°$$
$$\approx -19$$

For the vertical component:

$$b = |\mathbf{v}|\sin\theta$$
$$= 21\sin 205°$$
$$\approx -8.9$$

With $\mathbf{v}$ in QIII, its component form is approximately $\langle -19, -8.9 \rangle$. As a check, we apply the Pythagorean theorem: $\sqrt{(-19)^2 + (-8.9)^2} \approx 21$ ✓. See Figure 7.32.

Now try Exercises 25 through 30 ▶

✓ **B.** You've just learned how to represent a vector quantity graphically

C. Operations on Vectors and Vector Properties

The operations defined for vectors have a close knit graphical representation. Consider a local park having a large pond with pathways around both sides, so that a park visitor can enjoy the view from either side. Suppose $\mathbf{v} = \langle 8, 2 \rangle$ is the position vector representing a person who decides to turn to the right at the pond, while $\mathbf{u} = \langle 2, 6 \rangle$ represents a person who decides to first turn left. At (8, 2) the first person changes direction and walks to (10, 8) on the other side of the pond, while the second person arrives at (2, 6) and turns to head for (10, 8) as well. This is shown graphically in Figure 7.33 and demonstrates that (1) a parallelogram is formed (opposite sides equal and parallel), (2) the path taken is unimportant relative to the destination, and (3) the coordinates of the destination represent the *sum of corresponding coordinates* from the terminal points of $\mathbf{u}$ and $\mathbf{v}$: $(2, 6) + (8, 2) = (2 + 8, 6 + 2) = (10, 8)$. In other words, the result of adding $\mathbf{u}$ and $\mathbf{v}$ gives the new position vector $\mathbf{u} + \mathbf{v} = \mathbf{w}$, called the **resultant** or the **resultant vector.** Note the resultant vector is a diagonal of the parallelogram formed. Geometrically or graphically, the addition of vectors can be viewed as a "tail-to-tip" combination of one with another, by shifting one vector (without changing its direction) so that its tail (initial point) is at the tip (terminal point) of the other vector. This is illustrated in Figures 7.34 through 7.36.

Figure 7.33

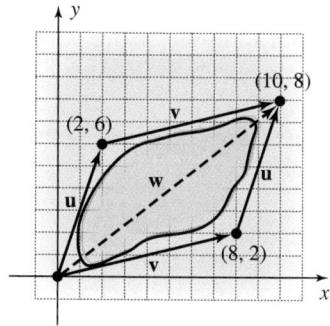

Figure 7.34
Given vectors u and v

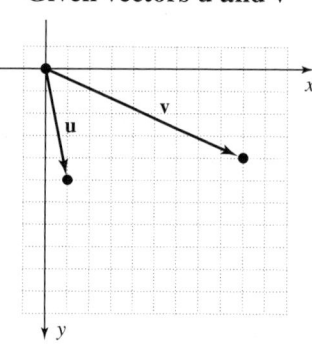

Figure 7.35
Shift vector v

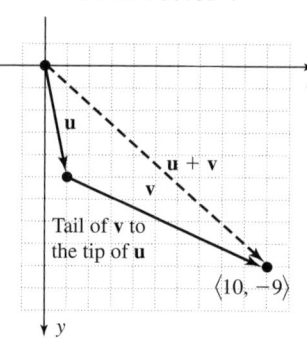

Figure 7.36
Shift vector u

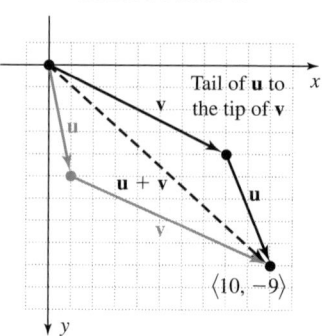

The subtraction of vectors can be understood as either as $\mathbf{u} - \mathbf{v}$ or $\mathbf{u} + (-\mathbf{v})$. Since the location of a vector is unimportant relative to the information it carries, vector subtraction can be interpreted as the *tip-to-tip diagonal* of the parallelogram from vector addition. In Figures 7.34 to 7.36, assume $\mathbf{u} = \langle 1, -5 \rangle$ and $\mathbf{v} = \langle 9, -4 \rangle$. Then $\mathbf{u} - \mathbf{v} = \langle 1, -5 \rangle - \langle 9, -4 \rangle = \langle 1 - 9, -5 + 4 \rangle$ giving the position vector $\langle -8, -1 \rangle$. By repositioning this vector with its tail at the tip of $\mathbf{v}$, we note the new vector points directly at $\mathbf{u}$, forming the diagonal (see Figure 7.37). Scalar multiplication of vectors also has a graphical representation that corresponds to the geometric description given earlier.

Figure 7.37

Operations on Vectors

Given vectors $\mathbf{u} = \langle a, b \rangle$, $\mathbf{v} = \langle c, d \rangle$, and a scalar k,

1. $\mathbf{u} + \mathbf{v} = \langle a + c, b + d \rangle$
2. $\mathbf{u} - \mathbf{v} = \langle a - c, b - d \rangle$
3. $k\mathbf{u} = \langle ka, kb \rangle$ for $k \in \mathbb{R}$

If $k > 0$, the new vector points in the same direction as $\mathbf{u}$.
If $k < 0$, the new vector points in the opposite direction as $\mathbf{u}$.

EXAMPLE 5 ▶ **Representing Operations on Vectors Graphically**

Given $\mathbf{u} = \langle -3, -2 \rangle$ and $\mathbf{v} = \langle 4, -6 \rangle$ compute each of the following and represent the result graphically:

a. $-2\mathbf{u}$ **b.** $\dfrac{1}{2}\mathbf{v}$ **c.** $-2\mathbf{u} + \dfrac{1}{2}\mathbf{v}$

Note the relationship between part (c) and parts (a) and (b).

Solution ▶   **a.** $-2\mathbf{u} = -2\langle -3, -2 \rangle$ **b.** $\dfrac{1}{2}\mathbf{v} = \dfrac{1}{2}\langle 4, -6 \rangle$ **c.** $-2\mathbf{u} + \dfrac{1}{2}\mathbf{v} = \langle 6, 4 \rangle + \langle 2, -3 \rangle$

$= \langle 6, 4 \rangle$ $= \langle 2, -3 \rangle$ $= \langle 8, 1 \rangle$

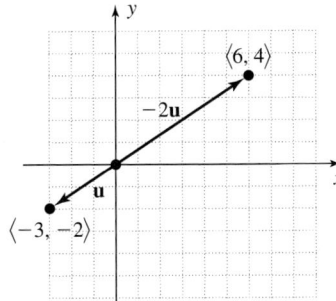
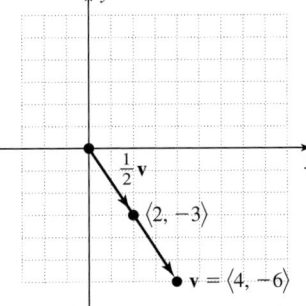
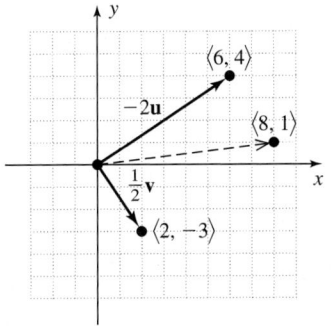

Now try Exercises 31 through 48 ▶

The properties that guide operations on vectors closely resemble the familiar properties of real numbers. Note we define the zero vector $\mathbf{0} = \langle 0, 0 \rangle$ as one having no magnitude or direction.

Properties of Vectors

For vector quantities **u**, **v**, and **w** and real numbers c and k,

1. $1\mathbf{u} = \mathbf{u}$
2. $0\mathbf{u} = \mathbf{0} = k\mathbf{0}$
3. $\mathbf{u} + \mathbf{v} = \mathbf{v} + \mathbf{u}$
4. $\mathbf{u} - \mathbf{v} = \mathbf{u} + (-\mathbf{v})$
5. $(\mathbf{u} + \mathbf{v}) + \mathbf{w} = \mathbf{u} + (\mathbf{v} + \mathbf{w})$
6. $(ck)\mathbf{u} = c(k\mathbf{u}) = k(c\mathbf{u})$
7. $\mathbf{u} + \mathbf{0} = \mathbf{u}$
8. $\mathbf{u} + (-\mathbf{u}) = \mathbf{0}$
9. $k(\mathbf{u} + \mathbf{v}) = k\mathbf{u} + k\mathbf{v}$
10. $(c + k)\mathbf{u} = c\mathbf{u} + k\mathbf{u}$

Proof of Property 3

For $\mathbf{u} = \langle a, b \rangle$ and $\mathbf{v} = \langle c, d \rangle$, we have

$$\begin{aligned}
\mathbf{u} + \mathbf{v} &= \langle a, b \rangle + \langle c, d \rangle & \text{sum of } \mathbf{u} \text{ and } \mathbf{v} \\
&= \langle a + c, b + d \rangle & \text{vector addition} \\
&= \langle c + a, d + b \rangle & \text{commutative property} \\
&= \langle c, d \rangle + \langle a, b \rangle & \text{vector addition} \\
&= \mathbf{v} + \mathbf{u} & \text{result}
\end{aligned}$$

☑ **C.** You've just learned how to perform defined operations on vectors

Proofs of the other properties are similarly derived (see **Exercises 89 through 97**).

D. Algebraic Vectors, Unit Vectors, and i, j Form

While the bold, small case **v** and the $\langle a, b \rangle$ notation for vectors has served us well, we now introduce an alternative form that is somewhat better suited to the **algebra of vectors,** and is used extensively in some of the physical sciences. Consider the vector $\langle 1, 0 \rangle$, a vector 1 unit in length extending along the x-axis. It is called the **horizontal unit vector** and given the special designation **i** (not to be confused with the imaginary unit $i = \sqrt{-1}$). Likewise, the vector $\langle 0, 1 \rangle$ is called the **vertical unit vector** and given the designation **j** (see Figure 7.38). Using scalar multiplication, the unit vector along the negative x-axis is $-\mathbf{i}$ and along the negative y-axis is $-\mathbf{j}$. Similarly, the vector 4**i** represents a position vector 4 units long along the x-axis, and $-5\mathbf{j}$ represents a position vector 5 units long along the negative y-axis. Using these conventions, any nonquadrantal vector $\langle a, b \rangle$ can be written as a **linear combination** of **i** and **j,** with a and b expressed as multiples of **i** and **j,** respectively: $a\mathbf{i} + b\mathbf{j}$. These ideas can easily be generalized and applied to any vector.

Figure 7.38

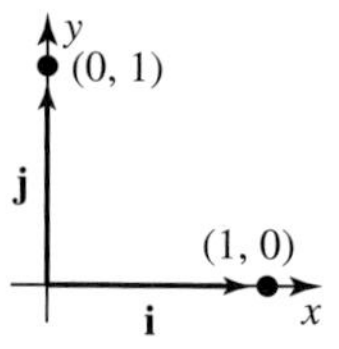

WORTHY OF NOTE

Earlier we stated, "Two vectors were equal if they have the same magnitude and direction." Note that this means two vectors are equal if *their components are equal.*

Algebraic Vectors and i, j Form

For the unit vectors $\mathbf{i} = \langle 1, 0 \rangle$ and $\mathbf{j} = \langle 0, 1 \rangle$, any arbitrary vector $\mathbf{v} = \langle a, b \rangle$ can be written as a linear combination of **i** and **j**:

$$\mathbf{v} = a\mathbf{i} + b\mathbf{j}$$

Graphically, **v** is being expressed as the resultant of a vector sum.

EXAMPLE 6 ▶ **Finding the Horizontal and Vertical Components of Algebraic Vectors**

Vector **u** is in QII, has a magnitude of 15, and makes an angle of 20° with the negative x-axis.

   a. Graph the vector.

   b. Find the horizontal and vertical components (round to one decimal place) then write **u** in component form.

   c. Write **u** in terms of **i** and **j**.

Solution ▶ **a.** The vector is graphed in Figure 7.39.   **Figure 7.39**

Figure 7.40

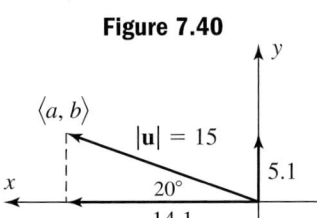

b. Horizontal Component   Vertical Component

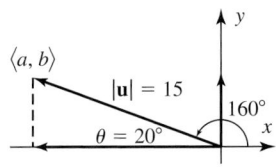

$$a = |\mathbf{v}|\cos\theta \qquad\qquad b = |\mathbf{v}|\sin\theta$$
$$= 15\cos 160° \qquad\qquad = 15\sin 160°$$
$$\approx -14.1 \qquad\qquad\qquad \approx 5.1$$

With the vector in QII, $\mathbf{u} = \langle -14.1, 5.1 \rangle$ in component form.

c. In terms of **i** and **j** we have $\mathbf{u} = -14.1\mathbf{i} + 5.1\mathbf{j}$. See Figure 7.40

Now try Exercises 49 through 62 ▶

Some applications require that we find a nonhorizontal, nonvertical vector one unit in length, having the same direction as a given vector **v**. To understand how this is done, consider vector $\mathbf{v} = \langle 6, 8 \rangle$. Using the Pythagorean theorem we find $|\mathbf{v}| = 10$, and can form a 6-8-10 triangle using the horizontal and vertical components (Figure 7.41). Knowing that similar triangles have sides that are proportional, we can find a unit vector in the same direction as **v** by dividing all three sides by 10, giving a triangle with sides $\frac{3}{5}, \frac{4}{5}$, and 1. The new vector "**u**" (along the hypotenuse) indeed points in the same direction since we have merely shortened **v**, and is a unit vector

Figure 7.41

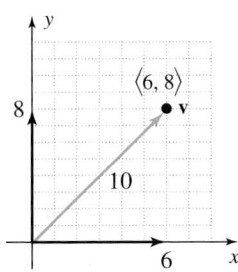

since $\left(\dfrac{3}{5}\right)^2 + \left(\dfrac{4}{5}\right)^2 \rightarrow \dfrac{9}{25} + \dfrac{16}{25} = 1$. In retrospect, we have divided the components of vector **v** by its magnitude $|\mathbf{v}|$ (or multiplied components by the reciprocal of $|\mathbf{v}|$) to obtain the desired unit vector: $\dfrac{\mathbf{v}}{|\mathbf{v}|} = \dfrac{\langle 6, 8 \rangle}{10} = \left\langle \dfrac{6}{10}, \dfrac{8}{10} \right\rangle = \left\langle \dfrac{3}{5}, \dfrac{4}{5} \right\rangle$. In general we have the following:

Unit Vectors

For any nonzero vector $\mathbf{v} = \langle a, b \rangle = a\mathbf{i} + b\mathbf{j}$, the vector

$$\mathbf{u} = \frac{\mathbf{v}}{|\mathbf{v}|} = \frac{a}{\sqrt{a^2 + b^2}}\mathbf{i} + \frac{b}{\sqrt{a^2 + b^2}}\mathbf{j}$$

is a unit vector in the same direction as **v**.

You are asked to verify this relationship in **Exercise 100.** In summary, for vector $\mathbf{v} = 6\mathbf{i} + 8\mathbf{j}$, we find $|\mathbf{v}| = \sqrt{6^2 + 8^2} = 10$, so the unit vector pointing in the same direction is $\dfrac{\mathbf{v}}{|\mathbf{v}|} = \dfrac{3}{5}\mathbf{i} + \dfrac{4}{5}\mathbf{j}$. **See Exercises 63 through 74.**

EXAMPLE 7 ▶ **Using Unit Vectors to Find Coincident Vectors**

Vectors **u** and **v** form the 37° angle illustrated in the figure. Find the vector **w** (in red), which points in the same direction as **v** (is coincident with **v**) and forms the base of the right triangle shown.

Solution ▶ Using the Pythagorean theorem we find $|\mathbf{u}| \approx 7.3$ and $|\mathbf{v}| = 10$. Using the cosine of 37° the magnitude of **w** is then $|\mathbf{w}| \approx 7.3\cos 37°$ or about 5.8. To ensure that **w** will point in the same direction as **v**, we simply multiply the 5.8 magnitude by the unit

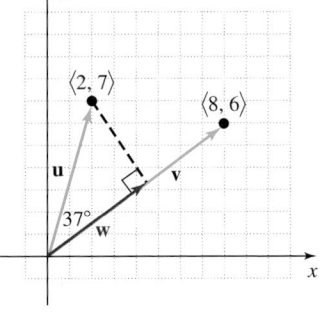

WORTHY OF NOTE

In this context **w** is called the projection of **u** on **v**, an idea applied more extensively in Section 7.4

vector for **v**: $|\mathbf{w}|\dfrac{\mathbf{v}}{|\mathbf{v}|} \approx (5.8)\dfrac{\langle 8, 6 \rangle}{10} = (5.8)\langle 0.8, 0.6 \rangle$,

and we find that $\mathbf{w} \approx \langle 4.6, 3.5 \rangle$. As a check we use the Pythagorean theorem: $\sqrt{4.6^2 + 3.5^2} = \sqrt{33.41} \approx 5.8$.

☑ D. You've just learned how to represent a vector quantity algebraically and find unit vectors

Now try Exercises 75 through 78 ▶

E. Vector Diagrams and Vector Applications

Applications of vectors are virtually unlimited, with many of these in the applied sciences. Here we'll look at two applications that are an extension of our work in this section. In Section 7.4 we'll see how vectors can be applied in a number of other creative and useful ways.

In Example 1, two tugboats were pulling on a barge to dislodge it from a sand bar, with the pulling force of each represented by a vector. Using our knowledge of vector components, vector addition, and **resultant forces** (a force exerted along the resultant), we can now determine the direction and magnitude of the resultant force if we know the angle formed by one of the vector forces and the barge.

EXAMPLE 8 ▶   **Solving an Application of Vectors—Force Vectors Acting on a Barge**

Two tugboats are attempting to free a barge that is stuck on a sand bar, and are exerting the forces shown in Figure 7.42. Find the magnitude and direction of the resultant force.

Figure 7.42

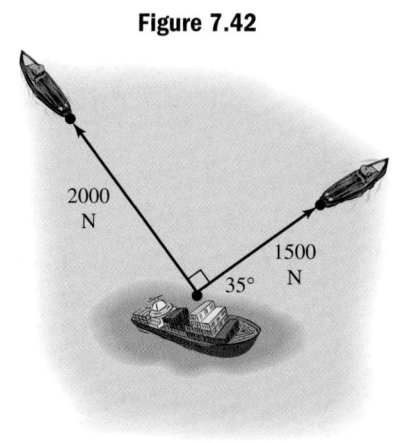

Solution ▶   Begin by orienting the diagram on a coordinate grid (see Figure 7.43). Since the angle between the vectors is 90°, we know the acute angle formed by the first tugboat and the x-axis is 55°. With this information, we can write each vector in "$\mathbf{i}, \mathbf{j}$" form and add the vectors to find the resultant.

For vector $\mathbf{v}_1$ (in QII):

Horizontal Component	Vertical Component
$a = |\mathbf{v}_1|\cos\theta$	$b = |\mathbf{v}_1|\sin\theta$
$= 2000 \cos 125°$	$= 2000 \sin 125°$
≈ -1147	≈ 1638

$$\mathbf{v}_1 \approx -1147\mathbf{i} + 1638\mathbf{j}.$$

For vector $\mathbf{v}_2$ (in QI):

Figure 7.43

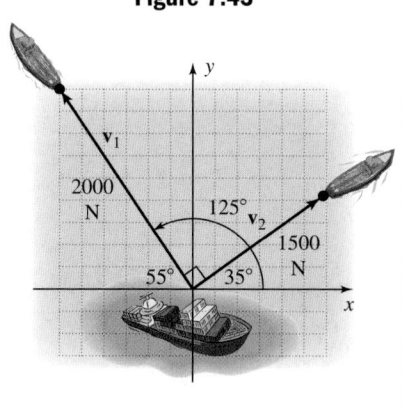

Horizontal Component	Vertical Component
$a = |\mathbf{v}_2|\cos\theta$	$b = |\mathbf{v}_2|\sin\theta$
$= 1500 \cos 35°$	$= 1500 \sin 35°$
≈ 1229	≈ 860

$$\mathbf{v}_2 \approx 1229\mathbf{i} + 860\mathbf{j}.$$

This gives a resultant of $\mathbf{v}_1 + \mathbf{v}_2 \approx (-1147\mathbf{i} + 1638\mathbf{j}) + (1229\mathbf{i} + 860\mathbf{j}) = 82\mathbf{i} + 2498\mathbf{j}$, with magnitude $|\mathbf{v}_1 + \mathbf{v}_2| \approx \sqrt{82^2 + 2498^2} \approx 2499$ N. To find the direction of the force, we have $\theta_r = \tan^{-1}\left(\dfrac{2498}{82}\right)$, or about 88°.

Now try Exercises 81 and 82 ▶

It's worth noting that a single tugboat pulling at 88° with a force of 2499 N would have the same effect as the two tugs in the original diagram. In other words, the resultant vector $82\mathbf{i} + 2498\mathbf{j}$ truly represents the "result" of the two forces.

Knowing that the location of a vector is unimportant enables us to model and solve a great number of seemingly unrelated applications. Although the final example concerns aviation, **headings,** and crosswinds, the solution process has a striking similarity to the "tugboat" example just discussed. In navigation, headings involve a single angle, which is understood to be the amount of rotation from due north in the clock-wise direction. Several headings are illustrated in Figures 7.44 through 7.47.

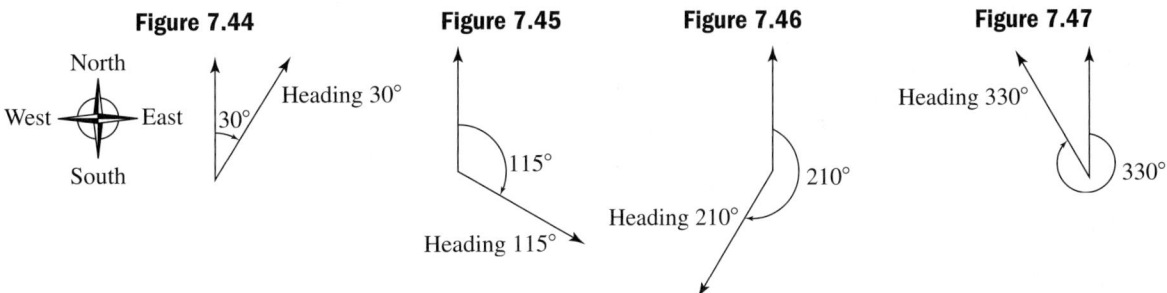

Figure 7.44   **Figure 7.45**   **Figure 7.46**   **Figure 7.47**

In order to keep an airplane on course, the captain must consider the direction and speed of any wind currents, since the plane's true course (relative to the ground) will be affected. Both the plane and the wind can be represented by vectors, with the plane's true course being the resultant vector.

EXAMPLE 9 ▶ **Solving an Application of Vectors—Airplane Navigation**

An airplane is flying at 240 mph, heading 75°, when it suddenly encounters a strong, 60 mph wind blowing *from* the southwest, heading 10°. What is the actual course and speed of the plane (relative to the ground) as it flies through this wind?

Solution ▶ Begin by drawing a vector **p** to represent the speed and direction of the airplane (Figure 7.48). Since the heading is 75°, the angle between the vector and the *x*-axis must be 15°. For convenience (and because location is unimportant) we draw it as a position vector. Note the vector **w** representing the wind will be $\frac{60}{240} = \frac{1}{4}$ as long, and can also be drawn as a position vector—with an acute 80° angle. To find the resultant, we first find the components of each vector, then add. For vector **w** (in QI):

Figure 7.48

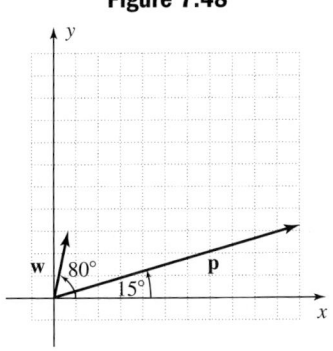

Horizontal Component	Vertical Component
$a = |\mathbf{w}|\cos\theta$	$b = |\mathbf{w}|\sin\theta$
$= 60\cos 80°$	$= 60\sin 80°$
≈ 10.4	≈ 59.1

$$\mathbf{w} \approx 10.4\mathbf{i} + 59.1\mathbf{j}.$$

For vector **p** (in QI):

Horizontal Component	Vertical Component
$a = |\mathbf{p}|\cos\theta$	$b = |\mathbf{p}|\sin\theta$
$= 240\cos 15°$	$= 240\sin 15°$
$= 231.8$	≈ 62.1

$$\mathbf{p} \approx 231.8\mathbf{i} + 62.1\mathbf{j}.$$

Figure 7.49

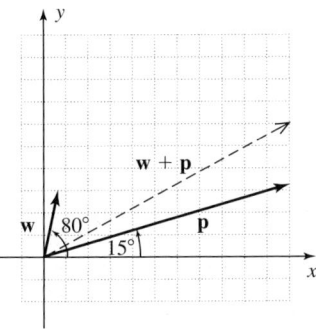

The resultant is $\mathbf{w} + \mathbf{p} \approx (10.4\mathbf{i} + 59.1\mathbf{j}) + (231.8\mathbf{i} + 62.1\mathbf{j}) = 242.2\mathbf{i} + 121.2\mathbf{j}$, with magnitude $|\mathbf{w} + \mathbf{p}| \approx \sqrt{(242.2)^2 + (121.2)^2} \approx 270.8$ mph (see Figure 7.49). To find the heading of the plane relative to the ground we use $\theta_r = \tan^{-1}\left(\frac{121.2}{242.2}\right)$, which shows $\theta_r \approx 26.6°$. The plane is flying on a course heading of $90° - 26.6° = 63.4°$ at a speed of about 270.8 mph relative to the ground. Note the airplane has actually "increased speed" due to the wind.

Now try Exercises 83 through 86 ▶

Applications like those in Examples 8 and 9 can also be solved using what is called the **parallelogram method,** which takes its name from the tail-to-tip vector addition noted earlier (See Figure 7.50). The resultant will be a diagonal of the parallelogram, whose magnitude can be found using the law of cosines. For Example 9, we note the parallelogram has two acute angles of $(80 - 15)° = 65°$, and since the adjacent angles must sum to $180°$, the obtuse angles must be $115°$. Using the law of cosines,

Figure 7.50

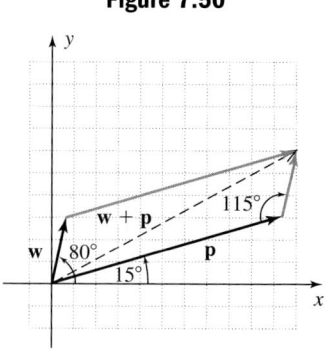

$$|\mathbf{w} + \mathbf{p}|^2 = \mathbf{p}^2 + \mathbf{w}^2 - 2\mathbf{pw} \cos 115° \qquad \text{law of cosines}$$
$$= 240^2 + 60^2 - 2(240)(60) \cos 115° \qquad \text{substitute 240 for } \mathbf{p}, \text{ 60 for } \mathbf{w}$$
$$= 73371.40594 \qquad \text{compute result}$$
$$|\mathbf{w} + \mathbf{p}| \approx 270.9 \qquad \text{take square roots}$$

Note this answer is slightly more accurate, since there was no rounding until the final stage.

✓ **E.** You've just learned how to use vector diagrams to solve applications

---

TECHNOLOGY HIGHLIGHT

Vector Components Given the Magnitude and the Angle θ

The TABLE feature of a graphing calculator can help us find the horizontal and vertical components of any vector with ease. Consider the vector **v** shown in Figure 7.51, which has a magnitude of 9.5 with $\theta = 15°$. Knowing this magnitude is used in both computations, first store 9.5 in storage location A: 9.5 [STO →] [ALPHA] [MATH] . Next, enter the expressions for the horizontal and vertical components as Y_1 and Y_2 on the [Y=] screen (see Figure 7.52). Note that storing the magnitude 9.5 in memory will prevent our having to alter Y_1 and Y_2 as we apply these ideas to other values of θ. As an additional check, note that Y_3 recomputes the magnitude of the vector using the components generated in Y_1 and Y_2. To access the function variables we press: [VARS] [▶] [ENTER] and select the desired function. Although our primary interest is the components for $\theta = 15°$, we use the TBLSET screen to begin at TblStart $= 0°$, ΔTbl $= 5$, and have it count **AUTO**matically, so we can make additional observations. Pressing [2nd] [GRAPH] (TABLE) brings up the screen shown in Figure 7.53. As expected, at $\theta = 0°$ the horizontal component is the same as the magnitude and the vertical component is zero. At $\theta = 15°$ we have the components of the vector pictured in Figure 7.51, approximately $\langle 9.18, 2.46 \rangle$. If the angle were increased to $\theta = 30°$, a 30-60-90 triangle could be formed and one component should be $\sqrt{3}$ times the other. Sure enough, $\sqrt{3}(4.75) \approx 8.2272$.

Figure 7.51

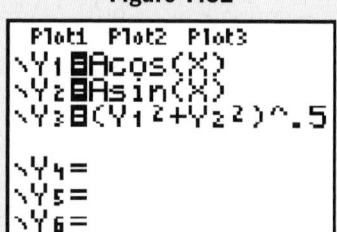

Figure 7.52

```
Plot1  Plot2  Plot3
\Y1■Acos(X)
\Y2■Asin(X)
\Y3■(Y1²+Y2²)^.5
\Y4=
\Y5=
\Y6=
```

Figure 7.53

X	Y1	Y2
0	9.5	0
5	9.4638	.82798
10	9.3557	1.6497
15	9.1763	2.4588
20	8.9271	3.2492
25	8.6099	4.0149
30	8.2272	4.75

X=0

Exercise 1: If $\theta = 45°$, what would you know about the lengths of the horizontal and vertical components? Scroll down to $\theta = 45°$ to verify.

Exercise 2: If $\theta = 60°$, what would you know about the lengths of the horizontal and vertical components? Scroll down to $\theta = 60°$ to verify.

Exercise 3: We used column Y_3 as a double check on the magnitude of **v** for any given θ. What would this value be for $\theta = 45°$ and $\theta = 60°$? Press the right arrow ▶ to verify. What do you notice?

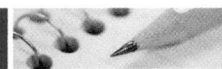

7.3 EXERCISES

▶ CONCEPTS AND VOCABULARY

Fill in each blank with the appropriate word or phrase. Carefully reread the section if needed.

1. Measurements that can be described using a single number are called _____ quantities.

2. _____ quantities require more than a single number to describe their attributes. Examples are force, velocity, and displacement.

3. To represent a vector quantity geometrically we use a _____ _____ segment.

4. Two vectors are equal if they have the same _____ and _____.

5. Discuss/Explain the geometric interpretation of vector addition. Give several examples and illustrations.

6. Describe the process of finding a resultant vector given the magnitude and direction of two arbitrary vectors **u** and **v**. Follow-up with an example.

▶ DEVELOPING YOUR SKILLS

Draw the comparative geometric vectors indicated.

7. Three oceanic research vessels are traveling on a parallel course in the same direction, mapping the ocean floor. One ship is traveling at 12 knots (nautical miles per hour), one at 9 knots, and the third at 6 knots.

8. As part of family reunion activities, the Williams Clan is at a bowling alley and using three lanes. Being amateurs they all roll the ball straight on, aiming for the 1 pin. Grand Dad in Lane 1 rolls his ball at 50 ft/sec. Papa in Lane 2 lets it rip at 60 ft/sec, while Junior in Lane 3 can muster only 30 ft/sec.

9. Vector v_1 is a geometric vector representing a boat traveling at 20 knots. Vectors v_2, v_3, and v_4 are geometric vectors representing boats traveling at 10 knots, 15 knots, and 25 knots, respectively. Draw these vectors given that v_2 and v_3 are traveling the same direction and parallel to v_1, while v_4 is traveling in the opposite direction and parallel to v_1.

10. Vector F_1 is a geometric vector representing a force of 50 N. Vectors F_2, F_3, and F_4 are geometric vectors representing forces of 25 N, 35 N, and 65 N, respectively. Draw these vectors given that F_2 and F_3 are applied in the same direction and parallel to F_1, while F_4 is applied in the opposite direction and parallel to F_1.

Represent each situation described using geometric vectors.

11. Two tractors are pulling at a stump in an effort to clear land for more crops. The Massey-Ferguson is pulling with a force of 250 N, while the John Deere is pulling with a force of 210 N. The chains attached to the stump and each tractor form a 25° angle.

12. In an effort to get their mule up and plowing again, Jackson and Rupert are pulling on ropes attached to the mule's harness. Jackson pulls with 200 lb of force, while Rupert, who is really upset, pulls with 220 lb of force. The angle between their ropes is 16°.

Draw the vector **v** indicated, then graph the equivalent position vector.

13. initial point $(-3, 2)$; terminal point $(4, 5)$

14. initial point $(-4, -4)$; terminal point $(2, 3)$

15. initial point $(5, -3)$; terminal point $(-1, 2)$

16. initial point $(1, 4)$; terminal point $(-2, 2)$

For each vector $\mathbf{v} = \langle a, b \rangle$ and initial point (x, y) given, find the coordinates of the terminal point and the magnitude $|\mathbf{v}|$ of the vector.

17. $\mathbf{v} = \langle 7, 2 \rangle$; initial point $(-2, -3)$

18. $\mathbf{v} = \langle -6, 1 \rangle$; initial point $(5, -2)$

19. $\mathbf{v} = \langle -3, -5 \rangle$; initial point $(2, 6)$

20. $\mathbf{v} = \langle 8, -2 \rangle$; initial point $(-3, -5)$

For each position vector given, (a) graph the vector and name the quadrant, (b) compute its magnitude, and (c) find the acute angle θ formed by the vector and the nearest x-axis.

21. $\langle 8, 3 \rangle$ 22. $\langle -7, 6 \rangle$

23. $\langle -2, -5 \rangle$ 24. $\langle 8, -6 \rangle$

For Exercises 25 through 30, the magnitude of a vector is given, along with the quadrant of the terminal point and the angle it makes with the nearest x-axis. Find the horizontal and vertical components of each vector and write the result in component form.

25. $|\mathbf{v}| = 12$; $\theta = 25°$; QII

26. $|\mathbf{u}| = 25$; $\theta = 32°$; QIII

27. $|\mathbf{w}| = 140.5$; $\theta = 41°$; QIV

28. $|\mathbf{p}| = 15$; $\theta = 65°$; QI

29. $|\mathbf{q}| = 10$; $\theta = 15°$; QIII

30. $|\mathbf{r}| = 4.75$; $\theta = 62°$; QII

For each pair of vectors **u** and **v** given, compute (a) through (d) and illustrate the indicated operations graphically.

  a. $\mathbf{u} + \mathbf{v}$ **b.** $\mathbf{u} - \mathbf{v}$

  c. $2\mathbf{u} + 1.5\mathbf{v}$ **d.** $\mathbf{u} - 2\mathbf{v}$

31. $\mathbf{u} = \langle 2, 3 \rangle$; $\mathbf{v} = \langle -3, 6 \rangle$

32. $\mathbf{u} = \langle -3, -4 \rangle$; $\mathbf{v} = \langle 0, 5 \rangle$

33. $\mathbf{u} = \langle 7, -2 \rangle$; $\mathbf{v} = \langle 1, 6 \rangle$

34. $\mathbf{u} = \langle -5, -3 \rangle$; $\mathbf{v} = \langle 6, -4 \rangle$

35. $\mathbf{u} = \langle -4, 2 \rangle$; $\mathbf{v} = \langle 1, 4 \rangle$

36. $\mathbf{u} = \langle 7, 3 \rangle$; $\mathbf{v} = \langle -7, 3 \rangle$

Use the graphs of vectors **a, b, c, d, e, f, g,** and **h** given to determine if the following statements are true or false.

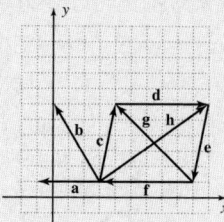

37. $\mathbf{a} + \mathbf{c} = \mathbf{b}$ 38. $\mathbf{f} - \mathbf{e} = \mathbf{g}$

39. $\mathbf{c} + \mathbf{f} = \mathbf{h}$ 40. $\mathbf{b} + \mathbf{h} = \mathbf{c}$

41. $\mathbf{d} - \mathbf{e} = \mathbf{h}$ 42. $\mathbf{d} + \mathbf{f} = \mathbf{0}$

For the vectors **u** and **v** shown, compute $\mathbf{u} + \mathbf{v}$ and $\mathbf{u} - \mathbf{v}$ and represent each result graphically.

43. 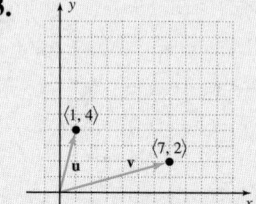   44.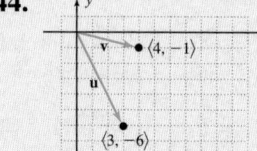

45. 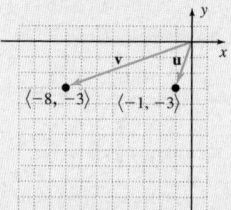   46.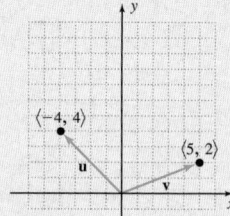

47. 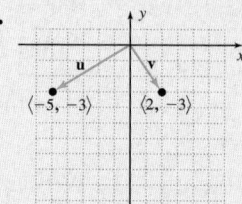   48.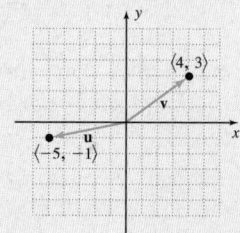

Graph each vector and write it as a linear combination of **i** and **j**. Then compute its magnitude.

49. $\mathbf{u} = \langle 8, 15 \rangle$ 50. $\mathbf{v} = \langle -5, 12 \rangle$

51. $\mathbf{p} = \langle -3.2, -5.7 \rangle$ 52. $\mathbf{q} = \langle 7.5, -3.4 \rangle$

For each vector here, θ_r represents the acute angle formed by the vector and the x-axis. (a) Graph each vector, (b) find the horizontal and vertical components and write the vector in component form, and (c) write the vector in **i, j** form. Round to the nearest tenth.

53. **v** in QIII, $|\mathbf{v}| = 12$, $\theta_r = 16°$

54. **u** in QII, $|\mathbf{u}| = 10.5$, $\theta_r = 25°$

55. **w** in QI, $|\mathbf{w}| = 9.5$, $\theta_r = 74.5°$

56. **v** in QIV, $|\mathbf{v}| = 20$, $\theta_r = 32.6°$

For vectors $\mathbf{v}_1$ and $\mathbf{v}_2$ given, compute the vector sums (a) through (d) and find the magnitude and direction of each resultant.

a. $\mathbf{v}_1 + \mathbf{v}_2 = \mathbf{p}$ **b.** $\mathbf{v}_1 - \mathbf{v}_2 = \mathbf{q}$

c. $2\mathbf{v}_1 + 1.5\mathbf{v}_2 = \mathbf{r}$ **d.** $\mathbf{v}_1 - 2\mathbf{v}_2 = \mathbf{s}$

57. $\mathbf{v}_1 = 2\mathbf{i} - 3\mathbf{j}; \mathbf{v}_2 = -4\mathbf{i} + 5\mathbf{j}$

58. $\mathbf{v}_1 = 7.8\mathbf{i} + 4.2\mathbf{j}; \mathbf{v}_2 = 5\mathbf{j}$

59. $\mathbf{v}_1 = 5\sqrt{2}\mathbf{i} + 7\mathbf{j}; \mathbf{v}_2 = -3\sqrt{2}\mathbf{i} - 5\mathbf{j}$

60. $\mathbf{v}_1 = 6.8\mathbf{i} - 9\mathbf{j}; \mathbf{v}_2 = -4\mathbf{i} + 9\mathbf{j}$

61. $\mathbf{v}_1 = 12\mathbf{i} + 4\mathbf{j}; \mathbf{v}_2 = -4\mathbf{i}$

62. $\mathbf{v}_1 = 2\sqrt{3}\mathbf{i} - 6\mathbf{j}; \mathbf{v}_2 = -4\sqrt{3}\mathbf{i} + 2\mathbf{j}$

Find a unit vector pointing in the same direction as the vector given. Verify that a unit vector was found.

63. $\mathbf{u} = \langle 7, 24 \rangle$ **64.** $\mathbf{v} = \langle -15, 36 \rangle$

65. $\mathbf{p} = \langle -20, 21 \rangle$ **66.** $\mathbf{q} = \langle 12, -35 \rangle$

67. $20\mathbf{i} - 21\mathbf{j}$ **68.** $-4\mathbf{i} - 7.5\mathbf{j}$

69. $3.5\mathbf{i} + 12\mathbf{j}$ **70.** $-9.6\mathbf{i} + 18\mathbf{j}$

71. $\mathbf{v}_1 = \langle 13, 3 \rangle$ **72.** $\mathbf{v}_2 = \langle -4, 7 \rangle$

73. $6\mathbf{i} + 11\mathbf{j}$ **74.** $-2.5\mathbf{i} + 7.2\mathbf{j}$

Vectors $\mathbf{p}$ and $\mathbf{q}$ form the angle indicated in each diagram. Find the vector $\mathbf{r}$ that points in the same direction as $\mathbf{q}$ and forms the base of the right triangle shown.

75.
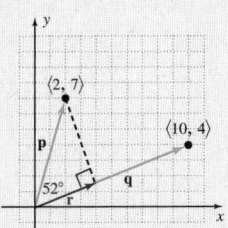

76.
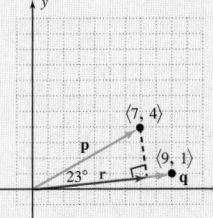

77.
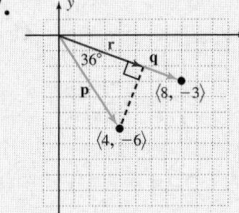

78.
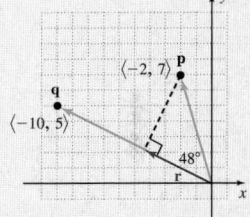

▶ WORKING WITH FORMULAS

The magnitude of a vector in three dimensions: $|\mathbf{v}| = \sqrt{a^2 + b^2 + c^2}$

79. The magnitude of a vector in three dimensional space is given by the formula shown, where the components of the position vector $\mathbf{v}$ are $\langle a, b, c \rangle$. Find the magnitude of $\mathbf{v}$ if $\mathbf{v} = \langle 5, 9, 10 \rangle$.

80. Find a cardboard box of any size and carefully measure its length, width, and height. Then use the given formula to find the magnitude of the box's diagonal. Verify your calculation by direct measurement.

▶ APPLICATIONS

81. Tow forces: A large van has careened off of the road into a ditch, and two tow trucks are attempting to winch it out. The cable from the first winch exerts a force of 900 lb, while the cable from the second exerts a force of 700 lb. Determine the angle θ for the first tow truck that will bring the van directly out of the ditch and along the line indicated.

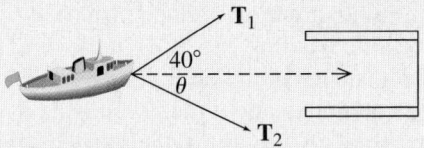

82. Tow forces: Two tugboats are pulling a large ship into dry dock. The first is pulling with a force of 1250 N and the second with a force of 1750 N. Determine the angle θ for the second tugboat that will keep the ship moving straight forward and into the dock.

83. Projectile components: An arrow is shot into the air at an angle of 37° with an initial velocity of 100 ft/sec. Compute the horizontal and vertical components of the representative vector.

84. Projectile components: A football is punted (kicked) into the air at an angle of 42° with an initial velocity of 20 m/sec. Compute the horizontal and vertical components of the representative vector.

85. Headings and cross-winds: An airplane is flying at 250 mph on a heading of 75°. There is a strong, 35 mph wind blowing from the southwest on a heading of 10°. What is the true course and speed of the plane (relative to the ground)?

86. Headings and currents: A cruise ship is traveling at 16 knots on a heading of 300°. There is a strong water current flowing at 6 knots from the northwest on a heading of 120°. What is the true course and speed of the cruise ship?

The lights used in a dentist's office are multijointed so they can be configured in multiple ways to accommodate various needs. As a simple model, consider such a light that has the three joints, as illustrated. The first segment has a length of 45 cm, the second is 40 cm in length, and the third is 35 cm.

87. If the joints of the light are positioned so a straight line is formed and the angle made with the horizontal is 15°, determine the approximate coordinates of the joint nearest the light.

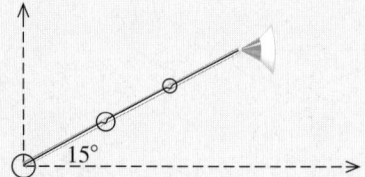

88. If the first segment is rotated 75° above horizontal, the second segment −30° (below the horizontal), and the third segment is parallel to the horizontal, determine the approximate coordinates of the joint nearest the light.

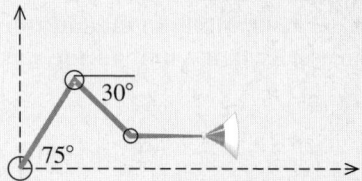

▶ EXTENDING THE CONCEPT

For the arbitrary vectors $\mathbf{u} = \langle a, b \rangle$, $\mathbf{v} = \langle c, d \rangle$, and $\mathbf{w} = \langle e, f \rangle$ and the scalars c and k, prove the following vector properties using the properties of real numbers.

89. $1\mathbf{u} = \mathbf{u}$

90. $0\mathbf{u} = \mathbf{0} = k\mathbf{0}$

91. $\mathbf{u} - \mathbf{v} = \mathbf{u} + (-\mathbf{v})$

92. $(\mathbf{u} + \mathbf{v}) + \mathbf{w} = \mathbf{u} + (\mathbf{v} + \mathbf{w})$

93. $(ck)\mathbf{u} = c(k\mathbf{u}) = k(c\mathbf{u})$

94. $\mathbf{u} + \mathbf{0} = \mathbf{u}$

95. $\mathbf{u} + (-\mathbf{u}) = \mathbf{0}$

96. $k(\mathbf{u} + \mathbf{v}) = k\mathbf{u} + k\mathbf{v}$ **97.** $(c + k)\mathbf{u} = c\mathbf{u} + k\mathbf{u}$

98. Consider an airplane flying at 200 mph at a heading of 45°. Compute the groundspeed of the plane under the following conditions. A strong, 40-mph wind is blowing (a) in the same direction; (b) in the direction of due north (0°); (c) in the direction heading 315°; (d) in the direction heading 270°; and (e) in the direction heading 225°. What did you notice about the groundspeed for (a) and (b)? Explain why the plane's speed is greater than 200 mph for (a) and (b), but less than 200 mph for the others.

99. Show that the sum of the vectors given, which form the sides of a closed polygon, is the zero vector. Assume all vectors have integer coordinates and each tick mark is 1 unit.

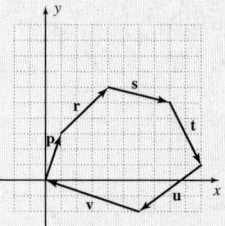

100. Verify that for $\mathbf{v} = a\mathbf{i} + b\mathbf{j}$ and
$$|\mathbf{v}| = \sqrt{a^2 + b^2}, \frac{\mathbf{v}}{|\mathbf{v}|} = 1.$$

(*Hint:* Create the vector $\mathbf{u} = \dfrac{\mathbf{v}}{|\mathbf{v}|}$ and find its magnitude.)

101. Referring to Exercises 87 and 88, suppose the dentist needed the pivot joint at the light (the furthest joint from the wall) to be at (80, 20) for a certain patient or procedure. Find at least one set of "joint angles" that will make this possible.

▶ **MAINTAINING YOUR SKILLS**

102. (6.1) Derive the other two common versions of the Pythagorean identities, given $\sin^2 x + \cos^2 x = 1$.

103. (2.5) Evaluate each expression for $x = 3$ (if possible):

    a. $y = \ln(2x - 7)$ **b.** $y = \dfrac{5}{x - 3}$

    c. $y = \sqrt{\dfrac{1}{3}x - 5}$

104. (6.5) Evaluate the expression $\csc\left[\tan^{-1}\left(\dfrac{55}{48}\right)\right]$ by drawing a representative triangle.

105. (3.4) Graph the function $g(x) = x^3 - 7x$ and find its zeroes.

MID-CHAPTER CHECK

1. Beginning with $\dfrac{\sin A}{a} = \dfrac{\sin B}{b}$, solve for $\sin B$.

2. Given $b^2 = a^2 + c^2 - 2ac\cos B$, solve for $\cos B$.

Solve the triangles shown below using any appropriate method.

3.

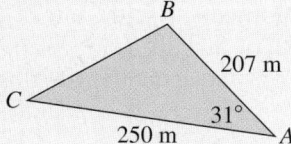

4.

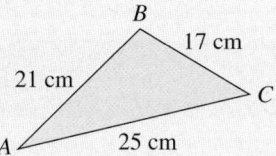

Solve the triangles described below using the law of sines. If more than one triangle exists, solve both.

5. $A = 44°$, $a = 2.1$ km, $c = 2.8$ km

6. $C = 27°$, $a = 70$ yd, $c = 100$ yd

7. A large highway sign is erected on a steep hillside that is inclined 45° from the horizontal. At 9:00 A.M.

the sign casts a 75 ft shadow. Find the height of the sign if the angle of elevation (measured from a horizontal line) from the tip of the shadow to the top of the sign is 65°.

8. Modeled after an Egyptian obelisk, the Washington Monument (Washington, D.C.) is one of the tallest masonry buildings in the world. Find the height of the monument given the measurements shown (see the figure).

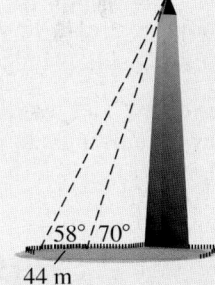

9. The circles shown here have radii of 4 cm, 9 cm, and 12 cm, and are tangent to each other. Find the angles formed by the line segments joining their centers.

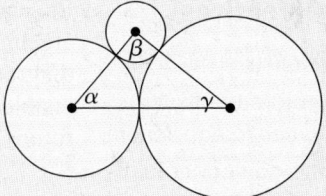

10. On her delivery route, Judy drives 23 miles to *C*olumbus, then 17 mi to *D*rake, then back home to *B*alboa. Use the diagram given to find the distance from *D*rake to *B*alboa.

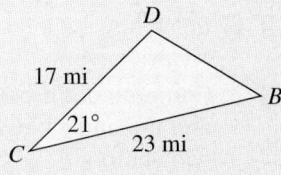

REINFORCING BASIC CONCEPTS

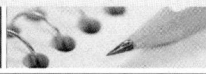

Scaled Drawings and the Laws of Sine and Cosine

In mathematics, there are few things as satisfying as the tactile verification of a concept or computation. In this *Reinforcing Basic Concepts*, we'll use scaled drawings to

verify the relationships stated by the law of sines and the law of cosines. First, gather a blank sheet of paper, a ruler marked in centimeters/millimeters, and a protractor. When working with scale models, always measure and mark as carefully as possible. The greater the care, the better the results. For the first illustration (see Figure 7.54), we'll draw a 20-cm horizontal line segment near the bottom of the paper, then use the left endpoint to mark off a 35°

Figure 7.54

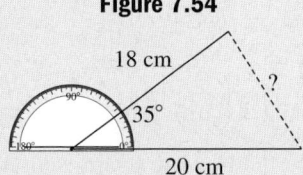

angle. Draw the second side a length of 18 cm. Our first goal is to compute the length of the side needed to complete the triangle, then verify our computation by measurement. Since the current "triangle" is SAS, we use the law of cosines. Label the 35° as $\angle A$, the top vertex as $\angle B$, and the right endpoint as $\angle C$.

$a^2 = b^2 + c^2 - 2bc \cos A$ law of cosines with respect to a

$\qquad = (20)^2 + (18)^2 - 2(20)(18)\cos 35$ substitute known values

$\qquad \approx 724 - 589.8$ simplify (round to 10)

$\qquad = 134.2$ combine terms

$a \approx 11.6$ solve for a

The computed length of side a is 11.6 cm, and if you took great care in drawing your diagram, you'll find the missing side is indeed very close to this length.

Exercise 1: Finish solving the triangle above using the law of sines. Once you've computed $\angle B$ and $\angle C$,

measure these angles from the diagram using your protractor. How close was the computed measure to the actual measure?

For the second illustration (see Figure 7.55), draw *any arbitrary triangle* on a separate blank sheet, noting that the larger the triangle, the easier it is to measure the angles. After you've drawn it, measure the length of

Figure 7.55

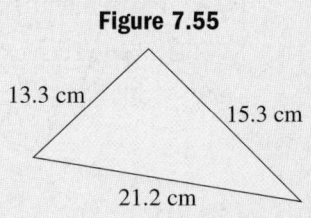

each side to the nearest millimeter (our triangle turned out to be 21.2 cm × 13.3 cm × 15.3 cm). Now use the law of cosines to find one angle, then the law of sines to solve the triangle. The computations for our triangle gave angles of 95.4°, 45.9°, and 38.7°. What angles did your computations give? Finally, use your protractor to measure the angles of the triangle you drew. With careful drawings, the measured results are often remarkably accurate!

Exercise 2: Using sides of 18 cm and 15 cm, draw a 35° angle, a 50° angle, and a 70° angle, then complete each triangle by connecting the endpoints. Use the law of cosines to compute the length of this third side, then actually measure each one. Was the actual length close to the computed length?

7.4 Vector Applications and the Dot Product

Learning Objectives

In Section 7.4 you will learn how to:

☐ **A.** Use vectors to investigate forces in equilibrium

☐ **B.** Find the components of one vector along another

☐ **C.** Solve applications involving work

☐ **D.** Compute dot products and the angle between two vectors

☐ **E.** Find the projection of one vector along another and resolve a vector into orthogonal components

☐ **F.** Use vectors to develop an equation for nonvertical projectile motion, and solve related applications

In Section 7.3 we introduced the concept of a vector, with its geometric, graphical, and algebraic representations. We also looked at operations on vectors and employed vector diagrams to solve basic applications. In this section we introduce additional ideas that enable us to solve a variety of new applications, while laying a strong foundation for future studies.

A. Vectors and Equilibrium

Much like the intuitive meaning of the word, vector forces are in **equilibrium** when they "counterbalance" each other. The simplest example is two vector forces of equal magnitude acting on the same point but in opposite directions. Similar to a tug-of-war with both sides equally matched, no one wins. If vector $\mathbf{F}_1$ has a magnitude of 500 lb in the positive direction, $\mathbf{F}_1 = \langle 500, 0 \rangle$ would need vector $\mathbf{F}_2 = \langle -500, 0 \rangle$ to counter it. If the forces are nonquadrantal, we intuitively sense the components must still sum to zero, and that $\mathbf{F}_3 = \langle 600, -200 \rangle$ would need $\mathbf{F}_4 = \langle -600, 200 \rangle$ for equilibrium to occur (see Figure 7.56). In other words, two vectors are in equilibrium when their sum is

Figure 7.56

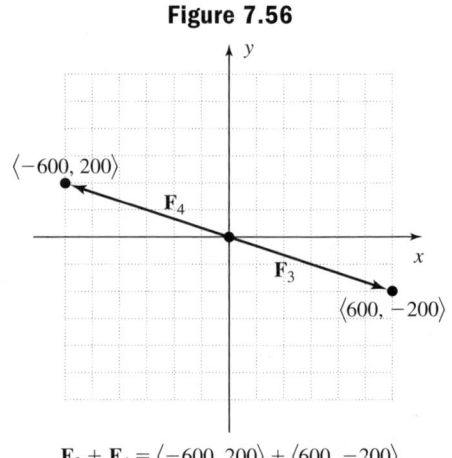

$\mathbf{F}_3 + \mathbf{F}_4 = \langle -600, 200 \rangle + \langle 600, -200 \rangle$
$\qquad\qquad = \langle 0, 0 \rangle = \mathbf{0}$

the zero vector **0**. If the forces have unequal magnitudes or do not pull in opposite directions, recall a resultant vector $\mathbf{F} = \mathbf{F}_a + \mathbf{F}_b$ can be found that represents the combined force. Equilibrium will then occur by adding the vector $-1\,(\mathbf{F})$ and this vector is sometimes called the **equilibriant.**

These ideas can be extended to include any number of vector forces acting on the same point. In general, we have the following:

Vectors and Equilibrium

Given vectors $\mathbf{F}_1, \mathbf{F}_2, \ldots, \mathbf{F}_n$ acting on a point P,
1. The resultant vector is $\mathbf{F} = \mathbf{F}_1 + \mathbf{F}_2 + \cdots + \mathbf{F}_n$.
2. Equilibrium for these forces requires the vector $-1\mathbf{F}$, where $\mathbf{F} + (-1)\mathbf{F} = 0$

EXAMPLE 1 ▶ **Finding the Equilibriant for Vector Forces**

Two force vectors $\mathbf{F}_1$ and $\mathbf{F}_2$ act on the point P as shown. Find a force $\mathbf{F}_3$ so equilibrium will occur, and sketch it on the grid.

Solution ▶ Begin by finding the horizontal and vertical components of each vector. For $\mathbf{F}_1$ we have $\langle -4.5 \cos 64°, 4.5 \sin 64° \rangle \approx \langle -2.0, 4.0 \rangle$, and for $\mathbf{F}_2$ we have $\langle 6.3 \cos 18°, 6.3 \sin 18° \rangle \approx \langle 6.0, 1.9 \rangle$. The resultant vector is $\mathbf{F} = \mathbf{F}_1 + \mathbf{F}_2 = \langle 4.0, 5.9 \rangle$, meaning equilibrium will occur by applying the force $-1\mathbf{F} = \langle -4.0, -5.9 \rangle$ (see figure).

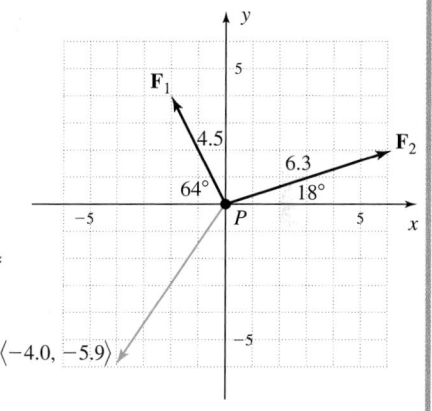

☑ **A.** You've just learned how to use vectors to investigate forces in equilibrium

---

Now try Exercises 7 through 20 ▶

B. The Component of u along v: comp$_v$u

As in Example 1, many simple applications involve position vectors where the angle and horizontal/vertical components are known or can easily be found. In these situations, the components are often quadrantal, that is, they lie along the x- and y-axes and meet at a right angle. Many other applications require us to find components of a vector that are nonquadrantal, with one of the components parallel to, or lying along a second vector. Given vectors **u** and **v**, as shown in Figure 7.57, we symbolize the component of **u** that lies along **v** as comp$_v$u, noting its value is simply $|\mathbf{u}|\cos\theta$ since

 $\cos\theta = \dfrac{\text{adj}}{\text{hyp}} = \dfrac{\text{comp}_v\text{u}}{|\mathbf{u}|}$. As the diagrams further indicate, comp$_v$u $= |\mathbf{u}|\cos\theta$ regardless

Figure 7.57

$$0 < \theta < \frac{\pi}{2} \qquad\qquad \frac{\pi}{2} < \theta < \pi$$

of how the vectors are oriented. Note that even when the components of a vector do not lie along the x- or y-axes, they are still **orthogonal** (meet at a 90° angle).

It is important to note that comp$_v$u is a *scalar quantity* (not a vector), giving only the magnitude of this component (the **vector projection** of **u** along **v** is studied later in this section). From these developments we make the following observations regarding the angle θ at which vectors **u** and **v** meet:

Vectors and the Component of u Along v

Given vectors **u** and **v**, which meet at an angle θ,
1. $\text{comp}_v u = |u| \cos \theta$.
2. If $0 < \theta < 90°$, $\text{comp}_v u > 0$; if $90° < \theta < 180°$, $\text{comp}_v u < 0$.
3. If $\theta = 0$, **u** and **v** have the same direction and $\text{comp}_v u = |u|$.
4. If $\theta = 90°$, **u** and **v** are orthogonal and $\text{comp}_v u = 0$.
5. If $\theta = 180°$, **u** and **v** have opposite directions and $\text{comp}_v \mathbf{u} = -|u|$.

EXAMPLE 2 ▶ **Finding the Component of Vector G Along Vector v**

Given the vectors **G** and **v** with $|G| = 850$ lb as shown in the figure, find $\text{comp}_v G$.

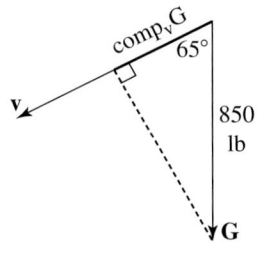

Solution ▶ Using $|G| \cos \theta = \text{comp}_v G$ we have $850 \cos 65° \approx 359$ lb. The component of **G** along **v** is about 359 pounds.

Now try Exercises 21 through 26 ▶

One interesting application of equilibrium and $\text{comp}_v u$ involves the force of gravity acting on an object placed on a ramp or an inclined plane. The greater the incline, the greater the tendency of the object to slide down the plane (for this study, we assume there is no friction between the object and the plane). While the force of gravity continues to pull straight downward (represented by the vector **G** in Figure 7.58), **G** is now the resultant of a force acting parallel to the plane along vector **v** (causing the object to slide) and a force acting perpendicular to the plane along vector **p** (causing the object to press against the plane). If we knew the component of **G** along **v** (indicated by the shorter, bold segment), we would know the force required to keep the object stationary as the two forces must be opposites. Note that **G** forms a right angle with the base of the inclined plane (see Figure 7.59), meaning that α and β must be complementary angles. Also note that since the location of a vector is unimportant, vector **p** has been repositioned for clarity.

Figure 7.58

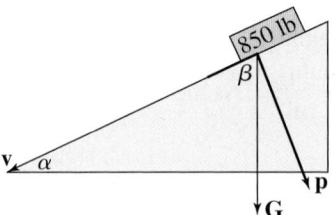

Figure 7.59

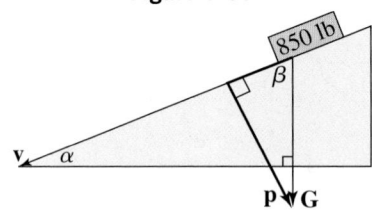

EXAMPLE 3A ▶ **Finding Components of Force for an Object on a Ramp**

A 850-lb object is sitting on a ramp that is inclined at 25°. Find the force needed to hold the object stationary (in equilibrium).

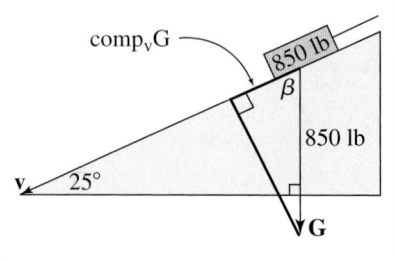

Solution ▶ Given $\alpha = 25°$, we know $\beta = 65°$. This means the component of **G** along the inclined plane is $\text{comp}_v G = 850 \cos 65°$ or about 359 lb. A force of 359 lb is required to keep the object from sliding down the incline (compare to Example 2).

EXAMPLE 3B ▶ A winch is being used to haul a 2000-lb block of granite up a ramp that is inclined at 15°. If the winch has a maximum tow rating of 500 lb, will it be successful?

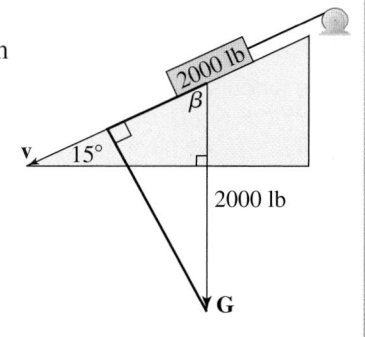

Solution ▶ We again need the component of **G** along the inclined plane: $\text{comp}_v G = 2000 \cos 75 \approx 518$ lb. Since the capacity of the winch is exceeded, the attempt will likely not be successful.

☑ **B.** You've just learned how to find the components of one vector along another

Now try Exercises 27 through 30 ▶

C. Vector Applications Involving Work

Figure 7.60

Figure 7.61

In common, everyday usage, **work** is understood to involve the exertion of energy or force to move an object a certain distance. For example, digging a ditch is hard work and involves moving dirt (exerting a force) from the trench to the bankside (over a certain distance). In an office, moving a filing cabinet likewise involves work. If the filing cabinet is heavier, or the distance it needs to be moved is greater, more work is required to move it (Figures 7.60 and 7.61).

To determine how much work was done by each person, we need to quantify the concept. Consider a constant force **F**, applied to move an object a distance D *in the same direction as the force*. In this case, work is defined as the product of the force applied and the distance the object is moved: Work = Force × Distance or $W = |\mathbf{F}|D$. If the force is given in pounds and the distance in feet, the amount of work is measured in a unit called **foot-pounds** (ft-lb). If the force is in newtons and the distance in meters, the amount of work is measured in **newton-meters** (N-m).

EXAMPLE 4 ▶ **Solving Applications of Vectors — Work and Force Parallel to the Direction of Movement**

While rearranging the office, Carrie must apply a force of 55.8 N to relocate a filing cabinet 4.5 m, while Bernard applies a 77.5 N force to move a second cabinet 3.2 m. Who did the most work?

Solution ▶
For Carrie: $W = |\mathbf{F}|D$ For Bernard: $W = |\mathbf{F}|D$
$\qquad\quad = (55.8)(4.5)$ $\qquad\qquad = (77.5)(3.2)$
$\qquad\quad = 251.1$ N-m $\qquad\qquad = 248$ N-m

Carrie did $251.1 - 248 = 3.1$ N-m more work than Bernard.

Now try Exercises 31 and 32 ▶

In many applications of work, the force **F** is not applied parallel to the direction of movement, as illustrated in Figures 7.62 and 7.63.

In calculating the amount of work done, the general concept of force × distance is preserved, *but only the component of force in the direction of movement is used*. In

Figure 7.62

Figure 7.63

terms of the component forces discussed earlier, if **F** is a constant force applied at angle θ to the direction of movement, the amount of work done is *the component of force along D times the distance the object is moved*.

<table>
<tr><td>

WORTHY OF NOTE

In the formula
$W = |\mathbf{F}|\cos\theta \times D$, observe that if $\theta = 0$, we have the old formula for work when the force is applied in the direction of movement
$W = \mathbf{F}D$. If $\theta \neq 0$, $\cos\theta \neq 1$ and the "effective force" on the object becomes $|\mathbf{F}|\cos\theta$.

</td></tr>
</table>

Force Vectors and Work W

Given a force **F** applied in the direction of movement at the acute angle θ to an object, and D the distance it is moved,

$$W = |\mathbf{F}|\cos\theta \times D$$

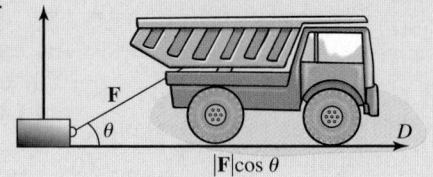

EXAMPLE 5 ▶ **Solving an Application of Vectors—Work and Force Applied at Angle θ to the Direction of Movement**

To help move heavy pieces of furniture across the floor, movers sometime employ a body harness similar to that used for a plow horse. A mover applies a constant 200-lb force to drag a piano 100 ft down a long hallway and into another room. If the straps make a 40° angle with the direction of movement, find the amount of work performed.

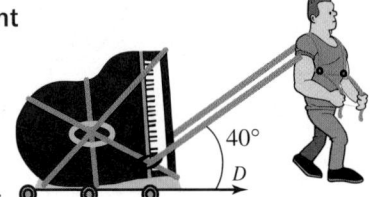

Solution ▶ The component of force in the direction of movement is 200 cos 40° or about 153 lb. The amount of work done is $W \approx 153(100) = 15{,}300$ ft-lb.

Now try Exercises 35 through 40 ▶

These ideas can be generalized to include work problems where the component of force in the direction of motion is along a *nonhorizontal* vector **v**. Consider Example 6.

EXAMPLE 6 ▶ **Solving an Application of Vectors—Forces Along a Nonhorizontal Vector**

The force vector $\mathbf{F} = \langle 5, 12 \rangle$ moves an object along the vector $\mathbf{v} = \langle 15.44, 2 \rangle$ as shown. Find the amount of work required to move the object along the entire length of **v**. Assume force is in pounds and distance in feet.

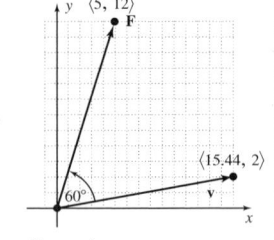

Solution ▶ To begin, we first determine the angle between the vectors. In this case we have $\theta = \tan^{-1}\left(\dfrac{12}{5}\right) - \tan^{-1}\left(\dfrac{2}{15.44}\right) \approx 60°$.

✓ **C.** You've just learned how to solve applications involving work

For $|\mathbf{F}| = 13$ (5-12-13 triangle), the component of force in the direction of motion is $\text{comp}_\mathbf{v}\mathbf{F} = 13\cos 60° = 6.5$. With $|\mathbf{v}| = \sqrt{(15.44)^2 + (2)^2} \approx 15.57$, the work required is $W = \text{comp}_\mathbf{v}\mathbf{F} \times |\mathbf{v}|$ or $(6.5)(15.57) \approx 101.2$ ft-lb.

Now try Exercises 41 through 44 ▶

D. Dot Products and the Angle Between Two Vectors

When the component of force in the direction of motion lies along a *nonhorizontal* vector (as in Example 6), the work performed can actually be computed more efficiently using an operation called the **dot product**. For any two vectors **u** and **v**, the dot product $\mathbf{u} \cdot \mathbf{v}$ is equivalent to $\text{comp}_\mathbf{v}\mathbf{u} \times |\mathbf{v}|$, yet is much easier to compute (for the proof of $\mathbf{u} \cdot \mathbf{v} = \text{comp}_\mathbf{v}\mathbf{u} \times |\mathbf{v}|$, see Appendix V). The operation is defined as follows:

The Dot Product u · v

Given vectors $\mathbf{u} = \langle a, b \rangle$ and $\mathbf{v} = \langle c, d \rangle$, $\mathbf{u} \cdot \mathbf{v} = \langle a, b \rangle \cdot \langle c, d \rangle = ac + bd$. In words, it is the *real number* found by taking the sum of corresponding component products.

EXAMPLE 7 ▶ **Using the Dot Product to Determine Force Along a Nonhorizontal Vector**

Verify the answer to Example 6 using the dot product $\mathbf{u} \cdot \mathbf{v}$.

Solution ▶ For $\mathbf{u} = \langle 5, 12 \rangle$ and $\mathbf{v} = \langle 15.44, 2 \rangle$, we have $\mathbf{u} \cdot \mathbf{v} = \langle 5, 12 \rangle \cdot \langle 15.44, 2 \rangle$ giving $5(15.44) + 12(2) = 101.2$. The result is 101.2, as in Example 6.

Now try Exercises 45 through 48 ▶

Note that dot products can also be used in the simpler case where the direction of motion is along a horizontal distance (Examples 4 and 5). While the dot product offers a powerful and efficient way to compute the work performed, it has many other applications; for example, to find the angle between two vectors. Consider that for any two vectors $\mathbf{u}$ and $\mathbf{v}$, $\mathbf{u} \cdot \mathbf{v} = |\mathbf{u}|\cos\theta \times |\mathbf{v}|$, leading directly to $\cos\theta = \dfrac{\mathbf{u}}{|\mathbf{u}|} \cdot \dfrac{\mathbf{v}}{|\mathbf{v}|}$ (solve for $\cos\theta$).

In summary,

Figure 7.64

The Angle θ Between Two Vectors

Given the nonzero vectors $\mathbf{u}$ and $\mathbf{v}$:

$$\cos\theta = \frac{\mathbf{u}}{|\mathbf{u}|} \cdot \frac{\mathbf{v}}{|\mathbf{v}|} \quad \text{and} \quad \theta = \cos^{-1}\left(\frac{\mathbf{u}}{|\mathbf{u}|} \cdot \frac{\mathbf{v}}{|\mathbf{v}|}\right)$$

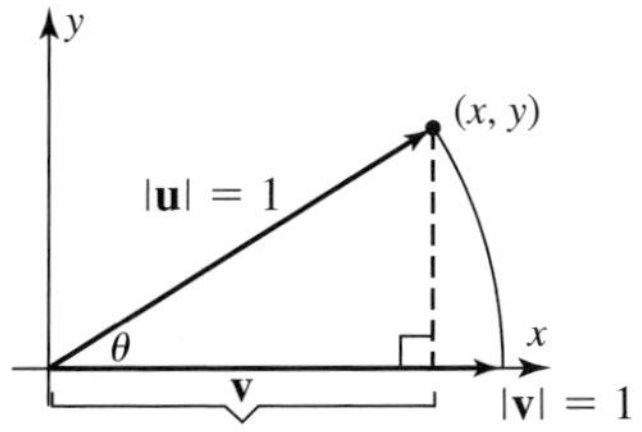

In the special case where $\mathbf{u}$ *and* $\mathbf{v}$ *are unit vectors*, this simplifies to $\cos\theta = \mathbf{u} \cdot \mathbf{v}$ since $|\mathbf{u}| = |\mathbf{v}| = 1$. This relationship is shown in Figure 7.64. The dot product $\mathbf{u} \cdot \mathbf{v}$ gives $\text{comp}_\mathbf{v}\mathbf{u} \times |\mathbf{v}|$, but $|\mathbf{v}| = 1$ and the component of $\mathbf{u}$ along $\mathbf{v}$ is simply the adjacent side of a right triangle whose hypotenuse is 1. Hence $\mathbf{u} \cdot \mathbf{v} = \cos\theta$.

EXAMPLE 8 ▶ **Determining the Angle Between Two Vectors**

Find the angle between the vectors given.

a. $\mathbf{u} = \langle -3, 4 \rangle$; $\mathbf{v} = \langle 5, 12 \rangle$ b. $\mathbf{v}_1 = 2\mathbf{i} - 3\mathbf{j}$; $\mathbf{v}_2 = 6\mathbf{i} + 4\mathbf{j}$

Solution ▶
a. $\cos\theta = \dfrac{\mathbf{u}}{|\mathbf{u}|} \cdot \dfrac{\mathbf{v}}{|\mathbf{v}|}$

$= \left\langle \dfrac{-3}{5}, \dfrac{4}{5} \right\rangle \cdot \left\langle \dfrac{5}{13}, \dfrac{12}{13} \right\rangle$

$= \dfrac{-15}{65} + \dfrac{48}{65}$

$= \dfrac{33}{65}$

$\theta = \cos^{-1}\left(\dfrac{33}{65}\right)$

$\approx 59.5°$

b. $\cos\theta = \dfrac{\mathbf{v}_1}{|\mathbf{v}_1|} \cdot \dfrac{\mathbf{v}_2}{|\mathbf{v}_2|}$

$= \left\langle \dfrac{2}{\sqrt{13}}, \dfrac{-3}{\sqrt{13}} \right\rangle \cdot \left\langle \dfrac{6}{\sqrt{52}}, \dfrac{4}{\sqrt{52}} \right\rangle$

$= \dfrac{12}{\sqrt{676}} + \dfrac{-12}{\sqrt{676}}$

$= \dfrac{0}{26} = 0$

$\theta = \cos^{-1}0$

$= 90°$

Now try Exercises 49 through 66 ▶

Note we have implicitly shown that if $\mathbf{u} \cdot \mathbf{v} = 0$, then $\mathbf{u}$ is orthogonal to $\mathbf{v}$. As with other vector operations, recognizing certain properties of the dot product will enable us to work with them more efficiently.

Properties of the Dot Product

Given vectors $\mathbf{u}$, $\mathbf{v}$, and $\mathbf{w}$ and a constant k,

1. $\mathbf{u} \cdot \mathbf{v} = \mathbf{v} \cdot \mathbf{u}$ 2. $\mathbf{u} \cdot \mathbf{u} = |\mathbf{u}|^2$

3. $\mathbf{w} \cdot (\mathbf{u} + \mathbf{v}) = \mathbf{w} \cdot \mathbf{u} + \mathbf{w} \cdot \mathbf{v}$ 4. $k(\mathbf{u} \cdot \mathbf{v}) = k\mathbf{u} \cdot \mathbf{v} = \mathbf{u} \cdot k\mathbf{v}$

5. $\mathbf{0} \cdot \mathbf{u} = \mathbf{u} \cdot \mathbf{0} = 0$ 6. $\dfrac{\mathbf{u}}{|\mathbf{u}|} \cdot \dfrac{\mathbf{v}}{|\mathbf{v}|} = \dfrac{\mathbf{u} \cdot \mathbf{v}}{|\mathbf{u}||\mathbf{v}|}$

Property 6 offers an alternative to unit vectors when finding $\cos \theta$—the dot product of the vectors can be computed first, and the result divided by the product of their magnitudes: $\cos \theta = \dfrac{\mathbf{u} \cdot \mathbf{v}}{|\mathbf{u}||\mathbf{v}|}$. Proofs of the first two properties are given here. Proofs of the others have a similar development **(see Exercises 79 through 82)**. For any two nonzero vectors $\mathbf{u} = \langle a, b \rangle$ and $\mathbf{v} = \langle c, d \rangle$:

Property 1: $\mathbf{u} \cdot \mathbf{v} = \langle a, b \rangle \cdot \langle c, d \rangle$ Property 2: $\mathbf{u} \cdot \mathbf{u} = \langle a, b \rangle \cdot \langle a, b \rangle$

$\qquad\qquad = ac + bd$ $\qquad\qquad = a^2 + b^2$

$\qquad\qquad = ca + db$ $\qquad\qquad = |\mathbf{u}|^2$

$\qquad\qquad = \langle c, d \rangle \cdot \langle a, b \rangle$ (since $|\mathbf{u}| = \sqrt{a^2 + b^2}$)

$\qquad\qquad = \mathbf{v} \cdot \mathbf{u}$

☑ **D. You've just learned how to compute dot products and the angle between two vectors**

Using $\text{comp}_{\mathbf{v}}\mathbf{u} = |\mathbf{u}|\cos \theta$ and $\mathbf{u} \cdot \mathbf{v} = \text{comp}_{\mathbf{v}}\mathbf{u} \times |\mathbf{v}|$, we can also state the following relationships, which give us some flexibility on how we approach applications of the dot product.

For any two vectors $\mathbf{u} = \langle a, b \rangle$ and $\mathbf{v} = \langle c, d \rangle$:

(1) $\mathbf{u} \cdot \mathbf{v} = ac + bd$ standard computation of the dot product

(2) $\mathbf{u} \cdot \mathbf{v} = |\mathbf{u}|\cos \theta \times |\mathbf{v}|$ alternative computation of the dot product

(3) $\mathbf{u} \cdot \mathbf{v} = \text{comp}_{\mathbf{v}}\mathbf{u} \times |\mathbf{v}|$ replace $|\mathbf{u}|\cos \theta$ in (2) with $\text{comp}_{\mathbf{v}}\mathbf{u}$

(4) $\dfrac{\mathbf{u} \cdot \mathbf{v}}{|\mathbf{u}||\mathbf{v}|} = \cos \theta$ divide (2) by scalars $|\mathbf{u}|$ and $|\mathbf{v}|$

(5) $\dfrac{\mathbf{u} \cdot \mathbf{v}}{|\mathbf{v}|} = \text{comp}_{\mathbf{v}}\mathbf{u}$ divide (3) by $|\mathbf{v}|$

E. Vector Projections and Orthogonal Components

In work problems and other simple applications, it is enough to find and apply $\text{comp}_{\mathbf{v}}\mathbf{u}$ (Figure 7.65). However, applications involving thrust and drag forces, tension and stress limits in a cable, electronic circuits, and cartoon animations often require that we also find the *vector form of* $\text{comp}_{\mathbf{v}}\mathbf{u}$. This is called the **projection** of $\mathbf{u}$ along $\mathbf{v}$ or **proj$_{\mathbf{v}}\mathbf{u}$**, and is a vector in the same direction of $\mathbf{v}$ with magnitude $\text{comp}_{\mathbf{v}}\mathbf{u}$ (Figures 7.66 and 7.67).

Figure 7.65

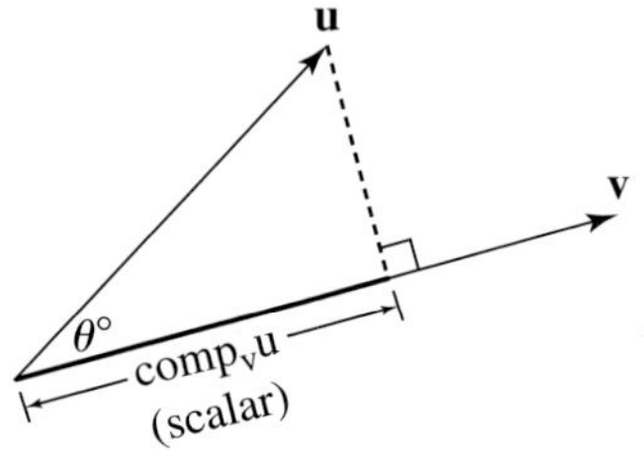

Figure 7.66

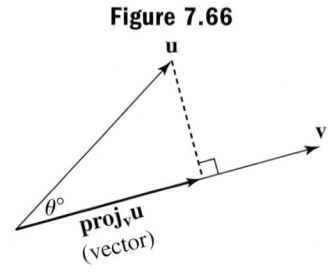

Figure 7.67

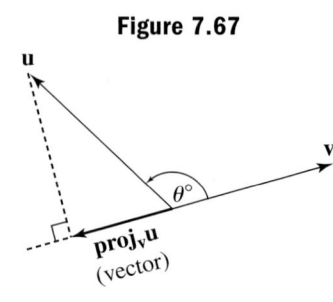

By its design, the unit vector $\dfrac{\mathbf{v}}{|\mathbf{v}|}$ has a length of one and points in the same direction as

$\mathbf{v}$, so $\mathbf{proj_v u}$ can be computed as $\mathrm{comp_v u} \times \dfrac{\mathbf{v}}{|\mathbf{v}|}$ (see Example 7, Section 7.3). Using

equation (5) above and the properties shown earlier, an alternative formula for $\mathbf{proj_v u}$ can be found that is usually easier to simplify:

$$\mathbf{proj_v u} = \mathrm{comp_v u} \times \frac{\mathbf{v}}{|\mathbf{v}|} \qquad \text{definition of a projection}$$

$$= \frac{\mathbf{u} \cdot \mathbf{v}}{|\mathbf{v}|} \times \frac{\mathbf{v}}{|\mathbf{v}|} \qquad \text{substitute } \frac{\mathbf{u} \cdot \mathbf{v}}{|\mathbf{v}|} \text{ for } \mathrm{comp_v u}$$

$$= \frac{\mathbf{u} \cdot \mathbf{v}}{|\mathbf{v}|^2} \times \mathbf{v} \qquad \text{rewrite factors}$$

Vector Projections

Given vectors $\mathbf{u}$ and $\mathbf{v}$, the projection of $\mathbf{u}$ along $\mathbf{v}$ is the *vector*

$$\mathbf{proj_v u} = \left(\frac{\mathbf{u} \cdot \mathbf{v}}{|\mathbf{v}|^2}\right)\mathbf{v}$$

EXAMPLE 9A ▶ **Finding the Projection of One Vector Along Another**

Given $\mathbf{u} = \langle -7, 1 \rangle$ and $\mathbf{v} = \langle 6, 6 \rangle$, find $\mathbf{proj_v u}$.

Solution ▶ To begin, find $\mathbf{u} \cdot \mathbf{v}$ and $|\mathbf{v}|$.

$$\mathbf{u} \cdot \mathbf{v} = \langle -7, 1 \rangle \cdot \langle 6, 6 \rangle \qquad\qquad |\mathbf{v}| = \sqrt{6^2 + 6^2}$$
$$= -42 + 6 \qquad\qquad\qquad\qquad = \sqrt{72}$$
$$= -36 \qquad\qquad\qquad\qquad\quad\; = 6\sqrt{2}$$

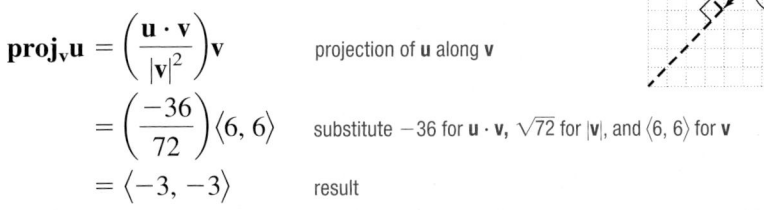

$$\mathbf{proj_v u} = \left(\frac{\mathbf{u} \cdot \mathbf{v}}{|\mathbf{v}|^2}\right)\mathbf{v} \qquad \text{projection of } \mathbf{u} \text{ along } \mathbf{v}$$

$$= \left(\frac{-36}{72}\right)\langle 6, 6 \rangle \qquad \text{substitute } -36 \text{ for } \mathbf{u} \cdot \mathbf{v}, \sqrt{72} \text{ for } |\mathbf{v}|, \text{ and } \langle 6, 6 \rangle \text{ for } \mathbf{v}$$

$$= \langle -3, -3 \rangle \qquad \text{result}$$

A useful consequence of computing $\mathbf{proj_v u}$ is we can then **resolve** the vector $\mathbf{u}$ into **orthogonal components** *that need not be quadrantal.* One component will be parallel to $\mathbf{v}$ and the other perpendicular to $\mathbf{v}$ (the dashed line in the diagram in Example 9A). In general terms, this means we can write $\mathbf{u}$ as the vector sum $\mathbf{u}_1 + \mathbf{u}_2$, where $\mathbf{u}_1 = \mathbf{proj_v u}$ and $\mathbf{u}_2 = \mathbf{u} - \mathbf{u}_1$ (note $\mathbf{u}_1 \parallel \mathbf{v}$).

Resolving a Vector into Orthogonal Components

Given vectors $\mathbf{u}$, $\mathbf{v}$, and $\mathbf{proj_v u}$, $\mathbf{u}$ can be resolved into the orthogonal components $\mathbf{u}_1$ and $\mathbf{u}_2$, where $\mathbf{u} = \mathbf{u}_1 + \mathbf{u}_2$, $\mathbf{u}_1 = \mathbf{proj_v u}$, and $\mathbf{u}_2 = \mathbf{u} - \mathbf{u}_1$.

WORTHY OF NOTE

Note that $\mathbf{u}_2 = \mathbf{u} - \mathbf{u}_1$ is the shorter diagonal of the parallelogram formed by the vectors $\mathbf{u}$ and $\mathbf{u}_1 = \mathbf{proj_v u}$. This can also be seen in the graph supplied for Example 9B.

EXAMPLE 9B ▶ **Resolving a Vector into Orthogonal Components**

Given $\mathbf{u} = \langle 2, 8 \rangle$ and $\mathbf{v} = \langle 8, 6 \rangle$, resolve $\mathbf{u}$ into orthogonal components $\mathbf{u}_1$ and $\mathbf{u}_2$, where $\mathbf{u}_1 \parallel \mathbf{v}$ and $\mathbf{u}_2 \perp \mathbf{v}$. Also verify $\mathbf{u}_1 \perp \mathbf{u}_2$.

Solution ▶ Once again, begin by finding $\mathbf{u} \cdot \mathbf{v}$ and $|\mathbf{v}|$.

$$\mathbf{u} \cdot \mathbf{v} = \langle 2, 8 \rangle \cdot \langle 8, 6 \rangle \qquad |\mathbf{v}| = \sqrt{8^2 + 6^2}$$
$$= 16 + 48 \qquad\qquad = \sqrt{100}$$
$$= 64 \qquad\qquad\qquad = 10$$

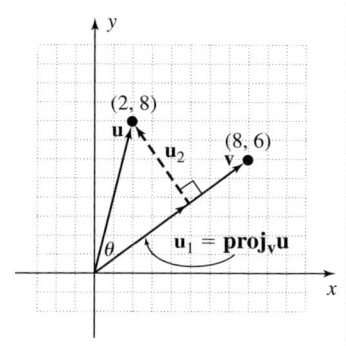

$$\mathbf{proj_v u} = \left(\frac{\mathbf{u} \cdot \mathbf{v}}{|\mathbf{v}|^2} \right) \mathbf{v} \qquad \text{projection of } \mathbf{u} \text{ along } \mathbf{v}$$

$$= \left(\frac{64}{100} \right) \langle 8, 6 \rangle \qquad \begin{array}{l}\text{substitute 64 for } \mathbf{u} \cdot \mathbf{v}, \text{ 10 for}\\ |\mathbf{v}|, \text{ and } \langle 8, 6 \rangle \text{ for } \mathbf{v}\end{array}$$

$$= \langle 5.12, 3.84 \rangle \qquad \text{result}$$

☑ **E.** You've just learned how to find the projection of one vector along another and resolve a vector into orthogonal components

For $\mathbf{proj_v u} = \mathbf{u}_1 = \langle 5.12, 3.84 \rangle$, we have $\mathbf{u}_2 = \mathbf{u} - \mathbf{u}_1 = \langle 2, 8 \rangle - \langle 5.12, 3.84 \rangle = \langle -3.12, 4.16 \rangle$. To verify $\mathbf{u}_1 \perp \mathbf{u}_2$, we need only show $\mathbf{u}_1 \cdot \mathbf{u}_2 = 0$:

$$\mathbf{u}_1 \cdot \mathbf{u}_2 = \langle 5.12, 3.84 \rangle \cdot \langle -3.12, 4.16 \rangle$$
$$= (5.12)(-3.12) + (3.84)(4.16)$$
$$= 0 ✓$$

Now try Exercises 67 through 72 ▶

F. Vectors and the Height of a Projectile

Our final application of vectors involves **projectile motion.** A projectile is any object that is thrown or projected upward, with no source of propulsion to sustain its motion. In this case, the only force acting on the projectile is gravity (air resistance is neglected), so the maximum height and the range of the projectile depend solely on its initial velocity and the angle θ at which it is projected. In a college algebra course, the equation $y = v_0 t - 16t^2$ is developed to model the height in feet (at time t) of a projectile thrown vertically upward with initial velocity of v_0 feet per second. Here, we'll modify the equation slightly to take into account that the object is now moving horizontally as well as vertically. As you can see in Figure 7.68, the vector $\mathbf{v}$ representing the initial velocity, as well as the velocity vector at other times, can easily be decomposed into horizontal and vertical components. This will enable us to find a more general relationship for the position of the projectile. For now, we'll let $\mathbf{v}_y$ represent the component of velocity in the vertical (y) direction, and $\mathbf{v}_x$ represent the component of velocity in the horizontal (x) direction. Since gravity acts only in the vertical (and negative) direction, the horizontal component of the velocity remains constant at $\mathbf{v}_x = |\mathbf{v}|\cos\theta$. Using $D = RT$, the x-coordinate of the projectile at time t is $x = (|\mathbf{v}|\cos\theta)t$. For the vertical component $\mathbf{v}_y$ we use the projectile equation developed earlier, *substituting* $|\mathbf{v}|\sin\theta$ *for* v_0, since the angle of projection is no longer 90°. This gives the y-coordinate at time t as $y = v_0 t - 16t^2 = (|\mathbf{v}|\sin\theta)t - 16t^2$.

Figure 7.68

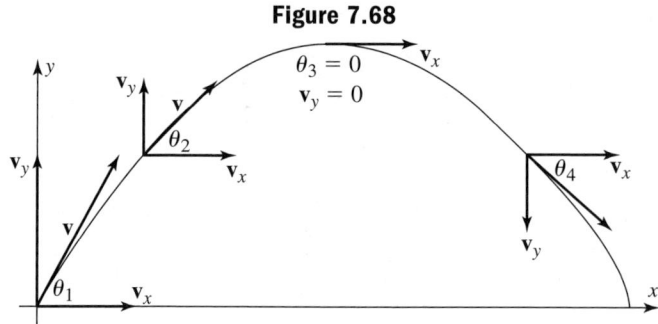

> **Projectile Motion**
>
> Given an object is projected upward from the origin with initial velocity $|\mathbf{v}|$ at angle $\theta°$.
> The x-coordinate of its position at time t is $x = (|\mathbf{v}|\cos\theta)t$.
> The y-coordinate of its position at time t is $y = (|\mathbf{v}|\sin\theta)t - 16t^2$.

EXAMPLE 10 ▶ **Solving an Application of Vectors—Projectile Motion**

An arrow is shot upward with an initial velocity of 150 ft/sec at an angle of 50°.
  a. Find the position of the arrow after 2 sec.
  b. How many seconds does it take to reach a height of 190 ft?

Solution ▶ **a.** Using the preceding equations yields these coordinates for its position at $t = 2$:

$$x = (|\mathbf{v}|\cos\theta)t \qquad\qquad y = (|\mathbf{v}|\sin\theta)t - 16t^2$$
$$= (150\cos 50°)(2) \qquad\qquad = (150\sin 50°)(2) - 16(2)^2$$
$$\approx 193 \qquad\qquad\qquad \approx 166$$

The arrow has traveled a horizontal distance of about 193 ft and is 166 ft high.

  b. To find the time required to reach 190 ft in height, set the equation for the y coordinate equal to 190, which yields a quadratic equation in t:

$$y = (|\mathbf{v}|\sin\theta)t - 16t^2 \qquad \text{equation for } y$$
$$190 = (150\sin 50°)t - 16t^2 \qquad \text{substitute 150 for } |\mathbf{v}| \text{ and 50° for } \theta$$
$$0 \approx -16(t)^2 + 115t - 190 \qquad 150\sin 50 \approx 115$$

✓ **F. You've just learned how to use vectors to develop an equation for nonvertical, projectile motion and solve related applications**

Using the quadratic formula we find that $t \approx 2.6$ sec and $t \approx 4.6$ sec are solutions. This makes sense, since the arrow reaches a given height once on the way up and again on the way down, as long as it hasn't reached its maximum height.

Now try Exercises 73 through 78 ▶

For more on projectile motion, see the *Calculator Exploration and Discovery* feature at the end of this chapter.

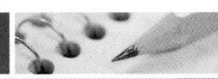

7.4 EXERCISES

▶ **CONCEPTS AND VOCABULARY**

Fill in each blank with the appropriate word or phrase. Carefully reread the section if needed.

1. Vector forces are in _____ when they counterbalance each other. Such vectors have a sum of _____.

2. The component of a vector $\mathbf{u}$ along another vector $\mathbf{v}$ is written notationally as _____, and is computed as _____.

3. Two vectors that meet at a right angle are said to be _____.

4. The component of $\mathbf{u}$ along $\mathbf{v}$ is a _____ quantity. The projection of $\mathbf{u}$ along $\mathbf{v}$ is a _____.

5. Explain/Discuss exactly what information the dot product of two vectors gives us. Illustrate with a few examples.

6. Compare and contrast the projectile equations $y = v_0 t - 16t^2$ and $y = (v_0\sin\theta)t - 16t^2$. Discuss similarities/differences using illustrative examples.

► **DEVELOPING YOUR SKILLS**

The force vectors given are acting on a common point *P*. Find an additional force vector so that equilibrium takes place.

7. $\mathbf{F}_1 = \langle -8, -3 \rangle$; $\mathbf{F}_2 = \langle 2, -5 \rangle$

8. $\mathbf{F}_1 = \langle -2, 7 \rangle$; $\mathbf{F}_2 = \langle 5, 3 \rangle$

9. $\mathbf{F}_1 = \langle -2, -7 \rangle$; $\mathbf{F}_2 = \langle 2, -7 \rangle$; $\mathbf{F}_3 = \langle 5, 4 \rangle$

10. $\mathbf{F}_1 = \langle -3, 10 \rangle$; $\mathbf{F}_2 = \langle -10, 3 \rangle$; $\mathbf{F}_3 = \langle -9, -2 \rangle$

11. $\mathbf{F}_1 = 5\mathbf{i} - 2\mathbf{j}$; $\mathbf{F}_2 = \mathbf{i} + 10\mathbf{j}$

12. $\mathbf{F}_1 = -7\mathbf{i} + 6\mathbf{j}$; $\mathbf{F}_2 = -8\mathbf{i} - 3\mathbf{j}$

13. $\mathbf{F}_1 = 2.5\mathbf{i} + 4.7\mathbf{j}$; $\mathbf{F}_2 = -0.3\mathbf{i} + 6.9\mathbf{j}$; $\mathbf{F}_3 = -12\mathbf{j}$

14. $\mathbf{F}_1 = 3\sqrt{2}\mathbf{i} - 2\sqrt{3}\mathbf{j}$; $\mathbf{F}_2 = -2\mathbf{i} + 7\mathbf{j}$; $\mathbf{F}_3 = 5\mathbf{i} + 2\sqrt{3}\mathbf{j}$

15.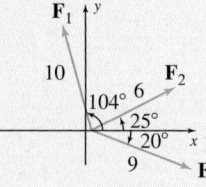

16.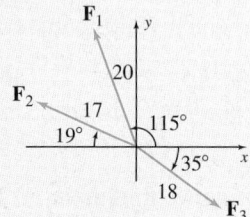

17. The force vectors $\mathbf{F}_1$ and $\mathbf{F}_2$ are simultaneously acting on a point *P*. Find a third vector $\mathbf{F}_3$ so that equilibrium takes place if $\mathbf{F}_1 = \langle 19, 10 \rangle$ and $\mathbf{F}_2 = \langle 5, 17 \rangle$.

18. The force vectors $\mathbf{F}_1$, $\mathbf{F}_2$, and $\mathbf{F}_3$ are simultaneously acting on a point *P*. Find a fourth vector $\mathbf{F}_4$ so that equilibrium takes place if $\mathbf{F}_1 = \langle -12, 2 \rangle$, $\mathbf{F}_2 = \langle -6, 17 \rangle$, and $\mathbf{F}_3 = \langle 3, 15 \rangle$.

19. A new "Survivor" game involves a three-team tug-of-war. Teams 1 and 2 are pulling with the magnitude and at the angles indicated in the diagram. If the teams are currently in a stalemate, find the magnitude and angle of the rope held by team 3.

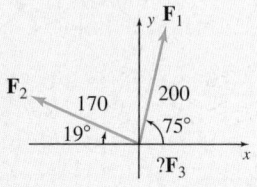

20. Three cowhands have roped a wild stallion and are attempting to hold him steady. The first and second cowhands are pulling with the magnitude and at the angles indicated in the diagram. If the stallion is held fast by the three cowhands, find the magnitude and angle of the rope from the third cowhand.

Find the component of **u** along **v** (compute comp$_v$**u**) for the vectors **u** and **v** given.

21.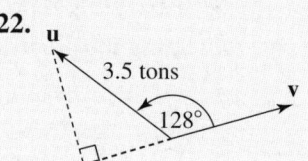

22.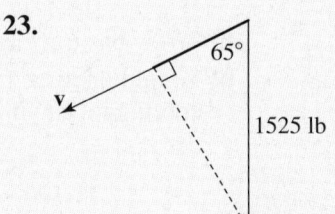

23.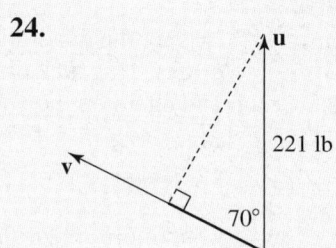

24.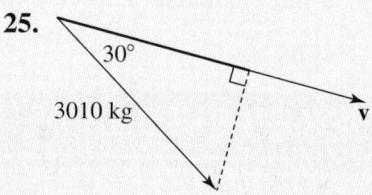

25.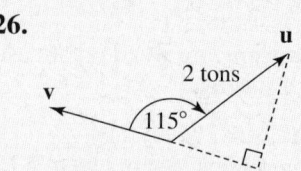

26.

27. **Static equilibrium:** A 500-lb crate is sitting on a ramp that is inclined at 35°. Find the force needed to hold the object stationary.

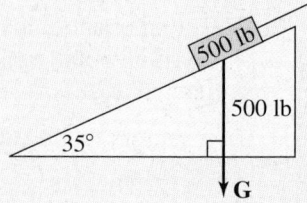

28. Static equilibrium:
A 1200-lb skiff is being pulled from a lake, using a boat ramp inclined at 20°. Find the minimum force needed to dock the skiff.

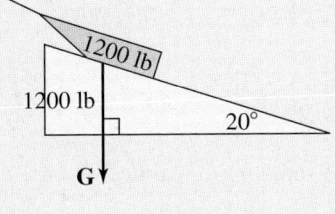

29. Static equilibrium: A 325-kg carton is sitting on a ramp, held stationary by 225 kg of tension in a restraining rope. Find the ramps's angle of incline.

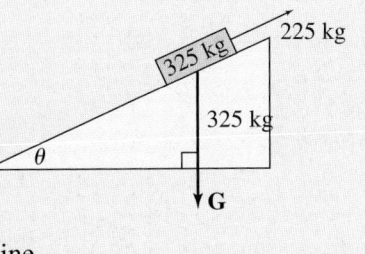

30. Static equilibrium:
A heavy dump truck is being winched up a ramp with an 18° incline. Approximate the weight of the truck if the winch is working at its maximum capacity of 1.75 tons and the truck is barely moving.

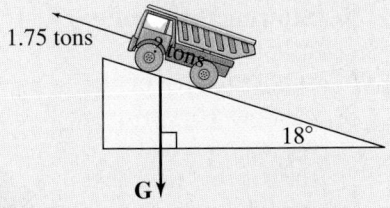

31. While rearranging the patio furniture, Rick has to push the weighted base of the umbrella stand 15 m. If he uses a constant force of 75 N, how much work did he do?

32. Vinny's car just broke down in the middle of the road. Luckily, a buddy is with him and offers to steer if Vinny will get out and push. If he pushes with a constant force of 185 N to move the car 30 m, how much work did he do?

► **WORKING WITH FORMULAS**

The range of a projectile: $R = \dfrac{v^2 \sin\theta \cos\theta}{16}$

33. The range of a projected object (total horizontal distance traveled) is given by the formula shown, where v is the initial velocity and θ is the angle at which it is projected. If an arrow leaves the bow traveling 175 ft/sec at an angle of 45°, what horizontal distance will it travel?

34. A collegiate javelin thrower releases the javelin at a 40° angle, with an initial velocity of about 95 ft/sec. If the NCAA record is 280 ft, will this throw break the record? What is the smallest angle of release that will break this record? If the javelin were released at the optimum 45°, by how many feet would the record be broken?

► **APPLICATIONS**

35. Plowing a field: An old-time farmer is plowing his field with a mule. How much work does the mule do in plowing one length of a field 300 ft long, if it pulls the plow with a constant force of 250 lb and the straps make a 30° angle with the horizontal.

36. Pulling a sled:
To enjoy a beautiful snowy day, a mother is pulling her three children on a sled along a level

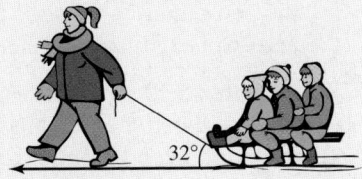

street. How much work (play) is done if the street is 100 ft long and she pulls with a constant force of 55 lb with the tow-rope making an angle of 32° with the street?

37. Tough-man contest: As part of a "tough-man" contest, participants are required to pull a bus along a level street for 100 ft. If one contestant did 45,000 ft-lb of work to accomplish the task and the straps used made an angle of 5° with the street, find the tension in the strap during the pull.

38. Moving supplies: An arctic explorer is hauling supplies from the supply hut to her tent, a distance of 150 ft, in a sled she is dragging behind her. If 9000 ft-lb of work was done and the straps used made an angle of 25° with the snow-covered ground, find the tension in the strap during the task.

39. Wheelbarrow rides: To break up the monotony of a long, hot, boring Saturday, a father decides to (carefully) give his kids a ride in a wheelbarrow. He applies a force of 30 N to move the "load" 100 m, then stops to rest. Find the amount of work done if the wheelbarrow makes an angle of 20° with level ground while in motion.

40. Mowing the lawn:
A home owner applies a force of 40 N to push her lawn mower back and forth across the back yard. Find the amount of work done if the yard is 50 m long, requires 24 passes to get the lawn mowed, and the mower arm makes an angle of 39° with the level ground.

39°

Force vectors: For the force vector **F** and vector **v** given, find the amount of work required to move an object along the entire length of **v**. Assume force is in pounds and distance in feet.

41. $F = \langle 15, 10 \rangle$; $v = \langle 50, 5 \rangle$

42. $F = \langle -5, 12 \rangle$; $v = \langle -25, 10 \rangle$

43. $F = \langle 8, 2 \rangle$; $v = \langle 15, -1 \rangle$

44. $F = \langle 15, -3 \rangle$; $v = \langle 24, -20 \rangle$

45. Use the dot product to verify the solution to Exercise 41.

46. Use the dot product to verify the solution to Exercise 42.

47. Use the dot product to verify the solution to Exercise 43.

48. Use the dot product to verify the solution to Exercise 44.

For each pair of vectors given, (a) compute the dot product **p · q** and (b) find the angle between the vectors to the nearest tenth of a degree.

49. $p = \langle 5, 2 \rangle$; $q = \langle 3, 7 \rangle$

50. $p = \langle -3, 6 \rangle$; $q = \langle 2, -5 \rangle$

51. $p = -2i + 3j$; $q = -6i - 4j$

52. $p = -4i + 3j$; $q = -6i - 8j$

53. $p = 7\sqrt{2}i - 3j$; $q = 2\sqrt{2}i + 9j$

54. $p = \sqrt{2}i - 3j$; $q = 3\sqrt{2}i + 5j$

Determine if the pair of vectors given are orthogonal.

55. $u = \langle 7, -2 \rangle$; $v = \langle 4, 14 \rangle$

56. $u = \langle -3.5, 2.1 \rangle$; $v = \langle -6, -10 \rangle$

57. $u = \langle -6, -3 \rangle$; $v = \langle -8, 15 \rangle$

58. $u = \langle -5, 4 \rangle$; $v = \langle -9, -11 \rangle$

59. $u = -2i - 6j$; $v = 9i - 3j$

60. $u = 3\sqrt{2}i - 2j$; $v = 2\sqrt{2}i + 6j$

Find comp$_v$u for the vectors **u** and **v** given.

61. $u = \langle 3, 5 \rangle$; $v = \langle 7, 1 \rangle$

62. $u = \langle 3, 5 \rangle$; $v = \langle -7, 1 \rangle$

63. $u = -7i + 4j$; $v = -10j$

64. $u = 8i$; $v = 10i + 3j$

65. $u = 7\sqrt{2}i - 3j$; $v = 6i + 5\sqrt{3}j$

66. $u = -3\sqrt{2}i + 6j$; $v = 2i + 5\sqrt{5}j$

For each pair of vectors given, (a) find the projection of **u** along **v** (compute **proj$_v$u**) and (b) resolve **u** into vectors **u$_1$** and **u$_2$**, where **u$_1$**‖**v** and **u$_2$**⊥**v**.

67. $u = \langle 2, 6 \rangle$; $v = \langle 8, 3 \rangle$

68. $u = \langle -3, 8 \rangle$; $v = \langle -12, 3 \rangle$

69. $u = \langle -2, -8 \rangle$; $v = \langle -6, 1 \rangle$

70. $u = \langle -4.2, 3 \rangle$; $v = \langle -5, -8.3 \rangle$

71. $u = 10i + 5j$; $v = 12i + 2j$

72. $u = -3i - 9j$; $v = 5i - 3j$

Projectile motion: A projectile is launched from a catapult with the initial velocity v_0 and angle θ indicated. Find (a) the position of the object after 3 sec and (b) the time required to reach a height of 250 ft.

73. $v_0 = 250$ ft/sec; $\theta = 60°$

74. $v_0 = 300$ ft/sec; $\theta = 55°$

75. $v_0 = 200$ ft/sec; $\theta = 45°$

76. $v_0 = 500$ ft/sec; $\theta = 70°$

77. At the circus, a "human cannon ball" is shot from a large cannon with an initial velocity of 90 ft/sec at an angle of 65° from the horizontal. How high is the acrobat after 1.2 sec? How long until the acrobat is again at this same height?

78. A center fielder runs down a long hit by an opposing batter and whirls to throw the ball to the infield to keep the hitter to a double. If the initial velocity of the throw is 130 ft/sec and the ball is released at an angle of 30° with level ground, how high is the ball after 1.5 sec? How long until the ball again reaches this same height?

▶ **EXTENDING THE CONCEPT**

For the arbitrary vectors $u = \langle a, b \rangle$, $v = \langle c, d \rangle$, and $w = \langle e, f \rangle$ and the scalar k, prove the following vector properties using the properties of real numbers.

79. $w \cdot (u + v) = w \cdot u + w \cdot v$

80. $k(u \cdot v) = ku \cdot v = u \cdot kv$

81. $0 \cdot u = u \cdot 0 = 0$ **82.** $\dfrac{u}{|u|} \cdot \dfrac{v}{|v|} = \dfrac{u \cdot v}{|u||v|}$

83. As alternative to $\cos \theta = \dfrac{u \cdot v}{|u||v|}$ for finding the angle between two vectors, the equation

$\tan \theta = \dfrac{m_2 - m_1}{1 + m_2 m_1}$ can be used, where m_1 and m_2

represent the slopes of the vectors. Find the angle between the vectors $1i + 5j$ and $5i + 2j$ using each equation and comment on which you found more efficient. Then see if you can find a geometric connection between the two equations.

84. Use the equations for the horizontal and vertical components of the projected object's position to obtain the equation of trajectory

$y = (\tan \theta)x - \dfrac{16}{v^2 \cos^2 \theta} x^2$. This is a quadratic

equation in x. What can you say about its graph? Include comments about the concavity, x-intercepts, maximum height, and so on.

▶ **MAINTAINING YOUR SKILLS**

85. (4.4) Solve for t: $2.9e^{-0.25t} + 7.6 = 438$

86. (5.5) Graph the function using a reference rectangle and the *rule of fourths*:

$y = 3 \cos\left(2\theta - \dfrac{\pi}{4}\right)$

87. (7.2) Solve the triangle shown, then compute its perimeter and area.

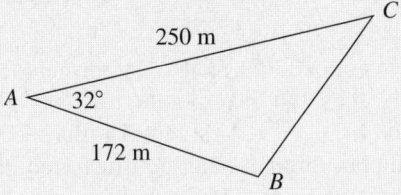

88. (7.3) A plane is flying 200 mph at heading 30°, with a 40 mph wind blowing from due west. Find the true course and speed of the plane.

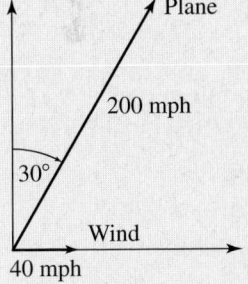

7.5 | Complex Numbers in Trigonometric Form

Learning Objectives

In Section 7.5 you will learn how to:

☐ **A.** Graph a complex number

☐ **B.** Write a complex number in trigonometric form

☐ **C.** Convert from trigonometric form to rectangular form

☐ **D.** Interpret products and quotients geometrically

☐ **E.** Compute products and quotients in trigonometric form

☐ **F.** Solve applications involving complex numbers (optional)

Once the set of complex numbers became recognized and defined, the related basic operations matured very quickly. With little modification—sums, differences, products, quotients, and powers all lent themselves fairly well to the algebraic techniques used for real numbers. But roots of complex numbers did not yield so easily and additional tools and techniques were needed. Writing complex numbers in trigonometric form enables us to find complex roots (Section 7.6) and in some cases, makes computing products, quotients, and powers more efficient.

A. Graphing Complex Numbers

In previous sections we defined a vector quantity as one that required more than a single component to describe its attributes. The complex number $z = a + bi$ certainly fits this description, since both a real number "component" and an imaginary "component" are needed to define it. In many respects, we can treat complex numbers in the same way we treated vectors and in fact, there is much we can learn from this connection.

Since both axes in the xy-plane have real number values, it's not possible to graph a complex number in $\mathbb{R}$ (the real plane). However, in the same way we used

WORTHY OF NOTE

Surprisingly, the study of complex numbers matured much earlier than the study of vectors, and representing complex numbers as directed line segments actually preceded their application to a vector quantity.

the x-axis for the horizontal component of a vector and the y-axis for the vertical, we can let the x-axis represent the real valued part of a complex number and the y-axis the imaginary part. The result is called the **complex plane** $\mathbb{C}$. Every point (a, b) in $\mathbb{C}$ can be associated with a complex number $a + bi$, and any complex number $a + bi$ can be associated with a point (a, b) in $\mathbb{C}$ (Figure 7.69). The point (a, b) can also be regarded as the terminal point of a position vector representing the complex number, generally named using the letter z.

Figure 7.69

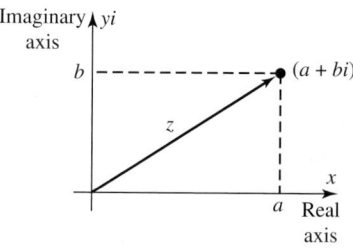

EXAMPLE 1 ▶ **Graphing Complex Numbers**

Graph the complex numbers below on the same complex plane.

    a. $z_1 = -2 - 6i$ **b.** $z_2 = 5 + 4i$

    c. $z_3 = 5$ **d.** $z_4 = 4i$

Solution ▶ The graph of each complex number is shown in the figure.

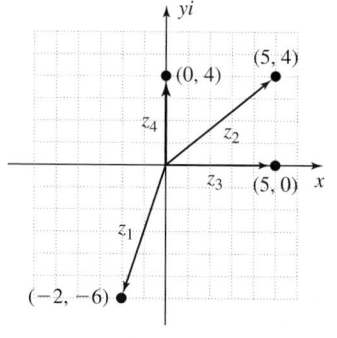

Now try Exercises 7 through 10 ▶

In Example 1, you likely noticed that from a vector perspective, z_2 is the "resultant vector" for the sum $z_3 + z_4$. To investigate further, consider $z_1 = (2 + 3i)$, $z_2 = (-5 + 2i)$, and the sum $z_1 + z_2 = z$ shown in Figure 7.70. The figure helps to confirm that the sum of complex numbers can be illustrated geometrically using the parallelogram (tail-to-tip) method employed for vectors in Section 7.4.

☑ **A.** You've just learned how to graph a complex number

Figure 7.70

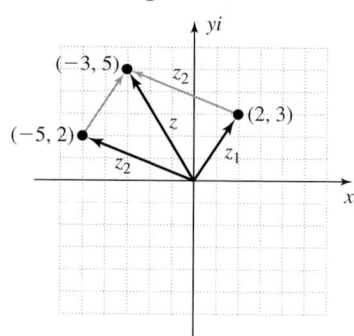

B. Complex Numbers in Trigonometric Form

The complex number $z = a + bi$ is said to be in **rectangular form** since it can be graphed using the rectangular coordinates of the complex plane. Complex numbers can also be written in **trigonometric form**. Similar to how $|x|$ represents the distance between the real number x and zero, $|z|$ represents the distance between (a, b) and the origin in the complex plane, and is computed as $|z| = \sqrt{a^2 + b^2}$. With any nonzero z, we can also associate an angle θ, which is the angle in standard position whose terminal side coincides with the graph of z. If we let r represent $|z|$, Figure 7.71 shows $\cos \theta = \dfrac{a}{r}$ and $\sin \theta = \dfrac{b}{r}$, yielding $r \cos \theta = a$ and $r \sin \theta = b$. The appropriate substitutions into $a + bi$ give the trigonometric form:

Figure 7.71

$$z = a + bi$$
$$= r \cos \theta + r \sin \theta \cdot i$$

Factoring out r and writing the imaginary unit as the lead factor of $\sin \theta$ gives the relationship in its more common form, $z = r(\cos \theta + i \sin \theta)$, where $\tan \theta = \dfrac{b}{a}$.

<table>
<tr>
<td>

WORTHY OF NOTE

While it is true the trigonometric form can more generally be written as $z = r[\cos(\theta + 2\pi k) + i \sin(\theta + 2\pi k)]$ for $k \in \mathbb{Z}$, the result is identical for any integer k and we will select θ so that $0 \le \theta < 2\pi$ or $0° \le \theta < 360°$, depending on whether we are working in radians or degrees.

</td>
<td>

The Trigonometric Form of a Complex Number

For the complex number $z = a + bi$ and angle θ shown, $z = r(\cos \theta + i \sin \theta)$ is the trigonometric form of z, where $r = \sqrt{a^2 + b^2}$, and $\tan \theta = \dfrac{b}{a}$; $a \ne 0$.

- $r = |z|$ represents the magnitude of z (also called the **modulus**).
- θ is often referred to as the **argument** of z.

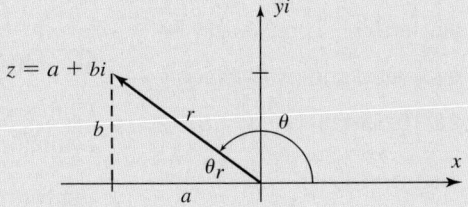

</td>
</tr>
</table>

Be sure to note that for $\tan \theta = \dfrac{b}{a}$, $\tan^{-1}\left(\dfrac{b}{a}\right)$ is equal to θ_r (the reference angle for θ) and the value of θ will ultimately *depend on the quadrant of z.*

EXAMPLE 2 ▶ **Converting a Complex Number from Rectangular to Trigonometric Form**

State the quadrant of the complex number, then write each in trigonometric form.

a. $z_1 = -2 - 2i$ **b.** $z_2 = 6 + 2i$

Solution ▶ Knowing that modulus r and angle θ are needed for the trigonometric form, we first determine these values. Once again, to find the correct value of θ, *it's important to note the quadrant of the complex number.*

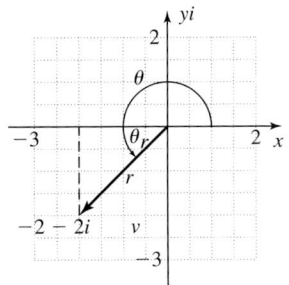

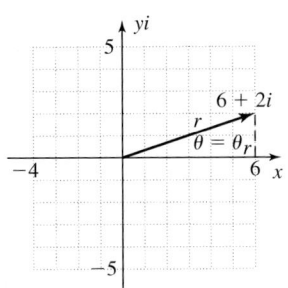

a. $z_1 = -2 - 2i$; QIII
$$r = \sqrt{(-2)^2 + (-2)^2}$$
$$= \sqrt{8} = 2\sqrt{2}$$
$$\theta_r = \tan^{-1}\left(\frac{-2}{-2}\right)$$
$$= \tan^{-1}(1)$$
$$= \frac{\pi}{4}$$

with z_1 in QIII, $\theta = \dfrac{5\pi}{4}$.

$$z_1 = 2\sqrt{2}\left[\cos\left(\frac{5\pi}{4}\right) + i \sin\left(\frac{5\pi}{4}\right)\right]$$

See the figure.

b. $z = 6 + 2i$; QI
$$r = \sqrt{(6)^2 + (2)^2}$$
$$= \sqrt{40} = 2\sqrt{10}$$
$$\theta_r = \tan^{-1}\left(\frac{2}{6}\right)$$
$$= \tan^{-1}\left(\frac{1}{3}\right)$$

z is in QI, so $\theta = \tan^{-1}\left(\dfrac{1}{3}\right)$

$$z = 2\sqrt{10}\left(\cos\left[\tan^{-1}\left(\frac{1}{3}\right)\right] + i \sin\left[\tan^{-1}\left(\frac{1}{3}\right)\right]\right)$$

Now try Exercises 11 through 26 ▶

☑ **B.** You've just learned how to write a complex number in trigonometric form

WORTHY OF NOTE

Using the triangle diagrams from Section 6.5,

$\cos\left[\tan^{-1}\left(\frac{1}{3}\right)\right]$ and

$\sin\left[\tan^{-1}\left(\frac{1}{3}\right)\right]$ can easily be

evaluated and used to verify

$2\sqrt{10}\,\text{cis}\left[\tan^{-1}\left(\frac{1}{3}\right)\right] = 6 + 2i.$

Since the angle θ is repeated for both cosine and sine, we often use an abbreviated notation for the trigonometric form, called "cis" (sis) notation: $z = r(\cos\theta + i\sin\theta) = r\,\text{cis}\,\theta$. The results of Example 2(a) and 2(b) would then be written $z = 2\sqrt{2}\,\text{cis}\left(\frac{5\pi}{4}\right)$ and $z = 2\sqrt{10}\,\text{cis}\left[\tan^{-1}\left(\frac{1}{3}\right)\right]$, respectively.

As in Example 2b, when $\theta_r = \tan^{-1}\left(\frac{b}{a}\right)$ is not a standard angle we either answer in exact form as shown, or use a four-decimal-place approximation: $2\sqrt{10}\,\text{cis}(0.3218)$.

C. Converting from Trigonometric Form to Rectangular Form

Converting from trigonometric form back to rectangular form is simply a matter of evaluating $r\,\text{cis}\,\theta$. This can be done regardless of whether θ is a standard angle or in the form $\tan^{-1}\left(\frac{b}{a}\right)$, since in the latter case we can construct a right triangle with side b opposite θ and side a adjacent θ, and find the needed values as in Section 6.5.

EXAMPLE 3 ▶ **Converting a Complex Number from Trigonometric to Rectangular Form**

Graph the following complex numbers, then write them in rectangular form.

a. $z = 12\,\text{cis}\left(\frac{\pi}{6}\right)$ **b.** $z = 13\,\text{cis}\left[\tan^{-1}\left(\frac{5}{12}\right)\right]$

Solution ▶ **a.** We have $r = 12$ and $\theta = \frac{\pi}{6}$, which yields the graph in Figure 7.72. In the nonabbreviated form we have $z = 12\left[\cos\left(\frac{\pi}{6}\right) + i\sin\left(\frac{\pi}{6}\right)\right]$. Evaluating within the brackets gives

$$z = 12\left[\frac{\sqrt{3}}{2} + \frac{1}{2}i\right] = 6\sqrt{3} + 6i.$$

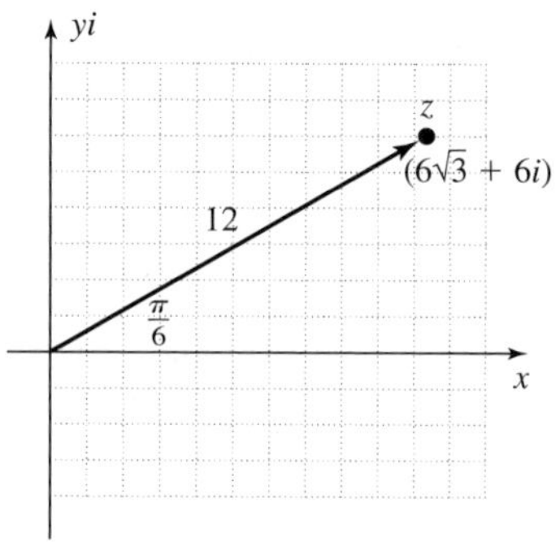

b. For $r = 13$ and $\theta = \tan^{-1}\left(\frac{5}{12}\right)$, we have the graph shown in Figure 7.73. Here we obtain the rectangular form directly from the diagram with $z = 12 + 5i$. Verify by noting that for $\theta = \tan^{-1}\left(\frac{5}{12}\right)$,

$\cos\theta = \frac{12}{13}$, and $\sin\theta = \frac{5}{13}$, meaning

$$z = 13(\cos\theta + i\sin\theta) =$$

$$13\left[\frac{12}{13} + \frac{5}{13}i\right] = 12 + 5i.$$

Figure 7.72

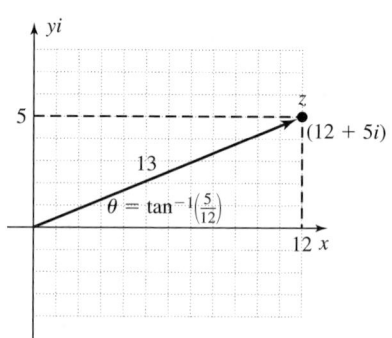

Figure 7.73

☑ **C.** You've just learned how to convert from trigonometric form to rectangular form

Now try Exercises 27 through 34 ▶

D. Interpreting Products and Quotients Geometrically

The multiplication and division of complex numbers has some geometric connections that can help us understand their computation in trigonometric form. Note the relationship between the modulus and argument of the following product, with the moduli (plural of modulus) and arguments from each factor.

EXAMPLE 4 ▶ **Noting Graphical Connections for the Product of Two Complex Numbers**

For $z_1 = 3 + 3i$ and $z_2 = 0 + 2i$,

    a. Graph the complex numbers and compute their moduli and arguments.

    b. Compute and graph the product z_1z_2 and find its modulus and argument. Discuss any connections you see between the factors and the resulting product.

Solution ▶   **a.** The graphs of z_1 and z_2 are shown in the figure. For the modulus and argument we have:

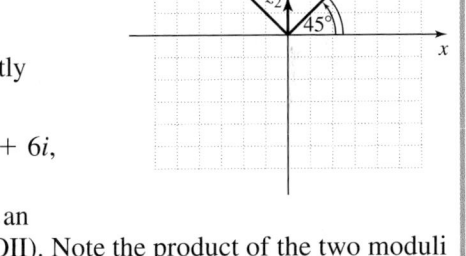

$$z_1 = 3 + 3i; \quad \text{QI} \qquad z_2 = 0 + 2i;$$
$$r = \sqrt{(3)^2 + (3)^2} \qquad \text{(quadrantal)}$$
$$= \sqrt{18} = 3\sqrt{2} \qquad r = 2 \text{ directly}$$
$$\theta = \tan^{-1}1 \qquad\qquad \theta = 90° \text{ directly}$$
$$\Rightarrow \theta = 45°$$

    b. The product z_1z_2 is $(3 + 3i)(2i) = -6 + 6i$, which is in QII. The modulus is $\sqrt{(-6)^2 + (6)^2} = \sqrt{72} = 6\sqrt{2}$, with an argument of $\theta_r = \tan^{-1}(-1)$ or 135° (QII). Note the product of the two moduli is *equal to the modulus of the final product:* $2 \cdot 3\sqrt{2} = 6\sqrt{2}$. Also note that the sum of the arguments for z_1 and z_2 *is equal to the argument of the product:* $45° + 90° = 135°$!

☑ **D.** You've just learned how to interpret products and quotients geometrically

---

                                                          Now try Exercises 35 and 36 ▶

A similar geometric connection exists for the division of complex numbers. This connection is explored in **Exercises 37 and 38** of the exercise set.

E. Products and Quotients in Trigonometric Form

The connections in Example 4 are not a coincidence, and can be proven to hold for all complex numbers. Consider any two nonzero complex numbers $z_1 = r_1(\cos \alpha + i \sin \alpha)$ and $z_2 = r_2(\cos \beta + i \sin \beta)$. For the product z_1z_2 we have

$$z_1z_2 = r_1(\cos \alpha + i \sin \alpha) \, r_2(\cos \beta + i \sin \beta) \qquad \text{product in trig form}$$
$$= r_1r_2[(\cos \alpha + i \sin \alpha)(\cos \beta + i \sin \beta)] \qquad \text{rearrange factors}$$
$$= r_1r_2[\cos \alpha \cos \beta + i \sin \beta \cos \alpha + i \sin \alpha \cos \beta + i^2\sin \alpha \sin \beta] \qquad \text{F-O-I-L}$$
$$= r_1r_2[(\cos \alpha \cos \beta - \sin \alpha \sin \beta) + i(\sin \beta \cos \alpha + \sin \alpha \cos \beta)] \qquad \text{commute}$$
$$\qquad\qquad\qquad\qquad\qquad\qquad\qquad\qquad\qquad\qquad\qquad\qquad\qquad\qquad \text{terms; } i^2 = -1$$
$$= r_1r_2[\cos(\alpha + \beta) + i \sin(\alpha + \beta)] \qquad \text{use sum/difference identities for sine/cosine}$$

In words, the proof says that to *multiply* complex numbers in trigonometric form, we *multiply* the moduli and *add* the arguments. For *division*, we *divide* the moduli and *subtract* the arguments. The proof for division resembles that for multiplication and is asked for in **Exercise 71.**

Products and Quotients of Complex Numbers in Trigonometric Form

For the complex numbers $z_1 = r_1(\cos \alpha + i \sin \alpha)$ and
$$z_2 = r_2(\cos \beta + i \sin \beta),$$
$$z_1 z_2 = r_1 r_2 [\cos(\alpha + \beta) + i \sin(\alpha + \beta)]$$

and

$$\frac{z_1}{z_2} = \frac{r_1}{r_2} [\cos(\alpha - \beta) + i \sin(\alpha - \beta)], z_2 \neq 0.$$

EXAMPLE 5 ▶ **Multiplying Complex Numbers in Trigonometric Form**

For $z_1 = -3 + \sqrt{3}i$ and $z_2 = \sqrt{3} + 1i$,

a. Write z_1 and z_2 in trigonometric form and compute $z_1 z_2$.

b. Compute the quotient $\dfrac{z_1}{z_2}$ in trigonometric form.

c. Verify the product using the rectangular form.

Solution ▶ **a.** For z_1 in QII we find $r = 2\sqrt{3}$ and $\theta = 150°$, for z_2 in QI, $r = 2$ and $\theta = 30°$. In trigonometric form,

$z_1 = 2\sqrt{3}(\cos 150° + i \sin 150°)$ and
$z_2 = 2(\cos 30° + i \sin 30°)$:
$$z_1 z_2 = 2\sqrt{3}(\cos 150° + i \sin 150°) \cdot 2(\cos 30° + i \sin 30°)$$
$$= 2\sqrt{3} \cdot 2[\cos(150° + 30°) + i \sin(150° + 30°)] \quad \text{multiply moduli, add arguments}$$
$$= 4\sqrt{3}(\cos 180° + i \sin 180°)$$
$$= 4\sqrt{3}\,(-1 + 0i)$$
$$= -4\sqrt{3}$$

b. $\dfrac{z_1}{z_2} = \dfrac{2\sqrt{3}(\cos 150° + i \sin 150°)}{2(\cos 30° + i \sin 30°)}$
$$= \sqrt{3}[\cos(150° - 30°) + i \sin(150° - 30°)] \quad \text{divide moduli, subtract arguments}$$
$$= \sqrt{3}(\cos 120° + i \sin 120°)$$
$$= \sqrt{3}\left(-\frac{1}{2} + \frac{\sqrt{3}}{2}i\right)$$
$$= -\frac{\sqrt{3}}{2} + \frac{3}{2}i$$

c. $z_1 z_2 = (-3 + \sqrt{3}i)(\sqrt{3} + 1i)$
$$= -3\sqrt{3} - 3i + 3i + \sqrt{3}i^2$$
$$= -4\sqrt{3}$$

Now try Exercises 39 through 46 ▶

Converting to trigonometric form for multiplication and division seems too clumsy for practical use, as we can often compute these results more efficiently in rectangular form. However, this approach leads to powers and roots of complex numbers, *an indispensable part of advanced equation solving*, and these are not easily found in rectangular form. In any case, note that the power and simplicity of computing products/quotients in trigonometric form is highly magnified when the complex numbers are *given in trig form*:

☑ **E.** You've just learned how to compute products and quotients in trigonometric form

$$(12 \text{ cis } 50°)(3 \text{ cis } 20°) = 36 \text{ cis } 70° \qquad \frac{12 \text{ cis } 50°}{3 \text{ cis } 20°} = 4 \text{ cis } 30°.$$

See Exercises 47 through 50.

F. (Optional) Applications of Complex Numbers

Somewhat surprisingly, complex numbers have several applications in the real world. Many of these involve a study of electricity, and in particular **AC (alternating current) circuits.**

In simplistic terms, when an armature (molded wire) is rotated in a uniform magnetic field, a voltage V is generated that depends on the strength of the field. As the armature is rotated, the voltage varies between a maximum and a minimum value, with the amount of voltage modeled by $V(\theta) = V_{\text{max}}\sin(B\theta)$, with θ in degrees. Here, V_{max} represents the maximum voltage attained, and the input variable θ represents the angle the armature makes with the **magnetic flux,** indicated in Figure 7.74 by the dashed arrows between the magnets.

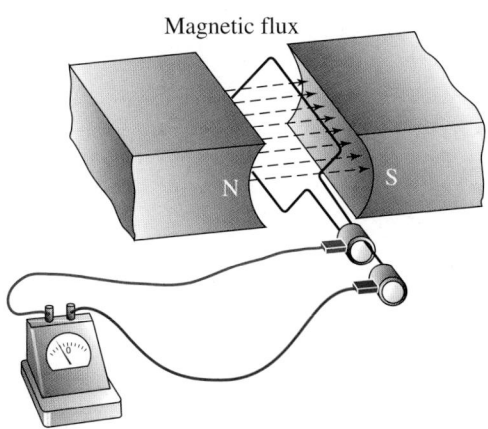
Figure 7.74

Magnetic flux

When the armature is perpendicular to the flux, we say $\theta = 0°$. At $\theta = 0°$ and $\theta = 180°$, no voltage is produced, while at $\theta = 90°$ and $\theta = 270°$, the voltage reaches its maximum and minimum values respectively (hence the name *alternating current*). Many electric dryers and other large appliances are labeled as 220 volt (V) appliances, but use an alternating current that varies from 311 V to -311 V (see *Worthy of Note*). This means when $\theta = 52°$, $V(52°) = 311 \sin(52°) = 245$ V is being generated. In practical applications, we use time t as the independent variable, rather the angle of the armature. These large appliances usually operate with a frequency of 60 cycles per second, or 1 cycle every $\dfrac{1}{60}$ of a second $\left(P = \dfrac{1}{60}\right)$. Using $B = \dfrac{2\pi}{P}$, we obtain $B = 120\pi$ and our equation model becomes $V(t) = 311 \sin(120\pi t)$ with t in radians. This variation in voltage is an excellent example of a simple harmonic model.

EXAMPLE 6 ▶ **Analyzing Alternating Current Using Trigonometry**

Use the equation $V(t) = 311 \sin(120\pi t)$ to:

a. Create a table of values illustrating the voltage produced every thousandth of a second for the first half-cycle $\left(t = \dfrac{1}{120} \approx 0.008 \right)$.

b. Use a graphing calculator to find the times t in this half-cycle when 160 V is being produced.

Solution ▶ **a.** Starting at $t = 0$ and using increments of 0.001 sec produces the table shown.

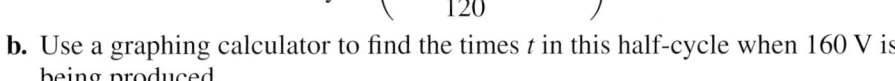

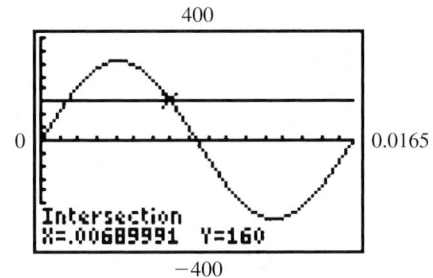

Time t	Voltage
0	0
0.001	114.5
0.002	212.9
0.003	281.4
0.004	310.4
0.005	295.8
0.006	239.6
0.007	149.8
0.008	39.9

b. From the table we note $V(t) = 160$ when $t \in (0.001, 0.002)$ and $t \in (0.006, 0.007)$. Using the intersection of graphs method places these values at $t \approx 0.0014$ and $t \approx 0.0069$ (see graph).

Now try Exercises 53 and 54 ▶

Figure 7.77

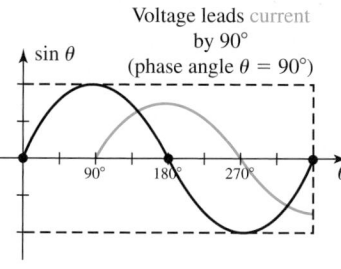

Voltage leads current by 90°
(phase angle $\theta = 90°$)

Figure 7.78

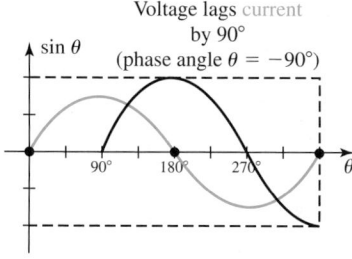

Voltage lags current by 90°
(phase angle $\theta = -90°$)

Figure 7.79

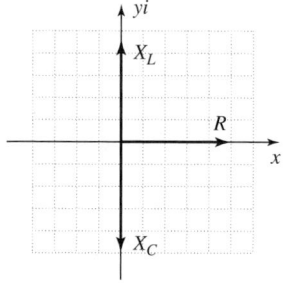

The chief components of AC circuits are **voltage** (V) and **current** (I). Due to the nature of how the current is generated, V and I can be modeled by sine functions. Other characteristics of electricity include pure **resistance** (R), **inductive reactance** (X_L), and **capacitive reactance** (X_C) (see Figure 7.75). Each of these is measured in a unit called ohms (Ω), while current I is measured in amperes (A), and voltages are measured in volts (V). These components of electricity *are related by fixed and inherent traits*, which include the following: (1) voltage across a resistor is always *in phase* with the current, meaning the phase shift or **phase angle** between them is 0° (Figure 7.76);(2) voltage across an inductor *leads the current* by 90° (Figure 7.77); (3) voltage across a capacitor *lags the current* by 90° (Figure 7.78); and (4) voltage is equal to the product of the current times the resistance or reactance: $V = IR$, $V = IX_L$, and $V = IX_C$.

Figure 7.75

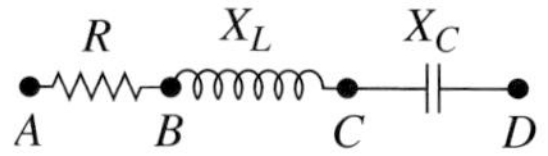

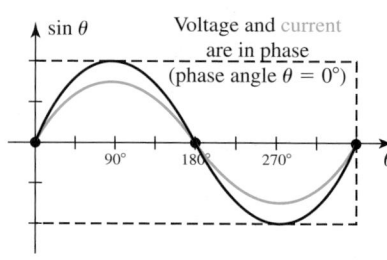

Figure 7.76

Voltage and current are in phase
(phase angle $\theta = 0°$)

Different combinations of R, X_L, and X_C in a combined (series) circuit alter the phase angle and the resulting voltage. Since voltage across a resistance is always in phase with the current (trait 1), we can model the resistance as a vector along the positive real axis (since the phase angle is 0°). For traits (2) and (3), X_L is modeled on the positive imaginary axis since voltage leads current by 90°, and X_C on the negative imaginary axis since voltage lags current by 90° (see Figure 7.79). These natural characteristics make the complex plane *a perfect fit for describing the characteristics of the circuit.*

Consider a series circuit (Figure 7.75), where $R = 12\ \Omega$, $X_L = 9\ \Omega$, and $X_C = 4\ \Omega$. For a current of $I = 2$ amps through this circuit, the voltage across each individual element would be $V_R = (2)(12) = 24$ V (A to B), $V_L = (2)(9) = 18$ V (B to C), and $V_C = (2)(4) = 8$ V (C to D). However, the resulting voltage across this circuit *cannot be an arithmetic sum*, since R is real while X_L and X_C are represented by imaginary numbers. The joint effect of resistance (R) and reactance (X_L, X_C) in a circuit is called the **impedance,** denoted by the letter Z, and is a measure of the total resistance to the flow of electrons. It is computed $Z = R + X_L j - X_C j$ (see *Worthy of Note*), due to the phase angle relationship of the voltage in each element (X_L and X_C point in opposite directions, hence the subtraction). The expression for Z is more commonly written $R + (X_L - X_C)j$, where we more clearly note Z *is a complex number* whose magnitude and angle with the x-axis can be found as before:

$$|Z| = \sqrt{R^2 + (X_L - X_C)^2} \text{ and } \theta_r = \tan^{-1}\left(\frac{X_L - X_C}{R}\right).$$ The angle θ represents the phase angle between the voltage and current brought about by this combination of elements. The resulting voltage of the circuit is then calculated as the product of the current with the magnitude of the impedance, or $V_{RLC} = I|Z|$ (Z is also measured in ohms, Ω).

EXAMPLE 7 ▶ **Finding the Impedance and Phase Angle of the Current in a Circuit**

For the circuit diagrammed in the figure, (a) find the magnitude of Z, the phase angle between current and voltage, and write the result in trigonometric form; and (b) find the total voltage across this circuit.

Solution ▶   **a.** Using the values given, we find $Z = R + (X_L - X_C)j =$
$12 + (9 - 4)j = 12 + 5j$ (QI). This gives a magnitude of

$$|Z| = \sqrt{(12)^2 + (5)^2} = \sqrt{169} = 13 \ \Omega, \text{ with a phase angle of}$$

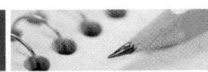

$R \quad X_L \qquad X_C$

$A \ 12\Omega \ B \quad 9\Omega \quad C \ 4\Omega \ D$

$$\theta = \tan^{-1}\left(\frac{5}{12}\right) \approx 22.6° \text{ (voltage leads the current by about 22.6°)}.$$

In trigonometric form $Z \approx 13 \text{ cis } 22.6°$.

☑ **F. You have just learned how to solve applications involving complex numbers**

b. With $I = 2$ amps, the total voltage across this circuit is
$V_{RLC} = I|Z| = 2(13) = 26$ V.

Now try Exercises 55 through 68 ▶

7.5 EXERCISES

▶ CONCEPTS AND VOCABULARY

Fill in each blank with the appropriate word or phrase. Carefully reread the section if needed.

1. For a complex number written in the form $z = r(\cos\theta + i\sin\theta)$, r is called the _____ and θ is called the _____.

2. The complex number $z = 2\left[\cos\left(\dfrac{\pi}{4}\right) + i\sin\left(\dfrac{\pi}{4}\right)\right]$ can be written as the abbreviated "cis" notation as _____.

3. To multiply complex numbers in trigonometric form, we _____ the moduli and _____ the arguments.

4. To divide complex numbers in trigonometric form, we _____ the moduli and _____ the arguments.

5. Write $z = -1 - \sqrt{3}i$ in trigonometric form and explain why the argument is $\theta = 240°$ instead of $60°$ as indicated by your calculator.

6. Discuss the similarities between finding the components of a vector and writing a complex number in trigonometric form.

▶ DEVELOPING YOUR SKILLS

Graph the complex numbers z_1, z_2, and z_3 given, then express one as the sum of the other two.

7. $z_1 = 7 + 2i$
$z_2 = 8 + 6i$
$z_3 = 1 + 4i$

8. $z_1 = 2 + 7i$
$z_2 = 3 + 4i$
$z_3 = -1 + 3i$

9. $z_1 = -2 - 5i$
$z_2 = 1 - 7i$
$z_3 = 3 - 2i$

10. $z_1 = -2 + 6i$
$z_2 = 7 - 2i$
$z_3 = 5 + 4i$

State the quadrant of each complex number, then write it in trigonometric form. For Exercises 11 through 14, answer in degrees. For 15 through 18, answer in radians.

11. $-2 - 2i$

12. $7 - 7i$

13. $-5\sqrt{3} - 5i$

14. $2 - 2\sqrt{3}i$

15. $-3\sqrt{2} + 3\sqrt{2}i$

16. $5\sqrt{7} - 5\sqrt{7}i$

17. $4\sqrt{3} - 4i$

18. $-6 + 6\sqrt{3}i$

Write each complex number in trigonometric form. For Exercises 19 through 22, answer in degrees using both an exact form and an approximate form, rounding to tenths. For 23 through 26, answer in radians using both an exact form and an approximate form, rounding to four decimal places.

19. $8 + 6i$

20. $-9 + 12i$

21. $-5 - 12i$

22. $-8 + 15i$

23. $6 + 17.5i$

24. $30 - 5.5i$

25. $-6 + 10i$

26. $12 - 4i$

Graph each complex number using its trigonometric form, then convert each to rectangular form.

27. $2 \operatorname{cis}\left(\dfrac{\pi}{4}\right)$

28. $12 \operatorname{cis}\left(\dfrac{\pi}{6}\right)$

29. $4\sqrt{3} \operatorname{cis}\left(\dfrac{\pi}{3}\right)$

30. $5\sqrt{3} \operatorname{cis}\left(\dfrac{7\pi}{6}\right)$

31. $17 \operatorname{cis}\left[\tan^{-1}\left(\dfrac{15}{8}\right)\right]$

32. $10 \operatorname{cis}\left[\tan^{-1}\left(\dfrac{3}{4}\right)\right]$

33. $6 \operatorname{cis}\left[\pi - \tan^{-1}\left(\dfrac{5}{\sqrt{11}}\right)\right]$

34. $4 \operatorname{cis}\left[\pi + \tan^{-1}\left(\dfrac{\sqrt{7}}{3}\right)\right]$

For the complex numbers z_1 and z_2 given, find their moduli r_1 and r_2 and arguments θ_1 and θ_2. Then compute their *product* in rectangular form. For modulus r and argument θ of the product, verify that $r_1 r_2 = r$ and $\theta_1 + \theta_2 = \theta$.

35. $z_1 = -2 + 2i$; $z_2 = 3 + 3i$

36. $z_1 = 1 + \sqrt{3}i$; $z_2 = 3 + \sqrt{3}i$

For the complex numbers z_1 and z_2 given, find their moduli r_1 and r_2 and arguments θ_1 and θ_2. Then compute their *quotient* in rectangular form. For modulus r and argument θ of the quotient, verify that $\dfrac{r_1}{r_2} = r$ and $\theta_1 - \theta_2 = \theta$.

37. $z_1 = \sqrt{3} + i$; $z_2 = 1 + \sqrt{3}i$

38. $z_1 = -\sqrt{3} + i$; $z_2 = 3 + 0i$

Compute the product $z_1 z_2$ and quotient $\dfrac{z_1}{z_2}$ using the trigonometric form. Answer in exact rectangular form where possible, otherwise round all values to two decimal places.

39. $z_1 = -4\sqrt{3} + 4i$

$z_2 = \dfrac{3\sqrt{3}}{2} + \dfrac{3}{2}i$

40. $z_1 = \dfrac{5\sqrt{3}}{2} + \dfrac{5}{2}i$

$z_2 = 0 + 6i$

41. $z_1 = -2\sqrt{3} + 0i$

$z_2 = -\dfrac{21}{2} + \dfrac{7i\sqrt{3}}{2}$

42. $z_1 = 0 - 6i\sqrt{2}$

$z_2 = \dfrac{3\sqrt{2}}{2} + \dfrac{3i\sqrt{6}}{2}$

43. $z_1 = 9\left[\cos\left(\dfrac{\pi}{15}\right) + i \sin\left(\dfrac{\pi}{15}\right)\right]$

$z_2 = 1.8\left[\cos\left(\dfrac{2\pi}{3}\right) + i \sin\left(\dfrac{2\pi}{3}\right)\right]$

44. $z_1 = 2\left[\cos\left(\dfrac{3\pi}{5}\right) + i \sin\left(\dfrac{3\pi}{5}\right)\right]$

$z_2 = 8.4\left[\cos\left(\dfrac{\pi}{5}\right) + i \sin\left(\dfrac{\pi}{5}\right)\right]$

45. $z_1 = 10(\cos 60° + i \sin 60°)$

$z_2 = 4(\cos 30° + i \sin 30°)$

46. $z_1 = 7(\cos 120° + i \sin 120°)$

$z_2 = 2(\cos 300° + i \sin 300°)$

47. $z_1 = 5\sqrt{2} \operatorname{cis} 210°$ **48.** $z_1 = 5\sqrt{3} \operatorname{cis} 240°$

$z_2 = 2\sqrt{2} \operatorname{cis} 30°$ $z_2 = \sqrt{3} \operatorname{cis} 90°$

49. $z_1 = 6 \operatorname{cis} 82°$ **50.** $z_1 = 1.6 \operatorname{cis} 59°$

$z_2 = 1.5 \operatorname{cis} 27°$ $z_2 = 8 \operatorname{cis} 275°$

▶ **WORKING WITH FORMULAS**

51. Equilateral triangles in the complex plane:
$u^2 + v^2 + w^2 = uv + uw + vw$

If the line segments connecting the complex numbers u, v, and w form the vertices of an equilateral triangle, the formula shown above holds true. Verify that $u = 2 + \sqrt{3}i$, $v = 10 + \sqrt{3}i$, and $w = 6 + 5\sqrt{3}i$ form the vertices of an equilateral triangle using the distance formula, then verify the formula given.

52. The cube of a complex number:
$(A + B)^3 = A^3 + 3A^2B + 3AB^2 + B^3$

The cube of any binomial can be found using the formula here, where A and B are the terms of the binomial. Use the formula to compute the cube of $1 - 2i$ (note $A = 1$ and $B = -2i$).

▶ **APPLICATIONS**

53. Electric current: In the United States, electric power is supplied to homes and offices via a "120 V circuit," using an alternating current that varies from 170 V to -170 V, at a frequency of 60 cycles/sec. (a) Write the voltage equation for U.S. households, (b) create a table of values illustrating the voltage produced every thousandth of a second for the first half-cycle, and (c) find the first time t in this half-cycle when exactly 140 V is being produced.

54. Electric current: While the electricity supplied in Europe is still not quite uniform, most countries employ 230-V circuits, using an alternating current that varies from 325 V to -325 V. However, the frequency is only *50 cycles per second.* (a) Write the voltage equation for these European countries, (b) create a table of values illustrating the voltage produced every thousandth of a second for the first half-cycle, and (c) find the first time t in this half-cycle when exactly 215 V is being produced.

AC circuits: For the circuits indicated in Exercises 55 through 60, (a) find the magnitude of Z, the phase angle between current and voltage, and write the result in trigonometric form; and (b) find the total voltage across this circuit. Recall $Z = R + (X_L - X_C)j$ and $|Z| = \sqrt{R^2 + (X_L - X_C)^2}$.

Exercises 55 through 58

55. $R = 15\ \Omega,\ X_L = 12\ \Omega,$ and $X_C = 4\ \Omega,$ with $I = 3$ A.

$$R \quad X_C \quad X_L$$
$$A \quad B \quad C \quad D$$

56. $R = 24\ \Omega,\ X_L = 12\ \Omega,$ and $X_C = 5\ \Omega,$ with $I = 2.5$ A.

57. $R = 7\ \Omega,\ X_L = 6\ \Omega,$ and $X_C = 11\ \Omega,$ with $I = 1.8$ A.

Exercises 59 and 60

58. $R = 9.2\ \Omega,\ X_L = 5.6\ \Omega,$ and $X_C = 8.3\ \Omega,$ with $I = 2.0$ A.

$$R \quad X_L$$
$$A \quad B \quad C$$

59. $R = 12\ \Omega$ and $X_L = 5\ \Omega,$ with $I = 1.7$ A.

60. $R = 35\ \Omega$ and $X_L = 12\ \Omega,$ with $I = 4$ A.

AC circuits—voltage: The current I and the impedance Z for certain AC circuits are given. Write I and Z in trigonometric form and find the voltage in each circuit. Recall $V = IZ.$

61. $I = \sqrt{3} + 1j$ A and $Z = 5 + 5j\ \Omega$

62. $I = \sqrt{3} - 1j$ A and $Z = 2 + 2j\ \Omega$

63. $I = 3 - 2j$ A and $Z = 2 + 3.75j\ \Omega$

64. $I = 4 + 3j$ A and $Z = 2 - 4j\ \Omega$

AC circuits—current: If the voltage and impedance are known, the current I in the circuit is calculated as the quotient $I = \dfrac{V}{Z}.$ Write V and Z in trigonometric form to find the current in each circuit.

65. $V = 2 + 2\sqrt{3}j$ and $Z = 4 - 4j\ \Omega$

66. $V = 4\sqrt{3} - 4j$ and $Z = 1 - 1j\ \Omega$

67. $V = 3 - 4j$ and $Z = 4 + 7.5j\ \Omega$

68. $V = 2.8 + 9.6j$ and $Z = 1.4 - 4.8j\ \Omega$

Parallel circuits: For AC circuits *wired in parallel*, the total impedance is given by $Z = \dfrac{Z_1 Z_2}{Z_1 + Z_2},$ where Z_1 and Z_2 represent the impedance in each branch. Find the total impedance for the values given. Compute the product in the numerator using trigonometric form, and the sum in the denominator in rectangular form.

69. $Z_1 = 1 + 2j$ and $Z_2 = 3 - 2j$

70. $Z_1 = 3 - j$ and $Z_2 = 2 + j$

▶ **EXTENDING THE CONCEPT**

71. Verify/prove that for the complex numbers $z_1 = r_1(\cos\alpha + i\sin\alpha)$ and $z_2 = r_2(\cos\beta + i\sin\beta),$

$$\frac{z_1}{z_2} = \frac{r_1}{r_2}\left[\cos(\alpha - \beta) + i\sin(\alpha - \beta)\right].$$

72. Using the Internet, a trade manual, or some other resource, find the voltage and frequency at which electricity is supplied to most of Japan (oddly enough—two different frequencies are in common use). As in Example 6, the voltage given will likely be the root-mean-square (rms) voltage. Use the information to find the true voltage and the equation model for voltage in most of Japan.

73. Recall that two lines are perpendicular if their slopes have a product of $-1.$ For the directed line segment representing the complex number $z_1 = 7 + 24i,$ find complex numbers z_2 and z_3 whose directed line segments are perpendicular to z_1 and have a magnitude one-fifth as large.

74. The magnitude of the impedance is $|Z| = \sqrt{R^2 + (X_L - X_C)^2}.$ If $R,\ X_L,$ and X_C are all nonzero, what conditions would make the magnitude of Z as small as possible?

▶ **MAINTAINING YOUR SKILLS**

75. (6.7) Solve for $x \in [0, 2\pi)$:

$$350 = 750 \sin\left(2x - \frac{\pi}{4}\right) - 25$$

76. (3.6) Name all asymptotes of the function

$$h(x) = \frac{1 + x^3}{x^2}$$

77. (2.7) Graph the piecewise-defined function given:

$$f(x) = \begin{cases} 2 & x < -2 \\ x^2 & -2 \le x < 1 \\ x & x \ge 1 \end{cases}$$

78. (7.2) A ship is spotted by two observation posts that are 4 mi apart. Using the line between them for reference, the first post reports the ship is at an angle of 41°, while the second reports an angle of 63°, as shown. How far is the ship from the closest post?

Exercise 78

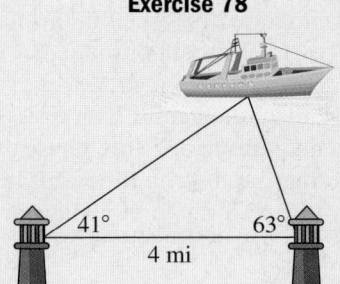

7.6 | De Moivre's Theorem and the Theorem on *n*th Roots

Learning Objectives

In Section 7.6 you will learn how to:

☐ **A.** Use De Moivre's theorem to raise complex numbers to any power

☐ **B.** Use De Moivre's theorem to check solutions to polynomial equations

☐ **C.** Use the *n*th roots theorem to find the *n*th roots of a complex number

The material in this section represents some of the most significant developments in the history of mathematics. After hundreds of years of struggle, mathematical scientists had not only come to recognize the existence of complex numbers, but were able to make operations on them commonplace and routine. This allowed for the unification of many ideas related to the study of polynomial equations, and answered questions that had puzzled scientists from many different fields for centuries. In this section, we will look at two fairly simple theorems that actually represent over 1000 years in the evolution of mathematical thought.

A. De Moivre's Theorem

Having found acceptable means for applying the four basic operations to complex numbers, our attention naturally shifts to the computation of powers and roots. Without them, we'd remain wholly unable to offer complete solutions to polynomial equations and find solutions for many applications. The computation of powers, squares, and cubes offer little challenge, as they can be computed easily using the formula for binomial squares $[(A + B)^2 = A^2 + 2AB + B^2]$ or by applying the **binomial theorem.** For larger powers, the binomial theorem becomes too time consuming and a more efficient method is desired. The key here is to use the trigonometric form of the complex number. In Section 7.5, we noted the product of two complex numbers involved multiplying their moduli and adding their arguments:

For $z_1 = r_1(\cos\theta_1 + i\sin\theta_1)$ and $z_2 = r_2(\cos\theta_2 + i\sin\theta_2)$ we have
$$z_1 z_2 = r_1 r_2 \left[\cos(\theta_1 + \theta_2) + i\sin(\theta_1 + \theta_2)\right]$$

For the square of a complex number, $r_1 = r_2$ and $\theta_1 = \theta_2$. Using θ itself yields
$$z^2 = r^2[\cos(\theta + \theta) + i\sin(\theta + \theta)]$$
$$= r^2[\cos(2\theta) + i\sin(2\theta)]$$

WORTHY OF NOTE

Sometimes the argument of cosine and sine becomes very large after applying De Moivre's theorem. In these cases, we use the fact that $\theta = \theta \pm 360°k$ and $\theta = \theta \pm 2\pi k$ represent coterminal angles for integers k, and use the coterminal angle θ where $0 \le \theta < 360°$ or $0 \le \theta < 2\pi$.

Multiplying this result by $z = r(\cos \theta + i \sin \theta)$ to compute z^3 gives

$$r^2[\cos(2\theta) + i \sin(2\theta)] \, r(\cos \theta + i \sin \theta) = r^3[\cos(2\theta + \theta) + i \sin(2\theta + \theta)]$$
$$= r^3[\cos(3\theta) + i \sin(3\theta)].$$

The result can be extended further and generalized into **De Moivre's theorem.**

De Moivre's Theorem

For any positive integer n, and $z = r(\cos \theta + i \sin \theta)$,
$$z^n = r^n[\cos(n\theta) + i \sin(n\theta)]$$

For a proof of the theorem where n is an integer and $n \ge 1$, see Appendix V.

EXAMPLE 1 ▶ **Using De Moivre's Theorem to Compute the Power of a Complex Number**

Use De Moivre's theorem to compute z^9, given $z = -\frac{1}{2} - \frac{1}{2}i$.

Solution ▶ Here we have $r = \sqrt{\left(-\dfrac{1}{2}\right)^2 + \left(-\dfrac{1}{2}\right)^2} = \dfrac{\sqrt{2}}{2}$. With z in QIII, $\tan \theta = 1$ yields

$\theta = \dfrac{5\pi}{4}$. The trigonometric form is $z = \dfrac{\sqrt{2}}{2}\left[\cos\left(\dfrac{5\pi}{4}\right) + i \sin\left(\dfrac{5\pi}{4}\right)\right]$ and applying

the theorem with $n = 9$ gives

$$z^9 = \left(\frac{\sqrt{2}}{2}\right)^9\left[\cos\left(9 \cdot \frac{5\pi}{4}\right) + i \sin\left(9 \cdot \frac{5\pi}{4}\right)\right] \qquad \text{De Moivre's theorem}$$

$$= \frac{\sqrt{2}}{32}\left[\cos\left(\frac{45\pi}{4}\right) + i \sin\left(\frac{45\pi}{4}\right)\right] \qquad \text{simplify}$$

$$= \frac{\sqrt{2}}{32}\left[\cos\left(\frac{5\pi}{4}\right) + i \sin\left(\frac{5\pi}{4}\right)\right] \qquad \text{coterminal angles}$$

$$= \frac{\sqrt{2}}{32}\left(-\frac{\sqrt{2}}{2} - \frac{\sqrt{2}}{2}i\right) \qquad \text{evaluate functions}$$

$$= -\frac{1}{32} - \frac{1}{32}i \qquad \text{result}$$

☑ **A.** You've just learned how to use De Moivre's theorem to raise complex numbers to any power

Now try Exercises 7 through 14 ▶

As with products and quotients, if the complex number is *given* in trigonometric form, computing any power of the number is both elegant and efficient. For instance, if $z = 2 \text{ cis } 40°$, then $z^4 = 16 \text{ cis } 160°$. **See Exercises 15 through 18.**

For cases where θ is not a standard angle, De Moivre's theorem requires an intriguing application of the skills developed in Chapter 6, including the use of multiple angle identities and working from a right triangle drawn relative to $\theta_r = \tan^{-1}\left(\dfrac{b}{a}\right)$.

See Exercises 57 and 58.

B. Checking Solutions to Polynomial Equations

One application of De Moivre's theorem is checking the complex roots of a polynomial, as in Example 2.

EXAMPLE 2 ▶ **Using De Moivre's Theorem to Check Solutions to a Polynomial Equation**

Use De Moivre's theorem to show that $z = -2 - 2i$ is a solution to
$z^4 - 3z^3 - 38z^2 - 128z - 144 = 0$.

Solution ▶ We will apply the theorem to the third and fourth degree terms, and compute the square directly. Since z is in QIII, the trigonometric form is $z = 2\sqrt{2}$ cis 225°. In the following illustration, note that 900° and 180° are coterminal, as are 675° and 315°.

$$(-2 - 2i)^4$$
$$= (2\sqrt{2})^4\text{cis}(4 \cdot 225°)$$
$$= (2\sqrt{2})^4\text{cis } 900°$$
$$= 64 \text{ cis } 180°$$
$$= 64(-1 + 0i)$$
$$= -64$$

$$(-2 - 2i)^3$$
$$= (2\sqrt{2})^3\text{cis}(3 \cdot 225°)$$
$$= (2\sqrt{2})^3\text{cis } 675°$$
$$= (2\sqrt{2})^3\text{cis } 315°$$
$$= 16\sqrt{2}\left(\frac{\sqrt{2}}{2} - \frac{\sqrt{2}}{2}i\right)$$
$$= 16 - 16i$$

$$(-2 - 2i)^2$$
$$= 4 + 8i + (2i)^2$$
$$= 4 + 8i + 4i^2$$
$$= 4 + 8i - 4$$
$$= 0 + 8i$$
$$= 8i$$

Substituting back into the original equation gives

$$1z^4 - 3z^3 - 38z^2 - 128z - 144 = 0$$
$$1(-64) - 3(16 - 16i) - 38(8i) - 128(-2 - 2i) - 144 = 0$$
$$-64 - 48 + 48i - 304i + 256 + 256i - 144 = 0$$
$$(-64 - 48 + 256 - 144) + (48 - 304 + 256)i = 0$$
$$0 = 0 ✓$$

☑ **B.** You've just learned to use De Moivre's theorem to check solutions to polynomial equations

Now try Exercises 19 through 26 ▶

Regarding Example 2, we know from a study of algebra that complex roots must occur in conjugate pairs, meaning $-2 + 2i$ is also a root. This equation actually has two real and two complex roots, with $z = 9$ and $z = -2$ being the two real roots.

C. The *n*th Roots Theorem

Having looked at De Moivre's theorem, which raises a complex number to any power, we now consider the **nth roots theorem,** which will compute the *n*th roots of a complex number. If we allow that De Moivre's theorem also holds for rational values $\frac{1}{n}$, instead of only the integers n illustrated previously, the formula for computing an *n*th root would be a direct result:

$$z^{\frac{1}{n}} = r^{\frac{1}{n}}\left[\cos\left(\frac{1}{n}\theta\right) + i \sin\left(\frac{1}{n}\theta\right)\right] \quad \text{De Moivre's theorem}$$

$$= \sqrt[n]{r}\left[\cos\left(\frac{\theta}{n}\right) + i \sin\left(\frac{\theta}{n}\right)\right] \quad \text{simplify}$$

However, this formula would *find only the principal nth root!* In other words, periodic solutions would be ignored. As in Section 7.5, it's worth noting the most general form of a complex number is $z = r[\cos(\theta + 360°k) + i \sin(\theta + 360°k)]$, for $k \in \mathbb{Z}$. When De Moivre's theorem is applied to this form for *integers n*, we obtain $z^n = r^n[\cos(n\theta + 360°kn) + i \sin(n\theta + 360°kn)]$, which returns a result identical to $r^n[\cos(n\theta) + i \sin(n\theta)]$. However, for the rational exponent $\frac{1}{n}$, the general form takes additional solutions into account and will return all *n*, *n*th roots.

$$z^{\frac{1}{n}} = r^{\frac{1}{n}}\left\{\cos\left[\frac{1}{n}(\theta + 360°k)\right] + i \sin\left[\frac{1}{n}(\theta + 360°k)\right]\right\} \quad \begin{array}{l}\text{De Moivre's}\\\text{theorem for rational}\\\text{exponents}\end{array}$$

$$= \sqrt[n]{r}\left[\cos\left(\frac{\theta}{n} + \frac{360°k}{n}\right) + i \sin\left(\frac{\theta}{n} + \frac{360°k}{n}\right)\right] \quad \text{simplify}$$

The nth Roots Theorem

For $z = r(\cos\theta + i\sin\theta)$, a positive integer n, and $r \in \mathbb{R}$, z has exactly n distinct nth roots determined by

$$\sqrt[n]{z} = \sqrt[n]{r}\left[\cos\left(\frac{\theta}{n} + \frac{360°k}{n}\right) + i\sin\left(\frac{\theta}{n} + \frac{360°k}{n}\right)\right]$$

where $k = 0, 1, 2, \ldots, n - 1$.

For ease of computation, it helps to note that once the argument for the principal root is found using $k = 0$, $\dfrac{\theta}{n} + \dfrac{360°k}{n}$ simply adds $\dfrac{360}{n}\left(\text{or }\dfrac{2\pi}{n}\right)$ to the previous argument for $k = 1, 2, 3, \ldots, n - 1$.

In Example 3 you're asked to find the three cube roots of 1, also called the **cube roots of unity,** and graph the results. The nth roots of unity play a significant role in the solution of many polynomial equations. For an in-depth study of this connection, visit www.mhhe.com/coburn and go to **Section 7.8: Trigonometry, Complex Numbers and Cubic Equations.**

EXAMPLE 3 ▶ Finding nth Roots

Use the nth roots theorem to solve the equation $x^3 - 1 = 0$. Write the results in rectangular form and graph.

Solution ▶ From $x^3 - 1 = 0$, we have $x^3 = 1$ and must find the three cube roots of unity. As before, we begin in trigonometric form: $1 + 0i = 1(\cos 0° + i\sin 0°)$. With $n = 3$, $r = 1$, and $\theta = 0°$, we have $\sqrt[3]{r} = \sqrt[3]{1} = 1$,

and $\dfrac{0°}{3} + \dfrac{360°k}{3} = 0° + 120°k$. The principal

root ($k = 0$) is $z_0 = 1(\cos 0° + i\sin 0°) = 1$. Adding $120°$ to each previous argument, we find the other roots are

$$z_1 = 1(\cos 120° + i\sin 120°)$$
$$z_2 = 1(\cos 240° + i\sin 240°).$$

In rectangular form these are $-\dfrac{1}{2} + \dfrac{\sqrt{3}}{2}i$, and

$-\dfrac{1}{2} - \dfrac{\sqrt{3}}{2}i$, as shown in the figure.

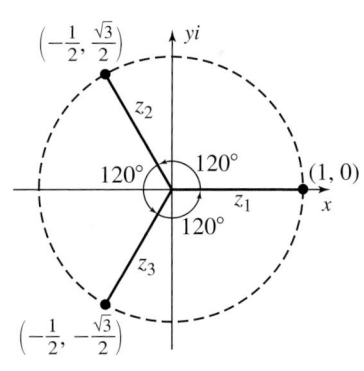

Now try Exercises 27 through 40 ▶

EXAMPLE 4 ▶ Finding nth Roots

Use the nth roots theorem to find the five fifth roots of $z = 16\sqrt{3} + 16i$.

Solution ▶ In trigonometric form, $16\sqrt{3} + 16i = 32(\cos 30° + i\sin 30°)$. With $n = 5$, $r = 32$, and $\theta = 30°$, we have $\sqrt[5]{r} = \sqrt[5]{32} = 2$, and

$\dfrac{30°}{5} + \dfrac{360°k}{5} = 6° + 72°k$. The principal root is $z_0 = 2(\cos 6° + i\sin 6°)$.

Adding $72°$ to each previous argument, we find the other four roots are

$$z_1 = 2(\cos 78° + i\sin 78°) \qquad z_2 = 2(\cos 150° + i\sin 150°)$$
$$z_3 = 2(\cos 222° + i\sin 222°) \qquad z_4 = 2(\cos 294° + i\sin 294°)$$

Now try Exercises 41 through 44 ▶

Of the five roots in Example 4, only $z_2 = 2(\cos 150° + i \sin 150°)$ uses a standard angle. Applying De Moivre's theorem with $n = 5$ gives $(2 \text{ cis } 150°)^5 = 32 \text{ cis } 750° = 32 \text{ cis } 30°$ or $16\sqrt{3} + 16i.$ ✓ **See Exercise 54.**

Figure 7.80

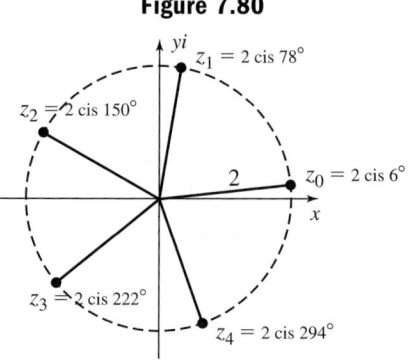

As a consequence of the arguments in a solution being uniformly separated by $\dfrac{360°}{n}$, the graphs of complex roots are equally spaced about a circle of radius r. The five fifth roots from Example 3 are shown in Figure 7.80 (note each argument differs by 72°).

For additional insight into roots of complex numbers, we reason that the nth roots of a complex number must also be complex. To find the four fourth roots of $z = 8 + 8\sqrt{3}i = 16(\cos 60° + i \sin 60°)$, we seek a number of the form $r(\cos \alpha + i \sin \alpha)$ such that $[r(\cos \alpha + i \sin \alpha)]^4 = 16(\cos 60° + i \sin 60°)$. Applying De Moivre's theorem to the left-hand side and equating equivalent parts we obtain

$$r^4[\cos(4\alpha) + i \sin(4\alpha)] = 16(\cos 60° + i \sin 60°), \text{ which leads to}$$

$$r^4 = 16 \text{ and}$$

$$4\alpha = 60°$$

From this it is obvious that $r = 2$, but as with similar equations solved in Chapter 6, the equation $4\alpha = 60°$ has multiple solutions. To find them, we first add $360°k$ to $60°$, *then* solve for α.

$$4\alpha = 60° + 360°k \quad \text{add } 360°k$$

$$\alpha = \frac{60° + 360°k}{4} \quad \text{divide by 4}$$

$$= 15° + 90°k \quad \text{result}$$

For convenience, we start with $k = 0, 1, 2$, and so on, which leads to

For $k = 0$: $\alpha = 15° + 90°(0)$ For $k = 1$: $\alpha = 15° + 90°(1)$
   $= 15°$ $= 105°$

For $k = 2$: $\alpha = 15° + 90°(2)$ For $k = 3$: $\alpha = 15° + 90°(3)$
   $= 195°$ $= 285°$

At this point it should strike us that we have four roots—exactly the number required. Indeed, using $k = 4$ gives $\alpha = 15° + 90°(4) = 375°$, which is coterminal with the $15°$ obtained when $k = 0$. Hence, the four fourth roots are

$$z_0 = 2(\cos 15° + i \sin 15°) \qquad z_1 = 2(\cos 105° + i \sin 105°)$$
$$z_2 = 2(\cos 195° + i \sin 195°) \qquad z_3 = 2(\cos 285° + i \sin 285°).$$

The check for these solutions is asked for in **Exercise 53.**

☑ **C.** You've just learned how to use the nth roots theorem to find the nth roots of a complex number

As a final note, it must have struck the mathematicians who pioneered these discoveries with some amazement that complex numbers and the trigonometric functions should be so closely related. The amazement must have been all the more profound upon discovering an additional connection between complex numbers and *exponential functions*. For more on these connections, visit www.mhhe.com/coburn and review **Section 7.7: Complex Numbers in Exponential Form.**

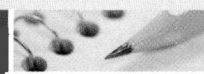

7.6 EXERCISES

▶ CONCEPTS AND VOCABULARY

Fill in each blank with the appropriate word or phrase. Carefully reread the section if needed.

1. For $z = r(\cos\theta + i\sin\theta)$, z^5 is computed as _____ according to _____ theorem.

2. If $z = 6i$, then z raised to an _____ power will be real and z raised to an _____ power will be _____ since $\theta = $ _____.

3. One application of De Moivre's theorem is to check _____ solutions to a polynomial equation.

4. The nth roots of a complex number are equally spaced on a circle of radius r, since their arguments all differ by _____ degrees or _____ radians.

5. From Example 4, go ahead and compute the value of z_5, z_6, and z_7. What do you notice? Discuss how this reaffirms that there are exactly n, nth roots.

6. Use a calculator to find $(1 - 3i)^4$. Then use it again to find the fourth root of the result. What do you notice? Explain the discrepancy and then resolve it using the nth roots theorem to find all four roots.

▶ DEVELOPING YOUR SKILLS

Use De Moivre's theorem to compute the following. Clearly state the value of r, n, and θ before you begin.

7. $(3 + 3i)^4$

8. $(-2 + 2i)^6$

9. $(-1 + \sqrt{3}i)^3$

10. $(\sqrt{3} - i)^3$

11. $\left(\dfrac{1}{2} - \dfrac{\sqrt{3}}{2}i\right)^5$

12. $\left(-\dfrac{\sqrt{3}}{2} + \dfrac{1}{2}i\right)^6$

13. $\left(\dfrac{\sqrt{2}}{2} - \dfrac{\sqrt{2}}{2}i\right)^6$

14. $\left(-\dfrac{\sqrt{2}}{2} + \dfrac{\sqrt{2}}{2}i\right)^5$

15. $(4\,\text{cis}\,330°)^3$

16. $(4\,\text{cis}\,300°)^3$

17. $\left(\dfrac{\sqrt{2}}{2}\,\text{cis}\,135°\right)^5$

18. $\left(\dfrac{\sqrt{2}}{2}\,\text{cis}\,135°\right)^8$

Use De Moivre's theorem to verify the solution given for each polynomial equation.

19. $z^4 + 3z^3 - 6z^2 + 12z - 40 = 0$; $z = 2i$

20. $z^4 - z^3 + 7z^2 - 9z - 18 = 0$; $z = -3i$

21. $z^4 + 6z^3 + 19z^2 + 6z + 18 = 0$; $z = -3 - 3i$

22. $2z^4 + 3z^3 - 4z^2 + 2z + 12 = 0$; $z = 1 - i$

23. $z^5 + z^4 - 4z^3 - 4z^2 + 16z + 16 = 0$; $z = \sqrt{3} - i$

24. $z^5 + z^4 - 16z^3 - 16z^2 + 256z + 256 = 0$; $z = 2\sqrt{3} + 2i$

25. $z^4 - 4z^3 + 7z^2 - 6z - 10 = 0$; $z = 1 + 2i$

26. $z^4 - 2z^3 - 7z^2 + 28z + 52 = 0$; $z = 3 - 2i$

Find the nth roots indicated by writing and solving the related equation.

27. five fifth roots of unity

28. six sixth roots of unity

29. five fifth roots of 243

30. three cube roots of 8

31. three cube roots of $-27i$

32. five fifth roots of $32i$

Solve each equation using the nth roots theorem.

33. $x^5 - 32 = 0$

34. $x^5 - 243 = 0$

35. $x^3 - 27i = 0$

36. $x^3 + 64i = 0$

37. $x^5 - \sqrt{2} - \sqrt{2}i = 0$

38. $x^5 - 1 + \sqrt{3}i = 0$

39. Solve the equation $x^3 - 1 = 0$ by factoring it as the difference of cubes and applying the quadratic formula. Compare results to those obtained in Example 3.

40. Use the nth roots theorem to find the four fourth roots of unity, then find all solutions to $x^4 - 1 = 0$ by factoring it as a difference of squares. What do you notice?

Use the nth roots theorem to find the *n*th roots. Clearly state *r*, *n*, and θ (from the trigonometric form of *z*) as you begin. Answer in exact form when possible, otherwise use a four decimal place approximation.

41. four fourth roots of $-8 + 8\sqrt{3}i$

42. five fifth roots of $16 - 16\sqrt{3}i$

43. four fourth roots of $-7 - 7i$

44. three cube roots of $9 + 9i$

▶ **WORKING WITH FORMULAS**

The discriminant of a cubic equation: $D = \dfrac{4p^3 + 27q^2}{108}$

For cubic equations of the form $z^3 + pz + q = 0$, where p and q are real numbers, one solution has the form

$z = \sqrt[3]{-\dfrac{q}{2} + \sqrt{D}} + \sqrt[3]{-\dfrac{q}{2} - \sqrt{D}}$, where D is called the

discriminant. Compute the value of D for the cubic equations given, then use the *n*th roots theorem to find the three cube roots of $-\dfrac{q}{2} + \sqrt{D}$ and $-\dfrac{q}{2} - \sqrt{D}$ in trigonometric form (also see Exercises 61 and 62).

45. $z^3 - 6z + 4 = 0$ **46.** $z^3 - 12z - 8 = 0$

▶ **APPLICATIONS**

 47. Powers and roots: Just after Example 4, the four fourth roots of $z = 8 + 8\sqrt{3}i$ were given as

$z_0 = 2(\cos 15° + i \sin 15°)$

$z_1 = 2(\cos 105° + i \sin 105°)$

$z_2 = 2(\cos 195° + i \sin 195°)$

$z_3 = 2(\cos 285° + i \sin 285°)$.

Verify these are the four fourth roots of $z = 8 + 8\sqrt{3}i$ using a calculator and De Moivre's theorem.

 48. Powers and roots: In Example 4 we found the five fifth roots of $z = 16\sqrt{3} + 16i$ were

$z_0 = 2(\cos 6° + i \sin 6°)$

$z_1 = 2(\cos 78° + i \sin 78°)$

$z_2 = 2(\cos 150° + i \sin 150°)$

$z_3 = 2(\cos 222° + i \sin 222°)$

$z_4 = 2(\cos 294° + i \sin 294°)$

Verify these are the five fifth roots of $16\sqrt{3} + 16i$ using a calculator and De Moivre's theorem.

Electrical circuits: For an AC circuit with three branches wired in parallel, the total impedance is given by

$Z_T = \dfrac{Z_1 Z_2 Z_3}{Z_1 Z_2 + Z_1 Z_3 + Z_2 Z_3}$, where Z_1, Z_2, and Z_3 represent the impedance in each branch of the circuit. If the impedance in each branch is identical, $Z_1 = Z_2 = Z_3 = Z$, and the numerator becomes Z^3 and the denominator becomes $3Z^2$, (a) use De Moivre's theorem to calculate the numerator and denominator for each value of Z given, (b) find the total impedance by computing the quotient $\dfrac{Z^3}{3Z^2}$, and (c) verify your result is identical to $\dfrac{Z}{3}$.

49. $Z = 3 + 4j$ in all three branches

50. $Z = 5\sqrt{3} + 5j$ in all three branches

▶ **EXTENDING THE CONCEPT**

In Chapter 6, you were asked to verify that $\sin(3\theta) = 3\sin\theta - 4\sin^3\theta$ and $\cos(4\theta) = 8\cos^2\theta - 8\cos^2\theta + 1$ were identities (Section 6.4, Exercises 21 and 22). For $z = 3 + \sqrt{7}i$, verify $|z| = 4$ and $\theta = \tan^{-1}\left(\dfrac{\sqrt{7}}{3}\right)$, then draw a right triangle with $\sqrt{7}$ opposite θ and 3 adjacent to θ. Discuss how this right triangle and the identities given can be used in conjunction with De Moivre's theorem to find the exact value of the powers given (also see Exercises 53 and 54).

For cases where θ is not a standard angle, working toward an exact answer using De Moivre's theorem requires the use of multiple angle identities and drawing the right triangle related to $\theta = \tan^{-1}\left(\dfrac{b}{a}\right)$. For Exercises 53 and 54, use De Moivre's theorem to compute the complex powers by (a) constructing the related right triangle for θ, (b) evaluating $\sin(4\theta)$ using two applications of double-angle identities, and (c) evaluating $\cos(4\theta)$ using a Pythagorean identity and the computed value of $\sin(4\theta)$.

51. $(3 + \sqrt{7}i)^3$ **52.** $(3 + \sqrt{7}i)^4$

53. $z = (1 + 2i)^4$ **54.** $(2 + \sqrt{5}i)^4$

The solutions to the cubic equations in Exercises 45 and 46 (repeated in Exercises 55 and 56) can be found by adding the cube roots of $-\frac{q}{2} + \sqrt{D}$ and $-\frac{q}{2} - \sqrt{D}$ that have arguments summing to 360°.

55. Find the roots of $z^3 - 6z + 4 = 0$

56. Find the roots of $z^3 - 12z - 8 = 0$

▶ MAINTAINING YOUR SKILLS

57. (6.2) Prove the following is a identity:
$$\frac{\tan^2 x}{\sec x + 1} = \frac{1 - \cos x}{\cos x}$$

58. (3.3) Given $f(x) = 2x^2 - 3x$, determine:
$$f(-1), f\left(\frac{1}{3}\right), f(a) \text{ and } f(a + h).$$

59. (2.3) Find the equation of the line whose graph is given.

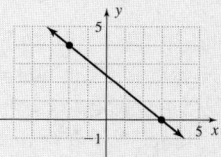

60. (5.2) Solve the triangle given. Round lengths to hundredths of a meter.

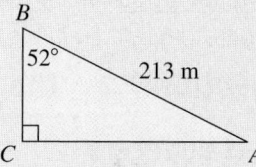

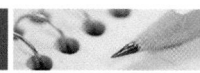

SUMMARY AND CONCEPT REVIEW

SECTION 7.1   Oblique Triangles and the Law of Sines

KEY CONCEPTS

- In any triangle, the ratio of the sine of an angle to its opposite side is constant: $\dfrac{\sin A}{a} = \dfrac{\sin B}{b} = \dfrac{\sin C}{c}$.
- The law of sines requires a known angle, a side opposite this angle and an additional side or angle, hence cannot be applied for SSS and SAS triangles.
- For AAS and ASA triangles, the law of sines yields a unique solution.
- When given two sides of a triangle and an angle opposite one of these sides (SSA), the number of solutions is *in doubt,* giving rise to the designation, "the ambiguous case."
- SSA triangles may have no solution, one solution, or two solutions, depending on the length of the side opposite the given angle.
- When solving triangles, always remember:
  - The sum of all angles must be 180°: $\angle A + \angle B + \angle C = 180°$.
  - The sum of any two sides must exceed the length of the remaining side.
  - Longer sides are opposite larger angles.
  - $k = \sin^{-1}\theta$ has no solution for $k > 1$.
  - $k = \sin^{-1}\theta$ has two solutions in $[0, 360°)$ for $0 < |k| < 1$.

EXERCISES

Solve the following triangles.

1.

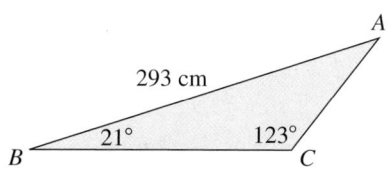

2.

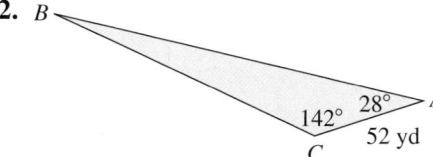

3. A tree is growing vertically on a hillside. Find the height of the tree if it makes an angle of 110° with the hillside and the angle of elevation from the base of the hill to the top of the tree is 25° at a distance of 70 ft.

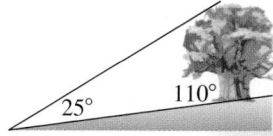

4. Find two values of θ that will make the equation true: $\dfrac{\sin \theta}{14} = \dfrac{\sin 50}{31}$.

5. Solve using the law of sines. If two solutions exist, find both (figure not drawn to scale).

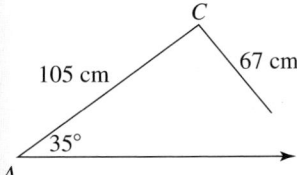

6. Jasmine is flying her tethered, gas-powered airplane at a local park, where a group of bystanders is watching from a distance of 60 ft, as shown. If the tether has a radius of 35 ft and one of the bystanders walks away at an angle of 40°, will he get hit by the plane? What is the smallest angle of exit he could take (to the nearest whole) without being struck by Jasmine's plane?

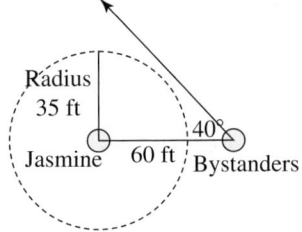

SECTION 7.2   The Law of Cosines; the Area of a Triangle

KEY CONCEPTS

- The law of cosines is used to solve SSS and SAS triangles.
- The law of cosines states that in any triangle, the square of any side is equal to the sums of the squares of the other two sides, minus twice their product times the cosine of the included angle:

$$a^2 = b^2 + c^2 - 2bc \cos A$$

- When using the law of cosines to solve a SSS triangle, always begin with the largest angle or the angle opposite the largest side.
- The area of a nonright triangle can be found using the following formulas. The choice of formula depends on the information given.

  - two sides a and b with included angle c
    $$A = \frac{1}{2}ab \sin C$$

  - two angles A and B with included side c
    $$A = \frac{c^2 \sin A \sin B}{2 \sin C}$$

  - three sides a, b, and c with $S = \dfrac{a + b + c}{2}$
    $$A = \sqrt{s(s - a)(s - b)(s - c)}$$

EXERCISES

7. Solve for B: $9^2 = 12^2 + 15^2 - 2(12)(15) \cos B$

8. Use the law of cosines to find the missing side.

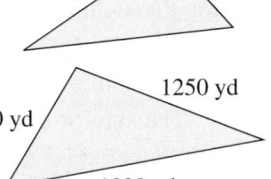

9. While preparing for the day's orienteering meet, Rick finds that the distances between the first three markers he wants to pick up are 1250 yd, 1820 yd, and 720 yd. Find the measure of each angle in the triangle formed so that Rick is sure to find all three markers.

10. The Great Pyramid of Giza, also known Khufu's pyramid, is the sole remaining member of the Seven Wonders of the Ancient World. It was built as a tomb for the Egyptian pharaoh Khufu from the fourth dynasty. This square pyramid is made up of four isosceles triangles, each with a base of 230.0 m and a slant height of about 218.7 m. Approximate the total surface area of Khufu's pyramid (excluding the base).

SECTION 7.3   Vectors and Vector Diagrams

KEY CONCEPTS

- Quantities/concepts that can be described using a single number are called scalar quantities. Examples are time, perimeter, area, volume, energy, temperature, weight, and so on.
- Quantities/concepts that require more than a single number to describe their attributes are called vector quantities. Examples are force, velocity, displacement, pressure, and so on.

- Vectors can be represented using directed line segments to indicate magnitude and direction. The origin of the segment is called the initial point, with the arrowhead pointing to the terminal point. When used solely for comparative analysis, they are called geometric vectors.
- Two vectors are equal if they have the same magnitude and direction.
- Vectors can be represented graphically in the xy-plane by naming the initial and terminal points of the vector or by giving the magnitude and angle of the related position vector [initial point at (0, 0)].
- For a vector with initial point (x_1, y_1) and terminal point (x_2, y_2), the related position vector can be written in the component form $\langle a, b \rangle$, where $a = x_2 - x_1$ and $b = y_2 - y_1$.
- For a vector written in the component form $\langle a, b \rangle$, a is called the horizontal component and b is called the vertical component of the vector.
- For vector $\mathbf{v} = \langle a, b \rangle$, the magnitude of $\mathbf{v}$ is $|\mathbf{v}| = \sqrt{a^2 + b^2}$.
- Vector components can also be written in trigonometric form. See page 661.
- For $\mathbf{u} = \langle a, b \rangle$, $\mathbf{v} = \langle c, d \rangle$, and any scalar k, we have the following operations defined:

$$\mathbf{u} + \mathbf{v} = \langle a + c, b + d \rangle \qquad \mathbf{u} - \mathbf{v} = \langle a - c, b - d \rangle \qquad k\mathbf{u} = \langle ka, kb \rangle \text{ for } k \in R$$

  If $k > 0$, the new vector has the same direction as $\mathbf{u}$; $k < 0$, the opposite direction.
- Vectors can be written in algebraic form using $\mathbf{i}$, $\mathbf{j}$ notation, where $\mathbf{i}$ is an x-axis unit vector and $\mathbf{j}$ is a y-axis unit vector. The vector $\langle a, b \rangle$ is written as a linear combination of $\mathbf{i}$ and $\mathbf{j}$: $\langle a, b \rangle = a\mathbf{i} + b\mathbf{j}$.
- For any nonzero vector $\mathbf{v}$, vector $\mathbf{u} = \dfrac{\mathbf{v}}{|\mathbf{v}|}$ is a unit vector in the same direction as $\mathbf{v}$.
- In aviation and shipping, the heading of a ship or plane is understood to be the amount of rotation from due north in the clockwise direction.

EXERCISES

11. Graph the vector $\mathbf{v} = \langle 9, 5 \rangle$, then compute its magnitude and direction angle.

12. Write the vector $\mathbf{u} = \langle -8, 3 \rangle$ in $\mathbf{i}$, $\mathbf{j}$ form and compute its magnitude and direction angle.

13. Approximate the horizontal and vertical components of the vector $\mathbf{u}$, where $|\mathbf{u}| = 18$ and $\theta = 52°$.

14. Compute $2\mathbf{u} + \mathbf{v}$, then find the magnitude and direction of the resultant: $\mathbf{u} = \langle -3, -5 \rangle$ and $\mathbf{v} = \langle 2, 8 \rangle$.

15. Find a unit vector that points in the same direction as $\mathbf{u} = 7\mathbf{i} + 12\mathbf{j}$.

16. Without computing, if $\mathbf{u} = \langle -9, 2 \rangle$ and $\mathbf{v} = \langle 2, 8 \rangle$, will the resultant sum lie in Quadrant I or II? Why?

17. It's once again time for the Great River Race, a $\frac{1}{2}$-mi swim across the Panache River. If Karl fails to take the river's 1-mph current into account and he swims the race at 3 mph, how far from the finish marker does he end up when he makes it to the other side?

18. Two Coast Guard vessels are towing a large yacht into port. The first is pulling with a force of 928 N and the second with a force of 850 N. Determine the angle θ for the second Coast Guard vessel that will keep the ship moving safely in a straight line.

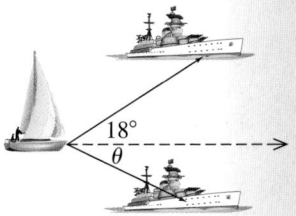

SECTION 7.4　Vector Applications and the Dot Product

KEY CONCEPTS

- Vector forces are in equilibrium when the sum of their components is the zero vector.
- When the components of vector $\mathbf{u}$ are nonquadrantal, with one of its components lying along vector $\mathbf{v}$, we call this component the "component of $\mathbf{u}$ along $\mathbf{v}$" or $\text{comp}_\mathbf{v}\mathbf{u}$.
- For vectors $\mathbf{u}$ and $\mathbf{v}$, $\text{comp}_\mathbf{v}\mathbf{u} = |\mathbf{u}|\cos\theta$, where θ is the angle between $\mathbf{u}$ and $\mathbf{v}$.
- Work done is computed as the product of the constant force $\mathbf{F}$ applied, times the distance D the force is applied: $W = \mathbf{F} \cdot D$.
- If force is not applied parallel to the direction of movement, only the component of the force in the direction of movement is used in the computation of work. If $\mathbf{u}$ is a force vector not parallel to the direction of vector $\mathbf{v}$, the equation becomes $W = \text{comp}_\mathbf{v}\mathbf{u} \cdot |\mathbf{v}|$.

- For vectors $\mathbf{u} = \langle a, b \rangle$ and $\mathbf{v} = \langle c, d \rangle$, the dot product $\mathbf{u} \cdot \mathbf{v}$ is defined as the scalar $ac + bd$.
- The dot product $\mathbf{u} \cdot \mathbf{v}$ is equivalent to $\text{comp}_{\mathbf{u}}\mathbf{v} \cdot |\mathbf{v}|$ and to $|\mathbf{u}||\mathbf{v}|\cos \theta$.
- The angle between two vectors can be computed using $\cos \theta = \dfrac{\mathbf{u}}{|\mathbf{u}|} \cdot \dfrac{\mathbf{v}}{|\mathbf{v}|} = \dfrac{\mathbf{u} \cdot \mathbf{v}}{|\mathbf{u}||\mathbf{v}|}$.
- Given vectors $\mathbf{u}$ and $\mathbf{v}$, the projection of $\mathbf{u}$ along $\mathbf{v}$ is the *vector* $\text{proj}_{\mathbf{v}}\mathbf{u}$ defined by $\text{proj}_{\mathbf{v}}\mathbf{u} = \left(\dfrac{\mathbf{u} \cdot \mathbf{v}}{|\mathbf{v}|^2} \right) \mathbf{v}$.
- Given vectors $\mathbf{u}$ and $\mathbf{v}$ and $\text{proj}_{\mathbf{v}}\mathbf{u}$, $\mathbf{u}$ can be resolved into the orthogonal components $\mathbf{u}_1$ and $\mathbf{u}_2$ where $\mathbf{u} = \mathbf{u}_1 + \mathbf{u}_2$, $\mathbf{u}_1 = \text{proj}_{\mathbf{v}}\mathbf{u}$, and $\mathbf{u}_2 = \mathbf{u} - \mathbf{u}_1$.
- The horizontal distance x a projectile travels in t seconds is $x = (|\mathbf{v}|\cos \theta)t$.
- The vertical height y of a projectile after t seconds is $y = (|\mathbf{v}|\sin \theta)t - 16t^2$, where $|\mathbf{v}|$ is the magnitude of the initial velocity, and θ is the angle of projection.

EXERCISES

19. For the force vectors $\mathbf{F}_1$ and $\mathbf{F}_2$ given, find the resultant and an additional force vector so that equilibrium takes place: $\mathbf{F}_1 = \langle -20, 70 \rangle$; $\mathbf{F}_2 = \langle 45, 53 \rangle$.

20. Find $\text{comp}_{\mathbf{v}}\mathbf{u}$ for $\mathbf{u} = -12\mathbf{i} - 16\mathbf{j}$ and $\mathbf{v} = 19\mathbf{i} - 13\mathbf{j}$.

21. Find the component d that ensures vectors $\mathbf{u}$ and $\mathbf{v}$ are orthogonal: $\mathbf{u} = \langle 2, 9 \rangle$ and $\mathbf{v} = \langle -18, d \rangle$.

22. Compute $\mathbf{p} \cdot \mathbf{q}$ and find the angle between them: $\mathbf{p} = \langle -5, -2 \rangle$; $\mathbf{q} = \langle 4, -7 \rangle$.

23. Given force vector $\mathbf{F} = \langle 50, 15 \rangle$ and $\mathbf{v} = \langle 85, 6 \rangle$, find the work required to move an object along the entire length of $\mathbf{v}$. Assume force is in pounds and distance in feet.

24. A 650-lb crate is sitting on a ramp that is inclined at $40°$. Find the force needed to hold the crate stationary.

25. An arctic explorer is hauling supplies from the supply hut to her tent, a distance of 120 ft, in a sled she is dragging behind her. If the straps used make an angle of $25°$ with the snow-covered ground and she pulls with a constant force of 75 lb, find the amount of work done.

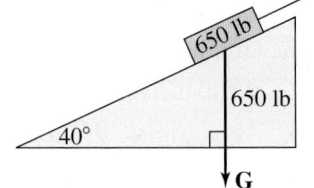

650 lb

650 lb

$40°$

G

26. A projectile is launched from a sling-shot with an initial velocity of $v_0 = 280$ ft/sec at an angle of $\theta = 50°$. Find (a) the position of the object after 1.5 sec and (b) the time required to reach a height of 150 ft.

SECTION 7.5   Complex Numbers in Trigonometric Form

KEY CONCEPTS

- A complex number $a + bi = (a, b)$ can be written in trigonometric form by noting (from its graph) that $a = r \cos \theta$ and $b = r \sin \theta$: $a + bi = r(\cos \theta + i \sin \theta)$.

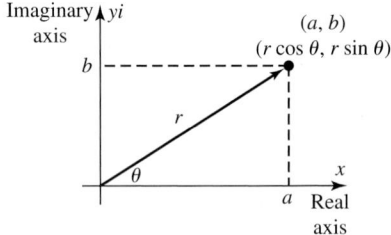

- The angle θ is called the argument of z and r is called the modulus of z.
- The argument of a complex number z is not unique, since any rotation of $\theta + 2\pi k$ (k an integer) will yield a coterminal angle.
- To convert from trigonometric to rectangular form, evaluate $\cos \theta$ and $\sin \theta$ and multiply by the modulus.
- To multiply complex numbers in trig form, multiply the moduli and add the arguments. To divide complex numbers in trig form, divide the moduli and subtract the arguments.
- Complex numbers have numerous real-world applications, particularly in a study of AC electrical circuits.
- The impedance of an AC circuit is given as $Z = R + j(X_L - X_C)$, where R is a pure resistance, X_C is the capacitive reactance, X_L is the inductive reactance, and $j = \sqrt{-1}$.
- Z is a complex number with magnitude $|Z| = \sqrt{R^2 + (X_L - X_C)^2}$ and phase angle $\theta = \tan^{-1}\left(\dfrac{X_L - X_C}{R} \right)$

  (θ represents the angle between the voltage and current).

- In an AC circuit, voltage $V = IZ$; current $I = \dfrac{V}{Z}$.

EXERCISES

27. Write in trigonometric form: $z = -1 - \sqrt{3}i$

28. Write in rectangular form: $z = 3\sqrt{2}\left[\operatorname{cis}\left(\dfrac{\pi}{4}\right)\right]$

29. Graph in the complex plane: $z = 5(\cos 30° + i \sin 30°)$

30. For $z_1 = 8 \operatorname{cis}\left(\dfrac{\pi}{4}\right)$ and $z_2 = 2 \operatorname{cis}\left(\dfrac{\pi}{6}\right)$, compute $z_1 z_2$ and $\dfrac{z_1}{z_2}$.

31. Find the current I in a circuit where $V = 4\sqrt{3} - 4j$ and $Z = 1 - \sqrt{3}j\ \Omega$.

32. In the V_{RLC} series circuit shown, $R = 10\ \Omega$, $X_L = 8\ \Omega$, and $X_C = 5\ \Omega$. Find the magnitude of Z and the phase angle between current and voltage. Express the result in trigonometric form.

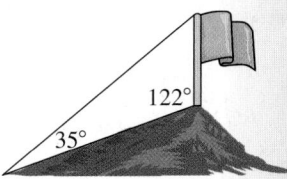

SECTION 7.6   De Moivre's Theorem and the Theorem on *n*th Roots

KEY CONCEPTS

* For complex number $z = r(\cos \theta + i \sin \theta)$, $z^n = r^n[\cos(n\theta) + i \sin(n\theta)]$ (De Moivre's theorem).
* De Moivre's theorem can be used to check complex solutions of polynomial equations.

* For complex number $z = r(\cos \theta + i \sin \theta)$, $\sqrt[n]{z} = \sqrt[n]{r}\left[\cos\left(\dfrac{\theta}{n} + \dfrac{2\pi k}{n}\right) + i \sin\left(\dfrac{\theta}{n} + \dfrac{2\pi k}{n}\right)\right]$, for $k = 1, 2, 3,\ldots, n - 1$ (*n*th roots theorem).

* The *n*th roots of a complex number are equally spaced around a circle of radius r in the complex plane.

EXERCISES

33. Use De Moivre's theorem to compute the value of $(-1 + i\sqrt{3})^5$.

34. Use De Moivre's theorem to verify that $z = 1 - i$ is a solution of $z^4 + z^3 - 2z^2 + 2z + 4 = 0$.

35. Use the *n*th roots theorem to find the three cube roots of $125i$.

36. Solve the equation using the *n*th roots theorem: $x^3 - 216 = 0$.

37. Given that $z = 2 + 2i$ is a fourth root of -64, state the other three roots.

38. Solve using the quadratic formula and the *n*th roots theorem: $z^4 + 6z^2 + 25 = 0$.

39. Use De Moivre's theorem to verify the three roots of $125i$ found in Exercise 35.

MIXED REVIEW

Solve each triangle using either the law of sines or law of cosines (whichever is appropriate) then find the area of each.

1.
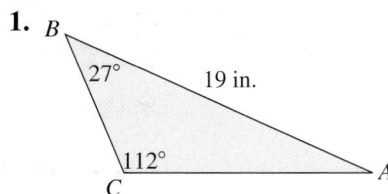

2.

3. Find the horizontal and vertical components of the vector **u**, where $|\mathbf{u}| = 21$ and $\theta = 40°$.

4. Compute $2\mathbf{u} + \mathbf{v}$, then find the magnitude and direction of the resultant: $\mathbf{u} = \langle 6, -3 \rangle$, $\mathbf{v} = \langle -2, 8 \rangle$.

5. Find the height of a flagpole that sits atop a hill, if it makes an angle of 122° with the hillside, and the angle of elevation between the side of the hill to the top of the flagpole is 35° at a distance of 120 ft.

6. A 900-lb crate is sitting on a ramp that is inclined at 28°. Find the force needed to hold the object stationary.

Exercise 6

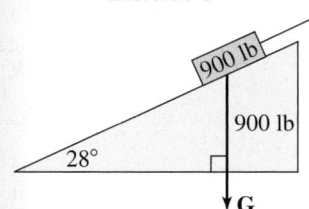

7. A jet plane is flying at 750 mph on a heading of 30°. There is a strong, 50-mph wind blowing from due south (heading 0°). What is the true course and speed of the plane (relative to the ground)?

8. Graph the vector $\mathbf{v} = \langle -8, 5 \rangle$, then compute its magnitude and direction.

9. Solve using the law of sines. If two solutions exist, find both.

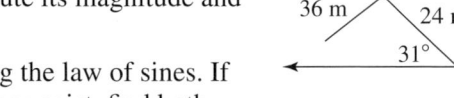

10. A local Outdoors Club sponsors a treasure hunt activity for its members, and has placed surprise packages at the corners of the triangular park shown. Find the measure of each angle to help club members find their way to the treasure.

Exercise 10

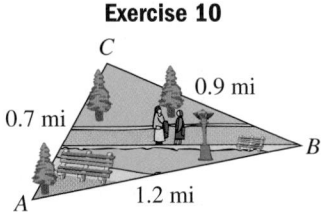

11. As part of a lab demonstrating centrifugal and centripetal forces, a physics teacher is whirling a tethered weight above her head while a group of students looks on from a distance of 20 ft as shown. If the tether has a radius of 10 ft and a student departs at the 35° angle shown, will the student be struck by the weight? What is the smallest angle of exit the student could take (to the nearest whole) without being struck by the whirling weight?

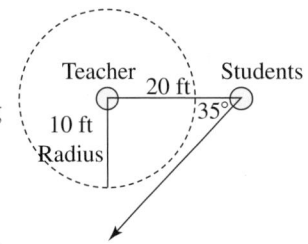

12. Given the vectors $\mathbf{p} = \langle -5, 2 \rangle$ and $\mathbf{q} = \langle 4, 7 \rangle$, use the dot product $\mathbf{p} \cdot \mathbf{q}$ to find the angle between them.

13. a. Graph the complex number using the rectangular form, then convert to trigonometric form: $z = 4 - 4i$.

   b. Graph the complex number using the trigonometric form, then convert to rectangular form: $z = 6(\cos 120° + i \sin 120°)$.

14 a. Verify that $z = 4 - 5i$ *and* its conjugate are solutions to $z^2 - 8z + 41 = 0$.

   b. Solve using the quadratic formula: $z^2 - 6iz + 7 = 0$

15. Two tractors are dragging a large, fallen tree into the brush pile that's being prepared for a large Fourth of July bonfire. The first is pulling with a force of 418 N and the second with a force of 320 N. Determine the angle θ for the second tractor that will keep the tree headed straight for the brush pile.

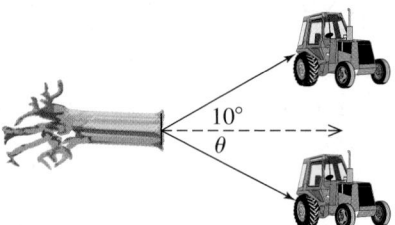

16. Given $z_1 = 8(\cos 45° + i \sin 45°)$ and $z_2 = 4(\cos 15° + i \sin 15°)$ compute:

   a. the product $z_1 z_2$ **b.** the quotient $\dfrac{z_1}{z_2}$

17. Given the vectors $\mathbf{u} = -12\mathbf{i} - 16\mathbf{j}$ and $\mathbf{v} = 19\mathbf{i} - 13\mathbf{j}$, find $\text{comp}_v\mathbf{u}$ and $\text{proj}_v\mathbf{u}$.

18. Find the result using De Moivre's theorem: $(2\sqrt{3} - 2i)^6$.

19. Use the *n*th roots theorem to find the four fourth roots of $-2 + 2i\sqrt{3}$.

20. The impedance of an AC circuit is $Z = R + j(X_L - X_C)$. The voltage across the circuit is $V_{RLC} = I|Z|$. Given $R = 12\ \Omega$, $X_L = 15.2\ \Omega$, and $X_C = 9.4\ \Omega$, write Z in trigonometric form and find the voltage in the circuit if the current is $I = 6.5$ A.

PRACTICE TEST

1. Within the Kilimanjaro Game Reserve, a fire is spotted by park rangers stationed in two towers known to be 10 mi apart. Using the line between them as a baseline, tower A reports the fire is at an angle of 39°, while tower B reports an angle of 68°. How far is the fire from the closer tower?

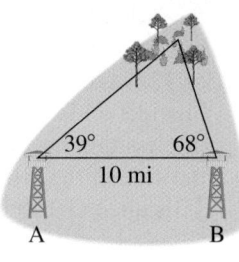

2. At the circus, Mac and Joe are watching a high-wire act from first-row seats on opposite sides of the center ring. Find the height of the performing acrobat at the instant Mac measures an angle of elevation of 68° while Joe measures an angle of 72°. Assume Mac and Joe are sitting 100 ft apart.

Exercise 2

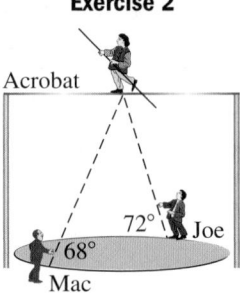

3. Three rods are attached via two joints and shaped into a triangle. How many triangles can be formed if the angle at the joint B must measure 20°? If two triangles can be formed, solve both.

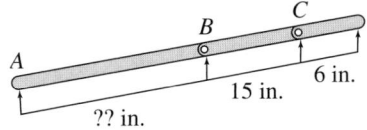

4. Jackie and Sam are rounding up cattle in the brush country, and are communicating via walkie-talkie. Jackie is at the water hole and Sam is at Dead Oak, which are 6 mi apart. Sam finds some strays and heads them home at the 32° indicated. (a) If the maximum range of Jackie's unit is 3 mi, will she be able to communicate with Sam as he heads home? (b) If the maximum range were 4 mi, how far from Dead Oak is Sam when he is first contacted by Jackie?

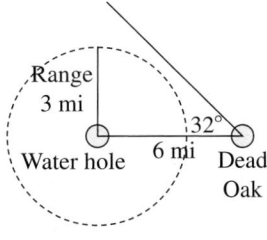

5. As part of an All-Star competition, a group of soccer players (forwards) stand where shown in the diagram and attempt to hit a moving target with a two-handed overhead pass. If a player has a maximum effective range of approximately (a) 25 yd, can the target be hit? (b) about 28 yd, how many "effective" throws can be made? (c) 35 yd and the target is moving at 5 yd/sec, how many seconds is the target within range?

Exercise 5

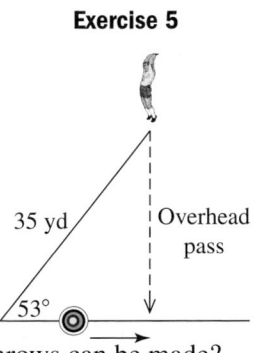

6. The summit of Triangle Peak can only be reached from one side, using a trail straight up the side that is approximately 3.5 mi long. If the mountain is 5 mi wide at its base and the trail makes a 24° angle with the horizontal, (a) what is the approximate length of the opposing side? (b) How tall is the peak (in feet)?

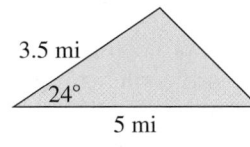

7. The Bermuda Triangle is generally thought to be the triangle formed by Miami, Florida, San Juan, Puerto Rico, and Bermuda itself. If the distances between these locations are the 1025 mi, 1020 mi, and 977 mi indicated, find the measure of each angle and the area of the Bermuda Triangle.

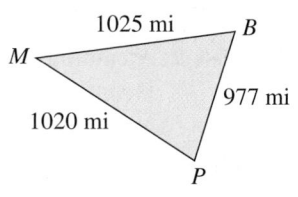

8. A helicopter is flying at 90 mph on a heading of 40°. A 20-mph wind is blowing from the NE on a heading of 190°. What is the true course and speed of the helicopter relative to the ground? Draw a diagram as part of your solution.

9. Two mules walking along a river bank are pulling a heavy barge up river. The first is pulling with a force of 250 N and the second with a force of 210 N. Determine the angle θ for the second mule that will ensure the barge stays midriver and does not collide with the shore.

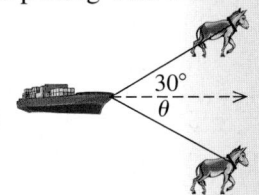

Exercise 10

10. Along a production line, various tools are attached to the ceiling with a multijointed arm so that workers can draw one down, position it for use, then move it up out of the way for the next tool (see the diagram). If the first segment is 100 cm, the second is 75 cm, and the third is 50 cm, determine the approximate coordinates of the last joint.

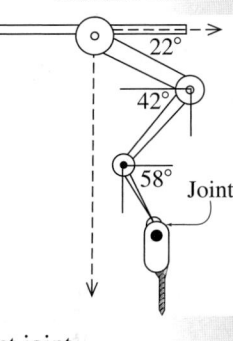

11. Three ranch hands have roped a run-away steer and are attempting to hold him steady. The first and second ranch hands are pulling with the magnitude and at the angles indicated in the diagram. If the steer is held fast by the efforts of all three, find the magnitude of the tension and angle of the rope from the third cowhand.

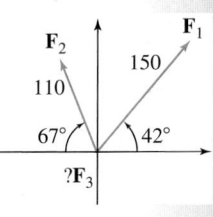

12. For $\mathbf{u} = \langle -9, 5 \rangle$ and $\mathbf{v} = \langle -2, 6 \rangle$, (a) compute the angle between $\mathbf{u}$ and $\mathbf{v}$; (b) find the projection of $\mathbf{u}$ along $\mathbf{v}$ (find $\mathbf{proj_v u}$; and (c) resolve $\mathbf{u}$ into vectors $\mathbf{u}_1$ and $\mathbf{u}_2$, where $\mathbf{u}_1 \| \mathbf{v}$ and $\mathbf{u}_2 \perp \mathbf{v}$.

13. A lacrosse player flips a long pass to a teammate way down field who is near the opponent's goal. If the initial velocity of the pass is 110 ft/sec and the ball is released at an angle of 50° with level ground, how high is the ball after 2 sec? How long until the ball again reaches this same height?

14. Compute the quotient $\dfrac{z_1}{z_2}$, given
$$z_1 = 6\sqrt{5}\operatorname{cis}\left(\frac{\pi}{8}\right) \text{ and } z_2 = 3\sqrt{5}\operatorname{cis}\left(\frac{\pi}{12}\right).$$

15. Compute the product $z = z_1 z_2$ in trigonometric form, then verify $|z_1||z_2| = |z|$ and $\theta_1 + \theta_2 = \theta$:
$z_1 = -6 + 6i; z_2 = 4 - 4\sqrt{3}i$

16. Use De Moivre's theorem to compute the value of $(\sqrt{3} - i)^4$.

17. Use De Moivre's theorem to verify $2 + 2\sqrt{3}i$ is a solution to $z^5 + 3z^3 + 64z^2 + 192 = 0$.

18. Use the nth roots theorem to solve $x^3 - 125i = 0$.

19. Solve using *u*-substitution, the quadratic formula, and the *n*th roots theorem: $z^4 - 6z^2 + 58 = 0$.

20. Due to its huge biodiversity, preserving Southeast Asia's Coral Triangle has become a top priority for conservationists. Stretching from the northern Philippines (*P*), south along the coast of Borneo (*B*) to the Lesser Sunda Islands (*L*), then eastward to the Solomon Islands (*S*), this area is home to over 75% of all coral species known. Use Heron's formula to help find the total area of this natural wonderland, given the dimensions shown.

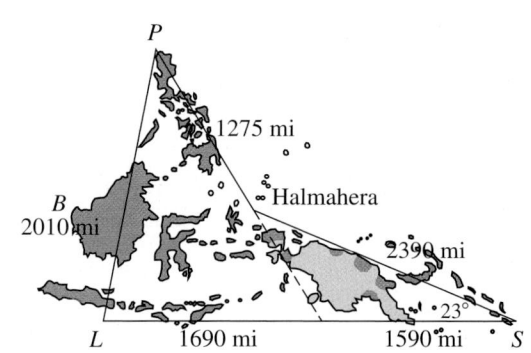

CALCULATOR EXPLORATION AND DISCOVERY

Investigating Projectile Motion

There are two important aspects of projectile motion that were not discussed earlier, the **range** of the projectile and the **optimum angle** θ that will maximize this range. Both can be explored using the equations for the horizontal and vertical components of the projectile's position: horizontal $\rightarrow (|\mathbf{v}|\cos\theta)t$ and vertical $\rightarrow (|\mathbf{v}|\sin\theta)$ $t - 16t^2$. In Example 10 of Section 7.4, an arrow was shot from a bow with initial velocity $|\mathbf{v}| = 150$ ft/sec at an angle of $\theta = 50°$. Enter the equations above on the [Y=] screen as Y_1 and Y_2, using these values (Figure 7.81). Then set up the TABLE using TblStart = 0, ΔTbl = 0.5 and the AUTO mode. The resulting table is shown in Figure 7.82, where Y_1 represents the horizontal distance the arrow has traveled, and Y_2 represents the height of the arrow. To find the *range* of the arrow, scroll downward [▼] until the height (Y_2) shows a value that is less than or equal to zero (the arrow has hit the ground). As Figure 7.83 shows, this happens somewhere between $t = 7$ and $t = 7.5$ sec. We could now change the TBLSET settings to TblStart = 0 and ΔTbl = 0.1 to get a better approximation of the time the arrow is in flight (it's just less than 7.2 sec) and the horizontal range of the arrow (about 692.4 ft), but our main interest is how to *compute these values exactly*. We begin with the equation for the arrow's vertical position $y = (|\mathbf{v}|\sin\theta)t - 16t^2$. Since the object returns to Earth when $y = 0$, we substitute 0 for y and factor out

Figure 7.81

```
Plot1 Plot2 Plot3
\Y1◘150cos(50)X
\Y2◘150sin(50)X-
16X²
\Y3=
\Y4=
\Y5=
\Y6=
```

Figure 7.82

X	Y1	Y2
0	0	0
.5	48.209	53.453
1	96.418	98.907
1.5	144.63	136.36
2	192.84	165.81
2.5	241.05	187.27
3	289.25	200.72

X=0

Figure 7.83

X	Y1	Y2
5	482.09	174.53
5.5	530.3	147.99
6	578.51	113.44
6.5	626.72	70.893
7	674.93	20.347
7.5	723.14	-38.2
8	771.35	-104.7

X=7.5

$t: 0 = t(|\mathbf{v}|\sin\theta - 16t)$. Solving for t gives $t = 0$ or $t = \dfrac{|\mathbf{v}|\sin\theta}{16}$. Since the component of velocity in the horizontal direction is $|\mathbf{v}|\cos\theta$, the basic distance relationship $D = \mathbf{r} \cdot \mathbf{t}$ gives the horizontal range of $R = |\mathbf{v}|\cos\theta \cdot \dfrac{|\mathbf{v}|\sin\theta}{16}$ or $\dfrac{|\mathbf{v}|^2\sin\theta\cos\theta}{16}$. Checking the values given for the arrow ($|\mathbf{v}| = 150$ ft/sec and $\theta = 50°$) verifies the range is $R \approx 692.4$. But what about the *maximum possible range* for the arrow? Using $|\mathbf{v}| = 150$ for R results in an equation in theta only: $R(\theta) = \dfrac{150^2\sin\theta\cos\theta}{16}$, which we can enter as Y_3 and investigate for various θ. After carefully entering $R(\theta)$ as Y_3 and resetting TBLSET to TblStart = 30 and ΔTbl = 5, the TABLE in Figure 7.84 shows a maximum range of about 703 ft at 45°. Resetting TBLSET to TblStart = 40 and ΔTbl = 1 verifies this fact.

For each of the following exercises, find (a) the height of the projectile after 1.75 sec, (b) the maximum height of the projectile, (c) the range of the projectile, and (d) the number of seconds the projectile is airborne.

Figure 7.84

X	Y3
30	608.92
35	660.72
40	692.44
45	703.13
50	692.44
55	660.72
60	608.92

X=30

Exercise 1: A javelin is thrown with an initial velocity of 85 ft/sec at an angle of 42°.

Exercise 2: A cannon ball is shot with an initial velocity of 1120 ft/sec at an angle of 30°.

Exercise 3: A baseball is hit with an initial velocity of 120 ft/sec at an angle of 50°. Will it clear the center field fence, 10 ft high and 375 ft away?

Exercise 4: A field goal (American football) is kicked with an initial velocity of 65 ft/sec at an angle of 35°. Will it clear the crossbar, 10 ft high and 40 yd away?

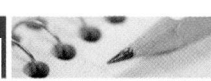

STRENGTHENING CORE SKILLS

Vectors and Static Equilibrium

In Sections 7.3 and 7.4, the concepts of vector forces, resultant forces, and equilibrium were studied extensively. A nice extension of these concepts involves what is called **static equilibrium.** Assuming that only coplanar forces are acting on an object, the object is said to be in static equilibrium if *the sum of all vector forces acting on it is 0.* This implies that the object is stationary, since the forces all counterbalance each other. The methods involved are simple and direct, with a wonderful connection to the systems of equations you've likely seen previously. Consider the following example.

Illustration 1 ▶ As part of their training, prospective FBI agents must move hand-over-hand across a rope strung between two towers. An agent-in-training weighing 180 lb is two-thirds of the way across, causing the rope to deflect from the horizontal at the angles shown. What is the tension in each part of the rope at this point?

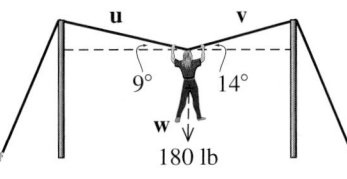

Solution ▶ We have three concurrent forces acting on the point where the agent grasps the rope. Begin by drawing a vector diagram and computing the components of each force, using the **i, j** notation. Note that **w** = −180**j**.

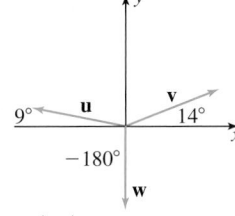

$$\mathbf{u} = -|\mathbf{u}|\cos(9°)\mathbf{i} + |\mathbf{u}|\sin(9°)\mathbf{j}$$
$$\approx -0.9877|\mathbf{u}|\mathbf{i} + 0.1564|\mathbf{u}|\mathbf{j}$$
$$\mathbf{v} = |\mathbf{v}|\cos(14°)\mathbf{i} + |\mathbf{v}|\sin(14°)\mathbf{j}$$
$$\approx 0.9703|\mathbf{v}|\mathbf{i} + 0.2419|\mathbf{v}|\mathbf{j}$$

For equilibrium, all vector forces must sum to the zero vector: **u** + **v** + **w** = 0, which results in the following

equation: $-0.9877|\mathbf{u}|\mathbf{i} + 0.1564|\mathbf{u}|\mathbf{j} + 0.9703|\mathbf{v}|\mathbf{i} +$ $0.2419|\mathbf{v}|\mathbf{j} - 180\mathbf{j} = 0\mathbf{i} + 0\mathbf{j}$. Factoring out **i** and **j** from the left-hand side yields $(-0.9877|\mathbf{u}| + 0.9703|\mathbf{v}|)\mathbf{i} +$ $(0.1564|\mathbf{u}| + 0.2419|\mathbf{v}| - 180)\mathbf{j} = 0\mathbf{i} + 0\mathbf{j}$. Since any two vectors are equal only when corresponding components are equal, we obtain a system in the two variables

$$|\mathbf{u}| \text{ and } |\mathbf{v}|: \begin{cases} -0.9877|\mathbf{u}| + 0.9703|\mathbf{v}|) = 0 \\ 0.1564|\mathbf{u}| + 0.2419|\mathbf{v}| - 180 = 0 \end{cases}.$$

Solving the system using matrix equations and a calculator (or any desired method), gives $|\mathbf{u}| \approx 447$ lb and $|\mathbf{v}| \approx 455$ lb.

At first it may seem surprising that the vector forces (tension) in each part of the rope are so much greater than the 180-lb the agent weighs. But with a 180-lb object hanging from the middle of the rope, the tension required to keep the rope taut (with small angles of deflection) must be very great. This should become more obvious to you after you work Exercise 2.

Exercise 1: A 500-lb crate is suspended by two ropes attached to the ceiling rafters. Find the tension in each rope.

Exercise 2: Two people team up to carry a 150-lb weight by passing a rope through an eyelet in the object. Find the tension in each rope.

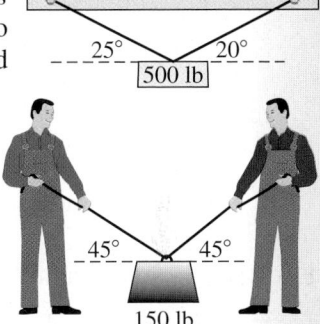

Exercise 3: Referring to Illustration 1, if the rope has a tension limit of 600-lb (before it snaps), can a 200-lb agent make it across?

CUMULATIVE REVIEW CHAPTERS 1–7

1. Solve using a standard triangle.
$a = 20, b =$ ____ , $c =$ ____
$\alpha = 30°, \beta =$ ____ , $\gamma =$ ____

2. Solve using trigonometric ratios.
$a \approx$ ____ , $b \approx$ ____ , $c = 82$
$\alpha =$ ____ , $\beta = 63°, \gamma =$ ____

3. A torus is a donut-shaped solid figure. Its surface area is given by the formula

$A = \pi^2(R^2 - r^2)$, where R is the outer radius of the donut, and r is the inner radius. Solve the formula for R in terms of r and A.

4. For a complex number $a + bi$, (a) verify the sum of a complex number and its conjugate is a real number, and (b) verify the product of a complex number and its conjugate is a real number.

5. State the value of all six trig functions given
$\tan \alpha = -\dfrac{3}{4}$ with $\cos \alpha > 0$.

6. Sketch the graph of $y = 4\cos\left(\dfrac{\pi}{6}x - \dfrac{\pi}{3}\right)$ using transformations of $y = \cos x$.

7. Solve using the quadratic formula:
$5x^2 + 8x + 2 = 0$.

8. Solve by completing the square: $3x^2 - 72x + 427 = 0$.

9. Given $\cos 53° \approx 0.6$ and $\cos 72° \approx 0.3$, approximate the value of $\cos 19°$ and $\cos 125°$ without using a calculator.

10. Find all real values of x that satisfy the equation $\sqrt{3} + 2\sin(2x) = 2\sqrt{3}$. State the answer in degrees.

11. a. Given that 1 acre $= 43{,}560$ ft^2, find the cost of a lot with the dimensions shown (to the nearest dollar) if land in this area is going for $4500 per acre.

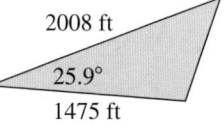

2008 ft
25.9°
1475 ft

b. After an accident at sea, a search and rescue team decides to focus their efforts on the area shown due to prevailing winds and currents. Find the distances between each vertex (use Pythagorean triples and a special triangle) and the number of square miles in the search area.

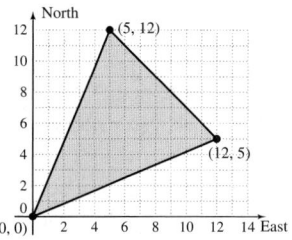

North
(5, 12)
(12, 5)
(0, 0)   2   4   6   8   10   12   14   East

12. State the domain of each function:
a. $f(x) = \sqrt{2x - 3}$
b. $g(x) = \log_b(x + 3)$
c. $h(x) = \dfrac{x + 3}{x^2 - 5}$
d. $v(x) = \sqrt{x^2 - x - 6}$

13. Write the following formulas from memory:
a. slope formula
b. midpoint formula
c. quadratic formula
d. distance formula
e. interest formula (compounded continuously)

Solve each triangle using the law of sines or the law of cosines, whichever is appropriate.

14.

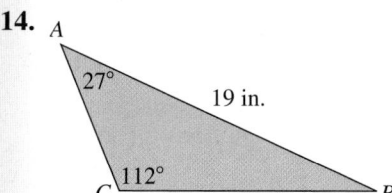

A
27°
19 in.
112°
C   B

15.

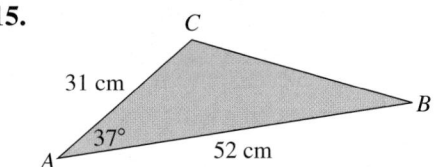

C
31 cm
37°
A   52 cm   B

16. A commercial fishery stocks a lake with 250 fish. Based on previous experience, the population of fish is expected to grow according to the model
$$P(t) = \dfrac{12{,}000}{1 + 25e^{-0.2t}},$$ where t is the time in months. From on this model, (a) how many months are required for the population to grow to 7500 fish? (b) If the fishery expects to harvest three-fourths of the fish population in 2 yr, approximately how many fish will be taken?

17. A 900-lb crate is sitting on a ramp which is inclined at 28°. Find the force needed to hold the object stationary.

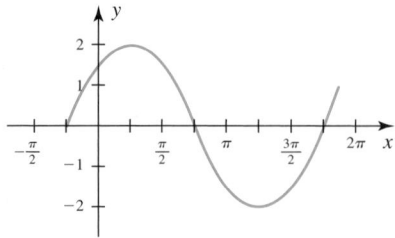

900 lb
900 lb
28°
G

18. A jet plane is flying at 750 mph on a heading of 30°. There is a strong, 50-mph wind blowing from due south (heading 0°). What is the true course and speed of the plane (relative to the ground)?

19. Use the *Guidelines for Graphing* to sketch the graph of function f given, then use it to solve $f(x) < 0$.
$$f(x) = x^3 - 4x^2 + x + 6$$

20. Use the *Guidelines for Graphing* to sketch the graph of function g given, then use it to name the intervals where $g(x)\downarrow$ and $g(x)\uparrow$.
$$g(x) = \dfrac{x^2 - 4}{x^2 - 1}$$

21. Find $(1 - \sqrt{3}i)^8$ using De Moivre's theorem.

22. Solve $\ln(x + 2) + \ln(x - 3) = \ln(4x)$.

23. If I saved $200 each month in an annuity program that paid 8% annual interest compounded monthly, how long would it take to save $10,000?

24. Mount Tortolas lies on the Argentine-Chilean border. When viewed from a distance of 5 mi, the angle of elevation to the top of the peak is 38°. How tall is Mount Tortolas? State the answer in feet.

25. The graph given is of the form $y = A\sin(Bx + C)$. Find the values of A, B, and C.

Chapter 7 unites our study of trigonometry with many of the skills you likely developed in previous coursework. Certainly *algebraic skills* continue to play a significant role, but here we've chosen to review and highlight certain *problem-solving skills* as well. From the vector diagrams seen in Chapter 7 to the triangle diagrams used in numerous applications, a careful sketch can highlight relationships and suggest a solution process, and this will certainly serve us well throughout the calculus sequence.

Trigonometry and Problem Solving

Perhaps to a higher degree in a calculus course, the quality of the diagrams we draw for problem solving have a greater impact on our ability to understand and solve applications. Draw the diagrams very carefully and to scale, if possible. Label the various parts of the diagram, make a list of all given information, assign variable names to unknown quantities, and build any needed relationships—in short, use the total accumulation of your problem-solving skills.

EXAMPLE 1 ▶ **Determining Sight Angles**

A 20-ft-tall road sign is to be placed along a busy highway by attaching it to a support pole, with the bottom edge of the sign 13 ft above the ground.

  a. Find a simplified expression for the "sight angle" (the angle at a person's eyes subtended by the sign) in terms of x, when the driver is x ft away from the base of the pole (assume that eye level for an average driver is 3 ft above the ground).

  b. Find the sight angle at the moment a driver is 50 ft away.

Solution ▶ **a.** Draw a car that is x ft away from a road sign with the dimensions described. Note that the triangle from eye level to the top of the sign, and the triangle from eye level to the bottom of the sign are right triangles, *while the sight angle is not.* Naming the acute angle of the larger right triangle (at the eyes of the driver) α, and the acute angle of the smaller right triangle β (see the figure), we see that the sight angle is then $\alpha - \beta$. Since we need to involve the distance x in our equation model, it appears the tangent function should be used, giving $\tan \alpha = \dfrac{30}{x}$ and $\tan \beta = \dfrac{10}{x}$. Knowing the identity for $\tan(\alpha - \beta)$ is expressed *in terms of tan α and tan β individually,* we have $\tan(\alpha - \beta) = \dfrac{\tan \alpha - \tan \beta}{1 + \tan \alpha \tan \beta}$. Our algebraic expression is then

$$\tan(\alpha - \beta) = \frac{\dfrac{30}{x} - \dfrac{10}{x}}{1 + \left(\dfrac{30}{x}\right)\left(\dfrac{10}{x}\right)} \qquad \text{make substitutions}$$

$$= \frac{\dfrac{20}{x}}{1 + \dfrac{300}{x^2}} \qquad \text{simplify}$$

$$= \frac{\dfrac{20}{x}}{\dfrac{x^2 + 300}{x^2}} \qquad \text{combine terms}$$

$$= \frac{20x}{x^2 + 300} \qquad \text{result}$$

The expression for the sight angle is $\alpha - \beta = \tan^{-1}\left(\dfrac{20x}{x^2 + 300}\right)$.

b. When $x = 50$ ft, $\alpha - \beta = \tan^{-1}\left(\dfrac{1000}{2800}\right)$, or about $19.65°$. For comparison, the sight angle at 25 ft is about $28.39°$, and at 100 ft the angle is about $10.99°$.

> **Now try Exercises 1 through 4 ▶**

Vectors in Three Dimensions

Many elements of a study of vectors in two dimensions transfer directly to the study of vectors in three dimensions. To help visualize a three-dimensional vector, consider a large $3 \text{ ft} \times 4 \text{ ft} \times 2 \text{ ft}$ cardboard box placed snugly in the corner of a room. The floor of the room forms the positive xy-plane (the axes are where the walls meet the floor, the x-axis to the left), with the third dimension (the z-axis) formed where the two walls meet

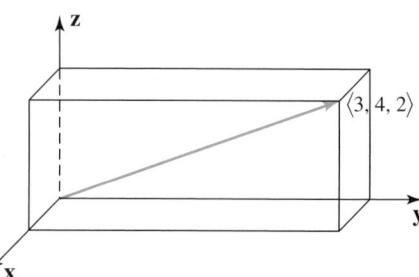

(see the figure). The origin is then $(0, 0, 0)$, where the three axes meet, and the position vector $\langle 3, 4, 2 \rangle$ forms an interior diagonal of the box. For three-dimensional vectors $\mathbf{u} = \langle a, b, c \rangle$ and $\mathbf{v} = \langle d, e, f \rangle$, the formulas for the magnitude of $\mathbf{v}$ and the dot product $\mathbf{u} \cdot \mathbf{v}$ are an extension of the two-dimensional cases, and are offered here without proof:

$$|\mathbf{u}| = \sqrt{a^2 + b^2 + c^2}, \text{ and } \mathbf{u} \cdot \mathbf{v} = ad + be + cf.$$

EXAMPLE 2 ▶ Finding the Angle between Two Vectors

For vector $\mathbf{u} = \langle 3, 4, 2 \rangle$ forming the interior diagonal of the box and $\mathbf{v} = \langle 3, 4, 0 \rangle$ forming the diagonal of the base,

a. Find $|\mathbf{u}|$, then verify the result using the Pythagorean theorem and $|\mathbf{v}|$.

b. Use the formula $\cos \theta = \dfrac{\mathbf{u} \cdot \mathbf{v}}{|\mathbf{u}||\mathbf{v}|}$ to find the angle θ between $\mathbf{u}$ and $\mathbf{v}$, then verify the result using right-triangle trig.

Solution ▶ a. First find $|\mathbf{u}| = \sqrt{3^2 + 4^2 + 2^2} = \sqrt{29}$. To verify, we use $|\mathbf{v}| = \sqrt{3^2 + 4^2 + 0^2} = 5$ and the height of the box in the Pythagorean theorem (since the triangle formed by $(0, 0, 0)$, $(3, 4, 0)$, and $(3, 4, 2)$ is a right triangle): $|\mathbf{v}|^2 + 2^2 = 25 + 4 = 29 = |\mathbf{u}|^2$ ✓.

b. For $\cos \theta = \dfrac{\mathbf{u} \cdot \mathbf{v}}{|\mathbf{u}||\mathbf{v}|}$, $\cos \theta = \dfrac{9 + 16 + 0}{5\sqrt{29}} = \dfrac{5}{\sqrt{29}}$, so $\theta = \cos^{-1}\left(\dfrac{5}{\sqrt{29}}\right) \approx 21.8°$.

To verify, we use $|\mathbf{v}| = 5$ and the height of the box, giving $\theta = \tan^{-1}\left(\dfrac{2}{5}\right) \approx 21.8°$ ✓.

> **Now try Exercises 5 through 10 ▶**

Connections to Calculus Exercises

For the following applications, draw a diagram and develop an appropriate equation model. Note that the emphasis on these exercises is shared equally between the *quality of diagram* and *developing the correct equation model*.

1. A jet plane is flying at a constant altitude of 9000 ft, straight toward a searchlight on the ground that is tracking it. (a) Find an expression for the angle the searchlight makes with the ground in terms of the horizontal distance d from the light to a point on the ground directly below the plane. (b) Find the angle θ at the moment $d = 12{,}850$ ft.

2. A light is hung from the rafters at a height h m above the floor, providing a circle of illumination. The illuminance E (in lumens/square meter or lm/m^2) at any point P on the floor varies directly as the cosine of the angle θ at which the rays leave the light source toward P, and inversely as the square of the height h of the light source. Using k as the constant of variation, (a) write an expression for the illuminance in terms of k, h, and θ. (b) If the distance from the light source to point P is 13 m, write an expression for the illuminance in terms of k and h alone. (c) For $k = 15{,}600$, what is the height of the light source if $E = 100$ lm/m^2? (d) For k and h as in part (c), at what angle θ is the illuminance 83 lm/m^2.

3. In a large factory, two automated carts run on concentric circular tracks. The inner track has a radius of 24 m, while the outer track has a radius of 32 m. (a) Write an expression for the straight-line distance x between the two carts in terms of the angle θ formed by the line segments from the center to each cart. (b) Determine the straight-line distance between the carts at the moment $\theta = 150°$. (c) At what angle θ are the carts 45 m apart?

4. A rectangle is inscribed in a semicircle of radius r. A straight line from the center of the circle to the point where a vertex of the rectangle meets the circle, forms an angle θ. (a) Find an expression for the area of the rectangle in terms of r and a sine function. (*Hint:* Write the length and width of the rectangle in terms of sine and cosine.) (b) Find the area of the circle if $r = 5$ cm and $\theta = 26°$. (c) At what angle θ does the rectangle have an area of 22 cm^2?

Solve each application of three-dimensional vectors using a careful sketch drawn from the information given. The emphasis on these exercises continues to be shared equally between the *quality of diagram* and *developing the correct equation model*.

5. A rectangular playground is 64 ft long and 48 ft wide. There is a light pole for evening play in one corner of the playground, with the light placed 18 ft high. (a) Find $|v|$ the length of the vector formed from the top of the light pole to the opposite corner of the playground. (b) If u is the vector formed by the opposite end of the playground and the base of the pole, find the angle between these two vectors using $\cos\theta = \dfrac{u \cdot v}{|u||v|}$ and verify the result using right triangle trig.

6. A gallery is about to open an exhibit in a room that is 22 ft long, 24 ft wide, and 16 ft high. The edge of the door to the exhibit is on the 22-ft side, 4 ft from one wall. For special effects lighting as a guest enters the door, let $\mathbf{u}$ be the vector representing the distance from the edge of the door (at the floor) to the farthest bottom corner of the room, and let $\mathbf{v}$ be the vector from the same point to the farthest upper corner. (a) Find $|\mathbf{v}|$, then (b) find the angle between $\mathbf{u}$ and $\mathbf{v}$ using $\cos\theta = \dfrac{\mathbf{u} \cdot \mathbf{v}}{|\mathbf{u}||\mathbf{v}|}$ and verify the result using right triangle trig.

7. At a small jungle airport, bush pilots are practicing their approach to drop zones for food aid. There is an observation post 525 ft from the center of one end of an 1800-ft straight dirt runway. At the moment the pilot of the first drop plane is over the opposite end of the runway, her altitude is 1000 ft. Let $\mathbf{d}$ be the vector representing the distance from the observation post to the opposite end of the runway, and let $\mathbf{D}$ represent the vector for the straight-line distance from the post to the plane at this moment. (a) Find $|\mathbf{D}|$, then (b) find the angle between $\mathbf{d}$ and $\mathbf{D}$ using $\cos\theta = \dfrac{\mathbf{d} \cdot \mathbf{D}}{|\mathbf{d}||\mathbf{D}|}$ and verify the result using right triangle trig.

8. An ornithologist is in the field studying what ornithologists study (in this case, a formation of geese), and is 180 m south and 385 m west of where she parked her car. Let $\mathbf{u}$ be the vector representing the distance between the ornithologist and her car, and let $\mathbf{v}$ be the vector representing the straight-line distance from the ornithologist to the formation. If the lead goose flies directly over her car at an altitude of 319 m, (a) find $|\mathbf{v}|$ then, (b) find the angle between $\mathbf{u}$ and $\mathbf{v}$ using $\cos\theta = \dfrac{\mathbf{u} \cdot \mathbf{v}}{|\mathbf{u}||\mathbf{v}|}$ and verify the result using right triangle trig.

9. A person is sitting in the center seat, in the center row, on the inclined seating of a new movie theater. The movie screen is 25 ft tall with the bottom of the screen 8 ft above a level floor. The person's eye level is 15 ft above the floor and the person's seat is a horizontal distance of x ft from the movie screen. (a) Write an expression for the person's sight angle to the screen in terms of x. (b) Find the sight angle if the person is 46 ft from the screen.

10. The main road in front of a school runs north/south along the east side. A sidewalk runs east/west along the south side of the school, leading to a crosswalk. A secondary road intersects the west side of the main road near the crosswalk, forming a 30° angle with the sidewalk. For the safety of the school children using the sidewalk, engineers are interested in the sight angle that drivers have as they drive on the secondary road approaching this intersection. (a) Find an expression for the sight angle of a driver in terms of x, the distance between their car and a point on the sidewalk 100 ft from the intersection. (b) Find the sight angle at the moment $x = 125$ ft.

8

Systems of Equations and Inequalities

CHAPTER OUTLINE

CHAPTER CONNECTIONS

The applications of systems of equations are virtually limitless, ranging from the extremely large systems used by governments to make economic forecasts, to the systems used in operations research to help companies minimize cost and risk. The idea of using a system to solve problems has been around for centuries, with mathematicians through the years often posing puzzle-like questions to their peers for fun and challenge. For instance, if you add the gestation periods for an elephant, camel, and rhino, the result is 1520 days. The gestation period for a rhino exceeds that of a camel by 58 days, while twice the camel's decreased by 162 gives that of an elephant. How long from conception to birth for each animal? This application appears as Exercise 57 of Section 8.2

$\frac{\Delta y}{\Delta x}$

Connections to Calculus

Recall that for exponential functions, the rate of growth is proportional to the current population as shown in (1) below. For logistics functions, the rate of growth is proportional to the current population times the *difference* between the carrying capacity c and the current population (2). For further analysis, this relationship can be rewritten as in (3), with a technique known as *partial fraction decomposition* then applied. This technique is presented in Section 8.3, and further explored in the *Connections to Calculus* feature for Chapter 8.

$$(1) \quad \frac{\Delta P}{\Delta t} = kP \qquad (2) \quad \frac{\Delta P}{\Delta t} = kP(c - P) \qquad (3) \quad \frac{\Delta P}{P(c - P)} = k\Delta t$$

Linear Systems in Two Variables with Applications

Learning Objectives

In Section 8.1 you will learn how to:

☐ **A.** Verify ordered pair solutions

☐ **B.** Solve linear systems by graphing

☐ **C.** Solve linear systems by substitution

☐ **D.** Solve linear systems by elimination

☐ **E.** Recognize inconsistent systems and dependent systems

☐ **F.** Use a system of equations to model and solve applications

In earlier chapters, we used linear equations in two variables to model a number of real-world situations. Graphing these equations gave us a visual image of how the variables were related, and helped us better understand this relationship. In many applications, two different measures of the independent variable must be considered simultaneously, leading to a **system of two linear equations in two unknowns.** Here, a graphical presentation once again supports a better understanding, as we explore systems and their many applications.

A. Solutions to a System of Equations

A **system of equations** is a set of two or more equations for which a common solution is sought. Systems are widely used to model and solve applications when the information given enables the relationship between variables to be stated in different ways. For example, consider an amusement park that brought in $3100 in revenue by charging $9.00 for adults and $5.00 for children, while selling 500 tickets. Using a for adult and c for children, we could write one equation modeling the number of tickets sold: $a + c = 500$, and a second modeling the amount of revenue brought in: $9a + 5c = 3100$. To show that we're considering both equations simultaneously, a large "left brace" is used and the result is called a **system of two equations in two variables:**

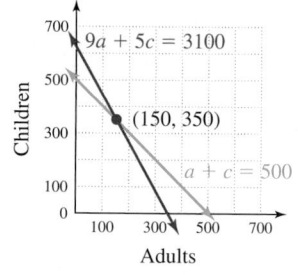

$$\begin{cases} a + c = 500 & \text{number of tickets} \\ 9a + 5c = 3100 & \text{amount of revenue} \end{cases}$$

We note that both equations are linear and will have different slope values, so their graphs must intersect at some point. Since every point on a line satisfies the equation of that line, this point of intersection must satisfy *both* equations simultaneously and is the solution to the system. The figure that accompanies Example 1 shows the point of intersecion for this system is (150, 350).

EXAMPLE 1 ▶ **Verifying Solutions to a System**

Verify that (150, 350) is a solution to $\begin{cases} a + c = 500 \\ 9a + 5c = 3100 \end{cases}$.

Solution ▶ Substitute the **150** for a and **350** for c in each equation.

$$a + c = 500 \quad \text{first equation} \qquad\qquad 9a + 5c = 3100 \quad \text{second equation}$$
$$(150) + (350) = 500 \qquad\qquad 9(150) + 5(350) = 3100$$
$$500 = 500 \checkmark \qquad\qquad\qquad 3100 = 3100 \checkmark$$

☑ **A.** You've just learned how to verify ordered pair solutions

Since (150, 350) satisfies both equations, it is the solution to the system and we find the park sold 150 adult tickets and 350 tickets for children.

Now try Exercises 7 through 18 ▶

B. Solving Systems Graphically

To **solve a system of equations** means we apply various methods in an attempt to find ordered pair solutions. As Example 1 suggests, one method for finding solutions is to graph the system. Any method for graphing the lines can be employed, but to keep important concepts fresh, the slope-intercept method is used here.

EXAMPLE 2 ▶ **Solving a System Graphically**

Solve the system by graphing: $\begin{cases} 4x - 3y = 9 \\ -2x + y = -5 \end{cases}$.

Solution ▶ First write each equation in slope-intercept form (solve for y):

$$\begin{cases} 4x - 3y = 9 \\ -2x + y = -5 \end{cases} \rightarrow \begin{cases} y = \dfrac{4}{3}x - 3 \\ y = 2x - 5 \end{cases}$$

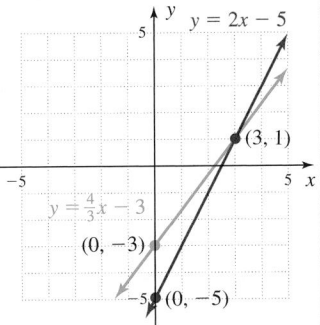

For the first line, $\frac{\Delta y}{\Delta x} = \frac{4}{3}$ with y-intercept $(0, -3)$. The second equation yields $\frac{\Delta y}{\Delta x} = \frac{2}{1}$ with $(0, -5)$ as the y-intercept. Both are then graphed on the grid as shown. The point of intersection appears to be $(3, 1)$, and checking this point in both equations gives

$$4x - 3y = 9 \qquad\qquad\qquad -2x + y = -5$$
$$4(3) - 3(1) = 9 \quad \text{substitute 3} \quad -2(3) + (1) = -5$$
$$9 = 9 \checkmark \quad \text{for } x \text{ and 1 for } y \qquad -5 = -5 \checkmark$$

☑ **B. You've just learned how to solve linear systems by graphing**

This verifies that $(3, 1)$ is the solution to the system.

Now try Exercises 19 through 22 ▶

C. Solving Systems by Substitution

While a graphical approach best illustrates *why* the solution must be an ordered pair, it does have one obvious drawback—noninteger solutions are difficult to spot. The ordered pair $\left(\frac{2}{5}, \frac{12}{5}\right)$ is the solution to $\begin{cases} 4x + y = 4 \\ y = x + 2 \end{cases}$, but this would be difficult to "pinpoint" as a precise location on a hand-drawn graph. To overcome this limitation, we next consider a method known as **substitution.** The method involves converting a system of two equations in two variables into a single equation in one variable by using an appropriate substitution. For $\begin{cases} 4x + y = 4 \\ y = x + 2 \end{cases}$, the second equation says "y is two more than x." We reason that *all* points on this line are related this way, *including the point where this line intersects the other.* For this reason, we can substitute $x + 2$ for y in the first equation, obtaining a single equation in x.

EXAMPLE 3 ▶ **Solving a System Using Substitution**

Solve using substitution: $\begin{cases} 4x + y = 4 \\ y = x + 2 \end{cases}$.

Solution ▶ Since $y = x + 2$, we can replace y with $x + 2$ in the first equation.

$$4x + y = 4 \qquad \text{first equation}$$
$$4x + (x + 2) = 4 \qquad \text{substitute } x + 2 \text{ for } y$$
$$5x + 2 = 4 \qquad \text{simplify}$$
$$x = \frac{2}{5} \qquad \text{result}$$

The x-coordinate is $\frac{2}{5}$. To find the y-coordinate, substitute $\frac{2}{5}$ for x into either of the original equations. Substituting in the second equation gives

$$y = x + 2 \qquad \text{second equation}$$
$$= \frac{2}{5} + 2 \qquad \text{substitute } \frac{2}{5} \text{ for } x$$
$$= \frac{12}{5} \qquad \frac{2}{1} = \frac{10}{5}, \frac{10}{5} + \frac{2}{5} = \frac{12}{5}$$

The solution to the system is $\left(\frac{2}{5}, \frac{12}{5}\right)$. Verify by substituting $\frac{2}{5}$ for x and $\frac{12}{5}$ for y into both equations.

Now try Exercises 23 through 32 ▶

If neither equation allows an immediate substitution, we first solve for one of the variables, either x or y, and *then* substitute. The method is summarized here, and can actually be used with either like variables or like variable expressions. **See Exercises 57 to 60.**

Solving Systems Using Substitution

1. Solve one of the equations for x in terms of y or y in terms of x.
2. Substitute for the appropriate variable in the *other* equation and solve for the variable that remains.
3. Substitute the value from step 2 into either of the original equations and solve for the other unknown.
4. Write the answer as an ordered pair and check the solution in both original equations.

☑ **C.** You've just learned how to solve linear systems by substitution

D. Solving Systems Using Elimination

Now consider the system $\begin{cases} -2x + 5y = 13 \\ 2x - 3y = -7 \end{cases}$, where solving for any one of the variables will result in fractional values. The substitution method can still be used, but often the **elimination method** is more efficient. The method takes its name from what happens when you add certain equations in a system (by adding the like terms from each). If the coefficients of either x or y are additive inverses—they sum to zero and are *eliminated*. For the system shown, adding the equations produces $2y = 6$, giving $y = 3$, then $x = 1$ using back-substitution (verify).

When neither variable term meets this condition, we can multiply one or both equations by a nonzero constant to "match up" the coefficients, so an elimination will take place. In doing so, we create an **equivalent system of equations,** meaning one that has the same solution as the original system. For $\begin{cases} 7x - 4y = 16 \\ -3x + 2y = -6 \end{cases}$, multiplying the second equation by 2 produces $\begin{cases} 7x - 4y = 16 \\ -6x + 4y = -12 \end{cases}$, giving $x = 4$ after "adding the equations." Note the three systems produced are equivalent, and have the solution $(4, 3)$ ($y = 3$ was found using back-substitution).

1. $\begin{cases} 7x - 4y = 16 \\ -3x + 2y = -6 \end{cases}$ 2. $\begin{cases} 7x - 4y = 16 \\ -6x + 4y = -12 \end{cases}$ 3. $\begin{cases} 7x - 4y = 16 \\ x = 4 \end{cases}$

In summary,

Operations that Produce an Equivalent System

1. Changing the order of the equations.
2. Replacing an equation by a nonzero constant multiple of that equation.
3. Replacing an equation with the sum of two equations from the system.

Before beginning a solution using elimination, check to make sure the equations are written in the **standard form** $Ax + By = C$, so that like terms will appear above/below each other. Throughout this chapter, we will use R1 to represent the equation in *row 1* of the system, R2 to represent the equation in *row 2*, and so on. These designations are used to help describe and document the steps being used to solve a system, as in Example 4 where $2R1 + R2$ indicates the first equation has been multiplied by two, with the result added to the second equation.

EXAMPLE 4 ▶ **Solving a System by Elimination**

Solve using elimination: $\begin{cases} 2x - 3y = 7 \\ 6y + 5x = 4 \end{cases}$

Solution ▶ The second equation is not in standard form, so we re-write the system as $\begin{cases} 2x - 3y = 7 \\ 5x + 6y = 4 \end{cases}$. If we "add the equations" now, we would get $7x + 3y = 11$, with neither variable eliminated. However, if we multiply *both sides* of the first equation by 2, the y-coefficients will be additive inverses. The sum then results in an equation with x as the only unknown.

$$\begin{array}{r} 2\text{R1} \\ + \\ \underline{\text{R2}} \\ \text{sum} \end{array} \begin{cases} 4x - 6y = 14 \\ 5x + 6y = 4 \\ \overline{9x + 0y = 18} \end{cases} \text{add}$$

$$9x = 18$$
$$x = 2 \quad \text{solve for } x$$

Substituting 2 for x back into either of the original equations yields $y = -1$. The ordered pair solution is $(2, -1)$. Verify using the original equations.

Now try Exercises 33 through 38 ▶

The elimination method is summarized here. If either equation has fraction or decimal coefficients, we can "clear" them using an appropriate constant multiplier.

Solving Systems Using Elimination

1. Write each equation in standard form: $Ax + By = C$.
2. Multiply one or both equations by a constant that will create coefficients of x (or y) that are additive inverses.
3. Combine the two equations using vertical addition and solve for the variable that remains.
4. Substitute the value from step 3 into either of the original equations and solve for the other unknown.
5. Write the answer as an ordered pair and check the solution in both original equations.

EXAMPLE 5 ▶ **Solving a System Using Elimination**

Solve using elimination: $\begin{cases} \frac{5}{8}x - \frac{3}{4}y = \frac{1}{4} \\ \frac{1}{2}x - \frac{2}{3}y = 1 \end{cases}$.

Solution ▶ Multiplying the first equation by 8(8R1) and the second equation by 6(6R2) will clear the fractions from each.

$$\begin{array}{l} 8\text{R1} \\ 6\text{R2} \end{array} \begin{cases} \frac{8}{1}(\frac{5}{8})x - \frac{8}{1}(\frac{3}{4})y = \frac{8}{1}(\frac{1}{4}) \\ \frac{6}{1}(\frac{1}{2})x - \frac{6}{1}(\frac{2}{3})y = 6(1) \end{cases} \rightarrow \begin{cases} 5x - 6y = 2 \\ 3x - 4y = 6 \end{cases}$$

The x-terms can now be eliminated if we use $3\text{R1} + (-5\text{R2})$.

$$\begin{array}{r} 3\text{R1} \\ + \\ \underline{-5\text{R2}} \\ \text{sum} \end{array} \begin{cases} 15x - 18y = 6 \\ -15x + 20y = -30 \\ \overline{0x + 2y = -24} \end{cases} \text{add}$$

$$y = -12 \quad \text{solve for } y$$

Substituting $y = -12$ in either of the original equations yields $x = -14$, and the solution is $(-14, -12)$. Verify by substituting in both equations.

Now try Exercises 39 through 44 ▶

> ⚠️ **CAUTION** ▶ Be sure to multiply *all* terms (on both sides) of the equation when using a constant multiplier. Also, note that for Example 5, we could have eliminated the *y*-terms using 2R1 with −3R2.

E. Inconsistent and Dependent Systems

A system having *at least one* solution is called a **consistent system.** As seen in Example 2, if the lines have different slopes, they intersect at a single point and the system has exactly one solution. Here, the lines are *independent* of each other and the system is called an **independent system.** If the lines have equal slopes *and* the same *y*-intercept, they are identical or **coincident lines.** Since one is right atop the other, they *intersect at all points,* and the system has an infinite number of solutions. Here, one line *depends* on the other and the system is called a **dependent system.** Using substitution or elimination on a dependent system results in the elimination of all variable terms and leaves a statement that is *always true,* such as 0 = 0 or some other simple identity.

EXAMPLE 6 ▶ **Solving a Dependent System**

Solve using elimination: $\begin{cases} 3x + 4y = 12 \\ 6x = 24 - 8y \end{cases}$.

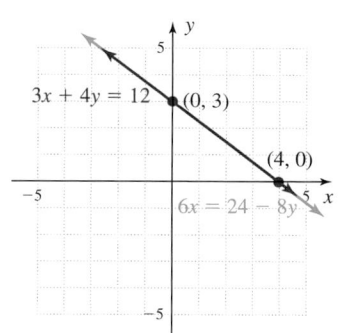

Solution ▶ Writing the system in standard form gives
$\begin{cases} 3x + 4y = 12 \\ 6x + 8y = 24 \end{cases}$. By applying −2R1, we can eliminate the variable *x*:

$$
\begin{array}{l}
-2R1 \\
+ \\
\underline{\quad R2 \quad}
\end{array}
\begin{cases}
-6x - 8y = -24 \\
6x + 8y = 24 \quad \text{add}
\end{cases}
$$

$$\text{sum} \qquad 0x + 0y = \quad 0 \quad \text{variables are eliminated}$$
$$0 = \quad 0 \quad \text{true statement}$$

Although we didn't expect it, both variables were eliminated and the final statement is true (0 = 0). This indicates the system is dependent, which the graph verifies (the lines are coincident). Writing both equations in slope-intercept form shows they represent the same line.

$$\begin{cases} 3x + 4y = 12 \\ 6x + 8y = 24 \end{cases} \longrightarrow \begin{cases} 4y = -3x + 12 \\ 8y = -6x + 24 \end{cases} \longrightarrow \begin{cases} y = -\dfrac{3}{4}x + 3 \\ y = -\dfrac{3}{4}x + 3 \end{cases}$$

> **WORTHY OF NOTE**
>
> When writing the solution to a dependent system using a parameter, the solution can be written in many different ways. For instance, if we let $p = 4b$ for the first coordinate of the solution to Example 6, we have $\dfrac{-3(4b)}{4} + 3 = -3b + 3$ as the second coordinate, and the solution becomes $(4b, -3b + 3)$ for any constant *b*.

The solutions of a dependent system are often written in set notation as the set of ordered pairs (x, y), where *y* is a specified function of *x*. For Example 6 the solution would be $\{(x, y)\,|\,y = -\frac{3}{4}x + 3\}$. Using an ordered pair with an arbitrary variable, called a **parameter,** is also common: $\left(p, \dfrac{-3p}{4} + 3\right)$.

Now try Exercises 45 through 56 ▶

> ✅ **E.** You've just learned how to recognize inconsistent and dependent systems

Finally, if the lines have equal slopes and *different y-intercepts,* they are parallel and the system will have no solution. A system with no solutions is called an **inconsistent system.** An "inconsistent system" produces an "inconsistent answer," such as 12 = 0 or some other false statement when substitution or elimination is applied. In other words, all variable terms are once again eliminated, but the remaining statement is *false.* A summary of the three possibilities is shown here for arbitrary slope *m* and *y*-intercept (0, *b*).

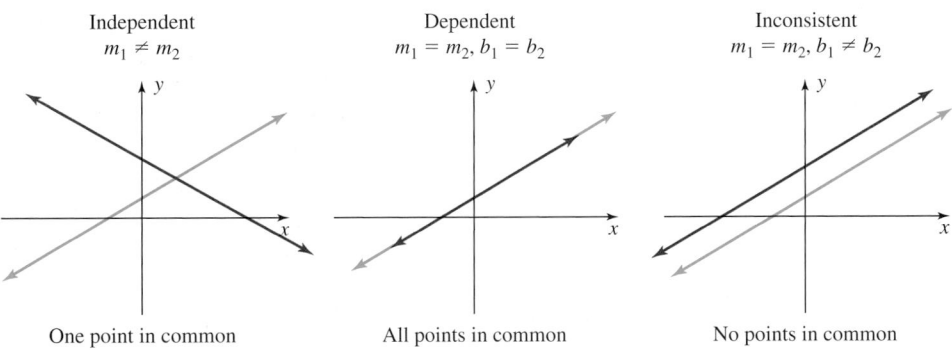

Independent
$m_1 \neq m_2$

One point in common

Dependent
$m_1 = m_2, b_1 = b_2$

All points in common

Inconsistent
$m_1 = m_2, b_1 \neq b_2$

No points in common

F. Systems and Modeling

In previous chapters, we solved numerous real-world applications by writing all given relationships in terms of a single variable. Many situations are easier to model using a system of equations with each relationship modeled independently using *two* variables. We begin here with a **mixture** application. Although they appear in many different forms (coin problems, metal alloys, investments, merchandising, and so on), mixture problems all have a similar theme. Generally one equation is related to *quantity* (how much of each item is being combined) and one equation is related to *value* (what is the value of each item being combined).

EXAMPLE 7 ▶ **Solving a Mixture Application**

A jeweler is commissioned to create a piece of artwork that will weigh 14 oz and consist of 75% gold. She has on hand two alloys that are 60% and 80% gold, respectively. How much of each should she use?

Solution ▶ Let x represent ounces of the 60% alloy and y represent ounces of the 80% alloy. The first equation must be $x + y = 14$, since the piece of art must weigh exactly 14 oz (this is the *quantity* equation). The x ounces are 60% gold, the y ounces are 80% gold, and the 14 oz will be 75% gold. This gives the *value* equation: $0.6x + 0.8y = 0.75(14)$. The system is $\begin{cases} x + y = 14 \\ 6x + 8y = 105 \end{cases}$ (after clearing decimals).

Solving for y in the first equation gives $y = 14 - x$. Substituting $14 - x$ for y in the second equation gives

$$
\begin{aligned}
6x + 8y &= 105 && \text{second equation} \\
6x + 8(14 - x) &= 105 && \text{substitute } 14 - x \text{ for } y \\
-2x + 112 &= 105 && \text{simplify} \\
x &= \frac{7}{2} && \text{solve for } x
\end{aligned}
$$

Substituting $\frac{7}{2}$ for x in the first equation gives $y = \frac{21}{2}$. She should use 3.5 oz of the 60% alloy and 10.5 oz of the 80% alloy.

WORTHY OF NOTE

As an estimation tool, note that if equal amounts of the 60% and 80% alloys were used (7 oz each), the result would be a 70% alloy (halfway in between). Since a 75% alloy is needed, more of the 80% gold will be used.

Now try Exercises 63 through 70 ▶

Systems of equations also play a significant role in *cost-based pricing* in the business world. The costs involved in running a business can broadly be understood as either a **fixed cost k** or a **variable cost v.** Fixed costs might include the monthly rent paid for facilities, which remains the same regardless of how many items are produced and sold. Variable costs would include the cost of materials needed to produce the item, which depends on the number of items made. The total cost can then be modeled by

$C(x) = vx + k$ for x number of items. Once a **selling price** p has been determined, the revenue equation is simply $R(x) = px$ (price times number of items sold). We can now set up and solve a system of equations that will determine how many items must be sold to break even, performing what is called a **break-even analysis.**

EXAMPLE 8 ▶ **Solving an Application of Systems: Break-Even Analysis**

In home businesses that produce items to sell on Ebay®, fixed costs are easily determined by rent and utilities, and variable costs by the price of materials needed to produce the item. Karen's home business makes large, decorative candles for all occasions. The cost of materials is $3.50 per candle, and her rent and utilities average $900 per month. If her candles sell for $9.50, how many candles must be sold each month to break even?

Solution ▶ Let x represent the number of candles sold. Her total cost is $C(x) = 3.5x + 900$ (variable cost plus fixed cost), and projected revenue is $R(x) = 9.5x$. This gives the system $\begin{cases} C(x) = 3.5x + 900 \\ R(x) = 9.5x \end{cases}$. To break even, Cost = Revenue which gives

$$9.5x = 3.5x + 900$$
$$6x = 900$$
$$x = 150$$

The analysis shows that Karen must sell 150 candles each month to break even.

Now try Exercises 71 through 74 ▶

WORTHY OF NOTE

This break-even concept can also be applied in studies of supply and demand, as well as in the decision to buy a new car or appliance that will enable you to break even over time due to energy and efficiency savings.

Our final example involves an application of uniform motion (distance = rate · time), and explores concepts of great importance to the navigation of ships and airplanes. As a simple illustration, if you've ever walked at your normal rate r on the "moving walkways" at an airport, you likely noticed an increase in your total speed. This is because the resulting speed combines your walking rate r with the speed w of the walkway: *total speed* $= r + w$. If you walk in the *opposite direction* of the walkway, your total speed is much slower, as now *total speed* $= r - w$.

This same phenomenon is observed when an airplane is flying with or against the wind, or a ship is sailing with or against the current.

EXAMPLE 9 ▶ **Solving an Application of Systems — Uniform Motion**

An airplane flying due south from St. Louis, Missouri, to Baton Rouge, Louisiana, uses a strong, steady tailwind to complete the trip in only 2.5 hr. On the return trip, the same wind slows the flight and it takes 3 hr to get back. If the flight distance between these cities is 912 km, what is the cruising speed of the airplane (speed with no wind)? How fast is the wind blowing?

Solution ▶ Let r represent the rate of the plane and w the rate of the wind. Since $D = RT$, the flight to Baton Rouge can be modeled by $912 = (r + w)(2.5)$, and the return flight by $912 = (r - w)(3)$. This produces the system $\begin{cases} 912 = 2.5r + 2.5w \\ 912 = 3r - 3w \end{cases}$. Using $\dfrac{R1}{2.5}$ and $\dfrac{R2}{3}$ gives the equivalent system $\begin{cases} 364.8 = r + w \\ 304 = r - w \end{cases}$, which is easily solved using elimination with R1 + R2.

$$364.8 = r + w$$
$$304 = r - w$$

$$668.8 = 2r \quad \text{R1 + R2}$$
$$334.4 = r \quad \text{divide by 2}$$

The cruising speed of the plane (with no wind) is 334.4 kph. Using $r - w = 304$ shows the wind is blowing at 30.4 kph.

☑ **F. You've just learned how to use a system of equations to model and solve applications**

Now try Exercises 75 through 78 ▶

TECHNOLOGY HIGHLIGHT

Solving Systems Graphically

When used with care, graphing calculators offer an accurate way to solve linear systems and to check solution(s) obtained by hand. We'll illustrate using the system from Example 3: $\begin{cases} 4x + y = 4 \\ y = x + 2 \end{cases}$, where we found the solution was $\left(\frac{2}{5}, \frac{12}{5}\right)$.

Figure 8.1

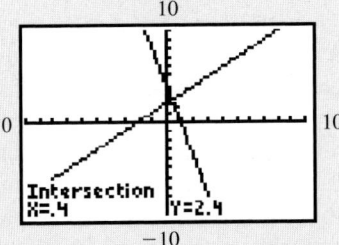

1. Solve for y in both equations:

$$\begin{cases} y = -4x + 4 \\ y = \quad x + 2 \end{cases}$$

2. Enter the equations as

$$Y_1 = -4x + 4$$
$$Y_2 = x + 2$$

3. Graph using **ZOOM** 6

$$Y_1 = -4x + 4$$
$$Y_2 = x + 2$$

4. Press **2nd** **TRACE** (CALC) 5

ENTER **ENTER** **ENTER** to have the calculator compute the point of intersection.

The coordinates of the intersection appear as decimal fractions at the bottom of the screen (Figure 8.1). In step 4, The first **ENTER** selects Y_1, the second **ENTER** selects Y_2 and the third **ENTER** bypasses the GUESS option (this option is most often used if the graphs intersect at more than one point). The calculator automatically registers the x-coordinate as its most recent entry, and from the home screen, converting it to a standard fraction (using **MATH** 1: ▶ **Frac** **ENTER**) shows $x = \frac{2}{5}$. You can also get an *approximate solution* by tracing along either line towards the point of intersection using the **TRACE** key and the left or right arrows.

Solve each system graphically, using a graphing calculator.

Exercise 1: $\begin{cases} 3x - y = -7 \\ y + 5x = -1 \end{cases}$ **Exercise 2:** $\begin{cases} 2x - 3y = 3 \\ 6 = 8x - 3y \end{cases}$

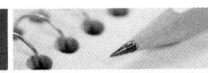

8.1 EXERCISES

▶ CONCEPTS AND VOCABULARY

Fill in the blank with the appropriate word or phrase. Carefully reread the section if needed.

1. Systems that have no solution are called _____ systems.

2. Systems having at least one solution are called _____ systems.

3. If the lines in a system intersect at a single point, the system is said to be _____ and _____.

4. If the lines in a system are coincident, the system is referred to as _____ and _____.

5. The given systems are equivalent. How do we obtain the second system from the first?

$$\begin{cases} \dfrac{2}{3}x + \dfrac{1}{2}y = \dfrac{5}{3} \\ 0.2x + 0.4y = 1 \end{cases} \quad \begin{cases} 4x + 3y = 10 \\ 2x + 4y = 10 \end{cases}$$

6. For $\begin{cases} 2x + 5y = 8 \\ 3x + 4y = 5 \end{cases}$, which solution method would be more efficient, substitution or elimination? Discuss/Explain why.

▶ DEVELOPING YOUR SKILLS

Show the lines in each system would intersect in a single point by writing the equations in slope-intercept form.

7. $\begin{cases} 7x - 4y = 24 \\ 4x + 3y = 15 \end{cases}$

8. $\begin{cases} 0.3x - 0.4y = 2 \\ 0.5x + 0.2y = -4 \end{cases}$

An ordered pair is a solution to an equation if it makes the equation true. Given the graph shown here, determine which equation(s) have the indicated point as a solution. If the point satisfies more than one equation, write the system for which it is a solution.

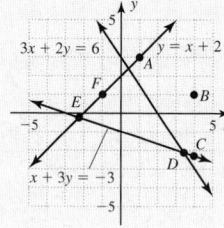

9. *A*   10. *B*

11. *C*   12. *D*

13. *E*   14. *F*

Substitute the x- and y-values indicated by the ordered pair to determine if it solves the system.

15. $\begin{cases} 3x + y = 11 \\ -5x + y = -13; \end{cases} (3, 2)$

16. $\begin{cases} 3x + 7y = -4 \\ 7x + 8y = -21; \end{cases} (-6, 2)$

17. $\begin{cases} 8x - 24y = -17 \\ 12x + 30y = 2; \end{cases} \left(-\frac{7}{8}, \frac{5}{12}\right)$

18. $\begin{cases} 4x + 15y = 7 \\ 8x + 21y = 11; \end{cases} \left(\frac{1}{2}, \frac{1}{3}\right)$

Solve each system by *graphing*. If the coordinates do not appear to be integers, estimate the solution to the nearest tenth (indicate that your solution is an estimate).

19. $\begin{cases} 3x + 2y = 12 \\ x - y = 9 \end{cases}$ 20. $\begin{cases} 5x + 2y = -2 \\ -3x + y = 10 \end{cases}$

21. $\begin{cases} 5x - 2y = 4 \\ x + 3y = -15 \end{cases}$ 22. $\begin{cases} 3x + y = 2 \\ 5x + 3y = 12 \end{cases}$

Solve each system using *substitution*. Write solutions as an ordered pair.

23. $\begin{cases} x = 5y - 9 \\ x - 2y = -6 \end{cases}$ 24. $\begin{cases} 4x - 5y = 7 \\ 2x - 5 = y \end{cases}$

25. $\begin{cases} y = \frac{2}{3}x - 7 \\ 3x - 2y = 19 \end{cases}$ 26. $\begin{cases} 2x - y = 6 \\ y = \frac{3}{4}x - 1 \end{cases}$

Identify the equation and variable that makes the substitution method easiest to use. Then solve the system.

27. $\begin{cases} 3x - 4y = 24 \\ 5x + y = 17 \end{cases}$ 28. $\begin{cases} 3x + 2y = 19 \\ x - 4y = -3 \end{cases}$

29. $\begin{cases} 0.7x + 2y = 5 \\ x - 1.4y = 11.4 \end{cases}$ 30. $\begin{cases} 0.8x + y = 7.4 \\ 0.6x + 1.5y = 9.3 \end{cases}$

31. $\begin{cases} 5x - 6y = 2 \\ x + 2y = 6 \end{cases}$ 32. $\begin{cases} 2x + 5y = 5 \\ 8x - y = 6 \end{cases}$

Solve using *elimination*. In some cases, the system must first be written in standard form.

33. $\begin{cases} 2x - 4y = 10 \\ 3x + 4y = 5 \end{cases}$ 34. $\begin{cases} -x + 5y = 8 \\ x + 2y = 6 \end{cases}$

35. $\begin{cases} 4x - 3y = 1 \\ 3y = -5x - 19 \end{cases}$ 36. $\begin{cases} 5y - 3x = -5 \\ 3x + 2y = 19 \end{cases}$

37. $\begin{cases} 2x = -3y + 17 \\ 4x - 5y = 12 \end{cases}$ 38. $\begin{cases} 2y = 5x + 2 \\ -4x = 17 - 6y \end{cases}$

39. $\begin{cases} 0.5x + 0.4y = 0.2 \\ 0.3y = 1.3 + 0.2x \end{cases}$ 40. $\begin{cases} 0.2x + 0.3y = 0.8 \\ 0.3x + 0.4y = 1.3 \end{cases}$

41. $\begin{cases} 0.32m - 0.12n = -1.44 \\ -0.24m + 0.08n = 1.04 \end{cases}$

42. $\begin{cases} 0.06g - 0.35h = -0.67 \\ -0.12g + 0.25h = 0.44 \end{cases}$

43. $\begin{cases} -\frac{1}{6}u + \frac{1}{4}v = 4 \\ \frac{1}{2}u - \frac{2}{3}v = -11 \end{cases}$ 44. $\begin{cases} \frac{3}{4}x + \frac{1}{3}y = -2 \\ \frac{3}{2}x + \frac{1}{5}y = 3 \end{cases}$

Solve using any method and identify the system as consistent, inconsistent, or dependent.

45. $\begin{cases} 4x + \frac{3}{4}y = 14 \\ -9x + \frac{5}{8}y = -13 \end{cases}$ 46. $\begin{cases} \frac{2}{3}x + y = 2 \\ 2y = \frac{5}{6}x - 9 \end{cases}$

47. $\begin{cases} 0.2y = 0.3x + 4 \\ 0.6x - 0.4y = -1 \end{cases}$ 48. $\begin{cases} 1.2x + 0.4y = 5 \\ 0.5y = -1.5x + 2 \end{cases}$

49. $\begin{cases} 6x - 22 = -y \\ 3x + \frac{1}{2}y = 11 \end{cases}$ 50. $\begin{cases} 15 - 5y = -9x \\ -3x + \frac{5}{3}y = 5 \end{cases}$

51. $\begin{cases} -10x + 35y = -5 \\ y = 0.25x \end{cases}$ 52. $\begin{cases} 2x + 3y = 4 \\ x = -2.5y \end{cases}$

53. $\begin{cases} 7a + b = -25 \\ 2a - 5b = 14 \end{cases}$ 54. $\begin{cases} -2m + 3n = -1 \\ 5m - 6n = 4 \end{cases}$

55. $\begin{cases} 4a = 2 - 3b \\ 6b + 2a = 7 \end{cases}$ **56.** $\begin{cases} 3p - 2q = 4 \\ 9p + 4q = -3 \end{cases}$ **57.** $\begin{cases} 2x + 4y = 6 \\ x + 12 = 4y \end{cases}$ **58.** $\begin{cases} 8x = 3y + 24 \\ 8x - 5y = 36 \end{cases}$

The substitution method can be used for like variables *or for like expressions.* Solve the following systems, *using the expression* common to both equations (do not solve for *x* or *y* alone).

59. $\begin{cases} 5x - 11y = 21 \\ 11y = 5 - 8x \end{cases}$ **60.** $\begin{cases} -6x = 5y - 16 \\ 5y - 6x = 4 \end{cases}$

► **WORKING WITH FORMULAS**

61. Uniform motion with current: $\begin{cases} (R + C)T_1 = D_1 \\ (R - C)T_2 = D_2 \end{cases}$

The formula shown can be used to solve uniform motion problems involving a *current,* where *D* represents distance traveled, *R* is the rate of the object with no current, *C* is the speed of the current, and *T* is the time. Chan-Li rows 9 mi up river (against the current) in 3 hr. It only took him 1 hr to row 5 mi downstream (with the current). How fast was the current? How fast can he row in still water?

62. Fahrenheit and Celsius temperatures:
$\begin{cases} y = \frac{9}{5}x + 32 & °F \\ y = \frac{5}{9}(x - 32) & °C \end{cases}$

Many people are familiar with temperature measurement in degrees Celsius and degrees Fahrenheit, but few realize that the equations are linear and there is one temperature at which the two scales agree. Solve the system using the method of your choice and find this temperature.

► **APPLICATIONS**

Solve each application by modeling the situation with a linear system. Be sure to clearly indicate what each variable represents.

Mixture

63. Theater productions: At a recent production of *A Comedy of Errors,* the Community Theater brought in a total of $30,495 in revenue. If adult tickets were $9 and children's tickets were $6.50, how many tickets of each type were sold if 3800 tickets in all were sold?

64. Milk-fat requirements: A dietician needs to mix 10 gal of milk that is $2\frac{1}{2}$% milk fat for the day's rounds. He has some milk that is 4% milk fat and some that is $1\frac{1}{2}$% milk fat. How much of each should be used?

65. Filling the family cars: Cherokee just filled both of the family vehicles at a service station. The total cost for 20 gal of regular unleaded and 17 gal of premium unleaded was $144.89. The premium gas was $0.10 more per gallon than the regular gas. Find the price per gallon for each type of gasoline.

66. Household cleaners: As a cleaning agent, a solution that is 24% vinegar is often used. How much pure (100%) vinegar and 5% vinegar must be mixed to obtain 50 oz of a 24% solution?

67. Alumni contributions: A wealthy alumnus donated $10,000 to his alma mater. The college used the funds to make a loan to a science major at 7% interest and a loan to a nursing student at 6% interest. That year the college earned $635 in interest. How much was loaned to each student?

68. Investing in bonds: A total of $12,000 is invested in two municipal bonds, one paying 10.5% and the other 12% simple interest. Last year the annual interest earned on the two investments was $1335. How much was invested at each rate?

69. Saving money: Bryan has been doing odd jobs around the house, trying to earn enough money to buy a new Dirt-Surfer©. He saves all quarters and dimes in his piggy bank, while he places all nickels and pennies in a drawer to spend. So far, he has 225 coins in the piggy bank, worth a total of $45.00. How many of the coins are quarters? How many are dimes?

70. Coin investments: In 1990, Molly attended a coin auction and purchased some rare "Flowing Hair" fifty-cent pieces, and a number of very rare two-cent pieces from the Civil War Era. If she bought 47 coins with a face value of $10.06, how many of each denomination did she buy?

71. Lawn service: Dave and his sons run a lawn service, which includes mowing, edging, trimming, and aerating a lawn. His fixed cost includes insurance, his salary, and monthly payments on equipment, and amounts to $4000/mo. The variable costs include gas, oil, hourly wages for his employees, and miscellaneous expenses, which run about $75 per lawn. The average charge for full-service lawn care is $115 per visit. Do a break-even analysis to (a) determine how many lawns Dave must service each month to break even and (b) the revenue required to break even.

72. Production of mini-microwave ovens: Due to high market demand, a manufacturer decides to introduce a new line of mini-microwave ovens for personal and office use. By using existing factory space and retraining some employees, fixed costs are estimated at $8400/mo. The components to assemble and test each microwave are expected to run $45 per unit. If market research shows consumers are willing to pay at least $69 for this product, find (a) how many units must be made and sold each month to break even and (b) the revenue required to break even.

In a market economy, the availability of goods is closely related to the market price. Suppliers are willing to produce more of the item at a higher price (the supply), with consumers willing to buy more of the item at a lower price (the demand). This is called the law of supply and demand. When supply and demand are equal, both the buyer and seller are satisfied with the current price and we have *market equilibrium*.

73. Farm commodities: One area where the law of supply and demand is clearly at work is farm commodities. Both growers and consumers watch this relationship closely, and use data collected by government agencies to track the relationship and make adjustments, as when a farmer decides to convert a large portion of her farmland from corn to soybeans to improve profits. Suppose that for x billion bushels of soybeans, supply is modeled by $y = 1.5x + 3$, where y is the current market price (in dollars per bushel). The related demand equation might be $y = -2.20x + 12$. (a) How many billion bushels will be supplied at a market price of $5.40? What will the demand be at this price? Is supply less than demand? (b) How many billion bushels will be supplied at a market price of $7.05? What will the demand be at this price? Is demand less than supply? (c) To the nearest cent, at what price does the market reach equilibrium? How many bushels are being supplied/demanded?

74. Digital music: Market research has indicated that by 2010, sales of MP3 portables will mushroom into a $70 billion dollar market. With a market this large, competition is often fierce—with suppliers fighting to earn and hold market shares. For x million MP3 players sold, supply is modeled by $y = 10.5x + 25$, where y is the current market price (in dollars). The related demand equation might be $y = -5.20x + 140$. (a) How many million MP3 players will be supplied at a market price of $88? What will the demand be at this price? Is supply less than demand? (b) How many million MP3 players will be supplied at a market price of $114? What will the demand be at this price? Is demand less than supply? (c) To the nearest cent, at what price does the market reach equilibrium? How many units are being supplied/demanded?

Uniform Motion

75. Canoeing on a stream: On a recent camping trip, it took Molly and Sharon 2 hr to row 4 mi upstream from the drop in point to the campsite. After a leisurely weekend of camping, fishing, and relaxation, they rowed back downstream to the drop in point in just 30 min. Use this information to find (a) the speed of the current and (b) the speed Sharon and Molly would be rowing in still water.

76. Taking a luxury cruise: A luxury ship is taking a Caribbean cruise from Caracas, Venezuela, to just off the coast of Belize City on the Yucatan Peninsula, a distance of 1435 mi. En route they encounter the Caribbean Current, which flows to the northwest, parallel to the coastline. From Caracas to the Belize coast, the trip took 70 hr. After a few days of fun in the sun, the ship leaves for Caracas, with the return trip taking 82 hr. Use this information to find (a) the speed of the Caribbean Current and (b) the cruising speed of the ship.

77. **Airport walkways:** As part of an algebra field trip, Jason takes his class to the airport to use their moving walkways for a demonstration. The class measures the longest walkway, which turns out to be 256 ft long. Using a stop watch, Jason shows it takes him just 32 sec to complete the walk going in the same direction as the walkway. Walking in a direction opposite the walkway, it takes him 320 sec—10 times as long! The next day in class, Jason hands out a two-question quiz: (1) What was the speed of the walkway in feet per second? (2) What is my (Jason's) normal walking speed? Create the answer key for this quiz.

78. **Racing pigeons:** The American Racing Pigeon Union often sponsors opportunities for owners to fly their birds in friendly competitions. During a recent competition, Steve's birds were liberated in Topeka, Kansas, and headed almost due north to their loft in Sioux Falls, South Dakota, a distance of 308 mi. During the flight, they encountered a steady wind from the north and the trip took 4.4 hr. The next month, Steve took his birds to a competition in Grand Forks, North Dakota, with the birds heading almost due south to home, also a distance of 308 mi. This time the birds were aided by the same wind from the north, and the trip took only 3.5 hr. Use this information to (a) find the racing speed of Steve's birds and (b) find the speed of the wind.

Descriptive Translation

79. **Important dates in U.S. history:** If you sum the year that the Declaration of Independence was signed and the year that the Civil War ended, you get 3641. There are 89 yr that separate the two events. What year was the Declaration signed? What year did the Civil War end?

80. **Architectual wonders:** When it was first constructed in 1889, the Eiffel Tower in Paris, France, was the tallest structure in the world. In 1975, the CN Tower in Toronto, Canada, became the world's tallest structure. The CN Tower is 153 ft less than twice the height of the Eiffel Tower, and the sum of their heights is 2799 ft. How tall is each tower?

81. **Pacific islands land area:** In the South Pacific, the island nations of Tahiti and Tonga have a combined land area of 692 mi^2. Tahiti's land area is 112 mi^2 more than Tonga's. What is the land area of each island group?

82. **Card games:** On a cold winter night, in the lobby of a beautiful hotel in Sante Fe, New Mexico, Marc and Klay just barely beat John and Steve in a close game of Trumps. If the sum of the team scores was 990 points, and there was a 12-point margin of victory, what was the final score?

▶ **EXTENDING THE CONCEPT**

83. Answer using observations only—no calculations. Is the given system consistent/independent, consistent/dependent, or inconsistent?
    Explain/Discuss your answer. $\begin{cases} y = 5x + 2 \\ y = 5.01x + 1.9 \end{cases}$

84. Federal income tax reform has been a hot political topic for many years. Suppose tax plan A calls for a flat tax of 20% tax on all income (no deductions or loopholes). Tax plan B requires taxpayers to pay

$5000 plus 10% of all income. For what income level do both plans require the same tax?

85. Suppose a certain amount of money was invested at 6% per year, and another amount at 8.5% per year, with a total return of $1250. If the amounts invested at each rate were switched, the yearly income would have been $1375. To the nearest whole dollar, how much was invested at each rate?

▶ **MAINTAINING YOUR SKILLS**

86. (2.6) Given the parent function $f(x) = |x|$, sketch the graph of $F(x) = -|x + 3| - 2$.

87. (5.1) Find two positive and two negative angles that are coterminal with $\theta = 112°$.

88. (4.4) Solve for x (rounded to the nearest thousandth): $33 = 77.5e^{-0.0052x} - 8.37$

89. (6.2) Verify that $\dfrac{\sin x - \csc x}{\csc x} = -\cos^2 x$ is an identity.

Learning Objectives

In Section 8.2 you will learn how to:

☐ **A.** Visualize a solution in three dimensions

☐ **B.** Check ordered triple solutions

☐ **C.** Solve linear systems in three variables

☐ **D.** Recognize inconsistent and dependent systems

☐ **E.** Use a system of three equations in three variables to solve applications

The transition to systems of three equations in three variables requires a fair amount of "visual gymnastics" along with good organizational skills. Although the techniques used are identical and similar results are obtained, the third equation and variable give us more to track, and we must work more carefully toward the solution.

A. Visualizing Solutions in Three Dimensions

The solution to an equation in one variable is the single number that satisfies the equation. For $x + 1 = 3$, the solution is $x = 2$ and its graph is a single *point* on the number line, a **one-dimensional graph.** The solution to an equation in two variables, such as $x + y = 3$, is an ordered pair (x, y) that satisfies the equation. When we graph this solution set, the result is a *line* on the xy-coordinate grid, a **two-dimensional graph.** The solutions to an equation in three variables, such as $x + y + z = 6$, are the **ordered triples** (x, y, z) that satisfy the equation. When we graph this solution set, the result is a **plane** in **space**, a *graph in three dimensions.* Recall a plane is a flat surface having infinite length and width, but no depth. We can graph this plane using the intercept method and the result is shown in Figure 8.2. For graphs in three dimensions, the xy-plane is parallel to the ground (the y-axis points to the right) and z is the **vertical axis.** To find an additional point on this plane, we use any three numbers whose sum is 6, such as $(2, 3, 1)$. Move 2 units along the x-axis, 3 units parallel to the y-axis, and 1 unit parallel to the z-axis, as shown in Figure 8.3.

WORTHY OF NOTE

We can visualize the location of a point in space by considering a large rectangular box 2 ft long × 3 ft wide × 1 ft tall, placed snugly in the corner of a room. The floor is the xy-plane, one wall is the xz-plane, and the other wall is the yz-plane. The z-axis is formed where the two walls meet and the corner of the room is the origin $(0, 0, 0)$. To find the corner of the box located at $(2, 3, 1)$, first locate the point $(2, 3)$ in the xy-plane (the floor), then move up 1 ft.

Figure 8.2

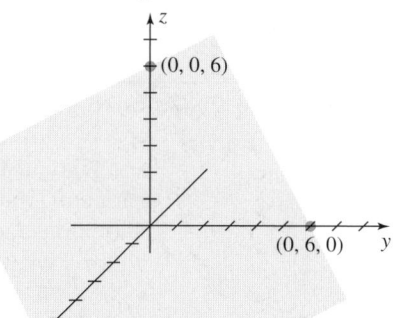

Figure 8.3

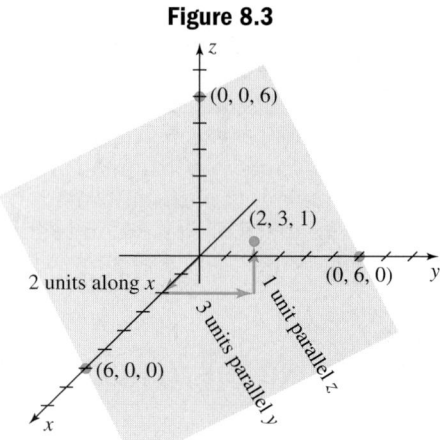

EXAMPLE 1 ▶ **Finding Solutions to an Equation in Three Variables**

Use a guess-and-check method to find four additional points on the plane determined by $x + y + z = 6$.

Solution ▶ We can begin by letting $x = 0$, then use any combination of y and z that sum to 6. Two examples are $(0, 2, 4)$ and $(0, 5, 1)$. We could also select any two values for x and y, then determine a value for z that results in a sum of 6. Two examples are $(-2, 9, -1)$ and $(8, -3, 1)$.

☑ **A.** You've just learned how to visualize a solution in three dimensions

Now try Exercises 7 through 10 ▶

B. Solutions to a System of Three Equations in Three Variables

When solving a system of three equations in three variables, remember each equation represents a plane in space. These planes can intersect in various ways, creating different possibilities for a solution set (see Figures 8.4 to 8.7). The system could have a **unique solution** (a, b, c), if the planes intersect at a single point (Figure 8.4) (the point satisfies all three equations simultaneously). If the planes intersect in a line (Figure 8.5), the system is **linearly dependent** and there are an infinite number of solutions. Unlike the two-dimensional case, the equation of a line in three dimensions is somewhat complex, and the coordinates of all points on this line are usually *represented* by a specialized ordered triple, which we use to state the solution set. If the planes intersect at all points, the system has **coincident dependence** (see Figure 8.6). This indicates the equations of the system differ by only a constant multiple—they are all "disguised forms" of the *same equation*. The solution set is any ordered triple (a, b, c) satisfying this equation. Finally, the system may have no solutions. This can happen a number of different ways, most notably if the planes intersect as shown in Figure 8.7 (other possibilities are discussed in the exercises). In the case of "no solutions," an ordered triple may satisfy none of the equations, only one of the equations, only two of the equations, but not all three equations.

Figure 8.4	Figure 8.5	Figure 8.6	Figure 8.7

Unique solution	Linear dependence	Coincident dependence	No solutions

---

EXAMPLE 2 ▶ **Determining If an Ordered Triple Is a Solution**

Determine if the ordered triple $(1, -2, 3)$ is a solution to the systems shown.

a. $\begin{cases} x + 4y - z = -10 \\ 2x + 5y + 8z = 4 \\ x - 2y - 3z = -4 \end{cases}$ **b.** $\begin{cases} 3x + 2y - z = -4 \\ 2x - 3y - 2z = 2 \\ x - y + 2z = 9 \end{cases}$

Solution ▶ Substitute 1 for x, -2 for y, and 3 for z in the first system.

a. $\begin{cases} x + 4y - z = -10 \\ 2x + 5y + 8z = 4 \\ x - 2y - 3z = -4 \end{cases} \rightarrow \begin{cases} (1) + 4(-2) - (3) = -10 \\ 2(1) + 5(-2) + 8(3) = 4 \\ (1) - 2(-2) - 3(3) = -4 \end{cases} \rightarrow \begin{cases} -10 = -10 \text{ true} \\ 16 = 4 \text{ false} \\ -4 = -4 \text{ true} \end{cases}$

No, the ordered triple $(1, -2, 3)$ is not a solution to the first system. Now use the same substitutions in the second system.

b. $\begin{cases} 3x + 2y - z = -4 \\ 2x - 3y - 2z = 2 \\ x - y + 2z = 9 \end{cases} \rightarrow \begin{cases} 3(1) + 2(-2) - (3) = -4 \\ 2(1) - 3(-2) - 2(3) = 2 \\ (1) - (-2) + 2(3) = 9 \end{cases} \rightarrow \begin{cases} -4 = -4 \text{ true} \\ 2 = 2 \text{ true} \\ 9 = 9 \text{ true} \end{cases}$

☑ **B.** You've just learned how to check ordered triple solutions

The ordered triple $(1, -2, 3)$ is a solution to the second system only.

Now try Exercises 11 and 12 ▶

C. Solving Systems of Three Equations in Three Variables Using Elimination

From Section 8.1, we know that two systems of equations are **equivalent** if they have the same solution set. The systems

$$\begin{cases} 2x + y - 2z = -7 \\ x + y + z = -1 \\ -2y - z = -3 \end{cases} \text{ and } \begin{cases} 2x + y - 2z = -7 \\ y + 4z = 5 \\ z = 1 \end{cases}$$

are equivalent, as both have the unique solution $(-3, 1, 1)$. In addition, it is evident that the second system can be solved more easily, since R2 and R3 have fewer variables than the first system. In the simpler system, mentally substituting 1 for z into R2 immediately gives $y = 1$, and these values can be back-substituted into the first equation to find that $x = -3$. This observation guides us to a general approach for solving larger systems—we would like to *eliminate variables in the second and third equations, until we obtain an equivalent system that can easily be solved by back-substitution.* To begin, let's review the three operations that "transform" a given system, and produce an equivalent system.

Operations That Produce an Equivalent System

1. Changing the order of the equations.
2. Replacing an equation by a nonzero constant multiple of that equation.
3. Replacing an equation with the sum of two equations from the system.

Building on the ideas from Section 8.1, we develop the following approach for solving a system of three equations in three variables.

Solving a System of Three Equations in Three Variables

1. Write each equation in standard form: $Ax + By + Cz = D$.
2. If the "x" term in any equation has a coefficient of 1, interchange equations (if necessary) so this equation becomes R1.
3. Use the x-term in R1 to eliminate the x-terms from R2 and R3. The original R1, with the new R2 and R3, form an equivalent system that contains a smaller "subsystem" of two equations in two variables.
4. Solve the subsystem and keep the result as the new R3. The result is an equivalent system that can be solved using back-substitution.

We'll begin by solving the system $\begin{cases} 2x + y - 2z = -7 \\ x + y + z = -1 \\ -2y - z = -3 \end{cases}$ using the elimination method and the procedure outlined. In Example 3, the notation $-2R1 + R2 \rightarrow R2$ indicates the equation in row 1 has been multiplied by -2 and added to the equation in row 2, with the result placed in the system as the new row 2.

EXAMPLE 3 ▶ Solving a System of Three Equations in Three Variables

Solve using elimination: $\begin{cases} 2x + y - 2z = -7 \\ x + y + z = -1 \\ -2y - z = -3 \end{cases}$.

Solution ▶   **1.** The system is in standard form.

2. If the x-term in any equation has a coefficient of 1, interchange equations so this equation becomes R1.

$$\begin{cases} 2x + y - 2z = -7 \\ x + y + z = -1 \\ -2y - z = -3 \end{cases} \xrightarrow{\text{R2} \leftrightarrow \text{R1}} \begin{cases} x + y + z = -1 \\ 2x + y - 2z = -7 \\ -2y - z = -3 \end{cases}$$

3. Use R1 to eliminate the x-term in R2 and R3. Since R3 has no x-term, the only elimination needed is the x-term from R2. Using $-2\text{R1} + \text{R2}$ will eliminate this term:

$$\begin{array}{rl} -2\text{R1} & -2x - 2y - 2z = 2 \\ + & \\ \text{R2} & \underline{2x + y - 2z = -7} \\ & 0x - 1y - 4z = -5 \quad \text{sum} \\ & y + 4z = 5 \quad \text{simplify} \end{array}$$

The new R2 is $y + 4z = 5$. The original R1 and R3, along with the new R2 form an equivalent system that contains a smaller **subsystem**

$$\begin{cases} x + y + z = -1 \\ 2x + y - 2z = -7 \\ -2y - z = -3 \end{cases} \xrightarrow[\text{R3} \to \text{R3}]{-2\text{R1} + \text{R2} \to \text{R2}} \begin{cases} x + y + z = -1 \quad \text{new} \\ y + 4z = 5 \quad \text{equivalent} \\ -2y - z = -3 \quad \text{system} \end{cases}$$

4. Solve the subsystem for either y or z, and keep the result as a *new* R3. We choose to eliminate y using $2\text{R2} + \text{R3}$:

$$\begin{array}{rl} 2\text{R2} & 2y + 8z = 10 \\ + & \\ \text{R3} & \underline{-2y - z = -3} \\ & 0y + 7z = 7 \quad \text{sum} \\ & z = 1 \quad \text{simplify} \end{array}$$

The new R3 is $z = 1$.

$$\begin{cases} x + y + z = -1 \\ y + 4z = 5 \\ -2y - z = -3 \end{cases} \xrightarrow{2\text{R2} + \text{R3} \to \text{R3}} \begin{cases} x + y + z = -1 \quad \text{new} \\ y + 4z = 5 \quad \text{equivalent} \\ z = 1 \quad \text{system} \end{cases}$$

The new R3, along with the original R1 and R2 from step 3, form an equivalent system that can be solved using back-substitution. Substituting 1 for z in R2 yields $y = 1$. Substituting 1 for z and 1 for y in R1 yields $x = -3$. The solution is $(-3, 1, 1)$.

Now try Exercises 13 through 18 ▶

     While not absolutely needed for the elimination process, there are two reasons for wanting the coefficient of x to be "1" in R1. First, it makes the elimination method more efficient since we can more easily see what to use as a multiplier. Second, it lays the foundation for developing other methods of solving larger systems. If no equation has an x-coefficient of 1, we simply use the y- or z-variable instead (see Example 7). Since solutions to larger systems generally are worked out in stages, we will sometimes track the transformations used by writing them *between* the original system and the equivalent system, rather than to the left as we did in Section 8.1.

     Here is an additional example illustrating the elimination process, but in *abbreviated form*. Verify the calculations indicated using a separate sheet.

EXAMPLE 4 ▶ **Solving a System of Three Equations in Three Variables**

Solve using elimination: $\begin{cases} -5y + 2x - z = -8 \\ -x + 3z + 2y = 13 \\ -z + 3y + x = 5 \end{cases}$

Solution ▶ **1.** Write the equations in standard form: $\begin{cases} 2x - 5y - z = -8 \\ -x + 2y + 3z = 13 \\ x + 3y - z = 5 \end{cases}$

2. $\begin{cases} 2x - 5y - z = -8 \\ -x + 2y + 3z = 13 \\ x + 3y - z = 5 \end{cases}$ $\xrightarrow{\text{R3} \leftrightarrow \text{R1}}$ $\begin{cases} x + 3y - z = 5 \\ -x + 2y + 3z = 13 \\ 2x - 5y - z = -8 \end{cases}$ equivalent system

3. Using R1 + R2 will eliminate the x-term from R2, yielding $5y + 2z = 18$.
Using -2R1 + R3 eliminates the x-term from R3, yielding $-11y + z = -18$.

$\begin{cases} x + 3y - z = 5 \\ -x + 2y + 3z = 13 \\ 2x - 5y - z = -8 \end{cases}$ $\xrightarrow[{-2\text{R1} + \text{R3} \to \text{R3}}]{\text{R1} + \text{R2} \to \text{R2}}$ $\begin{cases} x + 3y - z = 5 \\ 5y + 2z = 18 \\ -11y + z = -18 \end{cases}$ equivalent system

4. Using -2R3 + R2 will eliminate z from the subsystem, leaving $27y = 54$.

$\begin{cases} x + 3y - z = 5 \\ 5y + 2z = 18 \\ -11y + z = -18 \end{cases}$ $\xrightarrow{-2\text{R3} + \text{R2} \to \text{R3}}$ $\begin{cases} x + 3y - z = 5 \\ 5y + 2z = 18 \\ 27y = 54 \end{cases}$ equivalent system

☑ **C.** You've learned just how to solve linear systems in three variables

Solving for y in R3 shows $y = 2$. Substituting 2 for y in R2 yields $z = 4$, and substituting 2 for y and 4 for z in R1 shows $x = 3$. The solution is $(3, 2, 4)$.

Now try Exercises 19 through 24 ▶

D. Inconsistent and Dependent Systems

As mentioned, it is possible for larger systems to have no solutions or an infinite number of solutions. As with our work in Section 8.1, an inconsistent system (no solutions) will produce inconsistent results, ending with a statement such as $0 = -3$ or some other **contradiction.**

EXAMPLE 5 ▶ **Attempting to Solve an Inconsistent System**

Solve using elimination: $\begin{cases} 2x + y - 3z = -3 \\ 3x - 2y + 4z = 2 \\ 4x + 2y - 6z = -7 \end{cases}$

Solution ▶ **1.** This system has no equation where the coefficient of x is 1.

2. We can still use R1 to begin the solution process, but this time we'll use the variable y since it *does* have coefficient 1.

Using 2R1 + R2 eliminates the y-term from R2, leaving $7x - 2z = -4$. But using -2R1 + R3 to eliminate the y-term from R3 results in a contradiction:

$\begin{array}{ll} 2\text{R1} & 4x + 2y - 6z = -6 \\ + & \\ \underline{\text{R2}} & \underline{3x - 2y + 4z = 2} \\ & 7x - 2z = -4 \end{array}$
$\qquad$
$\begin{array}{ll} -2\text{R1} & -4x - 2y + 6z = 6 \\ + & \\ \underline{\text{R3}} & \underline{4x + 2y - 6z = -7} \\ & 0x + 0y + 0z = -1 \end{array}$

$$0 = -1 \text{ contradiction}$$

We conclude the system is inconsistent. The answer is the empty set $\varnothing$, and we need work no further.

Now try Exercises 25 and 26 ▶

Unlike our work with systems having only two variables, systems in three variables can have two forms of dependence—*linear dependence* or *coincident dependence*. To help understand linear dependence, consider a system of two equations in three variables: $\begin{cases} -2x + 3y - z = 5 \\ x - 3y + 2z = -1 \end{cases}$. Each of these equations represents a plane, and unless the planes are parallel, their intersection will be a line (see Figure 8.5). As in Section 8.1, we can state solutions to a dependent system using set notation with two of the variables written in terms of the third, or as an ordered triple using a parameter. The relationships named can then be used to generate specific solutions to the system.

Systems with two equations and two variables or three equations and three variables are called **square systems,** meaning there are exactly as many equations as there are variables. A system of linear equations cannot have a unique solution unless there are at least as many equations as there are variables in the system.

EXAMPLE 6 ▶ **Solving a Dependent System**

Solve using elimination: $\begin{cases} -2x + 3y - z = 5 \\ x - 3y + 2z = -1 \end{cases}$.

Solution ▶ Using R1 + R2 eliminates the y-term from R2, yielding $-x + z = 4$. This means (x, y, z) will satisfy both equations only when $x = z - 4$ (the x-coordinate must be 4 less than the z-coordinate). Since x is written in terms of z, we substitute $z - 4$ for x *in either equation* to find how y is related to z. Using R2 we have: $(z - 4) - 3y + 2z = -1$, which yields $y = z - 1$ (verify). This means the y-coordinate of the solution must be 1 less than z. In set notation the solution is $\{(x, y, z,) \mid x = z - 4, y = z - 1, z \in \mathbb{R}\}$. For $z = -2, 0$, and 3, the solutions would be $(-6, -3, -2)$, $(-4, -1, 0)$, and $(-1, 2, 3)$, respectively. Verify that these satisfy both equations. Using p as our parameter, the solution could be written $(p - 4, p - 1, p)$ in parameterized form.

Now try Exercises 27 through 30 ▶

The system in Example 6 was nonsquare, and we knew ahead of time the system would be dependent. The system in Example 7 *is* square, but only by applying the elimination process can we determine the nature of its solution(s).

EXAMPLE 7 ▶ **Solving a Dependent System**

Solve using elimination: $\begin{cases} 3x - 2y + z = -1 \\ 2x + y - z = 5 \\ 10x - 2y = 8 \end{cases}$.

Solution ▶ This system has no equation where the coefficient of x is 1. We will still use R1, but we'll try to eliminate z in R2 (there is no z-term in R3).

Using R1 + R2 eliminates the z-term from R2, yielding $5x - y = 4$.

$$\begin{cases} 3x - 2y + z = -1 \\ 2x + y - z = 5 \\ 10x - 2y = 8 \end{cases} \quad \begin{array}{c} \text{R1 + R2} \rightarrow \text{R2} \\ \hline \text{R3} \rightarrow \text{R3} \end{array} \quad \begin{cases} 3x - 2y + z = -1 \\ 5x - y = 4 \\ 10x - 2y = 8 \end{cases}$$

We next solve the subsystem. Using -2R2 + R3 eliminates the y-term in R3, but also all other terms:

$$\begin{array}{rl} -2\text{R2} & -10x + 2y = -8 \\ + & \\ \underline{\text{R3}} & \underline{10x - 2y = 8} \\ & 0x + 0y = 0 \quad \text{\small sum} \\ & 0 = 0 \quad \text{\small result} \end{array}$$

Since R3 is the same as 2R2, the system is linearly dependent and equivalent to $\begin{cases} 3x - 2y + z = -1 \\ 5x - y = 4 \end{cases}$. We can solve for y in R2 to write y in terms of x: $y = 5x - 4$.

Substituting $5x - 4$ for y in R1 enables us to also write z in terms of x:

$$
\begin{array}{lll}
3x - 2y \quad\quad + z = -1 & \text{R1} \\
3x - 2(5x - 4) + z = -1 & \text{substitute } 5x - 4 \text{ for } y \\
3x - 10x + 8 + z = -1 & \text{distribute} \\
\quad\quad -7x + z = -9 & \text{simplify} \\
\quad\quad\quad\quad z = 7x - 9 & \text{solve for } z
\end{array}
$$

The solution set is $\{(x, y, z) \mid x \in \mathbb{R}, y = 5x - 4, z = 7x - 9\}$. Three of the infinite number of solutions are $(0, -4, -9)$ for $x = 0$, $(2, 6, 5)$ for $x = 2$, and $(-1, -9, -16)$ for $x = -1$. Verify these triples satisfy all three equations. Again using the parameter p, the solution could be written as $(p, 5p - 4, 7p - 9)$ in parameterized form.

✓ D. You've just learned how to recognize inconsistent and dependent systems

Now try Exercises 31 through 34 ▶

Solutions to linearly dependent systems can actually be written in terms of either x, y, or z, depending on which variable is eliminated in the first step and the variable we elect to solve for afterward.

For **coincident dependence** the equations in a system differ by only a constant multiple. After applying the elimination process—all variables are eliminated from the other equations, leaving statements that are always true (such as $2 = 2$ or some other). See **Exercises 35 and 36.** For additional practice solving various kinds of systems, see **Exercises 37 to 51.**

E. Applications

Applications of larger systems are simply an extension of our work with systems of two equations in two variables. Once again, the applications come in a variety of forms and from many fields. In the world of business and finance, systems can be used to diversify investments or spread out liabilities, a financial strategy hinted at in Example 8.

EXAMPLE 8 ▶ **Modeling the Finances of a Business**

A small business borrowed $225,000 from three different lenders to expand their product line. The interest rates were 5%, 6%, and 7%. Find how much was borrowed at each rate if the annual interest came to $13,000 and twice as much was borrowed at the 5% rate than was borrowed at the 7% rate.

Solution ▶ Let x, y, and z represent the amount borrowed at 5%, 6%, and 7%, respectively. This means our first equation is $x + y + z = 225$ (in thousands). The second equation is determined by the total interest paid, which was $13,000: $0.05x + 0.06y + 0.07z = 13$. The third is found by carefully reading the problem.

"twice as much was borrowed at the 5% rate than was borrowed at the 7% rate", or $x = 2z$.

These equations form the system: $\begin{cases} x + y + z = 225 \\ 0.05x + 0.06y + 0.07z = 13 \\ x = 2z \end{cases}$. The x-term of the first equation has a coefficient of 1. Written in standard form we have:

$$\begin{cases} x + y + z = 225 & \text{R1} \\ 5x + 6y + 7z = 1300 & \text{R2} \quad \text{(multiplied by 100)} \\ x \qquad - 2z = 0 & \text{R3} \end{cases}$$

Using $-5\text{R1} + \text{R2}$ will eliminate the x term in R2, while $-\text{R1} + \text{R3}$ will eliminate the x-term in R3.

$$\begin{array}{rl} -5\text{R1} & -5x - 5y - 5z = -1125 \\ + & \\ \text{R2} & \underline{5x + 6y + 7z = 1300} \\ & y + 2z = 175 \end{array} \qquad \begin{array}{rl} -\text{R1} & -x - y - z = -225 \\ + & \\ \text{R3} & \underline{x - 2z = 0} \\ & -y - 3z = -225 \end{array}$$

The new R2 is $y + 2z = 175$, and the new R3 (after multiplying by -1) is

$y + 3z = 225$, yielding the equivalent system $\begin{cases} x + y + z = 225 \\ y + 2z = 175. \\ y + 3z = 225 \end{cases}$

 **E. You've just learned how to use a system of three equations in three variables to solve applications**

Solving the 2×2 subsystem using $-\text{R2} + \text{R3}$ yields $z = 50$. Back-substitution shows $y = 75$ and $x = 100$, yielding the solution $(100, 75, 50)$. This means $\$50,000$ was borrowed at the 7% rate, $\$75,000$ was borrowed at 6%, and $\$100,000$ at 5%.

Now try Exercises 54 through 63 ▶

TECHNOLOGY HIGHLIGHT

More on Parameterized Solutions

For linearly dependent systems, a graphing calculator can be used to both find and check possible solutions using the parameters Y_1, Y_2, and Y_3. This is done by assigning the chosen parameter to Y_1, then using Y_2 and Y_3 to form the other coordinates of the solution. We can then build the equations in the system using Y_1, Y_2, and Y_3 in place of x, y, and z. The system from Example 7 is

$$\begin{cases} 3x - 2y + z = -1 \\ 2x + y - z = 5 \\ 10x - 2y = 8 \end{cases}$$, which we found had solutions of the form $(x, 5x - 4, 7x - 9)$. We first form the

solution using $Y_1 = X$, $Y_2 = 5Y_1 - 4$ (for y), and $Y_3 = 7Y_1 - 9$ (for z). Then we form the equations in the system using $Y_4 = 3Y_1 - 2Y_2 + Y_3$, $Y_5 = 2Y_1 + Y_2 - Y_3$, and $Y_6 = 10Y_1 - 2Y_2$ (see Figure 8.8). After setting up the table (set on **AUTO**), solutions can be found by enabling only Y_1, Y_2, and Y_3, which gives values of x, y, and z, respectively (see Figure 8.9—use the right arrow ▶ to view Y_3). By enabling Y_4, Y_5, and Y_6 you can verify that for any value of the parameter, the first equation is equal to -1, the second is equal to 5, and the third is equal to 8 (see Figure 8.10—use the right arrow ▶ to view Y_6).

Figure 8.8

```
Plot1  Plot2  Plot3
\Y1■X
\Y2■5Y1-4
\Y3■7Y1-9
\Y4=3Y1-2Y2+Y3
\Y5=2Y1+Y2-Y3
\Y6=10Y1-2Y2
\Y7=
```

Figure 8.9

X	Y1	Y2
-3	-3	-19
-2	-2	-14
-1	-1	-9
0	0	-4
1	1	1
2	2	6
3	3	11

X= -3

Figure 8.10

X	Y4	Y5
-3	-1	5
-2	-1	5
-1	-1	5
0	-1	5
1	-1	5
2	-1	5
3	-1	5

X= -3

Exercise 1:   Use the ideas from this Technology Highlight to (a) find four specific solutions to Example 6, (b) check multiple variations of the solution given, and (c) determine if $(-9, -6, -5)$, $(-2, 1, 2)$, and $(6, 2, 4)$ are solutions.

8.2 EXERCISES

▶ CONCEPTS AND VOCABULARY

Fill in the blank with the appropriate word or phrase. Carefully reread the section if needed.

1. The solution to an equation in three variables is an ordered _____.

2. The graph of the solutions to an equation in three variables is a(n) _____.

3. Systems that have the same solution set are called _____ _____.

4. If a 3×3 system is linearly dependent, the ordered triple solutions can be written in terms of a single variable called a(n) _____.

5. Find a value of z that makes the ordered triple $(2, -5, z)$ a solution to $2x + y + z = 4$. Discuss/Explain how this is accomplished.

6. Explain the difference between linear dependence and coincident dependence, and describe how the equations are related.

▶ DEVELOPING YOUR SKILLS

Find any four ordered triples that satisfy the equation given.

7. $x + 2y + z = 9$

8. $3x + y - z = 8$

9. $-x + y + 2z = -6$

10. $2x - y + 3z = -12$

Determine if the given ordered triples are solutions to the system.

11. $\begin{cases} x + y - 2z = -1 \\ 4x - y + 3z = 3 \\ 3x + 2y - z = 4 \end{cases}$; $(0, 3, 2)$ $(-3, 4, 1)$

12. $\begin{cases} 2x + 3y + z = 9 \\ 5x - 2y - z = -32; \\ x - y - 2z = -13 \end{cases}$ $(-4, 5, 2)$ $(5, -4, 11)$

Solve each system using elimination and back-substitution.

13. $\begin{cases} x - y - 2z = -10 \\ x - z = 1 \\ z = 4 \end{cases}$

14. $\begin{cases} x + y + 2z = -1 \\ 4x - y = 3 \\ 3x = 6 \end{cases}$

15. $\begin{cases} x + 3y + 2z = 16 \\ -2y + 3z = 1 \\ 8y - 13z = -7 \end{cases}$

16. $\begin{cases} -x + y + 5z = 1 \\ 4x + y = 1 \\ -3x - 2y = 8 \end{cases}$

17. $\begin{cases} 2x - y + 4z = -7 \\ x + 2y - 5z = 13 \\ y - 4z = 9 \end{cases}$

18. $\begin{cases} 2x + 3y + 4z = -18 \\ x - 2y + z = 4 \\ 4x + z = -19 \end{cases}$

19. $\begin{cases} -x + y + 2z = -10 \\ x + y - z = 7 \\ 2x + y + z = 5 \end{cases}$

20. $\begin{cases} x + y - 2z = -1 \\ 4x - y + 3z = 3 \\ 3x + 2y - z = 4 \end{cases}$

21. $\begin{cases} 3x + y - 2z = 3 \\ x - 2y + 3z = 10 \\ 4x - 8y + 5z = 5 \end{cases}$

22. $\begin{cases} 2x - 3y + 2z = 0 \\ 3x - 4y + z = -20 \\ x + 2y - z = 16 \end{cases}$

23. $\begin{cases} 3x - y + z = 6 \\ 2x + 2y - z = 5 \\ 2x - y + z = 5 \end{cases}$

24. $\begin{cases} 2x - 3y - 2z = 7 \\ x - y + 2z = -5 \\ 2x - 2y + 3z = -7 \end{cases}$

Solve using the elimination method. If a system is inconsistent or dependent, so state. For systems with linear dependence, write solutions in set notation and as an ordered triple in terms of a parameter.

25. $\begin{cases} 3x + y + 2z = 3 \\ x - 2y + 3z = 1 \\ 4x - 8y + 12z = 7 \end{cases}$

26. $\begin{cases} 2x - y + 3z = 8 \\ 3x - 4y + z = 4 \\ -4x + 2y - 6z = 5 \end{cases}$

27. $\begin{cases} 4x + y + 3z = 8 \\ x - 2y + 3z = 2 \end{cases}$

28. $\begin{cases} 4x - y + 2z = 9 \\ 3x + y + 5z = 5 \end{cases}$

29. $\begin{cases} 6x - 3y + 7z = 2 \\ 3x - 4y + z = 6 \end{cases}$

30. $\begin{cases} 2x - 4y + 5z = -2 \\ 3x - 2y + 3z = 7 \end{cases}$

Solve using elimination. If the system is linearly dependent, state the general solution in terms of a parameter. Different forms of the solution are possible.

31. $\begin{cases} 3x - 4y + 5z = 5 \\ -x + 2y - 3z = -3 \\ 3x - 2y + z = 1 \end{cases}$

32. $\begin{cases} 5x - 3y + 2z = 4 \\ -9x + 5y - 4z = -12 \\ -3x + y - 2z = -12 \end{cases}$

33. $\begin{cases} x + 2y - 3z = 1 \\ 3x + 5y - 8z = 7 \\ x + y - 2z = 5 \end{cases}$

34. $\begin{cases} -2x + 3y - 5z = 3 \\ 5x - 7y + 12z = -8 \\ x - y + 2z = -2 \end{cases}$

Solve using elimination. If the system has coincident dependence, state the solution in set notation.

35. $\begin{cases} -0.2x + 1.2y - 2.4z = -1 \\ 0.5x - 3y + 6z = 2.5 \\ x - 6y + 12z = 5 \end{cases}$

36. $\begin{cases} 6x - 3y + 9z = 21 \\ 4x - 2y + 6z = 14 \\ -2x + y - 3z = -7 \end{cases}$

Solve using the elimination method. If a system is inconsistent or dependent, so state. For systems with linear dependence, write the answer in terms of a parameter. For coincident dependence, state the solution in set notation.

37. $\begin{cases} x + 2y - z = 1 \\ x + z = 3 \\ 2x - y + z = 3 \end{cases}$
38. $\begin{cases} 3x + 5y - z = 11 \\ 2x + y - 3z = 12 \\ y + 2z = -4 \end{cases}$

39. $\begin{cases} 2x - 5y - 4z = 6 \\ x - 2.5y - 2z = 3 \\ -3x + 7.5y + 6z = -9 \end{cases}$

40. $\begin{cases} x - 2y + 2z = 6 \\ 2x - 6y + 3z = 13 \\ 3x + 4y - z = -11 \end{cases}$

41. $\begin{cases} 4x - 5y - 6z = 5 \\ 2x - 3y + 3z = 0 \\ x + 2y - 3z = 5 \end{cases}$

42. $\begin{cases} x - 5y - 4z = 3 \\ 2x - 9y - 7z = 2 \\ 3x - 14y - 11z = 5 \end{cases}$

43. $\begin{cases} 2x + 3y - 5z = 4 \\ x + y - 2z = 3 \\ x + 3y - 4z = -1 \end{cases}$

44. $\begin{cases} \dfrac{1}{6}x + \dfrac{1}{3}y - \dfrac{1}{2}z = 2 \\ \dfrac{3}{4}x - \dfrac{1}{3}y + \dfrac{1}{2}z = 9 \\ \dfrac{1}{2}x - y + \dfrac{1}{2}z = 2 \end{cases}$
45. $\begin{cases} \dfrac{x}{2} + \dfrac{y}{3} - \dfrac{z}{2} = 2 \\ \dfrac{2x}{3} - y - z = 8 \\ \dfrac{x}{6} + 2y + \dfrac{3z}{2} = 6 \end{cases}$

Some applications of systems lead to systems similar to those that follow. Solve using elimination.

46. $\begin{cases} -2A - B - 3C = 21 \\ B - C = 1 \\ A + B = -4 \end{cases}$

47. $\begin{cases} -A + 3B + 2C = 11 \\ 2B + C = 9 \\ B + 2C = 8 \end{cases}$

48. $\begin{cases} A + 2C = 7 \\ 2A - 3B = 8 \\ 3A + 6B - 8C = -33 \end{cases}$

49. $\begin{cases} A - 2B = 5 \\ B + 3C = 7 \\ 2A - B - C = 1 \end{cases}$

50. $\begin{cases} C = -2 \\ 5A - 2C = 5 \\ -4B - 9C = 16 \end{cases}$

51. $\begin{cases} C = 3 \\ 2A + 3C = 10 \\ 3B - 4C = -11 \end{cases}$

▶ **WORKING WITH FORMULAS**

52. **Dimensions of a rectangular solid:**
$\begin{cases} 2w + 2h = P_1 \\ 2l + 2w = P_2 \\ 2l + 2h = P_3 \end{cases}$

$P_2 = 16$ cm (top)
$P_1 = 14$ cm (small side)
$P_3 = 18$ cm (large side)
h, w, l

Using the formula shown, the dimensions of a rectangular solid can be found if the perimeters of the three distinct faces are known. Find the dimensions of the solid shown.

53. **Distance from a point (x, y, z) to the plane**
$$Ax + By + Cz = D: \quad \left| \frac{Ax + By + Cz - D}{\sqrt{A^2 + B^2 + C^2}} \right|$$

The perpendicular distance from a given point (x, y, z) to the plane defined by $Ax + By + Cz = D$ is given by the formula shown. Consider the plane given in Figure 8.2 ($x + y + z = 6$). What is the distance from this plane to the point $(3, 4, 5)$?

▶ APPLICATIONS

Solve the following applications by setting up and solving a system of three equations in three variables. Note that some equations may have only two of the three variables used to create the system.

Investment/Finance and Simple Interest Problems

54. Investing the winnings: After winning $280,000 in the lottery, Maurika decided to place the money in three different investments: a certificate of deposit paying 4%, a money market certificate paying 5%, and some Aa bonds paying 7%. After 1 yr she earned $15,400 in interest. Find how much was invested at each rate if $20,000 more was invested at 7% than at 5%.

55. Purchase at auction: At an auction, a wealthy collector paid $7,000,000 for three paintings: a Monet, a Picasso, and a van Gogh. The Monet cost $800,000 more than the Picasso. The price of the van Gogh was $200,000 more than twice the price of the Monet. What was the price of each painting?

Descriptive Translation

56. Major wars: The United States has fought three major wars in modern times: World War II, the Korean War, and the Vietnam War. If you sum the years that each conflict ended, the result is 5871. The Vietnam War ended 20 years after the Korean War and 28 years after World War II. In what year did each end?

57. Animal gestation periods: The average gestation period (in days) of an elephant, rhinoceros, and camel sum to 1520 days. The gestation period of a rhino is 58 days longer than that of a camel. Twice the camel's gestation period decreased by 162 gives the gestation period of an elephant. What is the gestation period of each?

58. Moments in U.S. history: If you sum the year the Declaration of Independence was signed, the year the 13th Amendment to the Constitution abolished slavery, and the year the Civil Rights Act was signed, the total would be 5605. Ninety-nine years separate the 13th Amendment and the Civil Rights Act. The Civil Rights Act was signed 188 years after the Declaration of Independence. What year was each signed?

59. Aviary wingspan: If you combine the wingspan of the California Condor, the Wandering Albatross (see photo), and the prehistoric Quetzalcoatlus, you get an astonishing 18.6 m (over 60 ft). If the wingspan of the Quetzalcoatlus is equal to five times that of the Wandering Albatross minus twice that of the California Condor, and six times the wingspan of the Condor is equal to five times the wingspan of the Albatross, what is the wingspan of each?

Mixtures

60. Chemical mixtures: A chemist mixes three different solutions with concentrations of 20%, 30%, and 45% glucose to obtain 10 L of a 38% glucose solution. If the amount of 30% solution used is 1 L more than twice the amount of 20% solution used, find the amount of each solution used.

61. Value of gold coins: As part of a promotion, a local bank invites its customers to view a large sack full of $5, $10, and $20 gold pieces, promising to give the sack to the first person able to state the number of coins for each denomination. Customers are told there are exactly 250 coins, with a total face value of $1875. If there are also seven times as many $5 gold pieces as $20 gold pieces, how many of each denomination are there?

62. Rewriting a rational function: It can be shown that the rational function $V(x) = \dfrac{3x + 11}{x^3 - 3x^2 + x - 3}$ can be written as a sum of the terms $\dfrac{A}{x - 3} + \dfrac{Bx + C}{x^2 + 1}$, where the coefficients A, B, and C are solutions to $\begin{cases} A + B & = 0 \\ -3B + C = 3 \\ A \qquad - 3C = 11 \end{cases}$. Find the missing coefficients and verify your answer by adding the terms.

63. Rewriting a rational function: It can be shown that the rational function $V(x) = \dfrac{x - 9}{x^3 - 6x^2 + 9x}$ can be written as a sum of the terms $\dfrac{A}{x} + \dfrac{B}{x - 3} + \dfrac{C}{(x - 3)^2}$, where the coefficients A, B, and C are solutions to $\begin{cases} A + B & = 0 \\ -6A - 3B + C = 1 \\ 9A & = -9 \end{cases}$. Find the missing coefficients and verify your answer by adding the terms.

▶ **EXTENDING THE CONCEPT**

64. The system $\begin{cases} x - 2y - z = 2 \\ x - 2y + kz = 5 \\ 2x - 4y + 4z = 10 \end{cases}$ is inconsistent if
$k =$ _____, and dependent if $k =$ ____.

a. 9 cm   b. 10 cm   c. 11 cm
d. 12 cm   e. 13 cm

65. One form of the equation of a circle is
$x^2 + y^2 + Dx + Ey + F = 0$. Use a system to
find the equation of the circle through the points
$(2, -1)$, $(4, -3)$, and $(2, -5)$.

66. The lengths of each side of the squares A, B, C, D,
E, F, G, H, and I (the smallest square) shown are
whole numbers. Square B has sides of 15 cm and
square G has sides of 7 cm. What are the
dimensions of square D?

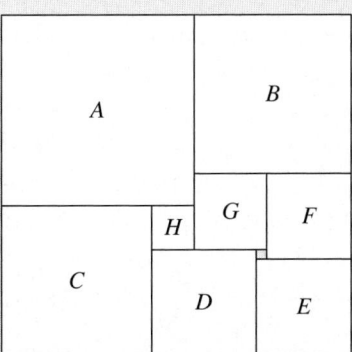

▶ **MAINTAINING YOUR SKILLS**

67. (7.3) Given $\mathbf{u} = \langle 1, -7 \rangle$ and $\mathbf{v} = \langle -3, \frac{1}{2} \rangle$, compute
$\mathbf{u} + 4\mathbf{v}$ and $3\mathbf{u} - \mathbf{v}$.

68. (5.2) Given $\cot A = 1.6831$, use a calculator to find
the acute angle A to the nearest tenth of a degree.

69. (4.4) Solve the logarithmic equation:
$\log(x + 2) + \log x = \log 3$

70. (2.5) Analyze the graph of g shown. Clearly state
the domain and range, the zeroes of g, intervals

where $g(x) > 0$, intervals
where $g(x) < 0$, local
maximums or minimums,
and intervals where the
function is increasing or
decreasing. Assume each
tick mark is one unit and
estimate endpoints to the
nearest tenths.

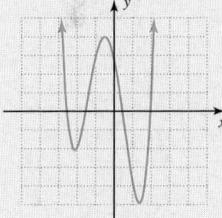

8.3 | Partial Fraction Decomposition

Learning Objectives

In Section 8.3 you will learn how to:

☐ **A.** Set up a decomposition template to help rewrite a rational expression as the sum of its partial fractions

☐ **B.** Decompose a rational expression using convenient values

☐ **C.** Decompose a rational expression by equating coefficients and using a system of equations

☐ **D.** Apply partial fraction decomposition to a telescoping sum

One application of linear systems often seen in higher mathematics involves rewriting a rational expression as a sum of its **partial fractions.** While most often used as a prelude to the application of other mathematical techniques, practical applications of the process range from a study of chemical reactions to the analysis of thermodynamic experiments.

A. Setting Up a Decomposition Template

Recall that a rational expression is one of the form $\dfrac{P(x)}{Q(x)}$, where P and Q are polynomials and $Q(x) \neq 0$. The addition of rational expressions is widely taught in courses prior to college algebra, and involves combining two rational expressions into a single term using a common denominator. For the *decomposition* of rational expressions, we seek to reverse this process. To begin, we make the following observations:

1. Consider the sum $\dfrac{7}{x + 2} + \dfrac{5}{x - 3}$, noting both terms are proper fractions (the degree of the numerator is less than the degree of the denominator) and have distinct linear denominators.

$$\frac{7}{x+2} + \frac{5}{x-3} = \frac{7(x-3)}{(x+2)(x-3)} + \frac{5(x+2)}{(x-3)(x+2)} \quad \text{common denominators}$$

$$= \frac{7(x-3) + 5(x+2)}{(x+2)(x-3)} \quad \text{combine numerators}$$

$$= \frac{12x - 11}{(x+2)(x-3)} \quad \text{result}$$

Assuming we didn't have the original sum to look at, reversing the process would require us to begin with the template

$$\frac{12x - 11}{(x+2)(x-3)} = \frac{A}{x+2} + \frac{B}{x-3}$$

and solve for the *constants A* and *B*. We know the numerators must be constant, otherwise the fraction(s) would be improper while the original expression is not.

2. Consider the sum $\dfrac{3}{x-1} + \dfrac{5}{x^2 - 2x + 1}$, again noting both terms are proper fractions.

$$\frac{3}{x-1} + \frac{5}{x^2 - 2x + 1} = \frac{3}{x-1} + \frac{5}{(x-1)(x-1)} \quad \text{factor denominators}$$

$$= \frac{3(x-1)}{(x-1)(x-1)} + \frac{5}{(x-1)(x-1)} \quad \text{common denominators}$$

$$= \frac{(3x-3) + 5}{(x-1)(x-1)} \quad \text{combine numerators}$$

$$= \frac{3x + 2}{(x-1)^2} \quad \text{result}$$

Note that while the new denominator is the repeated factor $(x-1)^2$, *both* $(x-1)$ and $(x-1)^2$ were denominators in the original sum. Assuming we didn't know the original sum, reversing the process would require us to begin with the template

$$\frac{3x + 2}{(x-1)^2} = \frac{A}{x-1} + \frac{B}{(x-1)^2}$$

and solve for the constants *A* and *B*. As before, the numerator of the first term must be constant. While the second term would still be a proper fraction if the numerator were linear (degree 1), the denominator is a *repeated* linear factor and using a single constant in the numerator of *all such fractions* will ensure we obtain unique values for *A* and *B* **(see Exercise 58).** Note that for any repeated linear factor $(x + a)^n$ in the original denominator, terms of the form

$$\frac{A_1}{x+a} + \frac{A_2}{(x+a)^2} + \frac{A_3}{(x+a)^3} + \cdots + \frac{A_{n-1}}{(x+a)^{n-1}} + \frac{A_n}{(x+a)^n} \text{ must appear in}$$

the decomposition template, although some of these numerators may turn out to be zero.

EXAMPLE 1 ▶ **Setting Up the Decomposition Template for Linear Factors**

Write the decomposition template for

a. $\dfrac{x - 8}{2x^2 + 5x + 3}$　　　　**b.** $\dfrac{x + 1}{x^2 - 6x + 9}$

Solution ▶   **a.** Factoring the denominator gives $\dfrac{x-8}{(2x+3)(x+1)}$. Since the denominator consists of two distinct linear factors, the decomposition template is

$$\frac{x-8}{(2x+3)(x+1)} = \frac{A}{2x+3} + \frac{B}{x+1} \qquad \text{decomposition template}$$

     b. After factoring the denominator we have $\dfrac{x+1}{(x-3)^2}$, where the denominator consists of a repeated linear factor. Using our previous observations, the decomposition template would be

$$\frac{x+1}{(x-3)^2} = \frac{A}{x-3} + \frac{B}{(x-3)^2} \qquad \text{decomposition template}$$

Now try Exercises 7 through 16 ▶

When both distinct and repeated linear factors are present in the denominator, the decomposition template maintains the elements illustrated above. Each distinct linear factor appearing in the denominator, and all powers of a repeated linear factor, will have a constant numerator.

EXAMPLE 2 ▶   **Setting Up the Decomposition Template for Repeated Linear Factors**

Write the decomposition template for $\dfrac{x^2-4x-15}{x^3-2x^2+x}$.

Solution ▶   Factoring the denominator gives $\dfrac{x^2-4x-15}{x(x^2-2x+1)}$ or $\dfrac{x^2-4x-15}{x(x-1)^2}$ after factoring completely. With a distinct linear factor of x, and the repeated linear factor $(x-1)^2$, the decomposition template becomes

$$\frac{x^2-4x-15}{x(x-1)^2} = \frac{A}{x} + \frac{B}{x-1} + \frac{C}{(x-1)^2} \qquad \text{decomposition template}$$

Now try Exercises 17 through 20 ▶

Returning to our observations:

3. Consider the sum $\dfrac{4}{x} + \dfrac{2x+3}{x^2+1}$, noting the denominator of the first term is linear, while the denominator of the second is an irreducible quadratic.

$$\frac{4}{x} + \frac{2x+3}{x^2+1} = \frac{4(x^2+1)}{x(x^2+1)} + \frac{(2x+3)x}{(x^2+1)x} \qquad \text{find common denominator}$$

$$= \frac{(4x^2+4)+(2x^2+3x)}{x(x^2+1)} \qquad \text{combine numerators}$$

$$= \frac{6x^2+3x+4}{x(x^2+1)} \qquad \text{result}$$

Here, reversing the process would require us to begin with the template

$$\frac{6x^2 + 3x + 4}{x(x^2 + 1)} = \frac{A}{x} + \frac{Bx + C}{x^2 + 1}$$

allowing that the numerator of the second term might be linear since it is quadratic but *not a repeated linear factor,* and noting the fraction would still be proper in cases where $B = 0$.

4. Finally, consider the sum $\dfrac{1}{x^2 + 3} + \dfrac{x - 2}{(x^2 + 3)^2}$, where we note the denominator of the first term is an irreducible quadratic, with the denominator of the second term being *the same factor* with multiplicity two.

$$
\begin{aligned}
\frac{1}{x^2 + 3} + \frac{x - 2}{(x^2 + 3)^2} &= \frac{1(x^2 + 3)}{(x^2 + 3)(x^2 + 3)} + \frac{x - 2}{(x^2 + 3)(x^2 + 3)} && \text{common denominators} \\
&= \frac{(x^2 + 3) + (x - 2)}{(x^2 + 3)(x^2 + 3)} && \text{combine numerators} \\
&= \frac{x^2 + x + 1}{(x^2 + 3)^2} && \text{result after simplifying}
\end{aligned}
$$

Reversing the process would require us to begin with the template

$$\frac{x^2 + x + 1}{(x^2 + 3)^2} = \frac{Ax + B}{x^2 + 3} + \frac{Cx + D}{(x^2 + 3)^2}$$

allowing that the numerator of either term might be nonconstant for the reasons in observation 3. Similar to our reasoning in observation 2, all powers of a repeated quadratic factor must be present in the decomposition template.

WORTHY OF NOTE

Note that the second term in the decomposition template would still be a proper fraction if the numerator were quadratic or cubic, but since the denominator is a *repeated* quadratic factor, using only a linear form ensures we obtain unique values for all coefficients.

EXAMPLE 3 ▶ **Setting Up the Decomposition Template for Quadratic Factors**

Write the decomposition template for

a. $\dfrac{x^2 + 10x + 1}{(x + 1)(x^2 + 3x + 4)}$ **b.** $\dfrac{x^2}{(x^2 + 2)^3}$

Solution ▶ **a.** With the denominator having one distinct linear factor and one irreducible quadratic factor, the decomposition template would be

$$\frac{x^2 + 10x + 1}{(x + 1)(x^2 + 3x + 4)} = \frac{A}{x + 1} + \frac{Bx + C}{x^2 + 3x + 4} \qquad \text{decomposition template}$$

b. The denominator consists of a repeated quadratic factor. Using our previous observations, the decomposition template would be

$$\frac{x^2}{(x^2 + 2)^3} = \frac{Ax + B}{x^2 + 2} + \frac{Cx + D}{(x^2 + 2)^2} + \frac{Ex + F}{(x^2 + 2)^3} \qquad \text{decomposition template}$$

Now try Exercises 21 through 24 ▶

When both distinct and repeated factors are present in the denominator, the decomposition template maintains the essential elements determined by observations 1 through 4. Using these observations, we can formulate a general approach to the decomposition template.

The Decomposition Template

For the rational expression $\dfrac{P(x)}{Q(x)}$ and constants $A, B, C, D, \ldots$,

1. If the degree of P is greater than or equal to the degree of Q, find the quotient and remainder using polynomial division. Only the remainder portion need be decomposed into partial fractions.
2. Factor Q completely into linear factors and irreducible quadratic factors.
3. For the linear factors, each distinct linear factor and each power of a repeated linear factor must appear in the decomposition template and have a constant numerator.
4. For the irreducible quadratic factors, each distinct quadratic factor and each power of a repeated quadratic factor must appear in the decomposition template and have a linear numerator.

☑ **A.** You've just learned how to set up a decomposition template

B. Decomposition Using Convenient Values

Once the decomposition template is obtained, we multiply both sides of the equation by the factored form of the original denominator and simplify. The resulting equation is an identity (a true statement for all real numbers x), and in many cases, all that's required is a choice of **convenient values** for x to identify the constants A, B, C, and so on.

EXAMPLE 4 ▶ **Decomposing a Rational Expression Using Convenient Values**

Decompose the expression $\dfrac{4x + 11}{x^2 + 7x + 10}$ into partial fractions.

Solution ▶ Factoring the denominator gives $\dfrac{4x + 11}{(x + 5)(x + 2)}$, and we note there are two distinct linear factors in the denominator. The decomposition template will be

$$\frac{4x + 11}{(x + 5)(x + 2)} = \frac{A}{x + 5} + \frac{B}{x + 2} \qquad \text{decomposition template}$$

Multiplying both sides by $(x + 5)(x + 2)$ clears all denominators and yields

$$4x + 11 = A(x + 2) + B(x + 5) \qquad \text{clear denominators}$$

Since the equation must be true for all x, using $x = -5$ will *conveniently* eliminate the term with B, and enable us to solve for A directly:

$$
\begin{aligned}
4(-5) + 11 &= A(-5 + 2) + B(-5 + 5) & \text{substitute } -5 \text{ for } x \\
-20 + 11 &= -3A + B(0) & \text{simplify} \\
-9 &= -3A & \text{term with } B \text{ is eliminated} \\
3 &= A & \text{solve for } A
\end{aligned}
$$

To find B, we repeat this procedure, using a value that *conveniently* eliminates the term with A, namely $x = -2$.

$$
\begin{aligned}
4x + 11 &= A(x + 2) + B(x + 5) & \text{original equation} \\
4(-2) + 11 &= A(-2 + 2) + B(-2 + 5) & \text{substitute } -2 \text{ for } x \\
-8 + 11 &= A(0) + 3B & \text{simplify} \\
3 &= 3B & \text{term with } A \text{ is eliminated} \\
1 &= B & \text{solve for } B
\end{aligned}
$$

With $A = 3$ and $B = 1$, the complete decomposition is

$$\frac{4x + 11}{(x + 5)(x + 2)} = \frac{3}{x + 5} + \frac{1}{x + 2}$$

which can be checked by adding the fractions on the right.

Now try Exercises 25 through 28 ▶

EXAMPLE 5 ▶ **Decomposing a Rational Expression Using Convenient Values**

Decompose the expression $\dfrac{9}{(x + 5)(x^2 + 7x + 10)}$ into partial fractions.

Solution ▶ Factoring the denominator gives $\dfrac{9}{(x + 5)(x + 2)(x + 5)}$ or $\dfrac{9}{(x + 2)(x + 5)^2}$ in simplified form. With one distinct linear factor and a repeated linear factor, the decomposition template will be $\dfrac{9}{(x + 2)(x + 5)^2} = \dfrac{A}{x + 2} + \dfrac{B}{x + 5} + \dfrac{C}{(x + 5)^2}$.

Multiplying both sides by $(x + 2)(x + 5)^2$ clears all denominators and yields

$$9 = A(x + 5)^2 + B(x + 2)(x + 5) + C(x + 2)$$

Using $x = -5$ will conveniently eliminate the terms with A and B, enabling us to solve for C directly:

$9 = A(-5 + 5)^2 + B(-5 + 2)(-5 + 5) + C(-5 + 2)$	substitute -5 for x
$9 = A(0) + B(-3)(0) - 3C$	simplify
$9 = -3C$	terms with A and B are eliminated
$-3 = C$	solve for C

Using $x = -2$ will conveniently eliminate the terms with B and C, enabling us to solve for A:

$9 = A(x + 5)^2 + B(x + 2)(x + 5) + C(x + 2)$	original equation
$9 = A(-2 + 5)^2 + B(-2 + 2)(-2 + 5) + C(-2 + 2)$	substitute -2 for x
$9 = A(3)^2 + B(0)(3) + C(0)$	simplify
$9 = 9A$	terms with B and C are eliminated
$1 = A$	solve for A

To find the value of B, we can substitute $A = 1$ and $C = -3$ into the previous equation, *and any value of x that does not eliminate B*. For efficiency's sake, we often elect to use $x = 0$ or $x = 1$ for this purpose (if possible).

$9 = A(x + 5)^2 + B(x + 2)(x + 5) + C(x + 2)$	original equation
$9 = 1(0 + 5)^2 + B(0 + 2)(0 + 5) - 3(0 + 2)$	substitute 1 for A, -3 for C, 0 for x
$9 = 25 + 10B - 6$	simplify
$-10 = 10B$	
$-1 = B$	solve for B

With $A = 1$, $B = -1$, and $C = -3$, the complete decomposition is

$$\frac{9}{(x + 2)(x + 5)^2} = \frac{1}{x + 2} + \frac{-1}{x + 5} + \frac{-3}{(x + 5)^2}, \text{ which can also be written as}$$

$$\frac{1}{x + 2} - \frac{1}{x + 5} - \frac{3}{(x + 5)^2}.$$

Now try Exercises 29 and 30 ▶

EXAMPLE 6 ▶ **Decomposing a Rational Expression Using Convenient Values**

Decompose the expression $\dfrac{3x + 11}{x^3 - 3x^2 + x - 3}$ into partial fractions.

Solution ▶ After inspection, we note the denominator can be factored by grouping, resulting in the expression $\dfrac{3x + 11}{(x - 3)(x^2 + 1)}$. With one distinct linear factor and one irreducible quadratic factor, the decomposition template will be

$$\frac{3x + 11}{(x - 3)(x^2 + 1)} = \frac{A}{x - 3} + \frac{Bx + C}{x^2 + 1}.$$

Multiplying both sides by $(x - 3)(x^2 + 1)$ yields

$$3x + 11 = A(x^2 + 1) + (Bx + C)(x - 3) \quad \text{clear denominators}$$

Using $x = 3$ will conveniently eliminate the term with B and C, giving

$$\begin{aligned} 3(3) + 11 &= A(3^2 + 1) + (B[3] + C)(3 - 3) && \text{substitute 3 for } x \\ 20 &= 10A + (B[3] + C)(0) && \text{simplify} \\ 20 &= 10A && \text{term with } B \text{ and } C \text{ is eliminated} \\ 2 &= A && \text{solve for } A \end{aligned}$$

Noting $x = 0$ will conveniently eliminate the term with B; we substitute 0 for x and 2 for A in order to solve for C:

$$\begin{aligned} 3x + 11 &= A(x^2 + 1) + (Bx + C)(x - 3) && \text{original equation} \\ 3(0) + 11 &= 2(0^2 + 1) + (B[0] + C)(0 - 3) && \text{substitute 2 for } A \text{ and 0 for } x \\ 11 &= 2 + C(-3) && \text{simplify} \\ 9 &= -3C && \text{solve for } C \\ -3 &= C && \text{result} \end{aligned}$$

To find the value of B, we substitute 2 for A, -3 for C, and any value of x that does not eliminate B. Here we chose $x = 1$.

$$\begin{aligned} 3x + 11 &= A(x^2 + 1) + (Bx + C)(x - 3) && \text{original equation} \\ 3(1) + 11 &= 2(1^2 + 1) + (B[1] + [-3])(1 - 3) && \text{substitute 2 for } A, -3 \text{ for } C, \text{ and 1 for } x \\ 14 &= 2(2) + (B - 3)(-2) && \text{simplify} \\ 14 &= 4 - 2B + 6 && \text{distribute} \\ 4 &= -2B && \text{solve for } B \\ -2 &= B && \text{result} \end{aligned}$$

With $A = 2$, $B = -2$, and $C = -3$, the complete decomposition is

$$\frac{3x + 11}{(x - 3)(x^2 + 1)} = \frac{2}{x - 3} + \frac{-2x - 3}{x^2 + 1}.$$

☑ **B.** You've just learned how to decompose a rational expression using convenient values

The result can be actually be written with fewer negative signs, as $\dfrac{2}{x - 3} - \dfrac{2x + 3}{x^2 + 1}$. Check the result by combining these fractions.

Now try Exercises 31 through 38 ▶

C. Decomposition Using a System of Equations

As an alternative to using convenient values, a system of equations can be set up by multiplying out the right-hand side (after clearing fractions) and equating coefficients of the terms with like degrees. Here we'll re-solve Example 6 using this method.

EXAMPLE 7 ▶ **Decomposing a Rational Expression Using a System of Equations**

Decompose the expression $\dfrac{3x + 11}{x^3 - 3x^2 + x - 3}$ into partial fractions by equating coefficients of like degree terms and using a system.

Solution ▶ From Example 6 we obtain

$$\frac{3x + 11}{(x - 3)(x^2 + 1)} = \frac{A}{x - 3} + \frac{Bx + C}{x^2 + 1} \qquad \text{decomposition template}$$

Multiplying both sides by $(x - 3)(x^2 + 1)$ yields

$$
\begin{aligned}
3x + 11 &= A(x^2 + 1) + (Bx + C)(x - 3) & \text{clear denominators}\\
&= Ax^2 + A + Bx^2 - 3Bx + Cx - 3C & \text{distribute/FOIL}\\
&= (A + B)x^2 + (-3B + C)x + (A - 3C) & \text{collect like terms}
\end{aligned}
$$

By comparing the like terms on the left and right, we find $A + B = 0$,
$-3B + C = 3$, and $A - 3C = 11$, resulting in the system $\begin{cases} A + B & = 0 \\ -3B + C & = 3 \\ A \quad\;\; - 3C & = 11 \end{cases}$.

Using $A = -B$ (from R1) in R3 results in the 2×2 subsystem $\begin{cases} -3B + C = 3 \\ -B - 3C = 11 \end{cases}$,
and $3R1 + R2$ of this system yields $-10B = 20$, giving $B = -2$ as before. Back-substitution again verifies $C = -3$ and $A = 2$.

Now try Exercises 39 and 40 ▶

In some cases, there are no "convenient values" to use and a system of equations is our best and only approach. More often than not, this happens when one or more of the denominators are irreducible quadratic factors.

EXAMPLE 8 ▶ **Decomposing a Rational Expression Using a System of Equations**

Decompose the expression $\dfrac{5x^2 + 2x + 12}{x^4 + 6x^2 + 9}$ into partial fractions by equating coefficients of like degree terms and using a system.

Solution ▶ By inspection or using a u-substitution, we find the denominator factors into a perfect square: $x^4 + 6x^2 + 9 = (x^2 + 3)^2$. The decomposition template is then

$$\frac{5x^2 + 2x + 12}{(x^2 + 3)^2} = \frac{Ax + B}{x^2 + 3} + \frac{Cx + D}{(x^2 + 3)^2}$$

After clearing fractions we obtain $5x^2 + 2x + 12 = (Ax + B)(x^2 + 3) + Cx + D$. The only "candidate" for a convenient value is $x = 0$, but this still leaves the unknowns B and D. Instead, we multiply out the right-hand side and equate coefficients of like terms as before.

$$
\begin{aligned}
5x^2 + 2x + 12 &= (Ax + B)(x^2 + 3) + Cx + D\\
&= Ax^3 + 3Ax + Bx^2 + 3B + Cx + D\\
&= Ax^3 + Bx^2 + (3A + C)x + (3B + D)
\end{aligned}
$$

By equating coefficients we find $A = 0$, since the left-hand side has no cubic term, and $B = 5$ by direct comparison. This yields the system

$$\begin{cases} A = 0 \\ B = 5 \\ 3A + C = 2 \\ 3B + D = 12 \end{cases}$$

☑ **C.** You've just learned how to decompose a rational expression by equating coefficients and using a system of equations

With $A = 0$ the third equation shows $C = 2$, and by substituting 5 for B in the fourth equation we find that $D = -3$. The final form of the decomposition is

$$\frac{5x^2 + 2x + 12}{(x^2 + 3)^2} = \frac{5}{x^2 + 3} + \frac{2x - 3}{(x^2 + 3)^2}$$

Now try Exercises 41 through 46 ▶

D. Partial Fractions and Telescoping Sums

From movies or the popular media (*Pirates of the Caribbean, Dances With Wolves,* etc), you might be aware that the telescopes of old were *retractable*. Since they were constructed as a series of nested tubes, you could compress the length with the lengths of the interior tubes being "negated" (see figure). In a similar fashion, some very extensive sums, called **telescoping sums,** can be rewritten in a more "compressed" form, where the interior terms are likewise negated and the resulting sum easily computed.

EXAMPLE 9 ▶ **Using Decomposition to Evaluate a Telescoping Sum**

For $t(x) = \dfrac{3}{x^2 + x}$

a. Evaluate the function for $x = 1, 2, 3,$ and 4, then find the sum of these four terms.
b. Decompose the expression into partial fractions.
c. Evaluate the decomposed form for $x = 1,\ 2,\ 3,$ and 4, leaving each result in unsimplified form.
d. Rewrite the sum from part (a) using the decomposed form of each term, and see if you can detect a pattern that enables you to compute the original sum more efficiently.
e. Use the pattern noted in part (d) to find the following sum:
$$\frac{3}{1 \cdot 2} + \frac{3}{2 \cdot 3} + \frac{3}{3 \cdot 4} + \cdots + \frac{3}{24 \cdot 25}.$$

Solution ▶ **a.** $\dfrac{3}{1 + 1} + \dfrac{3}{4 + 2} + \dfrac{3}{9 + 3} + \dfrac{3}{16 + 4} = \dfrac{3}{2} + \dfrac{1}{2} + \dfrac{1}{4} + \dfrac{3}{20}$

$$= \frac{30}{20} + \frac{10}{20} + \frac{5}{20} + \frac{3}{20}$$

$$= \frac{48}{20} = \frac{12}{5}$$

b. The denominator has two distinct linear factors and the decomposition template will be

$$\frac{3}{x(x + 1)} = \frac{A}{x} + \frac{B}{x + 1} \qquad \text{decomposition template}$$

Multiplying both sides by $x(x + 1)$ clears all denominators and gives

$$3 = A(x + 1) + Bx \qquad \text{clear denominators}$$

Since the equation must be true for all x, using $x = -1$ will eliminate the term with A, and enable us to solve for B directly:

$$3 = A(-1 + 1) + B(-1) \qquad \text{substitute } -1 \text{ for } x$$
$$3 = -B \qquad \text{simplify, term with } A \text{ is eliminated}$$
$$-3 = B \qquad \text{solve for } B$$

Repeat this procedure using $x = 1$ to solve for A.

$$3 = A(x + 1) + Bx \qquad \text{from template}$$
$$3 = A(1 + 1) - 3(1) \qquad \text{substitute } -3 \text{ for } B, 1 \text{ for } x$$
$$6 = 2A \qquad \text{add 3 to both sides}$$
$$3 = A \qquad \text{result}$$

With $A = 3$ and $B = -1$, the decomposition is

$$\frac{3}{x(x + 1)} = \frac{3}{x} - \frac{3}{x + 1} \qquad \text{decomposed form}$$

c. For $t(x) = \dfrac{3}{x} - \dfrac{3}{x + 1}$, we have

$$t(1) = \frac{3}{1} - \frac{3}{1 + 1} \qquad t(2) = \frac{3}{2} - \frac{3}{2 + 1} \qquad t(3) = \frac{3}{3} - \frac{3}{3 + 1} \qquad t(4) = \frac{3}{4} - \frac{3}{4 + 1}$$

$$= 3 - \frac{3}{2} \qquad\qquad = \frac{3}{2} - \frac{3}{3} \qquad\qquad = \frac{3}{3} - \frac{3}{4} \qquad\qquad = \frac{3}{4} - \frac{3}{5}$$

d. Replacing each term in the given sum by the equivalent term in decomposed form yields

$$\frac{3}{1 + 1} + \frac{3}{4 + 2} + \frac{3}{9 + 3} + \frac{3}{16 + 4} = \left(\frac{3}{1} - \frac{3}{2}\right) + \left(\frac{3}{2} - \frac{3}{3}\right) + \left(\frac{3}{3} - \frac{3}{4}\right) + \left(\frac{3}{4} - \frac{3}{5}\right)$$

and we note that all interior terms will cancel (add to zero) leaving only the first and last terms of the sum. The result is $\dfrac{3}{1} - \dfrac{3}{5} = \dfrac{15}{5} - \dfrac{3}{5} = \dfrac{12}{5}$.

e. From the pattern in part (d) we note that regardless of the number of terms in the sum, all interior terms will cancel (add to zero) leaving only the first and last terms.

$$\frac{3}{1 \cdot 2} + \frac{3}{2 \cdot 3} + \frac{3}{3 \cdot 4} + \cdots + \frac{3}{24 \cdot 25} = \left(\frac{3}{1} - \frac{3}{2}\right) + \left(\frac{3}{2} - \frac{3}{3}\right) + \left(\frac{3}{3} - \frac{3}{4}\right) + \cdots + \left(\frac{3}{24} - \frac{3}{25}\right)$$

$$= \frac{3}{1} - \frac{3}{25} = \frac{75}{25} - \frac{3}{25}$$

$$= \frac{72}{25} \quad \text{The result can be checked using a graphing calculator.}$$

Now try Exercises 51 through 54 ▶

As a final note, if the degree of the numerator is *greater than* the degree of the denominator in the original expression, divide using long division and apply the methods above to the remainder polynomial. For instance, you can check that

✓ **D. You've just learned how to apply partial fraction decomposition to a telescoping sum**

$$\frac{3x^3 + 6x^2 + 5x - 7}{x^2 + 2x + 1} = 3x + \frac{2x - 7}{(x + 1)^2}, \text{ and decomposing the remainder polynomial}$$

gives a final result of $3x + \dfrac{2}{x + 1} - \dfrac{9}{(x + 1)^2}$.

8.3 EXERCISES

▶ CONCEPTS AND VOCABULARY

Fill in the blank with the appropriate word or phrase. Carefully reread the section if needed.

1. In order to rewrite a rational expression as the sum of its partial fractions, we must set up a decomposition _____.

2. Before beginning the process of partial fraction decomposition, the rational expression must be _____. If not, use polynomial division.

3. If the denominator of a rational expression contains a _____ factor, each power of the factor must appear in the decomposition template and have a constant numerator.

4. If the denominator of a rational expression contains a distinct irreducible quadratic factor, it must appear in the decomposition template with a _____ numerator.

5. Discuss/Explain the process of writing $\dfrac{8x - 3}{x^2 - x}$ as a sum of partial fractions.

6. Discuss/Explain the first steps of the process of writing $\dfrac{x^2}{x^2 - 7x + 12}$ as a sum of partial fractions.

▶ DEVELOPING YOUR SKILLS

The exercises below are designed solely to reinforce the various possibilities for decomposing a rational expression. All are proper fractions whose denominators are completely factored. Set up the partial fraction decomposition using appropriate numerators, but *do not solve*.

7. $\dfrac{3x + 2}{(x + 3)(x - 2)}$

8. $\dfrac{-4x + 1}{(x - 2)(x - 5)}$

9. $\dfrac{2x + 5}{(x - 1)^2}$

10. $\dfrac{x - 7}{(x + 3)^2}$

11. $\dfrac{3x^2 - 2x + 5}{(x - 1)(x + 2)(x - 3)}$

12. $\dfrac{-2x^2 + 3x - 4}{(x + 3)(x + 1)(x - 2)}$

13. $\dfrac{x^2 + 5}{x(x - 3)(x + 1)}$

14. $\dfrac{x^2 - 7}{(x + 4)(x - 2)x}$

15. $\dfrac{x^2 + 2x - 4}{(x - 5)^3}$

16. $\dfrac{x^2 + 2x + 3}{(x + 4)^3}$

17. $\dfrac{x^2 + x - 1}{x^2(x + 2)}$

18. $\dfrac{x^2 - 3x + 5}{(x - 3)(x + 2)^2}$

19. $\dfrac{x^3 + 2x - 5}{x^2(x - 5)^2}$

20. $\dfrac{3x^2 - 5}{x^2(2x + 1)^2}$

21. $\dfrac{2x^2 + 3}{(x - 3)(x^2 + 5x + 7)}$

22. $\dfrac{7x^2 + 3x - 1}{(x + 1)(x^2 + 2x + 5)}$

23. $\dfrac{x^3 + 3x - 2}{(x + 1)(x^2 + 2)^2}$

24. $\dfrac{2x^3 + 3x^2 - 4x + 1}{x(x^2 + 3)^2}$

Decompose each rational expression into partial fractions using convenient values.

25. $\dfrac{2x - 27}{2x^2 + x - 15}$

26. $\dfrac{-11x + 6}{5x^2 - 4x - 12}$

27. $\dfrac{8x^2 - 3x - 7}{x^3 - x}$

28. $\dfrac{x^2 + 24x - 12}{x^3 - 4x}$

29. $\dfrac{3x^2 + 7x - 1}{x^3 + 2x^2 + x}$

30. $\dfrac{-2x^2 - 7x + 28}{x^3 - 4x^2 + 4x}$

31. $\dfrac{3x^3 + 3x^2 + 3x + 5}{x^4 + 3x^2 + 2}$

32. $\dfrac{2x^3 - 2x^2 + 6x}{x^4 + 4x^2 + 3}$

33. $\dfrac{6x^2 + x + 13}{x^3 + 2x^2 + 3x + 6}$

34. $\dfrac{3x^2 + 4x - 1}{x^3 - 1}$

35. $\dfrac{x^4 - 3x^2 - 2x + 1}{x^5 + 2x^3 + x}$

36. $\dfrac{-3x^4 - 11x^2 + x - 12}{x^5 - 4x^3 + 4x}$

37. $\dfrac{3x^3 + 2x^2 + 7x + 3}{x^4 + x^3 + 3x^2}$

38. $\dfrac{2x^3 - x + 6}{x^4 - 2x^3 + 3x^2}$

39. $\dfrac{3x^2 + 10x + 4}{8 - x^3}$

40. $\dfrac{2x^2 - 14x - 7}{x^3 - 2x^2 + 5x - 10}$

When the denominator of the rational expression contains repeated factors, using a system of equations sometimes offers a more efficient way to find the needed coefficients. Decompose each rational expression into partial fractions by equating coefficients and using a system of equations.

41. $\dfrac{5x + 13}{x^2 + 6x + 9}$

42. $\dfrac{14 - 3x}{x^2 - 8x + 16}$

43. $\dfrac{2x^3 + x^2 + 5x + 1}{x^4 + 2x^2 + 1}$

44. $\dfrac{x^3 + 5x^2 + 6x + 37}{x^4 + 14x^2 + 49}$

45. $\dfrac{2x^2 - 4x + 5}{x^3 - 3x^2 + 3x - 1}$

46. $\dfrac{5x^2 + 20x + 21}{x^3 + 6x^2 + 12x + 8}$

▶ WORKING WITH FORMULAS

Logistics equations and population size: Logistics equations are often used to model the growth of various populations. Initially growth is very near to exponential, but then due to certain limiting factors (food, limited resources, space, etc.), growth slows and reaches a limit called the carrying capacity c. In the process of solving the logistic equation for $P(t)$, we often need to decompose the expression shown into partial fractions, where $P(0)$ represents the initial population. Decompose the expression for the following values given.

$$\dfrac{1}{P\left(P(0) - \dfrac{P(0)}{c}P\right)}$$

47. $P(0) = 100, c = 10$ **48.** $P(0) = 80, c = 20$ **49.** $P(0) = 10, c = 100$ **50.** $P(0) = 20, c = 80$

▶ APPLICATIONS

Telescoping sums: For each function given, use the method illustrated in Example 9 to find a pattern that enables you to compute the sum shown quickly.

51. $t(x) = \dfrac{1}{x^2 + x}$;

$\dfrac{1}{1 \cdot 2} + \dfrac{1}{2 \cdot 3} + \dfrac{1}{3 \cdot 4} + \cdots + \dfrac{1}{49 \cdot 50}$

53. $t(x) = \dfrac{1}{(2x - 1)(2x + 1)}$;

$\dfrac{1}{1 \cdot 3} + \dfrac{1}{3 \cdot 5} + \dfrac{1}{5 \cdot 7} + \cdots + \dfrac{1}{123 \cdot 125}$

52. $t(x) = \dfrac{2}{x^2 + x}$;

$\dfrac{2}{1 \cdot 2} + \dfrac{2}{2 \cdot 3} + \dfrac{2}{3 \cdot 4} + \cdots + \dfrac{2}{19 \cdot 20}$

54. $t(x) = \dfrac{2x + 1}{(x^2 + 1)(x^2 + 2x + 2)}$;

$\dfrac{1}{1 \cdot 2} + \dfrac{3}{2 \cdot 5} + \dfrac{5}{5 \cdot 10} + \cdots + \dfrac{21}{101 \cdot 122}$

▶ EXTENDING THE CONCEPT

Decompose the following expressions into partial fractions.

55. $\dfrac{\ln x + 2}{(\ln x - 2)(\ln x - 1)^2}$

56. $\dfrac{1 + e^x}{(e^x - 3)(e^{2x} - 2e^x + 1)}$

57. As written, the rational expression given cannot be decomposed using the standard template (two distinct linear factors). Try to apply this template and see what happens. What can be done to decompose the expression?

$$\frac{x + 2}{(x - 1)(1 - x)}$$

58. Try to decompose the rational expression $\dfrac{3x + 1}{x^2}$ using the decomposition template $\dfrac{A}{x} + \dfrac{Bx + C}{x^2}$ (note the second rational expression in this template is a proper fraction). (a) How does this affect the decomposition process? Is the decomposition unique? (b) Next, use the standard template as outlined in this section. What do you notice?

▶ **MAINTAINING YOUR SKILLS**

59. (3.2) Use polynomial division to rewrite $\dfrac{2x^3 - 3x^2 + 13x - 5}{x^2 - x + 6}$ in the form $q(x) + \dfrac{r(x)}{d(x)}$.

60. (3.2) Use the remainder theorem to find $f(3)$ for $f(x) = 2x^3 - x^2 + 10$.

61. (6.2) Verify that $\dfrac{\cos^3\theta}{\sin\theta} = \cot\theta - \cos\theta\sin\theta$ is an identity.

62. (6.5) Evaluate the following expressions:

a. $\cos\left[\cos^{-1}\left(-\dfrac{\sqrt{3}}{2}\right)\right]$

b. $\cos^{-1}\left[\cos\left(-\dfrac{\pi}{6}\right)\right]$

8.4 | Systems of Inequalities and Linear Programming

Learning Objectives

In Section 8.4 you will learn how to:

☐ **A.** Solve a linear inequality in two variables

☐ **B.** Solve a system of linear inequalities

☐ **C.** Solve applications using a system of linear inequalities

☐ **D.** Solve applications using linear programming

In this section, we'll build on many of the ideas from Section 8.3, with a more direct focus on systems of linear inequalities. While systems of linear *equations* have an unlimited number of applications, there are many situations that can only be modeled using linear *inequalities*. For example, many decisions in business and industry are based on a large number of limitations or constraints, with many different ways these constraints can be satisfied.

A. Linear Inequalities in Two Variables

A linear equation in two variables is any equation that can be written in the form $Ax + By = C$, where A and B are real numbers, not simultaneously equal to zero. A **linear inequality** in two variables is similarly defined, with the " $=$ " sign replaced by the " $<$," " $>$," " $\leq$," or " $\geq$ " symbol:

$$Ax + By < C \qquad Ax + By > C$$
$$Ax + By \leq C \qquad Ax + By \geq C$$

Solving a linear inequality in two variables has many similarities with the one variable case. For one variable, we graph the *boundary point* on a number line, decide whether the endpoint is *included* or *excluded,* and *shade the appropriate half line.* For $x + 1 \leq 3$, we have the solution $x \leq 2$ with the endpoint included and the line shaded to the left (Figure 8.11):

Figure 8.11

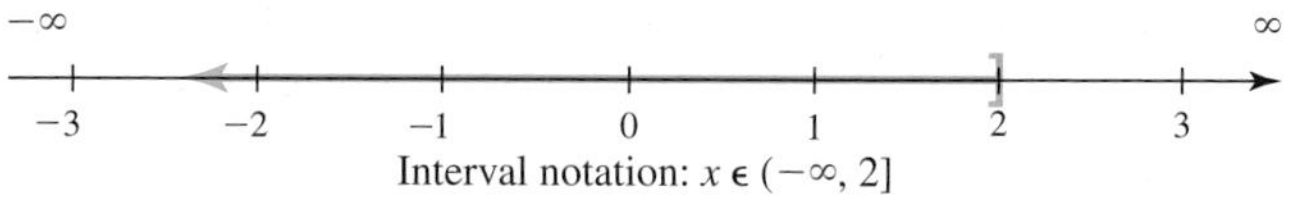

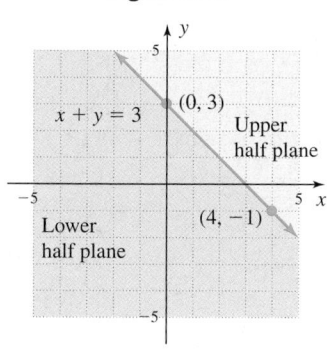

Figure 8.12

Interval notation: $x \in (-\infty, 2]$

For linear inequalities in two variables, we graph a *boundary line,* decide whether the boundary line is *included* or *excluded,* and *shade the appropriate half plane.* For $x + y \leq 3$, the boundary line $x + y = 3$ is graphed in Figure 8.12. Note it divides the coordinate plane into two regions called **half planes,** and it forms the **boundary** between the two regions. If the boundary is **included** in the solution set, we graph it using a *solid line.* If the boundary is **excluded,** a *dashed line* is used. Recall that solutions to a linear equation are ordered pairs that make the equation true. We use a similar idea to find or verify solutions to linear inequalities. If any one point in a half plane makes the inequality true, all points in that half plane will satisfy the inequality.

EXAMPLE 1 ▶   Checking Solutions to an Inequality in Two Variables

Determine whether the given ordered pairs are solutions to $-x + 2y < 2$:

a. $(4, -3)$ **b.** $(-2, 1)$

Solution ▶

a. Substitute 4 for x and -3 for y: $-(4) + 2(-3) < 2$ substitute 4 for x, -3 for y

$-10 < 2$ true

$(4, -3)$ is a solution.

b. Substitute -2 for x and 1 for y: $-(-2) + 2(1) < 2$ substitute -2 for x, 1 for y

$4 < 2$ false

$(-2, 1)$ is not a solution.

Now try Exercises 7 through 10 ▶

WORTHY OF NOTE

This relationship is often called the **trichotomy axiom** or the *"three-part truth."* Given any two quantities, they are either equal to each other, or the first is less than the second, or the first is greater than the second.

Earlier we graphed linear equations by plotting a small number of ordered pairs or by solving for y and using the slope-intercept method. The line represented all ordered pairs that made the equation true, meaning *the left-hand expression was equal to the right-hand expression.* To graph linear inequalities, we reason that if the line represents all ordered pairs that make the expressions *equal,* then any point *not on that line* must make the expressions *unequal*—either greater than or less than. These ordered pair solutions must lie in one of the half planes formed by the line, which we shade to indicate the **solution region.** Note this implies the boundary line for any inequality *is determined by the related equation,* temporarily replacing the inequality symbol with an "$=$" sign.

EXAMPLE 2 ▶   Solving an Inequality in Two Variables

Solve the inequality $-x + 2y \leq 2$.

Solution ▶   The related equation and boundary line is $-x + 2y = 2$. Since the inequality is inclusive (less than *or equal to*), we graph a solid line. Using the intercepts, we graph the line through $(0, 1)$ and $(-2, 0)$ shown in Figure 8.13. To determine the solution region and which side to shade, we select $(0, 0)$ as a test point, which results in a true statement: $-(0) + 2(0) \leq 2$✔. Since $(0, 0)$ is in the "lower" half plane, we shade this side of the boundary (see Figure 8.14).

Figure 8.13

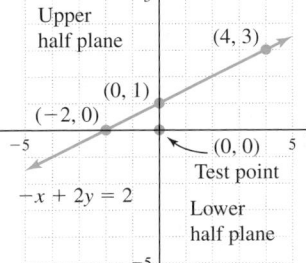

Figure 8.14

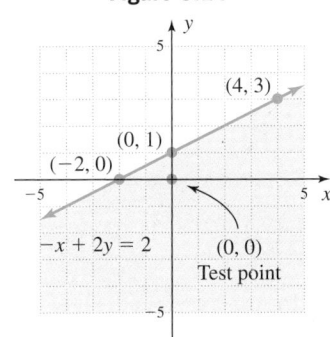

Now try Exercises 11 through 14 ▶

☑ **A. You've just learned how to solve a linear inequality in two variables**

The same solution would be obtained if we first solve for y and graph the boundary line using the slope-intercept method. However, using the slope-intercept method offers a distinct advantage — test points are no longer necessary since solutions to "less than" inequalities will always appear *below* the boundary line and solutions to "greater than" inequalities appear *above* the line. Written in slope-intercept form, the inequality from Example 2 is $y \le \frac{1}{2}x + 1$. Note that $(0, 0)$ still results in a true statement, but the "less than or equal to" symbol now indicates directly that solutions will be found in the lower half plane. This observation leads to our general approach for solving linear inequalities:

Solving a Linear Inequality

1. Graph the boundary line by solving for y and using the slope-intercept form.
   - Use a solid line if the boundary is included in the solution set.
   - Use a dashed line if the boundary is excluded from the solution set.
2. For "greater than" inequalities shade the upper half plane. For "less than" inequalities shade the lower half plane.

B. Solving Systems of Linear Inequalities

To solve a **system of inequalities,** we apply the procedure outlined above to all inequalities in the system, and note the ordered pairs that satisfy *all inequalities simultaneously.* In other words, we find *the intersection of all solution regions* (where they overlap), which then represents the solution for the system. In the case of vertical boundary lines, the designations *"above"* or *"below" the line* cannot be applied, and instead we simply note that for any vertical line $x = k$, points with x-coordinates larger than k will occur to the right.

EXAMPLE 3 ▶ Solving a System of Linear Inequalities

Solve the system of inequalities: $\begin{cases} 2x + y \ge 4 \\ x - y < 2 \end{cases}$.

Solution ▶ Solving for y, we obtain $y \ge -2x + 4$ and $y > x - 2$. The line $y = -2x + 4$ will be a solid boundary line (included), while $y = x - 2$ will be dashed (not included). Both inequalities are "greater than" and so we shade the upper half plane for each. The regions overlap and form the solution region (the lavender region shown). This sequence of events is illustrated here:

Shade above $y = -2x + 4$ (in blue)   Shade above $y = x - 2$ (in pink)   Overlapping region

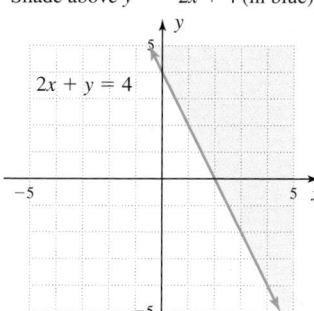
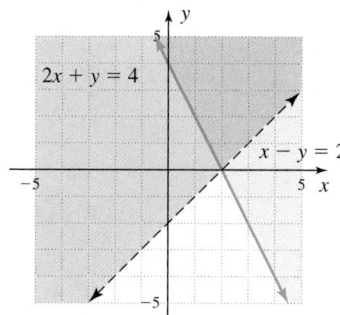
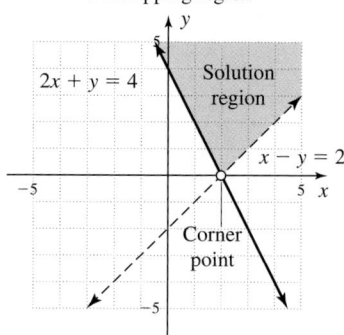

The solutions are all ordered pairs found in this region and its included boundaries. To verify the result, test the point $(2, 3)$ from inside the region, $(5, -2)$ from outside the region (the point $(2, 0)$ is not a solution since $x - y < 2$).

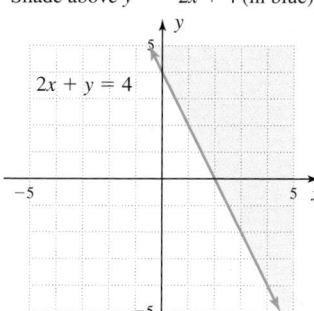 **B.** You've just learned how to solve a system of linear inequalities

Now try Exercises 15 through 42 ▶

For further reference, the point of intersection $(2, 0)$ is called a **corner point** or **vertex** of the solution region. If the point of intersection is not easily found from the graph, we can find it by solving a linear system using the two lines. For Example 3, the system is

$$\begin{cases} 2x + y = 4 \\ x - y = 2 \end{cases}$$

and solving by elimination gives $3x = 6$, $x = 2$, and $(2, 0)$ as the point of intersection.

C. Applications of Systems of Linear Inequalities

Systems of inequalities give us a way to model the decision-making process when certain *constraints* must be satisfied. A constraint is a fact or consideration that somehow limits or governs possible solutions, like the number of acres a farmer plants — which may be limited by time, size of land, government regulation, and so on.

EXAMPLE 4 ▶ **Solving Applications of Linear Inequalities**

As part of their retirement planning, James and Lily decide to invest up to $30,000 in two separate investment vehicles. The first is a bond issue paying 9% and the second is a money market certificate paying 5%. A financial adviser suggests they invest at least $10,000 in the certificate and not more than $15,000 in bonds. What various amounts can be invested in each?

Solution ▶ Consider the ordered pairs (B, C) where B represents the money invested in bonds and C the money invested in the certificate. Since they plan to invest no more than $30,000, the investment constraint would be $B + C \leq 30$ (in thousands). Following the adviser's recommendations, the constraints on each investment would be $B \leq 15$ and $C \geq 10$. Since they cannot invest less than zero dollars, the last two constraints are $B \geq 0$ and $C \geq 0$.

$$\begin{cases} B + C \leq 30 \\ B \leq 15 \\ C \geq 10 \\ B \geq 0 \\ C \geq 0 \end{cases}$$

The resulting system is shown in the figure, and indicates solutions will be in the first quadrant.

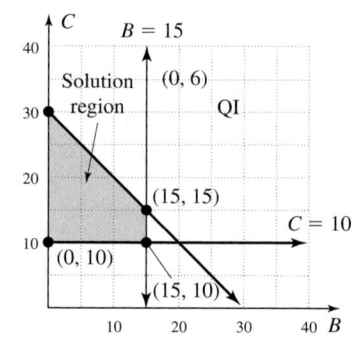

☑ C. You've just learned how to solve applications using a system of linear inequalities

There is a vertical boundary line at $B = 15$ with shading to the left (less than) and a horizontal boundary line at $C = 10$ with shading above (greater than). After graphing $C = 30 - B$, we see the solution region is a quadrilateral with vertices at $(0, 10)$, $(0, 30)$, $(15, 10)$, and $(15, 15)$, as shown.

Now try Exercises 53 and 54 ▶

D. Linear Programming

To become as profitable as possible, corporations look for ways to maximize their revenue and minimize their costs, while keeping up with delivery schedules and product demand. To operate at peak efficiency, plant managers must find ways to maximize productivity, while minimizing related costs and considering employee welfare, union agreements, and other factors. Problems where the goal is to **maximize** or **minimize** the value of a given quantity under certain **constraints** or restrictions are called programming problems. The quantity we seek to maximize or minimize is called the **objective function**. For situations where *linear* programming is used, the objective function is given as a linear function in two variables and is denoted $f(x, y)$. A function in two variables is evaluated in much the same way as a single variable function. To evaluate $f(x, y) = 2x + 3y$ at the point $(4, 5)$, we substitute 4 for x and 5 for y: $f(4, 5) = 2(4) + 3(5) = 23$.

EXAMPLE 5 ▶   **Determining Maximum Values**

Determine which of the following ordered pairs maximizes the value of $f(x, y) = 5x + 4y$: $(0, 6)$, $(5, 0)$, $(0, 0)$, or $(4, 2)$.

Solution ▶   Organizing our work in table form gives

Given Point	Evaluate $f(x, y) = 5x + 4y$
$(0, 6)$	$f(0, 6) = 5(0) + 4(6) = 24$
$(5, 0)$	$f(5, 0) = 5(5) + 4(0) = 25$
$(0, 0)$	$f(0, 0) = 5(0) + 4(0) = 0$
$(4, 2)$	$f(4, 2) = 5(4) + 4(2) = 28$

The function $f(x, y) = 5x + 4y$ is maximized at $(4, 2)$.

Now try Exercises 43 through 46 ▶

When the objective is stated as a linear function in two variables and the constraints are expressed as a system of linear inequalities, we have what is called a **linear programming** problem. The systems of inequalities solved earlier produced a solution region that was either **bounded** (as in Example 4) or **unbounded** (as in Example 3). We interpret the word *bounded* to mean we can enclose the solution region within a circle of appropriate size. If we cannot draw a circle around the region because it extends indefinitely in some direction, the region is said to be *unbounded*. In this study, we will consider only situations that produce a bounded solution region, meaning the region will have three or more vertices. The regions we study will also be **convex,** meaning that for any two points in the feasible region, the line segment between them is also in the region (Figure 8.15). Under these conditions, it can be shown that the optimal solution(s) *must occur at one of the corner points of the solution region*, also called the **feasible region.**

Figure 8.15

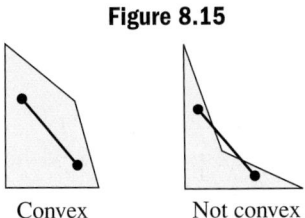

Convex   Not convex

EXAMPLE 6 ▶ Finding the Maximum of an Objective Function

Find the maximum value of the objective function $f(x, y) = 2x + y$ given the

constraints shown: $\begin{cases} x + y \leq 4 \\ 3x + y \leq 6 \\ x \geq 0 \\ y \geq 0 \end{cases}$.

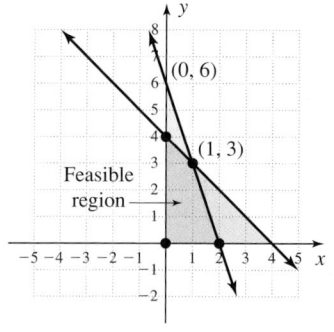

Solution ▶ Begin by noting that the solutions must be in QI, since $x \geq 0$ and $y \geq 0$. Graph the boundary lines $y = -x + 4$ and $y = -3x + 6$, shading the lower half plane in each case since they are "less than" inequalities. This produces the feasible region shown in lavender. There are four corner points to this region: $(0, 0)$, $(0, 4)$, $(2, 0)$, and $(1, 3)$. Three of these points are intercepts and can be found quickly. The point $(1, 3)$ was found by solving the system $\begin{cases} x + y = 4 \\ 3x + y = 6 \end{cases}$. Knowing that the objective function will be maximized at one of the corner points, we test them in the objective function, using a table to organize our work.

Corner Point	Objective Function $f(x, y) = 2x + y$
$(0, 0)$	$f(0, 0) = 2(0) + (0) = 0$
$(0, 4)$	$f(0, 4) = 2(0) + (4) = 4$
$(2, 0)$	$f(2, 0) = 2(2) + (0) = 4$
$(1, 3)$	$f(1, 3) = 2(1) + (3) = 5$

The objective function $f(x, y) = 2x + y$ is maximized at $(1, 3)$.

Now try Exercises 47 through 50 ▶

Figure 8.16

To help understand why solutions must occur at a vertex, note the objective function $f(x, y)$ is maximized using only (x, y) ordered pairs from the feasible region. If we let K represent this maximum value, the function from Example 6 becomes $K = 2x + y$ or $y = -2x + K$, which is a line with slope -2 and y-intercept K. The table in Example 6 suggests that K should range from 0 to 5 and graphing $y = -2x + K$ for $K = 1$, $K = 3$, and $K = 5$ produces the family of parallel lines shown in Figure 8.16. Note that values of K larger than 5 will cause the line to miss the solution region, and the maximum value of 5 occurs where the line intersects the feasible region at the vertex $(1, 3)$. These observations lead to the following principles, which we offer without a formal proof.

Linear Programming Solutions

1. If the feasible region is convex and bounded, a maximum and a minimum value exist.
2. If a unique solution exists, it will occur at a vertex of the feasible region.
3. If more than one solution exists, at least one of them occurs at a vertex of the feasible region with others on a boundary line.
4. If the feasible region is unbounded, a linear programming problem may have no solutions.

Solving linear programming problems depends in large part on two things: (1) identifying the **objective** and the **decision variables** (what each variable represents in context), and (2) using the decision variables to write the *objective function* and **constraint inequalities.** This brings us to our five-step approach for solving linear programming applications.

Solving Linear Programming Applications

1. Identify the main objective and the decision variables (descriptive variables may help) and write the objective function in terms of these variables.
2. Organize all information in a table, with the *decision variables* and *constraints* heading up the columns, and their *components* leading each row.
3. Complete the table using the information given, and write the constraint inequalities using the decision variables, constraints, and the domain.
4. Graph the constraint inequalities, determine the feasible region, and identify all corner points.
5. Test these points in the objective function to determine the optimal solution(s).

EXAMPLE 7 ▶ **Solving an Application of Linear Programming**

The owner of a snack food business wants to create two nut mixes for the holiday season. The regular mix will have 14 oz of peanuts and 4 oz of cashews, while the deluxe mix will have 12 oz of peanuts and 6 oz of cashews. The owner estimates he will make a profit of $3 on the regular mixes and $4 on the deluxe mixes. How many of each should be made in order to maximize profit, if only 840 oz of peanuts and 348 oz of cashews are available?

Solution ▶ Our *objective* is to maximize profit, and the *decision variables* could be r to represent the regular mixes sold, and d for the number of deluxe mixes. This gives $P(r, d) = \$3r + \$4d$ as our *objective function*. The information is organized in Table 8.1, using the variables r, d, and the constraints to head each column. Since the mixes are composed of peanuts and cashews, these lead the rows in the table.

Table 8.1

$$P(r, d) = \$3r \quad + \quad \$4d$$

	Regular r	Deluxe d	Constraints: Total Ounces Available
Peanuts	14	12	840
Cashews	4	6	348

After filling in the appropriate values, reading the table from left to right along the "peanut" row and the "cashew" row, gives the constraint inequalities $14r + 12d \le 840$ and $4r + 6d \le 348$. Realizing we won't be making a negative number of mixes, the remaining constraints are $r \ge 0$ and $d \ge 0$. The complete system is

$$\begin{cases} 14r + 12d \le 840 \\ 4r + 6d \le 348 \\ r \ge 0 \\ d \ge 0 \end{cases}$$

Note once again that the solutions must be in QI, since $r \ge 0$ and $d \ge 0$. Graphing the first two inequalities using slope-intercept form gives $d \le -\frac{7}{6}r + 70$ and $d \le -\frac{2}{3}r + 58$ producing the feasible region shown in lavender. The four corner

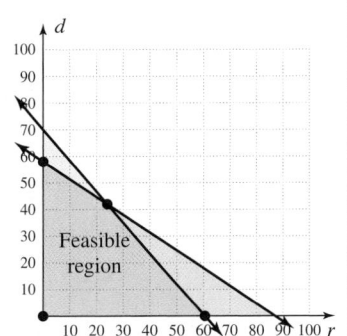

points are (0, 0), (60, 0), (0, 58), and (24, 42). Three of these points are intercepts and can be read from a table of values or the graph itself. The point (24, 42) was found by solving the system $\begin{cases} 14r + 12d = 840 \\ 4r + 6d = 348 \end{cases}$.

Knowing the objective function will be maximized at one of these points, we test them in the objective function (Table 8.2).

Table 8.2

Corner Point	Objective Function $P(r, d) = \$3r + \$4d$
(0, 0)	$P(0, 0) = \$3(0) + \$4(0) = 0$
(60, 0)	$P(60, 0) = \$3(60) + \$4(0) = \$180$
(0, 58)	$P(0, 58) = \$3(0) + \$4(58) = \$232$
(24, 42)	$P(24, 42) = \$3(24) + \$4(42) = \$240$

Profit will be maximized if 24 boxes of the regular mix and 42 boxes of the deluxe mix are made and sold.

Now try Exercises 55 through 60 ▶

Linear programming can also be used to minimize an objective function, as in Example 8.

EXAMPLE 8 ▶   Minimizing Costs Using Linear Programming

A beverage producer needs to minimize shipping costs from its two primary plants in Kansas City (KC) and St. Louis (STL). All wholesale orders within the state are shipped from one of these plants. An outlet in Macon orders 200 cases of soft drinks on the same day an order for 240 cases comes from Springfield. The plant in KC has 300 cases ready to ship and the plant in STL has 200 cases. The cost of shipping each case to Macon is $0.50 from KC, and $0.70 from STL. The cost of shipping each case to Springfield is $0.60 from KC, and $0.65 from STL. How many cases should be shipped from each warehouse to minimize costs?

Solution ▶   Our *objective* is to minimize costs, which depends on the number of cases shipped from each plant. To begin we use the following assignments:

$$A \rightarrow \text{cases shipped from KC to Macon}$$
$$B \rightarrow \text{cases shipped from KC to Springfield}$$
$$C \rightarrow \text{cases shipped from STL to Macon}$$
$$D \rightarrow \text{cases shipped from STL to Springfield}$$

From this information, the equation for total cost T is

$$T = 0.5A + 0.6B + 0.7C + 0.65D,$$

an equation in *four* variables. To make the cost equation more manageable, note since Macon ordered 200 cases, $A + C = 200$. Similarly, Springfield ordered 240 cases, so $B + D = 240$. After solving for C and D, respectively, these equations enable us to substitute for C and D, resulting in an equation with just two variables. For $C = 200 - A$ and $D = 240 - B$ we have

$$T(A, B) = 0.5A + 0.6B + 0.7(200 - A) + 0.65(240 - B)$$
$$= 0.5A + 0.6B + 140 - 0.7A + 156 - 0.65B$$
$$= 296 - 0.2A - 0.05B$$

The constraints involving the KC plant are $A + B \leq 300$ with $A \geq 0, B \geq 0$. The constraints for the STL plant are $C + D \leq 200$ with $C \geq 0, D \geq 0$. Since we want a system in terms of A and B only, we again substitute $C = 200 - A$ and $D = 240 - B$ in all the STL inequalities:

$C + D \leq 200$	STL inequalities	$C \geq 0$		$D \geq 0$
$(200 - A) + (240 - B) \leq 200$	substitute $200 - A$	$200 - A \geq 0$		$240 - B \geq 0$
	for C, $240 - B$ for D	$200 \geq A$		$240 \geq B$
$440 - A - B \leq 200$	simplify			
$240 \leq A + B$	result			

Combining the new STL constraints with those from KC produces the following system and solution. All points of intersection were read from the graph or located using the related system of equations.

$$\begin{cases} A + B \leq 300 \\ A + B \geq 240 \\ A \leq 200 \\ B \leq 240 \\ A \geq 0 \\ B \geq 0 \end{cases}$$

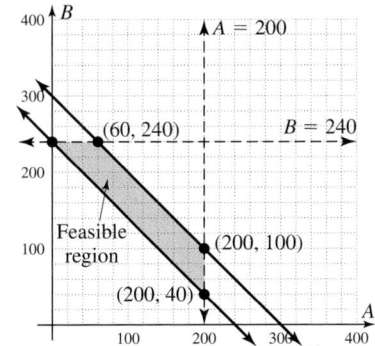

To find the minimum cost, we check each vertex in the objective function.

Vertices	Objective Function $T(A, B) = 296 - 0.2A - 0.05B$
$(0, 240)$	$P(0, 240) = 296 - 0.2(0) - 0.05(240) = \$284$
$(60, 240)$	$P(60, 240) = 296 - 0.2(60) - 0.05(240) = \$272$
$(200, 100)$	$P(200, 100) = 296 - 0.2(200) - 0.05(100) = \$251$
$(200, 40)$	$P(200, 40) = 296 - 0.2(200) - 0.05(40) = \$254$

The minimum cost occurs when $A = 200$ and $B = 100$, meaning the producer should ship the following quantities:

$A \rightarrow$ cases shipped from KC to Macon $= 200$
$B \rightarrow$ cases shipped from KC to Springfield $= 100$
$C \rightarrow$ cases shipped from STL to Macon $= 0$
$D \rightarrow$ cases shipped from STL to Springfield $= 140$

☑ **D.** You've just learned how to solve applications using linear programming

Now try Exercises 61 and 62 ▶

TECHNOLOGY HIGHLIGHT

Systems of Linear Inequalities

Solving systems of linear inequalities on the TI-84 Plus involves three steps, which are performed on both equations: (1) enter the related equations in Y_1 and Y_2 (solve for y in each equation) to create the boundary lines, (2) graph both lines and test the resulting half planes, and (3) shade the appropriate half plane. Since many real-world applications of linear inequalities preclude the use of negative numbers, we **set Xmin = 0 and Ymin = 0 for the WINDOW size.** Xmax and Ymax will depend on the equations given. We illustrate by solving the system $\begin{cases} 3x + 2y < 14 \\ x + 2y < 8 \end{cases}$.

—continued

1. *Enter the related equations.* For $3x + 2y = 14$, we have $y = -1.5 + 7$. For $x + 2y = 8$, we have $y = -0.5x + 4$. Enter these as Y_1 and Y_2 on the [Y=] screen.

2. *Graph the boundary lines.* Note the x- and y-intercepts of both lines are less than 10, so we can graph them using a friendly window where $x \in [0, 9.4]$ and $y \in [0, 6.2]$. After setting the window, press [GRAPH] to graph the lines.

Figure 8.17

```
Plot1  Plot2  Plot3
▶Y1◼-1.5X+7
▶Y2◼-0.5X+4
 Y3=
 Y4=
 Y5=
 Y6=
 Y7=
```

3. *Shade the appropriate half plane.* Since both equations are in slope-intercept form, we shade *below* both lines for the less than inequalities, using the "◣" feature located to the far left of Y_1 and Y_2. Simply overlay the diagonal line and press [ENTER] repeatedly until the symbol appears (Figure 8.17). After pressing the GRAPH key, the calculator draws both lines and shades the appropriate regions (Figure 8.18). Note the calculator uses two different kinds of shading. This makes it easy to identify the solution region—it will be the "checkerboard area" where the horizontal and vertical lines cross. As a final check, you could navigate the position marker into the solution region and test a few points in both equations.

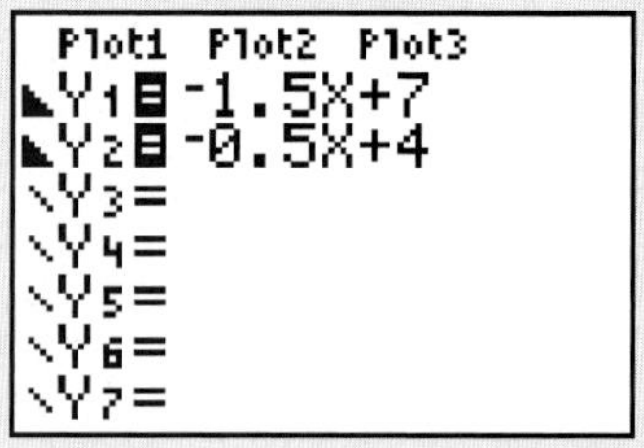

Figure 8.18

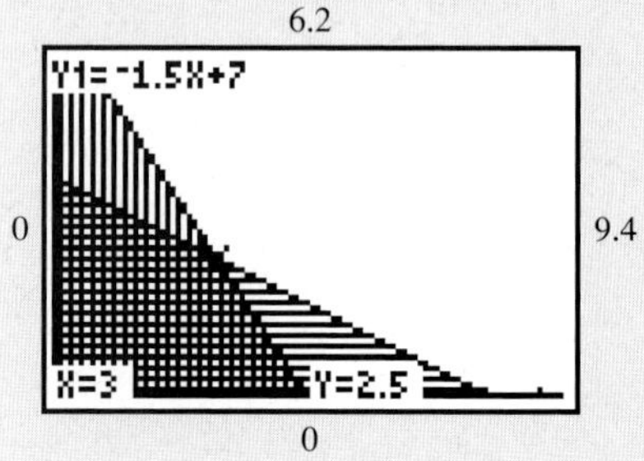

Use these ideas to solve the following systems of linear inequalities. Assume all solutions lie in Quadrant I.

Exercise 1: $\begin{cases} y + 2x < 8 \\ y + x < 6 \end{cases}$　　**Exercise 2:** $\begin{cases} 3x + y < 8 \\ x + y < 4 \end{cases}$　　**Exercise 3:** $\begin{cases} -4x - y > -9 \\ -3x - y > -7 \end{cases}$

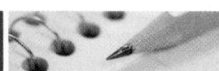

8.4 EXERCISES

▶ CONCEPTS AND VOCABULARY

Fill in the blank with the appropriate word or phrase. Carefully reread the section if needed.

1. Any line $y = mx + b$ drawn in the coordinate plane divides the plane into two regions called _____ _____.

2. For the line $y = mx + b$ drawn in the coordinate plane, solutions to $y > mx + b$ are found in the region _____ the line.

3. The overlapping region of two or more linear inequalities in a system is called the _____ region.

4. If a linear programming problem has a unique solution (x, y), it must be a _____ of the feasible region.

5. Suppose two boundary lines in a system of linear inequalities intersect, but the point of intersection is not a vertex of the feasible region. Describe how this is possible.

6. Describe the conditions necessary for a linear programming problem to have multiple solutions. (*Hint:* Consider the diagram in Figure 8.16, and the slope of the line from the objective function.)

▶ DEVELOPING YOUR SKILLS

Determine whether the ordered pairs given are solutions.

7. $2x + y > 3$; $(0, 0), (3, -5), (-3, -4), (-3, 9)$

8. $3x - y > 5$; $(0, 0), (4, -1), (-1, -5), (1, -2)$

9. $4x - 2y \leq -8$; $(0, 0), (-3, 5), (-3, -2), (-1, 1)$

10. $3x + 5y \geq 15$; $(0, 0), (3, 5), (-1, 6), (7, -3)$

Solve the linear inequalities by shading the appropriate half plane.

11. $x + 2y < 8$

12. $x - 3y > 6$

13. $2x - 3y \geq 9$

14. $4x + 5y \geq 15$

Determine whether the ordered pairs given are solutions to the accompanying system.

15. $\begin{cases} 5y - x \geq 10 \\ 5y + 2x \leq -5 \end{cases}$;
$(-2, 1), (-5, -4), (-6, 2), (-8, 2.2)$

16. $\begin{cases} 8y + 7x \geq 56 \\ 3y - 4x \geq -12 \\ y \geq 4 \end{cases}$; $(1, 5), (4, 6), (8, 5), (5, 3)$

Solve each system of inequalities by graphing the solution region. Verify the solution using a test point.

17. $\begin{cases} x + 2y \geq 1 \\ 2x - y \leq -2 \end{cases}$

18. $\begin{cases} -x + 5y < 5 \\ x + 2y \geq 1 \end{cases}$

19. $\begin{cases} 3x + y > 4 \\ x > 2y \end{cases}$

20. $\begin{cases} 3x \leq 2y \\ y \geq 4x + 3 \end{cases}$

21. $\begin{cases} 2x + y < 4 \\ 2y > 3x + 6 \end{cases}$

22. $\begin{cases} x - 2y < -7 \\ 2x + y > 5 \end{cases}$

23. $\begin{cases} x > -3y - 2 \\ x + 3y \leq 6 \end{cases}$

24. $\begin{cases} 2x - 5y < 15 \\ 3x - 2y > 6 \end{cases}$

25. $\begin{cases} 5x + 4y \geq 20 \\ x - 1 \geq y \end{cases}$

26. $\begin{cases} 10x - 4y \leq 20 \\ 5x - 2y > -1 \end{cases}$

27. $\begin{cases} 0.2x > -0.3y - 1 \\ 0.3x + 0.5y \leq 0.6 \end{cases}$

28. $\begin{cases} x > -0.4y - 2.2 \\ x + 0.9y \leq -1.2 \end{cases}$

29. $\begin{cases} y \leq \dfrac{3}{2}x \\ 4y \geq 6x - 12 \end{cases}$

30. $\begin{cases} 3x + 4y > 12 \\ y < \dfrac{2}{3}x \end{cases}$

31. $\begin{cases} \dfrac{-2}{3}x + \dfrac{3}{4}y \leq 1 \\ \dfrac{1}{2}x + 2y \geq 3 \end{cases}$

32. $\begin{cases} \dfrac{1}{2}x + \dfrac{2}{5}y \leq 5 \\ \dfrac{5}{6}x - 2y \geq -5 \end{cases}$

33. $\begin{cases} x - y \geq -4 \\ 2x + y \leq 4 \\ x \geq 1, y \geq 0 \end{cases}$

34. $\begin{cases} 2x - y \leq 5 \\ x + 3y \leq 6 \\ x \geq 1 \end{cases}$

35. $\begin{cases} y \leq x + 3 \\ x + 2y \leq 4 \\ y \geq 0 \end{cases}$

36. $\begin{cases} 4y < 3x + 12 \\ x \geq 0 \\ y \leq x + 1 \end{cases}$

37. $\begin{cases} 2x + 3y \leq 18 \\ x \geq 0 \\ y \geq 0 \end{cases}$

38. $\begin{cases} 8x + 5y \leq 40 \\ x \geq 0 \\ y \geq 0 \end{cases}$

Use the equations given to write the system of linear inequalities represented by each graph.

39.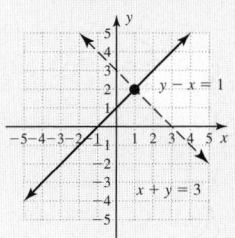

40.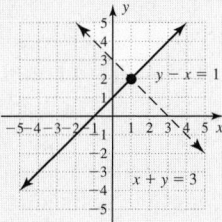

41.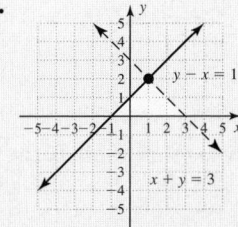

42.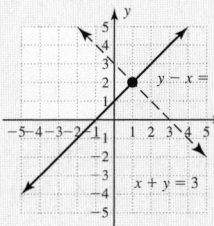

Determine which of the ordered pairs given produces the maximum value of $f(x, y)$.

43. $f(x, y) = 12x + 10y$; $(0, 0), (0, 8.5), (7, 0), (5, 3)$

44. $f(x, y) = 50x + 45y$; $(0, 0), (0, 21), (15, 0), (7.5, 12.5)$

Determine which of the ordered pairs given produces the minimum value of $f(x, y)$.

45. $f(x, y) = 8x + 15y$; $(0, 20), (35, 0), (5, 15), (12, 11)$

46. $f(x, y) = 75x + 80y$; $(0, 9), (10, 0), (4, 5), (5, 4)$

For Exercises 47 and 48, find the *maximum* value of the objective function $f(x, y) = 8x + 5y$ given the constraints shown.

47. $\begin{cases} x + 2y \leq 6 \\ 3x + y \leq 8 \\ x \geq 0 \\ y \geq 0 \end{cases}$

48. $\begin{cases} 2x + y \leq 7 \\ x + 2y \leq 5 \\ x \geq 0 \\ y \geq 0 \end{cases}$

For Exercises 49 and 50, find the *minimum* value of the objective function $f(x, y) = 36x + 40y$ given the constraints shown.

49. $\begin{cases} 3x + 2y \geq 18 \\ 3x + 4y \geq 24 \\ x \geq 0 \\ y \geq 0 \end{cases}$

50. $\begin{cases} 2x + y \geq 10 \\ x + 4y \geq 3 \\ x \geq 2 \\ y \geq 0 \end{cases}$

▶ WORKING WITH FORMULAS

Area Formulas

51. The area of a triangle is usually given as $A = \frac{1}{2}BH$, where B and H represent the base and height, respectively. The area of a rectangle can be stated as $A = BH$. If the base of both the triangle and rectangle is equal to 20 in., what are the possible values for H if the triangle must have an area *greater than* 50 in^2 and the rectangle must have an area *less than* 200 in^2?

Volume Formulas

52. The volume of a cone is $V = \frac{1}{3}\pi r^2 h$, where r is the radius of the base and h is the height. The volume of a cylinder is $V = \pi r^2 h$. If the radius of both the cone and cylinder is equal to 10 cm, what are the possible values for h if the cone must have a volume *greater than* 200 cm^3 and the volume of the cylinder must be *less than* 850 cm^3?

▶ APPLICATIONS

Write a system of linear inequalities that models the information given, then solve.

53. Gifts to grandchildren: Grandpa Augustus is considering how to divide a $50,000 gift between his two grandchildren, Julius and Anthony. After weighing their respective positions in life and family responsibilities, he decides he must bequeath at least $20,000 to Julius, but no more than $25,000 to Anthony. Determine the possible ways that Grandpa can divide the $50,000.

54. Guns versus butter: Every year, governments around the world have to make the decision as to how much of their revenue must be spent on national defense and domestic improvements (guns versus butter). Suppose total revenue for these two needs was $120 billion, and a government decides they need to spend at least $42 billion on butter and no more than $80 billion on defense. Determine the possible amounts that can go toward each need.

Solve the following linear programming problems.

55. Land/crop allocation: A farmer has 500 acres of land to plant corn and soybeans. During the last few years, market prices have been stable and the farmer anticipates a profit of $900 per acre on the corn harvest and $800 per acre on the soybeans. The farmer must take into account the time it takes to plant and harvest each crop, which is 3 hr/acre for corn and 2 hr/acre for soybeans. If the farmer has at most 1300 hr to plant, care for, and harvest each crop, how many acres of each crop should be planted in order to maximize profits?

56. Coffee blends: The owner of a coffee shop has decided to introduce two new blends of coffee in order to attract new customers—a *Deluxe Blend* and a *Savory Blend.* Each pound of the deluxe blend contains 30% Colombian and 20% Arabian coffee, while each pound of the savory blend contains 35% Colombian and 15% Arabian coffee (the remainder of each is made up of cheap and plentiful domestic varieties). The profit on the deluxe blend will be $1.25 per pound, while the profit on the savory blend will be $1.40 per pound. How many pounds of each should the owner make in order to maximize profit, if only 455 lb of Colombian coffee and 250 lb of Arabian coffee are currently available?

57. Manufacturing screws: A machine shop manufactures two types of screws—sheet metal screws and wood screws, using three different machines. Machine Moe can make a sheet metal screw in 20 sec and a wood screw in 5 sec. Machine Larry can make a sheet metal screw in 5 sec and a wood screw in 20 sec. Machine Curly, the newest machine (nyuk, nyuk) can make a sheet metal screw in 15 sec and a wood screw in 15 sec. (Shemp couldn't get a job because he failed the math portion of the employment exam.) Each machine can operate for only 3 hr each day before shutting down for maintenance. If sheet metal screws sell for 10 cents and wood screws sell for 12 cents, how many of each type should the machines be programmed to make in order to maximize revenue? (*Hint:* Standardize time units.)

58. Hauling hazardous waste: A waste disposal company is contracted to haul away some hazardous waste material. A full container of liquid waste weighs 800 lb and has a volume of 20 ft^3. A full container of solid waste weighs 600 lb and has a volume of 30 ft^3. The trucks used can carry at most 10 tons (20,000 lb) and have a carrying volume of 800 ft^3. If the trucking company makes $300 for disposing of liquid waste and $400 for disposing of solid waste, what is the maximum revenue per truck that can be generated?

59. Maximizing profit—food service: P. Barrett & Justin, Inc., is starting up a fast-food restaurant specializing in peanut butter and jelly sandwiches. Some of the peanut butter varieties are smooth, crunchy, reduced fat, and reduced sugar. The jellies will include those expected and common, as well as some exotic varieties such as kiwi and mango. Independent research has determined the two most popular sandwiches will be the traditional P&J (smooth peanut butter and grape jelly), and the Double-T (three slices of bread). A traditional P&J uses 2 oz of peanut butter and 3 oz of jelly. The Double-T uses 4 oz of peanut butter and 5 oz of jelly. The traditional sandwich will be priced at $2.00, and a Double-T at $3.50. If the restaurant has 250 oz of smooth peanut butter and 345 oz of grape jelly on hand for opening day, how many of each should they make and sell to maximize revenue?

60. Maximizing profit—construction materials: Mooney and Sons produces and sells two varieties of concrete mixes. The mixes are packaged in 50-lb bags. Type A is appropriate for finish work, and contains 20 lb of cement and 30 lb of sand. Type B is appropriate for foundation and footing work, and contains 10 lb of cement and 20 lb of sand. The remaining weight comes from gravel aggregate. The profit on type A is $1.20/bag, while the profit on type B is $0.90/bag. How many bags of each should the company make to maximize profit, if 2750 lb of cement and 4500 lb of sand are currently available?

61. Minimizing shipping costs: An oil company is trying to minimize shipping costs from its two primary refineries in Tulsa, Oklahoma, and Houston, Texas. All orders within the region are shipped from one of these two refineries. An order for 220,000 gal comes in from a location in Colorado, and another for 250,000 gal from a location in Mississippi. The Tulsa refinery has 320,000 gal ready to ship, while the Houston refinery has 240,000 gal. The cost of transporting each gallon to Colorado is $0.05 from Tulsa and $0.075 from Houston. The cost of transporting each gallon to Mississippi is $0.06 from Tulsa and $0.065 from Houston. How many gallons should be distributed from each refinery to minimize the cost of filling both orders?

62. Minimizing transportation costs: Robert's Las Vegas Tours needs to drive 375 people and 19,450 lb of luggage from Salt Lake City, Utah, to Las Vegas, Nevada, and can charter buses from two companies. The buses from company X carry 45 passengers and 2750 lb of luggage at a cost of $1250 per trip. Company Y offers buses that carry 60 passengers and 2800 lb of luggage at a cost of $1350 per trip. How many buses should be chartered from each company in order for Robert to minimize the cost?

▶ EXTENDING THE CONCEPT

63. Graph the feasible region formed by the system
$$\begin{cases} x \geq 0 \\ y \geq 0 \\ y \leq 3 \\ x \leq 3 \end{cases}.$$ How would you describe this region?

Select random points within the region or on any boundary line and evaluate the objective function $f(x, y) = 4.5x + 7.2y$. At what point (x, y) will this

function be maximized? How does this relate to optimal solutions to a linear programing problem?

64. Find the maximum value of the objective function $f(x, y) = 22x + 15y$ given the constraints
$$\begin{cases} 2x + 5y \leq 24 \\ 3x + 4y \leq 29 \\ x + 6y \leq 26 \\ x \geq 0 \\ y \geq 0 \end{cases}.$$

▶ MAINTAINING YOUR SKILLS

65. (5.3) Given the point $(-3, 4)$ is on the terminal side of angle θ with θ in standard position, find

   a. $\cos \theta$

   b. $\csc \theta$

   c. $\cot \theta$

66. (3.7) Solve the rational inequality. Write your answer in interval notation. $\dfrac{x + 2}{x^2 - 9} > 0$

67. (3.8) The resistance to current flow in copper wire varies directly as its length and inversely as the square of its diameter. A wire 8 m long with a 0.004-m diameter has a resistance of 1500 Ω. Find the resistance in a wire of like material that is 2.7 m long with a 0.005-m diameter.

68. (6.4) Use a half-angle identity to find an exact value for $\cos\left(\dfrac{7\pi}{12}\right)$.

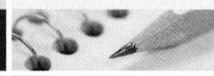

MID-CHAPTER CHECK

1. Solve using the substitution method. State whether the system is consistent, inconsistent, or dependent.
$$\begin{cases} x - 3y = -2 \\ 2x + y = 3 \end{cases}$$

2. Solve the system using elimination. State whether the system is consistent, inconsistent, or dependent:
$$\begin{cases} x - 3y = -4 \\ 2x + y = 13 \end{cases}$$

3. Solve using a system of linear equations and any method you choose. How many ounces of a 40% acid should be mixed with 10 oz of a 64% acid, to obtain a 48% acid solution?

4. Determine whether the ordered triple is a solution to the system.
$$\begin{cases} 5x + 2y - 4z = 22 \\ 2x - 3y + z = -1 \quad (2, 0, -3) \\ 3x - 6y + z = 2 \end{cases}$$

5. The system given is a dependent system. Without solving, state why.
$$\begin{cases} x + 2y - 3z = 3 \\ 2x + 4y - 6z = 6 \\ x - 2y + 5z = -1 \end{cases}$$

6. Solve the system of equations:
$$\begin{cases} x + 2y - 3z = -4 \\ 2y + z = 7 \\ 5y - 2z = 4 \end{cases}$$

7. Solve using elimination:
$$\begin{cases} 2x + 3y - 4z = -4 \\ x - 2y + z = 0 \\ -3 - 2y + 2z = -1 \end{cases}$$

8. Decompose the expression $\dfrac{15 - x^2 - 4x}{(x + 1)(x - 2)^2}$ into partial fractions.

9. **Child prodigies:** If you add Mozart's age when he wrote his first symphony, with the age of American chess player Paul Morphy when he began dominating the international chess scene, and the age of Blaise Pascal when he formulated his well-known *Essai pour les coniques* (Essay on Conics), the sum is 37. At the time of each event, Paul Morphy's age was 3 yr less than twice Mozart's, and Pascal was 3 yr older than Morphy. Set up a system of equations and find the age of each.

10. **Candle manufacturing:** David and Karen make table candles and holiday candles and sell them out of their home. Dave works 10 min and Karen works 30 min on each table candle, while Karen works 40 min and David works 20 min on each holiday candle. Dave can work at most 3 hr and 20 min (200 min) per day on the home business, while Karen can work at most 7 hr. If table candles sell for $4 and holiday candles sell for $6, how many of each should made to maximize their revenue?

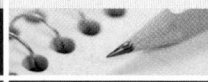

REINFORCING BASIC CONCEPTS

Window Size and Graphing Technology

Since most substantial applications involve noninteger values, technology can play an important role in applying mathematical models. However, with its use comes a heavy responsibility to use it carefully. A very real effort must be made to determine the best approach and to secure a reasonable estimate. This is the only way to guard against (the inevitable) keystroke errors, or ensure a window size that properly displays the results.

Rationale

On October 1, 1999, the newspaper *USA TODAY* ran an article titled, "Bad Math added up to Doomed Mars Craft."

The article told of how a $125,000,000.00 spacecraft was lost, apparently because the team of scientists that *plotted the course* for the craft used U.S. units of measurement, while the team of scientists *guiding* the craft were using metric units. NASA's space chief was later quoted, "The problem here was not the error, it was the failure of . . . the checks and balances in our process to detect the error."

No matter how powerful the technology, always try to begin your problem-solving efforts with an estimate. Begin by exploring the **context** of the problem, asking questions about the range of possibilities: How fast can a human run? How much does a new car cost? What is a reasonable price for a ticket? What is the total available to invest? There is no calculating involved in these estimates, they simply rely on "horse sense" and human experience. In many applied problems, the input and output values must be positive — which means the solution will appear in the first quadrant, narrowing the possibilities considerably. This information will be used to set the viewing window of your graphing calculator, in preparation for solving the problem using a system and graphing technology.

Illustration 1 ▶ Erin just filled both her boat and Blazer with gas, at a total cost of $211.14. She purchased 35.7 gallons of premium for her boat and 15.3 gal of regular for her Blazer. Premium gasoline cost $0.10 per gallon more than regular. What was the cost per gallon of each grade of gasoline?

Solution ▶ Asking how much *you* paid for gas the last time you filled up should serve as a fair estimate. Certainly (in 2008) a cost of $6.00 or more per gallon in the United

States is too high, and a cost of $1.50 per gallon or less would be too low. Also, we can estimate a solution by assuming that both kinds of gasoline cost the same. This would mean 51 gal were purchased for about $211, and a quick division would place the estimate at near $\frac{211}{51} \approx \$4.14$ per gallon. A good viewing window would be restricted to the first quadrant (since cost > 0) with maximum values of Xmax $= 6$ and Ymax $= 6$.

```
WINDOW
 Xmin=0
 Xmax=6
 Xscl=.5
 Ymin=0
 Ymax=6
 Yscl=.5
 Xres=1
```

Exercise 1: Solve Illustration 1 using graphing technology.

Exercise 2: Re-solve Exercises 63 and 64 from Section 8.1 using graphing technology. Verify results are identical.

8.5 | Solving Linear Systems Using Matrices and Row Operations

Learning Objectives

In Section 8.5 you will learn how to:

☐ **A.** State the size of a matrix and identify its entries

☐ **B.** Form the augmented matrix of a system of equations

☐ **C.** Solve a system of equations using row operations

☐ **D.** Recognize inconsistent and dependent systems

☐ **E.** Solve applications using linear systems

Just as synthetic division streamlines the process of polynomial division, matrices and row operations streamline the process of solving systems using elimination. With the equations of the system in standard form, the location of the variable terms and constant terms are set, and we simply apply the elimination process on the coefficients and constants.

A. Introduction to Matrices

In general terms, a **matrix** is simply a rectangular arrangement of numbers, called the **entries** of the matrix. **Matrices** (plural of matrix) are denoted by enclosing the entries between a left and right bracket, and named using a capital letter, such as

$$A = \begin{bmatrix} 1 & -3 & 2 \\ 5 & 1 & -1 \end{bmatrix} \text{ and } B = \begin{bmatrix} 2 & -1 & 3 \\ 4 & 6 & -2 \\ 1 & 0 & -1 \end{bmatrix}.$$ They occur in many different sizes

as defined by the number of **rows** and **columns** each has, with the number of rows always given first. Matrix A is said to be a 2×3 (two by three) matrix, since it has two rows and three columns. Matrix B is a 3×3 (three by three) matrix.

EXAMPLE 1A ▶ **Identifying the Size and Entries of a Matrix**

Determine the size of each matrix and identify the entry located in the second row and first column.

a. $C = \begin{bmatrix} 3 & -2 \\ 1 & 5 \\ -4 & 3 \end{bmatrix}$ **b.** $D = \begin{bmatrix} 0.2 & -0.5 & 0.7 & 3.3 \\ -0.4 & 0.3 & 1 & 2 \\ 2.1 & -0.1 & 0.6 & 4.1 \end{bmatrix}$

Solution ▶ **a.** Matrix C is 3×2. The row 2, column 1 entry is 1.

b. Matrix D is 3×4. The row 2, column 1 entry is -0.4.

If a matrix has the same number of rows and columns, it's called a **square matrix.** Matrix B above is a square matrix, while matrix A is not. For square matrices, the values on a diagonal line *from the upper left to the lower right* are called the **diagonal entries** and are said to be **on the diagonal** of the matrix. When solving systems using matrices, much of our focus is on these diagonal entries.

EXAMPLE 1B ▶ **Identifying the Diagonal Entries of a Square Matrix**

Name the diagonal entries of each matrix.

a. $E = \begin{bmatrix} 1 & 4 \\ -2 & -3 \end{bmatrix}$ **b.** $F = \begin{bmatrix} 0.2 & -0.5 & 0.7 \\ -0.4 & 0.3 & 1 \\ 2.1 & -0.1 & 0.6 \end{bmatrix}$

Solution ▶ **a.** The diagonal entries of matrix E are 1 and -3.

b. For matrix F, the diagonal entries are 0.2, 0.3, and 0.6.

✓ **A.** You've just learned how to state the size of a matrix and identify its entries

Now try Exercises 7 through 9 ▶

B. The Augmented Matrix of a System of Equations

A system of equations can be written in matrix form by augmenting or joining the **coefficient matrix,** formed by the variable coefficients, with the **matrix of constants.**

The coefficient matrix for the system $\begin{cases} 2x + 3y - z = 1 \\ x + \quad\quad z = 2 \\ x - 3y + 4z = 5 \end{cases}$ is $\begin{bmatrix} 2 & 3 & -1 \\ 1 & 0 & 1 \\ 1 & -3 & 4 \end{bmatrix}$ with

column 1 for the coefficients of x, column 2 for the coefficients of y, and so on. The

matrix of constants is $\begin{bmatrix} 1 \\ 2 \\ 5 \end{bmatrix}$. These two are joined to form the **augmented matrix,**

with a dotted line often used to separate the two as shown here: $\begin{bmatrix} 2 & 3 & -1 & \vdots & 1 \\ 1 & 0 & 1 & \vdots & 2 \\ 1 & -3 & 4 & \vdots & 5 \end{bmatrix}$.

It's important to note the use of a zero placeholder for the y-variable in the second row of the matrix, signifying there is no y-variable in the corresponding equation.

EXAMPLE 2 ▶ **Forming Augmented Matrices**

Form the augmented matrix for each system, and name the diagonal entries of each coefficient matrix.

a. $\begin{cases} 2x + y = 11 \\ -x + 3y = -2 \end{cases}$ **b.** $\begin{cases} x + 4y - z = -10 \\ 2x + 5y + 8z = 4 \\ x - 2y - 3z = -7 \end{cases}$ **c.** $\begin{cases} \frac{1}{2}x + y = -7 \\ x + \frac{2}{3}y + \frac{5}{6}z = \frac{11}{12} \\ -2y - z = -3 \end{cases}$

Solution ▶ **a.** $\begin{cases} 2x + y = 11 \\ -x + 3y = -2 \end{cases} \longrightarrow \begin{bmatrix} 2 & 1 & \vdots & 11 \\ -1 & 3 & \vdots & -2 \end{bmatrix}$

Diagonal entries: 2 and 3.

b. $\begin{cases} x + 4y - z = -10 \\ 2x + 5y + 8z = 4 \\ x - 2y - 3z = -7 \end{cases} \longrightarrow \begin{bmatrix} 1 & 4 & -1 & \vdots & -10 \\ 2 & 5 & 8 & \vdots & 4 \\ 1 & -2 & -3 & \vdots & -7 \end{bmatrix}$

Diagonal entries: 1, 5, and -3.

c. $\begin{cases} \frac{1}{2}x + \quad y \quad\quad = -7 \\ x + \quad \frac{2}{3}y + \frac{5}{6}z = \frac{11}{12} \\ \quad\quad -2y - \quad z = -3 \end{cases} \longrightarrow \begin{bmatrix} \frac{1}{2} & 1 & 0 & \vdots & -7 \\ 1 & \frac{2}{3} & \frac{5}{6} & \vdots & \frac{11}{12} \\ 0 & -2 & -1 & \vdots & -3 \end{bmatrix}$

Diagonal entries: $\frac{1}{2}, \frac{2}{3}$, and -1.

Now try Exercises 10 through 12 ▶

This process can easily be reversed to write a system of equations from a given augmented matrix.

EXAMPLE 3 ▶   **Writing the System Corresponding to an Augmented Matrix**

Write the system of equations corresponding to each matrix.

a. $\begin{bmatrix} 3 & -5 & \vdots & -14 \\ 0 & 1 & \vdots & 4 \end{bmatrix}$ **b.** $\begin{bmatrix} 1 & 4 & -1 & \vdots & -10 \\ 0 & -3 & 10 & \vdots & 7 \\ 0 & 0 & 1 & \vdots & 1 \end{bmatrix}$

Solution ▶   **a.** $\begin{bmatrix} 3 & -5 & \vdots & -14 \\ 0 & 1 & \vdots & 4 \end{bmatrix} \longrightarrow \begin{cases} 3x - 5y = -14 \\ 0x + 1y = 4 \end{cases}$

☑ **B. You've just learned how to form the augmented matrix of a system of equations**

b. $\begin{bmatrix} 1 & 4 & -1 & \vdots & -10 \\ 0 & -3 & 10 & \vdots & 7 \\ 0 & 0 & 1 & \vdots & 1 \end{bmatrix} \longrightarrow \begin{cases} 1x + 4y - \quad 1z = -10 \\ 0x - 3y + 10z = 7 \\ 0x + 0y + \quad 1z = 1 \end{cases}$

Now try Exercises 13 through 18 ▶

C. Solving a System Using Matrices

When a system of equations is written in augmented matrix form, we can solve the system by applying the same operations to each row of the matrix, that would be applied to the equations in the system. In this context, the operations are referred to as **elementary row operations.**

Elementary Row Operations

  1. Any two rows in a matrix can be interchanged.
  2. The elements of any row can be multiplied by a nonzero constant.
  3. Any two rows can be added together, and the sum used to replace one of the rows.

In this section, we'll use these operations to **triangularize the augmented matrix,** employing a solution method known as **Gaussian elimination.** A matrix is said to be in **triangular form** when all of the entries below the diagonal are zero. For example,

the matrix $\begin{bmatrix} 1 & 4 & -1 & \vdots & -10 \\ 0 & -3 & 10 & \vdots & 7 \\ 0 & 0 & 1 & \vdots & 1 \end{bmatrix}$ is in triangular form: $\begin{bmatrix} 1 & 4 & -1 & \vdots & -10 \\ 0 & -3 & 10 & \vdots & 7 \\ 0 & 0 & 1 & \vdots & 1 \end{bmatrix}$.

In system form we have $\begin{cases} x + 4y - \quad z = -10 \\ \quad\quad - 3y + 10z = 7 \\ \quad\quad\quad\quad z = 1 \end{cases}$, meaning a matrix written in triangular form can be used to solve the system using back-substitution. We'll illustrate by solving $\begin{cases} 1x + 4y - 1z = 4 \\ 2x + 5y + 8z = 15, \\ 1x + 3y - 3z = 1 \end{cases}$ using elimination to the left, and *row operations on the augmented matrix* to the right. As before, R1 represents the first equation in the system and the first row of the matrix, R2 represents equation 2 and row 2, and so on.

The calculations involved are shown for the first stage only and are designed to offer a careful comparison. In actual practice, the format shown in Example 4 is used.

<table>
<tr><td align="center">**Elimination**
(System of Equations)</td><td align="center">**Row Operations**
(Augmented Matrix)</td></tr>
</table>

$$\begin{cases} 1x + 4y - 1z = 4 \\ 2x + 5y + 8z = 15 \\ 1x + 3y - 3z = 1 \end{cases} \qquad \begin{bmatrix} 1 & 4 & -1 & | & 4 \\ 2 & 5 & 8 & | & 15 \\ 1 & 3 & -3 & | & 1 \end{bmatrix}$$

To eliminate the x-term in R2, we use $-2R1 + R2 \rightarrow R2$. For R3 the operations would be $-1R1 + R3 \rightarrow R3$. Identical operations are performed on the matrix, which begins the process of triangularizing the matrix.

<table>
<tr><td align="center">**System Form**</td><td align="center">**Matrix Form**</td></tr>
</table>

$$\begin{array}{ll} -2R1 & -2x - 8y + 2z = -8 \\ + & \\ \underline{R2} & \underline{2x + 5y + 8z = 15} \\ \text{New R2} & -3y + 10z = 7 \end{array} \qquad \begin{array}{lcccc} -2R1 & -2 & -8 & 2 & -8 \\ + & & & & \\ \underline{R2} & \underline{2} & \underline{5} & \underline{8} & \underline{15} \\ \text{New R2} & 0 & -3 & 10 & 7 \end{array}$$

$$\begin{array}{ll} -1R1 & -1x - 4y + 1z = -4 \\ + & \\ \underline{R3} & \underline{1x + 3y - 3z = 1} \\ \text{New R3} & -1y - 2z = -3 \end{array} \qquad \begin{array}{lcccc} -1R1 & -1 & -4 & 1 & -4 \\ + & & & & \\ \underline{R3} & \underline{1} & \underline{3} & \underline{-3} & \underline{1} \\ \text{New R3} & 0 & -1 & -2 & -3 \end{array}$$

As always, we should look for opportunities to simplify any equation in the system (and any row in the matrix). Note that $-1R3$ will make the coefficients and related matrix entries positive. Here is the new system and matrix.

<table>
<tr><td align="center">**New System**</td><td align="center">**New Matrix**</td></tr>
</table>

$$\begin{cases} 1x + 4y - 1z = 4 \\ -3y + 10z = 7 \\ 1y + 2z = 3 \end{cases} \qquad \begin{bmatrix} 1 & 4 & -1 & | & 4 \\ 0 & -3 & 10 & | & 7 \\ 0 & 1 & 2 & | & 3 \end{bmatrix}$$

On the left, we would finish by solving the 2×2 subsystem using $R2 + 3R3 \rightarrow R3$. In matrix form, we eliminate the corresponding entry (third row, second column) to triangularize the matrix.

$$\begin{bmatrix} 1 & 4 & -1 & | & 4 \\ 0 & -3 & 10 & | & 7 \\ 0 & 1 & 2 & | & 3 \end{bmatrix} \xrightarrow{R2 + 3R3 \rightarrow R3} \begin{bmatrix} 1 & 4 & -1 & | & 4 \\ 0 & -3 & 10 & | & 7 \\ 0 & 0 & 16 & | & 16 \end{bmatrix}$$

Dividing R3 by 16 gives $z = 1$ in the system, and entries of 0   0   1   1 in the augmented matrix. Completing the solution by back-substitution in the system gives the ordered triple (1, 1, 1). **See Exercises 19 through 27.**

The general solution process is summarized here.

Solving Systems by Triangularizing the Augmented Matrix

1. Write the system as an augmented matrix.
2. Use row operations to obtain zeroes below the first diagonal entry.
3. Use row operations to obtain zeroes below the second diagonal entry.
4. Continue until the matrix is triangularized (entries below diagonal are zero).
5. Divide to obtain a "1" in the last diagonal entry (if it is nonzero), then convert to equation form and solve using back-substitution.

Note: At each stage, look for opportunities to simplify row entries using multiplication or division. Also, to begin the process any equation with an x-coefficient of 1 can be made R1 by interchanging the equations.

WORTHY OF NOTE

The procedure outlined for solving systems using matrices is virtually identical to that for solving systems by elimination. Using a 3×3 system for illustration, the "zeroes below the first diagonal entry" indicates we've eliminated the x-term from R2 and R3, the "zeroes below the second entry" indicates we've eliminated the y-term from the subsystem, and the division "to obtain a '1' in the final entry" indicates we have just solved for z.

EXAMPLE 4 ▶ **Solving Systems Using the Augmented Matrix**

Solve by triangularizing the augmented matrix: $\begin{cases} 2x + y - 2z = -7 \\ x + y + z = -1 \\ -2y - z = -3 \end{cases}$

Solution ▶ $\begin{cases} 2x + y - 2z = -7 \\ x + y + z = -1 \\ -2y - z = -3 \end{cases}$ matrix form → $\begin{bmatrix} 2 & 1 & -2 & -7 \\ 1 & 1 & 1 & -1 \\ 0 & -2 & -1 & -3 \end{bmatrix}$ R1 ↔ R2 $\begin{bmatrix} 1 & 1 & 1 & -1 \\ 2 & 1 & -2 & -7 \\ 0 & -2 & -1 & -3 \end{bmatrix}$

$\begin{bmatrix} 1 & 1 & 1 & -1 \\ 2 & 1 & -2 & -7 \\ 0 & -2 & -1 & -3 \end{bmatrix}$ −2R1 + R2 → R2 $\begin{bmatrix} 1 & 1 & 1 & -1 \\ 0 & -1 & -4 & -5 \\ 0 & -2 & -1 & -3 \end{bmatrix}$ −1R2 → R2 $\begin{bmatrix} 1 & 1 & 1 & -1 \\ 0 & 1 & 4 & 5 \\ 0 & -2 & -1 & -3 \end{bmatrix}$

$\begin{bmatrix} 1 & 1 & 1 & -1 \\ 0 & 1 & 4 & 5 \\ 0 & -2 & -1 & -3 \end{bmatrix}$ 2R2 + R3 → R3 $\begin{bmatrix} 1 & 1 & 1 & -1 \\ 0 & 1 & 4 & 5 \\ 0 & 0 & 7 & 7 \end{bmatrix}$ $\dfrac{R3}{7} \to R3$ $\begin{bmatrix} 1 & 1 & 1 & -1 \\ 0 & 1 & 4 & 5 \\ 0 & 0 & 1 & 1 \end{bmatrix}$

Converting the augmented matrix back into equation form yields $\begin{cases} x + y + z = -1 \\ y + 4z = 5 \\ z = 1 \end{cases}$.

Back-substitution shows the solution is $(-3, 1, 1)$.

Now try Exercises 28 through 32 ▶

The process used in Example 4 is also called **Gaussian elimination** (Carl Friedrich Gauss, 1777–1855), with the last matrix written in **row-echelon form.** It's possible to solve a system entirely using only the augmented matrix, by continuing to use row operations until the diagonal entries are 1's, with 0's for all other entries: $\begin{bmatrix} 1 & 0 & 0 & a \\ 0 & 1 & 0 & b \\ 0 & 0 & 1 & c \end{bmatrix}$. The process is then called **Gauss-Jordan elimination** (Wilhelm Jordan, 1842–1899), with the final matrix written in **reduced row-echelon form** (see Appendix III).

☑ **C. You've just learned how to solve a system of equations using row operations**

Note that with Gauss-Jordan elimination, our *initial* focus is less on getting 1's along the diagonal, and more on obtaining zeroes for all entries *other than* the diagonal entries. This will enable us to work with integer values in the solution process.

EXAMPLE 5 ▶ **Solving a System Using Gauss-Jordan Elimination**

Solve using Gauss-Jordan elimination $\begin{cases} 2x + 5z - 15 = 2y \\ 2x + 3y = -1 + z. \\ 4y + z = -7 \end{cases}$

Solution ▶ standard form
$\begin{cases} 2x - 2y + 5z = 15 \\ 2x + 3y - 1z = -1 \\ 0x + 4y + 1z = -7 \end{cases}$ matrix form → $\begin{bmatrix} 2 & -2 & 5 & 15 \\ 2 & 3 & -1 & -1 \\ 0 & 4 & 1 & -7 \end{bmatrix}$ −R1 + R2 → R2 $\begin{bmatrix} 2 & -2 & 5 & 15 \\ 0 & 5 & -6 & -16 \\ 0 & 4 & 1 & -7 \end{bmatrix}$

$\begin{bmatrix} 2 & -2 & 5 & 15 \\ 0 & 5 & -6 & -16 \\ 0 & 4 & 1 & -7 \end{bmatrix}$ 2R2 + 5R1 → R1 / −4R2 + 5R3 → R3 $\begin{bmatrix} 10 & 0 & 13 & 43 \\ 0 & 5 & -6 & -16 \\ 0 & 0 & 29 & 29 \end{bmatrix}$ $\dfrac{R3}{29} \to R3$ $\begin{bmatrix} 10 & 0 & 13 & 43 \\ 0 & 5 & -6 & -16 \\ 0 & 0 & 1 & 1 \end{bmatrix}$

$$\begin{bmatrix} 10 & 0 & 13 & | & 43 \\ 0 & 5 & -6 & | & -16 \\ 0 & 0 & 1 & | & 1 \end{bmatrix} \begin{array}{l} -13R3 + R1 \to R1 \\ \\ 6R3 + R2 \to R2 \end{array} \begin{bmatrix} 10 & 0 & 0 & | & 30 \\ 0 & 5 & 0 & | & -10 \\ 0 & 0 & 1 & | & 1 \end{bmatrix} \begin{array}{l} \frac{R1}{10} \to R1 \\ \\ \frac{R2}{5} \to R2 \end{array} \begin{bmatrix} 1 & 0 & 0 & | & 3 \\ 0 & 1 & 0 & | & -2 \\ 0 & 0 & 1 & | & 1 \end{bmatrix}$$

The final matrix shows the solution is $(3, -2, 1)$.

Now try Exercises 33 through 36 ▶

D. Inconsistent and Dependent Systems

Due to the strong link between a linear system and its augmented matrix, inconsistent and dependent systems can be recognized just as in Sections 5.1 and 5.2. An inconsistent system will yield an inconsistent or contradictory statement such as $0 = -12$, meaning all entries in a row of the matrix of coefficients are zero, but the constant is not. A linearly dependent system will yield an identity statement such as $0 = 0$, meaning all entries in one row of the matrix are zero. If the system has coincident dependence, there will be only one nonzero row of coefficients.

EXAMPLE 6 ▶ **Solving a Dependent System**

Solve the system using Gauss-Jordan elimination: $\begin{cases} x + y - 5z = 3 \\ -x + 2z = -1 \\ 2x - y - z = 0 \end{cases}$

Solution ▶

$\begin{cases} x + y - 5z = 3 \\ -x + 2z = -1 \\ 2x - y - z = 0 \end{cases}$ standard form → $\begin{cases} x + y - 5z = 3 \\ -x + 0y + 2z = -1 \\ 2x - y - z = 0 \end{cases}$ matrix form → $\begin{bmatrix} 1 & 1 & -5 & | & 3 \\ -1 & 0 & 2 & | & -1 \\ 2 & -1 & -1 & | & 0 \end{bmatrix}$

$\begin{bmatrix} 1 & 1 & -5 & | & 3 \\ -1 & 0 & 2 & | & -1 \\ 2 & -1 & -1 & | & 0 \end{bmatrix} \begin{array}{l} R1 + R2 \to R2 \\ \\ -2R1 + R3 \to R3 \end{array} \begin{bmatrix} 1 & 1 & -5 & | & 3 \\ 0 & 1 & -3 & | & 2 \\ 0 & -3 & 9 & | & -6 \end{bmatrix} \begin{array}{l} -1R2 + R1 \to R1 \\ \\ 3R2 + R3 \to R3 \end{array} \begin{bmatrix} 1 & 0 & -2 & | & 1 \\ 0 & 1 & -3 & | & 2 \\ 0 & 0 & 0 & | & 0 \end{bmatrix}$

Since all entries in the last row are zeroes and it's the only row of zeroes, we conclude the system is linearly dependent and equivalent to $\begin{cases} x - 2z = 1 \\ y - 3z = 2 \end{cases}$. As in Chapter 5, we demonstrate this dependence by writing the (x, y, z) solution in terms of a parameter. Solving for y in R2 gives y in terms of z: $y = 3z + 2$. Solving for x in R1 gives x in terms of z: $x = 2z + 1$. As written, the solutions all depend on z: $x = 2z + 1$, $y = 3z + 2$, and $z = z$. Selecting p as the parameter (or some other "neutral" variable), we write the solution as $(2p + 1, 3p + 2, p)$. Two of the infinite number of solutions would be $(1, 2, 0)$ for $p = 0$, and $(-1, -1, -1)$ for $p = -1$. Test these triples in the original equations.

Now try Exercises 37 through 45 ▶

☑ **D. You've just learned how to recognize inconsistent and dependent systems**

E. Solving Applications Using Matrices

As in other areas, solving applications using systems relies heavily on the ability to mathematically model information given verbally or in context. As you work through the exercises, read each problem carefully. Look for relationships that yield a system of two equations in two variables, three equations in three variables and so on.

EXAMPLE 7 ▶ **Determining the Original Value of Collector's Items**

A museum purchases a famous painting, a ruby tiara, and a rare coin for its collection, spending a total of $30,000. One year later, the painting has tripled in value, while the tiara and the coin have doubled in value. The items now have a total value of $75,000. Find the purchase price of each if the original price of the painting was $1000 more than twice the coin.

Solution ▶ Let P represent the price of the painting, T the tiara, and C the coin.

Total spent was \$30,000: $\rightarrow P + T + C = 30{,}000$
One year later: $\rightarrow 3P + 2T + 2C = 75{,}000$
Value of painting versus coin: $\rightarrow P = 2C + 1000$

$$\begin{cases} P + T + C = 30000 \\ 3P + 2T + 2C = 75000 \\ P = 2C + 1000 \end{cases} \text{standard form} \rightarrow \begin{cases} 1P + 1T + 1C = 30000 \\ 3P + 2T + 2C = 75000 \\ 1P + 0T - 2C = 1000 \end{cases} \text{matrix form} \rightarrow \begin{bmatrix} 1 & 1 & 1 & \vdots & 30000 \\ 3 & 2 & 2 & \vdots & 75000 \\ 1 & 0 & -2 & \vdots & 1000 \end{bmatrix}$$

$$\begin{bmatrix} 1 & 1 & 1 & \vdots & 30000 \\ 3 & 2 & 2 & \vdots & 75000 \\ 1 & 0 & -2 & \vdots & 1000 \end{bmatrix} \begin{array}{l} -3R1 + R2 \rightarrow R2 \\ \\ -1R1 + R3 \rightarrow R3 \end{array} \begin{bmatrix} 1 & 1 & 1 & \vdots & 30000 \\ 0 & -1 & -1 & \vdots & -15000 \\ 0 & -1 & -3 & \vdots & -29000 \end{bmatrix} \begin{array}{l} -1R2 \rightarrow R2 \\ \\ -1R3 \rightarrow R3 \end{array} \begin{bmatrix} 1 & 1 & 1 & \vdots & 30000 \\ 0 & 1 & 1 & \vdots & 15000 \\ 0 & 1 & 3 & \vdots & 29000 \end{bmatrix}$$

$$\begin{bmatrix} 1 & 1 & 1 & \vdots & 30000 \\ 0 & 1 & 1 & \vdots & 15000 \\ 0 & 1 & 3 & \vdots & 29000 \end{bmatrix} \begin{array}{l} -1R2 + R3 \rightarrow R3 \end{array} \begin{bmatrix} 1 & 1 & 1 & \vdots & 30000 \\ 0 & 1 & 1 & \vdots & 15000 \\ 0 & 0 & 2 & \vdots & 14000 \end{bmatrix} \begin{array}{l} \dfrac{R3}{2} \rightarrow R3 \end{array} \begin{bmatrix} 1 & 1 & 1 & \vdots & 30000 \\ 0 & 1 & 1 & \vdots & 15000 \\ 0 & 0 & 1 & \vdots & 7000 \end{bmatrix}$$

From R3 of the triangularized form, $C = \$7000$ directly. Since R2 represents $T + C = 15{,}000$, we find the tiara was purchased for $T = \$8000$. Substituting these values into the first equation shows the painting was purchased for \$15,000. The solution is (15,000, 8,000, 7,000).

☑ **E.** You've just learned how to solve applications using linear systems

Now try Exercises 48 through 55 ▶

TECHNOLOGY HIGHLIGHT

Solving Systems Using Matrices and Calculating Technology

Graphing calculators offer a very efficient way to solve systems using matrices. Once the system has been written in matrix form, it can easily be entered and solved by asking the calculator to instantly perform the row operations needed. Pressing `2nd` `x⁻¹` **(MATRIX)** gives a screen similar to the one shown in Figure 8.19, where we begin by selecting the **EDIT** option (push the right arrow `▶` twice). Pressing `ENTER` places you on a screen where you can EDIT matrix A, changing the size as needed. Using the 3×4 matrix from Example 4, we press 3 and `ENTER`, then 4 and `ENTER`, giving the screen shown in Figure 8.20. The dash marks to the right indicate that there is a fourth column that cannot be seen, but that comes into view as you enter the elements of the matrix. Begin entering the first row of the matrix resulting from $R_1 \leftrightarrow R_2$, which has entries {1, 1, 1, −1}. Press `ENTER` after each entry and the cursor automatically goes to the next position in the matrix (note that the TI-84 Plus automatically shifts left and right to allow all four columns to be entered). After entering the second row {2, 1, −2, −7} and the third row {0, −2, −1, −3}, the completed matrix should look like the one shown in Figure 8.21 (the matrix is currently shifted to the right, showing the fourth column). To write this matrix in reduced row-echelon form **(rref)** we return to the home screen by pressing `2nd` `MODE` **(QUIT)**. Press the `CLEAR` key for a clean home screen. To access the **rref** function, press `2nd` `x⁻¹` **(MATRIX)** and select

Figure 8.19

```
NAMES MATH EDIT
1:[A]   1×1
2:[B]   1×1
3:[C]   1×1
4:[D]   1×1
5:[E]   1×1
6:[F]
7↓[G]
```

Figure 8.20

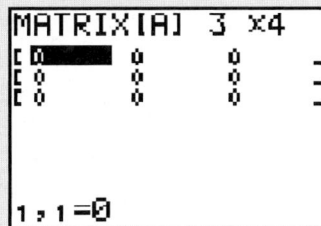

Figure 8.21

the **MATH** option, then scroll upward (or downward) until you get to **B:rref.** Pressing [ENTER] places this function on the home screen, where we must tell it to perform the **rref** operation on matrix [A]. Press [2nd] [x⁻¹] **(MATRIX)** to select a matrix (notice that matrix **NAMES** is automatically highlighted. Press [ENTER] to select matrix [A] as the object of the **rref** function. After pressing [ENTER] the calculator quickly computes the reduced row-echelon form and displays it on the screen as in Figure 8.22. The solution is easily read as $x = -3$, $y = 1$, and $z = 1$, as we found in Example 4. Use these ideas to complete the following.

Figure 8.22

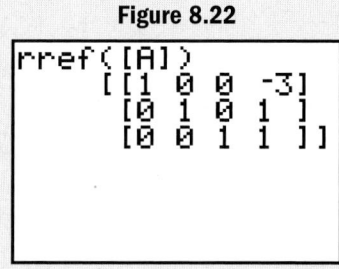

Exercise 1: Use this method to solve the 2 × 2 system from Exercise 30.

Exercise 2: Use this method to solve the 3 × 3 system from Exercise 32.

8.5 EXERCISES

▶ CONCEPTS AND VOCABULARY

Fill in the blank with the appropriate word or phrase. Carefully reread the section if needed.

1. A matrix with the same number of rows and columns is called a(n) _____ matrix.

2. When the coefficient matrix is used with the matrix of constants, the result is a(n) _____ matrix.

3. Matrix $A = \begin{bmatrix} 2 & 4 & -3 \\ 1 & -2 & 1 \end{bmatrix}$ is a ___ by ___ matrix. The entry in the second row and third column is ___.

4. Given matrix B shown here, the diagonal entries are ___, ___, and ___.

$$B = \begin{bmatrix} 1 & 4 & 3 \\ -1 & 5 & 2 \\ 3 & -2 & 1 \end{bmatrix}$$

5. The notation $-2R1 + R2 \rightarrow R2$ indicates that an equivalent matrix is formed by performing what operations/replacements?

6. Describe how to tell an inconsistent system apart from a dependent system when solving using matrix methods (row reduction).

▶ DEVELOPING YOUR SKILLS

Determine the size (order) of each matrix and identify the third row and second column entry. If the matrix given is a square matrix, identify the diagonal entries.

7. $\begin{bmatrix} 1 & 0 \\ 2.1 & 1 \\ -3 & 5.8 \end{bmatrix}$

8. $\begin{bmatrix} 1 & 0 & 4 \\ 1 & 3 & -7 \\ 5 & -1 & 2 \end{bmatrix}$

9. $\begin{bmatrix} 1 & 0 & 4 \\ 1 & 3 & -7 \\ 5 & -1 & 2 \\ 2 & -3 & 9 \end{bmatrix}$

Form the augmented matrix, then name the diagonal entries of the coefficient matrix.

10. $\begin{cases} 2x - 3y - 2z = 7 \\ x - y + 2z = -5 \\ 3x + 2y - z = 11 \end{cases}$

11. $\begin{cases} x + 2y - z = 1 \\ x + z = 3 \\ 2x - y + z = 3 \end{cases}$

12. $\begin{cases} 2x + 3y + z = 5 \\ 2y - z = 7 \\ x - y - 2z = 5 \end{cases}$

Write the system of equations for each matrix. Then use back-substitution to find its solution.

13. $\begin{bmatrix} 1 & 4 & \vdots & 5 \\ 0 & 1 & \vdots & \frac{1}{2} \end{bmatrix}$

14. $\begin{bmatrix} 1 & -5 & \vdots & -15 \\ 0 & -1 & \vdots & -2 \end{bmatrix}$

15. $\begin{bmatrix} 1 & 2 & -1 & \vdots & 0 \\ 0 & 1 & 2 & \vdots & 2 \\ 0 & 0 & 1 & \vdots & 3 \end{bmatrix}$

16. $\begin{bmatrix} 1 & 0 & 7 & \vdots & -5 \\ 0 & 1 & -5 & \vdots & 15 \\ 0 & 0 & 1 & \vdots & -26 \end{bmatrix}$

17. $\begin{bmatrix} 1 & 3 & -4 & \vdots & 29 \\ 0 & 1 & -\frac{3}{2} & \vdots & \frac{21}{2} \\ 0 & 0 & 1 & \vdots & 3 \end{bmatrix}$

18. $\begin{bmatrix} 1 & 2 & -1 & \vdots & 3 \\ 0 & 1 & \frac{1}{6} & \vdots & \frac{2}{3} \\ 0 & 0 & 1 & \vdots & \frac{22}{7} \end{bmatrix}$

Perform the indicated row operation(s) and write the new matrix.

19. $\begin{bmatrix} \frac{1}{2} & -3 & \vdots & -1 \\ -5 & 2 & \vdots & 4 \end{bmatrix}$ $\begin{matrix} 2R1 \rightarrow R1, \\ 5R1 + R2 \rightarrow R2 \end{matrix}$

20. $\begin{bmatrix} 7 & 4 & \vdots & 3 \\ 4 & -8 & \vdots & 12 \end{bmatrix}$ $\begin{matrix} \frac{1}{4}R2 \rightarrow R2, \\ R1 \leftrightarrow R2 \end{matrix}$

21. $\begin{bmatrix} -2 & 1 & 0 & \vdots & 4 \\ 5 & 8 & 3 & \vdots & -5 \\ 1 & -3 & 3 & \vdots & 2 \end{bmatrix}$ $\begin{matrix} R1 \leftrightarrow R3, \\ -5R1 + R2 \rightarrow R2 \end{matrix}$

22. $\begin{bmatrix} -3 & 2 & 0 & \vdots & 0 \\ 1 & 1 & 2 & \vdots & 6 \\ 4 & 1 & -3 & \vdots & 2 \end{bmatrix}$ $\begin{matrix} R1 \leftrightarrow R2, \\ -4R1 + R3 \rightarrow R3 \end{matrix}$

23. $\begin{bmatrix} 3 & 1 & 1 & \vdots & 8 \\ 6 & -1 & -1 & \vdots & 10 \\ 4 & -2 & -3 & \vdots & 22 \end{bmatrix}$ $\begin{matrix} -2R1 + R2 \rightarrow R2, \\ -4R1 + 3R3 \rightarrow R3 \end{matrix}$

24. $\begin{bmatrix} 2 & 1 & -1 & \vdots & -3 \\ 3 & 1 & 1 & \vdots & 0 \\ 4 & 3 & 2 & \vdots & 3 \end{bmatrix}$ $\begin{matrix} -3R1 + 2R2 \rightarrow R2, \\ -2R1 + R3 \rightarrow R3 \end{matrix}$

What row operations would produce zeroes beneath the first entry in the diagonal?

25. $\begin{bmatrix} 1 & 3 & 0 & \vdots & 2 \\ -2 & 4 & 1 & \vdots & 1 \\ 3 & -1 & -2 & \vdots & 9 \end{bmatrix}$

26. $\begin{bmatrix} 1 & 1 & -4 & \vdots & -3 \\ 3 & 0 & 1 & \vdots & 5 \\ -5 & 3 & 2 & \vdots & 3 \end{bmatrix}$

27. $\begin{bmatrix} 1 & 2 & 0 & \vdots & 10 \\ 5 & 1 & 2 & \vdots & 6 \\ -4 & 3 & -3 & \vdots & 2 \end{bmatrix}$

Solve each system by triangularizing the augmented matrix and using back-substitution. Simplify by clearing fractions or decimals before beginning.

28. $\begin{cases} 2y = 5x + 4 \\ -5x = 2 - 4y \end{cases}$

29. $\begin{cases} 0.15g - 0.35h = -0.5 \\ -0.12g + 0.25h = 0.1 \end{cases}$

30. $\begin{cases} -\frac{1}{5}u + \frac{1}{4}v = 1 \\ \frac{1}{10}u + \frac{1}{2}v = 7 \end{cases}$

31. $\begin{cases} x - 2y + 2z = 7 \\ 2x + 2y - z = 5 \\ 3x - y + z = 6 \end{cases}$

32. $\begin{cases} 2x - 3y - 2z = 7 \\ x - y + 2z = -5 \\ 3x + 2y - z = 11 \end{cases}$

33. $\begin{cases} x + 2y - z = 1 \\ x + z = 3 \\ 2x - y + z = 3 \end{cases}$

34. $\begin{cases} 2x + 3y + z = 5 \\ 2y - z = 7 \\ x - y - 2z = 5 \end{cases}$

35. $\begin{cases} -x + y + 2z = 2 \\ x + y - z = 1 \\ 2x + y + z = 4 \end{cases}$

36. $\begin{cases} x + y - 2z = -1 \\ 4x - y + 3z = 3 \\ 3x + 2y - z = 4 \end{cases}$

Solve each system by triangularizing the augmented matrix and using back-substitution. If the system is linearly dependent, give the solution in terms of a parameter. If the system has coincident dependence, answer in set notation as in Chapter 5.

37. $\begin{cases} 4x - 8y + 8z = 24 \\ 2x - 6y + 3z = 13 \\ 3x + 4y - z = -11 \end{cases}$

38. $\begin{cases} 3x + y + z = -2 \\ x - 2y + 3z = 1 \\ 2x - 3y + 5z = 3 \end{cases}$

39. $\begin{cases} x + 3y + 5z = 20 \\ 2x + 3y + 4z = 16 \\ x + 2y + 3z = 12 \end{cases}$

40. $\begin{cases} -x + 2y + 3z = -6 \\ x - y + 2z = -4 \\ 3x - 6y - 9z = 18 \end{cases}$

41. $\begin{cases} 3x - 4y + 2z = -2 \\ \frac{3}{2}x - 2y + z = -1 \\ -6x + 8y - 4z = 4 \end{cases}$

42. $\begin{cases} 2x - y + 3z = 1 \\ 4x - 2y + 6z = 2 \\ 10x - 5y + 15z = 5 \end{cases}$

43. $\begin{cases} 2x - y + 3z = 1 \\ 2y + 6z = 2 \\ x - \frac{1}{2}y + \frac{3}{2}z = 5 \end{cases}$

44. $\begin{cases} x + 2y + z = 4 \\ 3x - 4y + z = 4 \\ 6x - 8y + 2z = 8 \end{cases}$

45. $\begin{cases} -2x + 4y - 3z = 4 \\ 5x - 6y + 7z = -12 \\ x + 2y + z = -4 \end{cases}$

▶ WORKING WITH FORMULAS

Area of a triangle in the plane:

$$A = \pm\frac{1}{2}(x_1y_2 - x_2y_1 + x_2y_3 - x_3y_2 + x_3y_1 - x_1y_3)$$

The area of a triangle in the plane is given by the formula shown, where the vertices of the triangle are located at the points (x_1, y_1), (x_2, y_2), and (x_3, y_3), and the sign is chosen to ensure a positive value.

46. Find the area of a triangle whose vertices are $(-1, -3)$, $(5, 2)$, and $(1, 8)$.

47. Find the area of a triangle whose vertices are $(6, -2)$, $(-5, 4)$, and $(-1, 7)$.

▶ APPLICATIONS

Model each problem using a system of linear equations. Then solve using the augmented matrix.

Descriptive Translation

48. The distance (via air travel) from Los Angeles (LA), California, to Saint Louis (STL), Missouri, to Cincinnati (CIN), Ohio, to New York City (NYC), New York, is approximately 2480 mi. Find the distances between each city if the distance from LA to STL is 50 mi more than five times the distance between STL and CIN and 110 mi less than three times the distance from CIN to NYC.

49. In the 2006 NBA Championship Series, Dwayne Wade of the Miami Heat carried his team to the title after the first two games were lost to the Dallas Mavericks. If 187 points were scored in the

title game and the Heat won by 3 points, what was the final score?

50. Moe is lecturing Larry and Curly once again (Moe, Larry, and Curly of *The Three Stooges* fame) claiming he is twice as smart as Larry and three times as smart as Curly. If he is correct and the sum of their IQs is 165, what is the IQ of each stooge?

51. A collector of rare books buys a handwritten, autographed copy of Edgar Allan Poe's *Annabel Lee,* an original advance copy of L. Frank Baum's *The Wonderful Wizard of Oz,* and a first print copy of *The Caine Mutiny* by Herman Wouk, paying a total of $100,000. Find the cost of each one, given that the cost of *Annabel Lee* and twice the cost of *The Caine Mutiny* sum to the price paid for *The Wonderful Wizard of Oz,* and *The Caine Mutiny* cost twice as much as *Annabel Lee.*

Geometry

52. A right triangle has a hypotenuse of 39 m. If the perimeter is 90 m, and the longer leg is 6 m longer than twice the shorter leg, find the dimensions of the triangle.

53. In triangle *ABC*, the sum of angles *A* and *C* is equal to three times angle *B*. Angle *C* is 10 degrees more than twice angle *B*. Find the measure of each angle.

Investment and Finance

54. Suppose $10,000 is invested in three different investment vehicles paying 5%, 7%, and 9% annual interest. Find the amount invested at each rate if the interest earned after 1 yr is $760 and the amount invested at 9% is equal to the sum of the amounts invested at 5% and 7%.

55. The trustee of a union's pension fund has invested the funds in three ways: a savings fund paying 4% annual interest, a money market fund paying 7%, and government bonds paying 8%. Find the amount invested in each if the interest earned after one year is $0.178 million and the amount in government bonds is $0.3 million more than twice the amount in money market funds. The total amount invested is $2.5 million dollars.

▶ **EXTENDING THE CONCEPT**

56. In previous sections, we noted that one condition for a 3 × 3 system to be dependent was for the third equation to be a linear combination of the other two. To test this, write any two (different) equations using the same three variables, then form a third equation by performing some combination of elementary row operations. Solve the resulting 3 × 3 system. What do you notice?

57. Given the drawing shown, use a system of equations and the matrix method to find the measure of the angles labeled as *x* and *y*. Recall that vertical angles

are equal and that the sum of the angles in a triangle is 180°.

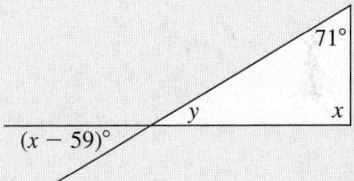

$(x - 59)°$ y x $71°$

58. The system given here has a solution of $(1, -2, 3)$. Find the value of *a* and *b*.

$$\begin{bmatrix} 1 & a & b & | & 1 \\ 2b & 2a & 5 & | & 13 \\ 2a & 7 & 3b & | & -8 \end{bmatrix}$$

▶ **MAINTAINING YOUR SKILLS**

59. (7.5) **a.** Convert $z_1 = -1 - 3i$ to trigonometric form.

   b. Convert $z_2 = 5 \operatorname{cis}\left(\dfrac{2\pi}{3}\right)$ to rectangular form.

60. (5.4) State the exact value of the following trig functions:

   a. $\sin\left(-\dfrac{\pi}{6}\right)$ **b.** $\cos\left(\dfrac{5\pi}{4}\right)$

   c. $\tan\left(\dfrac{2\pi}{3}\right)$ **d.** $\csc\left(\dfrac{3\pi}{2}\right)$

61. (4.4) Since 2005, cable installations for an Internet company have been modeled by the function $C(t) = 15 \ln (t + 1)$, where $C(t)$ represents cable installations in thousands, *t* yr after 2005. In what

year will the number of installations be greater than 30,000?

62. (4.5) If a set amount of money *p* is deposited regularly (daily, weekly, monthly, etc.) *n* times per year at a fixed interest rate *r*, the amount of money *A* accumulated in *t* years is given by the formula shown. If a parent deposits $250 per month for 18 yr at 4.6% beginning when her first child was born, how much has been accumulated to help pay for college expenses?

$$A = \dfrac{p\left[\left(1 + \dfrac{r}{n}\right)^{nt} - 1\right]}{\dfrac{r}{n}}$$

Learning Objectives

In Section 8.6 you will learn how to:

☐ **A.** Determine if two matrices are equal

☐ **B.** Add and subtract matrices

☐ **C.** Compute the product of two matrices

Matrices serve a much wider purpose than just a convenient method for solving systems. To understand their broader application, we need to know more about matrix theory, the various ways matrices can be combined, and some of their more practical uses. The common operations of addition, subtraction, multiplication, and division are all defined for matrices, as are other operations. Practical applications of matrix theory can be found in the social sciences, inventory management, genetics, operations research, engineering, and many other fields.

A. Equality of Matrices

To effectively study matrix algebra, we first give matrices a more general definition. For the *general* matrix A, all entries will be denoted using the lowercase letter "a," with their position in the matrix designated by the dual subscript a_{ij}. The letter "i" gives the *row* and the letter "j" gives the *column* of the entry's location. The general $m \times n$ matrix A is written

$$
\begin{array}{c}
\text{row } 1 \rightarrow \\
\text{row } 2 \rightarrow \\
\text{row } 3 \rightarrow \\
\\
\text{row } i \rightarrow \\
\\
\text{row } m \rightarrow
\end{array}
\begin{bmatrix}
a_{11} & a_{12} & a_{13} & \cdots & a_{1j} & \cdots & a_{1n} \\
a_{21} & a_{22} & a_{23} & \cdots & a_{2j} & \cdots & a_{2n} \\
a_{31} & a_{32} & a_{33} & \cdots & a_{3j} & \cdots & a_{3n} \\
\vdots & \vdots & \vdots & & \vdots & & \vdots \\
a_{i1} & a_{i2} & a_{i3} & \cdots & a_{ij} & \cdots & a_{in} \\
\vdots & \vdots & \vdots & & \vdots & & \vdots \\
a_{m1} & a_{m2} & a_{m3} & \cdots & a_{mj} & \cdots & a_{mn}
\end{bmatrix}
$$

with columns labeled col 1, col 2, col 3, col j, col n. a_{ij} is a general matrix element.

The size of a matrix is also referred to as its **order,** and we say the order of general matrix A is $m \times n$. Note that diagonal entries have the same row and column number, a_{ij}, where $i = j$. Also, where the general entry of matrix A is a_{ij}, the general entry of matrix B is b_{ij}, of matrix C is c_{ij}, and so on.

EXAMPLE 1 ▶ **Identifying the Order and Entries of a Matrix**

State the order of each matrix and name the entries corresponding to a_{22}, a_{31}; b_{22}, b_{31}; and c_{22}, c_{31}.

a. $A = \begin{bmatrix} 1 & 4 \\ -2 & -3 \end{bmatrix}$ **b.** $B = \begin{bmatrix} 3 & -2 \\ 1 & 5 \\ -4 & 3 \end{bmatrix}$ **c.** $C = \begin{bmatrix} 0.2 & -0.5 & 0.7 \\ -1 & 0.3 & 1 \\ 2.1 & -0.1 & 0.6 \end{bmatrix}$

Solution ▶ **a.** matrix A: order 2×2. Entry $a_{22} = -3$ (the row 2, column 2 entry is -3). There is no a_{31} entry (A is only 2×2).
b. matrix B: order 3×2. Entry $b_{22} = 5$, entry $b_{31} = -4$.
c. matrix C: order 3×3. Entry $c_{22} = 0.3$, entry $c_{31} = 2.1$.

Now try Exercises 7 through 12 ▶

Equality of Matrices

Two matrices are equal if they have the same order and their corresponding entries are equal. In symbols, this means that $A = B$ if $a_{ij} = b_{ij}$ for all i and j.

EXAMPLE 2 ▶ **Determining If Two Matrices Are Equal**

Determine whether the following statements are true, false, or conditional. If false, explain why. If conditional, find values that will make the statement true.

a. $\begin{bmatrix} 1 & 4 \\ -2 & -3 \end{bmatrix} = \begin{bmatrix} -3 & -2 \\ 4 & 1 \end{bmatrix}$
b. $\begin{bmatrix} 3 & -2 \\ 1 & 5 \\ -4 & 3 \end{bmatrix} = \begin{bmatrix} 3 & -2 & 1 \\ 5 & -4 & 3 \end{bmatrix}$

c. $\begin{bmatrix} 1 & 4 \\ -2 & -3 \end{bmatrix} = \begin{bmatrix} a-2 & 2b \\ c & -3 \end{bmatrix}$

Solution ▶ a. $\begin{bmatrix} 1 & 4 \\ -2 & -3 \end{bmatrix} = \begin{bmatrix} -3 & -2 \\ 4 & 1 \end{bmatrix}$ is false. The matrices have the same order and entries, but corresponding entries are not equal.

b. $\begin{bmatrix} 3 & -2 \\ 1 & 5 \\ -4 & 3 \end{bmatrix} = \begin{bmatrix} 3 & -2 & 1 \\ 5 & -4 & 3 \end{bmatrix}$ is false. Their orders are not equal.

☑ **A.** You've just learned how to determine if two matrices are equal

c. $\begin{bmatrix} 1 & 4 \\ -2 & -3 \end{bmatrix} = \begin{bmatrix} a-2 & 2b \\ c & -3 \end{bmatrix}$ is conditional. The statement is true when $a - 2 = 1$ ($a = 3$), $2b = 4$ ($b = 2$), $c = -2$, and is false otherwise.

Now try Exercises 13 through 16 ▶

B. Addition and Subtraction of Matrices

A sum or difference of matrices is found by combining the corresponding entries. This limits the operations to matrices of like orders, so that every entry in one matrix has a "corresponding entry" in the other. This also means the result is a new matrix of *like order*, whose entries are the corresponding sums or differences. Note that since a_{ij} represents a general entry of matrix A, $[a_{ij}]$ represents the entire matrix.

Addition and Subtraction of Matrices

Given matrices A, B, and C having like orders.

The sum $A + B = C$,   The difference $A - B = D$,
where $[a_{ij} + b_{ij}] = [c_{ij}]$.   where $[a_{ij} - b_{ij}] = [d_{ij}]$.

EXAMPLE 3 ▶ **Adding and Subtracting Matrices**

Compute the sum or difference of the matrices indicated.

$A = \begin{bmatrix} 2 & 6 \\ 1 & 0 \\ 1 & -3 \end{bmatrix}$ $B = \begin{bmatrix} -3 & 2 & -1 \\ -5 & 4 & 3 \end{bmatrix}$ $C = \begin{bmatrix} 3 & -2 \\ 1 & 5 \\ -4 & 3 \end{bmatrix}$

a. $A + C$ b. $A + B$ c. $C - A$

Solution ▶ a. $A + C = \begin{bmatrix} 2 & 6 \\ 1 & 0 \\ 1 & -3 \end{bmatrix} + \begin{bmatrix} 3 & -2 \\ 1 & 5 \\ -4 & 3 \end{bmatrix}$ sum of A and C

$= \begin{bmatrix} 2+3 & 6+(-2) \\ 1+1 & 0+5 \\ 1+(-4) & -3+3 \end{bmatrix} = \begin{bmatrix} 5 & 4 \\ 2 & 5 \\ -3 & 0 \end{bmatrix}$ add corresponding entries

b. $A + B = \begin{bmatrix} 2 & 6 \\ 1 & 0 \\ 1 & -3 \end{bmatrix} + \begin{bmatrix} -3 & 2 & -1 \\ -5 & 4 & 3 \end{bmatrix}$ Addition and subtraction are not defined for matrices of unlike order.

c. $C - A = \begin{bmatrix} 3 & -2 \\ 1 & 5 \\ -4 & 3 \end{bmatrix} - \begin{bmatrix} 2 & 6 \\ 1 & 0 \\ 1 & -3 \end{bmatrix}$ difference of C and A

$= \begin{bmatrix} 3-2 & -2-6 \\ 1-1 & 5-0 \\ -4-1 & 3-(-3) \end{bmatrix} = \begin{bmatrix} 1 & -8 \\ 0 & 5 \\ -5 & 6 \end{bmatrix}$ subtract corresponding entries

☑ **B. You've just learned how to add and subtract matrices**

Now try Exercises 17 through 20 ▶

Since the addition of two matrices is defined as the sum of corresponding entries, we find the properties of matrix addition closely resemble those of real number addition.

Properties of Matrix Addition

Given matrices A, B, C, and Z are $m \times n$ matrices, with Z the zero matrix. Then,

I.	$A + B = B + A$	matrix addition is commutative
II.	$(A + B) + C = A + (B + C)$	matrix addition is associative
III.	$A + Z = Z + A = A$	Z is the additive identity
IV.	$A + (-A) = (-A) + A = Z$	$-A$ is the additive inverse of A

C. Matrices and Multiplication

The algebraic terms $2a$ and ab have counterparts in matrix algebra. The product $2A$ represents a constant times a matrix and is called **scalar multiplication.** The product AB represents the product of two matrices.

Scalar Multiplication

Scalar multiplication is defined by taking the product of the constant with *each entry* in the matrix, forming a new matrix of like size. In symbols, for any real number k and matrix A, $kA = [ka_{ij}]$. Similar to standard algebraic properties, $-A$ represents the scalar product $-1 \cdot A$ and any subtraction can be rewritten as an algebraic sum: $A - B = A + (-B)$. As noted in the properties box, for any matrix A, the sum $A + (-A)$ will yield the **zero matrix Z,** a matrix of like size whose entries are all zeroes. Also note that matrix $-A$ is the **additive inverse** for A, while Z is the **additive identity.**

EXAMPLE 4 ▶ **Computing Operations on Matrices**

Given $A = \begin{bmatrix} 4 & 3 \\ \frac{1}{2} & 1 \\ 0 & -3 \end{bmatrix}$ and $B = \begin{bmatrix} 3 & -2 \\ 0 & 6 \\ -4 & 0.4 \end{bmatrix}$, compute the following:

a. $\frac{1}{2}B$ **b.** $-4A - \frac{1}{2}B$

Solution ▶ **a.** $\frac{1}{2}B = \left(\frac{1}{2}\right)\begin{bmatrix} 3 & -2 \\ 0 & 6 \\ -4 & 0.4 \end{bmatrix}$

$= \begin{bmatrix} \left(\frac{1}{2}\right)(3) & \left(\frac{1}{2}\right)(-2) \\ \left(\frac{1}{2}\right)(0) & \left(\frac{1}{2}\right)(6) \\ \left(\frac{1}{2}\right)(-4) & \left(\frac{1}{2}\right)(0.4) \end{bmatrix} = \begin{bmatrix} \frac{3}{2} & -1 \\ 0 & 3 \\ -2 & 0.2 \end{bmatrix}$

b. $-4A - \dfrac{1}{2}B = -4A + \left(-\dfrac{1}{2}\right)B$ rewrite using algebraic addition

$$= \begin{bmatrix} (-4)(4) & (-4)(3) \\ (-4)(\frac{1}{2}) & (-4)(1) \\ (-4)(0) & (-4)(-3) \end{bmatrix} + \begin{bmatrix} (-\frac{1}{2})(3) & (-\frac{1}{2})(-2) \\ (-\frac{1}{2})(0) & (-\frac{1}{2})(6) \\ (-\frac{1}{2})(-4) & (-\frac{1}{2})(0.4) \end{bmatrix}$$

$$= \begin{bmatrix} -16 & -12 \\ -2 & -4 \\ 0 & 12 \end{bmatrix} + \begin{bmatrix} -\frac{3}{2} & 1 \\ 0 & -3 \\ 2 & -0.2 \end{bmatrix} \quad \text{simplify}$$

$$= \begin{bmatrix} -16 + (-\frac{3}{2}) & -12 + 1 \\ -2 + 0 & -4 + (-3) \\ 0 + 2 & 12 + (-0.2) \end{bmatrix} = \begin{bmatrix} -\frac{35}{2} & -11 \\ -2 & -7 \\ 2 & 11.8 \end{bmatrix} \quad \text{result}$$

Now try Exercises 21 through 24 ▶

Matrix Multiplication

Consider a cable company offering three different levels of Internet service: Bronze—fast, Silver—very fast, and Gold—lightning fast. Table 8.3 shows the number and types of programs sold to households and businesses for the week. Each program has an incentive package consisting of a rebate and a certain number of free weeks, as shown in Table 8.4.

Table 8.3 Matrix A

	Bronze	Silver	Gold
Homes	40	20	**25**
Businesses	10	15	45

Table 8.4 Matrix B

	Rebate	Free Weeks
Bronze	$15	2
Silver	$25	4
Gold	**$35**	6

To compute the amount of rebate money the cable company paid to households for the week, we would take the first row (R1) in Table 8.3 and multiply by the corresponding entries (bronze with bronze, silver with silver, and so on) in the first column (C1) of Table 8.4 and add these products. In matrix form, we have

$$[40 \;\; 20 \;\; \mathbf{25}] \cdot \begin{bmatrix} 15 \\ 25 \\ \mathbf{35} \end{bmatrix} = 40 \cdot 15 + 20 \cdot 25 + \mathbf{25 \cdot 35} = \mathbf{\$1975}.$$ Using R1 of Table 8.3

with C2 from Table 8.4 gives the number of free weeks awarded to homes:

$$[40 \;\; 20 \;\; \mathbf{25}] \cdot \begin{bmatrix} 2 \\ 4 \\ \mathbf{6} \end{bmatrix} = 40 \cdot 2 + 20 \cdot 4 + \mathbf{25 \cdot 6} = 310.$$ Using the second row (R2) of

Table 8.3 with the two columns from Table 8.4 will give the amount of rebate money and the number of free weeks, respectively, awarded to business customers. When all computations are complete, the result is a product matrix P with order 2×2. This is because the product of R1 from matrix A, with C1 from matrix B, *gives the entry in position P_{11} of the product matrix:* $\begin{bmatrix} 40 & 20 & 25 \\ 10 & 15 & 45 \end{bmatrix} \cdot \begin{bmatrix} 15 & 2 \\ 25 & 4 \\ 35 & 6 \end{bmatrix} = \begin{bmatrix} 1975 & 310 \\ 2100 & 350 \end{bmatrix}.$

Likewise, the product R1 · C2 will give entry P_{12} (310), the product of R2 with C1 will give P_{21} (2100), and so on. This "row $\times$ column" multiplication can be generalized, and leads to the following. Given $m \times n$ matrix A and $s \times t$ matrix B,

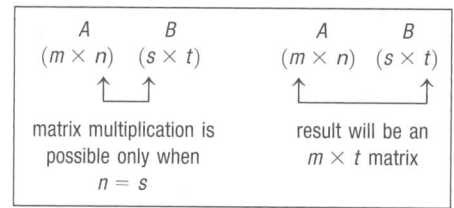

In more formal terms, we have the following definition of matrix multiplication.

Matrix Multiplication

Given the $m \times n$ matrix $A = [a_{ij}]$ and the $s \times t$ matrix $B = [b_{ij}]$. If $n = s$, then matrix multiplication is possible and the product AB is an $m \times t$ matrix $P = [p_{ij}]$, where p_{ij} is product of the ith row of A with the jth column of B.

In less formal terms, matrix multiplication involves multiplying the row entries of the first matrix with the corresponding column entries of the second, and adding them together.

In Example 5, two of the matrix products [parts (a) and (b)] are shown in full detail, with the first entry of the product matrix color-coded.

EXAMPLE 5 ▶ Multiplying Matrices

Given the matrices A through E shown here, compute the following products:

a. AB **b.** CD **c.** DC **d.** AE **e.** EA

$$A = \begin{bmatrix} -2 & 1 \\ 3 & 4 \end{bmatrix} \quad B = \begin{bmatrix} 4 & 3 \\ 6 & 1 \end{bmatrix} \quad C = \begin{bmatrix} -2 & 1 & 3 \\ 1 & 0 & 2 \\ 4 & 1 & -1 \end{bmatrix} \quad D = \begin{bmatrix} 2 & 5 & 1 \\ 4 & -1 & 1 \\ 0 & 3 & -2 \end{bmatrix} \quad E = \begin{bmatrix} -2 & -1 \\ 3 & 0 \\ 1 & 2 \end{bmatrix}$$

Solution ▶ a. $AB = \begin{bmatrix} -2 & 1 \\ 3 & 4 \end{bmatrix}\begin{bmatrix} 4 & 3 \\ 6 & 1 \end{bmatrix} = \begin{bmatrix} -2 & -5 \\ 36 & 13 \end{bmatrix}$

Computation: $\begin{bmatrix} (-2)(4) + (1)(6) & (-2)(3) + (1)(1) \\ (3)(4) + (4)(6) & (3)(3) + (4)(1) \end{bmatrix}$

A	B	A	B
(2×2)	(2×2)	(2×2)	(2×2)

multiplication is possible since 2 = 2

result will be a 2 × 2 matrix

b. $CD = \begin{bmatrix} -2 & 1 & \mathbf{3} \\ 1 & 0 & 2 \\ 4 & 1 & -1 \end{bmatrix}\begin{bmatrix} 2 & 5 & 1 \\ 4 & -1 & 1 \\ \mathbf{0} & 3 & -2 \end{bmatrix} = \begin{bmatrix} 0 & -2 & -7 \\ 2 & 11 & -3 \\ 12 & 16 & 7 \end{bmatrix}$

C	D	C	D
(3×3)	(3×3)	(3×3)	(3×3)

multiplication is possible since 3 = 3

result will be a 3 × 3 matrix

Computation: $\begin{bmatrix} (-2)(2) + (1)(4) + \mathbf{(3)(0)} & (-2)(5) + (1)(-1) + \mathbf{(3)(3)} & (-2)(1) + (1)(1) + \mathbf{(3)(-2)} \\ (1)(2) + (0)(4) + (2)\mathbf{(0)} & (1)(5) + (0)(-1) + (2)(3) & (1)(1) + (0)(1) + (2)(-2) \\ (4)(2) + (1)(4) + (-1)\mathbf{(0)} & (4)(5) + (1)(-1) + (-1)(3) & (4)(1) + (1)(1) + (-1)(-2) \end{bmatrix}$

c. $DC = \begin{bmatrix} 2 & 5 & 1 \\ 4 & -1 & 1 \\ 0 & 3 & -2 \end{bmatrix}\begin{bmatrix} -2 & 1 & 3 \\ 1 & 0 & 2 \\ 4 & 1 & -1 \end{bmatrix} = \begin{bmatrix} 5 & 3 & 15 \\ -5 & 5 & 9 \\ -5 & -2 & 8 \end{bmatrix}$

D	C	D	C
(3×3)	(3×3)	(3×3)	(3×3)

multiplication is possible since 3 = 3

result will be a 3 × 3 matrix

d. $AE = \begin{bmatrix} -2 & 1 \\ 3 & 4 \end{bmatrix} \begin{bmatrix} -2 & -1 \\ 3 & 0 \\ 1 & 2 \end{bmatrix}$

$$\begin{array}{cc} A & E \\ (2 \times 2) & (3 \times 2) \end{array}$$
$$\uparrow \qquad \uparrow$$
multiplication is not possible since $2 \neq 3$

e. $EA = \begin{bmatrix} -2 & -1 \\ 3 & 0 \\ 1 & 2 \end{bmatrix} \begin{bmatrix} -2 & 1 \\ 3 & 4 \end{bmatrix} = \begin{bmatrix} 1 & -6 \\ -6 & 3 \\ 4 & 9 \end{bmatrix}$

$$\begin{array}{cc} E & A \\ (3 \times 2) & (2 \times 2) \end{array} \qquad \begin{array}{cc} E & A \\ (3 \times 2) & (2 \times 2) \end{array}$$
$$\uparrow \qquad \uparrow \qquad\qquad \uparrow \qquad\qquad \uparrow$$
multiplication is possible   result will be a
since $2 = 2$ 3×2 matrix

Now try Exercises 25 through 36 ▶

Example 5 shows that in general, matrix multiplication is not commutative. Parts (b) and (c) show $CD \neq DC$ since we get different results, and parts (d) and (e) show $AE \neq EA$, since AE is not defined while EA is.

Operations on matrices can be a laborious process for larger matrices and for matrices with noninteger or large entries. For these, we can turn to available technology for assistance. This shifts our focus from a meticulous computation of entries, to carefully entering each matrix into the calculator, double-checking each entry, and appraising results to see if they're reasonable.

EXAMPLE 6 ▶ Using Technology for Matrix Operations

Use a calculator to compute the difference $A - B$ for the matrices given.

$$A = \begin{bmatrix} \frac{2}{11} & -0.5 & \frac{6}{5} \\ 0.9 & \frac{3}{4} & -4 \\ 0 & 6 & -\frac{5}{12} \end{bmatrix} \qquad B = \begin{bmatrix} \frac{1}{6} & \frac{-7}{10} & 0.75 \\ \frac{11}{25} & 0 & -5 \\ -4 & \frac{-5}{9} & \frac{-5}{12} \end{bmatrix}$$

Solution ▶ The entries for matrix A are shown in Figure 8.23. After entering matrix B, exit to the home screen [2nd MODE (**QUIT**)], call up matrix A, press the − (subtract) key, then call up matrix B and press ENTER . The calculator quickly finds the difference and displays the results shown in Figure 8.24. The last line on the screen shows the result can be stored for future use in a new matrix C by pressing the STO → key, calling up matrix C, and pressing ENTER .

Figure 8.23

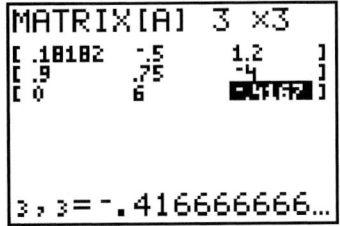

Figure 8.24

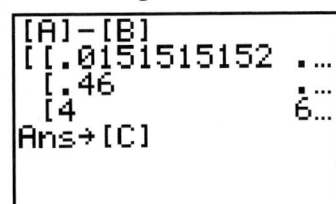

Now try Exercises 37 through 40 ▶

Figure 8.25

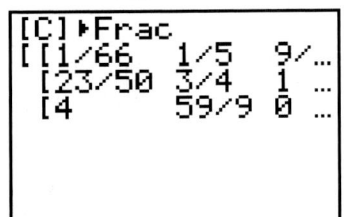

In Figure 8.24 the dots to the right on the calculator screen indicate there are additional digits or matrix columns that can't fit on the display, as often happens with larger matrices or decimal numbers. Sometimes, converting entries to fraction form will provide a display that's easier to read. Here, this is done by calling up the matrix C, and using the MATH **1: ▶ Frac** option. After pressing ENTER , all entries are converted to fractions in simplest form (where possible), as in Figure 8.25. The third column can be viewed by pressing the right arrow.

EXAMPLE 7 ▶ Using Technology for Matrix Operations

Use a calculator to compute the product AB.

$$A = \begin{bmatrix} 2 & -3 & 0 \\ -1 & 5 & 4 \\ 6 & 0 & 2 \\ 3 & 2 & -1 \end{bmatrix} \quad B = \begin{bmatrix} \frac{1}{2} & -0.7 & 1 \\ 0.5 & 3.2 & -3 \\ -2 & \frac{3}{4} & 4 \end{bmatrix}$$

Solution ▶ Carefully enter matrices A and B into the calculator, then press [2nd] [MODE] (**QUIT**) to get to the home screen. Use [A][B] [ENTER], and the calculator finds the product shown in the figure.

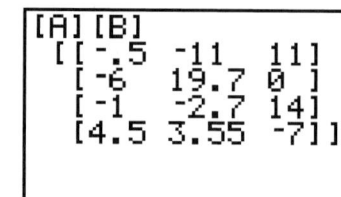

$$AB = \begin{bmatrix} 2 & -3 & 0 \\ -1 & 5 & 4 \\ 6 & 0 & 2 \\ 3 & 2 & -1 \end{bmatrix} \begin{bmatrix} \frac{1}{2} & -0.7 & 1 \\ 0.5 & 3.2 & -3 \\ -2 & \frac{3}{4} & 4 \end{bmatrix}$$

Now try Exercises 41 through 52 ▶

Properties of Matrix Multiplication

Earlier, Example 5 demonstrated that matrix multiplication is not commutative. Here is a group of properties that *do* hold for matrices. You are asked to check these properties in the exercise set using various matrices. See **Exercises 53 through 56.**

Properties of Matrix Multiplication

Given matrices A, B, and C for which the products are defined:

   I. $A(BC) = (AB)C$ matrix multiplication is associative

  II. $A(B + C) = AB + AC$ matrix multiplication is distributive from the left

 III. $(B + C)A = BA + CA$ matrix multiplication is distributive from the right

 IV. $k(A + B) = kA + kB$ a constant k can be distributed over addition

We close this section with an application of matrix multiplication. There are many other interesting applications in the exercise set.

EXAMPLE 8 ▶ Using Matrix Multiplication to Track Volunteer Enlistments

In a certain country, the number of males and females that will join the military depends on their age. This information is stored in matrix A (Table 8.5). The likelihood a volunteer will join a *particular branch* of the military also depends on their age, with this information stored in matrix B (Table 8.6). (a) Compute the product $P = AB$ and discuss/interpret what is indicated by the entries P_{11}, P_{13}, and P_{24} of the product matrix. (b) How many males are expected to join the Navy this year?

Table 8.5 Matrix A

A	Age Groups		
Sex	18–19	20–21	22–23
Female	1000	1500	500
Male	2500	3000	2000

Table 8.6 Matrix B

B	Likelihood of Joining			
Age Group	Army	Navy	Air Force	Marines
18–19	0.42	0.28	0.17	0.13
20–21	0.38	0.26	0.27	0.09
22–23	0.33	0.25	0.35	0.07

Solution ▶ **a.** Matrix A has order 2×3 and matrix B has order 3×4. The product matrix P can be found and is a 2×4 matrix. Carefully enter the matrices in your calculator. Figure 8.26 shows the entries of matrix B. Using $[A][B]$ **ENTER**, the calculator finds the product matrix shown in Figure 8.27. Pressing the right arrow shows the complete product matrix is

$$P = \begin{bmatrix} 1155 & 795 & 750 & 300 \\ 2850 & 1980 & 1935 & 735 \end{bmatrix}.$$

The entry P_{11} is the product of R1 from A and C1 from B, and indicates that for the year, 1155 females are projected to join the Army. In like manner, entry P_{13} shows that 750 females are projected to join the Air Force. Entry P_{24} indicates that 735 males are projected to join the Marines.

Figure 8.26   **Figure 8.27**

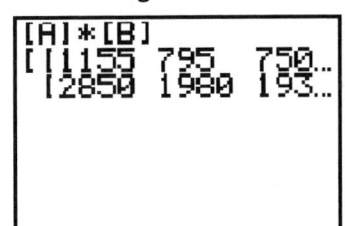

✓ **C.** You've just learned
how to compute the product
of two matrices

b. The product R2 (males) · C2 (Navy) gives $P_{22} = 1980$, meaning 1980 males are expected to join the Navy.

Now try Exercise 59 through 66 ▶

8.6 EXERCISES

CONCEPTS AND VOCABULARY

Fill in the blank with the appropriate word or phrase. Carefully reread the section if needed.

1. Two matrices are equal if they are like size and the corresponding entries are equal. In symbols, $A = B$ if _____ = _____.

2. The sum of two matrices (of like size) is found by adding the corresponding entries. In symbols, $A + B =$ _____.

3. The product of a constant times a matrix is called _____ multiplication.

4. The size of a matrix is also referred to as its _____.
   The order of $A = \begin{bmatrix} 1 & 2 & 3 \\ 4 & 5 & 6 \end{bmatrix}$ is _____.

5. Give two reasons why matrix multiplication is generally not commutative. Include several examples using matrices of various sizes.

6. Discuss the conditions under which matrix multiplication is defined. Include several examples using matrices of various sizes.

▶ **DEVELOPING YOUR SKILLS**

State the order of each matrix and name the entries in positions a_{12} and a_{23} if they exist. Then name the position a_{ij} of the 5 in each.

7. $\begin{bmatrix} 1 & -3 \\ 5 & -7 \end{bmatrix}$

8. $\begin{bmatrix} 19 \\ -11 \\ 5 \end{bmatrix}$

9. $\begin{bmatrix} 2 & -3 & 0.5 \\ 0 & 5 & 6 \end{bmatrix}$

10. $\begin{bmatrix} 2 & 0.4 \\ -0.1 & 5 \\ 0.3 & -3 \end{bmatrix}$

11. $\begin{bmatrix} -2 & 1 & -7 \\ 0 & 8 & 1 \\ 5 & -1 & 4 \end{bmatrix}$

12. $\begin{bmatrix} 89 & 55 & 34 & 21 \\ 13 & 8 & 5 & 3 \\ 2 & 1 & 1 & 0 \end{bmatrix}$

Determine if the following statements are true, false, or conditional. If false, explain why. If conditional, find values of $a, b, c, p, q,$ and r that will make the statement true.

13. $\begin{bmatrix} \sqrt{1} & \sqrt{4} & \sqrt{8} \\ \sqrt{16} & \sqrt{32} & \sqrt{64} \end{bmatrix} = \begin{bmatrix} 1 & 2 & 2\sqrt{2} \\ 4 & 4\sqrt{2} & 8 \end{bmatrix}$

14. $\begin{bmatrix} \dfrac{3}{2} & \dfrac{-7}{5} & \dfrac{13}{10} \\ \dfrac{-1}{2} & \dfrac{-2}{5} & \dfrac{1}{3} \end{bmatrix} = \begin{bmatrix} 1.5 & -1.4 & 1.3 \\ -0.5 & -0.4 & 0.\overline{3} \end{bmatrix}$

15. $\begin{bmatrix} -2 & 3 & a \\ 2b & -5 & 4 \\ 0 & -9 & 3c \end{bmatrix} = \begin{bmatrix} c & 3 & -4 \\ 6 & -5 & -a \\ 0 & -3b & -6 \end{bmatrix}$

16. $\begin{bmatrix} 2p+1 & -5 & 9 \\ 1 & 12 & 0 \\ q+5 & 9 & -2r \end{bmatrix} = \begin{bmatrix} 7 & -5 & 2-q \\ 1 & 3r & 0 \\ -2 & 3p & -8 \end{bmatrix}$

For matrices A through H as given, perform the indicated operation(s), if possible. Do not use a calculator. If an operation cannot be completed, state why.

$A = \begin{bmatrix} 2 & 3 \\ 5 & 8 \end{bmatrix}$ $B = \begin{bmatrix} 2 \\ 1 \\ -3 \end{bmatrix}$

$C = \begin{bmatrix} 2 & 0.5 \\ 0.2 & 5 \\ -1 & 3 \end{bmatrix}$ $D = \begin{bmatrix} 1 & 0 & 0 \\ 0 & 1 & 0 \\ 0 & 0 & 1 \end{bmatrix}$

$E = \begin{bmatrix} 1 & -2 & 0 \\ 0 & -1 & 2 \\ 4 & 3 & -6 \end{bmatrix}$ $F = \begin{bmatrix} 6 & -3 & 9 \\ 12 & 0 & -6 \end{bmatrix}$

$G = \begin{bmatrix} -1 & 2 & 0 \\ 0 & 1 & -2 \\ -4 & -3 & 6 \end{bmatrix}$ $H = \begin{bmatrix} 8 & -3 \\ -5 & 2 \end{bmatrix}$

17. $A + H$ **18.** $E + G$

19. $F + H$ **20.** $G + D$

21. $3H - 2A$ **22.** $2E + 3G$

23. $\dfrac{1}{2}E - 3D$ **24.** $F - \dfrac{2}{3}F$

25. ED **26.** DE

27. AH **28.** HA

29. FD **30.** FH

31. HF **32.** EB

33. H^2 **34.** F^2

35. FE **36.** EF

For matrices A through H as given, use a calculator to perform the indicated operation(s), if possible. If an operation cannot be completed, state why.

$A = \begin{bmatrix} -5 & 4 \\ 3 & 9 \end{bmatrix}$ $B = \begin{bmatrix} 1 & 0 \\ 0 & 1 \end{bmatrix}$

$C = \begin{bmatrix} \dfrac{\sqrt{3}}{2} & \dfrac{\sqrt{3}}{3} \\ \sqrt{3} & 2\sqrt{3} \end{bmatrix}$ $D = \begin{bmatrix} 1 & 0 & 0 \\ 0 & 1 & 0 \\ 0 & 0 & 1 \end{bmatrix}$

$E = \begin{bmatrix} 1 & -2 & 0 \\ 0 & -1 & 2 \\ 4 & 3 & -6 \end{bmatrix}$ $F = \begin{bmatrix} -0.52 & 0.002 & 1.032 \\ 1.021 & -1.27 & 0.019 \end{bmatrix}$

$G = \begin{bmatrix} 0 & \dfrac{3}{4} & \dfrac{1}{4} \\ \dfrac{-1}{2} & \dfrac{3}{8} & \dfrac{1}{8} \\ \dfrac{-1}{4} & \dfrac{11}{16} & \dfrac{1}{16} \end{bmatrix}$ $H = \begin{bmatrix} \dfrac{-3}{19} & \dfrac{4}{57} \\ \dfrac{1}{19} & \dfrac{5}{57} \end{bmatrix}$

37. $C + H$ **38.** $A - H$

39. $E + G$ **40.** $G - E$

41. AH **42.** HA

43. EG **44.** GE

45. HB **46.** BH

47. DG **48.** GD

49. C^2 **50.** E^2

51. FG **52.** AF

For Exercises 53 through 56, use a calculator and matrices A, B, and C to verify each statement.

$$A = \begin{bmatrix} -1 & 3 & 5 \\ 2 & 7 & -1 \\ 4 & 0 & 6 \end{bmatrix} \quad B = \begin{bmatrix} 0.3 & -0.4 & 1.2 \\ -2.5 & 2 & 0.9 \\ 1 & -0.5 & 0.2 \end{bmatrix}$$

$$C = \begin{bmatrix} 45 & -1 & 3 \\ -6 & 10 & -15 \\ 21 & -28 & 36 \end{bmatrix}$$

53. Matrix multiplication is not generally commutative: (a) $AB \neq BA$, (b) $AC \neq CA$, and (c) $BC \neq CB$.

54. Matrix multiplication is distributive from the left: $A(B + C) = AB + AC$.

55. Matrix multiplication is distributive from the right: $(B + C)A = BA + CA$.

56. Matrix multiplication is associative: $(AB)C = A(BC)$.

▶ **WORKING WITH FORMULAS**

$$\begin{bmatrix} 2 & 2 \\ W & 0 \end{bmatrix} \cdot \begin{bmatrix} L \\ W \end{bmatrix} = \begin{bmatrix} \text{Perimeter} \\ \text{Area} \end{bmatrix}$$

The perimeter and area of a rectangle can be simultaneously calculated using the matrix formula shown, where L represents the length and W represents the width of the rectangle. Use the matrix formula and your calculator to find the perimeter and area of the

rectangles shown, then check the results using $P = 2L + 2W$ and $A = LW$.

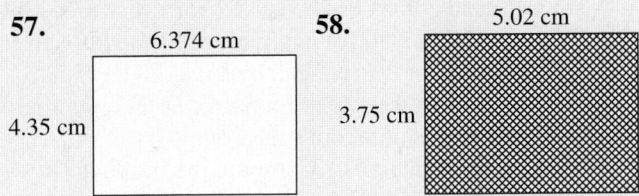

57. 6.374 cm / 4.35 cm

58. 5.02 cm / 3.75 cm

▶ **APPLICATIONS**

59. Custom T's designs and sells specialty T-shirts and sweatshirts, with plants in Verdi and Minsk. The company offers this apparel in three quality levels: standard, deluxe, and premium. Last fall the Verdi office produced 3820 standard, 2460 deluxe, and 1540 premium T-shirts, along with 1960 standard, 1240 deluxe, and 920 premium sweatshirts. The Minsk office produced 4220 standard, 2960 deluxe, and 1640 premium T-shirts, along with 2960 standard, 3240 deluxe, and 820 premium sweatshirts in the same time period.

   a. Write a 3 × 2 "production matrix" for each plant [$V \rightarrow$ Verdi, $M \rightarrow$ Minsk], with a *T-shirt* column, a *sweatshirt* column, and three rows showing how many of the different types of apparel were manufactured.

   b. Use the matrices from Part (a) to determine how many more or less articles of clothing were produced by Minsk than Verdi.

   c. Use scalar multiplication to find how many shirts of each type will be made at Verdi and Minsk next fall, if each is expecting a 4% increase in business.

   d. What will be Custom T's total production next fall (from both plants), for each type of apparel?

60. Terry's Tire Store sells automobile and truck tires through three retail outlets. Sales at the Cahokia store for the months of January, February, and March

broke down as follows: 350, 420, and 530 auto tires and 220, 180, and 140 truck tires. The Shady Oak branch sold 430, 560, and 690 auto tires and 280, 320, and 220 truck tires during the same 3 months. Sales figures for the downtown store were 864, 980, and 1236 auto tires and 535, 542, and 332 truck tires.

   a. Write a 2 × 3 "sales matrix" for each store [$C \rightarrow$ Cahokia, $S \rightarrow$ Shady Oak, $D \rightarrow$ Downtown], with *January, February,* and *March* columns, and two rows showing the sales of auto and truck tires respectively.

   b. Use the matrices from Part (a) to determine how many more or fewer tires of each type the downtown store sold (each month) over the other two stores combined.

   c. Market trends indicate that for the same three months in the following year, the Cahokia store will likely experience a 10% increase in sales, the Shady Oak store a 3% decrease, with sales at the downtown store remaining level (no change). What will be the combined monthly sales from all three stores next year, for each type of tire?

61. **Home improvements:** Dream-Makers Home Improvements specializes in replacement windows, replacement doors, and new siding. During the peak season, the number of contracts that came from various parts of the city (North, South, East, and West) are shown in matrix C. The average

profit per contract is shown in matrix P. Compute the product PC and discuss what each entry of the product matrix represents.

$$\begin{array}{c} \\ \text{Windows} \\ \text{Doors} \\ \text{Siding} \end{array} \begin{array}{cccc} N & S & E & W \\ \left[\begin{array}{cccc} 9 & 6 & 5 & 4 \\ 7 & 5 & 7 & 6 \\ 2 & 3 & 5 & 2 \end{array}\right] = C \end{array}$$

$$\begin{array}{ccc} \text{Windows} & \text{Doors} & \text{Siding} \\ [1500 & 500 & 2500] = P \end{array}$$

62. Classical music: Station 90.7—*The Home of Classical Music*—is having their annual fund drive. Being a loyal listener, Mitchell decides that for the next 3 days he will donate money according to his favorite composers, by the number of times their music comes on the air: $3 for every piece by Mozart (M), $2.50 for every piece by Beethoven (B), and $2 for every piece by Vivaldi (V). This information is displayed in matrix D. The number of pieces he heard from each composer is displayed in matrix C. Compute the product DC and discuss what each entry of the product matrix represents.

$$\begin{array}{c} \\ M \\ B \\ V \end{array} \begin{array}{ccc} \text{Mon.} & \text{Tue.} & \text{Wed.} \\ \left[\begin{array}{ccc} 4 & 3 & 5 \\ 3 & 2 & 4 \\ 2 & 3 & 3 \end{array}\right] = C \end{array}$$

$$\begin{array}{ccc} M & B & V \\ [3 & 2.5 & 2] = D \end{array}$$

63. Pizza and salad: The science department and math department of a local college are at a pre-semester retreat, and decide to have pizza, salads, and soft drinks for lunch. The quantity of food ordered by each department is shown in matrix Q. The cost of the food item at each restaurant is shown in matrix C using the published prices from three popular restaurants: Pizza Home (PH), Papa Jeff's (PJ), and Dynamos (D).

a. What is the total cost to the math department if the food is ordered from Pizza Home?

b. What is the total cost to the science department if the food is ordered from Papa Jeff's?

c. Compute the product QC and discuss the meaning of each entry in the product matrix.

$$\begin{array}{c} \\ \text{Science} \\ \text{Math} \end{array} \begin{array}{ccc} \text{Pizza} & \text{Salad} & \text{Drink} \\ \left[\begin{array}{ccc} 8 & 12 & 20 \\ 10 & 8 & 18 \end{array}\right] = Q \end{array}$$

$$\begin{array}{c} \\ \text{Pizza} \\ \text{Salad} \\ \text{Drink} \end{array} \begin{array}{ccc} \text{PH} & \text{PJ} & \text{D} \\ \left[\begin{array}{ccc} 8 & 7.5 & 10 \\ 1.5 & 1.75 & 2 \\ 0.90 & 1 & 0.75 \end{array}\right] = C \end{array}$$

64. Manufacturing pool tables: Cue Ball Incorporated makes three types of pool tables, for homes, commercial use, and professional use. The amount of time required to pack, load, and install each is summarized in matrix T, with all times in hours. The cost of these components in dollars per hour, is summarized in matrix C for two of its warehouses, one on the west coast and the other in the midwest.

a. What is the cost to package, load, and install a commercial pool table from the coastal warehouse?

b. What is the cost to package, load, and install a commercial pool table from the warehouse in the midwest?

c. Compute the product TC and discuss the meaning of each entry in the product matrix.

$$\begin{array}{c} \\ \text{Home} \\ \text{Comm} \\ \text{Prof} \end{array} \begin{array}{ccc} \text{Pack} & \text{Load} & \text{Install} \\ \left[\begin{array}{ccc} 1 & 0.2 & 1.5 \\ 1.5 & 0.5 & 2.2 \\ 1.75 & 0.75 & 2.5 \end{array}\right] = T \end{array}$$

$$\begin{array}{c} \\ \text{Pack} \\ \text{Load} \\ \text{Install} \end{array} \begin{array}{cc} \text{Coast} & \text{Midwest} \\ \left[\begin{array}{cc} 10 & 8 \\ 12 & 10.5 \\ 13.5 & 12.5 \end{array}\right] = C \end{array}$$

65. Joining a club: Each school year, among the students planning to join a club, the likelihood a student joins a particular club depends on their class standing. This information is stored in matrix C. The number of males and females from each class that are projected to join a club each year is stored in matrix J. Compute the product JC and use the result to answer the following:

a. Approximately how many females joined the chess club?

b. Approximately how many males joined the writing club?

c. What does the entry P_{13} of the product matrix tells us?

$$\begin{array}{c} \\ \text{Female} \\ \text{Male} \end{array} \begin{array}{ccc} \text{Fresh} & \text{Soph} & \text{Junior} \\ \left[\begin{array}{ccc} 25 & 18 & 21 \\ 22 & 19 & 18 \end{array}\right] = J \end{array}$$

$$\begin{array}{c} \\ \text{Fresh} \\ \text{Soph} \\ \text{Junior} \end{array} \begin{array}{ccc} \text{Spanish} & \text{Chess} & \text{Writing} \\ \left[\begin{array}{ccc} 0.6 & 0.1 & 0.3 \\ 0.5 & 0.2 & 0.3 \\ 0.4 & 0.2 & 0.4 \end{array}\right] = C \end{array}$$

66. Designer shirts: The SweatShirt Shoppe sells three types of designs on its products: stenciled (S), embossed (E), and applique (A). The quantity of each size sold is shown in matrix Q. The retail price of each sweatshirt depends on its size and whether it was finished by hand or machine. Retail prices are shown in matrix C. Assuming all stock is sold,

a. How much revenue was generated by the large sweatshirts?

b. How much revenue was generated by the extra-large sweatshirts?

c. What does the entry P_{11} of the product matrix QC tell us?

$$
\begin{array}{c}
\begin{array}{ccc} \text{S} & \text{E} & \text{A} \end{array} \\
\begin{array}{c} \text{med} \\ \text{large} \\ \text{x-large} \end{array}
\begin{bmatrix} 30 & 30 & 15 \\ 60 & 50 & 20 \\ 50 & 40 & 30 \end{bmatrix} = Q
\end{array}
$$

$$
\begin{array}{c}
\begin{array}{cc} \text{Hand} & \text{Machine} \end{array} \\
\begin{array}{c} \text{S} \\ \text{E} \\ \text{A} \end{array}
\begin{bmatrix} 40 & 25 \\ 60 & 40 \\ 90 & 60 \end{bmatrix} = C
\end{array}
$$

▶ **EXTENDING THE CONCEPT**

67. For the matrix A shown, use your calculator to compute A^2, A^3, A^4, and A^5. Do you notice a pattern? Try to write a "matrix formula" for A^n, where n is a positive integer, then use your formula to find A^6. Check results using a calculator.

$$
A = \begin{bmatrix} 1 & 0 & 1 \\ 1 & 1 & 1 \\ 1 & 0 & 1 \end{bmatrix}
$$

68. The matrix $M = \begin{bmatrix} 2 & 1 \\ -3 & -2 \end{bmatrix}$ has some very interesting properties. Compute the powers M^2,

M^3, M^4, and M^5, then discuss what you find. Try to find/create another 2×2 matrix that has similar properties.

69. For the "matrix equation"
$$
\begin{bmatrix} 2 & 1 \\ -3 & -2 \end{bmatrix} \cdot \begin{bmatrix} a & b \\ c & d \end{bmatrix} = \begin{bmatrix} 1 & 0 \\ 0 & 1 \end{bmatrix}, \text{ use matrix}
$$
multiplication and two systems of equations to find the entries a, b, c, and d that make the equation true.

▶ **MAINTAINING YOUR SKILLS**

70. (5.2) Solve the system using elimination.

$$
\begin{cases}
x + 2y - z = 3 \\
-2x - y + 3z = -5 \\
5x + 3y - 2z = 2
\end{cases}
$$

71. (6.5) Evaluate $\cos(\cos^{-1} 0.3211)$.

72. (7.6) Solve $z^4 - 81i = 0$ using the nth roots theorem. Leave your answer in trigonometric form.

73. (3.2) Find the quotient using synthetic division, then check using multiplication.

$$
\frac{x^3 - 9x + 10}{x - 2}
$$

8.7 | Solving Linear Systems Using Matrix Equations

Learning Objectives

In Section 8.7 you will learn how to:

☐ **A.** Recognize the identity matrix for multiplication

☐ **B.** Find the inverse of a square matrix

☐ **C.** Solve systems using matrix equations

☐ **D.** Use determinants to find whether a matrix is invertible

While using matrices and row operations offers a degree of efficiency in solving systems, we are still required to solve for each variable *individually*. Using matrix multiplication we can actually rewrite a given system as a single *matrix equation*, in which the solutions are computed *simultaneously*. As with other kinds of equations, the use of identities and inverses are involved, which we now develop in the context of matrices.

A. Multiplication and Identity Matrices

From the properties of real numbers, 1 is the identity for multiplication since $n \cdot 1 = 1 \cdot n = n$. A similar identity exists for matrix multiplication. Consider the 2×2 matrix $A = \begin{bmatrix} 1 & 4 \\ -2 & 3 \end{bmatrix}$. While matrix multiplication is not *generally* commutative, if we can find a matrix B where $AB = BA = A$, then B is a prime candidate for the identity matrix, which is denoted I. For the products AB and BA to be possible and have the same order as A, we note B must also be a 2×2 matrix. Using the arbitrary matrix $B = \begin{bmatrix} a & b \\ c & d \end{bmatrix}$, we have the following.

EXAMPLE 1A ▶ **Solving $AB = A$ to Find the Identity Matrix**

For $\begin{bmatrix} 1 & 4 \\ -2 & 3 \end{bmatrix} \begin{bmatrix} a & b \\ c & d \end{bmatrix} = \begin{bmatrix} 1 & 4 \\ -2 & 3 \end{bmatrix}$, use matrix multiplication, the equality of matrices, and systems of equations to find the value of a, b, c, and d.

Solution ▶ The product on the left gives $\begin{bmatrix} a + 4c & b + 4d \\ -2a + 3c & -2b + 3d \end{bmatrix} = \begin{bmatrix} 1 & 4 \\ -2 & 3 \end{bmatrix}$.

Since corresponding entries must be equal (shown by matching colors), we can find a, b, c, and d by solving the systems $\begin{cases} a + 4c = 1 \\ -2a + 3c = -2 \end{cases}$ and $\begin{cases} b + 4d = 4 \\ -2b + 3d = 3 \end{cases}$. For the first system, $2R1 + R2$ shows $a = 1$ and $c = 0$. Using $2R1 + R2$ for the second shows $b = 0$ and $d = 1$. It appears $\begin{bmatrix} 1 & 0 \\ 0 & 1 \end{bmatrix}$ is a candidate for the identity matrix.

▶

Before we name B as the identity matrix, we must show that $AB = BA = A$.

EXAMPLE 1B ▶ **Verifying $AB = BA = A$**

Given $A = \begin{bmatrix} 1 & 4 \\ -2 & 3 \end{bmatrix}$ and $B = \begin{bmatrix} 1 & 0 \\ 0 & 1 \end{bmatrix}$, determine if $AB = A$ and $BA = A$.

Solution ▶ $AB = \begin{bmatrix} 1 & 4 \\ -2 & 3 \end{bmatrix} \begin{bmatrix} 1 & 0 \\ 0 & 1 \end{bmatrix}$ $BA = \begin{bmatrix} 1 & 0 \\ 0 & 1 \end{bmatrix} \begin{bmatrix} 1 & 4 \\ -2 & 3 \end{bmatrix}$

$= \begin{bmatrix} 1(1) + 4(0) & 1(0) + 4(1) \\ -2(1) + 3(0) & -2(0) + 3(1) \end{bmatrix}$ $= \begin{bmatrix} 1(1) + 0(-2) & 1(4) + 0(3) \\ 0(1) + 1(-2) & 0(4) + 1(3) \end{bmatrix}$

$= \begin{bmatrix} 1 & 4 \\ -2 & 3 \end{bmatrix} = A ✓$ $= \begin{bmatrix} 1 & 4 \\ -2 & 3 \end{bmatrix} = A ✓$

Since $AB = A = BA$, B is the identity matrix I.

Now try Exercises 7 through 10 ▶

By replacing the entries of $A = \begin{bmatrix} 1 & 4 \\ -2 & -3 \end{bmatrix}$ with those of the general matrix $\begin{bmatrix} a_{11} & a_{12} \\ a_{21} & a_{22} \end{bmatrix}$, we can show that $I = \begin{bmatrix} 1 & 0 \\ 0 & 1 \end{bmatrix}$ is the identity for *all* 2×2 matrices. In considering the identity for larger matrices, we find that only *square matrices* have inverses, since $AI = IA$ is the primary requirement (the multiplication must be possible in both directions). This is commonly referred to as *multiplication from the right* and *multiplication from the left*. Using the same procedure as before we can show $\begin{bmatrix} 1 & 0 & 0 \\ 0 & 1 & 0 \\ 0 & 0 & 1 \end{bmatrix}$ is the identity for 3×3 matrices (denoted I_3). The $n \times n$ identity matrix I_n consists of 1's down the main diagonal and 0's for all other entries. Also, the identity I_n for a square matrix is unique.

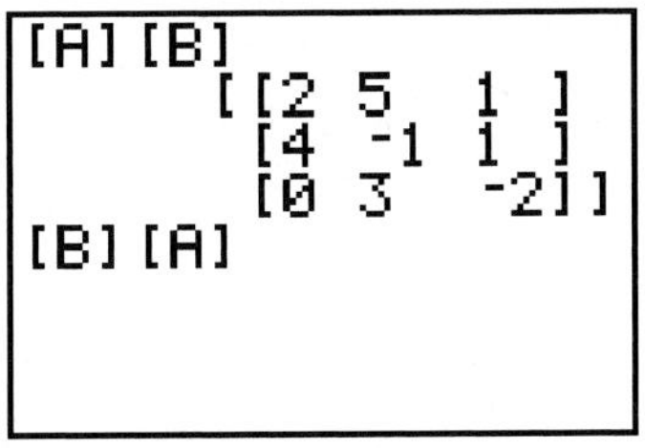

Figure 8.28

```
[A][B]
    [[2 5  1 ]
     [4 -1 1 ]
     [0 3  -2]]
[B][A]
```

☑ **A.** You've just learned how to recognize the identity matrix for multiplication

As in Section 9.2, a graphing calculator can be used to investigate operations on matrices and matrix properties. For the 3×3 matrix $A = \begin{bmatrix} 2 & 5 & 1 \\ 4 & -1 & 1 \\ 0 & 3 & -2 \end{bmatrix}$ and

$I_3 = \begin{bmatrix} 1 & 0 & 0 \\ 0 & 1 & 0 \\ 0 & 0 & 1 \end{bmatrix}$, a calculator will confirm that $AI_3 = A = I_3 A$. Carefully enter A

into your calculator as matrix A, and I_3 as matrix B. Figure 8.28 shows $AB = A$ and after pressing ENTER , the calculator will verify $BA = A$, although the screen cannot display the result without scrolling. **See Exercises 11 through 14.**

B. The Inverse of a Matrix

Again from the properties of real numbers, we know the multiplicative inverse for a is $a^{-1} = \dfrac{1}{a}$ ($a \neq 0$), since the products $a \cdot a^{-1}$ and $a^{-1} \cdot a$ yield the identity 1. To show that a similar inverse exists for matrices, consider the square matrix $A = \begin{bmatrix} 6 & 5 \\ 2 & 2 \end{bmatrix}$ and an arbitrary matrix $B = \begin{bmatrix} a & b \\ c & d \end{bmatrix}$. If we can find a matrix B, where $AB = BA = I$, then B is a prime candidate for the inverse matrix of A, which is denoted A^{-1}. Proceeding as in Examples 1A and 1B gives the result shown in Example 2.

EXAMPLE 2A ▶ Solving $AB = I$ to find A^{-1}

For $\begin{bmatrix} 6 & 5 \\ 2 & 2 \end{bmatrix} \begin{bmatrix} a & b \\ c & d \end{bmatrix} = \begin{bmatrix} 1 & 0 \\ 0 & 1 \end{bmatrix}$, use matrix multiplication, the equality of matrices, and systems of equations to find the entries of B.

Solution ▶ The product on the left gives $\begin{bmatrix} 6a + 5c & 6b + 5d \\ 2a + 2c & 2b + 2d \end{bmatrix} = \begin{bmatrix} 1 & 0 \\ 0 & 1 \end{bmatrix}$. Since corresponding entries must be equal (shown by matching colors), we find the values of a, b, c, and d by solving the systems $\begin{cases} 6a + 5c = 1 \\ 2a + 2c = 0 \end{cases}$ and $\begin{cases} 6b + 5d = 0 \\ 2b + 2d = 1 \end{cases}$. Using $-3R2 + R1$ for the first system shows $a = 1$ and $c = -1$, while $-3R2 + R1$ for the second system shows $b = -2.5$ and $d = 3$. Matrix $B = \begin{bmatrix} a & b \\ c & d \end{bmatrix} = \begin{bmatrix} 1 & -2.5 \\ -1 & 3 \end{bmatrix}$ is the prime candidate for A^{-1}.

To determine if A^{-1} has truly been found, we check to see if multiplication from the right and multiplication from the left yields the matrix I: $AB = BA = I$.

EXAMPLE 2B ▶ Verifying $B = A^{-1}$

For the matrices $A = \begin{bmatrix} 6 & 5 \\ 2 & 2 \end{bmatrix}$ and $B = \begin{bmatrix} 1 & -2.5 \\ -1 & 3 \end{bmatrix}$ from Example 2A, determine if $AB = BA = I$.

Solution ▶

$$AB = \begin{bmatrix} 6 & 5 \\ 2 & 2 \end{bmatrix} \begin{bmatrix} 1 & -2.5 \\ -1 & 3 \end{bmatrix}$$

$$= \begin{bmatrix} 6(1) + 5(-1) & 6(-2.5) + 5(3) \\ 2(1) + 2(-1) & 2(-2.5) + 2(3) \end{bmatrix}$$

$$= \begin{bmatrix} 1 & 0 \\ 0 & 1 \end{bmatrix} \checkmark$$

$$BA = \begin{bmatrix} 1 & -2.5 \\ -1 & 3 \end{bmatrix} \begin{bmatrix} 6 & 5 \\ 2 & 2 \end{bmatrix}$$

$$= \begin{bmatrix} 1(6) + (-2.5)(2) & 1(5) + (-2.5)(2) \\ -1(6) + 3(2) & -1(5) + 3(2) \end{bmatrix}$$

$$= \begin{bmatrix} 1 & 0 \\ 0 & 1 \end{bmatrix} \checkmark$$

Since $AB = BA = I$, we conclude $B = A^{-1}$.

Now try Exercises 15 through 22 ▶

These observations guide us to the following definition of an inverse matrix.

The Inverse of a Matrix

Given an $n \times n$ matrix A, if there exists an $n \times n$ matrix A^{-1} such that $AA^{-1} = A^{-1}A = I_n$, then A^{-1} is the inverse of matrix A.

We will soon discover that while only square matrices have inverses, not every square matrix has an inverse. If an inverse exists, the matrix is said to be **invertible.** For 2×2 matrices that are invertible, a simple formula exists for computing the inverse. The formula is derived in the *Strengthening Core Skills* feature at the end of Chapter 8.

The Inverse of a 2 × 2 Matrix

If $A = \begin{bmatrix} a & b \\ c & d \end{bmatrix}$, then $A^{-1} = \dfrac{1}{ad - bc} \begin{bmatrix} d & -b \\ -c & a \end{bmatrix}$ provided $ad - bc \neq 0$

To "test" the formula, again consider the matrix $A = \begin{bmatrix} 6 & 5 \\ 2 & 2 \end{bmatrix}$, where $a = 6, b = 5,$ $c = 2,$ and $d = 2$:

☑ **B. You've just learned how to find the inverse of a square matrix**

$$A^{-1} = \frac{1}{(6)(2) - (5)(2)} \begin{bmatrix} 2 & -5 \\ -2 & 6 \end{bmatrix}$$

$$= \frac{1}{2} \begin{bmatrix} 2 & -5 \\ -2 & 6 \end{bmatrix} = \begin{bmatrix} 1 & -2.5 \\ -1 & 3 \end{bmatrix} \checkmark$$

See Exercises 63 through 66 for more practice with this formula.

Almost without exception, real-world applications involve much larger matrices, with entries that are not integer-valued. Although the *equality of matrices* method from Example 2 can be extended to find the inverse of larger matrices, the process becomes very tedious and too time consuming to be useful. As an alternative, the **augmented matrix method** can be used. This process is discussed in the *Strengthening Core Skills* feature at the end Chapter 8 (see page 823). For practical reasons, we will rely on a calculator to produce these larger inverse matrices. This is done by (1) carefully entering a square matrix A into the calculator, (2) returning to the home screen and (3) calling up matrix A and pressing the ⌨️x⁻¹ key and ENTER to find A^{-1}. In the context of matrices, calculators are programmed to compute an inverse matrix, rather than to somehow find a reciprocal. **See Exercises 23 through 26.**

C. Solving Systems Using Matrix Equations

One reason matrix multiplication has its row $\times$ column definition is to assist in writing a linear system of equations as a single matrix equation. The equation consists of the matrix of constants B on the right, and a product of the coefficient matrix A with the matrix of variables X on the left: $AX = B$. For $\begin{cases} x + 4y - z = 10 \\ 2x + 5y - 3z = 7, \text{ the matrix} \\ 8x + y - 2z = 11 \end{cases}$

equation is $\begin{bmatrix} 1 & 4 & -1 \\ 2 & 5 & -3 \\ 8 & 1 & -2 \end{bmatrix} \begin{bmatrix} x \\ y \\ z \end{bmatrix} = \begin{bmatrix} 10 \\ 7 \\ 11 \end{bmatrix}$. Note that computing the product on the left will yield the original system.

Once written as a matrix equation, the system can be solved using an inverse matrix and the following sequence. If A represents the matrix of coefficients, X the matrix of variables, B the matrix of constants, and I the appropriate identity, the sequence is

$$(1) \qquad AX = B \qquad \text{matrix equation}$$
$$(2) \quad A^{-1}(AX) = A^{-1}B \qquad \text{multiply from the left by the inverse of } A$$
$$(3) \quad (A^{-1}A)X = A^{-1}B \qquad \text{associative property}$$
$$(4) \qquad IX = A^{-1}B \qquad A^{-1}A = I$$
$$(5) \qquad X = A^{-1}B \qquad IX = X$$

Lines 1 through 5 illustrate the steps that make the method work. In actual practice, after carefully entering the matrices, only step 5 is used when solving matrix equations using technology. Once matrix A is entered, the calculator will automatically *find* and *use* A^{-1} as we enter $A^{-1}B$.

EXAMPLE 3 ▶ **Using Technology to Solve a Matrix Equation**

Use a calculator and a matrix equation to solve the system
$$\begin{cases} x + 4y - z = 10 \\ 2x + 5y - 3z = 7. \\ 8x + y - 2z = 11 \end{cases}$$

Solution ▶ As before, the matrix equation is $\begin{bmatrix} 1 & 4 & -1 \\ 2 & 5 & -3 \\ 8 & 1 & -2 \end{bmatrix} \begin{bmatrix} x \\ y \\ z \end{bmatrix} = \begin{bmatrix} 10 \\ 7 \\ 11 \end{bmatrix}$.

Carefully enter (and double-check) the matrix of coefficients as matrix A in your calculator, and the matrix of constants as matrix B. The product $A^{-1}B$ shows the solution is $x = 2$, $y = 3$, $z = 4$. Verify by substitution.

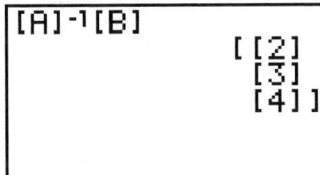

Now try Exercises 27 through 44 ▶

Figure 8.29

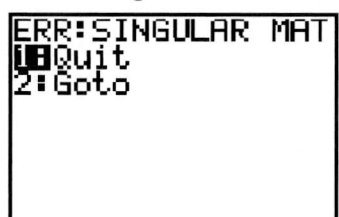

The matrix equation method does have a few shortcomings. Consider the system whose corresponding matrix equation is $\begin{bmatrix} 4 & -10 \\ -2 & 5 \end{bmatrix}\begin{bmatrix} x \\ y \end{bmatrix} = \begin{bmatrix} -8 \\ 13 \end{bmatrix}$. After entering the matrix of coefficients A and matrix of constants B, attempting to compute $A^{-1}B$ results in the error message shown in Figure 8.29. The calculator is unable to return a solution due to something called a **"singular matrix."** To investigate further, we attempt to find A^{-1} for $\begin{bmatrix} 4 & -10 \\ -2 & 5 \end{bmatrix}$ using the formula for a 2×2 matrix. With $a = 4, b = -10, c = -2,$ and $d = 5$, we have

$$A^{-1} = \frac{1}{ad - bc}\begin{bmatrix} d & -b \\ -c & a \end{bmatrix} = \frac{1}{(4)(5) - (-10)(-2)}\begin{bmatrix} 5 & 10 \\ 2 & 4 \end{bmatrix}$$

$$= \frac{1}{0}\begin{bmatrix} 5 & 10 \\ 2 & 4 \end{bmatrix}$$

☑ **C.** You've just learned how to solve systems using matrix equations

Since division by zero is undefined, we conclude that matrix A has no inverse. A matrix having no inverse is said to be **singular** or **noninvertible.** Solving systems using matrix equations is only possible when the matrix of coefficients is **nonsingular.**

D. Determinants and Singular Matrices

As a practical matter, it becomes important to know ahead of time whether a particular matrix has an inverse. To help with this, we introduce one additional operation on a square matrix, that of calculating its **determinant.** For a 1×1 matrix the determinant is the entry itself. For a 2×2 matrix $A = \begin{bmatrix} a_{11} & a_{12} \\ a_{21} & a_{22} \end{bmatrix}$, the determinant of A, written as $\det(A)$ or denoted with vertical bars as $|A|$, is computed as *a difference of diagonal products* beginning with the upper-left entry:

2nd diagonal
product

$$\det(A) = \begin{vmatrix} a_{11} & a_{12} \\ a_{21} & a_{22} \end{vmatrix} = a_{11}a_{22} - a_{21}a_{12}$$

1st diagonal
product

The Determinant of a 2 × 2 Matrix

Given any 2×2 matrix $A = \begin{bmatrix} a_{11} & a_{12} \\ a_{21} & a_{22} \end{bmatrix}$,

$$\det(A) = |A| = a_{11}a_{22} - a_{21}a_{12}$$

EXAMPLE 4 ▶ Calculating Determinants

Compute the determinant of each matrix given.

a. $B = \begin{bmatrix} 3 & 2 \\ 1 & -6 \end{bmatrix}$

b. $C = \begin{bmatrix} 5 & 2 & 1 \\ -1 & -3 & 4 \end{bmatrix}$

c. $D = \begin{bmatrix} 4 & -10 \\ -2 & 5 \end{bmatrix}$

Solution ▶   **a.** $\det(B) = \begin{vmatrix} 3 & 2 \\ 1 & -6 \end{vmatrix} = (3)(-6) - (1)(2) = -20$

       b. Determinants are only defined for square matrices.

       c. $\det(D) = \begin{vmatrix} 4 & -10 \\ -2 & 5 \end{vmatrix} = (4)(5) - (-2)(-10) = 20 - 20 = 0$

Now try Exercises 45 through 48 ▶

Notice from Example 4c, the determinant of $\begin{bmatrix} 4 & -10 \\ -2 & 5 \end{bmatrix}$ was zero, and this is the same matrix we earlier found had no inverse. This observation can be extended to larger matrices and offers the connection we seek between a given matrix, its inverse, and matrix equations.

Singular Matrices

If A is a square matrix and $\det(A) = 0$, the inverse matrix *does not exist* and A is said to be *singular* or *noninvertible*.

In summary, inverses exist only for square matrices, but not every square matrix has an inverse. If the determinant of a square matrix is zero, an inverse does not exist and the method of matrix equations cannot be used to solve the system.

    To use the determinant test for a 3×3 system, we need to compute a 3×3 determinant. At first glance, our experience with 2×2 determinants appears to be of little help. However, every entry in a 3×3 matrix is associated with a smaller 2×2 matrix, formed by *deleting the row and column* of that entry and using the entries that remain. These 2×2's are called the **associated minor matrices** or simply the **minors.** Using a general matrix of coefficients, we'll identify the minors associated with the entries in the first row.

WORTHY OF NOTE

For the determinant of a general $n \times n$ matrix using **cofactors,** see Appendix III.

$$\begin{bmatrix} \textcircled{a_{11}} & a_{12} & a_{13} \\ a_{21} & a_{22} & a_{23} \\ a_{31} & a_{32} & a_{33} \end{bmatrix} \qquad \begin{bmatrix} a_{11} & \textcircled{a_{12}} & a_{13} \\ a_{21} & a_{22} & a_{23} \\ a_{31} & a_{32} & a_{33} \end{bmatrix} \qquad \begin{bmatrix} a_{11} & a_{12} & \textcircled{a_{13}} \\ a_{21} & a_{22} & a_{23} \\ a_{31} & a_{32} & a_{33} \end{bmatrix}$$

Entry: a_{11}   **Entry: a_{12}**   **Entry: a_{13}**
associated minor   **associated minor**   **associated minor**

$$\begin{bmatrix} a_{22} & a_{23} \\ a_{32} & a_{33} \end{bmatrix} \qquad\qquad \begin{bmatrix} a_{21} & a_{23} \\ a_{31} & a_{33} \end{bmatrix} \qquad\qquad \begin{bmatrix} a_{21} & a_{22} \\ a_{31} & a_{32} \end{bmatrix}$$

    To illustrate, consider the system shown, and (1) form the matrix of coefficients, (2) identify the minor matrices associated with the entries in the first row, and (3) compute the determinant of each *minor*.

$$\begin{cases} 2x + 3y - z = 1 \\ x - 4y + 2z = -3 \\ 3x + y = -1 \end{cases} \qquad \text{(1) Matrix of coefficients} \begin{bmatrix} 2 & 3 & -1 \\ 1 & -4 & 2 \\ 3 & 1 & 0 \end{bmatrix}$$

$$\text{(2)} \begin{bmatrix} \textcircled{2} & 3 & -1 \\ 1 & -4 & 2 \\ 3 & 1 & 0 \end{bmatrix} \qquad \begin{bmatrix} 2 & \textcircled{3} & -1 \\ 1 & -4 & 2 \\ 3 & 1 & 0 \end{bmatrix} \qquad \begin{bmatrix} 2 & 3 & \textcircled{-1} \\ 1 & -4 & 2 \\ 3 & 1 & 0 \end{bmatrix}$$

Entry a_{11}: 2   **Entry a_{12}: 3**   **Entry a_{13}: -1**
associated minor   **associated minor**   **associated minor**

$$\begin{bmatrix} -4 & 2 \\ 1 & 0 \end{bmatrix} \qquad\qquad \begin{bmatrix} 1 & 2 \\ 3 & 0 \end{bmatrix} \qquad\qquad \begin{bmatrix} 1 & -4 \\ 3 & 1 \end{bmatrix}$$

|(3) **Determinant** | **Determinant of** | **Determinant of** |
|of minor | minor | minor |

$$(-4)(0) - (1)(2) = -2 \quad (1)(0) - (3)(2) = -6 \quad (1)(1) - (3)(-4) = 13$$

For computing a 3×3 determinant, we illustrate a technique called **expansion by minors.**

The Determinant of a 3 × 3 Matrix—Expansion by Minors

For the matrix M shown, $\det(M)$ is the unique number computed as follows:

matrix M

$$\begin{bmatrix} a_{11} & a_{12} & a_{13} \\ a_{21} & a_{22} & a_{23} \\ a_{31} & a_{32} & a_{33} \end{bmatrix}$$

1. Select any row or column and form the product of each entry with its minor matrix. The illustration here uses the entries in row 1:

$$\det(M) = +a_{11}\begin{vmatrix} a_{22} & a_{23} \\ a_{32} & a_{33} \end{vmatrix} - a_{12}\begin{vmatrix} a_{21} & a_{23} \\ a_{31} & a_{33} \end{vmatrix} + a_{13}\begin{vmatrix} a_{21} & a_{22} \\ a_{31} & a_{32} \end{vmatrix}$$

Sign Chart

$$\begin{bmatrix} + & - & + \\ - & + & - \\ + & - & + \end{bmatrix}$$

2. The *signs used between terms* of the expansion depends on the row or column chosen, according to the *sign chart* shown.

The determinant of a matrix is unique and *any* row or column can be used. For this reason, it's helpful to select the row or column having the most zero, positive, and/or smaller entries.

EXAMPLE 5 ▶ Calculating a 3 × 3 Determinant

Compute the determinant of $M = \begin{bmatrix} 2 & 1 & -3 \\ 1 & -1 & 0 \\ -2 & 1 & 4 \end{bmatrix}$.

Solution ▶ Since the second row has the "smallest" entries as well as a zero entry, we compute the determinant using this row. According to the sign chart, the signs of the terms will be negative–positive–negative, giving

$$\det(M) = -(1)\begin{vmatrix} 1 & -3 \\ 1 & 4 \end{vmatrix} + (-1)\begin{vmatrix} 2 & -3 \\ -2 & 4 \end{vmatrix} - (0)\begin{vmatrix} 2 & 1 \\ -2 & 1 \end{vmatrix}$$

$$= -1(4 + 3) + (-1)(8 - 6) - (0)(2 + 2)$$

$$= \quad -7 \quad + \quad (-2) \quad - \quad 0$$

$$= -9 \rightarrow \text{The value of } \det(M) \text{ is } -9.$$

Now try Exercises 49 through 54 ▶

Try computing the determinant of M two more times, using a different row or column each time. Since the determinant is unique, you should obtain the same result.

There are actually other alternatives for computing a 3×3 determinant. The first is called **determinants by column rotation,** and takes advantage of patterns generated from the expansion of minors. This method is applied to the matrix shown, which uses alphabetical entries for simplicity.

$$\det\begin{bmatrix} a & b & c \\ d & e & f \\ g & h & i \end{bmatrix} \quad \begin{aligned} &= a(ei - fh) - b(di - fg) + c(dh - eg) &&\text{expansion using R1} \\ &= aei - afh - bdi + bfg + cdh - ceg &&\text{distribute} \\ &= aei + bfg + cdh - afh - bdi - ceg &&\text{rewrite result} \end{aligned}$$

Although history is unsure of who should be credited, notice that if you repeat the first two columns to the right of the given matrix ("rotation of columns"), identical products are obtained using the six diagonals formed—three in the downward direction using addition, three in the upward direction using subtraction.

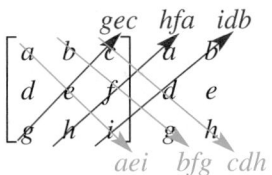

Adding the products in blue (regardless of sign) and subtracting the products in red (regardless of sign) gives the determinant. This method is more efficient than expansion by minors, *but can only be used for* 3×3 *matrices!*

EXAMPLE 6 ▶ **Calculating det(A) Using Column Rotation**

Use the column rotation method to find the determinant of $A = \begin{bmatrix} 1 & 5 & 3 \\ -2 & -8 & 0 \\ -3 & -11 & 1 \end{bmatrix}$.

Solution ▶ Rotate columns 1 and 2 to the right as above, and compute the diagonal products.

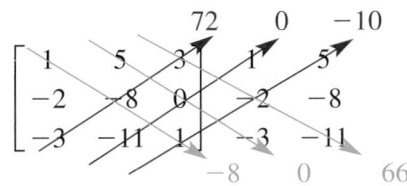

Adding the products in blue (regardless of sign) and subtracting the products in red (regardless of sign) shows $\det(A) = -4$:

$$-8 + 0 + 66 - 72 - 0 - (-10) = -4.$$

Now try Exercises 55 through 58 ▶

The final method is presented in the *Extending the Concept* feature of the Exercise Set, and shows that if certain conditions are met, the determinant of a matrix can be found using its triangularized form.

EXAMPLE 7 ▶ **Solving a System after Verifying A is Invertible**

Given the system shown here, (1) form the matrix equation $AX = B$; (2) compute the determinant of the coefficient matrix and determine if you can proceed; and (3) if so, solve the system using a matrix equation.

$$\begin{cases} 2x + 1y - 3z = 11 \\ 1x - 1y \qquad = 1 \\ -2x + 1y + 4z = -8 \end{cases}$$

Solution ▶ **1.** Form the matrix equation $AX = B$:

$$\begin{bmatrix} 2 & 1 & -3 \\ 1 & -1 & 0 \\ -2 & 1 & 4 \end{bmatrix} \begin{bmatrix} x \\ y \\ z \end{bmatrix} = \begin{bmatrix} 11 \\ 1 \\ -8 \end{bmatrix}$$

2. Since $\det(A)$ is nonzero (from Example 5), we can proceed.

3. For $X = A^{-1}B$, input $A^{-1}B$ on your calculator and press ENTER.

$$X = \begin{bmatrix} \frac{4}{9} & \frac{7}{9} & \frac{1}{3} \\ \frac{4}{9} & -\frac{2}{9} & \frac{1}{3} \\ \frac{1}{9} & \frac{4}{9} & \frac{1}{3} \end{bmatrix} \begin{bmatrix} 11 \\ 1 \\ -8 \end{bmatrix} = \begin{bmatrix} 3 \\ 2 \\ -1 \end{bmatrix}$$

calculator computes and uses A^{-1} in one step

The solution is the ordered triple $(3, 2, -1)$.

Now try Exercises 59 through 62 ▶

We close this section with an application involving a 4×4 system.

EXAMPLE 8 ▶ **Solving an Application Using Technology and Matrix Equations**

A local theater sells four sizes of soft drinks: 32 oz @ $2.25; 24 oz @ $1.90; 16 oz @ $1.50; and 12 oz @ $1.20/each. As part of a "free guest pass" promotion, the manager asks employees to try and determine the number of each size sold, given the following information: (1) the total revenue from soft drinks was $719.80; (2) there were 9096 oz of soft drink sold; (3) there was a total of 394 soft drinks sold; and (4) the number of 24-oz and 12-oz drinks sold was 12 more than the number of 32-oz and 16-oz drinks sold. Write a system of equations that models this information, then solve the system using a matrix equation.

Solution ▶ If we let x, l, m, and s represent the number of 32-oz, 24-oz, 16-oz, and 12-oz soft drinks sold, the following system is produced:

$$\begin{array}{rl} \text{revenue:} \\ \text{ounces sold:} \\ \text{quantity sold:} \\ \text{relationship between} \\ \text{amounts sold:} \end{array} \begin{cases} 2.25x + 1.90l + 1.50m + 1.20s = 719.8 \\ 32x + 24l + 16m + 12s = 9096 \\ x + l + m + s = 394 \\ l + s = x + m + 12 \end{cases}$$

☑ **D. You've just learned how to use determinants to find whether a matrix is invertible**

When written as a matrix equation the system becomes:

$$\begin{bmatrix} 2.25 & 1.9 & 1.5 & 1.2 \\ 32 & 24 & 16 & 12 \\ 1 & 1 & 1 & 1 \\ -1 & 1 & -1 & 1 \end{bmatrix} \begin{bmatrix} x \\ l \\ m \\ s \end{bmatrix} = \begin{bmatrix} 719.8 \\ 9096 \\ 394 \\ 12 \end{bmatrix}$$

To solve, carefully enter the matrix of coefficients as matrix A, and the matrix of constants as matrix B, then compute $A^{-1}B = X$ [verify $\det(A) \neq 0$]. This gives a solution of $(x, l, m, s) = (112, 151, 79, 52)$.

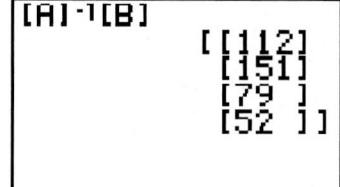

Now try Exercises 67 to 78 ▶

 **8.7 EXERCISES**

▶ **CONCEPTS AND VOCABULARY**

Fill in the blank with the appropriate word or phrase.
Carefully reread the section if needed.

1. The $n \times n$ identity matrix I_n consists of 1's down the _____ and _____ for all other entries.

2. Given square matrices A and B of like size, B is the inverse of A if ___ = ___ = ___. Notationally we write $B = $ ____.

3. The product of a square matrix A and its inverse A^{-1} yields the _____ matrix.

4. If the determinant of a matrix is zero, the matrix is said to be _____ or _____, meaning no inverse exists.

5. Explain why inverses exist only for square matrices, then discuss why some square matrices do not have an inverse. Illustrate each point with an example.

6. What is the connection between the determinant of a 2×2 matrix and the formula for finding its inverse? Use the connection to create a 2×2 matrix that is invertible, and another that is not.

▶ **DEVELOPING YOUR SKILLS**

Use matrix multiplication, equality of matrices, and the
arbitrary matrix given to show that $\begin{bmatrix} a & b \\ c & d \end{bmatrix} = \begin{bmatrix} 1 & 0 \\ 0 & 1 \end{bmatrix}$.

7. $A = \begin{bmatrix} 2 & 5 \\ -3 & -7 \end{bmatrix}\begin{bmatrix} a & b \\ c & d \end{bmatrix} = \begin{bmatrix} 2 & 5 \\ -3 & -7 \end{bmatrix}$

8. $A = \begin{bmatrix} 9 & -7 \\ -5 & 4 \end{bmatrix}\begin{bmatrix} a & b \\ c & d \end{bmatrix} = \begin{bmatrix} 9 & -7 \\ -5 & 4 \end{bmatrix}$

9. $A = \begin{bmatrix} 0.4 & 0.6 \\ 0.3 & 0.2 \end{bmatrix}\begin{bmatrix} a & b \\ c & d \end{bmatrix} = \begin{bmatrix} 0.4 & 0.6 \\ 0.3 & 0.2 \end{bmatrix}$

10. $A = \begin{bmatrix} \frac{1}{2} & \frac{1}{4} \\ \frac{1}{3} & \frac{1}{8} \end{bmatrix}\begin{bmatrix} a & b \\ c & d \end{bmatrix} = \begin{bmatrix} \frac{1}{2} & \frac{1}{4} \\ \frac{1}{3} & \frac{1}{8} \end{bmatrix}$

For $I_2 = \begin{bmatrix} 1 & 0 \\ 0 & 1 \end{bmatrix}, I_3 = \begin{bmatrix} 1 & 0 & 0 \\ 0 & 1 & 0 \\ 0 & 0 & 1 \end{bmatrix}$, **and**

$I_4 = \begin{bmatrix} 1 & 0 & 0 & 0 \\ 0 & 1 & 0 & 0 \\ 0 & 0 & 1 & 0 \\ 0 & 0 & 0 & 1 \end{bmatrix}$, **show** $AI = IA = A$ **for the**

matrices of like size. Use a calculator for Exercise 14.

11. $\begin{bmatrix} -3 & 8 \\ -4 & 10 \end{bmatrix}$

12. $\begin{bmatrix} 0.5 & -0.2 \\ -0.7 & 0.3 \end{bmatrix}$

13. $\begin{bmatrix} -4 & 1 & 6 \\ 9 & 5 & 3 \\ 0 & -2 & 1 \end{bmatrix}$

14. $\begin{bmatrix} 9 & 1 & 3 & -1 \\ 2 & 0 & -5 & 3 \\ 4 & 6 & 1 & 0 \\ 0 & -2 & 4 & 1 \end{bmatrix}$

Find the inverse of each 2×2 matrix using matrix
multiplication, equality of matrices, and a system of
equations.

15. $\begin{bmatrix} 5 & -4 \\ 2 & 2 \end{bmatrix}$

16. $\begin{bmatrix} 1 & -5 \\ 0 & -4 \end{bmatrix}$

17. $\begin{bmatrix} 1 & -3 \\ 4 & -10 \end{bmatrix}$

18. $\begin{bmatrix} -2 & 0.4 \\ 1 & 0.8 \end{bmatrix}$

Demonstrate that $B = A^{-1}$, by showing $AB = BA = I$.
Do not use a calculator.

19. $A = \begin{bmatrix} 1 & 5 \\ -2 & -9 \end{bmatrix}$

    $B = \begin{bmatrix} -9 & -5 \\ 2 & 1 \end{bmatrix}$

20. $A = \begin{bmatrix} -2 & -6 \\ 4 & 11 \end{bmatrix}$

    $B = \begin{bmatrix} 5.5 & 3 \\ -2 & -1 \end{bmatrix}$

21. $A = \begin{bmatrix} 4 & -5 \\ 0 & 2 \end{bmatrix}$

    $B = \begin{bmatrix} \frac{1}{4} & \frac{5}{8} \\ 0 & \frac{1}{2} \end{bmatrix}$

22. $A = \begin{bmatrix} -2 & 5 \\ 3 & -4 \end{bmatrix}$

    $B = \begin{bmatrix} \frac{4}{7} & \frac{5}{7} \\ \frac{3}{7} & \frac{2}{7} \end{bmatrix}$

Use a calculator to find $A^{-1} = B$, then confirm the inverse by showing $AB = BA = I$.

23. $A = \begin{bmatrix} -2 & 3 & 1 \\ 5 & 2 & 4 \\ 2 & 0 & -1 \end{bmatrix}$

24. $A = \begin{bmatrix} 0.5 & 0.2 & 0.1 \\ 0 & 0.3 & 0.6 \\ 1 & 0.4 & -0.3 \end{bmatrix}$

25. $A = \begin{bmatrix} -7 & 5 & -3 \\ 1 & 9 & 0 \\ 2 & -2 & -5 \end{bmatrix}$

26. $A = \dfrac{1}{12}\begin{bmatrix} 12 & -6 & 3 & 0 \\ 0 & -4 & 8 & -12 \\ 12 & -12 & 0 & 0 \\ 0 & 12 & 0 & -12 \end{bmatrix}$

Write each system in the form of a matrix equation. Do not solve.

27. $\begin{cases} 2x - 3y = 9 \\ -5x + 7y = 8 \end{cases}$

28. $\begin{cases} 0.5x - 0.6y = 0.6 \\ -0.7x + 0.4y = -0.375 \end{cases}$

29. $\begin{cases} x + 2y - z = 1 \\ x + z = 3 \\ 2x - y + z = 3 \end{cases}$

30. $\begin{cases} 2x - 3y - 2z = 4 \\ \frac{1}{4}x - \frac{2}{5}y + \frac{3}{4}z = \frac{-1}{3} \\ -2x + 1.3y - 3z = 5 \end{cases}$

31. $\begin{cases} -2w + x - 4y + 5z = -3 \\ 2w - 5x + y - 3z = 4 \\ -3w + x + 6y + z = 1 \\ w + 4x - 5y + z = -9 \end{cases}$

32. $\begin{cases} 1.5w + 2.1x - 0.4y + z = 1 \\ 0.2w - 2.6x + y = 5.8 \\ 3.2x + z = 2.7 \\ 1.6w + 4x - 5y + 2.6z = -1.8 \end{cases}$

Write each system as a matrix equation and solve (if possible) using inverse matrices and your calculator. If the coefficient matrix is singular, write *no solution*.

33. $\begin{cases} 0.05x - 3.2y = -15.8 \\ 0.02x + 2.4y = 12.08 \end{cases}$

34. $\begin{cases} 0.3x + 1.1y = 3.5 \\ -0.5x - 2.9y = -10.1 \end{cases}$

35. $\begin{cases} \frac{-1}{6}u + \frac{1}{4}v = 1 \\ \frac{1}{2}u - \frac{2}{3}v = -2 \end{cases}$

36. $\begin{cases} \sqrt{2}a + \sqrt{3}b = \sqrt{5} \\ \sqrt{6}a + 3b = \sqrt{7} \end{cases}$

37. $\begin{cases} \frac{-1}{8}a + \frac{3}{5}b = \frac{5}{6} \\ \frac{5}{16}a - \frac{3}{2}b = \frac{-4}{5} \end{cases}$

38. $\begin{cases} 3\sqrt{2}a + 2\sqrt{3}b = 12 \\ 5\sqrt{2}a - 3\sqrt{3}b = 1 \end{cases}$

39. $\begin{cases} 0.2x - 1.6y + 2z = -1.9 \\ -0.4x - y + 0.6z = -1 \\ 0.8x + 3.2y - 0.4z = 0.2 \end{cases}$

40. $\begin{cases} 1.7x + 2.3y - 2z = 41.5 \\ 1.4x - 0.9y + 1.6z = -10 \\ -0.8x + 1.8y - 0.5z = 16.5 \end{cases}$

41. $\begin{cases} x - 2y + 2z = 6 \\ 2x - 1.5y + 1.8z = 2.8 \\ \frac{-2}{3}x + \frac{1}{2}y - \frac{3}{5}z = -\frac{11}{30} \end{cases}$

42. $\begin{cases} 4x - 5y - 6z = 5 \\ \frac{1}{8}x - \frac{3}{5}y + \frac{5}{4}z = \frac{-2}{3} \\ -0.5x + 2.4y - 5z = 5 \end{cases}$

43. $\begin{cases} -2w + 3x - 4y + 5z = -3 \\ 0.2w - 2.6x + y - 0.4z = 2.4 \\ -3w + 3.2x + 2.8y + z = 6.1 \\ 1.6w + 4x - 5y + 2.6z = -9.8 \end{cases}$

44. $\begin{cases} 2w - 5x + 3y - 4z = 7 \\ 1.6w + 4.2y - 1.8z = 5.4 \\ 3w + 6.7x - 9y + 4z = -8.5 \\ 0.7x - 0.9z = 0.9 \end{cases}$

Compute the determinant of each matrix and state whether an inverse matrix exists. Do not use a calculator.

45. $\begin{bmatrix} 4 & -7 \\ 3 & -5 \end{bmatrix}$

46. $\begin{bmatrix} 0.6 & 0.3 \\ 0.4 & 0.5 \end{bmatrix}$

47. $\begin{bmatrix} 1.2 & -0.8 \\ 0.3 & -0.2 \end{bmatrix}$

48. $\begin{bmatrix} -2 & 6 \\ -3 & 9 \end{bmatrix}$

Compute the determinant of each matrix without using a calculator. If the determinant is zero, write *singular matrix*.

49. $A = \begin{bmatrix} 1 & 0 & -2 \\ 0 & -1 & -1 \\ 2 & 1 & -4 \end{bmatrix}$

50. $B = \begin{bmatrix} -2 & 2 & 1 \\ 0 & -1 & 2 \\ 4 & -4 & 0 \end{bmatrix}$

51. $C = \begin{bmatrix} -2 & 3 & 4 \\ 0 & 6 & 2 \\ 1 & -1.5 & -2 \end{bmatrix}$

52. $D = \begin{bmatrix} 1 & 2 & -0.8 \\ 2.5 & 5 & -2 \\ 3 & 0 & -2.5 \end{bmatrix}$

Use a calculator to compute the determinant of each matrix. If the determinant is zero, write *singular matrix*. If the determinant is nonzero, find A^{-1} and store the result as matrix B (STO→ 2nd X⁻¹ 2: [B] ENTER). Then verify each inverse by showing $AB = BA = I$.

53. $A = \begin{bmatrix} 1 & 0 & 3 & -4 \\ 2 & 5 & 0 & 1 \\ 8 & 15 & 6 & -5 \\ 0 & 8 & -4 & 1 \end{bmatrix}$

54. $M = \begin{bmatrix} 1 & 2 & 1 & 1 \\ 0 & 1 & -3 & 2 \\ -1 & 0 & 2 & -3 \\ 2 & -1 & 1 & 4 \end{bmatrix}$

Compute the determinant of each matrix using the column rotation method.

55. $\begin{bmatrix} 2 & -3 & 1 \\ 4 & -1 & 5 \\ 1 & 0 & -2 \end{bmatrix}$ **56.** $\begin{bmatrix} -3 & 2 & 4 \\ -1 & -2 & 0 \\ 3 & 1 & 5 \end{bmatrix}$

57. $\begin{bmatrix} 1 & -1 & 2 \\ 3 & -2 & 4 \\ 4 & 3 & 1 \end{bmatrix}$ **58.** $\begin{bmatrix} 5 & 6 & 2 \\ -2 & 1 & -2 \\ 3 & 4 & -1 \end{bmatrix}$

For each system shown, form the matrix equation $AX = B$; compute the determinant of the coefficient matrix and determine if you can proceed; and if possible, solve the system using the matrix equation.

59. $\begin{cases} x - 2y + 2z = 7 \\ 2x + 2y - z = 5 \\ 3x - y + z = 6 \end{cases}$ **60.** $\begin{cases} 2x - 3y - 2z = 7 \\ x - y + 2z = -5 \\ 3x + 2y - z = 11 \end{cases}$

61. $\begin{cases} x - 3y + 4z = -1 \\ 4x - y + 5z = 7 \\ 3x + 2y + z = -3 \end{cases}$ **62.** $\begin{cases} 5x - 2y + z = 1 \\ 3x - 4y + 9z = -2 \\ 4x - 3y + 5z = 6 \end{cases}$

▶ WORKING WITH FORMULAS

The inverse of a 2 × 2 matrix:

$A = \begin{bmatrix} a & b \\ c & d \end{bmatrix} \rightarrow A^{-1} = \dfrac{1}{ad - bc} \cdot \begin{bmatrix} d & -b \\ -c & a \end{bmatrix}$

The inverse of a 2 × 2 matrix can be found using the formula shown, as long as $ad - bc \neq 0$. Use the formula to find inverses for the matrices here (if possible), then verify by showing $A \cdot A^{-1} = A \cdot A^{-1} = I$.

63. $A = \begin{bmatrix} 3 & -5 \\ 2 & 1 \end{bmatrix}$ **64.** $B = \begin{bmatrix} 2 & 3 \\ -5 & -4 \end{bmatrix}$

65. $C = \begin{bmatrix} 0.3 & -0.4 \\ -0.6 & 0.8 \end{bmatrix}$ **66.** $\begin{bmatrix} 0.2 & 0.3 \\ -0.4 & -0.6 \end{bmatrix}$

▶ APPLICATIONS

Solve each application using a matrix equation.

Descriptive Translation

67. Convenience store sales: The local Moto-Mart sells four different sizes of Slushies—behemoth, 60 oz @ $2.59; gargantuan, 48 oz @ $2.29; mammoth, 36 oz @ $1.99; and jumbo, 24 oz @ $1.59. As part of a promotion, the owner offers free gas to any customer who can tell how many of each size were sold last week, given the following information: (1) The total revenue for the Slushies was $402.29; (2) 7884 ounces were sold; (3) a total of 191 Slushies were sold; and (4) the number of behemoth Slushies sold was one more than the number of jumbo. How many of each size were sold?

68. Cartoon characters: In America, four of the most beloved cartoon characters are Foghorn Leghorn, Elmer Fudd, Bugs Bunny, and Tweety Bird. Suppose that Bugs Bunny is four times as tall as Tweety Bird. Elmer Fudd is as tall as the combined height of Bugs Bunny and Tweety Bird. Foghorn Leghorn is 20 cm taller than the combined height of Elmer Fudd and Tweety Bird. The combined height of all four characters is 500 cm. How tall is each one?

69. Rolling Stones music: One of the most prolific and popular rock-and-roll bands of all time is the Rolling Stones. Four of their many great hits include: *Jumpin' Jack Flash, Tumbling Dice, You Can't Always Get What You Want,* and *Wild Horses.* The total playing time of all four songs is 20.75 min.

The combined playing time of *Jumpin' Jack Flash* and *Tumbling Dice* equals that of *You Can't Always Get What You Want*. *Wild Horses* is 2 min longer than *Jumpin' Jack Flash*, and *You Can't Always Get What You Want* is twice as long as *Tumbling Dice*. Find the playing time of each song.

70. **Mozart's arias:** Mozart wrote some of vocal music's most memorable arias in his operas, including *Tamino's Aria, Papageno's Aria,* the *Champagne Aria,* and the *Catalogue Aria.* The total playing time of all four arias is 14.3 min. *Papageno's Aria* is 3 min shorter than the *Catalogue Aria.* The *Champagne Aria* is 2.7 min shorter than *Tamino's Aria.* The combined time of *Tamino's Aria* and *Papageno's Aria* is five times that of the *Champagne Aria.* Find the playing time of all four arias.

Manufacturing

71. **Resource allocation:** Time Pieces Inc. manufactures four different types of grandfather clocks. Each clock requires these four stages: (1) assembly, (2) installing the clockworks, (3) inspection and testing, and (4) packaging for delivery. The time required for each stage is shown in the table, for each of the four clock types. At the end of a busy week, the owner determines that personnel on the assembly line worked for 262 hours, the installation crews for 160 hours, the testing department for 29 hours, and the packaging department for 68 hours. How many clocks of each type were made?

Dept.	Clock A	Clock B	Clock C	Clock D
Assemble	2.2	2.5	2.75	3
Install	1.2	1.4	1.8	2
Test	0.2	0.25	0.3	0.5
Pack	0.5	0.55	0.75	1.0

72. **Resource allocation:** Figurines Inc. makes and sells four sizes of metal figurines, mostly historical figures and celebrities. Each figurine goes through four stages of development: (1) casting, (2) trimming, (3) polishing, and (4) painting. The time required for each stage is shown in the table, for each of the four sizes. At the end of a busy week, the manager finds that the casting department put in 62 hr, and the trimming department worked for 93.5 hr, with the polishing and painting departments logging 138 hr and 358 hr respectively. How many figurines of each type were made?

Dept.	Small	Medium	Large	X-Large
Casting	0.5	0.6	0.75	1
Trimming	0.8	0.9	1.1	1.5
Polishing	1.2	1.4	1.7	2
Painting	2.5	3.5	4.5	6

73. **Thermal conductivity:** In lab experiments designed to measure the heat conductivity of a square metal plate of uniform density, the edges are held at four different (constant) temperatures. The *mean-value principle* from physics tells us that the temperature at a given point p_i on the plate, is equal to the average temperature of nearby points. Use this information to form a system of four equations in four variables, and determine the temperature at interior points p_1, p_2, p_3, and p_4 on the plate shown. (*Hint:* Use the temperature of the four points closest to each.)

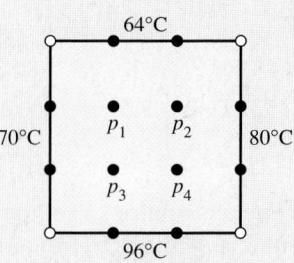

74. **Thermal conductivity:** Repeat Exercise 73 if (a) the temperatures at the top and bottom of the plate were *increased* by 10°, with the temperatures at the left and right edges *decreased* by 10° (what do you notice?); (b) the temperature at the top and the temperature to the left were *decreased* by 10°, with the temperatures at the bottom and right held at their original temperature.

Curve Fitting

75. **Cubic fit:** Find a cubic function of the form $y = ax^3 + bx^2 + cx + d$ such that $(-4, -6), (-1, 0), (1, -16),$ and $(3, 8)$ are on the graph of the function.

76. **Cubic fit:** Find a cubic function of the form $y = ax^3 + bx^2 + cx + d$ such that $(-2, 5), (0, 1), (2, -3),$ and $(3, 25)$ are on the graph of the function.

Nutrition

77. **Animal diets:** A zoo dietician needs to create a specialized diet that regulates an animal's intake of fat, carbohydrates, and protein during a meal. The table given shows three different foods and the amount of these nutrients (in grams) that each ounce of food provides. How many ounces of each should the dietician recommend to supply 20 g of fat, 30 g of carbohydrates, and 44 g of protein?

Nutrient	Food I	Food II	Food III
Fat	2	4	3
Carb.	4	2	5
Protein	5	6	7

78. **Training diet:** A physical trainer is designing a workout diet for one of her clients, and wants to supply him with 24 g of fat, 244 g of

carbohydrates, and 40 g of protein for the noontime meal. The table given shows three different foods and the amount of these nutrients (in grams) that each ounce of food provides. How many ounces of each should the trainer recommend?

Nutrient	Food I	Food II	Food III
Fat	2	5	0
Carb.	10	15	18
Protein	2	10	0.75

▶ EXTENDING THE CONCEPT

79. Some matrix applications require that you solve a matrix equation of the form $AX + B = C$, where A, B, and C are matrices with the appropriate number of rows and columns and A^{-1} exists. Investigate the solution process for such equations using $A = \begin{bmatrix} 2 & 3 \\ -5 & -4 \end{bmatrix}$, $B = \begin{bmatrix} 4 \\ 9 \end{bmatrix}$, $C = \begin{bmatrix} 12 \\ -4 \end{bmatrix}$, and $X = \begin{bmatrix} x \\ y \end{bmatrix}$, then solve $AX + B = C$ for X symbolically (using A^{-1}, I, and so on).

80. It is possible for the matrix of coefficients to be singular, yet for solutions to exist. If the system is dependent instead of inconsistent, there may be infinitely many solutions and the solution set must be written using a parameter or the set notation seen previously. Try solving the exercise given here using matrix equations. If this is not possible, discuss why, then solve using elimination. If the system is dependent, find at least *two* sets of three fractions that fit the criteria. The sum of the two smaller fractions equals the larger, the larger less the smaller equals the "middle" fraction, and four times the smaller fraction equals the sum of the other two.

81. Another alternative for finding determinants uses the triangularized form of a matrix and is offered without proof: *If nonsingular matrix A is written in triangularized form without exchanging any rows and without using the operations kR_i to replace any row (k a constant), then det(A) is equal to the product of resulting diagonal entries.* Compute the determinant of each matrix using this method. Be careful not to interchange rows and do not replace any row by a multiple of that row in the process.

a. $\begin{bmatrix} 1 & -2 & 3 \\ -4 & 5 & -6 \\ 2 & 5 & 3 \end{bmatrix}$ b. $\begin{bmatrix} 2 & 5 & -1 \\ -2 & -3 & 4 \\ 4 & 6 & 5 \end{bmatrix}$

c. $\begin{bmatrix} -2 & 4 & 1 \\ 5 & 7 & -2 \\ 3 & -8 & -1 \end{bmatrix}$ d. $\begin{bmatrix} 3 & -1 & 4 \\ 0 & -2 & 6 \\ -2 & 1 & -3 \end{bmatrix}$

82. Find 2×2 nonzero matrices A and B whose product gives the zero matrix $\begin{bmatrix} 0 & 0 \\ 0 & 0 \end{bmatrix}$.

▶ MAINTAINING YOUR SKILLS

83. (5.5) Find the amplitude and period of $y = -125 \cos(3t)$.

84. (2.5/4.3) Match each equation to its related graph. Justify your answers.

$y = \log_2(x - 2)$ $\qquad$ $y = \log_2 x - 2$

a. 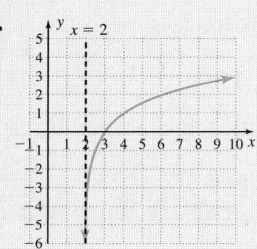 b.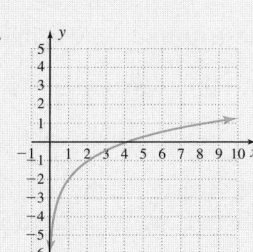

85. (1.3) Solve the absolute value inequality: $-3|2x + 5| - 7 \leq -19$.

86. (6.6) Find all solutions of $-7 \tan^2 x = -21$ in $[0, 2\pi)$.

8.8 | Applications of Matrices and Determinants: Cramer's Rule, Geometry, and More

Learning Objectives

In Section 8.8 you will learn how to:

☐ **A.** Solve a system using determinants and Cramer's Rule

☐ **B.** Use determinants in applications involving geometry in the coordinate plane

☐ **C.** Decompose a rational expression into partial fractions using matrices and technology

Introduction

In addition to their use in solving systems, matrices can be used to accomplish such diverse things as finding the volume of a three-dimensional solid or establishing certain geometrical relationships in the coordinate plane. Numerous uses are also found in higher mathematics, such as checking whether solutions to a differential equation are linearly independent.

A. Solving Systems Using Determinants and Cramer's Rule

In addition to identifying singular matrices, determinants can actually be used to *develop a formula approach* for the solution of a system. Consider the following illustration, in which we solve a *general* 2×2 system by modeling the process after a *specific* 2×2 system. With a view toward a solution involving determinants, the coefficients of x are written as a_{11} and a_{21} in the general system, and the coefficients of y are a_{12} and a_{22}.

Specific System	**General System**
$\begin{cases} 2x + 5y = 9 \\ 3x + 4y = 10 \end{cases}$	$\begin{cases} a_{11}x + a_{12}y = c_1 \\ a_{21}x + a_{22}y = c_2 \end{cases}$
eliminate the x-term in R2	eliminate the x-term in R2
$-3R1 + 2R2$	$-a_{21}R1 + a_{11}R2$

sums to zero

$\begin{cases} \boxed{-3 \cdot 2x} - 3 \cdot 5y = -3 \cdot 9 \\ \boxed{2 \cdot 3x} + 2 \cdot 4y = 2 \cdot 10 \end{cases}$

$2 \cdot 4y - 3 \cdot 5y = 2 \cdot 10 - 3 \cdot 9$

sums to zero

$\begin{cases} \boxed{-a_{21}a_{11}x} - a_{21}a_{12}y = -a_{21}c_1 \\ \boxed{a_{11}a_{21}x} + a_{11}a_{22}y = a_{11}c_2 \end{cases}$

$a_{11}a_{22}y - a_{21}a_{12}y = a_{11}c_2 - a_{21}c_1$

Notice the x-terms sum to zero in both systems. We are deliberately leaving the solution on the left unsimplified to show the pattern developing on the right. Next we solve for y.

Factor Out y

$(2 \cdot 4 - 3 \cdot 5)y = 2 \cdot 10 - 3 \cdot 9$

$$y = \frac{2 \cdot 10 - 3 \cdot 9}{2 \cdot 4 - 3 \cdot 5}$$

Factor Out y

$(a_{11}a_{22} - a_{21}a_{12})y = a_{11}c_2 - a_{21}c_1$

$$y = \frac{a_{11}c_2 - a_{21}c_1}{a_{11}a_{22} - a_{21}a_{12}}$$

On the left we find $y = \frac{-7}{-7} = 1$ and back-substitution shows $x = 2$. But more importantly, on the right we obtain a formula for the y-value of *any* 2×2 system: $y = \dfrac{a_{11}c_2 - a_{21}c_1}{a_{11}a_{22} - a_{21}a_{12}}$. If we had chosen to solve for x, the "formula" solution would be $x = \dfrac{a_{22}c_1 - a_{12}c_2}{a_{11}a_{22} - a_{21}a_{12}}$. Note these formulas are defined only if $a_{11}a_{22} - a_{21}a_{12} \neq 0$. You may have already noticed, but this denominator is the *determinant of the matrix of coefficients* $\begin{bmatrix} a_{11} & a_{12} \\ a_{21} & a_{22} \end{bmatrix}$ from the previous section! Since the numerator is also a difference of two products, we investigate the possibility that it too can be expressed as a determinant. Working backward, we're able to reconstruct the numerator for x in determinant form as $\begin{bmatrix} c_1 & a_{12} \\ c_2 & a_{22} \end{bmatrix}$, where it is apparent this matrix was formed by *replacing the coefficients of the x-variables with the constant terms.*

Forming the Numerator of the Solution for x

$$\begin{vmatrix} \boxed{a_{11}} & a_{12} \\ \boxed{a_{21}} & a_{22} \end{vmatrix} \qquad \begin{vmatrix} & a_{12} \\ & a_{22} \end{vmatrix} \qquad \begin{vmatrix} c_1 & a_{12} \\ c_2 & a_{22} \end{vmatrix}$$

$$\text{(removed)}$$

↘ remove coefficients of x replace with constants ↗

In a similar fashion, the numerator for y can be written in determinant form as $\begin{bmatrix} a_{11} & c_1 \\ a_{21} & c_2 \end{bmatrix}$, or the determinant formed by *replacing the coefficients of the y-variables with the constant terms.* If we use the notation D_y for this determinant, D_x for the determinant where x coefficients were replaced by the constants, and D as the determinant for the matrix of coefficients, the solutions can be written as shown next, with the result known as **Cramer's rule.**

Cramer's Rule applied to 2 × 2 Systems

Given a 2 × 2 system of linear equations

$$\begin{cases} a_{11} + a_{12} = c_1 \\ a_{21} + a_{22} = c_2 \end{cases}$$

the solution is the ordered pair (x, y), where

$$x = \frac{D_x}{D} = \frac{\begin{vmatrix} c_1 & a_{12} \\ c_2 & a_{22} \end{vmatrix}}{\begin{vmatrix} a_{11} & a_{12} \\ a_{21} & a_{22} \end{vmatrix}} \quad \text{and} \quad y = \frac{D_y}{D} = \frac{\begin{vmatrix} a_{11} & c_1 \\ a_{21} & c_2 \end{vmatrix}}{\begin{vmatrix} a_{11} & a_{12} \\ a_{21} & a_{22} \end{vmatrix}}$$

provided $D \neq 0$.

EXAMPLE 1 ▶ **Solving a 2 × 2 System Using Cramer's Rule**

Use Cramer's rule to solve the system $\begin{cases} 2x - 5y = 9 \\ -3x + 4y = -10 \end{cases}$.

Solution ▶ For $x = \dfrac{D_x}{D}$ and $y = \dfrac{D_y}{D}$, begin by finding the value of D, D_x, and D_y.

$$D = \begin{vmatrix} 2 & -5 \\ -3 & 4 \end{vmatrix} \qquad D_x = \begin{vmatrix} 9 & -5 \\ -10 & 4 \end{vmatrix} \qquad D_y = \begin{vmatrix} 2 & 9 \\ -3 & -10 \end{vmatrix}$$

$$(2)(4) - (-3)(-5) \qquad (9)(4) - (-10)(-5) \qquad (2)(-10) - (-3)(9)$$

$$= -7 \qquad\qquad = -14 \qquad\qquad = 7$$

This gives $x = \dfrac{D_x}{D} = \dfrac{-14}{-7} = 2$ and $y = \dfrac{D_y}{D} = \dfrac{7}{-7} = -1$. The solution is $(2, -1)$.

Check by substituting these values into the original equations.

Now try Exercises 7 through 14 ▶

Regardless of the method used to solve a system, always be aware that a consistent, inconsistent, or dependent system is possible. The system $\begin{cases} y - 2x = -3 \\ 4x + 6 = 2y \end{cases}$ yields $\begin{cases} -2x + y = -3 \\ 4x - 2y = -6 \end{cases}$ in standard form, with $D = \begin{vmatrix} -2 & 1 \\ 4 & -2 \end{vmatrix} = (-2)(-2) - (4)(1) = 0.$

We stop here since Cramer's rule cannot be applied, knowing the system is either inconsistent or dependent. To find out which, we write the equations in function form (solve for y). The result is $\begin{cases} y = 2x - 3 \\ y = 2x + 3 \end{cases}$, showing the system consists of two parallel lines and has no solutions.

Cramer's Rule for 3 × 3 Systems

Cramer's rule can be extended to a 3 × 3 system of linear equations, using the same pattern as for 2 × 2 systems. Given the general 3 × 3 system

$$\begin{cases} a_{11}x + a_{12}y + a_{13}z = c_1 \\ a_{21}x + a_{22}y + a_{23}z = c_2, \\ a_{31}x + a_{32}y + a_{33}z = c_3 \end{cases}$$

the solutions are $x = \dfrac{D_x}{D}$, $y = \dfrac{D_y}{D}$, and $z = \dfrac{D_z}{D}$, where D_x, D_y, and D_z are again formed by replacing the coefficients of the indicated variable with the constants, and D is the determinant of the matrix of coefficients.

Cramer's Rule Applied to 3 × 3 Systems

Given a 3 × 3 system of linear equations

$$\begin{cases} a_{11}x + a_{12}y + a_{13}z = c_1 \\ a_{21}x + a_{22}y + a_{23}z = c_2 \\ a_{31}x + a_{32}y + a_{33}z = c_3 \end{cases}$$

The solution is an ordered triple (x, y, z), where

$$x = \frac{D_x}{D} = \frac{\begin{vmatrix} c_1 & a_{12} & a_{13} \\ c_2 & a_{22} & a_{23} \\ c_3 & a_{32} & a_{33} \end{vmatrix}}{\begin{vmatrix} a_{11} & a_{12} & a_{13} \\ a_{21} & a_{22} & a_{23} \\ a_{31} & a_{32} & a_{33} \end{vmatrix}} \qquad y = \frac{D_y}{D} = \frac{\begin{vmatrix} a_{11} & c_1 & a_{13} \\ a_{21} & c_2 & a_{23} \\ a_{31} & c_3 & a_{33} \end{vmatrix}}{\begin{vmatrix} a_{11} & a_{12} & a_{13} \\ a_{21} & a_{22} & a_{23} \\ a_{31} & a_{32} & a_{33} \end{vmatrix}}$$

$$z = \frac{D_z}{D} = \frac{\begin{vmatrix} a_{11} & a_{12} & c_1 \\ a_{21} & a_{22} & c_2 \\ a_{31} & a_{32} & c_3 \end{vmatrix}}{\begin{vmatrix} a_{11} & a_{12} & a_{13} \\ a_{21} & a_{22} & a_{23} \\ a_{31} & a_{32} & a_{33} \end{vmatrix}}, \text{ provided } D \neq 0.$$

EXAMPLE 2 ▶ **Solving a 3 × 3 System Using Cramer's Rule**

Solve using Cramer's rule: $\begin{cases} x - 2y + 3z = -1 \\ -2x + y - 5z = 1 \\ 3x + 3y + 4z = 2 \end{cases}$

Solution ▶ Begin by computing the determinant of the matrix of coefficients, to ensure that Cramer's rule can be applied. Using the third row, we have

$$D = \begin{vmatrix} 1 & -2 & 3 \\ -2 & 1 & -5 \\ 3 & 3 & 4 \end{vmatrix} = +3\begin{vmatrix} -2 & 3 \\ 1 & -5 \end{vmatrix} - 3\begin{vmatrix} 1 & 3 \\ -2 & -5 \end{vmatrix} + 4\begin{vmatrix} 1 & -2 \\ -2 & 1 \end{vmatrix}$$

$$= 3(7) - 3(1) + 4(-3) = 6$$

Since $D \neq 0$ we continue, electing to compute the remaining determinants using a calculator.

$$D_x = \begin{vmatrix} -1 & -2 & 3 \\ 1 & 1 & -5 \\ 2 & 3 & 4 \end{vmatrix} = 12 \quad D_y = \begin{vmatrix} 1 & -1 & 3 \\ -2 & 1 & -5 \\ 3 & 2 & 4 \end{vmatrix} = 0 \quad D_z = \begin{vmatrix} 1 & -2 & -1 \\ -2 & 1 & 1 \\ 3 & 3 & 2 \end{vmatrix} = -6$$

The solution is $x = \dfrac{D_x}{D} = \dfrac{12}{6} = 2$, $y = \dfrac{D_y}{D} = \dfrac{0}{6} = 0$, and $z = \dfrac{D_z}{D} = \dfrac{-6}{6} = -1$, or $(2, 0, -1)$ in triple form. Check this solution in the original equations.

Now try Exercises 15 through 22 ▶

✓ **A.** You've just learned how to solve a system using determinants and Cramer's rule

B. Determinants, Geometry, and the Coordinate Plane

As mentioned in the introduction, the use of determinants extends far beyond solving systems of equations. Here, we'll demonstrate how determinants can be used to find the area of a triangle whose vertices are given as three points in the coordinate plane.

The Area of a Triangle in the xy-Plane

Given a triangle with vertices at (x_1, y_1), (x_2, y_2), and (x_3, y_3).

The area is given by the absolute value of one-half the determinant of T,

where $T = \begin{bmatrix} x_1 & y_1 & 1 \\ x_2 & y_2 & 1 \\ x_3 & y_3 & 1 \end{bmatrix}$: Area $= \left| \dfrac{\det(T)}{2} \right|$

EXAMPLE 3 ▶ **Finding the Area of a Triangle in the xy-Plane**

Find the area of a triangle with vertices at $(3, 1)$, $(-2, 3)$, and $(1, 7)$.

Solution ▶ Begin by forming matrix T and computing $\det(T)$:

$$\det(T) = \begin{vmatrix} x_1 & y_1 & 1 \\ x_2 & y_2 & 1 \\ x_3 & y_3 & 1 \end{vmatrix} = \begin{vmatrix} 3 & 1 & 1 \\ -2 & 3 & 1 \\ 1 & 7 & 1 \end{vmatrix}$$

$$= 3(3 - 7) - 1(-2 - 1) + 1(-14 - 3)$$

$$= -12 + 3 + (-17) = -26$$

Compute the Area: $A = \left| \dfrac{\det(T)}{2} \right| = \left| \dfrac{-26}{2} \right|$

$$= 13$$

The area of this triangle is 13 units^2.

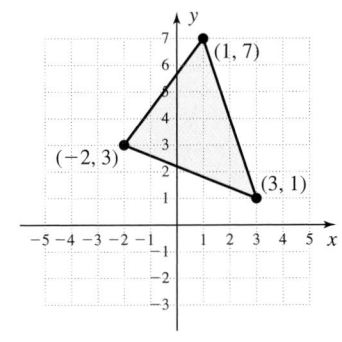

Now try Exercises 25 through 30 ▶

As an extension of this determinant formula, what if the three points were collinear? After a moment, it may occur to you that the formula would give an area of 0 units^2, since no triangle could be formed. This gives rise to a **test for collinear points.**

☑ **B.** You've just learned how to use determinants in applications involving geometry in the coordinate plane

Test for Collinear Points

Three points (x_1, y_1), (x_2, y_2), and (x_3, y_3) are collinear if the determinant of A is zero,

where $|A| = \begin{vmatrix} x_1 & y_1 & 1 \\ x_2 & y_2 & 1 \\ x_3 & y_3 & 1 \end{vmatrix}$.

See Exercises 31 through 36.

C. More on Partial Fraction Decomposition

Occasionally, the decomposition template for a rational expression becomes rather extensive. Instead of attempting a solution using convenient values or row operations as in Section 8.3, solutions can more readily be found using a matrix equation and technology.

EXAMPLE 4 ▶ **Decomposing a Rational Expression Using Matrices and Technology**

Use matrix equations and a graphing calculator to decompose the given expression into partial fractions: $\dfrac{12x^3 + 62x^2 + 102x + 56}{x^4 + 6x^3 + 12x^2 + 8x}$.

Solution ▶ The degree of the numerator is less than that of the denominator, so we begin by factoring the denominator. After removing the common factor of x and applying the rational roots theorem with synthetic division ($c = -2$), the completely factored form is $\dfrac{12x^3 + 62x^2 + 102x + 56}{x(x + 2)^3}$. The decomposed form will be

$$\frac{12x^3 + 62x^2 + 102x + 56}{x(x + 2)^3} = \frac{A}{x} + \frac{B}{(x + 2)^1} + \frac{C}{(x + 2)^2} + \frac{D}{(x + 2)^3}.$$

Clearing denominators and simplifying yields

$$12x^3 + 62x^2 + 102 + 56 = A(x + 2)^3 + Bx(x + 2)^2$$
$$+ Cx(x + 2) + Dx \qquad \text{clear denominators}$$

After expanding the powers on the right, grouping like terms, and factoring, we obtain

$$12x^3 + 62x^2 + 102x + 56 = (A + B)x^3 + (6A + 4B + C)x^2$$
$$+ (12A + 4B + 2C + D)x + 8A$$

By equating the coefficients of like terms, the following system and matrix equation are obtained:

$$\begin{cases} A + B = 12 \\ 6A + 4B + C = 62 \\ 12A + 4B + 2C + D = 102 \\ 8A = 56 \end{cases} \rightarrow \begin{bmatrix} 1 & 1 & 0 & 0 \\ 6 & 4 & 1 & 0 \\ 12 & 4 & 2 & 1 \\ 8 & 0 & 0 & 0 \end{bmatrix} \begin{bmatrix} A \\ B \\ C \\ D \end{bmatrix} = \begin{bmatrix} 12 \\ 62 \\ 102 \\ 56 \end{bmatrix}.$$

After carefully entering the matrices F (coefficients) and G (constants), we obtain the solution $A = 7, B = 5, C = 0$, and $D = -2$ as shown in the figure. The decomposed form is

```
[F]⁻¹[G]
        [[7 ]
         [5 ]
         [0 ]
         [-2]]
```

☑ **C.** You've just learned how to decompose a rational expression into partial fractions using matrices and technology

$$\frac{12x^3 + 62x^2 + 102x + 56}{x(x + 2)^3} = \frac{7}{x} + \frac{5}{(x + 2)^1} + \frac{-2}{(x + 2)^3}.$$

Now try Exercises 37 through 42

8.8 EXERCISES

▶ CONCEPTS AND VOCABULARY

Fill in the blank with the appropriate word or phrase. Carefully reread the section if needed.

1. The determinant $\begin{vmatrix} a_{11} & a_{12} \\ a_{21} & a_{22} \end{vmatrix}$ is evaluated as:

   _____.

2. _____ rule uses a ratio of determinants to solve for the unknowns in a system.

3. Given the matrix of coefficients D, the matrix D_x is formed by replacing the coefficients of x with the _____ terms.

4. The three points (x_1, y_1), (x_2, y_2), and (x_3, y_3) are collinear if $|T| = \begin{vmatrix} x_1 & y_1 & 1 \\ x_2 & y_2 & 1 \\ x_3 & y_3 & 1 \end{vmatrix}$ has a value of _____.

5. Discuss/Explain the process of writing $\dfrac{8x - 3}{x^2 - x}$ as a sum of partial fractions.

6. Discuss/Explain why Cramer's rule cannot be applied if $D = 0$. Use an example to illustrate.

▶ DEVELOPING YOUR SKILLS

Write the determinants D, D_x, and D_y for the systems given. Do not solve.

7. $\begin{cases} 2x + 5y = 7 \\ -3x + 4y = 1 \end{cases}$

8. $\begin{cases} -x + 5y = 12 \\ 3x - 2y = -8 \end{cases}$

Solve each system of equations using Cramer's rule, if possible. Do not use a calculator.

9. $\begin{cases} 4x + y = -11 \\ 3x - 5y = -60 \end{cases}$

10. $\begin{cases} x = -2y - 11 \\ y = 2x - 13 \end{cases}$

11. $\begin{cases} \dfrac{x}{8} + \dfrac{y}{4} = 1 \\ \dfrac{y}{5} = \dfrac{x}{2} + 6 \end{cases}$

12. $\begin{cases} \dfrac{2}{3}x - \dfrac{3}{8}y = \dfrac{7}{5} \\ \dfrac{5}{6}x + \dfrac{3}{4}y = \dfrac{11}{10} \end{cases}$

13. $\begin{cases} 0.6x - 0.3y = 8 \\ 0.8x - 0.4y = -3 \end{cases}$

14. $\begin{cases} -2.5x + 6y = -1.5 \\ 0.5x - 1.2y = 3.6 \end{cases}$

Write the determinants D, D_x, D_y, and D_z for the systems given, then determine if a solution using Cramer's rule is possible by computing the value of D without the use of a calculator (do not solve the system). Try to determine how the system from Part (b) is related to the system in Part (a).

15. a. $\begin{cases} 4x - y + 2z = -5 \\ -3x + 2y - z = 8 \\ x - 5y + 3z = -3 \end{cases}$

    b. $\begin{cases} 4x - y + 2z = -5 \\ -3x + 2y - z = 8 \\ x + y + z = -3 \end{cases}$

16. a. $\begin{cases} 2x + 3z = -2 \\ -x + 5y + z = 12 \\ 3x - 2y + z = -8 \end{cases}$

    b. $\begin{cases} 2x + 3z = -2 \\ -x + 5y + z = 12 \\ 3x - 5y + 2z = -8 \end{cases}$

Use Cramer's rule to solve each system of equations.

17. $\begin{cases} x + 2y + 5z = 10 \\ 3x + 4y - z = 10 \\ x - y - z = -2 \end{cases}$

18. $\begin{cases} x + 3y + 5z = 6 \\ 2x - 4y + 6z = 14 \\ 9x - 6y + 3z = 3 \end{cases}$

19. $\begin{cases} y + 2z = 1 \\ 4x - 5y + 8z = -8 \\ 8x - 9z = 9 \end{cases}$

20. $\begin{cases} x + 2y + 5z = 10 \\ 3x - z = 8 \\ -y - z = -3 \end{cases}$

21. $\begin{cases} w + 2x - 3y = -8 \\ x - 3y + 5z = -22 \\ 4w - 5x = 5 \\ -y + 3z = -11 \end{cases}$

22. $\begin{cases} w - 2x + 3y - z = 11 \\ 3w - 2y + 6z = -13 \\ 2x + 4y - 5z = 16 \\ 3x - 4z = 5 \end{cases}$

▶ WORKING WITH FORMULAS

Area of a Norman window: $A = \begin{vmatrix} L & r^2 \\ -\dfrac{\pi}{2} & W \end{vmatrix}$. The determinant shown can be used to find the area of a Norman

window (rectangle + half-circle) with length L, width W, and radius $r = \dfrac{W}{2}$. Find the area of the following windows.

23.

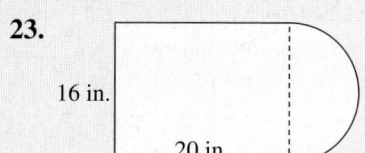

16 in.

20 in.

24.

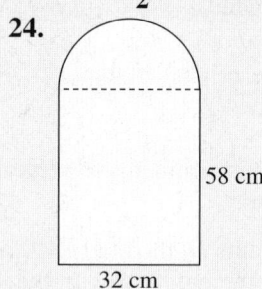

58 cm

32 cm

▶ APPLICATIONS

Area of a triangle: Find the area of the triangle with the vertices given. Assume units are in cm.

25. $(2, 1), (3, 7),$ and $(5, 3)$

26. $(-2, 3), (-3, -4),$ and $(-6, 1)$

Area of a parallelogram: Find the area of the parallelogram with vertices given (*Hint:* Use two triangles.) Assume units are in ft.

27. $(-4, 2), (-6, -1), (3, -1),$ and $(5, 2)$

28. $(-5, -6), (5, 0), (5, 4),$ and $(-5, -2)$

Volume of a pyramid: The volume of a triangular pyramid is given by the formula $V = \frac{1}{3}Bh$, where B represents the area of the triangular base and h is the height of the pyramid. Find the volume of a triangular pyramid whose height is given and whose base has the coordinates shown. Assume units are in m.

29. $h = 6$ m; vertices $(3, 5), (-4, 2),$ and $(-1, 6)$

30. $h = 7.5$ m; vertices $(-2, 3), (-3, -4),$ and $(-6, 1)$

Volume of a prism: The volume of a right prism is given by the formula $V = Bh$, where B represents the area of the base and h is the height of the prism. Find the volume of a triangular prism whose height is given and whose base has the coordinates shown. Assume units are in inches.

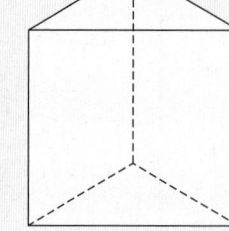

31. $(3, 0), (9, 0), (6, 4), h = 8$

32. $(2, 3), (9, 5), (0, 6), h = 5$

Test for collinear points: Determine if the following sets of points are collinear.

33. $(1, 5), (-2, -1),$ and $(4, 11)$

34. $(1, 1), (3, -5),$ and $(-2, 9)$

35. $(-2.5, 5.2), (1.2, -5.6),$ and $(2.2, -8.5)$

36. $(-0.5, 1.25), (-2.8, 3.75),$ and $(3, 6.25)$

Verifying points are collinear: For each linear equation given, substitute the first two points to verify they are solutions. Then use the test for collinear points to determine if the third point is also a solution. Verify by direct substitution.

37. $2x - 3y = 7; (2, -1), (-1.3, -3.2), (-3.1, -4.4)$

38. $5x + 2y = 4; (2, -3), (3.5, -6.75), (-2.7, 8.75)$

Decompose each rational expressing using a matrix equation and a graphing calculator.

39. $\dfrac{x^4 - x^2 - 2x + 1}{x^5 - 2x^3 + x}$

40. $\dfrac{-3x^4 + 13x^2 + x - 12}{x^5 - 4x^3 + 4x}$

41. $\dfrac{x^3 - 17x^2 + 76x - 98}{(x^2 - 6x + 9)(x^2 - 2x - 3)}$

42. $\dfrac{16x^3 - 66x^2 + 98x - 54}{(2x^2 - 3x)(4x^2 - 12x + 9)}$

Write a linear system that models each application. Then solve using Cramer's rule.

43. Return on investments: If \$15,000 is invested at a certain interest rate and \$25,000 is invested at another interest rate, the total return was \$2900. If the investments were reversed the return would be \$2700. What was the interest rate paid on each investment?

44. CD and DVD clearance sale: To generate interest in a music store clearance sale, the manger sets out a large box full of $2 CDs and $7 DVDs, with an advertised price of $800 for the lot. When asked how many of each are in the box, the manager will only say the box holds a total of 200 disks. How many CDs and DVDs are in the box?

45. Cost of fruit: Many years ago, two pounds of apples, 2 lb of kiwi, and 10 lb of pears cost $3.26. Three pounds of apples, 2 lb of kiwi, and 7 lb of pears cost $2.98. Two pounds of apples, 3 lb of kiwi, and 6 lb of pears cost $2.89. Find the cost of a pound of each fruit.

46. Vending machine receipts: The vending machines at an amusement park are stocked with various candies that sell for 5¢, 10¢, and 25¢. At week's end, the park collected $54.30 from one of the machines. How many of each type of candy were sold, if 484 total sales were made and the machine vended twice as many 5¢ candies as 25¢ candies?

47. Coffee blends: To make its morning coffees, a coffee shop uses three kinds of beans costing $1.90/lb, $2.25/lb, and $3.50/lb, respectively. By the end of the week, the shop went through 24 lb of coffee beans, having a total value of $58. Find how many pounds of each type of bean were used, given that the number of pounds used of the cheapest beans was four more than the most expensive beans.

48. Manufacturing ball bearings: A ball bearing producer makes three sizes of bearings, half-inch-diameter size weighing 30 grams (g), a three-quarter-inch-diameter size weighing 105 g, and a 1-in.-diameter size weighing 250 g. A large storage container holds ball bearings that have been rejected due to small defects. If the net weight of the container's contents is 95.6 kilograms (95,600 g), and automated tallies show 920 bearings have been rejected, how many of each size is in the reject bin, given that statistical studies show there are twice as many rejects of the smallest bearing, as compared to the largest?

▶ EXTENDING THE CONCEPT

49. Solve the given system four different ways: (1) elimination, (2) row reduction, (3) Cramer's rule, and (4) using a matrix equation. Which method seems to be the least error-prone? Which method seems most efficient (takes the least time)? Discuss the advantages and drawbacks of each method.

$$\begin{cases} x + 3y + 5z = 6 \\ 2x - 4y + 6z = 14 \\ 9x - 6y + 3z = 3 \end{cases}$$

50. Find the area of the pentagon whose vertices are: $(-5, -5)$, $(5, -5)$, $(8, 6)$, $(-8, 6)$, and $(0, 12.5)$.

51. The polynomial form for the equation of a circle is $x^2 + y^2 + Dx + Ey + F = 0$. Find the equation of the circle that contains the points $(-1, 7)$, $(2, 8)$, and $(5, -1)$.

52. For square matrix A, calculating A^2 is a simple matter of matrix multiplication. But what of those applications that require higher powers? For any matrix $A = \begin{bmatrix} a & b \\ c & d \end{bmatrix}$, we define the **characteristic polynomial** of A to be

$$p(\lambda) = \begin{vmatrix} \lambda - a & -b \\ -c & \lambda - d \end{vmatrix} = (\lambda - a)(\lambda - d) + cb.$$

The **Cayley-Hamilton theorem** states that if the original matrix A is substituted for λ, the result must be zero, or in other words, every square matrix satisfies its own characteristic equation. This fact is often used to compute higher powers of a square matrix, since all higher powers can then be expressed in terms of A. Specifically, for

$$A = \begin{bmatrix} 1 & 2 \\ 3 & 4 \end{bmatrix}, p(\lambda) = \begin{vmatrix} \lambda - 1 & -2 \\ -3 & \lambda - 4 \end{vmatrix} =$$

$(\lambda - 1)(\lambda - 4) - (-3)(-2) = \lambda^2 - 5\lambda - 2$,

with the theorem stating that $A^2 - 5A - 2I_2 = 0$ (note I_2 is used to ensure we remain in the context of 2×2 matrices). For $A^2 = 5A + 2I_2$, we have the follow sequence for any power of A:

$$\begin{aligned} A^3 &= A^2 A \\ &= (5A + 2I_2)A \\ &= 5A^2 + 2A \\ &= 5(5A + 2I_2) + 2A \\ &= 27A + 10I_2 \end{aligned} \qquad \begin{aligned} A^4 &= A^3 A \\ &= (27A + 10I_2)A \\ &= 27A^2 + 10A \\ &= 27(5A + 2I_2) + 10A \\ &= 145A + 54I_2 \end{aligned}$$

For $A = \begin{bmatrix} 2 & -1 \\ 4 & 1 \end{bmatrix}$, use the Cayley-Hamilton theorem to find A^3, A^4, and A^5. Verify A^3 using scalar multiplication and matrix addition.

▶ MAINTAINING YOUR SKILLS

53. (3.4) Graph the polynomial using information about end behavior, y-intercept, x-intercept(s), and midinterval points: $f(x) = x^3 - 2x^2 - 7x + 6$.

54. (3.3) Which is the graph (left or right) of $g(x) = -|x + 1| + 3$? Justify your answer.

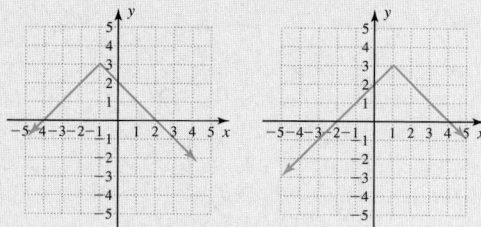

55. (5.6) Solve the triangle:
side $a = 8.7$ in.
side $b = 11.2$ in.
$\angle A = 49.0°$

56. (5.4) Graph $y = 3 \tan(2\pi t)$ over the interval $\left[-\frac{1}{2}, \frac{1}{2}\right]$. Note the period, asymptotes, zeroes and value of A.

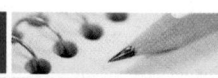

SUMMARY AND CONCEPT REVIEW

SECTION 8.1   Linear Systems in Two Variables with Applications

KEY CONCEPTS

- A *solution* to a linear system in two variables is any ordered pair (x, y) that makes all equations in the system true.
- Since every point on the graph of a line satisfies the equation of that line, a point where two lines intersect must satisfy both equations and is a solution of the system.
- A system with at least one solution is called a *consistent system*.
- If the lines have different slopes, there is a unique solution to the system (they intersect at a single point). The system is called a *consistent* and *independent system*.
- If the lines have equal slopes and the same y-intercept, they form identical or *coincident lines*. Since one line is right atop the other, they intersect at all points with an infinite number of solutions. The system is called a *consistent* and *dependent system*.
- If the lines have equal slopes but different y-intercepts, they will never intersect. The system has no solution and is called an *inconsistent system*.

EXERCISES

Solve each system by graphing. If the solution is not a lattice point on the graph (x- and y-values both integers), indicate your solution is an estimate. If the system is inconsistent or dependent, so state.

1. $\begin{cases} 3x - 2y = 4 \\ -x + 3y = 8 \end{cases}$

2. $\begin{cases} 0.2x + 0.5y = -1.4 \\ x - 0.3y = 1.4 \end{cases}$

3. $\begin{cases} 2x + y = 2 \\ x - 2y = 4 \end{cases}$

Solve using substitution. Indicate whether each system is consistent, inconsistent, or dependent. Write unique solutions as an ordered pair.

4. $\begin{cases} y = 5 - x \\ 2x + 2y = 13 \end{cases}$

5. $\begin{cases} x + y = 4 \\ 0.4x + 0.3y = 1.7 \end{cases}$

6. $\begin{cases} x - 2y = 3 \\ x - 4y = -1 \end{cases}$

Solve using elimination. Indicate whether each system is consistent, inconsistent, or dependent. Write unique solutions as an ordered pair.

7. $\begin{cases} 2x - 4y = 10 \\ 3x + 4y = 5 \end{cases}$

8. $\begin{cases} -x + 5y = 8 \\ x + 2y = 6 \end{cases}$

9. $\begin{cases} 2x = 3y + 6 \\ 2.4x + 3.6y = 6 \end{cases}$

10. When it was first constructed in 1968, the John Hancock building in Chicago, Ilinois, was the tallest structure in the world. In 1985, the Sears Tower in Chicago became the world's tallest structure. The Sears Tower is 323 ft taller than the John Hancock Building, and the sum of their heights is 2577 ft. How tall is each structure?

SECTION 8.2   Linear Systems in Three Variables with Applications

KEY CONCEPTS

- The graph of a linear equation in three variables is a *plane.*
- A linear system in three variables has the following possible solution sets:
  - If the planes intersect at a point, the system could have one *unique solution* (x, y, z).
  - If the planes intersect at a line, the system has *linear dependence* and the solution (x, y, z) can be written as linear combinations of a single variable (*a parameter*).
  - If the planes are *coincident,* the equations in the system differ by a constant multiple, meaning they are all "disguised forms" of the *same equation.* The solutions have *coincident dependence,* and the solution set can be represented by any one of the equations.
  - In all other cases, the system has *no solutions* and is an inconsistent system.

EXERCISES

Solve using elimination. If a system is inconsistent or dependent, so state. For systems with linear dependence, give the answer as an ordered triple using a parameter.

11. $\begin{cases} x + y - 2z = -1 \\ 4x - y + 3z = 3 \\ 3x + 2y - z = 4 \end{cases}$

12. $\begin{cases} -x + y + 2z = 2 \\ x + y - z = 1 \\ 2x + y + z = 4 \end{cases}$

13. $\begin{cases} 3x + y + 2z = 3 \\ x - 2y + 3z = 1 \\ 4x - 8y + 12z = 7 \end{cases}$

Solve using a system of three equations in three variables.

14. A large coin jar is full of nickels, dimes, and quarters. There are 217 coins in all. The number of nickels is 12 more than the number of quarters. The value of the dimes is \$4.90 more than the value of the nickels. How many coins of each type are in the bank?

15. If the point (x, y) is on the graph of a parabola, it must satisfy the equation $y = ax^2 + bx + c$. Use a system of three equations in three variables to find the equation of the parabola containing the points $(0, -9)$, $(2, 7)$, and $(6, 15)$.

SECTION 8.3   Partial Fraction Decomposition

KEY CONCEPTS

- Partial fraction decomposition is an attempt to reverse the process of adding rational expressions.
- The process begins by factoring the denominator and setting up a "decomposition template."
- For each linear factor $(x + c)$ of the denominator, the template will have a term of the form $\dfrac{A_1}{x + c}$, A_1 a constant.
- For each irreducible quadratic factor $(ax^2 + bx + c)$ of the denominator, the template will have a term of the form $\dfrac{A_1 x + B_1}{ax^2 + bx + c}$ for constants A_1 and B_1.
- If the factor $(x + c)$ is repeated n times the template will have n terms of the form $\dfrac{A_i}{(x + c)^i}$ for $i = 1$ to n.
- If the irreducible quadratic factor is repeated n times, the template will have n terms of the form $\dfrac{A_i x + B_i}{(ax^2 + bx + c)^i}$, for $i = 1$ to n.
- After the template has been set (original expression = decomposition template), multiply both sides by the LCD and (1) use convenient values or (2) simplify the right-hand side and equate coefficients to find the constants A_i and B_i needed. Note that some of the coefficients found may be zero.
- If the degree of the numerator is greater than the degree of the denominator, divide using long division and decompose the remainder portion.

EXERCISES

Decompose each of the following expressions into partial fractions.

16. $\dfrac{6x + 14}{x^2 + 2x - 15}$

17. $\dfrac{16x + 1}{2x^2 - 5x - 3}$

18. $\dfrac{x^2 + 2x + 9}{x^3 + 4x^2 + x + 4}$

19. $\dfrac{-2x^2 - 3x - 19}{x^3 - 5x^2 + 3x - 15}$

20. $\dfrac{-2x^2 + 13x - 24}{x^3 + 27}$

21. $\dfrac{6x^2 + 2x + 7}{x^3 - 1}$

22. $\dfrac{2x^2 - 8x + 33}{x^3 - x^2 - 8x + 12}$

23. $\dfrac{-x^2 - 15x + 22}{x^3 + 3x^2 - 9x + 5}$

24. $\dfrac{2x^2 - x + 1}{x^4 + 6x^2 + 9}$

SECTION 8.4   Systems of Inequalities and Linear Programming

KEY CONCEPTS

- To solve a *system of inequalities,* we find the intersecting or *overlapping regions* of the solution regions from the individual inequalities. The common area is called the *feasible region.*
- The process known as *linear programming* seeks to *maximize* or *minimize* the value of a given quantity under certain *constraints* or restrictions.
- The quantity we attempt to maximize or minimize is called the *objective function.*
- The solution(s) to a linear programming problem *occur at one of the corner points of the feasible region.*
- The process of solving a linear programming application contains these six steps:
  - Identify the main objective and the decision variables.
  - Write the objective function in terms of these variables.
  - Organize all information in a table, using the decision variables and constraints.
  - Fill in the table with the information given and write the constraint inequalities.
  - Graph the constraint inequalities and determine the feasible region.
  - Identify all corner points of the feasible region and test these points in the objective function.

EXERCISES

Graph the solution region for each system of linear inequalities and verify the solution using a test point.

25. $\begin{cases} -x - y > -2 \\ -x + y < -4 \end{cases}$

26. $\begin{cases} x - 4y \le 5 \\ -x + 2y \le 0 \end{cases}$

27. $\begin{cases} x + 2y \ge 1 \\ 2x - y \le -2 \end{cases}$

28. Carefully graph the feasible region for the system of inequalities shown, then maximize the

objective function: $f(x, y) = 30x + 45y$ $\begin{cases} x + y \le 7 \\ 2x + y \le 10 \\ 2x + 3y \le 18 \\ x \ge 0 \\ y \ge 0 \end{cases}$

29. After retiring, Oliver and Lisa Douglas buy and work a small farm (near Hooterville) that consists mostly of milk cows and egg-laying chickens. Although the price of a commodity is rarely stable, suppose that milk sales bring in an average of $85 per cow and egg sales an average of $50 per chicken over a period of time. During this time period, the new ranchers estimate that care and feeding of the animals took about 3 hr per cow and 2 hr per chicken, while maintaining the related equipment took 2 hr per cow and 1 hr per chicken. How many animals of each type should be maintained in order to maximize profits, if at most 1000 hr can be spent on care and feeding, and at most 525 hr on equipment maintenance?

SECTION 8.5   Solving Linear Systems Using Matrices and Row Operations

KEY CONCEPTS

- A *matrix* is a rectangular arrangement of numbers. An $m \times n$ matrix has m rows and n columns.
- The matrix derived from a system of linear equations is called the *augmented matrix*. It is created by augmenting the *coefficient matrix* (formed by the variable coefficients) with the *matrix of constants*.
- One matrix method for solving systems involves the augmented matrix and *row-reduction*.
  - If possible, interchange equations so that the coefficient of x is a "1" in R1.
  - Write the system in augmented matrix form (coefficient matrix with matrix of constants).
  - Use row operations to obtain zeroes below the first entry of the diagonal.
  - Use row operations to obtain zeroes below the second entry of the diagonal.
  - Continue until the matrix is triangularized (all entries below the diagonal are zero).
  - Convert the augmented matrix back into equation form and solve for z.

EXERCISES

Solve by triangularizing the matrix. Use a calculator for Exercise 32.

30. $\begin{cases} x - 2y = 6 \\ 4x - 3y = 4 \end{cases}$

31. $\begin{cases} x - 2y + 2z = 7 \\ 2x + 2y - z = 5 \\ 3x - y + z = 6 \end{cases}$

32. $\begin{cases} 2w + x + 2y - 3z = -19 \\ w - 2x - y + 4z = 15 \\ x + 2y - z = 1 \\ 3w - 2x - 5z = -60 \end{cases}$

SECTION 8.6   The Algebra of Matrices

KEY CONCEPTS

- The entries of a matrix are denoted a_{ij}, where i gives the row and j gives the column of its location.
- The $m \times n$ size of a matrix is also referred to as its *order*.
- Two matrices A and B are equal if corresponding entries are equal: $A = B$ if $a_{ij} = b_{ij}$.
- The sum or difference of two matrices is found by combining corresponding entries: $A + B = [a_{ij} + b_{ij}]$.
- The *identity matrix* for addition is an $m \times n$ matrix whose entries are all zeroes.
- The product of a constant times a matrix is called *scalar multiplication*, and is found by taking the product of the scalar with each entry, forming a new matrix of like size. For matrix A: $kA = ka_{ij}$.
- Matrix multiplication is performed as row entry $\times$ column entry according to the following procedure: For an $m \times n$ matrix $A = [a_{ij}]$ and a $p \times q$ matrix $B = [b_{ij}]$, matrix multiplication is possible if $n = p$, and the result will be an $m \times q$ matrix P. In symbols $A \cdot B = [p_{ij}]$, where p_{ij} is product of the ith row of A with the jth column of B.
- When technology is used to perform operations on matrices, the focus shifts from a meticulous computation of new entries, to carefully entering each matrix into the calculator, double checking that each entry is correct, and appraising the results to see if they are reasonable.

EXERCISES

Compute the operations indicated below (if possible), using the following matrices.

$A = \begin{bmatrix} \frac{-1}{4} & \frac{-3}{4} \\ \frac{-1}{8} & \frac{-7}{8} \end{bmatrix}$

$B = \begin{bmatrix} -7 & 6 \\ 1 & -2 \end{bmatrix}$

$C = \begin{bmatrix} -1 & 3 & 4 \\ 5 & -2 & 0 \\ 6 & -3 & 2 \end{bmatrix}$

$D = \begin{bmatrix} 2 & -3 & 0 \\ 0.5 & 1 & -1 \\ 4 & 0.1 & 5 \end{bmatrix}$

33. $A + B$

34. $B - A$

35. $C - B$

36. $8A$

37. BA

38. $C + D$

39. $D - C$

40. BC

41. $-4D$

42. CD

SECTION 8.7   Solving Linear Systems Using Matrix Equations

KEY CONCEPTS

- The *identity matrix I* for multiplication has 1s on the main diagonal and 0s for all other entries. For any $n \times n$ matrix A, the identity matrix is also $n \times n$, where $AI = IA = A$.
- For an $n \times n$ (square) matrix A, the *inverse matrix* for multiplication is a matrix B such that $AB = BA = I$. Only square matrices have an inverse. For matrix A the inverse is denoted A^{-1}.
- Any $n \times n$ system of equations can be written as a matrix equation and solved (if solutions exist) using an inverse

  matrix. For $\begin{cases} 2x + 3y - z = 5 \\ -3x + 2y + 4z = 13, \\ x - 5y + 2z = -3 \end{cases}$ the matrix equation is $\begin{bmatrix} 2 & 3 & -1 \\ -3 & 2 & 4 \\ 1 & -5 & 2 \end{bmatrix} \cdot \begin{bmatrix} x \\ y \\ z \end{bmatrix} = \begin{bmatrix} 5 \\ 13 \\ -3 \end{bmatrix}$.

- Every square matrix has a real number associated with it called its determinant. For a 2×2 matrix

  $A = \begin{vmatrix} a_{11} & a_{12} \\ a_{21} & a_{22} \end{vmatrix}$, the determinant $|A|$ is the difference of diagonal products: $a_{11}a_{22} - a_{21}a_{12}$.

- If the determinant of a matrix is zero, the matrix is said to be *singular* or *non-invertible*.
- For matrix equations, if the coefficient matrix is non-invertible, the system has no unique solution.

EXERCISES

Complete Exercises 43 through 45 using the following matrices:

$$A = \begin{bmatrix} 1 & 0 \\ 0 & 1 \end{bmatrix} \qquad B = \begin{bmatrix} 0.2 & 0.2 \\ -0.6 & 0.4 \end{bmatrix} \qquad C = \begin{bmatrix} 2 & -1 \\ 3 & 1 \end{bmatrix} \qquad D = \begin{bmatrix} 10 & -6 \\ -15 & 9 \end{bmatrix}$$

43. Exactly one of the matrices given is singular. Compute each determinant to identify it.

44. Show that $AB = BA = B$. What can you conclude about the matrix A?

45. Show that $BC = CB = I$. What can you conclude about the matrix C?

Complete Exercises 46 through 49 using these matrices:

$$E = \begin{bmatrix} 1 & -2 & 3 \\ -2 & 1 & -5 \\ -1 & -1 & -2 \end{bmatrix} \qquad F = \begin{bmatrix} 1 & -1 & 1 \\ 0 & 1 & 0 \\ -2 & 1 & -1 \end{bmatrix} \qquad G = \begin{bmatrix} 1 & 0 & 0 \\ 0 & 1 & 0 \\ 0 & 0 & 1 \end{bmatrix} \qquad H = \begin{bmatrix} -1 & 0 & -1 \\ 0 & 1 & 0 \\ 2 & 1 & 1 \end{bmatrix}$$

46. Exactly one of the matrices above is singular. Use a calculator to determine which one.

47. Show that $GF = FG = F$. What can you conclude about the matrix G?

48. Show that $FH = HF = I$. What can you conclude about the matrix H?

49. Verify that $EH \neq HE$ and $EF \neq FE$ and comment.

Solve manually using a matrix equation.

50. $\begin{cases} 2x - 5y = 14 \\ -3y + 4x = -14 \end{cases}$

Solve using a matrix equation and your calculator.

51. $\begin{cases} 0.5x - 2.2y + 3z = -8 \\ -0.6x - y + 2z = -7.2 \\ x + 1.5y - 0.2z = 2.6 \end{cases}$

SECTION 8.8   Applications of Matrices and Determinants, Cramer's Rule, Geometry, and More

KEY CONCEPTS

- Cramer's rule uses a ratio of determinants to solve systems of equations (if they exist).
- The determinant of the 2 × 2 matrix $\begin{vmatrix} a_{11} & a_{12} \\ a_{21} & a_{22} \end{vmatrix}$ is $a_{11}a_{22} - a_{21}a_{12}$.
- To compute the value of 3 × 3 and larger determinants, a calculator is generally used.
- Determinants can be used to find the area of a triangle in the plane if the vertices are known, and as a test to see if three points are collinear.
- A system of equations can be used to write a rational expression as a sum of its partial fractions.

EXERCISES

Solve using Cramer's rule.

52. $\begin{cases} 5x + 6y = 8 \\ 10x - 2y = -9 \end{cases}$

53. $\begin{cases} 2x + y = -2 \\ -x + y + 5z = 12 \\ 3x - 2y + z = -8 \end{cases}$

54. $\begin{cases} 2x + y - z = -1 \\ x - 2y + z = 5 \\ 3x - y + 2z = 8 \end{cases}$

55. Find the area of a triangle whose vertices have the coordinates $(6, 1)$, $(-1, -6)$, and $(-6, 2)$.

56. Use a matrix equation and a graphing calculator to find the partial fraction decomposition for $\dfrac{7x^2 - 5x + 17}{x^3 - 2x^2 + 3x - 6}$.

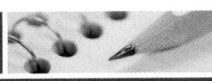

MIXED REVIEW

1. Write the equations in each system in slope-intercept form, then state whether the system is consistent/independent, consistent/dependent, or inconsistent. Do not solve.

  a. $\begin{cases} -3x + 5y = 10 \\ 6x + 20 = 10y \end{cases}$

  b. $\begin{cases} 4x - 3y = 9 \\ -2x + 5y = -10 \end{cases}$

  c. $\begin{cases} x - 3y = 9 \\ -6y + 2x = 10 \end{cases}$

2. Solve by graphing. $\begin{cases} x - 2y = 6 \\ -2x + y = -9 \end{cases}$

3. Solve using a substitution. $\begin{cases} 2x + 3y = 5 \\ -x + 5y = 17 \end{cases}$

4. Solve using elimination. $\begin{cases} 7x - 4y = -5 \\ 3x + 2y = 9 \end{cases}$

Solve using elimination.

5. $\begin{cases} x + 2y - 3z = -4 \\ -3x + 4y + z = 1 \\ 2x - 6y + z = 1 \end{cases}$

6. $\begin{cases} 0.1x - 0.2y + z = 1.7 \\ 0.3x + y - 0.1z = 3.6 \\ -0.2x - 0.1y + 0.2z = -1.7 \end{cases}$

Solve using row operations to triangularize the matrix.

7. $\begin{cases} \dfrac{1}{2}x + \dfrac{2}{3}y = 3 \\ \dfrac{-2}{5}x - \dfrac{1}{4}y = 1 \end{cases}$

8. $\begin{cases} -2x + y - 4z = -11 \\ x + 3y - z = -4 \\ 3x - 2y + z = 7 \end{cases}$

Decompose the rational expressions into partial fractions.

9. $\dfrac{13 - x}{x^2 - x - 6}$

10. $\dfrac{7x^2 - 3x - 4}{x^3 - 3x^2 + x - 3}$

Compute as indicated for

$$A = \begin{bmatrix} 2 & -1 \\ 0 & 3 \end{bmatrix} \qquad B = \begin{bmatrix} 1 & -2 \\ 3 & 0 \\ 2 & 4 \end{bmatrix}$$

$$C = \begin{bmatrix} 1 & -4 & 2 \\ -2 & 0 & -1 \end{bmatrix} \qquad D = \begin{bmatrix} 3 & 0 & 1 \\ -1 & 2 & 0 \\ 1 & 1 & -4 \end{bmatrix}$$

11. a. $-2AC$ **b.** CD

12. a. BA **b.** $CB - 4A$

13. Solve using a matrix equation:
$$\begin{cases} -x - 2z = 5 \\ 2y + z = -4 \\ -x + 2y = 3 \end{cases}$$

14. Use a matrix equation and a calculator to solve:
$$\begin{cases} w + \dfrac{1}{2}x + \dfrac{2}{3}y - z = -3 \\ \dfrac{3}{4}x - y + \dfrac{5}{8}z = \dfrac{41}{8} \\ \dfrac{2}{5}w - x - \dfrac{3}{10}z = -\dfrac{27}{10} \\ w + 2x - 3y + 4z = 16 \end{cases}$$

15. Solve using Cramer's rule:
$$\begin{cases} -x + 5y - 2z = 1 \\ 2x + 3y - z = 3 \\ 3x - y + 3z = -2 \end{cases}$$

16. A triangle in the coordinate plane has vertices $(-1, -7)$, $(6, 2)$, and $(1, 5)$. Use the geometry of matrices to find the area of the triangle.

17. Graph the solution region for the system of
$$\text{inequalities.} \quad \begin{cases} 4x + 2y \le 14 \\ 2x + 3y \le 15 \\ y \ge 0 \\ x \ge 0 \end{cases}$$

18. Maximize $P(x, y) = 2.5x + 3.75y$, given
$$\begin{cases} x + y \le 8 \\ x + 2y \le 14 \\ 4x + 3y \le 30 \\ x, y \ge 0 \end{cases}$$

19. It's the end of another big day at the circus, and the clowns are putting away their riding equipment—a motley collection of unicycles, bicycles, and tricycles. As she loads them into the storage shed, Trixie counts 21 cycles in all with a total of 40 wheels. In addition, she notes the number of bicycles is one fewer than twice the number of tricycles. How many cycles of each type do the clowns use?

20. A local fitness center is offering incentives in an effort to boost membership. If you buy a year's membership (Y), you receive a \$50 rebate and six tickets to a St. Louis Cardinals home game. For a half-year membership (H), you receive a \$30 rebate and four tickets to a Cardinals home game. For a monthly trial membership (M), you receive a \$10 rebate and two tickets to a Cardinals home game. During the last month, male clients purchased 40 one-year, 52 half-year, and 70 monthly memberships, while female clients purchased 50 one-year, 44 half-year, and 60 monthly memberships. Write the number of sales of each type to males and females as a 2×3 matrix, and the amount of the rebates and number of Cards tickets awarded per type of membership as a 3×2 matrix. Use these matrices to determine (a) the total amount of rebate money paid to males and (b) the number of Cardinals tickets awarded to females.

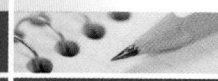

PRACTICE TEST

Solve each system and state whether the system is consistent, inconsistent, or dependent.

1. Solve graphically:
$$\begin{cases} 3x + 2y = 12 \\ -x + 4y = 10 \end{cases}$$

2. Solve using substitution:
$$\begin{cases} 3x - y = 2 \\ -7x + 4y = -6 \end{cases}$$

3. Solve using elimination:
$$\begin{cases} 5x + 8y = 1 \\ 3x + 7y = 5 \end{cases}$$

4. Solve using elimination:
$$\begin{cases} x + 2y - z = -4 \\ 2x - 3y + 5z = 27 \\ -5x + y - 4z = -27 \end{cases}$$

5. Given matrices A and B, compute:

   a. $A - B$ **b.** $\dfrac{2}{5}B$ **c.** AB

   d. A^{-1} **e.** $|A|$

$$A = \begin{bmatrix} -3 & -2 \\ 5 & 4 \end{bmatrix} \qquad B = \begin{bmatrix} 3 & 3 \\ -3 & -5 \end{bmatrix}$$

6. Given matrices C and D, use a calculator to find:

   a. $C - D$ **b.** $-0.6D$ **c.** DC

   d. D^{-1} **e.** $|D|$

$$C = \begin{bmatrix} 0.5 & 0 & 0.2 \\ 0.4 & -0.5 & 0 \\ 0.1 & -0.4 & -0.1 \end{bmatrix} \qquad D = \begin{bmatrix} 0.5 & 0.1 & 0.2 \\ -0.1 & 0.1 & 0 \\ 0.3 & 0.4 & 0.8 \end{bmatrix}$$

7. Solve using matrices and row reduction:

$$\begin{cases} 4x - 5y - 6z = 5 \\ 2x - 3y + 3z = 0 \\ x + 2y - 3z = 5 \end{cases}$$

8. Solve using Cramer's rule:

$$\begin{cases} 2x + 3y + z = 3 \\ x - 2y - z = 4 \\ x - y - 2z = -1 \end{cases}$$

9. Solve using a matrix equation and your calculator:

$$\begin{cases} 2x - 5y = 11 \\ 4x + 7y = 4 \end{cases}$$

10. Solve using a matrix equation and your calculator:

$$\begin{cases} x - 2y + 2z = 7 \\ 2x + 2y - z = 5 \\ 3x - y + z = 6 \end{cases}$$

Create a system of equations to model each exercise, then solve using the method of your choice.

11. The perimeter of a "legal-size" paper is 114.3 cm. The length of the paper is 7.62 cm less than twice the width. Find the dimensions of a legal-size sheet of paper.

12. The island nations of Tahiti and Tonga have a combined land area of 692 mi^2. Tahiti's land area is 112 mi^2 more than Tonga's. What is the land area of each island group?

13. Many years ago, two cans of corn (C), 3 cans of green beans (B), and 1 can of peas (P) cost $1.39.

Three cans of C, 2 of B, and 2 of P cost $1.73. One can of C, 4 of B, and 3 of P cost $1.92. What is the price of a single can of C, B, and P?

14. After inheriting $30,000 from a rich aunt, David decides to place the money in three different investments: a savings account paying 5%, a bond account paying 7%, and a stock account paying 9%. After 1 yr he earned $2080 in interest. Find how much was invested at each rate if $8000 less was invested at 9% than at 7%.

Using a radar gun placed in a protective cage beneath a falling baseball, the following data points are taken: (1, 128), (2, 80), (2.5, 44). Assume the data are of the form (time in seconds, distance from radar gun).

15. Use the data to set up a system of three equations in three variables to find the quadratic equation that models the ball's height at time t.

16. From what height is the baseball being released? In how many seconds will it strike the cage protecting the radar gun?

17. Solve the system of inequalities by graphing.

$$\begin{cases} x - y \le 2 \\ x + 2y \ge 8 \end{cases}$$

18. Maximize the objective function: $P = 50x - 12y$

$$\begin{cases} x + 2y \le 8 \\ 8x + 5y \ge 40 \\ x, y \ge 0 \end{cases}$$

Solve the linear programming problem.

19. A company manufactures two types of T-shirts, a plain T-shirt and a deluxe monogrammed T-shirt. To produce a plain shirt requires 1 hr of working time on machine A and 2 hr on machine B. To produce a deluxe shirt requires 1 hr on machine A and 3 hr on machine B. Machine A is available for at most 50 hr/week, while machine B is available for at most 120 hr/week. If a plain shirt can be sold at a profit of $4.25 each and a deluxe shirt can be sold at a profit of $5.00 each, how many of each should be manufactured to maximize the profit?

20. Decompose the expression into partial fractions:

$$\frac{4x^2 - 4x + 3}{x^3 - 27}.$$

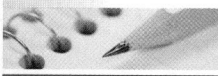

CALCULATOR EXPLORATION AND DISCOVERY

Optimal Solutions and Linear Programming

In Section 8.4 we learned, *"If a linear programming problem has a unique solution, it must occur at a vertex."* Although we explored the reason why using the feasible region and a family of lines from the objective function, it sometimes helps to have a good "old fashioned," point-by-point verification to support the facts. In this exercise, we'll use a graphing calculator to explore various areas of the feasible region, repeatedly evaluating the objective function to see where the maximal values (optimal solutions) seem to "congregate." If all goes as expected, ordered pairs nearest to a vertex should give relatively larger values. To demonstrate, we'll use Example 6 from Section 8.4, stated below.

Example 6 ▶ Find the maximum value of the objective function $f(x, y) = 2x + y$ given the constraints shown in

the system $\begin{cases} x + y \leq 4 \\ 3x + y \leq 6 \\ x \geq 0 \\ y \geq 0 \end{cases}$.

Solution ▶ Begin by noting the solutions must be in QI, since $x \geq 0$ and $y \geq 0$. Graph the boundary lines $y = -x + 4$ and $y = -3x + 6$, shading the lower half plane in each case since they are "less than" inequalities. This produces the feasible region shown in lavender. There are four corner points to this region: $(0, 0)$, $(0, 4)$, $(2, 0)$, and $(1, 3)$, the latter found by solving

the system: $\begin{cases} x + y = 4 \\ 3x + y = 6 \end{cases}$.

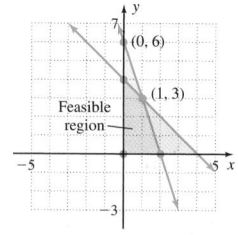

To explore this feasible region in terms of the objective function $f(x, y) = 2x + y$, enter the boundary lines $Y_1 = -x + 4$ and $Y_2 = -3x + 6$ on the $\boxed{Y=}$ screen. However, instead of shading below the lines to show the feasible region (using the ◣ feature to the extreme left), we shade above both lines (using the ◤ feature) so that the feasible region remains clear. Setting the window size at $x \in [0, 3]$ and $y \in [-1.5, 4]$ produces Figure 8.30. Using Ymin = -1.5 will leave a blank area just below QI that allows you to cleanly explore the feasible region as the x- and y-values are displayed. Next we place the

Figure 8.30

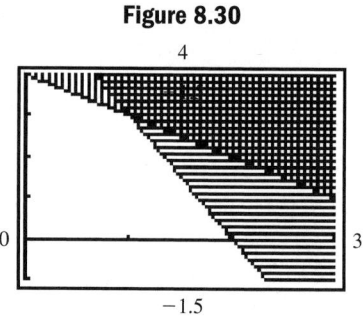

calculator in "split-screen" mode so that we can view the graph and the home screen simultaneously. Press the $\boxed{\text{MODE}}$ key and notice the second-to-last line reads **Full Horiz G-T.** The **Full** (screen) mode is the default operating mode. The **Horiz** mode splits the screen horizontally, placing the graph directly above a shorter home screen. The **G-T** mode splits the screen vertically, with the graph to the left and a table to the right. Highlight **Horiz,** then press $\boxed{\text{ENTER}}$ and $\boxed{\text{GRAPH}}$ to have the calculator reset the screen in this mode. The TI-84 Plus has a free-moving cursor (a cursor you can move around without actually tracing a curve). Pressing the left $\boxed{◄}$ or right $\boxed{►}$ arrow brings it into view (Figure 8.31). A useful feature of the free-moving cursor is that it automatically stores the current X value as the

Figure 6.31

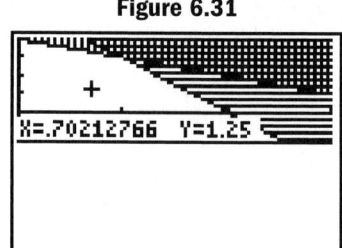

variable X ($\boxed{\text{X,T,θ,n}}$ or $\boxed{\text{ALPHA}}$ $\boxed{\text{STO →}}$) and the current Y value as the variable Y ($\boxed{\text{ALPHA}}$ 1), which enables us to evaluate the objective function $f(x, y) = 2x + y$ right on the home screen as we explore various corners of the feasible region. To access the graph and free-moving cursor you must press $\boxed{\text{GRAPH}}$ each time, and to access the home screen you must press $\boxed{\text{2nd}}$ $\boxed{\text{MODE}}$ **(QUIT)** each time. Begin by moving the cursor to the upper-left corner of the region, near the y-intercept [we stopped at $(\sim0.0957, 3.2\overline{6})$]. Once you have the cursor "tucked up into the corner," press $\boxed{\text{2nd}}$ $\boxed{\text{MODE}}$ **(QUIT)** to get to the home screen, then enter the objective function: 2X + Y. Pressing $\boxed{\text{ENTER}}$ automatically evaluates the function for the values indicated by the cursor's location (Figure 8.32). It appears the value of the objective function for points (x, y) in this corner are close to 4, and it's no accident that at the corner point $(0, 4)$ the maximum value is in fact 4. Now let's explore the area in the lower-right of the feasible region. Press $\boxed{\text{GRAPH}}$ and move the cursor using the arrow keys until it is "tucked" over into the lower-right corner [we stopped at $(\sim1.72, 0.\overline{3})$]. Pressing $\boxed{\text{2nd}}$ $\boxed{\text{ENTER}}$ recalls 2X + Y and pressing $\boxed{\text{ENTER}}$ reevaluates the objective function at these new X and Y values (Figure 8.33). It appears that values of the

Figure 8.32

Figure 8.33

objective function are also close to 4 in this corner, and it's again no accident that *at the corner* point (2, 0) the maximum value *is* 4. Finally, press **GRAPH** to explore the region in the upper-right corner, where the lines intersect. Move the cursor to this vicinity, locate it very near the point of intersection [we stopped at (~0.957, 2.71$\overline{6}$)] and return to the home screen and evaluate (Figure 8.34). The value of the objective function is near 5 in this corner of the region, and at the corner point (1, 3) the maximum value is 5. This investigation can be repeated for any feasible region and for any number of points within the region, and serves to support the statement that, "*If a linear programming problem has a unique solution, it must occur at a vertex.*"

Figure 8.34

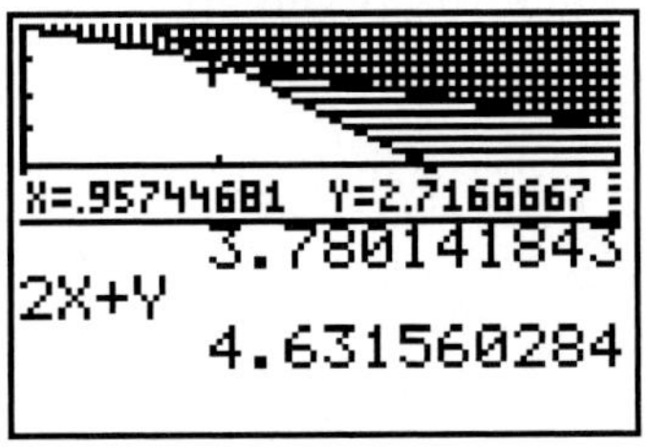

```
X=.95744681   Y=2.7166667
          3.780141843
2X+Y
          4.631560284
```

Exercise 1: Use the ideas discussed here to explore the solution to Example 7 from Section 8.4.

Exercise 2: The feasible region for the system given to the right has five corner points. Use the ideas here to explore the area near each corner point of the feasible region to determine which point is the likely candidate to produce the *maximum* value of the objective function $f(x, y) = -3.5x + y$. Then solve the linear programming problem to verify your guess.

$$\begin{cases} 8x + 3y \le 30 \\ 5x + 4y \le 23 \\ x + 2y \le 10 \\ x, y \ge 0 \end{cases}$$

Exercise 3: The feasible region for the system given to the right has four corner points. Use the ideas here to explore the area near each corner point of the feasible region to determine which point is the likely candidate to produce the *minimum* value of the objective function $f(x, y) = 2x + 4y$. Then solve the linear programming problem to verify your guess.

$$\begin{cases} 2x + 2y \le 15 \\ x + y \ge 6 \\ x + 4y \ge 9 \\ x, y \ge 0 \end{cases}$$

STRENGTHENING CORE SKILLS

Augmented Matrices and Matrix Inverses

The formula for finding the inverse of a 2×2 matrix has its roots in the more general method of computing the inverse of an $n \times n$ matrix. This involves augmenting a square matrix M with its corresponding identity I_n on the right (forming an $n \times 2n$ matrix), and using row operations to *transform M into the identity*. In some sense, as the original matrix is transformed, the "identity part" keeps track of the operations we used to convert M and we can use the results to "get back home," so to speak. We'll illustrate with the 2×2 matrix from Section 8.7, Example 3, where we found that $\begin{bmatrix} 1 & -2.5 \\ -1 & 3 \end{bmatrix}$ was the inverse matrix for $\begin{bmatrix} 6 & 5 \\ 2 & 2 \end{bmatrix}$. We begin by augmenting $\begin{bmatrix} 6 & 5 \\ 2 & 2 \end{bmatrix}$ with the 2×2 identity.

$$\left[\begin{array}{cc|cc} 6 & 5 & 1 & 0 \\ 2 & 2 & 0 & 1 \end{array}\right] \xrightarrow{-2R1 + 6R2 \longrightarrow R2} \left[\begin{array}{cc|cc} 6 & 5 & 1 & 0 \\ 0 & 2 & -2 & 6 \end{array}\right] \xrightarrow{\frac{R2}{2} \longrightarrow R2} \left[\begin{array}{cc|cc} 6 & 5 & 1 & 0 \\ 0 & 1 & -1 & 3 \end{array}\right]$$

$$\left[\begin{array}{cc|cc} 6 & 5 & 1 & 0 \\ 0 & 1 & -1 & 3 \end{array}\right] \xrightarrow{-5R2 + R1 \longrightarrow R1} \left[\begin{array}{cc|cc} 6 & 0 & 6 & -15 \\ 0 & 1 & -1 & 3 \end{array}\right] \xrightarrow{\frac{R1}{6} \longrightarrow R1} \left[\begin{array}{cc|cc} 1 & 0 & 1 & -2.5 \\ 0 & 1 & -1 & 3 \end{array}\right]$$

As you can see, the identity is automatically transformed into the inverse matrix when this method is applied. Performing similar row operations on the general matrix $\begin{bmatrix} a & b \\ c & d \end{bmatrix}$ results in the formula given earlier.

$$\left[\begin{array}{cc|cc} a & b & 1 & 0 \\ c & d & 0 & 1 \end{array}\right] \xrightarrow{-cR1 + aR2 \longrightarrow R2} \left[\begin{array}{cc|cc} a & b & 1 & 0 \\ 0 & ad - bc & -c & a \end{array}\right]$$

$$\xrightarrow{\frac{R2}{ad - bc} \longrightarrow R2} \left[\begin{array}{cc|cc} a & b & 1 & 0 \\ 0 & 1 & \dfrac{-c}{ad - bc} & \dfrac{a}{ad - bc} \end{array}\right]$$

$$\left[\begin{array}{cc|cc} a & b & 1 & 0 \\ 0 & 1 & \dfrac{-c}{ad - bc} & \dfrac{a}{ad - bc} \end{array}\right] \xrightarrow{-bR2 + R1 \longrightarrow R1} \left[\begin{array}{cc|cc} a & 0 & \dfrac{bc}{ad - bc} + 1 & \dfrac{-ba}{ad - bc} \\ 0 & 1 & \dfrac{-c}{ad - bc} & \dfrac{a}{ad - bc} \end{array}\right]$$

Finding a common denominator for $\dfrac{bc}{ad-bc}+1$ and combining like terms gives $\dfrac{bc}{ad-bc}+\dfrac{ad-bc}{ad-bc}=\dfrac{ad}{ad-bc}$.

$$\left[\begin{array}{cc|cc} a & 0 & \dfrac{ad}{ad-bc} & \dfrac{-ba}{ad-bc} \\ 0 & 1 & \dfrac{-c}{ad-bc} & \dfrac{a}{ad-bc} \end{array}\right] \ \dfrac{R1}{a}\longrightarrow R1 \ \left[\begin{array}{cc|cc} 1 & 0 & \dfrac{d}{ad-bc} & \dfrac{-b}{ad-bc} \\ 0 & 1 & \dfrac{-c}{ad-bc} & \dfrac{a}{ad-bc} \end{array}\right].$$

This shows $A^{-1}=\begin{bmatrix} \dfrac{d}{ad-bc} & \dfrac{-b}{ad-bc} \\ \dfrac{-c}{ad-bc} & \dfrac{a}{ad-bc} \end{bmatrix}$

and factoring out $\dfrac{1}{ad-bc}$ produces the familiar formula. As you might imagine, attempting this on a general 3×3 matrix is problematic at best, and instead we simply apply the procedure to any given 3×3 matrix. Here we'll use the augmented matrix method to find A^{-1}, given $A=\begin{bmatrix} 2 & 1 & 0 \\ -1 & 3 & -2 \\ 3 & -1 & 2 \end{bmatrix}$.

$$\left[\begin{array}{ccc|ccc} 2 & 1 & 0 & 1 & 0 & 0 \\ -1 & 3 & -2 & 0 & 1 & 0 \\ 3 & -1 & 2 & 0 & 0 & 1 \end{array}\right] \begin{array}{c} \\ R1+2R2\longrightarrow R2 \\ -3R1+2R3\longrightarrow R3 \end{array} \left[\begin{array}{ccc|ccc} 2 & 1 & 0 & 1 & 0 & 0 \\ 0 & 7 & -4 & 1 & 2 & 0 \\ 0 & -5 & 4 & -3 & 0 & 2 \end{array}\right]$$

$$\begin{array}{c} R2+7R1\longrightarrow R1 \\ \\ 5R2+7R3\longrightarrow R3 \end{array} \left[\begin{array}{ccc|ccc} -14 & 0 & -4 & -6 & 2 & 0 \\ 0 & 7 & -4 & 1 & 2 & 0 \\ 0 & 0 & 8 & -16 & 10 & 14 \end{array}\right]$$

$$\dfrac{R3}{8}\longrightarrow R3 \left[\begin{array}{ccc|ccc} -14 & 0 & -4 & -6 & 2 & 0 \\ 0 & 7 & -4 & 1 & 2 & 0 \\ 0 & 0 & 1 & -2 & 1.25 & 1.75 \end{array}\right]$$

$$\begin{array}{c} 4R3+R2\longrightarrow R2 \\ \\ 4R3+R1\longrightarrow R1 \end{array} \left[\begin{array}{ccc|ccc} -14 & 0 & 0 & -14 & 7 & 7 \\ 0 & 7 & 0 & -7 & 7 & 7 \\ 0 & 0 & 1 & -2 & 1.25 & 1.75 \end{array}\right]$$

Using $\dfrac{R1}{-14}\longrightarrow R1$ and $\dfrac{R2}{7}\longrightarrow R2$ produces the inverse matrix $A^{-1}=\begin{bmatrix} 1 & -0.5 & -0.5 \\ -1 & 1 & 1 \\ -2 & 1.25 & 1.75 \end{bmatrix}$.

To verify, we show $AA^{-1}=I$:

$$\begin{bmatrix} 2 & 1 & 0 \\ -1 & 3 & -2 \\ 3 & -1 & 2 \end{bmatrix}\begin{bmatrix} 1 & -0.5 & -0.5 \\ -1 & 1 & 1 \\ -2 & 1.25 & 1.75 \end{bmatrix}=\begin{bmatrix} 1 & 0 & 0 \\ 0 & 1 & 0 \\ 0 & 0 & 1 \end{bmatrix}\checkmark.$$

Exercise 1: Use the preceding inverse and a matrix equation to solve the system

$$\begin{cases} 2x+y=-2 \\ -x+3y-2z=-15. \\ 3x-y+2z=9 \end{cases}$$

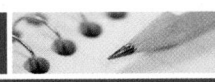

CUMULATIVE REVIEW CHAPTERS 1–8

1. Solve each equation by factoring.

    a. $9x^2 - 12x = -4$

    b. $x^2 - 7x = 0$

    c. $3x^3 - 15x^2 + 6x = 30$

    d. $x^3 = 4x + 3x^2$

2. Solve for x.
    $$\frac{-3}{x - 3} + \frac{3x}{x^2 - x - 6} = \frac{1}{x + 2}$$

3. A torus is a donut-shaped solid figure. Its surface area is given by the formula $A = \pi^2(R^2 - r^2)$, where R is the outer radius of the donut, and r is the inner radius. Solve the formula for R in terms of r and A.

4. State the value of all six trig functions given $(21, -28)$ is a point on the terminal side of θ.

5. Sketch the graph of $y = 4\cos\left(\dfrac{\pi}{6}x - \dfrac{\pi}{3}\right)$ using transformations of $y = \cos x$.

6. A jai alai player in an open-air court becomes frustrated and flings the pelota out and beyond the court. If the initial velocity of the ball is 136 mph and it is released at height of 5 ft, (a) how high is the ball after 3 sec? (b) What is the maximum height of the ball? (c) How long until the ball hits the ground (hopefully in an unpopulated area)? Recall the projectile equation is $h(t) = -16t^2 + v_0t + k$.

7. For a complex number $a + bi$, (a) verify the sum of a complex number and its conjugate is a real number. (b) Verify the product of a complex number and its conjugate is a real number.

8. Solve using the quadratic formula:
    $5x^2 + 8x + 2 = 0$.

9. Solve by completing the square:
    $3x^2 - 72x + 427 = 0$.

10. Given $\cos 53° \approx 0.6$ and $\cos 72° \approx 0.3$, approximate the value of $\cos 19°$ and $\cos 125°$ without using a calculator.

11. Given $\left(\dfrac{\sqrt{3}}{4}, y\right)$ is a point on the unit circle, find the value of $\sin\theta$, $\cos\theta$, and $\tan\theta$ if $y > 0$.

12. State the domain of each function:

    a. $f(x) = \sqrt{2x - 3}$

    b. $g(x) = \log_b(x + 3)$

    c. $h(x) = \dfrac{x + 3}{x^2 - 5}$

13. Write the following formulas from memory:

    a. slope formula

    b. midpoint formula

    c. quadratic formula

    d. distance formula

    e. interest formula (compounded continuously)

14. Find the equation of the line perpendicular to the line $4x + 5y = -20$, that contains the point $(0, 1)$.

15. For the force vectors $\mathbf{F}_1$ and $\mathbf{F}_2$ given, find a third force vector $\mathbf{F}_3$ that will bring these vectors into equilibrium: $\mathbf{F}_1\langle -5, 12\rangle$ $\mathbf{F}_2\langle 8, 6\rangle$

16. A commercial fishery stocks a lake with 250 fish. Based on previous experience, the population of fish is expected to grow according to the model
    $$P(t) = \frac{12{,}000}{1 + 25e^{-0.2t}}, \text{ where } t \text{ is the time in months.}$$
    From on this model, (a) how many months are required for the population to grow to 7500 fish? (b) If the fishery expects to harvest three-fourths of the fish population in 2 yr, approximately how many fish will be taken?

17. Evaluate each expression by drawing a right triangle and labeling the sides.

    a. $\sec\left[\sin^{-1}\left(\dfrac{x}{\sqrt{121 + x^2}}\right)\right]$

    b. $\sin\left[\csc^{-1}\left(\dfrac{\sqrt{9 + x^2}}{x}\right)\right]$

18. An luxury ship is traveling at 15 mph on a heading of 10°. There is a strong, 12 mph ocean current flowing from the southeast, at a heading of 330°. What is the true course and speed of the ship?

19. Use the *Guidelines for Graphing* to sketch the graph of function f given, then use it to solve $f(x) < 0$: $f(x) = x^3 - 4x^2 + x + 6$

20. Use the *Guidelines for Graphing* to sketch the graph of function g given, then use it to name the intervals where $g(x)\downarrow$ and $g(x)\uparrow$: $g(x) = \dfrac{x^2 - 4}{x^2 - 1}$

21. Find $(1 - \sqrt{3}i)^8$ using De Moivre's theorem.

22. Solve $\ln(x + 2) + \ln(x - 3) = \ln(4x)$.

23. If I saved $200 each month in an annuity program that paid 8% annual interest compounded monthly, how long would it take to save $10,000?

24. Mount Tortolas lies on the Argentine-Chilean border. When viewed from a distance of 5 mi, the angle of elevation to the top of the peak is 38°. How tall is Mount Tortolas? State the answer in feet.

25. The graph given is of the form $y = A \sin(Bx + C)$. Find the values of A, B, and C.

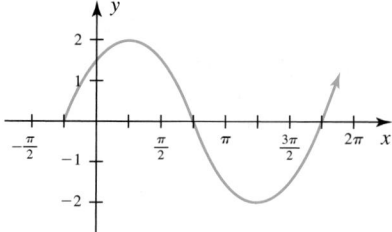

CONNECTIONS TO CALCULUS

As mentioned in the chapter opener, in calculus as well as algebra we often look for expressions that fit a special form, then apply a general formula to all expressions of that form. Here this general idea is applied to partial fraction decomposition, a technique we studied in Section 8.3 and an important part of integral calculus.

More on Partial Fraction Decomposition

To rewrite the expression $\dfrac{10}{x^2 - 25}$ as a sum of its partial fractions, we would use the template $\dfrac{10}{(x + 5)(x - 5)} = \dfrac{A}{x + 5} + \dfrac{B}{x - 5}$ and proceed as in Section 8.3. Using the ideas developed there, we'll attempt to develop a decomposition formula that works for *all* rational expressions of the form $\dfrac{2k}{x^2 - k^2}$, a form that certainly applies to the expression at hand.

EXAMPLE 1 ▶ **Decomposing Special Forms**

Use the techniques from Section 8.3 to rewrite the expression $\dfrac{2k}{x^2 - k^2}$ as the sum of its partial fractions.

Solution ▶ After factoring the denominator, we obtain the following template:

$$\frac{2k}{(x + k)(x - k)} = \frac{A}{x + k} + \frac{B}{x - k} \qquad \text{decomposition template}$$
$$2k = A(x - k) + B(x + k) \qquad \text{clear denominators}$$
$$= Ax - Ak + Bx + Bk \qquad \text{distribute}$$
$$= (A + B)x + (B - A)k \qquad \text{collect like terms}$$

This gives the system $\begin{cases} A + B = 0 \\ B - A = 2 \end{cases}$, yielding the solutions $B = 1$ and $A = -1$. As it turns out, any rational expression of the form $\dfrac{2k}{x^2 - k^2}$ can be written in decomposed form as $\dfrac{-1}{x + k} + \dfrac{1}{x - k}$.

Now try Exercises 1 and 2 ▶

EXAMPLE 2 ▶ **Decomposing Special Forms**

Use the "formula" from Example 1 to decompose the expression $\dfrac{10}{x^2 - 25}$ as the sum of its partial fractions, then verify the result.

Solution ▶ By comparing $\dfrac{2k}{x^2 - k^2}$ with $\dfrac{10}{x^2 - 25}$, we note $2k = 10$ and $k = 5$, so $\dfrac{10}{x^2 - 25}$ must be equal to $\dfrac{-1}{x + 5} + \dfrac{1}{x - 5}$. To check, we combine the fractions using the LCD $(x + 5)(x - 5)$, giving $\dfrac{-1(x - 5) + 1(x + 5)}{(x + 5)(x - 5)} = \dfrac{-x + 5 + x + 5}{(x + 5)(x - 5)}$ or $\dfrac{10}{x^2 - 25}$ ✓.

Now try Exercises 3 through 8 ▶

The Geometry of Vectors and Determinants

Although vectors (Section 7.3) and determinants (Section 8.8) are studied in two separate chapters in *Precalculus,* in a calculus course you'll see there is an intriguing connection between the two. In Section 7.3 we noted the tip-to-tail addition of vectors **u** and **v** formed a parallelogram, and used the parallelogram to illustrate a vector sum. In a calculus course, we say the vectors *span* the parallelogram, and in certain applications we need to know the area of this figure. Using the vectors $\mathbf{u} = \langle a, b \rangle$ and $\mathbf{v} = \langle c, d \rangle$ and relation $\cos \theta = \dfrac{\mathbf{u} \cdot \mathbf{v}}{|\mathbf{u}||\mathbf{v}|}$ for the angle θ between them, it can shown that the area of the parallelogram is the absolute value of the determinant $\begin{vmatrix} a & b \\ c & d \end{vmatrix}$ (see Figure C2C 8.1). For a complete proof, see Appendix V.

Figure C2C 8.1

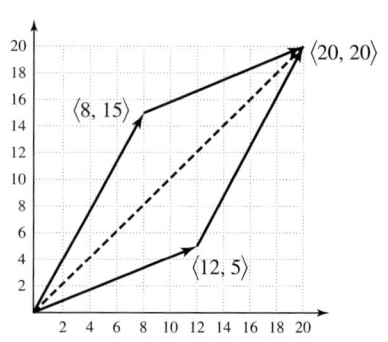

EXAMPLE 3 ▶ **Using Determinants to Find Area**

For the vectors $\mathbf{u} = \langle 12, 5 \rangle$ and $\mathbf{v} = \langle 8, 15 \rangle$,

a. Compute the vector sum $\mathbf{w} = \mathbf{u} + \mathbf{v}$ using the parallelogram (tail-to-tip) method.

b. Compute $|\mathbf{w}|$ and find the angle θ between **u** and **w**.

c. Use the formula $A = \dfrac{1}{2}|\mathbf{u}||\mathbf{w}|\sin \theta$ from Section 7.2 to find the area of the triangle formed.

d. Use the determinant $\begin{vmatrix} a & b \\ c & d \end{vmatrix} = \begin{vmatrix} 12 & 5 \\ 8 & 15 \end{vmatrix}$ to find the area of the parallelogram spanned by these vectors, then verify the area is twice as large as the triangle found in part (c).

Solution ▶ **a.** The vector sum is $\langle 12, 5 \rangle + \langle 8, 15 \rangle = \langle 20, 20 \rangle = \mathbf{w}$.

b. $|\mathbf{w}| = \sqrt{20^2 + 20^2} = \sqrt{800} = 20\sqrt{2}$. For $\mathbf{u} \cdot \mathbf{w} = 12(20) + 5(20) = 340$, and $|\mathbf{u}| = 13$ (verify this), we have $\cos \theta = \dfrac{340}{(13)(20\sqrt{2})}$ and $\theta = \cos^{-1}\left(\dfrac{17}{13\sqrt{2}}\right)$. This indicates $\theta \approx 22.38°$.

c. $A = \dfrac{1}{2}|\mathbf{u}||\mathbf{w}|\sin \theta$ area formula

$\approx \dfrac{1}{2}(13)(20\sqrt{2})\sin 22.38°$ substitute

$\approx 70 \text{ units}^2$ result

d. For $\mathbf{u} = \langle 12, 5 \rangle$ and $\mathbf{v} = \langle 8, 15 \rangle$, the determinant $\begin{vmatrix} a & b \\ c & d \end{vmatrix} = \begin{vmatrix} 12 & 5 \\ 8 & 15 \end{vmatrix} = (12)(15) - (8)(5) = 140 \text{ units}^2 ✓$, and the area of the parallelogram is indeed twice the area of the triangle.

Now try Exercises 9 through 14 ▶

Connections to Calculus Exercises

Use the techniques from Section 8.3 to complete Exercises 1 and 2.

1. A common form of partial fraction decomposition involves a constant over a factorable quadratic expression. (a) Rewrite the expression

$$\frac{k}{(x + a)(x + b)}$$ as the sum of partial fractions to

show the system obtained is $\begin{cases} A + B = 0 \\ Ab + Ba = k \end{cases}$. (b) Use

the fact that $A + B = 0$ to show the system can be

rewritten as $\begin{cases} A + B = 0 \\ A(b - a) = k \end{cases}$. (c) Solve for A in the

second equation and use the result to quickly write

$$\frac{16}{(x + 3)(x - 5)}$$ as the sum of its partial fractions.

2. Another common form involves a linear expression $x + k$ over a factorable quadratic expression.

(a) Rewrite the expression $\dfrac{x + k}{(x + a)(x + b)}$ as the

sum of partial fractions to show the system obtained

is $\begin{cases} A + B = 1 \\ Ab + Ba = k \end{cases}$. (b) Use the fact that $A + B = 1$

to show that system can be written as

$\begin{cases} A + B = 1 \\ A(b - a) + a = k \end{cases}$. (c) Solve for A in the

second equation, and use the result to write

$$\frac{x + 10}{(x + 3)(x + 4)}$$ as the sum of its partial fractions.

Use the "formulas" developed in Example 1 and Exercises 1 and 2 to decompose each expression into a sum of partial fractions.

3. $\dfrac{6}{x^2 - 9}$

4. $\dfrac{18}{x^2 - 81}$

5. $\dfrac{-22}{x^2 + 3x - 28}$

6. $\dfrac{15}{x^2 - 3x - 4}$

7. $\dfrac{x + 11}{x^2 + 13x + 40}$

8. $\dfrac{x - 17}{x^2 - 9x + 14}$

Find the area of the parallelogram spanned by the vectors given, then verify your answer using the method shown in Example 3.

9. $\mathbf{u} = \langle 2, 8 \rangle$ and $\mathbf{v} = \langle 15, 3 \rangle$

10. $\mathbf{u} = \langle -4, 1 \rangle$ and $\mathbf{v} = \langle 4, 4 \rangle$

11. $6\mathbf{i} + 3\mathbf{j}$ and $6\mathbf{i} - 3\mathbf{j}$

12. $9\mathbf{i}$ and $7\mathbf{j}$

Similar to how the absolute value of $\begin{vmatrix} a & b \\ c & d \end{vmatrix}$ **gives the** *area* **of a parallelogram spanned by** $\langle a, b \rangle$ **and** $\langle c, d \rangle$, **the absolute**

value of the determinant $\begin{vmatrix} a & b & c \\ d & e & f \\ g & h & i \end{vmatrix}$ **gives the** *volume* **of a parallelepiped spanned by** $\langle a, b, c, \rangle$, $\langle d, e, f \rangle$, **and** $\langle g, h, i \rangle$ **in**

three dimensions.

13. Find the volume of the parallelepiped spanned by $\langle 5, 0, 0 \rangle$, $\langle 0, 6, 0 \rangle$, and $\langle 0, 0, 8 \rangle$ using the determinant formula, then verify your answer using a more familiar formula.

14. Find the volume of the parallelepiped spanned by $\mathbf{u} = \langle 2, 5, -8 \rangle$, $\langle 3, 0, 4 \rangle$, and $\langle 0, 10, -7 \rangle$.

9

Analytical Geometry and the Conic Sections

CHAPTER OUTLINE

CHAPTER CONNECTIONS

Cathedral windows are just one place where we can appreciate the beauty and intricacy of polar graphs. The characteristics and features of these graphs are explored in this chapter, and introduce a historic alternative to graphing relations by simply computing points (x, y). While graphing relations using a rectangular coordinate system is much more widely known, both polar graphs and rectangular graphs seemed to have matured simultaneously, in the early 1600s. The graphs shown here appear as Exercises 57 and 65 in Section 9.6.

Connections to Calculus

When using polar curves in the design of windows or other forms of architechture, finding where two graphs intersect aids in the calculation of the area between curves and in determining the arclength of certain portions of the graph. These are important considerations in a study of calculus, and techniques for working with systems of polar equations are explored in the *Connections to Calculus* for Chapter 9.

Learning Objectives

In Section 9.1 you will learn how to:

☐ **A.** Verify theorems from basic geometry involving the distance between two points

☐ **B.** Verify that points (x, y) are an equal distance from a given point and a given line

☐ **C.** Use the defining characteristics of a conic section to find its equation

Generally speaking, **analytical geometry** is a study of geometry using the tools of algebra and a coordinate system. These tools include the midpoint and distance formulas; the algebra of parallel, perpendicular, and intersecting lines; and other tools that help establish geometric concepts. In this section, we'll use these tools to verify certain relationships, then use these relationships to introduce a family of curves known as the **conic sections.**

A. Verifying Relationships from Plane Geometry

For the most part, the algebraic tools used in this study were introduced in previous chapters. As the midpoint and distance formulas play a central role, they are restated here for convenience.

Algebraic Tools Used in Analytical Geometry

Given two points $P_1 = (x_1, y_1)$ and $P_2 = (x_2, y_2)$ in the xy-plane.

Midpoint Formula

The midpoint of line segment P_1P_2 is

$$(x, y) = \left(\frac{x_1 + x_2}{2}, \frac{y_1 + y_2}{2}\right)$$

Distance Formula

The distance from P_1 to P_2 is

$$d = \sqrt{(x_2 - x_1)^2 + (y_2 - y_1)^2}$$

These formulas can be used to verify the conclusion of many theorems from Euclidean geometry, while providing important links to an understanding of the conic sections.

EXAMPLE 1 ▶ Verifying a Theorem from Basic Geometry

A theorem from basic geometry states: *The midpoint of the hypotenuse of a right triangle is an equal distance from all three vertices.* Verify this statement for the right triangle formed by $(-4, -2)$, $(4, -2)$, and $(4, 4)$.

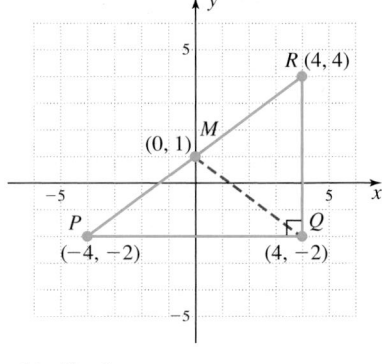

Solution ▶ After the plotting points and drawing a triangle, we note the hypotenuse has endpoints $(-4, -2)$ and $(4, 4)$, with midpoint

$$\left(\frac{4 + (-4)}{2}, \frac{4 + (-2)}{2}\right) = (0, 1).$$ Using the

distance formula to find the distance from $(0, 1)$ to $(4, 4)$ gives

$$\begin{aligned}
d &= \sqrt{(0 - 4)^2 + (1 - 4)^2} \\
&= \sqrt{(-4)^2 + (-3)^2} \\
&= \sqrt{25} \\
&= 5
\end{aligned}$$

From the definition of midpoint, $(0, 1)$ is also 5 units from $(-4, -2)$. Checking the distance from $(0, 1)$ to the vertex $(4, -2)$ gives

$$\begin{aligned}
d &= \sqrt{(4 - 0)^2 + (-2 - 1)^2} \\
&= \sqrt{4^2 + (-3)^2} \\
&= \sqrt{25} \\
&= 5
\end{aligned}$$

The midpoint of the hypotenuse *is* an equal distance from all three vertices (see the figure).

Now try Exercises 7 through 12 ▶

Recall from Section 2.1 that a circle is the set of all points that are an equal distance (called the radius) from a given point (called the center). If all three vertices of a triangle lie on the circumference of a circle, we say the circle **circumscribes** the triangle. Based on our earlier work, it appears we could also state the theorem in Example 1 as *For any circle in the xy-plane whose center (h, k) is the midpoint of the hypotenuse of a right triangle, all vertices lie on the circle defined by* $(x - h)^2 + (y - k)^2 = \left(\frac{d}{2}\right)^2$, *where d is the length of the hypotenuse.* See Figure 9.1 and **Exercises 13 through 20.**

Figure 9.1

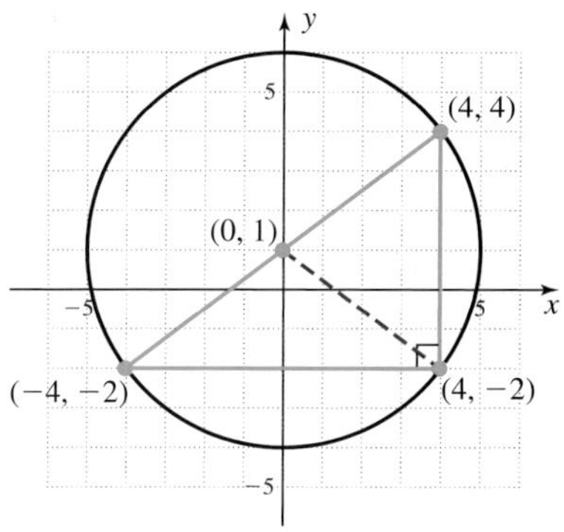

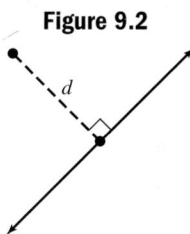

☑ **A.** You've just learned how to verify theorems from basic geometry involving the distance between two points

B. The Distance between a Point and a Line

In a study of analytical geometry, we are also interested in the distance d between a point and a *line*. This is always defined as the **perpendicular distance,** or the length of a line segment perpendicular to the given line, with the given point and the point of intersection as endpoints (see Figure 9.2).

Figure 9.2

EXAMPLE 2 ▶ **Locating Points That Are an Equal Distance from a Given Point and Line**

In Figure 9.3, the origin $(0, 0)$ is seen to be an equal distance from the point $(0, 2)$ and the line $y = -2$. Show that the following points are also an equal distance from $(0, 2)$ and $y = -2$:

　a. $\left(2, \frac{1}{2}\right)$　　**b.** $(4, 2)$　　**c.** $(8, 8)$

Solution ▶ Since the given line is horizontal, the perpendicular distance from the line to each point can be found by vertically counting the units. It remains to show that this is also the distance from the given point to $(0, 2)$ (see Figure 9.4).

a. The distance from $\left(2, \frac{1}{2}\right)$ to $y = -2$ is **2.5 units**. The distance from $\left(2, \frac{1}{2}\right)$ to $(0, 2)$ is

$$d = \sqrt{(0 - 2)^2 + (2 - 0.5)^2}$$
$$= \sqrt{(-2)^2 + (1.5)^2}$$
$$= \sqrt{6.25}$$
$$= 2.5 \checkmark$$

Figure 9.3

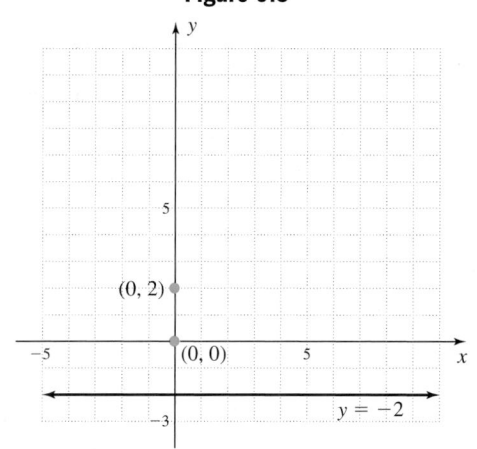

b. The distance from $(4, 2)$ to $y = -2$ is **4 units**. The distance from $(4, 2)$ to $(0, 2)$ is

$$d = \sqrt{(0 - 4)^2 + (2 - 2)^2}$$
$$= \sqrt{(-4)^2 + (0)^2}$$
$$= \sqrt{16}$$
$$= 4 \checkmark$$

c. The distance from $(8, 8)$ to $y = -2$ is **10 units**. The distance from $(8, 8)$ to $(0, 2)$ is

$$d = \sqrt{(0 - 8)^2 + (2 - 8)^2}$$
$$= \sqrt{(-8)^2 + (-6)^2}$$
$$= \sqrt{100}$$
$$= 10 \; \checkmark$$

Figure 9.4

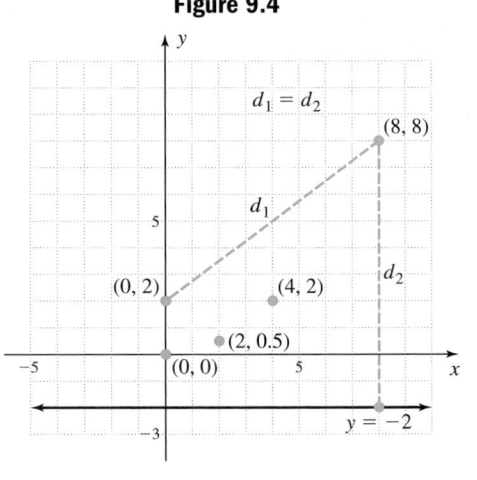

✓ **B.** You've just learned how to verify that points (x, y) are an equal distance from a given point and a given line

Now try Exercises 23 through 26 ▶

C. Characteristics of the Conic Sections

Figure 9.5

Axis

Nappe

Generator

Vertex

Nappe

Examples 1 and 2 bring us one step closer to the wider application of these ideas in a study of the conic sections. But before the connection is clearly made, we'll introduce some background on this family of curves. In common use, a cone might bring to mind the conical paper cups found at a water cooler. The point of the cone is called the **vertex** and the sheet of paper forming the sides is called a **nappe**. In mathematical terms, a cone has two nappes, formed by rotating a nonvertical line (called the generator), about a vertical line (called the axis), at their point of intersection—the vertex (see Figure 9.5). The conic sections are so named because all curves in the family can be formed by a *section* of the *cone,* or more precisely the intersection of a plane and a cone. Figure 9.6 shows that if the plane does not go through the vertex, the intersection will produce a circle, ellipse, parabola, or hyperbola.

Figure 9.6

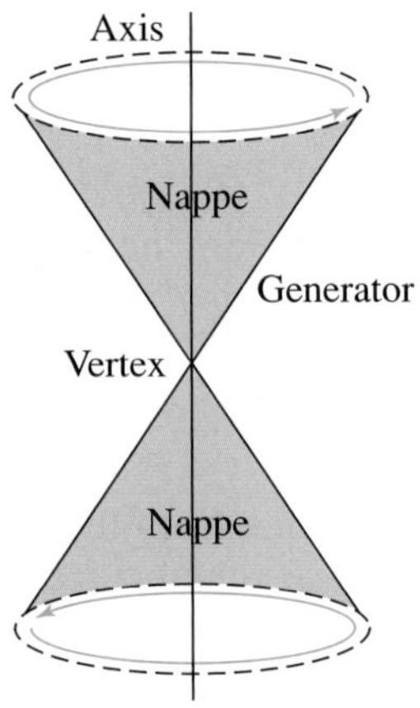

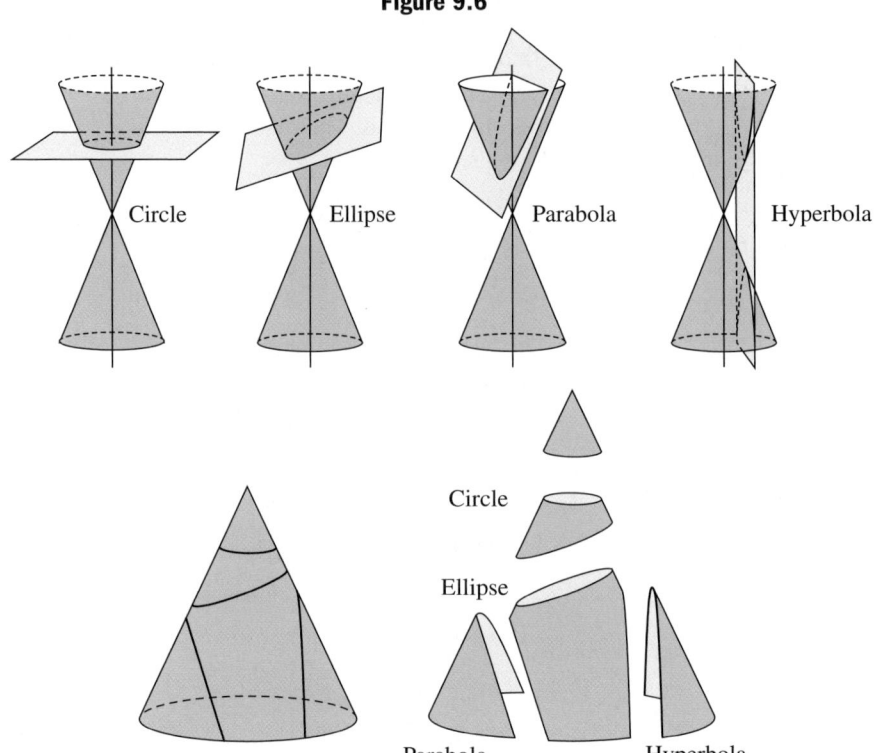

WORTHY OF NOTE

If the plane *does* go through the vertex, the result is a single point, a single line (if the plane contains the generator), or a pair of intersecting lines (if the plane contains the axis).

The connection we seek to make is that each conic section can be defined in terms of the distance between points in the plane, as in Example 1, or the distance between a given point and a line, as in Example 2. In Example 1, we noted the points $(-4, -2)$, $(4, -2)$, and $(4, 4)$ were all on a circle of radius 5 with center $(0, 1)$, in line with the analytic definition of a circle: *A circle is the set of all points that are an equal distance (called the radius) from a given point (called the center).*

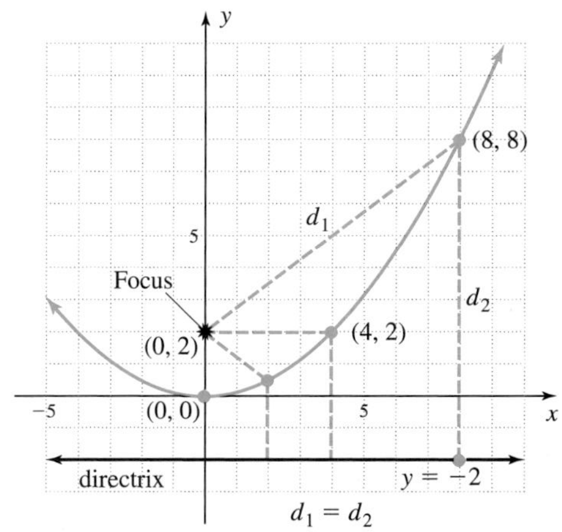

In Example 2, you may have noticed that the points seemed to form the right branch of a parabola (see Figure 9.7), and in fact, this example illustrates the analytic definition of a parabola: *A parabola is the set of all points that are an equal distance from a given point (called the **focus**), and a given line (called the **directrix**).*

Figure 9.7

The focus and directrix are not actually part of the graph, they are simply used to locate points on the graph. For this reason all foci (plural of focus) will be represented by a "*" symbol rather than a point.

EXAMPLE 3 ▶ **Finding an Equation for All Points That Form a Certain Parabola**

With Example 2 as a pattern, use the analytic definition to find a formula (equation) for the set of all points that form the parabola.

Solution ▶ Use the ordered pair (x, y) to represent an arbitrary point on the parabola. Since any point on the line $y = -2$ has coordinates $(x, -2)$, we set the distance from $(x, -2)$ to (x, y) equal to the distance from $(0, 2)$ to (x, y). The result is

$$\sqrt{(x - x)^2 + [y - (-2)]^2} = \sqrt{(x - 0)^2 + (y - 2)^2} \qquad \text{distances are equal}$$

$$\sqrt{(y + 2)^2} = \sqrt{x^2 + (y - 2)^2} \qquad \text{simplify}$$

$$(y + 2)^2 = x^2 + (y - 2)^2 \qquad \text{power property}$$

$$y^2 + 4y + 4 = x^2 + y^2 - 4y + 4 \qquad \text{expand binomials}$$

$$8y = x^2 \qquad \text{simplify}$$

$$y = \frac{1}{8}x^2 \qquad \text{result}$$

All points satisfying these conditions are on the parabola defined by $y = \frac{1}{8}x^2$.

Now try Exercises 27 and 28 ▶

At this point, it seems reasonable to ask what happens when the distance from the focus to (x, y) is *less than* the distance from the directrix to (x, y). For example, what if the distance is only two-thirds as long? As you might guess, the result is one of the other conic sections, in this case an ellipse. If the distance from the focus to a point (x, y) is *greater than* the distance from the directrix to (x, y), one branch of a hyperbola is formed. While we will defer a development of their general equations until later in the chapter, the following diagrams serve to illustrate this relationship for the ellipse, and show why we refer to the conic sections as a *family of curves*. In Figure 9.8, the line segment from the focus to each point on the graph (shown in blue), is exactly two-thirds the length of the line segment from the directrix to the same point (shown in red). Note the graph of these points forms the right half of an ellipse. In Figure 9.9, the lines and points forming the first half are moved to the background to more clearly show the remaining points that form the complete graph.

Figure 9.8   **Figure 9.9**

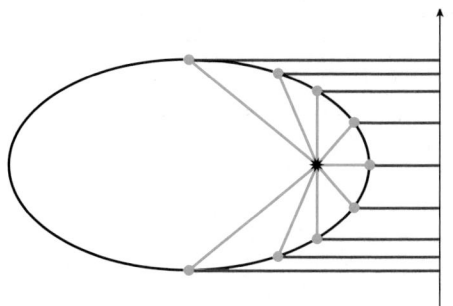
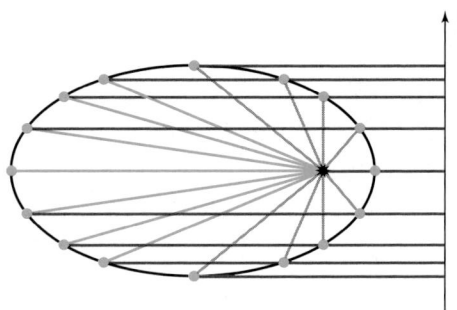

EXAMPLE 4 ▶   **Finding an Equation for All Points That Form a Certain Ellipse**

Suppose we arbitrarily select the point $(1, 0)$ as a focus and the (vertical) line $x = 4$ as the directrix. Use these to find an equation for the set of all points where the distance from the focus to a point (x, y) is $\frac{1}{2}$ the distance from the directrix to (x, y).

Solution ▶   Since any point on the line $x = 4$ has coordinates $(4, y)$, we have

$$\text{Distance from } (1, 0) \text{ to } (x, y) = \frac{1}{2}\left[\text{distance from } (4, y) \text{ to } (x, y)\right] \quad \text{in words}$$

$$\sqrt{(x - 1)^2 + [y - (0)]^2} = \frac{1}{2}\sqrt{(x - 4)^2 + (y - y)^2} \quad \text{resulting equation}$$

$$\sqrt{(x - 1)^2 + y^2} = \frac{1}{2}\sqrt{(x - 4)^2} \quad \text{simplify}$$

$$(x - 1)^2 + y^2 = \frac{1}{4}(x - 4)^2 \quad \text{power property}$$

$$x^2 - 2x + 1 + y^2 = \frac{1}{4}(x^2 - 8x + 16) \quad \text{expand binomials}$$

$$x^2 - 2x + 1 + y^2 = \frac{1}{4}x^2 - 2x + 4 \quad \text{distribute}$$

$$\frac{3}{4}x^2 + y^2 = 3 \quad \text{simplify: } 1x^2 - \frac{1}{4}x^2 = \frac{3}{4}x^2$$

$$3x^2 + 4y^2 = 12 \quad \text{polynomial form}$$

All points satisfying these conditions are on the ellipse defined by $3x^2 + 4y^2 = 12$.

Now try Exercises 29 and 30 ▶

Actually, any given ellipse has two foci (see Figure 9.10) and the equation from Example 4 could also have been developed using the left focus (with the directrix also on the left). This symmetrical relationship leads us to an *alternative definition* for the ellipse, which we will explore further in Section 9.2.

For foci f_1 and f_2, an ellipse is the set of all points (x, y) where the sum of the distances from f_1 to (x, y) and f_2 to (x, y) is constant.

See Figure 9.11 and **Exercises 31 and 32.**

Both the focus/directrix definition and the two foci definition have merit, and simply tend to call out different characteristics and applications of the ellipse. The hyperbola also has a focus/directrix definition and a two foci definition. **See Exercises 33 and 34.**

☑ **C.** You've just learned how to use the defining characteristics of a conic section to find its equation

Figure 9.10

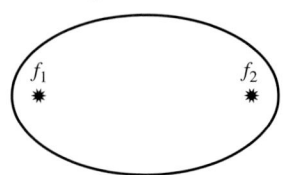

Figure 9.11

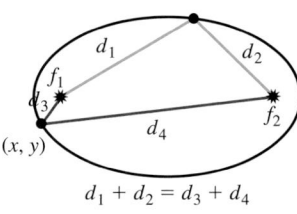

$d_1 + d_2 = d_3 + d_4$

 **9.1 EXERCISES**

▶ **CONCEPTS AND VOCABULARY**

Fill in the blank with the appropriate word or phrase. Carefully reread the section if needed.

1. Analytical geometry is a study of _____ using the tools of _____.

2. The distance formula is $d =$ _____; the midpoint formula is $M =$ _____.

3. The distance between a point and a line always refers to the _____ distance.

4. The conic sections are formed by the intersection of a _____ and a _____.

5. If a plane intersects a cone at its vertex, the result is a _____, a line, or a pair of _____ lines.

6. A circle is defined relative to an equal distance between two _____. A parabola is defined relative to an equal distance between a _____ and a _____.

▶ **DEVELOPING YOUR SKILLS**

The three points given form a right triangle. Find the midpoint of the hypotenuse and verify that the midpoint is an equal distance from all three vertices.

7. $P_1 = (-5, 2)$
   $P_2 = (1, 2)$
   $P_3 = (-5, -6)$

8. $P_1 = (3, 2)$
   $P_2 = (3, 14)$
   $P_3 = (8, 2)$

9. $P_1 = (-2, 1)$
   $P_2 = (6, -5)$
   $P_3 = (2, -7)$

10. $P_1 = (0, -5)$
    $P_2 = (-6, 4)$
    $P_3 = (6, -1)$

11. $P_1 = (10, -21)$
    $P_2 = (-6, -9)$
    $P_3 = (3, 3)$

12. $P_1 = (6, -6)$
    $P_2 = (-12, 18)$
    $P_3 = (20, 42)$

13. Find the equation of the circle that circumscribes the triangle in Exercise 7.

14. Find the equation of the circle that circumscribes the triangle in Exercise 8.

15. Find the equation of the circle that circumscribes the triangle in Exercise 9.

16. Find the equation of the circle that circumscribes the triangle in Exercise 10.

17. Find the equation of the circle that circumscribes the triangle in Exercise 11.

18. Find the equation of the circle that circumscribes the triangle in Exercise 12.

19. Of the following six points, four are an equal distance from the point $A(2, 3)$ and two are not. (a) Identify which four, and (b) find any two additional points that are this same (nonvertical, nonhorizontal) distance from $(2, 3)$:

    $B(7, 15)$ $C(-10, 8)$ $D(9, 14)$ $E(-3, -9)$
    $F(5, 4 + 3\sqrt{10})$ $G(2 - 2\sqrt{30}, 10)$

20. Of the following six points, four are an equal distance from the point $P(-1, 4)$ and two are not. (a) Identify which four, and (b) Find any two additional points that are the same (nonvertical, nonhorizontal) distance from $(-1, 4)$.

    $Q(-9, 10)$ $R(5, 12)$ $S(-7, 11)$ $T(4, 4 + 5\sqrt{3})$
    $U(-1 + 4\sqrt{6}, 6)$ $V(-7, 4 + \sqrt{51})$

▶ **WORKING WITH FORMULAS**

The Perpendicular Distance from a Point to a Line:
$$d = \left| \frac{Ax_1 + By_1 + C}{\sqrt{A^2 + B^2}} \right|$$

The perpendicular distance from a point (x_1, y_1) to a given line can be found using the formula shown, where $Ax + By + C = 0$ is the equation of the line in standard form (A, B, and C are integers).

21. Use the formula to verify that $P(-6, 2)$ and $Q(6, 4)$ are an equal distance from the line $y = -\frac{1}{2}x + 3$.

Exercise 21

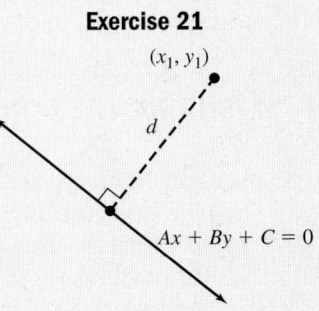

22. Find the value(s) for y that ensure $(1, y)$ is this same distance from $y = -\frac{1}{2}x + 3$.

▶ **APPLICATIONS**

23. Of the following four points, three are an equal distance from the point $A(0, 1)$ and the line $y = -1$. (a) Identify which three, and (b) find any two additional points that satisfy these conditions.

$B(-6, 9)$ $C(4, 4)$ $D(-2\sqrt{2}, 6)$ $E(4\sqrt{2}, 8)$

24. Of the following four points, three are an equal distance from the point $P(2, 4)$ and the line $y = -4$. (a) Identify which three, and (b) find any two additional points that satisfy these conditions.

$Q(-10, 9)$ $R(2 + 4\sqrt{2}, 3)$ $S(10, 4)$
$T(2 - 4\sqrt{5}, 5)$

25. Consider a fixed *point* $(0, -4)$ and a fixed *line* $y = 4$. Verify that the distance from each point to $(0, -4)$, is equal to the distance from the point to the line $y = 4$.

$A(4, -1)$ $B\left(10, -\dfrac{25}{4}\right)$ $C(4\sqrt{2}, -2)$

$D(8\sqrt{5}, -20)$

26. Consider a fixed *point* $(0, -2)$ and a fixed *line* $y = 2$. Verify that the distance from each point to $(0, -2)$, is equal to the distance from the point to the line $y = 2$.

$P(12, -18)$ $Q\left(6, -\dfrac{9}{2}\right)$ $R(4\sqrt{5}, -10)$

$S(4\sqrt{6}, -12)$

27. The points from Exercise 25 are on the graph of a parabola. Find the equation of the parabola.

28. The points from Exercise 26 are on the graph of a parabola. Find the equation of the parabola.

29. Using $(0, -2)$ as the focus and the horizontal line $y = -8$ as the directrix, find an equation for the set of all points (x, y) where the distance from the focus to (x, y) is one-half the distance from the directrix to (x, y).

30. Using $(4, 0)$ as the focus and the vertical line $x = 9$ as the directrix, find an equation for the set of all

points (x, y) where the distance from the focus to (x, y) is two-thirds the distance from the directrix to (x, y).

31. From Exercise 29, verify the points $(-3, 2)$ and $(\sqrt{12}, 0)$ are on the ellipse defined by $4x^2 + 3y^2 = 48$. Then verify that $d_1 + d_2 = d_3 + d_4$.

Exercise 31

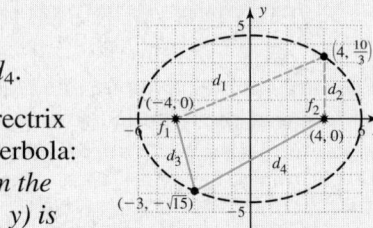

32. From Exercise 30, verify the points $\left(4, \dfrac{10}{3}\right)$ and $(-3, -\sqrt{15})$ are on the ellipse defined by $5x^2 + 9y^2 = 180$. Then verify that $d_1 + d_2 = d_3 + d_4$.

Exercise 32

33. From the focus/directrix definition of a hyperbola: *If the distance from the focus to a point (x, y) is greater than* the distance from the directrix to (x, y), one branch of a hyperbola is formed. Using $(2, 0)$ as the focus and the vertical line $x = \frac{1}{2}$ as the directrix, find an equation for the set of all points (x, y) where the distance from the focus to (x, y), is twice the distance from the directrix to (x, y).

34. From the two foci definition of a hyperbola: *For foci f_1 and f_2, a hyperbola is the set of all points (x, y) where the difference of the distances from f_1 to (x, y) and f_2 to (x, y) is constant.* Verify the points $(2, 3)$ and $(-3, -2\sqrt{6})$ are on the graph of the hyperbola from Exercise 33. Then verify $d_1 - d_2 = d_3 - d_4$.

▶ **EXTENDING THE CONCEPT**

35. Do some reading or research on the **orthocenter** of a triangle, and the **centroid** of a triangle. How are they found? What are their properties? Use the ideas and skills from this section to find the (a) orthocenter and (b) centroid of the triangle formed by the points $A(-8, 2)$, $B(-2, -6)$, and $C(4, 0)$.

36. **Properties of a circle:** A theorem from elementary geometry states: *If a radius is perpendicular to a chord, it bisects the chord.* Verify this is true for the circle, radii, and chords shown.

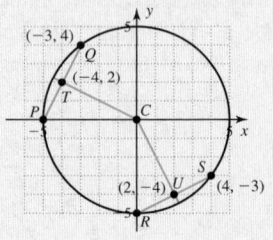

37. Verify that points $C(-2, 3)$ and $D(2\sqrt{2}, \sqrt{6})$ are points on the ellipse with foci at $A(-2, 0)$ and $B(2, 0)$, by verifying $d(AC) + d(BC) = d(AD) + d(BD)$. The expression that results has the form

$\sqrt{A + B} + \sqrt{A - B}$, which prior to the common use of technology had to be simplified using the formula $\sqrt{A + B} + \sqrt{A - B} = \sqrt{a} + \sqrt{b}$, where $a = 2A$ and $b = 4(A^2 - B^2)$. Use this relationship to verify the equation above.

▶ MAINTAINING YOUR SKILLS

38. (6.4) Verify the following is an identity:
$$\frac{\cos(2x) + \sin^2 x}{1 - \cos^2 x} = \cot^2(x)$$

39. (6.7) Find all solutions in $[0, 2\pi)$
$$-225 = 600 + 825 \sin\left(x + \frac{\pi}{6}\right)$$

40. (4.4) Solve for x in both exact and approximate form:

   a. $5 = \dfrac{10}{1 + 9e^{-0.5x}}$

   b. $345 = 5e^{0.4x} + 75$

41. (3.5) Sketch a complete graph of $h(x) = \dfrac{x^2 - 9}{x^2 - 4}$.

   Clearly label all intercepts and asymptotes.

9.2 | The Circle and the Ellipse

Learning Objectives

In Section 9.2 you will learn how to:

☐ **A.** Use the characteristics of a circle and its graph to understand the equation of an ellipse

☐ **B.** Use the equation of an ellipse to graph central and noncentral ellipses

☐ **C.** Locate the foci of an ellipse and use the foci and other features to write the equation

☐ **D.** Solve applications involving the foci

In Section 9.1, we introduced the equation of an ellipse using analytical geometry and the focus-directrix definition. Here we'll take a different approach, and use the equation of a circle to demonstrate that a circle is simply a special ellipse. In doing so, we'll establish a relationship between the foci and vertices of the ellipse, that enables us to apply these characteristics in context.

A. The Equation and Graph of a Circle

Recall that the equation of a circle with radius r and center at (h, k) is
$$(x - h)^2 + (y - k)^2 = r^2.$$

   As in Section 2.1, the standard form can be used to construct the equation of the circle given the center and radius as in Example 1, or to graph the circle as in Example 2.

EXAMPLE 1 ▶ **Determining the Equation of a Circle Given Its Center and Radius**

Find the equation of a circle with radius 5 and center at $(2, -1)$.

Solution ▶ With a center of $(2, -1)$, we have $h = 2$, $k = -1$, and $r = 5$. Making the corresponding substitutions into the standard form we obtain

$$(x - h)^2 + (y - k)^2 = r^2 \quad \text{standard form}$$
$$(x - 2)^2 + [y - (-1)]^2 = 5^2 \quad \text{substitute 2 for } h, -1 \text{ for } k, \text{ and 5 for } r$$
$$(x - 2)^2 + (y + 1)^2 = 25 \quad \text{simplify}$$

The equation of this circle is $(x - 2)^2 + (y + 1)^2 = 25$.

Now try Exercises 7 through 12 ▶

If the equation is given in polynomial form, recall that we first complete the square in x and y to identify the center and radius.

EXAMPLE 2 ▶   **Completing the Square to Graph a Circle**

Find the center and radius of the circle whose equation is given, then sketch its graph: $x^2 + y^2 - 6x + 4y - 3 = 0$.

Solution ▶   Begin by completing the square in both x and y.

$$(x^2 - 6x + \_\_) + (y^2 + 4y + \_\_) = 3 \qquad \text{group } x\text{- and } y\text{-terms; add 3}$$

$$\underbrace{(x^2 - 6x + 9)}_{\text{adds 9 to left side}} + \underbrace{(y^2 + 4y + 4)}_{\text{adds 4 to left side}} = \underbrace{3 + 9 + 4}_{\text{add } 9 + 4 \text{ to right side}} \qquad \text{complete the square}$$

$$(x - 3)^2 + (y + 2)^2 = 16 \qquad \text{factor and simplify}$$

The center is at $(3, -2)$, with radius is $r = \sqrt{16} = 4$.

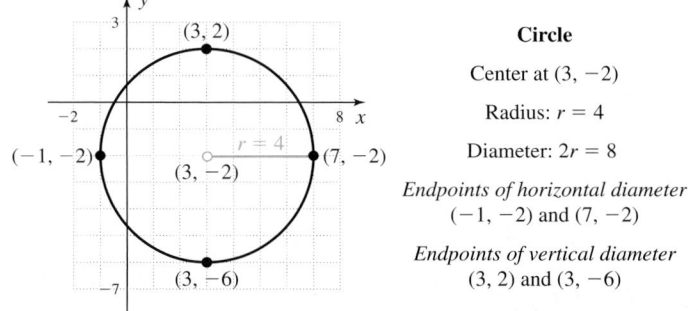

Circle

Center at $(3, -2)$

Radius: $r = 4$

Diameter: $2r = 8$

Endpoints of horizontal diameter
$(-1, -2)$ and $(7, -2)$

Endpoints of vertical diameter
$(3, 2)$ and $(3, -6)$

Now try Exercises 13 through 18 ▶

The equation of a circle in **standard form** provides a useful link to some of the other conic sections, and is obtained by *setting the equation equal to 1*. In the case of a circle, this means we simply divide by r^2.

$$(x - h)^2 + (y - k)^2 = r^2 \qquad \text{standard form}$$

$$\frac{(x - h)^2}{r^2} + \frac{(y - k)^2}{r^2} = 1 \qquad \text{divide by } r^2$$

In this form, the value of r in each denominator gives the *horizontal* and *vertical* distances, respectively, from the center to the graph. This is not so important in the case of a circle, since this distance is the same in *any* direction. But for other conics, these horizontal and vertical distances are *not* the same, making the new form a valuable tool for graphing. To distinguish the horizontal from the vertical distance, r^2 is replaced by a^2 in the "x-term" (horizontal distance), and by b^2 in the "y-term" (vertical distance).

☑ **A.** You've just learned how to use the characteristics of a circle and its graph to understand the equation of an ellipse

B. The Equation of an Ellipse

It then seems reasonable to ask, "What happens to the graph when $a \neq b$?" To answer, consider the equation from Example 2. We have $\dfrac{(x - 3)^2}{4^2} + \dfrac{(y + 2)^2}{4^2} = 1$ (after

dividing by 16), which we now compare to $\dfrac{(x - 3)^2}{4^2} + \dfrac{(y + 2)^2}{3^2} = 1$, where $a = 4$

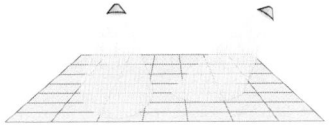

and $b = 3$. The center of the graph is still at $(3, -2)$, since $h = 3$ and $k = -2$ remain unchanged. Substituting $y = -2$ to find additional points, eliminates the y-term and gives two values for x:

$$\frac{(x - 3)^2}{4^2} + \frac{(-2 + 2)^2}{3^2} = 1 \qquad \text{substitute } -2 \text{ for } y$$

$$\frac{(x - 3)^2}{4^2} + 0 = 1 \qquad \text{simplify}$$

$$(x - 3)^2 = 16 \qquad \text{multiply by } 4^2 = 16$$

$$x - 3 = \pm 4 \qquad \text{property of square roots}$$

$$x = 3 \pm 4 \qquad \text{add 3}$$

$$x = 7 \text{ and } x = -1$$

This shows the horizontal distance from the center to the graph is still $a = 4$, and the points $(-1, -2)$ and $(7, -2)$ are on the graph (see Figure 9.12). Similarly, for $x = 3$ we have $(y + 2)^2 = 9$, giving $y = -5$ and $y = 1$, and showing the vertical distance from the center to the graph is $b = 3$, with points $(3, 1)$ and $(3, -5)$ also on the graph. Using this information to sketch the curve reveals the "circle" is elongated and has become an **ellipse**.

For an ellipse, the longest distance across the graph is called the **major axis**, with the endpoints of the major axis called **vertices**. The segment perpendicular to and bisecting the major axis (with its endpoints on the ellipse) is called the **minor axis**, as shown in see Figure 9.13.

Figure 9.12

Figure 9.13

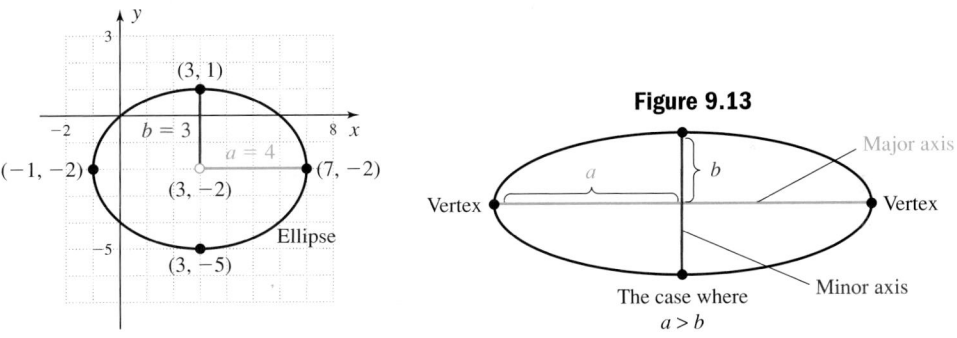

- If $a > b$, the major axis is horizontal (parallel to the x-axis) with length $2a$, and the minor axis is vertical with length $2b$ (see Example 3).
- If $b > a$ the major axis is vertical (parallel to the y-axis) with length $2b$, and the minor axis is horizontal with length $2a$ (see Example 4).

Generalizing this observation we obtain the equation of an ellipse in standard form.

The Equation of an Ellipse in Standard Form

Given $\dfrac{(x - h)^2}{a^2} + \dfrac{(y - k)^2}{b^2} = 1.$

If $a \neq b$ the equation represents the graph of an ellipse with center at (h, k).
- $|a|$ gives the horizontal distance from center to graph.
- $|b|$ gives the vertical distance from center to graph.

EXAMPLE 3 ▶ **Graphing a Horizontal Ellipse**

Sketch the graph of
$$\frac{(x-2)^2}{25} + \frac{(y+1)^2}{9} = 1.$$

Solution ▶ Noting $a \neq b$, we have an ellipse with center $(h, k) = (2, -1)$. The horizontal distance from the center to the graph is $a = 5$, and the vertical distance from the center to the graph is $b = 3$. After plotting the corresponding points and connecting them with a smooth curve, we obtain the graph shown.

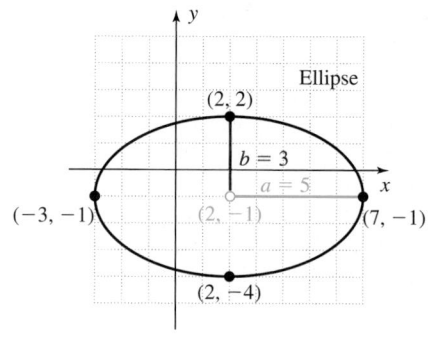

Now try Exercises 19 through 24 ▶

WORTHY OF NOTE

In general, for the equation $Ax^2 + By^2 = F$ ($A, B, F > 0$), the equation represents a circle if $A = B$, and an ellipse if $A \neq B$.

As with the circle, the equation of an ellipse can be given in polynomial form, and here our knowledge of circles is helpful. For the equation $25x^2 + 4y^2 = 100$, we know the graph cannot be a circle since the coefficients are unequal, and the center of the graph must be at the origin since $h = k = 0$. To actually draw the graph, we convert the equation to standard form.

EXAMPLE 4 ▶ **Graphing a Vertical Ellipse**

For $25x^2 + 4y^2 = 100$, (a) write the equation in standard form and identify the center and the values of a and b, (b) identify the major and minor axes and name the vertices, and (c) sketch the graph.

Solution ▶ The coefficients of x^2 and y^2 are unequal, and 25, 4, and 100 have like signs. The equation represents an ellipse with center at $(0, 0)$. To obtain standard form:

a. $25x^2 + 4y^2 = 100$ given equation

$$\frac{25x^2}{100} + \frac{4y^2}{100} = 1$$ divide by 100

$$\frac{x^2}{4} + \frac{y^2}{25} = 1$$ standard form

$$\frac{x^2}{2^2} + \frac{y^2}{5^2} = 1$$ write denominators in squared form; $a = 2$, $b = 5$

b. The result shows $a = 2$ and $b = 5$, indicating the major axis will be vertical and the minor axis will be horizontal. With the center at the origin, the x-intercepts will be $(2, 0)$ and $(-2, 0)$, with the vertices (and y-intercepts) at $(0, 5)$ and $(0, -5)$.

c. Plotting these intercepts and sketching the ellipse results in the graph shown.

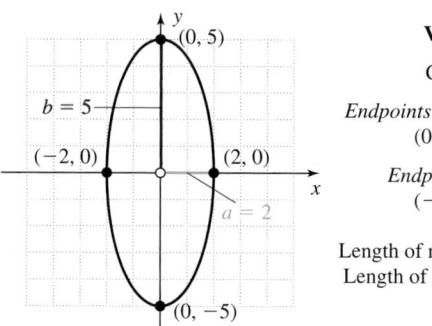

Vertical ellipse

Center at $(0, 0)$

Endpoints of major axis (vertices)
$(0, -5)$ and $(0, 5)$

Endpoints of minor axis
$(-2, 0)$ and $(2, 0)$

Length of major axis $2b$: $2(5) = 10$
Length of minor axis $2a$: $2(2) = 4$

Now try Exercises 25 through 36 ▶

If the center of the ellipse is not at the origin, the polynomial form has additional linear terms and we must first complete the square in x and y, then write the equation in standard form to sketch the graph (see the Reinforcing Basic Concepts feature for more on completing the square). Figure 9.14 illustrates how the central ellipse and the shifted ellipse are related.

Figure 9.14

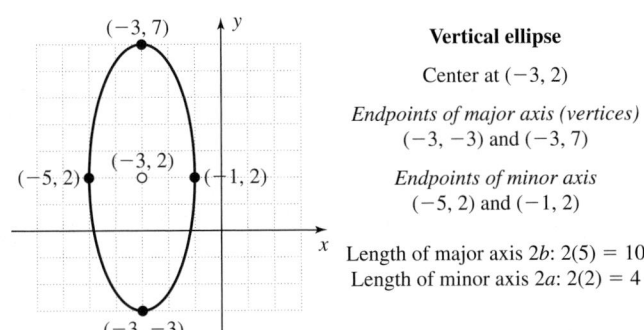

$$\frac{(x \pm h)^2}{a^2} + \frac{(y \pm k)^2}{b^2} = 1$$
$$a > b$$

$$\frac{x^2}{a^2} + \frac{y^2}{b^2} = 1$$
$$a > b$$

EXAMPLE 5 ▶ **Completing the Square to Graph an Ellipse**

Sketch the graph of $25x^2 + 4y^2 + 150x - 16y + 141 = 0$.

Solution ▶ The coefficients of x^2 and y^2 are unequal and have like signs, and we assume the equation represents an ellipse but wait until we have the factored form to be certain.

$$25x^2 + 4y^2 + 150x - 16y + 141 = 0 \qquad \text{given equation (polynomial form)}$$

$$25x^2 + 150x + 4y^2 - 16y = -141 \qquad \text{group like terms; subtract 141}$$

$$25(x^2 + 6x + \underline{\ \ }) + 4(y^2 - 4y + \underline{\ \ }) = -141 \qquad \text{factor out leading coefficient from each group}$$

$$25(x^2 + 6x + 9) + 4(y^2 - 4y + 4) = -141 + 225 + 16 \qquad \text{complete the square}$$
add 225 + 16 to right

adds 25(9) = 225　　　adds 4(4) = 16

$$25(x + 3)^2 + 4(y - 2)^2 = 100 \qquad \text{factor}$$

$$\frac{25(x + 3)^2}{100} + \frac{4(y - 2)^2}{100} = \frac{100}{100} \qquad \text{divide both sides by 100}$$

$$\frac{(x + 3)^2}{4} + \frac{(y - 2)^2}{25} = 1 \qquad \text{simplify (standard form)}$$

$$\frac{(x + 3)^2}{2^2} + \frac{(y - 2)^2}{5^2} = 1 \qquad \text{write denominators in squared form}$$

The result is a vertical ellipse with center at $(-3, 2)$, with $a = 2$ and $b = 5$. The vertices are a vertical distance of 5 units from center, and the endpoints of the minor axis are a horizontal distance of 2 units from center.

Note this is the same ellipse as in Example 4, but shifted 3 units left and 2 up.

Vertical ellipse

Center at $(-3, 2)$

Endpoints of major axis (vertices)
$(-3, -3)$ and $(-3, 7)$

Endpoints of minor axis
$(-5, 2)$ and $(-1, 2)$

Length of major axis $2b$: $2(5) = 10$
Length of minor axis $2a$: $2(2) = 4$

Now try Exercises 37 through 44 ▶

✓ **B. You've just learned how to use the equation of an ellipse to graph central and noncentral ellipses**

C. The Foci of an Ellipse

In Section 9.1, we noted that an ellipse could also be defined in terms of two special points called the **foci**. The Museum of Science and Industry in Chicago, Illinois (http://www.msichicago.org), has a permanent exhibit called the *Whispering Gallery*. The construction of the room is based on some of the reflective properties of an ellipse. If two people stand at designated points in the room and one of them whispers very softly, the other person can hear the whisper quite clearly—even though they are over 40 ft apart! The point where each person stands is a **focus** of the ellipse. This reflective property also applies to light and radiation, giving the ellipse some powerful applications in science, medicine, acoustics, and other areas. To understand and appreciate these applications, we introduce the analytic definition of an ellipse.

WORTHY OF NOTE

You can easily draw an ellipse that satisfies the definition. Press two pushpins (these form the foci of the ellipse) halfway down into a piece of heavy cardboard about 6 in. apart. Take an 8-in. piece of string and loop each end around the pins. Use a pencil to draw the string taut and keep it taut as you move the pencil in a circular motion—and the result is an ellipse! A different length of string or a different distance between the foci will produce a different ellipse.

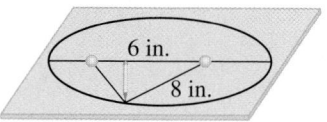

Definition of an Ellipse

Given two fixed points f_1 and f_2 in a plane, an ellipse is the set of all points (x, y) where the distance from f_1 to (x, y) added to the distance from f_2 to (x, y) remains constant.

$$d_1 + d_2 = k$$

The fixed points f_1 and f_2 are called the *foci* of the ellipse, and the points $P(x, y)$ are on the graph of the ellipse.

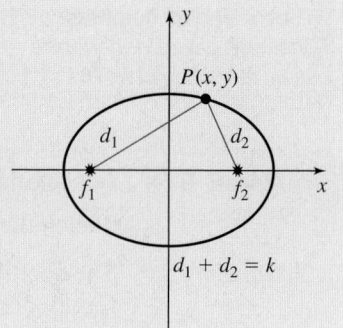

Figure 9.15

To find the equation of an ellipse in terms of a and b we combine the definition just given with the distance formula. Consider the ellipse shown in Figure 9.15 (for calculating ease we use a central ellipse). Note the vertices have coordinates $(-a, 0)$ and $(a, 0)$, and the endpoints of the minor axis have coordinates $(0, -b)$ and $(0, b)$ as before. It is customary to assign foci the coordinates $f_1 \rightarrow (-c, 0)$ and $f_2 \rightarrow (c, 0)$. We can calculate the distance between $(c, 0)$ and any point $P(x, y)$ on the ellipse using the distance formula:

$$\sqrt{(x - c)^2 + (y - 0)^2}$$

Likewise the distance between $(-c, 0)$ and any point (x, y) is

$$\sqrt{(x + c)^2 + (y - 0)^2}$$

According to the definition, the sum must be constant:

$$\sqrt{(x - c)^2 + y^2} + \sqrt{(x + c)^2 + y^2} = k$$

EXAMPLE 6 ▶ Finding the Value of k from the Definition of an Ellipse

Use the definition of an ellipse and the diagram given to determine the constant k used for this ellipse following (also see the following *Worthy of Note*). Note that $a = 5, b = 3$, and $c = 4$.

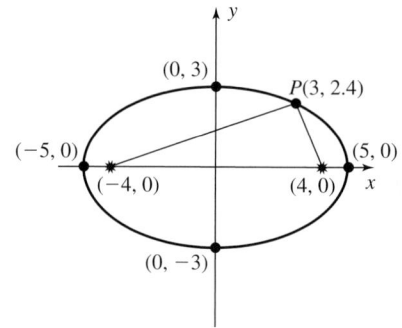

Solution ▶

$$\sqrt{(x - c)^2 + (y - 0)^2} + \sqrt{(x + c)^2 + (y - 0)^2} = k \quad \text{given}$$
$$\sqrt{(3 - 4)^2 + (2.4 - 0)^2} + \sqrt{(3 + 4)^2 + (2.4 - 0)^2} = k \quad \text{substitute}$$
$$\sqrt{(-1)^2 + 2.4^2} + \sqrt{7^2 + 2.4^2} = k \quad \text{add}$$
$$\sqrt{6.76} + \sqrt{54.76} = k \quad \text{simplify radicals}$$
$$2.6 + 7.4 = k \quad \text{compute square roots}$$
$$10 = k \quad \text{result}$$

The constant used for this ellipse is 10 units.

Now try Exercises 45 through 48 ▶

In Example 6, the sum of the distances could also be found by moving the point (x, y) to the location of a vertex $(a, 0)$, then using the symmetry of the ellipse. The sum is identical to the length of the major axis, since the overlapping part of the string from $(c, 0)$ to $(a, 0)$ is the same length as from $(-a, 0)$ to $(-c, 0)$ (see Figure 9.16). This shows the constant k is equal to *2a regardless of the distance between foci.*

As we noted, the result is

Figure 9.16

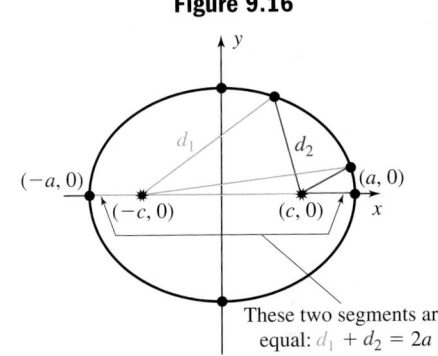

These two segments are equal: $d_1 + d_2 = 2a$

$$\sqrt{(x - c)^2 + y^2} + \sqrt{(x + c)^2 + y^2} = 2a \quad \text{substitute } 2a \text{ for } k$$

The details for simplifying this expression are given in Appendix IV, and the result is very close to the standard form seen previously:

$$\frac{x^2}{a^2} + \frac{y^2}{a^2 - c^2} = 1$$

By comparing the standard form $\frac{x^2}{a^2} + \frac{y^2}{b^2} = 1$ with $\frac{x^2}{a^2} + \frac{y^2}{a^2 - c^2} = 1$, we might suspect that $b^2 = a^2 - c^2$, and this is indeed the case. Note from Example 6 the relationship yields

$$b^2 = a^2 - c^2$$
$$3^2 = 5^2 - 4^2$$
$$9 = 25 - 16$$

Additionally, when we consider that $(0, b)$ is a point on the ellipse, the distance from $(0, b)$ to $(c, 0)$ must be equal to a due to symmetry (the "constant distance" used to form the ellipse is always $2a$). We then see in Figure 9.17, that $b^2 + c^2 = a^2$ (Pythagorean Theorem), yielding $b^2 = a^2 - c^2$ as above.

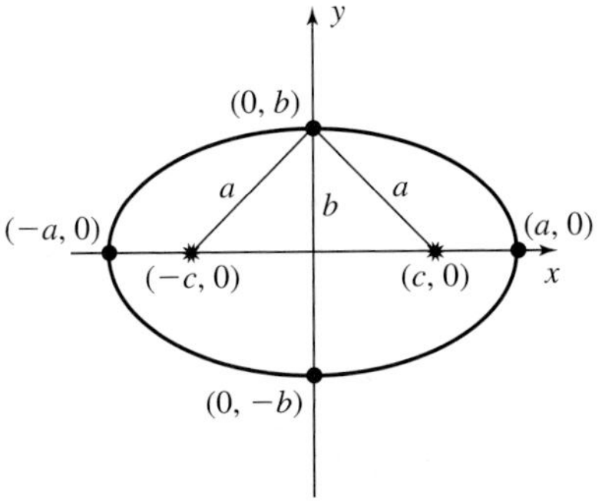

With this development, we now have the ability to *locate the foci of any ellipse*—an important step toward using the ellipse in practical applications. Because we're often asked to find the location of the foci, it's best to rewrite the relationship in terms of c^2, using absolute value bars to allow for a major axis that is vertical: $c^2 = |a^2 - b^2|$.

Figure 9.17

EXAMPLE 7 ▶ **Completing the Square to Graph an Ellipse and Locate the Foci**

For the ellipse defined by $25x^2 + 9y^2 - 100x - 54y - 44 = 0$, find the coordinates of the center, vertices, foci, and endpoints of the minor axis. Then sketch the graph.

Solution ▶

$$25x^2 + 9y^2 - 100x - 54y - 44 = 0 \qquad \text{given}$$

$$25x^2 - 100x + 9y^2 - 54y = 44 \qquad \text{group terms; add 44}$$

$$25(x^2 - 4x + \underline{}) + 9(y^2 - 6y + \underline{}) = 44 \qquad \text{factor out lead coefficients}$$

$$25(x^2 - 4x + 4) + 9(y^2 - 6y + 9) = 44 + 100 + 81 \qquad \text{add } 100 + 81 \text{ to right-hand side}$$

$$\underbrace{}_{\text{adds } 25(4) = 100} \qquad \underbrace{}_{\text{adds } 9(9) = 81}$$

$$25(x - 2)^2 + 9(y - 3)^3 = 225 \qquad \text{factored form}$$

$$\frac{25(x - 2)^2}{225} + \frac{9(y - 3)^2}{225} = \frac{225}{225} \qquad \text{divide by 225}$$

$$\frac{(x - 2)^2}{9} + \frac{(y - 3)^2}{25} = 1 \qquad \text{simplify (standard form)}$$

$$\frac{(x - 2)^2}{3^2} + \frac{(y - 3)^2}{5^2} = 1 \qquad \text{write denominators in squared form}$$

The result shows a vertical ellipse with $a = 3$ and $b = 5$. The center of the ellipse is at $(2, 3)$. The vertices are a vertical distance of $b = 5$ units from center at $(2, 8)$ and $(2, -2)$. The endpoints of the minor axis are a horizontal distance of $a = 3$ units from center at $(-1, 3)$ and $(5, 3)$. To locate the foci, we use the foci formula for an ellipse: $c^2 = |a^2 - b^2|$, giving $c^2 = |3^2 - 5^2| = 16$. This shows the foci "✻" are located a vertical distance of 4 units from center at $(2, 7)$ and $(2, -1)$.

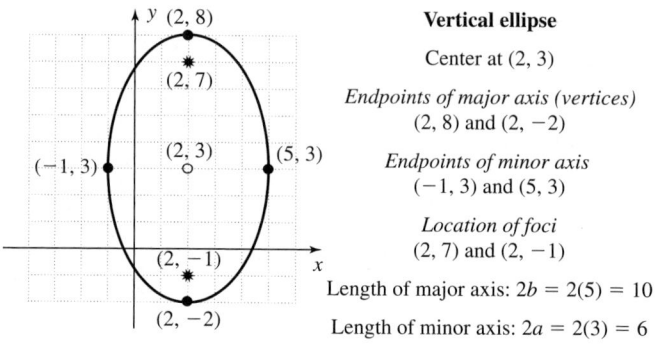

Vertical ellipse

Center at $(2, 3)$

Endpoints of major axis (vertices)
$(2, 8)$ and $(2, -2)$

Endpoints of minor axis
$(-1, 3)$ and $(5, 3)$

Location of foci
$(2, 7)$ and $(2, -1)$

Length of major axis: $2b = 2(5) = 10$

Length of minor axis: $2a = 2(3) = 6$

Now try Exercises 49 through 54 ▶

For future reference, remember the foci of an ellipse always occur on the major axis, with $a > c$ and $a^2 > c^2$ for a horizontal ellipse. This makes it easier to remember the **foci formula** for ellipses: $c^2 = |a^2 - b^2|$. Since a^2 is larger, it must be decreased by b^2 to equal c^2.

If any two of the values for a, b, and c are known, the relationship between them can be used to construct the equation of the ellipse.

EXAMPLE 8 ▶ **Finding the Equation of an Ellipse**

Find the equation of the ellipse (in standard form) that has foci at $(0, -2)$ and $(0, 2)$, with a minor axis 6 units in length.

Solution ▶ Since the foci must be on the major axis, we know this is a vertical and central ellipse with $c = 2$ and $c^2 = 4$. The minor axis has a length of $2a = 6$ units, meaning $a = 3$ and $a^2 = 9$. To find b^2, use the foci equation and solve.

LOOKING AHEAD

For the hyperbola, we'll find that $c > a$, and the formula for the foci of a hyperbola will be $c^2 = a^2 + b^2$.

✓ **C.** You've just learned how to locate the foci of an ellipse and use the foci and other features to write the equation

$$
\begin{aligned}
c^2 &= |a^2 - b^2| & & \text{foci equation (ellipse)} \\
4 &= |9 - b^2| & & \text{substitute} \\
-4 &= 9 - b^2 \qquad 4 = 9 - b^2 & & \text{solve} \\
b^2 &= 13 \qquad\quad\ b^2 = 5 & & \text{result}
\end{aligned}
$$

Since we know b^2 must be greater than a^2 (the major axis is always longer), $b^2 = 5$ can be discarded. The standard form is $\dfrac{x^2}{3^2} + \dfrac{y^2}{(\sqrt{13})^2} = 1$.

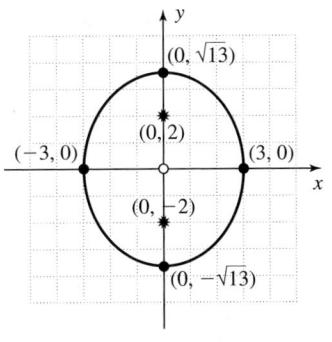

Now try Exercises 55 through 64 ▶

D. Applications Involving Foci

Applications involving the foci of a conic section can take various forms. In many cases, only partial information about the conic section is available and the ideas from Example 8 must be used to "fill in the gaps." In other applications, we must rewrite a known or given equation to find information related to the values of a, b, and c.

EXAMPLE 9 ▶ **Solving Applications Using the Characteristics of an Ellipse**

In Washington, D.C., there is a park called the *Ellipse* located between the White House and the Washington Monument. The park is surrounded by a path that forms an ellipse with the length of the major axis being about 1502 ft and the minor axis having a length of 1280 ft. Suppose the park manager wants to install water fountains at the location of the foci. Find the distance between the fountains rounded to the nearest foot.

Solution ▶ Since the major axis has length $2a = 1502$, we know $a = 751$ and $a^2 = 564{,}001$. The minor axis has length $2b = 1280$, meaning $b = 640$ and $b^2 = 409{,}600$. To find c, use the foci equation:

$$
\begin{aligned}
c^2 &= a^2 - b^2 \\
&= 564{,}001 - 409{,}600 \\
&= 154{,}401 \\
c &\approx -393 \text{ and } c \approx 393
\end{aligned}
$$

✓ **D.** You've just learned how to solve applications involving the foci

The distance between the water fountains would be $2(393) = 786$ ft.

Now try Exercises 65 through 76 ▶

9.2 EXERCISES

▶ **CONCEPTS AND VOCABULARY**

Fill in the blank with the appropriate word or phrase. Carefully reread the section if needed.

1. For an ellipse, the relationship between a, b, and c is given by the foci equation _____, since $c < a$ or $c < b$.

2. The greatest distance across an ellipse is called the _____ _____ and the endpoints are called _____.

3. For a vertical ellipse, the length of the minor axis is _____ and the length of the major axis is _____.

4. To write the equation $2x^2 + y^2 - 6x = 7$ in standard form, _____ the _____ in x.

5. Explain/Discuss how the relations $a > b$, $a = b$ and $a < b$ affect the graph of a conic section with equation $\dfrac{(x-h)^2}{a^2} + \dfrac{(y-k)^2}{b^2} = 1$.

6. Suppose foci are located at $(-3, 2)$ and $(5, 2)$. Discuss/Explain the conditions necessary for the graph to be an ellipse.

▶ DEVELOPING YOUR SKILLS

Find the equation of a circle satisfying the conditions given.

7. center $(0, 0)$, radius 7

8. center $(0, 0)$, radius 9

9. center $(5, 0)$, radius $\sqrt{3}$

10. center $(0, 4)$, radius $\sqrt{5}$

11. diameter has endpoints $(4, 9)$ and $(-2, 1)$

12. diameter has endpoints $(-2, -3)$, and $(3, 9)$

Identify the center and radius of each circle, then sketch its graph.

13. $x^2 + y^2 - 12x - 10y + 52 = 0$

14. $x^2 + y^2 + 8x - 6y - 11 = 0$

15. $x^2 + y^2 - 4x + 10y + 4 = 0$

16. $x^2 + y^2 + 4x + 6y - 3 = 0$

17. $x^2 + y^2 + 6x - 5 = 0$

18. $x^2 + y^2 - 8y - 5 = 0$

Sketch the graph of each ellipse.

19. $\dfrac{(x-1)^2}{9} + \dfrac{(y-2)^2}{16} = 1$

20. $\dfrac{(x-3)^2}{4} + \dfrac{(y-1)^2}{25} = 1$

21. $\dfrac{(x-2)^2}{25} + \dfrac{(y+3)^2}{4} = 1$

22. $\dfrac{(x+5)^2}{1} + \dfrac{(y-2)^2}{16} = 1$

23. $\dfrac{(x+1)^2}{16} + \dfrac{(y+2)^2}{9} = 1$

24. $\dfrac{(x+1)^2}{36} + \dfrac{(y+3)^2}{9} = 1$

For each exercise, (a) write the equation in standard form, then identify the center and the values of a and b, (b) state the coordinates of the vertices and the coordinates of the endpoints of the minor axis, and (c) sketch the graph.

25. $x^2 + 4y^2 = 16$ **26.** $9x^2 + y^2 = 36$

27. $16x^2 + 9y^2 = 144$ **28.** $25x^2 + 9y^2 = 225$

29. $2x^2 + 5y^2 = 10$ **30.** $3x^2 + 7y^2 = 21$

Identify each equation as that of an ellipse or circle, then sketch its graph.

31. $(x + 1)^2 + 4(y - 2)^2 = 16$

32. $9(x - 2)^2 + (y + 3)^2 = 36$

33. $2(x - 2)^2 + 2(y + 4)^2 = 18$

34. $(x - 6)^2 + y^2 = 49$

35. $4(x - 1)^2 + 9(y - 4)^2 = 36$

36. $25(x - 3)^2 + 4(y + 2)^2 = 100$

Complete the square in both x and y to write each equation in standard form. Then draw a complete graph of the relation and identify all important features.

37. $4x^2 + y^2 + 6y + 5 = 0$

38. $x^2 + 3y^2 + 8x + 7 = 0$

39. $x^2 + 4y^2 - 8y + 4x - 8 = 0$

40. $3x^2 + y^2 - 8y + 12x - 8 = 0$

41. $5x^2 + 2y^2 + 20y - 30x + 75 = 0$

42. $4x^2 + 9y^2 - 16x + 18y - 11 = 0$

43. $2x^2 + 5y^2 - 12x + 20y - 12 = 0$

44. $6x^2 + 3y^2 - 24x + 18y - 3 = 0$

Use the definition of an ellipse to find the constant k used for each ellipse (figures are not drawn to scale).

45.

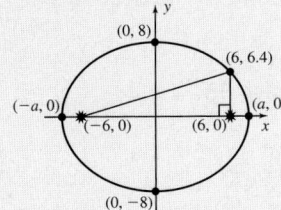

46.

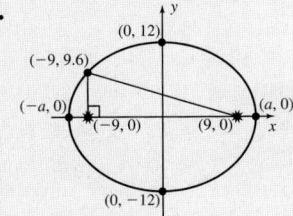

47.

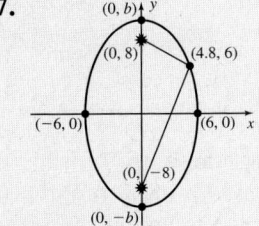

48.

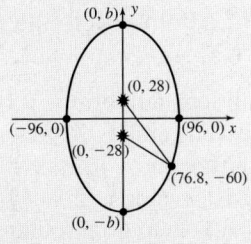

Find the coordinates of the (a) center, (b) vertices, (c) foci, and (d) endpoints of the minor axis. Then (e) sketch the graph.

49. $4x^2 + 25y^2 - 16x - 50y - 59 = 0$

50. $9x^2 + 16y^2 - 54x - 64y + 1 = 0$

51. $25x^2 + 16y^2 - 200x + 96y + 144 = 0$

52. $49x^2 + 4y^2 + 196x - 40y + 100 = 0$

53. $6x^2 + 24x + 9y^2 + 36y + 6 = 0$

54. $5x^2 - 50x + 2y^2 - 12y + 93 = 0$

Find the equation of an ellipse (in standard form) that satisfies the following conditions:

55. vertices at $(-6, 0)$ and $(6, 0)$; foci at $(-4, 0)$ and $(4, 0)$

56. vertices at $(-8, 0)$ and $(8, 0)$; foci at $(-5, 0)$ and $(5, 0)$

57. foci at $(3, -6)$ and $(3, 2)$; length of minor axis: 6 units

58. foci at $(-4, -3)$ and $(8, -3)$; length of minor axis: 8 units

Use the characteristics of an ellipse and the graph given to write the related equation and find the location of the foci.

59.

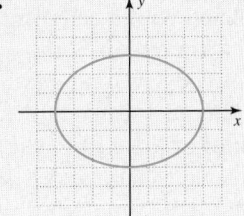

60.

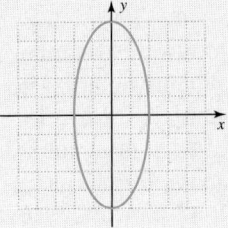

61.

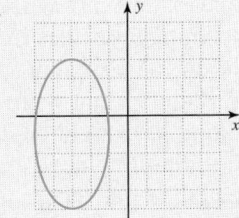

62.

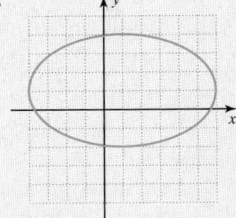

▶ WORKING WITH FORMULAS

63. Area of an Ellipse: $A = \pi ab$
The area of an ellipse is given by the formula shown, where a is the distance from the center to the graph in the horizontal direction and b is the distance from center to graph in the vertical direction. Find the area of the ellipse defined by $16x^2 + 9y^2 = 144$.

64. The Perimeter of an Ellipse: $P = 2\pi\sqrt{\dfrac{a^2 + b^2}{2}}$

The perimeter of an ellipse can be *approximated* by the formula shown, where a represents the length of the semimajor axis and b represents the length of the semiminor axis. Find the perimeter of the ellipse defined by the equation $\dfrac{x^2}{49} + \dfrac{y^2}{4} = 1$.

▶ **APPLICATIONS**

65. Decorative fireplaces: A bricklayer intends to build an elliptical fireplace 3 ft high and 8 ft wide, with two glass doors that open at the middle. The hinges to these doors are to be screwed onto a spine that is perpendicular to the hearth and goes through the foci of the ellipse. How far from center will the spines be located? How tall will each spine be?

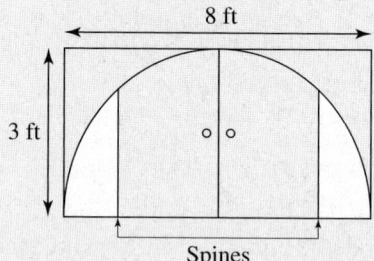

66. Decorative gardens: A retired math teacher decides to present her husband with a beautiful elliptical garden to help celebrate their 50th anniversary. The ellipse is to be 8 m long and 5 m across, with decorative fountains located at the foci. How far from the center of the ellipse should the fountains be located (round to the nearest 100th of a meter)? How far apart are the fountains?

67. Attracting attention to art: As part of an art show, a gallery owner asks a student from the local university to design a unique exhibit that will highlight one of the more significant pieces in the collection, an ancient sculpture. The student decides to create an elliptical showroom with reflective walls, with a rotating laser light on a stand at one focus, and the sculpture placed at the other focus on a stand of equal height. The laser light then points continually at the sculpture as it rotates. If the elliptical room is 24 ft long and 16 ft wide, how far from the center of the ellipse should the stands be located (round to the nearest 10th of a foot)? How far apart are the stands?

68. Medical procedures: The medical procedure called *lithotripsy* is a noninvasive medical procedure that is used to break up kidney and bladder stones in the body. A machine called a *lithotripter* uses its three-dimensional semielliptical shape and the foci properties of an ellipse to concentrate shock waves generated at one focus, on a kidney stone located at the other focus (see diagram — not drawn to scale). If the lithotripter has a length (semimajor axis) of 16 cm and a radius (semiminor axis) of 10 cm, how far from the vertex should a kidney stone be located

for the best result? Round to the nearest hundredth.

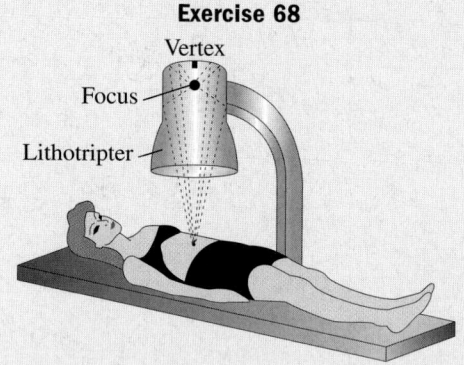

Exercise 68

69. Elliptical arches: In some situations, bridges are built using uniform elliptical archways as shown in the figure given. Find the equation of the ellipse forming each arch if it has a total width of 30 ft and a maximum center height (above level ground) of 8 ft. What is the height of a point 9 ft to the right of the center of each arch?

70. Elliptical arches: An elliptical arch bridge is built across a one lane highway. The arch is 20 ft across and has a maximum center height of 12 ft. Will a farm truck hauling a load 10 ft wide with a clearance height of 11 ft be able to go through the bridge without damage? (*Hint*: See Exercise 69.)

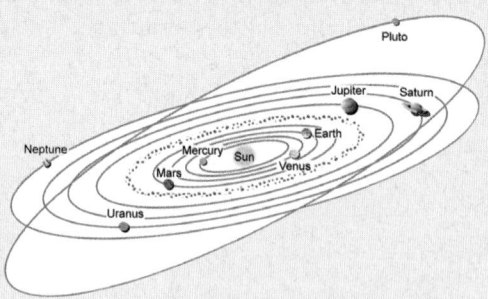

As a planet orbits around the Sun, it traces out an ellipse. If the center of the ellipse were placed at (0, 0) on a coordinate grid, the Sun would be actually off-centered (located at the *focus* of the ellipse). Use this information and the graphs provided to complete Exercises 71 through 74.

71. Orbit of Mercury: The approximate orbit of the planet Mercury is shown in the figure given. Find an equation that models this orbit.

Exercise 71

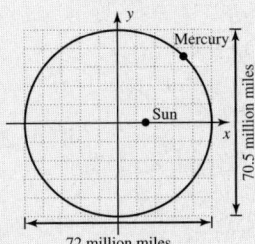

72 million miles

72. Orbit of Pluto: The approximate orbit of the dwarf planet Pluto is shown in the figure given. Find an equation that models this orbit.

Exercise 72

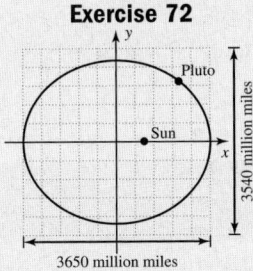

3650 million miles

73. Planetary orbits: Except for small variations, a planet's orbit around the Sun is elliptical with the Sun at one focus. The aphelion (maximum distance from the Sun) of the planet Mars is approximately 156 million miles, while the perihelion (minimum distance from the Sun) of Mars is about 128 million miles. Use this information to find the lengths of the semimajor and semiminor axes, rounded to the nearest million. If Mars has an orbital velocity of 54,000 miles per hour

(1.296 million miles per day), how many days does it take Mars to orbit the Sun? (*Hint:* Use the formula from Exercise 64).

74. Planetary orbits: The aphelion (maximum distance from the Sun) of the planet Saturn is approximately 940 million miles, while the perihelion (minimum distance from the Sun) of Saturn is about 840 million miles. Use this information to find the lengths of the semimajor and semiminor axes, rounded to the nearest million. If Saturn has an orbital velocity of 21,650 miles per hour (about 0.52 million miles per day), how many days does it take Saturn to orbit the Sun? How many years?

75. Area of a race track: Suppose the *Toronado 500* is a car race that is run on an elliptical track. The track is bounded by two ellipses with equations of $4x^2 + 9y^2 = 900$ and $9x^2 + 25y^2 = 900$, where x any y are in hundreds of yards. Use the formula given in Exercise 63 to find the area of the race track.

76. Area of a border: The table cloth for a large oval table is elliptical in shape. It is designed with two concentric ellipses (one within the other) as shown in the figure. The equation of the outer ellipse is $9x^2 + 25y^2 = 225$, and the equation of the inner ellipse is $4x^2 + 16y^2 = 64$ with x and y in feet. Use the formula given in Exercise 63 to find the area of the border of the tablecloth.

Exercise 76

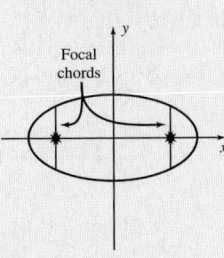

▶ **EXTENDING THE THOUGHT**

77. When graphing the conic sections, it is often helpful to use what is called a **focal chord,** as it gives additional points on the graph with very little effort. A focal chord is a line segment through a focus (perpendicular to the major or transverse axis), with the endpoints on the graph. For an ellipse, the length of the focal chord is given by $L = \frac{2m^2}{n}$, where m is the length of the semiminor axis, and n is the length of the semimajor axis. The focus will always be the midpoint of this line segment. Find the length of the focal chord for the ellipse

Exercise 77

Focal chords

$\frac{x^2}{81} + \frac{y^2}{36} = 1$ and the coordinates of the endpoints. Verify (by substituting into the equation) that these endpoints are indeed points on the graph, then use them to help complete the graph.

78. For the equation $6x^2 + 36x + 3y^2 - 24y + 74 = -28$, does the equation appear to be that of a circle, ellipse, or parabola? Write the equation in factored form. What do you notice? What can you say about the graph of this equation?

79. Verify that for the ellipse $\frac{x^2}{a^2} + \frac{y^2}{b^2} = 1$, the length of the focal chord is $\frac{2b^2}{a}$.

► **MAINTAINING YOUR SKILLS**

80. (4.4) Evaluate the expression using the change-of-base formula: $\log_3 20$.

81. (3.8) The resistance R to current flow in an electrical wire varies directly as the length L of the wire and inversely as the square of its diameter d. (a) Write the equation of variation; (b) find the constant of variation if a wire 2 m long with diameter $d = 0.005$ m has a resistance of 240 ohms (Ω); and (c) find the resistance in a similar wire 3 m long and 0.006 m in diameter.

82. (7.6) Use De Moivre's theorem to compute the value of $z = (1 - \sqrt{3}i)^6$.

83. (8.4) Find the true direction and groundspeed of the airplane shown, given the direction and speed of the wind (indicated in blue).

Exercise 83

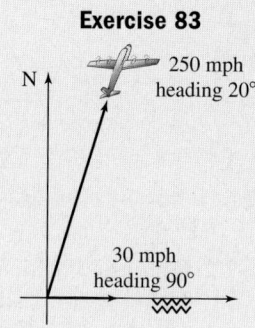

9.3 | The Hyperbola

Learning Objectives

In Section 9.3 you will learn how to:

☐ **A.** Use the equation of a hyperbola to graph central and noncentral hyperbolas

☐ **B.** Distinguish between the equations of a circle, ellipse, and hyperbola

☐ **C.** Locate the foci of a hyperbola and use the foci and other features to write its equation

☐ **D.** Solve applications involving foci

As seen in Section 9.1 (see Figure 9.18), a hyperbola is a conic section formed by a plane that cuts both nappes of a right circular cone. A hyperbola has two symmetric parts called **branches,** which open in opposite directions. Although the branches appear to resemble parabolas, we will soon discover they are actually a very different curve.

Figure 9.18

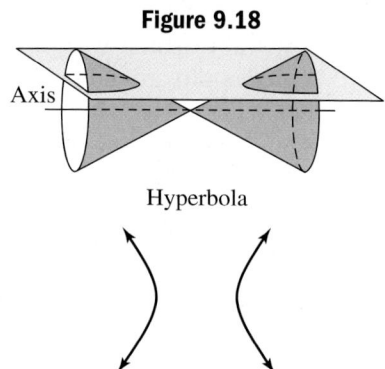

Hyperbola

A. The Equation of a Hyperbola

In Section 9.2, we noted that for the equation $Ax^2 + By^2 = F$, if $A = B$, the equation is that of a circle, if $A \neq B$, the equation represents an ellipse. Both cases contain a *sum* of second-degree terms. Perhaps driven by curiosity, we might wonder what happens if the equation has a *difference* of second-degree terms. Consider the equation $9x^2 - 16y^2 = 144$. It appears the graph will be centered at $(0, 0)$ since no shifts are applied (h and k are both zero). Using the intercept method to graph this equation reveals an entirely new curve, called a *hyperbola*.

EXAMPLE 1 ► **Graphing a Central Hyperbola**

Graph the equation $9x^2 - 16y^2 = 144$ using intercepts and additional points as needed.

Solution ►

$$9x^2 - 16y^2 = 144 \quad \text{given}$$
$$9(0)^2 - 16y^2 = 144 \quad \text{substitute 0 for } x$$
$$-16y^2 = 144 \quad \text{simplify}$$
$$y^2 = -9 \quad \text{divide by } -16$$

Since y^2 can never be negative, we conclude that the graph has *no y-intercepts.* Substituting $y = 0$ to find the *x*-intercepts gives

$$9x^2 - 16y^2 = 144 \qquad \text{given}$$
$$9x^2 - 16(0)^2 = 144 \qquad \text{substitute 0 for } y$$
$$9x^2 = 144 \qquad \text{simplify}$$
$$x^2 = 16 \qquad \text{divide by 9}$$
$$x = \sqrt{16} \quad \text{and} \quad x = -\sqrt{16} \qquad \text{square root property}$$
$$x = 4 \quad \text{and} \quad x = -4 \qquad \text{simplify}$$
$$(4, 0) \quad \text{and} \quad (-4, 0) \qquad \text{x-intercepts}$$

Knowing the graph has no *y*-intercepts, we select inputs greater than 4 and less than -4 to help sketch the graph. Using $x = 5$ and $x = -5$ yields

$$9x^2 - 16y^2 = 144 \qquad \text{given} \qquad\qquad 9x^2 - 16y^2 = 144$$
$$9(5)^2 - 16y^2 = 144 \qquad \text{substitute for } x \qquad 9(-5)^2 - 16y^2 = 144$$
$$9(25) - 16y^2 = 144 \qquad 5^2 = (-5)^2 = 25 \qquad 9(25) - 16y^2 = 144$$
$$225 - 16y^2 = 144 \qquad \text{simplify} \qquad\qquad 225 - 16y^2 = 144$$
$$-16y^2 = -81 \qquad \text{subtract 225} \qquad\qquad -16y^2 = -81$$
$$y^2 = \frac{81}{16} \qquad \text{divide by } -16 \qquad\qquad y^2 = \frac{81}{16}$$
$$y = \frac{9}{4} \quad y = -\frac{9}{4} \qquad \text{square root property} \qquad y = \frac{9}{4} \quad y = -\frac{9}{4}$$
$$y = 2.25 \quad y = -2.25 \qquad \text{decimal form} \qquad y = 2.25 \quad y = -2.25$$
$$(5, 2.25) \quad (5, -2.25) \qquad \text{ordered pairs} \qquad (-5, 2.25) \quad (-5, -2.25)$$

Plotting these points and connecting them with a smooth curve, while *knowing there are no y-intercepts,* produces the graph in the figure. The point at the origin (in blue) is not a part of the graph, and is given only to indicate the "center" of the hyperbola.

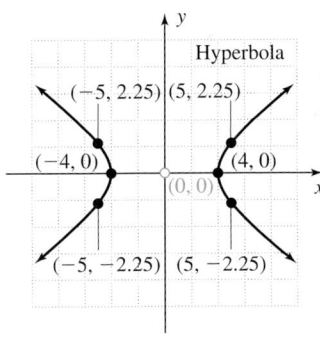

Now try Exercises 7 through 22 ▶

Since the hyperbola crosses a horizontal line of symmetry, it is referred to as a **horizontal hyperbola.** The points $(-4, 0)$ and $(4, 0)$ are called **vertices,** and the **center** of the hyperbola is always the point halfway between them. If the center is at the origin, we have a **central hyperbola.** The line passing through the center and both vertices is called the **transverse axis** (vertices are always on the transverse axis), and the line passing through the center and perpendicular to this axis is called the **conjugate axis** (see Figure 9.19).

In Example 1, the coefficient of x^2 was positive and we were subtracting $16y^2$: $9x^2 - 16y^2 = 144$. The result was a horizontal hyperbola. If the y^2-term is positive and we subtract the term containing x^2, the result is a **vertical hyperbola** (Figure 9.20).

Figure 9.19

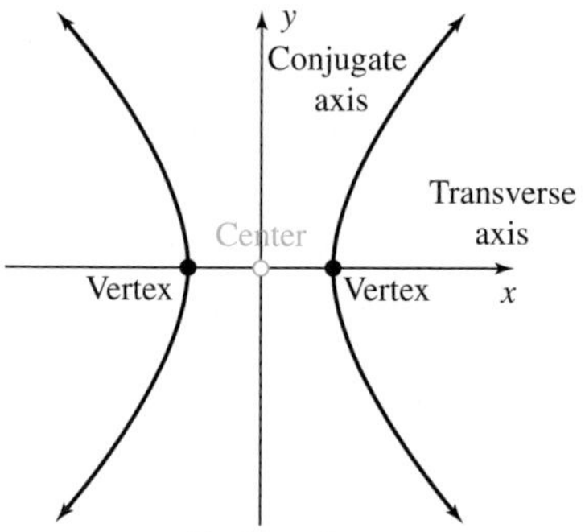

Figure 9.20

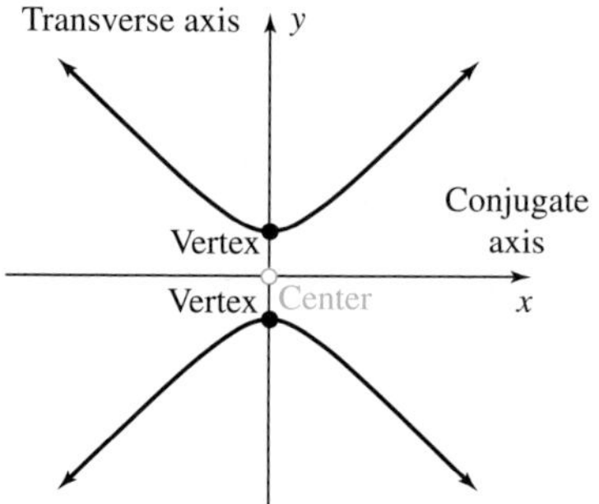

Horizontal hyperbola

Vertical hyperbola

EXAMPLE 2 ▶ **Identifying the Axes, Vertices, and Center of a Hyperbola from Its Graph**

For the hyperbola shown, state the location of the vertices and the equation of the transverse axis. Then identify the location of the center and the equation of the conjugate axis.

Solution ▶ By inspection we locate the vertices at (0, 0) and (0, 4). The equation of the transverse axis is $x = 0$. The center is halfway between the vertices at (0, 2), meaning the equation of the conjugate axis is $y = 2$.

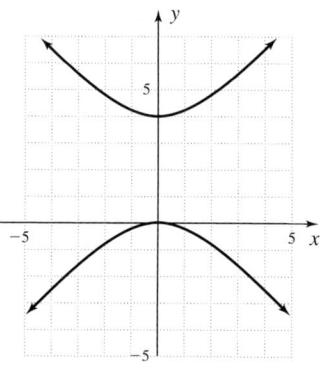

Now try Exercises 23 through 26 ▶

Standard Form

As with the ellipse, the polynomial form of the equation is helpful for *identifying* hyperbolas, but not very helpful when it comes to *graphing* a hyperbola (since we still must go through the laborious process of finding additional points). For graphing, standard form is once again preferred. Consider the hyperbola $9x^2 - 16y^2 = 144$ from Example 1. To write the equation in standard form, we divide by 144 and obtain $\dfrac{x^2}{4^2} - \dfrac{y^2}{3^2} = 1$. By comparing the standard form to the graph, we note $a = 4$ represents the distance from center to vertices, similar to the way we used a previously. But since the graph has no y-intercepts, what could $b = 3$ represent? The answer lies in the fact that branches of a hyperbola are **asymptotic,** meaning they will approach and become very close to imaginary lines that can be used to sketch the graph.

For a central hyperbola, the slopes of the asymptotic lines are given by the ratios $\dfrac{b}{a}$ and $-\dfrac{b}{a}$, with the related equations being $y = \dfrac{b}{a}x$ and $y = -\dfrac{b}{a}x$. The graph from Example 1 is repeated in Figure 9.21, with the asymptotes drawn. For a clearer understanding of how the equations for the asymptotes were determined, see **Exercise 88.**

A second method of drawing the asymptotes involves drawing a **central rectangle** with dimensions $2a$ by $2b$, as shown in Figure 9.22. The asymptotes will be the *extended diagonals* of this rectangle. This brings us to the equation of a hyperbola in standard form.

Figure 9.21

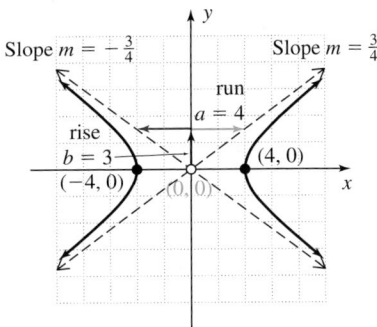

Slope method

Figure 9.22

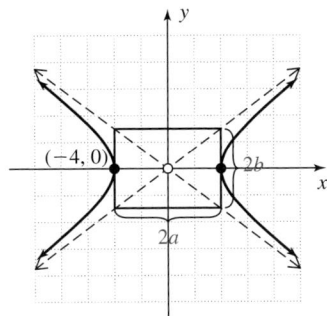

Central rectangle method

The Equation of a Hyperbola in Standard Form

The equation

$$\frac{(x - h)^2}{a^2} - \frac{(y - k)^2}{b^2} = 1$$

represents a *horizontal* hyperbola with center (h, k)
- *transverse* axis $y = k$
- *conjugate* axis $x = h$
- $|a|$ gives the distance from center to vertices.

The equation

$$\frac{(y - k)^2}{b^2} - \frac{(x - h)^2}{a^2} = 1$$

represents a *vertical* hyperbola with center (h, k)
- *transverse* axis $x = h$
- *conjugate* axis $y = k$
- $|b|$ gives the distance from center to vertices.

- Asymptotes can be drawn by starting at (h, k) and using slopes $m = \pm\dfrac{b}{a}$.

EXAMPLE 3 ▶ **Graphing a Hyperbola Using Its Equation in Standard Form**

Sketch the graph of $16(x - 2)^2 - 9(y - 1)^2 = 144$. Label the center, vertices, and asymptotes.

Solution ▶ Begin by noting a difference of the second-degree terms, with the x^2-term occurring first. This means we'll be graphing a horizontal hyperbola whose center is at $(2, 1)$. Continue by writing the equation in standard form.

$$16(x - 2)^2 - 9(y - 1)^2 = 144 \qquad \text{given equation}$$

$$\frac{16(x - 2)^2}{144} - \frac{9(y - 1)^2}{144} = \frac{144}{144} \qquad \text{divide by 144}$$

$$\frac{(x - 2)^2}{9} - \frac{(y - 1)^2}{16} = 1 \qquad \text{simplify}$$

$$\frac{(x - 2)^2}{3^2} - \frac{(y - 1)^2}{4^2} = 1 \qquad \text{write denominators in squared form}$$

Since $a = 3$ the vertices are a horizontal distance of 3 units from the center $(2, 1)$, giving $(2 + 3, 1) \rightarrow (5, 1)$ and $(2 - 3, 1) \rightarrow (-1, 1)$. After plotting the center and vertices, we can begin at the center and count off slopes of $m = \pm\dfrac{b}{a} = \pm\dfrac{4}{3}$, or draw a rectangle centered at $(2, 1)$ with dimensions $2(3) = 6$ (horizontal dimension) by $2(4) = 8$ (vertical dimension) to sketch the asymptotes. The complete graph is shown here.

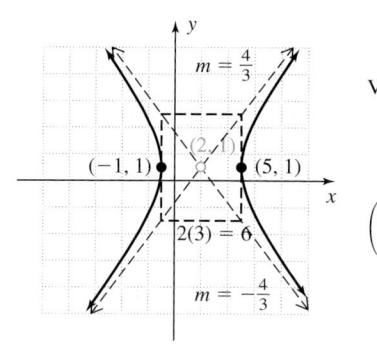

Horizontal hyperbola

Center at $(2, 1)$

Vertices at $(-1, 1)$ and $(5, 1)$

Transverse axis: $y = 1$
Conjugate axis: $x = 2$

Width of rectangle
$\left(\begin{array}{c}\text{horizontal dimension and} \\ \text{distance between vertices}\end{array}\right)$
$2a = 2(3) = 6$

Length of rectangle
(vertical dimension)
$2b = 2(4) = 8$

> **Now try Exercises 27 through 38 ▶**

Polynomial Form

If the equation is given as a polynomial in expanded form, complete the square in x and y, then write the equation in standard form.

EXAMPLE 4 ▶ **Graphing a Hyperbola by Completing the Square**

Graph the equation $9y^2 - x^2 + 54y + 4x + 68 = 0$.

Solution ▶ Since the y^2-term occurs first, we assume the equation represents a vertical hyperbola, but wait for the factored form to be sure (see **Exercise 87**).

$$9y^2 - x^2 + 54y + 4x + 68 = 0 \qquad \text{given}$$

$$9y^2 + 54y - x^2 + 4x = -68 \qquad \text{collect like-variable terms; subtract 68}$$

$$9(y^2 + 6y + \underline{\quad}) - 1(x^2 - 4x + \underline{\quad}) = -68 \qquad \text{factor out 9 from } y\text{-terms and } -1 \text{ from } x\text{-terms}$$

$$\underbrace{9(y^2 + 6y + 9)}_{\text{adds } 9(9) = 81} - \underbrace{1(x^2 - 4x + 4)}_{\text{adds } -1(4) = -4} = -68 + 81 + (-4) \qquad \substack{\text{complete the square} \\ \text{add } 81 + (-4) \text{ to right}}$$

$$9(y + 3)^2 - 1(x - 2)^2 = 9 \qquad \text{factor} \rightarrow \text{vertical hyperbola}$$

$$\frac{(y + 3)^2}{1} - \frac{(x - 2)^2}{9} = 1 \qquad \text{divide by 9 (standard form)}$$

$$\frac{(y + 3)^2}{1^2} - \frac{(x - 2)^2}{3^2} = 1 \qquad \text{write denominators in squared form}$$

The center of the hyperbola is $(2, -3)$ with $a = 3$, $b = 1$, and a transverse axis of $x = 2$. The vertices are at $(2, -3 + 1)$ and $(2, -3 - 1) \rightarrow (2, -2)$ and $(2, -4)$. After plotting the center and vertices, we draw a rectangle centered at $(2, -3)$ with a horizontal "width" of $2(3) = 6$ and a vertical "length" of $2(1) = 2$ to sketch the asymptotes. The completed graph is given in the figure.

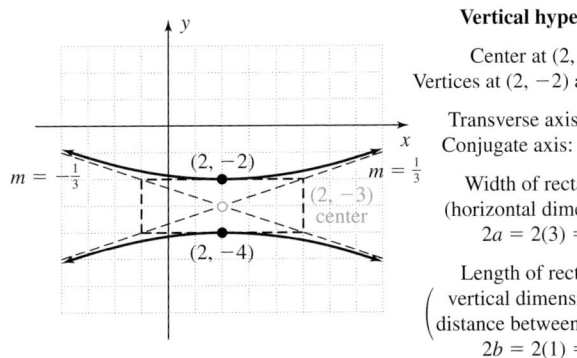

Vertical hyperbola

Center at $(2, -3)$
Vertices at $(2, -2)$ and $(2, -4)$

Transverse axis: $x = 2$
Conjugate axis: $y = -3$

Width of rectangle
(horizontal dimension)
$2a = 2(3) = 6$

Length of rectangle
$\left(\begin{array}{c} \text{vertical dimension and} \\ \text{distance between vertices} \end{array} \right)$
$2b = 2(1) = 2$

✓ **A.** You've just learned how to use the equation of a hyperbola to graph central and noncentral hyperbolas

Now try Exercises 39 through 48 ▶

B. Distinguishing between the Equations of a Circle, Ellipse, and Hyperbola

So far we've explored numerous graphs of circles, ellipses, and hyperbolas. In Example 5 we'll attempt to identify a given conic section from its equation alone (without graphing the equation). As you've seen, the corresponding equations have unique characteristics that can help distinguish one from the other.

EXAMPLE 5 ▶ **Identifying a Conic Section from Its Equation**

Identify each equation as that of a circle, ellipse, or hyperbola. Justify your choice and name the center, but do not draw the graphs.

a. $y^2 = 36 + 9x^2$ **b.** $4x^2 = 16 - 4y^2$

c. $x^2 = 225 - 25y^2$ **d.** $25x^2 = 100 + 4y^2$

e. $3(x - 2)^2 + 4(y + 3)^2 = 12$ **f.** $4(x + 5)^2 = 36 + 9(y - 4)^2$

Solution ▶ **a.** Writing the equation in factored form gives $y^2 - 9x^2 = 36$ $(h = 0, k = 0)$. Since the equation contains a difference of second-degree terms, it is the equation of a (vertical) hyperbola. The center is at $(0, 0)$.

b. Rewriting the equation as $4x^2 + 4y^2 = 16$ and dividing by 4 gives $x^2 + y^2 = 4$. The equation represents a circle of radius 2, with the center at $(0, 0)$.

c. Writing the equation as $x^2 + 25y^2 = 225$ we note a sum of second-degree terms with unequal coefficients. The equation is that of an ellipse, with the center at $(0, 0)$.

d. Rewriting the equation as $25x^2 - 4y^2 = 100$ we note the equation contains a difference of second-degree terms. The equation represents a central (horizontal) hyperbola, whose center is at $(0, 0)$.

e. The equation is in factored form and contains a sum of second-degree terms with unequal coefficients. This is the equation of an ellipse with the center at $(2, -3)$.

✓ **B.** You've just learned how to distinguish between the equations of a circle, ellipse, and hyperbola

f. Rewriting the equation as $4(x + 5)^2 - 9(y - 4)^2 = 36$ we note a difference of second-degree terms. The equation represents a horizontal hyperbola with center $(-5, 4)$.

Now try Exercises 49 through 60 ▶

C. The Foci of a Hyperbola

Like the ellipse, the foci of a hyperbola play an important part in their application. A long distance radio navigation system (called LORAN for short), can be used to determine the location of ships and airplanes and is based on the characteristics of a hyperbola (see **Exercises 85 and 86**). Hyperbolic mirrors are also used in some telescopes, and have the property that a beam of light directed at one focus will be reflected to the second focus. To understand and appreciate these applications, we use the analytic definition of a hyperbola:

Definition of a Hyperbola

Given two fixed points f_1 and f_2 in a plane, a hyperbola is the set of all points (x, y) such that the distance from f_2 to (x, y) subtracted from the distance from f_1 to (x, y) is a positive constant. In symbols,

$$d_1 - d_2 = k, k > 0$$

The fixed points f_1 and f_2 are called the foci of the hyperbola, and the points (x, y) are on the graph of the hyperbola.

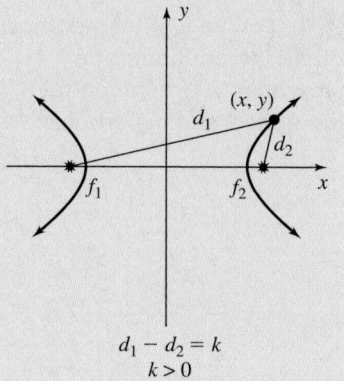

$$d_1 - d_2 = k$$
$$k > 0$$

Figure 9.23

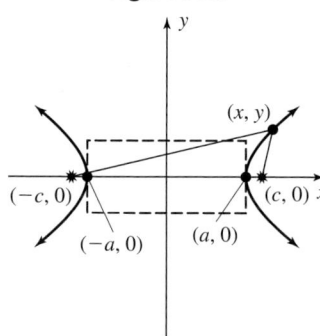

As with the analytic definition of the ellipse, it can be shown that the constant k is again equal to $2a$ (for horizontal hyperbolas). To find the equation of a hyperbola in terms of a and b, we use an approach similar to that of the ellipse (see Appendix IV), and the result is identical to that seen earlier: $\dfrac{x^2}{a^2} - \dfrac{y^2}{b^2} = 1$ where $b^2 = c^2 - a^2$ (see Figure 9.23).

We now have the ability to *find the foci of any hyperbola*—and can use this information in many significant applications. Since the location of the foci play such an important role, it is best to remember the relationship as $c^2 = a^2 + b^2$ (called the **foci formula** for hyperbolas), noting that for a hyperbola, $c > a$ and $c^2 > a^2$ (also $c > b$ and $c^2 > b^2$).

EXAMPLE 6 ▶ **Graphing a Hyperbola and Identifying Its Foci by Completing the Square.**

For the hyperbola defined by $7x^2 - 9y^2 - 14x + 72y - 200 = 0$, find the coordinates of the center, vertices, foci, and the dimensions of the central rectangle. Then sketch the graph.

Solution ▶

$$7x^2 - 9y^2 - 14x + 72y - 200 = 0 \qquad \text{given}$$

$$7x^2 - 14x - 9y^2 + 72y = 200 \qquad \text{group terms; add 200}$$

$$7(x^2 - 2x + \underline{\quad}) - 9(y^2 - 8y + \underline{\quad}) = 200 \qquad \text{factor out leading coefficients}$$

$$7(x^2 - 2x + 1) - 9(y^2 - 8y + 16) = 200 + 7 + (-144) \qquad \text{complete the square}$$
$$\rightarrow \text{add } 7 + (-144)$$
$$\text{to right-hand side}$$

$$\text{adds } 7(1) = 7 \qquad \text{adds } -9(16) = -144$$

$$7(x - 1)^2 - 9(y - 4)^2 = 63 \qquad \text{factored form}$$

$$\frac{(x - 1)^2}{9} - \frac{(y - 4)^2}{7} = 1 \qquad \text{divide by 63 and simplify}$$

$$\frac{(x - 1)^2}{3^2} - \frac{(y - 4)^2}{(\sqrt{7})^2} = 1 \qquad \text{write denominators in squared form}$$

This is a horizontal hyperbola with $a = 3$ ($a^2 = 9$) and $b = \sqrt{7}$ ($b^2 = 7$). The center is at $(1, 4)$, with vertices $(-2, 4)$ and $(4, 4)$. Using the foci formula $c^2 = a^2 + b^2$ yields $c^2 = 9 + 7 = 16$, showing the foci are $(-3, 4)$ and $(5, 4)$ (4 units from center). The central rectangle is $2\sqrt{7} \approx 5.29$ by $2(3) = 6$.

Drawing the rectangle and sketching the asymptotes to complete the graph, results in the graph shown.

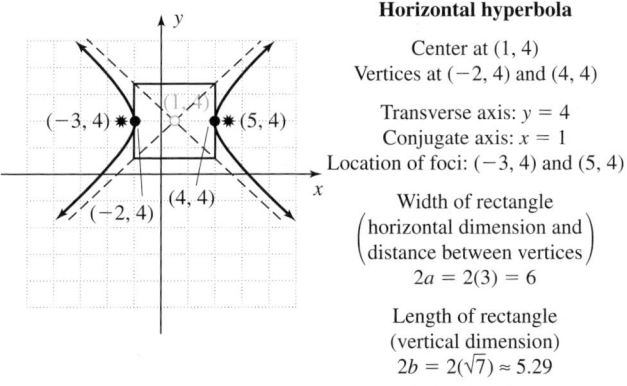

Horizontal hyperbola

Center at $(1, 4)$
Vertices at $(-2, 4)$ and $(4, 4)$

Transverse axis: $y = 4$
Conjugate axis: $x = 1$
Location of foci: $(-3, 4)$ and $(5, 4)$

Width of rectangle
$\begin{pmatrix} \text{horizontal dimension and} \\ \text{distance between vertices} \end{pmatrix}$
$2a = 2(3) = 6$

Length of rectangle
(vertical dimension)
$2b = 2(\sqrt{7}) \approx 5.29$

☑ **C.** You've just learned how to locate the foci of a hyperbola and use the foci and other features to write its equation

Now try Exercises 61 through 70 ▶

As with the ellipse, if any two of the values for a, b, and c are known, the relationship between them can be used to construct the equation of the hyperbola. See **Exercises 71 through 78.**

D. Applications Involving Foci

Applications involving the foci of a conic section can take many forms. As before, only partial information about the hyperbola may be available, and we'll determine a solution by manipulating a given equation, or constructing an equation from given facts.

EXAMPLE 7 ▶ Applying the Properties of a Hyperbola—The Path of a Comet

Comets with a high velocity cannot be captured by the Sun's gravity, and are slung around the Sun in a hyperbolic path with the Sun at one focus. If the path illustrated by the graph shown is modeled by the equation $2116x^2 - 400y^2 = 846{,}400$, how close did the comet get to the Sun? Assume units are in millions of miles and round to the nearest million.

Solution ▶ We are essentially asked to find the distance between a vertex and focus. Begin by writing the equation in standard form:

$2116x^2 - 400y^2 = 846{,}400$ given

$\dfrac{x^2}{400} - \dfrac{y^2}{2116} = 1$ divide by 846,400

$\dfrac{x^2}{20^2} - \dfrac{y^2}{46^2} = 1$ write denominators in squared form

This is a horizontal hyperbola with $a = 20$ $(a^2 = 400)$ and $b = 46$ $(b^2 = 2116)$. Use the foci formula to find c^2 and c.

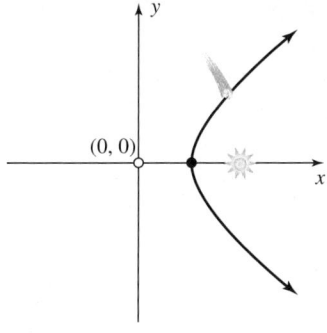

$$c^2 = a^2 + b^2$$
$$c^2 = 400 + 2116$$
$$c^2 = 2516$$
$$c \approx -50 \text{ and } c \approx 50$$

Since $a = 20$ and $|c| \approx 50$, the comet came within $50 - 20 = 30$ million miles of the Sun.

Now try Exercises 81 through 84 ▶

EXAMPLE 8 ▶ **Applying the Properties of a Hyperbola—The Location of a Storm**

Two amateur meteorologists, living 4 km apart (4000 m), see a storm approaching. The one farthest from the storm hears a loud clap of thunder 9 sec after the one nearest. Assuming the speed of sound is 340 m/sec, determine an equation that models possible locations for the storm at this time.

Solution ▶ Let M_1 represent the meteorologist nearest the storm and M_2 the farthest. Since M_2 heard the thunder 9 sec after M_1, M_2 must be $9 \cdot 340 = 3060$ m farther away from the storm S. In other words, $|M_2S| - |M_1S| = 3060$. The set of all points that satisfy this description fit the definition of a hyperbola, and we'll use this fact to develop an equation model for possible locations of the storm. Let's place the information on a coordinate grid. For convenience, we'll use the straight line distance between M_1 and M_2 as the x-axis, with the origin an equal distance from each. With the constant difference equal to 3060, we have $2a = 3060$, $a = 1530$ from the definition of a hyperbola,

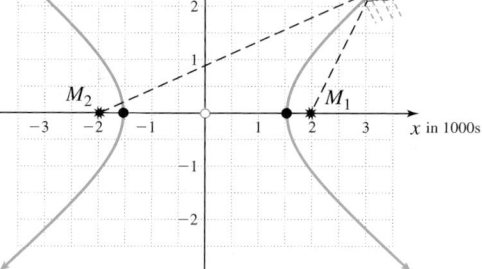

giving $\dfrac{x^2}{1530^2} - \dfrac{y^2}{b^2} = 1$. With

$c = 2000$ m (the distance from the origin to M_1 or M_2), we find the value of b using the equation $c^2 = a^2 + b^2$: $2000^2 = 1530^2 + b^2$ or $b^2 = (2000)^2 - (1530)^2 = 1,659,100 \approx 1288^2$. The equation that models possible locations of the storm is

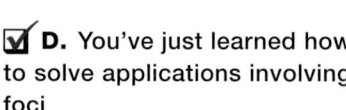

D. You've just learned how to solve applications involving foci

$\dfrac{x^2}{1530^2} - \dfrac{y^2}{1288^2} \approx 1$.

Now try Exercises 85 and 86 ▶

TECHNOLOGY HIGHLIGHT

Studying Hyperbolas

As with the circle and ellipse, the hyperbola must also be defined in two pieces in order to use a graphing calculator to study its graph. Consider the *relation* $4x^2 - 9y^2 = 36$. From our work in this section, we know this is the equation of a horizontal hyperbola centered at (0, 0). Solving for y gives

$$4x^2 - 9y^2 = 36 \qquad \text{original equation}$$

$$-9y^2 = 36 - 4x^2 \qquad \text{isolate } y^2\text{-term}$$

$$y^2 = \frac{36 - 4x^2}{-9} \qquad \text{divide by } -9$$

$$y = \pm\sqrt{\frac{36 - 4x^2}{-9}} \qquad \text{solve for } y$$

Figure 9.24

We can again separate this result into two parts: $Y_1 = \sqrt{\dfrac{36 - 4x^2}{-9}}$

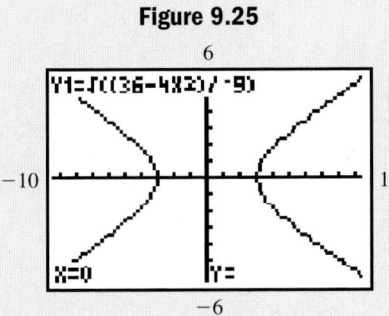

gives the "upper half" of the hyperbola, and $Y_2 = -\sqrt{\dfrac{36 - 4x^2}{-9}}$ gives

the "lower half." In Figure 9.24, note the use of parentheses on the
[Y=] screen to ensure we're taking the square root of the entire expression.

Entering these on the [Y=] screen, graphing them with the window shown, and pressing the [TRACE] key gives the graph shown in Figure 9.25. Note the location of the cursor at $x = 0$, but no y-value is displayed. This is because the hyperbola is not defined at $x = 0$. Press the right arrow key [▶] and walk the cursor to the right until the y-values begin appearing. In fact, they begin to appear at (3, 0), which is one of the vertices of the hyperbola. We could also graph the asymptotes ($y = \pm\frac{2}{3}x$) by entering the lines as Y_3 and Y_4 on the [Y=] screen. The resulting graph is shown in Figure 9.26 using the standard window (the [TRACE] key has been pushed and the down arrow used to highlight Y_2). Use these ideas to complete the following exercises.

Figure 9.25

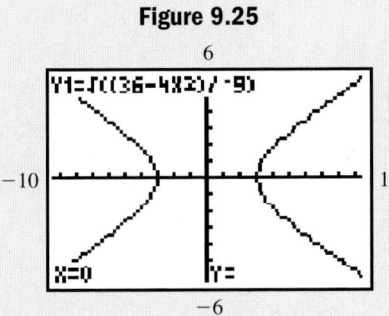

Figure 9.26

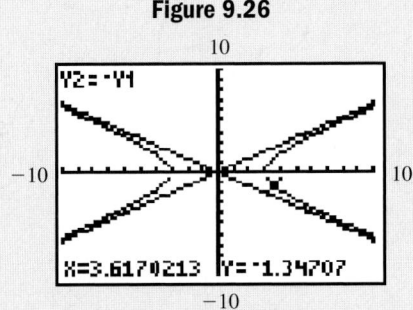

Exercise 1: Graph the hyperbola $25y^2 - 4x^2 = 100$ using a friendly window. What are the coordinates of the vertices? Use the [TRACE] feature to find the value(s) of y when $x = 4$. Determine (from the graph) the value(s) of y when $x = -4$, then verify your response using the [TABLE] feature.

Exercise 2: Graph the hyperbola $9x^2 - 16y^2 = 144$ using the standard window. Then determine the equations of the asymptotes and graph these as well. Why do the asymptotes intersect at the origin? When will the asymptotes *not* intersect at the origin?

9.3 EXERCISES

► CONCEPTS AND VOCABULARY

Fill in the blank with the appropriate word or phrase. Carefully reread the section if needed.

1. The line that passes through the vertices of a hyperbola is called the _____ axis.

2. The conjugate axis is _____ to the _____ axis and contains the _____ of the hyperbola.

3. The center of a hyperbola is located _____ between the vertices.

4. The center of the hyperbola defined by
$$\frac{(x-2)^2}{4^2} - \frac{(y-3)^2}{5^2} = 1 \text{ is at } \underline{\quad}.$$

5. Compare/Contrast the two methods used to find the asymptotes of a hyperbola. Include an example illustrating both methods.

6. Explore/Explain why $A(x-h)^2 - B(y-k)^2 = F$ results in a hyperbola regardless of whether $A = B$ or $A \neq B$. Illustrate with an example.

► DEVELOPING YOUR SKILLS

Graph each hyperbola. Label the center, vertices, and any additional points used.

7. $\dfrac{x^2}{9} - \dfrac{y^2}{4} = 1$

8. $\dfrac{x^2}{16} - \dfrac{y^2}{9} = 1$

9. $\dfrac{x^2}{4} - \dfrac{y^2}{9} = 1$

10. $\dfrac{x^2}{25} - \dfrac{y^2}{16} = 1$

11. $\dfrac{x^2}{49} - \dfrac{y^2}{16} = 1$

12. $\dfrac{x^2}{25} - \dfrac{y^2}{9} = 1$

13. $\dfrac{x^2}{36} - \dfrac{y^2}{16} = 1$

14. $\dfrac{x^2}{81} - \dfrac{y^2}{16} = 1$

15. $\dfrac{y^2}{9} - \dfrac{x^2}{1} = 1$

16. $\dfrac{y^2}{1} - \dfrac{x^2}{4} = 1$

17. $\dfrac{y^2}{12} - \dfrac{x^2}{4} = 1$

18. $\dfrac{y^2}{9} - \dfrac{x^2}{18} = 1$

19. $\dfrac{y^2}{9} - \dfrac{x^2}{9} = 1$

20. $\dfrac{y^2}{4} - \dfrac{x^2}{4} = 1$

21. $\dfrac{y^2}{36} - \dfrac{x^2}{25} = 1$

22. $\dfrac{y^2}{16} - \dfrac{x^2}{4} = 1$

For the graphs given, state the location of the vertices and the equation of the transverse axis. Then identify the location of the center and the equation of the conjugate axis. Note the scale used on each axis.

23.
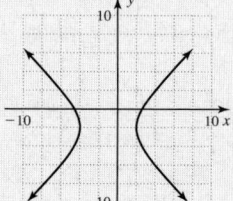

24.
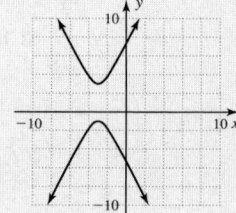

25.
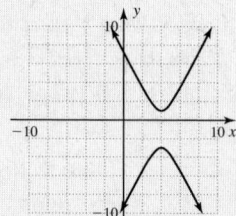

26.
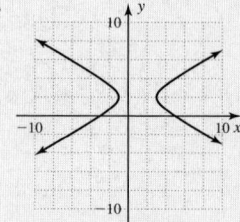

Sketch a complete graph of each equation, including the asymptotes. Be sure to identify the center and vertices.

27. $\dfrac{(y+1)^2}{4} - \dfrac{x^2}{25} = 1$

28. $\dfrac{y^2}{4} - \dfrac{(x-2)^2}{9} = 1$

29. $\dfrac{(x-3)^2}{36} - \dfrac{(y+2)^2}{49} = 1$

30. $\dfrac{(x-2)^2}{9} - \dfrac{(y-1)^2}{4} = 1$

31. $\dfrac{(y + 1)^2}{7} - \dfrac{(x + 5)^2}{9} = 1$

32. $\dfrac{(y - 3)^2}{16} - \dfrac{(x + 2)^2}{5} = 1$

33. $(x - 2)^2 - 4(y + 1)^2 = 16$

34. $9(x + 1)^2 - (y - 3)^2 = 81$

35. $2(y + 3)^2 - 5(x - 1)^2 = 50$

36. $9(y - 4)^2 - 5(x - 3)^2 = 45$

37. $12(x - 4)^2 - 5(y - 3)^2 = 60$

38. $8(x - 4)^2 - 3(y - 3)^2 = 24$

39. $16x^2 - 9y^2 = 144$

40. $16x^2 - 25y^2 = 400$

41. $9y^2 - 4x^2 = 36$

42. $25y^2 - 4x^2 = 100$

43. $12x^2 - 9y^2 = 72$

44. $36x^2 - 20y^2 = 180$

45. $4x^2 - y^2 + 40x - 4y + 60 = 0$

46. $x^2 - 4y^2 - 12x - 16y + 16 = 0$

47. $x^2 - 4y^2 - 24y - 4x - 36 = 0$

48. $-9x^2 + 4y^2 - 18x - 24y - 9 = 0$

Classify each equation as that of a circle, ellipse, or hyperbola. Justify your response.

49. $-4x^2 - 4y^2 = -24$

50. $9y^2 = -4x^2 + 36$

51. $x^2 + y^2 = 2x + 4y + 4$

52. $x^2 = y^2 + 6y - 7$

53. $2x^2 - 4y^2 = 8$

54. $36x^2 + 25y^2 = 900$

55. $x^2 + 5 = 2y^2$

56. $x + y^2 = 3x^2 + 9$

57. $2x^2 = -2y^2 + x + 20$

58. $2y^2 + 3 = 6x^2 + 8$

59. $16x^2 + 5y^2 - 3x + 4y = 538$

60. $9x^2 + 9y^2 - 9x + 12y + 4 = 0$

Use the definition of a hyperbola to find the distance between the vertices and the dimensions of the rectangle centered at (h, k). Figures are not drawn to scale. Note that Exercises 63 and 64 are *vertical hyperbolas*.

61.
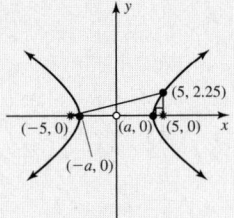

62.
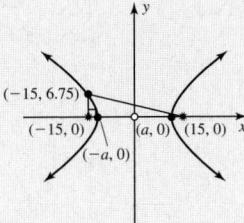

63.
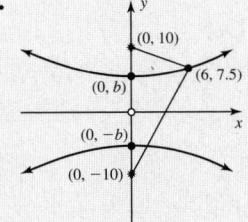

64.
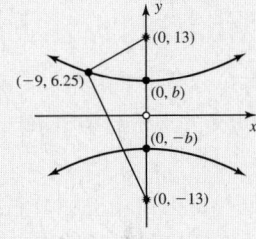

Find and list the coordinates of the (a) center, (b) vertices, (c) foci, and (d) dimensions of the central rectangle. Then (e) sketch the graph, including the asymptotes.

65. $4x^2 - 9y^2 - 24x + 72y - 144 = 0$

66. $4x^2 - 36y^2 - 40x + 144y - 188 = 0$

67. $16x^2 - 4y^2 + 24y - 100 = 0$

68. $81x^2 - 162x - 4y^2 - 243 = 0$

69. $9x^2 - 3y^2 - 54x - 12y + 33 = 0$

70. $10x^2 + 60x - 5y^2 + 20y - 20 = 0$

Find the equation of the hyperbola (in standard form) that satisfies the following conditions:

71. vertices at $(-6, 0)$ and $(6, 0)$; foci at $(-8, 0)$ and $(8, 0)$

72. vertices at $(-4, 0)$ and $(4, 0)$; foci at $(-6, 0)$ and $(6, 0)$

73. foci at $(-2, -3\sqrt{2})$ and $(-2, 3\sqrt{2})$; length of conjugate axis: 6 units

74. foci at $(-5, 2)$ and $(7, 2)$; length of conjugate axis: 8 units

Use the characteristics of a hyperbola and the graph given to write the related equation and state the location of the foci (75 and 76) or the dimensions of the central rectangle (77 and 78).

75.

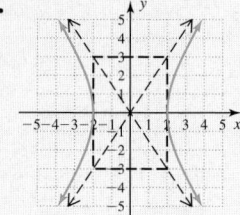

76.

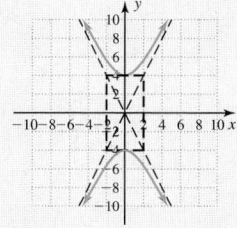

77.

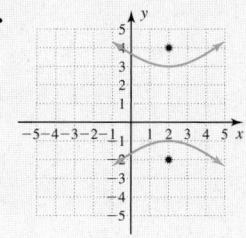

78.

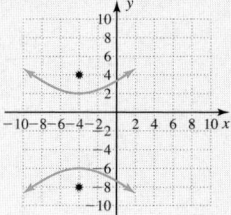

▶ WORKING WITH FORMULAS

79. Equation of a semi-hyperbola: $y = \sqrt{\dfrac{36 - 4x^2}{-9}}$

The "upper half" of a certain hyperbola is given by the equation shown. (a) Simplify the radicand, (b) state the domain of the expression, and (c) enter the expression as Y_1 on a graphing calculator and graph. What is the equation for the "lower half" of this hyperbola?

80. Focal chord of a hyperbola: $L = \dfrac{2b^2}{a}$

The focal chords of a hyperbola are line segments parallel to the conjugate axis with endpoints on the hyperbola, and containing certain points f_1 and f_2 called the *foci* (see grid). The length of the chord is given by the formula shown. Use it to find the length of the focal chord for the hyperbola indicated, then compare the calculated value with the length estimated from the given graph:

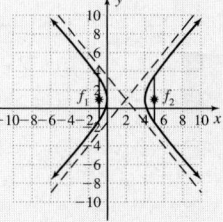

$$\frac{(x - 2)^2}{4} - \frac{(y - 1)^2}{5} = 1.$$

▶ APPLICATIONS

81. Stunt pilots: At an air show, a stunt plane dives along a hyperbolic path whose vertex is directly over the grandstands. If the plane's flight path can be modeled by the hyperbola $25y^2 - 1600x^2 = 40{,}000$, what is the minimum altitude of the plane as it passes over the stands? Assume x and y are in yards.

82. Flying clubs: To test their skill as pilots, the members of a flight club attempt to drop sandbags on a target placed in an open field, by diving along a hyperbolic path whose vertex is directly over the target area. If the flight path of the plane flown by the club's president is modeled by $9y^2 - 16x^2 = 14{,}400$, what is the minimum altitude of her plane as it passes over the target? Assume x and y are in feet.

83. Nuclear cooling towers: The natural draft cooling towers for nuclear power stations are called *hyperboloids of one sheet*. The

perpendicular cross sections of these hyperboloids form two branches of a hyperbola. Suppose the central cross section of one such tower is modeled by the hyperbola $1600x^2 - 400(y - 50)^2 = 640{,}000$. What is the minimum distance between the sides of the tower? Assume x and y are in feet.

84. Charged particles: It has been shown that when like particles with a common charge are hurled at each other, they deflect and travel along paths that are hyperbolic. Suppose the path of two such particles is modeled by the hyperbola $x^2 - 9y^2 = 36$. What is the minimum distance between the particles as they approach each other? Assume x and y are in microns.

85. Locating a ship using radar: Under certain conditions, the properties of a hyperbola can be used to help locate the position of a ship. Suppose two radio stations are located 100 km apart along a straight shoreline. A ship is sailing parallel to the shore and is 60 km out to sea. The ship sends out a distress call that is picked up by the closer station in 0.4 milliseconds (msec—one-thousandth of a second), while it takes 0.5 msec to reach the station that is farther away. Radio waves travel at a speed of approximately 300 km/msec. Use this information to find the equation of a hyperbola that will help you find the location of the ship, then find the coordinates of the ship. (*Hint:* Draw the hyperbola on a coordinate system with the radio stations on the *x*-axis at the foci, then use the definition of a hyperbola.)

86. **Locating a plane using radar:** Two radio stations are located 80 km apart along a straight shoreline, when a "mayday" call (a plea for immediate help) is received from a plane that is about to ditch in the ocean (attempt a water landing). The plane was flying at low altitude, parallel to the shoreline, and 20 km out when it ran into trouble. The plane's distress call is picked up by the closer station in 0.1 msec, while it takes 0.3 msec to reach the other. Use this information to construct the equation of a hyperbola that will help you find the location of the ditched plane, then find the coordinates of the plane. Also see Exercise 85.

87. It is possible for the plane to intersect only the vertex of the cone or to be tangent to the sides. These are called **degenerate cases** of a conic section. Many times we're unable to tell if the equation represents a degenerate case until it's written in standard form. Write the following equations in standard form and comment.

   a. $4x^2 - 32x - y^2 + 4y + 60 = 0$
   b. $x^2 - 4x + 5y^2 - 40y + 84 = 0$

88. For a greater understanding as to *why* the branches of a hyperbola are asymptotic, solve $\dfrac{x^2}{a^2} - \dfrac{y^2}{b^2} = 1$ for *y*, then consider what happens as $x \to \infty$ (note that $x^2 - k \approx x^2$ for large *x*).

89. Which has a greater area: (a) The central rectangle of the hyperbola given by $(x - 5)^2 - (y + 4)^2 = 57$, (b) the circle given by $(x - 5)^2 + (y + 4)^2 = 57$, or (c) the ellipse given by $9(x - 5)^2 + 10(y + 4)^2 = 570$?

90. Find the equation of the circle shown, given the equation of the hyperbola:
$9(x - 2)^2 - 25(y - 3)^2 = 225$

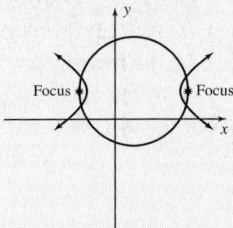

91. Find the equation of the ellipse shown, given the equation of the hyperbola and (2, 0) is on the graph of the ellipse. The hyperbola and ellipse share the same foci: $9(x - 2)^2 - 25(y - 3)^2 = 225$

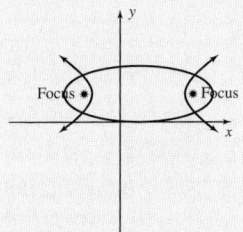

92. Verify that for a central hyperbola, a circle that circumscribes the central rectangle must also go through both foci.

▶ **MAINTAINING YOUR SKILLS**

93. **(7.3)** In weight-lifting competitions, Ursula Unger has shown she can lift up to 350 lb. Use a vector analysis to determine whether she will be able to pull the crate up the frictionless ramp shown.

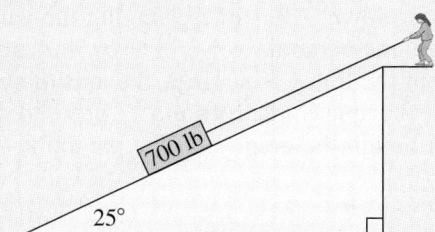

94. **(5.1)** The wheels on a motorized scooter are rotating at 403 rpm. If the wheels have a 2.5 in. radius, how fast is the scooter traveling in miles per hour?

95. **(1.4)** The number $z = 1 + i\sqrt{2}$ is a solution to two out of the three equations given. Which two?
   a. $x^4 + 4 = 0$
   b. $x^3 - 6x^2 + 11x - 12 = 0$
   c. $x^2 - 2x + 3 = 0$

96. **(5.4)** A government-approved company is licensed to haul toxic waste. Each container of solid waste weighs 800 lb and has a volume of 100 ft³. Each container of liquid waste weighs 1000 lb and is 60 ft³ in volume. The revenue from hauling solid waste is $300 per container, while the revenue from liquid waste is $375 per container. The truck used by this company has a weight capacity of 39.8 tons and a volume capacity of 6960 ft³. What combination of solid and liquid weight containers will produce the maximum revenue?

Learning Objectives

In Section 9.4 you will learn how to:

☐ **A.** Graph parabolas with a horizontal axis of symmetry

☐ **B.** Identify and use the focus-directrix form of the equation of a parabola

☐ **C.** Solve applications of the analytic parabola

In previous coursework, you likely learned that the graph of a quadratic function was a parabola. Parabolas are actually the fourth and final member of the family of conic sections, and as we saw in Section 9.1, the graph can be obtained by observing the intersection of a plane and a cone. If the plane is parallel to the generator of the cone (shown as a dark line in Figure 9.27), the intersection of the plane with one nappe forms a parabola. In this section we develop the general equation of a parabola from its analytic definition, opening a new realm of applications that extends far beyond those involving only zeroes and extreme values.

Figure 9.27

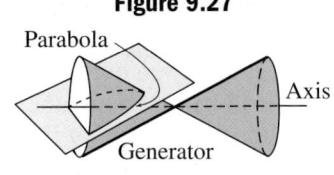

A. Parabolas with a Horizontal Axis

An introductory study of parabolas generally involves those with a vertical axis, defined by the equation $y = ax^2 + bx + c$. Unlike the previous conic sections, this equation has *only one second-degree (squared) term in x* and defines a function. As a review of our work in Section 3.1, the primary characteristics are listed here and illustrated in Figure 9.28.

Figure 9.28

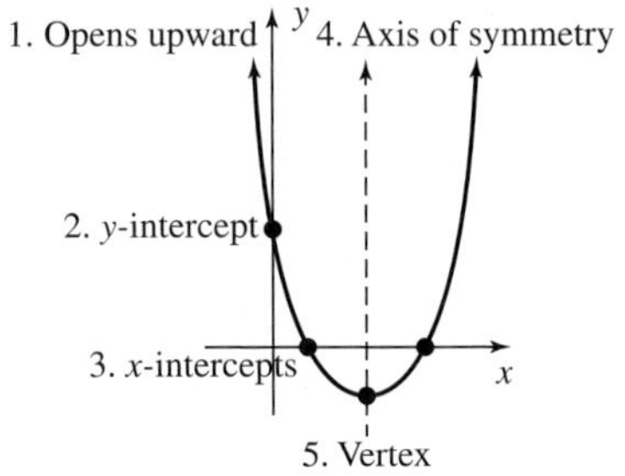

1. Opens upward
2. *y*-intercept
3. *x*-intercepts
4. Axis of symmetry
5. Vertex

Vertical Parabolas

For a second-degree equation of the form $y = ax^2 + bx + c$, the graph is a vertical parabola with these characteristics:

1. opens upward if $a > 0$; downward if $a < 0$.

2. *y*-intercept: $(0, c)$ (substitute 0 for *x*)

3. *x*-intercept(s): substitute 0 for *y* and solve.

4. axis of symmetry: $x = \dfrac{-b}{2a}$;

5. vertex: $\left(\dfrac{-b}{2a}, y\right)$

See **Exercises 7 through 12** for additional review and practice.

Horizontal Parabolas

Similar to our study of horizontal and vertical hyperbolas, the graph of a parabola can open *to the right or left,* as well as up or down. After interchanging the variables *x* and *y* in the standard equation, we obtain the parabola $x = ay^2 + by + c$, noting the resulting graph will be a reflection about the line $y = x$. Here, the axis of symmetry is a horizontal line and factoring or the quadratic formula is used to find the *y-intercepts* (if they exist). Note that although the graph is still a parabola—*it is not the graph of a function.*

Horizontal Parabolas

For a second-degree equation of the form $x = ay^2 + by + c$, the graph is a horizontal parabola with these characteristics:

1. opens right if $a > 0$,
   left if $a < 0$.

2. x-intercept: $(c, 0)$
   (substitute 0 for y)

3. y-intercept(s): substitute
   0 for x and solve.

4. axis of symmetry: $y = \dfrac{-b}{2a}$,

5. vertex: $\left(x, \dfrac{-b}{2a}\right)$

EXAMPLE 1 ▶ **Graphing a Horizontal Parabola**

Graph the relation whose equation is $x = y^2 + 3y - 4$, then state the domain and range of the relation.

Solution ▶ Since the equation has a single squared term in y, the graph will be a horizontal parabola. With $a > 0$ $(a = 1)$, the parabola opens to the right. The x-intercept is $(-4, 0)$. Factoring shows the y-intercepts are $y = -4$ and $y = 1$. The axis of symmetry is $y = \frac{-3}{2} = -1.5$, and substituting this value into the original equation gives $x = -6.25$. The coordinates of the vertex are $(-6.25, -1.5)$. Using horizontal and vertical boundary lines we find the domain for this relation is $x \in [-6.25, \infty)$ and the range is $y \in (-\infty, \infty)$. The graph is shown.

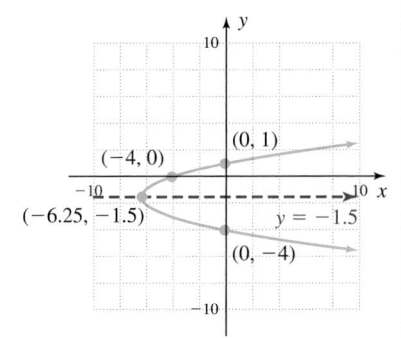

Now try Exercises 13 through 18 ▶

As with the vertical parabola, the equation of a horizontal parabola can be written as a transformation: $x = a(y \pm k)^2 \pm h$ by completing the square. Note that in this case, the vertical shift is k units *opposite the sign*, with a horizontal shift of h units in the same direction as the sign.

EXAMPLE 2 ▶ **Graphing a Horizontal Parabola by Completing the Square**

Graph by completing the square: $x = -2y^2 - 8y - 9$.

Solution ▶ Using the original equation, we note the graph will be a horizontal parabola opening to the left $(a = -2)$ and have an x-intercept of $(-9, 0)$. Completing the square gives $x = -2(y^2 + 4y + 4) - 9 + 8$, so $x = -2(y + 2)^2 - 1$. The vertex is at $(-1, -2)$ and $y = -2$ is the axis of symmetry. This means there are no y-intercepts, a fact that comes to light when we attempt to solve the equation after substituting 0 for x:

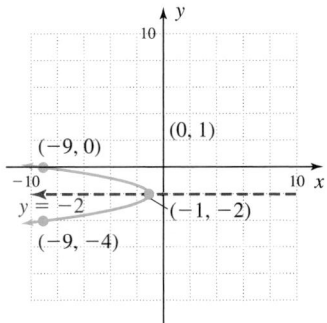

$$-2(y + 2)^2 - 1 = 0 \qquad \text{substitute 0 for } x$$

$$(y + 2)^2 = -\frac{1}{2} \qquad \text{isolate squared term}$$

☑ **A.** You've just learned how to graph parabolas with a horizontal axis of symmetry

The equation has no real roots. Using symmetry, the point $(-9, -4)$ is also on the graph. After plotting these points we obtain the graph shown.

Now try Exercises 19 through 36 ▶

B. The Focus-Directrix Form of the Equation of a Parabola

As with the ellipse and hyperbola, many significant applications of the parabola rely on its analytical definition rather than its algebraic form. From the construction of radio telescopes to the manufacture of flashlights, the location of the focus of a parabola is critical. To understand these and other applications, we use the analytic definition of a parabola first introduced in Section 9.1.

Definition of a Parabola

Given a fixed point f and fixed line D in the plane, a parabola is the set of all points (x, y) such that the distance from f to (x, y) is equal to the distance from line D to (x, y). The fixed point f is the **focus** of the parabola, and the fixed line is the **directrix**.

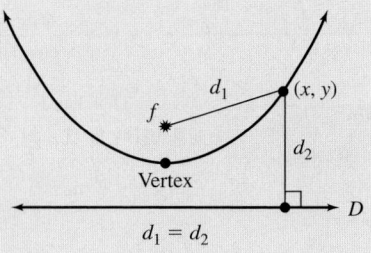

$d_1 = d_2$

WORTHY OF NOTE

For the analytic parabola, we use p to designate the focus, since c is so commonly used as the constant term in $y = ax^2 + bx + c$.

Figure 9.29

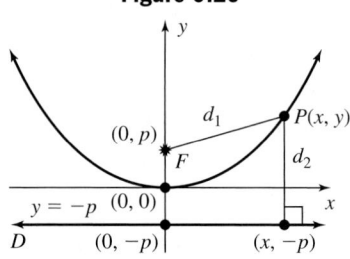

The general equation of a parabola can be obtained by combining this definition with the distance formula. With no loss of generality, we can assume the parabola shown in the definition box is oriented in the plane with the vertex at $(0, 0)$ and the focus at $(0, p)$. As the diagram in Figure 9.29 indicates, this gives the directrix an equation of $y = -p$ with all points on D having coordinates of $(x, -p)$.

Using $d_1 = d_2$, the distance formula yields

$$\sqrt{(x - 0)^2 + (y - p)^2} = \sqrt{(x - x)^2 + (y + p)^2} \quad \text{from definition}$$
$$(x - 0)^2 + (y - p)^2 = (x - x)^2 + (y + p)^2 \quad \text{square both sides}$$
$$x^2 + y^2 - 2py + p^2 = 0 + y^2 + 2py + p^2 \quad \text{simplify; expand binomials}$$
$$x^2 - 2py = 2py \quad \text{subtract } p^2 \text{ and } y^2$$
$$x^2 = 4py \quad \text{isolate } x^2$$

The resulting equation is called the **focus-directrix form** of a *vertical parabola* with vertex at $(0, 0)$. If we had begun by orienting the parabola so it opened to the right, we would have obtained the equation of a *horizontal parabola* with vertex $(0, 0)$: $y^2 = 4px$.

The Equation of a Parabola in Focus-Directrix Form

Vertical Parabola	**Horizontal Parabola**
$x^2 = 4py$	$y^2 = 4px$
focus $(0, p)$, directrix: $y = -p$	focus at $(p, 0)$, directrix: $x = -p$
If $p > 0$, opens upward.	If $p > 0$, opens to the right.
If $p < 0$, opens downward.	If $p < 0$, opens to the left.

For a parabola, note there is only one second-degree term.

EXAMPLE 3 ▶ **Locating the Focus and Directrix of a Parabola**

Find the vertex, focus, and directrix for the parabola defined by $x^2 = -12y$. Then sketch the graph, including the focus and directrix.

Solution ▶ Since the x-term is squared and no shifts have been applied, the graph will be a vertical parabola with a vertex of $(0, 0)$. Use a direct comparison between the given equation and the focus-directrix form to determine the value of p:

$$x^2 = -12y \quad \text{given equation}$$
$$\downarrow$$
$$x^2 = 4py \quad \text{focus-directrix form}$$

This shows:

$$4p = -12$$
$$p = -3$$

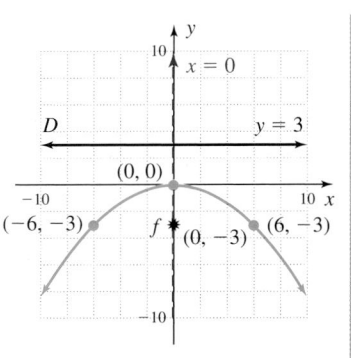

Since $p = -3$ ($p < 0$), the parabola opens downward, with the focus at $(0, -3)$ and directrix $y = 3$. To complete the graph we need a few additional points. Since 36 (6^2) is divisible by 12, we can use inputs of $x = 6$ and $x = -6$, giving the points $(6, -3)$ and $(-6, -3)$. Note the axis of symmetry is $x = 0$. The graph is shown.

Now try Exercises 37 through 48 ▶

Figure 9.30

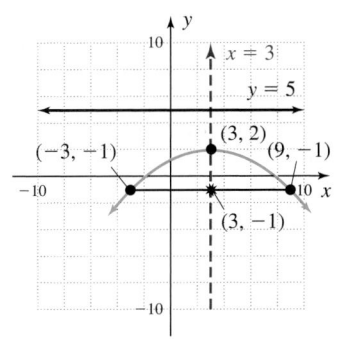

As an alternative to calculating additional points to sketch the graph, we can use what is called the **focal chord** of the parabola. Similar to the ellipse and hyperbola, the focal chord is a line segment that contains the focus, is parallel to the directrix, and has its endpoints on the graph. Using the definition of a parabola and the diagram in Figure 9.30, we see the horizontal distance from f to (x, y) is $2p$. Since $d_1 = d_2$, a line segment parallel to the directrix from the focus to the graph will also have a length of $|2p|$, and the focal chord of any parabola has a total length of $|4p|$. Note that in Example 3, the points we happened to choose were actually the endpoints of the focal chord.

Finally, if the vertex of a vertical parabola is shifted to (h, k), the equation will have the form $(x \pm h)^2 = 4p(y \pm k)$. As with the other conic sections, both the horizontal and vertical shifts are "opposite the sign."

EXAMPLE 4 ▶   **Locating the Focus and Directrix of a Parabola**

Find the vertex, focus, and directrix for the parabola whose equation is given, then sketch the graph, including the focus and directrix: $x^2 - 6x + 12y - 15 = 0$.

Solution ▶   Since only the x-term is squared, the graph will be a vertical parabola. To find the concavity, vertex, focus, and directrix, we complete the square in x and use a direct comparison between the shifted form and the focus-directrix form:

$$x^2 - 6x + 12y - 15 = 0 \qquad \text{given equation}$$
$$x^2 - 6x + \underline{} = -12y + 15 \qquad \text{complete the square in } x$$
$$x^2 - 6x + 9 = -12y + 24 \qquad \text{add 9 to both sides}$$
$$(x - 3)^2 = -12(y - 2) \qquad \text{factor}$$

Notice the parabola has been shifted 3 right and 2 up, so *all features of the parabola will likewise be shifted*. Since we have $4p = -12$ (the coefficient of the linear term), we know $p = -3$ ($p < 0$) and the parabola opens downward. If the parabola were in standard position, the vertex would be at $(0, 0)$, the focus at $(0, -3)$ and the directrix a horizontal line at $y = 3$. But since the parabola is shifted 3 right and 2 up, we add 3 to all x-values and 2 to all y-values to locate the features of the shifted parabola. The vertex is at $(0 + 3, 0 + 2) = (3, 2)$. The focus is $(0 + 3, -3 + 2) = (3, -1)$ and the directrix is $y = 3 + 2 = 5$. Finally, the horizontal distance from the focus to the graph is $|2p| = 6$ units (since $|4p| = 12$), giving us the additional points $(-3, -1)$ and $(9, -1)$. The graph is shown.

Now try Exercises 49 through 60 ▶

In many cases, we need to construct the equation of the parabola when only partial information in known, as illustrated in Example 5.

EXAMPLE 5 ▶ **Constructing the Equation of a Parabola**

Find the equation of the parabola with vertex (4, 4) and focus (4, 1). Then graph the parabola using the equation and focal chord.

Solution ▶ As the vertex and focus are on a vertical line, we have a vertical parabola with general equation $(x \pm h)^2 = 4p(x \pm k)$. The distance p from vertex to focus is 3 units, and with the focus below the vertex, the parabola opens downward so $p = -3$. Using the focal chord, the horizontal distance from (4, 1) to the graph is $|2p| = |2(-3)| = 6$, giving points $(-2, 1)$ and $(10, 1)$. The vertex is shifted 4 units right and 4 units up from (0, 0), showing $h = 4$ and $k = 4$, and the equation of the parabola must be $(x - 4)^2 = -12(y - 4)$, with directrix $y = 7$. The graph is shown.

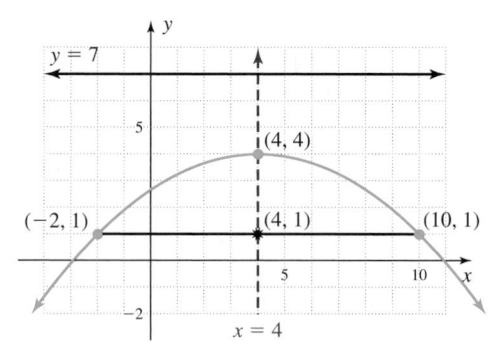

☑ **B. You've just learned how to identify and use the focus directrix form of the equation of a parabola**

Now try Exercises 61 through 76 ▶

C. Applications of the Analytic Parabola

Here is just one of the many ways the analytic definition of a parabola can be applied. There are several others in the exercise set.

EXAMPLE 6 ▶ **Locating the Focus of a Parabolic Receiver**

The diagram shows the cross section of a radio antenna dish. Engineers have located a point on the cross section that is 0.75 m above and 6 m to the right of the vertex. At what coordinates should the engineers build the focus of the antenna?

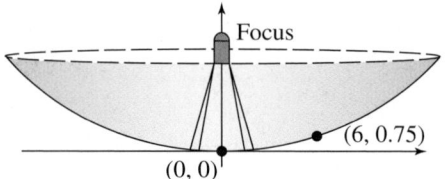

Solution ▶ By inspection we see this is a vertical parabola with center at (0, 0). This means its equation must be of the form $x^2 = 4py$. Because we know (6, 0.75) is a point on this graph, we can substitute (6, 0.75) in this equation and solve for p:

$$x^2 = 4py \qquad \text{equation for vertical parabola, vertex at (0, 0)}$$
$$(6)^2 = 4p(0.75) \qquad \text{substitute 6 for } x \text{ and 0.75 for } y$$
$$36 = 3p \qquad \text{simplify}$$
$$p = 12 \qquad \text{result}$$

☑ **C. You've just learned how to solve an application of the analytic parabola**

With $p = 12$, we see that the focus must be located at (0, 12), or 12 m directly above the vertex.

Now try Exercises 79 through 86 ▶

Note that in many cases, the focus of a parabolic dish may be taller than the rim of the dish.

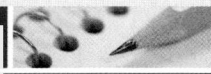

9.4 EXERCISES

▶ CONCEPTS AND VOCABULARY

Fill in the blank with the appropriate word or phrase. Carefully reread the section if needed.

1. The equation $x = ay^2 + by + c$ is that of a(n) _____ parabola, opening to the _____ if $a > 0$ and to the left if _____.

2. If point P is on the graph of a parabola with directrix D, the distance from P to line D is equal to the distance between P and the _____ of the parabola.

3. Given $y^2 = 4px$, the focus is at _____ and the equation of the directrix is _____.

4. Given $x^2 = -16y$, the value of p is _____ and the coordinates of the focus are _____.

5. Discuss/explain how to find the vertex, directrix, and focus from the equation $(x - h)^2 = 4p(y - k)$.

6. If a horizontal parabola has a vertex of $(2, -3)$ with $a > 0$, what can you say about the y-intercepts? Will the graph always have an x-intercept? Explain.

▶ DEVELOPING YOUR SKILLS

Find the x- and y-intercepts (if they exist) and the vertex of the parabola. Then sketch the graph by using symmetry and a few additional points or completing the square and shifting a parent function. Scale the axes as needed to comfortably fit the graph and state the domain and range.

7. $y = x^2 - 2x - 3$ 8. $y = x^2 + 6x + 5$

9. $y = 2x^2 - 8x - 10$ 10. $y = 3x^2 + 12x - 15$

11. $y = 2x^2 + 5x - 7$ 12. $y = 2x^2 - 7x + 3$

Find the x- and y-intercepts (if they exist) and the vertex of the graph. Then sketch the graph using symmetry and a few additional points (scale the axes as needed). Finally, state the domain and range of the relation.

13. $x = y^2 - 2y - 3$ 14. $x = y^2 - 4y - 12$

15. $x = -y^2 + 6y + 7$ 16. $x = -y^2 + 8y - 12$

17. $x = -y^2 + 8y - 16$ 18. $x = -y^2 + 6y - 9$

Sketch using symmetry and shifts of a basic function. Be sure to find the x- and y-intercepts (if they exist) and the vertex of the graph, then state the domain and range of the relation.

19. $x = y^2 - 6y$ 20. $x = y^2 - 8y$

21. $x = y^2 - 4$ 22. $x = y^2 - 9$

23. $x = -y^2 + 2y - 1$ 24. $x = -y^2 + 4y - 4$

25. $x = y^2 + y - 6$ 26. $x = y^2 + 4y - 5$

27. $x = y^2 - 10y + 4$ 28. $x = y^2 + 12y - 5$

29. $x = 3 - 8y - 2y^2$ 30. $x = 2 - 12y + 3y^2$

31. $y = (x - 2)^2 + 3$ 32. $y = (x + 2)^2 - 4$

33. $x = (y - 3)^2 + 2$ 34. $x = (y + 1)^2 - 4$

35. $x = 2(y - 3)^2 + 1$ 36. $x = -2(y + 3)^2 - 5$

Find the vertex, focus, and directrix for the parabolas defined by the equations given, then use this information to sketch a complete graph (illustrate and name these features). For Exercises 43 to 60, also include the focal chord.

37. $x^2 = 8y$ 38. $x^2 = 16y$

39. $x^2 = -24y$ 40. $x^2 = -20y$

41. $x^2 = 6y$ 42. $x^2 = 18y$

43. $y^2 = -4x$ 44. $y^2 = -12x$

45. $y^2 = 18x$ 46. $y^2 = 20x$

47. $y^2 = -10x$ 48. $y^2 = -14x$

49. $x^2 - 8x - 8y + 16 = 0$

50. $x^2 - 10x - 12y + 25 = 0$

51. $x^2 - 14x - 24y + 1 = 0$

52. $x^2 - 10x - 12y + 1 = 0$

53. $3x^2 - 24x - 12y + 12 = 0$

54. $2x^2 - 8x - 16y - 24 = 0$

55. $y^2 - 12y - 20x + 36 = 0$

56. $y^2 - 6y - 16x + 9 = 0$

57. $y^2 - 6y + 4x + 1 = 0$

58. $y^2 - 2y + 8x + 9 = 0$

59. $2y^2 - 20y + 8x + 2 = 0$

60. $3y^2 - 18y + 12x + 3 = 0$

For Exercises 61–72, find the equation of the parabola in standard form that satisfies the conditions given:

61. focus: $(0, 2)$
    directrix: $y = -2$

62. focus: $(0, -3)$
    directrix: $y = 3$

63. focus: $(4, 0)$
    directrix: $x = -4$

64. focus: $(-3, 0)$
    directrix: $x = 3$

65. focus: $(0, -5)$
    directrix: $y = 5$

66. focus: $(5, 0)$
    directrix: $x = -5$

67. vertex: $(2, -2)$
    focus: $(-1, -2)$

68. vertex: $(4, 1)$
    focus: $(1, 1)$

69. vertex: $(4, -7)$
    focus: $(4, -4)$

70. vertex: $(-3, -4)$
    focus: $(-3, -1)$

71. focus: $(3, 4)$
    directrix: $y = 0$

72. focus: $(-1, 2)$
    directrix: $x = -5$

For the graphs in Exercises 73–76, only two of the following four features are displayed: vertex, focus, directrix, and endpoints of the focal chord. Find the remaining two features and the equation of the parabola.

73.
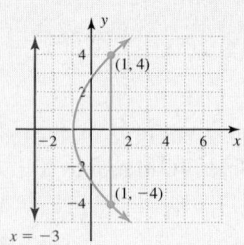

74.
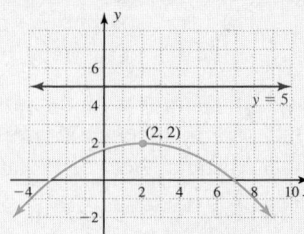

75.
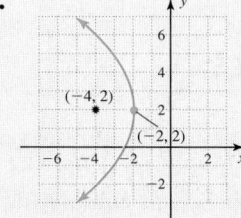

76.
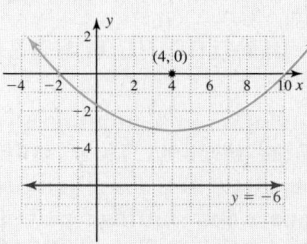

► **WORKING WITH FORMULAS**

77. The area of a right parabolic segment: $A = \frac{2}{3}ab$

A *right parabolic segment* is that part of a parabola formed by a line perpendicular to its axis, which cuts the parabola. The area of this segment is given by the formula shown, where b is the length of the chord cutting the parabola and a is the perpendicular distance from the vertex to this chord. What is the area of the parabolic segment shown in the figure?

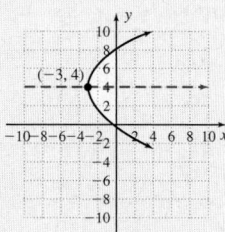

78. The arc length of a right parabolic segment:

$$\frac{1}{2}\sqrt{b^2 + 16a^2} + \frac{b^2}{8a}\ln\left(\frac{4a + \sqrt{b^2 + 16a^2}}{b}\right)$$

Although a fairly simple concept, finding the length of the parabolic arc traversed by a projectile requires a good deal of computation. To find the length of the arc ABC shown, we use the formula given where a is the maximum height attained by the projectile, b is the horizontal distance it traveled, and "ln" represents the natural log function. Suppose a baseball thrown from centerfield reaches a maximum height of 20 ft and traverses an arc length of 340 ft. Will the ball reach the catcher 310 ft away without bouncing?

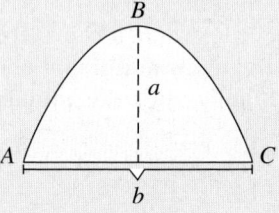

► **APPLICATIONS**

79. Parabolic car headlights: The cross section of a typical car headlight can be modeled by an equation similar to $25x = 16y^2$, where x and y are in inches and $x \in [0, 4]$. Use this information to graph the relation for the indicated domain.

80. Parabolic flashlights: The cross section of a typical flashlight reflector can be modeled by an equation similar to $4x = y^2$, where x and y are in centimeters and $x \in [0, 2.25]$. Use this information to graph the relation for the indicated domain.

Exercise 80

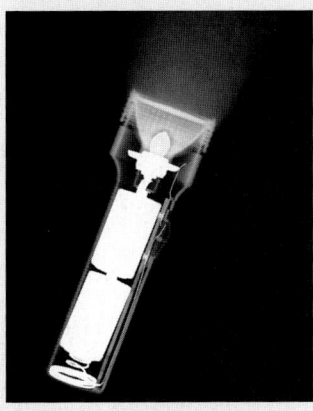

81. **Parabolic sound receivers:** Sound technicians at professional sports events often use parabolic receivers as they move along the sidelines. If a two-dimensional cross section of the receiver is modeled by the equation $y^2 = 54x$, and is 36 in. in *diameter,* how deep is the parabolic receiver? What is the location of the focus? [*Hint:* Graph the parabola on the coordinate grid (scale the axes).]

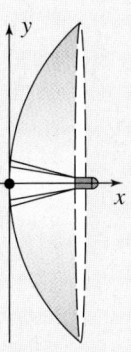

82. **Parabolic sound receivers:** Private investigators will often use a smaller and less expensive parabolic receiver (see Exercise 81) to gather information for their clients. If a two-dimensional cross section of the receiver is modeled by the equation $y^2 = 24x$, and the receiver is 12 in. in *diameter,* how deep is the parabolic dish? What is the location of the focus?

83. **Parabolic radio wave receivers:** The program known as S.E.T.I. (Search for Extra-Terrestrial Intelligence) identifies a group of scientists using radio

telescopes to look for radio signals from possible intelligent species in outer space. The radio

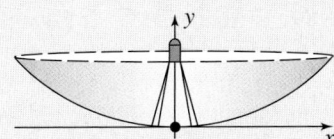

telescopes are actually parabolic dishes that vary in size from a few feet to hundreds of feet in diameter. If a particular radio telescope is 100 ft in diameter and has a cross section modeled by the equation $x^2 = 167y$, how deep is the parabolic dish? What is the location of the focus? [*Hint:* Graph the parabola on the coordinate grid (scale the axes).]

84. **Solar furnace:** Another form of technology that uses a parabolic dish is called a solar furnace. In general, the rays of the Sun are reflected by the dish and concentrated at the focus, producing extremely high temperatures. Suppose the dish of one of these parabolic reflectors had a 30-ft diameter and a cross section modeled by the equation $x^2 = 50y$. How deep is the parabolic dish? What is the location of the focus?

85. The reflector of a large, commercial flashlight has the shape of a parabolic dish, with a diameter of 10 cm and a depth of 5 cm. What equation will the engineers and technicians use for the manufacture of the dish? How far from the vertex (the lowest point of the dish) will the bulb be placed? (*Hint:* Analyze the information using a coordinate system.)

86. The reflector of an industrial spot light has the shape of a parabolic dish with a diameter of 120 cm. What is the depth of the dish if the correct placement of the bulb is 11.25 cm above the vertex (the lowest point of the dish)? What equation will the engineers and technicians use for the manufacture of the dish? (*Hint:* Analyze the information using a coordinate system.)

▶ **EXTENDING THE CONCEPT**

87. In a study of quadratic graphs from the equation $y = ax^2 + bx + c$, no mention is made of a parabola's focus and directrix. Generally, when $a \geq 1$, the focus of a parabola is very near its vertex. Complete the square of the function $y = 2x^2 - 8x$ and write the result in the form $(x - h)^2 = 4p(y - k)$. What is the value of p? What are the coordinates of the vertex?

88. Like the ellipse and hyperbola, the focal chord of a parabola (also called the **latus rectum**)

can be used to help sketch its graph. From our earlier work, we know the endpoints of the focal chord are $2p$ units from the focus. Write the equation $-12y + 15 = x^2 - 6x$ in the form $4p(y \pm k) = (x \pm h)^2$, and use the endpoints of the focal chord to help graph the parabola.

Exercise 88

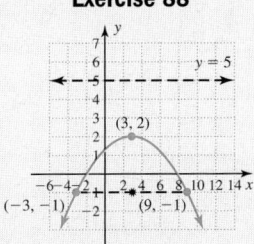

89. In Exercise 77, a formula was given for the area of a right parabolic segment. The area of an *oblique* parabolic segment (the line segment cutting the parabola is *not* perpendicular to the axis) is more complex, as it involves locating the point where a line parallel to this

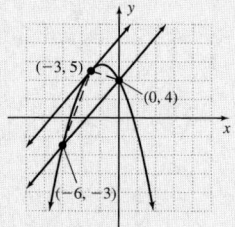

segment is tangent (touches at only one point) to the parabola. The formula is $A = \frac{4}{3}T$, where T represents the area of the triangle formed by the endpoints of the segment and this point of tangency. What is the area of the parabolic segment shown (assuming the lines are parallel)? See Section 8.5, Exercises 46 and 47 and Section 8.8, Example 3.

▶ **MAINTAINING YOUR SKILLS**

90. (3.3/3.4) Use the function
$$f(x) = x^5 + 2x^4 + 17x^3 + 34x^2 - 18x - 36$$
to comment and give illustrations of the tools available for working with polynomials: (a) synthetic division, (b) rational roots theorem, (c) the remainder and factor theorems, (d) the test for $x = -1$ and $x = 1$, (e) the upper/lower bounds property, (f) Descartes' rule of signs, and (g) roots of multiplicity (bounces, cuts, alternating intervals).

91. (1.6) Find all roots (real and complex) to the equation $x^6 - 64 = 0$. (*Hint:* Begin by factoring the expression as the difference of two perfect squares.)

92. (6.6) Find all real solutions to $\sec \theta = 1.1547$ (round to the nearest degree).

93. (6.5) The graph shown displays the variation in daylight from an average of 12 hours per day (i.e., the maximum is 15 hours and the minimum is 9). Use the graph to *approximate* the number of days in a year there are 10.5 or less hours of daylight. Answers may vary.

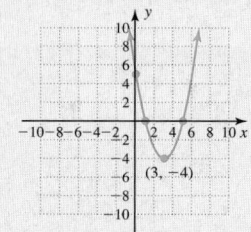

MID-CHAPTER CHECK

Sketch the graph of each conic section.

1. $(x - 4)^2 + (y + 3)^2 = 9$

2. $x^2 + y^2 - 10x + 4y + 4 = 0$

3. $\dfrac{(x - 2)^2}{16} + \dfrac{(y + 3)^2}{1} = 1$

4. $9x^2 + 4y^2 + 18x - 24y + 9 = 0$

5. $\dfrac{(x + 3)^2}{9} - \dfrac{(y - 4)^2}{4} = 1$

6. $9x^2 - 4y^2 + 18x - 24y - 63 = 0$

7. Find the equation of each relation and state its domain and range.

a.

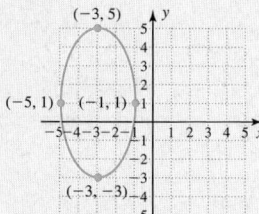

b.

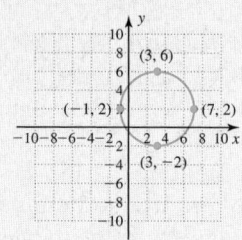

c.

8. Solve the following system of inequalities by graphing.
$$\begin{cases} \dfrac{x^2}{100} + \dfrac{y^2}{25} \le 1 \\ x^2 + (y - 4)^2 \le 36 \end{cases}$$

9. Find the equation of the ellipse (in standard form) if the vertices are $(-4, 0)$ and $(4, 0)$ and the distance between the foci is $4\sqrt{3}$ units.

10. The radio signal emanating from a tall radio tower spreads evenly in all directions with a range of 50 mi. If the tower is located at coordinates $(20, 30)$ and my home is at coordinates $(10, 78)$, will I be able to pick up this station on my home radio? Assume coordinates are in miles.

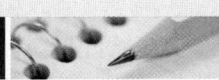

REINFORCING BASIC CONCEPTS

Ellipses and Hyperbolas with Rational/Irrational Values of *a* and *b*

Using the process known as completing the square, we were able to convert from the polynomial form of a conic section to the standard form. However, for some equations, values of *a* and *b* are somewhat difficult to identify, since the coefficients are not factors. Consider the equation $20x^2 + 120x + 27y^2 - 54y + 192 = 0$ the equation of an ellipse.

$$20x^2 + 120x + 27y^2 - 54y + 192 = 0 \qquad \text{original equation}$$

$$20(x^2 + 6x + \underline{\quad}) + 27(y^2 - 2y + \underline{\quad}) = -192 \qquad \text{subtract 192, begin process}$$

$$20(x^2 + 6x + 9) + 27(y^2 - 2y + 1) = -192 + 27 + 180 \qquad \text{complete the square in } x \text{ and } y$$

$$20(x + 3)^2 + 27(y - 1)^2 = 15 \qquad \text{factor and simplify}$$

$$\frac{4(x + 3)^2}{3} + \frac{9(y - 1)^2}{5} = 1 \qquad \text{standard form}$$

Unfortunately, we cannot easily identify the values of *a* and *b*, since the coefficients of each binomial square were not "1." In these cases, we can write the equation in standard form by using a simple property of fractions—the numerator and denominator of any fraction can be divided by the same quantity to obtain an equivalent fraction. Although the result may look odd, it can nevertheless be applied here, giving a result of $\frac{(x + 3)^2}{3/4} + \frac{(y - 1)^2}{5/9} = 1$. We can now identify *a* and *b* by writing these denominators in squared form, which gives the following expression: $\frac{(x + 3)^2}{\left(\frac{\sqrt{3}}{2}\right)^2} + \frac{(y - 1)^2}{\left(\frac{\sqrt{5}}{3}\right)^2} = 1$. The values of *a* and *b* are now easily seen as $a \approx 0.866$ and $b \approx 0.745$. Use this idea to complete the following exercises.

Exercise 1: Identify the values of *a* and *b* by writing the equation $100x^2 - 400x - 18y^2 - 108y + 230 = 0$ in standard form.

Exercise 2: Identify the values of *a* and *b* by writing the equation $28x^2 - 56x + 48y^2 + 192y + 195 = 0$ in standard form.

Exercise 3: Write the equation in standard form, then identify the values of *a* and *b* and use them to graph the ellipse.

$$\frac{4(x + 3)^2}{49} + \frac{25(y - 1)^2}{36} = 1$$

Exercise 4: Write the equation in standard form, then identify the values of *a* and *b* and use them to graph the hyperbola.

$$\frac{9(x + 3)^2}{80} - \frac{4(y - 1)^2}{81} = 1$$

9.5 | Nonlinear Systems of Equations and Inequalities

Learning Objectives

In Section 9.5 you will learn how to:

☐ **A.** Visualize possible solutions

☐ **B.** Solve nonlinear systems using substitution

☐ **C.** Solve nonlinear systems using elimination

☐ **D.** Solve nonlinear systems of inequalities

☐ **E.** Solve applications of nonlinear systems

Equations where the variables have exponents other than 1 or that are transcendental (like logarithmic and exponential equations) are all nonlinear equations. A nonlinear system of equations has at least one nonlinear equation, and these occur in a great variety.

A. Possible Solutions for a Nonlinear System

When solving nonlinear systems, it is often helpful to *visualize* the graphs of each equation in the system. This can help determine the number of possible intersections and further assist the solution process.

EXAMPLE 1 ▶   **Visualizing the Number of Possible Intersections for the Graphs in a System**

Identify each equation in the system as the equation of a line, parabola, circle, ellipse, or hyperbola. Then determine the number of solutions possible by considering the different ways the graphs might intersect: $\begin{cases} 4x^2 + 9y^2 = 36 \\ 2x + 3y = 6 \end{cases}$. Finally, solve the system by graphing.

Solution ▶   The first equation contains a sum of second-degree terms with unequal coefficients, and we recognize this as a central ellipse. The second equation is obviously linear. This means the system may have no solution, one solution, or two solutions, as shown in Figure 9.31. The graph of the system is shown in Figure 9.32 and the two points of intersection appear to be $(3, 0)$ and $(0, 2)$. After checking these in the original equations we find that both are solutions to the system.

Figure 9.31

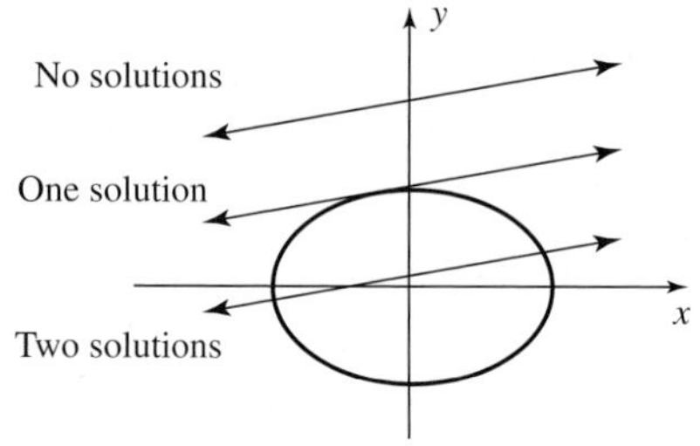

No solutions

One solution

Two solutions

Figure 9.32

$(0, 2)$

$(-3, 0)$ $(3, 0)$

$(0, -2)$

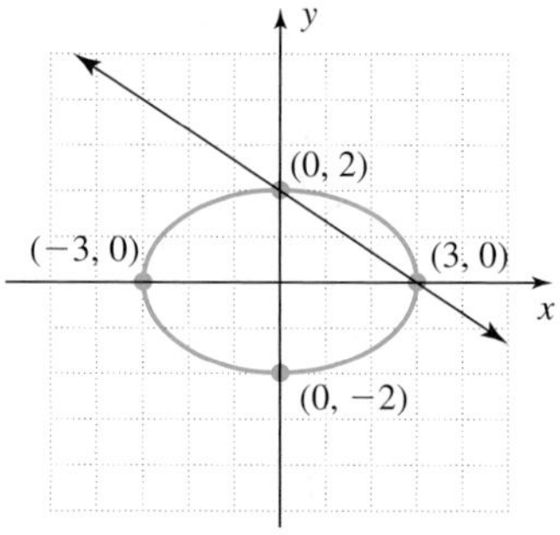

☑ **A.** You've just learned how to visualize possible solutions

Now try Exercises 7 through 12 ▶

B. Solving a Nonlinear System by Substitution

Since graphical methods at best offer an estimate for the solution (points of intersection may not have integer values), we more often turn to algebraic methods. Recall the substitution method involves solving one of the equations for a variable or expression that can be substituted in the other equation to eliminate one of the variables.

EXAMPLE 2 ▶   **Solving a Nonlinear System Using Substitution**

Solve the system using substitution: $\begin{cases} y = x^2 - 2x - 3 \\ 2x - y = 7 \end{cases}$.

Solution ▶   The first equation contains a single second-degree term in x, and is the equation of a parabola. The second equation is linear. Since the first equation is already written with y in terms of x, we can substitute $x^2 - 2x - 3$ for y in the second equation to solve.

$$2x - y = 7 \qquad \text{second equation}$$
$$2x - (x^2 - 2x - 3) = 7 \qquad \text{substitute } x^2 - 2x - 3 \text{ for } y$$
$$2x - x^2 + 2x + 3 = 7 \qquad \text{distribute}$$
$$-x^2 + 4x + 3 = 7 \qquad \text{simplify}$$
$$x^2 - 4x + 4 = 0 \qquad \text{set equal to zero}$$
$$(x - 2)^2 = 0 \qquad \text{factor}$$

We find that $x = 2$ is a repeated root.

☑ **B.** You've just learned how to solve nonlinear systems using substitution

Since the second equation is simpler than the first, we substitute 2 for x in this equation and find $y = -3$. The system has only one (repeated) solution at $(2, -3)$, which is shown in the figure.

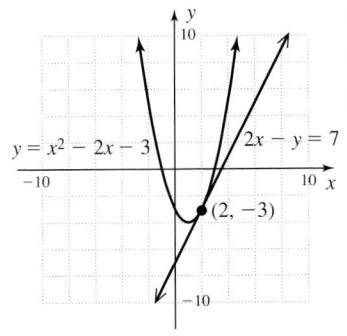

Now try Exercises 13 through 18 ▶

C. Solving Nonlinear Systems by Elimination

When both equations in the system have second-degree terms with like variables, it is generally easier to use the elimination method, rather than substitution. As in Chapter 8, watch for systems that have no solutions.

EXAMPLE 3 ▶ **Solving a Nonlinear System Using Elimination**

Solve the system using elimination: $\begin{cases} 2y^2 - 5x^2 = 13 \\ 3x^2 + 4y^2 = 39 \end{cases}$.

Solution ▶ The first equation contains a *difference of second-degree terms* and is the equation of a central hyperbola. The second has a *sum of second-degree terms* with unequal coefficients, and represents a central ellipse. By mentally visualizing the possibilities, there could be zero, one, two, three, or four points of intersection (see Example 1). However, with both centered at $(0, 0)$, we find there can only be zero, two, or four solutions. After writing the system so that x- and y-terms are in the same order, we find that multiplying the first equation by -2 will help eliminate the variable y:

$$\begin{cases} 10x^2 - 4y^2 = -26 \quad \text{rewrite first equation and multiply by } -2 \\ \underline{3x^2 + 4y^2 = 39} \quad \text{original second equation} \end{cases}$$

$$\begin{array}{ll} 13x^2 + 0 \;=\; 13 & \text{add} \\ \qquad\quad x^2 = 1 & \text{divide by 13} \\ x = -1 \quad \text{or} \quad x = 1 & \text{square root property} \end{array}$$

Substituting $x = 1$ and $x = -1$ into the second equation we obtain:

$$\begin{array}{ll}
3(1)^2 + 4y^2 = 39 & \qquad 3(-1)^2 + 4y^2 = 39 \\
3 + 4y^2 = 39 & \qquad\quad 3 + 4y^2 = 39 \\
4y^2 = 36 & \qquad\qquad\quad 4y^2 = 36 \\
y^2 = 9 & \qquad\qquad\quad\; y^2 = 9 \\
y = \pm 3 & \qquad\qquad\quad\; y = \pm 3
\end{array}$$

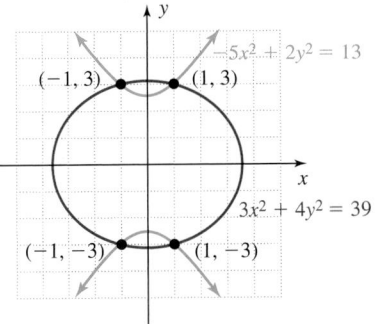

Since -1 and 1 each generated *two outputs*, there are a total of *four* ordered pair solutions: $(1, -3), (1, 3), (-1, -3),$ and $(-1, 3)$. Once again, the graph shown supports these calculations.

Now try Exercises 19 through 24 ▶

Nonlinear systems may involve other relations as well, including power, polynomial, logarithmic, or exponential functions. These are solved using the same methods.

EXAMPLE 4 ▶ **Solving a System of Logarithmic Equations**

Solve the system using the method of your choice: $\begin{cases} y = -\log(x + 7) + 2 \\ y = \log(x + 4) + 1 \end{cases}$.

Solution ▶ Since both equations have y written in terms of x, substitution appears to be the better choice. The result is a logarithmic equation, which we can solve using the techniques from Chapter 4.

$$\log(x + 4) + 1 = -\log(x + 7) + 2 \quad \text{substitute } \log(x+4) + 1 \text{ for } y \text{ in first equation}$$
$$\log(x + 4) + \log(x + 7) = 1 \quad \text{add } \log(x + 7); \text{ subtract 1}$$
$$\log(x + 4)(x + 7) = 1 \quad \text{product property of logarithms}$$
$$(x + 4)(x + 7) = 10^1 \quad \text{exponential form}$$
$$x^2 + 11x + 18 = 0 \quad \text{eliminate parentheses and set equal to zero}$$
$$(x + 9)(x + 2) = 0 \quad \text{factor}$$
$$x + 9 = 0 \quad \text{or} \quad x + 2 = 0 \quad \text{zero factor theorem}$$
$$x = -9 \quad \text{or} \quad x = -2 \quad \text{possible solutions}$$

By inspection, we see that $x = -9$ is not a solution, since $\log(-9 + 4)$ and $-\log(-9 + 7)$ are not real numbers. Substituting -2 for x in the second equation we find one form of the (exact) solution is $(-2, \log 2 + 1)$. If we substitute -2 for x in the first equation the exact solution is $(-2, -\log 5 + 2)$. Use a calculator to verify the answers are equivalent and approximately $(-2, 1.3)$.

WORTHY OF NOTE

Since the domain of the functions are $x > -7$ and $x > -4$ respectively, $x = -9$ cannot be a solution, as illustrated in the graphical representation of the system.

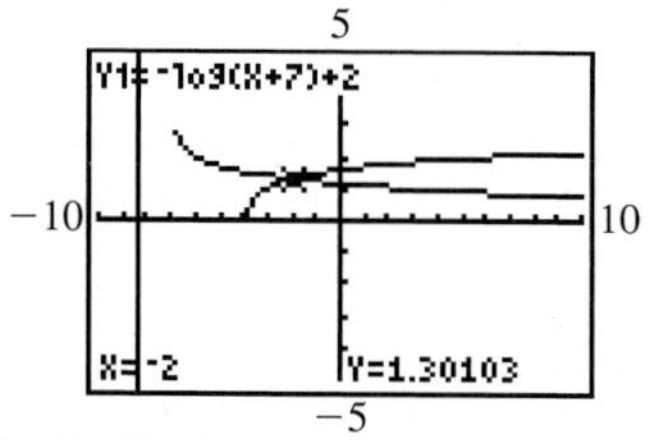

☑ **C.** You've just learned how to solve nonlinear systems using elimination

Now try Exercises 25 through 36 ▶

For practice solving more complex systems using a graphing calculator, **see Exercises 37 through 42.**

D. Solving Systems of Nonlinear Inequalities

As with our previous work with inequalities in two variables, nonlinear inequalities can be solved by graphing the boundary given by the related equation, and checking the regions that result using a test point. For example, the inequality $x^2 + 4y^2 < 25$ is solved by graphing $x^2 + 4y^2 = 25$ [a central ellipse with vertices at $(-5, 0)$ and $(5, 0)$], deciding if the boundary is included or excluded (in this case it is not), and using a test point from either the "outside" or "inside" region formed. The test point $(0, 0)$ results in a true statement since $(0)^2 + 4(0)^2 < 25$, so the inside of the ellipse is shaded to indicate the solution region (Figure 9.33). For a *system* of nonlinear inequalities, we identify regions where the solution set for each inequality overlap, paying special attention to points of intersection.

Figure 9.33

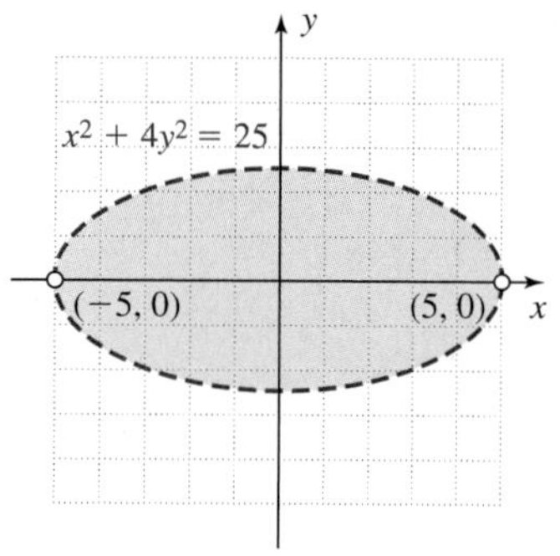

EXAMPLE 5 ▶ **Solving a System of Nonlinear Inequalities**

Solve the system: $\begin{cases} x^2 + 4y^2 < 25 \\ -x + 4y \geq 5 \end{cases}$.

Solution ▶ We recognize the first inequality from Figure 9.33, an ellipse with vertices at $(-5, 0)$ and $(5, 0)$, and a solution region in the interior. The second inequality is linear and after solving for x in the related equation we use a substitution to find points of intersection (if they exist). For $-x + 4y = 5$, we have $x = 4y - 5$ and substitute $4y - 5$ for x in $x^2 + 4y^2 = 25$:

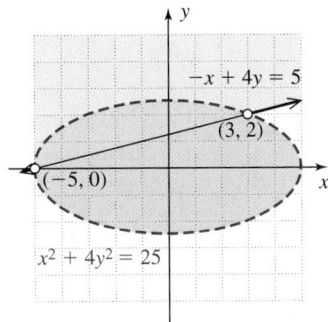

$$
\begin{aligned}
x^2 + 4y^2 &= 25 && \text{given} \\
(4y - 5)^2 + 4y^2 &= 25 && \text{substitute } 4y - 5 \text{ for } x \\
20y^2 - 40y + 25 &= 25 && \text{expand and simplify} \\
y^2 - 2y &= 0 && \text{subtract 25; divide by 20} \\
y(y - 2) &= 0 && \text{factor} \\
y = 0 \quad \text{or} \quad y &= 2 && \text{result}
\end{aligned}
$$

Back-substitution shows the graphs intersect at $(-5, 0)$ and $(3, 2)$. Graphing a line through these points and using $(0, 0)$ as a test point shows the upper half plane is the solution region for the linear inequality [$-(0) + 4(0) \geq 5$ is *false*]. The overlapping (solution) region for *both* inequalities is the elliptical sector shown. Note the points of intersection are graphed using "open dots," (see figure) since points on the graph of the ellipse are excluded from the solution set.

☑ **D.** You've just learned how to solve nonlinear systems of inequalities

Now try Exercises 43 through 50 ▶

E. Solving Applications of Nonlinear Systems

In the business world, a fast growing company can often reduce the average price of its products using what are called the **economies of scale.** These would include the ability to buy necessary materials in larger quantities, integrating new technology into the production process, and other means. However, there are also countering forces called the **diseconomies of scale,** which may include the need to hire additional employees, rent more production space, and the like.

EXAMPLE 6 ▶ **Solving an Application of Nonlinear Systems—Economies of Scale**

Suppose the cost to produce a new and inexpensive shoe made from molded plastic is modeled by the function $C(x) = x^2 - 5x + 18$, where $C(x)$ represents the cost to produce x thousand of these shoes. The revenue from the sales of these shoes is modeled by $R(x) = -x^2 + 10x - 4$. Use a break-even analysis to find the quantity of sales that will cause the company to break even.

Solution ▶ Essentially we are asked to solve the system formed by the two equations:
$\begin{cases} C(x) = x^2 - 5x + 18 \\ R(x) = -x^2 + 10x - 4 \end{cases}$. Since we want to know the point where the company breaks even, we set $C(x) = R(x)$ and solve.

$$
\begin{aligned}
C(x) &= R(x) && \text{break even: cost} = \text{revenue} \\
x^2 - 5x + 18 &= -x^2 + 10x - 4 && \text{substitute for } C(x) \text{ and } R(x) \\
2x^2 - 15x + 22 &= 0 && \text{set equal to zero} \\
(2x - 11)(x - 2) &= 0 && \text{factored form} \\
x = \frac{11}{2} \text{ or } x &= 2 && \text{result}
\end{aligned}
$$

☑ **E.** You've just learned how to solve applications of nonlinear systems

With x in thousands, it appears the company will break even if either 2000 shoes or 5500 shoes are made and sold.

Now try Exercises 53 through 56 ▶

For additional applications of nonlinear systems, see **Exercises 57 through 62.**

9.5 EXERCISES

▶ CONCEPTS AND VOCABULARY

Fill in each blank with the appropriate word or phrase. Carefully reread the section if needed.

1. Draw sketches showing the different ways each pair of relations can intersect and give one, two, three, and/or four points of intersection. If a given number of intersections is not possible, so state.

   a. circle and line   **b.** parabola and line

   c. circle and parabola   **d.** circle and hyperbola

   e. hyperbola and ellipse   **f.** circle and ellipse

2. By inspection only, identify the systems having *no solutions* and justify your choices.

   a. $\begin{cases} y^2 - x^2 = 16 \\ x^2 + y^2 = 9 \end{cases}$ **b.** $\begin{cases} y = x^2 + 4 \\ x^2 + 4y^2 = 4 \end{cases}$

   c. $\begin{cases} y = x + 1 \\ 3x^2 + 4y^2 = 12 \end{cases}$

3. The solution to a system of nonlinear inequalities is a(n) _____ of the plane where the _____ for each individual inequality overlap.

4. When both equations in the system have at least one _____ -degree term, it is generally easier to use the _____ method to find a solution.

5. Suppose a nonlinear system contained a central hyperbola and an exponential function. Are three solutions possible? Are four solutions possible? Explain/Discuss.

6. Solve the system twice, once using elimination, then again using substitution. Compare/Contrast each process and comment on which is more efficient in this case: $\begin{cases} 4x^2 + y^2 = 25 \\ 2x^2 + y = 5 \end{cases}$.

▶ DEVELOPING YOUR SKILLS

Identify each equation in the system as that of a line, parabola, circle, ellipse, or hyperbola, and solve the system by graphing.

7. $\begin{cases} x^2 + y = 6 \\ x + y = 4 \end{cases}$ 8. $\begin{cases} -2x + y = 4 \\ 4x^2 + y^2 = 16 \end{cases}$

9. $\begin{cases} y - x^2 = -1 \\ 4x^2 + y^2 = 100 \end{cases}$ 10. $\begin{cases} x^2 + y^2 = 25 \\ x^2 + y = 13 \end{cases}$

11. $\begin{cases} x^2 - y^2 = 9 \\ x^2 + y^2 = 41 \end{cases}$ 12. $\begin{cases} 4x^2 - y^2 = 36 \\ y^2 + 9x^2 = 289 \end{cases}$

Solve using substitution. [*Hint:* Substitute for x^2 (not y) in Exercises 17 and 18.]

13. $\begin{cases} x^2 + y^2 = 25 \\ y - x = 1 \end{cases}$ 14. $\begin{cases} x + 7y = 50 \\ x^2 + y^2 = 100 \end{cases}$

15. $\begin{cases} x^2 + 4y^2 = 25 \\ x + 2y = 7 \end{cases}$ 16. $\begin{cases} x^2 - 2y^2 = 8 \\ x + y = 6 \end{cases}$

17. $\begin{cases} x^2 + y = 13 \\ 9x^2 - y^2 = 81 \end{cases}$ 18. $\begin{cases} y - x^2 = -10 \\ 4x^2 + y^2 = 100 \end{cases}$

Solve using elimination.

19. $\begin{cases} x^2 + y^2 = 25 \\ 2x^2 - 3y^2 = 5 \end{cases}$ 20. $\begin{cases} y^2 - x^2 = 12 \\ x^2 + y^2 = 20 \end{cases}$

21. $\begin{cases} x^2 - y = 4 \\ x^2 - y^2 = 16 \end{cases}$ 22. $\begin{cases} 4x^2 + y^2 = 13 \\ x^2 + y^2 = 1 \end{cases}$

23. $\begin{cases} 5x^2 - 2y^2 = 75 \\ 2x^2 + 3y^2 = 125 \end{cases}$ 24. $\begin{cases} 3x^2 - 7y^2 = 20 \\ 4x^2 + 9y^2 = 45 \end{cases}$

Solve using the method of your choice. Answer in exact form.

25. $\begin{cases} y - 5 = \log x \\ y = 6 - \log(x - 3) \end{cases}$ 26. $\begin{cases} y = \log(x + 4) + 1 \\ y - 2 = -\log(x + 7) \end{cases}$

27. $\begin{cases} y = \ln(x^2) + 1 \\ y - 1 = \ln(x + 12) \end{cases}$ 28. $\begin{cases} \log(x + 1.1) = y + 3 \\ y + 4 = \log(x^2) \end{cases}$

29. $\begin{cases} y - 9 = e^{2x} \\ 3 = y - 7e^x \end{cases}$ 30. $\begin{cases} y - 2e^{2x} = 5 \\ y - 1 = 6e^x \end{cases}$

31. $\begin{cases} y = 4^{x+3} \\ y - 2^{x^2+3x} = 0 \end{cases}$ 32. $\begin{cases} y - 3^{x^2+2x} = 0 \\ y = 9^{x+2} \end{cases}$

33. $\begin{cases} x^3 - y = 2x \\ y - 5x = -6 \end{cases}$

34. $\begin{cases} y - x^3 = -2 \\ y + 4 = 3x \end{cases}$

35. $\begin{cases} x^2 - 6x = y - 4 \\ y - 2x = -8 \end{cases}$

36. $\begin{cases} y + x = -2 \\ y + 4x = x^2 \end{cases}$

Solve each system using a graphing calculator. Round solutions to hundredths (as needed).

37. $\begin{cases} x^2 + y^2 = 34 \\ y^2 + (x - 3)^2 = 25 \end{cases}$

38. $\begin{cases} 5x^2 + 5y^2 = 40 \\ y + 2x = x^2 - 6 \end{cases}$

39. $\begin{cases} y = 2^x - 3 \\ y + 2x^2 = 9 \end{cases}$

40. $\begin{cases} y = -2\log(x + 8) \\ y + x^3 = 4x - 2 \end{cases}$

41. $\begin{cases} y = \dfrac{1}{(x - 3)^2} + 2 \\ (x - 3)^2 + y^2 = 10 \end{cases}$

42. $\begin{cases} y^2 + x^2 = 5 \\ y = \dfrac{1}{x - 1} - 2 \end{cases}$

Solve each system of inequalities

43. $\begin{cases} y - x^2 \geq 1 \\ x + y \leq 3 \end{cases}$

44. $\begin{cases} x^2 + y^2 \leq 25 \\ x + 2y \leq 5 \end{cases}$

45. $\begin{cases} x^2 + y^2 > 16 \\ x^2 + y^2 \leq 64 \end{cases}$

46. $\begin{cases} y + 4 \geq x^2 \\ x^2 + y^2 \leq 34 \end{cases}$

47. $\begin{cases} y - x^2 \leq -16 \\ y^2 + x^2 < 9 \end{cases}$

48. $\begin{cases} x^2 + y^2 \leq 16 \\ x + 2y > 10 \end{cases}$

49. $\begin{cases} y^2 + x^2 \leq 25 \\ |x| - 1 > -y \end{cases}$

50. $\begin{cases} y^2 + x^2 \leq 4 \\ x + y < 4 \end{cases}$

▶ WORKING WITH FORMULAS

51. **Tunnel clearance:** $h = b\sqrt{1 - \left(\dfrac{d}{a}\right)^2}$

The maximum rectangular clearance allowed by an elliptical tunnel can be found using the formula shown, where $\begin{cases} \dfrac{x^2}{a^2} + \dfrac{y^2}{b^2} = 1 \end{cases}$ models the tunnel's elliptical cross section and h is the height of the

tunnel at a distance d from the center. If $a = 50$ and $b = 30$, find the maximum clearance at distances of $d = 20, 30,$ and 40 ft from center.

52. **Manufacturing cylindrical vents:** $\begin{cases} A = 2\pi rh \\ V = \pi r^2 h \end{cases}$

In the manufacture of cylindrical vents, a rectangular piece of sheet metal is rolled, riveted, and sealed to form the vent. The radius and height required to form a vent with a specified volume, using a piece of sheet metal with a given area, can be found by solving the system shown. Use the system to find the radius and height if the volume required is 4071 cm³ and the area of the rectangular piece is 2714 cm².

▶ APPLICATIONS

Solve the following applications involving economies of scale.

53. **Revenue from sales:** Early in 2008, The Tata Company (India) unveiled the Tata Nano, the world's most inexpensive car. With its low price and 54 miles per gallon, the car may prove to be very popular. *Assume* the cost to produce these cars is modeled by the function $C(x) = 2.5x^2 - 120x + 3500$, where $C(x)$ represents the cost to produce x thousand cars. Suppose the revenue from the sales of these cars is

modeled by $R(x) = -2x^2 + 180x - 500$. Use a break-even analysis to find the quantity of sales (to the nearest hundred) that will cause the company to break even.

54. **Document reproduction:** In a world of technology, document reproduction has become a billion dollar business. With very stiff competition, the price of a single black and white copy has varied greatly is recent years. Suppose the cost to produce these copies is modeled by the function $C(x) = 0.1x^2 - 1.2x + 7$, where $C(x)$ represents

the cost to produce x hundred thousand copies. If the revenue from the sales of these copies is modeled by $R(x) = -0.1x^2 + 1.8x - 2$, use a break-even analysis to find the quantity of copies sold (to the nearest thousand) that will cause the copy company to break even.

Market equilibrium: In a free-enterprise (supply and demand) economy, the amount buyers are willing to pay for an item and the number of these items manufacturers are willing to produce depend on the price of the item. As the price increases, demand for the item decreases since buyers are less willing to pay the higher price. On the other hand, an increase in price increases the supply of the item since manufacturers are now more willing to supply it. When the **supply and demand curves** are graphed, their point of intersection is called the **market equilibrium** for the item.

55. Suppose the monthly market demand D (in ten-thousands of gallons) for a new synthetic oil is related to the price P in dollars by the equation $10P^2 + 6D = 144$. For the market price P, assume the amount D that manufacturers are willing to supply is modeled by $8P^2 - 8P - 4D = 12$. (a) What is the minimum price at which manufacturers are willing to begin supplying the oil? (b) Use this information to create a system of nonlinear equations, then solve the system to find the market equilibrium price (per gallon) and the quantity of oil supplied and sold at this price.

56. The weekly demand D for organically grown carrots (in thousands of pounds) is related to the price per pound P by the equation $8P^2 + 4D = 84$. At this market price, the amount that growers are willing to supply is modeled by the equation $8P^2 + 6P - 2D = 48$. (a) What is the minimum price at which growers are willing to supply the organically grown carrots? (b) Use this information

to create a system of nonlinear equations, then solve the system to find the market equilibrium price (per pound) and the quantity of carrots supplied and sold at this price.

Solve by setting up and solving a system of nonlinear equations.

57. **Dimensions of a flag:** A large American flag has an area of 85 m² and a perimeter of 37 m. Find the dimensions of the flag.

58. **Dimensions of a sail:** The sail on a boat is a right triangle with a perimeter of 36 ft and a hypotenuse of 15 ft. Find the height and width of the sail.

Exercise 58

59. **Dimensions of a tract:** The area of a rectangular tract of land is 45 km². The length of a diagonal is $\sqrt{106}$ km. Find the dimensions of the tract.

60. **Dimensions of a deck:** A rectangular deck has an area of 192 ft² and the length of the diagonal is 20 ft. Find the dimensions of the deck.

61. **Dimensions of a trailer:** The surface area of a rectangular trailer with square ends is 928 ft². If the sum of all edges of the trailer is 164 ft, find its dimensions.

62. **Dimensions of cylindrical tank:** The surface area of a closed cylindrical tank is 192π m². Find the dimensions of the tank if the volume is 320π m³ and the radius is as small as possible.

▶ **EXTENDING THE CONCEPT**

63. The area of a vertical parabolic segment is given by $A = \frac{2}{3}BH$, where B is the length of the horizontal base of the segment and H is the height from the base to the vertex. Investigate how this formula can be used to find the *area* of the solution region for the general system of inequalities shown. $\begin{cases} y \geq x^2 - bx + c \\ y \leq c + bx - x^2 \end{cases}$ (*Hint:* Begin by investigating with $b = 6$ and $c = 8$, then use other values and try to generalize what you find.)

64. For what values of r will the volume of a sphere be numerically equal to its surface area? For what values of r will the volume of a cylinder be numerically equal to its lateral surface area? Can a similar relationship be found for the volume and lateral surface area of a cone? Why or why not?

65. A rectangular fish tank has a bottom and four sides made out of glass. Use a system of equations to help find the dimensions of the tank if the height is 18 in., surface area is 4806 in², the tank must hold 108 gal (1 gal = 231 in³), and all three dimensions are integers.

▶ **MAINTAINING YOUR SKILLS**

66. (7.4) Determine the angle θ between the vectors
$\mathbf{u} = 12\mathbf{i} - 35\mathbf{j}$ and $\mathbf{v} = -20\mathbf{i} + 21\mathbf{j}$.

67. (7.1) Estimate the length L to the base of the water
tank, if the angle of elevation to the base of the
tower is 10°, the angle of elevation to the base of
the tank is 35°, and the tower makes a 112° angle
with the hillside.

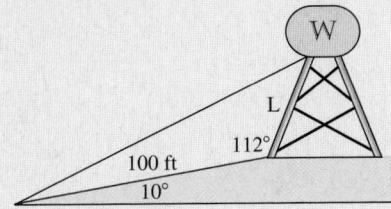

68. (2.6/4.2) Sketch each transformation:

    a. $y = 2|x + 3| - 1$　　　**b.** $y = \sqrt[3]{x - 2} + 3$

    c. $y = -(x + 4)^2 + 3$　　**d.** $y = 2^{x+1} - 3$

69. (2.3) In 2001, a small business purchased a copier
for $4500. In 2004, the value of the copier had
decreased to $3300. Assuming the depreciation is
linear: (a) find the rate-of-change $m = \dfrac{\Delta \text{value}}{\Delta \text{time}}$ and
discuss its meaning in this context; (b) find the
depreciation equation; and (c) use the equation to
predict the copier's value in 2008. (d) If the copier is
traded in for a new model when its value is less than
$700, how long will the company use this copier?

9.6 Polar Coordinates, Equations, and Graphs

Learning Objectives

In Section 9.6 you will learn how to:

☐ **A.** Plot points given in
polar form

☐ **B.** Convert from rectangular
form to polar form

☐ **C.** Convert from polar form
to rectangular form

☐ **D.** Sketch basic polar
graphs using an *r*-value
analysis

☐ **E.** Use symmetry and
families of curves to
write a polar equation
given a polar graph or
information about the
graph

One of the most enduring goals of mathematics is to express relations with the greatest
possible simplicity and ease of use. For $\dfrac{\tan \theta - \cot \theta}{\tan^2\theta - \cot^2\theta} = \sin \theta \cos\theta$, we would
definitely prefer working with $\sin \theta \cos \theta$, although the expressions are equivalent. Sim-
ilarly, we would prefer computing $(3 + \sqrt{3}i)^6$ in trigonometric form rather than algebraic form—and would quickly find the result is -1728. In just this way, many equations
and graphs are easier to work with in **polar form** rather than rectangular form. In rectangular form, a circle of radius 2 centered at (0, 2) has the equation $x^2 + (y - 2)^2 = 4$.
In polar form, the equation of the same circle is simply $r = 4 \sin \theta$. As you'll see, polar
coordinates offer an alternative method for plotting points and graphing relations.

A. Plotting Points Using Polar Coordinates

Suppose a Coast Guard station receives a distress call from a stranded boat. The boater
could attempt to give the location in rectangular form, but this might require imposing
an arbitrary coordinate grid on an uneven shoreline, using uncertain points of reference.
However, if the radio message said, "We're stranded 4 miles out, bearing 60°," the Coast
Guard could immediately locate the boat and send help. In **polar coordinates**, "4 miles
out, bearing 60°" would simply be written $(r, \theta) = (4, 30°)$, with r representing the
distance from the station and $\theta > 0$ measured from a horizontal axis in the counterclockwise direction as before (see Figure 9.34). If we placed the scenario on a
rectangular grid (assuming a straight shoreline), the coordinates of the boat would be
$(2\sqrt{3}, 2)$ using basic trigonometry. As you see, the **polar coordinate system** uses
angles and distances to locate a point in the plane. In this example, the Coast Guard
station would be considered the **pole** or origin, with the *x*-axis as the **polar axis** or axis
of reference (Figure 9.35). A distinctive feature of polar coordinates is that *we allow
r to be negative*, in which case $P(r, \theta)$ is the point $|r|$ units from the pole in a direction
opposite (180°) to that of θ (Figure 9.36). For convenience, polar graph paper is often
used when working with polar coordinates. It consists of a series of concentric circles
that share the same center and have integer radii. The standard angles are marked off
in multiples of $\dfrac{\pi}{12} = 15°$ depending on whether you're working in radians or degrees

Figure 9.34

(Figure 9.37). To plot the point $P(r, \theta)$, go a distance of $|r|$ at $0°$ then move $\theta°$ counterclockwise along a circle of radius r. If $r > 0$, plot a point at that location (you're finished). If $r < 0$, the point is plotted on a circle of the same radius, but $180°$ in the opposite direction.

Figure 9.35

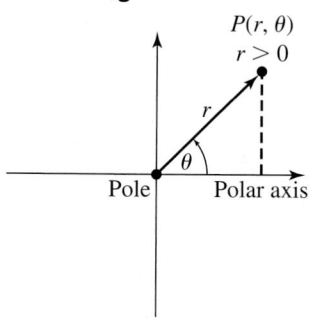

Figure 9.36

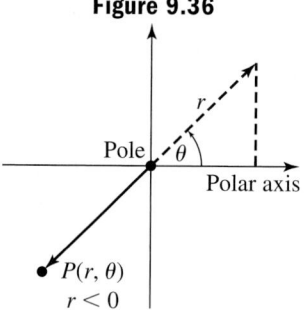

Figure 9.37

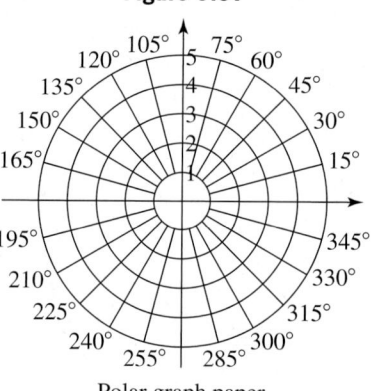

Polar graph paper

EXAMPLE 1 ▶ Plotting Points in Polar Coordinates

Plot each point $P(r, \theta)$ given $A(4, 45°)$; $B(-5, 135°)$; $C(-3, -30°)$; $D\left(2, \dfrac{2\pi}{3}\right)$; $E\left(-5, \dfrac{\pi}{3}\right)$; and $F\left(3, -\dfrac{\pi}{6}\right)$.

Solution ▶ For $A(4, 45°)$ go 4 units at $0°$, then rotate $45°$ counterclockwise and plot point A. For $B(-5, 135°)$, move $|-5| = 5$ units at $0°$, rotate $135°$, then actually plot point B $180°$ in the opposite direction, as shown. Point $C(-3, -30°)$ is plotted by moving $|-3| = 3$ units at $0°$, rotating $-30°$, then plotting point C $180°$ in the opposite direction (since $r < 0$). See Figure 9.38. The points $D\left(2, \dfrac{2\pi}{3}\right)$, $E\left(-5, \dfrac{\pi}{3}\right)$, and $F\left(3, -\dfrac{\pi}{6}\right)$ are plotted on the grid in Figure 9.39.

Figure 9.38

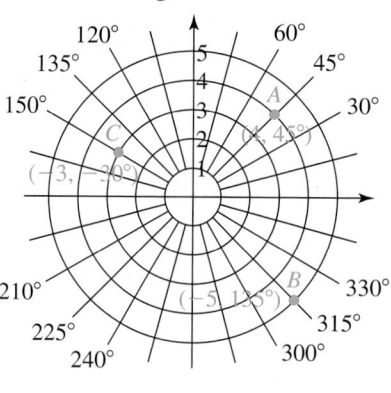

Figure 9.39

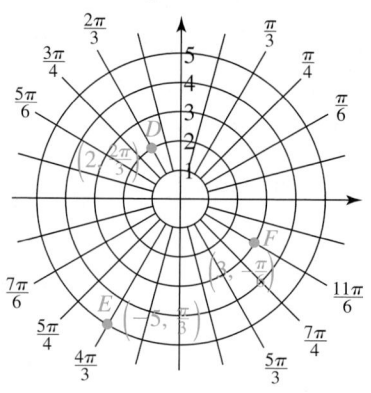

Now try Exercises 7 through 22 ▶

While plotting the points $B(-5, 135°)$ and $F\left(3, -\dfrac{\pi}{6}\right)$, you likely noticed that the coordinates of a point in polar coordinates are not unique. For $B(-5, 135°)$ it appears more natural to name the location $(5, 315°)$; while for $F\left(3, -\dfrac{\pi}{6}\right)$, the expression

☑ **A.** You've just learned
how to plot points given in
polar form

$\left(3, \dfrac{11\pi}{6}\right)$ is just as reasonable. In fact, for any point $P(r, \theta)$ in polar coordinates, $P(r, \theta \pm 2\pi)$ and $P(-r, \theta \pm \pi)$ name the same location. **See Exercises 23 through 36.**

☑ **A.** You've just learned how to plot points given in polar form

B. Converting from Rectangular Coordinates to Polar Coordinates

Conversions between rectangular and polar coordinates is a simple application of skills from previous sections, and closely resembles the conversion from the rectangular form to the trigonometric form of a complex number. To make the connection, we first assume $r > 0$ with θ in Quadrant II (see Figure 9.40). In rectangular form, the coordinates of the point are simply (x, y), with the lengths of x and y forming the sides of a right triangle. The distance r from the origin to point P resembles the modulus of a complex number and is computed in the same way:

Figure 9.40

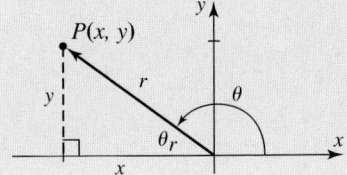

$r = \sqrt{x^2 + y^2}$. As long as $x \neq 0$, we have $\theta_r = \tan^{-1}\left(\dfrac{y}{x}\right)$, noting θ_r is a reference angle if the terminal side is not in Quadrant I. If needed, refer to Section 5.3 for a review of reference arcs and reference angles.

Converting from Rectangular to Polar Coordinates

Any point $P(x, y)$ in rectangular coordinates can be represented as $P(r, \theta)$ in polar coordinates, where $r = \sqrt{x^2 + y^2}$ and $\theta_r = \tan^{-1}\left(\dfrac{y}{x}\right)$, $x \neq 0$.

EXAMPLE 2 ▶ **Converting a Point from Rectangular Form to Polar Form**

Convert from rectangular to polar form, with $r > 0$ and $0 \leq \theta \leq 360°$ (round values to one decimal place as needed).

  a. $P(-5, 12)$ **b.** $P(3\sqrt{2}, -3\sqrt{2})$

Solution ▶   **a.** Point $P(-5, 12)$ is in Quadrant II.

$$r = \sqrt{(-5)^2 + 12^2} \qquad \theta = \tan^{-1}\left(\dfrac{12}{-5}\right)$$
$$= \sqrt{169} \qquad\qquad \theta_r \approx -67.4°$$
$$= 13 \qquad\qquad\quad \theta \approx 112.6°$$
$$P(-5, 12) \rightarrow P(13, 112.6°)$$

  b. Point $P(3\sqrt{2}, -3\sqrt{2})$ is in Quadrant IV.

$$r = \sqrt{(3\sqrt{2})^2 + (-3\sqrt{2})^2} \qquad \theta = \tan^{-1}\left(\dfrac{-3\sqrt{2}}{3\sqrt{2}}\right)$$
$$= \sqrt{36} \qquad\qquad\qquad \theta_r = -45°$$
$$= 6 \qquad\qquad\qquad\quad \theta = 315°$$
$$P(3\sqrt{2}, -3\sqrt{2}) \rightarrow P(6, 315°)$$

☑ **B.** You've just learned how to convert from rectangular form to polar form

Now try Exercises 37 through 44 ▶

C. Converting from Polar Coordinates to Rectangular Coordinates

The conversion from polar form to rectangular form is likewise straightforward. From Figure 9.41 we again note $\cos\theta = \dfrac{x}{r}$ and $\sin\theta = \dfrac{y}{r}$, giving $x = r\cos\theta$ and $y = r\sin\theta$. The conversion simply consists of making these substitutions and simplifying.

Figure 9.41

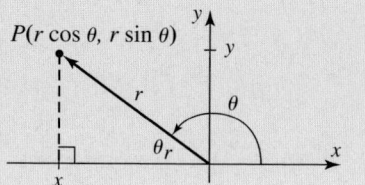

Converting from Polar to Rectangular Coordinates

Any point $P(r, \theta)$ in polar coordinates can be represented as $P(x, y)$ in rectangular coordinates, where $x = r\cos\theta$ and $y = r\sin\theta$.

EXAMPLE 3 ▶ **Converting a Point from Polar Form to Rectangular Form**

Convert from polar to rectangular form (round values to one decimal place as needed).

a. $P\left(12, \dfrac{5\pi}{3}\right)$ **b.** $P(6, 240°)$

Solution ▶   **a.** Point $P\left(12, \dfrac{5\pi}{3}\right)$ is in Quadrant IV.

$$x = r\cos\theta \qquad\qquad y = r\sin\theta$$
$$= 12\cos\left(\frac{5\pi}{3}\right) \qquad = 12\sin\left(\frac{5\pi}{3}\right)$$
$$= 12\left(\frac{1}{2}\right) \qquad\qquad = 12\left(\frac{-\sqrt{3}}{2}\right)$$
$$= 6 \qquad\qquad\qquad = -6\sqrt{3}$$
$$P\left(12, \frac{5\pi}{3}\right) \rightarrow P(6, -6\sqrt{3}) \approx P(6, -10.4)$$

b. Point $P(6, 240°)$ is in Quadrant III.

$$x = 6\cos 240° \qquad y = 6\sin 240°$$
$$= 6\left(-\frac{1}{2}\right) \qquad = 6\left(\frac{-\sqrt{3}}{2}\right)$$
$$= -3 \qquad\qquad = -3\sqrt{3}$$
$$P(6, 240°) \rightarrow P(-3, -3\sqrt{3}) \approx P(-3, -5.2)$$

☑ **C.** You've just learned how to convert from polar form to rectangular form

Now try Exercises 45 through 52 ▶

Using the relationships $x = r\cos\theta$, $y = r\sin\theta$, and $x^2 + y^2 = 1$, we can actually convert an equation given in polar form, to the equivalent equation in rectangular form. **See Exercises 105 and 106.**

D. Basic Polar Graphs and *r*-Value Analysis

To really understand polar graphs, an intuitive sense of how they're developed is needed. Polar equations are generally stated in terms of *r* and trigonometric functions of θ, with θ being the input value and *r* being the output value. First, it helps to view the length *r* as the long second hand of a clock, but extending an equal distance in both directions from center (Figure 9.42). This "second hand" ticks around the face of the clock in the counterclockwise direction, with the angular measure of each tick being $\dfrac{\pi}{12}$ radians $= 15°$. As each angle "ticks by,"

Figure 9.42

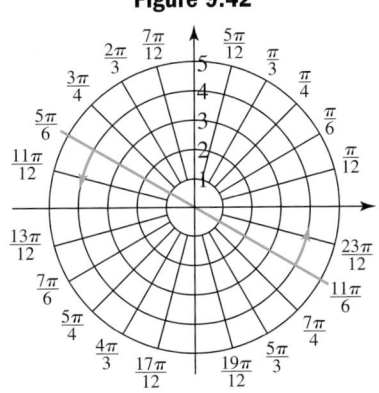

we locate a point somewhere along the radius, depending on whether *r* is positive or negative, and plot it on the face of the clock before going on to the next tick. For the purposes of this study, we will allow that all polar graphs are continuous and smooth curves, without presenting a formal proof.

EXAMPLE 4 ▶ Graphing Basic Polar Equations

Graph the polar equations.

    a. $r = 4$ **b.** $\theta = \dfrac{\pi}{4}$

Solution ▶   **a.** For $r = 4$, we're plotting all points of the form $(4, \theta)$ where *r* has a constant value and θ varies. As the second hand "ticks around the polar grid," we plot all points a distance of 4 units from the pole. As you might imagine, the graph is a circle with radius 4.

    b. For $\theta = \dfrac{\pi}{4}$, all points have the form $\left(r, \dfrac{\pi}{4}\right)$ with $\dfrac{\pi}{4}$ constant and *r* varying. In this case, the "second hand" is frozen at $\dfrac{\pi}{4}$, and we plot any selection of *r*-values, producing the straight line shown in the figure.

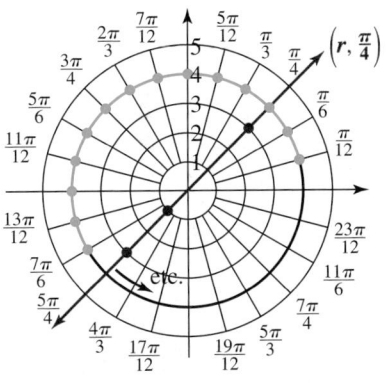

Now try Exercises 53 through 56 ▶

To develop an "intuitive sense" that allows for the efficient graphing of more sophisticated equations, we use a technique called *r*-**value analysis.** This technique basically takes advantage of the predictable patterns in $r = \sin \theta$ and $r = \cos \theta$ taken from their graphs, including the zeros and maximum/minimum values.

We begin with the *r*-value analysis for $r = \sin \theta$, using the graph shown in Figure 9.43. Note the analysis occurs in the four colored parts corresponding to Quadrants I, II, III, and IV, and that the maximum value of $|\sin \theta| = 1$.

Figure 9.43

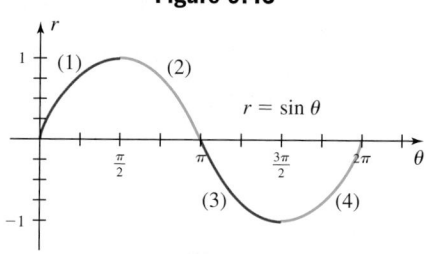

1. As θ moves from 0 to $\dfrac{\pi}{2}$, $\sin\theta$ is positive and $|\sin\theta|$ increases from 0 to 1.

   $\Rightarrow$ for $r = \sin\theta$, r is increasing

2. As θ moves from $\dfrac{\pi}{2}$ to π, $\sin\theta$ is positive and $|\sin\theta|$ decreases from 1 to 0.

   $\Rightarrow$ for $r = \sin\theta$, r is decreasing

3. As θ moves from π to $\dfrac{3\pi}{2}$, $\sin\theta$ is negative and $|\sin\theta|$ increases from 0 to 1.

   $\Rightarrow$ for $r = \sin\theta$, r is increasing

4. As θ moves from $\dfrac{3\pi}{2}$ to 2π, $\sin\theta$ is negative and $|\sin\theta|$ decreases from 1 to 0.

   $\Rightarrow$ for $r = \sin\theta$, r is decreasing

In summary, note that the value of $|r|$ goes through four cycles, two where it is increasing from 0 to 1 (in red), and two where it is decreasing from 1 to 0 (in blue).

> **WORTHY OF NOTE**
>
> It is important to remember that if $r < 0$, the related point on the graph is $|r|$ units from center, *180° in the opposite direction*: $(-r, \theta) \rightarrow (r, \theta + 180°)$. In addition, students are encouraged not to use a table of values, a conversion to rectangular coordinates, or a graphing calculator until after the r-value analysis.

EXAMPLE 5 ▶ **Graphing Polar Equations Using an *r*-Value Analysis**

Sketch the graph of $r = 4\sin\theta$ using an r-value analysis.

Solution ▶　Begin by noting that $r = 0$ at $\theta = 0$, and will increase from 0 to 4 as the clock "ticks" from 0 to $\dfrac{\pi}{2}$, since $\sin\theta$ is increasing from 0 to 1. (1) For $\theta = \dfrac{\pi}{6}, \dfrac{\pi}{4}$, and $\dfrac{\pi}{3}$, $r = 2$, $r \approx 2.8$, and $r \approx 3.5$, respectively (at $\theta = \dfrac{\pi}{2}$, $r = 4$). See Figure 9.44.

(2) As θ continues "ticking" from $\dfrac{\pi}{2}$ to π, $|r|$ decreases from 4 to 0, since $\sin\theta$ is decreasing from 1 to 0. For $\theta = \dfrac{2\pi}{3}, \dfrac{3\pi}{4}$, and $\dfrac{5\pi}{6}$, $r \approx 3.5$, $r \approx 2.8$, and $r = 2$, respectively (at $\theta = \pi$, $r = 0$). See Figure 9.45. (3) From π to $\dfrac{3\pi}{2}$, $|r|$ increases from 0 to 4, but since $r < 0$, this portion of the graph is reflected back into Quadrant I, overlapping the portion already drawn from 0 to $\dfrac{\pi}{2}$. (4) From $\dfrac{3\pi}{2}$ to 2π, $|r|$ decreases from 4 to 0, overlapping the portion drawn from $\dfrac{\pi}{2}$ to π. We conclude the graph is a closed figure limited to Quadrants I and II as shown in Figure 9.45. This is a circle with radius 2, centered at (0, 2). In summary:

Figure 9.44

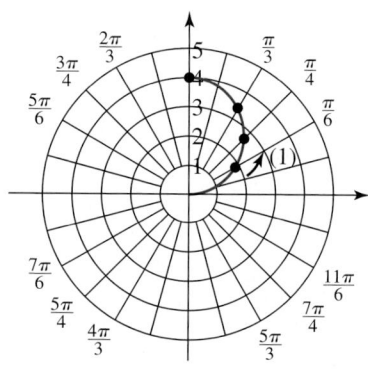

Figure 9.45

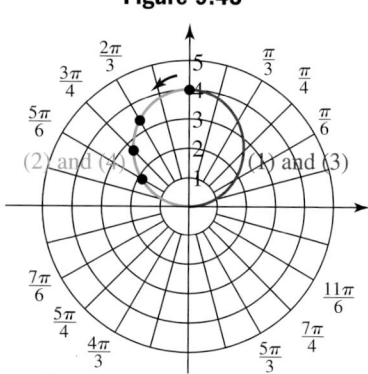

$r = 4\sin\theta$

θ	0 to $\dfrac{\pi}{2}$	$\dfrac{\pi}{2}$ to π	π to $\dfrac{3\pi}{2}$	$\dfrac{3\pi}{2}$ to 2π
$|r|$	0 to 4	4 to 0	0 to 4	4 to 0

Now try Exercises 57 and 58 ▶

Although it takes some effort, r-value analysis offers an efficient way to graph polar equations, and gives a better understanding of graphing in polar coordinates. In addition, it often enables you to sketch the graph with a minimum number of calculations and plotted points. As you continue using the technique, it will help to have Figure 9.43 in plain view for quick reference, as well as the corresponding analysis of $y = \cos\theta$ for polar graphs involving cosine (**see Exercise 98**).

| **EXAMPLE 6** ▶ | **Graphing Polar Equations Using an r-Value Analysis** |

Sketch the graph of $r = 2 + 2\sin\theta$ using an r-value analysis.

Solution ▶ Since the minimum value of $\sin\theta$ is -1, we note that r will always be greater than or equal to zero. At $\theta = 0$, r has a value of 2 ($\sin 0 = 0$), and will increase from 2 to 4 as the clock "ticks" from 0 to $\dfrac{\pi}{2}$ ($\sin\theta$ is positive and $|\sin\theta|$ is increasing).

From $\dfrac{\pi}{2}$ to π, r decreases from 4 to 2 ($\sin\theta$ is positive and $|\sin\theta|$ is decreasing).

From π to $\dfrac{3\pi}{2}$, r decreases from 2 to 0 ($\sin\theta$ is negative and $|\sin\theta|$ is increasing); and from $\dfrac{3\pi}{2}$ to 2π, r increases from 0 to 2 ($\sin\theta$ is negative and $|\sin\theta|$ is decreasing). We conclude the graph is a closed figure containing the points $(2, 0)$, $\left(4, \dfrac{\pi}{2}\right)$, $(2, \pi)$, and $\left(0, \dfrac{3\pi}{2}\right)$. Noting that $\theta = \dfrac{\pi}{6}$ and $\theta = \dfrac{5\pi}{6}$ will produce integer values, we evaluate $r = 2 + 2\sin\theta$ and obtain the additional points $\left(3, \dfrac{\pi}{6}\right)$ and $\left(3, \dfrac{5\pi}{6}\right)$. Using these points and the r-value analysis produces the graph shown here, called a **cardioid** (from the limaçon family of curves). In summary we have:

θ	$r = 4\sin\theta$
0	0
30	2
45	$2\sqrt{2} \approx 2.8$
60	$2\sqrt{3} \approx 3.5$
90	4
120	$2\sqrt{3} \approx 3.5$
135	$2\sqrt{2} \approx 2.8$
150	2
180	0

☑ **D.** You've just learned how to sketch basic polar graphs using an r-value analysis

	θ	$r = 2 + 2\sin\theta$
(1)	0 to $\dfrac{\pi}{2}$	2 to 4
(2)	$\dfrac{\pi}{2}$ to π	4 to 2
(3)	π to $\dfrac{3\pi}{2}$	2 to 0
(4)	$\dfrac{3\pi}{2}$ to 2π	0 to 2

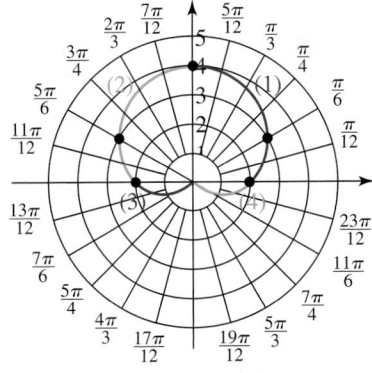

Now try Exercises 59 through 62 ▶

E. Symmetry and Families of Polar Graphs

Even with a careful r-value analysis, some polar graphs require a good deal of effort to produce. In many cases, symmetry can be a big help, as can recognizing certain families of equations and their related graphs. As with other forms of graphing, gathering this information beforehand will enable you to graph relations with a smaller number of plotted points. Figures 9.46 to 9.49 offer some examples of symmetry for polar graphs.

Figure 9.46
Vertical-axis symmetry:
$r = -2 - 2\sin\theta$

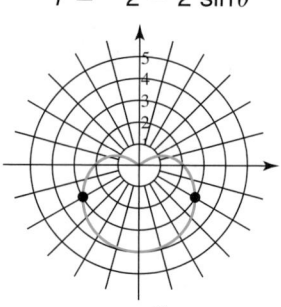

$\theta = \dfrac{\pi}{2}$

Figure 9.47
Polar-axis symmetry:
$r = 5\sin\theta$

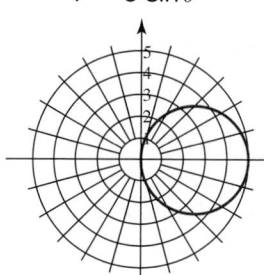

Figure 9.48
Polar symmetry:
$r = 5\sin(2\theta)$

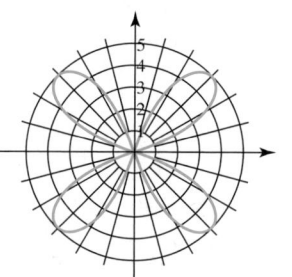

Figure 9.49
Polar symmetry:
$r^2 = 25\sin(2\theta)$

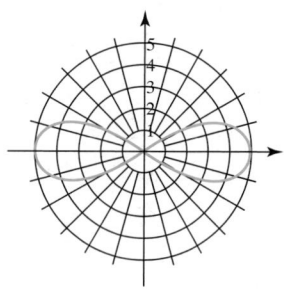

WORTHY OF NOTE

In mathematics we refer to the tests for polar symmetry as *sufficient but not necessary conditions*. The tests are *sufficient* to show symmetry (if the test is satisfied, the graph must be symmetric), but the tests are *not necessary* to show symmetry (the graph may be symmetric even if the test is not satisfied).

The tests for symmetry in polar coordinates bear a strong resemblance to those for rectangular coordinates, but there is a major difference. Since there are many different ways to name a point in polar coordinates, a polar graph may actually exhibit a form of symmetry without satisfying the related test. In other words, the tests are *sufficient* to establish symmetry, but not *necessary*.

The formal tests for symmetry are explored in **Exercises 100 to 102.** For our purposes, we'll rely on a somewhat narrower view, one that is actually a synthesis of our observations here and our previous experience with the sine and cosine.

Symmetry for Graphs of Certain Polar Equations

Given the polar equation $r = f(\theta)$,

1. If $f(\theta)$ represents an expression in terms of sine(s), the graph will be symmetric to $\theta = \dfrac{\pi}{2}$: (r, θ) and $(r, \pi - \theta)$ are on the graph.

2. If $f(\theta)$ represents an expression in terms of cosine(s), the graph will be symmetric to $\theta = 0$: (r, θ) and $(r, -\theta)$ are on the graph.

While the fundamental ideas from Examples 5 and 6 go a long way toward graphing other polar equations, our discussion would not be complete without a review of the *period* of sine and cosine. Many polar equations have factors of $\sin(n\theta)$ or $\cos(n\theta)$ in them, and it helps to recall the period formula $P = \dfrac{2\pi}{n}$. Comparing $r = 4\sin\theta$ from

Figure 9.50

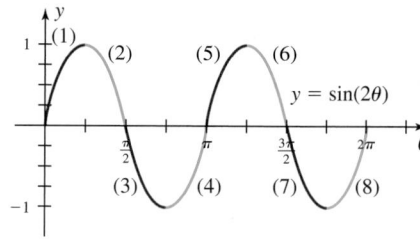

$y = \sin(2\theta)$

Example 5 with $r = 4\sin(2\theta)$, we note the period of sine changes from $P = 2\pi$ to $P = \dfrac{2\pi}{2} = \pi$, *meaning there will be twice as many cycles* and $|r|$ will now go through *eight* cycles—four where $|\sin(2\theta)|$ is increasing from 0 to 1 (in red), and four where it is decreasing from 1 to 0 (in blue). See Figure 9.50.

EXAMPLE 7 ▶   **Sketching Polar Graphs Using Symmetry and *r*-Values**

Sketch the graph of $r = 4\sin(2\theta)$ using symmetry and an *r*-value analysis.

Solution ▶   Since *r* is expressed in terms of sine, the graph will be symmetric to $\theta = \dfrac{\pi}{2}$. We note that $r = 0$ at $\theta = \dfrac{n\pi}{2}$, where *n* is even, and the graph will go through the pole at these points. This also tells us the graph will be a closed figure. From the graph of $\sin(2\theta)$ in Figure 9.50, we see $|\sin(2\theta)| = 1$ at $\theta = \dfrac{\pi}{4}, \dfrac{3\pi}{4}, \dfrac{5\pi}{4}$, and $\dfrac{7\pi}{4}$, so the graph will include the points $\left(4, \dfrac{\pi}{4}\right)$, $\left(4, \dfrac{3\pi}{4}\right)$, $\left(4, \dfrac{5\pi}{4}\right)$, and $\left(4, \dfrac{7\pi}{4}\right)$. Only the analysis of the first four cycles is given next, since the remainder of the graph can be drawn using symmetry.

	Cycle	*r*-Value Analysis	Location of Graph
(1)	0 to $\dfrac{\pi}{4}$	$\lvert r \rvert$ increases from 0 to 4	QI ($r > 0$)
(2)	$\dfrac{\pi}{4}$ to $\dfrac{\pi}{2}$	$\lvert r \rvert$ decreases from 4 to 0	QI ($r > 0$)
(3)	$\dfrac{\pi}{2}$ to $\dfrac{3\pi}{4}$	$\lvert r \rvert$ increases from 0 to 4	QIV ($r < 0$)
(4)	$\dfrac{3\pi}{4}$ to π	$\lvert r \rvert$ decreases from 4 to 0	QIV ($r < 0$)

Plotting the points and applying the *r*-value analysis with the symmetry involved produces the graph in the figure, called a **four-leaf rose.** At any time during this process, additional points can be calculated to "round-out" the graph.

$r = 4\sin(2\theta)$

θ	$\lvert r \rvert$
0 to $\dfrac{\pi}{4}$	0 to 4
$\dfrac{\pi}{4}$ to $\dfrac{\pi}{2}$	4 to 0
$\dfrac{\pi}{2}$ to $\dfrac{3\pi}{4}$	0 to 4
$\dfrac{3\pi}{4}$ to π	4 to 0

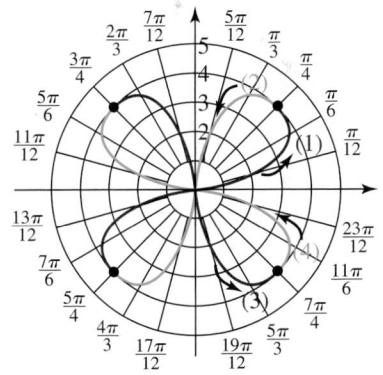

Now try Exercises 63 through 70 ▶

Graphing Polar Equations

To assist the process of graphing polar equations:
1. Carefully note any symmetries you can use.
2. Have graphs of $y = \sin(n\theta)$ and $y = \cos(n\theta)$ in view for quick reference.
3. Use these graphs to analyze the value of *r* as the "clock ticks" around the polar grid: (a) determine the max/min *r*-values and write them in polar form, and (b) determine the polar-axis intercepts and write them in polar form.
4. Plot the points, then use the *r*-value analysis and any symmetries to complete the graph.

Similar to polynomial graphs, polar graphs come in numerous shapes and varieties, yet many of them share common characteristics and can be organized into certain families. Some of the more common families are illustrated in Appendix VI, and give the general equation and related graph for common family members. Also included are characteristics of certain graphs that will enable you to develop the polar equation given its graph or information about its graph. For further investigations using a graphing calculator, **see Exercises 71 through 76.**

EXAMPLE 8 ▶ **Graphing a Limaçon Using Stated Conditions**

Find the equation of the polar curve satisfying the given conditions, then sketch the graph: limaçon, symmetric to $\theta = 90°$, with $a = 2$ and $b = -3$.

Solution ▶ The general equation of a limaçon symmetric to $\theta = 90°$ is $r = a + b \sin \theta$, so our desired equation is $r = 2 - 3 \sin \theta$. Since $|a| < |b|$, the limaçon has an inner loop of length $3 - 2 = 1$ and a maximum distance from the origin of $2 + 3 = 5$. The polar-axis intercepts are $(2, 0)$ and $(2, 180°)$. With $b < 0$, the graph is reflected across the polar axis (facing "downward"). The complete graph is shown in the figure.

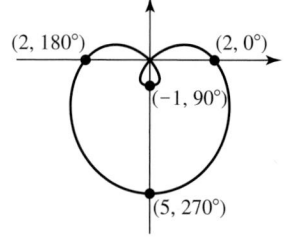

Now try Exercises 79 through 94 ▶

EXAMPLE 9 ▶ **Modeling the Flight Path of a Scavenger Bird**

Scavenger birds sometimes fly over dead or dying animals (called carrion) in a "figure-eight" formation, closely resembling the graph of a lemniscate. Suppose the flight path of one of these birds was plotted and found to contain the polar coordinates $(81, 0°)$ and $(0, 45°)$. Find the equation of the lemniscate. If the bird lands at the point $(r, 136°)$, how far is it from the carrion? Assume r is in yards.

Solution ▶ Since $(81, 0°)$ is a point on the graph, the lemniscate is symmetric to the polar axis and the general equation is $r^2 = a^2 \cos(2\theta)$. The point $(81, 0°)$ indicates $a = 81$, hence the equation is $r^2 = 6561 \cos(2\theta)$. At $\theta = 136°$ we have $r^2 = 6561 \cos 272°$, and the bird has landed $r \approx 15$ yd away.

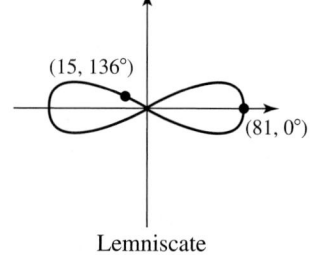

Lemniscate

Now try Exercises 95 through 97 ▶

✓ **E. You've just learned how to use symmetry and families of curves to write a polar equation given a polar graph or information about the graph**

You've likely been wondering how the different families of polar graphs were named. The roses are easy to figure as each graph has a flower-like appearance. The limaçon (pronounced li-ma-sawn) family takes its name from the Latin words *limax* or *lamacis,* meaning "snail." With some imagination, these graphs do have the appearance of a snail shell. The cardioids are a subset of the limaçon family and are so named due to their obvious resemblance to the human heart. In fact, the name stems from the Greek *kardia* meaning heart, and many derivative words are still in common use (a cardiologist is one who specializes in a study of the heart). Finally, there is the lemniscate family, a name derived from the Latin *lemniscus,* which describes a certain kind of ribbon. Once again, a little creativity enables us to make the connection between ribbons, bows, and the shape of this graph.

9.6 EXERCISES

▶ CONCEPTS AND VOCABULARY

Fill in each blank with the appropriate word or phrase. Carefully reread the section if needed.

1. The point (r, θ) is said to be written in _____ coordinates.

2. In polar coordinates, the origin is called the _____ and the horizontal axis is called the _____ axis.

3. The point $(4, 135°)$ is located in Q _____, while $(-4, 135°)$ is located in Q _____.

4. If a polar equation is given in terms of cosine, the graph will be symmetric to _____.

5. Write out the procedure for plotting points in polar coordinates, as though you were explaining the process to a friend.

6. Discuss the graph of $r = 6 \cos \theta$ in terms of an r-value analysis, using $y = \cos \theta$ and a color-coded graph.

▶ DEVELOPING YOUR SKILLS

Plot the following points using polar graph paper.

7. $\left(4, \dfrac{\pi}{2}\right)$ 8. $\left(3, \dfrac{3\pi}{2}\right)$ 9. $\left(2, \dfrac{5\pi}{4}\right)$

10. $\left(4.5, -\dfrac{\pi}{3}\right)$ 11. $\left(-5, \dfrac{5\pi}{6}\right)$ 12. $\left(-4, \dfrac{7\pi}{4}\right)$

13. $\left(-3, -\dfrac{2\pi}{3}\right)$ 14. $\left(-4, -\dfrac{\pi}{4}\right)$

Express the points shown using polar coordinates with θ in radians, $0 \le \theta < 2\pi$ and $r > 0$.

15.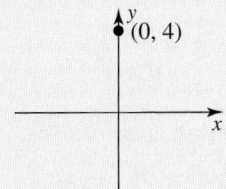

16.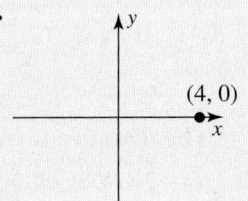

17.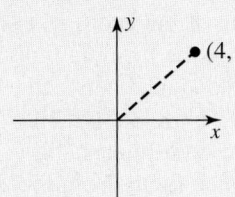

18.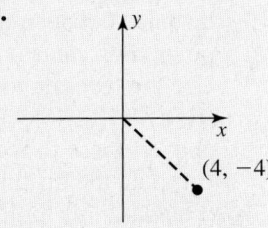

19.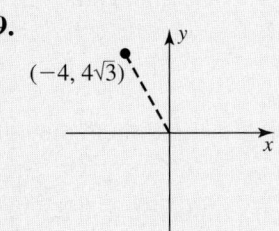

20.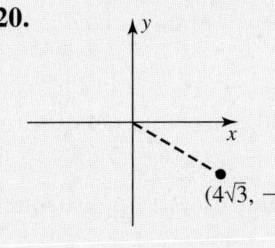

21.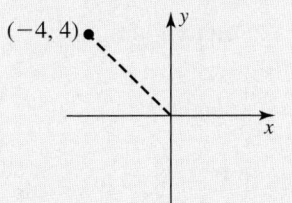

22.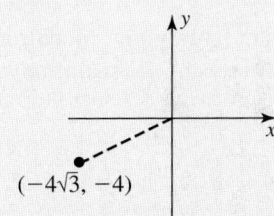

List three alternative ways the given points can be expressed in polar coordinates using $r > 0$, $r < 0$, and $\theta \in [-2\pi, 2\pi)$.

23. $\left(3\sqrt{2}, \dfrac{3\pi}{4}\right)$ 24. $\left(4\sqrt{3}, -\dfrac{5\pi}{3}\right)$

25. $\left(-2, \dfrac{11\pi}{6}\right)$ 26. $\left(-3, -\dfrac{7\pi}{6}\right)$

Match each (r, θ) given to one of the points A, B, C, or D shown.

Exercise 27–36

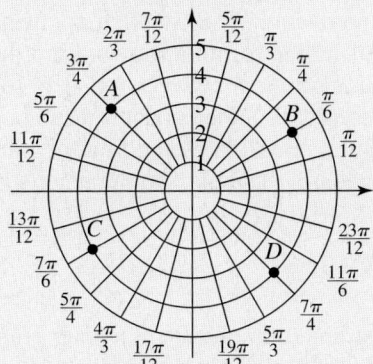

27. $\left(4, -\dfrac{5\pi}{6}\right)$ 28. $\left(4, -\dfrac{5\pi}{4}\right)$

29. $\left(-4, \dfrac{\pi}{6}\right)$ **30.** $\left(-4, \dfrac{3\pi}{4}\right)$

31. $\left(-4, -\dfrac{5\pi}{4}\right)$ **32.** $\left(-4, -\dfrac{\pi}{4}\right)$

33. $\left(4, \dfrac{13\pi}{6}\right)$ **34.** $\left(4, \dfrac{19\pi}{6}\right)$

35. $\left(-4, -\dfrac{21\pi}{4}\right)$ **36.** $\left(4, -\dfrac{35\pi}{6}\right)$

Convert from rectangular coordinates to polar coordinates. A diagram may help.

37. $(-8, 0)$ **38.** $(0, -7)$

39. $(4, 4)$ **40.** $(4\sqrt{3}, 4)$

41. $(5\sqrt{2}, 5\sqrt{2})$ **42.** $(6, -6\sqrt{3})$

43. $(-5, -12)$ **44.** $(-3.5, 12)$

Convert from polar coordinates to rectangular coordinates. A diagram may help.

45. $(8, 45°)$ **46.** $(6, 60°)$

47. $\left(4, \dfrac{3\pi}{4}\right)$ **48.** $\left(5, \dfrac{5\pi}{6}\right)$

49. $\left(-2, \dfrac{7\pi}{6}\right)$ **50.** $\left(-10, \dfrac{4\pi}{3}\right)$

51. $(-5, -135°)$ **52.** $(-4, -30°)$

Sketch each polar graph using an r-value analysis (a table may help), symmetry, and any convenient points.

53. $r = 5$ **54.** $r = 6$

55. $\theta = \dfrac{\pi}{6}$ **56.** $\theta = -\dfrac{3\pi}{4}$

57. $r = 4 \cos \theta$ **58.** $r = 2 \sin \theta$

59. $r = 3 + 3 \sin \theta$ **60.** $r = 2 + 2 \cos \theta$

61. $r = 2 - 4 \sin \theta$ **62.** $r = 1 - 2 \cos \theta$

63. $r = 5 \cos(2\theta)$ **64.** $r = 3 \sin(4\theta)$

65. $r = 4 \sin 2\theta$ **66.** $r = 6 \cos(5\theta)$

67. $r^2 = 9 \sin(2\theta)$ **68.** $r^2 = 16 \cos(2\theta)$

69. $r = 4 \sin\left(\dfrac{\theta}{2}\right)$ **70.** $r = 6 \cos\left(\dfrac{\theta}{2}\right)$

Use a graphing calculator in polar mode to produce the following polar graphs.

71. $r = 4\sqrt{1 - \sin^2\theta}$, *a hippopede*

72. $r = 3 + \csc \theta$, *a conchoid*

73. $r = 2 \cos \theta \cot \theta$, *a cissoid*

74. $r = \cot \theta$, *a kappa curve*

75. $r = 8 \sin \theta \cos^2\theta$, *a bifoliate*

76. $r = 8 \cos \theta(4 \sin^2\theta - 2)$, *a folium*

▶ WORKING WITH FORMULAS

77. The midpoint formula in polar coordinates:
$$M = \left(\frac{r \cos \alpha + R \cos \beta}{2}, \frac{r \sin \alpha + R \sin \beta}{2}\right)$$

The midpoint of a line segment connecting the points (r, α) and (R, β) in polar coordinates can be found using the formula shown. Find the midpoint of the line segment between $(r, \alpha) = (6, 45°)$ and $(R, \beta) = (8, 30°)$, then convert these points to rectangular coordinates and find the midpoint using the "standard" formula. Do the results match?

78. The distance formula in polar coordinates:
$$d = \sqrt{R^2 + r^2 - 2Rr \cos(\alpha - \beta)}$$

Using the law of cosines, it can be shown that the distance between the points (R, α) and (r, β) in polar coordinates is given by the formula indicated. Use the formula to find the distance between $(R, \alpha) = (6, 45°)$ and $(r, \beta) = (8, 30°)$, then convert these to rectangular coordinates and compute the distance between them using the "standard" formula. Do the results match?

▶ APPLICATIONS

Polar graphs: Find the equation of a polar graph satisfying the given conditions, then sketch the graph.

79. limaçon, symmetric to polar axis, $a = 4$ and $b = 4$

80. rose, four petals, two petals symmetric to the polar axis, $a = 6$

81. rose, five petals, one petal symmetric to the polar axis, $a = 4$

82. limaçon, symmetric to $\theta = \dfrac{\pi}{2}$, $a = 2$ and $b = 4$

83. lemniscate, $a = 4$ through $(\pi, 4)$

84. lemniscate, $a = 8$ through $\left(8, \dfrac{\pi}{4}\right)$

85. circle, symmetric to $\theta = \dfrac{\pi}{2}$, center at $\left(2, \dfrac{\pi}{2}\right)$, containing $\left(2, \dfrac{\pi}{6}\right)$

86. circle, symmetric to polar axis, through $(6, \pi)$

Matching: Match each graph to its equation a through h, which follow. Justify your answers.

87.
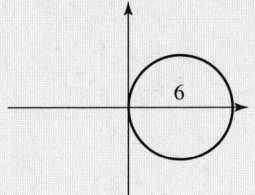

88.
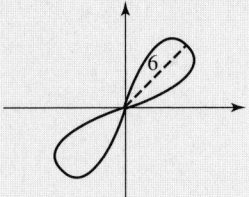

89.
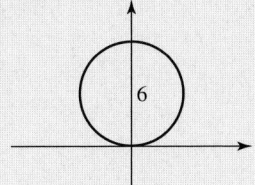

90.

91.
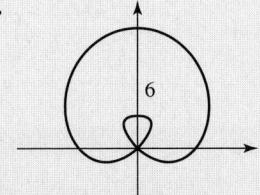

92.
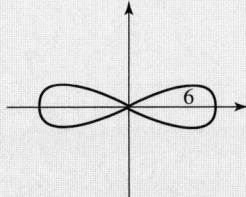

93.
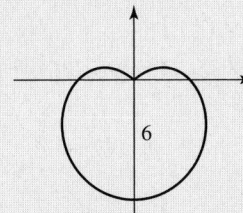

94.

a. $r = 6 \cos \theta$ **b.** $r = 3 - 3 \sin \theta$

c. $r = 6 \cos(4\theta)$ **d.** $r^2 = 36 \cos(2\theta)$

e. $r^2 = 36 \sin(2\theta)$ **f.** $r = 2 + 4 \sin \theta$

g. $r = 6 \sin \theta$ **h.** $r = 6 \sin(5\theta)$

95. Figure eights: Waiting for help to arrive on foot, a light plane is circling over some stranded hikers using a "figure eight" formation, closely resembling the graph of a lemniscate. Suppose the flight path of the plane was plotted (using the hikers as the origin) and found to contain the polar coordinates (7200, 45°) and (0, 90°) with r in meters. Find the equation of the lemniscate.

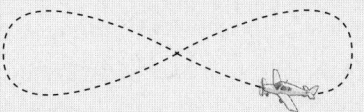

96. Animal territories: Territorial animals often prowl the borders of their territory, marking the boundaries with various bodily excretions. Suppose the territory of one such animal was limaçon shaped, with the pole representing the den of the animal. Find the polar equation defining the animal's territory if markings are left at (750, 0°), (1000, 90°), and (750, 180°). Assume r is in meters.

97. Prop manufacturing: The propellers for a toy boat are manufactured by stamping out a rose with n petals and then bending each blade. If the manufacturer wants propellers with five blades and a radius of 15 mm, what two polar equations will satisfy these specifications?

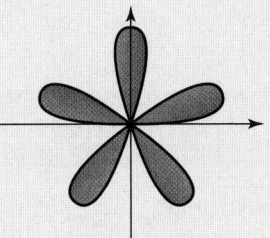

98. Polar curves and cosine: Do a complete r-value analysis for graphing polar curves involving cosine. Include a color-coded graph showing the relationship between r and θ, similar to the analysis for sines that preceded Example 6.

▶ **EXTENDING THE CONCEPT**

99. The polar graph $r = a\theta$ is called the *Spiral of Archimedes*. Consider the spiral $r = \dfrac{1}{2}\theta$. As this graph spirals around the origin, what is the distance between each positive, polar intercept? In QI, what is the distance between consecutive branches of the spiral each time it intersects $\theta = \dfrac{\pi}{4}$? What is the distance between consecutive branches of the spiral at $\theta = \dfrac{\pi}{2}$? What can you conclude?

As mentioned in the exposition, tests for symmetry of polar graphs are sufficient to show symmetry (if the test is satisfied, the graph must be symmetric), but the tests are not necessary to show symmetry (the graph may be symmetric even if the test is not satisfied). For $r = f(\theta)$, the formal tests for the symmetry are: (1) the graph will be symmetric to the polar axis if $f(\theta) = f(-\theta)$; (2) the graph will be symmetric to the line $\theta = \dfrac{\pi}{2}$ if $f(\pi - \theta) = f(\theta)$; and (3) the graph will be symmetric to the pole if $f(\theta) = -f(\theta)$.

100. Sketch the graph of $r = 4 \sin(2\theta)$. Show the equation fails the first test, yet the graph is still symmetric to the polar axis.

101. Why is the graph of every lemniscate symmetric to the pole?

102. Verify that the graph of every limaçon of the form $r = a + b \cos \theta$ is symmetric to the polar axis.

103. The graphs of $r = a \sin(n\theta)$ and $r = a \cos(n\theta)$ are from the rose family of polar graphs. If n is odd, there are n petals in the rose, and if n is even, there are $2n$ petals. An interesting extension of this fact is that the n petals enclose exactly 25% of the area of the circumscribed circle, and the $2n$ petals enclose exactly 50%. Find the area within the boundaries of the rose defined by $r = 6 \sin(5\theta)$.

To develop an understanding of polar equations, we used the following facts $x^2 + y^2 = r^2$, $x = r \cos \theta$, and $y = r \sin \theta$. Using these relationships, we can actually convert polar equations to rectangular equations and vice versa, showing that a particular equation *can be graphed in either form*. Use these relationships to write these polar equations in rectangular form. (*Hint:* Isolate the term kr (k a constant) on one side, then square.)

104. $r = \dfrac{1}{1 + \sin \theta}$ **105.** $r = \dfrac{6}{2 + 4 \sin \theta}$

▶ **MAINTAINING YOUR SKILLS**

106. (6.2) Verify the following is an identity:
$\cos^2 x - \sin^2 x = 1 - \sin(2x) \tan x$.

107. (6.7) Solve for $t \in [0, 2\pi)$:
$20 = 5 - 30 \sin\left(2t - \dfrac{\pi}{6}\right)$.

108. (1.3) Solve the absolute value inequality. Answer in interval notation:
$-3|2x + 5| - 7 > -19$

109. (2.7) Graph the piecewise function shown and state its domain and range.
$$f(x) = \begin{cases} x + 2 & -5 \le x < -1 \\ x & -1 \le x < 2 \\ 4 & 2 < x \le 5 \end{cases}$$

9.7 | More on the Conic Sections: Rotation of Axes and Polar Form

Learning Objectives

In Section 9.7 you will learn how to:

☐ **A.** Graph conic sections that have nonvertical and nonhorizontal axes (rotated conics)

☐ **B.** Identify conics using the discriminant of the polynomial form—the invariant $B^2 - 4AC$

☐ **C.** Write the equation of a conic section in polar form

☐ **D.** Solve applications involving the conic sections in polar form

Our study of conic sections would not be complete without considering conic sections whose graphs are not symmetric to a vertical or horizontal axis. The axis of symmetry still exists, but is rotated by some angle. We'll first study these **rotated conics** using the equation in its polynomial form, then investigate some interesting applications of the polar form.

A. Rotated Conics and the Rotation of Axes

It's always easier to understand a new idea in terms of a known idea, so we begin our study with a review of the reciprocal function $y = \dfrac{1}{x}$. From the equation we note:

1. The denominator is zero when $x = 0$, and the y-axis is a vertical asymptote (the vertical line $x = 0$).
2. Since the degree of the numerator is less than the degree of the denominator, the x-axis is a horizontal asymptote (the horizontal line $y = 0$).
3. Since $x < 0$ implies $y < 0$ and $x > 0$ implies $y > 0$, the graph will have two branches—one in the first quadrant and one in the third.

Figure 9.51

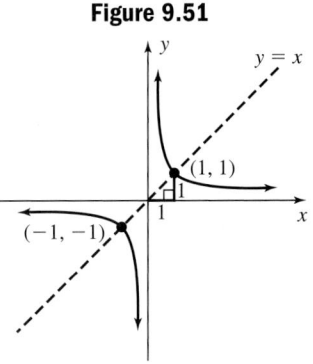

Note the polynomial form of this equation is $xy = 1$. The resulting graph is shown in Figure 9.51, *and is actually the graph of a hyperbola with a transverse axis of $y = x$.* Using the 45-45-90 triangle indicated, we find the distance from the origin to each vertex is $\sqrt{2}$. If we rotated the hyperbola 45° clockwise, we would obtain a more "standard" graph with a horizontal transverse axis and vertices at $(\pm a, 0) \rightarrow (\pm\sqrt{2}, 0)$. The asymptotes would be $y = \pm 1x$, and since $y = \pm\dfrac{b}{a}x$ is the general form we know $b = \pm\sqrt{2}$. This information can be used to find the equation of the rotated hyperbola.

EXAMPLE 1 ▶   **Finding the Equation of a Rotated Conic from Its Graph**

The hyperbola $xy = 1$ is rotated clockwise 45°, with new vertices at $(\pm\sqrt{2}, 0)$, asymptotes at $y = \pm 1x$ and $b = \pm\sqrt{2}$. Find the equation and graph the hyperbola.

Solution ▶   Using the standard form $\dfrac{x^2}{a^2} - \dfrac{y^2}{b^2} = 1$ and

substituting $\pm\sqrt{2}$ for a and b, the

equation of the rotated hyperbola is

$\dfrac{x^2}{2} - \dfrac{y^2}{2} = 1$ or $x^2 - y^2 = 2$ in polynomial

form. The resulting graph is the central hyperbola shown.

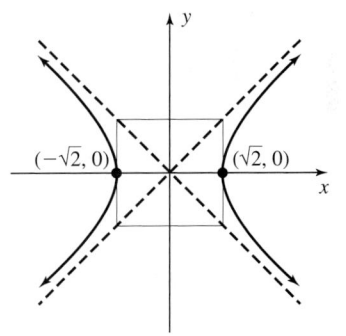

Now try Exercises 7 and 8 ▶

Figure 9.52

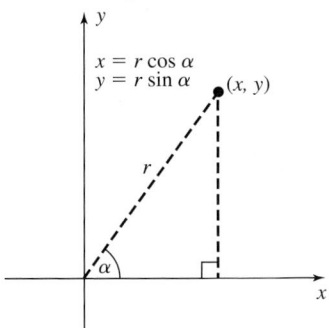

It's important to note the equation of the rotated hyperbola *is devoid of the mixed "xy" term.* In nondegenerate cases, the equation $Ax^2 + Cy^2 + Dx + Ey + F = 0$ is the polynomial form of a conic with axes that are vertical/horizontal. However, the most general form of the equation is $Ax^2 + Bxy + Cy^2 + Dx + Ey + F = 0$, and includes this Bxy term. As noted in Example 1, the inclusion of this term will rotate the graph through some angle β. Based on these observations, we reason that one approach to graphing these conics is to find the angle of rotation β with respect to the xy-axes. We can then use β to rewrite the equation so that it corresponds to a new set of XY-axes, *which are parallel to the axes of the conic.* The mixed xy-term will be absent from the new equation and we can graph the conic on the new axes using the same ideas as before (identifying a, b, foci, and so on). To find β, recall that a point (x, y) in the xy-plane can be written $x = r\cos\alpha$, $y = r\sin\alpha$, as in Figure 9.52. The diagram in Figure 9.53 shows the axes of a new XY-plane, rotated counterclockwise by angle β. In this new plane, the coordinates of the point (x, y) become $X = r\cos(\alpha - \beta)$ and $Y = r\sin(\alpha - \beta)$ as shown. Using the difference identities for sine and cosine and substituting $x = r\cos\alpha$ and $y = r\sin\alpha$ leads to

Figure 9.53

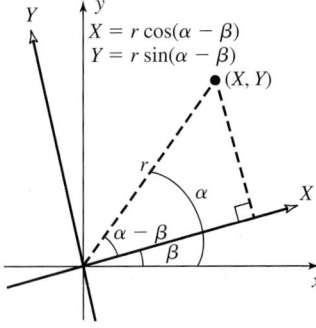

$$X = r\cos(\alpha - \beta) \qquad\qquad Y = r\sin(\alpha - \beta)$$
$$= r(\cos\alpha\cos\beta + \sin\alpha\sin\beta) \qquad = r(\sin\alpha\cos\beta - \cos\alpha\sin\beta)$$
$$= r\cos\alpha\cos\beta + r\sin\alpha\sin\beta \qquad = r\sin\alpha\cos\beta - r\cos\alpha\sin\beta$$
$$= x\cos\beta + y\sin\beta \qquad\qquad = y\cos\beta - x\sin\beta$$

The last two equations can be written as a system, which we will use to solve for x and y in terms of X and Y.

$$\begin{cases} X = x\cos\beta + y\sin\beta \\ Y = y\cos\beta - x\sin\beta \end{cases} \quad \text{original system}$$

$$\begin{cases} X\cos\beta = x\cos^2\beta + y\sin\beta\cos\beta & \text{multiply first equation by } \cos\beta \\ Y\sin\beta = y\sin\beta\cos\beta - x\sin^2\beta & \text{multiply second equation by } \sin\beta \end{cases}$$

$$X\cos\beta - Y\sin\beta = x\cos^2\beta + x\sin^2\beta \qquad \text{first equation} - \text{second equation}$$

$$X\cos\beta - Y\sin\beta = x \qquad \text{factor out } x(\cos^2\beta + \sin^2\beta = 1)$$

Re-solving the system for y results in $y = X\sin\beta + Y\cos\beta$, yielding what are called the **rotation of axes formulas (see Exercise 79).**

WORTHY OF NOTE

If you are familiar with matrices, it may be easier to remember the rotation formulas in their matrix form, since the pattern of functions is the same, with only a difference in sign:

$$\begin{bmatrix} x \\ y \end{bmatrix} = \begin{bmatrix} \cos\beta & -\sin\beta \\ \sin\beta & \cos\beta \end{bmatrix}\begin{bmatrix} X \\ Y \end{bmatrix}$$

$$\begin{bmatrix} X \\ Y \end{bmatrix} = \begin{bmatrix} \cos\beta & \sin\beta \\ -\sin\beta & \cos\beta \end{bmatrix}\begin{bmatrix} x \\ y \end{bmatrix}$$

See Exercises 86 and 87.

Rotation of Axes Formulas

If the x- and y-axes of the xy-plane are rotated counterclockwise by the (acute) angle β to form the X- and Y-axes of an XY-plane, the coordinates of the points (x, y) and (X, Y) are related by the formulas

$$x = X\cos\beta - Y\sin\beta \qquad\qquad X = x\cos\beta + y\sin\beta$$
$$y = X\sin\beta + Y\cos\beta \qquad\qquad Y = -x\sin\beta + y\cos\beta$$

EXAMPLE 2 ▶ **Naming the Location of a Point After Rotating the Axes**

Given the point $(1, \sqrt{3})$ in the xy-plane, find the coordinates of this point in the XY-plane given the angle β between the xy-axes and the XY-axes is $60°$.

Solution ▶ Using the formulas with $x = 1$, $y = \sqrt{3}$, and $\beta = 60°$, we obtain

$$X = x\cos\beta + y\sin\beta \qquad\qquad Y = -x\sin\beta + y\cos\beta$$
$$= 1\cos 60° + \sqrt{3}\sin 60° \qquad = -1\sin 60° + \sqrt{3}\cos 60°$$
$$= \left(\frac{1}{2}\right) + \sqrt{3}\left(\frac{\sqrt{3}}{2}\right) \qquad = -\frac{\sqrt{3}}{2} + \frac{\sqrt{3}}{2}$$
$$= 2 \qquad\qquad = 0$$

The coordinates of $P(X, Y)$ would be $(2, 0)$.

Now try Exercises 9 through 16 ▶

The diagram in Figure 9.54 provides a more intuitive look at the rotation from Example 2. As you can see, a 30-60-90 triangle is formed with a hypotenuse of 2, giving coordinates $(2, 0)$ in the XY-plane.

Figure 9.54

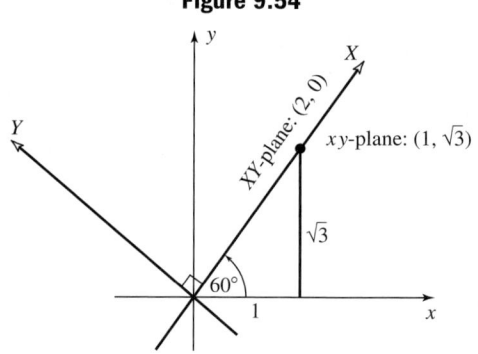

EXAMPLE 3 ▶ **Writing the Equation of a Conic After Rotating the Axes**

The ellipse $X^2 + 4Y^2 = 16$ is rotated clockwise 45°. What is the corresponding equation in the xy-plane?

Solution ▶ We proceed as before, using the rotation formulas $X = x \cos \beta + y \sin \beta$ and $Y = y \cos \beta - x \sin \beta$. With $\beta = 45°$ we have $\cos \beta = \sin \beta = \dfrac{\sqrt{2}}{2}$, yielding

$$X^2 + 4Y^2 = 16$$

$(x \cos \beta + y \sin \beta)^2 + 4(y \cos \beta - x \sin \beta)^2 = 16 \quad \text{use rotation formulas}$

$\left(\dfrac{\sqrt{2}}{2}x + \dfrac{\sqrt{2}}{2}y\right)^2 + 4\left(\dfrac{\sqrt{2}}{2}y - \dfrac{\sqrt{2}}{2}x\right)^2 = 16 \quad \text{substitute } \dfrac{\sqrt{2}}{2} \text{ for } \sin \beta \text{ and } \cos \beta$

$\left(\dfrac{1}{2}x^2 + xy + \dfrac{1}{2}y^2\right) + 4\left(\dfrac{1}{2}x^2 - xy + \dfrac{1}{2}y^2\right) = 16 \quad \text{square binomials}$

$\dfrac{1}{2}x^2 + xy + \dfrac{1}{2}y^2 + 2x^2 - 4xy + 2y^2 = 16 \quad \text{distribute}$

$\dfrac{5}{2}x^2 - 3xy + \dfrac{5}{2}y^2 = 16 \quad \text{result}$

Now try Exercises 17 through 20 ▶

Note the equation of the conic in the standard xy-plane contains the "mixed" Bxy-term. In practice, we seek to reverse this procedure by starting in the xy-plane, and finding the angle β needed to *eliminate* the Bxy-term. Using the rotation formulas and the appropriate angle β, the equation $Ax^2 + Bxy + Cy^2 + Dx + Ey + F = 0$ becomes $aX^2 + cY^2 + dX + eY + f = 0$, where the xy-term is absent. To find the angle β, note that without loss of generality, we can assume $D = E = 0$ since only the second-degree terms are used to identify a conic. Starting with the simplified equation $Ax^2 + Bxy + Cy^2 + F = 0$ and using the rotation formulas we obtain

$$Ax^2 \qquad + \qquad Bx \quad \cdot \quad y \qquad + \qquad Cy^2 \qquad + F = 0$$

$$A(X \cos \beta - Y \sin \beta)^2 + B(X \cos \beta - Y \sin \beta)(X \sin \beta + Y \cos \beta) + C(X \sin \beta + Y \cos \beta)^2 + F = 0$$

Expanding this expression and collecting like terms **(see Exercise 80),** gives the following expressions for coefficients a, b, and c of the corresponding equation $aX^2 + bXY + cY^2 + f = 0$:

$a \rightarrow A \cos^2\beta + B \sin \beta \cos \beta + C \sin^2\beta \qquad a \text{ is the coefficient of } X^2$

$b \rightarrow -2A \sin \beta \cos \beta + B(\cos^2\beta - \sin^2\beta) + 2C \sin \beta \cos \beta \quad b \text{ is the coefficient of } XY$

$c \rightarrow A \sin^2\beta - B \sin \beta \cos \beta + C \cos^2\beta \qquad c \text{ is the coefficient of } Y^2$

$f \rightarrow F \qquad\qquad f = F \text{ (the constant remains unchanged)}$

To accomplish our purpose, we require the coefficient b to be zero. While this expression looks daunting, the double-angle identities for sine and cosine simplify it very nicely:

$$b \rightarrow -A(2 \sin \beta \cos \beta) + B(\cos^2\beta - \sin^2\beta) + C(2 \sin \beta \cos \beta) = 0 \quad (1)$$

$$-A \sin(2\beta) + B \cos(2\beta) + C \sin(2\beta) = 0 \quad\quad (2)$$

$$(C - A)\sin(2\beta) = -B \cos(2\beta) \quad\quad (3)$$

$$\tan(2\beta) = \dfrac{-B}{C - A} \quad\quad (4)$$

$$\tan(2\beta) = \dfrac{B}{A - C}; A \neq C \quad\quad (5)$$

Note from line (3) that $A = C$ would imply $\cos(2\beta) = 0$, giving $2\beta = 90°$ or $-90°$, with $\beta = 45°$ or $-45°$ (for the sake of convenience, we select the angle in QI). This fact can many times be used to great advantage. If $A \neq C$, $\tan(2\beta) = \dfrac{B}{A - C}$ and we choose 2β between 0 and $180°$ so that β will be in the first quadrant $[0 < \beta < 90°]$.

The Equation of a Conic After Rotating the Axes

For a conic defined by $Ax^2 + Bxy + Cy^2 + Dx + Ey + F = 0$ and its graph in the xy-plane, an angle β can be determined using $\tan(2\beta) = \dfrac{B}{A - C}$ and used in the rotation formulas to find a polynomial $aX^2 + cY^2 + dX + eY + f = 0$ in XY-plane, where the conic is either vertical or horizontal.

EXAMPLE 4 ▶ **Rotating the Axes to Eliminate the *Bxy*-Term**

For $x^2 - 2\sqrt{3}xy + 3y^2 - \sqrt{3}x - y - 16 = 0$, eliminate the xy-term using a rotation of axes and identify the conic associated with the resulting equation. Then sketch the graph of the rotated conic in the XY-plane.

Solution ▶ Since $A \neq C$, we find β using $\tan(2\beta) = \dfrac{B}{A - C}$, giving $\tan(2\beta) = \dfrac{-2\sqrt{3}}{1 - 3} = \sqrt{3}$.

This shows $2\beta = \tan^{-1}\sqrt{3}$, yielding $2\beta = 60°$ so $\beta = 30°$. Using $\cos 30° = \dfrac{\sqrt{3}}{2}$ and $\sin 30° = \dfrac{1}{2}$ along with the rotation formulas we obtain the following XY-equation, with corresponding terms shown side-by-side for clarity:

Given Term in *xy*-Plane	**Corresponding Term in *xy*-Plane**

$$x^2 \longrightarrow \left(\frac{\sqrt{3}}{2}X - \frac{1}{2}Y\right)^2 = \frac{3}{4}X^2 - \frac{\sqrt{3}}{2}XY + \frac{1}{4}Y^2$$

$$-2\sqrt{3}xy \longrightarrow -2\sqrt{3}\left(\frac{\sqrt{3}}{2}X - \frac{1}{2}Y\right)\left(\frac{1}{2}X + \frac{\sqrt{3}}{2}Y\right) = -\frac{3}{2}X^2 - \sqrt{3}XY + \frac{3}{2}Y^2$$

$$3y^2 \longrightarrow 3\left(\frac{1}{2}x + \frac{\sqrt{3}}{2}Y\right)^2 = \frac{3}{4}X^2 + 3\frac{\sqrt{3}}{2}XY + \frac{9}{4}Y^2$$

$$-\sqrt{3}x \longrightarrow -\sqrt{3}\left(\frac{\sqrt{3}}{2}X - \frac{1}{2}Y\right) = -\frac{3}{2}X + \frac{\sqrt{3}}{2}Y$$

$$-y \longrightarrow -\left(\frac{1}{2}X + \frac{\sqrt{3}}{2}Y\right) = -\frac{1}{2}X - \frac{\sqrt{3}}{2}Y$$

$$-16 \longrightarrow -16$$

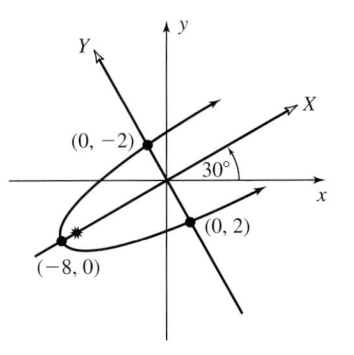

Adding the like terms to the far right, the X^2-terms (in red), the Y-terms (**in bold**), and the mixed XY-terms (in blue) sum to zero, leaving the equation $-2X + 4Y^2 - 16 = 0$, which is the parabola defined by $Y^2 = \frac{1}{2}(X + 8)$. This parabola is symmetric to the X-axis and opens to the right, with a vertex at $(-8, 0)$, Y-intercepts at $(0, -2)$ and $(0, 2)$, focus at $\left(-\frac{63}{8}, 0\right)$ and directrix through $\left(-\frac{65}{8}, 0\right)$. The graph is shown in the figure.

Now try Exercises 21 through 30 ▶

In Example 4, the angle β was a **standard angle** and easily found. In general, this is not the case and finding exact values of $\cos \beta$ and $\sin \beta$ for use in the rotation formulas requires using $\tan(2\beta) = \dfrac{\sin(2\beta)}{\cos(2\beta)}$, the corresponding (triangle) diagram, and the iden-

☑ **A.** You've just learned how to graph conic sections that have nonvertical and nonhorizontal axes (rotated conics)

tities $\cos \beta = \sqrt{\dfrac{1 + \cos(2\beta)}{2}}$ and $\sin \beta = \sqrt{\dfrac{1 - \cos(2\beta)}{2}}$. **See Exercises 31, 32, 84, and 85** for further study.

B. Identifying Conics Using the Discriminant

In addition to rotating the axes, the inclusion of the "xy-term" makes it impossible to identify the conic section using the tests seen earlier. For example, having $A = C$ no longer guarantees a circle, and $A = 0$ or $C = 0$ does not guarantee a parabola. Rather than continuing to look at what the mixed term and the resulting rotation *changes,* we now look at what the rotation *does not change,* called **invariants** of the transformation. These invariants can be used to double-check the algebra involved and to identify the conic using the **discriminant.** These are given here without proof.

Invariants of a Rotation and Classification Using the Discriminant

By rotating the coordinate axes through a predetermined angle β,

the equation $Ax^2 + Bxy + Cy^2 + Dx + Ey + F = 0$
can be transformed into $aX^2 + cY^2 + dX + eY + f = 0$

in which the xy-term is absent. This rotation has the following invariants:

$(1)\; F = f \qquad (2)\; A + C = a + c \qquad (3)\; B^2 - 4AC = b^2 - 4ac.$

The discriminant of a conic equation in polynomial form is $B^2 - 4AC$. Except in degenerate cases, the graph of the equation can be classified as follows:

If $B^2 - 4AC = 0$, the graph will be a parabola.
If $B^2 - 4AC < 0$, the graph will be a circle or an ellipse.
If $B^2 - 4AC > 0$, the graph will be a hyperbola.

EXAMPLE 5A ▶ **Verifying the Invariants of a Rotation of Axes**

Verify the invariants just given using the equations from Example 4. Also verify the discriminant test.

Solution ▶ From the equation $x^2 - 2\sqrt{3}xy + 3y^2 - \sqrt{3}x - y - 16 = 0$, we have $A = 1, B = -2\sqrt{3}, C = 3, D = -\sqrt{3}, E = -1$, and $F = -16$. After applying the rotation the equation became $-2X + 4Y^2 - 16 = 0$, with $a = 0, b = 0, c = 4, d = -2, e = 0$, and $f = -16$. Checking each invariant gives (1) $-16 = -16$✓, (2) $1 + 3 = 0 + 4$✓, and (3) $(-2\sqrt{3})^2 - 4(1)(3) = (0)^2 - 4(0)(4)$✓. With $B^2 - 4AC = 0$, the discriminant test indicates the conic is a parabola ✓.

EXAMPLE 5B ▶ **Identifying the Equation of a Conic Using the Discriminant**

Use the discriminant to identify each equation as that of a circle, ellipse, parabola, or hyperbola, but do not graph the equation.

  a. $3x^2 - 4xy + 3y^2 + 6x + 12y - 2 = 0$

  b. $4x^2 + 9xy + 4y^2 - 8x + 24y + 9 = 0$

  c. $6x^2 - 7xy + y^2 - 5 = 0$

  d. $x^2 - 6xy + 9y^2 + 6x = 0$

Solution ▶

a. $A = 3; B = -4; C = 3$
$$B^2 - 4AC = (-4)^2 - 4(3)(3)$$
$$= -20$$
circle or ellipse

b. $A = 4; B = 9; C = 4$
$$B^2 - 4AC = (9)^2 - 4(4)(4)$$
$$= 17$$
hyperbola

c. $A = 6; B = -7; C = 1$
$$B^2 - 4AC = (-7)^2 - 4(6)(1)$$
$$= 25$$
hyperbola

d. $A = 1; B = -6; C = 9$
$$B^2 - 4AC = (-6)^2 - 4(1)(9)$$
$$= 0$$
parabola

☑ **B.** You've just learned how to identify conics using the discriminant of the polynomial form—the invariant $B^2 - 4AC$

Now try Exercises 33 through 36 ▶

C. Conic Equations in Polar Form

You might recall that earlier in this chapter we defined ellipses and hyperbolas in terms of a distance between two points, but a parabola in terms of a distance between a point and a line (the focus and directrix). Actually, all conic sections can be defined using a focus/directrix development *and written in polar form*. This serves to unify and greatly simplify their study. We begin by revisiting the focus/directrix development of a parabola, using a directrix $\mathcal{L}$ and placing the focus at the origin. With the polar axis as the axis of symmetry and the point $P(r, \theta)$ in polar coordinates, we obtain the graph shown in Figure 9.55. Given D and A are points on $\mathcal{L}$ (with A on the polar axis), we note the following:

Figure 9.55

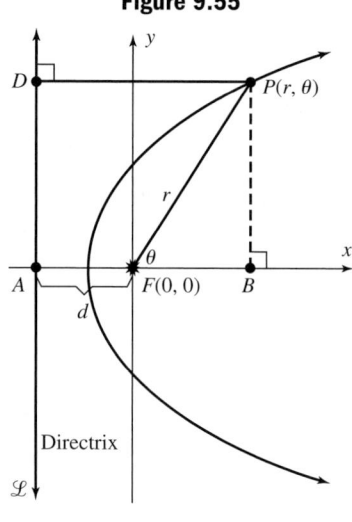

(1) $\overline{DP} = \overline{FP}$	definition of a parabola
(2) $\overline{DP} = \overline{AB}$	equal line segments
(3) $\overline{FB} = r \cos \theta$	$\cos \theta = \dfrac{\overline{FB}}{r}$
(4) $\overline{AB} = \overline{AF} + \overline{FB}$	sum of line segments

Using the preceding equations and representing the distance $\overline{AF}$ by the constant d, we obtain this sequence:

$$\overline{AB} = d + r \cos \theta \quad \text{substitute } d \text{ for } \overline{AF} \text{ and } r \cos \theta \text{ for } \overline{FB}$$

$$\overline{FP} = d + r \cos \theta \quad \text{substitute } \overline{FP} \text{ for } \overline{AB} \text{ since } \overline{FP} = \overline{DP} = \overline{AB}$$

$$r = d + r \cos \theta \quad \text{substitute } r \text{ for } \overline{FP}$$

Solving the last equation for r we have $r - r \cos \theta = d$, then $r = \dfrac{d}{1 - \cos \theta}$, which is the equation of a parabola in polar form with its focus at the origin, vertex at $\left(-\dfrac{d}{2}, \pi\right)$, and y-intercepts at $\left(d, \dfrac{\pi}{2}\right)$ and $\left(d, \dfrac{3\pi}{2}\right)$. Note the constant "1" in the denominator is a key characteristic of polar equations, and helps define the standard form.

EXAMPLE 6A ▶ **Identifying a Conic from Its Polar Equation**

Verify the equation $r = \dfrac{6}{3 - 3 \cos \theta}$ represents a parabola, then describe and sketch the graph.

Solution ▶ Write the equation in standard form by dividing the numerator and denominator by 3, obtaining $r = \dfrac{2}{1 - \cos \theta}$. From this we see $d = 2$ and the represents a parabola symmetric to the polar axis, with vertex at $(-1, \pi)$ and y-intercepts at $\left(2, \dfrac{\pi}{2}\right)$ and $\left(2, \dfrac{3\pi}{2}\right)$, as shown in the figure.

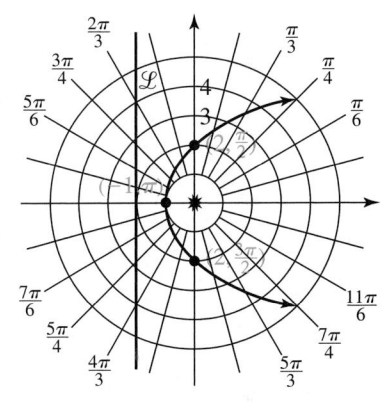

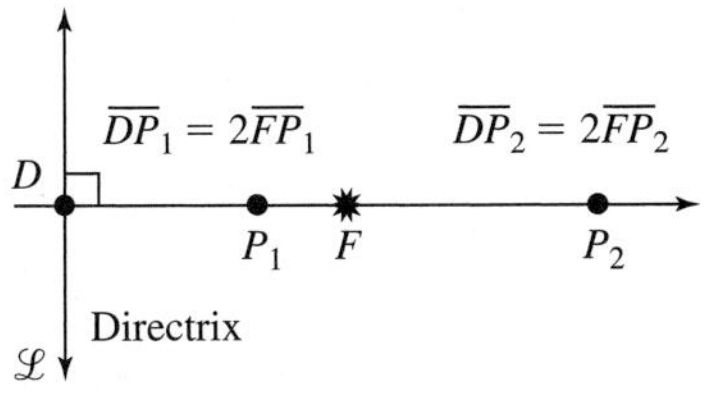

The polar equation for a parabola depended on $\overline{DP}$ and $\overline{FP}$ being equal in length, with ratio $\dfrac{\overline{FP}}{\overline{DP}} = 1$. But what if this ratio is not equal to 1? Similar to our introduction to conics in Section 9.1, we assume $\dfrac{\overline{FP}}{\overline{DP}} = \dfrac{1}{2}$ and investigate the graph that

Figure 9.56

$\overline{DP}_1 = 2\overline{FP}_1 \qquad \overline{DP}_2 = 2\overline{FP}_2$

D

$P_1 \quad F \qquad\qquad P_2$

Directrix

$\mathscr{L}$

results. Cross-multiplying gives $2\overline{FP} = \overline{DP}$, which states that the distance from D to P is twice the distance from F to P. Note that we are able to locate *two points P_1 and P_2 on the polar axis* that satisfy this relation, rather than only one as in the case of the parabola. Figure 9.56 illustrates the location of these points. Using the focal chord for convenience, two additional points P_3 and P_4 can be located that also satisfy the stated condition (see Figure 9.57). In fact, we can locate an infinite number of these

Figure 9.57

D_1

P_3

$\overline{D_1P_3} = 2\overline{FP_3}$

F

$P_1 \qquad \overline{D_2P_4} = 2\overline{FP_4} \qquad P_2$

D_2

P_4

$\mathscr{L}$

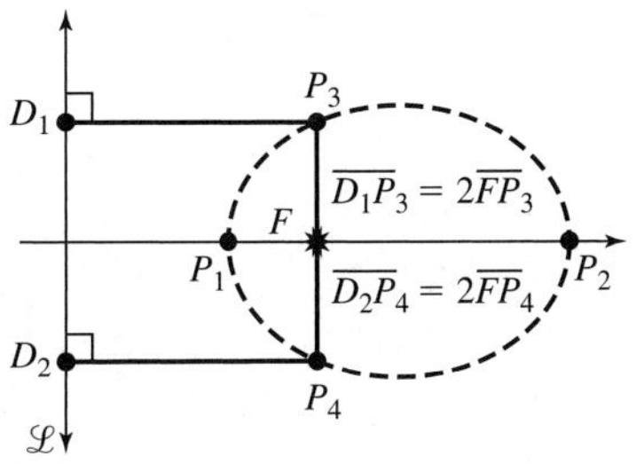

points using $\dfrac{\overline{FP}}{\overline{DP}} = \dfrac{1}{2}$, and the resulting graph appears to be an ellipse (and is definitely *not* a parabola). These illustrations provide the basis for stating a general focus/directrix definition of the conic sections. The ratio $\dfrac{\overline{FP}}{\overline{DP}}$ is often represented by the letter e, and represents the **eccentricity** of the conic. Using $\overline{FP} = r$ and $\overline{DP} = d + r \cos \theta$ from our initial development, $\dfrac{\overline{FP}}{\overline{DP}} = \dfrac{r}{d + r \cos \theta} = e$, which enables us to state the general equation of a conic in polar form. Solving for r leads to the equation $r = \dfrac{de}{1 - e \cos \theta}$, where the type of conic depends solely on e. Depending on the orientation of the conic, the general form may involve sine instead of cosine, and have a sum of terms in the denominator rather than a difference. Note once again that if $e = 1$, the relation simplifies into the parabolic equation seen earlier.

The Standard Equation of a Conic in Polar Form

Given a conic section with eccentricity e, one foci at the pole of the $r\theta$-plane, and directrix $\mathscr{L}$ located d units from this focus. Then the polar equations

$$r = \frac{de}{1 \pm e \cos \theta} \quad \text{and} \quad r = \frac{de}{1 \pm e \sin \theta}$$

represent one of the conic sections as determined by the value of e.

- If $e = 1$, the graph is a parabola.
- If $0 < e < 1$, the graph is an ellipse.
- If $e > 1$, the graph is a hyperbola.

For the ellipse and hyperbola, the major axis and transverse axis (respectively) are both perpendicular to the directrix and contain the vertices and foci. Our earlier development of eccentricity can then be expressed in terms of a and c, as the ratio $e = \dfrac{c}{a}$.

As in our previous study of polar equations, if the equation involves cosine the graph will be symmetric to the polar axis. If the graph involves sine, the line $\theta = \dfrac{\pi}{2}$ is the axis of symmetry. In addition, if the denominator contains a difference of terms (as in Example 6A), the graph will be above or to the right of the directrix (depending on whether the equation involves sine or cosine). If the denominator contains a sum of terms, the graph will be below or to the left of the directrix.

EXAMPLE 6B ▶ **Using the Standard Equation to Graph a Conic in Polar Form**

Determine if the equation $r = \dfrac{10}{5 - 3 \sin \theta}$ represents a parabola, ellipse, or hyperbola. Then describe and sketch the graph.

Solution ▶ To write the equation in standard form, we divide both numerator and denominator by 5, obtaining the equation $r = \dfrac{2}{1 - \dfrac{3}{5} \sin \theta}$. From the standard form we note $e = \dfrac{3}{5}$ so the equation represents an ellipse. With a difference of terms and the sine function involved, the graph is symmetric to $\theta = \dfrac{\pi}{2}$ and is above the directrix. Given so much information by the equation, we require very few points to sketch the graph and settle for those generated by $\theta = 0, \dfrac{\pi}{2}, \pi,$ and $\dfrac{3\pi}{2}$, yielding the points $(2, 0), \left(5, \dfrac{\pi}{2}\right), (2, \pi),$ and $\left(\dfrac{5}{4}, \dfrac{3\pi}{2}\right)$. The graph is shown in the figure.

✓ **C.** You've just learned how to write the equation of a conic section in polar form

Now try Exercises 37 through 56 ▶

D. Applications of Conics in Polar Form

For centuries it has been known that the orbits of the planets around the Sun are elliptical, with the Sun at one focus. In addition, comets may approach our Sun in an elliptical, hyperbolic, or parabolic path with the Sun again at the foci. This makes planetary studies a very natural application of the conic sections in polar form. To aid this study, it helps to know that in an elliptical orbit, the maximum distance of a planet from the Sun is called its **aphelion,** and the shortest distance is the **perihelion** (Figure 9.58). This means the length of the major axis is "aphelion + perihelion," enabling us to find the value of c if the aphelion and perihelion are known (Figure 9.59). Using $e = \dfrac{c}{a}$, we can then find the eccentricity of the planet's orbit.

Figure 9.58

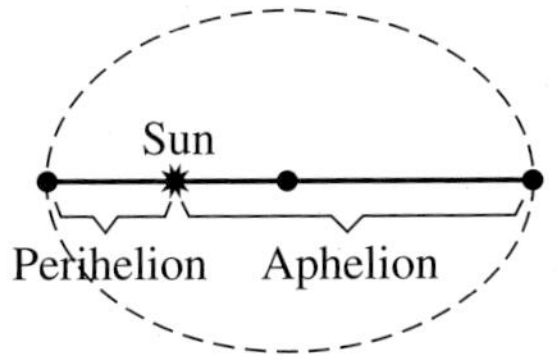

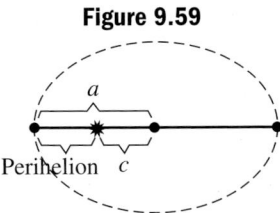

Figure 9.59

EXAMPLE 7 ▶   Determining the Eccentricity of a Planet's Orbit

In its elliptical orbit around the Sun, Mars has an aphelion of 154.9 million miles and a perihelion of 128.4 million miles. What is the eccentricity of its orbit?

Solution ▶   The length of the major axes would be $2a = (154.9 + 128.4)$ mi, yielding a semimajor axis of $a = 141.65$ million miles. Since $a = c +$ perihelion (Figure 9.59), we have $141.65 = c + 128.4$ so $c = 13.25$. The eccentricity of the orbit is

$$e = \frac{c}{a} = \frac{13.25}{141.65} \text{ or about } 0.0935.$$

Now try Exercises 59 and 60 ▶

We can also find the perihelion and aphelion directly in terms of a (semimajor axis) and e (eccentricity) if these quantities are known. Using $a = c +$ perihelion, we obtain: perihelion $= a - c$. For $e = \dfrac{c}{a}$, we have $ea = c$ and by direct substitution we obtain: perihelion $= a - ea = a(1 - e)$. For Example 8, recall that "AU" designates an *astronomical unit,* and represents the mean distance from the Earth to the Sun, approximately 92.96 million miles.

EXAMPLE 8 ▶   Determining the Perihelion of a Planet's Orbit

The orbit of the planet Jupiter has a semimajor axis of 5.2 AU (1 AU ≈ 92.96 million miles) and an eccentricity of 0.0489. What is the closest distance from Jupiter to the Sun?

Solution ▶   With perihelion $= a(1 - e)$, we have $5.2(1 - 0.0489) \approx 4.946$. At its closest approach, Jupiter is 4.946 AU from the Sun (about 460 million miles).

Now try Exercises 61 through 64 ▶

To find the polar equation of a planetary orbit, it's helpful to write the general polar equation in terms of the semimajor axis a, which is often known or easily found, rather than in terms of the distance d from directrix to focus, which is often unknown. Consider the diagram in Figure 9.60, which shows an elliptical orbit with the Sun at one focus, vertices

Figure 9.60

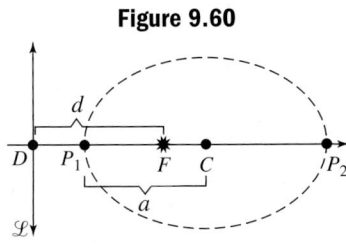

P_1 and P_2 (perihelion and aphelion), and the center C of the ellipse. Assume the point P used to define the conic sections is at position P_1, giving $\dfrac{\overline{FP_1}}{\overline{DP_1}} = e$. From Example 8 we have $\overline{FP_1} = a(1 - e)$. Substituting $a(1 - e)$ for $\overline{FP_1}$ and solving for $\overline{DP_1}$ gives $\overline{DP_1} = \dfrac{a(1 - e)}{e}$. Using $d = \overline{DP_1} + \overline{FP_1}$, we obtain the following sequence:

$$
\begin{aligned}
d &= \overline{DP_1} + \overline{FP_1} \\[4pt]
&= \frac{a(1 - e)}{e} + a(1 - e) &&\text{substitute } \frac{a(1 - e)}{e} \text{ for } \overline{DP_1} \text{ and } a(1 - e) \text{ for } \overline{FP_1} \\[4pt]
&= \frac{a(1 - e)}{e} + \frac{ae(1 - e)}{e} &&\text{common denominator} \\[4pt]
&= \frac{a(1 - e)(1 + e)}{e} &&\text{combine terms, factor out } a(1 - e) \\[4pt]
&= \frac{a(1 - e^2)}{e} &&(1 - e)(1 + e) = 1 - e^2 \\[4pt]
de &= a(1 - e^2) &&\text{multiply by } e
\end{aligned}
$$

Substituting $a(1 - e^2)$ for de in the standard equation $r = \dfrac{de}{1 - e\cos\theta}$ gives the equation of the orbit entirely in terms of a and e: $r = \dfrac{a(1 - e^2)}{1 - e\cos\theta}$.

EXAMPLE 9 ▶ Writing the Polar Equation of an Ellipse from Given Information

At its aphelion, the dwarf planet Pluto is the most distant from the Sun at 4538 million miles. It has a perihelion of 2756 million miles. Use this information to find the polar equation that models the orbit of Pluto, then find the length of the focal chord for this ellipse.

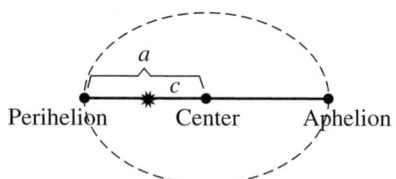

Perihelion   Center   Aphelion

Solution ▶ With all figures in millions of miles, the major axis is $2a = 4538 + 2756 = 7294$, so the semimajor axis has length $a = 3647$. With $a = c + \text{perihelion}$, we obtain $3647 = c + 2756$ or $c = 891$. The eccentricity of the orbit is $e = \dfrac{891}{3647} \approx 0.2443$.

The polar equation for the orbit of Pluto is $r \approx \dfrac{(3647)(1 - [0.244]^2)}{1 - [0.2443]\cos\theta}$ or

☑ **D.** you've just learned how to solve applications involving the conic sections in polar form

$r \approx \dfrac{3430}{1 - 0.2443\cos\theta}$. Substituting $\theta = \dfrac{\pi}{2}$ (since the left-most focus is at the pole), we obtain $r = 3430$, so the length of the focal chord is $2(3430) = 6860$ million miles.

Now try Exercises 65 through 70 ▶

TECHNOLOGY HIGHLIGHT

Investigating the Eccentricity *e*

One meaning of the word eccentric is "to deviate from a circular pattern." In a very real sense, this is the role that eccentricity plays as it helps to describe the conic sections. For an ellipse we've learned that $0 < e < 1$. If the eccentricity is near zero, there is little deviation and the ellipse appears nearly circular. If e is near 1, the ellipse is very elongated. To explore the eccentricity of an ellipse, enter the equation

$$r = \frac{a(1 - e^2)}{1 - e\cos\theta}$$ on the Y= screen, using $a = 2$ (arbitrarily chosen)

Figure 9.61

```
Plot1  Plot2  Plot3
\r1▤2(1-E²)/(1-E
cos(θ))
\r2=
\r3=
\r4=
\r5=
\r6=
```

and ALPHA SIN "E" for the eccentricity. The result is shown in Figure 9.61. We will enter and store values for E on the home screen and graph the resulting ellipse (see Exercise 2 for an alternative method). Return to the home screen and enter 0.1 STO►   ALPHA   SIN and graph the result on the ZOOM **4:ZDecimal** screen. Repeat the procedure using $e = 0.25$, 0.5, 0.75, and 0.9. The graphs for $e = 0.1$ and $e = 0.9$ are shown in Figures 9.62 and 9.63. As you can see, when $e = 0.1$ the ellipse is nearly circular, while $e = 0.9$ produces a graph that is cigar shaped.

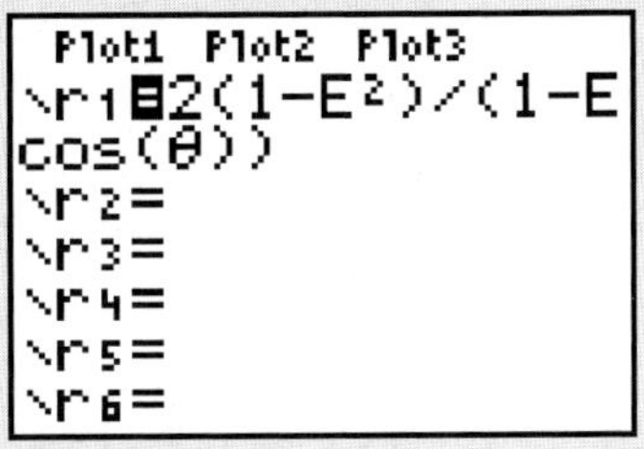

Figure 9.62

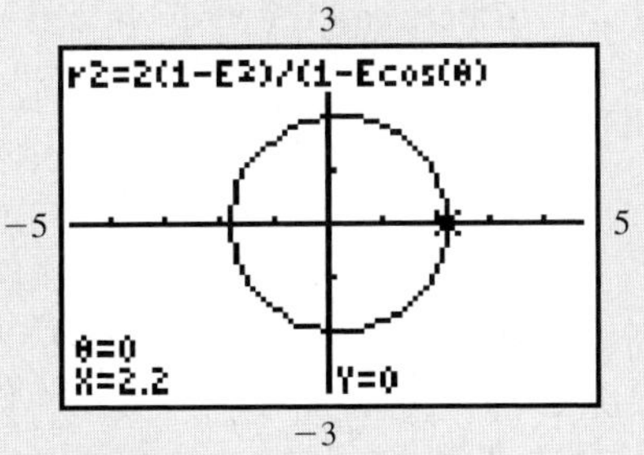

Figure 9.63

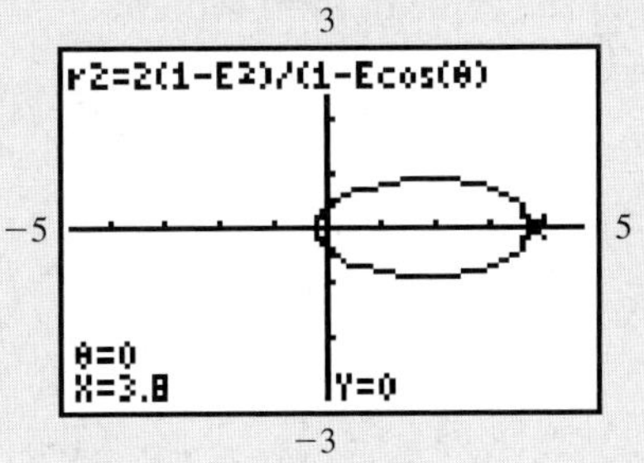

Exercise 1: Try entering a value of $e = 0$, then use your graphing calculator and basic knowledge to verify the resulting graph is a circle.

Exercise 2: Try the same exercise using the set/list option. In other words, enter the equation as shown here, with the values of e in braces { }: $r_1 = \dfrac{2(1 - \{0.1, 0.25, 0.5, 0.75, 0.9\}^2)}{(1 - \{0.1, 0.25, 0.5, 0.75, 0.9\}\cos(\theta))}$. This will enable you to view all five ellipses on the same \screen. Discuss the similarities and differences of this family of graphs.

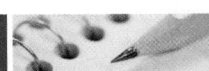

9.7 EXERCISES

► CONCEPTS AND VOCABULARY

**Fill in each blank with the appropriate word or phrase.
Carefully reread the section if needed.**

1. The set of points (x, y) in the xy-plane are related to points (X, Y) in the XY-plane by the _____ _____ formulas. To find the angle β between the original axes and the rotated axes, we use $\tan(2\beta) =$ _____.

2. For a point P on the graph of a conic with focus F and D a point on the directrix, the ratio $\dfrac{\overline{FP}}{\overline{DP}}$ gives the _____ of the graph. For the eccentricity e, if $e = 1$ the graph is a _____, if $e > 1$ the graph is a _____, and if $0 < e < 1$ the graph will be an ellipse.

3. Features or relationships that do not change when certain transformations are applied are called _____ of the transformation.

4. The _____ form of the equation of a conic is $r = \dfrac{de}{1 \pm e \cos \theta}$ if the graph is symmetric to

the _____ axis, and $r = \dfrac{de}{1 \pm e \sin \theta}$ if symmetric to the line _____.

5. Discuss the advantages of graphing a rotated conic using the rotation of axes, over graphing by simply plotting points.

6. Discuss the primary advantages of using $r = \dfrac{a(1 - e^2)}{1 - e \cos \theta}$ rather than $r = \dfrac{de}{1 - e \cos \theta}$ to develop the equation of planetary orbit.

▶ DEVELOPING YOUR SKILLS

The graph of a conic rotated in the xy-plane is given. Use the graph (not the rotation of axes formulas) to find the equation of the conic in the XY-plane.

7.

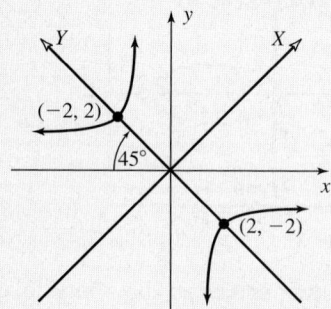

8.

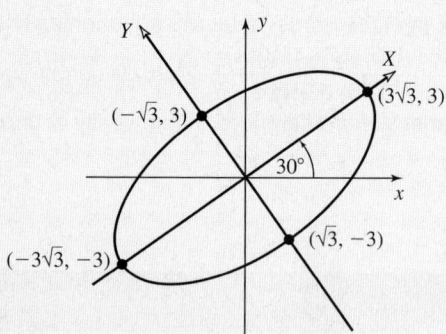

Given the point (x, y) in the xy-plane, find the coordinates of this point in the XY-plane given the angle β between the xy-axes and the XY-axes is 45°.

9. $(6\sqrt{2}, 6)$ 10. $(4, 3\sqrt{2})$

11. $(0, 5)$ 12. $(8, 0)$

Given the point (X, Y) in the XY-plane, find the coordinates of this point in the xy-plane given the angle β between the xy-axes and the XY-axes is 30°.

13. $(2, 2\sqrt{3})$ 14. $(\sqrt{3}, 3)$

15. $(3, 4)$ 16. $(12, 5)$

The conic sections whose equations are given in the XY-plane are rotated clockwise by the indicated angle. Find the corresponding equation in the xy-plane.

17. $X^2 - Y^2 = 9$; 60°   18. $X^2 + Y = 4$; 60°

The conic sections whose equations are given in the xy-plane are rotated by the indicated angle. What is the corresponding equation in the XY-plane?

19. $3x^2 + 2xy + 3y^2 = 9$; 45°

20. $x^2 + \sqrt{3}xy + 2y^2 = 8$; 60°

For the given conics in the xy-plane, (a) use a rotation of axes to find the corresponding equation in the XY-plane (clearly state the angle of rotation β), and (b) sketch its graph. Be sure to indicate the characteristic features of each conic in the XY-plane.

21. $x^2 + 4xy + y^2 - 2 = 0$

22. $x^2 + 2xy + y^2 - 12 = 0$

23. $5x^2 + 6xy + 5y^2 = 16$

24. $5x^2 - 26xy + 5y^2 = -72$

25. $x^2 + 10\sqrt{3}xy + 11y^2 = -64$

26. $37x^2 + 42\sqrt{3}xy + 79y^2 - 400 = 0$

27. $3x^2 - 2\sqrt{3}xy + y^2 - 8x - 8\sqrt{3}y = 0$

28. $6x^2 - 4\sqrt{3}xy + 2y^2 + 2x + 2\sqrt{3}y = 0$

29. $13x^2 - 6\sqrt{3}xy + 7y^2 - 100 = 0$

30. $x^2 + 4xy + y^2 + \sqrt{2}x + \sqrt{2}y = -11$

Identify the graph of each equation using the discriminant, then find the value of $\cos(2\beta)$ using

$\tan(2\beta) = \dfrac{\sin(2\beta)}{\cos(2\beta)}$ **and the related triangle diagram.**

Finally, find $\sin\beta$ and $\cos\beta$ using the half-angle

identities $\cos\beta = \sqrt{\dfrac{1 + \cos(2\beta)}{2}}$ and

$\sin\beta = \sqrt{\dfrac{1 - \cos(2\beta)}{2}}$.

31. $12x^2 + 24xy + 5y^2 - 40x - 30y = 25$

32. $25x^2 + 840xy - 16y^2 - 400 = 0$

For the following equations, (a) use the discriminant to identify the equation as that of a circle, ellipse, parabola, or hyperbola; (b) find the angle of rotation β and use it to find the corresponding equation in the XY-plane; and (c) verify all invariants of the transformation.

33. $x^2 - 2xy + y^2 - 5 = 0$

34. $2x^2 - 3xy + 2y^2 = 0$

35. $3x^2 + \sqrt{3}xy + 4y^2 + 4x = 1$

36. $3x^2 + 8\sqrt{3}xy - 5y^2 + 12y = -2$

Match each graph to its corresponding equation. Justify your answers (two equations have no match).

37.

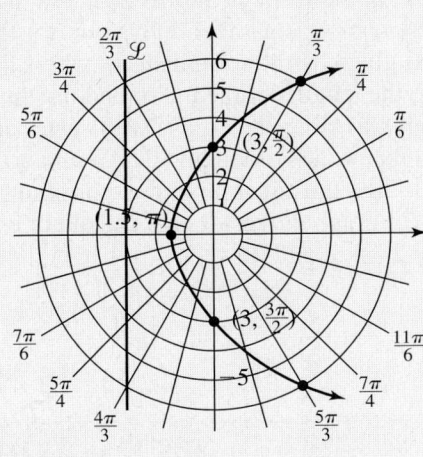

38.

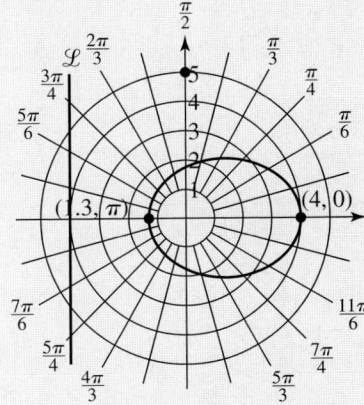

39.

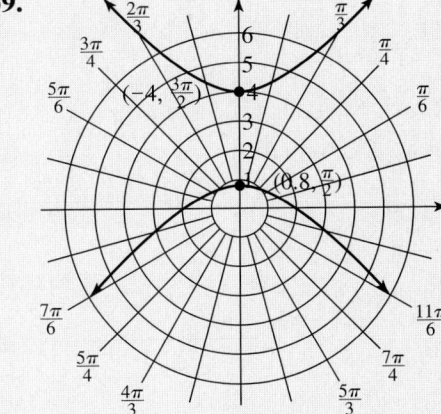

40.

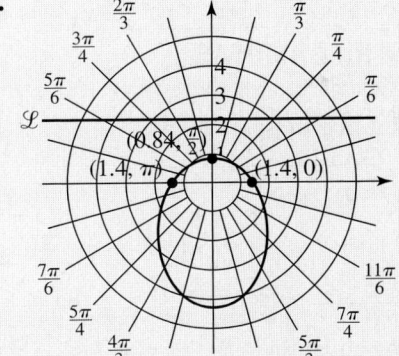

41.

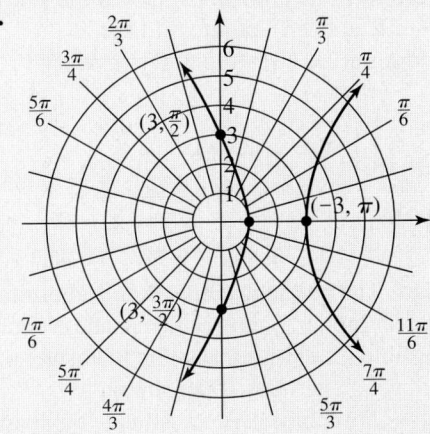

42.

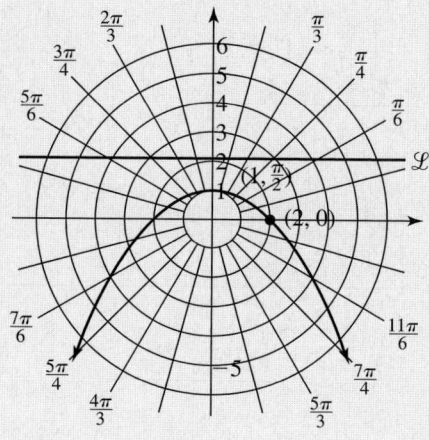

a. $r = \dfrac{10}{5 + 5 \sin \theta}$ **b.** $r = \dfrac{8}{4 - 2 \cos \theta}$

c. $r = \dfrac{5.4}{3 - 2 \sin \theta}$ **d.** $r = \dfrac{4.2}{3 + 2 \sin \theta}$

e. $r = \dfrac{12}{4 + 6 \sin \theta}$ **f.** $r = \dfrac{6}{2 - 2 \cos \theta}$

g. $r = \dfrac{4}{2 + 3 \sin \theta}$ **h.** $r = \dfrac{9}{3 + 6 \cos \theta}$

For the conic equations given, determine if the equation represents a parabola, ellipse, or hyperbola. Then describe and sketch the graphs using polar graph paper.

43. $r = \dfrac{4}{2 + 2 \sin \theta}$ **44.** $r = \dfrac{10}{5 - 5 \sin \theta}$

45. $r = \dfrac{12}{6 - 3 \sin \theta}$ **46.** $r = \dfrac{6}{4 + 3 \cos \theta}$

47. $r = \dfrac{6}{2 + 4 \cos \theta}$ **48.** $r = \dfrac{2}{2 - 3 \sin \theta}$

49. $r = \dfrac{5}{5 + 4 \cos \theta}$ **50.** $r = \dfrac{2}{4 - 5 \sin \theta}$

Write the equation of a conic that satisfies the conditions given. Assume each has one focus at the pole.

51. ellipse, $e = 0.8$, directrix to focus: $d = 4$

52. hyperbola, $e = 1.25$, directrix to focus: $d = 6$

53. parabola, vertex at $(2, \pi)$

54. ellipse, $e = 0.35$, vertex at $(4, 0)$

55. hyperbola, $e = 1.5$, vertex at $\left(3, \dfrac{\pi}{2} \right)$

56. parabola, directrix to focus: $d = 5.4$

▶ **WORKING WITH FORMULAS**

57. Equation of a line in polar form:

$$r = \dfrac{C}{A \cos \theta + B \sin \theta}$$

For the line $Ax + By = C$ in the xy-plane with slope $m = -\dfrac{A}{B}$ and y-intercept $\left(0, \dfrac{C}{B} \right)$, the corresponding equation in the $r\theta$-plane is given by the formula shown. (a) Given the line $2x + 3y = 12$ in the xy-plane, find the corresponding polar equation and (b) verify that $-\dfrac{A}{B} = -\dfrac{r(\pi/2)}{r(0)}$.

58. Polar form of an ellipse with center at the pole:

$$r^2 = \dfrac{a^2 b^2}{a^2 \sin^2 \theta + b^2 \cos^2 \theta}$$

If an ellipse in the $r\theta$-plane has its center at the pole (with major axis parallel to the x-axis), its equation is given by the formula here, where $2a$ and $2b$ are the lengths of the major and minor axes, respectively. (a) Given an ellipse with center at the pole has a major axis of length 8 and a minor axis of length 4, find the equation of the ellipse in polar form and (b) graph the result on a calculator and verify that $2a = 8$ and $2b = 4$.

▶ **APPLICATIONS**

Planetary motion: The perihelion, aphelion, and orbital period of the planets Jupiter, Saturn, Uranus, and Neptune are shown in the table. Use the information to answer or complete the following exercises. The formula $L = 2\pi \sqrt{0.5(a^2 + b^2)}$ can be used to estimate the length of the orbital path. Recall for an ellipse, $c^2 = a^2 - b^2$.

Planet	Perihelion (10^6 mi)	Aphelion (10^6 mi)	Period (yr)
Jupiter	460	507	11.9
Saturn	840	941	29.5
Uranus	1703	1866	84
Neptune	2762	2824	164.8

59. Find the eccentricity of the planets Jupiter and Saturn.

60. Find the eccentricity of the planets Uranus and Neptune.

61. The orbit of Pluto (a dwarf planet) has a semimajor axis of 3647 million miles and an eccentricity of $e = 0.2443$. Find the perihelion of Pluto.

62. The orbit of Ceres (a large asteroid) has a semimajor axis 257 million miles and an eccentricity of $e = 0.097$. Find the perihelion of Ceres.

63. Which of the four planets in the table given has the greatest orbital eccentricity?

64. Which of these four planets has the greatest orbital velocity?

65. Find the polar equation modeling the orbit of Jupiter.

66. Find the polar equation modeling the orbit of Saturn.

67. Find the polar equation modeling the orbit of Uranus.

68. Find the polar equation modeling the orbit of Neptune.

69. Suppose all four major planets arrived at the focal chord of their orbit $\left(\theta = \dfrac{\pi}{2}\right)$ simultaneously. Use the equations in Exercises 65 to 68 to determine the distance between each of the planets at this moment.

70. The polar equation for the orbit of Pluto (a dwarf planet) was developed in Example 9. From an earlier exercise, the polar equation for the orbit of Neptune is $r \approx \dfrac{2793}{1 - 0.0111 \cos \theta}$. Using the TABLE of your graphing calculator, determine if Pluto is *always* the farthest planet from the Sun. If not, how much further from the Sun is Neptune than Pluto at their perihelion?

Mirror manufacturing: A modern manufacturer of oval (elliptical) mirrors for consumer use has programmed the equipment to automatically cut the glass for each mirror (major axis horizontal). The most popular mirrors are those that fit within a golden rectangle (ratio of L to W is approximately 1 to 0.618). Find the polar equation the manufacturer should use to program the equipment for mirror orders of the following lengths. Recall that $c^2 = a^2 - b^2$ and $e = \dfrac{c}{a}$ and assume one focus is at the pole.

71. $L = 4$ ft   **72.** $L = 3.5$ ft

73. $L = 1.5$ m   **74.** $L = 0.5$ m

75. Referring to Exercises 71 to 74, find the total cost of each mirror (to the consumer) if they sell for $75 per square foot ($807 per square meter). The area of an ellipse is given by $A = \pi ab$.

76. Referring to Exercises 71 to 74, find the total cost of an elliptical frame for each mirror (to the consumer) if the frame sells for $12.50 per linear foot ($41.01 per meter). The circumference of an ellipse is approximated by $C = \pi \sqrt{2(a^2 + b^2)}$.

77. Home location: Candice is an enthusiastic golfer and an avid swimmer. After being transferred to a new city, she decides to buy a house that is an equal distance from the local golf course and the river running through the city. If the distance

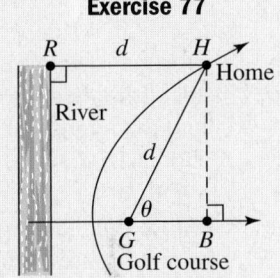

Exercise 77

between the river and the golf course at the closest point is 3 mi, find the polar equation of the parabola that will trace through the possible locations for her new home. Assume the golf course is at the focus of the parabola.

78. Home location: Referring to Exercise 77, assume Candice finds the perfect dream house in a subdivision located at $\left(6, \dfrac{\pi}{3}\right)$. Does this home fit the criteria (is it an equal distance from the river and golf course)?

79. Solve the system below for y to verify the rotation formula for y given on page 898.
$$\begin{cases} X = x \cos \beta + y \sin \beta \\ Y = y \cos \beta - x \sin \beta \end{cases}$$

80. Rotation of a conic section: Expand the following, collect like terms, and simplify. Show the result is the equation $aX^2 + bXY + cY^2 + f = 0$, where the coefficients a, b, c, and f are as given on page 899.
$A(X \cos \beta - Y \sin \beta)^2 + B(X \cos \beta - Y \sin \beta)$
$(X \sin \beta + Y \cos \beta) + C(X \sin \beta + Y \cos \beta)^2 + F = 0$

▶ **EXTENDING THE CONCEPT**

81. Using the rotation of axes formulas in the general equation $Ax^2 + Bxy + Cy^2 + F = 0(D = E = 0)$, we were able to obtain the equation $aX^2 + bXY + cY^2 + f = 0$ (see page 899), where

$$a \to A\cos^2\beta + B\sin\beta\cos\beta + C\sin^2\beta$$

$$b \to -2A\sin\beta\cos\beta + B(\cos^2\beta - \sin^2\beta) + 2C\sin\beta\cos\beta$$

$$c \to A\sin^2\beta - B\sin\beta\cos\beta + C\cos^2\beta \text{ and } f \to F$$

    a. Use these to verify   **b.** Use these to verify   **c.** Explain why the invariant
        $b^2 - 4ac = B^2 - 4AC$. $a + c = A + C$. $f = F$ must always hold.

82. A short-period comet is one that orbits the Sun in 200 yr or less. Two of the best known are Halley's Comet and Encke's Comet. Using any of the resources available to you, find the perihelion and aphelion of each comet and use the information to find the lengths of the semimajor and semiminor axes. Also find the period of each comet. If the length of an elliptical (orbital) path is approximated by $L = 2\pi\sqrt{0.5(a^2 + b^2)}$, find the approximate average speed of each comet in miles per hour. Finally, determine the polar equation of each orbit.

83. In the $r\theta$-plane, the equation of a circle having radius R, center at (R, β), and going through the pole is given by $r = 2R\cos(\theta - \beta)$. Consider the circle defined by $x^2 + y^2 - 6\sqrt{2}x - 6\sqrt{2}y = 0$ in the xy-plane. Verify this circle goes through the origin, then find the equation of the circle in polar form.

For the given conics in the xy-plane, use a rotation of axes to find the corresponding equation in the XY-plane. See Exercises 31 and 32.

84. $12x^2 + 24xy + 5y^2 - 40x - 30y = 25$

85. $25x^2 + 840xy - 16y^2 - 400 = 0$

86. A right triangle in the xy-plane had vertices at $(0, 0)$, $(8, 0)$, and $(8,6)$. Use the matrix equation $\begin{bmatrix} X \\ Y \end{bmatrix} = \begin{bmatrix} \cos\beta & \sin\beta \\ -\sin\beta & \cos\beta \end{bmatrix} \cdot \begin{bmatrix} x \\ y \end{bmatrix}$ to find the vertices in the XY-plane after the triangle is rotated $60°$.

87. A square in the XY-plane has vertices at $(0, 0)$, $(2\sqrt{3}, 2)$, $(2\sqrt{3} - 2, 2 + 2\sqrt{3})$ and $(-2, 2\sqrt{3})$. Use the matrix equation $\begin{bmatrix} x \\ y \end{bmatrix} = \begin{bmatrix} \cos\beta & -\sin\beta \\ \sin\beta & \cos\beta \end{bmatrix} \cdot \begin{bmatrix} X \\ Y \end{bmatrix}$ to find the vertices in the xy-plane after the triangle is rotated $-30°$.

▶ **MAINTAINING YOUR SKILLS**

88. (8.2) Solve the system using elimination.

$$\begin{cases} x + 2y - z = -3 \\ -2x - 6y + z = 4 \\ 5x + 4y - 2z = -3 \end{cases}$$

89. (4.5) Solve for x (to the nearest tenth):
$21.7 = 77.5e^{-0.0052x} - 44.95$

90. (5.5) Use the graph shown to write an equation of the form $y = A\sec(Bx + C)$. Clearly state the values of A, B, and C.

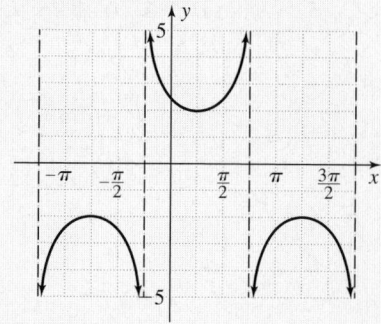

91. (7.3) A ship is moving at 12 mph on a heading of 325°, with a 5 mph current flowing at a 100° heading. Find the true course and speed of the ship.

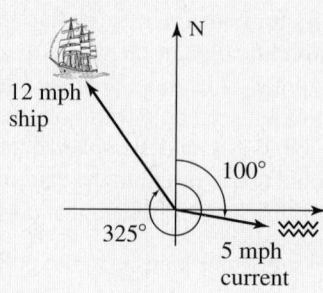

Learning Objectives

In Section 9.8 you will learn how to:

☐ **A.** Sketch the graph of a parametric equation

☐ **B.** Write parametric equations in rectangular form

☐ **C.** Graph curves from the cycloid family

☐ **D.** Solve applications involving parametric equations

A large portion of the mathematics curriculum is devoted to functions, due to their overall importance and widespread applicability. But there are a host of applications for which nonfunctions are a more natural fit. In this section, we show that many *nonfunctions* can be expressed as **parametric equations,** where each is actually a *function.* These equations can be appreciated for the diversity and versatility they bring to the mathematical spectrum.

A. Sketching a Curve Defined Parametrically

Suppose you were given the set of points in the table here, and asked to come up with an equation model for the data. To begin, you might plot the points to see if any patterns or clues emerge, but in this case the result seems to be a curve we've never seen before (see Figure 9.64).

| x | 0 | $\dfrac{\sqrt{3}}{2}$ | $\dfrac{\sqrt{3}}{2}$ | 0 | $-\dfrac{\sqrt{3}}{2}$ | $-\dfrac{\sqrt{3}}{2}$ | 0 |
| y | 1 | $\dfrac{\sqrt{3}}{2}$ | $\dfrac{1}{2}$ | 0 | $-\dfrac{1}{2}$ | $-\dfrac{\sqrt{3}}{2}$ | -1 |

You also might consider running a regression on the data, but it's not possible since the graph is obviously not a function. However, a closer look at the data reveals the *y*-values could be modeled *independently of the x-values* by a cosine function, $y = \cos t$ for $t \in [0, \pi]$. This observation leads to a closer look at the *x*-values, which we find could be modeled by a sine function over the same interval, namely, $x = \sin(2t)$ for $t \in [0, \pi]$. These two functions combine to name all points on this curve, and both use the independent variable t called a **parameter.** The functions $x = \sin(2t)$ and $y = \cos t$ are called the parametric equations for this curve. The complete curve, shown in Figure 9.65, is called a **Lissajous figure,** or a closed graph (coincident beginning and ending points) that crosses itself to form two or more loops. Note that since the maximum value of x and y is 1 (the amplitude of each function), the entire figure will fit within a 1×1 rectangle centered at the origin. This observation can often be used to help sketch parametric graphs with trigonometric parameters. In general, parametric equations can take many forms, including polynomial, exponential, trigonometric, and other forms.

Figure 9.64

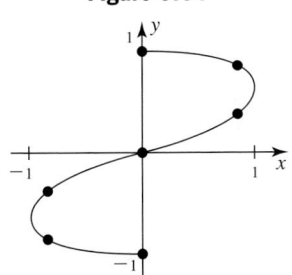

Figure 9.65

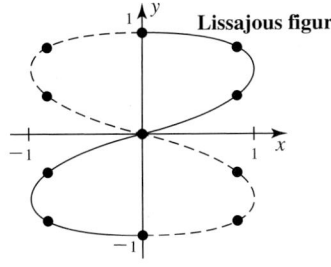

Lissajous figure

Parametric Equations

Given the set of points $P(x, y)$ such that $x = f(t)$ and $y = g(t)$, where f and g are both defined on an interval of the domain, the equations $x = f(t)$ and $y = g(t)$ are called parametric equations, with parameter t.

EXAMPLE 1 ▶ **Graphing a Parametric Curve Where *f* and *g* Are Algebraic**

Graph the curve defined by the parametric equations $x = t^2 - 3$ and $y = 2t + 1$.

Solution ▶ Begin by creating a table of values using $t \in [-3, 3]$. After plotting ordered pairs (x, y), the result appears to be a parabola, opening to the right.

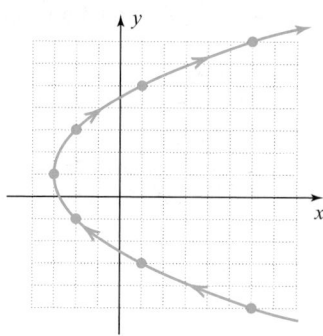

t	$x = t^2 - 3$	$y = 2t + 1$
-3	6	-5
-2	1	-3
-1	-2	-1
0	-3	1
1	-2	3
2	1	5
3	6	7

Now try Exercises 7 through 12, Part a ▶

If the parameter is a trig function, we'll often use standard angles as inputs to simplify calculations and the period of the function(s) to help sketch the resulting graph. Also note that successive values of t give rise to a directional evolution of the graph, meaning the curve is traced out in a direction dictated by the points that correspond to the next value of t. The arrows drawn along the graph illustrate this direction, also known as the **orientation** of the graph.

EXAMPLE 2 ▶ Graphing a Parametric Curve Where *f* and *g* Are Trig Functions

Graph the curve defined by the parametric equations $x = 2 \cos t$ and $y = 4 \sin t$.

Solution ▶ Using standard angle inputs and knowing the maximum value of any x- and y-coordinate will be 2 and 4, respectively, we begin computing and graphing a few points. After going from 0 to π, we note the graph appears to be a vertical ellipse. This is verified using standard values from π to 2π. Plotting the points and connecting them with a smooth curve produces the ellipse shown in the figure.

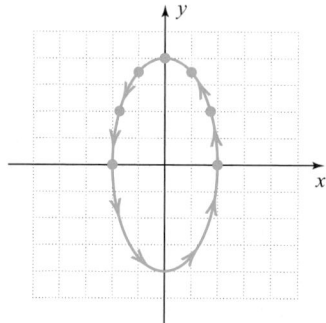

t	$x = 2 \cos t$	$y = 4 \sin t$
0	2	0
$\dfrac{\pi}{6}$	$\sqrt{3}$	2
$\dfrac{\pi}{3}$	1	$2\sqrt{3}$
$\dfrac{\pi}{2}$	0	4
$\dfrac{2\pi}{3}$	-1	$2\sqrt{3}$
$\dfrac{5\pi}{6}$	$-\sqrt{3}$	2
π	-2	0

☑ **A.** You've just learned how to sketch the graph of a parametric equation

Now try Exercises 13 through 18, Part a ▶

Note the ellipse has a counterclockwise orientation.

B. Writing Parametric Equations in Rectangular Form

When graphing parametric equations, there are sometimes alternatives to simply plotting points. One alternative is to try and *eliminate the parameter*, writing the parametric equations in standard, rectangular form. To accomplish this we use some connection that allows us to "rejoin" the parameterized equations, such as variable t itself, a trigonometric identity, or some other connection.

EXAMPLE 3 ▶ **Eliminating the Parameter to Obtain the Rectangular Form**

Eliminate the parameter from the equations in Example 1: $x = t^2 - 3$ and $y = 2t + 1$.

Solution ▶ Solving for t in the second equation gives $t = \dfrac{y - 1}{2}$, which we then substitute into

the first. The result is $x = \left(\dfrac{y - 1}{2}\right)^2 - 3 = \dfrac{1}{4}(y - 1)^2 - 3$. Notice this is indeed a

horizontal parabola, opening to the right, with vertex at $(-3, 1)$.

Now try Exercises 7 through 12, Part b ▶

EXAMPLE 4 ▶ **Eliminating the Parameter to Obtain the Rectangular Form**

Eliminate the parameter from the equations in Example 2: $x = 2 \cos t$ and $y = 4 \sin t$.

Solution ▶ Instead of trying to solve for t, we note the parametrized equations involve sine and cosine functions with the same argument (t), and opt to use the identity $\cos^2 t + \sin^2 t = 1$. Squaring both equations and solving for $\cos^2 t$ and $\sin^2 t$ yields $\dfrac{x^2}{4} = \cos^2 t$ and $\dfrac{y^2}{16} = \sin^2 t$. This shows $\cos^2 t + \sin^2 t = \dfrac{x^2}{4} + \dfrac{y^2}{16} = 1$, and as we suspected—the result is a vertical ellipse with vertices at $(0, \pm 4)$ and endpoints of the minor axis at $(\pm 2, 0)$.

Now try Exercises 13 through 16, Part b ▶

It's important to realize that a given curve can be represented parametrically in infinitely many ways. This flexibility sometimes enables us to simplify the given form, or to write a given polynomial form in an equivalent nonpolynomial form. The easiest way to write the function $y = f(x)$ in parametric form is $x = t$; $y = f(t)$, which is valid *as long as t is in the domain of f(t).*

EXAMPLE 5 ▶ **Writing an Equation in Terms of Various Parameters**

Write the equation $y = 4(x - 3)^2 + 1$ in three different parametric forms.

Solution ▶ **1.** If we let $x = t$, we have $y = 4(t - 3)^2 + 1$.

2. Letting $x = t + 3$ simplifies the related equation for y, and we begin to see some of the advantages of using a parameter: $x = t + 3$; $y = 4t^2 + 1$.

3. As a third alternative, we can let $x = \dfrac{1}{2} \tan t + 3$, which gives

☑ **B.** You've just learned how to write parametric equations in rectangular form

$$x = \dfrac{1}{2} \tan t + 3; \quad y = 4\left(\dfrac{1}{2} \tan t\right)^2 + 1 = \tan^2 t + 1 \text{ or } y = \sec^2 t.$$

Now try Exercises 19 through 26 ▶

C. Graphing Curves from the Cycloid Family

The **cycloids** are an important family of curves, and are used extensively to solve what are called **brachistochrone** applications. The name comes from the Greek *brakhus,* meaning short, and *khronos,* meaning time, and deal with finding the path along which a weight will fall in the shortest time possible. Cycloids are an excellent example of why parametric equations are important, as it's very difficult to name them in rectangular form. Consider a point fixed to the circumference of a wheel as it rolls from left to right. If we trace the path of the point as the wheel rolls, the resulting curve is a cycloid. Figure 9.66 shows the location of the point every one-quarter turn.

Figure 9.66

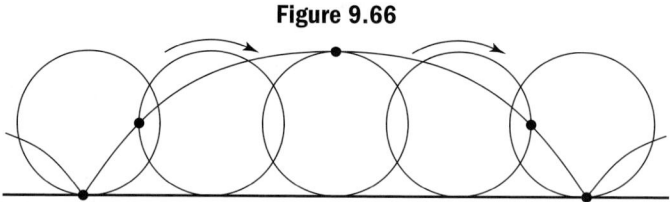

By superimposing a coordinate grid on the diagram in Figure 9.66, we can construct parametric equations that will produce the graph. This is done by developing equations for the location of a point $P(x, y)$ on the circumference of a circle with center (h, k), as the circle rotates through angle t. After a rotation of t rad, the x-coordinate of $P(x, y)$ is $x = h - a$ (Figure 9.67), and the y-coordinate is $y = k - b$. Using a right triangle with the radius as the hypotenuse, we find $\sin t = \dfrac{a}{r}$ and $\cos t = \dfrac{b}{r}$, giving $a = r \sin t$ and $b = r \cos t$. Substituting into $x = h - a$ and $y = k - b$ yields $x = h - r \sin t$ and $y = k - r \cos t$. Since the circle has radius r, we know $k = r$ (the "height" of the center is constantly $k = r$). The arc length subtended by t is the same as the distance h (see Figure 9.68), meaning $h = rt$ (t in radians) Substituting rt for h and r for k in the equations $x = h - r \sin t$ and $y = k - r \cos t$, gives the equation of the cycloid in parametric form: $x = rt - r \sin t$ and $y = r - r \cos t$, sometimes written $x = r(t - \sin t)$ and $y = r(1 - \cos t)$.

Figure 9.67

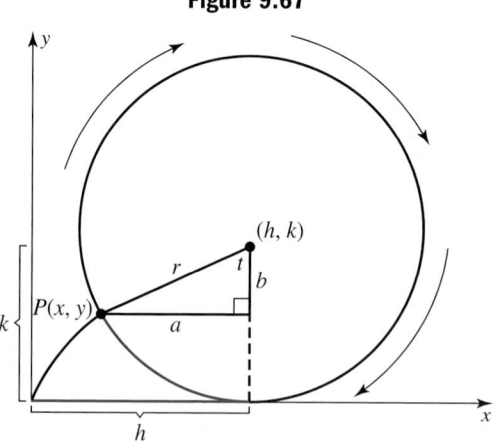

Figure 9.68

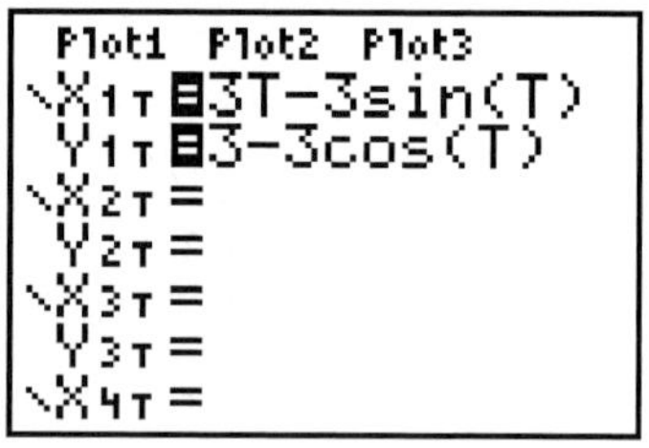

```
Plot1  Plot2  Plot3
\X₁ᴛ■3T-3sin(T)
 Y₁ᴛ■3-3cos(T)
\X₂ᴛ=
 Y₂ᴛ=
\X₃ᴛ=
 Y₃ᴛ=
\X₄ᴛ=
```

Most graphers have a parametric **MODE** that enables you to enter the equations for x and y separately, and graph the resulting points as a single curve. After pressing the **Y =** key (in parametric mode), the screen in Figure 9.68 comes into view using a TI-84 Plus, and we enter the equation of the cycloid formed by a circle of radius $r = 3$. To set the viewing window (including a frame), press **WINDOW** and set Ymin $= -1$ and Ymax at slightly more than 6 (since $r = 3$). Since the cycloid completes one cycle every $2\pi r$, we set Xmax at $2\pi rn$, where n is the number of cycles we'd like to see. In

Figure 9.69

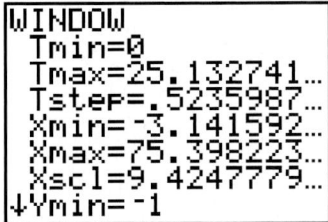

this case, we set it for four cycles $(2\pi)(3)(4) = 24\pi$ (Figure 9.69). With $r = 3$ we conveniently set Xscl at $3(2\pi) = 6\pi \approx 18.8$ to tick each cycle, and Xscl $= 3\pi \approx 9.4$ to tick each half cycle (Figure 9.69). For parametric equations, we must also specify a range of values for t, which we set at Tmin $= 0$, Tmax $= 8\pi \approx 25.1$ for the four cycles, and Tstep $= \dfrac{\pi}{6} \approx 0.52$ (Tstep controls the number of points plotted and joined to form the curve). The window settings and resulting graph are shown in Figure 9.70, which doesn't look much like a cycloid because the current settings do not produce a square viewing window. Using ZOOM **5:ZSquare** (and changing Yscl) produces the graph shown in Figure 9.71, which looks much more like the cycloid we expected.

Figure 9.70   **Figure 9.71**

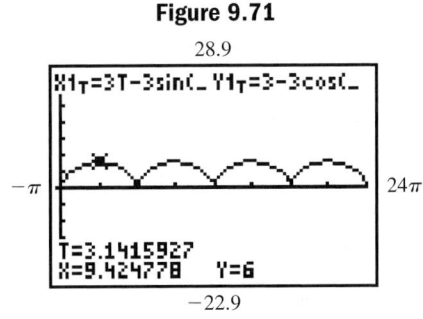

EXAMPLE 6 ▶ **Using Technology to Graph a Cycloid**

Use a graphing calculator to graph the curve defined by the equations $x = 3\cos^3 t$ and $y = 3\sin^3 t$, called a **hypocycloid with four cusps.**

Solution ▶ A hypocycloid is a curve traced out by the path of a point on the circumference of a circle as it rolls *inside a larger circle* of radius r (see Figure 9.72). Here $r = 3$ and we set Xmax and Ymax accordingly. Knowing ahead of time the hypocycloid will have four cusps, we set Tmax $= 4(2\pi) \approx 25.13$ to show all four. The window settings used and the resulting graph are shown in Figure 9.73 and 9.74.

Figure 9.72

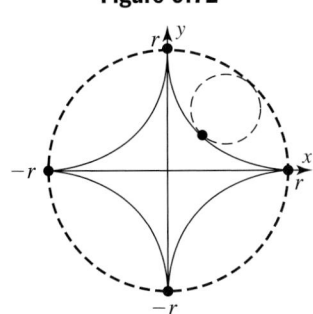

☑ **C.** You've just learned how to graph curves from the cycloid family

Figure 9.73   **Figure 9.74**

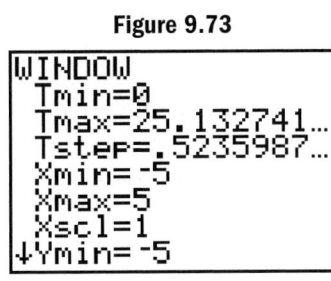   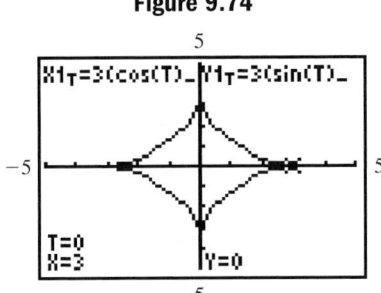

Now try Exercises 27 through 35 ▶

D. Common Applications of Parametric Equations

In Example 1 the parameter was simply the *real number t*, which enabled us to model the x- and y-values of an ordered pair (x, y) independently. In Examples 2 and 6, the parameter t represented an *angle*. Here we introduce yet another kind of parameter, that of *time t*.

A **projectile** is any object thrown, dropped, or projected in some way with no continuing source of propulsion. The parabolic path traced out by the projectile (assuming negligible air resistance) will be fully developed in Section 7.4. It is stated here in parametric terms. For the projectile's location $P(x, y)$ and any time t in seconds, the x-coordinate (horizontal distance from point of projection) is given by $x = v_0 t \cos \theta$, where v_0 is the initial velocity in feet per second and t is the time in seconds. The y-coordinate (vertical height) is $y = v_0 t \sin \theta - 16t^2$.

EXAMPLE 7 ▶ **Using Parametric Equations in Projectile Applications**

As part of a circus act, Karl the Human Cannonball is shot out of a specially designed cannon at an angle of 40° with an initial velocity of 120 ft/sec. Use a graphing calculator to graph the resulting parametric curve. Then use the graph to determine how high the Ring Master must place a circular ring for Karl to be shot through at the maximum height of his trajectory, and how far away the net must be placed to catch Karl.

Solution ▶ The information given leads to the equations $x = 120t \cos 40°$ and $y = 120t \sin 40° - 16t^2$. Enter these equations on the $\boxed{\text{Y =}}$ screen of your calculator, remembering to reset the $\boxed{\text{MODE}}$ to degrees (circus clowns may not know or understand radians). To set the window size, we can use trial and error, or estimate using $\theta = 45°$ (instead of 40°) and an estimate for t (the time that Karl will stay aloft). With $t = 6$ we get estimates of $x = 120(6)\left(\dfrac{\sqrt{2}}{2}\right) = 360\sqrt{2}$ for the horizontal distance. To find a range for y, use $t = 3$ since the maximum height of the parabolic path will occur halfway through the flight. This gives an estimate of $120(3)\left(\dfrac{\sqrt{2}}{2}\right) - 16(9) = 180\sqrt{2} - 144$ for y. The results are shown in Figures 9.75 and 9.76. Using the $\boxed{\text{TRACE}}$ feature or $\boxed{\text{2nd}}$ $\boxed{\text{GRAPH}}$ (**TABLE**) feature, we find the center of the net used to catch Karl should be set at a distance of about 450 ft from the cannon, and the ring should be located 220 ft from the cannon at a height of about 93 ft.

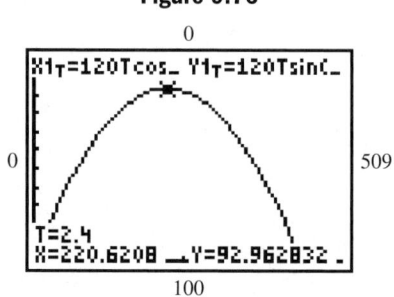

Figure 9.75

```
WINDOW
 Tmin=0
 Tmax=6
 Tstep=.2
 Xmin=0
 Xmax=509.11688…
 Xscl=50
↓Ymin=0
```

Figure 9.76

Now try Exercises 46 through 49 ▶

It is well known that planets orbit the Sun in elliptical paths. While we're able to model their orbits in both rectangular and polar form, neither of these forms can give a true picture of the *direction they travel*. This gives parametric forms a great advantage, in that they can model the shape of the orbit, *while also indicating the direction of travel*. We illustrate in Example 8 using a "planet" with a very simple orbit.

EXAMPLE 8 ▶ **Modeling Elliptical Orbits Parametrically**

The elliptical orbit of a certain planet is defined parametrically as $x = 4 \sin t$ and $y = -3 \cos t$. Graph the orbit and verify that for increasing values of t, the planet orbits in a counterclockwise direction.

Solution ▶ Eliminating the parameter as in Example 4, we obtain the equation $\dfrac{x^2}{16} + \dfrac{y^2}{9} = 1$, or the equation of an ellipse with center at $(0, 0)$, major axis of length 8, and minor axis of length 6. The path of the planet is traced out by the ordered pairs (x, y) generated by the parametric equations, shown in the table for $t \in [0, \pi]$. Starting at $t = 0$, $P(x, y)$ begins at $(0, -3)$ with x and y both increasing until $t = \dfrac{\pi}{2}$. Then from $t = \dfrac{\pi}{2}$ to $t = \pi$, y continues to increase as x decreases, indicating a counterclockwise orbit in this case. The orbit is illustrated in the figure.

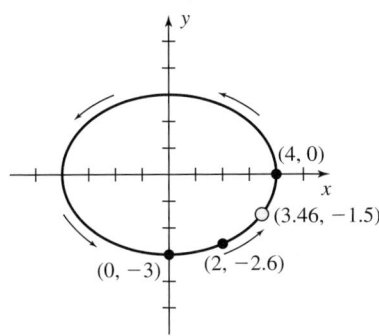

t	$x = 4 \sin t$	$y = -3 \cos t$
0	0	-3
$\dfrac{\pi}{6}$	2	-2.6
$\dfrac{\pi}{3}$	3.46	-1.5
$\dfrac{\pi}{2}$	4	0
$\dfrac{2\pi}{3}$	3.46	1.5
$\dfrac{5\pi}{6}$	2	2.6
π	0	3

Now try Exercises 50 and 51 ▶

Finally, you may recall from your previous work with linear 3×3 systems, that a dependent system occurs when one of the three equations is a linear combination of the other two. The result is a system with more variables than equations, with solutions expressed in terms of a parameter, or in *parametric form*. These solutions can be explored on a graphing calculator using ordered triples of the form $(t, f(t), g(t))$, where $Y_1 = f(t)$ and $Y_2 = g(t)$ (**see Exercises 52 through 55**). For more information, see the *Calculator Exploration and Discovery* feature on page 931.

☑ **D.** You've just learned how to solve applications involving parametric equations

---

TECHNOLOGY HIGHLIGHT

Exploring Parametric Graphs

Most graphing calculators have features that make it easy (and fun) to explore parametric equations. For example, the TI-84 Plus can use a circular cursor to trace the path of the plotted points, as they are generated by the equations. This can be used to illustrate the path of a projectile, the distance of a runner, or the orbit of a planet. Operations can also be applied to the parameter T to give the effect of "speed" (the points from one set of equations are plotted faster than the points of a second set). To help illustrate their use, consider again the simple, elliptical orbit of a planet in Example 8. Physics tells us the closer a planet is to the Sun, the faster its orbit. In fact, the orbital speed of Mercury is about twice that of Mars and about 10 times as fast as the dwarf planet Pluto (29.8, 15, and 2.9 mi/sec, respectively). With this information, we can explore a number of interesting questions. On the $\boxed{Y=}$ screen, let the orbits of Planet 1 and Planet 2 be modeled parametrically by the equations shown in Figure 9.77. Since the orbit of Planet 1 is "smaller" (closer to the Sun), we have T-values growing at a rate that is *four times as fast* as for Planet 2. Notice to the far left of X_{1T}, there is a symbol that looks like an old key "-0." By moving the cursor to the far left of the equation, you can change how the graph will look by repeatedly pressing $\boxed{\text{ENTER}}$. With this symbol in view, the calculator will trace out the curve with a circular cursor, which in this case represents the planets as they orbit (be sure you are in simultaneous $\boxed{\text{MODE}}$). Setting the window as in Figure 9.78 and pressing $\boxed{\text{GRAPH}}$ produces Figure 9.79, which displays their elliptical paths as they race around the Sun. Notice the inner planet has already completed one orbit while the outer planet has just completed one-fourth of an orbit.

Figure 9.77

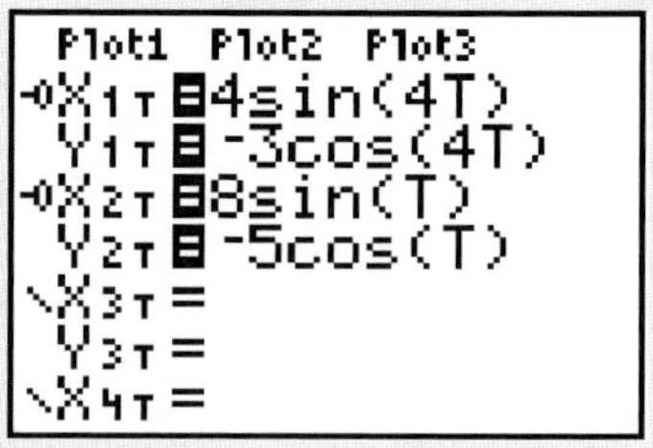

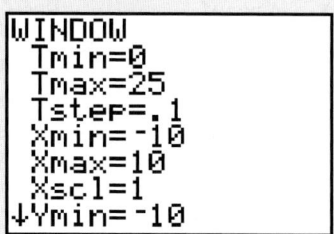

Figure 9.78

```
WINDOW
 Tmin=0
 Tmax=25
 Tstep=.1
 Xmin=-10
 Xmax=10
 Xscl=1
↓Ymin=-10
```

Figure 9.79

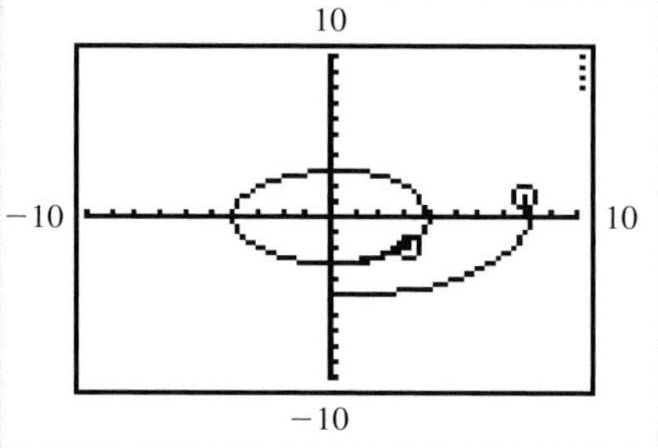

Exercise 1: Verify that the inner planet completes four orbits for every single orbit of the outer planet.

Exercise 2: Suppose that due to some cosmic interference, the orbit of the faster planet begins to decay at a rate of $T^{0.84}$ (replace T with $T^{0.84}$ in both equations for the inner planet). By observation, about how many orbits did the inner planet make for the first revolution of the outer planet? What is the ratio of orbits for the next complete orbit of the outer planet?

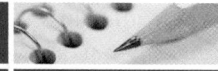

9.8 EXERCISES

▶ CONCEPTS AND VOCABULARY

Fill in each blank with the appropriate word or phrase. Carefully reread the section if needed.

1. When the coordinates of a point (x, y) are generated independently using $x = f(t)$ and $y = g(t)$, t is called a(n) _____.

2. The equations $x = f(t)$ and $y = g(t)$ used to generate the ordered pairs (x, y) are called _____ equations.

3. Parametric equations can both graph a curve *and* indicate the _____ traveled by a point on the curve.

4. To write parametric equations in rectangular form, we must _____ the parameter to write a single equation.

5. Discuss the connection between solutions to dependent systems and the parametric equations studied in this section.

6. In your own words, explain and illustrate the process used to develop the equation of a cycloid. Illustrate with a specific example.

▶ DEVELOPING YOUR SKILLS

For Exercises 7 through 18, (a) graph the curves defined by the parametric equations using the specified interval and identify the graph (if possible) and (b) eliminate the parameter (Exercises 7 to 16 only) and write the corresponding rectangular form.

7. $x = t + 2; t \in [-3, 3]$
$y = t^2 - 1$

8. $x = t - 3; t \in [-5, 5]$
$y = 2 - 0.5t^2$

9. $x = (2 - t)^2; t \in [0, 5]$
$y = (t - 3)^2$

10. $x = t^3 - 3; t \in [-2, 2.5]$
$y = t^2 + 1$

11. $x = \dfrac{5}{t}, t \neq 0; t \in [-3.5, 3.5]$
$y = t^2$

12. $x = \dfrac{t^3}{10}; t \in [-5, 5]$
$y = |t|$

13. $x = 4 \cos t; t \in [0, 2\pi)$
$y = 3 \sin t$

14. $x = 2 \sin t; t \in [0, 2\pi)$
$y = -3 \cos t$

15. $x = 4 \sin(2t); t \in [0, 2\pi)$
$y = 6 \cos t$

16. $x = 4 \cos(2t); t \in \left[\dfrac{\pi}{2}, \dfrac{3\pi}{2}\right]$
$y = 6 \sin t$

17. $x = \dfrac{-3}{\tan t}; t \in (0, \pi)$
$y = 5 \sin(2t)$

18. $x = \tan^2 t; t \neq \dfrac{\pi}{2}, t \in [0, \pi]$
$y = 3 \cos t$

Write each function in three different parametric forms by altering the parameter. For Exercises 19–22 use at least one trigonometric form, restricting the domain as needed.

19. $y = 3x - 2$ **20.** $y = 0.5x + 6$

21. $y = (x + 3)^2 + 1$ **22.** $y = 2(x - 5)^2 - 1$

23. $y = \tan^2(x - 2) + 1$ **24.** $y = \sin(2x - 1)$

25. Use a graphing calculator or computer to verify that the parametric equations from Example 5 all produce the same graph.

26. Use a graphing calculator or computer to verify that your parametric equations from Exercise 21 all produce the same graph.

The curves defined by the following parametric equations are from the cycloid family. (a) Use a graphing calculator or computer to draw the graph and (b) use the graph to approximate all x- and y-intercepts, and maximum and minimum values to one decimal place.

27. $x = 8 \cos t + 2 \cos(4t), y = 8 \sin t - 2 \sin(4t)$, hypocycloid (5-cusp)

28. $x = 8 \cos t + 4 \cos(2t), y = 8 \sin t - 4 \sin(2t)$, hypocycloid (3-cusp)

29. $x = \dfrac{2}{\tan t}, y = 8 \sin t \cos t$, serpentine curve

30. $x = 8 \sin^2 t, y = \dfrac{8 \sin^3 t}{\cos t}$, cissoid of Diocles

31. $x = 2(\cos t + t \sin t), y = 2(\sin t - t \cos t)$, involute of a circle

32. $4x = (16 - 36)\cos^3 t, 6y = (16 - 36)\sin^3 t$, evolute of an ellipse

33. $x = 3t - \sin t, y = 3 - \cos t$, curtate cycloid

34. $x = t - 3 \sin t, y = 1 - 3 \cos t$, prolate cycloid

35. $x = 2[3 \cos t - \cos(3t)], y = 2[3 \sin t - \sin(3t)]$, nephroid

Use a graphing calculator or computer to draw the following parametrically defined graphs, called Lissajous figures (Exercise 37 is a scaled version of the initial example from this section). Then find the dimensions of the rectangle necessary to frame the figure and state the number of times the graph crosses itself.

36. $x = 6 \sin(3t)$ **37.** $x = 6 \sin(2t)$
$y = 8 \cos t$ $y = 8 \cos t$

38. $x = 8 \sin(4t)$ **39.** $x = 5 \sin(7t)$
$y = 10 \cos t$ $y = 7 \cos(4t)$

40. $x = 8 \sin(4t)$ **41.** $x = 10 \sin(1.5t)$
$y = 10 \cos(3t)$ $y = 10 \cos(2.5t)$

42. Use a graphing calculator to experiment with parametric equations of the form $x = A \sin(mt)$ and $y = B \cos(nt)$. Try different values of A, B, m, and n, then discuss their effect on the Lissajous figures.

43. Use a graphing calculator to experiment with parametric equations of the form $x = \dfrac{a}{\tan t}$ and $y = b \sin t \cos t$. Try different values of a and b, then discuss their effect on the resulting graph, called a serpentine curve. Also see Exercise 29.

▶ WORKING WITH FORMULAS

44. The Folium of Descartes:
$$x(t) = \frac{3kt}{1 + t^3}; y(t) = \frac{3kt^2}{1 + t^3}$$

The Folium of Descartes is a parametric curve developed by Descartes in order to test the ability of Fermat to find its maximum and minimum values.

  a. Graph the curve on a graphing calculator with $k = 1$ using a reduced window (ZOOM 4), with Tmin = −6, Tmax = 6, and Tstep = 0.1. Locate the coordinates of the tip of the folium (the loop).

  b. This graph actually has a discontinuity (a break in the graph). At what value of t does this occur?

  c. Experiment with different values of k and generalize its effect on the basic graph.

45. The Witch of Agnesi: $x(t) = 2kt; y(t) = \dfrac{2k}{1 + t^2}$

The Witch of Agnesi is a parametric curve named by Maria Agnesi in 1748. Some believe she confused the Italian word for *witch* (*versiera*), with a similar word that meant *free to move*. In any case, the name stuck. The curve can also be stated in trigonometric form: $x(t) = 2k \cot t$ and $y = 2k \sin^2 t$.

  a. Graph the curve with $k = 1$ on a calculator or computer on a reduced window (ZOOM 4) using both of the forms shown with Tmin = −6, Tmax = 6, and Tstep = 0.1. Try to determine the maximum value.

  b. Explain why the x-axis is a horizontal asymptote.

  c. Experiment with different values of k and generalize its effect on the basic graph.

▶ APPLICATIONS

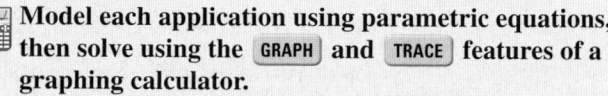 **Model each application using parametric equations, then solve using the** GRAPH **and** TRACE **features of a graphing calculator.**

46. Archery competition: At an archery contest, a large circular target 5 ft in diameter is laid flat on the ground with the bull's-eye exactly 180 yd (540 ft) away from the archers. Marion draws her bow and shoots an arrow at an angle of 25° above horizontal with an initial velocity of 150 ft/sec (assume the archers are standing in a depression and the arrow is shot from ground level). (a) What was the maximum height of the arrow? (b) Does the arrow hit the target? (c) What is the distance between Marion's arrow and the bull's-eye after the arrow hits?

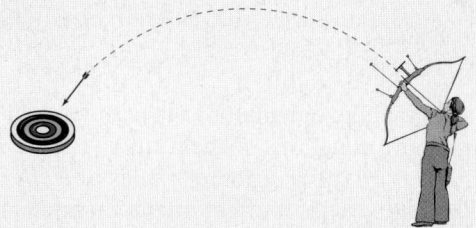

47. Football competition: As part of their contribution to charity, a group of college quarterbacks participate in a contest. The object is to throw a football through a hoop whose center is 30 ft high and 25 yd (75 ft) away, trying to hit a stationary (circular) target laid on the ground with the center

56 yd (168 ft) away. The hoop and target both have a diameter of 4 ft. On his turn, Lance throws the football at an angle of 36° with an initial velocity of 75 ft/sec. (a) Does the football make it through the hoop? (b) Does the ball hit the target? (c) What is the approximate distance between the football and the center of the target when the ball hits the ground?

48. Walk-off home run: It's the bottom of the ninth, two outs, the count is full, and the bases are loaded with the opposing team ahead 5 to 2. The home team has Heavy Harley, their best hitter at  the plate; the opposition has Raymond the Rocket on the mound. Here's the pitch . . . it's hit . . . a long fly ball to left-center field! If the ball left the bat at an angle of 30° with an initial velocity of 112 ft/sec, will it clear the home run fence, 9 ft high and 320 ft away?

49. Last-second win: It's fourth-and-long, late in the fourth quarter of the homecoming football game, with the home team trailing 29 to 27. The coach elects to kick a field goal, even though the goal posts are 50 yd (150 ft) away from the spot of the kick. If the ball leaves the kicker's foot at an angle of 29° with an initial velocity of 80 ft/sec, and the kick is "true," will the home team win (does the ball clear the 10-ft high cross bar)?

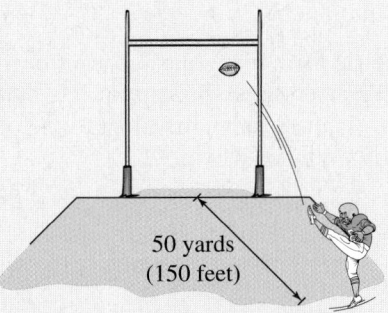

50 yards
(150 feet)

50. Particle motion: The motion of a particle is modeled by the parametric equations

$$\begin{cases} x = 5t - 2t^2 \\ y = 3t - 2 \end{cases}$$. Between $t = 0$ and $t = 1$, is the

particle moving to the right or to the left? Is the particle moving upward or downward?

51. Electron motion: The motion of an electron as it orbits the nucleus is modeled by the parametric

equations $\begin{cases} x = 6 \cos t \\ y = 2 \sin t \end{cases}$ with t in radians. Between

$t = 2$ and $t = 3$, is the electron moving to the right or to the left? Is the electron moving upward or downward?

Systems applications: Solve the following systems using elimination. If the system is dependent, write the general solution in parametric form and use a calculator to generate several solutions.

52. $\begin{cases} 2x - y + 3z = -3 \\ 3x + 2y - z = 4 \\ 8x + 3y + z = 5 \end{cases}$ **53.** $\begin{cases} x - 5y + z = 3 \\ 5x + y - 7z = -9 \\ 2x + 3y - 4z = -6 \end{cases}$

54. $\begin{cases} -5x - 3z = -1 \\ x + 2y - 2z = -3 \\ -2x + 6y - 9z = -10 \end{cases}$

55. $\begin{cases} x + y - 5z = -4 \\ 2y - 3z = -1 \\ x - 3y + z = -3 \end{cases}$

56. Regressions and parameters: Draw a scatter-plot of the data given in the table. Note that connecting the points with a smooth curve will not result in a function, so a standard regression cannot be run on the data. Now consider the x-values alone—what do you notice? Find a sinusoidal model for the x-values, using $T = 0, 1, 2, 3, \ldots, 8$. Use the same inputs to run some form of regression on the y-values, then use the results to form the "best-fit" parametric equations for this data (use L1 for T, L2 for the x-values, and L3 for the y-values). With your calculator in parametric **MODE**, enter the equations as X_{1T} and Y_{1T}, then graph these along with the scatterplot (L2, L3) to see the finished result. Use the **TABLE** feature of your calculator to comment on the accuracy of the model.

x	y
0	0
$\sqrt{2}$	0.25
2	2
$\sqrt{2}$	6.75
0	16
$-\sqrt{2}$	31.25
-2	54
$-\sqrt{2}$	85.75
0	128

57. Regressions and parameters: Draw a scatter-plot of the data given in the table, and connect the points with a smooth curve. The result is a function, but no standard regression seems to give an accurate model. The x-values alone are actually generated by an exponential function. Run a regression on these values using $T = 0, 1, 2, 3, \ldots, 8$ as inputs to find the exponential model. Then use the same inputs to run some form of regression on the y-values and use the results to form the "best-fit" parametric equations for this data (use L1 for T, L2 for the x-values, and L3 for the y-values). With your calculator in parametric **MODE**, enter the equations as X_{1T} and Y_{1T}, then graph these along with the scatterplot (L2, L3) to see the finished result. Use the **TABLE** feature of your calculator to comment on the accuracy of the model.

x	y
1	0
1.2247	-1.75
1.5	-3
1.8371	-3.75
2.25	-4
2.7557	-3.75
3.375	-3
4.1335	-1.75
5.0625	0

▶ **EXTENDING THE CONCEPT**

58. What is the difference between an *epicycloid*, a *hypercycloid*, and a *hypocycloid*? Do a word study on the prefixes *epi-*, *hyper-*, and *hypo-*, and see how their meanings match with the mathematical figures graphed in Exercises 27 to 35. To what other shapes or figures are these prefixes applied?

59. The motion of a particle in a certain medium is modeled by the parametric equations $\begin{cases} x = 6 \sin(4t) \\ y = 8 \cos t \end{cases}$.

Initially, use only the ⏹2nd ⏹GRAPH **(TABLE)** feature of your calculator (not the graph) to name the intervals for which the particle is moving (a) to the left and upward and (b) to the left and downward. Answer to the nearest tenth (set ΔTbl = 0.1). Is it *possible* for this particle to collide with another particle in this medium whose movement is modeled by $\begin{cases} x = 3 \cos t + 7 \\ y = 2 \sin t + 2 \end{cases}$? Discuss why or why not.

60. Write the function $y = \dfrac{1}{2}(x + 3)^2 - 1$ in parametric form using the substitution $x = 2 \cos t - 3$ and the appropriate double-angle identity. Is the result equivalent to the original function? Why or why not?

▶ **MAINTAINING YOUR SKILLS**

61. (1.1) The price of a popular video game is reduced by 20% and is selling for $39.96. By what percentage must the sale price be increased to return the item to its original price?

62. (5.2) When the tip of the antenna atop the Eiffel Tower is viewed at a distance of 265 ft from its base, the angle of elevation is 76°. Is the Eiffel Tower taller or shorter than the Chrysler Building (New York City) at 1046 ft?

63. (3.4) Graph $f(x) = x^3 + 2x^2 - 5x - 6$ using information about end behavior, y-intercept, x-intercept(s), and midinterval points:

64. (6.6) The maximum height a projectile will attain depends on the angle it is projected and its initial velocity. This phenomena is modeled by the function $H = \dfrac{v^2 \sin^2\theta}{64}$, where v is the initial velocity (in feet/sec) of the projectile and θ is the angle of projection. Find the angle of projection if the projectile attained a maximum height of 151 ft, and the initial velocity was 120 ft/sec.

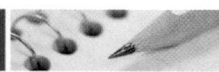

SUMMARY AND CONCEPT REVIEW

SECTION 9.1   A Brief Introduction to Analytical Geometry

KEY CONCEPTS

- The midpoint and distance formulas play an important role in the study of analytical geometry:

$$\text{midpoint: } (x, y) = \left(\frac{x_2 + x_1}{2}, \frac{y_2 + y_1}{2} \right) \qquad \text{distance: } d = \sqrt{(x_2 - x_1)^2 + (y_2 - y_1)^2}$$

- The perpendicular distance from a point to a line is the length of a line segment perpendicular to a given line with the given point and the point of intersection as endpoints.
- Using these tools, we can verify or construct relationships between points, lines, and curves in the plane; verify properties of geometric figures; prove theorems from Euclidean geometry; and construct relationships that define the conic sections.

EXERCISES

1. Verify the closed figure with vertices $(-3, -4)$, $(-5, 4)$, $(3, 6)$, and $(5, -2)$ is a square.

2. Find the equation of the circle that circumscribes the square in Exercise 1.

3. A theorem from Euclidean geometry states: *If any two points are equidistant from the endpoints of a line segment, they are on the perpendicular bisector of the segment.* Determine if the line through $(-3, 6)$ and $(6, -9)$ is a perpendicular bisector of the segment through $(-5, -2)$ and $(5, 4)$.

4. Four points are given below. Verify that the distance from each point to the line $y = -1$ is the same as the distance from the given point to the fixed point $(0, 1)$: $(-6, 9)$, $(-2, 1)$, $(4, 4)$, and $(8, 16)$.

SECTION 9.2   The Circle and the Ellipse

KEY CONCEPTS

- The equation of a circle centered at (h, k) with radius r is $(x - h)^2 + (y - k)^2 = r^2$.

- Dividing both sides by r^2, we obtain the standard form $\dfrac{(x - h)^2}{r^2} + \dfrac{(y - k)^2}{r^2} = 1$, showing the horizontal and vertical distance from center to graph is r.

- The equation of an ellipse in standard form is $\dfrac{(x - h)^2}{a^2} + \dfrac{(y - k)^2}{b^2} = 1$. The center of the ellipse is (h, k), with horizontal distance a and vertical distance b from center to graph.

- Given two fixed points f_1 and f_2 in a plane (called the foci), an ellipse is the set of all points (x, y) such that the distance from the first focus to (x, y), plus the distance from the second focus to (x, y), remains constant.

- For an ellipse, the distance a from center to vertex is *greater than* the distance c from center to one focus.

- To find the foci of an ellipse: $a^2 = b^2 + c^2$ (since $a > c$).

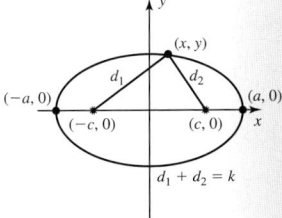

EXERCISES

Sketch the graph of each equation in Exercises 5 through 9.

5. $x^2 + y^2 = 16$ **6.** $x^2 + 4y^2 = 36$ **7.** $9x^2 + y^2 - 18x - 27 = 0$

8. $x^2 + y^2 + 6x + 4y + 12 = 0$ **9.** $\dfrac{(x + 3)^2}{16} + \dfrac{(y - 2)^2}{9} = 1$

10. Find the equation of the ellipse with minor axis of length 6 and foci at $(-4, 0)$ and $(4, 0)$.

11. Find the equation of the ellipse with vertices at (a) $(-13, 0)$ and $(13, 0)$, foci at $(-12, 0)$ and $(12, 0)$; (b) foci at $(0, -16)$ and $(0, 16)$, major axis: 40 units.

12. Write the equation in standard form and sketch the graph, noting all of the characteristic features of the ellipse. $4x^2 + 25y^2 - 16x - 50y - 59 = 0$

SECTION 9.3   The Hyperbola

KEY CONCEPTS

- The equation of a *horizontal* hyperbola in standard form is $\dfrac{(x - h)^2}{a^2} - \dfrac{(y - k)^2}{b^2} = 1$. The center of the hyperbola is (h, k) with horizontal distance a from center to vertices and vertical distance b from center to the midpoint of one side of the central rectangle.

- Given two fixed points f_1 and f_2 in a plane (called the foci), a hyperbola is the set of all points (x, y) such that the distance from the first focus to point (x, y), less the distance from the second focus to (x, y), remains constant.

- For a hyperbola, the distance from center to one of the vertices is *less than* the distance from center to one focus.

- To find the foci of a hyperbola: $c^2 = a^2 + b^2$ (since $c > a$).

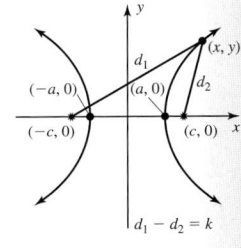

EXERCISES

Sketch the graph of each equation, indicating the center, vertices, and asymptotes. For Exercise 18, also give the equation of the hyperbola in standard form.

13. $4y^2 - 25x^2 = 100$ **14.** $\dfrac{(y - 3)^2}{16} - \dfrac{(x + 2)^2}{9} = 1$ **15.** $\dfrac{(x + 2)^2}{9} - \dfrac{(y - 1)^2}{4} = 1$

16. $9y^2 - x^2 - 18y - 72 = 0$ **17.** $x^2 - 4y^2 - 12x - 8y + 16 = 0$

18. vertices at $(-3, 0)$ and $(3, 0)$, asymptotes of $y = \pm\dfrac{4}{3}x$

19. Find the equation of the hyperbola with (a) vertices at $(\pm 15, 0)$, foci at $(\pm 17, 0)$, and (b) foci at $(0, \pm 5)$ with vertical dimension of central rectangle 8 units.

20. Write the equation in standard form and sketch the graph, noting all of the characteristic features of the hyperbola. $4x^2 - 9y^2 - 40x + 36y + 28 = 0$

SECTION 9.4   The Analytic Parabola

KEY CONCEPTS

- Horizontal parabolas have equations of the form $x = ay^2 + by + c;\ a \neq 0$.

- A horizontal parabola will open to the right if $a > 0$, and to the left if $a < 0$. The axis of symmetry is $y = \dfrac{-b}{2a}$, with the vertex (h, k) found by evaluating at $y = \dfrac{-b}{2a}$ or by completing the square and writing the equation in shifted form: $x = a(y - k)^2 + h$.

- Given a fixed point f (called the focus) and fixed line D in the plane, a parabola is the set of all points (x, y) such that the distance from f to (x, y) is equal to the distance from (x, y) to line D.

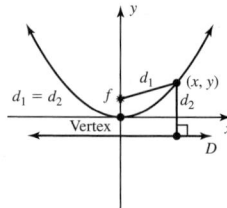

- The equation $x^2 = 4py$ describes a vertical parabola, opening upward if $p > 0$, and opening downward if $p < 0$.

- The equation $y^2 = 4px$ describes a horizontal parabola, opening to the right if $p > 0$, and opening to the left if $p < 0$.

- The focal chord of a parabola is a line segment that contains the focus and is parallel the directrix, with its endpoints on the graph. It has a total length of $|4p|$, meaning the distance from the focus to a point of the graph is $|2p|$. It is commonly used to assist in drawing a graph of the parabola.

EXERCISES

For Exercises 21 and 22, find the vertex and x- and y-intercepts if they exist. Then sketch the graph using symmetry and a few points or by completing the square and shifting a parent function.

21. $x = y^2 - 4$ **22.** $x = y^2 + y - 6$

For Exercises 23 and 24, find the vertex, focus, and directrix for each parabola. Then sketch the graph using this information and the focal chord. Also graph the directrix.

23. $x^2 = -20y$ **24.** $x^2 - 8x - 8y + 16 = 0$

SECTION 9.5   Nonlinear Systems of Equations and Inequalities

KEY CONCEPTS

- Nonlinear systems of equations can be solved using substitution or elimination.
- First identify the graphs of the equations in the system to help determine the number of solutions possible.
- For nonlinear systems of inequalities, graph the related equation for each inequality given, then use a test point to decide what region to shade as the solution.
- The solution for the system is the overlapping region (if it exists) created by the areas shaded for the individual inequalities.
- If the boundary is included, graph it using a solid line; if the boundary is not included (for strict inequalities) use a dashed line.

EXERCISES

Solve Exercises 25–30 using substitution or elimination. Identify the graph of each relation before you begin.

25. $\begin{cases} x^2 + y^2 = 25 \\ y - x = -1 \end{cases}$

26. $\begin{cases} x = y^2 - 1 \\ x + 4y = -5 \end{cases}$

27. $\begin{cases} -x^2 + y = -1 \\ x^2 + y^2 = 7 \end{cases}$

28. $\begin{cases} x^2 + y^2 = 10 \\ y - 3x^2 = 0 \end{cases}$

29. $\begin{cases} y \le x^2 - 2 \\ x^2 + y^2 \le 16 \end{cases}$

30. $\begin{cases} x^2 + y^2 > 9 \\ x^2 + y \le -3 \end{cases}$

SECTION 9.6   Polar Coordinates, Equations, and Graphs

KEY CONCEPTS

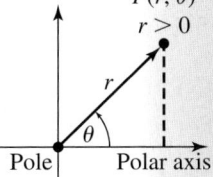

- In polar coordinates, the location of a point in the plane is denoted (r, θ), where r is the distance to the point from the origin or *pole*, and θ is the angle between a stipulated polar axis and a ray containing P.
- In the polar coordinate system, the location (r, θ) of a point is not unique for two reasons: (1) the angles θ and $\theta + 2\theta n$ are coterminal (n an integer), and (2) r may be negative.
- The point $P(r, \theta)$ can be converted to $P(x, y)$ in rectangular coordinates where $x = r \cos \theta$ and $y = r \sin \theta$.
- The point $P(x, y)$ in rectangular coordinates can be converted to $P(r, \theta)$ in polar coordinates, where $r = \sqrt{x^2 + y^2}$ and $\theta_r = \tan^{-1}\left(\dfrac{y}{x}\right)$.
- To sketch a polar graph, we view the length r as being along the second hand of a clock, ticking in a counterclockwise direction. Each "tick" is $\dfrac{\pi}{12}$ rad or 15°. For each tick we locate a point on the radius and plot it on the face of the clock before going on.
- For graphing, we also apply an "r-value" analysis, which looks where r is increasing, decreasing, zero, maximized, and/or minimized.
- If the polar equation is given in terms of sines, the graph will be symmetric to $\theta = \dfrac{\pi}{2}$.
- If the polar equation is given in terms of cosines, the graph will be symmetric to the polar axis.
- The graphs of several common polar equations are given in Appendix V.

EXERCISES

Sketch using an r-value analysis (include a table), symmetry, and any convenient points.

31. $r = 5 \sin \theta$ **32.** $r = 4 + 4 \cos \theta$ **33.** $r = 2 + 4 \cos \theta$ **34.** $r = 8 \sin(2\theta)$

SECTION 9.7   More on the Conic Sections: Rotation of Axes and Polar Form

KEY CONCEPTS

- Using a rotation, the conic equation $Ax^2 + Bxy + Cy^2 + Dx + Ey + F = 0$ in the xy-plane can be transformed into $aX^2 + cY^2 + dX + eY + f = 0$ in the XY-plane, in which the mixed xy-term is absent.
- The required angle of rotation β is found using $\tan(2\beta) = \dfrac{B}{A - C}$; $0 < 2\beta < 180°$.
- The change in coordinates from the xy-plane to the XY-plane is accomplished using the rotation formulas:

$$x = X \cos \beta - Y \sin \beta \qquad y = X \sin \beta + Y \cos \beta$$

- In the process of this conversion, certain quantities, called invariants, remain unchanged and can be used to check that the conversion was correctly performed. These invariants are (1) $F = f$, (2) $A + C = a + c$, and (3) $B^2 - 4AC = b^2 - 4ac$.

- The invariants $B^2 - 4AC = b^2 - 4ac$ are called discriminants and can be used to classify the type of graph the equation will give, except in degenerate cases:
  - If $B^2 - 4AC = 0$, the equation is that of a parabola.
  - If $B^2 - 4AC < 0$, the equation is that of a circle or an ellipse.
  - If $B^2 - 4AC > 0$, the equation is that of a hyperbola.
- All conics (not only the parabola) can be stated in terms of a focus/directrix definition. This is done using the concept of eccentricity, symbolized by the letter e.
- If F is a fixed point and $\mathcal{L}$ a fixed line in the plane with the point D on $\mathcal{L}$, the set of all points P such that $\dfrac{\overline{FP}}{\overline{DP}} = e$ (e a constant) is the graph of a conic section. If $e = 1$, the graph is a parabola. If $0 < e < 1$, the graph is an ellipse. If $e > 1$, the graph is a hyperbola.
- Given a conic section with eccentricity e, one focus at the pole of the $r\theta$-plane, and directrix $\mathcal{L}$ located d units from this focus, then the polar equations $r = \dfrac{de}{1 \pm e \cos \theta}$ and $r = \dfrac{de}{1 \pm e \sin \theta}$ represent one of the conic sections as determined by the value of e.

EXERCISES

For the given conics in the xy-plane, use a rotation of axes to find the corresponding equation in the XY-plane, then sketch its graph.

35. $2x^2 - 4xy + 2y^2 - 8\sqrt{2}y - 24 = 0$

36. $x^2 + 6\sqrt{3}xy + 7y^2 - 160 = 0$

For the conic equations given, determine if the equation represents a parabola, ellipse, or hyperbola. Then describe and sketch the graphs using polar graph paper.

37. $r = \dfrac{9}{3 - 2\cos\theta}$

38. $r = \dfrac{8}{4 - 6\cos\theta}$

39. $r = \dfrac{4}{3 + 3\sin\theta}$

40. Mars has a perihelion of 128.4 million miles and an aphelion of 154.9 million miles. Use this information to find a polar equation that models the elliptical orbit, then find the length of the focal chord.

SECTION 9.8 Parametric Equations and Graphs

KEY CONCEPTS

- If we consider the set of points $P(x, y)$ such that the x-values are generated by $f(t)$ and the y-values are generated by $g(t)$ (assuming f and g are both defined on an interval of the domain), the equations $x = f(t)$ and $y = g(t)$ are called parametric equations, with parameter t.
- Parametric equations can be converted to rectangular form by eliminating the parameter. This can sometimes be done by solving for t in one equation and substituting in the other, or by using trigonometric forms.
- A function can be written in parametric form many different ways, by altering the parameter or using trigonometric identities.
- The cycloids are an important family of curves, with equations $x = r(t - \sin t)$ and $y = r(1 - \cos t)$.
- The solutions to dependent systems of equations are often expression in parametric form, with the points $P(x, y)$ given by the parametric equations generating solutions to the system.

EXERCISES

Graph the curves defined by the parametric equations over the specified interval and identify the graph. Then eliminate the parameter and write the corresponding rectangular form.

41. $x = t - 4 : t \in [-3, 3]$:
$y = -2t^2 + 3$

42. $x = (2 - t)^2 : t \in [0, 5]$:
$y = (t - 3)^2$

43. $x = -3\sin t : t \in [0, 2\pi)$:
$y = 4\cos t$

44. Write the function in three different forms by altering the parameter: $y = 2(x - 5)^2 - 1$

45. Use a graphing calculator to graph the Lissajous figure indicated, then state the size of the rectangle needed to frame it: $x = 4\sin(5t)$; $y = 8\cos t$

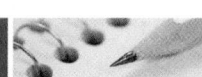

MIXED REVIEW

For Exercises 1 through 16, graph the conic section and locate the center, vertices, directrix, foci, focal chords, asymptotes, and other important features as these apply to a particular equation and conic.

1. $9x^2 + 9y^2 = 54$

2. $16x^2 + 25y^2 = 400$

3. $9y^2 - 25x^2 = 225$

4. $\dfrac{(x-3)^2}{9} + \dfrac{(y+1)^2}{25} = 1$

5. $4(x-1)^2 - 36(y+2)^2 = 144$

6. $16(x+2)^2 + 4(y-1)^2 = 64$

7. $y = -2x^2 - 10x + 15$

8. $x = -y^2 - 8y - 11$

9. $x = y^2 + 2y + 3$

10. $x = (y+2)^2 - 3$

11. $x^2 - 8x - 8y + 16 = 0$

12. $x^2 = -24y$

13. $4x^2 - 25y^2 - 24x + 150y - 289 = 0$

14. $4x^2 + 16y^2 - 12x - 48y - 19 = 0$

15. $49(x+2)^2 + (y-3)^2 = 49$

16. $x^2 + y^2 - 8x + 12y + 16 = 0$

17. Graph the curve defined by the parametric equations given, using the interval $t \in [0, 10]$. Then identify the graph: $x = (t-2)^2$, $y = (t-4)^2$

18. Plot the polar coordinates given, then convert to rectangular coordinates.

a. $\left(3.5, \dfrac{2\pi}{3}\right)$ **b.** $\left(-4, \dfrac{5\pi}{4}\right)$

19. Solve using elimination:

a. $\begin{cases} 4x^2 - y^2 = -9 \\ x^2 + 3y^2 = 79 \end{cases}$ **b.** $\begin{cases} 4x^2 + 9y^2 = 36 \\ x^2 + 3y = 6 \end{cases}$

20. Match each equation to its corresponding graph. Justify each response.

  (i) $r = 3.5 + \cos\theta$

  (ii) $r^2 = 20.25\sin(-2\theta)$

  (iii) $r = 4.5\cos\theta$

a.

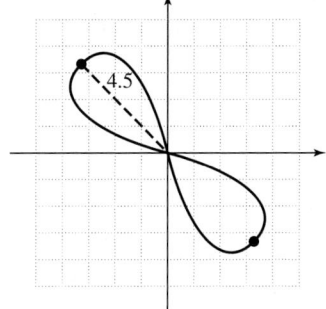

b.

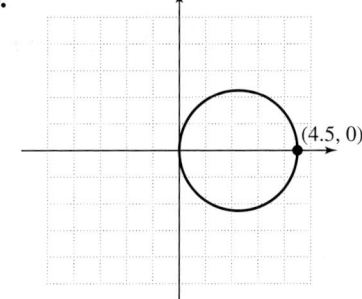

(4.5, 0)

c.

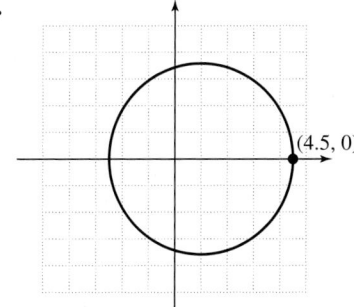

(4.5, 0)

21. A go-cart travels around an elliptical track with a 100-m major axis that is horizontal. The minor axis measures 60 m. Write an equation model for the track in parametric form.

22. Except for small variations, a planet's orbit around the Sun is elliptical, with the Sun at one focus. The *perihelion* or minimum distance from the planet Mercury to the Sun is about 46 million kilometers. Its *aphelion* or maximum distance from the Sun is approximately 70 million kilometers. Use this information to find the length of the major and minor axes, then determine the equation model for the orbit of Mercury in the standard form

$$\dfrac{x^2}{a^2} + \dfrac{y^2}{b^2} = 1.$$

23. The orbit of a comet can also be modeled by one of the conic sections, with the Sun at one focus. Assuming the equations given model a comet's path, (1) determine if the path is circular, elliptic, hyperbolic, or parabolic; and (2) determine the closest distance the comet will come to the Sun (in millions of miles).

a. $r = \dfrac{84}{100 + 70 \cos \theta}$ **b.** $r = \dfrac{31}{5 - 5 \sin \theta}$

24. In the design of their corporate headquarters, Centurion Computing includes a seven-leaf rose in a large foyer, with a fountain in the center. Each of the leaves is 5 m long (when measured from the center of the fountain), and will hold flower beds for carefully chosen perennials. The rose is to be symmetric to a vertical axis, with the leaf bisected by $\theta = \dfrac{\pi}{2}$ pointing directly to the elevators. Find the equation of the rose in polar form.

25. The hyperbola defined by $\dfrac{X^2}{80^2} - \dfrac{Y^2}{400^2} = 1$ in the XY-plane is rotated clockwise by $45°$. What is the corresponding equation in the xy-plane?

 **PRACTICE TEST**

By inspection only (no graphing), match each equation to its correct description.

1. $x^2 + y^2 - 6x + 4y + 9 = 0$ _____

2. $4y^2 + x^2 - 4x + 8y + 20 = 0$ _____

3. $x^2 - 4y^2 - 4x + 12y + 20 = 0$ _____

4. $y - x^2 - 4x + 20 = 0$ _____

    a. Parabola   **b.** Hyperbola   **c.** Circle   **d.** Ellipse

Identify and then graph each of the following conic sections. State the center, vertices, foci, asymptotes, and other important points when applicable.

5. $x^2 + y^2 - 4x + 10y + 20 = 0$

6. $25(x + 2)^2 + 4(y - 1)^2 = 100$

7. $r = \dfrac{10}{5 - 4 \cos \theta}$ **8.** $r = \dfrac{12}{5 - 5 \cos \theta}$

9. $\dfrac{(y + 3)^2}{9} - \dfrac{(x - 2)^2}{16} = 1$

10. $4(x - 1)^2 - 25(y + 2)^2 = 100$

Use the equation $80x^2 + 120xy + 45y^2 - 100y - 44 = 0$ to complete Exercises 11 and 12.

11. Use the discriminant $B^2 - 4AC$ to identify the graph, and $\tan(2\beta) = \dfrac{B}{A - C}$ to find $\cos \beta$ and $\sin \beta$.

12. Find the equation in the xy-plane and use a rotation of axes to draw a neat sketch of the graph in the XY-plane.

Graph each polar equation.

13. $r = 3 + 3 \cos \theta$ **14.** $r = 4 + 8 \cos \theta$

15. $r = 6 \sin(2\theta)$

For Exercises 16 and 17, identify and graph each conic section from the parametric equations given. Then remove the parameter and convert to rectangular form.

16. $x = 4 \sin t$ **17.** $x = (t - 3)^2 + 1$
    $y = 5 \cos t$ $y = t + 2$

 18. Use a graphing calculator to graph the cycloid, then identify the maximum and minimum values, and the period. $x = 4T - 4 \sin T$ $y = 4 - 4 \cos T$

19. Solve each nonlinear system using the technique of your choice.

    a. $\begin{cases} 4x^2 - y^2 = 16 \\ y - x = 2 \end{cases}$ **b.** $\begin{cases} 4y^2 - x^2 = 4 \\ x^2 + y^2 = 4 \end{cases}$

20. Halley's comet has a perihelion of 54.5 million miles and an aphelion of 3253 million miles. Use this information to find a polar equation that models its elliptical orbit. How does its eccentricity compare with that of the planets in our solar system?

21. The soccer match is tied, with time running out. In a desperate attempt to win, the opposing coach pulls his goalie and substitutes a forward. Suddenly, Marques gets a break-away and has an open shot at the empty net, 165 ft away. If the kick is on-line and leaves his foot at an angle of $28°$ with an initial velocity of 80 ft/sec, is the ball likely to go in the net and score the winning goal?

22. The orbit of Mars around the Sun is elliptical, with the Sun at one foci. When the orbit is expressed as a central ellipse on the coordinate grid, its equation is $\dfrac{x^2}{(141.65)^2} + \dfrac{y^2}{(141.03)^2} = 1$. Use this information to find the *aphelion* of Mars and the *perihelion* of Mars in millions of miles.

Determine the equation of each relation and state its domain and range. For the parabola and the ellipse, also give the location of the foci.

23.

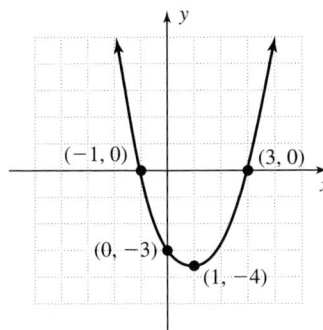

24.

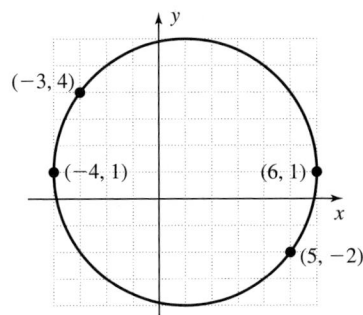

25.

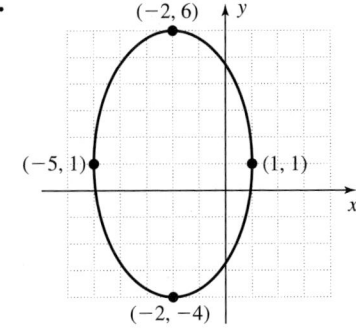

CALCULATOR EXPLORATION AND DISCOVERY

Conic Rotations in Polar Form

While all planets orbit around the sun in an elliptical path, their **ecliptic planes,** or the 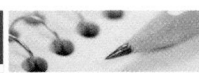 planes containing the orbits, differ considerably. For example, using the ecliptic plane of the Earth for reference, the plane containing Mercury's orbit is inclined by 7° and the plane of the dwarf planet Pluto by 17°! In addition, if we use the major axis of Earth's orbit for reference, the major axes of the other planets, assuming they are transformed to the ecliptic plane, are rotated by some angle θ. We can gain a basic understanding of the rotations of an elliptical path (relative to some point of reference) using skills developed in this chapter. Here we've seen that the equation of a conic can be given in rectangular form, polar form, and parametric form. Each form seems to have its advantages. When it comes to the rotations of a conic section, it's hard to match the ease and versatility of the polar form. To illustrate, recall that in polar form the general equation of a horizontal ellipse with one focus (the Sun) at the origin is $r = \dfrac{a(1 - e^2)}{1 - e \cos \theta}$. The constant a gives the length of the semimajor axis and e represents the eccentricity of the orbit. With the exception of Mercury and Pluto (a dwarf planet), the orbits of most planets are close to circular (e is very near zero). This makes the rota-

tions difficult to see. Instead we will explore the concept of axes rotation using "planets" with higher eccentricities. Consider the following planets and their orbital equations. The planet Agnesi has an eccentricity of $e = 0.5$, while the planet Erdös is the most eccentric at $e = 0.75$.

$$\text{Agnesi: } \frac{2.9}{1 - 0.5 \cos \theta}$$

$$\text{Galois: } \frac{5.75}{1 - 0.7 \cos \theta}$$

$$\text{Erdös: } \frac{7.875}{1 - 0.75 \cos \theta}$$

We'll investigate the concept of conic rotations in polar form by rotating these ellipses. With your calculator in polar **MODE**, enter these three equations on the **Y =** screen and use the settings shown in Figure 9.80 to set the window size (use θmax = 7).

Figure 9.80

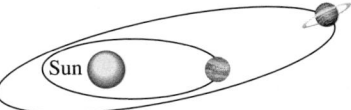

```
WINDOW
↑θstep=.1
Xmin=-10
Xmax=32
Xscl=4
Ymin=-15
Ymax=15
Yscl=3
```

The resulting graph is displayed in Figure 9.81, showing the very hypothetical case where all planets share the same major axis. To show a more realistic case where the planets approach the Sun along orbits with differing major axes, we'll use Galois as a reference and rotate Agnesi $\frac{\pi}{4}$ rad clockwise and Erdös $\frac{\pi}{12}$ rad counterclockwise. This is done by *simply adjusting the argument of cosine* in each equation, using $\cos\left(\theta - \frac{\pi}{4}\right)$ for Agnesi and $\cos\left(\theta + \frac{\pi}{12}\right)$ for Erdös. The adjusted [Y=] screen is shown in Figure 9.82, and new graphs in Figure 9.83.

Use these ideas to explore and investigate other rotations by completing the following exercises.

Figure 9.81

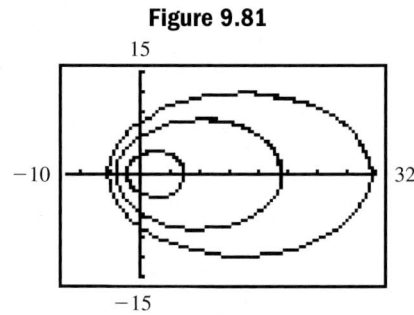

Figure 9.82

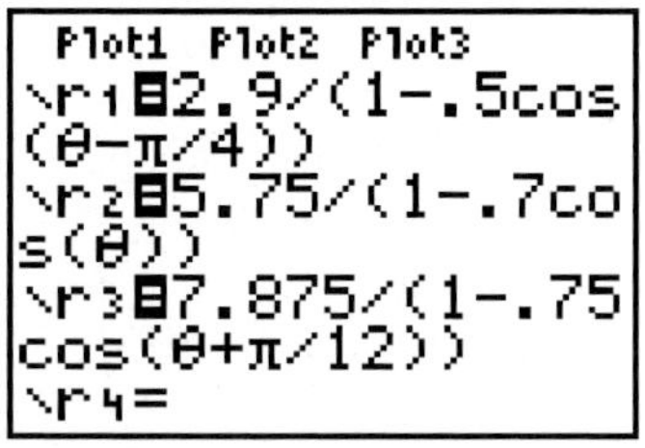

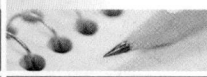

Exercise 1: What happens if the angle of rotation is π? Is the orbit identical if you rotate by $-\pi$?

Exercise 2: If the denominator in the equation is changed to a sum, what effect does it have on the graph?

Exercise 3: If the sign in the numerator is changed, what effect does it have on how the graph is generated?

Exercise 4: After resetting the orbits as originally given, use trial and error to approximate the smallest angle of rotation required for the orbit of Galois to intersect the orbit of Erdös.

Exercise 5: What minimum rotation is required for the orbit of Galois to intersect the orbit of *both* Agnesi and Erdös?

Exercise 6: What is the minimum rotation required for the orbit of Agnesi to intersect the orbit of Galois?

Figure 9.83

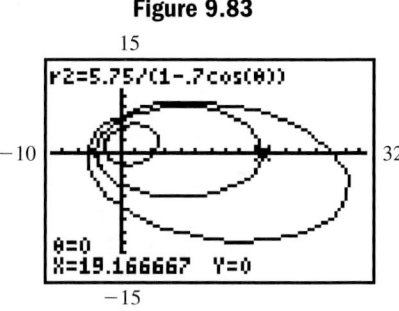

STRENGTHENING CORE SKILLS

Simplifying and Streamlining Computations for the Rotation of Axes

While the calculations involved for eliminating the mixed xy-term require a good deal of concentration, there are a few things we can do to simplify the overall process. Basically this involves two things. First, in Figure 9.84 we've organized the process in flowchart form to help you "see" the sequence involved in finding $\cos\beta$ and $\sin\beta$ (for use in the rotation formulas). Second, calculating x^2, y^2, and xy (from the equations $x = X\cos\beta - Y\sin\beta$ and $y = X\sin\beta + Y\cos\beta$) *as single terms and apart from their actual substitution* is somewhat less restrictive and seems to help to streamline the algebra.

Figure 9.85

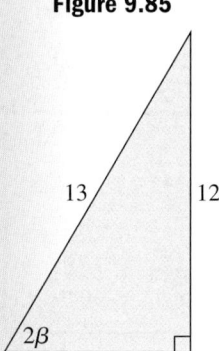

Illustration 1 ▶ For $2x^2 + 12xy - 3y^2 - 42 = 0$, use a rotation of axes to eliminate the xy-term, then identify the conic and its characteristic features.

Solution ▶ Since $A \neq C$, we find β using $\tan(2\beta) = \dfrac{B}{A - C}$, giving $\tan(2\beta) = \frac{12}{5}$. Using the triangle shown in Figure 9.85 we find $\cos(2\beta) = \frac{5}{13}$. We then find the values of $\cos\beta$ and $\sin\beta$ (choosing 2β in QII), using the double-angle identities as follows:

Figure 9.84

$$\cos \beta = \sqrt{\frac{1 + \cos(2\beta)}{2}}$$

$$= \sqrt{\frac{1 + \frac{5}{13}}{2}} \rightarrow \sqrt{\frac{\frac{18}{13}}{2}}$$

$$= \frac{3}{\sqrt{13}}$$

$$\sin \beta = \sqrt{\frac{1 - \cos(2\beta)}{2}}$$

$$= \sqrt{\frac{1 - \frac{5}{13}}{2}} \rightarrow \sqrt{\frac{\frac{8}{13}}{2}}$$

$$= \frac{2}{\sqrt{13}}$$

Is $A = C$?
($B \neq 0$) — **yes** → $\beta = 45°$

no

Is $\dfrac{B}{A - C}$ $\frac{1}{\sqrt{3}}$ or $\sqrt{3}$? — **yes** → $\beta = 30°$ or $\beta = 60°$

no

Find $\sin(2\beta)$ and $\cos(2\beta)$ from the triangle corresponding to $\tan(2\beta)$, then $\cos \theta$ and $\sin \theta$ from the double-angle identities.

Use $\cos \beta$ and $\sin \beta$ in the rotation formulas to compute x^2, xy, and y^2, *writing each as a single term.*

Substitute, simplify, and use the invariants to double-check your work.

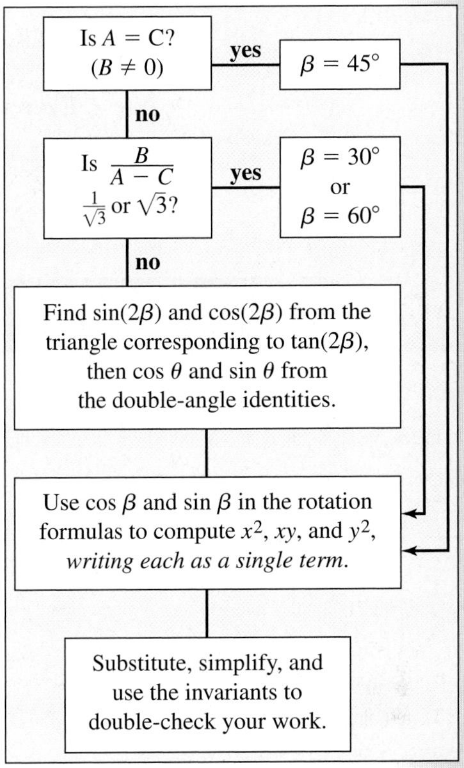

We now compute x^2, xy, and y^2 prior to substitution in the original equation, *writing each as a single term:*

- $x = \dfrac{3}{\sqrt{13}}X - \dfrac{2}{\sqrt{13}}Y = \dfrac{3X - 2Y}{\sqrt{13}}$
- $x^2 = \left(\dfrac{3X - 2Y}{\sqrt{13}}\right)^2$
$= \dfrac{9X^2 - 12XY + 4Y^2}{13}$

- $y = \dfrac{2}{\sqrt{13}}X + \dfrac{3}{\sqrt{13}}Y = \dfrac{2X + 3Y}{\sqrt{13}}$
- $y^2 = \left(\dfrac{2X + 3Y}{\sqrt{13}}\right)^2$
$= \dfrac{4X^2 + 12XY + 9Y^2}{13}$

- $xy = \dfrac{(3X - 2Y)(2X + 3Y)}{\sqrt{13}}$
$= \dfrac{6X^2 + 5XY - 6Y^2}{13}$

Next, we substitute into the original equation, clearing denominators *prior to* using the distributive property.

$$42 = 2x^2 + 12xy - 3y^2$$

$$42 = 2\left(\frac{9X^2 - 12XY + 4Y^2}{13}\right) + 12\left(\frac{6X^2 + 5XY - 6Y^2}{13}\right) - 3\left(\frac{4x^2 + 12XY + 9Y^2}{13}\right)$$

multiply both sides by 13, *then* distribute

$$546 = 18X^2 - 24XY + 8Y^2 + 72X^2 + 60XY - 72Y^2 - 12X^2 - 36XY - 27Y^2$$

$$546 = 78X^2 - 91Y^2 \qquad \text{combine like terms}$$

$$42 = 6X^2 - 7Y^2 \qquad \text{simplify and check invariants:} \quad F = f \checkmark \quad A + C = a + c \checkmark$$
$$B^2 - 4AC = b^2 - 4ac \checkmark$$

$$1 = \frac{X^2}{(\sqrt{7})^2} - \frac{Y^2}{(\sqrt{6})^2} \qquad \text{standard form}$$

The graph is a central hyperbola along the X-axis, with vertices at $(\pm\sqrt{7}, 0)$ and asymptotes $Y = \pm\sqrt{\dfrac{6}{7}}X$.

Exercise 1: Return to Section 9.6 and resolve Exercises 31 and 32 using these methods. Do the new ideas make a difference?

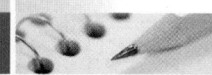

CUMULATIVE REVIEW CHAPTERS 1–9

Solve each equation.

1. $\sqrt{x + 2} + 2 = \sqrt{3x + 4}$

2. $x^2 - 6x + 13 = 0$

3. $4 \cdot 2^{x+1} = \frac{1}{8}$

4. $3^{x-2} = 7$

5. $\log_3 81 = x$

6. $\log_3 x + \log_3(x - 2) = 1$

7. $-6 \tan x = 2\sqrt{3}$

8. $25 \sin\left(\dfrac{\pi}{3}x - \dfrac{\pi}{6}\right) + 3 = 15.5$

9. $\dfrac{\sin 27°}{18} = \dfrac{\sin x}{35}$

10. Use De Moivre's theorem to find the three cube roots of $-8i$. Write the roots in $a + bi$ form.

11. The price of beef in Argentina varies directly with demand and inversely with supply. In the small town of Chascomus, the tender-cut lomito was selling for 18 pesos/kg last week. There were 1000 kg available, and 850 kg were bought. Next week there is a 3-day weekend, so the demand is expected to be closer to 1400 kg, but the butchers will only be able to supply 1200 kg. What will a kilogram of tender-cut lomito cost next week?

12. Find the inverse of $f(x) = 3 \sin(2x + 1)$.

13. A surveyor needs to estimate the width of a large rock formation in Canyonlands National Park. From her current position she is 540 yd from one edge of the formation and 850 yd from the other edge. If the included angle is 110°, how wide is the formation?

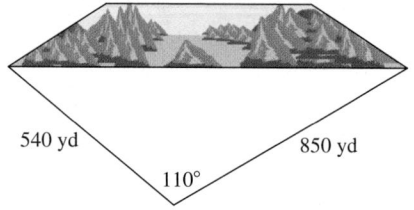

540 yd   850 yd   110°

Graph each relation. Include vertices, x- and y-intercepts, asymptotes, and other features.

14. $f(x) = |x - 2| + 3$ **15.** $y = \sqrt{x - 3} + 1$

16. $g(x) = (x - 3)(x + 1)(x + 4)$

17. $h(x) = \dfrac{x - 2}{x^2 - 9}$ **18.** $y = 2^x + 3$

19. $f(x) = \log_2(x + 1)$

20. $x^2 + y^2 + 10x - 4y + 20 = 0$

21. $4(x - 1)^2 - 36(y + 2)^2 = 144$

22. $y = -2 \cos\left(x - \dfrac{\pi}{4}\right) + 1$

23. $r = 4 \cos(2\theta)$ **24.** $x = 2 \sin t$
                                               $y = \tan t$

25. Use the dot product to find the angle between the vectors $\mathbf{u} = \langle -4, 5 \rangle$ and $\mathbf{v} = \langle 3, 7 \rangle$.

Solve each system of equations.

26. $\begin{cases} 4x + 3y = 13 \\ -9y + 5z = 19 \\ x - 4z = -4 \end{cases}$

27. $\begin{cases} x^2 + y^2 = 25 \\ 64x^2 + 12y^2 = 768 \end{cases}$

28. Find the equation of the parabola with vertex at $(2, 3)$ and directrix $x = 0$.

29. Decompose $y = \dfrac{3x^3 - 2x^2 + x - 3}{x^4 + x^2}$ into partial fractions.

30. In the summer, Hollywood releases its big budget, big star, big money movies. Suppose the weekly summer revenue generated by ticket sales was modeled by the function $R(w) = -w^4 + 25w^3 - 200w^2 + 560w - 234$, where $R(w)$ represents the revenue generated in week w and $1 \leq w \leq 12$. Use the remainder theorem to determine the amount of revenue generated in week 5.

CONNECTIONS TO CALCULUS

As with other relations and functions we've studied in precalculus, there is a high level of interest in finding rates of change in polar functions, and in solving systems of polar equations. From their use in architecture to their application in studies of planetary motion, both skills play a fundamental role in our continuing study.

Polar Graphs and Instantaneous Rates of Change

Using the tools of calculus, it can be shown that the slope of the tangent line for many different polar graphs can be generated from a specific template. It can also be shown that under certain conditions, the zeroes of the numerator will give the location of horizontal tangent lines, while the zeroes of the denominator will given the location of vertical tangent lines.

EXAMPLE 1 ▶ **Finding Slopes of Tangent Lines**

For the cardioid $r = 1 + \sin \theta$, it can be shown that $\dfrac{\cos \theta \sin \theta + (1 + \sin \theta)\cos \theta}{\cos \theta \cos \theta - (1 + \sin \theta)\sin \theta}$ gives the slope of the tangent line at θ.

  a. Simplify the expression using double-angle identities.

  b. Use the result to find the slope of the tangent line at $\theta = \dfrac{2\pi}{3}$.

Then, determine the value(s) of θ for which a tangent line is

  c. Horizontal.

  d. Vertical.

Solution ▶ **a.** After using the distributive property in the numerator and denominator, we have

$$\frac{\cos \theta \sin \theta + (1 + \sin \theta)\cos \theta}{\cos \theta \cos \theta - (1 + \sin \theta)\sin \theta} = \frac{\cos \theta \sin \theta + \cos \theta + \sin \theta \cos \theta}{\cos^2\theta - \sin \theta - \sin^2\theta} \quad \text{distribute}$$

$$= \frac{2 \sin \theta \cos \theta + \cos \theta}{\cos^2\theta - \sin^2\theta - \sin \theta} \quad \text{combine terms}$$

$$= \frac{\sin(2\theta) + \cos \theta}{\cos(2\theta) - \sin \theta} \quad \text{double angle identities}$$

  b. For $\theta = \dfrac{2\pi}{3}$, the expression gives $\dfrac{\sin\left(\dfrac{4\pi}{3}\right) + \cos\left(\dfrac{2\pi}{3}\right)}{\cos\left(\dfrac{4\pi}{3}\right) - \sin\left(\dfrac{2\pi}{3}\right)} = \dfrac{-\dfrac{\sqrt{3}}{2} - \dfrac{1}{2}}{-\dfrac{1}{2} - \dfrac{\sqrt{3}}{2}} = 1.$

  For the cardioid $r = 1 + \sin \theta$, the slope of the tangent line at $\theta = \dfrac{2\pi}{3}$ is 1 (Figure C2C 9.1).

Figure C2C 9.1

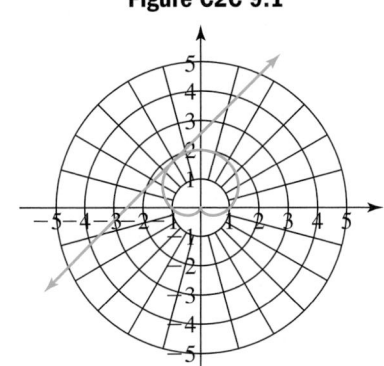

  c. For horizontal tangents, set the numerator equal to zero and solve.

$$\sin(2\theta) + \cos \theta = 0 \quad \text{zeroes of numerator}$$

$$2 \sin \theta \cos \theta + \cos \theta = 0 \quad \text{double angle identity}$$

$$\cos \theta \,(2 \sin \theta + 1) = 0 \quad \text{factor } \cos \theta$$

$$\cos \theta = 0 \quad \text{or} \quad \sin \theta = -\frac{1}{2}$$

Horizontal tangent lines will occur at $\theta = \dfrac{\pi}{2}, \dfrac{7\pi}{6},$ and $\dfrac{11\pi}{6}$. Note that a horizontal tangent does not occur at $\theta = \dfrac{3\pi}{2}$, as this value also makes the denominator of the expression equal to zero.

d. For vertical tangents, set the denominator equal to zero and solve.

Figure C2C 9.2

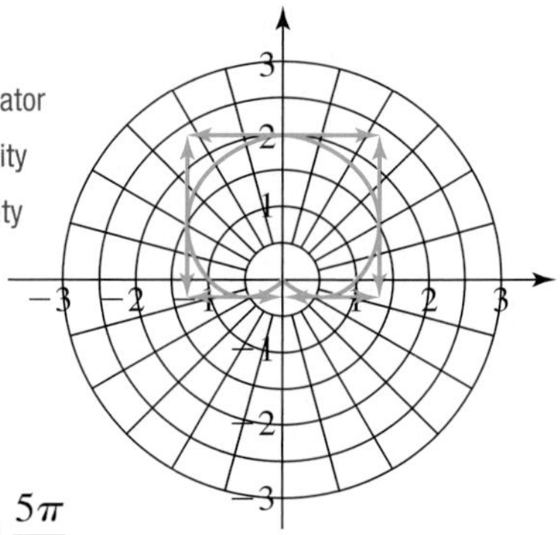

$$\cos(2\theta) - \sin\theta = 0 \quad \text{zeroes of denominator}$$
$$\cos^2\theta - \sin^2\theta - \sin\theta = 0 \quad \text{double angle identity}$$
$$1 - \sin^2\theta - \sin^2\theta - \sin\theta = 0 \quad \text{Pythagorean identity}$$
$$1 - \sin\theta - 2\sin^2\theta = 0 \quad \text{combine terms}$$
$$(1 - 2\sin\theta)(1 + \sin\theta) = 0 \quad \text{factor}$$
$$\sin\theta = \frac{1}{2} \quad \text{or} \quad \sin\theta = -1$$

Vertical tangent lines will occur at $\theta = \dfrac{\pi}{6}$ and $\dfrac{5\pi}{6}$

(see enlargement in Figure C2C 9.2). A vertical tangent does not occur at $\theta = \dfrac{3\pi}{2}$, as this value makes both the numerator and denominator of the expression equal to zero.

Now try Exercises 1 through 4 ▶

Systems of Polar Equations

In the "Connections" feature from Chapter 2, we investigated the area bounded by a given curve, the x-axis, and given vertical lines. In many applications, we need to find the area bounded *between two curves,* which requires that we find where the curves intersect. This can be accomplished by solving a system consisting of the two given functions. For polar equations, the system takes the form $\begin{cases} r = f(\theta) \\ r = g(\theta) \end{cases}$.

EXAMPLE 2 ▶ **Finding Intersections of Polar Curves**

To find the area of the region inside the circle $r = 4\sin\theta$ and outside the limacon $r = 1 + 2\sin\theta$, we must find where the curves intersect. Set up and solve the required system (see figure).

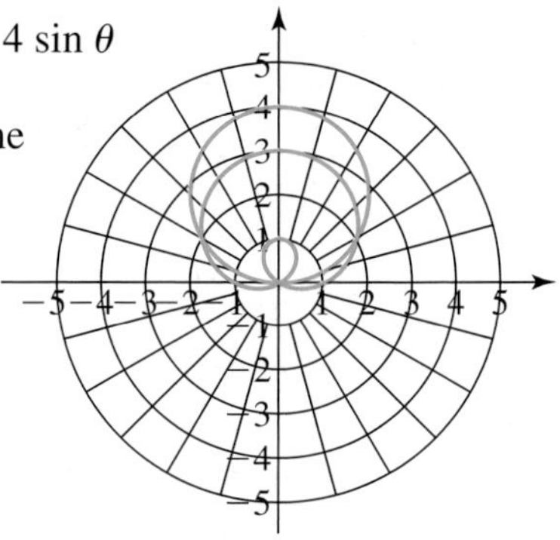

Solution ▶ For $r = 4\sin\theta$ and $r = 1 + 2\sin\theta$, the system is $\begin{cases} r = 4\sin\theta \\ r = 1 + 2\sin\theta \end{cases}$. Substituting $4\sin\theta$ for r in the second equation gives

$$4\sin\theta = 1 + 2\sin\theta, \quad \text{substitute } 4\sin\theta \text{ for } r$$
$$2\sin\theta = 1 \quad \text{subtract } 2\sin\theta$$
$$\sin\theta = \frac{1}{2} \quad \text{divide by 2}$$
$$\theta = \sin^{-1}\left(\frac{1}{2}\right) \quad \text{apply inverse sine}$$

This shows the graphs intersect at $\theta = \dfrac{\pi}{6}$ and $\theta = \dfrac{5\pi}{6}$. It appears we're finished, but an inspection of the equations (or the graph) shows there must be a third point of intersection at the pole $(0, 0)$. The reason we missed it is that in polar coordinates each point can be represented many different ways. Solutions using a system of polar equations finds only the intersections *represented by the identical ordered pair,* and a separate analysis is needed for the case where $r = 0$. Solving the equation

$0 = 4 \sin \theta$ gives $\theta = \pi k$, while $0 = 1 + 2 \sin \theta$ gives $\theta = \dfrac{7\pi}{6} + 2\pi k$ and

$\theta = \dfrac{11\pi}{6} + 2\pi k$. Although the ordered pairs are different, both represent the pole

$(0, 0)$ and the graphs must intersect there.

> **Now try Exercises 5 through 10 ▶**

Connections to Calculus Exercises

For each polar function given, the expression for the slope of the tangent line is also given. (a) Simplify the expression. (b) Find the slope of the tangent line at the values of θ shown. Then determine the value(s) of θ in $[0, 2\pi)$ for which a tangent line is (c) horizontal, and (d) vertical.

1. $r = 1 + \cos \theta$ (cardioid), slope of the tangent line:

$$\dfrac{\cos \theta + \sin \theta(-\sin \theta) + \cos \theta \cos \theta}{-\sin \theta + 2 \cos \theta(-\sin \theta)}; \theta = \dfrac{\pi}{2}, \dfrac{11\pi}{6}$$

2. $r = 6 \cos \theta$ (circle), slope of the tangent line:

$$\dfrac{-6 \sin \theta \sin \theta + 6 \cos \theta \cos \theta}{-6 \sin \theta \cos \theta - 6 \cos \theta \sin \theta}; \theta = \dfrac{\pi}{6}, \dfrac{2\pi}{3}$$

3. $r = \cos \theta - \sin \theta$ (circle), slope of the tangent line:

$$\dfrac{\cos \theta \cos \theta + \sin \theta(-\sin \theta) - 2 \sin \theta \cos \theta}{2 \cos \theta(-\sin \theta) - [\sin \theta(-\sin \theta) + \cos \theta \cos \theta]}$$

$$\theta = \dfrac{\pi}{4}, \dfrac{3\pi}{4}$$

4. $r = \sin \theta - \cos \theta$ (circle), slope of the tangent line:

$$\dfrac{\cos \theta \cos \theta + \sin \theta(-\sin \theta) - 2 \cos \theta(-\sin \theta)}{2 \sin \theta(\cos \theta) - [\cos \theta \cos \theta + \sin \theta(-\sin \theta)]}$$

$$\theta = \dfrac{\pi}{2}, \dfrac{3\pi}{2}$$

Solve each system. For Exercises 7 and 8, identify the graph of each function before you begin. Verify results using a graphing calculator.

5. $\begin{cases} r = \dfrac{3}{2} \\ r = 1 + \cos^2\theta \end{cases}$

6. $\begin{cases} r = 2 \sin^2\theta \\ r = 1 + \sin \theta \end{cases}$

7. $\begin{cases} r = \sin(2\theta) \\ r = \cos \theta \end{cases}$

8. $\begin{cases} r = 2 \sin(2\theta) \\ r = 2 \sin \theta \end{cases}$

9. Find the points of intersection for the circle $r = 4 \cos \theta$ and the curve $r = 4 \sin(2\theta)$, by solving the system $\begin{cases} r = 4 \cos \theta \\ r = 8 \sin \theta \cos \theta \end{cases}$. State your answer in both (r, θ) form and (x, y) form. Note from the graph there are three points of intersection, and that a polar grid has been superimposed on a rectangular grid.

10. Find the points of intersection of the two four-leaf roses formed by $r = -4 \sin(2\theta)$ and $r = 4 \cos(2\theta)$, by solving the system $\begin{cases} r = -4 \sin(2\theta) \\ r = 4 \cos(2\theta) \end{cases}$ (there are nine in all). Some of these points can be found by solving $4 \cos(2\theta) = -4 \sin(2\theta)$, others can be found using symmetry [since r can be positive or negative, we consider the possibility $r_1 = -r_2$, or $4 \cos(2\theta) = 4 \sin(2\theta)$]. Be sure to include an analysis of the case where $r = 0$. Answer in (r, θ) form.

Exercise 9

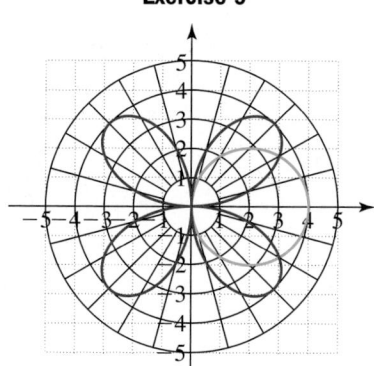

Exercise 10

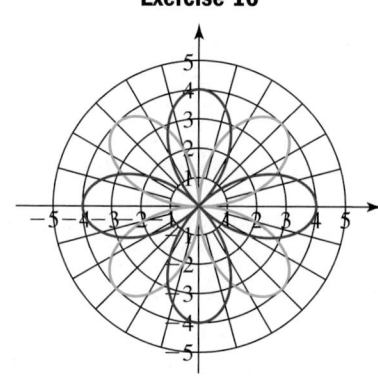

10

Additional Topics in Algebra

CHAPTER OUTLINE

CHAPTER CONNECTIONS

As part of a science experiment, a baseball, golf ball, superball, and bowling ball are dropped on a concrete surface from a height of 2 m. Why do the golf ball and superball seem to bounce almost as high as the drop, while the bowling ball hardly bounces at all? The answer lies in a study of *elastic rebound,* and depends on the density and elasticity of the material used to make the ball. Questions like these are an important part of a fast growing field of study called *sports science.* Further inquiries seek to know the total distance traveled by the balls before they come to rest, as an additional measure of their suitability for use in sports. This application appears as Exercise 117 in Section 10.3.

Connections to Calculus

While the notation and properties of summation were introduced as tools to work with finite sums, in Section 10.3 we will note that under certain conditions, consideration of an infinite sum is possible. The process will involve an ever larger sum of ever smaller pieces, a process further illustrated in the *Connections to Calculus* feature following Chapter 10. This feature also offers practice with the algebraic tools that make the ultimate conclusion possible.

Learning Objectives

In Section 10.1 you will learn how to:

☐ **A.** Write out the terms of a sequence given the general or *n*th term

☐ **B.** Work with recursive sequences and sequences involving a factorial

☐ **C.** Find the partial sum of a series

☐ **D.** Use summation notation to write and evaluate series

☐ **E.** Use sequences to solve applied problems

A *sequence* can be thought of as a pattern of numbers listed in a prescribed order. A *series* is the sum of the numbers in a sequence. Sequences and series come in countless varieties, and we'll introduce some general forms here. In following sections we'll focus on two special types: arithmetic and geometric sequences. These are used in a number of different fields, with a wide variety of significant applications.

A. Finding the Terms of a Sequence Given the General Term

Suppose a person had $10,000 to invest, and decided to place the money in government bonds that guarantee an annual return of 7%. From our work in Chapter 4, we know the amount of money in the account after x years can be modeled by the function $f(x) = 10,000(1.07)^x$. If you reinvest your earnings each year, the amount in the account would be (rounded to the nearest dollar):

Year:	$f(1)$	$f(2)$	$f(3)$	$f(4)$	$f(5)\ldots$
	↓	↓	↓	↓	↓
Value:	$10,700	$11,449	$12,250	$13,108	$14,026\ldots

Note the relationship (year, value) is a function that pairs 1 with $10,700, 2 with $11,449, 3 with $12,250 and so on. This is an example of a **sequence**. To distinguish sequences from other algebraic functions, we commonly name the functions a instead of f, use the variable n instead of x, and employ a subscript notation. The function $f(x) = 10,000(1.07)^x$ would then be written $a_n = 10,000(1.07)^n$. Using this notation $a_1 = 10,700$, $a_2 = 11,449$, and so on.

The values $a_1, a_2, a_3, a_4, \ldots$ are called the **terms** of the sequence. If the account were closed after a certain number of years (for example, after the fifth year) we have a **finite sequence.** If we let the investment grow indefinitely, the result is called an **infinite sequence.** The expression a_n that defines the sequence is called the **general** or **nth term** and the terms immediately preceding it are called the $(n-1)$st term, the $(n-2)$nd term, and so on.

WORTHY OF NOTE

Sequences can actually start with any natural number. For instance, the sequence

$a_n = \dfrac{2}{n-1}$ must start at

$n = 2$ to avoid division by zero. In addition, we will sometimes use a_0 to indicate a preliminary or inaugural element, as in $a_0 = \$10,000$ for the amount of money initially held, prior to investing it.

Sequences

A *finite sequence* is a function a_n whose domain is the set of natural numbers from 1 to n. The terms of the sequence are labeled

$$a_1, a_2, a_3, \ldots, \; a_k, a_{k+1}, \; \ldots, a_{n-1}, a_n$$

where a_k represents an arbitrary "interior" term and a_n also represents the last term of the sequence.

An *infinite sequence* is a function a_n whose domain is the set of <u>all</u> natural numbers.

EXAMPLE 1A ▶ **Computing Specified Terms of a Sequence**

For $a_n = \dfrac{n+1}{n^2}$, find a_1, a_3, a_6, and a_7.

Solution ▶ $a_1 = \dfrac{1+1}{1^2} = 2$ $\qquad\qquad$ $a_3 = \dfrac{3+1}{3^2} = \dfrac{4}{9}$

$a_6 = \dfrac{6+1}{6^2} = \dfrac{7}{36}$ $\qquad\qquad$ $a_7 = \dfrac{7+1}{7^2} = \dfrac{8}{49}$

EXAMPLE 1B ▶ **Computing the First *k* Terms of a Sequence**

Find the first four terms of the sequence $a_n = (-1)^n 2^n$. Write the terms of the sequence as a list.

Solution ▶ $a_1 = (-1)^1 2^1 = -2$ $a_2 = (-1)^2 2^2 = 4$

$a_3 = (-1)^3 2^3 = -8$ $a_4 = (-1)^4 2^4 = 16$

The sequence can be written $-2, 4, -8, 16, \ldots$, or more generally as $-2, 4, -8, 16, \ldots, (-1)^n 2^n, \ldots$ to show how each term was generated.

Now try Exercises 7 through 32 ▶

☑ **A.** You've just learned how to write out the terms of a sequence given the general or *n*th term

B. Recursive Sequences and Factorial Notation

Sometimes the formula defining a sequence uses the preceding term or terms to generate those that follow. These are called **recursive sequences** and are particularly useful in writing computer programs. Because of how they are defined, recursive sequences must give an inaugural term or **seed element,** to begin the recursion process.

Perhaps the most famous recursive sequence is associated with the work of Leonardo of Pisa (A.D. 1180–1250), better known to history as *Fibonacci*. In fact, it is commonly called the Fibonacci sequence in which each successive term is the sum of the previous two, beginning with $1, 1, \ldots$.

EXAMPLE 2 ▶   **Computing the Terms of a Recursive Sequence**

Write out the first eight terms of the recursive (Fibonacci) sequence defined by $c_1 = 1, c_2 = 1,$ and $c_n = c_{n-1} + c_{n-2}$.

Solution ▶   The first two terms are given, so we begin with $n = 3$.

$c_3 = c_{3-1} + c_{3-2}$ $c_4 = c_{4-1} + c_{4-2}$ $c_5 = c_{5-1} + c_{5-2}$
$\quad = c_2 + c_1$ $\quad = c_3 + c_2$ $\quad = c_4 + c_3$
$\quad = 1 + 1$ $\quad = 2 + 1$ $\quad = 3 + 2$
$\quad = 2$ $\quad = 3$ $\quad = 5$

At this point we can simply use the fact that each successive term is simply the sum of the preceding two, and find that $c_6 = 3 + 5 = 8, c_7 = 5 + 8 = 13,$ and $c_8 = 13 + 8 = 21$. The first eight terms are 1, 1, 2, 3, 5, 8, 13, and 21.

Now try Exercises 33 through 38 ▶

Sequences can also be defined using a **factorial,** which is the product of a given natural number with all those that precede it. The expression **5!** is read, "five factorial," and is evaluated as: $5! = 5 \cdot 4 \cdot 3 \cdot 2 \cdot 1 = 120$.

Factorials

For any natural number n,
$$n! = n \cdot (n - 1) \cdot (n - 2) \cdot \cdots \cdot 3 \cdot 2 \cdot 1$$

Rewriting a factorial in equivalent forms often makes it easier to simplify certain expressions. For example, we can rewrite $5! = 5 \cdot 4!$ or $5! = 5 \cdot 4 \cdot 3!$. Consider Example 3.

EXAMPLE 3 ▶   **Simplifying Expressions Using Factorial Notation**

Simplify by writing the numerator in an equivalent form.

a. $\dfrac{9!}{7!}$ **b.** $\dfrac{11!}{8!2!}$ **c.** $\dfrac{6!}{3!5!}$

Solution ▶ **a.** $\dfrac{9!}{7!} = \dfrac{9 \cdot 8 \cdot \cancel{7!}}{\cancel{7!}}$ **b.** $\dfrac{11!}{8!2!} = \dfrac{11 \cdot 10 \cdot 9 \cdot \cancel{8!}}{\cancel{8!}2!}$ **c.** $\dfrac{6!}{3!5!} = \dfrac{6 \cdot \cancel{5!}}{3!\cancel{5!}}$

$= 9 \cdot 8$ $= \dfrac{990}{2}$ $= \dfrac{6}{6}$

$= 72$ $= 495$ $= 1$

Now try Exercises 39 through 44 ▶

> **WORTHY OF NOTE**
>
> Most calculators have a factorial option or key. On the TI-84 Plus it is located on a submenu of the **MATH** key:
>
> **MATH** **PRB** (option) **4: !**

EXAMPLE 4 ▶ **Computing a Specified Term from a Sequence Defined Using Factorials**

Find the third term of each sequence.

a. $a_n = \dfrac{n!}{2^n}$ **b.** $c_n = \dfrac{(-1)^n(2n - 1)!}{n!}$

Solution ▶ **a.** $a_3 = \dfrac{3!}{2^3}$ **b.** $c_3 = \dfrac{(-1)^3[2(3) - 1]!}{3!}$

$= \dfrac{6}{8} = \dfrac{3}{4}$ $= \dfrac{(-1)(5)!}{3!} = \dfrac{(-1)[5 \cdot 4 \cdot \cancel{3!}]}{\cancel{3!}}$

$= -20$

☑ **B.** You've just learned how to work with recursive sequences and sequences involving a factorial

Now try Exercises 45 through 50 ▶

Figure 10.1

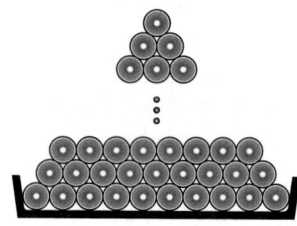

C. Series and Partial Sums

Sometimes the terms of a sequence are dictated by context rather than a formula. Consider the stacking of large pipes in a storage yard. If there are 10 pipes in the bottom row, then 9 pipes, then 8 (see Figure 10.1), how many pipes are in the stack if there is a single pipe at the top? The sequence generated is 10, 9, 8, . . . , 3, 2, 1 and to answer the question we would have to *compute the sum of all terms in the sequence.* When the terms of a finite sequence are added, the result is called a **finite series.**

> **Finite Series**
>
> Given the sequence $a_1, a_2, a_3, a_4, \ldots , a_n$, the sum of the terms is called a **finite series** or **partial sum** and is denoted S_n:
>
> $$S_n = a_1 + a_2 + a_3 + a_4 + \cdots + a_n$$

EXAMPLE 5 ▶ **Computing a Partial Sum**

Given $a_n = 2n$, find the value of
a. S_4 and   **b.** S_7.

Solution ▶ Since we eventually need the sum of the first seven terms (for Part b), begin by writing out these terms: 2, 4, 6, 8, 10, 12, and 14.

a. $S_4 = a_1 + a_2 + a_3 + a_4$

$= 2 + 4 + 6 + 8$

$= 20$

b. $S_7 = a_1 + a_2 + a_3 + a_4 + a_5 + a_6 + a_7$

$= 2 + 4 + 6 + 8 + 10 + 12 + 14$

$= 56$

☑ **C.** You've just learned how to find the partial sum of a series

Now try Exercises 51 through 56 ▶

D. Summation Notation

When the general term of a sequence is known, the Greek letter *sigma* Σ can be used to write the related series as a formula. For instance, to indicate the sum of the first four terms of $a_n = 3n + 2$, we write $\sum_{i=1}^{4}(3i + 2)$. This result is called **summation** or **sigma notation** and the letter i is called the **index of summation.** The letters $j, k, l,$ and m are also used as index numbers, and the summation need not start at 1.

EXAMPLE 6 ▶ **Computing a Partial Sum**

Compute each sum:

    a. $\displaystyle\sum_{i=1}^{4}(3i + 2)$ **b.** $\displaystyle\sum_{j=1}^{5}\frac{1}{j}$ **c.** $\displaystyle\sum_{k=3}^{6}(-1)^k k^2$

Solution ▶   **a.** $\displaystyle\sum_{i=1}^{4}(3i + 2) = (3 \cdot 1 + 2) + (3 \cdot 2 + 2) + (3 \cdot 3 + 2) + (3 \cdot 4 + 2)$

                                $= 5 + 8 + 11 + 14 = 38$

  b. $\displaystyle\sum_{j=1}^{5}\frac{1}{j} = \frac{1}{1} + \frac{1}{2} + \frac{1}{3} + \frac{1}{4} + \frac{1}{5}$

            $= \frac{60}{60} + \frac{30}{60} + \frac{20}{60} + \frac{15}{60} + \frac{12}{60} = \frac{137}{60}$

  c. $\displaystyle\sum_{k=3}^{6}(-1)^k k^2 = (-1)^3 \cdot 3^2 + (-1)^4 \cdot 4^2 + (-1)^5 \cdot 5^2 + (-1)^6 \cdot 6^2$

                      $= -9 + 16 + -25 + 36 = 18$

Now try Exercises 57 through 68 ▶

If a definite pattern is noted in a given series expansion, this process can be reversed, with the expanded form being expressed in summation notation using the *n*th term.

EXAMPLE 7 ▶ **Writing a Sum in Sigma Notation**

Write each of the following sums in summation (sigma) notation.

    a. $1 + 3 + 5 + 7 + 9$ **b.** $6 + 9 + 12 + \cdots$

Solution ▶   **a.** The series has five terms and each term is an odd number, or 1 less than a multiple of 2. The general term is $a_n = 2n - 1$, and the series is $\displaystyle\sum_{n=1}^{5}(2n - 1)$.

  b. This is an infinite sum whose terms are multiples of 3. The general term is $a_n = 3n$, but the series starts at 2 and not 1. The series is $\displaystyle\sum_{j=2}^{\infty}3j$.

WORTHY OF NOTE

By varying the function given and/or where the sum begins, more than one acceptable form is possible.

For example 7(b) $\displaystyle\sum_{n=1}^{\infty}(3 + 3n)$ also works.

Now try Exercises 69 through 78 ▶

Since the commutative and associative laws hold for the addition of real numbers, summations have the following properties:

Properties of Summation

Given any real number c and natural number n,

(I) $\displaystyle\sum_{i=1}^{n} c = cn$

If you add a constant c "n" times the result is cn.

(II) $\displaystyle\sum_{i=1}^{n} ca_i = c\sum_{i=1}^{n} a_i$

A constant can be factored out of a sum.

(III) $\displaystyle\sum_{i=1}^{n} (a_i \pm b_i) = \sum_{i=1}^{n} a_i \pm \sum_{i=1}^{n} b_i$

A summation can be distributed to two (or more) sequences.

(IV) $\displaystyle\sum_{i=1}^{m} a_i + \sum_{i=m+1}^{n} a_i = \sum_{i=1}^{n} a_i;\ 1 \le m < n$

A summation is cumulative and can be written as a sum of smaller parts.

The verification of property II depends solely on the distributive property.

Proof: $\displaystyle\sum_{i=1}^{n} ca_i = ca_1 + ca_2 + ca_3 + \cdots + ca_n$ expand sum

$\qquad\qquad = c(a_1 + a_2 + a_3 + \cdots + a_n)$ factor out c

$\qquad\qquad = c\displaystyle\sum_{i=1}^{n} a_i$ write series in summation form

The verification of properties III and IV simply uses the commutative and associative properties. You are asked to prove property III in **Exercise 91.**

EXAMPLE 8 ▶ **Computing a Sum Using Summation Properties**

Recompute the sum $\displaystyle\sum_{i=1}^{4} (3i + 2)$ from Example 6(a) using summation properties.

Solution ▶ $\displaystyle\sum_{i=1}^{4} (3i + 2) = \sum_{i=1}^{4} 3i + \sum_{i=1}^{4} 2$ property III

$\qquad\qquad\qquad = 3\displaystyle\sum_{i=1}^{4} i + \sum_{i=1}^{4} 2$ property II

☑ **D.** You've just learned how to use summation notation to write and evaluate series

$\qquad\qquad\qquad = 3(10) + 2(4)$ $1 + 2 + 3 + 4 = 10$; property I

$\qquad\qquad\qquad = 38$ result

Now try Exercises 79 through 82 ▶

E. Applications of Sequences

To solve applications of sequences, (1) identify where the sequence begins (the initial term) and (2) write out the first few terms to help identify the nth term.

EXAMPLE 9 ▶ Solving an Application — Accumulation of Stock

Hydra already owned 1420 shares of stock when her company began offering employees the opportunity to purchase 175 discounted shares per year. If she made no purchases other than these discounted shares each year, how many shares will she have 9 yr later? If this continued for the 25 yr she will work for the company, how many shares will she have at retirement?

Solution ▶ To begin, it helps to simply write out the first few terms of the sequence. Since she already had 1420 shares before the company made this offer, we let $a_0 = 1420$ be the inaugural element, showing $a_1 = 1595$ (after 1 yr, she owns 1595 shares). The first few terms are 1595, 1770, 1945, 2120, and so on. This supports a general term of $a_n = 1595 + 175(n - 1)$.

After 9 years

$$a_9 = 1595 + 175(8)$$
$$= 2995$$

After 25 years

$$a_{25} = 1595 + 175(24)$$
$$= 5795$$

☑ **E. You've just learned to use sequences to solve applied problems**

After 9 yr she would have 2995 shares. Upon retirement she would own 5795 shares of company stock.

Now try Exercises 85 through 90 ▶

TECHNOLOGY HIGHLIGHT

Studying Sequences and Series

To support a study of sequences and series, we can use a graphing calculator to generate the desired terms. This can be done either on the home screen or directly into the LIST feature of the calculator. On the TI-84 Plus this is accomplished using the **"seq("** and **"sum("** commands, which are accessed using the keystrokes `2nd` `STAT` (LIST) and the screen shown in Figure 10.2. The **"seq("** feature is option 5 under the **OPS** submenu (press `▶` 5) and the **"sum("** feature is option 5 under the **MATH** submenu (press `◀` 5).

To generate the first four terms of the sequence $a_n = n^2 + 1$, and to find the sum, `CLEAR` the home screen and press `2nd` `STAT` `▶` 5 to place **"seq("** on the home screen. This command requires four inputs: a_n (the nth term), variable used (the calculator can work with any letter), initial term and the last term. For this example the screen reads **"seq($x^2 + 1, x, 1, 4$),"** with the result being the four terms shown in Figure 10.3. To find the *sum* of these terms, we simply precede the **seq($x^2 + 1, x, 1, 4$)** command by **"sum(,"** and two methods are shown in Figure 10.4.

Each of the following sequences have some interesting properties or mathematical connections. Use your graphing calculator to generate the first 10 terms of each sequence and the sum of the first 10 terms. Next, generate the first 20 terms of each sequence and the sum of these terms. What conclusion (if any) can you reach about the sum of each sequence?

Figure 10.2

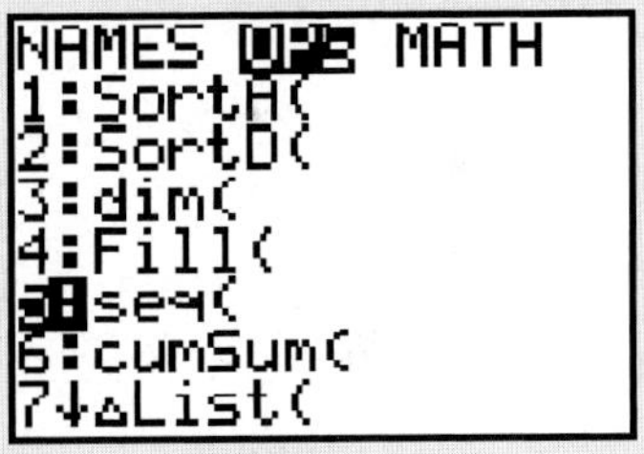

```
NAMES OPS MATH
1:SortA(
2:SortD(
3:dim(
4:Fill(
5▮seq(
6:cumSum(
7↓ΔList(
```

Figure 10.3

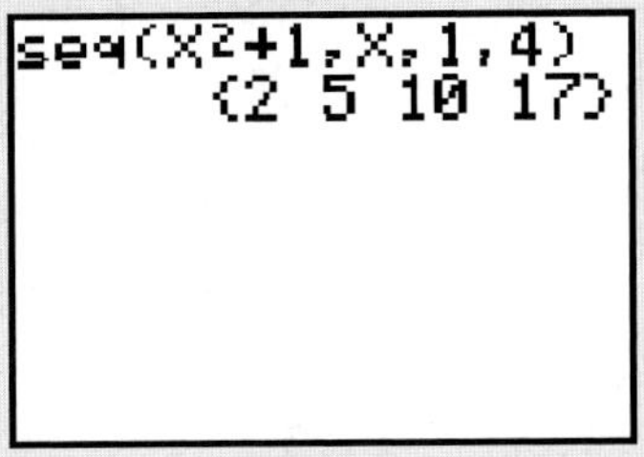

```
seq(X²+1,X,1,4)
        {2 5 10 17}
```

Figure 10.4

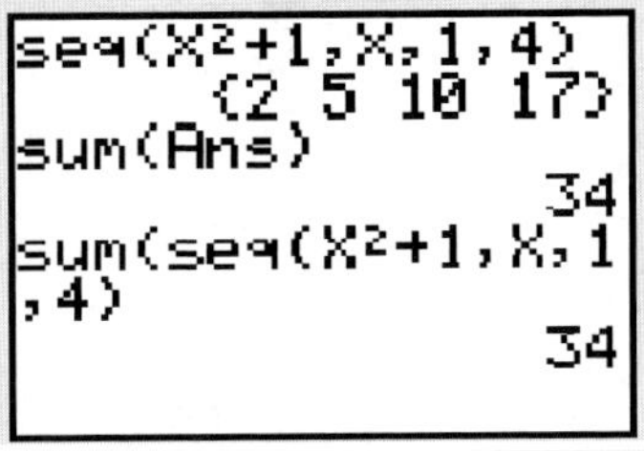

```
seq(X²+1,X,1,4)
        {2 5 10 17}
sum(Ans)
                34
sum(seq(X²+1,X,1
,4)
                34
```

Exercise 1: $a_n = \dfrac{1}{3^n}$ **Exercise 2:** $a_n = \dfrac{2}{n(n+1)}$ **Exercise 3:** $a_n = \dfrac{1}{(2n-1)(2n+1)}$

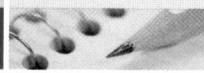

10.1 EXERCISES

▶ CONCEPTS AND VOCABULARY

Fill in the blank with the appropriate word or phrase. Carefully reread the section if needed.

1. A sequence is a(n) _____ of numbers listed in a specific _____.

2. A series is the _____ of the numbers from a given sequence.

3. When each term of a sequence is larger than the preceding term, the sequence is said to be _____.

4. When each term of a sequence is smaller than the preceding term, the sequence is said to be _____.

5. Describe the characteristics of a recursive sequence and give one example.

6. Describe the characteristics of an alternating sequence and give one example.

▶ DEVELOPING YOUR SKILLS

Find the first four terms, then find the 8th and 12th term for each nth term given.

7. $a_n = 2n - 1$

8. $a_n = 2n + 3$

9. $a_n = 3n^2 - 3$

10. $a_n = 2n^3 - 12$

11. $a_n = (-1)^n n$

12. $a_n = \dfrac{(-1)^n}{n}$

13. $a_n = \dfrac{n}{n + 1}$

14. $a_n = \left(1 + \dfrac{1}{n}\right)^n$

15. $a_n = \left(\dfrac{1}{2}\right)^n$

16. $a_n = \left(\dfrac{2}{3}\right)^n$

17. $a_n = \dfrac{1}{n}$

18. $a_n = \dfrac{1}{n^2}$

19. $a_n = \dfrac{(-1)^n}{n(n + 1)}$

20. $a_n = \dfrac{(-1)^{n+1}}{2n^2 - 1}$

21. $a_n = (-1)^n 2^n$

22. $a_n = (-1)^n 2^{-n}$

Find the indicated term for each sequence.

23. $a_n = n^2 - 2;\ a_9$

24. $a_n = (n - 2)^2;\ a_9$

25. $a_n = \dfrac{(-1)^{n+1}}{n};\ a_5$

26. $a_n = \dfrac{(-1)^{n+1}}{2n - 1};\ a_5$

27. $a_n = 2\left(\dfrac{1}{2}\right)^{n-1};\ a_7$

28. $a_n = 3\left(\dfrac{1}{3}\right)^{n-1};\ a_7$

29. $a_n = \left(1 + \dfrac{1}{n}\right)^n;\ a_{10}$

30. $a_n = \left(n + \dfrac{1}{n}\right)^n;\ a_9$

31. $a_n = \dfrac{1}{n(2n + 1)};\ a_4$

32. $a_n = \dfrac{1}{(2n - 1)(2n + 1)};\ a_5$

Find the first five terms of each recursive sequence.

33. $\begin{cases} a_1 = 2 \\ a_n = 5a_{n-1} - 3 \end{cases}$

34. $\begin{cases} a_1 = 3 \\ a_n = 2a_{n-1} - 3 \end{cases}$

35. $\begin{cases} a_1 = -1 \\ a_n = (a_{n-1})^2 + 3 \end{cases}$

36. $\begin{cases} a_1 = -2 \\ a_n = a_{n-1} - 16 \end{cases}$

37. $\begin{cases} c_1 = 64,\ c_2 = 32 \\ c_n = \dfrac{c_{n-2} - c_{n-1}}{2} \end{cases}$

38. $\begin{cases} c_1 = 1,\ c_2 = 2 \\ c_n = c_{n-1} + (c_{n-2})^2 \end{cases}$

Simplify each factorial expression.

39. $\dfrac{8!}{5!}$

40. $\dfrac{12!}{10!}$

41. $\dfrac{9!}{7!2!}$

42. $\dfrac{6!}{3!3!}$

43. $\dfrac{8!}{2!6!}$

44. $\dfrac{10!}{3!7!}$

Write out the first four terms in each sequence.

45. $a_n = \dfrac{n!}{(n+1)!}$

46. $a_n = \dfrac{n!}{(n+3)!}$

47. $a_n = \dfrac{(n+1)!}{(3n)!}$

48. $a_n = \dfrac{(n+3)!}{(2n)!}$

49. $a_n = \dfrac{n^n}{n!}$

50. $a_n = \dfrac{2^n}{n!}$

Find the indicated partial sum for each sequence.

51. $a_n = n;\ S_5$

52. $a_n = n^2;\ S_7$

53. $a_n = 2n - 1;\ S_8$

54. $a_n = 3n - 1;\ S_6$

55. $a_n = \dfrac{1}{n};\ S_5$

56. $a_n = \dfrac{n}{n+1};\ S_4$

Expand and evaluate each series.

57. $\displaystyle\sum_{i=1}^{4} (3i - 5)$

58. $\displaystyle\sum_{i=1}^{5} (2i - 3)$

59. $\displaystyle\sum_{k=1}^{5} (2k^2 - 3)$

60. $\displaystyle\sum_{k=1}^{5} (k^2 + 1)$

61. $\displaystyle\sum_{k=1}^{7} (-1)^k k$

62. $\displaystyle\sum_{k=1}^{5} (-1)^k 2^k$

63. $\displaystyle\sum_{i=1}^{4} \dfrac{i^2}{2}$

64. $\displaystyle\sum_{i=2}^{4} i^2$

65. $\displaystyle\sum_{j=3}^{7} 2j$

66. $\displaystyle\sum_{j=3}^{7} \dfrac{j}{2^j}$

67. $\displaystyle\sum_{k=3}^{8} \dfrac{(-1)^k}{k(k-2)}$

68. $\displaystyle\sum_{k=2}^{6} \dfrac{(-1)^{k+1}}{k^2 - 1}$

Write each sum using sigma notation. Answers are not necessarily unique.

69. $4 + 8 + 12 + 16 + 20$

70. $5 + 10 + 15 + 20 + 25$

71. $-1 + 4 - 9 + 16 - 25 + 36$

72. $1 - 8 + 27 - 64 + 125 - 216$

For the given general term a_n, write the indicated sum using sigma notation.

73. $a_n = n + 3;\ S_5$

74. $a_n = \dfrac{n^2 + 1}{n + 1};\ S_4$

75. $a_n = \dfrac{n^2}{3};$ third partial sum

76. $a_n = 2n - 1;$ sixth partial sum

77. $a_n = \dfrac{n}{2^n};$ sum for $n = 3$ to 7

78. $a_n = n^2;$ sum for $n = 2$ to 6

Compute each sum by applying properties of summation.

79. $\displaystyle\sum_{i=1}^{5} (4i - 5)$

80. $\displaystyle\sum_{i=1}^{6} (3 + 2i)$

81. $\displaystyle\sum_{k=1}^{4} (3k^2 + k)$

82. $\displaystyle\sum_{k=1}^{4} (2k^3 + 5)$

▶ **WORKING WITH FORMULAS**

83. Sum of $a_n = 3n - 2$: $S_n = \dfrac{n(3n - 1)}{2}$

The sum of the first n terms of the sequence defined by $a_n = 3n - 2 = 1, 4, 7, 10, \ldots,$ $(3n - 2), \ldots$ is given by the formula shown. Find S_5 using the formula, then verify by direct calculation.

84. Sum of $a_n = 3n - 1$: $S_n = \dfrac{n(3n + 1)}{2}$

The sum of the first n terms of the sequence defined by $a_n = 3n - 1 = 2, 5, 8, 11, \ldots, (3n - 1), \ldots$ is given by the formula shown. Find S_8 using the formula, then verify by direct calculation. Observing the results of Exercises 83 and 84, can you now state the sum formula for $a_n = 3n - 0$?

▶ **APPLICATIONS**

Use the information given in each exercise to determine the *n*th term a_n for the sequence described. Then use the *n*th term to list the specified number of terms.

85. **Blue-book value:** Steve's car has a blue-book value of $6000. Each year it loses 20% of its value (its value each year is 80% of the year before). List the value of Steve's car for the next 5 yr. (*Hint:* For $a_1 = 6000$, we need the *next* five terms.)

86. **Effects of inflation:** Suppose inflation (an increase in value) will average 4% for the next 5 yr. List the growing cost (year by year) of a DVD that costs $15 right now. (*Hint:* For $a_1 = 15$, we need the *next* five terms.)

87. **Wage increases:** Latisha gets $5.20 an hour for filling candy machines for Archtown Vending. Each year she receives a $0.50 hourly raise. List Latisha's wage for the first 5 yr. How much will she make in the fifth year if she works 8 hr per day for 240 working days?

88. **Average birth weight:** The average birth weight of a certain animal species is 900 g, with the baby gaining 125 g each day for the first 10 days. List the infant's weight for the first 10 days. How much does the infant weigh on the 10th day?

89. **Stocking a lake:** A local fishery stocks a large lake with 1500 bass and then adds an additional 100 mature bass per month until the lake nears maximum capacity. If the bass population grows at a rate of 5% per month through natural reproduction, the number of bass in the pond after *n* months is given by the recursive sequence $b_0 = 1500$, $b_n = 1.05b_{n-1} + 100$. How many bass will be in the lake after 6 months?

90. **Species preservation:** The Interior Department introduces 50 wolves (male and female) into a large wildlife area in an effort to preserve the species. Each year about 12 additional adult wolves are added from capture and relocation programs. If the wolf population grows at a rate of 10% per year through natural reproduction, the number of wolves in the area after *n* years is given by the recursive sequence $w_0 = 50$, $w_n = 1.10w_{n-1} + 12$. How many wolves are in the wildlife area after 6 years?

▶ **EXTENDING THE CONCEPT**

91. Verify that a summation may be distributed to two (or more) sequences. That is, verify that the following statement is true:

$$\sum_{i=1}^{n} (a_i \pm b_i) = \sum_{i=1}^{n} a_i \pm \sum_{i=1}^{n} b_i.$$

Surprisingly, some of the most celebrated numbers in mathematics can be represented or approximated by a series expansion. Use your calculator to find the partial

sums for $n = 4$, $n = 8$, and $n = 12$ for the summations given, and attempt to name the number the summation approximates:

92. $\displaystyle\sum_{k=0}^{n} \frac{1}{k!}$

93. $\displaystyle\sum_{k=1}^{n} \frac{1}{3^k}$

94. $\displaystyle\sum_{k=1}^{n} \frac{1}{2^k}$

▶ **MAINTAINING YOUR SKILLS**

95. (6.7) Solve $\csc x \sin\left(\dfrac{\pi}{2} - x\right) = -1$

96. (2.5) Set up the difference quotient for $f(x) = \sqrt{x}$, then rationalize the numerator.

97. (7.2) Given a triangle where $a = 0.4$ m, $b = 0.3$ m, and $c = 0.5$ m, find the three corresponding angles.

98. (6.3) Solve the system using a matrix equation. $\begin{cases} 25x + y - 2z = -14 \\ 2x - y + z = 40 \\ -7x + 3y - z = -13 \end{cases}$

Learning Objectives

In Section 10.2 you will learn how to:

☐ **A.** Identify an arithmetic sequence and its common difference

☐ **B.** Find the *n*th term of an arithmetic sequence

☐ **C.** Find the *n*th partial sum of an arithmetic sequence

☐ **D.** Solve applications involving arithmetic sequences

Similar to the way polynomials fall into certain groups or families (linear, quadratic, cubic, etc.), sequences and series with common characteristics are likewise grouped. In this section, we focus on sequences where each successive term is generated by adding a constant value, as in the sequence 1, 8, 15, 22, 29, . . . , where 7 is added to a given term in order to produce the next term.

A. Identifying an Arithmetic Sequence and Finding the Common Difference

An **arithmetic sequence** is one where each successive term is found by adding a fixed constant to the preceding term. For instance 3, 7, 11, 15, . . . is an arithmetic sequence, since adding 4 to any given term produces the next term. This also means if you take the difference of any two consecutive terms, the result will be 4 and in fact, 4 is called the **common difference** *d* for this sequence. Using the notation developed earlier, we can write $d = a_{k+1} - a_k$, where a_k represents any term of the sequence and a_{k+1} represents the term that follows a_k.

Arithmetic Sequences

Given a sequence $a_1, a_2, a_3, \ldots, a_k, a_{k+1}, \ldots, a_n$, where $k, n \in \mathbb{N}$ and $k < n$,

if there exists a common difference *d* such that $a_{k+1} - a_k = d$,

then the sequence is an *arithmetic sequence*.

The difference of successive terms can be rewritten as $a_{k+1} = a_k + d$ (for $k \geq 1$) to highlight that each following term is found by adding *d* to the previous term.

EXAMPLE 1 ▶ **Identifying an Arithmetic Sequence**

Determine if the given sequence is arithmetic.

a. 2, 5, 8, 11, . . .   **b.** $\frac{1}{2}, \frac{5}{6}, \frac{4}{3}, \frac{7}{6}, \ldots$

Solution ▶ **a.** Begin by looking for a common difference $d = a_{k+1} - a_k$. Checking each pair of consecutive terms we have

$$5 - 2 = 3 \quad 8 - 5 = 3 \quad 11 - 8 = 3 \quad \text{and so on.}$$

This is an arithmetic sequence with common difference $d = 3$.

b. Checking each pair of consecutive terms yields

$$\frac{5}{6} - \frac{1}{2} = \frac{5}{6} - \frac{3}{6} \qquad\qquad \frac{4}{3} - \frac{5}{6} = \frac{8}{6} - \frac{5}{6}$$
$$= \frac{2}{6} = \frac{1}{3} \qquad\qquad\qquad = \frac{3}{6} = \frac{1}{2}$$

Since the difference is not constant, this is not an arithmetic sequence.

Now try Exercises 7 through 18 ▶

EXAMPLE 2 ▶ **Writing the First *k* Terms of an Arithmetic Sequence**

Write the first five terms of the arithmetic sequence, given the first term a_1 and the common difference *d*.

a. $a_1 = 12$ and $d = -4$ **b.** $a_1 = \frac{1}{2}$ and $d = \frac{1}{3}$

Solution ▶

a. $a_1 = 12$ and $d = -4$. Starting at $a_1 = 12$, add -4 to each new term to generate the sequence: 12, 8, 4, 0, -4

b. $a_1 = \frac{1}{2}$ and $d = \frac{1}{3}$. Starting at $a_1 = \frac{1}{2}$ and adding $\frac{1}{3}$ to each new term will generate the sequence: $\frac{1}{2}, \frac{5}{6}, \frac{7}{6}, \frac{3}{2}, \frac{11}{6}$

Now try Exercises 19 through 30 ▶

B. Finding the *n*th Term of an Arithmetic Sequence

If the values a_1 and d from an arithmetic sequence are known, we could generate the terms of the sequence by adding *multiples of d to the first term,* instead of adding d to each new term. For example, we can generate the sequence 3, 8, 13, 18, 23 by adding multiples of 5 to the first term $a_1 = 3$:

$$3 = 3 + (0)5 \qquad a_1 = a_1 + 0d$$
$$8 = 3 + (1)5 \qquad a_2 = a_1 + 1d$$
$$13 = 3 + (2)5 \qquad a_3 = a_1 + 2d$$
$$18 = 3 + (3)5 \qquad a_4 = a_1 + 3d$$
$$23 = 3 + (4)5 \qquad a_5 = a_1 + 4d$$

current term ——↑   initial   ↑—— coefficient of common
term   difference

It's helpful to note the coefficient of d is 1 less than the subscript of the current term (as shown): $5 - 1 = 4$. This observation leads us to a formula for the *n*th term.

The *n*th Term of an Arithmetic Sequence

The *n*th term of an *arithmetic sequence* is given by

$$a_n = a_1 + (n - 1)d$$

where d is the common difference.

EXAMPLE 3 ▶ **Finding a Specified Term in an Arithmetic Sequence**

Find the 24th term of the sequence 0.1, 0.4, 0.7, 1, ...

Solution ▶ Instead of creating all terms up to the 24th, we determine the constant d and use the *n*th term formula. By inspection we note $a_1 = 0.1$ and $d = 0.3$.

$$a_n = a_1 + (n - 1)d \qquad \text{\textit{n}th term formula}$$
$$= 0.1 + (n - 1)0.3 \qquad \text{substitute 0.1 for } a_1 \text{ and 0.3 for } d$$
$$= 0.1 + 0.3n - 0.3 \qquad \text{eliminate parentheses}$$
$$= 0.3n - 0.2 \qquad \text{simplify}$$

To find the 24th term we substitute 24 for n:

$$a_{24} = 0.3(24) - 0.2 \qquad \text{substitute 24 for } n$$
$$= 7.0 \qquad \text{result}$$

Now try Exercises 31 through 42 ▶

EXAMPLE 4 ▶ **Finding the Number of Terms in an Arithmetic Sequence**

Find the number of terms in the arithmetic sequence 2, -5, -12, -19, . . . , -411.

Solution ▶ By inspection we see that $a_1 = 2$ and $d = -7$. As before,

$$a_n = a_1 + (n-1)d \qquad \text{\small nth term formula}$$
$$= 2 + (n-1)(-7) \qquad \text{\small substitute 2 for } a_1 \text{ and } -7 \text{ for } d$$
$$= -7n + 9 \qquad \text{\small simplify}$$

Although we don't know the number of terms in the sequence, we *do* know the last or *n*th term is -411. Substituting -411 for a_n gives

$$-411 = -7n + 9 \qquad \text{\small substitute } -411 \text{ for } a_n$$
$$60 = n \qquad \text{\small solve for } n$$

There are 60 terms in this sequence.

Now try Exercises 43 through 50 ▶

If the term a_1 is unknown but a term a_k is given, the *n*th term can be written

$$a_n = a_k + (n-k)d \text{ since } n = k + (n-k)$$

(the subscript of the term a_k and coefficient of d sum to n).

EXAMPLE 5 ▶ **Finding the First Term of an Arithmetic Sequence**

Given an arithmetic sequence where $a_6 = 0.55$ and $a_{13} = 0.9$, find the common difference d and the value of a_1.

Solution ▶ At first it seems that not enough information is given, but recall we can express a_{13} as the sum of any earlier term and the appropriate multiple of d. Since a_6 is known, we write $a_{13} = a_6 + 7d$ (note $13 = 6 + 7$ as required).

$$a_{13} = a_6 + 7d \qquad \text{\small } a_1 \text{ is unknown}$$
$$0.9 = 0.55 + 7d \qquad \text{\small substitute 0.9 for } a_{13} \text{ and 0.55 for } a_6$$
$$0.35 = 7d \qquad \text{\small subtract 0.55}$$
$$d = 0.05 \qquad \text{\small solve for } d$$

Having found d, we can now solve for a_1.

$$a_{13} = a_1 + 12d \qquad \text{\small nth term formula for } n = 13$$
$$0.9 = a_1 + 12(0.05) \qquad \text{\small substitute 0.9 for } a_{13} \text{ and 0.05 for } d$$
$$0.9 = a_1 + 0.6 \qquad \text{\small simplify}$$
$$a_1 = 0.3 \qquad \text{\small solve for } a_1$$

☑ **B. You've just learned how to find the *n*th term of an arithmetic sequence**

The first term is $a_1 = 0.3$ and the common difference is $d = 0.05$.

Now try Exercises 51 through 56 ▶

C. Finding the *n*th Partial Sum of an Arithmetic Sequence

Using sequences and series to solve applications often requires computing the sum of a given number of terms. Consider the sequence $a_1, a_2, a_3, a_4, \ldots, a_n$ with common difference d. Use S_n to represent the sum of the first n terms and write the original series, then the series in reverse order underneath. Since one row increases at the same rate the other decreases, the sum of each column remains constant, and for simplicity's sake we choose $a_1 + a_n$ to represent this sum.

$$
\begin{array}{lllllllll}
S_n = & a_1 & + & a_2 & + & a_3 & + \cdots + & a_{n-2} & + & a_{n-1} & + & a_n & \text{\small add} \\
S_n = & a_n & + & a_{n-1} & + & a_{n-2} & + \cdots + & a_3 & + & a_2 & + & a_1 & \text{\small columns} \\
\hline
2S_n = & (a_1 + a_n) & + & (a_1 + a_n) & + & (a_1 + a_n) & + \cdots + & (a_1 + a_n) & + & (a_1 + a_n) & + & (a_1 + a_n) & \text{\small vertically}
\end{array}
$$

To understand why each column adds to $a_1 + a_n$, consider the sum in the second column: $a_2 + a_{n-1}$. From $a_2 = a_1 + d$ and $a_{n-1} = a_n - d$, we obtain $a_2 + a_{n-1} = (a_1 + d) + (a_n - d)$ by direct substitution, which gives a result of $a_1 + a_n$. Since there are n columns, we end up with $2S_n = n(a_1 + a_n)$, and solving for S_n gives the formula for the first n terms of an arithmetic sequence.

The *n*th Partial Sum of an Arithmetic Sequence

Given an arithmetic sequence with first term a_1, the nth partial sum is given by

$$S_n = n\left(\frac{a_1 + a_n}{2}\right).$$

In words: The sum of an arithmetic sequence is the number of terms times the average of the first and last term.

EXAMPLE 6 ▶ **Computing the Sum of an Arithmetic Sequence**

Find the sum of the first 75 positive, odd integers: $\displaystyle\sum_{k=1}^{75}(2k - 1)$.

Solution ▶ The initial terms of the sequence are 1, 3, 5, ... and we note $a_1 = 1$, $d = 2$, and $n = 75$. To use the sum formula, we need the value of $a_n = a_{75}$. The nth term formula shows $a_{75} = a_1 + 74d = 1 + 74(2)$, so $a_{75} = 149$.

$$S_n = \frac{n(a_1 + a_n)}{2} \qquad \text{sum formula}$$

$$S_{75} = \frac{75(a_1 + a_{75})}{2} \qquad \text{substitute 75 for } n$$

$$= \frac{75(1 + 149)}{2} \qquad \text{substitute 1 for } a_1, \text{ 149 for } a_{75}$$

$$= 5625 \qquad \text{result}$$

☑ **C. You've just learned how to find the *n*th partial sum of an arithmetic sequence**

The sum of the first 75 positive, odd integers is 5625.

Now try Exercises 57 through 62 ▶

Figure 10.5

spiral fern

By substituting the nth term formula directly into the formula for partial sums, we're able to find a partial sum without actually having to find the nth term:

$$S_n = \frac{n(a_1 + a_n)}{2} \qquad \text{sum formula}$$

$$= \frac{n(a_1 + [a_1 + (n-1)d])}{2} \qquad \text{substitute } a_1 + (n-1)d \text{ for } a_n$$

$$= \frac{n}{2}[2a_1 + (n-1)d] \qquad \textit{alternative formula for the nth partial sum}$$

Figure 10.6

nautilus

See Exercises 63 through 68 for more on this alternative formula.

D. Applications

In the evolution of certain plants and shelled animals, sequences and series seem to have been one of nature's favorite tools (see Figures 10.5 and 10.6). Sequences and series also provide a good mathematical model for a variety of other situations as well.

EXAMPLE 7 ▶ **Solving an Application of Arithmetic Sequences: Seating Capacity**

Cox Auditorium is an amphitheater that has 40 seats in the first row, 42 seats in the second row, 44 in the third, and so on. If there are 75 rows in the auditorium, what is the auditorium's seating capacity?

Solution ▶ The number of seats in each row gives the terms of an arithmetic sequence with $a_1 = 40$, $d = 2$, and $n = 75$. To find the seating capacity, we need to find the total number of seats, which is the sum of this arithmetic sequence. Since the value of a_{75} is unknown, we opt for the alternative formula $S_n = \dfrac{n}{2}[2a_1 + (n - 1)d]$.

$$S_n = \frac{n}{2}[2a_1 + (n - 1)d] \qquad \text{sum formula}$$

$$S_{75} = \frac{75}{2}[2(40) + (75 - 1)(2)] \qquad \text{substitute 40 for } a_1, 2 \text{ for } d \text{ and 75 for } n$$

$$= \frac{75}{2}(228) \qquad \text{simplify}$$

$$= 8550 \qquad \text{result}$$

☑ **D. You've just learned how to solve applications involving arithmetic sequences**

The seating capacity for Cox Auditorium is 8550.

Now try Exercises 71 through 76 ▶

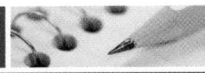

10.2 EXERCISES

▶ **CONCEPTS AND VOCABULARY**

Fill in the blank with the appropriate word or phrase. Carefully reread the section if needed.

1. Consecutive terms in an arithmetic sequence differ by a constant called the _____ _____.

2. The sum of the first n terms of an arithmetic sequence is called the nth _____ _____.

3. The formula for the nth partial sum of an arithmetic sequence is $s_n =$ _____, where a_n is the _____ term.

4. The nth term formula for an arithmetic sequence is $a_n =$ _____, where a_1 is the _____ term and d is the _____ _____.

5. Discuss how the terms of an arithmetic sequence can be written in various ways using the relationship $a_n = a_k + (n - k)d$.

6. Describe how the formula for the nth partial sum was derived, and illustrate its application using a sequence from the exercise set.

▶ **DEVELOPING YOUR SKILLS**

Determine if the sequence given is arithmetic. If yes, name the common difference. If not, try to determine the pattern that forms the sequence.

7. $-5, -2, 1, 4, 7, 10, \ldots$

8. $1, -2, -5, -8, -11, -14, \ldots$

9. $0.5, 3, 5.5, 8, 10.5, \ldots$

10. $1.2, 3.5, 5.8, 8.1, 10.4, \ldots$

11. $2, 3, 5, 7, 11, 13, 17, \ldots$

12. $1, 4, 8, 13, 19, 26, 34, \ldots$

13. $\dfrac{1}{24}, \dfrac{1}{12}, \dfrac{1}{8}, \dfrac{1}{6}, \dfrac{5}{24}, \ldots$

14. $\dfrac{1}{12}, \dfrac{1}{15}, \dfrac{1}{20}, \dfrac{1}{30}, \dfrac{1}{60}, \ldots$

15. 1, 4, 9, 16, 25, 36, . . .

16. −125, −64, −27, −8, −1, . . .

17. $\pi, \dfrac{5\pi}{6}, \dfrac{2\pi}{3}, \dfrac{\pi}{2}, \dfrac{\pi}{3}, \dfrac{\pi}{6}, \ldots$ **18.** $\pi, \dfrac{7\pi}{8}, \dfrac{3\pi}{4}, \dfrac{5\pi}{8}, \dfrac{\pi}{2}, \ldots$

Write the first four terms of the arithmetic sequence with the given first term and common difference.

19. $a_1 = 2, d = 3$

20. $a_1 = 8, d = 3$

21. $a_1 = 7, d = -2$

22. $a_1 = 60, d = -12$

23. $a_1 = 0.3, d = 0.03$

24. $a_1 = 0.5, d = 0.25$

25. $a_1 = \frac{3}{2}, d = \frac{1}{2}$

26. $a_1 = \frac{1}{5}, d = \frac{1}{10}$

27. $a_1 = \frac{3}{4}, d = -\frac{1}{8}$

28. $a_1 = \frac{1}{6}, d = -\frac{1}{3}$

29. $a_1 = -2, d = -3$

30. $a_1 = -4, d = -4$

Identify the first term and the common difference, then write the expression for the general term a_n and use it to find the 6th, 10th, and 12th terms of the sequence.

31. 2, 7, 12, 17, . . .

32. 7, 4, 1, −2, −5, . . .

33. 5.10, 5.25, 5.40, . . .

34. 9.75, 9.40, 9.05, . . .

35. $\frac{3}{2}, \frac{9}{4}, 3, \frac{15}{4}, \ldots$

36. $\frac{5}{7}, \frac{3}{14}, -\frac{2}{7}, -\frac{11}{14}, \ldots$

Find the indicated term using the information given.

37. $a_1 = 5, d = 4$; find a_{15}

38. $a_1 = 9, d = -2$; find a_{17}

39. $a_1 = \frac{3}{2}, d = -\frac{1}{12}$; find a_7

40. $a_1 = \frac{12}{25}, d = -\frac{1}{10}$; find a_9

41. $a_1 = -0.025, d = 0.05$; find a_{50}

42. $a_1 = 3.125, d = -0.25$; find a_{20}

Find the number of terms in each sequence.

43. $a_1 = 2, a_n = -22, d = -3$

44. $a_1 = 4, a_n = 42, d = 2$

45. $a_1 = 0.4, a_n = 10.9, d = 0.25$

46. $a_1 = -0.3, a_n = -36, d = -2.1$

47. −3, −0.5, 2, 4.5, 7, . . . , 47

48. −3.4, −1.1, 1.2, 3.5, . . . , 38

49. $\frac{1}{12}, \frac{1}{8}, \frac{1}{6}, \frac{5}{24}, \frac{1}{4}, \ldots, \frac{9}{8}$ **50.** $\frac{1}{12}, \frac{1}{15}, \frac{1}{20}, \frac{1}{30}, \ldots, -\frac{1}{4}$

Find the common difference d and the value of a_1 using the information given.

51. $a_3 = 7, a_7 = 19$

52. $a_5 = -17, a_{11} = -2$

53. $a_2 = 1.025, a_{26} = 10.025$

54. $a_6 = -12.9, a_{30} = 1.5$

55. $a_{10} = \frac{13}{18}, a_{24} = \frac{27}{2}$ **56.** $a_4 = \frac{5}{4}, a_8 = \frac{9}{4}$

Evaluate each sum. For Exercises 61 and 62, use the summation properties from Section 10.1.

57. $\displaystyle\sum_{n=1}^{30}(3n - 4)$

58. $\displaystyle\sum_{n=1}^{29}(4n - 1)$

59. $\displaystyle\sum_{n=1}^{37}\left(\frac{3}{4}n + 2\right)$

60. $\displaystyle\sum_{n=1}^{20}\left(\frac{5}{2}n - 3\right)$

61. $\displaystyle\sum_{n=4}^{15}(3 - 5n)$

62. $\displaystyle\sum_{n=7}^{20}(7 - 2n)$

Use the alternative formula for the nth partial sum to compute the sums indicated.

63. The sum S_{15} for the sequence
   $-12 + (-9.5) + (-7) + (-4.5) + \cdots$

64. The sum S_{20} for the sequence $\frac{9}{2} + \frac{7}{2} + \frac{5}{2} + \frac{3}{2} + \cdots$

65. The sum S_{30} for the sequence
   $0.003 + 0.173 + 0.343 + 0.513 + \cdots$

66. The sum S_{50} for the sequence
   $(-2) + (-7) + (-12) + (-17) + \cdots$

67. The sum S_{20} for the sequence
   $\sqrt{2} + 2\sqrt{2} + 3\sqrt{2} + 4\sqrt{2} + \cdots$

68. The sum S_{10} for the sequence
   $12\sqrt{3} + 10\sqrt{3} + 8\sqrt{3} + 6\sqrt{3} + \cdots$

▶ **WORKING WITH FORMULAS**

69. Sum of the first n natural numbers: $S_n = \dfrac{n(n + 1)}{2}$

The sum of the first n natural numbers can be found using the formula shown, where n represents the number of terms in the sum. Verify the formula by adding the first six natural numbers by hand, and then evaluating S_6. Then find the sum of the first 75 natural numbers.

70. Sum of the squares of the first n natural numbers: $S_n = \dfrac{n(n + 1)(2n + 1)}{6}$

If the first n natural numbers are squared, the sum of these squares can be found using the formula shown, where n represents the number of terms in the sum. Verify the formula by computing the sum of the squares of the first six natural numbers by hand, and then evaluating S_6. Then find the sum of the squares of the first 20 natural numbers: $(1^2 + 2^2 + 3^2 + \cdots + 20^2)$.

► **APPLICATIONS**

71. Temperature fluctuation: At 5 P.M. in Coldwater, the temperature was a chilly 36°F. If the temperature decreased by 3°F every half-hour for the next 7 hr, at what time did the temperature hit 0°F?

72. Arc of a baby swing: When Mackenzie's baby swing is started, the first swing (one way) is a 30-in. arc. As the swing slows down, each successive arc is $\frac{3}{2}$ in. less than the previous one. Find (a) the length of the tenth swing and (b) how far Mackenzie has traveled during the 10 swings.

73. Computer animations: The animation on a new computer game initially allows the hero of the game to jump a (screen) distance of 10 in. over booby traps and obstacles. Each successive jump is limited to $\frac{3}{4}$ in. less than the previous one. Find (a) the length of the seventh jump and (b) the total distance covered after seven jumps.

74. Seating capacity: The Fox Theater creates a "theater in the round" when it shows any of Shakespeare's plays. The first row has 80 seats, the second row  has 88, the third row has 96, and so on. How many seats are in the 10th row? If there is room for 25 rows, how many chairs will be needed to set up the theater?

75. Sales goals: *At the time that I was newly hired, 100 sales per month was what I required. Each following month—the last plus 20 more, as I work for the goal of top sales award. When 2500 sales are thusly made, it's Tahiti, Hawaii, and pina coladas in the shade.* How many sales were made by this person in the seventh month? What were the total sales after the 12th month? Was the goal of 2500 total sales met after the 12th month?

76. Bequests to charity: *At the time our mother left this Earth, she gave $9000 to her children of birth. This we kept and each year added $3000 more, as a lasting memorial from the children she bore. When $42,000 is thusly attained, all goes to charity that her memory be maintained.* What was the balance in the sixth year? In what year was the goal of $42,000 met?

► **EXTENDING THE THOUGHT**

77. From a study of numerical analysis, a function is known to be linear if its "first differences" (differences between each output) are constant. Likewise, a function is known to be quadratic if its "first differences" form an *arithmetic sequence*. Use this information to determine if the following sets of output come from a linear or quadratic function:

   a. 19, 11.8, 4.6, −2.6, −9.8, −17, −24.2, . . .
   b. −10.31, −10.94, −11.99, −13.46, −15.35, . . .

78. From elementary geometry it is known that the interior angles of a triangle sum to 180°, the interior angles of a quadrilateral sum to 360°, the interior angles of a pentagon sum to 540°, and so on. Use the pattern created by the relationship between the number of sides to the number of angles to develop a formula for the sum of the interior angles of an *n*-sided polygon. The interior angles of a decagon (10 sides) sum to how many degrees?

► **MAINTAINING YOUR SKILLS**

79. (5.7) Identify the amplitude (*A*), period (*P*), horizontal shift (HS), vertical shift (VS) and endpoints of the primary interval (PI) for

$$f(t) = 7 \sin\left(\frac{\pi}{3}t - \frac{\pi}{6}\right) + 10.$$

80. (3.1) Graph by completing the square. Label all important features: $y = x^2 - 2x - 3$.

81. (2.3) In 2000, the deer population was 972. By 2005 it had grown to 1217. Assuming the growth is linear, find the function that models this data and use it to estimate the deer population in 2008.

82. (6.1) Verify $\dfrac{\tan x}{\csc x} - \dfrac{\sin x}{\cos x} = \dfrac{\sin x - 1}{\cot x}$ is an identity.

Learning Objectives

In Section 10.3 you will learn how to:

☐ **A.** Identify a geometric sequence and its common ratio

☐ **B.** Find the nth term of a geometric sequence

☐ **C.** Find the nth partial sum of a geometric sequence

☐ **D.** Find the sum of an infinite geometric series

☐ **E.** Solve application problems involving geometric sequences and series

Recall that arithmetic sequences are those where each term is found by *adding* a constant value to the preceding term. In this section, we consider **geometric sequences,** where each term is found by *multiplying* the preceding term by a constant value. Geometric sequences have many interesting applications, as do **geometric series.**

A. Geometric Sequences

A geometric sequence is one where each successive term is found by multiplying the preceding term by a fixed constant. Consider growth of a bacteria population, where a single cell splits in two every hour over a 24-hr period. Beginning with a single bacterium ($a_0 = 1$), after 1 hr there are 2, after 2 hr there are 4, and so on. Writing the number of bacteria as a sequence we have:

hours:	a_1	a_2	a_3	a_4	a_5	. . .
	↓	↓	↓	↓	↓	
bacteria:	2	4	8	16	32	. . .

The sequence 2, 4, 8, 16, 32, . . . is a geometric sequence since each term is found by multiplying the previous term by the constant factor 2. This also means that the ratio of any two consecutive terms must be 2 and in fact, 2 is called the **common ratio r** for this sequence. Using the notation from Section 10.1 we can write $r = \dfrac{a_{k+1}}{a_k}$, where a_k represents any term of the sequence and a_{k+1} represents the term that follows a_k.

EXAMPLE 1 ▶ **Testing a Sequence for a Common Ratio**

Determine if the given sequence is geometric.

a. 1, 0.5, 0.25, 0.125, . . .   **b.** $\frac{1}{8}, \frac{1}{4}, \frac{3}{4}, 3, 15, \ldots$

Solution ▶ Apply the definition to check for a common ratio $r = \dfrac{a_{k+1}}{a_k}$.

a. For 1, 0.5, 0.25, 0.125, . . . , the ratio of consecutive terms gives

$$\frac{0.5}{1} = 0.5, \qquad \frac{0.25}{0.5} = 0.5, \qquad \frac{0.125}{0.25} = 0.5, \qquad \text{and so on.}$$

This is a geometric sequence with common ratio $r = 0.5$.

b. For $\frac{1}{8}, \frac{1}{4}, \frac{3}{4}, 3, 15, \ldots$, we have:

$$\frac{1}{4} \div \frac{1}{8} = \frac{1}{4} \cdot \frac{8}{1} \qquad \frac{3}{4} \div \frac{1}{4} = \frac{3}{4} \cdot \frac{4}{1} \qquad 3 \div \frac{3}{4} = \frac{3}{1} \cdot \frac{4}{3} \qquad \text{and so on.}$$
$$= 2 \qquad\qquad\qquad = 3 \qquad\qquad\qquad = 4$$

Since the ratio is not constant, this is not a geometric sequence.

Now try Exercises 7 through 24 ▶

EXAMPLE 2 ▶ **Writing the Terms of a Geometric Sequence**

Write the first five terms of the geometric sequence, given the first term $a_1 = -16$ and the common ratio $r = 0.25$.

Solution ▶ Given $a_1 = -16$ and $r = 0.25$. Starting at $a_1 = -16$, multiply each term by 0.25 to generate the sequence.

$$a_2 = -16 \cdot 0.25 = -4 \qquad a_3 = -4 \cdot 0.25 = -1$$
$$a_4 = -1 \cdot 0.25 = -0.25 \qquad a_5 = -0.25 \cdot 0.25 = -0.0625$$

The first five terms of this sequence are $-16, -4, -1, -0.25,$ and -0.0625.

☑ **A.** You've just learned how to identify a geometric sequence and its common ratio

Now try Exercises 25 through 38 ▶

B. Find the *n*th Term of a Geometric Sequence

If the values a_1 and r from a geometric sequence are known, we could generate the terms of the sequence by applying *additional factors of r to the first term,* instead of multiplying each new term by r. If $a_1 = 3$ and $r = 2$, we simply begin at a_1, and continue applying additional factors of r for each successive term.

$$3 = 3 \cdot 2^0 \qquad a_1 = a_1 r^0$$
$$6 = 3 \cdot 2^1 \qquad a_2 = a_1 r^1$$
$$12 = 3 \cdot 2^2 \qquad a_3 = a_1 r^2$$
$$24 = 3 \cdot 2^3 \qquad a_4 = a_1 r^3$$
$$48 = 3 \cdot 2^4 \qquad a_5 = a_1 r^4$$

current term —— initial term —— exponent on common ratio

From this pattern, we note the exponent on r is always 1 less than the subscript of the current term: $5 - 1 = 4$, which leads us to the formula for the *n*th term of a geometric sequence.

The *n*th Term of a Geometric Sequence

The *n*th term of a *geometric sequence* is given by

$$a_n = a_1 r^{n-1}$$

where r is the common ratio.

EXAMPLE 3 ▶ **Finding a Specific Term in a Sequence**

Find the 10th term of the sequence $3, -6, 12, -24, \ldots$.

Solution ▶ Instead of writing out all 10 terms, we determine the constant ratio r and use the *n*th term formula. By inspection we note that $a_1 = 3$ and $r = -2$.

$$a_n = a_1 r^{n-1} \qquad \text{\small{n}th term formula}$$
$$= 3(-2)^{n-1} \qquad \text{substitute 3 for } a_1 \text{ and } -2 \text{ for } r$$

To find the 10th term we substitute $n = 10$:

$$a_{10} = 3(-2)^{10-1} \qquad \text{substitute 10 for } n$$
$$= 3(-2)^9 = -1536 \qquad \text{simplify}$$

Now try Exercises 39 through 46 ▶

EXAMPLE 4 ▶ **Determining the Number of Terms in a Geometric Sequence**

Find the number of terms in the geometric sequence $4, 2, 1, \ldots, \frac{1}{64}$.

Solution ▶ Observing that $a_1 = 4$ and $r = \frac{1}{2}$, we have

$$a_n = a_1 r^{n-1} \qquad \text{\small{n}th term formula}$$
$$= 4\left(\frac{1}{2}\right)^{n-1} \qquad \text{substitute 4 for } a_1 \text{ and } \frac{1}{2} \text{ for } r$$

Although we don't know the number of terms in the sequence, we *do* know the last or *n*th term is $\frac{1}{64}$. Substituting $a_n = \frac{1}{64}$ gives

$$\frac{1}{64} = 4\left(\frac{1}{2}\right)^{n-1} \qquad \text{substitute } \frac{1}{64} \text{ for } a_n$$
$$\frac{1}{256} = \left(\frac{1}{2}\right)^{n-1} \qquad \text{divide by 4} \left(\text{multiply by } \frac{1}{4}\right)$$

From our work in Chapter 4, we attempt to write both sides as exponentials with a like base, or apply logarithms. Since $256 = 2^8$, we equate bases.

$$\left(\frac{1}{2}\right)^8 = \left(\frac{1}{2}\right)^{n-1} \qquad \text{write } \frac{1}{256} \text{ as } \left(\frac{1}{2}\right)^8$$

$$\rightarrow 8 = n - 1 \qquad \text{like bases imply exponents must be equal}$$

$$9 = n \qquad \text{solve for } n$$

This shows there are nine terms in the sequence.

Now try Exercises 47 through 58 ▶

If the term a_1 is unknown but a term a_k is given, the nth term can be written

$$a_n = a_k r^{n-k}, \text{ since } n = k + (n - k)$$

(the subscript on the term a_k and the exponent on r sum to n).

EXAMPLE 5 ▶ **Finding the First Term of a Geometric Sequence**

Given a geometric sequence where $a_4 = 0.075$ and $a_7 = 0.009375$, find the common ratio r and the value of a_1.

Solution ▶ Since a_1 is not known, we express a_7 as the product of a known term and the appropriate number of common ratios: $a_7 = a_4 r^3$ ($7 - 4 = 3$, as required).

$$a_7 = a_4 \cdot r^3 \qquad a_1 \text{ is unknown}$$

$$0.009375 = 0.075r^3 \qquad \text{substitute 0.009375 for } a_7 \text{ and 0.075 for } a_4$$

$$0.125 = r^3 \qquad \text{divide by 0.075}$$

$$r = 0.5 \qquad \text{solve for } r$$

Having found r, we can now solve for a_1

$$a_7 = a_1 r^6 \qquad n\text{th term formula}$$

$$0.009375 = a_1(0.5)^6 \qquad \text{substitute 0.009375 for } a_7 \text{ and 0.5 for } r$$

$$0.009375 = a_1(0.015625) \qquad \text{simplify}$$

$$a_1 = 0.6 \qquad \text{solve for } a_1$$

☑ **B.** You've just learned how to find the nth term of a geometric sequence

The first term is $a_1 = 0.6$ and the common ratio is $r = 0.5$.

Now try Exercises 59 through 64 ▶

C. Find the *n*th Partial Sum of a Geometric Sequence

As with arithmetic series, applications of geometric series often involve computing a sum of consecutive terms. We can adapt the method for finding the sum of an arithmetic sequence to develop a formula for adding the first n terms of a geometric sequence.

For the nth term $a_n = a_1 r^{n-1}$, we have $S_n = a_1 + a_1 r + a_1 r^2 + a_1 r^3 + \cdots + a_1 r^{n-1}$. If we multiply S_n by $-r$ then add the original series, the "interior terms" sum to zero.

$$- rS_n = -a_1 r + (-a_1 r^2) + (-a_1 r^3) + \cdots + (-a_1 r^{n-1}) + (-a_1 r^n)$$

$$+ S_n = a_1 + a_1 r + a_1 r^2 + \cdots + a_1 r^{n-2} + a_1 r^{n-1}$$

$$S_n - rS_n = a_1 + 0 + 0 + 0 + 0 + 0 + (-a_1 r^n)$$

We then have $S_n - rS_n = a_1 - a_1 r^n$, and can now solve for S_n:

$$S_n - rS_n = a_1 - a_1 r^n \qquad \text{difference of } S_n \text{ and } rS_n$$

$$S_n(1 - r) = a_1 - a_1 r^n \qquad \text{factor out } S_n$$

$$S_n = \frac{a_1 - a_1 r^n}{1 - r} \qquad \text{solve for } S_n$$

The result is a formula for the nth partial sum of a geometric sequence.

The nth Partial Sum of a Geometric Sequence

Given a geometric sequence with first term a_1 and common ratio r, the nth partial sum (the sum of the first n terms) is

$$S_n = \frac{a_1 - a_1 r^n}{1 - r} = \frac{a_1(1 - r^n)}{1 - r}, r \neq 1$$

In words: The sum of a geometric sequence is the difference of the first and $(n + 1)$st term, divided by 1 minus the common ratio.

EXAMPLE 6 ▶ **Computing a Partial Sum**

Find the sum: $\displaystyle\sum_{i=1}^{9} 3^i$ (the first nine powers of 3).

Solution ▶ The initial terms of this series are $3 + 9 + 27 + \cdots$, and we note $a_1 = 3$, $r = 3$, and $n = 9$. We could find the first nine terms and add, but using the partial sum formula gives

$$S_n = \frac{a_1(1 - r^n)}{1 - r} \qquad \text{sum formula}$$

$$S_9 = \frac{3(1 - 3^9)}{1 - 3} \qquad \text{substitute 3 for } a_1, \text{ 9 for } n, \text{ and 3 for } r$$

$$= \frac{3(-19{,}682)}{-2} \qquad \text{simplify}$$

$$= 29{,}523 \qquad \text{result}$$

☑ **C. You've just learned how to find the nth partial sum of a geometric sequence**

Now try Exercises 65 through 88 ▶

D. The Sum of an Infinite Geometric Series

To this point we've considered only partial sums of a geometric series. While it is impossible to add an infinite number of these terms, some of these "infinite sums" appear to have a limiting value. The sum appears to get ever closer to this value but never exceeds it—much like the asymptotic behavior of some graphs. We will define the sum of this **infinite geometric series** to be this limiting value, if it exists. Consider the illustration in Figure 10.7, where a standard sheet of typing paper is cut in half. One of the halves is again cut in half and the process is continued indefinitely, as shown. Notice the "halves" create an infinite sequence $\frac{1}{2}, \frac{1}{4}, \frac{1}{8}, \frac{1}{16}, \frac{1}{32}, \cdots$ with $a_1 = \frac{1}{2}$ and $r = \frac{1}{2}$. The corresponding infinite series is $\frac{1}{2} + \frac{1}{4} + \frac{1}{8} + \frac{1}{16} + \frac{1}{32} + \cdots + \frac{1}{2^n} + \cdots$.

Figure 10.7

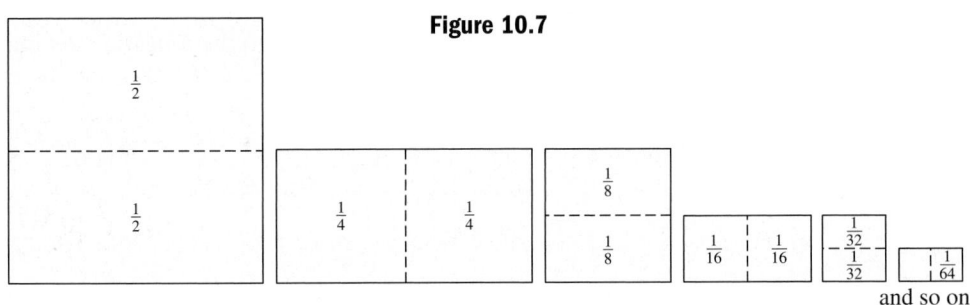

and so on

Figure 10.8

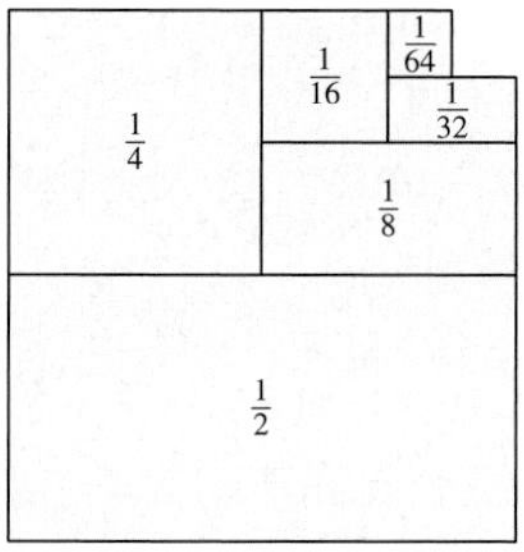

If we arrange one of the halves from each stage as shown in Figure 10.8, we would be rebuilding the original sheet of paper. As we add more and more of these halves together, we get closer and closer to the size of the original sheet. We gain an intuitive sense that this series must add to 1, because the *pieces* of the original sheet of paper must add to 1 whole sheet. To explore this idea further, consider what happens to $\left(\frac{1}{2}\right)^n$ as n becomes large.

$$n = 4: \left(\frac{1}{2}\right)^4 = 0.0625 \qquad n = 8: \left(\frac{1}{2}\right)^8 \approx 0.004 \qquad n = 12: \left(\frac{1}{2}\right)^{12} \approx 0.0002$$

Further exploration with a calculator seems to support the idea that as $n \to \infty$, $\left(\frac{1}{2}\right)^n \to 0$, although a definitive proof is left for a future course. In fact, it can be shown that for any $|r| < 1$, r^n becomes very close to zero as n becomes large. In symbols: as $n \to \infty$, $r^n \to 0$. For $S_n = \dfrac{a_1 - a_1 r^n}{1 - r} = \dfrac{a_1}{1 - r} - \dfrac{a_1 r^n}{1 - r}$, note that if $|r| < 1$ and "we sum an infinite number of terms," the second term becomes zero, leaving only the first term. In other words, the limiting value (represented by S_∞) is

$$S_\infty = \frac{a_1}{1 - r}.$$

WORTHY OF NOTE

The formula for the sum of an infinite geometric series can also be derived by noting that $S_\infty = a_1 + a_1 r + a_1 r^2 + a_1 r^3 + \cdots$ can be rewritten as $S_\infty = a_1 + r(a_1 + a_1 r + a_1 r^2 + a_1 r^3 + \cdots) = a_1 + r S_\infty$.

$$S_\infty - r S_\infty = a_1$$
$$S_\infty(1 - r) = a_1$$
$$S_\infty = \frac{a_1}{1 - r}.$$

Infinite Geometric Series

Given a geometric sequence with first term a_1 and $|r| < 1$, the sum of the related infinite series is given by

$$S_\infty = \frac{a_1}{1 - r}; r \neq 1$$

If $|r| > 1$, no finite sum exists.

EXAMPLE 7 ▶ **Computing an Infinite Sum**

Find the limiting value of each infinite geometric series (if it exists).

a. $1 + 2 + 4 + 8 + \cdots$ **b.** $3 + 2 + \frac{4}{3} + \frac{8}{9} + \cdots$

c. $0.185 + 0.000185 + 0.000000185 + \cdots$

Solution ▶ Begin by determining if the infinite series is geometric. If so, use $S_\infty = \dfrac{a_1}{1 - r}$.

a. Since $r = 2$ (by inspection), a finite sum does not exist.

b. Using the ratio of consecutive terms we find $r = \frac{2}{3}$ and the infinite sum exists. With $a_1 = 3$, we have

$$S_\infty = \frac{3}{1 - \frac{2}{3}} = \frac{3}{\frac{1}{3}} = 9$$

c. This series is equivalent to the repeating decimal $0.185185185\ldots = 0.\overline{185}$. The common ratio is $r = \frac{0.000185}{0.185} = 0.001$ and the infinite sum exists:

$$S_\infty = \frac{0.185}{1 - 0.001} = \frac{5}{27}$$

☑ **D.** You've just learned how to find the sum of an infinite geometric series

Now try Exercises 89 through 104 ▶

E. Applications Involving Geometric Sequences and Series

Here are a few of the ways these ideas can be put to use.

EXAMPLE 8 ▶ Solving an Application of Geometric Sequences: Pendulums

A pendulum is any object attached to a fixed point and allowed to swing freely under the influence of gravity. Suppose each swing is 0.9 the length of the previous one. Gradually the swings become shorter and shorter and at some point the pendulum will appear to have stopped (although *theoretically* it never does).

  a. How far does the pendulum travel on its eighth swing, if the first swing was 2 m?

  b. What is the total distance traveled by the pendulum for these eight swings?

  c. How many swings until the length of each swing falls below 0.5 m?

  d. What total distance does the pendulum travel before coming to rest?

Solution ▶ **a.** The lengths of each swing form the terms of a geometric sequence with $a_1 = 2$ and $r = 0.9$. The first few terms are 2, 1.8, 1.62, 1.458, and so on. For the 8th term we have:

$$a_n = a_1 r^{n-1} \qquad \text{\footnotesize nth term formula}$$
$$a_8 = 2(0.9)^{8-1} \qquad \text{\footnotesize substitute 8 for } n, \text{ 2 for } a_1, \text{ and 0.9 for } r$$
$$\approx 0.956$$

The pendulum travels about 0.956 m on its 8th swing.

  b. For the total distance traveled after eight swings, we compute the value of S_8.

$$S_n = \frac{a_1(1 - r^n)}{1 - r} \qquad \text{\footnotesize nth partial sum formula}$$
$$S_8 = \frac{2(1 - 0.9^8)}{1 - 0.9} \qquad \text{\footnotesize substitute 2 for } a_1, \text{ 0.9 for } r, \text{ and 8 for } n$$
$$\approx 11.4$$

The pendulum has traveled about 11.4 m by the end of the 8th swing.

  c. To find the number of swings until the length of each swing is less than 0.5 m, we solve for n in the equation $0.5 = 2(0.9)^{n-1}$. This yields

$$0.25 = (0.9)^{n-1} \qquad \text{\footnotesize divide by 2}$$
$$\ln 0.25 = (n - 1)\ln 0.9 \qquad \text{\footnotesize take the natural log, apply power property}$$
$$\frac{\ln 0.25}{\ln 0.9} + 1 = n \qquad \text{\footnotesize solve for } n \text{ (exact form)}$$
$$14.16 \approx n \qquad \text{\footnotesize solve for } n \text{ (approximate form)}$$

After the 14th swing, each successive swing will be less than 0.5 m.

  d. For the total distance traveled before coming to rest, we consider the related infinite geometric series, with $a_1 = 2$ and $r = 0.9$.

$$S_\infty = \frac{a_1}{1 - r} \qquad \text{\footnotesize infinite sum formula}$$
$$S_\infty = \frac{2}{1 - 0.9} \qquad \text{\footnotesize substitute 2 for } a_1 \text{ and 0.9 for } r$$
$$= 20 \qquad \text{\footnotesize result}$$

The pendulum would travel 20 m before coming to rest.

☑ **E. You've just learned how to solve application problems involving geometric sequences and series**

Now try Exercises 107 through 119 ▶

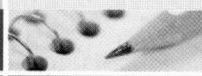

10.3 EXERCISES

▶ **CONCEPTS AND VOCABULARY**

Fill in the blank with the appropriate word or phrase. Carefully reread the section if needed.

1. In a geometric sequence, each successive term is found by _____ the preceding term by a fixed value r.

2. In a geometric sequence, the common ratio r can be found by computing the _____ of any two consecutive terms.

3. The nth term of a geometric sequence is given by $a_n =$ _____ , for any $n \geq 1$.

4. For the general sequence $a_1, a_2, a_3, \ldots, a_k, \ldots$, the fifth partial sum is given by $S_5 =$ _____ .

5. Describe/Discuss how the formula for the nth partial sum is related to the formula for the sum of an infinite geometric series.

6. Describe the difference(s) between an arithmetic and a geometric sequence. How can a student prevent confusion between the formulas?

▶ **DEVELOPING YOUR SKILLS**

Determine if the sequence given is geometric. If yes, name the common ratio. If not, try to determine the pattern that forms the sequence.

7. $4, 8, 16, 32, \ldots$

8. $2, 6, 18, 54, 162, \ldots$

9. $3, -6, 12, -24, 48, \ldots$

10. $128, -32, 8, -2, \ldots$

11. $2, 5, 10, 17, 26, \ldots$

12. $-13, -9, -5, -1, 3, \ldots$

13. $3, 0.3, 0.03, 0.003, \ldots$

14. $12, 0.12, 0.0012, 0.000012, \ldots$

15. $-1, 3, -12, 60, -360, \ldots$

16. $-\frac{2}{3}, 2, -8, 40, -240, \ldots$

17. $25, 10, 4, \frac{8}{5}, \ldots$

18. $-36, 24, -16, \frac{32}{3}, \ldots$

19. $\frac{1}{2}, \frac{1}{4}, \frac{1}{8}, \frac{1}{16}, \ldots$

20. $\frac{2}{3}, \frac{4}{9}, \frac{8}{27}, \frac{16}{81}, \ldots$

21. $3, \dfrac{12}{x}, \dfrac{48}{x^2}, \dfrac{192}{x^3}, \ldots$

22. $5, \dfrac{10}{a}, \dfrac{20}{a^2}, \dfrac{40}{a^3}, \ldots$

23. $240, 120, 40, 10, 2, \ldots$

24. $-120, -60, -20, -5, -1, \ldots$

Write the first four terms of the sequence, given a_1 and r.

25. $a_1 = 5, r = 2$

26. $a_1 = 2, r = -4$

27. $a_1 = -6, r = -\frac{1}{2}$

28. $a_1 = \frac{2}{3}, r = \frac{1}{5}$

29. $a_1 = 4, r = \sqrt{3}$

30. $a_1 = \sqrt{5}, r = \sqrt{5}$

31. $a_1 = 0.1, r = 0.1$

32. $a_1 = 0.024, r = 0.01$

Find the indicated term for each sequence.

33. $a_1 = -24, r = \frac{1}{2}$; find a_7

34. $a_1 = 48, r = -\frac{1}{3}$; find a_6

35. $a_1 = -\frac{1}{20}, r = -5$; find a_4

36. $a_1 = \frac{3}{20}, r = 4$; find a_5

37. $a_1 = 2, r = \sqrt{2}$; find a_7

38. $a_1 = \sqrt{3}, r = \sqrt{3}$; find a_8

Identify a_1 and r, then write the expression for the nth term $a_n = a_1 r^{n-1}$ and use it to find $a_6, a_{10},$ and a_{12}.

39. $\frac{1}{27}, -\frac{1}{9}, \frac{1}{3}, -1, 3, \ldots$

40. $-\frac{7}{8}, \frac{7}{4}, -\frac{7}{2}, 7, -14, \ldots$

41. $729, 243, 81, 27, 9, \ldots$

42. $625, 125, 25, 5, 1, \ldots$

43. $\frac{1}{2}, \frac{\sqrt{2}}{2}, 1, \sqrt{2}, 2, \ldots$

44. $36\sqrt{3}, 36, 12\sqrt{3}, 12, 4\sqrt{3}, \ldots$

45. $0.2, 0.08, 0.032, 0.0128, \ldots$

46. $0.5, -0.35, 0.245, -0.1715, \ldots$

Find the number of terms in each sequence.

47. $a_1 = 9, a_n = 729, r = 3$

48. $a_1 = 1, a_n = -128, r = -2$

49. $a_1 = 16, a_n = \frac{1}{64}, r = \frac{1}{2}$

50. $a_1 = 4, a_n = \frac{1}{512}, r = \frac{1}{2}$

51. $a_1 = -1, a_n = -1296, r = \sqrt{6}$

52. $a_1 = 2, a_n = 1458, r = -\sqrt{3}$

53. $2, -6, 18, -54, \ldots, -4374$

54. $3, -6, 12, -24, \ldots, -6144$

55. $64, 32\sqrt{2}, 32, 16\sqrt{2}, \ldots, 1$

56. $243, 81\sqrt{3}, 81, 27\sqrt{3}, \ldots, 1$

57. $\frac{3}{8}, -\frac{3}{4}, \frac{3}{2}, -3, \ldots, 96$

58. $-\frac{5}{27}, \frac{5}{9}, -\frac{5}{3}, -5, \ldots, -135$

Find the common ratio r and the value of a_1 using the information given (assume $r > 0$).

59. $a_3 = 324, a_7 = 64$ **60.** $a_5 = 6, a_9 = 486$

61. $a_4 = \frac{4}{9}, a_8 = \frac{9}{4}$ **62.** $a_2 = \frac{16}{81}, a_5 = \frac{2}{3}$

63. $a_4 = \frac{32}{3}, a_8 = 54$ **64.** $a_3 = \frac{16}{25}, a_7 = 25$

Find the indicated sum. For Exercises 81 and 82, use the summation properties from Section 10.1.

65. $a_1 = 8, r = -2$; find S_{12}

66. $a_1 = 2, r = -3$; find S_8

67. $a_1 = 96, r = \frac{1}{3}$; find S_5

68. $a_1 = 12, r = \frac{1}{2}$; find S_8

69. $a_1 = 8, r = \frac{3}{2}$; find S_7

70. $a_1 = -1, r = -\frac{3}{2}$; find S_{10}

71. $2 + 6 + 18 + \cdots$; find S_6

72. $2 + 8 + 32 + \cdots$; find S_7

73. $16 - 8 + 4 - \cdots$; find S_8

74. $4 - 12 + 36 - \cdots$; find S_8

75. $\frac{4}{3} + \frac{2}{9} + \frac{1}{27} + \cdots$; find S_9

76. $\frac{1}{18} - \frac{1}{6} + \frac{1}{2} - \cdots$; find S_7

77. $\displaystyle\sum_{j=1}^{5} 4^j$ **78.** $\displaystyle\sum_{k=1}^{10} 2^k$

79. $\displaystyle\sum_{k=1}^{8} 5\left(\frac{2}{3}\right)^{k-1}$ **80.** $\displaystyle\sum_{j=1}^{7} 3\left(\frac{1}{5}\right)^{j-1}$

81. $\displaystyle\sum_{i=4}^{10} 9\left(-\frac{1}{2}\right)^{i-1}$ **82.** $\displaystyle\sum_{i=3}^{8} 5\left(-\frac{1}{4}\right)^{i-1}$

Find the indicated partial sum using the information given. Write all results in simplest form.

83. $a_2 = -5, a_5 = \frac{1}{25}$; find S_5

84. $a_3 = 1, a_6 = -27$; find S_6

85. $a_3 = \frac{4}{9}, a_7 = \frac{9}{64}$; find S_6

86. $a_2 = \frac{16}{81}, a_5 = \frac{2}{3}$; find S_8

87. $a_3 = 2\sqrt{2}, a_6 = 8$; find S_7

88. $a_2 = 3, a_5 = 9\sqrt{3}$; find S_7

Determine whether the infinite geometric series has a finite sum. If so, find the limiting value.

89. $3 + 6 + 12 + 24 + \cdots$

90. $4 + 8 + 16 + 32 + \cdots$

91. $9 + 3 + 1 + \cdots$

92. $36 + 24 + 16 + \cdots$

93. $25 + 10 + 4 + \frac{8}{5} + \cdots$

94. $10 + 2 + \frac{2}{5} + \frac{2}{25} + \cdots$

95. $6 + 3 + \frac{3}{2} + \frac{3}{4} + \cdots$

96. $-49 + (-7) + \left(-\frac{1}{7}\right) + \cdots$

97. $6 - 3 + \frac{3}{2} - \frac{3}{4} + \cdots$

98. $10 - 5 + \frac{5}{2} - \frac{5}{4} + \cdots$

99. $0.3 + 0.03 + 0.003 + \cdots$

100. $0.63 + 0.0063 + 0.000063 + \cdots$

101. $\displaystyle\sum_{k=1}^{\infty} \frac{3}{4}\left(\frac{2}{3}\right)^k$ **102.** $\displaystyle\sum_{i=1}^{\infty} 5\left(\frac{1}{2}\right)^i$

103. $\displaystyle\sum_{j=1}^{\infty} 9\left(-\frac{2}{3}\right)^j$ **104.** $\displaystyle\sum_{k=1}^{\infty} 12\left(\frac{4}{3}\right)^k$

▶ WORKING WITH FORMULAS

105. Sum of the cubes of the first n natural numbers:
$$S_n = \frac{n^2(n + 1)^2}{4}$$

Compute $1^3 + 2^3 + 3^3 + \cdots + 8^3$ using the formula given. Then confirm the result by direct calculation.

106. Student loan payment: $A_n = P(1 + r)^n$

If P dollars is borrowed at an annual interest rate r with interest compounded annually, the amount of money to be paid back after n years is given by the indicated formula. Find the total amount of money that the student must repay to clear the loan, if $8000 is borrowed at 4.5% interest and the loan is paid back in 10 yr.

▶ APPLICATIONS

107. Pendulum movement: On each swing, a pendulum travels only 80% as far as it did on the previous swing. If the first swing is 24 ft, how far does the pendulum travel on the 7th swing? What total distance is traveled before the pendulum comes to rest?

108. Pendulum movement: Ernesto is swinging to and fro on his backyard tire swing. Using his legs and body, he pumps each swing until reaching a maximum height, then suddenly relaxes until the swing comes to a stop. With each swing, Ernesto travels 75% as far as he did on the previous swing. If the first arc (or swing) is 30 ft, find the distance Ernesto travels on the 5th arc. What total distance will he travel before coming to rest?

109. Depreciation: A certain new SUV depreciates in value about 20% per year (meaning it holds 80% of its value each year). If the SUV is purchased for $46,000, how much is it worth 4 yr later? How many years until its value is less than $5000?

110. Depreciation: A new photocopier under heavy use will depreciate about 25% per year (meaning it holds 75% of its value each year). If the copier is purchased for $7000, how much is it worth 4 yr later? How many years until its value is less than $1246?

111. Equipment aging: Tests have shown that the pumping power of a heavy-duty oil pump decreases by 3% per month. If the pump can move 160 gallons per minute (gpm) new, how many gpm can the pump move 8 months later? If the pumping rate falls below 118 gpm, the pump must be replaced. How many months until this pump is replaced?

112. Equipment aging: At the local mill, a certain type of saw blade can saw approximately 2 log-feet/sec when it is new. As time goes on, the blade becomes worn, and loses 6% of its cutting speed each week. How many log-feet/sec can the saw blade cut after 6 weeks? If the cutting speed falls below 1.2 log-feet/sec, the blade must be replaced. During what week of operation will this blade be replaced?

113. Population growth: At the beginning of the year 2000, the population of the United States was approximately 277 million. If the population is growing at a rate of 2.3% per year, what will the population be in 2010, 10 yr later?

114. Population growth: The population of the Zeta Colony on Mars is 1000 people. Determine the population of the Colony 20 yr from now, if the population is growing at a constant rate of 5% per year.

115. Population growth: A biologist finds that the population of a certain type of bacteria doubles *each half-hour*. If an initial culture has 50 bacteria, what is the population after 5 hr? How long will it take for the number of bacteria to reach 204,800?

116. Population growth: Suppose the population of a "boom town" in the old west doubled *every 2 months* after gold was discovered. If the initial population was 219, what was the population 8 months later? How many months until the population exceeds 28,000?

117. Elastic rebound: Megan discovers that a rubber ball dropped from a height of 2 m rebounds four-fifths of the distance it has previously fallen. How high does it rebound on the 7th bounce? How far does the ball travel before coming to rest?

118. Elastic rebound: The screen saver on my computer is programmed to send a colored ball vertically down the middle of the screen so that it rebounds 95% of the distance it last traversed. If the ball always begins at the top and the screen is 36 cm tall,

how high does the ball bounce after its 8th rebound? How far does the ball travel before coming to rest (and a new screen saver starts)?

119. Creating a vacuum: To create a vacuum, a hand pump is used to remove the air from an air-tight cube with a volume of 462 in^3. With each stroke of the pump, two-fifths of the air that remains in the cube is removed. How much air remains inside after the 5th stroke? How many strokes are required to remove all but 12.9 in^3 of the air?

▶ EXTENDING THE CONCEPT

120. As part of a science experiment, identical rubber balls are dropped from a certain height on these surfaces: slate, cement, and asphalt. When dropped on slate, the ball rebounds 80% of the height from which it last fell. On cement the figure is 75% and on asphalt the figure is 70%. The ball is dropped from 130 m on the slate, 175 m on the cement, and 200 m on the asphalt. Which ball has traveled the shortest total distance at the time of the fourth bounce? Which ball will travel farthest before coming to rest?

121. Consider the following situation. A person is hired at a salary of $40,000 per year, with a guaranteed raise of $1750 per year. At the same time, inflation is running about 4% per year. How many years until this person's salary is overtaken and eaten up by the actual cost of living?

122. A standard piece of typing paper is approximately 0.001 in. thick. Suppose you were able to fold this

piece of paper in half 26 times. How thick would the result be? (a) As tall as a hare, (b) as tall as a hen, (c) as tall as a horse, (d) as tall as a house, or (e) over 1 mi high? Find the actual height by computing the 27th term of a geometric sequence. Discuss what you find.

123. Find an alternative formula for the sum

$$S_n = \sum_{k=1}^{n} \log k,$$ that does not use the sigma notation.

124. Verify the following statements:

a. If $a_1, a_2, a_3, \ldots, a_n$ is a geometric sequence with r and a_1 greater than zero, then $\log a_1$, $\log a_2, \log a_3, \ldots, \log a_n$ is an arithmetic sequence.

b. If $a_1, a_2, a_3, \ldots, a_n$ is an arithmetic sequence, then $10^{a_1}, 10^{a_2}, \ldots, 10^{a_n}$, is a geometric sequence.

▶ MAINTAINING YOUR SKILLS

125. (1.5) Find the zeroes of f using the quadratic formula: $f(x) = x^2 + 5x + 9$.

126. (7.3) Find a unit vector in the same direction as $3\mathbf{i} - 7\mathbf{j}$.

127. (4.6) Graph the rational function:

$$h(x) = \frac{x^2}{x - 1}$$

128. (5.1) The cars on the Millenium Ferris Wheel are 100 ft from the center axle. If the top speed of the wheel is 1.5 revolutions per minute, find the linear velocity of a passenger in a car. Round your answer to the nearest whole number. Also, give the velocity in miles per hour.

Learning Objectives

In Section 10.4 you will learn how to:

☐ **A.** Use subscript notation to evaluate and compose functions

☐ **B.** Apply the principle of mathematical induction to sum formulas involving natural numbers

☐ **C.** Apply the principle of mathematical induction to general statements involving natural numbers

Since middle school (or even before) we have accepted that, "The product of two negative numbers is a positive number." But have you ever been asked to *prove* it? It's not as easy as it seems. We may think of several patterns that yield the result, analogies that indicate its truth, or even number line illustrations that lead us to believe the statement. But most of us have never seen a *proof* (see www.mhhe.com/coburn). In this section, we introduce one of mathematics' most powerful tools for proving a statement, called **proof by induction.**

A. Subscript Notation and Composition of Functions

One of the challenges in understanding a proof by induction is working with the notation. Earlier in the chapter, we introduced subscript notation as an alternative to function notation, since it is more commonly used when the functions are defined by a sequence. But regardless of the notation used, the functions can still be simplified, evaluated, composed, and even graphed. Consider the function $f(x) = 3x^2 - 1$ and the sequence defined by $a_n = 3n^2 - 1$. Both can be evaluated and graphed, with the only difference being that $f(x)$ is continuous with domain $x \in \mathbb{R}$, while a_n is discrete (made up of distinct points) with domain $n \in \mathbb{N}$.

EXAMPLE 1 ▶ **Using Subscript Notation for a Composition**

For $f(x) = 3x^2 - 1$ and $a_n = 3n^2 - 1$, find $f(k + 1)$ and a_{k+1}.

Solution ▶
$$f(k + 1) = 3(k + 1)^2 - 1 \qquad\qquad a_{k+1} = 3(k + 1)^2 - 1$$
$$= 3(k^2 + 2k + 1) - 1 \qquad\qquad = 3(k^2 + 2k + 1) - 1$$
$$= 3k^2 + 6k + 2 \qquad\qquad\qquad = 3k^2 + 6k + 2$$

Now try Exercises 7 through 18 ▶

☑ **A.** You've just learned how to use subscript notation to evaluate and compose functions

No matter which notation is used, every occurrence of the input variable is replaced by the new value or expression indicated by the composition.

B. Mathematical Induction Applied to Sums

Consider the sum of odd numbers $1 + 3 + 5 + 7 + 9 + 11 + 13 + \cdots$. The sum of the first four terms is $1 + 3 + 5 + 7 = 16$, or $S_4 = 16$. If we now add a_5 (the next term in line), would we get the same answer as if we had simply computed S_5? Common sense would say, "Yes!" since $S_5 = 1 + 3 + 5 + 7 + 9 = 25$ and $S_4 + a_5 = 16 + 9 = 25✓$. In diagram form, we have

add next term $a_5 = 9$ to S_4

$$1 + 3 + 5 + 7 + 9 + 11 + 13 + 15 + \cdots$$

S_4 —— sum of 4 terms

S_5 —— sum of 5 terms

Our goal is to develop this same degree of clarity in the *notational scheme* of things. For a given series, if we find the kth partial sum S_k (shown next) and then add the next term a_{k+1}, would we get the same answer if we had simply computed S_{k+1}? In other words, is $S_k + a_{k+1} = S_{k+1}$ true?

add next term a_{k+1}

$$a_1 \; + \; a_2 \; + \; a_3 \; + \; \cdots \; + \; a_{k-1} \; + \; a_k \; + \; a_{k+1} \; + \; \cdots \; + \; a_{n-1} \; + \; a_n$$

S_k ———————— sum of k terms ————

S_{k+1} ———————— sum of $k + 1$ terms ————

Now, let's return to the sum $1 + 3 + 5 + 7 + \cdots + 2n - 1$. This is an arithmetic series with $a_1 = 1$, $d = 2$, and nth term $a_n = 2n - 1$. Using the sum formula for an arithmetic sequence, an alternative formula for *this sum* can be established.

$$S_n = \frac{n(a_1 + a_n)}{2} \qquad \text{summation formula for an arithmetic sequence}$$

$$= \frac{n(1 + \mathbf{2n - 1})}{2} \qquad \text{substitute 1 for } a_1 \text{ and } 2n-1 \text{ for } a_n$$

$$= \frac{n(2n)}{2} \qquad \text{simplify}$$

$$= n^2 \qquad \text{result}$$

This shows that the sum of the first n positive odd integers is given by $S_n = n^2$. As a check we compute $S_5 = 1 + 3 + 5 + 7 + 9 = 25$ and compare to $S_5 = 5^2 = 25✓$. We also note $S_6 = 6^2 = 36$, and $S_5 + a_6 = 25 + 11 = 36$, showing $S_6 = S_5 + a_6$. For more on this relationship, **see Exercises 19 through 24.**

While it may seem simplistic now, showing $S_5 + a_6 = S_6$ and $S_k + a_{k+1} = S_{k+1}$ (in general) is a critical component of a proof by induction. Unfortunately, general summation formulas for many sequences cannot be established from known formulas. In addition, just because a formula works for the first few values of n, we cannot assume that it will hold true for *all* values of n (there are infinitely many). As an illustration, the formula $a_n = n^2 - n + 41$ yields a prime number for *every natural number n from 1 to 40,* but fails to yield a prime for $n = 41$. This helps demonstrate the need for a more conclusive proof, particularly when a relationship appears to be true, and can be "verified" in a finite number of cases, but whether it is true in *all* cases remains in question.

Proof by induction is based on a relatively simple idea. To help understand how it works, consider n relay stations that are used to transport electricity from a generating plant to a distant city. If we know the generating plant is operating, and if we assume that the kth relay station (any station in the series) is making the transfer to the $(k + 1)$st station (the next station in the series), then we're sure the city will have electricity.

WORTHY OF NOTE

No matter how distant the city or how many relay stations are involved, if the generating plant is working and the kth station relays to the $(k + 1)$st station, the city will get its power.

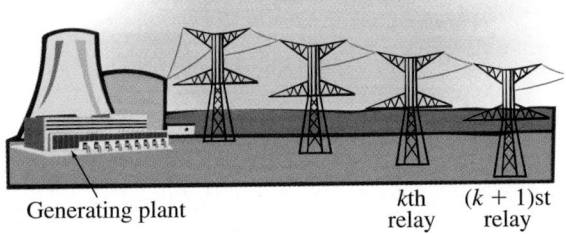

Generating plant

kth relay $(k + 1)$st relay

This idea can be applied mathematically as follows. Consider the statement, "The sum of the first n positive, even integers is $n^2 + n$." In other words, $2 + 4 + 6 + 8 + \cdots + 2n = n^2 + n$. We can certainly verify the statement for the first few even numbers:

The first even number is 2 and ...	$(1)^2 + 1 = 2$
The sum of the first *two* even numbers is $2 + 4 = 6$ and ...	$(2)^2 + 2 = 6$
The sum of the first *three* even numbers is $2 + 4 + 6 = 12$ and ...	$(3)^2 + 3 = 12$
The sum of the first *four* even numbers is $2 + 4 + 6 + 8 = 20$ and ...	$(4)^2 + 4 = 20$

While we could continue this process for a very long time (or even use a computer), *no finite number of checks can prove a statement is universally true.* To prove the statement true for *all* positive integers, we use a reasoning similar to that applied in the relay stations example. If we are sure the formula works for $n = 1$ (the generating station is operating), and assume that if the formula is true for $n = k$, it must also be true for $n = k + 1$ [the kth relay station is transferring electricity to the $(k + 1)$st station], then the statement is true for all n (the city will get its electricity). The case where $n = 1$ is called the **base case** of an inductive proof, and the assumption that the formula is true for $n = k$ is called the **induction hypothesis.** When the induction hypothesis is applied to a sum formula, we attempt to show that $S_k + a_{k+1} = S_{k+1}$. Since k and $k + 1$ are arbitrary, the statement must be true for all n.

Mathematical Induction Applied to Sums

Let S_n be a sum formula involving positive integers.

If   **1.** S_1 is true, and

      2. the truth of S_k implies that S_{k+1} is true,

then S_n must be true for all positive integers n.

Both parts 1 and 2 must be verified for the proof to be complete. Since the process requires the terms S_k, a_{k+1}, and S_{k+1}, we will usually compute these first.

EXAMPLE 2 ▶ **Proving a Statement Using Mathematical Induction**

Use induction to prove that *the sum of the first n perfect squares is given by*

$$1 + 4 + 9 + 16 + 25 + \cdots + n^2 = \frac{n(n + 1)(2n + 1)}{6}.$$

Solution ▶ Given $a_n = n^2$ and $S_n = \dfrac{n(n + 1)(2n + 1)}{6}$, the needed components are . . .

For $a_n = n^2$: $a_k = k^2$ and $a_{k+1} = (k + 1)^2$

For $S_n = \dfrac{n(n + 1)(2n + 1)}{6}$: $S_k = \dfrac{k(k + 1)(2k + 1)}{6}$ and $S_{k+1} = \dfrac{(k + 1)(k + 2)(2k + 3)}{6}$

1. Show S_n is true for $n = 1$.

$$S_n = \frac{n(n + 1)(2n + 1)}{6} \qquad \text{sum formula}$$

$$S_1 = \frac{1(2)(3)}{6} \qquad \text{base case: } n = 1$$

$$= 1✓ \qquad \text{result checks, the first term is 1}$$

2. Assume S_k is true,

$$1 + 4 + 9 + 16 + \cdots + k^2 = \frac{k(k + 1)(2k + 1)}{6} \qquad \text{induction hypothesis: } S_k \text{ is true}$$

and use it to show the truth of S_{k+1} follows. That is,

$$\underbrace{1 + 4 + 9 + 16 + \cdots + k^2}_{S_k} + \underbrace{(k + 1)^2}_{a_{k+1}} = \underbrace{\frac{(k + 1)(k + 2)(2k + 3)}{6}}_{S_{k+1}}$$

Working with the left-hand side, we have

$$\underbrace{1 + 4 + 9 + 16 + \cdots + k^2} + (k + 1)^2$$

$$= \frac{k(k + 1)(2k + 1)}{6} + (k + 1)^2 \qquad \text{use the induction hypothesis: substitute}$$
$$\qquad\qquad\qquad\qquad\qquad\qquad \frac{k(k + 1)(2k + 1)}{6} \text{ for } 1 + 4 + 9 + 16 + 25 + \cdots + k^2$$

$$= \frac{k(k + 1)(2k + 1) + 6(k + 1)^2}{6} \qquad \text{common denominator}$$

$$= \frac{(k + 1)[k(2k + 1) + 6(k + 1)]}{6} \qquad \text{factor out } k + 1$$

☑ **B.** You've just learned how to apply the principle of mathematical induction to sum formulas involving natural numbers

$$= \frac{(k + 1)[2k^2 + 7k + 6]}{6} \qquad \text{multiply and combine terms}$$

$$= \frac{(k + 1)(k + 2)(2k + 3)}{6} \qquad \text{factor the trinomial, result is } S_{k+1}✓$$

Since the truth of S_{k+1} follows from S_k, the formula is true for all n.

Now try Exercises 27 through 38 ▶

C. The General Principle of Mathematical Induction

Proof by induction can be used to verify many other kinds of relationships involving a natural number n. In this regard, the basic principles remain the same but are stated more broadly. Rather than having S_n represent a sum, we take it to represent *any statement or relationship* we might wish to verify. This broadens the scope of the proof and makes it more widely applicable, while maintaining its value to the sum formulas verified earlier.

The General Principle of Mathematical Induction

Let S_n be a statement involving natural numbers.

    If   **1.** S_1 is true, and
           2. the truth of S_k implies that S_{k+1} is also true

then S_n must be true for all natural numbers n.

EXAMPLE 3 ▶ **Proving a Statement Using the General Principle of Mathematical Induction**

Use the general principle of mathematical induction to show the statement S_n is true for all natural numbers n. S_n: $2^n \geq n + 1$

Solution ▶ The statement S_n is defined as $2^n \geq n + 1$. This means that S_k is represented by $2^k \geq k + 1$ and S_{k+1} by $2^{k+1} \geq k + 2$.

    1. Show S_n is true for $n = 1$:

$$S_n: \quad 2^n \geq n + 1 \quad \text{given statement}$$
$$S_1: \quad 2^1 \geq 1 + 1 \quad \text{base case: } n = 1$$
$$2 \geq 2✓ \quad \text{true}$$

Although not a part of the formal proof, a table of values can help to illustrate the relationship we're trying to establish. It *appears* that the statement is true.

n	1	2	3	4	5
2^n	2	4	8	16	32
$n+1$	2	3	4	5	6

2. Assume that S_k is true,

$$S_k: \quad 2^k \geq k + 1 \quad \text{induction hypothesis}$$

and use it to show that the truth of S_{k+1}. That is,

$$S_{k+1}: \quad 2^{k+1} \geq k + 2.$$

Begin by working with the left-hand side of the inequality, 2^{k+1}.

$$
\begin{aligned}
2^{k+1} &= 2(2^k) & \text{properties of exponents} \\
&\geq 2(\boldsymbol{k + 1}) & \text{\textit{induction hypothesis:} substitute } k + 1 \text{ for } 2^k \\
& & \text{(symbol changes since } k + 1 \text{ is less than } 2^k) \\
&\geq 2k + 2 & \text{distribute}
\end{aligned}
$$

Since k is a positive integer, $2k + 2 \geq k + 2$,

showing $2^{k+1} \geq k + 2$.

Since the truth of S_{k+1} follows from S_k, the formula is true for all n.

> **WORTHY OF NOTE**
>
> Note there is no reference to a_n, a_k, or a_{k+1} in the statement of the general principle of mathematical induction.

Now try Exercises 39 through 42 ▶

EXAMPLE 4 ▶ **Proving Divisibility Using Mathematical Induction**

Let S_n be the statement, "*$4^n - 1$ is divisible by 3 for all positive integers n.*" Use mathematical induction to prove that S_n is true.

Solution ▶ If a number is evenly divisible by three, it can be written as the product of 3 and some positive integer we will call p.

1. Show S_n is true for $n = 1$:

$$
\begin{aligned}
S_n: \quad & 4^n - 1 = 3p & 4^n - 1 = 3p, \ p \in \mathbb{Z} \\
S_1: \quad & 4^{(1)} - 1 = 3p & \text{substitute 1 for } n \\
& 3 = 3p\checkmark & \text{statement is true for } n = 1
\end{aligned}
$$

2. Assume that S_k is true . . .

$$
\begin{aligned}
S_k: \quad & 4^k - 1 = 3p & \text{induction hypothesis} \\
& 4^k = 3p + 1
\end{aligned}
$$

and use it to show the truth of S_{k+1}. That is,

$$S_{k+1}: \quad 4^{k+1} - 1 = 3q \text{ for } q \in \mathbb{Z} \text{ is also true.}$$

Beginning with the left-hand side we have:

$$
\begin{aligned}
4^{k+1} - 1 &= 4 \cdot 4^k - 1 & \text{properties of exponents} \\
&= 4 \cdot (\boldsymbol{3p + 1}) - 1 & \text{induction hypothesis: substitute } 3p + 1 \text{ for } 4^k \\
&= 12p + 3 & \text{distribute and simplify} \\
&= 3(4p + 1) = 3q & \text{factor}
\end{aligned}
$$

The last step shows $4^{k+1} - 1$ is divisible by 3. Since the original statement is true for $n = 1$, and the truth of S_k implies the truth of S_{k+1}, the statement, "$4^n - 1$ is divisible by 3" is true for all positive integers n.

Now try Exercises 43 through 47 ▶

We close this section with some final notes. Although the base step of a proof by induction seems trivial, both the base step and the induction hypothesis are necessary

☑ **C.** You've just learned how to apply the principle of mathematical induction to general statements involving natural numbers

parts of the proof. For example, the statement $\dfrac{1}{3^n} < \dfrac{1}{3n}$ is false for $n = 1$, but true for all other positive integers. Finally, for a fixed natural number p, some statements are false for all $n < p$, but true for all $n \geq p$. By modifying the base case to begin at p, we can use the induction hypothesis to prove the statement is true for all n greater than p. For example, $n < \frac{1}{3} n^2$ is false for $n < 4$, but true for all $n \geq 4$.

10.4 EXERCISES

▶ CONCEPTS AND VOCABULARY

Fill in the blank with the appropriate word or phrase. Carefully reread the section if needed.

1. No _____ number of verifications can prove a statement _____ true.

2. Showing a statement is true for $n = 1$ is called the _____ _____ of an inductive proof.

3. Assuming that a statement/formula is true for $n = k$ is called the _____ _____.

4. The graph of a sequence is _____, meaning it is made up of distinct points.

5. Explain the equation $S_k + a_{k+1} = S_{k+1}$. Begin by saying, "Since the kth term is arbitrary . . ." (continue from here).

6. Discuss the similarities and differences between mathematical induction applied to sums and the general principle of mathematical induction.

▶ DEVELOPING YOUR SKILLS

For the given nth term a_n, find a_4, a_5, a_k, and a_{k+1}.

7. $a_n = 10n - 6$

8. $a_n = 6n - 4$

9. $a_n = n$

10. $a_n = 7n$

11. $a_n = 2^{n-1}$

12. $a_n = 2(3^{n-1})$

For the given sum formula S_n, find S_4, S_5, S_k, and S_{k+1}.

13. $S_n = n(5n - 1)$

14. $S_n = n(3n - 1)$

15. $S_n = \dfrac{n(n + 1)}{2}$

16. $S_n = \dfrac{7n(n + 1)}{2}$

17. $S_n = 2^n - 1$

18. $S_n = 3^n - 1$

Verify that $S_4 + a_5 = S_5$ for each exercise. These are identical to Exercises 13 through 18.

19. $a_n = 10n - 6$; $S_n = n(5n - 1)$

20. $a_n = 6n - 4$; $S_n = n(3n - 1)$

21. $a_n = n$; $S_n = \dfrac{n(n + 1)}{2}$

22. $a_n = 7n$; $S_n = \dfrac{7n(n + 1)}{2}$

23. $a_n = 2^{n-1}$; $S_n = 2^n - 1$

24. $a_n = 2(3^{n-1})$; $S_n = 3^n - 1$

► **WORKING WITH FORMULAS**

25. Sum of the first n cubes (alternative form): $(1 + 2 + 3 + 4 + \cdots + n)^2$

Earlier we noted the formula for the sum of the first n cubes was $\dfrac{n^2(n + 1)^2}{4}$. An alternative is given by the formula shown.

  a. Verify the formula for $n = 1, 5,$ and 9.

  b. Verify the formula using
$$1 + 2 + 3 + \cdots + n = \frac{n(n + 1)}{2}.$$

26. Powers of the imaginary unit: $i^{n+4} = i^n$, where $i = \sqrt{-1}$

Use a proof by induction to prove that powers of the imaginary unit are cyclic. That is, that they cycle through the numbers $i, -1, -i,$ and 1 for consecutive powers.

► **APPLICATIONS**

Use mathematical induction to prove the indicated sum formula is true for all natural numbers n.

27. $2 + 4 + 6 + 8 + 10 + \cdots + 2n$;
$a_n = 2n, S_n = n(n + 1)$

28. $3 + 7 + 11 + 15 + 19 + \cdots + (4n - 1)$;
$a_n = 4n - 1, S_n = n(2n + 1)$

29. $5 + 10 + 15 + 20 + 25 + \cdots + 5n$;
$a_n = 5n, S_n = \dfrac{5n(n + 1)}{2}$

30. $1 + 4 + 7 + 10 + 13 + \cdots + (3n - 2)$;
$a_n = 3n - 2, S_n = \dfrac{n(3n - 1)}{2}$

31. $5 + 9 + 13 + 17 + \cdots + (4n + 1)$;
$a_n = 4n + 1, S_n = n(2n + 3)$

32. $4 + 12 + 20 + 28 + 36 + \cdots + (8n - 4)$;
$a_n = 8n - 4, S_n = 4n^2$

33. $3 + 9 + 27 + 81 + 243 + \cdots + 3^n$;
$a_n = 3^n, S_n = \dfrac{3(3^n - 1)}{2}$

34. $5 + 25 + 125 + 625 + \cdots + 5^n$;
$a_n = 5^n, S_n = \dfrac{5(5^n - 1)}{4}$

35. $2 + 4 + 8 + 16 + 32 + 64 + \cdots + 2^n$;
$a_n = 2^n, S_n = 2^{n+1} - 2$

36. $1 + 8 + 27 + 64 + 125 + 216 + \cdots + n^3$;
$a_n = n^3, S_n = \dfrac{n^2(n + 1)^2}{4}$

37. $\dfrac{1}{1(3)} + \dfrac{1}{3(5)} + \dfrac{1}{5(7)} + \cdots + \dfrac{1}{(2n - 1)(2n + 1)}$;
$a_n = \dfrac{1}{(2n - 1)(2n + 1)}, S_n = \dfrac{n}{2n + 1}$

38. $\dfrac{1}{1(2)} + \dfrac{1}{2(3)} + \dfrac{1}{3(4)} + \cdots + \dfrac{1}{n(n + 1)}$;
$a_n = \dfrac{1}{n(n + 1)}, S_n = \dfrac{n}{n + 1}$

Use the principle of mathematical induction to prove that each statement is true for all natural numbers n.

39. $3^n \geq 2n + 1$ **40.** $2^n \geq n + 1$

41. $3 \cdot 4^{n-1} \leq 4^n - 1$ **42.** $4 \cdot 5^{n-1} \leq 5^n - 1$

43. $n^2 - 7n$ is divisible by 2

44. $n^3 - n + 3$ is divisible by 3

45. $n^3 + 3n^2 + 2n$ is divisible by 3

46. $5^n - 1$ is divisible by 4

47. $6^n - 1$ is divisible by 5

► **EXTENDING THE THOUGHT**

 48. You may have noticed that the sum formula for the first n integers was *quadratic,* and the formula for the first n integer squares was *cubic.* Is the formula for the first n integer cubes, if it exists, a quartic (degree four) function? Use your calculator to run a quartic regression on the first five perfect cubes (enter 1

through 5 in L_1 and the cumulative sums in L_2). What did you find? Use proof by induction to show that the sum of the first n cubes is:

$$1 + 8 + 27 + \cdots + n^3 = \frac{n^4 + 2n^3 + n^2}{4} = \frac{n^2(n + 1)^2}{4}.$$

49. Use mathematical induction to prove that
$$\frac{x^n - 1}{x - 1} = (1 + x + x^2 + x^3 + \cdots + x^{n-1}).$$

50. Use mathematical induction to prove that for $1^4 + 2^4 + 3^4 + \cdots + n^4$, where $a_n = n^4$,
$$S_n = \frac{n(n + 1)(2n + 1)(3n^2 + 3n - 1)}{30}.$$

▶ MAINTAINING YOUR SKILLS

51. (6.2) Verify the identity
$(\sin \theta + \cos \theta)^2 + (\sin \theta - \cos \theta)^2 = 2.$

52. (2.7) State the domain and range of the piecewise function shown here.

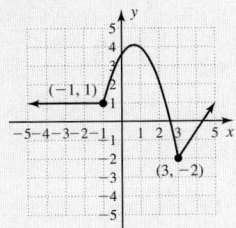

53. (2.1) State the equation of the circle whose graph is shown here.

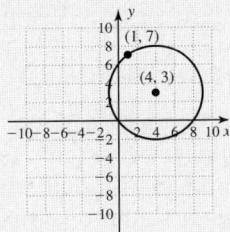

54. (7.4) Given $\mathbf{p} = \langle \sqrt{3}, -1 \rangle$ and $\mathbf{q} = \langle 1, 1 \rangle$, find
  a. the dot product $\mathbf{p} \cdot \mathbf{q}$
  b. the angle between the vectors

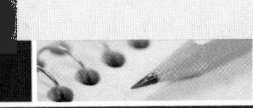

MID-CHAPTER CHECK

In Exercises 1 to 3, the nth term is given. Write the first three terms of each sequence and find a_9.

1. $a_n = 7n - 4$ **2.** $a_n = n^2 + 3$

3. $a_n = (-1)^n(2n - 1)$

4. Evaluate the sum $\displaystyle\sum_{n=1}^{4} 3^{n+1}$

5. Rewrite using sigma notation.
  $1 + 4 + 7 + 10 + 13 + 16$

Match each formula to its correct description.

6. $S_n = \dfrac{n(a_1 + a_n)}{2}$

7. $a_n = a_1 r^{n-1}$

8. $S_\infty = \dfrac{a_1}{1 - r}$

9. $a_n = a_1 + (n - 1)d$

10. $S_n = \dfrac{a_1(1 - r^n)}{1 - r}$

  a. sum of an infinite geometric series

  b. nth term formula for an arithmetic series

  c. sum of a finite geometric series

  d. summation formula for an arithmetic series

  e. nth term formula for a geometric series

11. Identify a_1 and the common difference d. Then find an expression for the general term a_n.
  a. $2, 5, 8, 11, \ldots$ **b.** $\frac{3}{2}, \frac{9}{4}, 3, \frac{15}{4}, \ldots$

Find the number of terms in each series and then find the sum.

12. $2 + 5 + 8 + 11 + \cdots + 74$

13. $\frac{1}{2} + \frac{3}{2} + \frac{5}{2} + \frac{7}{2} + \cdots + \frac{31}{2}$

14. For an arithmetic series, $a_3 = -8$ and $a_7 = 4$. Find S_{10}.

15. For a geometric series, $a_3 = -81$ and $a_7 = -1$. Find S_{10}.

16. Identify a_1 and the common ratio r. Then find an expression for the general term a_n.
  a. $2, 6, 18, 54, \ldots$ **b.** $\frac{1}{2}, \frac{1}{4}, \frac{1}{8}, \frac{1}{16}, \ldots$

17. Find the number of terms in the series then compute the sum.
  $\frac{1}{54} + \frac{1}{18} + \frac{1}{6} + \cdots + \frac{81}{2}$

18. Find the infinite sum (if it exists).

$-49 + (-7) + (-1) + (-\frac{1}{7}) + \cdots$

19. Barrels of toxic waste are stacked at a storage facility in pyramid form, with 60 barrels in the first row, 59 in the second row, and so on, until there are 10 barrels in the top row. How many barrels are in the storage facility?

20. As part of a conditioning regimen, a drill sergeant orders her platoon to do 25 continuous standing broad jumps. The best of these recruits was able to jump 96% of the distance from the previous jump, with a first jump distance of 8 ft. Use a sequence/series to determine the distance the recruit jumped on the 15th try, and the total distance traveled by the recruit after all 25 jumps.

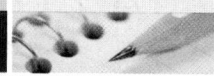

REINFORCING BASIC CONCEPTS

Applications of Summation

The properties of summation play a large role in the development of key ideas in a first semester calculus course, and the following summation formulas are an integral part of these ideas. The first three formulas were verified in Section 10.4, while proof of the fourth was part of Exercise 48 on page 972.

(1) $\sum_{i=1}^{n} c = cn$

(2) $\sum_{i=1}^{n} i = \dfrac{n(n+1)}{2}$

(3) $\sum_{i=1}^{n} i^2 = \dfrac{n(n+1)(2n+1)}{6}$

(4) $\sum_{i=1}^{n} i^3 = \dfrac{n^2(n+1)^2}{4}$

To see the various ways they can be applied consider the following.

Illustration 1 ▶ Over several years, the owner of Morgan's LawnCare has noticed that the company's monthly profits (in thousands) can be approximated by the sequence $a_n = 0.0625n^3 - 1.25n^2 + 6n$, with the points plotted in Figure 10.9 (the continuous graph is shown for effect only). Find the company's approximate annual profit.

Figure 10.9

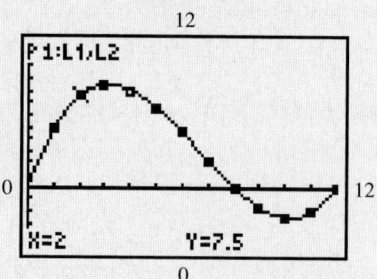

Solution ▶ The most obvious approach would be to simply compute terms a_1 through a_{12} (January through December) and find their sum: **sum(seq(Y1, X, 1, 12)** (see Section 10.1 Technology Highlight), which gives a result of 35.75 or $35,750.

As an alternative, we could add the amount of profit earned by the company in the first 8 months, then add the amount the company lost (or broke even) during the last 4 months. In other words, we could apply summation property

IV: $\sum_{i=1}^{12} a_n = \sum_{i=1}^{8} a_n + \sum_{i=9}^{12} a_n$ (see Figure 10.10), which gives the same result: $42 + (-6.25) = 35.75$ or $35,750.

Figure 10.10

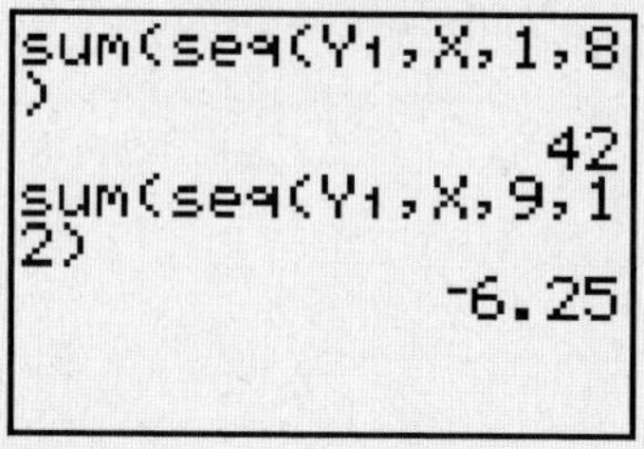

```
sum(seq(Y1,X,1,8
)
              42
sum(seq(Y1,X,9,1
2)
            -6.25
```

As a third option, we could use summation properties along with the appropriate summation formulas, and compute the result manually. Note the function is now written in terms of "i." *Distribute summations and factor out constants (properties II and III):*

$$\sum_{i=1}^{12} (0.0625i^3 - 1.25i^2 + 6i) =$$

$$0.0625 \sum_{i=1}^{12} i^3 - 1.25 \sum_{i=1}^{12} i^2 + 6 \sum_{i=1}^{12} i$$

Replace each summation with the appropriate summation formula, substituting 12 for n:

$$= 0.0625\left[\frac{n^2(n+1)^2}{4}\right] - 1.25\left[\frac{n(n+1)(2n+1)}{6}\right]$$
$$+ 6\left[\frac{n(n+1)}{2}\right]$$

$$= 0.0625\left[\frac{(12)^2(13)^2}{4}\right] - 1.25\left[\frac{(12)(13)(25)}{6}\right]$$
$$+ 6\left[\frac{(12)(13)}{2}\right]$$

$$= 0.0625(6084) - 1.25(650) + 6(78) \text{ or } 35.75$$

As we expected, the result shows profit was $35,750. While some approaches seem "easier" than others, all have great value, are applied in different ways at different times, and are necessary to adequately develop key concepts in future classes.

Exercise 1: Repeat Illustration 1 if the profit sequence is $a_n = 0.125x^3 - 2.5x^2 + 12x$.

Learning Objectives

In Section 10.5 you will learn how to:

☐ **A.** Count possibilities using lists and tree diagrams

☐ **B.** Count possibilities using the fundamental principle of counting

☐ **C.** Quick-count distinguishable permutations

☐ **D.** Quick-count nondistinguishable permutations

☐ **E.** Quick-count using combinations

How long would it take to estimate the number of fans sitting shoulder-to-shoulder at a sold-out basketball game? Well, it depends. You could actually begin counting 1, 2, 3, 4, 5, . . . , which would take a very long time, or you could try to simplify the process by counting the number of fans in the first row and multiplying by the number of rows. Techniques for "quick-counting" the objects in a set or various subsets of a large set play an important role in a study of probability.

A. Counting by Listing and Tree Diagrams

Consider the simple spinner shown in Figure 10.11, which is divided into three equal parts. What are the different possible outcomes for two spins, spin 1 followed by spin 2? We might begin by organizing the possibilities using a **tree diagram.** As the name implies, each choice or possibility appears as the branch of a tree, with the total possibilities being equal to the number of (unique) paths from the beginning point to the end of a branch. Figure 10.12 shows how the spinner exercise would appear (possibilities for two spins). Moving from top to bottom we can trace nine possible paths: *AA, AB, AC, BA, BB, BC, CA, CB,* and *CC.*

Figure 10.11

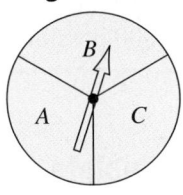

Figure 10.12

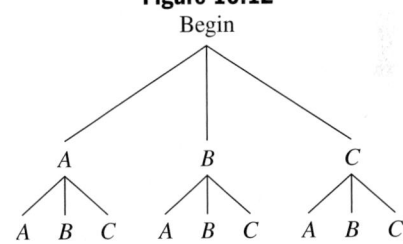

EXAMPLE 1 ▶ Listing Possibilities Using a Tree Diagram

A basketball player is fouled and awarded three free throws. Let H represent the possibility of a hit (basket is made), and M the possibility of a miss. Determine the possible outcomes for the three shots using a tree diagram.

Solution ▶ Each shot has two possibilities, hit (H) or miss (M), so the tree will branch in two directions at each level. As illustrated in the figure, there are a total of eight possibilities: HHH, HHM, HMH, HMM, MHH, MHM, MMH, and MMM.

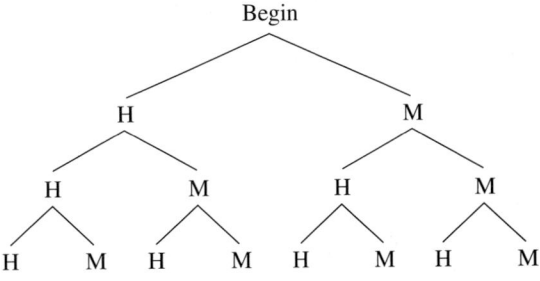

Now try Exercises 7 through 10 ▶

To assist our discussion, an **experiment** is any task that can be done repeatedly and has a well-defined set of possible outcomes. Each repetition of the experiment is called a **trial.** A **sample outcome** is any potential outcome of a trial, and a **sample space** is a set of all possible outcomes.

In our first illustration, the *experiment* was spinning a spinner, there were *three sample outcomes* (*A, B,* or *C*), the experiment had *two trials* (spin 1 and spin 2), and there were *nine* elements in the *sample space*. Note that after the first trial, each of the three sample outcomes will again have three possibilities (*A*, *B*, and *C*). For two trials we have $3^2 = 9$ possibilities, while three trials would yield a sample space with $3^3 = 27$ possibilities. In general, we have

---

A "Quick-Counting" Formula for a Sample Space

If an experiment has N sample outcomes that are equally likely and the experiment is repeated t times, the number of elements in the sample space is N^t.

EXAMPLE 2 ▶ **Counting the Outcomes in a Sample Space**

Many combination locks have the digits 0 through 39 arranged along a circular dial. Opening the lock requires stopping at a sequence of three numbers within this range, going counterclockwise to the first number, clockwise to the second, and counterclockwise to the third. How many three-number combinations are possible?

Solution ▶ There are 40 sample outcomes ($N = 40$) in this experiment, and three trials ($t = 3$). The number of possible combinations is identical to the number of elements in the sample space. The quick-counting formula gives $40^3 = 64{,}000$ possible combinations.

☑ **A. You've just learned how to count possibilities using lists and tree diagrams**

Now try Exercises 11 and 12 ▶

B. Fundamental Principle of Counting

The number of possible outcomes may differ depending on how the event is defined. For example, some security systems, license plates, and telephone numbers exclude certain numbers. For example, phone numbers cannot begin with 0 or 1 because these are reserved for operator assistance, long distance, and international calls. Constructing a three-digit area code is like filling in three blanks ___ ___ ___ with three
                                                           digit digit digit
digits. Since the area code must start with a number between 2 and 9, there are eight choices for the first blank. Since there are 10 choices for the second digit and 10 choices for the third, there are $8 \cdot 10 \cdot 10 = 800$ possibilities in the sample space.

EXAMPLE 3 ▶ **Counting Possibilities for a Four-Digit Security Code**

A digital security system requires that you enter a four-digit PIN (personal identification number), using only the digits 1 through 9. How many codes are possible if
  a. Repetition of digits is allowed?
  b. Repetition is not allowed?
  c. The first digit must be even and repetitions are not allowed?

Solution ▶ **a.** Consider filling in the four blanks ___ ___ ___ ___ with the number of
                                                    digit digit digit digit
  ways the digit can be chosen. If repetition is allowed, the experiment is similar to that of Example 2 and there are $N^t = 9^4 = 6561$ possible PINs.

  b. If repetition is not allowed, there are only eight possible choices for the second digit of the PIN, then seven for the third, and six for the fourth. The number of possible PIN numbers decreases to $9 \cdot 8 \cdot 7 \cdot 6 = 3024$.

  c. There are four choices for the first digit (2, 4, 6, 8). Once this choice has been made there are eight choices for the second digit, seven for the third, and six for the last: $4 \cdot 8 \cdot 7 \cdot 6 = 1344$ possible codes.

Now try Exercises 13 through 20 ▶

WORTHY OF NOTE

In Example 4, we could also reason that since there are $6! = 720$ random seating arrangements and 240 of them consist of Bob and Carol sitting together [Example 4(a)], the remaining $720 - 240 = 480$ must consist of Bob and Carol *not* sitting together. More will be said about this type of reasoning in Section 10.6.

Given *any* experiment involving a sequence of tasks, if the first task can be completed in p possible ways, the second task has q possibilities, and the third task has r possibilities, a tree diagram will show that the number of possibilities in the sample space for task_1–task_2–task_3 is $p \cdot q \cdot r$. Even though the examples we've considered to this point have varied a great deal, this idea was fundamental to counting all possibilities in a sample space and is, in fact, known as the **fundamental principle of counting** (FPC).

Fundamental Principle of Counting (Applied to Three Tasks)

Given any experiment with three defined tasks, if there are p possibilities for the first task, q possibilities for the second, and r possibilities for the third, the total number of ways the experiment can be completed is $p \cdot q \cdot r$.

This fundamental principle can be extended to include any number of tasks.

EXAMPLE 4 ▶ Counting Possibilities for Seating Arrangements

Adrienne, Bob, Carol, Dax, Earlene, and Fabian bought tickets to see *The Marriage of Figaro.* Assuming they sat together in a row of six seats, how many different seating arrangements are possible if

a. Bob and Carol are sweethearts and must sit together?

b. Bob and Carol are enemies and must not sit together?

Solution ▶

Figure 10.13

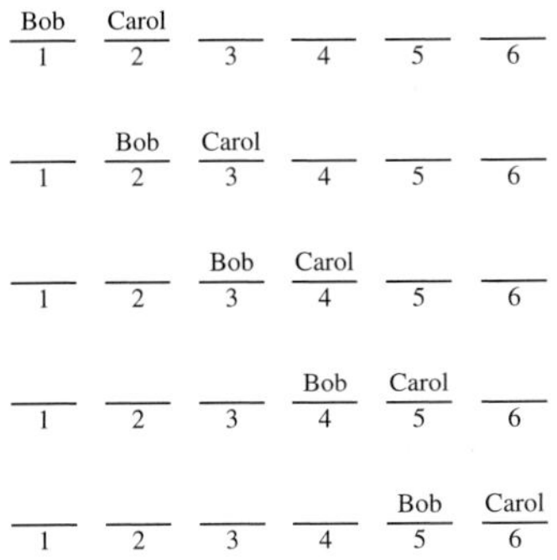

a. Since a restriction has been placed on the seating arrangement, it will help to divide the experiment into a sequence of tasks: *task 1:* they sit together; *task 2:* either Bob is on the left or Bob is on the right; and *task 3:* the other four are seated. Bob and Carol can sit together in five different ways, as shown in Figure 10.13, so there are five possibilities for task 1. There are two ways they can be side-by-side: Bob on the left and Carol on the right, as shown, or Carol on the left and Bob on the right. The remaining four people can be seated randomly, so task 3 has $4! = 24$ possibilities. Under these conditions they can be seated $5 \cdot 2 \cdot 4! = 240$ ways.

b. This is similar to Part (a), but now we have to count the number of ways they can be separated by *at least one seat: task 1:* Bob and Carol are in nonadjacent seats; *task 2:* either Bob is on the left or Bob is on the right; and *task 3:* the other four are seated. For task 1, be careful to note there is no multiplication involved, just a simple counting. If Bob sits in seat 1, there are four nonadjacent seats. If Bob sits in seat 2, there are three nonadjacent seats, and so on. This gives $4 + 3 + 2 + 1 = 10$ possibilities for Bob and Carol not sitting together. Task 2 and task 3 have the same number of possibilities as in Part (a), giving $10 \cdot 2 \cdot 4! = 480$ possible seating arrangements.

Now try Exercises 21 through 28 ▶

☑ **B. You've just learned how to count possibilities using the fundamental principle of counting**

C. Distinguishable Permutations

In the game of Scrabble® (Milton Bradley), players attempt to form words by rearranging letters. Suppose a player has the letters P, S, T, and O at the end of the game. These letters could be rearranged or *permuted* to form the words POTS, SPOT, TOPS, OPTS, POST, or STOP. These arrangements are called permutations of the four letters. A permutation is any new arrangement, listing, or sequence of objects obtained by changing an existing order. A **distinguishable permutation** is a permutation that produces a result different from the original. For example, a distinguishable permutation of the digits in the number 1989 is 8199.

Example 4 considered six people, six seats, and the various ways they could be seated. But what if there were fewer seats than people? By the FPC, with six people and four seats there could be $6 \cdot 5 \cdot 4 \cdot 3 = 360$ different arrangements, with six people and three seats there are $6 \cdot 5 \cdot 4 = 120$ different arrangements, and so on. These rearrangements are called distinguishable permutations. You may have noticed that for six people and six seats, we used all six factors of 6!, while for six people and four seats we used the first four, six people and three seats required only the first three, and so on. Generally, for n people and r seats, the first r factors of $n!$ will be used. The notation and formula for *distinguishable permutations of n objects taken r at a time* is

$_nP_r = \dfrac{n!}{(n-r)!}$. By defining $0! = 1$, the formula includes the case where all n objects

are selected, which of course results in $_nP_n = \dfrac{n!}{(n-n)!} = \dfrac{n!}{0!} = \dfrac{n!}{1} = n!$.

Distinguishable Permutations: Unique Elements

If r objects are selected from a set containing n unique elements $(r \le n)$ and placed in an ordered arrangement, the number of distinguishable permutations is

$$_nP_r = \frac{n!}{(n-r)!} \quad \text{or} \quad _nP_r = n(n-1)(n-2)\cdots(n-r+1)$$

EXAMPLE 5 ▶ Computing a Permutation

Compute each value of $_nP_r$ using the methods just described.

a. $_7P_4$ **b.** $_{10}P_3$

Solution ▶ Begin by evaluating each expression using the formula $_nP_r = \dfrac{n!}{(n-r)!}$, noting the third line (in bold) gives the first r factors of $n!$.

a. $_7P_4 = \dfrac{7!}{(7-4)!}$

$= \dfrac{7 \cdot 6 \cdot 5 \cdot 4 \cdot \cancel{3!}}{\cancel{3!}}$

$\mathbf{= 7 \cdot 6 \cdot 5 \cdot 4}$

$= 840$

b. $_{10}P_3 = \dfrac{10!}{(10-3)!}$

$= \dfrac{10 \cdot 9 \cdot 8 \cdot \cancel{7!}}{\cancel{7!}}$

$\mathbf{= 10 \cdot 9 \cdot 8}$

$= 720$

Now try Exercises 29 through 36 ▶

EXAMPLE 6 ▶ Counting the Possibilities for Finishing a Race

As part of a sorority's initiation process, the nine new inductees must participate in a 1-mi race. Assuming there are no ties, how many first- through fifth-place finishes are possible if it is well known that Mediocre Mary will finish fifth and Lightning Louise will finish first?

Solution ▶ To help understand the situation, we can diagram the possibilities for finishing first through fifth. Since Louise will finish first, this slot can be filled in only one way, by Louise herself. The same goes for Mary and her fifth-place finish: $\dfrac{\text{Louise}}{\text{1st}}$

$\underset{\text{2nd}}{\underline{\qquad}} \ \underset{\text{3rd}}{\underline{\qquad}} \ \underset{\text{4th}}{\underline{\qquad}} \ \underset{\text{5th}}{\underline{\text{Mary}}}$. The remaining three slots can be filled in $_7P_3 = 7 \cdot 6 \cdot 5$

different ways, indicating that under these conditions, there are $1 \cdot 7 \cdot 6 \cdot 5 \cdot 1 = 210$ different ways to finish.

☑ **C. You've just learned how to quick-count distinguishable permutations**

Now try Exercises 37 through 42 ▶

D. Nondistinguishable Permutations

As the name implies, certain permutations are nondistinguishable, meaning you cannot tell one apart from another. Such is the case when the original set contains elements or sample outcomes that are identical. Consider a family with four children, Lyddell, Morgan, Michael, and Mitchell, who are at the photo studio for a family picture. Michael and Mitchell are identical twins and cannot be told apart. In how many ways can they be lined up for the picture? Since this is an ordered arrangement of four children taken from a group of four, there are $_4P_4 = 24$ ways to line them up. A few of them are

Lyddell	Morgan	Michael	Mitchell	Lyddell	Morgan	Mitchell	Michael
Lyddell	Michael	Morgan	Mitchell	Lyddell	Mitchell	Morgan	Michael
Michael	Lyddell	Morgan	Mitchell	Mitchell	Lyddell	Morgan	Michael

But of these six arrangements, half will appear to be the same picture, since the difference between Michael and Mitchell cannot be distinguished. In fact, of the 24 total permutations, every picture where Michael and Mitchell have switched places will be nondistinguishable. To find the distinguishable permutations, we need to take the total permutations ($_4P_4$) *and divide by* 2!, *the number of ways the twins can be permuted:* $\frac{_4P_4}{(2)!} = \frac{24}{2} = 12$ distinguishable pictures.

These ideas can be generalized and stated in the following way.

Nondistinguishable Permutations: Nonunique Elements

In a set containing n elements where one element is repeated p times, another is repeated q times, and another is repeated r times ($p + q + r = n$), the number of nondistinguishable permutations is

$$\frac{_nP_n}{p!q!r!} = \frac{n!}{p!q!r!}$$

The idea can be extended to include any number of repeated elements.

> **WORTHY OF NOTE**
>
> In Example 7, if a Scrabble player is able to play all seven letters in one turn, he or she "bingos" and is awarded 50 extra points. The player in Example 7 did just that. Can you determine what word was played?

EXAMPLE 7 ▶ **Counting Distinguishable Permutations**

A Scrabble player has the seven letters S, A, O, O, T, T, and T in his rack. How many distinguishable arrangements can be formed as he attempts to play a word?

Solution ▶ Essentially the exercise asks for the number of distinguishable permutations of the seven letters, given T is repeated three times and O is repeated twice. There are $\frac{_7P_7}{3!2!} = 420$ distinguishable permutations.

☑ **D.** You've just learned how to quick-count nondistinguishable permutations

Now try Exercises 43 through 54 ▶

E. Combinations

Similar to nondistinguishable permutations, there are other times the total number of permutations must be reduced to quick-count the elements of a desired subset. Consider a vending machine that offers a variety of 40¢ candies. If you have a quarter (Q), dime (D), and nickel (N), the machine wouldn't care about the order the coins were deposited. Even though QDN, QND, DQN, DNQ, NQD, and NDQ give the $_3P_3 = 6$ possible permutations, the machine considers them as equal and will vend your snack. Using sets, this is similar to saying the set $A = \{X, Y, Z\}$ has only one subset with three elements, since $\{X, Z, Y\}$, $\{Y, X, Z\}$, $\{Y, Z, X\}$, and so on, all represent the same set. Similarly, there are six, two-letter permutations of X, Y, and Z ($_3P_2 = 6$): XY, XZ, YX,

YZ, ZX, and *ZY,* but only three two-letter subsets: {*X, Y*}, {*X, Z*} and {*Y, Z*}. When permutations having the same elements are considered identical, the result is the number of possible **combinations** and is denoted $_nC_r$. Since the *r* objects can be selected in *r*! ways, we divide $_nP_r$ by *r*! to "quick-count" the number of possibilities: $_nC_r = \dfrac{_nP_r}{r!}$, which can be thought of as *the first r factors of n!, divided by r!.* By substituting $\dfrac{n!}{(n-r)!}$ for $_nP_r$ in this formula, we find an alternative method for computing $_nC_r$ is $\dfrac{n!}{r!(n-r)!}$. Take special note that when *r* objects are selected from a set with *n* elements and the order they're listed is unimportant (because you end up with the same subset), the result is a *combination,* not a permutation.

Combinations

The number of combinations of *n* objects taken *r* at a time is given by

$$_nC_r = \frac{_nP_r}{r!} \qquad \text{or} \qquad _nC_r = \frac{n!}{r!(n-r)!}$$

EXAMPLE 8 ▶ **Computing Combinations Using a Formula**
Compute each value of $_nC_r$ given.

  a. $_7C_4$ **b.** $_8C_3$ **c.** $_5C_2$

Solution ▶   **a.** $_7C_4 = \dfrac{7 \cdot 6 \cdot 5 \cdot 4}{4!}$ **b.** $_8C_3 = \dfrac{8 \cdot 7 \cdot 6}{3!}$ **c.** $_5C_2 = \dfrac{5 \cdot 4}{2!}$

  $= 35$ $= 56$ $= 10$

Now try Exercises 55 through 64 ▶

EXAMPLE 9 ▶ **Applications of Combinations-Lottery Results**
A small city is getting ready to draw five Ping-Pong balls of the nine they have numbered 1 through 9 to determine the winner(s) for its annual raffle. If a ticket holder has the same five numbers, they win. In how many ways can the winning numbers be drawn?

Solution ▶ Since the winning numbers can be drawn in any order, we have a combination of 9 things taken 5 at a time. The five numbers can be drawn in $_9C_5 = \dfrac{9 \cdot 8 \cdot 7 \cdot 6 \cdot 5}{5!} = 126$ ways.

Now try Exercises 65 and 66 ▶

Somewhat surprisingly, there are many situations where the order things are listed is not important. Such situations include

- The formation of committees, since the order people volunteer is unimportant
- Card games with a standard deck, since the order cards are dealt is unimportant
- Playing BINGO, since the order the numbers are called is unimportant

When the order in which people or objects are selected from a group is unimportant, the number of possibilities is a *combination,* not a permutation.

Another way to tell the difference between permutations and combinations is the following memory device: Permutations have *Priority* or *Precedence;* in other words, the *Position* of each element matters. By contrast, a *Combination* is like a *Committee* of *Colleagues* or *Collection* of *Commoners*; all members have equal rank. For permutations, *a-b-c* is different from *b-a-c*. For combinations, *a-b-c* is the same as *b-a-c*.

EXAMPLE 10 ▶ **Applications of Quick-Counting—Committees and Government**

The Sociology Department of Lakeside Community College has 12 dedicated faculty members. (a) In how many ways can a three-member textbook selection committee be formed? (b) If the department is in need of a Department Chair, Curriculum Chair, and Technology Chair, in how many ways can the positions be filled?

Solution ▶ **a.** Since textbook selection depends on a *Committee* of *Colleagues*, the order members are chosen is not important. This is a *Combination* of 12 people taken 3 at a time, and there are $_{12}C_3 = 220$ ways the committee can be formed.

☑ **E.** You've just learned how to quick-count using combinations

b. Since those selected will have *Position* or *Priority*, this is a *Permutation* of 12 people taken 3 at a time, giving $_{12}P_3 = 1320$ ways the positions can be filled.

Now try Exercises 67 through 78 ▶

The Exercise Set contains a wide variety of additional applications. See **Exercises 81 through 107.**

TECHNOLOGY HIGHLIGHT

Calculating Permutations and Combinations

Both the $_nP_r$ and $_nC_r$ functions are accessed using the [MATH] key and the **PRB** submenu (see Figure 10.14). To compute the permutations of 12 objects taken 9 at a time ($_{12}P_9$), clear the home screen and enter a 12, then press [MATH] [◄] **2:$_nP_r$** to access the $_nP_r$ operation, which is automatically pasted on the home screen after the 12. Now enter a 9, press [ENTER] and a result of 79833600 is displayed (Figure 10.15). Repeat the sequence to compute the value of $_{12}C_9$ ([MATH] [◄] **3:$_nC_r$**). Note that the value of $_{12}P_9$ is much larger than $_{12}C_9$ and that they differ by a factor of 9! since $_nC_r = \dfrac{_nP_r}{r!}$.

Figure 10.14

```
MATH NUM CPX PRB
1:rand
2:nPr
3:nCr
4:!
5:randInt(
6:randNorm(
7:randBin(
```

Figure 10.15

```
12 nPr 9
              79833600
12 nCr 9
                   220
220*9!
              79833600
```

Exercise 1: The Department of Humanities has nine faculty members who must serve on at least one committee per semester. How many different committees can be formed that have (a) two members, (b) three members, (c) four members, and (d) five members?

Exercise 2: A certain state places 45 Ping-Pong balls numbered 1 through 45 in a container, then draws out five to form the winning lottery numbers. How many different ways can the five numbers be picked?

Exercise 3: Dairy King maintains six different toppings at a self-service counter, so that customers can top their ice cream sundaes with as many as they like. How many different sundaes can be created if a customer were to select any three ingredients?

10.5 EXERCISES

▶ CONCEPTS AND VOCABULARY

Fill in the blank with the appropriate word or phrase. Carefully reread the section if needed.

1. A(n) _____ is any task that can be repeated and has a(n) _____ set of possible outcomes.

2. If an experiment has N equally likely outcomes and is repeated t times, the number of elements in the sample space is given by ____.

3. When unique elements of a set are rearranged, the result is called a(n) _____ permutation.

4. If some elements of a group are identical, certain rearrangements are identical and the result is a(n) _____ permutation.

5. A three-digit number is formed from digits 1 to 9. Explain how forming the number with repetition differs from forming it without repetition.

6. Discuss/Explain the difference between a permutation and a combination. Try to think of new ways to help remember the distinction.

▶ DEVELOPING YOUR SKILLS

7. For the spinner shown here, (a) draw a tree diagram illustrating all possible outcomes for two spins and (b) create an ordered list showing all possible outcomes for two spins.

8. For the fair coin shown here, (a) draw a tree diagram illustrating all possible outcomes for four flips and (b) create an ordered list showing the possible outcomes for four flips.

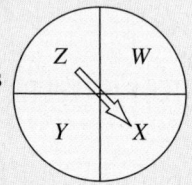

9. A fair coin is flipped five times. If you extend the tree diagram from Exercise 8, how many elements are in the sample space?

10. A spinner has the two equally likely outcomes A or B and is spun four times. How is this experiment related to the one in Exercise 8? How many elements are in the sample space?

11. An inexpensive lock uses the numbers 0 to 24 for a three-number combination. How many different combinations are possible?

12. Grades at a local college consist of A, B, C, D, F, and W. If four classes are taken, how many different report cards are possible?

License plates. In a certain (English-speaking) country, license plates for automobiles consist of two letters followed by one of four symbols (■, ♦, ○, or ●), followed by three digits. How many license plates are possible if

13. Repetition is allowed?

14. Repetition is not allowed?

15. A remote access door opener requires a five-digit (1–9) sequence. How many sequences are possible if (a) repetition is allowed? (b) repetition is not allowed?

16. An instructor is qualified to teach Math 020, 030, 140, and 160. How many different four-course schedules are possible if (a) repetition is allowed? (b) repetition is not allowed?

Use the fundamental principle of counting and other quick-counting techniques to respond.

17. **Menu items:** At Joe's Diner, the manager is offering a dinner special that consists of one choice of entree (chicken, beef, soy meat, or pork), two vegetable servings (corn, carrots, green beans, peas, broccoli, or okra), and one choice of pasta, rice, or potatoes. How many different meals are possible?

18. **Getting dressed:** A frugal businessman has five shirts, seven ties, four pairs of dress pants, and three pairs of dress shoes. Assuming that all possible arrangements are appealing, how many different shirt-tie-pants-shoes outfits are possible?

19. **Number combinations:** How many four-digit numbers can be formed using the even digits 0, 2, 4, 6, 8, if (a) no repetitions are allowed; (b) repetitions are allowed; (c) repetitions are not allowed and the number must be less than 6000 and divisible by 10.

20. **Number combinations:** If I was born in March, April, or May, after the 19th but before the 30th,

and after 1949 but before 1981, how many different MM–DD–YYYY dates are possible for my birthday?

Seating arrangements: William, Xayden, York, and Zelda decide to sit together at the movies. How many ways can they be seated if

21. They sit in random order?

22. York must sit next to Zelda?

23. York and Zelda must be on the outside?

24. William must have the aisle seat?

Course schedule: A college student is trying to set her schedule for the next semester and is planning to take five classes: English, art, math, fitness, and science. How many different schedules are possible if

25. The classes can be taken in any order.

26. She wants her science class to immediately follow her math class.

27. She wants her English class to be first and her fitness class to be last.

28. She can't decide on the best order and simply takes the classes in alphabetical order.

Find the value of $_nP_r$ in two ways: (a) compute r factors of $n!$ and (b) use the formula $_nP_r = \dfrac{n!}{(n-r)!}$.

29. $_{10}P_3$ **30.** $_{12}P_2$ **31.** $_9P_4$

32. $_5P_3$ **33.** $_8P_7$ **34.** $_8P_1$

Determine the number of three-letter permutations of the letters given, then use an organized list to write them all out. How many of them are actually words or common names?

35. T, R, and A   **36.** P, M, and A

37. The regional manager for an office supply store needs to replace the manager and assistant manager at the downtown store. In how many ways can this be done if she selects the personnel from a group of 10 qualified applicants?

38. The local chapter of Mu Alpha Theta will soon be electing a president, vice-president, and treasurer. In how many ways can the positions be filled if the chapter has 15 members?

39. The local school board is going to select a principal, vice-principal, and assistant vice-principal from a pool of eight qualified candidates. In how many ways can this be done?

40. From a pool of 32 applicants, a board of directors must select a president, vice-president, labor relations liaison, and a director of personnel for the company's day-to-day operations. Assuming all applicants are qualified and willing to take on any of these positions, how many ways can this be done?

41. A hugely popular chess tournament now has six finalists. Assuming there are no ties, (a) in how many ways can the finalists place in the final round? (b) In how many ways can they finish first, second, and third? (c) In how many ways can they finish if it's sure that Roberta Fischer is going to win the tournament and that Geraldine Kasparov will come in sixth?

42. A field of 10 horses has just left the paddock area and is heading for the gate. Assuming there are no ties in the big race, (a) in how many ways can the horses place in the race? (b) In how many ways can they finish in the win, place, or show positions? (c) In how many ways can they finish if it's sure that John Henry III is going to win, Seattle Slew III will come in second (place), and either Dumb Luck II or Calamity Jane I will come in tenth?

Assuming all multiple births are identical and the children cannot be told apart, how many distinguishable photographs can be taken of a family of six, if they stand in a single row and there is

43. one set of twins

44. one set of triplets

45. one set of twins and one set of triplets

46. one set of quadruplets

47. How many distinguishable numbers can be made by rearranging the digits of 105,001?

48. How many distinguishable numbers can be made by rearranging the digits in the palindrome 1,234,321?

How many distinguishable permutations can be formed from the letters of the given word?

49. logic   **50.** leave

51. lotto   **52.** levee

A Scrabble player (see Example 7) has the six letters shown remaining in her rack. How many distinguishable, six-letter permutations can be formed? (If all six letters are played, what was the word?)

53. A, A, A, N, N, B

54. D, D, D, N, A, E

Find the value of $_nC_r$: (a) using $_nC_r = \dfrac{_nP_r}{r!}$ (r factors of $n!$ over $r!$) and (b) using $_nC_r = \dfrac{n!}{r!(n-r)!}$.

55. $_9C_4$ **56.** $_{10}C_3$ **57.** $_8C_5$

58. $_6C_3$ **59.** $_6C_6$ **60.** $_6C_0$

Use a calculator to verify that each pair of combinations is equal.

61. $_9C_4,\ _9C_5$ **62.** $_{10}C_3,\ _{10}C_7$

63. $_8C_5,\ _8C_3$ **64.** $_7C_2,\ _7C_5$

65. A platoon leader needs to send four soldiers to do some reconnaissance work. There are 12 soldiers in the platoon and each soldier is assigned a number between 1 and 12. The numbers 1 through 12 are placed in a helmet and drawn randomly. If a soldier's number is drawn, then that soldier goes on the mission. In how many ways can the reconnaissance team be chosen?

66. Seven colored balls (red, indigo, violet, yellow, green, blue, and orange) are placed in a bag and three are then withdrawn. In how many ways can the three colored balls be drawn?

67. When the company's switchboard operators went on strike, the company president asked for three volunteers from among the managerial ranks to temporarily take their place. In how many ways can the three volunteers "step forward," if there are 14 managers and assistant managers in all?

68. Becky has identified 12 books she wants to read this year and decides to take four with her to read while on vacation. She chooses *Pastwatch* by Orson Scott Card for sure, then decides to randomly choose any three of the remaining books. In how many ways can she select the four books she'll end up taking?

69. A new garage band has built up their repertoire to 10 excellent songs that really rock. Next month they'll be playing in a *Battle of the Bands* contest, with the winner getting some guaranteed gigs at the city's most popular hot spots. In how many ways can the band select 5 of their 10 songs to play at the contest?

70. Pierre de Guirré is an award-winning chef and has just developed 12 delectable, new main-course recipes for his restaurant. In how many ways can he select three of the recipes to be entered in an international culinary competition?

For each exercise, determine whether a permutation, a combination, counting principles, or a determination of the number of subsets is the most appropriate tool for obtaining a solution, then solve. Some exercises can be completed using more than one method.

71. In how many ways can eight second-grade children line up for lunch?

72. If you flip a fair coin five times, how many different outcomes are possible?

73. Eight sprinters are competing for the gold, silver, and bronze medals. In how many ways can the medals be awarded?

74. Motorcycle license plates are made using two letters followed by three numbers. How many plates can be made if repetition of letters (only) is allowed?

75. A committee of five students is chosen from a class of 20 to attend a seminar. How many different ways can this be done?

76. If onions, cheese, pickles, and tomatoes are available to dress a hamburger, how many different hamburgers can be made?

77. A caterer offers eight kinds of fruit to make various fruit trays. How many different trays can be made using four different fruits?

78. Eighteen females try out for the basketball team, but the coach can only place 15 on her roster. How many different teams can be formed?

▶ **WORKING WITH FORMULAS**

79. Stirling's Formula: $n! \approx \sqrt{2\pi} \cdot (n^{n+0.5}) \cdot e^{-n}$

Values of $n!$ grow very quickly as n gets larger (13! is already in the billions). For some applications, scientists find it useful to use the approximation for $n!$ shown, called Stirling's Formula.

  a. Compute the value of 7! on your calculator, then use Stirling's Formula with $n = 7$. By what percent does the approximate value differ from the true value?

  b. Compute the value of 10! on your calculator, then use Stirling's Formula with $n = 10$. By what percent does the approximate value differ from the true value?

80. Factorial formulas: For $n, k \in \mathbb{W}$, where $n > k$,

$$\frac{n!}{(n-k)!} = n(n-1)(n-2)\cdots(n-k+1)$$

  a. Verify the formula for $n = 7$ and $k = 5$.

  b. Verify the formula for $n = 9$ and $k = 6$.

▶ **APPLICATIONS**

81. Yahtzee: In the game of "Yahtzee"® (Milton Bradley) five dice are rolled simultaneously on the first turn in an attempt to obtain various arrangements (worth various point values). How many different arrangements are possible?

82. Twister: In the game of "Twister"® (Milton Bradley) a simple spinner is divided into four quadrants designated Left Foot (LF), Right Hand (RH), Right Foot (RF), and Left Hand (LH), with four different color possibilities in each quadrant (red, green, yellow, blue). Determine the number of possible outcomes for three spins.

83. Clue: In the game of "Clue"® (Parker Brothers) a crime is committed in one of nine rooms, with one of six implements, by one of six people. In how many different ways can the crime be committed?

Phone numbers in North America have 10 digits: a three-digit area code, a three-digit exchange number, and the four final digits that make each phone number unique. Neither area codes nor exchange numbers can start with 0 or 1. Prior to 1994 the second digit of the area code *had to be* a 0 or 1. Sixteen area codes are reserved for special services (such as 911 and 411). In 1994, the last area code was used up and the rules were changed to allow the digits 2 through 9 as the middle digit in area codes.

84. How many different area codes were possible prior to 1994?

85. How many different exchange numbers were possible prior to 1994?

86. How many different phone numbers were possible *prior to* 1994?

87. How many different phone numbers were possible *after* 1994?

Aircraft N-numbers: In the United States, private aircraft are identified by an "N-Number," which is generally the letter "N" followed by five characters and includes these restrictions: (1) the N-Number can consist of five digits, four digits followed by one letter, or three digits followed by two letters; (2) the first digit cannot be a zero; (3) to avoid confusion with the numbers zero and one, the letters O and I cannot be used; and (4) repetition of digits and letters is allowed. How many unique N-Numbers can be formed

88. that have four digits and one letter?

89. that have three digits and two letters?

90. that have five digits?

91. that have three digits, two letters with no repetitions of any kind allowed?

Seating arrangements: Eight people would like to be seated. Assuming some will have to stand, in how many ways can the seats be filled if the number of seats available is

92. eight

93. five

94. three

95. one

Seating arrangements: In how many different ways can eight people (six students and two teachers) sit in a row of eight seats if

96. the teachers must sit on the ends

97. the teachers must sit together

Television station programming: A television station needs to fill eight half-hour slots for its Tuesday evening schedule with eight programs. In how many ways can this be done if

98. there are no constraints

99. *Seinfeld* must have the 8:00 P.M. slot

100. *Seinfeld* must have the 8:00 P.M. slot and *The Drew Carey Show* must be shown at 6:00 P.M.

101. *Friends* can be aired at 7:00 or 9:00 P.M. and *Everybody Loves Raymond* can be aired at 6:00 or 8:00 P.M.

Scholarship awards: Fifteen students at Roosevelt Community College have applied for six available scholarship awards. How many ways can the awards be given if

102. there are six different awards given to six different students

103. there are six identical awards given to six different students

Committee composition: The local city council has 10 members and is trying to decide if they want to be governed by a committee of three people or by a president, vice-president, and secretary.

104. If they are to be governed by committee, how many unique committees can be formed?

105. How many different president, vice-president, and secretary possibilities are there?

106. Team rosters: A soccer team has three goalies, eight defensive players, and eight forwards on its roster. How many different starting line-ups can be formed (one goalie, three defensive players, and three forwards)?

107. e-mail addresses: A business wants to standardize the e-mail addresses of its employees. To make them easier to remember and use, they consist of two letters and two digits (followed by @esmtb.com), with zero being excluded from use as the first digit and no repetition of letters or digits allowed. Will this provide enough unique addresses for their 53,000 employees worldwide?

▶ **EXTENDING THE CONCEPT**

108. In Exercise 79, we learned that an approximation for $n!$ can be found using Stirling's Formula: $n! \approx \sqrt{2\pi}(n^{n+0.5})e^{-n}$. As with other approximations, mathematicians are very interested in whether the approximation gets better or worse for larger values of n (does their ratio get closer to 1 or farther from 1). Use your calculator to investigate and answer the question.

109. Verify that the following equations are true, then generalize the patterns and relationships noted to create your own equation. Afterward, write each of the four factors from Part (a) (the two combinations on each side) in expanded form and discuss/explain why the two sides are equal.

   a. $_{10}C_3 \cdot {}_7C_2 = {}_{10}C_2 \cdot {}_8C_5$

   b. $_9C_3 \cdot {}_6C_2 = {}_9C_2 \cdot {}_7C_4$

   c. $_{11}C_4 \cdot {}_7C_5 = {}_{11}C_5 \cdot {}_6C_4$

   d. $_8C_3 \cdot {}_5C_2 = {}_8C_2 \cdot {}_6C_3$

110. Tic-Tac-Toe: In the game *Tic-Tac-Toe*, players alternately write an "X" or an "O" in one of nine squares on a 3 × 3 grid. If either player gets three in a row horizontally, vertically, or diagonally, that player wins. If all nine squares are played with neither person winning, the game is a draw. Assuming "X" always goes first,

   a. How many different "boards" are possible if the game ends after five plays?

   b. How many different "boards" are possible if the game ends after six plays?

▶ **MAINTAINING YOUR SKILLS**

111. (5.4) Solve the given system of linear inequalities by graphing. Shade the feasible region.
$$\begin{cases} 2x + y < 6 \\ x + 2y < 6 \\ x \geq 0 \\ y \geq 0 \end{cases}$$

112. (5.2) Given $\sin \theta = \dfrac{12}{13}$, determine the other five trig functions of the acute angle θ.

113. (6.3) Rewrite $\cos(2\alpha)\cos(3\alpha) - \sin(2\alpha)\sin(3\alpha)$ as a single expression.

114. (7.3) Graph the hyperbola that is defined by $\dfrac{(x-2)^2}{4} - \dfrac{(y+3)^2}{9} = 1$.

Learning Objectives

In Section 10.6 you will learn how to:

☐ **A.** Define an event on a sample space

☐ **B.** Compute elementary probabilities

☐ **C.** Use certain properties of probability

☐ **D.** Compute probabilities using quick-counting techniques

☐ **E.** Compute probabilities involving nonexclusive events

There are few areas of mathematics that give us a better view of the world than **probability** and **statistics.** Unlike statistics, which seeks to analyze and interpret data, probability (for our purposes) attempts to use observations and data to make statements concerning the likelihood of future events. Such predictions of what *might* happen have found widespread application in such diverse fields as politics, manufacturing, gambling, opinion polls, product life, and many others. In this section, we develop the basic elements of probability.

A. Defining an Event

In Section 10.5 we defined the following terms: experiment and sample outcome. Flipping a coin twice in succession is an *experiment,* and two sample outcomes are HH and HT. An **event** E is *any designated set of sample outcomes,* and is a subset of the sample space. One event might be E_1: (two heads occur), another possibility is E_2: (at least one tail occurs).

EXAMPLE 1 ▶ **Stating a Sample Space and Defining an Event**

Consider the experiment of rolling one standard, six-sided die (plural is dice). State the sample space S and define any two events relative to S.

Solution ▶ S is the set of all possible outcomes, so $S = \{1, 2, 3, 4, 5, 6\}$. Two possible events are E_1: (a 5 is rolled) and E_2: (an even number is rolled).

Now try Exercises 7 through 10 ▶

☑ **A.** You've just learned how to define an event on a sample space

B. Elementary Probability

When rolling the die, we know the result can be any of the six equally likely outcomes in the sample space, so the chance of E_1:(a five is rolled) is $\frac{1}{6}$. Since three of the elements in S are even numbers, the chance of E_2:(an even number is rolled) is $\frac{3}{6} = \frac{1}{2}$. This suggests the following definition.

> **The Probability of an Event E**
>
> Given S is a sample space of equally likely events and E is an event relative to S, the probability of E, written $P(E)$, is computed as
>
> $$P(E) = \frac{n(E)}{n(S)}$$
>
> where $n(E)$ represents the number of elements in E, and $n(S)$ represents the number of elements in S.

A standard deck of playing cards consists of 52 cards divided in four groups or *suits.* There are 13 hearts (♥), 13 diamonds (♦), 13 spades (♠), and 13 clubs (♣). As you can see in the illustration, each of the 13 cards in a suit is labeled 2, 3, 4, 5, 6, 7, 8, 9, 10, J, Q, K, and A. Also notice that 26 of the cards are red (hearts and diamonds), 26 are black (spades and clubs) and 12 of the cards are "face cards" (J, Q, K of each suit).

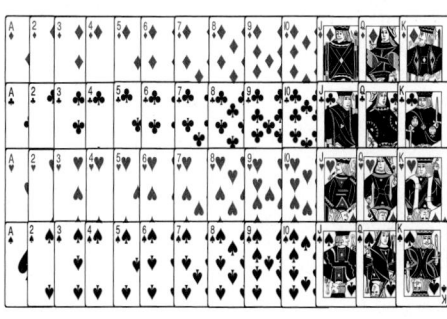

EXAMPLE 2 ▶   Stating a Sample Space and the Probability of a Single Outcome

A single card is drawn from a well-shuffled deck. Define S and state the probability of any single outcome. Then define E as *a King is drawn* and find $P(E)$.

Solution ▶   Sample space: $S = \{$the 52 cards$\}$. There are 52 equally likely outcomes, so the probability of any one outcome is $\frac{1}{52}$. Since S has four Kings,

$$P(E) = \frac{n(E)}{n(S)} = \frac{4}{52} \text{ or about } 0.077.$$

Now try Exercises 11 through 14 ▶

EXAMPLE 3 ▶   Stating a Sample Space and the Probability of a Single Outcome

A family of five has two girls and three boys named Sophie, Maria, Albert, Isaac, and Pythagoras. Their ages are 21, 19, 15, 13, and 9, respectively. One is to be selected randomly. Find the probability a teenager is chosen.

Solution ▶   The sample space is $S = \{9, 13, 15, 19, 21\}$. Three of the five are teenagers, meaning the probability is $\frac{3}{5}$, 0.6, or 60%.

☑ **B.** You've just learned how to compute elementary probabilities

Now try Exercises 15 and 16 ▶

C. Properties of Probability

A study of probability necessarily includes recognizing some basic and fundamental properties. For example, when a fair die is rolled, what is $P(E)$ if E is defined as a *1, 2, 3, 4, 5,* or *6* is rolled? The event E will occur 100% of the time, since 1, 2, 3, 4, 5, 6 are the only possibilities. In symbols we write P(outcome is in the sample space) or simply $P(S) = 1$ (100%).

What percent of the time will a result *not* in the sample space occur? Since the die has only the six sides numbered 1 through 6, the probability of rolling something else is zero. In symbols, P(outcome is not in sample space) $= 0$ or simply $P(\sim S) = 0$.

WORTHY OF NOTE

In probability studies, the tilde "~" acts as a negation symbol. For any event E defined on the sample space, $\sim E$ means the event does not occur.

Properties of Probability
Given sample space S and any event E defined relative to S.
1. $P(S) = 1$ **2.** $P(\sim S) = 0$ **3.** $0 \leq P(E) \leq 1$

EXAMPLE 4 ▶   Determining the Probability of an Event

A game is played using a spinner like the one shown. Determine the probability of the following events:

E_1: A nine is spun. E_2: An integer greater than 0 and less than 9 is spun.

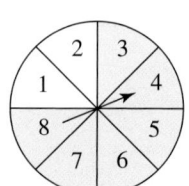

Solution ▶   The sample space consists of eight equally likely outcomes.

$$P(E_1) = \frac{0}{8} = 0 \qquad P(E_2) = \frac{8}{8} = 1.$$

Technically, E_1: A nine is spun is not an "event," since it is not in the sample space and cannot occur, while E_2 contains the entire sample space and must occur.

Now try Exercises 17 and 18 ▶

Because we know $P(S) = 1$ and all sample outcomes are equally likely, the probabilities of all single events defined on the sample space must sum to 1. For the experiment of rolling a fair die, the sample space has six outcomes that are equally likely. Note that $P(1) = P(2) = P(3) = P(4) = P(5) = P(6) = \frac{1}{6}$, and $\frac{1}{6} + \frac{1}{6} + \frac{1}{6} + \frac{1}{6} + \frac{1}{6} + \frac{1}{6} = 1$.

Probability and Sample Outcomes

Given a sample space S with n equally likely sample outcomes $s_1, s_2, s_3, \ldots, s_n$.

$$\sum_{i=1}^{n} P(s_i) = P(s_1) + P(s_2) + P(s_3) + \cdots + P(s_n) = 1$$

The **complement** of an event E is the set of sample outcomes in S not contained in E. Symbolically, $\sim E$ is the complement of E.

Probability and Complementary Events

Given sample space S and any event E defined relative to S, the complement of E, written $\sim E$, is the set of all outcomes not in E and:

1. $P(E) = 1 - P(\sim E)$ **2.** $P(E) + P(\sim E) = 1$

EXAMPLE 5 ▶ Stating a Probability Using Complements

Use complementary events to answer the following questions:
  a. A single card is drawn from a well-shuffled deck. What is the probability that it is not a diamond?
  b. A single letter is picked at random from the letters in the word "divisibility." What is the probability it is not an "i"?

Solution ▶

WORTHY OF NOTE

Probabilities can be written in fraction form, decimal form, or as a percent. For $P(E_2)$ from Example 1, the probability is $\frac{3}{4}$, 0.75, or 75%.

a. Since there are 13 diamonds in a standard 52-card deck, there are 39 nondiamonds: $P(\sim D) = 1 - P(D) = 1 - \frac{13}{52} = \frac{39}{52} = 0.75$.

b. Of the 12 letters in d-i-v-i-s-i-b-i-l-i-t-y, 5 are "i's." This means $P(\sim i) = 1 - P(i)$, or $1 - \frac{5}{12} = \frac{7}{12}$. The probability of choosing a letter other than i is $0.58\overline{3}$.

Now try Exercises 19 through 22 ▶

EXAMPLE 6 ▶ Stating a Probability Using Complements

Inter-Island Waterways has just opened hydrofoil service between several islands. The hydrofoil is powered by two engines, one forward and one aft, and will operate if either of its two engines is functioning. Due to testing and past experience, the company knows the probability of the aft engine failing is $P(\text{aft engine fails}) = 0.05$, the probability of the forward engine failing is $P(\text{forward engine fails}) = 0.03$, and the probability that both fail is $P(\text{both engines simultaneously fail}) = 0.012$. What is the probability the hydrofoil completes its next trip?

Solution ▶ Although the answer may *seem* complicated, note that $P(\text{trip is completed})$ and $P(\text{both engines simultaneously fail})$ are complements.

$$P(\text{trip is completed}) = 1 - P(\text{both engines simultaneously fail})$$
$$= 1 - 0.012$$
$$= 0.988$$

There is close to a 99% probability the trip will be completed.

Now try Exercises 23 and 24 ▶

The chart in Figure 10.16 shows all 36 possible outcomes (the sample space) from the experiment of rolling two fair dice.

Figure 10.16

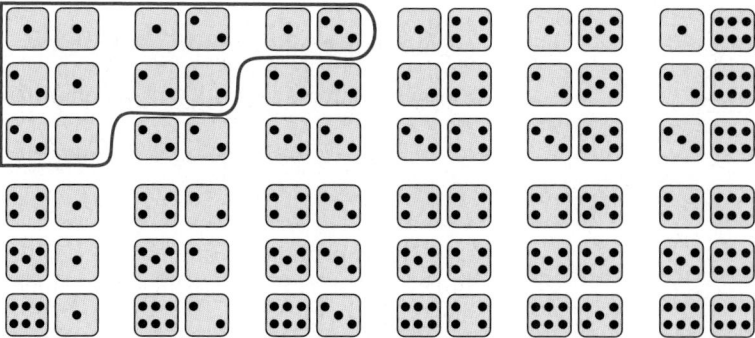

EXAMPLE 7 ▶ **Stating a Probability Using Complements**

Two fair dice are rolled. What is the probability the sum of both dice is greater than or equal to 5, $P(\text{sum} \geq 5)$?

Solution ▶ See Figure 10.16. For $P(\text{sum} \geq 5)$ it may be easier to use complements as there are far fewer possibilities: $P(\text{sum} \geq 5) = 1 - P(\text{sum} < 5)$, which gives

☑ **C.** You've just learned how to use certain properties of probability

$$1 - \frac{6}{36} = 1 - \frac{1}{6} = \frac{5}{6} = 0.8\overline{3}.$$

Now try Exercises 25 and 26 ▶

D. Probability and Quick-Counting

Quick-counting techniques were introduced earlier to help count the number of elements in a large or more complex sample space, and the number of sample outcomes in an event.

EXAMPLE 8A ▶ **Stating a Probability Using Combinations**

Five cards are drawn from a shuffled 52-card deck. Calculate the probability of E_1:(*all five cards are face cards*) or E_2:(*all five cards are hearts*)?

Solution ▶ The sample space for both events consists of all five-card groups that can be formed from the 52 cards or $_{52}C_5$. For E_1 we are to select five face cards from the 12 that are available (three from each suit), or $_{12}C_5$. The probability of five face

WORTHY OF NOTE

It seems reasonable that the probability of 5 hearts is slightly higher, as 13 of the 52 cards are hearts, while only 12 are face cards.

cards is $\dfrac{n(E)}{n(S)} = \dfrac{_{12}C_5}{_{52}C_5}$, which gives $\dfrac{792}{2{,}598{,}960} \approx 0.0003$. For E_2 we are to select five

hearts from the 13 available, or $_{13}C_5$. The probability of five hearts is $\dfrac{n(E)}{n(S)} = \dfrac{_{13}C_5}{_{52}C_5}$,

which is $\dfrac{1287}{2{,}598{,}960} \approx 0.0005$.

EXAMPLE 8B ▶ **Stating a Probability Using Combinations and the Fundamental Principle of Counting**

Of the 42 seniors at Jacoby High School, 23 are female and 19 are male. A group of five students is to be selected at random to attend a conference in Reno, Nevada. What is the probability the group will have exactly three females?

Solution ▶ The sample space consists of all five-person groups that can be formed from the 42 seniors or $_{42}C_5$. The event consists of selecting 3 females from the 23 available $(_{23}C_3)$ and 2 males from the 19 available $(_{19}C_2)$. Using the fundamental principle of counting $n(E) = {_{23}C_3} \cdot {_{19}C_2}$ and the probability the group has 3 females is

☑ **D.** You've just learned how to compute probabilities using quick-counting techniques

$$\frac{n(E)}{n(S)} = \frac{_{23}C_3 \cdot {_{19}C_2}}{_{42}C_5}, \text{ which gives } \frac{302{,}841}{850{,}668} \approx 0.356. \text{ There is approximately a}$$

35.6% probability the group will have exactly 3 females.

Now try Exercises 27 through 34 ▶

E. Probability and Nonexclusive Events

Sometimes the way events are defined causes them to share sample outcomes. Using a standard deck of playing cards once again, if we define the events E_1:(a club is drawn) and E_2:(a face card is drawn), they share the outcomes J♣, Q♣, and K♣ as shown in Figure 10.17. This overlapping region is the intersection of the events, or $E_1 \cap E_2$. If we compute $n(E_1 \cup E_2)$ as $n(E_1) + n(E_2)$ as before, this intersecting region gets counted *twice*!

Figure 10.17

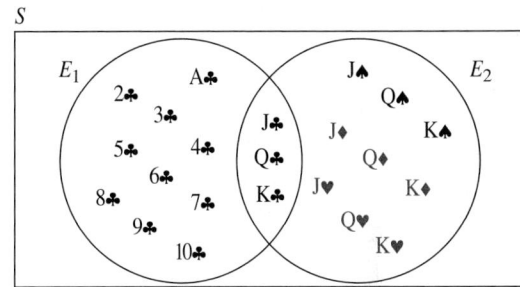

In cases where the events are **nonexclusive** (not mutually exclusive), we maintain the correct count by subtracting one of the two intersections, obtaining $n(E_1 \cup E_2) = n(E_1) + n(E_2) - n(E_1 \cap E_2)$. This leads to the following calculation for the probability of nonexclusive events:

$$P(E_1 \cup E_2) = \frac{n(E_1) + n(E_2) - n(E_1 \cap E_2)}{n(S)} \quad \text{definition of probability}$$

$$= \frac{n(E_1)}{n(S)} + \frac{n(E_1)}{n(S)} - \frac{n(E_1 \cap E_2)}{n(S)} \quad \text{property of rational expressions}$$

$$= P(E_1) + P(E_2) - P(E_1 \cap E_2) \quad \text{definition of probability}$$

Probability and Nonexclusive Events

Given sample space S and *nonexclusive events* E_1 and E_2 defined relative to S, the probability of E_1 *or* E_2 is given by

$$P(E_1 \cup E_2) = P(E_1) + P(E_2) - P(E_1 \cap E_2)$$

EXAMPLE 9A ▶ Stating the Probability of Nonexclusive Events

What is the probability that a club or a face card is drawn from a standard deck of 52 well-shuffled cards?

Solution ▶ As before, define the events E_1:(a club is drawn) and E_2:(a face card is drawn). Since there are 13 clubs and 12 face cards, $P(E_1) = \frac{13}{52}$ and $P(E_2) = \frac{12}{52}$. But three of the face cards are clubs, so $P(E_1 \cap E_2) = \frac{3}{52}$. This leads to

$$P(E_1 \cup E_2) = P(E_1) + P(E_2) - P(E_1 \cap E_2) \quad \text{nonexclusive events}$$

$$= \frac{13}{52} + \frac{12}{52} - \frac{3}{52} \quad \text{substitute}$$

$$= \frac{22}{52} \approx 0.423 \quad \text{combine terms}$$

There is about a 42% probability that a club or face card is drawn.

EXAMPLE 9B ▶   **Stating the Probability of Nonexclusive Events**

A survey of 100 voters was taken to gather information on critical issues and the demographic information collected is shown in the table. One out of the 100 voters is to be drawn at random to be interviewed on the 5 O'Clock News. What is the probability the person is a woman (W) or a Republican (R)?

	Women	Men	Totals
Republican	17	20	37
Democrat	22	17	39
Independent	8	7	15
Green Party	4	1	5
Tax Reform	2	2	4
Totals	53	47	100

Solution ▶   Since there are 53 women and 37 Republicans, $P(W) = 0.53$ and $P(R) = 0.37$. The table shows 17 people are both female and Republican so $P(W \cap R) = 0.17$.

$$\begin{aligned}
P(W \cup R) &= P(W) + P(R) - P(W \cap R) &&\text{nonexclusive events} \\
&= 0.53 + 0.37 - 0.17 &&\text{substitute} \\
&= 0.73 &&\text{combine}
\end{aligned}$$

☑ **E.** You've just learned how to compute probabilities involving nonexclusive events

There is a 73% probability the person is a woman or a Republican.

Now try Exercises 35 through 48 ▶

Two events that have no common outcomes are called **mutually exclusive** events (one excludes the other and vice versa). For example, in rolling one die, E_1:(*a 2 is rolled*) and E_2:(*an odd number is rolled*) are mutually exclusive, since 2 is not an odd number. For the probability of E_3:(*a 2 is rolled* or *an odd number is rolled*), we note that $n(E_1 \cap E_2) = 0$ and the previous formula simply reduces to $P(E_1) + P(E_2)$. See **Exercises 49 and 50.**

There is a large variety of additional applications in the Exercise Set. See **Exercises 53 through 68.**

TECHNOLOGY HIGHLIGHT

Principles of Quick-Counting, Combinations, and Probability

At this point you are likely using the [Y=] screen and tables ([TBLSET], [2nd] [GRAPH] **TABLE,** and so on) with relative ease. When probability calculations require a repeated use of permutations and combinations, these features can make the work more efficient and help to explore the patterns they generate. For choosing r children from a group of six children ($n = 6$), set the [TBLSET] to **AUTO,** then press [Y=] and enter 6 $_nC_r$ X as Y_1 (Figure 10.18). Access the **TABLE** ([2nd] [GRAPH]) and note that the calculator has automatically computed the value of $_6C_0, {}_6C_1, {}_6C_2, \ldots, {}_6C_6$ (Figure 10.19) and the pattern of outputs is symmetric. For calculations similar to those required in Example 8B ($_{23}C_3 \cdot {}_{19}C_2$), enter

Figure 10.18

Figure 10.19

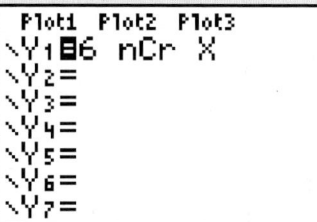

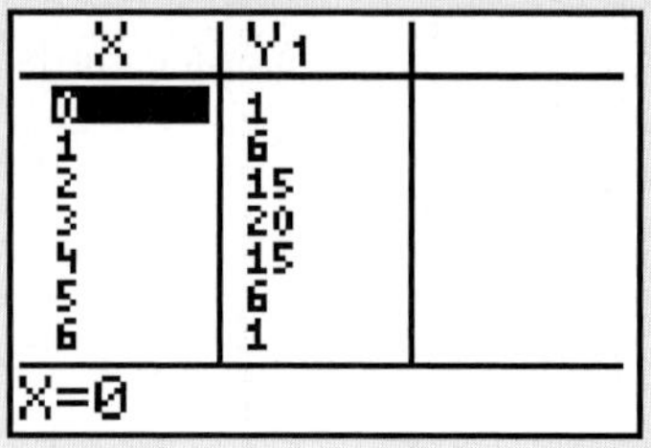

$Y_1 = 23 \, _nC_r \, X$, $Y_2 = 19 \, _nC_r \, (X - 1)$, and $Y_3 = Y_1 \cdot Y_2$, or any variation of these. Use these ideas to work the following exercises.

Exercise 1: Use your calculator to display the values of $_5C_0, {}_5C_1, \ldots, {}_5C_5$. Is the result a pattern similar to that for $_6C_0, {}_6C_1, {}_6C_2, \ldots, {}_6C_6$? Repeat for $_7C_r$. Why are the "middle values" repeated for $n = 7$ and $n = 5$, but not for $n = 6$?

Exercise 2: A committee consists of 10 Republicans and eight Democrats. In how many ways can a committee of four Republicans and three Democrats be formed?

10.6 EXERCISES

▶ CONCEPTS AND VOCABULARY

Fill in the blank with the appropriate word or phrase. Carefully reread the section if needed.

1. Given a sample space S and an event E defined relative to S, $P(E) = \dfrac{}{n(S)}$.

2. In elementary probability, we consider all events in the sample space to be _____ likely.

3. Given a sample space S and an event E defined relative to S: ___ $\le P(E) \le$ ___, $P(S) =$ ___, and $P(\sim S) =$ ___.

4. The _____ of an event E is the set of sample outcomes in S which are not contained in E.

5. Discuss/Explain the difference between mutually exclusive events and nonmutually exclusive events. Give an example of each.

6. A single die is rolled. With no calculations, explain why the probability of rolling an even number is greater than rolling a number greater than four.

▶ DEVELOPING YOUR SKILLS

State the sample space S and the probability of a single outcome. Then define any two events E relative to S (many answers possible).

7. Two fair coins (heads and tails) are flipped.

8. The simple spinner shown is spun.

Exercise 8

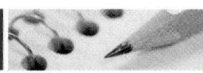

9. The head coaches for six little league teams (the Patriots, Cougars, Angels, Sharks, Eagles, and Stars) have gathered to discuss new changes in the rule book. One of them is randomly chosen to ask the first question.

10. Experts on the planets Mercury, Venus, Mars, Jupiter, Saturn, Uranus, Neptune, and the dwarf planet Pluto have gathered at a space exploration conference. One group of experts is selected at random to speak first.

Find $P(E)$ for the events defined.

11. Nine index cards numbered 1 through 9 are shuffled and placed in an envelope, then one of the cards is randomly drawn. Define event E as *the number drawn is even*.

12. Eight flash cards used for studying basic geometric shapes are shuffled and one of the cards is drawn at random. The eight cards include information on circles, squares, rectangles, kites, trapezoids, parallelograms, pentagons, and triangles. Define event E as *a quadrilateral is drawn*.

13. One card is drawn at random from a standard deck of 52 cards. What is the probability of
    a. drawing a Jack   b. drawing a spade
    c. drawing a black card   d. drawing a red three

14. Pinochle is a card game played with a deck of 48 cards consisting of 2 Aces, 2 Kings, 2 Queens, 2 Jacks, 2 Tens, and 2 Nines in each of the four

standard suits [hearts (♥), diamonds (♦), spades (♠), and clubs (♣)]. If one card is drawn at random from this deck, what is the probability of

a. drawing an Ace   b. drawing a club

c. drawing a red card

d. drawing a face card (Jack, Queen, King)

15. A group of finalists on a game show consists of three males and five females. Hank has a score of 520 points, with Harry and Hester having 490 and 475 points, respectively. Madeline has 532 points, with Mackenzie, Morgan, Maggie, and Melanie having 495, 480, 472, and 470 points, respectively. One of the contestants is randomly selected to start the final round. Define E_1 as *Hester is chosen*, E_2 as *a female is chosen*, and E_3 as *a contestant with less than 500 points is chosen*. Find the probability of each event.

16. Soccer coach Maddox needs to fill the last spot on his starting roster for the opening day of the season and has to choose between three forwards and five defenders. The forwards have jersey numbers 5, 12, and 17, while the defenders have jersey numbers 7, 10, 11, 14, and 18. Define E_1 as *a forward is chosen*, E_2 as *a defender is chosen*, and E_3 as *a player whose jersey number is greater than 10 is chosen*. Find the probability of each event.

17. A game is played using a spinner like the one shown. For each spin,

a. What is the probability the arrow lands in a shaded region?

b. What is the probability your spin is less than 5?

c. What is the probability you spin a 2?

d. What is the probability the arrow lands on a prime number?

18. A game is played using a spinner like the one shown here. For each spin,

a. What is the probability the arrow lands in a lightly shaded region?

b. What is the probability your spin is greater than 2?

c. What is the probability the arrow lands in a shaded region?

d. What is the probability you spin a 5?

Use the complementary events to complete Exercises 19 through 22.

19. One card is drawn from a standard deck of 52. What is the probability it is not a club?

20. Four standard dice are rolled. What is the probability the sum is less than 24?

21. A single digit is randomly selected from among the digits of 10!. What is the probability the digit is not a 2?

22. A corporation will be moving its offices to Los Angeles, Miami, Atlanta, Dallas, or Phoenix. If the site is randomly selected, what is the probability Dallas is not chosen?

23. A large manufacturing plant can remain at full production as long as one of its two generators is functioning. Due to past experience and the age difference between the systems, the plant manager estimates the probability of the main generator failing is 0.05, the probability of the secondary generator failing is 0.01, and the probability of both failing is 0.009. What is the probability the plant remains in full production today?

24. A fire station gets an emergency call from a shopping mall in the mid-afternoon. From a study of traffic patterns, Chief Nozawa knows the probability the most direct route is clogged with traffic is 0.07, while the probability of the secondary route being clogged is 0.05. The probability both are clogged is 0.02. What is the probability they can respond to the call unimpeded using one of these routes?

25. Two fair dice are rolled (see Figure 10.16). What is the probability of

a. a sum less than four

b. a sum less than eleven

c. the sum is not nine

d. a roll is not a "double" (both dice the same)

"Double-six" dominos is a game played with the 28 numbered tiles shown in the diagram.

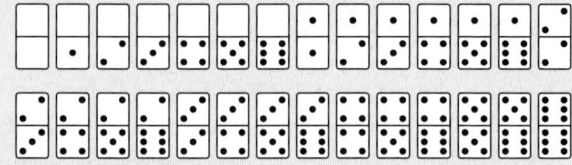

26. The 28 dominos are placed in a bag, shuffled, and then one domino is randomly drawn. What is the probability the total number of dots on the domino

a. is three or less

b. is greater than three

c. does not have a blank half

d. is not a "double" (both sides the same)

Find $P(E)$ given the values for $n(E)$ and $n(S)$ shown.

27. $n(E) = {}_6C_3 \cdot {}_4C_2$; $n(S) = {}_{10}C_5$

28. $n(E) = {}_{12}C_9 \cdot {}_8C_7$; $n(S) = {}_{20}C_{16}$

29. $n(E) = {}_9C_6 \cdot {}_5C_3$; $n(S) = {}_{14}C_9$

30. $n(E) = {}_7C_6 \cdot {}_3C_2$; $n(S) = {}_{10}C_8$

31. Five cards are drawn from a well-shuffled, standard deck of 52 cards. Which has the greater probability: (a) all five cards are red or (b) all five cards are numbered cards? How much greater?

32. Five cards are drawn from a well-shuffled pinochle deck of 48 cards (see Exercise 14). Which has the greater probability (a) all five cards are face cards (King, Queen, or Jack) or (b) all five cards are black? How much greater?

33. A dietetics class has 24 students. Of these, 9 are vegetarians and 15 are not. The instructor receives enough funding to send six students to a conference. If the students are selected randomly, what is the probability the group will have
    a. exactly two vegetarians
    b. exactly four nonvegetarians
    c. at least three vegetarians

34. A large law firm has a support staff of 15 employees: six paralegals and nine legal assistants. Due to recent changes in the law, the firm wants to send five of them to a forum on the new changes. If the selection is done randomly, what is the probability the group will have
    a. exactly three paralegals
    b. exactly two legal assistants
    c. at least two paralegals

Find the probability indicated using the information given.

35. Given $P(E_1) = 0.7$, $P(E_2) = 0.5$, and $P(E_1 \cap E_2) = 0.3$, compute $P(E_1 \cup E_2)$.

36. Given $P(E_1) = 0.6$, $P(E_2) = 0.3$, and $P(E_1 \cap E_2) = 0.2$, compute $P(E_1 \cup E_2)$.

37. Given $P(E_1) = \frac{3}{8}$, $P(E_2) = \frac{3}{4}$, and $P(E_1 \cup E_2) = \frac{15}{18}$; compute $P(E_1 \cap E_2)$.

38. Given $P(E_1) = \frac{1}{2}$, $P(E_2) = \frac{3}{5}$, and $P(E_1 \cup E_2) = \frac{17}{20}$; compute $P(E_1 \cap E_2)$.

39. Given $P(E_1 \cup E_2) = 0.72$, $P(E_2) = 0.56$, and $P(E_1 \cap E_2) = 0.43$; compute $P(E_1)$.

40. Given $P(E_1 \cup E_2) = 0.85$, $P(E_1) = 0.4$, and $P(E_1 \cap E_2) = 0.21$; compute $P(E_2)$.

41. Two fair dice are rolled. What is the probability the sum of the dice is

a. a multiple of 3 and an odd number
b. a sum greater than 5 and a 3 on one die
c. an even number and a number greater than 9
d. an odd number and a number less than 10

42. *Eight Ball* is a game played on a pool table with 15 balls numbered 1 through 15 and a cue ball that is solid white. Of the 15 numbered balls, 8 are a solid (nonwhite) color and numbered 1 through 8, and seven are striped balls numbered 9 through 15. The fifteen numbered pool balls (no cueball) are placed in a large bowl and mixed, then one is drawn out. What is the probability of drawing

a. the eight ball
b. a number greater than fifteen
c. an even number
d. a multiple of three
e. a solid color and an even number
f. a striped ball and an odd number
g. an even number and a number divisible by three
h. an odd number and a number divisible by 4

43. A survey of 50 veterans was taken to gather information on their service career and what life is like out of the military. A breakdown of those surveyed is shown in the table. One out of the 50 will be selected at random for an interview and a biographical sketch. What is the probability the person chosen is

	Women	Men	Totals
Private	6	9	15
Corporal	10	8	18
Sergeant	4	5	9
Lieutenant	2	1	3
Captain	2	3	5
Totals	24	26	50

a. a woman and a sergeant
b. a man and a private
c. a private and a sergeant
d. a woman and an officer
e. a person in the military

44. Referring to Exercise 43, what is the probability the person chosen is
    a. a woman or a sergeant
    b. a man or a private
    c. a woman or a man
    d. a woman or an officer
    e. a captain or a lieutenant

A computer is asked to randomly generate a three-digit number. What is the probability the

45. ten's digit is odd or the one's digit is even

46. first digit is prime and the number is a multiple of 10

A computer is asked to randomly generate a four-digit number. What is the probability the number is

47. at least 4000 or a multiple of 5

48. less than 7000 and an odd number

49. Two fair dice are rolled. What is the probability of rolling

    a. boxcars (a sum of 12) or snake eyes (a sum of 2)

    b. a sum of 7 or a sum of 11

    c. an even-numbered sum or a prime sum

d. an odd-numbered sum or a sum that is a multiple of 4

e. a sum of 15 or a multiple of 12

f. a sum that is a prime number

50. Suppose all 16 balls from a game of pool (see Exercise 42) are placed in a large leather bag and mixed, then one is drawn out. Consider the cue ball as "0." What is the probability of drawing

    a. a striped ball

    b. a solid-colored ball

    c. a polka-dotted ball

    d. the cue ball

    e. the cue ball or the eight ball

    f. a striped ball or a number less than five

    g. a solid color or a number greater than 12

    h. an odd number or a number divisible by 4

▶ WORKING WITH FORMULAS

51. Games involving a fair spinner (with numbers 1 through 4): $P(n) = \left(\frac{1}{4}\right)^n$

Games that involve moving pieces around a board using a fair spinner are fairly common. If a fair spinner has the numbers 1 through 4, the probability that any one number is spun n times in succession is given by the formula shown, where n represents the number of spins. What is the probability (a) the first player spins a two? (b) all four players spin a two? (c) Discuss the graph of $P(n)$ and explain the connection between the graph and the probability of consistently spinning a two.

52. Games involving a fair coin (heads and tails): $P(n) = \left(\frac{1}{2}\right)^n$

When a fair coin is flipped, the probability that heads (or tails) is flipped n times in a row is given by the formula shown, where n represents the number of flips. What is the probability (a) the first flip is heads? (b) the first four flips are heads? (c) Discuss the graph of $P(n)$ and explain the connection between the graph and the probability of consistently flipping heads.

▶ APPLICATIONS

53. To improve customer service, a company tracks the number of minutes a caller is "on hold" and waiting for a customer service representative. The table shows the probability that a caller will wait m minutes. Based on the table, what is the probability a caller waits

    a. at least 2 minutes

    b. less than 2 minutes

    c. 4 minutes or less

    d. over 4 minutes

    e. less than 2 or more than 4 minutes

    f. 3 or more minutes

Wait Time (minutes m)	Probability
0	0.07
$0 < m < 1$	0.28
$1 \leq m < 2$	0.32
$2 \leq m < 3$	0.25
$3 \leq m < 4$	0.08

54. To study the impact of technology on American families, a researcher first determines the probability that a family has n computers at home. Based on the table, what is the probability a home

a. has at least one computer
b. has two or more computers
c. has less than four computers
d. has five computers
e. has one, two, or three computers
f. does not have two computers

Number of Computers	Probability
0	9%
1	51%
2	28%
3	9%
4	3%

55. Jolene is an experienced markswoman and is able to hit a 10 in. by 20 in. target 100% of the time at a range of 100 yd. Assuming the probability she hits a target is related to its area, what is the probability she hits the shaded portions shown?

10 in.

20 in.

a. isosceles triangle   b. right triangle

   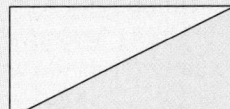

c. equilateral triangle

56. a. square   b. circle

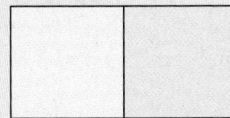   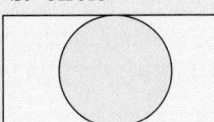

c. isosceles trapezoid with $b = \frac{B}{2}$

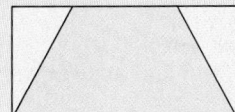

57. A circular dartboard has a total radius of 8 in., with circular bands that are 2 in. wide, as shown. You are skilled enough to hit this board 100% of the time so you always score at least two points each

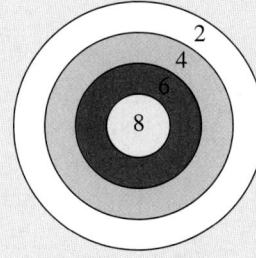

time you throw a dart. Assuming the probabilities are related to area, on the next dart that you throw what is the probability you

a. score at least a 4?   b. score at least a 6?
c. hit the bull's-eye?   d. score exactly 4 points?

58. Three red balls, six blue balls, and four white balls are placed in a bag. What is the probability the first ball you draw out is

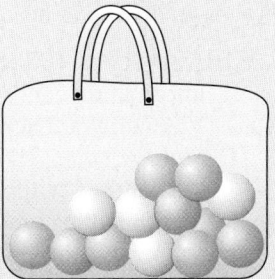

a. red   b. blue
c. not white
d. purple
e. red or white
f. red and white

59. Three red balls, six blue balls, and four white balls are placed in a bag, then two are drawn out and placed in a rack. What is the probability the balls drawn are

a. first red, second blue
b. first blue, second red
c. both white
d. first blue, second not red
e. first white, second not blue
f. first not red, second not blue

60. Consider the 210 discrete points found in the first and second quadrants where $-10 \le x \le 10$, $1 \le y \le 10$, and x and y are integers. The coordinates of each point is written on a slip of paper and placed in a box. One of the slips is then randomly drawn. What is the probability the point (x, y) drawn

a. is on the graph of $y = |x|$
b. is on the graph of $y = 2|x|$
c. is on the graph of $y = 0.5|x|$
d. has coordinates $(x, y > -2)$
e. has coordinates $(x \le 5, y > -2)$
f. is between the branches of $y = x^2$

61. Your instructor surprises you with a True/False quiz for which you are totally unprepared and must guess randomly. What is the probability you pass the quiz with an 80% or better if there are

a. three questions   b. four questions
c. five questions

62. A robot is sent out to disarm a timed explosive device by randomly changing some switches from a neutral position to a *positive flow* or *negative flow* position. The problem is, the switches are independent and unmarked, and it is unknown

which direction is positive and which direction is negative. The bomb is harmless if a majority of the switches yield a positive flow. All switches must be thrown. What is the probability the device is disarmed if there are

a. three switches   b. four switches

c. five switches

63. A survey of 100 retirees was taken to gather information concerning how they viewed the Vietnam War back in the early 1970s. A breakdown of those surveyed is shown in the table. One out of the hundred will be selected at random for a personal, taped interview. What is the probability the person chosen had a

a. career of any kind and opposed the war

b. medical career and supported the war

c. military career and opposed the war

d. legal or business career and opposed the war

e. academic or medical career and supported the war

Career	Support	Opposed	T
Military	9	3	12
Medical	8	16	24
Legal	15	12	27
Business	18	6	24
Academics	3	10	13
Totals	53	47	100

64. Referring to Exercise 63, what is the probability the person chosen

a. had a career of any kind or opposed the war

b. had a medical career or supported the war

c. supported the war or had a military career

d. had a medical or a legal career

e. supported or opposed the war

▶ **EXTENDING THE CONCEPT**

69. The function $f(x) = (\frac{1}{2})^x$ gives the probability that x number of flips will all result in heads (or tails). Compute the probability that 20 flips results in *20 heads in a row,* then use the Internet or some other resource to find the probability of winning a state lottery. Which is more likely to happen (which has the greater probability)? Were you surprised?

70. Recall that a function is a relation in which each element of the domain is paired with only one element from the range. Is the relation defined by $C(x) = {}_nC_x$ (n is a constant) a function? To

65. The Board of Directors for a large hospital has 15 members. There are six doctors of nephrology (kidneys), five doctors of gastroenterology (stomach and intestines), and four doctors of endocrinology (hormones and glands). Eight of them will be selected to visit the nation's premier hospitals on a 3-week, expenses-paid tour. What is the probability the group of eight selected consists of exactly

a. four nephrologists and four gastroenterologists

b. three endocrinologists and five nephrologists

66. A support group for hodophobics (an irrational fear of travel) has 32 members. There are 15 aviophobics (fear of air travel), eight siderodrophobics (fear of train travel), and nine thalassophobics (fear of ocean travel) in the group. Twelve of them will be randomly selected to participate in a new therapy. What is the probability the group of 12 selected consists of exactly

a. two aviophobics, six siderodrophobics, and four thalassophobics

b. five thalassophobics, four aviophobics, and three siderodrophobics

67. A trained chimpanzee is given a box containing eight wooden cubes with the letters p, a, r, a, l, l, e, l printed on them (one letter per block). Assuming the chimp can't read or spell, what is the probability he draws the eight blocks in order and actually forms the word "parallel"?

68. A number is called a "perfect number" if the sum of its proper factors is equal to the number itself. Six is the first perfect number since the sum of its proper factors is six: $1 + 2 + 3 = 6$. Twenty-eight is the second since: $1 + 2 + 4 + 7 + 14 = 28$. A young child is given a box containing eight wooden blocks with the following numbers (one per block) printed on them: four 3's, two 5's, one 0, and one 6. What is the probability she draws the eight blocks in order and forms the fifth perfect number: 33,550,336?

investigate, plot the points generated by $C(x) = {}_6C_x$ for $x = 0$ to $x = 6$ and answer the following questions:

a. Is the resulting graph continuous or discrete (made up of distinct points)?

b. Does the resulting graph pass the vertical line test?

c. Discuss the features of the relation and its graph, including the domain, range, maximum or minimum values, and symmetries observed.

► **MAINTAINING YOUR SKILLS**

71. (5.3) Given $\csc \theta = 3$ and $\cos \theta < 0$, find the values of the remaining five trig functions of θ.

72. (4.4) Complete the following logarithmic properties:

$\log_b b = $ ___ $\log_b 1 = $ ___

$\log_b b^n = $ ___ $b^{\log_b n} = $ ___

73. (6.4) Find exact values for $\sin(2\theta)$, $\cos(2\theta)$, and $\tan(2\theta)$ given $\cos \theta = -\dfrac{21}{29}$ and θ is in Quadrant II.

74. (8.3) A rubber ball is dropped from a height of 25 ft onto a hard surface. With each bounce, it rebounds 60% of the height from which it last fell. Use sequences/series to find (a) the height of the sixth bounce, (b) the total distance traveled up to the sixth bounce, and (c) the distance the ball will travel before coming to rest.

10.7 | The Binomial Theorem

Learning Objectives

In Section 10.7 you will learn how to:

- ☐ **A.** Use Pascal's triangle to find $(a + b)^n$
- ☐ **B.** Find binomial coefficients using $\dbinom{n}{k}$ notation
- ☐ **C.** Use the binomial theorem to find $(a + b)^n$
- ☐ **D.** Find a specific term of a binomial expansion

Strictly speaking, a binomial is a polynomial with two terms. This limits us to terms with real number coefficients and whole number powers on variables. In this section, we will loosely regard a binomial as the sum or difference of *any* two terms. Hence $3x^2 - y^4$, $\sqrt{x} + 4$, $x + \dfrac{1}{x}$, and $-\dfrac{1}{2} + \dfrac{\sqrt{3}}{2}i$ are all "binomials." Our goal is to develop an ability to raise a binomial to any natural number power, with the results having important applications in genetics, probability, polynomial theory, and other areas. The tool used for this purpose is called the *binomial theorem*.

A. Binomial Powers and Pascal's Triangle

Much of our mathematical understanding comes from a study of patterns. One area where the study of patterns has been particularly fruitful is **Pascal's triangle** (Figure 10.20), named after the French scientist Blaise Pascal (although the triangle was well known before his time). It begins with a "1" at the vertex of the triangle, with 1's extending diagonally downward to the left and right as shown. The entries on the interior of the triangle are found by adding the two entries directly above and to the left and right of each new position.

There are a variety of patterns hidden within the triangle. In this section, we'll use the *horizontal rows* of the triangle to help us raise a binomial to various powers. To begin, recall that $(a + b)^0 = 1$ and $(a + b)^1 = 1a + 1b$ (unit coefficients are included for emphasis). In our earlier work, we saw that a binomial square (a binomial raised to the second power) always

Figure 10.20

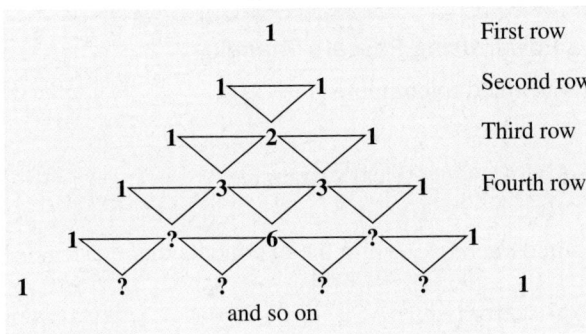

First row
Second row
Third row
Fourth row

and so on

followed the pattern $(a + b)^2 = 1a^2 + 2ab + 1b^2$. Observe the overall pattern that is developing as we include $(a + b)^3$:

$(a + b)^0$	1	row 1
$(a + b)^1$	$1a + 1b$	row 2
$(a + b)^2$	$1a^2 + 2ab + 1b^2$	row 3
$(a + b)^3$	$1a^3 + 3a^2b + 3ab^2 + 1b^3$	row 4

Apparently the coefficients of $(a + b)^n$ will occur in row $n + 1$ of Pascal's triangle. Also observe that in each term of the expansion, the exponent of the first term a *decreases by 1* as the exponent on the second term b *increases by 1*, keeping the degree of each term constant (recall the degree of a term with more than one variable is the sum of the exponents).

$$1a^3b^0 + 3a^2b^1 + 3a^1b^2 + 1a^0b^3$$

$$\begin{array}{cccc} 3 + 0 & 2 + 1 & 1 + 2 & 0 + 3 \\ \text{degree 3} & \text{degree 3} & \text{degree 3} & \text{degree 3} \end{array}$$

These observations help us to quickly expand a binomial power.

EXAMPLE 1 ▶ **Expanding a Binomial Using Pascal's Triangle**

Use Pascal's triangle and the patterns noted to expand $(x + \frac{1}{2})^4$.

Solution ▶ Working step-by-step we have

1. The coefficients will be in the fifth row of Pascal's triangle.

   $$1 \qquad 4 \qquad 6 \qquad 4 \qquad 1$$

2. The exponents on x begin at 4 and *decrease,* while the exponents on $\frac{1}{2}$ begin at 0 and *increase.*

   $$1x^4\left(\frac{1}{2}\right)^0 + 4x^3\left(\frac{1}{2}\right)^1 + 6x^2\left(\frac{1}{2}\right)^2 + 4x^1\left(\frac{1}{2}\right)^3 + 1x^0\left(\frac{1}{2}\right)^4$$

3. Simplify each term.

   The result is $x^4 + 2x^3 + \frac{3}{2}x^2 + \frac{1}{2}x + \frac{1}{16}$.

Now try Exercises 7 through 10 ▶

If the exercise involves a difference rather than a sum, we simply rewrite the expression using algebraic addition and proceed as before.

EXAMPLE 2 ▶ **Raising a Complex Number to a Power Using Pascal's Triangle**

Use Pascal's triangle and the patterns noted to compute $(3 - 2i)^5$.

Solution ▶ Begin by rewriting $(3 - 2i)^5$ as $[3 + (-2i)]^5$.

1. The coefficients will be in the sixth row of Pascal's triangle.

   $$1 \qquad 5 \qquad 10 \qquad 10 \qquad 5 \qquad 1$$

2. The exponents on 3 begin at 5 and *decrease,* while the exponents on $(-2i)$ begin at 0 and *increase.*

$$\mathbf{1}(3^5)(-2i)^0 + \mathbf{5}(3^4)(-2i)^1 + \mathbf{10}(3^3)(-2i)^2 + \mathbf{10}(3^2)(-2i)^3 + \mathbf{5}(3^1)(-2i)^4 + \mathbf{1}(3^0)(-2i)^5$$

3. Simplify each term.

$$243 - 810i - 1080 + 720i + 240 - 32i$$

☑ **A. You've just learned how to use Pascal's triangle to find $(a + b)^n$**

The result is $-597 - 122i$.

Now try Exercises 11 and 12 ▶

> **Expanding Binomial Powers $(a + b)^n$**
>
> 1. The coefficients will be in row $n + 1$ of Pascal's triangle.
> 2. The exponents on *the first term* begin at n and *decrease*, while the exponents on *the second term* begin at 0 and *increase*.
> 3. For any binomial difference $(a - b)^n$, rewrite the base as $[a + (-b)]^n$ using algebraic addition and proceed as before, then simplify each term.

B. Binomial Coefficients and Factorials

Pascal's triangle can easily be used to find the coefficients of $(a + b)^n$, as long as the exponent is relatively small. If we needed to expand $(a + b)^{25}$, writing out the first 26 rows of the triangle would be rather tedious. To overcome this limitation, we introduce a *formula* for the binomial coefficients that enables us to find the coefficients of any expansion.

> **The Binomial Coefficients**
>
> For natural numbers n and r where $n \geq r$, the expression $\binom{n}{r}$, read "n choose r," is called the **binomial coefficient** and evaluated as:
> $$\binom{n}{r} = \frac{n!}{r!(n - r)!}$$

In Example 1, we found the coefficients of $(a + b)^4$ using the fifth or $(n + 1)$st row of Pascal's triangle. In Example 3, these coefficients are found using the formula for binomial coefficients.

EXAMPLE 3 ▶ **Computing Binomial Coefficients**

Evaluate $\binom{n}{r} = \dfrac{n!}{r!(n - r)!}$ as indicated:

a. $\binom{4}{1}$ b. $\binom{4}{2}$ c. $\binom{4}{3}$

Solution ▶

a. $\binom{4}{1} = \dfrac{4!}{1!(4 - 1)!} = \dfrac{4 \cdot 3!}{1!3!} = 4$

b. $\binom{4}{2} = \dfrac{4!}{2!(4 - 2)!} = \dfrac{4 \cdot 3 \cdot 2!}{2!2!} = \dfrac{4 \cdot 3}{2} = 6$

c. $\binom{4}{3} = \dfrac{4!}{3!(4 - 3)!} = \dfrac{4 \cdot 3!}{3!1!} = 4$

Now try Exercises 13 through 20 ▶

Note $\binom{4}{1} = 4$, $\binom{4}{2} = 6$, and $\binom{4}{3} = 4$ give the *interior entries* in the fifth row of Pascal's triangle: 1 4 6 4 1. For consistency and symmetry, we define $0! = 1$, which enables the formula to generate all entries of the triangle, including the "1's."

$\binom{4}{0} = \dfrac{4!}{0!(4 - 0)!}$ apply formula $\binom{4}{4} = \dfrac{4!}{4!(4 - 4)!} = \dfrac{4!}{4! \cdot 0!}$ apply formula

$= \dfrac{4!}{1 \cdot 4!} = 1$ $0! = 1$ $= \dfrac{4!}{4! \cdot 1} = 1$ $0! = 1$

The formula for $\binom{n}{r}$ with $0 \leq r \leq n$ now gives all coefficients in the $(n + 1)$st row. For $n = 5$, we have

$$\binom{5}{0} \quad \binom{5}{1} \quad \binom{5}{2} \quad \binom{5}{3} \quad \binom{5}{4} \quad \binom{5}{5}$$

$$1 \qquad 5 \qquad 10 \qquad 10 \qquad 5 \qquad 1$$

EXAMPLE 4 ▶ **Computing Binomial Coefficients**

Compute the binomial coefficients:

a. $\binom{9}{0}$ **b.** $\binom{9}{1}$ **c.** $\binom{6}{5}$ **d.** $\binom{6}{6}$

Solution ▶ **a.** $\binom{9}{0} = \dfrac{9!}{0!(9 - 0)!}$ **b.** $\binom{9}{1} = \dfrac{9!}{1!(9 - 1)!}$

$\qquad\qquad = \dfrac{9!}{9!} = 1$ $\qquad\qquad = \dfrac{9!}{8!} = 9$

c. $\binom{6}{5} = \dfrac{6!}{5!(6 - 5)!}$ **d.** $\binom{6}{6} = \dfrac{6!}{6!(6 - 6)!}$

$\qquad = \dfrac{6!}{5!} = 6$ $\qquad = \dfrac{6!}{6!} = 1$

Now try Exercises 21 through 24 ▶

You may have noticed that the formula for $\binom{n}{r}$ is identical to that of $_nC_r$, and both yield like results for given values of n and r. For future use, it will help to commit the general results from Example 4 to memory: $\binom{n}{0} = 1$, $\binom{n}{1} = n$, $\binom{n}{n-1} = n$, and $\binom{n}{n} = 1$.

☑ **B.** You've just learned how to find binomial coefficients using $\binom{n}{k}$ notation

C. The Binomial Theorem

Using $\binom{n}{r}$ notation and the observations made regarding binomial powers, we can now state the **binomial theorem.**

Binomial Theorem

For any binomial $(a + b)$ and natural number n,

$$(a + b)^n = \binom{n}{0}a^nb^0 + \binom{n}{1}a^{n-1}b^1 + \binom{n}{2}a^{n-2}b^2 + \cdots$$

$$+ \binom{n}{n-1}a^1b^{n-1} + \binom{n}{n}a^0b^n$$

The theorem can also be stated in summation form as

$$(a + b)^n = \sum_{r=0}^{n} \binom{n}{r}a^{n-r}b^r$$

The expansion actually looks overly impressive in this form, and it helps to summarize the process in words, as we did earlier. The exponents on the first term a begin at n and decrease, while the exponents on the second term b begin at 0 and increase, keeping the degree of each term constant. The $\binom{n}{r}$ notation simply gives the coefficients of each term. As a final note, observe that the r in $\binom{n}{r}$ gives the exponent on b.

EXAMPLE 5 ▶ **Expanding a Binomial Using the Binomial Theorem**

Expand $(a + b)^6$ using the binomial theorem.

Solution ▶ $(a + b)^6 = \binom{6}{0}a^6b^0 + \binom{6}{1}a^5b^1 + \binom{6}{2}a^4b^2 + \binom{6}{3}a^3b^3 + \binom{6}{4}a^2b^4 + \binom{6}{5}a^1b^5 + \binom{6}{6}a^0b^6$

$= \dfrac{6!}{0!6!}a^6 + \dfrac{6!}{1!5!}a^5b^1 + \dfrac{6!}{2!4!}a^4b^2 + \dfrac{6!}{3!3!}a^3b^3 + \dfrac{6!}{4!2!}a^2b^4 + \dfrac{6!}{5!1!}a^1b^5 + \dfrac{6!}{6!0!}b^6$

$= 1a^6 + 6a^5b + 15a^4b^2 + 20a^3b^3 + 15a^2b^4 + 6ab^5 + 1b^6$

Now try Exercises 25 through 32 ▶

EXAMPLE 6 ▶ **Using the Binomial Theorem to Find the Initial Terms of an Expansion**

Find the first three terms of $(2x + y^2)^{10}$.

Solution ▶ Use the binomial theorem with $a = 2x$, $b = y^2$, and $n = 10$.

$(2x + y^2)^{10} = \binom{10}{0}(2x)^{10}(y^2)^0 + \binom{10}{1}(2x)^9(y^2)^1 + \binom{10}{2}(2x)^8(y^2)^2 + \cdots$ first three terms

$= (1)1024x^{10} + (10)512x^9y^2 + \dfrac{10!}{2!8!}256x^8y^4 + \cdots$ $\binom{10}{0} = 1, \binom{10}{1} = 10$

☑ **C. You've just learned how to use the binomial theorem to find $(a + b)^n$**

$= 1024x^{10} + 5120x^9y^2 + (45)256x^8y^4 + \cdots$ $\dfrac{10!}{2!8!} = 45$

$= 1024x^{10} + 5120x^9y^2 + 11{,}520x^8y^4 + \cdots$ result

Now try Exercises 33 through 36 ▶

D. Finding a Specific Term of the Binomial Expansion

In some applications of the binomial theorem, our main interest is a *specific term* of the expansion, rather than the expansion as a whole. To find a specified term, it helps to consider that the expansion of $(a + b)^n$ has $n + 1$ terms: $(a + b)^0$ has one term, $(a + b)^1$ has two terms, $(a + b)^2$ has three terms, and so on. Because the notation $\binom{n}{r}$ always begins at $r = 0$ for the first term, the value of r will be *1 less than the term we are seeking*. In other words, for the seventh term of $(a + b)^9$, we use $r = 6$.

The kth Term of a Binomial Expansion

For the binomial expansion $(a + b)^n$, the kth term is given by

$$\binom{n}{r}a^{n-r}b^r, \text{ where } r = k - 1.$$

EXAMPLE 7 ▶ **Finding a Specific Term of a Binomial Expansion**

Find the eighth term in the expansion of $(x + 2y)^{12}$.

Solution ▶ By comparing $(x + 2y)^{12}$ to $(a + b)^n$ we have $a = x$, $b = 2y$, and $n = 12$. Since we want the eighth term, $k = 8 \rightarrow r = 7$. The eighth term of the expansion is

$$\binom{12}{7}x^5(2y)^7 = \frac{12!}{7!5!}128x^5y^7 \qquad 2^7 = 128$$

$$= (792)(128x^5y^7) \qquad \binom{12}{7} = 792$$

$$= 101{,}376x^5y^7 \qquad \text{result}$$

Now try Exercises 37 through 42 ▶

One application of the binomial theorem involves a **binomial experiment** and **binomial probability.** For binomial probabilities, the following must be true: (1) The experiment must have only two possible outcomes, typically called success and failure, and (2) if the experiment has n trials, the probability of success must be constant for all n trials. If the probability of success for each trial is p, the formula $\binom{n}{k}(1 - p)^{n-k}p^k$ gives the probability that exactly k trials will be successful.

Binomial Probability

Given a binomial experiment with n trials, where the probability for success in each trial is p. The probability that exactly k trials are successful is given by

$$\binom{n}{k}(1 - p)^{n-k}p^k.$$

EXAMPLE 8 ▶ **Applying the Binomial Theorem–Binomial Probability**

Paula Rodrigues has a free-throw shooting average of 85%. On the last play of the game, with her team behind by three points, she is fouled at the three-point line, and is awarded two additional free throws via technical fouls on the opposing coach (a total of five free-throws). What is the probability she makes *at least three* (meaning they at least tie the game)?

Solution ▶ Here we have $p = 0.85$, $1 - p = 0.15$, and $n = 5$. The key idea is to recognize the phrase *at least three* means "3 or 4 or 5." So $P(\text{at least } 3) = P(3 \cup 4 \cup 5)$.

$$P(\text{at least } 3) = P(3 \cup 4 \cup 5) \qquad \text{"or" implies a union}$$

$$= P(3) + P(4) + P(5) \qquad \text{sum of probabilities (mutually exclusive events)}$$

$$= \binom{5}{3}(0.15)^2(0.85)^3 + \binom{5}{4}(0.15)^1(0.85)^4 + \binom{5}{5}(0.15)^0(0.85)^5$$

$$\approx 0.1382 + 0.3915 + 0.4437$$

$$= 0.9734$$

Paula's team has an excellent chance ($\approx 97.3\%$) of at least tying the game.

☑ **D.** You've just learned how to find a specific term of a binomial expansion

New try Exercises 45 and 46 ▶

10.7 EXERCISES

▶ CONCEPTS AND VOCABULARY

Fill in the blank with the appropriate word or phrase. Carefully reread the section if needed.

1. In any binomial expansion, there is always _____ more term than the power applied.

2. In all terms in the expanded form of $(a + b)^n$, the exponents on a and b must sum to _____.

3. To expand a binomial *difference* such as $(a - 2b)^5$, we rewrite the binomial as _____ and proceed as before.

4. In a binomial experiment with n trials, the probability there are exactly k successes is given by the formula _____.

5. Discuss why the expansion of $(a + b)^n$ has $n + 1$ terms.

6. For any defined binomial experiment, discuss the relationships between the phrases, "exactly k success," and "at least k successes."

▶ DEVELOPING YOUR SKILLS

Use *Pascal's triangle* and the patterns explored to write each expansion.

7. $(x + y)^5$

8. $(a + b)^6$

9. $(2x + 3)^4$

10. $(x^2 + \frac{1}{3})^3$

11. $(1 - 2i)^5$

12. $(2 - 5i)^4$

Evaluate each of the following

13. $\binom{7}{4}$

14. $\binom{8}{2}$

15. $\binom{5}{3}$

16. $\binom{9}{5}$

17. $\binom{20}{17}$

18. $\binom{30}{26}$

19. $\binom{40}{3}$

20. $\binom{45}{3}$

21. $\binom{6}{0}$

22. $\binom{5}{0}$

23. $\binom{15}{15}$

24. $\binom{10}{10}$

Use the *binomial theorem* to expand each expression. Write the general form first, then simplify.

25. $(c + d)^5$

26. $(v + w)^4$

27. $(a - b)^6$

28. $(x - y)^7$

29. $(2x - 3)^4$

30. $(a - 2b)^5$

31. $(1 - 2i)^3$

32. $(2 + \sqrt{3}i)^5$

Use the *binomial theorem* to write the first three terms.

33. $(x + 2y)^9$

34. $(3p + q)^8$

35. $(v^2 - \frac{1}{2}w)^{12}$

36. $(\frac{1}{2}a - b^2)^{10}$

Find the indicated term for each binomial expansion.

37. $(x + y)^7$; 4th term

38. $(m + n)^6$; 5th term

39. $(p - 2)^8$; 7th term

40. $(a - 3)^{14}$; 10th term

41. $(2x + y)^{12}$; 11th term

42. $(3n + m)^9$; 6th term

▶ WORKING WITH FORMULAS

43. **Binomial probability:** $P(k) = \binom{n}{k}\left(\frac{1}{2}\right)^k \left(\frac{1}{2}\right)^{n-k}$

    The theoretical probability of getting exactly k heads in n flips of a fair coin is given by the formula above. What is the probability that you would get 5 heads in 10 flips of the coin?

44. **Binomial probability:** $P(k) = \binom{n}{k}\left(\frac{1}{5}\right)^k \left(\frac{4}{5}\right)^{n-k}$

    A multiple choice test has five options per question. The probability of guessing correctly k times out of n questions is found using the formula shown. What is the probability a person scores a 70% by guessing randomly (7 out of 10 questions correct)?

► **APPLICATIONS**

45. Batting averages: Tony Gwynn (San Diego Padres) had a lifetime batting average of 0.347, ranking him as one of the greatest hitters of all time. Suppose he came to bat five times in any given game.

  a. What is the probability that he will get exactly three hits?

  b. What is the probability that he will get at least three hits?

46. Pollution testing: Erin suspects that a nearby iron smelter is contaminating the drinking water over a large area. A statistical study reveals that 83% of the wells in this area are likely contaminated. If the figure is accurate, find the probability that if another 10 wells are tested

  a. exactly 8 are contaminated

  b. at least 8 are contaminated

47. Late rental returns: The manager of Victor's DVD Rentals knows that 6% of all DVDs rented are returned *late*. Of the eight videos rented in the last hour, what is the probability that

  a. exactly five are returned on time

  b. exactly six are returned on time

  c. at least six are returned on time

  d. none of them will be returned late

48. Opinion polls: From past experience, a research firm knows that 20% of telephone respondents will agree to answer an opinion poll. If 20 people are contacted by phone, what is the probability that

  a. exactly 18 refuse to be polled

  b. exactly 19 refuse to be polled

  c. at least 18 refuse to be polled

  d. none of them agree to be polled

► **EXTENDING THE CONCEPT**

49. Prior to calculators and computers, the binomial theorem was frequently used to approximate the value of compound interest given by the expression $\left(1 + \dfrac{r}{n}\right)^{nt}$ by expanding the first three terms. For example, if the interest rate were 8% ($r = 0.08$) and the interest was compounded quarterly ($n = 4$) for 5 yr ($t = 5$), we have $\left(1 + \frac{0.08}{4}\right)^{(4)(5)} = (1 + 0.02)^{20}$. The first three terms of the expansion give a value of: $1 + 20(0.02) + 190(0.0004) = 1.476$.

  a. Calculate the percent error:
  $$\%\,\text{error} = \frac{\text{approximate value}}{\text{actual value}}$$

  b. What is the percent error if only two terms are used.

50. If you sum the entries in each row of Pascal's triangle, a pattern emerges. Find a formula that generalizes the result for any row of the triangle, and use it to find the sum of the entries in the 12th row of the triangle.

51. Show that $\dbinom{n}{k} = \dbinom{n}{n-k}$ for $n = 6$ and $k \leq 6$.

52. The *derived polynomial* of $f(x)$ is $f(x + h)$ or the original polynomial evaluated at $x + h$. Use Pascal's triangle or the binomial theorem to find the derived polynomial for $f(x) = x^3 + 3x^2 + 5x - 11$. Simplify the result completely.

► **MAINTAINING YOUR SKILLS**

53. (2.7) Graph the function shown and find
$f(3)$: $f(x) = \begin{cases} x + 2 & x \leq 2 \\ (x - 4)^2 & x > 2 \end{cases}$

54. (5.4) Given the point $(-0.6, y)$ is a point on the unit circle in the third quadrant, find y.

55. (3.4) Graph the function $g(x) = x^3 - x^2 - 6x$. Clearly indicate all intercepts and intervals where $g(x) > 0$.

56. (6.5) Evaluate $\arcsin\left[\sin\left(\dfrac{5\pi}{6}\right)\right]$.

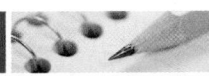

SUMMARY AND CONCEPT REVIEW

SECTION 10.1   Sequences and Series

KEY CONCEPTS

- A *finite sequence* is a function a_n whose domain is the set of natural numbers from 1 to n.
- The terms of the sequence are labeled $a_1, a_2, a_3, \ldots, a_{k-1}, a_k, a_{k+1}, \ldots, a_{n-2}, a_{n-1}, a_n$.
- The expression a_n, which defines the sequence (generates the terms in order), is called the nth term.
- An *infinite sequence* is a function whose domain is the set of natural numbers.
- When each term of a sequence is larger than the preceding term, it is called an increasing sequence.
- When each term of a sequence is smaller than the preceding term, it is called a decreasing sequence.
- When successive terms of a sequence alternate in sign, it is called an alternating sequence.
- When the terms of a sequence are generated using previous term(s), it is called a recursive sequence.
- Sequences are sometimes defined using factorials, which are the product of a given natural number with all natural numbers that precede it: $n! = n \cdot (n-1) \cdot (n-2) \cdot \cdots \cdot 3 \cdot 2 \cdot 1$.
- Given the sequence $a_1, a_2, a_3, a_4, \ldots, a_n$ the sum is called a finite series and is denoted S_n.
- $S_n = a_1 + a_2 + a_3 + a_4 + \cdots + a_n$. The sum of the first n terms is called a partial sum.
- In sigma notation, the expression $\displaystyle\sum_{i=1}^{k} a_i = a_1 + a_2 + \cdots + a_k$ represents a finite series.
- When sigma notation is used, the letter "i" is called the index of summation.

EXERCISES

Write the first four terms that are defined and the value of a_{10}.

1. $a_n = 5n - 4$

2. $a_n = \dfrac{n+1}{n^2+1}$

Find the general term a_n for each sequence, and the value of a_6.

3. $1, 16, 81, 256, \ldots$

4. $-17, -14, -11, -8, \ldots$

Find the eighth partial sum (S_8).

5. $\frac{1}{2}, \frac{1}{4}, \frac{1}{8}, \ldots$

6. $-21, -19, -17, \ldots$

Evaluate each sum.

7. $\displaystyle\sum_{n=1}^{7} n^2$

8. $\displaystyle\sum_{n=1}^{5} (3n - 2)$

Write the first five terms that are defined.

9. $a_n = \dfrac{n!}{(n-2)!}$

10. $\begin{cases} a_1 = \frac{1}{2} \\ a_{n+1} = 2a_n - \frac{1}{4} \end{cases}$

Write as a single summation and evaluate.

11. $\displaystyle\sum_{n=1}^{7} n^2 + \sum_{n=1}^{7} (3n - 2)$

SECTION 10.2   Arithmetic Sequences

KEY CONCEPTS

- In an arithmetic sequence, successive terms are found by adding a fixed constant to the preceding term.
- In a sequence, if there exists a number d, called the common difference, such that $a_{k+1} - a_k = d$, then the sequence is arithmetic. Alternatively, $a_{k+1} = a_k + d$ for $k \geq 1$.

- The nth term n of an arithmetic sequence is given by $a_n = a_1 + (n - 1)d$, where a_1 is the first term and d is the common difference.
- If the initial term is unknown or is not a_1 the nth term can be written $a_n = a_k + (n - k)d$, where the subscript of the term a_k and the coefficient of d sum to n.
- For an arithmetic sequence with first term a_1, the nth partial sum (the sum of the first n terms) is given by
$$S_n = \frac{n(a_1 + a_n)}{2}.$$

EXERCISES

Find the general term (a_n) for each arithmetic sequence. Then find the indicated term.

12. $2, 5, 8, 11, \ldots$; find a_{40}

13. $3, 1, -1, -3, \ldots$; find a_{35}

Find the sum of each series.

14. $-1 + 3 + 7 + 11 + \cdots + 75$

15. $1 + 4 + 7 + 10 + \cdots + 88$

16. $3 + 6 + 9 + 12 + \cdots$; S_{20}

17. $1 + \frac{3}{4} + \frac{1}{2} + \frac{1}{4} + \cdots$; S_{15}

18. $\sum_{n=1}^{25} (3n - 4)$

19. $\sum_{n=1}^{40} (4n - 1)$

SECTION 10.3 Geometric Sequences

KEY CONCEPTS

- In a geometric sequence, successive terms are found by multiplying the preceding term by a nonzero constant.
- In other words, if there exists a number r, called the common ratio, such that $\frac{a_{k+1}}{a_k} = r$, then the sequence is geometric. Alternatively, we can write $a_{k+1} = a_k r$ for $k \geq 1$.
- The nth term a_n of a geometric sequence is given by $a_n = a_1 r^{n-1}$, where a_1 is the first term and a_n represents the general term of a finite sequence.
- If the initial term is unknown or is not a_1, the nth term can be written $a_n = a_k r^{n-k}$, where the subscript of the term a_k and the exponent on r sum to n.
- The nth partial sum of a geometric sequence is $S_n = \dfrac{a_1(1 - r^n)}{1 - r}$.
- If $|r| < 1$, the sum of an infinite geometric series is $S_\infty = \dfrac{a_1}{1 - r}$.

EXERCISES

Find the indicated term for each geometric sequence.

20. $a_1 = 5, r = 3$; find a_7

21. $a_1 = 4, r = \sqrt{2}$; find a_7

22. $a_1 = \sqrt{7}, r = \sqrt{7}$; find a_8

Find the indicated sum, if it exists.

23. $16 - 8 + 4 - \cdots$ find S_7

24. $2 + 6 + 18 + \cdots$; find S_8

25. $\frac{4}{5} + \frac{2}{5} + \frac{1}{5} + \frac{1}{10} + \cdots$; find S_{12}

26. $4 + 8 + 12 + 24 + \cdots$

27. $5 + 0.5 + 0.05 + 0.005 + \cdots$

28. $6 - 3 + \frac{3}{2} - \frac{3}{4} + \cdots$

29. $\sum_{n=1}^{8} 5\left(\frac{2}{3}\right)^n$

30. $\sum_{n=1}^{\infty} 12\left(\frac{4}{3}\right)^n$

31. $\sum_{n=1}^{\infty} 5\left(\frac{1}{2}\right)^n$

32. Charlene began to work for Grayson Natural Gas in January of 1990 with an annual salary of $26,000. Her contract calls for a $1220 raise each year. Use a sequence/series to compute her salary after nine years, and her total earnings up to and including that year. (*Hint:* For $a_1 = 26,000$, her salary after *9 yrs* will be what term of the sequence?)

33. Sumpter reservoir contains 121,500 ft³ of water and is being drained in the following way. Each day one-third of the water is *drained* (and not replaced). Use a sequence/series to compute how much water *remains in the pond* after 7 days.

34. Credit-hours taught at Cody Community College have been increasing at 7% per year since it opened in 2000 and taught 1225 credit-hours. For the new faculty, the college needs to predict the number of credit-hours that will be taught in 2009. Use a sequence/series to compute the credit-hours for 2009 and to find the total number of credit hours taught through the 2009 school year.

SECTION 10.4 Mathematical Induction

KEY CONCEPTS

- Functions written in subscript notation can be evaluated, graphed, and composed with other functions.
- A sum formula involving only natural numbers n as inputs can be proven valid using a proof by induction. Given that S_n represents a sum formula involving natural numbers, if (1) S_1 is true and (2) $S_k + a_{k+1} = S_{k+1}$, then S_n must be true for all natural numbers.
- Proof by induction can also be used to validate other relationships, using a more general statement of the principle. Let S_n be a statement involving the natural numbers n. If (1) S_1 is true (S_n for $n = 1$) and (2) the truth of S_k implies that S_{k+1} is also true, then S_n must be true for all natural numbers n.

EXERCISES

Use the principle of mathematical induction to prove the indicated sum formula is true for all natural numbers n.

35. $1 + 2 + 3 + 4 + 5 + \cdots + n$;
$a_n = n$ and $S_n = \dfrac{n(n+1)}{2}$.

36. $1 + 4 + 9 + 16 + 25 + 36 + \cdots + n^2$;
$a_n = n^2$ and $S_n = \dfrac{n(n+1)(2n+1)}{6}$.

Use the principle of mathematical induction to prove that each statement is true for all natural numbers n.

37. $4^n \geq 3n + 1$

38. $6 \cdot 7^{n-1} \leq 7^n - 1$

39. $3^n - 1$ is divisible by 2

SECTION 10.5 Counting Techniques

KEY CONCEPTS

- An experiment is any task that can be repeated and has a well-defined set of possible outcomes.
- Each repetition of an experiment is called a trial.
- Any potential outcome of an experiment is called a sample outcome.
- The set of all sample outcomes is called the sample space.
- An experiment with N (equally likely) sample outcomes that is repeated t times, has a sample space with N^t elements.
- If a sample outcome can be used more than once, the counting is said to be with repetition. If a sample outcome can be used only once the counting is said to be without repetition.
- The fundamental principle of counting states: If there are p possibilities for a first task, q possibilities for the second, and r possibilities for the third, the total number of ways the experiment can be completed is pqr. This fundamental principle can be extended to include any number of tasks.
- If the elements of a sample space have precedence or priority (order or rank is important), the number of elements is counted using a permutation, denoted $_nP_r$ and read, "the distinguishable permutations of n objects taken r at a time."
- To expand $_nP_r$, we can write out the first r factors of $n!$ or use the formula $_nP_r = \dfrac{n!}{(n-r)!}$.
- If any of the sample outcomes are identical, certain permutations will be nondistinguishable. In a set containing n elements where one element is repeated p times, another is repeated q times, and another r times ($p + q + r \leq n$), the number of distinguishable permutations is given by $\dfrac{_nP_n}{p!q!r!} = \dfrac{n!}{p!q!r!}$.

- If the elements of a set have no rank, order, or precedence (as in a committee of colleagues) permutations with the same elements are considered identical. The result is the number of combinations, $_nC_r = \dfrac{n!}{r!(n-r)!}$.

EXERCISES

40. Three slips of paper with the letters A, B, and C are placed in a box and randomly drawn one at a time. Show all possible ways they can be drawn using a tree diagram.

41. The combination for a certain bicycle lock consists of three digits. How many combinations are possible if (a) repetition of digits is not allowed and (b) repetition of digits is allowed.

42. Jethro has three work shirts, four pairs of work pants, and two pairs of work shoes. How many different ways can he dress himself (shirt, pants, shoes) for a day's work?

43. From a field of 12 contestants in a pet show, three cats are chosen at random to be photographed for a publicity poster. In how many different ways can the cats be chosen?

44. How many subsets can be formed from the elements of this set: { ■, ⊡, ☐, ◼, ▣ }?

45. Compute the following values by hand, showing all work:

    a. $7!$ **b.** $_7P_4$ **c.** $_7C_4$

46. Six horses are competing in a race at the McClintock Ranch. Assuming there are no ties, (a) how many different ways can the horses finish the race? (b) How many different ways can the horses finish first, second, and third place? (c) How many finishes are possible if it is well known that Nellie-the-Nag will finish last and Sea Biscuit will finish first?

47. How many distinguishable permutations can be formed from the letters in the word "tomorrow"?

48. Quality Construction Company has 12 equally talented employees. (a) How many ways can a three-member crew be formed to complete a small job? (b) If the company is in need of a Foreman, Assistant Foreman, and Crew Chief, in how many ways can the positions be filled?

SECTION 10.6   Introduction to Probability

KEY CONCEPTS

- An event E is any designated set of sample outcomes.
- Given S is a sample space of equally likely sample outcomes and E is an event relative to S, the probability of E, written $P(E)$, is computed as $P(E) = \dfrac{n(E)}{n(S)}$, where $n(E)$ represents the number of elements in E, and $n(S)$ represents the number of elements in S.
- The complement of an event E is the set of sample outcomes in S, but not in E and is denoted $\sim E$.
- Given sample space S and any event E defined relative to S:
  (1) $P(\sim S) = 0$,   (2) $0 \le P(E) \le 1$,   (3) $P(S) = 1$,   (4) $P(E) = 1 - P(\sim E)$, and
  (5) $P(E) + P(\sim E) = 1$.
- Two events that have no outcomes in common are said to be mutually exclusive.
- If two events are not mutually exclusive, $P(E_1 \text{ or } E_2) \rightarrow P(E_1 \cup E_2) = P(E_1) + P(E_2) - P(E_1 \cap E_2)$.
- If two events are mutually exclusive, $P(E_1 \text{ or } E_2) \rightarrow P(E_1 \cup E_2) = P(E_1) + P(E_2)$.

EXERCISES

49. One card is drawn from a standard deck. What is the probability the card is a ten or a face card?

50. One card is drawn from a standard deck. What is the probability the card is a Queen or a face card?

51. One die is rolled. What is the probability the result is not a three?

52. Given $P(E_1) = \frac{3}{8}$, $P(E_2) = \frac{3}{4}$, and $P(E_1 \cup E_2) = \frac{5}{6}$, compute $P(E_1 \cap E_2)$.

53. Find $P(E)$ given that $n(E) = {_7C_4} \cdot {_5C_3}$ and $n(S) = {_{12}C_7}$.

54. To determine if more physicians should be hired, a medical clinic tracks the number of days between a patient's request for an appointment and the actual appointment date. The table given shows the probability that a patient must wait "d" days. Based on the table, what is the probability a patient must wait

Wait (days d)	Probability
0	0.002
$0 < d < 10$	0.07
$10 \leq d < 20$	0.32
$20 \leq d < 30$	0.43
$30 \leq d < 40$	0.178

a. at least 20 days   **b.** less than 20 days

c. 40 days or less   **d.** over 40 days

e. less than 40 and more than 10 days   **f.** 30 or more days

SECTION 10.7 The Binomial Theorem

KEY CONCEPTS

- To expand $(a + b)^n$ for n of "moderate size," we can use Pascal's triangle and observed patterns.
- For any natural numbers n and r, where $n \geq r$, the expression $\binom{n}{r}$ (read "n choose r") is called the *binomial coefficient* and evaluated as $\binom{n}{r} = \dfrac{n!}{r!(n-r)!}$.
- If n is large, it is more efficient to expand using the binomial coefficients and binomial theorem.
- The following binomial coefficients are useful/common and should be committed to memory:

$$\binom{n}{0} = 1 \quad \binom{n}{1} = n \quad \binom{n}{n-1} = n \quad \binom{n}{n} = 1$$

- We define $0! = 1$; for example $\binom{n}{n} = \dfrac{n!}{n!(n-n)!} = \dfrac{1}{0!} = \dfrac{1}{1} = 1$.
- The binomial theorem: $(a+b)^n = \binom{n}{0}a^n b^0 + \binom{n}{1}a^{n-1}b^1 + \binom{n}{2}a^{n-2}b^2 + \cdots + \binom{n}{n-1}a^1 b^{n-1} + \binom{n}{n}a^0 b^n$.
- The kth term of $(a+b)^n$ can be found using the formula $\binom{n}{r}a^{n-r}b^r$, where $r = k - 1$.

EXERCISES

55. Evaluate each of the following:

    a. $\binom{7}{5}$ **b.** $\binom{8}{3}$

56. Use Pascal's triangle to expand the binomials:

    a. $(x - y)^4$ **b.** $(1 + 2i)^5$

Use the binomial theorem to:

57. Write the first four terms of

    a. $(a + \sqrt{3})^8$ **b.** $(5a + 2b)^7$

58. Find the indicated term of each expansion.

    a. $(x + 2y)^7$; fourth   **b.** $(2a - b)^{14}$; 10th

MIXED REVIEW

1. Identify each sequence as arithmetic, geometric, or neither. If neither, try to identify the pattern that forms the sequence.

    a. 120, 163, 206, 249, . . .

    b. 4, 4, 4, 4, 4, 4, . . .

    c. 1, 2, 6, 24, 120, 720, 5040, . . .

    d. 2.00, 1.95, 1.90, 1.85, . . .

    e. $\frac{5}{8}, \frac{5}{64}, \frac{5}{512}, \frac{5}{4096}, \cdots$

    f. $-5.5, 6.05, -6.655, 7.3205, \ldots$

    g. $0.\overline{1}, 0.\overline{2}, 0.\overline{3}, 0.\overline{4}, \ldots$

    h. 525, 551.25, 578.8125, . . .

    i. $\frac{1}{2}, \frac{1}{4}, \frac{1}{6}, \frac{1}{8}, \cdots$

2. Compute by hand (show your work).

a. $10!$ **b.** $\dfrac{10!}{6!}$ **c.** $_{10}P_4$

d. $_{10}P_9$ **e.** $_{10}C_6$ **f.** $_{10}C_4$

3. The call letters for a television station must consist of four letters and begin with either a K or a W. How many distinct call letters are possible if repeating any letter is not allowed?

4. Given $a_1 = 9$ and $r = \frac{1}{3}$, write out the first five terms and the 15th term.

5. Given $a_1 = 0.1$ and $r = 5$, write out the first five terms and the 15th term.

6. One card is drawn from a well-shuffled deck of standard cards. What is the probability the card is a Queen or an Ace?

7. Two fair dice are rolled. What is the probability the result is not doubles (doubles $=$ same number on both die)?

8. A house in a Boston suburb cost $185,000 in 1985. Each year its value increased by 8%. If this appreciation were to continue, find the value of the house in 2005 and 2015 using a sequence.

9. Evaluate each sum using summation formulas.

a. $\displaystyle\sum_{n=1}^{\infty}\left(\frac{2}{3}\right)^n$ **b.** $\displaystyle\sum_{n=1}^{10}(9 + 2n)$

c. $\displaystyle\sum_{n=1}^{5} 12n + \sum_{n=1}^{5}(-5) + \sum_{n=1}^{5} n^2$

10. Expand each binomial using the binomial theorem. Simplify each term.

a. $(2x + 5)^5$ **b.** $(1 - 2i)^4$

11. For $(a + b)^n$, determine:

a. the first three terms for $n = 20$

b. the last three terms for $n = 20$

c. the fifth term for $n = 35$

d. the fifth term for $n = 35$, $a = 0.2$, and $b = 0.8$

12. On average, bears older than 3 yr old increase their weight by 0.87% per day from July to November. If a bear weighed 110 kg on June 30th: (a) identify the type of sequence that gives the bear's weight each day; (b) find the general term for the sequence; and (c) find the bear's weight on July 1, July 2, July 3, July 31, August 31, and September 30.

13. Use a proof by induction to show that
$$3 + 6 + 9 + \cdots + 3n = \frac{3n(n + 1)}{2}.$$

14. The owner of an arts and crafts store makes specialty key rings by placing five colored beads on a nylon cord and tying it to the ring that will hold the keys. If there are eight different colors to choose from, (a) how many distinguishable key rings are possible if no colors are repeated? (b) How many distinguishable key rings are possible if a repetition of colors is allowed?

15. Donell bought 15 raffle tickets from the Inner City Children's Music School, and five tickets from the Arbor Day Everyday raffle. The Music School sold a total of 2000 tickets and the Arbor Day foundation sold 550 tickets. For E_1: Donell wins the Music School raffle and E_2: Donell wins the Arbor Day raffle, find $P(E_1 \text{ or } E_2)$.

Find the sum if it exists.

16. $\frac{1}{3} + \frac{2}{3} + 1 + \frac{4}{3} + \cdots + \frac{20}{3}$

17. $0.36 + 0.0036 + 0.000036 + 0.00000036 + \cdots$

Find the first five terms of the sequences in Exercises 18 and 19.

18. $a_n = \dfrac{12!}{(12 - n)!}$ **19.** $\begin{cases} a_1 = 10 \\ a_{n+1} = a_n(\frac{1}{5}) \end{cases}$

20. A random survey of 200 college students produces the data shown. One student from this group is randomly chosen for an interview. Use the data to find

a. P(student works more than 10 hr)

b. P(student takes less than 13 credit-hours)

c. P(student works more than 20 hr and takes more than 12 credit-hours)

d. P(student works between 11 and 20 hr or takes 6 to 12 credit-hours)

	0-10 hr	11-20 hr	Over 20 hr	Total
1-5 credits	3	7	10	20
6-12 credits	21	55	48	124
over 13 credits	8	28	20	56
Total:	32	90	78	200

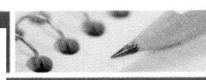

PRACTICE TEST

1. The general term of a sequence is given. Find the first four terms, the 8th term, and the 12th term.

a. $a_n = \dfrac{2n}{n+3}$ 　　　　**b.** $a_n = \dfrac{(n+2)!}{n!}$

c. $a_n = \begin{cases} a_1 = 3 \\ a_{n+1} = \sqrt{(a_n)^2 - 1} \end{cases}$

2. Expand each series and evaluate.

a. $\displaystyle\sum_{k=2}^{6} (2k^2 - 3)$ 　　　**b.** $\displaystyle\sum_{j=2}^{6} (-1)^j \left(\dfrac{j}{j+1}\right)$

c. $\displaystyle\sum_{j=1}^{5} (-2)\left(\dfrac{3}{4}\right)^j$ 　　　**d.** $\displaystyle\sum_{k=1}^{\infty} 7\left(\dfrac{1}{2}\right)^k$

3. Identify the first term and the common difference or common ratio. Then find the general term a_n.

a. $7, 4, 1, -2, \ldots$ 　　　　**b.** $-8, -6, -4, -2, \ldots$

c. $4, -8, 16, -32, \ldots$ 　　　**d.** $10, 4, \frac{8}{5}, \frac{16}{25}, \ldots$

4. Find the indicated value for each sequence.

a. $a_1 = 4, d = 5$; find a_{40}

b. $a_1 = 2, a_n = -22, d = -3$; find n

c. $a_1 = 24, r = \frac{1}{2}$; find a_6

d. $a_1 = -2, a_n = 486, r = -3$; find n

5. Find the sum of each series.

a. $7 + 10 + 13 + \cdots + 100$

b. $\displaystyle\sum_{k=1}^{37} (3k + 2)$

c. For $4 - 12 + 36 - 108 + \cdots$, find S_7

d. $6 + 3 + \frac{3}{2} + \frac{3}{4} + \cdots$

6. Each swing of a pendulum (in one direction) is 95% of the previous one. If the first swing is 12 ft, (a) find the length of the seventh swing and (b) determine the distance traveled by the pendulum for the first seven swings.

7. A rare coin that cost $3000 appreciates in value 7% per year. Find the value of after 12 yr.

8. A car that costs $50,000 decreases in value by 15% per year. Find the value of the car after 5 yr.

9. Use mathematical induction to prove that for $a_n = 5n - 3$, the sum formula $S_n = \dfrac{5n^2 - n}{2}$ is true for all natural numbers n.

10. Use the principle of mathematical induction to prove that S_n: $2 \cdot 3^{n-1} \leq 3^n - 1$ is true for all natural numbers n.

11. Three colored balls (Aqua, Brown, and Creme) are to be drawn without replacement from a bag. List all possible ways they can be drawn using (a) a tree diagram and (b) an organized list.

12. Suppose that license plates for motorcycles must consist of three numbers followed by two letters. How many license plates are possible if zero and "Z" cannot be used and no repetition is allowed?

13. How many subsets can be formed from the elements in this set: $\{\,\square, \square, \square, \square, \square, \square\,\}$.

14. Compute the following values by hand, showing all work: (a) 6! (b) $_6P_3$ (c) $_6C_3$

15. An English major has built a collection of rare books that includes two identical copies of *The Canterbury Tales* (Chaucer), three identical copies of *Romeo and Juliet* (Shakespeare), four identical copies of *Faustus* (Marlowe), and four identical copies of *The Faerie Queen* (Spenser). If these books are to be arranged on a shelf, how many distinguishable permutations are possible?

16. A company specializes in marketing various *cornucopia* (traditionally a curved horn overflowing with fruit, vegetables, gourds, and ears of grain) for Thanksgiving table settings. The company has seven fruit, six vegetable, five gourd, and four grain varieties available. If two from each group (without repetition) are used to fill the horn, how many different cornucopia are possible?

17. Use Pascal's triangle to expand/simplify:

a. $(x - 2y)^4$ 　　　　**b.** $(1 + i)^4$

18. Use the binomial theorem to write the first three terms of (a) $(x + \sqrt{2})^{10}$ and (b) $(a - 2b^3)^8$.

19. Michael and Mitchell are attempting to make a nonstop, 100-mi trip on a tandem bicycle. The probability that Michael cannot continue pedaling for the entire trip is 0.02. The probability that Mitchell cannot continue pedaling for the entire trip is 0.018. The probability that neither one can pedal the entire trip is 0.011. What is the probability that they complete the trip?

20. The spinner shown is spun once. What is the probability of spinning

a. a striped wedge

b. a shaded wedge

c. a clear wedge

d. an even number

e. a two or an odd number

f. a number greater than nine

g. a shaded wedge *or* a number greater than 12

h. a shaded wedge *and* a number greater than 12

21. To improve customer service, a cable company tracks the number of days a customer must wait until their cable service is installed. The table shows the probability that a customer must wait d days. Based on the table, what is the probability a customer waits

Wait (days d)	Probability
0	0.02
$0 < d < 1$	0.30
$1 \leq d < 2$	0.60
$2 \leq d < 3$	0.05
$3 \leq d < 4$	0.03

a. at least 2 days   **b.** less than 2 days

c. 4 days or less   **d.** over 4 days

e. less than 2 or at least 3 days

f. three or more days

22. An experienced archer can hit the rectangular target shown 100% of the time at a range of 75 m. Assuming the probability the target is hit is related to its area, what is the probability the archer hits within the

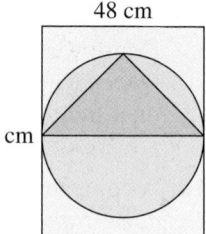

48 cm

64 cm

a. triangle   **b.** circle

c. circle but outside the triangle

d. lower half-circle

e. rectangle but outside the circle

f. lower half-rectangle, outside the circle

23. A survey of 100 union workers was taken to register concerns to be raised at the next bargaining session. A breakdown of those surveyed is shown in the table

Expertise Level	Women	Men	Total
Apprentice	16	18	34
Technician	15	13	28
Craftsman	9	9	18
Journeyman	7	6	13
Contractor	3	4	7
Totals	50	50	100

in the right column. One out of the hundred will be selected at random for a personal interview. What is the probability the person chosen is a

a. woman or a craftsman

b. man or a contractor

c. man and a technician

d. journeyman or an apprentice

24. Cheddar is a 12-year-old male box turtle. Provolone is an 8-year-old female box turtle. The probability that Cheddar will live another 8 yr is 0.85. The probability that Provolone will live another 8 yr is 0.95. Find the probability that

a. both turtles live for another 8 yr

b. neither turtle lives for another 8 yr

c. at least one of them will live another 8 yr

25. Use a proof by induction to show that the sum of the first n natural numbers is $\dfrac{n(n + 1)}{2}$. That is, prove

$$1 + 2 + 3 + \cdots + n = \frac{n(n + 1)}{2}.$$

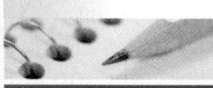

CALCULATOR EXPLORATION AND DISCOVERY

Infinite Series, Finite Results

Although there were many earlier flirtations with infinite processes, it may have been the paradoxes of Zeno of Elea (~450 B.C.) that crystallized certain questions that simultaneously frustrated and fascinated early mathematicians. The first paradox, called the dichotomy paradox, can be summarized by the following question: How can one ever finish a race, seeing that one-half the distance must first be traversed, then one-half the remaining distance, then one-half the distance that then remains, and so on an infinite number of times? Although we easily accept that races can be finished, the subtleties involved in this question stymied mathematicians for centuries and were not satisfactorily resolved until the eighteenth century. In modern notation, Zeno's first paradox says $\frac{1}{2} + \frac{1}{4} + \frac{1}{8} + \frac{1}{16} + \cdots < 1$. This is a geometric series with $a_1 = \frac{1}{2}$ and $r = \frac{1}{2}$.

Illustration 1 ▶ For the geometric sequence with $a_1 = \frac{1}{2}$ and $r = \frac{1}{2}$, the nth term is $a_n = \frac{1}{2^n}$. Use the "**sum(**" and "**seq(**" features of your calculator to compute S_5, S_{10}, and S_{15} (see *Technology Highlight* from Section 10.1). Does the sum appear to be approaching some "limiting value"? If so, what is this value? Now compute S_{20}, S_{25}, and S_{30}. Does there still appear to be a limit to the sum? What happens when you have the calculator compute S_{35}?

Solution ▶ [CLEAR] the calculator and enter sum(seq (0.5^X, X, 1, 5)) on the home screen. Pressing [ENTER] gives

$S_5 = 0.96875$ (Figure 10.21). Press [2nd] [ENTER] to recall the expression and overwrite the 5, changing it to a 10. Pressing [ENTER] shows $S_{10} = 0.9990234375$. Repeating these steps gives $S_{15} = 0.9999694824$, and it seems that "1" may be a limiting value. Our conjecture receives further support as S_{20}, S_{25}, and S_{30} are closer and closer to 1, but do not exceed it.

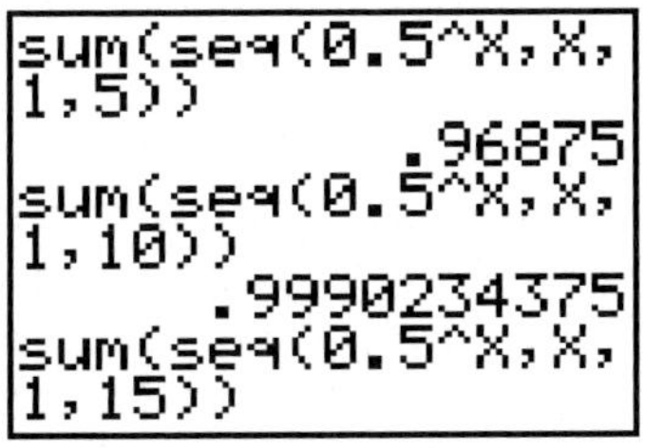

Note that the sum of additional terms will create a longer string of 9's. That the sum of an infinite number of these terms *is* **1** can be understood by converting the repeating decimal $0.\overline{9}$ to its fractional form (as shown). For $x = 0.\overline{9}$, $10x = 9.\overline{9}$ and it follows that

$$10x = 9.\overline{9}$$
$$-x = -0.\overline{9}$$
$$9x = 9$$
$$x = 1$$

Figure 10.21

```
sum(seq(0.5^X,X,
1,5))
            .96875
sum(seq(0.5^X,X,
1,10))
       .9990234375
sum(seq(0.5^X,X,
1,15))
```

For a geometric sequence, the result of an infinite sum can be verified using $S_\infty = \dfrac{a_1}{1-r}$. However, there are many nongeometric, infinite series that also have a limiting value. In some cases these require many, many more terms before the limiting value can be observed.

Use a calculator to write the first five terms and to find S_5, S_{10}, and S_{15}. Decide if the sum appears to be approaching some limiting value, then compute S_{20} and S_{25}. Do these sums support your conjecture?

Exercise 1: $a_1 = \frac{1}{3}$ and $r = \frac{1}{3}$

Exercise 2: $a_1 = 0.2$ and $r = 0.2$

Exercise 3: $a_n = \dfrac{1}{(n-1)!}$

Additional Insight: Zeno's first paradox can also be "resolved" by observing that the "half-steps" needed to complete the race require increasingly shorter (infinitesimally short) amounts of time. Eventually the race is complete.

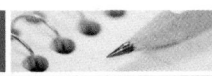

STRENGTHENING CORE SKILLS

Probability, Quick-Counting, and Card Games

The card game known as *Five Card Stud* is often played for fun and relaxation, using toothpicks, beans, or pocket change as players attempt to develop a winning "hand" from the five cards dealt. The various "hands" are given here with the higher value hands listed first (e.g., a full house is a better/higher hand than a flush).

Five Card Hand	Description	Probability of Being Dealt
royal flush	five cards of the same suit in sequence from Ace to 10	0.000 001 540
straight flush	any five cards of the same suit in sequence (exclude royal)	0.000 013 900
four of a kind	four cards of the same rank, any fifth card	
full house	three cards of the same rank, with one pair	
flush	five cards of the same suit, no sequence required	0.001 970
straight	five cards in sequence, regardless of suit	
three of a kind	three cards of the same rank, any two other cards	
two pairs	two cards of the one rank, two of another rank, one other card	0.047 500
one pair	two cards of the same rank, any three others	0.422 600

For this study, we will consider the hands that are based on suit (the flushes) and the sample space to be five cards dealt from a deck of 52, or $_{52}C_5$.

A flush consists of five cards in the same suit, a straight consists of five cards in sequence. Let's consider that an Ace can be used as either a high card (as in 10, J, Q, K, A) or a low card (as in A, 2, 3, 4, 5). Since the dominant characteristic of a flush is its *suit,* we first consider choosing one suit of the four, then the number of ways that the straight can be formed (if needed).

Illustration 1 ▶ What is the probability of being dealt a royal flush?

Solution ▶ For a royal flush, all cards must be of one suit. Since there are four suits, it can be chosen in $_4C_1$ ways. A royal flush must have the cards A, K, Q, J, and 10 and once the suit has been decided, it can happen in only this (one) way or $_1C_1$. This means

$$P \text{ (royal flush)} = \frac{_4C_1 \cdot {_1C_1}}{_{52}C_5} \approx 0.000\ 001\ 540.$$

Illustration 2 ▶ What is the probability of being dealt a straight flush?

Solution ▶ Once again all cards must be of one suit, which can be chosen in $_4C_1$ ways. A straight flush is any five cards in sequence and once the suit has been decided, this can happen in 10 ways (Ace on down, King on down, Queen on down, and so on). By the FCP, there are $_4C_1 \cdot {_{10}C_1} = 40$ ways this can happen, but *four of these will be royal flushes that are of a higher value* and must be subtracted from this total. So in the intended context we have

$$P \text{(straight flush)} = \frac{_4C_1 \cdot {_{10}C_1} - 4}{_{52}C_5} \approx 0.000\ 013\ 900$$

Using these examples, determine the probability of being dealt

Exercise 1: a simple flush (no royal or straight flushes)

Exercise 2: three cards of the same suit and any two other (nonsuit) cards

Exercise 3: four cards of the same suit and any one other (nonsuit) card

Exercise 4: a flush having no face cards

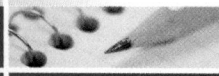

CUMULATIVE REVIEW CHAPTERS 1–10

1. Robot Moe is assembling memory cards for computers. At 9:00 A.M., 52 cards had been assembled. At 11:00 A.M., a total of 98 had been made. Assuming the production rate is linear
   a. Find the slope of this line and explain what it means in this context.
   b. Determine how many boards Moe can assemble in an eight-hour day.
   c. Find a linear equation model for this data.
   d. Determine the approximate time that Moe began work this morning.

2. When using a calculator to find $\sin 120°$, you get $\dfrac{\sqrt{3}}{2}$, yet $\sin^{-1}\left(\dfrac{\sqrt{3}}{2}\right) \neq 120°$. Explain why.

3. Complete this table of special values for $y = \cos x$ without using a calculator.

Table for Exercise 3

x	y
0	
$\dfrac{\pi}{6}$	
$\dfrac{\pi}{4}$	
$\dfrac{\pi}{3}$	
$\dfrac{\pi}{2}$	
$\dfrac{2\pi}{3}$	
$\dfrac{5\pi}{6}$	
π	

4. Sketch the graph of $y = \sqrt{x + 4} - 3$ using transformations of a parent function. Label the x- and y-intercepts and state what transformations were used.

5. Solve using the quadratic formula: $3x^2 + 5x - 7 = 0$. State your answer in exact and approximate form.

6. The orbit of Venus around the Sun is nearly circular, with a radius of 67 million miles. The planet completes one revolution in about 225 days. Calculate the planet's (a) angular velocity in radians per hour and (b) the planet's orbital velocity in miles per hour.

7. For the graph of $g(x)$ shown, state where
   a. $g(x) = 0$ b. $g(x) < 0$
   c. $g(x) > 0$ d. $g(x)\uparrow$
   e. $g(x)\downarrow$ f. local max
   g. local min   h. $g(x) = 2$
   i. $g(4)$ j. $g(-1)$
   k. as $x \to -1^+$, $g(x) \to$ _____
   l. as $x \to \infty$, $g(x) \to$ _____
   m. the domain of $g(x)$

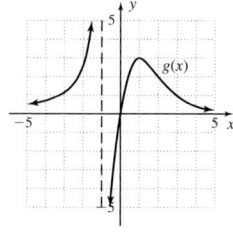

8. Match each equation to its corresponding graph.

   a. $y = \sin(\pi x)$ **b.** $y = \sin(\pi x - \pi)$

   c. $y = \sin(2\pi x - \pi)$ **d.** $y = \sin\left(\pi x - \dfrac{\pi}{2}\right)$

   e. $y = \sin(2\pi x)$ **f.** $y = \sin\left(\pi x + \dfrac{\pi}{2}\right)$

(1)
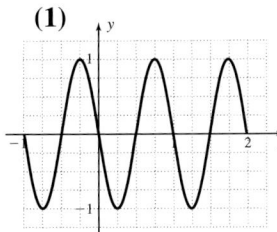

(2)
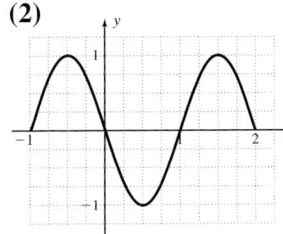

(3)
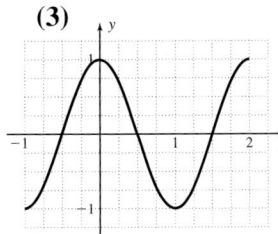

(4)
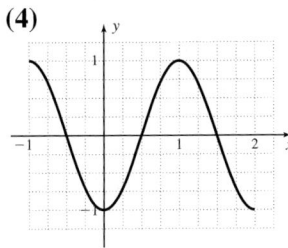

(5)
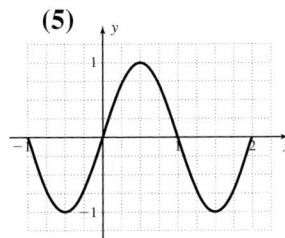

(6)
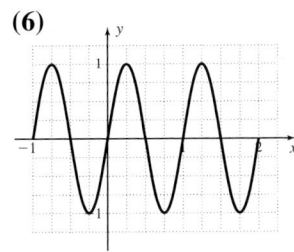

9. Graph the piecewise function and state the domain and range.

$$y = \begin{cases} -2 & -3 \le x \le -1 \\ x & -1 < x < 2 \\ x^2 & 2 \le x \le 3 \end{cases}$$

10. For $\mathbf{u} = 3\mathbf{i} + 4\mathbf{j}$ and $\mathbf{v} = -12\mathbf{i} + 8\mathbf{j}$, find the resultant vector $\mathbf{w} = \mathbf{u} + \mathbf{v}$, then use the dot product to compute the angle between $\mathbf{u}$ and $\mathbf{w}$.

11. Compute the difference quotient for each function given.

   a. $f(x) = 2x^2 - 3x$ **b.** $h(x) = \dfrac{1}{x - 2}$

12. Graph the polynomial function given. Clearly indicate all intercepts. $f(x) = x^3 + x^2 - 4x - 4$

13. Graph the rational function $h(x) = \dfrac{2x^2 - 8}{x^2 - 1}$. Clearly indicate all asymptotes and intercepts.

14. Write each expression in logarithmic form:

   a. $x = 10^y$ **b.** $\dfrac{1}{81} = 3^{-4}$

15. Write each expression in exponential form:

   a. $3 = \log_x(125)$ **b.** $\ln(2x - 1) = 5$

16. What interest rate is required to ensure that $2000 will double in 10 yr if interest is compounded continuously?

17. Solve for x.

   a. $e^{2x-1} = 217$ **b.** $\log(3x - 2) + 1 = 4$

18. Solve using matrices and row reduction:

$$\begin{cases} 2a + 3b - 6c = 15 \\ 4a - 6b + 5c = 35 \\ 3a + 2b - 5c = 24 \end{cases}$$

19. Solve using a calculator and inverse matrices.

$$\begin{cases} 0.7x + 1.2y - 3.2z = -32.5 \\ 1.5x - 2.7y + 0.8z = -7.5 \\ 2.8x + 1.9y - 2.1z = 1.5 \end{cases}$$

20. Find the equation of the hyperbola with foci at $(-6, 0)$ and $(6, 0)$ and vertices at $(-4, 0)$ and $(4, 0)$.

21. Write $x^2 + 4y^2 - 24y + 6x + 29 = 0$ by completing the square, then identify the center, vertices, and foci.

22. Use properties of sequences to determine a_{20} and S_{20}.

   a. $262144, 65536, 16384, 4096, \ldots$

   b. $\dfrac{7}{8}, \dfrac{27}{40}, \dfrac{19}{40}, \dfrac{11}{40}, \ldots$

23. Use the difference identity for cosine to

   a. verify that $\cos\left(\dfrac{\pi}{2} - \theta\right) = \sin\theta$ and

   b. find the value of $\cos 15°$ in exact form.

24. Caleb's grandparents live in a small town that lies 125 mi away at a heading of 110°. Having just received his pilot's license, he sets out on a heading of 110° in a rented plane, traveling at 125 mph (total flight time 1 hr). Unfortunately, he forgets to account for the wind, which is coming from the northeast at 20 mph on a heading of 190°. (a) If Caleb starts out at coordinates $(0, 0)$, what are the coordinates of his grandparent's town? (b) What are the vector coordinates of the plane 1 hr later? (c) How many miles is he actually away from his grandparent's town?

25. Empty 55-gal drums are stacked at a storage facility in the form of a pyramid with 52 barrels in the bottom row, 51 barrels in the next row, and so on, until there are 10 barrels in the top row. Use properties of sequences to determine how many barrels are in this stack.

26. Three $20 bills, six $50 bills, and four $100 bills are placed in a large box and mixed thoroughly, then two bills are drawn out and placed in a savings account. What is the probability the bills drawn are

   a. first $20, second $50

   b. first $50, second $20

   c. both $100

   d. first $100, second not $20

   e. first $100, second not $50

   f. first not $20, second $20

27. The manager of Tom's Tool and Equipment Rentals knows that 4% of all tools rented are returned late. Of the 12 tools rented in the last hour, what is the probability that

   a. exactly ten will be returned on time

   b. at least eleven will be returned on time

   c. at least ten will be returned on time

   d. none of them will be returned on time

28. Use a proof by induction to show
$$3 + 7 + 11 + 15 + \cdots + (4n - 1) = n(2n + 1)$$
for all natural numbers n.

29. State the three double angle formulas for cosine. If $\cos(2\theta) = \dfrac{1}{2}$, what is the value of $\sin \theta$?

30. A park ranger tracks the number of campers at a remote national park from January ($m = 1$) to December ($m = 12$) and collects the following data (month, number of campers): (3, 6), (5, 110), (7, 134), and (9, 78). Assuming the data is quadratic ($y = ax^2 + bx + c$), (a) select any of the three points and create a system of three equations in three variables to obtain a parabolic equation model for the data and (b) determine the month that brought the maximum number of campers. (c) What was this maximum number? (d) How many campers might be expected in September? (e) Based on your model, what month(s) is the park apparently closed to campers (number of campers is zero or negative)?

The properties of summation play an important role in the development of many key ideas in a first semester calculus course. Here, we'll see how the summation properties (Section 10.1) are combined with the summation formulas (Section 10.4) and the concept of the area under a curve (*Connections to Calculus* Chapter 2), to produce some very interesting results. For convenience, the summation formulas are restated here:

$$(1) \ \sum_{i=1}^{n} c = cn \qquad\qquad (2) \ \sum_{i=1}^{n} i = \frac{n^2 + n}{2}$$

$$(3) \ \sum_{i=1}^{n} i^2 = \frac{2n^3 + 3n^2 + n}{6} \qquad (4) \ \sum_{i=1}^{n} i^3 = \frac{n^4 + 2n^3 + n^2}{4}$$

Applications of Summation

For this study, consider the area under the line $f(x) = x + 5$, in the interval $[0, 4]$ (between the y-axis and the vertical line $x = 4$, Figure C2C 10.1). To find the total area, we could add the area of the rectangle and triangle shown (see *Reinforcing Basic Concepts,* page 974), or use $f(0) = 5$ and $f(4) = 9$ as the bases of a trapezoid and simply apply the formula $A = \frac{h}{2}(b_1 + b_2)$.

Instead of being so direct, we'll use this simple geometric shape to develop some powerful ideas, to be used when the desired area is much more complex. Using what we'll call the **rectangle method,** we first *approximate* the total area by adding the area of the four rectangles shown in Figure C2C 10.2. Note this will give an area greater than the true value, but yields a reasonable estimate.

Figure C2C 10.1

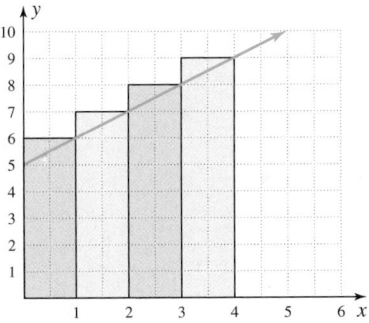

Each rectangle will have a width of 1, since there are four rectangles in an interval four units wide ($\frac{4}{4} = 1$), and the length of each rectangle will be $f(x) = x + 5$ for $x = 1, 2, 3,$ and 4. Using the formula $A = LW$, we can write this as $\sum LW = \sum_{i=1}^{4} f(i)(1)$, where the function is written in terms of i [$f(i) = i + 5$] for the sake of the summation notation.

Figure C2C 10.2

$$\sum_{i=1}^{4} f(i)(1) = f(1)(1) + f(2)(1) + f(3)(1) + f(4)(1)$$
$$= 6 + 7 + 8 + 9 = 30 \text{ units}^2$$

To refine this estimate, consider the <u>eight</u> rectangles shown in Figure C2C 10.3. The original four-unit interval is now divided into eight parts, so each part is $\frac{4}{8} = \frac{1}{2}$ unit wide. The length of each rectangle is still given by $f(x) = x + 5$, but we now evaluate the function in increments of $\frac{1}{2}$:

Figure C2C 10.3

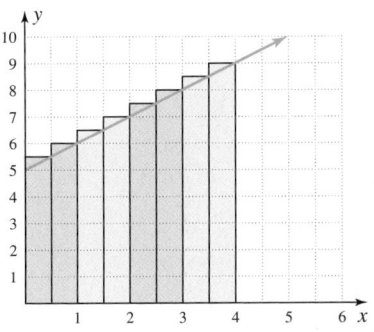

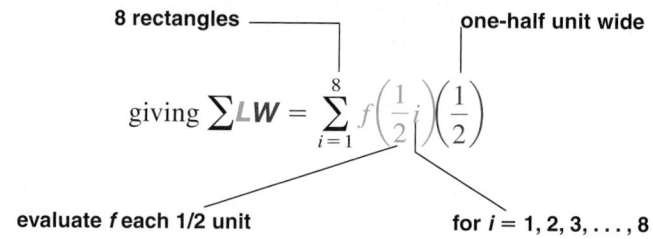

giving $\sum LW = \sum\limits_{i=1}^{8} f\left(\dfrac{1}{2}i\right)\left(\dfrac{1}{2}\right)$

8 rectangles — one-half unit wide

evaluate f each 1/2 unit   for $i = 1, 2, 3, \ldots, 8$

Since the length of each rectangle is multiplied by $W = \frac{1}{2}$, we can factor out $\frac{1}{2}$ and write the expression as $\dfrac{1}{2}\sum\limits_{i=1}^{8} f\left(\dfrac{1}{2}i\right)$ for convenience.

EXAMPLE 1 ▶ **Computing a Sum of Rectangular Areas**

Find the sum $\dfrac{1}{2}\sum\limits_{i=1}^{8} f\left(\dfrac{1}{2}i\right)$ by writing out each term.

Solution ▶ $\dfrac{1}{2}\sum\limits_{i=1}^{8} f\left(\dfrac{1}{2}i\right) = \dfrac{1}{2}\left[f\left(\dfrac{1}{2}\right) + f(1) + f\left(\dfrac{3}{2}\right) + f(2) + f\left(\dfrac{5}{2}\right) + f(3) + f\left(\dfrac{7}{2}\right) + f(4) \right]$

$= \dfrac{1}{2}\left(\dfrac{11}{2} + 6 + \dfrac{13}{2} + 7 + \dfrac{15}{2} + 8 + \dfrac{17}{2} + 9 \right)$

$= \dfrac{11}{4} + 3 + \dfrac{13}{4} + \dfrac{7}{2} + \dfrac{15}{4} + 4 + \dfrac{17}{4} + \dfrac{9}{2} = 29 \text{ units}^2$

Now try Exercises 1 and 2 ▶

This approximation is still too large, but closer to the true value of 28 units². At this point, we could get an even better estimate using 16 rectangles, with each width being $\frac{4}{16} = \frac{1}{4}$ unit wide, and the length still computed by $f(x) = x + 5$ but using increments of $\frac{1}{4}$. This would yield $\sum LW = \sum\limits_{i=1}^{16} f\left(\dfrac{1}{4}i\right)\left(\dfrac{1}{4}\right) = \dfrac{1}{4}\sum\limits_{i=1}^{16} f\left(\dfrac{1}{4}i\right)$. As you might imagine, finding the sum $\dfrac{1}{4}\sum\limits_{i=1}^{16} f\left(\dfrac{1}{4}i\right)$ by

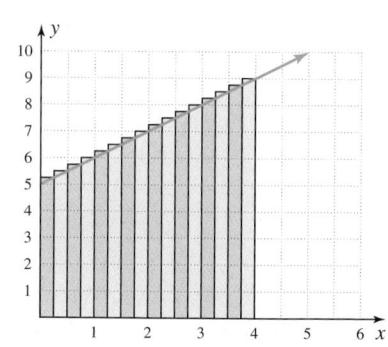

writing out each term would be tedious and time consuming. Instead, the properties of summation and the summation formulas can be used to develop an expression where we can find *the sum for n = 16 directly*. The key idea is to note that since $f\left(\dfrac{1}{4}i\right) = \dfrac{1}{4}i + 5$, we can write the summation as $\dfrac{1}{4}\sum\limits_{i=1}^{n}\left(\dfrac{1}{4}i + 5\right)$ and work as in Example 2.

EXAMPLE 2 ▶ Computing a Sum of Rectangular Areas

For the sum $\dfrac{1}{4}\displaystyle\sum_{i=1}^{n}\left(\dfrac{1}{4}i + 5\right)$, use the summation properties and summation formulas to develop an expression where the sum can be found by substituting $n = 16$ directly.

Solution ▶

$$\frac{1}{4}\sum_{i=1}^{n}\left(\frac{1}{4}i + 5\right) = \frac{1}{4}\left(\sum_{i=1}^{n}\frac{1}{4}i + \sum_{i=1}^{n}5\right) \qquad \text{summation properties (distribute)}$$

$$= \frac{1}{4}\left(\frac{1}{4}\sum_{i=1}^{n}i + \sum_{i=1}^{n}5\right) \qquad \text{factor } \frac{1}{4} \text{ from first summation}$$

$$= \frac{1}{4}\left[\frac{1}{4}\left(\frac{n^2 + n}{2}\right) + 5n\right] \qquad \text{use summation formulas}$$

$$= \frac{1}{4}\left[\frac{1}{4}\left(\frac{16^2 + 16}{2}\right) + 5(16)\right] \qquad \text{substitute 16 for } n$$

$$= \frac{1}{16}\left(\frac{272}{2}\right) + 20 \qquad \text{distribute and simplify}$$

$$= 8.5 + 20 = 28.5 \qquad \text{result}$$

The approximate area is 28.5 units^2.

Now try Exercises 3 and 4 ▶

The significance of this result cannot be overstated. As we'll see in Chapter 11, similar calculations can be used if n rectangles are assumed, with the area easily calculated for *any value of n*. Further, you might imagine the amazement of the early students of mathematics, when they realized that these ideas could still be applied for areas where no elementary formula existed, as for the area under a nonlinear graph.

Connections to Calculus Exercises

1. For $f(x) = -x + 6$ and $x \in [0, 4]$, (a) graph the function, then approximate the area under the graph in this interval using four rectangles of width 1 and length $f(x)$ for $x = 1, 2, 3,$ and 4. Begin by developing the summation formula for $\sum LW$, then expand the sum. (b) Repeat part (a) using eight rectangles of width one-half and length $f(x)$. Draw both the graph and the rectangles in each case, and verify the increased number of rectangles gives a better estimate.

2. For $f(x) = x + 4$ and $x \in [0, 6]$, (a) graph the function, then approximate the area under the graph in this interval using six rectangles of width 1 and length $f(x)$ for $x = 1, 2, 3, 4, 5,$ and 6. Begin by developing the summation formula for $\sum LW$, then expand the sum. (b) Repeat part (a) using 12 rectangles of width one-half and length $f(x)$. Draw both the graph and the rectangles in each case, and verify the increased number of rectangles gives a better estimate.

For Exercises 3 and 4, show all work requested using a methodical step-by-step process, similar to that illustrated in Example 2.

3. For $f(x) = -x + 6$ and $x \in [0, 4]$ from Exercise 1, (a) discuss how the area under the graph of f in this interval can be approximated using the rectangle method and $n = 32$ rectangles without expanding the sum. (b) Find the area for $n = 32$ by applying summation properties and formulas.

4. For $f(x) = x + 4$ and $x \in [0, 6]$ from Exercise 2, (a) discuss how the area under the graph of f in this interval can be approximated using the rectangle method and $n = 36$ rectangles without expanding the sum. (b) Find the area for $n = 36$ by applying summation properties and formulas.

11

Bridges to Calculus: An Introduction to Limits

CHAPTER OUTLINE

CHAPTER CONNECTIONS

This chapter introduces the most central and powerful idea in calculus, the concept of a limit. While the study of limits isn't a recent phenomenon (Archimedes inscribed a polygon within a circle and considered an ever increasing number of sides to estimate the circumference), a complete development of the idea had to wait for the talents of Augustin-Louis Cauchy (c. 1821) and a maturation of the notation needed to work with the concept effectively. Using a rational function to illustrate, essentially the concept involves recognizing the distinction between the *value* of $f(x) = \dfrac{x^2 + 2x - 3}{x - 1}$ when $x = 1$ (the function is not defined at 1), and the *limit* of $f(x)$ as x becomes very close to 1. This application appears as Exercise 33 in Section 11.1.

Learning Objectives

In Section 11.1 you will learn how to:

☐ **A.** Distinguish between a limit and an approximation

☐ **B.** Estimate limits using tables

☐ **C.** Evaluate one-sided limits

☐ **D.** Determine when a limit does not exist

The concept of a limit is the foundation of all calculus. If the concept is faulty, the conclusions of calculus can be neither supported nor sustained, and the whole becomes a great leap of faith. The inventors of calculus (Isaac Newton and Gottfried Leibniz) tried mightily to bridge the gap between the finite and the infinite, but it was only after a long evolution that the ideas would mature to a point where infinity could be tamed and calculus placed on an unassailable footing.

A. Distinguishing between Limits and Approximations

To begin our study, consider the area of a regular polygon inscribed in a circle of radius r (inscribed: all vertices of the polygon are on the circumference). The radii drawn from the center to each vertex divide the polygons into triangles, enabling us to compute the area of each polygon by summing the area of these triangles (Figure 11.1).

Figure 11.1

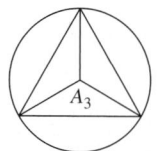 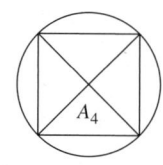   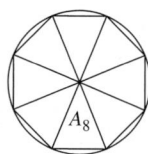

If we let A_n represent the area of an inscribed polygon with n sides, we note that as the number of sides increases, the area of the polygon gives a better and better approximation for the area of the circle. In fact, it seems intuitive that as n becomes infinitely large, the area of the polygon becomes infinitely close to the area of the circle. Using the notation seen in previous chapters we would say, as $n \to \infty$, $A_n \to \pi r^2$. In a study of limits, **limit notation** is used, and the same statement is written

$$\lim_{n \to \infty} A_n = \pi r^2$$

In words, "the limit of A_n as n becomes infinitely large is πr^2." For practice using the new notation, **see Exercises 7 through 12.**

In a study of limits it becomes extremely important to accept that no finite number of sides for our polygon will give the exact area of the circle (even with 100 sides we still have only a very good approximation). But that's not what the concept and notation are saying. They only say that πr^2 is the **limiting value,** and that we can become as close as we like to this limit, by choosing n sufficiently large.

EXAMPLE 1 ▶ **Finding the Area of an Inscribed Polygon**

If an n-sided regular polygon is inscribed in a circle of radius r, its area is given by $A_n = \dfrac{nr^2}{2} \sin\left(\dfrac{2\pi}{n}\right)$. Assume the circle has a radius of $r = 10$ cm. Use a table of values to determine the number of sides needed (to the nearest 100) for the area of the polygon to approximate the area of the circle rounded to two decimal places ($A \approx 314.16$ units² rounded).

x	A_n
100	313.95
200	314.11
300	314.14
400	314.15
500	314.15
600	314.15
700	314.16
$y = 314.155046835$	

WORTHY OF NOTE

From the area formula

$A = \dfrac{1}{2} ab \sin C$, note that for each triangle within the circle, the angle at the center C is $\dfrac{2\pi}{n}$, and $a = b = r$. This shows each triangle has area

$A = \dfrac{1}{2} r^2 \sin\left(\dfrac{2\pi}{n}\right)$ and there are n of these triangles.

Solution ▶ With $r = 10$, the formula becomes $A_n = 50n \sin\left(\dfrac{2\pi}{n}\right)$, where n represents the number of sides. Evaluating the equation using multiples of 100 produces the table shown, and we note that when $n = 700$, the area of the polygon approximates the area of the circle rounded to two decimal places.

☑ **A.** You've just learned how to distinguish between a limit and an approximation

Now try Exercises 13 through 16 ▶

B. Estimating Limits Using Tables and Graphs

The limit in the discussion prior to Example 1 ($\lim\limits_{n\to\infty} A_n$) is called a **limit at infinity,** or a limit as $n \to \infty$. These are discussed in more detail in Section 11.3. Here we turn our attention to limits at a constant c, or a limit as $x \to c$ (using the more common variable x). Consider the function $f(x) = \dfrac{\sin x}{x}$ (x in radians), noting that $x = 0$ is not in the domain of f. However, the fact that we could not evaluate $\lim\limits_{n\to\infty} A_n$ for $n = \infty$ in Example 1, did not stop us from evaluating A_n as $n \to \infty$. Similarly, we cannot evaluate $f(x) = \dfrac{\sin x}{x}$ at $x = 0$, but we can evaluate f for values of x slightly less than zero, or slightly greater than zero, to investigate the limit of $\dfrac{\sin x}{x}$ as $x \to 0$. For additional practice with the limit notation, **see Exercises 17 through 22.**

EXAMPLE 2 ▶	Determining Whether a Limiting Value Exists

Use a table of values to evaluate $f(x) = \dfrac{\sin x}{x}$ as $x \to 0$. If the function seems to approach a limiting value, write the relationship in words and using the limit notation.

Solution ▶ For this exercise it seems appropriate to begin a short distance from zero, and approach it in small increments. Noting that 0 can be "approached" from either side, we use values less than zero (from the left-hand side) and values greater than zero (from the right-hand side). Beginning at $x = -0.5$ and approaching 0 from the left produces Table 11.1. Starting at $x = 0.5$ and approaching from the right yields Table 11.2. The tables seems to indicate that as x becomes very close to 0, $\dfrac{\sin x}{x}$ becomes very close to 1, or seems to have a limiting value of 1. In limit notation we write $\lim\limits_{x\to 0} \dfrac{\sin x}{x} = 1$.

Table 11.1

x	$f(x)$
-0.5	0.95885
-0.4	0.97355
-0.3	0.98507
-0.2	0.99335
-0.1	0.99833
-0.01	0.99998
-0.001	1
$y = 0.999999833333$	

Table 11.2

x	$f(x)$
0.5	0.95885
0.4	0.97355
0.3	0.98507
0.2	0.99335
0.1	0.99833
0.01	0.99998
0.001	1
$y = 0.999999833333$	

Now try Exercises 23 through 28 ▶

Similar to our work with the polygon and circle, note that we can make $\dfrac{\sin x}{x}$ as close as we like to 1 by using values of x that are sufficiently close to 0. This suggests the following definition of a limit, a definition, however, that will be refined and made more precise as we gain a greater understanding of the limit concept.

The Definition of a Limit

For a fixed real number L and constant c,

$$\lim_{x \to c} f(x) = L$$

means that values of $f(x)$ can be made arbitrarily close (infinitely close) to L, by taking values of x sufficiently close to c, $x \neq c$.

The graph of (1) $f(x) = \dfrac{\sin x}{x}$ is shown in Figure 11.2. As we did in Section 2.7, we could create a piecewise function that defines $f(x)$ at $x = 0$, and two possibilities are listed here, followed by their graphs (Figures 11.3 and 11.4, respectively).

$$(2)\ f(x) = \begin{cases} \dfrac{\sin x}{x} & x \neq 0 \\ 2 & x = 0 \end{cases} \qquad\qquad (3)\ f(x) = \begin{cases} \dfrac{\sin x}{x} & x \neq 0 \\ 1 & x = 0 \end{cases}$$

Figure 11.2   **Figure 11.3**   **Figure 11.4**

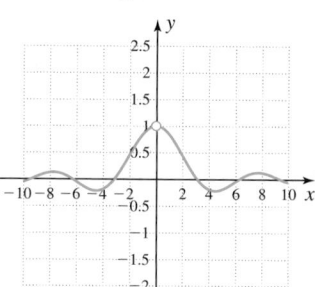   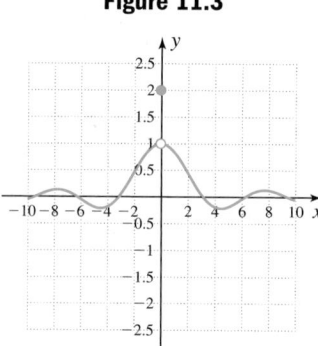   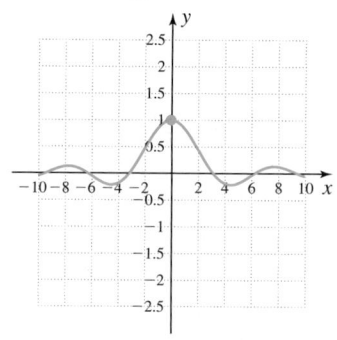

In (2), $f(x)$ defines the function at $(0, 2)$ but leaves a hole in the graph at $(0, 1)$, while in (3), $f(x)$ is a continuous function as we can now draw the entire graph without lifting our pencil **(see Exercises 29 and 30)**. Note that in all three cases, $\displaystyle\lim_{x \to 0} \dfrac{\sin x}{x} = 1$ remains *true*, as the definition of a limit is still satisfied in every respect. In other words, the limit of a function as $x \to c$ can exist even if

(1) the function is not defined   (2) the function is defined at c but $f(c) \neq L$

at c as in $f(x) = \dfrac{\sin x}{x}$,   as in $f(x) = \begin{cases} \dfrac{\sin x}{x}, & x \neq 0 \\ 2 & x = 0 \end{cases}$,

or (3) the function is defined at c and $f(c) = L$ as in $f(x) = \begin{cases} \dfrac{\sin x}{x}, & x \neq 0 \\ 1 & x = 0 \end{cases}$.

EXAMPLE 3 ▶ Determining Whether a Limiting Value Exists

Use a table of values to evaluate $g(x) = \dfrac{\frac{1}{2}x^2 - x}{x - 2}$, as $x \to 2$. If the function seems to approach a limiting value, write the relationship in words and using the limit notation.

Solution ▶ Similar to Example 2, it seems appropriate to start a short distance from 2 on the left- and right-hand sides, then approach 2 in small increments. Starting from the left at 1.6 produces Table 11.3. Starting from the right at 2.4 yields Table 11.4.

Figure 11.5

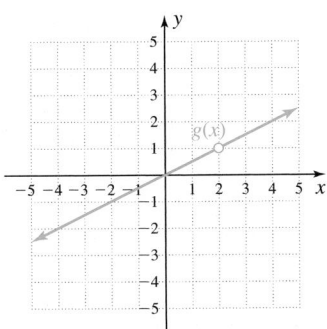

Table 11.3

x	g(x)
1.6	0.8
1.7	0.85
1.8	0.9
1.9	0.95
1.99	0.995
1.999	0.9995
1.9999	0.99995
y = 0.99995	

Table 11.4

x	g(x)
2.4	1.2
2.3	1.15
22	11
2.1	1.05
2.01	1.005
2.001	1.0005
2.0001	1.00005
y = 1.00005	

☑ **B. You've just learned how to estimate limits using tables**

The tables seem to indicate that as x becomes very close to 2, $g(x)$ becomes very close to 1, or seems to have a limiting value of 1. In limit notation,

$\lim\limits_{x \to 2} \dfrac{\frac{1}{2}x^2 - x}{x - 2} = 1$. The graph is shown in Figure 11.5.

Now try Exercises 31 through 36 ▶

C. One-Sided Limits

In Examples 2 and 3 our discussion focused on what are called **two-sided limits,** as we approached the limiting value by selecting inputs to the left *and* to the right of c ($x \neq c$). But what of functions like $y = \sqrt{2 - x}$, $y = \dfrac{x^2 - 2x}{20\sqrt{x - 2}}$ and many others, that may only be defined to the left or right of $c = 2$? Can these functions approach a limiting value as x approaches c from only one side?

EXAMPLE 4 ▶ **Determining Whether a Limit Exists for a One-Sided Approach**

Use a table of values to help determine if $f(x) = \dfrac{x^2 - 2x}{20\sqrt{x - 2}}$

has a limiting value as $x \to 2$ from the right.

Solution ▶ With the radical $\sqrt{x - 2}$ in the denominator, the domain of f is $x \in (2, \infty)$. For $y = f(x)$, we input values close to 2 but greater than 2, since 2 can only be "approached" from the right-hand side. The result is the table shown, and seems to indicate that as x becomes very close to 2 from the right, $f(x)$ becomes very close to 0. Further, it appears we can make $f(x)$ as close to zero as we like, by choosing values of x sufficiently close to 2 but greater than 2. This suggests the limit of $f(x)$ as x approaches 2 from the right is 0.

x	f(x)
2.4	0.07589
2.3	0.06299
2.2	0.04919
2.1	0.0332
2.01	0.01005
2.001	0.00316
2.0001	0.001
y = 0.00100005	

Now try Exercises 37 through 42 ▶

For functions like those in Example 4, we can define a **one-sided limit,** using the general definition as a template and introducing a modified notation to represent the idea.

Right-Hand Limits

For a fixed real number L and constant c,

$$\lim_{x \to c^+} f(x) = L$$

means that values of $f(x)$ can be made arbitrarily close (infinitely close) to L, by taking values of x sufficiently close to c and to the right of c ($x > c$).

Left-Hand Limits

For a fixed real number L and constant c,

$$\lim_{x \to c^-} f(x) = L$$

means that values of $f(x)$ can be made arbitrarily close (infinitely close) to L, by taking values of x sufficiently close to c and to the left of c ($x < c$).

Note how the superscript on c indicates the direction from which x is approaching c. For $x \to c^-$, x is approaching c from the left, for $x \to c^+$, x is approaching c from the right (c can be any real number). For practice with this notation, **see Exercises 43 through 46.**

EXAMPLE 5 ▶ **Determining Left- and Right-Handed Limits**

For $h(x) = \dfrac{|x - 2|}{x - 2}$, use a table to evaluate the expressions

a. $\lim\limits_{x \to 2^-} h(x)$ **b.** $\lim\limits_{x \to 2^+} h(x)$

If a limiting value appears to exist from either side, write the relationship in words and using the limit notation.

Solution ▶ Begin by noting the domain of h is $x \in \mathbb{R}, x \neq 2$.

a. For $\lim\limits_{x \to 2^-} h(x)$, we use a table to track the value of h as x approaches 2 from the

left ($x \to 2^-$). The result is shown in Table 11.5, and suggests $\lim\limits_{x \to 2^-} \dfrac{|x - 2|}{x - 2} = -1$.

In words, the limit of $h(x)$ as x approaches 2 from the left is -1.

Figure 11.6

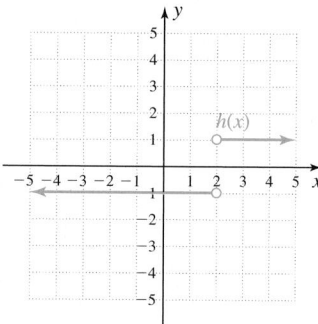

Table 11.5

x	$h(x)$
1.6	−1
1.7	−1
1.8	−1
1.9	−1
1.99	−1
1.999	−1
1.9999	−1
$y = -1$	

Table 11.6

x	$h(x)$
2.4	1
2.3	1
2.2	1
2.1	1
2.01	1
2.001	1
2.0001	1
$y = 1$	

✓ **C. You've just learned how to evaluate one-sided limits**

b. Similarly, Table 11.6 suggests $\lim\limits_{x \to 2^+} \dfrac{|x - 2|}{x - 2} = 1$, or the limit of $h(x)$ as x approaches 2 from the right is 1. The graph of h is shown in Figure 11.6 and seems to support this conclusion.

Now try Exercises 47 through 54 ▶

D. Limits That Fail to Exist

Unlike the function in Example 4 (where x could approach 2 only from the right), the functions in Examples 3 and 5 allowed an approach from both the left and right sides. Note that the statement, "taking values of x sufficiently close to c" from the general definition of a limit *implies a two-sided approach to c*, and we must approach c from both sides wherever possible. Since this is not possible in Example 4, we use the notation for one-sided limits to state: $\lim\limits_{x \to 2^+} \dfrac{x^2 - 2x}{\sqrt{x - 2}} = 0$. For the function $g(x)$ in Example 3, a "two-sided" approach was possible and we saw the approach from each side was equal to 1. Using the general definition we were able to write $\lim\limits_{x \to 2} g(x) = 1$ (the limit is $L = 1$). For the function $h(x)$ in Example 5, x could also approach c from both sides, but the left-hand limit was not equal to the right, and there is no single number L that $h(x)$ approaches as $x \to 2$. This leads us to the following statement:

The Existence of a Limit

For a fixed real number L and constant c,

$$\lim_{x \to c} f(x) = L \quad \text{if and only if} \quad \lim_{x \to c^-} f(x) = \lim_{x \to c^+} f(x) = L.$$

In words, "the limit of $f(x)$ as x approaches c is L, if and only if the left-hand limit as x approaches c is equal to the right-hand limit as x approaches c, and both are also equal to L."

EXAMPLE 6 ▶ Determining Whether a Limit Exists

Use a table of values to evaluate the expression $\lim\limits_{x \to 3} \dfrac{x^2 - 3x}{\sqrt{x^2 - 6x + 9}}$.

Solution ▶ As a matter of choice we first evaluate the left-hand limit, and Table 11.7 suggests

$$\lim_{x \to 3^-} \dfrac{x^2 - 3x}{\sqrt{x^2 - 6x + 9}} = -3. \text{ For the right-hand limit, Table 11.8 suggests}$$

$$\lim_{x \to 3^+} \dfrac{x^2 - 3x}{\sqrt{x^2 - 6x + 9}} = 3.$$

Table 11.7

x	y
2.5	−2.5
2.6	−2.6
2.7	−2.7
2.8	−2.8
2.9	−2.9
2.99	−2.99
2.999	−2.999
$y = -2.999$	

Table 11.8

x	y
3.4	3.4
3.3	3.3
3.2	3.2
3.1	3.1
3.01	3.01
3.001	3.001
3.0001	3.0001
$y = 3.0001$	

Figure 11.7

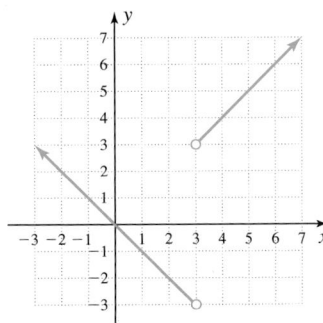

Since the left-hand limit $\neq$ right-hand limit, $\lim\limits_{x \to 3} \dfrac{x^2 - 3x}{\sqrt{x^2 - 6x + 9}}$ does not exist.

The graph of $y = \dfrac{x^2 - 3x}{\sqrt{x^2 - 6x + 9}}$ is shown in Figure 11.7.

Now try Exercises 55 through 60 ▶

There are actually other ways that a limit can fail to exist, and as a matter of order we will adopt the following notation for the case where the left-hand limit is not equal to the right-hand limit. Using dne as an abbreviation for "does not exist," we will write $\lim\limits_{x \to c} f(x) = \left(\begin{smallmatrix} \text{dne} \\ \text{LH} \neq \text{RH} \end{smallmatrix}\right)$. In Example 7 we note a second way that a limit can fail to exist.

EXAMPLE 7 ▶ **Determining Whether a Limit Exists**

Given $f(x) = \dfrac{9x}{(x-2)^2}$, evaluate $\lim\limits_{x \to 2} f(x)$.

x	$f(x)$
1.5	54
1.6	90
1.7	170
1.8	405
1.9	1710
1.99	179100
1.999	1.8E7
$y = 17991000$	

Solution ▶ For the left-hand limit, we obtain the table shown and note as $x \to 2^-$, the value of $f(x)$ increases without bound [for $x = 1.999$, $f(x)$ is almost $18{,}000{,}000$]. A similar check shows that as $x \to 2^+$, the values of $f(x)$ likewise become infinitely large. Since $f(x)$ becomes infinitely large and positive as $x \to 2$, the limit does not exist and we write $\lim\limits_{x \to 2} f(x) = \left(\begin{smallmatrix} \text{dne} \\ \infty \end{smallmatrix}\right)$.

Now try Exercises 61 through 66 ▶

From our work with rational functions in Section 3.5, we know the graph of $f(x) = \dfrac{9x}{(x-2)^2}$ has a vertical asymptote at $x = 2$, verifying it is impossible for the function to have a limit as $x \to 2$.

Finally, we consider a third way that a limit can fail to exist, as when a function does not approach a fixed real number L, but does not grow infinitely large and positive, nor infinitely large and negative.

EXAMPLE 8 ▶ **Determining Whether a Limit Exists**

For $h(x) = 2\left|\sin\left(\dfrac{1}{x}\right)\right|$, try to find a fixed number L such that $\lim\limits_{x \to 0} f(x) = L$.

Solution ▶ Using an approach from the right gives the result shown in the table. This time there is no apparent pattern in the output values, and a graph of the function (Figure 11.8) reveals why. As $x \to 0$, the function begins to oscillate wildly between 0 and 2, with no approach to any fixed number L. A similar thing happens for a left-handed approach. As per our definition, the limit does not exist because L does not exist. In this case, we write $\lim\limits_{x \to 2} h(x) = \left(\begin{smallmatrix} \text{dne} \\ \not{L} \end{smallmatrix}\right)$.

WORTHY OF NOTE

For students and instructors with a sense of humor, Example 8 illustrates what's sometimes called the Christmas Case, because there's No-el.

Figure 11.8

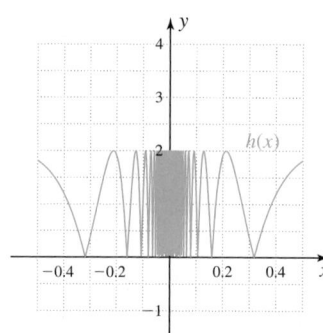

x	$h(x)$
0.4	1.1969
0.3	0.38114
0.2	1.9178
0.1	1.088
0.01	1.0127
0.001	1.6538
1E-4	0.61123
$x = 0.0001$	

☑ **D.** You've just learned how to determine when a limit does not exist

Now try Exercises 67 through 70 ▶

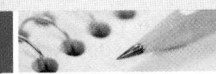

11.1 EXERCISES

▶ CONCEPTS AND VOCABULARY

Fill in each blank with the appropriate word or phrase. Carefully reread the section, if necessary.

1. The limit of $f(x)$ as $x \to \infty$, is called a limit at _____.

2. $\lim\limits_{x \to c} g(x) = L$ means that values of $g(x)$ can be made _____ to L, by taking values of _____ sufficiently close to _____.

3. To evaluate the limit of a function as $x \to c$, we must find the _____ limit using values of x less than c, and the _____ limit using values of x _____ than c.

4. If a rational function $v(x)$ has a vertical asymptote at $x = c$, the limit as $x \to c$ does not exist and we write $\lim\limits_{x \to c} v(x) =$ _____.

5. Discuss/Explain why $\lim\limits_{x \to 3} g(x) = 5$ for $g(x) = \dfrac{x^2 - x - 6}{x - 3}$, but $g(3) \neq 5$. Can we redefine $g(x)$ to create a piecewise-defined function where $\lim\limits_{x \to 3} g(x) = g(3)$?

6. Discuss/Explain why $\lim\limits_{x \to 1} f(x)$ does not exist for $f(x) = \dfrac{x^2 - x}{\sqrt{(x - 1)^2}}$, even though the left-hand limit and the right-hand limit do exist.

▶ DEVELOPING YOUR SKILLS

Write each of the following statements using limit notation.

7. As n approaches ∞, V_n approaches $\dfrac{4}{3}\pi r^3$.

8. As n approaches ∞, A_n approaches $\dfrac{a}{3}x^3 + \dfrac{b}{2}x^2 + cx + d$.

9. As t approaches $-\infty$, $e^{f(t)}$ approaches 0.

10. As t approaches $-\infty$, $\sin[g(t)]$ approaches 1.

11. As x increases without bound, $\cos\left(\dfrac{1}{x}\right)$ approaches 1.

12. As x decreases without bound, $\dfrac{5x^2 - 3}{2x^2 - x - 1}$ approaches $\dfrac{5}{2}$.

Use a table of values for the following exercises.

13. If an n-sided regular polygon *circumscribes* a circle of radius r (see the figure), its area is given by $A_n = nr^2 \tan\left(\dfrac{\pi}{n}\right)$. For a circle with radius $r = 10$ cm,

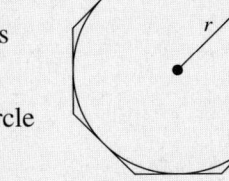

determine the number of sides needed (to the nearest 100) for the area of the polygon to approximate the area of the circle correct to two decimal places when rounded. Compare your result with that of Example 1.

14. If an n-sided regular polygon is *inscribed* in a circle of radius r, its perimeter is given by $P = 2nr \sin\left(\dfrac{\pi}{n}\right)$. For a circle with radius $r = 50$ mm, determine the number of sides needed (to the nearest 25) for the perimeter of the polygon to approximate the circumference of the circle correct to two decimal places when rounded.

15. If an n-sided regular polygon *circumscribes* a circle of radius r, its perimeter is given by $P = 2nr \tan\left(\dfrac{\pi}{n}\right)$. For a circle with radius $r = 50$ mm, determine the number of sides needed (to the nearest 25) for the perimeter of the polygon to approximate the circumference of the circle correct to two decimal places when rounded. Compare your results with those in Exercise 13.

16. One of the most famous and useful numbers in all of mathematics is the one defined as $\lim_{x \to \infty} \left(1 + \frac{1}{x}\right)^x = e$. Determine the smallest x to (the nearest 250) that can be used to approximate the value of e to three decimal places.

Write each of the following statements using limit notation.

17. As t approaches 5, s_t approaches $5r$.

18. As t approaches -3, d_t approaches $\sqrt{9 + r^2}$.

19. As x approaches a, $\tan^{-1}[g(x)]$ approaches $\frac{\pi}{3}$.

20. As x approaches b, $\csc^2[f(x)]$ approaches $\frac{4}{3}$.

21. As $x \to -3$, $\frac{x + 3}{x^2 - 9}$ approaches $-\frac{1}{6}$.

22. As $x \to \pi$, $\dfrac{\sin x}{\cos\left(\dfrac{x}{2}\right)}$ approaches 2.

Use a table of values to evaluate each function as x approaches the value indicated. If the function seems to approach a limiting value, write the relationship in words and using the limit notation.

23. $p(x) = \cos x + \sin\left(\dfrac{3x}{2}\right)$; $x \to \pi$

24. $q(x) = \tan x + 3 \sin 2x$; $x \to \dfrac{\pi}{4}$

25. $v(x) = \dfrac{\cos\left(\dfrac{\pi}{x}\right)}{\sin(\pi x)}$; $x \to 2$

26. $w(x) = \dfrac{\tan(x - 2)}{x - 2}$; $x \to 2$

27. $s(x) = \dfrac{2 \cos x - 2}{x}$; $x \to 0$

28. $t(x) = \dfrac{\sin x - 1}{\sin(2x)}$; $x \to \dfrac{\pi}{2}$

The functions in Exercises 29 and 30 have a hole (discontinuity) in their graphs at $x = 2$. Write a related piecewise-defined function that creates a continuous graph for each.

29. $r(x) = \dfrac{2x^2 - 7x + 6}{\sin(x - 2)}$

30. $s(x) = \dfrac{\sin(x^2 - 4x + 4)}{x - 2}$

Use a table of values to evaluate each function as x approaches the value indicated. If the function seems to approach a limiting value, write the relationship in words and using the limit notation.

31. $f(x) = \dfrac{x^2 - 2x}{x^2 - 4}$; $x \to 2$

32. $v(x) = \dfrac{x^2 - 3x - 10}{2x + 4}$; $x \to -2$

33. $g(x) = \dfrac{x^2 + 2x - 3}{x - 1}$; $x \to 1$

34. $w(x) = \dfrac{x(x - 2) - 8}{4(x - 4)}$; $x \to 4$

35. $f(x) = 3x^2 - x - 2$; $x \to 1$

36. $g(x) = 10x - x^2$; $x \to -2$

See if a table of values suggests a limit exists for the functions and approaches indicated.

37. $f(x) = 2x^2 - 3x$ as $x \to -3$ from the right.

38. $g(x) = 6x^2 - x$ as $x \to -\dfrac{1}{2}$ from the right.

39. $f(x) = \dfrac{x^3 + 8}{\frac{1}{2}x + 1}$ as $x \to -2$ from the left.

40. $g(x) = \dfrac{x^4 - 1}{x + 1}$ as $x \to -1$ from the left.

41. $f(x) = \dfrac{\sin \sqrt{x}}{\sqrt{x}}$ as $x \to 0$ from the right.

42. $g(x) = \dfrac{x}{\ln x}$ as $x \to 0$ from the right.

Write each statement using limit notation.

43. As x approaches 3 from the left, I_x approaches $3 \cos^2(R_1 + R_2)$.

44. As x approaches -1 from the right, S_x approaches $\log(\varphi + 1)$.

45. As x approaches m from the left, f approaches L.

46. As x approaches n from the right, g approaches L.

Evaluate the following limits using a table of values.

47. Given $f(x) = \dfrac{\sin x}{|\sin x|}$, find

    a. $\lim_{x \to \pi^-} f(x)$ **b.** $\lim_{x \to \pi^+} f(x)$

48. Given $g(x) = -\dfrac{\cos x}{|\cos x|}$, find

    a. $\displaystyle\lim_{x \to \frac{\pi}{2}^-} g(x)$ **b.** $\displaystyle\lim_{x \to \frac{\pi}{2}^+} g(x)$

49. Given $f(x) = \dfrac{x^2 - 10x + 24}{\sqrt{x^2 - 12x + 36}}$, find

    a. $\displaystyle\lim_{x \to 6^-} f(x)$ **b.** $\displaystyle\lim_{x \to 6^+} f(x)$

50. Given $g(x) = \dfrac{x^2 + 2x - 3}{\sqrt{x^2 + 6x + 9}}$, find

    a. $\displaystyle\lim_{x \to -3^-} g(x)$ **b.** $\displaystyle\lim_{x \to -3^+} g(x)$

51. Given $f(x) = \dfrac{2}{x} - 3x^{\frac{1}{2}}$, find

    a. $\displaystyle\lim_{x \to 4^-} f(x)$ **b.** $\displaystyle\lim_{x \to 4^+} f(x)$

52. Given $g(x) = e^x - 5$, find

    a. $\displaystyle\lim_{x \to \ln 7.3^-} g(x)$ **b.** $\displaystyle\lim_{x \to \ln 7.3^+} g(x)$

53. For $f(x) = \begin{cases} x^2 - 4 & x < 2 \\ \sin(x^2 - 4) & x \geq 2 \end{cases}$, find

    a. $\displaystyle\lim_{x \to 2^-} f(x)$ **b.** $\displaystyle\lim_{x \to 2^+} f(x)$

54. For $g(x) = \begin{cases} 3\tan\left[\dfrac{\pi}{4}(x + 2)\right] & x \leq -1 \\ \sqrt{x^2 + 8} & x > -1 \end{cases}$, find

    a. $\displaystyle\lim_{x \to -1^-} g(x)$ **b.** $\displaystyle\lim_{x \to -1^+} g(x)$

55. Given $f(x) = \begin{cases} 2x^2 - 7 & x < -5 \\ 3 - 2x & x \geq -5 \end{cases}$, find

    a. $\displaystyle\lim_{x \to -5^-} f(x)$ **b.** $\displaystyle\lim_{x \to -5^+} f(x)$

    c. $\displaystyle\lim_{x \to 5} f(x)$

56. Given $g(x) = \begin{cases} -(x + 1) & x \leq 10 \\ \log x & x > 10 \end{cases}$, find

    a. $\displaystyle\lim_{x \to 10^-} g(x)$ **b.** $\displaystyle\lim_{x \to 10^+} g(x)$

    c. $\displaystyle\lim_{x \to 10} g(x)$

57. Given $f(x) = \dfrac{\sqrt{1 - \cos(2x)}}{\sqrt{2}\sin x}$, find

    a. $\displaystyle\lim_{x \to \pi^-} f(x)$ **b.** $\displaystyle\lim_{x \to \pi^+} f(x)$

    c. $\displaystyle\lim_{x \to \pi} f(x)$

58. Given $g(x) = \dfrac{\sin[\pi(x - 1)]}{|x|}$, find

    a. $\displaystyle\lim_{x \to 0^-} g(x)$ **b.** $\displaystyle\lim_{x \to 0^+} g(x)$

    c. $\displaystyle\lim_{x \to 0} g(x)$

59. Given $f(x) = \begin{cases} \sin x & x < \dfrac{\pi}{4} \\ \tan x & x = \dfrac{\pi}{4} \\ \cos x & x > \dfrac{\pi}{4} \end{cases}$, find

    a. $\displaystyle\lim_{x \to \frac{\pi}{4}^-} f(x)$ **b.** $\displaystyle\lim_{x \to \frac{\pi}{4}^+} f(x)$

    c. $\displaystyle\lim_{x \to \frac{\pi}{4}} f(x)$

60. Given $g(x) = \begin{cases} \sec x & x < \dfrac{5\pi}{4} \\ \cot x & x = \dfrac{5\pi}{4} \\ \csc x & x > \dfrac{5\pi}{4} \end{cases}$, find

    a. $\displaystyle\lim_{x \to \frac{5\pi}{4}^-} g(x)$ **b.** $\displaystyle\lim_{x \to \frac{5\pi}{4}^+} g(x)$

    c. $\displaystyle\lim_{x \to \frac{5\pi}{4}} g(x)$

61. Given $f(x) = x^2 - \dfrac{3}{x + 2}$, find $\displaystyle\lim_{x \to -2} f(x)$.

62. Given $g(x) = 2x^2 - 3x + 4x^{-2}$, find $\displaystyle\lim_{x \to 0} g(x)$.

63. Given $f(x) = \dfrac{-4\cos\left(\dfrac{\pi}{4}x\right)}{x - 2}$, find $\displaystyle\lim_{x \to 2} f(x)$.

64. Given $g(x) = \csc(x^2 - 17x + 52)$, find $\displaystyle\lim_{x \to 13} g(x)$.

65. Given $f(x) = \dfrac{2x}{x^3 + 216}$, find $\displaystyle\lim_{x \to -6} f(x)$.

66. Given $g(x) = \dfrac{6x^2 - x - 1}{2x - 1}$, find $\displaystyle\lim_{x \to \frac{1}{2}} g(x)$.

67. Given $f(x) = \sin\left(\dfrac{x + 1}{x - 1}\right)$, find $\displaystyle\lim_{x \to 1} f(x)$.

68. Given $g(x) = \cos\left(\dfrac{x - 1}{x + 1}\right)$, find $\displaystyle\lim_{x \to 1} g(x)$.

69. Given $f(x) = \tan(\cos x)$, find $\displaystyle\lim_{x \to \frac{\pi}{2}} f(x)$.

70. Given $f(x) = \cos(\tan x)$, find $\displaystyle\lim_{x \to \frac{\pi}{2}} f(x)$.

MAINTAINING YOUR SKILLS

71. (3.3) Use the rational zeroes theorem to write the polynomial in completely factored form: $3x^4 - 19x^3 + 15x^2 + 27x - 10$.

72. (1.4) Find the sum, difference, product, and quotient of $1 + 3i$ and $1 - 3i$.

73. (7.2) Use Heron's formula to find the area of a triangle with sides $a = 5$ in., $b = 8$ in., $c = 9$ in., rounded to two decimal places.

74. (5.7) Write a sinusoidal equation given the maximum is 100, the minimum is -20 and the period is 365.

11.2 | The Properties of Limits

Learning Objectives

In Section 11.2 you will learn how to:

☐ **A.** Distinguish between a declarative statement and a proof

☐ **B.** Establish limit properties using a table

☐ **C.** Evaluate limits using the limit properties

After the invention of calculus, but before the concept of a limit had been fully developed, the ideas of Newton (the method of fluxions) and Leibniz (the method of differentials) were often met with skepticism and sarcasm even within the academic world. In a tract called *The Analyst* (1794), mathematician and philosopher George Berkeley once said, "he who can digest a second or third fluxion . . . need not, methinks, be squeamish about any point in divinity." In this section, we introduce the properties of limits, which helped to place the study of calculus on a sure footing. The plausibility of these properties will be illustrated using a table or graph, with their formal proof left for a future course.

A. Distinguishing between a Declarative Statement and a Proof

Consider the function $p(x) = x^2 - x + 41$. Evaluating p for the natural numbers 1 through 7 reveals that all outputs are prime numbers (Table 11.9). Made curious by this discovery, we investigate further and find that values of x from 8 through 14 also generate prime numbers (Table 11.10), as do the inputs 15 through 40. At this point it might seem reasonable to declare that p generates a prime number for all natural numbers n. This exercise points out the dangers of making a declarative statement without formal proof, because when $x = 41$, the function returns a value of 1681, which is not a prime number ($41 \cdot 41 = 1681$).

Table 11.9

x	$p(x)$
1	41
2	43
3	47
4	53
5	61
6	71
7	83
$x = 7$	

Table 11.10

x	$p(x)$
8	97
9	113
10	131
11	151
12	173
13	197
14	223
$x = 14$	

Also, in Section 10.4 we noted $1 + 3 + 5 + 7 = 16$, $1 + 3 + 5 + 7 + 9 + 11 = 36$, $1 + 3 + 5 + 7 + 9 + 11 + 13 + 15 + 17 + 19 + 21 + 23 = 144$ and it appeared the sum of the first n odd numbers was equal to n^2. While this declaration turns out to be true, no number of finite checks can *prove* the statement is true for all n, and a formal proof had to be rendered using mathematical induction. In exactly the same way, we can view a table of values to "declare" $\lim_{x \to 3} \dfrac{x^2 - 9}{x - 3} = 6$, using values closer and closer to 3, then even closer still, but how do we *prove* the limit is 6?

It was this type of question that prompted additional sarcasm from George Berkeley. "And what are these fluxions? The velocities of evanescent increments? And what are these same increments? They are neither finite quantities, nor quantities infinitely small, nor yet nothing. May we not then call them the ghosts of departed quantities?" Example 1 demonstrates that in the absence of a formal proof, it appears Mr. Berkeley's skepticism was justified.

EXAMPLE 1 ▶ **Evaluating Limits Using a Table**

Evaluate $\lim\limits_{x \to 0}\left(x^2 + \dfrac{\cos x}{1000} + 1 \right)$ using a table of values.

Solution ▶ For the left- and right-hand limits we obtain Tables 11.11 and 11.12 respectively, and it appears we can declare that $\lim\limits_{x \to 0} f(x) = 1$.

Table 11.11	
x	$f(x)$
−0.6	1.361
−0.5	1.251
−0.4	1.161
−0.3	1.091
−0.2	1.041
−0.1	1.011
−0.01	1.0011
$y = 1.00109999998$	

Table 11.12	
x	$f(x)$
0.6	1.361
0.5	1.251
0.4	1.161
0.3	1.091
0.2	1.041
0.1	1.011
0.01	1.0011
$y = 1.00109999998$	

☑ **A. You've just learned how to distinguish between a declarative statement and a proof**

However, if we continue the approach using values of x that are even closer to $c = 0$, we find $f(0.001) = 1.001001, f(0.0001) = 1.00100001$, and it turns out the limit is actually 1.001. Our declarative statement then has no standing and our work falls far short of a proof.

Now try Exercises 7 through 14 ▶

B. Establishing the Limit Properties

From Example 1 and our work in Section 11.1, it seems we need a more definitive method for evaluating limits. As it turns out, limits possess certain properties that, once established, enable us to evaluate even complex limits with complete confidence. The properties we're interested in involve the various ways that limits can be combined using basic operations. For instance, from our study of operations on functions, we know $(f + g)(x) = f(x) + g(x)$. Is it similarly true that $\lim\limits_{x \to c} [f(x) + g(x)] = \lim\limits_{x \to c} f(x) + \lim\limits_{x \to c} g(x)$? While a formal proof of the properties that follow will be offered in a first calculus course, for the time being we will accept them as true after demonstrating their plausibility using tables and our intuition (despite their shortcomings).

EXAMPLE 2 ▶ **Combining Limits to Establish Limit Properties**

Using a table of values and the techniques from Section 11.1, it can be shown that for $f(x) = x$ and $g(x) = -x^2 + 7$, $\lim\limits_{x \to 3} f(x) = 3$ and $\lim\limits_{x \to 3} g(x) = -2$. Use these results to create a new function $h(x) = (f + g)(x)$ and investigate whether $\lim\limits_{x \to c} [f(x) + g(x)] = \lim\limits_{x \to c} f(x) + \lim\limits_{x \to c} g(x)$.

Solution ▶ For f and g as given, we have $h(x) = x + (-x^2 + 7)$. This means we need to investigate $\lim_{x \to 3} h(x) = \lim_{x \to 3} [x + (-x^2 + 7)]$. The results are shown in Tables 11.13 and 11.14 and suggest $\lim_{x \to 3} h(x) = 1$.

Table 11.13

x	$h(x)$
2.6	2.84
2.7	2.41
2.8	1.96
2.9	1.49
2.99	1.0499
2.999	1.005
2.9999	1.0005
$y = 1.00049999$	

Table 11.14

x	$h(x)$
3.4	-1.16
3.3	-0.59
3.2	-0.04
3.1	0.49
3.01	0.9499
3.001	0.995
3.0001	0.9995
$y = 0.99949999$	

It appears that such a property exists, as the tables suggest

$$\lim_{x \to 3} [f(x) + g(x)] = \lim_{x \to 3} f(x) + \lim_{x \to 3} g(x)$$
$$1 = 3 + (-2)$$
$$1 = 1 \checkmark$$

Now try Exercises 15 through 34 ▶

Other investigations might involve whether limits have a "multiplicative" property. In other words, is $\lim_{x \to c} [f(x)g(x)] = \lim_{x \to c} f(x) \lim_{x \to c} g(x)$? Is $\lim_{x \to c} \dfrac{g(x)}{f(x)} = \dfrac{\lim_{x \to c} g(x)}{\lim_{x \to c} f(x)}$ a true statement? We explore these questions in Example 3.

EXAMPLE 3 ▶ **Combining Limits to Establish Limit Properties**

Using a table of values and the techniques from Section 11.1 suggests that for $f(x) = x^2$ and $g(x) = -3x$, $\lim_{x \to -2} f(x) = 4$ and $\lim_{x \to -2} g(x) = 6$. Note this yields

$\lim_{x \to -2} f(x) \cdot \lim_{x \to -2} g(x) = 4(6) = 24$ and $\dfrac{\lim_{x \to -2} g(x)}{\lim_{x \to -2} f(x)} = \dfrac{6}{4} = \dfrac{3}{2}$. Use this information to

a. Create a new function $m(x) = f(x)g(x)$ to investigate whether $\lim_{x \to -2} [f(x)g(x)] = \lim_{x \to -2} f(x) \lim_{x \to -2} g(x)$.

b. Create a new function $v(x) = \dfrac{g(x)}{f(x)}$ to investigate

whether $\lim_{x \to -2} \dfrac{g(x)}{f(x)} = \dfrac{\lim_{x \to -2} g(x)}{\lim_{x \to -2} f(x)}$.

Table 11.15

x	$m(x)$
-1.9	20.577
-1.99	23.642
-1.999	23.964
-2	
-2.001	24.036
-2.01	24.362
-2.1	27.783

Solution ▶ **a.** For the product of f and g we have $m(x) = (x^2)(-3x) = -3x^3$. Now using a single table to view the approach from the left and from the right (indicated by the arrows), Table 11.15 suggests $\lim_{x \to -2} m(x) = \lim_{x \to -2} (-3x^3) = 24$.

b. For the quotient of g and f we have

$v(x) = \dfrac{-3x}{x^2} = -\dfrac{3}{x}$ $(x \neq 0)$. Table 11.16 indicates

$\displaystyle\lim_{x \to -2} v(x) = \lim_{x \to -2} \dfrac{-3}{x} = \dfrac{3}{2}$. The tables once again seem

to indicate that such properties exist, as they suggest

$\displaystyle\lim_{x \to -2} [f(x)g(x)] = \lim_{x \to -2} f(x) \cdot \lim_{x \to -2} g(x) = 24$, and

$\displaystyle\lim_{x \to -2} \dfrac{g(x)}{f(x)} = \dfrac{\displaystyle\lim_{x \to -2} g(x)}{\displaystyle\lim_{x \to -2} f(x)} = \dfrac{3}{2}.$

Table 11.16

x	$v(x)$
-1.9	1.5789
-1.99	1.5075
-1.999	1.5008
-2	
-2.001	1.4993
-2.01	1.4925
-2.1	1.4286

Now try Exercises 35 through 44 ▶

The preceding observations seem to imply that the limit of a product can be found by taking the limit of each factor, and the limit of a quotient is equal to the quotient of the limits (if the limit of the denominator is not 0). In fact, each of these observations (and others) can be generalized and formally established using the definition of a limit.

C. Finding Limits Using the Limit Properties

Building on our previous work, we might expect that limit properties exist for the other basic operations, and this is indeed the case:

Properties of Limits

Given that the limits $\displaystyle\lim_{x \to c} f(x)$ and $\displaystyle\lim_{x \to c} g(x)$ exist,

(I) $\displaystyle\lim_{x \to c} [f(x) + g(x)] = \lim_{x \to c} f(x) + \lim_{x \to c} g(x)$

the limit of a sum is the sum of the limits

(II) $\displaystyle\lim_{x \to c} [f(x) - g(x)] = \lim_{x \to c} f(x) - \lim_{x \to c} g(x)$

the limit of a difference is the difference of the limits

(III) $\displaystyle\lim_{x \to c} [f(x)g(x)] = \lim_{x \to c} f(x)\lim_{x \to c} g(x)$

the limit of a product is the product of the limits

(IV) $\displaystyle\lim_{x \to c} [kg(x)] = k \lim_{x \to c} g(x)$, k a constant

the limit of a constant times a function is the constant times the limit of the function

(V) $\displaystyle\lim_{x \to c} \dfrac{f(x)}{g(x)} = \dfrac{\displaystyle\lim_{x \to c} f(x)}{\displaystyle\lim_{x \to c} g(x)}$, *provided* $\displaystyle\lim_{x \to c} g(x) \neq 0$

the limit of a quotient is the quotient of the limits

By repeatedly applying property III with $f(x) = g(x)$, we obtain the property related to the limit of a power.

(VI) $\displaystyle\lim_{x \to c} [f(x)]^n = \left[\lim_{x \to c} f(x)\right]^n$, for n a natural number

the limit of a power is the power of the limit

A similar property holds for the limit of an nth root.

(VII) $\displaystyle\lim_{x \to c} \sqrt[n]{f(x)} = \sqrt[n]{\lim_{x \to c} f(x)}$, for $n \in \mathbb{N}$, $n > 1$ [if n is even, then $f(x) > 0$]

the limit of an nth root is the nth root of the limit

Before we begin applying these properties to evaluate limits, we will accept the validity of the following basic limits, most of which are supported by the previous examples. These are used extensively in our application of the limit properties.

Basic Limits

1. $\lim\limits_{x \to c} k = k$ **2.** $\lim\limits_{x \to c} x = c$

3. $\lim\limits_{x \to c} x^n = c^n$, n a natural number   **4.** $\lim\limits_{x \to c} \sqrt[n]{x} = \sqrt[n]{c}$, n a natural number, $n > 1$ (if n is even, then $c > 0$)

EXAMPLE 4 ▶ **Using Limit Properties to Determine a Limit**

Find the following limits using the limit properties. State the property that justifies each step.

a. $\lim\limits_{x \to -3} (x^3 + 3x - 4)$ **b.** $\lim\limits_{x \to 2} \dfrac{x^4 + 3x^2 - 5}{7 - 4x}$

Solution ▶ **a.** $\lim\limits_{x \to -3} (x^3 + 3x - 4) = \lim\limits_{x \to -3} x^3 + \lim\limits_{x \to -3} 3x - \lim\limits_{x \to -3} 4$ properties I and II

$= \lim\limits_{x \to -3} x^3 + 3 \lim\limits_{x \to -3} x - \lim\limits_{x \to -3} 4$ property IV

$= (-3)^3 + 3(-3) - 4$ basic limits 3, 2, and 1

$= -40$ result

b. $\lim\limits_{x \to 2} \dfrac{x^4 + 3x^2 - 5}{7 - 4x} = \dfrac{\lim\limits_{x \to 2} (x^4 + 3x^2 - 5)}{\lim\limits_{x \to 2} (7 - 4x)}$ property V

$= \dfrac{\lim\limits_{x \to 2} x^4 + \lim\limits_{x \to 2} 3x^2 - \lim\limits_{x \to 2} 5}{\lim\limits_{x \to 2} 7 - \lim\limits_{x \to 2} 4x}$ properties I and II

$= \dfrac{\lim\limits_{x \to 2} x^4 + 3 \lim\limits_{x \to 2} x^2 - \lim\limits_{x \to 2} 5}{\lim\limits_{x \to 2} 7 - 4 \lim\limits_{x \to 2} x}$ property IV

$= \dfrac{2^4 + 3(2^2) - 5}{7 - 4(2)}$ basic limits 3, 2, and 1

$= \dfrac{23}{-1} = -23$ result

Note the general strategy is to "break-up" the expressions using limit properties, then use the basic limits to evaluate the result.

Now try Exercises 45 through 56 ▶

EXAMPLE 5 ▶ **Using the Limit Properties to Determine a Limit**

Evaluate the following limits using the limit properties.

a. $\lim\limits_{x \to -4} \sqrt[3]{x^3 - 7x + 28}$ **b.** $\lim\limits_{x \to 0} \dfrac{9\sqrt{x^2 + 4}}{x^2 + 5}$

Solution ▶   **a.** $\lim\limits_{x \to -4} \sqrt[3]{x^3 - 7x + 28} = \sqrt[3]{\lim\limits_{x \to -4} (x^3 - 7x + 28)}$ property VII

$$= \sqrt[3]{\lim\limits_{x \to -4} x^3 - \lim\limits_{x \to -4} 7x + \lim\limits_{x \to -4} 28} \qquad \text{properties I and II}$$

$$= \sqrt[3]{\lim\limits_{x \to -4} x^3 - 7 \lim\limits_{x \to -4} x + \lim\limits_{x \to -4} 28} \qquad \text{property IV}$$

$$= \sqrt[3]{(-4)^3 - 7(-4) + 28} \qquad \text{basic limits 3, 2, and 1}$$

$$= \sqrt[3]{-8} = -2 \qquad \text{result}$$

b. $\lim\limits_{x \to 0} \dfrac{9\sqrt{x^2 + 4}}{x^2 + 5} = \dfrac{\lim\limits_{x \to 0} (9\sqrt{x^2 + 4})}{\lim\limits_{x \to 0} (x^2 + 5)}$ property V

$$= \dfrac{9 \lim\limits_{x \to 0} \sqrt{x^2 + 4}}{\lim\limits_{x \to 0} (x^2 + 5)} \qquad \text{property IV}$$

$$= \dfrac{9\sqrt{\lim\limits_{x \to 0} (x^2 + 4)}}{\lim\limits_{x \to 0} (x^2 + 5)} \qquad \text{property VII}$$

$$= \dfrac{9\sqrt{\lim\limits_{x \to 0} x^2 + \lim\limits_{x \to 0} 4}}{\lim\limits_{x \to 0} x^2 + \lim\limits_{x \to 0} 5} \qquad \text{properties I and II}$$

$$= \dfrac{9\sqrt{0^2 + 4}}{0^2 + 5} \qquad \text{basic limits 3, 2, and 1}$$

$$= \dfrac{9(2)}{5} = \dfrac{18}{5} \qquad \text{result}$$

Now try Exercises 57 through 64 ▶

EXAMPLE 6 ▶   **Using Limit Properties to Determine Limits**

Find the following limits using the limit properties. If a limit does not exist, state why.

a. $\lim\limits_{x \to -1} \dfrac{x^2}{x^2 + 2x + 1}$ **b.** $\lim\limits_{x \to -2} \dfrac{x^2 + 2x}{\sqrt{x^2 + 4x + 4}}$

a. $\lim\limits_{x \to -1} \dfrac{x^2}{x^2 + 2x + 1} = \dfrac{\lim\limits_{x \to -1} x^2}{\lim\limits_{x \to -1} (x^2 + 2x + 1)}$ property V

$$= \dfrac{\lim\limits_{x \to -1} x^2}{\lim\limits_{x \to -1} x^2 + \lim\limits_{x \to -1} 2x + \lim\limits_{x \to -1} 1} \qquad \text{property I}$$

$$= \dfrac{(-1)^2}{(-1)^2 + 2(-1) + 1} \qquad \text{basic limits 3, 2, and 1}$$

$$= \dfrac{1}{0} \qquad \text{result}$$

In this case, applying the limit properties results in an undefined expression, and we resort to a table of values to see what's happening as $x \to -1$.

Table 11.17

x	y
-1.5	9
-1.4	12.25
-1.3	18.778
-1.2	36
-1.1	121
-1.01	10201
-1.001	1E6
$y = 1002001$	

Table 11.18

x	y
-0.5	1
-0.6	2.25
-0.7	5.4444
-0.8	16
-0.9	81
-0.99	9801
-0.999	998001
$y = 998001$	

Tables 11.17 and 11.18 show that as $x \to -1$, function values grow infinitely large and positive, and we conclude $\lim\limits_{x \to -1} \dfrac{x^2}{x^2 + 2x + 1} = \left(\substack{\text{dne} \\ \infty}\right)$.

b. $\lim\limits_{x \to -2} \dfrac{x^2 + 2x}{\sqrt{x^2 + 4x + 4}}$

$= \dfrac{\lim\limits_{x \to -2} (x^2 + 2x)}{\lim\limits_{x \to -2} \sqrt{x^2 + 4x + 4}}$ property V

$= \dfrac{\lim\limits_{x \to -2} x^2 + \lim\limits_{x \to -2} 2x}{\sqrt{\lim\limits_{x \to -2}(x^2 + 4x + 4)}}$ properties I and VII

$= \dfrac{\lim\limits_{x \to -2} x^2 + \lim\limits_{x \to -2} 2x}{\sqrt{\lim\limits_{x \to -2} x^2 + \lim\limits_{x \to -2} 4x + \lim\limits_{x \to -2} 4}}$ property I

$= \dfrac{\left(\lim\limits_{x \to -2} x\right)^2 + 2\lim\limits_{x \to -2} x}{\sqrt{\left(\lim\limits_{x \to -2} x\right)^2 + 4\lim\limits_{x \to -2} x + \lim\limits_{x \to -2} 4}}$ properties VI and IV

$= \dfrac{(-2)^2 + 2(-2)}{\sqrt{(-2)^2 + 4(-2) + 4}} = \dfrac{0}{\sqrt{0}}$ basic limits 3, 2, and 1

$= \dfrac{0}{0}$ result

> **WORTHY OF NOTE**
>
> Recall that the equation $\dfrac{1}{0} = x$ (with an undefined expression) has no solution, while the equation $\dfrac{0}{0} = x$ (with an indeterminate expression) has infinitely many solutions—a unique solution cannot be "determined."

Here, the limit properties result in an indeterminate form, and we again resort to a table of values to check the behavior of the function as $x \to -2$.

Table 11.19

x	y
-2.5	2.5
-2.4	2.4
-2.3	2.3
-2.2	2.2
-2.1	2.1
-2.01	2.01
-2.001	2.001
$y = 2.001$	

Table 11.20

x	y
-1.5	-1.5
-1.6	-1.6
-1.7	-1.7
-1.8	-1.8
-1.9	-1.9
-1.99	-1.99
-1.999	-1.999
$y = -1.999$	

Tables 11.19 and 11.20 imply that $\lim\limits_{x \to -2^-} \dfrac{x^2 + 2x}{\sqrt{x^2 + 4x + 4}} = 2$ (the left-hand

limit is 2), while $\lim\limits_{x \to -2^+} \dfrac{x^2 + 2x}{\sqrt{x^2 + 4x + 4}} = -2$ (the right-hand limit is -2).

Since these do not agree, the limit does not exist and we write

$$\lim\limits_{x \to -2} \dfrac{x^2 + 2x}{\sqrt{x^2 + 4x + 4}} = \left(\underset{\text{LH} \neq \text{RH}}{\text{dne}}\right).$$

 **C.** You've just learned how to evaluate limits using the limit properties

Now try Exercises 65 through 76 ▶

11.2 EXERCISES

▶ **CONCEPTS AND VOCABULARY**

Fill in each blank with the appropriate word or phrase.
Carefully reread the section, if necessary.

1. The limit of a sum is the _____ of the _____ .

2. The limit of a product is the _____ of the _____ .

3. The limit of an nth _____ is the _____ root of the limit: $\lim\limits_{x \to c} \sqrt[n]{f(x)} = \sqrt[n]{\lim\limits_{x \to c} f(x)}$, provided _____ if n is even.

4. The limit of a _____ is the quotient of the _____, $\lim\limits_{x \to c} \dfrac{f(x)}{g(x)} = \dfrac{\lim\limits_{x \to c} f(x)}{\lim\limits_{x \to c} g(x)}$, provided _____.

5. Discuss/Explain how applying the limit properties can result in an undefined or indeterminate expression. What should be done in these cases?

6. Discuss/Explain the relationship between limit properties VI and VII, and basic limits 3 and 4. How is basic limit 2 involved here?

▶ **DEVELOPING YOUR SKILLS**

For Exercises 7 and 8, (a) complete the table given for each function and state the apparent limit as $x \to 0$, then (b) continue the approach using values even closer to 0 and comment.

7. $f(x) = x^2 + \dfrac{\sin x}{1000x}$

x	$f(x)$	x	$f(x)$
0.5		-0.5	
0.4		-0.4	
0.3		-0.3	
0.2		-0.2	
0.1		-0.1	
0.01		-0.01	
0.001		-0.001	

8. $f(x) = \dfrac{1000x^{\frac{2}{3}} - 1}{1000}$

x	$f(x)$	x	$f(x)$
0.5		-0.5	
0.4		-0.4	
0.3		-0.3	
0.2		-0.2	
0.1		-0.1	
0.01		-0.01	
0.001		-0.001	

9. Complete the table shown (also see Exercise 11).

x	$y = x + \dfrac{\sqrt{x} - 2}{10^x}$
2.7	
2.8	
2.9	
3	
3.1	
3.2	
3.3	

10. Complete the table shown (also see Exercise 12).

x	$= 2x + \dfrac{x + 3}{40(x - 3)^2}$
-1.7	
-1.8	
-1.9	
-2	
-2.1	
-2.2	
-2.3	

11. Complete the table shown (also see Exercise 9).

x	$y = x + \dfrac{\sqrt{x} - 2}{10^x}$
2.99	
2.999	
2.9999	
3	
3.0001	
3.001	
3.01	

12. Complete the table shown (also see Exercise 10).

x	$y = 2x + \dfrac{x + 3}{40(x - 3)^2}$
-1.999	
-1.9999	
-1.99999	
-2	
-2.00001	
-2.0001	
-2.001	

13. a. Based on the table in Exercise 9, estimate
$$\lim_{x \to 3}\left(x + \frac{\sqrt{x} - 2}{10^x}\right).$$

b. Based on the table in Exercise 11, what appears to be the actual value of $\lim\limits_{x \to 3}\left(x + \dfrac{\sqrt{x} - 2}{10^x}\right)$?

14. a. Based on the table in Exercise 10, estimate
$$\lim_{x \to -2}\left[2x + \frac{x + 3}{40(x - 3)^2}\right].$$

b. Based on the table in Exercise 12, what appears to be the actual value of $\lim\limits_{x \to -2}\left[2x + \dfrac{x + 3}{40(x - 3)^2}\right]$?

The following functions will be used for Exercises 15 through 44.

$a(x) = 3 - 2x$	$b(x) = \lvert x - 2 \rvert$
$c(x) = 1 - x^2$	$d(x) = 5^x$
$f(x) = \sqrt{x + 7}$	$g(x) = \cos\left(\dfrac{\pi}{2x}\right)$
$h(x) = \sin(x + 3)$	$j(x) = \log[(5x)^2]$
$k(x) = \lvert x^2 - 16 \rvert$	$l(x) = \sqrt{x^2 - 10x + 25}$

For Exercises 15 through 24, use a table of values to evaluate each limit.

15. $\lim\limits_{x \to -3} a(x)$ **16.** $\lim\limits_{x \to 2} b(x)$

17. $\lim\limits_{x \to -3} c(x)$ **18.** $\lim\limits_{x \to 2} d(x)$

19. $\lim\limits_{x \to -3} f(x)$ **20.** $\lim\limits_{x \to 2} g(x)$

21. $\lim\limits_{x \to -3} h(x)$ **22.** $\lim\limits_{x \to 2} j(x)$

23. $\lim\limits_{x \to -3} k(x)$ **24.** $\lim\limits_{x \to 2} l(x)$

For Exercises 25 through 34, use a table of values to evaluate each limit. Compare each result with that of the corresponding Exercises in 15 through 24.

25. $\lim\limits_{x \to -3}[a(x) + c(x)]$ **26.** $\lim\limits_{x \to 2}[d(x) + g(x)]$

27. $\lim\limits_{x \to -3}[f(x) + h(x) + k(x)]$

28. $\lim\limits_{x \to 2}[b(x) + d(x) + j(x)]$

29. $\lim\limits_{x \to -3}[h(x) - c(x)]$ **30.** $\lim\limits_{x \to 2}[b(x) - g(x)]$

31. $\lim\limits_{x \to -3}[f(x) + k(x) - a(x)]$

32. $\lim\limits_{x \to 2}[j(x) - d(x) + l(x)]$

33. $\lim\limits_{x \to -3}[k(x) + k(x)]$ **34.** $\lim\limits_{x \to 2}[d(x) + d(x)]$

For Exercises 35 through 44, use a table of values to evaluate each limit. Compare each result with that of the corresponding Exercises in 15 through 24.

35. $\lim\limits_{x\to-3}\left[c(x)f(x)\right]$ **36.** $\lim\limits_{x\to2}\left[j(x)d(x)\right]$

37. $\lim\limits_{x\to-3}\dfrac{k(x)}{f(x)}$ **38.** $\lim\limits_{x\to2}\dfrac{d(x)}{g(x)}$

39. $\lim\limits_{x\to-3}\left[a(x)h(x)k(x)\right]$ **40.** $\lim\limits_{x\to2}\left[b(x)g(x)l(x)\right]$

41. $\lim\limits_{x\to-3}\left[c(x)c(x)\right]$ **42.** $\lim\limits_{x\to2}\left[d(x)d(x)\right]$

43. $\lim\limits_{x\to-3}\dfrac{a(x)c(x)}{f(x)}$ **44.** $\lim\limits_{x\to2}\dfrac{d(x)}{g(x)j(x)}$

For Exercises 45 through 62, evaluate the limits using the limit properties.

45. $\lim\limits_{x\to-4}(x^3-5)$ **46.** $\lim\limits_{x\to16}(\sqrt{x}-16)$

47. $\lim\limits_{x\to-4}(2x^2-x-7)$ **48.** $\lim\limits_{x\to2}(x^4-2x+5)$

49. $\lim\limits_{x\to4}(x^2-5\sqrt{x}+3)$

50. $\lim\limits_{x\to-8}\left(\dfrac{1}{2}x^2-\sqrt[3]{x}+3\right)$

51. $\lim\limits_{x\to2}\dfrac{2x^2-5}{x+3}$ **52.** $\lim\limits_{x\to-3}\dfrac{x^2-2x+3}{x-3}$

53. $\lim\limits_{x\to1}\dfrac{\dfrac{1}{x+1}-x^2}{\dfrac{1}{x}+2x^2-x}$ **54.** $\lim\limits_{x\to3}\dfrac{\dfrac{1}{x-2}+3x^2}{x^2-6+\dfrac{1}{x-1}}$

55. $\lim\limits_{x\to3}(x^2-3)^3$ **56.** $\lim\limits_{x\to2}(3x^2-2x)^2$

57. $\lim\limits_{x\to10}5\sqrt[3]{2x+7}$ **58.** $\lim\limits_{x\to-2}-2\sqrt[3]{x-6}$

59. $\lim\limits_{x\to-3}(\sqrt{x+7}-7x)$ **60.** $\lim\limits_{x\to-5}(2x+\sqrt{4-x})$

61. $\lim\limits_{x\to2}\dfrac{x^3-2x-10}{2\sqrt[3]{5x^2+2x+3}}$ **62.** $\lim\limits_{x\to2}\dfrac{11-3x^2}{\sqrt{x^2+3x-1}}$

63. $\lim\limits_{x\to3}\dfrac{\dfrac{2}{x+1}-3x}{\sqrt{x-2}+1}$ **64.** $\lim\limits_{x\to-1}\dfrac{\left(\dfrac{x}{2x+1}\right)^2-3x}{\sqrt{x^2+3}+1}$

For Exercises 65 through 76, evaluate the limits using limit properties. If a limit does not exist, state why.

65. $\lim\limits_{x\to-2}\dfrac{3x^2-11x-4}{x^2-2x-8}$ **66.** $\lim\limits_{x\to1}\dfrac{2x^2-x-3}{x^2-1}$

67. $\lim\limits_{x\to7}\dfrac{x^2-5x-14}{x-7}$ **68.** $\lim\limits_{x\to-4}\dfrac{x^2+7x+12}{2x+8}$

69. $\lim\limits_{x\to-2}\dfrac{x^2+5x+6}{x^2+4x+4}$ **70.** $\lim\limits_{x\to1}\dfrac{x^2-4x+3}{x^2-2x+1}$

71. $\lim\limits_{x\to-7}\dfrac{\sqrt{x^2+14x+49}}{x^2+8x+7}$ **72.** $\lim\limits_{x\to-1}\dfrac{2x^2-x-3}{\sqrt{x^2+2x+1}}$

73. $\lim\limits_{x\to0}\dfrac{(x+3)^2-9}{x}$ **74.** $\lim\limits_{x\to0}\dfrac{(x-1)^3+1}{x}$

75. $\lim\limits_{x\to1}\dfrac{x^3-x^2}{\sqrt{x^2-2x+1}}$ **76.** $\lim\limits_{x\to1}\dfrac{x^3|x-1|}{x-1}$

▶ **MAINTAINING YOUR SKILLS**

77. (2.5) If $p(x)=2x^2-x-3$, in what intervals is $p(x)\le0$?

78. (5.4) Solve the logarithmic equation $\log(x+2)+\log x=\log3$.

79. (6.2) Verify the identity: $\sec y-\cos y=\tan y\sin y$

80. (6.7) Use any appropriate method to solve in $[0,2\pi)$: $\sin2x=\cos x$.

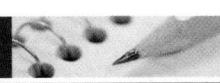

MID-CHAPTER CHECK

1. Express the following statement using limit notation:

As x decreases without bound, $\dfrac{6x^2-3}{12x^3-x-1}$ approaches 0.

Use a table of values to complete Exercises 2 through 7.

2. Evaluate each limit.

   a. $\lim\limits_{h\to0}\dfrac{\sin h}{h}$ **b.** $\lim\limits_{h\to0}\dfrac{\cos h-1}{h}$

3. The formula for interest compounded n times per year is $A = P\left(1 + \dfrac{r}{n}\right)^{nt}$, where P represents the principal deposit, r is the annual interest rate, t is the number of years the funds are on deposit, and n is the number of compoundings per year. If we let $n \to \infty$, this formula can be transformed into the formula for interest compounded continuously: $A = Pe^{rt}$. Verify that $1000 invested at 3% compounded continuously for 1 yr grows to $1030.45. Then find the minimum number of compoundings per year needed to yield the same amount with the same $1000 principal.

4. The function $f(x)$ shown has a pointwise discontinuity. Create a related piecewise-defined function $F(x)$ that is continuous for $x \in \mathbb{R}$.

$$f(x) = \frac{6x^2 - 19x - 7}{2x - 7}$$

5. Given $g(x) = \begin{cases} \dfrac{-x}{2x + 5} & x < -5 \\ \cos(x + 5) & x \geq -5 \end{cases}$, find the following:

   a. $\displaystyle\lim_{x \to -5^-} g(x)$ b. $\displaystyle\lim_{x \to -5^+} g(x)$

   c. $\displaystyle\lim_{x \to 5} g(x)$

6. Given $h(x) = \cot\left(\dfrac{x^2 - 4}{x + 2}\right)$, find $\displaystyle\lim_{x \to 2} h(x)$.

7. a. Complete the table of values shown.
   b. Based on your table, it appears that

   $$\lim_{x \to 0} \cos\left(\frac{\pi}{x}\right)$$

   is approaching what value? Graph $y = \cos\left(\dfrac{\pi}{x}\right)$ on a graphing calculator, and "Zoom In" a few times. Is your prior observation still valid?

x	$y = \cos\left(\dfrac{\pi}{x}\right)$
0.1	
0.01	
0.001	
0	
−0.001	
−0.01	
−0.1	

For Exercises 8 through 10, evaluate each limit using the limit properties. If a limit does not exist, state why.

8. $\displaystyle\lim_{x \to 2} \frac{4x^2 + x - 5}{x^5 - 3x}$

9. $\displaystyle\lim_{x \to -3} \frac{2x + 10}{2\sqrt[3]{11 - x^2}}$

10. $\displaystyle\lim_{x \to 2} \frac{\sqrt{x^2 - 4x + 4}}{x - 2}$

11.3 | Continuity and More on Limits

Learning Objectives

In Section 11.3 you will learn how to:

☐ **A.** Determine if a function is continuous at c and find limits by direct substitution

☐ **B.** Evaluate limits using algebra and the properties of limits

☐ **C.** Evaluate limits at infinity

☐ **D.** Use the definition of a limit to evaluate limits graphically

Historically, the careful development of the limit concept also impacted the concept of continuity and the two seem to have matured simultaneously. In this section, we'll see how limits help define the continuity of a function more precisely, and once established, how continuity makes finding certain limits an easier task. We'll also see how the algebra skills developed in previous courses (adding rational expressions, simplifying complex fractions, multiplying by conjugates, etc.) are used to help determine some very important limits.

A. Continuity and Finding Limits by Direct Substitution

In previous chapters, we stated that a function was **continuous** if you could draw the entire graph without lifting your pencil. We also noted there were several ways that continuity could be interrupted, such as the holes and gaps seen in a study of piecewise-defined functions (Section 2.7), and the vertical asymptotes in some rational and trigonometric graphs. These "interruptions" are called **discontinuities** and can occur with other types of functions as well. The graph of $f(x)$ given in Figure 11.9 shows a vertical asymptote at $x = -1$, which interrupts the continuity of the graph and indicates that f is not defined at -1.

Figure 11.9

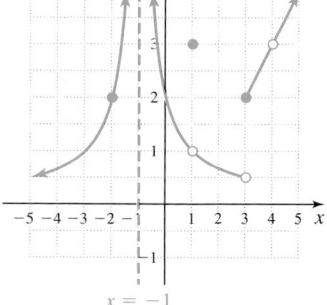

This is called an **asymptotic discontinuity.** The graph also shows that continuity is interrupted at $x = 1$ (there is a "hole" in the graph), even though f is still defined there: $f(1) = 3$. This is called a **removable discontinuity,** because the discontinuity can be removed/repaired by redefining the function as $f(x) = 1$ when $x = 1$. The discontinuity at $x = 4$ is also removable, even though the function is not defined at 4. A break in the graph also occurs at $x = 3$, but this time it leaves a gap in the output values, and there is no way to redefine f to "fill the gap." This is called a **jump discontinuity** or a **nonremovable discontinuity.** In fact, of all the points *called out* by this graph, the function is only continuous at $x = -2$. By comparing the graph at $x = -2$ with the graph of f at other points, we find the following conditions are required for continuity.

Continuity

A function f is continuous at c if

1. f is defined at c **2.** $\lim\limits_{x \to c^-} f(x) = \lim\limits_{x \to c^+} f(x)$ **3.** $\lim\limits_{x \to c} f(x) = f(c)$

In Figure 11.9, the discontinuities at $x = -1$ and $x = 4$ violate condition 1, the discontinuity at $x = 3$ violates condition 2, and the discontinuity at $x = 1$ violates condition 3.

EXAMPLE 1 ▶   **Analyzing the Continuity of a Function Graphically**

The graph of a function f is given. Use the graph to comment on the continuity of the function for all integer values from -2 to 3. If the function is discontinuous at any value, state which of the three conditions are violated.

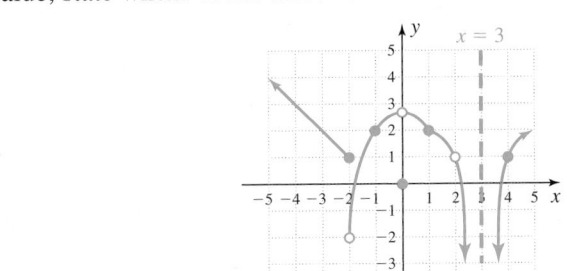

> **WORTHY OF NOTE**
>
> The continuity of a function really boils down to condition 3, since if this condition is satisfied, conditions 1 and 2 are satisfied.

Solution ▶ $x = -2$: discontinuous, condition 2 is violated $x = -1$: continuous
$x = 0$: discontinuous, condition 3 is violated $x = 1$: continuous
$x = 2$: discontinuous, condition 1 is violated $x = 3$: discontinuous,
                                                   condition 1 is violated

Now try Exercises 7 through 16 ▶

The fact that condition 3, $\lim\limits_{x \to c} f(x) = f(c)$, is one of the requirements for continuity offers a great advantage when evaluating the limit of many common functions. For example, using properties of limits it can be shown that polynomials are continuous for all real numbers, and rational, root, and trigonometric functions are continuous wherever they are defined. This means as long as c is in the domain of these functions, we can find the limit by **direct substitution** since $\lim\limits_{x \to c} f(x) = f(c)$ can be assumed.

Finding Limits Using Direct Substitution

If f is a polynomial, rational, root, or trigonometric function
and c is in the domain of f, then

$$\lim_{x \to c} f(x) = f(c)$$

> **EXAMPLE 2** ▶ **Finding Limits Using Direct Substitution**
>
> Evaluate the following limits using direct substitution, if possible.
>
> **a.** $\lim\limits_{x \to -3} x^3 + 3x - 4$ **b.** $\lim\limits_{x \to 2} \dfrac{x^4 + 3x^2 - 5}{7 - 4x}$ **c.** $\lim\limits_{x \to 2} \dfrac{x^2 + 3x - 10}{x - 2}$
>
> **Solution** ▶ **a.** Since $x^3 + 3x - 4$ is a polynomial, the limit can be found by direct substitution:
>
> $$\lim\limits_{x \to -3} x^3 + 3x - 4 = (-3)^3 + 3(-3) - 4 \quad \text{substitute } -3 \text{ for } x$$
>
> $$= -27 + (-9) - 4 \quad \text{simplify}$$
>
> $$= -40 \quad \text{result}$$
>
> See Example 4 of Section 11.2 where this limit was found using limit properties.
>
> **b.** Since $\lim\limits_{x \to 2} \dfrac{x^4 + 3x^2 - 5}{7 - 4x}$ is a rational function and 2 is in the domain, the limit can be found by direct substitution:
>
> $$\lim\limits_{x \to 2} \dfrac{x^4 + 3x^2 - 5}{7 - 4x} = \dfrac{2^4 + 3(2)^2 - 5}{7 - 4(2)} \quad \text{substitute 2 for } x$$
>
> $$= \dfrac{16 + 12 - 5}{7 - 8} \quad \text{simplify}$$
>
> $$= -23 \quad \text{result}$$
>
> See Example 4 of Section 11.2 where this limit was found using limit properties.
>
> **c.** Since $\lim\limits_{x \to 2} \dfrac{x^2 + 3x - 10}{x - 2}$ is a rational function and 2 is *not* in the domain, the
>
> ☑ **A.** You've just learned how to determine if a function is continuous at c and find limits by direct substitution
>
> limit cannot be found by direct substitution. This doesn't mean the limit doesn't exist, only that direct substitution cannot be used. A table of values seems to indicate $\lim\limits_{x \to 2} \dfrac{x^2 + 3x - 10}{x - 2} = 7$ (also see Example 3a).
>
> **Now try Exercises 17 through 26** ▶

B. Evaluating Limits Using Algebra and Limit Properties

Recall that for $\lim\limits_{x \to c} f(x) = L$, the difference between x and c is made very small by taking values of x very close to c but not equal to c. In many cases, knowing that $\dfrac{x - c}{x - c} = 1$ (even if $x - c$ is very small), can help evaluate limits where a direct substitution is not initially possible. This is done by using algebra to rewrite the function prior to taking the limit.

> **EXAMPLE 3** ▶ **Finding Limits Using Algebra and Limit Properties**
>
> Evaluate the following limits.
>
> **a.** $\lim\limits_{x \to 2} \dfrac{x^2 + 3x - 10}{x - 2}$ **b.** $\lim\limits_{x \to 5} \dfrac{\sqrt{x + 4} - 3}{x - 5}$

Solution ▶   **a.** Although the function given is rational, $x = 2$ is not in the domain and the limit cannot be found using direct substitution. But noting the numerator is factorable, we can work as follows.

$$\lim_{x \to 2} \frac{x^2 + 3x - 10}{x - 2} = \lim_{x \to 2} \frac{(x + 5)(x - 2)}{x - 2} \quad \text{factor the numerator}$$

$$= \lim_{x \to 2} \frac{(x + 5)\cancel{(x - 2)}}{\cancel{x - 2}} \quad \text{factors reduce since } x - 2 \neq 0$$

$$= \lim_{x \to 2} (x + 5) \quad \text{result is a polynomial}$$

$$= 2 + 5 = 7 \quad \text{evaluate by direct substitution}$$

As in Example 2c, a table of values will support this result.

b. Since $x = 5$ is not in the domain, we again attempt to rewrite the expression in a form that would make direct substitution possible. Note that multiplying the numerator and denominator by $\sqrt{x + 4} + 3$ (the conjugate of $\sqrt{x + 4} - 3$) will eliminate the radical in the numerator since $(A - B)(A + B) = A^2 - B^2$. This yields

$$\lim_{x \to 5} \frac{\sqrt{x + 4} - 3}{x - 5} = \lim_{x \to 5} \frac{(\sqrt{x + 4} - 3)(\sqrt{x + 4} + 3)}{(x - 5)(\sqrt{x + 4} + 3)} \quad \text{multiply by } \frac{\sqrt{x + 4} + 3}{\sqrt{x + 4} + 3}$$

$$= \lim_{x \to 5} \frac{[(x + 4) - 9]}{(x - 5)(\sqrt{x + 4} + 3)} \quad (A - B)(A + B) = A^2 - B^2$$

$$= \lim_{x \to 5} \frac{\cancel{(x - 5)}}{\cancel{(x - 5)}(\sqrt{x + 4} + 3)} \quad \text{simplify numerator}$$

$$= \lim_{x \to 5} \frac{1}{\sqrt{x + 4} + 3} \quad \text{factors reduce since } x - 5 \neq 0$$

$$= \frac{\lim\limits_{x \to 5} 1}{\lim\limits_{x \to 5} (\sqrt{x + 4} + 3)} \quad \text{property V}$$

$$= \frac{1}{\sqrt{5 + 4} + 3} \quad \text{evaluate by direct substitution}$$

$$= \frac{1}{\sqrt{9} + 3} = \frac{1}{6} \quad \text{result}$$

The result can be supported using a table of values.

Now try Exercises 27 through 36 ▶

In Example 4, we need to find the limit as $h \to 0$ rather than $x \to c$, but the concepts are identical $\left(\dfrac{h}{h} = 1 \text{ as long as } h \text{ is not } 0 \right)$. If h is not in the domain, the limit cannot be found by direct substitution, and we again attempt to write the expression in an alternative form.

EXAMPLE 4 ▶   **Finding Limits Using Algebra and Limit Properties**

Evaluate the following limits.

a. $\displaystyle \lim_{h \to 0} \frac{[(x + h)^2 + 3(x + h)] - (x^2 + 3x)}{h}$ **b.** $\displaystyle \lim_{h \to 0} \frac{\dfrac{1}{x + h} - \dfrac{1}{x}}{h}$

Solution ▶   **a.** Here we begin by simplifying the numerator.

$$\lim_{h \to 0} \frac{[(x + h)^2 + 3(x + h)] - (x^2 + 3x)}{h}$$

$$= \lim_{h \to 0} \frac{x^2 + 2xh + h^2 + 3x + 3h - x^2 - 3x}{h} \qquad \text{square binomial, distribute 3, distribute } -1$$

$$= \lim_{h \to 0} \frac{2xh + h^2 + 3h}{h} \qquad \text{combine like terms}$$

$$= \lim_{h \to 0} \frac{\cancel{h}(2x + h + 3)}{\cancel{h}} \qquad \text{factor out } h$$

$$= \lim_{h \to 0} (2x + h + 3) \qquad \text{factors reduce since } h \neq 0$$

$$= 2x + 3 \qquad \text{evaluate by direct substitution}$$

b. Here we begin by simplifying the complex fraction, hoping to rewrite the expression in a form that allows a direct substitution.

$$\lim_{h \to 0} \frac{\dfrac{1}{x + h} - \dfrac{1}{x}}{h} = \lim_{h \to 0} \frac{\dfrac{x}{x(x + h)} - \dfrac{(x + h)}{x(x + h)}}{h} \qquad \text{LCD for the numerator is } x(x + h)$$

$$= \lim_{h \to 0} \frac{\dfrac{x - (x + h)}{x(x + h)}}{h} \qquad \text{combine terms in the numerator}$$

$$= \lim_{h \to 0} \frac{\dfrac{-h}{x(x + h)}}{h} \qquad \text{simplify}$$

$$= \lim_{h \to 0} \frac{-\cancel{h}}{x(x + h)} \cdot \frac{1}{\cancel{h}} \qquad \text{invert and multiply}$$

$$= \lim_{h \to 0} \frac{-1}{x(x + h)} \qquad \text{factors reduce since } h \neq 0$$

$$= \frac{-1}{x^2} \qquad \text{evaluate by direct substitution}$$

☑ **B.** You've just learned how to evaluate limits using algebra and the properties of limits

Now try Exercises 37 through 44 ▶

C. Evaluating Limits at Infinity

The limit of a function as $x \to c$ necessarily has x moving *infinitely close* to a number c from the left or right. In contrast, limits at infinity are concerned with the value of a function as x moves *infinitely away from* an arbitrary number c to the left or right, meaning as $x \to -\infty$ or $x \to \infty$.

Figure 11.10

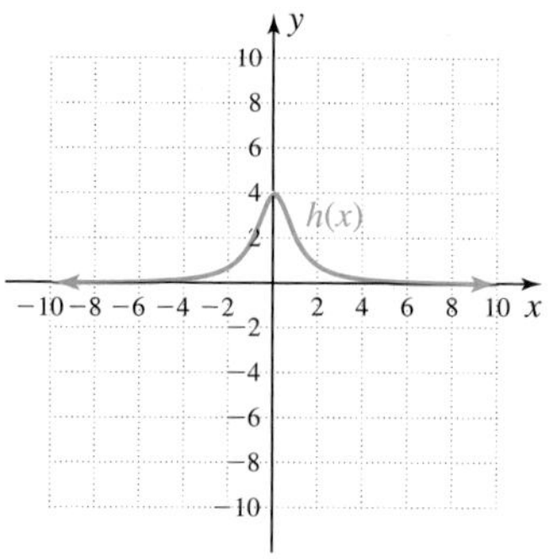

The graph of $h(x) = \dfrac{4}{x^2 + 1}$ is shown in Figure 11.10, and we note as $x \to \infty$, values of $h(x)$ become closer and closer to 0. Similar to our previous work, we could say that values of $h(x)$ can be made arbitrarily close to 0, by taking values of x sufficiently large and positive. This suggests the following definition.

Limits at Positive Infinity

For a function f defined on an interval (c, ∞),

$$\lim_{x \to \infty} f(x) = L$$

means that values of $f(x)$ can be made arbitrarily close to L, by taking values of x sufficiently large and positive.

In the present case, we have $\lim\limits_{x \to \infty} \dfrac{4}{x^2 + 1} = 0$.

EXAMPLE 5 ▶ **Evaluating Limits at Infinity**

For $f(x) = \dfrac{3x^2 - x - 12}{2x^2 - 8}$, use a table of values to evaluate $\lim\limits_{x \to \infty} f(x)$.

x	y
250	1.498
500	1.499
750	1.4993
1000	1.4995
1250	1.4996
1500	1.4997
1750	1.4997
$y = 1.49971428534$	

Solution ▶ From our work with rational functions in Chapter 3 (ratio of leading terms), we suspect $\lim\limits_{x \to \infty} f(x) = \dfrac{3}{2}$. This is supported by the table shown.

Now try Exercises 45 through 52 ▶

WORTHY OF NOTE

Since ∞ is not a number, the notation $x \to \infty$ should no longer be read, "as x *approaches* infinity," and instead we say, "as x becomes infinitely large," or "as x increases without bound."

The limit of a function as $x \to -\infty$ is likewise defined.

Limits at Negative Infinity

For a function f defined on an interval $(-\infty, c)$

$$\lim_{x \to -\infty} f(x) = L$$

means that values of $f(x)$ can be made arbitrarily close to L, by taking values of x sufficiently large and negative.

EXAMPLE 6 ▶ **Finding Limits at Infinity**

For $f(x) = \dfrac{1}{x}$, evaluate $\lim\limits_{x \to -\infty} f(x)$ and $\lim\limits_{x \to \infty} f(x)$.

Solution ▶ You might recognize this function from our work in Chapter 3, where we observed that as x becomes very large, $\dfrac{1}{x}$ becomes very small and close to zero (see figure).

The result was a horizontal asymptote at $y = 0$ (the x-axis). In fact, we can make $f(x)$ as close to 0 as we like, by taking x to be a sufficiently large positive number, or a sufficiently large negative number.

This shows $\lim\limits_{x \to -\infty} \dfrac{1}{x} = 0$ and $\lim\limits_{x \to \infty} \dfrac{1}{x} = 0$.

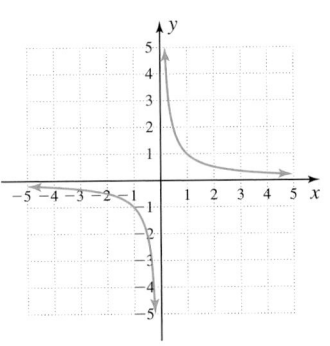

Now try Exercises 53 through 60 ▶

From Example 6 we learn two important things. First, horizontal asymptotes can be defined in terms of limits at infinity.

Horizontal Asymptotes

The line $y = L$ is a horizontal asymptote if

$$\lim_{x \to -\infty} f(x) = L \qquad \text{or} \qquad \lim_{x \to \infty} f(x) = L$$

Second, by repeatedly applying limit property VI (the limit properties also hold for limits at infinity), we can state the following result.

Limits of Reciprocal Powers

For any positive integer k,

$$\lim_{x \to -\infty} \frac{1}{x^k} = 0 \qquad \text{and} \qquad \lim_{x \to \infty} \frac{1}{x^k} = 0$$

These limits are a great asset to finding limits at infinity, as we often need to rewrite an expression and apply *this* limit in order to find the desired limit. For rational functions, this involves dividing numerator and denominator by the highest power of x occurring in the denominator.

EXAMPLE 7 ▶ **Finding Limits at Infinity for a Rational Function**

Evaluate $\displaystyle\lim_{x \to \infty} \frac{5x^2 - 6x + 3}{4x^2 - 9x + 12}$.

Solution ▶ Begin by dividing the numerator and denominator by x^2, the highest power of x in the denominator.

$$\lim_{x \to \infty} \frac{5x^2 - 6x + 3}{4x^2 - 9x + 12} = \lim_{x \to \infty} \frac{\dfrac{5x^2}{x^2} - \dfrac{6x}{x^2} + \dfrac{3}{x^2}}{\dfrac{4x^2}{x^2} - \dfrac{9x}{x^2} + \dfrac{12}{x^2}} \qquad \text{divide numerator and denominator by } x^2$$

$$= \lim_{x \to \infty} \frac{5 - \dfrac{6}{x} + \dfrac{3}{x^2}}{4 - \dfrac{9}{x} + \dfrac{12}{x^2}} \qquad \text{simplify}$$

$$= \frac{\displaystyle\lim_{x \to \infty} \left(5 - \dfrac{6}{x} + \dfrac{3}{x^2} \right)}{\displaystyle\lim_{x \to \infty} \left(4 - \dfrac{9}{x} + \dfrac{12}{x^2} \right)} \qquad \text{property V}$$

$$= \frac{\displaystyle\lim_{x \to \infty} 5 - \lim_{x \to \infty} \dfrac{6}{x} + \lim_{x \to \infty} \dfrac{3}{x^2}}{\displaystyle\lim_{x \to \infty} 4 - \lim_{x \to \infty} \dfrac{9}{x} + \lim_{x \to \infty} \dfrac{12}{x^2}} \qquad \text{properties I and II}$$

$$= \frac{5 - 0 + 0}{4 - 0 + 0} = \frac{5}{4} \qquad \lim_{x \to \infty} \dfrac{k}{x^n} = 0$$

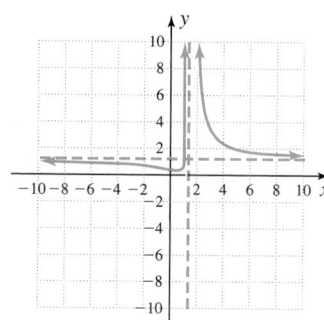

A similar calculation for $x \to -\infty$ shows $\displaystyle\lim_{x \to -\infty} \frac{5x^2 - 6x + 3}{4x^2 - 9x + 12}$ is also $\dfrac{5}{4}$. The graph is shown in the figure.

Now try Exercises 61 through 64 ▶

If the function contains a radical term in the numerator or denominator, we attempt to rewrite the expression as a single radical term. We can then apply limit property VII and treat the radicand as a rational function. When doing so, you must particularly note whether you're taking the limit as $x \to -\infty$ or as $x \to \infty$, since this may affect the sign of the final result.

EXAMPLE 8 ▶ Finding Limits at Infinity for a Radical Function

Evaluate each limit:

a. $\lim\limits_{x\to\infty} \dfrac{8x}{\sqrt{4x^2 + 1}}$ **b.** $\lim\limits_{x\to-\infty} \dfrac{8x}{\sqrt{4x^2 + 1}}$

Solution ▶ **a.** Since x is becoming an infinitely large *positive* number, we know $8x$ can be written as $+\sqrt{(8x)^2} = +\sqrt{64x^2}$ (signs added for emphasis). The original expression can then be rewritten as follows.

$$\lim\limits_{x\to\infty} \frac{8x}{\sqrt{4x^2 + 1}} = \lim\limits_{x\to\infty} \frac{\sqrt{64x^2}}{\sqrt{4x^2 + 1}} \qquad 8x = \sqrt{64x^2}$$

$$= \lim\limits_{x\to\infty} \sqrt{\frac{64x^2}{4x^2 + 1}} \qquad \text{quotient property of radicals}$$

$$= \sqrt{\lim\limits_{x\to\infty}\left(\frac{64x^2}{4x^2 + 1}\right)} \qquad \text{property VII}$$

We are now taking the limit of a rational expression, so we next divide the numerator and denominator by x^2.

$$= \sqrt{\lim\limits_{x\to\infty}\left(\dfrac{\dfrac{64x^2}{x^2}}{\dfrac{4x^2}{x^2} + \dfrac{1}{x^2}}\right)} \qquad \text{divide by } x^2$$

$$= \sqrt{\lim\limits_{x\to\infty} \dfrac{64}{4 + \dfrac{1}{x^2}}} \qquad \text{simplify}$$

$$= \sqrt{\dfrac{\lim\limits_{x\to\infty} 64}{\lim\limits_{x\to\infty}\left(4 + \dfrac{1}{x^2}\right)}} \qquad \text{property V}$$

$$= \sqrt{\dfrac{\lim\limits_{x\to\infty} 64}{\lim\limits_{x\to\infty} 4 + \lim\limits_{x\to\infty}\dfrac{1}{x^2}}} \qquad \text{property I}$$

☑ **C.** You've just learned how to evaluate limits at infinity

$$= \sqrt{\dfrac{64}{4 + 0}} \qquad \lim\limits_{x\to\infty}\dfrac{k}{x^n} = 0$$

$$= \sqrt{16} = 4 \qquad \text{result}$$

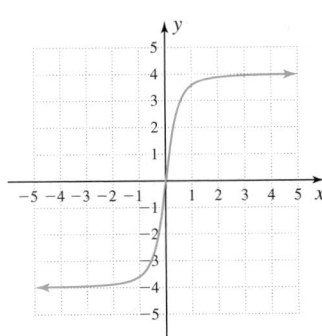

b. For part (b), x is becoming an infinitely large *negative* number, so the sign of $\dfrac{8x}{\sqrt{4x^2 + 1}}$ will be negative for all such values. For this reason $8x$ must be written as $-\sqrt{(8x)^2} = -\sqrt{64x^2}$. This is the only change to be made, and after applying algebra and the limit properties we find $\lim\limits_{x\to-\infty} \dfrac{8x}{\sqrt{4x^2 + 1}} = -4$.

The graph is shown in the figure and shows that this function has two unique horizontal asymptotes.

Now try Exercises 65 through 70 ▶

D. Evaluating Limits Graphically

As a complement to the equations and tables we've used to explore the limit concept, we now look at how these concepts appear graphically. This will offer a heightened understanding of important ideas, and an excellent summary of the definitions, properties, and relationships involved. The graphs of two general functions f and g are given (Figures 11.11 and 11.12, respectively) for use in the following examples.

Figure 11.11

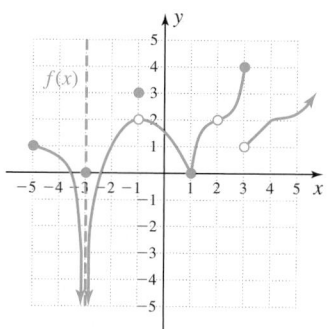

Figure 11.12

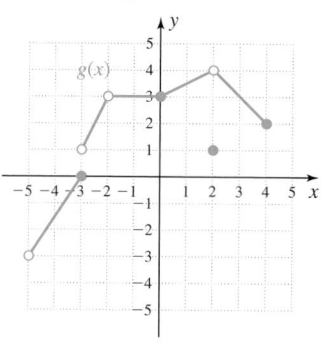

EXAMPLE 9 ▶ **Evaluating and Combining Function Values Using a Graph**

Use the graphs of f and g to perform the operations indicated. If any computation is not possible, state why.

   a. $f(-1) - g(4)$ **b.** $g(2) + f(4)$ **c.** $f(-3) - f(2)$ **d.** $g(2)\,g(-3)$

Solution ▶ **a.** $f(-1) - g(4) = 3 - 2$ **b.** $g(2) + f(4) = 1 + 2$

                 $= 1$ $= 3$

   c. $f(-3) - f(2) = $ X   **d.** $g(2)\,g(-3) = (1)(0)$

     not possible since $f(2)$ is not defined $= 0$

Now try Exercises 71 through 76 ▶

EXAMPLE 10 ▶ **Evaluating One-Sided Limits Graphically**

Use the graphs of f and g to evaluate the one-sided limits necessary to complete each calculation. If any computation is not possible, state why.

   a. $\displaystyle\lim_{x \to 1^+} f(x) + \lim_{x \to 4^-} g(x)$ **b.** $\displaystyle\lim_{x \to 3^+} f(x) - \lim_{x \to -5^+} g(x)$

   c. $\displaystyle\lim_{x \to -2^+} g(x) \lim_{x \to -3^+} f(x)$ **d.** $\displaystyle\lim_{x \to -1^+} f(x) + \lim_{x \to 5^+} g(x)$

Solution ▶ **a.** $\displaystyle\lim_{x \to 1^+} f(x) + \lim_{x \to 4^-} g(x) = 0 + 2$ **b.** $\displaystyle\lim_{x \to 3^+} f(x) - \lim_{x \to -5^+} g(x) = 1 - (-3)$

                             $= 2$ $= 4$

   c. $\displaystyle\lim_{x \to -2^+} g(x) \lim_{x \to -3^+} f(x) = $ X   **d.** $\displaystyle\lim_{x \to -1^+} f(x) + \lim_{x \to 5^+} g(x) = $ X

     not possible since   not possible since 5 is not

     $\displaystyle\lim_{x \to -3^+} f(x) = \left(\begin{matrix} \text{dne} \\ -\infty \end{matrix}\right)$ in the domain of g

Now try Exercises 77 through 82 ▶

EXAMPLE 11 ▶ **Evaluating Limits Graphically**

Use the graphs of f and g to evaluate the limits necessary to complete each calculation. If any computation is not possible, state why.

   a. $\displaystyle\lim_{x \to -2} [f(x) + g(x)]$ **b.** $\displaystyle\lim_{x \to -1} [3f(x) - g(x)]$

   c. $\displaystyle\lim_{x \to 1} \frac{g(x)}{f(x)}$ **d.** $\displaystyle\lim_{x \to 2} \left[(f(x))^2 + \sqrt{g(x)} \right]$

Solution ▶   **a.** $\displaystyle\lim_{x\to-2}[f(x)+g(x)] = \lim_{x\to-2}f(x) + \lim_{x\to-2}g(x)$ property I

$$= 1 + 3 = 4 \qquad \text{result}$$

b. $\displaystyle\lim_{x\to-1}[3f(x)-g(x)] = \lim_{x\to-1}3f(x) - \lim_{x\to-1}g(x)$ property II

$$= 3\lim_{x\to-1}f(x) - \lim_{x\to-1}g(x) \qquad \text{property IV}$$

$$= 3(2) - 3 = 3 \qquad \text{result}$$

c. $\displaystyle\lim_{x\to1}\frac{g(x)}{f(x)} = \frac{\displaystyle\lim_{x\to1}g(x)}{\displaystyle\lim_{x\to1}f(x)}$ property V

not possible since $\displaystyle\lim_{x\to1}f(x) = 0$ limit of the denominator is 0

d. $\displaystyle\lim_{x\to2}[(f(x))^2 + \sqrt{g(x)}] = \lim_{x\to2}[f(x)]^2 + \lim_{x\to2}\sqrt{g(x)}$ property I

$$= \left[\lim_{x\to2}f(x)\right]^2 + \sqrt{\lim_{x\to2}g(x)} \qquad \text{properties VI and VII}$$

$$= (2)^2 + \sqrt{4} = 6 \qquad \text{result}$$

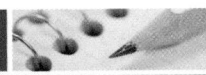 **D.** You've just learned how to use the definition of a limit to evaluate limits graphically

> Now try Exercises 83 through 88 ▶

11.3 EXERCISES

▶ CONCEPTS AND VOCABULARY

Fill in each blank with the appropriate word or phrase. Carefully reread the section, if necessary.

1. There are three types of discontinuities we may encounter. These are (1) _____, (2) _____, and (3) _____ discontinuities.

2. In order for a function $f(x)$ to be continuous at c, it must be _____ at c. Furthermore, $\displaystyle\lim_{x\to c^-}f(x)$ must equal _____ with $\displaystyle\lim_{x\to c^-}f(x) = $ _____.

3. If $f(x)$ is continuous at c, we can evaluate $\displaystyle\lim_{x\to c}f(x)$ by _____ _____.

4. The notation $x\to\infty$ is read, "as x becomes _____ _____," or "as x increases without _____."

5. Discuss/Explain the relationship between the horizontal asymptote of a rational function f, and $\displaystyle\lim_{x\to\infty}f(x)$.

6. Discuss/Explain how the three different types of discontinuities appear on the graph of a function.

▶ DEVELOPING YOUR SKILLS

For Exercises 7 through 16, use the given graphs to comment on the continuity of the function at the given x-values. If the function is discontinuous at any value, state which of the three conditions for continuity are violated.

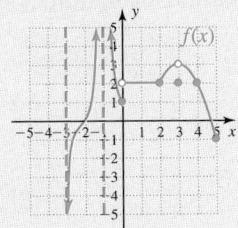

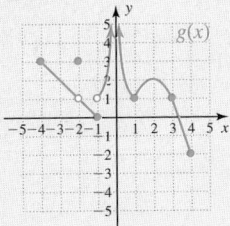

7. $f(x)$ at $x = -1$ 8. $f(x)$ at $x = 3$

9. $f(x)$ at $x = 1$ 10. $f(x)$ at $x = 2$

11. $f(x)$ at $x = 0$ 12. $g(x)$ at $x = -1$

13. $g(x)$ at $x = -3$ 14. $g(x)$ at $x = 1$

15. $g(x)$ at $x = -2$ 16. $g(x)$ at $x = 0$

Evaluate the following limits using direct substitution, if possible. If not possible, state why.

17. $\displaystyle\lim_{x\to-3}2x^2 - 5x + 3$ 18. $\displaystyle\lim_{x\to-2}3x^2 - 5x - 2$

19. $\lim\limits_{x \to 5} \sqrt{3x - 19}$

20. $\lim\limits_{x \to -6} \sqrt{x^2 + 7x}$

21. $\lim\limits_{x \to 2} \dfrac{x^2}{5x - 2}$

22. $\lim\limits_{x \to 8} \dfrac{2x - 5}{x^2 - 5x}$

23. $\lim\limits_{x \to -1} \dfrac{x + 1}{x^2 - 1}$

24. $\lim\limits_{x \to -2} \dfrac{4 - x^2}{x^2 + 5x + 6}$

25. $\lim\limits_{x \to -5} \sqrt{x^2 - 6x}$

26. $\lim\limits_{x \to 6} \sqrt{3x - 5}$

Evaluate the following limits by rewriting the given expression as needed.

27. $\lim\limits_{x \to -1} \dfrac{x + 1}{x^2 - 1}$

28. $\lim\limits_{x \to -2} \dfrac{4 - x^2}{x^2 + 5x + 6}$

29. $\lim\limits_{x \to 4} \dfrac{x - 4}{\sqrt{x} - 2}$

30. $\lim\limits_{x \to 16} \dfrac{x - 16}{4 - \sqrt{x}}$

31. $\lim\limits_{x \to 3} \dfrac{2x^2 - 3x - 9}{x - 3}$

32. $\lim\limits_{x \to -2} \dfrac{3x^2 + 7x + 2}{x + 2}$

33. $\lim\limits_{x \to -3} \dfrac{\sqrt{x + 7} - 2}{x + 3}$

34. $\lim\limits_{x \to 12} \dfrac{\sqrt{2x + 1} - 5}{x - 12}$

35. $\lim\limits_{x \to -4} \dfrac{x^3 + 8x^2 + 16x}{x^2 + 7x + 12}$

36. $\lim\limits_{x \to 3} \dfrac{2x^3 - 12x^2 + 18x}{x^2 - 7x + 12}$

Evaluate the following limits. Write your answer in simplest form.

37. $\lim\limits_{h \to 0} \dfrac{[2(x + h)^2 - (x + h)] - (2x^2 - x)}{h}$

38. $\lim\limits_{h \to 0} \dfrac{[3(x + h) - (x + h)^2] - (3x - x^2)}{h}$

39. $\lim\limits_{h \to 0} \dfrac{\dfrac{3}{x + h + 2} - \dfrac{3}{x + 2}}{h}$

40. $\lim\limits_{h \to 0} \dfrac{\dfrac{2}{x + h - 1} - \dfrac{2}{x - 1}}{h}$

41. $\lim\limits_{h \to 0} \dfrac{\sqrt{x + h + 2} - \sqrt{x + 2}}{h}$

42. $\lim\limits_{h \to 0} \dfrac{\sqrt{x + h - 1} - \sqrt{x - 1}}{h}$

43. $\lim\limits_{h \to 0} \dfrac{(x + h + 2)^3 - (x + 2)^3}{h}$

44. $\lim\limits_{h \to 0} \dfrac{(x + h - 4)^3 - (x - 4)^3}{h}$

Use a table of values to evaluate the following limits as x increases without bound.

45. $\lim\limits_{x \to \infty} \dfrac{5x^3 + 2}{10x^3 - 2x + 1}$

46. $\lim\limits_{x \to \infty} \dfrac{8x - 3x^2}{7x^2 - x + 1}$

47. $\lim\limits_{x \to \infty} \dfrac{6x^2 - x + 2}{2x^2 + 1}$

48. $\lim\limits_{x \to \infty} \dfrac{3x - 5x^3}{x^3 + 2x^2 - 3x + 7}$

49. $\lim\limits_{x \to \infty} \dfrac{7x^3}{5x^2 + 3x}$

50. $\lim\limits_{x \to \infty} \dfrac{x^2 + 1}{2x - 11}$

51. $\lim\limits_{x \to \infty} \dfrac{x^2 + 6x + 9}{2x^3}$

52. $\lim\limits_{x \to \infty} \dfrac{10 - 3x^2}{10 - 3x^3}$

Use a table of values to evaluate the following limits as x decreases without bound.

53. $\lim\limits_{x \to -\infty} \dfrac{5x^3 + 2}{10x^3 - 2x + 1}$

54. $\lim\limits_{x \to -\infty} \dfrac{6x^2 - x + 2}{2x^2 + 1}$

55. $\lim\limits_{x \to -\infty} \dfrac{x^2 + 1}{2x - 11}$

56. $\lim\limits_{x \to -\infty} \dfrac{7x^3}{5x^2 + 3x}$

57. $\lim\limits_{x \to -\infty} \dfrac{10 - 3x^2}{10 - 3x^3}$

58. $\lim\limits_{x \to -\infty} \dfrac{x^2 + 6x + 9}{2x^3}$

59. Given $\lim\limits_{x \to \infty} \dfrac{1}{x} = 0$, find the smallest positive value of x such that $\dfrac{1}{x} \leq 0.001$.

60. Given $\lim\limits_{x \to -\infty} \dfrac{1}{x} = 0$, find the largest negative value of x such that $\dfrac{1}{x} \geq -0.01$.

For Exercises 61 through 64, evaluate the limits by dividing the numerator and denominator by the highest power of x occurring in the denominator.

61. $\lim\limits_{x \to \infty} \dfrac{3x^2 - 2x + 1}{8x^2 + 5}$

62. $\lim\limits_{x \to \infty} \dfrac{5x^2 + 11x - 3}{12x^2 + 7x + 5}$

63. $\lim\limits_{x \to -\infty} \dfrac{2x^2 - 1}{x^3 + 2x + 12}$

64. $\lim\limits_{x \to -\infty} \dfrac{8x^3 - 27}{x^4 - 1}$

For Exercises 65 through 70, evaluate each limit.

65. $\lim\limits_{x \to \infty} \dfrac{\sqrt{36x^2 - 11}}{3x}$

66. $\lim\limits_{x \to -\infty} \dfrac{\sqrt{36x^2 - 11}}{3x}$

67. $\lim\limits_{x\to\infty} \dfrac{\sqrt{12x^2 - 6x + 1}}{7x}$

68. $\lim\limits_{x\to -\infty} \dfrac{\sqrt{12x^2 - 6x + 1}}{7x}$

69. $\lim\limits_{x\to\infty} \dfrac{\sqrt[3]{216x^3 + 36x^2 - 6x + 1}}{2x}$

70. $\lim\limits_{x\to -\infty} \dfrac{\sqrt[3]{216x^3 + 36x^2 - 6x + 1}}{2x}$

The graphs of two functions f and g are given here, for use in Exercises 71 through 88.

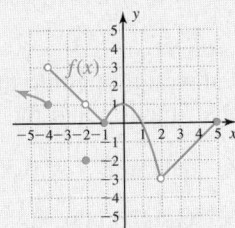
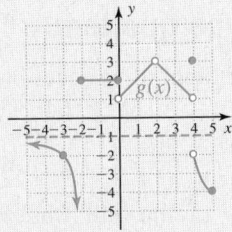

For Exercises 71 through 76, use the graphs of f and g to perform the operations indicated. If any computation is not possible, state why.

71. $f(0) + g(-1)$

72. $f(5) - g(-3)$

73. $g(4) - f(2)$

74. $f(-2) + g(2)$

75. $g(0) + f(-4)$

76. $g(5) + f(-1)$

For Exercises 77 through 82, use the graphs of f and g to evaluate the one-sided limits necessary to complete each calculation. If any computation is not possible, state why.

77. $\lim\limits_{x\to -4^-} f(x) - \lim\limits_{x\to 0^+} g(x)$

78. $\lim\limits_{x\to 4^+} g(x) + \lim\limits_{x\to -2^-} f(x)$

79. $\lim\limits_{x\to -4^+} f(x) - \lim\limits_{x\to 0^-} g(x)$

80. $\lim\limits_{x\to 4^-} g(x) + \lim\limits_{x\to -2^+} f(x)$

81. $\lim\limits_{x\to 5^-} f(x) + \lim\limits_{x\to -2^-} g(x)$

82. $\lim\limits_{x\to 2^+} g(x) + \lim\limits_{x\to 6^+} f(x)$

For Exercises 83 through 88, use the graphs of f and g to evaluate the limits necessary to complete each calculation. If any computation is not possible, state why.

83. $\lim\limits_{x\to 0}[f(x) - g(x)]$ **84.** $\lim\limits_{x\to 2}[g(x) - f(x)]$

85. $\lim\limits_{x\to -3}\left[\dfrac{3f(x)}{g(x)}\right]$

86. $\lim\limits_{x\to -1}\left\{\dfrac{3[g(x)]^2}{f(x)}\right\}$

87. $\lim\limits_{x\to -\infty}[f(x) - \sqrt[3]{g(x)}]$

88. $\lim\limits_{x\to 3}[2g(x)f(x)]$

► **MAINTAINING YOUR SKILLS**

89. (1.3) Solve each equation:

   a. $3x^2 + 4x - 12 = 0$

   b. $\sqrt{3x + 1} - \sqrt{2x} = 1$

   c. $\dfrac{1}{x + 2} + \dfrac{3}{x^2 + 5x + 6} = \dfrac{2}{x + 3}$

90. (6.6) Find $\cos^{-1}\left[\cos\left(-\dfrac{\pi}{6}\right)\right]$.

91. (5.6) Solve the right triangle shown. Round answers to the nearest tenth.

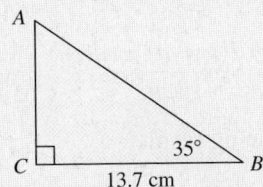

92. (2.3) In 2005, a small business purchased a copier for $4500. By 2008, the value of the copier had decreased to $3300. Assuming the depreciation is linear, (a) find the rate-of-change $m = \dfrac{\Delta \text{ value}}{\Delta \text{ time}}$ and discuss its meaning in this context. (b) Find the depreciation equation and (c) use the equation to predict the copier's value in 2012. (d) If the copier is traded in for a new model when its value is less than $700, how long will the company use this copier?

Learning Objectives

In Section 11.4 you will learn how to:

☐ **A.** Evaluate the limit of a difference quotient to find instantaneous rates of change

☐ **B.** Evaluate the limit of a sum to find the area under a curve

At this point, the concept of rates of change is a familiar theme. From their first introduction in a linear context $\left(\dfrac{\Delta y}{\Delta x} = \dfrac{y_2 - y_1}{x_2 - x_1}\right)$ to investigations of nonlinear functions using the difference quotient, the idea has proven powerful and effective. In Example 10 from Section 2.5, we used the difference quotient to approximate the velocity of a falling wrench with great accuracy. While this study certainly answers Galileo's historic questions regarding the speed of a falling body, business, science, industry, education, and government present us with equally compelling questions that can likewise be answered using the difference quotient.

A. The Limit of a Difference Quotient

In Example 8 of Section 2.5, the height of a soccer ball kicked straight up into the air was modeled by the function $d(t) = -16t^2 + 64t$. Using this function and the rate-of-change formula $\left[\dfrac{\Delta d}{\Delta t} = \dfrac{d(t_2) - d(t_1)}{t_2 - t_1}\right]$, we found that between $t = 0.5$ and $t = 1.0$ sec, the average velocity of the ball was 40 ft/sec. Graphically, this was represented by the slope of the secant line through (0.5, 28) and (1.0, 48). For the time interval [1.0, 1.5], the average velocity had slowed to 24 ft/sec (the slope of this secant line is $\frac{24}{1}$). Using smaller and smaller increments of time enables a better and better estimate of the velocity over a given time interval, but requires a new calculation for each interval. Instead, we consider applying the difference quotient, as it allows for a fixed point t and a second point $(t + h)$ *that becomes arbitrarily close to t as $h \to 0$*: $\dfrac{d(t + h) - d(t)}{h}$.

Graphically, this means the two points used for the secant line are becoming arbitrarily close, and in fact, can be made as close as we like by applying the concept of a limit.

In Figure 11.13, several secant lines are shown, each with a left endpoint at $t = 1$. Selecting right endpoints that become closer and closer to 1 ($1 + h$ as $h \to 0$), we obtain the sequence of secant lines shown in black, then purple, then red. As $1 + h$ gets *infinitely close* to 1, the secant line reaches a "limiting position" (the yellow line) as the two points forming the line become infinitely close. The end result is a **tangent line** to the graph of d at $x = 1$ (tangent: touching the curve at only one point). As the slope of each secant line gives the average rate of change between two points, in this case the velocity of the ball, the slope of the tangent line gives the **instantaneous velocity** at precisely $x = 1$. As the graph indicates, the slope of the tangent line is the limit of the slopes of the secant lines and we write,

Figure 11.13

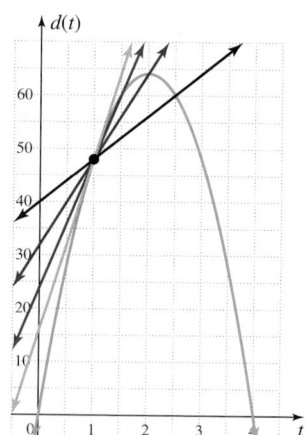

$$m_{\tan} = \lim_{h \to 0} m_{\sec} = \lim_{h \to 0} \frac{d(t + h) - d(t)}{h}$$

For a general function $f(x)$, we have the following result.

The Slope of a Tangent Line

Given a function $f(x)$ that is smooth and continuous over the interval containing x and $x + h$, the slope of a line drawn tangent to the graph of f at x is given by the function $f'(x)$, where

$$f'(x) = \lim_{h \to 0} \frac{f(x + h) - f(x)}{h}$$

EXAMPLE 1 ▶   Computing the Limit of a Difference Quotient

For the function $d(t) = -16t^2 + 64t$ (see page 146)

a. Find the limit of the difference quotient for d, to obtain a function $d'(t)$ that represents the instantaneous velocity at time t.

b. Use $d'(t)$ to find the velocity at $t = 1$, $t = 2$, and $t = 3.5$.

Solution ▶   a. For $d(t) = -16t^2 + 64t$, $d(t + h) = -16(t + h)^2 + 64(t + h)$. After squaring the binomial and applying the distributive property, we obtain

$$d(t + h) = -16(t^2 + 2th + h^2) + 64t + 64h$$
$$= -16t^2 - 32th - 16h^2 + 64t + 64h$$

To find $d'(t)$, we then have

$$\lim_{h \to 0} \frac{d(t + h) - d(t)}{h} = \lim_{h \to 0} \frac{-16t^2 - 32th - 16h^2 + 64t + 64h - (-16t^2 + 64t)}{h} \quad \text{apply difference quotient}$$

$$= \lim_{h \to 0} \frac{-16t^2 - 32th - 16h^2 + 64t + 64h + 16t^2 - 64t}{h} \quad \text{distribute } -1$$

$$= \lim_{h \to 0} \frac{-32th - 16h^2 + 64h}{h} \quad \text{combine like terms}$$

$$= \lim_{h \to 0} \frac{h(-32t - 16h + 64)}{h} \quad \text{factor out } h$$

$$= \lim_{h \to 0} (-32t - 16h + 64) \quad \frac{h}{h} = 1, h \neq 0$$

$$d'(t) = -32t + 64 \quad \text{result}$$

b. For the velocity of the ball at $t = 1$, $t = 2$, and $t = 3.5$, we evaluate $d'(t)$.

$$t = 1: \qquad d'(1) = -32(1) + 64 \qquad \text{substitute 1 for } t$$
$$= -32 + 64 = 32 \qquad \text{simplify}$$

At the moment $t = 1$ sec, the velocity is 32 ft/sec.

$$t = 2: \qquad d'(2) = -32(2) + 64 \qquad \text{substitute 2 for } t$$
$$= -64 + 64 = 0 \qquad \text{simplify}$$

At the moment $t = 2$ sec, the velocity is 0 ft/sec. This seems reasonable, as we expect the ball will slow as it approaches its maximum height, where it momentarily stops (velocity is 0), then begins its return to the ground.

$$t = 3.5: \quad d'(3.5) = -32(3.5) + 64 \qquad \text{substitute 3.5 for } t$$
$$= -112 + 64 = -48 \quad \text{simplify}$$

At the moment $t = 3.5$ sec, the velocity is -48 ft/sec. This indicates the ball is *falling back to Earth* (velocity is negative) at a rate of 48 ft/sec.

Now try Exercises 7 through 12 ▶

In Example 10 of Section 2.5 (page 148), we modeled the distance a wrench had fallen from a skyscraper with the function $d(t) = 16t^2$. Using a very small time interval and the difference quotient, we obtained a close approximation for the velocity of the wrench in various intervals. Using the concept of a limit we can now find the *instantaneous velocity* at any time t.

EXAMPLE 2 ▶ **Finding an Instantaneous Rate of Change**

For the function $d(t) = 16t^2$,

   a. Find the limit of the difference quotient for d to obtain a function $\mathcal{v}(t)$ that represents the instantaneous velocity of the falling wrench.

   b. Use $\mathcal{v}(t)$ to find the velocity at $t = 2$, $t = 7$, and $t = 9$.

Solution ▶   **a.** In the referenced example we obtained the expression $32t + 16h$ after applying the difference quotient. Using the ideas developed here, we know the velocity of the wrench will be

$$\mathcal{v}(t) = \lim_{h \to 0}(32t + 16h)$$

$$= 32t \quad \text{by direct substitution.}$$

   b. For the instantaneous velocity at $t = 2$, $t = 7$, and $t = 9$, we evaluate $\mathcal{v}(t)$.

$$t = 2: \quad \mathcal{v}(2) = 32(2) = 64 \quad \text{substitute 2 for } t$$

At $t = 2$ sec, the velocity of the wrench is 64 ft/sec.

$$t = 7: \quad \mathcal{v}(7) = 32(7) = 224 \quad \text{substitute 7 for } t$$

At $t = 7$ sec, the velocity of the wrench is 224 ft/sec.

$$t = 9: \quad \mathcal{v}(9) = 32(9) = 288 \quad \text{substitute 9 for } t$$

At the instant $t = 9$ sec, the velocity of the wrench has increased to 288 ft/sec. In fact, in the absence of air resistance the velocity of the wrench will continue to increase until it hits the ground (with a wham!).

Now try Exercises 13 through 16 ▶

Mathematical models of real-world phenomena come in many different forms, and in a calculus course these ideas are applied to a very wide variety of functions. For practice with the algebra required to make these ideas work, **see Exercises 17 through 20.** The following applications demonstrate how these ideas are applied to root and rational function models.

EXAMPLE 3 ▶ **Computing a Rate of Change for the Distance to the Horizon**

Many new high-rise buildings have external elevators, allowing an unobstructed view of the city as it extends toward the horizon. The distance that Melody can see as the elevator takes her higher is modeled by the function $d(x) = 1.15\sqrt{x}$, where $d(x)$ represents her sight distance in nautical miles, at a height of x feet.

   a. Find the limit of the difference quotient for $d(x)$, to obtain a function $\mathcal{v}(x)$ that represents the instantaneous rate of change in the sight distance at x.

   b. Find the instantaneous rate of change at heights of 225, 400, and 625 ft.

Solution ▶   **a.** For $d(x) = 1.15\sqrt{x}$, $d(x+h) = 1.15\sqrt{x+h}$. To find $d'(x)$, we then have

$$\lim_{h\to 0}\frac{d(x+h)-d(x)}{h} = \lim_{h\to 0}\frac{1.15\sqrt{x+h}-1.15\sqrt{x}}{h} \qquad \text{apply difference quotient}$$

$$= 1.15\lim_{h\to 0}\frac{\sqrt{x+h}-\sqrt{x}}{h} \qquad \text{factor out the constant}$$

Noting that multiplying the numerator by $\sqrt{x+h}+\sqrt{x}$ will "free up" the h in the first radicand, we multiply numerator and denominator by $\sqrt{x+h}+\sqrt{x}$.

$$= 1.15\lim_{h\to 0}\frac{(\sqrt{x+h}-\sqrt{x})(\sqrt{x+h}+\sqrt{x})}{h\,(\sqrt{x+h}+\sqrt{x})}$$

$$= 1.15\lim_{h\to 0}\frac{x+h-x}{h(\sqrt{x+h}+\sqrt{x})} \qquad (A-B)(A+B)=A^2-B^2$$

$$= 1.15\lim_{h\to 0}\frac{\cancel{h}}{\cancel{h}(\sqrt{x+h}+\sqrt{x})} \qquad \text{combine like terms}$$

$$= 1.15\lim_{h\to 0}\frac{1}{\sqrt{x+h}+\sqrt{x}} \qquad \frac{h}{h}=1,\, h\neq 0$$

$$= 1.15\left(\frac{1}{\sqrt{x}+\sqrt{x}}\right) \qquad \text{apply limit}$$

$$d'(x) = \frac{1.15}{2\sqrt{x}} \qquad \text{result}$$

b. To find the instantaneous rate of change at $x = 225$, 400, and 625 ft, we evaluate $d'(t)$.

$$d'(225) = \frac{1.15}{2\sqrt{225}} = \frac{1.15}{2(15)} \qquad \text{substitute 225 for } x$$

$$\approx 0.038$$

At $x = 225$ ft, the sight distance is increasing at a rate of about 0.038 mi/ft (about 3.8 nautical miles per 100 ft).

$$d'(400) = \frac{1.15}{2\sqrt{400}} = \frac{1.15}{2(20)} \qquad \text{substitute 400 for } x$$

$$\approx 0.029$$

At $x = 400$ ft, the sight distance is increasing at a rate of about 0.029 mi/ft (about 2.9 nautical miles per 100 ft).

$$d'(625) = \frac{1.15}{2\sqrt{625}} = \frac{1.15}{2(25)} \qquad \text{substitute 625 for } x$$

$$= 0.023$$

At $x = 625$ ft, the sight distance is increasing at a rate of 0.023 mi/ft (2.3 nautical miles per 100 ft).

Now try Exercises 21 through 24 ▶

Note that for square root functions, the algebra involves rationalizing the numerator to "free up" the constant h in hopes that the limit can eventually be found by direct substitution. For rational functions, the difference quotient yields a difference of rational expressions in the numerator, which we combine for the same reason.

EXAMPLE 4 ▶ **Computing Instantaneous Rates of Change for a Cost Function**

A county government finds the cost of removing pollutants from a stream can be modeled by the function $C(x) = \dfrac{2x}{1-x}$, where $C(x)$ represents the cost in millions of dollars, to remove x percent of the pollutants $[C(x) > 0, x$ is a decimal less than 1$]$.

a. Find the limit of the difference quotient for C, to obtain a function $C(x)$ that represents the instantaneous rate of change in the cost of removing pollutants.

b. Find the rate the cost is increasing at the moment 60%, 70%, and 80% of the pollutants are removed.

c. Use the information from Part (b) to *estimate* the rate the cost will be increasing at the moment 90% of the pollutants are removed, then compute the actual rate. Were you surprised?

Solution ▶ **a.** For $C(x) = \dfrac{2x}{1-x}$, $C(x+h) = \dfrac{2(x+h)}{1-(x+h)}$. To find $C(x)$ we then have

$$\lim_{h \to 0} \frac{C(x+h) - C(x)}{h} = \lim_{h \to 0} \frac{\dfrac{2(x+h)}{1-(x+h)} - \dfrac{2x}{1-x}}{h} \qquad \text{apply difference quotient}$$

$$= 2 \lim_{h \to 0} \frac{\dfrac{(x+h)}{1-x-h} - \dfrac{x}{1-x}}{h} \qquad \text{factor out 2, distribute } -1$$

$$= 2 \lim_{h \to 0} \frac{\dfrac{(x+h)(1-x) - x(1-x-h)}{(1-x-h)(1-x)}}{h} \qquad \frac{A}{B} - \frac{C}{D} = \frac{AD - BC}{BD}$$

$$= 2 \lim_{h \to 0} \frac{\dfrac{x - x^2 + h - xh - x + x^2 + xh}{(1-x-h)(1-x)}}{h} \qquad \text{F-O-I-L, distribute}$$

$$= 2 \lim_{h \to 0} \frac{\dfrac{h}{(1-x-h)(1-x)}}{h} \qquad \text{simplify numerator}$$

$$= 2 \lim_{h \to 0} \frac{\cancel{h}}{(1-x-h)(1-x)\cancel{h}} \qquad \text{invert and multiply}$$

$$= 2 \lim_{h \to 0} \frac{1}{(1-x-h)(1-x)} \qquad \frac{h}{h} = 1, h \neq 0$$

$$C(x) = \frac{2}{(1-x)^2} \qquad \text{apply limit, multiply by 2}$$

b. Evaluating $C(x)$ for $x = 0.6, 0.7, 0.8$ (60%, 70%, and 80%) yields

$$C(0.6) = \frac{2}{(1-0.6)^2} = \frac{2}{0.16} \qquad \text{substitute 0.6 for } x \text{ and simplify}$$

$$= 12.5$$

At $x = 0.6$, the cost is increasing at a rate of \$12,500,000 for each additional percentage point of pollution removed.

$$C(0.7) = \frac{2}{(1-0.7)^2} = \frac{2}{0.09} \qquad \text{substitute 0.7 for } x \text{ and simplify}$$

$$= 22.\overline{2}$$

At $x = 0.7$, the cost is increasing at a rate of about $22,222,222 for each additional percentage point.

$$C(0.8) = \frac{2}{(1 - 0.8)^2} = \frac{2}{0.04} \qquad \text{substitute 0.8 for } x \text{ and simplify}$$
$$= 50$$

At $x = 0.8$, the cost is increasing at a rate of $50,000,000 for each additional percentage point.

☑ **A.** You've just learned how to evaluate the limit of a difference quotient to find instantaneous rates of change

c. Evaluating $C(x)$ for $x = 0.9$ shows the cost rises dramatically to a rate of $200,000,000 per percentage point. Removing anywhere near 100% of the pollutants would become prohibitively expensive.

Now try Exercises 25 through 32 ▶

B. Limits and the Area under a Curve

In the *Connections to Calculus* from Chapter 10, the area under the graph of $f(x) = x + 5$ for x in the interval $[0, 4]$ was approximated using the rectangle method with $n = 4$ and $n = 8$ rectangles (see Example 1). While we agreed that using 16 rectangles would give a better estimate (see Figure 11.14), expanding the related

summation $\sum LW = \sum_{i=1}^{16} f\left(\frac{1}{4}i\right)\left(\frac{1}{4}\right)$ seemed too

tedious and instead we opted to use the properties of summation to write the expression in a form where the sum could be found by substituting $n = 16$ directly. Here we extend this idea to a situation where n rectangles are used, as we can then take the limit of this sum as $n \to \infty$. The key to this extension is a result from Section 11.3, where we noted for any positive integer k,

Figure 11.14

$$\lim_{n \to -\infty} \frac{1}{n^k} = 0 \text{ and } \lim_{n \to \infty} \frac{1}{n^k} = 0 \text{ (note the change}$$

of variable from x to n, which is more commonly used for summations). The original

4-unit interval would then be divided into n parts, so each rectangle is $\frac{4}{n}$ units wide.

The length of each rectangle is still given by $f(x) = x + 5$, but we now evaluate f in

increments of $\frac{4}{n}$, giving $\sum LW = \sum_{i=1}^{n} f\left(\frac{4}{n}i\right)\left(\frac{4}{n}\right)$. Since $f\left(\frac{4}{n}i\right) = \frac{4}{n}i + 5$, we can

write the sum as $\sum_{i=1}^{n} \left(\frac{4}{n}i + 5\right)\frac{4}{n}$ and proceed as before.

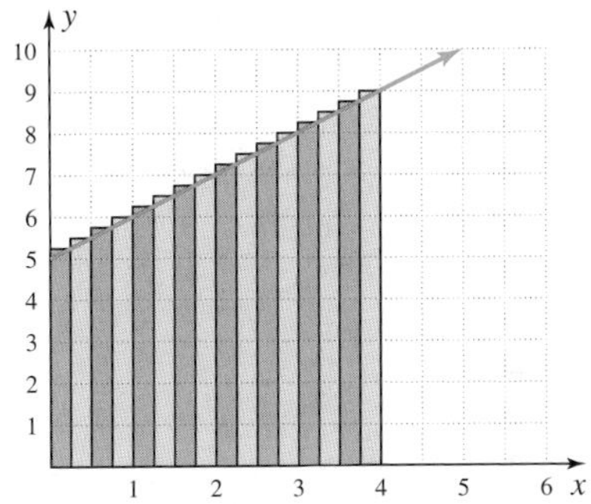

EXAMPLE 5 ▶ **Finding the Limit of a Sum**

Evaluate the following limit: $\lim_{n \to \infty} \sum_{i=1}^{n} \left(\frac{4}{n}i + 5\right)\frac{4}{n}$.

Solution ▶ Begin by applying *summation properties* and formulas to write the expression in a

form where the *limit properties* can be applied. Since $\frac{4}{n}$ is a constant, we begin by

factoring it out of the summation (summation property II).

$$\frac{4}{n} \sum_{i=1}^{n} \left(\frac{4}{n}i + 5\right) = \frac{4}{n}\left(\sum_{i=1}^{n}\frac{4}{n}i + \sum_{i=1}^{n}5\right) \qquad \text{summation properties (distribute)}$$

$$= \frac{4}{n}\left(\frac{4}{n}\sum_{i=1}^{n}i + \sum_{i=1}^{n}5\right) \qquad \text{factor } \frac{4}{n} \text{ from first summation}$$

$$= \frac{4}{n}\left[\frac{4}{n}\left(\frac{n^2 + n}{2}\right) + 5n\right] \qquad \text{apply summation formulas}$$

$$= \frac{16}{n^2}\left(\frac{n^2 + n}{2}\right) + 20 \qquad \text{distribute } \frac{4}{n}$$

$$= \frac{16}{2}\left(\frac{n^2 + n}{n^2}\right) + 20 \qquad \text{rewrite denominators (commute)}$$

$$= 8\left(\frac{n^2}{n^2} + \frac{n}{n^2}\right) + 20 \qquad \text{decompose rational expression}$$

$$= 8\left(1 + \frac{1}{n}\right) + 20 = 28 + \frac{8}{n} \qquad \text{result}$$

We can now take the limit of this result and find

$$\lim_{n \to \infty}\left(28 + \frac{8}{n}\right) = 28.$$

This shows

$$\lim_{n \to \infty} \sum_{i=1}^{n}\left(\frac{4}{n}i + 5\right)\frac{4}{n} = 28,$$

and that the area under the graph of $f(x) = x + 5$, between $x = 0$ and $x = 4$, *is exactly 28 units^2!*

Now try Exercises 37 through 40 ▶

This result may seem unimpressive, as we could have quickly found the area using simple geometry. But as mentioned earlier, this method is easily applied to areas under a *general function f(x)*. In the *Connections to Calculus* feature from Chapter 2, we noted how the area under a graph could represent the distance covered by a runner. Our work here now takes on added meaning, as the area under a general curve has many other real-world implications in a study of calculus.

EXAMPLE 6 ▶ Using Limits to Find the Area under a Curve

During a training session for the Boston Marathon, Georgiana plans to begin at a leisurely walk of 2 mi/hr and gradually increase her speed over a 5-hr period, finishing up at a clip of over 8 mi/hr. Her speed might then be modeled by the function

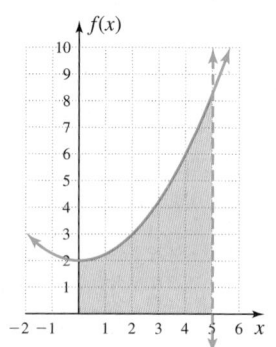

$f(x) = \frac{1}{4}x^2 + 2$, where $f(x)$ represents the velocity

in miles per hour, and x is the time in hours. The area under the curve in the interval [0, 5] then represents the distance Georgiana has run in these 5 hr.

a. Show that using the rectangle method results in the expression

$$\frac{5}{n} \sum_{i=1}^{n}\left[\frac{1}{4}\left(\frac{5}{n}i\right)^2 + 2\right].$$

b. Determine if Georgiana has run a "marathon distance" (about 26.2 mi) in these 5 hr, by applying summation properties and formulas, and taking the limit of this sum as $n \to \infty$.

Solution ▶ **a.** The area described is shown in the figure. The given interval is 5 units wide, and after dividing it into n parts, each rectangle will have a width of $\dfrac{5}{n}$. The length of each one is given by $f(x) = \dfrac{1}{4}x^2 + 2$, evaluated at each of the $\dfrac{5}{n}$ subintervals. The total area is then

$$\sum_{i=1}^{n} LW = \sum_{i=1}^{n} f\left(\frac{5}{n}i\right)\left(\frac{5}{n}\right) = \frac{5}{n}\sum_{i=1}^{n} f\left(\frac{5}{n}i\right) = \frac{5}{n}\sum_{i=1}^{n}\left[\frac{1}{4}\left(\frac{5}{n}i\right)^2 + 2\right].$$

b. Proceed as in Example 5.

$$\frac{5}{n}\sum_{i=1}^{n}\left[\frac{1}{4}\left(\frac{5}{n}i\right)^2 + 2\right] = \frac{5}{n}\left[\sum_{i=1}^{n}\frac{1}{4}\left(\frac{5}{n}i\right)^2 + \sum_{i=1}^{n} 2\right] \qquad \text{summation properties (distribute)}$$

$$= \frac{5}{n}\left[\sum_{i=1}^{n}\frac{1}{4}\left(\frac{25}{n^2}\right)i^2 + \sum_{i=1}^{n} 2\right] \qquad \left(\frac{5}{n}i\right)^2 = \frac{25}{n^2}i^2$$

$$= \frac{5}{n}\left[\sum_{i=1}^{n}\frac{25}{4n^2}i^2 + \sum_{i=1}^{n} 2\right] \qquad \text{multiply}$$

$$= \frac{5}{n}\left[\left(\frac{25}{4n^2}\right)\sum_{i=1}^{n} i^2 + \sum_{i=1}^{n} 2\right] \qquad \text{factor } \frac{25}{n^2} \text{ from first summation}$$

$$= \frac{5}{n}\left[\frac{25}{4n^2}\left(\frac{2n^3 + 3n^2 + n}{6}\right) + 2n\right] \qquad \text{apply summation formulas}$$

$$= \frac{125}{4n^3}\left(\frac{2n^3 + 3n^2 + n}{6}\right) + 10 \qquad \text{distribute } \frac{5}{n}$$

$$= \frac{125}{24}\left(\frac{2n^3 + 3n^2 + n}{n^3}\right) + 10 \qquad \text{rewrite denominators (commute)}$$

$$= \frac{125}{24}\left(2 + \frac{3}{n} + \frac{1}{n^2}\right) + 10 \qquad \text{decompose rational expression}$$

We can now take the limit of this result, and after applying the needed limit properties we obtain

$$= \frac{125}{24}\lim_{n\to\infty}\left(2 + \frac{3}{n} + \frac{1}{n^2}\right) + \lim_{n\to\infty} 10 \quad \text{apply properties of limits}$$

$$= \left(\frac{125}{24}\right)(2) + 10 = 20.41\overline{6}$$

The area under the graph of $f(x) = \dfrac{1}{4}x^2 + 2$ between $x = 0$ and $x = 5$ is $20.41\overline{6}$ units^2, which is numerically equivalent to the distance run. Georgiana is still about 5.8 mi short of a marathon.

☑ **B. You've just learned how to evaluate the limit of a sum to find the area under a curve**

Now try Exercises 41 through 44 ▶

Note: For Learning Objective B, the functions and intervals offered in the Exercise Set were deliberately chosen for their simplicity. In particular, all are polynomials where $f(x) > 0$ in the interval selected. In a calculus course, the same ideas are applied to a greater variety of functions, not all intervals begin at 0, and the function may vary from positive to negative within the interval. Regardless, the general ideas demonstrated here remain consistent.

11.4 EXERCISES

▶ CONCEPTS AND VOCABULARY

Fill in each blank with the appropriate word or phrase. Carefully reread the section, if necessary.

1. The slope of a line drawn tangent to the graph of any function $f(x)$ can be found by taking the limit of the _____ quotient for f, as h approaches zero.

2. The instantaneous rate of change of a function $f(x)$ can be found by evaluating the expression _____ .

3. In order to obtain a better approximation for the area under a curve, we can use narrower (and thus more) _____ in our computation.

4. As we let the number of approximating rectangles increase without bound, the resulting limit gives a(n) _____ value for the area under the curve.

5. Discuss/Explain the relationship between the difference quotient and the formula for finding the slope of a line given two points.

6. Do some research on geometric shapes other than rectangles that can be used to approximate the area under a curve. Return to the graph on page 1061, and comment on why approximating rectangles consistently give an overestimate of the area.

▶ DEVELOPING YOUR SKILLS

Use the following information to answer Exercises 7 through 12.

Two model rockets are launched at a gathering of the National Association of Rocketry (NAR: www.nar.org). Frank's Apollo II motor burns out at a height of 500 m, at which point the rocket has a velocity of 88.2 meters per second (m/sec). His rocket's height in meters, t sec after engine burnout, is given by $f(t) = 500 + 88.2t - 4.9t^2$. Gwen's Icarus Alpha motor burns out at a height of 600 m, at which point the rocket has a velocity of 78.4 m/sec. Her rocket's height in meters, t sec after burnout, is given by $g(t) = 600 + 78.4t - 4.9t^2$.

7. Find the limit of the difference quotient for f, to obtain a function $f(t)$ that represents the instantaneous velocity at time t.

8. Use the result from Exercise 7 to find the instantaneous velocity of Frank's rocket at
   a. $t = 5$ **b.** $t = 9$ **c.** $t = 13$

9. Find the limit of the difference quotient for g, to obtain a function $g(t)$ that represents the instantaneous velocity at time t.

10. Use the result from Exercise 9 to find the instantaneous velocity of Gwen's rocket at
    a. $t = 2$ **b.** $t = 8$ **c.** $t = 10$

11. Use the result from Exercise 7 to find the maximum height of Frank's rocket. This occurs when $v = 0$.

12. Use the result from Exercise 9 to find the maximum height of Gwen's rocket. This occurs when $v = 0$.

13. To replicate Galileo's famous test as to whether the velocity of a falling body depends on its weight, a science class is dropping bowling balls of different weights from an 11-story building onto the lawn below. The ball's height in meters, t sec after it is released, is modeled by $d(t) = -4.9t^2 + 44.1$. Find the limit of the difference quotient for d, to obtain a function $d(t)$ that represents the instantaneous velocity of the bowling ball at time t.

14. Use the results of Exercise 13 to find the instantaneous velocity of the bowling ball at
    a. $t = 1$ **b.** $t = 2$ **c.** $t = 3$ (time of impact)

15. A rock climber's carabineer falls off her harness 256 ft above the floor of the Grand Canyon. It's height in feet, t sec after it falls, can be modeled by $d(t) = -16t^2 + 256$. Find the limit of the difference quotient for d, to obtain a function $d(t)$ that represents the instantaneous velocity of the "biner" at time t.

16. Use the results of Exercise 15 to find the instantaneous velocity of the "biner" at

    a. $t = 1$ **b.** $t = 3$

    c. $t = 4$ (time of impact)

Find the limit of the difference quotient for each function $f(x)$ **given, to obtain a function** $\hat{f}(x)$ **that represents the instantaneous rate of change at** x **for each function.**

17. $f(x) = \dfrac{1}{2}x + 5$ **18.** $f(x) = x^2 - 3x$

19. $f(x) = x^3$ **20.** $f(x) = \dfrac{1}{x}$

21. The population of a small town can be modeled by the function $p(t) = 1.2\sqrt{t} + 40$, where p is measured in thousands and t is the number of years after 2008. Find the limit of the difference quotient for p to obtain a function $\hat{p}(t)$ that represents the instantaneous rate of change of population at time t.

22. Use the results of Exercise 21 to find the instantaneous rate of change of population of the town in

    a. 2009   **b.** 2012   **c.** 2024

23. The number of bacteria in a person's body, after they begin a regimen of antibiotics, can be modeled by the function $b(t) = 6 - \sqrt{t}$, where b is measured in tens of thousands and t is the number of hours after the first dose. Find the limit of the difference quotient for b to obtain a function $\hat{b}(t)$ that represents the instantaneous rate of change of number of bacteria at time t.

24. Use the results of Exercise 23 to find the instantaneous rate of change of number of bacteria after

    a. 1 hr   **b.** 16 hr   **c.** 25 hr

For Exercises 25 through 28, find the limit of the difference quotient of the given function to obtain a function that represents the slope of a line drawn tangent to the curve at x**.**

25. $f(x) = \dfrac{2}{x - 1}$ **26.** $g(x) = \dfrac{3}{x + 2}$

27. $h(x) = \dfrac{5x}{x + 5}$ **28.** $j(x) = \dfrac{x}{2(x + 1)}$

Use the results from the corresponding Exercises 25 through 28 for the following exercises.

29. Find the slope of the line drawn tangent to the graph of $f(x)$ at $x = 3$.

30. Find the slope of the line drawn tangent to the graph of $g(x)$ at $x = -1$.

31. Find the slope of the line drawn tangent to the graph of $h(x)$ at $x = 0$.

32. Find the slope of the line drawn tangent to the graph of $j(x)$ at $x = -\dfrac{1}{2}$.

For Exercises 33 through 36, graph each function over the interval [0, 7]. Then use geometry to find the area of the region below the graph, and above the x-axis in the interval [0, 6].

33. $q(x) = \dfrac{1}{2}x$ **34.** $r(x) = \dfrac{1}{3}x$

35. $t(x) = \dfrac{1}{2}x + 1$ **36.** $s(x) = \dfrac{1}{3}x + 2$

For Exercises 37 through 40, evaluate the following limits and compare your result to the corresponding exercise in 33 through 36.

37. $\displaystyle\lim_{n\to\infty} \sum_{i=1}^{n} \left(\dfrac{1}{2}\cdot\dfrac{6}{n}i\right)\dfrac{6}{n}$ **38.** $\displaystyle\lim_{n\to\infty} \sum_{i=1}^{n} \left(\dfrac{1}{3}\cdot\dfrac{6}{n}i\right)\dfrac{6}{n}$

39. $\displaystyle\lim_{n\to\infty} \sum_{i=1}^{n} \left(\dfrac{1}{2}\cdot\dfrac{6}{n}i + 1\right)\dfrac{6}{n}$

40. $\displaystyle\lim_{n\to\infty} \sum_{i=1}^{n} \left(\dfrac{1}{3}\cdot\dfrac{6}{n}i + 2\right)\dfrac{6}{n}$

41. For a new machine shop employee, the rate of production for a specialized part is modeled by the function

$$p(x) = \dfrac{1}{2}x^2 + 3$$

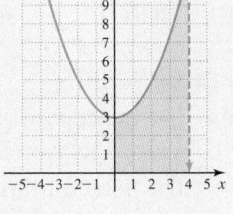

(production increases quickly with experience), where $p(x)$ represents the number of parts completed per day after x days on the job. The area under this curve in the interval $x \in [0, 4]$ then represents the total number of parts produced in the first 4 days. Using the rectangle method results in the expression $\dfrac{4}{n}\displaystyle\sum_{i=1}^{n}\left[\dfrac{1}{2}\left(\dfrac{4}{n}i\right)^2 + 3\right]$. Find the total number of parts produced by applying the summation properties/formulas and taking the limit as $n \to \infty$.

42. Water has been leaking undetected from a reservoir formed by an earthen dam. Due to the accelerating erosion, the rate of water loss can be modeled by the function $g(x) = -\frac{1}{4}x^2 + 8$, $x > 0$, where $g(x)$ represents hundreds of gallons of water lost per hour. The area under the curve in the interval $[0, 5]$ then represents the total water loss in 5 hr. Using the rectangle method results in the expression $\frac{5}{n}\sum_{i=1}^{n}\left[-\frac{1}{4}\left(\frac{5}{n}i\right)^2 + 8\right]$. Find the total amount of water lost by applying summation properties and formulas and taking the limit as $n \to \infty$.

Exercise 42

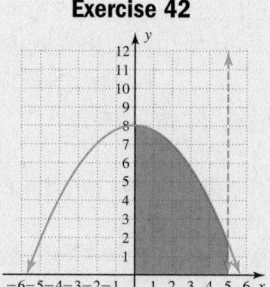

Find the area under the curve for each function and interval given, using the rectangle method and n subintervals of equal width.

43. $f(x) = -\frac{1}{2}x^2 + 4x$, $x \in [0, 6]$

44. $f(x) = -\frac{1}{2}x^3 + 6x$, $x \in [0, 3]$

▶ **MAINTAINING YOUR SKILLS**

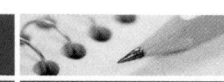

45. (4.4) Solve for x: $-350 = 211\, e^{-0.025x} - 450$

46. (7.4) Given $\mathbf{u} = \langle -2, -3 \rangle$ and $\mathbf{v} = \langle -3, -5 \rangle$, find $3\mathbf{u} - \mathbf{v}$.

47. (1.5/1.6) Find all solutions by factoring: $x^3 - 5x^2 + 3x - 15 = 0$.

48. (5.2) Evaluate all six trigonometric functions at $t = -\frac{2\pi}{3}$

SUMMARY AND CONCEPT REVIEW

SECTION 11.1   An Introduction to Limits Using Tables and Graphs

KEY CONCEPTS

• The concept of a limit involves the incremental approach of a variable x to a fixed constant c (the cause), with a focus on the behavior the function f during this approach (the effect).

• The notation $\lim_{x \to c} f(x) = L$ means values of $f(x)$ can be made arbitrarily close to L by taking values of x sufficiently close to c, but not equal to c. L is called the limit of f as x approaches c.

• Limits can be investigated using a table of values or the graph of a function. Choosing values of x that are closer and closer to c, we observe the resulting values of $f(x)$ to see if it appears a limiting value exists.

• The $\lim_{x \to c} f(x)$ may exist even if $f(x)$ is not defined at c, if $f(x)$ is defined at c but $f(c) \neq L$, or if $f(x)$ is defined at c and $f(c) = L$.

• We can also define a left-hand limit and a right-hand limit: $\lim_{x \to c^-} f(x) = L$ means values of $f(x)$ can be made arbitrarily close to L by taking values of x sufficiently close to c and to the left of c. The right-hand limit is similarly defined.

• A limit can fail to exist in one of three ways.

   (1) the left-hand limit is not equal to the right-hand limit: $\lim_{x \to c} f(x) = \left(\genfrac{}{}{0pt}{}{\text{dne}}{\text{LH} \neq \text{RH}}\right)$,

   (2) as $x \to c$, $f(x)$ grows without bound: $\lim_{x \to c} f(x) = \left(\genfrac{}{}{0pt}{}{\text{dne}}{\pm\infty}\right)$, or

   (3) as $x \to c$, there is no fixed number L that $f(x)$ approaches: $\lim_{x \to c} f(x) = \left(\genfrac{}{}{0pt}{}{\text{dne}}{\cancel{L}}\right)$.

EXERCISES

1. Estimate the following limits (if they exist) using the tables shown.

   a. $\lim\limits_{x \to 2^-} f(x)$ **b.** $\lim\limits_{x \to 2^+} f(x)$ **c.** $\lim\limits_{x \to 2} f(x)$

x	$f(x)$
1.5	−2.5
1.6	−2.6
1.7	−2.7
1.8	−2.8
1.9	−2.9
1.99	−2.99
1.999	−2.999
$y = -2.999$	

x	$f(x)$
2.5	5.5
2.4	5.4
2.3	5.3
2.2	5.2
2.1	5.1
2.01	5.01
2.001	5.001
$y = 5.001$	

Use a table of values to evaluate each limit.

2. $\lim\limits_{x \to 5} \dfrac{x^2 - 2x - 15}{x^2 - 6x + 5}$ **3.** $\lim\limits_{x \to \frac{3}{2}} \dfrac{\sin(2x - 3)}{2x - 3}$ **4.** $\lim\limits_{x \to -2} f(x),\ f(x) = \begin{cases} \dfrac{x - 4}{|x|} & x \le -2 \\ x^2 - 1 & x > -2 \end{cases}$

SECTION 11.2   The Properties of Limits

KEY CONCEPTS

- When evaluating limits using a table, great care must be exercised as results can sometimes be misleading. Using a graph and some common sense will sometimes help.
- Limits possess certain properties that seem plausible using a table of values. These include the sum, difference, product, and quotient properties, as well as the power and root properties. For a review of these properties, see page 1037.
- Once the limit properties have been established, they can be used to evaluate the limit of many different kinds of functions.
- These four basic limits play an important role in the evaluation of general limits.

   (1) $\lim\limits_{x \to c} k = k$ (2) $\lim\limits_{x \to c} x = c$ (3) $\lim\limits_{x \to c} x^k = c^k$ (4) $\lim\limits_{x \to c} \sqrt[k]{x} = \sqrt[k]{c}$, if $c \ge 0$ when k is even

- If the limit properties lead to an undefined or indeterminate expression, a limit may still exist and other means should be used to investigate.

EXERCISES

Determine the following limits using limit properties.

5. $\lim\limits_{x \to -3} 2x^3 - 5x + 1$ **6.** $\lim\limits_{x \to 7} \dfrac{\sqrt{11 - x}}{\sqrt[3]{x^2 - 3x - 1}}$ **7.** $\lim\limits_{x \to 1} \dfrac{\sqrt{x}}{(x^2 - 12x + 9)^5}$ **8.** $\lim\limits_{x \to 1} \dfrac{\sqrt{2 - x} - 1}{1 - x}$

SECTION 11.3   Continuity and More on Limits

KEY CONCEPTS

- Generally, a function is continuous if you can draw the entire graph without lifting your pencil.
- Specifically, a function f is continuous at c if these three conditions are met:

   (1) c is in the domain of f (2) $\lim\limits_{x \to c^-} f(x) = \lim\limits_{x \to c^+} f(x)$ (3) $\lim\limits_{x \to c} f(x) = f(c)$.

- The continuity of a function can be interupted by (1) asymptotic, (2) jump (non-removable), or (3) removable discontinuities.
- If a function is continuous at c, the limit $\lim\limits_{x \to c} f(x)$ can be found by direct substitution.
- Polynomial functions are continuous for $x \in \mathbb{R}$, while rational, root, and trigonometric functions are continuous everywhere they are defined.

• Using algebraic methods, a limit that cannot be evaluated by direct substitution can sometimes be rewritten in a form where a direct substitution becomes possible.
• Limits can also occur as x becomes infinitely large: $\lim\limits_{x \to \pm\infty} f(x) = L$, provided $f(x)$ can be made arbitrarily close to L, by taking values of x sufficiently large.

EXERCISES

9. Name the x-value(s) where the graph has

   a. an asymptotic discontinuity

   b. a jump discontinuity

   c. a removable discontinuity

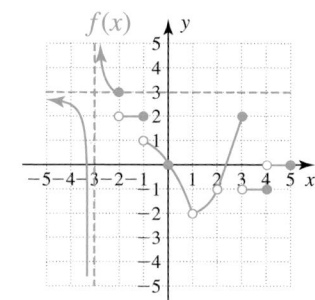

Use the graph of f to evaluate the following limits.

10. $\lim\limits_{x \to -\infty} f(x)$ 11. $\lim\limits_{x \to 3^+} f(x)$ 12. $\lim\limits_{x \to 3^-} f(x)$

13. $\lim\limits_{x \to -3^-} f(x)$ 14. $\lim\limits_{x \to 3} f(x)$ 15. $\lim\limits_{x \to 1} f(x)$

SECTION 11.4 Applications of Limits: Instantaneous Rates of Change and the Area under a Curve

KEY CONCEPTS

• The rate of change of $f(x)$ in the interval $[x_1, x_2]$ can be approximated using the rate-of-change formula:
$$\frac{\Delta f}{\Delta x} = \frac{f(x_2) - f(x_1)}{x_2 - x_1}.$$

• The rate of change of $f(x)$ in the arbitrary interval $[x, x + h]$ can be approximated using the difference quotient:
$$\frac{f(x + h) - f(x)}{h}.$$

• The instantaneous rate of change of f at point x is the limit of the difference quotient (if it exists):
$$f(x) = \lim\limits_{h \to 0} \frac{f(x + h) - f(x)}{h}.$$

• Graphically, the slope of the secant line gives a better and better approximation for the instantaneous rate of change as $h \to 0$, and we define the instantaneous rate of change as the limiting value of this slope:
$$m_{\text{tan}} = \lim\limits_{h \to 0} m_{\text{sec}} = \lim\limits_{h \to 0} \frac{f(x + h) - f(x)}{h}.$$

• For $f(x) > 0$ in the first quadrant, the area under the graph of f in the interval $[0, b]$ can be found using the rectangle method and n rectangles of equal width. The width of each rectangle will be $\dfrac{b}{n}$, the height of each is found by evaluating f in increments of $\dfrac{b}{n}$:
$$A = \lim\limits_{n \to \infty} \sum_{i=1}^{n} LW = \lim\limits_{n \to \infty} \sum_{i=1}^{n} f\left(\frac{b}{n}i\right)\frac{b}{n}.$$

EXERCISES

Find the limit of the difference quotient as $h \to 0$ to determine the instantaneous rate of change for the functions given.

16. $f(x) = x^2 + 5x - 2$ 17. $g(x) = \sqrt{2x - 1}$ 18. $v(x) = \dfrac{1}{x + 3}$

19. Take the limit of a difference quotient to find the slope of a line drawn tangent to $f(x) = -x^2 + 3x$ at $x = 4$.

20. Use the rectangle method to find the area in QI, under the graph of $f(x) = x^2 - 6x + 9$ in the interval $[0, 3]$.

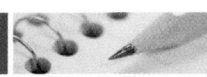

MIXED REVIEW

1. Express the following statements using limit notation:

   a. As x approaches a from the right, $f(x)$ approaches b.

   b. As x increases without bound, $f(x)$ approaches b.

   c. As x approaches a from both sides, $f(x)$ approaches b.

   d. As x approaches a from the left, $f(x)$ increases without bound.

Use a table of values to evaluate the following limits.

2. $\lim\limits_{x \to 0} \dfrac{\sin x}{x + \tan x}$

3. $\lim\limits_{x \to 0} \dfrac{2x^2 - 7x}{\sin x}$

4. Determine the limiting value of $g(x) = \dfrac{x^3 - 8}{x - 2}$

   as $x \to 2$ from the right.

5. Given $f(x) = \sin |\csc x|$, find $\lim\limits_{x \to \pi} f(x)$.

For Exercises 6 and 7, use the following information to evaluate the limits:

$$\lim\limits_{x \to c} f(x) = 5 \quad \lim\limits_{x \to c} g(x) = -13 \quad \lim\limits_{x \to c} h(x) = \frac{2}{3}$$

6. $\lim\limits_{x \to c} \dfrac{f(x) - g(x)}{h(x)}$

7. $\lim\limits_{x \to c} \sqrt{g(x) + \dfrac{6f(x)}{h(x)}}$

Evaluate each limit using the limit properties. If a limit does not exist, state why.

8. $\lim\limits_{x \to \frac{1}{2}} (x^{-2} - 6x + 2)$

9. $\lim\limits_{x \to 2} \dfrac{\dfrac{1}{3 - x} - x^2}{5x^2 - x}$

10. $\lim\limits_{x \to \sqrt{6}} \dfrac{x^2 + 6}{x^2 - 6}$

Evaluate the following limits.

11. Evaluate $\lim\limits_{x \to -2} \dfrac{x^3 + 8}{x^2 - 4}$

12. Evaluate $\lim\limits_{x \to 10} \dfrac{\sqrt{x - 1} - 3}{x - 10}$

13. Evaluate the limits by dividing the numerator and denominator by the highest power of x occurring in the denominator: $\lim\limits_{x \to \infty} \dfrac{-x^3 - 2x + 1}{-2x^3 + 5}$

For Exercises 14 and 15, use the graphs shown to evaluate each limit. If any computation is not possible, state why.

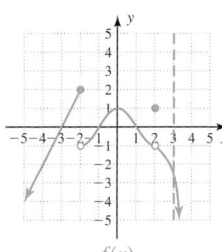

$f(x)$

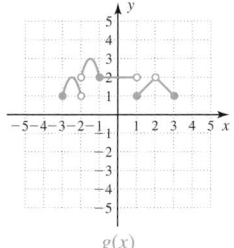

$g(x)$

14. $\lim\limits_{x \to -2^+} f(x) - \lim\limits_{x \to 1^-} g(x) + \lim\limits_{x \to -2^-} 2f(x)$

15. $\lim\limits_{x \to 2} \left[\dfrac{f(x)}{\sqrt{g(x)}} \right]$

16. A little boy throws a rock straight down off a bridge into a river. The rock's height in feet, t sec after it is released, is modeled by the function $d(t) = -16t^2 - 10t + 85$. Compute the limit of the difference quotient for d, to obtain a function $d(t)$ that gives the instantaneous velocity of the rock at time t.

17. What is the rock's instantaneous velocity at

    a. $t = 1$ sec?   **b.** $t = 2$ sec?

    c. $t = 0$ sec (when it was released)?

18. The spreading of a particularly juicy bit of gossip at a local college can be modeled by the function $p(t) = 0.1\sqrt{10t}$, where p is the percentage of faculty/staff "in the know" after t days. Find the limit of the difference quotient for p, to obtain a function $p(t)$ that represents the instantaneous rate of change of percentage of faculty who have heard the gossip at time t.

19. Referring to Exercise 18, what is the instantaneous rate of change in this percentage after

    a. $\dfrac{2}{5}$ day?   **b.** 1.6 days?   **c.** 5 days

20. Find the area in QI, under the graph of $g(x) = 10x - 2x^2$ in the interval $[0, 5]$. Use the rectangle method and n subintervals of equal width.

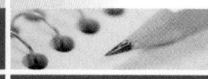

PRACTICE TEST

1. Write the expression in words, as though you were reading it out loud: $\lim_{x \to 5} f(x) = 10$

2. Fill in the blanks with the correct word: $\lim_{x \to c} f(x) = L$, means that values of _____ can be made arbitrarily close to _____, by taking values of x _____ close to _____.

State true or false and justify your answer.

3. For $f(x) = \dfrac{x^2 - 1}{x - 1}$, $\lim_{x \to 1} f(x)$ cannot exist since $x = 1$ is not in the domain.

4. For $f(x) = \begin{cases} x + 1, & x \geq 1 \\ x - 1, & x < 1 \end{cases}$, $\lim_{x \to 1} f(x)$ must exist since $x = 1$ is in the domain.

5. For $g(x) = \sqrt{x - 1} + 2$, state which of the following one-sided limits exist. Justify your answer and then find the limit.
 a. $\lim_{x \to 1^-} (\sqrt{x - 1} + 2)$
 b. $\lim_{x \to 1^+} (\sqrt{x - 1} + 2)$

6. Match each expression to its conclusion.
 a. $\lim_{x \to 0} \cos\left(\dfrac{1}{x}\right)$ **I.** $\left(\begin{matrix} \text{dne} \\ \infty \end{matrix}\right)$
 b. $\lim_{x \to 2} \dfrac{5x}{x^2 - 4x + 4}$ **II.** $\left(\begin{matrix} \text{dne} \\ \not L \end{matrix}\right)$
 c. $\lim_{x \to 0} \dfrac{\sin x}{x}$ **III.** $\left(\begin{matrix} \text{dne} \\ \text{LH} \neq \text{RH} \end{matrix}\right)$
 d. $\lim_{x \to 1} \dfrac{2x^2 - 2x}{\sqrt{x^2 - 2x + 1}}$ **IV.** 1

Determine the following using the graphs of $f(x)$ and $g(x)$ shown.

7. a. $\lim_{x \to -1^+} f(x)$ **b.** $\lim_{x \to 3^+} f(x)$

8. a. $\lim_{x \to -4} g(x)$ **b.** $\lim_{x \to 3^-} g(x)$

9. a. $f(0) - g(0)$
 b. $g(-5)f(-4)$

10. a. $[g(1)]^2$ **b.** $\dfrac{f(-4)}{g(3)}$

11. a. $\lim_{x \to -4} [f(x) + g(x)]$
 b. $\lim_{x \to 1^+} f(x)g(x)$

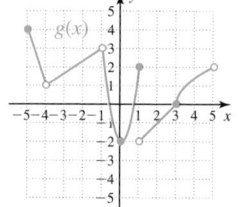

12. a. $\lim_{x \to -5^+} \sqrt{g(x)}$ **b.** $\lim_{x \to -1^+} [g(x)]^2$

13. Evaluate the following limits using a table. If the limit does not exist, state why.
 a. $\lim_{x \to 0} \dfrac{\cos x - 1}{x}$ **b.** $\lim_{x \to -1} \dfrac{x^3 + 1}{\sqrt{(x + 1)^2}}$
 c. $\lim_{x \to 0} \dfrac{\sin x}{2x + \tan x}$ **d.** $\lim_{x \to 0} \dfrac{\sin(3x)}{3x}$

14. Each of the following limits does not exist. State why.
 a. $\lim_{x \to 1} \dfrac{|x^2 - 1|}{x - 1}$ **b.** $\lim_{x \to 0} \tan\left(\dfrac{1}{x}\right)$
 c. $\lim_{x \to -2^+} \dfrac{x^2 - 1}{x + 2}$ **d.** $\lim_{x \to 3} \sqrt{x - 4}$

Evaluate each limit.

15. $\lim_{x \to -1} \dfrac{x^3 + 1}{x + 1}$ **16.** $\lim_{x \to -\infty} \dfrac{3x^2 + 5x - 4}{2x^2 - x - 3}$

17. $\lim_{x \to 25} \dfrac{\sqrt{x} - 5}{x - 25}$ **18.** $\lim_{x \to 0} \dfrac{\cos(2x)}{\cos^2 x}$

19. A controlled explosion at a strip mine causes a huge shower of dirt and rock. The height of any debris that has been projected vertically upward by the blast can be modeled by the function $d(t) = -16t^2 + 224t$, where $d(t)$ is the height in feet after t sec. (a) Compute the limit of the difference quotient for d to find a function $\mathcal{d}(t)$ that gives the instantaneous velocity of the debris at time t. (b) Find the velocity at $t = 2, 6, 7,$ and 11, and explain the significance of each result.

20. In preparation for another sold-out performance by the Danish pianist and comedian Victor Borge (Phonetic Punctuation), stage hands move a large grand piano 15 ft across the main stage and then carefully into position. The force applied can be modeled by the function $f(x) = 225 - x^2$ (the force is much greater as they start than when they finish). Since **Work = Force · Distance**, the area under the curve from 0 to x represents the amount of work done in foot-pounds. Use the rectangle method to (a) find an expression representing the amount of work done in moving the piano the first 6 ft, and (b) find the amount of work done in the interval [0, 6].

CUMULATIVE REVIEW CHAPTERS 1–11

1. Find all zeroes, real and complex:
$f(x) = 2x^3 - 3x^2 = 9 - 6x$

2. Graph the function $f(x) = 2x^2 - 12x + 15$ by completing the square.

3. Solve the system using elimination.
$$\begin{cases} x + 3y - 2z = 6 \\ 2x + y + z = 2 \\ -3x + 4y - 2z = 3 \end{cases}$$

4. Write as a sum/difference of logarithmic terms.
$$\ln\left[\frac{x^{\frac{2}{3}} \sqrt{x^2 + 3}}{(2x + 1)^4} \right]$$

5. Standing 134 ft from the St. Louis Arch, the angle of elevation to the top of the arch is 78°. How tall is the arch?

6. Verify that $\left(\dfrac{8}{17}, -\dfrac{15}{17} \right)$ is a point on the unit circle, then use it to find the value of $\sin \theta$, $\cos \theta$, and $\tan \theta$.

7. Find the sum of the infinite series
$$\frac{1}{3} + \frac{1}{9} + \frac{1}{27} + \frac{1}{81} + \cdots$$

8. Use the Guidelines for Graphing Rational Functions to graph $v(x) = \dfrac{x}{x^2 - 9}$.

9. Write $f(x) = x^3 - 2x^2 - 5x + 6$ in completely factored form, then graph the function.

10. Evaluate the following limits:
    a. $\displaystyle\lim_{x \to 1^-} \frac{|x - 1|}{x^2 - 1}$ and $\displaystyle\lim_{x \to 1^+} \frac{|x - 1|}{x^2 - 1}$
    b. $\displaystyle\lim_{x \to 1} \frac{|x - 1|}{x^2 - 1}$

11. For $z = 2 + 2i$, use DeMoivre's theorem to find $z^6 = (2 + 2i)^6$.

12. Solve the equation $2 + \sqrt{3} \tan (\alpha - 1) = 3$. Find all solutions in $[0, 2\pi)$.

13. I am a heavy pencil user. My new pencils are 19 cm long, but after 30 days of heavy use they now average only 9 cm in length. Assuming their length decreases linearly, (a) find a function $L(x)$ that models the length of my pencils after x days of use. (b) Use the model to find the length of a pencil I've been using for 15 days. (c) If one of my pencils is 11 cm long, how many days has it been in use?

14. An airline pilot is having to fight a stiff crosswind to stay on course. If the plane is flying at 250 mph on a heading of 20°, and the wind is from the west at 50 mph, what would be the course and heading of the plane if no course corrections are made?

15. Graph the piecewise-defined function given, then state whether its graph is continuous.
$$f(x) = \begin{cases} (x + 2)^2 & x \le 0 \\ x + 2 & x > 0 \end{cases}$$

16. Solve the following system using inverse matrices and your calculator.
$$\begin{cases} w + x + y + z = -2 \\ 2w - x + 3y + z = 9 \\ w - x - y + z = -4 \\ 3w + 2x - 2y + 3z = -19 \end{cases}$$

17. Solve for x: $-3e^{2x-1} = -28.08$.

18. Graph the function $y = 2 \cos\left(x - \dfrac{\pi}{4} \right)$.

19. Evaluate the limit:
$$\lim_{x \to -\infty} \frac{6x}{\sqrt{4x^2 + 5}}$$

20. Verify the following is an identity:
$\sin^4 x - \cos^4 x = 2 \sin^2 x - 1$

21. Decompose the rational expression $\dfrac{x^2 + x + 4}{x^2 + x - 2}$ into a sum of partial fractions.

22. Write each expression in exponential form:
    a. $-3 = \log 0.001$ b. $\ln x = 2$

23. Find all real solutions for
$$3500 = 7000 \sin\left(4\theta - \frac{\pi}{6} \right) + 10.$$

24. Identify and sketch the graph of the relation defined by $r = \dfrac{8}{4 + 3 \sin \theta}$.

25. Find the equation of the ellipse with vertices at $(-4, 1)$ and $(6, 1)$, and foci at $(-3, 1)$ and $(5, 1)$.

26. Graph the hyperbola by completing the square in x and y, then writing the equation in standard form:
$x^2 - 4y^2 + 24y + 6x - 43 = 0$.

Determine the following using the graph of $f(x)$ shown:

27. a. $\lim\limits_{x \to -3^-} f(x)$

b. $\lim\limits_{x \to -3^+} f(x)$

c. $\lim\limits_{x \to 2} f(x)$

d. $\lim\limits_{x \to 5} f(x)$

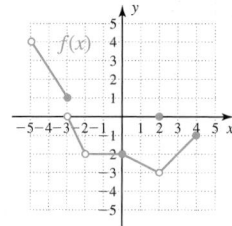

28. a. $f(-3)f(0)$

b. $f(2) - f(4)$

c. $f(-1) + f(3)$

d. $\dfrac{f(0)}{f(2)}$

29. Use the rectangle method to find the area under the graph of $f(x) = -\dfrac{1}{4}x^2 + 9$ for $x \in [0, 4]$.

30. Hi'ilani has to reset her five-digit PIN number. How many PINs are possible if no repetitions are allowed and two common vowels must be followed by three nonzero digits?

A Review of Basic Concepts and Skills

APPENDIX OUTLINE

A | The Language, Notation, and Numbers of Mathematics

In Section A you will review:

- ☐ **A.** Sets of numbers, graphing real numbers, and set notation
- ☐ **B.** Inequality symbols and order relations
- ☐ **C.** The absolute value of a real number
- ☐ **D.** The order of operations

A. Sets of Numbers, Graphing Real Numbers, and Set Notation

To effectively use mathematics as a problem-solving tool, we must first be familiar with the **sets of numbers** used to quantify (give a numeric value to) the things we investigate. Only then can we make comparisons and develop the equation models that lead to informed decisions. Our primary focus here is the set of **real numbers,** symbolized by the letter $\mathbb{R}$ written in the special font shown. This set is made up of several smaller sets called **subsets,** which are described in Table AI.1. Note the members or **elements** of a set are grouped within braces "{ }," and separated by commas. The three dots "..." seen in some subsets is called an **ellipsis,** and indicates the list of elements is infinite.

Table AI.1 Subsets of $\mathbb{R}$

Subset	Description	Additional Examples
Natural numbers $\mathbb{N}$	numbers used to count physical objects: $\{1, 2, 3, 4, 5, \dots\}$	19, 52, 1089
Whole numbers $\mathbb{W}$	natural numbers with the 0 element: $\{0, 1, 2, 3, 4, 5, \dots\}$	0, 31, 107, 3215
Integers $\mathbb{Z}$	whole numbers and their opposites: $\{\dots, -3, -2, -1, 0, 1, 2, 3, \dots\}$	$-725, -48, -7,$ $0, 7, 48, 725$
Rational numbers $\mathbb{Q}$	numbers that can be wriiten in fraction form, as the ratio of two integers $\frac{p}{q}, q \neq 0$	$-\frac{19}{2}, -7, 0.4, \frac{1}{3}, 29$
Irrational numbers $\mathbb{H}$	numbers with a nonrepeating and nonterminating decimal form; numbers that cannot be written as a ratio of two integers	$\pi \approx 3.141529\dots$ $-\frac{\sqrt{3}}{2} \approx -0.866025\dots$ $0.02002000200002\dots$

EXAMPLE 1 ▶ **Graphing Rational and Irrational Numbers**

Graph the each number by writing the decimal form and estimating its location between two integers (round to hundredths as needed).

a. $-\dfrac{2}{3}$ **b.** $\dfrac{3}{2}$ **c.** $\sqrt{19}$ **d.** π

Solution ▶ **a.** $-\dfrac{2}{3} = -0.67$ **b.** $\dfrac{3}{2} = 1.5$ **c.** $\sqrt{19} \approx 4.36$ **d.** $\pi \approx 3.14$

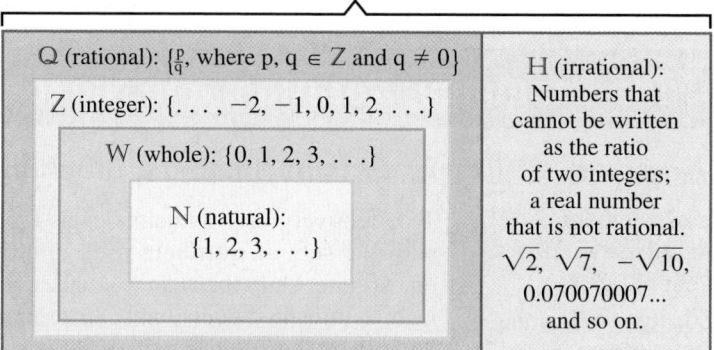

Now try Exercises 1 through 8 ▶

Figure AI.1 illustrates the relationship between these sets of numbers. Notice how each subset appears "nested" in a larger set.

$\mathbb{R}$ (real): All rational and irrational numbers

$\mathbb{Q}$ (rational): $\left\{\dfrac{p}{q}, \text{ where } p, q \in \mathbb{Z} \text{ and } q \neq 0\right\}$

$\mathbb{Z}$ (integer): $\{\ldots, -2, -1, 0, 1, 2, \ldots\}$

$\mathbb{W}$ (whole): $\{0, 1, 2, 3, \ldots\}$

$\mathbb{N}$ (natural): $\{1, 2, 3, \ldots\}$

$\mathbb{H}$ (irrational): Numbers that cannot be written as the ratio of two integers; a real number that is not rational. $\sqrt{2}$, $\sqrt{7}$, $-\sqrt{10}$, 0.070070007... and so on.

Figure AI.1

EXAMPLE 2 ▶ **Identifying Numbers**

List the numbers in set $A = \{-2, 0, 5, \sqrt{7}, 12, \frac{2}{3}, 4.5, \sqrt{21}, \pi, -0.75\}$ that belong to

a. $\mathbb{Q}$ **b.** $\mathbb{H}$ **c.** $\mathbb{W}$ **d.** $\mathbb{Z}$

Solution ▶ **a.** $-2, 0, 5, 12, \frac{2}{3}, 4.5, -0.75 \in \mathbb{Q}$ **b.** $\sqrt{7}, \sqrt{21}, \pi \in \mathbb{H}$

c. $0, 5, 12 \in \mathbb{W}$ **d.** $-2, 0, 5, 12 \in \mathbb{Z}$

☑ **A.** You've just reviewed sets of numbers, graphing real numbers, and set notation

Now try Exercises 9 through 12 ▶

B. Inequality Symbols and Order Relations

We compare numbers of different size using **inequality notation,** known as the **greater than** ($>$) and **less than** ($<$) symbols. Note that $-4 < 3$ is the same as saying -4 is to the left of 3 on the number line. In fact, on a number line, any given number is smaller than any number to the right of it (see Figure AI.2).

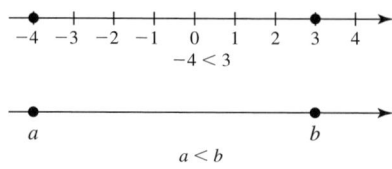

Figure AI.2

> **Order Property of Real Numbers**
>
> Given any two real numbers a and b.
>   1. $a < b$ if a is to the left of b on the number line.
>   2. $a > b$ if a is to the right of b on the number line.

Inequality notation is used with numbers and variables to write mathematical statements. A **variable** is a symbol, commonly a letter of the alphabet, used to represent an unknown quantity. Over the years x, y, and n have become most common, although any letter (or symbol) can be used. Often we'll use variables that remind us of the quantities they represent, like L for length, and D for distance.

EXAMPLE 3 ▶ **Writing Mathematical Models Using Inequalities**

Use a variable and an inequality symbol to represent the statement: "To hit a home run out of Jacobi Park, the ball must travel over three hundred twenty-five feet."

Solution ▶ Let D represent distance: $D > 325$ ft.

Now try Exercises 13 through 16 ▶

In Example 3, note the number 325 itself is not a possible value for D. If the ball traveled *exactly* 325 ft, it would hit the fence and stay in play. Numbers that mark the limit or boundary of an inequality are called **endpoints.** If the endpoint(s) are *not* included, the less than ($<$) or greater than ($>$) symbols are used. When the endpoints *are* included, the *less than or equal to symbol* ($\leq$) or the *greater than or equal to symbol* ($\geq$) is used. The decision to *include* or *exclude* an endpoint is often an important one, and many mathematical decisions (and real-life decisions) depend on a clear understanding of the distinction.

☑ **B.** You've just reviewed inequality symbols and order relations

C. The Absolute Value of a Real Number

In some applications, our primary interest is simply the *size* of n, rather than its sign. This is called the **absolute value** of n, denoted $|n|$, and can be thought of as its *distance from zero on the number line,* regardless of the direction (see Figure AI.3). Since distance is always positive or zero, $|n| \geq 0$.

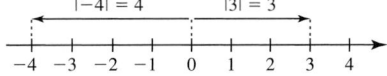

Figure AI.3

EXAMPLE 4 ▶ **Absolute Value Reading and Reasoning**

In the table shown, the absolute value of a number is given in column 1. Complete the remaining columns.

Solution ▶

Column 1 (In Symbols)	Column 2 (Spoken)	Column 3 (Result)	Column 4 (Reason)
$\lvert 7.5 \rvert$	"the absolute value of seven and five-tenths"	7.5	the distance between 7.5 and 0 is 7.5 units
$\lvert -2 \rvert$	"the absolute value of negative two"	2	the distance between -2 and 0 is 2 units
$-\lvert -6 \rvert$	"the opposite of the absolute value of negative six"	-6	the distance between -6 and 0 is 6 units, the opposite of 6 is -6

Now try Exercises 17 through 20 ▶

Example 4 shows the absolute value of a positive number is the number itself, while the absolute value of a negative number is the *opposite of that number* (recall that $-n$ is positive if n itself is negative). For this reason the formal definition of absolute value is stated as follows.

Absolute Value

For any real number n,

$$|n| = \begin{cases} n & \text{if} \quad n \geq 0 \\ -n & \text{if} \quad n < 0 \end{cases}$$

The concept of absolute value is often used to find the distance between two numbers on a number line. For instance, we know the distance between 2 and 8 is 6. Using absolute values, we write $|8 - 2| = |6| = 6$, or $|2 - 8| = |-6| = 6$. Generally, if a and b are two numbers on the real number line, the distance between them is $|a - b|$ or $|b - a|$.

EXAMPLE 5 ▶ **Using Absolute Value to Find the Distance between Points**

Find the distance between -5 and 3 on the number line.

Solution ▶ $|-5 - 3| = |-8| = 8$ or $|3 - (-5)| = |8| = 8$.

Now try Exercises 21 through 30 ▶

☑ **C.** You've just reviewed the absolute value of a real number

D. The Order of Operations

When a number is repeatedly multiplied by itself as in $(10)(10)(10)(10)$, we write it using **exponential notation** as 10^4. The number used for repeated multiplication (in this case 10) is called the **base,** and the superscript number is called an **exponent.** The exponent tells how many times the base occurs as a factor, and we say 10^4 is written in **exponential form.** Numbers that result from squaring an integer are called **perfect squares,** while numbers that result from cubing an integer are called **perfect cubes** (see Table AI.2).

Table AI.2

Perfect Squares				Perfect Cubes	
N	N^2	N	N^2	N	N^3
1	1	7	49	1	1
2	4	8	64	2	8
3	9	9	81	3	27
4	16	10	100	4	64
5	25	11	121	5	125
6	36	12	144	6	216

EXAMPLE 6 ▶ **Evaluating Numbers in Exponential Form**

Write each exponential in expanded form, then determine its value.

a. 4^3 b. $(-6)^2$ c. -6^2 d. $\left(\frac{2}{3}\right)^3$

Solution ▶   a. $4^3 = 4 \cdot 4 \cdot 4 = 64$ b. $(-6)^2 = (-6) \cdot (-6) = 36$
c. $-6^2 = -(6 \cdot 6) = -36$ d. $\left(\frac{2}{3}\right)^3 = \frac{2}{3} \cdot \frac{2}{3} \cdot \frac{2}{3} = \frac{8}{27}$

Now try Exercises 31 and 32 ▶

Examples 6(b) and 6(c) illustrate an important distinction. The expression $(-6)^2$ is read, "the square of negative six" and the negative sign is included in both factors. The expression -6^2 is read, "the opposite of six squared," and the square of six is calculated first, then made negative.

Square Roots and Cube Roots

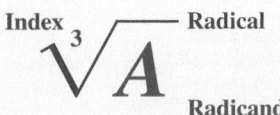

Index   Radical
Radicand

For the square root operation, either the $\sqrt{}$ or $\sqrt[2]{}$ notation can be used. The $\sqrt{}$ symbol is called a **radical,** the number under the radical is called the **radicand,** and the small case number used is called the **index.** The index tells how many factors are needed to obtain the radicand. In general, $\sqrt{a} = b$ only if $b^2 = a$. All numbers greater than zero have one positive and one negative square root. The *positive* or **principal square root** of 49 is 7 ($\sqrt{49} = 7$) since $7^2 = 49$. The *negative* square root of 49 is -7 ($-\sqrt{49} = -7$). The cube root of a number has the form $\sqrt[3]{a} = b$, where $b^3 = a$. This means $\sqrt[3]{27} = 3$ since $3^3 = 27$, and $\sqrt[3]{-8} = -2$ since $(-2)^3 = -8$. The cube root of a real number has one unique real value. In general, we have the following:

Square Roots	Cube Roots
$\sqrt{a} = b$ if $b^2 = a$	$\sqrt[3]{a} = b$ if $b^3 = a$
$(a \geq 0)$	$(a \in \mathbb{R})$
This indicates that	This indicates that
$\sqrt{a} \cdot \sqrt{a} = a$	$\sqrt[3]{a} \cdot \sqrt[3]{a} \cdot \sqrt[3]{a} = a$
or $(\sqrt{a})^2 = a$	or $(\sqrt[3]{a})^3 = a$

EXAMPLE 7 ▶ **Evaluating Square Roots and Cube Roots**

Determine the value of each expression.

a. $\sqrt{49}$ b. $\sqrt[3]{125}$ c. $\sqrt{\frac{9}{16}}$ d. $-\sqrt{16}$ e. $\sqrt{-25}$

Solution ▶   a. 7 since $7 \cdot 7 = 49$ b. 5 since $5 \cdot 5 \cdot 5 = 125$
c. $\frac{3}{4}$ since $\frac{3}{4} \cdot \frac{3}{4} = \frac{9}{16}$ d. -4 since $\sqrt{16} = 4$
e. not a real number since $5 \cdot 5 = (-5)(-5) = 25$

Now try Exercises 33 through 38 ▶

When basic operations are combined into a larger mathematical expression, we use a specified **priority** or **order of operations** to evaluate them.

The Order of Operations

1. Simplify within grouping symbols (parentheses, brackets, braces, etc.). If there are "nested" symbols of grouping, begin with the innermost group. If a fraction bar is used, simplify the numerator and denominator separately.
2. Evaluate all exponents and roots.
3. Compute all multiplications or divisions *in the order they occur from left to right.*
4. Compute all additions or subtractions *in the order they occur from left to right.*

EXAMPLE 8 ▶ **Evaluating Expressions Using the Order of Operations**

Simplify using the order of operations:

a. $5 + 2 \cdot 3$

b. $8 + 36 \div 4(12 - 3^2)$

c. $7500\left(1 + \dfrac{0.075}{12}\right)^{12 \cdot 15}$

d. $\dfrac{-4.5(8) - 3}{\sqrt[3]{125} + 2^3}$

Solution ▶

a. $5 + 2 \cdot 3 = 5 + 6$ multiplication before addition

            $= 11$ result

b. $8 + 36 \div 4(12 - 3^2)$

    $= 8 + 36 \div 4(12 - 9)$ simplify within parentheses

    $= 8 + 36 \div 4(3)$ $12 - 9 = 3$

    $= 8 + 9(3)$ division before multiplication

    $= 8 + 27$ multiply

    $= 35$ result

c. $7500\left(1 + \dfrac{0.075}{12}\right)^{12 \cdot 15}$ original expression

    $= 7500(1.00625)^{12 \cdot 15}$ simplify within the parenthesis (division before addition)

    $= 7500(1.00625)^{180}$ simplify the exponent

    $\approx 7500(3.069451727)$ exponents before multiplication

    $\approx 23{,}020.89$ result (rounded to hundredths)

d. $\dfrac{-4.5(8) - 3}{\sqrt[3]{125} + 2^3}$ original expression

    $= \dfrac{-36 - 3}{5 + 8}$ simplify the numerator and denominator

    $= \dfrac{-39}{13}$ combine terms

    $= -3$ result

☑ **D.** You've just reviewed the order of operations

Now try Exercises 39 through 54 ▶

A EXERCISES

▶ DEVELOPING YOUR SKILLS

Convert to decimal form and graph by estimating the number's location between two integers.

1. $\frac{4}{3}$ **2.** $-\frac{7}{8}$ **3.** $2\frac{5}{9}$ **4.** $-1\frac{5}{6}$

Use a calculator to find the principal square root of each number (round to hundredths as needed). Then graph each number by estimating its location between two integers.

5. 7 **6.** $\frac{75}{4}$ **7.** 3 **8.** $\frac{25\pi}{2}$

For the sets in Exercises 9 through 12:

a. List all numbers that are elements of (i) $\mathbb{N}$, (ii) $\mathbb{W}$, (iii) $\mathbb{Z}$, (iv) $\mathbb{Q}$, (v) $\mathbb{H}$, and (vi) $\mathbb{R}$.

b. Reorder the elements of each set from smallest to largest.

c. Graph the elements of each set on a number line.

9. $\left\{-1, 8, 0.75, \frac{9}{2}, 5.\overline{6}, 7, \frac{3}{5}, 6\right\}$

10. $\left\{-7, 2.\overline{1}, 5.73, -3\frac{5}{6}, 0, -1.12, \frac{7}{8}\right\}$

11. $\left\{-5, \sqrt{49}, 2, -3, 6, -1, \sqrt{3}, 0, 4, \pi\right\}$

12. $\left\{-8, 5, -2\frac{3}{5}, 1.75, -\sqrt{2}, -0.6, \pi, \frac{7}{2}, \sqrt{64}\right\}$

Use a descriptive variable and an inequality symbol ($<, >, \leq, \geq$) to write a model for each statement.

13. To spend the night at a friend's house, Kylie must be at least 6 yr old.

14. Monty can spend at most $2500 on the purchase of a used automobile.

15. If Jerod gets no more than two words incorrect on his spelling test he can play in the soccer game this weekend.

16. Andy must weigh less than 112 lb to be allowed to wrestle in his weight class at the meet.

Evaluate/simplify each expression.

17. $|-2.75|$

18. $|-7.24|$

19. $-|-4|$

20. $-|-6|$

21. $|3-(-6)|$

22. $|-4-7|$

Use the concept of absolute value to complete Exercises 23 to 30.

23. Write the statement two ways, then simplify. "The distance between -7.5 and 2.5 is . . ."

24. Write the statement two ways, then simplify. "The distance between $13\frac{2}{5}$ and $-2\frac{3}{5}$ is . . ."

25. What two numbers on the number line are five units from negative three?

26. What two numbers on the number line are three units from two?

27. If n is positive, then $-n$ is _____.

28. If n is negative, then $-n$ is _____.

29. If $n < 0$, then $|n| =$ _____.

30. If $n > 0$, then $|n| =$ _____.

Without computing the actual answer, state whether the result will be positive or negative. Be careful to note what power is used and whether the negative sign is included in parentheses.

31. a. $(-7)^2$ **b.** -7^2
    c. $(-7)^5$ **d.** -7^5

32. a. $(-7)^3$ **b.** -7^3
    c. $(-7)^4$ **d.** -7^4

Evaluate without the aid of a calculator.

33. $-\sqrt{\dfrac{121}{36}}$

34. $-\sqrt{\dfrac{25}{49}}$

35. $\sqrt[3]{-8}$

36. $\sqrt[3]{-64}$

37. What perfect square is closest to 78?

38. What perfect cube is closest to -71?

Perform the operation indicated without the aid of a calculator.

39. $-24-(-31)$

40. $-45-(-54)$

41. $4\frac{5}{6}+(-\frac{1}{2})$

42. $1\frac{1}{8}+(-\frac{3}{4})$

43. $\frac{4}{5}\div(-8)$

44. $-15\div\frac{1}{2}$

45. $-\frac{2}{3}\div\frac{16}{21}$

46. $-\frac{3}{4}\div\frac{7}{8}$

Evaluate without a calculator, using the order of operations.

47. $12-10\div2\times5+(-3)^2$

48. $(5-2)^2-16\div4\cdot2-1$

49. $\sqrt{\dfrac{9}{16}-\dfrac{3}{5}\cdot\left(\dfrac{5}{3}\right)^2}$

50. $\left(\dfrac{3}{2}\right)^2\div\left(\dfrac{9}{4}\right)-\sqrt{\dfrac{25}{64}}$

51. $\dfrac{4(-7)-6^2}{6-\sqrt{49}}$

52. $\dfrac{5(-6)-3^2}{9-\sqrt{64}}$

Evaluate using a calculator (round to hundredths).

53. $2475\left(1+\dfrac{0.06}{4}\right)^{4\cdot10}$

54. $5100\left(1+\dfrac{0.078}{52}\right)^{52\cdot20}$

▶ **APPLICATIONS**

Use positive and negative numbers **to model each situation, then compute.**

55. At 6:00 P.M., the temperature was 50°F. A cold front moves through that causes the temperature to *drop* 3°F each hour until midnight. What is the temperature at midnight?

56. Most air conditioning systems are designed to create a 2° *drop* in the air temperature each hour. How long would it take to reduce the air temperature from 86° to 71°?

57. The state of California holds the record for the greatest temperature swing between a record high and a record low. The record high was 134°F and the record low was −45°F. How many degrees *difference* are there between the record high and the record low?

58. In Juneau, Alaska, the temperature was 17°F early one morning. A cold front later moved in and the temperature *dropped* 32°F by lunch time. What was the temperature at lunch time?

▶ **EXTENDING THE CONCEPT**

59. Here are some historical approximations for π. Which one is closest to the true value?

Archimedes: $3\frac{1}{7}$ Tsu Ch'ung-chih: $\frac{355}{113}$
Aryabhata: $\frac{62,832}{20,000}$ Brahmagupta: $\sqrt{10}$

60. If $A > 0$ and $B < 0$, is the product $A \cdot (-B)$ positive or negative?

61. If $A < 0$ and $B < 0$, is the quotient $-(A \div B)$ positive or negative?

B | Algebraic Expressions and the Properties of Real Numbers

In Section B you will review how to:

☐ **A.** Identify algebraic expressions and create mathematical models

☐ **B.** Evaluate algebraic expressions

☐ **C.** Identify and use properties of real numbers

☐ **D.** Simplify algebraic expressions

A. Algebraic Expressions and Mathematical Models

An **algebraic term** is a *collection of factors* that may include numbers, variables, or expressions within parentheses. Here are some examples:

(1) 3   (2) $-6P$ (3) $5xy$ (4) $-8n^2$ (5) n (6) $2(x + 3)$

If a term consists of a single nonvariable number, it is called a **constant** term. In (1), 3 is a constant term. Any term that contains a variable is called a **variable term.** We call the constant factor of a term the **numerical coefficient** or simply the **coefficient.**

An **algebraic expression** can be a single term or a sum or difference of terms. To avoid confusion when identifying the coefficient of each term, the expression can be rewritten using algebraic addition if desired: $A - B = A + (-B)$. To identify the coefficient of a rational term, it sometimes helps to **decompose** the term, rewriting it using a unit fraction: $\frac{n-2}{5} = \frac{1}{5}(n - 2)$ and $\frac{x}{2} = \frac{1}{2}x$.

The key to solving many applied problems is finding an algebraic expression that accurately models the situation given. First, we assign a variable to represent an unknown quantity, then build related expressions using words that suggest a mathematical operation.

EXAMPLE 1 ▶ **Translating English Phrases into Algebraic Expressions**

Assign a variable to the unknown number, then translate each phrase into an algebraic expression.

  a. twice a number, increased by five

  b. six less than three times the width

Solution ▶   **a.** Let n represent the number. Then $2n$ represents twice the number, and $2n + 5$ represents twice a number, increased by five.

  b. Let W represent the width. Then $3W$ represents three times the width, and $3W - 6$ represents six less than three times the width.

Now try Exercises 1 through 14 ▶

☑ **A.** You've just reviewed how to identify algebraic expressions and how to create mathematical models

Identifying and translating such phrases *when they occur in context* is an important problem-solving skill. Note how this is done in Example 2.

> **EXAMPLE 2 ▶** **Creating a Mathematical Model**
>
> *The cost for a rental car is $35 plus 15 cents per mile.* Express the cost of renting a car in terms of the number of miles driven.
>
> **Solution ▶** Let m represent the number of miles driven. Then $0.15m$ represents the cost for each mile and $C = 35 + 0.15m$ represents the total cost for renting the car.

Now try Exercises 15 through 26 ▶

B. Evaluating Algebraic Expressions

We often need to **evaluate** expressions to investigate patterns and note relationships.

Evaluating a Mathematical Expression

1. Replace each variable with open parentheses ( ).
2. Substitute the given values for each variable.
3. Simplify using the order of operations.

In this evaluation, it's best to use a **vertical format,** with the original expression written first, the substitutions shown next, followed by the simplified forms and the final result. The numbers substituted or "plugged into" the expression are often called the **input values,** with the resulting values called **outputs.**

> **EXAMPLE 3 ▶** **Evaluating an Algebraic Expression**
>
> Evaluate the expression $x^3 - 2x^2 + 5$ for $x = -3$.
>
> **Solution ▶** For $x = -3$: $\begin{aligned} x^3 - 2x^2 + 5 &= (-3)^3 - 2(-3)^2 + 5 \quad &\text{substitute } -3 \text{ for } x \\ &= -27 - 2(9) + 5 \quad &\text{simplify: } (-3)^3 = -27, (-3)^2 = 9 \\ &= -27 - 18 + 5 \quad &\text{simplify: } 2(9) = 18 \\ &= -40 \quad &\text{result} \end{aligned}$

Now try Exercises 27 through 42 ▶

☑ **B.** You've just reviewed how to evaluate algebraic expressions

For exercises that combine the skills from Examples 2 and 3, **see Exercises 59 to 62.**

C. Properties of Real Numbers

While the phrase, "an unknown number times five," is accurately modeled by the expression $n5$ for some number n, in algebra we prefer to write numerical coefficients before variable factors. When we reorder the factors as $5n$, we're using the **commutative property of multiplication.** A reordering of terms involves the **commutative property of addition.**

The Commutative Properties

Given that a and b represent real numbers:

ADDITION: $a + b = b + a$ MULTIPLICATION: $a \cdot b = b \cdot a$

| Terms can be combined in any order without changing the sum. | Factors can be multiplied in any order without changing the product. |

Each property can be extended to include any number of terms or factors. While the commutative property implies a *reordering* or *movement* of terms (to commute implies back-and-forth movement), the **associative property** implies a *regrouping* or reassociation of terms. For example, the sum $\left(\frac{3}{4} + \frac{3}{5}\right) + \frac{2}{5}$ is easier to compute if we regroup the addends as $\frac{3}{4} + \left(\frac{3}{5} + \frac{2}{5}\right)$. This illustrates the **associative property of addition.** Multiplication is also associative.

The Associative Properties

Given that a, b, and c represent real numbers:

ADDITION:

$$(a + b) + c = a + (b + c)$$

Terms can be regrouped.

MULTIPLICATION:

$$(a \cdot b) \cdot c = a \cdot (b \cdot c)$$

Factors can be regrouped.

EXAMPLE 4 ▶ **Simplifying Expressions Using Properties of Real Numbers**

Use the commutative and associative properties to simplify each calculation.

a. $\frac{3}{8} - 19 + \frac{5}{8}$ **b.** $[-2.5 \cdot (-1.2)] \cdot 10$

Solution ▶ **a.** $\frac{3}{8} - 19 + \frac{5}{8} = -19 + \frac{3}{8} + \frac{5}{8}$ commutative property

$$= -19 + \left(\frac{3}{8} + \frac{5}{8}\right)$$ associative property

$$= -19 + 1$$ simplify

$$= -18$$ result

b. $[-2.5 \cdot (-1.2)] \cdot 10 = -2.5 \cdot [(-1.2) \cdot 10]$ associative property

$$= -2.5 \cdot (-12)$$ simplify

$$= 30$$ result

Now try Exercises 43 and 44 ▶

For any real number x, $x + 0 = x$ and 0 is called the **additive identity** since the original number was returned or "identified." Similarly, 1 is called the **multiplicative identity** since $1 \cdot x = x$.

For any real number x, there is a real number $-x$ such that $x + (-x) = 0$. The number $-x$ is called the **additive inverse** of x, since their sum results in the additive identity. Similarly, the **multiplicative inverse** of any nonzero number x is $\frac{1}{x}$, since $x \cdot \frac{1}{x} = 1$ (the multiplicative identity). This property can also be stated as $\frac{p}{q} \cdot \frac{q}{p} = 1$ ($p, q \neq 0$) for any rational number $\frac{p}{q}$. Note that $\frac{p}{q}$ and $\frac{q}{p}$ are **reciprocals.**

The **distributive property of multiplication over addition** is widely used in a study of algebra, because it enables us to rewrite a product as an equivalent sum and vice versa.

The Distributive Property of Multiplication over Addition

Given that a, b, and c represent real numbers:

$$a(b + c) = ab + ac$$

A factor outside a sum can be distributed to each addend in the sum.

$$ab + ac = a(b + c)$$

A factor common to each addend in a sum can be "undistributed" and written outside a group.

EXAMPLE 5 ▶ Simplifying Expressions Using the Distributive Property

Apply the distributive property as appropriate. Simplify if possible.

a. $7(p + 5.2)$ **b.** $-(2.5 - x)$ **c.** $7x^3 - x^3$ **d.** $\dfrac{5}{2}n + \dfrac{1}{2}n$

Solution ▶

a. $7(p + 5.2) = 7p + 7(5.2)$
$= 7p + 36.4$

b. $-(2.5 - x) = -1(2.5 - x)$
$= -1(2.5) - (-1)(x)$
$= -2.5 + x$

c. $7x^3 - x^3 = 7x^3 - 1x^3$
$= (7 - 1)x^3$
$= 6x^3$

d. $\dfrac{5}{2}n + \dfrac{1}{2}n = \left(\dfrac{5}{2} + \dfrac{1}{2}\right)n$
$= \left(\dfrac{6}{2}\right)n$
$= 3n$

Now try Exercises 45 through 50 ▶

☑ **C.** You've just reviewed how to identify and use properties of real numbers

---

D. Simplifying Algebraic Expressions

Two terms are **like terms** only if they have the *same variable factors* (the coefficient is not used to identify like terms). For instance, $3x^2$ and $-\frac{1}{7}x^2$ are like terms, while $5x^3$ and $5x^2$ are not. We simplify expressions by **combining like terms** using the distributive property, along with the commutative and associative properties.

EXAMPLE 6 ▶ Combining Like Terms Using the Distributive Property

Simplify the expression completely: $7(2p^2 + 1) - (p^2 + 3)$.

Solution ▶

$7(2p^2 + 1) - 1(p^2 + 3)$	original expression; note coefficient of −1
$= 14p^2 + 7 - 1p^2 - 3$	distributive property
$= (14p^2 - 1p^2) + (7 - 3)$	commutative and associative properties (collect like terms)
$= (14 - 1)p^2 + 4$	distributive property
$= 13p^2 + 4$	result

Now try Exercises 51 through 58 ▶

The steps for simplifying an algebraic expression are summarized here:

To Simplify an Expression

☑ **D.** You've just reviewed how to simplify algebraic expressions

1. Eliminate parentheses by applying the distributive property.
2. Use the commutative and associative properties to group like terms.
3. Use the distributive property to combine like terms.

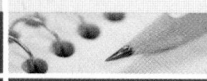

B EXERCISES

▶ DEVELOPING YOUR SKILLS

Translate each phrase into an algebraic expression.

1. seven fewer than a number

2. x decreased by six

3. the sum of a number and four

4. a number increased by nine

5. the difference between a number and five is squared

6. the sum of a number and two is cubed

7. thirteen less than twice a number

8. five less than double a number

9. a number squared plus the number doubled

10. a number cubed less the number tripled

11. five fewer than two-thirds of a number

12. fourteen more than one-half of a number

13. three times the sum of a number and five, decreased by seven

14. five times the difference of a number and two, increased by six

Create a mathematical model using descriptive variables.

15. The length of the rectangle is three meters less than twice the width.

16. The height of the triangle is six centimeters less than three times the base.

17. The speed of the car was fifteen miles per hour more than the speed of the bus.

18. It took Romulus three minutes more time than Remus to finish the race.

19. **Hovering altitude:** The helicopter was hovering 150 ft above the top of the building. Express the altitude of the helicopter in terms of the building's height.

20. **Stacks on a cruise liner:** The smoke stacks of the luxury liner cleared the bridge by 25 ft as it passed beneath it. Express the height of the stacks in terms of the bridge's height.

21. **Dimensions of a city park:** The length of a rectangular city park is 20 m more than twice its width. Express the length of the park in terms of the width.

22. **Dimensions of a parking lot:** In order to meet the city code while using the available space, a contractor planned to construct a parking lot with a length that was 50 ft less than three times its width. Express the length of the lot in terms of the width.

23. **Cost of milk:** In 2008, a gallon of milk cost two and one-half times what it did in 1990. Express the cost of a gallon of milk in 2008 in terms of the 1990 cost.

24. **Cost of gas:** In 2008, a gallon of gasoline cost one and one-half times what it did in 1990. Express the cost of a gallon of gas in 2008 in terms of the 1990 cost.

25. **Pest control:** In her pest control business, Judy charges $50 per call plus $12.50 per gallon of insecticide for the control of spiders and other insects. Express the total charge in terms of the number of gallons of insecticide used.

26. **Computer repairs:** As his reputation and referral business grew, Keith began to charge $75 per service call plus an hourly rate of $50 for the repair and maintenance of home computers. Express the cost of a service call in terms of the number of hours spent on the call.

Evaluate each algebraic expression given $x = 2$ and $y = -3$.

27. $4x - 2y$

28. $5x - 3y$

29. $-2x^2 + 3y^2$

30. $-5x^2 + 4y^2$

31. $2y^2 + 5y - 3$

32. $3x^2 + 2x - 5$

33. $(3x - 2y)^2$

34. $(2x - 3y)^2$

35. $\dfrac{-12y + 5}{-3x + 1}$

36. $\dfrac{12x + (-3)}{-3y + 1}$

37. $\sqrt{-12y} \cdot 4$

38. $7 \cdot \sqrt{-27y}$

Evaluate each expression for integers from -3 to 3 inclusive. What input(s) give an output of zero?

39. $x^2 - 3x - 4$ **40.** $x^2 - 2x - 3$

41. $x^3 - 6x + 4$ **42.** $x^3 + 5x + 18$

Rewrite each expression using the given property and simplify if possible.

43. Commutative property of addition

     a. $-5 + 7$ **b.** $-2 + n$

     c. $-4.2 + a + 13.6$ **d.** $7 + x - 7$

44. Associative property of multiplication

     a. $2 \cdot (3 \cdot 6)$ **b.** $3 \cdot (4 \cdot b)$

     c. $-1.5 \cdot (6 \cdot a)$ **d.** $-6 \cdot (-\frac{5}{6} \cdot x)$

Simplify by removing all grouping symbols (as needed) and combining like terms.

45. $-5(x - 2.6)$ **46.** $-12(v - 3.2)$

47. $\frac{2}{3}(-\frac{1}{5}p + 9)$

48. $\frac{5}{6}(-\frac{2}{15}q + 24)$

49. $\frac{2}{3}x + \frac{3}{4}x$

50. $\frac{5}{12}y - \frac{3}{8}y$

51. $3(a^2 + 3a) - (5a^2 + 7a)$

52. $2(b^2 + 5b) - (6b^2 + 9b)$

53. $x^2 - (3x - 5x^2)$

54. $n^2 - (5n - 4n^2)$

55. $(3a + 2b - 5c) - (a - b - 7c)$

56. $(x - 4y + 8z) - (8x - 5y - 2z)$

57. $\frac{3}{4}(5n - 4) + \frac{5}{8}(n + 16)$

58. $\frac{2}{3}(2x - 9) + \frac{3}{4}(x + 12)$

▶ APPLICATIONS

Translate each key phrase into an algebraic expression, then evaluate as indicated.

59. Cruising speed: A turbo-prop airliner has a cruising speed that is one-half the cruising speed of a 767 jet aircraft. (a) Express the speed of the turbo-prop in terms of the speed of the jet, and (b) determine the speed of the airliner if the cruising speed of the jet is 550 mph.

60. Softball toss: Macklyn can throw a softball two-thirds as far as her father. (a) Express the distance that Macklyn can throw a softball in terms of the distance her father can throw. (b) If her father can throw the ball 210 ft, how far can Macklyn throw the ball?

61. Dimensions of a lawn: The length of a rectangular lawn is 3 ft more than twice its width. (a) Express the length of the lawn in terms of the width. (b) If the width is 52 ft, what is the length?

62. Pitch of a roof: To obtain the proper pitch, the crossbeam for a roof truss must be 2 ft less than three-halves the rafter. (a) Express the length of the crossbeam in terms of the rafter. (b) If the rafter is 18 ft, how long is the crossbeam?

63. Postage costs: In 2010, a first class stamp cost 27¢ more than it did in 1978. Express the cost of a 2010 stamp in terms of the 1978 cost. If a stamp cost 15¢ in 1978, what was the cost in 2004?

64. Minimum wage: In 2009, the federal minimum wage was $4.95 per hour more than it was in 1976. Express the 2009 wage in terms of the 1976 wage. If the hourly wage in 1976 was $2.30, what was it in 2009?

65. Repair costs: The TV repairman charges a flat fee of $43.50 to come to your house and $25 per hour for labor. Express the cost of repairing a TV in terms of the time it takes to repair it. If the repair took 1.5 hr, what was the total cost?

66. Repair costs: At the local car dealership, shop charges are $79.50 to diagnose the problem and $85 per shop hour for labor. Express the cost of a repair in terms of the labor involved. If a repair takes 3.5 hr, how much will it cost?

67. If C must be a positive odd integer and D must be a negative even integer, then $C^2 + D^2$ must be a:

     a. positive odd integer.

     b. positive even integer.

     c. negative odd integer.

     d. negative even integer.

     e. cannot be determined.

68. Historically, several attempts have been made to create metric time using factors of 10, but our current system won out. If 1 day was 10 metric hours, 1 metric hour was 10 metric minutes, and 1 metric minute was 10 metric seconds, what time would it really be if a metric clock read 4:3:5? Assume that each new day starts at midnight.

A. The Properties of Exponents

The Product and Power Properties

Recall that b^n indicates the number b is multiplied by itself n times. There are two properties that follow immediately from this definition. When b^3 is multiplied by b^2, we have an uninterrupted string of five factors: $b^3 \cdot b^2 = (b \cdot b \cdot b) \cdot (b \cdot b)$, which can be written as b^5. This is an example of the **product property of exponents.**

Product Property Of Exponents

For any base b and positive integers m and n:
$$b^m \cdot b^n = b^{m+n}$$

A special application of the product property uses repeated factors of the *same* exponential term, as in $(x^2)^3$. Using the product property, we have $(x^2)(x^2)(x^2) = x^6$. Notice the same result can be found more quickly by multiplying the inner exponent by the outer exponent: $(x^2)^3 = x^{2 \cdot 3} = x^6$. We generalize this idea to state the **power property of exponents.**

Power Property of Exponents

For any base b and positive integers m and n:
$$(b^m)^n = b^{m \cdot n}$$

WORTHY OF NOTE

In this statement of the product property and the exponential properties that follow, it is assumed that for any expression of the form 0^m, $m > 0$ hence $0^m = 0$.

EXAMPLE 1 ▶ **Multiplying Terms Using Exponential Properties**

Compute each product.

a. $-4x^3 \cdot \frac{1}{2}x^2$ **b.** $(p^3)^2 \cdot (p^4)^2$

Solution ▶ **a.** $-4x^3 \cdot \frac{1}{2}x^2 = (-4 \cdot \frac{1}{2})(x^3 \cdot x^2)$ commutative and associative properties

$\qquad\qquad\qquad = (-2)(x^{3+2})$ product property; simplify

$\qquad\qquad\qquad = -2x^5$ result

b. $(p^3)^2 \cdot (p^4)^2 = p^6 \cdot p^8$ power property

$\qquad\qquad\qquad = p^{6+8}$ product property

$\qquad\qquad\qquad = p^{14}$ result

Now try Exercises 1 through 6 ▶

When the power property is extended to include more than one factor within the parentheses, we obtain the **product to a power property.** We can also raise a quotient of exponential terms to a power. The result is called the **quotient to a power property.**

Product to a Power Property

For any bases a and b, and positive integers m, n, and p:
$$(a^m b^n)^p = a^{mp} \cdot b^{np}$$

Quotient to a Power Property

For any bases a and $b \neq 0$, and positive integers m, n, and p:

$$\left(\frac{a^m}{b^n}\right)^p = \frac{a^{mp}}{b^{np}}$$

EXAMPLE 2 ▶ **Simplifying Terms Using the Power Properties**

Simplify using the power property (if possible):

a. $(-3a)^2$

b. $\left(-\frac{5}{2}a^3 b\right)^2$

Solution ▶ a. $(-3a)^2 = (-3)^2 \cdot (a^1)^2$

$\qquad = 9a^2$

b. $\left(-\frac{5}{2}a^3 b\right)^2 = \left(-\frac{5}{2}\right)^2 (a^3)^2 b^2$

$\qquad\qquad = \frac{25}{4}a^6 b^2$

Now try Exercises 7 through 12 ▶

Applications of exponents sometimes involve linking one exponential term with another using a substitution. The result is then simplified using exponential properties.

EXAMPLE 3 ▶ **Applying the Power Property after a Substitution**

The formula for the volume of a cube is $V = S^3$, where S is the length of one edge. If the length of each edge is $2x^2$:

a. Find a formula for volume in terms of x.

b. Find the volume if $x = 2$.

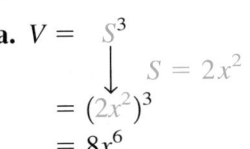

Solution ▶ a. $V = \quad S^3$

$\qquad \downarrow \quad S = 2x^2$

$\qquad = (2x^2)^3$

$\qquad = 8x^6$

b. For $V = 8x^6$,

$\qquad V = 8(2)^6 \quad$ substitute 2 for x

$\qquad = 8 \cdot 64$ or $512 \quad (2)^6 = 64$

The volume of the cube would be 512 units^3.

Now try Exercises 13 and 14 ▶

The Quotient Property

For $\dfrac{x^5}{x^2} = \dfrac{x \cdot x \cdot x \cdot x \cdot x}{x \cdot x}$ or x^3, the exponent of the final result appears to be the *difference between the exponent in the numerator and the exponent in the denominator.* This seems reasonable since the subtraction would indicate a removal of the factors that reduce to 1. Regardless of how many factors are used, we can generalize the idea and state the **quotient property of exponents.**

Quotient Property of Exponents

For any base $b \neq 0$ and positive integers m and n: $\dfrac{b^m}{b^n} = b^{m-n}$

Zero and Negative Numbers as Exponents

If the exponent of the denominator is *greater* than the exponent in the numerator, the quotient property yields a negative exponent: $\dfrac{x^2}{x^5} = x^{2-5} = x^{-3}$. To help understand what a negative exponent *means*, let's look at the expanded form of the expression: $\dfrac{x^2}{x^5} = \dfrac{x \cdot x}{x \cdot x \cdot x \cdot x \cdot x} = \dfrac{1}{x^3}$. A negative exponent can literally be interpreted as "write the factors as a reciprocal." A good way to remember this is

$$2^{-3} \overset{\text{three factors of 2}}{\underset{\text{written as a reciprocal}}{\curvearrowright}} \qquad \dfrac{2^{-3}}{1} = \dfrac{1}{2^3} = \dfrac{1}{8}$$

Since the result would be similar regardless of the base used, we can generalize this idea and state the **property of negative exponents.**

<table>
<tr><td>

WORTHY OF NOTE

The use of zero as an exponent should not strike you as strange or odd; it's simply a way of saying that *no factors of the base remain,* since all terms have been reduced to 1. For $\dfrac{2^3}{2^3}$, we have $\dfrac{8}{8} = 1$, or $\dfrac{\overset{1}{2} \cdot \overset{1}{2} \cdot \overset{1}{2}}{2 \cdot 2 \cdot 2} = 1$, or $2^{3-3} = 2^0 = 1$.

</td><td>

Property of Negative Exponents

For any base $b \neq 0$ and integer n:

$$b^{-n} = \dfrac{1}{b^n} \qquad \dfrac{1}{b^{-n}} = b^n \qquad \left(\dfrac{a}{b}\right)^{-n} = \left(\dfrac{b}{a}\right)^n; a \neq 0$$

Finally, when we consider that $\dfrac{x^3}{x^3} = 1$ by division, and $\dfrac{x^3}{x^3} = x^{3-3} = x^0$ using the quotient property, we conclude that $x^0 = 1$ as long as $x \neq 0$. We can also generalize this observation and state the meaning of zero as an exponent. In words the property says, *any nonzero quantity raised to an exponent of zero is equal to 1.*

</td></tr>
</table>

Zero Exponent Property

For any base $b \neq 0$: $b^0 = 1$

EXAMPLE 4 ▶ Simplifying Expressions Using Exponential Properties

Simplify using exponential properties. Answer using positive exponents only.

a. $\left(\dfrac{2a^3}{b^2}\right)^{-2}$

b. $(3hk^{-2})^3(6h^{-2}k^{-3})^{-2}$

c. $(3x)^0 + 3x^0 + 3^{-2}$

d. $\dfrac{(-2m^2n^3)^5}{(4mn^2)^3}$

Solution ▶

a. $\left(\dfrac{2a^3}{b^2}\right)^{-2} = \left(\dfrac{b^2}{2a^3}\right)^2$ property of negative exponents

$= \dfrac{(b^2)^2}{2^2(a^3)^2}$ power property

$= \dfrac{b^4}{4a^6}$ result

b. $(3hk^{-2})^3(6h^{-2}k^{-3})^{-2} = (3^3h^3k^{-6})(6^{-2}h^4k^6)$ power property

$= 3^3 \cdot 6^{-2} \cdot h^{3+4} \cdot k^{-6+6}$ product property

$= \dfrac{27h^7k^0}{36}$ simplify $\left(6^{-2} = \dfrac{1}{6^2} = \dfrac{1}{36}\right)$

$= \dfrac{3h^7}{4}$ result ($k^0 = 1$)

WORTHY OF NOTE

Notice in Example 4(c), we have $(3x)^0 = (3 \cdot x)^0 = 1$, while $3x^0 = 3 \cdot x^0 = 3(1)$. This is another example of operations and grouping symbols working together: $(3x)^0 = 1$ because any *quantity* to the zero power is 1. However, for $3x^0$ there are no grouping symbols, so the exponent 0 acts only on the x and not the 3.

c. $(3x)^0 + 3x^0 + 3^{-2} = 1 + 3(1) + \dfrac{1}{3^2}$ zero exponent property; property of negative exponents

$$= 4 + \dfrac{1}{9}$$ simplify

$$= 4\dfrac{1}{9} = \dfrac{37}{9}$$ result

d. $\dfrac{(-2m^2n^3)^5}{(4mn^2)^3} = \dfrac{(-2)^5(m^2)^5(n^3)^5}{4^3m^3(n^2)^3}$ power property

$$= \dfrac{-32m^{10}n^{15}}{64m^3n^6}$$ simplify

$$= -\dfrac{m^7n^9}{2}$$ quotient property

Now try Exercises 15 through 46 ▶

Summary of Exponential Properties

For real numbers a and b, and integers m, n, and p (excluding 0 raised to a nonpositive power)

Product property: $b^m \cdot b^n = b^{m+n}$

Power property: $(b^m)^n = b^{m \cdot n}$

Product to a power: $(a^m b^n)^p = a^{mp} \cdot b^{np}$

Quotient to a power: $\left(\dfrac{a^m}{b^n}\right)^p = \dfrac{a^{mp}}{b^{np}}\ (b \neq 0)$

Quotient property: $\dfrac{b^m}{b^n} = b^{m-n}\ (b \neq 0)$

Zero exponents: $b^0 = 1\ (b \neq 0)$

Negative exponents: $b^{-n} = \dfrac{b^{-n}}{1} = \dfrac{1}{b^n},\ \dfrac{1}{b^{-n}} = b^n,\ \left(\dfrac{a}{b}\right)^{-n} = \left(\dfrac{b}{a}\right)^n\ (a, b \neq 0)$

☑ **A.** You've just reviewed how to apply properties of exponents

B. Exponents and Scientific Notation

In many technical and scientific applications, we encounter numbers that are either extremely large or very, very small. For example, the mass of the moon is over 73 quintillion kilograms (73 followed by 18 zeroes), while the constant for universal gravitation contains 10 zeroes before the first nonzero digit. When computing with numbers of this size, scientific notation has a distinct advantage over the common decimal notation (base-10 place values).

WORTHY OF NOTE

Recall that multiplying by 10's (or multiplying by 10^k, $k > 0$) shifts the decimal to the right k places, making the number larger. Dividing by 10's (or multiplying by 10^{-k}, $k > 0$) shifts the decimal to the left k places, making the number smaller.

Scientific Notation

A nonzero number written in scientific notation has the form

$$N \times 10^k$$

where $1 \leq |N| < 10$ and k is an integer.

To convert a number from decimal notation into scientific notation, move the decimal point to the immediate right of the first nonzero digit (creating a number less than 10 but greater than or equal to 1) and multiply by 10^k. Here k represents the number of decimal places needed to return the decimal to its original position.

EXAMPLE 5 ▶ Converting from Decimal Notation to Scientific Notation

The mass of the moon is about 73,000,000,000,000,000,000 kg. Write this number in scientific notation.

Solution ▶ Place decimal to the right of first nonzero digit (7) and multiply by 10^k.

$$73,000,000,000,000,000,000 = 7.3 \times 10^k$$

To return the decimal to its original position would require 19 shifts to the *right*, so k must be *positive* 19.

$$73,000,000,000,000,000,000 = 7.3 \times 10^{19}$$

The mass of the moon is 7.3×10^{19} kg.

Now try Exercises 47 and 48 ▶

Converting a number from scientific notation to decimal notation is simply an application of multiplication or division with powers of 10.

EXAMPLE 6 ▶ Converting from Scientific Notation to Decimal Notation

The constant of gravitation is 6.67×10^{-11}. Write this number in common decimal form.

Solution ▶ Since the exponent is *negative* 11, shift the decimal 11 *places to the left*, using placeholder zeroes as needed to return the decimal to its original position:

$$6.67 \times 10^{-11} = 0.000\ 000\ 000\ 066\ 7$$

☑ **B.** You've just reviewed how to perform operations in scientific notation

Now try Exercises 49 through 52 ▶

C. Identifying and Classifying Polynomial Expressions

A **monomial** is a term using *only whole number exponents* on variables, with no variables in the denominator. A **polynomial** is a monomial or any sum or difference of monomial terms. For instance, $\frac{1}{2}x^2 - 5x + 6$ is a polynomial, while $3n^{-2} + 2n - 7$ is not. We classify polynomials according to their *degree* and *number of terms*. The **degree of a polynomial** in one variable is the largest exponent occurring on the variable. The degree of a polynomial in more than one variable is the largest sum of exponents in any one term. A polynomial with two terms is called a **binomial** (*bi* means two) and a polynomial with three terms is called a **trinomial** (*tri* means three). There are special names for polynomials with four or more terms, but for these, we simply use the general name *polynomial* (*poly* means many).

EXAMPLE 7 ▶ Classifying and Describing Polynomials

For each expression:
   a. Classify as a monomial, binomial, trinomial, or polynomial.
   b. State the degree of the polynomial.
   c. Name the coefficient of each term.

Solution ▶

Expression	Classification	Degree	Coefficients
$5x^2y - 2xy$	binomial	three	$5, -2$
$x^2 - 0.81$	binomial	two	$1, -0.81$
$z^3 - 3z^2 + 9z - 27$	polynomial (four terms)	three	$1, -3, 9, -27$
$\frac{-3}{4}x + 5$	binomial	one	$\frac{-3}{4}, 5$
$2x^2 + x - 3$	trinomial	two	$2, 1, -3$

Now try Exercises 53 through 58 ▶

A polynomial expression is in **standard form** when the terms of the polynomial are written in *descending order of degree,* beginning with the highest-degree term. The coefficient of the highest-degree term is called the **leading coefficient.**

EXAMPLE 8 ▶ Writing Polynomials in Standard Form

Write each polynomial in standard form, then identify the leading coefficient.

Solution ▶

Polynomial	Standard Form	Leading Coefficient
$9 - x^2$	$-x^2 + 9$	-1
$5z + 7z^2 + 3z^3 - 27$	$3z^3 + 7z^2 + 5z - 27$	3
$2 - \frac{3}{4}x$	$\frac{-3}{4}x + 2$	$\frac{-3}{4}$
$-3 + 2x^2 + x$	$2x^2 + x - 3$	2

☑ **C.** You've just reviewed how to identify and classify polynomial expressions

Now try Exercises 59 through 64 ▶

D. Adding and Subtracting Polynomials

To add polynomials, use the distributive, commutative, and associative properties to combine like terms. As with real numbers, the subtraction of polynomials involves adding the opposite of the second polynomial using algebraic addition.

EXAMPLE 9 ▶ Adding and Subtracting Polynomials

Perform the indicated operations:

$(0.7n^3 + 4n^2 + 8) + (0.5n^3 - n^2 - 6n) - (3n^2 + 7n - 10).$ eliminate parentheses (distributive property)

Solution ▶ $0.7n^3 + 4n^2 + 8 + 0.5n^3 - n^2 - 6n - 3n^2 - 7n + 10$ use real number properties to collect like terms

$= 0.7n^3 + 0.5n^3 + 4n^2 - 1n^2 - 3n^2 - 6n - 7n + 8 + 10$ combine like terms

$= 1.2n^3 - 13n + 18$

☑ **D.** You've just reviewed how to add and subtract polynomials

Now try Exercises 65 through 70 ▶

E. The Product of Two Polynomials

To multiply polynomials, we use the distributive property together with the product property of exponents.

EXAMPLE 10 ▶ Multiplying Polynomials

Find the product: **a.** $-2a^2(a^2 - 2a + 1)$ **b.** $(2z + 1)(z - 2)$

Solution ▶

a. $-2a^2(a^2 - 2a + 1) = -2a^2(a^2) - (-2a^2)(2a^1) + (-2a^2)(1)$ distribute

$\qquad\qquad\qquad\qquad = -2a^4 + 4a^3 - 2a^2$ simplify

b. $(2z + 1)(z - 2) = 2z(z - 2) + 1(z - 2)$ distribute to every term in the first binomial

$\qquad\qquad\qquad = 2z^2 - 4z + 1z - 2$ eliminate parentheses (distribute again)

$\qquad\qquad\qquad = 2z^2 - 3z - 2$ simplify

Now try Exercises 71 through 78 ▶

WORTHY OF NOTE

Consider the product $(x + 3)(x + 2)$ in the context of *area*. If we view $x + 3$ as the length of a rectangle (an unknown length plus 3 units), and $x + 2$ as its width (the same unknown length plus 2 units), a diagram of the total area would look like the following, with the result $x^2 + 5x + 6$ clearly visible.

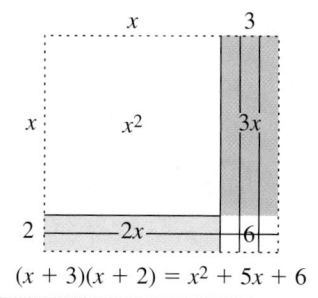

$(x + 3)(x + 2) = x^2 + 5x + 6$

The F-O-I-L Method

By observing the product of two binomials in Example 10(b), we note a pattern that can make the process more efficient. We illustrate here using the product $(2x - 1)(3x + 2)$.

The F-O-I-L Method for Multiplying Binomials

The product of two binomials can quickly be computed by multiplying:

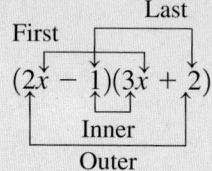

$$6x^2 + 4x - 3x - 2$$
First Outer Inner Last

and combining like terms

$$6x^2 + x - 2$$

The first term of the result will always be the product of the first terms from each binomial, and the last term of the result is the product of their last terms. We also note that here, the middle term is found by adding the *outermost product* with the *innermost product*.

EXAMPLE 11 ▶ Multiplying Binomials Using F-O-I-L

Compute each product mentally:

a. $(5n - 1)(n + 2)$

b. $(2b + 3)(5b - 6)$

Solution ▶

a. $(5n - 1)(n + 2)$: $5n^2 + 9n - 2$

                 $\boxed{10n + (-1n) = 9n}$

product of   sum of   product of
first two terms   outer and inner   last two terms

b. $(2b + 3)(5b - 6)$: $10b^2 + 3b - 18$

                 $\boxed{-12b + 15b = 3b}$

product of   sum of   product of
first two terms   outer and inner   last two terms

☑ **E.** You've just reviewed how to compute the product of two polynomials

Now try Exercises 79 through 86 ▶

F. Special Polynomial Products

Certain polynomial products are considered "special" for two reasons: (1) their products follow a predictable pattern, and (2) their results are frequently used to simplify expressions, graph functions, solve equations, and/or develop other skills. Recall that for an expression of the form $A + B$, the conjugate of the expression is $A - B$.

Special Products

Binomial Conjugates	Binomial Squares
$(A + B)(A - B) = A^2 - B^2$	$(A + B)^2 = A^2 + 2AB + B^2$
	$(A - B)^2 = A^2 - 2AB + B^2$

These special products can be verified using the F-O-I-L method, and should be committed to memory.

EXAMPLE 12 ▶ **Using Patterns for Special Products**

Compute each special product using the patterns outlined previously.

a. $(2x + 5)(2x - 5)$ **b.** $\left(x + \dfrac{2}{5}\right)\left(x - \dfrac{2}{5}\right)$

c. $(a + 9)^2$ **d.** $(3x - 5)^2$

Solution ▶ **a.** $(2x + 5y)(2x - 5y) = (2x)^2 - (5y)^2$ $(A + B)(A - B) = A^2 - B^2$
$$= 4x^2 - 25y^2 \quad \text{result}$$

b. $\left(x + \dfrac{2}{5}\right)\left(x - \dfrac{2}{5}\right) = x^2 - \left(\dfrac{2}{5}\right)^2$ $(A + B)(A - B) = A^2 - B^2$
$$= x^2 - \dfrac{4}{25} \quad \text{result}$$

c. $(a + 9)^2 = a^2 + 2(a \cdot 9) + 9^2$ special product $A^2 + 2AB + B^2$
$$= a^2 + 18a + 81 \quad \text{result}$$

d. $(3x - 5)^2 = (3x)^2 - 2(3x \cdot 5) + 5^2$ special product $A^2 + 2AB + B^2$
$$= 9x^2 - 30x + 25 \quad \text{result}$$

 F. You've just reviewed how to compute special products

Now try Exercises 87 through 102 ▶

⚠ **CAUTION** ▶ Note the square of a binomial always results in a trinomial (three terms). Specifically $(A + B)^2 \neq A^2 + B^2$.

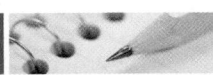

C EXERCISES

▶ DEVELOPING YOUR SKILLS

Determine each product using the product and/or power properties.

1. $\dfrac{2}{3}n^2 \cdot 21n^5$ **2.** $24g^5 \cdot \dfrac{3}{8}g^9$

3. $(-6p^2q)(2p^3q^3)$ **4.** $(-1.5vy^2)(-8v^4y)$

5. $(a^2)^4 \cdot (a^3)^2 \cdot b^2 \cdot b^5$ **6.** $d^2 \cdot d^4 \cdot (c^5)^2 \cdot (c^3)^2$

Simplify each expression using the product to a power property.

7. $(6pq^2)^3$ **8.** $(-3p^2q)^2$

9. $(-0.7c^4)^2(10c^3d^2)^2$ **10.** $(-2.5a^3)^2(3a^2b^2)^3$

11. $\left(\dfrac{3}{4}x^3y\right)^2$ **12.** $\left(\dfrac{4}{5}x^3\right)^2$

13. Volume of a cube: The formula for the volume of a cube is $V = S^3$, where S is the length of one edge. If the length of each edge is $3x^2$,

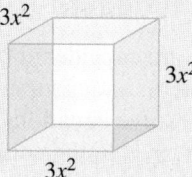

    a. Find a formula for volume in terms of the variable x.

    b. Find the volume of the cube if $x = 2$.

14. Area of a circle: The formula for the area of a circle is $A = \pi r^2$, where r is the length of the radius. If the radius is given as $5x^3$,

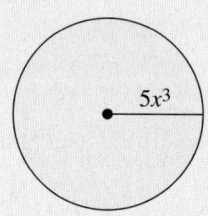

    a. Find a formula for area in terms of the variable x.

    b. Find the area of the circle if $x = 2$.

Simplify using the quotient property or the property of negative exponents. Write answers using positive exponents only.

15. $\dfrac{-6w^5}{-2w^2}$

16. $\dfrac{8z^7}{16z^5}$

17. $\dfrac{-12a^3b^5}{4a^2b^4}$

18. $\dfrac{5m^3n^5}{10mn^2}$

19. $\left(\frac{2}{3}\right)^{-3}$

20. $\left(\frac{5}{6}\right)^{-1}$

21. $\dfrac{2}{h^{-3}}$

22. $\dfrac{3}{m^{-2}}$

23. $(-2)^{-3}$

24. $(-4)^{-2}$

Simplify each expression using the quotient to a power property.

25. $\left(\dfrac{2p^4}{q^3}\right)^2$

26. $\left(\dfrac{-5v^4}{7w^3}\right)^2$

27. $\left(\dfrac{0.2x^2}{0.3y^3}\right)^3$

28. $\left(\dfrac{-0.5a^3}{0.4b^2}\right)^2$

29. $\left(\dfrac{5m^2n^3}{2r^4}\right)^2$

30. $\left(\dfrac{4p^3}{3x^2y}\right)^3$

Use properties of exponents to simplify the following. Write the answer using positive exponents only.

31. $\dfrac{9p^6q^4}{-12p^4q^6}$

32. $\dfrac{5m^5n^2}{10m^5n}$

33. $\dfrac{20h^{-2}}{12h^5}$

34. $\dfrac{5k^3}{20k^{-2}}$

35. $\dfrac{a^3b^{-2}}{(a^{-2}b^3)^4}$

36. $\dfrac{(p^{-4}q^8)^2}{p^5q^{-2}}$

37. $\dfrac{-6(2x^{-3})^2}{10x^{-2}}$

38. $\dfrac{18n^{-3}}{-8(3n^{-2})^3}$

39. $4^0 + 5^0$

40. $(-3)^0 + (-7)^0$

41. $2^{-1} + 5^{-1}$

42. $4^{-1} + 8^{-1}$

43. $3^0 + 3^{-1} + 3^{-2}$

44. $2^{-2} + 2^{-1} + 2^0$

45. $-5x^0 + (-5x)^0$

46. $-2n^0 + (-2n)^0$

Convert the following numbers to scientific notation.

47. In mid-2007, the U.S. Census Bureau estimated the world population at nearly 6,600,000,000 people.

48. The mass of a proton is generally given as 0.000 000 000 000 000 000 000 000 001 670 kg.

Convert the following numbers to decimal notation.

49. As of 2006, the smallest microprocessors in common use measured 6.5×10^{-9} m across.

50. In 2007, the estimated net worth of Bill Gates, the founder of Microsoft, was 5.6×10^{10} dollars.

Compute using scientific notation. Show all work.

51. The average distance between the Earth and the planet Jupiter is 465,000,000 mi. How many hours would it take a satellite to reach the planet if it traveled an average speed of 17,500 mi per hour? How many days? Round to the nearest whole.

52. In fiscal terms, a nation's debt-per-capita is the ratio of its total debt to its total population. In the year 2007, the total U.S. debt was estimated at $9,010,000,000,000, while the population was estimated at 303,000,000. What was the U.S. debt-per-capita ratio for 2007? Round to the nearest whole dollar.

Identify each expression as a polynomial or nonpolynomial (if a nonpolynomial, state why); classify each as a monomial, binomial, trinomial, or none of these; and state the degree of the polynomial.

53. $-35w^3 + 2w^2 + (-12w) + 14$

54. $-2x^3 + \frac{2}{3}x^2 - 12x + 1.2$

55. $5n^{-2} + 4n + \sqrt{17}$

56. $\dfrac{4}{r^3} + 2.7r^2 + r + 1$

57. $p^3 - \frac{2}{5}$

58. $q^3 + 2q^{-2} - 5q$

Write the polynomial in standard form and name the leading coefficient.

59. $7w + 8.2 - w^3 - 3w^2$

60. $-2k^2 - 12 - k$

61. $c^3 + 6 + 2c^2 - 3c$

62. $-3v^3 + 14 + 2v^2 + (-12v)$

63. $12 - \frac{2}{3}x^2$

64. $8 + 2n^2 + 7n$

Find the indicated sum or difference.

65. $(3p^3 - 4p^2 + 2p - 7) + (p^2 - 2p - 5)$

66. $(5q^2 - 3q + 4) + (-3q^2 + 3q - 4)$

67. $(5.75b^2 + 2.6b - 1.9) + (2.1b^2 - 3.2b)$

68. $(0.4n^2 + 5n - 0.5) + (0.3n^2 - 2n + 0.75)$

69. $(\frac{3}{4}x^2 - 5x + 2) - (\frac{1}{2}x^2 + 3x - 4)$

70. $(\frac{5}{9}n^2 + 4n - \frac{1}{2}) - (\frac{2}{3}n^2 - 2n + \frac{3}{4})$

Compute each product.

71. $-3x(x^2 - x - 6)$

72. $-2v^2(v^2 + 2v - 15)$

73. $(3r - 5)(r - 2)$

74. $(s - 3)(5s + 4)$

75. $(x - 3)(x^2 + 3x + 9)$

76. $(z + 5)(z^2 - 5z + 25)$

77. $(b^2 - 3b - 28)(b + 2)$

78. $(2h^2 - 3h + 8)(h - 1)$

79. $(7v - 4)(3v - 5)$ **80.** $(6w - 1)(2w + 5)$

81. $(3 - m)(3 + m)$ **82.** $(5 + n)(5 - n)$

83. $(m + \frac{3}{4})(m - \frac{3}{4})$ **84.** $(n - \frac{2}{5})(n + \frac{2}{5})$

85. $(2x^2 + 5)(x^2 - 3)$ **86.** $(3y^2 - 2)(2y^2 + 1)$

For each binomial, determine its conjugate and then find the product of the binomial with its conjugate.

87. $4m - 3$ **88.** $6n + 5$

89. $7x - 10$ **90.** $c + 3$

91. $6 + 5k$ **92.** $11 - 3r$

93. $x + \sqrt{6}$ **94.** $p - \sqrt{2}$

Find each binomial square.

95. $(x + 4)^2$ **96.** $(a - 3)^2$

97. $(4g + 3)^2$ **98.** $(5x - 3)^2$

99. $(4p - 3q)^2$ **100.** $(5c + 6d)^2$

101. $(4 - \sqrt{x})^2$ **102.** $(\sqrt{x} + 7)^2$

▶ **APPLICATIONS**

103. Attraction between particles: In electrical theory, the force of attraction between two particles P and Q with opposite charges is modeled by $F = \dfrac{kPQ}{d^2}$, where d is the distance between them and k is a constant that depends on certain conditions. This is known as Coulomb's law. Rewrite the formula using a negative exponent.

104. Intensity of light: The intensity of illumination from a light source depends on the distance from the source according to $I = \dfrac{k}{d^2}$, where I is the intensity measured in footcandles, d is the distance from the source in feet, and k is a constant that depends on the conditions. Rewrite the formula using a negative exponent.

Rewriting an expression: In advanced mathematics, negative exponents are widely used because they are easier to work with than rational expressions.

105. Rewrite the expression $\dfrac{5}{x^3} + \dfrac{3}{x^2} + \dfrac{2}{x^1} + 4$ using negative exponents.

106. Rewrite the expression $\dfrac{6}{a^4} - \dfrac{2}{a^2} + \dfrac{1}{a}$ using negative exponents.

107. Maximizing revenue: A sporting goods store finds that if they price their video games at $20, they make 200 sales per day. For each decrease of $1, 20 additional video games are sold. This means the store's revenue can be modeled by the formula $R = (20 - 1x)(200 + 20x)$, where x is the number of $1 decreases. Multiply out the binomials and use a table of values to determine what price will give the most revenue.

108. Maximizing revenue: Due to past experience, a jeweler knows that if they price jade rings at $60, they will sell 120 each day. For each decrease of $2, five additional sales will be made. This means the jeweler's revenue can be modeled by the formula $R = (60 - 2x)(120 + 5x)$, where x is the number of $2 decreases. Multiply out the binomials and use a table of values to determine what price will give the most revenue.

► **EXTENDING THE CONCEPT**

109. If $(3x^2 + kx + 1) - (kx^2 + 5x - 7) + (2x^2 - 4x - k) = -x^2 - 3x + 2$, what is the value of k?

110. If $\left(2x + \dfrac{1}{2x}\right)^2 = 5$, then the expression $4x^2 + \dfrac{1}{4x^2}$ is equal to what number?

D | Factoring Polynomials

In Section D you will review:

☐ **A.** Factoring out the greatest common factor and factoring by grouping

☐ **B.** Factoring quadratic polynomials

☐ **C.** Factoring special forms and quadratic forms

A. The Greatest Common Factor and Factoring by Grouping

To **factor** an expression means to *rewrite the expression as an equivalent product.* The distributive property is an example of factoring in action. To factor $2x^2 + 6x$, first rewrite each term using the common factor $2x$: $2x^2 + 6x = 2x \cdot x + 2x \cdot 3$, then apply the distributive property to obtain $2x(x + 3)$. The **greatest common factor** (or GCF) is the largest factor common to *all* terms in the polynomial.

EXAMPLE 1 ► **Factoring Polynomials**

Factor each polynomial:

   a. $12x^2 + 18xy - 30y$ **b.** $x^5 + x^2$ **c.** $(x + 3)x^2 + (x + 3)5$

Solution ►   **a.** 6 is common to all three terms:

$$12x^2 + 18xy - 30y \qquad \text{mentally: } 6 \cdot 2x^2 + 6 \cdot 3xy - 6 \cdot 5y$$
$$= 6(2x^2 + 3xy - 5y)$$

   b. x^2 is common to both terms:

$$x^5 + x^2 \qquad \text{mentally: } x^2 \cdot x^3 + x^2 \cdot 1$$
$$= x^2(x^3 + 1)$$

   c. $(x + 3)$ is common to both terms:

$$(x + 3)x^2 + (x + 3)5$$
$$= (x + 3)(x^2 + 5)$$

Now try Exercises 1 through 4 ►

One application of removing a binomial factor involves **factoring by grouping.** At first glance it appears, the expression $3t^3 + 15t^2 - 6t - 30 = 3(t^3 + 5t^2 - 2t - 10)$ cannot be factored further. But by grouping the terms (applying the associative property), we can remove a monomial factor from each subgroup, which then reveals a common binomial factor.

EXAMPLE 2 ► **Factoring by Grouping**

Factor $3t^3 + 15t^2 - 6t - 30$.

Solution ►   Notice that all four terms have a common factor of 3. Begin by factoring it out.

$$3t^3 + 15t^2 - 6t - 30 \qquad \text{original polynomial}$$
$$= 3(t^3 + 5t^2 - 2t - 10) \qquad \text{factor out 3}$$
$$= 3(t^3 + 5t^2 - 2t - 10) \qquad \text{group remaining terms}$$
$$= 3[t^2(t + 5) - 2(t + 5)] \qquad \text{factor common } \textit{monomial}$$
$$= 3(t + 5)(t^2 - 2) \qquad \text{factor common } \textit{binomial}$$

☑ **A.** You've just reviewed how to factor out the greatest common factor and factor by grouping

Now try Exercises 5 and 6 ►

When asked to factor an expression, first look for common factors. The resulting expression will be easier to work with and help ensure the final answer is written in **completely factored form.** If a four-term polynomial cannot be factored as written, try rearranging the terms to find a combination that enables factoring by grouping.

B. Factoring Quadratic Polynomials

A quadratic polynomial is one that can be written in the form $ax^2 + bx + c$, where a, b, $c \in \mathbb{R}$, and $a \neq 0$. One common form of factoring involves quadratic trinomials such as $x^2 + 7x + 10$ and $2x^2 - 13x + 15$. For this study, it will help to place the trinomials in two families—those with a leading coefficient of 1 and those with a leading coefficient other than 1.

$ax^2 + bx + c$, where $a = 1$

When $a = 1$, the only factor pair for x^2 (other than $1 \cdot x^2$) is $x \cdot x$ and the first term in each binomial will be x: $(x \quad)(x \quad)$. The following observation helps guide us to the complete factorization. Consider the product $(x + b)(x + a)$:

$$(x + m)(x + n) = x^2 + nx + mx + mn \quad \text{F-O-I-L}$$
$$= x^2 + (m + n)x + mn \quad \text{distributive property}$$

This shows that to factor the expression $x^2 + bx + c$, we're looking for two numbers m and n that multiply to c ($mn = c$) and add to b ($m + n = b$). It is also helpful to note that if the constant term is positive, the binomials will have *like* signs, since only *the product of like signs is positive*. If the constant term is negative, the binomials will have *unlike* signs, since only *the product of unlike signs is negative*. This means we can use the sign of the linear term (the term with degree 1) to guide our choice of factors.

> #### Template for Factoring Trinomials with a Leading Coefficient of 1
>
> If the constant term is positive, the binomials will have *like* signs:
> $$(x + \quad)(x + \quad) \text{ or } (x - \quad)(x - \quad),$$
> to match the sign of the linear (middle) term.
>
> If the constant term is negative, the binomials will have *unlike* signs:
> $$(x + \quad)(x - \quad),$$
> with the larger factor placed in the binomial whose sign *matches* the linear (middle) term.

EXAMPLE 3 ▶ **Factoring Trinomials**

Factor these expressions:
a. $-x^2 + 11x - 24$ **b.** $x^2 - 10 - 3x$

Solution ▶ **a.** First rewrite the trinomial in standard form as $-1(x^2 - 11x + 24)$. For $x^2 - 11x + 24$, the constant term is positive so the binomials will have like signs. Since the linear term is negative,

$$-1(x^2 - 11x + 24) = -1(x - \quad)(x - \quad) \quad \text{like signs, both negative}$$
$$= -1(x - 8)(x - 3) \quad (-8)(-3) = 24; -8 + (-3) = -11$$

b. First rewrite the trinomial in standard form as $x^2 - 3x - 10$. The constant term is negative so the binomials will have unlike signs. Since the linear term is negative,

$$x^2 - 3x - 10 = (x + \quad)(x - \quad) \quad \text{unlike signs, one positive and one negative}$$
$$5 > 2, 5 \text{ is placed in the second binomial;}$$
$$= (x + 2)(x - 5) \quad (2)(-5) = -10; 2 + (-5) = -3$$

Now try Exercises 7 and 8 ▶

WORTHY OF NOTE

Similarly, a cubic polynomial is one of the form $ax^3 + bx^2 + cx + d$. It's helpful to note that a cubic polynomial can be factored by grouping only when $ad = bc$, where a, b, c, and d are the coefficients shown. This is easily seen in Example 2, where $(3)(-30) = (15)(-6)$ gives $-90 = -90$✓.

Sometimes we encounter **prime polynomials,** such as $x^2 + 9x + 15$. The factor pairs of 15 are $1 \cdot 15$ and $3 \cdot 5$, with neither pair having a sum of $+9$.

$ax^2 + bx + c$, where $a \neq 1$

If the leading coefficient is not one, the possible combinations of outers and inners are more numerous and we generally employ a trial-and-error approach. To factor $2x^2 - 13x + 15$, note the constant term is positive so the binomials *must have like signs.* The negative linear term indicates these signs will be negative. We then list possible factors for the first and last terms of each binomial, then sum the outer and inner products in a search for the correct linear term.

Possible First and Last Terms for $2x^2$ and 15	Sum of Outers and Inners
1. $(2x - 1)(x - 15)$	$-30x - 1x = -31x$
2. $(2x - 15)(x - 1)$	$-2x - 15x = -17x$
3. $(2x - 3)(x - 5)$	$-10x - 3x = -13x$ ←
4. $(2x - 5)(x - 3)$	$-6x - 5x = -11x$

As you can see, only possibility 3 yields a linear term of $-13x$, and the correct factorization is then $(2x - 3)(x - 5)$. With practice, this **trial-and-error** process can be completed very quickly.

If the constant term is negative, the number of possibilities can be reduced by finding a factor pair with a sum *or* difference equal to the *absolute value* of the linear coefficient, as we can then arrange the sign of each binomial to obtain the needed result (see Example 4).

EXAMPLE 4 ▶ **Factoring a Trinomial Using Trial and Error**

Factor $6z^2 - 11z - 35$.

Solution ▶ Note the constant term is negative (binomials will have unlike signs), $|-11| = 11$, and the factors of 35 are $1 \cdot 35$ and $5 \cdot 7$. Two possible first terms are: $(6z \quad)(z \quad)$ and $(3z \quad)(2z \quad)$, and we begin with 5 and 7 as factors of 35.

$(6z \quad)(z \quad)$		Outers/Inners		$(3z \quad)(2z \quad)$		Outers/Inners	
		Sum	Diff			Sum	Diff
1. $(6z \; 5)(z \; 7)$		$47z$	$37z$	3. $(3z \; 5)(2z \; 7)$		$31z$	$11z$ ←
2. $(6z \; 7)(z \; 5)$		$37z$	$23z$	4. $(3z \; 7)(2z \; 5)$		$29z$	$1z$

Since possibility 3 yields the linear term of $11z$, we need not consider other factors of 35 and write the factored form as $6z^2 - 11z - 35 = (3z \; 5)(2z \; 7)$. The signs can then be arranged to obtain a middle term of $-11z$: $(3z + 5)(2z - 7)$, $-21z + 10z = -11z$ ✓.

☑ **B. You've just reviewed how to factor quadratic polynomials**

Now try Exercises 9 and 10 ▶

C. Factoring Special Forms and Quadratic Forms

Multiplying and factoring are reverse processes. This means that to factor the special products from Section C, we simply view the multiplication process "in reverse." When doing so, it may help to rewrite certain terms in the original expression as a perfect square: $(\quad)^2$.

> **WORTHY OF NOTE**
>
> In an attempt to factor a *sum* of two perfect squares, say $v^2 + 49$, let's list all possible binomial factors. These are (1) $(v + 7)(v + 7)$, (2) $(v - 7)(v - 7)$, and (3) $(v + 7)(v - 7)$. Note that (1) and (2) are the binomial squares $(v + 7)^2$ and $(v - 7)^2$, with each product resulting in a "middle" term, whereas (3) is a binomial times its conjugate, resulting in a *difference* of squares: $v^2 - 49$. With all possibilities exhausted, we conclude that *the sum of two squares is prime!*

Factoring Special Forms

Difference of Squares	Perfect Square Trinomials
$A^2 - B^2 = (A + B)(A - B)$	$A^2 + 2AB + B^2 = (A + B)^2$
	$A^2 - 2AB + B^2 = (A - B)^2$

Note that the *sum* of two perfect squares $A^2 + B^2$ *cannot be factored* using real numbers (the expression is prime). As a reminder, always check for a common factor first and be sure to write all results in completely factored form. See Example 5(c).

EXAMPLE 5 ▶ **Factoring a Difference of Squares**

Factor each expression completely.

    a. $4w^2 - 81$ **b.** $v^2 + 49$ **c.** $-3n^2 + 48$ **d.** $z^4 - \frac{1}{81}$ **e.** $x^2 - 7$

Solution ▶   **a.** $4w^2 - 81 = (2w)^2 - 9^2$ write as a difference of squares

                 $= (2w + 9)(2w - 9)$ $A^2 - B^2 = (A + B)(A - B)$

    b. $v^2 + 49$ is prime.

    c. $-3n^2 + 48 = -3(n^2 - 16)$ factor out -3

                 $= -3\left[n^2 - (4)^2\right]$ write as a difference of squares

                 $= -3(n + 4)(n - 4)$ $A^2 - B^2 = (A + B)(A - B)$

    d. $z^4 - \frac{1}{81} = (z^2)^2 - (\frac{1}{9})^2$ write as a difference of squares

               $= (z^2 + \frac{1}{9})(z^2 - \frac{1}{9})$ $A^2 - B^2 = (A + B)(A - B)$

               $= (z^2 + \frac{1}{9})\left[z^2 - (\frac{1}{3})^2\right]$ write as a difference of squares

               $= (z^2 + \frac{1}{9})(z + \frac{1}{3})(z - \frac{1}{3})$ result

    e. $x^2 - 7 = (x)^2 - (\sqrt{7})^2$ write as a difference of squares

                $= (x + \sqrt{7})(x - \sqrt{7})$ $A^2 - B^2 = (A + B)(A - B)$

Now try Exercises 11 and 12 ▶

EXAMPLE 6 ▶ **Factoring a Perfect Square Trinomial**

Factor $12m^3 - 12m^2 + 3m$.

Solution ▶ $12m^3 - 12m^2 + 3m$ check for common factors: GCF = $3m$

        $= 3m(4m^2 - 4m + 1)$ factor out $3m$

For the remaining trinomial $4m^2 - 4m + 1$ . . .

    1. Are the first and last terms perfect squares?

$$4m^2 = (2m)^2 \text{ and } 1 = (1)^2 \checkmark \quad \text{Yes.}$$

    2. Is the linear term twice the product of $2m$ and 1?

$$2 \cdot 2m \cdot 1 = 4m \checkmark \quad \text{Yes.}$$

    Factor as a binomial square: $4m^2 - 4m + 1 = (2m - 1)^2$

This shows $12m^3 - 12m^2 + 3m = 3m(2m - 1)^2$.

Now try Exercises 13 and 14 ▶

> ⚠ **CAUTION** ▶ As shown in Example 6, be sure to include the GCF in your final answer. It is a common error to "leave the GCF behind."

In actual practice, the tests for a perfect square trinomial are performed mentally, with only the factored form being written down.

There is one additional special form that merits attention, and that is factoring the sum or difference of two cubes.

Factoring the Sum or Difference of Two Cubes: $A^3 \pm B^3$

$$A^3 + B^3 = (A + B)(A^2 - AB + B^2)$$
$$A^3 - B^3 = (A - B)(A^2 + AB + B^2)$$

EXAMPLE 7 ▶ **Factoring the Sum and Difference of Two Cubes**

Factor completely:

a. $x^3 + 125$ **b.** $-5m^3n + 40n^4$

Solution ▶   **a.** $x^3 + 125 = x^3 + 5^3$ write terms as perfect cubes

Use $A^3 + B^3 = (A + B)(A^2 - AB + B^2)$ factoring template

$x^3 + 5^3 = (x + 5)(x^2 - 5x + 25)$ $A \to x$ and $B \to 5$

b. $-5m^3n + 40n^4 = -5n(m^3 - 8n^3)$ check for common factors (GCF = $-5n$)

$= -5n[m^3 - (2n)^3]$ write terms as perfect cubes

Use $A^3 - B^3 = (A - B)(A^2 + AB + B^2)$ factoring template

$m^3 - (2n)^3 = (m - 2n)[m^2 + m(2n) + (2n)^2]$ $A \to m$ and $B \to 2n$

$= (m - 2n)(m^2 + 2mn + 4n^2)$ simplify

$\Rightarrow -5m^3n + 40n^4 = -5n(m - 2n)(m^2 + 2mn + 4n^2).$ factored form

The results for parts (a) and (b) can be checked using multiplication.

Now try Exercises 15 and 16 ▶

Using *u*-Substitution to Factor Quadratic Forms

For any quadratic expression $ax^2 + bx + c$ in standard form, the degree of the leading term is twice the degree of the middle term. Generally, a trinomial is in **quadratic form** if it can be written as $a(\underline{\quad})^2 + b(\underline{\quad}) + c$, where the parentheses "hold" the same factors. The equation $x^4 - 13x^2 + 36 = 0$ is in quadratic form since $(x^2)^2 - 13(x^2) + 36 = 0$. In many cases, we can factor these expressions using a **placeholder substitution** that transforms these expressions into a more recognizable form. In a study of algebra, the letter "*u*" often plays this role. If we let *u* represent x^2, the expression $(x^2)^2 - 13(x^2) + 36$ becomes $u^2 - 13u + 36$, which can be factored into $(u - 9)(u - 4)$. After "unsubstituting" (replace *u* with x^2), we have $(x^2 - 9)(x^2 - 4) = (x + 3)(x - 3)(x + 2)(x - 2)$.

EXAMPLE 8 ▶ **Factoring a Quadratic Form**

Write in completely factored form: $(x^2 - 2x)^2 - 2(x^2 - 2x) - 3$.

Solution ▶ Expanding the binomials would produce a fourth-degree polynomial that would be very difficult to factor. Instead we note the expression is in *quadratic form*. Letting u represent $x^2 - 2x$ (the variable part of the "middle" term), $(x^2 - 2x)^2 - 2(x^2 - 2x) - 3$ becomes $u^2 - 2u - 3$.

$$u^2 - 2u - 3 = (u - 3)(u + 1) \quad \text{factor}$$

To finish up, write the expression in terms of x, substituting $x^2 - 2x$ for u.

$$= (x^2 - 2x - 3)(x^2 - 2x + 1) \quad \text{substitute } x^2 - 2x \text{ for } u$$

The resulting trinomials can be further factored.

✓ **C.** You've just reviewed how to factor special forms and quadratic forms

$$= (x - 3)(x + 1)(x - 1)^2 \quad x^2 - 2x + 1 = (x - 1)^2$$

Now try Exercises 17 through 20 ▶

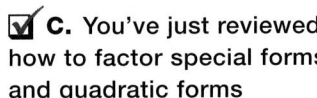

D EXERCISES

▶ **DEVELOPING YOUR SKILLS**

Factor each expression using the method indicated.

Greatest Common Factor

1. a. $-17x^2 + 51$ **b.** $21b^3 - 14b^2 + 56b$
   c. $-3a^4 + 9a^2 - 6a^3$

2. a. $-13n^2 - 52$ **b.** $9p^2 + 27p^3 - 18p^4$
   c. $-6g^5 + 12g^4 - 9g^3$

Common Binomial Factor

3. a. $2a(a + 2) + 3(a + 2)$
   b. $(b^2 + 3)3b + (b^2 + 3)2$
   c. $4m(n + 7) - 11(n + 7)$

4. a. $5x(x - 3) - 2(x - 3)$
   b. $(v - 5)2v + (v - 5)3$
   c. $3p(q^2 + 5) + 7(q^2 + 5)$

Grouping

5. a. $9q^3 + 6q^2 + 15q + 10$
   b. $h^5 - 12h^4 - 3h + 36$
   c. $k^5 - 7k^3 - 5k^2 + 35$

6. a. $6h^3 - 9h^2 - 2h + 3$
   b. $4k^3 + 6k^2 - 2k - 3$
   c. $3x^2 - xy - 6x + 2y$

Trinomial Factoring where $|a| = 1$

7. a. $-p^2 + 5p + 14$ **b.** $q^2 - 4q - 45$
   c. $n^2 + 20 - 9n$

8. a. $-m^2 + 13m - 42$ **b.** $x^2 + 12 + 13x$
   c. $v^2 + 10v + 15$

Trinomial Factoring where $a \neq 1$

9. a. $3p^2 - 13p - 10$ **b.** $4q^2 + 7q - 15$
   c. $10u^2 - 19u - 15$

10. a. $6v^2 + v - 35$ **b.** $20x^2 + 53x + 18$
   c. $15z^2 - 22z - 48$

Difference of Perfect Squares

11. a. $4s^2 - 25$ **b.** $9x^2 - 49$
   c. $50x^2 - 72$ **d.** $121h^2 - 144$
   e. $b^2 - 5$

12. a. $9v^2 - \frac{1}{25}$ **b.** $25w^2 - \frac{1}{49}$
   c. $v^4 - 1$ **d.** $16z^4 - 81$
   e. $x^2 - 17$

Perfect Square Trinomials

13. a. $a^2 - 6a + 9$ **b.** $b^2 + 10b + 25$
   c. $4m^2 - 20m + 25$ **d.** $9n^2 - 42n + 49$

14. a. $x^2 + 12x + 36$ **b.** $z^2 - 18z + 81$
   c. $25p^2 - 60p + 36$ **d.** $16q^2 + 40q + 25$

   c. $x^2 + x + 6$ **d.** $x^2 - x - 6$
   e. $x^2 - 5x + 6$ **f.** $x^2 + 5x - 6$

Sum/Difference of Perfect Cubes

15. a. $8p^3 - 27$ **b.** $m^3 + \frac{1}{8}$
   c. $g^3 - 0.027$ **d.** $-2t^4 + 54t$

16. a. $27q^3 - 125$ **b.** $n^3 + \frac{8}{27}$
   c. $b^3 - 0.125$ **d.** $3r^4 - 24r$

u-Substitution

17. a. $x^4 - 10x^2 + 9$ **b.** $x^4 + 13x^2 + 36$
   c. $x^6 - 7x^3 - 8$

18. a. $x^6 - 26x^3 - 27$
   b. $3(n + 5)^2 + (2n + 10) - 21$
   c. $2(z + 3)^2 + (3z + 9) - 54$

19. Completely factor each of the following (recall that "1" is its own perfect square and perfect cube).

   a. $n^2 - 1$ **b.** $n^3 - 1$
   c. $n^3 + 1$ **d.** $28x^3 - 7x$

20. Carefully factor each of the following trinomials, if possible. Note differences and similarities.

   a. $x^2 - x + 6$ **b.** $x^2 + x - 6$

Factor each expression completely, if possible. Rewrite the expression in standard form (factor out "−1" if needed) and factor out the GCF if one exists. If you believe the expression will not factor, write "prime."

21. $a^2 + 7a + 10$ **22.** $b^2 + 9b + 20$

23. $2x^2 - 24x + 40$ **24.** $10z^2 - 140z + 450$

25. $64 - 9m^2$ **26.** $25 - 16n^2$

27. $-9r + r^2 + 18$ **28.** $28 + s^2 - 11s$

29. $2h^2 + 7h + 6$ **30.** $3k^2 + 10k + 8$

31. $9k^2 - 24k + 16$ **32.** $4p^2 - 20p + 25$

33. $-6x^3 + 39x^2 - 63x$ **34.** $-28z^3 + 16z^2 + 80z$

35. $12m^2 - 40m + 4m^3$ **36.** $-30n - 4n^2 + 2n^3$

37. $a^2 - 7a - 60$ **38.** $b^2 - 9b - 36$

39. $8x^3 - 125$ **40.** $27r^3 + 64$

41. $m^2 + 9m - 24$ **42.** $n^2 - 14n - 36$

43. $x^3 - 5x^2 - 9x + 45$ **44.** $x^3 + 3x^2 - 4x - 12$

▶ APPLICATIONS

In many cases, factoring an expression can make it easier to evaluate as in the following applications.

45. Conical shells: The volume of a conical shell (like the shell of an ice cream cone) is given by the formula $V = \frac{1}{3}\pi R^2 h - \frac{1}{3}\pi r^2 h$, where R is the outer radius and r is the inner radius of the cone. Write the formula in completely factored form, then find the volume of a shell when $R = 5.1$ cm, $r = 4.9$ cm, and $h = 9$ cm. Answer in exact form and in approximate form rounded to the nearest tenth.

46. Spherical shells: The volume of a spherical shell (like the outer shell of a cherry cordial) is given by the formula $V = \frac{4}{3}\pi R^3 - \frac{4}{3}\pi r^3$, where R is the outer radius and r is the inner radius of the shell. Write the right-hand side in completely factored form, then find the volume of a shell where $R = 1.8$ cm and $r = 1.5$ cm.

47. Volume of a box: The volume of a rectangular box x inches in height is given by the relationship $V = x^3 + 8x^2 + 15x$. Factor the right-hand side to determine: (a) The number of inches that the width exceeds the height, (b) the number of inches the length exceeds the height, and (c) the volume given the height is 2 ft.

48. Shipping textbooks: A publisher ships paperback books stacked x copies high in a box. The total number of books shipped per box is given by the relationship $B = x^3 - 13x^2 + 42x$. Factor the right-hand side to determine (a) how many more or fewer books fit the width of the box (than the height), (b) how many more or fewer books fit the length of the box (than the height), and (c) the number of books shipped per box if they are stacked 10 high in the box.

49. Space-Time relationships: Due to the work of Albert Einstein and other physicists who labored on space-time relationships, it is known that the faster an object moves the shorter it appears to become. This phenomenon is modeled by the

Lorentz transformation $L = L_0\sqrt{1 - \left(\dfrac{v}{c}\right)^2}$,

where L_0 is the length of the object at rest, L is the relative length when the object is moving at velocity v, and c is the speed of light. Factor the radicand and use the result to determine the relative length of a 12-in. ruler if it is shot past a stationary observer at 0.75 times the speed of light ($v = 0.75c$).

50. Tubular fluid flow: As a fluid flows through a tube, it is flowing faster at the center of the tube than at the sides, where the tube exerts a backward drag. **Poiseuille's law** gives the velocity of the flow at any point of the cross section: $v = \dfrac{G}{4\eta}(R^2 - r^2)$, where R is the inner radius of the tube, r is the distance from the center of the tube to a point in the flow, G represents what is called the pressure gradient, and η is a constant that depends on the viscosity of the fluid. Factor the right-hand side and find v given $R = 0.5$ cm, $r = 0.3$ cm, $G = 15$, and $\eta = 0.25$.

▶ **EXTENDING THE CONCEPT**

51. Factor out a constant that leaves integer coefficients for each term:
  a. $\frac{1}{2}x^4 + \frac{1}{8}x^3 - \frac{3}{4}x^2 + 4$
  b. $\frac{2}{3}b^5 - \frac{1}{6}b^3 + \frac{4}{9}b^2 - 1$

52. If $x = 2$ is substituted into $2x^3 + hx + 8$, the result is zero. What is the value of h?

53. Factor the expression: $192x^3 - 164x^2 - 270x$.

54. As an alternative to evaluating polynomials by direct substitution, **nested factoring** can be used. The method has the advantage of using only products and sums—no powers. For $P = x^3 + 3x^2 + 1x + 5$, we begin by grouping all variable terms and factoring x: $P = [x^3 + 3x^2 + 1x] + 5 = x[x^2 + 3x + 1] + 5$. Then we group the inner terms with x and factor again: $P = x[x^2 + 3x + 1] + 5 = x[x(x + 3) + 1] + 5$. The expression can now be evaluated using any input and the order of operations. If $x = 2$, we quickly find that $P = 27$. Use this method to evaluate $H = x^3 + 2x^2 + 5x - 9$ for $x = -3$.

Factor each expression completely.

55. $x^4 - 81$

56. $16n^4 - 1$

57. $p^6 - 1$

58. $m^6 - 64$

59. $q^4 - 28q^2 + 75$

60. $a^4 - 18a^2 + 32$

E | Rational Expressions

In Section E you will learn how to:

☐ **A.** Write a rational expression in simplest form

☐ **B.** Multiply and divide rational expressions

☐ **C.** Add and subtract rational expressions

☐ **D.** Simplify compound fractions

☐ **E.** Rewrite formulas and algebraic models

A. Writing a Rational Expression in Simplest Form

A rational number is one that can be written as the quotient of two integers. Similarly, a *rational expression* is one that can be written as the quotient of two polynomials. We can apply the skills developed in a study of fractions (how to reduce, add, subtract, multiply, and divide) to **rational expressions,** sometimes called **algebraic fractions.** A rational expression is in **simplest form** when the numerator and denominator have no common factors (other than 1). After factoring the numerator and denominator, we apply the **fundamental property of rational expressions.**

Fundamental Property of Rational Expressions

If P, Q, and R are polynomials, with $Q, R \neq 0$,

$$(1)\quad \frac{P \cdot R}{Q \cdot R} = \frac{P}{Q} \quad \text{and} \quad (2)\quad \frac{P}{Q} = \frac{P \cdot R}{Q \cdot R}$$

1. A rational expression can be simplified by canceling common factors in the numerator and denominator.
2. An equivalent expression can be formed by multiplying numerator and denominator by the same nonzero polynomial.

EXAMPLE 1 ▶ Simplifying a Rational Expression

Write the expression in simplest form: $\dfrac{x^2 - 1}{x^2 - 3x + 2}$.

Solution ▶

$$\frac{x^2 - 1}{x^2 - 3x + 2} = \frac{(x - 1)(x + 1)}{(x - 1)(x - 2)} \qquad \text{factor numerator and denominator}$$

$$= \frac{\cancel{(x - 1)}(x + 1)}{\cancel{(x - 1)}(x - 2)} \qquad \text{cancel common factors}$$

$$= \frac{x + 1}{x - 2} \qquad \text{simplest form}$$

Now try Exercises 1 through 4 ▶

⚠ CAUTION ▶ When reducing rational numbers or expressions, only common *factors* can be reduced. It is incorrect to reduce (or divide out) individual terms: $\dfrac{-6 + 4\sqrt{3}}{2} \neq -3 + 4\sqrt{3}$, and

$\dfrac{x + 1}{x + 2} \neq \dfrac{1}{2}$ (except for $x = 0$)

When simplifying rational expressions, we sometimes encounter expressions of the form $\dfrac{a - b}{b - a}$. If we factor -1 from the numerator, we see that $\dfrac{a - b}{b - a} = \dfrac{-1\cancel{(b - a)}}{\cancel{b - a}} = -1$.

EXAMPLE 2 ▶ Simplifying a Rational Expression

Write the expression in simplest form: $\dfrac{(6 - 2x)}{x^2 - 9}$.

Solution ▶

$$\frac{(6 - 2x)}{x^2 - 9} = \frac{2(3 - x)}{(x - 3)(x + 3)} \qquad \text{factor numerator and denominator}$$

$$= \frac{(2)(-1)}{x + 3} \qquad \text{reduce:} \frac{(3 - x)}{(x - 3)} = -1$$

$$= \frac{-2}{x + 3} \qquad \text{simplest form}$$

☑ A. You've just reviewed how to write a rational expression in simplest form

Now try Exercises 5 through 10 ▶

B. Multiplication and Division of Rational Expressions

Operations on rational expressions use the factoring skills reviewed earlier, along with much of what we know about rational numbers.

Multiplying Rational Expressions

Given that P, Q, R, and S are polynomials with $Q, S \neq 0$,

$$\frac{P}{Q} \cdot \frac{R}{S} = \frac{PR}{QS}$$

1. Factor all numerators and denominators completely.
2. Reduce common factors.
3. Multiply numerator $\times$ numerator and denominator $\times$ denominator.

EXAMPLE 3 ▶ **Multiplying Rational Expressions**

Compute the product: $\dfrac{2a + 2}{3a - 3a^2} \cdot \dfrac{3a^2 - a - 2}{9a^2 - 4}$.

Solution ▶
$$\frac{2a + 2}{3a - 3a^2} \cdot \frac{3a^2 - a - 2}{9a^2 - 4} = \frac{2(a + 1)}{3a(1 - a)} \cdot \frac{(3a + 2)(a - 1)}{(3a - 2)(3a + 2)} \qquad \text{factor}$$

$$= \frac{2(a + 1)}{3a\underset{1}{(1 - a)}} \cdot \frac{\overset{(-1)}{\cancel{(3a + 2)}}\cancel{(a - 1)}}{(3a - 2)\underset{1}{\cancel{(3a + 2)}}} \qquad \text{reduce: } \frac{a - 1}{1 - a} = -1$$

$$= \frac{-2(a + 1)}{3a(3a - 2)} \qquad \text{simplest form}$$

Now try Exercises 11 through 14 ▶

To divide fractions, we multiply the first expression by the *reciprocal of the second*. The quotient of two rational expressions is computed in the same way.

Dividing Rational Expressions

Given that P, Q, R, and S are polynomials with $Q, R, S \neq 0$,

$$\frac{P}{Q} \div \frac{R}{S} = \frac{P}{Q} \cdot \frac{S}{R} = \frac{PS}{QR}$$

Invert the divisor and multiply.

EXAMPLE 4 ▶ **Dividing Rational Expressions**

Compute the quotient: $\dfrac{4m^3 - 12m^2 + 9m}{m^2 - 49} \div \dfrac{10m^2 - 15m}{m^2 + 4m - 21}$.

Solution ▶
$$\frac{4m^3 - 12m^2 + 9m}{m^2 - 49} \div \frac{10m^2 - 15m}{m^2 + 4m - 21}$$

$$= \frac{4m^3 - 12m^2 + 9m}{m^2 - 49} \cdot \frac{m^2 + 4m - 21}{10m^2 - 15m} \qquad \text{invert and multiply}$$

$$= \frac{m(4m^2 - 12m + 9)}{(m + 7)(m - 7)} \cdot \frac{(m + 7)(m - 3)}{5m(2m - 3)} \qquad \text{factor}$$

$$= \frac{\overset{1}{\cancel{m}}(2m - 3)\overset{1}{\cancel{(2m - 3)}}}{\cancel{(m + 7)}(m - 7)} \cdot \frac{\cancel{(m + 7)}(m - 3)}{5\underset{1}{\cancel{m}}\underset{1}{\cancel{(2m - 3)}}} \qquad \text{factor and reduce}$$

$$= \frac{(2m - 3)(m - 3)}{5(m - 7)} \qquad \text{lowest terms}$$

Note that we sometimes refer to simplest form as *lowest terms*.

Now try Exercises 15 through 30 ▶

> ⚠ **CAUTION** ▶ For products like $\dfrac{(w + 7)(w - 7)}{(w - 7)(w - 2)} \cdot \dfrac{(w - 2)}{(w + 7)}$, it is a common mistake to think that all factors "cancel," leaving an answer of zero. Actually, all factors *reduce to 1,* and the result is a value of 1 for all inputs where the product is defined.
>
> $$\dfrac{(w \overset{1}{+} 7)(w \overset{1}{-} 7)}{(w \underset{1}{-} 7)(w \underset{1}{-} 2)} \cdot \dfrac{(w \overset{1}{-} 2)}{(w \underset{1}{+} 7)} = 1$$

☑ **B.** You've just reviewed how to multiply and divide rational expressions

C. Addition and Subtraction of Rational Expressions

Recall that the addition and subtraction of *fractions* requires finding the lowest common denominator (LCD) and building equivalent fractions. The sum or difference of the numerators is then placed over this denominator. The procedure for the addition and subtraction of *rational expressions* is very much the same.

> **Addition and Subtraction of Rational Expressions**
>
> 1. Find the LCD of all rational expressions.
> 2. Build equivalent expressions using the LCD.
> 3. Add or subtract numerators as indicated.
> 4. Write the result in lowest terms.

EXAMPLE 5 ▶ **Adding and Subtracting Rational Expressions**

Compute as indicated:

a. $\dfrac{7}{10x} + \dfrac{3}{25x^2}$ **b.** $\dfrac{10x}{x^2 - 9} - \dfrac{5}{x - 3}$

Solution ▶ **a.** The LCD for $10x$ and $25x^2$ is $50x^2$.   find the LCD

$$\dfrac{7}{10x} + \dfrac{3}{25x^2} = \dfrac{7}{10x} \cdot \dfrac{(5x)}{(5x)} + \dfrac{3}{25x^2} \cdot \dfrac{(2)}{(2)} \qquad \text{write equivalent expressions}$$

$$= \dfrac{35x}{50x^2} + \dfrac{6}{50x^2} \qquad \text{simplify}$$

$$= \dfrac{35x + 6}{50x^2} \qquad \text{add the numerators and write the result over the LCD}$$

The result is in simplest form.

b. The LCD for $x^2 - 9$ and $x - 3$ is $(x - 3)(x + 3)$.   find the LCD

$$\dfrac{10x}{x^2 - 9} - \dfrac{5}{x - 3} = \dfrac{10x}{(x - 3)(x + 3)} - \dfrac{5}{(x - 3)} \cdot \dfrac{(x + 3)}{(x + 3)} \qquad \text{write equivalent expressions}$$

$$= \dfrac{10x - 5(x + 3)}{(x - 3)(x + 3)} \qquad \text{subtract numerators, write the result over the LCD}$$

$$= \dfrac{10x - 5x - 15}{(x - 3)(x + 3)} \qquad \text{distribute}$$

$$= \dfrac{5x - 15}{(x - 3)(x + 3)} \qquad \text{combine like terms}$$

$$= \dfrac{5(x \overset{1}{-} 3)}{(x \underset{1}{-} 3)(x + 3)} = \dfrac{5}{x + 3} \qquad \text{factor and reduce}$$

Now try Exercises 31 through 36 ▶

EXAMPLE 6 ▶ **Adding and Subtracting Rational Expressions**

Perform the operations indicated:

a. $\dfrac{5}{n+2} - \dfrac{n-3}{n^2-4}$ **b.** $\dfrac{b^2}{4a^2} - \dfrac{c}{a}$

Solution ▶ **a.** The LCD for $n+2$ and n^2-4 is $(n+2)(n-2)$.

$$\dfrac{5}{n+2} - \dfrac{n-3}{n^2-4} = \dfrac{5}{(n+2)}\cdot\dfrac{(n-2)}{(n-2)} - \dfrac{n-3}{(n+2)(n-2)} \quad \text{write equivalent expressions}$$

$$= \dfrac{5(n-2)-(n-3)}{(n+2)(n-2)} \quad \text{subtract numerators, write the result over the LCD}$$

$$= \dfrac{5n-10-n+3}{(n+2)(n-2)} \quad \text{distribute}$$

$$= \dfrac{4n-7}{(n+2)(n-2)} \quad \text{result}$$

b. The LCD for a and $4a^2$ is $4a^2$: $\dfrac{b^2}{4a^2} - \dfrac{c}{a} = \dfrac{b^2}{4a^2} - \dfrac{c}{a}\cdot\dfrac{(4a)}{(4a)} \quad \text{write equivalent expressions}$

$$= \dfrac{b^2}{4a^2} - \dfrac{4ac}{4a^2} \quad \text{simplify}$$

$$= \dfrac{b^2-4ac}{4a^2} \quad \text{subtract numerators, write the result over the LCD}$$

☑ **C.** You've just reviewed how to add and subtract rational expressions

Now try Exercises 37 through 48 ▶

⚠ **CAUTION** ▶ When the second term in a subtraction has a binomial numerator as in Example 6(a), be sure the subtraction *is applied to both terms*. It is a common error to write $\dfrac{5(n-2)}{(n+2)(n-2)} - \dfrac{n-3}{(n+2)(n-2)} = \dfrac{5n-10\boxed{-n-3}}{(n+2)(n-2)}$ ✗ in which the subtraction is applied to the first term only. This is incorrect!

D. Simplifying Compound Fractions

Rational expressions whose numerator or denominator contain a fraction are called **compound fractions**. The expression $\dfrac{\frac{2}{3m}-\frac{3}{2}}{\frac{3}{4m}-\frac{1}{3m^2}}$ is a compound fraction with a numerator of $\dfrac{2}{3m}-\dfrac{3}{2}$ and a denominator of $\dfrac{3}{4m}-\dfrac{1}{3m^2}$. The two methods commonly used to simplify compound fractions are summarized in the following boxes.

Simplifying Compound Fractions (Method I)

1. Add/subtract fractions in the numerator, writing them as a single expression.
2. Add/subtract fractions in the denominator, also writing them as a single expression.
3. Multiply the numerator by the reciprocal of the denominator and simplify if possible.

> **Simplifying Compound Fractions (Method II)**
>
> 1. Find the LCD of all fractions in the numerator and denominator.
> 2. Multiply the numerator and denominator by this LCD and simplify.
> 3. Simplify further if possible.

Method II is illustrated in Example 7 (for Method I, see Example 9).

EXAMPLE 7 ▶ **Simplifying a Compound Fraction**

Simplify the compound fraction:

$$\frac{\dfrac{2}{3m} - \dfrac{3}{2}}{\dfrac{3}{4m} - \dfrac{1}{3m^2}}$$

Solution ▶ The LCD for all fractions is $12m^2$.

$$\frac{\dfrac{2}{3m} - \dfrac{3}{2}}{\dfrac{3}{4m} - \dfrac{1}{3m^2}} = \frac{\left(\dfrac{2}{3m} - \dfrac{3}{2}\right)\left(\dfrac{12m^2}{1}\right)}{\left(\dfrac{3}{4m} - \dfrac{1}{3m^2}\right)\left(\dfrac{12m^2}{1}\right)}$$

multiply numerator and denominator by $12m^2 = \dfrac{12m^2}{1}$

$$= \frac{\left(\dfrac{2}{3m}\right)\left(\dfrac{12m^2}{1}\right) - \left(\dfrac{3}{2}\right)\left(\dfrac{12m^2}{1}\right)}{\left(\dfrac{3}{4m}\right)\left(\dfrac{12m^2}{1}\right) - \left(\dfrac{1}{3m^2}\right)\left(\dfrac{12m^2}{1}\right)}$$

distribute

$$= \frac{8m - 18m^2}{9m - 4}$$

simplify

$$= \frac{2m(4 \overset{-1}{\cancel{-9m}})}{\cancel{9m - 4}} = -2m$$

factor and write in lowest terms

☑ **D.** You've just reviewed how to simplify compound fractions

Now try Exercises 49 through 58 ▶

E. Rewriting Formulas and Algebraic Models

In many fields of study, formulas and algebraic models involve rational expressions and we often need to write them in an alternative form.

EXAMPLE 8 ▶ **Rewriting a Formula**

In an electrical circuit with two resistors in parallel, the total resistance R is related to resistors R_1 and R_2 by the formula $\dfrac{1}{R} = \dfrac{1}{R_1} + \dfrac{1}{R_2}$. Rewrite the right-hand side as a single term.

Solution ▶ $\dfrac{1}{R} = \dfrac{1}{R_1} + \dfrac{1}{R_2}$ LCD for the right-hand side is R_1R_2

$= \dfrac{R_2}{R_1R_2} + \dfrac{R_1}{R_1R_2}$ build equivalent expressions using LCD

$= \dfrac{R_2 + R_1}{R_1R_2}$ write as a single expression

Now try Exercises 59 and 60 ▶

EXAMPLE 9 ▶ **Simplifying an Algebraic Model**

When studying rational expressions and rates of change, we encounter the

expression $\dfrac{\dfrac{1}{x+h} - \dfrac{1}{x}}{h}$. Simplify the compound fraction.

Solution ▶ Using Method I gives:

$$\dfrac{\dfrac{1}{x+h} - \dfrac{1}{x}}{h} = \dfrac{\dfrac{1}{x+h}\cdot\dfrac{x}{x} - \dfrac{1}{x}\cdot\dfrac{(x+h)}{(x+h)}}{h} \qquad \text{LCD for the numerator is } x(x+h)$$

$$= \dfrac{\dfrac{x-(x+h)}{x(x+h)}}{h} \qquad \text{write numerator as a single expression}$$

$$= \dfrac{\dfrac{-h}{x(x+h)}}{h} \qquad \text{simplify}$$

$$= \dfrac{-h}{x(x+h)}\cdot\dfrac{1}{h} \qquad \text{invert and multiply}$$

$$= \dfrac{-1}{x(x+h)} \qquad \text{result}$$

☑ **E.** You've just reviewed how to rewrite formulas and algebraic models

Now try Exercises 61 through 64 ▶

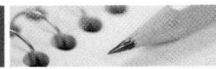

E EXERCISES

▶ DEVELOPING YOUR SKILLS

Reduce to lowest terms.

1. a. $\dfrac{a-7}{-3a+21}$ **b.** $\dfrac{2x+6}{4x^2-8x}$

2. a. $\dfrac{x-4}{-7x+28}$ **b.** $\dfrac{3x-18}{6x^2-12x}$

3. a. $\dfrac{x^2-5x-14}{x^2+6x-7}$ **b.** $\dfrac{a^2+3a-28}{a^2-49}$

4. a. $\dfrac{r^2+3r-10}{r^2+r-6}$ **b.** $\dfrac{m^2+3m-4}{m^2-4m}$

5. a. $\dfrac{x-7}{7-x}$ **b.** $\dfrac{5-x}{x-5}$

6. a. $\dfrac{v^2-3v-28}{49-v^2}$ **b.** $\dfrac{u^2-10u+25}{25-u^2}$

7. a. $\dfrac{-12a^3b^5}{4a^2b^{-4}}$ **b.** $\dfrac{7x+21}{63}$

   c. $\dfrac{y^2-9}{3-y}$ **d.** $\dfrac{m^3n-m^3}{m^4-m^4n}$

8. a. $\dfrac{5m^{-3}n^5}{-10mn^2}$ **b.** $\dfrac{-5v+20}{25}$

   c. $\dfrac{n^2-4}{2-n}$ **d.** $\dfrac{w^4-w^4v}{w^3v-w^3}$

9. a. $\dfrac{2n^3+n^2-3n}{n^3-n^2}$ **b.** $\dfrac{6x^2+x-15}{4x^2-9}$

   c. $\dfrac{x^3+8}{x^2-2x+4}$ **d.** $\dfrac{mn^2+n^2-4m-4}{mn+n+2m+2}$

10. a. $\dfrac{x^3+4x^2-5x}{x^3-x}$ **b.** $\dfrac{5p^2-14p-3}{5p^2+11p+2}$

   c. $\dfrac{12y^2-13y+3}{27y^3-1}$ **d.** $\dfrac{ax^2-5x^2-3a+15}{ax-5x+5a-25}$

Compute as indicated. Write final results in lowest terms.

11. $\dfrac{a^2-4a+4}{a^2-9}\cdot\dfrac{a^2-2a-3}{a^2-4}$

12. $\dfrac{b^2+5b-24}{b^2-6b+9}\cdot\dfrac{b}{b^2-64}$

13. $\dfrac{x^2 - 7x - 18}{x^2 - 6x - 27} \cdot \dfrac{2x^2 + 7x + 3}{2x^2 + 5x + 2}$

14. $\dfrac{6v^2 + 23v + 21}{4v^2 - 4v - 15} \cdot \dfrac{4v^2 - 25}{3v + 7}$

15. $\dfrac{p^3 - 64}{p^3 - p^2} \div \dfrac{p^2 + 4p + 16}{p^2 - 5p + 4}$

16. $\dfrac{a^2 + 3a - 28}{a^2 + 5a - 14} \div \dfrac{a^3 - 4a^2}{a^3 - 8}$

17. $\dfrac{3x - 9}{4x + 12} \div \dfrac{3 - x}{5x + 15}$

18. $\dfrac{5b - 10}{7b - 28} \div \dfrac{2 - b}{5b - 20}$

19. $\dfrac{8}{a^2 - 25} \cdot (a^2 - 2a - 35)$

20. $(m^2 - 16) \cdot \dfrac{m^2 - 5m}{m^2 - m - 20}$

21. $\dfrac{xy - 3x + 2y - 6}{x^2 - 3x - 10} \div \dfrac{xy - 3x}{xy - 5y}$

22. $\dfrac{2a - ab + 7b - 14}{b^2 - 14b + 49} \div \dfrac{ab - 2a}{ab - 7a}$

23. $\dfrac{m^2 + 2m - 8}{m^2 - 2m} \div \dfrac{m^2 - 16}{m^2}$

24. $\dfrac{18 - 6x}{x^2 - 25} \div \dfrac{2x^2 - 18}{x^3 - 2x^2 - 25x + 50}$

25. $\dfrac{x^2 - 0.49}{x^2 + 0.5x - 0.14} \div \dfrac{x^2 - 0.10x + 0.21}{x^2 - 0.09}$

26. $\dfrac{x^2 - 0.25}{x^2 + 0.1x - 0.2} \div \dfrac{x^2 - 0.8x + 0.15}{x^2 - 0.16}$

27. $\dfrac{n^2 - \dfrac{4}{9}}{n^2 - \dfrac{13}{15}n + \dfrac{2}{15}} \div \dfrac{n^2 + \dfrac{4}{3}n + \dfrac{4}{9}}{n^2 - \dfrac{1}{25}}$

28. $\dfrac{q^2 - \dfrac{9}{25}}{q^2 - \dfrac{1}{10}q - \dfrac{3}{10}} \div \dfrac{q^2 + \dfrac{17}{20}q + \dfrac{3}{20}}{q^2 - \dfrac{1}{16}}$

29. $\dfrac{3a^3 - 24a^2 - 12a + 96}{a^2 - 11a + 24} \div \dfrac{6a^2 - 24}{3a^3 - 81}$

30. $\dfrac{p^3 + p^2 - 49p - 49}{p^2 + 6p - 7} \div \dfrac{p^2 + p + 1}{p^3 - 1}$

Compute as indicated. Write answers in lowest terms [recall that $a - b = -1(b - a)$].

31. $\dfrac{3}{8x^2} + \dfrac{5}{2x}$

32. $\dfrac{15}{16y} - \dfrac{7}{2y^2}$

33. $\dfrac{7}{4x^2y^3} - \dfrac{1}{8xy^4}$

34. $\dfrac{3}{6a^3b} + \dfrac{5}{9ab^3}$

35. $\dfrac{4p}{p^2 - 36} - \dfrac{2}{p - 6}$

36. $\dfrac{3q}{q^2 - 49} - \dfrac{3}{2q - 14}$

37. $\dfrac{m}{m^2 - 16} + \dfrac{4}{4 - m}$

38. $\dfrac{2}{m - 7} - 5$

39. $\dfrac{y + 1}{y^2 + y - 30} - \dfrac{2}{y + 6}$

40. $\dfrac{4n}{n^2 - 5n} - \dfrac{3}{4n - 20}$

41. $\dfrac{1}{a + 4} + \dfrac{a}{a^2 - a - 20}$

42. $\dfrac{2x - 1}{x^2 + 3x - 4} - \dfrac{x - 5}{x^2 + 3x - 4}$

43. $\dfrac{3y - 4}{y^2 + 2y + 1} - \dfrac{2y - 5}{y^2 + 2y + 1}$

44. $\dfrac{-2}{3a + 12} - \dfrac{7}{a^2 + 4a}$

45. $\dfrac{2}{m^2 - 9} + \dfrac{m - 5}{m^2 + 6m + 9}$

46. $\dfrac{m + 2}{m^2 - 25} - \dfrac{m + 6}{m^2 - 10m + 25}$

Write each term as a rational expression. Then compute the sum or difference indicated.

47. a. $p^{-2} - 5p^{-1}$ **b.** $x^{-2} + 2x^{-3}$

48. a. $3a^{-1} + (2a)^{-1}$ **b.** $2y^{-1} - (3y)^{-1}$

Simplify each compound rational expression. Use either method.

49. $\dfrac{\dfrac{5}{a} - \dfrac{1}{4}}{\dfrac{25}{a^2} - \dfrac{1}{16}}$

50. $\dfrac{\dfrac{8}{x^3} - \dfrac{1}{27}}{\dfrac{2}{x} - \dfrac{1}{3}}$

51. $\dfrac{p + \dfrac{1}{p - 2}}{1 + \dfrac{1}{p - 2}}$

52. $\dfrac{1 + \dfrac{3}{y - 6}}{y + \dfrac{9}{y - 6}}$

53. $\dfrac{\dfrac{2}{3-x}+\dfrac{3}{x-3}}{\dfrac{4}{x}+\dfrac{5}{x-3}}$

54. $\dfrac{\dfrac{1}{y-5}-\dfrac{2}{5-y}}{\dfrac{3}{y-5}-\dfrac{2}{y}}$

58. a. $\dfrac{4-9a^{-2}}{3a^{-2}}$ **b.** $\dfrac{3+2n^{-1}}{5n^{-2}}$

Rewrite each expression as a single term.

55. $\dfrac{\dfrac{2}{y^2-y-20}}{\dfrac{3}{y+4}-\dfrac{4}{y-5}}$

56. $\dfrac{\dfrac{2}{x^2-3x-10}}{\dfrac{6}{x+2}-\dfrac{4}{x-5}}$

59. $\dfrac{1}{f_1}+\dfrac{1}{f_2}$ **60.** $\dfrac{1}{w}+\dfrac{1}{x}-\dfrac{1}{y}$

61. $\dfrac{\dfrac{a}{x+h}-\dfrac{a}{x}}{h}$ **62.** $\dfrac{\dfrac{a}{h-x}-\dfrac{a}{-x}}{h}$

Rewrite each expression as a compound fraction. Then simplify using either method.

57. a. $\dfrac{1+3m^{-1}}{1-3m^{-1}}$ **b.** $\dfrac{1+2x^{-2}}{1-2x^{-2}}$

63. $\dfrac{\dfrac{1}{2(x+h)^2}-\dfrac{1}{2x^2}}{h}$ **64.** $\dfrac{\dfrac{a}{(x+h)^2}-\dfrac{a}{x^2}}{h}$

▶ APPLICATIONS

65. Stock prices: When a hot new stock hits the market, its price will often rise dramatically and then taper off over time. The equation
$$P=\dfrac{50(7d^2+10)}{d^3+50}$$ models the price
of stock XYZ d days after it has "hit the market." Create a table of values showing the price of the stock for the first 10 days and comment on what you notice. Find the opening price of the stock— does the stock ever return to its original price?

66. Population growth: The Department of Wildlife introduces 60 elk into a new game reserve. It is projected that the size of the herd will grow according to the equation $N=\dfrac{10(6+3t)}{1+0.05t}$, where
N is the number of elk and t is the time in years. Approximate the population of elk after 14 yr.

67. Typing speed: The number of words per minute that a beginner can type is approximated by the equation $N=\dfrac{60t-120}{t}$, where N is the number of words per minute after t weeks, $2 < t < 12$. Use a table to determine how many weeks it takes for a student to be typing an average of forty-five words per minute.

68. Memory retention: A group of students is asked to memorize 50 Russian words that are unfamiliar to them. The number N of these words that the average student remembers D days later is modeled by the equation $N=\dfrac{5D+35}{D}$ $(D\geq 1)$. How many words are remembered after (a) 1 day? (b) 5 days? (c) 12 days? (d) 35 days? (e) 100 days? According to this model, is there a certain number of words that the average student never forgets? How many?

▶ EXTENDING THE CONCEPT

69. One of these expressions is *not* equal to the others. Identify which and explain why.

a. $\dfrac{20n}{10n}$ **b.** $20\cdot n\div 10\cdot n$

c. $20n\cdot\dfrac{1}{10n}$ **d.** $\dfrac{20}{10}\cdot\dfrac{n}{n}$

70. The average of A and B is x. The average of C, D, and E is y. The average of A, B, C, D, and E is

a. $\dfrac{3x+2y}{5}$ **b.** $\dfrac{2x+3y}{5}$

c. $\dfrac{2(x+y)}{5}$ **d.** $\dfrac{3(x+y)}{5}$

Square roots and cube roots come from a much larger family called **radical expressions**. Expressions containing radicals can be found in virtually every field of mathematical study, and are an invaluable tool for modeling many real-world phenomena.

A. Simplifying Radical Expressions of the Form $\sqrt[n]{a^n}$

In Section A we noted $\sqrt{a} = b$ only if $b^2 = a$. The expression $\sqrt{-16}$ does not represent a real number because there is no number b such that $b^2 = -16$ ($\sqrt{a}$ is a real number only if $a \geq 0$). Of particular interest to us now is an inverse operation for a^2. In other words, what operation can be applied to a^2 to return a? Consider the following.

EXAMPLE 1 ▶ **Evaluating a Radical Expression**

Evaluate $\sqrt{a^2}$ for the values given:

   a. $a = 3$ **b.** $a = 5$ **c.** $a = -6$

Solution ▶   **a.** $\sqrt{3^2} = \sqrt{9}$ **b.** $\sqrt{5^2} = \sqrt{25}$ **c.** $\sqrt{(-6)^2} = \sqrt{36}$

               $= 3$ $= 5$ $= 6$

                                Now try Exercises 1 and 2 ▶

The pattern seemed to indicate that $\sqrt{a^2} = a$ and that our search for an inverse operation was complete—until Example 1(c), where we found that $\sqrt{(-6)^2} \neq -6$. Using the absolute value concept, we can repair this apparent discrepancy and state a general rule for simplifying these expressions: $\sqrt{a^2} = |a|$. For expressions like $\sqrt{49x^2}$ and $\sqrt{y^6}$, the radicands can be rewritten as perfect squares and simplified in the same manner: $\sqrt{49x^2} = \sqrt{(7x)^2} = 7|x|$ and $\sqrt{y^6} = \sqrt{(y^3)^2} = |y^3|$.

The Square Root of a^2: $\sqrt{a^2}$

For any real number a, $\sqrt{a^2} = |a|$.

EXAMPLE 2 ▶ **Simplifying Square Root Expressions**

Simplify each expression.

   a. $\sqrt{169x^2}$ **b.** $\sqrt{x^2 - 10x + 25}$

Solution ▶   **a.** $\sqrt{169x^2} = |13x|$

                   $= 13|x|$ since *x* could be negative

    b. $\sqrt{x^2 - 10x + 25} = \sqrt{(x - 5)^2}$

                         $= |x - 5|$ since *x* − 5 could be negative

                                Now try Exercises 3 and 4 ▶

⚠ **CAUTION** ▶ In Section C, we noted that $(A + B)^2 \neq A^2 + B^2$. In a similar way, $\sqrt{A^2 + B^2} \neq A + B$, and you cannot take the square root of individual terms. There is a big difference between the expressions $\sqrt{A^2 + B^2}$ and $\sqrt{(A + B)^2} = |A + B|$. Try evaluating each when $A = 3$ and $B = 4$.

To investigate expressions like $\sqrt[3]{x^3}$, note the radicand in both $\sqrt[3]{8}$ and $\sqrt[3]{-64}$ can be written as a perfect cube. From our earlier definition of cube roots we know $\sqrt[3]{8} = \sqrt[3]{(2)^3} = 2$, $\sqrt[3]{-64} = \sqrt[3]{(-4)^3} = -4$, and that every real number has only one real cube root. For this reason, absolute value notation is not used or needed when taking cube roots.

The Cube Root of a^3: $\sqrt[3]{a^3}$

For any real number a, $\sqrt[3]{a^3} = a$.

> **WORTHY OF NOTE**
>
> Just as $\sqrt[2]{-16}$ is not a real number, $\sqrt[4]{-16}$ or $\sqrt[6]{-16}$ do not represent real numbers. An even number of repeated factors is always positive!

We can extend these ideas to fourth roots, fifth roots, and so on. For example, the fifth root of a is b only if $b^5 = a$. In symbols, $\sqrt[5]{a} = b$ implies $b^5 = a$. Since an odd number of negative factors is always negative: $(-2)^5 = -32$, and an even number of negative factors is always positive: $(-2)^4 = 16$, we must take the index into account when evaluating expressions like $\sqrt[n]{a^n}$. If n is even and the radicand is unknown, absolute value notation must be used.

The nth Root of a^n: $\sqrt[n]{a^n}$

For any real number a,

1. $\sqrt[n]{a^n} = |a|$ when n is even.   2. $\sqrt[n]{a^n} = a$ when n is odd.

EXAMPLE 3 ▶ **Simplifying Radical Expressions**

Simplify each expression.

    a. $\sqrt[3]{-27x^3}$ **b.** $\sqrt[3]{-64n^6}$ **c.** $\sqrt[4]{81}$ **d.** $\sqrt[4]{-81}$

    e. $\sqrt[4]{16m^4}$ **f.** $\sqrt[5]{32p^5}$ **g.** $\sqrt[6]{(m+5)^6}$ **h.** $\sqrt[7]{(x-2)^7}$

Solution ▶   **a.** $\sqrt[3]{-27x^3} = \sqrt[3]{(-3x)^3}$ **b.** $\sqrt[3]{-64n^6} = \sqrt[3]{(-4n^2)^3}$

                $= -3x$ $= -4n^2$

    c. $\sqrt[4]{81} = 3$ **d.** $\sqrt[4]{-81}$ is not a real number

    e. $\sqrt[4]{16m^4} = \sqrt[4]{(2m)^4}$ **f.** $\sqrt[5]{32p^5} = \sqrt[5]{(2p)^5}$

                  $= |2m|$ or $2|m|$ $= 2p$

    g. $\sqrt[6]{(m+5)^6} = |m+5|$ **h.** $\sqrt[7]{(x-2)^7} = x-2$

> ☑ **A.** You've just reviewed how to simplify radical expressions of the form $\sqrt[n]{a^n}$

Now try Exercises 5 through 8 ▶

B. Radical Expressions and Rational Exponents

As an alternative to radical notation, a rational (fractional) exponent can be used, along with the power property of exponents. For $\sqrt[3]{a^3} = a$, notice that an exponent of one-third can replace the cube root notation and produce the same result: $\sqrt[3]{a^3} = (a^3)^{\frac{1}{3}} = a^{\frac{3}{3}} = a$. In the same way, an exponent of one-half can replace the square root notation: $\sqrt{a^2} = (a^2)^{\frac{1}{2}} = a^{\frac{2}{2}} = |a|$. In general, we have the following:

Rational Exponents

If a is a real number and n is an integer greater than 1,

$$\text{then } \sqrt[n]{a} = \sqrt[n]{a^1} = a^{\frac{1}{n}}$$

provided $\sqrt[n]{a}$ represents a real number.

EXAMPLE 4 ▶ **Simplifying Radical Expressions Using Rational Exponents**

Simplify by rewriting each radicand as a perfect nth power and converting to rational exponent notation.

a. $\sqrt[3]{-125}$ **b.** $-\sqrt[4]{16x^{20}}$

Solution ▶ **a.** $\sqrt[3]{-125} = \sqrt[3]{(-5)^3}$ **b.** $-\sqrt[4]{16x^{20}} = -\sqrt[4]{(2x^5)^4}$

$= \left[(-5)^3\right]^{\frac{1}{3}}$ $= -\left[(2x^5)^4\right]^{\frac{1}{4}}$

$= (-5)^{\frac{3}{3}}$ $= -(2x^5)^{\frac{4}{4}}$

$= -5$ $= -2|x|^5$

Now try Exercises 9 and 10 ▶

WORTHY OF NOTE

Any rational number can be decomposed into the product of a unit fraction and an integer: $\dfrac{m}{n} = \dfrac{1}{n} \cdot m$.

When a rational exponent is used, as in $\sqrt[n]{a} = \sqrt[n]{a^1} = a^{\frac{1}{n}}$, the denominator of the exponent represents the index number, while the numerator of the exponent represents the original power on a. *This is true even when the exponent on a is something other than one!* In other words, the radical expression $\sqrt[4]{16^3}$ can be rewritten as $\left(16^3\right)^{\frac{1}{4}} = 16^{\frac{3}{4}}$. This is further illustrated in Figure AI.4 where we see the rational exponent has the form, "power over root." To evaluate this expression without the aid of a calculator, we use the commutative property to rewrite $\left(16^{\frac{3}{1}}\right)^{\frac{1}{4}}$ as $\left(16^{\frac{1}{4}}\right)^{\frac{3}{1}}$ and begin with the fourth root of 16: $\left(16^{\frac{1}{4}}\right)^{\frac{3}{1}} = 2^3 = 8$.

In general, if m and n have no common factors (other than 1) the expression $a^{\frac{m}{n}}$ can be interpreted in the following two ways.

$\left(\sqrt[n]{a^m}\right) \quad \begin{matrix}\rightarrow \\ \end{matrix} a^{\frac{m}{n}}$

Figure AI.4

WORTHY OF NOTE

While the expression $(-8)^{\frac{1}{3}} = \sqrt[3]{-8}$ represents the real number -2, the expression $(-8)^{\frac{2}{6}} = \left(\sqrt[6]{-8}\right)^2$ is not a real number, even though $\dfrac{1}{3} = \dfrac{2}{6}$. Note that the second exponent is not in lowest terms.

Rational Exponents

If $\dfrac{m}{n}$ is a rational number expressed in lowest terms with $n \geq 2$, then

(1) $a^{\frac{m}{n}} = \left(\sqrt[n]{a}\right)^m$ or (2) $a^{\frac{m}{n}} = \sqrt[n]{a^m}$

(compute $\sqrt[n]{a}$, then take the mth power) (compute a^m, then take the nth root)

provided $\sqrt[n]{a}$ represents a real number.

EXAMPLE 5 ▶ **Simplifying Expressions with Rational Exponents**

Find the value of each expression without a calculator, by rewriting the exponent as the product of a unit fraction and an integer.

a. $27^{\frac{2}{3}}$ **b.** $(-8)^{\frac{4}{3}}$ **c.** $\left(\dfrac{4x^6}{9}\right)^{\frac{5}{2}}$ **d.** $-49^{\frac{3}{2}}$

Solution ▶ **a.** $27^{\frac{2}{3}} = 27^{\frac{1}{3} \cdot 2}$ **b.** $(-8)^{\frac{4}{3}} = (-8)^{\frac{1}{3} \cdot 4}$

$= \left(27^{\frac{1}{3}}\right)^2$ $= \left[(-8)^{\frac{1}{3}}\right]^4$

$= 3^2 \text{ or } 9$ $= [-2]^4 = 16$

c. $\left(\dfrac{4x^6}{9}\right)^{\frac{5}{2}} = \left(\dfrac{4x^6}{9}\right)^{\frac{1}{2} \cdot 5}$ **d.** $-49^{\frac{3}{2}} = -\left(49^{\frac{1}{2}}\right)^3$

$= \left[\left(\dfrac{4x^6}{9}\right)^{\frac{1}{2}}\right]^5$ $= -\left(\sqrt{49}\right)^3$

$= -(7)^3 \text{ or } -343$

B. You've just reviewed how to rewrite and simplify radical expressions using rational exponents

$= \left[\dfrac{2|x|^3}{3}\right]^5 = \dfrac{32|x|^{15}}{243}$

Now try Exercises 11 through 16 ▶

C. Properties of Radicals and Simplifying Radical Expressions

The properties used to simplify radical expressions are closely connected to the properties of exponents. For instance, the product to a power property: $(xy)^n = x^n y^n$ holds true, even when n is a rational number. This means $(xy)^{\frac{1}{2}}$ is equal to $x^{\frac{1}{2}}y^{\frac{1}{2}}$ and $(4 \cdot 25)^{\frac{1}{2}}$ is equal to $4^{\frac{1}{2}} \cdot 25^{\frac{1}{2}}$. When this statement is expressed in radical form, we have $\sqrt{4 \cdot 25} = \sqrt{4} \cdot \sqrt{25}$, with both having a value of 10. It is likewise true that $\dfrac{\sqrt{100}}{\sqrt{25}} = \sqrt{\dfrac{100}{25}}$, which suggests the following **properties of radicals.** These can be extended to include cube roots, fourth roots, and so on.

Properties of Radicals

If $\sqrt[n]{A}$ and $\sqrt[n]{B}$ represent real valued expressions, and $B \neq 0$,

$$(1) \quad \sqrt[n]{AB} = \sqrt[n]{A} \cdot \sqrt[n]{B}; \qquad\qquad (2) \quad \sqrt[n]{\dfrac{A}{B}} = \dfrac{\sqrt[n]{A}}{\sqrt[n]{B}};$$

$$\sqrt[n]{A} \cdot \sqrt[n]{B} = \sqrt[n]{AB} \qquad\qquad\qquad \dfrac{\sqrt[n]{A}}{\sqrt[n]{B}} = \sqrt[n]{\dfrac{A}{B}}$$

⚠ CAUTION   Note that this property applies only to a *product* of two terms, not to a sum or difference. In other words, while $\sqrt{9x^2} = |3x|$, $\sqrt{9 + x^2} \neq |3 + x|$!

One application of the product property is to simplify radical expressions. In general, the expression $\sqrt[n]{a}$ is in simplified form if a has no factors (other than 1) that are perfect nth roots.

EXAMPLE 6 ▶ **Simplifying Radical Expressions**

Write each expression in simplest form using the product property.

$\quad$ **a.** $\sqrt{18}$ $\qquad$ **b.** $5\sqrt[3]{125x^4}$ $\qquad$ **c.** $1.2\sqrt[3]{16n^4} \; \sqrt[3]{4n^5}$

Solution ▶ $\quad$ **a.** $\sqrt{18} = \sqrt{9 \cdot 2}$ $\qquad\qquad$ **b.** $5\sqrt[3]{125x^4} = 5 \cdot \sqrt[3]{125 \cdot x^4}$

$\qquad\qquad\qquad = \sqrt{9}\sqrt{2}$ $\qquad$ *These steps can* $\quad \left\{ \begin{array}{l} = 5 \cdot \sqrt[3]{125} \cdot \sqrt[3]{x^3} \cdot \sqrt[3]{x^1} \\ = 5 \cdot 5 \cdot x \cdot \sqrt[3]{x} \end{array} \right.$

$\qquad\qquad\qquad = 3\sqrt{2}$ $\qquad$ *be done mentally.*

$\qquad\qquad\qquad\qquad\qquad\qquad\qquad\qquad\qquad = 25x\sqrt[3]{x}$

$\quad$ **c.** $1.2\sqrt[3]{16n^4} \; \sqrt[3]{4n^5} = 1.2\sqrt[3]{64 \cdot n^9}$ $\qquad$ product property

$\quad$ Since the index is 3, we look for perfect cube factors in the radicand.

$$= 1.2\sqrt[3]{64} \; \sqrt[3]{n^9} \qquad \text{product property}$$
$$= 1.2\sqrt[3]{64} \; \sqrt[3]{(n^3)^3} \qquad \text{rewrite } n^9 \text{ as a perfect cube}$$
$$= 1.2(4)n^3 \qquad \text{simplify}$$
$$= 4.8n^3 \qquad \text{result}$$

Now try Exercises 17 through 20 ▶

WORTHY OF NOTE

Rational exponents also could have been used to simplify the expression from Example 9, since $1.2\sqrt[3]{64} \; \sqrt[3]{n^9} = 1.2(4)n^{\frac{9}{3}} = 4.8n^3$. Also see Example 8.

When radicals are *combined* using the product property, the result may contain a perfect nth root, which should be simplified. Note that the *index numbers must be the same* in order to use this property.

EXAMPLE 7 ▶ **Simplifying Radical Expressions**

Simplify each expression:

a. $\dfrac{\sqrt{18a^5}}{\sqrt{2a}}$

b. $\sqrt[3]{\dfrac{81}{125x^3}}$

Solution ▶ **a.** $\dfrac{\sqrt{18a^5}}{\sqrt{2a}} = \sqrt{\dfrac{18a^5}{2a}}$

$= \sqrt{9a^4}$

$= 3a^2$

b. $\sqrt[3]{\dfrac{81}{125x^3}} = \dfrac{\sqrt[3]{81}}{\sqrt[3]{125x^3}}$

$= \dfrac{\sqrt[3]{27 \cdot 3}}{5x}$

$= \dfrac{3\sqrt[3]{3}}{5x}$

Now try Exercises 21 and 22 ▶

Radical expressions can also be simplified using rational exponents.

EXAMPLE 8 ▶ **Using Rational Exponents to Simplify Radical Expressions**

Simplify using rational exponents:

a. $\sqrt{36p^4q^5}$ **b.** $v\sqrt[3]{v^4}$ **c.** $\sqrt[3]{\sqrt{x}}$ **d.** $\sqrt[3]{m}\sqrt{m}$

Solution ▶ **a.** $\sqrt{36p^4q^5} = (36p^4q^5)^{\frac{1}{2}}$

$= 36^{\frac{1}{2}}p^{\frac{4}{2}}q^{\frac{5}{2}}$

$= 6p^2q^{(\frac{4}{2}+\frac{1}{2})}$

$= 6p^2q^2q^{\frac{1}{2}}$

$= 6p^2q^2\sqrt{q}$

b. $v\sqrt[3]{v^4} = v^1 \cdot v^{\frac{4}{3}}$

$= v^{\frac{3}{3}} \cdot v^{\frac{4}{3}}$

$= v^{\frac{7}{3}}$

$= v^{\frac{6}{3}}v^{\frac{1}{3}}$

$= v^2\sqrt[3]{v}$

c. $\sqrt[3]{\sqrt{x}} = \sqrt[3]{x^{\frac{1}{2}}}$

$= (x^{\frac{1}{2}})^{\frac{1}{3}}$

$= x^{\frac{1}{2}\cdot\frac{1}{3}}$

$= x^{\frac{1}{6}}$ or $\sqrt[6]{x}$

d. $\sqrt[3]{m}\sqrt{m} = m^{\frac{1}{3}}m^{\frac{1}{2}}$

$= m^{\frac{1}{3}+\frac{1}{2}}$

$= m^{\frac{5}{6}}$

$= \sqrt[6]{m^5}$

☑ **C.** You've just reviewed how to use properties of radicals to simplify radical expressions

Now try Exercises 23 and 24 ▶

D. Addition and Subtraction of Radical Expressions

Since $3x$ and $5x$ are like terms, we know $3x + 5x = 8x$. If $x = \sqrt[3]{7}$, the sum becomes $3\sqrt[3]{7} + 5\sqrt[3]{7} = 8\sqrt[3]{7}$, illustrating how *like* radical expressions can be combined. Like radicals are those that have *the same index and radicand.* In some cases, we can identify like radicals only after radical terms have been simplified.

EXAMPLE 9 ▶ **Adding and Subtracting Radical Expressions**

Simplify and combine (if possible).

a. $\sqrt{45} + 2\sqrt{20}$ **b.** $\sqrt[3]{16x^5} - x\sqrt[3]{54x^2}$

Solution ▶ **a.** $\sqrt{45} + 2\sqrt{20} = 3\sqrt{5} + 2(2\sqrt{5})$ simplify radicals: $\sqrt{45} = \sqrt{9\cdot5}$; $\sqrt{20} = \sqrt{4\cdot5}$

$= 3\sqrt{5} + 4\sqrt{5}$ like radicals

$= 7\sqrt{5}$ result

b. $\sqrt[3]{16x^5} - x\sqrt[3]{54x^2} = \sqrt[3]{8 \cdot 2 \cdot x^3 \cdot x^2} - x\sqrt[3]{27 \cdot 2 \cdot x^2}$

$= 2x\sqrt[3]{2x^2} - 3x\sqrt[3]{2x^2}$ simplify radicals

$= -x\sqrt[3]{2x^2}$ result

☑ **D.** You've just reviewed how to add and subtract radical expressions

Now try Exercises 25 through 28 ▶

E. Multiplication and Division of Radical Expressions; Radical Expressions in Simplest Form

Multiplying radical expressions is simply an extension of our earlier work. The multiplication can take various forms, from the distributive property to any of the special products reviewed in Section C. For instance, $(A \pm B)^2 = A^2 \pm 2AB + B^2$, even if A or B is a radical term.

EXAMPLE 10 ▶ **Evaluating a Quadratic Expression**

Show that when $x^2 - 4x + 1$ is evaluated at $x = 2 + \sqrt{3}$, the result is zero.

Solution ▶

$$x^2 - 4x + 1 \quad \text{original expression}$$
$$(2 + \sqrt{3})^2 - 4(2 + \sqrt{3}) + 1 \quad \text{substitute } 2 + \sqrt{3} \text{ for } x$$
$$4 + 4\sqrt{3} + 3 - 8 - 4\sqrt{3} + 1 \quad \text{multiply}$$
$$(4 + 3 - 8 + 1) + (4\sqrt{3} - 4\sqrt{3}) \quad \text{commutative and associative properties}$$
$$0 \checkmark$$

Now try Exercises 29 through 36 ▶

When we applied the quotient property in Example 7, we obtained a denominator free of radicals. Sometimes the denominator is not automatically free of radicals, and the need to write radical expressions in *simplest form* comes into play. This process is called **rationalizing the denominator.**

Radical Expressions in Simplest Form

A radical expression is in simplest form if:
1. The radicand has no perfect nth root factors.
2. The radicand contains no fractions.
3. No radicals occur in a denominator.

As with other types of simplification, the desired form can be achieved in various ways. If the denominator is a single radical term, we multiply the numerator and denominator by the factors required to eliminate the radical in the denominator [see Example 11(a)]. If the radicand is a rational expression, it is generally easier to build an equivalent fraction *within the radical* having perfect nth root factors in the denominator [see Example 11(b)]. In some applications, the denominator may be a sum or difference containing a radical term. In this case, we multiply by a conjugate since $(A + B)(A - B) = A^2 - B^2$. If either A or B is a square root, the result will be a denominator free of radicals.

EXAMPLE 11 ▶ **Simplifying Radical Expressions**

Simplify by rationalizing the denominator. Assume $a \neq 0$.

a. $\dfrac{2}{5\sqrt{3}}$ 　　　 **b.** $\sqrt[3]{\dfrac{3}{4a^4}}$ 　　　 **c.** $\dfrac{1}{\sqrt{6} - \sqrt{2}}$

Solution ▶

a. $\dfrac{2}{5\sqrt{3}} = \dfrac{2}{5\sqrt{3}} \cdot \dfrac{\sqrt{3}}{\sqrt{3}}$ multiply numerator and denominator by $\sqrt{3}$

$= \dfrac{2\sqrt{3}}{5(\sqrt{3})^2} = \dfrac{2\sqrt{3}}{15}$ simplify—denominator is now rational

b. $\sqrt[3]{\dfrac{3}{4a^4}} = \sqrt[3]{\dfrac{3}{4a^4} \cdot \dfrac{2a^2}{2a^2}}$ $4 \cdot 2 = 8$ is the smallest perfect cube with 4 as a factor; $a^4 \cdot a^2 = a^6$ is the smallest perfect cube with a^4 as a factor

$= \sqrt[3]{\dfrac{6a^2}{8a^6}}$ the denominator is now a perfect cube—simplify

$= \dfrac{\sqrt[3]{6a^2}}{2a^2}$ result

c. $\dfrac{1}{\sqrt{6} - \sqrt{2}} = \dfrac{1}{\sqrt{6} - \sqrt{2}} \cdot \dfrac{(\sqrt{6} + \sqrt{2})}{(\sqrt{6} + \sqrt{2})}$

$= \dfrac{\sqrt{6} + \sqrt{2}}{(\sqrt{6})^2 - (\sqrt{2})^2}$

$= \dfrac{\sqrt{6} + \sqrt{2}}{6 - 2}$

$= \dfrac{\sqrt{6} + \sqrt{2}}{4}$

☑ **E. You've just reviewed how to multiply and divide radical expressions and write a radical expression in simplest form**

Now try Exercises 37 through 42 ▶

F. Formulas and Radicals

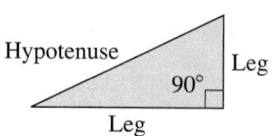

Hypotenuse   Leg   90°   Leg

One application that often involves radical terms is the Pythagorean theorem. A right triangle is one that has a 90° angle. The longest side (opposite the right angle) is called the **hypotenuse,** while the other two sides are simply called "legs." The **Pythagorean theorem** is a formula that says if you add the square of each leg, the result will be equal to the square of the hypotenuse. Furthermore, we note the converse of this theorem is also true.

Pythagorean Theorem

1. For any right triangle with legs a, b and hypotenuse c,
$$a^2 + b^2 = c^2,$$

2. For any triangle with sides a, b, and c, if $a^2 + b^2 = c^2$, then the triangle is a right triangle.

A geometric interpretation of the theorem is given in the figure, which shows $3^2 + 4^2 = 5^2$.

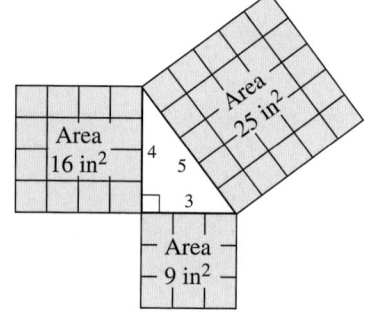

Area 16 in²   4   5   3   Area 25 in²   Area 9 in²

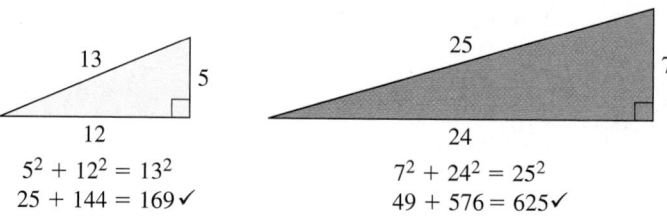

13   5   12
$5^2 + 12^2 = 13^2$
$25 + 144 = 169\checkmark$

25   7   24
$7^2 + 24^2 = 25^2$
$49 + 576 = 625\checkmark$

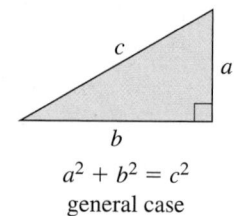

c a b
$a^2 + b^2 = c^2$
general case

EXAMPLE 12 ▶ Applying the Pythagorean Theorem

An extension ladder is placed 9 ft from the base of a building in an effort to reach a third-story window that is 27 ft high. What is the minimum length of the ladder required? Answer in exact form using radicals, and approximate form by rounding to one decimal place.

Solution ▶ We can assume the building makes a 90° angle with the ground, and use the Pythagorean theorem to find the required length. Let c represent this length.

$$c^2 = a^2 + b^2 \qquad \text{Pythagorean theorem}$$
$$c^2 = (9)^2 + (27)^2 \qquad \text{substitute 9 for } a \text{ and 27 for } b$$
$$c^2 = 81 + 729 \qquad 9^2 = 81, 27^2 = 729$$
$$c^2 = 810 \qquad \text{add}$$
$$c = \sqrt{810} \qquad \text{definition of square root; } c > 0$$
$$c = 9\sqrt{10} \qquad \text{exact form: } \sqrt{810} = \sqrt{81 \cdot 10} = 9\sqrt{10}$$
$$c \approx 28.5 \text{ ft} \qquad \text{approximate form}$$

The ladder must be at least 28.5 ft tall.

Now try Exercises 43 and 44 ▶

c 27 ft

9 ft

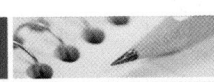 **F.** You've just reviewed how to evaluate formulas involving radicals

F EXERCISES

▶ DEVELOPING YOUR SKILLS

Evaluate the expression $\sqrt{x^2}$ for the values given.

1. a. $x = 9$ **b.** $x = -10$

2. a. $x = 7$ **b.** $x = -8$

Simplify each expression, assuming that variables can represent any real number.

3. a. $\sqrt{49p^2}$ **b.** $\sqrt{(x-3)^2}$
   c. $\sqrt{81m^4}$ **d.** $\sqrt{x^2 - 6x + 9}$

4. a. $\sqrt{25n^2}$ **b.** $\sqrt{(y+2)^2}$
   c. $\sqrt{v^{10}}$ **d.** $\sqrt{4a^2 + 12a + 9}$

5. a. $\sqrt[3]{64}$ **b.** $\sqrt[3]{-125x^3}$
   c. $\sqrt[3]{216z^{12}}$ **d.** $\sqrt[3]{\dfrac{v^3}{-8}}$

6. a. $\sqrt[3]{-8}$ **b.** $\sqrt[3]{-125p^3}$
   c. $\sqrt[3]{27q^9}$ **d.** $\sqrt[3]{\dfrac{w^3}{-64}}$

7. a. $\sqrt[6]{64}$ **b.** $\sqrt[6]{-64}$
   c. $\sqrt[5]{243x^{10}}$ **d.** $\sqrt[5]{-243x^5}$
   e. $\sqrt[5]{(k-3)^5}$ **f.** $\sqrt[6]{(h+2)^6}$

8. a. $\sqrt[4]{216}$ **b.** $\sqrt[4]{-216}$
   c. $\sqrt[5]{1024z^{15}}$ **d.** $\sqrt[5]{-1024z^{20}}$
   e. $\sqrt[5]{(q-9)^5}$ **f.** $\sqrt[6]{(p+4)^6}$

9. a. $\sqrt[3]{-125}$ **b.** $-\sqrt[4]{81n^{12}}$
   c. $\sqrt{-36}$ **d.** $\sqrt{\dfrac{49v^{10}}{36}}$

10. a. $\sqrt[3]{-216}$ **b.** $-\sqrt[4]{16m^{24}}$
   c. $\sqrt{-121}$ **d.** $\sqrt{\dfrac{25x^6}{4}}$

11. a. $8^{\frac{2}{3}}$ **b.** $\left(\dfrac{16}{25}\right)^{\frac{3}{2}}$
   c. $\left(\dfrac{4}{25}\right)^{-\frac{3}{2}}$ **d.** $\left(\dfrac{-27p^6}{8q^3}\right)^{\frac{2}{3}}$

12. a. $9^{\frac{3}{2}}$

b. $\left(\dfrac{4}{9}\right)^{\frac{3}{2}}$

c. $\left(\dfrac{16}{81}\right)^{-\frac{3}{4}}$

d. $\left(\dfrac{-125v^9}{27w^6}\right)^{\frac{2}{3}}$

13. a. $-144^{\frac{3}{2}}$

b. $\left(-\dfrac{4}{25}\right)^{\frac{3}{2}}$

c. $(-27)^{-\frac{2}{3}}$

d. $-\left(\dfrac{27x^3}{64}\right)^{-\frac{4}{3}}$

14. a. $-100^{\frac{3}{2}}$

b. $\left(-\dfrac{49}{36}\right)^{\frac{3}{2}}$

c. $(-125)^{-\frac{2}{3}}$

d. $-\left(\dfrac{x^9}{8}\right)^{-\frac{4}{3}}$

Use properties of exponents to simplify. Answer in exponential form without negative exponents.

15. a. $\left(2n^2p^{-\frac{2}{5}}\right)^5$

b. $\left(\dfrac{8y^{\frac{3}{4}}}{64y^{\frac{3}{2}}}\right)^{\frac{1}{3}}$

16. a. $\left(\dfrac{24x^{\frac{3}{8}}}{4x^{\frac{1}{2}}}\right)^2$

b. $\left(2x^{-\frac{1}{4}}y^{\frac{3}{4}}\right)^4$

Simplify each expression. Assume all variables represent nonnegative real numbers.

17. a. $\sqrt{18m^2}$

b. $-2\sqrt[3]{-125p^3q^7}$

c. $\dfrac{3}{8}\sqrt[3]{64m^3n^5}$

d. $\sqrt{32p^3q^6}$

e. $\dfrac{-6+\sqrt{28}}{2}$

f. $\dfrac{27-\sqrt{72}}{6}$

18. a. $\sqrt{8x^6}$

b. $3\sqrt[3]{128a^4b^2}$

c. $\dfrac{2}{9}\sqrt[3]{27a^2b^6}$

d. $\sqrt{54m^6n^8}$

e. $\dfrac{12-\sqrt{48}}{8}$

f. $\dfrac{-20+\sqrt{32}}{4}$

19. a. $2.5\sqrt{18a}\sqrt{2a^3}$

b. $-\dfrac{2}{3}\sqrt{3b}\sqrt{12b^2}$

c. $\sqrt{\dfrac{x^3y}{3}}\sqrt{\dfrac{4x^5y}{12y}}$

d. $\sqrt[3]{9v^2u}\sqrt[3]{3u^5v^2}$

20. a. $5.1\sqrt{2p}\sqrt{32p^5}$

b. $-\dfrac{4}{5}\sqrt{5q}\sqrt{20q^3}$

c. $\sqrt{\dfrac{ab^2}{3}}\sqrt{\dfrac{25ab^4}{27}}$

d. $\sqrt[3]{5cd^2}\sqrt[3]{25cd}$

21. a. $\dfrac{\sqrt{8m^5}}{\sqrt{2m}}$

b. $\dfrac{\sqrt[3]{108n^4}}{\sqrt[3]{4n}}$

c. $\sqrt{\dfrac{45}{16x^2}}$

d. $12\sqrt[3]{\dfrac{81}{8z^9}}$

22. a. $\dfrac{\sqrt{27y^7}}{\sqrt{3y}}$

b. $\dfrac{\sqrt[3]{72b^5}}{\sqrt[3]{3b^2}}$

c. $\sqrt{\dfrac{20}{4x^4}}$

d. $-9\sqrt[3]{\dfrac{125}{27x^6}}$

23. a. $\sqrt[5]{32x^{10}y^{15}}$

b. $x\sqrt[4]{x^5}$

c. $\sqrt[4]{\sqrt[3]{b}}$

d. $\dfrac{\sqrt[3]{6}}{\sqrt{6}}$

e. $\sqrt{b}\sqrt[4]{b}$

24. a. $\sqrt[4]{81a^{12}b^{16}}$

b. $a\sqrt[5]{a^6}$

c. $\sqrt{\sqrt[4]{a}}$

d. $\dfrac{\sqrt[3]{3}}{\sqrt[4]{3}}$

e. $\sqrt[3]{c}\sqrt[4]{c}$

Simplify and add (if possible).

25. a. $12\sqrt{72}-9\sqrt{98}$

b. $8\sqrt{48}-3\sqrt{108}$

c. $7\sqrt{18m}-\sqrt{50m}$

d. $2\sqrt{28p}-3\sqrt{63p}$

26. a. $-3\sqrt{80}+2\sqrt{125}$

b. $5\sqrt{12}+2\sqrt{27}$

c. $3\sqrt{12x}-5\sqrt{75x}$

d. $3\sqrt{40q}+9\sqrt{10q}$

27. a. $3x\sqrt[3]{54x}-5\sqrt[3]{16x^4}$

b. $\sqrt{4}+\sqrt{3x}-\sqrt{12x}+\sqrt{45}$

c. $\sqrt{72x^3}+\sqrt{50}-\sqrt{7x}+\sqrt{27}$

28. a. $5\sqrt[3]{54m^3}-2m\sqrt[3]{16m^3}$

b. $\sqrt{10b}+\sqrt{200b}-\sqrt{20}+\sqrt{40}$

c. $\sqrt{75r^3}+\sqrt{32}-\sqrt{27r}+\sqrt{38}$

Use a substitution to verify the solutions to the quadratic equation given.

29. $x^2-4x+1=0$

   a. $x=2+\sqrt{3}$ **b.** $x=2-\sqrt{3}$

30. $x^2-10x+18=0$

   a. $x=5-\sqrt{7}$ **b.** $x=5+\sqrt{7}$

31. $x^2+2x-9=0$

   a. $x=-1+\sqrt{10}$ **b.** $x=-1-\sqrt{10}$

32. $x^2-14x+29=0$

   a. $x=7-2\sqrt{5}$ **b.** $x=7+2\sqrt{5}$

Compute each product and simplify the result.

33. a. $(7\sqrt{2})^2$ **b.** $\sqrt{3}(\sqrt{5}+\sqrt{7})$

   c. $(n+\sqrt{5})(n-\sqrt{5})$ **d.** $(6-\sqrt{3})^2$

34. a. $(0.3\sqrt{5})^2$ **b.** $\sqrt{5}(\sqrt{6} - \sqrt{2})$
   c. $(4 + \sqrt{3})(4 - \sqrt{3})$ **d.** $(2 + \sqrt{5})^2$

35. a. $(3 + 2\sqrt{7})(3 - 2\sqrt{7})$
   b. $(\sqrt{5} - \sqrt{14})(\sqrt{2} + \sqrt{13})$
   c. $(2\sqrt{2} + 6\sqrt{6})(3\sqrt{10} + \sqrt{7})$

36. a. $(5 + 4\sqrt{10})(1 - 2\sqrt{10})$
   b. $(\sqrt{3} + \sqrt{2})(\sqrt{10} + \sqrt{11})$
   c. $(3\sqrt{5} + 4\sqrt{2})(\sqrt{15} + \sqrt{6})$

Rationalize each expression by building perfect *n*th root factors for each denominator. Assume all variables represent positive quantities.

37. a. $\dfrac{3}{\sqrt{12}}$ **b.** $\sqrt{\dfrac{20}{27x^3}}$

   c. $\sqrt{\dfrac{27}{50b}}$ **d.** $\sqrt[3]{\dfrac{1}{4p}}$ **e.** $\dfrac{5}{\sqrt[3]{a}}$

38. a. $\dfrac{-4}{\sqrt{20}}$ **b.** $\sqrt{\dfrac{125}{12n^3}}$

   c. $\sqrt{\dfrac{5}{12x}}$ **d.** $\sqrt[3]{\dfrac{3}{2m^2}}$ **e.** $\dfrac{-8}{3\sqrt[3]{5}}$

Simplify the following expressions by rationalizing the denominators. Where possible, state results in exact form and approximate form, rounded to hundredths.

39. a. $\dfrac{8}{3 + \sqrt{11}}$ **b.** $\dfrac{1}{2 - \sqrt{3}}$

40. a. $\dfrac{7}{\sqrt{7} + 3}$ **b.** $\dfrac{12}{\sqrt{x} + \sqrt{3}}$

41. a. $\dfrac{\sqrt{10} - 3}{\sqrt{3} + \sqrt{2}}$ **b.** $\dfrac{7 + \sqrt{6}}{3 - 3\sqrt{2}}$

42. a. $\dfrac{1 + \sqrt{2}}{\sqrt{6} + \sqrt{14}}$ **b.** $\dfrac{1 + \sqrt{6}}{5 + 2\sqrt{3}}$

▶ **APPLICATIONS**

43. Length of a cable: A radio tower is secured by cables that are anchored in the ground 8 m from its base. If the cables are attached to the tower 24 m above the ground, what is the length of each cable? Answer in (a) exact form using radicals, and (b) approximate form by rounding to one decimal place.

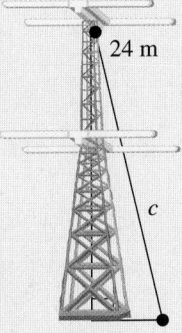
24 m
c
8 m

44. Height of a kite: Benjamin Franklin is flying his kite in a storm once again. John Adams has walked to a position directly under the kite and is 75 m from Ben. If the kite is 50 m above John Adams' head, how much string *S* has Ben let out? Answer in (a) exact form using radicals, and (b) approximate form by rounding to one decimal place.

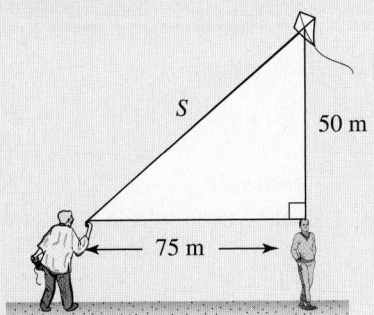
S
50 m
75 m

The time *T* (in days) required for a planet to make one revolution around the sun is modeled by the function $T = 0.407R^{\frac{3}{2}}$, where *R* is the maximum radius of the planet's orbit (in millions of miles). This is known as *Kepler's third law of planetary motion*. Use the equation given to approximate the number of days required for one complete orbit of each planet, given its maximum orbital radius.

45. a. Earth: 93 million mi
   b. Mars: 142 million mi
   c. Mercury: 36 million mi

46. a. Venus: 67 million mi
   b. Jupiter: 480 million mi
   c. Saturn: 890 million mi

47. Accident investigation: After an accident, police officers will try to determine the approximate velocity *V* that a car was traveling using the formula $V = 2\sqrt{6L}$, where *L* is the length of the skid marks in feet and *V* is the velocity in miles per hour. (a) If the skid marks were 54 ft long, how fast was the car traveling? (b) Approximate the speed of the car if the skid marks were 90 ft long.

48. Wind-powered energy: If a wind-powered generator is delivering *P* units of power, the velocity *V* of the wind (in miles per hour) can be determined using $V = \sqrt[3]{\dfrac{P}{k}}$, where *k* is a constant that depends on the size and efficiency of the generator. Rationalize the radical expression and use the new version to find the velocity of the wind if $k = 0.004$ and the generator is putting out 13.5 units of power.

49. **Surface area:** The lateral surface area (surface area excluding the base) S of a cone is given by the formula $S = \pi r \sqrt{r^2 + h^2}$, where r is the radius of the base and h is the height

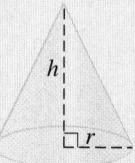

of the cone. Find the surface area of a cone that has a radius of 6 m and a height of 10 m. Answer in simplest form.

50. **Surface area:** The lateral surface area S of a frustum (a truncated cone) is given by the formula $S = \pi(a + b)\sqrt{h^2 + (b - a)^2}$, where a is the radius of the upper base, b is the radius of the lower base, and h is the height. Find the surface area of a frustum where $a = 6$ m, $b = 8$ m, and $h = 10$ m. Answer in simplest form.

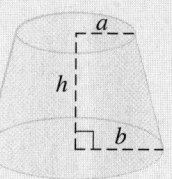

The expression $x^2 - 7$ is not factorable using *integer values*. But the expression *can be written* in the form $x^2 - (\sqrt{7})^2$, enabling us to factor it as a "binomial" and its conjugate: $(x + \sqrt{7})(x - \sqrt{7})$. Use this idea to factor the following expressions.

51. a. $x^2 - 5$ b. $n^2 - 19$

52. a. $4v^2 - 11$ b. $9w^2 - 11$

▶ **EXTENDING THE CONCEPT**

53. The following terms $\sqrt{3x} + \sqrt{9x} + \sqrt{27x} + \ldots$ form a pattern that continues until the sixth term is found. (a) Compute the sum of all six terms; (b) develop a system (investigate the pattern further) that will enable you to find the sum of 12 such terms *without actually writing out the terms*.

54. Find a quick way to simplify the expression without the aid of a calculator.

$$\left(\left(\left(\left(\left(\left(3^{\frac{5}{6}}\right)^{\frac{3}{2}}\right)^{\frac{4}{5}}\right)^{\frac{3}{4}}\right)^{\frac{2}{5}}\right)^{\frac{10}{3}}\right.$$

55. If $\left(x^{\frac{1}{2}} + x^{-\frac{1}{2}}\right)^2 = \frac{9}{2}$, find the value of $x^{\frac{1}{2}} + x^{-\frac{1}{2}}$.

56. Rewrite by rationalizing the *numerator:*
$$\frac{\sqrt{x + h} - \sqrt{x}}{h}$$

OVERVIEW OF APPENDIX I

Important Definitions, Properties, Formulas, and Relationships

A Notation and Relations

concept	notation	description	example
• Set notation:	{*members*}	braces enclose the members of a set	set of even whole numbers $A = \{0, 2, 4, 6, 8, \ldots\}$
• Is an element of	$\in$	indicates membership in a set	$14 \in A$
• Empty set	{ }	a set having no elements	odd numbers in A
• Is a proper subset of	$\subset$	indicates the elements of one set are entirely contained in another	$S = \{0, 6, 12, 18, 24, \ldots\}$ $S \subset A$
• Defining a set	$\{x \mid x \ldots\}$	the set of all x, such that $x \ldots$	$S = \{x \mid x = 6n \text{ for } n \in \mathbb{W}\}$

A Sets of Numbers

- Natural: $\mathbb{N} = \{1, 2, 3, 4, \ldots\}$

- Integers: $\mathbb{Z} = \{\ldots, -3, -2, -1, 0, 1, 2, 3, \ldots\}$

- Irrational: $\mathbb{H} = \{$numbers with a nonterminating, nonrepeating decimal form$\}$

- Whole: $\mathbb{W} = \{0, 1, 2, 3, \ldots\}$

- Rational: $\mathbb{Q} = \left\{\dfrac{p}{q}, \text{where } p, q \in \mathbb{Z}; q \neq 0\right\}$

- Real: $\mathbb{R} = \{$all rational and irrational numbers$\}$

Absolute Value of a Number

$$|n| = \begin{cases} n & \text{if } n \geq 0 \\ -n & \text{if } n < 0 \end{cases}$$

Distance Between a and b on a Number Line

$$|a - b| \text{ or } |b - a|$$

B Properties of Real Numbers: For real numbers a, b, and c,

Commutative Property
- Addition: $a + b = b + a$
- Multiplication: $a \cdot b = b \cdot a$

Identities
- Additive: $0 + a = a$
- Multiplicative: $1 \cdot a = a$

Associative Property
- Addition: $(a + b) + c = a + (b + c)$
- Multiplication: $(a \cdot b) \cdot c = a \cdot (b \cdot c)$

Inverses
- Additive: $a + (-a) = 0$
- Multiplicative: $\dfrac{p}{q} \cdot \dfrac{q}{p} = 1; p, q \neq 0$

C Properties of Exponents: For real numbers a and b, and integers m, n, and p (excluding 0 raised to a nonpositive power),

- Product property: $b^m \cdot b^n = b^{m+n}$

- Product to a power: $(a^m b^n)^p = a^{mp} \cdot b^{np}$

- Quotient property: $\dfrac{b^m}{b^n} = b^{m-n} \ (b \neq 0)$

- Negative exponents: $b^{-n} = \dfrac{1}{b^n}; \left(\dfrac{a}{b}\right)^{-n} = \left(\dfrac{b}{a}\right)^n$
  $(a, b \neq 0)$

- Power property: $(b^m)^n = b^{mn}$

- Quotient to a power: $\left(\dfrac{a^m}{b^n}\right)^p = \dfrac{a^{mp}}{b^{np}} \ (b \neq 0)$

- Zero exponents: $b^0 = 1 \ (b \neq 0)$

- Scientific notation: $N \times 10^k; 1 \leq |N| < 10, k \in \mathbb{Z}$

Special Products

- $(A + B)(A - B) = A^2 - B^2$

- $(A - B)(A^2 + AB + B^2) = A^3 - B^3$

- $(A + B)^2 = A^2 + 2AB + B^2$;
  $(A - B)^2 = A^2 - 2AB + B^2$

- $(A + B)(A^2 - AB + B^2) = A^3 + B^3$

D Special Factorizations

- $A^2 - B^2 = (A + B)(A - B)$
- $A^3 - B^3 = (A - B)(A^2 + AB + B^2)$

- $A^2 \pm 2AB + B^2 = (A \pm B)^2$
- $A^3 + B^3 = (A + B)(A^2 - AB + B^2)$

E Rational Expressions: For polynomials P, Q, R, and S with no denominator of zero,

- Lowest terms: $\dfrac{P \cdot R}{Q \cdot R} = \dfrac{P}{Q}$

- Multiplication: $\dfrac{P}{Q} \cdot \dfrac{R}{S} = \dfrac{P \cdot R}{Q \cdot S} = \dfrac{PR}{QS}$

- Addition: $\dfrac{P}{R} + \dfrac{Q}{R} = \dfrac{P + Q}{R}$

- Equivalence: $\dfrac{P}{Q} = \dfrac{P \cdot R}{Q \cdot R}$

- Division: $\dfrac{P}{Q} \div \dfrac{R}{S} = \dfrac{P}{Q} \cdot \dfrac{S}{R} = \dfrac{PS}{QR}$

- Subtraction: $\dfrac{P}{R} - \dfrac{Q}{R} = \dfrac{P - Q}{R}$

- Addition/subtraction with unlike denominators:
    1. Find the LCD of all rational expressions.
    2. Build equivalent expressions using LCD.
    3. Add/subtract numerators as indicated.
    4. Write the result in lowest terms.

F Properties of Radicals

- $\sqrt{a}$ is a real number only for $a \geq 0$
- $\sqrt[n]{a} = b$, only if $b^n = a$

- For any real number a, $\sqrt[n]{a^n} = |a|$ when n is even

- If a is a real number and n is an integer greater than 1, then $\sqrt[n]{a} = a^{\frac{1}{n}}$ provided $\sqrt[n]{a}$ represents a real number

- If $\sqrt[n]{A}$ and $\sqrt[n]{B}$ represent real numbers, $\sqrt[n]{AB} = \sqrt[n]{A} \cdot \sqrt[n]{B}$

- A radical expression is in simplest form when:

- $\sqrt{a} = b$, only if $b^2 = a$
- If n is even, $\sqrt[n]{a}$ represents a real number only if $a \geq 0$
- For any real number a, $\sqrt[n]{a^n} = a$ when n is odd

- If $\frac{m}{n}$ is a rational number written in lowest terms with $n \geq 2$, then $a^{\frac{m}{n}} = (\sqrt[n]{a})^m$ and $a^{\frac{m}{n}} = \sqrt[n]{a^m}$ provided $\sqrt[n]{a}$ represents a real number.

- If $\sqrt[n]{A}$ and $\sqrt[n]{B}$ represent real numbers and $B \neq 0$, $\sqrt[n]{\dfrac{A}{B}} = \dfrac{\sqrt[n]{A}}{\sqrt[n]{B}}$

1. the radicand has no factors that are perfect nth roots,
2. the radicand contains no fractions, and
3. no radicals occur in a denominator.

Pythagorean Theorem

- For any right triangle with legs a and b and hypotenuse c: $a^2 + b^2 = c^2$.

- For any triangle with sides a, b, and c, if $a^2 + b^2 = c^2$, then the triangle is a right triangle.

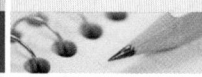

PRACTICE TEST

1. State true or false. If false, state why.
   a. $\mathbb{H} \subset \mathbb{R}$ b. $\mathbb{N} \subset \mathbb{Q}$
   c. $\sqrt{2} \in \mathbb{Q}$ d. $\frac{1}{2} \notin \mathbb{W}$

2. State the value of each expression.
   a. $\sqrt{121}$ b. $\sqrt[3]{-125}$
   c. $\sqrt{-36}$ d. $\sqrt{400}$

3. Evaluate each expression:
   a. $\frac{7}{8} - \left(-\frac{1}{4}\right)$ b. $-\frac{1}{3} - \frac{5}{6}$
   c. $-0.7 + 1.2$ d. $1.3 + (-5.9)$

4. Evaluate each expression:
   a. $(-4)\left(-2\frac{1}{3}\right)$ b. $(-0.6)(-1.5)$
   c. $\frac{-2.8}{-0.7}$ d. $4.2 \div (-0.6)$

5. Evaluate using a calculator: $2000\left(1 + \frac{0.08}{12}\right)^{12 \cdot 10}$

6. State the value of each expression, if possible.
   a. $0 \div 6$ b. $6 \div 0$

7. State the number of terms in each expression and identify the coefficient of each.
   a. $-2v^2 + 6v + 5$ b. $\frac{c + 2}{3} + c$

8. Evaluate each expression given $x = -0.5$ and $y = -2$. Round to hundredths as needed.
   a. $2x - 3y^2$ b. $\sqrt{2} - x(4 - x^2) + \frac{y}{x}$

9. Translate each phrase into an algebraic expression.
   a. Nine less than twice a number is subtracted from the number cubed.
   b. Three times the square of half a number is subtracted from twice the number.

10. Create a mathematical model using descriptive variables.

    a. The radius of the planet Jupiter is approximately 119 mi less than 11 times the radius of the Earth. Express the radius of Jupiter in terms of the Earth's radius.

    b. Last year, Video Venue Inc. earned \$1.2 million more than four times what it earned this year. Express last year's earnings of Video Venue Inc. in terms of this year's earnings.

11. Simplify by combining like terms.

    a. $8v^2 + 4v - 7 + v^2 - v$

    b. $-4(3b - 2) + 5b$

    c. $4x - (x - 2x^2) + x(3 - x)$

12. Factor each expression completely.

    a. $9x^2 - 16$ **b.** $4v^3 - 12v^2 + 9v$

    c. $x^3 + 5x^2 - 9x - 45$

13. Simplify using the properties of exponents.

    a. $\dfrac{5}{b^{-3}}$ **b.** $(-2a^3)^2(a^2b^4)^3$

    c. $\left(\dfrac{m^2}{2n}\right)^3$ **d.** $\left(\dfrac{5p^2q^3r^4}{-2pq^2r^4}\right)^2$

14. Simplify using the properties of exponents.

    a. $\dfrac{-12a^3b^5}{3a^2b^4}$

    b. $(3.2 \times 10^{-17}) \times (2.0 \times 10^{15})$

    c. $\left(\dfrac{a^{-3} \cdot b}{c^{-2}}\right)^{-4}$ **d.** $-7x^0 + (-7x)^0$

15. Compute each product.

    a. $(3x^2 + 5y)(3x^2 - 5y)$

    b. $(2a + 3b)^2$

16. Add or subtract as indicated.

    a. $(-5a^3 + 4a^2 - 3) + (7a^4 + 4a^2 - 3a - 15)$

    b. $(2x^2 + 4x - 9) - (7x^4 - 2x^2 - x - 9)$

Simplify or compute as indicated.

17. a. $\dfrac{x - 5}{5 - x}$ **b.** $\dfrac{4 - n^2}{n^2 - 4n + 4}$

    c. $\dfrac{x^3 - 27}{x^2 + 3x + 9}$ **d.** $\dfrac{3x^2 - 13x - 10}{9x^2 - 4}$

    e. $\dfrac{x^2 - 25}{3x^2 - 11x - 4} \div \dfrac{x^2 + x - 20}{x^2 - 8x + 16}$

    f. $\dfrac{m + 3}{m^2 + m - 12} - \dfrac{2}{5(m + 4)}$

18. a. $\sqrt{(x + 11)^2}$ **b.** $\sqrt[3]{\dfrac{-8}{27v^3}}$

    c. $\left(\dfrac{25}{16}\right)^{\frac{-3}{2}}$ **d.** $\dfrac{-4 + \sqrt{32}}{8}$

    e. $7\sqrt{40} - \sqrt{90}$ **f.** $(x + \sqrt{5})(x - \sqrt{5})$

    g. $\sqrt{\dfrac{2}{5x}}$ **h.** $\dfrac{8}{\sqrt{6} - \sqrt{2}}$

19. Maximizing revenue: Due to past experience, the manager of a video store knows that if a popular video game is priced at \$30, the store will sell 40 each day. For each decrease of \$0.50, one additional sale will be made. The formula for the store's revenue is then $R = (30 - 0.5x)(40 + x)$, where x represents the number of times the price is decreased. Multiply the binomials and use a table of values to determine (a) the number of 50¢ decreases that will give the most revenue and (b) the maximum amount of revenue.

20. Use the Pythagorean theorem to determine the length of the diagonal of the rectangular prism shown in the figure. (*Hint:* First find the diagonal of the base.)

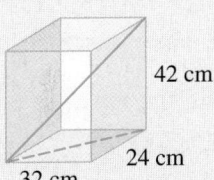

42 cm

24 cm

32 cm

More on Synthetic Division

As the name implies, synthetic division simulates the long division process, but in a condensed and more efficient form. It's based on a few simple observations of long division, as noted in the division $(x^3 - 2x^2 - 13x - 17) \div (x - 5)$ shown in Figure AII.1.

Figure AII.1

$$
\begin{array}{r}
x^2 + 3x + 2 \\
x - 5 \overline{) x^3 - 2x^2 - 13x - 17} \\
- (x^3 - 5x^2) \\
\hline
3x^2 - 13x \\
- (3x^2 - 15x) \\
\hline
2x - 17 \\
- (2x - 10) \\
\hline
-7 \quad \text{remainder}
\end{array}
$$

Figure AII.2

$$
\begin{array}{r}
1 \quad 3 \quad 2 \\
x - 5 \overline{) 1 \quad -2 \quad -13 \quad -17} \\
5 \\
\hline
3 \\
15 \\
\hline
2 \\
10 \\
\hline
-7 \quad \text{remainder}
\end{array}
$$

A careful observation reveals a great deal of repetition, as any term in red is a duplicate of the term above it. In addition, since the dividend and divisor must be written in decreasing order of degree, the variable part of each term is unnecessary as we can let the *position of each coefficient* indicate the degree of the term. In other words, we'll agree that

$$1 \quad -2 \quad -13 \quad -17 \quad \text{represents the polynomial} \quad 1x^3 - 2x^2 - 13x - 17.$$

Finally, we know in advance that we'll be subtracting each partial product, so we can "distribute the negative," shown at each stage. Removing the repeated terms and variable factors, then distributing the negative to the remaining terms produces Figure AII.2. The entire process can now be condensed by vertically compressing the rows of the division so that a minimum of space is used (Figure AII.3).

Figure AII.3

$$
\begin{array}{r}
1 \quad\quad 3 \quad\quad 2 \quad \text{quotient}\\
x - 5 \overline{) 1 \quad -2 \quad -13 \quad -17} \quad \text{dividend}\\
5 \quad\quad 15 \quad\quad 10 \quad \text{products}\\
\hline
3 \quad\quad 2 \quad\quad 7 \quad \text{sums}
\end{array}
$$

Figure AII.4

$$
\begin{array}{r}
1 \quad\quad 3 \quad\quad 2 \\
x - 5 \overline{) 1 \quad -2 \quad -13 \quad -17} \quad \text{dividend}\\
5 \quad\quad 15 \quad\quad 10 \quad \text{products}\\
\hline
1 \quad 3 \quad 2 \quad -7 \quad \textbf{remainder}\\
\text{quotient}
\end{array}
$$

Further, if we include the lead coefficient in the bottom row (Figure AII.4), the coefficients in the top row (in blue) are duplicated and no longer necessary, since the quotient and remainder now appear in the last row. Finally, note all entries in the product row (in **red**) are five times the sum from the prior column. There is a direct connection between this multiplication by 5 and the divisor $x - 5$, and in fact, it is the *zero of the divisor* that is used in synthetic division ($x = 5$ from $x - 5 = 0$). A simple change in format makes this method of division easier to use, and highlights the location of the divisor and remainder (the blue brackets in Figure AII.5). Note the process begins by "dropping the lead coefficient into place" (shown in **bold**). The full process of synthetic division is shown in Figure AII.6 for the same exercise.

Figure AII.5

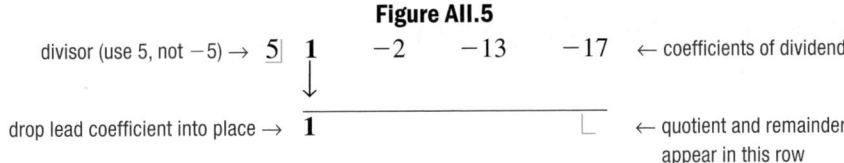

We then multiply this coefficient by the "divisor," place the result in the next column and add. In a sense, we "multiply in the diagonal direction," and "add in the vertical direction." Continue the process until the division is complete.

Figure AII.6

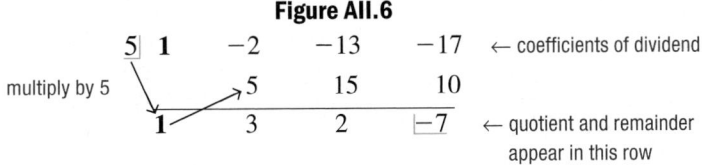

The result is $x^2 + 3x + 2 + \dfrac{-7}{x - 5}$, read from the last row.

More on Matrices

Reduced Row-Echelon Form

A matrix is in reduced row-echelon form if it satisfies the following conditions:

1. All null rows (zeroes for all entries) occur at the bottom of the matrix.
2. The first non-zero entry of any row must be a 1.
3. For any two consecutive, nonzero rows, the leading 1 in the higher row is to the left of the 1 in the lower row.
4. Every column with a leading 1 has zeroes for all other entries in the column.

Matrices A through D are in reduced row-echelon form.

$$A = \begin{bmatrix} 0 & 1 & 0 & 0 & 0 & 5 \\ 0 & 0 & 0 & 1 & 0 & 3 \\ 0 & 0 & 0 & 0 & 1 & 2 \end{bmatrix} \quad B = \begin{bmatrix} 1 & 0 & 0 & 5 \\ 0 & 1 & 0 & 3 \\ 0 & 0 & 0 & 0 \end{bmatrix} \quad C = \begin{bmatrix} 1 & 0 & 0 & 5 \\ 0 & 1 & 3 & -2 \\ 0 & 0 & 0 & 0 \end{bmatrix} \quad D = \begin{bmatrix} 1 & 0 & 5 & 0 \\ 0 & 1 & 2 & 0 \\ 0 & 0 & 0 & 1 \end{bmatrix}$$

Where *Gaussian elimination* places a matrix in *row-echelon form* (satisfying the first three conditions), *Gauss-Jordan elimination* places a matrix in *reduced row-echelon form*. To obtain this form, continue applying row operations to the matrix until the fourth condition above is also satisfied. For a 3 × 3 system having a unique solution, the diagonal entries of the coefficient matrix will be 1's, with 0's for all other entries. To illustrate, we'll extend Example 4 from Section 8.5 until reduced row-echelon form is obtained.

EXAMPLE 4 ▶ **Solving Systems Using the Augmented Matrix**

Solve using Gauss-Jordan elimination: $\begin{cases} 2x + y - 2z = -7 \\ x + y + z = -1 \\ -2y - z = -3 \end{cases}$

$$\begin{cases} 2x + y - 2z = -7 \\ x + y + z = -1 \\ -2y - z = -3 \end{cases} \xrightarrow{R1 \leftrightarrow R2} \begin{cases} x + y + z = -1 \\ 2x + y - 2z = -7 \\ -2y - z = -3 \end{cases} \text{ matrix form} \rightarrow \begin{bmatrix} 1 & 1 & 1 & -1 \\ 2 & 1 & -2 & -7 \\ 0 & -2 & -1 & -3 \end{bmatrix}$$

$$\begin{bmatrix} 1 & 1 & 1 & -1 \\ 2 & 1 & -2 & -7 \\ 0 & -2 & -1 & -3 \end{bmatrix} \xrightarrow{-2R1 + R2 \to R2} \begin{bmatrix} 1 & 1 & 1 & -1 \\ 0 & -1 & -4 & -5 \\ 0 & -2 & -1 & -3 \end{bmatrix} \xrightarrow{-1R2 \to R2} \begin{bmatrix} 1 & 1 & 1 & -1 \\ 0 & 1 & 4 & 5 \\ 0 & -2 & -1 & -3 \end{bmatrix}$$

$$\begin{bmatrix} 1 & 1 & 1 & -1 \\ 0 & 1 & 4 & 5 \\ 0 & -2 & -1 & -3 \end{bmatrix} \xrightarrow{2R2 + R3 \to R3} \begin{bmatrix} 1 & 1 & 1 & -1 \\ 0 & 1 & 4 & 5 \\ 0 & 0 & 7 & 7 \end{bmatrix} \xrightarrow{\frac{R3}{7} \to R3} \begin{bmatrix} 1 & 1 & 1 & -1 \\ 0 & 1 & 4 & 5 \\ 0 & 0 & 1 & 1 \end{bmatrix}$$

$$\begin{bmatrix} 1 & 1 & 1 & -1 \\ 0 & 1 & 4 & 5 \\ 0 & 0 & 1 & 1 \end{bmatrix} \xrightarrow{-R2 + R1 \to R1} \begin{bmatrix} 1 & 0 & -3 & -6 \\ 0 & 1 & 4 & 5 \\ 0 & 0 & 1 & 1 \end{bmatrix} \begin{smallmatrix} 3R3 + R1 \to R1 \\ \\ -4R3 + R2 \to R2 \end{smallmatrix} \begin{bmatrix} 1 & 0 & 0 & -3 \\ 0 & 1 & 0 & 1 \\ 0 & 0 & 1 & 1 \end{bmatrix}$$

The final matrix is in reduced row-echelon form with solution $(-3, 1, 1)$ just as in Section 8.5.

The Determinant of a General Matrix

To compute the determinant of a general square matrix, we introduce the idea of a **cofactor.** For an $n \times n$ matrix A, $A_{ij} = (-1)^{i+j} |M_{ij}|$ is the cofactor of matrix element a_{ij}, where $|M_{ij}|$ represents the determinant of the corresponding minor matrix. Note that $i + j$ is the sum of the row and column of the entry, and if this sum is even, $(-1)^{i+j} = 1$, while if the sum is odd, $(-1)^{i+j} = -1$ (this is how the sign table for a 3×3 determinant was generated). To compute the determinant of an $n \times n$ matrix, multiply each element in any row or column by its cofactor and add. The result is a tier-like process in which the determinant of a larger matrix requires computing the determinant of smaller matrices. In the case of a 4×4 matrix, each of the minor matrices will be size 3×3, whose determinant then requires the computation of other 2×2 determinants. In the following illustration, two of the entries in the first row are zero for convenience. For

$$
A = \begin{bmatrix} -2 & 0 & 3 & 0 \\ 1 & 2 & 0 & -2 \\ 3 & -1 & 4 & 1 \\ 0 & -3 & 2 & 1 \end{bmatrix},
$$

we have: $\det(A) = -2 \cdot (-1)^{1+1} \begin{vmatrix} 2 & 0 & -2 \\ -1 & 4 & 1 \\ -3 & 2 & 1 \end{vmatrix} + (3) \cdot (-1)^{1+3} \begin{vmatrix} 1 & 2 & -2 \\ 3 & -1 & 1 \\ 0 & -3 & 1 \end{vmatrix}$

Computing the first 3×3 determinant gives -16, the second 3×3 determinant is 14. This gives:

$$
\det(A) = -2(-16) + 3(14)
$$
$$
= 74
$$

Deriving the Equation of a Conic

The Equation of an Ellipse

In Section 9.2, the equation $\sqrt{(x + c)^2 + y^2} + \sqrt{(x - c)^2 + y^2} = 2a$ was developed using the distance formula and the definition of an ellipse. To find the standard form of the equation, we treat this result as a radical equation, isolating one of the radicals and squaring both sides.

$$\sqrt{(x + c)^2 + y^2} = 2a - \sqrt{(x - c)^2 + y^2} \qquad \text{isolate one radical}$$

$$(x + c)^2 + y^2 = 4a^2 - 4a\sqrt{(x - c)^2 + y^2} + (x - c)^2 + y^2 \qquad \text{square both sides}$$

We continue by simplifying the equation, isolating the remaining radical, and squaring again.

$$x^2 + 2cx + c^2 + y^2 = 4a^2 - 4a\sqrt{(x - c)^2 + y^2} + x^2 - 2cx + c^2 + y^2 \quad \begin{array}{l}\text{expand}\\\text{binomials}\end{array}$$

$$4cx = 4a^2 - 4a\sqrt{(x - c)^2 + y^2} \qquad \text{simplify}$$

$$a\sqrt{(x - c)^2 + y^2} = a^2 - cx \qquad \text{isolate radical; divide by 4}$$

$$a^2[(x - c)^2 + y^2] = a^4 - 2a^2cx + c^2x^2 \qquad \text{square both sides}$$

$$a^2x^2 - 2a^2cx + a^2c^2 + a^2y^2 = a^4 - 2a^2cx + c^2x^2 \qquad \text{expand and distribute } a^2 \text{ on left}$$

$$a^2x^2 - c^2x^2 + a^2y^2 = a^4 - a^2c^2 \qquad \text{add } 2a^2cx \text{ and rewrite equation}$$

$$x^2(a^2 - c^2) + a^2y^2 = a^2(a^2 - c^2) \qquad \text{factor}$$

$$\frac{x^2}{a^2} + \frac{y^2}{a^2 - c^2} = 1 \qquad \text{divide by } a^2(a^2 - c^2)$$

Since $a > c$, we know $a^2 > c^2$ and $a^2 - c^2 > 0$. For convenience, we let $b^2 = a^2 - c^2$ and it also follows that $a^2 > b^2$ and $a > b$ (since $c > 0$). Substituting b^2 for $a^2 - c^2$ we obtain the standard form of the equation of an ellipse (major axis horizontal, since we stipulated $a > b$): $\frac{x^2}{a^2} + \frac{y^2}{b^2} = 1$. Note once again the x-intercepts are $(\pm a, 0)$, while the y-intercepts are $(0, \pm b)$.

The Equation of a Hyperbola

Similar to the development of the equation of an ellipse, the equation $\sqrt{(x + c)^2 + y^2} - \sqrt{(x - c)^2 + y^2} = 2a$ could have been developed using the distance formula and the definition of a hyperbola. To find the standard form of this equation, we apply the same procedures as before.

$$\sqrt{(x + c)^2 + y^2} = 2a + \sqrt{(x - c)^2 + y^2} \qquad \text{isolate one radical}$$

$$(x + c)^2 + y^2 = 4a^2 + 4a\sqrt{(x - c)^2 + y^2} + (x - c)^2 + y^2 \qquad \text{square both sides}$$

$$x^2 + 2cx + c^2 + y^2 = 4a^2 + 4a\sqrt{(x - c)^2 + y^2} + x^2 - 2cx + c^2 + y^2 \quad \begin{array}{l}\text{expand}\\\text{binomials}\end{array}$$

$$4cx = 4a^2 + 4a\sqrt{(x - c)^2 + y^2} \qquad \text{simplify}$$

$$cx - a^2 = a\sqrt{(x - c)^2 + y^2} \qquad \text{isolate radical; divide by 4}$$

$$c^2x^2 - 2a^2cx + a^4 = a^2[(x - c)^2 + y^2] \qquad \text{square both sides}$$

$$c^2x^2 - 2a^2cx + a^4 = a^2x^2 - 2a^2cx + a^2c^2 + a^2y^2 \qquad \text{expand and distribute } a^2 \text{ on the right}$$

$$c^2x^2 - a^2x^2 - a^2y^2 = a^2c^2 - a^4 \qquad \text{add } 2a^2cx \text{ and rewrite equation}$$

$$x^2(c^2 - a^2) - a^2y^2 = a^2(c^2 - a^2) \qquad \text{factor}$$

$$\frac{x^2}{a^2} - \frac{y^2}{c^2 - a^2} = 1 \qquad \text{divide by } a^2(c^2 - a^2)$$

From the definition of a hyperbola we have $0 < a < c$, showing $c^2 > a^2$ and $c^2 - a^2 > 0$. For convenience, we let $b^2 = c^2 - a^2$ and substitute to obtain the standard form of the equation of a hyperbola (transverse axis horizontal): $\dfrac{x^2}{a^2} - \dfrac{y^2}{b^2} = 1$. Note the x-intercepts are $(0, \pm a)$ and there are no y-intercepts.

The Asymptotes of a Central Hyperbola

From our work in Section 9.3, a central hyperbola with a horizontal axis will have asymptotes at $y = \pm \dfrac{b}{a} x$. To understand why, recall that for asymptotic behavior we investigate what happens to the relation for large values of x, meaning as $|x| \to \infty$. Starting with $\dfrac{x^2}{a^2} - \dfrac{y^2}{b^2} = 1$, we have

$$b^2 x^2 - a^2 y^2 = a^2 b^2 \qquad \text{clear denominators}$$

$$a^2 y^2 = b^2 x^2 - a^2 b^2 \qquad \text{isolate term with } y$$

$$a^2 y^2 = b^2 x^2 \left(1 - \frac{a^2}{x^2} \right) \qquad \text{factor out } b^2 x^2 \text{ from right side}$$

$$y^2 = \frac{b^2}{a^2} x^2 \left(1 - \frac{a^2}{x^2} \right) \qquad \text{divide by } a^2$$

$$y = \pm \frac{b}{a} x \sqrt{1 - \frac{a^2}{x^2}} \qquad \text{square root both sides}$$

As $|x| \to \infty$, $\dfrac{a^2}{x^2} \to 0$, and we find that for large values of x, $y \approx \pm \dfrac{b}{a} x$.

Selected Proofs

Proof of the Complex Conjugates Corollary

Given $p(x)$ is a polynomial with real number coefficients, complex solutions must occur in conjugate pairs. If $a + bi$, $b \neq 0$ is a solution, then $a - bi$ must also be a solution.

To prove this for polynomials of degree $n > 2$, we let $z_1 = a + bi$ and $z_2 = c + di$ be complex numbers, and $\bar{z}_1 = a - bi$ and $\bar{z}_2 = c - di$ represent their conjugates, and observe the following properties:

1. The conjugate of a sum is equal to the sum of the conjugates.

sum: $z_1 + z_2$

$(a + bi) + (c + di)$

$(a + c) + (b + d)i$ $\rightarrow$ conjugate of sum $\rightarrow$

sum of conjugates: $\bar{z}_1 + \bar{z}_2$

$(a - bi) + (c - di)$

$(a + c) - (b + d)i$ ✓

2. The conjugate of a product is equal to the product of the conjugates.

product: $z_1 \cdot z_2$

$(a + bi) \cdot (c + di)$

$ac + adi + bci + bdi^2$

$(ac - bd) + (ad + bc)i$ $\rightarrow$ conjugate of product $\rightarrow$

product of conjugates: $\bar{z}_1 \cdot \bar{z}_2$

$(a - bi) \cdot (c - di)$

$ac - adi - bci + bdi^2$

$(ac - bd) - (ad + bc)i$ ✓

Since polynomials involve only sums and products, and the complex conjugate of any real number is the number itself, we have the following:

Given polynomial $p(x) = a_n x^n + a_{n-1} x^{n-1} + \cdots + a_1 x^1 + a_0$, where a_n, $a_{n-1}, \ldots, a_1, a_0$ are real numbers and $z = a + bi$ is a zero of p, we must show that $\bar{z} = a - bi$ is also a zero.

$$a_n z^n + a_{n-1} z^{n-1} + \cdots + a_1 z^1 + a_0 = p(z) \quad \text{evaluate } p(x) \text{ at } z$$

$$a_n z^n + a_{n-1} z^{n-1} + \cdots + a_1 z^1 + a_0 = 0 \quad p(z) = 0 \text{ given}$$

$$\overline{a_n z^n + a_{n-1} z^{n-1} + \cdots + a_1 z^1 + a_0} = \bar{0} \quad \text{conjugate both sides}$$

$$\overline{a_n z^n} + \overline{a_{n-1} z^{n-1}} + \cdots + \overline{a_1 z^1} + \overline{a_0} = \bar{0} \quad \text{property 1}$$

$$\bar{a}_n (\overline{z^n}) + \bar{a}_{n-1} (\overline{z^{n-1}}) + \cdots + \bar{a}_1 (\overline{z^1}) + \overline{a_0} = \bar{0} \quad \text{property 2}$$

$$a_n (\overline{z^n}) + a_{n-1} (\overline{z^{n-1}}) + \cdots + a_1 (\overline{z^1}) + a_0 = 0 \quad \text{conjugate of a real number is the number}$$

$$p(\bar{z}) = 0 \text{ ✓} \quad \text{result}$$

Proof of the Determinant Formula for the Area of a Parallelogram

Since the area of a triangle is $A = \dfrac{1}{2} ab \sin \theta$, the area of the corresponding parallelogram is twice as large: $A = ab \sin \theta$. In terms of the vectors $\mathbf{u} = \langle a, b \rangle$ and $\mathbf{v} = \langle c, d \rangle$, we have $A = |\mathbf{u}||\mathbf{v}| \sin \theta$, and it follows that

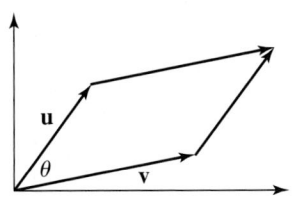

$$\text{Area} = |\mathbf{u}||\mathbf{v}|\sqrt{1 - \cos^2\theta}, \qquad \text{Pythagorean identity}$$

$$= |\mathbf{u}||\mathbf{v}|\sqrt{1 - \frac{(\mathbf{u} \cdot \mathbf{v})^2}{|\mathbf{u}|^2|\mathbf{v}|^2}} \qquad \text{substitute for } \cos\theta$$

$$= |\mathbf{u}||\mathbf{v}|\sqrt{\frac{|\mathbf{u}|^2|\mathbf{v}|^2 - (\mathbf{u} \cdot \mathbf{v})^2}{|\mathbf{u}|^2|\mathbf{v}|^2}} \qquad \text{common denominator}$$

$$= \sqrt{|\mathbf{u}|^2|\mathbf{v}|^2 - (\mathbf{u} \cdot \mathbf{v})^2} \qquad \text{simplify}$$

$$= \sqrt{(a^2 + b^2)(c^2 + d^2) - (ac + bd)^2} \qquad \text{substitute}$$

$$= \sqrt{a^2c^2 + a^2d^2 + b^2c^2 + b^2d^2 - (a^2c^2 + 2acbd + b^2d^2)} \qquad \text{expand}$$

$$= \sqrt{a^2d^2 - 2acbd + b^2c^2} \qquad \text{simplify}$$

$$= \sqrt{(ad - bc)^2} \qquad \text{factor}$$

$$= |ad - bc| \qquad \text{result}$$

Proof of Heron's Formula Using Algebra

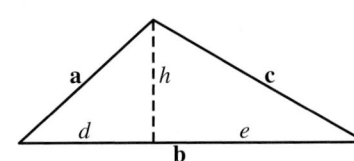

Note that $\sqrt{a^2 - d^2} = h = \sqrt{c^2 - e^2}$. It follows that

$$\sqrt{a^2 - d^2} = \sqrt{c^2 - e^2}$$

$$a^2 - d^2 = c^2 - e^2$$

$$a^2 - (b - e)^2 = c^2 - e^2$$

$$a^2 - b^2 + 2be - e^2 = c^2 - e^2$$

$$a^2 - b^2 - c^2 = -2be$$

$$\frac{b^2 + c^2 - a^2}{2b} = e$$

This shows:
$$A = \frac{1}{2}bh$$

$$= \frac{1}{2}b\sqrt{c^2 - e^2}$$

$$= \frac{1}{2}b\sqrt{c^2 - \left(\frac{b^2 + c^2 - a^2}{2b}\right)^2}$$

$$= \frac{1}{2}b\sqrt{c^2 - \left(\frac{b^4 + 2b^2c^2 - 2a^2b^2 - 2a^2c^2 + a^4 + c^4}{4b^2}\right)}$$

$$= \frac{1}{2}b\sqrt{\frac{4b^2c^2}{4b^2} - \left(\frac{b^4 + 2b^2c^2 - 2a^2b^2 - 2a^2c^2 + a^4 + c^4}{4b^2}\right)}$$

$$= \frac{1}{2}b\sqrt{\frac{4b^2c^2 - b^4 - 2b^2c^2 + 2a^2b^2 + 2a^2c^2 - a^4 - c^4}{4b^2}}$$

$$= \frac{1}{4}\sqrt{2a^2b^2 + 2a^2c^2 + 2b^2c^2 - a^4 - b^4 - c^4}$$

$$= \frac{1}{4}\sqrt{[(a + b)^2 - c^2][c^2 - (a - b)^2]}$$

$$= \frac{1}{4}\sqrt{(a + b + c)(a + b - c)(c + a - b)(c - a + b)}$$

From this point, the conclusion of the proof is the same as the trigonometric development found on page 652.

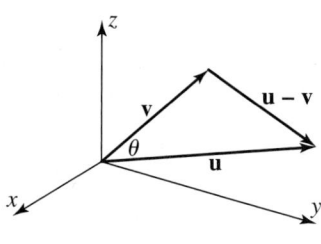

Proof that $u \cdot v = \text{comp}_v\, u \times |v|$

Consider the vectors in the figure shown, which form a triangle. Applying the Law of Cosines to this triangle yields:

$$|\mathbf{u} - \mathbf{v}|^2 = |\mathbf{u}|^2 + |\mathbf{v}|^2 - 2|\mathbf{u}||\mathbf{v}| \cos \theta$$

Using properties of the dot product (page 680), we can rewrite the left-hand side as follows:

$$|\mathbf{u} - \mathbf{v}|^2 = (\mathbf{u} - \mathbf{v}) \cdot (\mathbf{u} - \mathbf{v})$$
$$= \mathbf{u} \cdot \mathbf{u} - \mathbf{u} \cdot \mathbf{v} - \mathbf{v} \cdot \mathbf{u} + \mathbf{v} \cdot \mathbf{v}$$
$$= |\mathbf{u}|^2 - 2\,\mathbf{u} \cdot \mathbf{v} + |\mathbf{v}|^2$$

Substituting the last expression for $|\mathbf{u} - \mathbf{v}|^2$ from the Law of Cosines gives

$$|\mathbf{u}|^2 - 2\,\mathbf{u} \cdot \mathbf{v} + |\mathbf{v}|^2 = |\mathbf{u}|^2 + |\mathbf{v}|^2 - 2\,|\mathbf{u}||\mathbf{v}| \cos \theta$$
$$-2\,\mathbf{u} \cdot \mathbf{v} = -2\,|\mathbf{u}||\mathbf{v}| \cos \theta$$
$$\mathbf{u} \cdot \mathbf{v} = |\mathbf{u}||\mathbf{v}| \cos \theta$$
$$= |\mathbf{u}| \cos \theta \times |\mathbf{v}|.$$

Substituting $\text{comp}_v\, u$ for $|\mathbf{u}| \cos \theta$ completes the proof:

$$\mathbf{u} \cdot \mathbf{v} = \text{comp}_v\, u \times |\mathbf{v}|$$

Proof of DeMoivre's Theorem: $(\cos x + i \sin x)^n = \cos(nx) + i \sin(nx)$

For $n > 0$, we proceed using mathematical induction.

1. Show the statement is true for $n = 1$ (base case):

$$(\cos x + i \sin x)^1 = \cos(1x) + i \sin(1x)$$
$$\cos x + i \sin x = \cos x + i \sin x \quad \checkmark$$

2. Assume the statement is true for $n = k$ (induction hypothesis):

$$(\cos x + i \sin x)^k = \cos(kx) + i \sin(kx)$$

3. Show the statement is true for $n = k + 1$:

$$(\cos x + i \sin x)^{k+1} = (\cos x + i \sin x)^k (\cos x + i \sin x)^1$$
$$= [\cos(kx) + i \sin(kx)](\cos x + i \sin x) \qquad \text{induction hypothesis}$$
$$= \cos(kx)\cos x - \sin(kx)\sin x + i[\cos(kx)\sin x + \sin(kx)\cos x] \qquad \text{F-O-I-L}$$
$$= \cos[(k + 1)x] + i \sin[(k + 1)x] \quad \checkmark \qquad \text{sum/difference identities}$$

By the principle of mathematical induction, the statement is true for all positive integers. For $n < 0$ (the theorem is obviously true for $n = 0$), consider a positive integer m, where $n = -m$.

$$(\cos x + i \sin x)^n = (\cos x + i \sin x)^{-m}$$

$$= \frac{1}{(\cos x + i \sin x)^m} \qquad \text{negative exponent property}$$

$$= \frac{1}{\cos(mx) + i \sin(mx)} \qquad \text{DeMoivre's theorem for } n > 0$$

$$= \cos(mx) - i \sin(mx) \qquad \begin{array}{l}\text{multiply numerator and denom by}\\ \cos(mx) - i\sin(mx) \text{ and simplify}\end{array}$$

$$= \cos(-mx) + i \sin(-mx) \qquad \text{even/odd identities}$$
$$= \cos(nx) + i \sin(nx) \qquad n = -m$$

Families of Polar Curves

Circles and Spiral Curves

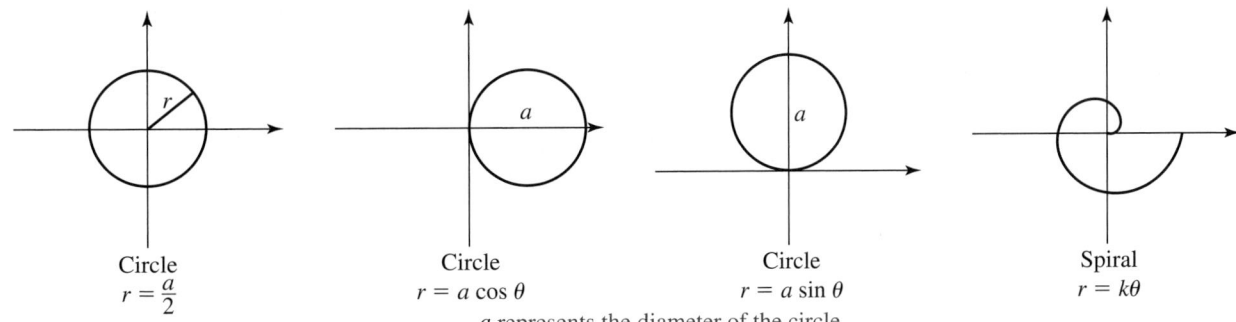

Circle
$r = \dfrac{a}{2}$

Circle
$r = a \cos \theta$

Circle
$r = a \sin \theta$

a represents the diameter of the circle

Spiral
$r = k\theta$

Roses: $r = a \sin(n\theta)$ (illustrated here) and $r = a \cos(n\theta)$

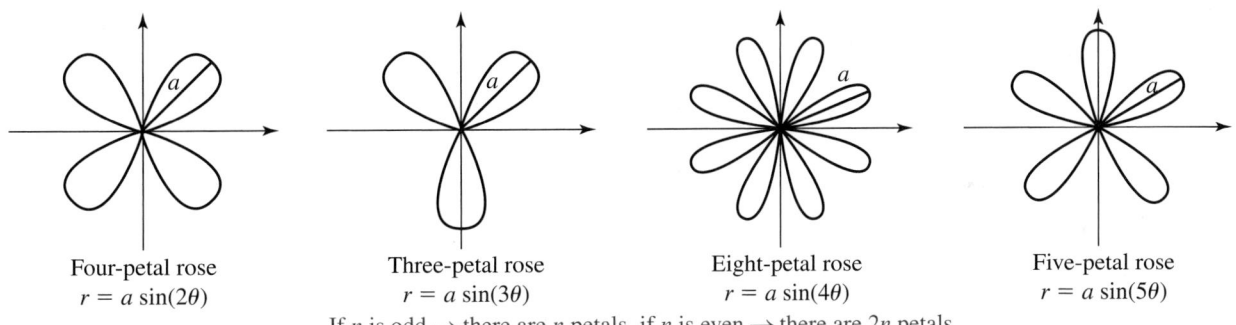

Four-petal rose
$r = a \sin(2\theta)$

Three-petal rose
$r = a \sin(3\theta)$

Eight-petal rose
$r = a \sin(4\theta)$

Five-petal rose
$r = a \sin(5\theta)$

If n is odd → there are n petals, if n is even → there are $2n$ petals.
$|a|$ represents the maximum distance from the origin
(the radius of a circumscribed circle)

Limaçons: $r = a + b \sin \theta$ (illustrated here) and $r = a + b \cos \theta$

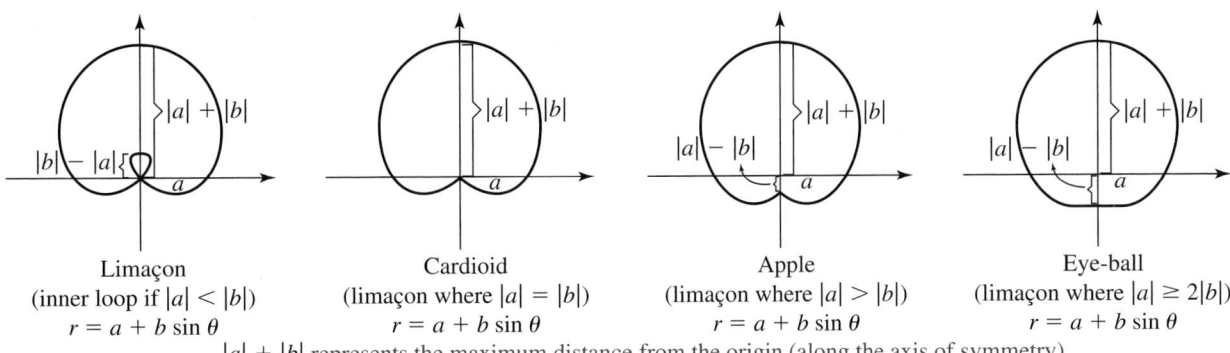

Limaçon
(inner loop if $|a| < |b|$)
$r = a + b \sin \theta$

Cardioid
(limaçon where $|a| = |b|$)
$r = a + b \sin \theta$

Apple
(limaçon where $|a| > |b|$)
$r = a + b \sin \theta$

Eye-ball
(limaçon where $|a| \geq 2|b|$)
$r = a + b \sin \theta$

$|a| + |b|$ represents the maximum distance from the origin (along the axis of symmetry)

Lemniscates: $r^2 = a^2\sin(2\theta)$ and $r^2 = a^2\cos(2\theta)$

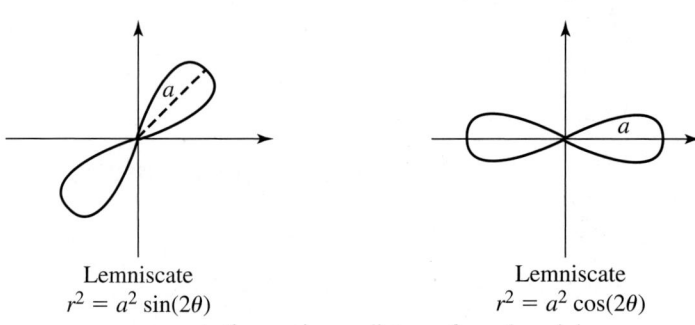

Lemniscate
$r^2 = a^2 \sin(2\theta)$

Lemniscate
$r^2 = a^2 \cos(2\theta)$

a represents the maximum distance from the origin
(the radius of a circumscribed circle)

Student Answer Appendix

CHAPTER 1

Exercises 1.1, pp. 10–13

1. identity; unknown **3.** literal; two **5.** Answers will vary. **7.** $x = 3$

9. $v = -11$ **11.** $b = \dfrac{6}{5}$ **13.** $b = -15$ **15.** $m = -\dfrac{27}{4}$ **17.** $x = 12$

19. $x = 12$ **21.** $p = -56$ **23.** $a = -3.6$ **25.** $v = -0.5$

27. $n = \dfrac{20}{21}$ **29.** $p = \dfrac{12}{5}$ **31.** contradiction; { }

33. conditional; $n = -\dfrac{11}{10}$ **35.** identity; $\{x \mid x \in \mathbb{R}\}$ **37.** $C = \dfrac{P}{1 + M}$

39. $r = \dfrac{C}{2\pi}$ **41.** $T_2 = \dfrac{T_1 P_2 V_2}{P_1 V_1}$ **43.** $h = \dfrac{3V}{4\pi r^2}$ **45.** $n = \dfrac{2S_n}{a_1 + a_n}$

47. $P = \dfrac{2(S - B)}{S}$ **49.** $y = -\dfrac{A}{B}x + \dfrac{C}{B}$ **51.** $y = \dfrac{-20}{9}x + \dfrac{16}{3}$

53. $y = \dfrac{-4}{5}x - 5$ **55.** $a = 3; b = 2; c = -19; x = -7$

57. $a = -6; b = 1; c = 33; x = \dfrac{-16}{3}$

59. $a = 7; b = -13; c = -27; x = -2$ **61.** $h = 17$ cm **63.** 510 ft

65. 56 in. **67.** 3084 ft **69.** 48; 50 **71.** 5; 7 **73.** 11: 30 A.M.

75. 36 min **77.** 4 quarts; 50% O.J. **79.** 16/lb; $1.80/lb **81.** 12 lb

83. 16 lb **85.** Answers will vary **87.** 69 **89.** -3

91. a. $(2x + 3)(2x - 3)$ **b.** $(x - 3)(x^2 + 3x + 9)$

Exercises 1.2, pp. 20–23

1. set; interval **3.** intersection; union **5.** Answers will vary.

7. $w \geq 45$ **9.** $250 < T < 450$

11.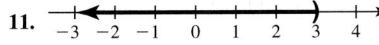

13.

15.

17.

19. $\{x \mid x \geq -2\}; [-2, \infty)$ **21.** $\{x \mid -2 \leq x \leq 1\}; [-2, 1]$

23. $\{a \mid a \geq 2\};$

$; a \in [2, \infty)$

25. $\{n \mid n \geq 1\};$

$; n \in [1, \infty)$

27. $\{x \mid x < \tfrac{-32}{5}\};$

$; x \in (-\infty, \tfrac{-32}{5})$

29. { } **31.** $\{x \mid x \in \mathbb{R}\}$ **33.** $\{x \mid x \in \mathbb{R}\}$

35. $\{2\}; \{-3, -2, -1, 0, 1, 2, 3, 4, 6, 8\}$

37. $\{\}; \{-3, -2, -1, 0, 1, 2, 3, 4, 5, 6, 7\}$

39. $\{4, 6\}; \{2, 4, 5, 6, 7, 8\}$

41. $x \in (-\infty, -2) \cup (1, \infty);$

43. $x \in [-2, 5);$

45. no solution

47. $x \in (-\infty, \infty);$

49. $x \in [-5, 0];$

51. $x \in (\tfrac{-1}{3}, \tfrac{-1}{4});$

53. $x \in (-\infty, \infty);$

55. $x \in [-4, 1);$

57. $x \in [-1.4, 0.8];$

59. $x \in (-16, 8);$

61. $m \in (-\infty, 0) \cup (0, \infty)$ **63.** $y \in (-\infty, -7) \cup (-7, \infty)$

65. $a \in (-\infty, \tfrac{1}{2}) \cup (\tfrac{1}{2}, \infty)$ **67.** $x \in (-\infty, 4) \cup (4, \infty)$

69. $x \in [2, \infty)$ **71.** $n \in [4, \infty)$ **73.** $b \in [\tfrac{4}{3}, \infty)$ **75.** $y \in (-\infty, 2]$

77. a. $W = \dfrac{BH^2}{704}$ **b.** $W < 177.34$ lb **79.** $x \geq 81\%$ **81.** $b \geq \$2000$

83. $0 < w < 7.5$ m **85.** $7.2° < C < 29.4°$ **87.** $h > 6$

89. Answers may vary. **91.** $<$ **93.** $<$ **95.** $<$ **97.** $>$ **99.** $2n - 8$

101. $\tfrac{17}{18}x - 5$

Exercises 1.3, pp. 29–31

1. reverse **3.** $-7; 7$ **5.** no solution; answers will vary. **7.** $\{-4, 6\}$

9. $\{2, -12\}$ **11.** $\{-3.35, 0.85\}$ **13.** $\left\{-\dfrac{8}{7}, 2\right\}$ **15.** $\left\{-\dfrac{1}{2}, \dfrac{1}{2}\right\}$

17. { } **19.** $\{-10, -6\}$ **21.** $\{3.5, 11.5\}$ **23.** $\{-1.6, 1.6\}$

25. $[-5, 9]$ **27.** $\varnothing$ **29.** $\left(-1, \dfrac{3}{5}\right)$ **31.** $(-5, -3)$ **33.** $\left[\dfrac{8}{3}, \dfrac{14}{3}\right]$

35. $\varnothing$ **37.** $\left[-\dfrac{7}{4}, 0\right]$ **39.** $(-\infty, -10) \cup (4, \infty)$

41. $(-\infty, -3] \cup [3, \infty)$ **43.** $\left(-\infty, -\dfrac{7}{3}\right] \cup \left[\dfrac{7}{3}, \infty\right)$

45. $\left(-\infty, \dfrac{3}{7}\right] \cup [1, \infty)$ **47.** $(-\infty, \infty)$ **49.** $(-\infty, 0) \cup (5, \infty)$

51. $(-\infty, -0.75] \cup [3.25, \infty)$ **53.** $\left(-\infty, -\dfrac{7}{15}\right] \cup (1, \infty)$

55. $45 \leq d \leq 51$ in. **57.** in feet: $[32,500, 37,600]$; yes

59. in feet: $d < 210$ or $d > 578$ **61. a.** $|s - 37.58| \leq 3.35$

b. $[34.23, 40.93]$ **63. a.** $|s - 125| \leq 23$ **b.** $[102, 148]$

65. a. $|d - 42.7| < 0.03$ **b.** $|d - 73.78| < 1.01$

c. $|d - 57.150| < 0.127$ **d.** $|d - 2171.05| < 12.05$

e. golf: $t \approx 0.0014$ **67. a.** $x = 4$ **b.** $\left[\dfrac{4}{3}, 4\right]$ **c.** $x = 0$ **d.** $\left(-\infty, \dfrac{3}{5}\right]$

e. { }   **69.** $3x(2x + 5)(3x - 4)$ **71.** $\dfrac{-3 + \sqrt{3}}{6} \approx -0.21$

Mid-Chapter Check, pp. 31–32

1. a. $r = -9$ **b.** $x = -6$ **c.** identity; $m \in \mathbb{R}$ **d.** $y = \dfrac{50}{13}$

e. contradiction: { }   **f.** $x = 5.5$ **2.** $v_0 = \dfrac{H + 16t^2}{t}$

3. $x = \sqrt{\dfrac{S}{\pi(2 + y)}}$

4. a. $x \geq 1$ or $x \leq -2$

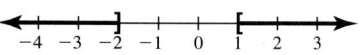

b. $16 < x \leq 19$

5. a. $x \in \left(-\infty, \dfrac{5}{2}\right) \cup \left(\dfrac{5}{2}, \infty\right)$ **b.** $x \in \left(-\infty, \dfrac{17}{6}\right]$

6. a. $\{-4, 14\}$ **b.** { }   **7. a.** $q \in (-8, 0)$ **b.** $\{-6\}$

8. a. $d \in (-\infty, 0] \cup [4, \infty)$ **b.** $y \in \left(-\infty, -\dfrac{19}{2}\right) \cup \left(\dfrac{23}{2}, \infty\right)$

c. $k \in (-\infty, \infty)$ **9.** 1 hr, 20 min   **10.** $w \in [8, 26]$; yes

Reinforcing Basic Concepts, p. 32

Exercise 1: $x = -3$ or $x = 7$
Exercise 2: $x \in [-5, 3]$
Exercise 3: $x \in (-\infty, -1] \cup [4, \infty)$

Exercises 1.4, pp. 39–42

1. $3 - 2i$ **3.** $2; 3\sqrt{2}$ **5.** (b) is correct.   **7. a.** $4i$ **b.** $7i$ **c.** $3\sqrt{3}$
d. $6\sqrt{2}$ **9. a.** $-3i\sqrt{2}$ **b.** $-5i\sqrt{2}$ **c.** $15i$ **d.** $6i$ **11. a.** $i\sqrt{19}$
b. $i\sqrt{31}$ **c.** $\dfrac{2\sqrt{3}}{5}i$ **d.** $\dfrac{3\sqrt{2}}{8}i$ **13. a.** $1 + i; a = 1, b = 1$
b. $2 + \sqrt{3}i; a = 2, b = \sqrt{3}$ **15. a.** $4 + 2i; a = 4, b = 2$
b. $2 - \sqrt{2}i; a = 2, b = -\sqrt{2}$ **17. a.** $5 + 0i; a = 5, b = 0$
b. $0 + 3i; a = 0, b = 3$ **19. a.** $18i; a = 0, b = 18$
b. $\dfrac{\sqrt{2}}{2}i; a = 0, b = \dfrac{\sqrt{2}}{2}$ **21. a.** $4 + 5\sqrt{2}i; a = 4, b = 5\sqrt{2}$
b. $-5 + 3\sqrt{3}i; a = -5, b = 3\sqrt{3}$

23. a. $\dfrac{7}{4} + \dfrac{7\sqrt{2}}{8}i; a = \dfrac{7}{4}, b = \dfrac{7\sqrt{2}}{8}$ **b.** $\dfrac{1}{2} + \dfrac{\sqrt{10}}{2}i; a = \dfrac{1}{2}, b = \dfrac{\sqrt{10}}{2}$
25. a. $19 + i$ **b.** $2 - 4i$ **c.** $9 + 10\sqrt{3}i$ **27. a.** $-3 + 2i$ **b.** 8
c. $2 - 8i$ **29. a.** $2.7 + 0.2i$ **b.** $15 + \dfrac{1}{12}i$ **c.** $-2 - \dfrac{1}{8}i$
31. a. 15   **b.** 16   **33. a.** $-21 - 35i$ **b.** $-42 - 18i$
35. a. $-12 - 5i$ **b.** $1 + 5i$ **37. a.** $4 - 5i; 41$ **b.** $3 + i\sqrt{2}; 11$
39. a. $-7i; 49$ **b.** $\dfrac{1}{2} + \dfrac{2}{3}i; \dfrac{25}{36}$ **41. a.** 41   **b.** 74   **43. a.** 11   **b.** $\dfrac{17}{36}$
45. a. $-5 + 12i$ **b.** $-7 - 24i$ **47. a.** $-21 - 20i$ **b.** $7 + 6\sqrt{2}i$
49. no   **51.** yes   **53.** yes   **55.** yes   **57.** yes   **59.** Answers will vary.

61. a. 1   **b.** -1 **c.** $-i$ **d.** i **63. a.** $\dfrac{2}{7}i$ **b.** $\dfrac{-4}{5}i$

65. a. $\dfrac{21}{13} - \dfrac{14}{13}i$ **b.** $\dfrac{-10}{13} - \dfrac{15}{13}i$ **67. a.** $1 - \dfrac{3}{4}i$ **b.** $-1 - \dfrac{2}{3}i$
69. a. $\sqrt{13}$ **b.** 5   **c.** $\sqrt{11}$ **71.** $A + B = 10$ $AB = 40$
73. $7 - 5i \, \Omega$ **75.** $25 + 5i$ V   **77.** $\dfrac{7}{4} + i \, \Omega$ **79. a.** $(x + 6i)(x - 6i)$
b. $(m + i\sqrt{3})(m - i\sqrt{3})$ **c.** $(n + 2i\sqrt{3})(n - 2i\sqrt{3})$
d. $(2x + 7i)(2x - 7i)$ **81.** $-8 - 6i$ **83. a.** $P = 4s; A = s^2$

b. $P = 2L + 2W; A = LW$ **c.** $P = a + b + c; A = \dfrac{bh}{2}$
d. $C = \pi d; A = \pi r^2$ **85.** John

Exercises 1.5, pp. 52–56

1. descending; 0   **3.** quadratic; 1   **5.** GCF factoring: $x = 0, x = \dfrac{5}{4}$
7. $a = -1; b = 2; c = -15$ **9.** not quadratic
11. $a = \dfrac{1}{4}; b = -6; c = 0$ **13.** $a = 2; b = 0; c = 7$ **15.** not quadratic
17. $a = 1; b = -1; c = -5$ **19.** $x = 5$ or $x = -3$ **21.** $m = 4$
23. $p = 0$ or $p = 2$ **25.** $h = 0$ or $h = \dfrac{-1}{2}$ **27.** $a = 3$ or $a = -3$
29. $g = -9$ **31.** $m = -5$ or $m = -3$ or $m = 3$ **33.** $c = -3$ or $c = 15$
35. $r = 8$ or $r = -3$ **37.** $t = -13$ or $t = 2$ **39.** $x = 5$ or $x = -3$
41. $w = -\dfrac{1}{2}$ or $w = 3$ **43.** $m = \pm 4$ **45.** $y = \pm 2\sqrt{7}; y \approx \pm 5.29$
47. no real solutions   **49.** $x = \pm \dfrac{\sqrt{21}}{4}; x \approx \pm 1.15$ **51.** $n = 9; n = -3$
53. $w = -5 \pm \sqrt{3}; w \approx -3.27$ or $w \approx -6.73$ **55.** no real solutions
57. $m = 2 \pm \dfrac{3\sqrt{2}}{7}; m \approx 2.61$ or $m \approx 1.39$ **59.** $9; (x + 3)^2$
61. $\dfrac{9}{4}; \left(n + \dfrac{3}{2}\right)^2$ **63.** $\dfrac{1}{9}; \left(p + \dfrac{1}{3}\right)^2$ **65.** $x = -1; x = -5$
67. $p = 3 \pm \sqrt{6}; p \approx 5.45$ or $p \approx 0.55$
69. $p = -3 \pm \sqrt{5}; p \approx -0.76$ or $p \approx -5.24$
71. $m = \dfrac{-3}{2} \pm \dfrac{\sqrt{13}}{2}; m \approx 0.30$ or $m \approx -3.30$
73. $n = \dfrac{5}{2} \pm \dfrac{3\sqrt{5}}{2}; n \approx 5.85$ or $n \approx -0.85$
75. $x = \dfrac{1}{2}$ or $x = -4$ **77.** $n = 3$ or $n = \dfrac{-3}{2}$
79. $p = \dfrac{3}{8} \pm \dfrac{\sqrt{41}}{8}; p \approx 1.18$ or $p \approx -0.43$
81. $m = \dfrac{7}{2} \pm \dfrac{\sqrt{33}}{2}; m \approx 6.37$ or $m \approx 0.63$
83. $x = 6$ or $x = -3$ **85.** $m = \pm \dfrac{5}{2}$
87. $n = \dfrac{2 \pm \sqrt{5}}{2}; n \approx 2.12$ or $n \approx -0.12$ **89.** $w = \dfrac{2}{3}$ or $w = \dfrac{-1}{2}$
91. $m = \dfrac{3}{2} \pm \dfrac{\sqrt{6}}{2}i; m \approx 1.5 \pm 1.12i$ **93.** $n = \pm \dfrac{3}{2}$
95. $w = \dfrac{-4}{5}$ or $w = 2$ **97.** $a = \dfrac{1}{6} \pm \dfrac{\sqrt{23}}{6}i; a \approx 0.1\overline{6} \pm 0.80i$
99. $p = \dfrac{3 \pm 2\sqrt{6}}{5}; p \approx 1.58$ or $p \approx -0.38$
101. $w = \dfrac{1 \pm \sqrt{21}}{10}; w \approx 0.56$ or $w \approx -0.36$
103. $a = \dfrac{3}{4} \pm \dfrac{\sqrt{31}}{4}i; a \approx 0.75 \pm 1.39i$
105. $p = 1 \pm \dfrac{3\sqrt{2}}{2}i; p \approx 1 \pm 2.12i$
107. $w = \dfrac{-1}{3} \pm \dfrac{\sqrt{2}}{3}i; w \approx 0.14$ or $w \approx -0.80$
109. $a = \dfrac{-6 \pm 3\sqrt{2}}{2}; a \approx -0.88$ or $a \approx -5.12$
111. $p = \dfrac{4 \pm \sqrt{394}}{6}; p \approx 3.97$ or $p \approx -2.64$
113. two rational; factorable   **115.** two complex   **117.** two rational;
factorable   **119.** two complex   **121.** two irrational   **123.** one repeated;
factorable   **125.** $x = \dfrac{3}{2} \pm \dfrac{1}{2}i$ **127.** $x = -\dfrac{1}{2} \pm \dfrac{i\sqrt{3}}{2}$

129. $x = \dfrac{5}{4} \pm \dfrac{3i\sqrt{7}}{4}$ **131.** $t = \dfrac{v \pm \sqrt{v^2 - 64h}}{32}$

133. $t = \dfrac{6 + \sqrt{138}}{2}$ sec, $t \approx 8.87$ sec   **135.** 30,000 ovens
137. a. $P = -x^2 + 120x - 2000$ **b.** 10,000   **139.** $t = 2.5$ sec, 6.5 sec
141. $x \approx 13.5$, or the year 2008   **143.** 36 ft, 78 ft
145. a. $7x^2 + 6x - 16 = 0$ **b.** $6x^2 + 5x - 14 = 0$

c. $5x^2 - x - 6 = 0$ **147.** $x = -2i; x = 5i$ **149.** $x = \dfrac{-3}{4}i; x = 2i$

151. $x = -1 - i; x = -13 - i$ **153. a.** $P = 2L + 2W, A = LW$
b. $P = 2\pi r, A = \pi r^2$ **c.** $A = \dfrac{1}{2}h(b_1 + b_2), P = c + h + b_1 + b_2$
d. $A = \dfrac{1}{2}bh, P = a + b + c$ **155.** 700 \$30 tickets; 200 \$20 tickets

Exercises 1.6, pp. 65–70

1. excluded   **3.** extraneous   **5.** Answers will vary.
7. $x = -2, x = 0, x = 11$ **9.** $x = -3, x = 0, x = \dfrac{2}{3}$
11. $x = -\dfrac{3}{2}, x = 0, x = 3$ **13.** $x = 0, x = 2, x = -1 \pm i\sqrt{3}$
15. $x = \pm 2, x = 5$ **17.** $x = 3, x = \pm 2i$ **19.** $x = \pm \sqrt{5}, x = 6$
21. $x = 0, x = 7, x = \pm 2i$ **23.** $x = \pm 3, x = \pm 3i$
25. $x = \pm 4; x = \pm 4i$ **27.** $x = \pm \sqrt{2}, x = \pm 1, \pm i$
29. $x = \pm 1, x = 2, x = -1 \pm i\sqrt{3}$
31. $x = -\dfrac{1}{2} \pm \dfrac{i\sqrt{3}}{2}, x = \dfrac{1}{2} \pm \dfrac{i\sqrt{3}}{2}, x = \pm 1$ **33.** $x = 1$ **35.** $a = \dfrac{3}{2}$
37. $y = 12$ **39.** $x = 3; x = 7$ is extraneous   **41.** $n = 7$

43. $a = -1, a = -8$ **45.** $f = \dfrac{f_1 f_2}{f_1 + f_2}$ **47.** $r = \dfrac{E - IR}{I}$ or $\dfrac{E}{I} - R$

49. $h = \dfrac{3V}{\pi r^2}$ **51.** $r^3 = \dfrac{3V}{4\pi}$ **53. a.** $x = \frac{14}{3}$

b. $x = 8, x = 1$ is extraneous **55. a.** $m = 3$ **b.** $x = 5$ **c.** $m = -64$
d. $x = -16$ **57. a.** $x = 25$ **b.** $x = 7; x = -2$ is extraneous
c. $x = 2, x = 18$ **d.** $x = 6; x = 0$ is extraneous **59.** $x = -32$
61. $x = 9$ **63.** $x = -32, x = 22$ **65.** $x = -27, 125$
67. $x = \pm 5, x = \pm i$ **69.** $x = \pm 1, \pm 2$ **71.** $x = -1, \frac{1}{4}$
73. $x = \pm \frac{1}{3}, \pm \frac{1}{2}$ **75.** $x = -4, 45$

77. $x = -6; x = \frac{-74}{9}$ is extraneous **79. a.** $h = \sqrt{\left(\dfrac{S}{\pi r}\right)^2 - r^2}$

b. $S = 12\pi\sqrt{34}\ \text{m}^2$ **81.** $x = \pm 3, x = -2$ **83.** $x = 2, 4, 6$ or
$x = -2, 0, 2$ **85.** 11 in. by 13 in. **87.** $r = 3$ m; $r = 0$ and $r = 12$ m
do not fit the context **89.** either \$50 or \$30 **91. a.** 32 ft, $(h = -32)$
b. 11 sec **c.** pebble is at canyon's rim **93.** 12 min **95.** $v = 6$ mph
97. $P \approx 52.1\%$ **99. a.** 36 million mi **b.** 67 million mi
c. 93 million mi **d.** 142 million mi **e.** 484 million mi
f. 887 million mi **101.** The constant "3" was not multiplied by the LCD,
$3x(x + 3) - 8x = x + 3; x = -1, 1$ **103.** $x \in [1, 2) \cup (2, \infty)$
105. a. $x = -5, -3, 5, 7$ **b.** $x = -2, -1, 6, 7$ **c.** $x = -2, 1, 3$
d. $x = -4, -2, 3$ **e.** $x = -1, 1, 7$ **f.** $x = -1, 1, 2, 7$
107. $2\sqrt{11}$ cm **109.** $-1 < x < 5$;

$$\xleftarrow{\quad} \overset{-2\ -1\ \ 0\ \ 1\ \ 2\ \ 3\ \ 4\ \ 5\ \ 6}{\underset{(\qquad\qquad\qquad)}{\rule{0pt}{0pt}}} \xrightarrow{\quad}$$

Summary and Concept Review, pp. 70–74

1. a. yes **b.** yes **c.** yes **2.** $b = 6$ **3.** $n = 4$ **4.** $m = -1$

5. $x = \frac{1}{6}$ **6.** no solution **7.** $g = 10$ **8.** $h = \dfrac{V}{\pi r^2}$ **9.** $L = \dfrac{P - 2W}{2}$

10. $x = \dfrac{c - b}{a}$ **11.** $y = \frac{2}{3}x - 2$ **12.** 8 gal **13.** $12 + \frac{9}{8}\pi\ \text{ft}^2 \approx 15.5\ \text{ft}^2$
14. $\frac{2}{3}$ hr $= 40$ min **15.** $a \geq 35$ **16.** $a < 2$ **17.** $s \leq 65$
18. $c \geq 1200$ **19.** $(5, \infty)$ **20.** $(-10, \infty)$ **21.** $(-\infty, 2]$
22. $(-9, 9]$ **23.** $(-6, \infty)$ **24.** $\left(-\infty, \frac{-8}{5}\right) \cup \left(\frac{23}{10}, \infty\right)$
25. a. $(-\infty, 3) \cup (3, \infty)$ **b.** $\left(-\infty, \frac{3}{2}\right) \cup \left(\frac{3}{2}, \infty\right)$ **c.** $[-5, \infty)$
d. $(-\infty, 6]$ **26.** $x \geq 96\%$ **27.** $\{-4, 10\}$ **28.** $\{-7, 3\}$ **29.** $\{-5, 8\}$
30. $\{-4, -1\}$ **31.** $(-\infty, -6) \cup (2, \infty)$ **32.** $[4, 32]$ **33.** $\{\ \}$ **34.** $\{\ \}$
35. $(-\infty, \infty)$ **36.** $[-2, 6]$ **37.** $(-\infty, -2] \cup [\frac{10}{3}, \infty)$
38. a. $|r - 2.5| \leq 1.7$ **b.** highest: 4.2 in., lowest: 0.8 in. **39.** $6\sqrt{2}i$
40. $24\sqrt{3}i$ **41.** $-2 + \sqrt{2}i$ **42.** $3\sqrt{2}i$ **43.** i **44.** $21 + 20i$
45. $-2 + i$ **46.** $-5 + 7i$ **47.** 13 **48.** $-20 - 12i$
49. $(5i)^2 - 9 = -34 \qquad (-5i)^2 - 9 = -34$
$\ 25i^2 - 9 = -34 \qquad\ 25i^2 - 9 = -34$
$\ -25 - 9 = -34\checkmark \qquad -25 - 9 = -34\checkmark$
50. $(2 + i\sqrt{5})^2 - 4(2 + i\sqrt{5}) + 9 = 0$
$\ 4 + 4i\sqrt{5} + 5i^2 - 8 - 4i\sqrt{5} + 9 = 0$
$\ 5 + (-5) = 0\checkmark$
$\ (2 - i\sqrt{5})^2 - 4(2 - i\sqrt{5}) + 9 = 0$
$\ 4 - 4i\sqrt{5} + 5i^2 - 8 + 4i\sqrt{5} + 9 = 0$
$\ 5 + (-5) = 0\checkmark$
51. a. $2x^2 + 3 = 0; a = 2, b = 0, c = 3$ **b.** not quadratic
c. $x^2 - 8x - 99 = 0; a = 1, b = -8, c = -99$
d. $x^2 + 16 = 0; a = 1, b = 0, c = 16$ **52. a.** $x = 5$ or $x = -2$
b. $x = -5$ or $x = 5$ **c.** $x = -\frac{5}{3}$ or $x = 3$ **d.** $x = -2$ or $x = 2$ or $x = 3$
53. a. $x = \pm 3$ **b.** $x = 2 \pm \sqrt{5}$ **c.** $x = \pm\sqrt{5}i$ **d.** $x = \pm 5$
54. a. $x = 3$ or $x = -5$ **b.** $x = -8$ or $x = 2$
c. $x = 1 \pm \frac{\sqrt{10}}{2}; x \approx 2.58$ or $x \approx -0.58$ **d.** $x = 2$ or $x = \frac{1}{3}$
55. a. $x = 2 \pm \sqrt{5}i; x \approx 2 \pm 2.24i$ **b.** $x = \frac{3 \pm \sqrt{2}}{2}; x \approx 2.21$ or
$x \approx 0.79$ **c.** $x = \frac{3}{2} \pm \frac{1}{2}i$ **56. a.** 1.3 sec **b.** 4.67 sec **c.** 6 sec
57. a. 0.8 sec **b.** 3.2 sec **c.** 5 sec **58.** \$3.75; 3000 **59.** 6 hr
60. $x = \pm\sqrt{3}, x = 7$ **61.** $x = -2, x = 0, x = \frac{1}{3}$
62. $x = 0, x = 2, x = -1 \pm \sqrt{3}i$ **63.** $x = \pm\frac{1}{2}, x = \pm\frac{1}{2}i$ **64.** $x = \frac{-1}{2}$
65. $h = -\frac{5}{3}, h = 2$ **66.** $n = 13; n = -2$ is extraneous
67. $x = -3; x = 3$ **68.** $x = -4; x = 5$

69. $x = -1; x = 7$ is extraneous **70.** $x = \frac{5}{2}$ **71.** $x = -5.8; x = 5$
72. $x = -2, x = -1, x = 4, x = 5$
73. $x = -3, x = 3, x = -i\sqrt{2}, x = i\sqrt{2}$ **74. a.** 12,000 kilocalories
b. 810 kg **75.** width, 6 in.; length, 9 in. **76.** 1 sec; 244 ft; 8 sec
77. \$24 per load; \$42 per load

Mixed Review, p. 75

1. a. $x \in (8, \infty)$ **b.** $x \in \left(-\infty, \frac{-4}{3}\right) \cup \left(\frac{-4}{3}, \infty\right)$ **3. a.** $x = 2, x = \pm 5i$
b. $x = 0, x = -5, x = \frac{5}{2} \pm \frac{5i\sqrt{3}}{2}$ **c.** $x = -\frac{7}{3}, x = \frac{5}{3}$
d. $(-\infty, 3] \cup [27, \infty)$ **e.** $v = \pm 27$ **f.** $x = 80$ **5.** $y = \frac{-3}{4}x - 3$
7. a. $x = -2$ **b.** $n = 5$ **9.** $x = 7, 11$ **11.** $x = -\sqrt{6}, \sqrt{6}$
13. $x = \frac{4}{5}$ **15.** $x = \pm\sqrt{5}, \pm i\sqrt{5}$ **17. a.** $v = 6, 2$ is extraneous
b. $x = -5; x = 4$ **c.** $x = 2; x = 18$ is extraneous **19.** 6'10"

Practice Test, pp. 75–76

1. a. $x = 27$ **b.** $x = 2$ **c.** $C = \frac{P}{1 + k}$ **d.** $x = -4, x = -1$
2. 30 gal **3. a.** $x > -30$ **b.** $-5 \leq x < 4$ **c.** $x \in \mathbb{R}$
d. $x = 2, x = 4$ **e.** $x < -4$ or $x > 2$ **4.** $S \geq 177$
5. $z = -3, z = 10$ **6.** $x = \pm 5i$ **7.** $x = 1 \pm i\sqrt{3}$
8. $x = \pm 1, x = \pm 4$ **9.** $x = \frac{2}{3}, x = 6$ **10.** $x = -2, x = \pm\frac{3}{2}$
11. $x = 6, x = -2$ is extraneous **12.** $x = -\frac{3}{2}, x = 2$
13. $x = 16, x = 4$ is extraneous **14.** $x = -11, x = 5$
15. a. \$4.50 per tin **b.** 90 tins **16. a.** $t = 5$ (May) **b.** $t = 9$ (Sept.)
c. July; \$3000 more **17.** $-\frac{4}{3} \pm \frac{i\sqrt{5}}{3}$ **18.** $-i$ **19. a.** 1 **b.** $i\sqrt{3}$
c. 1 **20.** $-\frac{3}{2} + \frac{3}{2}i$ **21.** 34
22. $(2 - 3i)^2 - 4(2 - 3i) + 13 = 0$
$\ -5 - 12i - 8 + 12i + 13 = 0$
$\qquad\qquad\qquad\qquad 0 = 0\checkmark$
23. a. $x = 5 \pm \frac{\sqrt{2}}{2}$ **b.** $x = \frac{5}{4} \pm \frac{i\sqrt{7}}{4}$ **24. a.** $x = \frac{3 \pm \sqrt{3}}{3}$
b. $x = 1 \pm 3i$ **25. a.** $F \approx 64.8$ g **b.** $W \approx 256$ g

Strengthening Core Skills, pp. 77–78

Exercise 1: $\dfrac{7}{2} + (-1) = \dfrac{5}{2} = -\dfrac{b}{a}\checkmark \quad \dfrac{7}{2} \cdot (-1) = \dfrac{-7}{2} = \dfrac{c}{a}\checkmark$

Exercise 2: $\dfrac{2 + 3\sqrt{2}}{2} + \dfrac{2 - 3\sqrt{2}}{2} = \dfrac{4}{2} = \dfrac{-b}{a}\checkmark \quad \dfrac{2 + 3\sqrt{2}}{2} \cdot \dfrac{2 - 3\sqrt{2}}{2}$

$= \dfrac{-14}{4} = \dfrac{-7}{2} = \dfrac{c}{a}\checkmark$

Exercise 3: $(5 + 2\sqrt{3}i) + (5 - 2\sqrt{3}i) = 10 = \dfrac{-b}{a}\checkmark$

$(5 + 2\sqrt{3}i)(5 - 2\sqrt{3}i) = 25 + 12 = 37 = \dfrac{c}{a}\checkmark$

Connections to Calculus Exercises, p. 82

1. $x = \dfrac{3}{\sqrt{2}}, 2x = \dfrac{6}{\sqrt{2}} \approx 4.24$ in. **3.** $t = \dfrac{6}{5}$ sec, $t = \dfrac{3}{2}$ sec

5. If the difference between x and $\dfrac{3}{2}$ is very small, then the difference

between $h(x)$ and 6 is very small: *if* $\left|x - \dfrac{3}{2}\right| < \delta$, *then* $|h(x) - 6| < \epsilon$.

7. If the difference between x and 2 is very small, then the difference
between $w(x)$ and 14 is very small: *if* $|x - 2| < \delta$, *then* $|w(x) - 14| < \epsilon$.

CHAPTER 2
Exercises 2.1, pp. 93–96

1. first, second **3.** radius, center **5.** Answers will vary.
7.

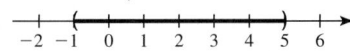

Year in college → GPA

$D = \{1, 2, 3, 4, 5\}$
$R = \{2.75, 3.00, 3.25, 3.50, 3.75\}$

9. $D = \{1, 3, 5, 7, 9\}; R = \{2, 4, 6, 8, 10\}$
11. $D = \{4, -1, 2, -3\}; R = \{0, 5, 4, 2, 3\}$

13.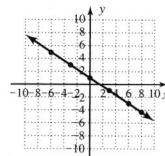

15.

x	y
-6	5
-3	3
0	1
3	-1
6	-3
8	$\frac{-13}{3}$

x	y
-2	0
0	$2, -2$
1	$3, -3$
3	$5, -5$
6	$8, -8$
7	$9, -9$

17.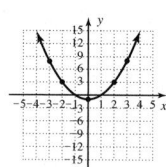

19.

x	y
-3	8
-2	3
0	-1
2	3
3	8
4	15

x	y
-4	3
-3	4
0	5
2	$\sqrt{21}$
3	4
4	3

21.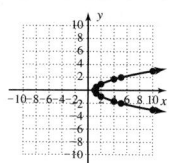

23.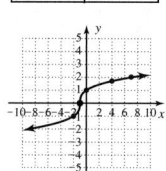

x	y
10	$3, -3$
5	$2, -2$
4	$\sqrt{3}, -\sqrt{3}$
2	$1, -1$
1.25	$0.5, -0.5$
1	0

x	y
-9	-2
-2	-1
-1	0
0	1
4	$\sqrt[3]{5}$
7	2

25. $(3, 1)$ **27.** $(-0.7, -0.3)$ **29.** $\left(\frac{1}{20}, \frac{1}{24}\right)$ **31.** $(0, -1)$
33. $(-1, 0)$ **35.** $2\sqrt{34}$ **37.** 10 **39.** not a right triangle
41. not a right triangle   **43.** right triangle
45. $x^2 + y^2 = 9$ **47.** $(x - 5)^2 + y^2 = 3$

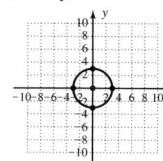   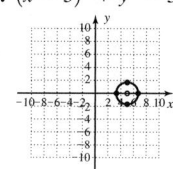

49. $(x - 4)^2 + (y + 3)^2 = 4$ **51.** $(x + 7)^2 + (y + 4)^2 = 7$

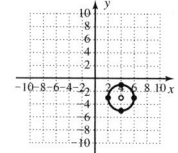   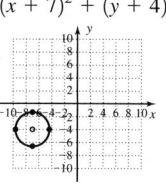

53. $(x - 1)^2 + (y + 2)^2 = 9$ **55.** $(x - 4)^2 + (y - 5)^2 = 12$

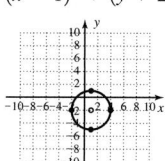   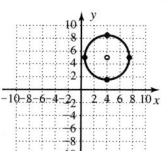

57. $(x - 7)^2 + (y - 1)^2 = 100$ **59.** $(x - 3)^2 + (y - 4)^2 = 41$

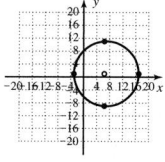   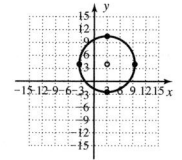

61. $(x - 5)^2 + (y - 4)^2 = 9$ **63.** $(2, 3), r = 2, x \in [0, 4], y \in [1, 5]$

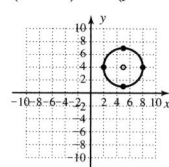   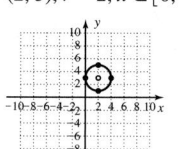

65. $(-1, 2), r = 2\sqrt{3}, x \in [-1 - 2\sqrt{3}, -1 + 2\sqrt{3}],$
$y \in [2 - 2\sqrt{3}, 2 + 2\sqrt{3}]$

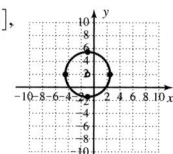

67. $(-4, 0), r = 9, x \in [-13, 5], y \in [-9, 9]$

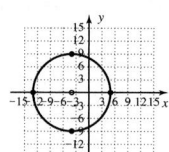

69. $(x - 5)^2 + (y - 6)^2 = 57, (5, 6), r = \sqrt{57}$

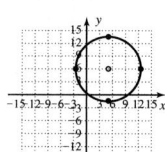

71. $(x - 5)^2 + (y + 2)^2 = 25, (5, -2), r = 5$

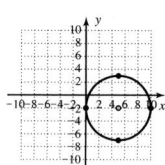

73. $x^2 + (y + 3)^2 = 14, (0, -3), r = \sqrt{14}$

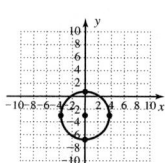

75. $(x + 2)^2 + (y + 5)^2 = 11, (-2, -5), r = \sqrt{11}$

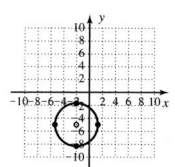

77. $(x + 7)^2 + y^2 = 37, (-7, 0), r = \sqrt{37}$

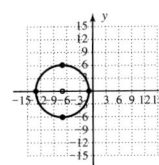

79. $(x - 3)^2 + (y + 5)^2 = 32, (3, -5), r = 4\sqrt{2}$

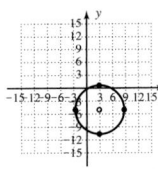

81. a. $(1, 71.5), (2, 84), (3, 96.5), (5, 121.5), (7, 146.5);$ yes **b.** $159
c. 2011 **d.**

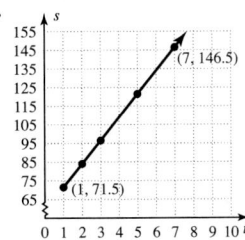

83. a. $(x - 5)^2 + (y - 12)^2 = 625$ **b.** no
85. Red: $(x - 2)^2 + (y - 2)^2 = 4$; Blue: $(x - 2)^2 + y^2 = 16$;
Area blue $= 12\pi$ units^2
87.

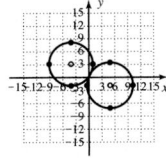

No, distance between centers is less than sum of radii.

89. Answers will vary.   **91. a.** center: $(6, -2); r = 0$ (degenerate case)
b. center: $(1, 4); r = 5$ **c.** $r^2 = -1$; degenerate case
93. a. 0   **b.** not possible   **c.** 0.3; many answers possible
d. not possible   **e.** not possible   **f.** $\sqrt{3}$; many answers possible
95. $n = 1$ is a solution, $n = -2$ is extraneous

Exercises 2.2, pp. 106–110

1. 0, 0   **3.** negative, downward   **5.** yes $m_1 \neq m_2$ no $m_1 \cdot m_2 \neq -1$

7.

x	y
-6	6
-3	4
0	2
3	0

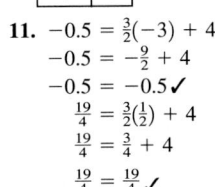

9.

x	y
-2	1
0	4
2	7
4	10

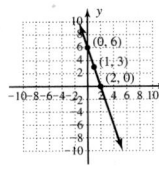

11. $-0.5 = \frac{3}{2}(-3) + 4$ **13.**
$-0.5 = -\frac{9}{2} + 4$
$-0.5 = -0.5 ✓$
$\frac{19}{4} = \frac{3}{2}(\frac{1}{2}) + 4$
$\frac{19}{4} = \frac{3}{4} + 4$
$\frac{19}{4} = \frac{19}{4} ✓$

15. **17.** **19.**

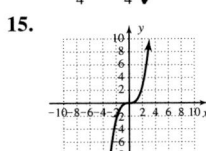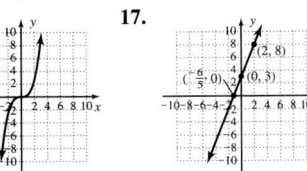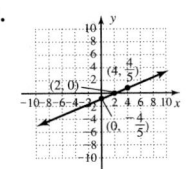

21. **23.** **25.**

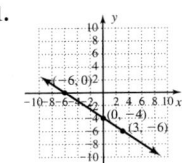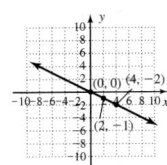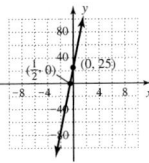

27. **29.** **31.**

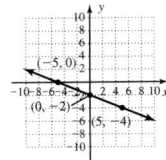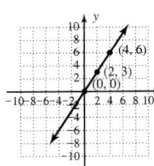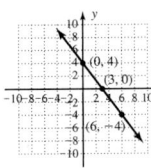

33. $m = 1;$
$(2, 4)$ and $(1, 3)$

35. $m = \frac{4}{3};$
$(7, -1)$ and $(1, -9)$

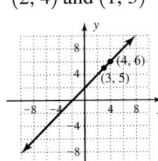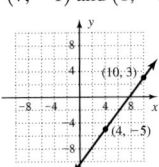

37. $m = \frac{-15}{4};$
$(1, -8)$ and $(-1, -\frac{1}{2})$

39. $m = \frac{-4}{7};$
$(-10, 10)$ and $(11, -2)$

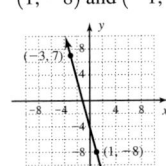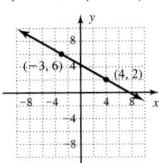

41. a. $m = 125$, cost increased $125,000 per 1000 sq ft   **b.** $375,000
43. a. $m = 22.5$, distance increases 22.5 mph   **b.** about 186 mi
45. a. $m = \frac{23}{6}$, a person weighs 23 lb more for each additional 6 in. in height   **b.** 3.8
47. In inches: $(0, -6)$ and $(576, -18)$: $m = \frac{-1}{48}$. The sewer line is 1 in. deeper for each 48 in. in length.
49. **51.**

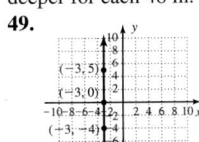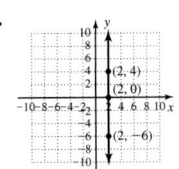

53. $L_1: x = 2; L_2: y = 4$; point of intersection $(2, 4)$
55. a. For any two points chosen $m = 0$, indicating there has been no increase or decrease in the number of supreme court justices.   **b.** For any two points chosen $m = \frac{1}{10}$, which indicates that over the last 5 decades, one nonwhite or nonfemale justice has been added to the court every 10 yr.   **57.** parallel

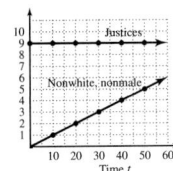

59. neither   **61.** parallel   **63.** not a right triangle
65. not a right triangle   **67.** right triangle
69. a. 76.4 yr   **b.** 2010   **71.** $v = -1250t + 8500$ **a.** $3500   **b.** 5 yr
73. $h = -3t + 300$ **a.** 273 in.   **b.** 20 months
75. Yes they will meet, the two roads are not parallel: $\frac{38}{12} \neq \frac{30}{9.5}$.
77. a. $3789   **b.** 2012   **79. a.** 23%   **b.** 2005
81. $a = -6$ **83. a.** 142   **b.** -83 **c.** 9   **d.** $\frac{27}{2}$
85. perimeter of a rectangle, volume of a rectangular prism, volume of a right circular cylinder, circumference of a circle   **87.** 2 hr

Exercises 2.3, pp. 118–122

1. $\frac{-7}{4}$; $(0, 3)$ **3.** 2.5   **5.** Answers will vary.
7. $y = \frac{-4}{5}x + 2$ **9.** $y = 2x + 7$ **11.** $y = \frac{-5}{3}x - 5$

x	y
−5	6
−2	$\frac{18}{5}$
0	2
1	$\frac{6}{5}$
3	$\frac{-2}{5}$

x	y
−5	−3
−2	3
0	7
1	9
3	13

x	y
−5	$\frac{10}{3}$
−2	$\frac{-5}{3}$
0	−5
1	$\frac{-20}{3}$
3	−10

13. $y = 2x - 3$: $2, -3$ **15.** $y = \frac{-5}{3}x - 7$: $\frac{-5}{3}, -7$
17. $y = \frac{-35}{6}x - 4$: $\frac{-35}{6}, -4$

19. 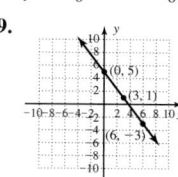   **21.** 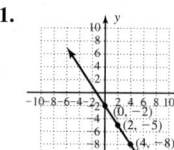   **23.**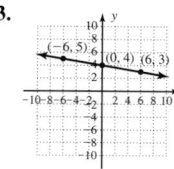

25. a. $\frac{-3}{4}$ **b.** $y = \frac{-3}{4}x + 3$ **c.** The coeff. of x is the slope and the constant is the y-intercept.
27. a. $\frac{2}{5}$ **b.** $y = \frac{2}{5}x - 2$ **c.** The coeff. of x is the slope and the constant is the y-intercept.
29. a. $\frac{4}{5}$ **b.** $y = \frac{4}{5}x + 3$ **c.** The coeff. of x is the slope and the constant is the y-intercept.   **31.** $y = \frac{-2}{3}x + 2$, $m = \frac{-2}{3}$, y-intercept $(0, 2)$
33. $y = \frac{-5}{4}x + 5$, $m = \frac{-5}{4}$, y-intercept $(0, 5)$ **35.** $y = \frac{1}{3}x$, $m = \frac{1}{3}$, y-intercept $(0, 0)$ **37.** $y = \frac{-3}{4}x + 3$, $m = \frac{-3}{4}$, y-intercept $(0, 3)$
39. $y = \frac{2}{3}x + 1$ **41.** $y = 3x + 3$ **43.** $y = 3x + 2$
45. $y = 250x + 500$ **47.** $y = \frac{75}{2}x + 150$ **49.** $y = 2x - 13$
51. $y = -\frac{3}{5}x + 4$ **53.** $y = \frac{2}{3}x - 5$ **55.**

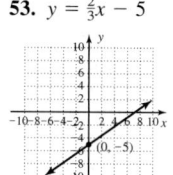   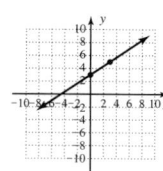   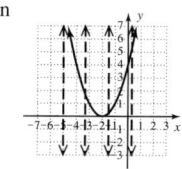

57.   **59.**   **61.**

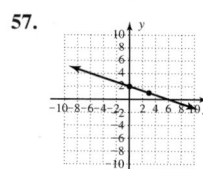   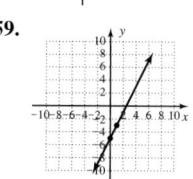   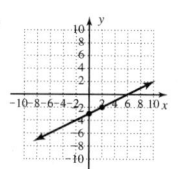

63. $y = \frac{2}{5}x + 4$ **65.** $y = \frac{-5}{3}x + 7$ **67.** $y = \frac{-12}{5}x - \frac{29}{5}$

69. $y = 5$ **71.** perpendicular   **73.** neither   **75.** neither
77. a. $y = \frac{-3}{4}x - \frac{5}{2}$ **b.** $y = \frac{4}{3}x - \frac{20}{3}$
79. a. $y = \frac{4}{9}x + \frac{31}{9}$ **b.** $y = \frac{-9}{4}x + \frac{3}{4}$
81. a. $y = \frac{-1}{2}x - 2$ **b.** $y = 2x - 2$
83. $y + 5 = 2(x - 2)$ **85.** $y + 4 = \frac{3}{8}(x - 3)$

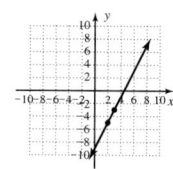   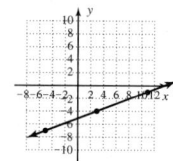

87. $y + 3.1 = 0.5(x - 1.8)$

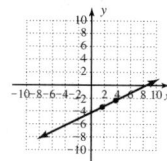

89. $y - 2 = \frac{6}{5}(x - 4)$; For each 5000 additional sales, income rises \$6000.
91. $y - 100 = \frac{-20}{1}(x - 0.5)$; For every hour of television, a student's final grade falls 20%.   **93.** $y - 10 = \frac{35}{2}(x - \frac{1}{2})$; Every 2 in. of rainfall increases the number of cattle raised per acre by 35.   **95.** C   **97.** A
99. B   **101.** D   **103.** $m = \frac{-a}{b}$, y-intercept $= \frac{c}{b}$ **a.** $m = \frac{-3}{4}$, y-intercept $(0, 2)$ **b.** $m = \frac{-2}{5}$, y-intercept $(0, -3)$ **c.** $m = \frac{5}{6}$, y-intercept $(0, 2)$
d. $m = \frac{5}{3}$, y-intercept $(0, 3)$ **105. a.** As the temperature increases 5°C, the velocity of sound waves increases 3 m/s. At a temperature of 0°C, the velocity is 331 m/s.   **b.** 343 m/s   **c.** 50°C   **107. a.** $V = \frac{20}{3}t + 150$ **b.** Every 3 yr the value of the coin increases by \$20; the initial value was \$150.
c. \$223.33   **d.** 15 years, in 2013   **e.** 3 yr   **109. a.** $N = 7t + 9$
b. Every 1 yr the number of homes with Internet access increases by 7 million.   **c.** 1993   **d.** 86 million   **e.** 13 yr   **f.** 2010
111. a. $P = 58,000t + 740,000$ **b.** Each year, the prison population increases by 58,000.   **c.** 1,726,000   **113.** Answers will vary.
115. (1) d   (2) a   (3) c   (4) b   (5) f   (6) h
117. $x = \frac{5 \pm 2\sqrt{13}}{3}$; $x \approx -0.74$ or $x \approx 4.07$ **119.** 113.10 yd^2

Exercises 2.4, pp. 132–137

1. first   **3.** range   **5.** Answers will vary.   **7.** function   **9.** Not a function. The Shaq is paired with two heights.   **11.** Not a function; 4 is paired with 2 and −5.   **13.** function   **15.** function   **17.** Not a function; −2 is paired with 3 and −4.   **19.** function   **21.** function   **23.** Not a function; 0 is paired with 4 and −4.   **25.** function   **27.** Not a function; 5 is paired with −1 and 1.   **29.** function

31. function   **33.** function

35. function, $x \in [-4, 5]$ $y \in [-2, 3]$ **37.** function, $x \in [-4, \infty)$ $y \in [-4, \infty)$ **39.** function, $x \in [-4, 4]$ $y \in [-5, -1]$ **41.** function, $x \in (-\infty, \infty)$ $y \in (-\infty, \infty)$ **43.** Not a function, $x \in [-3, 5]$ $y \in [-3, 3]$ **45.** Not a function, $x \in (-\infty, 3]$ $y \in (-\infty, \infty)$
47. $x \in (-\infty, 5) \cup (5, \infty)$ **49.** $x \in [\frac{-5}{3}, \infty)$
51. $x \in (-\infty, -5) \cup (-5, 5) \cup (5, \infty)$
53. $v \in (-\infty, -3\sqrt{2}) \cup (-3\sqrt{2}, 3\sqrt{2}) \cup (3\sqrt{2}, \infty)$
55. $x \in (-\infty, \infty)$ **57.** $x \in (-\infty, \infty)$ **59.** $x \in (-\infty, \infty)$
61. $x \in (-\infty, -2) \cup (-2, 5) \cup (5, \infty)$ **63.** $x \in [2, \frac{5}{2}) \cup (\frac{5}{2}, \infty)$
65. $x \in (2, \infty)$ **67.** $x \in (-4, \infty)$ **69.** $f(c + 1) = \frac{1}{2}c + \frac{7}{2}$
71. $f(c + 1) = 3c^2 + 2c - 1$ **73.** $h(a - 2) = \frac{3}{a-2}$
75. $h(a - 2) = 5\left(\frac{|a - 2|}{a - 2}\right)$
77. $g(4) = 8\pi$, $g\left(\frac{3}{2}\right) = 3\pi$, $g(2c) = 4\pi c$, $g(c + 3) = 2\pi(c + 3)$
79. $g(4) = 16\pi$, $g\left(\frac{3}{2}\right) = \frac{9}{4}\pi$, $g(2c) = 4\pi c^2$, $g(c + 3) = (c^2 + 6c + 9)\pi$
81. $p(5) = \sqrt{13}$, $p\left(\frac{3}{2}\right) = \sqrt{6}$, $p(3a) = \sqrt{6a + 3}$, $p(a - 1) = \sqrt{2a + 1}$
83. $p(5) = \frac{14}{5}$, $p\left(\frac{3}{2}\right) = \frac{7}{9}$, $p(3a) = \frac{27a^2 - 5}{9a^2}$, $p(a - 1) = \frac{3a^2 - 6a - 2}{a^2 - 2a + 1}$
85. a. $D = \{-1, 0, 1, 2, 3, 4, 5\}$ **b.** $R = \{-2, -1, 0, 1, 2, 3, 4\}$ **c.** 1
d. −1   **87. a.** $D = [-5, 5]$ **b.** $y \in [-3, 4]$ **c.** −2   **d.** −4 and 0
89. a. $D = [-3, \infty)$ **b.** $y \in (-\infty, 4]$ **c.** 2   **d.** −2 and 2
91. a. 186.5 lb   **b.** 37 lb   **93.** $A = \frac{1}{2}(8) + 22 - 1 = 25$ units^2
95. a. $N(g) = 2.5g$ **b.** $g \in [0, 5]$; $N \in [0, 12.5]$ **97. a.** $[0, \infty)$
b. 750π **c.** 800   **99. a.** $c(t) = 42.50t + 50$ **b.** \$156.25   **c.** 5 hr

d. $t \in [0, 10.6]$; $c \in [0, 500]$ **101. a.** Yes. Each x is paired with exactly one y.   **b.** 10 P.M.   **c.** 0.9 m   **d.** 7 P.M. and 1 A.M.
103. a. $\frac{\Delta \text{fertility}}{\Delta \text{time}} = \frac{-1}{20}$, negative, fertility is decreasing by one child every 20 yr   **b.** 1940 to 1950: $\frac{\Delta f}{\Delta t} = \frac{0.8}{10}$; positive, fertility is increasing by less than one child every 10 yr   **c.** 1940 to 1950: $\frac{\Delta f}{\Delta t} = \frac{0.8}{10}$; 1980 to 1990: $\frac{\Delta f}{\Delta t} = \frac{0.2}{10}$, the fertility rate was increasing four times as fast from 1940 to 1950.
105. negative outputs become positive

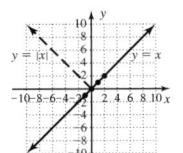 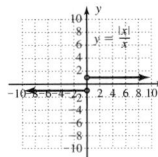

107. a. $x \in (-\infty, -2) \cup (2, \infty)$; $x = \frac{2y+3}{1-y}$; $y \in (-\infty, 1) \cup (1, \infty)$
b. $x \in \mathbb{R}$ $x = \pm\sqrt{y+3}$; $y \in [-3, \infty)$ **109. a.** $19\sqrt{6}$ **b.** 1
111. a. $(x-3)(x-5)(x+5)$ **b.** $(2x+3)(x-8)$
c. $(2x-5)(4x^2+10x+25)$

Mid-Chapter Check, p. 137

1.

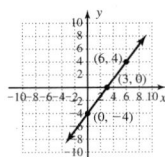

2. $\frac{-18}{7}$ **3.** positive, loss is decreasing (profit is increasing); $m = \frac{3}{2}$, yes; $\frac{1.5}{1}$, each year Data.com's loss decreases by 1.5 million.
4.

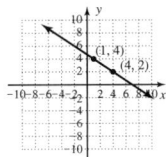

$y = \frac{3}{2}x + \frac{5}{2}$

5. $x = -3$; no; input -3 is paired with more than one output.
6. $y = \frac{-4}{3}x + 4$; yes   **7. a.** 0   **b.** $x \in [-3, 5]$ **c.** -1
d. $y \in [-4, 5]$ **8.** from $x = 1$ to $x = 2$; steeper line $\rightarrow$ greater slope
9. $F(p) = \frac{3}{4}p + \frac{5}{4}$, For every 4000 pheasants, the fox population increases by 300: 1625.   **10. a.** $x \in \{-3, -2, -1, 0, 1, 2, 3, 4\}$
$y \in \{-3, -2, -1, 0, 1, 2, 3, 4\}$ **b.** $x \in [-3, 4]$ $y \in [-3, 4]$
c. $x \in (-\infty, \infty)$ $y \in (-\infty, \infty)$

Reinforcing Basic Concepts, p. 138

1. a. $\frac{1}{3}$, increasing   **b.** $y - 5 = \frac{1}{3}(x - 0)$ **c.** $y = \frac{1}{3}x + 5$
d. $x - 3y = -15$ **e.** $(0, 5), (-15, 0)$

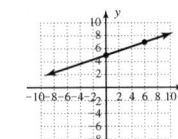

2. a. $\frac{-7}{3}$, decreasing   **b.** $y - 9 = \frac{-7}{3}(x - 0)$ **c.** $y = \frac{-7}{3}x + 9$
d. $7x + 3y = 27$ **e.** $(0, 9), (\frac{22}{7}, 0)$

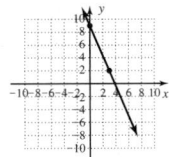

3. a. $\frac{1}{2}$, increasing   **b.** $y - 2 = \frac{1}{2}(x - 3)$ **c.** $y = \frac{1}{2}x + \frac{1}{2}$
d. $x - 2y = -1$ **e.** $(0, \frac{1}{2}), (-1, 0)$

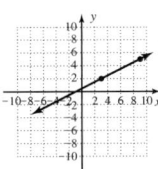

4. a. $\frac{3}{4}$, increasing   **b.** $y + 4 = \frac{3}{4}(x + 5)$ **c.** $y = \frac{3}{4}x - \frac{1}{4}$
d. $3x - 4y = 1$ **e.** $(0, \frac{-1}{4}), (\frac{1}{3}, 0)$

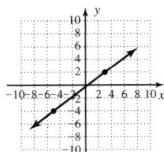

5. a. $\frac{-3}{4}$, decreasing   **b.** $y - 5 = \frac{-3}{4}(x + 2)$ **c.** $y = \frac{-3}{4}x + \frac{7}{2}$
d. $3x + 4y = 14$ **e.** $(0, \frac{7}{2}), (\frac{14}{3}, 0)$

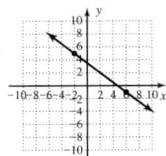

6. a. $\frac{-1}{2}$, decreasing   **b.** $y + 7 = \frac{-1}{2}(x - 2)$ **c.** $y = \frac{-1}{2}x - 6$
d. $x + 2y = -12$ **e.** $(0, -6), (-12, 0)$

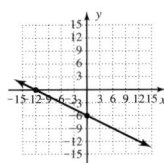

Exercises 2.5, pp. 150–156

1. linear; bounce   **3.** increasing   **5.** Answers will vary.
7.   **9.** even   **11.** even

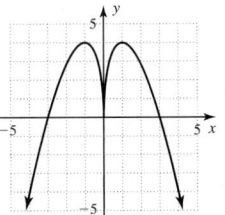

13.

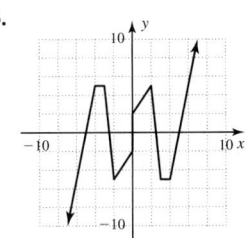

15. odd   **17.** not odd   **19.** neither   **21.** odd   **23.** neither
25. $x \in [-1, 1] \cup [3, \infty)$ **27.** $x \in (-\infty, -1) \cup (-1, 1) \cup (1, \infty)$
29. $p(x) \geq 0$ for $x \in [2, \infty)$ **31.** $f(x) \leq 0$ for $x \in (-\infty, 2]$
33. $V(x)\uparrow$: $x \in (-3, 1) \cup (4, 6)$ $V(x)\downarrow$: $x \in (-\infty, -3) \cup (1, 4)$
constant: none   **35.** $f(x)\uparrow$: $x \in (1, 4)$ $f(x)\downarrow$: $x \in (-2, 1) \cup (4, \infty)$
constant: $x \in (-\infty, -2)$ **37. a.** $p(x)\uparrow$: $x \in (-\infty, \infty)$ $p(x)\downarrow$: none
b. down, up   **39. a.** $f(x)\uparrow$: $x \in (-3, 0) \cup (3, \infty)$
$f(x)\downarrow$: $x \in (-\infty, -3) \cup (0, 3)$ **b.** up, up

41. a. $x \in (-\infty, \infty), y \in (-\infty, 5)$ **b.** $x = 1, 3$
c. $H(x) \geq 0: x \in [1, 3]$ $H(x) \leq 0: x \in (-\infty, 1] \cup [3, \infty)$
d. $H(x)\uparrow: x \in (-\infty, 2)$ $H(x)\downarrow: x \in (2, \infty)$ **e.** local max: $y = 5$ at $(2, 5)$
43. a. $x \in (-\infty, \infty), y \in (-\infty, \infty)$ **b.** $x = -1, 5$
c. $g(x) \geq 0: x \in [-1, \infty)$ $g(x) \leq 0: x \in (-\infty, -1] \cup [0, 3.5]$
d. $g(x)\uparrow: x \in (-\infty, 1) \cup (5, \infty)$ $g(x)\downarrow: x \in (1, 5)$ **e.** local max: $y = 6$ at
$(1, 6)$; local min: $y = 0$ at $(5, 0)$ **45. a.** $x \in [-4, \infty), y \in (-\infty, 3]$
b. $x = -4, 2$ **c.** $Y_1 \geq 0: x \in [-4, 2]$ $Y_1 \leq 0: x \in [2, \infty)$
d. $Y_1\uparrow: x \in (-4, -2)$ $Y_1\downarrow: x \in (-2, \infty)$ **e.** local max: $y = 3$ at $(-2, 3)$
47. a. $x \in \mathbb{R}, y \in \mathbb{R}$ **b.** $x = -4$ **c.** $p(x) \geq 0: x \in [-4, \infty); p(x) \leq 0:$
$x \in (-\infty, -4]$ **d.** $p(x)\uparrow: x \in (-\infty, -3) \cup (-3, \infty); p(x)\downarrow:$ never
decreasing   **e.** local max: none; local min: none
49. a. $x \in (-\infty, -3] \cup [3, \infty), y \in [0, \infty)$ **b.** $(-3, 0), (3, 0)$
c. $f(x)\uparrow: x \in (3, \infty)$ $f(x)\downarrow: x \in (-\infty, -3)$ **d.** even
51. a. $x \in [0, 260], y \in [0, 80]$ **b.** 80 ft   **c.** 120 ft   **d.** yes   **e.** $(0, 120)$
f. $(120, 260)$ **53. a.** $x \in (-\infty, \infty); y \in [-1, \infty)$ **b.** $(-1, 0), (1, 0)$
c. $f(x) \geq 0: x \in (-\infty, -1] \cup [1, \infty); f(x) < 0: x \in (-1, 1);$
d. $f(x)\uparrow: x \in (0, \infty), f(x)\downarrow: x \in (-\infty, 0)$ **e.** min: $(0, -1)$
55. a. $t \in [72, 96], I \in [7.25, 16]$
b. $I(t)\uparrow: t \in (72, 74) \cup (77, 81) \cup (83, 84) \cup (93, 94)$
$I(t)\downarrow: t \in (74, 75) \cup (81, 83) \cup (84, 86) \cup (90, 93) \cup (94, 95)$ $I(t)$
constant: $t \in (75, 77) \cup (86, 90) \cup (95, 96)$ **c.** max: (74, 9.25), (81, 16)
(global max), (84, 13), (94, 8.5), min: (72, 7.5), (83, 12.75), (93, 7.25)
d. Increase: 80 to 81; Decrease: 82 to 83 or 85 to 86
57. zeroes: $(-8, 0), (-4, 0), (0, 0), (4, 0);$
min: $(-2, -1), (4, 0)$; max: $(-6, 2), (2, 2)$

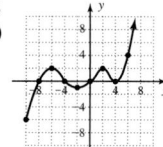

59. a. 7   **b.** 7   **c.** They are the same.
d. Slopes are equal.

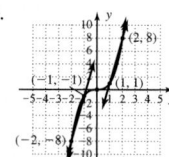

61. a. 176 ft   **b.** 320 ft   **c.** 144 ft/sec   **d.** -144 ft/sec; The arrow is
going down.   **63. a.** 17.89 ft/sec; 25.30 ft/sec   **b.** 30.98 ft/sec;
35.78 ft/sec   **c.** Between 5 and 10.   **d.** 1.482 ft/sec, 0.96 ft/sec   **65.** 2

67. $2x + h$ **69.** $2x + 2 + h$ **71.** $\dfrac{-2}{x(x + h)}$

73. a. $\dfrac{\Delta g}{\Delta x} = 2x + 2 + h$ **b.** $\dfrac{\Delta g}{\Delta x} = -3.9$ **c.** $\dfrac{\Delta g}{\Delta x} = 3.01$
d.
The rates of change have opposite sign, with the
secant line to the left being slightly more steep.

75. a. $\dfrac{\Delta g}{\Delta x} = 3x^2 + 3xh + h^2$ **b.** $\dfrac{\Delta g}{\Delta x} \approx 12.61$ **c.** $\dfrac{\Delta g}{\Delta x} \approx 0.49$
d.
Both lines have a positive slope, but the line at
$x = -2$ is much steeper.

77. a. $\dfrac{\Delta d}{\Delta h} \approx 0.25$ **b.** $\dfrac{\Delta d}{\Delta h} \approx 0.05$

c.
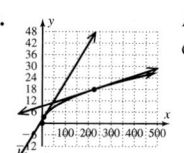
As height increases you can see farther, the sight
distance is increasing much slower.

79. no; no; Answers will vary.   **81.** Answers will vary.

83. $x = -2, x = 10$ **85.** $y = \dfrac{2}{3}x - 1$

Exercises 2.6, pp. 166–171

1. stretch; compression   **3.** $(-5, -9)$; upward   **5.** Answers will vary.
7. a. quadratic;   **b.** up/up, $(-2, -4)$, $x = -2$, $(-4, 0)$, $(0, 0)$, $(0, 0)$;
c. $D: x \in \mathbb{R}, R: y \in [-4, \infty)$ **9. a.** quadratic;   **b.** up/up, $(1, -4)$,
$x = 1$, $(-1, 0)$, $(3, 0)$, $(0, -3)$; **c.** $D: x \in \mathbb{R}, R: y \in [-4, \infty)$
11. a. quadratic;   **b.** up/up, $(2, -9)$, $x = 2$, $(-1, 0)$, $(5, 0)$, $(0, -5)$;
c. $D: x \in \mathbb{R}, R: y \in [-9, \infty)$ **13. a.** square root;   **b.** up to the right,
$(-4, -2)$, $(-3, 0)$, $(0, 2)$; **c.** $D: x \in [-4, \infty), R: y \in [-2, \infty)$
15. a. square root;   **b.** down to the left, $(4, 3)$, $(3, 0)$, $(0, -3)$;
c. $D: x \in (-\infty, 4], R: y \in (-\infty, 3]$ **17. a.** square root;   **b.** up to the
left, $(4, 0)$, $(4, 0)$, $(0, 4)$;   **c.** $D: x \in (-\infty, 4], R: y \in [0, \infty)$
19. a. absolute value;   **b.** up/up, $(-1, -4)$, $x = -1$, $(-3, 0)$, $(1, 0)$, $(0, -2)$;
c. $D: x \in \mathbb{R}, R: y \in [-4, \infty)$ **21. a.** absolute value;   **b.** down/down,
$(-1, 6)$, $x = -1$, $(-4, 0)$, $(2, 0)$, $(0, 4)$;   **c.** $D: x \in \mathbb{R}, R: y \in (-\infty, 6]$
23. a. absolute value;   **b.** down/down, $(0, 6)$, $x = 0$, $(-2, 0)$, $(2, 0)$, $(0, 6)$;
c. $D: x \in \mathbb{R}, R: y \in (-\infty, 6]$ **25. a.** cubic;   **b.** up/down, $(1, 0)$, $(1, 0)$,
$(0, 1)$;   **c.** $D: x \in \mathbb{R}, R: y \in \mathbb{R}$ **27. a.** cubic;   **b.** down/up, $(0, 1)$,
$(-1, 0)$, $(0, 1)$;   **c.** $D: x \in \mathbb{R}, R: y \in \mathbb{R}$ **29. a.** cube root;   **b.** down/up,
$(1, -1)$, $(2, 0)$, $(0, -2)$;   **c.** $D: x \in \mathbb{R}, R: y \in \mathbb{R}$ **31.** square root function;
y-int $(0, 2)$; x-int $(-3, 0)$; initial point $(-4, -2)$; up on right;
$D: x \in [-4, \infty), R: y \in [-2, \infty)$ **33.** cubic function; y-int $(0, -2)$;
x-int $(-2, 0)$; inflection point $(-1, -1)$; up, down; $D: x \in \mathbb{R}, R: y \in \mathbb{R}$

35.
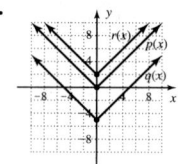

37.
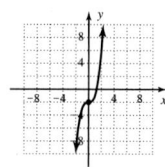

39.

41.
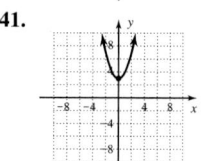

43.
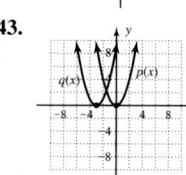

45.
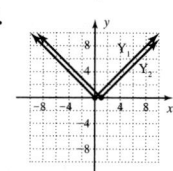

47.
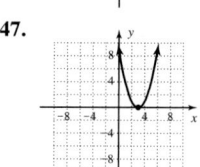

49.
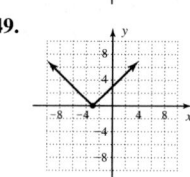

51.
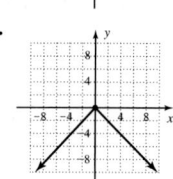

53.
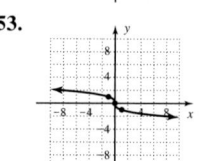

55.
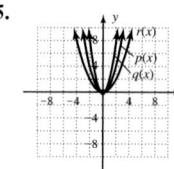

57.
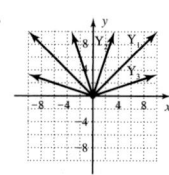

59.
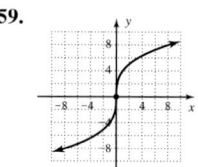

61.
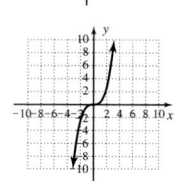

63. g   **65.** i   **67.** e   **69.** j   **71.** l   **73.** c

75. left 2, down 1

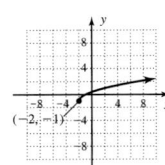

77. left 3, reflected across x-axis, down 2

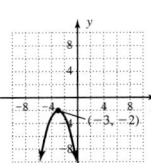

79. left 3, down 1

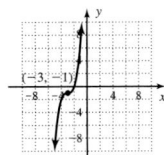

81. left 1, down 2

83. left 3, reflected across x-axis, down 2

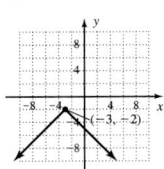

85. left 1, reflected across x-axis, stretched vertically, down 3

87. left 2, reflected across x-axis, compressed vertically down 1,

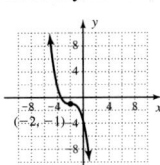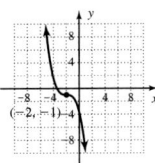

89. left 1, reflected across y-axis, reflected across x-axis, stretched vertically, up 3,

91. right 3, compressed vertically, up 1

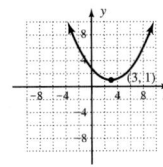

93. a.

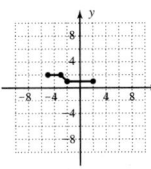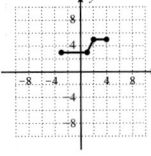

b.   **c.**   **d.**

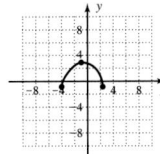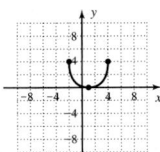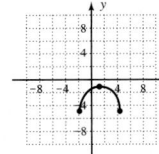

95. a.   **b.**   **c.**

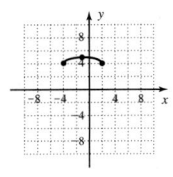

d.

97. $f(x) = -(x - 2)^2$ **99.** $p(x) = 1.5\sqrt{x + 3}$

101. $f(x) = \frac{4}{5}|x + 4|$ **103.**

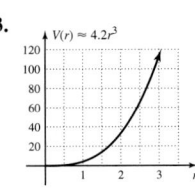

≈ 4.2, about 65 units², 65.4 units³, yes

105.

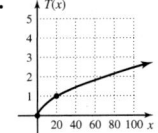

compressed vertically, 2.25 sec

107. **a.** compressed vertically, **b.** 216 W, **c.** ≈ 15.6, 161.5, power increases dramatically at higher windspeeds

109. **a.** vertical stretch by a factor of 2, **b.** 12.5 ft, **c.** 5, 13, distance fallen by unit time increases very fast

111. $x \in (0, 4)$; yes, $x \in (4, \infty)$; yes

113. Any points in Quadrants III and IV will reflect across the x-axis and move to Quadrants I and II.

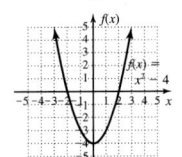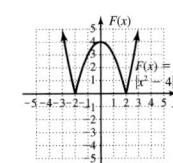

115. $p = 140$ in. $A = 1168$ in²

117. $f(x) \downarrow : x \in (-\infty, 4)$ $f(x) \uparrow : x \in (4, \infty)$

Exercises 2.7, pp. 180–185

1. continuous   **3.** smooth   **5.** Each piece must be continuous on the corresponding interval, and the function values at the endpoints of each interval must be equal. Answers will vary.

7. a. $f(x) = \begin{cases} x^2 - 6x + 10 & 0 \le x \le 5 \\ \frac{3}{2}x - \frac{5}{2} & 5 < x \le 9 \end{cases}$ **b.** $y \in [1, 11]$

9. $-2, 2, \frac{1}{2}, 0, 2.999, 5$ **11.** $5, 5, 0, -4, 5, 11$

13. $D: x \in [-6, \infty); R: y \in [-4, \infty)$

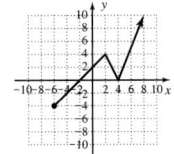

15. $D: x \in [-2, \infty); R: y \in [-4, \infty)$

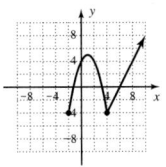

17. $D: x \in (-\infty, \infty)$; $R: y \in (-\infty, 3), \cup (3, \infty)$

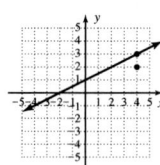

19. $x \in (-\infty, 9)$; $y \in [2, \infty)$ **21.** $x \in (-\infty, \infty)$; $y \in [0, \infty)$

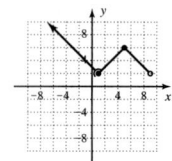   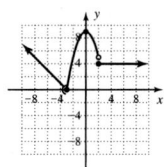

23. $x \in (-\infty, \infty)$; $y \in (-\infty, -6) \cup (-6, \infty)$; discontinuity at $x = -3$, redefine $f(x) = -6$ at $x = -3$; $c = -6$

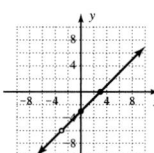

25. $x \in (-\infty, \infty)$; $y \in [0.75, \infty)$; discontinuity at $x = 1$, redefine $f(x) = 3$ at $x = 1$; $c = 3$

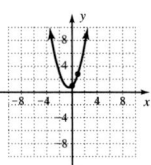

27. $f(x) = \begin{cases} \frac{1}{2}x - 1 & -4 \le x < 2 \\ 3x - 6 & x \ge 2 \end{cases}$ **29.** $p(x) = \begin{cases} x^2 + 2x - 3 & x \le 1 \\ x + 1 & x > 1 \end{cases}$

31. Graph is discontinuous at $x = 0$; $f(x) = 1$ for $x > 0$; $f(x) = -1$ for $x < 0$.

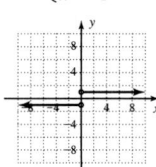

33. a. $S(t) = \begin{cases} -t^2 + 6t & 0 \le t \le 5 \\ 5 & t > 5 \end{cases}$ **b.** $S(t) \in [0, 9]$

35. a.

Year (0 → 1950)	Percent
5	7.33
15	14.13
25	14.93
35	22.65
45	41.55
55	60.45

b. Each piece gives a slightly different value due to rounding of coefficients in each model. At $t = 30$, we use the "first" piece: $P(30) = 13.08$.

37. $C(h) = \begin{cases} 0.09h & 0 \le h \le 1000 \\ 0.18h - 90 & h > 1000 \end{cases}$ $C(1200) = \$126$

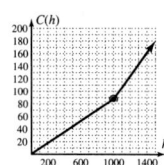

39. $C(t) = \begin{cases} 0.75t & 0 \le t \le 25 \\ 1.5t - 18.75 & t > 25 \end{cases}$ $C(45) = \$48.75$

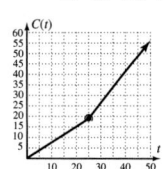

41. $S(t) = \begin{cases} -1.35t^2 + 31.9t + 152; & 0 \le t \le 12 \\ 2.5t^2 - 80.6t + 950; & 12 < t \le 22 \end{cases}$

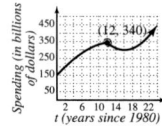

$\$498$ billion, $\$653$ billion, $\$782$ billion

43. $c(m) = \begin{cases} 3.3m & 0 \le m \le 30 \\ 7m - 111 & m > 30 \end{cases}$; $\$2.11$

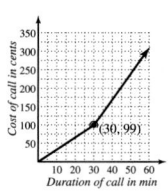

45. $C(a) = \begin{cases} 0 & a < 2 \\ 2 & 2 \le a < 13 \\ 5 & 13 \le a < 20 \\ 7 & 20 \le a < 65 \\ 5 & a \ge 65 \end{cases}$ $\$38$

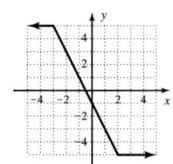

47. a. $C(w - 1) = 17[w - 1] + 80$, **b.** $0 < w \le 13$; **c.** 80¢, **d.** 165¢, **e.** 165¢, **f.** 165¢, **g.** 182¢

49. yes; $h(x) = \begin{cases} 5 & x \le -3 \\ -2x - 1 & -3 < x < 2 \\ -5 & x \ge 2 \end{cases}$

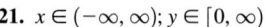

51. Y_1 has a removable discontinuity at $x = -2$; Y_2 has a discontinuity at $x = -2$ **53.** $x = -7, x = 4$ **55. a.** $4\sqrt{5}$ cm **b.** $16\sqrt{5}$ cm^2 **c.** $V = 320\sqrt{5}$ cm^3

Exercises 2.8, pp. 196–202

1. $(f + g)(x)$; $A \cap B$ **3.** intersection; $g(x)$ **5.** Answers will vary.

7. a. $x \in \mathbb{R}$ **b.** $f(-2) - g(-2) = 13$ **9. a.** $h(x) = x^2 - 6x - 3$ **b.** $h(-2) = 13$ **c.** they are identical **11. a.** $x \in [3, \infty)$ **b.** $h(x) = \sqrt{x - 3} + 2x^3 - 54$ **c.** $h(4) = 75$, 2 is not in the domain of h. **13. a.** $x \in [-5, 3]$ **b.** $r(x) = \sqrt{x + 5} + \sqrt{3 - x}$ **c.** $2(7) = \sqrt{7} + 1$, 4 is not in the domain of r. **15. a.** $x \in [-4, \infty)$ **b.** $h(x) = \sqrt{x + 4}(2x + 3)$ **c.** $h(-4) = 0, h(21) = 225$

17. a. $x \in [-1, 7]$ **b.** $r(x) = \sqrt{-x^2 + 6x + 7}$ **c.** 15 is not in the domain of r, $r(3) = 4$ **19. a.** $x \in (-\infty, -4) \cup (-4, \infty)$ **b.** $h(x) = x - 4, x \ne -4$ **21. a.** $x \in (-\infty, -4) \cup (-4, \infty)$ **b.** $h(x) = x^2 - 2, x \ne -4$ **23. a.** $x \in (-\infty, 1) \cup (1, \infty)$ **b.** $h(x) = x^2 - 6x, x \ne 1$ **25. a.** $x \in (-\infty, 5) \cup (5, \infty)$ **b.** $h(x) = \dfrac{x + 1}{x - 5}, x \ne 5$ **27. a.** $x \in (-\infty, -2)$ **b.** $r(x) = \dfrac{2x - 3}{\sqrt{-2 - x}}$ **c.** 6 is not in the domain of r. $r(-6) = -\dfrac{15}{2}$ **29. a.** $x \in (5, \infty)$ **b.** $r(x) = \dfrac{x - 5}{\sqrt{x - 5}}$ **c.** $r(6) = 1$; -6 is not in the domain of r.

31. a. $x \in \left(-\dfrac{13}{2}, \infty\right)$ **b.** $r(x) = \dfrac{x^2 - 36}{\sqrt{2x + 13}}$ **c.** $r(6) = 0, r(-6) = 0$

33. a. $h(x) = \dfrac{2x + 4}{x - 3}$ **b.** $x \in (-\infty, 3) \cup (3, \infty)$ **c.** $x \ne -2, x \ne 0$

35. sum: $3x + 1, x \in (-\infty, \infty)$; difference: $x + 5, x \in (-\infty, \infty)$; product: $2x^2 - x - 6, x \in (-\infty, \infty)$; quotient: $\dfrac{2x + 3}{x - 2}, x \in (-\infty, 2) \cup (2, \infty)$

37. sum: $x^2 + 3x + 5, x \in (-\infty, \infty)$; difference: $x^2 - 3x + 9$, $x \in (-\infty, \infty)$; product: $3x^3 - 2x^2 + 21x - 14, x \in (-\infty, \infty)$; quotient: $\dfrac{x^2 + 7}{3x - 2}, x \in \left(-\infty, \dfrac{2}{3}\right) \cup \left(\dfrac{2}{3}, \infty\right)$

39. sum: $x^2 + 3x - 4, x \in (-\infty, \infty)$; difference: $x^2 + x - 2$, $x \in (-\infty, \infty)$; product: $x^3 + x^2 - 5x + 3, x \in (-\infty, \infty)$; quotient: $x + 3, x \in (-\infty, 1) \cup (1, \infty)$

41. sum: $3x + 1 + \sqrt{x - 3}, x \in [3, \infty)$; difference: $3x + 1 - \sqrt{x - 3}$, $x \in [3, \infty)$; product: $(3x + 1)\sqrt{x - 3}, x \in [3, \infty)$; quotient: $\dfrac{3x + 1}{\sqrt{x - 3}}, x \in (3, \infty)$ **43.** sum: $2x^2 + \sqrt{x + 1}, x \in [-1, \infty)$;

difference: $2x^2 - \sqrt{x+1}$, $x \in [-1, \infty)$; product:

$2x^2\sqrt{x+1}$, $x \in [-1, \infty)$; quotient: $\dfrac{2x^2}{\sqrt{x+1}}$, $x \in (-1, \infty)$

45. sum: $\dfrac{7x-11}{(x-3)(x+2)}$, $x \in (-\infty, -2) \cup (-2, 3) \cup (3, \infty)$;

difference: $\dfrac{-3x+19}{(x-3)(x+2)}$, $x \in (-\infty, -2) \cup (-2, 3) \cup (3, \infty)$;

product: $\dfrac{10}{(x^2-x-6)}$, $x \in (-\infty, -2) \cup (-2, 3) \cup (3, \infty)$;

quotient: $\dfrac{2x+4}{(5x-15)}$, $x \in (-\infty, -2) \cup (-2, 3) \cup (3, \infty)$

47. 0; 0; $4a^2 - 10a - 14$ $a^2 - 9a$ **49. a.** $h(x) = \sqrt{2x-2}$
b. $H(x) = 2\sqrt{x+3} - 5$ **c.** D of $h(x)$: $x \in [1, \infty)$; D of $H(x)$:
$x \in [-3, \infty)$ **51. a.** $h(x) = \sqrt{3x+1}$ **b.** $H(x) = 3\sqrt{x-3} + 4$
c. D of $h(x)$: $x \in [-\frac{1}{3}, \infty)$ D of $H(x)$: $x \in [3, \infty)$
53. a. $h(x) = x^2 + x - 2$ **b.** $H(x) = x^2 - 3x + 2$ **c.** D of $h(x)$:
$x \in (-\infty, \infty)$ D of $H(x)$: $x \in (-\infty, \infty)$ **55. a.** $h(x) = x^2 + 7x + 8$
b. $H(x) = x^2 + x - 1$ **c.** D of $h(x)$: $x \in (-\infty, \infty)$ D of $H(x)$:
$x \in (-\infty, \infty)$ **57. a.** $h(x) = |-3x+1| - 5$ **b.** $H(x) = -3|x| + 16$
c. D of $h(x)$: $x \in (-\infty, \infty)$ D of $H(x)$: $x \in (-\infty, \infty)$
59. a. $(f \circ g)(x)$: For $g(x)$ to be defined, $x \neq 0$.

For $f[g(x)] = \dfrac{2g(x)}{g(x)+3}$, $g(x) \neq -3$ so $x \neq -\dfrac{5}{3}$.

domain: $\left\{x \mid x \neq 0, x \neq -\dfrac{5}{3}\right\}$

b. $(g \circ f)(x)$: For $f(x)$ to be defined, $x \neq -3$.

For $g[f(x)] = \dfrac{5}{f(x)}$, $f(x) \neq 0$ so $x \neq 0$.

domain: $\{x \mid x \neq 0, x \neq -3\}$

c. $(f \circ g)(x) = \dfrac{10}{5+3x}$; $(g \circ f)(x) = \dfrac{5x+15}{2x}$;

the domain of a composition cannot always be determined from the composed form
61. a. $(f \circ g)(x)$: For $g(x)$ to be defined, $x \neq 5$.

For $f[g(x)] = \dfrac{4}{g(x)}$, $g(x) \neq 0$ and $g(x)$ is never zero

domain: $\{x \mid x \neq 5\}$

b. $(g \circ f)(x)$: For $f(x)$ to be defined, $x \neq 0$.

For $g[f(x)] = \dfrac{1}{f(x)-5}$, $f(x) \neq 5$ so $x \neq \dfrac{4}{5}$.

domain: $\left\{x \mid x \neq 0, x \neq \dfrac{4}{5}\right\}$

c. $(f \circ g)(x) = 4x - 20$; $(g \circ f)(x) = \dfrac{x}{4-5x}$; the domain of a composition

cannot always be determined from the composed form
63. a. 41 **b.** 41 **65.** $g(x) = \sqrt{x-2} + 1$, $f(x) = x^3 - 5$
67. $p(x) = 2(x+4)^2 - 3$, $q(x) = (2x+7)^2 - 1$ **69. a.** 6000
b. 3000 **c.** 8000 **d.** $C(9) - T(9)$; 4000 **71. a.** \$1 billion
b. \$5 billion **c.** 2003, 2007, 2010 **d.** $t \in (2000, 2003) \cup (2007, 2010)$
e. $t \in (2003, 2007)$ **f.** $R(5) - C(5)$; \$4 billion **73. a.** 4 **b.** 0 **c.** 2

d. 3 **e.** $-\dfrac{1}{3}$ **f.** 6 **g.** -3 **h.** 1 **i.** 1 **j.** undefined **k.** 0.5 **l.** 2

75. $h(x) = -\dfrac{2}{3}x + 4$ **77.** $h(x) = 4x - x^2$

79. $A = 2\pi r(20 + r)$; $f(r) = 2\pi r$, $g(r) = 20 + r$; $A(5) = 250\pi$ units^2
81. a. $P(x) = 12{,}000x - 108{,}000$; **b.** nine boats must be sold
83. a. $P(n) = 11.45n - 0.1n^2$ **b.** \$123 **c.** \$327
d. $C(115) > R(115)$ **85.** $h(x) = x - 2.5$; 10.5 **87. a.** 4160
b. 45,344 **c.** $M(x) = 453.44x$; yes **89. a.** 6 ft **b.** 36π ft^2
c. $A(t) = 9\pi t^2$; yes **91. a.** 1995 to 1996; 1999 to 2004 **b.** 30; 1995

c. 20 seats; 1997 **d.** The total number in the senate (50); the number of additional seats held by the majority **93.** Answers will vary.

no, yes

95.

x	$f(x)$	$g(x)$	$(f-g)(x)$
-2	27	15	12
-1	18	11	7
0	11	7	4
1	6	3	3
2	3	-1	4
3	2	-5	7
4	3	-1	4
5	6	3	3
6	11	7	4
7	18	11	7
8	27	15	12

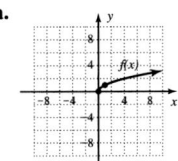

97. a.

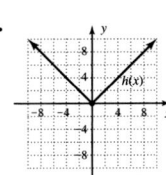

b.

c.

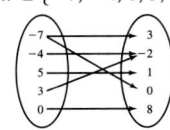

99. $y = -\dfrac{3}{2}x$

Summary and Concept Review, pp. 202–209

1. $x \in \{-7, -4, 0, 3, 5\}$ $y \in \{-2, 0, 1, 3, 8\}$

2. $x \in [-5, 5]$ $y \in [0, 5]$

x	y
-5	0
-4	3
-2	$\sqrt{21} \approx 4.58$
0	5
2	$\sqrt{21} \approx 4.58$
4	3
5	0

3. 65 mi **4.** $\left(\frac{5}{2}, -3\right)$ **5.**

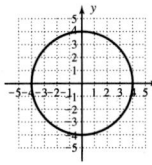

6.

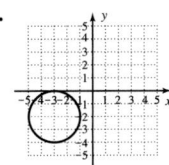

7. $(x+1.5)^2 + (y-2)^2 = 6.25$
8. a.

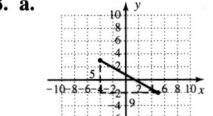

b.

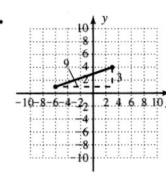

$\dfrac{-5}{9}$, $(14, -7)$ 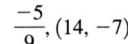 $\frac{1}{3}$, $(0, 3)$

9. a. parallel   **b.** perpendicular

10. a.

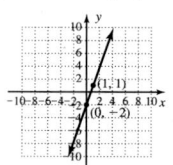

b.

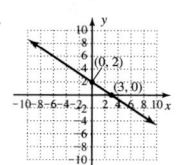

11. a.

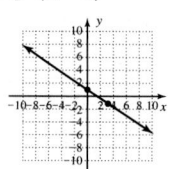

b.

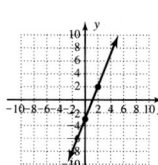

12. a. vertical
   b. horizontal
   c. neither

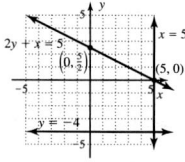

13. yes   **14.** $m = \frac{2}{3}$, y-intercept $(0, 2)$; when the rodent population increases by 3000, the hawk population increases by 200.

15. a. $y = \frac{-4}{3}x + 4$, $m = \frac{-4}{3}$, y-intercept $(0, 4)$ **b.** $y = \frac{5}{3}x - 5$, $m = \frac{5}{3}$, y-intercept $(0, -5)$

16. a.

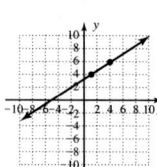

b.

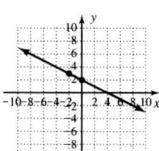

17. a.

b.

18. $y = 5, x = -2$; $y = 5$ **19.** $y = \frac{-3}{4}x + \frac{11}{4}$ **20.** $f(x) = \frac{4}{3}x$

21. $m = \frac{2}{5}$, y-intercept $(0, 2)$, $y = \frac{2}{5}x + 2$. When the rabbit population increases by 500, the wolf population increases by 200.

22. a. $y - 90 = \frac{-15}{2}(x - 2)$ **b.** $(14, 0), (0, 105)$ **c.** $f(x) = \frac{-15}{2}x + 105$

d. $f(20) = -45, x = 12$ **23. a.** $x \in [-\frac{5}{4}, \infty)$

b. $x \in (-\infty, -2) \cup (-2, 3) \cup (3, \infty)$ **24.** $14; \frac{26}{9}; 18a^2 - 9a$ **25.** It is a function.   **26. I. a.** $D = \{-1, 0, 1, 2, 3, 4, 5\}$

$R = \{-2, 1, 0, 1, 2, 3, 4\}$ **b.** 1   **c.** 2   **II. a.** $x \in (-\infty, \infty)$

$y \in (-\infty, \infty)$ **b.** -1 **c.** 3   **III. a.** $x \in [-3, \infty$ $y \in [-4, \infty)$

b. -1 **c.** -3 or 3   **27.** $D: x \in (-\infty, \infty)$ $R: y \in [-5, \infty)$

$f(x)\uparrow: x \in (2, \infty)$ $f(x)\downarrow: x \in (-\infty, 2)$ $f(x) > 0: x \in (-\infty, -1) \cup (5, \infty)$

$f(x) < 0: x \in (-1, 5)$ **28.** $D: x \in [-3, \infty)$ $R: y \in (-\infty, 0)$ $f(x)\uparrow$: none

$f(x)\downarrow: x \in (-3, \infty)$ $f(x) > 0$: none $f(x) < 0: x \in (-3, \infty)$

29. $D: x \in (-\infty, \infty), R: y \in (-\infty, \infty)$ $f(x)\uparrow: x \in (-\infty, -3) \cup (1, \infty)$

$f(x)\downarrow: x \in (-3, 1)$ $f(x) > 0: x \in (-5, -1) \cup (4, \infty)$

$f(x) < 0: x \in (-\infty, -5) \cup (-1, 4)$

30. a. odd   **b.** even   **c.** neither   **d.** odd   **31. a.** $\frac{1}{4}$; the graph is rising to the right.   **b.** $2x - 1 + h$; 3.01

32.

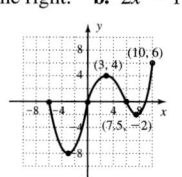

zeroes: $(-6, 0), (0, 0), (6, 0)$ $(9, 0)$
min: $(-3, -8), (7.5 -2)$
max: $(-6, 0), (3, 4)$

33. squaring function   **a.** up on left/up on the right;   **b.** x-intercepts: $(-4, 0), (0, 0)$; y-intercept: $(0, 0)$ **c.** vertex $(-2, -4)$

d. $x \in (-\infty, \infty), y \in [-4, \infty)$ **34.** square root function   **a.** down on the right;   **b.** x-intercept: $(0, 0)$; y-intercept: $(0, 0)$ **c.** initial point $(-1, 2)$;

d. $x \in [-1, \infty), y \in (-\infty, 2]$ **35.** cubing function   **a.** down on left/up on the right   **b.** x-intercepts: $(-2, 0), (1, 0), (4, 0)$; y-intercept: $(0, 2)$

c. inflection point: $(1, 0)$ **d.** $x \in (-\infty, \infty), y \in (-\infty, \infty)$

36. absolute value function   **a.** down on left/down on the right

b. x-intercepts: $(-1, 0), (3, 0)$; y-intercept: $(0, 1)$ **c.** vertex: $(1, 2)$;

d. $x \in (-\infty, \infty), y \in (-\infty, 2]$ **37.** cube root   **a.** up on left, down on right   **b.** x-intercept: $(1, 0)$; y-intercept: $(0, 1)$ **c.** inflection point: $(1, 0)$

d. $x \in (-\infty, \infty), y \in (-\infty, \infty)$

38. quadratic

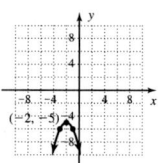

39. absolute value

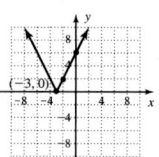

40. cubic

41. square root

42. cube root

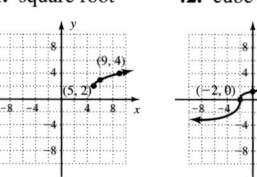

43. a.

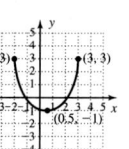

b.

c.

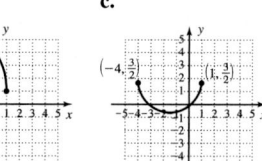

44. a. $f(x) = \begin{cases} 5 & x \le -3 \\ -x + 1 & -3 < x \le 3 \\ 3\sqrt{x - 3} - 1 & x > 3 \end{cases}$ **b.** $R: y \in [-2, \infty)$

45.

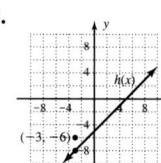

$D: x \in (-\infty, \infty)$,
$R: y \in (-\infty, -8) \cup (-8, \infty)$,
discontinuity at $x = -3$;
define $h(x) = -8$ at $x = -3$

46. $-4, -4, -4.5, -4.99, 3\sqrt{3} - 9, 3\sqrt{3.5} - 9$

47. $D: x \in (-\infty, \infty) R: y \in [-4, \infty)$

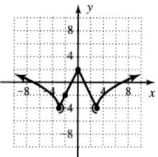

48. $\begin{cases} 20x & x \le 2 \\ 30x - 20 & 2 < x \le 4 \\ 40x - 60 & x > 4 \end{cases}$

For 5 hr the total cost is $140.

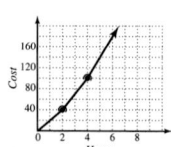

49. $a^2 + 7a - 2$ **50.** 147   **51.** $x \in (-\infty, \frac{2}{3}) \cup (\frac{2}{3}, \infty)$

52. $4x^2 + 8x - 3$ **53.** 99   **54.** $x; x$

55. $f(x) = \sqrt{x} + 1; g(x) = 3x - 2$

56. $f(x) = x^2 - 3x - 10; g(x) = x^{\frac{1}{3}}$ **57.** $A(t) = \pi(2t + 3)^2$

58. a. 4   **b.** 7   **c.** 6   **d.** $\frac{-1}{5}$ **e.** 14

Mixed Review, pp. 209–210

1. $y = -\dfrac{4}{3}x + 4$ **3. a.** $(-\infty, 1) \cup (1, 4) \cup (4, \infty)$ **b.** $\left(\dfrac{3}{2}, \infty\right)$

5. $y = -\dfrac{3}{2}x - 2$ **7.** $(2, 2); (x - 2)^2 + (y - 2)^2 = 50$

9.

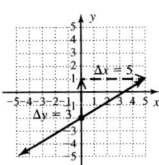

11. a.

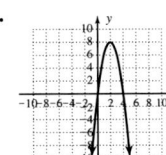

rate of change is positive in $[-2, -1]$ since p is increasing in $(-\infty, 2)$; less; $\dfrac{\Delta y}{\Delta x} = \dfrac{14}{1}$ in $[-2, -1]$; $\dfrac{\Delta y}{\Delta x} = \dfrac{2}{1}$ in $[1, 2]$

b. In the interval $[15, 15.01]$, $\dfrac{\Delta A}{\Delta t} \approx 200.1$

13. $\dfrac{1}{3x^2 - 4x + 1}$; $\left(-\infty, \dfrac{1}{3}\right) \cup \left(\dfrac{1}{3}, 1\right) \cup (1, \infty)$

15. $\dfrac{\Delta f}{\Delta x} = 2x + h$, $\dfrac{\Delta g}{\Delta x} = 3$; For small h, $2x + h = 3$ when $x \approx \dfrac{3}{2}$.

17. $D: x \in (-\infty, 6]; R: y \in (-\infty, 3]$ $g(x)\uparrow: x \in (-\infty, -6) \cup (3, 6)$ $g(x)\downarrow: x \in (-3, 3)$ $g(x)$ constant: $x \in (-6, -3)$ $g(x) > 0: x \in (-7, -1)$ $g(x) < 0: x \in (-\infty, -7) \cup (-1, 6)$ max: $y = 3$ for $x \in (-6, -3)$; $y = 0$ at $(6, 0)$ min: $y = -3$ at $(3, -3)$ **19.** $f(x) = -2x^2 + x + 3$

Practice Test, pp. 211–212

1. a. a and c are nonfunctions, they do not pass the vertical line test

2. neither   **3.**

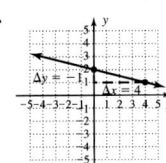

4.

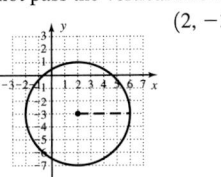

$(2, -3); r = 4$

5. $y = -\dfrac{6}{5}x + \dfrac{2}{5}$ **6. a.** $(7.5, 1.5)$,   **b.** ≈ 61.27 mi

7. $L_1: x = -3$ $L_2: y = 4$ **8. a.** $x \in \{-4, -2, 0, 2, 4, 6\}$ $y \in \{-2, -1, 0, 1, 2, 3\}$ **b.** $x \in [-2, 6]$ $y \in [1, 4]$

9. a. 300   **b.** 30   **c.** $W(h) = \dfrac{25}{2}h$ **d.** Wages are $12.50 per hr.

e. $h \in [0, 40]; w \in [0, 500]$ **10. I. a.** square root

b. $x \in [-4, \infty), y \in [-3, \infty)$ **c.** $(-2, 0), (0, 1)$

d. up on right   **e.** $x \in (-2, \infty)$ **f.** $x \in [-4, -2)$

II. a. cubic   **b.** $x \in (-\infty, \infty)$ $y \in (-\infty, \infty)$ **c.** $(2, 0), (0, -1)$

d. down on left, up on right   **e.** $x \in (2, \infty)$ **f.** $x \in (-\infty, 2)$

III. a. absolute value   **b.** $x \in (-\infty, \infty)$ $y \in (-\infty, 4]$

c. $(-1, 0), (3, 0), (0, 2)$ **d.** down/down   **e.** $x \in (-1, 3)$

f. $x \in (-\infty, -1) \cup (3, \infty)$ **IV. a.** quadratic   **b.** $x \in (-\infty, \infty)$; $y \in [-5.5, \infty)$ **c.** $(0, 0), (5, 0)$ **d.** up/up

e. $x \in (-\infty, 0) \cup (5, \infty)$ **f.** $x \in (0, 5)$

11. a. $\dfrac{7}{2}$ **b.** $\dfrac{-a^2 - 6a - 7}{a^2 + 6a + 9}$ **c.** $\dfrac{31}{25} - \dfrac{8}{25}i$

12. $3x + 1; x \in [\frac{1}{3}, \infty)$ **13. a.** No, new company and sales should be growing   **b.** 19 for $[5, 6]$; 23 for $[6, 7]$

c. $\dfrac{\Delta s}{\Delta t} = 4t - 3 + 2h$. For small h, sales volume is approximately

$\dfrac{37{,}000 \text{ units}}{1 \text{ mo}}$ in month 10, $\dfrac{69{,}000 \text{ units}}{1 \text{mo}}$ in month 18, and $\dfrac{93{,}000 \text{ units}}{1 \text{ mo}}$ in month 24

14.

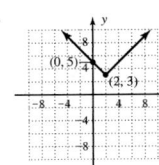

15.

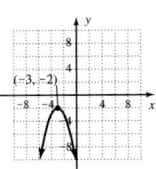

16. a. $V(t) = \dfrac{4}{3}\pi(\sqrt{t})^3$ **b.** 36π in^3 **17. a.** $D: x \in [-4, \infty)$; $R: y \in [-3, \infty)$ **b.** $f(-1) \approx 2.2$ **c.** $f(x) < 0: x \in (-4, -3)$ $f(x) > 0: x \in (-3, \infty)$ **d.** $f(x)\uparrow: x \in (-4, \infty)$ $f(x)\downarrow$: none

e. $f(x) = 3\sqrt{x + 4} - 3$ **18. a.** 4, -4, 6.25

b.

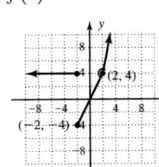

19.

20.

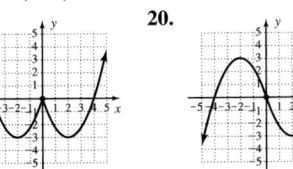

Strengthening Core Skills, p. 213

Exercise 1: $h(x) = x^2 - 28; x = 4 \pm 2\sqrt{7}$

Exercise 2: $h(x) = x^2 + 1; x = -2 \pm i$

Exercise 3: $h(x) = 2x^2 - \frac{3}{2}; x = \dfrac{5}{2} \pm \dfrac{\sqrt{3}}{2}$

Cumulative Review Chapters 1–2, p. 214

1. $x^2 + 2$ **3.** 29.45 cm   **5.** $x = 1$ **7. a.** $\dfrac{-1}{3}$ **b.** $\dfrac{3}{5}$

9.

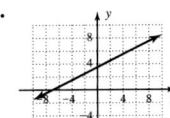

$y = \frac{1}{2}x + \frac{7}{2}$ **11.** $(f \cdot g)(x) = 3x^3 - 12x^2 + 12x$; $\left(\dfrac{f}{g}\right)(x) = 3x, x \neq 2; (g \circ f) = 22$

13. a. $D: x \in (-\infty, 8], R: y \in [-4, \infty)$ **b.** 5, -3, -3, 1, 2

c. $(-2, 0)$ **d.** $f(x) < 0: x \in (-2, 2)$ $f(x) > 0: x \in (-\infty, -2) \cup [2, 8]$

e. min: $(0, -4)$, max: $(8, 7)$ **f.** $f(x)\uparrow: x \in (0, 8)$ $f(x)\downarrow: x \in (-\infty, 0)$

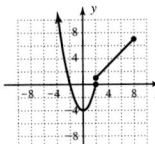

15. a. $\dfrac{x - 7}{(x - 5)(x + 2)}$ **b.** $\dfrac{b^2 - 4ac}{4a^2}$

17. a. False; $\mathbb{Z} \not\subset \mathbb{W}$ **b.** False; $\mathbb{W} \not\subset \mathbb{N}$ **c.** True   **d.** False; $\mathbb{R} \not\subset \mathbb{Z}$

19. $x = -5 \pm \dfrac{\sqrt{2}}{2}; x \approx -5.707; x \approx -4.293$

21. $W = 31$ cm, $L = 47$ cm   **23. a.** $x = \dfrac{-4}{3}, \dfrac{5}{2}$ **b.** $x = -5, -\sqrt{3}, \sqrt{3}$

25. $p = 15 + \sqrt{97}$ units ≈ 24.8 units. No, it is not a right triangle. $5^2 + (\sqrt{97})^2 \neq 10^2$

Connections to Calculus Exercises, pp. 217–218

1. a. -3 **b.** as $h \to 0$, -3 remains constant

3. a. $2x - 3 + h$ **b.** as $h \to 0$, $2(2) - 3 + h \to 1$

5. a. $\dfrac{-1}{(x + h)x}$ **b.** as $h \to 0$, $\dfrac{-1}{(2 + h)^2} \to \dfrac{-1}{4}$

7. a. $\dfrac{-2x - h}{2x^2(x + h)^2}$ **b.** as $h \to 0$, $\dfrac{-2(2) - h}{2(2)^2(2 + h)^2} - c \to \dfrac{-1}{8}$

9.

11.

$A = 24$, distance $= 24$ ft $A = 40$, distance $= 40$ ft

CHAPTER 3

Exercises 3.1, pp. 226–230

1. $\frac{25}{2}$ **3.** $0; f(x)$ **5.** Answers will vary.

7. left 2, down 9

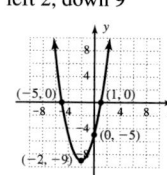

9. reflected across x-axis; right 1, up 4

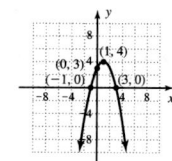

11. stretched vertically, left 1, down 8

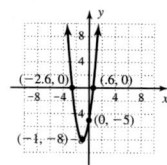

13. stretched vertically reflected across x-axis; right 2, up 15

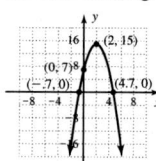

15. right $\frac{7}{4}$, down $\frac{-25}{8}$, stretched vertically

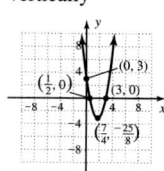

17. reflected across x-axis; left $\frac{7}{6}$, up $\frac{121}{12}$; stretched vertically

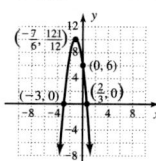

19. right $\frac{5}{2}$, down $\frac{17}{4}$

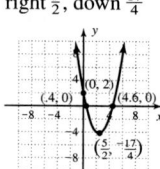

21. left 1, down 7

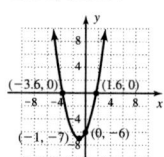

23. reflected across x-axis; right 2, up 6

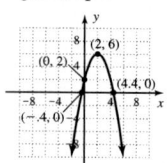

25. left 3, up $\frac{5}{2}$; compressed vertically

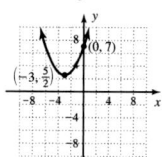

27. reflected across x-axis; right $\frac{5}{2}$, up $\frac{11}{2}$; stretched vertically

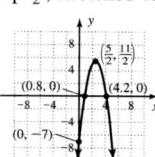

29. right $\frac{3}{2}$, down 6; stretched vertically

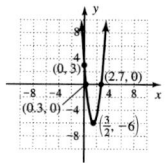

31. left 3, down $\frac{19}{2}$; compressed vertically

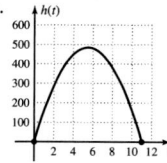

33. $y = 1(x - 2)^2 - 1$ **35.** $y = -1(x + 2)^2 + 4$

37. $y = -\frac{3}{2}(x + 2)^2 + 3$ **39. i.** $x = -3 \pm \sqrt{5}$ **ii.** $x = 4 \pm \sqrt{3}$

iii. $x = -4 \pm \frac{\sqrt{14}}{2}$ **iv.** $x = 2 \pm \sqrt{2}$ **v.** $t = -2.7, t = 1.3$

vi. $t = -1.4, t = 2.6$ **41. a.** $(0, -66,000)$; when no cars are produced, there is a loss of \$66,000.   **b.** $(20, 0), (330, 0)$; no profit will be made if less than 20 or more than 330 cars are produced.   **c.** 175   **d.** \$240,250

43. a. 6 mi   **b.** 3600 ft   **c.** 3200 ft   **d.** 12 mi   **45. a.** $(0, -3300)$; if no appliances are sold, the loss will be \$3300.   **b.** $(20, 0), (330, 0)$; if less than 20 or more than 330 appliances are made and sold, there will be no profit.   **c.** $0 \le x \le 200$; maximum capacity is 200   **d.** 175, \$12,012.50

47. a. 288 ft   **b.**

(graph of $h(t)$ with vertical axis labeled 100–600 and horizontal axis labeled 2 4 6 8 10 12 t)

c. 484 ft; 5.5 sec   **d.** 11 sec

49. a. $h(t) = -16t^2 + 32t + 5$ **b.** (i) 17 ft (ii) 17 ft

c. it must occur between $t = 0.5$ and $t = 1.5$ **d.** $t = 1$ sec

e. $h(1) = 21$ ft   **f.** 2 sec   **51.** 155,000; \$16,625   **53. a.** 96 ft $\times$ 48 ft

b. 32 ft $\times$ 48 ft   **55.** $f(x) = x^2 - 4x + 13$

57. a. radicand will be negative—two complex zeroes.   **b.** radicand will be positive—two real zeroes.   **c.** radicand is zero—one real zero.

d. two real, rational zeroes.   **e.** two real, irrational zeroes.

59. $\dfrac{x - 2}{x - 5}$ **61.** $x \in \left[-3, \frac{2}{3}\right]$

Exercises 3.2, pp. 238–242

1. synthetic; zero   **3.** $P(c)$; remainder   **5.** Answers will vary.

7. $x^3 - 5x^2 - 4x + 21 = (x - 2)(x^2 - 3x - 10) + 3$

9. $2x^3 + 5x^2 + 4x + 17 = (x + 3)(2x^2 - x + 7) - 4$

11. $x^3 - 8x^2 + 11x + 20 = (x - 5)(x^2 - 3x - 4) + 0$

13. a. $\dfrac{2x^2 - 5x - 3}{x - 3} = (2x + 1) + \dfrac{0}{x - 3}$

b. $2x^2 - 5x - 3 = (x - 3)(2x + 1) + 0$

15. a. $\dfrac{x^3 - 3x^2 - 14x - 8}{x - 2} = (x^2 - 5x - 4) + \dfrac{0}{x + 2}$

b. $x^3 - 3x^2 - 14x - 8 = (x + 2)(x^2 - 5x - 4) + 0$

17. a. $\dfrac{x^3 - 5x^2 - 4x + 23}{x - 2} = (x^2 - 3x - 10) + \dfrac{3}{x - 2}$

b. $x^3 - 5x^2 - 4x + 23 = (x - 2)(x^2 - 3x - 10) + 3$

19. a. $\dfrac{2x^3 - 5x^2 - 11x - 17}{x - 4} = (2x^2 + 3x + 1) + \dfrac{-13}{x - 4}$

b. $2x^3 - 5x^2 - 11x - 17 = (x - 4)(2x^2 + 3x + 1) - 13$

21. $x^3 + 5x^2 + 7 = (x + 1)(x^2 + 4x - 4) + 11$

23. $x^3 - 13x - 12 = (x - 4)(x^2 + 4x + 3) + 0$

25. $3x^3 - 8x + 12 = (x - 1)(3x^2 + 3x - 5) + 7$

27. $n^3 + 27 = (n + 3)(n^2 - 3n + 9) + 0$

29. $x^4 + 3x^3 - 16x - 8 = (x - 2)(x^3 + 5x^2 + 10x + 4) + 0$

31. $(2x + 7) + \dfrac{-7x + 5}{x^2 + 3}$ **33.** $-(x^2 - 4) + \dfrac{-4x + 3}{x^2 - 1}$

35. a. -30 **b.** 12   **37. a.** -2 **b.** -22 **39. a.** -1 **b.** 3

41. a. 31   **b.** 0   **43. a.** -10 **b.** 0   **45. a.** yes   **b.** yes   **47. a.** no

b. yes   **49. a.** yes   **b.** yes

51.
$$\begin{array}{r|rrr} -3 & 1 & 2 & -5 & -6 \\ & & -3 & 3 & 6 \\ \hline & 1 & -1 & -2 & 0 \end{array}$$

53.
$$\begin{array}{r|rrr} 2 & 1 & 0 & -7 & 6 \\ & & 2 & 4 & -6 \\ \hline & 1 & 2 & -3 & 0 \end{array}$$

55.
$$\begin{array}{r|rrr} \frac{2}{3} & 9 & 18 & -4 & -8 \\ & & 6 & 16 & 8 \\ \hline & 9 & 24 & 12 & 0 \end{array}$$

57. $P(x) = (x + 2)(x - 3)(x + 5)$, $P(x) = x^3 + 4x^2 - 11x - 30$

59. $P(x) = (x + 2)(x - \sqrt{3})(x + \sqrt{3})$, $P(x) = x^3 + 2x^2 - 3x - 6$

61. $P(x) = (x + 5)(x - 2\sqrt{3})(x + 2\sqrt{3})$, $P(x) = x^3 + 5x^2 - 12x - 60$

63. $P(x) = (x - 1)(x + 2)(x - \sqrt{10})(x + \sqrt{10})$,

$P(x) = x^4 + x^3 - 12x^2 - 10x + 20$ **65.** $P(x) = (x + 2)(x - 3)(x - 4)$

67. $p(x) = (x + 3)^2(x - 3)(x - 1)$ **69.** $f(x) = 2\left(x - \frac{3}{2}\right)(x + 2)(x + 5)$

71. $p(x) = (x + 3)(x - 3)^2$ **73.** $p(x) = (x - 2)^3$
75. $p(x) = (x + 3)(x - 3)^3$ **77.** $p(x) = (x + 3)(x - 3)^2(x + 4)^2$
79. 4-in. squares; 16 in. $\times$ 10 in. $\times$ 4 in. **81. a.** week 10, 22.5 thousand
b. one week before closing, 36 thousand **c.** week 9
83. a. 198 ft^3 **b.** 2 ft **c.** about 7 ft **85.** $k = 10$ **87.** $k = -3$
89. The theorems also apply to complex zeroes of polynomials.
91. $S_3 = 36; S_5 = 225$ **93.** yes, John wins.
95. $G(t) = 1400t + 5000$

b. $C(z) = (z - 9i)(z + 4)(z + 1)$
c. $C(z) = (z - 3i)(z - 1 - 2i)(z - 1 + 2i)$
d. $C(z) = (z - i)(z - 2 - 5i)(z - 2 + 5i)$
e. $C(z) = (z - 6i)(z - 1 - \sqrt{3}\, i)(z - 1 + \sqrt{3}\, i)$
f. $C(z) = (z + 4i)(z - 3 - \sqrt{2}\, i)(z - 3 + \sqrt{2}\, i)$
g. $C(z) = (z - 2 + i)(z - 3i)(z + i)$
h. $C(z) = (z - 2 + 3i)(z - 5i)(z + 2i)$ **117. a.** $w = 150$ ft, $l = 300$;
b. $A = 15{,}000$ ft^2 **119.** $r(x) = 2\sqrt{x + 4} - 2$

Exercises 3.3, pp. 252–257

1. coefficients **3.** $a - bi$ **5.** b; 4 is not a factor of 6
7. $P(x) = (x + 2)(x - 2)(x + 3i)(x - 3i)$
$x = -2, x = 2, x = 3i, x = -3i$
9. $Q(x) = (x + 2)(x - 2)(x + 2i)(x - 2i)$
$x = -2, x = 2, x = 2i, x = -2i$ **11.** $P(x) = (x + 1)(x + 1)(x - 1)$
$x = -1, x = -1, x = 1$ **13.** $P(x) = (x - 5)(x + 5)(x - 5)$
$x = 5, x = -5, x = 5$
15. $(x - 5)^3(x + 9)^2$; $x = 5$, multiplicity 3; $x = -9$, multiplicity 2
17. $(x - 7)^2(x + 2)^2(x + 7)$; $x = 7$, multiplicity 2; $x = -2$,
multiplicity 2; $x = -7$, multiplicity 1
19. $P(x) = x^3 - 3x^2 + 4x - 12$ **21.** $P(x) = x^4 - x^3 - x^2 - x - 2$
23. $P(x) = x^4 - 6x^3 + 13x^2 - 24x + 36$
25. $P(x) = x^4 + 2x^2 + 8x + 5$ **27.** $P(x) = x^4 + 4x^3 + 27$
29. a. yes **b.** yes **31. a.** yes **b.** yes
33. $\{\pm 1, \pm 15, \pm 3, \pm 5, \pm\frac{1}{4}, \pm\frac{15}{4}, \pm\frac{3}{4}, \pm\frac{5}{4}, \pm\frac{1}{2}, \pm\frac{15}{2}, \pm\frac{3}{2}, \pm\frac{5}{2}\}$
35. $\{\pm 1, \pm 15, \pm 3, \pm 5, \pm\frac{1}{2}, \pm\frac{15}{2}, \pm\frac{3}{2}, \pm\frac{5}{2}\}$
37. $\{\pm 1, \pm 28, \pm 2, \pm 14, \pm 4, \pm 7, \pm\frac{1}{6}, \pm\frac{14}{3}, \pm\frac{1}{3}, \pm\frac{7}{3}, \pm\frac{2}{3}, \pm\frac{7}{6}, \pm\frac{1}{2}, \pm\frac{7}{2}, \pm\frac{28}{3}, \pm\frac{4}{3}\}$
39. $\{\pm 1, \pm 3, \pm\frac{1}{32}, \pm\frac{1}{2}, \pm\frac{1}{16}, \pm\frac{1}{4}, \pm\frac{1}{8}, \pm\frac{3}{32}, \pm\frac{3}{2}, \pm\frac{3}{16}, \pm\frac{3}{4}, \pm\frac{3}{8}\}$
41. $(x + 4)(x - 1)(x - 3), x = -4, 1, 3$
43. $(x + 3)(x + 2)(x - 5), x = -3, -2, 5$
45. $(x + 3)(x - 1)(x - 4), x = -3, 1, 4$
47. $(x + 2)(x - 3)(x - 5), x = -2, 3, 5$
49. $(x + 4)(x + 1)(x - 2)(x - 3), x = -4, -1, 2, 3$
51. $(x + 7)(x + 2)(x + 1)(x - 3), x = -7, -2, -1, 3$
53. $(2x + 3)(2x - 1)(x - 1); x = -\frac{3}{2}, \frac{1}{2}, 1$
55. $(2x + 3)^2(x - 1); x = -\frac{3}{2}, 1$
57. $(x + 2)(x - 1)(2x - 5); x = -2, 1, \frac{5}{2}$
59. $(x + 1)(2x + 1)(x - \sqrt{5})(x + \sqrt{5}); x = -1, -\frac{1}{2}, \sqrt{5}, -\sqrt{5}$
61. $(x - 1)(3x - 2)(x - 2i)(x + 2i); x = 1, \frac{2}{3}, 2i, -2i$
63. $x = 1, 2, 3, \frac{-3}{2}$ **65.** $x = -2, 1, \frac{-2}{3}$ **67.** $x = -2, -\frac{3}{2}, 4$
69. $x = 3, -1, \frac{5}{3}$ **71.** $x = 1, 2, -3, \pm\sqrt{7}\, i$ **73.** $x = -2, \frac{2}{3}, 1, \pm\sqrt{3}\, i$
75. $x = 1, 2, 4, -2$ **77.** $x = -3, 1, \pm\sqrt{2}$ **79.** $x = -1, \frac{3}{2}, \pm\sqrt{3}\, i$
81. $x = \frac{1}{2}, 1, 2, \pm\sqrt{3}\, i$ **83. a.** possible roots: $\{\pm 1, \pm 8, \pm 2, \pm 4\}$;
b. neither -1 nor 1 is a root; **c.** 3 or 1 positive roots, 1 negative root;
d. roots must lie between -2 and 2 **85. a.** possible roots: $\{\pm 1, \pm 2\}$;
b. -1 is a root; **c.** 2 or 0 positive roots, 3 or 1 negative roots; **d.** roots
must lie between -3 and 2 **87. a.** possible roots: $\{\pm 1, \pm 12, \pm 2, \pm 6,$
$\pm 3, \pm 4\}$; **b.** $x = 1$ and $x = -1$ are roots; **c.** 4, 2, or 0 positive roots,
1 negative root; **d.** roots must lie between -1 and 4 **89. a.** possible
roots: $\pm 1, \pm 20, \pm 2, \pm 10, \pm 4, \pm 5, \pm\frac{1}{2}, \pm\frac{5}{2}$; **b.** $x = 1$ is a root;
c. 1 positive root, 1 negative root; **d.** roots must lie between -2 and 1
91. $(x - 4)(2x - 3)(2x + 3); x = 4, \frac{3}{2}, -\frac{3}{2}$
93. $(2x + 1)(3x - 2)(x - 12); x = -\frac{1}{2}, \frac{2}{3}, 12$
95. $(x - 2)(2x - 1)(2x + 1)(x + 12); x = 2, \frac{1}{2}, -\frac{1}{2}, -12$
97. a. 5 **b.** 13 **c.** 2 **99.** yes **101.** yes
103. a. 4 cm $\times$ 4 cm $\times$ 4 cm **b.** 5 cm $\times$ 5 cm $\times$ 5 cm
105. length 10 in., width 5 in., height 3 in.
107. 1994, 1998, 2002, about 5 yr **109. a.** 8.97 m, 11.29 m, 12.05 m,
12.94 m; **b.** 9.7 m, $+3.7$ **111. a.** yes, **b.** no, **c.** about 14.88
113A. a. $(x + 5i)(x - 5i)$ **b.** $(x + 3i)(x - 3i)$
c. $(x + i\sqrt{7})(x - i\sqrt{7})$ **113B. a.** $x = -\sqrt{7}, \sqrt{7}$
b. $x = -2\sqrt{3}, 2\sqrt{3}$ **c.** $x = -3\sqrt{2}, 3\sqrt{2}$
115. a. $C(z) = (z - 4i)(z + 3)(z - 2)$

Exercises 3.4, pp. 267–270

1. zero; m **3.** bounce; flatter **5.** Answers will vary.
7. polynomial, degree 3 **9.** not a polynomial, sharp turns
11. polynomial, degree 2 **13.** up/down **15.** down/down
17. down/up; $(0, -2)$ **19.** down/down; $(0, -6)$ **21.** up/down; $(0, -6)$
23. a. even **b.** -3 odd, -1 even, 3 odd **c.** $f(x) = (x + 3)$
$(x + 1)^2(x - 3)$, deg 4 **d.** $x \in \mathbb{R}, y \in [-9, \infty)$
25. a. even **b.** -3 odd, -1 odd, 2 odd, 4 odd
c. $f(x) = -(x + 3)(x + 1)(x - 2)(x - 4)$, deg 4
d. $x \in \mathbb{R}, y \in (-\infty, 25]$ **27. a.** odd **b.** -1 even, 3 odd
c. $f(x) = -(x + 1)^2(x - 3)$, deg 3 **d.** $x \in \mathbb{R}, y \in \mathbb{R}$
29. degree 6; up/up; $(0, -12)$ **31.** degree 5; up/down; $(0, -24)$
33. degree 6; up/up; $(0, -192)$ **35.** degree 5; up/down; $(0, 2)$
37. b **39.** e **41.** c
43. 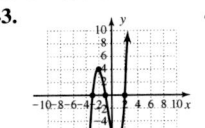 **45.** 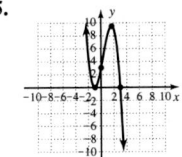 **47.**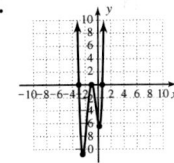

49. 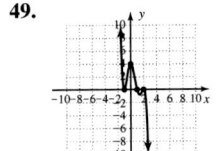 **51.** 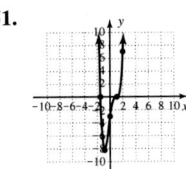 **53.**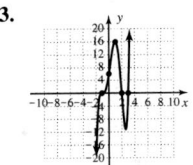

55. 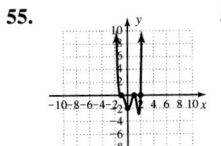 **57.** 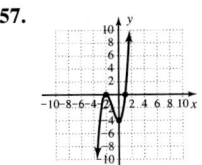 **59.**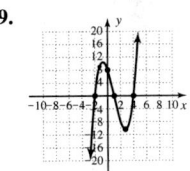

61. 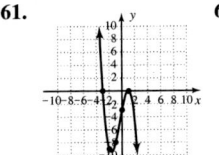 **63.** 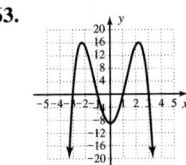 **65.**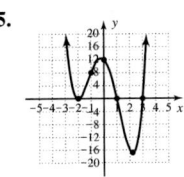

67. 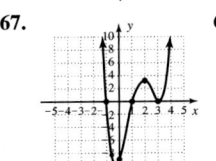 **69.** 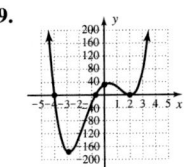 **71.**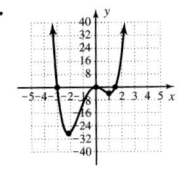

73. 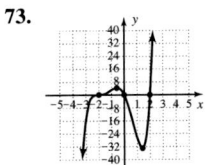 **75.**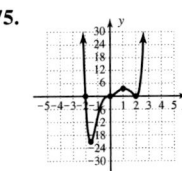

77. $h(x) = (x + 4)(x - \sqrt{3})(x + \sqrt{3})(x - \sqrt{3}i)(x + \sqrt{3}i)$
79. $f(x) = 2(x + \frac{5}{2})(x - \sqrt{2})(x + \sqrt{2})(x - \sqrt{3})(x + \sqrt{3})$

81. $P(x) = \frac{1}{6}(x + 4)(x - 1)(x - 3)$, $P(x) = \frac{1}{6}(x^3 - 13x + 12)$
83. $P(x) = x^4 - 2x^3 - 13x^2 + 14x + 24$
85. a. 280 vehicles above average, 216 vehicles below average, 154 vehicles below average   **b.** 6:00 A.M. $(t = 0)$, 10:00 A.M. $(T = 4)$, 3:00 P.M. $(t = 9)$, 6:00 P.M. $(t = 12)$
c. max: about 300 vehicles above average at 7:30 A.M.;
    min: about 220 vehicles below average at 12 noon

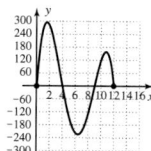

87. a. 3   **b.** 5   **c.** $B(x) = \frac{1}{4}x(x - 4)(x - 9)$, $-\$80,000$
89. a. $f(x) \to \infty, f(x) - \infty$ **b.** $g(x) \to \infty, g(x) \to \infty; x^4 \geq 0$ for all x
91. verified   **93.** $h(x) = \dfrac{1 - 2x}{x^2}; D : x \in \{x | x \neq 0\}; H(x) = \dfrac{1}{x^2 - 2x};$
$D : x \in \{x | x \neq 0, x \neq 2\}$ **95. a.** $x = 2$ **b.** $x = 8$ **c.** $x = 4, x = -6$

Mid-Chapter Check, p. 271

1. a. $x^3 + 8x^2 + 7x - 14 = (x^2 + 6x - 5)(x + 2) - 4$
b. $\dfrac{x^3 + 8x^2 + 7x - 14}{x + 2} = x^2 + 6x - 5 - \dfrac{4}{x + 2}$
2. $f(x) = (2x + 3)(x + 1)(x - 1)(x - 2)$ **3.** $f(-2) = 7$
4. $f(x) = x^3 - 2x + 4$ **5.** $g(2) = -8$ and $g(3) = 5$ have opposite signs
6. $f(x) = (x - 2)(x + 1)(x + 2)(x + 4)$
7. $x = -2, x = 1, x = -1 \pm 3i$
8. 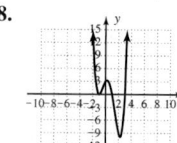   **9.**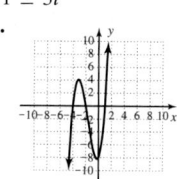

10. a. degree 4; three turning points   **b.** 2 sec
c. $A(t) = (t - 1)^2(t - 3)(t - 5)$ $A(t) = t^4 - 10t^3 + 32t^2 - 38t + 15$
$A(2) = 3$; altitude is 300 ft above hard-deck $A(4) = -9$; altitude is 900 ft below hard-deck

Reinforcing Basic Concepts, pp. 271–272

Exercise 1: 1.532
Exercise 2: $-2.152, 1.765$

Exercises 3.5, pp. 283–289

1. as $x \to -\infty, y \to 2$ **3.** denominator; numerator   **5.** about $x = 98$
7. a. as $x \to -\infty, y \to 2$ **9. a.** as $x \to -\infty, y \to 1$
      as $x \to \infty, y \to 2$ as $x \to \infty, y \to 1$
b. as $x \to 1^-, y \to -\infty$ **b.** as $x \to -2^-, y \to \infty$
     as $x \to 1^+, y \to \infty$ as $x \to -2^+, y \to \infty$
11. reciprocal quadratic, $S(x) = \dfrac{1}{(x + 1)^2} - 2$
13. reciprocal function, $Q(x) = \dfrac{1}{x + 1} - 2$
15. reciprocal quadratic, $v(x) = \dfrac{1}{(x + 2)^2} - 5$
17. $\to -2$ **19.** $\to -\infty$ **21.** $-1; \pm\infty$
23. $x = 3, x \in (-\infty, 3) \cup (3, \infty)$
25. $x = 3, x = -3, x \in (-\infty, -3) \cup (-3, 3) \cup (3, \infty)$
27. $x = \frac{-5}{2}, x = 1, x \in (-\infty, -\frac{5}{2}) \cup (-\frac{5}{2}, 1) \cup (1, \infty)$
29. No V.A., $x \in (-\infty, \infty)$ **31.** $x = 3$, yes; $x = -2$, yes
33. $x = 3$, no   **35.** $x = 2$, yes; $x = -2$, no   **37.** $y = 0$, crosses at $(\frac{3}{2}, 0)$
39. $y = 4$, crosses at $(-\frac{21}{4}, 4)$ **41.** $y = 3$, does not cross

43. $(0, 0)$ cross, $(3, 0)$ cross   **45.** $(-4, 0)$ cross, $(0, 4)$
47. $(0, 0)$ cross, $(3, 0)$ bounce
49. 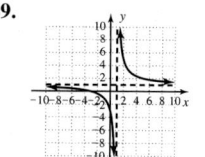   **51.** 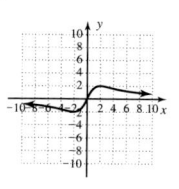   **53.**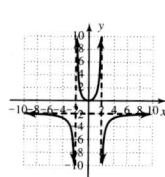

55. 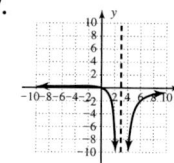   **57.** 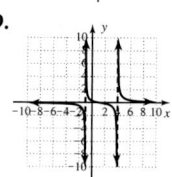   **59.**

61. 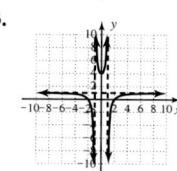   **63.** 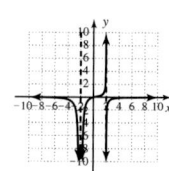   **65.**

67. $f(x) = \dfrac{(x - 4)(x + 1)}{(x + 2)(x - 3)}$ **69.** $f(x) = \dfrac{x^2 - 4}{9 - x^2}$
71. a. Population density approaches zero far from town.
b. 10 mi, 20 mi   **c.** 4.5 mi, 704 people per square mi
73. a. $\$20,000, \$80,000, \$320,000$; cost increases dramatically
b. 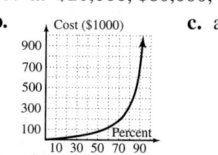   **c.** as $p \to 100^-, C \to \infty$

75. a. 5 hr; about 0.28   **b.** $-0.019, -0.005$; As the number of hours increases, the rate of change decreases.   **c.** $h \to \infty, C \to 0^+$; horizontal asymptote
77. a. 2; 10   **b.** 10; 20   **79. a.**
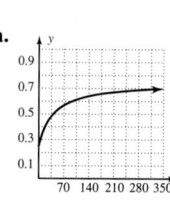   **c.** On average, 6 words will be remembered for life.

b. 35%; 62.5%; 160 gal;   **c.** 160 gal; 200 gal;   **d.** 70%; 75%
81. a. $\$225; \$175$ **b.** 2000 heaters   **c.** 4000   **d.** The horizontal asymptote at $y = 125$ means the average cost approaches $\$125$ as monthly production gets very large. Due to limitations on production (maximum of 5000 heaters) the average cost will never fall below $A(5000) = 135$.
83. a. 5   **b.** 18   **c.** The horizontal asymptote at $y = 95$ means her average grade will approach 95 as the number of tests taken increases; no **d.** 6   **85. a.** 16.0 28.7 65.8 277.8   **b.** 12.7, 37.1, 212.0   **c. a.** 22.4, 40.2, 92.1, 388.9   **b.** 17.8, 51.9, 296.8; answers will vary.
87. a. $q(x) = 3$, horizontal asymptote at $y = 3$; $r(x) = -7x + 10$, graph crosses HA at $x = \dfrac{10}{7}$ **b.** $q(x) = -2$, horizontal asymptote at $y = -2$; $r(x) = 7$, no zeroes—graph will not cross
89. $y = \dfrac{-4}{3}x - \dfrac{1}{3}$ **91.** $39, \frac{3}{2}, 1$

Exercises 3.6, pp. 298–302

1. nonremovable   **3.** two   **5.** Answers will vary.

7. $F(x) = \begin{cases} \dfrac{x^2 - 4}{x + 2} & x \neq -2 \\ -4 & x = -2 \end{cases}$ 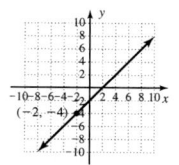

9. $G(x) = \begin{cases} \dfrac{x^2 - 2x - 3}{x + 1} & x \neq -1 \\ -4 & x = -1 \end{cases}$

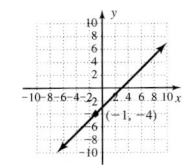

11. $H(x) = \begin{cases} \dfrac{3x - 2x^2}{2x - 3} & x \neq \dfrac{3}{2} \\ \dfrac{-3}{2} & x = \dfrac{3}{2} \end{cases}$

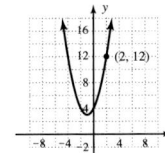

13. $P(x) = \begin{cases} \dfrac{x^3 - 8}{x - 2} & x \neq 2 \\ 12 & x = 2 \end{cases}$

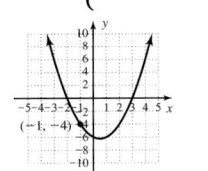

15. $q(x) = \begin{cases} \dfrac{x^3 - 7x - 6}{x + 1} & x \neq -1 \\ -4 & x = -1 \end{cases}$

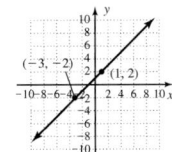

17. $R(x) = \begin{cases} \dfrac{x^3 + 3x^2 - x - 3}{x^2 + 2x - 3} & x \neq -3, x \neq 1 \\ -2 & x = -3 \\ 2 & x = 1 \end{cases}$

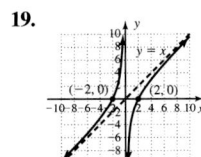

19. 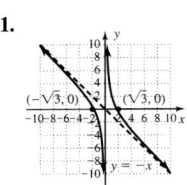   **21.** 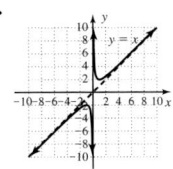   **23.**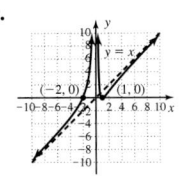

25. 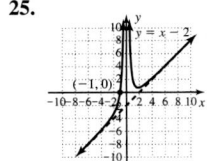   **27.** 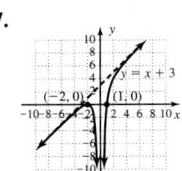   **29.**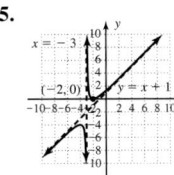

31. 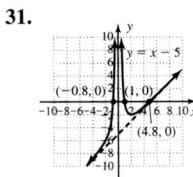   **33.** 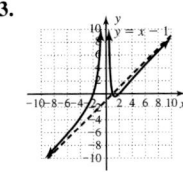   **35.**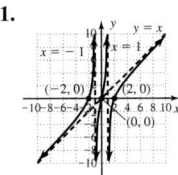

37. 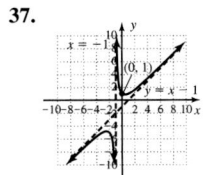   **39.** 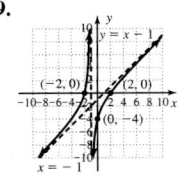   **41.**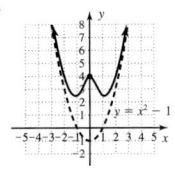

43. 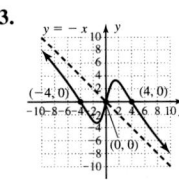   **45.** 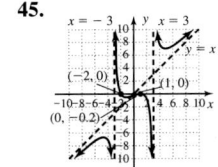   **47.**

49. 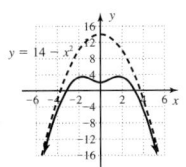   **51.** 119.1

53. a. $a = 5$, $y = 3a + 15$ **b.** 60.5   **c.** 10

55. a. $A(x) = \dfrac{4x^2 + 53x + 250}{x}$; $x = 0$, $g(x) = 4x + 53$

b. cost: \$307, \$372, \$445, Avg. cost: \$307, \$186, \$148.33   **c.** 8, \$116.25

d.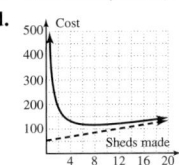

57. a. $S(x, y) = 2x^2 + 4xy$; $V(x, y) = x^2 y$ **b.** $S(x) = \dfrac{2x^3 + 48}{x}$

c. $S(x)$ is asymptotic to $y = 2x^2$.   **d.** $x = 2$ ft 3.5 in.; $y = 2$ ft 3.5 in.

59. a. $A(x, y) = xy$; $R(x, y) = (x - 2.5)(y - 2)$ **b.** $y = \dfrac{2x + 55}{x - 2.5}$

$A(x) = \dfrac{2x^2 + 55x}{x - 2.5}$ **c.** $A(x)$ is asymptotic to $y = 2x + 60$

d. $x \approx 11.16$ in.; $y = 8.93$ in.   **61. a.** $h = \dfrac{V}{\pi r^2}$ **b.** $S = 2\pi r^2 + \dfrac{2V}{r}$

c. $S = \dfrac{2\pi r^3 + 2V}{r}$ **d.** $r \approx 5.76$ cm, $h \approx 11.51$ cm; $S \approx 625.13$ cm^2

63. Answers will vary.   **65.** $S = \dfrac{\pi r^3 + 2V}{r}$; $r = 3.1$ in., $h = 3$ in.

67. $y = \frac{3}{4}x - 4$, $m = \frac{3}{4}$, $(0, -4)$ **69. a.** $P = 30$ cm,   **b.** $\overline{CD} = \frac{60}{13}$ cm,

c. 30 cm^2,   **d.** $A = \frac{750}{169}$ cm^2, and $A = \frac{4320}{169}$ cm^2

Exercises 3.7, pp. 311–315

1. vertical; multiplicity   **3.** empty   **5.** Answers will vary.
7. $x \in (0, 4)$ **9.** $x \in (-\infty, -5] \cup [1, \infty)$ **11.** $x \in \left(-1, \frac{7}{2}\right)$
13. $x \in [-\sqrt{7}, \sqrt{7}]$ **15.** $x \in \left[-\frac{3}{2} - \frac{\sqrt{33}}{2}, -\frac{3}{2} + \frac{\sqrt{33}}{2}\right]$
17. $x \in \left(-\infty, -\frac{5}{3}\right] \cup [1, \infty)$ **19.** $x \in (-\infty, \infty)$ **21.** { }
23. $x \in (-\infty, 5) \cup (5, \infty)$ **25.** { }   **27.** $x \in (-\infty, \infty)$
29. $x \in (-\infty, \infty)$ **31.** $x \in (-\infty, -5] \cup [5, \infty)$
33. $x \in (-\infty, 0] \cup [5, \infty)$ **35.** { }   **37.** $x \in (-3, 5)$
39. $x \in [4, \infty) \cup \{-1\}$ **41.** $x \in (-\infty, -2] \cup \{2\} \cup [4, \infty)$
43. $x \in (-2 - \sqrt{3}, -2 + \sqrt{3})$ **45.** $x \in [-\infty, -3] \cup \{1\}$
47. $x \in (-3, 1) \cup (2, \infty)$ **49.** $x \in (-\infty, -3) \cup (-1, 1) \cup (3, \infty)$
51. $x \in (-\infty, -2) \cup (-2, 1) \cup (3, \infty)$ **53.** $x \in [-1, 1] \cup \{3\}$
55. $x \in [-3, 2)$ **57.** $x \in (-\infty, -2) \cup (-2, -1)$
59. $x \in (-\infty, -2) \cup [2, 3)$ **61.** $x \in (-\infty, -5) \cup (0, 1) \cup (2, \infty)$
63. $x \in (-4, -2] \cup (1, 2] \cup (3, \infty)$ **65.** $x \in (-7, -3) \cup (2, \infty)$
67. $x \in (-\infty, -2] \cup (0, 2)$ **69.** $x \in (-\infty, -17) \cup (-2, 1) \cup (7, \infty)$
71. $x \in \left(-3, \frac{-7}{4}\right] \cup (2, \infty)$ **73.** $x \in (-2, \infty)$ **75.** $x \in (-1, \infty)$
77. $(-\infty, -3) \cup (3, \infty)$ **79.** $x \in (-\infty, -3] \cup [5, \infty)$
81. $x \in [-3, 0] \cup [3, \infty)$ **83.** $x \in (-\infty, -2) \cup (2, 3)$
85. $x \in (-\infty, -2] \cup (-1, 1) \cup [3, \infty)$ **87.** b   **89.** b   **91. a.** verified
b. $D = -4\left(p + \frac{3}{4}\right)(p + 3)^2$, $p = -3$, $q = -2$; $p = \frac{-3}{4}$, $q = \frac{1}{4}$
c. $\left(-\infty, -3\right) \cup \left(-3, \frac{-3}{4}\right)$ **d.** verified
93. $d(x) = k(x^3 - 192x + 1024)$ **a.** $x \in (5, 8]$ **b.** 320 units
c. $x \in [0, 3)$ **d.** 2 ft   **95. a.** verified   **b.** horizontal: $r_2 = 20$, as r_1
increases, r_2 decreases to maintain $R = 40$ vertical: $r_1 = 20$, as r_1 decreases,
r_2 increases to maintain $R = 40$ **c.** $r_1 \in (20, 40)$
97. $R(t) = 0.01t^2 + 0.1t + 30$ **a.** $[0°, 30°)$ **b.** $(20°, \infty)$
c. $(50°, \infty)$ **99. a.** $n \geq 4$ **b.** $n \leq 9$ **c.** 13

101. a. yes, $x^2 \geq 0$ **b.** yes, $\dfrac{x^2}{x^2 + 1} \geq 0$

103. $x(x + 2)(x - 1)^2 > 0; \dfrac{x(x + 2)}{(x - 1)} > 0$

105. $R(x) < 0$ for $x \in (2, 8) \cup (8, 14)$

107. $F(x) = \begin{cases} f(x) & x \neq -4 \\ -6 & x = -4 \end{cases}$

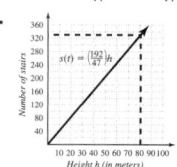

109.

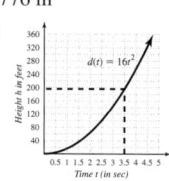

Exercises 3.8, pp. 321–326

1. constant   **3.** $y = \dfrac{k}{x^2}$ **5.** Answers will vary.   **7.** $d = kr$ **9.** $F = ka$

11. $y = 0.025x$

x	y
500	12.5
650	16.25
750	18.75

13. $w = 9.18$; h \$321.30; the hourly wage; $k = \$9.18/\text{hr}$

15. a. $k = \frac{192}{47}$ $S = \frac{192}{47} h$

b.

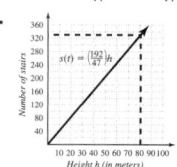

c. 330 stairs   **d.** $S = 331$; yes

17. $A = kS^2$ **19.** $P = kc^2$

21. $k = 0.112$; $p = 0.112 q^2$

q	p
45	226.8
55	338.8
70	548.8

23. $k = 6$, $A = 6s^2$; 55,303,776 m^2

25. a. $k = 16$ $d = 16t^2$ **b.**

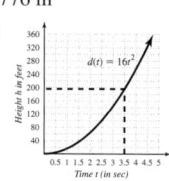

c. about 3.5 sec   **d.** 3.5 sec; yes   **e.** 2.75 sec   **27.** $F = \dfrac{k}{d^2}$ **29.** $S = \dfrac{k}{L}$

31. $Y = \dfrac{12{,}321}{Z^2}$

Z	Y
37	9
74	2.25
111	1

33. $w = \dfrac{3{,}072{,}000{,}000}{r^2}$; 48 kg   **35.** $l = krt$

37. $A = kh(B + b)$ **39.** $V = ktr^2$

41. $C = \dfrac{6.75R}{S^2}$

R	S	C
120	6	22.5
200	12.5	8.64
350	15	10.5

43. $E = 0.5mv^2$; 612.50 J

45. cube root family; answers will vary; 0.054 or 5.4%

Amount A	Rate R
1.0	0.000
1.05	0.016
1.10	0.032
1.15	0.048
1.20	0.063
1.25	0.077

47. $T = \dfrac{48}{V}$; 32 volunteers   **49.** $M = \dfrac{1}{6}E$; ≈ 41.7 kg

51. $D = 21.6\sqrt{S}$; ≈ 144.9 ft   **53.** $C = 8.5LD$; \$76.50

55. $C \approx (4.4 \times 10^{-4})\dfrac{p_1 p_2}{d^2}$; about 223 calls   **57. a.** about 23.39 cm^3,

b. about 191%   **59. a.** $M = kwh^2(\frac{1}{L})$ **b.** 180 lb

61. For f: $\dfrac{\Delta y}{\Delta x} = \dfrac{-10}{3}$ For g: $\dfrac{\Delta y}{\Delta x} = \dfrac{-110}{9}$; less; for both f and g, as

$x \to \infty$, $y \to 0$ **63. a.** about 3.5 ft   **b.** about 6.9 ft

65. $x = 0, x = -2 \pm 2i$

67.

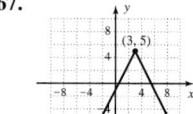

Summary and Concept Review, pp. 326–331

1.

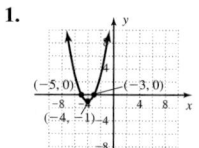

2.

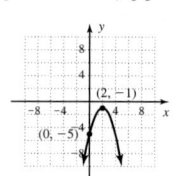

3.

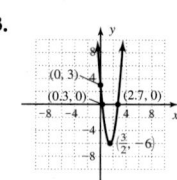

4. a. 0 ft   **b.** 108 ft   **c.** 2.25 sec   **d.** 144 ft, $t = 3$ sec

5. $q(x) = x^2 + 6x + 7$; $R = 8$ **6.** $q(x) = x + 1$; $R = 3x - 4$

7. $\underline{-7}$ 2   13 -6 9   14

         -14 7 -7 -14

      2   -1 1   2 $\underline{|0}$

Since $R = 0$, -7 is a root and $x + 7$ is a factor.

8. $x^3 - 4x + 5 = (x - 2)(x^2 + 2x) + 5$ **9.** $(x + 4)(x + 1)(x - 3)$

10. $h(x) = (x - 1)(x - 4)(x^2 + 2x + 2)$

11. $\frac{1}{2}$ 4   8 -3 -1

        2   5   1

    4   10   2 $\underline{|0}$

Since $R = 0$, $\frac{1}{2}$ is a root and $(x - \frac{1}{2})$ is a factor.

12. $3i$ 1 -2 9 -18

        $3i$ $-9 - 6i$ 18

    1    $-2 + 3i$ $-6i$ $\underline{|0}$

Since $R = 0$, $3i$ is a zero

13. $\underline{-7}$ 1   9   13 -10

        -7 -14 7

    1   2 -1 $\underline{|-3}$

$h(-7) = -3$

14. $P(x) = x^3 - x^2 - 5x + 5$ **15.** $C(x) = x^4 - 2x^3 + 5x^2 - 8x + 4$

16. a. $C(0) = 350$ customers   **b.** more at 2 P.M., 170

c. busier at 1 P.M., $760 > 710$

17. $\{\pm 1, \pm\frac{1}{2}, \pm\frac{1}{4}, \pm 5, \pm 10, \pm\frac{5}{2}, \pm\frac{5}{4}, \pm 2\}$ **18.** $x = -\frac{1}{2}, 2, \frac{5}{2}$

19. $p(x) = (2x + 3)(x - 4)(x + 1)$ **20.** only possibilities are $\pm 1, \pm 3$,

none give a remainder of zero   **21.** [1, 2], [4, 5]; verified

22. one sign change for $g(x) \to 1$ positive zero; three sign changes for $g(-x) \to 3$ or 1 negative zeroes; 1 positive, 3 negative, 0 complex, or 1 positive, 1 negative, 2 complex; verified   **23.** degree 5; up/down; $(0, -4)$
24. degree 4; up/up; $(0, 8)$

25. 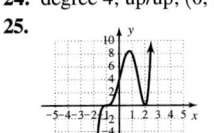   **26.** 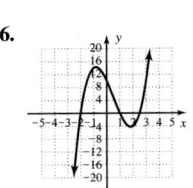   **27.**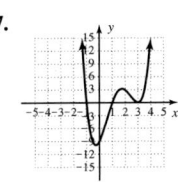

28. a. even   **b.** $x = -2$, odd; $x = -1$, even; $x = 1$, odd
c. deg 6: $P(x) = (x + 2)(x + 1)^2(x - 1)^3$
29. a. $\{x | x \in \mathbb{R}; x \neq -1, 4\}$ **b.** HA: $y = 1$; VA: $x = -1, x = 4$
c. $V(0) = \frac{9}{4}$ (y-intercept); $x = -3, 3$ (x-intercepts)   **d.** $V(1) = \frac{4}{3}$
30. No—even multiplicity; yes—odd multiplicity

31. 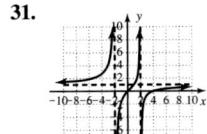   **32.**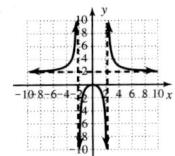

33. $V(x) = \dfrac{x^2 - x - 12}{x^2 - x - 6}$; $V(0) = 2$

34. a. $y = 15$; as $|x| \to \infty$ $A(x) \to 15^+$. As production increases, average cost decreases and approaches 15.   **b.** $x > 2000$
35. removable discontinuity at $(2, -5)$;

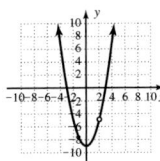

36. $H(x) = \begin{cases} \dfrac{x^2 - 3x - 4}{x + 1} & x \neq -1 \\ -5 & x = -1 \end{cases}$

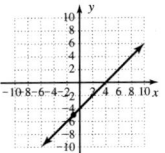

37. 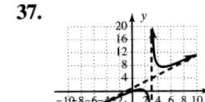   **38.** 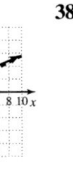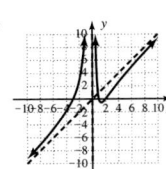   **39. a.**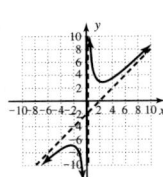

b. about 2450 favors **c.** about \$2.90 ea.
40. factored form $(x + 4)(x - 1)(x - 2) > 0$

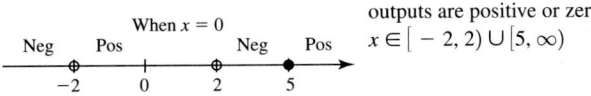

When $x = 0$: outputs are positive for $x \in (-4, 1) \cup (2, \infty)$

41. $\dfrac{x^2 - 3x - 10}{x - 2} = \dfrac{(x - 5)(x + 2)}{x - 2} \geq 0$

When $x = 0$: outputs are positive or zero for $x \in [-2, 2) \cup [5, \infty)$

42. $\dfrac{(x + 2)(x - 1)}{x(x - 2)} \leq 0$

When $x = -1$: outputs are negative or zero for $x \in [-2, 0) \cup [1, 2)$

43. $k = 17.5$; $y = 17.5\sqrt[3]{x}$ **44.** $k = 0.72$; $z = \dfrac{0.72v}{w^2}$

x	y
216	105
0.343	12.25
729	157.5

v	w	z
196	7	2.88
38.75	1.25	17.856
24	0.6	48

45. $t = 160$ **46.** 4.5 sec

Mixed Review, pp. 331–332

1. $y = -2(x - \frac{1}{2})^2 + \frac{9}{2}$ **3.** 80 GB, \$40.00
5. $q(x) = x^3 - 2x^2 + x + 3$; $R = -7$ **7. a.** $P(-1) = 42$
b. $P(1) = -26$ **c.** $P(5) = 6$ **9. a.** $x = 9$; $x = \frac{8}{3}$
b. $P(x) = (x - 2)(x + 1)(x^2 + 9)$; $x = 2, x = -1, x = -3i, x = 3i$
11. 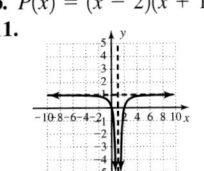   **13.**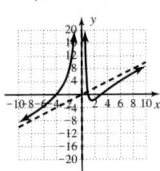

15. $x \in (-\infty, 3) \cup (-2, 2)$
17. a. $V(x) = (24 - 2x)(16 - 2x)(x) = 4x^3 - 80x^2 + 384x$
b. $512 = 4x^3 - 80x^2 + 384x$ $0 = x^3 - 20x^2 + 96x - 128$
c. for $0 < x < 8$, possible rational zeroes are 1, 2, and 4
d. $x = 4$ **e.** $x = 8 - 4\sqrt{2} \approx 2.34$ in.   **19.** $R = kL(\frac{1}{A})$

Practice Test, pp. 332–333

1. a. $f(x) = -(x - 5)^2 + 9$ **b.** $g(x) = \frac{1}{2}(x + 4)^2 + 8$

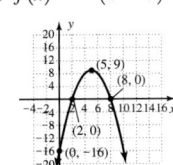   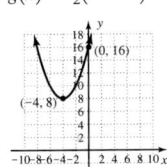

2. $(-2, 0)$, $y = 2x^2 + 4x$ **3. a.** 40 ft, 48 ft   **b.** 49 ft   **c.** 14 sec
4. $x - 5 + \dfrac{14x + 3}{x^2 + 2x + 1}$ **5.** $x^2 + 2x - 9 + \dfrac{-2}{x + 2}$

6. $\underline{-3|}$
```
-3|  1    0   -15  -10   24
        -3    9    18  -24
    1   -3   -6     8    0   R = 0 ✓
```
7. -1 **8.** $P(x) = x^3 - 2x^2 + 9x - 18$
9. $Q(x) = (x - 2)^2(x - 1)^2(x + 1)$, 2 mult 2, 1 mult 2, -1 mult 1
10. a. $\pm 1, \pm 18, \pm 2, \pm 9, \pm 3, \pm 6$ **b.** 1 positive zero, 3 or 1 negative zeroes; 2 or 0 complex zeroes   **c.** $C(x) = (x + 2)(x - 1)(x - 3i)(x + 3i)$
11. a. 1992, 1994, 1998   **b.** 4 yr   **c.** surplus of \$2.5 million
12. 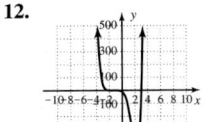   **13.** 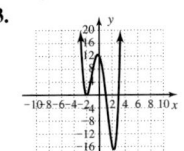   **14.**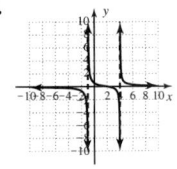

15. a. removal of 100% of the contaminants   **b.** \$500,000; \$3,000,000; dramatic increase   **c.** 88%
16. a. 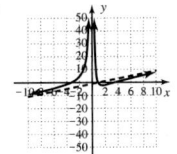   **b.**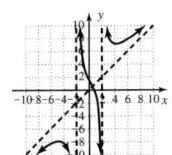

17. 800   **18. a.** $x \in (-\infty, -3] \cup [-1, 4]$ **b.** $x \in (-\infty, -4) \cup (0, 2)$
19. a.

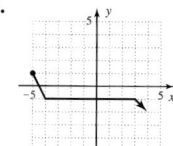

b. $h = -\sqrt[3]{55}$; no   **c.** 28.6% 29.6%
d. $\approx$11.7 hr   **e.** 4 hr 43.7%
f. The amount of the chemical in the blood-stream becomes neglible.

20. 520 lb

Strengthening Core Skills, pp. 334–335

Exercise 1: $x \in (-\infty, 3]$
Exercise 2: $x \in (-2, -1) \cup (2, \infty)$
Exercise 3: $x \in (-\infty, -4) \cup (1, 3)$
Exercise 4: $x \in [-2, \infty)$
Exercise 5: $x \in (-\infty, -2) \cup (2, \infty)$
Exercise 6: $x \in [-3, 1] \cup [3, \infty)$

Cumulative Review Chapters 1–3, pp. 335–336

1. $R = \dfrac{R_1 R_2}{R_1 + R_2}$ **3. a.** $(x - 1)(x^2 + x + 1)$

b. $(x - 3)(x + 2)(x - 2)$ **5.** all reals   **7.** verified

9. $y = \dfrac{11}{60}x + \dfrac{1009}{60}$; 39 min, driving time increases 11 min every 60 days

11. Month 9   **13.** $f^{-1}(x) = \dfrac{x^3 + 3}{2}$

15.

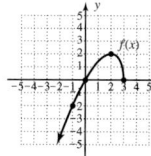

17. $X = 63$ **19.**

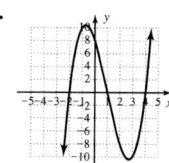

Connections to Calculus Exercises, pp. 339–340

1. $x \in (-\infty, 3]$,
as $x \to -\infty, y \to -\infty$, $(0, 0)$;
$(0, 0), (3, 0)$

3. $x \in [-3, 3]$, as $x \to -3, y \to -3$,
as $x \to 3, y \to 3$; $(0, 0)$;
$(-3, 0), (0, 0), (3, 0)$

5. $x \in (-\infty, \infty)$, down/up,
$(0, -6), (-3, 0), (-1, 0), (2, 0)$

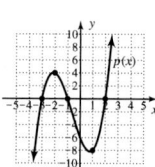

7. max $y = 2$ at $x = 2$ **9.** max at $y \approx 4.5$ at $x = 1.5\sqrt{2}$; min at $y \approx -4.5$

at $x = -1.5\sqrt{2}$ **11.** $\left(\dfrac{-2 - \sqrt{19}}{3}, p\left(\dfrac{-2 - \sqrt{19}}{3}\right)\right)$ or about

$(-2.11, 4.06)$. The skater has a maximum anxiety level of near 4,

about 2 min before starting his routine; $\left(\dfrac{-2 + \sqrt{19}}{3}, p\left(\dfrac{-2 + \sqrt{19}}{3}\right)\right)$ or

about $(0.79, -8.21)$. The skater has a minimum anxiety level of near -8,
shortly after starting his routine.

CHAPTER 4

Exercises 4.1, pp. 350–354

1. second; one   **3.** $(-11, -2), (-5, 0), (1, 2), (19, 4)$ **5.** False, answers
will vary.   **7.** one-to-one   **9.** one-to-one   **11.** not a function   **13.** one-to-one   **15.** not one-to-one, fails horizontal line test: $x = -3$ and $x = 3$ are

paired with $y = 1$ **17.** not one-to-one, $y = 7$ is paired with $x = -2$ and
$x = 2$ **19.** one-to-one   **21.** one-to-one   **23.** not one-to-one; $p(t) > 5$,
corresponds to two x-values   **25.** one-to-one   **27.** one-to-one
29. $f^{-1}(x) = \{(1, -2), (4, -1), (5, 0), (9, 2), (15, 5)\}$
31. $v^{-1}(x) = \{(3, -4), (2, -3), (1, 0), (0, 5), (-1, 12), (-2, 21), (-3, 32)\}$

33. $f^{-1}(x) = x - 5$ **35.** $p^{-1}(x) = \dfrac{-5}{4}x$ **37.** $f^{-1}(x) = \dfrac{x - 3}{4}$

39. $Y_1^{-1} = x^3 + 4$ **41.** $f^{-1}(x) = x^3 + 2$ **43.** $f^{-1}(x) = \sqrt[3]{x - 1}$

45. $f^{-1}(x) = \dfrac{8}{x} - 2$ **47.** $f^{-1}(x) = \dfrac{x}{1 - x}$ **49. a.** $x \geq -5, y \geq 0$

b. $f^{-1}(x) = \sqrt{x} - 5, x \geq 0, y \geq -5$ **51. a.** $x > 3, y > 0$

b. $v^{-1}(x) = \sqrt{\dfrac{8}{x}} + 3, x > 0, y > 3$ **53 a.** $x \geq -4, y \geq -2$

b. $p^{-1}(x) = \sqrt{x + 2} - 4, x \geq -2, y \geq -4$
55. $(f \circ g)(x) = x, (g \circ f)(x) = x$ **57.** $(f \circ g)(x) = x, (g \circ f)(x) = x$
59. $(f \circ g)(x) = x, (g \circ f)(x) = x$ **61.** $(f \circ g)(x) = x, (g \circ f)(x) = x$

63. $f^{-1}(x) = \dfrac{x + 5}{3}$ **65.** $f^{-1}(x) = 2x + 5$ **67.** $f^{-1}(x) = 2x + 6$

69. $f^{-1}(x) = \sqrt[3]{x - 3}$ **71.** $f^{-1}(x) = \dfrac{x^3 - 1}{2}$ **73.** $f^{-1}(x) = 2\sqrt[3]{x} + 1$

75. $f^{-1}(x) = \dfrac{x^2 - 2}{3}, x \geq 0; y \in \left[-\dfrac{2}{3}, \infty\right)$

77. $p^{-1}(x) = \dfrac{x^2}{4} + 3, x \geq 0; y \in [3, \infty)$

79. $v^{-1}(x) = \sqrt{x - 3}, x \geq 3; y \in [0, \infty)$
81.   **83.**

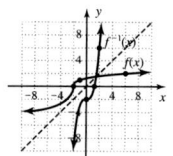

85.   **87.**

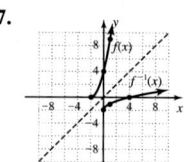

89.

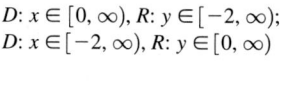

$D: x \in [0, \infty), R: y \in [-2, \infty)$;
$D: x \in [-2, \infty), R: y \in [0, \infty)$

91.

$D: x \in (0, \infty), R: y \in (-\infty, \infty)$;
$D: x \in (-\infty, \infty), R: y \in (0, \infty)$

93.

$D: x \in (-\infty, 4], R: y \in (-\infty, 4]$;
$D: x \in (-\infty, 4], R: y \in (-\infty, 4]$

95. a. 31.5 cm   **b.** The result is 80 cm. It gives the distance of the projector from the screen.   **97. a.** $-63.5°F$ **b.** $f^{-1}(x) = \frac{-2}{7}(x - 59)$; independent: temperature, dependent: altitude   **c.** 22,000 ft   **99. a.** 144 ft

b. $f^{-1}(x) = \dfrac{\sqrt{x}}{4}$, independent: distance fallen, dependent: time fallen

c. 7 sec   **101. a.** 28,260 ft^3 **b.** $f^{-1}(x) = \sqrt[3]{\dfrac{3x}{\pi}}$, independent: volume,

dependent: height **c.** 9 ft **103.** Answers will vary. **105.** d
107. $x \in [-1, 2]$ **109. a.** $P = 2l + 2w$ **b.** $A = \pi r^2$ **c.** $V = \pi r^2 h$
d. $V = \frac{1}{3}\pi r^2 h$ **e.** $C = 2\pi r$ **f.** $A = \frac{1}{2}bh$ **g.** $A = \frac{1}{2}(b_1 + b_2)h$
h. $V = \frac{4}{3}\pi r^3$ **i.** $a^2 + b^2 = c^2$

Exercises 4.2, pp. 362–366

1. b^x; b; b; x **3.** a; 1 **5.** False; for $|b| < 1$ and $x_2 > x_1$, $b^{x_2} < b^{x_1}$, so function is decreasing **7.** 40,000; 5000; 20,000; 27,589.162 **9.** 500; 1.581; 2.321; 221.168 **11.** 10,000; 1975.309; 1487.206; 1316.872

13. increasing

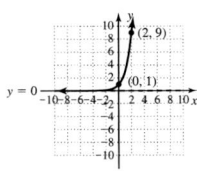

increasing

15. decreasing

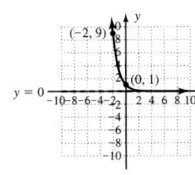

decreasing

17. up 2

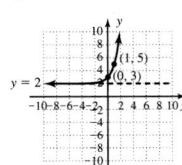

19. left 3

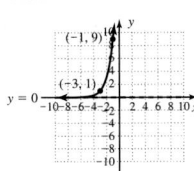

21. reflect across y-axis

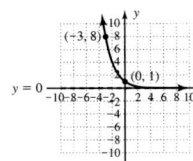

23. reflect across y-axis, up 3

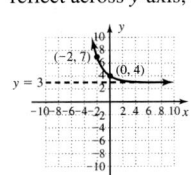

25. left 1, down 3

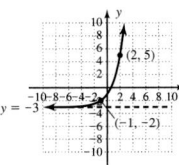

27. up 1

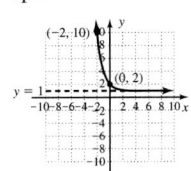

29. right 2

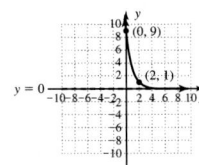

31. down 2

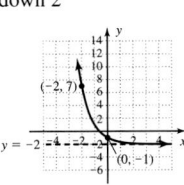

33. e **35.** a **37.** b **39.** 2.718282 **41.** 7.389056 **43.** 4.481689
45. 4.113250

47.

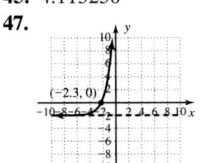

49.

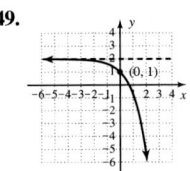

51.

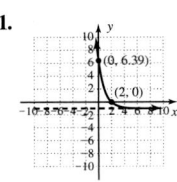

53. 3 **55.** $\frac{3}{2}$ **57.** $-\frac{1}{3}$ **59.** 4 **61.** -3 **63.** 3 **65.** 2 **67.** -2
69. 2 **71.** 3
73. a. 1732, 3000, 5196, 9000
b. yes **c.** as $t \to \infty$, $P \to \infty$

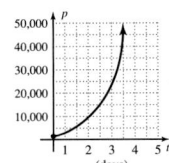

75. no, they will have to wait about 10 min **77. a.** $100,000 **b.** 3 yr
79. a. $\approx$ $86,806 **b.** 3 yr **81. a.** $40 million **b.** 7 yr **83.** 32% transparent **85.** 17% transparent **87.** $\approx$$32,578 **89. a.** 8 g
b. 48 min **91.** 9.5×10^{-7}; answers will vary **93.** 9 **95.** $\frac{3}{2}$

97. a. $\frac{\Delta y}{\Delta x} = 0.3842, 0.056, 0.011, 0.003$; the rate of growth seems to be approaching zero **b.** 16,608 **c.** yes, the secant lines are becoming virtually horizontal
99. $x \in [-2, \infty), y \in [-1, \infty)$

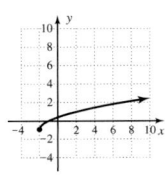

101. a. volume of a sphere **b.** area of a triangle **c.** volume of a rectangular prism **d.** Pythagorean theorem

Exercises 4.3, pp. 375–379

1. $\log_b x$; b; b; greater **3.** $(1, 0)$; 0 **5.** 5; answers will vary **7.** $2^3 = 8$
9. $7^{-1} = \frac{1}{7}$ **11.** $9^0 = 1$ **13.** $8^{\frac{1}{3}} = 2$ **15.** $2^1 = 2$ **17.** $7^2 = 49$
19. $10^2 = 100$ **21.** $e^4 \approx 54.598$ **23.** $\log_4 64 = 3$ **25.** $\log_3 \frac{1}{9} = -2$
27. $0 = \log_e 1$ **29.** $\log_3 27 = -3$ **31.** $\log 1000 = 3$ **33.** $\log_{10} \frac{1}{100} = -2$
35. $\log_4 8 = \frac{3}{2}$ **37.** $\log_4 \frac{1}{8} = \frac{-3}{2}$ **39.** 1 **41.** 2 **43.** 1 **45.** $\frac{1}{2}$
47. -2 **49.** -2 **51.** 1.6990 **53.** 0.4700 **55.** 5.4161 **57.** 0.7841
59. shift up 3

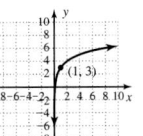

61. shift right 2, up 3

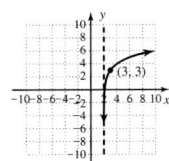

63. shift left 1

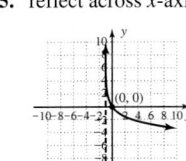

65. reflect across x-axis, shift left 1

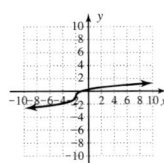

67. II **69.** VI **71.** V **73.** $x \in (-\infty, -1) \cup (3, \infty)$ **75.** $x \in (\frac{3}{2}, \infty)$
77. $x \in (-3, 3)$ **79.** pH $\approx$ 4.1; acid **81. a.** $\approx$4.7 **b.** $\approx$4.9
83. about 3.2 times **85. a.** $\approx$2.4 **b.** $\approx$1.2 **87. a.** 20 dB **b.** 120 dB
89. about 3162 times **91.** 6,194 m **93. a.** about 5434 m
b. 4000 m **95. a.** 2225 items **b.** 2732 items **c.** $117,000
d. verified **97. a.** about 58.6 cfm **b.** about 1605 ft^2 **99. a.** 95%
b. 67% **c.** 39% **101.** $\approx$4.3; acid **103.** Answers will vary. **a.** 0 dB
b. 90 dB **c.** 15 dB **d.** 120 dB **e.** 100 dB **f.** 140 dB **105. a.** $\frac{-2}{3}$
b. $\frac{-3}{2}$ **c.** $\frac{-5}{2}$ **107.** D: $x \in \mathbb{R}$ R: $y \in \mathbb{R}$

109. $x \in (-\infty, -5)$; $f(x) = (x + 5)(x - 4)^2 = x^3 - 3x^2 - 24x + 80$

Mid-Chapter Check, pp. 379–380

1. a. $\frac{2}{3} = \log_{27} 9$ **b.** $\frac{5}{4} = \log_{81} 243$ **2. a.** $8^{\frac{5}{3}} = 32$ **b.** $1296^{0.25} = 6$
3. a. $x = 5$ **b.** $b = \frac{5}{4}$ **4. a.** $x = 3$ **b.** $b = 5$ **5. a.** $71,191.41
b. 6 yr **6.** $F(x) = 4 \cdot 5^{x-3} + 2$ **7.** $f^{-1}(x) = (x - 1)^2 + 3$,
D: $x \in [1, \infty)$; R: $y \in [3, \infty)$; verified **8. a.** $4 = \log_3 81$, verified
b. $4 = \ln 54.598$, verified **9. a.** $27^{\frac{2}{3}} = 9$, verified **b.** $e^{1.4} \approx 4.0552$, verified **10.** $\approx$7.9 times more intense

Reinforcing Basic Concepts, p. 380

Exercise 1: about 158 times   **Exercise 4:** about 398 times
Exercise 2: about 501 times   **Exercise 5:** about 39,811 times
Exercise 3: about 12,589 times

Exercises 4.4, pp. 392–396

1. e **3.** extraneous   **5.** 2.316566275   **7.** $x \approx 29.964$ **9.** $x \approx 1.778$

11. $x \approx 2.200$ **13.** $x \approx 1.260$ **15.** $x = \ln\frac{65}{4} + 2, x \approx 4.7881$

17. $x = \log(78) - 5, x \approx -3.1079$ **19.** $x = -\frac{\ln 2.32}{0.75}, x \approx -1.1221$

21. $x = e^{\frac{8}{3}} - 4, x \approx 10.3919$ **23.** $x = 5 - 10^{1.25}, x \approx -12.7828$

25. $x = \frac{e^{0.4} - 5}{2}, x \approx -1.7541$ **27.** $\ln(2x^2 - 14x)$ **29.** $\log(x^2 - 1)$

31. $\log_3 4$ **33.** $\log\left(\frac{x}{x+1}\right)$ **35.** $\ln\left(\frac{x-5}{x}\right)$ **37.** $\ln(x-2)$ **39.** $\log_2 42$

41. $\log_5(x - 2)$ **43.** $(x+2)\log 8$ **45.** $(2x-1)\ln 5$ **47.** $\frac{1}{2}\log 22$

49. $4\log_5 3$ **51.** $3\log a + \log b$ **53.** $\ln x + \frac{1}{4}\ln y$ **55.** $2\ln x - \ln y$

57. $\frac{1}{2}[\log(x-2) - \log x]$

59. $\ln 7 + \ln x + \frac{1}{2}\ln(3 - 4x) - \ln 2 - 3\ln(x - 1)$

61. $\frac{\ln 60}{\ln 7}$; 2.104076884   **63.** $\frac{\ln 152}{\ln 5}$; 3.121512475

65. $\frac{\log 1.73205}{\log 3}$; 0.499999576   **67.** $\frac{\log 0.125}{\log 0.5}$; 3

69. $f(x) = \frac{\log(x)}{\log(3)}; f(5) \approx 1.4650; f(15) \approx 2.4650; f(45) \approx 3.4650;$

outputs increase by 1; $f(3^3 \cdot 5) = 4.465$

71. $h(x) = \frac{\log(x)}{\log(9)}; h(2) \approx 0.3155; h(4) \approx 0.6309; h(8) \approx 0.9464;$

outputs are multiples of 0.3155; $h(2^4) = 4(0.3155) \approx 1.2619$

73. $x = 32$ **75.** $x = 6.4$ **77.** $x = 20, -5$ is extraneous

79. $x = 2, -\frac{5}{2}$ is extraneous   **81.** $x = 0$ **83.** $x = \frac{5}{2}$ **85.** $x - \frac{2}{3}$

87. $x = \frac{3}{2}$ **89.** $x = \frac{-19}{9}$ **91.** $x = \frac{e^2 - 63}{9}$

93. $x = 2; -9$ is extraneous   **95.** $x = 3e^3 - \frac{1}{2}; x \approx 59.75661077$

97. no solution   **99.** $t = -\frac{1}{2}; -4$ is extraneous

101. $x = 2 + \sqrt{3}, x = 2 - \sqrt{3}$ is extraneous

103. $x = \frac{\ln 231}{\ln 7} - 2; x \approx 0.7968$ **105.** $x = \frac{\ln 128,965}{3\ln 5} + \frac{2}{3}; x \approx 3.1038$

107. $x = \frac{\ln 2}{\ln 3 - \ln 2}; x \approx 1.7095$ **109.** $x = \frac{\ln 9 - \ln 5}{2\ln 5 - \ln 9}; x \approx 0.5753$

111. $x \approx 46.2$ **113.** $t = \frac{\ln\left(\frac{C}{a} - 1\right)}{-k}, t \approx 55.45$

115. a. 30 fish   **b.** about 37 months   **117.** about 3.2 cmHg
119. about 50.2 min   **121.** \$15,641   **123.** 6 hr, 18.0%
125. $M_f = 52.76$ tons   **127. a.** 26 planes   **b.** 9 days
129. a. $\log_3 4 + \log_3 5 = 2.7268$ **b.** $\log_3 4 - \log_3 5 = -0.203$
c. $2\log_3 5 = 2.9298$ **131. a.** d   **b.** e   **c.** b   **d.** f   **e.** a   **f.** c
133. $x = 0.69314718$ **135. a.** $(f \circ g)(x) = 3^{(\log_3 x + 2) - 2} = 3^{\log_3 x} = x;$
$(g \circ f)(x) = \log_3(3^{x-2}) + 2 = x - 2 + 2 = x$
b. $(f \circ g)(x) = e^{(\ln x + 1) - 1} = e^{\ln x} = x;$
$(g \circ f)(x) = \ln e^{x-1} + 1 = x - 1 + 1 = x$
137. a. $y = e^{x\ln 2} = e^{\ln 2^x} = 2^x;$
$y = 2^x \Rightarrow \ln y = x\ln 2, e^{\ln y} = e^{x\ln 2} \Rightarrow y = e^{x\ln 2}$
b. $y = b^x, \ln y = x\ln b, e^{\ln y} = e^{x\ln b}, y = e^{xr}$ for $r = \ln b$
139. Answers will vary.   **141.** b   **143.**

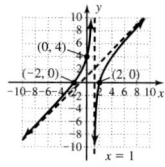

Exercises 4.5, pp. 405–410

1. Compound   **3.** $Q_0 e^{-rt}$ **5.** Answers will vary.   **7.** \$4896   **9.** 250%
11. \$2152.47   **13.** 5.25 yr   **15.** 80%   **17.** 4 yr   **19.** 16 yr
21. \$7561.33   **23.** about 5 yr   **25.** 7.5 yr   **27.** no   **29. a.** no
b. 9.12%   **31.** 7.9 yr   **33.** 7.5 yr   **35. a.** no   **b.** 9.4%   **37. a.** no
b. approx 13,609 euros   **39.** No; \$234,612.01   **41.** about 7 yr

43. 23 yr   **45. a.** no   **b.** \$302.25   **47. a.** $t = \frac{A - P}{pr}$ **b.** $p = \frac{A}{1 + rt}$

49. a. $r = n\left(\sqrt[nt]{\frac{A}{p}} - 1\right)$ **b.** $t = \frac{\ln\left(\frac{A}{p}\right)}{n\ln\left(1 + \frac{r}{n}\right)}$ **51. a.** $Q_0 = \frac{Q(t)}{e^{rt}}$

b. $t = \frac{\ln\left(\frac{Q(t)}{Q_0}\right)}{r}$ **53.** \$709.74   **55. a.** 5.78%   **b.** 91.67 hr   **57.** 0.65 g

59. 816 yr   **61.** about 12.4%   **63.** \$17,027,502.21   **65.** 7.93%
67. 2548.8 m   **69.** $P(x) = x^4 - 4x^3 + 6x^2 - 4x - 15$

Summary and Concept Review, pp. 410–414

1. no   **2.** no   **3.** yes   **4.** $f^{-1}(x) = \frac{x-2}{-3}$ **5.** $f^{-1}(x) = \sqrt{x + 2}$

6. $f^{-1}(x) = x^2 + 1; x \geq 0$ **8.** $f(x): D: x \in (-\infty, \infty),$
7. $f(x): D: x \in [-4, \infty),$ $R: y \in (-\infty, \infty);$
    $R: y \in [0, \infty);$ $f^{-1}(x): D: (-\infty, \infty),$
    $f^{-1}(x): D: x \in [0, \infty);$ $R: y \in (-\infty, \infty)$
    $R: y \in [-4, \infty)$

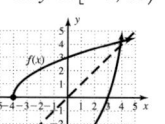 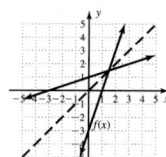

9. $f(x): D: x \in (-\infty, \infty),$
    $R: y \in (0, \infty);$
    $f^{-1}(x): D: x \in (0, \infty),$
    $R: y \in (-\infty, \infty)$

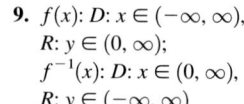 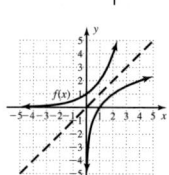

10. a. \$3.05   **b.** $f^{-1}(t) = \frac{t-2}{0.15}, f^{-1}(3.05) = 7$ **c.** 12 days
11.   **12.**   **13.**

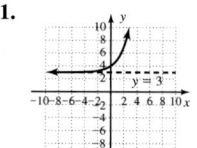 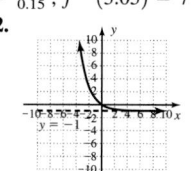 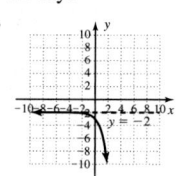

14. 2   **15.** -2 **16.** $\frac{5}{2}$ **17.** 12.1 yr   **18.** $3^2 = 9$ **19.** $5^{-3} = \frac{1}{125}$
20. $e^{3.7612} \approx 43$ **21.** $\log_5 25 = 2$ **22.** $\ln 0.7788 \approx -0.25$
23. $\log_3 81 = 4$ **24.** 5   **25.** -1 **26.** $\frac{1}{2}$
27.   **28.**   **29.**

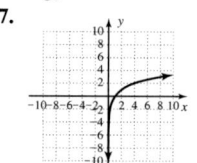 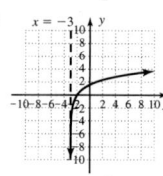 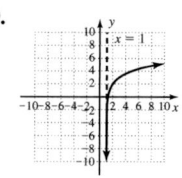

30. $x \in (-\infty, 0) \cup (6, \infty)$ **31.** $x \in \left(-\frac{3}{2}, \infty\right)$ **32. a.** 4.79   **b.** $10^{7.3} I_0$

33. a. $x = e^{32}$ **b.** $x = 10^{2.38}$ **c.** $x = \ln 9.8$ **d.** $x = \frac{1}{2}\log 7$

34. a. $x = \frac{\ln 4}{0.5}, x \approx 2.7726$ **b.** $x = \frac{\ln 19}{0.2}, x \approx 6.3938$

c. $x = \frac{10^3}{3}, x \approx 33.3333$ **d.** $x = e^{-2.75}, x \approx 0.0639$ **35. a.** $\ln 42$

b. $\log_9 30$ **c.** $\ln\left(\frac{x+3}{x-1}\right)$ **d.** $\log(x^2 + x)$ **36. a.** $2\log_5 9$ **b.** $2\log_7 4$

c. $(2x - 1)\ln 5$ **d.** $(3x + 2)\ln 10$ **37. a.** $\ln x + \frac{1}{4}\ln y$
b. $\frac{1}{3}\ln p + \ln q$ **c.** $\frac{5}{3}\log x + \frac{4}{3}\log y - \frac{5}{2}\log x - \frac{3}{2}\log y$
d. $\log 4 + \frac{5}{3}\log p + \frac{4}{3}\log q - \frac{3}{2}\log p - \log q$ **38. a.** $\frac{\log 45}{\log 6} \approx 2.215$
b. $\frac{\log 128}{\log 3} \approx 4.417$ **c.** $\frac{\ln 124}{\ln 2} \approx 6.954$ **d.** $\frac{\ln 0.42}{\ln 5} \approx -0.539$
39. $x = \frac{\ln 7}{\ln 2}$ **40.** $x = \frac{\ln 5}{\ln 3} - 1$ **41.** $x = \frac{2}{1 - \ln 3}$ **42.** $x \approx 6.389$
43. $x = 5$; -2 is extraneous **44.** $x = 4.25$ **45. a.** 17.77%
b. 23.98 days **46.** 38.6 cmHg **47.** 18.5% **48.** Almost, she needs
$42.15 more. **49. a.** no **b.** $268.93 **50.** 55.0%

Mixed Review, pp. 414–415

1. a. $\frac{\log 30}{\log 2} \approx 4.9069$ **b.** -1.5 **c.** $\frac{1}{3}$ **3. a.** $2\log_{10}20$ **b.** $0.05x$
c. $(x - 3)\ln 2$
5. 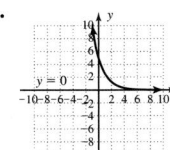 **7.** 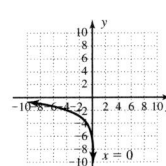 **9. a.** $5^4 = 625$

b. $e^{0.45} = 0.15x$ **c.** $10^7 = 0.1 \times 10^8$ **11. a.** $x \in [1, \infty), y \in [2, \infty)$
b. $g^{-1}(x) = (x - 2)^2 + 1, x \in [2, \infty), y \in [1, \infty)$ **c.** Answers will
vary. **13.** $6 + \log 2$ **15.** $\frac{9}{4} + \frac{\sqrt{129}}{4}$ **17.** $I \approx 6.3 \times 10^{17}$
19. 1.6 m, 1.28 m, 1.02 m, 0.82 m, 0.66 m, 0.52 m

Practice Test, pp. 415–416

1. $3^4 = 81$ **2.** $\log_{25}5 = \frac{1}{2}$ **3.** $\frac{5}{2}\log_b x + 3\log_b y - \log_b z$
4. $\log_b \frac{m\sqrt{n^3}}{\sqrt{p}}$ **5.** $x = 10$ **6.** $x = \frac{-5}{3}$ **7.** 2.68 **8.** -1.24
9. 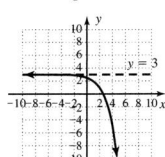 **10.** 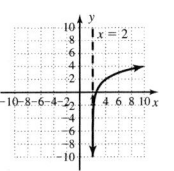 **11. a.** 4.19 **b.** -0.81

12. f is a parabola (hence not one-to-one), $x \in \mathbb{R}, y \in [-3, \infty)$; vertex is
at $(2, -3)$, so restricted domain could be $x \in [2, \infty)$ to create a one-to-one
function; $f^{-1}(x) = \sqrt{x + 3} + 2, x \in [-3, \infty), y \in [2, \infty)$.
13. $x = 1 + \frac{\ln 89}{\ln 3}$ **14.** $x = 1, x = -5$ is extraneous **15.** ≈ 5 yr
16. ≈ 8.7 yr **17.** 19.1 months **18.** 7% compounded semi-annually
19. a. no **b.** $54.09 **20. a.** 10.2 lb **b.** 19 weeks

Strengthening Core Skills, p. 418

Exercise 1: Answers will vary.
Exercise 2: a. $\log(x^2 + 3x)$ **b.** $\ln(x^2 - 4)$ **c.** $\log_{\frac{x}{x + 3}}$
Exercise 3: Answers will vary.
Exercise 4: a. $x\log 3$ **b.** $5\ln x$ **c.** $(3x - 1)\ln 2$

Cumulative Review Chapters 1–4, pp. 418–419

1. $x = 2 \pm 7i$ **3.** $(4 + 5i)^2 - 8(4 + 5i) + 41 = 0$
$-9 + 40i - 32 - 40i + 41 = 0$ $0 = 0$ ✓ **5.** $f(g(x)) = x$ $g(f(x)) = x$
Since $(f \circ g)(x) = (g \circ f)(x)$, they are inverse functions.
7. a. $T(t) = 455t + 2645$ (1991 → year 1) **b.** $\frac{\Delta T}{\Delta t} = \frac{455}{1}$, triple births
increase by 455 each year **c.** $T(6) = 5375$ sets of triplets,
$T(17) = 10,380$ sets of triplets
9. 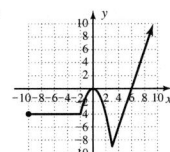 $D: x \in [-10, \infty), R: y \in [-9, \infty)$
$h(x)\uparrow: x \in (-2, 0) \cup (3, \infty)$ $h(x)\downarrow: x \in (0, 3)$

11. $x = 3, x = 2$ (multiplicity 2); $x = -4$ **13.** $\sqrt{\frac{2V}{\pi a}} = b$
15. $f^{-1}(x) = \frac{5x - 3}{2}$

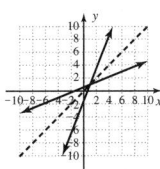

17. $x = 5, x = -6$ is an extraneous root **19. a.** ≈ 88 hp for sport
wagon, ~81 hp for minivan **b.** ≈ 3294 rpm
c. minivan, 208 hp at 5800 rpm

Connections to Calculus Exercises, p. 423

1. $5x + \frac{3}{2}\ln x$ **3.** $5\log x - 3\log y - \frac{1}{2}\log z$
5. $A(x) = \frac{1}{2}x^2$

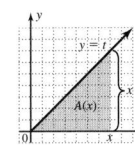

7. verified (factor out $\frac{1}{2}$, then combine like terms and simplify)
9. $t = \frac{1}{2}\ln\left(\frac{x + 1}{1 - x}\right)$

CHAPTER 5
Exercises 5.1, pp. 435–440

1. Complementary; 180; less; greater **3.** $r\theta, \frac{1}{2}r^2\theta$, radians
5. Answers will vary. **7. a.** $77.5°$ **b.** $30.8°$ **9.** $53°$ **11.** $42.5°$
13. $67.555°$ **15.** $285.0025°$ **17.** $45.7625°$ **19.** $20° 15' 00''$
21. $67° 18' 25.2''$ **23.** $275° 19' 48''$ **25.** $5° 27' 9''$ **27.** No,
$19 + 16 < 40$ **29.** $69°$ **31.** $25°$ **33.** 62.5 m
35. $41\sqrt{2}$ ft ≈ 58 ft $+ 10$ ft $= 68$ ft **37.** $-645°, -285°, 435°, 795°$
39. $-765°, -405°, 315°, 675°$ **41.** $s = 980$ m **43.** $\theta = 0.75$ rad
45. $r \approx 1760$ yd **47.** $s = \frac{8\pi}{3}$ mi **49.** $\theta = 0.2575$ rad
51. $r \approx 9.4$ km **53.** $A = 115.6$ km^2 **55.** $\theta = 0.6$ rad
57. $r \approx 3$ m **59.** $\theta = 1.5$ rad; $s = 7.5$ cm; $r = 5$ cm; $A = 18.75$ cm^2
61. $\theta = 4.3$ rad $s = 43$ m; $r = 10$ m; $A = 215$ m^2
63. $\theta = 3$ rad; $A = 864$ mm^2; $s = 72$ mm; $r = 24$ mm
65. 2π rad **67.** $\frac{\pi}{4}$ rad **69.** $\frac{7\pi}{6}$ rad **71.** $\frac{-2\pi}{3}$ rad **73.** 0.4712 rad
75. 3.9776 rad **77.** $60°$ **79.** $30°$ **81.** $120°$ **83.** $720°$ **85.** $165°$
87. $186.4°$ **89.** $171.9°$ **91.** $-143.2°$
93. $h \approx 7.06$ cm; $m \approx 3.76$ cm; $n \approx 13.24$ cm
95. 960.7 mi apart **97. a.** ≈ 50.3 m^2 **b.** $\approx 80°$ **c.** ≈ 17 m
99. a. 1.5π rad/sec **b.** about 15 mi/hr
101. a. 40π rad/min **b.** $\frac{\pi}{6}$ ft/sec ≈ 0.52 ft/sec **c.** about 11.5 sec
103. a. 1000 m **b.** 1000 m
c. 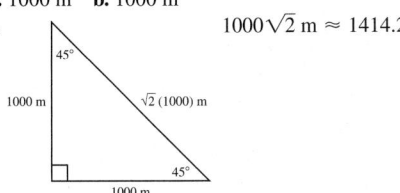 $1000\sqrt{2}$ m ≈ 1414.2 m

105. $50\sqrt{2}$ or about 10.7 mi apart
107. a. $\approx 50.3°$/day; ≈ 0.8788 rad/day **b.** ≈ 0.0366 rad/hr **c.** ≈ 6.67 mi/sec
109. Answers will vary. **111. a.** ≈ 192 yd **b.** ≈ 86.6 rpm
113. $\approx 8.14\%$ **115.** $y = \frac{1}{4}(x - 2)^2 - 4$

Exercises 5.2, pp. 449–454

1. x, y, origin   **3.** x, y, $\frac{y}{x}$, sec t, csc t, cot t **5.** Answers will vary.

7. $(-0.6, -0.8)$ **9.** $\left(\frac{5}{13}, \frac{-12}{13}\right)$ **11.** $\left(\frac{\sqrt{11}}{6}, \frac{5}{6}\right)$ **13.** $\left(\frac{-\sqrt{11}}{4}, \frac{\sqrt{5}}{4}\right)$

15. $(-0.9769, -0.2137)$ **17.** $(-0.9928, 0.1198)$

19. $\left(\frac{-\sqrt{3}}{2}, \frac{-1}{2}\right), \left(\frac{\sqrt{3}}{2}, \frac{1}{2}\right), \left(\frac{\sqrt{3}}{2}, \frac{-1}{2}\right)$

21. $\left(\frac{-\sqrt{11}}{6}, \frac{-5}{6}\right), \left(\frac{-\sqrt{11}}{6}, \frac{5}{6}\right), \left(\frac{\sqrt{11}}{6}, \frac{5}{6}\right)$

23. $(-0.3325, 0.9431), (-0.3325, -0.9431), (0.3325, -0.9431)$

25. $(0.9937, 0.1121), (-0.9937, 0.1121), (-0.9937, -0.1121)$

27. $\left(\frac{1}{2}, \frac{\sqrt{3}}{2}\right)$ is on unit circle   **29.** $\frac{\pi}{4}; \left(\frac{-\sqrt{2}}{2}, \frac{-\sqrt{2}}{2}\right)$

31. $\frac{\pi}{6}; \left(\frac{-\sqrt{3}}{2}, \frac{-1}{2}\right)$ **33.** $\frac{\pi}{4}; \left(\frac{-\sqrt{2}}{2}, \frac{\sqrt{2}}{2}\right)$ **35.** $\frac{\pi}{6}; \left(\frac{\sqrt{3}}{2}, \frac{1}{2}\right)$

37. a. $\frac{\sqrt{2}}{2}$ **b.** $\frac{\sqrt{2}}{2}$ **c.** $\frac{-\sqrt{2}}{2}$ **d.** $\frac{-\sqrt{2}}{2}$ **e.** $\frac{\sqrt{2}}{2}$ **f.** $\frac{-\sqrt{2}}{2}$ **g.** $\frac{\sqrt{2}}{2}$

h. $\frac{-\sqrt{2}}{2}$ **39. a.** -1 **b.** 1 **c.** 1 **d.** 0 **41. a.** $\frac{\sqrt{3}}{2}$ **b.** $\frac{-\sqrt{3}}{2}$

c. $\frac{-\sqrt{3}}{2}$ **d.** $\frac{\sqrt{3}}{2}$ **e.** $\frac{\sqrt{3}}{2}$ **f.** $\frac{\sqrt{3}}{2}$ **g.** $\frac{-\sqrt{3}}{2}$ **h.** $\frac{\sqrt{3}}{2}$

43. a. 0 **b.** 0 **c.** undefined   **d.** undefined

45. $\sin t = 0.6$, $\cos t = -0.8$, $\tan t = -0.75$, $\csc t = 1.\overline{6}$, $\sec t = -1.25$, $\cot t = -1.3$

47. $\sin t = -\frac{12}{13}$, $\cos t = -\frac{5}{13}$, $\tan t = \frac{12}{5}$, $\csc t = -\frac{13}{12}$, $\sec t = -\frac{13}{5}$, $\cot t = \frac{5}{12}$

49. $\sin t = \frac{\sqrt{11}}{6}$, $\cos t = \frac{5}{6}$, $\tan t = \frac{\sqrt{11}}{5}$, $\csc t = \frac{6\sqrt{11}}{11}$, $\sec t = \frac{6}{5}$, $\cot t = \frac{5\sqrt{11}}{11}$

51. $\sin t = \frac{\sqrt{21}}{5}$, $\cos t = \frac{-2}{5}$, $\tan t = \frac{-\sqrt{21}}{2}$, $\csc t = \frac{5\sqrt{21}}{21}$, $\sec t = \frac{-5}{2}$, $\cot t = \frac{-2\sqrt{21}}{21}$

53. $\sin t = \frac{-2\sqrt{2}}{3}$, $\cos t = \frac{-1}{3}$, $\tan t = 2\sqrt{2}$, $\csc t = \frac{-3\sqrt{2}}{4}$, $\sec t = -3$, $\cot t = \frac{\sqrt{2}}{4}$

55. $\sin t = \frac{\sqrt{3}}{2}$, $\cos t = \frac{1}{2}$, $\tan t = \sqrt{3}$, $\csc t = \frac{2\sqrt{3}}{3}$, $\sec t = 2$, $\cot t = \frac{\sqrt{3}}{3}$

57. $\sin t = \frac{\sqrt{2}}{2}$, $\cos t = \frac{-\sqrt{2}}{2}$, $\tan t = -1$, $\csc t = \sqrt{2}$, $\sec t = -\sqrt{2}$, $\cot t = -1$

59. QI, 0.7 **61.** QIV, 0.7 **63.** QI, 1 **65.** QII, 1.1 **67.** QII, -0.4

69. QIV, -3.1 **71.** $\frac{2\pi}{3}$ **73.** $\frac{7\pi}{6}$ **75.** $\frac{2\pi}{3}$ **77.** $\frac{\pi}{2}$ **79.** $\frac{3\pi}{4}, \frac{5\pi}{4}$

81. $\frac{\pi}{2}, \frac{3\pi}{2}$ **83.** $\frac{3\pi}{4}, \frac{5\pi}{4}$ **85.** $0, \pi$ **87. a.** $\left(\frac{3}{4}, \frac{4}{5}\right)$ **b.** $\left(\frac{-3}{4}, \frac{4}{5}\right)$

89. 2.3416 **91.** 1.7832 **93.** 3.5416

95. a. $\left(\frac{5}{13}, \frac{12}{13}, 1\right), \left(\frac{5}{13}\right)^2 + \left(\frac{12}{13}\right)^2 = \frac{25}{169} + \frac{144}{169} = \frac{169}{169} = 1$; $\sin t = \frac{12}{13}$, $\cos t = \frac{5}{13}$, $\tan t = \frac{12}{5}$, $\csc t = \frac{13}{12}$, $\sec t = \frac{13}{5}$, $\cot t = \frac{5}{12}$

b. $\left(\frac{7}{25}, \frac{24}{25}, 1\right), \left(\frac{7}{25}\right)^2 + \left(\frac{24}{25}\right)^2 = \frac{49}{625} + \frac{576}{625} = \frac{625}{625} = 1$; $\sin t = \frac{24}{25}$, $\cos t = \frac{7}{25}$, $\tan t = \frac{24}{7}$, $\csc t = \frac{25}{24}$, $\sec t = \frac{25}{7}$, $\cot t = \frac{7}{24}$

c. $\left(\frac{12}{37}, \frac{35}{37}, 1\right), \left(\frac{12}{37}\right)^2 + \left(\frac{35}{37}\right)^2 = \frac{144}{1369} + \frac{1225}{1369} = \frac{1369}{1369} = 1$; $\sin t = \frac{35}{37}$, $\cos t = \frac{12}{37}$, $\tan t = \frac{35}{12}$, $\csc t = \frac{37}{35}$, $\sec t = \frac{37}{12}$, $\cot t = \frac{12}{35}$

d. $\left(\frac{9}{41}, \frac{40}{41}, 1\right), \left(\frac{9}{41}\right)^2 + \left(\frac{40}{41}\right)^2 = \frac{81}{1681} + \frac{1600}{1681} = \frac{1681}{1681} = 1$; $\sin t = \frac{40}{41}$, $\cos t = \frac{9}{41}$, $\tan t = \frac{40}{9}$, $\csc t = \frac{41}{40}$, $\sec t = \frac{41}{9}$, $\cot t = \frac{9}{40}$

97. a. 5 rad   **b.** 30 rad   **99. a.** 5 dm   **b.** ≈ 6.28 dm   **101. a.** 2.5 AU
b. ≈ 6.28 AU   **103.** yes   **105.** range of $\sin t$ and $\cos t$ is $[-1, 1]$
107. a. $2t \approx 2.2$ **b.** QI   **c.** $\cos t \approx 0.5$ **d.** No   **109. a.** $d = 10$
b. midpoint: $(1, -1)$ **c.** $m = \frac{3}{4}$ **111. a.** $x = -6, 4$ **b.** $x = 24$

Exercises 5.3, pp. 466–470

1. increasing   **3.** $(-\infty, \infty); [-1, 1]$ **5.** Answers will vary.

7.

t	$y = \cos t$
π	-1
$\frac{7\pi}{6}$	$-\frac{\sqrt{3}}{2}$
$\frac{5\pi}{4}$	$-\frac{\sqrt{2}}{2}$
$\frac{4\pi}{3}$	$-\frac{1}{2}$
$\frac{3\pi}{2}$	0
$\frac{5\pi}{3}$	$\frac{1}{2}$
$\frac{7\pi}{4}$	$\frac{\sqrt{2}}{2}$
$\frac{11\pi}{6}$	$\frac{\sqrt{3}}{2}$
2π	1

9. a. II   **b.** V   **c.** IV   **d.** I   **e.** III

11.

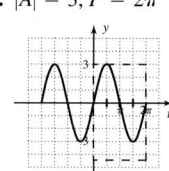

13.

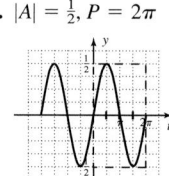

15. $|A| = 3$, $P = 2\pi$

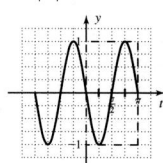

17. $|A| = 2$, $P = 2\pi$

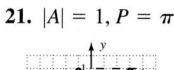

19. $|A| = \frac{1}{2}$, $P = 2\pi$

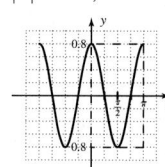

21. $|A| = 1$, $P = \pi$

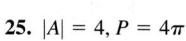
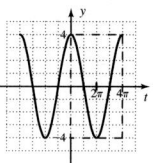

23. $|A| = 0.8$, $P = \pi$

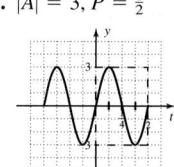

25. $|A| = 4$, $P = 4\pi$

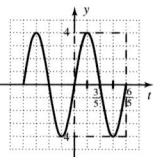

27. $|A| = 3$, $P = \frac{1}{2}$

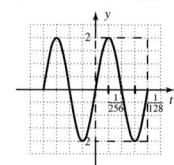

29. $|A| = 4$, $P = \frac{6}{5}$

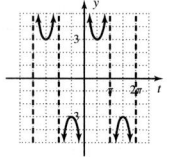

31. $|A| = 2$, $P = \frac{1}{128}$

33.

35.

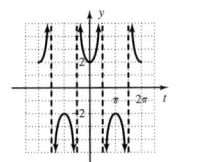

37. $|A| = 2$, $P = \dfrac{\pi}{2}$, k

39. $|A| = 3$, $P = \pi$, f   **41.** $P = 4\pi$, h   **43.** $|A| = \frac{3}{4}$, $P = 5\pi$, b

45. $P = \frac{1}{4}$, j   **47.** $|A| = 4$, $P = \frac{1}{72}$, d   **49.** $y = -\frac{3}{4}\cos(8t)$

51. $y = -0.2\csc(\frac{1}{2}t)$ **53.** $y = 6\cos\left(\dfrac{2\pi}{3}t\right)$

55. red: $y = -\cos x$; blue: $y = \sin x$; $x = \dfrac{3\pi}{4}, \dfrac{7\pi}{4}$

57. red: $y = -2\cos x$; blue: $y = 2\sin(3x)$;

$x = \dfrac{3\pi}{8}, \dfrac{3\pi}{4}, \dfrac{7\pi}{8}, \dfrac{11\pi}{8}, \dfrac{7\pi}{4}, \dfrac{15\pi}{8}$

59. $\cos t = \frac{112}{113}$, $(15, 112, 113)$

61. a. 3 ft   **b.** 80 mi   **c.** $h = 1.5\cos\left(\dfrac{\pi}{40}x\right)$

63. a. $D = -4\cos\left(\dfrac{\pi}{12}t\right)$ **b.** $D \approx 3.86$ **c.** 72°

65. a. $D = 15\cos(\pi t)$ **b.** at center   **c.** Swimming leisurely.
One complete cycle in 2 sec!   **67. a.** Graph a   **b.** 76 days   **c.** 96 days
69. a. 480 nm → blue   **b.** 620 nm → orange
71. $I = 30\sin(50\pi t)$, $I \approx 21.2$ amps

73. Since $m = -M$, 0;

t	y	
0	3	avg. value = 3; shifted up 3 units;
$\dfrac{\pi}{2}$	5	avg. value = 1; amplitude is
π	3	"centered" on average value.
$\dfrac{3\pi}{2}$	1	
2π	3	

75. $g(t)$ has the shortest period;

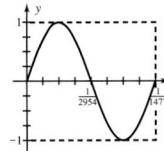

77. distance $= \dfrac{200}{\sqrt{3}}$ yd ≈ 115.5 yd

79. a. $3 - 4i$ **b.** $-1 - 6i$ **c.** $7 - 3i$ **d.** $\dfrac{-3 - 7i}{2}$

Exercises 5.4, pp. 479–484

1. π; $P = \dfrac{\pi}{|B|}$ **3.** odd; $-f(t)$; -0.268 **5. a.** use $\dfrac{\cos t}{\sin t}$

b. use reciprocals of $\tan t$ **7.** $0, \dfrac{1}{\sqrt{3}}, 1, \sqrt{3}$, und.

9. 1.6, 0.8, 0.5, 1.4, 0.7, 1.2   **11. a.** -1 **b.** $\sqrt{3}$ **c.** -1 **d.** $\sqrt{3}$

13. a. $\dfrac{7\pi}{4}$ **b.** $\dfrac{7\pi}{6}$ **c.** $\dfrac{5\pi}{3}$ **d.** $\dfrac{3\pi}{4}$ **15.** und., $\sqrt{3}, 1, \dfrac{1}{\sqrt{3}}, 0$

17. $\dfrac{-13\pi}{24}, \dfrac{35\pi}{24}, \dfrac{59\pi}{24}$ **19.** $-1.6, 4.6, 7.8$ **21.** $\dfrac{\pi}{10} + \pi k, k \in \mathbb{Z}$

23. $\dfrac{\pi}{12} + \pi k, k \in \mathbb{Z}$

25.

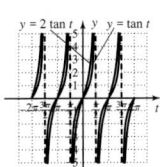

27.

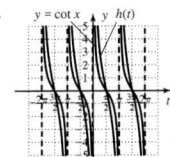

29.

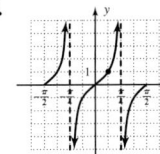

31.

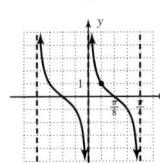

33.

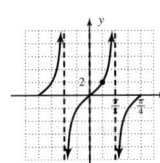

35.

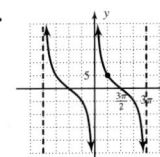

37.

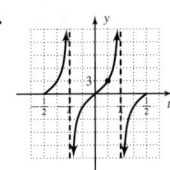

39.

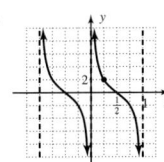

41. $y = 3\tan\left(\dfrac{1}{2}t\right)$ **43.** $y = 2\cot\left(\dfrac{2\pi}{3}t\right)$ **45.** $\dfrac{\pi}{8}, \dfrac{3\pi}{8}$

47. about 137.8 ft

49. $y = 5.2\tan\left(\dfrac{\pi}{12}x\right)$; $P = 12$; asymptotes at $x = 6 + 12k, k \in \mathbb{Z}$; using

(3, 5.2), $|A| = 5.2$; at $x = 2$, model gives $y \approx 3.002$; at $x = -2$,
model gives $y \approx -3.002$; answers will vary.

51. Answers will vary; $y = 11.95\tan\theta$; $P = 180°$; asymptotes at
$\theta = 90° + 180°k$; $|A| = 11.95$ from $(30°, 6.9$ cm$)$; pen is
≈ 12 cm long

53. a. 20π cm ≈ 62.8 cm   **b.** 80 cm; it is a square

c.

n	P
10	64.984
20	63.354
30	63.063
100	62.853

getting close to 20π

55.

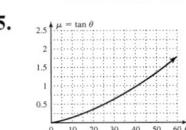

a. no; $\approx 35°$ **b.** 1.05
c. Angles will be greater than 68.2°; soft rubber on sandstone
57. a. 5.67 units   **b.** 86.5°   **c.** Yes. Range of $\tan\theta$ is $(-\infty, \infty)$.
d. The closer θ gets to 90°, the longer the line segment gets.
59. $[2, 3] \to \approx 7.1$ m/sec; $[3, 3.5] \to \approx 26.1$ m/sec;
$[3.5, 3.8] \to \approx 128$ m/sec. The velocity of the beam is increasing
dramatically, $[3.9, 3.99] \to \approx 12{,}733$ m/sec
61. a. x-intercepts: (0, 0), (3, 0); y-intercepts: (0, 0); vertical

asymptotes: $x = -2$, $x = 2$; horizontal asymptote: $y = \dfrac{3}{2}$

b. x-intercept: $(-1, 0)$; y-intercept: none; vertical asymptotes:
$x = 0$, $x = 4$; horizontal asymptote: $y = 0$

c. x-intercepts: $(-1, 0)$, $(1, 0)$; y-intercepts: $\left(0, -\dfrac{1}{2}\right)$; vertical asymptote:

$x = -2$ slant asymptote: $y = x - 2$

63. ≈ 7.37 hr

Mid-Chapter Check, pp. 484–485

1. a. 36.11°N, 115.08°W   **b.** 2495.7 mi.

2. $\theta = 4.3$; $A = 860$ cm^2

3. a. $\dfrac{1}{\sqrt{3}}$ **b.** $\dfrac{-\sqrt{2}}{2}$

4. a. ≈ 1.0353 **b.** ≈ 8.9152

5. $y = \dfrac{-2}{3}$; $\sin\theta = \dfrac{-2}{3}$, $\csc\theta = \dfrac{-3}{2}$, $\cos\theta = \dfrac{-\sqrt{5}}{3}$, $\sec\theta = \dfrac{-3}{\sqrt{5}}$,

$\tan\theta = \dfrac{2}{\sqrt{5}}$, $\cot\theta = \dfrac{\sqrt{5}}{2}$ **6.** 221.8°, 3.8711

7. asymptotes: $x = -5, -3, -1, 1, 3, 5$ **8.** $|A| = 3$, $P = 4$

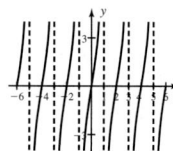
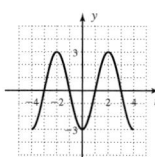

9. a. QIV   **b.** $2\pi - 5.94 \approx 0.343$ **c.** $\sin t$, $\tan t$

10. a. $|A| = 6$, $P = \dfrac{3\pi}{4}$ **b.** $f(t) = -6\cos\left(\dfrac{8}{3}t\right)$

c. $f(\pi) = 3$

Reinforcing Basic Concepts, pp. 485–486

1. $\left(-\dfrac{1}{2}, \dfrac{\sqrt{3}}{2}\right)$, $\cos t = -\dfrac{1}{2}$, $\sin t = \dfrac{\sqrt{3}}{2}$

2. $t = \dfrac{5\pi}{6}$, negative since $x < 0$

3. QIV, negative since $y < 0$

4. QI, $\cos t = \dfrac{1}{2}$, $\sin t = \dfrac{\sqrt{3}}{2}$, $t = \dfrac{\pi}{3}$

Exercises 5.5, pp. 494–499

1. $y = A\sin(Bt + C) + D$, $y = A\cos(Bt + C) + D$

3. $0 \le Bt + C < 2\pi$ **5.** Answers will vary.

7. a. $|A| = 50$, $P = 24$ **b.** ≈ -25 **c.** [1.6, 10.4]

9. a. $|A| = 200$, $P = 3$ **b.** -175 **c.** [1.75, 2.75]

11. $y = 40\sin\left(\dfrac{\pi}{15}t\right) + 60$

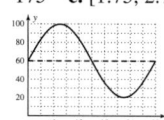

13. $y = 8\sin\left(\dfrac{\pi}{180}t\right) + 12$

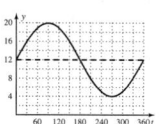

15. a. $y = 5\sin\left(\dfrac{\pi}{12}t\right) + 34$ **b.**

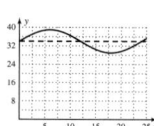

c. $\approx$ 1:30 A.M., 10:30 A.M.

17. a. $y = -6.4\cos\left(\dfrac{\pi}{6}t\right) + 12.4$

b. 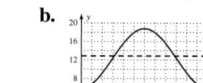   **c.** $\approx$ 134 days

19. a. $P = 11$ yr   **c.** max $= 1200$, min $= 700$

b.   **d.** about 2 yr.

21. $P(t) = 250\cos\left[\dfrac{2\pi}{11}(t - 2.75)\right] + 950$; $P(t) = 250\sin\left(\dfrac{2\pi}{11}t\right) + 950$

23. $|A| = 120$; $P = 24$; HS: 6 units right; VS: (none); PI: $6 \le t < 30$

25. $|A| = 1$; $P = 12$; HS: 2 units right; VS: (none); PI: $2 \le t < 14$

27. $|A| = 1$; $P = 8$; HS: $\frac{2}{3}$ unit right; VS: (none); PI: $\frac{2}{3} \le t < \frac{26}{3}$

29. $|A| = 24.5$; $P = 20$; HS: 2.5 units right;
VS: 15.5 units up; PI: $2.5 \le t < 22.5$

31. $|A| = 28$; $P = 12$; HS: $\frac{5}{2}$ units right; VS: 92 units up; PI: $\frac{5}{2} \le t < \frac{29}{2}$

33. $|A| = 2500$; $P = 8$; HS: $\frac{1}{3}$ unit left;
VS: 3150 units up; PI: $-\frac{1}{3} \le t < \frac{23}{3}$

35. $y = 250\sin\left(\dfrac{\pi}{12}t\right) + 350$ **37.** $y = 5\sin\left(\dfrac{\pi}{50}t + \dfrac{\pi}{2}\right) + 13$

39. $y = 4\sin\left(\dfrac{\pi}{180}t + \dfrac{\pi}{4}\right) + 7$ **41.**

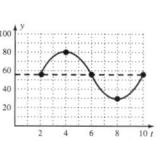

43.

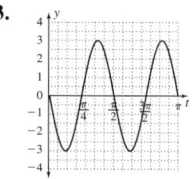

45. $P = \dfrac{2\pi}{B}$, $B = \dfrac{2\pi}{P}$; $f = \dfrac{1}{P}$, $P = \dfrac{1}{f}$; $B = \dfrac{2\pi}{1/f} = 2\pi f$.
$A\sin(Bt) = A\sin[(2\pi f)t]$

47. a. $P = 4$ sec, $f = \dfrac{1}{4}$ cycle/sec   **b.** -4.24 cm, moving away

c. -4.24 cm, moving toward   **d.** about 1.76 cm. avg. vel. $= 3.52$ cm/sec, greater, still gaining speed

49. $d(t) = 15\cos\left(\dfrac{5\pi}{4}t\right)$ **51.** red $\to$ D$_3$; blue $\to$ A#$_3$

53. D$_3$: $y = \sin[146.84 (2\pi t)]$; $P \approx 0.0068$ sec;
G$_4$: $y = \sin[392 (2\pi t)]$; $P \approx 0.00255$ sec

55. a. Caracas: $\approx$11.4 hr, Tokyo: $\approx$9.9 hr
   b. (i) Same # of hours on 79th day & 261st day
      (ii) Caracas: $\approx$81 days, Tokyo: $\approx$158 days

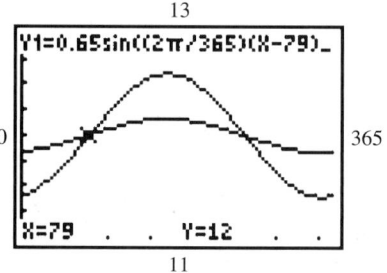

57. a. Adds 12 hr. The sinusoidal behavior is actually based on hours more/less than an average of 12 hr of light.
   b. Means 12 hr of light and dark on March 20, day 79 (Solstice!).
   c. Additional hours of deviation from average. In the north, the planet is tilted closer toward the Sun or farther from Sun, depending on date. Variations will be greater!

59. QIII; $3.7 - \pi \approx 0.5584$

61. sum: -2, difference: $2i\sqrt{5}$, product: 6, quotient: $\dfrac{-2}{3} - \dfrac{i\sqrt{5}}{3}$

Exercises 5.6, pp. 506–511

1. $\theta = \tan^{-1}x$ **3.** opposite; hypotenuse

5. To find the measure of all three angles and all three sides.

7. $\sin\theta = \frac{12}{13}$, $\csc\theta = \frac{13}{12}$, $\sec\theta = \frac{13}{5}$, $\tan\theta = \frac{12}{5}$, $\cot\theta = \frac{5}{12}$

9. $\cos \theta = \frac{13}{85}$, $\sec \theta = \frac{85}{13}$, $\cot \theta = \frac{13}{84}$, $\sin \theta = \frac{84}{85}$, $\csc \theta = \frac{85}{84}$

11. $\sin \theta = \frac{11}{5\sqrt{5}}$, $\tan \theta = \frac{11}{2}$, $\csc \theta = \frac{5\sqrt{5}}{11}$, $\cos \theta = \frac{2}{5\sqrt{5}}$, $\sec \theta = \frac{5\sqrt{5}}{2}$

13.

Angles	Sides
$A = 30°$	$a = 98$ cm
$B = 60°$	$b = 98\sqrt{3}$ cm
$C = 90°$	$c = 196$ cm

15.

Angles	Sides
$A = 45°$	$a = 9.9$ mm
$B = 45°$	$b = 9.9$ mm
$C = 90°$	$c = 9.9\sqrt{2}$ mm

17.

Angles	Sides
$A = 22°$	$a = 14$ m
$B = 68°$	$b \approx 34.65$ m
$C = 90°$	$c \approx 37.37$ m

verified

19.

Angles	Sides
$A = 32°$	$a = 5.6$ mi
$B = 58°$	$b \approx 8.96$ mi
$C = 90°$	$c \approx 10.57$ mi

verified

21.

Angles	Sides
$A = 65°$	$a = 625$ mm
$B = 25°$	$b \approx 291.44$ mm
$C = 90°$	$c \approx 689.61$ mm

23. 0.4540 **25.** 0.8391 **27.** 1.3230 **29.** 0.9063 **31.** 27° **33.** 40°

35. 40.9° **37.** 65° **39.** 44.7° **41.** 20.2° **43.** 18.4° **45.** 46.2°

47. 61.6° **49.** 21.98 mm **51.** 3.04 mi **53.** 177.48 furlongs

55. They have like values. **57.** They have like values.

59. 43° **61.** 21° **63.** $\frac{1}{2}, \frac{\sqrt{3}}{2}, \frac{\sqrt{3}}{3}, \frac{\sqrt{3}}{2}, \frac{1}{2}, \sqrt{3}, 2, \frac{2\sqrt{3}}{3}, \sqrt{3}$

65. $6 + 2\sqrt{3}$ **67.** $7 + 4\sqrt{3}$ **69.** $\theta \approx 11.0°, \beta \approx 23.9°, \gamma \approx 145.1°$

71. approx. 300.6 m **73.** approx. 1483.8 ft **75.** approx. 118.1 mph

77. a. approx. 250.0 yd **b.** approx. 351.0 yd **c.** approx. 23.1 yd

79. approx. 1815.2 ft; approx. 665.3 ft **81.** approx. 386.0 Ω

83. a. 875 m **b.** 1200 m
   c. 1485 m; 36.1°

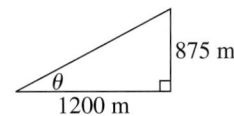

85. approx. 450 ft **87. a.** approx. 20.2 cm for each side **b.** approx. 35.3°

89. $\cot u = \dfrac{x}{h}$

$x = h \cot u$

$\cot v = \dfrac{x - d}{h}$

$\cot v = \dfrac{h \cot u - d}{h}$

$h \cot v = h \cot u - d$

$d = h \cot u - h \cot v$

$h = \dfrac{d}{\cot u - \cot v}$

91. a. approx. 3055.6 mi **b.** approx. 9012.8 mi **c.** approx. 7 hr, 13 min

93. a. local max: $(-5, 2), (2, 3)$;
   local min: $(-2, -1), (-7, -2), (6, -3)$
   b. zeroes: $x = -6, -3, -1, 4$
   c. $T(x)\downarrow$: $x \in (-5, -2) \cup (2, 6)$; $T(x)\uparrow$: $x \in (-7, -5) \cup (-2, 2)$
   d. $T(x) > 0$: $x \in (-6, -3) \cup (-1, 4)$;
   $T(x) < 0$: $x \in (-7, -6) \cup (-3, -1) \cup (4, 6)$

95. $d \approx 53.74$ in.
$D \approx 65.82$ in.

Exercises 5.7, pp. 519–522

1. origin; x-axis **3.** positive; clockwise **5.** Answers will vary.

7. slope $= \sqrt{3}$, equation: $y = \sqrt{3}x$,

$\sin 60° = \dfrac{\sqrt{3}}{2}$, $\cos 60° = \dfrac{1}{2}$, $\tan 60° = \sqrt{3}$

9.

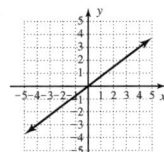

QI/III;

$(4, 3)$: $\sin \theta = \dfrac{3}{5}$; $(-4, -3)$: $\sin \theta = -\dfrac{3}{5}$

$\cos \theta = \dfrac{4}{5}$ $\qquad \cos \theta = -\dfrac{4}{5}$

$\tan \theta = \dfrac{3}{4}$ $\qquad \tan \theta = \dfrac{3}{4}$

11.

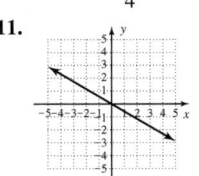

QII/QIV;

$(-3, \sqrt{3})$: $\sin \theta = \dfrac{1}{2}$; $(3, -\sqrt{3})$: $\sin \theta = -\dfrac{1}{2}$

$\cos \theta = -\dfrac{\sqrt{3}}{2}$ $\qquad \cos \theta = \dfrac{\sqrt{3}}{2}$

$\tan \theta = -\dfrac{1}{\sqrt{3}}$ $\qquad \tan \theta = -\dfrac{1}{\sqrt{3}}$

13. $\sin \theta = \dfrac{15}{17}$, $\csc \theta = \dfrac{17}{15}$, $\cos \theta = \dfrac{8}{17}$, $\sec \theta = \dfrac{17}{8}$, $\tan \theta = \dfrac{15}{8}$,
$\cot \theta = \dfrac{8}{15}$

15. $\sin \theta = \dfrac{21}{29}$, $\csc \theta = \dfrac{29}{21}$, $\cos \theta = \dfrac{-20}{29}$,
$\sec \theta = \dfrac{-29}{20}$, $\tan \theta = \dfrac{-21}{20}$, $\cot \theta = \dfrac{-20}{21}$

17. $\sin \theta = \dfrac{-\sqrt{2}}{2}$, $\csc \theta = \dfrac{-2}{\sqrt{2}}$, $\cos \theta = \dfrac{\sqrt{2}}{2}$,
$\sec \theta = \dfrac{2}{\sqrt{2}}$, $\tan \theta = -1$, $\cot \theta = -1$

19. $\sin \theta = \dfrac{1}{2}$, $\csc \theta = 2$, $\cos \theta = \dfrac{\sqrt{3}}{2}$,
$\sec \theta = \dfrac{2}{\sqrt{3}}$, $\tan \theta = \dfrac{1}{\sqrt{3}}$, $\cot \theta = \sqrt{3}$

21. $\sin \theta = \dfrac{4}{\sqrt{17}}$, $\csc \theta = \dfrac{\sqrt{17}}{4}$, $\cos \theta = \dfrac{1}{\sqrt{17}}$,
$\sec \theta = \sqrt{17}$, $\tan \theta = 4$, $\cot \theta = \dfrac{1}{4}$

23. $\sin \theta = \dfrac{-2}{\sqrt{13}}$, $\csc \theta = \dfrac{-\sqrt{13}}{2}$, $\cos \theta = \dfrac{-3}{\sqrt{13}}$,
$\sec \theta = \dfrac{-\sqrt{13}}{3}$, $\tan \theta = \dfrac{2}{3}$, $\cot \theta = \dfrac{3}{2}$

25. $\sin \theta = \dfrac{6}{\sqrt{61}}$, $\csc \theta = \dfrac{\sqrt{61}}{6}$, $\cos \theta = \dfrac{-5}{\sqrt{61}}$,
$\sec \theta = \dfrac{-\sqrt{61}}{5}$, $\tan \theta = \dfrac{-6}{5}$, $\cot \theta = \dfrac{-5}{6}$

27. $\sin \theta = \dfrac{-2\sqrt{5}}{\sqrt{21}}$, $\csc \theta = \dfrac{-\sqrt{21}}{2\sqrt{5}}$, $\cos \theta = \dfrac{1}{\sqrt{21}}$,
$\sec \theta = \sqrt{21}$, $\tan \theta = -2\sqrt{5}$, $\cot \theta = \dfrac{-1}{2\sqrt{5}}$

29. $x = 0$, $y = k$; $k > 0$; $r = |k|$;
$\sin 90° = \dfrac{k}{k}$, $\cos 90° = \dfrac{0}{k}$, $\tan 90° = \dfrac{k}{0}$,
$\sin 90° = 1$, $\cos 90° = 0$, $\tan 90°$ undefined
$\csc 90° = 1$, $\sec 90°$ undefined
$\cot 90° = 0$

31. 60°   **33.** 45°   **35.** 45°   **37.** 68°   **39.** 40°   **41.** 11.6°

43. QII   **45.** QII

47. $\sin\theta = -\dfrac{1}{2}; \cos\theta = \dfrac{\sqrt{3}}{2}; \tan\theta = -\dfrac{1}{\sqrt{3}}$

49. $\sin\theta = \dfrac{-\sqrt{2}}{2}; \cos\theta = \dfrac{\sqrt{2}}{2}; \tan\theta = -1$

51. $\sin\theta = \dfrac{-\sqrt{3}}{2}; \cos\theta = \dfrac{-1}{2}; \tan\theta = \sqrt{3}$

53. $\sin\theta = -\dfrac{1}{2}; \cos\theta = \dfrac{-\sqrt{3}}{2}; \tan\theta = \dfrac{1}{\sqrt{3}}$

55. $x = 4, y = -3, r = 5;$ QIV; $\sin\theta = \dfrac{-3}{5}, \csc\theta = \dfrac{-5}{3}, \cos\theta = \dfrac{4}{5},$
$\sec\theta = \dfrac{5}{4}, \tan\theta = \dfrac{-3}{4}, \cot\theta = \dfrac{-4}{3}$

57. $x = -12, y = -35, r = 37;$ QIII; $\sin\theta = \dfrac{-35}{37}, \csc\theta = \dfrac{-37}{35},$
$\cos\theta = \dfrac{-12}{37}, \sec\theta = \dfrac{-37}{12}, \tan\theta = \dfrac{35}{12}, \cot\theta = \dfrac{12}{35}$

59. $x = 2\sqrt{2}, y = 1, r = 3;$ QI; $\sin\theta = \dfrac{1}{3}, \csc\theta = 3, \cos\theta = \dfrac{2\sqrt{2}}{3},$
$\sec\theta = \dfrac{3}{2\sqrt{2}}, \tan\theta = \dfrac{1}{2\sqrt{2}}, \cot\theta = 2\sqrt{2}$

61. $x = -\sqrt{15}, y = -7, r = 8;$ QIII; $\sin\theta = \dfrac{-7}{8}, \csc\theta = \dfrac{-8}{7},$
$\cos\theta = \dfrac{-\sqrt{15}}{8}, \sec\theta = -\dfrac{8}{\sqrt{15}}, \tan\theta = \dfrac{7}{\sqrt{15}}, \cot\theta = \dfrac{\sqrt{15}}{7}$

63. $52° + 360°k$ **65.** $87.5° + 360°k$ **67.** $225° + 360°k$

69. $-107° + 360°k$ **71.** $\dfrac{\sqrt{3}}{2}, \dfrac{-1}{2}, -\sqrt{3}$ **73.** $-\dfrac{1}{2}, \dfrac{\sqrt{3}}{2}, \dfrac{1}{\sqrt{3}}$

75. $\sin\theta = \dfrac{-\sqrt{3}}{2}, \cos\theta = \dfrac{-1}{2}, \tan\theta = \sqrt{3}$

77. $\sin\theta = \dfrac{-\sqrt{3}}{2}, \cos\theta = -\dfrac{1}{2}, \tan\theta = \sqrt{3}$

79. $\sin\theta = \dfrac{-1}{2}, \cos\theta = \dfrac{-\sqrt{3}}{2}, \tan\theta = \dfrac{1}{\sqrt{3}}$

81. $\sin\theta = \dfrac{-1}{2}, \cos\theta = \dfrac{-\sqrt{3}}{2}, \tan\theta = \dfrac{1}{\sqrt{3}} = -1$

83. QIV, neg., -0.0175 **85.** QIV, neg., -1.6643

87. QIV, neg., -1.5890 **89.** QI, pos., 0.0872

91. a. approx. 144.78 units²   **b.** 53°   **c.** The parallelogram is a rectangle
whose area is $A = ab$.   **d.** $A = \dfrac{ab}{2}\sin\theta$

93. $\theta = 60° + 360°k$ and $\theta = 300° + 360°k$

95. $\theta = 240° + 360°k$ and $\theta = 300° + 360°k$

97. $\theta = 61.1° + 360°k$ and $\theta = 118.9° + 360°k$

99. $\theta = 113.0° + 360°k$ and $\theta = 293.0° + 360°k$

101. $1890°; 90° + 360°k$ **103.** head first; 900°

105. approx. 701.6°   **107.** 343.12 in²   **109.** Answers will vary.

111. a. 12,960°   **b.** 125.66 in.   **c.** 15,080 in.   **d.** 85.68 mph

113. about 555.4 ft   **115.** $y = -\frac{5}{4}x + 2$

Summary and Concept Review, pp. 522–530

1. $147.61\overline{3}$ **2.** $32°52'12''$ **3.** $10.125 \times 13.5 \times 16.875$

4. approx. 692.82 yd   **5.** 120°   **6.** $\dfrac{7\pi}{6}$ **7.** approx. 4.97 units   **8.** $-\dfrac{1}{2}$

9. $s = 25.5$ cm. $A = 191.25$ cm²   **10.** $r \approx 41.74$ in., $A \approx 2003.48$ in²

11. $\theta = 4.75$ rad, $s = 38$ m   **12. a.** approx. 9.4248 rad/sec
b. approx. 3.9 ft/sec   **c.** about 15.4 sec

13. $y = -\dfrac{6}{7}, \left(-\dfrac{\sqrt{13}}{7}, \dfrac{6}{7}\right), \left(-\dfrac{\sqrt{13}}{7}, -\dfrac{6}{7}\right),$ and $\left(\dfrac{\sqrt{13}}{7}, \dfrac{6}{7}\right)$

14. $\sin t = -\dfrac{\sqrt{7}}{4}, \csc t = -\dfrac{4}{\sqrt{7}},$
$\cos t = \dfrac{3}{4}, \sec t = \dfrac{4}{3}, \tan t = -\dfrac{\sqrt{7}}{3}, \cot t = -\dfrac{3}{\sqrt{7}}$

15. $\dfrac{\pi}{3}$ and $\dfrac{2\pi}{3}$ **16.** $t \approx 2.44$ **17. a.** approx. 19.6667 rad   **b.** 25 rad

18. $|A| = 3, P = 2\pi$

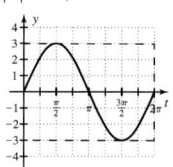

19. $P = 2\pi$

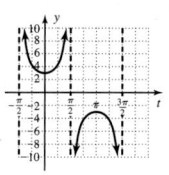

20. $|A| = 1, P = \pi$

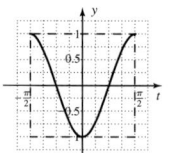

21. $|A| = 1.7, P = \dfrac{\pi}{2}$

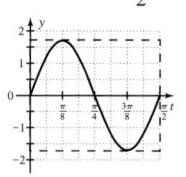

22. $|A| = 2, P = \dfrac{1}{2}$

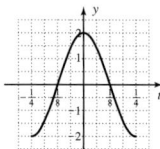

23. $|A| = 3, P = \dfrac{1}{199}$

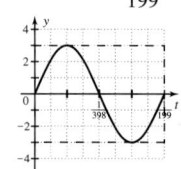

24. $y = 0.75\sin(6t)$ **25.** $y = 4\csc(3\pi t)$ **26.** green; red

27. $\tan\left(\dfrac{7\pi}{4}\right) = -1; \cot\left(\dfrac{\pi}{3}\right) = \dfrac{1}{\sqrt{3}}$ **28.** $\theta = \dfrac{2\pi}{3}; \theta = \dfrac{2\pi}{3}$

29.

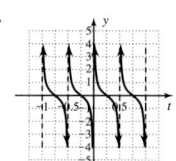

30.

31. $1.55 + k\pi$ radians; $k \in Z$ **32.** 3.5860   **33.** ≈ 151.14 m

34. $y = 5.2\tan\left(\dfrac{\pi}{12}x\right)$; period = 12; $A = 5.2$; asymptotes $x = -6, x = 6$

35. a. $|A| = 240, P = 12,$ HS: 3 units right, VS: 520 units up
b.

36. a. $|A| = 3.2, P = 8,$ HS: 6 units left, VS: 6.4 units up
b.

37. $|A| = 125, P = 24,$ HS: 3 units right, VS: 175 units up,
$y = 125\cos\left[\dfrac{\pi}{12}(t - 3)\right] + 175$

38. $A = 75, P = \dfrac{3\pi}{8},$ HS: (none),
VS: 105 units up, $y = 75\sin\left(\dfrac{16}{3}t\right) + 105$

39. a. $P(t) = 0.91 \sin\left(\dfrac{\pi}{6}t\right) + 1.35$ **b.** August: 1.81 in., Dec: 0.44 in.

40. a. $A \approx 0.80$, **b.** $A \approx 64.3°$ **41. a.** $\cot 32.6°$, **b.** $\cos(70°29'45'')$

42.

Angles	Sides
$A = 49°$	$a = 89$ in.
$B = 41°$	$b \approx 77.37$ in.
$C = 90°$	$c \approx 117.93$ in.

43.

Angles	Sides
$A \approx 43.6°$	$a = 20$ m
$B \approx 46.4°$	$b = 21$ m
$C = 90°$	$c = 29$ m

44. approx. 5.18 m   **45. a.** approx. 239.32 m
b. approx. 240.68 m apart   **46.** approx. 54.5° and 35.5°
47. $207° + 360°k$; answers will vary.   **48.** 28°, 19°, 30°
49. a. $\sin \theta = \dfrac{35}{37}$, $\csc \theta = \dfrac{37}{35}$, $\cos \theta = \dfrac{-12}{37}$, $\sec \theta = \dfrac{-37}{12}$,

$\tan \theta = \dfrac{-35}{12}$, $\cot \theta = \dfrac{-12}{35}$

b. $\sin \theta = \dfrac{-3}{\sqrt{13}}$, $\csc \theta = \dfrac{-\sqrt{13}}{3}$, $\cos \theta = \dfrac{2}{\sqrt{13}}$, $\sec \theta = \dfrac{\sqrt{13}}{2}$,

$\tan \theta = \dfrac{-3}{2}$, $\cot \theta = \dfrac{-2}{3}$

50. a. $x = 4$, $y = -3$, $r = 5$; QIV; $\sin \theta = -\dfrac{3}{5}$, $\csc \theta = -\dfrac{5}{3}$,

$\cos \theta = \dfrac{4}{5}$, $\sec \theta = \dfrac{5}{4}$, $\tan \theta = \dfrac{-3}{4}$, $\cot \theta = \dfrac{-4}{3}$

b. $x = 5$, $y = -12$, $r = 13$; QIV; $\sin \theta = \dfrac{-12}{13}$, $\csc \theta = \dfrac{-13}{12}$,

$\cos \theta = \dfrac{5}{13}$, $\sec \theta = \dfrac{13}{5}$, $\tan \theta = \dfrac{-12}{5}$, $\cot \theta = \dfrac{-5}{12}$
51. a. $\theta = 135° + 180°k$ **b.** $\theta = 30° + 360°k$ or $\theta = 330° + 360°k$
c. $\theta \approx 76.0° + 180°k$ **d.** $\theta \approx -27.0° + 360°k$ or $\theta = 207.0° + 360°k$

Mixed Review, pp. 530–531

1. a. $A = 10$ **b.** $D = 15$ **c.** $P = 6$ **d.** $f(4) = 20$

3. $t = \dfrac{2\pi}{3}$ and $t = \dfrac{4\pi}{3}$ **5.** $220°48'50''$

7. $12\sqrt{2}$ in.; $60\sqrt{2} \approx 84.9$ in.

9. arc length: $\dfrac{28}{3}\pi \approx 29.3$ units; area: $\dfrac{112\pi}{3} \approx 117.3$ units^2 **11.** $86.915°$

13. $\sin \theta = \dfrac{-8}{17}$, $\sec \theta = \dfrac{17}{15}$, $\cos \theta = \dfrac{15}{17}$,

$\csc \theta = \dfrac{-17}{8}$, $\tan \theta = \dfrac{-8}{15}$, $\cot \theta = \dfrac{-15}{8}$

15. 60°   **17. a.** 6π rad/sec   **b.** $20(6\pi)$ cm/sec ≈ 377 cm/sec

19. a. $|A| = 5$; $P = \pi$; HS: (none);   **b.** $|A| = \dfrac{7}{2}$; $P = 4$; HS: 1 unit right;

VS: 8 units down; PI: $0 \leq t < \pi$ VS: (none); PI: $1 \leq t < 5$

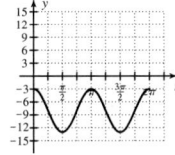   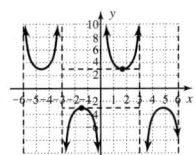

c. $|A|$: NA; $P = 4\pi$; HS: none;   **d.** $|A|$: NA; $P = 2\pi$; HS: $\dfrac{\pi}{2}$ to the

VS: none; PI: $(-2\pi, 2\pi)$ right; VS: none; PI: $(-\pi, \pi)$

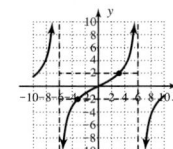

Practice Test, pp. 532–533

1. complement: 55°; supplement: 145°   **2. a.** 45°   **b.** 30°   **c.** $\dfrac{\pi}{6}$ **d.** $\dfrac{\pi}{3}$
3. $30° + 360°k$; $k \in Z$ **4. a.** $100.755°$ **b.** $48°12'45''$
5. a. 430 mi   **b.** $215\sqrt{3} \approx 372$ mi
6.

t	$\sin t$	$\cos t$	$\tan t$	$\csc t$	$\sec t$	$\cot t$
0	0	1	0	undefined	1	undefined
$\dfrac{2\pi}{3}$	$\dfrac{\sqrt{3}}{2}$	$-\dfrac{1}{2}$	$-\sqrt{3}$	$\dfrac{2\sqrt{3}}{3}$	-2	$\dfrac{-\sqrt{3}}{3}$
$\dfrac{7\pi}{6}$	$-\dfrac{1}{2}$	$-\dfrac{\sqrt{3}}{2}$	$\dfrac{\sqrt{3}}{3}$	-2	$\dfrac{-2\sqrt{3}}{3}$	$\sqrt{3}$
$\dfrac{5\pi}{4}$	$-\dfrac{\sqrt{2}}{2}$	$-\dfrac{\sqrt{2}}{2}$	1	$-\sqrt{2}$	$-\sqrt{2}$	1
$\dfrac{5\pi}{3}$	$-\dfrac{\sqrt{3}}{2}$	$\dfrac{1}{2}$	$-\sqrt{3}$	$\dfrac{2\sqrt{3}}{3}$	2	$\dfrac{-\sqrt{3}}{3}$
$\dfrac{13\pi}{6}$	$\dfrac{1}{2}$	$\dfrac{\sqrt{3}}{2}$	$\dfrac{\sqrt{3}}{3}$	2	$\dfrac{2\sqrt{3}}{3}$	$\sqrt{3}$

7. $\sec \theta = \dfrac{5}{2}$, $\sin \theta = \dfrac{-\sqrt{21}}{5}$, $\tan \theta = \dfrac{-\sqrt{21}}{2}$, $\csc \theta = \dfrac{-5}{\sqrt{21}}$,

$\cot \theta = \dfrac{-2}{\sqrt{21}}$

8. $\left(\dfrac{1}{3}\right)^2 + \left(\dfrac{-2\sqrt{2}}{3}\right)^2 = \dfrac{1}{9} + \dfrac{8}{9} = 1$

9. a. ≈ 225.8 ft or 225 ft 9.6 in.   **b.** $\dfrac{23\pi}{480} \approx 0.1505$ rad/sec

c. 11.29 ft/sec ≈ 7.7 mph

10.

Angles	Sides
$A = 33°$	$a \approx 8.2$ cm
$B = 57°$	$b \approx 12.6$ cm
$C = 90°$	$c = 15.0$ cm

11. about 67 cm, 49.6°   **12.** 57.9 m   **13. a.** $\dfrac{7\pi}{6}$ **b.** $\dfrac{11\pi}{6}$ **c.** $\dfrac{3\pi}{4}$

14. a. $W(t) = 18.4 \sin\left(\dfrac{\pi}{12}t\right) + 34.1$

b. 433,000 gal; 249,000 gal
15. a. $D: t \in R$, $R: y \in [-2, 2]$,
$P = 10$, $|A| = 2$;

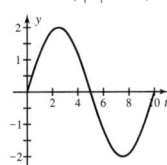

b. $D: t \neq \dfrac{\pi}{2}(2k + 1)$ for $k \in Z$,

$R: y \in (-\infty, -1] \cup [1, +\infty)$, $P = 2\pi$

c. $D: t \neq \dfrac{\pi}{6}(2k + 1)$ for $k \in Z$, $R: y \in R$, $P = \dfrac{\pi}{3}$

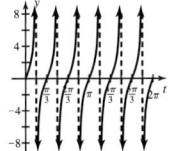   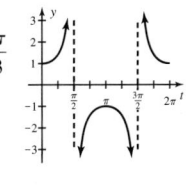

16. $|A| = 12$, $P = \dfrac{2\pi}{3}$, HS: $\dfrac{\pi}{12}$ units right, VS: 19 units up

PI: $\dfrac{\pi}{12} \le t < \dfrac{3\pi}{4}$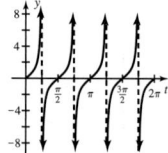

17. 1260°

18. a. $D: t \ne \dfrac{\pi}{4}(2k+1)$, $k \in Z$; $R: y \in R$; $P = \dfrac{\pi}{2}$;

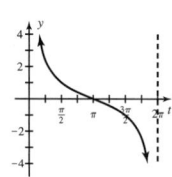

b. $D: t \ne 2\pi k$, $k \in Z$; $R: y \in R$, $P = 2\pi$,

19. $y = 7.5 \sin\left(\dfrac{\pi}{6}t - \dfrac{\pi}{2}\right) + 12.5$ **20. a.** ≈ 4 **b.** ≈ 2.3

Strengthening Core Skills, pp. 535–536

Exercise 1:

t	0	$\dfrac{\pi}{6}$	$\dfrac{\pi}{4}$	$\dfrac{\pi}{3}$	$\dfrac{\pi}{2}$
$\sin t = y$	0	$\dfrac{1}{2}$	$\dfrac{\sqrt{2}}{2}$	$\dfrac{\sqrt{3}}{2}$	1
$\cos t = x$	1	$\dfrac{\sqrt{3}}{2}$	$\dfrac{\sqrt{2}}{2}$	$\dfrac{1}{2}$	0
$\tan t = \dfrac{y}{x}$	0	$\dfrac{\sqrt{3}}{3}$	1	$\sqrt{3}$	—

$\dfrac{2\pi}{3}$	$\dfrac{3\pi}{4}$	$\dfrac{5\pi}{6}$	π	$\dfrac{7\pi}{6}$	$\dfrac{5\pi}{4}$
$\dfrac{\sqrt{3}}{2}$	$\dfrac{\sqrt{2}}{2}$	$\dfrac{1}{2}$	0	$\dfrac{-1}{2}$	$\dfrac{-\sqrt{2}}{2}$
$\dfrac{-1}{2}$	$\dfrac{-\sqrt{2}}{2}$	$\dfrac{-\sqrt{3}}{2}$	-1	$\dfrac{-\sqrt{3}}{2}$	$\dfrac{-\sqrt{2}}{2}$
$-\sqrt{3}$	-1	$\dfrac{-\sqrt{3}}{3}$	0	$\dfrac{\sqrt{3}}{3}$	1

Exercise 2:
a. $t = \dfrac{4\pi}{3}, \dfrac{5\pi}{3}$ **b.** $t = \dfrac{\pi}{4}, \dfrac{7\pi}{4}$ **c.** $t = \dfrac{\pi}{6}, \dfrac{7\pi}{6}$ **d.** $t = \dfrac{\pi}{4}, \dfrac{7\pi}{4}$

Exercise 3:
a. no solution   **b.** $t \approx 1.2310$, $t \approx 5.0522$ **c.** $t \approx 6.0382$, $t \approx 2.8966$
d. $t \approx 1.9823$, $t \approx 4.3009$

Cumulative Review Chapters 1–5, pp. 536–537

1. $-5 < x < 3$

3.
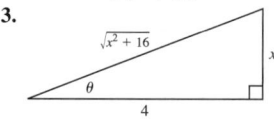

hyp $= 89$; $\theta \approx 64°$; $90 - \theta = 26°$

5. $\cos t = \dfrac{3}{4}$, $\sin t = \dfrac{-\sqrt{7}}{4}$, $\tan t = \dfrac{-\sqrt{7}}{3}$, $\sec t = \dfrac{4}{3}$,

$\csc t = \dfrac{-4}{\sqrt{7}} = \dfrac{-4\sqrt{7}}{7}$, $\cot t = \dfrac{-3}{\sqrt{7}} = \dfrac{-3\sqrt{7}}{7}$

7. a. $D: x \in \left[\dfrac{3}{2}, \infty\right)$, $R: y \in [0, \infty)$

b. $D: x \in (-\infty, -7) \cup (-7, 7) \cup (7, \infty)$, $R: y \in (-\infty, \infty)$

9. a. max: $(-2, 4)$, endpoint max: $(4, 0)$
min: $(2, -4)$, endpoint min: $(-4, 0)$

b. $f(x) \ge 0: x \in [-4, 0) \cup \{4\}$
$f(x) < 0: x \in (0, 4)$

c. $f(x)\uparrow: x \in (-4, -2) \cup (2, 4)$
$f(x)\downarrow: x \in (-2, 2)$

d. function is odd: $f(-x) = -f(x)$

11. ≈ 114.3 ft   **13.**
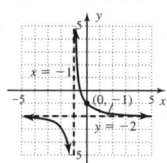

15. $x = -9$, $y = 40$, $r = 41$, QII;

$\cos \theta = \dfrac{-9}{41}$, $\sin \theta = \dfrac{40}{41}$, $\tan \theta = \dfrac{-40}{9}$, $\sec \theta = \dfrac{-41}{9}$, $\csc \theta = \dfrac{41}{40}$,

$\cot \theta = \dfrac{-9}{40}$, $\theta \approx 102.7°$

17. $S = 18$ m; $A = 135$ m²   **19.** $y = \dfrac{3}{2}\sin\left(4t - \dfrac{\pi}{2}\right) + \dfrac{1}{2}$

21.   **23.** $y = \dfrac{3}{4}x - 2$, $m = \dfrac{3}{4}$, y-intercept $(0, -2)$

25. about 6.85%

Connections to Calculus Exercises, pp. 541–542

1. hyp $= x + 4$

$\sin \theta = \dfrac{4}{x+4}$, $\cos \theta = \dfrac{\sqrt{x^2 + 8x}}{x+4}$

$\csc \theta = \dfrac{x+4}{4}$, $\sec \theta = \dfrac{x+4}{\sqrt{x^2 + 8x}}$

$\tan \theta = \dfrac{4}{\sqrt{x^2 + 8x}}$, $\cot \theta = \dfrac{\sqrt{x^2 + 8x}}{4}$

3.

$\sin \theta = \dfrac{x}{\sqrt{x^2 + 16}}$, $\cos \theta = \dfrac{4}{\sqrt{x^2 + 16}}$

$\csc \theta = \dfrac{\sqrt{x^2 + 16}}{x}$, $\sec \theta = \dfrac{\sqrt{x^2 + 16}}{4}$

$\cot \theta = \dfrac{4}{x}$

5. $\cot \theta = \dfrac{13}{u}$, $\sec \theta = \dfrac{\sqrt{u^2 + 169}}{13}$ **7.** $r = \dfrac{2}{\sin \theta}$

9. $r = \dfrac{3}{\sin \theta + 2 \cos \theta}$ **11.** $x^2 + y^2 - 5y = 0$; circle

13. $3x - 2y = 6$; line

CHAPTER 6
Exercises 6.1, pp. 548–551

1. $\sin\theta$; $\sec\theta$; $\cos\theta$ **3.** one; false   **5.** $\dfrac{1-\sin^2 x}{\sin x \sec x}$; Answers will vary.

7. Answers may vary;

$$\tan x = \frac{\sec x}{\csc x}; \frac{\sin x}{\cos x} = \frac{\sec x}{\csc x}; \frac{1}{\cot x} = \frac{\sec x}{\csc x}; \frac{1}{\cot x} = \frac{\sin x}{\cos x}$$

9. $1 = \sec^2 x - \tan^2 x$; $\tan^2 x = \sec^2 x - 1$;
$1 = (\sec x + \tan x)(\sec x - \tan x)$; $\tan x = \pm\sqrt{\sec^2 x - 1}$

11. $\sin x \cot x = \sin x \dfrac{\cos x}{\sin x} = \cos x$

13. $\sec^2 x \cot^2 x = \dfrac{1}{\cos^2 x}\dfrac{\cos^2 x}{\sin^2 x} = \dfrac{1}{\sin^2 x} = \csc^2 x$

15. $\cos x\,(\sec x - \cos x) = \cos x \sec x - \cos^2 x =$
$\cos x \dfrac{1}{\cos x} - \cos^2 x = 1 - \cos^2 x = \sin^2 x$

17. $\sin x(\csc x - \sin x) = 1 - \sin^2 x = \cos^2 x$

19. $\tan x(\csc x + \cot x) = \tan x \csc x + \tan x \cot x =$
$\dfrac{\sin x}{\cos x}\dfrac{1}{\sin x} + \dfrac{\sin x}{\cos x}\dfrac{\cos x}{\sin x} = \dfrac{1}{\cos x} + 1 = \sec x + 1$

21. $\tan^2 x \csc^2 x - \tan^2 x = \tan^2 x\,(\csc^2 x - 1) = 1$; $\tan^2 x\,(\cot^2 x) = 1$

23. $\dfrac{\sin x \cos x + \sin x}{\cos x + \cos^2 x} = \dfrac{\sin x\,(\cos x + 1)}{\cos x\,(1 + \cos x)} = \tan x$; $\dfrac{\sin x}{\cos x} = \tan x$

25. $\dfrac{1 + \sin x}{\cos x + \cos x \sin x} = \dfrac{(1)(1 + \sin x)}{(\cos x)(1 + \sin x)} = \dfrac{1}{\cos x} = \sec x$

27. $\dfrac{\sin x \tan x + \sin x}{\tan x + \tan^2 x} = \dfrac{\sin x\,(\tan x + 1)}{\tan x(1 + \tan x)} = \dfrac{\sin x}{\tan x}$
$\dfrac{\sin x}{\sin x/\cos x} = \dfrac{\sin x \cos x}{\sin x} = \cos x$

29. $\dfrac{(\sin x + \cos x)^2}{\cos x} = \dfrac{\sin^2 x + 2\sin x \cos x + \cos^2 x}{\cos x} =$
$\dfrac{\sin^2 x + \cos^2 x + 2\sin x \cos x}{\cos x} = \dfrac{1 + 2\sin x \cos x}{\cos x} =$
$\dfrac{1}{\cos x} + \dfrac{2\sin x \cos x}{\cos x} = \sec x + 2\sin x$

31. $(1 + \sin x)[1 + \sin(-x)] = (1 + \sin x)(1 - \sin x) = 1 - \sin^2 x = \cos^2 x$

33. $\dfrac{(\csc x - \cot x)(\csc x + \cot x)}{\tan x} = \dfrac{\csc^2 x - \cot^2 x}{\tan x} = \dfrac{1}{\tan x} = \cot x$

35. $\dfrac{\cos^2 x}{\sin x} + \dfrac{\sin x}{1} = \dfrac{\cos^2 x + \sin^2 x}{\sin x} = \dfrac{1}{\sin x} = \csc x$

37. $\dfrac{\tan x}{\csc x} - \dfrac{\sin x}{\cos x} = \dfrac{\tan x \cos x - \sin x \csc x}{\csc x \cos x} = \dfrac{\dfrac{\sin x}{\cos x}\cos x - 1}{\dfrac{1}{\sin x}\cos x} = \dfrac{\sin x - 1}{\cot x}$

39. $\dfrac{\sec x}{\sin x} - \dfrac{\csc x}{\sec x} = \dfrac{\sec^2 x - \sin x \csc x}{\sin x \sec x} = \dfrac{\sec^2 x - 1}{\sin x \dfrac{1}{\cos x}} = \dfrac{\tan^2 x}{\tan x} = \tan x$

41. $\dfrac{\sin x}{\pm\sqrt{1 - \sin^2 x}}$ **43.** $\pm\sqrt{\dfrac{1}{\cot^2 x} + 1}$ **45.** $\dfrac{\pm\sqrt{1 - \sin^2 x}}{\sin x}$

47. $\sin\theta = \dfrac{21}{29}$, $\tan\theta = -\dfrac{21}{20}$, $\sec\theta = -\dfrac{29}{20}$, $\csc\theta = \dfrac{29}{21}$, $\cot\theta = -\dfrac{20}{21}$

49. $\cos\theta = -\dfrac{8}{17}$, $\sin\theta = -\dfrac{15}{17}$, $\sec\theta = -\dfrac{17}{8}$, $\csc\theta = -\dfrac{17}{15}$, $\cot\theta = \dfrac{8}{15}$

51. $\cos\theta = \dfrac{x}{\sqrt{x^2 + 25}}$, $\sin\theta = \dfrac{5}{\sqrt{x^2 + 25}}$, $\tan\theta = \dfrac{5}{x}$,
$\sec\theta = \dfrac{\sqrt{x^2 + 25}}{x}$, $\csc\theta = \dfrac{\sqrt{x^2 + 25}}{5}$

53. $\cos\theta = -\dfrac{\sqrt{120}}{13} = -\dfrac{2\sqrt{30}}{13}$, $\tan\theta = \dfrac{7}{2\sqrt{30}}$, $\sec\theta = -\dfrac{13}{2\sqrt{30}}$,
$\csc\theta = -\dfrac{13}{7}$, $\cot\theta = \dfrac{2\sqrt{30}}{7}$

55. $\sin\theta = \dfrac{4\sqrt{2}}{9}$, $\cos\theta = -\dfrac{7}{9}$, $\tan\theta = -\dfrac{4\sqrt{2}}{7}$,
$\csc\theta = \dfrac{9}{4\sqrt{2}}$, $\cot\theta = -\dfrac{7}{4\sqrt{2}}$ **57.** Answers will vary.

59. Answers will vary.   **61.** Answers will vary.

63. a. $A = \dfrac{nx^2}{4}\cot\left(\dfrac{\pi}{n}\right)$ **b.** $A = \dfrac{4\,(8\text{ m})^2}{4}\cot\left(\dfrac{\pi}{4}\right) = 64\text{ m}^2 \cdot 1 = 64\text{ m}^2$

c. $A \approx 119.62\text{ in}^2$

65. $\cos^3 x = (\cos x)(\cos^2 x) = (\cos x)(1 - \sin^2 x)$

67. $\tan x + \tan^3 x = (\tan x)(1 + \tan^2 x) = (\tan x)(\sec^2 x)$

69. $\tan^2 x \sec x - 4\tan^2 x = (\tan^2 x)(\sec x - 4)$
$= (\sec x - 4)(\tan^2 x) = (\sec x - 4)(\sec^2 - 1)$
$= (\sec x - 4)(\sec x - 1)(\sec x + 1)$

71. $\cos^2 x \sin x - \cos^2 x = (\cos^2 x)(\sin x - 1)$
$= (1 - \sin^2 x)(\sin x - 1)$
$= (1 + \sin x)(1 - \sin x)(\sin x - 1)$
$= (1 + \sin x)(1 - \sin x)(-1)(1 - \sin x)$
$= (-1)(1 + \sin x)(1 - \sin x)^2$

73. a. $A = nr^2\tan\left(\dfrac{\pi}{n}\right)$ **b.** $A = 4 \cdot 4^2\tan\left(\dfrac{\pi}{4}\right) = 64\text{ m}^2$

c. $A \approx 51.45\text{ m}^2$ **75.** $\tan\theta = \dfrac{1 + m_1 m_2}{m_2 - m_1}$ **77.** $\theta = 45°$

79. $0, \dfrac{\pi}{3}, \dfrac{\pi}{2}, \dfrac{2\pi}{3}, \pi, \dfrac{3\pi}{2}$ **81.** about 1148 ft

83.

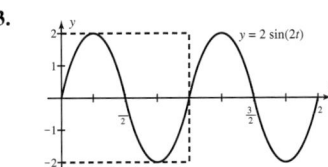

Exercises 6.2, pp. 555–558

1. substituted   **3.** complicated; simplify; build

5. Because we don't know if the equation is true.   **7.** $\dfrac{1 + \sin x}{\cos x}$

9. $\cos x$ **11.** $\dfrac{1 - \cos x}{\sin x}$

13. $\cos^2 x \tan^2 x = \cos^2 x \dfrac{\sin^2 x}{\cos^2 x}$
$= \sin^2 x$
$= 1 - \cos^2 x$

15. $\tan x + \cot x = \dfrac{\sin x}{\cos x} + \dfrac{\cos x}{\sin x}$
$= \dfrac{\sin^2 x + \cos^2 x}{\cos x \sin x}$
$= \dfrac{1}{\cos x \sin x}$
$= \dfrac{1}{\cos x \sin x}$
$= \dfrac{1}{\cos x}\dfrac{1}{\sin x}$
$= \sec x \csc x$

17. $\csc x - \sin x = \dfrac{1}{\sin x} - \sin x$
$= \dfrac{1 - \sin^2 x}{\sin x}$
$= \dfrac{\cos^2 x}{\sin x}$
$= \dfrac{\cos x}{\sin x/\cos x}$
$= \dfrac{\cos x}{\tan x}$

19. $\sec\theta + \tan\theta = \dfrac{1}{\cos\theta} + \dfrac{\sin\theta}{\cos\theta}$

$\qquad\qquad = \dfrac{1 + \sin\theta}{\cos\theta}$

$\qquad\qquad = \dfrac{(1 + \sin\theta)(1 - \sin\theta)}{\cos\theta(1 - \sin\theta)}$

$\qquad\qquad = \dfrac{1 - \sin^2\theta}{\cos\theta(1 - \sin\theta)}$

$\qquad\qquad = \dfrac{\cos^2\theta}{\cancel{\cos\theta}(1 - \sin\theta)}$

$\qquad\qquad = \dfrac{\cos\theta}{1 - \sin\theta}$

21. $\dfrac{1 - \sin x}{\cos x} = \dfrac{(1 - \sin x)(1 + \sin x)}{\cos x(1 + \sin x)}$

$\qquad\qquad = \dfrac{1 - \sin^2 x}{\cos x(1 + \sin x)}$

$\qquad\qquad = \dfrac{\cos^2 x}{\cancel{\cos x}(1 + \sin x)}$

$\qquad\qquad = \dfrac{\cos x}{1 + \sin x}$

23. $\dfrac{\csc x}{\cos x} - \dfrac{\cos x}{\csc x} = \dfrac{\csc^2 x - \cos^2 x}{\cos x \csc x}$

$\qquad\qquad = \dfrac{\csc^2 x - (1 - \sin^2 x)}{\cos x \dfrac{1}{\sin x}}$

$\qquad\qquad = \dfrac{\csc^2 x - 1 + \sin^2 x}{\cot x}$

$\qquad\qquad = \dfrac{\cot^2 x + \sin^2 x}{\cot x}$

25. $\dfrac{\sin x}{1 + \sin x} - \dfrac{\sin x}{1 - \sin x} = \dfrac{\sin x(1 - \sin x) - \sin x(1 + \sin x)}{(1 + \sin x)(1 - \sin x)}$

$\qquad\qquad = \dfrac{\cancel{\sin x} - \sin^2 x - \cancel{\sin x} - \sin^2 x}{1 - \sin^2 x}$

$\qquad\qquad = \dfrac{-2\sin^2 x}{\cos^2 x}$

$\qquad\qquad = -2\tan^2 x$

27. $\dfrac{\cot x}{1 + \csc x} - \dfrac{\cot x}{1 - \csc x} = \dfrac{\cot x(1 - \csc x) - \cot x(1 + \csc x)}{(1 + \csc x)(1 - \csc x)}$

$\qquad\qquad = \dfrac{\cancel{\cot x} - \cot x \csc x - \cancel{\cot x} - \cot x \csc x}{1 - \csc^2 x}$

$\qquad\qquad = \dfrac{2\cot x \csc x}{\cot^2 x}$

$\qquad\qquad = \dfrac{2\csc x}{\cot x}$

$\qquad\qquad = \dfrac{2\dfrac{1}{\cancel{\sin x}}}{\dfrac{\cos x}{\cancel{\sin x}}}$

$\qquad\qquad = \dfrac{2}{\cos x}$

$\qquad\qquad = 2\sec x$

29. $\dfrac{\sec^2 x}{1 + \cot^2 x} = \dfrac{\sec^2 x}{\csc^2 x}$

$\qquad\qquad = \dfrac{\dfrac{1}{\cos^2 x}}{\dfrac{1}{\sin^2 x}}$

$\qquad\qquad = \dfrac{\sin^2 x}{\cos^2 x}$

$\qquad\qquad = \tan^2 x$

31. $\sin^2 x(\cot^2 x - \csc^2 x) = \sin^2 x \cot^2 x - \sin^2 x \csc^2 x$

$\qquad\qquad = \cancel{\sin^2 x}\dfrac{\cos^2 x}{\cancel{\sin^2 x}} - \cancel{\sin^2 x}\dfrac{1}{\cancel{\sin^2 x}}$

$\qquad\qquad = \cos^2 x - 1$

$\qquad\qquad = -\sin^2 x$

33. $\cos x \cot x + \sin x = \cos x \dfrac{\cos x}{\sin x} + \sin x$

$\qquad\qquad = \dfrac{\cos^2 x}{\sin x} + \sin x$

$\qquad\qquad = \dfrac{\cos^2 x + \sin^2 x}{\sin x}$

$\qquad\qquad = \dfrac{1}{\sin x}$

$\qquad\qquad = \csc x$

35. $\dfrac{\sec x}{\cot x + \tan x} = \dfrac{\dfrac{1}{\cos x}(\sin x)(\cos x)}{\left(\dfrac{\cos x}{\sin x} + \dfrac{\sin x}{\cos x}\right)(\sin x)(\cos x)}$

$\qquad\qquad = \dfrac{\sin x}{\cos^2 x + \sin^2 x}$

$\qquad\qquad = \dfrac{\sin x}{1}$

$\qquad\qquad = \sin x$

37. $\dfrac{\sin x - \csc x}{\csc x} = \dfrac{\sin x}{\csc x} - \dfrac{\csc x}{\csc x}$

$\qquad\qquad = \sin^2 x - 1$

$\qquad\qquad = -\cos^2 x$

39. $\dfrac{1}{\csc x - \sin x} = \dfrac{1}{(\csc x - \sin x)}\dfrac{\sin x}{\sin x}$

$\qquad\qquad = \dfrac{\sin x}{1 - \sin^2 x}$

$\qquad\qquad = \dfrac{\sin x}{\cos^2 x}$

$\qquad\qquad = \dfrac{\sin x}{\cos x}\dfrac{1}{\cos x}$

$\qquad\qquad = \tan x \sec x$

41. $\dfrac{1 + \sin x}{1 - \sin x} = \dfrac{(1 + \sin x)}{(1 - \sin x)}\dfrac{(1 + \sin x)}{(1 + \sin x)}$

$\qquad\qquad = \dfrac{1 + 2\sin x + \sin^2 x}{1 - \sin^2 x}$

$\qquad\qquad = \dfrac{1 + 2\sin x + \sin^2 x}{\cos^2 x}$

$\qquad\qquad = \dfrac{1}{\cos^2 x} + 2\dfrac{\sin x}{\cos x}\dfrac{1}{\cos x} + \dfrac{\sin^2 x}{\cos^2 x}$

$\qquad\qquad = \sec^2 x + 2\tan x \sec x + \tan^2 x$

$\qquad\qquad = (\sec x + \tan x)^2$

$\qquad\qquad = (\tan x + \sec x)^2$

43. $\dfrac{\cos x - \sin x}{1 - \tan x} = \dfrac{(\cos x - \sin x)}{(1 - \tan x)}\dfrac{(\cos x + \sin x)}{(\cos x + \sin x)}$

$\qquad\qquad = \dfrac{(\cos x - \sin x)(\cos x + \sin x)}{\cos x + \sin x - \sin x - \dfrac{\sin^2 x}{\cos x}}$

$\qquad\qquad = \dfrac{(\cos x - \sin x)(\cos x + \sin x)}{\cos x\left(1 - \dfrac{\sin^2 x}{\cos^2 x}\right)}$

$\qquad\qquad = \dfrac{(\cos x - \sin x)(\cos x + \sin x)}{\cos x(1 - \tan^2 x)}$

$\qquad\qquad = \dfrac{(\cos x - \sin x)(\cos x + \sin x)}{\cos x(1 - \tan x)(1 + \tan x)}$

$\qquad\qquad = \dfrac{(\cancel{\cos x - \sin x})(\cos x + \sin x)}{(\cancel{\cos x - \sin x})(1 + \tan x)}$

$\qquad\qquad = \dfrac{\cos x + \sin x}{1 + \tan x}$

45. $\dfrac{\tan^2 x - \cot^2 x}{\tan x - \cot x} = \dfrac{(\tan x + \cot x)(\tan x - \cot x)}{(\tan x - \cot x)}$

$= \tan x + \cot x$

$= \dfrac{\sin x}{\cos x} + \dfrac{\cos x}{\sin x}$

$= \dfrac{\sin^2 x + \cos^2 x}{\cos x \sin x}$

$= \dfrac{1}{\cos x \sin x}$

$= \dfrac{1}{\cos x} \dfrac{1}{\sin x}$

$= \sec x \csc x$

$= \csc x \sec x$

47. $\dfrac{\cot x}{\cot x + \tan x} = \dfrac{\dfrac{\cos x}{\sin x}}{\dfrac{\cos x}{\sin x} + \dfrac{\sin x}{\cos x}} \dfrac{(\cos x)(\sin x)}{(\cos x)(\sin x)}$

$= \dfrac{\cos^2 x}{\cos^2 x + \sin^2 x}$

$= \dfrac{\cos^2 x}{1}$

$= 1 - \sin^2 x$

49. $\dfrac{\sec^4 x - \tan^4 x}{\sec^2 x + \tan^2 x} = \dfrac{(\sec^2 x - \tan^2 x)(\cancel{\sec^2 x + \tan^2 x})}{(\cancel{\sec^2 x + \tan^2 x})}$

$= \sec^2 x - \tan^2 x$

$= 1$

51. $\dfrac{\cos^4 x - \sin^4 x}{\cos^2 x} = \dfrac{(\cos^2 x - \sin^2 x)(\cos^2 x + \sin^2 x)}{\cos^2 x}$

$= \dfrac{(\cos^2 x - \sin^2 x)(1)}{\cos^2 x}$

$= \dfrac{\cos^2 x}{\cos^2 x} - \dfrac{\sin^2 x}{\cos^2 x}$

$= 1 - \tan^2 x$

$= 1 - (\sec^2 x - 1)$

$= 1 - \sec^2 x + 1$

$= 2 - \sec^2 x$

53. $(\sec x + \tan x)^2 = \sec^2 x + 2 \sec x \tan x + \tan^2 x$

$= \dfrac{1}{\cos^2 x} + \dfrac{2 \sin x}{\cos^2 x} + \dfrac{\sin^2 x}{\cos^2 x}$

$= \dfrac{1 + 2 \sin x + \sin^2 x}{\cos^2 x}$

$= \dfrac{(1 + \sin x)^2}{\cos^2 x}$

$= \dfrac{(\sin x + 1)^2}{\cos^2 x}$

55. $\dfrac{\cos x}{\sin x} + \dfrac{\sin x}{\cos x} + \dfrac{\csc x}{\sec x} = \dfrac{\cos^2 x \sec x + \sin^2 x \sec x + \csc x \sin x \cos x}{\sin x \cos x \sec x}$

$= \dfrac{\sec x(\cos^2 x + \sin^2 x) + (1)\cos x}{\sin x(1)}$

$= \dfrac{\sec x + \cos x}{\sin x}$

57. $\dfrac{\sin^4 x - \cos^4 x}{\sin^3 x + \cos^3 x} = \dfrac{(\sin^2 x + \cos^2 x)(\sin^2 x - \cos^2 x)}{(\sin x + \cos x)(\sin^2 x - \sin x \cos x + \cos^2 x)}$

$= \dfrac{(1)(\cancel{\sin x + \cos x})(\sin x - \cos x)}{(\cancel{\sin x + \cos x})(\sin^2 x + \cos^2 x - \sin x \cos x)}$

$= \dfrac{\sin x - \cos x}{1 - \sin x \cos x}$

59. a. $d^2 = (20 + x \cos \theta)^2 + (20 - x \sin \theta)^2$

$= 400 + 40x \cos \theta + x^2 \cos^2 \theta + 400 - 40 x \sin \theta + x^2 \sin^2 \theta$

$= 800 + 40x(\cos \theta - \sin \theta) + x^2(\cos^2 \theta + \sin^2 \theta)$

$= 800 + 40x(\cos \theta - \sin \theta) + x^2$

b. ≈ 42.2 ft

61. a. $h = \sqrt{\cot x + \tan x};$
$h \approx 3.76$

b. $\cot x + \tan x = \dfrac{\cos x}{\sin x} + \dfrac{\sin x}{\cos x}$

$= \dfrac{\cos^2 x + \sin^2 x}{\sin x \cos x}$

$= \dfrac{1}{\sin x \cos x}$

$= \csc x \sec x;$

$h = \sqrt{\csc x \sec x}$

$h \approx 3.76;$ yes

63. $D^2 = 400 + 40x \cos \theta + x^2$
$D \approx 40.5$ ft

65. $\sin \alpha = \cos \theta$ **67.** Answers will vary.

69. $(\sin^2 x + \cos^2 x)^2 = (1)^2 = 1$

71. $\sin t = \dfrac{3}{4}$, $\cos t = \dfrac{\sqrt{7}}{4}$, $\tan t = \dfrac{3}{\sqrt{7}}$

73.

Exercises 6.3, pp. 563–567

1. false; QII   **3.** repeat; opposite   **5.** Answers will vary.

7. $\dfrac{\sqrt{2} - \sqrt{6}}{4}$ **9.** $\dfrac{\sqrt{2} - \sqrt{6}}{4}$

11. a. $\cos(45° + 30°) = \cos 45° \cos 30° - \sin 45° \sin 30° = \dfrac{\sqrt{6} - \sqrt{2}}{4}$

b. $\cos(120° - 45°) = \cos 120° \cos 45° + \sin 120° \sin 45° =$

$\dfrac{-\sqrt{2} + \sqrt{6}}{4} = \dfrac{\sqrt{6} - \sqrt{2}}{4}$

13. $\cos(5\theta)$ **15.** $\dfrac{\sqrt{3}}{2}$ **17.** $\dfrac{-16}{65}$ **19.** $\sin 33°$ **21.** $\cot\left(\dfrac{\pi}{12}\right)$

23. $\cos\left(\dfrac{\pi}{3} + \theta\right)$ **25.** $\sin(8x)$ **27.** $\tan(3\theta)$ **29.** 1   **31.** $\sqrt{3}$

33. a. $\dfrac{-304}{425}$ **b.** $\dfrac{-304}{297}$ **35.** $\dfrac{\sqrt{6} + \sqrt{2}}{4}$ **37.** $\dfrac{\sqrt{6} + \sqrt{2}}{4}$

39. $-\dfrac{1}{\sqrt{3}} = -\dfrac{\sqrt{3}}{3}$ **41.** $-\sqrt{3}$

43. a. $\sin(45° - 30°) = \sin 45° \cos 30° - \cos 45° \sin 30° = \dfrac{\sqrt{6} - \sqrt{2}}{4}$

b. $\sin(135° - 120°) = \sin 135° \cos 120° - \cos 135° \sin 120°$

$= \left(\dfrac{\sqrt{2}}{2}\right)\left(-\dfrac{1}{2}\right) - \left(\dfrac{-\sqrt{2}}{2}\right)\left(\dfrac{\sqrt{3}}{2}\right)$

$= \dfrac{-\sqrt{2}}{4} + \dfrac{\sqrt{6}}{4}$

$= \dfrac{\sqrt{6} - \sqrt{2}}{4}$

45. $\dfrac{-\sqrt{2} - \sqrt{6}}{4}$ **47. a.** $\dfrac{319}{481}$ **b.** $\dfrac{480}{481}$ **c.** $-\dfrac{319}{360}$

49. a. $\dfrac{3416}{4505}$ **b.** $\dfrac{-1767}{4505}$ **c.** $\dfrac{3416}{2937}$

51. a. $\dfrac{12 + 5\sqrt{3}}{26}$ **b.** $\dfrac{12\sqrt{3} - 5}{26}$ **c.** $\dfrac{12 + 5\sqrt{3}}{12\sqrt{3} - 5}$

53. $(90° - \alpha) + \theta + (90° - \beta) = 180°$ **a.** $\dfrac{247}{265}$ **b.** $\dfrac{96}{265}$ **c.** $\dfrac{247}{96}$

55. $\sin(\pi - \alpha) = \sin \pi \cos \alpha - \cos \pi \sin \alpha$

$= 0 - (-1)\sin \alpha$

$= \sin \alpha$

57. $\cos\left(x + \dfrac{\pi}{4}\right) = \cos x \cos\left(\dfrac{\pi}{4}\right) - \sin x \sin\left(\dfrac{\pi}{4}\right) =$

$\cos x\left(\dfrac{\sqrt{2}}{2}\right) - \sin x\left(\dfrac{\sqrt{2}}{2}\right) = \dfrac{\sqrt{2}}{2}(\cos x - \sin x)$

59. $\tan\left(x + \dfrac{\pi}{4}\right) = \dfrac{\tan x + \tan\left(\dfrac{\pi}{4}\right)}{1 - \tan x \tan\left(\dfrac{\pi}{4}\right)} = \dfrac{\tan x + 1}{1 - \tan x} = \dfrac{1 + \tan x}{1 - \tan x}$

61. $\cos(\alpha + \beta) + \cos(\alpha - \beta) =$
$\cos\alpha\cos\beta - \sin\alpha\sin\beta + \cos\alpha\cos\beta + \sin\alpha\sin\beta =$
$2\cos\alpha\cos\beta$

63. $\cos(2t) = \cos(t + t)$
$= \cos t \cos t - \sin t \sin t$
$= \cos^2 t - \sin^2 t$

65. $\sin(3t) = \sin(2t + t)$
$= \sin(2t)\cos t + \cos(2t)\sin t$
$= 2\sin t \cos t \cos t + (\cos^2 t - \sin^2 t)\sin t$
$= 2\sin t \cos^2 t + \sin t \cos^2 t - \sin^3 t$
$= 3\sin t \cos^2 t - \sin^3 t$
$= 3\sin t(1 - \sin^2 t) - \sin^3 t$
$= 3\sin t - 3\sin^3 t - \sin^3 t$
$= -4\sin^3 t + 3\sin t$

67. $\cos\left(x - \dfrac{\pi}{4}\right) = \cos x \cos\left(\dfrac{\pi}{4}\right) + \sin x \sin\left(\dfrac{\pi}{4}\right)$

$= \cos x\left(\dfrac{\sqrt{2}}{2}\right) + \sin x\left(\dfrac{\sqrt{2}}{2}\right)$

$= \dfrac{\sqrt{2}}{2}(\cos x + \sin x)$

69. $F = \dfrac{Wk}{c}\dfrac{1 - \sqrt{3}}{1 + \sqrt{3}}$

71. $R = \dfrac{\cos s \cos t}{\overline{\omega}C \sin(s + t)}$

$= \dfrac{\cos s \cos t}{\overline{\omega}C(\sin s \cos t + \cos s \sin t)}$

$= \dfrac{\cos s \cos t \dfrac{1}{\cos s \cos t}}{\overline{\omega}C(\sin s \cos t + \cos s \sin t)\dfrac{1}{\cos s \cos t}}$

$= \dfrac{1}{\overline{\omega}C\left(\dfrac{\sin s \cos t}{\cos s \cos t} + \dfrac{\cos s \sin t}{\cos s \cos t}\right)}$

$= \dfrac{1}{\overline{\omega}C(\tan s + \tan t)}$

73. $\dfrac{A}{B} = \dfrac{\sin\theta\cos(90° - \theta)}{\cos\theta\sin(90° - \theta)}$

$\dfrac{A}{B} = \dfrac{\sin\theta(\cos 90°\cos\theta + \sin 90°\sin\theta)}{\cos\theta(\sin 90°\cos\theta - \cos 90°\sin\theta)}$

$= \dfrac{\sin\theta(0 + \sin\theta)}{\cos\theta(\cos\theta - 0)}$

$= \dfrac{\sin^2\theta}{\cos^2\theta}$

$= \tan^2\theta$

75. verified using sum identity for sine

77. $\dfrac{f(x + h) - f(x)}{h} = \dfrac{\sin(x + h) - \sin x}{h}$

$= \dfrac{\sin x \cos h + \cos x \sin h - \sin x}{h} = \dfrac{\sin x \cos h - \sin x + \cos x \sin h}{h}$

$= \dfrac{\sin x(\cos h - 1) + \cos x \sin h}{h} = \sin x\dfrac{\cos h - 1}{h} + \cos x\dfrac{\sin h}{h}$

79. $\dfrac{-\sqrt{2}}{2}$ **81.** $\dfrac{1}{2}$

83. $D = d$, so $D^2 = d^2$, and
$D^2 = (\cos\alpha - \cos\beta)^2 + (\sin\alpha - \sin\beta)^2$
$= \cos^2\alpha - 2\cos\alpha\cos\beta + \cos^2\beta + \sin^2\alpha -$
$2\sin\alpha\sin\beta + \sin^2\beta$
$= 2 - 2\cos\alpha\cos\beta - 2\sin\alpha\sin\beta$
$d^2 = \sin^2(\alpha - \beta) + [\cos(\alpha - \beta) - 1]^2$
$= \sin^2(\alpha - \beta) + \cos^2(\alpha - \beta) - 2\cos(\alpha - \beta) + 1$
$= 2 - 2\cos(\alpha - \beta)$
$D^2 = d^2$ so
$2 - 2\cos\alpha\cos\beta - 2\sin\alpha\sin\beta = 2 - 2\cos(\alpha - \beta)$
$\dfrac{-2\cos\alpha\cos\beta - 2\sin\alpha\sin\beta}{-2} = \dfrac{-2\cos(\alpha - \beta)}{-2}$
$\cos\alpha\cos\beta + \sin\alpha\sin\beta = \cos(\alpha - \beta)$

85. a. $P = 16$,   **b.** $P = \dfrac{\pi}{2}$ **87.** about 19.3 ft

Exercises 6.4, pp. 576–580

1. sum; $\alpha = \beta$ **3.** $2x; x$ **5.** Answers will vary

7. $\sin(2\theta) = \dfrac{-120}{169}, \cos(2\theta) = \dfrac{119}{169}, \tan(2\theta) = \dfrac{-120}{119}$

9. $\sin(2\theta) = \dfrac{-720}{1681}, \cos(2\theta) = \dfrac{-1519}{1681}, \tan(2\theta) = \dfrac{720}{1519}$

11. $\sin(2\theta) = \dfrac{2184}{7225}, \cos(2\theta) = \dfrac{6887}{7225}, \tan(2\theta) = \dfrac{2184}{6887}$

13. $\sin(2\theta) = \dfrac{-5280}{5329}, \cos(2\theta) = \dfrac{721}{5329}, \tan(2\theta) = \dfrac{-5280}{721}$

15. $\sin(2\theta) = \dfrac{-24}{25}, \cos(2\theta) = \dfrac{7}{25}, \tan(2\theta) = \dfrac{-24}{7}$

17. $\sin\theta = \dfrac{4}{5}, \cos\theta = \dfrac{3}{5}, \tan\theta = \dfrac{4}{3}$

19. $\sin\theta = \dfrac{21}{29}, \cos\theta = \dfrac{20}{29}, \tan\theta = \dfrac{21}{20}$

21. $\sin(3\theta) = \sin(2\theta + \theta)$
$= \sin(2\theta)\cos\theta + \cos(2\theta)\sin\theta$
$= (2\sin\theta\cos\theta)\cos\theta + (1 - 2\sin^2\theta)\sin\theta$
$= 2\sin\theta\cos^2\theta + \sin\theta - 2\sin^3\theta$
$= 2\sin\theta(1 - \sin^2\theta) + \sin\theta - 2\sin^3\theta$
$= 2\sin\theta - 2\sin^3\theta + \sin\theta - 2\sin^3\theta$
$= 3\sin\theta - 4\sin^3\theta$

23. $\dfrac{1}{4}$ **25.** $\dfrac{\sqrt{2}}{2}$ **27.** 1   **29.** $4.5\sin(6x)$ **31.** $\dfrac{1}{8} - \dfrac{1}{8}\cos(4x)$

33. $\dfrac{9}{8} + \dfrac{3}{2}\cos(2x) + \dfrac{3}{8}\cos(4x)$

35. $\dfrac{5}{8} - \dfrac{7}{8}\cos(2x) + \dfrac{3}{8}\cos(4x) - \dfrac{1}{8}\cos(2x)\cos(4x)$

37. $\sin\theta = \dfrac{\sqrt{2 - \sqrt{2}}}{2}, \cos\theta = \dfrac{\sqrt{2 + \sqrt{2}}}{2}, \tan\theta = \sqrt{2} - 1$

39. $\sin\theta = \dfrac{\sqrt{2 - \sqrt{3}}}{2}, \cos\theta = \dfrac{\sqrt{2 + \sqrt{3}}}{2}, \tan\theta = 2 - \sqrt{3}$

41. $\sin\theta = \dfrac{\sqrt{2 + \sqrt{2}}}{2}, \cos\theta = \dfrac{\sqrt{2 - \sqrt{2}}}{2}, \tan\theta = \sqrt{2} + 1$

43. $\sin\theta = \dfrac{\sqrt{2 + \sqrt{2}}}{2}, \cos\theta = \dfrac{\sqrt{2 - \sqrt{2}}}{2}, \tan\theta = \sqrt{2} + 1$

45. $\dfrac{\sqrt{2 - \sqrt{2 + \sqrt{2}}}}{2}$ **47.** $\dfrac{\sqrt{2 - \sqrt{2 + \sqrt{3}}}}{2}$ **49.** $\cos 15°$

51. $\tan 2\theta$ **53.** $\tan x$

55. $\sin\left(\dfrac{\theta}{2}\right) = \dfrac{3}{\sqrt{13}}, \cos\left(\dfrac{\theta}{2}\right) = \dfrac{2}{\sqrt{13}}, \tan\left(\dfrac{\theta}{2}\right) = \dfrac{3}{2}$

57. $\sin\left(\dfrac{\theta}{2}\right) = \dfrac{3}{\sqrt{10}}, \cos\left(\dfrac{\theta}{2}\right) = \dfrac{1}{\sqrt{10}}, \tan\left(\dfrac{\theta}{2}\right) = 3$

59. $\sin\left(\dfrac{\theta}{2}\right) = \dfrac{7}{\sqrt{74}}, \cos\left(\dfrac{\theta}{2}\right) = \dfrac{5}{\sqrt{74}}, \tan\left(\dfrac{\theta}{2}\right) = \dfrac{7}{5}$

61. $\sin\left(\dfrac{\theta}{2}\right) = \dfrac{1}{\sqrt{226}}, \cos\left(\dfrac{\theta}{2}\right) = \dfrac{15}{\sqrt{226}}, \tan\left(\dfrac{\theta}{2}\right) = \dfrac{1}{15}$

63. $\sin\left(\dfrac{\theta}{2}\right) = \dfrac{5}{\sqrt{29}}, \cos\left(\dfrac{\theta}{2}\right) = \dfrac{-2}{\sqrt{29}}, \tan\left(\dfrac{\theta}{2}\right) = -\dfrac{5}{2}$

65. $\dfrac{1}{2}[\cos(12\theta) - \cos(4\theta)]$ **67.** $\cos(2t) + \cos(5t)$

69. $\cos(1540\pi t) + \cos(2418\pi t)$ **71.** $\dfrac{1 + \sqrt{3}}{2}$ **73.** $\dfrac{-1}{4}$

75. $2\sin\left(\dfrac{55}{2}k\right)\cos\left(\dfrac{27}{2}k\right)$ **77.** $-2\sin x \sin\left(\dfrac{x}{6}\right)$

79. $2\cos\left(\dfrac{2061}{2}\pi t\right)\cos\left(\dfrac{357}{2}\pi t\right)$ **81.** $\dfrac{-\sqrt{2}}{2}$

83. $\dfrac{2\sin x \cos x}{\cos^2 x - \sin^2 x} = \dfrac{\sin(2x)}{\cos(2x)}$
$= \tan(2x)$

85. $(\sin x + \cos x)^2 = \sin^2 x + 2\sin x \cos x + \cos^2 x$
$= \sin^2 x + \cos^2 x + 2\sin x \cos x$
$= 1 + 2\sin x \cos x$
$= 1 + \sin(2x)$

87. $\cos(8\theta) = \cos(2 \cdot 4\theta)$
$= \cos^2(4\theta) - \sin^2(4\theta)$

89. $\dfrac{\cos(2\theta)}{\sin^2\theta} = \dfrac{\cos^2\theta - \sin^2\theta}{\sin^2\theta}$
$= \dfrac{\cos^2\theta}{\sin^2\theta} - \dfrac{\sin^2\theta}{\sin^2\theta}$
$= \cot^2\theta - 1$

91. $\tan(2\theta) = \dfrac{2\tan\theta}{1 - \tan^2\theta}$
$= \dfrac{(2\tan\theta)\dfrac{1}{\tan\theta}}{(1 - \tan^2\theta)\dfrac{1}{\tan\theta}}$
$= \dfrac{2}{\dfrac{1}{\tan\theta} - \tan\theta}$
$= \dfrac{2}{\cot\theta - \tan\theta}$

93. $2\csc(2x) = \dfrac{2}{\sin(2x)}$
$= \dfrac{2}{2\sin x \cos x}$
$= \dfrac{1}{\sin x \cos x}$
$= \dfrac{\sin^2 x + \cos^2 x}{\sin x \cos x}$
$= \dfrac{\sin^2 x}{\sin x \cos x} + \dfrac{\cos^2 x}{\sin x \cos x}$
$= \dfrac{\sin x}{\cos x} + \dfrac{\cos x}{\sin x}$
$= \tan x + \cot x$

95. $\cos^2\left(\dfrac{x}{2}\right) - \sin^2\left(\dfrac{x}{2}\right) = \cos\left(2 \cdot \dfrac{x}{2}\right)$
$= \cos x$

97. $1 - 4\sin^2\theta + 4\sin^4\theta = (1 - 2\sin^2\theta)^2$
$= [\cos(2\theta)]^2$
$= \cos^2(2\theta)$
$= 1 - \sin^2(2\theta)$

99. $\dfrac{\sin(120\pi t) + \sin(80\pi t)}{\cos(120\pi t) - \cos(80\pi t)} = \dfrac{2\sin(100\pi t)\cos(20\pi t)}{-2\sin(100\pi t)\sin(20\pi t)}$
$= -\dfrac{\cos(20\pi t)}{\sin(20\pi t)}$
$= -\cot(20\pi t)$

101. $\sin^2\alpha + (1 - \cos\alpha)^2 = \sin^2\alpha + 1 - 2\cos\alpha + \cos^2\alpha$
$= \sin^2\alpha + \cos^2\alpha + 1 - 2\cos\alpha = 1 + 1 - 2\cos\alpha = 2 - 2\cos\alpha$
$= 2(1 - \cos\alpha) = 4\left(\dfrac{1 - \cos\alpha}{2}\right) = 4\sin^2\left(\dfrac{\alpha}{2}\right) = \left[2\sin\left(\dfrac{\alpha}{2}\right)\right]^2$

103. $\sin(2\alpha) = \sin(\alpha + \alpha)$
$= \sin\alpha\cos\alpha + \cos\alpha\sin\alpha$
$= \sin\alpha\cos\alpha + \sin\alpha\cos\alpha$
$= 2\sin\alpha\cos\alpha$

$\tan(\alpha + \beta) = \tan(\alpha + \alpha)$
$= \dfrac{\tan\alpha + \tan\alpha}{1 - \tan\alpha\tan\alpha}$
$= \tan(2\alpha) = \dfrac{2\tan\alpha}{1 - \tan^2\alpha}$

105. $\dfrac{1}{2}[\cos(\alpha - \beta) - \cos(\alpha + \beta)] = \sin\alpha\sin\beta$

107. a. $\mathcal{M} = \dfrac{2}{\sqrt{2 - \sqrt{3}}}, \mathcal{M} \approx 3.9$

b. $\mathcal{M} = \dfrac{2}{\sqrt{2 - \sqrt{2}}}, \mathcal{M} \approx 2.6$ **c.** $\theta = 60°$

109. a. $288 - 144\sqrt{2}$ ft ≈ 84.3 ft   **b.** $288 - 144\sqrt{2}$ ft ≈ 84.3 ft

111. $\cos[2\pi(1209)t] + \cos[2\pi(941)t]$; the ⌗ key

113. $d(t) = \left|6\sin\left(\dfrac{\pi t}{60}\right)\right|$
$= \left|6\sin\left(\dfrac{1}{2} \cdot \dfrac{\pi t}{30}\right)\right|$
$= \left|6\left(\pm\sqrt{\dfrac{1 - \cos\left(\dfrac{\pi t}{30}\right)}{2}}\right)\right|$
$= 6\sqrt{\dfrac{1 - \cos\left(\dfrac{\pi t}{30}\right)}{2}}$
$= \sqrt{36\dfrac{1 - \cos\left(\dfrac{\pi t}{30}\right)}{2}}$
$= \sqrt{18\left[1 - \cos\left(\dfrac{\pi t}{30}\right)\right]}$

115. a. $\sin(2\theta - 90°) + 1$
$= \sin(2\theta)\cos 90° - \cos(2\theta)\sin 90° + 1$
$= 0 - \cos(2\theta) + 1$
$= 1 - \cos(2\theta)$

b. $2\sin^2\theta = \sin^2\theta + \sin^2\theta$
$= 1 - \cos^2\theta + \sin^2\theta$
$= 1 - (\cos^2\theta - \sin^2\theta)$
$= 1 - \cos(2\theta)$

c. $1 + \sin^2\theta - \cos^2\theta = 1 - (\cos^2\theta - \sin^2\theta)$
$= 1 - \cos(2\theta)$

d. $1 - \cos(2\theta) = 1 - \cos(2\theta)$

117. a. $\approx 0.9659; \approx 0.9659$

b. $\left(\dfrac{\sqrt{2 + \sqrt{3}}}{2}\right)^2 \overset{?}{=} \left(\dfrac{\sqrt{6} + \sqrt{2}}{4}\right)^2$
$\dfrac{2 + \sqrt{3}}{4} \overset{?}{=} \dfrac{6 + 2\sqrt{12} + 2}{16}$
$\dfrac{2 + \sqrt{3}}{4} \overset{?}{=} \dfrac{8 + 4\sqrt{3}}{16}$
$\dfrac{2 + \sqrt{3}}{4} = \dfrac{2 + \sqrt{3}}{4}$

119. Must be a unit circle with θ in radians. Must use a right triangle definition of tangent: $\text{tangent}\left(\dfrac{\theta}{2}\right) = \dfrac{\text{opposite side}}{\text{adjacent side}} = \dfrac{\sin\theta}{1 + \cos\theta}$.

121. $x = 1; x = -2; x = -\sqrt{6}; x = \sqrt{6}$

123. $\left(\dfrac{16}{65}\right)^2 + \left(\dfrac{63}{65}\right)^2 = \dfrac{256}{4224} + \dfrac{3969}{4225} = \dfrac{4225}{4225} = 1,$

$$\tan\theta = \frac{63}{16}; \sec\theta = \frac{65}{16}$$

$$1 + \left(\frac{63}{16}\right)^2 = \left(\frac{65}{16}\right)^2$$

$$1 + \frac{3969}{256} = \frac{4225}{256}$$

$$\frac{256}{256} + \frac{3969}{256} = \frac{4225}{256}$$

Mid-Chapter Check, pp. 580–581

1. $\sin x[\csc x - \sin x] = \sin x \csc x - \sin^2 x$

$$= \sin x \frac{1}{\sin x} - \sin^2 x$$
$$= 1 - \sin^2 x$$
$$= \cos^2 x$$

2. $\cos^2 x - \cot^2 x = \cos^2 x - \frac{\cos^2 x}{\sin^2 x}$

$$= \cos^2 x\left(1 - \frac{1}{\sin^2 x}\right)$$
$$= \cos^2 x(1 - \csc^2 x)$$
$$= \cos^2 x(-\cot^2 x)$$
$$= -\cos^2 x \cot^2 x$$

3. $\dfrac{2\sin x}{\sec x} - \dfrac{\cos x}{\csc x} = \dfrac{2\sin x \csc x - \cos x \sec x}{\sec x \csc x}$

$$= \frac{2(1) - 1}{\sec x \csc x}$$
$$= \frac{1}{\sec x \csc x}$$
$$= \cos x \sin x$$

4. $1 + \sec^2 x = \tan^2 x$
$$1 + \sec^2 0 = \tan^2 0$$
$$1 + 1^2 = 0^2$$
$$1 + 1 = 0$$
$$2 = 0 \text{ False}$$

5. a. $\dfrac{\sin^3 x + \cos^3 x}{\sin x + \cos x} = \dfrac{(\cancel{\sin x + \cos x})(\sin^2 x - \sin x \cos x + \cos^2 x)}{(\cancel{\sin x + \cos x})}$

$$= (\sin^2 x + \cos^2 x - \sin x \cos x)$$
$$= 1 - \sin x \cos x$$

b. $\dfrac{1 + \sec x}{\csc x} - \dfrac{1 + \cos x}{\cot x} = \dfrac{1 + \dfrac{1}{\cos x}}{\dfrac{1}{\sin x}} - \dfrac{1 + \cos x}{\dfrac{\cos x}{\sin x}}$

$$= \left(\sin x + \frac{\sin x}{\cos x}\right) - \left(\frac{\sin x}{\cos x} + \sin x\right)$$
$$= \sin x + \frac{\sin x}{\cos x} - \frac{\sin x}{\cos x} - \sin x$$
$$= 0$$

6. a. $\dfrac{\sec^2 x - \tan^2 x}{\sec^2 x} = \dfrac{\sec^2 x}{\sec^2 x} - \dfrac{\tan^2 x}{\sec^2 x}$

$$= 1 - \frac{\dfrac{\sin^2 x}{\cos^2 x}}{\dfrac{1}{\cos^2 x}}$$
$$= 1 - \sin^2 x$$
$$= \cos^2 x$$

b. $\dfrac{\cot x - \tan x}{\csc x \sec x} = \dfrac{\cot x}{\csc x \sec x} - \dfrac{\tan x}{\csc x \sec x}$

$$= \frac{\dfrac{\cos x}{\sin x}}{\dfrac{1}{\sin x} \cdot \dfrac{1}{\cos x}} - \frac{\dfrac{\sin x}{\cos x}}{\dfrac{1}{\sin x} \cdot \dfrac{1}{\cos x}}$$
$$= \frac{\cos^2 x \cancel{\sin x}}{\cancel{\sin x}} - \frac{\sin^2 x \cancel{\cos x}}{\cancel{\cos x}}$$
$$= \cos^2 x - \sin^2 x$$

7. a. $\dfrac{456}{5785}$ **b.** $\dfrac{-3193}{5785}$ **c.** $\dfrac{456}{5767}$

8. $\sin A = \dfrac{7 + 24\sqrt{3}}{50}$, $\cos A = \dfrac{24 - 7\sqrt{3}}{50}$, $\tan A = \dfrac{7 + 24\sqrt{3}}{24 - 7\sqrt{3}}$

9. $\sin\left(\dfrac{\theta}{2}\right) = \dfrac{4}{\sqrt{17}}$, $\cos\left(\dfrac{\theta}{2}\right) = \dfrac{1}{\sqrt{17}}$

10. $\sin(2\alpha) = \dfrac{336}{625}$, $\cos(2\alpha) = \dfrac{527}{625}$, $\tan(2\alpha) = \dfrac{336}{527}$

Reinforcing Basic Concepts, pp. 581–582

1. $\sin^2 x + \cos^2 x = 1$

$$\frac{\sin^2 x}{\sin^2 x} + \frac{\cos^2 x}{\sin^2 x} = \frac{1}{\sin^2 x}$$
$$1 + \cot^2 x = \csc^2 x \checkmark$$
$$\sin^2 x + \cos^2 x = 1$$
$$\frac{\sin^2 x}{\cos^2 x} + \frac{\cos^2 x}{\cos^2 x} = \frac{1}{\cos^2 x}$$
$$\tan^2 x + 1 = \sec^2 x \checkmark$$

2. $\cos(\alpha + \beta) = \cos(\alpha + \alpha) = \cos\alpha\cos\alpha - \sin\alpha\sin\alpha$

$$= \cos^2\alpha - \sin^2\alpha$$
$$= \cos^2\alpha - (1 - \cos^2\alpha)$$
$$= 2\cos^2\alpha - 1$$
$$= \cos^2\alpha - \sin^2\alpha$$
$$= (1 - \sin^2\alpha) - \sin^2\alpha$$
$$= 1 - 2\sin^2\alpha$$

Exercises 6.5, pp. 593–598

1. horizontal; line; one; one   **3.** $[-1, 1]$; $\left[-\dfrac{\pi}{2}, \dfrac{\pi}{2}\right]$ **5.** $\cos^{-1}\left(\frac{1}{5}\right)$

7. $0; \dfrac{1}{2}; -\dfrac{\pi}{6}; -\dfrac{\pi}{2}$ **9.** $\dfrac{\pi}{4}$ **11.** $\dfrac{\pi}{2}$ **13.** $1.0956, 62.8°$

15. $0.3876, 22.2°$ **17.** $\dfrac{\sqrt{2}}{2}$ **19.** $\dfrac{\pi}{3}$ **21.** $45°$ **23.** 0.8205

25. $0; \dfrac{\sqrt{3}}{2}; 120°; \pi$ **27.** $\dfrac{\pi}{3}$ **29.** π **31.** $1.4352; 82.2°$

33. $0.7297; 41.8°$ **35.** $\dfrac{\pi}{4}$ **37.** 0.5560 **39.** $-\dfrac{\sqrt{2}}{2}$ **41.** $\dfrac{3\pi}{4}$

43. $0; -\sqrt{3}; 30°; \sqrt{3}; \dfrac{\pi}{3}$ **45.** $-\dfrac{\pi}{6}$ **47.** $\dfrac{\pi}{3}$ **49.** $-1.1170, -64.0°$

51. $0.9441, 54.1°$ **53.** $-\dfrac{\pi}{6}$ **55.** $\dfrac{\sqrt{3}}{3}$ **57.** $\sqrt{2}$ **59.** $-30°$

61. cannot evaluate $\tan\left(\dfrac{\pi}{2}\right)$

63. $\csc\dfrac{\pi}{4} = \sqrt{2} > 1$, not in domain of $\sin^{-1} x$.

65. $\sin\theta = \dfrac{3}{5}, \cos\theta = \dfrac{4}{5}, \tan\theta = \dfrac{3}{4}$

67. $\sin\theta = \dfrac{\sqrt{x^2 - 36}}{x}, \cos\theta = \dfrac{6}{x}, \tan\theta = \dfrac{\sqrt{x^2 - 36}}{6}$

69. $\dfrac{24}{25}$ **71.** $\dfrac{\sqrt{5}}{3}$

73. $\dfrac{\sqrt{25 - 9x^2}}{3x}$ **75.** $\sqrt{\dfrac{12}{12 + x^2}}$

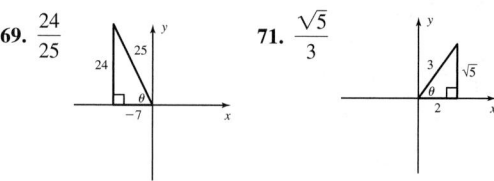

77. $0; 2; 30°; -1; \pi$ **79.** $\dfrac{\pi}{6}$ **81.** $\dfrac{\pi}{6}$ **83.** $80.1°$ **85.** $67.8°$

87. a. $F_N \approx 2.13$ N; $F_N \approx 1.56$ N   **b.** $\theta \approx 63°$ for $F_N = 1$ N, $\theta \approx 24.9°$ for $F_N = 2$ N   **89.** $\approx 30°$ **91.** $\theta \approx 72.3°$; straight line distance; ≈ 157.5 yd

93. a. $\alpha = \tan^{-1}\left(\dfrac{4}{x}\right), \beta = \tan^{-1}\left(\dfrac{1.5}{x}\right)$

b. $\theta = \tan^{-1}\left(\dfrac{4}{x}\right) - \tan^{-1}\left(\dfrac{1.5}{x}\right)$ **c.** $\theta \approx 27.0°$ at $x \approx 2.5$ ft

95. a. $\theta = \tan^{-1}\left(\dfrac{94}{x}\right) - \tan^{-1}\left(\dfrac{70}{x}\right)$ **b.** $\theta \approx 8.4°$ at $d \approx 81.1$ ft

97. a. $\theta \approx 15.5°; \theta \approx 0.2705$ rad   **b.** ≈ 29 mi   **99. a.** 413.6 ft away

b. -503 ft   **c.** ≈ 651.2 ft   **101.** $\sin(2\theta) = \dfrac{84}{85}$

103. $x \in (-\infty, -3] \cup [0, 3]$

Exercises 6.6, pp. 606–610

1. principal; $[0, 2\pi)$; real   **3.** $\dfrac{\pi}{4}; \dfrac{\pi}{4}; \dfrac{3\pi}{4}; \dfrac{\pi}{4} + 2\pi k; \dfrac{3\pi}{4} + 2\pi k$

5. Answers will vary.   **7. a.** QIV **b.** 2 roots   **9. a.** QIV **b.** 2 roots

11.

θ	$\sin\theta$	$\cos\theta$	$\tan\theta$
0	0	1	0
$\dfrac{\pi}{6}$	$\dfrac{1}{2}$	$\dfrac{\sqrt{3}}{2}$	$\dfrac{\sqrt{3}}{3}$
$\dfrac{\pi}{3}$	$\dfrac{\sqrt{3}}{2}$	$\dfrac{1}{2}$	$\sqrt{3}$
$\dfrac{\pi}{2}$	1	0	und.
$\dfrac{2\pi}{3}$	$\dfrac{\sqrt{3}}{2}$	$-\dfrac{1}{2}$	$-\sqrt{3}$
$\dfrac{5\pi}{6}$	$\dfrac{1}{2}$	$-\dfrac{\sqrt{3}}{2}$	$-\dfrac{\sqrt{3}}{3}$
π	0	-1	0
$\dfrac{7\pi}{6}$	$-\dfrac{1}{2}$	$-\dfrac{\sqrt{3}}{2}$	$\dfrac{\sqrt{3}}{3}$
$\dfrac{4\pi}{3}$	$-\dfrac{\sqrt{3}}{2}$	$-\dfrac{1}{2}$	$\sqrt{3}$

13. $\dfrac{\pi}{4}$ **15.** $-\dfrac{\pi}{4}$ **17.** $\dfrac{\pi}{6}$ **19.** $-\dfrac{\pi}{3}$ **21.** π **23.** $\dfrac{\pi}{3}$ **25.** $\dfrac{\pi}{6}$ **27.** $\dfrac{5\pi}{6}$

29. $\dfrac{\pi}{6}, \dfrac{5\pi}{6}$ **31.** $\dfrac{2\pi}{3}, \dfrac{5\pi}{3}$ **33.** $\dfrac{3\pi}{4}, \dfrac{7\pi}{4}$ **35.** $\dfrac{\pi}{6}, \dfrac{5\pi}{6}, \dfrac{7\pi}{6}, \dfrac{11\pi}{6}$

37. $\dfrac{\pi}{3}, \dfrac{2\pi}{3}, \dfrac{4\pi}{3}, \dfrac{5\pi}{3}$ **39.** $\dfrac{\pi}{4}, \dfrac{3\pi}{4}, \dfrac{5\pi}{4}, \dfrac{7\pi}{4}$ **41.** $\dfrac{\pi}{2}, \dfrac{3\pi}{2}$

43. $\theta = 1.2310 + 2\pi k$ or $5.0522 + 2\pi k$

45. $x = \dfrac{\pi}{2} + \pi k$ or $\dfrac{\pi}{6} + 2\pi k$ or $\dfrac{5\pi}{6} + 2\pi k$

47. $x = \dfrac{2\pi}{3} + 2\pi k$ or $\dfrac{4\pi}{3} + 2\pi k$ or $1.4455 + 2\pi k$ or $4.8377 + 2\pi k$

49. $x = \dfrac{\pi}{6} + \pi k$ or $\dfrac{5\pi}{6} + \pi k$ **51.** $x = \dfrac{5\pi}{4} + 2\pi k$ or $\dfrac{7\pi}{4} + 2\pi k$

53. $x = \dfrac{3\pi}{4} + 2\pi k$ or $\dfrac{5\pi}{4} + 2\pi k$ **55.** $x = \dfrac{3\pi}{4} + \pi k$

57. $x = \dfrac{\pi}{3} + \pi k$ or $\dfrac{2\pi}{3} + \pi k$ **59.** $x = \dfrac{3\pi}{8} + \dfrac{\pi}{2}k$ **61.** $x = 3\pi + 6\pi k$

63. $x = \dfrac{\pi}{2} + \pi k$ **65.** $x = \dfrac{\pi}{6} + \dfrac{\pi}{3}k$ or $\dfrac{\pi}{12} + \pi k$ or $\dfrac{5\pi}{12} + \pi k$

67. a. $x \approx 1.2310$ **b.** $x \approx 1.2310 + 2\pi k, 5.0522 + 2\pi k$
69. a. $x \approx 1.2094$ **b.** $x \approx 1.2094 + 2\pi k, 5.0738 + 2\pi k$
71. a. $\theta \approx 0.3649$ **b.** $\theta \approx 0.3649 + \pi k, 1.2059 + \pi k$
73. a. $\theta \approx 0.8861$ **b.** $\theta \approx 0.8861 + \pi k, 2.2555 + \pi k$

75. $x = \dfrac{\pi}{6} + \pi k$ or $\dfrac{5\pi}{6} + \pi k$ **77.** $x = \dfrac{2\pi}{9} + \dfrac{4\pi}{3}k$ or $\dfrac{10\pi}{9} + \dfrac{4\pi}{3}k$

79. $\theta = \dfrac{\pi}{2}k$ **81.** $\theta \approx 0.3398 + 2\pi k$ or $2.8018 + 2\pi k$ **83.** $x \approx 0.7290$

85. $x \approx 2.6649$ **87.** $x \approx 0.4566$ **89.** $22.1°$ and $67.9°$
91. $0°$; the ramp is horizontal.   **93.** $30.7°$; smaller
95. $\alpha = 35°, \beta \approx 25.5°$ **97.** $k \approx 1.36, \alpha \approx 20.6°$ **99. a.** 7 in.

b. ≈ 1.05 in. and ≈ 5.24 in.   **101.** 1.1547   **103.** $\dfrac{\pi}{2} + \pi k$, explanations will vary.

105. $f(2 + i) = (2 + i)^2 - 4(2 + i) + 5$ **107. a.** $-\dfrac{1}{\sqrt{3}}$ **b.** $-\dfrac{\sqrt{2}}{2}$
$= 4 + 4i + i^2 - 8 - 4i + 5$
$= \cancel{4} + \cancel{4i} - \cancel{1} - \cancel{8} - \cancel{4i} + \cancel{5} = 0$

Exercises 6.7, pp. 615–618

1. $\sin^2 x + \cos^2 x = 1; 1 + \tan^2 x = \sec^2 x; 1 + \cot^2 x = \csc^2 x$

3. factor; grouping   **5.** Answers will vary.   **7.** $\dfrac{\pi}{12}, \dfrac{5\pi}{12}$ **9.** 0

11. $0.4456, 1.1252$ **13.** $\dfrac{\pi}{4}, \dfrac{5\pi}{4}, \dfrac{\pi}{6}, \dfrac{5\pi}{6}$ **15.** $\dfrac{\pi}{4}, \dfrac{3\pi}{4}, \dfrac{5\pi}{4}, \dfrac{7\pi}{4}, 0.8411, 5.4421$

17. $\dfrac{\pi}{4}, \dfrac{3\pi}{4}, \dfrac{5\pi}{4}, \dfrac{7\pi}{4}$ **19.** $\dfrac{\pi}{6}, \dfrac{5\pi}{6}, 0.7297, 2.4119$ **21.** $\dfrac{2\pi}{3}$

23. $\dfrac{\pi}{9} + \dfrac{2\pi}{3}k, \dfrac{5\pi}{9} + \dfrac{2\pi}{3}k; k = 0, 1, 2$ **25.** $\dfrac{\pi}{4}, \dfrac{3\pi}{4}, \dfrac{5\pi}{4}, \dfrac{7\pi}{4},$

27. $P = 12; x = 3; x = 11$ **29.** $P = 24; x \approx 0.4909, x \approx 5.5091$

31. $\dfrac{\pi}{12}, \dfrac{17\pi}{12}$ **33.** $0.3747, 5.9085, 2.7669, 3.5163$

35. $\dfrac{\pi}{2}\left(\dfrac{3\pi}{2} \text{ is extraneous}\right)$ **37.** $\dfrac{3\pi}{4}, \dfrac{7\pi}{4}$ **39.** $\dfrac{\pi}{12}, \dfrac{5\pi}{12}, \dfrac{13\pi}{12}, \dfrac{17\pi}{12}$

41. I. a. $\left(\dfrac{5}{2}, \dfrac{5}{2}\right)$ **b.** $D = \sqrt{12.5}, \theta = \dfrac{\pi}{4}, y = \dfrac{\sqrt{12.5} - x\cos\left(\dfrac{\pi}{4}\right)}{\sin\left(\dfrac{\pi}{4}\right)}$

c. verified

II. a. $(2, 4)$ **b.** $D = 2\sqrt{5}, \theta \approx 1.1071, y = \dfrac{2\sqrt{5} - x\cos 1.1071}{\sin 1.1071}$

c. verified

III. a. $(1, \sqrt{3})$ **b.** $D = 2, \theta = \dfrac{\pi}{3}, y = \dfrac{2 - x\cos\left(\dfrac{\pi}{3}\right)}{\sin\left(\dfrac{\pi}{3}\right)}$ **c.** verified

43. a. 2500π ft³ ≈ 7853.98 ft³   **b.** ≈ 7824.09 ft³   **c.** $\theta \approx 78.5°$
45. a. ≈ 78.53 m³/sec   **b.** during the months of August, September, October, and November   **47. a.** $\approx \$3554.52$ **b.** during the months of May, June, July, and August   **49. a.** ≈ 12.67 in.   **b.** during the months of April, May, June, July, and August   **51. a.** ≈ 8.39 gal **b.** approx. day 214 to day 333   **53. a.** 68 bpm   **b.** ≈ 176.2 bpm
c. from about 4.6 min to 7.4 min

55. a. $y = 19\cos\left(\pi - \dfrac{\pi}{6}x\right) + 53$ **b.** $y = -21\sin\left(\dfrac{2\pi}{365}x\right) + 29$

57. a. $L \approx 25.5$ cm.   **b.** $\theta \approx 38.9°$ or $33.4°$, depending on what side you consider the base.
59. $(-1, 0), (0, 0), (2, 0)$ (multiplicity 2): up/up;

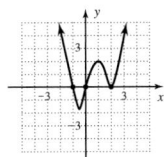

61. $\theta \approx 4.56°$

Summary and Concept Review, pp. 619–623

1. $\sin x(\csc x - \sin x) = \sin x \csc x - \sin x \sin x$
$= \sin x \dfrac{1}{\sin x} - \sin^2 x$
$= 1 - \sin^2 x$
$= \cos^2 x$

2. $\dfrac{\tan^2 x \csc x + \csc x}{\sec^2 x} = \dfrac{\csc x(\tan^2 x + 1)}{\sec^2 x}$

$\qquad\qquad = \dfrac{\csc x \sec^2 x}{\sec^2 x}$

$\qquad\qquad = \csc x$

3. $\dfrac{(\sec x - \tan x)(\sec x + \tan x)}{\csc x} = \dfrac{\sec^2 x + \sec x \tan x - \sec x \tan x - \tan^2 x}{\csc x}$

$\qquad\qquad = \dfrac{\sec^2 x - \tan^2 x}{\csc x}$

$\qquad\qquad = \dfrac{1 + \tan^2 x - \tan^2 x}{\csc x}$

$\qquad\qquad = \dfrac{1}{\csc x}$

$\qquad\qquad = \sin x$

4. $\dfrac{\sec^2 x}{\csc x} - \sin x = \dfrac{\sec^2 x - \sin x \csc x}{\csc x}$

$\qquad\qquad = \dfrac{\sec^2 x - 1}{\csc x}$

$\qquad\qquad = \dfrac{\tan^2 x}{\csc x}$

5. $\sin\theta = \dfrac{-35}{37},\ \csc\theta = \dfrac{-37}{35},\ \cot\theta = \dfrac{12}{35},\ \tan\theta = \dfrac{35}{12},\ \sec\theta = \dfrac{-37}{12}$

6. $\sin\theta = \dfrac{-4\sqrt{6}}{25},\ \csc\theta = \dfrac{-25}{4\sqrt{6}},\ \cot\theta = \dfrac{-23}{4\sqrt{6}},\ \tan\theta = -\dfrac{4\sqrt{6}}{23},$

$\cos\theta = \dfrac{23}{25}$

7. $\dfrac{1 + \cos x}{\sin x}$; answers will vary. **8.** $\sec x - \tan x$; answers will vary.

9. $\dfrac{\csc^2 x(1 - \cos^2 x)}{\tan^2 x} = \dfrac{\csc^2 x \sin^2 x}{\tan^2 x}$

$\qquad\qquad = \dfrac{1}{\tan^2 x}$

$\qquad\qquad = \cot^2 x$

10. $\dfrac{\cot x}{\sec x} - \dfrac{\csc x}{\tan x} = \cot x \dfrac{1}{\sec x} - \cot x \csc x$

$\qquad\qquad = \cot x \cos x - \cot x \csc x$

$\qquad\qquad = \cot x(\cos x - \csc x)$

11. $\dfrac{\sin^4 x - \cos^4 x}{\sin x \cos x} = \dfrac{(\sin^2 x - \cos^2 x)(\sin^2 x + \cos^2 x)}{\sin x \cos x}$

$\qquad\qquad = \dfrac{(\sin^2 x - \cos^2 x)(1)}{\sin x \cos x}$

$\qquad\qquad = \dfrac{\sin x \sin x}{\sin x \cos x} - \dfrac{\cos x \cos x}{\sin x \cos x}$

$\qquad\qquad = \dfrac{\sin x}{\cos x} - \dfrac{\cos x}{\sin x}$

$\qquad\qquad = \tan x - \cot x$

12. $\dfrac{(\sin x + \cos x)^2}{\sin x \cos x} = \dfrac{\sin^2 x + 2\sin x \cos x + \cos^2 x}{\sin x \cos x}$

$\qquad\qquad = \dfrac{\sin^2 x + \cos^2 x}{\sin x \cos x} + \dfrac{2\sin x \cos x}{\sin x \cos x}$

$\qquad\qquad = \dfrac{1}{\sin x \cos x} + 2$

$\qquad\qquad = \csc x \sec x + 2$

13. a. $\cos 75° = \dfrac{\sqrt{6} - \sqrt{2}}{4}$

b. $\tan\left(\dfrac{\pi}{12}\right) = \dfrac{\sqrt{3} - 1}{1 + \sqrt{3}} = \dfrac{(\sqrt{3} - 1)^2}{2} = 2 - \sqrt{3}$

14. a. $\tan 15° = \dfrac{\sqrt{3} - 1}{1 + \sqrt{3}} = \dfrac{(\sqrt{3} - 1)^2}{2} = 2 - \sqrt{3}$

b. $\sin\left(\dfrac{-\pi}{12}\right) = \dfrac{\sqrt{2} - \sqrt{6}}{4}$ **15. a.** $\cos 180° = -1$ **b.** $\sin 120° = \dfrac{\sqrt{3}}{2}$

16. a. $\cos x$ **b.** $\sin\left(\dfrac{5x}{8}\right)$ **17. a.** $\cos 1170° = \cos 90° = 0$

b. $\sin\left(\dfrac{57\pi}{4}\right) = \sin\left(\dfrac{\pi}{4}\right) = \dfrac{\sqrt{2}}{2}$

18. a. $\cos\left(\dfrac{x}{8}\right) = \sin\left(\dfrac{\pi}{2} - \dfrac{x}{8}\right)$ **b.** $\sin\left(x - \dfrac{\pi}{12}\right) = \cos\left(\dfrac{7\pi}{12} - x\right)$

19. $\tan(45° - 30°) = \dfrac{\tan 45° - \tan 30°}{1 + \tan 45° \tan 30°}$

$= \dfrac{1 - \dfrac{\sqrt{3}}{3}}{1 + 1 \cdot \dfrac{\sqrt{3}}{3}} = \dfrac{1 - \dfrac{\sqrt{3}}{3}}{1 + \dfrac{\sqrt{3}}{3}} = \dfrac{\dfrac{3 - \sqrt{3}}{3}}{\dfrac{3 + \sqrt{3}}{3}}$

$= \dfrac{3 - \sqrt{3}}{3} \cdot \dfrac{3}{3 + \sqrt{3}} = \dfrac{3 - \sqrt{3}}{3 + \sqrt{3}} = \dfrac{\sqrt{3}(\sqrt{3} - 1)}{\sqrt{3}(\sqrt{3} + 1)} = \dfrac{\sqrt{3} - 1}{\sqrt{3} + 1}$

$\tan(135° - 120°) = \dfrac{\tan 135° - \tan 120°}{1 + \tan 135° \tan 120°}$

$= \dfrac{-1 + \sqrt{3}}{1 + (-1)(-\sqrt{3})} = \dfrac{\sqrt{3} - 1}{1 + \sqrt{3}} = \dfrac{\sqrt{3} - 1}{\sqrt{3} + 1}$

20. $\cos\left(x + \dfrac{\pi}{6}\right) + \cos\left(x - \dfrac{\pi}{6}\right) = \sqrt{3}\cos x$

$= \cos x \cos\left(\dfrac{\pi}{6}\right) - \sin x \sin\left(\dfrac{\pi}{6}\right) + \cos x \cos\left(\dfrac{\pi}{6}\right) + \sin x \sin\left(\dfrac{\pi}{6}\right)$

$= 2\cos x \cos\left(\dfrac{\pi}{6}\right) + 0 = 2\cos x\left(\dfrac{\sqrt{3}}{2}\right) = \sqrt{3}\cos x$

21. a. $\sin(2\theta) = 2\left(\dfrac{-84}{85}\right)\left(\dfrac{13}{85}\right) = \dfrac{-2184}{7225}$

$\cos(2\theta) = \left(\dfrac{13}{85}\right)^2 - \left(\dfrac{84}{85}\right)^2 = \dfrac{-6887}{7225}$

$\tan(2\theta) = \dfrac{2184}{-7225}\left(\dfrac{7225}{-6887}\right) = \dfrac{2184}{6887}$

b. $\sin(2\theta) = 2\left(\dfrac{-20}{29}\right)\left(\dfrac{-21}{29}\right) = \dfrac{840}{841}$

$\cos(2\theta) = \left(\dfrac{-21}{29}\right)^2 - \left(\dfrac{-20}{29}\right)^2 = \dfrac{441 - 400}{841} = \dfrac{41}{841}$

$\tan(2\theta) = \dfrac{2\left(\dfrac{20}{21}\right)}{1 - \left(\dfrac{20}{21}\right)^2} = \dfrac{840}{41}$

22. a. $\sin\theta = \dfrac{21}{29},\ \cos\theta = \dfrac{-20}{29},\ \tan\theta = -\dfrac{21}{20},$

b. $\sin\theta = \dfrac{7}{25}$ or $\sin\theta = \dfrac{24}{25},\ \cos\theta = \dfrac{-24}{25}$ or $\cos\theta = \dfrac{-7}{25},\ \tan\theta = \dfrac{-7}{24}$

or $\tan\theta = \dfrac{-24}{7}$

23. a. $\cos 45° = \dfrac{\sqrt{2}}{2}$ **b.** $\cos\left(\dfrac{\pi}{6}\right) = \dfrac{\sqrt{3}}{2}$

24. a. $\sin 67.5 = \sqrt{\dfrac{1 - \cos 135°}{2}} = \sqrt{\dfrac{1 + \dfrac{\sqrt{2}}{2}}{2}} = \sqrt{\dfrac{2 + \sqrt{2}}{4}}$

$= \dfrac{\sqrt{2 + \sqrt{2}}}{2}$

$\cos 67.5 = \sqrt{\dfrac{1 + \cos 135°}{2}} = \sqrt{\dfrac{1 - \dfrac{\sqrt{2}}{2}}{2}} = \sqrt{\dfrac{2 - \sqrt{2}}{4}}$

$= \dfrac{\sqrt{2 - \sqrt{2}}}{2}$

b. $\sin\left(\dfrac{5\pi}{8}\right) = \sqrt{\dfrac{1 - \cos\left(\dfrac{5\pi}{4}\right)}{2}} = \sqrt{\dfrac{1 + \dfrac{\sqrt{2}}{2}}{2}} = \sqrt{\dfrac{2 + \sqrt{2}}{4}}$

$= \dfrac{\sqrt{2 + \sqrt{2}}}{2}$

$\cos\left(\dfrac{5\pi}{8}\right) = -\sqrt{\dfrac{1 + \cos\left(\dfrac{5\pi}{4}\right)}{2}} = -\sqrt{\dfrac{1 - \dfrac{\sqrt{2}}{2}}{2}} = -\sqrt{\dfrac{2 - \sqrt{2}}{4}}$

$= -\dfrac{\sqrt{2 - \sqrt{2}}}{2}$

25. a. $\sin\left(\dfrac{\theta}{2}\right) = \sqrt{\dfrac{1 - 24/25}{2}} = \sqrt{\dfrac{25 - 24}{50}} = +\dfrac{1}{5\sqrt{2}}, \dfrac{\theta}{2}$ in QII

$\cos\left(\dfrac{\theta}{2}\right) = -\sqrt{\dfrac{1 + 24/25}{2}} = -\sqrt{\dfrac{25 + 24}{50}}$

$= -\sqrt{\dfrac{49}{50}} = \dfrac{-7}{5\sqrt{2}}, \dfrac{\theta}{2}$ in QII

b. $\sin\left(\dfrac{\theta}{2}\right) = -\sqrt{\dfrac{1 - 56/65}{2}} = -\sqrt{\dfrac{65 - 56}{130}}$

$= -\sqrt{\dfrac{9}{130}} = \dfrac{-3}{\sqrt{130}}, \dfrac{\theta}{2}$ in QIV

$\cos\left(\dfrac{\theta}{2}\right) = \sqrt{\dfrac{1 + 56/65}{2}} = \sqrt{\dfrac{65 + 56}{130}} = \sqrt{\dfrac{121}{130}} = +\dfrac{11}{\sqrt{130}}, \dfrac{\theta}{2}$ in QIV

26. $\dfrac{\cos(3\alpha) - \cos\alpha}{\cos(3\alpha) + \cos\alpha} = \dfrac{-2\sin(2\alpha)\sin\alpha}{2\cos(2\alpha)\cos\alpha}$

$= \dfrac{-2\sin^2\alpha}{\cos^2\alpha - \sin^2\alpha} = \dfrac{2\sin^2\alpha}{\sin^2\alpha - \cos^2\alpha}$

$= \dfrac{2\sin^2\alpha}{1 - 2\cos^2\alpha} = \dfrac{2\tan^2\alpha}{\sec^2\alpha - 2}$

27. $\cos(3x) + \cos x = 0 \to 2\cos(2x)\cos x = 0$

$\cos(2x) = 0$: $x = \dfrac{\pi}{4} + \dfrac{\pi}{2}k;\ k \in \mathbb{Z}$

$\cos x = 0$: $x = \dfrac{\pi}{2} + \pi k;\ k \in \mathbb{Z}$

28. a. $A = 12^2\sin\left(\dfrac{30°}{2}\right)\cos\left(\dfrac{30°}{2}\right) = 144\sqrt{\dfrac{1 - \cos 30°}{2}}\sqrt{\dfrac{1 + \cos 30°}{2}}$

$= 144\sqrt{\dfrac{1 - \dfrac{\sqrt{3}}{2}}{2}}\sqrt{\dfrac{1 + \dfrac{\sqrt{3}}{2}}{2}} = 144\sqrt{\dfrac{2 - \sqrt{3}}{4}}\sqrt{\dfrac{2 + \sqrt{3}}{4}}$

$= \dfrac{144\sqrt{4 - 3}}{4} = 36\ \text{cm}^2$; yes

b. $x^2\sin\left(\dfrac{\theta}{2}\right)\cos\left(\dfrac{\theta}{2}\right)$

Let $u = \dfrac{\theta}{2}$, then $= x^2\sin u\cos u = \dfrac{1}{2}x^2(2\sin u\cos u) = \dfrac{1}{2}x^2\sin(2u)$

$= \dfrac{1}{2}x^2\sin\theta;\ A = \dfrac{1}{2}(12)^2\sin(30°) = 72\left(\dfrac{1}{2}\right) = 36\ \text{cm}^2$; yes

29. $\dfrac{\pi}{4}$ or 45°   **30.** $\dfrac{\pi}{6}$ or 30°   **31.** $\dfrac{5\pi}{6}$ or 150°   **32.** 1.3431 or 77.0°

33. 1.0956 or 62.8°   **34.** 0.5054 or 29.0°   **35.** $\dfrac{1}{2}$ **36.** $\dfrac{\pi}{4}$

37. undefined   **38.** 1.0245   **39.** 60°   **40.** $\dfrac{3\pi}{4}$

41. 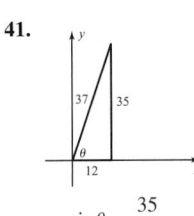   **42.**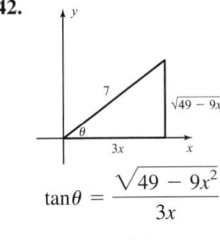

$\sin\theta = \dfrac{35}{37}$ $\tan\theta = \dfrac{\sqrt{49 - 9x^2}}{3x}$

43. 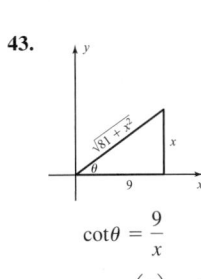   **44.** $\theta = \cos^{-1}\left(\dfrac{x}{5}\right)$ **45.** $\theta = \sec^{-1}\left(\dfrac{x}{7\sqrt{3}}\right)$

$\cot\theta = \dfrac{9}{x}$

46. $\theta = \sin^{-1}\left(\dfrac{x}{4}\right) + \dfrac{\pi}{6}$

47. a. $\dfrac{\pi}{4}$ **b.** $\dfrac{\pi}{4}, \dfrac{3\pi}{4}$ **c.** $x = \dfrac{\pi}{4} + 2\pi k$ or $\dfrac{3\pi}{4} + 2\pi k, k \in \mathbb{Z}$

48. a. $\dfrac{2\pi}{3}$ **b.** $\dfrac{2\pi}{3}, \dfrac{4\pi}{3}$ **c.** $\dfrac{2\pi}{3} + 2\pi k$ or $\dfrac{4\pi}{3} + 2\pi k, k \in \mathbb{Z}$

49. a. $-\dfrac{\pi}{3}$ **b.** $\dfrac{2\pi}{3}, \dfrac{5\pi}{3}$ **c.** $\dfrac{2\pi}{3} + \pi k, k \in \mathbb{Z}$

50. a. ≈ 1.1102 **b.** $\approx 1.1102, 5.1729$ **c.** $\approx 1.1102 + 2\pi k$ or $5.1729 + 2\pi k, k \in \mathbb{Z}$ **51. a.** ≈ 0.3376 **b.** $\approx 0.3376, 1.2332, 3.4792, 4.3748$ **c.** $\approx 0.3376 + \pi k$ or $1.2332 + \pi k, k \in \mathbb{Z}$ **52. a.** ≈ 0.3614 **b.** $\approx 0.3614, 2.7802$ **c.** $\approx 0.3614 + 2\pi k$ or $2.7802 + 2\pi k, k \in \mathbb{Z}$

53. $\theta \approx 1.1547$ **54.** $x = \dfrac{\pi}{12}, \dfrac{5\pi}{12}$ **55.** $x \approx 0.7297, 2.4119;\ x = \dfrac{\pi}{6}, \dfrac{5\pi}{6}$

56. $x = \dfrac{\pi}{6}, \dfrac{5\pi}{6}, \dfrac{11\pi}{6}$ **57.** $x = \dfrac{\pi}{2}$ **58.** $P = 12; x \approx 2.6931, x \approx 9.3069$

59. $P = 6; x = 0, x = \dfrac{9}{2}$ **60. a.** $\approx 43(1000) = \$43,000$

b. April through August

Mixed Review, pp. 623–624

1. $\sin\theta = \dfrac{6}{\sqrt{117}}, \sec\theta = \dfrac{-\sqrt{117}}{9}, \tan\theta = \dfrac{-6}{9} = \dfrac{-2}{3}, \cos\theta = \dfrac{-9}{\sqrt{117}},$
$\csc\theta = \dfrac{\sqrt{117}}{6}, \cot\theta = \dfrac{-3}{2}$ **3.** $\sqrt{3} + 2$

5. $\tan\theta = \dfrac{x}{\sqrt{100 - x^2}}$

7. $x = 0.4103; x = 4.9230$

9. a. $\left(\dfrac{1}{32}\right)\left(\dfrac{2}{1}\right)v^2\sin\theta\cos\theta$

$\left(\dfrac{1}{32}\right)v^2\,2\sin\theta\cos\theta$

$\left(\dfrac{1}{32}\right)v^2\sin 2\theta$

b. $\sin(2\theta) = \sin[2(90 - \theta)]$
$= \sin(180° - 2\theta)$
$= \sin(2\theta)$

11. $\cos\left[2\left(\dfrac{\pi}{12}\right)\right] = \cos\left(\dfrac{\pi}{6}\right) = \dfrac{\sqrt{3}}{2}$

13. $\dfrac{(\cos t + \sin t)^2}{\tan t} = \dfrac{\cos^2 t + 2\cos t\sin t + \sin^2 t}{\tan t}$

$= \dfrac{1 + 2\cos t\sin t}{\tan t}$

$= \dfrac{1}{\tan t} + 2\cos t\,\cancel{\sin t}\cdot\dfrac{\cos t}{\cancel{\sin t}}$

$= \cot t + 2\cos^2 t$

15. $\dfrac{3\pi}{4}$ or 135°   **17.** $\dfrac{\pi}{3}$ or 60°   **19.** $\theta = \tan^{-1}\left(\dfrac{x}{10}\right)$

21. a. 6 ft: 2 A.M., 2 P.M., 10 A.M., 10 P.M. 10 ft: 6 A.M., 6 P.M.

b. about 8.9 ft   **23. a.** $\dfrac{-\sqrt{3}\sqrt{2 - \sqrt{2}}}{2}$ **b.** $\dfrac{-\sqrt{2 - \sqrt{2}}}{2}$

25. a. $\dfrac{1 + \sqrt{2}}{4}$ **b.** $\dfrac{\sqrt{2} + \sqrt{3}}{4}$

Practice Test, pp. 625–626

1. $\dfrac{(\csc x - \cot x)(\csc x + \cot x)}{\sec x} = \dfrac{\csc^2 x + \csc x\cot x - \csc x\cot x - \cot^2 x}{\sec x}$

$= \dfrac{\csc^2 x - \cot^2 x}{\sec x}$

$= \dfrac{(1 + \cot^2 x) - \cot^2 x}{\sec x}$

$= \dfrac{1}{\sec x}$

$= \cos x$

2. $\dfrac{\sin^3 x - \cos^3 x}{1 + \cos x \sin x} = \dfrac{(\sin x - \cos x)(\sin^2 x + \sin x \cos x + \cos^2 x)}{1 + \cos x \sin x}$

$= \dfrac{(\sin x - \cos x)(1 + \sin x \cos x)}{1 + \cos x \sin x}$

$= \sin x - \cos x$

3. $\sin \theta = \dfrac{-55}{73}$, $\sec \theta = \dfrac{73}{48}$, $\cot \theta = \dfrac{-48}{55}$, $\tan \theta = \dfrac{-55}{48}$, $\csc \theta = \dfrac{-73}{55}$

4. $\dfrac{\sqrt{3} - 1}{\sqrt{3} + 1}$ **5.** $\dfrac{\sqrt{2}}{2}$ **6.** $\dfrac{-\sqrt{2}}{2}$

7. $\sin\left(x + \dfrac{\pi}{4}\right) - \sin\left(x - \dfrac{\pi}{4}\right)$

$= \sin x \cos\left(\dfrac{\pi}{4}\right) + \cos x \sin\left(\dfrac{\pi}{4}\right) - \sin x \cos\left(\dfrac{\pi}{4}\right) + \cos x \sin\left(\dfrac{\pi}{4}\right)$

$= \sin\left(\dfrac{\pi}{4}\right)\cos x + \sin\left(\dfrac{\pi}{4}\right)\cos x$

$= 2\sin\left(\dfrac{\pi}{4}\right)\cos x$

$= 2 \dfrac{\sqrt{2}}{2}\cos x$

$= \sqrt{2}\cos x$

8. $\sin \theta = \dfrac{15}{17}$, $\cos \theta = \dfrac{8}{17}$, $\tan \theta = \dfrac{15}{8}$ **9.** $\dfrac{-\sqrt{3}}{2}$ **10.** $\dfrac{1}{\sqrt{37}}$; $\dfrac{6}{\sqrt{37}}$

11. $20\sqrt{2 - \sqrt{2}}$ **12.** $\dfrac{\sqrt{6} - \sqrt{2}}{4} \approx 0.2588$; $\dfrac{\sqrt{6} + \sqrt{2}}{4} \approx 0.9659$

13. a. $y = 30°$ **b.** $f(x) = \dfrac{1}{2}$ **c.** $y = 30°$

14. a. $y = 0.8523$ rad or $y = 48.8°$ **b.** $y = 78.5°$ or $\dfrac{157\pi}{360}$ rad

c. $y = \dfrac{7\pi}{24}$ rad or $52.5°$

15. 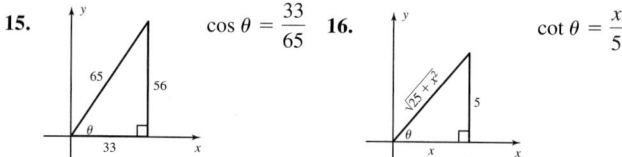 $\cos \theta = \dfrac{33}{65}$ **16.** $\cot \theta = \dfrac{x}{5}$

17. I. a. $\cos^{-1}\left(\dfrac{-\sqrt{2}}{2}\right) = \dfrac{3\pi}{4}$ **b.** $x = \dfrac{3\pi}{4}, \dfrac{5\pi}{4}$

c. $x = \dfrac{3\pi}{4} + 2\pi k$ or $\dfrac{5\pi}{4} + 2\pi k, k \in \mathbb{Z}$ **II. a.** $\dfrac{\pi}{6}$ **b.** $x = \dfrac{\pi}{6}, \dfrac{11\pi}{6}$

c. $x = \dfrac{\pi}{6} + 2\pi k$ or $\dfrac{11\pi}{6} + 2\pi k, k \in \mathbb{Z}$ **18. I. a.** $x \approx 0.1922$

b. $x \approx 0.1922, 1.3786, 3.3338, 4.5202$ **c.** $x \approx 0.1922 + \pi k$ or $1.3786 + \pi k, k \in \mathbb{Z}$ **II. a.** $x \approx 0.9204$ **b.** $x \approx 0.9204, 2.2212,$ $4.0620, 5.3628$ **c.** $x \approx 0.9204 + \pi k$ or $2.2212 + \pi k, k \in \mathbb{Z}$

19. a. $x \approx -1.6875, -0.3413, 1.1321, 2.8967$ **b.** $x \approx 0.9671, 2.6110,$ 3.4538 **20. a.** $x = 0, \pi, \dfrac{7\pi}{6}, \dfrac{11\pi}{6}$ **b.** $x = \dfrac{7\pi}{12}, \dfrac{11\pi}{12}, \dfrac{19\pi}{12}, \dfrac{23\pi}{12}$

21. $x = \dfrac{\pi}{2}, \dfrac{3\pi}{2}$; $x \approx 3.3090, 6.1157$ **22.** $x = \dfrac{5\pi}{6}, \dfrac{11\pi}{6}$

23. a. 6 or \$6,000 **b.** January through July
24. a. $y = 35.223 \sin(0.576x - 2.589) + 6.120$
b.

25. $\cos(2418\pi t) + \cos(1540\pi t)$

Month (Jan → 1)	Low Temp. (°F)
1	−26
3	−21
5	16
7	41
9	25
11	−14

Strengthening Core Skills p. 627

Exercise 1: $x \in (0.6025, 2.5391)$
Exercise 2: $x \in [0, 0.7945] \cup [4.4415, 2\pi]$
Exercise 3: $x \in [0, 2.6154] \cup [9.3847, 12]$
Exercise 4: $x \in (67.3927, 202.6073)$

Cumulative Review Chapters 1–6, pp. 628–629

1. $\sin \theta = \frac{84}{85}$, $\csc \theta = \frac{85}{84}$, $\cos \theta = \frac{-13}{85}$, $\sec \theta = \frac{-85}{13}$, $\tan \theta = \frac{-84}{13}$, $\cot \theta = \frac{-13}{84}$

3. $g(2 + \sqrt{3}) = (2 + \sqrt{3})^2 - 4(2 + \sqrt{3}) + 1$
$= 4 + 4\sqrt{3} + 3 - 8 - 4\sqrt{3} + 1$
$= 0$

5. about 474 ft **7.**

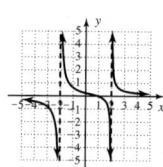

9. 50.89 km/hr **11.** $x \in \left[\dfrac{-9}{2}, \dfrac{11}{2}\right]$ **13. a.** $y = -\dfrac{1}{2}x + 31$ **b.** every 2 years, the amount of emissions decreases by 1 million tons. **c.** 23.5 million tons; 11 million tons **15.** $x \in (1, 5)$ **17.** \$7

19. $\dfrac{\cos x}{\sec x - 1} = \dfrac{\cos x(\sec x + 1)}{(\sec x - 1)(\sec x + 1)}$ **21.** $\dfrac{99}{101}$

$= \dfrac{1 + \cos x}{\sec^2 x - 1}$

$= \dfrac{1 + \cos x}{\tan^2 x}$

23. a. $y = 5.4 \sin\left(\dfrac{\pi}{6}x - \dfrac{2\pi}{3}\right) + 27.1$ **b.** from early May until late August

25. a. volume of a cylinder **b.** volume of a rectangular solid **c.** circumference of a circle **d.** area of a triangle

Connections to Calculus Exercises, p. 632

1. a. $= \csc \theta$, **b.** $y = \csc\left[\tan^{-1}\left(\dfrac{x}{13}\right)\right]$ **c.** verified

3. $4\tan \theta \sec \theta$ **5.** $\sin \theta$ **7.** $0, \dfrac{2\pi}{3}, \dfrac{4\pi}{3}, \pi$

9. $x = \dfrac{\pi}{4}, \dfrac{3\pi}{4}, \dfrac{5\pi}{4}, \dfrac{7\pi}{4}$

CHAPTER 7

Exercises 7.1, pp. 641–646

1. ambiguous **3.** I; II **5.** Answers will vary. **7.** $a \approx 8.98$
9. $C \approx 49.2°$ **11.** $C \approx 21.4°$ **13.** $\angle C = 78°$, $b \approx 109.5$ cm, $c \approx 119.2$ cm **15.** $\angle C = 90°$, $a = 10$ in., $c = 20$ in.
17. **19.** $\angle C = 90°$, $a = 15$ mi, $b = 15$ mi

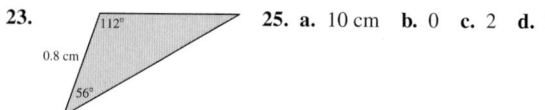

21. $\angle A = 57°$, $b \approx 49.5$ km, $c \approx 17.1$ km

23. **25. a.** 10 cm **b.** 0 **c.** 2 **d.** 1

27. not possible **29.** $B = 60°$, $C = 90°$, $b = 12.9\sqrt{3}$ mi
31. $A \approx 39°$, $B \approx 82°$, $a \approx 42.6$ mi or $A \approx 23°$, $B \approx 98°$, $a \approx 26.4$ mi
33. $A \approx 39°$, $B \approx 82°$, $a \approx 42.6$ ft or $A \approx 23°$, $B \approx 98°$, $a \approx 26.4$ ft
35. not possible **37.** $A \approx 80.0°$, $B \approx 38.0°$, $b \approx 1.8 \times 10^{25}$ mi
39. $A_1 \approx 19.3°$, $A_2 \approx 160.7°$, $48° + 160.7° > 180°$; no second solution possible **41.** $C_1 \approx 71.3°$, $C_2 \approx 108.7°$, $57° + 108.7° < 180°$; two

solutions possible   **43.** not possible, $\sin A > 1$ **45.** $\dfrac{\sqrt{2}}{2}$

47. 34.6 million miles or 119.7 million miles   **49. a.** No   **b.** ≈ 3.9 mi
51. $V \leftrightarrow S = 41.7$ km, $V \leftrightarrow P = 80.8$ km
53. a. No   **b.** about 201.5 ft   **c.** ≈ 15 sec
55. Two triangles

Angles	Sides		Angles	Sides
$A_1 \approx 41.1°$	$a = 12$ cm		$A_2 \approx 138.9°$	$a = 12$ cm
$B = 26°$	$b = 8$ cm		$B = 26°$	$b = 8$ cm
$C_1 \approx 112.9°$	$c_1 \approx 16.8$ cm		$C_2 \approx 15.1°$	$c_2 \approx 4.8$ cm

57.

Angles	Sides		Angles	Sides
$A_1 \approx 47.0°$	$a = 9$ cm		$A_2 \approx 133.0°$	$a = 9$ cm
$B_1 \approx 109.0°$	$b_1 \approx 11.6$ cm		$B_2 \approx 23.0°$	$b_2 \approx 4.8$ cm
$C \approx 24°$	$c = 5$ cm		$C \approx 24°$	$c = 5$ cm

59. $a \approx 33.7$ ft, $c \approx 22.3$ ft   **61.** Rhymes to Tarryson: 61.7 km, Sexton to Tarryson: 52.6 km   **63.** ≈ 3.2 mi   **65.** $h \approx 161.9$ yd
67. angle $= 90°$; sides ≈ 9.8 cm, 11 cm; diameter ≈ 11 cm; it is a right triangle.   **69. a.** about 3187 m   **b.** about 2613 m   **c.** about 2368 m
71.

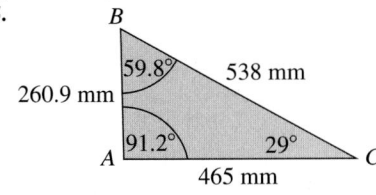

$\sqrt{3} = \dfrac{\sin 60°}{\sin 30°}$; $\sqrt{2} = \dfrac{\sin 90°}{\sin 45°}$

73. $A = 19°$, $B = 31°$, $C = 130°$, $a = 45$ cm, $b = 71.2$ cm, $c \approx 105.8$ cm

75. $\approx 12{,}564$ mph

77.
$$\tan^2 x - \sin^2 x = \dfrac{\sin^2 x}{\cos^2 x} - \sin^2 x$$
$$= \dfrac{\sin^2 x}{\cos^2 x} - \dfrac{\sin^2 x \cos^2 x}{\cos^2 x}$$
$$= \dfrac{\sin^2 x - \sin^2 x \cos^2 x}{\cos^2 x}$$
$$= \dfrac{\sin^2 x(1 - \cos^2 x)}{\cos^2 x}$$
$$= \dfrac{\sin^2 x \sin^2 x}{\cos^2 x}$$
$$= \sin^2 x \dfrac{\sin^2 x}{\cos^2 x}$$
$$= \sin^2 x \tan^2 x$$

79. a. $y = \dfrac{5}{9}x - \dfrac{2}{9}$ **b.** $\sqrt{106}$ units

Exercises 7.2, pp. 653–658
1. cosines   **3.** Pythagorean   **5.** $B \approx 33.1°$, $C \approx 129.9°$, $a \approx 19.8$ m; law of sines   **7.** yes   **9.** no   **11.** yes   **13.** verified   **15.** $B \approx 41.4°$
17. $a \approx 7.24$ **19.** $A \approx 41.6°$ **21.** $A \approx 120.4°$, $B \approx 21.6°$, $c \approx 53.5$ cm
23. $A \approx 23.8°$, $C \approx 126.2°$, $b \approx 16$ mi
25.

B, 59.8°, 538 mm, 260.9 mm, A 91.2°, 29° C, 465 mm

27. $A \approx 137.9°$, $B \approx 15.6°$, $C \approx 26.5°$
29. $A \approx 119.3°$, $B \approx 41.5°$, $C = 19.2°$
31.

A, 2.9×10^{25} mi, 103.3°, B 33.2°, 2.3×10^{25} mi, 4.1×10^{25} mi, 43.5°, C

33. $A \approx 139.7°$, $B \approx 23.7°$, $C \approx 16.6°$
35. $C \approx 86.3°$ **37.** about 1688 mi   **39.** $P \approx 27.7°$; heading 297.7°
41. It cannot be constructed (available length $\approx 10{,}703.6$ ft)
43. 1678.2 mi   **45.** $P \approx 22.4$ cm, $A = 135°$, $B \approx 23.2°$, $C \approx 21.8°$
47. $A \approx 20.6°$, $B \approx 15.3°$, $C \approx 144.1°$ **49.** 58.78 cm
51. $a = 13$ $A \approx 133.2°$ **53.** 33.7°; 150 ft^2
$b = 5$ $B \approx 16.3°$
$c = \sqrt{82}$ $C \approx 30.5°$
55. a. $0.65 = 65\%$ **b.** $\$1{,}950{,}000$ **57.** about 483,529 km^2
59. $387 + 502 = 889 < 902$ **61.** (1) $a^2 = b^2 + c^2 - 2bc \cos A$
(2) $b^2 = a^2 + c^2 - 2ac \cos B$, use substitution for a^2 and
(2) becomes $b^2 = (b^2 + c^2 - 2bc \cos A) + c^2 - 2ac \cos B$.
Then $0 = 2c^2 - 2bc \cos A - 2ac \cos B$, $2bc \cos A +$
$2ac \cos B = 2c^2$, $b \cos A + a \cos B = c$ **63.** 2
65. $\sin x = \dfrac{-5}{13}$, $\csc x = \dfrac{-13}{5}$, $\cos x = \dfrac{12}{13}$, $\sec x = \dfrac{13}{12}$,
$\tan x = \dfrac{-5}{12}$, $\cot x = \dfrac{-12}{5}$

Exercises 7.3, pp. 669–673
1. scalar   **3.** directed; line   **5.** Answers will vary.
7. 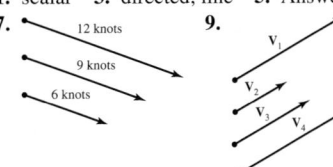   **9.** 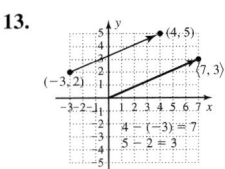   **11.**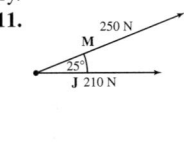

13. 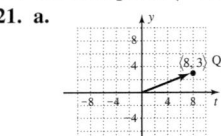   **15.**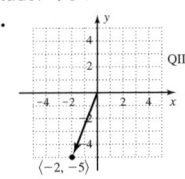

17. Terminal point: $(5, -1)$, magnitude: $\sqrt{53}$
19. Terminal point: $(-1, 1)$, magnitude: $\sqrt{34}$
21. a. 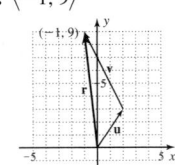   **23. a.**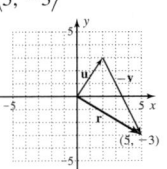

b. $\sqrt{73}$ **c.** 20.6°   **b.** $\sqrt{29}$ **c.** 68.2°
25. $\langle -10.9, 5.1 \rangle$ **27.** $\langle 106, -92.2 \rangle$ **29.** $\langle -9.7, -2.6 \rangle$
31. a. $\langle -1, 9 \rangle$ **b.** $\langle 5, -3 \rangle$

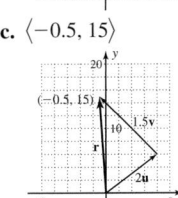   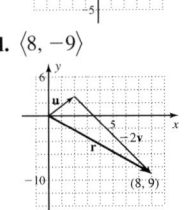

c. $\langle -0.5, 15 \rangle$ **d.** $\langle 8, -9 \rangle$

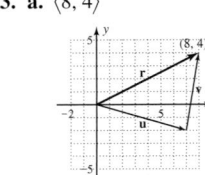   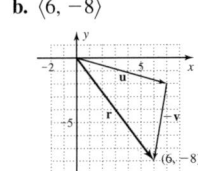

33. a. $\langle 8, 4 \rangle$ **b.** $\langle 6, -8 \rangle$

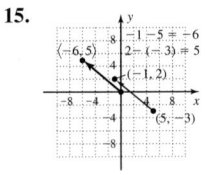

c. $\langle 15.5, 5 \rangle$ **d.** $\langle 5, -14 \rangle$

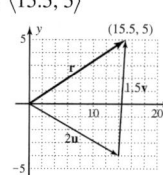 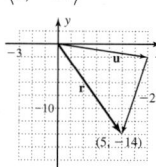

35. a. $\langle -3, 6 \rangle$ **b.** $\langle -5, -2 \rangle$

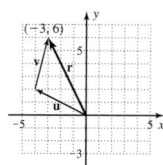 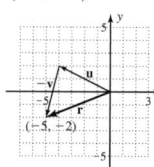

c. $\langle -6.5, 10 \rangle$ **d.** $\langle -6, -6 \rangle$

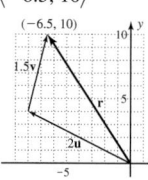 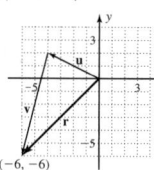

37. True   **39.** False   **41.** True

43. $\mathbf{u} + \mathbf{v} = \langle 8, 6 \rangle$ **45.** $\mathbf{u} + \mathbf{v} = \langle -9, -6 \rangle$
     $\mathbf{u} - \mathbf{v} = \langle -6, 2 \rangle$ $\mathbf{u} - \mathbf{v} = \langle 7, 0 \rangle$

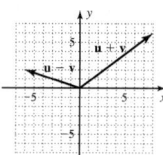 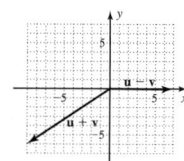

47. $\mathbf{u} + \mathbf{v} = \langle -3, -6 \rangle$ **49.** $\mathbf{u} = 8\mathbf{i} + 15\mathbf{j}$
     $\mathbf{u} - \mathbf{v} = \langle -7, 0 \rangle$ $|\mathbf{u}| = 17$

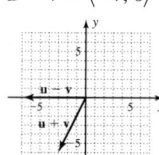 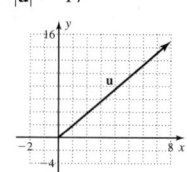

51. $\mathbf{p} = -3.2\mathbf{i} - 5.7\mathbf{j}$ **53. a.**
     $|\mathbf{p}| \approx 6.54$

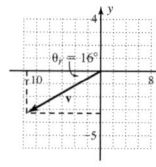

                      b. $\mathbf{v} = \langle -11.5, -3.3 \rangle$
                      c. $\mathbf{v} = -11.5\mathbf{i}, -3.3\mathbf{j}$

55. a.   **b.** $\mathbf{w} = \langle 2.5, 9.2 \rangle$
                c. $\mathbf{w} = 2.5\mathbf{i} + 9.2\mathbf{j}$

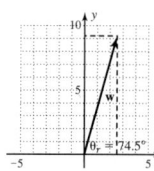

57. a. $\mathbf{p} = -2\mathbf{i} + 2\mathbf{j}; |\mathbf{p}| = 2\sqrt{2}, \theta = 135°$
    b. $\mathbf{q} = 6\mathbf{i} - 8\mathbf{j}; |\mathbf{q}| = 10, \theta = 306.9°$
    c. $\mathbf{r} = -2\mathbf{i} + 1.5\mathbf{j}; |\mathbf{r}| = 2.5, \theta = 143.1°$
    d. $\mathbf{s} = 10\mathbf{i} - 13\mathbf{j}; |\mathbf{s}| \approx 16.4, \theta \approx 307.6°$
59. a. $\mathbf{p} = 2\sqrt{2}\mathbf{i} + 2\mathbf{j}; |\mathbf{p}| \approx 3.5, \theta \approx 35.3°$
    b. $\mathbf{q} = 8\sqrt{2}\mathbf{i} + 12\mathbf{j}; |\mathbf{q}| \approx 16.5, \theta \approx 46.7°$
    c. $\mathbf{r} = 5.5\sqrt{2}\mathbf{i} + 6.5\mathbf{j}; |\mathbf{r}| \approx 10.1, \theta \approx 39.9°$
    d. $\mathbf{s} = 11\sqrt{2}\mathbf{i} + 17\mathbf{j}; |\mathbf{s}| \approx 23.0, \theta \approx 47.5°$
61. a. $\mathbf{p} = 8\mathbf{i} + 4\mathbf{j}; |\mathbf{p}| \approx 8.9, \theta \approx 26.6°$
    b. $\mathbf{q} = 16\mathbf{i} + 4\mathbf{j}; |\mathbf{q}| \approx 16.5, \theta \approx 14.0°$

c. $\mathbf{r} = 18\mathbf{i} + 8\mathbf{j}; |\mathbf{r}| \approx 19.7, \theta \approx 24.0°$
d. $\mathbf{s} = 20\mathbf{i} + 4\mathbf{j}; |\mathbf{s}| \approx 20.4, \theta \approx 11.3°$
63. $\left\langle \dfrac{7}{25}, \dfrac{24}{25} \right\rangle$, verified   **65.** $\left\langle \dfrac{-20}{29}, \dfrac{21}{29} \right\rangle$, verified
67. $\dfrac{20}{29}\mathbf{i} - \dfrac{21}{29}\mathbf{j}$, verified   **69.** $\dfrac{7}{25}\mathbf{i} + \dfrac{24}{25}\mathbf{j}$, verified
71. $\left\langle -\dfrac{13}{\sqrt{178}}, \dfrac{3}{\sqrt{178}} \right\rangle$, verified   **73.** $\dfrac{6}{\sqrt{157}}\mathbf{i} + \dfrac{11}{\sqrt{157}}\mathbf{j}$, verified
75. $\approx 4.48\left\langle \dfrac{5}{\sqrt{29}}, \dfrac{2}{\sqrt{29}} \right\rangle \approx \langle 4.16, 1.66 \rangle$
77. $\approx 5.83\left\langle \dfrac{8}{\sqrt{73}}, \dfrac{-3}{\sqrt{73}} \right\rangle \approx \langle 5.46, -2.05 \rangle$ **79.** ≈ 14.4 **81.** $\approx 24.3°$
83. hor. comp. ≈ 79.9 ft/sec; vert. comp. ≈ 60.2 ft/sec
85. heading $68.2°$ at 266.7 mph   **87.** $\approx (82.10$ cm, 22.00 cm$)$
89. $1\langle a, b \rangle = \langle 1a, 1b \rangle = \langle a, b \rangle$
91. $\langle a, b \rangle - \langle c, d \rangle = \langle a - c, b - d \rangle = \langle a + (-c), b + (-d) \rangle$
    $= \langle a, b \rangle + \langle -c, -d \rangle = \langle a, b \rangle + -1\langle c, d \rangle = \mathbf{u} + (-1\mathbf{v})$
93. $(ck)\mathbf{u} = \langle cka, ckb \rangle = c\langle ka, kb \rangle = c(k\mathbf{u})$
    $c(k\mathbf{u}) = \langle cka, ckb \rangle = \langle kca, kcb \rangle = k\langle ca, cb \rangle = k(c\mathbf{u})$
95. $\mathbf{u} + (-\mathbf{u}) = \langle a, b \rangle + \langle -a, -b \rangle = \langle a - a, b - b \rangle = \langle 0, 0 \rangle$
97. $(c + k)\mathbf{u} = (c + k)\langle a, b \rangle = \langle (c + k)a, (c + k)b \rangle =$
    $\langle ca + ka, cb + kb \rangle = \langle ca, cb \rangle + \langle ka, kb \rangle = c\mathbf{u} + k\mathbf{u}$
99. $\langle 1, 3 \rangle + \langle 3, 3 \rangle + \langle 4, -1 \rangle + \langle 2, -4 \rangle + \langle -4, -3 \rangle + \langle -6, 2 \rangle = \langle 0, 0 \rangle$
101. Answers will vary, one possibility: $0°, 81.4°, -34°$
103. a. not a real number   **b.** not possible   **c.** not a real number
105. $x = 0, \pm\sqrt{7}$; see graph

Mid-Chapter Check, p. 673

1. $\sin B = \dfrac{b \sin A}{a}$ **2.** $\cos B = \dfrac{a^2 + c^2 - b^2}{2ac}$
3. $a \approx 129$ m, $B \approx 86.5°$, $C \approx 62.5°$
4. $A \approx 42.3°$, $B \approx 81.5°$, $C \approx 56.2°$
5. $A = 44°$ $a = 2.1$ km   **6.** $A \approx 18.5°$ $a = 70$ yd
    $B \approx 68.1°$ $b \approx 2.8$ km $B \approx 134.5°$ $b \approx 157.1$ yd
    $C \approx 67.9°$ $c = 2.8$ km $C = 27°$ $c = 100$ yd
    or
    $A = 44°$ $a = 2.1$ km
    $B \approx 23.9°$ $b \approx 1.2$ km
    $C \approx 112.1°$ $c = 2.8$ km
7. about 60.7 ft   **8.** 169 m   **9.** $\alpha \approx 49.6°; \beta \approx 92.2°; \gamma \approx 38.2°$
10. 9.4 mi

Reinforcing Basic Concepts, pp. 673–674

1.

Angles	Sides
$A = 35°$	$a = 11.6$ cm
$B \approx 81.5°$	$a = 20$ cm
$C \approx 63.5°$	$c = 18$ cm

Very close.

2. For $\angle A = 35°, a \approx 10.3$
For $\angle A = 50°, a \approx 14.2$
For $\angle A = 70°, a \approx 19.1$;
yes, very close

Exercises 7.4, pp. 683–687

1. equilibrium; zero   **3.** orthogonal   **5.** Answers will vary.
7. $\langle 6, 8 \rangle$ **9.** $\langle -5, 10 \rangle$ **11.** $-6\mathbf{i} - 8\mathbf{j}$ **13.** $-2.2\mathbf{i} + 0.4\mathbf{j}$
15. $\langle -11.48, -9.16 \rangle$ **17.** $\langle -24, -27 \rangle$ **19.** $|\mathbf{F}_3| \approx 3336.8; \theta \approx 268.5°$
21. 37.16 kg   **23.** 644.49 lb   **25.** 2606.74 kg   **27.** approx. 286.79 lb
29. approx. 43.8°   **31.** 1125 N-m   **33.** approx. 957.0 ft   **35.** approx.
64,951.90 ft-lb   **37.** approx. 451.72 lb   **39.** approx. 2819.08 N-m
41. 800 ft-lb   **43.** 118 ft-lb   **45.** verified   **47.** verified   **49. a.** 29
b. 45°   **51. a.** 0   **b.** 90°   **53. a.** 1   **b.** 89.4°   **55.** yes   **57.** no
59. yes   **61.** 3.68   **63.** -4 **65.** 3.17   **67. a.** $\langle 3.73, 1.40 \rangle$
b. $\mathbf{u}_1 = \langle 3.73, 1.40 \rangle, \mathbf{u}_2 = \langle -1.73, 4.60 \rangle$ **69. a.** $\langle -0.65, 0.11 \rangle$
b. $\mathbf{u}_1 = \langle -0.65, 0.11 \rangle, \mathbf{u}_2 = \langle -1.35, -8.11 \rangle$ **71. a.** $10.54\mathbf{i} + 1.76\mathbf{j}$

b. $u_1 = 10.54i + 1.76j$, $u_2 = -0.54i + 3.24j$ **73. a.** projectile is about 375 ft away, and 505.52 ft high **b.** approx. 1.27 sec and 12.26 sec
75. a. projectile is about 424.26 ft away, and 280.26 ft high **b.** approx. 2.44 sec and 6.40 sec   **77.** about 74.84 ft; $t \approx 3.9 - 1.2 = 2.7$ sec
79. $w \cdot (u + v) = \langle e, f \rangle \cdot \langle a + c, b + d \rangle$
$= e(a + c) + f(b + d) = ea + ec + fb + fd$
$= (ea + fb) + (ec + fd)$
$= \langle e, f \rangle \cdot \langle a, b \rangle + \langle e, f \rangle \cdot \langle c, d \rangle$
$= w \cdot u + w \cdot v$
81. $0 \cdot u = \langle 0, 0 \rangle \cdot \langle a, b \rangle = 0(a) + 0(b) = 0$
$u \cdot 0 = \langle a, b \rangle \cdot \langle 0, 0 \rangle = a(0) + b(0) = 0$
83. $\theta \approx 56.9°$; answers will vary.   **85.** $x \approx -20$
87. $a \approx 138.4$,
$B \approx 106.8°$
$C \approx 41.2°$;
$P \approx 560.4$ m,
$A \approx 11,394.3$ m^2

Exercises 7.5, pp. 695–698
1. modulus; argument   **3.** multiply; add
5. $2(\cos 240° + i \sin 240°)$, z is in QIII
7. $z_2 = z_1 + z_3$ **9.** $z_2 = z_1 + z_3$

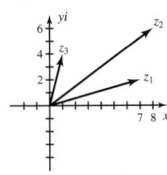   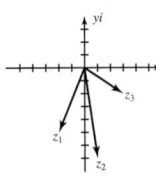

11. $2\sqrt{2}(\cos 225° + i \sin 225°)$ **13.** $10(\cos 210° + i \sin 210°)$

15. $6\left[\cos\left(\dfrac{3\pi}{4}\right) + i \sin\left(\dfrac{3\pi}{4}\right)\right]$ **17.** $8\left[\cos\left(\dfrac{11\pi}{6}\right) + i \sin\left(\dfrac{11\pi}{6}\right)\right]$

19. $10 \operatorname{cis}\left[\tan^{-1}\left(\dfrac{6}{8}\right)\right]$; $10 \operatorname{cis} 36.9°$

21. $13 \operatorname{cis}\left[180° + \tan^{-1}\left(\dfrac{12}{5}\right)\right]$; $13 \operatorname{cis} 247.4°$

23. $18.5 \operatorname{cis}\left[\tan^{-1}\left(\dfrac{17.5}{6}\right)\right]$; $18.5 \operatorname{cis} 1.2405$

25. $2\sqrt{34} \operatorname{cis}\left[\pi + \tan^{-1}\left(-\dfrac{5}{3}\right)\right]$; $2\sqrt{34} \operatorname{cis} 2.1112$

27. $r = 2, \theta = \dfrac{\pi}{4}$
$z = 2 \operatorname{cis}\left(\dfrac{\pi}{4}\right)$
$= \sqrt{2} + \sqrt{2}i$

29. $r = 4\sqrt{3}, \theta = \dfrac{\pi}{3}$
$z = 4\sqrt{3} \operatorname{cis}\left(\dfrac{\pi}{3}\right)$
$= 2\sqrt{3} + 6i$

31. $r = 17, \theta = \tan^{-1}\left(\dfrac{15}{8}\right)$
$z = 17 \operatorname{cis}\left[\tan^{-1}\left(\dfrac{15}{8}\right)\right]$
$= 17\left(\dfrac{8}{17} + \dfrac{15}{17}i\right) = 8 + 15i$

33. $r = 6, \theta = \pi - \tan^{-1}\left(\dfrac{5}{\sqrt{11}}\right)$
$z = 6 \operatorname{cis}\left[\pi - \tan^{-1}\dfrac{5}{\sqrt{11}}\right]$
$= 6\left(-\dfrac{\sqrt{11}}{6} + \dfrac{5}{6}i\right) = -\sqrt{11} + 5i$

35. $r_1 = 2\sqrt{2}$, $r_2 = 3\sqrt{2}$, $\theta_1 = 135°$, $\theta_2 = 45°$;
$z = z_1 z_2 = -12 + 0i \Rightarrow r = 12$, $\theta = 180°$;
$r_1 r_2 = 2\sqrt{2}(3\sqrt{2}) = 12$✓
$\theta_1 + \theta_2 = 135° + 45° = 180°$✓

37. $r_1 = 2, r_2 = 2, \theta_1 = 30°, \theta_2 = 60°$;
$z = \dfrac{z_1}{z_2} = \dfrac{\sqrt{3}}{2} - \dfrac{1}{2}i \Rightarrow r = 1, \theta = -30°; \dfrac{r_1}{r_2} = \dfrac{2}{2} = 1$✓
$\theta_1 - \theta_2 = 30° - 60° = -30°$✓

39. $z_1 z_2 = -24 + 0i, \dfrac{z_1}{z_2} = -\dfrac{4}{3} + \dfrac{4\sqrt{3}}{3}i$

41. $z_1 z_2 = 21\sqrt{3} - 21i, \dfrac{z_1}{z_2} = \dfrac{\sqrt{3}}{7} + \dfrac{1}{7}i$

43. $z_1 z_2 = -10.84 + 12.04i, \dfrac{z_1}{z_2} = -1.55 - 4.76i$

45. $z_1 z_2 = 0 + 40i, \dfrac{z_1}{z_2} = \dfrac{5\sqrt{3}}{4} + \dfrac{5}{4}i$

47. $z_1 z_2 = -10 - 10\sqrt{3}i, \dfrac{z_1}{z_2} = \dfrac{-5}{2} + 0i$

49. $z_1 z_2 = -2.93 + 8.5i, \dfrac{z_1}{z_2} = 2.29 + 3.28i$

51. verified; verified, $u^2 + v^2 + w^2 = uv + uw + vw$
$(1 + 4\sqrt{3}i) + (97 + 20\sqrt{3}i) + (-39 + 60\sqrt{3}i)$
$= (17 + 12\sqrt{3}i) + (-3 + 16\sqrt{3}i) + (45 + 56\sqrt{3}i)$,
$59 + 84\sqrt{3}i = 59 + 84\sqrt{3}i$

53. a. $V(t) = 170 \sin(120\pi t)$
b.

t	$V(t)$
0	0
0.001	62.6
0.002	116.4
0.003	153.8
0.004	169.7
0.005	161.7
0.006	131.0
0.007	81.9
0.008	21.3

c. $t \approx 0.00257$ sec
55. a. $17 \operatorname{cis} 28.1°$ **b.** 51 V   **57. a.** $8.60 \operatorname{cis} 324.5°$ **b.** 15.48 V
59. a. $13 \operatorname{cis} 22.6°$ **b.** 22.1 V
61. $I = 2 \operatorname{cis} 30°; Z = 5\sqrt{2} \operatorname{cis} 45°; V = 10\sqrt{2} \operatorname{cis} 75°$
63. $I = \sqrt{13} \operatorname{cis} 326.3°; Z = \dfrac{17}{4} \operatorname{cis} 61.9°; V = \dfrac{17\sqrt{13}}{4} \operatorname{cis} 28.2°$
65. $V = 4 \operatorname{cis} 60°; Z = 4\sqrt{2} \operatorname{cis} 315°; I = \dfrac{\sqrt{2}}{2} \operatorname{cis} 105°$
67. $V = 5 \operatorname{cis} 306.9°; Z = 8.5 \operatorname{cis} 61.9°; I = \dfrac{10}{17} \operatorname{cis} 245°$
69. $\dfrac{\sqrt{65} \operatorname{cis} 29.7°}{4}$ **71.** verified
73. $z_2 = \dfrac{24}{5} - \dfrac{7}{5}i, z_3 = -\dfrac{24}{5} + \dfrac{7}{5}i$ **75.** $\dfrac{5\pi}{24}, \dfrac{13\pi}{24}, \dfrac{29\pi}{24}, \dfrac{37\pi}{24}$
77.

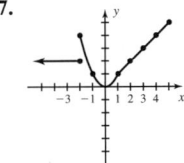

Exercises 7.6, pp. 703–705

1. $r^5[\cos(5\theta) + i\sin(5\theta)]$; De Moivre's   **3.** complex

5. $z_5 = 2\,\text{cis}\,366° = 2\,\text{cis}\,6°,\ z_6 = 2\,\text{cis}\,438° = 2\,\text{cis}\,78°,$
$z_7 = 2\,\text{cis}\,510° = 2\,\text{cis}\,150°$; Answers will vary.

7. $r = 3\sqrt{2}; n = 4; \theta = 45°; -324$ **9.** $r = 2; n = 3; \theta = 120°;\ 8$

11. $r = 1; n = 5; \theta = -60°; \dfrac{1}{2} + \dfrac{\sqrt{3}}{2}i$ **13.** $r = 1; n = 6; \theta = -45°;\ i$

15. $r = 4; n = 3; \theta = 330°; -64i$

17. $r = \dfrac{\sqrt{2}}{2}; n = 5; \theta = 135°; \dfrac{1}{8} - \dfrac{1}{8}i$

19. verified   **21.** verified   **23.** verified   **25.** verified

27. $r = 1; n = 5; \theta = 0°$; roots: $1, 0.3090 \pm 0.9511i, -0.8090 \pm 0.5878i$

29. $r = 243; n = 5; \theta = 0°$; roots: $3, 0.9271 \pm 2.8532i, -2.4271 \pm 1.7634i$

31. $r = 27; n = 3; \theta = 270°$; roots: $3i, \dfrac{-3\sqrt{3}}{2} - \dfrac{3}{2}i, \dfrac{3\sqrt{3}}{2} - \dfrac{3}{2}i$

33. $2, 0.6180 \pm 1.9021i, -1.6180 \pm 1.1756i$

35. $\dfrac{3\sqrt{3}}{2} + \dfrac{3}{2}i, -\dfrac{3\sqrt{3}}{2} + \dfrac{3}{2}i, -3i$

37. $1.1346 + 0.1797i, 0.1797 + 1.1346i, -1.0235 + 0.5215i,$
$-0.8123 - 0.8123i, 0.5215 - 1.0235i$

39. $x = 1, -\dfrac{1}{2} \pm \dfrac{\sqrt{3}}{2}i$. These are the same results as in Example 3.

41. $r = 16; n = 4; \theta = 120°$; roots: $\sqrt{3} + i, -1 + \sqrt{3}i, -\sqrt{3} - i, 1 - \sqrt{3}i$

43. $r = 7\sqrt{2}; n = 4; \theta = 225°$; roots: $0.9855 + 1.4749i, -1.4749 + 0.9855i, -0.9855 - 1.4749i, 1.4749i - 0.9855i$

45. $D = -4, z_0 = 8^{\frac{1}{6}}\text{cis}\,45°, z_1 = 8^{\frac{1}{6}}\text{cis}\,165°, z_2 = 8^{\frac{1}{6}}\text{cis}\,285°,$
$z_0 = 8^{\frac{1}{6}}\text{cis}\,75°, z_1 = 8^{\frac{1}{6}}\text{cis}\,195°, z_2 = 8^{\frac{1}{6}}\text{cis}\,315°$ **47.** verified

49. a. numerator: $-117 + 44j$, denominator: $-21 + 72j$ **b.** $1 + \dfrac{4}{3}j$

c. verified   **51.** Answers will vary.   **53.** $-7 - 24i$

55. $z \approx -2.7320, z \approx 0.7320, z = 2.$
Note: Using sum and difference identities, all three solutions can actually be given in exact form: $-1 - \sqrt{3}, -1 + \sqrt{3}, 2.$

57.
$$\frac{\tan^2 x}{\sec x + 1} = \frac{\sec^2 x - 1}{\sec x + 1}$$
$$= \frac{(\sec x + 1)(\sec x - 1)}{\sec x + 1}$$
$$= \sec x - 1$$
$$= \frac{1}{\cos x} - \frac{\cos x}{\cos x}$$
$$= \frac{1 - \cos x}{\cos x}$$

59. $y = -\dfrac{4}{5}x + \dfrac{12}{5}$

Summary and Concept Review pp. 705–709

1.

Angles	Sides
$A = 36°$	$a \approx 205.35$ cm
$B = 21°$	$b \approx 125.20$ cm
$C = 123°$	$c = 293$ cm

2.

Angles	Sides
$A = 28°$	$a \approx 140.59$ yd
$B = 10°$	$b = 52$ yd
$C = 142°$	$c \approx 184.36$ yd

3. approx. 41.84 ft

4. approx. 20.2° and 159.8°

5.

Angles	Sides		Angles	Sides
$A = 35°$	$a = 67$ cm		$A = 35°$	$a = 67$ cm
$B_1 \approx 64.0°$	$b = 105$ cm		$B_2 \approx 116.0°$	$b = 105$ cm
$C_1 \approx 81.0°$	$c_1 \approx 115.37$ cm		$C_2 \approx 29.0°$	$c_2 \approx 56.63$ cm

6. no; 36°   **7.** approx. 36.9°   **8.** approx. 385.5 m

9. 133.2°, 30.1°, and 16.7°   **10.** 85,570.7 m²

11.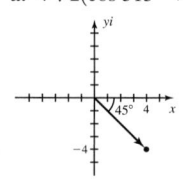

12. $-8i + 3j$; $|\mathbf{u}| \approx 8.54$; $\theta \approx 159.4°$

13. horiz. comp. ≈ 11.08, vertical comp. ≈ 14.18

14. $\langle -4, -2 \rangle$; $|2\mathbf{u} + \mathbf{v}| \approx 4.47$, $\theta \approx 206.6°$ **15.** $\dfrac{7}{\sqrt{193}}\mathbf{i} + \dfrac{12}{\sqrt{193}}\mathbf{j}$

16. QII; since the x-component is negative and the y-component is positive.

17. $\frac{1}{6}$ mi   **18.** approx. 19.7°   **19.** $\langle -25, -123 \rangle$ **20.** approx. -0.87

21. 4   **22.** $\mathbf{p} \cdot \mathbf{q} = -6$; $\theta \approx 97.9°$ **23.** 4340 ft-lb   **24.** approx. 417.81 lb

25. approx. 8156.77 ft-lb   **26. a.** $x \approx 269.97$ ft; $y \approx 285.74$ ft

b. approx. 0.74 sec   **27.** $2(\cos 240° + i \sin 240°)$ **28.** $3 + 3i$

29.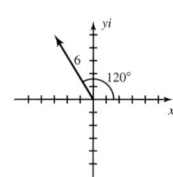

30. $z_1 z_2 = 16\,\text{cis}\left(\dfrac{5\pi}{12}\right); \dfrac{z_1}{z_2} = 4\,\text{cis}\left(\dfrac{\pi}{12}\right)$

31. $2\sqrt{3} + 2j$ **32.** $|Z| \approx 10.44$; $\theta \approx 16.7°$, $10.44\,\text{cis}\,16.7°$

33. $-16 - 16\sqrt{3}i$ **34.** verified   **35.** $\dfrac{5\sqrt{3}}{2} + \dfrac{5}{2}i, -\dfrac{5\sqrt{3}}{2} + \dfrac{5}{2}i, -5i$

36. $6, -3 \pm 3i\sqrt{3}$ **37.** $2 - 2i, -2 \pm 2i$ **38.** $1 \pm 2i, -1 \pm 2i$

39. verified

Mixed Review pp. 709–710

1.

Angles	Sides
$A = 41°$	$a \approx 13.44$ in.
$B = 27°$	$b \approx 9.30$ in.
$C = 112°$	$c = 19$ in.

Area ≈ 57.9 in²

3. $x \approx 16.09$, $y \approx 13.50$ **5.** approx. 176.15 ft   **7.** approx. 793.70 mph; heading 28.2°

9. One solution possible since side $a >$ side b

Angles	Sides
$A = 31°$	$a = 36$ m
$B \approx 20.1°$	$b = 24$ m
$C \approx 128.9°$	$c \approx 54.4$ m

11. No; barely touches ("tangent") at 30°

13. a. $4\sqrt{2}(\cos 315° + i \sin 315°)$ **b.** $-3 + 3\sqrt{3}i$

15. $\approx 13.1°$ **17.** $\text{comp}_v\mathbf{u} \approx -0.87$, $\text{proj}_v\mathbf{u} \approx \dfrac{-38}{53} + \dfrac{26}{53}\mathbf{i}$

19. $z_0 = \dfrac{\sqrt{6}}{2} + \dfrac{\sqrt{2}}{2}i, z_1 = \dfrac{-\sqrt{2}}{2} + \dfrac{\sqrt{6}}{2}i,$
$z_2 = \dfrac{-\sqrt{6}}{2} - \dfrac{\sqrt{2}}{2}i, z_3 = \dfrac{\sqrt{2}}{2} - \dfrac{\sqrt{6}}{2}i$

Practice Test pp. 710–712

1. 6.58 mi   **2.** 137.18 ft

3.

Angles	Sides (in.)		Angles	Sides (in.)
$A_1 \approx 58.8°$	$a = 15$		$A_2 \approx 121.2°$	$a = 15$
$B = 20°$	$b = 6$		$B = 20°$	$b = 6$
$C_1 \approx 101.2°$	$c_1 \approx 17.21$		$C_2 \approx 38.8°$	$c_2 \approx 11.0$

4. a. No   **b.** 2.66 mi   **5. a.** No   **b.** 1   **c.** 8.43 sec

6. a. 2.30 mi   **b.** 7516.5 ft   **7.** $A \approx 438,795$ mi², $P \approx 61.7°$,
$B \approx 61.2°$, $M \approx 57.1°$ **8.** speed ≈ 73.36 mph, bearing $\approx 47.8°$

9. $\theta \approx 36.5°$ **10.** 63.48 cm to the right and 130.05 cm down from the initial point on the ceiling   **11.** $|\mathbf{F}_3| \approx 212.94$ N, $\theta \approx 251.2°$
12. a. $\theta \approx 42.5°$ **b.** $\mathbf{proj}_v\, \mathbf{u} = \langle -2.4, 7.2 \rangle$
c. $\mathbf{u}_1 = \langle -2.4, 7.2 \rangle$, $\mathbf{u}_2 = \langle -6.6, -2.2 \rangle$
13. 104.53 ft; 3.27 sec   **14.** $2 \operatorname{cis}\left(\dfrac{\pi}{24}\right)$ **15.** $48\sqrt{2} \operatorname{cis} 75°$; verified
16. $-8 - 8\sqrt{3}i$ **17.** verified   **18.** $\dfrac{5\sqrt{3}}{2} + \dfrac{5}{2}i, -\dfrac{5\sqrt{3}}{2} + \dfrac{5}{2}i, -5i$
19. $2.3039 \pm 1.5192i, -2.3039 \pm 1.5192i$ **20.** $\approx 2,414,300$ mi^2

Strengthening Core Skills p. 713

Exercise 1: 664.46 lb, 640.86 lb   **Exercise 2:** 106.07 lb, 106.07 lb
Exercise 3: yes

Cumulative Review Chapters 1–7 pp. 713–714

1. $20\sqrt{3}$; 40; 60°; 90°   **3.** $R = \dfrac{1}{\pi}\sqrt{A + (\pi r)^2}$
5. QIV $\sin \theta = \frac{-3}{5}$; $\cos \theta = \frac{4}{5}$; $\tan \theta = \frac{-3}{4}$; $\csc \theta = -\frac{5}{3}$;
$\sec \theta = \frac{5}{4}$; $\cot \theta = \frac{-4}{3}$ **7.** $x = \dfrac{-4}{5} \pm \dfrac{\sqrt{6}}{5}$
9. $\cos = 19° \approx 0.94$, $\cos 125° \approx -0.58$ **11. a.** about \$66,825
b. $13, 13, 7\sqrt{2}$; $A = 59.5$ mi^2 **13. a.** $m = \dfrac{y_2 - y_1}{x_2 - x_1}$
b. $\left(\dfrac{x_2 + x_1}{2}, \dfrac{y_2 + y_1}{2}\right)$ **c.** $x = \dfrac{-b \pm \sqrt{b^2 - 4ac}}{2a}$
d. $d = \sqrt{(x_2 - x_1)^2 + (y_2 - y_1)^2}$ **e.** $A = Pe^{rt}$
15. $\angle A = 37°, a \approx 33$ cm,
$\angle B = 34.4°, b = 31$ cm,
$\angle C = 108.6°, c = 52$ cm
17. about 422.5 lb
19.

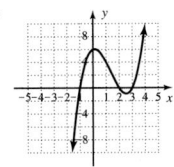

$x \in (-\infty, -1) \cup (2, 3)$
21. $-128 - 128i\sqrt{3}$ **23.** about 3.6 yr   **25.** $A = 2, B = 1, C = \dfrac{\pi}{4}$

Connections to Calculus Exercises, pp. 717–718

1.

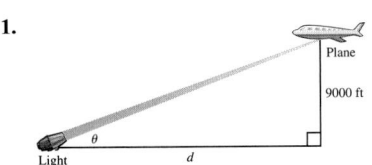

a. $\theta = \tan^{-1}\left(\dfrac{9000}{d}\right)$
b. $\theta \approx 35°$

3.

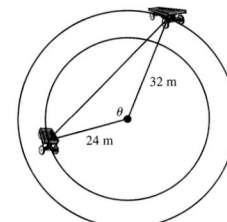

a. $d = \sqrt{24^2 + 32^2 - 2(24)(32)\cos \theta}$
b. $d \approx 54.13$ m   **c.** about 106.1°

5.

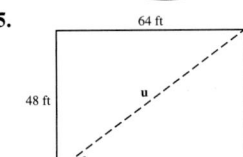

a. $|\mathbf{v}| = \sqrt{48^2 + 64^2 + 18^2} = 82$ ft
b. $\mathbf{u} = \langle 48, 64, 0 \rangle$, $\mathbf{v} = \langle 48, 64, 18 \rangle$
$\cos \theta_2 = \dfrac{\langle 48, 64, 0 \rangle \cdot \langle 48, 64, 18 \rangle}{(80)(82)}$
$\theta_2 = \cos^{-1}\left(\dfrac{40}{41}\right) \approx 12.68°$; verified

7.

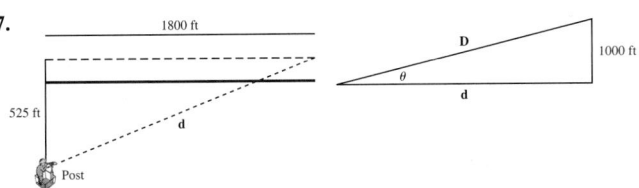

a. $|\mathbf{D}| = \sqrt{525^2 + 1800^2 + 1000^2} = 2125$ ft
b. $\mathbf{d} = \langle 525, 1800, 0 \rangle$, $\mathbf{D} = \langle 525, 1800, 1000 \rangle$
$\cos \theta = \dfrac{\langle 525, 1800, 0 \rangle \cdot \langle 525, 1800, 1000 \rangle}{(1875)(2125)}$
$\theta = \cos^{-1}\left(\dfrac{15}{17}\right) \approx 28.07°$; verified

9.

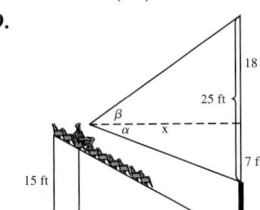

a. $\theta = \alpha + \beta$
$= \tan^{-1}\left(\dfrac{7}{x}\right) + \tan^{-1}\left(\dfrac{18}{x}\right)$
b. $\theta = \tan^{-1}\left(\dfrac{7}{46}\right) + \tan^{-1}\left(\dfrac{18}{46}\right)$
$\approx 30.02°$

CHAPTER 8

Exercises 8.1, pp. 727–731

1. inconsistent   **3.** consistent; independent
5. Multiply the first equation by 6 and the second equation by 10.
7. $y = \frac{7}{4}x - 6$, $y = \frac{-4}{3}x + 5$ **9.** $y = x + 2$ **11.** $x + 3y = -3$
13. $y = x + 2$, $x + 3y = -3$ **15.** yes   **17.** yes
19.

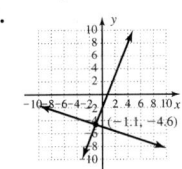

21.

23. $(-4, 1)$ **25.** $(3, -5)$ **27.** second equation, y, $(4, -3)$
29. second equation, x, $(10, -1)$ **31.** second equation, x, $(\frac{5}{2}, \frac{7}{4})$
33. $(3, -1)$ **35.** $(-2, -3)$ **37.** $(\frac{11}{2}, 2)$ **39.** $(-2, 3)$ **41.** $(-3, 4)$
43. $(-6, 12)$ **45.** $(2, 8)$; consistent/independent   **47.** $\varnothing$; inconsistent
49. $\{(x, y)\,|\,6x + y = 22\}$; consistent/dependent   **51.** $(4, 1)$;
consistent/independent   **53.** $(-3, -4)$; consistent/independent
55. $(\frac{-1}{2}, \frac{4}{3})$; consistent/independent   **57.** $(-2, \frac{5}{2})$ **59.** $(2, -1)$
61. 1 mph   4 mph   **63.** 2318 adult tickets; 1482 child tickets
65. premium: \$3.97, regular: \$3.87   **67.** nursing student \$6500; science major \$3500   **69.** 150 quarters, 75 dimes   **71. a.** 100 lawns/mo,
b. \$11,500/mo   **73. a.** 1.6 billion bu, 3 billion bu, yes;   **b.** 2.7 billion bu, 2.25 billion bu, yes;   **c.** \$6.65, 2.43 billion bu   **75. a.** 3 mph,   **b.** 5 mph
77. a. 3.6 ft/sec,   **b.** 4.4 ft/sec   **79.** 1776; 1865   **81.** Tahiti: 402 mi^2, Tonga: 290 mi^2 **83.** $m_1 \neq m_2$; consistent/independent   **85.** \$6552 at 8.5%; \$11,551 at 6%   **87.** 472°, 832°, -248°, -608°   **89.** verified

Exercises 8.2, pp. 740–743

1. triple   **3.** equivalent; systems   **5.** $z = 5$ **7.** Answers will vary.
9. Answers will vary.   **11.** yes;   no   **13.** $(5, 7, 4)$ **15.** $(-2, 4, 3)$
17. $(1, 1, -2)$ **19.** $(4, 0, -3)$ **21.** $(3, 4, 5)$ **23.** $(1, 6, 9)$
25. no solution, inconsistent   **27.** $(p, 2 - p, 2 - p)$
29. $\left(-\dfrac{5}{3}p - \dfrac{2}{3}, -p - 2, p\right)$, other solutions possible
31. $(p, 2p, p + 1)$ **33.** $(p + 9, p - 4, p)$
35. $\{(x, y, z)\,|\,x - 6y + 12z = 5\}$
37. $(1, 1, 2)$ **39.** $\left\{(x, y, z)\,\Big|\,x - \dfrac{5}{2}y - 2z = 3\right\}$ **41.** $\left(2, 1, \dfrac{-1}{3}\right)$

43. $(p + 5, p - 2, p)$ **45.** $(18, -6, 10)$ **47.** $\left(\dfrac{11}{3}, \dfrac{10}{3}, \dfrac{7}{3}\right)$

49. $(1, -2, 3)$ **51.** $\left(\dfrac{1}{2}, \dfrac{1}{3}, 3\right)$ **53.** ≈ 3.464 units

55. Monet \$1,900,000; Picasso \$1,100,000; van Gogh \$4,000,000

57. elephant, 650 days; rhino, 464 days; camel, 406 days

59. Albatross: 3.6 m, Condor: 3.0 m, Quetzalcoatlus: 12.0 m

61. 175 \$5 gold pieces; 50 \$10 gold pieces; 25 \$20 gold pieces

63. $A = -1, B = 1, C = -2$; verified   **65.** $x^2 + y^2 - 4x + 6y + 9 = 0$

67. $\langle -11, -5 \rangle$; $\langle 6, -\dfrac{43}{2} \rangle$ **69.** $x = 1$

Exercises 8.3, pp. 753–755

1. template   **3.** repeated linear   **5.** Answers will vary

7. $\dfrac{A}{x + 3} + \dfrac{B}{x - 2}$ **9.** $\dfrac{A}{x - 1} + \dfrac{B}{(x - 1)^2}$

11. $\dfrac{A}{x - 1} + \dfrac{B}{x + 2} + \dfrac{C}{x - 3}$ **13.** $\dfrac{A}{x} + \dfrac{B}{x - 3} + \dfrac{C}{x + 1}$

15. $\dfrac{A}{x - 5} + \dfrac{B}{(x - 5)^2} + \dfrac{C}{(x - 5)^3}$ **17.** $\dfrac{A}{x} + \dfrac{B}{x^2} + \dfrac{C}{x + 2}$

19. $\dfrac{A}{x} + \dfrac{B}{x^2} + \dfrac{C}{x - 5} + \dfrac{D}{(x - 5)^2}$ **21.** $\dfrac{A}{x - 3} + \dfrac{Bx + C}{x^2 + 5x + 7}$

23. $\dfrac{A}{x + 1} + \dfrac{Bx + C}{x^2 + 2} + \dfrac{Dx + E}{(x^2 + 2)^2}$ **25.** $\dfrac{-4}{2x - 5} + \dfrac{3}{x + 3}$

27. $\dfrac{7}{x} + \dfrac{2}{x + 1} - \dfrac{1}{x - 1}$ **29.** $\dfrac{-1}{x} + \dfrac{4}{x + 1} + \dfrac{5}{(x + 1)^2}$

31. $\dfrac{2}{x^2 + 1} + \dfrac{3x + 1}{x^2 + 2}$ **33.** $\dfrac{5}{x + 2} + \dfrac{x - 1}{x^2 + 3}$

35. $\dfrac{1}{x} - \dfrac{5x + 2}{(x^2 + 1)^2}$ **37.** $\dfrac{2}{x} + \dfrac{1}{x^2} + \dfrac{x - 1}{x^2 + x + 3}$

39. $\dfrac{3}{2 - x} - \dfrac{4}{4 + 2x + x^2}$ **41.** $\dfrac{5}{x + 3} + \dfrac{-2}{(x + 3)^2}$

43. $\dfrac{2x + 1}{x^2 + 1} + \dfrac{3x}{(x^2 + 1)^2}$ **45.** $\dfrac{2}{x - 1} + \dfrac{3}{(x - 1)^3}$

47. $\dfrac{\frac{1}{100}}{P} + \dfrac{\frac{1}{10}}{100 - 10P}$ **49.** $\dfrac{\frac{1}{10}}{P} + \dfrac{\frac{1}{100}}{10 - \frac{1}{10}P}$

51. $\dfrac{49}{50}$ **53.** $\dfrac{62}{125}$ **55.** $\dfrac{4}{\ln x - 2} - \dfrac{4}{\ln x - 1} - \dfrac{3}{(\ln x - 1)^2}$

57. Factor out -1 from the denominator:

$\dfrac{x + 2}{(x - 1)(1 - x)} = \dfrac{x + 2}{(x - 1)(-1)(x - 1)} = \dfrac{-x - 2}{(x - 1)^2} = \dfrac{-1}{x - 1} + \dfrac{-3}{(x - 1)^2}$

59. $2x - 1 + \dfrac{1}{x^2 - x + 6}$ **61.** Verified

Exercises 8.4, pp. 764–767

1. half; planes   **3.** solution

5. The feasible region may be bordered by three or more oblique lines, with two of them intersecting outside and away from the feasible region.

7. No, No, No, No   **9.** No, Yes, Yes, No

11. 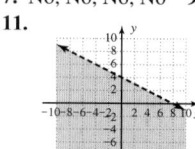   **13.**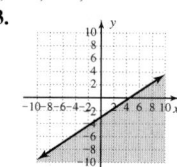

15. No, No, No, Yes   **17.**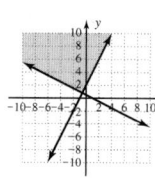

19. 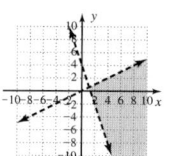   **21.** 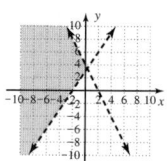   **23.**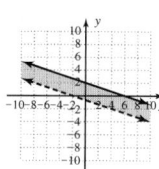

25. 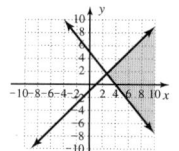   **27.** 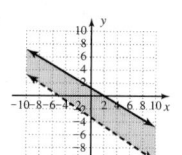   **29.**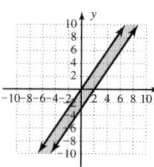

31. 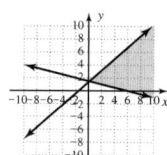   **33.** 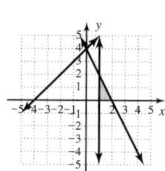   **35.**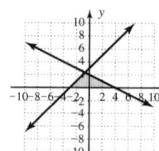

37. 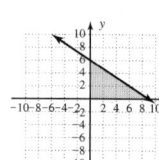   **39.** $\begin{cases} y - x \le 1 \\ x + y > 3 \end{cases}$

41. $\begin{cases} y - x \le 1 \\ x + y < 3 \\ y \ge 0 \end{cases}$ **43.** $(5, 3)$ **45.** $(12, 11)$ **47.** $(2, 2)$

49. $(4, 3)$ **51.** $5 < H < 10$

53. 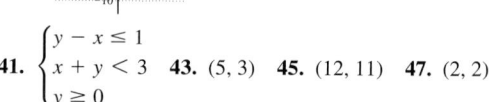   **55.** 300 acres of corn; 200 acres of soybeans

$J + A \le 50,000$

$J \ge 20,000$

$A \le 25,000$

57. 240 sheet metal screws; 480 wood screws   **59.** 65 traditionals, 30 Double-T's   **61.** 220,000 gallons from Tulsa to Colorado; 100,000 gal from Tulsa to Mississippi; 0 thousand gal from Houston to Colorado; 150,000 gal from Houston to Mississippi

63. 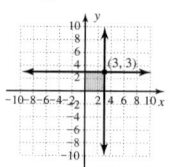   **65. a.** $-\dfrac{3}{5}$ **b.** $\dfrac{5}{4}$ **c.** $-\dfrac{3}{4}$ **67.** 324 Ω

$(3, 3)$; optimal solutions occur at vertices

Mid-Chapter Check, p. 768

1. $(1, 1)$ consistent   **2.** $(5, 3)$ consistent   **3.** 20 oz   **4.** No

5. The second equation is a multiple of the first equation.   **6.** $(1, 2, 3)$

7. $(1, 2, 3)$ **8.** $\dfrac{2}{x + 1} - \dfrac{3}{x - 2} + \dfrac{1}{(x - 2)^2}$ **9.** Mozart = 8 yr;

Morphy = 13 yr; Pascal = 16 yr   **10.** 2 table candles, 9 holiday candles

Reinforcing Basic Concepts, pp. 768–769

Exercise 1. Premium: \$4.17/gal, Regular: \$4.07/gal

$\begin{cases} 15.3R + 35.7P = 211.14 \\ P = R + 0.10 \end{cases}$

Exercise 2. Verified

Exercises 8.5, pp. 776–779

1. square **3.** 2; 3; 1 **5.** Multiply R1 by -2 and add that result to R2. This sum will be the new R2. **7.** 3×2, 5.8 **9.** 4×3, -1

11. $\begin{bmatrix} 1 & 2 & -1 & | & 1 \\ 1 & 0 & 1 & | & 3 \\ 2 & -1 & 1 & | & 3 \end{bmatrix}$; diagonal entries 1, 0, 1

13. $\begin{cases} x + 4y = 5 \\ y = \frac{1}{2} \end{cases} \to (3, \frac{1}{2})$ **15.** $\begin{cases} x + 2y - z = 0 \\ y + 2z = 2 \\ z = 3 \end{cases} \to (11, -4, 3)$

17. $\begin{cases} x + 3y - 4z = 29 \\ y - \frac{3}{2}z = \frac{21}{2} \\ z = 3 \end{cases} \to (-4, 15, 3)$ **19.** $\begin{bmatrix} 1 & -6 & -2 \\ 0 & -28 & -6 \end{bmatrix}$

21. $\begin{bmatrix} 1 & -3 & 3 & 2 \\ 0 & 23 & -12 & -15 \\ -2 & 1 & 0 & 4 \end{bmatrix}$ **23.** $\begin{bmatrix} 3 & 1 & 1 & 8 \\ 0 & -3 & -3 & -6 \\ 0 & -10 & -13 & 34 \end{bmatrix}$

25. $2R_1 + R_2 \to R_2$ **27.** $-5R_1 + R_2 \to R_2$
$-3R_1 + R_3 \to R_3$ $4R_1 + R_3 \to R_3$

29. (20, 10) **31.** (1, 6, 9) **33.** (1, 1, 2) **35.** (1, 1, 1) **37.** $(-1, \frac{-3}{2}, 2)$
39. linear dependence $(p - 4, -2p + 8, p)$ **41.** coincident dependence $\{(x, y, z)|3x - 4y + 2z = -2\}$ **43.** no solution **45.** linear dependence, $(-\frac{5}{4}p - 3, \frac{1}{8}p - \frac{1}{2}, p)$ **47.** 28.5 units^2 **49.** Heat: 95, Mavericks: 92
51. Poe, \$12,500; Baum, \$62,500; Wouk, \$25,000
53. $A = 35°, B = 45°, C = 100°$ **55.** \$.4 million at 4%; \$.6 million at 7%; \$1.5 million at 8% **57.** $x = 84°; y = 25°$

59. a. $z_1 = \sqrt{10} \text{ cis}[\pi + \tan^{-1}(3)]$ **b.** $z_2 = -\frac{5}{2} + \frac{5\sqrt{3}}{2}i$

61. $C > 30,000$ in the year 2011 $(t \approx 6.39)$

Exercises 8.6, pp. 787–791

1. $a_{ij}; b_{ij}$ **3.** scalar **5.** Answers will vary. **7.** 2×2, $a_{12} = -3$, $a_{21} = 5$
9. 2×3, $a_{12} = -3$, $a_{23} = 6$, $a_{22} = 5$ **11.** 3×3, $a_{12} = 1$, $a_{23} = 1$, $a_{31} = 5$
13. true **15.** conditional, $c = -2$, $a = -4$, $b = 3$

17. $\begin{bmatrix} 10 & 0 \\ 0 & 10 \end{bmatrix}$ **19.** different orders, sum not possible

21. $\begin{bmatrix} 20 & -15 \\ -25 & -10 \end{bmatrix}$ **23.** $\begin{bmatrix} \frac{5}{2} & -1 & 0 \\ 0 & \frac{-7}{2} & 1 \\ 2 & \frac{3}{2} & -6 \end{bmatrix}$ **25.** $\begin{bmatrix} 1 & 2 & 0 \\ 0 & -1 & 2 \\ 4 & 3 & -6 \end{bmatrix}$

27. $\begin{bmatrix} 1 & 0 \\ 0 & 1 \end{bmatrix}$ **29.** $\begin{bmatrix} 6 & -3 & 9 \\ 12 & 0 & -6 \end{bmatrix}$ **31.** $\begin{bmatrix} 12 & -24 & 90 \\ -6 & 15 & -57 \end{bmatrix}$

33. $\begin{bmatrix} 79 & -30 \\ -50 & 19 \end{bmatrix}$ **35.** $\begin{bmatrix} 42 & 18 & -60 \\ -12 & -42 & 36 \end{bmatrix}$

37. $\begin{bmatrix} 0.71 & 0.65 \\ 1.78 & 3.55 \end{bmatrix}$ **39.** $\begin{bmatrix} 1 & -1.25 & 0.25 \\ -0.5 & -0.63 & 2.13 \\ 3.75 & 3.69 & -5.94 \end{bmatrix}$

41. $\begin{bmatrix} 1 & 0 \\ 0 & 1 \end{bmatrix}$ **43.** $\begin{bmatrix} 1 & 0 & 0 \\ 0 & 1 & 0 \\ 0 & 0 & 1 \end{bmatrix}$ **45.** $\begin{bmatrix} \frac{-3}{19} & \frac{4}{57} \\ \frac{1}{19} & \frac{5}{57} \end{bmatrix}$ **47.** $\begin{bmatrix} 0 & \frac{3}{4} & \frac{1}{4} \\ \frac{-1}{2} & \frac{3}{8} & \frac{1}{8} \\ \frac{-1}{4} & \frac{11}{16} & \frac{1}{16} \end{bmatrix}$

49. $\begin{bmatrix} 1.75 & 2.5 \\ 7.5 & 13 \end{bmatrix}$ **51.** $\begin{bmatrix} -0.26 & 0.32 & -0.07 \\ 0.63 & 0.30 & 0.10 \end{bmatrix}$

53. verified **55.** verified **57.** $P = 21.448$ cm; $A = 27.7269$ cm^2
59. a.

$$V = \begin{array}{c} \\ S \\ D \\ P \end{array} \begin{bmatrix} T & S \\ 3820 & 1960 \\ 2460 & 1240 \\ 1540 & 920 \end{bmatrix} \quad M = \begin{array}{c} \\ S \\ D \\ P \end{array} \begin{bmatrix} T & S \\ 4220 & 2960 \\ 2960 & 3240 \\ 1640 & 820 \end{bmatrix}$$

b. 3900 more by Minsk
c. $V = \begin{bmatrix} 3972.8 & 2038.4 \\ 2558.4 & 1289.6 \\ 1601.6 & 956.8 \end{bmatrix}$ **d.** $\begin{bmatrix} 8361.6 & 5116.8 \\ 5636.8 & 4659.2 \\ 3307.2 & 1809.6 \end{bmatrix}$

$M = \begin{bmatrix} 4388.8 & 3078.4 \\ 3078.4 & 3369.6 \\ 1705.6 & 852.8 \end{bmatrix}$

61. $[22{,}000 \ 19{,}000 \ 23{,}500 \ 14{,}000]$;
total profit
North: \$22,000
South: \$19,000
East: \$23,500
West: \$14,000
63. a. \$108.20 **b.** \$101
c. $\begin{array}{c} \text{Science} \\ \text{Math} \end{array} \begin{bmatrix} 100 & 101 & 119 \\ 108.2 & 107 & 129.5 \end{bmatrix}$
First row, total cost for science from each restaurant; Second row, total cost for math from each restaurant.
65. a. 10 **b.** 20
c.
$$\begin{array}{c} \\ \text{Female} \\ \text{Male} \end{array} \begin{array}{ccc} \text{Spanish} & \text{Chess} & \text{Writing} \\ \begin{bmatrix} 32.4 & 10.3 & 21.3 \\ 29.9 & 9.6 & 19.5 \end{bmatrix} \end{array},$$
the approximate number of females expected to join the writing club
67. $\begin{bmatrix} 2^{n-1} & 0 & 2^{n-1} \\ 2^n - 1 & 1 & 2^n - 1 \\ 2^{n-1} & 0 & 2^{n-1} \end{bmatrix}$
69. $a = 2, b = 1, c = -3, d = -2$ **71.** 0.3211 **73.** $x^2 + 2x - 5$

Exercises 8.7, pp. 801–805

1. diagonal; zeroes **3.** identity **5.** Answers will vary.
7. verified **9.** verified **11.** verified **13.** verified
15. $\begin{bmatrix} \frac{1}{9} & \frac{2}{9} \\ \frac{-1}{9} & \frac{5}{18} \end{bmatrix}$ **17.** $\begin{bmatrix} -5 & 1.5 \\ -2 & 0.5 \end{bmatrix}$ **19.** verified **21.** verified
23. $\begin{bmatrix} \frac{-2}{39} & \frac{1}{13} & \frac{10}{39} \\ \frac{1}{3} & 0 & \frac{1}{3} \\ \frac{-4}{39} & \frac{2}{13} & \frac{-19}{39} \end{bmatrix}$ **25.** $\begin{bmatrix} \frac{-9}{80} & \frac{31}{400} & \frac{27}{400} \\ \frac{1}{80} & \frac{41}{400} & \frac{-3}{400} \\ \frac{-1}{20} & \frac{-1}{100} & \frac{-17}{100} \end{bmatrix}$ **27.** $\begin{bmatrix} 2 & -3 \\ -5 & 7 \end{bmatrix}\begin{bmatrix} x \\ y \end{bmatrix} = \begin{bmatrix} 9 \\ 8 \end{bmatrix}$

29. $\begin{bmatrix} 1 & 2 & -1 \\ 1 & 0 & 1 \\ 2 & -1 & 1 \end{bmatrix}\begin{bmatrix} x \\ y \\ z \end{bmatrix} = \begin{bmatrix} 1 \\ 3 \\ 3 \end{bmatrix}$

31. $\begin{bmatrix} -2 & 1 & -4 & 5 \\ 2 & -5 & 1 & -3 \\ -3 & 1 & 6 & 1 \\ 1 & 4 & -5 & 1 \end{bmatrix}\begin{bmatrix} w \\ x \\ y \\ z \end{bmatrix} = \begin{bmatrix} -3 \\ 4 \\ 1 \\ -9 \end{bmatrix}$

33. (4, 5) **35.** (12, 12) **37.** no solution **39.** $(1.5, -0.5, -1.5)$
41. no solution **43.** $(-1, -0.5, 1.5, 0.5)$ **45.** 1, yes **47.** 0, no **49.** 1
51. singular matrix **53.** singular matrix **55.** -34 **57.** 7
59. $\det(A) = -5$; (1, 6, 9) **61.** $\det(A) = 0$ **63.** $A^{-1} = \begin{bmatrix} \frac{1}{13} & \frac{5}{13} \\ \frac{-2}{13} & \frac{3}{13} \end{bmatrix}$
65. singular **67.** 31 behemoth, 52 gargantuan, 78 mammoth, 30 jumbo
69. Jumpin' Jack Flash: 3.75 min
Tumbling Dice: 3.75 min
You Can't Always Get: 7.5 min
Wild Horses: 5.75 min
71. 30 of clock A; 20 of clock B; 40 of clock C; 12 of clock D
73. $p_1 = 72.25°, p_2 = 74.75°, p_3 = 80.25°, p_4 = 82.75°$
75. $y = x^3 + 2x^2 - 9x - 10$ **77.** 2 oz food I, 1 oz Food II, 4 oz Food III
79. Answers will vary. **81. a.** -45, **b.** 52 **c.** -19 **d.** -4
83. $A = 125$, period $= \frac{2\pi}{3}$ **85.** $x \in \left(-\infty, -\frac{9}{2}\right) \cup \left(-\frac{1}{2}, \infty\right)$

Exercises 8.8, pp. 811–814

1. $a_{11}a_{22} - a_{21}a_{12}$ **3.** constant **5.** Answers will vary.
7. $D = \begin{vmatrix} 2 & 5 \\ -3 & 4 \end{vmatrix}; D_x = \begin{vmatrix} 7 & 5 \\ 1 & 4 \end{vmatrix}; D_y = \begin{vmatrix} 2 & 7 \\ -3 & 1 \end{vmatrix}$
9. $(-5, 9)$ **11.** $\left(\frac{-26}{3}, \frac{25}{3}\right)$ **13.** no solution

15. a. $D = \begin{vmatrix} 4 & -1 & 2 \\ -3 & 2 & -1 \\ 1 & -5 & 3 \end{vmatrix}$, $D_x = \begin{vmatrix} -5 & -1 & 2 \\ 8 & 2 & -1 \\ -3 & -5 & 3 \end{vmatrix}$,

$D_y = \begin{vmatrix} 4 & -5 & 2 \\ -3 & 8 & -1 \\ 1 & -3 & 3 \end{vmatrix}$, $D_z = \begin{vmatrix} 4 & -1 & -5 \\ -3 & 2 & 8 \\ 1 & -5 & -3 \end{vmatrix}$

$D = 22$, solutions possible   **b.** $D = 0$, Cramer's rule cannot be used; The sum of the first two rows of D gives row 3

17. $(1, 2, 1)$

19. $\left(\dfrac{3}{4}, \dfrac{5}{3}, \dfrac{-1}{3}\right)$ **21.** $(0, -1, 2, -3)$ **23.** $320 + 32\pi \approx 420.5$ in^2

25. 8 cm^2 **27.** 27 ft^2 **29.** 19 m^3 **31.** $V = 96$ in^3 **33.** yes

35. no   **37.** yes yes yes verified

39. $\dfrac{1}{x} + \dfrac{x-2}{(x^2-1)^2}$ **41.** $\dfrac{3}{x+1} - \dfrac{2}{x-3} + \dfrac{1}{(x-3)^3}$

43. $\begin{cases} 15{,}000x + 25{,}000y = 2900 \\ 25{,}000x + 15{,}000y = 2700 \end{cases}$; 6%, 8%

45. $\begin{cases} 2x + 2y + 10z = 3.26 \\ 3x + 2y + 7z = 2.98 \\ 2x + 3y + 6z = 2.89 \end{cases}$; apples, 29¢/lb   kiwi, 39¢/lb   pears, 19¢/lb

47. 10 lb of $1.90, 8 lb of $2.25, 6 lb of $3.50

49. Answers will vary.   **51.** $x^2 + y^2 - 4x - 6y - 12 = 0$

53.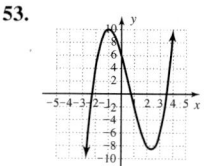

55. $\angle B \approx 76.3°$, $\angle C \approx 54.7°$, side $c = 9.4$ in.

Summary and Concept Review, pp. 814–819

1. 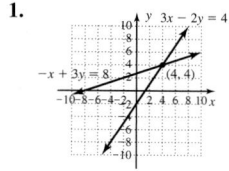   **2.**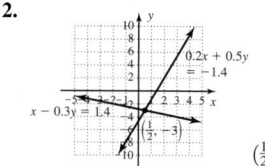

$(4, 4)$ $\left(\dfrac{1}{2}, -3\right)$

3.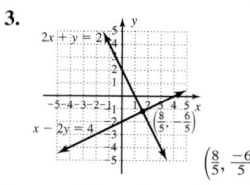

$\left(\dfrac{8}{5}, \dfrac{-6}{5}\right)$

4. no solution; inconsistent   **5.** $(5, -1)$; consistent

6. $(7, 2)$; consistent   **7.** $(3, -1)$; consistent   **8.** $(2, 2)$; consistent

9. $\left(\dfrac{11}{4}, \dfrac{-1}{6}\right)$; consistent   **10.** Sears Tower is 1450 ft; Hancock Building is 1127 ft.   **11.** $(0, 3, 2)$ **12.** $(1, 1, 1)$ **13.** no solution, inconsistent

14. 72 nickels, 85 dimes, 60 quarters   **15.** $y = -x^2 + 10x - 9$

16. $\dfrac{4}{x-3} + \dfrac{2}{x+5}$ **17.** $\dfrac{7}{x-3} + \dfrac{2}{2x+1}$ **18.** $\dfrac{1}{x+4} + \dfrac{2}{x^2+1}$

19. $\dfrac{-3}{x-5} + \dfrac{x+2}{x^2+3}$ **20.** $\dfrac{-3}{x+3} + \dfrac{x+1}{x^2-3x+9}$

21. $\dfrac{5}{x-1} + \dfrac{x-2}{x^2+x+1}$ **22.** $\dfrac{5}{(x-2)^2} - \dfrac{1}{x-2} - \dfrac{3}{x+3}$

23. $\dfrac{1}{(x-1)^2} - \dfrac{3}{x-1} + \dfrac{2}{x+5}$ **24.** $\dfrac{2}{x^2+3} - \dfrac{x+5}{(x^2+3)^2}$

25. 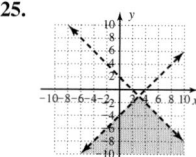   **26.** 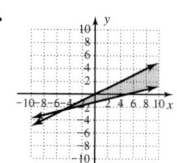   **27.**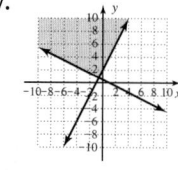

28. 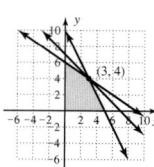   Maximum of 270 occurs at both $(0, 6)$ and $(3, 4)$

29. 50 cows, 425 chickens   **30.** $(-2, -4)$

31. $(1, 6, 9)$ **32.** $(-2, 7, 1, 8)$

33. $\begin{bmatrix} -7.25 & 5.25 \\ 0.875 & -2.875 \end{bmatrix}$ **34.** $\begin{bmatrix} -6.75 & 6.75 \\ 1.125 & -1.125 \end{bmatrix}$ **35.** not possible

36. $\begin{bmatrix} -2 & -6 \\ -1 & -7 \end{bmatrix}$ **37.** $\begin{bmatrix} 1 & 0 \\ 0 & 1 \end{bmatrix}$ **38.** $\begin{bmatrix} 1 & 0 & 4 \\ 5.5 & -1 & -1 \\ 10 & -2.9 & 7 \end{bmatrix}$

39. $\begin{bmatrix} 3 & -6 & -4 \\ -4.5 & 3 & -1 \\ -2 & 3.1 & 3 \end{bmatrix}$ **40.** not possible   **41.** $\begin{bmatrix} -8 & 12 & 0 \\ -2 & -4 & 4 \\ -16 & -0.4 & -20 \end{bmatrix}$

42. $\begin{bmatrix} 15.5 & 6.4 & 17 \\ 9 & -17 & 2 \\ 18.5 & -20.8 & 13 \end{bmatrix}$

43. D **44.** It's an identity.   **45.** It's the inverse of B.   **46.** E

47. It's an identity matrix.   **48.** It's the inverse of F.

49. matrix multiplication is not generally commutative

50. $(-8, -6)$ **51.** $(2, 0, -3)$

52. $\left(\dfrac{-19}{35}, \dfrac{25}{14}\right)$ **53.** $\left(\dfrac{-37}{19}, \dfrac{36}{19}, \dfrac{31}{19}\right)$ **54.** $(1, -1, 2)$

55. $\dfrac{91}{2}$ units^2 **56.** $\dfrac{5}{x-2} + \dfrac{2x-1}{x^2+3}$

Mixed Review, pp. 819–820

1. consistent/dependent; consistent; inconsistent

3 $(-2, 3)$ **5.** $\left(1, \dfrac{1}{2}, 2\right)$ **7.** $(-10, 12)$

9. $\dfrac{-3}{x+2} + \dfrac{2}{x-3}$ **11. a.** $\begin{bmatrix} -8 & 16 & -10 \\ 12 & 0 & 6 \end{bmatrix}$ **b.** $\begin{bmatrix} 9 & -6 & -7 \\ -7 & -1 & 2 \end{bmatrix}$

13. $(-9, -3, 2)$ **15.** $\left(\dfrac{33}{31}, \dfrac{-10}{31}, \dfrac{-57}{31}\right)$

17. 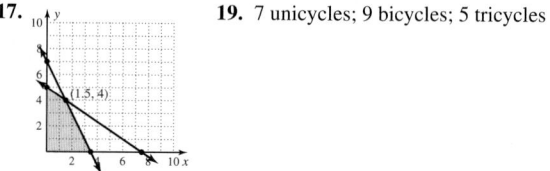   **19.** 7 unicycles; 9 bicycles; 5 tricycles

Practice Test, pp. 820–821

1. 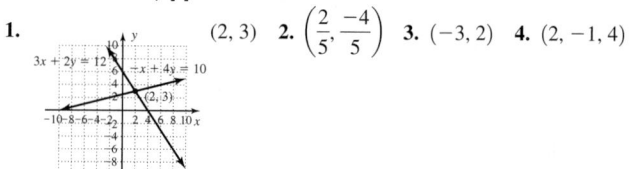 $(2, 3)$ **2.** $\left(\dfrac{2}{5}, \dfrac{-4}{5}\right)$ **3.** $(-3, 2)$ **4.** $(2, -1, 4)$

5. a. $\begin{bmatrix} -6 & -5 \\ 8 & 9 \end{bmatrix}$ **b.** $\begin{bmatrix} 1.2 & 1.2 \\ -1.2 & -2 \end{bmatrix}$ **c.** $\begin{bmatrix} -3 & 1 \\ 3 & -5 \end{bmatrix}$

d. $\begin{bmatrix} -2 & -1 \\ 2.5 & 1.5 \end{bmatrix}$ **e.** -2

6. a. $\begin{bmatrix} 0 & -0.1 & 0 \\ 0.5 & -0.6 & 0 \\ -0.2 & -0.8 & -0.9 \end{bmatrix}$ **b.** $\begin{bmatrix} -0.3 & -0.06 & -0.12 \\ 0.06 & -0.06 & 0 \\ -0.18 & -0.24 & -0.48 \end{bmatrix}$

c. $\begin{bmatrix} 0.31 & -0.13 & 0.08 \\ -0.01 & -0.05 & -0.02 \\ 0.39 & -0.52 & -0.02 \end{bmatrix}$ **d.** $\begin{bmatrix} \frac{40}{17} & 0 & \frac{-10}{17} \\ \frac{40}{17} & 10 & \frac{-10}{17} \\ \frac{-35}{17} & -5 & \frac{30}{17} \end{bmatrix}$ **e.** $\dfrac{17}{500}$

7. $\left(2, 1, \dfrac{-1}{3}\right)$ **8.** $(3, -2, 3)$ **9.** $\left(\dfrac{97}{34}, \dfrac{-18}{17}\right)$ **10.** $(1, 6, 9)$

11. 21.59 cm by 35.56 cm   **12.** Tahiti 402 mi²; Tonga 290 mi²
13. Corn 25¢   Beans 20¢   Peas 29¢   **14.** $15,000 at 7% $8000 at 5%
$7000 at 9%   **15.** $h(t) = -16t^2 + 144$ **16.** 144 ft, 3 sec
17.   **18.** $(5, 0)$ **19.**   30 plain;
                                                                                              20 deluxe

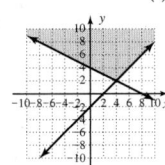   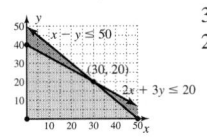

20. $\dfrac{1}{x - 3} + \dfrac{3x + 2}{x^2 + 3x + 9}$

Strengthening Core Skills, p. 824

1. Exercise 1 $(1, -4, 1)$

Cumulative Review Chapters 1–8, pp. 825–826

1. a. $x = \dfrac{2}{3}$ **b.** $x = 0, 7$ **c.** $x = 5, \pm i\sqrt{2}$ **d.** $x = -1, 0, 4$

3. $R = \pm\dfrac{1}{\pi}\sqrt{A + (\pi r)^2}$ **5.**

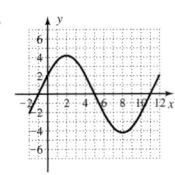

7. a. $(a + bi) + (a - bi) = 2a$ **b.** $(a + bi)(a - bi) = a^2 - (bi)^2$
$= a^2 + b^2$ **9.** $x - 12 \pm \dfrac{\sqrt{15}}{3}$ **11.** $\sin\theta = \dfrac{\sqrt{13}}{4}, \cos\theta = \dfrac{\sqrt{3}}{4}$,
$\tan\theta = \dfrac{\sqrt{39}}{3}$ **13. a.** $m = \dfrac{y_2 - y_1}{x_2 - x_1}$ **b.** $\left(\dfrac{x_2 + x_1}{2}, \dfrac{y_2 + y_1}{2}\right)$
c. $x = \dfrac{-b \pm \sqrt{b^2 - 4ac}}{2a}$ **d.** $d = \sqrt{(x_2 - x_1)^2 + (y_2 - y_1)^2}$
e. $A = Pe^{rt}$ **15.** $\langle -3, -18\rangle$ **17. a.** $\dfrac{\sqrt{121 + x^2}}{11}$ **b.** $\dfrac{x}{\sqrt{9 + x^2}}$

19. $x \in (-\infty, -1) \cup (2, 3)$ **21.** $-128 - 128i\sqrt{3}$

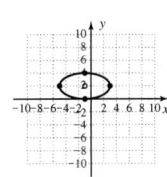

23. about 3.6 yr   **25.** $A = 2, B = 1, C = \dfrac{\pi}{4}$

Connections to Calculus Exercises, p. 829

1. a. $\dfrac{k}{(x + a)(x + b)} = \dfrac{A}{x + a} + \dfrac{B}{x + b}; (A + B)x + (AB + Ba) = k$
b. $B = -A \Rightarrow Ab + (-Aa) = k; A(b - a) = k$
c. $A = \dfrac{k}{b - a}; \dfrac{16}{(x + 3)(x - 5)} = \dfrac{-2}{x + 3} + \dfrac{2}{x - 5}$
3. $\dfrac{-1}{x + 3} + \dfrac{1}{x - 3}$ **5.** $\dfrac{2}{x + 7} + \dfrac{-2}{x - 4}$ **7.** $\dfrac{-1}{x + 8} + \dfrac{2}{x + 5}$
9. $A = 114$ units²   **11.** $A = 36$ units²
13. $V = 240$ units³
   $V = LWH; (5)(6)(8) = 240$ units³

CHAPTER 9
Exercises 9.1, pp. 837–839

1. geometry, algebra   **3.** perpendicular   **5.** point, intersecting
7. $(-2, -2)$; verified   **9.** $(2, -2)$; verified   **11.** $\left(\frac{13}{2}, -9\right)$; verified
13. $(x + 2)^2 + (y + 2)^2 = 5^2$ **15.** $(x - 2)^2 + (y + 2)^2 = 5^2$
17. $\left(x - \dfrac{13}{2}\right)^2 + (y + 9)^2 = \left(\dfrac{25}{2}\right)^2$
19. a. $d = 13; B, C, E, G;$ **b.** $(13, 3 + 4\sqrt{3}), (14, 8);$ Many others
21. Verified, $d = \dfrac{8\sqrt{5}}{5}$ **23. a.** $B, C, E;$ **b.** Answers will vary.

25. Verified   **27.** $y = -\dfrac{1}{16}x^2$ **29.** $4x^2 + 3y^2 = 48$
31. Verified, verified   **33.** $3x^2 - y^2 = 3$
35. a. $\left(-\dfrac{12}{7}, -\dfrac{30}{7}\right),$ **b.** $\left(-2, -\dfrac{4}{3}\right)$ **37.** Verified (both add to 8)
39. $x = \dfrac{4\pi}{3}$ **41.** $h(x) = \dfrac{(x + 3)(x - 3)}{(x + 2)(x - 2)}$

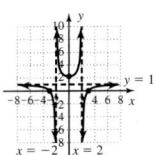

Exercises 9.2, pp. 847–852

1. $c^2 = |a^2 - b^2|$ **3.** $2a; 2b$ **5.** answers will vary.   **7.** $x^2 + y^2 = 49$
9. $(x - 5)^2 + y^2 = 3$ **11.** $(x - 1)^2 + (y - 5)^2 = 25$
13. $(x - 6)^2 + (y - 5)^2 = 9$ **15.** $(x - 2)^2 + (y + 5)^2 = 25$
   center: $(6, 5), r = 3$ center: $(2, -5), r = 5$

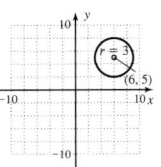   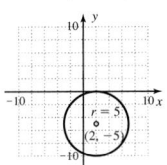

17. $(x + 3)^2 + y^2 = 14$
   center: $(-3, 0), r = \sqrt{14}$

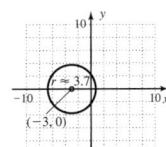

19.   **21.**   **23.**

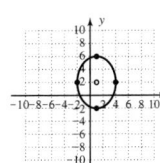   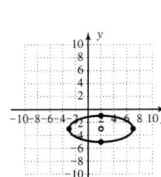   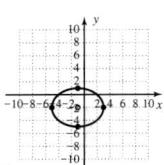

25. a. $\dfrac{x^2}{16} + \dfrac{y^2}{4} = 1, (0, 0), a = 4, b = 2$
   b. $(-4, 0), (4, 0), (0, -2), (0, 2)$ c.

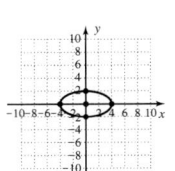

27. a. $\dfrac{x^2}{9} + \dfrac{y^2}{16} = 1, (0, 0), a = 3, b = 4$
   b. $(0, -4), (0, 4), (-3, 0), (3, 0)$ c.

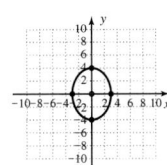

29. a. $\dfrac{x^2}{5} + \dfrac{y^2}{2} = 1, (0, 0), a = \sqrt{5}, b = \sqrt{2}$
   b. $(-\sqrt{5}, 0), (\sqrt{5}, 0), (0, -\sqrt{2}), (0, \sqrt{2})$ c.

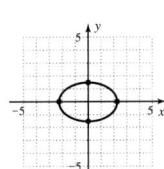

31. ellipse   **33.** circle

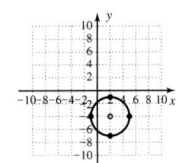

35. ellipse

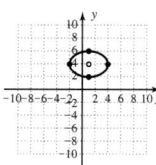

37. $x^2 + \dfrac{(y+3)^2}{4} = 1$ **39.** $\dfrac{(x+2)^2}{16} + \dfrac{(y-1)^2}{4} = 1$

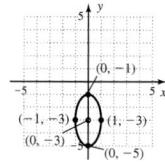

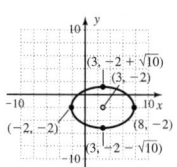

41. $\dfrac{(x-3)^2}{4} + \dfrac{(y+5)^2}{10} = 1$ **43.** $\dfrac{(x-3)^2}{25} + \dfrac{(y+2)^2}{10} = 1$

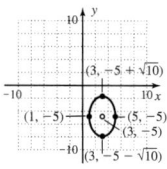

45. 20   **47.** 20

49. a. $(2, 1)$ **b.** $(-3, 1)$ and $(7, 1)$ **c.** $(2 - \sqrt{21}, 1)$ and $(2 + \sqrt{21}, 1)$
    d. $(2, 3)$ and $(2, -1)$ **e.**

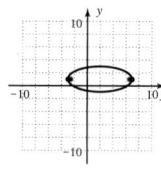

51. a. $(4, -3)$ **b.** $(4, 2)$ and $(4, -8)$ **c.** $(4, 0)$ and $(4, -6)$
    d. $(0, -3)$ and $(8, -3)$ **e.**

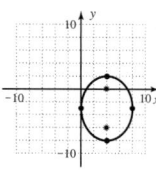

53. a. $(-2, -2)$ **b.** $(-5, -2)$ and $(1, -2)$ **c.** $(-2 + \sqrt{3}, -2)$ and
    $(-2 - \sqrt{3}, -2)$ **d.** $(-2, -2 + \sqrt{6})$ and $(-2, -2 - \sqrt{6})$
    e.

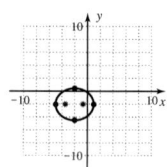

55. $\dfrac{x^2}{36} + \dfrac{y^2}{20} = 1$ **57.** $\dfrac{(x-3)^2}{9} + \dfrac{(y+2)^2}{25} = 1$

59. $\dfrac{x^2}{16} + \dfrac{y^2}{9} = 1, (\pm\sqrt{7}, 0)$ **61.** $\dfrac{(x+3)^2}{4} + \dfrac{(y+1)^2}{16} = 1,$

$(-3, -1 \pm 2\sqrt{3})$ **63.** $A = 12\pi$ units^2 **65.** $\sqrt{7} \approx 2.65$ ft; 2.25 ft

67. 8.9 ft; 17.9 ft   **69.** $\dfrac{x^2}{15^2} + \dfrac{y^2}{8^2} = 1$; 6.4 ft   **71.** $\dfrac{x^2}{36^2} + \dfrac{y^2}{(35.25)^2} = 1$

73. $a \approx 142$ million miles, $b \approx 141$ million miles, orbit time ≈ 686 days

75. $90{,}000\pi$ yd^2

77. $L = 8$ units; $(3\sqrt{5}, 4), (3\sqrt{5}, -4), (-3\sqrt{5}, 4), (-3\sqrt{5}, -4)$; verified

79. Verified   **81.** $R = \dfrac{kL}{d^2}$ $k = 0.003$ 250 Ω

83. 261.8 mph, heading 26.2°

Exercises 9.3, pp. 862–865

1. transverse   **3.** midway   **5.** Answers will vary.

7.
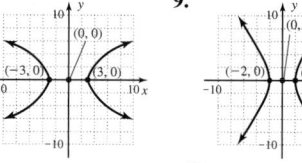

9.
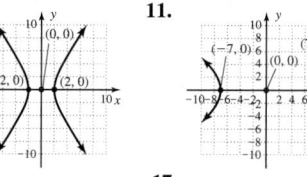

11.

13.
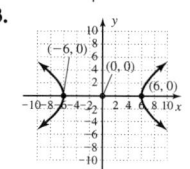

15.
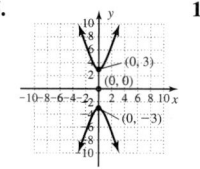

17.
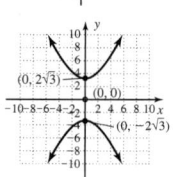

19.
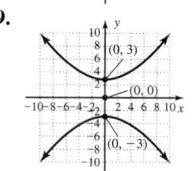

21.
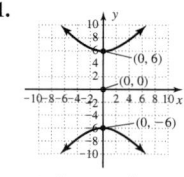

23. $(-4, -2), (2, -2), y = -2, (-1, -2), x = -1$
25. $(4, 1), (4, -3), x = 4, (4, -1), y = -1$

27.
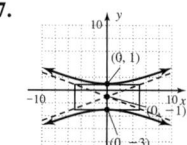

29.
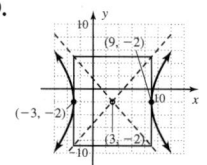

31.
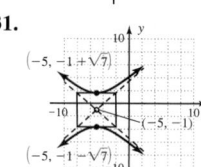

33.
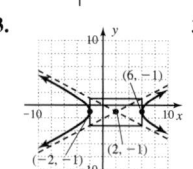

35.
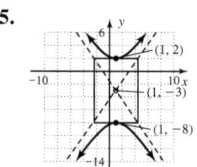

37.
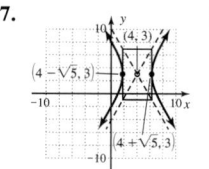

39.
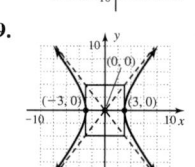

41.
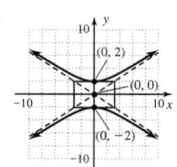

43.
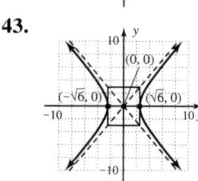

45.
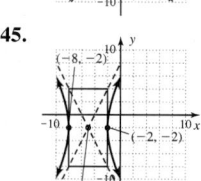

47.
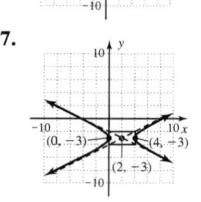

49. circle   **51.** circle   **53.** hyperbola   **55.** hyperbola
57. circle   **59.** ellipse   **61.** 8, $2a = 8, 2b = 6$
63. 12, $2a = 16, 2b = 12$
65. a. $(3, 4)$ **b.** $(0, 4)$ and $(6, 4)$ **c.** $(3 - \sqrt{13}, 4)$ and $(3 + \sqrt{13}, 4)$
    d. $2a = 6, 2b = 4$ **e.**

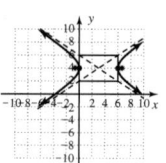

67. a. $(0, 3)$ **b.** $(-2, 3)$ and $(2, 3)$ **c.** $(-2\sqrt{5}, 3$ and $(2\sqrt{5}, 3)$
    d. $2a = 4, 2b = 8$ **e.**

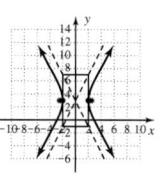

69. a. $(3, -2)$ **b.** $(1, -2)$ and $(5, -2)$ **c.** $(-1, -2)$ and $(7, -2)$
d. $2a = 4, 2b = 4\sqrt{3}$ **e.**

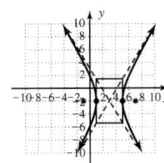

71. $\dfrac{x^2}{36} - \dfrac{y^2}{28} = 1$ **73.** $\dfrac{y^2}{9} - \dfrac{(x+2)^2}{9} = 1$

75. $\dfrac{x^2}{4} - \dfrac{y^2}{9} = 1, (\pm\sqrt{13}, 0)$ **77.** $\dfrac{(y-1)^2}{4} - \dfrac{(x-2)^2}{5} = 1$, 4 by $2\sqrt{5}$

79. a. $y = \frac{2}{3}\sqrt{x^2 - 9}$ **b.** $x \in (-\infty, -3] \cup [3, \infty)$ **c.** $y = \frac{-2}{3}\sqrt{x^2 - 9}$

81. 40 yd **83.** 40 ft

85. $\dfrac{x^2}{225} - \dfrac{y^2}{2275} = 1$, about $(24.1, 60)$ or $(-24.1, 60)$

87. a. $\dfrac{(x-4)^2}{\frac{1}{4}} - (y-2)^2 = 0$ **b.** $(x-2)^2 + \dfrac{(x-4)^2}{\frac{1}{5}} = 0$

89. a **91.** $\dfrac{(x-2)^2}{43} + \dfrac{(y-3)^2}{9} = 1$ **93.** $700 \cos 65° \approx 295.8$, yes

95. b and c

Exercises 9.4, pp. 871–874

1. horizontal; right; $a < 0$ **3.** $(p, 0)$, $x = -p$
5. Answers will vary.
7. $x \in (-\infty, \infty), y \in [-4, \infty)$ **9.** $x \in (-\infty, \infty), y \in [-18, \infty)$

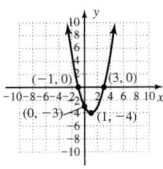 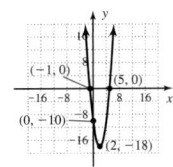

11. $x \in (-\infty, \infty), y \in [-10.125, \infty)$ **13.** $x \in [-4, \infty), y \in (-\infty, \infty)$

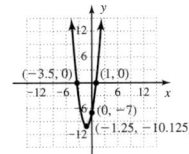 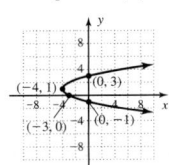

15. $x \in (-\infty, 16], y \in (-\infty, \infty)$ **17.** $x \in (-\infty, 0], y \in (-\infty, \infty)$

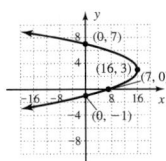 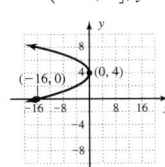
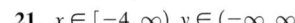

19. $x \in [-9, \infty), y \in (-\infty, \infty)$ **21.** $x \in [-4, \infty), y \in (-\infty, \infty)$

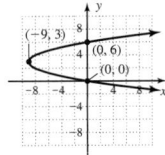 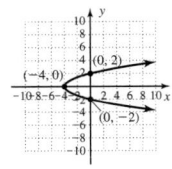
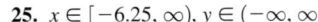

23. $x \in (-\infty, 0], y \in (-\infty, \infty)$ **25.** $x \in [-6.25, \infty), y \in (-\infty, \infty)$

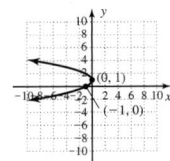 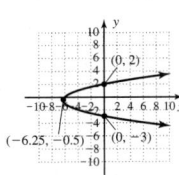

27. $x \in [-21, \infty), y \in (-\infty, \infty)$ **29.** $x \in (-\infty, 11], y \in (-\infty, \infty)$

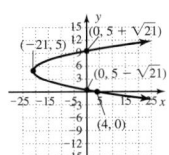 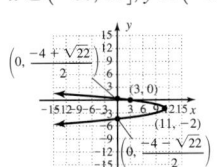

31. $x \in (-\infty, \infty), y \in [3, \infty)$ **33.** $x \in [2, \infty), y \in (-\infty, \infty)$

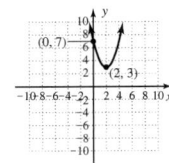 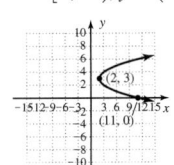

35. $x \in [1, \infty), y \in (-\infty, \infty)$

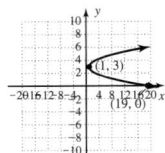

37. **39.** **41.**

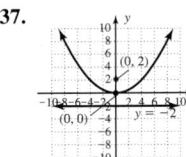 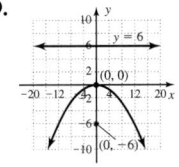 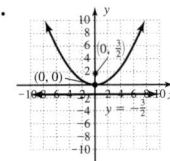

43. **45.** **47.**

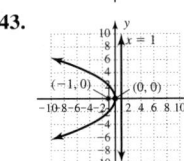 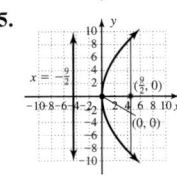 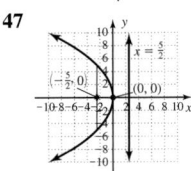

49. **51.** **53.**

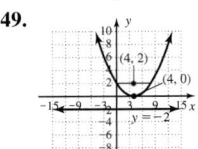 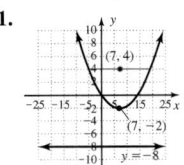 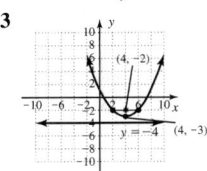

55. **57.** **59.**

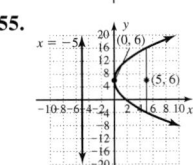 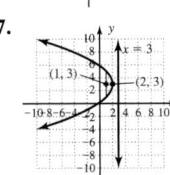 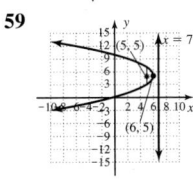

61. $x^2 = 8y$ **63.** $y^2 = 16x$ **65.** $x^2 = -20y$
67. $(y+2)^2 = -12(x-2)$ **69.** $(x-4)^2 = 12(y+7)$
71. $(x-3)^2 = 8(y-2)$ **73.** $y^2 = 8(x+1)$ vertex $(-1, 0)$ focus $(1, 0)$
75. $(y-2)^2 = -8(x+2)$ directrix: $x = 0$ endpoints $(-4, 4)$ and $(-4, 0)$
77. 16 units^2
79.

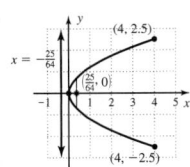

81. 6 in.; $(13.5, 0)$ **83.** 14.97 ft, $(0, 41.75)$
85. $y^2 = 5x$ or $x^2 = 5y$, 1.25 cm
87. $(x-2)^2 = \frac{1}{2}(y+8); p = \frac{1}{8}; (2, -8)$ **89.** 18 units^2
91. $-2, 2, 1 + \sqrt{3}i, 1 - \sqrt{3}i, -1 + \sqrt{3}i, -1 - \sqrt{3}i$
93. about 120 days

Mid-Chapter Check, p. 874

1. 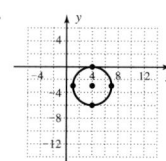 **2.** 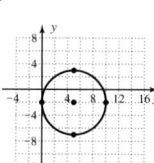 **3.**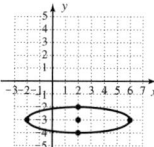

4. 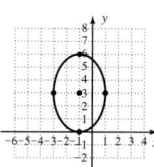 **5.** 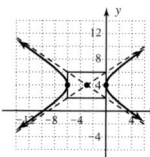 **6.**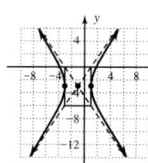

7. a. $\dfrac{(x+3)^2}{4} + \dfrac{(y-1)^2}{16} = 1$; $D: x \in [-5, -1]$; $R: y \in [-3, 5]$

b. $(x-3)^2 + (y-2)^2 = 16$; $D: x \in [-1, 7]$; $R: y \in [-2, 6]$

c. $y = (x-3)^2 - 4$ $D: x \in (-\infty, \infty)$; $R: y \in [-4, \infty)$

8.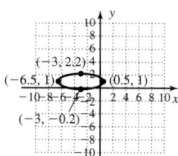

9. $\dfrac{x^2}{16} + \dfrac{y^2}{4} = 1$ **10.** yes, distance $d \approx 49$ mi

Reinforcing Basic Concepts, p. 875

Exercise 1. $\dfrac{(x-2)^2}{\left(\dfrac{\sqrt{2}}{5}\right)^2} + \dfrac{(y+3)^2}{\left(\dfrac{2}{3}\right)^2} = 1$ $a = \dfrac{\sqrt{2}}{5}, b = \dfrac{2}{3}$

Exercise 2. $\dfrac{(x-1)^2}{\left(\dfrac{5\sqrt{7}}{14}\right)^2} + \dfrac{(y+2)^2}{\left(\dfrac{5\sqrt{3}}{12}\right)^2} = 1$ $a = \dfrac{5\sqrt{17}}{14}, b = \dfrac{5\sqrt{3}}{12}$

Exercise 3. $\dfrac{(x+3)^2}{\left(\dfrac{7}{2}\right)^2} + \dfrac{(y-1)^2}{\left(\dfrac{6}{5}\right)^2} = 1$; $a = \dfrac{7}{2}, b = \dfrac{6}{5}$

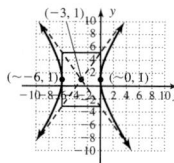

Exercise 4. $\dfrac{(x+3)^2}{\left(\dfrac{4\sqrt{5}}{3}\right)^2} - \dfrac{(y-1)^2}{\left(\dfrac{9}{2}\right)^2} = 1$; $a = \dfrac{4\sqrt{5}}{3} \approx 3, b = \dfrac{9}{2}$

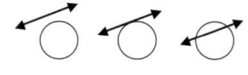

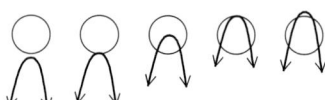

Exercises 9.5, pp. 880–883

1. a. 0, 1, and 2 solutions possible   **b.** 0, 1, and 2 solutions possible

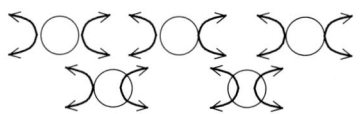

c. 0, 1, 2, 3, and 4 solutions possible

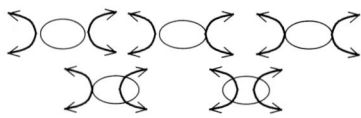

d. 0, 1, 2, 3, and 4 solutions possible

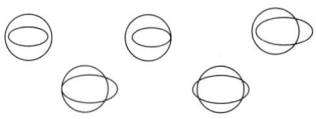

e. 0, 1, 2, 3, and 4 solutions possible

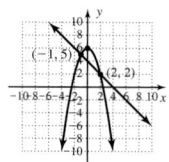

f. 0, 1, 2, 3, and 4 solutions possible

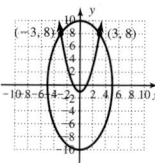

3. region; solutions   **5.** Answers will vary.

7. first: parabola; second: line

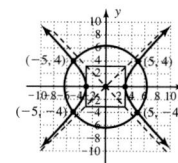

9. first: parabola; second: ellipse   **11.** first: hyperbola; second: circle

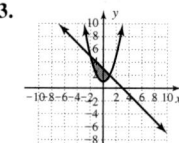
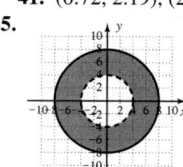

13. $(-4, -3), (3, 4)$ **15.** $\left(4, \dfrac{3}{2}\right), (3, 2)$

17. $(\sqrt{10}, 3), (-\sqrt{10}, 3), (5, -12), (-5, -12)$

19. $(4, 3), (4, -3), (-4, 3), (-4, -3)$ **21.** no solutions

23. $(5, -5), (5, 5), (-5, 5), (-5, -5)$ **25.** $(5, \log 5 + 5)$

27. $(-3, \ln 9 + 1), (4, \ln 16 + 1)$ **29.** $(0, 10), (\ln 6, 45)$

31. $(-3, 1), (2, 1024)$ **33.** $(-3, -21), (1, -1), (2, 4)$

35. $(2, -4), (6, 4)$ **37.** $(3, 5), (3, -5)$

39. $(-2.43, -2.81), (2, 1)$ **41.** $(0.72, 2.19), (2, 3), (4, 3), (5.28, 2.19)$

43. 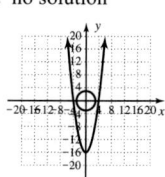   **45.** 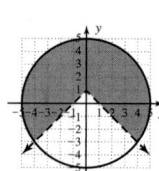   **47.** no solution

49.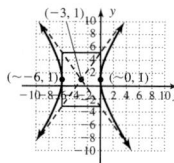

51. $h \approx 27.5$ ft; $h = 24$ ft; $h = 18$ ft

53. The company breaks even if either 18,400 or 48,200 Nanos are sold.

55. $1.83; $3

90,000 gal

$\begin{cases} 10P^2 + 6D = 144 \\ 8P^2 - 8P - 4D = 12 \end{cases}$

57. 8.5 m × 10 m   **59.** 5 km, 9 km   **61.** 8 × 8 × 25 ft

63. Answers will vary.   **65.** 18 in. by 18 in. by 77 in.   **67.** $h \approx 62$ ft

69. a. $m = \dfrac{-400}{1}$, the copier depreciates by $400 a year.

b. $y = -400x + 4500$ **c.** $1700   **d.** 9.5 yr

Exercises 9.6, pp. 893–896

1. polar   **3.** II; IV   **5.** To plot the point (r, θ) start at the origin or pole and move $|r|$ units out along the polar axis. Then move counterclockwise an angle measure of θ. You should be r units straight out from the pole in a direction of θ from the positive polar axis. If r is negative, final resting place for the point (r, θ) will be 180° from θ.

7. 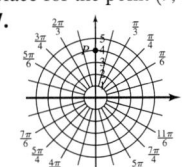   **9.** 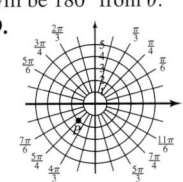   **11.**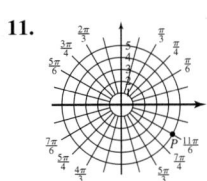

13. 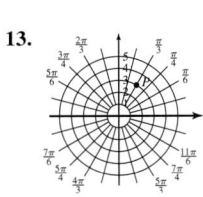   **15.** $\left(4, \dfrac{\pi}{2}\right)$ **17.** $\left(4\sqrt{2}, \dfrac{\pi}{4}\right)$

19. $\left(8, \dfrac{2\pi}{3}\right)$ **21.** $\left(4\sqrt{2}, \dfrac{3\pi}{4}\right)$

23. $\left(3\sqrt{2}, \dfrac{-5\pi}{4}\right), \left(-3\sqrt{2}, \dfrac{7\pi}{4}\right), \left(3\sqrt{2}, \dfrac{11\pi}{4}\right), \left(-3\sqrt{2}, \dfrac{-\pi}{4}\right)$

25. $\left(2, \dfrac{5\pi}{6}\right), \left(2, \dfrac{-7\pi}{6}\right), \left(2, \dfrac{17\pi}{6}\right), \left(-2, \dfrac{-\pi}{6}\right)$ **27.** C   **29.** C   **31.** D

33. B   **35.** D   **37.** $(8, 180°)$ or $(8, \pi)$ **39.** $(4\sqrt{2}, 45°)$ or $\left(4\sqrt{2}, \dfrac{\pi}{4}\right)$

41. $(10, 45°)$ or $\left(10, \dfrac{\pi}{4}\right)$ **43.** $(13, 247.4°)$ or $(13, 4.3176)$

45. $(4\sqrt{2}, 4\sqrt{2})$ **47.** $(-2\sqrt{2}, 2\sqrt{2})$ **49.** $(\sqrt{3}, 1)$ **51.** $\left(\dfrac{5\sqrt{2}}{2}, \dfrac{5\sqrt{2}}{2}\right)$

53. 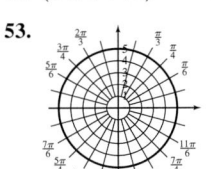   **55.** 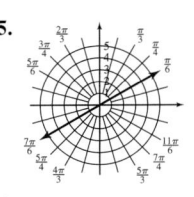   **57.**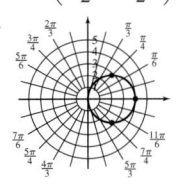

59. 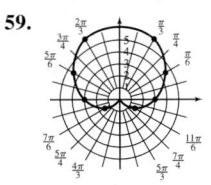   **61.** 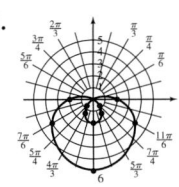   **63.**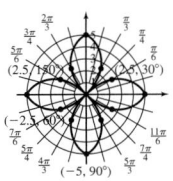

65. 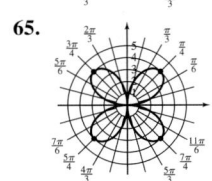   **67.** 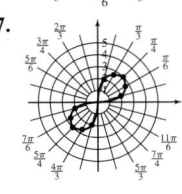   **69.**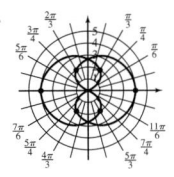

71. 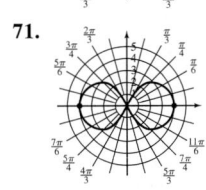   **73.**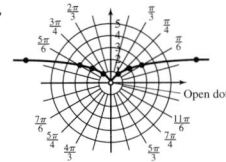

75.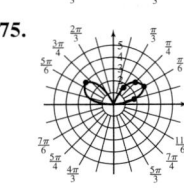

77. 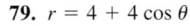 $\left(\dfrac{4\sqrt{3} + 3\sqrt{2}}{2}, \dfrac{4 + 3\sqrt{2}}{2}\right)$; $(3\sqrt{2}, 3\sqrt{2})$; $(4\sqrt{3}, 4)$; yes

$M = \left(\dfrac{3\sqrt{2} + 4\sqrt{3}}{2}, \dfrac{3\sqrt{2} + 4}{2}\right)$

79. $r = 4 + 4\cos\theta$ **81.** $r = 4\cos(5\theta)$

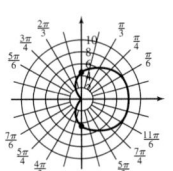   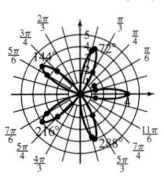

83. $r^2 = 16\cos(2\theta)$ **85.** $r = 4\sin\theta$

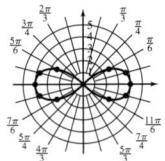   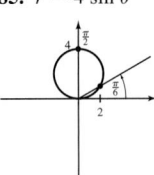

87. a; this is a circle through $(6, 0°)$ symmetric about the polar axis

89. g; this is a circle through $\left(6, \dfrac{\pi}{2}\right)$ symmetric about $\theta = \dfrac{\pi}{2}$.

91. f; this is a limaçon symmetric about $\theta = \dfrac{\pi}{2}$ with an inner loop. Thus $a < b$.

93. b; this is a cardioid symmetric about $\theta = \dfrac{\pi}{2}$ through $\left(6, \dfrac{3\pi}{2}\right)$.

95. $r^2 = 7200^2\sin(2\theta)$ **97.** $r = 15\cos(5\theta)$ or $r = 15\sin(5\theta)$

99. π; π; π; Answers will vary.

101. Consider $r = a\sqrt{\cos(2\theta)}$ and $r = -a\sqrt{\cos(2\theta)}$; both satisfy $r^2 = a^2\cos(2\theta)$. Thus, (r, θ) and $(-r, \theta)$ will both be on the curve. The same is true with $a\sqrt{\sin(2\theta)}$ and $-a\sqrt{\sin(2\theta)}$.

103. 9π units²   **105.** $3y^2 - x^2 - 12y + 9 = 0$

107. $t = 0, \dfrac{2\pi}{3}, \pi, \dfrac{5\pi}{3}$

109. $D: x \in [-5, 2) \cup (2, 5]$
$R: y \in [-3, 2) \cup \{4\}$

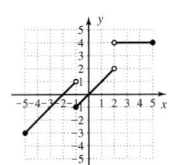

Exercises 9.7, pp. 907–912

1. rotation of axes; $\dfrac{B}{A - C}$ **3.** invariants   **5.** Answers will vary.

7. $\dfrac{Y^2}{8} - \dfrac{X^2}{8} = 1$ **9.** $6 + 3\sqrt{2} = X, -6 + 3\sqrt{2} = Y$

11. $\dfrac{5\sqrt{2}}{2} = X, \dfrac{5\sqrt{2}}{2} = Y$ **13.** $0 = x, 4 = y$

15. $\dfrac{3\sqrt{3}}{2} - 2 = x; \dfrac{3}{2} + 2\sqrt{3} = y$ **17.** $\dfrac{-x^2}{2} + xy\sqrt{3} + \dfrac{y^2}{2} = 9$

19. $4X^2 + 2Y^2 = 9$

21. a. $3X^2 - Y^2 = 2$ **23. a.** $4X^2 + Y^2 = 8$

b. 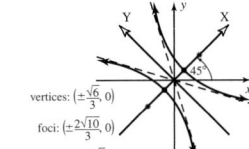   **b.**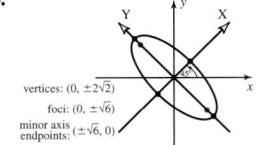

vertices: $\left(\pm\sqrt{6}, 0\right)$ vertices: $(0, \pm2\sqrt{2})$

foci: $\left(\pm\dfrac{2\sqrt{10}}{3}, 0\right)$ foci: $(0, \pm\sqrt{6})$

asymptotes: $Y = \pm\sqrt{3}X$ minor axis endpoints: $(\pm\sqrt{6}, 0)$

25. a. $Y^2 - 4X^2 = 16$ **27. a.** $Y^2 - 4X = 0$
b.

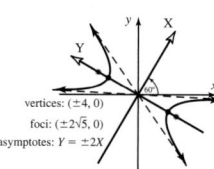

vertices: $(\pm 4, 0)$
foci: $(\pm 2\sqrt{5}, 0)$
asymptotes: $Y = \pm 2X$

b.

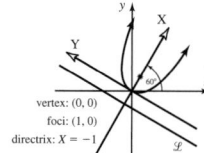

vertex: $(0, 0)$
foci: $(1, 0)$
directrix: $X = -1$

29. a. $X^2 + 4Y^2 = 25$
b.

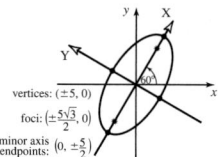

vertices: $(\pm 5, 0)$
foci: $\left(\pm \frac{5\sqrt{3}}{2}, 0\right)$
minor axis endpoints: $\left(0, \pm \frac{5}{2}\right)$

31. $336 > 0$; hyperbola; $\cos(2\beta) = \frac{7}{25}$; $\frac{4}{5} = \cos \beta$; $\frac{3}{5} = \sin \beta$

33. a. parabola   **b.** $\beta = 45°$; $2Y^2 = 5$ **c.** verified

35. a. circle or ellipse   **b.** $\beta = 60°$; $\frac{9}{2}X^2 + \frac{5Y^2}{2} + 2X - 2\sqrt{3}Y = 1$

(ellipse)   **c.** verified   **37.** f   **39.** g   **41.** h

43. parabola   **45.** ellipse   **47.** hyperbola

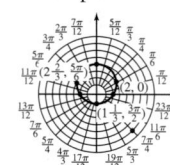
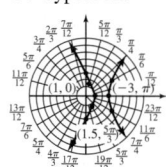

49. ellipse

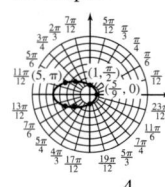

51. $r = \dfrac{3.2}{1 - 0.8 \cos \theta}$

53. $r = \dfrac{4}{1 - \cos \theta}$ **55.** $r = \dfrac{7.5}{1 + 1.5 \sin \theta}$

57. a. $r = \dfrac{12}{2 \cos \theta + 3 \sin \theta}$ **b.** $-\dfrac{r(\pi/2)}{r(0)} = \dfrac{-2}{3}$ and $\dfrac{-A}{B} = \dfrac{-2}{3}$

59. Jupiter: $e \approx 0.0486$, Saturn: $e \approx 0.0567$

61. about 2757.1 million miles   **63.** Saturn: $e \approx 0.0567$

65. $r \approx \dfrac{482.36}{1 - 0.0486 \cos\theta}$ **67.** $r \approx \dfrac{1780.77}{1 - 0.0457 \cos\theta}$

69. In millions of miles (approx): $\overline{JS}$: 405.3, $\overline{JU}$: 1298.4, $\overline{JN}$: 2310.3, $\overline{SU}$: 893.1, $\overline{SN}$: 1905.0, $\overline{UN}$: 1011.9

71. $r = \dfrac{0.7638}{1 \pm 0.7862 \cos \theta}$ **73.** $r = \dfrac{0.2864}{1 \pm 0.7862 \cos \theta}$

75. \$582.45; \$445.94; \$881.32; \$97.92   **77.** $y = \dfrac{3}{1 - \cos \theta}$

79. verified   **81.** Answers will vary

83. $r = 12 \cos\left(\theta - \dfrac{\pi}{4}\right) = 6\sqrt{2}(\cos \theta + \sin \theta)$

85. $425X^2 - 416Y^2 - 400 = 0$ **87.** $(0, 0)$, $(4, 0)$, $(4, 4)$, $(0, 4)$

89. $x \approx 29.0$ **91.** 9.2 mph at heading 347.7°

Exercises 9.8, pp. 920–924

1. parameter   **3.** direction   **5.** Answers will vary.
7. a. parabola with vertex at $(2, -1)$ **9. a.** parabola
b. $y = x^2 - 4x + 3$ **b.** $y = x \pm 2\sqrt{x} + 1$

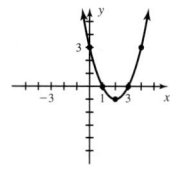
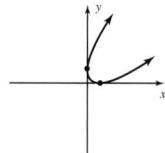

11. a. power function with $p = -2$ **13. a.** ellipse

b. $y = \dfrac{25}{x^2}, x \neq 0$ **b.** $\dfrac{x^2}{16} + \dfrac{y^2}{9} = 1$

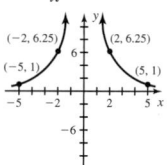
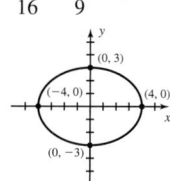

15. a. Lissajous figure   **17.**

b. $y = 6 \cos\left[\dfrac{1}{2} \sin^{-1}\left(\dfrac{x}{4}\right)\right]$

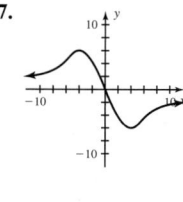

19. $x = t, y = 3t - 2; x = \dfrac{1}{3}t, y = t - 2; x = \cos t, y = 3 \cos t - 2$

21. $x = t, y = (t + 3)^2 + 1; x = t - 3, y = t^2 + 1; x = \tan t - 3,$
$y = \sec^2 t, t \neq \dfrac{(2k + 1)\pi}{2}, k \in \mathbb{Z}$

23. $x = t, y = \tan^2(t - 2) + 1, t \neq \pi k + \dfrac{\pi}{2} + 2, k \in \mathbb{Z}; x = t + 2,$
$y = \sec^2 t, t \neq \left(k + \dfrac{1}{2}\right)\pi, k \in \mathbb{Z}; x = \tan^{-1} t + 2, y = t^2 + 1$

25. verified

27. a.

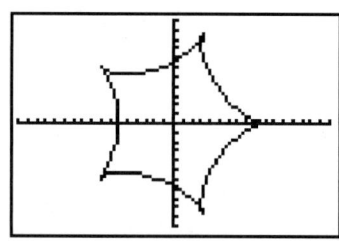

b. x-intercepts: $t = 0, x = 10, y = 0$ and $t = \pi, x = -6, y = 0$; y-intercepts: $t \approx 1.757, x = 0, y \approx 6.5$ and $t \approx 4.527, x = 0, y \approx -6.5$; minimum x-value is -8.1; maximum x-value is 10; minimum y-value is -9.5; the maximum y-value is 9.5

29. a.

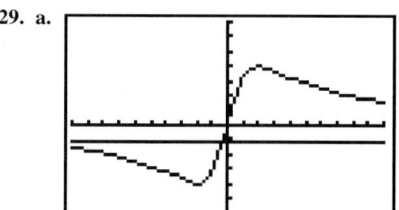

b. x-intercepts none, y-intercepts none; no minimum or maximum x-values; minimum y-value is -4 and maximum y-value is 4

31. a.

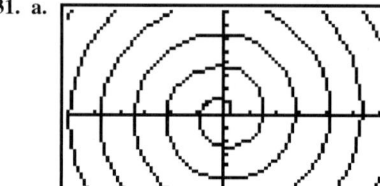

b. *x*-intercepts: $t = 0, x = 2, y = 0$ and $t \approx 4.493, x \approx -9.2, y = 0$; infinitely many others; *y*-intercepts: $t \approx 2.798, x = 0, y \approx 5.9$ and $t \approx 6.121, x = 0, y \approx -12.4$; infinitely many others; no minimum or maximum values for *x* or *y*

33. a.

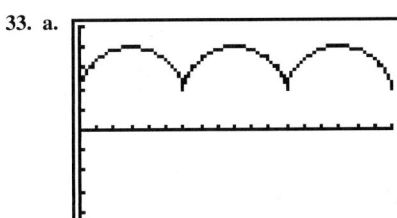

b. no *x*-intercepts; *y*-intercept is $t = 0, x = 0, y = 2$; no minimum or maximum *x*-values; minimum *y*-value is 2; maximum *y*-value is 4

35. a.

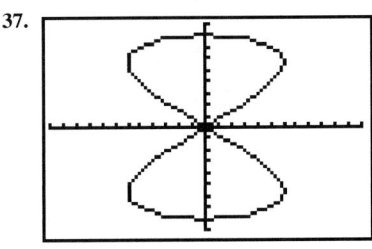

b. *x*-intercepts: $t = 0, x = 4, y = 0$ and $t = \pi, x = -4, y = 0$; *y*-intercepts: $t = \dfrac{\pi}{2}, x = 0, y = 8$ and $t = \dfrac{3\pi}{2}, x = 0, y = -8$; minimum and maximum *x*-values are approx. ± 5.657; minimum and maximum *y*-values are ± 8

37.

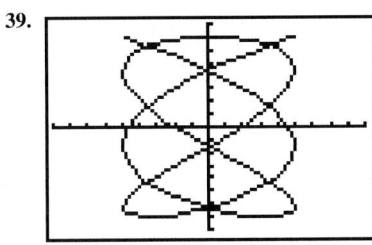

width 12 and length 16; including the endpoint $t = 2\pi$, the graph crosses itself two times from 0 to 2π.

39.

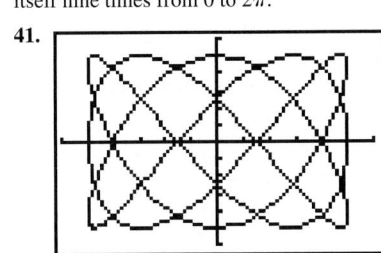

width 10 and length 14; including the endpoint $t = 2\pi$, the graph crosses itself nine times from 0 to 2π.

41.

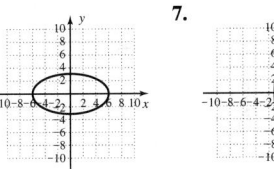

width 20 and length 20; including the endpoint $t = 4\pi$, the graph crosses itself 23 times from 0 to 4π.

43. The maximum value (as the graph swells to a peak) is at $(x, y) = \left(a, \dfrac{b}{2}\right)$. The minimum value (as the graph dips to the valley) is at $(x, y) = \left(-a, \dfrac{-b}{2}\right)$.

45. a. The curve is approaching $y = 2$ as *t* approaches $\dfrac{3\pi}{2}$, but $\cot\left(\dfrac{3\pi}{2}\right)$ is undefined, and the trig form seems to indicate a hole at $t = \dfrac{3\pi}{2}, x = 0, y = 2$. The algebraic form does not have this problem and shows a maximum defined at $t = 0, x = 0, y = 2$.
b. As $|t| \to \infty, y(t) \to 0$ **c.** The maximum value occurs at $(0, 2k)$.

47. a. Yes   **b.** Yes   **c.** ≈ 0.82 ft   **49.** No, the kick is short.

51. The electron is moving left and downward.
53. $\left(t, \dfrac{6t}{17} - \dfrac{6}{17}, \dfrac{13t}{17} + \dfrac{21}{17}\right)$ **55.** Inconsistent, no solutions

57. $x = 1.22475^t$
$y = 0.25t^2 - 2t$

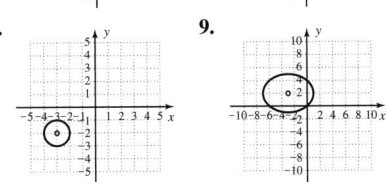

The parametric equations fit the data very well.

59. Answers will vary.   **61.** by 25%   **63.**

Summary and Concept Review, pp. 924–928

1. verified (segments are perpendicular and equal length)
2. $x^2 + (y - 1)^2 = 34$ **3.** verified   **4.** verified
5.   **6.**   **7.**

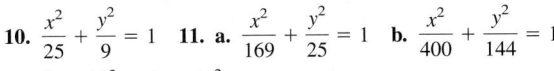

8.   **9.**

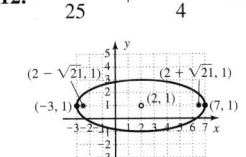 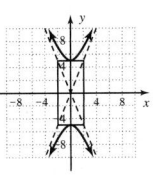

10. $\dfrac{x^2}{25} + \dfrac{y^2}{9} = 1$ **11. a.** $\dfrac{x^2}{169} + \dfrac{y^2}{25} = 1$ **b.** $\dfrac{x^2}{400} + \dfrac{y^2}{144} = 1$
12. $\dfrac{(x - 2)^2}{25} + \dfrac{(y - 1)^2}{4} = 1$ **13.**

14.   **15.**   **16.**

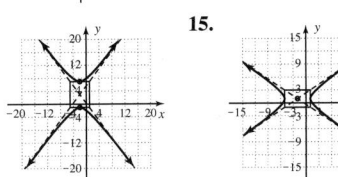 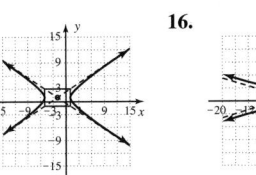

17.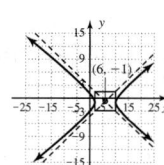

18. 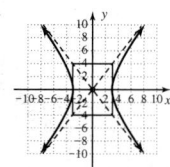 $\dfrac{x^2}{9} - \dfrac{y^2}{16} = 1$

19. a. $\dfrac{x^2}{225} - \dfrac{y^2}{64} = 1$ **b.** $\dfrac{y^2}{16} - \dfrac{x^2}{9} = 1$

20. $\dfrac{(x-5)^2}{9} - \dfrac{(y-2)^2}{4} = 1$ **21.**

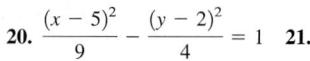

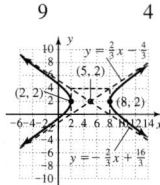   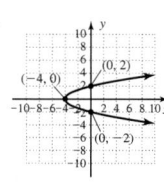

22. 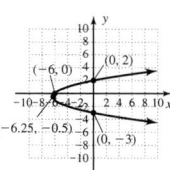 **23.** 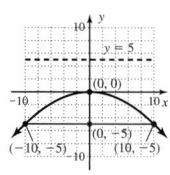 **24.**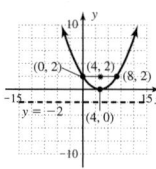

25. circle, line, $(4, 3), (-3, -4)$ **26.** parabola, line, $(3, -2)$
27. parabola, circle, $(\sqrt{3}, 2), (-\sqrt{3}, 2)$ **28.** circle, parabola, $(1, 3), (-1, 3)$
29. Parabola, circle   **30.** Circle, parabola

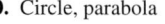

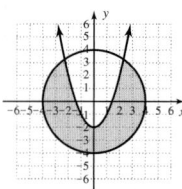 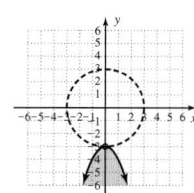 **31.**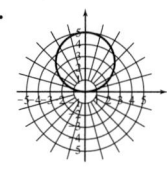

32. 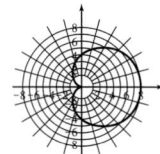 **33.** 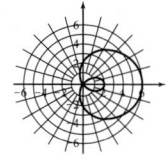 **34.**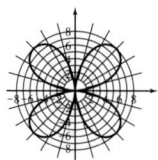

35. $Y^2 - 2Y - 2X - 6 = 0$ **36.** $5X^2 - Y^2 - 80 = 0$
$(Y - 1)^2 = 2\left(X + \dfrac{7}{2}\right)$ $\dfrac{X^2}{16} - \dfrac{Y^2}{80} = 1$

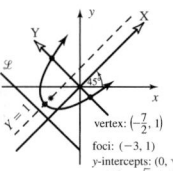

vertex: $\left(-\dfrac{7}{2}, 1\right)$
foci: $(-3, 1)$
y-intercepts: $(0, \sqrt{7})$
and $(0, -\sqrt{7} + 1)$

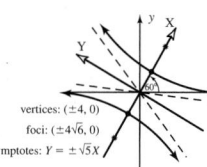

vertices: $(\pm 4, 0)$
foci: $(\pm 4\sqrt{6}, 0)$
asymptotes: $Y = \pm\sqrt{5}X$

37. ellipse, $e = \dfrac{2}{3}$;   **38.** hyperbola, $e = \dfrac{3}{2}$;   **39.** parabola, $e = 1$;

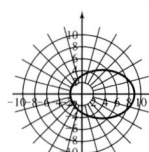 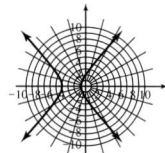 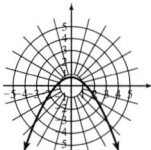

40. $r = \dfrac{de}{1 - e\cos\theta}$ with $e \approx 0.0935$ and $d \approx 1501.1$; focal cord:
≈ 280.82 million miles

41. $y = -2(x + 4)^2 + 3$ **42.** $y = (-1 \pm \sqrt{x})^2$ **43.** $\dfrac{x^2}{9} + \dfrac{y^2}{16} = 1$

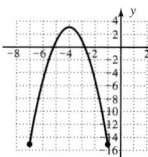 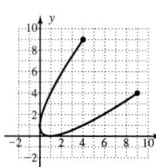 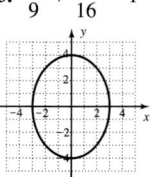

44. Answers will vary. **45.** $x \in [-4, 4]; y \in [-8, 8]$

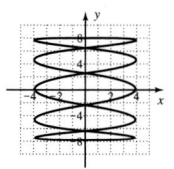

Mixed Review, pp. 929–930

1. circle, center: $(0, 0); r = \sqrt{6}$ **3.** hyperbola; center $(0, 0)$;
$a = 5, b = 3, c = \sqrt{34}$;
vertices $(0, 5), (0, -5)$;
foci $(0, \sqrt{34}), (0, -\sqrt{34})$;
asymptotes $y = \dfrac{5}{3}x, y = -\dfrac{5}{3}x$

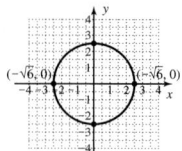

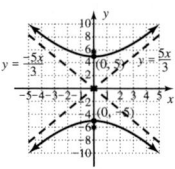

5. hyperbola; center $(1, -2)$;
$a = 6, b = 2, c = 2\sqrt{10}$
vertices $(-5, -2), (7, -2)$;
foci $(1 - 2\sqrt{10}, -2), (1 + 2\sqrt{10}, -2)$;
asymptotes: $y = \dfrac{1}{3}x - \dfrac{7}{3}$,
$y = -\dfrac{1}{3}x - \dfrac{5}{3}$

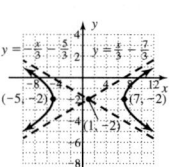

7. parabola; $p = 0.125$
vertex $(-2.5, 27.5)$;
focus $(-2.5, 27.375)$;
directrix $y = 27.625$;
y-intercepts: approx. $(-6.2, 0), (1.2, 0)$

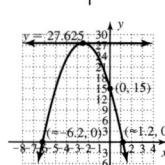

9. parabola; $p = 0.25$
vertex $(2, -1)$;
focus $(2.25, -1)$;
directrix $x = 1.75$;
y-intercepts: none

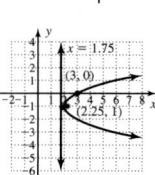

11. parabola; $p = 2$ vertex $(4, 0)$;
focus $(4, 2)$; directrix $y = -2$;
x-intercept $(4, 0)$

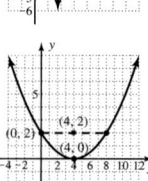

13. hyperbola; center $(3, 3)$;
$a = 5, b = 2, c = \sqrt{29}$
vertices $(8, 3), (-2, 3)$;
foci $(3 - \sqrt{29}, 3), (3 + \sqrt{29}, 3)$,
$\approx (-2.39, 3), (8.39, 3)$;
asymptotes $y = \dfrac{2}{5}x + \dfrac{9}{5}$,
$y = -\dfrac{2}{5}x + \dfrac{21}{5}$

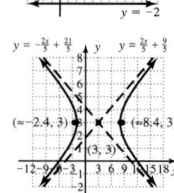

15. ellipse; center $(-2, 3)$;
$a = 1, b = 7, c = 4\sqrt{3}$
vertices $(-2, -4), (-2, 10)$;
endpoints of minor axis $(-3, 3), (-1, 3)$;
foci $(-2, 3 - 4\sqrt{3}), (-2, 3 + 4\sqrt{3})$,
$\approx (-2, -3.93), \approx (-2, 9.93)$

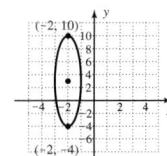

17. parabola

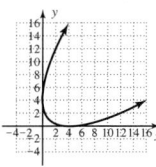

19. a. $(2, 5), (-2, 5), (-2, -5), (2, -5)$

b. $(0, 2), \left(-2\sqrt{2}, -\frac{2}{3}\right), \left(2\sqrt{2}, -\frac{2}{3}\right)$ **21.** $x = 50 \cos t;\ y = 30 \sin t$

23. a. elliptic; 0.494 million miles   **b.** parabolic; 3.1 million miles
25. $12x^2 + 26xy + 12y^2 = 160{,}000$

Practice Test, pp. 930–931

1. c   **2.** d   **3.** b   **4.** a
5. circle; center $(2, -5)$; radius 3

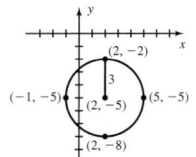

6. ellipse; center $(-2, 1)$; vertices
$(-2, -4), (-2, 6)$; foci $(-2, 1 - \sqrt{21})$,
$(-2, 1 + \sqrt{21})$

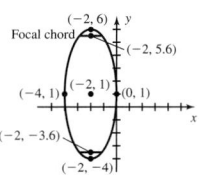

7. ellipse; center $(\frac{40}{9}, 0)$; vertices
$(\frac{-10}{9}, 0), (10, 0)$; foci $(0, 0), (\frac{80}{9}, 0)$

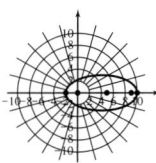

8. parabola; vertex $(-1.2, 0)$;
focus $(0, 0)$; directrix at $y = -2.4$

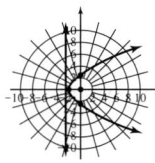

9. hyperbola; center: $(2, -3)$;
vertices: $(2, 0), (2, -6)$; foci:
$(2, -8), (2, 2)$; asymptotes:
$y = \frac{3}{4}x - \frac{9}{2}, y = \frac{-3}{4}x - \frac{3}{2}$

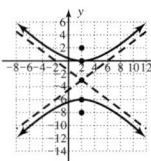

10. hyperbola; center $(1, -2)$;
vertices $(-4, -2), (6, -2)$;
foci $(1 - \sqrt{29}, -2), (1 + \sqrt{29}, -2)$,
$\approx (-4.39, -2), (6.39, -2)$;
asymptotes: $y = -\frac{2}{5}x - \frac{8}{5}$,
$y = \frac{2}{5}x - \frac{12}{5}$

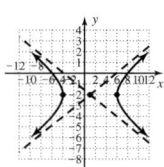

11. parabola; $\beta = 36.87°$; $\cos \beta = \frac{4}{5}$, $\sin \beta = \frac{3}{5}$
12. $Y = \frac{25}{16}X^2 - \frac{3}{4}X - \frac{11}{20}$

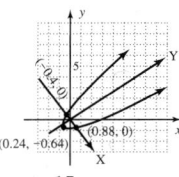

13.   **14.**   **15.**

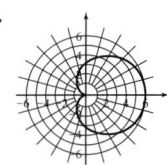   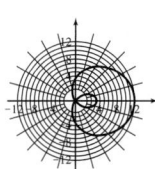   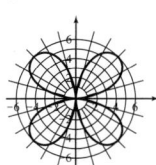

16. ellipse; $\dfrac{x^2}{16} + \dfrac{y^2}{25} = 1$ **17.** parabola; $x = (y - 5)^2 + 1$

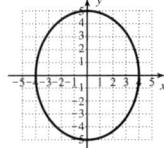   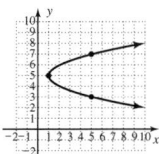

18. max: $y = 8$; min: $y = 0$; $P = 8\pi$

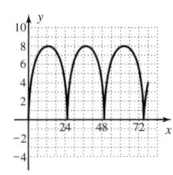

19. a. $(3\frac{1}{3}, 5\frac{1}{3}), (-2, 0)$ **b.** $\left(\sqrt{\dfrac{12}{5}}, \sqrt{\dfrac{8}{5}}\right), \left(-\sqrt{\dfrac{12}{5}}, -\sqrt{\dfrac{8}{5}}\right)$
$\left(-\sqrt{\dfrac{12}{5}}, \sqrt{\dfrac{8}{5}}\right), \left(\sqrt{\dfrac{12}{5}}, -\sqrt{\dfrac{8}{5}}\right)$ **20.** $r \approx \dfrac{1654(1 - 0.967^2)}{1 - 0.967 \cos \theta}$
e is very close to 1. This makes its orbit a very elongated ellipse, where the
orbit of most planets is nearly circular.   **21.** The ball is 0.43 ft above the
ground at $x = 165$ ft, and will likely go into the goal.   **22.** Perihelion:
128.41 million miles Aphelion: 154.89 million miles
23. $y = (x - 1)^2 - 4$; D: $x \in$ R; R: $y \in [-4, \infty)$; focus: $(1, -3.75)$
24. $(x - 1)^2 + (y - 1)^2 = 25$; D: $x \in [-4, 6]$; R: $y \in [-4, 6]$
25. $\dfrac{(x + 2)^2}{9} + \dfrac{(y - 1)^2}{25} = 1$; D: $x \in [-5, 1]$; R: $y \in [-4, 6]$

Strengthening Core Skills, pp. 932–934

Exercise 1:   Yes, the calculations are much "cleaner" with the right triangle definitions.

Cumulative Review Chapters 1–9, p. 934

1. $x = 7, x = -1$ is extraneous   **3.** $x = -6$ **5.** $x = 4$
7. $\dfrac{5\pi}{6} + k\pi, k \in$ Z
9. $x \approx 61.98° + 360°k; k \in$ Z
  $x \approx 118.02° + 360°k; k \in$ Z
11. about 24.7 pesos/kg   **13.** The formation is 1152.4 yd wide
15.   **17.** horizontal asymptote: $y = 0$
vertical asymptotes: $x = 3, x = -3$,
x-intercept: $(2, 0)$; y-intercept: $(0, \frac{2}{9})$

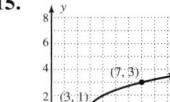   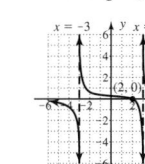

19.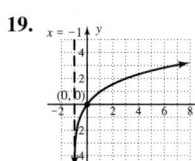

21. center $(1, -2)$;

foci $(1 + 2\sqrt{10}, -2) \approx (7.32, -2)$,

$(1 - 2\sqrt{10}, -2) \approx (-5.32, -2)$; asymptotes

$y = \frac{1}{3}x - \frac{7}{3}, y = -\frac{1}{3}x - \frac{5}{3}$

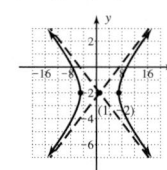

23.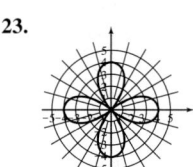

25. $61.9°$ **27.** $(-3, 4), (3, 4),$

$(-3, -4), (3, -4)$ **29.** $\frac{1}{x} - \frac{3}{x^2} + \frac{2x + 1}{x^2 + 1}$

Connections to Calculus Exercises, pp. 937–938

1. a. $-\dfrac{\cos(2\theta) + \cos\theta}{\sin(2\theta) + \sin\theta}$ **b.** $1, 1$ **c.** $\dfrac{\pi}{3}, \dfrac{5\pi}{3}$ **d.** $0, \dfrac{2\pi}{3}, \dfrac{4\pi}{3}$

3. a. $\dfrac{\cos(2\theta) - \sin(2\theta)}{\sin(2\theta) + \cos(2\theta)}$ **b.** $1, 1$ **c.** $\dfrac{\pi}{8}, \dfrac{5\pi}{8}, \dfrac{9\pi}{8}, \dfrac{13\pi}{8}$

d. $\dfrac{3\pi}{8}, \dfrac{7\pi}{8}, \dfrac{11\pi}{8}, \dfrac{15\pi}{8}$ **5.** $\dfrac{\pi}{4}, \dfrac{3\pi}{4}, \dfrac{5\pi}{4}, \dfrac{7\pi}{4}$; verified

7. four-leave rose; circle; $\dfrac{\pi}{6}, \dfrac{\pi}{2}, \dfrac{5\pi}{6}, \dfrac{3\pi}{2}$; verified   **9.** $(r, \theta): \left(0, \dfrac{\pi}{2}\right),$

$\left(2\sqrt{3}, \dfrac{\pi}{6}\right), \left(-2\sqrt{3}, \dfrac{5\pi}{6}\right), \left(0, \dfrac{3\pi}{2}\right)$; (x, y): $(0, 0), (3, \sqrt{3}), (3, -\sqrt{3})$

CHAPTER 10

Exercises 10.1, pp. 946–948

1. pattern; order   **3.** increasing   **5.** formula defining the sequence uses the preceding term(s); answers will vary.   **7.** $1, 3, 5, 7$; $a_8 = 15$; $a_{12} = 23$ **9.** $0, 9, 24, 45$; $a_8 = 189$; $a_{12} = 429$ **11.** $-1, 2, -3, 4$; $a_8 = 8$; $a_{12} = 12$ **13.** $\dfrac{1}{2}, \dfrac{2}{3}, \dfrac{3}{4}, \dfrac{4}{5}$; $a_8 = \dfrac{8}{9}$; $a_{12} = \dfrac{12}{13}$

15. $\dfrac{1}{2}, \dfrac{1}{4}, \dfrac{1}{8}, \dfrac{1}{16}$; $a_8 = \dfrac{1}{256}$; $a_{12} = \dfrac{1}{4096}$ **17.** $1, \dfrac{1}{2}, \dfrac{1}{3}, \dfrac{1}{4}$; $a_8 = \dfrac{1}{8}$; $a_{12} = \dfrac{1}{12}$

19. $\dfrac{-1}{2}, \dfrac{1}{6}, \dfrac{-1}{12}, \dfrac{1}{20}$; $a_8 = \dfrac{1}{72}$; $a_{12} = \dfrac{1}{156}$

21. $-2, 4, -8, 16$; $a_8 = 256$; $a_{12} = 4096$ **23.** 79 **25.** $\dfrac{1}{5}$ **27.** $\dfrac{1}{32}$

29. $\left(\dfrac{11}{10}\right)^{10}$ **31.** $\dfrac{1}{36}$ **33.** $2, 7, 32, 157, 782$ **35.** $-1, 4, 19, 364, 132, 499$

37. $64, 32, 16, 8, 4$ **39.** 336 **41.** 36 **43.** 28 **45.** $\dfrac{1}{2}, \dfrac{1}{3}, \dfrac{1}{4}, \dfrac{1}{5}$

47. $\dfrac{1}{3}, \dfrac{1}{120}, \dfrac{1}{15,120}, \dfrac{1}{3,991,680}$ **49.** $1, 2, \dfrac{9}{2}, \dfrac{32}{3}$ **51.** 15 **53.** 64 **55.** $\dfrac{137}{60}$

57. 10 **59.** 95 **61.** -4 **63.** 15 **65.** 50 **67.** $\dfrac{-27}{112}$ **69.** $\displaystyle\sum_{n=1}^{5} (4n)$

71. $\displaystyle\sum_{n=1}^{6} (-1)^n n^2$ **73.** $\displaystyle\sum_{n=1}^{5} (n + 3)$ **75.** $\displaystyle\sum_{n=1}^{3} \dfrac{n^2}{3}$ **77.** $\displaystyle\sum_{n=3}^{7} \dfrac{n}{2^n}$ **79.** 35

81. 100 **83.** 35 **85.** $a_n = 6000(0.8)^{n-1}$; $6000, 4800, 3840, 3072,$ $2457.60, 1966.08$ **87.** $5.20, 5.70, 6.20, 6.70, 7.20, \$13{,}824$ **89.** ≈ 2690

91. verified   **93.** approaches $\dfrac{1}{2}$ **95.** $\dfrac{3\pi}{4}, \dfrac{7\pi}{4}$

97. $\angle A \approx 53.1°, \angle B = 90°, \angle C \approx 36.9°$

Exercises 10.2, pp. 953–955

1. common; difference   **3.** $\dfrac{n(a_1 + a_n)}{2}$; nth   **5.** Answers will vary.

7. arithmetic; $d = 3$ **9.** arithmetic; $d = 2.5$

11. not arithmetic; all prime   **13.** arithmetic; $d = \dfrac{1}{24}$

15. not arithmetic; $a_n = n^2$ **17.** arithmetic; $d = \dfrac{-\pi}{6}$ **19.** $2, 5, 8, 11$

21. $7, 5, 3, 1$ **23.** $0.3, 0.33, 0.36, 0.39$ **25.** $\dfrac{3}{2}, 2, \dfrac{5}{2}, 3$ **27.** $\dfrac{3}{4}, \dfrac{5}{8}, \dfrac{1}{2}, \dfrac{3}{8}$

29. $-2, -5, -8, -11$ **31.** $a_1 = 2, d = 5, a_n = 5n - 3,$ $a_6 = 27, a_{10} = 47, a_{12} = 57$ **33.** $a_1 = 5.10, d = 0.15,$ $a_n = 0.15n + 4.95, a_6 = 5.85, a_{10} = 6.45, a_{12} = 6.75$

35. $a_1 = \dfrac{3}{2}, d = \dfrac{3}{4}, a_n = \dfrac{3}{4}n + \dfrac{3}{4}, a_6 = \dfrac{21}{4}, a_{10} = \dfrac{33}{4}, a_{12} = \dfrac{39}{4}$

37. 61 **39.** 1 **41.** 2.425 **43.** 9 **45.** 43 **47.** 21 **49.** 26

51. $d = 3, a_1 = 1$ **53.** $d = 0.375, a_1 = 0.65$ **55.** $d = \dfrac{115}{126}, a_1 = \dfrac{-472}{63}$

57. 1275 **59.** 601.25 **61.** -534 **63.** 82.5 **65.** 74.04 **67.** $210\sqrt{2}$

69. $S_6 = 21; S_{75} = 2850$ **71.** at 11 P.M.   **73.** 5.5 in.; 54.25 in.

75. $220; 2520$; yes   **77. a.** linear function   **b.** quadratic

79. $A = 7, P = 6$, HS: $\dfrac{1}{2}$ unit right, VS: 10 units up, PI: $\dfrac{1}{2} \le t < \dfrac{13}{2}$.

81. $f(x) = 4ax + 972, 1364$

Exercises 10.3, pp. 962–965

1. multiplying   **3.** $a_1 r^{n-1}$ **5.** Answers will vary.   **7.** $r = 2$

9. $r = -2$ **11.** $a_n = n^2 + 1$ **13.** $r = 0.1$ **15.** not geometric; ratio of terms decreases by 1   **17.** $r = \dfrac{2}{5}$ **19.** $r = \dfrac{1}{2}$ **21.** $r = \dfrac{4}{x}$

23. not geometric; $a_n = \dfrac{240}{n!}$ **25.** $5, 10, 20, 40$ **27.** $-6, 3, \dfrac{-3}{2}, \dfrac{3}{4}$

29. $4, 4\sqrt{3}, 12, 12\sqrt{3}$ **31.** $0.1, 0.01, 0.001, 0.0001$ **33.** $-\dfrac{3}{8}$ **35.** $\dfrac{25}{4}$

37. 16 **39.** $a_1 = \dfrac{1}{27}, r = -3, a_n = \dfrac{1}{27}(-3)^{n-1}, a_6 = -9, a_{10} = -729,$ $a_{12} = -6561$

41. $a_1 = 729, r = \dfrac{1}{3}, a_n = 729(\frac{1}{3})^{n-1}, a_6 = 3, a_{10} = \dfrac{1}{27}, a_{12} = \dfrac{1}{243}$

43. $a_1 = \dfrac{1}{2}, r = \sqrt{2}, a_n = \dfrac{1}{2}(\sqrt{2})^{n-1}, a_6 = 2\sqrt{2}, a_{10} = 8\sqrt{2},$ $a_{12} = 16\sqrt{2}$ **45.** $a_1 = 0.2, r = 0.4, a_n = 0.2(0.4)^{n-1}$ $a_6 = 0.002048, a_{10} = 0.0000524288, a_{12} = 0.000008388608$ **47.** 5

49. 11 **51.** 9 **53.** 8 **55.** 13 **57.** 9 **59.** $r = \dfrac{2}{3}, a_1 = 729$

61. $r = \dfrac{3}{2}, a_1 = \dfrac{32}{243}$ **63.** $r = \dfrac{3}{2}, a_1 = \dfrac{256}{81}$ **65.** $-10,920$

67. $\dfrac{3872}{27} \approx 143.41$ **69.** $\dfrac{2059}{8} = 257.375$ **71.** 728 **73.** $\dfrac{85}{8} = 10.625$

75. ≈ 1.60 **77.** 1364 **79.** $\dfrac{31{,}525}{2187} \approx 14.41$ **81.** $\dfrac{-387}{512} \approx -0.76$

83. $\dfrac{521}{25}$ **85.** $\dfrac{3367}{1296}$ **87.** $14 + 15\sqrt{2}$ **89.** no   **91.** $\dfrac{27}{7}$ **93.** $\dfrac{125}{3}$

95. 12 **97.** 4 **99.** $\dfrac{1}{3}$ **101.** $\dfrac{3}{2}$ **103.** $-\dfrac{18}{5}$ **105.** 1296

107. about 6.3 ft; 120 ft   **109.** $\$18{,}841.60$; 10 yr

111. 125.4 gpm; 10 months   **113.** about 347.7 million

115. 51,200 bacteria; 12 half-hours later (6 hr)   **117.** ≈ 0.42 m; 8 m

119. 35.9 in^3; 7 strokes   **121.** 6 yr   **123.** $S_n = \log n!$

125. $x = \dfrac{-5}{2} \pm \dfrac{\sqrt{11}}{2}i$ **127.**

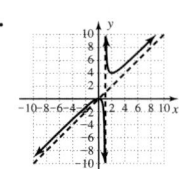

Exercises 10.4, pp. 971–973

1. finite; universally   **3.** induction; hypothesis   **5.** Answers will vary.

7. $a_n = 10n - 6$
$a_4 = 10(4) - 6 = 40 - 6 = 34;$
$a_5 = 10(5) - 6 = 50 - 6 = 44;$
$a_k = 10k - 6;$
$a_{k+1} = 10(k + 1) - 6 = 10k + 10 - 6 = 10k + 4$

9. $a_n = n$
$a_4 = 4;$
$a_5 = 5;$
$a_k = k;$
$a_{k+1} = k + 1$

11. $a_n = 2^{n-1}$
$a_4 = 2^{4-1} = 2^3 = 8;$
$a_5 = 2^{5-1} = 2^4 = 16;$
$a_k = 2^{k-1};$
$a_{k+1} = 2^{k+1-1} = 2^k$

13. $S_n = n(5n - 1)$
$S_4 = 4(5(4) - 1) = 4(20 - 1) = 4(19) = 76;$
$S_5 = 5(5(5) - 1) = 5(25 - 1) = 5(24) = 120;$
$S_k = k(5k - 1);$
$S_{k+1} = (k + 1)(5(k + 1) - 1) = (k + 1)(5k + 5 - 1) = (k + 1)(5k + 4)$

15. $S_n = \dfrac{n(n + 1)}{2}$

$S_4 = \dfrac{4(4 + 1)}{2} = \dfrac{4(5)}{2} = 10;$

$S_5 = \dfrac{5(5 + 1)}{2} = \dfrac{5(6)}{2} = 15;$

$S_k = \dfrac{k(k + 1)}{2};$

$S_{k+1} = \dfrac{(k + 1)(k + 1 + 1)}{2} = \dfrac{(k + 1)(k + 2)}{2}$

17. $S_n = 2^n - 1$
$S_4 = 2^4 - 1 = 16 - 1 = 15;$
$S_5 = 2^5 - 1 = 32 - 1 = 31;$
$S_k = 2^k - 1;$
$S_{k+1} = 2^{k+1} - 1$

19. $a_n = 10n - 6; S_n = n(5n - 1)$
$S_4 = 4(5(4) - 1) = 4(20 - 1) = 4(19) = 76;$
$a_5 = 10(5) - 6 = 50 - 6 = 44;$
$S_5 = 5(5(5) - 1) = 5(25 - 1) = 5(24) = 120;$
$S_4 + a_5 = S_5$
$76 + 44 = 120$
$120 = 120$ Verified

21. $a_n = n; S_n = \dfrac{n(n + 1)}{2}$

$S_4 = \dfrac{4(4 + 1)}{2} = \dfrac{4(5)}{2} = 10;$

$a_5 = 5;$

$S_5 = \dfrac{5(5 + 1)}{2} = \dfrac{5(6)}{2} = 15;$

$S_4 + a_5 = S_5$
$10 + 5 = 15$
$15 = 15$ Verified

23. $a_n = 2^{n-1}; S_n = 2^n - 1$
$S_4 = 2^4 - 1 = 16 - 1 = 15;$
$a_5 = 2^{5-1} = 2^4 = 16;$
$S_5 = 2^5 - 1 = 32 - 1 = 31;$
$S_4 + a_5 = S_5$
$15 + 16 = 31$
$31 = 31$ Verified

25. $a_n = n^3; S_n = (1 + 2 + 3 + 4 + \cdots + n)^2$
$S_1 = 1^2 = 1^3$
$S_5 = (1 + 2 + 3 + 4 + 5)^2$
$= 15^2$
$= 225$
$1 + 8 + 27 + 64 + 125 = 225$
$S_9 = (1 + 2 + \cdots + 9)^2$
$= 45^2$
$= 2025$
$1 + 8 + \cdots + 729 = 2025$
$\left[\dfrac{n(n + 1)}{2}\right]^2 = \dfrac{n^2(n + 1)^2}{4}$

27. 1. Show S_n is true for $n = 1.$
$S_1 = 1(1 + 1) = 1(2) = 2$ Verified
2. Assume S_k is true: $2 + 4 + 6 + 8 + 10 + \cdots + 2k = k(k + 1)$
and use it to show the truth of S_{k+1} follows. That is:
$2 + 4 + 6 + \cdots + 2k + 2(k + 1) = (k + 1)(k + 2)$
$S_k + a_{k+1} = S_{k+1}$
Working with the left hand side:
$2 + 4 + 6 + \cdots + 2k + 2(k + 1)$
$= k(k + 1) + 2(k + 1)$
$= k^2 + k + 2k + 2$
$= k^2 + 3k + 2$
$= (k + 1)(k + 2)$
$= S_{k+1}$
Since the truth of S_{k+1} follows from S_k, the formula is true for all n.

29. 1. Show S_n is true for $n = 1.$
$S_1 = \dfrac{5(1)(1 + 1)}{2} = \dfrac{5(2)}{2} = 5$ Verified
2. Assume S_k is true:
$5 + 10 + 15 + \cdots + 5k = \dfrac{5k(k + 1)}{2}$
and use it to show the truth of S_{k+1} follows. That is:
$5 + 10 + 15 + \cdots + 5k + 5(k + 1) = \dfrac{5(k + 1)(k + 1 + 1)}{2}$
$S_k + a_{k+1} = S_{k+1}$
Working with the left hand side:
$5 + 10 + 15 + \cdots + 5k + 5(k + 1)$
$= \dfrac{5k(k + 1)}{2} + 5(k + 1)$
$= \dfrac{5k(k + 1) + 10(k + 1)}{2}$
$= \dfrac{(k + 1)(5k + 10)}{2}$
$= \dfrac{5(k + 1)(k + 2)}{2}$
$= S_{k+1}$
Since the truth of S_{k+1} follows from S_k, the formula is true for all n.

31. 1. Show S_n is true for $n = 1.$
$S_1 = 1(2(1) + 3) = 5$ Verified
2. Assume S_k is true:
$5 + 9 + 13 + 17 + \cdots + 4k + 1 = k(2k + 3)$
and use it to show the truth of S_{k+1} follows. That is:
$5 + 9 + 13 + 17 + \cdots + 4k + 1 + 4(k + 1) + 1$
$= (k + 1)(2(k + 1) + 3)$
$S_k + a_{k+1} = S_{k+1}$ Working with the left hand side:
$5 + 9 + 13 + 17 + \cdots + 4k + 1 + 4k + 5$
$= k(2k + 3) + 4k + 5$
$= 2k^2 + 3k + 4k + 5 = 2k^2 + 7k + 5$
$= (k + 1)(2k + 5) = S_{k+1}$
Since the truth of S_{k+1} follows from S_k, the formula is true for all n.

33. 1. Show S_n is true for $n = 1.$
$S_1 = \dfrac{3(3^1 - 1)}{2} = \dfrac{3(3 - 1)}{2} = \dfrac{3(2)}{2} = 3$ Verified
2. Assume S_k is true:
$3 + 9 + 27 + \cdots + 3^k = \dfrac{3(3^k - 1)}{2}$
and use it to show the truth of S_{k+1} follows. That is:
$3 + 9 + 27 + \cdots + 3^k + 3^{k+1}$
$= \dfrac{3(3^{k+1} - 1)}{2}$
$S_k + a_{k+1} = S_{k+1}$
Working with the left hand side:
$3 + 9 + 27 + \cdots + 3^k + 3^{k+1}$
$= \dfrac{3(3^k - 1)}{2} + 3^{k+1}$
$= \dfrac{3(3^k - 1) + 2(3^{k+1})}{2}$
$= \dfrac{3^{k+1} - 3 + 2(3^{k+1})}{2}$
$= \dfrac{3(3^{k+1}) - 3}{2}$
$= \dfrac{3(3^{k+1} - 1)}{2}$
$= S_{k+1}$
Since the truth of S_{k+1} follows from S_k, the formula is true for all n.

35. 1. Show S_n is true for $n = 1.$
$S_n = 2^{n+1} - 2$
$S_1 = 2^{1+1} - 2 = 2^2 - 2 = 4 - 2 = 2$ Verified

2. Assume S_k is true:
$$2 + 4 + 8 + \cdots + 2^k = 2^{k+1} - 2$$
and use it to show the truth of S_{k+1} follows. That is:
$$2 + 4 + 8 + \cdots + 2^k + 2^{k+1} = 2^{k+1} - 2$$
$$S_k + a_{k+1} = S_{k+1}$$
Working with the left hand side:
$$2 + 4 + 8 + \cdots + 2^k + 2^{k+1}$$
$$= 2^{k+1} - 2 + 2^{k+1}$$
$$= 2(2^{k+1}) - 2$$
$$= 2^{k+2} - 2$$
$$= S_{k+1}$$
Since the truth of S_{k+1} follows from S_k, the formula is true for all n.

37. 1. Show S_n is true for $n = 1$.
$$S_n = \frac{n}{2n + 1}$$
$$S_1 = \frac{1}{2(1) + 1} = \frac{1}{2 + 1} = \frac{1}{3} \quad \text{Verified}$$
2. Assume S_k is true:
$$\frac{1}{3} + \frac{1}{15} + \frac{1}{35} + \cdots + \frac{1}{(2k - 1)(2k + 1)} = \frac{k}{2k + 1}$$
and use it to show the truth of S_{k+1} follows. That is:
$$\frac{1}{3} + \frac{1}{15} + \frac{1}{35} + \cdots + \frac{1}{(2k - 1)(2k + 1)}$$
$$+ \frac{1}{(2(k + 1) - 1)(2(k + 1) + 1)} = \frac{k + 1}{2(k + 1) + 1}$$
$$S_k + a_{k+1} = S_{k+1}$$
Working with the left hand side:
$$\frac{1}{3} + \frac{1}{15} + \frac{1}{35} + \cdots + \frac{1}{(2k - 1)(2k + 1)} + \frac{1}{(2k + 1)(2k + 3)}$$
$$= \frac{k}{2k + 1} + \frac{1}{(2k + 1)(2k + 3)}$$
$$= \frac{k(2k + 3) + 1}{(2k + 1)(2k + 3)}$$
$$= \frac{2k^2 + 3k + 1}{(2k + 1)(2k + 3)}$$
$$= \frac{(2k + 1)(k + 1)}{(2k + 1)(2k + 3)}$$
$$= \frac{k + 1}{2k + 3}$$
$$= S_{k+1}$$
Since the truth of S_{k+1} follows from S_k, the formula is true for all n.

39. 1. Show S_n is true for $n = 1$.
$$S_1 : 3^1 \geq 2(1) + 1$$
$$3 \geq 2 + 1$$
$$3 \geq 3 \quad \text{Verified}$$
2. Assume $S_k : 3^k \geq 2k + 1$ is true and use it to show the truth of S_{k+1} follows. That is: $3^{k+1} \geq 2k + 3$.
Working with the left hand side:
$$3^{k+1} = 3(3^k)$$
$$\geq 3(2k + 1)$$
$$\geq 6k + 3$$
Since k is a positive integer, $6k + 3 \geq 2k + 3$
Showing $S_{k+1} : 3^{k+1} \geq 2k + 3 \quad \text{Verified}$

41. 1. Show S_n is true for $n = 1$.
$$S_1 : 3 \cdot 4^{1-1} \leq 4^1 - 1$$
$$3 \cdot 4^0 \leq 4 - 1$$
$$3 \cdot 1 \leq 3$$
$$3 \leq 3 \quad \text{Verified}$$
2. Assume $S_k : 3 \cdot 4^{k-1} \leq 4^k - 1$ is true and use it to show the truth of S_{k+1} follows. That is: $3 \cdot 4^k \leq 4^{k+1} - 1$.
Working with the left hand side:
$$3 \cdot 4^k = 3 \cdot 4(4^{k-1})$$
$$= 4 \cdot 3(4^{k-1})$$
$$\leq 4(4^k - 1)$$
$$\leq 4^{k+1} - 4$$
Since k is a positive integer, $4^{k+1} - 4 \leq 4^{k+1} - 1$
Showing that $3 \cdot 4^k \leq 4^{k+1} - 1$

43. $n^2 - 7n$ is divisible by 2
1. Show S_n is true for $n = 1$.
$$S_n : n^2 - 7n = 2m$$
$$S_1 : (1)^2 - 7(1) = 2m$$
$$1 - 7 = 2m$$
$$-6 = 2m \quad \text{Verified}$$
2. Assume $S_k : k^2 - 7k = 2m$ for $m \in Z$ and use it to show the truth of S_{k+1} follows. That is: $(k + 1)^2 - 7(k + 1) = 2p$ for $p \in Z$.
Working with the left hand side:
$$= (k + 1)^2 - 7(k + 1)$$
$$= k^2 + 2k + 1 - 7k - 7$$
$$= k^2 - 7k + 2k - 6$$
$$= 2m + 2k - 6$$
$$= 2(m + k - 3) \quad \text{is divisible by 2.}$$

45. $n^3 + 3n^2 + 2n$ is divisible by 3
1. Show S_n is true for $n = 1$. $S_1 : n^3 + 3n^2 + 2n = 3m$
$$S_1 : (1)^3 + 3(1)^2 + 2(1) = 3m$$
$$1 + 3 + 2 = 3m$$
$$6 = 3m$$
$$2 = m \quad \text{Verified}$$
2. Assume $S_k : k^3 + 3k^2 + 2k = 3m$ for $m \in Z$ and use it to show the truth of S_{k+1} follows.
That is: $S_{k+1} : (k + 1)^3 + 3(k + 1)^2 + 2(k + 1) = 3p$ for $p \in Z$.
Working with the left hand side:
$(k + 1)^3 + 3(k + 1)^2 + 2(k + 1)$ is true.
$$= k^3 + 3k^2 + 3k + 1 + 3(k^2 + 2k + 1) + 2k + 2$$
$$= k^3 + 3k^2 + 2k + 3(k^2 + 2k + 1) + 3k + 3$$
$$= k^3 + 3k^2 + 2k + 3(k^2 + 2k + 1) + 3(k + 1)$$
$$= 3m + 3(k^2 + 2k + 1) + 3(k + 1) \text{ is divisible by 3.}$$

47. $6^n - 1$ is divisible by 5
1. Show S_n is true for $n = 1$. $S_n : 6^n - 1 = 5m$
$$S_1 : 6^1 - 1 = 5m$$
$$6 - 1 = 5m$$
$$5 = 5m$$
$$1 = m \quad \text{Verified}$$
2. Assume $S_k : 6^k - 1 = 5m$ for $m \in Z$ and use it to show the truth of S_{k+1} follows.
That is: $S_{k+1} : 6^{k+1} - 1 = 5p$ for $p \in Z$.
Working with the left hand side:
$$= 6^k - 1$$
$$= 6(6^k) - 1$$
$$= 6(5m + 1) - 1$$
$$= 30m + 6 - 1$$
$$= 30m + 5$$
$$= 5(6m + 1) \quad \text{is divisible by 5, Verified}$$

49. Verified **51.** Verified **53.** $(x - 3)^2 + (x - 4)^2 = 25$

Mid-Chapter Check, pp. 973–974

1. 3, 10, 17, $a_9 = 59$ **2.** 4, 7, 12, $a_9 = 84$ **3.** $-1, 3, -5, a_9 = -17$

4. 360 **5.** $\sum_{k=1}^{6}(3k - 2)$ **6.** d **7.** e **8.** a **9.** b **10.** c

11. a. $a_1 = 2, d = 3, a_n = 3n - 1$ **b.** $a_1 = \frac{3}{2}, d = \frac{3}{4}, a_n = \frac{3}{4}n + \frac{3}{4}$
12. $n = 25, S_{25} = 950$ **13.** $n = 16, S_{16} = 128$ **14.** $S_{10} = -5$
15. $S_{10} = \frac{-29{,}524}{27}$ **16. a.** $a_1 = 2, r = 3, a_n = 2(3)^{n-1}$
b. $a_1 = \frac{1}{2}, r = \frac{1}{2}, a_n = (\frac{1}{2})^n$ **17.** $n = 8, S_8 = \frac{1640}{27}$ **18.** $\frac{-343}{6}$ **19.** 1785
20. ≈ 4.5 ft; ≈ 127.9 ft

Reinforcing Basic Concepts, p. 974

Exercise 1: $71,500

Exercises 10.5, pp. 982–986

1. experiment; well-defined **3.** distinguishable **5.** Answers will vary.
7. a. 16 possible

b. WW, WX, WY, WZ, XW, XX, XY, XZ, YW, YX, YY, YZ, ZW, ZX, ZY, ZZ
9. 32   **11.** 15,625   **13.** 2,704,000   **15. a.** 59,049   **b.** 15,120
17. 360 if double veggies are not allowed, 432 if double veggies are allowed.   **19. a.** 120   **b.** 625   **c.** 12   **21.** 24   **23.** 4   **25.** 120
27. 6   **29.** 720   **31.** 3024   **33.** 40,320   **35.** 6; 3   **37.** 90   **39.** 336
41. a. 720   **b.** 120   **c.** 24   **43.** 360   **45.** 60   **47.** 60   **49.** 120
51. 30   **53.** 60, BANANA   **55.** 126   **57.** 56   **59.** 1   **61.** verified
63. verified   **65.** 495   **67.** 364   **69.** 252   **71.** 40,320   **73.** 336
75. 15,504   **77.** 70   **79. a.** $\approx 1.2\%$ **b.** $\approx 0.83\%$ **81.** 7776   **83.** 324
85. 800   **87.** 6,272,000,000   **89.** 518,400   **91.** 357,696   **93.** 6720
95. 8   **97.** 10,080   **99.** 5040   **101.** 2880   **103.** 5005   **105.** 720
107. 52,650, no   **109. a.** $\dfrac{10!}{2!3!5!}$ **b.** $\dfrac{9!}{2!3!4!}$ **c.** $\dfrac{11!}{4!5!2!}$ **d.** $\dfrac{8!}{2!3!3!}$
111.

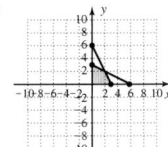

113. $\cos(5\alpha)$

Exercises 10.6, pp. 993–999

1. $n(E)$ **3.** 0; 1; 1; 0   **5.** Answers will vary.
7. $S = \{HH, HT, TH, TT\}, \frac{1}{4}$
9. $S = \{$coach of Patriots, Cougars, Angels, Sharks, Eagles, Stars$\}, \frac{1}{6}$
11. $P(E) = \frac{4}{9}$ **13. a.** $\frac{1}{13}$ **b.** $\frac{1}{4}$ **c.** $\frac{1}{2}$ **d.** $\frac{1}{26}$
15. $P(E_1) = \frac{1}{8}, P(E_2) = \frac{5}{8}, P(E_3) = \frac{3}{4}$ **17. a.** $\frac{3}{4}$ **b.** 1   **c.** $\frac{1}{4}$ **d.** $\frac{1}{2}$
19. $\frac{3}{4}$ **21.** $\frac{6}{7}$ **23.** 0.991   **25. a.** $\frac{1}{12}$ **b.** $\frac{11}{12}$ **c.** $\frac{8}{9}$ **d.** $\frac{5}{6}$ **27.** $\frac{10}{21}$
29. $\frac{60}{143}$ **31.** b, about 12%   **33. a.** 0.3651   **b.** 0.3651   **c.** 0.3969
35. 0.9   **37.** $\frac{7}{24}$ **39.** 0.59   **41. a.** $\frac{1}{6}$ **b.** $\frac{7}{36}$ **c.** $\frac{1}{9}$ **d.** $\frac{4}{9}$
43. a. $\frac{2}{25}$ **b.** $\frac{9}{50}$ **c.** 0   **d.** $\frac{2}{25}$ **e.** 1   **45.** $\frac{3}{4}$ **47.** $\frac{11}{15}$
49. a. $\frac{1}{18}$ **b.** $\frac{2}{9}$ **c.** $\frac{8}{9}$ **d.** $\frac{3}{4}$ **e.** $\frac{1}{36}$ **f.** $\frac{5}{12}$ **51.** $\frac{1}{4}; \frac{1}{256}$; answers will vary.
53. a. 0.33   **b.** 0.67   **c.** 1   **d.** 0   **e.** 0.67   **f.** 0.08
55. a. $\frac{1}{2}$ **b.** $\frac{1}{2}$ **c.** 0.2165   **57. a.** $\frac{9}{16}$ **b.** $\frac{1}{4}$ **c.** $\frac{1}{16}$ **d.** $\frac{5}{16}$
59. a. $\frac{3}{26}$ **b.** $\frac{3}{26}$ **c.** $\frac{1}{13}$ **d.** $\frac{9}{26}$ **e.** $\frac{2}{13}$ **f.** $\frac{11}{26}$ **61. a.** $\frac{1}{8}$ **b.** $\frac{1}{16}$ **c.** $\frac{3}{16}$
63. a. $\frac{47}{100}$ **b.** $\frac{2}{25}$ **c.** $\frac{1}{100}$ **d.** $\frac{9}{50}$ **e.** $\frac{11}{100}$ **65. a.** $\frac{5}{429}$ **b.** $\frac{8}{2145}$ **67.** $\frac{1}{3360}$
69. $\frac{1}{1,048,576}$; answers will vary; 20 heads in a row.
71. $\sin\theta = \dfrac{1}{3}, \cos\theta = -\dfrac{2\sqrt{2}}{3}, \tan\theta = -\dfrac{1}{2\sqrt{2}}, \sec\theta = -\dfrac{3}{2\sqrt{2}},$
$\cot\theta = -2\sqrt{2}$ **73.** $\sin(2\theta) = -\dfrac{840}{841}, \cos(2\theta) = \dfrac{41}{841}, \tan(2\theta) = -\dfrac{840}{41}$

Exercises 10.7, pp. 1005–1006

1. one   **3.** $(a + (-2b))^5$ **5.** Answers will vary.
7. $x^5 + 5x^4y + 10x^3y^2 + 10x^2y^3 + 5xy^4 + y^5$
9. $16x^4 + 96x^3 + 216x^2 + 216x + 81$ **11.** $41 + 38i$ **13.** 35   **15.** 10
17. 1140   **19.** 9880   **21.** 1   **23.** 1
25. $c^5 + 5c^4d + 10c^3d^2 + 10c^2d^3 + 5cd^4 + d^5$
27. $a^6 - 6a^5b + 15a^4b^2 - 20a^3b^3 + 15a^2b^4 - 6ab^5 + b^6$
29. $16x^4 - 96x^3 + 216x^2 - 216x + 81$
31. $-11 + 2i$ **33.** $x^9 + 18x^8y + 144x^7y^2 + \cdots$
35. $v^{24} - 6v^{22}w + \frac{33}{2}v^{20}w^2 - \cdots$ **37.** $35x^4y^3$ **39.** $1792p^2$ **41.** $264x^2y^{10}$
43. ≈ 0.25 **45 a.** $\approx 17.8\%$ **b.** $\approx 23.0\%$ **47 a.** $\approx 0.88\%$
b. $\approx 6.9\%$ **c.** $\approx 99.0\%$ **d.** $\approx 61.0\%$ **49. a.** 99.33%   **b.** 94.22%
51. $\binom{6}{6} = \binom{6}{0} = 1; \binom{6}{5} = \binom{6}{1} = 6; \binom{6}{4} = \binom{6}{2} = 15; \binom{6}{3} = 20$
53.

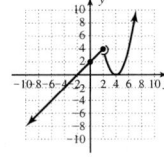

$f(3) = 1$

55. $g(x) > 0: x \in (-2, 0) \cup (3, \infty)$

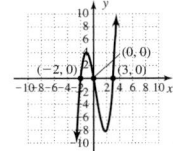

Summary and Concept Review, pp. 1007–1011

1. 1, 6, 11, 16; $a_{10} = 46$ **2.** $1, \frac{3}{5}, \frac{2}{5}, \frac{5}{17}; a_{10} = \frac{11}{101}$
3. $a_n = n^4; a_6 = 1296$ **4.** $a_n = -17 + (n-1)(3); a_6 = -2$ **5.** $\frac{255}{256}$
6. -112 **7.** 140   **8.** 35   **9.** not defined, 2, 6, 12, 20, 30
10. $\frac{1}{2}, \frac{3}{4}, \frac{5}{4}, \frac{9}{4}, \frac{17}{4}$ **11.** $\displaystyle\sum_{n=1}^{7}(n^2 + 3n - 2); 210$
12. $a_n = 2 + 3(n-1); 119$ **13.** $a_n = 3 + (-2)(n-1); -65$
14. 740   **15.** 1335   **16.** 630   **17.** -11.25 **18.** 875   **19.** 3240
20. 3645   **21.** 32   **22.** 2401   **23.** 10.75   **24.** 6560   **25.** $\frac{819}{512}$
26. does not exist   **27.** $\frac{50}{9}$ **28.** 4   **29.** $\frac{63,050}{6561}$ **30.** does not exist
31. 5   **32.** $a_9 = \$36,980; S_9 = \$314,900$ **33.** ≈ 7111.1 ft^3
34. $a_9 \approx 2105$ credit hrs; $S_9 \approx 14,673$ credit hours   **35.** verified
(1) Show S_n is true for $n = 1$: $S_1 = \dfrac{1(1+1)}{2} = 1$ ✔
(2) Assume S_k is true:
$$1 + 2 + 3 + \cdots + k = \dfrac{k(k+1)}{2}$$
Use it to show the truth of S_{k+1}:
$$1 + 2 + 3 + \cdots + k + (k+1) = \dfrac{(k+1)(k+2)}{2}$$
left-hand side: $1 + 2 + 3 + \cdots + k + (k+1)$
$$= \dfrac{\mathbf{k(k+1)}}{\mathbf{2}} + \dfrac{2(k+1)}{2} = \dfrac{k(k+1) + 2(k+1)}{2}$$
$$= \dfrac{(k+1)(k+2)}{2}$$

36. verified
(1) Show S_n is true $n = 1$: $S_1 = \dfrac{1[2(1) + 1](1+1)}{6} = 1$ ✔
(2) Assume S_k is true:
$$1 + 4 + 9 + \cdots + k^2 = \dfrac{k(2k+1)(k+1)}{6}$$
Use it to show the truth of S_{k+1}:
$$1 + 4 + 9 + \cdots + k^2 + (k+1)^2 = \dfrac{(k+1)(2k+3)(k+2)}{6}$$
left-hand side: $1 + 4 + 9 + \cdots + k^2 + (k+1)^2$
$$= \dfrac{\mathbf{k(k+1)(2k+1)}}{\mathbf{6}} + \dfrac{6(k+1)^2}{6} = \dfrac{(k+1)[(2k^2 + k + 6k + 6]}{6}$$
$$= \dfrac{(k+1)(2k^2 + 7k + 6)}{6} = \dfrac{(k+1)(2k+3)(k+2)}{6}$$

37. verified
(1) Show S_n is true for $n = 1$: $S_1: 4^1 \geq 3(1) + 1$ ✔
(2) Assume S_k is true: $4^k \geq 3k + 1$
Use it to show the truth of S_{k+1}:
$4^{k+1} \geq 3(k+1) + 1 = 3k + 4$
left-hand side: $4^{k+1} = 4(4^k)$
$$\geq 4(\mathbf{3k+1}) = 12k + 4$$
Since k is a positive integer, $12k + 4 \geq 3k + 4$ showing
$4^{k+1} \geq 3k + 4$

38. verified
(1) Show S_n is true for $n = 1$: $S_1: 6 \cdot 7^{1-1} \leq 7^1 - 1$ ✔
(2) Assume S_k is true: $6 \cdot 7^{k-1} \leq 7^k - 1$
Use it to show the truth of S_{k+1}:
$6 \cdot 7^k \leq 7^{k+1} - 1$
left-hand side: $6 \cdot 7^k = 7 \cdot 6 \cdot 7^{k-1}$
$$\leq 7 \cdot \mathbf{7^k} - \mathbf{1}$$
$$\leq 7^{k+1} - 1$$

39. verified
(1) Show S_n is true for $n = 1$: $S_1: 3^1 - 1 = 2$ or $2(1)$ ✔
(2) Assume S_k is true: $3^k - 1 = 2p$ for $p \in \mathbb{Z}$
Use it to show the truth of S_{k+1}:
$3^{k+1} - 1 = 2q$ for $q \in \mathbb{Z}$
left-hand side: $3^{k+1} - 1 = 3 \cdot 3^k - 1$
$$= 3 \cdot \mathbf{2p}$$
$$= 2(3p) = 2q \text{ is divisible by 2}$$

40. 6 ways

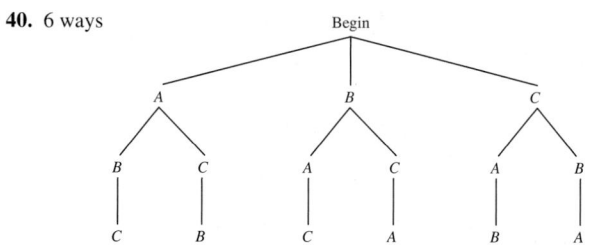

41. 720; 1000   **42.** 24   **43.** 220   **44.** 32   **45. a.** 5040   **b.** 840
c. 35   **46. a.** 720   **b.** 120   **c.** 24   **47.** 3360   **48. a.** 220   **b.** 1320
49. $\frac{4}{13}$ **50.** $\frac{3}{13}$ **51.** $\frac{5}{6}$ **52.** $\frac{7}{24}$ **53.** $\frac{175}{396}$ **54. a.** 0.608   **b.** 0.392
c. 1   **d.** 0   **e.** 0.928   **f.** 0.178   **55. a.** 21   **b.** 56
56. a. $x^4 - 4x^3y + 6x^2y^2 - 4xy^3 + y^4$ **b.** $41 - 38i$
57. a. $a^8 + 8\sqrt{3}a^7 + 84a^6 + 168\sqrt{3}a^5$
b. $78{,}125a^7 + 218{,}750a^6b + 262{,}500a^5b^2 + 175{,}000a^4b^3$
58. a. $280x^4y^3$ **b.** $-64{,}064a^5b^9$

Mixed Review, pp. 1011–1012

1. a. arithmetic   **b.** $a_n = 4$ **c.** $a_n = n!$ **d.** arithmetic   **e.** geometric
f. geometric   **g.** arithmetic   **h.** geometric   **i.** $a_n = \dfrac{1}{2n}$ **3.** 27,600
5. 0.1, 0.5, 2.5, 12.5, 62.5; $a_{15} = 610{,}351{,}562.5$ **7.** $\frac{5}{6}$ **9. a.** 2   **b.** 200
c. 210   **11. a.** $a^{20} + 20a^{19}b + 190a^{18}b^2$ **b.** $190a^2b^{18} + 20ab^{19} + b^{20}$
c. $52{,}360a^{31}b^4$ **d.** 4.6×10^{-18} **13.** verified   **15.** 0.01659   **17.** $\frac{4}{11}$
19. $10, 2, \frac{2}{5}, \frac{2}{25}, \frac{2}{125}$

Practice Test, pp. 1013–1014

1. a. $\frac{1}{2}, \frac{4}{5}, 1, \frac{8}{7}; a_8 = \frac{16}{11}, a_{12} = \frac{8}{5}$ **b.** 6, 12, 20, 30; $a_8 = 90, a_{12} = 182$
c. $3, 2\sqrt{2}, \sqrt{7}, \sqrt{6}; a_8 = \sqrt{2}, a_{12} = i\sqrt{2}$
2. a. 165   **b.** $\frac{311}{420}$ **c.** $\frac{-2343}{512}$ **d.** 7
3. a. $a_1 = 7, d = -3, a_n = 10 - 3n$
b. $a_1 = -8, d = 2, a_n = 2n - 10$ **c.** $a_1 = 4, r = -2, a_n = 4(-2)^{n-1}$
d. $a_1 = 10, r = \frac{2}{5}, a_n = 10(\frac{2}{5})^{n-1}$
4. a. 199   **b.** 9   **c.** $\frac{3}{4}$ **d.** 6   **5. a.** 1712   **b.** 2183   **c.** 2188   **d.** 12
6. a. ≈8.82 ft   **b.** ≈72.4 ft   **7.** $6756.57   **8.** $22,185.27   **9.** verified
10. verified
11. a.

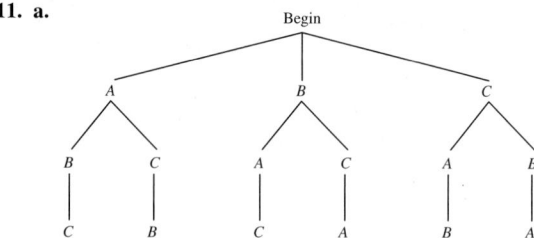

b. ABC, ACB, BAC, BCA, CAB, CBA
12. 302,400   **13.** 64   **14.** 720, 120, 20   **15.** 900,900   **16.** 302,400
17. a. $x^4 - 8x^3y + 24x^2y^2 - 32xy^3 + 16y^4$ **b.** -4
18. a. $x^{10} + 10\sqrt{2}x^9 + 90x^8$ **b.** $a^8 - 16a^7b^3 + 112a^6b^6$
19. 0.989   **20. a.** $\frac{1}{4}$ **b.** $\frac{5}{12}$ **c.** $\frac{1}{3}$ **d.** $\frac{1}{2}$ **e.** $\frac{7}{12}$ **f.** $\frac{1}{4}$ **g.** $\frac{5}{12}$ **h.** 0
21. a. 0.08   **b.** 0.92   **c.** 1   **d.** 0   **e.** 0.95   **f.** 0.03   **22. a.** 0.1875
b. 0.589   **c.** 0.4015   **d.** 0.2945   **e.** 0.4110   **f.** 0.2055   **23. a.** $\frac{59}{100}$
b. $\frac{53}{100}$ **c.** $\frac{13}{100}$ **d.** $\frac{47}{100}$ **24. a.** 0.8075   **b.** 0.0075   **c.** 0.9925
25. verified

Strengthening Core Skills, pp. 1015–1016

Exercise 1. $\dfrac{{}_4C_1 \cdot {}_{13}C_5 - 40}{{}_{52}C_5} \approx 0.001\,970$

Exercise 2. $\dfrac{4 \cdot {}_{13}C_3 \cdot {}_{39}C_2}{{}_{52}C_5} \approx 0.326\,170$

Exercise 3. $\dfrac{4 \cdot {}_{13}C_4 \cdot {}_{39}C_1}{{}_{52}C_5} \approx 0.042\,917$

Exercise 4. $\dfrac{4 \cdot {}_{10}C_5}{{}_{52}C_5} \approx 0.000\,388$

Cumulative Review Chapters 1–10, pp. 1016–1018

1. a. 23 cards are assembled each hour.   **b.** 184 cards
c. $y = 23x - 155$ **d.** ≈ 6:45 A.M.

3.

x	0	$\dfrac{\pi}{6}$	$\dfrac{\pi}{4}$	$\dfrac{\pi}{3}$	$\dfrac{\pi}{2}$	$\dfrac{2\pi}{3}$	$\dfrac{5\pi}{6}$	π
y	1	$\dfrac{\sqrt{3}}{2}$	$\dfrac{\sqrt{2}}{2}$	$\dfrac{1}{2}$	0	$-\dfrac{1}{2}$	$-\dfrac{\sqrt{3}}{2}$	-1

5. $x = \dfrac{-5 \pm \sqrt{109}}{6}$; $x \approx 0.91$; $x \approx -2.57$

7. a. $x = 0$ **b.** $x \in (-1, 0)$ **c.** $x \in (-\infty, -1) \cup (0, \infty)$
d. $x \in (-\infty, -1) \cup (-1, 1)$ **e.** $x \in (1, \infty)$ **f.** $y = 3$ at $(1, 3)$
g. none   **h.** $x \approx -2.3, 0.4, 2$ **i.** $g(4) \approx 0.25$ **j.** does not exist
k. $-\infty$ **l.** 0   **m.** $x \in (-\infty, 1) \cup (-1, \infty)$

9.

$D: x \in [-3, 3]$
$R: y \in [-2, -2] \cup (-1, 2) \cup [4, 9]$

11. a. $4x + 2h - 3$ **b.** $\dfrac{-1}{(x + h - 2)(x - 2)}$

13.

15. a. $x^3 = 125$ **b.** $e^5 = 2x - 1$ **17. a.** $x \approx 3.19$ **b.** $x = 334$
19. (5, 10, 15)
21. $(-3, 3)$; $(-7, 3)$, $(1, 3)$; $(-3 - 2\sqrt{3}, 3)$, $(-3 + 2\sqrt{3}, 3)$
23. a. verified   **b.** $\dfrac{\sqrt{6} + \sqrt{2}}{4}$
25. 1333   **27. a.** ≈7.0%   **b.** ≈91.9%   **c.** ≈98.9%
d. $\dbinom{12}{0}(0.04)^{12}(0.96)^0$; virtually nil

29. $\cos(2\theta) = \cos^2\theta - \sin^2\theta$
$= 1 - 2\sin^2\theta$
$= 2\cos^2\theta - 1;$
$\dfrac{1}{2} = 1 - 2\sin^2\theta$
$\dfrac{1}{4} = \sin^2\theta$
$\pm\dfrac{1}{2} = \sin\theta$
$\theta = \dfrac{\pi}{6}, \dfrac{5\pi}{6}, \dfrac{7\pi}{6}, \dfrac{13\pi}{6}$

Connections to Calculus Exercises, p. 1021

1. a.

$\sum LW = \sum_{i=1}^{4} f(i)(1) = 1\sum_{i=1}^{4} f(i)$ sum of rectangles

$1\sum_{i=1}^{4} f(i) = 1[f(1) + f(2) + f(3) + f(4)]$ definition of summation

$= 5 + 4 + 3 + 2$ evaluate

$= 14$ units^2 result

b. $\sum LW = \sum_{i=1}^{8} f\left(\frac{1}{2}i\right)\left(\frac{1}{2}\right) = \frac{1}{2}\sum_{i=1}^{8} f\left(\frac{1}{2}i\right)$ sum of rectangles

$\frac{1}{2}\sum_{i=1}^{8} f\left(\frac{1}{2}i\right) = \frac{1}{2}\left[f\left(\frac{1}{2}\right) + f(1) + f\left(\frac{3}{2}\right) + f(2) + f\left(\frac{5}{2}\right)\right.$

$\left. + f(3) + f\left(\frac{7}{2}\right) + f(4)\right]$ definition of summation

$$= \frac{1}{2}\left[\frac{11}{2} + 5 + \frac{9}{2} + 4 + \frac{7}{2} + 3 + \frac{5}{2} + 2\right] \text{ evaluate}$$

$$= \frac{1}{2}[30] = 15 \text{ units}^2 \quad \text{result}$$

True area is 16 units², more rectangles → better estimate.

3. a. Since the interval [0, 4] is 4 units wide and we're using 32 subintervals of equal length, the width of each interval (the width of each rectangle) will be $\frac{4}{32} = \frac{1}{8}$. The length of each rectangle is determined by a point of the graph of $f(x) = -x + 6$, so the length of the first rectangle is $f(\frac{1}{8})$, the second length is $f(\frac{2}{8})$, the third is $f(\frac{3}{8})$, and so on up to the 32nd rectangle. Since $A = LW$, we multiply each length $f(\frac{1}{8}i)$ by width $\frac{1}{8}$ and sum the areas of all such rectangles. Using i for a counter, this can be written as

$$A = \sum_{i=1}^{32} LW = \sum_{i=1}^{32} f\left(\frac{1}{8}i\right)\left(\frac{1}{8}\right). \text{ Since all lengths are multiplied by } \frac{1}{8}$$

(the counter i does not affect the constant $\frac{1}{8}$), we can factor out this term

and evaluate $f(x) = -x + 6$ at $x = \frac{1}{8}i$. The result is $\frac{1}{8}\sum_{i=1}^{32}\left(-\frac{1}{8}i + 6\right)$.

b. $\frac{1}{8}\sum_{i=1}^{32}\left(-\frac{1}{8}i + 6\right) = \frac{1}{8}\left[\sum_{i=1}^{32}\left(-\frac{1}{8}i\right) + \sum_{i=1}^{32}6\right]$ summation properties (distribute)

$$= \frac{1}{8}\left[-\frac{1}{8}\sum_{i=1}^{32}i + \sum_{i=1}^{32}6\right] \text{ factor } -\frac{1}{8} \text{ from first summation}$$

$$= \frac{1}{8}\left[-\frac{1}{8}\left(\frac{32^2 + 32}{2}\right) + 6(32)\right] \text{ apply summation formulas}$$

$$= -\frac{1}{64}\left(\frac{1056}{2}\right) + 24 \text{ distribute and simplify}$$

$$= -8.25 + 24 = 15.75 \text{ result}$$

For $n = 32$ the approximate area under the graph is 15.75 units², even closer to the known area of 16 units².

CHAPTER 11
Exercises 11.1, pp. 1031–1034

1. infinity   **3.** left-hand; right-hand; greater   **5.** Answers will vary.

7. $\lim_{n\to\infty} V_n = \frac{4}{3}\pi r^3$ **9.** $\lim_{t\to-\infty} e^{f(t)} = 0$

11. $\lim_{x\to\infty} \cos\left(\frac{1}{x}\right) = 0$ **13.** 500 sides

15. 425 sides   **17.** $\lim_{t\to5} s_t = 5r$

19. $\lim_{x\to a} \tan^{-1}[g(x)] = \frac{\pi}{3}$ **21.** $\lim_{x\to-3} \frac{x+3}{x^2-9} = -\frac{1}{6}$

23. As x approaches π, $p(x)$ approaches -2: $\lim_{x\to\pi} p(x) = -2$
25. As x approaches 2, $v(x)$ approaches $\frac{1}{4}$: $\lim_{x\to2} v(x) = \frac{1}{4}$

27. As x approaches 0, $s(x)$ approaches 0: $\lim_{x\to0} s(x) = 0$

29. $R(x) = \begin{cases} \dfrac{2x^2 - 7x + 6}{\sin(x-2)} & x \neq 2 \\ 1 & x = 2 \end{cases}$

31. As x approaches 2, $f(x)$ approaches $\frac{1}{2}$: $\lim_{x\to2} f(x) = \frac{1}{2}$

33. As x approaches 1, $g(x)$ approaches 4: $\lim_{x\to1} g(x) = 4$

35. As x approaches 1, $f(x)$ approaches 0: $\lim_{x\to1} f(x) = 0$
37. 27   **39.** 24   **41.** 1

43. $\lim_{x\to3} I_x = 3\cos^2(R_1 + R_2)$ **45.** $\lim_{x\to m} f = L$ **47. a.** 1 **b.** -1

49. a. -2 **b.** 2   **51. a.** $-\frac{11}{2}$ **b.** $-\frac{11}{2}$ **53. a.** 0 **b.** 0

55. a. 43 **b.** 13 **c.** $\left(\text{dne}\atop\text{LH}\neq\text{RH}\right)$ **57. a.** 1 **b.** -1 **c.** $\left(\text{dne}\atop\text{LH}\neq\text{RH}\right)$

59. a. $\frac{\sqrt{2}}{2}$ **b.** $\frac{\sqrt{2}}{2}$ **c.** $\frac{\sqrt{2}}{2}$ **61.** $\left(\text{dne}\atop\infty\right)$ **63.** π **65.** $\left(\text{dne}\atop-\infty\right)$

67. $\left(\text{dne}\atop\mathcal{L}\right)$ **69.** 0   **71.** $(x-5)(x-2)(x+1)(3x-1)$ **73.** 19.90 in²

Exercises 11.2, pp. 1041–1043

1. sum; limits
3. root; nth; $f(x) > 0$
5. Answers will vary.
7. a. limit appears to be 0   **b.** limit is actually 0.001

x	$f(x)$	x	$f(x)$
0.5	0.2510	-0.5	0.2510
0.4	0.1610	-0.4	0.1610
0.3	0.0910	-0.3	0.0910
0.2	0.0410	-0.2	0.0410
0.1	0.0110	-0.1	0.0110
0.01	0.0011	-0.01	0.0011
0.001	0.001001	-0.001	0.001001

9.

x	$y = x + \dfrac{\sqrt{x-2}}{10^x}$
2.7	≈2.7017
2.8	≈2.8014
2.9	≈2.9012
3	—
3.1	≈3.1008
3.2	≈3.2007
3.3	≈3.3006

11.

x	$y = x + \dfrac{\sqrt{x-2}}{10^x}$
2.99	≈2.991
2.999	≈3
2.9999	≈3.0009
3	—
3.0001	≈3.0011
3.001	≈3.002
3.01	≈3.011

13. a. 3 **b.** 3.001   **15.** 9   **17.** -8 **19.** 2   **21.** 0   **23.** 7   **25.** 1

27. 9   **29.** 8   **31.** 0   **33.** 14   **35.** -16 **37.** $\frac{7}{2}$ **39.** 0   **41.** 64   **43.** -36

45. -69 **47.** 29   **49.** 9   **51.** $\frac{3}{5}$ **53.** $-\frac{1}{4}$ **55.** 216   **57.** 15   **59.** 23

61. -1 **63.** $-\frac{17}{4}$ **65.** $\left(\text{dne}\atop-\infty\right)$ **67.** 9   **69.** $\left(\text{dne}\atop\infty\right)$ **71.** $\left(\text{dne}\atop\text{LH}\neq\text{RH}\right)$ **73.** 6

75. $\left(\text{dne}\atop\text{LH}\neq\text{RH}\right)$ **77.** $[-1, \frac{3}{2}]$ **79.** verified

Mid-Chapter Check, pp. 1043–1044

1. $\lim_{x\to-\infty} \frac{6x^2-3}{12x^3-x-1} = 0$ **2. a.** 1 **b.** 0   **3.** 49 per year (almost weekly)

4. $F(x) = \begin{cases} \dfrac{6x^2-19x-7}{2x-7} & x \neq \dfrac{7}{2} \\ \dfrac{23}{2} & x = \dfrac{7}{2} \end{cases}$ **5. a.** -1 **b.** 1 **c.** $\left(\text{dne}\atop\text{LH}\neq\text{RH}\right)$ **6.** $\left(\text{dne}\atop\infty\right)$

7. a.

x	$y = \cos\left(\dfrac{\pi}{x}\right)$
0.1	1
0.01	1
0.001	1
0	—
-0.001	1
-0.01	1
-0.1	1

b. Answers may vary.

8. $\frac{1}{2}$ **9.** $\sqrt[3]{4}$ **10.** $\left(\text{dne}\atop\text{LH}\neq\text{RH}\right)$

Exercises 11.3, pp. 1053–1055

1. asymptotic; removable; jump   **3.** direct substitution
5. Answers may vary.   **7.** not continuous, condition 1 is violated
9. continuous   **11.** not continuous, condition 2 is violated
13. continuous   **15.** not continuous, condition 3 is violated
17. 36   **19.** Direct substitution not possible, 5 is not in the domain.

21. $\dfrac{1}{2}$ **23.** Direct substitution not possible, -1 is not in the domain.

25. $\sqrt{55}$ **27.** $-\dfrac{1}{2}$ **29.** 4   **31.** 9   **33.** $\dfrac{1}{4}$ **35.** 0   **37.** $4x - 1$

39. $\dfrac{-3}{(x+2)^2}$ **41.** $\dfrac{1}{2\sqrt{x+2}}$ **43.** $3(x+2)^2$ **45.** $\dfrac{1}{2}$ **47.** 3

49. $\left(\dfrac{\text{dne}}{\infty}\right)$ **51.** 0   **53.** $\dfrac{1}{2}$ **55.** $\left(\dfrac{\text{dne}}{-\infty}\right)$ **57.** 0   **59.** $x = 1000$

61. $\dfrac{3}{8}$ **63.** 0   **65.** 2   **67.** $\dfrac{2\sqrt{3}}{7}$ **69.** 3   **71.** 3

73. not possible since $f(2)$ not defined   **75.** 3   **77.** 0
79. 1   **81.** not possible, $\lim\limits_{x \to -2^-} g(x)$ does not exist

83. not possible, $\lim\limits_{x \to 0} g(x)$ does not exist   **85.** -3 **87.** 3

89. a. $x = \dfrac{-2 \pm 2\sqrt{10}}{3}$ **b.** $x = 0, x = 8$ **c.** $x = 2$

91. $A = 55°, C = 90°, b = 9.6$ cm, $c = 16.7$ cm

Exercises 11.4, pp. 1064–1066

1. difference   **3.** rectangles   **5.** Answers may vary.
7. $f(t) = 88.2 - 9.8t$ **9.** $g(t) = 78.4 - 9.8t$ **11.** 896.9 m
13. $d(t) = -9.8t$ **15.** $d(t) = -32t$

17. $f(x) = \dfrac{1}{2}$ **19.** $f(x) = 3x^2$ **21.** $p(t) = \dfrac{0.6}{\sqrt{t}}$ **23.** $b(t) = -\dfrac{1}{2\sqrt{t}}$

25. $f(x) = \dfrac{-2}{(x-1)^2}$ **27.** $h(x) = \dfrac{25}{(x+5)^2}$ **29.** $-\dfrac{1}{2}$

31. 1   **33.** 9 units^2 **35.** 15 units^2 **37.** 9 units^2 **39.** 15 units^2

41. $\dfrac{4}{n} \sum\limits_{i=1}^{n}\left[\dfrac{1}{2}\left(\dfrac{4}{n}i\right)^2 + 3\right]$

$= \dfrac{4}{n}\left[\sum\limits_{i=1}^{n}\dfrac{1}{2}\left(\dfrac{4}{n}i\right)^2 + \sum\limits_{i=1}^{n}3\right]$ summation properties (distribute)

$= \dfrac{4}{n}\left[\dfrac{1}{2}\sum\limits_{i=1}^{n}\dfrac{16}{n^2}i^2 + \sum\limits_{i=1}^{n}3\right]$ simplify

$= \dfrac{4}{n}\left[\dfrac{16}{2n^2}\sum\limits_{i=1}^{n}i^2 + \sum\limits_{i=1}^{n}3\right]$ factor $\dfrac{16}{n^2}$ from first summation

$= \dfrac{4}{n}\left[\dfrac{8}{n^2}\left(\dfrac{2n^3 + 3n^2 + n}{6}\right) + 3n\right]$ apply summation formula

$= \left[\dfrac{32}{n^3}\left(\dfrac{2n^3 + 3n^2 + n}{6}\right) + 12\right]$ distribute $\dfrac{4}{n}$

$= \left[\dfrac{32}{6}\left(\dfrac{2n^3 + 3n^2 + n}{n^3}\right) + 12\right]$ rewrite denominators

$= \dfrac{16}{3}\left(2 + \dfrac{3}{n} + \dfrac{1}{n^2}\right) + 12$ decompose rational expression

Applying the limit properties gives $\dfrac{16}{3}\lim\limits_{n\to\infty}\left(2 + \dfrac{3}{n} + \dfrac{1}{n^2}\right) + \lim\limits_{n\to\infty}12$, and

the area under the curve is $\left(\dfrac{16}{3}\right)(2) + 12 = \dfrac{68}{3}$ units^2. The new employee has produced 22 complete parts.

43. $A = \sum\limits_{i=1}^{n} LW$

$= \sum\limits_{i=1}^{n} f\left(\dfrac{6}{n}i\right)\left(\dfrac{6}{n}\right)$ area formula, rectangle method

$= \dfrac{6}{n}\sum\limits_{i=1}^{n} f\left(\dfrac{6}{n}i\right)$ factor $\dfrac{6}{n}$

$= \dfrac{6}{n}\sum\limits_{i=1}^{n}\left[-\dfrac{1}{2}\left(\dfrac{6}{n}i\right)^2 + 4\left(\dfrac{6}{n}i\right)\right]$ evaluate f at $\dfrac{6}{n}i$

$= \dfrac{6}{n}\left[-\dfrac{1}{2}\sum\limits_{i=1}^{n}\left(\dfrac{6}{n}i\right)^2 + 4\sum\limits_{i=1}^{n}\left(\dfrac{6}{n}i\right)\right]$ distribute summation

$= \dfrac{6}{n}\left[-\dfrac{1}{2}\sum\limits_{i=1}^{n}\dfrac{36}{n^2}i^2 + 4\sum\limits_{i=1}^{n}\dfrac{6}{n}i\right]$ simplify

$= \dfrac{6}{n}\left[-\dfrac{36}{2n^2}\sum\limits_{i=1}^{n}i^2 + \dfrac{24}{n}\sum\limits_{i=1}^{n}i\right]$ factor $\dfrac{36}{n^2}$ and $\dfrac{6}{n}$

$= \dfrac{6}{n}\left[-\dfrac{18}{n^2}\left(\dfrac{2n^3 + 3n^2 + n}{6}\right) + \dfrac{24}{n}\left(\dfrac{n^2 + n}{2}\right)\right]$ summation formulas

$= -\dfrac{108}{n^3}\left(\dfrac{2n^3 + 3n^2 + n}{6}\right) + \dfrac{144}{n^2}\left(\dfrac{n^2 + n}{2}\right)$ distribute $\dfrac{6}{n}$

$= -\dfrac{108}{6}\left(\dfrac{2n^3 + 3n^2 + n}{n^3}\right) + \dfrac{144}{2}\left(\dfrac{n^2 + n}{n^2}\right)$ rewrite denominators

$= -18\left(2 + \dfrac{3}{n} + \dfrac{1}{n^2}\right) + 72\left(1 + \dfrac{1}{n}\right)$ decompose rational expression

As $n \to \infty$, $\dfrac{3}{n}, \dfrac{1}{n^2} \to 0$, and the area is $-(18)(2) + 72 = 36$ units^2.

45. $x \approx 29.87$ **47.** $x = 5, \pm\sqrt{3}i$

Summary and Concept Review, pp. 1066–1068

1. a. -3, **b.** 5, **c.** $\left(\dfrac{\text{dne}}{\text{LH}\neq\text{RH}}\right)$ **2.** 2   **3.** 1   **4.** $\left(\dfrac{\text{dne}}{\text{LH}\neq\text{RH}}\right)$

5. -38 **6.** $\dfrac{2}{3}$ **7.** $-\dfrac{1}{32}$ **8.** $\dfrac{1}{2}$

9. a. $x = -3$, **b.** $x = -2, -1, 3, 4$, **c.** $x = 1, 2$
10. 3   **11.** -1 **12.** 2

13. $\left(\dfrac{\text{dne}}{-\infty}\right)$ **14.** $\left(\dfrac{\text{dne}}{\text{LH}\neq\text{RH}}\right)$ **15.** -2 **16.** $f(x) = 2x + 5$

17. $g(x) = \dfrac{1}{\sqrt{2x-1}}$ **18.** $v(x) = \dfrac{-1}{(x+3)^2}$

19. $m_{\text{tan}} = -2x + 3$; at $x = 4, m_{\text{tan}} = -5$
20. 9 units^2

Mixed Review, p. 1069

1. a. $\lim\limits_{x\to a^+} f(x) = b$ **b.** $\lim\limits_{x\to\infty} f(x) = b$

   c. $\lim\limits_{x\to a} f(x) = b$ **d.** $\lim\limits_{x\to a} f(x) = \left(\dfrac{\text{dne}}{\infty}\right)$

3. -7 **5.** (dne)   **7.** $4\sqrt{2}$ **9.** $-\dfrac{1}{6}$ **11.** -3 **13.** $\dfrac{1}{2}$ **15.** $-\dfrac{\sqrt{2}}{2}$
17. a. -42 ft/sec   **b.** -74 ft/sec   **c.** -10 ft/sec
19. a. 25% of the staff per day   **b.** 12.5% of the staff per day

   c. $\dfrac{\sqrt{2}}{20} \approx 7\%$ of the staff per day

Practice Test, p. 1070

1. The limit of $f(x)$ as x approaches 5 is 10.

2. $f(x)$; L; sufficiently; c

3. False, a limit can exist even if c is not in the domain.

4. False, a limit can fail to exists even if a function is defined at c.

5. As the domain of g is $x \geq 1$, the limit in **b.** exists and the limit in
   a. does not. $\lim\limits_{x\to 1^+}(\sqrt{x-1} + 2) = 2$.

6. a. $\to$ II   **b.** $\to$ I   **c.** $\to$ IV   **d.** $\to$ III

7. a. 2   **b.** $\left(\dfrac{\text{dne}}{\infty}\right)$ **8. a.** 1   **b.** 0   **9. a.** 3   **b.** 12
10. a. 4   **b.** undefined, $g(3) = 0$ **11. a.** 4   **b.** -4

12. a. 2   **b.** 9   **13. a.** 0   **b.** $\left(\dfrac{\text{dne}}{\text{LH}\neq\text{RH}}\right)$, **c.** $\dfrac{1}{3}$ **d.** 1

14. a. $\left(\dfrac{\text{dne}}{\text{LH}\neq\text{RH}}\right)$ **b.** $\left(\dfrac{\text{dne}}{\infty}\right)$, **c.** $\left(\dfrac{\text{dne}}{\infty}\right)$ **d.** 3 is not in the domain

15. 3   **16.** $\dfrac{3}{2}$ **17.** $\dfrac{1}{10}$ **18.** 1

19. a. $d(t) = -32t + 224$ **b.** $d(2) = 160$, the debris is rising at a
velocity of 160 ft/sec; $d(6) = 32$, the upward velocity of the debris has
slowed to 32 ft/sec; $d(7) = 0$, the debris has reached its maximum height
(velocity is 0 ft/sec); $d(11) = -128$, the velocity of the debris in now in
the downward direction ($v < 0$) at 128 ft/sec.

20. $\lim\limits_{x\to\infty}\sum\limits_{i=1}^{n} LW = \lim\limits_{n\to\infty}\sum\limits_{i=1}^{n} f\left(\dfrac{6}{n}i\right)\dfrac{6}{n} = \lim\limits_{n\to\infty}\dfrac{6}{n}\sum\limits_{i=1}^{n}\left[225 - \left(\dfrac{6}{n}i\right)^2\right]$; 1278 ft-lb

Cumulative Review Chapters 1–11, pp. 1071–1072

1. $x = \dfrac{3}{2}, x = \pm i\sqrt{3}$ **3.** $(1, 1, -1)$ **5.** about 630 ft   **7.** $\dfrac{1}{2}$

9. $f(x) = (x - 3)(x - 1)(x + 2)$

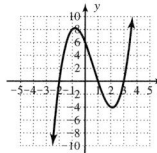

11. $-512i$ **13. a.** $L(x) = -\dfrac{1}{3}x + 19$ **b.** 14 cm   **c.** 24 days

15. no, $\lim\limits_{x \to 0^-} f(x) \neq \lim\limits_{x \to 0^+} f(x)$

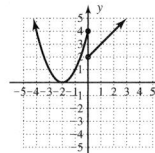

17. $x = \dfrac{\ln 9.36 + 1}{2}$, $x \approx 1.618$ **19.** -3

21. $1 + \dfrac{-2}{x + 2} + \dfrac{2}{x - 1}$

23. $\theta = \dfrac{\pi}{12} + \dfrac{\pi}{2}k; \dfrac{\pi}{4} + \dfrac{\pi}{2}k$

25. $\dfrac{(x - 1)^2}{25} + \dfrac{(y - 1)^2}{9} = 1$

27. a. 1   **b.** 0   **c.** -3 **d.** 5 not in domain

29. $30.\overline{6}$ units^2

APPENDIX I

Exercises AI-A, pp. A-6–A-8

1. $1.\overline{3}$

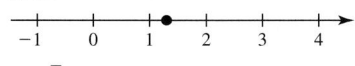

3. $2.\overline{5}$

5. ≈ 2.65

7. ≈ 1.73

9. a. i. $\{8, 7, 6\}$ **ii.** $\{8, 7, 6\}$ **iii.** $(-1, 8, 7, 6)$
iv. $\{-1, 8, 0, 75, \frac{9}{2}, 5, \overline{.6}, 7, \frac{3}{5}, 6\}$ **v.** $\{\ \}$
vi. $\{-1, 8, 0.75, \frac{9}{2}, 5.\overline{6}, 7, \frac{3}{5}, 6\}$ **b.** $\{-1, \frac{3}{5}, 0.75, \frac{9}{2}, 5.\overline{6}, 6, 7, 8\}$
c.

11. a. i. $\{\sqrt{49}, 2, 6, 4\}$ **ii.** $\{\sqrt{49}, 2, 6, 0, 4\}$
iii. $\{-5, \sqrt{49}, 2, -3, 6, -1, 0, 4\}$ **iv.** $\{-5, \sqrt{49}, 2, -3, 6, -1, 0, 4\}$
v. $\{\sqrt{3}, \pi\}$ **vi.** $\{-5, \sqrt{49}, 2, -3, 6, -1, \sqrt{3}, 0, 4, \pi\}$
b. $\{-5, -3, -1, 0, \sqrt{3}, 2, \pi, 4, 6, \sqrt{49}\}$
c.

13. Let a represent Kylie's age: $a \geq 6$ years.
15. Let n represent the number of incorrect words: $n \leq 2$ incorrect.
17. 2.75   **19.** -4 **21.** 9   **23.** 10   **25.** $-8, 2$ **27.** negative   **29.** $-n$
31. a. positive   **b.** negative   **c.** negative   **d.** negative   **33.** $-\dfrac{11}{6}$
35. -2 **37.** $9^2 = 81$ is closest   **39.** 7   **41.** $4\frac{1}{3}$ **43.** $-\frac{1}{10}$ **45.** $-\frac{7}{8}$

47. -4 **49.** $\dfrac{-11}{12}$ **51.** 64   **53.** 4489.70   **55.** 32°F   **57.** 179°F
59. Tsu Ch'ung-chih: $\frac{355}{113}$ **61.** negative

Exercises AI-B, pp. A-12–A-13

1. $n - 7$ **3.** $n + 4$ **5.** $(n - 5)^2$ **7.** $2n - 13$ **9.** $n^2 + 2n$
11. $\frac{2}{3}n - 5$ **13.** $3(n + 5) - 7$
15. Let w represent the width. Then $2w$ represents twice the width and $2w - 3$ represents three meters less than twice the width.   **17.** Let b represent the speed of the bus. Then $b + 15$ represents 15 mph more than the speed of the bus.   **19.** $h = b + 150$ **21.** $L = 2W + 20$
23. $M = 2.5N$ **25.** $T = 12.50g + 50$ **27.** 14   **29.** 19   **31.** 0
33. 144   **35.** $\dfrac{-41}{5}$ **37.** 24

39.

x	Output
-3	14
-2	6
-1	0
0	-4
1	-6
2	-6
3	-4

-1 has an output of 0.

41.

x	Output
-3	-5
-2	8
-1	9
0	4
1	-1
2	0
3	13

2 has an output of 0.

43. a. $7 + (-5) = 2$ **b.** $n + (-2)$ **c.** $a + (-4.2) + 13.6 = a + 9.4$
d. $x + 7 - 7 = x$ **45.** $-5x + 13$ **47.** $-\frac{2}{15}p + 6$ **49.** $\frac{17}{12}x$
51. $-2a^2 + 2a$ **53.** $6x^2 - 3x$ **55.** $2a + 3b + 2c$ **57.** $\frac{35}{8}n + 7$
59. a. $t = \frac{1}{2}j$ **b.** $t = 225$ mph   **61. a.** $L = 2W + 3$ **b.** 107 ft
63. $t = c + 27; 42¢$ **65.** $C = 25t + 43.50; \$81$ **67. a.** positive odd integer

Exercises AI-C, pp. A-21–A-24

1. $14n^7$ **3.** $-12p^5q^4$ **5.** $a^{14}b^7$ **7.** $216p^3q^6$ **9.** $49c^{14}d^4$ **11.** $\frac{9}{16}x^6y^2$
13. a. $V = 27x^6$ **b.** 1728 units^3 **15.** $3w^3$ **17.** $-3ab$ **19.** $\dfrac{27}{8}$
21. $2h^3$ **23.** $\dfrac{-1}{8}$ **25.** $\dfrac{4p^8}{q^6}$ **27.** $\dfrac{8x^6}{27y^9}$ **29.** $\dfrac{25m^4n^6}{4r^8}$ **31.** $\dfrac{3p^2}{-4q^2}$
33. $\dfrac{5}{3h^7}$ **35.** $\dfrac{a^{11}}{b^{14}}$ **37.** $\dfrac{-12}{5x^4}$ **39.** 2   **41.** $\frac{7}{10}$ **43.** $\frac{13}{9}$ **45.** -4
47. 6.6×10^9 **49.** 0.000 000 006 5   **51.** 26,571 hr; 1,107 days
53. polynomial, none of these, degree 3   **55.** nonpolynomial because exponents are not whole numbers, NA, NA   **57.** polynomial, binomial, degree 3   **59.** $-w^3 - 3w^2 + 7w + 8.2; -1$ **61.** $c^3 + 2c^2 - 3c + 6; 1$
63. $\frac{-2}{3}x^2 + 12; \frac{-2}{3}$ **65.** $3p^3 - 3p^2 - 12$ **67.** $7.85b^2 - 0.6b - 1.9$
69. $\frac{1}{4}x^2 - 8x + 6$ **71.** $-3x^3 + 3x^2 + 18x$ **73.** $3r^2 - 11r + 10$
75. $x^3 - 27$ **77.** $b^3 - b^2 - 34b - 56$ **79.** $21v^2 - 47v + 20$
81. $9 - m^2$ **83.** $m^2 - \frac{9}{16}$ **85.** $2x^4 - x^2 - 15$ **87.** $4m + 3; 16m^2 - 9$
89. $7x + 10; 49x^2 - 100$ **91.** $6 - 5k; 36 - 25k^2$ **93.** $x - \sqrt{6}; x^2 - 6$
95. $x^2 + 8x + 16$ **97.** $16g^2 + 24g + 9$ **99.** $16p^2 - 24pq + 9q^2$
101. $16 - 8\sqrt{x} + x$ **103.** $F = kPQd^{-2}$
105. $5x^{-3} + 3x^{-2} + 2x^{-1} + 4$ **107.** \$15   **109.** 6

Exercises AI-D, pp. A-29–A-31

1. a. $-17(x^2 - 3)$ **b.** $7b(3b^2 - 2b + 8)$ **c.** $-3a^2(a^2 + 2a - 3)$
3. a. $(a + 2)(2a + 3)$ **b.** $(b^2 + 3)(3b + 2)$ **c.** $(n + 7)(4m - 11)$
5. a. $(3q + 2)(3q^2 + 5)$ **b.** $(h - 12)(h^4 - 3)$ **c.** $(k^2 - 7)(k^3 - 5)$
7. a. $-1(p - 7)(p + 2)$ **b.** $(q - 9)(q + 5)$ **c.** $(n - 4)(n - 5)$
9. a. $(3p + 2)(p - 5)$ **b.** $(4q - 5)(q + 3)$ **c.** $(5u + 3)(2u - 5)$
11. a. $(2s + 5)(2s - 5)$ **b.** $(3x + 7)(3x - 7)$ **c.** $2(5x + 6)(5x - 6)$
d. $(11h + 12)(11h - 12)$ **e.** $(b + \sqrt{5})(b - \sqrt{5})$ **13. a.** $(a - 3)^2$
b. $(b + 5)^2$ **c.** $(2m - 5)^2$ **d.** $(3n - 7)^2$
15. a. $(2p - 3)(4p^2 + 6p + 9)$ **b.** $(m + \frac{1}{2})(m^2 - \frac{1}{2}m + \frac{1}{4})$
c. $(g - 0.3)(g^2 + 0.3g + 0.09)$ **d.** $-2t(t - 3)(t^2 + 3t + 9)$
17. a. $(x + 3)(x - 3)(x + 1)(x - 1)$ **b.** $(x^2 + 9)(x^2 + 4)$
c. $(x - 2)(x^2 + 2x + 4)(x + 1)(x^2 - x + 1)$ **19. a.** $(n + 1)(n - 1)$
b. $(n - 1)(n^2 + n + 1)$ **c.** $(n + 1)(n^2 - n + 1)$

d. $7x(2x+1)(2x-1)$ **21.** $(a+5)(a+2)$ **23.** $2(x-2)(x-10)$
25. $-1(3m+8)(3m-8)$ **27.** $(r-3)(r-6)$ **29.** $(2h+3)(h+2)$
31. $(3k-4)^2$ **33.** $-3x(2x-7)(x-3)$ **35.** $4m(m+5)(m-2)$
37. $(a+5)(a-12)$ **39.** $(2x-5)(4x^2+10x+25)$ **41.** prime

43. $(x-5)(x+3)(x-3)$ **45.** $V=\frac{1}{3}\pi h(R+r)(R-r)$; 6π cm³; 18.8 cm³

47. $V=x(x+5)(x+3)$ **a.** 3 in. **b.** 5 in. **c.** $V=24(29)(27)=18{,}792$ in³

49. $L=L_0\sqrt{\left(1+\frac{v}{c}\right)\left(1-\frac{v}{c}\right)}$, $L=12\sqrt{(1+0.75)(1-0.75)}$ $=3\sqrt{7}$ in. ≈ 7.94 in. **51. a.** $\frac{1}{8}(4x^4+x^3-6x^2+32)$
b. $\frac{1}{18}(12b^5-3b^3+8b^2-18)$ **53.** $2x(16x-27)(6x+5)$
55. $(x+3)(x-3)(x^2+9)$
57. $(p+1)(p^2-p+1)(p-1)(p^2+p+1)$
59. $(q+5)(q-5)(q+\sqrt{3})(q-\sqrt{3})$

Exercises AI-E, pp. A-37–A-39

1. a. $-\frac{1}{3}$ **b.** $\frac{x+3}{2x(x-2)}$ **3. a.** simplified **b.** $\frac{a-4}{a-7}$ **5. a.** -1
b. -1 **7. a.** $-3ab^9$ **b.** $\frac{x+3}{9}$ **c.** $-1(y+3)$ **d.** $\frac{-1}{m}$
9. a. $\frac{2n+3}{n}$ **b.** $\frac{3x+5}{2x+3}$ **c.** $x+2$ **d.** $n-2$ **11.** $\frac{(a-2)(a+1)}{(a+3)(a+2)}$
13. 1 **15.** $\frac{(p-4)^2}{p^2}$ **17.** $\frac{-15}{4}$ **19.** $\frac{8(a-7)}{a-5}$ **21.** $\frac{y}{x}$ **23.** $\frac{m}{m-4}$
25. $\frac{x+0.3}{x-0.2}$ **27.** $\frac{n+\frac{1}{5}}{n+\frac{2}{3}}$ **29.** $\frac{3(a^2+3a+9)}{2}$ **31.** $\frac{3+20x}{8x^2}$
33. $\frac{14y-x}{8x^2y^4}$ **35.** $\frac{2}{p+6}$ **37.** $\frac{-3m-16}{(m+4)(m-4)}$ **39.** $\frac{-y+11}{(y+6)(y-5)}$
41. $\frac{2a-5}{(a+4)(a-5)}$ **43.** $\frac{1}{y+1}$ **45.** $\frac{m^2-6m+21}{(m+3)^2(m-3)}$
47. a. $\frac{1}{p^2}-\frac{5}{p}$; $\frac{1-5p}{p^2}$ **b.** $\frac{1}{x^2}+\frac{2}{x^3}$; $\frac{x+2}{x^3}$ **49.** $\frac{4a}{a+20}$ **51.** $p-1$
53. $\frac{x}{9x-12}$ **55.** $\frac{-2}{y+31}$ **57. a.** $\frac{1+\frac{3}{m}}{1-\frac{3}{m}}$; $\frac{m+3}{m-3}$ **b.** $\frac{1+\frac{2}{x^2}}{1-\frac{2}{x^2}}$; $\frac{x^2+2}{x^2-2}$
59. $\frac{f_2+f_1}{f_1f_2}$ **61.** $\frac{-a}{x(x+h)}$ **63.** $\frac{-(2x+h)}{2x^2(x+h)^2}$

65. Price rises rapidly for first four days, then begins a gradual decrease. Yes, on the 35th day of trading.

Day	Price
0	10
1	16.67
2	32.76
3	47.40
4	53.51
5	52.86
6	49.25
7	44.91
8	40.75
9	37.03
10	33.81

67. $t=8$ weeks **69. b.** $20\cdot n\div 10\cdot n=2n^2$, all others equal 2

Exercises AI-F, pp. A-47–A-50

1. a. 9 **b.** 10 **3. a.** $7|p|$ **b.** $|x-3|$ **c.** $9m^2$ **d.** $|x-3|$
5. a. 4 **b.** $-5x$ **c.** $6z^4$ **d.** $\frac{v}{-2}$ **7. a.** 2 **b.** not a real number

c. $3x^2$ **d.** $-3x$ **e.** $k-3$ **f.** $|h+2|$ **9. a.** -5 **b.** $-3|n^3|$
c. not a real number **d.** $\frac{7|v^5|}{6}$ **11. a.** 4 **b.** $\frac{64}{125}$ **c.** $\frac{125}{8}$ **d.** $\frac{9p^4}{4q^2}$
13. a. -1728 **b.** not a real number **c.** $\frac{1}{9}$ **d.** $\frac{-256}{81x^4}$
15. a. $\frac{32n^{10}}{p^2}$ **b.** $\frac{1}{2y^{\frac{1}{4}}}$ **17. a.** $3m\sqrt{2}$ **b.** $10pq^2\sqrt[3]{q}$ **c.** $\frac{3}{2}mn\sqrt[3]{n^2}$
d. $4pq^3\sqrt{2p}$ **e.** $-3+\sqrt{7}$ **f.** $\frac{9}{2}-\sqrt{2}$ **19. a.** $15a^2$ **b.** $-4b\sqrt{b}$
c. $\frac{x^4\sqrt{y}}{3}$ **d.** $3u^2v\sqrt[3]{v}$ **21. a.** $2m^2$ **b.** $3n$ **c.** $\frac{3\sqrt{5}}{4x}$ **d.** $\frac{18\sqrt[3]{3}}{z^3}$
23. a. $2x^2y^3$ **b.** $x^2\sqrt[4]{x}$ **c.** $\sqrt[12]{b}$ **d.** $\frac{1}{\sqrt[6]{6}}=\frac{\sqrt[6]{6^5}}{6}$ **e.** $b^{\frac{3}{4}}$
25. a. $9\sqrt{2}$ **b.** $14\sqrt{3}$ **c.** $16\sqrt{2m}$ **d.** $-5\sqrt{7p}$
27. a. $-x^3\sqrt{2x}$ **b.** $2-\sqrt{3x}+3\sqrt{5}$
c. $6x\sqrt{2x}+5\sqrt{2}-\sqrt{7x}+3\sqrt{3}$ **29.** Verified **31.** Verified
33. a. 98 **b.** $\sqrt{15}+\sqrt{21}$ **c.** n^2-5 **d.** $39-12\sqrt{3}$
35. a. -19 **b.** $\sqrt{10}+\sqrt{65}-2\sqrt{7}-\sqrt{182}$
c. $12\sqrt{5}+2\sqrt{14}+36\sqrt{15}+6\sqrt{42}$
37. a. $\frac{\sqrt{3}}{2}$ **b.** $\frac{2\sqrt{15x}}{9x^2}$ **c.** $\frac{3\sqrt[3]{6b}}{10b}$ **d.** $\frac{\sqrt[3]{2p^2}}{2p}$ **e.** $\frac{5\sqrt[3]{a^2}}{a}$
39. a. $-12+4\sqrt{11}$; 1.27 **b.** $2+\sqrt{3}$
41. a. $\sqrt{30}-2\sqrt{5}-3\sqrt{3}+3\sqrt{2}$; 0.05
b. $\frac{7+7\sqrt{2}+\sqrt{6}+2\sqrt{3}}{-3}$; -7.60
43. a. $8\sqrt{10}$ m; **b.** about 25.3 m **45. a.** 365.02 days
b. 688.69 days **c.** 87.91 days **47. a.** 36 mph **b.** 46.5 mph
49. $12\pi\sqrt{34}\approx 219.82$ m² **51. a.** $(x+\sqrt{5})(x-\sqrt{5})$
b. $(n+\sqrt{19})(n-\sqrt{19})$
53. a. $13\sqrt{3x}+39\sqrt{x}$ **b.** Answers will vary. **55.** $\frac{3\sqrt{2}}{2}$

Practice Test, pp. A-52–A-53

1. a. True **b.** True **c.** False; $\sqrt{2}$ cannot be expressed as a ratio of two integers. **d.** True **2. a.** 11 **b.** -5 **c.** not a real number **d.** 20
3. a. $\frac{9}{8}$ **b.** $\frac{-7}{6}$ **c.** 0.5 **d.** -4.6 **4. a.** $\frac{28}{3}$ **b.** 0.9 **c.** 4 **d.** -7
5. ≈ 4439.28 **6. a.** 0 **b.** undefined **7. a.** 3; $-2,6,5$ **b.** 2; $\frac{1}{3}$, 1
8. a. -13 **b.** ≈ 7.29 **9. a.** $x^3-(2x-9)$ **b.** $2n-3\left(\frac{n}{2}\right)^2$
10. a. Let r represent Earth's radius. Then $11r-119$ represents Jupiter's radius. **b.** Let e represent this year's earnings. Then $4e+1.2$ million represents last year's earnings. **11. a.** $9v^2+3v-7$ **b.** $-7b+8$
c. x^2+6x **12. a.** $(3x+4)(3x-4)$ **b.** $v(2v-3)^2$
c. $(x+5)(x+3)(x-3)$ **13. a.** $5b^3$ **b.** $4a^{12}b^{12}$ **c.** $\frac{m^6}{8n^3}$ **d.** $\frac{25}{4}p^2q^2$
14. a. $-4ab$ **b.** $6.4\times 10^{-2}=0.064$ **c.** $\frac{a^{12}}{b^4c^8}$ **d.** -6
15. a. $9x^4-25y^2$ **b.** $4a^2+12ab+9b^2$
16. a. $7a^4-5a^3+8a^2-3a-18$ **b.** $-7x^4+4x^2+5x$
17. a. -1 **b.** $\frac{2+n}{2-n}$ **c.** $x-3$ **d.** $\frac{x-5}{3x-2}$ **e.** $\frac{x-5}{3x+1}$
f. $\frac{3(m+7)}{5(m+4)(m-3)}$ **18. a.** $|x+11|$ **b.** $\frac{-2}{3v}$ **c.** $\frac{64}{125}$ **d.** $-\frac{1}{2}+\frac{\sqrt{2}}{2}$
e. $11\sqrt{10}$ **f.** x^2-5 **g.** $\frac{\sqrt{10x}}{5x}$ **h.** $2(\sqrt{6}+\sqrt{2})$
19. $-0.5x^2+10x+1200$; **a.** 10 decreases of 0.50 or $5.00
b. Maximum revenue is $1250. **20.** 58 cm

Index